SOLID STATE PHYSICS LITERATURE GUIDES
Volume 7

SCATTERING OF
THERMAL NEUTRONS
A Bibliography (1932-1974)

Solid State Physics Literature Guides

Prepared under the auspices of the Research Materials Information Center
Oak Ridge National Laboratory

General Editor: T. F. Connolly

Solid State Division
Oak Ridge National Laboratory*
Oak Ridge, Tennessee

*Oak Ridge National Laboratory is operated by Union Carbide Corporation for the U.S. Atomic Energy Commission.

SOLID STATE PHYSICS LITERATURE GUIDES
Volume 7

SCATTERING OF THERMAL NEUTRONS
A Bibliography (1932-1974)

Compiled by

André Larose
and
Jake Vanderwal

McMaster University
Hamilton, Ontario, Canada

IFI/PLENUM • NEW YORK-WASHINGTON-LONDON

Library of Congress Cataloging in Publication Data

Larose, André.
  Scattering of thermal neutrons.

  (Solid state physics literature guides; v. 7)
  Includes index.
  1. Thermal neutrons — Scattering — Bibliography. I. Vanderwal, Jake, joint
author. II. Title. III. Series.
Z7144.S58S65 vol. 7 [Z7144.N8] 016.5304′1′08s
ISBN 978-1-4684-6215-9   ISBN 978-1-4684-6213-5 (eBook)
DOI 10.1007/978-1-4684-6213-5   [016.5304]   74-23310

© 1974 IFI/Plenum Data Company
Softcover reprint of the hardcover 1st edition 1974

A Division of Plenum Publishing Corporation
227 West 17th Street, New York, N.Y. 10011

United Kingdom edition published by Plenum Press, London
A Division of Plenum Publishing Company, Ltd.
4a Lower John Street, London W1R 3PD, England

INTRODUCTION

Solid state physicists have long appreciated the usefulness of thermal neutron scattering in the investigation of condensed matter. This technique was first made possible by the advent of the nuclear reactor and has, since then, undergone many refinements. The developments in this field of research have, we felt, necessitated the making of a comprehensive compilation of the published thermal neutron papers. The large number of titles collected in this book, as well as their diversity and their yearly distribution, reflects the continued contribution of the neutron probe to our understanding of physical systems.

This bibliography is an updated and improved version of the one first published by us in March of 1973 under a similar title. Many of the omissions and inconsistencies of the first edition, such as occurred, for example, in the initialing of authors' names, have been corrected. The literature search has been carried back to 1932, the year when the existence of the neutron was experimentally confirmed. Several additional journals have also been searched and brought up to date together with those listed in our first publication. The number of entries is now 8543, an increase of 65 per cent relative to the first edition.

This volume is divided into 25 sections and an author index, the first four sections being as follows:

1: Introduction
2: List of the Surveyed Periodicals
3: Special References
4: Books, Treatises, and Proceedings of Conferences

Section 3 contains the list of papers which deal with large numbers of substances. In these cases the detailed information is given in Section 3 while the individual substances are listed in Sections 5 or 6 as one-line entries. In Section 4 we have gathered the complete information about books and proceedings. These works are usually referred to in an abbreviated form in the main body of the bibliography.

Sections 5 and 6 constitute a major part of the bibliography; here one will find references describing elastic and inelastic scattering studies, respectively, performed on various substances. In these two sections ordering is based on the chemical formulas. In some cases, however, we have listed papers under generic names, e.g., ferromagnets, rare earths, etc., when these dealt with general categories of substances. In order to avoid artificial swelling of the bibliography, we have listed only single entries in the case of alloys and compounds; that is, no multiple entries were made by permutation of the element symbols within a formula.

Sections 7 to 24 list the theoretical and technical references. Here the emphasis is not so much on particular substances as on interpretations of neutron scattering results or on topics of a technical nature. The extensive subsectioning will, we hope, facilitate the user's search for any desired articles. In each of these sections, the references are displayed in an order based on the name of the journals and the year of publication.

Section 25 groups all the publications prior to 1945. It seemed appropriate to collect these into a single section because of their historical significance, even though we appreciate the fact that many pioneering contributions were made after this period. All references in this section are also listed elsewhere in the bibliography under various headings.

A special convention had to be adopted since computer characters do not allow one to write chemical formulas in the conventional way. Hence adjacent elements within a given formula were separated by an asterisk (*) in the case of chemical bonds and by a hyphen (-) in the case of mixtures and alloys. Names of elements found in brackets will indicate that these elements are minority constituents. In general, each record contains the title of the reference with its list of authors. Exceptions occur in the case of very short notes or in the case of some papers published in Russian. In the latter instance, the code "C.C.C.P." is placed in lieu of the author list.

The majority of records have a key indicating the type of work performed in a given reference. The scheme used goes as follows:

- A: Coherent Elastic Scattering Studies
    - A1: Crystal Structure Determination
    - A2: Magnetic Structure Determination
    - A3: Concerning Liquids and Gases
    - A4: Miscellaneous

- B: Coherent Nuclear Inelastic Scattering Studies
    - B1: Dispersion Relations for Crystals
    - B2: Anharmonic Properties of Crystals
    - B3: Dynamics of Crystals with Defects
    - B4: Dynamics of Liquids
    - B5: Dynamics of Gases
    - B6: Concerning Critical Phenomena
    - B7: Miscellaneous

- C: Incoherent Scattering Studies
    - C1: Using Subthermal Neutrons on Crystals
    - C2: Measurement of the Density of States
    - C3: Using Subthermal Neutrons on Liquids and Gases
    - C4: Determination of $S(\mathbf{Q}, \omega)$; Liquids and Gases
    - C5: Determination of $S(\mathbf{Q}, \omega)$; Solids

- D: Magnetic Scattering Studies
    - D1: Spin Waves
    - D2: Critical Scattering
    - D3: Paramagnetic Scattering
    - D4: Miscellaneous

- E: Experimental Considerations

- P: Measurement of Parameters

- T: Theoretical Communications

The author index contains the list of publications of each author accompanied by an address number. This number is actually composite: the first part refers to the section and the second part to the rank within that section. It should be noted that the author index lists only the different publications of an author and ignores any multiple references that may be made to a given publication. A multiple reference would possibly occur because of a plurality of substances studied and/or because of relevance to various sections. For operational reasons, composite authors' names were written with hyphens;

2

e.g., "Van Laar" becomes "Van-Laar." In addition, the author index procedure can only handle names of less than 17 characters; this has resulted in some particularly long names being amputated in the index. We hope that no author whose name has been so modified will be offended. Finally, the ordering of authors' names is not strictly alphabetical in the cases where names of different lengths have common initial letters; e.g., "Popovici" is listed before "Popov." Here, a word of caution would be appropriate. It is possible that different authors having the same name and initial(s) are listed herein. If such is the case, their publications will be grouped together in the index as if produced by the very same author. On the other hand, it will be noticed that occasionally a given individual is listed with differing numbers of initials. This apparent anomaly results from the fact that the names are not always written in a uniform way in the journals.

The large amount of keypunching required for this project was accomplished efficiently by two staff members of the University Computing Centre, Mrs. C. Stiles and Mrs. J. Posner; their contribution has been indispensable to the completion of this volume. Also, at several stages of the literature search we have benefited from the valuable contributions of Dr. B. N. Brockhouse and Mr. R. R. Dymond. Finally, generous consultation services were provided by Drs. M. F. Collins and W. A. Kamitakahara. We take the opportunity to thank these persons as well as those who have written to us about the first edition; needless to say, we again encourage feedback on this edition so as to be informed of any possible errors or omissions.

This project was first suggested by Dr. B. N. Brockhouse. Financial assistance has been provided under a grant from the National Research Council of Canada. The compilation was carried out on the CDC-6400 computer at McMaster University.

Finally, it would not be inappropriate to say that one of us (A.L.) owes a debt of gratitude to his co-editor, who joined the project when hopes of publishing this work were all but lost.

André Larose
Jake Vanderwal

McMaster University
August, 1974

Correspondence concerning the content of this book should be addressed to:

> Neutron Bibliography
> c/o Dr. B. N. Brockhouse
> Institute for Materials Research
> McMaster University
> Hamilton, Ontario
> Canada, L8S 4M1

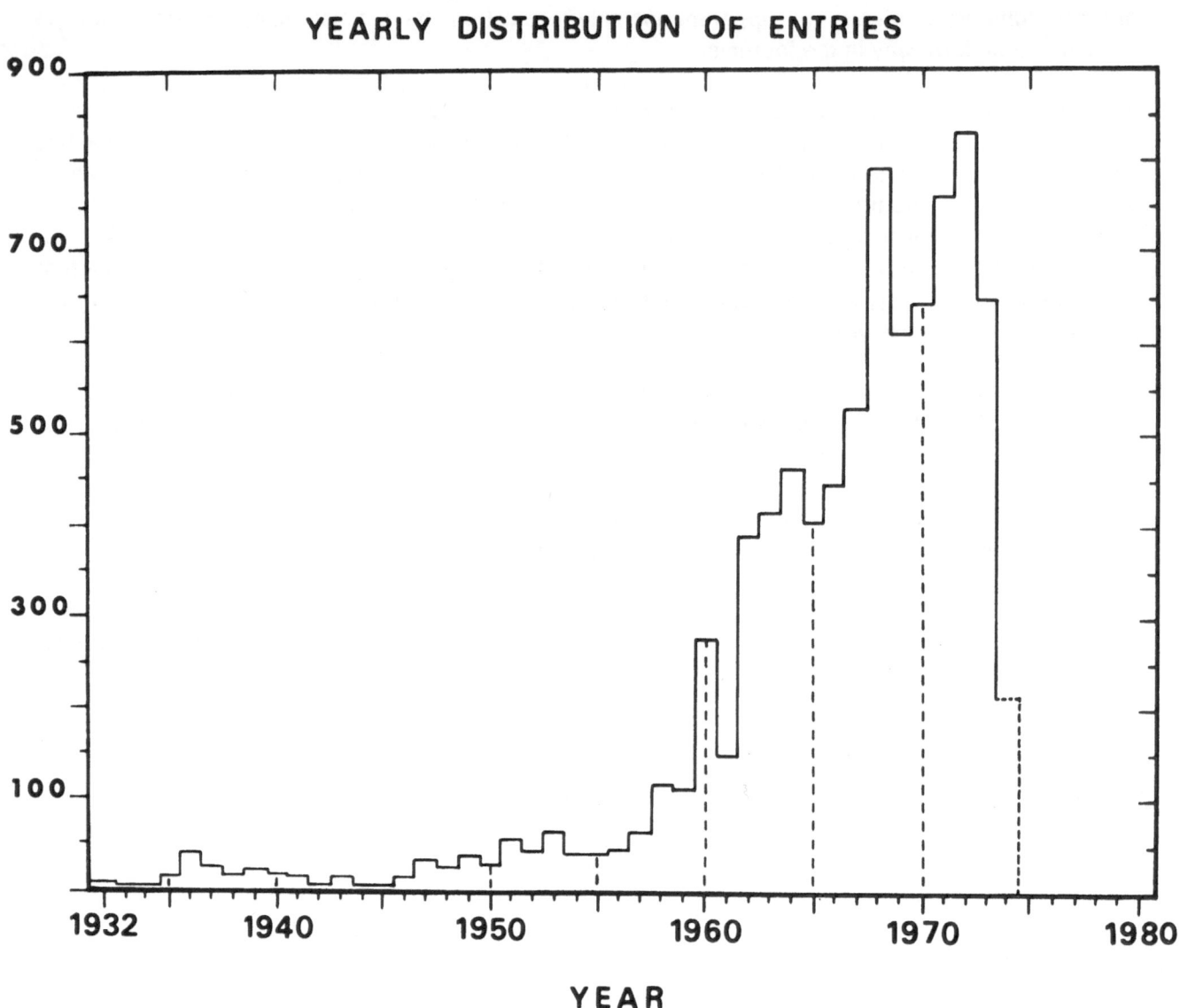

YEARLY DISTRIBUTION OF ENTRIES

YEAR

CONTENTS

SECTION   2 -LIST OF THE SURVEYED PERIODICALS . . . .

        THE BIBLIOGRAPHY WAS PREPARED ALMOST ENTIRELY BY CHECKING AGAINST
ORIGINAL JOURNALS OR BOOKS. THE JOURNALS LISTED BELOW WERE SEARCHED IN
DETAIL.   GUIDANCE   AS   TO   OTHER   PAPERS FOR INCLUSION WAS OBTAINED FROM
≠CURRENT PAPERS IN PHYSICS≠, ≠PHYSICS ABSTRACTS≠, ≠CHEMICAL ABSTRACTS≠,
OTHER BIBLIOGRAPHIES AS LISTED IN SECTION 4, AND FROM PRIVATE COMMUNI-
CATIONS. IN PARTICULAR, ≠CHEMICAL ABSTRACTS≠ WAS SEARCHED FOR THE YEARS
1932-1963, AND ≠PHYSICS ABSTRACTS≠ FOR 1958 TO MAY, 1974.

ACTA CRYSTALLOGRAPHICA . . . . MUNKSGAARD: . . . . . . .COPENHAGEN
        FROM: VOL.1, 1948. . . . . TO: . . VOL.A30, NO.3 (INCL.), 1974
                                             AND VOL.B30, NO.5 (INCL.), 1974

ANNALES DE PHYSIQUE. . . . . MASSON, CIE: . . . . PARIS
        FROM: SER.13, TOME 8, 1963 . . . . . TO: . TOME 8, NO.2 (INCL.), 1973-74

ANNALS OF PHYSICS. . . . . ACADEMIC: . . . NEW YORK AND LONDON
        FROM: VOL.21, 1963 . . . . TO: . . . . . . .VOL.83 (INCL.), 1974

CANADIAN JOURNAL OF PHYSICS. . . . . NATIONAL RESEARCH COUNCIL: .OTTAWA
        FROM: VOL.41, 1963 . . . . TO: . . .VOL.52, NO.9 (INCL.), 1974

COMPTES RENDUS/SEANCES DE L≠ACAD./SCIENCES B GAUTHIER-VILLARS: . . . . . PARIS
        FROM: TOME 256, 1963 . . . . TO: . TOME 278, NO.11 (INCL.), 1974

CRYSTAL STRUCTURE COMMUNICATIONS . . . . . . UNIV. OF PARMA: . . . . . . ITALY
        FROM: VOL.1, 1972. . . . . TO: . . VOL.3, NO.2 (INCL.), 1974

DISSERTATION ABSTRACTS . . . . XEROX (UNIV. MICROFILMS): ANN ARBOR
        FROM: VOL.1, 1938. . . . . TO: . VOL.B34, NO.10 (INCL.), 1974

JOURNAL DE PHYSIQUE. . . . . SOCIETE FRANCAISE/PHYSIQUE: . PARIS
        FROM: TOME 24, 1963. . . . . TO: . TOME 35, NO.4 (INCL.), 1974

JOURNAL OF APPLIED CRYSTALLOGRAPHY . . . . . MUNKSGAARD: . . . . . . .COPENHAGEN
        FROM: VOL.1, 1968. . . . . TO: . . VOL.7, NO.2 (INCL.), 1974

JOURNAL OF APPLIED PHYSICS . . . . A.I.P.: . .VOL.45, NO. .PENNSYLVANIA
        FROM: VOL.34, 1963 . . . . TO: . . .VOL.45, NO.6 (INCL.), 1974

JOURNAL OF CHEMICAL PHYSICS. . . . . A.I.P.: . .VOL.60, NO. .PENNSYLVANIA
        FROM: VOL.34, 1961 . . . . TO: . . .VOL.60, NO.8 (INCL.), 1974

JOURNAL OF PHYSICS C: SOLID STATE PHYSICS. . INSTITUTE OF PHYSICS: . . . .LONDON
        FROM: VOL.3, 1970 . . . . TO: . . . VOL.7, NO.8 (INCL.), 1974
        SEE ALSO: PROCEEDINGS OF THE PHYSICAL SOCIETY.

JOURNAL OF PHYSICS F: METAL PHYSICS. . . . . INSTITUTE OF PHYSICS: . . . .LONDON
        FROM: VOL.1, 1971 . . . . TO: . . . VOL.4, NO.4 (INCL.), 1974

JOURNAL OF/PHYSICS AND CHEMISTRY OF SOLIDS . PERGAMON: . . . . NEW YORK AND LONDON
        FROM: VOL.24, 1963 . . . . TO: . . .VOL.35, NO.7 (INCL.), 1974
        EXCEPT: VOL.26, NO.7.

NUCLEAR INSTRUMENTS AND METHODS. . . . . . . NORTH-HOLLAND: . . . . . . .AMSTERDAM
        FROM: VOL.20, 1963 . . . . TO: . . . . . . VOL.116 (INCL.), 1974

PHYSICA. . . . . NORTH-HOLLAND: . . . . . . .AMSTERDAM
        FROM: VOL.29, 1963 . . . . TO: . . .VOL.72, NO.2 (INCL.), 1974

PHYSICA STATUS SOLIDI. . . . . AKADEMIE-VERLAG: . . . . . . .BERLIN
        FROM: VOL.1, 1962 . . . . TO: . .VOL.A-20, NO.2 (INCL.), 1973
                                             AND VOL.B-61, NO.1 (INCL.), 1974

PHYSICAL REVIEW. . . . . A.I.P.: . . . . . .PENNSYLVANIA
        FROM: VOL.39, 1932 . . . . TO: . . VOL.A-9, NO.5 (INCL.), 1974
                                             AND VOL.B-9, NO.10(INCL.), 1974

PHYSICAL REVIEW LETTERS. . . . . AMERICAN PHYSICAL SOCIETY: NEW YORK
        FROM: VOL.10, 1963 . . . . TO: . . VOL.32, NO.22 (INCL.), 1974

PHYSICAL SOCIETY OF JAPAN (JOURNAL). . . . . PHYSICAL SOCIETY/JAPAN: . . . TOKYO
        FROM: VOL.18, 1963 . . . . TO: . . .VOL.36, NO.3 (INCL.), 1974

PHYSICS LETTERS. . . . . NORTH-HOLLAND: . . . . . . .AMSTERDAM
        FROM: VOL.1, 1962. . . . . TO: . . . . . . VOL.47A (INCL.), 1974

2 -LIST OF THE SURVEYED PERIODICALS . . . .

PROCEEDINGS OF THE PHYSICAL SOCIETY. . . . . INST. OF PHYSICS/PHYS. SOC.: LONDON
        FROM: VOL.81, 1963 . . . . TO: . . .VOL.2, SER.2 (INCL.), 1969
        SEE ALSO: J. PHYSICS C

PROCEEDINGS OF THE ROYAL SOCIETY SERIES A. . THE ROYAL SOCIETY . . . . LONDON
        FROM: VOL.271, 1963. . . . . TO: . VOL.337, NO.1608(INCL.), 1974

REVIEW OF SCIENTIFIC INSTRUMENTS . . . . . . A.I.P.: . VOL.45, NO.5 : PENNSYLVANIA
        FROM: VOL.34, 1963 . . . . TO: . . .VOL.45, NO.5 (INCL.), 1974

SOLID STATE COMMUNICATIONS . . . . PERGAMON: . . . OXFORD AND NEW YORK
        FROM: VOL.1, 1963. . . . . TO: . . .VOL.14, NO.9 (INCL.), 1974

SOVIET PHYSICS DOKLADY . . . . A.I.P., INC.: . . . . .PENNSYLVANIA
        FROM: VOL.14, 1969 . . . . TO: . . .VOL.18, NO.11 (INCL.), 1974

SOVIET PHYSICS J.E.T.P. . . . . A.I.P., INC.: . . . . .PENNSYLVANIA
        FROM: VOL.16, 1963 . . . . TO: . . .VOL.37, NO.2 (INCL.), 1973

SOVIET PHYSICS SOLID STATE . . . . A.I.P., INC.: . . . . .PENNSYLVANIA
        FROM: VOL.4, 1962 . . . . TO: . . .VOL.15, NO.9 (INCL.), 1974

SPRINGER TRACTS IN MODERN PHYSICS. . . . . . SPRINGER-VERLAG:BERLIN AND NEW YORK
        FROM: VOL.37, 1965 . . . . TO: . . . . . .VOL.69 (INCL.), 1973

        WE   HAVE   ALSO   INCLUDED IN OUR SEARCH THE MAJOR CONFERENCES SUCH AS
THE   I.A.E.A.   SYMPOSIA   ON   INELASTIC NEUTRON SCATTERING AND THE ANNUAL
A.I.P. MAGNETISM CONFERENCES. FULL DETAILS ARE PROVIDED IN SECTION 4.

SECTION   3 -SPECIAL REFERENCES . . . .

47/HAVENS. . . . . . .PHYS. REV. VOL.71. . . . . .165, (1947)
        SLOW NEUTRON VELOCITY SPECTROMETER STUDIES. II. AU, IN, TA, W, PT, ZR.
        HAVENS-JR(W.W.), WU(C.S.), RAINWATER(L.J.), MEAKER(C.L.).

47/WU. . . . . . .PHYS. REV. VOL.71. . . . . .174, (1947)
        SLOW NEUTRON VELOCITY SPECTROMETER STUDIES. III. I, OS, CO, TL, CB, GE.
        WU(C.S.), RAINWATER(L.J.), HAVENS-JR(W.W.).

48/HAVENS. . . . . . .PHYS. REV. VOL.73. . . . . .963, (1948)
        SLOW NEUTRON VELOCITY SPECTROMETER STUDIES OF CU, NI, BI, FE, SN, AND
        CALCITE.
        HAVENS-JR(W.W.), RAINWATER(L.J.), WU(C.S.), DUNNING(J.R.).

48/RAINWATER . . . . . . .PHYS. REV. VOL.73. . . . . .733, (1948)
        SLOW NEUTRON VELOCITY SPECTROMETER STUDIES OF H, D, F, MG, S, SI, AND
        QUARTZ.
        RAINWATER(L.J.), HAVENS-JR(W.W.), DUNNING(J.R.), WU(C.S.).

60/WOODS. . . . . . .IAEA SYMPOSIUM (VIENNA). 487, (1960)
        ENERGY DISTRIBUTIONS OF NEUTRONS SCATTERED FROM GRAPHITE, LIGHT AND HEA-
        VY WATER, ICE, ZR*H2, LI*H, NA*H AND N*H4*CL BY THE BE DETECTOR METHOD.
        WOODS(A.D.B.), BROCKHOUSE(B.N.), SAKAMOTO(M.), SINCLAIR(R.W.).

61/ATOJI . . . . . . .J. CHEM. PHYS. VOL.35. . . . . .1950, (1961)
        NEUTRON DIFFRACTION STUDIES OF CA*C2, Y*C2, LA*C2, CE*C2, TB*C2, YB*C2,
        LU*C2, AND U*C2.
        ATOJI(M.).

63/BRAJOVIC. . . . . . .J. PHYS. CHEM. SOL. VOL.24. . . . . . . .617, (1963)
        A STUDY OF ROTATIONAL FREEDOM IN SEVERAL AMMONIUM SALTS BY SLOW NEUTRON
        INELASTIC SCATTERING.
        BRAJOVIC(V.), BOUTIN(H.), SAFFORD(G.J.), PALEVSKY(H.).

        COMPOUNDS STUDIED: N*H4*P*F6, N*H4*(K-I), (N*H4)2*S2*O8, (N*H4)*S*O3*F.

63/CHILD . . . . . . .PHYS. REV. VOL.131. . . . . .922, (1963)
        NEUTRON DIFFRACTION INVESTIGATION OF THE MAGNETIC PROPERTIES OF COM-
        POUNDS OF RARE-EARTH METALS WITH GROUP V ANIONS.
        CHILD(H.R.), WILKINSON(M.K.), CABLE(J.W.), KOEHLER(W.C.), WOLLAN(E.O.).

        COMPOUNDS STUDIED: TB*(AS, N, P, SB), DY*N, TM*N, HO*N, HO*P, HO*SB,
        ER*(N, P, SB).

64/BERTAUT . . . . . . .PROC. INT. CONF. (NOTTINGHAM). 275, (1964)
        INVESTIGATIONS AT THE NUCLEAR CENTRE IN GRENOBLE: THE MAGNETIC STRUCTURE
        OF CR*HO*O3, MN*Y*O3, GE*M2*O4 (M=NI, CO), AND ND*CO5.
        BERTAUT(E.F.), BUISSON(G.), DELAPALME(A.), VAN-LAAR(B.), LEMAIRE(R.);
        MARESCHAL(J.), ROULT(G.), SCHWETZER(J.), VU-VAN-QUI, BARTHOLIN(H.);
        MERCIER(M.), PAUTHENET(R.).

64/CABLE . . . . . . .PHYS. REV. VOL.136. . . . . .A240, (1964)
        MAGNETIC ORDER IN RARE-EARTH INTERMETALLIC COMPOUNDS. . . . .
        CABLE(J.W.), KOEHLER(W.C.), WOLLAN(E.O.).

        COMPOUNDS STUDIED: TB*(CU, ZN, GA, AG, HG, IN)

64/JANIK . . . . . . .J. PHYS. CHEM. SOL. VOL.25. . . . . . . .1091, (1964)
        STUDY OF MOLECULAR ROTATIONS IN SOLIDS AND LIQUIDS BY THE INELASTIC
        SCATTERING OF COLD NEUTRONS.
        JANIK(J.A.), JANIK(J.M.), MELLOR(J.), PALEVSKY(H.).

        COMPOUNDS STUDIED: C*H4, C*H3*I, N*H4*CL*O4, N*H4*P*F6, H3*O*CL*O4,
        DIMETHYLACETYLENE, DIMETHOXYAZOXYBENZENE.

64/KOEHLER(1). . . . .PHYS. LETT. VOL.9. . . . . .93, (1964)
        A NOTE ON THE MAGNETIC STRUCTURES OF RARE EARTH MANGANESE OXIDES.
        KOEHLER(W.C.), YAKEL(H.L.), WOLLAN(E.O.), CABLE(J.W.).

        COMPOUNDS STUDIED: (HO,ER,TM,LU,SC)*MN*O3

64/KOEHLER(2). . . . .PROC. INT. CONF. (NOTTINGHAM). 271, (1964)
        NEUTRON MAGNETIC SCATTERING STUDIES AT THE OAK RIDGE NATIONAL LABORATORY
        KOEHLER(W.C.), CABLE(J.W.), CHILD(H.R.), MOON(R.M.), WOLLAN(E.O.).

        COMPOUNDS STUDIED: TB*(CU,ZN,AG,HG), ND*CO2, ER*CO2, (ER,DY,TB,GD)-Y,
        (TM,ER,HO,DY,TB,GD)-LA.

65/CABLE . . . . . . .PHYS. REV. VOL.138. . . . . .A755, (1965)
        DISTRIBUTION OF MAGNETIC MOMENTS IN PD-3D AND NI-3D ALLOYS.
        CABLE(J.W.), WOLLAN(E.O.), KOEHLER(W.C.).

        COMPOUNDS STUDIED: NI-CO, NI3*CO, PD-FE, PD3*FE, PD-CO, PD3*CO.

65/CHILD . . . . . . .PHYS. REV. VOL.138. . . . . .A1655, (1965)
        MAGNETIC PROPERTIES OF HEAVY RARE EARTHS DILUTED BY YTTRIUM AND LUTETIUM
        CHILD(H.R.), KOEHLER(W.C.), WOLLAN(E.O.), CABLE(J.W.).

        COMPOUNDS STUDIED: Y-(TB,DY,HO,ER,TM), TB-LU.

3 -SPECIAL REFERENCES . . . .

65/COLLINS . . . . . .PROC. PHYS. SOC. VOL.86. . . . . .535, (1965)
      THE MAGNETIC MOMENT DISTRIBUTION AROUND TRANSITION ELEMENT IMPURITIES IN
      IRON AND NICKEL.
      COLLINS(M.F.), LOW(G.G.E.).

      COMPOUNDS STUDIED: NI-(CR,V,MN,FE), FE-(NI,CO,MN,CR,V,TI,PD,RH,RU,PT,IR,
      OS,RE).

65/KOEHLER . . . . . .J. APPL. PHYS. VOL.36. . . . . .1078, (1965)
      MAGNETIC PROPERTIES OF RARE-EARTH METALS AND ALLOYS.
      KOEHLER(W.C.).

      COMPOUNDS STUDIED: CE, DY, ER, EU, GD, HO, ND, PR, SM, TB, TM.

66/BERTAUT . . . . . .J. APPL. PHYS. VOL.37. . . . . .1038, (1966)
      SOME NEUTRON-DIFFRACTION INVESTIGATIONS AT THE NUCLEAR CENTER/GRENOBLE.
      BERTAUT(E.F.),   BASSI(G.),   BUISSON(G.),   BURLET(P.),   CHAPPERT(J.),
      DELAPALME(A.), MARESCHAL(J.), ROULT(G.), ALEONARD(R.), PAUTHENET(R.),
      REBOUILLAT(J.P.).

      COMPOUNDS STUDIED: (LA,CE,PR,ND,SM,EU,GD,TB,DY,HO,ER,TM,YB,LU,Y)*C*O3,
      (FE-GA)*O3, FE*S, (CR-MN)*S.

66/KOEHLER(1). . . . .J. APPL. PHYS. VOL.37. . . . . .1259, (1966)
      NEUTRON DIFFRACTION STUDY OF DILUTE CHROMIUM-BASE ALLOYS.
      KOEHLER(W.C.), TREGO(A.L.), MOON(R.M.), MACKINTOSH(A.R.).

      COMPOUNDS STUDIED: CR-(NB,MO,TC,RU,RH,TA,W,RE).

66/KOEHLER(2). . . . .PHYS. REV. VOL.151 . . . . .405, (1966)
      ANTIFERROMAGNETISM IN CHROMIUM ALLOYS. 1. NEUTRON DIFFRACTION.
      KOEHLER(W.C.), MOON(R.M.), TREGO(A.L.), MACKINTOSH(A.R.).

      COMPOUNDS STUDIED: CR-(MN,MO,NB,RE,RH,RU,TA,TC,V,W).

66/SIDOROV . . . . . .PHYS. STAT. SOL. VOL.16. . . . . .737, (1966)
      ON THE MAGNETIC STRUCTURE OF SOME ALLOYS OF TRANSITION METALS.
      SIDOROV(S.K.), DOROSHENKO(A.V.).

      COMPOUNDS STUDIED: NI-MN, NI-FE, PD-FE.

66/VENKATARAMAN. . .J. PHYS. CHEM. SOL. VOL.27 . . . . .1103, (1966)
      STUDY OF THE ROTATIONAL BEHAVIOUR OF THE AMMONIUM ION IN SEVERAL SALTS
      BY NEUTRON SPECTROMETRY.
      VENKATARAMAN(G.), USHA-DENIZ(K.), IYENGAR(P.K.), ROY(A.P.),
      VIJAYARAGHAVAN(P.R.).

      COMPOUNDS STUDIED: N*H4*(I,CL,BR), N*D4*(CL,BR), (N*H4)2*SN*(CL6,BR6).

67/HOLDEN. . . . . . .PROC. PHYS. SOC. VOL.92. . . . . .726, (1967)
      MAGNETIZATION DISTRIBUTION ASSOCIATED WITH NON-TRANSITION METAL IMPUR-
      ITIES IN IRON.
      HOLDEN(T.M.), COMLY(J.B.), LOW(G.G.E.).

      COMPOUNDS STUDIED: FE-(AL,GA,SI,GE,SN,SB).

68/BAJOREK . . . . . .IAEA SYMPOSIUM (COPENHAGEN), VOL.II. 143, (1968)
      INVESTIGATION OF THE DYNAMICS OF WATER MOLECULES IN CRYSTALLO-HYDRATES
      BY NEUTRON INELASTIC SCATTERING.
      BAJOREK(A.), JANIK(J.A.), JANIK(J.M.), NATKANIEC(I.), PARLINSKI(K.),
      POKOTILOVSKY(YU.N.), SUDNIK-HRYNKIEWICZ(M.).

      COMPOUNDS STUDIED: H2*O   (ICE),   LI2*S*O4.H2*O,   LI*CL*O4.(H2*O)3,
      NI*S*O4.(H2*O)7, NI*S*O4.(H2*O)6, LI*CL.H2*O, LA2*MG3*(N*O3)12.(H2*O)24.

68/MARESCHAL . . . . .J. APPL. PHYS. VOL.39. . . . . .1364, (1968)
      MAGNETIC ORDERING OF TERBIUM IN SOME PEROVSKITE COMPOUNDS.
      MARESCHAL(J.), SIVARDIERE(J.), DE-VRIES(G.F.), BERTAUT(E.F.).

      COMPOUNDS STUDIED: TB*(FE,CR,AL,CO,V)*O3.

SECTION 4 -BOOKS, TREATISES, AND PROCEEDINGS OF CONFERENCES . . . . . .

1. ELASTIC NUCLEAR SCATTERING.

NEUTRON OPTICS . . . .(1954)
      HUGHES(D.J.).
      PUBL. INTERSCIENCE: NEW YORK 1954 (136 PP.)

APPLICATIONS OF NEUTRON DIFFRACTION TO SOLID STATE PROBLEMS. . . . .(1956)
      SHULL(C.G.), WOLLAN(E.O.). (IN: SOLID STATE PHYSICS VOL.2, PG.138. SEITZ
      AND TURNBULL (EDS.); PUBL. ACADEMIC: NEW YORK 1956).

I.A.E.A. SYMP. (PILE RESEARCH), VIENNA . . . .(OCT.17-21, 1960)
      PROCEEDINGS OF THE SYMPOSIUM ON PILE NEUTRON RESEARCH IN PHYSICS
      PUBL. I.A.E.A.: VIENNA 1962 (654PP.).

NEUTRON DIFFRACTION. . . . .(1962)
      BACON(G.E.).
      PUBL. CLARENDON PRESS: OXFORD 1962, 2ND EDITION (426PP.)

THE STRUCTURE OF SOLIDS. . . . .(1962)
      HAMILTON(W.C.).
      SEE: AM. REV. PHYS. CHEM. VOL.13, PG.19 (1962)

APPLICATIONS OF NEUTRON DIFFRACTION IN CHEMISTRY . . . .(1963)
      BACON(G.E.).
      PUBL. PERGAMON PRESS: NEW YORK 1963

SLOW NEUTRONS. . . . .(1965)
      TURCHIN(V.F.).
      PUBL. OLDBOURNE PRESS: LONDON 1965

SINGLE CRYSTAL DIFFRACTOMETRY. . . . .(1966)
      ARNDT(U.W.), WILLIS(B.T.M.).
      HERZENBERG, ZIMAN (EDS.); PUBL. CAMBRIDGE UNIVERSITY PRESS 1966 (331PP.)

X-RAYS AND NEUTRON DIFFRACTION . . . .(1966)
      BACON(G.E.)-ED.; PUBL. PERGAMON PRESS: NEW YORK 1966 (368PP.).

DYNAMICS AND STRUCTURE IN MOLECULAR CRYSTALS . . . .(1967)
      HAMILTON(W.C.).
      PUBL. BROOKHAVEN NATIONAL LABORATORY 11836

MOL. DYN. AND STRUCT. OF SOLIDS. . . . . (OCT.16-19, 1967)
      MOLECULAR DYNAMICS AND STRUCTURE OF SOLIDS
      NBS SPECIAL PUBLICATION 301: WASHINGTON, D.C. 1969

HARWELL SUMMER SCHOOL. . . . .(JULY 1-5, 1968)
      PROCEEDINGS OF THE INTERNATIONAL SUMMER SCHOOL AT HARWELL (1968) ON THE
      ACCURATE DETERMINATION OF NEUTRON INTENSITIES AND STRUCTURE FACTORS. (IN
      BOOK: THERMAL NEUTRON DIFFRACTION. B.T.M. WILLIS (ED.); PUBL. OXFORD
      UNIVERSITY PRESS: OXFORD 1970 (229PP.)).

HYDROGEN BONDING IN SOLIDS . . . .(1968)
      HAMILTON(W.C.), IBERS(J.A.)
      PUBL. W.A. BENJAMIN, INC.: NEW YORK 1968 (284 PP.)

LOW ENERGY NEUTRON PHYSICS . . . .(1968)
      GUREVICH(I.I.), TARASOV(L.V.).
      PUBL. NORTH-HOLLAND: AMSTERDAM; WILEY AND SONS: NEW YORK 1968 (607 PP.)

PROC/CONF/NEUTRON C.S. + TECH. . . . . (MAR.4-7, 1968)
      PROCEEDINGS OF A CONFERENCE ON NEUTRON CROSS SECTION AND TECHNOLOGY 1, 2
      NBS SPECIAL PUBLICATION 299: WASHINGTON, D.C..

STRUCTURAL CHEMISTRY IN THE NUCLEAR AGE. . . . .(1968)
      HAMILTON(W.C.).
      SEE: J. CHEM. EDUC. VOL.45, PG.296 (1968)

THE THEORY OF NEUTRON RESONANCE REACTIONS. . . . .(1968)
      LYNN(J.E.).
      PUBL. CLARENDON PRESS: OXFORD 1968 (504PP.)

THEORY OF X-RAY AND THERMAL-NEUTRON SCATTERING BY REAL CRYSTALS. . . . . .(1969)
      KRIVOGLAZ(M.A.).
      PUBL. PLENUM PRESS: NEW YORK 1969 (405PP.)

SLOW NEUTRON SCATTERING AND THERMALIZATION (WITH REACTOR APPLICATIONS) . .(1970)
      PARKS(D.E.), BEYSTER(J.R.), NELKIN(M.S.), WIKNER(N.F.).
      PUBL. W.A. BENJAMIN, INC.: NEW YORK 1970 (825PP.).

THE THEORY OF THERMAL NEUTRON SCATTERING . . . .(1971)
      SUB-TITLE: THE USE OF NEUTRONS FOR THE INVESTIGATION OF CONDENSED MATT.
      MARSHALL(W.), LOVESEY(S.W.).
      PUBL. CLARENDON PRESS: OXFORD 1971 (599PP.)

SPRINGER TRACTS IN MODERN PHYSICS VOL.64 . . . .(1972)
      QUASIELASTIC NEUTRON SCATTERING FOR THE INVESTIGATION OF DIFFUSIVE
      MOTIONS IN SOLIDS AND LIQUIDS.
      SPRINGER(T.).   PUBL.   SPRINGER-VERLAG: BERLIN,   NEW YORK (100PP.)

CHEMICAL APPLICATIONS OF THERMAL NEUTRON SCATTERING. . . . .(1973)
      WILLIS(B.T.M.)-ED.; PUBL. OXFORD UNIV. PRESS: LONDON 1973 (312 PP.).

4 -BOOKS, TREATISES, AND PROCEEDINGS OF CONFERENCES . . . . . .

2. INELASTIC NUCLEAR SCATTERING.

NEUTRON OPTICS . . . .(1954)
HUGHES(D.J.).
PUBL. INTERSCIENCE: NEW YORK 1954 (136 PP.)

USE OF SLOW NEUTRONS TO INVESTIGATE THE SOLID STATE. . . . . . .(OCT. 3-4, 1957)
PROCEEDINGS OF THE MEETING ON THE USE OF SLOW NEUTRONS TO INVESTIGATE
THE SOLID STATE, STOCKHOLM, 1957.
R. PAULI AND R. STEDMAN (EDS.).

PROC. NEUTRON THERMALIZATION CONF. . . . . (APRIL 28-30, 1958)
PROCEEDINGS OF THE NEUTRON THERMALIZATION CONFERENCE AT GATLINBURG,
TENNESSEE. PUBL. AT OAK RIDGE NATIONAL LABORATORY. REPORT ORNL-2739.

INTERACTIONS OF THERMAL NEUTRONS WITH SOLIDS . . . .(1959)
KOTHARI(L.S.),   SINGWI(K.S.). (IN: SOLID STATE PHYSICS VOL.8, PG.110.
SEITZ AND TURNBULL (EDS.); PUBL. ACADEMIC: NEW YORK 1959).

LECTURES ON NEUTRON SCATTERING/HARVARD. . . . .(1959)
MARSHALL(W.). (NOT PUBLISHED).

I.A.E.A. SYMP. (PILE RESEARCH), VIENNA . . . .(OCT.17-21, 1960)
PROCEEDINGS OF THE SYMPOSIUM ON PILE NEUTRON RESEARCH IN PHYSICS
PUBL. I.A.E.A.: VIENNA 1962 (654PP.).

I.A.E.A. SYMPOSIUM, VIENNA . . . . (OCT.11-14, 1960)
INELASTIC SCATTERING OF NEUTRONS IN SOLIDS AND LIQUIDS
PUBL. I.A.E.A.: VIENNA 1961 (647PP.)

RENSSELAER POL. INST. SYMP. . . . . (MAY 5-6, 1961)
PROCEEDINGS OF THE SYMPOSIUM AT RENSSELAER POLYTECHNIC (IN
BOOK: NEUTRON PHYSICS. M.L. YEATER (ED.): PUBL. ACADEMIC: NEW YORK AND
LONDON 1962).

I.A.E.A. SYMPOSIUM, CHALK RIVER VOLS.1AND 2. . . . .(SEPT.10-14, 1962)
INELASTIC SCATTERING OF NEUTRONS IN SOLIDS AND LIQUIDS
PUBL. I.A.E.A.: VIENNA 1963 VOL.1 (469 PP.); VOL.2 (341 PP.)

AARHUS SUMMER SCHOOL . . . .(1963)
AARHUS SUMMER SCHOOL LECTURES. (IN BOOK: PHONONS AND PHONON INTERACTIONS
THOR A. BAK (ED.); PUBL. W.A. BENJAMIN, INC.: NEW YORK 1964 (640PP.).

COLLOQUE INTERNATIONAL, GRENOBLE . . . .(SEPT.3-5, 1963)
NO.126: LA DIFFRACTION ET LA DIFFUSION DES NEUTRONS
EDITIONS DU CENTRE NATIONAL DE LA RECHERCHE SCIENTIFIQUE: PARIS (1964)
[SEE ALSO: J. PHYSIQUE, TOME 25, NO 5, PP. 425-656 (1964)]

COPENHAGEN CONF.- LATT. DYNAMICS . . . .(AUG.5-9, 1963)
PROCEEDINGS OF THE INTERNATIONAL CONFERENCE AT COPENHAGEN, DENMARK
LATTICE DYNAMICS. WALLIS (ED.); PUBL. PERGAMON PRESS: LONDON 1965

INTERN. SUMMER SCHOOL, MOL . . . . (AUG.12-31, 1963)
PROCEEDINGS OF THE INTERNATIONAL SUMMER SCHOOL ON SOLID STATE PHYSICS AT
MOL, BELGIUM. (IN BOOK: THE INTERACTION OF RADIATION WITH SOLIDS. STRU-
MANE, NIHOUL, GEVERS, AMELINCKX (EDS.), PUBL. NORTH-HOLLAND: AMST. 1964)

LATTICE VIBRATIONS . . . .(1963)
COCHRAN(W.). (SEE: REP. PROGRESS PHYS. VOL.26, NO.1).

THEORY OF LATTICE DYNAMICS IN THE HARMONIC APPROXIMATION . . . .(1963)
MARADUDIN(A.A.), MONTROLL(E.W.), WEISS(G.H.).
SEE: SOLID STATE PHYSICS (SUPPLEMENT 3). PUBL. ACADEMIC: NEW YORK AND
LONDON 1963 (319PP.).

I.A.E.A. SYMPOSIUM, BOMBAY VOLS.1 AND 2. . . . . (DEC.15-19, 1964)
INELASTIC SCATTERING OF NEUTRONS
PUBL. I.A.E.A.: VIENNA 1965 VOL.1 (460 PP.); VOL.2 (574 PP.)

INELASTIC SCATTERING OF SLOW NEUTRONS. . . . .(1964)
BROCKHOUSE(B.N.), HAUTECLER(S.), STILLER(H.). (IN BOOK: THE INTERACTION
OF RADIATION WITH SOLIDS. STRUMANE ET AL. (EDS.): NORTH-HOLLAND: AMST.).

PHONONS AND NEUTRON SCATTERING . . . .(1964)
BROCKHOUSE(B.N.). (IN: PHONONS AND PHONON INTERACTIONS, PG.221. SEE:
AARHUS SUMMER SCHOOL).

PROC. INT. CONF. (NOTTINGHAM). . . . . (SEPT., 1964)
PROCEEDINGS OF THE INTERNATIONAL CONFERENCE ON MAGNETISM/NOTTINGHAM
PUBL. INSTITUTE OF PHYSICS AND THE PHYSICAL SOCIETY IN ASSOCIATION WITH
PROCEEDINGS OF THE PHYSICAL SOCIETY: LONDON, S.W.1

BROOKHAVEN SYMPOSIUM . . . .(SEPT.20-22, 1965)
SYMPOSIUM ON INELASTIC SCATTERING OF NEUTRONS BY CONDENSED SYSTEMS
PUBL. ASSOCIATED UNIVERSITIES, INC., U.S.A. BNL 940(C-45).

NEUTRON SCATTERING BY PHONONS. . . . .(1965)
BROCKHOUSE(B.N.). (IN: PHONONS IN PERFECT LATTICES AND IN LATTICES WITH
POINT IMPERFECTIONS, PG.110. SEE: SCOTTISH UNIV. SUMMER SCHOOL).

SCOTTISH UNIV. SUMMER SCHOOL . . . .(1965)
SCOTTISH UNIVERSITIES SUMMER SCHOOL/1965. (IN BOOK: PHONONS IN PERFECT
LATTICES AND IN LATTICES WITH POINT IMPERFECTIONS. STEVENSON (ED.);
PUBL. OLIVER AND BOYD: EDINBURGH, LONDON 1966).

SLOW NEUTRONS. . . . .(1965)
TURCHIN(V.F.).
PUBL. OLDBOURNE PRESS: LONDON 1965

TH. NEUTRON SCATT. . . . .(1965)
THERMAL NEUTRON SCATTERING.
EGELSTAFF (ED.); PUBL. ACADEMIC: LONDON AND NEW YORK 1965 (523PP.).

NEUTRON SPECTROSCOPY OF SOLIDS . . . .(1966)
LOMER(W.M.). (SEE: CONTEMP. PHYSICS VOL.7, PG.278).

AN INTRODUCTION TO THE LIQUID STATE. . . . .(1967)
EGELSTAFF(P.A.).
PUBL. ACADEMIC PRESS: NEW YORK AND LONDON 1967 (236PP.)

4 -BOOKS, TREATISES, AND PROCEEDINGS OF CONFERENCES . . . . . .

ATLANTA SYMP. GEORGIA INST. TECH. . . . .(1967)
    PROCEEDINGS OF THE SYMPOSIUM ON ≠THERMAL NEUTRON SCATTERING≠ APPLIED TO
    CHEMICAL AND SOLID STATE PHYSICS≠
    TRANSACTIONS OF THE AMERICAN CRYSTALLOGRAPHIC ASSOCIATION VOL.3, 1967
    PUBL. BY: THE AMERICAN CRYSTALLOGRAPHIC ASSOCIATION.

DYNAMICS AND STRUCTURE IN MOLECULAR CRYSTALS . . . .(1967)
    HAMILTON(W.C.).
    PUBL. BROOKHAVEN NATIONAL LABORATORY 11836

EXPERIMENTAL INVESTIGATIONS OF CRITICAL PHENOMENA. . . . .(1967)
    HELLER(P.). (IN: REPORTS ON PROGRESS IN PHYSICS VOL.XXX PART II, 1967,
    PG.731).

METAL. SOC. CONF., LOS ANGELES . . . .(1967)
    METALLURGICAL SOCIETY CONFERENCES VOL.43
    MAGNETIC AND INELASTIC SCATTERING OF NEUTRONS BY METALS. ROWLAND, BECK
    (EDS.); PUBL. GORDON AND BREACH: LONDON, NEW YORK, PARIS 1968

MOL. DYN. AND STRUCT. OF SOLIDS. . . . . (OCT.16-19, 1967)
    MOLECULAR DYNAMICS AND STRUCTURE OF SOLIDS
    NBS SPECIAL PUBLICATION 301: WASHINGTON, D.C. 1969

THE THEORY OF EQUILIBRIUM CRITICAL PHENOMENA . . . .(1967)
    FISHER(M.E.). (IN: REPORTS ON PROGRESS IN PHYSICS VOL.XXX PART II, 1967,
    PG.615).

TRIESTE/INTERN. COURSE . . . .(1967)
    LECTURES PRESENTED AT AN INTERNATIONAL COURSE (OCT.3-DEC.16, 1967) AT
    TRIESTE, ITALY. (IN BOOK: THEORY OF CONDENSED MATTER. PUBL. I.A.E.A.:
    VIENNA 1968).

ANHARMONIC CRYSTALS. . . . .(1968)
    COWLEY(R.A.). (IN: REPORTS ON PROGRESS IN PHYSICS VOL.31, PART I, PG.123
    PUBL. THE INSTITUTE OF PHYSICS AND THE PHYSICAL SOCIETY: LONDON S.W.1).

ATOMIC VIBRATIONS IN METALS AND ALLOYS STUDIED BY NEUTRON SPECTROSCOPY . .(1968)
    BROCKHOUSE(B.N.), HALLMAN(E.D.), NG(S.C.). (IN BOOK: MAGNETIC AND IN-
    ELASTIC SCATTERING OF NEUTRONS BY METALS. ROWLAND, BECK (EDS.); PUBL.
    GORDON AND BREACH: NEW YORK)

C.N.E.N. SYMP., CASACCIA . . . .(SEPT.24-27, 1968)
    PROCEEDINGS OF THE SYMPOSIUM HELD AT C.N.E.N., CASACCIA CENTRE
    CURRENT PROBLEMS IN NEUTRON SCATTERING
    PUBL. COMITATO NAZIONALE ENERGIA NUCLEARE: ROME 1970

DENSITY OF STATES MEASUREMENTS FOR OPTICAL AND ACOUSTIC PHONONS BY NEUTRON-SCAT-
    TERING SPECTROSCOPY. . . . .(1968)
    WHITE(J.W.). (IN BOOK: EXCITONS, MAGNONS AND PHONONS IN MOLECULAR CRYS-
    TALS. ZAHLAN (ED.); PUBL. CAMBRIDGE PRESS: CAMBRIDGE 1968, PG.43.).

DETERMINATION OF PHONON SPECTRA. . . . .(1968)
    COWLEY(R.A.). (IN: SIMON FRASER SUMMER SCHOOL ON SOLID STATE PHYSICS,
    VOL.II- PHONONS AND THEIR INTERACTIONS. ENNS, HAERING-(EDS.); PUBL.
    GORDON AND BREACH: NEW YORK 1968).

HYDROGEN BONDING IN SOLIDS. . . . .(1968)
    HAMILTON(W.C.), IBERS(J.A.)
    PUBL. W.A. BENJAMIN, INC.: NEW YORK 1968   (284 PP.)

I.A.E.A. SYMPOSIUM, COPENHAGEN VOLS.1 AND 2. . . . . (MAY 20-25, 1968)
    NEUTRON INELASTIC SCATTERING
    PUBL. I.A.E.A.: VIENNA 1968   VOL.1 (641 PP.); VOL.2 (457 PP.)

KYOTO CONF. STAT. MECHANICS. . . . .(SEPT.9-14, 1968)
    INTERNATIONAL CONFERENCE ON STATISTICAL MECHANICS/KYOTO
    SUPPLEMENT TO JOURNAL OF THE PHYSICAL SOCIETY OF JAPAN, VOL.26, 1969.

LOW ENERGY NEUTRON PHYSICS . . . .(1968)
    GUREVICH(I.I.), TARASOV(L.V.).
    PUBL. NORTH-HOLLAND: AMSTERDAM; WILEY AND SONS: NEW YORK 1968 (607 PP.)

MOLECULAR SPECTROSCOPY WITH NEUTRONS . . . .(1968)
    BOUTIN(H.), YIP(S.).
    PUBL. M.I.T. PRESS: MASSACHUSETTS 1968

2ND SUMMER SCHOOL S.S. PHYS. . . . . (JUNE 3-13, 1968)
    THE 2ND SUMMER SCHOOL ON SOLID STATE PHYSICS AT HERCEG-NOVI, YUGOSLAVIA.
    (IN BOOK: DYNAMIC AND MAGNETIC PROPERTIES OF SOLIDS AND LIQUIDS.
    D.M. JOVIC-(ED.); PUBL. ≠BORIS KIDRIC≠ INSTITUTE OF NUCLEAR SCIENCES:
    BEOGRAD 1970   (345 PP.)).

THE THEORY OF NEUTRON RESONANCE REACTIONS. . . . .(1968)
    LYNN(J.E.).
    PUBL. CLARENDON PRESS: OXFORD 1968   (504PP.)

DETERMINATION OF PHONON AND MAGNON DISPERSION CURVES BY NEUTRON SPECTROS..(1969)
    COLLINS(M.F.). (IN BOOK: ELEMENTARY EXCITATIONS IN SOLIDS. MARADUDIN,
    NARDELLI (EDS.); PUBL. PLENUM PRESS: NEW YORK).

NEUTRON PHYSICS. . . . .(1969)
    BACON(G.E.).
    THE WYKEHAM SCIENCE SERIES FOR SCHOOL AND UNIVERSITIES
    PUBL. WYKEHAM PUBLICATIONS LTD.: LONDON 1969   (140PP.)

THE INTERACTIONS OF NEUTRONS WITH SOLIDS . . . .(1969)
    MARSHALL(W.), LOVESEY(S.W.). (IN: COMMENTS ON SOLID STATE PHYSICS VOL.2,
    NO.3, PG.88).

THEORY OF X-RAY AND THERMAL-NEUTRON SCATTERING BY REAL CRYSTALS. . . . .(1969)
    KRIVOGLAZ(M.A.).
    PUBL. PLENUM PRESS: NEW YORK 1969   (405PP.)

DYN. ASPECTS OF CRITICAL PHEN. . . . .(1970)
    DYNAMICAL ASPECTS OF CRITICAL PHENOMENA. BUDNICK, KAWATRA (EDS.).
    PUBL. GORDON AND BREACH: LONDON, ENGLAND

NEUTRON BEAM AND MAGNETIC STUDIES. . . . .(1970)
    GARDNER-(ED.). (RESEARCH REPORTS OF U.K.A.E.A. RESEARCH GROUP, APR. 1969
    TO MARCH 1970. ATOMIC ENERGY RESEARCH ESTABLISHMENT, HARWELL. REPORT
    AERE- P.R/MPD.2).

SLOW NEUTRON SCATTERING AND THERMALIZATION (WITH REACTOR APPLICATIONS) . .(1970)
    PARKS(D.E.), BEYSTER(J.R.), NELKIN(M.S.), WIKNER(N.F.).
    PUBL. W.A. BENJAMIN, INC.: NEW YORK 1970   (825PP.).

4 -BOOKS, TREATISES, AND PROCEEDINGS OF CONFERENCES . . . . . .

2ND INTERNAT. CONF./NUCL. DATA . . . . (JUNE15-19, 1970)
        SECOND INTERNATIONAL CONFERENCE ON NUCLEAR DATA FOR REACTORS
        PUBL. I.A.E.A.: VIENNA, AUSTRIA

CONF. INTERN., RENNES. . . . .(1971)
        COMPTES RENDUS DE LA CONFERENCE INTERNATIONALE- RENNES, FRANCE
        PHONONS. NUSIMOVICI (ED.); PUBL. FLAMMARION SCIENCES: PARIS

THE THEORY OF THERMAL NEUTRON SCATTERING . . . .(1971)
        SUB-TITLE: THE USE OF NEUTRONS FOR THE INVESTIGATION OF CONDENSED MATT.
        MARSHALL(W.), LOVESEY(S.W.)
        PUBL. CLARENDON PRESS: OXFORD 1971   (599PP.)

THEORY OF LATTICE DYNAMICS IN THE HARMONIC APPROXIMATION . . . .(1971)
        MARADUDIN(A.A.), MONTROLL(E.W.), WEISS(G.H.), IPATOVA(I.P.).
        SEE: SOLID STATE PHYSICS (SUPPLEMENT 3), 2ND EDITION. PUBL. ACADEMIC:
        NEW YORK AND LONDON 1971   (708PP.).

CHEMICAL APPLICATIONS OF NEUTRON SCATTERING. . . . .(1972)
        B.T.M. WILLIS-(ED.); PUBL. OXFORD UNIV. PRESS: LONDON 1972.

I.A.E.A. SYMPOSIUM, GRENOBLE . . . .(MARCH 6-10, 1972)
        NEUTRON INELASTIC SCATTERING
        PUBL. I.A.E.A.: VIENNA 1972   (888 PP.).

NUCL./S.S. PHYS. SYMP. ABSTR.. . . . .(1972)
        NUCLEAR PHYSICS AND SOLID STATE PHYSICS SYMPOSIUM ABSTRACTS
        PUBL. AT: BHABHA ATOMIC RES. CENTER, BOMBAY, INDIA.

SPRINGER TRACTS IN MODERN PHYSICS VOL.64 . . . .(1972)
        QUASIELASTIC NEUTRON SCATTERING FOR THE INVESTIGATION OF DIFFUSIVE
        MOTIONS IN SOLIDS AND LIQUIDS.
        SPRINGER(T.).   PUBL.   SPRINGER-VERLAG: BERLIN,   NEW YORK (100PP.)

CHEMICAL APPLICATIONS OF THERMAL NEUTRON SCATTERING. . . . .(1973)
        WILLIS(B.T.M.)-ED.; PUBL. OXFORD UNIV. PRESS: LONDON 1973 (312 PP.).

PROPERTIES LIQUID METALS . . . .(1973)
        THE   PROPERTIES   OF LIQUID METALS. (PROCEEDINGS OF THE 2ND INTERNATIONAL
        CONFERENCE HELD AT TOKYO, JAPAN, SEPT.3-8, 1972)
        TAKEUCHI (ED.); PUBL. TAYLOR AND FRANCIS LTD.: LONDON 1973 (640 PP.).

SPECTROSCOPY BIOL. CHEM. . . . .(1974)
        SPECTROSCOPY IN BIOLOGY AND CHEMISTRY: NEUTRON, X-RAY, LASER
        S.H.   CHEN,   S. YIP (EDS.); PUBL. ACADEMIC PRESS, INC.: NEW YORK AND
        LONDON 1974   (410 PP.)

4 -BOOKS, TREATISES, AND PROCEEDINGS OF CONFERENCES . . . . . .

3. MAGNETIC SCATTERING.

NEUTRON OPTICS . . . .(1954)
    HUGHES(D.J.).
    PUBL. INTERSCIENCE: NEW YORK 1954 (136 PP.)

APPLICATIONS OF NEUTRON DIFFRACTION TO SOLID STATE PROBLEMS . . .(1956)
    SHULL(C.G.), WOLLAN(E.O.). (IN: SOLID STATE PHYSICS VOL.2; PG.138: SEITZ
    AND TURNBULL (EDS.); PUBL. ACADEMIC: NEW YORK 1956).

USE OF SLOW NEUTRONS TO INVESTIGATE THE SOLID STATE . . . .(OCT. 3-4, 1957)
    PROCEEDINGS OF THE MEETING ON THE USE OF SLOW NEUTRONS TO INVESTIGATE
    THE SOLID STATE, STOCKHOLM, 1957.
    R. PAULI AND R. STEDMAN (EDS.).

NEUTRON DIFFRACTION. . . . .(1962)
    BACON(G.E.).
    PUBL. CLARENDON PRESS: OXFORD 1962, 2ND EDITION (426PP.)

SPIN ARRANGEMENTS IN METALS. . . . .(1963)
    NATHANS(R.), PICKART(S.J.). (IN: MAGNETISM VOL.3; PG.211: RADO AND SUHL
    (EDS.); PUBL. ACADEMIC: NEW YORK AND LONDON 1963).

THEORY OF NEUTRON SCATTERING BY MAGNETIC CRYSTALS. . . . .(1963)
    DE-GENNES(P.G.). (IN: MAGNETISM VOL.3, PG.115. RADO AND SUHL (EDS.)).

THEORY OF SCATTERING OF SLOW NEUTRONS IN MAGNETIC CRYSTALS . . .(1963)
    IZYUMOV(YU.A.). (IN: SOVIET PHYSICS USPEKHI VOL.6, PG.359.).

AN ADVANCED COURSE ON MAGNETIC EXCHANGE INTERACTIONS . . . . . .(1965)
    LOWDE(R.D.)
    PUBL. AT: INSTITUT FOR ATOMENERGIE, KJELLER, NORWAY.

TH. NEUTRON SCATT. . . . .(1965)
    THERMAL NEUTRON SCATTERING,
    EGELSTAFF (ED.); PUBL. ACADEMIC: LONDON AND NEW YORK 1965 (523PP.).

FERROMAGNETISM . . . .(1966)
    HANDBUCH DER PHYSIK VOL. XVIII, PART 2
    H.P.J. WIJN (ED.); PUBL. SPRINGER-VERLAG: BERLIN, NEW YORK 1966 (560PP.)

NEUTRON SPECTROSCOPY OF SOLIDS . . . .(1966)
    LOMER(W.M.). (SEE: CONTEMP. PHYSICS VOL.7; PG.278).

THE INVESTIGATION OF MAGNETIC STRUCTURES BY NEUTRON DIFFRACTION. .(1966)
    BACON(G.E.). (IN: ADVANCES IN STRUCTURE RESEARCH BY DIFFRACTION METHODS
    VOL.2, PG.1. BRILL, MASON (EDS.); PUBL. WILEY AND SONS: NEW YORK.).

ATLANTA SYMP. GEORGIA INST. TECH. . . . .(1967)
    PROCEEDINGS OF THE SYMPOSIUM ON ≠THERMAL NEUTRON SCATTERING≠ APPLIED TO
    CHEMICAL AND SOLID STATE PHYSICS≠
    TRANSACTIONS OF THE AMERICAN CRYSTALLOGRAPHIC ASSOCIATION VOL.3, 1967
    PUBL. BY: THE AMERICAN CRYSTALLOGRAPHIC ASSOCIATION.

EXPERIMENTAL INVESTIGATIONS OF CRITICAL PHENOMENA. . . . .(1967)
    HELLER(P.). (IN: REPORTS ON PROGRESS IN PHYSICS VOL.XXX PART II; 1967,
    PG.731).

METAL. SOC. CONF., LOS ANGELES . . . .(1967)
    METALLURGICAL SOCIETY CONFERENCES VOL.43
    MAGNETIC AND INELASTIC SCATTERING OF NEUTRONS BY METALS. ROWLAND, BECK
    (EDS.); PUBL. GORDON AND BREACH: LONDON, NEW YORK, PARIS 1968

THE THEORY OF EQUILIBRIUM CRITICAL PHENOMENA . . . .(1967)
    FISHER(M.E.). (IN: REPORTS ON PROGRESS IN PHYSICS VOL.XXX PART II; 1967,
    PG.615).

C.N.E.N. SYMP., CASACCIA . . . .(SEPT.24-27, 1968)
    PROCEEDINGS OF THE SYMPOSIUM HELD AT C.N.E.N.; CASACCIA CENTRE
    CURRENT PROBLEMS IN NEUTRON SCATTERING
    PUBL. COMITATO NAZIONALE ENERGIA NUCLEARE: ROME 1970

HARWELL SUMMER SCHOOL. . . . .(JULY 1-5, 1968)
    PROCEEDINGS OF THE INTERNATIONAL SUMMER SCHOOL AT HARWELL (1968) ON THE
    ACCURATE DETERMINATION OF NEUTRON INTENSITIES AND STRUCTURE FACTORS. (IN
    BOOK: THERMAL NEUTRON DIFFRACTION. B.T.M. WILLIS (ED.); PUBL. OXFORD
    UNIVERSITY PRESS: OXFORD 1970 (229PP.)).

LOW ENERGY NEUTRON PHYSICS . . . .(1968)
    GUREVICH(I.I.), TARASOV(L.V.).
    PUBL. NORTH-HOLLAND: AMSTERDAM; WILEY AND SONS: NEW YORK 1968 (607 PP.)

MAGNETIC CORRELATIONS AND NEUTRON SCATTERING . . . .(1968)
    MARSHALL(W.), LOWDE(R.D.). (IN: REPORTS ON PROGRESS IN PHYSICS VOL.31,
    PART II, PG.705.).

MAGNETIC PROPERTIES OF RARE EARTH METALS . . . .(1968)
    COOPER(B.R.). (IN: SOLID STATE PHYSICS VOL.21; PG.393).

PROC/CONF/NEUTRON C.S. + TECH. . . . .(MAR.4-7, 1968)
    PROCEEDINGS OF A CONFERENCE ON NEUTRON CROSS SECTION AND TECHNOLOGY 1, 2
    NBS SPECIAL PUBLICATION 299: WASHINGTON, D.C..

2ND SUMMER SCHOOL S.S. PHYS. . . . .(JUNE 3-13, 1968)
    THE 2ND SUMMER SCHOOL ON SOLID STATE PHYSICS AT HERCEG-NOVI, YUGOSLAVIA.
    (IN BOOK: DYNAMIC AND MAGNETIC PROPERTIES OF SOLIDS AND LIQUIDS.
    D.M. JOVIC-(ED.); PUBL. ≠BORIS KIDRIC≠ INSTITUTE OF NUCLEAR SCIENCES:
    BEOGRAD 1970 (345 PP.)).

DETERMINATION OF PHONON AND MAGNON DISPERSION CURVES BY NEUTRON SPECTROS..(1969)
    COLLINS(M.F.). (IN BOOK: ELEMENTARY EXCITATIONS IN SOLIDS. MARADUDIN,
    NARDELLI (EDS.); PUBL. PLENUM PRESS: NEW YORK).

NEUTRON PHYSICS. . . . .(1969)
    BACON(G.E.).
    THE WYKEHAM SCIENCE SERIES FOR SCHOOL AND UNIVERSITIES
    PUBL. WYKEHAM PUBLICATIONS LTD.: LONDON 1969 (140PP.)

TABLE OF ANTIFERROMAGNETIC MATERIALS STUDIED BY NEUTRON DIFFRACTION. . . .(1969)
    COX(D.E.).
    PUBL. BROOKHAVEN NATIONAL LABORATORY 13822 (1969).

4 -BOOKS, TREATISES, AND PROCEEDINGS OF CONFERENCES . . . . . .

THE THEORY OF ELASTIC SCATTERING OF NEUTRONS BY MAGNETIC SALTS . . . .(1969)
        LOVESEY(S.W.), RIMMER(D.E.). (IN: REPORTS ON PROGRESS IN PHYSICS VOL.32,
        PART I, PG.333.).

DYN. ASPECTS OF CRITICAL PHEN. . . . .(1970)
        DYNAMICAL   ASPECTS   OF   CRITICAL   PHENOMENA.   BUDNICK,   KAWATRA   (EDS.).
        PUBL. GORDON AND BREACH: LONDON, ENGLAND

MAGNETIC NEUTRON DIFFRACTION . . . .(1970)
        IZYUMOV(YU.A.), OZEROV(R.P.).
        TRANSL. BY ABRAHAMS(S.C.); PUBL. PLENUM PRESS: NEW YORK 197J (593PP.)

CONF./RARE EARTHS/ACTINIDES. . . . .(1971)
        CONFERENCE   ON   RARE   EARTHS   AND   ACTINIDES   AT   DURHAM,   ENGLAND,   JULY,   1971
        PUBL. INST. PHYS.: LONDON, ENGLAND 1971

THE THEORY OF THERMAL NEUTRON SCATTERING . . . .(1971)
        SUB-TITLE: THE USE OF NEUTRONS FOR THE INVESTIGATION OF CONDENSED MATT.
        MARSHALL(W.), LOVESEY(S.W.),
        PUBL. CLARENDON PRESS: OXFORD 1971   (599PP.)

AIP CONF. PROC. NO.5, PARTS 1,2. . . . .(1972)
        MAGNETISM   AND MAGNETIC MATERIALS- 1971 (17TH ANNUAL CONFERENCE-CHICAGO)
        GRAHAM, RHYNE (EDS.); PUBL. A.I.P.: NEW YORK 1972

NEUTRON DIFFRACTION DETERMINATION OF MAGNETIC STRUCTURES . . . .(1972)
        COX(D.E.). (IN: IEEE TRANS. MAGN. VOL.8, PG.161 1972).

NUCL./S.S. PHYS. SYMP. ABSTR. . . . .(1972)
        NUCLEAR PHYSICS AND SOLID STATE PHYSICS SYMPOSIUM ABSTRACTS
        PUBL. AT: BHABHA ATOMIC RES. CENTER, BOMBAY, INDIA.

A.I.P. CONF. PROC. NO.10, PARTS 1,2. . . . .(1973)
        MAGNETISM   AND MAGNETIC MATERIALS- 1972 (18TH ANNUAL CONFERENCE- DENVER)
        GRAHAM, RHYNE (EDS.); PUBL. A.I.P.: NEW YORK 1973

4 -BOOKS, TREATISES, AND PROCEEDINGS OF CONFERENCES . . . . . .

4. TECHNIQUES.

PILE NEUTRON RESEARCH. . . . .(1953)
HUGHES(D.J.).
PUBL. ADDISON-WESLEY, INC.: CAMBRIDGE, MASS. 1953

NEUTRON OPTICS . . . .(1954)
HUGHES(D.J.).
PUBL. INTERSCIENCE: NEW YORK 1954 (136 PP.)

METHODS FOR NEUTRON SPECTROMETRY . . . . : I.A.E.A. SYMPOSIUM, VIENNA .(1960)
BROCKHOUSE(B.N.). (IN: PROCEEDINGS I.A.E.A. SYMPOSIUM, VIENNA (1960),
PG. 113.).

I.A.E.A. SYMP. (PILE RESEARCH), VIENNA . . . .(OCT.17-21, 1960)
PROCEEDINGS OF THE SYMPOSIUM ON PILE NEUTRON RESEARCH IN PHYSICS
PUBL. I.A.E.A.: VIENNA 1962 (654PP.)

RENSSELAER POL. INST. SYMP. . . . .(MAY 5-6, 1961)
PROCEEDINGS OF THE SYMPOSIUM AT RENSSELAER POLYTECHNIC INSTITUTE. (IN
BOOK: NEUTRON PHYSICS. M.L. YEATER (ED.); PUBL. ACADEMIC: NEW YORK AND
LONDON 1962).

INELASTIC SCATTERING OF SLOW NEUTRONS . . . .(1964)
BROCKHOUSE(B.N.), HAUTECLER(S.), STILLER(H.). (IN BOOK: THE INTERACTION
OF RADIATION WITH SOLIDS. STRUMANE ET AL. (EDS.); NORTH-HOLLAND: AMST.).

RECENT METHODS IN CRYSTAL SPECTROMETRY . . . .(1964)
IYENGAR(P.K.). (IN: PROCEEDINGS I.A.E.A. SYMPOSIUM, BOMBAY (1964),
VOL.II, PG.483.).

TH. NEUTRON SCATT. . . . .(1965)
THERMAL NEUTRON SCATTERING.
EGELSTAFF (ED.); PUBL. ACADEMIC: LONDON AND NEW YORK 1965 (523PP.).

THE INVESTIGATION OF MAGNETIC STRUCTURES BY NEUTRON DIFFRACTION. . . . . .(1966)
BACON(G.E.). (IN: ADVANCES IN STRUCTURE RESEARCH 3Y DIFFRACTION METHODS
VOL.2, PG.1. BRILL, MASON (EDS.); PUBL. WILEY AND SONS: NEW YORK.).

LOW ENERGY NEUTRON PHYSICS . . . .(1968)
GUREVICH(I.I.), TARASOV(L.V.);
PUBL. NORTH-HOLLAND: AMSTERDAM; WILEY AND SONS: NEW YORK 1968 (607 PP.)

I.A.E.A. SYMP. (INSTRUMENT.), VIENNA . . . .(DEC.1-5, 1969)
INSTRUMENTATION FOR NEUTRON INELASTIC SCATTERING RESEARCH
PUBL. I.A.E.A.: VIENNA 1970 (299PP.).

NEUTRON PHYSICS. . . . .(1969)
BACON(G.E.)
THE WYKEHAM SCIENCE SERIES FOR SCHOOL AND UNIVERSITIES
PUBL. WYKEHAM PUBLICATIONS LTD.: LONDON 1969 (140PP.)

NEUTRON BEAM AND MAGNETIC STUDIES. . . . .(1970)
GARDNER-(ED.). (RESEARCH REPORTS OF U.K.A.E.A. RESEARCH GROUP, APR. 1969
TO MARCH 1970. ATOMIC ENERGY RESEARCH ESTABLISHMENT, HARWELL. REPORT
AERE- P.R/MPD.2).

SLOW NEUTRON SCATTERING AND THERMALIZATION (WITH REACTOR APPLICATIONS) . .(1970)
PARKS(D.E.), BEYSTER(J.R.), NELKIN(M.S.), WIKNER(N.F.).
PUBL. W.A. BENJAMIN, INC.: NEW YORK 1970 (825PP.).

2ND INTERNAT. CONF./NUCL. DATA . . . .(JUNE15-19, 1970)
SECOND INTERNATIONAL CONFERENCE ON NUCLEAR DATA FOR REACTORS
PUBL. I.A.E.A.: VIENNA, AUSTRIA

5. BIBLIOGRAPHIES.

NEUTRON CROSS SECTIONS (2ND ED.) . . . .(1958)
BROOKHAVEN NATIONAL LABORATORY, REPORT BNL 325: COMPILED BY HUGHES(D.J.)
AND SCHWARTZ(R.B.). (AVAIL. SUPERINTENDENT OF DOCUMENTS, U.S. GOVERNMENT
PRINTING OFFICE, WASHINGTON 25, D.C.)

NEUTRON CROSS SECTIONS (2ND ED.) . . . .(1966)
BNL 325, SUPPLEMENT NO.2: VOLS. IIA,IIB,IIC. (AVAIL. CLEARINGHOUSE FOR
SCIENTIFIC AND TECHNICAL INFO., NATIONAL BUREAU OF STANDARDS, U.S. DEPT.
OF COMMERCE, SPRINGFIELD, VIRGINIA)

METHODS OF OBTAINING MONOCHROMATIC X-RAYS AND NEUTRONS . . . .(1967)
INTERNATIONAL UNION OF CRYSTALLOGRAPHY- COMMISSION ON CRYSTALLOGRAPHIC
APPARATUS, BIBLIOGRAPHY NO.3: ED. BY HERBSTEIN(F.H.). (AVAIL. OOSTHOEK:S
UITGEVERS MIJ N.V., DOMSTRAAT 11-13, UTRECHT, THE NETHERLANDS)

TABLES OF MAGNETIC STRUCTURES DETERMINED BY NEUTRON DIFFRACTION. . . .(1970-1971)
INST. OF NUCLEAR TECHNIQUES- CRACOW; PARTS 1-5, REPORT NOS. 17PS-11/PS.
(AVAIL. NUCLEAR ENERGY INFO. CENTRE, PALACE OF CULTURE AND SCIENCE,
WARSAW, POLAND)

BIBL. FOR THERMAL NEUTRON SCATTERING (3RD ED.) . . . .(1971)
THE THERMALIZATION GROUP, JAPANESE NUCLEAR DATA COMMITTEE, JAPAN ATOMIC
ENERGY RESEARCH INSTITUTE. REPORT JAERI 4043. (AVAIL. DIVISION OF TECH.
INFORMATION, JAERI, TOKAI-MURA, NAKA-GUN, IBARAKI-KEN, JAPAN).

BIBL. FOR THERMAL NEUTRON SCATTERING (3RD ED.) . . . .(1971)
SUPPLEMENT 1. REPORT JAERI-M 4666; IBID.

BIBL. FOR THERMAL NEUTRON SCATTERING (4TH ED.) . . . .(1973)
REPORT JAERI-M 5395; IBID.

MAGNETIC STRUCTURE DATA SHEETS VOL.1 . . . .(1972)
NEUTRON DIFFRACTION COMMISSION, PHYSICS DEPT.; BROOKHAVEN NATIONAL
LABORATORY, UPTON, NEW YORK 11973, U.S.A..

SECTION 5 -STRUCTURE AND CROSS-SECTION DETERMINATIONS . . . .

1 ACID/ACETIC, ...A1 . . . . . . .ACTA CRYST. VOL.B27, ... 893, (1971).
 *HYDROGEN BOND STUDIES.XLIV. NEUTRON DIFFRACTION STUDY OF ACETIC ACID.
 *JONSSON(P.G.).

2 ACID/ACETIC-PHOSPHORIC, ...A1 . . . . . . .ACTA CHEM. SCAND. VOL.26, ... 1599, (1972).
 *HYDROGEN BOND STUDIES. LVI.NEUTRON AND X-RAY DIFFRACTION STUDIES OF THE
 1:1 ADDITION COMPOUND OF ACETIC ACID WITH PHOSPHORIC ACID.
 *JONSSON(P.G.).

3 ACID/ALPHA-OXALIC DIHYDRATE, ...A1 . . . . . . .ACTA CRYST. VOL.B25, ... 2437, (1969).
 *A NEUTRON DIFFRACTION STUDY OF ALPHA-OXALIC ACID DIHYDRATE.
 *SABINE(T.M.), COX(G.W.), CRAVEN(B.M.).

4 ACID/AMINO, ...A1 . . . . . . .ACTA CRYST. A28 PART S-4, ... 193, (1972).
 *PRECISION NEUTRON DIFFRACTION STRUCTURE DETERMINATION OF ALPHA-AMINO
 ACIDS.
 *KOETZLE(T.F.), FREY(M.N.), GOLIC(L.), HAMILTON(W.C.), JONSSON(P.G.),
 KVICK(A.), LEHMANN(M.S.), VERBIST(J.J.).

5 ACID/CYANURIC, ...A4 . . . . . . .ACTA CRYST. VOL.B27, ... 146, (1971).
 *ELECTRON DENSITY DISTRIBUTION IN CYANURIC ACID. II. NEUTRON DIFFRACTION
 STUDY AT 78K AND COMPARISON OF X-RAY AND NEUTRON DIFFRACTION RESULTS.
 *COPPENS(P.), VOS(A.).

6 ACID/D-TARTARIC, ...A1 . . . . . . .ACTA CRYST. 21, ... 237, (1966).
 *REFINEMENT OF THE STRUCTURE OF D-TARTARIC ACID BY X-RAY AND NEUTRON
 DIFFRACTION.
 *OKAYA(Y.), STEMPLE(N.R.), KAY(M.I.).

7 ACID/GLYCOLIC, ...A1 . . . . . . .ACTA CRYST. VOL.B27, ... 333, (1971).
 *GLYCOLIC ACID: DIRECT NEUTRON DIFFRACTION DETERMINATION OF CRYSTAL
 STRUCTURE AND THERMAL MOTION ANALYSIS.
 *ELLISON(R.D.), JOHNSON(C.K.), LEVY(H.A.).

8 ACID/GLYCINE(ALPHA) . . . . .A1 . . . . . . .ACTA CRYST. VOL.B28, ... 1827, (1972).
 *PRECISION NEUTRON DIFFRACTION STRUCTURE DETERMINATION OF PROTEIN AND
 NUCLEIC ACID COMPONENTS. III.THE CRYST./MOLEC. STRUC. OF/ALPHA-GLYCINE.
 *JONSSON(P.G.), KVICK(A.).

9 ACID/GLYCINE(ALPHA), ...A1 . . . . . . .J. CHEM. PHYS. VOL.59, ... 3901, (1973).
 *HYDROGEN BOND STUDIES. 77.ELECTRON DENSITY DISTRIBUTION IN ALPHA-
 GLYCINE: X-N DIFFERENCE FOURIER SYNTHESIS VS. AB INITIO CALCULATIONS.
 *ALMLOF(J.), KVICK(A.), THOMAS(J.O.).

10 ACID/L-ALANINE, ...A1 . . . . . . .J. AMER. CHEM. SOC. VOL.94, ... 2657, (1972).
 *PRECISION NEUTRON DIFFRACTION STRUCTURE DETERMINATION OF PROTEIN/NUCLEIC
 ACID COMP. I.THE CRYST./MOLECULAR STRUCTURE OF/AMINO ACID L-ALANINE.
 *LEHMANN(M.S.), KOETZLE(T.F.), HAMILTON(W.C.).

11 ACID/L-ARGININE DIHYDRATE, ...A1 . . . . . . .J.C.S. PERKIN TRANS. II, ... 133, (1973).
 *PRECISION NEUTRON DIFFRACTION STRUC. DETERMINATION OF PROTEIN/NUCLEIC ACID
 COMP. V.CRYSTAL AND MOLECULAR STRUCTURE OF/AMINO ACID L-ARGININE DIHYDP.
 *LEHMANN(M.S.), VERBIST(J.J.), HAMILTON(W.C.), KOETZLE(T.F.).

12 ACID/L-ASCORBIC, ...A1 . . . . . . .ACTA CRYST. B24, ... 1431, (1968).
 *THE CRYSTAL STRUCTURE OF L-ASCORBIC ACID, VITAMIN C. II. THE NEUTRON
 DIFFRACTION ANALYSIS.
 *HVOSLEF(J.).

13 ACID/L-ASPARAGINE MONOHYDRATE, ...A1 . . . . . . .ACTA CRYST. VOL.B28, ... 3006, (1972).
 *PRECISION NEUTRON DIFFRACTION STRUC. DETERMINATION OF PROTEIN/NUCLEIC
 ACID COMP. VI.THE CRYST./MOLEC. STRUC. OF/L-ASPARAGINE MONOHYDRATE.
 *VERBIST(J.J.), LEHMANN(M.S.), KOETZLE(T.F.), HAMILTON(W.C.).

14 ACID/L-ASPARAGINE MONOHYDRATE, ...A1 . . . . . . .ACTA CRYST. VOL.B28, ... 3000, (1972).
 *STRUCTURE OF L-ASPARAGINE MONOHYDRATE BY NEUTRON DIFFRACTION.
 *RAMANADHAM(M.), SIKKA(S.K.), CHIDAMBARAM(R.).

15 ACID/L-CYSTEIC.H2*O, ...A4 . . . . . . .ACTA CRYST. VOL.B29, ... 1167, (1973).
 *REFINEMENT OF HYDROGEN-ATOM POSITIONS IN L-CYSTEIC ACID.H2*O FROM
 NEUTRON DIFFRACTION DATA.
 *RAMANADHAM(M.), SIKKA(S.K.), CHIDAMBARAM(R.).

16 ACID/L-CYSTINE.2H*CL, ...A1 . . . . . . .ACTA CRYST. VOL.B30, ... 562, (1974).
 *A NEUTRON DIFFRACTION STUDY OF THE STRUCTURE OF L-CYSTINE.2H*CL.
 *GUPTA(S.C.), SEQUEIRA(A.), CHIDAMBARAM(R.).

17 ACID/L-CYSTINE DIHYDROCHLORIDE, ...A1 . . . . . . .ACTA CRYST. VOL.B30, ... 1220, (1974).
 *PRECISION NEUTRON DIFFRACTION STRUCTURE DETERMINATION OF PROTEIN/NUCLEIC
 ACID COMPONENTS. XIV. CRYSTAL/MOLECULAR STRUCTURE OF THE AMINO ACID ...
 *JONES(D.D.), BERNAL(I.), FREY(M.N.), KOETZLE(T.F.).

18 ACID/L-GLUTAMIC HYDROCHLORIDE, ...A1 . . . . . . .ACTA CRYST. VOL.A28 PART S-4, ... 193, (1972).
 *A NEUTRON DIFFRACTION STUDY OF THE STRUCTURE OF L-GLUTAMIC ACID HYDRO-
 CHLORIDE.
 *SEQUEIRA(A.), RAJAGOPAL(H.), CHIDAMBARAM(R.). (ABSTRACT ONLY).

19 ACID/L-GLUTAMIC HYDROCHLORIDE, ...A1 . . . . . . .ACTA CRYST. VOL.B28, ... 2514, (1972).
 *A NEUTRON DIFFRACTION STUDY OF THE STRUCTURE OF L-GLUTAMIC ACID.H*CL.
 *SEQUEIRA(A.), RAJAGOPAL(H.), CHIDAMBARAM(R.).

20 ACID/L-GLUTAMIC (BETA) . . . . .A1 . . . . . . .J. CRYST. MOL. STRUC. VOL.2, ... 225, (1972).
 *PRECISION NEUTRON DIFFRACTION STRUC. DETERMINATION OF PROTEIN/NUCLEIC
 ACID COMP. VIII: CRYST./MOL. STRUC./BETA-FORM/AMINO ACID L-GLUTAMIC ACID
 *LEHMANN(M.S.), KOETZLE(T.F.), HAMILTON(W.C.).

21 ACID/L-GLUTAMINE, ...A1 . . . . . . .ACTA CRYST. VOL.B29, ... 2571, (1973).
 *PRECISION NEUTRON DIFFRACTION STRUC. DETERMINATION OF PROTEIN/NUCLEIC
 ACID COMP. XIII. MOLECULAR/CRYSTAL STRUCTURE/AMINO ACID L-GLUTAMINE.
 *KOETZLE(T.F.), FREY(M.N.), LEHMANN(M.S.), HAMILTON(W.C.).

22 ACID/L-HISTIDINE, ...A1 . . . . . . .INT. J. PEPTIDE PROTEIN RES. 4, . . . 229, (1972).
 *PRECISION NEUTRON DIFFRACTION STRUCTURE DETERMINATION OF PROTEIN/NUCLEIC
 ACID COMP. IV. CRYST./MOLECULAR STRUCTURE OF/AMINO ACID L-HISTIDINE.
 *LEHMANN(M.S.), KOETZLE(T.F.), HAMILTON(W.C.).

23 ACID/L-LYSINE MONOHYDROCHLORIDE, ...A1 . . . . . . .ACTA CRYST. VOL.B28, ... 3214, (1972).
 *A NEUTRON DIFFRACTION STUDY OF THE STRUCTURE OF L-LYSINE MONOHYDRO-
 CHLORIDE DIHYDRATE.
 *BUGAYONG(R.R.), SEQUEIRA(A.), CHIDAMBARAM(R.).

SECTION   5 -STRUCTURE AND CROSS-SECTION DETERMINATIONS . . . .

24 ACID/L-LYSINE MONOHYDROCHLORIDE,....A1........ACTA CRYST. VOL.B28,........ 3207, (1972).
    *PRECISION NEUTRON DIFFRACTION STRUCTURE DETERMINATION OF PROTEIN AND
    NUCLEIC ACID COMP. VII.CRYST./MOL. STRUC. OF/AMINO ACID...DIHYDRATE.
    *KOETZLE(T.F.), LEHMANN(M.S.), VERBIST(J.J.), HAMILTON(W.C.).

25 ACID/L-TYROSINE, ...A1........J. CHEM. PHYS. VOL.58 NO.6,.... 2547, (1973).
    *PRECISION NEUTRON DIFF. STRUCT. DETERMINATION/PROTEIN/NUCLEI ACID COMP.:
    X.A COMP. BETWEEN/CRYST/MOLEC. STRUCT. OF L-TYROSINE/L-TYROSINE-H*CL:
    *FREY(M.N.), KOETZLE(T.F.), LEHMANN(M.S.), HAMILTON(W.C.).

26 ACID/NITRANILIC HEXAHYDRATE,...A1.....ACTA CRYST. (INTERACT.) VOL.A25, S113, (1969).
    *AN UNSYMMETRIC (H2*O*H*O*H2)+ ION: A NEUTRON DIFFRACTION STUDY OF
    NITRANILIC ACID HEXAHYDRATE.
    *WILLIAMS(J.M.), PETERSON(S.W.).

27 ACID/ORTHANILIC, ...A1.......ACTA CRYST. 22,.......... 216, (1967).
    *THE CRYSTAL STRUCTURE OF ORTHANILIC ACID.
    *HALL(S.R.), MASLEN(E.N.).

28 ACID/OXALIC DIHYDRATE, ...A4.......ACTA CRYST. VOL.B25,....... 2451, (1969).
    *AN EXPERIMENTAL DETERMINATION OF THE ASPHERICITY OF THE ATOMIC CHARGE
    DISTRIBUTION IN OXALIC ACID DIHYDRATE.
    *COPPENS(P.), SABINE(T.M.), DELAPLANE(R.G.), IBERS(J.A.).

29 ACID/OXALIC DIHYDRATE,.......A4.......ACTA CRYST. VOL.B25,....... 2442, (1969).
    *NEUTRON DIFFRACTION STUDY OF HYDROGEN BONDING AND THERMAL MOTION IN
    DEUTERATED ALPHA AND BETA OXALIC ACID DIHYDRATE.
    *COPPENS(P.), SABINE(T.M.).

30 ACIDS/AMINO, .. *.........MATER. RES. BULL. VOL.7,.... 1225, (1972).
    *STEREOSCOPIC ATLAS OF AMINO ACID STRUCTURES.
    *HAMILTON(W.C.), FREY(M.N.), GOLIC(L.), KOETZLE(T.F.), LEHMANN(M.S.),
    VERBIST(J.J.).

31 ACID/SALICYLIC,.. ....A1.......ACTA CRYST. VOL.A28 PART S-4,.... 193, (1972).
    *NEUTRON DIFFRACTION STUDIES OF SALICYLIC ACID AND ALPHA-RESORCINOL.
    *BACON(G.E.), JUDE(R.J.). (ABSTRACT ONLY).

32 ACID/SALICYLIC,.. ....A1.......PURE APPL. CHEM. VOL.18,.... 517, (1969).
    *NEUTRON DIFFRACTION STUDIES OF ORGANIC MOLECULES.
    *BACON(G.E.).

33 ACID/SERINE (C3*H7*N*O3),.....A1.....ACTA CRYST. VOL.B29,........ 876, (1973).
    *PRECISION NEUTRON DIFF. STRUC. DETERMINATION OF PROTEIN/NUCLEIC ACID
    COMP. XI.MOL. CONFIG./H-BONDING OF SERINE IN/L-SERINE.H2*O AND DL-SERINE
    *FREY(M.N.), LEHMANN(M.S.), KOETZLE(T.F.), HAMILTON(W.C.).

34 ACID/SULFAMIC, ...A1.......ACTA CRYST. VOL.13,........ 320, (1960).
    *A NEUTRON DIFFRACTION STUDY ON THE CRYSTAL STRUCTURE OF SULFAMIC ACID.
    *SASS(R.L.).

35 ACID/TRICHLOROACETIC,.....A4.......J. CHEM. PHYS. VOL.56,....... 4433, (1972).
    *HYDROGEN BOND STUDIES. LX. A SINGLE CRYSTAL NEUTRON DIFFRACTION STUDY
    OF TRICHLOROACETIC ACID DIMER.
    *JONSSON(P.G.), HAMILTON(W.C.).

36 ACID/UREA-PHOSPHORIC,........A1.......ACTA CRYST. VOL.B28,...... 2454, (1972).
    *A SINGLE-CRYSTAL NEUTRON-DIFFRACTION STUDY OF UREA-PHOSPHORIC ACID.
    *KOSTANSEK(E.C.), BUSING(W.R.).

37 ACID/VIOLURIC-ACID-MONOHYDRATE,....A1.......ACTA CRYST. 17,...... 415, (1964).
    *THE CRYSTAL STRUCTURE OF PERDEUTERATED VIOLURIC ACID MONOHYDRATE:
    THE NEUTRON DIFFRACTION ANALYSIS.
    *CRAVEN(B.M.), TAKEI(W.J.).

38 ACID/VITAMIN B12 DERIVATIVE, ...A1.......NATURE VOL.214,..... 130, (1967).
    *CRYSTAL AND MOLECULAR STRUCTURE FROM NEUTRON DIFFRACTION ANALYSIS:
    *MOORE(F.M.), WILLIS(B.T.M.), HODGKIN(D.C.).

39 AG,. . . . .P.......PROC. PHYS.-MATH. SOC. JAPAN 22, 391, (1940).
    *SCATTERING OF SLOW NEUTRONS BY SOME ELEMENTS.
    *KIMURA(M.).

40 AG*BI*SE2/AG*BI*S2,......A4.......J. MATERIALS SCI. VOL.3,........ 498, (1968).
    *X-RAY AND NEUTRON DIFFRACTION STUDIES OF THE HIGH-TEMPERATURE BETA-PHASE
    OF THE AG*BI*SE2/AG*BI*S2 SYSTEM.
    *GLATZ(A.C.), PINELLA(A.).

41 AG*CR*SE2, ...A1,A2....J. SOLID STATE CHEM. VOL.6,.... 574, (1973).
    *CRYSTAL STRUCTURES AND MAGNETIC STRUCTURES OF SOME METAL(I) CHROMIUM
    (III) SULFIDES AND SELENIDES.
    *ENGELSMAN(F.M.R.), WIEGERS(G.A.), JELLINEK(F.), VAN-LAAR(B.).

42 AG*F2, ...A1,A2....J. PHYS. CHEM. SOLIDS VOL.32,.... 1641, (1971).
    *CRYSTAL AND MAGNETIC STRUCTURE OF SILVER DIFLUORIDE-II. WEAK 4D-FERRO-
    MAGNETISM OF AG*F2.
    *FISCHER(P.), ROULT(G.), SCHWARZENBACH(D.).

43 AG*F2, ...A1,A2....J. PHYS. CHEM. SOLIDS VOL.32,.... 543, (1971).
    *CRYSTAL AND MAGNETIC STRUCTURE OF SILVER DIFLUORIDE I. DETERMINATION OF
    THE AG*F2 STRUCTURE.
    *FISCHER(P.), SCHWARZENBACH(D.), RIETVELD(H.M.).

44 AG*F2, ...A1,A2....BULL. SOC. FRA. MINER. CRIST. 93 7, (1970).
    *A NEUTRON DIFFRACTION STUDY OF THE CRYSTAL AND MAGNETIC STRUCTURE OF
    AG*F2.
    *CHARPIN(P.), PLURIEN(P.), MERIEL(P.). (ED. NOTE: IN FRENCH).

45 AG2*H3*I*O6, ...A1.......DISS. ABS. XXV,........ 5588, (1965).
    *THE CRYSTAL STRUCTURES OF SILVER PARAPERIODATE AG2*H3*I*O6 AND OF
    POTASSIUM BOROHYDRIDE (K*B*H4): A NEUTRON AND X-RAY DIFFRACTION STUDY.
    *PETERSON(E.R.).

46 (AG-IN)*CR2*S4,.......A2.......SOL. STATE COMM. VOL.9 .... 413, (1971).
    *MISE EN EVIDENCE PAR DIFFRACTION DES NEUTRONS D≠UN TERME D≠ECHANGE
    BIQUADRATIQUE DANS L ≠HELIMAGNETIQUE (AG-IN)*CR2*S4.
    *PLUMIER(R.), SOUGI(M.).

47 AG7*N*O11, ...A1.......ACTA CRYST. 19,........ 180, (1965).
    *THE STRUCTURE OF THE AG(I,III) OXIDE PHASES.
    *NARAY-SZABO(I.), ARGAY(G.), SZABO(P.).

48 AL,. . . . .A1......PHYS. STAT. SOL. 5,........ K23, (1964).
    *X-RAY AND NEUTRON STUDIES OF THE LINEAGE STRUCTURE IN LARGE ALUMINUM
    SINGLE CRYSTALS.
    *MODRZEJEWSKI(A.), SZARRAS(S.).

SECTION   5 -STRUCTURE AND CROSS-SECTION DETERMINATIONS . . . .

49   AL,. . . . . .P. . . . . DISS. ABS. VOL.32, ...35538, (1971).
                         *STRUCTURE FACTOR FOR LIQUID ALUMINUM.
                         *STALLARD(J.M.).

50   AL,. . . . . .P. . . . . PHYS. SCRIPTA(SWEDEN) VOL.4, ...125, (1971).
                         *THE ATOMIC SCATTERING FACTORS OF CU AND AL FROM THERMAL DIFFUSE
                         SCATTERING MEASUREMENTS.
                         *LINKOAHO(M.), SERVOMAA(A.).

51   AL (ISO. A=27),. .P. . . . . J. NUCL. ENERGY VOL.24,. . . . . . . 419, (1970).
                         *THERMAL NEUTRON CAPTURE CROSS-SECTION MEASUREMENTS FOR NA(ISOTOPE A=23),
                         AL(ISOTOPE A=27), CL(ISOTOPE A=37) AND V(ISOTOPE A=51).
                         *RYVES(T.B.), PERKINS(D.R.).

52   AL,. . . . . .A1. . . . . PHYS. STAT. SOL. 4, ...349, (1964).
                         *A NEW METHOD FOR NEUTRON DIFFRACTION CRYSTAL STRUCTURE INVESTIGATION.
                         *BURAS(B.), LECIEJEWICZ(J.).

53   AL,. . . . . .P. . . . . PHYS. LETT. 21, ... 286, (1966).
                         *THE STRUCTURE FACTOR FOR LIQUID METALS AT LOW ANGLES.
                         *EGELSTAFF(P.A.), DUFFILL(C.), RAINEY(V.S.), ENDERBY(J.E.), NORTH(D.M.).

54   AL,. . . . . .A4. . . . . ACTA CRYST. 23, ... 185, (1967).
                         *NEUTRON DIFFRACTION STUDY OF DEBYE-WALLER FACTOR FOR ALUMINUM.
                         *MCDONALD(D.L.).

55   AL,. . . . . .A1. . . . . SOV. PHYS. SOL. STATE 6, ...1070, (1964).
                         *STRUCTURE INVESTIGATION BY NEUTRON DIFFRACTION USING PULSED FAST REACTOR
                         (IBR).
                         *NITTS(V.V.), PAPULOVA(Z.G.), SOSNOVSKAYA(I.), SOSNOVSKII(E.).

56   AL,. . . . . .P. . . . . PHYS. REV. A VOL.8,. . . . . 368, (1973).
                         *LIQUID-ALUMINUM STRUCTURE FACTOR BY NEUTRON DIFFRACTION.
                         *STALLARD(J.M.), DAVIS-JR(C.M.).

57   AL,. . . . . .P. . . . . NUCL. SCI. ENGNG. VOL.12,. . . . . 157, (1962).
                         *THE TOTAL CROSS SECTION OF ALUMINUM FOR NEUTRONS OF ENERGIES FROM
                         0.003 EV TO 0.009 EV.
                         *BALLY(D.), TODIREANU(S.), RIPEANU(S.).

58   AL,. . . . . .P. . . . . Z. PHYSIK VOL.250, ... 166, (1972).
                         *TOTAL CROSS-SECTIONS OF VARIOUS HOMOGENEOUS SUBSTANCES FOR ULTRACOLD
                         NEUTRONS.
                         *STEYERL(A.), VONACH(H.).

59   AL,. . . . . .A4. . . . . J. APPL. PHYS. VOL.45, ... 43, (1974).
                         *NEUTRON SMALL-ANGLE SCATTERING INVESTIGATION OF VOIDS IN IRRADIATED
                         MATERIALS.
                         *MOOK(H.A.).

60   AL,. . . . . .P. . . . . PHYS. REV. VOL.72, ... 408, (1947).
                         *SPIN DEPENDENCE OF SCATTERING OF SLOW NEUTRONS BY BE,AL, AND BI.
                         *FERMI(E.), MARSHALL(L.).

61   AL,. . . . . .P. . . . . PROC. PHYS. MATH. SOC. JAPAN 25, 495, (1943).
                         *SCATTERING OF THERMAL NEUTRONS BY SOLIDS. I.VARIATION IN TOTAL CROSS
                         SECTION OF METALS BY COLD-WORKING.
                         *KIMURA(M.), HASHIGUCHI(R.).

62   AL,. . . . . .A4. . . . . PHYS. REV. VOL.73, ... 830, (1948).
                         *THE DIFFRACTION OF NEUTRONS BY CRYSTALLINE POWDERS.
                         *WOLLAN(E.O.), SHULL(C.G.).

63   AL,. . . . . .P. . . . . PHYS. REV. VOL.73, ... 1385, (1948).
                         *CRYSTAL ORIENTATION IN AL BY SLOW NEUTRON DIFFRACTION.
                         *ARNOLD(G.P.), WEBER(A.H.).

64   AL,. . . . . .A4. . . . . PHIL. MAG. VOL.3, ...476, (1958).
                         *SMALL ANGLE SCATTERING FROM COLD-WORKED AND FATIGUED METALS.
                         *ATKINSON(H.H.).

65   AL2*CA3*(SI*O4)3, .A1. . . . . COLL. INTER. N.126 (GRENOBLE),. . 16, (1963).
                         *REFINEMENT OF THE GARNET STRUCTURE BY NEUTRON DIFFRACTION.
                         *PRANDL(W.).

66   AL*CL3*6(H2*O),. .A1. . . . . DISS. ABS. XXIV, ... 90, (1964).
                         *A NEUTRON AND X-RAY DIFFRACTION DETERMINATION OF THE STRUCTURE AND AM-
                         PLITUDES OF THERMAL MOTION IN CRYSTALLINE ALUMINUM CHLORIDE HEXAHYDRATE.
                         *BUCHANAN(D.R.).

67   AL*CL3*6(H2*O),. .A1. . . . . ACTA CRYST VOL.B24,. . . . . 954, (1968).
                         *A NEUTRON AND X-RAY DIFFRACTION INVESTIGATION OF ALUMINUM CHLORIDE
                         HEXAHYDRATE.
                         *BUCHANAN(D.R.), HARRIS(P.M.).

68   AL2*CO*O4, ...A2. . . . . COLL. INTER. N.126 (GRENOBLE), . . 83, (1963).
                         *MAGNETIC PROPERTIES OF NORMAL SPINELS WITH ONLY A-A INTERACTIONS.
                         *ROTH(W.L.).

69   AL-CR2,. . . . . .A2. . . . . J. CHEM. PHYS. VOL.43, ... 222, (1965).
                         *ANTIFERROMAGNETIC STRUCTURE OF AL-CR2.
                         *ATOJI(M.).

70   AL*(CR-FE),. . . .A2. . . . . COMP. REND. 268, ... 455, (1969).
                         *ORDRE ANTIFERROMAGNETIQUE DANS LES ALLIAGES CR(2-X)*FE(X)*AL.
                         *KALLEL(A.).

71   AL-(CU), ...A4. . . . . J. PHYS. SOC. JAPAN VOL.20,. . . .1723, (1965).
                         *SCATTERING OF LONG WAVELENGTH NEUTRONS BY PRECIPITATIONS IN AL(99.5)-
                         CU(0.5) ALLOYS.
                         *SAKAMOTO(M.), KUNITOMI(N.).

72   AL-FE, ...A2. . . . . COLL. INTER. N.126 (GRENOBLE), . 180, (1963).
                         *MAGNETISM RESEARCH WITH POLARIZED NEUTRONS AT THE CENTRO DI STUDI
                         NEUCLEARI DELLA CASACCIA DEL C.N.E.N., ROMA, ITALY.
                         *ANTONINI(B.), FELCHER(G.P.), MENZINGER(F.), PAOLETTI(A.), RICCI(F.P.),
                         PASSARI(L.).

73   AL2*FE*O4, ...A2. . . . . COLL. INTER. N.126 (GRENOBLE), . 83, (1963).
                         *MAGNETIC PROPERTIES OF NORMAL SPINELS WITH ONLY A-A INTERACTION.
                         *ROTH(W.L.).

74   ALKALI-METALS, . .A3. . . . . J. CHEM. PHYS. VOL.34, ... 873, (1961).
                         *STRUCTURE OF ALKALI METALS IN THE LIQUID STATE.
                         *GINGRICH(N.S.), HEATON(L.).

SECTION   5 -STRUCTURE AND CROSS-SECTION DETERMINATIONS . . . .

75 AL-LI, ...NEUTRON-DIFFRACTION STUDY OF AL-LI ALLOY. A4 . . . . . . LATV. PSR. . .FIZ. TEHN. SER. NO.1   88, (1970).
                         .NEUTRON-DIFFRACTION STUDY OF AL-LI ALLOY.
                         .NOZIK(YU.Z.), LIPIN(YU.V.), MAKSIMYUK(P.A.), MIKHALKO(V.D.).

76 AL-MG, ...ETUDE PAR DIFFUSION CENTRALE DES NEUTRONS DES PHENOMENES DE   A4 . . . . . . C.R. ACAD. SCI. B TOME 277, . . . 225, (1973).
                         .ETUDE PAR DIFFUSION CENTRALE DES NEUTRONS DES PHENOMENES DE
                         PREPRECIPITATION DANS LES ALLIAGES AL-MG RICHES EN ALUMINIUM.
                         .RAYNAL(J.M.), ROTH(P.), BERNOLE(M.), GUYOT(P.), GRAF(R.).

77 AL-MG, ...SMALL-ANGLE NEUTRON SCATTERING BY GUINIER-PRESTON ZONES IN AL-MG ALLOYS. A4 . . . . . . J. APPL. CRYST. VOL.7, ... 219, (1974).
                         .SMALL-ANGLE NEUTRON SCATTERING BY GUINIER-PRESTON ZONES IN AL-MG ALLOYS.
                         .ROTH(M.), RAYNAL(J.M.).

78 AL2*MG*O4, ...REDETERMINATION OF THE CATION DISTRIBUTION OF SPINEL MG*AL2*O4 BY MEANS A4 . . . . . . COLL. INTER. N.126 (GRENOBLE) . . 23, (1963).
                         .REDETERMINATION OF THE CATION DISTRIBUTION OF SPINEL MG*AL2*O4 BY MEANS
                         OF NEUTRON DIFFRACTION.
                         .STOLL(E.), FISCHER(P.), HALG(W.), MAIER(G.).

79 AL(0.89)*MN(1.11), . .   .AN X-RAY AND NEUTRON-DIFFRACTION INVESTIGATION OF THE MAGNETIC PHASE A2 . . . . . . ACTA CRYST. 16, ... 737, (1963).
                         .AN X-RAY AND NEUTRON-DIFFRACTION INVESTIGATION OF THE MAGNETIC PHASE
                         AL(0.89)*MN(1.11).
                         .BRAUN(E.), GOEDKOOP(J.A.).

80 AL*MN*GE, ...MAGNETIC STRUCTURE OF MN-AL-GE. A2 . . . . . . J. APP. PHYS. VOL.40, ... 1870, (1969).
                         .MAGNETIC STRUCTURE OF MN-AL-GE.
                         .MURTHY(N.S.S.),   BEGUM(R.J.),   SOMANATHAN(C.S.),   MURTHY(M.R.L.N.).

81 AL2*MN*O4, ...MAGNETIC PROPERTIES OF NORMAL SPINELS WITH ONLY A-A INTERACTION. A2 . . . . . . COLL. INTER. N.126 (GRENOBLE) . . 83, (1963).
                         .MAGNETIC PROPERTIES OF NORMAL SPINELS WITH ONLY A-A INTERACTION.
                         .ROTH(W.L.).

82 AL2*MN3*(SI*O4)3, ...RHOMBOHEDRAL MAGNETIC STRUCTURE IN SPESSARTITE TYPE GARNETS. A2 . . . . . . PHYS. STAT. SOLIDI B VOL.55 NO.2 K159, (1973).
                         .RHOMBOHEDRAL MAGNETIC STRUCTURE IN SPESSARTITE TYPE GARNETS.
                         .PRANDL(W.).

83 AL2*O3.3CA*O.6H2*O, ...ETUDE DE LA POSITION DES ATOMES D≠HYDROGENE DANS L≠HYDROGRENAT A1 . . . . . . SOL. STATE COMM. 1, ... 85, (1963).
                         .ETUDE DE LA POSITION DES ATOMES D≠HYDROGENE DANS L≠HYDROGRENAT
                         AL2*O3,3CA*O,6(H2*O).
                         .COHEN-ADDAD(C.), DUCROS(P.), DURIF(A.), BERTAUT(E.F.), DELAPALME(A.).

84 AL2*O3.3CA*O.6H2*O, ...DETERMINATION DE LA POSITION DES ATOMES D≠HYDROGENE DANS L≠HYDROGRENAT A4 . . . . . . COLL. INTER. N.126 (GRENOBLE) . . 54, (1963).
                         .DETERMINATION DE LA POSITION DES ATOMES D≠HYDROGENE DANS L≠HYDROGRENAT
                         AL2*O3,3CA*O,6H2*O: RESONNANCE MAG. NUCLEAIRE ET DIFFRACTION NEUTRONIQUE
                         .COHEN-ADDAD(C.), DUCROS(P.), DURIF(A.), BERTAUT(E.F.), DELAPALME(A.).

85 AL2*O3-(CR+++), .   .THE MAGNETIC DIFFUSE SCATTERING OF NEUTRONS BY ISOLATED METAL IONS: A4 . . . . . . J. PHYS. C VOL.4, ... 1279, (1971).
                         .THE MAGNETIC DIFFUSE SCATTERING OF NEUTRONS BY ISOLATED METAL IONS:
                         CR(+3) IN AL2*O3.
                         .TOFIELD(B.C.), FENDER(B.E.F.).

86 5(AL2*O3)*3(DY2*O3), . .   .ETUDE PAR DIFFRACTION DE NEUTRONS DE LA STRUCTURE MAGNETIQUE DU GRENAT A2 . . . . . . COMP. REND. 259, ... 2416, (1964).
                         .ETUDE PAR DIFFRACTION DE NEUTRONS DE LA STRUCTURE MAGNETIQUE DU GRENAT
                         D≠ALUMINUM ET DE DYSPROSIUM.
                         .HERPIN(A.), MERIEL(P.).

87 AL*O*(O*H), ...A SINGLE CRYSTAL NEUTRON DIFFRACTION STUDY OF DIASPORE, AL*O*(O*H). A1 . . . . . . ACTA CRYST. VOL.11, ... 798, (1958).
                         .A SINGLE CRYSTAL NEUTRON DIFFRACTION STUDY OF DIASPORE, AL*O*(O*H).
                         .BUSING(W.R.), LEVY(H.A.).

88 AL2-TH, ...NEUTRON-DIFFRACTION STUDY OF AL2-TH. A1 . . . . . . ACTA CRYST. VOL.8, ... 118, (1955).
                         .NEUTRON-DIFFRACTION STUDY OF AL2-TH.
                         .ANDRESEN(A.F.), GOEDKOOP(J.A.).

89 AL*TH2-D(N) (N=0,2,3,4)   .NEUTRON DIFFRACTION INVESTIGATION OF SOLID SOLUTIONS AL*TH2-D(N). A1 . . . . . . ACTA CRYST. VOL.14, ... 223, (1961).
                         .NEUTRON DIFFRACTION INVESTIGATION OF SOLID SOLUTIONS AL*TH2-D(N).
                         .BERGSMA(J.), GOEDKOOP(J.A.), VAN-VUCHT(J.H.N.).

90 AL-ZN, ...A STUDY OF PRE-PRECIPITATION IN AL-ZN ALLOYS WITH X-RAYS AND NEUTRONS. A4 . . . . . . J. APPL. CRYST. VOL.4, ... 511, (1971).
                         .A STUDY OF PRE-PRECIPITATION IN AL-ZN ALLOYS WITH X-RAYS AND NEUTRONS.
                         .RAYNAL(J.M.), SCHELTEN(J.), SCHMATZ(W.).

91 ANTHRACENE(DEUTERATED), . .   .THE STRUCTURE OF PERDEUTERIOANTHRACENE BY NEUTRON DIFFRACTION. A1 . . . . . . ACTA CHEM. SCAND. VOL.26, ... 1996, (1972).
                         .THE STRUCTURE OF PERDEUTERIOANTHRACENE BY NEUTRON DIFFRACTION.
                         .LEHMANN(M.S.), PAWLEY(G.S.).

92 AR, ...DETERMINATION OF THE ELASTIC CONSTANTS OF ARGON. P . . . . . . PHYS LETT 27A, ... 695, (1968).
                         .DETERMINATION OF THE ELASTIC CONSTANTS OF ARGON.
                         .SANGER(M.G.), EGGER(H.), LUSCHER(E.).

93 AR, ...A STATISTICAL EVALUATION OF ELASTIC SCATTERING DATA WITH APPLICATION TO A4 . . . . . . BROOKHAVEN SYMPOSIUM, ... 183, (1965).
                         .A STATISTICAL EVALUATION OF ELASTIC SCATTERING DATA WITH APPLICATION TO
                         THERMAL NEUTRON SCATTERING IN LIQUID ARGON.
                         .SCHLUP(W.A.).

94 AR, ...NON-GAUSSIAN EFFECTS IN THERMAL-NEUTRON SCATTERING IN LIQUID ARGON. A3 . . . . . . J. PHYS. SOC. JAP. VOL.31, . . . 1162, (1971).
                         .NON-GAUSSIAN EFFECTS IN THERMAL-NEUTRON SCATTERING IN LIQUID ARGON.
                         .CHATURVEDI(D.K.), BAIJAL(J.S.).

95 AR, ...SLOW-NEUTRON CROSS SECTIONS OF HE, NE, AR AND KR. A3 . . . . . . NUCL. PHYS. VOL.A133, ... 410, (1969).
                         .SLOW-NEUTRON CROSS SECTIONS OF HE, NE, AR AND KR.
                         .RORER(D.C.), ECKER(B.M.), AKYUZ(R.O.).

96 AR, ...THERMAL NEUTRON CAPTURE IN NATURAL ARGON. P . . . . . . PHYS. SCRIPTA VOL.1, ... 85, (1970).
                         .THERMAL NEUTRON CAPTURE IN NATURAL ARGON.
                         .HARDELL(R.), BEER(C.).

97 AR, ...COLLECTIVE MOTIONS IN CLASSICAL SYSTEMS.3. THEORETICAL STUDY OF THERMAL T . . . . . . ANN. PHYS. (GER.) VOL.23, ... 49, (1969).
                         .COLLECTIVE MOTIONS IN CLASSICAL SYSTEMS.3. THEORETICAL STUDY OF THERMAL
                         NEUTRON SCATTERING IN LIQUID ARGON.
                         .RICHTER(J.), VOSS(K.).

98 AR, ...CORRECTION TO: PAIR POTENTIAL FOR ARGON FROM NEUTRON DIFFRACTION AT LOW A4 . . . . . . PHYSICA VOL.62, ... 474, (1972).
                         .CORRECTION TO: PAIR POTENTIAL FOR ARGON FROM NEUTRON DIFFRACTION AT LOW
                         DENSITY.
                         .FREDRIKZE(H.), ANDRIESSE(C.D.), LEGRAND(E.).

99 AR, ...STRUCTURE FACTOR AND RADIAL DISTRIBUTION FUNCTION FOR LIQUID ARGON AT P . . . . . . PHYS. REV. A VOL.7, ... 2130, (1973).
                         .STRUCTURE FACTOR AND RADIAL DISTRIBUTION FUNCTION FOR LIQUID ARGON AT
                         85 DEGREES K.
                         .YARNELL(J.L.), KATZ(M.J.), WENZEL(R.G.), KOENIG(S.H.).

100 AR, ...NORMALIZATION OF DIFFRACTION DATA FROM LIQUIDS. A3 . . . . . . J. CHEM. PHYS. VOL.42, ... 3540, (1965).
                         .NORMALIZATION OF DIFFRACTION DATA FROM LIQUIDS.
                         .RAHMAN(A.).

SECTION   5 -STRUCTURE AND CROSS-SECTION DETERMINATIONS . . . .

101 AR,. . . . .   *PAIR POTENTIAL FOR ARGON FROM NEUTRON DIFFRACTION AT LOW DENSITY.
     PHYSICA VOL.57, ... 191, (1972).
     *ANDRIESSE(C.D.), LEGRAND(E.).

102 AR,. . . . .   *ATOMIC OVERLAP IN THE LIQUID STATE.   PHYS. LETT. VOL.29A, ... 296, (1969).
     *PAGE(D.I.), EGELSTAFF(P.A.), ENDERBY(J.E.), WINGFIELD(B.R.).

103 AR,. . . . .   *NEUTRON REFRACTION IN O2, N2, HE, AR GASES.   PHYS. REV. VOL.84, ... 969, (1951).
     *MCREYNOLDS(A.W.).

104 AR,. . . . .   *STRUCTURE OF LIQUID NITROGEN, OXYGEN, AND ARGON BY NEUTRON DIFFRACTION.
     PHYS. REV. VOL.92, . . . 1229, (1953).
     *HENSHAW(D.G.), HURST(D.G.), POPE(N.K.).

105 AR,. . . . .   *SLOW NEUTRON SPECTROMETER STUDIES OF OXYGEN, NITROGEN, AND ARGON.
     PHYS. REV. VOL.73, . . . 1399, (1948).
     *MELKONIAN(E.), RAINWATER(L.J.), HAVENS-JR(W.W.), DUNNING(J.R.).

106 AR,. . . . .   *ON THE STRUCTURE FACTOR OF GASEOUS ARGON.   PHYS. LETT. VOL.34A, ... 112, (1971).
     *HASMAN(A.), ZANDVELD(P.).

107 AR,. . . . .   *ATOMIC DISTRIBUTION IN LIQUID AND SOLID NEON AND SOLID ARGON BY
     NEUTRON DIFFRACTION.   PHYS. REV. VOL.111, ... 1470, (1958).
     *HENSHAW(D.G.).

108 AR,. . . . .   *THE MEASUREMENT OF THE TOTAL CROSS-SECTION OF SOLID NEON AND ARGON FOR
     SLOW NEUTRONS.   Z. NATURFORSCH. VOL.15A, ... 828, (1960).
     *SPRINGER(T.), WIEDEMANN(W.). (IN GERMAN).

109 AR (A=36,40),. .   *ATOMIC DISTRIBUTION IN LIQUID ARGON BY NEUTRON DIFFRACTION AND THE
     CROSS SECTIONS OF A(ISO. A=36) AND A(ISO. A=40).   PHYS. REV. VOL.105, ... 976, (1957).
     *HENSHAW(D.G.).

110 AS,. . . . .   *TOTAL CROSS-SECTION OF ARSENIC FOR NEUTRONS BETWEEN 0.05 AND 40 EV.
     REV. ROUMAINE PHYS. VOL.9, . . . 121, (1964).
     *APOSTOLESCU(S.), CONSTANTINESCU(M.)... ET AL. (IN FRENCH).

111 AS,. . . . .   *NEUTRON COHERENT-SCATTERING AMPLITUDES OF GA, IN, AS AND SB.
     PHYS. REV. 131, ... 2098, (1963).
     *ARNOLD(G.P.), NERESON(N.G.).

112 AS2*SE3, ...   *NEUTRON SCATTERING IN AMORPHOUS SOLIDS.   AMORPHOUS MATERIALS, ... 423, (1972).
     *LEADBETTER(A.J.), WRIGHT(A.C.), APLING(A.J.).   (IN BOOK: AMORPHOUS
     MATERIALS ED. BY DOUGLAS, ELLIS: PUBL. WILEY: LONDON 1972).

113 AU (ISO. A=198),   *AVERAGE THERMAL NEUTRON CAPTURE CROSS SECTIONS OF AU(ISO. A=198), NI
     (ISO. A=65), AND CU(ISO. A=66).   NUCL. APPL. TECHNOL. VOL.9, . . . 662, (1970).
     *SERMENT(V.), ABU-SAMRA(A.), EMMONS(A.H.).

114 AU,. . . . .   *THE ABSORPTION CROSS SECTION OF COPPER FOR THERMAL NEUTRONS.
     NUCL. SCI. ENGNG. VOL.7, ... 184, (1960).
     *DONAHUE(D.J.), BENNETT(R.A.), LANNING(D.D.).

115 AU,. . . . .   *LONG WAVELENGTH CRYSTAL SPECTROMETER AND THE NEUTRON ABSORPTION CROSS
     SECTIONS OF GOLD AND BORON.   NUCL. SCI. ENGNG. VOL.8, ... 453, (1960).
     *GOULD(F.T.), TAYLOR(T.I.), HAVENS-JR(W.W.), RUSTAD(B.M.), MELKONIAN(E.).

116 AU,. . . . .   *TOTAL CROSS-SECTIONS OF VARIOUS HOMOGENEOUS SUBSTANCES FOR ULTRACOLD
     NEUTRONS.   Z. PHYSIK VOL.250, ... 166, (1972).
     *STEYERL(A.), VONACH(H.).

117 AU,. . . . .   *PRECISION NEUTRON TOTAL CROSS SECTION MEASUREMENTS ON GOLD AND COBALT
     IN THE .00004 TO .005 EV RANGE.   Z. PHYSIK VOL.264, ... 427, (1973).
     *DILG(W.), MANNHART(W.), STEICHELE(E.), ARNOLD(P.).

118 AU,. . . . .   *THERMAL NEUTRON ABSORPTION CROSS SECTIONS OF BORON AND GOLD.
     PHYS. REV. VOL.92, ... 716, (1953).
     *CARTER(R.S.), PALEVSKY(H.), MYERS(V.W.), HUGHES(D.J.).

119 AU,. . . . .   *SLOW NEUTRON CROSS SECTIONS FOR HE3, B AND AU.   PHYS. REV. VOL.133, ... B925, (1964).
     *ALS-NIELSEN(J.), DIETRICH(O.W.).

120 AU-CR, ...   *MAGNETIC FORM FACTOR OF GOLD-CHROMIUM ALLOYS.   J. PHYS. SOC. JAPAN VOL.29,. . . 978, (1970).
     *NAKAI(Y.), KUNITOMI(N.), YAMADA(Y.).

121 AU-FE, ...   *MAGNETIC-MOMENT DISTRIBUTION IN NI-FE AND AU-FE ALLOYS.   PHYS. REV. B VOL.7,. . . . . . 2005, (1973).
     *CABLE(J.W.), WOLLAN(E.O.).

122 AU-MN, ...   *ANTIFERROMAGNETIC ORDERING IN SOME AU-MN ALLOYS.   J. PHYS. F VOL.3,. . . . . . 1054, (1973).
     *MAKHURANE(P.M.).

123 AU-MN, ...   *THE MAGNETIC STRUCTURE OF AU-MN.   PROC. PHYS. SOC. VOL.79, ... 938, (1962).
     *BACON(G.E.).

124 AU5-MN2, ...   *ANTIFERROMAGNETISM IN AU5*MN2.   J. OF PHYS. C VOL.2, ... 356, (1969).
     *SMITH(J.H.), WELLS(P.).

125 AU4-MN,. . . . .   *THE MAGNETIC STRUCTURE OF DISORDERED AU4*MN.   AIP CONF. PROC. VOL.5, ... 497, (1971).
     *CHAKRABARTI(D.J.).

126 AU2-MN,. . . . .   *A NEUTRON DIFFRACTION INVESTIGATION OF AU2*MN UNDER PRESSURE.
     J. PHYS. CHEM. SOL. 27,. . . . 925, (1966).
     *SMITH(F.A.), BRADLEY(J.G.), BACON(G.E.).

127 AU3-MN,. . . . .   *LONG PERIOD MODULATION OF STACKING ORDER IN AU3*MN.   J. PHYS. CHEM. SOL. 27,. . . . 413, (1966).
     *SATO(H.), TOTH(R.S.), SHIRANE(G.), COX(D.E.).

SECTION   5 -STRUCTURE AND CROSS-SECTION DETERMINATIONS . . . .

128 AU-MN, ... A1 . . . . PROC. PHYS. SOC. VOL.72, ...   470, (1958).
     *GOLD-MANGANESE ALLOYS; SOME PRELIMINARY STUDIES BY NEUTRON DIFFRACTION.
     *BACON(G.E.), STREET(R.).

129 AU2-(MN,AL)2, ... A2 . . . . . . PROC. PHYS. SOC. 92, ...   713, (1967).
     *THE MAGNETIC STRUCTURES OF THE ALLOYS AU2(MN,AL)2.
     *BACON(G.E.), MASON(E.W.).

130 AU-MN, AU2-(MN-AL, CU, GA, IN, ZN)2, ... A2 . . . . J. PHYS. F VOL.3, ...   2003, (1973).
     *MAGNETIC ORDER IN THE AU-MN ALLOY DERIVATIVES AU2-(MN-Z)2 WHERE
     Z IS AL, CU, GA, IN OR ZN.
     *BACON(G.E.), PLANT(J.S.).

131 AUSTENITE, . . .   *THE ARRANGEMENT OF THE CARBON ATOMS IN THE AUSTENITE LATTICE.
     *BYKOV(V.N.), VINOGRADOV(S.I.).   A1 . . . . . . SOV. PHYS. CRYST. VOL.3, ...308, (1958).

132 AU2-TB, AU2-TM, . *ON THE MAGNETIC STRUCTURE OF R*AU2 COMPOUNDS (R:TB-TM). . 1390, (1971).
     *FEDRO(A.J.), SHAFFER(J.C.).   A2 . . . . . . AIP CONF. PROC. VOL.5,

133 AU4-X, ... A1 . . . . . . J. APP. PHYS. VOL.40, ...   1373, (1969).
     *STUDIES OF AU4-X- ORDERED ALLOYS: ELECTRON AND NEUTRON DIFFRACTION,
     RESISTIVITY AND SPECIFIC HEAT.
     *TOTH(R.S.),   ARROTT(A.),   SHINOZAKI(S.S.),   WERNER(S.A.),   SATO(H.).

134 AZOXYANISOLE (PARA), ... A4, A3 . . . . SOL. STATE COMM. VOL.11, ... 1365, (1972).
     *COHERENT NEUTRON SCATTERING BY A NEMATIC LIQUID CRYSTAL.
     *PYNN(R.), OTNES(K.), RISTE(T.).

135 AZOXYANISOLE(PARA-DEUTERATED), ... A4 . . . . SOL. STATE COMM. VOL.12 NO.5, . . 409, (1973).
     *PRETRANSITIONAL EFFECTS ASSOCIATED WITH THE MELTING OF A MATERIAL WITH
     A NEMATIC MESOPHASE.
     *RISTE(T.), PYNN(R.).

136 AZOXYANISOLE (PARA-DEUTERATED), ... A1, A3 . . . PHYS. NORV. VOL.6 NO.3-4, . . .   205, (1972).
     *COHERENT NEUTRON SCATTERING BY A NEMATIC LIQUID CRYSTAL.
     *PYNN(R.), OTNES(K.), RISTE(T.).

137 AZOXYANISOLE (PARA-DEUTERATED), ... T . . . . J. PHYS. CHEM. SOLIDS VOL.34, . . 735, (1973).
     *CALCULATIONS OF NEUTRON DIFFRACTION PATTERNS OBTAINED WITH A NEMATIC
     LIQUID CRYSTAL.
     *PYNN(R.).

138 AZOXYANISOLE, ... A3 . . . . . ACTA PHYS. POLON. VOL.17, . . .   483, (1958).
     *THE INFLUENCE OF POLARIZATION OF LIQUID CRYSTAL MOLECULES ON THE
     SCATTERING OF SLOW NEUTRONS.
     *JANIK(J.A.), KRASNICKI(S.), MURASIK(A.).

139 B (ISOTOPE A=10), ... P . . . . J. NUCL. ENERGY VOL.24, . . .   85, (1970).
     *THE LOW ENERGY SCATTERING CROSS SECTION OF B (ISOTOPE A=10).
     *ASAMI(A.), MOXON(M.C.).

140 B, ... P . . . . NUCL. SCI. ENGNG. VOL.8, ...   453, (1960).
     *LONG WAVELENGTH CRYSTAL SPECTROMETER AND THE NEUTRON ABSORPTION CROSS
     SECTIONS OF GOLD AND BORON.
     *GOULD(F.T.), TAYLOR(T.I.), HAVENS-JR(W.W.), RUSTAD(B.M.), MELKONIAN(E.).

141 B, ... P . . . . NUCL. SCI. ENGNG. VOL.9, ...   98, (1961).
     *PRECISION MEASUREMENTS OF THE SLOW NEUTRON ABSORPTION CROSS-SECTIONS OF
     NORMAL BORON STANDARDS.
     *SAFFORD(G.J.), TAYLOR(T.I.), RUSTAD(B.M.), HAVENS-JR(W.W.).

142 B (ISO. A=10), ... P . . . . NUCL. SCI. ENGNG. VOL.40, . .   12, (1970).
     *THE THERMAL-NEUTRON-ABSORPTION CROSS SECTIONS OF LI(ISOTOPE A=6) AND
     B(ISOTOPE A=10).
     *MEADOWS(J.W.), WHALEN(J.F.).

143 B, ... P . . . . PHYS. REV. VOL.92, ...716, (1953).
     *THERMAL NEUTRON ABSORPTION CROSS SECTIONS OF BORON AND GOLD.
     *CARTER(R.S.), PALEVSKY(H.), MYERS(V.W.), HUGHES(D.J.).

144 B, ... P . . . . NUCL. PHYS. VOL.17, ...   109, (1960).
     *TOTAL NEUTRON CROSS-SECTION OF B(A=10) IN THE THERMAL NEUTRON ENERGY
     RANGE.
     *SCHMITT(H.W.), BLOCK(R.C.), BAILEY(R.L.).

145 B, ... P . . . . PHYS. REV. VOL.69, . .   443, (1946).
     *EXPERIMENTS WITH A SLOW-NEUTRON VELOCITY SPECTROMETER.
     *BACHER(R.F.), BAKER(C.P.), MCDANIEL(B.D.).

146 B, ... P . . . . PHYS. REV. VOL.119, ...   1291, (1960).
     *PRECISION DETERMINATION OF THE SLOW NEUTRON ABSORPTION CROSS SECTION OF
     B(A=10).
     *SAFFORD(G.J.), TAYLOR(T.I.), RUSTAD(B.M.), HAVENS-JR(W.W.).

147 B, ... P . . . . PHYS. REV. VOL.133, ... B925, (1964).
     *SLOW NEUTRON CROSS SECTIONS FOR HE3, B AND AU.
     *ALS-NIELSEN(J.), DIETRICH(O.W.).

148 B, ... P . . . . NUCL. INST. MET. 39, ...   350, (1966).
     *THE THERMAL NEUTRON ABSORPTION CROSS SECTION OF THE AERE BORON STANDARD.
     *WYNCHANK(S.A.R.), COX(A.E.), COLLIE(C.H.).

149 B, ... P . . . . PHYS. REV. VOL.138, ... B1116, (1965).
     *DETERMINATION OF THE COHERENT NEUTRON SCATTERING AMPLITUDES OF BORON,
     NITROGEN, AND OXYGEN BY MIRROR REFLECTION.
     *DONALDSON(R.E.), PASSELL(L.), BARTOLINI(W.), GROVES(D.).

150 B10*H14, ... A4 . . . . . ACTA CRYST. VOL.B27, ...   2003, (1971).
     *DIE VERTEILUNG DER BINDUNGSELEKTRONEN IM DEKABORAN-MOLEKUL (B1J*H14).
     *BRILL(R.), DIETRICH(H.), DIERKS(H.).

151 B2*03, ... P . . . . PHYS. REV. VOL.146, ...   660, (1966).
     *COHERENT NEUTRON SCATTERING AMPLITUDES OF B (A=10) AND B (A=11).
     *DONALDSON(R.E.), GROVES(D.J.), PEARSON(R.K.).

152 B*P, ... P . . . . J. PHYS. SOC. JAPAN VOL.17 B-II,   335, (1962).
     *COHERENT BRAGG SCATTERING OF RESONANT NUCLIDES.
     *PETERSON(S.W.), SMITH(H.G.).

153 BA, ... P . . . . NUCL. PHYS. A VOL.A177, ...   393, (1971).
     *NEUTRON RESONANCES IN BARIUM ISOTOPES.
     *VAN-DE-VYVER(R.E.), PATTENDEN(N.J.).

SECTION   5 -STRUCTURE AND CROSS-SECTION DETERMINATIONS . . . .

154 BA,. . . . . .P. . . . . . .Z. KRISTALLOGR. VOL.135,. . . . . .316, (1972).
                         *THE NEUTRON SCATTERING AMPLITUDES OF POTASSIUM, STRONTIUM AND BARIUM.
                         *COOPER(M.J.), ROUSE(K.D.).

155 BA*AL2*04, ...A4. . . . . . .BULL. SOC. FRANC. MIN. CRIST. 88 413, (1965).
                         *ATOMIC PARAMETERS OF BA*AL2*04 AND AN INVESTIGATION OF SOLID SOLUTIONS
                         OF BA*(FE-AL)*04 AND BA*(GA-AL)*04.
                         *DO-DINH(C.), BERTAUT(E.F.). (IN FRENCH).

156 BA*(CE,PR,TB)*03, ...A1. . . . . . .ACTA CRYST. VOL.B28,. . . . . . . 956, (1972).
                         *THE STRUCTURES OF BA*CE*03, BA*PR*03 AND BA*TB*03 BY NEUTRON DIFFRAC-
                         TION: LATTICE PARAMETER RELATIONS AND IONIC RADII IN O-PEROVSKITES.
                         *JACOBSON(A.J.), TOFIELD(B.C.), FENDER(B.E.F.).

157 BA*CL2*2(H2*0),.   . . . . .A1. . . . . . .COLL. INTER. N.126 (GRENOBLE), 45 (1963),
                         *RECENT CRYSTAL STRUCTURE DETERMINATION BY NEUTRON DIFFRACTION AT
                         OAK RIDGE.
                         *BROWN(G.M.), LEVY(H.A.).

158 BA*(CL*03)2*(H2*0), ...A1. . . . . . .J. CHEM. PHYSICS VOL.48, ... 1883, (1968).
                         *NEUTRON-DIFFRACTION REFINEMENT OF THE CRYSTAL STRUCTURE OF BARIUM
                         CHLORATE MONOHYDRATE BA*(CL*03)2*(H2*0).
                         *SIKKA(S.K.), MOMIN(S.N.), RAJAGOPAL(H.), CHIDAMBARAM(R.).

159 BA*CO*F4,. . . .   . .A2. . . . . . .PHYS. REV. B VOL.6,. . . . . . . 2677, (1972).
                         *MAGNETIC STRUCTURE OF THE TWO-DIMENSIONAL ANTIFERROMAGNET BA*CO*F4.
                         *EIBSCHUTZ(M.), HOLMES(L.), GUGGENHEIM(H.J.), COX(D.E.).

160 BA*(CO-FE)*027,.   . . . . .A2. . . . . . .SOV. PHYS. J.E.T.P. VOL.23,. . . 395, (1966).
                         *SPIN ORDERING AND MAGNETOCRYSTALLINE ANISOTROPY IN SINGLE CRYSTALS OF
                         BA*(CO-FE)*027 FERRITES.
                         *YAMZIN(I.I.), SIZOV(R.A.), ZHELUDEV(I.S.), PEREKALINA(T.M.),
                         ZALESSKII(A.V.).

161 BA2*(CO,NI)*W*06, ...A2. . . . . . .J. APPL. PHYS. VOL.38,. . . . . . 1459, (1967).
                         *NEUTRON-DIFFRACTION STUDY OF ANTIFERROMAGNETIC BA2*CO*W*06 AND
                         BA2*NI*W*06.
                         *COX(D.E.), SHIRANE(G.), FRAZER(B.C.).

162 BA2*CO*RE*06,. .   . . . .A2. . . . . . .A.I.P. CONF. PROC. NO.10 PART 1, 674, (1973).
                         *SPIRAL STRUCTURE OF BA2*CO*RE*06.
                         *KHATTAK(C.P.), COX(D.E.), WANG(F.F.Y.).

163 BA*F2, ...A4. . . . . . .HARWELL SUMMER SCHOOL, ... 101, (1968).
                         *NEUTRON STUDIES OF NUCLEAR CHARGE DISTRIBUTIONS IN BARIUM FLUORIDE AND
                         HEXAMETHYLENETETRAMINE.
                         *DAWSON(B.).

164 BA*F2, ...A3. . . . . . .ACTA CRYST. A24, ... 484, (1968).
                         *NEUTRON DIFFRACTION STUDIES OF ANHARMONIC TEMPERATURE FACTORS IN BA*F2.
                         *COOPER(M.J.), ROUSE(K.D.), WILLIS(B.T.M.).

165 BA*FE*F4,. . . .   . .A2. . . . . . .J. PHYSIQUE TOME 32 COL.1 VOL.II 759, (1971).
                         *MAGNETIC BEHAVIOUR OF THE TWO DIMENSIONAL ANTIFERROMAGNET BA*FE*F4.
                         *EIBSCHUTZ(M.), HOLMES(L.), GUGGENHEIM(H.J.).

166 BA*(FE(X)-AL(1-X))2*04,. . . . . . .A4. . . . . . .J. PHYSIQUE TOME 31,. . . . . . . 401, (1970).
                         *MOSSBAUER AND NEUTRON DIFFRACTION STUDY OF INFINITE CLUSTERS IN
                         DISORDERED SOLID SOLUTIONS OF BA*(FE(X)-AL(1-X))2*04.
                         *DO-DINH(C.), CHEVALIER(R.), BURLET(P.), BERTAUT(E.F.).

167 BA*GA2*(SC-FE)10*019,. . . . . . . .A2. . . . . . .SOV. PHYS. CRYST. VOL.18,. . . . 393, (1973).
                         *BLOCK CONICAL FERROMAGNETIC SPIRAL IN THE HEXAGONAL FERRITE (M).
                         *ALESHKO-OZHEVSKII(O.P.), CHEPARIN(V.P.), CHERKASOV(A.P.), YAMZIN(I.I.).

168 BA*(IN-FE)*019,.   . . . . .A2. . . . . . .SOV. PHYS. CRYST. VOL.16,. . . . 711, (1972).
                         ANGLED BLOCK MAGNETIC STRUCTURE IN THE HEXAGONAL FERRITE BA(IN(3.4)-
                         FE(8.6))*019(M).
                         *ALESHKO-OZHEVSKII(O.P.), LYUBIMTSEV(V.A.), NAMTALISHVILI(M.I.),
                         YAMZIN(I.I.).

169 BA*LA2*FE2*07, ...A1. . . . . . .J. SOLID STATE CHEM. VOL.7,. . . 337, (1973).
                         *DETERMINATION DES STRUCTURES DE DEUX FERRITES MIXTES NOUVEAUX DE
                         FORMULE BA*LA2*FE2*07 ET SR*TB2*FE2*07.
                         *SAMARAS(D.), COLLOMB(A.), JOUBERT(J.C.).

170 BA*LI*H3, BA*LI*D3, ...A1. . . . . . .J. CHEM. PHYSICS VOL.48,. . . . . 4660, (1968).
                         *TERNARY HYDRIDES POSSESSING THE CUBIC PEROVSKITE STRUCTURE. I. A NEUTRON
                         DIFFRACTION STUDY OF BA*LI*H3 AND BA*LI*D3.
                         *MAELAND(A.J.), ANDRESEN(A.F.).

171 BA*(N3)2,. . . .   . .A4. . . . . . .ACTA CRYST. VOL.B25, ... 2638, (1969).
                         NEUTRON DIFFRACTION STUDY OF BA(N3)2.
                         *CHOI(C.S.).

172 BA*NI*F4,. . . .   . .A2. . . . . . .J. APPL. PHYS. 41,. . . . . 943, (1970).
                         *NEUTRON DIFFRACTION STUDY OF THE MAGNETIC STRUCTURE OF BA*NI*F4.
                         *COX(D.E.), EIBSCHUTZ(M.), GUGGENHEIM(H.J.), HOLMES(L.).

173 BA*NI*02,. . . .   . .A1,2. . . . . .COLL. INTER. N.126 (GRENOBLE),. . 18, (1963).
                         *NEUTRON DIFFRACTION STUDIES AT THE PUERTO RICO NUCLEAR CENTRE.
                         *ALMODOVAR(I.), BIELEN(H.J.), FRAZER(B.C.), KAY(M.I.).

174 BA*0*FE2*03, ...A2. . . . . . .J. PHYSIQUE TOME 30,. . . . . . . 566, (1969).
                         *X-RAY, NEUTRON DIFFRACTION AND MOSSBAUER EFFECT STUDIES OF BARIUM MONO-
                         FERRITE BA*0*FE2*03.
                         *DO-DINH(C.), BERTAUT(E.F.), CHAPPERT(J.).

175 BA*(PR,TB)*03, ...A2. . . . . . .J. PHYS. C VOL.5,. . . . . 2887, (1972).
                         *SOME MAGNETIC PROPERTIES OF BA*PR*03 AND BA*TB*03: THE MAGNETIC FORM
                         FACTOR FOR TB(4+) BY NEUTRON DIFFRACTION.
                         *TOFIELD(B.C.), JACOBSON(A.J.), FENDER(B.E.F.).

176 BA*(SC-FE)12*019(M),. . . . . .A2. . . . . . .SOV. PHYS. JETP LETT. VOL.7, . . 158, (1968).
                         *HELICOIDAL ANTIPHASE SPIN ORDERING IN HEXAGONAL FERRITES WITH MAGNETO-
                         PLUMBITE STRUCTURE.
                         *ALESHKO-OZHEVSKII(O.P.), SIZOV(R.A.), CHEPARIN(V.P.), YAMZIN(I.I.).

177 (BA-SR)2*ZN2*FE12*022,. . . . . . . .A2. . . . . . .SOV. PHYS. J.E.T.P. VOL.33,. . . 737, (1971).
                         *DELOCALIZATION OF THE MAGNETIC MOMENT OF FE+++ IONS IN TYPE Y HEXAGONAL
                         FERRITE AT 293 DEG. K.
                         *SIZOV(R.A.).

178 (BA-SR)*ZN2*FE12*022*(Y),. . . . . . .A2. . . . . . .SOV. PHYS. SOL. STATE VOL.12,. . 2316, (1971).
                         *DELOCALIZATION OF THE MAGNETIC MOMENT OF FE+++ IONS IN THE LATTICE OF A
                         HEXAGONAL TYPE Y FERRITE.
                         *SIZOV(R.A.).

SECTION   5 -STRUCTURE AND CROSS-SECTION DETERMINATIONS . . . .

179 (BA-SR)*ZN2*FE12*022,. . . . . .T. . . . . . . .SOV. PHYS. SOL. STATE VOL.11,. . 2177, (1969).
     ↦MAGNETIC   PROPERTIES   OF   (BA-SR)*ZN2*FE12*022   HEXAFERRITES   WITH   A
      HELICOIDAL MAGNETIC STRUCTURE.
     ↦PETROVA(I.I.), VINNIK(M.A.).

180 (BA-SR)*ZN2*FE12*022(Y) FERRITES,. . .A2. . . . . . . .SOV. PHYS. J.E.T.P. VOL.26,. . . 736, (1968).
     ↦A NEW TYPE OF SPIN ORDERING IN THE SYSTEM OF HEXAGONAL
      (BA-SR)*ZN2*FE12*022(Y) FERRITES.
     ↦SIZOV(V.A.), SIZOV(R.A.), YAMZIN(I.I.).

181 BA-SR-ZN-FE-O, ...A2. . . . . . . .SOV. PHYS. SOL. STATE VOL.10,. . 2537, (1969).
     ↦NEUTRON DIFFRACTION INVESTIGATION OF THE SPIN ORDERING IN HEXAGONAL
      Y-TYPE FERRITES UNDER HIGH HYDROSTATIC PRESSURES.
     ↦SIZOV(R.A.), BOKHENKOV(E.L.), SIZOV(V.A.).

182 (BA-SR)*ZN2*FE24*041,. . . . .A2. . . .SOV. PHYS. JETP VOL.35,. . . . 370, (1972).
     ↦NONCOLLINEAR MAGNETIC STRUCTURES IN HEXAGONAL FERRITES OF THE
      BA(3-X)-SR(3-X)*ZN2*FE24*041(Z) SYSTEM.
     ↦NAMTALISHVILI(M.I.), ALESHKO-OZHEVSKII(O.P.), YAMZIN(I.I.).

183 BA*TB*03,. . . . . .A1,A2. . . . .ACTA CRYST. VOL.B28,. . . . . 3429, (1972).
     ↦A NOTE ON THE CRYSTAL AND MAGNETIC STRUCTURE OF BA*TB*03.
     ↦BANKS(E.), LA-PLACA(S.J.), KUNNMANN(W.), CORLISS(L.M.), HASTINGS(J.M.).

184 BA*TI*03,. . . . . .A1. . . . . . .ACTA CRYST. VOL.A26,. . . . . 336, (1970).
     ↦X-RAY AND NEUTRON DIFFRACTION STUDY OF TETRAGONAL BARIUM TITANATE.
     ↦HARADA(J.), PEDERSEN(T.), BARNEA(Z.).

185 BA*TI*03,. . . . . .A1. . . . . . .PHYS. STAT. SOL. 31,. . . . . 121, (1969).
     ↦INCORPORATION OF ANTIMONY INTO THE BARIUM TITANATE LATTICE.
     ↦SCHMELZ(H.).

186 BA*TI*03,. . . . . .A4. . . . . . .J. PHYS. SOC. JAPAN VOL.17 B-II, 376, (1962).
     ↦NEUTRON DIFFRACTION STUDIES OF FERROELECTRICS.
     ↦FRAZER(B.C.).

187 BA*TI*03,. . . . . .A1. . . . . . .PHYS. REV. VOL.100,. . . . . 745, (1955).
     ↦SINGLE-CRYSTAL NEUTRON ANALYSIS OF TETRAGONAL BA*TI*03.
     ↦FRAZER(B.C.), DANNER(H.R.), PEPINSKY(R.).

188 BA*TI*03,. . . . . .A1. . . . . . .PHYS. REV. VOL.105,. . . . 856, (1957).
     ↦NEUTRON DIFFRACTION STUDY OF ORTHORHOMBIC BA*TI*03.
     ↦SHIRANE(G.), DANNER(H.), PEPINSKY(R.).

189 BA*U*04, ...A1. . . . . . .ACTA CRYST. B25,. . . . . 787, (1969).
     ↦THE STRUCTURE OF SOME ALKALINE-EARTH METAL URANATES.
     ↦LOOPSTRA(B.O.), RIETVELD(H.M.).

190 BA5*W3*LI2*015,. . . . . .A1. . . . . . .ACTA CRYST. VOL.B30,. . . . . 816, (1974).
     ↦A POWDER NEUTRON-DIFFRACTION DETERMINATION OF THE STRUCTURE OF
      BA5*W3*LI2*015.
     ↦JACOBSON(A.J.), COLLINS(B.M.), FENDER(B.E.F.).

191 BA*W*04, ...A1. . . . . . .J. CHEM. PHYSICS VOL.55,. . . . . 1093, (1971).
     ↦CRYSTAL STRUCTURE REFINEMENT OF SR*MO*04, SR*W*04, CA*MO*04, AND BA*W*04
      BY NEUTRON DIFFRACTION.
     ↦GURMEN(E.), DANIELS(E.), KING(J.S.).

192 BA-ZN-AL-FE-O, ...A1. . . . . . .SOV. PHYS. SOL. STATE VOL.10,. . 2258, (1969).
     ↦NEUTRON DIFFRACTION STUDY ON THE HEXAGONAL FERRITE
      BA2*ZN2*AL(2.5)*FE(9.5)*022.
     ↦AGANOVA(N.N.), SIZOV(V.A.), YARNZIN(I.I.).

193 BA2*ZN2*FE12*022,. . . . . .A1. . . . . . .SOV. PHYS. CRYST. VOL.15,. . . . . 280, (1970).
     ↦NEUTRON DIFFRACTION INVESTIGATION OF THE HEXAGONAL FERRITES
      BA2*ZN2*FE12*022 AND 22(Y).
     ↦YAMZIN(I.I.), LETSIEVICH(YA.).

194 BA2*ZN2*FE12*022,. . . . . .A2. . . . . . .ACTA CRYST. VOL.12,. . . . . 476, (1959).
     ↦SPIN ORIENTATION AND EXTINCTION IN FERRIMAGNETIC BA2*ZN2*FE12*022 BY
      NEUTRON DIFFRACTION.
     ↦GOEDKOOP(J.A.), HVOSLEF(J.), ZIVADINOVIC(M.).

195 BE,. . . . . .A3. . . . . . .IAEA SYMPOSIUM VIENNA,. . . . 277, (1960).
     ↦SLOW-NEUTRON INELASTIC SCATTERING MEASUREMENTS AT THE MATERIALS TESTING
      REACTOR.
     ↦BRUGGER(R.M.), EVANS(J.E.).

196 BE,. . . . . .A1,T. . . . . . .NUCL. SCI. ENG. VOL.42,. . . . . 137, (1970).
     ↦A CALCULATION OF THE COHERENT NEUTRON SCATTERING FROM POLYCRYSTALLINE
      BERYLLIUM.
     ↦BORGONOVI(G.M.), SPREVAK(D.).

197 BE,. . . . . .P. . . . . . .PHYS. REV. VOL.72,. . . . . 408, (1947).
     ↦SPIN DEPENDENCE OF SCATTERING OF SLOW NEUTRONS BY BE,AL, AND BI.
     ↦FERMI(E.), MARSHALL(L.).

198 BE,. . . . . .E. . . . . . .ACTA CRYST. VOL.11,. . . . . 228, (1958).
     ↦THE SCATTERING OF 4 ANGSTROM NEUTRONS BY A BERYLLIUM CRYSTAL.
     ↦HAY(H.J.), PATTENDEN(N.J.), EGELSTAFF(P.A.).

199 BE,. . . . . .T,P. . . . . . .J. NUCL. ENERGY VOL.6,. . . . . . 104, (1957).
     ↦SCATTERING OF THERMAL NEUTRONS IN BERYLLIUM.
     ↦BHANDARI(R.C.).

200 BE (ISO. A=9), ...P. . . . . . .PROC/CONF/NEUTRON C. S. + TECH., 851, (1968).
     ↦TOTAL NEUTRON CROSS SECTIONS OF BE (ISO. A=9), N (ISO. A=14), AND
      O (ISO. A=16).
     ↦JOHNSON(C.H.), HAAS(F.X.), FOWLER(J.L.), MARTIN(F.D.), KERNELL(R.L.),
      COHN(H.O.).

201 BE,BE*O, ...P. . . . . . .PHYS. REV. VOL.71,. . . . . 589, (1947).
     ↦THE TRANSMISSION OF SLOW NEUTRONS THROUGH MICROCRYSTALLINE MATERIALS.
     ↦FERMI(E.), STURM(W.J.), SACHS(R.G.).

202 BE*F2, ...P. . . . . . .AMORPHOUS MATERIALS,. . . . . 423, (1972).
     ↦NEUTRON SCATTERING IN AMORPHOUS SOLIDS.
     ↦LEADBETTER(A.J.),   WRIGHT(A.C.),   APLING(A.J.).   (IN BOOK: AMORPHOUS
      MATERIALS ED. BY DOUGLAS, ELLIS; PUBL. WILEY: LONDON 1972).

203 BE*F2, ...A1. . . . . . .J. NON-CRYST. SOLIDS VOL.7,. . . . . 156, (1972).
     ↦DIFFRACTION STUDIES OF GLASS STRUCTURE IV.THE STRUCTURE OF VITREOUS
      BE*F2 BY X-RAY AND NEUTRON DIFFRACTION.
     ↦LEADBETTER(A.J.), WRIGHT(A.C.).

SECTION   5 -STRUCTURE AND CROSS-SECTION DETERMINATIONS . . . .

204 BE*O, ... P . . . . .SOV. AT. ENERGY VOL.13, ... 852, (1962).
     ⌐EFFECT OF TEMPERATURE AND MICROSTRUCTURE OF BAKED BERYLLIUM OXIDE ON
     THE SCATTERING CROSS-SECTION FOR THERMAL NEUTRONS.
     ⌐ZHEZHERUN(I.F.), SADIKOV(I.P.), CHERNYSHOV(A.A.).

205 BE*O, ... A4 . . . . .ACTA CRYST. VOL.B25 . . . . . . .2254, (1969).
     ⌐THE WURTZITE Z PARAMETER FOR BERYLLIUM OXIDE AND ZINC OXIDE.
     ⌐SABINE(T.M.), HOGG(S.).

206 BE*S*O4*4(H2*O), . . A1 . . . . .ACTA CRYST. VOL.B25 . . . . 310, (1969).
     ⌐A NEUTRON DIFFRACTION DETERMINATION OF THE STRUCTURE OF BERYLLIUM SUL-
     PHATE TETRAHYDRATE, BE*S*O4*4(H2*O).
     ⌐SIKKA(S.K.), CHIDAMBARAM(R.).

207 BI, ... P . . . . .J. OF PHYS. C VOL.1, ... 1075, (1968).
     ⌐THE STRUCTURE FACTOR FOR LIQUID METALS. II. RESULTS FOR LIQUID ZN, TL,
     PB, SN AND BI.
     ⌐NORTH(D.M.), ENDERBY(J.E.), EGELSTAFF(P.A.).

208 BI, ... A3 . . . . .PHYS. MET. METALLOGR. VOL.27, 6, 1011, (1969).
     ⌐NEUTRON DIFFRACTION BY LIQUID BISMUTH AT TEMPERATURES BETWEEN ITS
     MELTING AND BOILING POINTS.
     ⌐KHRUSHCHEV(B.I.), BOGOMOLOV(A.M.).

209 BI, ... A1 . . . . .PHYS. LETT. 24A, ... 714, (1967).
     ⌐THE CRYSTAL STRUCTURE OF BISMUTH-II AT 26 KBAR.
     ⌐BRUGGER(R.M.), BENNION(R.B.), WORLTON(T.G.).

210 BI, ... P . . . . .PHYS. LETT. 21, ... 286, (1966).
     ⌐THE STRUCTURE FACTOR FOR LIQUID METALS AT LOW ANGLES.
     ⌐EGELSTAFF(P.A.), DUFFILL(C.), RAINEY(V.S.), ENDERBY(J.E.), NORTH(D.M.).

211 BI, ... A3 . . . . .J. CHEM. PHYS. VOL.21, ... 228, (1953).
     ⌐NEUTRON DIFFRACTION AND ATOMIC DISTRIBUTION IN LIQUID LEAD AND LIQUID
     BISMUTH AT TWO TEMPERATURES.
     ⌐SHARRAH(P.C.), SMITH(G.P.).

212 BI, ... A3 . . . . .PHYS. REV. VOL.77, ... 305, (1950).
     ⌐NEUTRON DIFFRACTION IN LIQUID SULFUR, LEAD, AND BISMUTH.
     ⌐CHAMBERLAIN(O.).

213 BI, ... P . . . . .PHYS. REV. VOL.72, ... 408, (1947).
     ⌐SPIN DEPENDENCE OF SCATTERING OF SLOW NEUTRONS BY BE,AL, AND BI.
     ⌐FERMI(E.), MARSHALL(L.).

214 (BI-CA)*MN*O3, . . A1 . . . . .SOV. PHYS. SOL. STATE VOL.10, . . . 754, (1968).
     ⌐NEUTRON-DIFFRACTION INVESTIGATION OF THE BI*MN*O3-CA*MN*O3 SYSTEM:
     BI(0.15)*CA(0.85)*MN*O3.
     ⌐TURKEVICH(E.I.), PLAKHTII(V.P.).

215 BI*FE*O3, ... A1 . . . . .J. PHYS. CHEM. SOLIDS VOL.32, . . 1315, (1971).
     ⌐FERROELECTRIC BI*FE*O3 X-RAY AND NEUTRON DIFFRACTION STUDY.
     ⌐MOREAU(J.M.), MICHEL(C.), GERSON(R.), JAMES(W.J.).

216 BI*FE*O3, ... A2 . . . . .BULL. ACAD. SCI. USSR VOL.28, . . 350, (1964).
     ⌐NEUTRON DIFFRACTION STUDIES OF SOME COMPOUNDS WITH THE PEROVSKITE
     STRUCTURE.
     ⌐PLAKHTII(V.P.), MAL≠TSEV(E.I.), KAMINKER(D.M.).

217 BI*FE*O3, ... A2 . . . . .SOV. PHYS. DOKLADY VOL.7, . . . 742, (1962).
     ⌐NEUTRON DIFFRACTION DETECTION OF MAGNETIC ORDERING IN THE FERROELECTRIC
     BI*FE*O3.
     ⌐KISELEV(S.V.), OZEROV(R.P.), ZHDANOV(G.S.).

218 BI*FE*O3, ... A2 . . . . .PROC. INT. CONF. (NOTTINGHAM), . . 354, (1964).
     ⌐INVESTIGATIONS OF NEW ANTIFERROMAGNETS AT THE INSTITUTE OF SEMICONDUC-
     TORS IN THE U.S.S.R. (LENINGRAD).
     ⌐SMOLENSKY(G.A.), BOKOV(V.A.), KIZHAEV(S.A.), MAL≠TSEV(E.I.),
     NEDLIN(G.M.), PLAKHTII(V.P.), TUTOV(A.G.), JUDIN(V.M.).

219 BI*MN, ... A1,A2 . . . .PHYS. REV. VOL.104, ... 607, (1956).
     ⌐NEUTRON DIFFRACTION STUDY OF THE STRUCTURES AND MAGNETIC PROPERTIES
     OF MANGANESE BISMUTHIDE.
     ⌐ROBERTS(B.W.).

220 BI*MN2*O5, . . . A2,T . . . . .ACTA CRYST A24, ... 217, (1968).
     ⌐REPRESENTATION ANALYSIS OF MAGNETIC STRUCTURE.
     ⌐BERTAUT(E.F.).

221 BI*MN2*O5, . . . A2 . . . . .SOLID STATE COMM. 5, ... 25, (1967).
     ⌐STRUCTURE MAGNETIQUE ET PROPRIETES MAGNETIQUES DE BI*MN2*O5.
     ⌐BERTAUT(E.F.), BUISSON(G.), QUEZEL-AMBRUNAZ(S.), QUEZEL(G.).

222 BI4*SI3*O12 (EULYTITE), ... A1 . . . . .Z. KRIST. VOL.123, ... 73, (1966).
     ⌐NEUTRON-DIFFRACTION STUDY OF BI4*SI3*O12.
     ⌐SEGAL(D.J.), SANTORO(R.P.), NEWNHAM(R.E.).

223 BI2*TE3-(BI2*SE3) 5-PERCENT, , ...A1 . . . . .J. OF PHYS.-C-, SER. 2, VOL.2, . 870, (1969).
     ⌐A PULSED NEUTRON DIFFRACTION MEASUREMENT OF PREFERRED ORIENTATION IN
     PRESSED POWDER COMPACTS OF BISMUTH SELENO-TELLURIDE.
     ⌐DAY(D.H.), SINCLAIR(R.N.).

224 BI4*TI3*O12, . . A1 . . . . .DISS. ABS. VOL.32, ...6587B, (1972).
     ⌐THE CRYSTAL STRUCTURES OF BI4*TI3*O12 AND LI2*TI*O3.
     ⌐DORRIAN(J.F.).

225 BR, ... P . . . . .ACTA CRYST. VOL.A28, ... 663, (1972).
     ⌐COHERENT NEUTRON SCATTERING AMPLITUDES OF BR AND I (ISO. A=127).
     ⌐ATOJI(M.).

226 BR, ... A3 . . . . .NUOVO CIMENTO VOL.24, ... 103, (1962).
     ⌐THE STRUCTURE OF LIQUID BROMINE.
     ⌐CAGLIOTI(G.), RICCI(F.P.).

227 BR, ... A3 . . . . .J. PHYS. SOC. JAPAN VOL.17 B-II, 348, (1962).
     ⌐NEUTRON DIFFRACTION WORK AT THE ISPRA CENTER.
     ⌐CAGLIOTI(G.), RICCI(F.P.), SANTORO(A.), SCATTURIN(V.).

228 BR, ... A3,P . . . . . . .PHYS. LETT. VOL.45A NO.4, . . . 273, (1973).
     ⌐HIGH MOMENTUM TRANSFER STRUCTURE FACTOR OF LIQUID BROMINE BY T-O-F
     NEUTRON DIFFRACTION.
     ⌐MISAWA(M.), FUKUSHIMA(Y.), SUZUKI(K.), TAKEUCHI(S.).

229 BUERGERITE, . . . A1,A2 . . . .AMER. MINERAL. VOL.56, ... 101, (1971).
     ⌐A NEUTRON-DIFFRACTION STUDY OF THE FERRIC TOURMALINE, BUERGERITE.
     ⌐TIPPE(A.), HAMILTON(W.C.).

SECTION   5 -STRUCTURE AND CROSS-SECTION DETERMINATIONS . . . .

230 C (ISOTOPE A=12),.....T.........TRANS. AM. NUCL. SOC. VOL.13,.... 728, (1970).
     *THE APPLICATION OF EXTENDED R-MATRIX THEORY TO ELASTIC SCATTERING OF
     NEUTRONS BY C (ISOTOPE A=12).
     *YOST(K.J.), PITKANEN(P.H.).

231 C (ISOTOPE A=12),.........P.......NUCL. PHYS. A VOL.A181,....... 177, (1972).
     *POLARIZATION AND DIFFERENTIAL CROSS SECTION FOR NEUTRONS ELASTICALLY
     SCATTERED FROM C(ISOTOPE A=12).
     *DRIGO(L.), MANDUCHI(C.), MOSCHINI(G.), MANDUCHI(M.T.R.), TORNIELLI(G.),
     ZANNONI(G.).

232 C (GRAPHITE),.... ....A4.......NUCL. SCI. ENGNG. VOL.37,..... 198, (1969).
     *COHERENT NEUTRON SCATTERING FROM POLYCRYSTALLINE GRAPHITE.
     *TAKAHASHI(H.).

233 C (ISO. A=12), ...P.......PHYS. REV. VOL.188,....... 1618, (1969).
     *POLARIZATION AND DIFFERENTIAL CROSS SECTION FOR NEUTRONS SCATTERED FROM
     C (ISO. A=12).
     *LANE(R.O.), KOSHEL(R.D.), MONAHAN(J.E.).

234 C (GRAPHITE),. . ...P.......ACTA CRYST. VOL.A28,....... 59, (1972).
     *BESTIMMUNG DER ANDERUNG DER GITTERKONSTANTEN/ANISOTROPEN DEBYE-WALLER-
     FAKTORS/GRAPHIT MITTELS NEUTRONENBEUGUNG/TEMPERATURBEREICH VON 25-1850C.
     *LUDSTECK(A.).

235 C (A=12),. . . . .T,P....ANNALS OF PHYS. VOL.75,....... 132, (1973).
     *A STUDY OF THE NEUTRON PLUS C (A=12) SYSTEM USING THE UNIFIED-REACTION
     FORMALISM. I.ELASTIC SCATTERING.
     *LEUNG(T.T.), KOSHEL (R.D.).

236 C, ...A4.......REPORT AERE-R 6052,. . . . . , (1969).
     *A NEUTRON DIFFRACTION STUDY OF CARBON FIBRES.
     *SINCLAIR(R.N.),   WEDGWOOD(A.),   HARRIS(D.H.C.),   EGELSTAFF(P.A.). (ED.
     NOTE# AVAIL. ATOMIC ENERGY RES. ESTABL., HARWELL, DIDCOT, BERKSHIRE,ENG)

237 C (GRAPHITE),.. ....A1.......J. NUCL. MATERIALS VOL.7,..... 92, (1962).
     *STUDY OF THE CRYSTALLINE ANISOTROPY OF ARTIFICIAL GRAPHITE BY NEUTRON
     DIFFRACTION.
     *LECOMTE(M.). (IN FRENCH).

238 C (GRAPHITE),. . .....P.......ACTA CRYST. VOL.5,....... 492, (1952).
     *THE POWDER DIFFRACTION INTENSITIES OF GRAPHITE FOR X-RAYS AND NEUTRONS.
     *BACON(G.E.).

239 C (GRAPHITE),. . .....E,T.....PHYS. REV. VOL.75,....... 217, (1949).
     *THE EFFECT OF CRYSTAL ORIENTATION ON THE SCATTERING OF SLOW NEUTRONS.
     *ARNOLD(G.P.), MYERS(V.W.), WEBER(A.H.).

240 C (A=13),. . . . .P.......PHYS. REV. VOL.85,....... 491, (1952).
     *THE COHERENT NEUTRON SCATTERING CROSS SECTION OF C(A=13).
     *KOEHLER(W.C.), WOLLAN(E.O.).

241 C (DIAMOND, GRAPHITE),. . . . . . .A4.......PHYS. REV. VOL.73,....... 830, (1948).
     *THE DIFFRACTION OF NEUTRONS BY CRYSTALLINE POWDERS.
     *WOLLAN(E.O.), SHULL (C.G.).

242 C (GRAPHITE),. . .....P.......PHYS. REV. VOL.75,....... 1098, (1949).
     *THE EFFECT OF SMALL ANGLE SCATTERING ON NEUTRON CROSS-SECTION MEASURE-
     MENTS.
     *KRUEGER(H.H.A.), MENEGHETTI(D.), RINGO(G.R.), WINSBERG(L.).

243 C (GRAPHITE),. . .....P.......J. NUCL. ENERGY VOL.5,. . . . . 203, (1957).
     *THE SLOW-NEUTRON CROSS-SECTION OF GRAPHITE.
     *EGELSTAFF(P.A.).

244 C, ...P.......Z. PHYSIK VOL.198,....... 187, (1967).
     *ABSOLUTE MEASUREMENT OF COHERENT SCATTERING LENGTHS OF H, C AND CL AND
     DETERMINATION OF GPAV. ACC. FOR FREE NEUTRONS WITH A GRAVITY REFRACTOM.
     *KOESTER(L.). (IN GERMAN).

245 C (GRAPHITE),. . .....P.......J. APPL. CRYST. VOL.7,. . . . . 38, (1974).
     *IS PYROLYTIC GRAPHITE AN IDEAL MOSAIC CRYSTAL.
     *DORNER(B.), KOLLMAR(A.).

246 C (GLASSY),. . . ..P.......J. NUCL. SCI. TECHNOL. VOL.5,... 649, (1968).
     *TOTAL CROSS SECTION OF GLASSY CARBON FROM 0.0C1 TO 0.1 EV.
     *IIZUMI(M.), AYAO(S.).

247 C*(C*H2*O*H)4, ...A1.......ACTA CRYST. VOL.11,....... 383, (1958).
     *A NEUTRON DIFFRACTION STUDY OF PENTAERYTHRITOL.
     *HVOSLEF(J.).

248 C2*(C*N)4-OXIDE, ...A1.......J. AMER. CHEM. SOC. VOL.93,.... 5945, (1971).
     *BONDING AND VALENCE ELECTRON DISTRIBUTIONS IN MOLECULES. AN X-RAY AND
     NEUTRON DIFFRACTION STUDY OF ... TETRACYANOETHYLENE OXIDE.
     *MATTHEWS(D.A.), SWANSON(J.), MUELLER(M.H.), STUCKY(G.D.).

249 C*CL4, ...A3.......J. CHEM. PHYSICS VOL.48,....... 2395, (1968).
     *NEUTRON DIFFRACTION FROM LIQUID CARBON TETRACHLORIDE.
     *RAO(K.R.).

250 C*CL3*C*H*(O*H)2,.......A1.......CRYST. STRUC. COMM. VOL.2 NO.1,. 107, (1973).
     *CHLORAL HYDRATE,.... (FROM NEUTRON DATA).
     *BROWN(G.M.), LEVY(H.A.).

251 C6*CL4*(O*H)2, ...A1.......ACTA CRYST. 23,....... 107, (1967).
     *NEUTRON DIFFRACTION STUDY OF THE HYDROGEN BOND SYSTEM IN TETRACHLOROHY-
     DROQUINONE.
     *SIKKA(S.K.), CHIDAMBARAM(R.).

252 C*D4,. . . . .A1.......NATURWISSENSCHAFTEN VOL.52,. . . 512, (1965).
     *THE CRYSTAL STRUCTURE OF METHANE.
     *GISSLER(W.), STILLER(H.).

253 C*D4,. . . . .A1.......PHYS. LETT. VOL.31A,....... 253, (1970).
     *A NEUTRON DIFFRACTION STUDY ON SOLID METHANE C*D4 IN PHASE I.
     *PRESS(W.), DORNER(B.), WILL(G.).

254 C*D4,. . . . .A1,A4.....ACTA CRYST. VOL.A28 PART S-4,.. 188, (1972).
     *CRYSTAL STRUCTURE AND PHASE TRANSITION IN DEUTERO-METHANE.
     *ARZI(E.), SANDOR(E.). (ABSTRACT ONLY).

255 C*D4,. . . . .A1.......J. CHEM. PHYS. VOL.56,....... 2597, (1972).
     *STRUCTURE AND PHASE TRANSITIONS IN SOLID HEAVY METHANE (C*D4).
     *PRESS(W.).

SECTION   5 -STRUCTURE AND CROSS-SECTION DETERMINATIONS . . . .

256   C10*D8,. . . . . . .A1. . . . . . .ACTA CRYST. B25,. . . . . 2009, (1969).
      ⊹A NEUTRON-DIFFRACTION STUDY OF PERDEUTERONAPHTHALENE.
      ⊹PAWLEY(G.S.), YEATS(E.A.).

257   C14*D10 (ANTHRACENE),. . . . .A1. . . . . .ACTA CHEM. SCAND. VOL.26,. . . . . . 1996, (1972).
      ⊹THE STRUCTURE OF PERDEUTERIOANTHRACENE BY NEUTRON DIFFRACTION.
      ⊹LEHMANN(M.S.), PAWLEY(G.S.).

258   (C*D3)4*N*MN*CL3,. . . . . .A2. . . . .PHYS. REV. LETT. VOL.26,. . . . . .718, (1971).
      ⊹SPIN CORRELATIONS IN A ONE-DIMENSIONAL HEISENBERG ANTIFERROMAGNET.
      ⊹BIRGENEAU(R.J.), DINGLE(R.), HUTCHINGS(M.T.), SHIRANE(G.), HOLT(S.L.).

259   C*F4,. . . . . . . .A3. . . . . .PHYS. REV. VOL.83,. . . . . . 1100, (1951).
      ⊹NEUTRON DIFFRACTION BY THE GASES N2, C*F4, AND C*H4.
      ⊹ALCOCK(N.Z.), HURST(D.G.).

260   C*H4,. . . . . . . .A3. . . . . .PHYS. REV. VOL.83,. . . . . . 1100, (1951).
      ⊹NEUTRON DIFFRACTION BY THE GASES N2, C*F4, AND C*H4.
      ⊹ALCOCK(N.Z.), HURST(D.G.).

261   C*H4,. . . . . . . .T,P. . . . .PHYS. REV. VOL.138,. . . . . .A692, (1965).
      ⊹EFFECT OF SPIN CORRELATION UPON THE NEUTRON CROSS SECTION OF METHANE.
      ⊹MICHAEL(P.).

262   C*H4,. . . . . . . .A3. . . . . .PROC. PHYS. SOC. 89,. . . . . . 379, (1966).
      ⊹NEUTRON SCATTERING FROM GASEOUS METHANE AND AMMONIA.
      ⊹VENKATARAMAN(G.), RAO(K.R.). . .ET AL.

263   C*H4, C2*H4, ...P. . . . . .PHYSICA VOL.33,. . . . . . 523, (1967).
      ⊹THE NEUTRON DOUBLE DIFFERENTIAL CROSS SECTIONS OF METHANE AND ETHYLENE.
      ⊹BALLY(D.), TODIREANU(S.), TREPADUS(V.), TARINA(V.).

264   C*H4,. . . . . . . .P. . . . . .J. CHEM. PHYS. VOL.57,. . . . . 5007, (1972).
      ⊹NUCLEAR SPIN CONVERSION IN C*H4 BY NEUTRON SCATTERING.
      ⊹JOHNSTON(N.T.), COLLINS(M.F.).

265   C*H4,. . . . . . . .P. . . . . .NUCL./S.S. PHYS. SYMP. ABSTR.,. . . ., (1972).
      ⊹NEUTRON SCATTERING CROSS SECTION FOR LIQUID METHANE.
      ⊹RAO(K.R.), DASANNACHARYA(B.A.).

266   C2*H2, C*H4, ...P. . . . . .PHYSICA VOL.30,. . . . . .237, (1964).
      ⊹TOTAL CROSS SECTION FOR SCATTERING OF COLD NEUTRONS BY PROTONS IN SOME
      COOLED HYDROGENOUS COMPOUNDS.
      ⊹VAN-DINGENEN(W.), NEVE-DE-MEVERGNIES(M.).

267   C2*H4, ...A3. . . . .PHYS. LETT. 24A,. . . . . . 544, (1967).
      ⊹THE SCATTERING OF SLOW NEUTRONS BY ETHYLENE ADSORBED ON ACTIVATED
      CHARCOAL.
      ⊹TODIREANU(S.).

268   C5*H12,. . . . . . .A4. . . . .CHEM. PHYS. LETTERS VOL.4,. . . . . .444, (1969).
      ⊹STUDY OF MOLECULAR REORIENTATION IN SOLID NEOPENTANE BY QUASIELASTIC
      NEUTRON SCATTERING.
      ⊹LECHNER(R.E.), ROWE(J.M.), SKOLD(K.), RUSH(J.J.).

269   C6*H6, ...P. . . . . .NUKLEONIKA VOL.8,. . . . . . . 581, (1963).
      ⊹SLOW NEUTRON SCATTERING BY BENZENE MOLECULES.
      ⊹MANIAWSKI(F.), PARLINSKI(K.).

270   C6*H6, ...A1. . . . . .PROC. ROY. SOC. A VOL.279,. . . . . 98, (1964).
      ⊹A CRYSTALLOGRAPHIC STUDY OF SOLID BENZENE BY NEUTRON DIFFRACTION.
      ⊹BACON(G.E.), CURRY(N.A.), WILSON(S.A.).

271   C6*H6, C12*H10,. . . . .P. . . . . .PHYSICA VOL.30,. . . . . .1647, (1964).
      ⊹NEUTRON TOTAL CROSS SECTION OF BENZENE AND DIPHENYL BETWEEN 0.1 AND
      0.001 EV.
      ⊹ANTONINI(B.), PAOLETTI(A.).

272   C6*H6, C7*H8 (TOLUENE),. . . .P. . . . . .SOV. AT. ENERGY VOL.18,. . . . . .350, (1965).
      ⊹TOTAL NEUTRON CROSS-SECTIONS FOR BENZENE, TOLUENE AND SODIUM ACETATE
      IN THE ENERGY RANGE 0.03-0.5 EV.
      ⊹ANISIMOV(I.S.), NIKITIN(V.I.), SAUKOV(A.I.), UGODENKO(A.A.).

273   C6*H6, C12*H10 (DIPHENYL),. . . . .P. . . . . .PHYSICA VOL.32,. . . . . .119, (1966).
      ⊹TOTAL NEUTRON CROSS SECTION OF SOME POLYPHENYLS IN SOLID AND LIQUID
      PHASE.
      ⊹ANTONINI(B.), PAOLETTI(A.), GAMBETTA(V.).

274   C10*H14 (DURENE),. . . . .A1. . . . .ACTA CRYST. VOL.B29,. . . . . . 184, (1973).
      ⊹A CONSTRAINED REFINEMENT OF THE STRUCTURE OF DURENE.
      ⊹PRINCE(E.), SCHROEDER(L.W.), RUSH(J.J.).

275   C12*H18, ...A1,A4. . . . .DISC. FARADAY SOC. NO.48,. . . . . .192, (1969).
      ⊹METHYL GROUP ROTATION AND THE LOW TEMPERATURE TRANSITION IN HEXAMETHYL-
      BENZENE. NEUTRON DIFFRACTION STUDY.
      ⊹HAMILTON(W.C.), EDMONDS(J.W.), TIPPE(A.), RUSH(J.J.).

276   C14*H10, ...A1. . . . .ACTA CRYST. VOL.B27,. . . . . . 26, (1971).
      ⊹A REFINEMENT OF THE STRUCTURE OF THE ROOM TEMPERATURE PHASE OF
      PHENANTHRENE, C14*H10 FROM X-RAY AND NEUTRON DIFFRACTION DATA.
      ⊹KAY(M.I.), OKAYA(Y.), COX(D.E.).

277   C16*H10 (PYRENE),. . . . .A1. . . . . .ACTA CRYST. VOL.B28,. . . . . .2977, (1972).
      ⊹A NEUTRON DIFFRACTION STUDY OF THE CRYSTAL STRUCTURE OF PYRENE, C16*H10.
      ⊹HAZELL(A.C.), LARSEN(F.K.), LEHMANN(M.S.).

278   C18*H10 (TRIPHENYLENE),. . . . .A1. . . . . .ACTA CRYST. VOL.A28 PART S-4,. . . . . 27, (1972).
      ⊹OVERCROWDING IN POLYCYCLIC HYDROCARBONS:NEUTRON DIFFRACTION STUDIES.
      ⊹JONES(D.W.), YERKESS(J.), FERRARIS(G.). (ABSTRACT ONLY).

279   C18*H14 (P-DIPHENYLBENZENE),. . . . .A1. . . . . .NATURE VOL.192,. . . . . .154, (1961).
      ⊹X-RAY AND NEUTRON DIFFRACTION EXAMINATION OF P-DIPHENYLBENZENE.
      ⊹CLEWS(C.J.B.), MASLEN(E.N.), RIETVELD(H.M.).

280   C8*H12*BA*05,. . . . .A1. . . . . .CRYST. STRUCT. COMM. VOL.1 N.3,. 193, (1972).
      ⊹BARIUM METHACRYLATEMONOHYDRATE,. . .
      ⊹ISAACS(N.W.), VAN-DER-ZEE(J.J.), SHIELDS(K.G.), TILLACK(J.V.),
      WHEELER(D.H.), MOORE(F.H.), KENNARD(C.H.L.).

281   C*H3(C*H2)4*O*H,. . . . .A3. . . . .PHYS. LETT. 26A,. . . . . . 72, (1967).
      ⊹QUASI ELASTIC SCATTERING OF SLOW NEUTRONS IN PENTANOL.
      ⊹RAPEANU(S.), ILIESCU(N.).

282   C*H3*C*N*2(H*CL),. . . . .A4. . . . .DISS. ABS. XXVII,. . . . .1871B, (1967).
      ⊹THE CRYSTAL STRUCTURES OF HYPOPHOSPHOROUS ACID H3*P*04 AND ACETONITRILE
      -2-(HYDROGEN CHLORIDE) (CH3*C*N*2(H*CL): A NEUTRON AND X-RAY. . .STUDY.
      ⊹WILLIAMS(J.M.).

SECTION   5 -STRUCTURE AND CROSS-SECTION DETERMINATIONS . . . .

283 C*H3*C*O*O*H,. . . . . .A1. . . . . . .ACTA CRYST. VOL.B27, ... 893, (1971).
     *HYDROGEN BOND STUDIES.XLIV. NEUTRON DIFFRACTION STUDY OF ACETIC ACID.
     *JONSSON(P.G.).

284 C6*H3*CL3, ...A1. . . . . . .ACTA CRYST. VOL.B28, ... 1388, (1972).
     *THE CRYSTAL STRUCTURE OF 1,2,3-TRICHLOROBENZENE: NEUTRON DIFFRACTION AND
     CONSTRAINED REFINEMENTS.
     *HAZELL(R.G.), LEHMANN(M.S.), PAWLEY(G.S.).

285 C6*H6*CR*(C*O)3, ...A4. . . . . . .ACTA CRYST. VOL.B29, ... 2516, (1973).
     *ELECTRONIC STRUCTURE OF BENZENE CHROMIUM TRICARBONYL BY X-RAY AND
     NEUTRON DIFFRACTION AT 78 DEGREES K.
     *REES(B.), COPPENS(P.).

286 C6*H2*D2*N*O2*N*O2,. . . . . .A1. . . . . . .CZECH. J. PHYS. B VOL.19,. . . . . . 857, (1969).
     *THE TWO-DIMENSIONAL NEUTRON DIFFRACTION STUDY OF THE CRYSTAL STRUCTURE
     OF DEUTERATED 4-NITROANILINE C6*H2*D2*N*O2*N*O2.
     *TICHY(K.), PRELESNIK(B.).

287 C3*H4*N2 (PYRAZOLE), ...A1. . . . . . .ACTA CHEM. SCAND. VOL.24,. . . . . 3248, (1970).
     *A NEUTRON DIFFRACTION STUDY OF THE CRYSTAL AND MOLECULAR STRUCTURE OF
     PYRAZOLE, C3*H4*N2.
     *LARSEN(F.K.), LEHMANN(M.S.), SOTOFTE(I.), RASMUSSEN(S.E.).

288 C32*H18*N8 (PHTHALOCYANINE), ...A1. . . . . . .CHEM. COMMUNICATIONS,. . . . . 554, (1969).
     *THE LOCATION OF THE INNER HYDROGEN ATOMS OF PHTHALOCYANINE: A NEUTRON
     DIFFRACTION STUDY.
     *HOSKINS(B.F.), MASON(S.A.), WHITE(J.C.B.).

289 (C*H3*N*H3)2*MN*CL4, ...A1,A2. . . .SOL. STATE COMM. VOL.12, ... 1157, (1973).
     *INVESTIGATIONS OF A QUASI TWO-DIMENSIONAL HEISENBERG ANTIFERROMAGNETIC
     SYSTEM. NONDEUTERATED/DEUTERATED ALKYL-AMMONIUM-TETRACHLOROMANGANATE.
     *HEGER(G.), HENRICH(E.), KANELLAKOPULOS(B.).

290 C5*H6*N*O*CL.H2*O, ...A1. . . . . . .TETRAHEDRON LETT. NO.59, ... 5219, (1969).
     *THE MOLECULAR STRUCTURES OF TWO HYDROXYPYRIDINE HYDROCHLORIDES.
     *MASON(S.A.), WHITE(J.C.B.), WOODLOCK(A.).

291 C5*H9*N*O3(4-HYDROXY-L-PROLINE),. . . .A1. . . . . . .ACTA CRYST. VOL.B29, ... 231, (1973).
     *PRECISION NEUTRON DIFFRACTION STRUCTURE DETERMINATION OF PROTEIN AND
     NUCLEIC ACID COMP..IX.THE CRYST./MOLEC. STRUCTURE OF 4-HYDROXY-L-PROLINE
     *KOETZLE(T.F.), LEHMANN(M.S.), HAMILTON(W.C.).

292 C6*H3*N3*O6 (TRINITROBENZENE), ...A1. . . . . . .ACTA CRYST. VOL.B28, ... 193, (1972).
     *THE CRYSTAL STRUCTURE OF 1,3,5-TRINITROBENZENE BY NEUTRON DIFFRACTION.
     *CHOI(C.S.), ABEL(J.E.).

293 C9*H11*N*O3.H*CL,. . . . . .A1. . . .J. CHEM. PHYS. VOL.58 NO.6, ... 2547, (1973).
     *PRECISION NEUTRON DIFF. STRUCT. DETERMINATION/PROTEIN/NUCLEI ACID COMP..
     X..A COMP. BETWEEN/CRYST/MOLEC. STRUCT. OF L-TYROSINE/L-TYROSINE-H*CL.
     *FREY(M.N.), KOETZLE(T.F.), LEHMANN(M.S.), HAMILTON(W.C.).

294 C6*H14*O6, ...A1. . . . . . .ACTA CRYST. VOL.B27, ... 2393, (1971).
     *DETERMINATION OF THE CRYSTAL STRUCTURE OF THE A FORM OF D-GLUCITOL BY
     NEUTRON AND X-RAY DIFFRACTION.
     *YOUNG-JA-PARK, JEFFREY(G.A.), HAMILTON(W.C.).

295 C10*H20*O2,. . . . . .A1. . . . . . .ACTA CRYST. VOL.B29, ... 2278, (1973).
     *THE STRUCTURES OF MEDIUM-RING COMPOUNDS. XVIII: X-RAY AND NEUTRON
     DIFFRACTION ANALYSIS OF CYCLODECANE-1,6-TRANS-DIOL.
     *ERMER(O.), DUNITZ(J.D.), BERNAL(I.).

296 C12*H12*O3 (TRIACETYLBENZENE), ...A1. . . . . . .ACTA CRYST. VOL.B29, ... 1903, (1973).
     *NEUTRON-DIFFRACTION REFINEMENT OF THE CRYSTAL STRUCTURE OF 1,3,5-TRI-
     ACETYLBENZENE.
     *O≠CONNOR(B.H.), MOORE(F.H.).

297 C12*H22*O11 (SUCROSE), ...A1. . . . . . .SCIENCE VOL.141, ... 921, (1963).
     *SUCROSE: PRECISE DETERMINATION OF CRYSTAL AND MOLECULAR STRUCTURE BY
     NEUTRON DIFFRACTION.
     *BROWN(G.M.), LEVY(H.A.).

298 C12*H22*O11 (SUCROSE), ...A1. . . . . . .ACTA CRYST. VOL.B29, ... 790, (1973).
     *FURTHER REFINEMENT OF THE STRUCTURE OF SUCROSE BASED ON NEUTRON-
     DIFFRACTION DATA.
     *BROWN(G.M.), LEVY(H.A.).

299 C12*H22*O11 (SUCROSE), ...A1. . . . . . .ACTA CRYST. VOL.B29, ... 797, (1973).
     *SUCROSE: X-RAY REFINEMENT AND COMPARISON WITH NEUTRON REFINEMENT.
     *HANSON(J.C.), SIEKER(L.C.), JENSEN(L.H.).

300 C*H2*O2,C6*H6, ...P. . . . . . .Z. PHYSIK VOL.163, ... 218, (1961).
     *THE SCATTERING OF SUBTHERMAL NEUTRONS BY H2*O, C*H2*O2 AND C6*H6.
     *HEINLOTH(K.). (IN GERMAN).

301 C*H3*O*H,C*H3*I, ...A3. . . . . .ACTA PHYS. POLON. VOL.12, ... 45, (1953).
     *INVESTIGATION OF THE MOLECULAR STRUCTURE OF METHYL ALCOHOL BY THE
     SCATTERING OF THERMAL NEUTRONS.
     *JANIK(J.A.).

302 C*H3*O*H,C*H3*I, ...A3. . . . . .ACTA PHYS. POLON. VOL.13, ... 167, (1954).
     *SCATTERING OF 80 DEGREE K NEUTRONS BY C*H3*O*H AND C*H3*I MOLECULES.
     *JANIK(J.A.).

303 C*H3*O*H,C*H3*S*H, ...P. . . . . . .NUKLEONIKA VOL.10, ... 201, (1965).
     *SLOW NEUTRON SCATTERING BY C*H3*O*H AND C*H3*S*H MOLECULES.
     *BOROWSKI(F.), RZANY(A.).

304 C6*H4*(O*H)2 (RESORCINOL), ...A1. . . . . . .PROC. ROY. SOC. A VOL.235, ... 552, (1956).
     *A STUDY OF ALPHA-RESORCINOL BY NEUTRON DIFFRACTION.
     *BACON(G.E.), CURRY(N.A.).

305 C4*H4*O4*N2-H2*O, ...A1. . . . . . .ACTA CRYST. VOL.B25, ... 1970, (1969).
     *THE 5-HYDROXYL CONFIGURATION IN DIALURIC ACID MONOHYDRATE BY NEUTRON
     CRYSTAL STRUCTURE DETERMINATION.
     *CRAVEN(B.M.), SABINE(T.M.).

306 (C*H3)-X,. . . . . .A3. . . . . . .IAEA SYMPOSIUM VIENNA, ... 293, (1960).
     *SCATTERING OF THERMAL NEUTRONS BY POLARIZED MOLECULES.
     *JANIK(J.), KRASNICKI(SZ.), MURASIK(A.).

307 C2*N4*H4,. . . . . .A1. . . . . . .SOV. PHYS. CRYST. VOL.11, ... 177, (1966).
     *A NEUTRON-DIFFRACTION STUDY OF THE CRYSTAL STRUCTURE OF DICYANDIAMIDE.
     *RANNEV(N.V.), OZEROV(R.P.), DATT(I.D.), KSHNYAKINA(A.N.).

SECTION   5 -STRUCTURE AND CROSS-SECTION DETERMINATIONS . . . .

308 C6*N4*H12, . . .
. . . . A1 . . . . . ACTA CRYST. VOL.10, ... 107, (1957).
*INVESTIGATION OF HEXAMETHYLENE TETRAMINE BY NEUTRON DIFFRACTION.
*ANDRESEN(A.F.).

309 C3*N3*H3, ...
. . . . A4 . . . . . SCIENCE VOL.158, ... 1577, (1967).
*COMPARATIVE X-RAY AND NEUTRON DIFFRACTION STUDY OF BONDING EFFECTS
IN S-TRIAZINE.
*COPPENS(P.).

310 C5*N8*H12*2(H*CL)*H2*O, .
. . . . A1 . . . . . ACTA CRYST. VOL.B24, ... 1147, (1968).
*CRYSTAL/MOLECULAR STRUCTURE OF AN ANTI-LEUKEMIA DRUG: METHYLGLYOXAL
BISGUANYLHYDRAZONE DIHYDROCHLORIDE MONOHYDRATE:NEUTRON/X-RAY DIFFRACTION
*HAMILTON(W.C.), LA-PLACA(S.J.).

311 C4*N2*O3*H9*CL.H2*O, .
. . . . A1 . . . . . ACTA CRYST. VOL.B28, ... 2083, (1972).
*PRECISION NEUTRON DIFFR. STRUC. DETERMINATION OF PROTEIN/NUCLEIC ACID
COMPONENTS. II.THE CRYST./MOLEC. STRUC. OF GLYCYLGLYCINE ...MONOHYDRATE.
*KOETZLE(T.F.), HAMILTON(W.C.), PARTHASARATHY(R.).

312 C4*N2*O2*H8, . . .
. . . . A1 . . . . . ACTA CRYST. VOL.14, ... 95, (1961).
*A NEUTRON DIFFRACTION REFINEMENT OF THE CRYSTAL STRUCTURE OF DIMETHYL-
GLYOXIME.
*HAMILTON(W.C.).

313 C*O2, ...
. . . . A3 . . . . . PHYS. REV. VOL.75, ... 1609, (1949).
*NEUTRON DIFFRACTION BY GASES.
*ALCOCK(N.Z.), HURST(D.G.).

314 C2*O4*H2.2H2*O (OXALIC ACID). .
. . . . A1 . . . . . DISS. ABSTR. VOL.14, ... 1152, (1954).
*THE CRYSTAL STRUCTURES OF OXALIC ACID DIHYDRATE AND ALPHA IODIC ACID AS
DETERMINED BY NEUTRON DIFFRACTION.
*GARRETT-SR(B.S.).

315 (C*O*O*D)2*2(D2*O), .
. . . . A1 . . . . . ACTA CRYST. 23, ... 64, (1967).
*THE CRYSTAL STRUCTURE OF DEUTERATED OXALIC ACID DIHYDRATED
(C*O*O*D)2*2(D2*O).
*WASAKI(F.F.), IWASAKI(H.), SAITO(Y.).

316 C*S2, C*SE2, . .
. . . . A3 . . . . . CAN. J. PHYS. VOL.52, ... 241, (1974).
*MOLECULAR CORRELATION FUNCTIONS IN LIQUID CARBON DISULFIDE AND CARBON
DISELENIDE.
*SUZUKI(K.), EGELSTAFF(P.A.).

317 CA, ...
. . . . P . . . . . NUCL. PHYS. A VOL.A169, ... 95, (1971).
*THERMAL NEUTRON CAPTURE CROSS SECTIONS IN CALCIUM.
*CRANSTON(F.P.), WHITE(D.H.).

318 CA (ISO. A=44), .
. . . . P . . . . . J. INORG. NUCL. CHEM. VOL.32, . 2839, (1970).
*THE THERMAL NEUTRON CAPTURE CROSS-SEC. AND RES. CAPTURE INTEG. OF CA(IS.
A=44), NI(IS. A=62), YB(IS. A=168, 174), TM(IS. A=169) AND TL(IS. A=203)
*SIMS(G.H.E.), JUHNKE(D.G.).

319 CA3*AL2*(O4*D4)3, .
. . . . A1 . . . . . J. CHEM. PHYSICS VOL.48, ... 3037, (1968).
*NEUTRON AND X-RAY DIFFRACTION STUDY OF CA3*AL2*(O4*D4)3, A GARNETOID.
*FOREMAN-JR(D.W.).

320 CA3*AL2*(SI*O4)3, .
. . . . A1 . . . . . COLL. INTER. N.126 (GRENOBLE), . 16, (1963).
*REFINEMENT OF THE GARNET STRUCTURE BY NEUTRON DIFFRACTION.
*PRANDL(W.).

321 CA*(AL2*SI4*O12)*2H2*O, .
. . . . A1 . . . . . ACTA CRYST. (INTERACT.) VOL.A25, S119, (1969).
*NEUTRON DIFFRACTION MEASUREMENTS ON THE STRUCTURE OF THE ZEOLITE
MINERAL LAUMONTITE, CA*(AL2*SI4*O12)*2H2*O.
*BARTL(H.). (ED. NOTE: IN GERMAN).

322 (CA2-BI-FE4-V)*O12, .
. . . . A1 . . . . . LATV. PSR. FIZ. TEHN. SER. NO.4 87, (1970).
*A NEUTRON-DIFFRACTION STUDY OF FERRITE-GARNET (CA2-BI-FE4-V)*O12.
*DUKHOVSKAYA(E.L.), LIPIN(YU.V.), NOZIK(YU.Z.).

323 CA-BI-V, ...
. . . . A2 . . . . . PHYS. STAT. SOLIDI A VOL.13, . . K119, (1972).
*NEUTRON-DIFFRACTION INVESTIGATION OF CA-BI-V FERRITE GARNETS.
*NOZIK(YU.Z.).

324 CA*C2, ...
. . . . A1 . . . . . J. CHEM. PHYSICS VOL.54, ... 3514, (1971).
*NEUTRON STRUCTURE ANALYSIS OF CUBIC CA*C2 AND K*C*N.
*ATOJI(M.).

325 CA*C2, ...
. . . . A1,A4 . . . J. CHEM. PHYS. VOL.35, ... 1950, (1961).
*NEUTRON DIFFRACTION STUDIES OF CA*C2, Y*C2, LA*C2, CE*C2, TB*C2, YB*C2,
LU*C2, AND U*C2.
*ATOJI(M.).

326 CA*C2 TYPE, . . .
. . . . A4 . . . . . J. PHYS. SOC. JAPAN VOL.17 B-II, 395, (1962).
*NEUTRON DIFFRACTION STUDIES OF HIGHER CARBIDES OF HEAVY METALS.
*ATOJI(M.).

327 CA*C2, ...
. . . . A1 . . . . . J. CHEM. PHYS. VOL.31, ... 332, (1959).
*STRUCTURES OF CALCIUM DICARBIDE AND URANIUM DICARBIDE BY NEUTRON
DIFFRACTION.
*ATOJI(M.), MEDRUD(R.C.).

328 CA*C*O3, ...
. . . . P . . . . . PHYS. REV. VOL.58, ... 321, (1940).
*SCATTERING OF THERMAL NEUTRONS BY CRYSTALS.
*RASETTI(F.).

329 CA*(CO,FE)*SI*O4, .
. . . . A2 . . . . . J. AMER. CERAM. SOC. VOL.49, . . 284, (1966).
*MAGNETIC PROPERTIES OF CA*CO*SI*O4 AND CA*FE*SI*O4.
*NEWNHAM(R.E.), CARON(L.G.), SANTORO(R.P.).

330 CA*(CR-FE)2*O4, .
. . . . A2 . . . . . J. APP. PHYS. 38, ... 946, (1967).
*MAGNETIC PHASE EQUILIBRIUM IN CR-SUBSTITUTED CA*FE2*O4.
*CORLISS(L.M.), HASTINGS(J.M.), KUNNMANN(W.).

331 CA*(CR-FE)2*O4, .
. . . . A2 . . . . . PHYS. REV. 160, ... 408, (1967).
*MAGNETIC PHASE EQUILIBRIUM IN CHROMIUM-SUBSTITUTED CALCIUM FERRITE.
*CORLISS(L.M.), HASTINGS(J.M.), KUNNMANN(W.).

332 CA3*CR2*(GE*O4)3, .
. . . . A2 . . . . . SOL. STATE COMM. VOL.11 NO.5, . 645, (1972).
*MAGNETIC STRUCTURE AND SPACE GROUP OF THE GARNET CA3*CR2*(GE*O4)3.
*PRANDL(W.).

333 CA*F2, ...
. . . . A4 . . . . . ACTA CRYST. VOL.A27, ... 622, (1971).
*A NEUTRON DIFFRACTION STUDY OF SR*F2 AND CA*F2.
*COOPER(M.J.), ROUSE(K.D.).

SECTION   5 -STRUCTURE AND CROSS-SECTION DETERMINATIONS . . . .

334 CA*F2, ...A1. . . . . . . ACTA CRYST. 18, ... 75, (1965).
      *THE' 'ANOMALOUS' 'BEHAVIOUR' OF' 'THE' 'NEUTRON 'REFLECTIONS' OF 'FLUORITE.
      *WILLIS(B.T.M.).

335 CA*F2, ... T.P. . . . PROC. ROY. SOC. 298, ... 289, (1967).
      *ANHARMONIC VIBRATION IN FLUORITE STRUCTURE.
      *DAWSON(B.), HURLEY(A.C.), MASLEN(V.W.).

336 CA*F2, ... A4. . . . . PHYS. LETT. 1 . . . . 187, (1962).
      *DETERMINATION EXPERIMENTALE DES FREQUENCES ≠OPTIQUES≠ PRINCIPALES! CA*F2
      *CRIBIER(D.), FARNOUX(B.), JACROT(B.).

337 CA2*F2*O5, ... A1. . . . . PHYS. LETT. 25A, ... 9, (1967).
      *NEUTRON DIFFRACTION STUDY OF DICALCIUM FERRITE.
      *FRIEDMAN(Z.), SHAKED(H.), SHTRIKMAN(S.).

338 CA*F2-(Y*F3), ... A1. . . . . J. PHYS. C VOL.4, ... 3107, (1971).
      *DEFECT STRUCTURE OF CALCIUM FLUORIDE CONTAINING EXCESS'ANIONS! I. BRAGG
      SCATTERING.
      *CHEETHAM(A.K.), FENDER(B.E.F.), COOPER(M.J.).

339 CA2*(FE-AL)*O5, ... A2. . . . . . J. PHYS. CHEM. SOLIDS VOL.31, . . 793, (1970).
      MAGNETIC STRUCTURES IN THE CA2(FE-AL)*O5 SYSTEM.
      *GELLER(S.), GRANT(R.W.), FULLMER(L.O.).

340 CA3*(FE2,CR2)*(GE*O4)3, ... A2 . . . . ACTA CRYST. VOL.A28 PART.S-4. . . 194, (1972).
      *MAGNETIC STRUCTURE OF GARNETS WITH SINGLE 16A OR 24C MAGNETIC SUBLATTICE
      *PRANDL(W.). (ABSTRACT ONLY).

341 CA3*FE2*GE3*O12, ... A2 . . . . SOL. STATE COMM. VOL.10 . . . . . 5, (1972).
      *DETERMINATION' PAR DIFFRACTION DES NEUTRONS DE LA'STRUCTURE ANTIFERROMA-
      GNETIQUE DU GRENAT CA3*FE2*GE3*O12.
      *PLUMIER(R.).

342 CA3*FE2*(GE*O4)3, ... A2. . . . . . SOL. STATE COMM. VOL.10 . . . . 529, (1972).
      *MAGNETIC 'STRUCTURE' 'AND' 'SPACE' 'GROUP' OF' THE GARNET CA3*FE2*(GE*O4)3.
      *PRANDL(W.).

343 CA*FE2*O4, ... A2. . . . . . COMP. REND. 263, ... 98, (1966).
      *STRUCTURE' ET PROPRIETES MAGNETIQUES DE L≠ANTIFERROMAGNETIQUE CA*FE2*O4.
      *ALLAIN(Y.), BOUCHER(B.), IMBERT(P.), PERRIN(M.).

344 CA2*FE2*O5, ... A2. . . . . . J. PHYS. SOC. JAPAN VOL.24, . . . 446, (1968).
      *MAGNETIC STRUCTURE OF CA2*FE2*O5.
      *TAKEDA(T.), YAMAGUCHI(Y.), TOMIYOSHI(S.), FUKASE(M.), SUGIMOTO(M.),
      WATANABE(H.).

345 CA*FE2*O4, ... A2. . . . . . J. PHYS. SOC. JAPAN VOL.22, . . . 939, (1967).
      *NEUTRON'DIFFRACTION STUDY OF CA*FE2*O4.
      *WATANABE(H.), YAMAGUCHI(H.), OHASHI(M.). . . ET AL.

346 CA3*FE2*((SI*O2)1.15)*((O*H)7.4) . . A2. . . . . ACTA CRYST. VOL.A27 . . . . . . 68, (1971).
      *ETUDE DU COMPOSE CA3*FE2*((SI*O2)1.15)*((O*H)7.4) ' PAR ABSORPTION
      INFRAROUGE ET DIFFRACTION DES RAYONS X ET DES NEUTRONS.
      *COHEN-ADDAD(C.). (ED. NOTE! REFERENCE IS ERRONEOUSLY GIVEN AS
      ACTA CRYST.(1970), A26, 68 ON THE FIRST PAGE OF THIS PAPER.).

347 CA3*FE2*SI3*O12, ... A2 . . . . . SOV. PHYS. SOL. STATE VOL.14, . . 2387, (1973).
      *MAGNETIC STRUCTURE OF THE GARNET ANDRADITE.
      *PLAKHTII(V.P.), GOLOSOVSKII(I.V.).

348 CA3*(FE(3.5)-V(1.5))*O12, . . . A4 . . . . . LATV. PSR. . . FIZ. TEHN. SER. NO.2 124, (1970).
      NEUTRON-DIFFRACTION 'STUDY' OF CA3(FE(3.5)-V(1.5))*O12 FERRITE-GARNET.
      *DUKHOVSKAYA(E.L.), LIPIN(YU.V.), NOZIK(YU.Z.).

349 (CA-FE-V)*O12, ... A2. . . . . . SOV. PHYS. SOL. STATE VOL.9, . . 1762, (1968).
      *NEUTRON' DIFFRACTION STUDY OF ATOMIC AND MAGNETIC STRUCTURES OF'CALCIUM-
      VANADIUM-FERRITE GARNET.
      *PLAKHTII(V.P.), GORDEEV(G.P.), KUZINA(A.A.), TURKEVICH(E.I.).

350 CA*H*AS*O4.H2*O, ... A1. . . . . . ACTA CRYST. VOL.B28 . . . . . . 209, (1972).
      *A' NEUTRON AND X-RAY REFINEMENT OF THE CRYSTAL STRUCTURE OF CA*H*AS*O4.-
      H2*O (HAIDINGERITE).
      *FERRARIS(G.), JONES(D.W.), YERKESS(J.).

351 CA*(H2*AS*O4)2, ... A1. . . . . . ACTA CRYST. VOL.B28 . . . . . . 2430, (1972).
      *A' NEUTRON DIFFRACTION'STUDY OF THE CRYSTAL STRUCTURE'OF CALCIUM BIS(DI-
      HYDROGEN ARSENATE), CA*(H2*AS*O4)2.
      *FERRARIS(G.), JONES(D.W.), YERKESS(J.).

352 CA*H*AS*O4.2H2*O, ... A1. . . . . . ACTA CRYST. VOL.B27 . . . . . . 349, (1971).
      *DETERMINATION' OF HYDROGEN ATOM POSITIONS IN THE CRYSTAL STRUCTURE OF
      PHARMACOLITE, CA*H*AS*O4*2H2*O, BY NEUTRON DIFFRACTION.
      *FERRARIS(G.), JONES(D.W.), YERKESS(J.).

353 CA*H2,CA*O2, ... A1. . . . . . ACTA CRYST. VOL.15, ... 92, (1962).
      *THE'CRYSTAL'STRUCTURE OF CALCIUM HYDRIDE.
      *BERGSMA(J.), LOOPSTRA(B.O.).

354 CA*H*P*O4, ... A1. . . . . . J. CRYST. MOL. STRUCT.(GB) VOL.1 347, (1971).
      *NEUTRON' DIFFRACTION INVESTIGATION OF' THE HYDROGEN POSITIONS IN THE
      CRYSTAL STRUCTURE OF MONETITE CA*H*P*O4.
      *DENNE(W.A.), JONES(D.W.).

355 CA*H*P*O4.2H2*O, ... A1. . . . . . J. CHEM. SOC. A (1971) . . . . 3725, (1971).
      *CRYSTAL STRUCTURE OF BRUSHITE, CALCIUM HYDROGEN 'ORTHOPHOSPHATE DI-
      HYDRATE! A NEUTRON-DIFFRACTION INVESTIGATION.
      *CURRY(N.A.), JONES(D.W.).

356 CA*H*P*O4, ... A1. . . . . BULL. SOC. CHIM. FR. 1968, . . . 1748, (1968).
      *HYDROGEN LOCATION IN DICALCIUM PHOSPHATE STRUCTURES.
      *CURRY(N.A.), DENNE(W.A.), JONES(D.W.).

357 CA3*MN2*GE3*O12, ... A1. . . . . LATV. PSR. . . FIZ. TEHN. SER. NO.6 120, (1971).
      *NEUTRON DIFFRACTION STUDY OF A CA3*MN2*GE3*O12 GARNET.
      *LIPIN(YU.), NOZIK(YU.). (ED. NOTE! IN RUSSIAN ONLY).

358 CA3*MN2*GE3*O12, ... A2. . . . . SOL. STATE COMM. VOL.9 . . . . . . 1723, (1971).
      *DETERMINATION' PAR DIFFRACTION DES NEUTRONS DE LA'STRUCTURE ANTIFERROMA-
      GNETIQUE DU GRENAT CA3*MN2*GE3*O12.
      *PLUMIER(R.).

359 CA2*MN*O4, ... A2. . . . . J. PHYS. CHEM. SOLIDS VOL.32, . . 1189, (1971).
      *INFLUENCE' DE ' LA PREPARATION CHIMIQUE SUR LES STRUCTURES'MAGNETIQUES DE
      CA2*MN*O4.
      *OLLIVIER(G.), BUISSON(G.).

SECTION   5 -STRUCTURE AND CROSS-SECTION DETERMINATIONS . . . .

360 CA2*MN*04, ...A2 . . . . . .PHYS. REV. VOL.188, ... 930, (1969).
     *NEUTRON-DIFFRACTION STUDY OF MAGNETIC ORDERING IN CA2*MN*04.
     *COX(D.E.), SHIRANE(G.), BIRGENEAU(R.J.), MACCHESNEY(J.B.).

361 CA*MN2*04, ...A1,A2 . . . .J. PHYSIQUE TOME 26, ... 789, (1965).
     *MAGNETIC MEASUREMENTS AND NEUTRON CRYSTALLOGRAPHY OF ANTIFERROMAGNETIC
     CA*MN2*04.
     *ALLAIN(Y.), BOUCHER(B.). (IN FRENCH).

362 CA*MN*03, ...A2 . . . . . .PHYS. REV. VOL.100, ... 545, (1955).
     *NEUTRON DIFFRACTION STUDY OF THE MAGNETIC PROPERTIES OF THE SERIES OF
     PEROVSKITE-TYPE COMPOUNDS ((1-X)LA,(X)CA)*MN*03.
     *WOLLAN(E.O.), KOEHLER(W.C.).

363 CA*MN*03, ...A2 . . . . . .J. PHYS. CHEM. SOLIDS VOL.31, . . 2741, (1970).
     *COVALENCY PARAMETERS FOR CR(3+), FE(3+), AND MN(4+) IN AN OXIDE
     ENVIRONMENT.
     *TOFIELD(B.C.), FENDER(B.E.F.).

364 CA*MN*SI*04, ...A2 . . . . . .J. PHYS. CHEM. SOL. 26, ... 927, (1965).
     *MAGNETIC STRUCTURE OF CA*MN*SI*04.
     *CARON(L.G.), SANTORO(R.P.), NEWNHAM(R.E.).

365 CA*MO*04, ...A1 . . . . . . .J. CHEM. PHYSICS VOL.55, ...1093, (1971).
     *CRYSTAL STRUCTURE REFINEMENT OF SR*MO*04, SR*W*04, CA*MO*04, AND BA*W*04
     BY NEUTRON DIFFRACTION.
     *GURMEN(E.), DANIELS(E.), KING(J.S.).

366 CA*O, ...P . . . . . . .PHYS. REV. VOL.75, ... 975, (1949).
     *TOTAL NEUTRON CROSS SECTIONS OF COMPOUNDS WITH DIFFERENT CRYSTALLINE
     STRUCTURES.
     *WINSBERG(L.), MENEGHETTI(D.), SIDHU(S.S.).

367 3CA*O.AL2*03.6D2*O, ...A1 . . . . . .DISS. ABS. XXVII, ... 2459B, (1967).
     *THE TETRAHEDRON IN TRICALCIUM ALUMINATE HEXADEUTERATE DETERMINED
     BY MEANS OF NEUTRON AND X-RAY DIFFRACTION.
     *FOREMAN-JR(D.W.).

368 CA*(O*H)2, ...A1 . . . . . .J. CHEM. PHYS. VOL.26, ... 563, (1957).
     *NEUTRON DIFFRACTION STUDY OF CALCIUM HYDROXIDE.
     *BUSING(W.R.), LEVY(H.A.).

369 CA10*(P*04)6*(O*H)2, ...A1 . . . . . .NATURE VOL.204, ...1050, (1964).
     *CRYSTAL STRUCTURE OF HYDROXY APATITE.
     *KAY(M.I.), YOUNG(R.A.), POSNER(A.S.).

37C CA*S*04.2H2*0 (GYPSUM), ...A1 . . . . . .J. CHEM. PHYS. VOL.29, ... 1306, (1958).
     *NEUTRON DIFFRACTION STUDY OF GYPSUM, CA*S*04.2H2*0.
     *ATOJI(M.), RUNDLE(R.E.).

371 CA*SN*(O*H)6, ...A1 . . . . . .ACTA CRYST. B24, ... 1358, (1968).
     *APPLICATIONS DE LA METHODE DU SIMPLEX A LA DETERMINATION DIRECTE DES
     STRUCTURES CRISTALLINES.
     *BASSI(G.).

372 CA2*SR(C*H3*C*H2*C*O*O)6, ...A4 . . . . . .PHYS. LETT. 25A, ... 123, (1967).
     *SLOW NEUTRON SCATTERING ON CA2*SR(C*H3*C*H2*C*O*O)6.
     *DIMIC(V.), OSREDKAR(M.), PETKOVSEK(J.).

373 CA3*U*06, ...A4 . . . . . .ACTA CRYST. 20, ... 508, (1966).
     *THE CRYSTAL STRUCTURE OF SOME ALKALINE EARTH METALS URANATES OF THE TYPE
     M3*U*06.
     *RIETVELD(H.M.).

374 CA*U*04,CA2*U*05,CA3*U*06, ...A1 . . . . . .ACTA CRYST. B25, ... 787, (1969).
     *THE STRUCTURE OF SOME ALKALINE-EARTH METAL URANATES.
     *LOOPSTRA(B.O.), RIETVELD(H.M.).

375 CA*V2*04, ...A2 . . . . . .J. PHYS. CHEM. SOL. 28, ...1089, (1967).
     *MAGNETIC STRUCTURE OF CALCIUM VANADITE.
     *HASTINGS(J.M.), CORLISS(L.M.), KUNNMANN(W.), LA-PLACA(S.J.).

376 CA*W*04, ...A1 . . . . . .J. CHEM. PHYSICS VOL.40, ... 504, (1964).
     *NEUTRON DIFFRACTION REFINEMENT OF CA*W*04.
     *KAY(M.I.), FRAZER(B.C.), ALMODOVAR(I.).

377 CA*W*04, ...A1,2 . . . .COLL. INTER. N.126 (GRENOBLE), . . 18, (1963).
     *NEUTRON DIFFRACTION STUDIES AT THE PUERTO RICO NUCLEAR CENTER.
     *ALMODOVAR(I.), BIELEN(H.J.), FRAZER(B.C.), KAY(M.I.).

378 (CA-Y)*F(2.10), ...A4 . . . . . .SOL. STATE COMM. VOL.8, ... 171, (1970).
     *DEFECT STRUCTURE OF FLUORITE COMPOUNDS CONTAINING EXCESS ANIONS.
     *CHEETHAM(A.K.), FENDER(B.E.F.), STEELE(D.), TAYLOR(R.I.), WILLIS(B.T.M.)

379 CD, ...P . . . . . . .UKR. FIZ. ZH.(USSR) VOL.17, . . 38, (1972).
     *STUDY OF THERMAL NEUTRON INTERACTION WITH CADMIUM ISOTOPES.
     *VERTEBNYI(V.P.), VLASOV(M.F.), GNIDAK(N.L.), GRISHANIN(E.I.),
     ZATSERKOVSKII(R.A.), KIZILYUK(A.L.), LEPENDIN(V.I.), TROFIMOVA(N.A.).

380 CD, ...P . . . . . . .NUKLEONIK VOL.11, ... 297, (1968).
     *TOTAL CROSS SECTION OF CADMIUM FOR NEUTRONS OF ENERGY BETWEEN 0.01 AND
     10 EV.
     *WIDDER(F.), BRUNNER(J.).

381 CD, ...P . . . . . . .ACTA POLYTECH. SCANDIN. SER.52, . 3, (1968).
     *MEASUREMENT OF THE CROSS SECTION OF CADMIUM FOR NEUTRONS IN THE
     10 MICRO EV RANGE.
     *PALMGREN(A.).

382 CD, ...P . . . . . . .PHYS. REV. VOL.124, ...1848, (1961).
     *NEUTRON COHERENT SCATTERING AMPLITUDES FOR CD AND EU.
     *ARNOLD(G.P.), NERESON(N.G.).

383 CD, ...P . . . . . . .PHYS. REV. VOL.72, ... 109, (1947).
     *A BENT CRYSTAL NEUTRON SPECTROMETER AND ITS APPLICATION TO NEUTRON
     CROSS-SECTION MEASUREMENTS.
     *SAWYER(R.B.), WOLLAN(E.O.), BERNSTEIN(S.), PETERSON(K.C.).

384 CD, ...P . . . . . . .PHYS. REV. VOL.83, ... 840, (1951).
     *RESONANT SCATTERING OF SLOW NEUTRONS BY CADMIUM.
     *BROCKHOUSE(B.N.), HURST(D.G.), BLOOM(M.).

385 CD*(N*03)2-4D2*0, ...A1 . . . . . .ACTA CRYST. B25, ...1804, (1969).
     *THE DETERMINATION OF THE CRYSTAL STRUCTURE OF CADMIUM NITRATE TETRA-
     DEUTERATE BY MEANS OF NEUTRON ANAMALOUS DISPERSION MEASUREMENT.
     *MACDONALD(A.C.), SIKKA(S.K.).

SECTION   5 -STRUCTURE AND CROSS-SECTION DETERMINATIONS . . . .

386 CO*(N*O3)2*4(D2*O)......A1.......ACTA CRYST. VOL.A25..........S76, (1969).
     SOLUTION OF THE STRUCTURE OF CD(N*O3)2*4(D2*O) BY NEUTRON ANOMALOUS
     DISPERSION MEASUREMENT
     *MACDONALD(A.C.), SIKKA(S.K.).

387 CD*S,. . . . .A4.......PHYS. REV. LETT. VOL.6,. . . . . .   7, (1961).
     *ANOMALOUS NEUTRON DIFFRACTION IN ALPHA CADMIUM SULFIDE.
     *PETERSON(S.W.), SMITH(H.G.).

388 CD*S,. . . . .P.......J. PHYS. SOC. JAPAN VOL.17 B-II, 335, (1962).
     *COHERENT BRAGG SCATTERING OF RESONANT NUCLIDES.
     *PETERSON(S.W.), SMITH(H.G.).

389 CE,. . . . .T.......UKR. FIZ. ZH.(USSR) VOL.16,. . . 1735, (1971).
     *ANALYSIS OF ELASTIC SCATTERING OF NEUTRONS BY K, CO AND CE NUCLEI USING
     A LOCAL OPTICAL NUCLEAR MODEL.
     *KASHUBA(I.E.), KOSTYUK(T.A.), KOTISHEVSKAYA(E.YA.).

390 CE,. . . . .A4.......PHYS. REV. VOL.122.........1409, (1961).
     *NEUTRON DIFFRACTION INVESTIGATIONS OF METALLIC CERIUM AT LOW TEMP.
     *WILKINSON(M.K.), CHILD(H.R.), MCHARGUE(C.J.), KOEHLER(W.C.),
     WOLLAN(E.O.).

391 CE,. . . . .A2.......J. APP. PHYS. 36,. . . . . . . 1078, (1965).
     *MAGNETIC PROPERTIES OF RARE-EARTH METALS AND ALLOYS.
     *KOEHLER(W.C.). (ED. NOTE: REF. SECTION 3, 65/KOEHLER).

392 CE,. . . . .P.......PHYS. REV. VOL.91,. . . . . . . 597, (1953).
     *SLOW NEUTRON SCATTERING CROSS SECTIONS FOR RARE EARTH NUCLIDES.
     *KOEHLER(W.C.), WOLLAN(E.O.).

393 CE-AL3,. . . . ...A2.......PHYS. REV. B VOL.9,. . . . . . . 154, (1974).
     *EXPERIMENTAL EVIDENCE FOR THE FORMATION OF A SINGLET GROUND STATE AT
     LOW TEMPERATURES IN THE DENSE KONDO SYSTEM CE-AL3.
     *MAHONEY(J.V.), RAO(V.U.S.), WALLACE(W.E.), CRAIG(R.S.), NERESON(N.G.).

394 CE-BI, ...A2.......PHYS. STAT. SOL. VOL.44,. . . . K25, (1971).
     *CRYSTAL FIELD AND MAGNETIC PROPERTIES OF CE-BI.
     *TSUCHIDA(T.), SUZAWA(T.), NAKAMURA(Y.).

395 CE*BI, ...A2.......AIP CONF. PROC. VOL.5, ... 1381, (1971).
     *THE MAGNETIC STRUCTURES OF CE*BI.
     *CABLE(J.W.), KOEHLER(W.C.).

396 CE*BI, ...A2.......J. APPL. PHYS. VOL.42, ... 1625, (1971).
     *MAGNETIC PROPERTIES OF CE*BI, ND*BI, TB*BI, AND DY*BI.
     *NERESON(N.), ARNOLD(G.).

397 CE*C2, ...A1,2.....PHYS. LETT. 22,. . . . . . .   21, (1966).
     *CRYSTALS AND MAGNETIC STRUCTURES OF CE*C2, PR*C2 AND ND*C2.
     *ATOJI(M.).

398 CE2*C3,. . . . ...A1,A2.....J. CHEM. PHYS. VOL.46,. . . 4148, (1967).
     *NEUTRON DIFFRACTION STUDY OF CE2*C3 AT LOW TEMPERATURES.
     *ATOJI(M.).

399 CE*C2, ...A1,A2.....J. CHEM. PHYS. VOL.46,. . . 1891, (1967).
     *MAGNETIC AND CRYSTAL STRUCTURES OF CE*C2,PR*C2,ND*C2,TB*C2, AND HO*C2
     AT LOW TEMPERATURES.
     *ATOJI(M.).

400 CE2*C3,. . . . ...A1,A4.....J. CHEM. PHYS. VOL.35,. . . 1960, (1961).
     *NEUTRON-DIFFRACTION STUDIES OF LA2*C3, CE2*C3, PR2*C3, AND TB2*C3.
     *ATOJI(M.), WILLIAMS(D.E.).

401 CE*C2, ...A1,A4....REF. SECTION 3, 61/ATOJI,. . . . .

402 CE*C*O3, ...A2,A4....REF. SECTION 3, 66/BERTAUT,. . . . .

403 CE*CO5,. . . . ...A2.......J. PHYSIQUE TOME 28,. . . . . 216, (1967).
     *MAGNETIC STRUCTURES OF THE INTERMETALLIC COMPOUNDS CE*CO5 AND TB*CO5.
     *LEMAIRE(R.), SCHWEIZER(J.). (IN FRENCH).

404 CE*CO5-D,. . . ...A1.......J. PHYS. CHEM. SOLIDS VOL.35, 301, (1974).
     *A NEUTRON-DIFFRACTION STUDY ON THE STRUCTURAL RELATIONSHIPS OF R*CO5
     HYDRIDES.
     *KUIJPERS(F.A.), LOOPSTRA(B.O.).

405 CE*D,. . . . .A1.......J. PHYS. C VOL.5,. . . . . . . L35, (1972).
     *NEUTRON DIFFRACTION STUDY OF NONSTOICHIOMETRIC CERIUM HYDRIDE.
     *CHEETHAM(A.K.), FENDER(B.E.F.).

406 CE2*FE17,. . . ...A2.......C.R. ACAD. SCI. B TOME 274,. 1166, (1972).
     *TRANSITION FERROMAGNETIQUE-HELIMAGNETIQUE DANS LES COMPOSES LU*FE(9.5)
     ET CE2*FE17.
     *GIVORD(D.), LEMAIRE(R.).

407 CE2*S3,. . . . ...A1.......J. CHEM. PHYSICS VOL.54,. . . 3226, (1971).
     *NEUTRON DIFFRACTION STUDY OF CE2*S3 AT 300-5 K.
     *ATOJI(M.).

408 CE*SB, ...A1.......AIP CONF. PROC. VOL.5, . . . 1385, (1971).
     *NEUTRON DIFFRACTION STUDIES ON DY*SB, ND*SB, AND CE*SB.
     *NERESON(N.), STRUEBING(V.).

409 CE*SB, ...A2.......CONF./RARE EARTHS/ACTINIDES, . 204, (1971).
     *THE MAGNETIC STRUCTURES OF CE*SB.
     *LEBECH(B.), FISCHER(P.).

410 (CE-TB)-RU2, ...A2.......J. PHYS. C VOL.6,. . . . . . 3465, (1973).
     *NEUTRON SCATTERING EXPERIMENT ON THE COEXISTENCE OF SUPERCONDUCTIVITY
     AND FERROMAGNETISM.
     *ROTH(S.), IBEL(K.), JUST(W.).

411 (CE-TB)*RU2, ...A2.......J. APPL. CRYST. VOL.7,. . . . 230, (1974).
     *SUPERCONDUCTIVITY AND CORRELATION OF MAGNETIC MOMENTS IN (CE-TB)*RU2.
     *ROTH(S.), IBEL(K.), JUST(W.).

412 CE*ZN2,. . . . ...A2.......CONF./RARE EARTHS/ACTINIDES, . 218, (1971).
     *MAGNETIC STRUCTURES OF CE*ZN2 AND TB*ZN2.
     *DEBRAY(D.), SOUGI(M.), MERIEL(P.).

413 CE*ZN2,. . . . ...A2.......J. CHEM. PHYS. VOL.56, ... 4325, (1972).
     *MAGNETIC STRUCTURES OF CE*ZN2 AND TB*ZN2.
     *DEBRAY(D.), SOUGI(M.), MERIEL(P.).

SECTION   5 -STRUCTURE AND CROSS-SECTION DETERMINATIONS . . . .

414 CF (ISO. A=252), ...P. . . . . . . .NUCL. SCI. ENG. VOL.37, ... 228, (1969),
     *THE THERMAL-NEUTRON CAPTURE CROSS SECTION AND RESONANCE INTEGRAL OF
     CF (ISOTOPE A=252).
     *HALPERIN(J.), BEMIS-JR(C.E.), DRUSCHEL(R.E.), STOKELY(J.R.).

415 CHOLINE IODIDE (DIMETHYLACETYL), ...A1. . . . . . .J. PHARM. PHARMACOL. VOL.22, . . 724, (1970).
     *NEUTRON DIFFRACTION STUDY OF AN ACETYLCHOLINE ANALOG- ERYTHRO-(ALPHA,
     BETA)-DIMETHYLACETYLCHOLINE IODIDE.
     *BRENNAN(T.), ROSS(F.K.), HAMILTON(W.C.), SHEFTER(E.).

416 CHROMITES(ORTHO), ...A4. . . . . .J. PHYSIQUE TOME 31, ... 803, (1970).
     *STUDY OF SOME RARE EARTH ORTHOFERRITES AND ORTHOCHROMITES BY NEUTRON
     DIFFRACTION.
     *PATAUD(P.), SIVARDIERE(J.).

417 CL (ISO. A=37), , ...P. . . . . . . .J. NUCL. ENERGY VOL.24, ... 419, (1970).
     *THERMAL NEUTRON CAPTURE CROSS-SECTION MEASUREMENTS FOR NA(ISOTOPE A=23),
     AL(ISOTOPE A=27), CL(ISOTOPE A=37) AND V(ISOTOPE A=51).
     *RYVES(T.B.), PERKINS(O.R.).

418 CL (ISO. A=35), , ...P. . . . . . . .J. INORG. NUCL. CHEM. VOL.31, . . 3721, (1969).
     *THE THERMAL NEUTRON CAPTURE CROSS SECTION AND RESONANCE CAPTURE INTEGRAL
     OF CL(ISOTOPE A=35) FOR (NEUTRON, GAMMA) AND (NEUTRON, PROTON) REACTIONS
     *SIMS(G.H.E.), JUHNKE(O.G.).

419 CL, ...P. . . . . . .Z. PHYSIK VOL.198, ... 187, (1967).
     *ABSOLUTE MEASUREMENT OF COHERENT SCATTERING LENGTHS OF H, C AND CL AND
     DETERMINATION OF GRAV. ACC. FOR FREE NEUTRONS WITH A GRAVITY REFRACTOM.
     *KOESTER(L.). (IN GERMAN).

420 CL-CU, ...A3,P. . . .J. OF PHYS.-C-, SER.2 VOL.4, ... 3034, (1971).
     *THE PARTIAL STRUCTURE FACTORS OF MOLTEN CUPROUS CHLORIDE FROM NEUTRON
     DIFFRACTION MEASUREMENTS.
     *PAGE(D.I.), MIKA(K.).

421 CM (ISO. A=244+248), ...P. . . . . . .J. INORG. NUCL. CHEM. VOL.33, . . 1553, (1971).
     *THERMAL NEUTRON CROSS SECTIONS AND RESONANCE INTEGRALS FOR CM(ISOTOPES
     A=244 THROUGH 248).
     *THOMPSON(M.C.), HYDER(M.L.), REULAND(P.J.).

422 CM, ...P. . . . . . .PHYS. REV. VOL.134, ...B1281, (1964).
     *TOTAL NEUTRON CROSS SECTION OF CM (ISOTOPE A=244).
     *COTE(R.E.), BARNES(R.F.), DIAMOND(H.).

423 CO ++ (IN CO*O), ...A2,P. . . .PHYS. REV. B-1, ... 2243, (1970).
     *MAGNETIC FORM FACTOR OF CO++ ION IN COBALTOUS OXIDE.
     *KHAN(D.C.), ERICKSON(R.A.).

424 CO, ...A2. . . . . . .SOL. STATE COMM. VOL.10, ... 667, (1972).
     *MAGNETIC FORM FACTOR OF PURE COBALT IN THE HIGH TEMPERATURE F.C.C. PHASE
     *MENZINGER(F.), SACCHETTI(F.).

425 CO, ...A2. . . . . . .METAL. SOC. CONF. LOS ANGELES, . 15, (1967).
     *SPIN DENSITY DISTRIBUTION IN FE, CO, AND NI.
     *SHULL(C.G.).

426 CO, ...A2. . . . . . .J. APP. PHYS. VOL.40, ... 1454, (1969).
     *SEARCH OF NONCOLLINEAR SPIN DENSITY IN HEXAGONAL CO.
     *MOON(R.M.), KOEHLER(W.C.).

427 CO, ...A2. . . . . . .COLL. INTER. N.126 (GRENOBLE), . 186, (1963).
     *ROTATION DES SPINS DANS LE COBALT HEXAGONAL.
     *BERTAUT(E.F.), DELAPALME(A.), PAUTHENET(R.).

428 CO, ...A2. . . . . . .COLL. INTER. N.126 (GRENOBLE), . 180, (1963).
     *MAGNETISM RESEARCH WITH POLARIZED NEUTRONS AT THE CENTRO DE STUDI
     NUCLEARI DELLA CASACCIA DEL C.N.E.N., ROMA, ITALY.
     *ANTONINI(B.), FELCHER(G.P.), MENZINGER(F.), PAOLETTI(A.), RICCI(F.P.),
     PASSARI(L.).

429 CO, ...T. . . . . . .UKR. FIZ. ZH.(USSR) VOL.16, . . . 1735, (1971).
     *ANALYSIS OF ELASTIC SCATTERING OF NEUTRONS BY K, CO AND CE NUCLEI USING
     A LOCAL OPTICAL NUCLEAR MODEL.
     *KASHUBA(I.E.), KOSTYUK(T.A.), KOTISHEVSKAYA(E.YA.).

430 CO (ISOTOPE A=59), ...P. . . . . . . .REPORT HEPL-630 (14PP.), ... (1970).
     *EXPERIMENTAL OBSERVATION OF A SPIN-SPIN EFFECT IN THE NEUTRON TOTAL
     CROSS SECTION OF CO(ISOTOPE A=59).
     *HEALEY(D.C.), MCCARTHY(J.S.), PARKS(D.), FISHER(T.R.). (ED. NOTE: AVAIL.
     CFSTI, SPRINGFIELD, VA.22151, USA.).

431 CO (ISO. A=59), , ...P. . . . . . . .J. NUCL. ENERGY VOL.24, ... 43, (1970).
     *A PULSED NEUTRON DETERMINATION OF THE THERMAL NEUTRON ABSORPTION CROSS-
     SECTION OF COBALT-59.
     *SILK(M.G.), WADE(B.O.).

432 CO (ISO. A=59), , ...P. . . . . . . .PHYS. REV. LETT. VOL.25, ... 117, (1970).
     *EXPERIMENTAL OBSERVATION OF A SPIN-SPIN EFFECT IN THE NEUTRON TOTAL
     CROSS SECTION OF CO (ISO. A=59).
     *HEALEY(D.C.), MCCARTHY(J.S.), PARKS(D.), FISHER(T.R.).

433 CO, ...A4. . . . . . .ATOMKERNENERGIE VOL.11, ... 286, (1966).
     *INVESTIGATION ON FERROMAGNETIC DOMAINS IN COBALT BY POLARIZED NEUTRONS.
     *NAGIB(M.N.). (IN GERMAN).

434 CO, ...A2. . . . . . .PHYS. REV. 181, ... 883, (1969).
     *SEARCH FOR NONCOLLINEAR SPIN DENSITY IN HEXAGONAL COBALT.
     *MOON(R.M.), KOEHLER(W.C.).

435 CO, ...A2. . . . . . .SOL. STATE COMM. 1, ... 81, (1963).
     *ROTATION DES SPINS DANS LE COBALT HEXAGONAL.
     *BERTAUT(E.F.), DELAPALME(A.), PAUTHENET(R.).

436 CO, ...A2. . . . . . .PHYS. REV. 136, ... A195, (1964).
     *DISTRIBUTION OF MAGNETIC MOMENT IN HEXAGONAL COBALT.
     *MOON(R.M.).

437 CO, ...A4. . . . . . .REV. ROUMAINE PHYS. VOL.9, . . . 245, (1964).
     *TOTAL REFLECTION OF NEUTRONS ON COBALT LAYERS.
     *BALLY(D.), GRABCHEV(B.), POPESKU(M.), POPOVICI(M.). (IN RUSSIAN).

438 CO, ...P. . . . . . .Z. PHYSIK VOL.264, ... 427, (1973).
     *PRECISION NEUTRON TOTAL CROSS SECTION MEASUREMENTS ON GOLD AND COBALT
     IN THE .00004 TO .005 EV RANGE.
     *DILG(W.), MANNHART(W.), STEICHELE(E.), ARNOLD(P.).

SECTION   5 -STRUCTURE AND CROSS-SECTION DETERMINATIONS . . . .

439 CO, ... A2 . . . . . . PHYS. REV. VOL. 84, ... 912, (1951).
     ↪NEUTRON SCATTERING AND POLARIZATION BY FERROMAGNETIC MATERIALS.
     ↪SHULL(C.G.), WOLLAN(E.O.), KOEHLER(W.C.).

440 CO, ... A2 . . . . . . PHYS. REV. LETT. VOL.2, ... 254, (1959).
     ↪MAGNETIC FORM FACTOR OF COBALT.
     ↪NATHANS(R.), PAOLETTI(A.).

441 CO, ... P . . . . . . ATOMKERNENERGIE VOL.22, ... 87, (1973).
     ↪MEASUREMENTS OF THE TOTAL NEUTRON CROSS-SECTION OF CHROMIUM, COBALT
     AND ZIRCONIUM IN THE ENERGY RANGE .003 EV TO 1 EV.
     ↪SALAMA(M.), ADIB(M.), ABDEL-KAWI(A.), EL-MINIAWY(S.), EL-KHOSHT(M.),
     HAMOUDA(I.).

442 CO*AL2*O4, ... A2 . . . . . . COLL. INTER. N.126 (GRENOBLE), . 83, (1963).
     ↪MAGNETIC PROPERTIES OF NORMAL SPINELS WITH ONLY A-A INTERACTION.
     ↪ROTH(W.L.).

443 CO3*B2*O6, ... A2 . . . . . . PHYS. STAT. SOL. 16, ... K17, (1966).
     ↪ANTIFERROMAGNETISM IN MN3*B3*O6, CO3*B2*O6, AND NI3*B2*O6.
     ↪NEWNHAM(R.E.), SANTORO(R.P.), SEAL(P.F.), STALLINGS(G.R.).

444 CO3*B2*O6, ... A2 . . . . . . Z. KRIST. VOL.121, ... 418, (1965).
     ↪NEUTRON-DIFFRACTION STUDY OF CO3*B2*O6.
     ↪NEWNHAM(R.E.), REDMAN(M.J.), SANTORO(R.P.).

445 CO*BR2,CO*CL2, ... A2 . . . . . . PHYS. REV. VOL.113, ... 497, (1959).
     ↪NEUTRON DIFFRACTION INVESTIGATIONS OF THE MAGNETIC ORDERING IN FE*BR2,
     CO*BR2, FE*CL2, AND CO*CL2.
     ↪WILKINSON(M.K.), CABLE(J.W.), WOLLAN(E.O.), KOEHLER(W.C.).

446 CO*C*O3, ... A2,D4 . . . . J. PHYS. C VOL.6, ... 1405, (1973).
     ↪MAGNETIZATION DENSITY AND THE MAGNETIC STRUCTURE OF COBALT CARBONATE.
     ↪BROWN(P.J.), WELFORD(P.J.), FORSYTH(J.B.).

447 CO*C*O3, ... A2 . . . . . . SOV. PHYS. JETP VOL.12, ... 1029, (1961).
     ↪THE ANTIFERROMAGNETISM OF CO*C*O3.
     ↪ALIKHANOV(R.A.).

448 CO*CL2.2D2*O, ... A2 . . . . . . PHYS. LETT. 17, ... 103, (1965).
     ↪THE MAGNETIC STRUCTURE OF CO*CL2*2(D2*O).
     ↪COX(D.E.), FRAZER(B.C.), SHIRANE(G.).

449 CO*CL2*(6H2*O), ... A1,A2 . . . . J. CHEM. PHYSICS VOL.53, ... 2660, (1970).
     ↪NEUTRON-DIFFRACTION STUDY OF CO*CL2*(6H2*O).
     ↪KLEINBERG(R.).

450 CO*CL2.2H2*O, ... A2 . . . . . . J. APP. PHYS. 37, ... 1126, (1966).
     ↪NEUTRON DIFFRACTION DETERMINATION OF THE INTERMEDIATE-FIELD MAGNETIC
     STRUCTURE OF CO*CL2*2(H2*O).
     ↪COX(D.E.), SHIRANE(G.), FRAZER(B.C.), NARATH(A.).

451 CO-CR, ... A2 . . . . . . PHYS. REV. VOL.97, ... 304, (1955).
     ↪NEUTRON DIFFRACTION STUDIES OF THE MAGNETIC STRUCTURE OF ALLOYS OF
     TRANSITION ELEMENTS.
     ↪SHULL(C.G.), WILKINSON(M.K.).

452 CO*CR2*O4, ... A2 . . . . . . COLL. INTER. N.126 (GRENOBLE), . 104, (1963).
     ↪FERRIMAGNETIC SPIRAL CONFIGURATIONS IN COBALT CHROMITE.
     ↪MENYUK(N.), DWIGHT(K.), WOLD(A.).

453 CO*CR*O4, ... A2 . . . . . . BULL. SOC. FRANC. MIN. CRIST. 92 264, (1969).
     ↪MAGNETIC STRUCTURES OF COBALT AND NICKEL CHROMATES.
     ↪PERNET(M.), QUEZEL(G.), COING-BOYAT(J.), BERTAUT(E.F.).

454 CO*CR2*O4, ... A2 . . . . . . COMP. REND. 265, ... 672B, (1967).
     ↪ETUDE MAGNETIQUE PAR DIFFRACTION DES NEUTRONS DU COMPOSE SPINELLE
     CO*CR2*O4.
     ↪PLUMIER(R.).

455 CO*CR2*O4, ... A2 . . . . . . J. APPL. PHYS. VOL.39, ... 635, (1968).
     ↪REINVESTIGATION OF MAGNETIC STRUCTURES OF CO*CR2*O4 AND MN*CR2*O4
     OBTAINED BY NEUTRON DIFFRACTION.
     ↪PLUMIER(R.).

456 CO*CR2*S4, ... A2 . . . . . . BULL. SOC. FRANC. MIN. CRIST. 90 169, (1967).
     ↪MAGNETIC STRUCTURE OF CO*CR2*S4.
     ↪COLOMINAS-BROQUETAS(C.), VU-VAN-QUI, BERTAUT(E.F.). (IN FRENCH).

457 CO*CS3*CL5, ... A2 . . . . . . PHYSICA VOL.43, ... 277, (1969).
     ↪NEUTRON DIFFRACTION STUDY OF THE ANTIFERROMAGNETISM OF CO*CS3*CL5.
     ↪HAMMANN(J.). (ED. NOTE: IN FRENCH).

458 CO*F2, ... A2 . . . . . . PHYS. REV. VOL.90, ... 779, (1953).
     ↪NEUTRON DIFFRACTION STUDIES OF ANTIFERROMAGNETISM IN MANGANOUS FLUORIDE
     AND SOME ISOMORPHOUS COMPOUNDS.
     ↪ERICKSON(R.A.).

459 CO*F3, ... A2 . . . . . . PHYS. REV. VOL.112, ... 1132, (1958).
     ↪ANTIFERROMAGNETIC PROPERTIES OF THE IRON GROUP TRIFLUORIDES.
     ↪WOLLAN(E.O.), CHILD(H.R.), KOEHLER(W.C.), WILKINSON(M.K.).

460 CO-FE, ... A2 . . . . . . PHYS. STAT. SOL. VOL.38, ... 103, (1970).
     ↪ASPHERICITY OF THE 3D MAGNETIC MOMENT DISTRIBUTION AND CONDUCTION
     ELECTRON POLARIZATION IN CO-FE ALLOYS.
     ↪DOBRZYNSKI(L.), MANIAWSKI(F.), MODRZEJEWSKI(A.), SIKORSKA(D.).

461 CO-FE, ... A4 . . . . . . PHIL. MAG. 8, ... 401, (1963).
     ↪THE MAGNETIC MOMENT DISTRIBUTION IN SOME TRANSITION METAL ALLOYS.
     ↪COLLINS(M.F.), FORSYTH(J.B.).

462 CO-FE, ... A2 . . . . . . J. APPL. PHYS. VOL.44, ... 4181, (1973).
     ↪POLARIZED NEUTRON TECHNIQUES FOR THE OBSERVATION OF FERROMAGNETIC
     DOMAINS.
     ↪SCHLENKER(M.), SHULL(C.G.).

463 CO-FE, ... A2 . . . . . . NUOVO CIMENTO VOL.10B, ... 565, (1972).
     ↪TEMPERATURE DEPENDENCE OF MAGNETIC-MOMENT DISTRIBUTION IN F.C.C.
     CO(.91)-FE(.09) ALLOY.
     ↪MENZINGER(F.), PAOLETTI(A.).

464 CO-FE,CO-NI, ... A4 . . . . . . SOL. STATE COMM. VOL.12, ... 909, (1973).
     ↪CONCENTRATION AND TEMPERATURE DEPENDENCE OF ASPHERICITY OF 3D-ELECTRONS
     IN F.C.C. CO-NI AND CO-FE ALLOYS.
     ↪KADAR(G.), MENZINGER(F.).

SECTION 5 -STRUCTURE AND CROSS-SECTION DETERMINATIONS . . . .

465 (CO-FE)3*O4, ...A4. . . . . .PHYS. REV. VOL.102, ... 674, (1956).
*NEUTRON DIFFRACTION OBSERVATION OF HEAT TREATMENT IN COBALT FERRITE.
*PRINCE(E.).

466 CO2-HF-(AL,GA,SN),CO2-NB-SN, ...A1,A2. . . .J. PHYS. CHEM. SOLIDS VOL.35, . . . 1, (1974).
*A NEUTRON DIFFRACTION AND MAGNETIZATION STUDY OF HEUSLER ALLOYS
CONTAINING CO AND ZR, HF, V OR NB.
*ZIEBECK(K.R.A.), WEBSTER(P.J.).

467 CO2*MN*(AL,SI,GA,GE,SN,SB),. . . . . .A1,A2. . . .J. PHYS. CHEM. SOLIDS VOL. 32, . . 1221, (1971).
*MAGNETIC AND CHEMICAL ORDER IN HEUSLER ALLOYS CONTAINING COBALT AND
MANGANESE.
*WEBSTER(P.J.).

468 CO-MN-FE,. . . . . .A1,A2. . . .J. PHYS. SOC. JAPAN VOL.35,. . . 426, (1973).
*NEUTRON DIFFRACTION INVESTIGATIONS OF (CO-MN)-FE.
*ADACHI(K.), SATO(K.), MATSUI(M.), MITANI(S.).

469 CO*(MN-FE)*O4, ...A1. . . . . .J. SOLID STATE CHEM. VOL.8, ... 50, (1973).
*DISTORSION CRISTALLINE ET DISTRIBUTION DES CATIONS DANS LES SPINELLES
DE LA SERIE CO*(MN(X)-FE(2-X))*O4.
*BERNARD(J.L.), BAFFIER(N.), HUBER(M.).

470 CO*MN2*O4, ...A1,2. . . . .J. PHYS. CHEM. SOLIDS 30, ... 805, (1969).
*MANGANITES SPINELLES PURS D≠ELEMENTS DE TRANSITION: PREPARATIONS ET
STRUCTURES CRISTALLOGRAPHIQUES.
*BUHL(R.).

471 (CO-MN)3*O4, ...A1,A2. . . .J. PHYSIQUE TOME 31, ... 113, (1970).
*NEUTRON DIFFRACTION AND MAGNETIC STUDY OF THE CRYSTAL AND MAGNETIC
PROPERTIES OF CUBIC SPINEL COMPOUNDS CO(3-X)-MN(X)*O4 (0.6≤X≤1.2).
*BOUCHER(B.), BUHL(R.), DI-BELLA(R.), PERRIN(M.).

472 CO*MN2*O4, ...A2. . . . . .J. APP. PHYS. 39, ... 632, (1968).
*MAGNETIC STRUCTURE OF COBALT MANGANITE BY NEUTRON DIFFRACTION.
*BOUCHER(B.), BUHL(R.), PERRIN(M.).

473 (CO-MN)*O, ...A1. . . . . .PROC. ROY. SOC. A VOL.217, . . . 252, (1953).
*THE ANTIFERROMAGNETIC PROPERTIES OF MIXED COBALT AND MANGANESE OXIDES.
*BACON(G.E.), STREET(R.), TREDGOLD(R.H.).

474 CO-MN-SB,. . . . . .A2. . . . . .PHYS. STAT. SOLIDI A VOL.9 NO.1, . 97, (1972).
*ATOMIC AND MAGNETIC STRUCTURE OF THE HEUSLER ALLOYS NI-MN-SB AND
CO-MN-SB.
*SZYTULA(A.), DIMITRIJEVIC(Z.), TODOROVIC(J.), KOLODZIEJCZYK(A.),
SZELAG(J.), WANIC(A.).

475 CO-MN-SB,. . . . . .A1,A2. . . .PHYS. STAT. SOLIDI A VOL.3, . . . 959, (1970).
*ATOMIC AND MAGNETIC STRUCTURE OF THE HEUSLER ALLOYS PD2-MN-GE, PD2-MN-SN
CU2-MN-IN, AND CO-MN-SB.
*NATERA(M.G.), MURTHY(M.R.L.), BEGUM(R.J.), MURTHY(N.S.S.).

476 CO2-MN-SN, ...A1,A2. . . .PHYS. STAT. SOLIDI A VOL.11 NO.1 . 57, (1972).
*ATOMIC AND MAGNETIC STRUCTURE OF THE HEUSLER ALLOYS NI2-MN-SB, NI2-MN-SN
AND CO2-MN-SN.
*SZYTULA(A.), KOLODZIEJCZYK(A.), RZANY(H.), TODOROVIC(J.), WANIC(A.).

477 CO*NB2*O6, ...A2. . . . . .SOL. STATE COMM. VOL.12 NO.2, . . 113, (1973).
*MAGNETISCHE STRUKTUR VON COLUMBITEN MN*TA2*O6 UND CO*NB2*O6.
*WEITZEL(H.), KLEIN(S.).

478 CO-NI, ...A2. . . . . .SOL. STATE COMM. VOL.8,. . . . . 1, (1970).
*MAGNETIC MOMENT DISTRIBUTION IN CO-NI ALLOY.
*ANTONINI(B.), LUCARI(F.), MENZINGER(F.).

479 CO-NI, ...A2. . . . . .INT. J. MAGN. VOL.1 NO.2,. . . . 183, (1971).
*CONCENTRATION DEPENDENCE OF MAGNETIC MOMENT DISTRIBUTION IN CO-NI ALLOYS
*ANTONINI(B.), MENZINGER(F.), PAOLETTI(A.), SACCHETTI(F.).

480 CO*O, ...A2,P. . . . .PHYS. REV. B-1, ... 2243, (1970).
*MAGNETIC FORM FACTOR OF CO++ ION IN COBALTOUS OXIDE.
*KHAN(D.C.), ERICKSON(R.A.).

481 CO*O, ...A1. . . . . .DISS. ABS. XXVII, ... 276B, (1967).
*NEUTRON DIFFRACTION STUDY OF COBALTOUS OXIDE.
*CHARAN-KHAN(D.).

482 CO*O, ...T. . . . . . .PHYS. REV. B VOL.4, ... 3901, (1971).
*THEORETICAL MAGNETIC FORM FACTORS OF CO++ AND FE++ IN THEIR MONOXIDES.
*MAHENDRA(A.), KHAN(D.C.).

483 CO*O, ...A2. . . . . .PHYS. REV. B VOL.5, ... 2693, (1972).
*SHORT-RANGE MAGNETIC ORDER IN CO*O.
*RECHTIN(M.D.), AVERBACH(B.L.).

484 CO*O, ...A2,T. . . . .ACTA CRYST A24, ... 217, (1968).
*REPRESENTATION ANALYSIS OF MAGNETIC STRUCTURES.
*BERTAUT(E.F.).

485 CO*O, ...A2,4. . . . .J. PHYS. CHEM. SOLIDS 29, . . . 2087, (1968).
*TEMPERATURE DEPENDENCE OF THE SUBLATTICE MAGNETIZATION IN COBALTOUS
OXIDE.
*KHAN(D.C.), ERICKSON(R.A.).

486 CO3*O4, ...A2. . . . . .COLL. INTER. N.126 (GRENOBLE), . 83, (1963).
*MAGNETIC PROPERTIES OF NORMAL SPINELS WITH ONLY A-A INTERACTION.
*ROTH(W.L.).

487 CO*O, ...A4. . . . . .BULL. AMER. PHYS. SOC. VOL.17, . 667, (1972).
*NEUTRON DIFFRACTION STUDIES OF CO*O UNDER PRESSURE.
*WORLTON(T.G.), BEYERLEIN(R.), DECKER(D.L.), ROULT(G.).

488 CO*O, ...A2. . . . . .PHYS. REV. B VOL.6, ... 4294, (1972).
*LONG-RANGE MAGNETIC ORDER IN CO*O.
*RECHTIN(M.D.), AVERBACH(B.L.).

489 CO3*O4, ...A2. . . . . .J. PHYS. CHEM. SOL. 25, ... 1, (1964).
*THE MAGNETIC STRUCTURE OF CO3*O4.
*ROTH(W.L.).

490 CO*O, ...A2. . . . . .PHYS. REV. 138, A, ... 584, (1965).
*MULTI-SPIN-AXIS STRUCTURE OF CO*O.
*VAN-LAAR(B.).

SECTION   5 -STRUCTURE AND CROSS-SECTION DETERMINATIONS . . . .

491 CO*O,. . . . . .A2. . . . . . .PHYS. REV. 141,. . . . .   538, (1966).
     *NEUTRON DIFFRACTION INVESTIGATION OF CO*O SINGLE CRYSTAL.
     *VAN-LAAR(B.), SCHWEIZER(J.), LEMAIRE(R.).

492 CO*O,. . . . . .A4. . . . . . .DISS. ABSTR. B VOL.34,. . . . . 2247, (1973).
     *A NEUTRON DIFFRACTION STUDY OF ANTIFERROMAGNETIC COBALTOUS OXIDE WITH
     NUCLEAR POLARIZATION FROM THE HFS INTERACTION, IN THE REGION .35-4.2 K.
     *GOER(D.A.).

493 CO*O,. . . . . . . .A2. . . . . . .PHYS. REV. VOL.110,. . . . . . . 1333, (1958).
     *MAGNETIC STRUCTURES OF MN*O, FE*O, CO*O, AND NI*O.
     *ROTH(W.L.).

494 (CO*O)-(P2*O5),. . . . .A4. . . . . . .J. PHYS. C VOL.5,. . . . . . L261, (1972).
     *AMORPHOUS ANTIFEPROMAGNETISM IN IRON AND COBALT PHOSPORUS PENTOXIDE
     GLASSES.
     *EGAMI(T.),   SACLI(O.A.),   SIMPSON(A.W.),   TERRY(A.L.),   WEDGWOOD(F.A.).

495 CO-PT, ...A2. . . . . . .PHYS. LETT. 25A,. . . . .   372, (1967).
     *MAGNETIZATION DISTRIBUTION IN DISORDERED CO-PT ALLOY.
     *ANTONINI(B.), MENZINGER(F.), PAOLETTI(A.).

496 CO-PT3,. . . . . . .A4. . . . . . .PROC. INT. CONF. (NOTTINGHAM), .   288, (1964).
     *POLARIZED NEUTRON INVESTIGATION OF THE CO-PT3 AND FE-PD ALLOYS.
     *ANTONINI(B.), FELCHER(G.P.),   MAZZONE(G.), MENZINGER(F.), PAOLETTI(A.).

497 CO-PT, ...A2. . . . . . .COLL. INTER. N.126 (GRENOBLE), .   176, (1963).
     *THE MAGNETIC STRUCTURE OF CO-PT.
     *VAN-LAAR(B.).

498 CO*S2, ...A2. . . . . . .ACTA CHEM. SCAND. VOL.21,. . . .   833, (1967).
     *ON THE FERROMAGNETISM OF CO*S2.
     *ANDRESEN(A.F.), FURUSETH(S.), KJEKSHUS(A.).

499 CO*S*O4 (BETA),. . . . . .A2,T. . . . . . .ACTA CRYST A24,. . . . . . .   217, (1968).
     *REPRESENTATION ANALYSIS OF MAGNETIC STRUCTURES.
     *BERTAUT(E.F.).

500 CO*S*O4, ...A2. . . . . . .PHYS. LETT. 3, ...   178, (1962).
     *STRUCTURE MAGNETIQUE DE CO*S*O4.
     *BERTAUT(E.F.), COING-BOYAT(J.), DELAPALME(A.).

501 CO*S*O4, ...A2. . . . . . .PHYS. REV. 129,. . . . . . . 1145, (1963).
     *MAGNETIC STRUCTURE OF BETA-CO*S*O4.
     *BROWN(P.J.), FRAZER(B.C.).

502 CO*S*O4, ...A2. . . . . . .J. APP. PHYS. 34,. . . . . 1333, (1963).
     *MAGNETIC STRUCTURE INVESTIGATIONS AT THE NUCLEAR CENTRE.
     *BALLESTRACCI(R.),   BERTAUT(E.F.),   COING-BOYAT(J.),   DELAPALME(A.),
     JAMES(W.J.), LEMAIRE(R.).

503 CO*S*O4, ...T,A2. . . . . . .J. PHYS. SOC. JAPAN VOL.18,. . . . 1641, (1963).
     *ANTIFERROMAGNETIC STRUCTURE OF THE ANHYDROUS COBALT SULFATE CO*S*O4.
     *URYU(N.).

504 CO*S*O4, ...A2. . . . . . .PHYS. REV. VOL.125,. . . . . . . 1283, (1962).
     *ANTIFERROMAGNETIC STRUCTURE OF CR*V*O4 AND THE ANHYDROUS SULFATES OF
     DIVALENT FE,NI AND CO.
     *FRAZER(B.C.), BROWN(P.J.).

505 CO*SE2,. . . . . .A1. . . . . . .ACTA CHEM. SCAND. VOL.23,. . . . 2325, (1969).
     *ON THE MAGNETIC PROPERTIES OF CO*SE2, NI*S2, AND NI*SE2.
     *FURUSETH(S.), KJEKSHUS(A.), ANDRESEN(A.F.).

506 CO*SE*O4,. . . . . .A2. . . . . . .Z. ANGEW. PHYS. VOL.27,. . . . . .   311, (1969).
     *MAGNETIC STRUCTURES OF NI*SE*O4 AND CO*SE*O4 AND DETERMINATION OF THE
     NEEL POINTS OF MN*SE*O4, CO*SE*O4 AND NI*SE*O4.
     *FUESS(H.).

507 CO*SE*O4,. . . . . .A1. . . . . . .Z. ANORG. ALLG. CHEM. BAND 358,. .   125, (1968).
     *BESTIMMUNG DER KRISTALLSTRUKTUR DER SELENATE M*SE*O4 (M=MN,CO,NI) DURCH
     RONTGEN- UND NEUTRONENBEUGUNG.
     *FUESS(H.), WILL(G.).

508 CO2*SI*O4, ...A2. . . . . . .J. PHYS. CHEM. SOL. 25,. . . . .   901, (1964).
     *ANTIFERROMAGNETISM IN COBALT ORTHOSILICATE.
     *NOMURA(S.), SANTORO(R.P.), FANG(J.H.), NEWNHAM(R.E.).

509 CO2-TI-(AL,SI,GA,GE,SN,SB),. . . . . .A1,A2. . . .J. PHYS. CHEM. SOLIDS VOL.34,. . 1647, (1973).
     *MAGNETIC AND CHEMICAL ORDER IN HEUSLER ALLOYS CONTAINING COBALT AND
     TITANIUM.
     *WEBSTER(P.J.), ZIEBECK(K.R.A.).

510 CO*TI*O3,. . . . . .T,A2. . . . . . .PHYS. REV. 164,. . . . . . .   768, (1967).
     *THEORY OF MAGNETIC PROPERTIES OF ILMENITES M*TI*O3.
     *GOODENOUGH(J.B.), STICKLER(J.J.).

511 CO*TI*O3,. . . . . .A1,2. . . . . .ACTA CRYST. 17,. . . . .   240, (1964).
     *CRYSTAL STRUCTURE AND MAGNETIC PROPERTIES OF CO*TI*O3.
     *NEWNHAM(R.E.), FANG(J.H.), SANTORO(R.P.).

512 CO-TI-SB,. . . . . .A1. . . . . . .ACTA PHYS. POLON. VOL.A44,. . . . .   147, (1973).
     *CRYSTAL STRUCTURE AND MAGNETIC PROPERTIES OF HEUSLER-TYPE ALLOYS
     M-TI-SB (M=NI,CO,FE) AND FE2-TI-SN.
     *SZYTULA(A.), TOMKOWICZ(Z.), TUROWSKI(M.).

513 CO*U*O4, ...A2. . . . . . .J. PHYS. RADIUM VOL.23,. . . . . .   477, (1962).
     *ETUDE DES URANATES DE COBALT ET DE MANGANESE.
     *BERTAUT(E.F.), DELAPALME(A.), FORRAT(F.), PAUTHENET(R.).

514 CO3*U2*O8, ...A1. . . . . . .ACTA CRYST. VOL.B29,. . . . . . 1570, (1973).
     *STRUCTURE CRISTALLINE DU NOUVEAU COMPOSE CO3*U2*O8.
     *BACHMANN(M.).

515 CO2-V-(AL,GA),CO2-ZR-(AL,SN),. . . . . .A1,A2. . . .J. PHYS. CHEM. SOLIDS VOL.35,. .   1, (1974).
     *A NEUTRON DIFFRACTION AND MAGNETIZATION STUDY OF HEUSLER ALLOYS
     CONTAINING CO AND ZR, HF, V OR NB.
     *ZIEBECK(K.R.A.), WEBSTER(P.J.).

516 CO-(V,CR,MN,NI), ...A2. . . . . . .PHYS. REV. B-2,. . . . .   176, (1970).
     *MAGNETIC-MOMENT DISTRIBUTION FOR 3D-TRANSITION-METAL IMPURITIES IN
     COBALT.
     *CABLE(J.W.), HICKS(T.J.).

SECTION 5 -STRUCTURE AND CROSS-SECTION DETERMINATIONS . . . .

517 CO*V2*O4,. . . . . *.MAGNETIC PROPERTIES OF V+++ IONS IN CUBIC SPINELS. A4 OF V..PROC. INT. CONF. (NOTTINGHAM), . 538, (1964).
 *.DWIGHT(K.), MENYUK(N.), ROGERS(D.B.), WOLD(A.).

518 CO-W,. . . . . . . *.NEUTRON DIFFRACTION STUDY OF MAGNETIC STRUCTURES IN CO(X)-W HEXAGONAL A2.....SOV. PHYS. DOKLADY VOL.11,.....379, (1966).
 FERRITES.
 *.SIZOV(R.A.), YAMZIN(I.I.).

519 CO*W*O4, ... *.MAGNETISCHE STRUKTUR VON CO*W*O4, NI*W*O UND CU*W*O4. A2.......SOL. STATE COMM. VOL.8,. . . . . 2071, (1970).
 *.WEITZEL(H.).

520 CO3*X (X=PR,ND,TB,HO,ER),. . . . . T,A2.......PHYS. STAT. SOL. VOL.50,. . . 747, (1972).
 *.INTERPRETATION DES STRUCTURES MAGNETIQUES DES COMPOSES T*CO3(T=PR,ND,TB,
 HO,ER).
 *.YAKINTHOS(J.), ROSSAT-MIGNOD(J.).

521 CO3*Y2*O8, . . . *.STRUCTURE AUX RAYONS X, NEUTRONS ET PROPRIETES MAGNETIQUES DES A1.......ACTA CRYST. VOL.B26,. . . . . 2036, (1970).
 ORTHOVANADATES DE NICKEL ET DE COBALT.
 *.FUESS(H.), BERTAUT(E.F.), PAUTHENET(R.), DURIF(A.).

522 CO-ZN FERRITES,. . *.NEUTRON DIFFRACTION STUDY OF COBALT-ZINC FERRITES. A2.......NUCL./S.S. PHYS. SYMP. ABSTR.,. . . . , (1972).
 *.RADHAKRISHNAN(N.K.), PARANOPE(S.K.), MURTHY(M.R.L.), BEGUM(R.J.),
 MADHAV-RAO(L.), MURTHY(N.S.S.).

523 CR,. . . . . . . *.THE MAGNETIC STRUCTURE OF CHROMIUM. A2.......IAEA SYMP. (PILE RESEARCH)VIENNA 429, (1960).
 *.C.C.C.P.

524 CR,. . . . . . . *.Q-SWITCH AND POLARIZATION DOMAINS IN ANTIFERROMAGNETIC CHROMIUM OBSERVED A2.......PHYS. REV. LETT. VOL.29,. . . . 281, (1972).
 WITH NEUTRON-DIFFRACTION TOPOGRAPHY.
 *.ANDO(M.), HOSOYA(S.).

525 CR,. . . . . . . *.NEUTRONENBEUGUNG AN HEXAGONALEM CHROM. A1.......PHYS. STAT. SOL. 3,. . . . . . . K249, (1963).
 *.ALBRECHT(G.), DOENITZ(F.D.), KLEINSTUCK(K.), BETZL(M.).

526 CR (ISO. A=50, 52, 53, 54),. . . . P.......THESIS (231PP.). . . . . (1970).
 *.KEY NEUTRON CAPTURE AND TRANSMISSION MEASUREMENTS ON CR(ISOTOPES A=50,
 52, 53, 54), NI(ISOTOPE A=60), AND V.
 *.STIEGLITZ(R.G.). (ED. NOTE: AVAIL. UNIV. MICROFILMS, ANN ARBOR, MICH.
 ORDER NO. 71-12829).

527 CR,. . . . . . . *.NEUTRON ELASTIC SCATTERING CROSS SECTIONS OF VANADIUM, CHROMIUM, IRON A4.......2ND INTERNAT. CONF./NUCL. DATA,. . . . , (1970).
 AND NICKEL.
 *.HOLMQVIST(B.), WIEDLING(T.).

528 CR,. . . . . . . *.NEUTRON DIFFRACTION STUDY OF CR UNDER HIGH PRESSURE. A2.......J. PHYS. SOC. JAPAN VOL.24,. . . 368, (1968).
 *.UMEBAYASHI(H.), SHIRANE(G.), FRAZER(B.C.).

529 CR,. . . . . . . *.MAGNETIC FORM FACTOR OF CHROMIUM. P.......J. APP. PHYS. 37,. . . . . 1036, (1966).
 *.MOON(R.M.), KOEHLER(W.C.), TREGO(A.L.).

530 CR,. . . . . . . *.ANOMALIES IN THE MAGNETIC STRUCTURE OF CHROMIUM. A2.......SOV. PHYS. J.E.T.P. VOL.22,. . . 754, (1965).
 *.GOLOVKIN(V.S.), BYKOV(V.N.), LEVDIK(V.A.).

531 CR,. . . . . . . *.EFFECTS OF PRESSURE AND A MAGNETIC FIELD ON CHROMIUM STUDIED BY NEUTRON A4.......J. APP. PHYS. 39,. . . . . . . 671, (1968).
 DIFFRACTION.
 *.WERNER(S.A.), ARROTT(A.), ATOJI(M.).

532 CR,. . . . . . . *.TEMPERATURE AND MAGNETIC-FIELD DEPENDENCE OF ANTIFERROMAGNETISM IN PURE A2.......PHYS. REV. 155,. . . . . 528, (1967).
 CHROMIUM.
 *.WERNER(S.A.), ARROTT(A.), KENDRICK(H.).

533 CR,. . . . . . . *.MAGNETIC STRUCTURES OF FIELD-COOLED AND STRESS-COOLED CHROMIUM. A2.......PHYS. REV. 141,. . . . . . . 510, (1966).
 *.BASTOW(T.J.), STREET(R.).

534 CR,. . . . . . . *.ELASTICITY AND ANELASTICITY OF CHROMIUM. A4.......PHYS. REV. LETT. 10,. . . . . . 210, (1963).
 *.STREET(R.).

535 CR,. . . . . . . *.NONELASTIC TRANSITIONS IN CHROMIUM. A4.......PHYS. REV. LETT. 10,. . . . . . 208, (1963).
 *.DE-MORTON(M.E.).

536 CR,. . . . . . . *.THE MAGNETIC STRUCTURE OF PLASTICALLY DEFORMED CHROMIUM. A2.......J. PHYS. CHEM. SOL. 27,. . . . . 1955, (1966).
 *.SABINE(T.M.), COX(G.W.).

537 CR,. . . . . . . *.CHANGES IN APPARENT SYMMETRY WITH MAGNETIC FIELD IN CHROMIUM. A2,4.......J. APP. PHYS. 37,. . . . . . 1260, (1966).
 *.WERNER(S.A.), ARROTT(A.), KENDRICK(H.).

538 CR,. . . . . . . *.NEUTRON DIFFRACTION STUDY OF CR AND CR ALLOYS. A2.......J. APP. PHYS. 38,. . . . . . . 1243, (1967).
 *.WERNER(S.A.), ARROTT(A.), KENDRICK(H.).

539 CR,. . . . . . . *.SOME OBSERVATIONS ON THE HIGHER TEMPERATURE ANTIFERROMAGNETIC PHASE: CR. A2,4.......PROC. PHYS. SOC. 85,. . . . . . 1185, (1965).
 *.BROWN(P.J.), WILKINSON(C.), FORSYTH(J.B.), NATHANS(R.).

540 CR,. . . . . . . *.THE ANTIFERROMAGNETIC PHASES OF FIELD-COOLED CHROMIUM. A2,4.......PROC. PHYS. SOC. 86,. . . . . . 1143, (1965).
 *.BASTOW(T.J.), STREET(R.).

541 CR,. . . . . . . *.THE ANTIFERROMAGNETIC STRUCTURE OF CHROMIUM. A2.......SOV. PHYS. CRYST. VOL.7,. . . . 639, (1963).
 *.GOMAN≠KOV(V.I.), LITVIN(D.F.), LOSHMANOV(A.A.), LYASHCHENKO(B.G.).

542 CR,. . . . . . . *.FIRST-ORDER MAGNETIC PHASE CHANGE IN CHROMIUM AT 38.5 DEGREES C. A2.......PHYS. REV. LETT. VOL.14,. . . 1022, (1965).
 *.ARROTT(A.), WERNER(S.A.), KENDRICK(H.).

SECTION   5 -STRUCTURE AND CROSS-SECTION DETERMINATIONS . . . .

543 CR,. . . . . .THE ANTIFERROMAGNETIC PHASES OF FIELD-COOLED CHROMIUM. . .   893, (1965).
             A2 . . . . PROC. PHYS. SOC. VOL.8,.
            *HEARN(C.J.).

544 CR,. . . . . .A NEUTRON-DIFFRACTION STUDY OF VERY PURE CHROMIUM. . . .   823, (1961).
            A2 . . . . ACTA CRYST. VOL.14,.
            *BACON(G.E.).

545 CR,. . . . . .NEUTRON DIFFRACTION STUDY OF CHROMIUM SINGLE CRYSTALS.   35, (1962).
            A2 . . . . J. PHYS. SOC. JAPAN VOL.17 B-III
            *SHIRANE(G.), TAKEI(W.J.).

546 CR,. . . . . .MAGNETIC STUDIES OF ANNEALED AND ALLOYED CHROMIUM BY NEUTRON DIFFRACTION
            A2 . . . . J. OF PHYS.-C- SER.2 VOL.2,.   238, (1969).
            *BACON(G.E.), COWLAM(N.).

547 CR,. . . . . .THE INFLUENCE OF PRESSURE ON THE ANTIFERROMAGNETIC TRANSITION   388, (1964).
            A4 . . . . SOV. PHYS. DOKLADY VOL.9,.
            TEMPERATURE OF CHROMIUM.
            *LITVIN(D.F.), PONYATOVSKII(E.G.).

548 CR,. . . . . .NEUTRON DIFFRACTION STUDIES OF VARIOUS TRANSITION ELEMENTS.   100, (1953).
            A2 . . . . REV. MOD. PHYS. VOL.25,.
            *SHULL(C.G.), WILKINSON(M.R.).

549 CR,. . . . . .MECHANISM OF ANTIFERROMAGNETISM IN CHROMIUM.   226, (1960).
            A4 . . . . PHYS. REV. LETT. VOL.4,.
            *OVERHAUSER(A.W.), ARROTT(A.).

550 CR,. . . . . .ANTIPHASE ANTIFERROMAGNETIC STRUCTURE OF CHROMIUM.   211, (1959).
            A2 . . . . PHYS. REV. LETT. VOL.3,.
            *CORLISS(L.M.), HASTINGS(J.M.), WEISS(R.J.).

551 CR,. . . . . .HELICAL SPIN ARRANGEMENT IN CHROMIUM METAL.   135, (1960).
            T . . . . PHYS. REV. VOL.118,.
            *COOPER(B.R.).

552 CR,. . . . . .POLARIZED-NEUTRON STUDY OF THE FIELD-INDUCED MAGNETIC MOMENT IN CHROMIUM
            A2 . . . . PHYS. REV. LETT. VOL.31,.   1498, (1973).
            *STASSIS(C.), KLINE(G.R.), SINHA(S.K.).

553 CR,. . . . . .ON THE MAGNETIC STRUCTURE OF CHROMIUM.   1070, (1960).
            A2 . . . . SOV. PHYS. DOKLADY VOL.4,.
            *BYKOV(V.N.), GOLOVKIN(V.S.), AGEEV(N.V.), LEVDIK(V.A.), VINOGRADOV(S.I.)

554 CR,. . . . . .MEASUREMENTS OF THE TOTAL NEUTRON CROSS-SECTION OF CHROMIUM, COBALT
            P . . . . ATOMKERNENERGIE VOL.22,.   87, (1973).
            AND ZIRCONIUM IN THE ENERGY RANGE .003 EV TO 1 EV.
            *SALAMA(M.), ADIB(M.), ABDEL-KAWI(A.), EL-MINIAWY(S.), EL-KHOSHT(M.),
            HAMOUDA(I.).

555 CR-AL, ...ANTIFERROMAGNETISM OF A SOLID SOLUTION OF ALUMINIUM IN CHROMIUM.   955, (1967).
            A2 . . . . SOL. STATE COMM. VOL.5,.
            *KALLEL(A.), DE-BERGEVIN(F.).

556 CR-ALLOYS, . . .NEUTRON DIFFRACTION STUDIES OF CR AND CR ALLOYS.   59, (1967).
            A2 . . . . METAL. SOC. CONF. LOS ANGELES, .
            *ARROTT(A.), WERNER(S.A.).

557 CR*AS, ...FIRST ORDER CRYSTALLOGRAPHIC AND MAGNETIC PHASE TRANSITION IN CR*AS.   1699, (1971).
            A2 . . . . SOL. STATE COMM. VOL.9,.
            *BOLLER(H.), KALLEL(A.).

558 CR2*AS,. . . . .THE NEUTRON DIFFRACTION STUDY OF CR2*AS.   2244, (1965).
            A2 . . . . J. PHYS. SOC. JAPAN VOL.20,.
            *WATANABE(H.), NAKAGAWA(Y.), SATO(K.).

559 CR*AS, ...MAGNETIC STRUCTURE CR*AS AND MN-SUBSTITUTED CR*AS.   1128, (1969).
            A2 . . . . J. APP. PHYS. VOL.40,.
            *WATANABE(H.), KAZENNA(N.), YAMAGUCHI(Y.), OHASHI(M.).

560 CR2*AS,. . . . .NEUTRON DIFFRACTION STUDY OF CR2*AS.   958, (1972).
            A4 . . . . J. PHYS. SOC. JAPAN VOL.32,.
            *YAMAGUCHI(Y.), WATANABE(H.), YAMAGUCHI(H.), TOMIGOSHI(S.).

561 CR*AS, ...MAGNETIC STRUCTURE AND PROPERTIES OF CR*AS.   1703, (1971).
            A2 . . . . ACTA CHEM. SCAND. VOL.25,.
            *SELTE(K.), KJEKSHUS(A.), JAMISON(W.E.), ANDRESEN(A.F.),
            ENGEBRETSEN(J.E.).

562 CR*(AS-SB),. . .CRYSTALLOGRAPHIC DISTORTION IN CR*(AS(.5)-SB(.5)).   665, (1973).
            A1,A2 . . . . SOL. STATE COMM. VOL.12 NO.7,.
            *KALLEL(A.), NASR-EDDINE(M.), BOLLER(H.).

563 CR2*BE*04, . . .NEUTRON DIFFRACTION INVESTIGATION OF THE SPIRAL MAGNETIC STRUCTURE IN
            A2 . . . . J. APPL. PHYS. VOL.40,.   1124, (1969).
            CR2*BE*04.
            *COX(D.E.), FRAZER(B.C.), NEWNHAM(R.E.), SANTORO(R.P.).

564 CR*BR3,. . . . .NOTE ON THE MAGNETIC STRUCTURE OF CR*BR3.   K112, (1962).
            A2 . . . . PHYS. STAT. SOL. VOL. 2,.
            *LEGRAND(E.), PLUMIER(R.).

565 CR23*C6, ...THE CRYSTAL STRUCTURE OF CR23*C6.   3102, (1972).
            A1 . . . . ACTA CRYST. VOL.B28, .
            *BOWMAN(A.L.), ARNOLD(G.P.), STORMS(E.K.), NERESON(N.G.).

566 CR3*C2, ...STRUCTURE INVESTIGATION OF CHROMIUM CARBIDE CR3*C2 BY THERMAL NEUTRONS.   880, (1960).
            A1 . . . . Z. NATURFORSCH. VOL.15A,.
            *MEINHARDT(D.), KRISEMENT(O.). (IN GERMAN).

567 CR*CL3,. . . . .NEUTRON DIFFRACTION INVESTIGATION OF ANTIFERROMAGNETISM IN CR*CL3.   29, (1961).
            A2 . . . . J. PHYS. CHEM. SOLIDS VOL.19,.
            *CABLE(J.W.), WILKINSON(M.K.), WOLLAN(E.O.).

568 CR(CO),CR(NI), .ANTIFERROMAGNETISM OF DILUTE CR ALLOYS WITH CO AND NI.   263, (1968).
            A2 . . . . J. PHYS. SOC. JAPAN VOL.24,.
            *ENDOH(Y.), ISHIKAWA(Y.), OHNO(H.).

569 CR2*CO*04, . . .FERRIMAGNETIC SPIRAL CONFIGURATIONS IN COBALT CHROMITE.   104, (1963).
            A2 . . . . COLL. INTER. N.126 (GRENOBLE), .
            *MENYUK(N.), DWIGHT(K.), WOLD(A.).

570 CR*CU*02,. . . .NEUTRON DIFFRACTION STUDY AND MAGNETOSTATIC MEASUREMENTS OF CR*CU*02.   165, (1971).
            A4 . . . . ANNU. UNIV. SOFIA FAC. PHYS. 63,.
            *APOSTOLOV(A.). (ED. NOTE! IN UKRAINIAN).

SECTION 5 -STRUCTURE AND CROSS-SECTION DETERMINATIONS . . . .

571 CR*F3, ...A4. . . . . . .J. PHYS. C VOL.7 . . . . 783, (1974).
. . . .*A NEUTRON DIFFRACTION DETERMINATION OF COVALENCY PARAMETERS FOR FE(3+)
. . . . .AND CR(3+) IN FE*F3 AND CR*F3.
. . . .*JACOBSON(A.J.), MCBRIDE(L.), FENDER(B.E.F.).

572 CR*F3, ...A2. . . . . . .PHYS. REV. VOL.112, ... 1132, (1958).
. . . .*ANTIFERROMAGNETIC PROPERTIES OF THE IRON GROUP TRIFLUORIDES.
. . . .*WOLLAN(E.O.), CHILD(H.R.), KOEHLER(W.C.), WILKINSON(M.K.).

573 CR*F2,CR*CL2, ...A2. . . . . . .PHYS. REV. VOL.118, ... 950, (1960).
. . . .*NEUTRON DIFFRACTION STUDIES OF ANTIFERROMAGNETISM IN CR*F2 AND CR*CL2.
. . . .*CABLE(J.W.), WILKINSON(M.K.), WOLLAN(E.O.).

574 CR-(FE), ...A1. . . . . . .PHYS. REV. 153, ... 624, (1967).
. . . .*NEUTRON-DIFFRACTION STUDY OF DILUTE CHROMIUM ALLOYS WITH IRON.
. . . .*ARROTT(A.), WERNER(S.A.), KENDRICK(H.).

575 CR-(FE), ...A2. . . . . . .J. PHYS. SOC. JAPAN VOL.22, . . . 1221, (1967).
. . . .*ANTIFERROMAGNETISM IN DILUTE IRON CHROMIUM ALLOYS.
. . . .*ISHIKAWA(Y.), HOSHINO(S.), ENDOH(Y.).

576 CR-(FE), ...A2. . . . . . .J. PHYS. SOC. JAPAN VOL.20, . . . 1729, (1965).
. . . .*NEUTRON DIFFRACTION STUDY OF CR-RICH CR-FE ALLOY.
. . . .*HOSHINO(S.), ISHIKAWA(Y.), YAMADA(Y.), YAMADA(T.).

577 CR-FE-CO, ...A4. . . . . . .J. APP. PHYS. 38, ... 1243, (1967).
. . . .*NEUTRON-DIFFRACTION STUDY OF CR AND CR ALLOYS.
. . . .*WERNER(S.A.), ARROTT(A.), KENDRICK(H.).

578 CR-GE, ...A2. . . . . . .PHYS. STAT. SOLIDI B VOL.57 NO.2 K107, (1973).
. . . .*NEUTRON DIFFRACTION STUDY OF CR-GE.
. . . .*KOLENDA(M.), LECIEJEWICZ(J.), SZYTULA(A.).

579 CR*HO*O3, ...A2. . . . . . .PROC. INT. CONF. (NOTTINGHAM), . . 275, (1964).
. . . .*INVESTIGATIONS AT THE NUCLEAR CENTRE IN GRENOBLE: THE MAGNETIC STRUCTURE
. . . . .OF CR*HO*O3, MN*Y*O3, GE*M2*O4 (M=NI,CO) AND ND*CO5.
. . . .*BERTAUT(E.F.), BUISSON(G.), DELAPALME(A.), VAN-LAAR(B.), LEMAIRE(R.),
. . . . .MARESCHAL(J.)... ET AL. (ED. NOTE: REF. SECTION 3, 64/BERTAUT).

580 CR*K*(S*O4)2.12H2*O, ...A1. . . . . . .PROC. ROY. SOC. A VOL.246, . . . 78, (1958).
. . . .*THE STRUCTURE OF CHROMIUM POTASSIUM ALUM.
. . . .*BACON(G.E.), GARDNER(W.E.).

581 CR*LA*O3, ...A4. . . . . . .PROC. INT. CONF. (NOTTINGHAM), . 327, (1964).
. . . .*NEUTRON DIFFRACTION DATA AND COVALENCY EFFECTS.
. . . .*NATHANS(R.), WILL(G.), COX(D.E.).

582 CR-MN, ...A1,2. . . . . .COLL. INTER. N.126 (GRENOBLE), . 38, (1963).
. . . .*NEUTRON DIFFRACTION AND SCATTERING STUDIES AT J.A.E.R.I.
. . . .*KUNITOMI(N.), HAMAGUCHI(Y.), SAKAMOTO(M.), DOI(R.), KOMURA(S.).

583 CR-MN, ...A2. . . . . . .REF. SECTION 3, 66/KOEHLER(2), ...

584 CR-MN, ...A2. . . . . . .SOL. STATE COMM. VOL.3, ... 137, (1965).
. . . .*ANTIFERROMAGNETISM IN CHROMIUM ALLOY SINGLE CRYSTALS.
. . . .*MOLLER(H.B.), TREGO(A.L.), MACKINTOSH(A.R.).

585 CR-MN, ...A2. . . . . . .PROC. PHYS. SOC. 88, ... 935, (1966).
. . . .*THE MAGNETIC STRUCTURES OF DILUTE CR-MN ALLOYS.
. . . .*BASTOW(T.J.).

586 CR-(MN), ...A1. . . . . . .PHYS. REV. 138, ...A737, (1965).
. . . .*NEUTRON-DIFFRACTION INVESTIGATION OF CHROMIUM WITH SMALL ADDITIONS OF
. . . . .MANGANESE AND VANADIUM.
. . . .*HAMAGUCHI(Y.), WOLLAN(E.O.), KOEHLER(W.C.).

587 CR-MN, ...A2. . . . . . .J. PHYS. SOC. JAPAN VOL.19, . . . 1849, (1964).
. . . .*ANTIFERROMAGNETISM IN DISORDERED B.C.C. CR-MN ALLOYS.
. . . .*HAMAGUCHI(Y.), KUNITOMI(N.).

588 CR-MN, ...A2. . . . . . .SOV. PHYS. CRYST. VOL.6, ... 628, (1962).
. . . .*ANTIFERROMAGNETISM IN THE CR-MN ALLOY.
. . . .*GOMAN≠KOV(V.I.), LOSHMANOV(A.A.).

589 (CR-MN)*AS, ...A2. . . . . . .J. PHYS. SOC. JAPAN VOL.30, . . . 1319, (1971).
. . . .*MAGNETIC PROPERTIES OF (CR(1-X)-MN(X))*AS SYSTEM.
. . . .*KAZAMA(N.), WATANABE(H.).

590 CR*MN2*O4, ...A1,2. . . . . .J. PHYS. CHEM. SOLIDS 30, ... 805, (1969).
. . . .*MANGANITES SPINELLES PURS D≠ELEMENTS DE TRANSITION: PREPARATIONS ET
. . . . .STRUCTURES CRISTALLOGRAPHIQUES.
. . . .*BUHL(R.).

591 (CR-MN)*S, ...A2,A4. . . . .J. APPL. PHYS. VOL.37, ... 1038, (1966).
. . . .*SOME NEUTRON-DIFFRACTION INVESTIGATIONS AT THE NUCLEAR CENTRE OF
. . . . .GRENOBLE. (ED. NOTE: REF. SECTION 3, 66/BERTAUT).
. . . .*BERTAUT(E.F.), BASSI(G.), BUISSON(G.), BURLET(P.), CHAPPERT(J.),
. . . . .DELAPALME(A.), MARESCHAL(J.), ROULT(G.), ALEONARD(R.), PAUTHENET(R.).

592 (CR-MN)*SB, ...A2. . . . . . .PHYS. REV. 131, ... 1511, (1963).
. . . .*MAGNETIC STRUCTURES OF CHROMIUM-MODIFIED MN2*SB.
. . . .*AUSTIN(A.E.), ADELSON(E.), CLOUD(H.).

593 CR-MO, ...A2. . . . . . .REF. SECTION 3, 66/KOEHLER(1,2), ...

594 CR*N, ...A2. . . . . . .SOL. STATE COMM. VOL. 9, ... 717, (1971).
. . . .*ETUDE DE LA TRANSITION DE PREMIER ORDRE DANS CR*N.
. . . .*NASR-EDDINE(M.), BERTAUT(E.F.).

595 CR*N, ...A1. . . . . . .COMP. REND. 269, ... 574, (1969).
. . . .*CONTRIBUTION A L≠ETUDE DE CR*N A BASSES TEMPERATURES PAR DIFFRACTION
. . . . .DE RAYONS X ET DE NEUTRONS.
. . . .*EDDINE(M.N.), SAYETAT(F.), BERTAUT(E.F.).

596 CR*N, ...A2. . . . . . .PHYS. REV. VOL.117, ... 929, (1960).
. . . .*ANTIFERROMAGNETIC STRUCTURE OF CR*N.
. . . .*CORLISS(L.M.), ELLIOTT(N.), HASTINGS(J.M.).

597 CR-NB, ...A2. . . . . . .PHYS. REV. VOL.151, ... 405, (1966).
. . . .*ANTIFERROMAGNETISM IN CHROMIUM ALLOYS. I. NEUTRON DIFFRACTION.
. . . .*KOEHLER(W.C.), MOON(R.M.), TREGO(A.L.), MACKINTOSH(A.R.). (ED. NOTE: REF
. . . . .SECTION 3, 66/KOEHLER(1,2)).

598 CR-NB, ...A2. . . . . . .J. APP. PHYS. 37, ... 1259, (1966).
. . . .*NEUTRON DIFFRACTION STUDY OF DILUTE CHROMIUM-BASE ALLOYS.
. . . .*KOEHLER(W.C.), TREGO(A.L.), MOON(R.M.), MACKINTOSH(A.R.). (ED. NOTE:
. . . . .REF. SECTION 3, 66/KOEHLER(1,2)).

SECTION   5 -STRUCTURE AND CROSS-SECTION DETERMINATIONS . . . .

599   CR2*NI*S4, ... A2. . . . . . J. PHYSIQUE TOME 27, ... 619, (1966).
      *NEUTRON DIFFRACTION STUDY OF CR2*NI*S4.
      *ANDRON(B.), BERTAUT(E.F.). (IN FRENCH).

600   CR2*O3, ... A2. . . . . . DISS. ABS. 28, ... 3C298, (1967).
      *NEUTRON DIFFRACTION STUDIES ON (ALPHA)-FE2*O3 AND CR2*O3 AT HIGH
      PRESSURE.
      *WORLTON(T.G.).

601   CR2*O3, ... A2. . . . . . COLL. INTER. N.126 (GRENOBLE) . . 133, (1963).
      *MAGNETIC STRUCTURE STUDIES AT BROOKHAVEN NATIONAL LABORATORY.
      *CORLISS(L.M.), HASTINGS(J.M.).

602   CR2*O3, ... A2. . . . . . J. SOLID STATE CHEM. VOL.4, ... 420, (1972).
      *NEUTRON DIFFRACTION AND OTHER STUDIES OF MAGNETIC ORDERING IN PHASES
      BASED ON CR2*O3, V2*O3, AND TI2*O3.
      *REID(A.F.), SABINE(T.M.), WHEELER(D.A.).

603   CR2*O3, ... A4. . . . . . J. PHYS. CHEM. SOLIDS VOL.29, . . 435, (1968).
      *PRESSURE DEPENDENCE OF THE NEEL TEMPERATURE OF CR2*O3.
      *WORLTON(T.G.), BRUGGER(R.M.), BENNION(R.B.).

604   CR2*O3, ... A2. . . . . . J. APP. PHYS. 36, ... 1099, (1965).
      *MAGNETIC STRUCTURE OF CR2*O3.
      *CORLISS(L.M.), HASTINGS(J.M.), NATHANS(R.), SHIRANE(G.).

605   CR2*O3, ... A2. . . . . . J. CHEM. PHYS. VOL.21, ... 961, (1953).
      *ANTIFERROMAGNETIC STRUCTURE IN CR2*O3.
      *BROCKHOUSE(B.N.).

606   (CR2*O3)-(FE2*O3), ... A1,2. . . . . J. PHYS. CHEM. SOL. 24, ... 405, (1963).
      *A MAGNETIC AND NEUTRON DIFFRACTION STUDY OF THE (CR2*O3)-(FE2*O3) SYSTEM
      *COX(D.E.), TAKEI(W.J.), SHIRANE(G.).

607   CR-PD, ... A2. . . . . . PHYS. LETT. VOL.43A NO.4, ... 381, (1973).
      *MAGNETIC PROPERTIES OF CR-PD ALLOYS.
      *ABOUL-NOOR(S.S.), BOOTH(J.G.).

608   CR-PT, ... A2. . . . . . J. APP. PHYS. 34, ... 1203, (1963).
      *NEUTRON DIFFRACTION INVESTIGATION OF PT-BASED ALLOYS OF THE FIRST
      TRANSITION SERIES.
      *PICKART(S.J.), NATHANS(R.).

609   CR-RE, ... A2. . . . . . J. PHYS. CHEM. SOLIDS VOL.33, . . 1651, (1972).
      *NEUTRON DIFFRACTION STUDIES OF DILUTE CR-RE SINGLE CRYSTAL ALLOYS.
      *LEBECH(B.), MIKKE(K.).

610   CR-RE, ... A2. . . . . . REF. SECTION 3, 66/KOEHLER(1,2), ...

611   CR-RE, ... A2. . . . . . SOL. STATE COMM. VOL.3, ... 137, (1965).
      *ANTIFERROMAGNETISM IN CHROMIUM ALLOY SINGLE CRYSTALS.
      *MOLLER(H.B.), TREGO(A.L.), MACKINTOSH(A.R.).

612   CR-RH, ... A2. . . . . . REF. SECTION 3, 66/KOEHLER(1,2), ...

613   CR-RU, ... A2. . . . . . REF. SECTION 3, 66/KOEHLER(1,2), ...

614   CR*S, ... A2. . . . . . COLL. INTER. N.126 (GRENOBLE), . 143, (1963).
      *ETUDES DE DIFFRACTION NEUTRONIQUE A LIVERMORE.
      *SPARKS(J.T.), KOMOTO(T.).

615   CR3*S4, ... A2. . . . . . COLL. INTER. N.126 (GRENOBLE), . 158, (1963).
      *STRUCTURES MAGNETIQUES DE CR3*X4 (X=S,SE,TE).
      *BERTAUT(E.F.), ROULT(G.), ALEONARD(R.), PAUTHENET(R.), CHEVRETON(M.),
      JANSEN(R.).

616   CR2*S3, ... A2. . . . . . PHYS. LETT. 25A, ...   27, (1967).
      *THE MAGNETIC STRUCTURE OF TRIGONAL CR2*S3.
      *VAN-LAAR(B.).

617   CR5*S6, ... A2. . . . . . PHYS. REV. 156, ... 654, (1967).
      *FERRIMAGNETIC AND ANTIFERROMAGNETIC STRUCTURES OF CR5*S6.
      *VAN-LAAR(B.).

618   CR2*S3, ... A2. . . . . . J. PHYSIQUE TOME 29, ... 813, (1968).
      *ETUDE DE CR2*S3 RHOMBOEDRIQUE PAR DIFFRACTION NEUTRONIQUE ET MESURES
      MAGNETIQUES.
      *BERTAUT(E.F.), COHEN(J.), LAMBERT-ANDRON(B.), MOLLARD(P.).

619   CR2*S3,CR5*S6, ... A2. . . . . . J. PHYS. CHEM. SOLIDS VOL.32, . . 581, (1971).
      *SPIN STRUCTURE AND MAGNETIC ANISOTROPY OF CR5*S6 AND RHOMBOHEDRAL CR2*S3
      *POPMA(T.J.A.), HAAS(C.), VAN-LAAR(B.).

620   CR*SB, ... A2. . . . . . PHYS. REV. VOL.85, ... 365, (1952).
      *NEUTRON DIFFRACTION INVESTIGATION OF THE ATOMIC MAGNETIC MOMENT
      ORIENTATION IN THE ANTIFERROMAGNETIC COMPOUND CR*SB.
      *SNOW(A.I.).

621   CR*SB2, ... A1,A2. . . . ACTA CHEM. SCAND. VOL.24, ... 3309, (1970).
      *COMPOUNDS WITH THE MARCASITE TYPE CRYSTAL STRUCTURE VI. NEUTRON
      DIFFRACTION STUDIES OF CR*SB2 AND FE*SB2.
      *HOLSETH(H.), KJEKSHUS(A.), ANDRESEN(A.F.).

622   (CR*SB)-(MN*SB), ... A2. . . . . . PROC. INT. CONF. (NOTTINGHAM), . . 291, (1964).
      *MAGNETIC STRUCTURES IN THE MN*SB-CR*SB AND CR*TE-CR*SB SYSTEMS.
      *COX(D.E.), SHIRANE(G.), TAKEI(W.J.).

623   CR3*SE4, ... A2. . . . . . COLL. INTER. N.126 (GRENOBLE), . 158, (1963).
      *STRUCTURES MAGNETIQUES DE CR3*X4 (X=S,SE,TE).
      *BERTAUT(E.F.), ROULT(G.), ALEONARD(R.), PAUTHENET(R.), CHEVRETON(M.),
      JANSEN(R.).

624   CR*SE, ... A2. . . . . . PHYS. REV. VOL.122, ... 1402, (1961).
      *MAGNETIC STRUCTURE OF CHROMIUM SELENIDE.
      *CORLISS(L.M.), ELLIOTT(N.), HASTINGS(J.M.), SASS(R.L.).

625   CR3*SE4, ... A2. . . . . . J. APPL. PHYS. VOL.33, SUPPL., . . 1123, (1962).
      *MAGNETIC STRUCTURE WORK AT THE NUCLEAR CENTER OF GRENOBLE.
      *BERTAUT(E.F.), DELAPALME(A.), FORRAT(F.), ROULT(G.), DE-BERGEVIN(F.),
      PAUTHENET(R.).

626   CR3*SI, ... A2. . . . . . PHYS. REV. VOL.95, ... 280, (1954).
      *SEARCH FOR ANTIFERROMAGNETISM IN THE SILICIDES V3*SI, CR3*SI, AND MO3*SI
      *KOEHLER(W.C.), WOLLAN(E.O.).

SECTION 5 -STRUCTURE AND CROSS-SECTION DETERMINATIONS . . . .

627 CR-TA, ...A2. . . . . . .REF. SECTION 3, 66/KOEHLER(1,2), ...

628 CR*TA*04,. . . .
*ATOMIC AND MOLECULAR ORDERING IN ME*TA*O4 (ME=TI,V,CR,FE) WITH A RUTILE ....A1........SOV. PHYS. CRYST. VOL.17,...... 1017, (1973).
STRUCTURE.
*ASTROV(D.N.), KRYUKOVA(N.A.), ZORIN(R.B.), MAKAROV(V.A.), OZEROV(R.P.),
ROZHDESTVENSKII(F.A.), SMIRNOV(V.P.), TURCHANINOV(A.M.), FADEEVA(N.V.).

629 CR-TC, ...A2. . . . . . .REF. SECTION 3, 66/KOEHLER(1,2), ...

630 CR3*TE4, ...
.........A2........COLL. INTER. N.126 (GRENOBLE), . 158, (1963).
*STRUCTURES MAGNETIQUES DE CR3*X4 (X=S,SE,TE).
*BERTAUT(E.F.), ROULT(G.), ALEONARD(R.), PAUTHENET(R.), CHEVRETON(M.),
JANSEN(R.).

631 (CR*TE)-(CR*SB), .
.......A2........PROC. INT. CONF. (NOTTINGHAM), . 291, (1964).
*MAGNETIC STRUCTURES IN THE MN*SB-CR*SB AND CR*TE-CR*SB SYSTEMS.
*COX(D.E.), SHIRANE(G.), TAKEI(W.J.).

632 CR*TE-CR*SB, . .
.......A2........J. APP. PHYS. 37,...... 973, (1966).
*MAGNETIC STRUCTURES IN CR*TE-CR*SB SOLID SOLUTIONS.
*TAKEI(W.J.), COX(D.E.), SHIRANE(G.).

633 CR2*TE*06, . . .
.......A2........J. PHYS. CHEM. SOLIDS 29,...... 1359, (1968).
*MAGNETIC STRUCTURES OF THE ORDERED TRIRUTILES CR2*W*06, CR2*TE*06 AND
FE2*TE*06.
*KUNNMANN(W.), LA-PLACA(S.J.), CORLISS(L.M.), HASTINGS(J.M.), BANKS(E.).

634 CR2*(TE,W)*06, .
.......A2........SOL. STATE COMM. VOL.6 NO.5,..... 323, (1968).
*STRUCTURES MAGNETIQUES DES COMPOSES TRIRUTILES CR2*TE*06 ET CR2*W*06.
*MONTMORY(M.C.), NEWNHAM(R.).

635 CR*TI*ND*05, . .
.......A1........J. PHYS. CHEM. SOLIDS VOL. 31,.... 1171, (1970).
*ETUDE PAR RAYONS X ET NEUTRONS DE LA SERIE ISOMORPHE A*TI*T*05.
(A=CR, MN, FE, T=TERRES RARES).
*BUISSON(G.).

636 (CR-TI)2*03, . .
.......A2........SOV. PHYS. SOL. STATE VOL. 13, . 644, (1971).
*MAGNETIC STRUCTURES OF (CR-TI)2*03 AND (CR-V)2*03 SOLID SOLUTIONS.
*ZUBKOV(V.G.), MATVEENKO(I.I.), SHEINKER(M.E.), GEL#D(P.V.).

637 CR-V, ...A2. . . . . . .REF. SECTION 3, 66/KOEHLER(2), ...

638 CR-V, ...
.......A2........SOL. STATE COMM. VOL.3,.... 137, (1965).
*ANTIFERROMAGNETISM IN CHROMIUM ALLOY SINGLE CRYSTALS.
*MOLLER(H.B.), TREGO(A.L.), MACKINTOSH(A.R.).

639 CR-(V), ...
.......A1........PHYS. REV. 138,......A737, (1965).
*NEUTRON-DIFFRACTION INVESTIGATION OF CHROMIUM WITH SMALL ADDITIONS OF
MANGANESE AND VANADIUM.
*HAMAGUCHI(Y.), WOLLAN(E.O.), KOEHLER(W.C.).

640 CR-(V), ...
.......A2........J. PHYS. SOC. JAPAN VOL.20,.... 103, (1965).
*NEUTRON DIFFRACTION STUDY ON CHROMIUM ALLOY WITH SMALL AMOUNTS OF
VANADIUM.
*KOMURA(S.), KUNITOMI(N.).

641 CR-V-MN, ...
.......A1........PHYS. LETT. 24A,...... 299, (1967).
*NEUTRON DIFFRACTION STUDIES ON CHROMIUM-BASED CR-V-MN TERNARY DILUTE
ALLOYS.
*KOMURA(S.), HAMAGUCHI(Y.), KUNITOMI(N.).

642 (CR-V)*N, ...
.......A2........SOL. STATE COMM. VOL.13, . 905, (1973).
*CONTRIBUTION A L#ETUDE DE LA TRANSITION MAGNETIQUE DANS CR(.75)-V(.25)*N
*NASR-EDDINE(M.), BERTAUT(E.F.), MOLLARD(P.), CHAUSSY(J.).

643 (CR-V)2*03, . . .
.......A2........SOV. PHYS. SOL. STATE VOL. 13, . 644, (1971).
*MAGNETIC STRUCTURES OF (CR-TI)2*03 AND (CR-V)2*03 SOLID SOLUTIONS.
*ZUBKOV(V.G.), MATVEENKO(I.I.), SHEINKER(M.E.), GEL#D(P.V.).

644 CR*V*04, ...
.......A2........PHYS. REV. VOL.125,..... 1283, (1962).
*ANTIFERROMAGNETIC STRUCTURE OF CR*V*04 AND THE ANHYDROUS SULFATES OF
DIVALENT FE,NI AND CO.
*FRAZER(B.C.), BROWN(P.J.).

645 CR-W, ...A2. . . . . . .REF. SECTION 3, 66/KOEHLER(1,2), ...

646 CR2*W*06, ...
.......A2........SOL. STATE COMM. 4,..... 249, (1966).
*STRUCTURE MAGNETIQUE DE CR2*W*06.
*MONTMORY(M.C.), BERTAUT(E.F.), MOLLARD(P.).

647 CR2*W*06, ...
.......A2........J. PHYS. CHEM. SOLIDS 29,...... 1359, (1968).
*MAGNETIC STRUCTURES OF THE ORDERED TRIRUTILES CR2*W*06, CR2*TE*06 AND
FE2*TE*06.
*KUNNMANN(W.), LA-PLACA(S.J.), CORLISS(L.M.), HASTINGS(J.M.), BANKS(E.).

648 CS, ...
.......P........HELV. PHYS. ACTA VOL.45,..... 46, (1972).
*NEUTRON-CAPTURE CROSS-SECTIONS OF V, MN, CS, DY AND LU IN THE ENERGY
RANGE FROM 0.01 TO 100 EV.
*WIDDER(F.).

649 CS, ...
.......P........Z. KRIST. VOL.134,...... 308, (1971).
*THE COHERENT NEUTRON SCATTERING AMPLITUDE OF CAESIUM.
*CHADWICK(B.M.), JONES(D.W.), SARNESKI(J.E.), WILDE(H.J.), YERKESS(J.).

650 CS, ...
.......P........ACTA CRYST. VOL.A27,...... 494, (1971).
*A REDETERMINATION OF THE COHERENT NEUTRON SCATTERING AMPLITUDE OF CS.
*COX(D.E.), MINKIEWICZ(V.J.).

651 CS*AL(S*04)2*12(H2*0)
.......A1........ACTA CRYST. 21,...... 383, (1966).
*REFINEMENT OF ALUM STRUCTURES. I. X-RAY AND NEUTRON DIFFRACTION STUDY OF
CS*AL(S*04)2*12(H2*0).
*CROMER(D.T.), KAY(M.I.), LARSON(A.C.).

652 CS*BR, ...
.......A3........ANN. N.Y. ACAD. SCI. VOL.79, . 762, (1960).
*X-RAY AND NEUTRON DIFFRACTION STUDIES OF MOLTEN ALKALI HALIDES.
*LEVY(H.A.), AGRON(P.A.), BREDIG(M.A.), DANFORD(M.D.).

653 CS*CL, ...
.......P........ACTA CRYST. VOL.20,...... 315, (1966).
*THE NEUTRON COHERENT SCATTERING AMPLITUDE OF CESIUM.
*ZIVADINOVIC(M.S.), PRELESNIK(B.).

654 CS*(CL,BR), . . .
.......P........PHYS LETT 28A, ... 301, (1968).
*DEBYE-WALLER FACTORS OF CESIUM HALIDES.
*HARIDASAN(T.M.), NANDINI(R.).

SECTION   5 -STRUCTURE AND CROSS-SECTION DETERMINATIONS . . . .

655   CS*CO*CL3.2H2*O,, ...A2. . . . . . .ACTA CRYST. VOL.A28 PART S-4,. . . 194, (1972).
      *MAGNETIC STRUCTURE, MAGNETIZATION, AND SHORT RANGE SPIN CORRELATIONS IN
      THE ISING LINEAR CHAIN ANTIFERROMAGNET CS*CO*CL3.2H2*O.
      *VAN-LAAR(B.), BONGAARTS(A.L.M.). (ABSTRACT ONLY).

656   CS*CO*CL3.2H2*O, ...A2. . . . . . .PHYS. REV. B VOL.6,. . . . . . 2669, (1972).
      *MAGNETIC STRUCTURE, MAGNETIZATION, AND SHORT-RANGE SPIN CORRELATIONS
      IN THE ISING LINEAR-CHAIN ANTIFERROMAGNET CS*CO*CL3.2H2*O.
      *BONGAARTS(A.L.M.), VAN-LAAR(B.).

657   CS2*CU*H12*O14*S2, ...A1. . . . . . .CRYST. STRUCT. COMM. VOL.1 N.3,. 189, (1972).
      *CESIUM HEXA-AQUACOPPER(II) SULPHATE,. . .
      *SHIELDS(K.G.), KENNARD(C.H.L.).

658   CS*FE*F3,. . . . . .A2. . . . . . .J. APPL. PHYS. VOL.42,. . . . . . 1617, (1971).
      *MAGNETIC STRUCTURES OF FERRIMAGNETIC RB*NI*F3 AND CS*FE*F3.
      *PICKART(S.J.), ALPERIN(H.A.).

659   CS*H3*(SE*O3)2,. . . . . .A1. . . . . . .J. PHYS. SOC. JAPAN VOL.32,. . . 1670, (1972).
      *NEUTRON STRUCTURE ANALYSIS OF CS*H3*(SE*O3)2 AT ROOM TEMPERATURE.
      *SATO(S.).

660   CS*I, ...P. . . . . . .PHYS LETT 28A, ... 301, (1968).
      *DEBYE-WALLER FACTORS OF CAESIUM HALIDES.
      *HARIDASAN(T.M.), NANDINI(R.).

661   CS2*LI*(CR,MN,FE,CO)*(C*N)6, ...A1. . . . .INORG. NUCL. CHEM. LETT. VOL.9,. 1025, (1973).
      *SINGLE CRYSTAL VIBRATIONAL SPECTROSCOPIC AND NEUTRON DIFFRACTION
      STUDIES OF THE HEXACYANOMETALLATES, CS2*LI*M*(C*N)6 (M=CR,MN,FE,CO).
      *ARMSTRONG(J.R.), CHADWICK(B.M.), JONES(D.W.), SARNESKI(J.E.),
      WILDE(H.J.), YERKESS(J.).

662   CS*MN*BR3, ...A2. . . . . . .A.I.P. CONF. PROC. NO.10 PART 1, 684, (1973).
      *MAGNETIC ORDERING OF THE LINEAR-CHAIN ANTIFERROMAGNET CS*MN*BR3.
      *EIBSCHUTZ(M.), SHERWOOD(R.C.), HSU(F.S.L.), COX(D.E.).

663   CS*MN*CL3, ...A2. . . . . . .PHYS. REV. B VOL.3,. . . . . . 3873, (1971).
      *NEUTRON DIFFRACTION AND MAGNETIC STRUCTURE OF CS*MN*CL3.
      *MELAMUD(M.), MAKOVSKY(J.), SHAKED(H.).

664   CS*MN*CL3, ...A2. . . . . . .BU. ISRAEL PHYS. SOC. 1971. . . 27, (1971).
      *NEUTRON DIFFRACTION AND MAGNETIC STRUCTURE OF CS*MN*CL3.
      *MELAMUD(M.), MAKOVSKY(J.), SHAKED(H.).

665   CS2*MN*CL4,. . . . . .A2. . . . . . .PHYS. REV. B VOL.2,. . . . . . 3703, (1970).
      *MAGNETIC STRUCTURE AND TWO-DIMENSIONAL BEHAVIOUR OF RB2*MN*CL4 AND
      CS2*MN*CL4.
      *EPSTEIN(A.), GUREWITZ(E.), MAKOVSKY(J.), SHAKED(H.).

666   CS*MN*CL3(2*H2*O),. . . . . .A2. . . . . . .PHYS. REV. B-2,. . . . . 1310, (1970).
      *NEUTRON SCATTERING IN THE LINEAR CHAIN ANTIFERROMAGNET CS*MN*CL3(2*H2*O).
      *SKALYO-JR(J.), SHIRANE(G.), FRIEDBERG(S.A.), KOBAYASHI(H.).

667   CS*MN*F3,. . . . . .A1. . . . . . .J. CHEM. PHYSICS VOL.37, ... 697, (1962).
      *CRYSTAL STRUCTURE OF CS*MN*F3.
      *ZALKIN(A.), LEE(K.), TEMPLETON(D.H.).

668   CS*MN*F3, ...A2. . . . . . .COLL. INTER. N.126 (GRENOBLE), . 141, (1963).
      *MAGNETIC STRUCTURE OF BINARY FLUORIDES CONTAINING MN(+2).
      *PICKART(S.J.), ALPERIN(H.A.).

669   (CS-MN-FE)*F6, ...A2. . . . . . .SOL. STATE COMM. VOL.11,. . . . 1471, (1972).
      *THE MAGNETIC STRUCTURE OF CS(0.4)-MN(0.4)-FE(1.6)*F6 AND OF CS(0.4)-
      NI(0.4)-FE(1.5)*F6.
      *STEINER(M.), BINDER(F.).

670   CS*NI*CL3, ...A2. . . . . . .SOL. STATE COMM. VOL.8,. . . . . 1001, (1970).
      *THE MAGNETIC STRUCTURES OF RB*NI*CL3 AND CS*NI*CL3.
      *MINKIEWICZ(V.J.), COX(D.E.), SHIRANE(G.).

671   CS*NI*CL3, ...A2. . . . . . .PHYS. REV. B VOL.4,. . . . . . 2209, (1971).
      *MAGNETIC ORDERING AND LOW NI(2+) MOMENT IN CS*NI*CL3.
      *COX(D.E.), MINKIEWICZ(V.J.).

672   CS*NI*CL3, ...A2. . . . . . .PHYS. REV. B VOL.7,. . . . . . . 2024, (1973).
      *MAGNETIC TRANSITIONS IN CS*NI*CL3.
      *YELON(W.B.), COX(D.E.).

673   CS*NI*F3,. . . . . .A2. . . . . . .SOL. STATE COMM. VOL.11. . . . . 73, (1972).
      *THE 3-D MAGNETIC STRUCTURE OF CS*NI*F3: COMPARISON OF THE MEASURED
      SHORT RANGE ORDER WITH DIFFERENT MODELS.
      *STEINER(M.).

674   CS*NI*F3,. . . . . .A2. . . . . . .A.I.P. CONF. PROC. NO.10 PART 1, 664, (1973).
      *THE THREE-DIMENSIONAL MAGNETIC STRUCTURE OF CS*NI*F3.
      *STEINER(M.), DACHS(H.), BABEL(D.).

675   CS*NI*F3, ...A2. . . . . . .SOL. STATE COMM. VOL.14 NO.9,. . 841, (1974).
      *THE MAGNETIC PHASE DIAGRAM OF CS*NI*F3 AS DETERMINED BY NEUTRON
      DIFFRACTION.
      *STEINER(M.), DACHS(H.).

676   (CS-NI-FE)*F6, ...A2. . . . . . .SOL. STATE COMM. VOL.11,. . . . 1471, (1972).
      *THE MAGNETIC STRUCTURE OF CS(0.4)-MN(0.4)-FE(1.6)*F6 AND OF CS(0.4)-
      NI(0.4)-FE(1.5)*F6.
      *STEINER(M.), BINDER(F.).

677   CU,. . . . . .A3. . . . . . .J. PHYS. CHEM. SOLIDS VOL.31,. . 549, (1970).
      *DETERMINATION DE LA FONCTION DE DISTRIBUTION DE PAIRE DU CUIVRE LIQUIDE
      PAR DIFFRACTION DE NEUTRONS.
      *BREUIL(M.), TOURAND(G.).

678   CU,. . . . . .A4. T. . . . .PHYS. STAT. SOL. 29,. . . . . . 259, (1968).
      *SMALL ANGLE SCATTERING OF SUBTHERMAL NEUTRONS FROM DEFORMED POLY-
      CRYSTALLINE COPPER.
      *TAGLAUER(E.).

679   CU,. . . . . .P. . . . . . .PHYS. SCRIPTA(SWEDEN) VOL.4,. . 125, (1971).
      *THE ATOMIC SCATTERING FACTORS OF CU AND AL FROM THERMAL DIFFUSE
      SCATTERING MEASUREMENTS.
      *LINKOAHO(M.), SERVOMAA(A.).

680   CU (ISO. A=66),. . . . . .P. . . . . . .NUCL. APPL. TECHNOL. VOL.9,. . . 662, (1970).
      *AVERAGE THERMAL NEUTRON CAPTURE CROSS SECTIONS OF AU(ISO. A=198), NI
      (ISO. A=65), AND CU(ISO. A=66).
      *SERMENT(V.), ABU-SAMRA(A.), EMMONS(A.H.).

SECTION 5 -STRUCTURE AND CROSS-SECTION DETERMINATIONS . . . .

681 CU,. . . . . .ACTA CRYST. (INTERACT.) VOL.A25, S218, (1969).
  *NEUTRON SMALL ANGLE SCATTERING AT DISLOCATIONS IN COPPER AND NICKEL
  SINGLE CRYSTALS DEFORMED IN TENSION.
  *HERGET(P.), SCHMATZ(W.).

682 CU,. . . . . . .A4. . . . . . . .THESIS (86 PP.), ..., (1970).
  *THE REFLECTIVITY OF NEUTRONS BY DISTORTED COPPER CRYSTALS.
  *DYMOND(R.R.). (ED. NOTE: M.SC. DISSERTATION, MCMASTER UNIVERSITY).

683 CU (SPINEL), ...A1. . . . . . .SOL. STATE COMM. 5,. . . . . . . 577, (1967).
  *TETRAHEDRAL-SITE COPPER IN CHALCOGENIDE SPINELS.
  *GOODENOUGH(J.B.).

684 CU,. . . . . . . .A4. . . . . . . .J. APPL. CRYST. VOL.4, ... 303, (1971).
  *THE DEVELOPMENT OF THE ROLLING TEXTURE IN COPPER MEASURED BY NEUTRON
  DIFFRACTION.
  *BUNGE(H.J.), TOBISCH(J.), SONNTAG(W.).

685 CU,. . . . . . . .P. . . . . . . .NUCL. SCI. ENGNG. VOL.7, ...184, (1960).
  *THE ABSORPTION CROSS SECTION OF COPPER FOR THERMAL NEUTRONS.
  *DONAHUE(D.J.), BENNETT(R.A.), LANNING(D.D.).

686 CU,. . . . . . . .P. . . . . . . .Z. PHYSIK VOL.250, ...166, (1972).
  *TOTAL CROSS-SECTIONS OF VARIOUS HOMOGENEOUS SUBSTANCES FOR ULTRACOLD
  NEUTRONS.
  *STEYERL(A.), VONACH(H.).

687 CU,. . . . . . .P. . . . . . . .PHYS. REV. VOL.68, ... 159, (1945).
  *THERMAL NEUTRON SCATTERING STUDIES IN METALS.
  *NIX(F.C.), CLEMENT(G.F.).

688 CU,. . . . . . .P. . . . . . . .PROC. PHYS. MATH. SOC. JAPAN 25, 495, (1943).
  *SCATTERING OF THERMAL NEUTRONS BY SOLIDS. I.VARIATION IN TOTAL CROSS
  SECTION OF METALS BY COLD-WORKING.
  *KIMURA(M.), HASHIGUCHI(R.).

689 CU,. . . . . . . .A4. . . . . . . .PHIL. MAG. VOL.3,. . . . . .476, (1958).
  *SMALL ANGLE SCATTERING FROM COLD-WORKED AND FATIGUED METALS.
  *ATKINSON(H.H.).

690 CU,. . . . . . . .P. . . . . . . .Z. PHYSIK VOL.266, ...157, (1974).
  *NEUTRON TOTAL CROSS SECTION OF SC, V, CU AND RH AT MICRO-EV ENERGIES.
  *DILG(W.), MANNHART(W.).

691 CU (A=63,65),. . . . . .P. . . . . . . .PHYS. REV. VOL.111,. . . . . . 261, (1958).
  *NEUTRON CROSS SECTIONS AND SCATTERING LENGTHS OF CU (A=63) AND CU (A=65).
  *KEATING(D.T.), NEIDHARDT(W.J.), GOLAND(A.N.).

692 CU3-AU,. . . . . . . .A1. . . . . . .PHYS. REV. VOL.75, ... 1008, (1949).
  *NEUTRON DIFFRACTION STUDIES OF ORDER-DISORDER IN ALLOYS.
  *SHULL(C.G.), SIEGEL (S.).

693 CU-(AU,NI,ZN), ...A4. . . . . . .MATERIALS RES. BULL. VOL.2,. . . . .69, (1967).
  *CHARACTERIZATION OF LARGE ALLOY SINGLE CRYSTALS BY NEUTRON DIFFRACTION.
  *NG(S.C.), BROCKHOUSE(B.N.), HALLMAN(E.D.).

694 CU2*(C*H3*C*O2)4.2H2*O,. . . . . . . .A1. . . . . . .ACTA CRYST. VOL.B29, ...2393, (1973).
  *DINUCLEAR COPPER(II) ACETATE MONOHYDRATE: A REDETERMINATION OF THE
  STRUCTURE BY NEUTRON-DIFFRACTION ANALYSIS.
  *BROWN(G.M.), CHIDAMBARAM(R.).

695 CU*CL, ...A3,P. . . . .J. OF PHYS.-C-, SER.2, VOL.4, ...3034, (1971).
  *THE PARTIAL STRUCTURE FACTORS OF MOLTEN CUPROUS CHLORIDE FROM NEUTRON
  DIFFRACTION MEASUREMENTS.
  *PAGE(D.I.), MIKA(K.).

696 CU*CL2.2D2*O,. . . . . . . .A2. . . . . . . .PHYS. REV. VOL.167,. . . . . . . 519, (1968).
  *SPIN-DENSITY DISTRIBUTION IN CU*CL2.2D2*O.
  *UMEBAYASHI(H.), FRAZER(B.C.), COX(D.E.), SHIRANE(G.).

697 CU*CL2.2D2*O,. . . . . . . .A2. . . . . . . .J. APPL. PHYS. VOL.38, ... 1461, (1967).
  *CANTED ANTIFERROMAGNETISM OF CU*CL2.2D2*O.
  *UMEBAYASHI(H.), SHIRANE(G.), FRAZER(B.C.), COX(D.E.).

698 CU*CL2.2H2*O,. . . . . . . .A2. . . . . . . .PHYS. LETT. 17,. . . . . 95, (1965).
  *MAGNETIC STRUCTURE OF CU*CL2*2(H2*O).
  *SHIRANE(G.), FRAZER(B.C.), FRIEDBERG(S.A.).

699 CU*CL2.2H2*O,. . . . . . . .T. . . . . . . .PHYS. STAT. SOL. VOL.38, ...643, (1970).
  *SYMMETRY PROPERTIES OF THE ANTIFERROMAGNETIC STRUCTURE OF CU*CL2*2H2*O.
  *JOSHUA(S.J.).

700 CU-(CO), ...P. . . . . . . .PHYS. STAT. SOL. A VOL.7,. . . . . .477, (1971).
  *NEUTRON SMALL-ANGLE SCATTERING STUDY OF THE TRANSITION FROM SINGLE- TO
  MULTI-DOMAIN BEHAVIOUR IN PRECIPITATIONS OF CU-CO(.01).
  *ERNST(M.), SCHELTEN(J.), SCHMATZ(W.).

701 CU-(CO), ...P. . . . . . . .PHYS. STAT. SOL. A VOL.7,. . . . .469, (1971).
  *SMALL-ANGLE SCATTERING OF NEUTRONS AT SINGLE-DOMAIN PRECIPITATIONS IN
  A CU-CO(.01) SINGLE CRYSTAL.
  *ERNST(M.), SCHELTEN(J.), SCHMATZ(W.).

702 CU*CR2*O4, ...A2. . . . . . . .ACTA CRYST. VOL.10,. . . . . . . 554, (1957).
  *CRYSTAL AND MAGNETIC STRUCTURE OF COPPER CHROMITE.
  *PRINCE(E.).

703 CU*CR*(S2,SE2),. . . . . . . .A1. . . . . . .J. SOLID STATE CHEM. VOL.6,. . . . . 574, (1973).
  *CRYSTAL STRUCTURES AND MAGNETIC STRUCTURES OF SOME METAL(I) CHROMIUM
  (III) SULFIDES AND SELENIDES.
  *ENGELSMAN(F.M.R.), WIEGERS(G.A.), JELLINEK(F.), VAN-LAAR(B.).

704 CU*CR2*SE4,. . . . . . . .A1,2. . .J. PHYS. CHEM. SOL. 28. . . . . .897, (1967).
  *NEUTRON DIFFRACTION AND ELECTRICAL TRANSPORT PROPERTIES OF CU*CR2*SE4.
  *ROBBINS(M.), LEHMANN(H.W.), WHITE(J.G.).

705 CU*CR2*SE4,. . . . . . . .A1. . . . . . .PHYS. REV. 153,. . . . . 558, (1967).
  *NEUTRON-DIFFRACTION INVESTIGATION OF CU*CR2*SE4.
  *COLOMINAS(C.).

706 CU*CR2*SE3*BR, ...A2. . . . . . .J. APP. PHYS. 39,. . . . . . .664, (1968).
  *NEUTRON-DIFFRACTION STUDY OF THE FERROMAGNETIC MIXED-ANION SPINEL
  CU*CR2*SE3*BR.
  *WHITE(J.G.), ROBBINS(M.).

SECTION 5 -STRUCTURE AND CROSS-SECTION DETERMINATIONS . . . .

707 CU*CR2*TE4, ... A1 . . . . PHYS. REV. 153, ... 558, (1967).
⌐NEUTRON DIFFRACTION INVESTIGATION OF CU*CR2*TE4.
⌐COLOMINAS(C.).

708 CU*(D*C*O*O)2.4D2*O, ... A1 . . . . ACTA CRYST. VOL.A28 PART.S-4, . . 186, (1972).
⌐NEUTRON DIFFRACTION STUDY OF THE ANTIFERROELECTRIC PHASE OF COPPER
FORMATE TETRAHYDRATE.
⌐KAY(M.I.), KLEINBERG(R.). (ABSTRACT ONLY).

709 CU*F2*2(H2*O), ... A1 . . . . J. CHEM. PHYSICS VOL.36, ... 56, (1962).
⌐CRYSTAL AND MAGNETIC STRUCTURE OF CUPRIC FLUORIDE DIHYDRATE AT 4.2 K.
⌐ABRAHAMS(S.C.).

710 CU*F2*2(H2*O), ... A1 . . . . J. CHEM. PHYSICS VOL.36, ... 50, (1962).
⌐CRYSTAL STRUCTURE OF CUPRIC FLUORIDE DIHYDRATE AT 298 K.
⌐ABRAHAMS(S.C.), PRINCE(E.).

711 CU-(FE), ... A4 . . . . PHYS. REV. B VOL.5, ... 1040, (1972).
⌐NEUTRON-DIFFRACTION STUDY OF THE KONDO EFFECT IN CU-(FE).
⌐STASSIS(C.), SHULL(C.G.).

712 CU-FE, ... A4 . . . . J. APPL. PHYS. VOL.41, ... 1146, (1970).
⌐NEUTRON DIFFRACTION STUDY OF THE KONDO EFFECT IN CU-FE.
⌐STASSIS(C.), SHULL(C.G.).

713 (CU(0.5)-FE(2.5))*O4, ... A2 . . . . LATV. PSR...FIZ. TEHN SER. NO.3, 64, (1972).
⌐NEUTRON-DIFFRACTION STUDY OF SPINEL FERRITE (CU(0.5)-FE(2.5))*O4.
⌐TSITSENOVSKAYA(S.E.).

714 CU*FE2*O4, ... A1 . . . . ACTA CRYST. VOL.9, ... 1025, (1956).
⌐THE STRUCTURE OF TETRAGONAL COPPER FERRITE.
⌐PRINCE(E.), TREUTING(R.G.).

715 CU*FE*S2, ... A2 . . . . PHYS. REV. VOL.112, ... 1917, (1958).
⌐SYMMETRY OF MAGNETIC STRUCTURES: MAGNETIC STRUCTURE OF CHALCOPYRITE.
⌐DONNAY(G.), CORLISS(L.M.), DONNAY(J.D.H.), ELLIOTT(N.), HASTINGS(J.M.).

716 CU-FORMATE, ... A1 . . . . J. CHEM. PHYSICS VOL.44, ... 1648, (1966).
⌐CRYSTAL STRUCTURE BY NEUTRON DIFFRACTION AND THE ANTIFERROELECTRIC PHASE
TRANSITION IN COPPER FORMATE TETRAHYDRATE.
⌐OKADA(K.), KAY(M.I.), CROMER(D.T.), ALMODOVAR(I.).

717 CU(H*C*O*O)2*4(D2*O), ... A1 . . . . SOL. STATE COMM. 5, ... 887, (1967).
⌐CELL DOUBLING IN COPPER FORMATE TETRAHYDRATE AT THE ANTIFERROELECTRIC
PHASE TRANSITION.
⌐TURBERFIELD(K.C.).

718 CU*H,CU*D, ... A1 . . . . ACTA CRYST. VOL.8, ... 118, (1955).
⌐THE CRYSTAL STRUCTURE OF COPPER HYDRIDE.
⌐GOEDKOOP(J.A.), ANDRESEN(A.F.).

719 CU-MN, ... A2 . . . . MAGN. AND MAG. MATERIALS-1971, . 508, (1972).
⌐SHORT RANGE ORDER IN CU-MN ALLOYS.
⌐SATO(H.), WERNER(S.A.), YESSIK(M.). (AIP CONF. PROC. NO.5 PART I).

720 CU-(MN), ... A2 . . . . J. APP. PHYS. 36, ... 1093, (1965).
⌐NEUTRON DIFFRACTION STUDIES OF PD, NI, FE*MN, AND CU(MN) SINGLE CRYSTAL.
⌐ARROTT(A.).

721 CU-MN, ... A1,A2 . . . . A.I.P. CONF. PROC. NO.10 PART 1, 679, (1973).
⌐EFFECTS OF TEMPERATURE AND HEAT TREATMENT ON THE SHORT-RANGE ORDER IN
CU-MN ALLOYS.
⌐WERNER(S.A.), SATO(H.), YESSIK(M.).

722 CU-MN, ... A2 . . . . PHYS. REV. VOL.105, ... 130, (1957).
⌐MAGNETIC STRUCTURES IN COPPER-MANGANESE ALLOYS.
⌐MENEGHETTI(D.), SIDHU(S.S.).

723 CU-MN-AL, ... A1,A2 . . . . J. PHYS. SOC. JAPAN VOL.18, . . . 93, (1963).
⌐NEUTRON DIFFRACTION STUDY OF KAPPA PHASE CU-MN-AL ALLOYS.
⌐KATSURAKI(H.), TAKADA(H.), SUZUKI(K.).

724 CU2-MN-AL, ... A2 . . . . J. PHYS. SOC. JAPAN VOL.20, . . . 1743, (1965).
⌐MAGNETIC FORM FACTOR OF CU2-MN-AL.
⌐TAKATA(H.).

725 CU2-MN-AL, ... A2 . . . . J. PHYS. CHEM. SOL. 24, ... 1663, (1963).
⌐THE MAGNETIC MOMENT DISTRIBUTION IN CU2*MN*AL.
⌐FELCHER(G.P.), CABLE(J.W.), WILKINSON(M.K.).

726 CU2-MN-AL, ... A2 . . . . PHYS. CHEM. SOLIDS 29, ... 193, (1968).
⌐STUDIES ON THE HEUSLER ALLOYS-I. CU2*MN*AL AND ASSOCIATED STRUCTURES.
⌐JOHNSTON(G.B.), HALL(E.O.).

727 CU3-MN2-AL, ... A1,2 . . . . J. PHYS. CHEM. SOLIDS 29, ... 201, (1968).
⌐STUDIES ON THE HEUSLER ALLOYS-II. CU3*MN2*AL AND ASSOCIATED STRUCTURES.
⌐JOHNSTON(G.B.), HALL(E.O.).

728 (CU-MN)*FE2*O4, ... A1,A2 . . . . INORG. MATER. VOL.8 NO.12, . . . 1916, (1973).
⌐NEUTRON-DIFFRACTION ANALYSIS OF THE SPINEL (CU(.5)-MN(.5))*FE2*O4.
⌐TSITSENOVSKAYA(S.E.), TARASYUK(YU.A.).

729 CU2-MN-IN, ... A1,A2 . . . . PHYS. STAT. SOLIDI A VOL.3, . . . 959, (1970).
⌐ATOMIC AND MAGNETIC STRUCTURE OF THE HEUSLER ALLOYS PD2-MN-GE, PD2-MN-SN
CU2-MN-IN, AND CO-MN-SB.
⌐NATERA(M.G.), MURTHY(M.R.L.), BEGUM(R.J.), MURTHY(N.S.S.).

730 CU*MN2*O4, ... A2 . . . . SOV. PHYS. SOL. STATE VOL.11, . . 672, (1969).
⌐CATION DISTRIBUTION IN A COPPER-MANGANESE SPINEL.
⌐ZASLAVSKII(A.I.), PLAKHTII(V.P.).

731 CU*MN2*O4, ... A1,2 . . . . J. PHYS. CHEM. SOLIDS 30, ... 805, (1969).
⌐MANGANITES SPINELLES PURS D*ELEMENTS DE TRANSITION: PREPARATIONS ET
STRUCTURES CRISTALLOGRAPHIQUES.
⌐BUHL(R.).

732 CU-MN-SB, ... A1,A2 . . . . J. PHYS. CHEM. SOLIDS 29, ... 855, (1968).
⌐STUDIES ON THE HEUSLER ALLOYS-III. THE ANTIFERROMAGNETIC PHASE IN THE
CU-MN-SB SYSTEM.
⌐FORSTER(R.H.), JOHNSTON(G.B.), WHEELER(D.A.).

733 CU*(N*H4)2*(S*O4)2*6(H2*O), ... A1 . . . . ACTA CRYST. VOL.B25, ... 676, (1969).
⌐STRUCTURE OF COPPER AMMONIUM SULFATE HEXAHYDRATE FROM
NEUTRON-DIFFRACTION DATA.
⌐BROWN(G.M.), CHIDAMBARAM(R.).

SECTION 5 -STRUCTURE AND CROSS-SECTION DETERMINATIONS . . . .

734 CU-NI, ...A2. . . . . .PHYS. REV. LETT. 22, ... 531, (1969).
⊬GIANT MOMENTS IN NI-CU ALLOYS NEAR THE CRITICAL COMPOSITION.
⊬HICKS(T.J.), RAINFORD(B.D.), KOUVEL(J.S.), LOW(G.G.), COMLY(J.B.).

735 CU*NI, ...A4. . . . . .PHYS. REV. 175, ... 868, (1968).
⊬NEUTRON MEASUREMENT OF CLUSTERING IN ALLOY CU-NI.
⊬MOZER(B.), KEATING(D.T.), MOSS(S.C.).

736 (CU,NI)*O, ...A1. . . . . .SOL. STATE COMM. VOL.8, ... 935, (1970).
⊬ETUDE DE L≠ORDRE DANS LES SOLUTIONS SOLIDES (CU,NI)*O.
⊬ZILBER(R.), BERTAUT(E.F.), BURLET(P.).

737 CU2-NI-ZN, ...A1. . . . . .J. PHYS. SOC. JAPAN VOL.20, . . . 381, (1965).
⊬NEUTRON DIFFRACTION STUDY OF THE LONG-RANGE ORDER IN A SINGLE CRYSTAL
CU2-NI-ZN.
⊬HIRABAYASHI(M.), HOSHINO(S.), SATO(K.).

738 CU-NI-ZN, ...A4. . . . . .TRANS. METALL. SOC. AIME VOL.236 1012, (1966).
⊬NEUTRON-DIFFRACTION EVIDENCE SUGGESTING CLUSTERING IN COMMERCIAL NICKEL
SILVER CLOSE TO THE CU2-NI-ZN COMPOSITION.
⊬PHILLIPS(V.A.), ROBERTS(B.W.).

739 CU2*O, ...A3. . . . . .J. PHYS. SOC. JAPAN VOL.17 B-II, 348, (1962).
⊬NEUTRON DIFFRACTION WORK AT THE ISPRA CENTER.
⊬CAGLIOTI(G.), RICCI(F.P.), SANTORO(A.), SCATTURIN(V.).

740 CU*O, ...A2. . . . . .PHYS. REV. VOL.94, ... 781, (1954).
⊬ANTIFERROMAGNETISM IN CUPRIC OXIDE.
⊬BROCKHOUSE(B.N.).

741 CU2*P2*O7, ...A2. . . . . .CAN. J. PHYS. VOL.50, ... 3079, (1972).
⊬MAGNETIC STRUCTURE OF MANGANESE PYROPHOSPHATE AND COPPER PYROPHOSPHATE.
⊬STILES(J.A.R.), STAGER(C.V.).

742 CU*S*O4, ...A1,2. . . . .COLL. INTER. N.126 (GRENOBLE), . . 18, (1963).
⊬NEUTRON DIFFRACTION STUDIES AT THE PUERTO RICO NUCLEAR CENTRE.
⊬ALMODOVAR(I.), BIELEN(H.J.), FRAZER(B.C.), KAY(M.I.).

743 CU*S*O4, ...A2. . . . . .PHYS. REV. VOL.181, ... 936, (1969).
⊬POLARIZED-NEUTRON STUDY OF THE MAGNETIC MOMENT DENSITY IN ANTIFERRO-
MAGNETIC CU*S*O4.
⊬MENZINGER(F.), COX(D.E.), FRAZER(B.C.), UMEBAYASHI(H.).

744 CU*S*O4, ...A2. . . . . .PHYS. REV. 138, ... A153, (1965).
⊬MAGNETIC STRUCTURE OF CU*S*O4.
⊬ALMODOVAR(I.), FRAZER(B.C.), HURST(H.J.), COX(D.E.), BROWN(P.J.).

745 CU*S*O4.5H2*O, ...A1,A4. . . .PROC. ROY. SOC. A VOL.266, . . . 95, (1962).
⊬THE WATER MOLECULES IN CU*S*O4.5H2*O.
⊬BACON(G.E.), CURRY(N.A.).

746 CU-SB, ...A3. . . . . .PHYS. CHEM. LIQ. VOL.4, ... 39, (1973).
⊬STRUCTURE OF MOLTEN COPPER-ANTIMONY ALLOYS BY COMBINATION OF NEUTRON
AND X-RAY DIFFRACTION.
⊬KNOLL(W.), STEEB(S.).

747 CU*SE*O4, ...A2. . . . . .J. PHYSIQUE TOME 32 COL.1 VOL.II 855, (1971).
⊬A NEUTRON DIFFRACTION STUDY OF CU*SE*O4.
⊬SCHARENBERG(W.), WILL(G.).

748 CU6-SN5, ...T,A3. . . . .ADV. IN PHYSICS VOL.16, ... 171, (1967).
⊬THE PARTIAL STRUCTURE FACTORS OF LIQUID ALLOYS.
⊬ENDERBY(J.E.), NORTH(D.M.), EGELSTAFF(P.A.).

749 CU6-SN5, ...A3. . . . . .PHIL. MAG. VOL.14, ... 961, (1966).
⊬THE PARTIAL STRUCTURE FACTORS OF LIQUID CU-SN.
⊬ENDERBY(J.E.), NORTH(D.M.), EGELSTAFF(P.A.).

750 CU*W*O4, ...A2. . . . . .SOL. STATE COMM. VOL.8, ... 2071, (1970).
⊬MAGNETISCHE STRUKTUR VON CO*W*O4, NI*W*O UND CU*W*O4.
⊬WEITZEL(H.).

751 CU-ZN, ...A4. . . . . .PHYS. REV. B-2, ... 277, (1970).
⊬LONG-RANGE ORDER IN BETA BRASS.
⊬NORVELL(J.C.), ALS-NIELSEN(J.).

752 CU-ZN, ...A4. . . . . .PHYS. REV. 153, ... 711, (1967).
⊬TEMPERATURE DEPENDENCE OF SHORT RANGE ORDER IN BETA-BRASS.
⊬ALS-NIELSEN(J.), DIETRICH(O.W.).

753 CU-ZN, ...A4. . . . . .PHYS. REV. 153, ... 706, (1967).
⊬PAIR CORRELATION FUNCTION IN DISORDERED BETA-BRASS AS STUDIED BY
NEUTRON DIFFRACTION.
⊬ALS-NIELSEN(J.), DIETRICH(O.W.).

754 CU-ZN, ...A4. . . . . .PHYS. REV. 130, ... 1726, (1963).
⊬NEUTRON DIFFRACTION STUDY OF SHORT RANGE ORDERS IN (BETA) CU-ZN.
⊬WALKER(C.B.), KEATING(D.T.).

755 CU-ZN, ...A1. . . . . .BULL. SOC. FRANC. MIN. CRIST. 85 57, (1962).
⊬MEASUREMENT OF THE LONG-RANGE ORDER PARAMETER IN BETA-BRASS BY NEUTRON
DIFFRACTION.
⊬ROUSSEL(J.). (IN FRENCH).

756 CU-ZN (ALPHA), ...A4. . . . . .J. APPL. CRYST. VOL.5, ... 27, (1972).
⊬THE TEXTURE TRANSITION IN ALPHA-BRASSES DETERMINED BY NEUTRON
DIFFRACTION.
⊬BUNGE(H.J.), TOBISCH(J.).

757 CU-ZN, ...A4. . . . . .J. APPL. PHYS. VOL.23, ... 1379, (1952).
⊬NEUTRON DIFFRACTION STUDIES OF COLD-WORKED BRASS.
⊬WEISS(R.J.), CLARK(J.R.), CORLISS(L.M.), HASTINGS(J.M.).

758 CU-ZN, ...A1. . . . . .PHYS. REV. B VOL.9, ... 3921, (1974).
⊬LONG-RANGE ORDER IN BETA-BRASS STUDIED BY NEUTRON DIFFRACTION.
⊬RATHMANN(O.), ALS-NIELSEN(J.).

759 CYCLOTETRAMETHYLENE TETRANITR. . . . .A1. . . . . .ACTA CRYST. VOL.B26 . . . . 1235, (1970).
⊬A STUDY OF THE CRYSTAL STRUCTURE OF CYCLOTETRAMETHYLENE TETRANITRAMINE
(BETA) BY NEUTRON DIFFRACTION.
⊬CHOI(C.S.), BOUTIN(H.P.).

760 D2, ...A1. . . . . .J. CHEM. PHYSICS VOL.49, ... 1922, (1968).
⊬STRUCTURE OF SOLID DEUTERIUM ABOVE AND BELOW THE LAMBDA TRANSITION AS
DETERMINED BY NEUTRON DIFFRACTION.
⊬MUCKER(K.F.), HARRIS(P.M.), WHITE(D.), ERICKSON(R.A.).

SECTION   5 -STRUCTURE AND CROSS-SECTION DETERMINATIONS . . . .

761 D2,. . . . . .SCATTERING OF THERMAL NEUTRONS BY LIQUID SOLUTIONS OF ORTHO- AND PARA-
A3....J. CHEM. PHYSICS VOL.48,....1273, (1968).
DEUTERIUM.
TALHOUK(S.J.), HARRIS(P.M.), WHITE(D.), ERICKSON(R.A.).

762 D2,. . . . . . . .STRUCTURES OF SOLID DEUTERIUM ABOVE AND BELOW THE (LAMDA) TRANSITION
A1......DISS. ABS. XXVII,. . . . . .3892B, (1967).
AS DETERMINED BY NEUTRON DIFFRACTION.
MUCKER(K.F.).

763 D2,. . . . . . . .AN INVESTIGATION OF STRUCTURE IN PARA- AND ORTHO-LIQUID DEUTERIUM BY
A1......DISS. ABS. XXVII,. . . . . . . .1447B, (1967).
NEUTRON SCATTERING.
TALHOUK(S.J.).

764 D2,. . . . . . . .OBSERVABLES OF THE NEUTRONS ELASTICALLY SCATTERED BY DEUTERONS AT LOW
A4......NUCL. PHYS. A(NETHERLANDS) A182, 541, (1972).
ENERGY.
JACCARD(S.), VIENNET(R.).

765 D2,. . . . . . . .REVIEW OF THE LOW ENERGY SCATTERING CROSS SECTIONS OF DEUTERIUM.
P......NUCL. TECHNOL. VOL.15,. . . . . . .49, (1972).
LEONARD-JR(B.R.).

766 D2,. . . . . . . .MEASUREMENT OF THE COHERENT NEUTRON SCATTERING AMPLITUDE OF DEUTERIUM,
P......PHYS. REV. VOL.174,. . . . . .313, (1968).
MERCURY, AND FLUORINE BY MIRROR REFLECTION.
BARTOLINI(W.), DONALDSON(R.E.), GROVES(D.J.).

767 D2,. . . . . . . .MEASUREMENT OF THE THERMAL NEUTRON SCATTERING CROSS-SECTIONS FOR LIQUID
P......Z. NATURFORSCH. A VOL.25,. . . . .967, (1970).
AND SOLID HYDROGEN, DEUTERIUM AND DEUTERIUM HYDRIDE.
SEIFFERT(W.D.), WECKERMANN(B.), MISENTA(R.).

768 D2,. . . . . . . .CRYSTAL STRUCTURE OF SOLID DEUTERIUM FROM NEUTRON-DIFFRACTION STUDIES.
A1......PHYS. REV. LETT. 15,. . . . . . . .586, (1965).
MUCKER(K.F.), TALHOUK(S.J.), HARRIS(P.M.), WHITE(D.), ERICKSON(R.A.).

769 D2,. . . . . . . .THE PHASE TRANSITIONS AND CRYSTAL STRUCTURES OF DEUTERIUM.
A1,A4....ACTA CRYST. VOL.A28 PART S-4,. .188, (1972).
SCHUCH(A.F.), YARNELL(J.L.), MILLS(R.L.). (ABSTRACT ONLY).

770 D2,. . . . . . . .CRYSTAL STRUCTURE OF PARA-ENRICHED SOLID DEUTERIUM BELOW THE (LAMDA)
A1......PHYS. REV. LETT. 16,. . . . . . .799, (1966).
TRANSITION.
MUCKER(K.F.),   TALHOUK(S.J.),   HARRIS(P.M.),   WHITE(D.), ERICKSON(R.A.).

771 D2,. . . . . . . .SPIN DEPENDENCE OF SLOW NEUTRON SCATTERING BY DEUTERONS.
P......PHYS. REV. VOL.75,. . . . . .578, (1949).
FERMI(E.), MARSHALL(L.).

772 D2,. . . . . . . .THE SCATTERING LENGTHS OF THE DEUTERON.
P,A3.....CAN. J. PHYS. VOL.29,. . . . . .36, (1951).
HURST(D.G.), ALCOCK(N.Z.).

773 D3*B*O3,. . . . .A NEUTRON DIFFRACTION STUDY OF ORTHOBORIC ACID D3*B*O3.
A1......ACTA CRYST. 20,. . . . . . .214, (1966).
CRAVEN(B.M.), SABINE(T.M.).

774 D*CL,. . . . . .MOLECULAR MOTION IN SOLID D*CL.
A4......ACTA CRYST. VOL.A28 PART S-4,. .192, (1972).
NIIMURA(N.),   FUJII(Y.),   MOTEGI(H.),   HOSHINO(S.). (ABSTRACT   ONLY).

775 D*CL,. . . . . .NEUTRON DIFFRACTION STUDY ON MOLECULAR LIBRATION IN SOLID D*CL.
A4......J. PHYS. SOC. JAPAN VOL.35,. . . . .842, (1973).
NIIMURA(N.), FUJII(Y.), MOTEGI(H.), HOSHINO(S.).

776 D3*CO*(C*N)6,. .NEUTRON-DIFFRACTION STUDY OF D3*CO*(C*N)6.
A1......J. CHEM. PHYSICS VOL.53,. . . . 1917, (1970).
GUDEL(H.U.), LUDI(A.), FISCHER(P.), HALG(G.).

777 D*CR*O2,. . . .STRUCTURES OF H*CR*O2 AND D*CR*O2.
A1......ACTA CRYST. 16,. . . . . . . .1209, (1963).
HAMILTON(W.C.), IBERS(J.A.).

778 D4*FE*(C*N)6,. .STRUCTURE PROTONIQUE ET DEUTONIQUE DE L≠ACIDE FERROCYANHYDRIQUE.
A1......ACTA CRYST. B25,. . . . . 1685, (1969).
PIERROT(M.), KERN(R.).

779 D*H,. . . . . .MEASUREMENT OF THE THERMAL NEUTRON SCATTERING CROSS-SECTIONS FOR LIQUID
P......Z. NATURFORSCH. A VOL.25,. . . . .967, (1970).
AND SOLID HYDROGEN, DEUTERIUM AND DEUTERIUM HYDRIDE.
SEIFFERT(W.D.), WECKERMANN(B.), MISENTA(R.).

780 D2,H2,. . . . .NEUTRON DIFFRACTION STUDY OF THE CRYSTAL STRUCTURE OF SOLID HYDROGEN
A1......J. PHYS. SOC. JAPAN VOL.17 B-II, 385, (1962).
AND DEUTERIUM.
OZEROV(R.P.), KOGAN(V.S.), ZHDANOV(G.S.), LAZAREV(B.G.).

781 D*I,. . . . . .CRYSTAL STRUCTURE AND PHASE TRANSITION IN DEUTERIUM IODIDE.
A1,A4.....ACTA CRYST. VOL.A28 PART S-4,. .188, (1972).
SANDOR(E.), CLARKE(J.H.). (ABSTRACT ONLY).

782 D2*O,. . . . .NEUTRON-DIFFRACTION STUDY OF ICE POLYMORPHS. III, ICE IC.
A1......J. CHEM. PHYSICS VOL.49,. . . . 4365, (1968).
ARNOLD(G.P.), FINCH(E.D.), RABIDEAU(S.W.), WENZEL(R.G.).

783 D2*O,. . . . .NEUTRON-DIFFRACTION STUDY OF ICE POLYMORPHS. II, ICE II.
A1......J. CHEM. PHYSICS VOL.49,. . . . 4361, (1968).
FINCH(E.D.), RABIDEAU(S.W.), WENZEL(R.G.), NERESON(N.G.).

784 D2*O,. . . . .NEUTRON DIFFRACTION STUDY OF ICE POLYMORPHS UNDER HELIUM PRESSURE.
A1......J. CHEM. PHYSICS VOL.55,. . . . .589, (1971).
ARNOLD(G.P.), WENZEL(R.G.), RABIDEAU(S.W.), NERESON(N.G.), BOWMAN(A.L.).

785 D2*O,. . . . .SLOW-NEUTRON SCATTERING BY LIGHT AND HEAVY WATER.
A3......NUCL. SCI. ENG. VOL.45,. . . . .126, (1971).
GOTOH(Y.), TAKAHASHI(H.).

786 D2*O,. . . . .ON A NEARLY PROTON-ORDERED STRUCTURE FOR ICE IX.
A1,A4....J. CHEM. PHYS. VOL.58,. . . . .567, (1973).
LA-PLACA(S.J.), HAMILTON(W.C.), KAMB(B.), PRAKASH(A.).

787 D2*O,. . . . .DIFFRACTION OF NEUTRONS ON HEAVY WATER.
A3......NUOVO CIMENTO B VOL.46,. . . . . .7, (1966).
FORTE(M.), MENARDI(S.).

SECTION   5 -STRUCTURE AND CROSS-SECTION DETERMINATIONS . . . .

788   D2*0, ...A1.....PHYS. REV. VOL. 75 . . . . . 1348, (1949).
                         *NEUTRON DIFFRACTION STUDY OF THE STRUCTURE OF ICE.
                         *WOLLAN(E.O.), DAVIDSON(W.L.), SHULL(C.G.).

789   D2*0, ...A1.....ACTA CRYST. VOL. 10 . . . . . 70, (1957).
                         *A SINGLE-CRYSTAL NEUTRON DIFFRACTION STUDY OF HEAVY ICE.
                         *PETERSON(S.W.), LEVY(H.A.). (SEE CORRECTION NOTE ON PG. 344 AS WELL).

790   D2*0, ...P.....Z. KRIST. VOL. 116 . . . . . 328, (1961).
                         *UNTERSUCHUNG DES NEUTRONENSTREUQUERSCHNITTES VON SCHWEREM EIS IN DER
                         UMGEBUNG DER BRAGGSCHEN GRENZWELLENLANGE.
                         *GISSLER(W.), REINSCH(C.), SPRINGER(T.), WIEDEMANN(W.).

791   D2*0, ...P.....Z. PHYSIK VOL. 219 . . . . . 300, (1969).
                         *MESSUNG KOHARENTER STREULANGEN DURCH KLEINWINKELSTREUUNG VON NEUTRONEN
                         IM SCHWERKRAFT-REFRAKTOMETER.
                         *KOESTER(L.), UNGERER(H.).

792   D2*0, ...A1.....PHYS. ICE . . . . . 59, (1969).
                         *STRUCTURAL STUDIES OF ICE POLYMORPHS BY NEUTRON DIFFRACTION, PROTON AND
                         DEUTERON NMR.
                         *RABIDEAU(S.W.), FINCH(E.D.). (PROC. INT. SYMP., SEPT., 1968. IN BOOK:
                         PHYSICS OF ICE: ED. BY RIEHL ET AL.).

793   D2*0, ...A1.....PHYS. ICE . . . . . 44, (1969).
                         *DEUTERON ARRANGEMENTS IN HIGH-PRESSURE FORMS OF ICE.
                         *HAMILTON(W.C.), KAMB(B.), LA-PLACA(S.J.), PRAKASH(A.). (PROC. INT. SYMP.
                         SEPT., 1968. IN BOOK: PHYSICS OF ICE: ED. BY RIEHL ET AL.).

794   D2*0, ...T.....J. CHEM. PHYS. VOL. 60 . . . . . 1545, (1974).
                         *IMPROVED SIMULATION OF LIQUID WATER BY MOLECULAR DYNAMICS.
                         *STILLINGER(F.H.), RAHMAN(A.).

795   D2*0, ...P.....Z. KRIST. VOL. 118 . . . . . 149, (1963).
                         *UNTERSUCHUNG DES NEUTRONENSTREUQUERSCHNITTES VON SCHWEREM EIS IN DER
                         UMGEBUNG DER BRAGGSCHEN GRENZWELLENLAENGE.
                         *GISSLER(W.).

796   D2*0, ...A1.....J. CHEM. PHYSICS VOL. 49 . . . . . 2514, (1968).
                         *NEUTRON DIFFRACTION STUDY OF ICE POLYMORPHS. I. ICE IH.
                         *RABIDEAU(S.W.), FINCH(E.D.), ARNOLD(G.P.), BOWMAN(A.L.).

797   D3*0(+)*C*H3*C6*H4*S*03(-) ......A1.....J. CHEM. PHYS. VOL. 59 . . . . . 5114, (1973).
                         THE HYDRATED PROTON H(+)(H2*O)N. III. A NEUTRON DIFFRACTION STUDY OF
                         DEUTERATED PARATOLUENE SULFONIC ACID MONOHYDRATE,...
                         *FINHOLT(J.E.), WILLIAMS(J.M.).

798   D2*0-(NI*CL2,NA*CL,BA*CL2) . . . . .A3.....CHEM. PHYS. LETT. VOL. 21 NO.1, . 109, (1973).
                         *THE STRUCTURE OF AQUEOUS SOLUTIONS.
                         *ENDERBY(J.E.), HOWELLS(W.S.), HOWE(R.A.).

799   D2*U3*010, ...A1.....ACTA CRYST. VOL. B30 . . . . . 151, (1974).
                         *THE DEUTERIUM LOCATION IN DEUTERIUM TRIURANATE; D2*U3*010, BY NEUTRON
                         DIFFRACTION.
                         *TAYLOR(J.C.), WILSON(P.W.).

800   DY, ...A5.....J. PHYS. CHEM. SOLIDS VOL. 30, . 2175, (1969).
                         *A NEUTRON DEPOLARIZATION STUDY ON DYSPROSIUM AND TERBIUM.
                         *LOFFLER(E.), RAUCH(H.).

801   DY, ...A4.....ATOMKERNENERGIE VOL. 19 . . . . . 167, (1972).
                         *NEUTRON DEPOLARIZATION MEASUREMENTS ON A DY-SINGLE CRYSTAL.
                         *RAUCH(H.), ZEILINGER(A.).

802   DY, ...A4.....Z. ANGEW PHYS. VOL. 32, ... 109, (1971).
                         *A NEUTRON DEPOLARIZATION EXPERIMENT ON DY NEAR THE FERROMAGNETIC
                         TRANSITION POINT.
                         *RAUCH(H.), SEIDL(E.), ZEILINGER(A.).

803   DY, ...P.....HELV. PHYS. ACTA VOL. 45 . . . . . 46, (1972).
                         *NEUTRON-CAPTURE CROSS-SECTIONS OF V, MN, CS, DY AND LU IN THE ENERGY
                         RANGE FROM 0.01 TO 100 EV.
                         *WIDDER(F.).

804   DY, ...A2.....REF. SECTION 3, 65/KOEHLER, ...

805   DY, ...A2.....Z. ANGEW. PHYS. VOL. 15, ... 371, (1963).
                         *METAMAGNETISM. II.
                         *VOGT(E.). (IN GERMAN).

806   DY, ...A2.....Z. PHYS. VOL. 210 . . . . . 265, (1968).
                         *NEUTRON DEPOLARIZATION MEASUREMENT ON DY NEAR MAGNETIC TRANSITION POINTS
                         *RAUCH(H.), LOFFLER(E.). (IN GERMAN).

807   DY, ...A2.....J. APPL. PHYS. VOL. 32, SUPPL. . 48S, (1961).
                         *NEUTRON DIFFRACTION INVESTIGATION OF MAGNETIC ORDERING IN DYSPROSIUM.
                         *WILKINSON(M.K.), KOEHLER(W.C.), WOLLAN(E.O.), CABLE(J.W.).

808   DY*AG2, ...A2.....J. CHEM. PHYSICS VOL. 51, ... 3877, (1969).
                         *MAGNETIC STRUCTURES OF DY*AU2 AND DY*AG2.
                         *ATOJI(M.).

809   DY*AG, ...A2.....J. CHEM. PHYS. VOL. 46, ... 4041, (1967).
                         *MAGNETIC STRUCTURE OF DY*AG.
                         *ARNOLD(G.P.), NERESON(N.G.), OLSEN(C.E.).

810   DY3*AL2, ...A2.....Z. ANGEW PHYS. VOL. 32, ... 113, (1971).
                         *MAGNETIC PROPERTIES AND MAGNETIC STRUCTURES OF TB3*AL2 AND DY3*AL2.
                         *BARBARA(B.), BECLE(C.), LEMAIRE(R.). (ED. NOTE: IN FRENCH).

811   DY*AL2, ...A2.....J. APP. PHYS. 37, ... 4575, (1966).
                         *MAGNETIC PROPERTIES OF DY*AL2.
                         *NERESON(N.G.), OLSEN(C.E.), ARNOLD(G.P.).

812   DY*AL-GARNET, ...A2.....PHYS. REV. LETT. VOL. 32, ... 544, (1974).
                         *OBSERVATION OF AN ANTIFERROMAGNET IN AN INDUCED STAGGERED MAGNETIC
                         FIELD: DYSPROSIUM ALUMINUM GARNET NEAR THE TRICRITICAL POINT.
                         *BLUME(M.), CORLISS(L.M.), HASTINGS(J.M.), SCHILLER(E.).

813   DY3*AL5*012, ...A2.....PHYS. REV. VOL. 186, ... 567, (1969).
                         *MAGNETIC NEUTRON SCATTERING IN DYSPROSIUM ALUMINUM GARNET. II. SHORT-
                         RANGE ORDER AND CRITICAL SCATTERING.
                         *NORVELL(J.C.), WOLF(W.P.), CORLISS(L.M.), HASTINGS(J.M.), NATHANS(R.).

SECTION   5 -STRUCTURE AND CROSS-SECTION DETERMINATIONS . . . .

814 DY3*AL5*O12, ...A2. . . . . . .PHYS. REV. 138, ... A176, (1965).
                     *ANTIFERROMAGNETIC STRUCTURE OF DYSPROSIUM ALUMINUM GARNET.
                     *HASTINGS(J.M.), CORLISS(L.M.), WINDSOR(C.G.).

815 DY*AL*O3,. . . . . .A1,A2. . . . .J. PHYSIQUE VOL.29 NO.2-3 . . . 220, (1968).
                     *NEUTRON DIFFRACTION STUDY OF THE NUCLEAR AND MAGNETIC STRUCTURE OF
                     DY*AL*O3.
                     *BIDAUX(R.), MERIEL(P.). (IN FRENCH).

816 DY3*AL5*O12, ...A2. . . . . . .PHYS. REV. VOL.186,. . . . . 557, (1969).
                     *MAGNETIC NEUTRON SCATTERING IN DYSPROSIUM ALUMINUM GARNET. I.LONG-RANGE
                     ORDER.
                     *NORVELL(J.C.),   WOLF(W.P.),   CORLISS(L.M.), HASTINGS(J.M.), NATHANS(R.).

817 DY*AS*O4,. . . . . .A2. . . . . .J. PHYS. C, SER.2 VOL.4,. . . . 3224, (1971).
                     *NEUTRON DIFFRACTION STUDY OF ANTIFERROMAGNETIC DY*AS*O4.
                     *SCHAFER(W.), WILL(G.).

818 DY*AU2,. . . . . .A2. . . . .J. CHEM. PHYSICS VOL.51, . . . 3877, (1969).
                     *MAGNETIC STRUCTURES OF DY*AU2 AND DY*AG2.
                     *ATOJI(M.).

819 DY*BI, ...A2. . . . .J. APPL. PHYS. VOL.42, ... 1625, (1971).
                     *MAGNETIC PROPERTIES OF CE*BI, ND*BI, TB*BI, AND DY*BI.
                     *NERESON(N.), ARNOLD(G.).

820 DY*C2, ...A2. . . . .J. CHEM. PHYS. VOL.48, ... 3384, (1968).
                     *MAGNETIC STRUCTURE OF DY*C2.
                     *ATOJI(M.).

821 DY*C*O3, ...A2,A4. . . .REF. SECTION 3, 66/BERTAUT,. . . . .

822 DY*CR*O3,. . . . . .A2,T. . . . .ACTA CRYST A24,. . . . . 217, (1968).
                     *REPRESENTATION ANALYSIS OF MAGNETIC STRUCTURES.
                     *BERTAUT(E.F.).

823 DY*CR*O3,. . . . . .A2. . . . . .J. PHYSIQUE TOME 32, ... 301, (1971).
                     *ON THE MAGNETIC STRUCTURE OF DY*CR*O3.
                     *VAN-LAAR(B.), ELEMANS(J.B.A.).

824 DY*CR*O3,. . . . . .A2,T. . . . .J. PHYSIQUE TOME 29, ...   67, (1968).
                     *STRUCTURE MAGNETIQUE DE DY*CR*O3.
                     *BERTAUT(E.F.), MARESCHAL(J.).

825 DY-(ER,HO),. . . . . .A4. . . . . .INT. J. MAGN.(G3) VOL.2, ... 389, (1971).
                     *SINGLE CRYSTAL NEUTRON DIFFRACTION STUDY OF BINARY RARE EARTH ALLOYS.
                     *MILLHOUSE(A.H.), KOEHLER(W.C.).

826 DY*FE-GARNET,. . . . . .A2. . . . . .J. APPL. PHYS. VOL.41, ... 1192, (1970).
                     *RARE-EARTH SUBLATTICE CANTING IN DY*IG, ER*IG, AND YB*IG.
                     *PICKART(S.J.), ALPERIN(H.A.), CLARK(A.E.).

827 DY*GA2,. . . . . .A2. . . . . .J. PHYSIQUE TOME 32 COL.1 VOL.II 1126, (1971).
                     *PROPRIETES MAGNETIQUES ET STRUCTURES MAGNETIQUES DE QUELQUES COMPOSES
                     T*GA2
                     *BARBARA(B.), BECLE(C.), SIAUD(E.).

828 DY*N,. . . . . .A2. . . . . .REF. SECTION 3, 63/CHILD,. . . . .

829 DY*NI3,. . . . . .T. . . . . . .PHYS. STAT. SOL. VOL.47, ... 247, (1971).
                     *CHAMP CRISTALLIN ET EFFET MOSSBAUER DANS LE COMPOSE DY*NI3.
                     *YAKINTHOS(J.), ROSSAT-MIGNOD(J.), BELAKHOVSKY(M.).

830 DY*NI, ...A2. . . . . .SOL. STATE COMM. VOL.11, ... 1709, (1972).
                     *MAGNETIC STRUCTURE OF THE COMPOUND DY*NI.
                     *GIGNOUX(D.), SHAH(J.S.).

831 DY-NI-CO,. . . . . .A2. . . . . .C. R. ACAD. SCI. B TOME 274,. . 319, (1972).
                     *STRUCTURE MAGNETIQUE DU COMPOSE INTERMETALLIQUE DY-NI-CO.
                     *BARBARA(B.), GIGNOUX(D.).

832 DY2*O3,. . . . . .A1. . . . . .PHYS. STAT. SOL. 3,. . . . . . . K446, (1963).
                     *NEUTRONOGRAPHISCHE BESTIMMUNG DER KRISTALLSTRUKTURPARAMETER VON DY2*O3
                     TM2*O3 UND (ALPHA)-MN2*O3.
                     *HASE(W.).

833 DY2*O3,. . . . . .P. . . . . . .PHYS. STAT. SOL. 2,. . . . . . . K164, (1962).
                     *MESSUNG DER KOHARENTEN STREUAMPLITUDEN VON DYSPROSIUM UND THULIUM FUR
                     THERMISCHE NEUTRONEN.
                     *BETZL(M.), HASE(W.), KLEINSTUCK(K.), TOBISCH(J.).

834 DY*O*CL, ...A2. . . . . .J. PHYSIQUE TOME 32 COL.1 VOL.II 741, (1971).
                     *ETUDE MAGNETIQUE ET DIFFRACTION NEUTRONIQUE DANS LES COMPOSES DY*O*CL ET
                     TB*O*CL.
                     *ELMALEH(D.), FRUCHART(D.), JOUBERT(J.C.).

835 DY*O*O*H,. . . . . .A2. . . . . .SOL. STATE COMM. VOL. 9, ... 925, (1971).
                     *TEMPERATURE MAGNETIC PROPERTIES OF THE MONOCLINIC MODIFICATION OF
                     DY*O*O*H.
                     *CHRISTENSEN(A.N.), QUEZEL(S.), BELAKHOVSKY(M.).

836 DY2*O2*S,. . . . . .A2. . . . . .J. PHYS. CHEM. SOLIDS 29,. . . . 1001, (1968).
                     *ETUDE PAR DIFFRACTION NEUTRONIQUE ET MESURES MAGNETIQUES DES OXYSULFURES
                     DE TERRES RARES T2*O2*S.
                     *BALLESTRACCI(R.), BERTAUT(E.F.), QUEREL(G.).

837 DY*P*O4, ...A2. . . . . .SOL. STATE COMM. VOL. 9, ... 1949, (1971).
                     *NEUTRON DIFFRACTION DETERMINATION OF THE MAGNETIC STURCTURE OF DY*P*O4.
                     *FUESS(H.), KALLEL(A.), TCHEOU(F.).

838 DY*SB, ...A2. . . . . .PHYS. REV. LETT. VOL.28, ... 746, (1972).
                     *MAGNETIC AND STRUCTURAL PHASE TRANSITION IN DY*SB.
                     *BUCHER(E.), BIRGENEAU(R.J.), MAITA(J.P.), FELCHER(G.P.), BRUN(T.O.).

839 DY*SB, ...A1. . . . . .AIP CONF. PROC. VOL.5, ... 1385, (1971).
                     *NEUTRON DIFFRACTION STUDIES ON DY*SB, ND*SB, AND CE*SB.
                     *NERESON(N.), STRUEBING(V.).

840 DY*SB, ...A2. . . . . .PHYS. REV. B VOL.8,. . . . . . . 260, (1973).
                     *MAGNETIC STRUCTURE OF DY*SB.
                     *FELCHER(G.P.), BRUN(T.O.), GAMBINO(R.J.), KUZNIETZ(M.).

841 DY-TH, ...A2. . . . . .J. PHYSIQUE TOME 32 COL.1 VOL.II 1128, (1971).
                     *SINGLE CRYSTAL NEUTRON DIFFRACTION STUDIES OF HCP RARE EARTH THORIUM
                     ALLOYS.
                     *CHILD(H.R.), KOEHLER(W.C.).

SECTION   5 -STRUCTURE AND CROSS-SECTION DETERMINATIONS . . . .

842 DY*(V,AS)*O4,. .   .A1,A2.....CONF./RARE EARTHS/ACTINIDES,   226, (1971).
                        *NEUTRON- AND X-RAY DIFFRACTION STUDY OF THE MAGNETIC AND CRYSTALLOGRAPH-
                        .IC PHASE TRANSITIONS IN DY*V*O4 AND DY*AS*O4.
                        *WILL(G.), SCHAFER(W.), GOEBEL(H.).

843 DY*V*O4, ...A2.......J. OF PHYS.-C-, SER.2, VOL.4,. . 811, (1971).
                        *THE MAGNETIC STRUCTURE OF ANTIFERROMAGNETIC DY*V*O4.
                        *WILL(G.), SCHAFER(W.).

844 DY*V*O4, ...A2.......PHYS. LETT. VOL.34A, ... 361, (1971).
                        *EXPERIMENTAL STUDY OF MAGNETIC AND CRYSTALLOGRAPHIC TRANSITIONS IN
                        DY*V*O4.
                        *SAYETAT(F.), BOUCHERLE(J.X.), BELAKHOVSKY(M.), KALLEL(A.), TCHEOU(F.),
                        FUESS(H.).

845 DY*V*O4, ...A1.......J. SOLID STATE CHEM. VOL.5,. .   11, (1972).
                        *REFINEMENT OF THE CRYSTAL STRUCTURE OF SOME RARE EARTH VANADATES R*V*O4
                        (R=DY,TB,HO,YB).
                        *FUESS(H.), KALLEL(A.).

846 ELECTRON,. . . .   .P........PHYS. REV. VOL.85 . . . .   483, (1952).
                        *A MEASUREMENT OF THE ELECTRON-NEUTRON INTERACTION.
                        *HAMERMESH(M.), RINGO(G.R.), WATTENBERG(A.).

847 ER (ISO. A=171),   .P........J. PHYS. SOC. JAP. VOL.31 . . . 1304, (1971).
                        *THE SLOW NEUTRON CAPTURE CROSS SECTIONS FOR THE LONG-LIVED ACTIVITIES;
                        ER(ISOTOPE A=171) AND TM(ISOTOPE A=171).
                        *MIYANO(K.).

848 ER,. . . . . . .   .A2.......MAGN. AND MAG. MATERIALS-1971, .1436, (1972).
                        *FIELD-DEPENDENT SPIN STRUCTURE OF ERBIUM.
                        *RHYNE(J.J.), PICKART(S.J.). (AIP CONF. PROC. NO.5 PART II).

849 ER,. . . . . . .   .A2.......REF. SECTION 3, 65/KOEHLER,. . . . .

850 ER,. . . . . . .   .A2.......PHYS. REV. 140,. . . . . . .A1896, (1965).
                        *MAGNETIC STRUCTURES OF METALLIC ERBIUM.
                        *CABLE(J.W.), WOLLAN(E.O.), KOEHLER(W.C.), WILKINSON(M.K.).

851 ER,. . . . . . .   .A2.......ACTA CRYST. VOL.A28 PART S-4,. . 197, (1972).
                        *MAGNETIC STRUCTURE OF ER SINGLE CRYSTAL.
                        *ATOJI(M.). (ABSTRACT ONLY).

852 ER,. . . . . . .   .A2.......J. APPL. PHYS. VOL.32, SUPPL.,. . 49S, (1961).
                        *NEUTRON DIFFRACTION STUDY OF METALLIC ERBIUM.
                        *CABLE(J.W.), WOLLAN(E.O.), KOEHLER(W.C.), WILKINSON(M.K.).

853 ER,. . . . . . .   .A2.......PHYS. REV. VOL.97, ... 1177, (1955).
                        *NEUTRON DIFFRACTION BY METALLIC ERBIUM.
                        *KOEHLER(W.C.), WOLLAN(E.O.).

854 ER*AG2,. . . . .   .A2.......J. CHEM. PHYS. VOL.57 NO.2,. . 851, (1972).
                        *MAGNETIC STRUCTURE OF ER*AG2.
                        *ATOJI(M.).

855 ER*AG, ...A2.......J. APPL. PHYS. VOL.44, ... 4727, (1973).
                        *MAGNETIC PROPERTIES OF ER*AG.
                        *NERESON(N.).

856 ER*AL3,. . . . .   .A2.......PHYS. LETT. VOL.36A, ...   50, (1971).
                        *THE MAGNETIC STRUCTURE OF HEXAGONAL ER*AL3.
                        *BARGOUTH(M.O.), WILL(G.).

857 ER*AL, ...A2.......PHYS LETT 27A, ... 541, (1968).
                        *PROPRIETES MAGNETIQUES ET STRUCTURE MAGNETIQUE DU COMPOSE ER*AL.
                        *BECLE(C.), LEMAIRE(R.).

858 ER-AL2,. . . . .   .A2,A1.....J. PHYS. CHEM. SOLIDS VOL.33,. .1031, (1972).
                        *STRUCTURAL, MAGNETIC AND NEUTRON DIFFRACTION STUDIES ON ER-MN2, ER-AL2
                        AND TB-MN-AL.
                        *OESTERREICHER(H.).

859 ER*AL3,. . . . .   .A2.......CONF./RARE EARTHS/ACTINIDES, . . 196, (1971).
                        *THE MAGNETIC STRUCTURES OF CUBIC AND HEXAGONAL ER*AL3.
                        *WILL(G.), BARGOUTH(M.O.).

860 ER*AL2,. . . . .   .A2.......Z. NATURFORSCH. VOL.23A . . . . 413, (1968).
                        *INVESTIGATION OF THE MAGNETIC STRUCTURE OF ER*AL2 WITH THERMAL NEUTRONS.
                        *WILL(G.).

861 ER*AL3,. . . . .   .A2.......J. PHYS. CHEM. SOLIDS VOL.35,. . 861, (1974).
                        *THE MAGNETIC STRUCTURE OF CUBIC ER*AL3.
                        *WILL(G.), COX(D.E.).

862 ER*AL2-ER*AL3, .   .A2.......INT. J. MAGN. VOL.3,. . . .   87, (1972).
                        *NEUTRON DIFFRACTION STUDIES IN THE SYSTEM ER*AL2-ER*AL3.
                        *WILL(G.), BARGOUTH(M.O.).

863 ER*AL*O3,. . . .   .A4.......C. R. ACAD. SCI. B TOME 273,. . 619, (1971).
                        *ETUDE PAR DIFFRACTION NEUTRONIQUE DES PEROVSKITES ER*AL*O3, HO*AL*O3,
                        TB*RH*O3, ETC.
                        *SIVARDIERE(J.), QUEZEL-AMBRUNAZ(S.).

864 ER*AU2,. . . . .   .A2.......J. CHEM. PHYS. VOL.57, ... 2407, (1972).
                        *MAGNETIC STRUCTURE OF ER*AU2.
                        *ATOJI(M.).

865 ER*C2, ...A1,A2.....J. CHEM. PHYS. VOL.57,. .   2410, (1972).
                        *MAGNETIC AND CRYSTAL STRUCTURES OF ER*C2 AT 297-2 DEGREES K.
                        *ATOJI(M.).

866 ER*C*O3, ...A2,A4....REF. SECTION 3, 66/BERTAUT,. . . . .

867 ER3*CO,. . . . .   .A2.......SOL. STATE COMM. VOL.8,. . . 391, (1970).
                        *ETUDE DES STRUCTURES MAGNETIQUES DES COMPOSES ER3*CO ET ER3*NI PAR
                        DIFFRACTION NEUTRONIQUE.
                        *GIGNOUX(D.), LEMAIRE(R.), PACCARD(D.).

868 ER*CO2,. . . . .   .A2.......J. APP. PHYS. 36,. . . . . 978, (1965).
                        *MAGNETIC STRUCTURE OF RARE-EARTH-COBALT (R*CO2) INTERMETALIC COMPOUNDS.
                        *MOON(R.M.), KOEHLER(W.C.), FARRELL(J.).

869 ER-CO-AL,. . . .   .A1.......J. LESS-COMMON METALS VOL.33,. .   25, (1973).
                        *X-RAY AND NEUTRON DIFFRACTION STUDY OF ORDERING ON CRYSTALLOGRAPHIC
                        SITES IN RARE-EARTH-BASE ALLOYS CONTAINING ALUMINIUM AND TRANSITION MET.
                        *OESTERREICHER(H.).

SECTION   5 -STRUCTURE AND CROSS-SECTION DETERMINATIONS . . . .

870 ER*CO2-ER*AL2, ...A4. . . . . . .J. APPL. PHYS. VOL.41, ... 2326, (1970).
    ⊷NEUTRON DIFFRACTION STUDY OF, THE PSEUDOBINARY SYSTEM ER*CO2-ER*AL2.
    ⊷OESTERREICHER(H.), CORLISS(L.M.), HASTINGS(J.M.).

871 ER*CR*O3,. . . . . .A2,T. . . . .ACTA CRYST A24, ... 217, (1968).
    ⊷REPRESENTATION ANALYSIS OF MAGNETIC STRUCTURES.
    ⊷BERTAUT(E.F.).

872 ER*CR*O3,. . . . . .A2. . . . . . .IEEE TRANS. MAGNETICS MAG-2, . . 453, (1966).
    ⊷ETUDES DES PROPRIETES MAGNETOSTATIQUES ET DES STRUCTURES MAGNETIQUES
    DES CHROMITES DES TERRES RARES ET D≠YTTRIUM.
    ⊷BERTAUT(E.F.), MARESCHAL(J.). . .ET AL.

873 ER*CR*O3,. . . . . .A2. . . . . . .SOL. STATE COMM. 5, ... 93, (1967).
    ⊷ETUDE DE LA STRUCTURE MAGNETIQUE DES CHROMITES D≠ERBIUM ET DE NEODYME
    PAR DIFFRACTION NEUTRONIQUE.
    ⊷BERTAUT(E.F.), MARESCHAL(J.).

874 ER*CR*O3,. . . . . .A4. . . . . . .J. PHYSIQUE TOME 31, ER*C*O3. . 607, (1970).
    ⊷PROPRIETES METAMAGNETIQUES D≠UN MONOCRISTAL DE ER*CR*O3.
    ⊷VEYRET(C.), AYASSE(J.B.), CHAUSSY(J.), MARESCHAL(J.), SIVARDIERE(J.).

875 ER-(DY,GD,TB,HO),. . . . . .A4. . . . . . .INT. J. MAGN.(GB) VOL.2, . . . 389, (1971).
    ⊷SINGLE CRYSTAL NEUTRON DIFFRACTION STUDY OF BINARY RARE EARTH ALLOYS.
    ⊷MILLHOUSE(A.H.), KOEHLER(W.C.).

876 ER*FE2,. . . . . .A2. . . . . . .PHYS. KOND. MATER.(GER.) VOL.13, . 137, (1971).
    ⊷MAGNETIC STRUCTURE AND BEHAVIOR OF ER*FE2: A NEUTRON-DIFFRACTION STUDY.
    ⊷WILL(G.), BARGOUTH(M.O.).

877 ER*FE2,. . . . . .A2. . . . . . .J. PHYSIQUE TOME 32 COL.1 VOL.II 675, (1971).
    ⊷NEUTRON DIFFRACTION STUDY OF ER*FE2.
    ⊷BARGOUTH(M.O.), WILL(G.).

878 ER*FE-GARNET,. . . . . .A2. . . . . . .J. APPL. PHYS. VOL.41. . . . . . 1192, (1970).
    ⊷RARE-EARTH SUBLATTICE CANTING IN DY*IG, ER*IG, AND YB*IG.
    ⊷PICKART(S.J.), ALPERIN(H.A.), CLARK(A.E.).

879 ER*FE*O3,. . . . . .A2. . . . . . .PHYS. REV. B VOL.3, ... 3861, (1971).
    ⊷SPIN REORIENTATION IN ER*FE*O3 SINGLE CRYSTALS OBSERVED BY NEUTRON
    DIFFRACTION.
    ⊷PINTO(H.), SHACHAR(G.), SHAKED(H.), SHTRIKMAN(S.).

880 ER*FE*O3,. . . . . .A2,T. . . . .ACTA CRYST A24, ... 217, (1968).
    ⊷REPRESENTATION ANALYSIS OF MAGNETIC STRUCTURES.
    ⊷BERTAUT(E.F.).

881 ER*FE*O3,. . . . . .A2. . . . . . .PHYS. REV. VOL.118, ... 58, (1960).
    ⊷NEUTRON DIFFRACTION STUDY OF THE MAGNETIC PROPERTIES OF RARE-EARTH-IRON
    PEROVSKITES.
    ⊷KOEHLER(W.C.), WOLLAN(E.O.), WILKINSON(M.K.).

882 ER*FE*O3,. . . . . .A2. . . . . . .PHYS. REV. B VOL.8, ... 3398, (1973).
    ⊷MAGNETIC STRUCTURE OF ER*FE*O3 BELOW 4.5 DEGREES K.
    ⊷GORODETSKY(G.), HORNREICH(R.M.), YAEGER(I.), PINTO(H.), SHACHAR(G.),
    SHAKED(H.).

883 ER*GA GARNET,. . . . . .A2. . . . . . .J. PHYSIQUE TOME 29, ... 495, (1968).
    ⊷ETUDE PAR DIFFRACTION DE NEUTRONS A 0,31 DEGRES K DE L≠ANTIFERROMAGNE-
    TISME DU GRENAT DE GALLIUM ET D≠ERBIUM.
    ⊷HAMMANN(J.).

884 ER-LU, ...A2. . . . . . .PHYS. LETT. VOL.45A NO.4, . . . 281, (1973).
    ⊷NEUTRON DIFFRACTION STUDY OF THE MAGNETIC STRUCTURE OF ER(.75)-LU(.25).
    ⊷HABENSCHUSS(M.), STASSIS(C.), SINHA(S.K.), SPEDDING(F.H.).

885 ER-MN2,. . . . . .A2,A1. . . .J. PHYS. CHEM. SOLIDS VOL.33, . . 1031, (1972).
    ⊷STRUCTURAL, MAGNETIC AND NEUTRON DIFFRACTION STUDIES ON ER-MN2, ER-AL2
    AND TB-MN-AL.
    ⊷OESTERREICHER(H.).

886 ER*MN2,. . . . . .A2. . . . . . .J. APP. PHYS. 36, ... 1001, (1965).
    ⊷INVESTIGATION OF THE MAGNETIC STRUCTURE OF ER*MN2, TM*MN2, TB*NI2 BY
    NEUTRON DIFFRACTION.
    ⊷FELCHER(G.P.), CORLISS(L.M.), HASTINGS(J.M.).

887 ER*MN*O3,. . . . . .A2. . . . . . .REF. SECTION 3, 64/KOEHLER(1), ...

888 ER*MN2*O5, ...A2. . . . . . .PHYS. STAT. SOLIDI A VOL.17 NO.1 191, (1973).
    ⊷STRUCTURES MAGNETIQUES SINUSOIDALES ET HELICOIDALES DE LA TERRE RARE
    DANS T*MN2*O5 (T=ND,TB,ER).
    ⊷BUISSON(G.).

889 ER*N,. . . . . .A2. . . . . . .REF. SECTION 3, 63/CHILD,. . . . .

890 ER3*NI,. . . . . .A2. . . . . . .SOL. STATE COMM. VOL.8, ... 391, (1970).
    ⊷ETUDE DES STRUCTURES MAGNETIQUES DES COMPOSES ER3*CO ET ER3*NI PAR
    DIFFRACTION NEUTRONIQUE.
    ⊷GIGNOUX(D.), LEMAIRE(R.), PACCARD(D.).

891 ER*NI5,. . . . . .A2. . . . . . .COLL. INTER. N.126 (GRENOBLE), . 133, (1963).
    ⊷MAGNETIC STRUCTURE STUDIES AT THE BROOKHAVEN NATIONAL LABORATORY.
    ⊷CORLISS(L.M.), HASTINGS(J.M.).

892 ER2*O3,. . . . . .A2,T. . . . .ACTA CRYST A24, ... 217, (1968).
    ⊷REPRESENTATION ANALYSIS OF MAGNETIC STRUCTURES.
    ⊷BERTAUT(E.F.).

893 ER2*O3,. . . . . .A2. . . . . . .COMP. REND. 262, ...1707B, (1966).
    ⊷STRUCTURE MAGNETIQUE DE ER2*O3.
    ⊷BERTAUT(E.F.), CHEVALIER(R.).

894 ER2*O3,. . . . . .A2. . . . . . .PHYS. REV. 176,. . . . . 722, (1968).
    ⊷MAGNETIC STRUCTURES ER2*O3 AND YB2*O3.
    ⊷MOON(R.M.), KOEHLER(W.C.), CHILD(H.R.), RAUBENHEIMER(L.J.).

895 ER*O*O*H,. . . . . .A2. . . . . . .SOL. STATE COMM. VOL.10, ... 765, (1972).
    ⊷THE MAGNETIC STRUCTURE OF THE MONOCLINIC MODIFICATION OF ER*O*O*H.
    ⊷CHRISTENSEN(A.N.), QUEZEL(S.).

896 ER*P,. . . . . .A2. . . . . . .REF. SECTION 3, 63/CHILD,. . . . .

897 ER*P*O4, ...A1. . . . . . .PHYS. STAT. SOL. B VOL.46, . . . 597, (1971).
    ⊷NEUTRON DIFFRACTION AND SUSCEPTIBILITY MEASUREMENTS ON ER*P*O4 AND
    ER*V*O4.
    ⊷WILL(G.), LUGSCHEIDER(W.), ZINN(W.), PATSCHEKE(E.).

SECTION   5 -STRUCTURE AND CROSS-SECTION DETERMINATIONS . . . .

898 ER*P*04,ER*V*04, ...A1 . . . . . . .CHEM. PHYS. LETT. VOL.2,NO.1,. .   47, (1968).
     #NEUTRON DIFFRACTION STUDY OF ER*P*04 AND ER*V*04.
     #PATSCHEKE(E.), FUESS(H.), WILL(G.).

899 ER*SB, ...A2 . . . . . . .REF. SECTION 3, 63/CHILD,. . . . .

900 ER-SC, ...A2 . . . . . . .PHYS. REV. 174,. . . . .   562, (1968).
     #MAGNETIC STRUCTURE PROPERTIES OF ALLOYS OF TB, HO AND ER WITH SC.
     #CHILD(H.R.), KOEHLER(W.C.).

901 ER-TH, ...A2 . . . . . . .J. APP. PHYS. 39,. . . . . 1329, (1968).
     #MAGNETIC PROPERTIES OF FCC RARE-EARTH-THORIUM ALLOYS.
     #CHILD(H.R.), KOEHLER(W.C.), MILLHOUSE(A.H.).

902 ER2*TI2*07,. . . . . .A1 . . . . . . .CANAD. J. CHEM. VOL.43,. . . . . 2819, (1965).
     #DETERMINATION OF THE CRYSTAL STRUCTURE OF ERBIUM TITANATE, ER2*TI2*07,
     BY X-RAY AND NEUTRON DIFFRACTION.
     #KNOP(O.), BRISSE(F.), CASTELLIZ(L.), SUTARNO.

903 ER*V*04, ...A1 . . . . . . .PHYS. STAT. SOL. B VOL.46,. . .   597, (1971).
     #NEUTRON DIFFRACTION AND SUSCEPTIBILITY MEASUREMENTS ON ER*P*04 AND
     ER*V*04.
     #WILL(G.), LUGSCHEIDER(W.), ZINN(W.), PATSCHEKE(E.).

904 (ER-Y)*AL2,. . . . . .A2 . . . . . . .Z. NATURFORSCH. A VOL.27A, . . . 1581, (1972).
     #NEUTRON DIFFRACTION MEASUREMENTS ON ER(X)-Y(1-X)*AL2.
     #WILL(G.), BARGOUTH(M.O.).

905 EU,. . . . . .A2 . . . . . . .REF. SECTION 3, 65/KOEHLER,. . . . .

906 EU,. . . . . .A2 . . . . . . .PHYS. REV. 135,. . . . . A176, (1964).
     #MAGNETIC STRUCTURES OF EUROPIUM.
     #NERESON(N.G.), OLSEN(C.E.), ARNOLD(G.P.).

907 EU,. . . . . .P . . . . .PHYS. REV. VOL.124,. . . . . . . 1848, (1961).
     #NEUTRON COHERENT SCATTERING AMPLITUDES FOR CD AND EU.
     #ARNOLD(G.P.), NERESON(N.G.).

908 EU,. . . . . .A2 . . . . . . .J. APPL. PHYS. VOL.33, SUPPL., . 1135, (1962).
     #NEUTRON DIFFRACTION STUDIES ON EUROPIUM METAL.
     #OLSEN(C.E.), NERESON(N.G.), ARNOLD(G.P.).

909 EU,. . . . . .A2 . . . . . . .SOL. STATE COMM. VOL.13,. . . .   339, (1973).
     #NEUTRON DIFFRACTION STUDY OF SINGLE CRYSTAL EUROPIUM IN AN APPLIED
     MAGNETIC FIELD.
     #MILLHOUSE(A.H.), MCEWEN(K.A.).

910 EU*C*03, ...A2,A4. . . .REF. SECTION 3, 66/BERTAUT,. . . . .

911 (EU-GD)*S, ...A2 . . . . . . .A.I.P. CONF. PROC. NO.10 PART 2, 1569, (1973).
     #MAGNETIC STRUCTURES OF (EU(1-X)-GD(X))*S.
     #PICKART(S.J.), ALPERIN(H.A.), HOLTZBERG(F.), MCGUIRE(T.R.).

912 EU*0,. . . . . .A2 . . . . . . .PHYS. REV. VOL.127,. . . . . . . 2101, (1962).
     #NEUTRON DIFFRACTION STUDIES ON EU*0.
     #NERESON(N.G.), OLSEN(C.E.), ARNOLD(G.P.).

913 EU*SE, ...A2 . . . . . . .J. PHYS. CHEM. SOLIDS 29,. . . .   414, (1968).
     #MAGNETIC STRUCTURE OF EU*SE.
     #PICKART(S.J.), ALPERIN(H.A.).

914 EU*SE, ...A2 . . . . . . .PHYS. KONDENS. MAT. (GER.) VOL.9 249, (1969).
     #NEUTRON DIFFRACTION EVIDENCE FOR MAGNETIC PHASE TRANSITION IN EUROPIUM
     SELENIDE.
     #FISCHER(P.), HALG(W.), VON-WARTBURG(W.), SCHWOB(P.), VOGT(O.).

915 EU*TE, ...A2 . . . . . . .J. PHYS. CHEM. SOL. 24,. . . . . 1679, (1963).
     #ANTIFERROMAGNETIC STRUCTURE OF EU*TE.
     #WILL(G.), PICKART(S.J.), ALPERIN(H.A.), NATHANS(R.).

916 EU*TE, ...A2 . . . . . . .J. APPL. PHYS. VOL.35, ...   984, (1964).
     #FERROMAGNETIC EUROPIUM COMPOUNDS.
     #MCGUIRE(T.R.), SHAFER(M.W.).

917 EU*TI*03,. . . . . .A2 . . . . . . .J. APP. PHYS. 37,. . . . .   981, (1966).
     #MAGNETIC STRUCTURE OF EU*TI*03.
     #MCGUIRE(T.R.), SHAFER(M.W.), JOENK(R.J.), ALPERIN(H.A.), PICKART(S.J.).

918 F (ISO. A=19), ...T . . . . . . .NUCL. PHYS. A VOL.A176,. . . . .   225, (1971).
     #THEORY OF POLARIZED THERMAL NEUTRON SCATTERING ON F(ISOTOPE A=19).1.
     #GILLET(V.), NORMAND(J.M.).

919 F (ISO. A=19), ...P . . . . . . .NUCL. PHYS. VOL.A123,. . . . . .   215, (1969).
     #THERMAL NEUTRON CAPTURE IN F (ISO. A=19).
     #HARDELL(R.), HASSELGREN(A.).

920 F2,. . . . . .P . . . . . . .PHYS. REV. VOL.174,. . . . . . .   313, (1968).
     #MEASUREMENT OF THE COHERENT NEUTRON SCATTERING AMPLITUDE OF DEUTERIUM,
     MERCURY, AND FLUORINE BY MIRROR REFLECTION.
     #BARTOLINI(W.), DONALDSON(R.E.), GROVES(D.J.).

921 FE,. . . . . .A2 . . . . . . .METAL. SOC. CONF. LOS ANGELES, .   15, (1967).
     #SPIN DENSITY DISTRIBUTION IN FE, CO, AND NI.
     #SHULL(C.G.).

922 FE,. . . . . .A3 . . . . . . .PHYS. STAT. SOL. VOL.39,. . . .   669, (1970).
     #ATOMIC DISTRIBUTION AND MAGNETIC MOMENT IN LIQUID IRON BY NEUTRON
     DIFFRACTION.
     #WASEDA(Y.), SUZUKI(K.).

923 FE,. . . . . .A4 . . . . . . .ACTA PHYS. POL. A VOL.A39, . . .   465, (1971).
     #SPIN-PAIR CORRELATION IN IRON BY NEUTRON DIFFRACTION TECHNIQUE.
     #CISZEWSKI(R.).

924 FE (ISOTOPE A=56), ...T . . . . . . .NUCL. SCI. ENG. VOL.46,. . . . .   255, (1971).
     #STATISTICAL MODEL CALCULATIONS OF IRON-56 NEUTRON CROSS SECTIONS.
     #BLOOM(S.D.), GREEN(J.M.), HUBBARD(H.W.), MOSZKOWSKI(S.A.).

925 FE,. . . . . .A4 . . . . . . .NUCL. PHYS. VOL.A123,. . . . . .   33, (1969).
     #POSSIBLE DOORWAY-STATE STRUCTURE IN THE SCATTERING OF NEUTRONS BY FE.
     #ELWYN(A.J.), MONAHAN(J.E.).

926 FE,. . . . . .P . . . . . . .NUCL. SCI. ENG. VOL.42,. . . . .   28, (1970).
     #HIGH RESOLUTION MEASUREMENTS OF THE TOTAL NEUTRON CROSS SECTIONS OF
     NITROGEN AND IRON.
     #CARLSON(A.D.), CERBONE(R.J.).

SECTION 5 -STRUCTURE AND CROSS-SECTION DETERMINATIONS . . . .

927 FE,. . . . . .A4. . . . . . . .2ND INTERNAT. CONF./NUCL. DATA,. . . . , (1970).
◦NEUTRON ELASTIC SCATTERING CROSS SECTIONS OF VANADIUM, CHROMIUM, IRON
AND NICKEL.
◦HOLMQVIST(B.), WIEDLING(T.).

928 FE,. . . . . .A2. . . . . . . .PROC. PHYS. SOC. 89, 71, (1966).
◦THE MAGNETIC MOMENT DISTRIBUTIONS FOR TRANSITION METAL IMPURITIES IN
IRON.
◦CAMPBELL(I.A.).

929 FE,. . . . . .A4. . . . . . . .PHYS. LETT. 29A, 679, (1969).
◦NEUTRON DIFFRACTION BY A SINGLE CRYSTAL OF IRON EXCITED BY MEANS OF
MAGNETOSTRICTIVE RESONATOR.
◦MICHALEC(R.), CHALUPA(B.), CECH(J.)...ET AL.

930 FE (TRANSITION EL.). . . . . . . .A2. . . . . . .PROC. PHYS. SOC. 86, 535, (1965).
◦THE MAGNETIC MOMENT DISTRIBUTION AROUND TRANSITION ELEMENT IMPURITIES
IN IRON AND NICKEL.
◦COLLINS(M.F.), LOW(G.G.E.).

931 FE,. . . . . .A1. . . . . . .PHYS. LETT. 20, 470, (1966).
◦ERHOLUNG VON STRAHLENSCHADEN IN KUBISCH-RAUMZENTRIERTEN METALLEN NACH
NEUTRONEN BESTRAHLUNG BEI 4.5 DEG. K.
◦BURGER(G.), ISEBECK(K.), KERLER(R.), VOLKL(J.), WENZL(H.).

932 FE,. . . . . .A2. . . . . . .PHYS. REV. VOL.127, 2052, (1962).
◦NEUTRON DIFFRACTION DETERMINATION OF ANTIFERROMAGNETISM IN FACE-
CENTERED CUBIC (GAMMA) IRON.
◦ABRAHAMS(S.C.), GUTTMAN(L.), KASPER(J.S.).

933 FE,. . . . . .A2,A4. . . . .J. PHYS. SOC. JAPAN VOL.17 B-III 7, (1962).
◦SPIN DENSITY DISTRIBUTIONS AND HARTREE-FOCK CALCULATIONS FOR THE IRON
GROUP SERIES.
◦NATHANS(R.), PICKART(S.J.), ALPERIN(H.A.).

934 FE,. . . . . .A2. . . . . . .J. PHYS. SOC. JAPAN VOL.17 B-III 1, (1962).
◦MAGNETIC ELECTRON CONFIGURATION IN IRON.
◦SHULL(C.G.), YAMADA(Y.).

935 FE,. . . . . .A4. . . . . . .PHYS. REV. LETT. VOL.16, 184, (1966).
◦DISTRIBUTION OF INTERNAL MAGNETIZATION IN IRON.
◦SHULL(C.G.), MOOK(H.A.).

936 FE,. . . . . .A4. . . . . . .J. APP. PHYS. 35, 678, (1964).
◦NEUTRON DIFFRACTION EFFECTS WITH MOVING LATTICES.
◦SHULL(C.G.), GINGRICH(N.S.).

937 FE,. . . . . .A4. . . . . . .COMP. REND. TOME 203, 73, (1936).
◦PREUVE EXPERIMENTALE DE LA DIFFRACTION DES NEUTRONS.
◦VON-HALBAN(H.), PREISWERK(P.).

938 FE,. . . . . .A2. . . . . . .REV. MOD. PHYS. VOL.25, 100, (1953).
◦NEUTRON DIFFRACTION STUDIES OF VARIOUS TRANSITION ELEMENTS.
◦SHULL(C.G.), WILKINSON(M.K.).

939 FE,. . . . . .P. . . . . . .PHYS. REV. VOL.68, 159, (1945).
◦THERMAL NEUTRON SCATTERING STUDIES IN METALS.
◦NIX(F.C.), CLEMENT(G.F.).

940 FE,. . . . . .E. . . . . . .PHYS. REV. VOL.70, 235, (1946).
◦TRANSMISSION OF VELOCITY-SELECTED NEUTRONS THROUGH MAGNETIZED IRON.
◦FRYER(E.M.).

941 FE,. . . . . .P. . . . . . .PROC. PHYS. MATH. SOC. JAPAN 25, 495, (1943).
◦SCATTERING OF THERMAL NEUTRONS BY SOLIDS. I.VARIATION IN TOTAL CROSS
SECTION OF METALS BY COLD-WORKING.
◦KIMURA(M.), HASHIGUCHI(R.).

942 FE,. . . . . .A2. . . . . . .PHYS. REV. VOL.103, 516, (1956).
◦NEUTRON DIFFRACTION STUDIES ON IRON AT HIGH TEMPERATURES.
◦WILKINSON(M.K.), SHULL(C.G.).

943 FE,. . . . . .P. . . . . . .PHYS. REV. VOL.54, 827, (1938).
◦THE MAGNETIC SCATTERING OF NEUTRONS.
◦POWERS(P.N.).

944 FE,. . . . . .A4. . . . . . .PROC. PHYS. SOC. A VOL.65, 857, (1952).
◦DIFFUSE REFLECTION OF NEUTRONS FROM A SINGLE CRYSTAL.
◦LOWDE(R.D.).

945 FE,. . . . . .A2. . . . . . .PHYS. REV. VOL.84, 912, (1951).
◦NEUTRON SCATTERING AND POLARIZATION BY FERROMAGNETIC MATERIALS.
◦SHULL(C.G.), WOLLAN(E.O.), KOEHLER(W.C.).

946 FE,. . . . . .A2. . . . . . .J. PHYS. RADIUM VOL.19, 617, (1958).
◦NEUTRON STUDIES OF MAGNETISM.
◦ERICSON(M.), DE-GENNES(P.G.), HERPIN(A.), JACROT(B.), MERIEL(P.). (IN
FRENCH).

947 FE,. . . . . .A2. . . . . . .J. PHYS. RADIUM VOL.20, 169, (1959).
◦RECENT MAGNETIC STRUCTURE STUDIES BY NEUTRON DIFFRACTION.
◦SHULL(C.G.).

948 FE,. . . . . .A4. . . . . . .J. PHYS. CHEM. SOLIDS VOL.10, 147, (1959).
◦X-RAY AND NEUTRON SCATTERING FROM ELECTRONS IN A CRYSTALLINE FIELD AND
THE DETERMINATION OF OUTER ELECTRON CONFIGURATIONS IN IRON AND NICKEL.
◦WEISS(R.J.), FREEMAN(A.J.).

949 FE,. . . . . .P. . . . . . .J. PHYS. CHEM. SOLIDS VOL.10, 138, (1959).
◦THE USE OF POLARIZED NEUTRONS IN DETERMINING THE MAGNETIC SCATTERING
BY IRON AND NICKEL.
◦NATHANS(R.), SHULL(C.G.), SHIRANE(G.), ANDRESEN(A.).

950 FE,. . . . . .P. . . . . . .PROC. PHYS. SOC. A VOL.67, 248, (1954).
◦THE SCATTERING OF SLOW NEUTRONS BY FERROMAGNETIC CRYSTALS.
◦SQUIRES(G.L.).

951 FE-AL,. . . . . .A2. . . . . . .COLL. INTER. N.126 (GRENOBLE), 180, (1963).
◦MAGNETISM RESEARCH WITH POLARIZED NEUTRONS AT THE CENTRO DI STUDI
NUCLEARI DELLA CASACCIA DEL C.N.E.N ROMA, ITALY.
◦ANTONINI(B.), FELCHER(G.P.), MENZINGER(F.), PAOLETTI(A.), RICCI(F.P.),
PASSARI(L.).

952 FE-AL,. . . . . .A2. . . . . . .ACTA PHYS. AUSTRALIA VOL.29, 342, (1969).
◦THE EFFECT OF FE3AL ISLANDS IN THE FE-AL SYSTEM ON THE MAGNETIC
SCATTERING OF SLOW NEUTRONS.
◦ELK(K.). (ED. NOTE: IN GERMAN).

SECTION   5 -STRUCTURE AND CROSS-SECTION DETERMINATIONS . . . .

953 FE-AL, ...A2. . . . . .ACTA PHYS. POLON. VOL.27. . . . . 343, (1965).
     *INVESTIGATION OF AN ANTIFERROMAGNETIC STRUCTURE IN THE ALLOY FE-AL.
     *OLES(A.). (IN FRENCH).

954 FE3-AL,. .A2. . . . . .PHYS. REV. VOL.123,. . . . . . . 1163, (1961).
     *UNPAIRED SPIN DENSITY IN ORDERED FE3-AL.
     *PICKART(S.J.), NATHANS(R.).

955 FE-(AL), .A2. . . . . .PROC. PHYS. SOC. VOL.92. . . . . . 726, (1967).
     *MAGNETIZATION DISTRIBUTION ASSOCIATED WITH NON-TRANSITION METAL
     IMPURITIES IN IRON.
     *HOLDEN(T.M.), COMLY(J.B.), LOW(G.G.E.). (REF. SECTION 3, 67/HOLDEN).

956 FE-AL, ...A2,T. . . . . .J. APPL. PHYS. VOL.29. . . . . . 515, (1958).
     *TRANSITIONS FROM FERROMAGNETISM TO ANTIFERROMAGNETISM IN IRON ALUMINUM
     ALLOYS.
     *SATO(H.), ARROTT(A.).

957 FE3-AL,. .A2. . . . . . .J. APPL. PHYS. VOL.31, SUPPL., . 372, (1960).
     *MAGNETIC-FORM FACTOR OF FE3-AL.
     *PICKART(S.J.), NATHANS(R.).

958 FE3-AL,. .A2. . . . . . .J. PHYS. CHEM. SOLIDS VOL.6, . . 38, (1958).
     *THE MAGNETIC STRUCTURE OF FE3-AL.
     *NATHANS(R.), PIGOTT(M.T.), SHULL(C.G.).

959 FE-AL, ...T,A2. . . . .PHYS. REV. VOL.114,. . . . . . . 1427, (1959).
     *TRANSITIONS FROM FERROMAGNETISM TO ANTIFERROMAGNETISM IN IRON-ALUMINUM
     ALLOYS. THEORETICAL INTERPRETATION.
     *SATO(H.), ARROTT(A.).

960 FE-AL, ...A2. . . . . .PHYS. REV. VOL.114,. . . . . . . 1420, (1959).
     *TRANSITIONS FROM FERROMAGNETISM TO ANTIFERROMAGNETISM IN IRON-ALUMINUM
     ALLOYS, EXPERIMENTAL RESULTS.
     *ARROTT(A.), SATO(H.).

961 FE-AL,FE,. . . . . .A4. . . . . . .J. METALS VOL.8, ... 1259, (1956).
     *NEUTRON DIFFRACTION STUDY OF ANNEALING TEXTURES IN DRAWN B.C.C. METALS.
     *SWALIN(R.A.), GEISLER(A.H.).

962 FE*AL2*O4, ...A2. . . . . .COLL. INTER. N.126 (GRENOBLE), . 83, (1963).
     *MAGNETIC PROPERTIES OF NORMAL SPINELS WITH ONLY A-A INTERACTIONS.
     *ROTH(W.L.).

963 FE2*AS,. .A2. . . . . .J. APP. PHYS. 36,. . . . . 1094, (1965).
     *MAGNETIC STRUCTURE OF FE2*AS.
     *KATSURAKI(H.), SUZUKI(K.).

964 FE2*AS,. .A2. . . . . .J. PHYS. SOC. JAPAN VOL.21,. . . 2238, (1966).
     *THE MAGNETIC STRUCTURE OF FE2*AS.
     *KATSURAKI(H.), ACHIWA(N.).

965 FE2*AS,. .A2. . . . . .J. PHYS. SOC. JAPAN VOL.19,. . . 1988, (1964).
     *NEUTRON DIFFRACTION STUDY OF FE2*AS SINGLE CRYSTALS.
     *KATSURAKI(H.).

966 FE*AS, ...A2. . . . . .ACTA CHEM. SCAND. VOL.26,. . . . 3101, (1972).
     *MAGNETIC STRUCTURE AND PROPERTIES OF FE*AS.
     *SELTE(K.), KJEKSHUS(A.), ANDRESEN(A.F.).

967 (FE+++)-(BA-SR)2*ZN2*FE12*O22, ...A2. . . . .SOV. PHYS. J.E.T.P. VOL.33,. . . . . 737, (1971).
     *DELOCALIZATION OF THE MAGNETIC MOMENT OF FE+++ IONS IN TYPE Y HEXAGONAL
     FERRITE AT 293 DEG. K.
     *SIZOV(R.A.).

968 FE*BR2,FE*CL2, ...A2. . . . . .PHYS. REV. VOL.113,. . . . . . . 497, (1959).
     *NEUTRON DIFFRACTION INVESTIGATIONS OF THE MAGNETIC ORDERING IN FE*BR2,
     CO*BR2, FE*CL2, AND CO*CL2.
     *WILKINSON(M.K.), CABLE(J.W.), WOLLAN(E.O.), KOEHLER(W.C.).

969 FE3*C, ...A1. . . . . .ACTA CRYST. 17,. . . . . 1331, (1964).
     *COMPARISON OF X-RAY AND NEUTRON DIFFRACTION REFINEMENTS OF THE STRUCTURE
     OF CEMENTITE FE3*C.
     *HERBSTEIN(F.H.), SMUTS(J.).

970 FE3*C, ...A4. . . . . .SOV. PHYS. CRYST. VOL.8,. . . . . 300, (1963).
     *DETERMINATION OF THE POSITION OF CARBON IN CEMENTITE BY THE NEUTRON
     DIFFRACTION METHOD.
     *LYASHCHENKO(B.G.), SOROKIN(L.M.).

971 FE3*C, ...T,A1. . . . .ARCH. EISENHUTTENW. VOL.30,. . . 51, (1959).
     *STRUCTURE AMPLITUDES FOR NEUTRON DIFFRACTION ON CEMENTITE.
     *MEINHARDT(D.).

972 FE-C,. . . . . .A3. . . . . .PHYS. KOND. MATER. VOL.17,. . . . 11, (1973).
     *X-RAY AND NEUTRON WIDE-ANGLE DIFFRACTION IN MOLTEN ALLOYS OF THE SYSTEM
     FE-C.
     *MAIER(U.), STEEB(S.). (IN GERMAN).

973 FE*C*O3, .A2. . . . . .SOV. PHYS. JETP VOL.9,. . . . . . 1204, (1959).
     *NEUTRON DIFFRACTION INVESTIGATION OF THE ANTIFERROMAGNETISM OF THE
     CARBONATES OF MANGANESE AND IRON.
     *ALIKHANOV(R.A.).

974 FE*C*O3, .A2. . . . . .A.I.P. CONF. PROC. NO.10 PART 2, 1163, (1973).
     *MAGNETIC LONG RANGE ORDER IN FE*C*O3.
     *ALTMAN(R.F.), SPOONER(S.), LANDAU(D.P.).

975 FE*C*O3,FE*F2, ...P,T. . . . . .J. PHYS. C VOL.6,. . . . . 3746, (1973).
     *THE ELASTIC MAGNETIC NEUTRON CROSS-SECTION FOR 3D TRANSITION COMPOUNDS:
     CALCULATIONS FOR V2*O3, FE*F2 AND FE*C*O3.
     *BALCAR(E.), LOVESEY(S.W.), WEDGWOOD(F.A.).

976 FE*CL2,. .A4. . . . . .PHYS. REV. B VOL.5,. . . . . . . 2615, (1972).
     *MAGNETIC PROPERTIES OF FE*CL2 IN ZERO FIELD: II. LONG-RANGE ORDER.
     *YELON(W.B.), BIRGENEAU(R.J.).

977 FE*CL3,. .A2. . . . . .PHYS. REV. VOL.127,. . . . . . . 714, (1962).
     *NEUTRON-DIFFRACTION STUDY OF ANTIFERROMAGNETIC FE*CL3.
     *CABLE(J.W.), WILKINSON(M.K.), WOLLAN(E.O.), KOEHLER(W.C.).

978 FE*CL2,. .A2. . . . . .J. PHYS. RADIUM VOL.19,. . . . . 617, (1958).
     *NEUTRON STUDIES OF MAGNETISM.
     *ERICSON(M.), DE-GENNES(P.G.), HERPIN(A.), JACROT(B.), MERIEL(P.). (IN
     FRENCH).

SECTION   5 -STRUCTURE AND CROSS-SECTION DETERMINATIONS . . . .

979 FE*CL3,. . . . . *THE MAGNETIC PHASE A2 TRANSITION IN FE*CL3. A.I.P. CONF. PROC. NO.10 PART 1,   98, (1973).
                      *ENDOH(Y.), SKALYO(J.), OOSTERHUIS(W.T.), STAMPFEL(J.P.).

980 FE*CL2,FE*BR2, . *SINGLE CRYSTAL NEUTRON A2 DIFFRACTION J. PHYS. RADIUM VOL.20 OF ANTIFERROMAGNETS 180, (1959) AT
                      LOW TEMPERATURES IN APPLIED MAGNETIC FIELDS.
                      *KOEHLER(W.C.), WILKINSON(M.K.), CABLE(J.H.), WOLLAN(E.O.).

981 FE*CL2.4H2*O,. . *NEUTRON DIFFRACTION A1 STUDY OF IRON J. CHEM. PHYS. VOL.56, 3257, (1972).
                      (FE*CL2)*4(H2*O).
                      *VERBIST(J.J.), HAMILTON(W.C.), KOETZLE(T.F.), LEHMANN(M.S.).

982 FE*CL2.2H2*O,. . *NEUTRON DIFFRACTION A2 DETERMINATION SOL. STATE COMM. VOL.13, 303, (1973).
                      *SCHNEIDER(W.), WEITZEL(H.).

983 FE-CO, ... *MAGNETIC MOMENT DISTRIBUTION A4 IN SOME PHIL. MAG. 8, TRANSITION METAL ALLOYS. 401, (1963).
                      *COLLINS(M.F.), FORSYTH(J.B.).

984 FE-CO, ... *INVESTIGATION OF THE A2 ORDERING MAGN. AND MAG. MATERIALS-1971 1415, (1972). AN
                      EQUIATOMIC IRON COBALT ALLOY.
                      *SPOONER(S.), LYNN(J.W.), CABLE(J.W.). (AIP CONF. PROC. NO.5 PART II).

985 FE-(CO), ... A4. . . . . .REF. SECTION 3, 65/COLLINS,. . . . . .

986 FE-CO, ... *NEUTRON-DIFFRACTION A1 STUDY OF SOV. PHYS. CRYST. VOL.7, 637, (1963).
                      ORDER IN AN FE-CO ALLOY.
                      *GOMAN≠KOV(V.I.),   LITVIN(D.F.),   LOSHMANOV(A.A.),   LYASHCHENKO(B.G.),
                      PUZEI(I.M.).

987 FE-CO, ... *NEUTRON DIFFRACTION A1 STUDY OF SOV. PHYS. CRYST. VOL.6, 443, (1962).
                      *LYASHCHENKO(B.G.), LITVIN(D.F.), ABOV(YU.B.).

988 FE-CO, ... *NEUTRON DIFFRACTION A2 STUDIES OF BULL. ACAD. SCI. USSR VOL.30, 1007, (1966).
                      TURES IN (FE,CR)CO, (FE,MN)CO AND FE(CO,NI) ALLOYS.
                      *DOROFEEV(YU.A.), LYASHCHENKO(B.G.), NOVAK(L.I.).

989 FE-CO, ... *NEUTRON DIFFRACTION A4 DETERMINATION UKR. FIZ. ZH. (USSR) VOL.8, 268, (1963).
                      ATOMIC ORDER IN (THE EQUIATOMIC) FE-CO ALLOY.
                      *GOMAN≠KOV(V.I.), LITVIN(D.F.)...ET AL. (IN RUSSIAN).

990 FE-CO, ... *NEUTRON DIFFRACTION A1 STUDIES OF PHYS. REV. VOL.75, ORDER-DISORDER IN ALLOYS. 1008, (1949).
                      *SHULL(C.G.), SIEGEL(S.).

991 FE-CO-(V), . . . *NEUTRON DIFFRACTION A4 STUDY OF ACTA PHYS. SINICA VOL.21, 1304, (1965).
                      ORDER OF 50 PCT. FE-CO ALLOYS.
                      *GWAN-TING(DU)... ET AL. (IN CHINESE).

992 FE-(CR), ... A4. . . . . .REF. SECTION 3, 65/COLLINS,. . . . .

993 FE-CR, ... *FIRST ORDER A2 TRANSITION SOL. STATE COMM. 4, 657, (1966).
                      STRUCTURE OF DILUTE FE-CR ALLOY.
                      *ISHIKAWA(Y.), ENDOH(Y.), HASHIURA(H.), OHNO(H.).

994 FE-(CR), ... *MAGNETIC PROPERTIES OF A2 CR RICH J. PHYS. CHEM. SOL. 26, 1727, (1965).
                      *ISHIKAWA(Y.), TOURNIER(R.), FILIPPI(J.).

995 FE-CR, ... *NEUTRON DIFFRACTION A2 STUDIES OF PHYS. REV. VOL.97, 304, (1955).
                      TRANSITION ELEMENTS.
                      *SHULL(C.G.), WILKINSON(M.K.).

996 FE-(CR), ... *NEUTRON DIFFRACTION A4 STUDIES OF J. PHYSIQUE TOME 25, 596, (1964).
                      METAL ALLOYS.
                      *COLLINS(M.F.), LOW(G.G.E.).

997 FE-CR-MO,. . . . *THE ORDERING OF A1 ATOMS IN ACTA MET. VOL.2, 456, (1954).
                      SYSTEM.
                      *KASPER(J.S.).

998 FE*CR2*O4, . . . *ETUDE SUR LA A2 STRUCTURE COLL. INTER. N.126 (GRENOBLE), 113, (1963).
                      *BACCHELLA(G.L.), PINOT(M.).

999 FE*CR2*S4, . . . *ETUDE PAR DIFFRACTION A2 NEUTRONIQUE COLL. INTER. N.126 (GRENOBLE), 102, (1963).
                      *BROQUETAS-COLOMINAS(C.), BALLESTRACCI(R.), ROULT(G.).

1000 FE*CR2*(S4,O4),. *MAGNETIC STRUCTURES A2 IN FE*CR2*S4 J. APPL. PHYS. VOL.35, 954, (1964).
                      *SHIRANE(G.), COX(D.E.), PICKART(S.J.).

1001 FE*CR2*SE4,. . . *ETUDE PAR DIFFRACTION A2 NEUTRONIQUE COMP. REND. 264, 3168, (1967).
                      *CHEVRETON(M.), ANDRON(B.).

1002 FE*CU*O2,. . . . *STUDY OF FE*CU*O2 A2 BY MOSSBAUER ANNU. UNIV. SOFIA FAC. PHYS. 59, 47, (1966).
                      *APOSTOLOV(A.). (IN RUSSIAN).

1003 (FE-CU)*RH2*S4,. *ANTIFERROMAGNETIC A2 INTERACTIONS SOL. STATE COMM. VOL.8, 477, (1970).
                      (FE-CU)*RH2*S4.
                      *PLUMIER(R.), LOTGERING(F.K.).

1004 FE*F2, ... *NONCOLLINEARITY EFFECT IN P THE NEUTRON-SCATTERING PHYS. REV. B VOL.6, STRUCTURE FACTOR: FE*F2 178, (1972).
                      CASE.
                      *LEONI(F.).

SECTION   5 -STRUCTURE AND CROSS-SECTION DETERMINATIONS . . . .

1005   FE*F2, ...   *NEUTRON DIFFRACTION STUDIES OF ANTIFERROMAGNETISM IN MANGANOUS FLUORIDE
　　　　　　　　　　　　　　　　　　　A2. . . . . . PHYS. REV. VOL.90, ... 779, (1953).
　　　　　　　　　　　　　　　　AND SOME ISOMORPHOUS COMPOUNDS.
　　　　　　　　　　　　　　　*ERICKSON(R.A.).

1006   FE*F3, ...   　　　　　　　　　　　A4. . . . . . J. PHYS. C VOL.7, ... 783, (1974).
　　　　　　　　　　　　　　　*A NEUTRON DIFFRACTION DETERMINATION OF COVALENCY PARAMETERS FOR FE(3+)
　　　　　　　　　　　　　　　AND CR(3+) IN FE*F3 AND CR*F3.
　　　　　　　　　　　　　　　*JACOBSON(A.J.), MCBRIDE(L.), FENDER(B.E.F.).

1007   FE*F3, ...   　　　　　　　　　　　A2. . . . . . PHYS. REV. VOL.112, ... 1132, (1958).
　　　　　　　　　　　　　　　*ANTIFERROMAGNETIC PROPERTIES OF THE IRON GROUP TRIFLUORIDES.
　　　　　　　　　　　　　　　*WOLLAN(E.O.), CHILD(H.R.), KOEHLER(W.C.), WILKINSON(M.K.).

1008   FE-GA, ...A2. . . . . . REF. SECTION 3, 67/HOLDEN, ...

1009   FE-GA-O, ...   　　　　　　　　A1,2. . . . . PHYS. STAT. SOL. 29, ... 323, (1968).
　　　　　　　　　　　　　　　*A NEUTRON DIFFRACTION STUDY ON GALLIUM-SUBSTITUTED MAGNETITE.
　　　　　　　　　　　　　　　*GAMARI-SEALE(H.), PAPAMANTELLOS(P.).

1010   (FE-GA)*O4,. . .   　　　　　　　　　　A2. . . . . . COMP. REND. 260, ... 6075, (1965).
　　　　　　　　　　　　　　　*ETUDE D*UNE SPINELLE FER GALLIUM PAR DIFFRACTION DE NEUTRONS.
　　　　　　　　　　　　　　　*OLES(A.).

1011   (FE-GA)*O3,. . .   　　　　　　　　　A2. . . . . . J. PHYSIQUE TOME 27, ... 433, (1966).
　　　　　　　　　　　　　　　*STUDY OF FE(1.15)-GA(0.85)*O3 BY MOSSBAUER EFFECT, X-RAYS, NEUTRON
　　　　　　　　　　　　　　　DIFFRACTION AND MAGNETIC MEASUREMENTS.
　　　　　　　　　　　　　　　*BERTAUT(E.F.), BASSI(G.)..ET AL. (IN FRENCH).

1012   (FE-GA)*O4,. . .   　　　　　　　　　A2. . . . . . ACTA PHYS. POLON. VOL.30, . . . 125, (1966).
　　　　　　　　　　　　　　　*NEUTRON DIFFRACTION STUDY OF THE GA2*O3*FE*O-FE2*O3*FE*O SYSTEM.
　　　　　　　　　　　　　　　*OLES(A.). (IN FRENCH).

1013   (FE-GA)*O3,. . .   　　　　　　　　A1. . . . . . J. PHYS. CHEM. SOL. 28, . . . 1451, (1967).
　　　　　　　　　　　　　　　*ETUDE CRISTALLOGRAPHIQUE DU COMPOSE FE(X)*GA(2-X)*O3.
　　　　　　　　　　　　　　　*DELAPALME(A.).

1014   (FE-GA)*O3,. . . . . .A2,A4. . . . REF. SECTION 3, 66/BERTAUT, ...

1015   FE-GARNETS (RARE-EARTH), ...A4. . . . . . J. PHYSIQUE TOME 32 COL.1 VOL.I, 202, (1971).
　　　　　　　　　　　　　　　*NEUTRON DIFFRACTION STUDY OF SOME RARE EARTH IRON GARNETS RIG (R=ER,DY,
　　　　　　　　　　　　　　　YB,TM)
　　　　　　　　　　　　　　　*TCHEOU(F.), BERTAUT(E.F.), FUESS(H.).

1016   FE-GE, ...   　　　　　　　　A1,2. . . . . PROC. INT. CONF. (NOTTINGHAM), . 524, (1964).
　　　　　　　　　　　　　　　*THE CRYSTALLOGRAPHIC AND MAGNETIC STRUCTURE OF FE(1.67)-GE.
　　　　　　　　　　　　　　　*FORSYTH(J.B.), BROWN(P.J.).

1017   FE*GE2,. . . . .   　　　　　　　　　A2. . . . . . SOL. STATE COMM. 3, ... 117, (1965).
　　　　　　　　　　　　　　　*COMMENTS ON THE MAGNETIC STRUCTURES OF DELTA-FE*GE2.
　　　　　　　　　　　　　　　*BERTAUT(E.F.), CHENAVAS(J.).

1018   FE-GE, ...   　　　　　　　　　A2. . . . . . SOL. STATE COMM. 3, ... 113, (1965).
　　　　　　　　　　　　　　　*MAGNETIC STRUCTURES IN THE IRON-GERMANIUM SYSTEM.
　　　　　　　　　　　　　　　*MURTHY(N.S.S.), BEGUM(R.J.), SOMANATHAN(C.S.), MURTHY(M.R.L.).

1019   FE*GE2,. . . . .   　　　　　　　　　A2. . . . . . SOL. STATE COMM. 4, ... 255, (1966).
　　　　　　　　　　　　　　　*ON THE MAGNETIC STRUCTURE OF FE*GE2.
　　　　　　　　　　　　　　　*SOLYOM(J.), KREN(E.).

1020   FE*GE2,. . . . .   　　　　　　　　　A2. . . . . . PHYS. LETT. 11, ... 215, (1964).
　　　　　　　　　　　　　　　*ANTIFERROMAGNETIC STRUCTURE OF FE*GE2.
　　　　　　　　　　　　　　　*KREN(E.), SZABO(P.).

1021   FE-GE, ...   　　　　　　　　　A2. . . . . . J. PHYS. CHEM. SOL. 26, ... 1795, (1965).
　　　　　　　　　　　　　　　*MAGNETIC STRUCTURES OF IRON GERMANIDES.
　　　　　　　　　　　　　　　*ADELSON(E.), AUSTIN(A.E.).

1022   FE-GE, ...   　　　　　　　　　A2. . . . . . REV. ELECT. COMMUN. LAB. VOL.12, 424, (1964).
　　　　　　　　　　　　　　　*NEUTRON DIFFRACTION STUDY OF FE(1.76)-GE SINGLE CRYSTAL.
　　　　　　　　　　　　　　　*KATSURAKI(H.), SUZUKI(K.).

1023   FE*GE, ...   　　　　　　　　　A2. . . . . . J. PHYS. SOC. JAPAN VOL.21, . . 1932, (1966).
　　　　　　　　　　　　　　　*ON THE NEUTRON DIFFRACTION STUDY OF FE*GE.
　　　　　　　　　　　　　　　*WATANABE(H.), KUNITOMI(N.).

1024   FE-GE, ...   　　　　　　　　　A2. . . . . . J. PHYS. SOC. JAPAN VOL.19, . . 863, (1964).
　　　　　　　　　　　　　　　*NEUTRON DIFFRACTION STUDY OF FE(1.76)-GE SINGLE CRYSTALS.
　　　　　　　　　　　　　　　*KATSURAKI(H.).

1025   FE*GE2,. . . . .   　　　　　　　　　A2. . . . . . PHIL. MAG. VOL.10, ... 713, (1964).
　　　　　　　　　　　　　　　*THE MAGNETIC STRUCTURE AND HYPERFINE FIELD OF FE*GE2.
　　　　　　　　　　　　　　　*FORSYTH(J.B.), JOHNSON(C.E.), BROWN(P.J.).

1026   FE-GE, ...A2. . . . . . REF. SECTION 3, 67/HOLDEN, ...

1027   FE2*GE*S4, . . .   　　　　　　　　　A2. . . . . . J. PHYS. CHEM. SOLIDS VOL.34, . 151, (1973).
　　　　　　　　　　　　　　　*ETUDE CRISTALLOGRAPHIQUE ET MAGNETIQUE DE FE2*GE*S4 STRUCTURES
　　　　　　　　　　　　　　　MAGNETIQUES A 85 ET 4.2 DEGRES.
　　　　　　　　　　　　　　　*VINCENT(H.), BERTAUT(E.F.).

1028   FE*(H*C*O2)2*2(H2*O), ...A2. . . . . . SOL. STATE COMM. VOL. 9, . . . 1633, (1971).
　　　　　　　　　　　　　　　*ETUDE PAR DIFFRACTION NEUTRONIQUE DU FORMIATE DE FER DIHYDRATE.
　　　　　　　　　　　　　　　*BURLET(MME.P.), BURLET(P.), BERTAUT(E.F.).

1029   FE*I2, ...   　　　　　　　　　A2. . . . . . SOL. STATE COMM. VOL.14 NO.2, . 187, (1974).
　　　　　　　　　　　　　　　*MAGNETIC STRUCTURE OF FE*I2 BY NEUTRON DIFFRACTION EXPERIMENTS.
　　　　　　　　　　　　　　　*GELARD(J.), FERT(A.R.), MERIEL(P.), ALLAIN(Y.).

1030   FE-(IR), ...A4. . . . . . REF. SECTION 3, 65/COLLINS, ...

1031   FE*LA*O3,. . . .   　　　　　　　　A4. . . . . . PROC. INT. CONF. (NOTTINGHAM), . 327, (1964).
　　　　　　　　　　　　　　　*NEUTRON DIFFRACTION DATA AND COVALENCY EFFECTS.
　　　　　　　　　　　　　　　*NATHANS(R.), WILL(G.), COX(D.E.).

1032   (FE-LI)*(LI-FE-CR)2*O4, ...A2. . . . . . ACTA PHYS. POLON. VOL.A44, . . . 587, (1973).
　　　　　　　　　　　　　　　*THE INFLUENCE OF THERMAL TREATMENT ON THE MAGNETIC STRUCTURE OF THE
　　　　　　　　　　　　　　　FERRITE FE(X)-LI(1-X)-(LI-FE-CR)2*O4.
　　　　　　　　　　　　　　　*DARGEL(L.), KUBEL(W.), MIGON(K.).

1033   FE5*LI*O8, . . .   　　　　　　　　　A2. . . . . . COLL. INTER. N.126 (GRENOBLE), . 79, (1963).
　　　　　　　　　　　　　　　*SUBLATTICE MAGNETIZATION IN LITHIUM FERRITE AS A FUNCTION OF TEMPERATURE
　　　　　　　　　　　　　　　*PRINCE(E.).

SECTION   5 -STRUCTURE AND CROSS-SECTION DETERMINATIONS . . . .

1034 FE-MG-MN-O,. . .   . . .A4. . . . . .PHYS. STAT. SOL. VOL.8,. . . . . . 271, (1965).
                         *NEUTRONOGRAPHISCHE UNTERSUCHUNGEN AN STOCHIOMETRISCHEM MANGANFERRIT UND
                         MAGNESIUM-MANGAN-FERRITEN ZUR DEUTUNGIHRER STRUKTUR.
                         *KLEINSTUCK(K.), WIESER(E.), KLEINERT(P.), PERTHEL(R.).

1035 FE-(MN), ...A4. . . . . .REF. SECTION 3, 65/COLLINS,. . . . .

1036 FE-MN, ...A2. . . . . .J. APP. PHYS. 36,. . . . . . . 1093, (1965).
                         *NEUTRON DIFFRACTION STUDIES OF PD, NI, FE*MN AND CU(MN) SINGLE CRYSTALS.
                         *ARROTT(A.).

1037 FE-MN, ...A2. . . . . .J. PHYS. SOC. JAPAN VOL.23,. . . 205, (1967).
                         *ANTIFERROMAGNETISM OF GAMMA-FE-MN ALLOYS. II.NEUTRON DIFFRACTION AND
                         MOSSBAUER EFFECT STUDIES.
                         *ISHIKAWA(Y.), ENDOH(Y.).

1038 FE-MN, ...A2. . . . . .J. PHYS. SOC. JAPAN VOL.21,. . . 1281, (1966).
                         *ANTIFERROMAGNETISM OF GAMMA FE-MN ALLOYS.
                         *UMEBAYASHI(H.), ISHIKAWA(Y.).

1039 FE-(MN), ...A4. . . . . .J. PHYSIQUE TOME 25,. . . . . . 596, (1964).
                         *NEUTRON DIFFRACTION STUDIES OF MAGNETIC MOMENTS IN DILUTE TRANSITION
                         METAL ALLOYS.
                         *COLLINS(M.F.), LOW(G.G.E.).

1040 FE-MN, ...A2. . . . . .J. APP. PHYS. VOL.39,. . . . . . 1318, (1968).
                         *ANTIFERROMAGNETISM OF GAMMA FE-MN ALLOYS.
                         *ISHIKAWA(Y.), ENDOH(Y.).

1041 FE*MN*AS,. . . .   . .A2. . . . . .J. PHYS. SOC. JAPAN VOL.22,. . . 674, (1967).
                         *NEUTRON DIFFRACTION STUDY OF FE*MN*AS.
                         *YOSHII(S.), KATSURAKI(H.).

1042 FE-MN-(CU),. . .   . . .A2. . . . . .J. PHYS. SOC. JAPAN VOL.30,. . . 1614, (1971).
                         *ANTIFERROMAGNETISM OF GAMMA IRON MANGANESE ALLOYS.
                         *ENDOH(Y.), ISHIKAWA(Y.).

1043 FE(0.9)*MN(0.9)*GE, ...A2. . . . . .ACTA CRYST. A24, ... 513, (1968).
                         *MAGNETIC STRUCTURE OF FE(0.9)*MN(0.9)*GE.
                         *SUZUOKA(T.), ADELSON(E.), AUSTIN(A.E.).

1044 FE*MN2*04, ...A2. . . . . .COLL. INTER. N.126 (GRENOBLE). . . 98, (1963).
                         *STRUCTURE MAGNETIQUE DU MANGANITE DE FER PAR DIFFRACTION NEUTRONIQUE.
                         *MURASIK(A.), ROULT(G.).

1045 FE*MN2*04, ...A4. . . . . .J. PHYSIQUE TOME 32 COL.1 VOL.I, 322, (1971).
                         *ETUDE DE L≠ANISOTROPIE DU MANGANITE DE FER PAR DIFFRACTION DE NEUTRONS
                         A 4.2 DEGRES K.
                         *BOUCHER(B.), BUHL(R.), PERRIN(M.).

1046 FE*MN2*04, ...A1,2. . . . .J. PHYS. CHEM. SOLIDS 30,. . . . 805, (1969).
                         *MANGANITES SPINELLES PURS D≠ELEMENTS DE TRANSITION: PREPARATIONS ET
                         STRUCTURES CRISTALLOGRAPHIQUES.
                         *BUHL(R.).

1047 FE*MN2*04, ...A2. . . . . .J. APP. PHYS. 40,. . . . . 1126, (1969).
                         *MAGNETIC STRUCTURE OF IRON MANGANITE BY NEUTRON DIFFRACTION.
                         *BOUCHER(B.), BUHL(R.), PERRIN(M.).

1048 FE-MN-P, ...A2. . . . . .J. PHYS. SOC. JAPAN VOL.34,. . . 911, (1973).
                         *MAGNETIC STRUCTURE OF FE-MN-P.
                         *SUZUKI(T.), YAMAGUCHI(Y.), YAMAMOTO(H.), WATANABE(H.).

1049 (FE-MN-ZN)*04, ...A2. . . . . .Z. ANGEW. PHYS. VOL.22,. . . . . 103, (1967).
                         *CATION DISTRIBUTION AND MAGNETIC MOMENTS OF THE IONS AT THE TETRAHEDRAL
                         AND OCTAHEDRAL SITES OF THE MANGANESE-ZINC FERRITES.
                         *KONIG(U.). (IN GERMAN).

1050 FE2*MO*04, ...A2. . . . . .J. PHYS. SOC. JAPAN VOL.33,. . . 1296, (1972).
                         *X-RAY AND NEUTRON DIFFRACTION STUDIES IN SPINEL FE2*MO*04.
                         *ABE(M.), KAWACHI(M.), NOMURA(S.).

1051 FE-N,. . . . . .A2. . . . . .J. PHYS. CHEM. SOL. 25,. . . . . 717, (1964).
                         *MAGNETIC PROPERTIES OF EPSILON-IRON NITRIDE.
                         *ROBBINS(M.), WHITE(J.G.).

1052 FE4*N, ...A2. . . . . .PHYS. REV. VOL.112,. . . . . . . 751, (1958).
                         *MAGNETIC STRUCTURE OF FE4*N.
                         *FRAZER(B.C.).

1053 FE*NA*02,. . . .   . .A2. . . . . .COMP. REND. 257,. . . . . 421, (1963).
                         *STRUCTURE MAGNETIQUE DE (BETA)-FE*NA*02. RAFFINEMENT DES PARAMETRES
                         ATOMIQUES.
                         *BERTAUT(E.F.), DELAPALME(A.), BASSI(G.).

1054 FE*NA*02,. . . .   . .A2. . . . . .COLL. INTER. N.126 (GRENOBLE). . 121, (1963).
                         *STRUCTURE MAGNETIQUE DE (BETA)-FE*NA*02 ET RAFFINEMENT DES POSITIONS
                         ATOMIQUES.
                         *BERTAUT(E.F.), DELAPALME(A.), BASSI(G.).

1055 FE*NA*02,. . . .   . .A2. . . . . .J. APPL. PHYS. VOL.35,. . . . . 952, (1964).
                         *MAGNETIC STRUCTURE INVESTIGATIONS AT THE (GRENOBLE) NUCLEAR CENTRE.
                         *BERTAUT(E.F.), ROULT(G.)...ET AL.

1056 (FE-NB)*04,. . .   . . .A2. . . . . .SOL. STATE COMM. VOL.11 NO.2,. . 313, (1972).
                         *MAGNETIC STRUCTURE OF (FE(0.9)-NB(1.1))*04.
                         *WEITZEL(H.). (ED. NOTE: IN GERMAN).

1057 FE*(NB-TA)*06 (TAPIOLITE), ...A1. . . . . .ARK. KEMI(SWEDEN)VOL.28 PAPER23, 375, (1968).
                         *NEUTRON AND X-RAY DIFFRACTION STUDIES ON TAPIOLITE AND SOME SYNTHETIC
                         SUBSTANCES OF TRIRUTILE STRUCTURE.
                         *VON-HEIDENSTAM(O.).

1058 FE-NI, ...A4. . . . . .PHIL. MAG. 8,. . . . . 401, (1963).
                         *THE MAGNETIC MOMENT DISTRIBUTION IN SOME TRANSITION METAL ALLOYS.
                         *COLLINS(M.F.), FORSYTH(J.B.).

1059 FE-NI, ...A4. . . . . .SOV. PHYS. JETP VOL.34,. . . . . 799, (1972).
                         *TEMPERATURE DEPENDENCE OF ELASTIC MAGNETIC DIFFUSE SCATTERING OF NEUT-
                         RONS IN IRON-NICKEL ALLOYS.
                         *ARKHIPOV(V.E.), MEN≠SHIKOV(A.Z.), SIDOROV(S.K.).

1060 FE-NI, ...A2. . . . . .REF. SECTION 3, 66/SIDOROV,. . . . .

SECTION 5 -STRUCTURE AND CROSS-SECTION DETERMINATIONS . . . .

1061 FE-(NI), ...A4. . . . . . .REF. SECTION 3, 65/COLLINS,. . . . .

1062 FE-NI, ...A2. . . . . . .J. PHYS. SOC. JAPAN VOL.17 B-III . 19, (1962).
*ON THE MAGNETIC MOMENTS AND THE DEGREE OF ORDER IN IRON-NICKEL ALLOYS.
*COLLINS(M.F.), JONES(R.V.), LOWDE(R.D.).

1063 FE-NI, ...A4. . . . . . .J. PHYS. SOC. JAPAN VOL.35,. . . 706, (1973).
*MAGNETIC DIFFUSE SCATTERING OF NEUTRONS FROM FE-NI INVAR ALLOY.
*KOMURA(S.), TAKEDA(T.), OHARA(S.).

1064 FE-NI, ...A2. . . . . . .NUOVO CIMENTO VOL.20B NO.1,. . . 1, (1974).
*MAGNETIC-MOMENT DISTRIBUTION IN FE-NI ALLOYS. EXPERIMENT AND CALCULATION
*MENZINGER(F.), SACCHETTI(F.), LEONI(F.).

1065 FE-NI, ...A2. . . . . . .J. APPL. CRYST. VOL.7,. . 233, (1974).
*MAGNETIC CLUSTER STRUCTURES IN AN IRON-NICKEL INVAR ALLOY.
*KOMURA(S.), LIPPMANN(G.), SCHMATZ(W.).

1066 (FE-NI)-AU,. . . . . .A2. . . . . . .INT. J. MAGN. VOL.2 NO.2,. . . 1, (1972).
*MAGNETIC MOMENT PROPERTIES OF THE (FE-NI)-AU SYSTEM.
*CABLE(J.W.), WOLLAN(E.O.).

1067 FE-NI,FE-CO, ...A2. . . . . . .J. PHYS. SOC. JAPAN VOL.17 B-III . 49, (1962).
*NEUTRON DIFFRACTION INVESTIGATION OF ORDER-DISORDER IN THE ALLOYS
#FERRUM-NICKEL# AND #FERRUM-COBALT#.
*LYASHCHENKO(B.G.), LITVIN(D.F.), PUZEI(I.M.), ABOV(YU.G.).

1068 FE-NI-MN,. . . . . .A2. . . . . . .J. PHYS. CHEM. SOL. 24,. . . 529, (1963).
*LONG-RANGE ANTIFERROMAGNETISM IN DISORDERED FE-NI-MN ALLOY.
*KOUVEL(J.S.), KASPER(J.S.).

1069 FE-NI-(MO,SI,CU),. . . . . .A2. . . . . . .BULL. ACAD. SCI. USSR VOL.28,. . 354, (1964).
*NEUTRON DIFFRACTION DETERMINATION OF THE ATOMIC MAGNETIC MOMENTS IN
IRON-NICKEL ALLOYS CONTAINING MO,SI OR CU.
*PUZEI(I.M.), GOMAN#KOV(V.I.), LOSHMANOV(A.A.).

1070 (FE)65-(NI(1-X)-MN(X))35,. . . . . .A4. . . . . . .J. PHYS. SOC. JAPAN VOL.27,. . . 1470, (1969).
*MOSSBAUER EFFECT AND NEUTRON DIFFRACTION OF (FE)65-(NI(1-X)-MN(X))35
ALLOYS.
*NAKAMURA(Y.), SHIGA(M.), TAKEDA(Y.).

1071 FE2*03,. . . . . .A2. . . . . . .DISS. ABS. 28,. . . 3029B, (1967).
*NEUTRON DIFFRACTION STUDIES ON (ALPHA)-FE2*03 AND CR2*03 AT HIGH
PRESSURE.
*WORLTON(T.G.).

1072 FE2*03,. . . . . .A2. . . . . . .SOV. PHYS. SOL. STATE VOL.13,. . 44, (1971).
*NEUTRON-DIFFRACTION STUDY OF HEMATITE IN MAGNETIC FIELDS UP TO 120 KOE
IN A FAST PULSED REACTOR.
*ANTSUPOV(P.S.), VOSKANYAN(R.A.), LEVITIN(R.Z.), NIZEL(S.), NITTS(V.V.),
OZEROV(R.P.), PAK-HWANG-O, SHAFRAN(S.).

1073 FE*0,. . . . . .T. . . . . . .PHYS. REV. B VOL.4,. . . 3901, (1971).
*THEORETICAL MAGNETIC FORM FACTORS OF CO++ AND FE++ IN THEIR MONOXIDES.
*MAHENDRA(A.), KHAN(D.C.).

1074 FE2*03,. . . . . .A2. . . . . . .PHYS REV 171,. . . 596, (1968).
*NEUTRON DIFFRACTION STUDY OF THE MAGNETIC STRUCTURE HEMATITE TO 41 KBAR.
*WORLTON(T.G.), DECKER(D.L.).

1075 FE2*03 (ALPHA),. . . . . .A2,A4. . . .PHYS. LETT. VOL.22,. . . 407, (1966).
*PRESSURE DEPENDENCE OF THE LOW-TEMPERATURE MAGNETIC TRANSITION IN ALPHA-
FE2*03.
*UMEBAYASHI(H.), FRAZER(B.C.), SHIRANE(G.), DANIELS(W.B.).

1076 FE3*04,. . . . . .T, A2. . . .PHYS. REV. 150,. . . . . 367, (1966).
*MAGNETIC PROPERTIES OF MAGNETITE.
*CALLEN(E.).

1077 FE2*03,. . . . . .A2. . . . . . .BULL. ACAD. SCI. USSR VOL.28,. . 359, (1964).
*NEUTRON DIFFRACTION STUDIES OF ALPHA-FE2*03 TYPE OXIDES.
*SHIRANE(G.).

1078 FE2*03,. . . . . .A2. . . . . . .SOL. STATE COMM. 7, NO22,. . . 1665, (1969).
*NEUTRON-DIFFRACTION STUDY OF THE MAGNETIC STRUCTURE OF HEMATITE IN
MAGNETIC FIELD UP TO 120 KOE.

SECTION   5 -STRUCTURE AND CROSS-SECTION DETERMINATIONS . . . .

1088 FE2*O3,. . . . . . .MAGNETIC 'STRUCTURE''AND''VACANCY''DISTRIBUTION''IN''GAMMA-FE2'*O3''BY''NEUTRON''(1958).
                           DIFFRACTION.
                          .FERGUSON-JR(G.A.), HASS(M.).

1089 FE3*O4,. . . . . .A2. . . . . . .PHYS. REV. VOL.110,. . . . . 1050, (1958).
                          .NEUTRON DIFFRACTION INVESTIGATION OF THE 119 DEGREES K TRANSITION IN
                           MAGNETITE.
                          .HAMILTON(W.C.).

1090 FE2*O3,. . . . . . .SPIN DENSITY OF THE CANTED MOMENT IN (ALPHA)-FE2*O3.   118, (1963).
                          .PICKART(S.J.), NATHANS(R.), ALPERIN(H.A.).

1091 FE3*O4,. . . . . .A1. . . . . . .J. PHYS. C VOL.7,. . . . . L115, (1974).
                          .NOTE ON THE SPACE GROUP OF MAGNETITE.
                          .SAMUELSEN(E.J.).

1092 FE2*O3*CA*O, . .   .MOSSBAUER EFFECT AND NEUTRON DIFFRACTION STUDY OF THE MONOCALCITE
                           FERRITE FE2*O3*CA*O.
                          .BERTAUT(E.F.), CHAPPERT(J.), APOSTOLOV(A.), SEMENOV(V.). (IN FRENCH).

1093 FE*O*F,. . . . . .A2. . . . . . .SOL. STATE COMM. 4,. . . . . 395, (1966).
                          .STRUCTURE ANTIFERROMAGNETIQUE DE FE*O*F.
                          .CHAPPERT(J.), PORTIER(J.).

1094 FE2*O3,FE2*O3(GA,AL,TI). . . . . . . .A4. . . . . . .PHIL. MAG. VOL.12,. . . . . 221, (1965).
                          .NEUTRON DIFFRACTION MEASUREMENTS ON PURE AND DOPED SYNTHETIC HEMATITE
                           CRYSTALS.
                          .CURRY(N.A.), JOHNSTON(G.B.), BESSER(P.J.), MORRISH(A.H.).

1095 FE*O*O*D,. . . . . .A1,2. . . . . .PHYS. STAT. SOL. 26,. . . . . . . 429, (1968).
                          .SEMICONDUCTIVITY IN PYRITE, MARCASITE AND ARSENOPYRITE PHASES.
                          .SZYTULA(A.), BUREWICZ(A.), DINITRIJEVIC(Z.), KRASNICKI(S.), RZANY(H.),
                           TODOROVIC(J.),.

1096 FE*O*O*H,. . . . . .A1,2. . . . . .PHYS. STAT. SOL. 26,. . . . . . . 429, (1968).
                          .NEUTRON DIFFRACTION STUDIES OF (ALPHA)-FE*O*O*H.
                          .SZYTULA(A.), BUREWICZ(A.), DINITRIJEVIC(Z.), KRASNICKI(S.), RZANY(H.),
                           TODOROVIC(J.), WANIC(A.), WOLSKI(W.).

1097 FE*O*O*H,. . . . . .A2. . . . . . .J. OF PHYS. C VOL.1,. . . . . . . 179, (1968).
                          .THE MAGNETIC STRUCTURE AND HYPERFINE FIELD OF GOETHITE (ALPHA-FE*O*O*H).
                          .FORSYTH(J.B.), HEDLEY(I.G.), JOHNSON(C.E.).

1098 FE*O*O*H (PHASE: BETA). . . . . .A1. . . . . . .PHYS. STAT. SOL. A VOL.3,. . . . 1033, (1970).
                          .NEUTRON DIFFRACTION STUDIES OF FE*O*O*H (BETA PHASE).
                          .SZYTULA(A.), BALANDA(M.), DIMITRIJEVIC(Z.).

1099 FE*O*O*H (PHASE: GAMMA). . . . . .A1. . . . . . .PHYS. STAT. SOL. 41,. . . . . . 173, (1970).
                          .NEUTRON DIFFRACTION STUDY OF GAMMA-FE*O*O*H.
                          .OLES(A.), SZYTULA(A.), WANIC(A.).

1100 (FE2*O3)-(P2*O5),. . . . . .A4. . . . . . .J. PHYS. C VOL.5,. . . . . L261, (1972).
                          .AMORPHOUS ANTIFERROMAGNETISM IN IRON AND COBALT PHOSPORUS PENTOXIDE
                           GLASSES.
                          .EGAMI(T.),   SACLI(O.A.), SIMPSON(A.W.), TERRY(A.L.), WEDGWOOD(F.A.).

1101 (FE2*O3)-(RH2*O3),. . . . . .A1,2. . . . . .PHYS. LETT. 19,. . . . . 103, (1965).
                          .NEUTRON DIFFRACTION STUDIES ON THE (1-X)FE2*O3-(X)RH2*O3.
                          .KREN(E.), SZABO(P.), KONCZOS(G.).

1102 FE2*O3-V2*O3,. . .   .A MAGNETIC AND NEUTRON DIFFRACTION STUDY OF THE FE2*O3-V2*O3 SYSTEM.
                          .COX(D.E.), TAKEI(W.J.), MILLER(R.C.), SHIRANE(G.).

1103 FE-(OS), ...A4. . . . . .REF. SECTION 3, 65/COLLINS,. . . . .

1104 FE*P,. . . . . .   .MAGNETIC STRUCTURE OF IRON MONOPHOSPHIDE.
                          .FELCHER(G.P.), SMITH(F.A.), BELLAVANCE(D.), WOLD(A.).

1105 FE3*P, ...STUDIES OF THE MAGNETIC STRUCTURE OF FE3*P.
                          .LISHER(E.J.), WILKINSON(C.), ERICSSON(T.), HAGGSTROM(L.), LUNDGREN(L.),
                           WAPPLING(R.).

1106 FE3*(P*O4)2.4H2*O,. . . . . .A1,A2. . . . . .J. CHEM. PHYS. VOL.44,. . . . . 2230, (1966).
                          .FERROMAGNETIC AND CRYSTAL STRUCTURE OF LUDLAMITE,... AT 4.2 DEGREES K.
                          .ABRAHAMS(S.C.).

1107 FE3*(P*O4)2-8H2*O,. . . . . .A2. . . . . . .J. OF PHYS. C VOL.3,. . . . . . . 1127, (1970).
                          .THE MAGNETIC STRUCTURE OF VIVIANITE, FE3*(P*O4)2-8H2*O.
                          .FORSYTH(J.B.), JOHNSON(C.E.), WILKINSON(C.).

1108 FE-PD, ...MAGNETIZATION DISTRIBUTION IN A PALLADIUM-RICH FE*PD ALLOY.
                          .PHILLIPS(W.C.).

1109 FE-PD, ...A4. . . . . . .PROC. INT. CONF. (NOTTINGHAM), . 133, (1964).
                          .NEUTRON DIFFRACTION STUDIES OF PD-RICH FERROMAGNETIC ALLOYS.
                          .LOW(G.G.).

1110 FE-PD, ...A2. . . . . . .REF. SECTION 3, 66/SIDOROV,. . . . .

1111 FE-(PD), ...A4. . . . . . .REF. SECTION 3, 65/COLLINS,. . . . .

1112 FE-PD, ...ATOMIC MAGNETIC MOMENTS IN DILUTE IRON-PALLADIUM ALLOYS.
                          .CABLE(J.W.), WOLLAN(E.O.), KOEHLER(W.C.).

1113 FE-PD3,. . . . .   .SPIN DENSITIES OF 4D ATOMS IN FERROMAGNETIC ALLOYS.
                          .SHIRANE(G.), NATHANS(R.), PICKART(S.J.), ALPERIN(H.A.).

1114 FE-PD, ...POLARIZED NEUTRON INVESTIGATION OF THE CO-PT3 AND FE-PD ALLOYS.
                          .ANTONINI(B.), FELCHER(G.P.), MAZZONE(G.), MENZINGER(F.), PAOLETTI(A.).

1115 FE(PD-PT)3,. . .   .NEUTRON-DIFFRACTION STUDY OF ANTIFERRO-FERROMAGNETIC TRANSITION IN A
                           SYSTEM OF ORDERED FE(PD(X)-PT(1-X))3 ALLOYS.
                          .KELAREV(V.V.), SIDOROV(S.K.), KLYUSHIN(V.V.), ABDULOV(R.Z.).

SECTION   5 -STRUCTURE AND CROSS-SECTION DETERMINATIONS . . . .

1116 FE-(PD-PT)3, ... NEUTRON DIFFRACTION STUDY OF ORDERED FE-(PD(.53)-PT(.47))3 SINGLE
      CRYSTAL.
      A2   J. PHYS. SOC. JAPAN VOL.35, 1554, (1973).
      KADOMATSU(H.), INOUE(C.), FUJII(H.), OKAMOTO(T.).

1117 FE-(PD-PT)3, ... FERROMAGNETIC   TO   CANTED-FERRIMAGNETIC TRANSITION IN FE-(PD-PT)3.
      A2   J. APP. PHYS. 40, 1359, (1969).
      KOUVEL(J.S.), FORSYTH(J.B.).

1118 FE-(PT), ... A4 . . . . REF. SECTION 3, 65/COLLINS, ...

1119 FE-PT, ... MAGNETIC MOMENTS IN THE ORDERED FE(2,8)PT(1,2)ALLOY.
      A2   SOL. STATE COMM. 3, 371, (1965).
      KREN(E.), SZABO(P.).

1120 FE-PT3, ... THE DEPENDENCE OF THE MAGNETIC STRUCTURE OF AN FE-PT3 ALLOY UPON THE
      DEGREE OF LONG-RANGE ORDER.
      A2   PHYS. MET. METALLOGR. VOL.17 N.5 136, (1964).
      KELAREV(V.V.), KLYUSHIN(V.V.), LYASHCHENKO(B.G.).

1121 FE-(RE), ... A4 . . . . REF. SECTION 3, 65/COLLINS, ...

1122 FE-RH, ... SPIN DENSITIES OF 4D ATOMS IN FERRIMAGNETIC ALLOYS.
      A4   PROC. INT. CONF. (NOTTINGHAM), . 223, (1964).
      SHIRANE(G.), NATHANS(R.), PICKART(S.J.), ALPERIN(H.A.).

1123 FE-RH, ... INVESTIGATION OF THE ANTIFERROMAGNETIC-FERROMAGNETIC TRANSFORMATION IN
      IRON-RHODIUM ALLOYS.
      A4   PROC. INT. CONF. (NOTTINGHAM), . 158, (1964).
      PAL(L.), TARNOCZI(T.), SZABO(P.), KREN(E.), TOTH(J.).

1124 FE-(RH), ... A4 . . . . REF. SECTION 3, 65/COLLINS, ...

1125 FE-RH, ... MAGNETIC MOMENTS AND UNPAIRED SPIN DENSITIES IN THE FE-RH ALLOYS.
      A2   PHYS. REV. VOL.134, A1547, (1964).
      SHIRANE(G.), NATHANS(R.), CHEN(C.W.).

1126 FE-RH, ... ETUDE PAR DIFFRACTION NEUTRONIQUE DE FE(O,47)-RH(O,53).
      A2   COMP. REND. 256, 1688, (1963).
      BERTAUT(E.F.), DE-BERGEVIN(F.), ROULT(G.).

1127 FE-RH, ... MAGNETIC AND MAGNOELASTIC PROPERTIES OF METAMAGNETIC IRON-RHODIUM ALLOY.
      A2   SOV. PHYS. J.E.T.P. VOL.19, 1348, (1964).
      ZAKHAROV(A.I.), KADOMTSEVA(A.M.), LEVITIN(R.Z.), PONYATOVSKII(E.G.).

1128 FE-RH, ... THE STUDY OF CERTAIN MAGNETIC STRUCTURES AT THE CENTRE FOR NUCLEAR
      STUDIES, GRENOBLE.
      A2   J. PHYS. SOC. JAPAN VOL.17 B-III 53, (1962).
      BERTAUT(E.F.), DELAPALME(A.) . . . ET AL. (IN FRENCH).

1129 FE*RH, ... MAGNETIC STRUCTURE WORK AT THE NUCLEAR CENTER OF GRENOBLE.
      A2   J. APPL. PHYS. VOL.33, SUPPL., . 1123, (1962).
      BERTAUT(E.F.), DELAPALME(A.), FORRAT(F.), ROULT(G.), DE-BERGEVIN(F.),
      PAUTHENET(R.).

1130 FE-RH, ... NEUTRON DIFFRACTION INVESTIGATION OF THE ANTIFERROMAGNETIC-FERROMAGNETIC
      TRANSFORMATION IN THE FE-RH ALLOY.
      A2   PHYS. LETT. 9, 297, (1964).
      KREN(E.), PAL(L.), SZABO(P.).

1131 FE-RH, ... HYPERFINE FIELDS AND MAGNETIC MOMENTS IN THE FE-RH SYSTEM.
      A2   J. APP. PHYS. 34, 1044, (1963).
      SHIRANE(G.), CHEN(C.W.), FLINN(P.A.), NATHANS(R.).

1132 FERRITE/HEXAGONAL E24(3+)O41(Z), NEUTRON-DIFFRACTION STUDY OF THE MAGNETIC STRUCTURE OF THE HEXAGONAL
      FERRITE E24(3+)O41(Z)
      A2   SOV. PHYS. CRYST. VOL.16, 935, (1971).
      NAMTALISHVILI(M.I.), ALESHKO-OZHEVSKII(O.P.), LIDER(V.V.), YAMZIN(I.I.).

1133 FERRITES, ... A NEUTRON DIFFRACTION STUDY OF THE TEMPERATURE VARIATION OF THE SPON-
      TANEOUS MAGNETIZATION OF FERRITES AND THE NEEL THEORY OF FERRIMAGNETISM.
      A4   J. PHYS. CHEM. SOLIDS VOL.19, 117, (1961).
      RISTE(T.), TENZER(L.).

1134 FERRITES(ORTHO), STUDY OF SOME RARE EARTH ORTHOFERRITES AND ORTHOCHROMITES BY NEUTRON
      DIFFRACTION.
      A4   J. PHYSIQUE TOME 31, 803, (1970).
      PATAUD(P.), SIVARDIERE(J.).

1135 FERRITIN-(H2*O,D2*O), NEUTRON SMALL-ANGLE SCATTERING OF BIOLOGICAL MACROMOLECULES IN SOLUTION.
      A3   J. APPL. CRYST. VOL.7, 173, (1974).
      STUHRMANN(H.B.).

1136 FERROMAGNETS, . . NEUTRON DEPOLARIZATION AS A METHOD TO DETERMINE THE MAGNETIZATION, THE
      MEAN DOMAIN SIZE . . . OF THE INNER MAGNETIZATION OF FERROMAGNETS.
      A2   J. PHYSIQUE TOME 32 COL.1 VOL.I, 579, (1971).
      REKVELDT(M.TH.).

1137 FERROMAGNETS, . . ANOMALOUS SCATTERING OF SLOW NEUTRONS AND ELECTROMAGNETIC WAVES IN
      FERROMAGNETS WITH SMALL MAGNETIC ANISOTROPY.
      A4   SOV. PHYS. JETP LETT. VOL.8, 128, (1968).
      AKHIEZER(I.A.), BELOZOROV(D.P.).

1138 FE-(RU), ... A4 . . . . REF. SECTION 3, 65/COLLINS, ...

1139 FE*S, ... SUR L'ABSENCE D'ORDRE MAGNETIQUE DANS LA FORME QUADRATIQUE DE FE*S.
      A2   SOL. STATE COMM. 3, 335, (1965).
      BERTAUT(E.F.), BURLET(P.), CHAPPERT(J.).

1140 FE-S, ... NEUTRON DIFFRACTION INVESTIGATION OF THE MAGNETIC AND STRUCTURAL
      PROPERTIES OF NEAR-STOICHIOMETRIC IRON SULFIDE.
      A1,A2   J. PHYS. SOC. JAPAN VOL.17 B-I, 249, (1962).
      SPARKS(J.T.), MEAD(W.), KOMOTO(T.).

1141 FE*S2, ... NEUTRON DIFFRACTION STUDIES IN SOLID STATE PHYSICS.
      A1,A2   NUKLEONIKA VOL.5, 414, (1960).
      BLINOWSKI(K.).

1142 FE*S, ... MAGNETIC PHASE TRANSITIONS IN STOICHIOMETRIC FE*S STUDIED BY MEANS OF
      NEUTRON DIFFRACTION.
      A2   ACTA CHEM. SCAND. VOL.14, 919, (1960).
      ANDRESEN(A.F.).

1143 FE-S, ... NEUTRON DIFFRACTION INVESTIGATION OF THE FE(1-X)-S SYSTEM.
      A2   J. APPL. PHYS. VOL.31, SUPPL., 356, (1960).
      SPARKS(J.T.), MEAD(W.), KIRSCHBAUM(A.J.), MARSHALL(W.).

SECTION   5 -STRUCTURE AND CROSS-SECTION DETERMINATIONS . . . .

1144 FE*S, ...A2,A4....REF. SECTION 3, 66/BERTAUT, ...

1145 FE-S, ...A2.......COLL. INTER. N.126 (GRENOBLE), .   143, (1963).
      ~ETUDES DE DIFFRACTION NEUTRONIQUE A LIVERMORE.
      ~SPARKS(J.T.), KOMOTO(T.).

1146 FE-S,FE*S, . . . A1,A2.....ACTA CHEM. SCAND. VOL.21,.   2841, (1967).
      ~PHASE TRANSITIONS IN FE(X)-S (X=.9~1.0) STUDIED BY NEUTRON DIFFRACTION.
      ~ANDRESEN(A.F.), TORBO(P.).

1147 FE*S*O4, ... A2.......PHYS. REV. VOL.125,.   1283, (1962).
      ~ANTIFERROMAGNETIC STRUCTURE OF CR*V*O4 AND THE ANHYDROUS SULFATES OF
      DIVALENT FE,NI AND CO.
      ~FRAZER(B.C.), BROWN(P.J.).

1148 FE-SB, ... A2.......J. PHYS. SOC. JAPAN VOL.34,. . .   58, (1973).
      ~MAGNETIC STRUCTURE OF FE(1+X)-S9.
      ~YASHIRO(T.), YAMAGUCHI(Y.), TOMIYOSHI(S.),   KAZAMA(N.),   WATANABE(H.).

1149 FE*SB2, ... A1,A2....ACTA CHEM. SCAND. VOL.24,. . .   3309, (1970).
      ~COMPOUNDS WITH THE MARCASITE TYPE CRYSTAL STRUCTURE VI. NEUTRON
      DIFFRACTION STUDIES OF CR*SB2 AND FE*SB2.
      ~HOLSETH(H.), KJEKSHUS(A.), ANDRESEN(A.F.).

1150 FE-SB, ... A2.......REF. SECTION 3, 67/HOLDEN, ...

1151 FE*SB2*O4, . . . A2.......PHYS. REV. 147,. . . . .   415, (1966).
      ~THE MAGNETIC STRUCTURE OF FE*SB2*O4.
      ~GONZALO(J.A.), COX(D.E.), SHIRANE(G.).

1152 FE7*SE8, ... A2.......COLL. INTER. N.126 (GRENOBLE), .   150, (1963).
      ~A NEUTRON DIFFRACTION STUDY OF FE7*SE8.
      ~ANDRESEN(A.F.), LECIEJEWICZ(J.).

1153 FE7*SE8, ... A4.......J. PHYS. SOC. JAPAN VOL.29,. . .   649, (1970).
      ~NEUTRON DIFFRACTION STUDY OF FE7*SE8. II.
      ~KAWAMINAMI(M.), OKAZAKI(A.).

1154 FE3*SE4, ... A2.......SOL. STATE COMM. VOL.7,. . . . .   623, (1969).
      ~ETUDE PAR DIFFRACTION NEUTRONIQUE DE FE3*SE4.
      ~LAMBERT-ANDRON(B.), BERODIAS(G.).

1155 FE3*SE4, M*FE2*SE4 (M=TI,V...). . .A2.......J. PHYSIQUE TOME 32 COL.1 VOL.II 985, (1971).
      ~STRUCTURES MAGNETIQUES DE M*FE2*SE4 AVEC M=TI,V,CR,FE,CO,NI.
      ~BABOT(D.), BERODIAS(G.), LAMBERT-ANDRON(B.).

1156 FE7*SE8, ... A2.......J. PHYS. SOC. JAPAN VOL.22,. . .   924, (1967).
      ~NEUTRON DIFFRACTION STUDY OF FE7*SE8.
      ~KAWAMINAMI(M.), OKAZAKI(A.).

1157 FE3*SE4, ... A2.......ACTA CHEM. SCAND. VOL.24,. . . . 2435, (1970).
      ~THE MAGNETIC STRUCTURE OF FE3*SE4.
      ~ANDRESEN(A.F.), VAN-LAAR(B.).

1158 FE3*SE4, ... A2.......ACTA CHEM. SCAND. VOL.22,. . . .   827, (1968).
      ~A NEUTRON DIFFRACTION INVESTIGATION OF FE3*SE4.
      ~ANDRESEN(A.F.).

1159 FE2*SE4*(TI,V,CR,FE,CO,NI). . . .A2.......J. PHYS. CHEM. SOLIDS VOL.33,. .   87, (1972).
      ~STRUCTURES MAGNETIQUES DES COMPOSES M*FE2*SE4 (M=TI,V,CR,FE,CO,NI).
      ~LAMBERT-ANDRON(B.), BERODIAS(G.), BABOT(D.).

1160 FE-SI, ... A2,D4....J. PHYS.F: METAL PHYS. VOL.2,. .   358, (1972).
      ~SPIN DENSITY DISTRIBUTION IN IRON-SILICON ALLOYS.
      ~MOSS(J.), BROWN(P.J.).

1161 FE-SI, ... A2.......COLL. INTER. N.126 (GRENOBLE), .   180, (1963).
      ~MAGNETISM RESEARCH WITH POLARIZED NEUTRON AT THE CENTRO DI STUDI
      NUCLEARI DELLA CASACCIA DEL C.N.E.N., ROMA, ITALY.
      ~ANTONINI(B.),   FELCHER(G.P.),   MENZINGER(F.), PAOLETTI(A.), RICCI(F.P.),
      PASSARI(L.).

1162 FE-SI, ... A2.......Z. PHYS. VOL.174,. . . . .   472, (1963).
      ~THE MAGNETIC STRUCTURE OF ORDERED ALLOYS OF THE IRON-SILICON SYSTEM.
      ~MEINHARDT(D.), KRISEMENT(O.). (IN GERMAN).

1163 FE3-SI, ... A2.......NUOVO CIMENTO VOL.32,. . . . . .   25, (1964).
      ~A POLARIZED NEUTRON INVESTIGATION OF THE FE3-SI ALLOY.
      ~PAOLETTI(A.), PASSARI(L.).

1164 FE*SI, ... A2.......J. PHYS. SOC. JAPAN VOL.18,. . .   995, (1963).
      ~NEUTRON DIFFRACTION STUDY OF THE INTERMETALLIC COMPOUND FE*SI.
      ~WATANABE(H.), YAMAMOTO(H.), ITO(K.).

1165 FE*SI, ... A4.......INT. J. MAGN. VOL.5,. . . . . .   223, (1973).
      ~THE SMALL-ANGLE SCATTERING OF THERMAL NEUTRONS BY BLOCH WALLS IN FE*SI
      (2.5 PCT.) AND NI-SINGLE CRYSTALS.
      ~SCHAERPF(O.), VEHOFF(H.), SCHWINK(CH.). (IN GERMAN).

1166 FE-SI, ... A2.......J. APPL. PHYS. VOL.44,. . . . . 4181, (1973).
      ~POLARIZED NEUTRON TECHNIQUES FOR THE OBSERVATION OF FERROMAGNETIC
      DOMAINS.
      ~SCHLENKER(M.), SHULL(C.G.).

1167 FE-SI, ... A4.......SOV. PHYS. JETP VOL.36,. . . . . 1170, (1973).
      ~NEUTRON REFRACTION AT INDIVIDUAL DOMAIN BOUNDARIES IN A FERROMAGNET.
      ~SHIL≠SHTEIN(S.SH.), SOMENKOV(V.A.), KALANOV(N.).

1168 FE-SI, ... A2.......REF. SECTION 3, 67/HOLDEN, ...

1169 FE*SI*F6.6H2*O, . A1.......ACTA CRYST. VOL.15,. . . . . .   353, (1962).
      ~BOND DISTANCES AND THERMAL MOTION IN FERROUS FLUOSILICATE HEXAHYDRATE:
      A NEUTRON DIFFRACTION STUDY.
      ~HAMILTON(W.C.).

1170 FE2*SI*O4, . . . A2.......J. PHYS. CHEM. SOL. 27,. . . . .   655, (1966).
      ~MAGNETIC PROPERTIES OF MN2*SI*O4 AND FE2*SI*O4.
      ~SANTORO(R.P.), NEWNHAM(R.E.), NOMURA(S.).

1171 FE2*SI*O4, . . . A1,2....COLL. INTER. N.126 (GRENOBLE),.   18, (1963).
      ~NEUTRON DIFFRACTION STUDIES AT THE PUERTO RICO NUCLEAR CENTER.
      ~ALMODOVAR(I.), BIELEN(H.J.), FRAZER(B.C.), KAY(M.I.).

1172 FE*SN, ... A2.......J. PHYS. SOC. JAPAN VOL.22,. . . 1210, (1967).
      ~NEUTRON DIFFRACTION STUDY OF FE*SN.
      ~YAMAGUCHI(K.), WATANABE(H.).

SECTION 5 -STRUCTURE AND CROSS-SECTION DETERMINATIONS . . . .

1173 FE-SN2,. . . . . .A2.......J. PHYS. SOC. JAPAN VOL.17,. . . 247, (1962).
&NEUTRON DIFFRACTION STUDY OF ANTIFERROMAGNETISM IN FE-SN2.
&IYENGAR(P.K.), DASANNACHARYA(B.A.), VIJAYARAGHAVAN(P.R.), ROY(A.P.).

1174 FE-SN2,. . . . . .A2.......J. PHYS. SOC. JAPAN VOL.17 B-III 41, (1962).
&NEUTRON DIFFRACTION STUDY OF ANTIFERROMAGNETISM IN FE-SN2.
&IYENGAR(P.K.), DASANNACHARYA(B.A.), VIJAYARAGHAVAN(P.R.), ROY(A.P.).

1175 FE-SN, ...A2.......REF. SECTION 3, 67/HOLDEN, ...

1176 FE*TA*04,. . . . . .A1.......SOV. PHYS. CRYST. VOL.17,. . 1017, (1973).
&ATOMIC AND MOLECULAR ORDERING IN ME*TA*04 (ME=TI,V,CR,FE) WITH A RUTILE
STRUCTURE.
&ASTROV(D.N.), KRYUKOVA(N.A.), ZORIN(R.B.), MAKAROV(V.A.), OZEROV(R.P.),
ROZHDESTVENSKII(F.A.), SMIRNOV(V.P.), TURCHANINOV(A.M.), FADEEVA(N.V.).

1177 FE*TA2*06, ...A2.......ACTA CRYST. VOL.A30, ... 380, (1974).
&MAGNETISCHE STRUKTUR DES TRIRUTILS FE*TA2*06.
&WEITZEL(H.), KLEIN(S.).

1178 FE2*TE*06, ...A2.......SOL. STATE COMM. VOL.6 NO.5,. . . 317, (1968).
&ETUDE PAR DIFFRACTION NEUTRONIQUE ET EFFET MOSSBAUER DU TELLURATE DE
FER FE2*TE*06.
&MONTMORY(M.C.), BELAKHOVSKY(M.), CHEVALIER(R.), NEWNHAM(R.).

1179 FE2*TE*06, ...A2.......J. PHYS. CHEM. SOLIDS 29,. . 1359, (1968).
&MAGNETIC STRUCTURES OF THE ORDERED TRIRUTILES CR2*W*06, CR2*TE*06 AND
FE2*TE*06.
&KUNNMANN(W.), LA-PLACA(S.J.), CORLISS(L.M.), HASTINGS(J.M.), BANKS(E.).

1180 FE-(TI), ...A4.......REF. SECTION 3, 65/COLLINS,. . . . .

1181 FE-(TI), ...A4.......J. PHYSIQUE TOME 25,. . . . . . 596, (1964).
&NEUTRON DIFFRACTION STUDIES OF MAGNETIC MOMENTS IN DILUTE TRANSITION
METAL ALLOYS.
&COLLINS(M.F.), LOW(G.G.E.).

1182 FE*TI*ND*05, ...A1.......J. PHYS. CHEM. SOLIDS VOL. 31,. . 1171, (1970).
&ETUDE PAR RAYONS X ET NEUTRONS DE LA SERIE ISOMORPHE A*TI*T*05.
(A=CR, MN, FE, T=TERRES RARES).
&BUISSON(G.).

1183 FE2*TI*04, ...A2.......J. PHYS. SOC. JAP. VOL.31,. . 452, (1971).
&NEUTRON AND MAGNETIC STUDIES OF A SINGLE CRYSTAL OF FE2*TI*04.
&ISHIKAWA(Y.), SATO(S.), SYONO(Y.).

1184 FE2*TI*04, ...A1.......ACTA CRYST. VOL. 18,. . 857, (1965).
&A NEUTRON AND X-RAY DIFFRACTION STUDY OF ULVOSPINEL, FE2*TI*04.
&FORSTER(R.H.), HALL(E.O.).

1185 FE*TI*03,. . . . . .T,A2.......PHYS. REV. 164,. . . . . . 768, (1967).
&THEORY OF MAGNETIC PROPERTIES OF ILMENITES M*TI*03.
&GOODENOUGH(J.B.), STICKLER(J.J.).

1186 FE*TI*03-FE2*03, ...A2.......J. PHYS. SOC. JAPAN VOL.17,. . . 1598, (1962).
&A STUDY OF THE MAGNETIC PROPERTIES OF THE FE*TI*03-FE2*03 SYSTEM BY
NEUTRON DIFFRACTION AND THE MOSSBAUER EFFECT.
&SHIRANE(G.), COX(D.E.), TAKEI(W.J.), RUBY(S.L.).

1187 FE*TI*03,FE*TI*03-FE2*03,. . . .A2.......J. PHYS. CHEM. SOLIDS VOL.10,. . 35, (1959).
&NEUTRON-DIFFRACTION STUDY OF ANTIFERROMAGNETIC FE*TI*03 AND ITS SOLID
SOLUTIONS WITH ALPHA-FE2*03.
&SHIRANE(G.), PICKART(S.J.), NATHANS(R.), ISHIKAWA(Y.).

1188 FE2-TI-SN, ...A1.......ACTA PHYS. POLON. VOL.A44,. . . 147, (1973).
&CRYSTAL STRUCTURE AND MAGNETIC PROPERTIES OF HEUSLER-TYPE ALLOYS
M-TI-SB (M=NI,CO,FE) AND FE2-TI-SN.
&SZYTULA(A.), TOMKOWICZ(Z.), TUROWSKI(M.).

1189 FE-(V),. . . . . .A4.......REF. SECTION 3, 65/COLLINS,. . . . .

1190 FE-(V),. . . . . .A4.......J. PHYSIQUE TOME 25,. . . . . . 596, (1964).
&NEUTRON DIFFRACTION STUDIES OF MAGNETIC MOMENTS IN DILUTE TRANSITION
METAL ALLOYS.
&COLLINS(M.F.), LOW(G.G.E.).

1191 FE*W*04, ...A2.......SOL. STATE COMM. VOL.12 NO.8,. . . 779, (1973).
&INVESTIGATIONS CONCERNING THE COEXISTENCE OF TWO MAGNETIC PHASES IN
MIXED CRYSTALS (FE,MN)*W*04.
&OBERMAYER(H.A.), DACHS(H.), SCHROCKE(H.).

1192 FE*W*04, ...A1,A2.......Z. KRIST. VOL.124,. . 192, (1967).
&UNTERSUCHUNGEN ZUR KRISTALLSTRUKTUR UND MAGNETISCHEN STRUKTUR DES
FERBERITS FE*W*04.
&ULKU(D.).

1193 FE-(X),. . . . . .A2.......J. APP. PHYS. VOL. 39,. . . . . . 1174, (1968).
&MAGNETIC NEUTRON SCATTERING FROM ATOMS DISSOLVED IN FERROMAGNETIC
IRON AND NICKEL.
&LOW(G.G.).

1194 GA,. . . . . .A4.......PHYS. REV. 143,. . . . . . 36, (1966).
&ATOMIC RADIAL DISTRIBUTIONS AND ION-ION POTENTIAL IN LIQUID GALLIUM.
&ASCARELLI(P.).

1195 GA,. . . . . .P.......SOV. PHYS. CRYST. VOL.8,. . . . . . 537, (1964).
&COHERENT-SCATTERING AMPLITUDES FOR NEUTRONS AND GALLIUM NUCLEI.
&KUZ#MINOV(YU.S.), BELOV(N.V.).

1196 GA,. . . . . .A3.......J. PHYS. C VOL.6,. . . . . . 212, (1973).
&NEUTRON SCATTERING FROM SUPERCOOLED GALLIUM.
&PAGE(D.I.), SAUNDERSON(D.H.), WINDSOR(C.G.).

1197 GA (LIQUID), ...T.......J. CHEM. PHYS. VOL.56,. . . . . . 1185, (1972).
&LIQUID GALLIUM: COMPARISON OF X-RAY AND NEUTRON-DIFFRACTION DATA.
&NARTEN(A.H.).

1198 GA,. . . . . .P.......PHYS. REV. 131,. . . . . . 2098, (1963).
&NEUTRON COHERENT-SCATTERING AMPLITUDES OF GA, IN AND SB.
&ARNOLD(G.P.), NERESON(N.G.).

1199 GA*AS, ...P.......SOV. PHYS. CRYST. VOL.8,. . . . . . 626, (1964).
&DETERMINATION OF THE COHERENT-SCATTERING CROSS-SECTION FOR NEUTRONS ON
GALLIUM.
&KONAKHOVICH(YU.YU.), SOMENKOV(V.A.).

SECTION   5 -STRUCTURE AND CROSS-SECTION DETERMINATIONS . . . .

1200 GA-HG, ...*. . . . . . .E. . . . . . .PHYS. REV. LETT. VOL.28, ...22, (1972).
                          *SHAPE OF THE COEXISTENCE CURVE OF THE GA-HG SYSTEM NEAR T(CRITICAL).
                          *D≠ABRAMO(G.), RICCI(F.P.), MENZINGER(F.).

1201 GA*MN3*C, ...*. . . . . . .A2. . . . . . .J. SOLID STATE CHEM. VOL.8, ...182, (1973).
                          *ETUDES PAR DIFFRACTION NEUTRONIQUE ET RMN DE ZN*MN3*C ET GA*MN3*C(.935).
                          *FRUCHART(D.), BERTAUT(E.F.), LE-CLERC(B.), LE-DANG-KHOI, VEILLET(P.)...

1202 GARNETS (RARE EARTH), . . .*. . . . . .A2. . . . . . .SOL. STATE COMM. VOL.8, ...1751, (1970).
                          *II.- NEUTRON DIFFRACTION STUDY OF SOME RARE EARTH IRON GARNETS RIG
                          (R=DY,ER,YB,TM) AT LOW TEMPERATURES.
                          *TCHEOU(F.), BERTAUT(E.F.), FUESS(H.).

1203 GARNETS (RARE EARTH), . . .*. PARAMETER . . . .A2. . . . . . .SOL. STATE COMM. VOL.8, ...1745, (1970).
                          *I.- PARAMETER REFINEMENT AT 400 K/MAGNETIC MOMENTS AT ROOM TEMPERATURE
                          OF SOME RARE EARTH IRON GARNETS RIG(R=DY,ER,YB,TM)BY NEUTRON DIFFRACTION
                          *TCHEOU(F.), FUESS(H.), BERTAUT(E.F.).

1204 GD (ISOTOPE A=160), ...A2. . . . . . .SOV. PHYS. J.E.T.P. 28, ... 649, (1969).
                          *NEUTRON DIFFRACTION INVESTIGATION OF SINGLE CRYSTAL GD(160).
                          *KUCHIN(V.M.),   SOMENKOV(V.A.),   SHIL≠SHTEIN(S.SH.),   PATRIKEEV(YU.B.).

1205 GD, ...*. . . . . . .A2. . . . . . .PHYS. REV. B VOL.5, ... 997, (1972).
                          *DISTRIBUTION OF MAGNETIC MOMENT IN METALLIC GADOLINIUM.
                          *MOON(R.M.), KOEHLER(W.C.), CABLE(J.W.), CHILD(H.R.).

1206 GD, ...*. . . . . . .A2. . . . . . .J. APPL. PHYS. VOL.35, ...1045, (1964).
                          *NEUTRON DIFFRACTION INVESTIGATION OF A GADOLINIUM SINGLE CRYSTAL.
                          *WILL(G.), NATHANS(R.), ALPERIN(H.A.).

1207 GD, ...A2. . . . . . .REF. SECTION 3, 65/KOEHLER, ...

1208 GD, ...*. . . . . . .A2,T. . . . . .INT. J. MAGN. VOL.3 NO.4, ... 311, (1972).
                          *NEUTRON MAGNETIC FORM FACTOR OF GADOLINIUM.
                          *FREEMAN(A.J.), DESCLAUX(J.P.).

1209 GD, ...*. . . . . . .A2. . . . . . .PHYS. REV. 165, ...733, (1968).
                          *NEUTRON DIFFRACTION STUDY OF THE MAGNETIC BEHAVIOR OF GADOLINIUM.
                          *CABLE(J.W.), WOLLAN(E.O.).

1210 GD, ...*. . . . . . .T. . . . . .A.I.P. CONF. PROC. NO.10 PART 2, 1309, (1973).
                          *SPIN-POLARIZED ENERGY BANK STRUCTURE, SPIN DENSITIES AND THE NEUTRON
                          MAGNETIC FORM FACTOR OF GADOLINIUM METAL.
                          *HARMON(B.N.), FREEMAN(A.J.).

1211 GD*C*03, ...A2,A4. . . . .REF. SECTION 3, 66/BERTAUT, ...

1212 GD2*03, ...*. . . . . . .A2. . . . . . .J. APP. PHYS. 38, ...1381, (1967).
                          *THE PARAMAGNETIC FORM FACTOR OF GADOLINIUM.
                          *CHILD(H.R.), MOON(R.M.), RAUBENHEIMER(L.J.), KOEHLER(W.C.).

1213 GD*(S,SE,SB,BI), ...*. . . . . .A2. . . . . . .J. APPL. PHYS. VOL.40, ...1009, (1969).
                          *MAGNETIC STRUCTURE AND EXCHANGE INTERACTIONS IN CUBIC GADOLINIUM
                          COMPOUNDS.
                          *MCGUIRE(T.R.), GAMBINO(R.J.), PICKART(S.J.), ALPERIN(H.A.).

1214 GD-SC,GD-Y, ...*. . . . . . .A2. . . . . . .J. APP. PHYS. VOL.40, ... 1003, (1969).
                          *MAGNETIC STRUCTURE PROPERTIES OF GD-Y AND GD-SC ALLOYS.
                          *CHILD(H.R.), CABLE(J.W.).

1215 GE, ...*. . . . . . .A4. . . . . . .ACTA CRYST. VOL.A27, ...219, (1971).
                          *A NEUTRON DIFFRACTION SEARCH FOR NON-CENTROSYMMETRIC THERMAL
                          OSCILLATIONS IN GERMANIUM AND SILICON.
                          *NUNES(A.C.).

1216 GE, ...*. . . . . . .P. . . . . . .PROC. ROY. SOC. 298, ...307, (1967).
                          *ANHARMONIC VIBRATION AND FORBIDDEN REFLECTIONS IN SILICON AND GERMANIUM.
                          *DAWSON(B.), WILLIS(B.T.M.).

1217 GE, ...*. . . . . . .A4. . . . . . .ACTA CRYST. VOL.A28 PART S-4, . . . 220, (1972).
                          *NEUTRON DIFFRACTION TOPOGRAPHY OF HOT-PRESSED GERMANIUM CRYSTAL.
                          *DOI(K.),   MINAKAWA(N.),   MOTOHASHI(H.),   MASAKI(N.),   TOMIMITSU(H.).
                          (ABSTRACT ONLY).

1218 GE, ...*. . . . . . .T,P. . . . .ACTA CRYST. A25, ...116, (1969).
                          *DEBYE-WALLER FACTOR AND ANOMALOUS ABSORPTION (GE, 293≠5 DEG. K).
                          *LUDEWIG(J.), BORRMANN(G.).

1219 GE, ...*. . . . . . .E. . . . . . .Z. NATURFORSCH. VOL.28A, ...657, (1973)
                          *NEUTRON PENDELLOSUNG FRINGE STRUCTURE IN THE LAUE DIFFRACTION BY GE.
                          *SHULL(C.G.), SHAW(W.M.).

1220 GE, ...*. . . . . . .A4,P. . . . .CRYSTAL LATTICE DEFECTS VOL.2, . . 105, (1971).
                          *LONG WAVELENGTH NEUTRON SCATTERING BY DEFECTS IN IRRADIATED GERMANIUM.
                          *CLARK(C.D.), MITCHELL(E.W.J.), STEWART(R.J.).

1221 GE, ...*. . . . . . .A4. . . . . . .J. APPL. CRYST. VOL.7, ...59, (1974).
                          *A NEUTRON DIFFRACTION TOPOGRAPHIC OBSERVATION OF STRAIN FIELD IN A
                          HOT-PRESSED GERMANIUM CRYSTAL.
                          *TOMIMITSU(H.), DOI(K.).

1222 GE, ...*. . . . . . .A4. . . . . . .PHYS. STAT. SOLIDI B VOL.59, ...K59, (1973).
                          *FORBIDDEN (222) NEUTRON REFLECTION IN GERMANIUM, ANHARMONICITY IN THE
                          NUCLEAR MOTION.
                          *ROBERTO(J.B.), BATTERMAN(B.W.), KEATING(D.T.).

1223 GE, ...*. . . . . . .A4. . . . . . .PHYS. REV. B VOL.9, ...2590, (1974).
                          *DIFFRACTION STUDIES OF THE (222) REFLECTION IN GE AND SI: ANHARMONICITY
                          AND THE BONDING ELECTRONS.
                          *ROBERTO(J.B.), BATTERMAN(B.W.), KEATING(D.T.).

1224 (GE-CO)*04, ...A2. . . . . . .REF. SECTION 3, 64/BERTAUT, ...

1225 GE*CO2*04, . . .*. . . . . . .A2. . . . . . .COMP. REND. 264, ...2788, (1967).
                          *ETUDE PAR DIFFRACTION DES NEUTRONS DU COMPOSE SPINELLE ANTIFERROMA-
                          GNETIQUE GE*CO2*04.
                          *PLUMIER(R.).

1226 GE-FE, ...*. . . . . . .A1,2. . . . .PROC. INT. CONF. (NOTTINGHAM), . . 524, (1964).
                          *THE CRYSTALLOGRAPHIC AND MAGNETIC STRUCTURE OF FE(1.67)*4GE.
                          *FORSYTH(J.B.), BROWN(P.J.).

1227 GE*FE2*04, . . .*. . . . . . .A2. . . . . . .COMP. REND. 263, ...1738, (1966).
                          *ETUDE PAR DIFFRACTION DES NEUTRONS DU COMPOSE SPINELLE ANTIFERROMA-
                          GNETIQUE GE*FE2*04.
                          *PLUMIER(R.).

SECTION   5 -STRUCTURE AND CROSS-SECTION DETERMINATIONS . . . .

1228 GE*MN*AL,. . . . . .MAGNETIC STRUCTURE....A2............J. APP. PHYS. VOL.40,. . . . . . . 1870, (1969).
                        .MAGNETIC STRUCTURE OF MN-AL-GE.
                        .MURTHY(N.S.S.),   BEGUM(R.J.),   SOMANATHAN(C.S.),   MURTHY(M.R.L.N.).

1229 GE*MN*03,. . . . .STRUCTURE ET.....A2,A4.....J. PHYSIQUE TOME 32 COL.1 VOL.II 853, (1971).
                        .STRUCTURE ET PROPRIETES MAGNETIQUES DE L#ANTIFERROMAGNETIQUE'A QUATRE
                        SOUS-RESEAUX GE*MN*03.
                        .BOUCHER(B.), SOUGI(M.), WHULER(A.).

1230 GE*NI2*04, ...STRUCTURE MAGNETIQUE....A2....et....COLL. INTER. N.126 (GRENOBLE).. . 92, (1963).
                        .STRUCTURE MAGNETIQUE ET PROPRIETES MAGNETIQUES DE GE*NI2*04.
                        .BERTAUT(E.F.), VU-VAN-QUI, PAUTHENET(R.), MURASIK(A.).

1231 (GE-NI)*04,. . . . . .A2.......REF. SECTION 3, 64/BERTAUT,. . . . .

1232 GE*NI2*04, ...MAGNETIC STRUCTURE....A2....J. APPL. PHYS. VOL.35,. . . . . . 952, (1964).
                        .MAGNETIC STRUCTURE INVESTIGATIONS AT THE (GRENOBLE) NUCLEAR CENTRE.
                        .BERTAUT(E.F.), ROULT(G.)...ET AL.

1233 GE*02, ...NEUTRON SCATTERING....P........AMORPHOUS MATERIALS, ... 423, (1972).
                        .NEUTRON SCATTERING IN AMORPHOUS SOLIDS.
                        .LEADBETTER(A.J.),   WRIGHT(A.C.),   APLING(A.J.).   (IN BOOK: AMORPHOUS
                        MATERIALS ED. BY DOUGLAS, ELLIS; PUBL. WILEY: LONDON 1972).

1234 GE*02, ...NEUTRON DIFFRACTION....A4.......J. AMER. CERAMIC SOC. VOL.53,. . 109, (1970).
                        .NEUTRON DIFFRACTION INVESTIGATION OF VITREOUS GERMANIA.
                        .FERGUSON(G.A.), HASS(M.).

1235 GERMANIA (VITREOUS),. . .NEUTRON....A4.......J. OF PHYS.-C, SER. 2, VOL.2,. . 229, (1969).
                        .NEUTRON DIFFRACTION BY GERMANIA, SILICA AND RADIATION-DAMAGED SILICA
                        GLASSES.
                        .LORCH(E.).

1236 GE-TE, ...STRUCTURAL INVESTIGATION....A3....J. NON-CRYST. SOLIDS VOL.11 NO.5, 417, (1973).
                        .STRUCTURAL INVESTIGATION OF AMORPHOUS AND LIQUID GE(.175)-TE(.825) BY
                        NEUTRON SCATTERING.
                        .NICOTERA(E.),   CORCHIA(M.),   DE-GIORGI(G.),   VILLA(F.),   ANTONINI(M.).

1237 GLASSES, ...STRUCTURE AND ATOMIC....CHEM. APPL. THERMAL NEUTRON SCAT 146, (1973).
                        .STRUCTURE AND ATOMIC MOTION IN GLASSES.
                        .LEADBETTER(A.J.).   (IN BOOK: CHEMICAL   APPLICATIONS OF THERMAL NEUTRON
                        SCATTERING. ED. BY B.T.M. WILLIS; OXFORD UNIV. PRESS: LONDON).

1238 GLASSES, ...DIFFRACTION STUDIES....A4,T.......J. NON-CRYST. SOLIDS VOL.7,. . . 141, (1972).
                        .DIFFRACTION STUDIES OF GLASS STRUCTURE III.LIMITATIONS OF THE FOURIER
                        METHOD FOR POLYATOMIC GLASSES.
                        .LEADBETTER(A.J.), WRIGHT(A.C.).

1239 GLASSES, ...DIFFRACTION STUDIES....T.......J. NON-CRYST. SOLIDS VOL.7,. . . . 23, (1972).
                        .DIFFRACTION STUDIES OF GLASS STRUCTURE I.THEORY AND QUASI-CRYSTALLINE
                        MODEL.
                        .LEADBETTER(A.J.), WRIGHT(A.C.).

1240 GLYCINE(TRI)-SULFATE,. . . . . . .A1.......FERROELECTRICS VOL.5,. . . . . . 45, (1973).
                        .THE CRYSTAL STRUCTURE OF TRIGLYCINE SULFATE.
                        .KAY(M.I.), KLEINBERG(R.).

1241 H2,. . . . . . .NEUTRON DIFFRACTION....A1.......SOV. PHYS. J.E.T.P. 27,. . . . . 210, (1968).
                        .NEUTRON DIFFRACTION INVESTIGATIONS OF HYDROGEN ISOTOPES.
                        .BULATOV(A.S.), KOGAN(V.S.).

1242 H2,. . . . . . .NEUTRON CAPTURE....P.......NUCL. SCI. ENGNG. VOL.38,. . . . 180, (1969).
                        .NEUTRON CAPTURE ON MOLECULAR HYDROGEN.
                        .RIETSCHEL(H.), SCHOTT(W.), FINK(J.), KAPULLA(H.).

1243 H2,. . . . . . .CALCULATION....T.......NUCL. PHYS. VOL.A139,. . . . . . 100, (1969).
                        .CALCULATION OF THE THERMAL NEUTRON CAPTURE CROSS SECTION OF MOLECULAR
                        HYDROGEN.
                        .RIETSCHEL(H.).

1244 H2 (TRITIUM),. . .MEASUREMENT....P.......PHYS. REV. C VOL.5,. . . . . 1952, (1972).
                        .MEASUREMENT OF THE COHERENT-SCATTERING AMPLITUDE OF TRITIUM.
                        .DONALDSON(R.E.), BARTOLINI(W.), OTSUKI(H.).

1245 H2,. . . . . . .MEASUREMENT....P.......Z. NATURFORSCH. A VOL.25,. . . . 967, (1970).
                        .MEASUREMENT OF THE THERMAL NEUTRON SCATTERING CROSS-SECTIONS FOR LIQUID
                        AND SOLID HYDROGEN, DEUTERIUM AND DEUTERIUM HYDRIDE.
                        .SEIFFERT(W.D.), WECKERMANN(B.), MISENTA(R.).

1246 H (BINARY SOLUTIONS),. . . . .P.......ATOMKERNENERGIE VOL.19 NO.4, . . 325, (1972).
                        .TOTAL NEUTRON CROSS SECTION OF HYDROGEN IN BINARY SOLUTIONS OF LIQUIDS.
                        .SUSZKIN(A.).

1247 H2,. . . . . . .HARMONIC....P.......PROC. PHYS. SOC., 85,. . . . . . 79, (1965).
                        .HARMONIC FREE DETERMINATION OF THE THERMAL NEUTRON ABSORPTION CROSS-
                        SECTION OF HYDROGEN BY A PULSED SOURCE TECHNIQUE.
                        .HARRIS(M.J.), KAY(R.E.).

1248 H2 (TRITIUM),. . .MEASUREMENTS....P.......BULL. ACAD. SCI. USSR/PHYS. 31,. . 334, (1967).
                        .MEASUREMENTS OF THE NEUTRON CROSS SECTIONS OF TRITIUM IN THE 0.007 TO
                        5 EV REGION.
                        .VERTEBNYI(V.P.), VLASOV(M.F.), KIRILYUK(A.L.), KOLOTYI(V.V.),
                        PASECHNIK(M.V.), STEPANENKO(V.A.).

1249 H2,. . . . . . .NEUTRON SCATTERING....P.......PHYS. REV. VOL.69,. . . . . . 236, (1946).
                        .NEUTRON SCATTERING IN ORTHO- AND PARAHYDROGEN AND THE RANGE OF NUCLEAR
                        FORCES.
                        .WU(C.S.), RAINWATER(L.J.), HAVENS-JR(W.W.), DUNNING(J.R.).

1250 H2,. . . . . . .NEUTRON SCATTERING....P,A3.......PHYS. REV. VOL.54,. . . . . . 266, (1938).
                        .NEUTRON SCATTERING CROSS SECTIONS OF PARA- AND ORTHOHYDROGEN, AND OF
                        N2, O2 AND H2*O.
                        .BRICKWEDDE(F.G.), DUNNING(J.R.), HOGE(H.J.), MANLEY(J.H.).

1251 H2,. . . . . . .THE SCATTERING....T.......PHYS. REV. VOL.52,. . . . . . 286, (1937).
                        .THE SCATTERING OF NEUTRONS BY ORTHO- AND PARAHYDROGEN.
                        .SCHWINGER(J.), TELLER(E.).

1252 H2,. . . . . . .THE SCATTERING....P,A3.......PHYS. REV. VOL.55,. . . . . . 339, (1939).
                        .THE SCATTERING OF SLOW NEUTRONS BY GASEOUS ORTHO- AND PARAHYDROGEN: SPIN
                        DEPENDENCE OF THE NEUTRON-PROTON FORCE.
                        .LIBBY(W.F.), LONG(E.A.).

1253 H2,. . . . . . .NEUTRON SCATTERING....P.......PHYS. REV. VOL.71,. . . . . . 678, (1947).
                        .NEUTRON SCATTERING IN ORTHO- AND PARAHYDROGEN.
                        .HAMERMESH(M.), SCHWINGER(J.).

SECTION   5 -STRUCTURE AND CROSS-SECTION DETERMINATIONS . . . .

1254 H2,. . . . . ⊕SCATTERING OF 20 DEGREE NEUTRONS IN ORTHO- AND PARAHYDROGEN. PHYS. REV. VOL.58, 1003, (1940).
       ⊕ALVAREZ(L.W.), PITZER(K.S.).

1255 H2,. . . . . ⊕THE INTERACTION OF NEUTRONS WITH NORMAL AND PARAHYDROGEN. PHYS. REV. VOL.52, 1076, (1937).
       ⊕DUNNING(J.R.), MANLEY(J.H.), HOGE(H.J.), BRICKWEDDE(F.G.).

1256 H2,. . . . . ⊕SCATTERING OF SLOW NEUTRONS BY ORTHO- AND PARAHYDROGEN. PHYS. REV. VOL.72, 1147, (1947).
       ⊕SUTTON(R.B.), HALL(T.), ANDERSON(E.E.), BRIDGE(H.S.)...ET AL.

1257 H, ... ⊕ABSOLUTE MEASUREMENT OF COHERENT SCATTERING LENGTHS OF H, C AND CL AND Z. PHYSIK VOL.198, 187, (1967).
       DETERMINATION OF GRAV. ACC. FOR FREE NEUTRONS WITH A GRAVITY REFRACTOM.
       ⊕KOESTER(L.). (IN GERMAN).

1258 H2,. . . . . . . ⊕THE SCATTERING OF NEUTRONS BY PARA- AND ORTHO-HYDROGEN. J. NUCL. ENERGY VOL.4, 115, (1957).
       ⊕OROZDOV(S.I.).

1259 H, ... ⊕DETERMINATION OF HYDROGEN POSITIONS IN CRYSTALS BY NEUTRON DIFFRACTION. REV. MOD. PHYS. VOL.30, 94, (1958).
       ⊕BACON(G.E.).

1260 H (PROTON),. . . ⊕COHERENT NEUTRON-PROTON SCATTERING BY LIQUID MIRROR REFLECTION. PHYS. REV. VOL.84, 1160, (1951).
       ⊕BURGY(M.T.), RINGO(G.R.), HUGHES(D.J.).

1261 H, ... ⊕EFFECTS OF CHEMICAL BINDING ON THE NEUTRON CROSS SECTION OF HYDROGEN. PHYS. REV. VOL.113, 806, (1959).
       ⊕WHITTEMORE(W.L.), MCREYNOLDS(A.W.).

1262 H2 (PARA), . . . ⊕CROSS SECTION OF SLOW NEUTRONS ON PARAHYDROGEN. PHYS. REV. LETT. VOL.26, 1581, (1971).
       ⊕HOUK(T.L.), SHAMBROOM(D.), WILSON(R.).

1263 H2,. . . . . . . ⊕THE SCATTERING OF SLOW NEUTRONS BY ORTHO- AND PARA-HYDROGEN. PHYS. REV. VOL.90, 1125, (1953).
       ⊕STEWART(A.T.), SQUIRES(G.L.).

1264 H-ATOM,. . . . . ⊕COMMENTS ON RECENT WORK ON HYDROGEN ATOM LOCATION AND THERMAL MOTION J. PHYS. SOC. JAPAN VOL.17 B-II, 374, (1962).
       DETERMINATION BY NEUTRON DIFFRACTION.
       ⊕HAMILTON(W.C.).

1265 H-ATOM,. . . . . ⊕PROTON MOTIONS IN SOLIDS BY SLOW NEUTRON SCATTERING CROSS SECTIONS. J. CHEM. PHYS. VOL.35, 2265, (1961).
       ⊕RUSH(J.J.), TAYLOR(T.I.), HAVENS-JR(W.W.).

1266 H-ATOM,. . . . . ⊕THE STRUCTURE OF SOLIDS (REVIEW PAPER). ANN. REV. PHYS. CHEM. VOL.13, 19, (1962).
       ⊕HAMILTON(W.C.).

1267 H-ATOM,. . . . . ⊕THE DETERMINATION OF CRYSTAL STRUCTURES BY NEUTRON-DIFFRACTION ADV. STRUC. RES. DIFFR. METHOD 1 1, (1964).
       MEASUREMENTS (REVIEW).
       ⊕BACON(G.E.).

1268 H-BOND,. . . . . ⊕HYDROGEN BONDING, AND SOME RESULTS OF ITS STUDY BY NEUTRON DIFFRACTION. CHEM. APPL.THERMAL NEUTRON SCAT. 201, (1973).
       ⊕SPEAKMAN(J.C.). (IN BOOK: CHEMICAL APPLICATIONS OF THERMAL NEUTRON
       SCATTERING. ED. BY B.T.M. WILLIS; OXFORD UNIV. PRESS: LONDON).

1269 H-BONDS, ... ⊕ON HYDROGEN BONDING IN INORGANIC CRYSTALS: SOME GENERALIZATIONS, SOME STRUC. CHEM. MOL. BIOL., 466, (1968).
       RECENT RESULTS, AND SOME NEW TECHNIQUES (REVIEW).
       ⊕HAMILTON(W.C.). (IN BOOK: STRUCTURAL CHEMISTRY AND MOLECULAR BIOLOGY;
       ED. BY A. RICH AND N. DAVIDSON.).

1270 H-BONDS, ... ⊕SOME NEW X-RAY AND NEUTRON STUDIES OF HYDROGEN BONDING. REV. MOD. PHYS. VOL.30, 100, (1958).
       ⊕PEPINSKY(R.).

1271 H-BONDS, ... ⊕NEUTRON DIFFRACTION STUDIES OF HYDROGEN BONDING IN ORGANIC AND BIOCHEM- SPECTROSCOPY BIOL. CHEM., 177, (1974).
       ICAL SYSTEMS.
       ⊕KOETZLE(T.F.). (REFER TO SECTION 4).

1272 H-C BONDS (C6*H6), ⊕RIDING MOTION AND HIGHER CUMULANTS IN CRYSTALLOGRAPHIC MODELS. PROC. ROYAL SOC. EDINBURGH A-70, 225, (1972).
       ⊕PAWLEY(G.S.).

1273 H*CL,. . . . . . ⊕CRYSTAL STRUCTURE AND PHASE TRANSITION OF HYDROGEN CHLORIDE. J. PHYS. SOC. JAPAN VOL.32, 1019, (1972).
       ⊕NIIMURA(N.), SHIMAOKA(K.), MOTEGI(H.), HOSHINO(S.).

1274 H-CONTAINING CRYSTALS, . . . ⊕NEUTRON DIFFRACTION STUDIES OF HYDROGEN CONTAINING CRYSTALS. REV. MOD. PHYS. VOL.30, 101, (1958).
       ⊕LEVY(H.A.), PETERSON(S.W.).

1275 H*CR*O2, ... ⊕STRUCTURES OF H*CR*O2 AND D*CR*O2. ACTA CRYST. 16, 1209, (1963).
       ⊕HAMILTON(W.C.), IBERS(J.A.).

1276 H*CR*O2,D*CR*O2, . ⊕A NEUTRON DIFFRACTION STUDY OF POLYCRYSTALLINE H*CR*O2 AND D*CR*O2. J. PHYS. SOC. JAPAN VOL.17 B-II, 383, (1962).
       ⊕HAMILTON(W.C.), IBERS(J.A.).

1277 H2,D2, ... ⊕NEUTRON DIFFRACTION STUDY OF THE CRYSTALLINE STRUCTURE OF SOLID SOV. PHYS. JETP VOL.13, 718, (1961).
       HYDROGEN AND DEUTERIUM.
       ⊕KOGAN(V.S.), LAZAREV(B.G.), OZEROV(R.P.), ZHDANOV(G.S.).

1278 H3*(FE(III),CO(III))*(C*N)6, . . . ⊕ETUDE STRUC./SERIE DES HEXACYANOFERRATES(II,III) D≠HYDROGENE:/I.STRUC. ACTA CRYST. VOL.B28, 2530, (1972).
       CRIST. DES PHASES HEXAGONALES H..., PAR DIFFR. DES RAYONS X ET/NEUTRONS
       ⊕HASER(R.), DE-BROIN(C.E.), PIERROT(M.).

1279 H2,HYDROCARBONS, ⊕SLOW NEUTRON VELOCITY SPECTROMETER STUDIES OF O2, N2, A, H2, H2≠O AND PHYS. REV. VOL.76, 1750, (1949).
       SEVEN HYDROCARBONS.
       ⊕MELKONIAN(E.).

SECTION   5 -STRUCTURE AND CROSS-SECTION DETERMINATIONS . . . .

1280 H5*I*06, ...   A1 . . . . . ACTA CRYST. 20, ... 765, (1966).
      *THE CRYSTAL STRUCTURES OF TWO OXY-ACIDS OF IODINE I: A STUDY OF ORTHO-
      PERIODIC ACID, H5*I*06.
      *FEIKEMA(Y.D.).

1281 H*I*03, ...   A1 . . . . . DISS. ABSTR. VOL.14, ... 1152, (1954).
      *THE CRYSTAL STRUCTURES OF OXALIC ACID DIHYDRATE AND ALPHA IODIC ACID AS
      DETERMINED BY NEUTRON DIFFRACTION.
      *GARRETT-SR(B.S.).

1282 H*MN*(C*O)5, . .   A4 . . . . . INORG. CHEM. VOL.8, ... 1928, (1969).
      *NATURE OF THE METAL-HYDROGEN BOND IN TRANSITION METAL-H COMPLEXES:
      NEUTRON AND X-RAY DIFFRACTION STUDIES OF BETA-PENTACARBONYLMANGANESE HYD
      *LA-PLACA(S.J.), HAMILTON(W.C.), IBERS(J.A.), DAVISON(A.).

1283 H2*02, ...   A1 . . . . . J. CHEM. PHYSICS VOL.42, ... 3054, (1965).
      *CRYSTAL AND MOLECULAR STRUCTURE OF HYDROGEN PEROXIDE: A NEUTRON-DIFFRAC-
      TION STUDY.
      *BUSING(W.R.), LEVY(H.A.).

1284 H2*O, ...   A1 . . . . . ACTA CRYST. 19, ... 909, (1965).
      *ON HYDROGEN BONDS IN CRYSTALLINE HYDRATES.
      *BAUR(W.H.).

1285 H2*O, ...   A3 . . . . . IAEA SYMPOSIUM VIENNA, ... 265, (1960).
      *A STUDY OF THE DIFFUSIVE MOTIONS OF LIQUIDS BY MEANS OF COLD-NEUTRON
      SCATTERING EXPERIMENTS.
      *PALEVSKY(H.).

1286 H2*O, ...   A3 . . . . . IAEA SYMPOSIUM VIENNA, ... 297, (1960).
      *TOTAL NEUTRON SCATTERING CROSS-SECTIONS OF SOME HYDROGENOUS MOLECULES:
      EXPERIMENTS AND INTERPRETATION.
      *JANIK(J.A.),   JANIK(J.),   MANIAWSKI(F.),   RZANY(H.),   SZKATULA(J.),
      SCIESINSKI(J.), WANIC(A.).

1287 H2*O, ...   P . . . . . NUCL. SCI. ENG. VOL.45, ... 308, (1971).
      *THERMAL-NEUTRON DIFFUSION PARAMETERS IN WATER BY THE POISONING METHOD.
      *MARTINHO(E.), COSTA-PAIVA(M.M.).

1288 H2*O, ...   A1 . . . . . NUCL. SCI. ENG. VOL.47, ... 153, (1972).
      *STEADY-STATE NEUTRON SPECTRA IN ICE IN THE TEMPERATURE RANGE 268 TO 4K.
      *GANGWANI(G.S.), TEWARI(S.P.), KOTHARI(L.S.).

1289 H2*O, ...   A3 . . . . . NUCL. SCI. ENG. VOL.45, ... 126, (1971).
      *SLOW-NEUTRON SCATTERING BY LIGHT AND HEAVY WATER.
      *GOTOH(Y.), TAKAHASHI(H.).

1290 H2*O, ...   T . . . . . RAD. ZAVODA FIZ. (YUGOSLAVIA), . 41, (1971).
      *AN ANALYTICAL IMPROVEMENT OF NELKIN#S FORMALISM FOR SLOW NEUTRON
      SCATTERING BY WATER MOLECULES. (IN ENGLISH).
      *STANCIC(V.).

1291 H2*O, ...   REPORT AEEW-R 701 (31PP.), ... (1970).
      *THERMAL NEUTRON DIFFUSION DATA AND THE WIMS SCATTERING MODEL FOR LIGHT
      WATER.
      *BUTLAND(A.T.D.), CHUDLEY(C.T.). (ED. NOTE: AVAIL. HMSO, 49 HIGH HOLBORN,
      LONDON, WC1, ENG.).

1292 H2*O (ICE II), .   J. CHEM. PHYS. VOL.55, ... 1934, (1971).
      *ORDERED PROTON CONFIGURATION IN ICE II, FROM SINGLE-CRYSTAL NEUTRON DIF-
      FRACTION.
      *KAMB(B.), HAMILTON(W.C.), LA-PLACA(S.J.), PRAKASH(A.).

1293 H2*O (CRYSTAL HYDRATES), ... SOV. PHYS. CRYST. VOL.17, NO.2, . 383, (1972).
      *CRYSTAL CHEMISTRY OF HYDROGEN-CONTAINING COMPOUNDS, ON THE BASIS OF
      NEUTRON DIFFRACTION DATA (REVIEW PAPER).
      *DATT(I.D.), OZEROV(R.P.).

1294 H2*O, ...   A3 . . . . . PHYS. REV. 153, ... 184, (1967).
      *QUASI-ELASTIC SCATTERING OF SLOW NEUTRONS BY WATER AND AN AQUEOUS
      SOLUTION OF SODIUM CHLORIDE.
      *GOLAND(A.N.), OTNES(K.).

1295 H2*O, ...   P . . . . . Z. PHYSIK VOL.163, ... 218, (1961).
      *THE SCATTERING OF SUBTHERMAL NEUTRONS BY H2*O, C*H2*O2 AND C6*H6.
      *HEINLOTH(K.). (IN GERMAN).

1296 H2*O, ...   P . . . . . Z. PHYSIK VOL.163, ... 424, (1961).
      *MEASUREMENT OF THE DIFFERENTIAL EFFECTIVE CROSS-SECTION AND THE AVERAGE
      LOGARITHMIC ENERGY LOSS ON THE SCAT. OF SLOW NEUTRONS BY WATER AND ICE.
      *REINSCH(C.). (IN GERMAN).

1297 H2*O, ...   P . . . . . Z. PHYSIK VOL.164, ... 111, (1961).
      *STUDY OF THE SCATTERING CROSS-SECTION IN ICE FOR VERY SLOW NEUTRONS AT
      4 DEGREES K.
      *SPRINGER(T.), WIEDEMANN(W.).

1298 H2*O, ...   P,T . . . . . REACTOR SCI. VOL.11, ... 89, (1960).
      *SCATTERING OF SLOW NEUTRONS BY THE WATER MOLECULE.
      *KHUBCHANDANI(P.G.), RAHMAN(A.).

1299 H2*O, ...   P . . . . . Z. NATURFORSCH. VOL.16A, ... 112, (1961).
      *MEASUREMENT OF THE DIFFERENTIAL SCATTERING CROSS-SECTION AND OF THE MEAN
      ENERGY CHANGE IN THE SCATTERING OF SLOW NEUTRONS IN WATER AND ICE.
      *REINSCH(C.), SPRINGER(T.).

1300 H2*O, ...   T,P . . . . . NUKLEONIK VOL.4, ... 110, (1962).
      *CALCULATION OF THE TOTAL CROSS-SECTION FOR THE SCATTERING OF SUB-THERMAL
      NEUTRONS BY WATER AND ICE ACCORDING TO A CRYSTAL MODEL.
      *GOSSMANN(G.).

1301 H2*O, ...   A3 . . . . . J. CHEM. PHYS. VOL.56, ... 5681, (1972).
      *LIQUID WATER: ATOM PAIR CORRELATION FUNCTIONS FROM NEUTRON AND X-RAY
      DIFFRACTION.
      *NARTEN(A.H.).

1302 H2*O, ...   A3 . . . . . MOL. PHYS. VOL.21, ... 901, (1971).
      *THE CORRELATION OF MOLECULAR ORIENTATION IN LIQUID WATER BY NEUTRON AND
      X-RAY SCATTERING.
      *PAGE(D.I.), POWLES(J.G.).

1303 H2*O (ICE-I), .   A1 . . . . . ADV. IN PHYS. VOL.7, ... 171, (1958).
      *THE STRUCTURE OF ICE-I, AS DETERMINED BY X-RAY AND NEUTRON DIFFRACTION
      ANALYSIS.
      *OWSTON(P.G.).

SECTION   5 -STRUCTURE AND CROSS-SECTION DETERMINATIONS . . . .

1304 H2*O,. . . . . *THE SCATTERING OF SLOW NEUTRONS BY WATER MOLECULES. . . . .   109, (1957).
      *GORYUNOV(A.F.).

1305 H2*O,. . . . . *DIE STREUUNG VON LANGSAMEN NEUTRONEN AN WASSER, EIS UND WASSERDAMPF.   110, (1961).
      *SPRINGER(T.).

1306 H2*O,. . . . . *TOTAL NEUTRON CROSS SECTION OF WATER. NUCL. SCI. ENG. VOL.33,. . . . .   265, (1968).
      *NEILL(J.M.), RUSSELL-JR(J.L.), BROWN(J.R.).

1307 H2*O,. . . . . *THE MEASUREMENT OF THE TOTAL CROSS-SECTION OF H2*O BETWEEN -150 AND +200   323, (1960).
      DEG. C WITH VERY SLOW NEUTRONS.
      *HEINLOTH(K.), SPRINGER(T.).

1308 H2*O,. . . . . *SCATTERING OF SLOW NEUTRONS BY WATER MOLECULES. SOV. PHYS. JETP. 27,. . . . . .   15, (1968).
      *SAMOSVAT(S.S.), SAYASOV(YU.S.), CHURASKOV(V.I.).

1309 H3*O(+)*(C*H3*C6*H4*S*O3)- *THE HYDRATED PROTON H+*(H2*O)N. I. A SINGLE CRYSTAL NEUTRON DIFFRACTION   788, (1973).
      STUDY OF THE OXONIUM ION IN P-TOLUENESULFONIC ACID MONOHYDRATE,...
      *LUNDGREN(J.O.), WILLIAMS(J.M.).

1310 H2*O-CRYSTALLINE HYDRATES,. . . *SURVEY OF THE GEOMETRY AND ENVIRONMENT OF WATER MOLECULES IN CRYSTALLINE   3572, (1972).
      HYDRATES STUDIED BY NEUTRON DIFFRACTION.
      *FERRARIS(G.), FRANCHINI-ANGELA(M.).

1311 H2*O-SOILS,. . . *NEUTRON SCATTERING AND SOIL MOISTURE. DISS. ABSTR. VOL.23,. . . . . .   967, (1962).
      *FISHER-JR(C.P.).

1312 H3*P*O2, ... *THE CRYSTAL STRUCTURES OF HYPOPHOSPHOROUS ACID H3*P*O2 AND ACETONITRILE   1871B, (1967).
      -2-(HYDROGEN CHLORIDE) (CH3*C*N*2(H*CL)) A NEUTRON AND X-RAY...STUDY.
      *WILLIAMS(J.M.).

1313 H3*P*O4, ... *THE CRYSTAL STRUCTURE OF PHOSPHORIC ACID- A NEUTRON DIFFRACTION STUDY;   1850B, (1967).
      THE CRYSTAL STRUCTURE OF STRONTIUM 2-ISOPROPYL MALATE- AN X-RAY...STUDY.
      *COLE(F.E.).

1314 H2*S,. . . . . . *TOTAL NEUTRON SCATTERING CROSS-SECTIONS OF SOME HYDROGENOUS MOLECULES   297, (1960).
      EXPERIMENTS AND INTERPRETATION.
      *JANIK(J.A.), JANIK(J.), MANIAWSKI(F.), RZANY(H.), SZKATULA(J.),
      SCIESINSKI(J.), WANIC(A.).

1315 H2*S,. . . . . . *THE TOTAL SCATTERING CROSS-SECTION OF SLOW NEUTRONS IN GASEOUS H2*S.   517, (1962).
      *TUBBS(N.), SAGAN(U.), RZANY(H.), JANIK(J.A.), JANIK(J.).

1316 H2*SE*O3,. . . . *A NEUTRON DIFFRACTION STUDY OF SELENIOUS ACID, H2*SE*O3. ACTA CHEM. SCAND. VOL.25,. .   1233, (1971).
      *LARSEN(F.K.), LEHMANN(M.S.), SOTOFTE(I.).

1317 HCP CRYSTALS,. . *DETERMINATION OF MAGNETIC STRUCTURES IN HCP CRYSTALS.3. ACTA PHYS. POLON. A VOL.A37,. .   625, (1970).
      *LEHMANN(A.M.).

1318 HCP CRYSTALS,. . *DETERMINATION OF MAGNETIC STRUCTURES IN HCP CRYSTALS. I. ACTA PHYS. POLON. VOL.35,. .   245, (1969).
      *LEHMANN(A.M.).

1319 HE,. . . . . . . *COLD NEUTRON SCATTERING EXPERIMENTS WITH LIQUID HELIUM. IAEA SYMPOSIUM VIENNA,   223, (1960).
      *PALEVSKY(H.).

1320 HE,. . . . . . . *SCATTERING OF SLOW NEUTRONS IN HE NEAR THE LAMBDA CURVE. SOV. PHYS. J.E.T.P. VOL.21,. . .   733, (1965).
      *POKROVSKII(V.L.), SARDUTOVICH(G.I.).

1321 HE,. . . . . . . *SLOW-NEUTRON CROSS SECTIONS OF HE, NE, AR AND KR. NUCL. PHYS. VOL.A133,. . . . .   410, (1969).
      *RORER(D.C.), ECKER(B.M.), AKYUZ(R.O.).

1322 HE,. . . . . . . *ON THE STRUCTURE OF LIQUID HE(A=4) FROM THE ELASTIC SCATTERING OF   374, (1961).
      NEUTRONS.
      *FRANCHETTI(S.).

1323 HE,. . . . . . . *DETERMINATION OF ABSORPTION AND SCATTERING CROSS-SECTIONS OF RARE-GASES   21, (1963).
      FOR THERMAL NEUTRONS.
      *GENIN(R.), BEIL(H.)... ET AL.

1324 HE,. . . . . . . *TRANSMISSION OF SLOW NEUTRONS BY LIQUID HELIUM. PHYS. REV. VOL.97,. . . . . . .   855, (1955).
      *SOMMERS-JR(H.S.), DASH(J.G.), GOLDSTEIN(L.).

1325 HE,. . . . . . . *NEUTRON REFRACTION IN O2, N2, HE, AR GASES. PHYS. REV. VOL.84, ...   969, (1951).
      *MCREYNOLDS(A.W.).

1326 HE,. . . . . . . *ATOMIC DISTRIBUTION IN LIQUID HELIUM BY NEUTRON DIFFRACTION. PHYS. REV. VOL.100,.   994, (1955).
      *HURST(D.G.), HENSHAW(D.G.).

1327 HE,. . . . . . . *TRANSMISSION OF SLOW NEUTRONS BY LIQUID HE(A=4). II. PHYS. REV. VOL.101,. . . .   1235, (1956).
      *GOLDSTEIN(L.), SOMMERS-JR(H.S.).

1328 HE,. . . . . . . *NEUTRON-DIFFRACTION STUDIES IN LIQUID HE(A=4). PHYS. REV. A VOL.9,. . . . . .   448, (1974).
      *MOZER(B.), DE-GRAAF(L.A.), LE-NEINDRE(B.).

1329 HE,. . . . . . . *STRUCTURE STUDIES IN LIQUID HE(A=4). PHYS. REV. A VOL.9,. . . . . .   435, (1974).
      *RAVECHE(H.J.), MOUNTAIN(R.D.).

1330 HE,. . . . . . . *STRUCTURE OF SOLID HELIUM BY NEUTRON DIFFRACTION. PHYS. REV. VOL.109,. . . . .   328, (1958).
      *HENSHAW(D.G.).

SECTION 5 -STRUCTURE AND CROSS-SECTION DETERMINATIONS . . . .

1331 HE,. . . . .
 PRESSURE EFFECT IN THE ATOMIC DISTRIBUTION IN LIQUID HELIUM BY NEUTRON
 DIFFRACTION.
 HENSHAW(D.G.).
 A3 PHYS. REV. VOL.119, 14, (1960).

1332 HE,. . . . .
 EFFECT OF THE LAMBDA TRANSITION ON THE ATOMIC DISTRIBUTION IN LIQUID
 HELIUM BY NEUTRON DIFFRACTION.
 HENSHAW(D.G.).
 A3 PHYS. REV. VOL.119, 9, (1960).

1333 HE,. . . . .
 THE LAMBDA-TRANSITION IN HELIUM.
 KUPER(C.G.).
 T,A3 NATURE VOL.183, 1544, (1959).

1334 HE,. . . . . . .
 STRUCTURE AND EXCITATIONS OF LIQUID HELIUM (REVIEW PAPER).
 WOODS(A.D.B.), COWLEY(R.A.).
 REP. PROGR. PHYS. VOL.36, 1135, (1973).

1335 HE,. . . . . . .
 CRYSTAL STRUCTURE OF HELIUM ISOTOPES.
 DONOHUE(J.).
 A1 PHYS. REV. VOL.114, 1009, (1959).

1336 HE (A=3),. . . .
 MEASUREMENT OF THE COHERENT NEUTRON SCATTERING LENGTH OF HE(A=3) BY
 REFLECTION FROM A QUARTZ-LIQUID-HELIUM INTERFACE.
 KITCHENS(T.A.), OVERSLUIZEN(T.), PASSELL(L.), SCHERMER(R.I.).
 P PHYS. REV. LETT. VOL.32, 791, (1974).

1337 HE,. . . . . . .
 NEUTRON DIFFRACTION BY LIQUID HELIUM.
 HENSHAW(D.G.), HURST(D.G.).
 A3 PHYS. REV. VOL.91, 1222, (1953).

1338 HE,NE,AR...,. .
 MEASUREMENT OF THE ELECTRON-NEUTRON INTERACTION BY THE ASYMMETRICAL
 SCATTERING OF THERMAL NEUTRONS BY NOBLE GASES.
 KROHN-JR(V.E.), RINGO(G.R.).
 P PHYS. REV. VOL.148, 1303, (1966).

1339 HEUSLER ALLOYS,.
 CHEMICAL ORDERING IN HEUSLER ALLOYS WITH THE GENERAL FORMULA A2*B*C
 OR A*B*C.
 BACON(G.E.), PLANT(J.S.).
 A1 J. PHYS.F: METAL PHYS. VOL.1, 524, (1971).

1340 HEXAMETHYLENETETRAMINE,.
 NEUTRON STUDIES OF NUCLEAR CHARGE DISTRIBUTIONS IN BARIUM FLUORIDE AND
 HEXAMETHYLENETETRAMINE.
 DAWSON(B.).
 A4 HARWELL SUMMER SCHOOL, 101, (1968).

1341 HEXAMETHYLENETETRAMINE,.
 JOINT REFINEMENT OF NEUTRON AND X-RAY DIFFRACTION DATA.
 DUCKWORTH(J.A.K.), WILLIS(B.T.M.), PAWLEY(G.S.).
 A4 ACTA CRYST. VOL.A25, 482, (1969).

1342 HEXAMETHYLENETETRAMINE,.
 NEUTRON DIFFRACTION STUDY OF THE ATOMIC AND MOLECULAR MOTION IN
 HEXAMETHYLENETETRAMINE.
 DUCKWORTH(J.A.K.), WILLIS(B.T.M.), PAWLEY(G.S.).
 A4 ACTA CRYST. VOL.A26, 263, (1970).

1343 HF,. . . . .
 SLOW-NEUTRON CROSS SECTIONS OF HAFNIUM.
 ATOJI(M.).
 P ACTA CRYST. 17, 1087, (1964).

1344 HF-H,HF-D,. . . .
 NEUTRON DIFFRACTION STUDIES OF HAFNIUM-HYDROGEN AND TITANIUM-HYDROGEN
 SYSTEMS.
 SIDHU(S.S.), HEATON(L.), ZAUBERIS(D.D.).
 A1 ACTA CRYST. VOL.9, 607, (1956).

1345 HF-O,. . . . . .
 SUPERSTRUCTURE AND ORDER-DISORDER TRANSFORMATION OF INTERSTITIAL OXYGEN
 IN HAFNIUM.
 HIRABAYASHI(M.), YAMAGUCHI(S.), ARAI(T.).
 A1 J. PHYS. SOC. JAPAN VOL.35, 473, (1973).

1346 HG,. . . . . . .
 MEASUREMENT OF THE COHERENT NEUTRON SCATTERING AMPLITUDE OF DEUTERIUM,
 MERCURY, AND FLUORINE BY MIRROR REFLECTION.
 BARTOLINI(W.), DONALDSON(R.E.), GROVES(D.J.).
 P PHYS. REV. VOL.174, 313, (1968).

1347 HG,. . . . . . .
 SCATTERING OF SLOW NEUTRONS BY SOME ELEMENTS.
 KIMURA(M.).
 P PROC. PHYS.-MATH. SOC. JAPAN 22, 391, (1940).

1348 HG,. . . . . . .
 SCATTERING OF THERMAL NEUTRONS BY MERCURY.
 HIBDON(C.T.), MUEHLHAUSE(C.O.), RINGO(G.R.), ROBILLARD(T.R.).
 P,A3 PHYS. REV. VOL.82, 560, (1951).

1349 HG,. . . . . . .
 NEUTRON DIFFRACTION STUDY OF LIQUID MERCURY.
 VINEYARD(G.H.).
 A3 J. CHEM. PHYS. VOL.22, 1665, (1954).

1350 HG,. . . . . . .
 ABSOLUTE MEASUREMENT OF THE COHERENT SCATTERING LENGTH OF MERCURY WITH
 THE NEUTRON GRAVITY-REFRACTOMETER AT THE MUNICH RESEARCH REACTOR.
 KOESTER(L.). (IN GERMAN).
 P Z. PHYSIK VOL.182, 328, (1964).

1351 HG*CR2*S4, . . .
 MAGNETIC STRUCTURE AND METAMAGNETISM OF HG*CR2*S4.
 HASTINGS(J.M.), CORLISS(L.M.).
 A2 J. PHYS. CHEM. SOLIDS 29, 9, (1968).

1352 HG*O,. . . . . .
 THE STRUCTURE OF MERCURIC OXIDE.
 ROTH(W.L.).
 A1 ACTA CRYST. VOL.9, 277, (1956).

1353 HG3*O*CL4, . . .
 A NEUTRON DIFFRACTION STRUCTURE DETERMINATION OF HG3*O*CL4.
 AURIVILLUS(K.).
 A1 ARK. KEMI VOL.22, 517, (1964).

1354 HG*(O*H)*N*O3, .
 CRYSTAL STRUCTURE OF HG*(O*H)*N*O3 BY NEUTRON DIFFRACTION.
 RIBAR(B.), MATKOVIC(B.), PRELESNIK(B.), HERAK(R.), GABELA(F.).
 (ABSTRACT ONLY).
 A1 ACTA CRYST. VOL.A28 PART S-4, 60, (1972).

1355 HG*S,. . . . . .
 NON-ADDITIVITY IN SCATTERING CROSS-SECTIONS OF SLOW NEUTRONS.
 KIMURA(M.).
 P PROC. IMP. ACAD. JAPAN VOL.15, 214, (1939).

1356 HO,. . . . . . .
 NEUTRON DIFFRACTION STUDY OF TB AND HO UNDER HIGH PRESSURE.
 UMEBAYASHI(H.), SHIRANE(G.), FRAZER(B.C.), DANIELS(W.B.).
 A2 PHYS. REV. VOL.165, 688, (1968).

SECTION 5 -STRUCTURE AND CROSS-SECTION DETERMINATIONS . . . .

1357 HO,. . . . . .A2. . . . . . .REF. SECTION 3, 65/KOEHLER,. . . . .

1358 HO,. . . . . .A2. . . . . . .PHYS. REV. 158, ... 450, (1967).
 *MAGNETIC STRUCTURES OF HOLMIUM.II. THE MAGNETIZATION PROCESS.
 *KOEHLER(W.C.), CABLE(J.W.), CHILD(H.R.), WILKINSON(M.K.), WOLLAN(E.O.).

1359 HO,. . . . . .A2. . . . . . .PHYS. REV. 151, ... 414, (1966).
 *MAGNETIC STRUCTURES OF HOLMIUM.I.THE VIRGIN STATE.
 *KOEHLER(W.C.), CABLE(J.W.), WILKINSON(M.K.), WOLLAN(E.O.).

1360 HO,. . . . . .A2,T. . . . .SOL. STATE COMM. VOL.12, ...1167, (1973).
 *ON THE DIFFRACTION OF NEUTRONS BY MAGNETIC SPIRAL STRUCTURES.
 *FELCHER(G.P.).

1361 HO,. . . . . .P. . . . . . . .J. PHYSIQUE TOME 34, ... 423, (1973).
 *DETERMINATION PAR DIFFRACTION DE NEUTRONS DU TERME DE L≠AMPLITUDE DE
 DIFFUSION DE HO(A=165) DEPENDANT DU SPIN NUCLEAIRE.
 *HERPIN(A.), MERIEL(P.).

1362 HO,. . . . . .A2,A4. . . . .DISS. ABSTR. B VOL.34. . . . . 3992, (1974).
 *A NEUTRON DIFFRACTION STUDY OF SINGLE CRYSTAL HOLMIUM TO 0.4 K.
 *LITTLE(G.R.).

1363 HO*AG2,. . . . . .A2. . . . . . .J. CHEM. PHYSICS VOL.51, ... 3882, (1969).
 *MAGNETIC STRUCTURE OF HO*AG2.
 *ATOJI(M.).

1364 HO-AG,. . . . . .A2,A4. . . .A.I.P. CONF. PROC. NO.10 PART 1, 669, (1973).
 *MAGNETIC PROPERTIES OF HO-AG.
 *NERESON(N.).

1365 HO*AL,. . . . . .A2. . . . . . .COMPTES RENDUS, SERIE B, 266,. . 994, (1968).
 *PROPRIETES MAGNETIQUES ET STRUCTURE MAGNETIQUE DU COMPOSE HO*AL.
 *BECLE(C.), LEMAIRE(R.), PAUTHENET(R.).

1366 HO-AL GARNETS,. . . . . .A1. . . . . . .ACTA CRYST. B25, ... 1853, (1969).
 *ETUDE PAR DIFFRACTION DE NEUTRONS A 0,31 DEG. K DE LA STRUCTURE ANTIFER-
 ROMAGNETIQUE DES GRENATS D≠ALUMINIUM-TERBIUM ET D≠ALUMINIUM-HOLMIUM.
 *HAMMANN(J.).

1367 HO*AL2,. . . . . .A2. . . . . . .SOL. STATE COMM. VOL.11 NO.5,. . 707, (1972).
 *ELASTIC NEUTRON DIFFRACTION STUDY OF TB*AL2 AND HO*AL2.
 *MILLHOUSE(A.H.), PURWINS(H.G.), WALKER(E.).

1368 HO*AL*03,. . . . . .A4. . . . . . .C. R. ACAD. SCI. B TOME 273, . . 619, (1971).
 *ETUDE PAR DIFFRACTION NEUTRONIQUE DES PEROVSKITES ER*AL*03, HO*AL*03,
 TB*RH*03, ETC.
 *SIVARDIERE(J.), QUEZEL-AMBRUNAZ(S.).

1369 HO*AU2,. . . . . .A2. . . . . . .J. CHEM. PHYS. VOL.57, ... 2402, (1972).
 *MAGNETIC STRUCTURE OF HO*AU2.
 *ATOJI(M.).

1370 HO2*C3 (PHASE ALPHA),. . . . .A2. . . . . . .J. CHEM. PHYSICS VOL.54, ... 3510, (1971).
 *MAGNETIC STRUCTURE OF ALPHA-HO2*C3.
 *ATOJI(M.), TSUNODA(Y.).

1371 HO2*C3,. . . . . .A2. . . . . . .J. APPL. PHYS. 42, ... 1630, (1971).
 *MAGNETIC STRUCTURES OF TB2*C3 AND HO2*C3.
 *ATOJI(M.).

1372 HO*C2,. . . . . .A2. . . . . . .PHYS. LETT. 23,. . . . . 208, (1966).
 *MAGNETIC STRUCTURES OF TB*C2 AND HO*C2.
 *ATOJI(M.).

1373 HO2*C,. . . . . .A2. . . . .BULL. SOC. FRANC. MIN. CRIST. 89 216, (1966).
 *NEUTRON DIFFRACTION STUDY OF HO2*C.
 *BACCHELLA(G.L.), LALLEMENT(R.), MERIEL(P.), PINOT(M.). (IN FRENCH).

1374 HO*C2,. . . . . .A1,A2. . . .J. CHEM. PHYS. VOL.46, ... 1891, (1967).
 *MAGNETIC AND CRYSTAL STRUCTURES OF CE*C2,PR*C2,ND*C2,TB*C2, AND HO*C2
 AT LOW TEMPERATURES.
 *ATOJI(M.).

1375 HO*C*03,. . . . . .A2,A4. . . .REF. SECTION 3, 66/BERTAUT,. . . . .

1376 HO*CO2,. . . . . .A2. . . . . . .J. APP. PHYS. 36,. . . . . 978, (1965).
 *MAGNETIC STRUCTURE OF RARE EARTH-COBALT (R*CO2) INTERMETALLIC COMPOUNDS.
 *MOON(R.M.), KOEHLER(W.C.), FARRELL(J.).

1377 HO-CO5,. . . . . .A2. . . . . . .C.R. ACAD. SCI. VOL.255, ... 896, (1962).
 *MAGNETIC STRUCTURE OF THE ALLOYS Y-CO5 AND HO-CO5.
 *JAMES(W.), LEMAIRE(R.), BERTAUT(E.F.). (IN FRENCH).

1378 HO-CO5,. . . . . .A2. . . . . . .J. APP. PHYS. 34,. . . . . 1333, (1963).
 *MAGNETIC STRUCTURE INVESTIGATIONS AT THE NUCLEAR CENTRE.
 *BALLESTRACCI(R.), BERTAUT(E.F.), COING-BOYAT(J.), DELAPALME(A.),
 JAMES(W.J.), LEMAIRE(R.).

1379 HO*CO2-HO*AL2,. . . . . . .A2. . . . . . .J. PHYS. CHEM. SOLIDS VOL.34,. . .1267, (1973).
 *STRUCTURAL, MAGNETIC AND NEUTRON DIFFRACTION STUDIES ON TB*FE2-TB*AL2,
 TB*CO2-TB*AL2 AND HO*CO2-HO*AL2.
 *OESTERREICHER(H.).

1380 HO*D2,. . . . . .A2. . . . . . .J. APP. PHYS. 34,. . . . . 1352, (1963).
 *MAGNETIC STRUCTURES OF TB*D2 AND HO*D2.
 *COX(D.E.), SHIRANE(G.), TAKEI(W.J.), WALLACE(W.E.).

1381 HO*D3,. . . . . .A4. . . . . . .COLL. INTER. N.126 (GRENOBLE), . 30, (1963).
 *THE STRUCTURE OF HO*D3.
 *MANSMANN(M), WALLACE(W.E.).

1382 HO-DY,. . . . . .A4. . . . .J. PHYSIQUE TOME 32 COL.1 VOL.1, 362, (1971).
 *NEUTRON DIFFRACTION STUDY OF HO(25)-DY(75) AND HO(50)-DY(50) IN AN
 EXTERNAL MAGNETIC FIELD.
 *KHAN(Q.H.), KOEHLER(W.C.).

1383 HO-(DY,ER),. . . . . .A4. . . . . . .INT. J. MAGN.(GB) VOL.2, ... 389, (1971).
 *SINGLE CRYSTAL NEUTRON DIFFRACTION STUDY OF BINARY RARE EARTH ALLOYS.
 *MILLHOUSE(A.H.), KOEHLER(W.C.).

1384 HO-ER,. . . . . .A2. . . . . . .J. APP. PHYS. 37, ... 1032, (1966).
 *NEUTRON-DIFFRACTION STUDIES OF MAGNETIC STRUCTURES IN HO-ER ALLOYS.
 *SHIRANE(G.), PICKART(S.J.).

SECTION   5 -STRUCTURE AND CROSS-SECTION DETERMINATIONS . . . .

1385 HO-FE, ...   A2 . . . . . J. PHYSIQUE TOME 32 COL.1 VOL.II   670, (1971).
                         NEUTRON DIFFRACTION STUDY OF THE HO-FE SYSTEM.
                         MOREAU(J.M.),   MICHEL(C.),   SIMMONS(M.),   O#KEEFE(T.J.),   JAMES(W.J.).

1386 HO*FE*03,. . . .   A2 . . . . . PHYS. REV. VOL.118,   58, (1960).
                         NEUTRON DIFFRACTION STUDY OF THE MAGNETIC PROPERTIES OF RARE-EARTH-IRON
                         PEROVSKITES.
                         KOEHLER(W.C.), WOLLAN(E.O.), WILKINSON(M.K.).

1387 HO*GA2,. . . . .   A2 . . . . . J. PHYSIQUE TOME 32 COL.1 VOL.II 1126, (1971).
                         PROPRIETES MAGNETIQUES ET STRUCTURES MAGNETIQUES DE QUELQUES COMPOSES
                         T*GA2.
                         BARBARA(B.), BECLE(C.), SIAUD(E.).

1388 HO-GARNET, . . .   A2 . . . . . COMP. REND. 251   1359, (1960)
                         MAGNETIC STRUCTURE OF THE HOLMIUM GARNET AT LOW TEMPERATURE (4.2 DEG. K)
                         HERPIN(A.), KOEHLER(W.C.), MERIEL(P.). (IN FRENCH).

1389 HO*IR2,. . . . .   A2 . . . . . PHYS. REV. 131   1518, (1963).
                         MAGNETIC STRUCTURE OF RARE-EARTH IRIDIUM COMPOUNDS R*IR2.
                         FELCHER(G.P.), KOEHLER(W.C.).

1390 HO*IR2,. . . . .   A2 . . . . . COLL. INTER. N.126 (GRENOBLE)   190, (1963).
                         STRUCTURE MAGNETIQUE DES COMPOSES R*IR2-TERRES RARES-IRIDIUM.
                         FELCHER(G.P.), KOEHLER(W.C.).

1391 HO*MN*03,. . . .   A2 . . . . . PHYS. LETT. VOL.9,   93, (1964).
                         A NOTE ON THE MAGNETIC STRUCTURES OF RARE EARTH MANGANESE OXIDES.
                         KOEHLER(W.C.),   YAKEL(H.L.),   WOLLAN(E.O.), CABLE(J.W.). (ED. NOTE: REF.
                         SECTION 3, 64/KOEHLER(1)).

1392 HO*N,. . . . . .   A2 . . . . . REF. SECTION 3, 63/CHILD . . . .

1393 HO*N,. . . . . .   A2 . . . . . J. APPL. PHYS. VOL.31, SUPPL.,   358, (1960).
                         NEUTRON DIFFRACTION INVESTIGATIONS OF THE MAGNETIC ORDERING IN RARE
                         EARTH NITRIDES.
                         WILKINSON(M.K.),   CHILD(H.R.), CABLE(J.W.), WOLLAN(E.O.), KOEHLER(W.C.).

1394 HO2*03,. . . . .   P . . . . . PHYS. REV. VOL.110,   37, (1958).
                         PARAMAGNETIC AND NUCLEAR SCATTERING CROSS SECTIONS OF HOLMIUM
                         SESQUIOXIDE.
                         KOEHLER(W.C.), WOLLAN(E.O.), WILKINSON(M.K.).

1395 HO2*02*S,. . . .   A2 . . . . . J. PHYS. CHEM. SOLIDS 29,   1001, (1968).
                         ETUDE PAR DIFFRACTION NEUTRONIQUE ET MESURES MAGNETIQUES DES OXYSULFURES
                         DE TERRES-RARES TC2*02*S.
                         BALLESTRACCI(R.), BERTAUT(E.F.), QUEREL(G.).

1396 HO2*02*SE, . . .   A2 . . . . . SOL. STATE COMM. VOL.10,   735, (1972).
                         PROPRIETES MAGNETIQUES DES OXYSELENIURES DE GD, TB, DY, HO, ER, TM ET
                         STRUCTURE MAGNETIQUE DE HO2*02*SE ET DE YB2*02*SE.
                         QUEZEL(G.), ROSSAT-MIGNOD(J.), LANG(H.Y.).

1397 HO*P,. . . . .   A2 . . . . . REF. SECTION 3, 63/CHILD, ...

1398 HO*SB, ...   A2 . . . . . REF. SECTION 3, 63/CHILD, ...

1399 HO-SC, ...   A2 . . . . . PHYS. REV. 174,   562, (1968).
                         MAGNETIC STRUCTURE PROPERTIES OF ALLOYS OF TB, HO AND ER WITH SC.
                         CHILD(H.R.), KOEHLER(W.C.).

1400 HO-TB, ...   A2 . . . . . J. CHEM. PHYSICS VOL.54,   1995, (1971).
                         MAGNETIC PROPERTIES OF TB-HO SINGLE CRYSTAL ALLOYS. III. NEUTRON DIF-
                         FRACTION STUDIES OF HO RICH ALLOYS.
                         SPEDDING(F.H.), ITO(Y.), JORDAN(R.G.), CROAT(J.).

1401 HO-TB, ...   A2 . . . . . J. CHEM. PHYSICS VOL.53,   1455, (1970).
                         MAGNETIC PROPERTIES OF TB-HO SINGLE-CRYSTAL ALLOYS. II. NEUTRON DIFFRAC-
                         TION STUDIES.
                         SPEDDING(F.H.), ITO(Y.), JORDAN(R.G.).

1402 HO-TH, ...   A2 . . . . . J. APP. PHYS. 39,   1329, (1968).
                         MAGNETIC PROPERTIES OF FCC RARE EARTH-THORIUM ALLOYS.
                         CHILD(H.R.), KOEHLER(W.C.), MILLHOUSE(A.H.).

1403 HO*V*04, ...   A1 . . . . . J. SOLID STATE CHEM. VOL.5,   11, (1972).
                         REFINEMENT OF THE CRYSTAL STRUCTURE OF SOME RARE EARTH VANADATES R*V*04
                         (R=DY,TB,HO,YB)
                         FUESS(H.), KALLEL(A.).

1404 HO*ZN2,. . . . .   A2 . . . . . J. CHEM. PHYS. VOL.57,   2156, (1972).
                         MAGNETIC STRUCTURE OF HO*ZN2.
                         DEBRAY(D.), SOUGI(M.).

1405 HO*ZN2,. . . . .   A2 . . . . . PHYS. STAT. SOLIDI A VOL.18 NO.1   227, (1973).
                         MAGNETOCRYSTALLINE ANISOTROPY AND EFFECT OF APPLIED FIELD ON THE
                         SINUSOIDAL MAGNETIC STRUCTURES OF TB*ZN2 AND HO*ZN2.
                         DEBRAY(D.).

1406 I (ISO. A=127),.   P . . . . . ACTA CRYST. VOL.A28,   663, (1972).
                         COHERENT NEUTRON SCATTERING AMPLITUDES OF BR AND I (ISO. A=127).
                         ATOJI(M.).

1407 I, ...   P . . . . . PHYS. REV. VOL.71,   174, (1947).
                         SLOW NEUTRON VELOCITY SPECTROMETER STUDIES. III. I, OS, CO, TL, CB, GE.
                         WU(C.S.), RAINWATER(L.J.), HAVENS-JR(W.W.).

1408 IN,. . . . . . .   P . . . . . PHYS. REV. 131,   2098, (1963).
                         NEUTRON COHERENT-SCATTERING AMPLITUDES OF GA, IN, AS AND SB.
                         ARNOLD(G.P.), NERESON(N.G.).

1409 IN,IN2*03, . . .   P,A1 . . . . . Z. KRIST. VOL.118,   473, (1963).
                         MESSUNG DER KOHAERENTEN STREUAMPLITUDE VON IN FUER THERMISCHE NEUTRONEN
                         UND BESTIMMUNG DER STRUKTUR PARAMETER VON IN2*03.
                         BETZL(M.), HASE(W.), KLEINSTUCK(K.), TOBISCH(J.).

1410 IN*0*0*H,. . . .   A1 . . . . . ACTA CHEM. SCAND. VOL.24,   1662, (1970).
                         NEUTRON AND X-RAY CRYSTALLOGRAPHIC STUDIES ON INDIUM OXIDE HYDROXIDE.
                         LEHMANN(M.S.), LARSEN(F.K.), POULSEN(F.R.), CHRISTENSEN(A.N.),
                         RASMUSSEN(S.E.).

1411 IR,. . . . . . .   P . . . . . PHYS. REV. VOL.72,   109, (1947).
                         A BENT CRYSTAL NEUTRON SPECTROMETER AND ITS APPLICATION TO NEUTRON
                         CROSS-SECTION MEASUREMENTS.
                         SAWYER(R.B.), WOLLAN(E.O.), BERNSTEIN(S.), PETERSON(K.C.).

SECTION 5 -STRUCTURE AND CROSS-SECTION DETERMINATIONS . . . .

1412 IR-MN, ... A2 . . . . ACTA CHEM. SCAND. VOL.22, ... 3039, (1968).
&EQUIATOMIC TRANSITION METAL ALLOYS OF MANGANESE VII: A NEUTRON
DIFFRACTION STUDY OF MAGNETIC ORDERING IN THE IR-MN PHASE.
&SELTE(K.), KJEKSHUS(A.), ANDRESEN(A.F.), PEARSON(W.B.).

1413 K, ... A4 . . . . PHYS. REV. 186, ... 705, (1969).
&NEUTRON-DIFFRACTION MEASUREMENT OF THE LATTICE PARAMETER OF POTASSIUM
METAL AT 5.2 K.
&WERNER(S.A.), GURMEN(E.), ARROTT(A.).

1414 K, ... T . . . . UKR. FIZ. ZH.(USSR) VOL.16, . . . 1735, (1971).
&ANALYSIS OF ELASTIC SCATTERING OF NEUTRONS BY K, CO AND CE NUCLEI USING
A LOCAL OPTICAL NUCLEAR MODEL.
&KASHUBA(I.E.), KOSTYUK(T.A.), KOTISHEVSKAYA(E.YA.).

1415 K, ... P . . . . Z. KRISTALLOGR. VOL.135, ... 316, (1972).
&THE NEUTRON SCATTERING AMPLITUDES OF POTASSIUM, STRONTIUM AND BARIUM.
&COOPER(M.J.), ROUSE(K.D.).

1416 K, ... A4,P . . . . . SOL. STATE COMM. VOL.7, ... 1681, (1969).
&A NEUTRON DIFFRACTION SEARCH FOR A CHARGE DENSITY WAVE IN POTASSIUM MET.
&ATOJI(M.), WERNER(S.A.).

1417 K, ... P . . . . Z. KRIST. VOL.126, ... 460, (1968).
&THE NEUTRON-SCATTERING AMPLITUDE OF POTASSIUM.
&BACON(G.E.), PLANT(J.S.).

1418 K*(AU*(C*N)4)*H2*O, ... A1 . . . . ACTA CRYST. VOL.B26 , ... 422, (1970).
&STRUCTURE CRISTALLINE DE L#AURICYANURE DE POTASSIUM MONOHYDRATE PAR LA
DIFFRACTION DES NEUTRONS.
&BERTINOTTI(C.), BERTINOTTI(A.).

1419 K*B*H4, ... A1 . . . . DISS. ABS. XXV . . . . 5588, (1965).
&THE CRYSTAL STRUCTURES OF SILVER PARAPERIODATE AG2*H3*I*06 AND OF
POTASSIUM BOROHYDRIDE (K*B*H4): A NEUTRON AND X-RAY DIFFRACTION STUDY.
&PETERSON(E.R.).

1420 (K*C4*H3*04)2*(C4*H4*04) . . A1 . . . . Z. KRIST. VOL.137, ... 173, (1973).
&THE CRYSTAL STRUCTURE OF DIPOTASSIUM TETRAHYDROGEN TRIFUMARATE:
2K*C4*H3*04.C4*H4*04, A NEUTRON-DIFFRACTION STUDY.
&GUPTA(M.P.), PRASAD(N.).

1421 K*C*N, ... A1 . . . . J. CHEM. PHYSICS VOL.54, ... 3514, (1971).
&NEUTRON STRUCTURE ANALYSIS OF CUBIC CA*C2 AND K*C*N.
&ATOJI(M.).

1422 K*C*N, ... A1 . . . . ACTA CRYST. VOL.14, ... 1018, (1961).
&NEUTRON DIFFRACTION INVESTIGATION OF K*C*N.
&ELLIOTT(N.), HASTINGS(J.).

1423 K*C*N, ... A4 . . . . J. CHEM. PHYS. VOL.56, ... 3697, (1972).
&SINGLE CRYSTAL NEUTRON DIFFRACTION STUDY OF POTASSIUM CYANIDE.
&PRICE(D.L.), ROWE(J.M.), RUSH(J.J.), PRINCE(E.), HINKS(D.G.), SUSMAN(S.)

1424 K4*(C*N)8*MO.2H2*O, ... A1 . . . . ACTA CRYST. VOL.A28 PART S-4, . . 93, (1972).
&NEUTRON DIFFRACTION STUDY OF THE OCTOCYANO MOLYBDATE (IV) ION IN
K4*(C*N)8*MO.2H2*O.
&BARTL(H.). (ABSTRACT ONLY).

1425 K2*C2*04*(H2*O), ... A1 . . . . J. CHEM. PHYSICS VOL.41, ... 3616, (1964).
&NEUTRON-DIFFRACTION STUDY OF THE STRUCTURE OF POTASSIUM OXALATE MONOHY-
DRATE: LONEPAIR COORDINATION OF THE H-BONDED WATER MOLECULE IN CRYSTALS.
&CHIDAMBARAM(R.), SEQUEIRA(A.), SIKKA(S.K.).

1426 K*CL, ... A4 . . . . ACTA CRYST. VOL.A29, ... 514, (1973).
&A NEUTRON DIFFRACTION STUDY OF K*CL.
&COOPER(M.J.), ROUSE(K.D.).

1427 K*CL, ... A3 . . . . ANN. N.Y. ACAD. SCI. VOL.79, . . 762, (1960).
&X-RAY AND NEUTRON DIFFRACTION STUDIES OF MOLTEN ALKALI HALIDES.
&LEVY(H.A.), AGRON(P.A.), BREDIG(M.A.), DANFORD(M.D.).

1428 K3*CO*(C*N)6, ... A1 . . . . ACTA CRYST. VOL.12, ... 674, (1959).
&A NEUTRON-DIFFRACTION STUDY OF POTASSIUM COBALTICYANIDE.
&CURRY(N.A.), RUNCIMAN(W.A.).

1429 K*(CO,CR,CU)*F3, ... A2 . . . . ACTA CRYST. VOL.14, ... 19, (1961).
&MAGNETIC STRUCTURES OF 3D TRANSITION METAL DOUBLE FLUORIDES, K*(ME)*F3.
&SCATTURIN(V.), CORLISS(L.), ELLIOTT(N.), HASTINGS(J.).

1430 K2*CO*F4, ... A2 . . . . SOL. STATE COMM. VOL.14, ... 529, (1974).
&NEUTRON SCATTERING STUDY OF TWO-DIMENSIONAL ISING NATURE OF K2*CO*F4.
&IKEDA(H.), HIRAKAWA(K.).

1431 K2*CO*F4, ... A2 . . . . J. PHYS. CHEM. SOLIDS VOL.35, . . 785, (1974).
&CRITICAL BEHAVIOR OF THE TWO-DIMENSIONAL ISING ANTIFERROMAGNETS K2*CO*F4
AND RB2*CO*F4.
&SAMUELSEN(E.J.).

1432 K*CR*S2, ... A2 . . . . J. SOLID STATE CHEM. VOL.6, . . . 384, (1973).
&THE MAGNETIC STRUCTURE OF K*CR*S2.
&VAN-LAAR(B.), ENGELSMAN(F.M.R.).

1433 K2*CU*CL4.2H2*O, ... A1 . . . . ACTA CRYST. VOL.B26 . . . . . . 827, (1970).
&NEUTRON DIFFRACTION REFINEMENT OF THE CRYSTAL STRUCTURE OF POTASSIUM
COPPER CHLORIDE DIHYDRATE, K2*CU*CL4*2H2*O.
&CHIDAMBARAM(R.), NAVARRO(Q.O.), GARCIA(A.), LINGGOATMODJO(K.),
LIN-SHI-CHIEN, IL-HWAN-SUH, SEQUEIRA(A.), SRIKANTA(S.).

1434 K*CU*F3, ... A2 . . . . PHYS. REV. VOL.188, ... 919, (1969).
&NEUTRON-DIFFRACTION DETERMINATION OF THE ANTIFERROMAGNETIC STRUCTURE OF
K*CU*F3.
&HUTCHINGS(M.T.), SAMUELSEN(E.J.), SHIRANE(G.), HIRAKAWA(K.).

1435 K2*CU*F4, ... A2 . . . . J. PHYS. SOC. JAPAN VOL.33, . . 1483, (1972).
&NEUTRON SCATTERING EXPERIMENTS ON THE TWO-DIMENSIONAL FERROMAGNET
K2*CU*F4.
&HIRAKAWA(K.), IKEDA(H.).

1436 K*CU*F3, ... A1,A2 . . . . . J. PHYS. SOC. JAPAN VOL.35, . . 722, (1973).
&NEUTRON DIFFRACTION STUDY IN ONE-DIMENSIONAL ANTIFERROMAGNET K*CU*F3.
&IKEDA(H.), HIRAKAWA(K.).

1437 K2*CU*H12*014*S2, ... A1 . . . . CRYST. STRUCT. COMM. VOL.1 N.3, . 185, (1972).
&POTASSIUM HEXA-AQUACOPPER(II) SULPHATE, . . .
&ROBINSON(D.J.), KENNARD(C.H.L.).

SECTION   5 -STRUCTURE AND CROSS-SECTION DETERMINATIONS . . . .

1438   K*D2*P*O4, ...A4. . . . . .SOLID STATE COMM. VOL. 5, ... 591, (1967).
~EFFECT OF HYDROSTATIC PRESSURE ON THE FERROELECTRIC CURIE TEMPERATURES
OF K*H2*P*O4 AND K*D2*P*O4.
~UMEBAYASHI(H.), FRAZER(B.C.), SHIRANE(G.), DANIELS(W.B.).

1439   K4*FE*(C*N)6*3(D2*O) . . . . .A1. . . . . . .ACTA CRYST. VOL.A26, ... 559, (1970).
~A NEUTRON DIFFRACTION STUDY OF FERROELECTRIC, KFCT, K4*FE*(C*N)6*3(D2*O)
ABOVE THE CURIE TEMPERATURE.
~TAYLOR(J.C.), MUELLER(M.H.), HITTERMAN(R.L.).

1440   K*FE*CL3, ...A2. . . . . . .PHYS. REV. B VOL.9, ... 1071, (1974).
~NEUTRON-DIFFRACTION STUDY OF THE MAGNETIC STRUCTURE OF K*FE*CL3.
~GUREWITZ(E.), MAKOVSKY(J.), SHAKED(H.).

1441   K*FE*F4, ...A2. . . . . . .CONF. ON LOW-TEMP. PHYS.(ABSTR.) . . 31, (1972).
~NEUTRON AND MOSSBAUER STUDIES OF THE PLANAR ANTIFERROMAGNET K*FE*F4 IN
THE CRITICAL REGION.
~HEGER(G.), GELLER(R.). (ED. NOTE: AVAIL. FROM EUROPEAN PHYS. SOC.,
GENEVA, SWITZERLAND).

1442   K*(FE,MN,NI)*F3, ...A2. . . . . .ACTA CRYST. VOL.14, ... 19, (1961).
~MAGNETIC STRUCTURES OF 3D TRANSITION METAL DOUBLE FLUORIDES, K*(ME)*F3.
~SCATTURIN(V.), CORLISS(L.), ELLIOTT(N.), HASTINGS(J.).

1443   K*H*(C6*H5*C*H2*C*O*O)2, ...A1. . . . . .ACTA CRYST. VOL.13, ... 717, (1960).
~A NEUTRON-DIFFRACTION STUDY OF POTASSIUM HYDROGEN BIS-PHENYLACETATE:
PART 2.
~BACON(G.E.), CURRY(N.A.).

1444   K*H*(C6*H5*C*H2*C*O*O)2, ...A1. . . . . .ACTA CRYST. VOL.10, ... 524, (1957).
~A NEUTRON-DIFFRACTION STUDY OF POTASSIUM HYDROGEN BIS-PHENYLACETATE.
~BACON(G.E.), CURRY(N.A.).

1445   K*H*C*O3, ...A4. . . . . .COLL. INTER. N.126 (GRENOBLE) . . 60, (1963).
~LOCALISATION D#HYDROGENE ET RAFFINEMENT DES POSITIONS ATOMIQUES DANS
K*H*C*O3.
~HERPIN(P.), MERIEL(P.).

1446   K*H3*C4*O8.2H2*O, ...A1. . . . . .J. CHEM. SOC. A (1967) . . . . . 1862, (1967).
~THE CRYSTAL STRUCTURES OF THE ACID SALTS OF SOME DIBASIC ACIDS. PART 1.
A NEUTRON DIFFRACTION STUDY OF AMMONIUM (AND POTASSIUM) TETROXALATE.
~CURRIE(M.), SPEAKMAN(J.C.), CURRY(N.A.).

1447   K-H-CHLOROMALEATE, ...A1. . . . . .ACTA CRYST. 19, ... 260, (1965).
~A CENTERED HYDROGEN BOND IN POTASSIUM CHLOROMALEATE: A NEUTRON
DIFFRACTION DETERMINATION.
~ELLISON(R.D.), LEVY(H.A.).

1448   K*H-DIASPIRINATE, ...A1. . . . . .J. MOLEC. STRUC. VOL.1, ... 283, (1968).
~STRUCTURE AND DYNAMICS IN HYDROGEN BONDING SYSTEMS. I. A NEUTRON DIF-
FRACTION STUDY OF POTASSIUM HYDROGEN DIASPIRINATE (BISACETYLSALICYLATE).
~SEQUEIRA(A.), BERKEBILE(C.A.), HAMILTON(W.C.).

1449   K*H*F2, ...A1. . . . . .J. CHEM. PHYSICS VOL.40, ... 402, (1964).
~REFINEMENT OF PETERSON AND LEVY≠S NEUTRON DIFFRACTION DATA ON K*H*F2.
~IBERS(J.A.).

1450   K*H*F2, ...A1. . . . . .J. CHEM. PHYS. VOL.20, ... 704, (1952).
~A SINGLE CRYSTAL NEUTRON DIFFRACTION DETERMINATION OF THE HYDROGEN
POSITION IN POTASSIUM BIFLUORIDE.
~PETERSON(S.W.), LEVY(H.A.).

1451   K*H*F2, ...A4. . . . . .COLL. INTER. N.126 (GRENOBLE), . . 50, (1963).
~THE NATURE OF THE HYDROGEN BOND IN THE BIFLUORIDE ION.
~IBERS(J.A.).

1452   K*H*(MALONATE), ...A1. . . . . .J. CHEM. SOC. A (1970) . . . . . 1923, (1970).
~THE CRYSTAL STRUCTURES OF THE ACID SALTS OF SOME DIBASIC ACIDS. III.
POTASSIUM HYDROGEN MALONATE: A NEUTRON DIFFRACTION STUDY.
~CURRIE(M.), SPEAKMAN(J.C.).

1453   K*(H3*O)2*(B5*O10*H2), ...A1. . . . . .CAN. J. PHYS. 48, ... 1091, (1970).
~HYDROGEN POSITIONS IN POTASSIUM PENTABORATE TETRAHYDRATE AS DETERMINED
BY NEUTRON DIFFRACTION.
~ASHMORE(J.P.), PETCH(H.E.).

1454   K*H*O*(C*H2*C*O*O)2, ...A1. . . . . .ACTA CRYST. VOL.B29, ... 2751, (1973).
~A NEUTRON DIFFRACTION STUDY OF POTASSIUM AND RUBIDIUM HYDROGEN OXYDI-
ACETATE. THE DYNAMICS OF THEIR HYDROGEN BONDS.
~ALBERTSSON(J.), GRENTHE(I.).

1455   K*H2*P*O4, ...A4. . . . . .SOLID STATE COMM. VOL. 5, ... 591, (1967).
~EFFECT OF HYDROSTATIC PRESSURE ON THE FERROELECTRIC CURIE TEMPERATURES
OF K*H2*P*O4 AND K*D2*P*O4.
~UMEBAYASHI(H.), FRAZER(B.C.), SHIRANE(G.), DANIELS(W.B.).

1456   K*H2*P*O4, ...A4. . . . . .PHYS. STAT. SOL. VOL.42, ... 207, (1970).
~NEUTRON MEASUREMENTS ON THE HYDROGEN BOND POTENTIAL IN PARAELECTRIC KDP.
~GRIMM(H.), STILLER(H.), PLESSER(T.).

1457   K*H2*P*O4, ...A1. . . . . .SOV. PHYS. SOL. STATE VOL.10, ... 675, (1968).
~SCATTERING OF SLOW NEUTRONS BY THE PROTON IN HYDROGEN BONDS USING THE
DOUBLE-POTENTIAL-WELL MODEL.
~SHAMENKOVICH(S.S.).

1458   K*H2*P*O4, ...A4. . . . . .1ST EUR. CONF./CONDENSED MATTER, . . 60, (1971).
~NEUTRON MEASUREMENTS ON THE FERROELECTRIC PHASE TRANSFORMATION IN
K*H2*P*O4.
~ARSIC-ESKINJA(M.), GRIMM(H.), STILLER(H.). (ED. NOTE: PUBL. EUROPEAN
PHYS. SOC. 1971, GENEVA, SWITZ.).

1459   K*H2*P*O4, ...P. . . . . .REPORT JUL-696-FF, ... 1, (1970).
~NEUTRON SCATTERING EXPERIMENT ON THE PROTON DENSITY DISTRIBUTION IN
PARAELECTRIC KDP.
~GRIMM(H.). (ED. NOTE: AVAIL. KERNFORSCHUNGSANLAGE JULICH, GERMANY).

1460   K*H2*P*O4-K*D2*P*O4, ...A1,A4. . . . .SOL. STATE COMM. VOL.11, ... 1261, (1972).
~STRUCTURAL STUDIES OF THE SYSTEM K*H2*P*O4-K*D2*P*O4.
~NELMES(R.J.), EIRIKSSON(V.R.), ROUSE(K.D.).

1461   K*H2*P*O4, ...A4. . . . . .IAEA SYMP. GRENOBLE, ... 825, (1972).
~NEUTRON MEASUREMENTS ON THE FERROELECTRIC PHASE TRANSFORMATION
IN K*H2*P*O4.
~ARSIC-ESKINJA(M.), GRIMM(H.), STILLER(H.).

SECTION 5 -STRUCTURE AND CROSS-SECTION DETERMINATIONS . . . .

1462 K*H2*P*04, . . . A4 . . . . . . SOL. STATE COMM. 5, . . . 41, (1967).
⇔ANISOTROPY AND PHASE TRANSITION EFFECTS IN NEUTRON SCATTERING FROM
⇔K*H2*P*04.
⇔IMRY(Y.), PELAH(I.), WIENER(E.), ZAFRIR(H.).

1463 K*H2*P*04, . . . A3 . . . . . . ACTA CRYST. A24, . . . 237, (1968).
⇔ANOMALE INKOHÄRENTE STREUUNG THERMISCHER NEUTRONEN BEI BILDUNG STEHENDER
⇔NEUTRONENWELLE/NAHEZU IDEALEN KRISTALLEN VON KALIUMDIHYDROGENPHOSPHAT.
⇔SIPPEL(D.), EICHHORN(F.).

1464 K*H2*P*04, K*D2*P*04 . . . A4 . . . . . . ACTA CRYST. VOL.A28 PART S-4, . . . 181, (1972).
⇔NEUTRON DIFFRACTION STUDIES OF FERROELECTRIC PHASE TRANSITIONS IN KDP
⇔AND DKDP.
⇔SHIBUYA(I.), IWATA(Y.), KOYANO(N.). (ABSTRACT ONLY).

1465 K*H5*(P*04)2,. . . A1 . . . . . . REV. CHIM. MINER. VOL.9, . . . 825, (1972).
⇔REFINEMENT OF THE K*H5*(P*04)2 CRYSTAL STRUCTURE BY NEUTRON DIFFRACTION.
⇔PHILIPPOT(E.), RICHARD(P.), ROUDAULT(R.), MAURIN(M.). (IN FRENCH).

1466 K*H2*P*04, . . . A1 . . . . . . PROC. ROY. SOC. A VOL.230, . . . 359, (1955).
⇔A NEUTRON-DIFFRACTION STUDY OF THE FERROELECTRIC TRANSITION OF
⇔POTASSIUM DIHYDROGEN PHOSPHATE.
⇔BACON(G.E.), PEASE(R.S.).

1467 K*H2*P*04, . . . A1 . . . . . . PROC. ROY. SOC. A VOL.220, . . . 397, (1953).
⇔A NEUTRON DIFFRACTION STUDY OF POTASSIUM DIHYDROGEN PHOSPHATE BY
⇔FOURIER SYNTHESIS.
⇔BACON(G.E.), PEASE(R.S.).

1468 K*H2*P*04, . . . A1 . . . . . . PHYS. REV. VOL.93, . . . 1120, (1954).
⇔NEUTRON DIFFRACTION STUDY OF THE FERROELECTRIC MODIFICATION OF POTASSIUM
⇔DIHYDROGEN PHOSPHATE.
⇔LEVY(H.A.), PETERSON(S.W.).

1469 K*H2*P*04, . . . P . . . . . . J. PHYSIQUE TOME 33 COL.2, . . . 83, (1972).
⇔INCOHERENT NEUTRON SCATTERING FROM HYDROGEN BOND IN KDP AND ADP.
⇔ANTONINI(M.), SOSNOWSKA(I.), VADACCHINO(M.).

1470 K*H2*P*04, . . . C1 . . . . . . PHYS. REV. VOL.172, . . . 576, (1968).
⇔DISAPPEARANCE OF A VIBRATIONAL MODE IN THE FERROELECTRIC PHASE
⇔TRANSITION OF K*H2*P*04.
⇔SCHENK(C.), WIENER(E.), WECKERMANN(B.), KLEY(W.).

1471 K*H2*P*04, . . . A4 . . . . . . ACTA CRYST. A25, . . . 514, (1969).
⇔ANOMALIES IN NEUTRON DIFFRACTION INTENSITIES OF K*H2*P*04.
⇔SCHENK(C.), WECKERMANN(B.).

1472 K*H2*P*04, . . . A1 . . . . . . J. CHEM. PHYS. VOL.21, . . . 2084, (1953).
⇔NEUTRON DIFFRACTION STUDY OF TETRAGONAL POTASSIUM DIHYDROGEN PHOSPHATE.
⇔PETERSON(S.W.), LEVY(H.A.).

1473 K*H3*(SE*03)2, K*D3*(SE*03)2, . . . A4 . . . . . . ACTA CHEM. SCAND. VOL.25, . . . 3859, (1971).
⇔THE HYDROGEN BOND SYSTEM IN . . . K*H3*(SE*03)2, AND IN . . . K*D3*(SE*03)2,
⇔AS DETERMINED BY NEUTRON DIFFRACTION.
⇔LEHMANN(M.S.), LARSEN(F.K.).

1474 K*H3*(SE*03)2, . . . A1 . . . . . . ACTA CRYST. VOL.B28, . . . 3164, (1972).
⇔A NEUTRON DIFFRACTION STUDY OF POTASSIUM TRIHYDROGEN SELENITE,
⇔K*H3*(SE*03)2.
⇔PRELESNIK(B.), HERAK(R.).

1475 K*I*03,. . . . . A1 . . . . . . SOV. PHYS. CRYST. VOL.7, ... 499, (1962).
⇔STRUCTURE OF K*I*03 AT ROOM TEMPERATURE.
⇔OZEROV(R.P.), RANNEV(N.V.), PAKHOMOV(V.I.), REZ(I.S.), ZHDANOV(G.S.).

1476 K2*IR*CL6, . . . A2 . . . . . . J. PHYS. CHEM. SOLIDS 29, . . . 881, (1968).
⇔NEUTRON DIFFRACTION STUDY OF MAGNETIC ORDERING IN K2*IR*CL6, K2*RE*BR6
⇔AND K2*RE*CL6.
⇔MINKIEWICZ(V.J.), SHIRANE(G.), FRAZER(B.C.), WHEELER(R.G.),
⇔DORAIN(P.B.).

1477 K2*IR*CL6, . . . A2 . . . . . . PROC. PHYS. SOC. 91, ... 928, (1967).
⇔THE MAGNETIC STRUCTURE OF K2*IR*CL6.
⇔HUTCHINGS(M.T.), WINDSOR(C.G.).

1478 K*MG3*(AL*SI3*03*010)*F2, . . . E . . . . . . INSTRUM. EXP. TECH. VOL.16 NO.2, 399, (1973).
⇔FLUORPHLOGOPITE SINGLE CRYSTAL AS A LONG-WAVE NEUTRON MONOCHROMATOR.
⇔ZELENYUK(F.M.), ZAITSEV(K.N.)... ET AL.

1479 K*MN*F3, ... A1 . . . . . . SOL. STATE COMM. VOL.8, . . . 1941, (1970).
⇔NEUTRON SCATTERING STUDY OF THE LATTICE DYNAMICAL PHASE TRANSITIONS IN
⇔K*MN*F3.
⇔SHIRANE(G.), MINKIEWICZ(V.J.), LINZ(A.).

1480 K2*MN*F4,. . . . D2 . . . . . . J. PHYS. SOC. JAPAN VOL.33, . . . 393, (1972).
⇔ELASTIC AND QUASI-ELASTIC NEUTRON SCATTERING FROM K2*MN*F4 NEAR THE
⇔CRITICAL POINT.
⇔IKEDA(H.), HIRAKAWA(K.).

1481 K2*MN*F4,. . . . A2 . . . . . . PHYS. LETT. VOL.26A NO.11, . . . 526, (1968).
⇔ZERO-POINT SPIN REDUCTION IN K2*MN*F4.
⇔LOOPSTRA(B.O.), VAN-LAAR(B.), BREED(D.J.).

1482 K2*MN*F4,. . . . A2 . . . . . . ACTA CRYST. VOL.A28 PART S-4, . . . 175, (1972).
⇔NEUTRON DIFFRACTION STUDY IN A TYPICAL TWO-DIMENSIONAL ANTIFERROMAGNET
⇔K2*MN*F4.
⇔IKEDA(H.), HIRAKAWA(K.). (ABSTRACT ONLY).

1483 K2*MN*F4,. . . . A2,D1. . . . PHYS. REV. B VOL.8, ... 304, (1973).
⇔SPIN WAVES AND MAGNETIC ORDERING IN K2*MN*F4.
⇔BIRGENEAU(R.J.), GUGGENHEIM(H.J.), SHIRANE(G.).

1484 K2*MN*F4,. . . . A1 . . . . . . J. CRYST. GROWTH VOL.21, . . . 82, (1974).
⇔GROWTH OF LARGE K2*MN*F4 SINGLE CRYSTALS AND CHARACTERIZATION BY X-RAY
⇔AND NEUTRON DIFFRACTION INVESTIGATIONS.
⇔BITTERMANN(K.), HEGER(G.).

1485 K2*MN*(S*04)2.4(H2*O) . . . A1 . . . . . . ACTA CRYST. VOL.B24, . . . 1176, (1968).
⇔NEUTRON DIFFRACTION STUDY OF THE SPACE GROUP AND STRUCTURE OF MANGANESE-
⇔LEONITE, K2*MN*(S*04)2*4(H2*O).
⇔SRIKANTA(S.), SEQUEIRA(A.), CHIDAMBARAM(R.).

1486 K*N*03,. . . . . A1 . . . . . . J. PHYS. C VOL.6, . . . 201, (1973).
⇔A NEUTRON DIFFRACTION DETERMINATION OF THE CRYSTAL STRUCTURE OF ALPHA-
⇔PHASE POTASSIUM NITRATE AT 25 DEGREES C AND 100 DEGREES C.
⇔NIMMO(J.K.), LUCAS(B.W.).

SECTION   5 -STRUCTURE AND CROSS-SECTION DETERMINATIONS . . . .

1487 K2*NB*F7,. . . .   *REFINEMENT OF THE STRUCTURE OF POTASSIUM HEPTAFLUORONIOBATE, K2*NB*F7
                          FROM NEUTRON-DIFFRACTION DATA.   ACTA CRYST. 20,. 220, (1966),
                          *BROWN(G.M.), WALKER(L.A.).

1488 K2*NB*F7,. . . .   *RECENT CRYSTAL STRUCTURE DETERMINATIONS BY NEUTRON DIFFRACTION AT
                          OAK RIDGE.   A1. . . . . . . COLL. INTER. N.126 (GRENOBLE), .   45, (1963).
                          *BROWN(G.M.), LEVY(H.A.).

1489 K*NB*O3, ...   *SOFT MODES AND THE STRUCTURE, SPONTANEOUS POLARIZATION AND CURIE
                          CONSTANTS OF PEROVSKITE FERROELECTRICS: TETRAGONAL POTASSIUM NIOBATE.
                          A1. . . . . J. PHYS. C VOL.6 NO.6,. 1074, (1973),
                          *HEWAT(A.W.).

1490 K*NB*O3, ...   *CUBIC-TETRAGONAL-ORTHORHOMBIC-RHOMBOHEDRAL FERROELECTRIC TRANSITIONS IN
                          PEROVSKITE POTASSIUM NIOBATE: NEUTRON POWDER PROFILE REFINEMENT OF/STRUC
                          A1. . . . . J. PHYS. C VOL.6,. . . . . . 2559, (1973).
                          *HEWAT(A.W.).

1491 K*(NB(0.37)-TA(0.63))*O3,. . . .   *NEUTRON SCATTERING AND DIELECTRIC PROPERTIES OF PEROVSKITE-TYPE CRYSTALS
                          A4. . . . . PHYS. STAT. SOL. 30,. . . . . . K157, (1968).
                          *COCHRAN(W.).

1492 K2*NI*F4,. . . .   *L≠INTERACTION MAGNETIQUE DANS LES STRUCTURES DE K2*NI*F4.
                          A2. . . . . COLL. INTER. N.126 (GRENOBLE), .   154, (1963).
                          *LEGRAND(E.), VERSCHUEREN(M.).

1493 K2*NI*F4,. . . .   *NEUTRON SCATTERING FROM K2*NI*F4: A TWO DIMENSIONAL HEISENBERG
                          ANTIFERROMAGNET.   A2. . . . . . PHYS. REV. LETT. 22,. . . . . . 720, (1969).
                          *BIRGENEAU(R.J.), GUGGENHEIM(H.J.), SHIRANE(G.).

1494 K2*NI*F4,. . . .   *NEUTRON DIFFRACTION INVESTIGATION OF ANTIFERROMAGNETIC K2*NI*F4.
                          A2. . . . . . PHYS. STAT. SOL. 2,. . . . . 317, (1962).
                          *LEGRAND(E.), PLUMIER(R.).

1495 K2*NI*F4,. . . .   *MAGNETIC STRUCTURE OF K2*NI*F4.
                          A2. . . . . . J. PHYS. RADIUM VOL.23,. . . . . 474, (1962).
                          *PLUMIER(R.), LEGRAND(E.). (IN FRENCH).

1496 K2*NI*F4,. . . .   *NEUTRON DIFFRACTION STUDY OF MAGNETIC STACKING FAULTS IN ANTIFERRO-
                          MAGNETIC K2*NI*F4.   A2. . . . . . J. APPL. PHYS. VOL.35,. . . . . 950, (1964).
                          *PLUMIER(R.).

1497 K2*NI*F4,. . . .   *ETUDE PAR DIFFRACTION DES NEUTRONS DU DESORDRE MAGNETIQUE D≠EMPILEMENT
                          DANS L≠ANTIFERROMAGNETIQUE K2*NI*F4.   A2. . . . . . J. PHYSIQUE TOME 24,. . . . . 741, (1963).
                          *PLUMIER(R.).

1498 K*NI*F3-(NI++),.   *NEUTRON DIFFRACTION MEASUREMENT OF COVALENCY AND SPIN DEVIATION: A TEST
                          OF THEORY AND EXPERIMENTAL METHOD USING (NI++)-K*NI*F3.
                          A4. . . . . . J. OF PHYS.-C-, SER.2, VOL.3,. . . 1303, (1970).
                          *HUTCHINGS(M.T.), GUGGENHEIM(H.J.).

1499 K*NI*F3-(NI++),.   *NEUTRON DIFFRACTION MEASUREMENT OF THE EFFECTIVE MAGNETIC MOMENT OF NI++
                          IN K*NI*F3.   A2. . . . . . J. APPL. PHYS. 41,. . . . . . 945, (1970).
                          *HUTCHINGS(M.T.), GUGGENHEIM(H.J.).

1500 K*O2,. . . . . .   *ANTIFERROMAGNETISM IN K*O2.   A1,2. . . . . J. APP. PHYS. 37,. . . . . . 1047, (1966).
                          *SMITH(H.G.), NICKLOW(R.M.), RAUBENHEIMER(L.J.), WILKINSON(M.K.).

1501 K*O2,. . . . . .   *MAGNETIC PROPERTIES OF NA*O2 AND K*O2.   A2. . . . . . J. APP. PHYS. 37,. . . . . . 1040, (1966).
                          *SPARKS(J.T.), KOMOTO(T.).

1502 K*O2*C*C*H:C*H*C*O2*H,   *STRUCTURE OF POTASSIUM HYDROGEN MALEATE BY NEUTRON DIFFRACTION.
                          A1. . . . . . J. CHEM. PHYS. VOL.29,. . . . . 948, (1958).
                          *PETERSON(S.W.), LEVY(H.A.).

1503 K-OXALATE MONOHYDRATE,. . . .   *NEUTRON DIFFRACTION REFINEMENT OF THE STRUCTURE OF POTASSIUM OXALATE
                          MONOHYDRATE.   A1. . . . . . ACTA CRYST. VOL.B26,. . . . . 77, (1970).
                          *SEQUEIRA(A.), SRIKANTA(S.), CHIDAMBARAM(R.).

1504 K2*PT*(C*N)4*BR(.3).3D2*O,. . . .   *NEUTRON-SCATTERING STUDY OF THE STRUCTURAL PHASE TRANSITION IN THE ONE-
                          DIMENSIONAL CONDUCTOR K2*PT*(C*N)4*BR(.3).3D2*O.
                          A1,A4. . . . . PHYS. REV. LETT. VOL.32,. . . . . 836, (1974).
                          *RENKER(B.), PINTSCHOVIUS(L.), GLASER(W.), RIETSCHEL(H.), COMES(R.),
                          LIEBERT(L.), DREXEL(W.).

1505 K2*RE*BR6, . . .   *NEUTRON DIFFRACTION STUDY OF MAGNETIC ORDERING IN K2*IR*CL6, K2*RE*BR6
                          AND K2*RE*CL6.   A2. . . . . . J. PHYS. CHEM. SOLIDS 29,. . . . . 881, (1968).
                          *MINKIEWICZ(V.J.), SHIRANE(G.), FRAZER(B.C.), WHEELER(R.G.),
                          DORAIN(P.B.).

1506 K2*RE*CL6, . . .   *NEUTRON DIFFRACTION STUDY OF MAGNETIC ORDERING IN K2*RE*CL6.
                          A2. . . . . . J. APP. PHYS. 37,. . . . . . 979, (1966).
                          *SMITH(H.G.), BACON(G.E.).

1507 K2*RE*CL6, . . .   *NEUTRON DIFFRACTION STUDY OF MAGNETIC ORDERING IN K2*IR*CL6, KR*RE*BR6
                          AND K2*RE*CL6.   A2. . . . . . J. PHYS. CHEM. SOLIDS 29,. . . . . 881, (1968).
                          *MINKIEWICZ(V.J.), SHIRANE(G.), FRAZER(B.C.), WHEELER(R.G.),
                          DORAIN(P.B.).

1508 K2*RE*H9,. . . .   *THE HYDROGEN ATOM ARRANGEMENT IN RE*H9(2-).
                          A4. . . . . . COLL. INTER. N.126 (GRENOBLE), .   37, (1963).
                          *ABRAHAMS(S.C.), KNOX(K.).

1509 K2*RE*H9,. . . .   *TRANSITION METAL-HYDROGEN COMPOUNDS. II.THE CRYSTAL AND MOLECULAR
                          STRUCTURE OF POTASSIUM RHENIUM HYDRIDE, K2*RE*H9.
                          A1. . . . . . INORG. CHEM. VOL.3,. . . . . . 558, (1964).
                          *ABRAHAMS(S.C.), GINSBERG(A.P.), KNOX(K.).

1510 K*S*O3*N*H2, . .   *A NEUTRON DIFFRACTION STUDY OF POTASSIUM SULPHAMATE K*S*O3*N*H9(-2).
                          A1. . . . . . ACTA CRYST. 23,. . . . . . 578, (1967).
                          *COX(G.W.), SABINE(T.M.), PADMANABHAN(V.M.), BAN(N.T.), CHUNG(M.K.),
                          SURJADI(A.J.).

SECTION 5 -STRUCTURE AND CROSS-SECTION DETERMINATIONS . . . .

1511 K2*SB2*(C4*H2*O6)2.3H2*O. . . . . . .A1. . . . . .DISS. ABSTR. B VOL.34, ...1958, (1973).
⌐THE CRYSTAL AND MOLECULAR STRUCTURES OF ... AND X-RAY AND NEUTRON
STUDY OF K2*SB2*(C4*H2*O6)2.3H2*O.
⌐GRESS(M.E.).

1512 K*(TA-NB)*O3,. . . . . .A1. . . . . .FERROELECT. VOL.4, ...153, (1972).
⌐SOFT MODES AND THE STRUCTURE OF FERROELECTRIC TETRAGONAL POTASSIUM
TANTALATE NIOBATE.
⌐HEWAT(A.W.), ROUSE(K.D.), ZACCAI(G.).

1513 K2*ZN*(C*N)4,. . . . . .A4. . . . . .ACTA CRYST. 20, ...910, (1966).
⌐CONFIGURATION OF THE CYANIDE ION IN POTASSIUM ZINC CYANIDE: A NEUTRON
DIFFRACTION STUDY.
⌐SEQUEIRA(A.), CHIDAMBARAM(R.).

1514 K2*ZN*(C*N)4,. . . . . .A4. . . . . .ACTA CRYST. 18, ...291, (1965).
⌐ON THE DETECTION OF CYANIDE-ION ROTATION IN POTASSIUM CYANIDE BY NEUTRON
DIFFRACTION.
⌐SEQUEIRA(A.).

1515 KR (ISOTOPE A=85), ...P. . . . . . .NUCL. SCI. ENG. VOL.47,. . . . . .371, (1972).
⌐THERMAL-NEUTRON CAPTURE CROSS SECTION AND RESONANCE INTEGRAL FOR 10.7-
YEAR KRYPTON-85.
⌐BEMIS-JR(C.E.), DRUSCHEL(R.E.), HALPERIN(J.), WALTON(J.R.).

1516 KR (ISOTOPE A=80), ...P. . . . . . .NUCL. SCI. ENG. (USA) VOL.47, . .151, (1972).
⌐THE THERMAL-NEUTRON CAPTURE CROSS SECTION AND RESONANCE CAPTURE INTEGRAL
OF KRYPTON-80.
⌐BRADLEY(J.G.), JOHNSON(W.H.).

1517 KR,. . . . . .A3. . . . . .NUCL. PHYS. VOL.A133,. . . . . .410, (1969).
⌐SLOW-NEUTRON CROSS SECTIONS OF HE, NE, AR AND KR.
⌐RORER(D.C.), ECKER(B.M.), AKYUZ(R.O.).

1518 KR,. . . . . .A3. . . . . .DISS. ABSTR. VOL.21, ...2759, (1961).
⌐A NEUTRON DIFFRACTION STUDY OF KRYPTON IN THE LIQUID STATE.
⌐CLAYTON(G.T.).

1519 KR,. . . . . .A3. . . . . .CAN. J. PHYS. VOL.51, ...1965, (1973).
⌐SHORT RANGE TRIPLET CORRELATIONS IN KRYPTON NEAR THE CRITICAL POINT.
⌐WINFIELD(D.J.), EGELSTAFF(P.A.).

1520 KR,. . . . . .A3. . . . . .PHYS. REV. VOL.121, ...649, (1961).
⌐NEUTRON DIFFRACTION STUDY OF KRYPTON IN THE LIQUID STATE.
⌐CLAYTON(G.T.), HEATON(L.).

1521 KR,. . . . . .P. . . . . . .PHYS. REV. VOL.102, ...1321, (1956).
⌐COHERENT NEUTRON SCATTERING AMPLITUDES OF KRYPTON AND XENON, AND THE
ELECTRON-NEUTRON INTERACTION.
⌐CROUCH(M.F.), KROHN(V.E.), RINGO(G.R.).

1522 LA,. . . . . .A3. . . . . .PHYS. LETT. 29A, ...506, (1969).
⌐ETUDE DE LA STRUCTURE DU LANTHANE LIQUIDE PAR DIFFRACTION DE NEUTRONS.
⌐BREUIL(M.), TOURAND(G.).

1523 LA2*C3,. . . . . .A1,A4. . . .J. CHEM. PHYS. VOL.35, ...1960, (1961).
⌐NEUTRON-DIFFRACTION STUDIES OF LA2*C3, CE2*C3, PR2*C3, AND TB2*C3.
⌐ATOJI(M.), WILLIAMS(D.E.).

1524 LA*C2,. . . . . .A1. . . . . .ACTA CRYST. VOL.B24, ...459, (1968).
⌐THE CRYSTAL STRUCTURE OF LA*C2.
⌐BOWMAN(A.L.), KRIKORIAN(N.H.), ARNOLD(G.P.), WALLACE(T.C.),
NERESON(N.G.).

1525 LA*C2, ...A1,A4. . .REF. SECTION 3, 61/ATOJI,. . . . .

1526 LA*C*O3, ...A2,A4. . .REF. SECTION 3, 66/BERTAUT,. . . . .

1527 LA*CO5-H,. . . . . .A1. . . . . .J. PHYS. CHEM. SOLIDS VOL.35,. . .301, (1974).
⌐A NEUTRON-DIFFRACTION STUDY ON THE STRUCTURAL RELATIONSHIPS OF R*CO5
HYDRIDES.
⌐KUIJPERS(F.A.), LOOPSTRA(B.O.).

1528 LA*CO*O3,. . . . . .A1,2. . . .J. PHYS. CHEM. SOL. 28, ...549, (1967).
⌐LOW TEMPERATURE CRYSTALLOGRAPHIC AND MAGNETIC STUDY OF LA*CO*O3.
⌐MENYUK(N.), DWIGHT(K.), RACCAH(P.M.).

1529 LA*(CR,MN,FE,CO,NI)*O3,. . . . . . .A2. . . . .J. PHYS. CHEM. SOLIDS VOL.2, . . .100, (1957).
⌐NEUTRON-DIFFRACTION STUDY OF THE MAGNETIC PROPERTIES OF PEROVSKITE-LIKE
COMPOUNDS LA*B*O3.
⌐KOEHLER(W.C.), WOLLAN(E.O.).

1530 LA*CR*O3,. . . . . .A4. . . . . .PROC. INT. CONF. (NOTTINGHAM), . .327, (1964).
⌐NEUTRON DIFFRACTION DATA AND COVALENCY EFFECTS.
⌐NATHANS(R.), WILL(G.), COX(D.E.).

1531 LA*CR*O3,. . . . . .A2. . . . . .J. PHYS. CHEM. SOLIDS VOL.31,. . .2741, (1970).
⌐COVALENCY PARAMETERS FOR CR(3+), FE(3+), AND MN(4+) IN AN OXIDE
ENVIRONMENT.
⌐TOFIELD(B.C.), FENDER(B.E.F.).

1532 LA*ER*O3,. . . . . .A2. . . . . .SOL. STATE COMM. VOL.6,. . . . . .751, (1968).
⌐ETUDE DE LA*ER*O3 PAR DIFFRACTION NEUTRONIQUE.
⌐MOREAU(J.M.), MARESCHAL(J.), BERTAUT(E.F.).

1533 LA*ER*O3,. . . . . .A2. . . . . .SOL. STATE COMM. VOL.7 NO.22,. . .1669, (1969).
⌐PROPRIETES METAMAGNETIQUES DE LA*ER*O3.
⌐MARESCHAL(J.), MOREAU(J.M.), OLLIVIER(G.), PATAUD(P.), SIVARDIERE(J.).

1534 LA*FE*O3,. . . . . .A4. . . . . .PROC. INT. CONF. (NOTTINGHAM), . .327, (1964).
⌐NEUTRON DIFFRACTION DATA AND COVALENCY EFFECTS.
⌐NATHANS(R.), WILL(G.), COX(D.E.).

1535 LA*H3, ...P. . . . . . .J. PHYS. RADIUM VOL.22, ...648, (1961).
⌐SCATTERING CROSS-SECTION FOR SLOW NEUTRONS FROM HYDROGEN IN METALLIC
HYDRIDES.
⌐GENIN(R.), RIBRAG(M.). (IN FRENCH).

1536 LA-LU-TB,. . . . . .A4. . . . . .DISS. ABS. VOL.31, ...8798, (1970).
⌐SPIN SCATTERING IN SUPERCONDUCTORS.
⌐WILLIAMS(L.J.).

1537 LA*MN*O3,. . . . . .A4. . . . . .PROC. INT. CONF. (NOTTINGHAM), . .327, (1964).
⌐NEUTRON DIFFRACTION DATA AND COVALENCY EFFECTS.
⌐NATHANS(R.), WILL(G.), COX(D.E.).

SECTION   5 -STRUCTURE AND CROSS-SECTION DETERMINATIONS . . . .

1538   LA*MN*03,(LA-CA)*MN*03, ... A2 . . . . . PHYS. REV. VOL.100, ... 545, (1955).
       *NEUTRON DIFFRACTION STUDY OF THE MAGNETIC PROPERTIES OF THE SERIES OF
       PEROVSKITE-TYPE COMPOUNDS ((1-X)LA,(X)CA)*MN*03.
       *WOLLAN(E.O.), KOEHLER(W.C.).

1539   LA2*03,. . . . . A1 . . . . ACTA CRYST. VOL.6, ... 741, (1953).
       *NEUTRON-DIFFRACTION STUDY OF THE STRUCTURE OF THE A-FORM OF THE RARE
       EARTH SESQUIOXIDES.
       *KOEHLER(W.C.), WOLLAN(E.O.).

1540   LA*(0*D)3, ... A1 . . . . . J. CHEM. PHYS. VOL.31, ... 329, (1959).
       *DEUTERIUM POSITIONS IN LANTHANUM DEUTEROXIDE BY NEUTRON DIFFRACTION.
       *ATOJI(M.), WILLIAMS(D.E.).

1541   LA-RARE-EARTH ALLOYS,. . . . . . A2 . . . . . . PHYS. REV. 176,. . . . . 712, (1968).
       *NEUTRON DIFFRACTION AND SUSCEPTIBILITY STUDY OF DILUTE LA-RARE-EARTH
       ALLOYS.
       *FINNEMORE(D.K.),   WILLIAMS(L.J.),   SPEDDING(F.H.),   HOPKINS(D.C.).

1542   LA2*TI*05, ... A1 . . . . COMP. REND. 262, ... 9628, (1966).
       *DETERMINATION DE LA STRUCTURE DE LA2*TI*05 AUX RAYONS X ET AUX NEUTRONS.
       *GUILLEN(R.), BERTAUT(E.F.).

1543   LA*V*03, ... A2 . . . . . SOV. PHYS. SOL. STATE VOL.15,. . 1079, (1973).
       *MAGNETIC STRUCTURE OF LA*V*03.
       *ZUBKOV(V.G.), BAZUEV(G.V.), PERELYAEV(V.A.), SHVEIKIN(G.P.).

1544   LI (ISOTOPE A=6),. . . . . P . . . . 2ND INTERNAT. CONF./NUCL. DATA,. . . . , (1970).
       *LOW ENERGY NEUTRON SCATTERING CROSS-SECTION OF LI(ISOTOPE A=6).
       *ASAMI(A.), MOXON(M.C.).

1545   LI (ISO. A=7), ... A4 . . . . PROC/CONF/NEUTRON C. S. + TECH.,   827, (1968).
       *MEASUREMENTS OF NEUTRON SCATTERING FROM LI (ISO. A=7).
       *KNITTER(H.H.), COPPOLA(M.).

1546   LI (ISO. A=6), ... P . . . . NUCL. SCI. ENGNG. VOL.40,. . . . . 12, (1970).
       *THE THERMAL-NEUTRON-ABSORPTION CROSS SECTIONS OF LI(ISOTOPE A=6) AND
       B(ISOTOPE A=10).
       *MEADOWS(J.W.), WHALEN(J.F.).

1547   LI-AL FERRITES,. . . . . A1 . . . . SOV. PHYS. J. NO.11,. . . . . . . 88, (1968).
       *NEUTRON DIFFRACTION INVESTIGATION OF LITHIUM ALUMINIUM FERRITES.
       I. ATOMIC STRUCTURE AND CATION DISTRIBUTION.
       *NAIDEN(E.P.).

1548   LI*CL, ... A3 . . . . ANN. N.Y. ACAD. SCI. VOL.79, . . . 762, (1960).
       *X-RAY AND NEUTRON DIFFRACTION STUDIES OF MOLTEN ALKALI HALIDES.
       *LEVY(H.A.), AGRON(P.A.), BREDIG(M.A.), DANFORD(M.D.).

1549   LI*CL.H2*0,. . . . . A3 . . . . . J. CHEM. PHYS. VOL.58, ... 5017, (1973).
       *DIFFRACTION PATTERN AND STRUCTURE OF AQUEOUS LITHIUM CHLORIDE SOLUTIONS.
       *NARTEN(A.H.), VASLOW(F.), LEVY(H.A.).

1550   LI*CL*04.3H2*0,. . . . . A1 . . . . . SOV. PHYS. CRYST. VOL.13 . . . . 204, (1968).
       *NEUTRON-DIFFRACTION STUDY OF THE STRUCTURE OF HYDRATED LITHIUM SALTS.II.
       LI*CL*04.3H2*0.
       *DATT(I.D.), RANNEV(N.V.), OZEROV(R.P.).

1551   LI*CL*04.3H2*0,. . . . . A4 . . . . . MOL. DYN. AND STRUCT. OF SOLIDS, 249, (1969).
       *A NEUTRON DIFFRACTION STUDY OF LITHIUM PERCHLORATE TRIHYDRATE.
       *SEQUEIRA(A.), BERNAL(I.).

1552   LI*(CO,NI)*P*04, ... A2 . . . . . J. PHYS. CHEM. SOL. 27,. . . . . 1192, (1966).
       *MAGNETIC PROPERTIES OF LI*CO*P*04 AND LI*NI*P*04.
       *SANTORO(R.P.), SEGAL(D.J.), NEWNHAM(R.E.).

1553   LI*CU*CL3.2H2*0, ... A1,A2 . . . . J. CHEM. PHYS. VOL.39, ... 2923, (1963).
       *ANTIFERROMAGNETIC AND CRYSTAL STRUCTURE OF LITHIUM CUPRIC CHLORIDE
       DIHYDRATE.
       *ABRAHAMS(S.C.), WILLIAMS(H.J.).

1554   LI*(D,T),. . . . . J. PHYS. CHEM. SOLIDS VOL.31,. . 1193, (1970).
       *LOW TEMPERATURE NEUTRON DIFFRACTION STUDIES.
       *ANDERSON(J.L.), BOWMAN(A.L.), ARNOLD(G.P.).

1555   (LI-FE-AL)*04, ... A2 . . . . . SOV. PHYS. J. VOL.12 NO.8, . . . 1018, (1972).
       *NEUTRON DIFFRACTION STUDY OF LITHIUM FERRITE ALUMINATES. II. MAGNETIC
       MOMENTS.
       *NAIDEN(E.P.).

1556   (LI-FE-AL)*04, ... A2 . . . . . SOV. PHYS. SOL. STATE VOL.12,. . 770, (1970).
       *NEUTRON-DIFFRACTION INVESTIGATION OF THE MAGNETIZATION IN LITHIUM
       FERRITE-ALUMINATES.
       *NAIDEN(E.P.), ZHILYAKOV(S.M.).

1557   LI*FE2*F6, ... A2 . . . . . PHYS. REV. B VOL.6,. . . . . . . 1968, (1972).
       *NEUTRON-DIFFRACTION STUDY OF THE MAGNETIC STRUCTURE OF THE TRIRUTILE
       LI*FE2*F6.
       *SHACHAR(G.), MAKOVSKY(J.), SHAKED(H.).

1558   LI*FE2*F6, ... A2 . . . . . SOL. STATE COMM. VOL.10, ... 739, (1972).
       *DETERMINATION DE LA STRUCTURE MAGNETIQUE DE LI*FE2*F6 PAR DIFFRACTION
       NEUTRONIQUE.
       *WINTENBERGER(M.), TRESSAUD(M.A.), MENIL(F.).

1559   LI*FE5*08, ... A2 . . . . . COLL. INTER. N.126 (GRENOBLE). . 79, (1963).
       *SUBLATTICE MAGNETIZATIONS IN LITHIUM FERRITE AS FUNCTION OF TEMPERATURE.
       *PRINCE(E.).

1560   LI*FE*02,. . . . . A1 . . . . . . PHYS. REV. 132,. . . . . 1547, (1963).
       *NEUTRON DIFFRACTION AND MOSSBAUER STUDY OF ORDERED AND DISORDERED
       LI*FE*02.
       *COX(D.E.), SHIRANE(G.), FLINN(P.A.), RUBY(S.L.), TAKEI(W.J.).

1561   LI*FE5*08, ... A4 . . . . . . J. APPL. PHYS. VOL.36, ... 161, (1965).
       *BIQUADRATIC EXCHANGE AND THE TEMPERATURE DEPENDENCE OF SUBLATTICE
       MAGNETIZATION IN LITHIUM FERRITE.
       *PRINCE(E.).

1562   LI*FE*P*04,. . . . . A2 . . . . . . ACTA CRYST. 22,. . . . . 344, (1967).
       *ANTIFERROMAGNETISM IN LI*FE*P*04.
       *SANTORO(R.P.), NEWNHAM(R.E.).

1563   LI*H,. . . . . A1 . . . . . J. PHYS. CHEM. SOLIDS VOL.23,. . 621, (1962).
       *AN X-RAY AND NEUTRON DIFFRACTION ANALYSIS OF LITHIUM HYDRIDE.
       *CALDER(R.S.), COCHRAN(W.), GRIFFITHS(D.), LOWDE(R.D.).

SECTION   5 -STRUCTURE AND CROSS-SECTION DETERMINATIONS . . . .

1564 LI*H3(SE*O3)2, ...A4. . . . . . .J. SOLID STATE CHEM. VOL.4,. . . . . 255, (1972).
      *HYDROGEN BOND STUDIES. 54. A NEUTRON DIFFRACTION STUDY OF THE FERRO-
      ELECTRIC LITHIUM TRIHYDROGEN SELENITE, LI*H3(SE*O3)2.
      *TELLGREN(R.), LIMINGA(R.).

1565 LI2-MG-PB, ...A1. . . . . . .J. PHYS. CHEM. SOLIDS VOL.24,. . 1066, (1963).
      *NEUTRON DIFFRACTION BY LI2-MG-PB.
      *RAMSEY(W.J.), SANOS(D.E.), MEAD(S.W.).

1566 LI*MN*P*O4,. . . . . .A1. . . . . . .J. PHYS. CHEM. SOL. 26,. . . . . . 445, (1965).
      *NEUTRON DIFFRACTION STUDY OF LI*MN*P*O4.
      *NEWNHAM(R.E.), SANTORO(R.P.), REDMAN(M.J.).

1567 LI*MN*P*O4,. . . . . .A2. . . . . . .ACTA CRYST. 22,. . . . . 344, (1967).
      *ANTIFFEROMAGNETISM IN LI*MN*P*O4.
      *SANTORO(R.P.), NEWNHAM(R.E.).

1568 (LI-MN)*SE, ...A2. . . . . . .PHYS. REV. VOL.121,. . . . . . . 707, (1961).
      *MAGNETIC STRUCTURE TRANSITIONS IN (LI-MN)*SE.
      *PICKART(S.J.), NATHANS(R.), SHIRANE(G.).

1569 LI*N*D2, ...A1. . . . . . .ATOMKERNENERGIE VOL.21,. . . . . 275, (1973).
      *NEUTRON DIFFRACTION BY LITHIUMDEUTEROAMIDE.
      *NAGIB(M.), JACOBS(H.). (IN GERMAN).

1570 LI*(N2*H5)*S*O4, ...A4. . . . . . .ACTA CRYST. VOL.A28,. . . . . . 663, (1972).
      *ANOMALOUS NEUTRON SCATTERING AND THE QUESTION OF FERROELECTRICITY IN
      LI*(N2*H5)*S*O4.
      *ANDERSON(M.R.), BROWN(I.D.).

1571 LI*N2*H5*S*O4, ...A1. . . . . . .ACTA CRYST. 22,. . . . . 532, (1967).
      *A NEUTRON DIFFRACTION STUDY OF THE CRYSTAL STRUCTURE OF LITHIUM
      HYDRAZONIUM SULPHATE.
      *PADMANABHAN(V.M.), BALASUBRAMANIAN(R.).

1572 LI*NB*O3,. . . . . .A1. . . . . . .J. PHYS. CHEM. SOL. 27,. . . . . 1013, (1966).
      *FERROELECTRIC LITHIUM NIOBATE. 4. SINGLE CRYSTAL NEUTRON DIFFRACTION
      STUDY AT 24 DEG. C.
      *ABRAHAMS(S.C.), HAMILTON(W.C.), REDDY(J.F.).

1573 LI*NB*O3,. . . . . .A1. . . . . . .J. PHYS. CHEM. SOL. 24,. . . . . 1057, (1963).
      *POWDER NEUTRON DIFFRACTION STUDY OF LI*NB*O3.
      *SHIOZAKI(Y.), MITSUI(T.).

1574 LI2*O-FE2*O3-CR2*O3, ...A2. . . . . . .ACTA PHYS. POLON. VOL.A43, . . . 673, (1973).
      *MAGNETIC STRUCTURE OF LITHIUM FERRITES-CHROMITES.
      *DARGEL(L.).

1575 LI*O*H,. . . . . .A1. . . . . . .Z. KRIST. VOL.112,. . . . . 60, (1959).
      *BESTIMMUNG DER LAGE DES WASSERSTOFFS IN LI*O*H DURCH NEUTRONENBEUGUNG.
      *DACHS(H.).

1576 LI-PB, ...A3. . . . . . .PHYS. LETT. VOL.46A NO.1,. . . . 75, (1973).
      *A NEUTRON DIFFRACTION INVESTIGATION OF LIQUID LITHIUM-LEAD ALLOYS.
      *RUPPERSBERG(H.).

1577 LI*RH, ...A1. . . . . . .ACTA CRYST. 18,. . . . . 906, (1965).
      *NEUTRON AND X-RAY DIFFRACTION STUDY OF LI*RH.
      *SIDHU(S.S.), ANDERSON(K.D.), ZAUBERIS(D.D.).

1578 LI2*(S*O4)*(H2*O), ...A1. . . . . . .J. CHEM. PHYSICS VOL.48, ... 5561, (1968).
      *NEUTRON-DIFFRACTION STUDY OF LITHIUM SULFATE MONOHYDRATE.
      *SMITH(H.G.), PETERSON(S.W.), LEVY(H.A.).

1579 LI2*S*O4.H2*O,. . . . . .A1. . . . . . .SOV. PHYS. DOKLADY VOL.8,. . . . 131, (1963).
      *NEUTRON DIFFRACTION LOCALIZATION OF HYDROGEN ATOMS IN THE STRUCTURE OF
      LITHIUM SULPHATE MONOHYDRATE, LI2*S*O4,H2*O.
      *OZEROV(R.P.), FYKIN(L.E.), RANNEV(N.V.), ZHDANOV(G.S.).

1580 LI2*S*O4.H2*O, ...P. . . . . . .J. PHYS. SOC. JAPAN VOL.17 B-II, 335, (1962).
      *COHERENT BRAGG SCATTERING OF RESONANT NUCLIDES.
      *PETERSON(S.W.), SMITH(H.G.).

1581 LI*TA*O3,. . . . . .A1. . . . . . .J. PHYS. CHEM. SOL. 28,. . . . . 1693, (1967).
      *FERROELECTRIC LITHIUM TANTALATE-2. SINGLE CRYSTAL NEUTRON DIFFRACTION
      STUDY AT 24 DEG. C.
      *ABRAHAMS(S.C.), HAMILTON(W.C.), SEQUEIRA(A.).

1582 LI*TA*O3,. . . . . .A4. . . . . . .J. PHYS. CHEM. SOLIDS VOL.34,. . 521, (1973).
      *FERROELECTRIC LITHIUM TANTALATE-III. TEMPERATURE DEPENDENCE OF THE
      STRUCTURE IN/FERROELECTRIC PHASE AND/PARAELECTRIC STRUCTURE AT 940 K.
      *ABRAHAMS(S.C.), BUEHLER(E.), HAMILTON(W.C.), LA-PLACA(S.J.).

1583 LI*TB*F4,. . . . . .D4. . . . . . .PHYS. REV. LETT. VOL.32,. . . . . 610, (1974).
      *WAVE-VECTOR-DEPENDENT SUSCEPTIBILITY AT T>T(CRIT.) IN A DIPOLAR ISING
      FERROMAGNET.
      *ALS-NIELSEN(J.), HOLMES(L.M.), GUGGENHEIM(H.J.).

1584 LI2*TI*O3, ...A1. . . . . . .DISS. ABS. VOL.32,. . . . . .6587B, (1972).
      *THE CRYSTAL STRUCTURES OF BI4*TI3*O12 AND LI2*TI*O3.
      *DORRIAN(J.F.).

1585 LN2*O2*S2, ...A1. . . . . . .MATERIALS RES. BULL. VOL.2 NO.7, 473, (1967).
      *STRUCTURE CRISTALLINE DES OXYDISULFURES DE TERRES RARES LN2*O2*S2.
      *BALLESTRACCI(R.).

1586 LU,. . . . . .P. . . . . . .HELV. PHYS. ACTA VOL.45, ... 46, (1972).
      *NEUTRON-CAPTURE CROSS-SECTIONS OF V, MN, CS, DY AND LU IN THE ENERGY
      RANGE FROM 0.01 TO 100 EV.
      *WIDDER(F.).

1587 LU,. . . . . .P. . . . . . .PHYS. REV. VOL.121,. . . . . . . 610, (1961).
      *SLOW-NEUTRON SCATTERING CROSS SECTIONS OF TERBIUM, YTTERBIUM, AND
      LUTETIUM.
      *ATOJI(M.).

1588 LU*C2, ...A1,A4. . . .REF. SECTION 3, 61/ATOJI,. . . . .

1589 LU*C*O3, ...A2,A4. . . .REF. SECTION 3, 66/BERTAUT,. . . . .

1590 LU-FE, ...A1. . . . . . .J. LESS-COMMON MET. VOL.29,. . . 361, (1972).
      *X-RAY AND NEUTRON DETERMINATION OF A SO-CALLED TH2*NI17-TYPE STRUCTURE
      IN THE LUTETIUM-IRON SYSTEM.
      *GIVORD(D.), LEMAIRE(R.).

SECTION   5 -STRUCTURE AND CROSS-SECTION DETERMINATIONS . . . .

1591 LU*FE(9.5),. . . . . A2. . . . . C.R. ACAD. SCI. B TOME 274, ... 1166, (1972).
      *TRANSITION FERROMAGNETIQUE-HELIMAGNETIQUE DANS LES COMPOSES LU*FE(9.5)
      ET CE2*FE17.
      *GIVORD(D.), LEMAIRE(R.).

1592 LU*MN*O3,. . . . . .A2. . . . . . . REF. SECTION 3, 64/KOEHLER(1), ...

1593 METALS-ALKALI, ...A3. . . . . . . J. CHEM. PHYS. VOL.34, ... 873, (1961).
      *STRUCTURE OF ALKALI METALS IN THE LIQUID STATE.
      *GINGRICH(N.S.), HEATON(L.).

1594 METHYLADENINE(9-)*1-METHYLTHYMIN...A1. . . . . . J. CHEM. PHYS. VOL.59, ... 915, (1973).
      *PRECISION NEUTRON DIFF. STRUC. DETERMINATION/PROTEIN/NUCLEIC ACID COMP.
      XII.A STUDY OF H-BONDING IN/PURINE-PYRIMIDINE BASE PAIR 9-METHYLAD.-1-ME
      *FREY(M.N.), KOETZLE(T.F.), LEHMANN(M.S.), HAMILTON(W.C.).

1595 MG (ISO. A=24,25,26),. . . . . P. . . . . ACTA CRYST. VOL.A28 (ISO.A=24); 473, (1972).
      *COHERENT NEUTRON-SCATTERING AMPLITUDES FOR MG (ISO.A=24); MG(ISO. A=25)
      AND MG(ISO. A=26) ISOTOPES.
      *ABUL-KHAIL(A.), AMIN(F.A.), AL-NAIMI(A.), AL-SAJI(A.), AL-SHAHERY(G.Y.),
      PETRUNIN(V.F.), ZEMLYANOV(M.G.).

1596 MG*AL2*O4, ... A4. . . . . . . COLL. INTER. N.126 (GRENOBLE), 23, (1963).
      *REDETERMINATION OF THE CATION DISTRIBUTION OF SPINEL (MG*AL2*O4)
      BY MEANS OF NEUTRON DIFFRACTION.
      *STOLL(E.), FISCHER(P.), HALG(W.), MAIER(G.).

1597 MG*AL2*O4, ... A1. . . . . . . ACTA CRYST. VOL.5, ... 684, (1952).
      *A NEUTRON-DIFFRACTION STUDY OF MAGNESIUM ALUMINIUM OXIDE.
      *BACON(G.E.).

1598 MG*AL2*O4, ... A1. . . . . . . Z. KRIST. VOL.124, ... 275, (1967).
      *NEUTRONENBEUGUNGSUNTERSUCHUNG DER STRUKTUREN VON MG*AL2*O4- UND
      ZN*AL2*O4- SPINELLEN, IN ABHANGIGKEIT VON DER VORGESCHICHTE.
      *FISCHER(P.).

1599 MG*CL2*6(H2*O),. . . . . A1. . . . . . . ACTA CRYST. VOL.A25, ... S118, (1969).
      *A NEUTRON DIFFRACTION STUDY OF MAGNESIUM CHLORIDE HEXAHYDRATE.
      *AGRON(P.A.), BUSING(W.R.).

1600 MG*CR2*O,. A2. . . . . . . PHYS. REV. B-1, ... 3116, (1970).
      *MAGNETIC STRUCTURE OF MAGNESIUM CHROMITE.
      *SHAKED(H.), HASTINGS(J.M.), CORLISS(L.M.).

1601 MG*CR2*O4, ... A2. . . . . . . COMP. REND. 268, ... 365, (1969).
      *ETUDE PAR DIFFRACTION DES NEUTRONS DU COMPOSE SPINELLE MG*CR2*O4.
      *PLUMIER(R.), SOUGI(M.).

1602 MG*CR2*O4, ... A1. . . . . . . J. PHYS. C VOL.6, ... L333, (1973).
      *OFF-CENTRE DISPLACEMENTS IN SPINELS: A NEUTRON DIFFRACTION EXAMINATION
      OF MG*CR2*O4.
      *INFANTE(C.), FENDER(B.E.F.).

1603 MG*CR2*O4, ... A2. . . . . . . COMP. REND. 267, ... 98, (1968).
      *ETUDE PAR DIFFRACTION DES NEUTRONS DU COMPOSE SPINELLE NORMAL MG*CR2*O4.
      *PLUMIER(R.).

1604 MG(D-H)2, ... A1. . . . . . . ACTA CRYST. 16, ... 352, (1963).
      *NEUTRON DIFFRACTION STUDY OF MAGNESIUM DEUTERIDE.
      *ZACHARIASEN(W.H.), HOLLEY-JR(C.E.), STAMPER-JR(J.F.).

1605 MG*(FE-AL)2*O4,MG*FE2*O4, ... A1. . . . . . . ACTA CRYST. VOL.6, ... 57, (1953).
      *NEUTRON DIFFRACTION STUDIES OF MAGNESIUM FERRITE-ALUMINATE POWDERS.
      *BACON(G.E.), ROBERTS(F.F.).

1606 MG*FE2*O4, ... A2. . . . . . . PHYS. STAT. SOL. A VOL.1, ... 749, (1970).
      *A CONTRIBUTION TO THE MAGNETIC STRUCTURE OF MG*FE2*O4 BY NEUTRON
      DIFFRACTION AND MOSSBAUER EFFECT.
      *WIESER(E.), SCHRODER(H.), KLEINSTUCK(K.).

1607 MG*FE2*O4, ... A2. . . . . . . PHYS. REV. VOL.90, ... 1013, (1953).
      *A NEUTRON DIFFRACTION STUDY OF MAGNESIUM FERRITE.
      *CORLISS(L.M.), HASTINGS(J.M.), BROCKMAN(F.G.).

1608 MG*(GA-MN)*O4,MG*(CR-MN)*O4, . . . A1. . . . . . . J. PHYS. CHEM. SOLIDS VOL.28, . . 2441, (1967).
      *STUDY OF THE JAHN-TELLER EFFECT TETRAGONAL DISTORTION BY X-RAY AND
      NEUTRON DIFFRACTION IN THE SPINELS MG*(GA(2-X)-MN(X))*O4 AND . . .
      *GRENOT(M.), HUBER(M.). (IN FRENCH).

1609 MG-MN-FE-O,. . . . . A4. . . . . . . PHYS. STAT. SOL. VOL.8, ... 271, (1965).
      *NEUTRONOGRAPHISCHE UNTERSUCHUNGEN AN STOICHIOMETRISCHEM MANGANFERRIT UND
      MAGNESIUM-MANGAN-FERRITEN ZUR DEUTUNGIHRER STRUKTUR.
      *KLEINSTUCK(K.), WIESER(E.), KLEINERT(P.), PERTHEL(R.).

1610 (MG-MN)*FE2*O4,. . . . . A1. . . . . . . PROC. I.E.E. VOL.104B SUP.4-7, 217, (1957).
      *NEUTRON DIFFRACTION STUDIES OF THE MANGANESE-MAGNESIUM FERRITE SYSTEM.
      *NATHANS(R.), PICKART(S.J.), HARRISON(S.E.), KRIESSMAN(C.J.).

1611 MG*O,. . . . . A4,P. . . . . . PROC. PHYS. SOC. SER.2, 1, ... 333C, (1968).
      *A STUDY OF VERY SMALL DEFECT CLUSTERS IN IRRADIATED MAGNESIUM OXIDE,
      USING LONG-WAVELENGTH NEUTRON SCATTERING MEASUREMENTS.
      *MARTIN(D.G.).

1612 MG*O-AL2*O3-SI*O2-TI*O2, ... A1. . . . . . . J. APPL. CRYST. VOL.7, ... 207, (1974).
      *SMALL-ANGLE NEUTRON SCATTERING ON SILICA GLASSES CONTAINING TITANIA.
      *LOSHMANOV(A.A.), SIGAEV(V.N.), KHODAKOVSKAYA(R.YA.), PAVLUSHKIN(N.M.),
      YAMZIN(I.I.).

1613 MG*O-(MN,NI)*O,. . . . . A2,A4. . . . ACTA CRYST. VOL.A28 PART S-4,. . 176, (1972).
      *NEUTRON DIFFRACTION STUDIES OF DILUTE MAGNETIC SYSTEMS.
      *SABINE(T.M.). (ABSTRACT ONLY).

1614 (MG*S*O4)*4(H2*O), ... A1. . . . . . . ACTA CRYST. 17, ... 863, (1964).
      *ON CRYSTAL CHEMISTRY OF SALT HYDRATES II. A NEUTRON DIFFRACTION STUDY
      OF (MG*S*O4)*4(H2*O).
      *BAUR(W.H.).

1615 MG*V2*O4,. . . . . A2. . . . . . . COMP. REND. 257, ... 3858, (1963).
      *ETUDE PAR DIFFRACTION DES NEUTRONS DU COMPOSE SPINELLE MG*V2*O4.
      *PLUMIER(R.), TARDIEU(A.).

1616 (MG-ZN)*FE2*O4,. . . . . A2. . . . . . . PHYS. STAT. SOL. VOL.36, ... K25, (1969).
      *NEUTRON DIFFRACTION STUDY OF MAGNESIUM-ZINC FERRITES.
      *SIROTA(N.N.), NECHAI(E.F.).

SECTION   5 -STRUCTURE AND CROSS-SECTION DETERMINATIONS . . . .

1617 MN (ALPHA), ...A1,A2. . . . .PHYS. REV. B-2, ... 670, (1970).
      ~NEUTRON AND X-RAY DIFFRACTION STUDY OF THE LOW-TEMPERATURE CHEMICAL
      AND MAGNETIC STRUCTURE OF ALPHA-MANGANESE.
      ~OBERTEUFFER(J.A.),   MARCUS(J.A.),   SCHWARTZ(L.H.),   FELCHER(G.P.).

1618 MN, ...A2. . . . . . .DISS. ABS. VOL.30, ...4756B, (1970).
      ~THE CHEMICAL AND MAGNETIC STRUCTURE OF SINGLE CRYSTAL ALPHA MANGANESE.
      ~OBERTEUFFER(J.A.).

1619 MN, ...A2. . . . . . .PHYS LETT 28A, ... 267, (1968).
      ~MAGNETIC STRUCTURE OF A SINGLE CRYSTAL OF ALPHA MANGANESE.
      ~OBERTEUFFER(J.A.),   MARCUS(J.A.),   SCHWARTZ(L.H.),   FELCHER(G.P.).

1620 MN (ALPHA), ...A2. . . . . . .J. PHYS. SOC. JAPAN VOL.28, . . . 615, (1970).
      ~MAGNETIC STRUCTURE OF ALPHA-MN.
      ~YAMADA(T.), KUNITOMI(N.), NAKAI(Y.), COX(D.E.), SHIRANE(G.).

1621 MN, ...P. . . . . . .HELV. PHYS. ACTA VOL.45, ... 46, (1972).
      ~NEUTRON-CAPTURE CROSS-SECTIONS OF V, MN, CS, DY AND LU IN THE ENERGY
      RANGE FROM 0.01 TO 100 EV.
      ~WIDDER(F.).

1622 MN (GAMMA), ...A2. . . . . . .J. OF PHYS. C VOL.2, ... 761, (1969).
      ~A NEUTRON DIFFRACTION STUDY OF ELECTROLYTIC GAMMA MANGANESE.
      ~SMITH(J.H.), VANCE(E.R.).

1623 MN, ...A2. . . . . . .REV. MOD. PHYS. VOL.25, ... 100, (1953).
      ~NEUTRON DIFFRACTION STUDIES OF VARIOUS TRANSITION ELEMENTS.
      ~SHULL(C.G.), WILKINSON(M.K.).

1624 MN (ALPHA,BETA), . . .A2. . . . . . .PHYS. REV. VOL.101, ... 537, (1956).
      ~ANTIFERROMAGNETIC STRUCTURE OF ALPHA-MANGANESE AND A MAGNETIC STRUCTURE
      STUDY OF BETA-MANGANESE.
      ~KASPER(J.S.), ROBERTS(B.W.).

1625 MN, ...P. . . . . . .PHYS. REV. VOL.134, ...B1047, (1964).
      ~TOTAL NEUTRON CROSS SECTION OF MANGANESE.
      ~COTE(R.E.), BOLLINGER(L.M.), THOMAS(G.E.).

1626 MN-AL, ...A2. . . . . . .SOLID STATE COMM. VOL.11, ... 179, (1972).
      ~EXPERIMENTAL EVIDENCE FOR MAGNETIC SCATTERING OF NEUTRONS FROM MN ATOMS
      DISSOLVED IN AL.
      ~BAUER(G.), SEITZ(E.).

1627 MN-AL, ...A1. . . . . . .SOV. PHYS. DOKLADY VOL.18, . . . 432, (1973).
      ~NEW ORDERED PHASE IN THE MN-AL ALLOY SYSTEM.
      ~VINTAIKIN(E.Z.),   UDOVENKO(V.A.), LUARSABISHVILI(N.N.),   LITVIN(L.F.).

1628 MN*AL*GE, ...A2. . . . . . .J. APP. PHYS. VOL.40, ... 1870, (1969).
      ~MAGNETIC STRUCTURE OF MN*AL*GE.
      ~MURTHY(N.S.S.),   BEGUM(R.J.),   SOMANATHAN(C.S.),   MURTHY(M.R.L.N.).

1629 MN3*AL2*GE3*012, ...A2. . . . . . .SOL. STATE COMM. VOL.12 NO.2, . . . 109, (1973).
      ~DETERMINATION PAR DIFFRACTION DES NEUTRONS DE LA STRUCTURE ANTIFERROMA-
      GNETIQUE DU GRENAT MN3*AL2*GE3*012.
      ~PLUMIER(R.).

1630 MN-ALLOYS (PHASE: GAMMA), ...A2. . . . . . .J. OF PHYS.-C, SER.2, VOL.3, . . . 675, (1970).
      ~A STUDY OF SOME ALLOYS OF GAMMA-MANGANESE BY NEUTRON DIFFRACTION.
      ~BACON(G.E.), COWLAM(N.).

1631 MN*AL2*04, ...A2. . . . . . .COLL. INTER. N.126 (GRENOBLE) . . 83, (1963).
      ~MAGNETIC PROPERTIES OF NORMAL SPINELS WITH ONLY A-A INTERACTIONS.
      ~ROTH(W.L.).

1632 MN*AS, ...A2. . . . . . .J. APPL. PHYS. 42, ... 1621, (1971).
      ~MAGNETIC STRUCTURE OF MN*AS AND MN*AS-(MN*P).
      ~SCHWARTZ(L.H.), HALL(E.L.), FELCHER(G.P.).

1633 MN*AS, ...A1,A2. . . . .J. PHYSIQUE TOME 32 COL.1 VOL.II 987, (1971).
      ~NEUTRON DIFFRACTION STUDY OF MAGNETIC AND CRYSTALLOGRAPHIC PHASE TRANS-
      FORMATION IN MANGANESE ARSENIDE AS A FUNCTION TEMPERATURE AND PRESSURE.
      ~SIROTA(N.N.), VASILEV(E.A.), GOVOR(G.A.).

1634 MN*AS, ...A1. . . . . . .ACTA CRYST. 17, ... 95, (1964).
      ~THE CRYSTAL STRUCTURE OF MN*AS ABOVE 40 DEG. C.
      ~WILSON(R.H.), KASPER(J.S.).

1635 MN*AS, ...A2. . . . . . .DISS. ABS. VOL.31, ...4278B, (1971).
      ~A NEUTRON DIFFRACTION STUDY OF THE MAGNETIC PROPERTIES OF MN*AS AND
      MN*(AS-P).
      ~HALL(E.L.).

1636 MN*AS, ...A2. . . . . . .NATURE VOL.175, ... 518, (1955).
      ~MAGNETIC STRUCTURE OF MANGANESE ARSENIDE.
      ~BACON(G.E.), STREET(R.).

1637 (MN*AS)-CR*AS, ...A2. . . . . . .J. APP. PHYS. VOL.40, ... 1128, (1969).
      ~MAGNETIC STRUCTURE OF CR*AS AND MN-SUBSTITUTED CR*AS.
      ~WATANABE(H.), KAZENNA(N.), YAMAGUCHI(Y.), OHASHI(M.).

1638 MN*AS-(MN*P), ...A2. . . . . . .J. APPL. PHYS. 42, ... 1621, (1971).
      ~MAGNETIC STRUCTURE OF MN*AS AND MN*AS-(MN*P).
      ~SCHWARTZ(L.H.), HALL(E.L.), FELCHER(G.P.).

1639 MN2*AS-MN2*SB, ...A2. . . . . . .SOV. PHYS. DOKLADY VOL.17, ... 370, (1972).
      ~NEUTRONOGRAPHIC INVESTIGATION OF MAGNETIC TRANSFORMATIONS OF THE MN2*AS-
      MN2*SB SYSTEM.
      ~SIROTA(N.N.), RYZHKOVSKII(V.M.).

1640 MN2*AS,MN2*(SB-AS), ...A2. . . . . . .J. APPL. PHYS. VOL.33, SUPPL., . 1356, (1962).
      ~MAGNETIC STRUCTURES OF MN2*AS AND MN2*(SB-AS).
      ~AUSTIN(A.E.), ADELSON(E.), CLOUD(W.H.).

1641 MN*(AS-P), ...A1,A2. . . . .J. APPL. PHYS. 41, ... 939, (1970).
      ~CRYSTAL AND MAGNETIC STRUCTURE OF MN*(AS-P).
      ~HALL(E.L.), SCHWARTZ(L.H.), FELCHER(G.P.), RIDGLEY(D.H.).

1642 MN*(AS-P), ...A2. . . . . . .DISS. ABS. VOL.31, ...4278B, (1971).
      ~A NEUTRON DIFFRACTION STUDY OF THE MAGNETIC PROPERTIES OF MN*AS AND
      MN*(AS-P).
      ~HALL(E.L.).

1643 MN2*AU, MN3*AU, ...A1. . . . . . .ACTA CRYST. VOL.A26, ... 379, (1970).
      ~THE STRUCTURE OF MN2*AU, MN3*AU.
      ~WELLS(P.), SMITH(J.H.).

SECTION   5 -STRUCTURE AND CROSS-SECTION DETERMINATIONS . . . .

1644 MN*AU2,. . . . . . .METAMAGNETISM. II. . . .A2. . . . . . .Z. ANGEW. PHYS. VOL.15,. . . . .   371, (1963).
                         .VOGT(E.). (IN GERMAN).

1645 MN*AU2,. . . . . . .HELICAL ANTIFERROMAGNETISM. . . .A2. . . . . . .J. PHYS. RADIUM VOL.21,. . . . . .   67, (1960).
                         .HERPIN(A.), MERIEL(P.), VILLAIN(J.). (IN FRENCH).

1646 MN*AU2,. . . . . .A2. . . . . . .J. PHYS. RADIUM VOL.22,. . . . . . 337, (1961).
                         .HELICOIDAL ANTIFERROMAGNETISM OF MN*AU2 STUDIED BY NEUTRON DIFFRACTION.
                         .HERPIN(A.), MERIEL(P.).

1647 MN*AU2,. . . . . .A1,A2. . . .COMP. REND. TOME 246,. . . . . . 3170, (1958).
                         .ETUDE PAR DIFFRACTION DE NEUTRONS DE L≠ALLIAGE MN*AU2.
                         .HERPIN(A.), MERIEL(P.), MEYER(A.P.).

1648 MN*AU2,. . . . . .A2. . . .COMP. REND. 250,. . . . . . . 1450, (1960).
                         .MAGNETIC BEHAVIOUR OF A SPIRAL ANTIFERROMAGNET.
                         .HERPIN(A.), MERIEL(P.).

1649 MN-AU2,. . . . . .A2. . . .COMP. REND. 249,. . . . . . . 1334, (1959).
                         .MAGNETIC STRUCTURE OF THE ALLOY MN-AU2.
                         .HERPIN(A.), MERIEL(P.), VILLAIN(J.).

1650 MN3*B4,. . . . . .A2. . . . . . .PHYS. STAT. SOL. VOL.49,. . . .   589, (1972).
                         .NEUTRON DIFFRACTION STUDY OF THE MAGNETIC STRUCTURE OF MN3*B4.
                         .NEOV(S.), LEGRAND(E.).

1651 MN*B2,. . . . . .A2. . . . . . .SOL. STATE COMM. VOL.10,. . . .   883, (1972).
                         .NEUTRON DIFFRACTION STUDY OF MN*B2.
                         .LEGRAND(E.), NEOV(S.).

1652 MN*B2,. . . . . .A2. . . . . . .A.I.P. CONF. PROC. NO.10 PART 1,   658, (1973).
                         .MAGNETIC PROPERTIES OF MN*B2.
                         .NERESON(N.), BOWMAN(A.), ARNOLD(G.).

1653 MN3*B2*06,. . . . . .A2. . . . . . .PHYS. STAT. SOL. 16,. . . . . . . K17, (1966).
                         .ANTIFERROMAGNETISM IN MN3*B3*06, CO3*B2*06, AND NI3*B2*06.
                         .NEWNHAM(R.E.), SANTORO(R.P.), SEAL(P.F.), STALLINGS(G.R.).

1654 MN*BI,. . . . . .A1,A2. . . .ACTA CHEM. SCAND. VOL.21,. . . . . 1543, (1967).
                         .THE MAGNETIC AND CRYSTALLOGRAPHIC PROPERTIES OF MN*BI STUDIED BY NEUTRON
                         DIFFRACTION.
                         .ANDRESEN(A.F.), HALG(W.), FISCHER(P.), STOLL(E.).

1655 MN*BI,MN*(BI-SB),. . . . .A2. . . .ACTA CRYST. VOL.A28 PART S-4,. . . . 196, (1972).
                         .NEUTRON DIFFRACTION STUDIES OF QUENCHED MN*BI AND MN*(BI(.9)-SB(.1)).
                         .ANDRESEN(A.F.),   ENGEBRETSEN(J.E.),   REFSNES(J.).   (ABSTRACT   ONLY).

1656 MN*BI,MN*(BI-SB),. . . . .A2. . . .ACTA CHEM. SCAND. VOL.26,. . . . . 175, (1972).
                         .NEUTRON DIFFRACTION INVESTIGATIONS ON QUENCHED MN*BI AND MN*(BI-SB).
                         .ANDRESEN(A.F.), ENGEBRETSEN(J.E.), REFSNES(J.).

1657 MN*BR2,. . . . . .A2. . . . . . .PHYS. REV. VOL.110,. . . . . . . 638, (1958).
                         .NEUTRON DIFFRACTION STUDY OF THE MAGNETIC PROPERTIES OF MN*BR2.
                         .WOLLAN(E.O.), KOEHLER(W.C.), WILKINSON(M.K.).

1658 MN*BR2,MN*CL2,. . . . .A2. . . . . . .J. PHYS. RADIUM VOL.20,. . . . . . 180, (1959).
                         .SINGLE CRYSTAL NEUTRON DIFFRACTION STUDIES OF ANTIFERROMAGNETS AT
                         LOW TEMPERATURES IN APPLIED MAGNETIC FIELDS.
                         .KOEHLER(W.C.), WILKINSON(M.K.), CABLE(J.W.), WOLLAN(E.O.).

1659 MN*(C*H3*C*O*O)2.4H2*O,. . . . .A2. . . . . .SOL. STATE COMM. VOL.14,. . . . . 665, (1974).
                         .STRUCTURE MAGNETIQUE EN CHAMP APPLIQUE NUL DE L≠ACETATE DE MANGANESE
                         TETRAHYDRATE MN*(C*H3*C*O*O)2.4H2*O.
                         .BURLET(MME.P.), BURLET(P.), BERTAUT(E.F.).

1660 MN*C*03,. . . . . .A2. . . . . . .PROC. PHYS. SOC. 92,. . . . . . . 125, (1967).
                         .THE SPATIAL DISTRIBUTION OF FERROMAGNETIC MOMENT IN MN*C*03.
                         .BROWN(P.J.), FORSYTH(J.B.).

1661 MN*C*03,. . . . . .A2. . . . . . .SOV. PHYS. JETP VOL.9,. . . . . . 1204, (1959).
                         .NEUTRON DIFFRACTION INVESTIGATION OF THE ANTIFERROMAGNETISM OF THE
                         CARBONATES OF MANGANESE AND IRON.
                         .ALIKHANOV(R.A.).

1662 MN*CL2*4H2*O,. . . . .A1. . . . . . .ACTA CRYST. VOL.B27,. . . . .   66, (1971).
                         .THE STRUCTURE OF MANGANESE DICHLORIDE TETRAHYDRATE:
                         A NEUTRON-DIFFRACTION STUDY.
                         .EL-SAFFAR(Z.M.), BROWN(G.M.).

1663 MN2*CO2*C,. . . . . .A1,2. . . .J. PHYS. CHEM. SOLIDS 30,. . . . . 939, (1969).
                         .FERRIMAGNETIC STRUCTURE OF MN2*CO2*C.
                         .MURTHY(N.S.S.),   BEGUM(R.J.),   SOMANATHAN(C.S.),   SRINIVASAN(B.S.),
                         MURTHY(M.R.L.N.).

1664 MN-CR,. . . . . .A1,2. . . .COLL. INTER. N.126 (GRENOBLE),.   38, (1963).
                         .NEUTRON DIFFRACTION AND SCATTERING STUDIES AT J.A.E.R.I.
                         .KUNITOMI(N.),   HAMAGUCHI(Y.),   SAKAMOTO(M.),   DOI(K.),   KOMURA(S.).

1665 MN-CR,. . . . . .A1. . . . . . .ACTA CRYST. VOL.9,. . . . . 289, (1956).
                         .ORDERING OF ATOMS IN THE SIGMA PHASE.
                         .KASPER(J.S.), WATERSTRAT(R.M.).

1666 MN-CR,. . . . . .A2. . . . . . .PHYS. REV. VOL.109,. . . . . . . 1551, (1958).
                         .ANTIFERROMAGNETISM IN THE BODY-CENTERED CUBIC MN-CR SOLID SOLUTION.
                         .KASPER(J.S.), WATERSTRAT(R.M.).

1667 MN*CR2*04,. . . . . .A2. . . . . . .COMP. REND. 265,. . . . . . . 7268, (1967).
                         .ETUDE MAGNETIQUE PAR DIFFRACTION DES NEUTRONS DU COMPOSE SPINELLE
                         MN*CR2*04.
                         .PLUMIER(R.).

1668 MN*CR2*04,. . . . . .A2. . . . . . .J. APP. PHYS. 37,. . . . . 962, (1966).
                         .REDUCED MANGANESE MOMENT IN MANGANESE CHROMITE.
                         .DWIGHT(K.), MENYUK(N.), FEINLEIB(J.), WOLD(A.).

1669 MN*CR2*04,. . . . . .A2. . . . . . .J. APPL. PHYS. VOL.39,. . . . . . 635, (1968).
                         .REINVESTIGATION OF MAGNETIC STRUCTURES OF CO*CR2*04 AND MN*CR2*04
                         OBTAINED BY NEUTRON DIFFRACTION.
                         .PLUMIER(R.).

1670 MN*CR2*04,. . . . . .A2. . . . . . .PHYS. REV. VOL.126,. . . . . . . 556, (1962).
                         .MAGNETIC STRUCTURE OF MANGANESE CHROMITE.
                         .HASTINGS(J.M.), CORLISS(L.M.).

SECTION   5 -STRUCTURE AND CROSS-SECTION DETERMINATIONS . . . .

1671   MN*CR2*O4, . . .   A2   J. PHYS. SOC. JAPAN VOL.17 B-III   43, (1962).
                          ⇒THE MAGNETIC STRUCTURE OF MN*CR2*O4.
                          ⇒HASTINGS(J.M.), CORLISS(L.M.).

1672   (MN-CR)*O2, . .   A2   APP. PHYS. VOL.39,   590, (1968).
                          ⇒MAGNETIC OXIDES CRYSTALLIZING IN THE RUTILE STRUCTURE.
                          ⇒VILLERS(G.), GIBART(P.), DRUILHE(R.), BURLET(P.).

1673   MN*CR2*S4, . .   A2   COMP. REND. 268,   1549, (1969).
                          ⇒ETUDE PAR DIFFRACTION DES NEUTRONS D'UNE TRANSITION MAGNETIQUE A BASSE
                            TEMPERATURE DANS LE THIOSPINELLE MN*CR2*S4.
                          ⇒PLUMIER(R.), SOUGI(M.).

1674   (MN-CR)*S, . . .   A2   COMP. REND. 264,   3238, (1967).
                          ⇒STRUCTURE MAGNETIQUE DES SOLUTIONS SOLIDES MN(X)*CR(1-X)*S.
                          ⇒BURLET(P.), BERTAUT(E.F.).

1675   (MN-CR)*S, . . .   A2   SOL. STATE COMM. 5,   279, (1967).
                          ⇒ORDRE MAGNETIQUE A COURTE DISTANCE DANS LES SOLUTIONS SOLIDES (MN-CR)*S.
                          ⇒BERTAUT(E.F.), BURLET(P.).

1676   MN*CR2*S4, . .   A2   J. APP. PHYS. 36,   1088, (1965).
                          ⇒MAGNETIC PROPERTIES OF MN*CR2*S4.
                          ⇒MENYUK(N.), DWIGHT(K.), WOLD(A.).

1677   MN-CR-SB, . .   A2   J. APPL. PHYS. VOL.30, SUPPL.   28C, (1959).
                          ⇒NEUTRON DIFFRACTION INVESTIGATION OF A POSSIBLE FERRO-ANTIFERROMAGNETIC
                            TRANSITION IN MN(0.2)-CR(0.8)-SB.
                          ⇒PICKART(S.J.), NATHANS(R.).

1678   MN*CS*F3, . . .   A2   COLL. INTER. N.126 (GRENOBLE),   141, (1963).
                          ⇒MAGNETIC STRUCTURE OF BINARY FLUORIDES CONTAINING MN(+2).
                          ⇒PICKART(S.J.), ALPERIN(H.A.).

1679   MN-CU, ...   A1   PROC. ROY. SOC. A VOL.241,   223, (1957).
                          ⇒THE ANTIFERROMAGNETISM OF MANGANESE COPPER ALLOYS.
                          ⇒BACON(G.E.), DUNMUR(I.W.), SMITH(J.H.), STREET(R.).

1680   MN*(D*C*O*O)2.2D2*O,   A2   SOL. STATE COMM. VOL.7 NO.2,   343, (1969).
                          ⇒ETUDE DE L'ANTIFERROMAGNETISME DANS MN*(D*C*O*O)2.2D2*O PAR DIFFRACTION
                            NEUTRONIQUE.
                          ⇒BERTAUT(E.F.), BURLET(MME.P.), BURLET(P.).

1681   MN*(D*C*O*O)2.2D2*O,   A2   SOL. STATE COMM. VOL.7 NO.19,   1403, (1969).
                          ⇒ETUDE DE L'ANTIFERROMAGNETISME A TRES BASSE TEMPERATURE DANS LE FORMIATE
                            DE MANGANESE DIHYDRATE.
                          ⇒BURLET(MME.P.), BURLET(P.), BERTAUT(E.F.), ROULT(G.), DE-COMBARIEU(A.),
                            PILLON(J.J.).

1682   MN*F2, ...   A4   PHYS. REV. 181,   920, (1969).
                          ⇒POLARIZATION ANALYSIS OF THERMAL-NEUTRON SCATTERING.
                          ⇒MOON(R.M.), RISTE(T.), KOEHLER(W.C.).

1683   MN*F2, ...   A2   PHYS. REV. LETT. VOL.8,   237, (1962).
                          ⇒POLARIZED NEUTRON STUDY OF ANTIFERROMAGNETIC DOMAINS IN MN*F2.
                          ⇒ALPERIN(H.A.), BROWN(P.J.), NATHANS(R.), PICKART(S.J.).

1684   MN*F2, ...   A2   PHYS. REV. VOL.90,   779, (1953).
                          ⇒NEUTRON DIFFRACTION STUDIES OF ANTIFERROMAGNETISM IN MANGANOUS FLUORIDE
                            AND SOME ISOMORPHOUS COMPOUNDS.
                          ⇒ERICKSON(R.A.).

1685   MN*F3, ...   A2   PHYS. REV. VOL.112,   1132, (1958).
                          ⇒ANTIFERROMAGNETIC PROPERTIES OF THE IRON GROUP TRIFLUORIDES.
                          ⇒WOLLAN(E.O.), CHILD(H.R.), KOEHLER(W.C.), WILKINSON(M.K.).

1686   MN*(FE-CR)*O3,   A2   J. PHYS. CHEM. SOL. 24,   1531, (1963).
                          ⇒FERRIMAGNETISM OF MANGANESE FERRITE-CHROMITE.
                          ⇒MCGUIRE(T.R.), PICKART(S.J.).

1687   MN*(FE-CR)*O4,   A2   DISS. ABSTR. VOL.20,   333, (1959).
                          ⇒A NEUTRON DIFFRACTION STUDY OF CHROMIUM-SUBSTITUTED FERRITES.
                          ⇒PICKART(S.J.).

1688   MN*(FE-CR)*O4,   A1,A2   PHYS. REV. VOL.116,   317, (1959).
                          ⇒NEUTRON STUDY OF THE CRYSTAL AND MAGNETIC STRUCTURES OF MN*(FE-CR)*O4.
                          ⇒PICKART(S.J.), NATHANS(R.).

1689   MN-FE-MG-O, . .   A4   PHYS. STAT. SOL. VOL.8,   271, (1965).
                          ⇒NEUTRONOGRAPHISCHE UNTERSUCHUNGEN AN STOCHIOMETRISCHEM MANGANFERRIT UND
                            MAGNESIUM-MANGAN-FERRITEN ZUR DEUTUNG IHRER STRUKTUR.
                          ⇒KLEINSTUCK(K.), WIESER(E.), KLEINERT(P.), PERTHEL(R.).

1690   MN2*FE*O4, . .   A2   COLL. INTER. N.126 (GRENOBLE),   98, (1963).
                          ⇒STRUCTURE MAGNETIQUE DU MANGANITE DE FER PAR DIFFRACTION NEUTRONIQUE.
                          ⇒MURASIK(A.), ROULT(G.).

1691   MN*FE2*O4, . .   A1   J. PHYS. SOC. JAPAN VOL.23,   1426, (1967).
                          ⇒NEUTRON DIFFRACTION OF MANGANESE FERRITE PREPARED FROM AQUEOUS SOLUTION.
                          ⇒SAKURAI(J.), SHINJO(T.).

1692   (MN-FE)*O4, . .   A1,A2   SOV. PHYS. CRYST. VOL.6,   744, (1962).
                          ⇒THE STRUCTURE OF MANGANESE FERRITE.
                          ⇒NOZIK(YU.Z.), YAMZIN(I.I.).

1693   (MN-FE2)*O4, . .   A4   J. PHYS. SOC. JAPAN VOL.17 B-III   55, (1962).
                          ⇒A NEUTRON-DIFFRACTION STUDY OF MANGANESE FERRITES.
                          ⇒YAMZIN(I.I.), BELOV(N.V.), NOZIK(YU.Z.).

1694   MN*FE2*O4, . .   A1   PHYS. REV. VOL.104,   328, (1956).
                          ⇒NEUTRON DIFFRACTION STUDY OF MANGANESE FERRITE.
                          ⇒HASTINGS(J.M.), CORLISS(L.M.).

1695   (MN-FE)*PT, . .   A2   PHYS. REV. 171,   574, (1968).
                          ⇒MAGNETIC STRUCTURES AND EXCHANGE INTERACTIONS IN THE MN-PT SYSTEM.
                          ⇒KREN(E.), KADAR(G.), PAL(L.), SOLYOM(J.), SZABO(P.), TARNOCZI(T.).

1696   (MN-FE)3-PT, . .   A2   PHYS. LETT. VOL.26A,   556, (1968).
                          ⇒EFFECT OF FE SUBSTITUTION ON THE MAGNETIC STRUCTURE OF MN3-PT.
                          ⇒KREN(E.), KADAR(G.), SZABO(P.).

1697   (MN-FE)5*SI3, . .   A1,A2   PHYS. STAT. SOLIDI A VOL.19 NO.1   K13, (1973).
                          ⇒ATOMIC AND MAGNETIC STRUCTURE OF (MN(5-X)-FE(X))*SI3.
                          ⇒BINCZYCKA(H.), DIMITRIJEVIC(Z.), GAJIC(B.), SZYTULA(A.).

SECTION   5 -STRUCTURE AND CROSS-SECTION DETERMINATIONS . . . .

1698 (MN,FE)3*SN, ...A2. . . . . .J. APP. PHYS. 36, ... 980, (1965).
      -EXCHANGE ANISOTROPY AND LONG RANGE MAGNETIC ORDER IN THE MIXED
      INTERMETALLIC COMPOUNDS (MN,FE)3*SN.
      -KOUVEL(J.S.).

1699 (MN-FE)*W*O4,. . . . . .A2. . . . . . .Z. KRIST. VOL.131, ... 289, (1970).
      -NEUTRONENBEUGUNG AN MISCHKRISTALLEN (MN,FE)*W*O4, WOLFRAMIT.
      -WEITZEL(H.).

1700 MN3*GA,. . . . . .A1,A2. . . .SOL. STATE COMM. VOL.8, ... 1653, (1970).
      -NEUTRON DIFFRACTION STUDY OF MN3*GA.
      -KREN(E.), KADAR(G.).

1701 MN3*(GA,GE,SN), ...A2. . . .A.I.P. CONF. PROC. NO.10 PART 2, 1379, (1973).
      -MAGNETIC STRUCTURE OF DO(19) TYPE COMPOUNDS.
      -ZIMMER(G.J.), KREN(E.).

1702 MN3*GA*N,. . . . . .A2. . . .SOL. STATE COMM. VOL.6 NO.5, . . 251, (1968).
      -DIFFRACTION NEUTRONIQUE DE MN3*GA*N.
      -BERTAUT(E.F.), FRUCHART(D.), BOUCHAUD(J.P.), FRUCHART(R.).

1703 MN*GA2*O4, ...A2. . . . . . .J. APP. PHYS. 37, ... 960, (1966).
      -ANTIFERROMAGNETISM OF SPINEL MN*GA2*O4.
      -BOUCHER(B.), HERPIN(A.G.), OLES(A.).

1704 MN*GA2*O4, ...A2. . . . . . .J. PHYSIQUE TOME 27, ... 51, (1966)
      -NEUTRON DIFFRACTION AND MAGNETIC STUDY OF THE NORMAL ANTIFERROMAGNETIC
      SPINEL MN*GA2*O4.
      -BOUCHER(B.), OLES(A.). (IN FRENCH).

1705 MN5*GE3, ...A1. . . . . .ACTA CRYST. VOL.B26, ... 2079, (1970).
      -AN INVESTIGATION OF THE CRYSTAL STRUCTURE OF MN5*GE3 USING
      SINGLE-CRYSTAL NEUTRON TIME-OF-FLIGHT TECHNIQUES.
      -DAY(D.H.), SINCLAIR(R.N.).

1706 MN3*GE,. . . . . .A1. . . . .INT. J. MAGN. VOL.1, ... 143, (1971).
      -NEUTRON DIFFRACTION STUDY OF MN3*GE.
      -KADAR(G.), KREN(E.).

1707 MN5*GE3, ...A2. . . . . .PHYS. STAT. SOL. 7, ... 1015, (1964).
      -NEUTRON DIFFRACTION BY PARAMAGNETIC MN5*GE3.
      -CISZEWSKI(R.).

1708 MN5*GE3, ...A2. . . . . .PHYS. STAT. SOL. 4,. . . . . . . 199, (1964).
      -MAGNETIC STRUCTURE OF MN5*GE3 NEAR CURIE POINT.
      -CISZEWSKI(R.).

1709 MN5-GE3, ...A2. . . . . .PHYS. STAT. SOL. 3,. . . . . . . 1999, (1963).
      -MAGNETIC STRUCTURE OF THE MN5-GE3 ALLOY.
      -CISZEWSKI(R.).

1710 MN*GE*N2,. . . . . .A1,A2. . . .SOL. STATE COMM. VOL.11 NO.11. . 1485, (1972).
      -ETUDE CRISTALLOGRAPHIQUE ET MAGNETIQUE DE MN*GE*N2 PAR DIFFRACTION NEUT.
      -WINTENBERGER(M.), GUYADER(J.), MAUNAYE(M.).

1711 MN2*GEO4,. . . . . .A1,A2. . . .SOL. STATE COMM. VOL.8, ... 1183, (1970).
      -THE CRYSTAL AND MAGNETIC STRUCTURES OF MN2*GE*O4.
      -CREER(J.G.), TROUP(G.J.F.).

1712 MN*GE*O3,. . . . . .A1,A2. . . .PHYS. STAT. SOL. VOL.44, ... 71, (1971).
      -ETUDE CRISTALLOGRAPHIQUE ET MAGNETIQUE DE MN*GE*O3.
      -HERPIN(P.), WHULER(A.), BOUCHER(B.), SOUGI(M.).

1713 MN2*GE*S4, ...A1,A2. . . .SOL. STATE COMM. VOL.7, ... 641, (1969).
      -ETUDE PAR DIFFRACTION NEUTRONIQUE DE LA STRUCTURE NUCLEAIRE ET
      MAGNETIQUE DE L*ORTHOTHIOGERMANATE DE MANGANESE.
      -DUC(T.), VINCENT(H.), BERTAUT(E.F.), VU-VAN-QUI.

1714 MN*(H*C*O*O)2*2(H2*O), ...A2. . . . . .PHYS. REV. VOL.188,. . . . . 1037, (1969).
      -TWO-DIMENSIONAL ANTIFERROMAGNETISM IN MN*(H*C*O*O)2*2(H2*O).
      -SKALYO-JR(J.), SHIRANE(G.), FRIEDBERG(S.A.).

1715 MN*(H*C*O2)2*2(H2*O) . . . . .A1. . . . . .ACTA CRYST B24,. . . . . 1312, (1968).
      -HYDROGEN ATOM POSITIONS IN MANGANOUS FORMATE DIHYDRATE AND A REFINEMENT
      OF COPPER FORMATE DIHYDRATE.
      -KAY(M.I.), ALMODOVAR(I.), KAPLAN(S.F.).

1716 MN*HG, ...A2. . . . . .PHYS. STAT. SOL. 8, ... K167, (1965).
      -MAGNETIC STRUCTURE OF MN*HG.
      -OLES(A.).

1717 MN*HG, ...A2. . . . . .PHYS. STAT. SOL. 14, ... K39, (1966).
      -MAGNETIC ANISOTROPY OF THE ALLOY MN*HG.
      -OLES(A.).

1718 MN2*HG5, ...A2. . . . . .PHYS. STAT. SOLIDI A VOL.15 NO.1 K37, (1973).
      -THE MAGNETIC STRUCTURE OF MN2*HG5.
      -LEGRAND(E.).

1719 MN*HG, ...A2. . . . . .J. PHYS. SOC. JAPAN VOL.19,. . . 2078, (1964).
      -MAGNETIC STRUCTURE OF THE INTERMETALLIC COMPOUND MN*HG.
      -NAKAGAWA(Y.), WATANABE(H.), HORI(T.).

1720 MN*I2, ...A2. . . . . .PHYS. REV. VOL.125,. . . . . . . 1860, (1962).
      -NEUTRON DIFFRACTION INVESTIGATION OF THE MAGNETIC ORDER IN MN*I2.
      -CABLE(J.W.), WILKINSON(M.K.), WOLLAN(E.O.), KOEHLER(W.C.).

1721 MN-IR, ...A2. . . . . .J. PHYS. SOC. JAPAN VOL.36,. . . . 438, (1974).
      -NEUTRON DIFFRACTION STUDY OF GAMMA-PHASE MN-IR SINGLE CRYSTALS.
      -YAMAOKA(T.), MEKATA(M.), TAKAKI(H.).

1722 MN-IR, ...A2. . . . . .J. PHYS. SOC. JAPAN VOL.31,. . . . 301, (1971)
      -NEUTRON DIFFRACTION STUDY OF ANTIFERROMAGNETISM IN FACE-CENTERED CUBIC
      MN-IR ALLOYS.
      -YAMAOKA(T.), MEKATA(M.), TAKAKI(H.).

1723 MN*LA*O3,. . . . . .A4. . . . . .PROC. INT. CONF. (NOTTINGHAM), . 327, (1964).
      -NEUTRON DIFFRACTION DATA AND COVALENCY EFFECTS.
      -NATHANS(R.), WILL(G.), COX(D.E.).

1724 MN4*N, ...A2. . . . . .PHYS. REV. VOL.125,. . . . . . . 1893, (1962).
      -MAGNETIC STRUCTURE OF MN4*N-TYPE COMPOUNDS.
      -TAKEI(W.J.), HEIKES(R.R.), SHIRANE(G.).

SECTION   5 -STRUCTURE AND CROSS-SECTION DETERMINATIONS . . . .

1725 MN2*N, ...A1,A2. . .J. PHYS. SOC. JAPAN VOL.25,. . .   234, (1968).
                          ⊶NEUTRON DIFFRACTION STUDY OF ANTIFERROMAGNETIC MN2*N.
                          ⊶MEKATA(M.), HARUNA(J.), TAKAKI(H.).

1726 MN4*N, ...A2. . . . . .PHYS. REV. VOL.119,. . . . . . .   122, (1960).
                          ⊶MAGNETIC STRUCTURE OF MN4*N.
                          ⊶TAKEI(W.J.), SHIRANE(G.), FRAZER(B.C.).

1727 MN*N*H4*F3,. . .   . . .A2. . . . . .COLL. INTER. N.126 (GRENOBLE),.   141, (1963).
                          ⊶MAGNETIC STRUCTURES OF BINARY FLUORIDES CONTAINING MN(+2).
                          ⊶PICKART(S.J.), ALPERIN(H.A.).

1728 MN*NA*F3,. . . .   . .A2. . . . . .COLL. INTER. N.126 (GRENOBLE),.   141, (1963).
                          ⊶MAGNETIC STRUCTURES OF BINARY FLUORIDES CONTAINING MN(+2).
                          ⊶PICKART(S.J.), ALPERIN(H.A.).

1729 MN*ND*03,. . . .   . .A1,A2. . . .BU. SOC. FRA. MIN. CRIST. VOL.91 339, (1968).
                          ⊶NEUTRON DIFFRACTION STUDY OF THE PARAMETERS OF RARE EARTH MANGANITES OF
                          PEROVSKITE-TYPE AND MAGNETIC STRUCTURES OF MN*PR*03 AND MN*ND*03.
                          ⊶QUEZEL-AMBRUNAZ(S.).

1730 MN-NI, ...A1,2. . . . .J. PHYS. CHEM. SOLIDS 29,. . . . .   101, (1968).
                          ⊶STRUCTURES AND PHASE TRANSFORMATIONS IN THE MN-NI SYSTEM NEAR
                          EQUIATOMIC CONCENTRATION.
                          ⊶KREN(E.), NAGY(E.), NAGY(I.), PAL(L.), SZABO(P.).

1731 MN-NI, ...A2. . . . . .COLL. INTER. N.126 (GRENOBLE), .   180, (1963).
                          ⊶MAGNETISM RESEARCH WITH POLARIZED NEUTRONS AT THE CENTRO DI STUDI
                          NUCLEARI DELLA CASACCIA DEL C.N.E.N., ROMA ITALY.
                          ⊶ANTONINI(B.), FELCHER(G.P.), MENZINGER(F.), PAOLETTI(A.), RICCI(F.P.),
                          PASSARI(L.).

1732 MN-NI, ...A1. . . . . .J. PHYS. F VOL.3,. . . . . . .   6, (1973).
                          ⊶MULTIPHASE STRUCTURES IN MANGANESE RICH MN-NI ALLOYS.
                          ⊶BACON(G.E.), COWLAM(N.), SELF(A.G.).

1733 MN-NI, ...A2. . . . . .REF. SECTION 3, 66/SIDOROV,. . . . .

1734 MN-NI, ...   ⊶. . . . . .A2,D4. . .J. OF PHYS. C VOL.1,. . . . . .1683, (1968).
                          ⊶ANTIFERROMAGNETISM IN GAMMA-PHASE MANGANESE-PALLADIUM AND MANGANESE-
                          NICKEL ALLOYS.
                          ⊶HICKS(T.J.), PEPPER(A.R.), SMITH(J.H.).

1735 MN-NI3,. . . . . .A2. . . . . .J. APP. PHYS. 37,. . . . .   3236, (1966).
                          ⊶MAGNETIZATION AND STATE OF ORDER IN MN-NI3.
                          ⊶PAOLETTI(A.), RICCI(F.P.), PASSARI(L.).

1736 MN-NI3,. . . . . .A4. . . . . .J. APP. PHYS. 34,. . . . .   1571, (1963).
                          ⊶A POLARIZED-NEUTRON INVESTIGATION OF MN-NI3 ALLOY.
                          ⊶PAOLETTI(A.), RICCI(F.P.).

1737 MN-NI3,. . . . . .A2. . . . . .SOL. STATE COMM. 5,. . . . . .   769, (1967).
                          ⊶ETUDE DE LA DENSITE DE SPIN DU MANGANESE DANS L≠ALLIAGE ORDONNE MN-NI3.
                          ⊶DELAPALME(A.).

1738 MN-NI, ...A2. . . . . .J. APP. PHYS. 39,. . . . . . .   538, (1968).
                          ⊶MAGNETIC STRUCTURES AND PHASE TRANSFORMATIONS IN MN-BASED CU-AU I-TYPE
                          ALLOYS.
                          ⊶PAL(L.), KREN(E.), KADAR(G.), SZABO(P.), TARNOCZI(T.).

1739 MN-NI3,. . . . . .A4. . . . . .PHYS. LETT. 24A,. . . . .   371, (1967).
                          ⊶COEXISTING PHASES IN PARTIALLY ORDERED MN-NI3.
                          ⊶PAOLETTI(A.), RICCI(F.P.).

1740 MN-NI (GAMMA), ...A4. . . . . .J. PHYSIQUE TOME 32 COL.1 VOL.I,   70, (1971).
                          ⊶NEUTRON DIFFRACTION STUDY OF SHORT RANGE ORDER IN GAMMA MN-NI.
                          ⊶WELLS(P.), SMITH(J.H.).

1741 MN3*NI*N,. . . .   . .A2. . . . . .SOL. STATE COMM. VOL. 9,. . . . .   1793, (1971).
                          ⊶STRUCTURE MAGNETIQUE ET ROTATION DE SPIN DE MN3*NI*N.
                          ⊶FRUCHART(D.), BERTAUT(E.F.), MADAR(R.), LORTHIOIR(G.), FRUCHART(R.).

1742 MN3*04,. . . . . .A2. . . . . .J. PHYS. CHEM. SOLIDS VOL.32,. .   2429, (1971).
                          ⊶PROPRIETES ET STRUCTURE MAGNETIQUE DE MN3*04.
                          ⊶BOUCHER(B.), BUHL(R.), PERRIN(M.).

1743 MN2*03,. . . . . .A1. . . . . .PHYS. STAT. SOL. 3,. . . . . . .   K446, (1963).
                          ⊶NEUTRONOGRAPHISCHE BESTIMMUNG DER KRISTALLSTRUKTURPARAMETER VON DY2*03
                          TM2*03 UND (ALPHA)-MN2*03.
                          ⊶HASE(W.).

1744 MN3*04,. . . . . .A1,2. . . .J. PHYS. CHEM. SOLIDS 30,. . . . .   805, (1969).
                          ⊶MANGANITES 'SPINELLES' PURES D≠ELEMENTS DE TRANSITION: PREPARATIONS ET
                          STRUCTURES CRISTALLOGRAPHIQUES.
                          ⊶BUHL(R.).

1745 MN*0,. . . . . .A4. . . . . .PROC. INT. CONF. (NOTTINGHAM), .   327, (1964).
                          ⊶NEUTRON DIFFRACTION DATA AND COVALENCY EFFECTS.
                          ⊶NATHANS(R.), WILL(G.), COX(D.E.).

1746 MN*02, ...A2. . . . . .AN. FIS. (SPAIN) VOL.66, ...   407, (1970).
                          ⊶NEUTRON DIFFRACTION STUDY OF THE MAGNETIC SPIRAL STRUCTURE OF MN*02.
                          ⊶GONZALO(J.A.), COX(D.). (ED. NOTE: IN ENGLISH).

1747 MN2*03,. . . . . .A2. . . . . .SOL. STATE COMM. 5,. . . . . .   7, (1967).
                          ⊶ETUDE PAR EFFECT MOSSBAUER DU SYSTEM MN(2-X)*FE(X)*03 ET TRANSITIONS
                          MAGNETIQUES DANS MN2*03 PAR DIFFRACTION NEUTRONIQUE.
                          ⊶CHEVALIER(R.), ROULT(G.), BERTAUT(E.F.).

1748 MN3*04,. . . . . .A2. . . . . .J. APPL. PHYS. 42,. . . . .   1615, (1971).
                          ⊶MAGNETIC STRUCTURE OF MN3*04 BY NEUTRON DIFFRACTION.
                          ⊶BOUCHER(B.), BUHL(R.), PERRIN(M.).

1749 MN*0,. . . . . .A2. . . . . .PHYS. REV. 147,. . . . .   418, (1966).
                          ⊶LOCAL ANTIFERROMAGNETIC ORDER IN SINGLE-CRYSTAL MN*0 ABOVE THE NEEL
                          TEMPERATURE.
                          ⊶RENNINGER(A.), MOSS(S.C.), AVERBACH(B.L.).

1750 MN*0,. . . . . .A2. . . . . .PHYS. REV. 142,. . . . .   287, (1966).
                          ⊶LONG-RANGE MAGNETIC ORDER IN MN*0.
                          ⊶BLECH(I.A.), AVERBACH(B.L.).

1751 MN*0,. . . . . .A2. . . . . .J. PHYS. C VOL.6,. . . . .   1615, (1973).
                          ⊶THE MAGNETIC FORM FACTOR OF MN(2+) IN MN*0 BY POWDER NEUTRON
                          DIFFRACTION.
                          ⊶JACOBSON(A.J.), TOFIELD(B.C.), FENDER(B.E.F.).

SECTION   5 -STRUCTURE AND CROSS-SECTION DETERMINATIONS . . . .

1752 MN*O,. . . . . . . *DETECTION OF ANTIFERROMAGNETISM BY NEUTRON DIFFRACTION.   E. . . . . . . .PHYS. REV. VOL.76,. . . . . 1256, (1949).
*SHULL(C.G.), SMART(J.S.).

1753 MN*O,. . . . . . . *NEUTRON DIFFRACTION BY PARAMAGNETIC AND ANTIFERROMAGNETIC SUBSTANCES.   A2. . . . . .PHYS. REV. VOL.83,. . . . . 333, (1951).
*SHULL(C.G.), STRAUSER(W.A.), WOLLAN(E.O.).

1754 MN3*O4,. . . . . . *THE MAGNETIC STRUCTURE OF MN3*O4 (HAUSMANNITE) BETWEEN 4.7 K AND THE   A2. . . . . .J. PHYS. C VOL.7,. . . . . . . 409, (1974).
NEEL POINT, 41 K.
*JENSEN(G.B.), NIELSEN(O.V.).

1755 MN*O,. . . . . . *MAGNETIC STRUCTURES OF MN*O, FE*O, CO*O, AND NI*O.   A2. . . . . .PHYS. REV. VOL.110,. . . . . . 1333, (1958).
*ROTH(W.L.).

1756 MN*O2, ... *A NEW TYPE OF ANTIFERROMAGNETIC STRUCTURE IN THE RUTILE TYPE CRYSTAL.   A2. . . . . .J. PHYS. SOC. JAPAN VOL.14,. . . . 807, (1959).
*YOSHIMORI(A.).

1757 (MN3*O4)-(CU*(FE,CR)2*O4), *ETUDE PAR DIFFRACTION DES RAYONS X ET DES NEUTRONS, DES RELATIONS ENTRE   A4. . . . . .J. PHYS. CHEM. SOLIDS VOL.33,. . . 737, (1972).
DISTR. CATIONIQUE ET DISTORSION CRIST. DANS/FERRO-MANGANITES SPINELLES..
*BAFFIER(N.), HUBER(M.).

1758 MN2*O3-FE2*O3, . *MAGNETIC STRUCTURES IN THE ALPHA-MN2*O3-FE2*O3 SYSTEM.   A2. . . . .J. APP. PHYS. VOL.40,. . . . . . 1136, (1969).
*GELLER(S.), GRANT(R.W.), CAPE(J.A.), ESPINOSA(G.P.).

1759 MN*(O*H)2, . . . *HYDROTHERMAL PREPARATION AND LOW TEMPERATURE MAGNETIC PROPERTIES OF   A2. . . . .SOL. STATE COMM. VOL.10,. . . . 609, (1972).
MN*(O*H)2.
*CHRISTENSEN(A.N.), OLLIVIER(G.).

1760 MN*O,MN*S, . . . *PROBLEM OF SPIN ARRANGEMENTS IN MN*O AND SIMILAR ANTIFERROMAGNETS.   A2. T. . . . .PHYS. REV. VOL.108,. . . . . . . 637, (1957).
*KEFFER(F.), O#SULLIVAN(W.).

1761 MN*O*O*H,. . . . *THE MAGNETIC STRUCTURE OF MANGANITE MN*O*O*H AT 4.2 DEG. K.   A2. . . . . .COLL. INTER. N.126 (GRENOBLE), . 139, (1963).
*DACHS(H.).

1762 MN*O*O*H,. . . . *DETERMINATION OF HYDROGEN POSITIONS IN MANGANITE BY NEUTRON DIFFRACTION.   A1. . . . . .J. PHYS. SOC. JAPAN VOL.17 B-II, 387, (1962).
*DACHS(H.). (IN GERMAN)

1763 MN*O*O*H,. . . . *THE MAGNETIC STRUCTURE AND CRYSTAL SYMMETRY OF MN*O*O*H.   A1,A2. . . .INT. J. MAGN. VOL.4 NO.1,. . . . 5, (1973).
*DACHS(H.).

1764 MN*P,. . . . . . *POLARIZATION EFFECTS IN NEUTRON SCATTERING FROM SPIRAL DOMAINS IN MN*P.   A2. . . . . .J. PHYSIQUE TOME 32 COL.1 VOL.I, 577, (1971).
*FELCHER(G.P.), LANDER(G.H.), BRUN(T.O.).

1765 MN*P,. . . . . . *MAGNETIC STRUCTURE OF MN*P.   A2. . . . . .J. APP. PHYS. 37,. . . . . 1056, (1966).
*FELCHER(G.P.).

1766 MN*P,. . . . . . *MAGNETIC STRUCTURE OF MN*P.   A2. . . . . .J. APP. PHYS. 37,. . . . . 1053, (1966).
*FORSYTH(J.B.), PICKART(S.J.), BROWN(P.J.).

1767 MN2*P, ... *THE MAGNETIC STRUCTURE OF MN2*P.   A2. . . . . .PHIL. MAG. VOL.17, ... 623, (1968).
*YESSIK(M.).

1768 MN2*P2*O7, . . . *MAGNETIC STRUCTURE OF MANGANESE PYROPHOSPHATE, MN2*P2*O7.   A2. . . . . .CAN. J. PHYS. 49,. . . . . 979, (1971).
*COLLINS(M.F.), GILL(G.S.), STAGER(C.V.).

1769 MN2*P2*O7, . . . *MAGNETIC STRUCTURE OF MANGANESE PYROPHOSPHATE AND COPPER PYROPHOSPHATE.   A2. . . . . .CAN. J. PHYS. VOL.50,. . . . . 3079, (1972).
*STILES(J.A.R.), STAGER(C.V.).

1770 MN-PD, ... *NEUTRON DIFFRACTION STUDY OF ORDERED MN-ALLOYS.   A2. . . . . .J. PHYSIQUE TOME 32 COL.1 VOL.II 980, (1971).
*KREN(E.), KADAR(G.), PAL(L.), ZSOLDOS(E.), BARBERON(M.), FRUCHART(R.).

1771 MN-PD, ... *MAGNETIC STRUCTURES AND PHASE TRANSFORMATIONS IN MN-BASED CU-AU I-TYPE   A2. . . . . .J. APP. PHYS. 39,. . . . . . . 538, (1968).
ALLOYS.
*PAL(L.), KREN(E.), KADAR(G.), SZABO(P.), TARNOCZI(T.).

1772 MN-PD, ... *NEUTRON DIFFRACTION STUDY OF MN*PD.   A1. . . . . .PHYS. LETT. 22,. . . . . 273, (1966).
*KREN(E.), SOLYOM(J.).

1773 MN-PD, ... *ANTIFERROMAGNETISM IN GAMMA-PHASE MANGANESE-PALLADIUM AND MANGANESE-   A2,D4. . . .J. OF PHYS. C VOL.1,. . . . . . 1683, (1968).
NICKEL ALLOYS.
*HICKS(T.J.), PEPPER(A.R.), SMITH(J.H.).

1774 MN2-PD3, ... *THE CRYSTAL AND MAGNETIC STRUCTURE OF AN ORDERED MN2-PD3 ALLOY.   A1,A2. . . .PHYS. MET. METALLOGR. VOL.16 N.5 145, (1963).
*KKUSCIN(W.W.), ZISCEWSKI(R.).

1775 MN-PD3,. . . . . *NEUTRON DIFFRACTION STUDY OF THE MN-PD3 PHASE.   A2. . . . . .SOL. STATE COMM. VOL.10, . . . 1195, (1972).
*KREN(E.), KADAR(G.), MARTON(M.).

1776 MN-PD3,. . . . . *EFFECT OF ATOMIC ORDERING ON THE MAGNETIC STRUCTURE OF THE MN-PD3 PHASE.   A2. . . . . .J. APPL. PHYS. 41,. . . . . . . 941, (1970).
*KREN(E.), KADAR(G.), PAL(L.).

1777 MN3-PD5, ... *CRYSTAL AND MAGNETIC STRUCTURE OF THE MN3-PD5 PHASE.   A1,A2. . . .SOL. STATE COMM. VOL.11,. . . . 933, (1972).
*KADAR(G.), KREN(E.).

1778 MN-PD2,. . . . . *NEW ANTIFERROMAGNETIC INTERMETALLIC COMPOUND IN THE MN-PD SYSTEM:MN-PD2.   A1,A2. . . .J. PHYS. CHEM. SOLIDS VOL.33,. . 212, (1972).
*KADAR(G.), KREN(E.), MARTON(M.).

1779 MN2-PD3, ... *CRYSTAL AND MAGNETIC STRUCTURE OF THE ORDERING ALLOY MN2-PD3.   A1,A2. . . .PHYS. MET. METALLOGR. VOL.16 N.5 145, (1963).
*KKUSCIN(W.W.), ZISCEWSKI(R.).

SECTION   5 -STRUCTURE AND CROSS-SECTION DETERMINATIONS . . . .

1780 MN-PD, ...   A1 A2   PHYS. LETT. VOL.29A   340, (1969).
      *CRYSTAL AND MAGNETIC STRUCTURES IN THE MN-PD SYSTEM NEAR MN-PD3.
      *KREN(E.), KADAR(G.).

1781 MN2-PD3, ...   A2   PHYS. LETT. VOL.25A   56, (1967).
      *ATOMIC AND MAGNETIC ORDER IN MN2-PD3.
      *KREN(E.), KADAR(G.), TARNOCZI(T.).

1782 MN*PR*O3,. . . .   A1 A2   BU. SOC. FRA. MIN. CRIST. VOL.91   339, (1968).
      *NEUTRON DIFFRACTION STUDY OF THE PARAMETERS OF RARE EARTH MANGANITES OF
      PEROVSKITE-TYPE AND MAGNETIC STRUCTURES OF MN*PR*O3 AND MN*ND*O3.
      *QUEZEL-AMBRUNAZ(S.).

1783 MN-PT, ...   A2   J. APP. PHYS. 39   538, (1968).
      *MAGNETIC STRUCTURES AND PHASE TRANSFORMATIONS IN MN-BASED CU-AU I-TYPE
      ALLOYS.
      *PAL(L.), KREN(E.), KADAR(G.), SZABO(P.), TARNOCZI(T.).

1784 MN-PT, ...   A2   J. APP. PHYS. 34   1203, (1963).
      *NEUTRON DIFFRACTION INVESTIGATION AT PT-BASED ALLOYS OF THE FIRST
      TRANSITION SERIES.
      *PICKART(S.J.), NATHANS(R.).

1785 MN-PT, ...   A2   PHYS. LETT. VOL.24A   198, (1967).
      *MAGNETIC STRUCTURE TRANSFORMATION IN MN-PT.
      *KREN(E.), CSELIK(M.), KADAR(G.), PAL(L.).

1786 MN-PT, ...   A2   CENT. RES INST PHYS. KFKI NO.2,   1, (1968).
      *MAGNETIC STRUCTURES AND EXCHANGE INTERACTIONS IN THE MN-PT SYSTEM.
      *KREN(E.), KADAR(G.), PAL(L.), SOLYOM(J.), SZABO(P.), TARNOCZI(T.).

1787 MN-PT, ...   A2   PHYS. REV. 171   574, (1968).
      *MAGNETIC STRUCTURES AND EXCHANGE INTERACTIONS IN THE MN-PT SYSTEM.
      *KREN(E.), KADAR(G.), PAL(L.), SOLYOM(J.), SZABO(P.), TARNOCZI(T.).

1788 MN3-PT,. . . . .   A2   J. APPL. PHYS. VOL.38   1265, (1967).
      *INVESTIGATION OF THE FIRST-ORDER MAGNETIC TRANSFORMATION IN MN3-PT.
      *KREN(E.), KADAR(G.), PAL(L.), SZABO(P.).

1789 MN-PT3,. . . . .   A2   PHYS. REV. B VOL.5.   3778, (1972).
      *LOCAL ANTIFERROMAGNETIC ORDERING IN FERROMAGNETIC MN-PT3 ALLOYS NEAR THE
      STOICHIOMETRIC COMPOSITION.
      *MENZINGER(F.), SACCHETTI(F.), ROMANAZZO(M.).

1790 MN-PT3,. . . . .   A2   PHYS. REV. VOL.187.   611, (1969).
      *MAGNETIZATION DISTRIBUTION IN FERROMAGNETIC MN-PT3 BY A POLARIZED-
      NEUTRON INVESTIGATION.
      *ANTONINI(B.), LUCARI(F.), MENZINGER(F.), PAOLETTI(A.).

1791 MN3*PT*N(.25), .   A4   INT. J. MAGN.(GB) VOL.1   341, (1971).
      *NEUTRON DIFFRACTION STUDY OF THE HEXAGONAL NITRIDES MN3*RH*N(.20) AND
      MN3*PT*N(.25).
      *KREN(E.), KADAR(G.), BARBERON(M.), FRUCHART(R.).

1792 MN3-PT-N,. . . .   A2   J. PHYSIQUE TOME 32 COL.1 VOL.II   980, (1971).
      *NEUTRON DIFFRACTION STUDY OF ORDERED MN-ALLOYS.
      *KREN(E.), KADAR(G.), PAL(L.), ZSOLDOS(E.), BARBERON(M.), FRUCHART(R.).

1793 MN3*PT*(N),. . .   A2   SOL. STATE COMM. VOL. 9   27, (1971).
      *MAGNETIC PROPERTIES OF THE MN3*PT*(N) SYSTEM.
      *KREN(E.), ZSOLDOS(E.), BARBERON(M.), FRUCHART(R.).

1794 MN3*(PT-RH), . .   A2   PHYS. REV. 171   574, (1968).
      *MAGNETIC STRUCTURES AND EXCHANGE INTERACTIONS IN THE MN-PT SYSTEM.
      *KREN(E.), KADAR(G.), PAL(L.), SOLYOM(J.), SZABO(P.), TARNOCZI(T.).

1795 MN3*(PT-RH), . .   A2   PHYS. LETT. 21,   383, (1966).
      *ANTIFERROMAGNETISM IN DISORDERED MU3*PT(1-X)*RH(X) ALLOYS.
      *KREN(E.).

1796 MN*RB*F3,. . . .   A2   COLL. INTER. N.126 (GRENOBLE),   141, (1963).
      *MAGNETIC STRUCTURE OF BINARY FLUORIDES CONTAINING MN(+2).
      *PICKART(S.J.), ALPERIN(H.A.).

1797 MN3*RH*N(.20), .   A4   INT. J. MAGN.(GB) VOL.1   341, (1971).
      *NEUTRON DIFFRACTION STUDY OF THE HEXAGONAL NITRIDES MN3*RH*N(.20) AND
      MN3*PT*N(.25).
      *KREN(E.), KADAR(G.), BARBERON(M.), FRUCHART(R.).

1798 MN3*(RH,PT), . .   A2   PHYS. LETT. 20,   331, (1966).
      MAGNETIC STRUCTURES AND MAGNETIC TRANSFORMATIONS IN ORDERED MN3(RH,PT).
      *KREN(E.), KADAR(G.), PAL(L.), SOLYOM(J.), SZABO(P.).

1799 MN*S,. . . . . .   A2   PHYS. REV. VOL.104.   924, (1956).
      *MAGNETIC STRUCTURES OF THE POLYMORPHIC FORMS OF MANGANOUS SULFIDE.
      *CORLISS(L.M.), ELLIOTT(N.), HASTINGS(J.M.).

1800 MN*S2,MN*SE2,. .   A2   J. APPL. PHYS. VOL.29   391, (1958).
      *ANTIFERROMAGNETIC STRUCTURES OF MN*S2, MN*SE2, AND MN*TE2.
      *CORLISS(L.M.), ELLIOTT(N.), HASTINGS(J.M.).

1801 MN*S2,MN*SE2,. .   A2   PHYS. REV. VOL.115.   13, (1959).
      *ANTIFERROMAGNETIC STRUCTURES OF MN*S2, MN*SE2, AND MN*TE2.
      *HASTINGS(J.M.), ELLIOTT(N.), CORLISS(L.M.).

1802 MN*S*O4, ...   A1   ACTA CRYST. 19,   854, (1965).
      *THE CRYSTAL STRUCTURE OF MN*S*O4.
      *WILL(G.), FRAZER(B.C.), COX(D.E.).

1803 MN*S*O4, ...   A2   PHYS. REV. 140,   A2139, (1965).
      *MAGNETIC STRUCTURE OF MN*S*O4.
      *WILL(G.), FRAZER(B.C.), SHIRANE(G.), COX(D.E.), BROWN(P.J.).

1804 MN*S*O4, ...   A2,4   J. APP. PHYS. 36,   1095, (1965).
      *CYCLOIDAL SPIN CONFIGURATION OF ORTHORHOMBIC MN*S*O4.
      *WILL(G.), FRAZER(B.C.), SHIRANE(G.), COX(D.E.), BROWN(P.J.).

1805 MN*SB, ...   A2   ACTA CRYST. VOL.A28 PART S-4,   196, (1972).
      *DISTRIBUTION OF UNPAIRED ELECTRONS IN MN*SB.
      *WATANABE(H.), YAMAGUCHI(Y.), SUZUKI(T.), KAZAMA(N.). (ABSTRACT ONLY).

1806 MN2*SB,. . . . .   A2   J. PHYS. CHEM. SOLIDS VOL.2,   289, (1957).
      *THE MAGNETIC STRUCTURE OF MN2*SB.
      *WILKINSON(M.K.), GINGRICH(N.S.), SHULL(C.G.).

SECTION 5 -STRUCTURE AND CROSS-SECTION DETERMINATIONS . . . .

1807 MN2-SB-(CR), ...A2 . . . . .PHYS. REV. VOL.120 . . . . 1969, (1960).
&NEUTRON DIFFRACTION STUDIES OF CHROMIUM-MODIFIED MN2-SB.
&CLOUD(W.H.), JARRETT(H.S.), AUSTIN(A.E.), ADELSON(E.).

1808 (MN*SB)-(CR*SB), ...A2. . . . . .PROC. INT. CONF. (NOTTINGHAM), . . 291, (1964).
&MAGNETIC STRUCTURES IN THE MN*SB-CR*SB AND CR*TE-CR*SB SYSTEMS.
&COX(D.E.), SHIRANE(G.), TAKEI(W.J.).

1809 (MN*SB)-(CR*SB), ...A2. . . . . .PHYS. REV. 129, ... 2008, (1963).
&MAGNETIC STRUCTURES IN THE (MN*SB)-(CR*SB) SYSTEM.
&TAKEI(W.J.), COX(D.E.), SHIRANE(G.).

1810 MN-SB-SN, ...A2,A4. . . .J. PHYSIQUE TOME 32 COL.1 VOL.I,   78, (1971).
&NEUTRON DIFFRACTION AND MAGNETIC PROPERTIES OF MN-SB-SN.
&BOUWMA(J.), VAN-BRUGGEN(C.F.), HAAS(C.), VAN-LAAR(B.).

1811 MN*SE, ...A4. . . . . .PHYS. NORVEG. VOL.3, ...203, (1969).
&A NEUTRON DIFFRACTION INVESTIGATION OF LATTICE TRANSFORMATIONS IN MN*SE.
&ANDRESEN(A.F.), ROTTERUD(H.).

1812 MN*SE, ...A4. . . . . .ACTA CRYST. (INTERACT.) VOL.A25, S250, (1969).
&PHASE TRANSITIONS IN MN*SE STUDIED BY NEUTRON DIFFRACTION.
&ANDRESEN(A.F.), ROTTERUD(H.).

1813 MN*SE2, ...A2. . . . . .COLL. INTER. N.126 (GRENOBLE), . .133, (1963).
&MAGNETIC STRUCTURE STUDIES AT BROOKHAVEN NATIONAL LABORATORY.
&CORLISS(L.M.), HASTINGS(J.M.).

1814 MN*SE*O4, ...A1. . . . . .Z. ANORG. ALLG. CHEM. BAND 358, . . 125, (1968).
&BESTIMMUNG DER KRISTALLSTRUKTUR DER SELENATE M*SE*O4 (M=MN,CO,NI) DURCH
RONTGEN- UND NEUTRONENBEUGUNG.
&FUESS(H.), WILL(G.).

1815 MN*SE*O4, ...A2. . . . . .J. APP. PHYS. 39, ... 628, (1968).
&MAGNETIC STRUCTURE OF MN*SE*O4.
&FUESS(H.), WILL(G.).

1816 MN*SE*O4, ...A2. . . . . .INT. J. MAGN. VOL.5, ...197, (1973).
&THE MAGNETIC STRUCTURE OF THE HIGH-PRESSURE PHASE OF MN*SE*O4.
&KIRFEL(A.), WILL(G.).

1817 MN5*SI3, ...A2. . . . . .PROC. PHYS. SOC. 91, ... 332, (1967).
&THE ANTIFERROMAGNETIC STRUCTURE OF MN5*SI3.
&LANDER(G.H.), BROWN(P.J.), FORSYTH(J.B.).

1818 MN*SI, ...A2. . . . . .J. APP. PHYS. 39, ... 1331, (1968).
&SPATIAL DISTRIBUTION OF THE MAGNETIC MOMENT MN*SI.
&BROWN(P.J.), FORSYTH(J.B.), LANDER(G.H.).

1819 MN*SI, ...A2. . . . . .A.I.P. CONF. PROC. NO.10 PART 2, 1138, (1973).
&ANOMALOUS MAGNETIC BEHAVIOR OF MN*SI.
&LEVINSON(L.M.), LANDER(G.H.), STEINITZ(M.O.).

1820 MN2*SI*O4, ...A2. . . . . .J. PHYS. CHEM. SOL. 27, ... 655, (1966).
&MAGNETIC PROPERTIES OF MN2*SI*O4 AND FE2*SI*O4.
&SANTORO(R.P.), NEWNHAM(R.E.), NOMURA(S.).

1821 MN3*SN, ...A2,A4. . . .MAGN. AND MAG. MATERIALS-1971, . .513, (1972).
&INVESTIGATION OF THE MAGNETIC PHASE TRANSFORMATION IN MN3*SN.
&ZIMMER(G.J.), KREN(E.). (AIP CONF. PROC. NO.5 PART I).

1822 MN*SN2, ...A1,2. . . . .J. APP. PHYS. 39, ... 461, (1968).
&NEUTRON-DIFFRACTION STUDY OF MN*SN2.
&CORLISS(L.M.), HASTINGS(J.M.).

1823 MN*SN2, ...A2. . . . . .J. APP. PHYS. 34, ... 1192, (1963).
&MAGNETIC STRUCTURE OF MN*SN2.
&CORLISS(L.M.), HASTINGS(J.M.).

1824 MN-SN, ...A1,A2. . . .PHYS. LETT. VOL.15, ... 225, (1965).
&ATOMIC AND MAGNETIC STRUCTURE OF MN-SN.
&MURTHY(N.S.S.),   BEGUM(R.J.),   SRINIVASAN(B.S.),   MURTHY(M.R.L.N.).

1825 MN*SN2, ...A2. . . . . .COLL. INTER. N.126 (GRENOBLE), . .133, (1963).
&MAGNETIC STRUCTURE STUDIES AT BROOKHAVEN NATIONAL LABORATORY.
&CORLISS(L.M.), HASTINGS(J.M.).

1826 MN*TA2*O6, ...A2. . . . . .SOL. STATE COMM. VOL.12 NO.2, . . .113, (1973).
&MAGNETISCHE STRUKTUR VON COLUMBITEN MN*TA2*O6 UND CO*NB2*O6.
&WEITZEL(H.), KLEIN(S.).

1827 MN*TE, ...A2. . . . . .COLL. INTER. N.126 (GRENOBLE), . 144, (1963).
&NEUTRON DIFFRACTION STUDY ON MANGANESE TELLURIDE.
&KUNITOMI(N.), HAMAGUCHI(Y.), ANZAI(S.).

1828 MN*TE, ...A2. . . . . .PHYS. MET. METALLOGR. VOL.12, . . 119, (1961).
&NEUTRON DIFFRACTION ANALYSIS OF MANGANESE TELLURIDE.
&DOROSHENKO(A.V.),   KLYUSHIN(V.V.),   LOSHMANOV(A.A.), GOMAN≠KOV(V.I.).

1829 MN*TE, ...A2. . . . . .SOV. PHYS. DOKLADY VOL.11, ...888, (1967).
&NEUTRON DIFFRACTION STUDY OF THE ANTIFERROMAGNETIC TRANSFORMATION IN
MANGANESE TELLURIDE.
&SIROTA(N.N.), MAKOVETSKII(G.I.).

1830 MN*TE2, ...A2. . . . . .J. APPL. PHYS. VOL.29, ...391, (1958).
&ANTIFERROMAGNETIC STRUCTURES OF MN*S2, MN*SE2, AND MN*TE2.
&CORLISS(L.M.), ELLIOTT(N.), HASTINGS(J.M.).

1831 MN*TE2, ...A2. . . . . .PHYS. REV. VOL.115, ... 13, (1959).
&ANTIFERROMAGNETIC STRUCTURES OF MN*S2, MN*SE2, AND MN*TE2.
&HASTINGS(J.M.), ELLIOTT(N.), CORLISS(L.M.).

1832 MN*(TI*MN)*O4, ...A1,A2. . . .SOL. STATE COMM. VOL.6 NO.5, . . .269, (1968).
&ETUDE PAR DIFFRACTION NEUTRONIQUE DE LA FORME ORDONNEE DE L≠ORTHO-
TITANATE DE MANGANESE-STRUCTURE CRISTALLINE ET STRUCTURE MAGNETIQUE.
&BERTAUT(E.F.), VINCENT(H.).

1833 MN*TI*ND*O5, ...A1. . . . . .J. PHYS. CHEM. SOLIDS VOL.31, . 1171, (1970).
&ETUDE PAR RAYONS X ET NEUTRONS DE LA SERIE ISOMORPHE A*TI*T*O5.
(A=CR, MN, FE, T=TERRES RARES).
&BUISSON(G.).

1834 MN*TI*O3, ...A2. . . . . .SOL. STATE COMM. VOL.8, ... 87, (1970).
&ON THE TWO-DIMENSIONAL ANTIFERROMAGNETIC CHARACTER OF MN*TI*O3.
&AKIMITSU(J.), ISHIKAWA(Y.), ENDOH(Y.).

SECTION   5 -STRUCTURE AND CROSS-SECTION DETERMINATIONS . . . .

1835 MN*TI*03, ...T,A2...PHYS. REV. 164,.........768, (1967).
      ↗THEORY OF THE MAGNETIC PROPERTIES OF THE ILMENITES MN*TI*03.
      ↗GOODENOUGH(J.B.), STICKLER(J.J.).

1836 MN*TI*03, ...A2.......J. PHYS. SOC. JAPAN VOL.14,. 1352, (1959).
      ↗NEUTRON DIFFRACTION STUDY OF ANTIFERROMAGNETIC MN*TI*03 AND NI*TI*03.
      ↗SHIRANE(G.), PICKART(S.J.), ISHIKAWA(Y.).

1837 MN*U*04, ...A2.......J. PHYS. RADIUM VOL.23,.....477, (1962).
      ↗ETUDE DES URANATES DE COBALT ET DE MANGANESE.
      ↗BERTAUT(E.F.), DELAPALME(A.), FORRAT(F.), PAUTHENET(R.).

1838 MN*U*04, ...A1,A2.....J. PHYSIQUE TOME 27,.......726, (1966).
      ↗PARAMETRES ATOMIQUES ET STRUCTURE MAGNETIQUE DE MN*U*04.
      ↗BACHMANN(M.), BERTAUT(E.F.).

1839 MN*V2*04, ...A4.......PROC. INT. CONF. (NOTTINGHAM), . 295, (1964).
      ↗A NEUTRON DIFFRACTION STUDY OF THE INFLUENCE OF TEMPERATURE ON SPIN
      CONFIGURATION IN SPINEL MN*V2*04.
      ↗PLUMIER(R.).

1840 MN*V2*04, ...A4.......PROC. INT. CONF. (NOTTINGHAM), . 538, (1964).
      ↗MAGNETIC PROPERTIES OF V+++ IONS IN CUBIC SPINELS.
      ↗DWIGHT(K.), MENYUK(N.), ROGERS(D.B.), WOLD(A.).

1841 MN*V2*04, ...A2.......C. R. ACAD. SCI. VOL.255,.... 2244, (1962).
      ↗NEUTRON DIFFRACTION STUDY OF THE SPINEL MN*V2*04.
      ↗PLUMIER(R.). (IN FRENCH).

1842 MN*V2*04, ...A2,A4...SOV. PHYS. CRYST. VOL.12,..... 811, (1968).
      ↗DETERMINATION OF THE OXYGEN PARAMETER AND DEGREE OF INVERSION OF
      MANGANESE VANADATE BY NEUTRON DIFFRACTION.
      ↗KUZ≠MINOV(YU.S.).

1843 MN*V2*04, ...A2.......COMP. REND. 267,......... 1057, (1968).
      ↗ETUDE PAR DIFFRACTION DES NEUTRONS DU MOUVEMENT DES PAROIS DE BLOCH DANS
      LE SPINELLE FERRIMAGNETIQUE MN*V2*04.
      ↗PLUMIER(R.).

1844 (MN-V)*TE, ...A2.......INT. J. MAGN. VOL.5,..... 175, (1973).
      ↗THE MAGNETIC STRUCTURE AND THE MAGNETIC BEHAVIOUR OF (MN-V)*TE: A
      NEUTRON DIFFRACTION STUDY.
      ↗SCHAFER(W.), WILL(G.). (IN GERMAN).

1845 MN*W*04, ...A2.......SOL. STATE COMM. 4,....... 473, (1966).
      ↗MAGNETIC STRUCTURE OF MANGANESE TUNGSTATE MN*W*04 AT 4.2 DEG. K.
      ↗DACHS(H.), WEITZEL(H.), STOLL(E.).

1846 MN*W*04, ...A2.......SOL. STATE COMM. VOL.12 NO.8,... 779, (1973).
      ↗INVESTIGATIONS CONCERNING THE COEXISTENCE OF TWO MAGNETIC PHASES IN
      MIXED CRYSTALS (FE,MN)*W*04.
      ↗OBERMAYER(H.A.), DACHS(H.), SCHROCKE(H.).

1847 MN*W*04, ...A1,A2....Z. KRIST. VOL.125,....... 120, (1967).
      ↗KRISTALLSTRUKTUR UND MAGNETISCHE ORDNUNG DES HUEBNERITS, MN*W*04.
      ↗DACHS(H.), STOLL(E.), WEITZEL(H.).

1848 MN3*X (X=SN,GE,RH)...........A2.......PROC. INT. CONF. (NOTTINGHAM), . 169, (1964).
      ↗TRIANGULAR SPIN CONFIGURATIONS IN THE ANTIFERROMAGNETIC INTERMETALLIC
      COMPOUNDS MN3*SN, MN3*GE AND MN3*RH.
      ↗KOUVEL(J.S.), KASPER(J.S.).

1849 MN*Y*03, ...A2.......COLL. INTER. N.126 (GRENOBLE), . 126, (1963).
      ↗ORDRE MAGNETIQUE ET PROPRIETES MAGNETIQUES DE MN*Y*03.
      ↗BERTAUT(E.F.), MERCIER(M.), PAUTHENET(R.).

1850 MN*Y*03, ...A2.......REF. SECTION 3, 64/BERTAUT,. . . . .

1851 MN*Y*03, ...A2,T....PHYS. LETT. 7,......... 110, (1963).
      ↗PROPRIETES MAGNETIQUES ET STRUCTURES DU MANGANITE D≠YTTRIUM.
      ↗BERTAUT(E.F.), PAUTHENET(R.), MERCIER(M.).

1852 MN*Y*03, ...A2.......PHYS. LETT. 5,......... 27, (1963).
      ↗STRUCTURE MAGNETIQUE DE MN*Y*03.
      ↗BERTAUT(E.F.), MERCIER(M.).

1853 MN*Y*03, ...A2.......PHYS. LETT. 18,........ 13, (1965).
      ↗SUR DES PROPRIETES MAGNETIQUES DU MANGANITE D≠YTTRIUM.
      ↗BERTAUT(E.F.), PAUTHENET(R.), MERCIER(M.).

1854 MN*Y*03, ...A4.......SOV. PHYS. SOL. STATE 8,.... 215, (1966).
      ↗MAGNETIC PROPERTIES OF Y*MN*03.
      ↗KIZHAEV(S.A.), BOROV(V.A.), KOCHALOV(O.V.).

1855 MN*Y*03, ...A2.......J. APPL. PHYS. VOL.35,..... 952, (1964).
      ↗MAGNETIC STRUCTURE INVESTIGATIONS AT THE (GRENOBLE) NUCLEAR CENTRE.
      ↗BERTAUT(E.F.), ROULT(G.)...ET AL.

1856 MN-ZN, ...A2.......J. PHYS. SOC. JAPAN VOL.19,. . . 2082, (1964).
      ↗NEUTRON DIFFRACTION STUDIES OF MN-ZN ALLOYS.
      ↗NAKAGAWA(Y.), HORI(T.).

1857 MN-ZN, ...A2.......J. PHYS. SOC. JAPAN VOL.21,. . . 2080, (1966).
      ↗CANTED SPIN STRUCTURE OF BETA 1-MN-ZN.
      ↗HORI(T.), NAKAGAWA(Y.), ISHIKAWA(Y.).

1858 MN-ZN, ...A2.......J. PHYS. SOC. JAPAN VOL.19,. . . 1255, (1964).
      ↗A NEW FERROMAGNETIC PHASE IN MN-ZN ALLOY SYSTEM.
      ↗HORI(T.), NAKAGAWA(Y.).

1859 (MN3-ZN)*C, ...A2.......CAN. J. PHYS. VOL.35,...... 313, (1957).
      ↗NEW TYPE OF MAGNETIC TRANSITION IN (MN3-ZN)*C.
      ↗BROCKHOUSE(B.N.), MYERS(H.P.).

1860 (MN-ZN)*FE2*04,.↗X-RAY AND NEUTRON DIFFRACTION INVESTIGATIONS OF FERRITES OF THE TYPE
      ....A1.......J. APPL. CRYST. VOL.1 PT.2,.....124, (1968).
      (MN(X)-ZN(1-X))*FE2*04.
      ↗KONIG(U.), CHOL(G.).

1861 MN3*ZN*N,. . . . .A2.......J. PHYSIQUE TOME 32 COL.1 VOL.II 876, (1971).
      ↗DIFFRACTION NEUTRONIQUE DE MN3*ZN*N(1).
      ↗FRUCHART(D.), BERTAUT(E.F.), MADAR(R.), FRUCHART(R.).

1862 MO,. . . . .A2.......J. PHYS. CHEM. SOLIDS VOL.23,.. 1348, (1962).
      ↗ABSENCE OF ANTIFERROMAGNETISM IN MOLYBDENUM.
      ↗ABRAHAMS(S.C.).

SECTION   5 -STRUCTURE AND CROSS-SECTION DETERMINATIONS . . . .

1863   MO,. . . . .A2.......REV. MOD. PHYS. VOL.25,. . . . . .   100, (1953).
                            *NEUTRON DIFFRACTION STUDIES OF VARIOUS TRANSITION ELEMENTS.
                            *SHULL(C.G.), WILKINSON(M.K.).

1864   MO,. . . . .A4.......J. METALS VOL.8,. . . . . . .   1259, (1956).
                            *NEUTRON DIFFRACTION STUDY OF ANNEALING TEXTURES IN DRAWN B.C.C. METALS.
                            *SWALIN(R.A.), GEISLER(A.H.).

1865   MO2*C, ...A1.......ACTA CRYST. 16,. . . . . . .   202, (1963).
                            *THE STRUCTURE OF DIMOLYBDENUM CARBIDE BY NEUTRON DIFFRACTION TECHNIQUE.
                            *PARTHE(E.), SADAGOPAN(V.).

1866   MO*F3, ...A2.......PHYS. REV. VOL.121,. . . . .   74, (1961).
                            *NEUTRON DIFFRACTION INVESTIGATION OF MAGNETIC ORDERING IN THE TRIFLUOR-
                            IDES OF 40 TRANSITION ELEMENTS.
                            *WILKINSON(M.K.), WOLLAN(E.O.), CHILD(H.R.), CABLE(J.W.).

1867   MO-FE-MN,. . . .   .A1.......ACTA CRYST. 18,. . . . . . .   37, (1965).
                            *NEUTRON DIFFRACTION STUDIES OF THE ORDER OF THE ATOMS IN THE P-PHASE
                            AND R-PHASE.
                            *SHOEMAKER(C.B.), SHOEMAKER(D.P.), MELLOR(J.).

1868   MO-NI-CR,. . . .   .A1.......ACTA CRYST. 18,. . . . . . .   37, (1965).
                            *NEUTRON DIFFRACTION STUDIES OF THE ORDER OF THE ATOMS IN THE P-PHASE
                            AND R-PHASE.
                            *SHOEMAKER(C.B.), SHOEMAKER(D.P.), MELLOR(J.).

1869   MO3*SI,. . . .   .A2.......PHYS. REV. VOL.95,. . . . .   280, (1954).
                            *SEARCH FOR ANTIFERROMAGNETISM IN THE SILICIDES V3*SI, CR3*SI, AND MO3*SI
                            *KOEHLER(W.C.), WOLLAN(E.O.).

1870   MO-SI3-C,. . . .   .A1.......ACTA CRYST. 19,. . . . . .   1031, (1965).
                            *A NEUTRON DIFFRACTION STUDY OF THE NOWOTNY PHASE MO-SI-C.
                            *PARTHE(E.), JEITSCHKO(W.), SADAGOPAN(V.).

1871   MUSKOVITE, ...A1.......ACTA CRYST. VOL. A25,. . . . . .   S119, (1969).
                            *NEUTRONENBEUGUNGSMESSUNGEN AM MUSKOVIT.
                            *ROTHBAUER(R.), O≠DANIEL(H.).

1872   MYELIN,. . . . .A1.......SPECTROSCOPY BIOL. CHEM.,. . . . .   203, (1974).
                            *COMPARATIVE X-RAY AND NEUTRON DIFFRACTION FROM NERVE MYELIN MEMBRANES.
                            *KIRSCHNER(D.A.). (REFER TO SECTION 4).

1873   MYOGLOBIN, ...A1.......BER. BUNSENGES. PHYS. CHEM. 74,. 1202, (1970).
                            *NEUTRON DIFFRACTION ANALYSIS OF BIOLOGICAL STRUCTURES.
                            *SCHOENBORN(B.P.), NUNES(A.C.), NATHANS(R.).

1874   MYOGLOBIN, ...A1.......NATURE VOL.224,. . . . .   143, (1969).
                            *NEUTRON DIFFRACTION ANALYSIS OF MYOGLOBIN.
                            *SCHOENBORN(B.P.).

1875   MYOGLOBIN, ...A1.......SYMP. QUANT. BIOL. VOL.36,. . .   569, (1971).
                            *A NEUTRON DIFFRACTION ANALYSIS OF MYOGLOBIN. III. HYDROGEN-DEUTERIUM
                            BONDING IN SIDE CHAINS.
                            *SCHOENBORN(B.P.). (COLD SPRING HARBOR SYMPOSIA ON QUANTITATIVE BIOLOGY).

1876   MYOGLOBIN-(H2*O,D2*O)   ....A3.......J. APPL. CRYST. VOL.7,. . . . . .   173, (1974).
                            *NEUTRON SMALL-ANGLE SCATTERING OF BIOLOGICAL MACROMOLECULES IN SOLUTION.
                            *STUHRMANN(H.B.).

1877   N, ...P.......NUCL. SCI. ENG. VOL.42,. . . . .   28, (1970).
                            *HIGH RESOLUTION MEASUREMENTS OF THE TOTAL NEUTRON CROSS SECTIONS OF
                            NITROGEN AND IRON.
                            *CARLSON(A.D.), CERBONE(R.J.).

1878   N (ISO. A=15), ...P.......ACTA CRYST. VOL.A28,. . . . .   655, (1972).
                            *THE COHERENT NEUTRON SCATTERING AMPLITUDE OF N (ISO. A=15).
                            *KUZNIETZ(M.).

1879   N (ISOTOPE A=15), ...T.......NUCL. PHYS. VOL.A166,. . . . .   461, (1971).
                            *NEUTRON SCATTERING FROM N(ISOTOPE A=15). 2. COUPLED CHANNEL CALCULATIONS
                            *DOVER(C.B.), CIERJACKS(S.), KIROUAC(G.J.), NEBE(J.), DUBENKROPP(H.),
                            PUTZKI(R.), ZEITNITZ(B.).

1880   N (ISOTOPE A=15), ...T.......NUCL. PHYS. VOL.A166,. . . . .   443, (1971).
                            *NEUTRON SCATTERING FROM N(ISOTOPE A=15). 1. R-MATRIX AND PHASE-SHIFT
                            ANALYSES.
                            *ZEITNITZ(B.), DUBENKROPP(H.), PUTZKI(R.), KIROUAC(G.J.), CIERJACKS(S.),
                            NEBE(J.), DOVER(C.B.).

1881   N2,. . . . .P,A3.......SOV. AT. ENERGY VOL.18,. . . . .   585, (1965).
                            *COHERENT EFFECTS IN THE INTERACTION OF SLOW NEUTRONS WITH LIQUIDS.
                            *VERTEBNYI(V.P.), DZYUB(I.P.), MAISTRENKO(A.N.), PASECHNIK(M.V.).

1882   N2,. . . . .P,A3.......PHYS. REV. VOL.76,. . . . . .   1750, (1949).
                            *SLOW NEUTRON VELOCITY SPECTROMETER STUDIES OF O2, N2, A, H2, H2*O AND
                            SEVEN HYDROCARBONS.
                            *MELKONIAN(E.).

1883   N (V*N,K*N3),. .   ...P.......PHYS. REV. VOL.87,. . . . . .   462, (1952).
                            *THE COHERENT NEUTRON SCATTERING CROSS SECTIONS OF NITROGEN AND VANADIUM.
                            *PETERSON(S.W.), LEVY(H.A.).

1884   N2,. . . . .A3,P.......HELV. PHYS. ACTA VOL.19,. . . .   493, (1946).
                            *INFLUENCE DES LIAISONS MOLECULAIRES DANS LA DIFFUSION DES NEUTRONS
                            THERMIQUES PAR L≠AZOTE.
                            *GIBERT(A.), KELLER(R.), ROSSEL(J.).

1885   N2,. . . . .P,A3.......PHYS. REV. VOL.84,. . . . . . .   969, (1951).
                            *NEUTRON REFRACTION IN O2, N2, HE, AR GASES.
                            *MCREYNOLDS(A.W.).

1886   N2,. . . . .A3.......PHYS. REV. VOL.83,. . . . .   1100, (1951).
                            *NEUTRON DIFFRACTION BY THE GASES N2, C*F4, AND C*H4.
                            *ALCOCK(N.Z.), HURST(D.G.).

1887   N2,. . . . .A3.......PHYS. REV. VOL.92,. . . . .   1229, (1953).
                            *STRUCTURE OF LIQUID NITROGEN, OXYGEN, AND ARGON BY NEUTRON DIFFRACTION.
                            *HENSHAW(D.G.), HURST(D.G.), POPE(N.K.).

1888   N2,. . . . .P,A3.......PHYS. REV. VOL.73,. . . . .   1399, (1948).
                            *SLOW NEUTRON SPECTROMETER STUDIES OF OXYGEN, NITROGEN, AND ARGON.
                            *MELKONIAN(E.), RAINWATER(L.J.), HAVENS-JR(W.W.), DUNNING(J.R.).

SECTION   5 -STRUCTURE AND CROSS-SECTION DETERMINATIONS . . . .

1889   N2,. . . . . P . . . . PHYS. REV. VOL.138 . . . . B1116, (1965).
                            ╶DETERMINATION OF THE COHERENT NEUTRON SCATTERING AMPLITUDES OF BORON,
                            NITROGEN, AND OXYGEN BY MIRROR REFLECTION.
                            ╶DONALDSON(R.E.), PASSELL(L.), BARTOLINI(W.), GROVES(D.).

1890   N (ISO. A=14), ... P . . . . . . PROC/CONF/NEUTRON C. S. + TECH., 851, (1968).
                            ╶TOTAL NEUTRON CROSS SECTIONS OF BE (ISO. A=9), N (ISO. A=14), AND
                            O (ISO. A=16).
                            ╶JOHNSON(C.H.), HAAS(F.X.), FOWLER(J.L.), MARTIN(F.D.), KERNELL(R.L.),
                            COHN(H.O.).

1891   N*D3,. . . . . .A1. . . . . . .J. CHEM. PHYS. VOL.35 . . . . . 1730, (1961).
                            ╶NEUTRON DIFFRACTION STUDY OF SOLID DEUTEROAMMONIA.
                            ╶REED(J.W.), HARRIS(P.M.).

1892   N*D4*AL(S*O4)2*12(D2*O), ...A1. . . . . . .ACTA CRYST. 22, ... 830, (1967).
                            ╶REFINEMENT OF THE ALUM STRUCTURES IV, NEUTRON DIFFRACTION STUDY OF
                            DEUTERATED AMMONIUM ALUM, N*D4*AL(S*O4)2*12(D2*O).
                            ╶CROMER(D.T.), KAY(M.I.).

1893   N*D4*BR, ...T.A1. . . . .ACTA CRYST. VOL.A29 . . . . . . .257, (1973).
                            ╶ANALYSIS OF ORIENTATIONALLY DISORDERED STRUCTURES.II. SOLID C*D4, P-D2
                            AND N*D4*BR.
                            ╶PRESS(W.).

1894   N*D4*BR, ...ACTA CRYST. VOL.A27 . . . . . . .348, (1971).
                            ╶SHORT-RANGE ORDER IN N*D4*BR STUDIED BY DIFFUSE NEUTRON SCATTERING.
                            ╶SEYMOUR(R.S.).

1895   N*D4*BR, ...A1. . . . .PHYS. REV. VOL.83 . . . . 1270, (1951).
                            ╶THE NATURE OF THE SECOND-ORDER TRANSITION IN N*D4*BR.
                            ╶LEVY(H.A.), PETERSON(S.W.).

1896   N*D4*BR, ...A1. . . . .J. AMER. CHEM. SOC. VOL.75 . . . 1536, (1953).
                            ╶NEUTRON-DIFFRACTION DETERMINATION OF THE CRYSTAL STRUCTURE OF AMMONIUM
                            BROMIDE IN FOUR PHASES.
                            ╶LEVY(H.A.), PETERSON(S.W.).

1897   N*D4*BR,N*D4*I,. .   . . . .A1. . . .J. CHEM. PHYS. VOL.21 . . . . . 366, (1953).
                            ╶NEUTRON DIFFRACTION STUDY OF THE NA*CL-TYPE MODIFICATION OF N*D4*BR
                            AND N*D4*I.
                            ╶LEVY(H.A.), PETERSON(S.W.).

1898   N*D4*CL, ...A1. . . .SOL. STATE COMM. VOL.11, . . . 1011, (1972).
                            ╶FIRST ORDER PHASE TRANSITION IN N*D4*CL.
                            ╶YELON(W.B.), COX(D.E.).

1899   N*D4*CL, ...A1. . . . . .PHYS. REV. VOL.83 . . . . 88, (1951).
                            ╶THE STRUCTURE OF AMMONIUM CHLORIDE BY NEUTRON DIFFRACTION.
                            ╶GOLDSCHMIDT(G.H.), HURST(D.G.).

1900   N*D4*CL, ...A1. . . . . .PHYS. REV. VOL.86, ... 797, (1952).
                            ╶THE STRUCTURE OF AMMONIUM CHLORIDE BY NEUTRON DIFFRACTION.
                            ╶GOLDSCHMIDT(G.H.), HURST(D.G.).

1901   N*D4*CL, ...A4. . . .PHYS. REV. B VOL.9 NO.11 . . . . ., (1974).
                            ╶NEUTRON-DIFFRACTION STUDY OF N*D4*CL IN THE TRICRITICAL REGION.
                            ╶YELON(W.B.), COX(D.E.), KORTMAN(P.J.), DANIELS(W.B.). (REF. OBTAINED
                            FROM PHYS. REV. ABSTRACTS).

1902   N*D4*D2*P*O4,. .   . . . .A1. . . . .NATURE VOL.246 NO.5428 . . . . . 90, (1973).
                            ╶LOCATION OF HYDROGEN ATOMS IN ADP BY NEUTRON POWDER PROFILE REFINEMENT.
                            ╶HEWAT(A.W.).

1903   N2-GRAPHITE, ...A3,A4. . . .PHYS. REV. LETT. VOL.32, . . . 724, (1974).
                            ╶NEUTRON SCATTERING FROM NITROGEN ADSORBED ON BASAL-PLANE-ORIENTED
                            GRAPHITE.
                            ╶KJEMS(J.K.), PASSELL(L.), TAUB(H.), DASH(J.G.).

1904   N*H3,. . . . . .A3. . . . . .IAEA SYMPOSIUM VIENNA, ... 297, (1960).
                            ╶TOTAL NEUTRONS SCATTERING CROSS-SECTIONS OF SOME HYDROGENOUS MOLECULES:
                            EXPERIMENTS AND INTERPRETATION.
                            ╶JANIK(J.A.), JANIK(J.), MANIAWSKI(F.), RZANY(H.), SZKATULA(J.),
                            SCIESINSKI(J.), WANIC(A.).

1905   N*H3,. . . . . .P. . . . . . .NUCL./S.S. PHYS. SYMP. ABSTR., . . . , (1972).
                            ╶NEUTRON SCATTERING KERNEL FOR SOLID AMMONIA.
                            ╶BANSAL(R.M.), TEWARI(S.P.).

1906   N*H3,. . . . . .A3. . . . . .PROC. PHYS. SOC. 89, ... 379, (1966).
                            ╶NEUTRON SCATTERING FROM GASEOUS METHANE AND AMMONIA.
                            ╶VENKATARAMAN(G.), RAO(K.R.), DASANNACHARYA(B.A.), DAYANIDHI(P.K.).

1907   N*H3,. . . . . .A3. . . . . .PHYSICA VOL.30 . . . . 237, (1964).
                            ╶TOTAL CROSS SECTION FOR SCATTERING OF COLD NEUTRONS BY PROTONS IN SOME
                            COOLED HYDROGENOUS COMPOUNDS.
                            ╶VAN-DINGENEN(W.), NEVE-DE-MEVERGNIES(M.).

1908   N*H3,. . . . . .P,A3. . . .PHYSICA VOL.26, ... 449, (1960).
                            ╶SCATTERING OF SLOW NEUTRONS BY N*H3 MOLECULES. I. SCATTERING BY GASEOUS
                            N*H3.
                            ╶JANIK(J.A.), JANIK(J.), WANIC(A.).

1909   N*H4*BR, ...A1. . . .ACTA CRYST. VOL.B26, ... 1487, (1970).
                            ╶NEUTRON DIFFRACTION STUDY OF N*H4*BR AND N*H4*I.
                            ╶SEYMOUR(R.S.), PRYOR(A.W.).

1910   N*H4*BR, ...E.A1. . . . .NUOVO CIMENTO B VOL.46 . . . . . 248, (1966).
                            ╶ENHANCEMENT OF THE SIGNAL-TO-BACKGROUND RATIO IN THE CRYSTAL STRUCTURE
                            ANALYSIS OF HYDROGENOUS COMPOUNDS BY/ELASTIC DIFFRACTION OF SLOW NEUTRON
                            ╶CAGLIOTI(G.), POMPA(F.).

1911   (N*H2*C*H2*C*O*O*H)3*H2*S*O4, ...A1. . . . . .SOV. PHYS. DOKLADY VOL.16 . . . 9, (1971).
                            ╶INVESTIGATION OF THE PHASE TRANSITION IN TRIGLYCINE SULFATE.
                            ╶LEONIDOVA(G.G.), BUZIN(V.N.), ALIKHANOV(R.A.).

1912   N*H3*C6*H4*S*O6, ...A1. . . . .ACTA CRYST. 22, ... 216, (1967).
                            ╶THE CRYSTAL STRUCTURE OF ORTHANILIC ACID.
                            ╶HALL(S.R.), MASLEN(E.N.).

1913   N*H2*C6*H4*S*O2*N*H2, ...A1. . . . .ACTA CRYST. 22, ... 134, (1967).
                            ╶X-RAY AND NEUTRON DIFFRACTION STUDIES OF BETA-SULPHANILAMIDE.
                            ╶O'CONNELL(A.M.), MASLEN(E.N.).

SECTION   5 -STRUCTURE AND CROSS-SECTION DETERMINATIONS . . . .

1914   (N*H2*C*H2*C*O2)3*S*O4,. . . . . .A1. . . . . .FERROELECTRICS VOL.5,. . . . . .   45, (1973).
       *THE CRYSTAL STRUCTURE OF TRIGLYCINE SULFATE.
       *KAY(M.I.), KLEINBERG(R.).

1915   (N*H2)2*(C*N)2 (DICYANDIAMIDE),. . . . .A1. . . . . .DOKL. AKAD. NAUK SSSR VOL.155, . 1415, (1964).
       *A NEUTRON DIFFRACTION DETERMINATION OF THE HYDROGEN ATOM POSITIONS
       AND THE STRUCTURE OF DICYANDIAMIDE.
       *RANNEV(N.B.), OZEROV(R.P.). (IN RUSSIAN).

1916   (N*H4)2*C2*O4.H2*O,. . . . . .A1. . . . . .ACTA CRYST. VOL.B28, ... 3343, (1972).
       *ISOTOPE AND BONDING EFFECTS IN AMMONIUM OXALATE MONOHYDRATE, DETERMINED
       BY THE COMBINED USE OF NEUTRON AND X-RAY DIFFRACTION ANALYSES.
       *TAYLOR(J.C.), SABINE(T.M.).

1917   (N*H4)2*C2*O4.H2*O,. . . . . .A1. . . . . .ACTA CRYST. VOL.18,. . . . . . .   567, (1965).
       *NEUTRON DIFFRACTION STUDY OF AMMONIUM OXALATE MONOHYDRATE,
       (N*H4)2*C2*O4.H2*O.
       *PADMANABHAN(V.M.), SRIKANTA(S.), MEDHI-ALI(S.).

1918   (N*H2)2*C*O*H*N*O3,. . . . . .A1. . . . . .ACTA CRYST. VOL.B25, ...   572, (1969).
       *THE CRYSTAL STRUCTURE OF URONIUM NITRATE (UREA NITRATE) BY NEUTRON DIFF.
       *WORSHAM(J.E.), BUSING(W.R.).

1919   N*H4*(CL,BR,I),. . . . . .A4. . . . . .IAEA SYMP. BOMBAY, VOL.2,. . . . .   347, (1964).
       *STUDY OF ELASTIC INCOHERENT SCATTERING BY AMMONIUM SALTS.
       *VENKATARAMAN(G.),   USHA-DENIZ(K.),   IYENGAR(P.K.),   VIJAYARAGHAVAN(P.R.),
       ROY(A.P.).

1920   N*H4*CL,N*D4*CL,. . . . . .A1. . . . . .PHYS. REV. VOL.86,. . . . . .   766, (1952).
       *NEUTRON DIFFRACTION STUDY OF THE CRYSTAL STRUCTURE OF AMMONIUM CHLORIDE.
       *LEVY(H.A.), PETERSON(S.W.).

1921   N*H4*CL*O4,. . . . . .A4. . . . . .PHYS. STAT. SOL. VOL.44,. . . . .   437, (1971).
       *QUASI-FREE ROTATION OF THE AMMONIUM ION IN CRYSTALLINE N*H4*CL*O4.
       *JANIK(J.A.), JANIK(J.M.), MAYER(J.).

1922   (N*H4)*CL*O4,. . . . . .A1. . . . . .ACTA CRYST. VOL.15,. . . . . . 1201, (1962).
       *NEUTRON DIFFRACTION STUDY OF AMMONIUM PERCHLORATE.
       *SMITH(H.G.), LEVY(H.A.).

1923   N*H4*F,. . . . . .A1. . . . . .ACTA CRYST. VOL.A25,. . . . . . .   438, (1969).
       *THE STRUCTURE OF N*H4*F AS DETERMINED BY NEUTRON AND X-RAY DIFFRACTION.
       *ADRIAN(H.W.W.), FEIL(D.).

1924   N*H4*H3*C4*O8.2H2*O,. . . . . .A1. . . . . .J. CHEM. SOC. A (1967),. . . . . . 1862, (1967).
       *THE CRYSTAL STRUCTURES OF THE ACID SALTS OF SOME DIBASIC ACIDS. PART 1.
       A NEUTRON DIFFRACTION STUDY OF AMMONIUM (AND POTASSIUM) TETROXALATE.
       *CURRIE(M.), SPEAKMAN(J.C.), CURRY(N.A.).

1925   N2*H5*H*C2*O4, ...A1. . . . . .ACTA CHEM. SCAND. VOL.22, ...   719, (1968).
       *HYDROGEN BOND STUDIES. 25.A NEUTRON DIFFRACTION STUDY OF HYDRAZINIUM
       HYDROGEN OXALATE, N2*H5*H*C2*O4.
       *NILSSON(A.), LIMINGA(R.), OLOVSSON(I.).

1926   N*H4*H2*P*O4,. . . . . .P. . . . . . . .J. PHYSIQUE TOME 33 COL.2,. . . .   83, (1972).
       *INCOHERENT NEUTRON SCATTERING FROM HYDROGEN BOND IN KDP AND ADP.
       *ANTONINI(M.), SOSNOWSKA(I.), VADACCHINO(M.).

1927   N2*H5*H2*P*O4, ...A1. . . . . .ACTA CHEM. SCAND. VOL.25,. . . . . 1729, (1971).
       *HYDROGEN BOND STUDIES. 48.NEUTRON DIFFRACTION STUDY OF HYDRAZINIUM
       DIHYDROGEN PHOSPHATE, N2*H5*H2*P*O4.
       *JONSSON(P.G.), LIMINGA(R.).

1928   N2*H6*(H2*P*O4)2,. . . . . .A1. . . . . .ACTA CHEM. SCAND. VOL.26,. . . . . 1087, (1972).
       *HYDROGEN BOND STUDIES. 52.NEUTRON DIFFRACTION STUDY OF HYDRAZINIUM
       BIS(DIHYDROGEN PHOSPHATE), N2*H6*(H2*P*O4)2.
       *KVICK(A.), JONSSON(P.G.), LIMINGA(R.).

1929   N*H4*H2*P*O4,. . . . . .A4. . . . . .J. APPL. PHYS. VOL.45,. . . . . 2021, (1974).
       *NEUTRON DIFFRACTION BY VIBRATING ADP CRYSTALS.
       *PARKINSON(T.F.), GURMEN(E.), LOYALKA(S.K.), MÜHLESTEIN(L.D.).

1930   N*H4*H2*P*O4,. . . . . .A1. . . . . .ACTA CRYST. VOL.11,. . . . . .   505, (1958).
       *A NEUTRON STRUCTURE ANALYSIS OF TETRAGONAL N*H4*H2*P*O4.
       *TENZER(L.), FRAZER(B.C.), PEPINSKY(R.).

1931   N*H4*I,. . . . . .A1. . . . . .ACTA CRYST. VOL.B26,. . . . . 1487, (1970).
       *NEUTRON DIFFRACTION STUDY OF N*H4*BR AND N*H4*I.
       *SEYMOUR(R.S.), PRYOR(A.W.).

1932   N*H4*(K-I),. . . . . .A4,P. . . . . .REF. SECTION 3, 63/BRAJOVIC,. . . . .

1933   (N*H4)*MN*CL3, ...A2. . . . . .SOL. STATE COMM. VOL. 9,. . . .   493, (1971).
       *NEUTRON DIFFRACTION AND MAGNETIC MEASUREMENTS OF POLYCRYSTALLINE
       (N*H4)*MN*CL3.
       *SHACHAR(G.), MAKOVSKY(J.), SHAKED(H.).

1934   N*H4*MN*F3,. . . . . .A2. . . . . .COLL. INTER. N.126 (GRENOBLE),. .   141, (1963).
       *MAGNETIC STRUCTURE OF BINARY FLUORIDES CONTAINING MN(+2).
       *PICKART(S.J.), ALPERIN(H.A.).

1935   N*H3*O*H*CL, ...A1. . . . . .ACTA CRYST. 22,. . . . . .   928, (1967).
       *NEUTRON DIFFRACTION STUDY OF HYDROXYLAMONIUM CHLORIDE, N*H3*O*H*CL.
       *PADMANABHAN(V.M.), SMITH(H.G.), PETERSON(S.W.).

1936   N*H3*O*H*CL*O4,. . . . . .A1. . . . . .ACTA CRYST. VOL.B30,. . . . . . 1167, (1974).
       *A STUDY OF ONE-DIMENSIONAL HINDERED ROTATION IN N*H3*O*H*CL*O4.
       *PRINCE(E.), DICKENS(B.), RUSH(J.J.).

1937   (N*H4)2*S*O4,. . . . . .A1. . . . . .J. CHEM. PHYSICS VOL.44,. . . . . 4498, (1966).
       *NEUTRON-DIFFRACTION STUDY OF THE STRUCTURE OF FERROELECTRIC AND PARA-
       ELECTRIC AMMONIUM SULFATE.
       *SCHLEMPER(E.O.), HAMILTON(W.C.).

1938   (N*H4)*S*O3*F,. . . . . .A4,P. . . . . .REF. SECTION 3, 63/BRAJOVIC,. . . . .

1939   (N*H4)2*S2*O8,. . . . . .A4,P. . . . . .REF. SECTION 3, 63/BRAJOVIC,. . . . .

1940   N2*H6*S*O4,. . . . . .A1. . . . . .ACTA CRYST. VOL.B26,. . . . . .   536, (1970).
       *NEUTRON AND X-RAY DIFFRACTION STUDIES OF HYDRAZINIUM SULFATE,
       N2*H6*S*O4.
       *JONSSON(P.G.), HAMILTON(W.C.).

1941   N*H2*S*O3*H,. . . . . .A1. . . . . .ACTA CRYST. VOL.13,. . . . . . .   320, (1960).
       *A NEUTRON DIFFRACTION STUDY ON THE CRYSTAL STRUCTURE OF SULFAMIC ACID.
       *SASS(R.L.).

SECTION   5 -STRUCTURE AND CROSS-SECTION DETERMINATIONS . . . .

1942 (N*H4)2*SI*F6, ...A1,C1. . . .J. CHEM. PHYSICS VOL.44,. . . . . 2499, (1966).
&STRUCTURE OF CUBIC AMMONIUM FLUOSILICATE: NEUTRON-DIFFRACTION AND
NEUTRON-INELASTIC-SCATTERING STUDIES.
&SCHLEMPER(E.O.), HAMILTON(W.C.), RUSH(J.J.).

1943 (N*H4)2*SI*F6, ...A1. . . . . .J. CHEM. PHYSICS VOL.45,. . . . . 408, (1966).
&ON THE STRUCTURE OF TRIGONAL AMMONIUM FLUOROSILICATE.
&SCHLEMPER(E.O.), HAMILTON(W.C.).

1944 (N*H4)2*SN*(CL,BR)6, ...A4. . . . . . .IAEA SYMP. BOMBAY, VOL.2,. . . . . 347, (1964).
&STUDY OF ELASTIC INCOHERENT SCATTERING BY AMMONIUM SALTS.
&VENKATARAMAN(G.), USHA-DENIZ(K.), IYENGAR(P.K.), VIJAYARAGHAVAN(P.R.),
ROY(A.P.).

1945 (N*H4)3*ZR*F7, ...A1. . . . . .ACTA CRYST. VOL.B26,. . . . . 2136, (1970).
&A NEUTRON DIFFRACTION ANALYSIS OF THE DISORDER IN AMMONIUM
HEPTAFLUOROZIRCONATE.
&HURST(H.J.), TAYLOR(J.C.).

1946 N2*O,. . . . . .A1. . . . . .J. PHYS. CHEM. VOL.65,. . . . . 1453, (1961).
&CONFIRMATION OF DISORDER IN SOLID NITROUS OXIDE BY NEUTRON DIFFRACTION.
&HAMILTON(W.C.), PETRIE(M.).

1947 NA (ISO. A=23),. . . . . .P. . . . . . .J. NUCL. ENERGY VOL.24,. . . . . 419, (1970).
&THERMAL NEUTRON CAPTURE CROSS-SECTION MEASUREMENTS FOR NA(ISOTOPE A=23),
AL(ISOTOPE A=27), CL(ISOTOPE A=37) AND V(ISOTOPE A=51).
&RYVES(T.B.), PERKINS(D.R.).

1948 NA,. . . . . .A4. . . . . .CAN. J. PHYS. 46,. . . . . . 1727, (1968).
&NATURE OF THE MARTENSITIC TRANSFORMATION IN SODIUM.
&DOLLING(G.), POWELL(B.M.), MARTEL(P.).

1949 NA,. . . . . .P. . . . . . .PHYS. LETT. VOL.47A NO.1,. . . . . 91, (1974).
&MEASUREMENT OF THE LATTICE PARAMETER OF SODIUM BY NEUTRON-BACK-
SCATTERING.
&ADLHART(W.), FRITSCH(G.), HEIDEMANN(A.), LUSCHER(E.).

1950 NA*AL(S*O4)2*12(H2*O), ...A1. . . . . .ACTA CRYST. 22,. . . . . . 182, (1967).
&REFINEMENT OF ALUM STRUCTURES. II. X-RAY AND NEUTRON DIFFRACTION OF
NA*AL(S*O4)2*12(H2*O).
&CROMER(D.T.), KAY(M.I.), LARSON(A.C.).

1951 NA*AL*SI2*O6.H2*O, ...A1. . . . . .Z. KRIST. VOL.135,. . . . . 240, (1972).
&A NEUTRON-DIFFRACTION STUDY OF THE CRYSTAL STUCTURE OF ANALCIME,. . .
&FERRARIS(G.), JONES(D.W.), YERKESS(J.).

1952 NA2*AL2*SI3*O10.2H2*O, ...A1. . . . . .CAN. J. PHYS. 42,. . . . . 229, (1964).
&NEUTRON DIFFRACTION DETERMINATION OF THE HYDROGEN POSITIONS IN NATROLITE
&TORRIE(B.H.), BROWN(I.D.), PETCH(H.E.).

1953 NA2*C2,. . . . . .A1. . . . . .J. CHEM. PHYS. VOL.60,. . . . . 3324, (1974).
&CRYSTAL STRUCTURE DETERMINATION OF DISODIUM ACETYLIDE, NA2*C2, BY
NEUTRON DIFFRACTION.
&ATOJI(M.).

1954 NA*C2*H, ...A1. . . . . .J. CHEM. PHYS. VOL.56,. . . . . 4947, (1972).
&NEUTRON STRUCTURE DETERMINATION OF MONOSODIUM ACETYLIDE NA*C2*H, AT
293 AND 5K.
&ATOJI(M.).

1955 NA*C2*H3*O2, ...P. . . . . . . .SOV. AT. ENERGY VOL.18,. . . . . 350, (1965).
&TOTAL NEUTRON CROSS-SECTIONS FOR BENZENE, TOLUENE AND SODIUM ACETATE
IN THE ENERGY RANGE 0.03-0.5 EV.
&ANISIMOV(I.S.), NIKITIN(V.I.), SAUKOV(A.I.), UGODENKO(A.A.).

1956 NA*C*N, ...A1. . . . . .J. KOREAN PHYS. SOC. VOL.1,. . . . . 108, (1968).
&NEUTRON DIFFRACTION INVESTIGATION OF SODIUM CYANIDE.
&MOON(Y.S.), YOON(B.G.), BAK(H.I.), KIM(H.J.).

1957 NA*C*N, ...A1. . . . . .J. CHEM. PHYS. VOL.58,. . . . . 2039, (1973).
&SINGLE CRYSTAL NEUTRON DIFFRACTION STUDY OF SODIUM CYANIDE.
&ROWE(J.M.), HINKS(D.G.), PRICE(D.L.), SUSMAN(S.), RUSH(J.J.).

1958 NA2*C*O3.NA*H*C*O3.2H2*O,. . .A1. . . . . .ACTA CRYST. VOL.9,. . . . . 82, (1956).
&A NEUTRON-DIFFRACTION STUDY OF SODIUM SESQUICARBONATE.
&BACON(G.E.), CURRY(N.A.).

1959 (NA-CA)3*CO2*(V*O4)3,. . . . . . .A2. . . . . .SOV. PHYS. JETP LETT. VOL.16,. . . . . 198, (1972).
&MAGNETIC STRUCTURE OF THE ANTIFERROMAGNETIC GARNET NA*CA2*CO2*V3*O12.
&OZEROV(R.P.), FADEEVA(N.V.).

1960 NA*CA2*MN2*V3*O12,. . . . . . . .A2. . . . . .SOL. STATE COMM. VOL.14 NO.4,. . . . . 309, (1974).
&MAGNETIC ORDERING IN THE GARNET NA*CA2*MN2*V3*O12.
&GOLOSOVSKII(I.V.), PLAKHTII(V.P.), MILL(B.V.), SOKOLOV(V.I.),
SHEVALEEVSKY(O.P.).

1961 NA*CA2*NI2*V3*O12,. . . . . . . .A1. . . . . .LATV. PSR...FIZ. TEHN. SER. NO.5 122, (1971).
&NEUTRON DIFFRACTION STUDY OF NA*CA2*NI2*V3*O12 GARNET WITH ONE SUB-
LATTICE.
&LIPIN(YU.), NOZIK(YU.).

1962 NA*CL, ...A3. . . . . .PHYS. REV. 153,. . . . . 184, (1967).
&QUASI-ELASTIC SCATTERING OF SLOW NEUTRONS BY WATER AND AN AQUEOUS
SOLUTION OF SODIUM CHLORIDE.
&GOLAND(A.N.), OTNES(K.).

1963 NA*CL, ...P. . . . . . .ACTA CRYST. VOL.A29,. . . . . 727, (1973).
&THE DEBYE-WALLER FACTORS OF SODIUM CHLORIDE.
&BUTT(N.M.), CHEETHAM(A.K.), WILLIS(B.T.M.).

1964 NA*CO*F3,. . . . . .A2. . . . . .PHYS. REV. B VOL.2,. . . . . 179, (1970).
&MAGNETIC STRUCTURE OF NA*CO*F3.
&FRIEDMAN(Z.), MELAMUD(M.), MAKOVSKY(J.), SHAKED(H.).

1965 NA2*CO*SI*O4,NA2*CO*GE*O4,. . .A2. . . . . .PHYS. STAT. SOLIDI A VOL.18 NO.1 209, (1973).
&STRUCTURES MAGNETIQUES DE NA2*CO*SI*O4 ET NA2*CO*GE*O4.
&WINTENBERGER(M.), LAMBERT-ANDRON(B.).

1966 NA*CR*(S2,SE2),. . . . . .A1,A2. . . .J. SOLID STATE CHEM. VOL.6,. . . . . 574, (1973).
&CRYSTAL STRUCTURES AND MAGNETIC STRUCTURES OF SOME METAL(I) CHROMIUM
(III) SULFIDES AND SELENIDES.
&ENGELSMAN(F.M.R.), WIEGERS(G.A.), JELLINEK(F.), VAN-LAAR(B.).

1967 NA*D,. . . . . .P. . . . . . .PHYS. REV. VOL.83,. . . . . 700, (1951).
&NEUTRON-DEUTERON SCATTERING AMPLITUDES.
&WOLLAN(E.O.), SHULL(C.G.), KOEHLER(W.C.).

SECTION   5 -STRUCTURE AND CROSS-SECTION DETERMINATIONS . . . .

1968 NA*D*F2, ...A4. . . . . . .COLL. INTER. N.126 (GRENOBLE), .   50, (1963).
      ↪THE NATURE OF THE HYDROGEN BOND IN THE BIFLUORIDE ION.
      ↪IBERS(J.A.).

1969 NA*FE*O2,. . . .   . .A2. . . . . . .COLL. INTER. N.126 (GRENOBLE), .   121, (1963).
      ↪STRUCTURE MAGNETIQUE DE (BETA)-FE*NA*O2 ET RAFFINEMENT DES POSITIONS
      ATOMIQUES.
      ↪BERTAUT(E.F.), DELAPALME(A.), BASSI(G.).

1970 NA2*H*AS*O4*7H2*O, ...A1. . . . . . .ACTA CRYST. VOL.B27. . . . .   354, (1971).
      ↪A NEUTRON DIFFRACTION STUDY OF THE CRYSTAL STRUCTURE OF SODIUM ARSENATE
      HEPTAHYDRATE, NA2*H*AS*O4*7H2*O.
      ↪FERRARIS(G.), JONES(D.W.), YERKESS(J.).

1971 NA*H*F2, ...A4. . . . . . .COLL. INTER. N.126 (GRENOBLE), .   50, (1963).
      ↪THE NATURE OF THE HYDROGEN BOND IN THE BIFLUORIDE ION.
      ↪IBERS(J.A.).

1972 NA*H,NA*D, ...A1. . . . . . .PHYS. REV. VOL.73. . . . .   842, (1948).
      ↪NEUTRON DIFFRACTION STUDIES OF NA*H AND NA*D.
      ↪SHULL(C.G.), WOLLAN(E.O.), MORTON(G.A.), DAVIDSON(W.L.).

1973 NA*H3*(SE*O3)2,.   . . . . .A1. . . . . . .FERROELECTRICS . . . .   31, (1970).
      ↪NEUTRON DIFFRACTION STUDY ON PARAELECTRIC NA*H3*(SE*O3)2.
      ↪KAPLAN(S.F.), KAY(M.I.), MOROSIN(B.).

1974 NA-K,. . . . . .A3. . . . . . .J. CHEM. PHYSICS VOL.44. . . . .   1758, (1966).
      ↪ATOMIC STRUCTURE AND CORRELATION IN LIQUID BINARIES BY X-RAY AND NEUTRON
      DIFFRACTION WITH APPLICATION TO NA-K.
      ↪HENNINGER(E.H.), BUSCHERT(R.C.), HEATON(L.).

1975 NA*K*C4*H4*O6*4(H2*O), ...A1. . . . . . .J. PHYS. CHEM. SOL. 24. . . . . .   1341, (1963).
      ↪STRUCTURAL EFFECTS OF IONIZING RADIATION IN FERROELECTRIC ROCHELLE SALT.
      ↪BOUTIN(H.), FRAZER(B.C.), JONA(F.).

1976 (NA,LI)7*TH6*F31,. . . . . .A1. . . . . . .ACTA CRYST. VOL.B25. . . . . . .   2519, (1969).
      ↪THE NEUTRON AND X-RAY CRYSTAL STRUCTURE OF (NA,LI)7*TH6*F31 WITH MORE
      THAN ONE LEAST-SQUARES MINIMUM.
      ↪BRUNTON(G.), SEARS(D.R.).

1977 NA*MN*F3,. . . .   . .A2. . . . . . .COLL. INTER. N.126 (GRENOBLE), .   141, (1963).
      ↪MAGNETIC STRUCTURE OF BINARY FLUORIDES CONTAINING MN(+2).
      ↪PICKART(S.J.), ALPERIN(H.A.).

1978 NA*N*O2, ...A4. . . . . . .2ND INTERNAT. MEET./FERROELECTR. . . , (1969).
      ↪NEUTRON DIFFRACTION STUDY OF NA*N*O2.
      ↪KAY(M.I.), GONZALO(J.A.). (ED. NOTE: PUBL. PHYS. SOC. JAPAN, TOKYO,
      JAPAN 1969).

1979 NA*N*O3, ...A1. . . . . . .ACTA CRYST. VOL.B28. . . . . . .   2700, (1972).
      ↪THE STUDY OF SODIUM NITRATE BY NEUTRON DIFFRACTION.
      ↪PAUL(G.L.), PRYOR(A.W.).

1980 NA*N*O2, ...A4. . . . . . .J. PHYS. SOC. JAPAN VOL.23. . . .   461, (1967).
      ↪NEUTRON DIFFRACTION STUDIES OF ANISOTROPIC ROTATIONAL MOTIONS OF N*O2
      RADICALS IN NA*N*O2.
      ↪IWATA(Y.), TOKUNAGA(M.), MITANI(S.), FUKUI(S.), KOYANO(N.), SHIBUYA(I.).

1981 NA*N*O2, ...A1. . . . . . .J. PHYS. SOC. JAPAN VOL.17 B-II,   389, (1962).
      ↪X-RAY AND NEUTRON STUDY ON THE PHASE TRANSFORMATION OF NA*N*O2.
      ↪KAY(M.I.), FRAZER(B.C.), UEDA(R.).

1982 NA*N*O2, ...A4. . . . . . .J. PHYS. SOC. JAPAN VOL.35. . . .   628, (1973).
      ↪DIRECT OBSERVATION OF TRANSIENT PHENOMENA BY THE NEUTRON TIME-OF-FLIGHT
      METHOD.
      ↪NIIMURA(N.), MUTO(M.).

1983 NA*N*O2, ...A1. . . . . . .ACTA CRYST. VOL.15. . . . .   506, (1962).
      ↪THE DISORDERED STRUCTURE OF NA*N*O2 AT 185 DEGREES C.
      ↪KAY(M.I.), FRAZER(B.C.), UEDA(R.).

1984 NA*N*O2, ...A1. . . . . . .ACTA CRYST. VOL.14. . . . .   56, (1961).
      ↪A NEUTRON DIFFRACTION REFINEMENT OF THE LOW TEMPERATURE PHASE OF NA*N*O2
      ↪KAY(M.I.), FRAZER(B.C.).

1985 NA,NA*(BR,CL,F), ...A4. . . . . . .PHYS. REV. VOL.73. . . . .   830, (1948).
      ↪THE DIFFRACTION OF NEUTRONS BY CRYSTALLINE POWDERS.
      ↪WOLLAN(E.O.), SHULL(C.G.).

1986 NA2*NI*(AL,FE)*F7, ...A2. . . . . . .INT. J. MAGN. VOL.5. . . . . . .   119, (1973).
      ↪MAGNETIC PROPERTIES OF THE FERRIMAGNET NA2*NI*FE*F7 AND THE LINEAR
      ANTIFERROMAGNET NA2*NI*AL*F7.
      ↪HEGER(G.).

1987 NA*NI*F4,. . . .   . .A2. . . . . . .PHYS. REV. 174,. . . . .   560, (1968).
      ↪MAGNETIC STRUCTURE OF NA*NI*F4.
      ↪EPSTEIN(A.), MAKOVSKY(J.), MELAMUD(M.), SHAKED(H.).

1988 NA2*NI*FE*F7,. .   . . . .A2. . . . . . .SOL. STATE COMM. VOL.11. . . . .   1119, (1972).
      ↪THE MAGNETIC STRUCTURE OF NA2*NI*FE*F7.
      ↪HEGER(G.), VIEBAHN-HANSLER(R.).

1989 NA*O2, ...A2. . . . . . .J. APP. PHYS. 37,. . . . .   1040, (1966).
      ↪MAGNETIC PROPERTIES OF NA*O2 AND K*O2.
      ↪SPARKS(J.T.), KOMOTO(T.).

1990 NA2*O-AL2*O3-SI*O2-TI*O2,. . . . . . .A1. . . . . . .J. APPL. CRYST. VOL.7. . . . . .   207, (1974).
      ↪SMALL-ANGLE NEUTRON SCATTERING ON SILICA GLASSES CONTAINING TITANIA.
      ↪LOSHMANOV(A.A.), SIGAEV(V.N.), KHODAKOVSKAYA(R.YA.), PAVLUSHKIN(N.M.),
      YAMZIN(I.I.).

1991 (NA2*O)*(SI*O2)*6(H2*O), ...A1. . . . . . .ACTA CRYST. VOL.B27. . . . . . .   2269, (1971).
      ↪SODIUM SILICATE HYDRATES.IV. LOCATION OF HYDROGEN ATOMS IN
      (NA2*O)*(SI*O2)*6(H2*O) BY NEUTRON DIFFRACTION.
      ↪WILLIAMS(P.P.), DENT-GLASSER(L.S.).

1992 NAPHTHALENE, ...A4. . . . . . .ADV. STRUC. RES. DIFFR. METHOD 4   1, (1972).
      ↪CONSTRAINED REFINEMENTS IN CRYSTALLOGRAPHY.
      ↪PAWLEY(G.S.).

1993 NA*S*H,. . . . . .A1. . . . . . .J. CHEM. PHYS. VOL.55. . . . . .   5363, (1971).
      ↪NEUTRON DIFFRACTION STUDY OF THE TRIGONAL AND CUBIC PHASES OF NA*S*H.
      ↪SCHROEDER(L.W.), DE-GRAAF(L.A.), RUSH(J.J.).

SECTION   5 -STRUCTURE AND CROSS-SECTION DETERMINATIONS . . . .

1994 NA2*S2*03*5H2*O, ...A1. . . . . . .ACTA CRYST. VOL.B27, ... 253, (1971).
      *NEUTRON DIFFRACTION STUDY OF SODIUM THIOSULPHATE PENTAHYDRATE,
      NA2*S2*03*5H2*O.
      *PADMANABHAN(V.M.), YADAVA(V.S.), NAVAPRO(Q.O.), GARCIA(A.), KARSONO(L.),
      IL-HWAN-SUH, LIN-SHI-CHIEN.

1995 NA2*SI2*05,. . . . . .A4. . . . . . .NATURWISSENSCHAFTEN VOL.53, ... 16, (1966).
      *NEUTRON DIFFRACTION IN NA2*SI2*05-, NA2*SI*O3- AND SI*O2-GLASSES AND THE
      ADDITIVE PROPERTIES OF THE INTENSITY DATA.
      *HOFFMAN(W.), FISCHER(P.), MAIER(G.). (IN GERMAN).

1996 (NA-W)*O3, ...A1. . . . . . .J. CHEM. PHYS. VOL.32, ... 627, (1960).
      *NEUTRON DIFFRACTION STUDY ON SODIUM TUNGSTEN BRONZES NA(X)*W*O3.
      *ATOJI(M.), RUNDLE(R.E.).

1997 NA7*ZR6*F31, ...A1. . . . . . .ACTA CRYST. B24, ... 230, (1968).
      *THE CRYSTAL STRUCTURE OF NA7*ZR6*F31.
      *BURNS(J.H.), ELLISON(R.D.), LEVY(H.A.).

1998 NB,. . . . . .A4. . . . . . .PHYS. STAT. SOL. VOL.48, ... 619, (1971).
      *NEUTRON DIFFRACTION BY VORTEX LATTICES IN SUPERCONDUCTING NB AND NB-TA.
      *SCHELTEN(J.), ULLMAIER(H.), SCHMATZ(W.).

1999 NB,. . . . . .A4. . . . . . .PROC. INT. CONF. (NOTTINGHAM), . 285, (1964).
      *NEUTRON DIFFRACTION VORTEX LINES IN SUPERCONDUCTING NIOBIUM.
      *CRIBIER(D.),   FARNOUX(B.),   JACROT(B.),   MADHAV-RAO(L.),   VIVET(B.)
      ANTONINI(M.).

2000 NB,. . . . . .A2. . . . . . .PHYS. LETT. 9, ... 106, (1964).
      *MISE EN EVIDENCE PAR DIFFRACTION DE NEUTRONS D#UNE STRUCTURE PERIODIQUE#
      CHAMP MAGNETIQUE DANS LE NIOBIUM SUPRACONDUCTEUR.
      *CRIBIER(D.), JACROT(B.), RAO(K.R.), FARNOUX(B.).

2001 NB,. . . . . .A1. . . . . . .PHYS. LETT. 20, ... 470, (1966).
      *ERHOLUNG VON STRAHLENSCHADEN IN KUBISCH-RAUMZENTRIERTEN METALLEN NACH
      NEUTRONEN BESTRAHLUNG BEI 4.5 DEG. K.
      *BURGER(G.), ISEBECK(K.), KERLER(R.), VOLKL(J.), WENZL(H.).

2002 NB,. . . . . .A4. . . . . . .PHYS. STAT. SOLIDI B VOL.58 NO.2 633, (1973).
      *MISALIGNMENT OF FLUX LINES IN A TYPE II SUPERCONDUCTOR STUDIED BY
      NEUTRON SMALL-ANGLE DIFFRACTION.
      *LIPPMANN(G.), SCHELTEN(J.), HENDRICKS(R.W.), SCHMATZ(W.).

2003 NB,. . . . . .A4. . . . . . .PHYS. STAT. SOLIDI B VOL.57 NO.2 515, (1973).
      *MICROSCOPIC MAGNETIC FIELD DISTRIBUTION IN SUPERCONDUCTING NIOBIUM
      SINGLE CRYSTALS IN THE #DIRTY LIMIT#.
      *WEBER(H.W.), SCHELTEN(J.), LIPPMANN(G.).

2004 NB,. . . . . .A4. . . . . . .J. PHYSIQUE TOME 34, ... 447, (1973).
      *FABRICATION ET ETUDE D#UN MONOCRISTAL DE VORTEX DANS LE NIOBIUM
      SUPRACONDUCTEUR.
      *THOREL(P.), KAHN(R.), SIMON(Y.), CRIBIER(D.).

2005 NB,. . . . . .A2. . . . . . .Z. PHYSIK VOL.253, ... 219, (1972).
      *LOCAL MAGNETIC FIELD DISTRIBUTIONS IN SUPERCONDUCTING NIOBIUM AT 4.2 K
      BY NEUTRON DIFFRACTION.
      *SCHELTEN(J.), ULLMAIER(H.), LIPPMANN(G.).

2006 NB,. . . . . .A4. . . . . . .PROG. LOW TEMP. PHYS. VOL.5, . . 161, (1967).
      *STUDY OF THE SUPERCONDUCTIVE MIXED STATE BY NEUTRON-DIFFRACTION.
      *CRIBIER(D.), JACROT(B.), MADHAV-RAO(L.), FARNOUX(B.).

2007 NB,. . . . . .A4. . . . . . .SOL. STATE COMM. VOL.13, ... 1839, (1973).
      *ANISOTROPIC FLUX-LINE LATTICE IN SUPERCONDUCTING NIOBIUM.
      *KAHN(R.), PARETTE(G.).

2008 NB,. . . . . .A4. . . . . . .PHYS. REV. LETT. VOL.28, ... 1370, (1972).
      *OBSERVATION OF A SINGLE CRYSTAL OF VORTEX LINES IN A TYPE-II SUPER-
      CONDUCTOR.
      *CRIBIER(D.), SIMON(Y.), THOREL(P.).

2009 NB,. . . . . .A4. . . . . . .J. LOW TEMP. PHYS. VOL.14, . . . 213, (1974).
      *NEUTRON DIFFRACTION STUDIES OF THE MORPHOLOGY OF FLUX LINE CRYSTALS IN
      TYPE II SUPERCONDUCTORS.
      *SCHELTEN(J.), LIPPMANN(G.), ULLMAIER(H.).

2010 NB,. . . . . .A4. . . . . . .J. APPL. CRYST. VOL.7, ... 236, (1974).
      *NEUTRON DIFFRACTION BY VORTEX LATTICES IN SUPERCONDUCTING NIOBIUM.
      *LIPPMANN(G.), SCHELTEN(J.).

2011 NB-D,. . . . . .A1. . . . . . .SOV. PHYS. SOL. STATE VOL.10,. . 1076, (1968).
      *NEUTRON SCATTERING STUDY OF STRUCTURE AND PHASE TRANSITIONS IN NIOBIUM
      HYDRIDES AND DEUTERIDES.
      *SOMENKOV(V.A.), GURSKAYA(A.V.), ZEMLYANOV(M.G.), KOST(M.E.),
      CHERNOPLEKOV(N.A.), CHERTKOV(A.A.).

2012 NB-D,. . . . . .A1. . . . . . .SOV. PHYS. CRYST. VOL.14, . . . 522, (1970).
      *NEUTRON-DIFFRACTION STUDY OF THE DECOMPOSITION AND ORDERING IN THE
      NIOBIUM-HYDROGEN SYSTEM.
      *SOMENKOV(V.A.),   PETRUNIN(V.F.),   SHIL#SHTEIN(S.SH.),   CHERTKOV(A.A.).

2013 NB-H,. . . . . .A4. . . . . . .J. PHYS. CHEM. SOLIDS VOL.31,. . 2361, (1970).
      *QUASIELASTIC NEUTRON SCATTERING BY HYDROGEN IN NIOBIUM.
      *GISSLER(W.), ALEFOLD(G.), SPRINGER(T.).

2014 NB-H,. . . . . .A1. . . . . . .SOV. PHYS. SOL. STATE VOL.10,. . 1076, (1968).
      *NEUTRON SCATTERING STUDY OF STRUCTURE AND PHASE TRANSITIONS IN NIOBIUM
      HYDRIDES AND DEUTERIDES.
      *SOMENKOV(V.A.), GURSKAYA(A.V.), ZEMLYANOV(M.G.), KOST(M.E.),
      CHERNOPLEKOV(N.A.), CHERTKOV(A.A.).

2015 NB*K2*F7,. . . . . .A1. . . . . . .COLL. INTER. N.126 (GRENOBLE), . 45, (1963).
      *RECENT CRYSTAL STRUCTURE DETERMINATIONS BY NEUTRON DIFFRACTION AT
      OAK RIDGE.
      *BROWN(G.M.), LEVY(H.A.).

2016 NB*LI3*O4, ...A1. . . . . . .BULL. SOC. FRANC. MIN. CRIST. 88 345, (1965).
      *ACCURATE STRUCTURE OF NB*LI3*O4.
      *GRENIER(J.C.), BASSI(G.). (IN FRENCH).

2017 NB2*M4*O9 (M=MG,MN,FE,CO,NI),. . . . . .A2. . . . . . .C.R. ACAD. SCI. VOL.251, ... 1733, (1960).
      *THE CRYSTAL STRUCTURE AND MAGNETIC STRUCTURE OF NIOBATES AND TANTALATES
      OF BIVALENT TRANSITION METALS.
      *BERTAUT(E.F.), CORLISS(L.), FORRAT(F.). (IN FRENCH).

SECTION   5 -STRUCTURE AND CROSS-SECTION DETERMINATIONS . . . .

2018 NB2*MN4*O9,NB2*CO4*O9, ... A2. . . . . J. PHYS. CHEM. SOLIDS VOL.21, ... 234, (1961).
      *A STUDY OF THE NIOBATES AND TANTALATES OF BIVALENT TRANSITION METALS.
      *BERTAUT(E.F.), CORLISS(L.), FORRAT(F.), ALEONARD(R.), PAUTHENET(R.).
      (IN FRENCH).

2019 NB*O, ... A1. . . . . . . ACTA CRYST. 21, ... 843, (1966).
      *THE CRYSTAL STRUCTURE OF NIOBIUM MONOXIDE.
      *BOWMAN(A.L.), WALLACE(T.C.), YARNELL(J.L.), WENZEL(R.G.).

2020 NB3*SN, ... A4. . . . . . . PHYS. REV. B VOL.4, ... 2957, (1971).
      *NEUTRON SCATTERING STUDY OF THE LATTICE-DYNAMICAL PHASE TRANSITION
      IN NB3*SN.
      *SHIRANE(G.), AXE(J.D.).

2021 NB-TA, ... A4. . . . . . PHYS. STAT. SOL VOL.48, ... 619, (1971).
      *NEUTRON DIFFRACTION BY VORTEX LATTICES IN SUPERCONDUCTING NB AND NB-TA.
      *SCHELTEN(J.), ULLMAIER(H.), SCHMATZ(W.).

2022 NB-TA, ... T. . . . PHYS. REV. B VOL.9, ... 130, (1974).
      *APPLICATIONS OF THE BOSON FORMALISM TO MAGNETIC PROPERTIES OF SUPER-
      CONDUCTORS.
      *MANCINI(F.), SCARPETTA(G.), SRINIVASAN(V.), UMEZAWA(H.).

2023 ND, ... A2. . . . . . . J. PHYSIQUE TOME 32 COL.1 VOL.I, 370, (1971).
      *THE MAGNETIC STRUCTURES OF PRASEODYMIUM AND NEODYMIUM.
      *LEBECH(B.), RAINFORD(B.D.).

2024 ND, ... A2. . . . . . . CONF./RARE EARTHS/ACTINIDES, . . 43, (1971).
      *MAGNETIC FIELD DEPENDENCE OF THE MAGNETIC STRUCTURE OF NEODYMIUM.
      *LEBECH(B.), RAINFORD(B.D.).

2025 ND, ... A2. . . . . . . REF. SECTION 3, 65/KOEHLER, ...

2026 ND, ... A2. . . . . . . PHYS. REV. LETT. VOL.25, ... 524, (1970).
      *CRYSTAL FIELDS AND THE MAGNETIC PROPERTIES OF PRASEODYMIUM AND NEODYMIUM
      *JOHANSSON(T.), LEBECH(B.), NIELSEN(M.), MOLLER(H.B.), MACKINTOSH(A.R.).

2027 ND, ... A2. . . . . . . J. APPL. PHYS. VOL.35, ... 1041, (1964).
      *MAGNETIC STRUCTURE OF NEODYMIUM.
      *MOON(R.M.), CABLE(J.W.), KOEHLER(W.C.).

2028 ND, ... P. . . . . . . PHYS. REV. VOL.91, ... 597, (1953).
      *SLOW NEUTRON SCATTERING CROSS SECTIONS FOR RARE EARTH NUCLIDES.
      *KOEHLER(W.C.), WOLLAN(E.O.).

2029 ND, ... P. . . . . . . J. NUCL. ENERGY VOL.7, ... 199, (1958).
      *THE SLOW NEUTRON TOTAL CROSS-SECTION OF ND(A=143).
      *HAY(H.J.).

2030 ND*AL2, ... A2. . . . . . . J. APP. PHYS. 37, ... 4575, (1966).
      *MAGNETIC PROPERTIES OF DY*AL2 AND ND*AL2.
      *NERESON(N.G.), OLSEN(C.E.), ARNOLD(G.P.).

2031 ND*BI, ... A2. . . . . . . J. APPL. PHYS. VOL.42, ... 1625, (1971).
      *MAGNETIC PROPERTIES OF CE*BI, ND*BI, TB*BI, AND DY*BI.
      *NERESON(N.), ARNOLD(G.).

2032 ND*C2, ... A1,2. . . . . PHYS. LETT. 22, ... 21, (1966).
      *CRYSTALS AND MAGNETIC STRUCTURES OF CE*C2, PR*C2 AND ND*C2.
      *ATOJI(M.).

2033 ND*C2, ... A1,A2. . . . . J. CHEM. PHYS. VOL.46, ... 1891, (1967).
      *MAGNETIC AND CRYSTAL STRUCTURES OF CE*C2,PR*C2,ND*C2,TB*C2, AND HO*C2
      AT LOW TEMPERATURES.
      *ATOJI(M.).

2034 ND*C*O3, ... A2,A4. . . . REF. SECTION 3, 66/BERTAUT, ...

2035 ND*CO5, ... A2. . . . . . . REF. SECTION 3, 64/BERTAUT, ...

2036 ND*CO2, ... A2. . . . . . . J. APP. PHYS. 36, ... 978, (1965).
      *MAGNETIC STRUCTURE RARE EARTH-COBALT (R*CO2) INTERMETALLIC COMPOUNDS.
      *MOON(R.M.), KOEHLER(W.C.), FARRELL(J.).

2037 ND*CO5, ... A2. . . . . . . J. PHYS. CHEM. SOL. 27, ... 1287, (1966).
      *ETUDE MAGNETIQUE DU COMPOSE INTERMETALLIQUE ND*CO5.
      *BARTHOLIN(H.), VAN-LAAR(B.), LEMAIRE(R.), SCHWEIZER(J.).

2038 ND*CO5-D, ... A1. . . . . . . J. PHYS. CHEM. SOLIDS VOL.35, . . 301, (1974).
      *A NEUTRON-DIFFRACTION STUDY ON THE STRUCTURAL RELATIONSHIPS OF R*CO5
      HYDRIDES.
      *KUIJPERS(F.A.), LOOPSTRA(B.O.).

2039 ND*CR*O3, ... A2. . . . . . . SOL. STATE COMM. 5, ... 93, (1967).
      *ETUDE DE LA STRUCTURE MAGNETIQUE DES CHROMITES D#ERBIUM ET DE NEODYME
      PAR DIFFRACTION NEUTRONIQUE.
      *BERTAUT(E.F.), MARESCHAL(J.).

2040 ND*FE*O3, ... A2. . . . . . . SOL. STATE COMM. VOL.10, ... 663, (1972).
      *LONG WAVELENGTH NEUTRON DIFFRACTION STUDY OF THE MAGNETIC STRUCTURES OF
      PR*FE*O3 AND ND*FE*O3.
      *PINTO(H.), SHAKED(H.).

2041 ND*FE*O3, ... A2. . . . . . . PHYS. LETT. 29A, ... 659, (1969).
      *MAGNETIC STRUCTURE OF ND*FE*O3 AT ROOM TEMPERATURE.
      *EPSTEIN(A.), SHAKED(H.).

2042 ND*FE*O3, ... A2. . . . . . . PHYS. REV. VOL.118, ... 58, (1960).
      *NEUTRON DIFFRACTION STUDY OF THE MAGNETIC PROPERTIES OF RARE-EARTH-IRON
      PEROVSKITES.
      *KOEHLER(W.C.), WOLLAN(E.O.), WILKINSON(M.K.).

2043 ND*FE2*SI2, ... A1,A2. . . . ACTA CRYST. VOL.A28 PART S-4, . . 194, (1972).
      *NEUTRON DIFFRACTION STUDY OF ND*FE2*SI2.
      *PINTO(H.), SHAKED(H.). (ABSTRACT ONLY).

2044 ND*FE2*SI2, ... A1,A2. . . . PHYS. REV. B VOL.7, ... 3261, (1973).
      *NEUTRON-DIFFRACTION STUDY OF ND*FE2*SI2.
      *PINTO(H.), SHAKED(H.).

2045 ND-GA GARNET, ... A2. . . . . . . PHYS. LETT. VOL.26A, ... 263, (1968).
      *ETUDE PAR DIFFRACTION DE NEUTRONS A 0.31 DEGRES K DE LA STRUCTURE
      MAGNETIQUE DU GRENAT DE NEODYME ET DE GALLIUM.
      *HAMMANN(J.).

SECTION   5 -STRUCTURE AND CROSS-SECTION DETERMINATIONS . . . .

2046   ND3*IN,. . . . . .A2. . . .J. SOLID STATE CHEM. VOL.9,. . .   152, (1974).
                          +MAGNETIC PROPERTIES OF LN3*IN INTERMETALLIC COMPOUNDS.
                          +HUTCHENS(R.D.), WALLACE(W.E.), NERESON(N.).

2047   ND*MN2*05, ...A2. . . . .PHYS. STAT. SOLIDI A VOL.17 NO.1 191, (1973).
                          +STRUCTURES MAGNETIQUES SINUSOIDALES ET HELICOIDALES DE LA TERRE RARE
                          DANS T*MN2*05 (T=ND,TB,ER).
                          +BUISSON(G.).

2048   ND*(N,P,AS,SB),.   . . . . .A2. . . . .J. PHYS. C VOL.6,. . . . . . .   725, (1973).
                          +MAGNETIC ORDERING OF NEODYMIUM MONOPNICTIDES DETERMINED BY NEUTRON
                          DIFFRACTION.
                          +SCHOBINGER-PAPAMANTELLOS(P.),   FISCHER(P.),   VOGT(O.),   KALDIS(E.).

2049   ND2*03,. . . . . .A1. . . . .ACTA CRYST. VOL.6,. . . . . . .   741, (1953).
                          +NEUTRON-DIFFRACTION STUDY OF THE STRUCTURE OF THE A-FORM OF THE RARE
                          EARTH SESQUIOXIDES.
                          +KOEHLER(W.C.), WOLLAN(E.O.).

2050   ND*(O*H)2*CL,. .   . . . .A1. . . . .SOV. PHYS. DOKLADY VOL.17,. . . 1131, (1973).
                          +X-RAY AND NEUTRON-DIFFRACTION RESULTS ON THE CRYSTAL STRUCTURE OF
                          ND*(O*H)2*CL.
                          +BUKIN(V.I.).

2051   ND*S,. . . . . .A2. . . . .PHYS. STAT. SOL. B VOL.56 NO.1,.   61, (1973).
                          +MAGNETIC ORDERING IN ND*S.
                          +GOLOSOVSKII(I.V.), PLAKHTII(V.P.).

2052   ND*SB, ...A1. . . . .AIP CONF. PROC. VOL.5, ...   1385, (1971).
                          +NEUTRON DIFFRACTION STUDIES ON DY*SB, ND*SB, AND CE*SB.
                          +NERESON(N.), STRUEBING(V.).

2053   ND-TH, ...A2. . . . .J. APP. PHYS. 39,. . . . . . .   1329, (1968).
                          +MAGNETIC PROPERTIES OF FCC RARE EARTH-THORIUM ALLOYS.
                          +CHILD(H.R.), KOEHLER(W.C.), MILLHOUSE(A.H.).

2054   ND-V,. . . . . .A2,A4. . .CONF./RARE EARTHS/ACTINIDES, . .   206, (1971).
                          +NEUTRON SCATTERING BY ND-V COMPOUNDS: MAGNETIC PROPERTIES AND CRYSTAL
                          FIELD EFFECTS.
                          +FISCHER(P.), FURRER(A.), HEER(H.), HALG(W.)...ET AL.

2055   ND*ZN2,. . . . . .A2. . . . .J. CHEM. PHYS. VOL.58, ...   1783, (1973).
                          +MAGNETIC STRUCTURE OF ND*ZN2.
                          +DEBRAY(D.), SOUGI(M.).

2056   NE,. . . . . .A3. . . . .PHYS. CAN. VOL.27,. . . . . .   68, (1971).
                          +OSCILLATORY BEHAVIOUR IN THE NEUTRON SCATTERING FROM LIQUID NEON AT HIGH
                          MOMENTUM TRANSFER.
                          +BUYERS(W.J.L.), DELONNGI(D.A.), LONNGI(P.A.).

2057   NE,. . . . . .A3. . . . .NUCL. PHYS. VOL.A133,. . . . .   410, (1969).
                          +SLOW-NEUTRON CROSS SECTIONS OF HE, NE, AR AND KR.
                          +RORER(D.C.), ECKER(B.M.), AKYUZ(R.O.).

2058   NE,. . . . . .A3. . . . .J. CHEM. PHYS. VOL.55,. . . .   4967, (1971).
                          +STRUCTURE STUDY OF LIQUID NEON BY NEUTRON DIFFRACTION.
                          +DE-GRAAF(L.A.), MOZER(B.).

2059   NE,. . . . . .A3. . . . .J. CHEM. PHYS. VOL.57, ...   3987, (1972).
                          +THREE ATOM CORRELATIONS IN LIQUID NEON.
                          +RAVECHE(H.J.), MOUNTAIN(R.D.).

2060   NE,. . . . . .P,A3. . . .J. PHYSIQUE TOME 24,. . . . . .   21, (1963).
                          +DETERMINATION OF ABSORPTION AND SCATTERING CROSS-SECTIONS OF RARE-GASES
                          FOR THERMAL NEUTRONS.
                          +GENIN(R.), BEIL(H.)... ET AL.

2061   NE,. . . . . .A1,A3. . .PHYS. REV. VOL.111,. . . . . . .   1470, (1958).
                          +ATOMIC DISTRIBUTION IN LIQUID AND SOLID NEON AND SOLID ARGON BY
                          NEUTRON DIFFRACTION.
                          +HENSHAW(D.G.).

2062   NE,. . . . . .P. . . . .Z. NATURFORSCH. VOL.15A,. . .   828, (1960).
                          +THE MEASUREMENT OF THE TOTAL CROSS-SECTION OF SOLID NEON AND ARGON FOR
                          SLOW NEUTRONS.
                          +SPRINGER(T.), WIEDEMANN(W.). (IN GERMAN).

2063   NE,. . . . . .T,A3. . .J. PHYS. CHEM. SOLIDS VOL.35,. .   585, (1974).
                          +TRIPLET CORRELATION AND PAIR POTENTIAL FUNCTIONS IN LIQUID NEON AND
                          SODIUM.
                          +WASEDA(Y.), OHTANI(M.), SUZUKI(K.).

2064   NI,. . . . . .A2. . . . .METAL. SOC. CONF. LOS ANGELES,   15, (1967).
                          +SPIN DENSITY DISTRIBUTION IN FE, CO, AND NI.
                          +SHULL(C.G.).

2065   NI,. . . . . .A2. . . . .SOV. PHYS. J.E.T.P. VOL.29,. .   261, (1969).
                          +INVESTIGATION OF A PHASE TRANSITION IN NICKEL WITH POLARIZED NEUTRONS.
                          +DRABKIN(G.M.), ZABIDAROV(E.I.), KASMAN(YA.A.), OKOROKOV(A.I.).

2066   NI,. . . . . .P. . . . .ACTA CRYST. VOL.A25,. . . . .   714, (1969).
                          +AN EXPERIMENTAL DETERMINATION OF THE DEBYE-WALLER TEMPERATURE FACTOR
                          FOR NICKEL.
                          +COOPER(M.J.), TAYLOR(R.I.).

2067   NI,. . . . . .A3. . . . .PHYS. STAT. SOL. VOL.39,. . .   181, (1970).
                          +NEUTRON DIFFRACTION STUDY OF NICKEL IN THE LIQUID STATE.
                          +WASEDA(Y.), SUZUKI(K.), TAMAKI(S.), TAKEUCHI(S.).

2068   NI (ISO. A=64),.   . . . . .P. . . . .PHYS. SCRIPTA(SWEDEN) VOL.4,.   89, (1971).
                          +THERMAL NEUTRON CAPTURE IN TI(ISOTOPE A=50) AND NI(ISOTOPE A=64).
                          +ARNELL(S.E.), HARDELL(R.), HASSELGREN(A.), MATTSSON(C.G.),
                          SKEPPSTEDT(O.).

2069   NI (ISO. A=60),.   . . . . .P. . . . .THESIS (231PP.). . . . . . .   (1970).
                          +KEY NEUTRON CAPTURE AND TRANSMISSION MEASUREMENTS ON CR(ISOTOPES A=50,
                          52, 53, 54), NI(ISOTOPE A=60), AND V.
                          +STIEGLITZ(R.G.). (ED. NOTE: AVAIL. UNIV. MICROFILMS, ANN ARBOR, MICH.
                          ORDER NO. 71-12829).

2070   NI (ISO. A=65),.   . . . . .P. . . . .NUCL. APPL. TECHNOL. VOL.9,.   662, (1970).
                          +AVERAGE THERMAL NEUTRON CAPTURE CROSS SECTIONS OF AU(ISO. A=198), NI
                          (ISO. A=65), AND CU(ISO. A=66).
                          +SERMENT(V.), ABU-SAMRA(A.), EMMONS(A.H.).

SECTION 5 -STRUCTURE AND CROSS-SECTION DETERMINATIONS . . . .

2071 NI,. . . . . .A4. . . . . .2ND INTERNAT. CONF./NUCL. DATA;. . . . .(1970).
⇌NEUTRON ELASTIC SCATTERING CROSS SECTIONS OF VANADIUM, CHROMIUM, IRON
AND NICKEL.
⇌HOLMQVIST(B.), WIEDLING(T.).

2072 NI,. . . . . .ACTA CRYST. (INTERACT.) VOL.A25, S218, (1969).
⇌NEUTRON SMALL ANGLE SCATTERING AT DISLOCATIONS IN COPPER AND NICKEL
SINGLE CRYSTALS DEFORMED IN TENSION.
⇌HERGET(P.), SCHMATZ(W.).

2073 NI (ISO. A=62),. . . . . .P. . . . . . .J. INORG. NUCL. CHEM. VOL.32,. . . 2839, (1970).
⇌THE THERMAL NEUTRON CAPTURE CROSS-SEC. AND RES. CAPTURE INTEG. OF CA(IS.
A=44), NI(IS. A=62), YB(IS. A=168, 174), TM(IS. A=169) AND TL(IS. A=203)
⇌SIMS(G.H.E.), JUHNKE(D.G.).

2074 NI,. . . . . .A4. . . . . . . .PHYS. REV. 181,. . . . . 920, (1969).
⇌POLARIZATION ANALYSIS OF THERMAL-NEUTRON SCATTERING.
⇌MOON(R.M.), RISTE(T.), KOEHLER(W.C.).

2075 NI,. . . . . .A2. . . . . . .PHYS. REV. 148,. . . . . 495, (1966).
⇌MAGNETIC MOMENT DISTRIBUTION OF NICKEL METAL.
⇌MOOK(H.A.).

2076 NI,. . . . . .A2. . . . . .J. APP. PHYS. 36,. . . . . . . 1093, (1965).
⇌NEUTRON DIFFRACTION STUDIES OF PD, NI, FE*MN, AND CU(MN) SINGLE CRYSTALS
⇌ARROTT(A.).

2077 NI,. . . . . .P. . . . . . .PHYS. REV. VOL.156,. . . . . . . 1225, (1967).
⇌THERMAL-NEUTRON COHERENT SCATTERING AMPLITUDES AND CROSS SECTIONS OF
NICKEL-61 AND NICKEL-64.
⇌SIDHU(S.S.), ANDERSON(K.D.).

2078 NI,. . . . . .A4. . . . . . .PHYS. LETT. VOL.27A,. . . . . . 69, (1968).
⇌NEUTRON DEPOLARIZATION MEASUREMENTS IN NICKEL NEAR THE CURIE POINT.
⇌BAKKER(H.K.), REKVELDT(M.TH.), VAN-LOEF(J.J.).

2079 NI,. . . . . .A4. . . . . . .INT. J. MAGN. VOL.5,. . . . . 223, (1973).
⇌THE SMALL-ANGLE SCATTERING OF THERMAL NEUTRONS BY BLOCH WALLS IN FE*SI
(2.5 PCT.) AND NI-SINGLE CRYSTALS.
⇌SCHAERPF(O.), VEHOFF(H.), SCHWINK(CH.). (IN GERMAN).

2080 NI,. . . . . .P. . . . . . .PHYS. REV. VOL.79,. . . . . . 395, (1950).
⇌THE COHERENT NEUTRON SCATTERING CROSS SECTIONS OF NICKEL AND ITS
ISOTOPES.
⇌KOEHLER(W.C.), WOLLAN(E.O.), SHULL(C.G.).

2081 NI,. . . . . .A2. . . . . .J. PHYS. RADIUM VOL.20,. . . . . 169, (1959).
⇌RECENT MAGNETIC STRUCTURE STUDIES BY NEUTRON DIFFRACTION.
⇌SHULL(C.G.).

2082 NI,. . . . . .A4. . . . . .J. PHYS. CHEM. SOLIDS VOL.10,. . . 147, (1959).
⇌X-RAY AND NEUTRON SCATTERING FROM ELECTRONS IN A CRYSTALLINE FIELD AND
THE DETERMINATION OF OUTER ELECTRON CONFIGURATIONS IN IRON AND NICKEL.
⇌WEISS(R.J.), FREEMAN(A.J.).

2083 NI,. . . . . .P. . . . . .J. PHYS. CHEM. SOLIDS VOL.10,. . . 138, (1959).
⇌THE USE OF POLARIZED NEUTRONS IN DETERMINING THE MAGNETIC SCATTERING
BY IRON AND NICKEL.
⇌NATHANS(R.), SHULL(C.G.), SHIRANE(G.), ANDRESEN(A.).

2084 NI,. . . . . .A2. . . . . .SOV. PHYS. SOL. STATE VOL.15,. . 919, (1973).
⇌INVESTIGATION OF THE STRUCTURE OF MAGNETIC DOMAINS IN NICKEL WITH THE
AID OF MONOCHROMATIC POLARIZED NEUTRONS.
⇌TRUNOV(V.A.), DMITRIEV(R.P.), UL*YANOV(V.A.).

2085 NI,. . . . . .P. . . . . . .PROC. PHYS. SOC. A VOL.67,. . . 248, (1954).
⇌THE SCATTERING OF SLOW NEUTRONS BY FERROMAGNETIC CRYSTALS.
⇌SQUIRES(G.L.).

2086 NI,. . . . . .A4. . . . . .CAN. J. PHYS. VOL.31,. . . . . 339, (1953).
⇌THE INITIAL MAGNETIZATION OF NICKEL UNDER TENSION.
⇌BROCKHOUSE(B.N.). (ED. NOTE: SEE APPENDIX TO THIS PAPER).

2087 NI,. . . . . .T.P. . . . . .J. APPL. PHYS. 37,. . . . . 1449, (1966).
⇌MAGNETIC FORM FACTOR OF NICKEL.
⇌HODGES(L.), LANG(N.D.), EHRENREICH(H.), FREEMAN(A.J.).

2088 NI,. . . . . .T.A2. . . . .PHYS. REV. VOL.152,. . . . . 505, (1966).
⇌INTERPOLATION SCHEME FOR BAND STRUCTURE OF NOBLE AND TRANSITION METALS:
FERROMAGNETISM AND NEUTRON DIFFRACTION IN NI.
⇌HODGES(L.), EHRENREICH(H.), LANG(N.D.).

2089 NI3-AL-(FE), ...A2. . . . . .J. PHYS. F VOL.3,. . . . . 697, (1973).
⇌GIANT MOMENTS IN IRON DOPED NI3-AL.
⇌LING(P.C.), HICKS(T.J.).

2090 NI3*B2*06, ...A2. . . . . .PHYS. STAT. SOL. 16,. . . . . K17, (1966).
⇌ANTIFERROMAGNETISM IN MN3*B3*06, CO3*B2*06, AND NI3*B2*06.
⇌NEWNHAM(R.E.), SANTORO(R.P.), SEAL(P.F.), STALLINGS(G.R.).

2091 NI3*B7*013*I (NI-I BORACITE),. . . . .A1. . . . . . .Z. KRIST. VOL.131,. . . . . 139, (1970).
⇌ROENTGEN- UND NEUTRONENBEUGUNGSUNTERSUCHUNGEN AN NI-J-BORACIT.
⇌BECKER(H.J.), WILL(G.).

2092 NI*BA*02,. . . . . .A1,2. . . . .COLL. INTER. N.126 (GRENOBLE),. . . . 18, (1963).
⇌NEUTRON DIFFRACTION STUDIES AT THE PUERTO RICO NUCLEAR CENTRE.
⇌ALMODOVAR(I.), BIELEN(H.J.), FRAZER(B.C.), KAY(M.I.).

2093 (NI*(C5*H11*N2*O)2*H)*CL.H2*O, ...A1,A4. . .J. CHEM. PHYS. VOL.54,. . . . . 3990, (1971).
⇌A SHORT, SLIGHTLY ASYMMETRICAL, INTRAMOLECULAR H BOND: A NEUTRON DIFFR.
STUDY OF BIS(2-AMINO-2-METHYL-3-BUTANONE OXIMATO)NI(II)*CL.H2*O.
⇌SCHLEMPER(E.O.), HAMILTON(W.C.), LA-PLACA(S.J.).

2094 NI*C*O3, ...A2. . . . . .J. PHYS. SOC. JAPAN VOL.17 B-III 58, (1962).
⇌ANTIFERROMAGNETISM OF NI*C*O3.
⇌ALIKHANOV(R.A.).

2095 NI*CL2-D2*O, ...A3. . . . . .J. PHYS. C VOL.7,. . . . . L111, (1974).
⇌ION DISTRIBUTION AND LONG-RANGE ORDER IN CONCENTRATED ELECTROLYTE
SOLUTIONS.
⇌HOWE(R.A.), HOWELLS(W.S.), ENDERBY(J.E.).

2096 NI*CL2*6(H2*O),. . . . . .A1. . . . . .J. CHEM. PHYSICS VOL.50,. . . . 4690, (1969).
⇌CRYSTAL STRUCTURE OF NI*CL2*6(H2*O) AT ROOM TEMPERATURE AND 4.2 K BY
NEUTRON DIFFRACTION.
⇌KLEINBERG(R.).

SECTION   5 -STRUCTURE AND CROSS-SECTION DETERMINATIONS . . . .

2097 NI*CL2.2H2*O,. . . A2 A4 PHYS. LETT. VOL.41A NO.5,. . . .   411, (1972).
            *PHASE TRANSITION IN NI*CL2.2H2*O.
            *BONGAARTS(A.L.M.),   VAN-LAAR(B.),   BOTTERMAN(A.C.),   DE-JONGE(W.J.M.).

2098 NI*CL2.6H2*O,. . . A2 J. APPL. PHYS. VOL.38, ... 1453, (1967).
            *MAGNETIC STRUCTURE OF NI*CL2.6H2*O.
            *KLEINBERG(R.).

2099 NI-CO, ... A2 J. APPL. PHYS. VOL.33, SUPPL., . 1340, (1962).
            *NEUTRON DIFFRACTION INVESTIGATIONS OF FERROMAGNETIC PALLADIUM AND IRON
            GROUP ALLOYS.
            *CABLE(J.W.), WOLLAN(E.O.), KOEHLER(W.C.), WILKINSON(M.K.).

2100 NI-CO, ... A2 J. PHYS. SOC. JAPAN VOL.17 B-III   38, (1962).
            *MAGNETIC MOMENT DISTRIBUTION IN PALLADIUM AND IRON GROUP ALLOYS.
            *WOLLAN(E.O.), CABLE(J.W.), KOEHLER(W.C.), WILKINSON(M.K.).

2101 NI-CO, ... A2 PHYS. REV. 138, ... A755, (1965).
            *DISTRIBUTION OF MAGNETIC MOMENTS IN PD-3D AND NI-3D ALLOYS.
            *CABLE(J.W.), WOLLAN(E.O.), KOEHLER(W.C.). (ED. NOTE: REF. SECTION 3,
            65/CABLE).

2102 NI-CO-P, ... A1 MATER. RES. BULL. VOL.8, ... 229, (1973).
            *SUBSTITUTION SELECTIVITY IN MM*P PHASES. ORDERING IN NI-CO-P BY NEUTRON
            DIFFRACTION STUDY.
            *SENATEUR(J.P.), ROUAULT(A.)... ET AL. (IN FRENCH).

2103 NI2*CR,. . . . . . A1 SOV. PHYS. DOKLADY VOL.14, . . . 263, (1969).
            *HEAT OF ORDERING FOR NI2*CR ALLOY.
            *VINTAIKIN(E.Z.), ITKIN(V.P.), MOGUTNOV(B.M.), URUSHADZE(G.G.).

2104 NI-CR, ... A1 SOV. PHYS. DOKLADY VOL.14, . . . 84, (1969).
            *FORMATION OF LONG-RANGE ORDER IN NICKEL-CHROMIUM ALLOYS.
            *VINTAIKIN(E.Z.), URUSHADZE(G.G.).

2105 NI-CR, ... A4 UKRAINIAN PHYS. VOL.15,. . . . . 133, (1970).
            *NEUTRON DIFFRACTION STUDY OF ATOM ORDERING IN NICKEL-CHROMIUM ALLOYS.
            *VINTAIKIN(E.Z.), URUSHADZE(G.G.).

2106 NI-CR, ... A2 J. PHYSIQUE TOME 32 COL.1 VOL.I, 575, (1971).
            *MAGNETIC POLARIZATION CLOUDS IN FERROMAGNETIC NICKEL-CHROMIUM ALLOYS.
            *RAINFORD(B.D.), ALDRED(A.T.), LOW(G.G.).

2107 NI(CR), NI(V),. . A2 ACTA CRYST. VOL.16,. . . . . . . A126, (1963).
            *MAGNETIC MOMENT DISTRIBUTIONS IN DILUTE FERROMAGNETIC ALLOYS.
            *COLLINS(M.F.), LOW(G.G.). (ED. NOTE: ABSTRACT ONLY. PAPERS PRESENTED
            AT: INT. UNION OF CRYSTALLOGRAPHY/ROME, SEPT., 1963).

2108 NI-CR, ... A2 PHYS. MET. METALLOGR. VOL.14,. . 25, (1962).
            *NEUTRON DIFFRACTION STUDY OF NI-CR ALLOYS.
            *GOMAN*KOV(V.I.),   LITVIN(D.F.),   LOSHMANOV(A.A.),   LYASHCHENKO(B.G.).

2109 NI-(CR), ... A4 PROC. PHYS. SOC. 86, ... 535, (1965).
            *THE MAGNETIC MOMENT DISTRIBUTION AROUND TRANSITION ELEMENT IMPURITIES
            IN IRON AND NICKEL.
            *COLLINS(M.F.), LOW(G.G.E.). (ED. NOTE: REF. SECTION 3, 65/COLLINS).

2110 NI-CR, ... A2 J. APPL. PHYS. VOL.34, ... 1195, (1963).
            *MAGNETIC MOMENT DISTRIBUTION IN DILUTE NICKEL ALLOYS.
            *LOW(G.G.E.), COLLINS(M.F.).

2111 NI-CR, ... A2 J. PHYS. SOC. JAPAN VOL.36,. . . 431, (1974).
            *LOCALIZED MOMENT IN NI(.95)-CR(.05) ALLOY.
            *ITO(Y.), AKIMITSU(J.).

2112 NI*CR*O4,. . . . . A2 BULL. SOC. FRANC. MIN. CRIST. 92 264, (1969).
            *MAGNETIC STRUCTURES OF COBALT AND NICKEL CHROMATES.
            *PERNET(M.), QUEZEL(G.), COING-BOYAT(J.), BERTAUT(E.F.).

2113 NI*CR2*O4, ... A2 ACTA CRYST. VOL.A28 PART S-4,. . 195, (1972).
            *APPLICATION OF REPRESENTATION ANALYSIS TO THE MAGNETIC STRUCTURE OF
            NICKEL CHROMITE SPINEL.
            *BERTAUT(E.F.). (ABSTRACT ONLY).

2114 NI*CR2*O4, ... A1 J. APPL. PHYS. VOL.32, SUPPL., . 68S, (1961).
            *STRUCTURE OF NICKEL CHROMITE.
            *PRINCE(E.).

2115 NI-CU, ... A2 PHYS. REV. B VOL.7,. . . . . . . 218, (1973).
            *MAGNETIC-MOMENT DISTRIBUTIONS IN FERROMAGNETIC NI-CU ALLOYS.
            *ALDRED(A.T.), RAINFORD(B.D.), HICKS(T.J.), KOUVEL(J.S.).

2116 NI-CU, ... A2 PHYS. REV. LETT. VOL.22,. . . . 1256, (1969).
            *THE DISTRIBUTION OF MAGNETIC MOMENT IN THE NI-CU ALLOY SYSTEM.
            *CABLE(J.W.), WOLLAN(E.O.), CHILD(H.R.).

2117 NI-CU, ... A2 ACTA CRYST. VOL.A28 PART S-4,. . 196, (1972).
            *MAGNETIC FORM FACTOR OF NI-CU ALLOY SYSTEM.
            *ITO(Y.), AKIMITSU(J.). (ABSTRACT ONLY).

2118 NI-CU, ... A2 J. PHYS. SOC. JAPAN VOL.35,. . . 1000, (1973).
            *THE BEHAVIOR OF LOCALIZED MOMENT OF NI IN NI-CU ALLOY SYSTEMS.
            *ITO(Y.), AKIMITSU(J.).

2119 NI-CU, ... T PHYS. REV. B VOL.9,. . . . . . . 2354, (1974).
            *CONTRIBUTION OF GIANT SPIN CLUSTERS TO THE RESISTIVITY, NEUTRON-
            SCATTERING CROSS SECTION, AND SPECIFIC HEAT IN ALLOYS: APPLIC. TO NI-CU.
            *LEVIN(K.), MILLS(D.L.).

2120 (NI-CU)-MN-SB, . . A2 ACTA PHYS. POLON. VOL.A43,. . . 787, (1973).
            *ATOMIC AND MAGNETIC STRUCTURE OF THE NI(X)-CU(1-X)-MN-SB.
            *SZYTULA(A.).

2121 NI5*ER,. . . . . . A2 COLL. INTER. N.126 (GRENOBLE). . 133, (1963).
            *MAGNETIC STRUCTURE STUDIES AT BROOKHAVEN NATIONAL LABORATORY.
            *CORLISS(L.M.), HASTINGS(J.M.).

2122 NI*F2, ... A2 PROC. PHYS. SOC. 85, ... 967, (1965).
            *THE MAGNETIC PROPERTIES OF NI*F2.
            *COOKE(A.H.), GEHRING(K.A.), LAZENBY(R.).

2123 NI*F2, ... A1 ACTA CRYST. VOL.B30,. . . . . . 554, (1974).
            *THE STRUCTURES OF FLUORIDES. V. THE X-PARAMETER IN NI*F2.
            *TAYLOR(J.C.), WILSON(P.W.).

SECTION   5 -STRUCTURE AND CROSS-SECTION DETERMINATIONS . . . .

2124 NI*F2, ...A2. . . . . .PHYS. REV. VOL.90, ... 779, (1953).
⌐NEUTRON DIFFRACTION STUDIES OF ANTIFERROMAGNETISM IN MANGANOUS FLUORIDE
AND SOME ISOMORPHOUS COMPOUNDS.
⌐ERICKSON(R.A.).

2125 NI*F2, ...A2. . . . . .SOV. PHYS. JETP VOL.10, ... 814, (1960).
⌐ANTIFERROMAGNETISM IN NI*F2.
⌐ALIKHANOV(R.A.).

2126 NI-FE, ...A4. . . . . .PHIL. MAG. 8, ... 401, (1963).
⌐THE MAGNETIC MOMENT DISTRIBUTION IN SOME TRANSITION METAL ALLOYS.
⌐COLLINS(M.F.), FORSYTH(J.B.).

2127 NI-FE, ...A4. . . . . .MATERIALS RES. BULL. VOL.2, . . . 69, (1967).
⌐CHARACTERIZATION OF LARGE ALLOY SINGLE CRYSTALS BY NEUTRON DIFFRACTION.
⌐NG(S.C.), BROCKHOUSE(B.N.), HALLMAN(E.D.).

2128 NI(FE), ...A2. . . . . .PROC. PHYS. SOC. VOL.89, ... 893, (1966).
⌐DISTRIBUTION OF MAGNETIZATION IN MIXED MAGNETIC SYSTEMS. III. MAGNETIC
IMPURITY.
⌐LOVESEY(S.W.).

2129 NI-FE, ...A2. . . . . .REF. SECTION 3, 66/SIDOROV, ...

2130 NI-(FE), ...A4. . . . . .REF. SECTION 3, 65/COLLINS, ...

2131 NI*FE, ...A2. . . . . .J. APPL. PHYS. VOL.34, ... 1195, (1963).
⌐MAGNETIC MOMENT DISTRIBUTIONS IN DILUTE NICKEL ALLOYS.
⌐LOW(G.G.E.), COLLINS(M.F.).

2132 NI3-FE, ...A1. . . . . .SOV. PHYS. CRYST. VOL.10, ... 338, (1965).
⌐SUPERLATTICE STRUCTURE IN NI3-FE.
⌐GOMAN≠KOV(V.I.), PUZEI(I.M.), LOSHMANOV(A.A.).

2133 NI-FE, ...A2. . . . . .SOV. PHYS. JETP VOL.34, ... 163, (1972).
⌐MAGNETIC STRUCTURE OF FACE-CENTERED CUBIC NICKEL-IRON ALLOYS.
⌐MEN≠SHIKOV(A.Z.), SIDOROV(S.K.), ARKHIPOV(V.E.).

2134 NI-FE, ...A2. . . . . .PHYS. REV. B VOL.7, ... 2005, (1973).
⌐MAGNETIC-MOMENT DISTRIBUTION IN NI-FE AND AU-FE ALLOYS.
⌐CABLE(J.W.), WOLLAN(E.O.).

2135 NI-FE, ...A2. . . . . .PHYS. REV. VOL.97, ... 304, (1955).
⌐NEUTRON DIFFRACTION STUDIES OF THE MAGNETIC STRUCTURE OF ALLOYS OF
TRANSITION ELEMENTS.
⌐SHULL(C.G.), WILKINSON(M.K.).

2136 NI-FE, ...A1. . . . . .PHYS. REV. VOL.58, ... 1031, (1940).
⌐NEUTRON STUDIES OF ORDER IN FE-NI ALLOYS.
⌐NIX(F.C.), BEYER(H.G.), DUNNING(J.R.).

2137 NI*(FE-AL)*O4, ...A1,A2. . . .PHYS. STAT. SOLIDI A VOL.17 NO.2 555, (1973).
⌐INVESTIGATION OF THE INFLUENCE OF THE ALUMINIUM CONTENT ON THE CRYSTAL-
LOGRAPHIC AND MAGNETIC STRUCTURES OF THE NI*(FE(2-X)-AL(X))*O4 FERRITE..
⌐NIZIOL(S.).

2138 NI*(FE-CR)*O4, ...A2. . . . . .DISS. ABSTR. VOL.20, ... 333, (1959).
⌐A NEUTRON DIFFRACTION STUDY OF CHROMIUM-SUBSTITUTED FERRITES.
⌐PICKART(S.J.).

2139 NI*(FE-MN)*O4, ...A1. . . . . .J. OF PHYS.-C, SER.2, VOL.4, . . 2266, (1971).
⌐NEUTRON AND X RAY DIFFRACTION STUDIES OF CERTAIN DOPED NICKEL FERRITES.
⌐SUBRAMANYAM(K.N.).

2140 NI-FE,NI-FE-MO-CU, ...A4. . . . . .SOV. PHYS. CRYST. VOL.3, ... 147, (1958).
⌐NEUTRON DIFFRACTION STUDY OF DEFECT STRUCTURE IN SINGLE CRYSTALS OF
METALS.
⌐LYASHCHENKO(B.G.), LITVIN(D.F.), PUZEI(I.M.), ABOV(YU.G.),
GOLOVKIN(V.S.).

2141 NI*FE2*O4, ...A1,A2. . . .REV. MOD. PHYS. VOL.25, ... 114, (1953).
⌐NEUTRON DIFFRACTION STUDIES OF ZINC FERRITE AND NICKEL FERRITE.
⌐HASTINGS(J.M.), CORLISS(L.M.).

2142 NI-FERRITE, ...A2. . . . . .J. PHYS. CHEM. SOLIDS VOL.30, . 1941, (1969).
⌐POLARIZED NEUTRON DIFFRACTION STUDY OF NICKEL FERRITE.
⌐YOUSSEF(S.I.), NATERA(M.G.), BEGUM(R.J.), SRINIVASAN(B.S.),
MURTHY(N.S.S.).

2143 NI3*FE-(X), ...A4. . . . . .SOV. PHYS. DOKLADY VOL.15, . . . 874, (1971).
⌐EFFECT OF ALLOYING ELEMENTS ON THE SUPERSTRUCTURE OF NI3*FE-(X),
X=SI, GE, V, CR, MN, CO, CU, MO, W.
⌐GOMAN≠KOV(V.I.), PUZEI(I.M.), MAL≠TSEV(E.I.).

2144 NI3*GA-(FE), ...A2. . . . . .A.I.P. CONF. PROC. NO.10 PART 2, 1623, (1973).
⌐THE SPATIAL DISTRIBUTION OF THE MAGNETIZATION AROUND FE IMPURITIES IN
NI3*GA.
⌐CABLE(J.W.), CHILD(H.R.).

2145 NI2*GE*O4, ...A2. . . . . .COLL. INTER. N.126 (GRENOBLE), . 92, (1963).
⌐STRUCTURE MAGNETIQUE ET PROPRIETES MAGNETIQUES DE GE*NI2*O4.
⌐BERTAUT(E.F.), VU-VAN-QUI, PAUTHENET(R.), MURASIK(A.).

2146 NI*H, ...A1. . . . . .COLL. INTER. N.126 (GRENOBLE), . 36, (1963).
⌐THE CRYSTAL STRUCTURE OF NICKEL HYDRIDE.
⌐CABLE(J.W.), WOLLAN(E.O.), KOEHLER(W.C.).

2147 NI*H, ...A1. . . . . .J. PHYS. CHEM. SOL. 24, ... 1141, (1963).
⌐THE HYDROGEN ATOM POSITIONS IN FACE CENTERED CUBIC NICKEL HYDRIDE.
⌐WOLLAN(E.O.), CABLE(J.W.), KOEHLER(W.C.).

2148 NI*(I*O3)2.2O2*O, ...A1,A2. . . .PHYSICA VOL.57, ... 215, (1972).
⌐THE CRYSTALLOGRAPHIC AND MAGNETIC STRUCTURE OF NI*(I*O3)2.2O2*O.
⌐ELEMANS(J.B.A.), VAN-LAAR(B.), LOOPSTRA(B.O.).

2149 NI*K2*F4, ...A2. . . . . .COLL. INTER. N.126 (GRENOBLE), . 154, (1963).
⌐L≠INTERACTION MAGNETIQUE DANS LES STRUCTURES DE K2*NI*F4.
⌐LEGRAND(E.), VERSCHUEREN(M.).

2150 (NI++)-K*NI*F3, ...A4. . . . . .J. OF PHYS.-C, SER.2, VOL.3, . . 1303, (1970).
⌐NEUTRON DIFFRACTION MEASUREMENT OF COVALENCY AND SPIN DEVIATION: A TEST
OF THEORY AND EXPERIMENTAL METHOD USING (NI++)-K*NI*F3.
⌐HUTCHINGS(M.T.), GUGGENHEIM(H.J.).

SECTION 5 -STRUCTURE AND CROSS-SECTION DETERMINATIONS . . . .

2151 NI3*MN,. . . . . .SOV. PHYS. DOKLADY VOL.16, ... 486, (1971).
&PHASE TRANSITIONS NEAR THE STOICHIOMETRIC COMPOSITION NI3*MN.
&LITVIN(D.F.), UDOVENKO(V.A.), VINTAIKIN(E.Z.).

2152 NI-MN, ...A2. . . . .COLL. INTER. N.126 (GRENOBLE). . . 180, (1963)
&MAGNETISM WITH POLARIZED NEUTRONS AT THE CENTRO DI STUDI NUCLEARI DELLA
CASACCIA DEL C.N.E.N., ROMA, ITALY.
&ANTONINI(B.), FELCHER(G.P.), MENZINGER(F.), PAOLETTI(A.), RICCI(F.P.),
PASSARI(L.).

2153 NI-MN, ...A2. . . . .J. PHYSIQUE TOME 32 COL.1 VOL.1. . 67, (1971).
&THE MAGNETIC MOMENT DISTRIBUTION IN NI RICH, NI-MN ALLOYS.
&CABLE(J.W.), CHILD(H.R.).

2154 NI-(MN), ...A4. . . . . .REF. SECTION 3, 65/COLLINS,. . . . .

2155 NI-MN, ...A2. . . . .PHYS. STAT. SOL. VOL.16. . . . . 737, (1966).
&ON THE MAGNETIC STRUCTURE OF SOME ALLOYS OF TRANSITION METALS.
&SIDOROV(S.K.), DOROSHENKO(A.V.). (ED. NOTE: REF. SECTION 3, 66/SIDOROV).

2156 NI*MN, ...A2. . . . .J. APPL. PHYS. VOL.34. . . . . 1195, (1963).
&MAGNETIC MOMENT DISTRIBUTIONS IN DILUTE NICKEL ALLOYS.
&LOW(G.G.E.), COLLINS(M.F.).

2157 NI3*MN,. . . . . .A1,2. . . . .SOV. PHYS. SOL. STATE 5,. . . 918, (1963).
&ATOMIC AND MAGNETIC ORDERING OF QUENCHED NI3*MN ALLOYS.
&DEKHTYAR(M.V.).

2158 NI3-MN,. . . . . .A2. . . . .PHYS. MET. METALLOGR. VOL.15 N.6 119, (1963).
&EFFECT OF ORDERING ON THE SATURATION MAGNETIZATION OF THE ALLOY NI3-MN.
&DOROSHENKO(A.V.).

2159 NI-MN,. . . . . .A2. . . . .PHYS. MET. METALLOGR. VOL.20 N.6 48, (1965).
&A NEUTRON-DIFFRACTION STUDY OF ORDERED NICKEL-MANGANESE ALLOYS.
&SIDOROV(S.K.), DOROSHENKO(A.V.).

2160 NI3-MN,. . . . . .A2. . . . .J. APPL. PHYS. VOL.32, ... 375, (1961).
&TRANSFORMATION-DISORDER TO ORDER IN NI3-MN.
&MARCINKOWSKI(M.J.), BROWN(N.).

2161 NI3-MN,. . . . . .A1. . . . .PHYS. REV. VOL.75. . . . . . . 1008, (1949).
&NEUTRON DIFFRACTION STUDIES OF ORDER-DISORDER IN ALLOYS.
&SHULL(C.G.), SIEGEL(S.).

2162 NI-MN, ...A2. . . . .J. PHYS. CHEM. SOLIDS VOL.11,. . 231, (1959).
&THE ANTIFERROMAGNETIC STRUCTURE OF NI*MN.
&KASPER(J.S.), KOUVEL(J.S.).

2163 NI-MN, ...A2. . . . .PHYS. STAT. SOLIDI A VOL.21 NO.1 K31, (1974).
&ON ANTIFERROMAGNETISM OF DISORDERED NI-MN ALLOYS.
&TEPLYUGOV(S.G.), SIDOROV(S.K.), DUBININ(S.F.), NIKULIN(YU.M.).

2164 NI3*MN-(CO,CR,FE),. . . . .A1. . . . .SOV. PHYS. DOKLADY VOL.14,. . 1119, (1970).
&ON THE INFLUENCE OF ALLOYING A NI3*MN ALLOY WITH A THIRD 3D-TRANSITION
ELEMENT ON ORDERING KINETICS.
&PANIN(V.E.), PRUSHINSKII(V.V.), FADIN(V.P.), NOVAK(L.I.).

2165 NI*MN2*O4, ...A1,2. . . .J. PHYS. CHEM. SOLIDS 30. . . . . 805, (1969).
&MANGANITES 'SPINELLES' PURES D*ELEMENTS DE TRANSITION: PREPARATIONS ET
STRUCTURES CRISTALLOGRAPHIQUES.
&BUHL(R.).

2166 NI*MN2*O4, ...A2. . . . .COMP. REND. 263,. . . . . . 344B, (1966).
&DETERMINATION DE LA STRUCTURE MAGNETIQUE DU SPINELLE CUBIQUE NI*MN2*O4.
&BOUCHER(B.), BUHL(R.), PERRIN(M.).

2167 NI*MN2*O4, ...A2. . . . .ACTA CRYST. B25,. . . . . . 2326, (1969).
&ETUDE CRISTALLOGRAPHIQUE DU MANGANITE SPINELLE CUBIQUE NI*MN2*O4 PAR
DIFFRACTION DE NEUTRONS.
&BOUCHER(B.), BULL(R.), PERRIN(M.).

2168 NI*MN2*O4, ...A2. . . . .COMP. REND. 249, ... 514, (1959).
&MAGNETIC STRUCTURE OF NICKEL MANGANITE.
&BOUCHER(B.). (IN FRENCH).

2169 NI-MN-SB,. . . . .A2. . . . .PHYS. STAT. SOLIDI A VOL.9 NO.1 97, (1972).
&ATOMIC AND MAGNETIC STRUCTURE OF THE HEUSLER ALLOYS NI-MN-SB AND
CO-MN-SB.
&SZYTULA(A.), DIMITRIJEVIC(Z.), TODOROVIC(J.), KOLODZIEJCZYK(A.),
SZELAG(J.), WANIC(A.).

2170 NI2-MN-(SB,SN),. . . . .A1,A2. . . .PHYS. STAT. SOLIDI A VOL.11 NO.1 57, (1972).
&ATOMIC AND MAGNETIC STRUCTURE OF THE HEUSLER ALLOYS NI2-MN-SB, NI2-MN-SN
AND CO2-MN-SN.
&SZYTULA(A.), KOLODZIEJCZYK(A.), RZANY(H.), TODOROVIC(J.), WANIC(A.).

2171 NI*O,. . . . . . .A2. . . . .J. PHYS. SOC. JAPAN VOL.17 B-III 12, (1962).
&THE MAGNETIC FORM FACTOR OF NICKEL OXIDE.
&ALPERIN(H.A.).

2172 NI*O,. . . . . . .A2. . . . .J. APPL. PHYS. VOL.31, SUPPL.,. 354, (1960).
&NEUTRON DIFFRACTION INVESTIGATION OF THE MAGNETIC STRUCTURE OF NICKEL
OXIDE.
&ALPERIN(H.A.).

2173 NI*O,. . . . . . .A2. . . . .J. APPL. PHYS. VOL.31, SUPPL.,. 352, (1960).
&ANTIFERROMAGNETIC STRUCTURE AND DOMAINS IN SINGLE CRYSTAL NI*O.
&ROTH(W.L.), SLACK(G.A.).

2174 NI*O,. . . . . . .A2. . . . .SOL. STATE COMM. VOL.13,. . . . 249, (1973).
&THE CRITICAL BEHAVIOUR OF SPONTANEOUS MAGNETIZATION IN THE ANTIFERRO-
MAGNETIC NI*O.
&NEGOVETIC(I.), KONSTANTINOVIC(J.).

2175 NI*O,. . . . . . .A2. . . . .J. APPL. PHYS. VOL.31, ... 2000, (1960).
&NEUTRON AND OPTICAL STUDIES OF DOMAINS IN NI*O.
&ROTH(W.L.).

2176 NI*O,. . . . . . .A2. . . . .DISS. ABSTR. VOL.21, ... 1979, (1961).
&A NEUTRON DIFFRACTION STUDY OF NICKEL OXIDE.
&ALPERIN(H.A.).

2177 NI*O,. . . . . . .A2. . . . .PHYS. REV. VOL.110,. . . . . . 1333, (1958).
&MAGNETIC STRUCTURES OF MN*O, FE*O, CO*O, AND NI*O.
&ROTH(W.L.).

SECTION   5 -STRUCTURE AND CROSS-SECTION DETERMINATIONS . . . .

2178 NI*O, ... A2 . . . . . . PHYS. REV. VOL.111, ... 772, (1958).
      #MULTISPIN AXIS STRUCTURES FOR ANTIFERROMAGNETS.
      #ROTH(W.L.).

2179 NI*O, ... A2 . . . . . DISS. ABSTR. VOL.23, ... 668, (1962).
      #A NEUTRON DIFFRACTION STUDY OF NICKELOUS OXIDE.
      #MURRAY(D.O.).

2180 NI*O, ... A2 . OF NI . FIZIKA (YUGOSLAVIA) VOL.5, . . . 179, (1973).
      #CRITICAL BEHAVIOUR OF NI*O AS AN ANTIFERROMAGNET WITH NONRIGID LATTICE.
      #KONSTANTINOVIC(J.), MILOSEVIC(S.), NEGOVETIC(I.).

2181 NI*(O*H)2, ... A1 . STUDY . PHYS. STAT. SOL. B VOL.43, . . . 125, (1971).
      #NEUTRON DIFFRACTION STUDY OF NI*(O*H)2.
      #SZYTULA(A.), MURASIK(A.), BALANDA(M.).

2182 NI-PD, ... A2 . . . . . . PHYS. REV. B-1, ... 3809, (1970).
      #MAGNETIC MOMENT DISTRIBUTION IN NI-PD ALLOYS.
      #CABLE(J.W.), CHILD(H.R.).

2183 NI3*(PR,ND,TB,DY,TM), ... A2 . . . . . . SOL. STATE COMM. VOL.10, . . . 989, (1972).
      #STRUCTURE MAGNETIQUE DES COMPOSES INTERMETALLIQUES TERRE RARE-NICKEL DE
      FORMULE T*NI3.
      #YAKINTHOS(J.), PACCARD(D.).

2184 NI*S, ... A4 . . . . . 1ST EUR. CONF./CONDENSED MATTER, 39, (1971).
      #NEUTRON SCATTERING INVESTIGATION OF NI*S. (METAL-INSULATOR PARAMAGNETIC
      ANTIFERROMAGNETIC TRANSITION).
      #HUTCHINGS(T.). (ED. NOTE: PUBL. EUROPEAN PHYS. SOC. 1971, GENEVA, SWI.).

2185 NI*S, ... A1 . . . . . J. APPL. PHYS. VOL.40, ... 1332, (1969).
      #NEUTRON DIFFRACTION STUDY OF NI*S UNDER PRESSURE.
      #SMITH(F.A.), SPARKS(J.T.).

2186 NI*S, ... A1 . . . . . J. APP. PHYS. 34, ... 1191, (1963).
      #NEUTRON DIFFRACTION STUDY NI*S.
      #SPARKS(J.T.), KOMOTO(T.).

2187 NI*S2, ... A2 . . . . . PHYS. LETT. VOL.44A NO.7, . . . 529, (1973).
      #MAGNETIC PROPERTIES OF PYRITE TYPE NI*S2.
      #MIYADAI(T.), MIYAHARA(S.), TAKIZAWA(K.), UCHINO(K.).

2188 NI*S, ... A2 . . . . . COLL. INTER. N.126 (GRENOBLE), . 143, (1963).
      #ETUDES DE DIFFRACTION NEUTRONIQUE A LIVERMORE.
      #SPARKS(J.T.), KOMOTO(T.).

2189 NI*S, ... A2 . . . . . PHYS. REV. LETT. VOL.32, . . . 1257, (1974).
      #NICKEL SULFIDE- AN ITINERANT-ELECTRON ANTIFERROMAGNET.
      #COEY(J.M.D.),   BRUSETTI(R.),   KALLEL(A.),   SCHWEIZER(J.),   FUESS(H.).

2190 NI*S*O4, ... A2 . . . . . . PHYS. REV. VOL.125, ... 1283, (1962).
      #ANTIFERROMAGNETIC STRUCTURE OF CR*V*O4 AND THE ANHYDROUS SULFATES OF
      DIVALENT FE,NI AND CO.
      #FRAZER(B.C.), BROWN (P.J.).

2191 (NI*S*O4)*6(D2*O), ... A1 . . . . . . ACTA CRYST. 21, ... 705, (1966).
      #A NEUTRON DIFFRACTION ANALYSIS OF THE CRYSTAL STRUCTURE OF TETRAGONAL
      NICKEL SULPHATE HEXADEUTERATED.
      #O#CONNOR(B.H.), DALE(D.H.).

2192 NI*(SC-FE)*O4, ... A4.P. . . . . . SOV. PHYS. CRYST. VOL.16, . . . 634, (1971).
      #NEUTRON DIFFR. DETERMINATION OF THE OXYGEN PARAM. AND CATION DISTRIB. IN
      THE FERRITE-SPINELS (NI-SC(.15)-FE(1.85))*O4 AND (NI-SC(.4)-FE(1.6))*O4.
      #KOCHAROV(A.G.), YAMZIN(I.I.), FAEK(M.K.).

2193 NI*SE2,NI*S2, ... A1 . . . . . . ACTA CHEM. SCAND. VOL.23, . . . 2325, (1969).
      #ON THE MAGNETIC PROPERTIES OF CO*SE2, NI*S2, AND NI*SE2.
      #FURUSETH(S.), KJEKSHUS(A.), ANDRESEN(A.F.).

2194 NI*SE*O4, ... A2 . . . . . . Z. ANGEW. PHYS. VOL.27, ... 311, (1969).
      #MAGNETIC STRUCTURES OF NI*SE*O4 AND CO*SE*O4 AND DETERMINATION OF THE
      NEEL POINTS OF MN*SE*O4, CO*SE*O4 AND NI*SE*O4.
      #FUESS(H.).

2195 NI*SE*O4, ... A1 . . . . . . Z. ANORG. ALLG. CHEM. BAND 358, . 125, (1968).
      #BESTIMMUNG DER KRISTALLSTRUKTUR DER SELENATE M*SE*O4 (M=MN,CO,NI) DURCH
      RONTGEN- UND NEUTRONENBEUGUNG.
      #FUESS(H.), WILL(G.).

2196 NI2*SI*O4, ... A2 . . . . . . ACTA CRYST. 19, ... 147, (1965).
      #ANTIFERROMAGNETISM IN NICKEL ORTHOSILICATE.
      #NEWNHAM(R.E.), SANTORO(R.P.), FANG(J.H.), NOMURA(S.).

2197 NI3*T (T=RARE EARTHS), ... T. . . . . . PHYS. STAT. SOL. VOL.47, . . . 239, (1971).
      #INTERPRETATION DES STRUCTURES MAGNETIQUES DES COMPOSES T*NI3 A L#AIDE
      DU CHAMP CRISTALLIN.
      #ROSSAT-MIGNOD(J.), YAKINTHOS(J.).

2198 NI*TI*O3, ... T.A2. . . . . . PHYS. REV. 164, ... 768, (1967).
      #THEORY OF THE MAGNETIC PROPERTIES OF THE ILMENITES OF THE M*(TI*O3).
      #GOODENOUGH(J.B.), STICKLER(J.J.).

2199 NI*TI*O3, ... A2 . . . . . . J. PHYS. SOC. JAPAN VOL.14, . . 1352, (1959).
      #NEUTRON DIFFRACTION STUDY OF ANTIFERROMAGNETIC MN*TI*O3 AND NI*TI*O3.
      #SHIRANE(G.), PICKART(S.J.), ISHIKAWA(Y.).

2200 NI-TI-SB, ... A1 . . . . . . ACTA PHYS. POLON. VOL.A44, . . . 147, (1973).
      #CRYSTAL STRUCTURE AND MAGNETIC PROPERTIES OF HEUSLER-TYPE ALLOYS
      M-TI-SB (M=NI,CO,FE) AND FE2-TI-SN.
      #SZYTULA(A.), TOMKOWICZ(Z.), TUROWSKI(M.).

2201 4-NITROANILINE (DEUTERATED), ... A1 . . . . . CZECH. J. PHYS. B VOL.19, . . . 857, (1969).
      #THE TWO-DIMENSIONAL NEUTRON DIFFRACTION STUDY OF THE CRYSTAL STRUCTURE
      OF DEUTERATED 4-NITROANILINE C6*H2*D2*N*O2*N*D2.
      #TICHY(K.), PRELESNIK(B.).

2202 2-NITROBENZALDEHYDE, ... A1 . . . . . . ACTA CRYST. 17, ... 573, (1964).
      #A NEUTRON DIFFRACTION STUDY OF 2-NITROBENZALDEHYDE AND THE C-H....O
      INTERACTION.
      #COPPENS(P.).

2203 NI-V, ... A2 . . . . PHYSICA VOL.51, ... 627, (1971).
      #POLARIZED-NEUTRON STUDY OF THE NI(.968)-V(.032) ALLOY.
      #MANIAWSKI(F.), DOBRZYNSKI(L.), SIKORSKA(D.).

SECTION   5 -STRUCTURE AND CROSS-SECTION DETERMINATIONS . . . .

2204 NI-(V),. . . . . .A4. . . . . . .REF. SECTION 3, 65/COLLINS,. . . . .

2205 NI*V,. . . . . .A2. . . . . . .J. APPL. PHYS. VOL.34. . . . . . . 1195, (1963).
      »MAGNETIC MOMENT DISTRIBUTIONS IN DILUTE NICKEL ALLOYS.
      »LOW(G.G.E.), COLLINS(M.F.).

2206 NI3*V2*O8, ...A1. . . . . . .ACTA CRYST. VOL.B26. . . . . 2036, (1970).
      »STRUCTURE AUX RAYONS X, NEUTRONS ET PROPRIETES MAGNETIQUES DES
      ORTHOVANADATES DE NICKEL ET DE COBALT.
      »FUESS(H.), BERTAUT(E.F.), PAUTHENET(R.), DURIF(A.).

2207 NI*W*O4, ...A2. . . . . . .SOL. STATE COMM. VOL.8. . . . . . 2071, (1970).
      »MAGNETISCHE STRUKTUR VON CO*W*O4, NI*W*O UND CU*W*O4.
      »WEITZEL(H.).

2208 NI-(X),. . . . . .A2. . . . . . .J. APP. PHYS. VOL. 39. . . . . . 1174, (1968).
      »MAGNETIC NEUTRON SCATTERING FROM ATOMS DISSOLVED IN FERROMAGNETIC
      IRON AND NICKEL.
      »LOW(G.G.).

2209 NI*ZR*H3,. . . . . .A4. . . . . . .COLL. INTER. N.126 (GRENOBLE), .   27, (1963).
      »NEUTRON DIFFRACTION STUDY OF NICKEL ZIRCONIUM HYDRIDE.
      »PETERSON(S.W.), SADANA(V.N.), KORST(W.L.).

2210 NOBLE GASES, ...P. . . . . . .PHYS. REV. VOL.148. . . . . . . 1303, (1966).
      »MEASUREMENT OF THE ELECTRON-NEUTRON INTERACTION BY THE ASYMMETRICAL
      SCATTERING OF THERMAL NEUTRONS BY NOBLE GASES.
      »KROHN-JR(V.E.), RINGO(G.R.).

2211 NP,. . . . . .A4,P. . . . . . .J. PHYS. CHEM. SOL. 28,. . . . . . 1651, (1967).
      »A NEUTRON DIFFRACTION DETERMINATION OF THE COHERENT SCATTERING AMPLITUDE
      OF NP AND THE POSSIBLE ANTIFERROMAGNETISM OF NEPTUNIUM DIOXIDE.
      »HEATON(L.), MUELLER(M.H.), WILLIAMS(J.M.).

2212 NP*(AS,N,P,SB),. . . . . .A2. . . . . . .PHYS. REV. B VOL.9,. . . . . . 3766, (1974).
      »MAGNETIC PROPERTIES OF THE NEPTUNIUM MONOPNICTIDES.
      »ALDRED(A.T.),   DUNLAP(B.D.),   HARVEY(A.R.),   LAM(D.J.),   LANDER(G.H.),
      MUELLER(M.H.).

2213 NP*C,. . . . . .A2. . . . . . .J. PHYS. CHEM. SOLIDS 30,. . . . .   733, (1969).
      »NEUTRON DIFFRACTION STUDY OF NP*C.
      »LANDER(G.H.), HEATON(L.), MUELLER(M.H.), ANDERSON(K.D.).

2214 NP*C,. . . . . .A2. . . . . . .J. PHYSIQUE TOME 32 COL.1 VOL.II   917, (1971).
      »THE EFFECT OF CARBON CONCENTRATION ON THE MAGNETIC PROPERTIES AND
      HYPERFINE INTERACTIONS OF NEPTUNIUM MONOCARBIDE.
      »LAM(D.J.), MUELLER(M.H.), PAULIKAS(A.P.), LANDER(G.H.).

2215 NP*O2,. . . . . .A2. . . . . . .J. PHYS. CHEM. SOL. 28,. . . . . . 1649, (1967).
      »A NEUTRON DIFFRACTION STUDY OF NP*O2.
      »COX(D.E.), FRAZER(B.C.).

2216 NP*P,. . . . . .A2. . . . . . .A.I.P. CONF. PROC. NO.10 PART 1,   88, (1973).
      »MAGNETIC PROPERTIES OF ANTIFERROMAGNETIC NP*P.
      »LANDER(G.H.),   DUNLAP(B.D.),   LAM(D.J.),   HARVEY(A.R.),   MUELLER(M.H.),
      ALDRED(A.T.), NOWIK(I.).

2217 NP*P,. . . . . .A2. . . . . . .INT. J. MAGN. VOL.4 NO.2,. . . . .   99, (1973).
      »A NEUTRON AND MOSSBAUER STUDY OF THE MAGNETIC ORDERING IN NP*P.
      »LANDER(G.H.),   DUNLAP(B.D.),   MUELLER(M.H.),   NOWIK(I.),   REDDY(J.F.).

2218 NP*PD3,. . . . . .A2. . . . . . .PHYS. REV. B VOL.9,. . . . . . 1041, (1974).
      »MAGNETIC PROPERTIES OF NP*PD3 AND PU*PD3 INTERMETALLIC COMPOUNDS.
      »NELLIS(W.J.),   HARVEY(A.R.),   LANDER(G.H.),   DUNLAP(B.D.),   BRODSKY(M.B.),
      MUELLER(M.H.), REDDY(J.F.), DAVIDSON(G.R.).

2219 O2,. . . . . .P. . . . . . .PHYS. STAT. SOL. VOL.50,. . . . .   385, (1972).
      »MAGNETIC FORM FACTOR OF MOLECULAR OXYGEN.
      »ALIKHANOV(R.A.), ILYINA(I.L.), SMIRNOV(L.S.).

2220 O2,. . . . . .A4. . . . . . .COLL. INTER. N.126 (GRENOBLE), .   25, (1963).
      »ETUDE PAR DIFFRACTION NEUTRONIQUE DE L≠OXYGENE SOLIDE.
      »ALIKHANOV(R.A.).

2221 O (ISOTOPE A=16),. . . . . .P. . . . . . .FIZIKA (YUGOSLAVIA) VOL.2,. . . .   97, (1970).
      »NEUTRON TOTAL ELASTIC SCATTERING CROSS SECTION MEASUREMENT ON OXYGEN
      (ISOTOPE A=16).
      »BORELI(F.).

2222 O2 (GAMMA),. . . . . .A2. . . . . . .J. PHYSIQUE TOME 32 COL.1 VOL.I,   582, (1971).
      »SPIN DENSITY DISTRIBUTION IN PARAMAGNETIC GAMMA-OXYGEN.
      »COX(D.E.), SAMUELSEN(E.J.), BECKURTS(K.H.).

2223 O2,. . . . . .A1,2. . . . . . .SOV. PHYS. J.E.T.P. VOL.18,. . .   556, (1964).
      »NEUTRON DIFFRACTION INVESTIGATIONS OF SOLID OXYGEN.
      »ALIKHANOV(R.A.).

2224 O2,. . . . . .A2. . . . . . .PROC. PHYS. SOC. 89,. . . . . . .   415, (1966).
      »MAGNETIC STRUCTURE OF SOLID OXYGEN.
      »COLLINS(M.F.).

2225 O2 (ALPHA),. . . . . .A1. . . . . . .JETP LETT. VOL.5,. . . . .   349, (1967).
      »STRUCTURE OF ALPHA MODIFICATION OF OXYGEN.
      »ALIKHANOV(R.A.).

2226 O2 (GAMMA),. . . . . .T,A2. . . . . . .PHYS. REV. B VOL.7,. . . . . 3112, (1973).
      »MAGNETIC-STRUCTURE-FACTOR CALCULATIONS IN PARAMAGNETIC GAMMA-OXYGEN.
      »LEONI(F.), SACCHETTI(F.).

2227 O2 (GAMMA),. . . . . .A1,A2. . . . . . .PHYS. REV. B VOL.7,. . . . . 3102, (1973).
      »NEUTRON-DIFFRACTION DETERMINATION OF THE CRYSTAL STRUCTURE AND MAGNETIC
      FORM FACTOR OF GAMMA-OXYGEN.
      »COX(D.E.), SAMUELSEN(E.J.), BECKURTS(K.H.).

2228 O2 (ISOTOPE A=18),. . . . . .P. . . . . . .ACTA CRYST. 22,. . . . .   927, (1967).
      »COHERENT NEUTRON SCATTERING LENGTH OF OXYGEN (A=18).
      »O≠CONNOR(B.H.).

2229 O2,. . . . . .P,A3. . . . . . .SOV. AT. ENERGY VOL.18,. . . . .   585, (1965).
      »COHERENT EFFECTS IN THE INTERACTION OF SLOW NEUTRONS WITH LIQUIDS.
      »VERTEBNYI(V.P.),   DZYUB(I.P.),   MAISTRENKO(A.N.),   PASECHNIK(M.V.).

2230 O2,. . . . . .P,A3. . . . . . .PHYS. REV. VOL.76,. . . . . . . 1750, (1949).
      »SLOW NEUTRON VELOCITY SPECTROMETER STUDIES OF O2, N2, A, H2, H2*O AND
      SEVEN HYDROCARBONS.
      »MELKONIAN(E.).

SECTION 5 -STRUCTURE AND CROSS-SECTION DETERMINATIONS . . . .

2231 O2,. . . . . .NEUTRON REFRACTION IN O2, N2, HE, AR GASES.P,A3.....PHYS. REV. VOL.84, ... 969, (1951).
.MCREYNOLDS(A.W.).

2232 O2,. . . . . .STRUCTURE OF LIQUID NITROGEN, OXYGEN, AND ARGON BY NEUTRON DIFFRACTION.A3.......PHYS. REV. VOL.92, ... 1229, (1953).
.HENSHAW(D.G.), HURST(D.G.), POPE(N.K.).

2233 O2,. . . . . .SLOW NEUTRON SPECTROMETER STUDIES OF OXYGEN, NITROGEN, AND ARGON.P,A3.....PHYS. REV. VOL.73, ... 1399, (1948).
.MELKONIAN(E.), RAINWATER(L.J.), HAVENS-JR(W.W.), DUNNING(J.R.).

2234 O2,. . . . . .NEUTRON DIFFRACTION BY GASES.A3.......PHYS. REV. VOL.75, ... 1609, (1949).
.ALCOCK(N.Z.), HURST(D.G.).

2235 O2,. . . . . .STRUCTURE OF LIQUID OXYGEN BY NEUTRON DIFFRACTION.A3.......PHYS. REV. VOL.119,. . . . . . 22, (1960).
.HENSHAW(D.G.).

2236 O2,. . . . . .DETERMINATION OF THE COHERENT NEUTRON SCATTERING AMPLITUDES OF BORON, NITROGEN, AND OXYGEN BY MIRROR REFLECTION.P......PHYS. REV. VOL.138,. . . . . . B1116, (1965).
.DONALDSON(R.E.), PASSELL(L.), BARTOLINI(W.), GROVES(D.).

2237 O (ISO. A=16),. .TOTAL NEUTRON CROSS SECTIONS OF BE (ISO. A=9), N (ISO. A=14), AND O (ISO. A=16).P......PROC/CONF/NEUTRON C. S. + TECH. 851, (1968).
.JOHNSON(C.H.), HAAS(F.X.), FOWLER(J.L.), MARTIN(F.D.), KERNELL(R.L.), COHN(H.O.).

2238 (O-H) BONDS, . .CRYSTAL CHEMISTRY OF HYDROGEN-CONTAINING COMPOUNDS, ON THE BASIS OF NEUTRON DIFFRACTION DATA (REVIEW PAPER).SOV. PHYS. CRYST. VOL.17, NO.2,. 383, (1972).
.DATT(I.D.), OZEROV(R.P.).

2239 O-H-O, ...ON SYMMETRICAL O-H-O HYDROGEN BONDS.A4.......COLL. INTER. N.126 (GRENOBLE), . 63, (1963).
.RUNDLE(R.E.).

2240 O2*PT*F6,. . .CRYSTAL STRUCTURE OF O2*PT*F6: A NEUTRON-DIFFRACTION STUDY.A1.......J. CHEM. PHYSICS VOL.44, ... 1748, (1966).
.IBERS(J.A.), HAMILTON(W.C.).

2241 ORTHOFERRITES,MAGNETIC STRUCTURES OF RARE EARTH ORTHOFERRITES.A2.......J. PHYSIQUE TOME 30,. . . . . 967, (1969).
.MARESCHAL(J.), SIVARDIERE(J.).

2242 OS (ISOTOPES A=187*189),SLOW NEUTRON TOTAL CROSS-SECTIONS AND RESONANCE LEVELS OF OSMIUM- 187 188 AND 189.A4.......UKRAINIAN PHYS. J. VOL.14, . . . 1968, (1969).
.VERTEBNYI(V.P.), VLASOV(M.F.), DADAKINA(A.F.), ZATSERKOVSKA(R.A.), KIRILYUK(A.L.), PASECHNIK(M.V.), TROFIMOVA(N.A.).

2243 OS,. . . . . . . .THERMAL-NEUTRON COHERENT SCATTERING AMPLITUDES OF THALLIUM AND OSMIUM.P......PHYS. REV. VOL.105,. . . . . 216, (1957).
.HEATON(L.), SIDHU(S.S.).

2244 OS*O2, ...NEUTRON DIFFRACTION INVESTIGATION OF OS*O2.A1.......J. LESS-COMMON METALS VOL.17,. . 459, (1969).
.THIELE(G.), WODITSCH(P.). (ED. NOTE: IN GERMAN).

2245 P*H4*BR, ...NEUTRON DIFFRACTION STUDY OF THE STRUCTURE AND THERMAL MOTION OF PHOSPHONIUM BROMIDE.A1.......J. CHEM. PHYS. VOL.54, ... 1968, (1971).
.SCHROEDER(L.W.), RUSH(J.J.).

2246 P*H4*I,. . . . .HYDROGEN BONDING IN PHOSPHONIUM IODIDE: A NEUTRON-DIFFRACTION STUDY.A1.......J. CHEM. PHYSICS VOL.47, . . . 1818, (1967).
.SEQUEIRA(A.), HAMILTON(W.C.).

2247 (P2*O5)-(CO*O,FE2*O3),AMORPHOUS ANTIFERROMAGNETISM IN IRON AND COBALT PHOSPORUS PENTOXIDE GLASSES.A4.......J. PHYS. C VOL.5,. . . . . L261, (1972).
.EGAMI(T.), SACLI(O.A.), SIMPSON(A.W.), TERRY(A.L.), WEDGWOOD(F.A.).

2248 PA (A=233),. . .TOTAL NEUTRON CROSS SECTION OF PA (A=233) FROM 0.01 TO 1.0 EV.P......NUCL. SCI. ENG. VOL.20,. . . . 235, (1964).
.SIMPSON(F.B.), CODDING(J.W.), BERRETH(J.R.).

2249 PARAMAGNETICS,THE MAGNETIC SCATTERING OF NEUTRONS BY LIQUID PARAMAGNETICS.D3.......UKRAINIAN PHYS. J. VOL.14, . . . 1909, (1969).
.ZATOVSKII(A.V.), KRASNYI(YU.P.).

2250 PB,. . . . . . .A STUDY OF THE DIFFUSIVE MOTIONS OF LIQUIDS BY MEANS OF COLD-NEUTRON SCATTERING.A3.......IAEA SYMPOSIUM VIENNA, . . . 265, (1960).
.PALEVSKY(H.).

2251 PB,. . . . . . .STRUCTURE FACTOR FOR LIQUID METALS.II. RESULTS FOR LIQUID ZN, TL, PB, SN AND BI.P......J. OF PHYS. C VOL.1, ... 1075, (1968).
.NORTH(D.M.), ENDERBY(J.E.), EGELSTAFF(P.A.).

2252 PB,. . . . . . .THE STRUCTURE FACTOR FOR LIQUID METALS AT LOW ANGLES.P......PHYS. LETT. 21, ... 286, (1966).
.EGELSTAFF(P.A.), DUFFILL(C.), RAINEY(V.S.), ENDERBY(J.E.), NORTH(D.M.).

2253 PB,. . . . . . .TOTAL CROSS SECTION OF LEAD FOR SLOW NEUTRONS.P......PHIL. MAG. VOL.6,. . . . . . 485, (1961).
.COLLINS(M.F.), DOLLING(G.).

2254 PB,. . . . . . .DETERMINATION OF THE TOTAL CROSS-SECTION OF LEAD FOR COLD NEUTRONS.P......STUD. CERCETARI FIZ. VOL.13,. 477, (1962).
.TEUTSCH(H.), MATEESCU(N.), TIMIS(P.). (IN ROUMANIAN).

2255 PB,. . . . . . .COHERENT SCATTERING OF SLOW NEUTRONS BY LIQUID LEAD.A3.......UKR. FIZ. ZH.(USSR) VOL.15,. . . 1772, (1970).
.GOLIKOV(V.V.), KOZLOV(ZH.A.).

2256 PB,. . . . . . .NEUTRON DIFFRACTION BY LIQUID METALS.A3.......PHYS. MET. METALLOGR. VOL.22 N.2 123, (1967).
.KHRUSHCHEV(B.I.), BOGOMOLOV(A.M.), SHARIPOVA(L.S.).

SECTION 5 -STRUCTURE AND CROSS-SECTION DETERMINATIONS . . . .

2257 PB,. . . . . .A3. . . . . .J. CHEM. PHYS. VOL.21. . . . . . .228, (1953).
&NEUTRON DIFFRACTION AND ATOMIC DISTRIBUTION IN LIQUID LEAD AND LIQUID BISMUTH AT TWO TEMPERATURES.
&SHARRAH(P.C.), SMITH(G.P.).

2258 PB,. . . . . .P. . . . . . . .NATURE VOL.161,. . . . . 282, (1948).
&SCATTERING OF NEUTRONS BY LEAD.
&LATHAM(R.), CASSELS(J.M.).

2259 PB,. . . . . .A3. . . . . . .PHYS. REV. VOL.77,. . . 305, (1950).
&NEUTRON DIFFRACTION IN LIQUID SULFUR, LEAD, AND BISMUTH.
&CHAMBERLAIN(O.).

2260 PB,. . . . . .E,P. . . . . . .PHYS. REV. VOL.81,. . . . . . .527, (1951).
&COHERENT SCATTERING AMPLITUDES AS DETERMINED BY NEUTRON DIFFRACTION.
&SHULL(C.G.), WOLLAN(E.O.).

2261 PB,. . . . . .P,A3. . . . . .PHYS. REV. VOL.72,. . . . . . . • 634, (1947).
&INTERACTION OF NEUTRONS WITH ELECTRONS IN LEAD.
&HAVENS-JR(W.W.), RABI(I.I.), RAINWATER(L.J.).

2262 PB-BI, ...ACTA CRYST. (INTERACT.) VOL.A25, S211, (1969).
&DISORDER SCATTERING OF LONG-WAVELENGTH NEUTRONS AT LEAD SINGLE CRYSTALS WITH 2 AND 4 AT. PCT. BISMUTH CONCENTRATION.
&SCHUMACHER(H.), SCHMATZ(W.), EISENRIEGLER(E.).

2263 PB-BI, ...A3. . . . . .J. CHEM. PHYS. VOL.32. . . . . . .241, (1960).
&DETERMINATION OF ATOMIC DISTRIBUTIONS IN LIQUID LEAD-BISMUTH ALLOYS BY NEUTRON AND X-RAY DIFFRACTION.
&SHARRAH(P.C.), PETZ(J.I.), KRUH(R.F.).

2264 PB-BI, ...A4. . . . . . . .PROG. LOW TEMP. PHYS. VOL.5,. . . . 161, (1967).
&STUDY OF THE SUPERCONDUCTIVE MIXED STATE BY NEUTRON-DIFFRACTION.
&CRIBIER(D.), JACROT(B.), MADHAV-RAO(L.), FARNOUX(B.).

2265 PB-(BI), ...A1. . . . . . .PHYS. STAT. SOLIDI A VOL.20, . . 109, (1973).
&THE STRUCTURE OF THE STRAIN FIELD OF SUBSTITUTIONAL BISMUTH IN LEAD INVESTIGATED BY DIFFUSE NEUTRON SCATTERING.
&SCHUMACHER(H.), SCHMATZ(W.), SEITZ(E.).

2266 PB*BI, ...A4. . . . . . . .J. LOW TEMP. PHYS. VOL.14,. . . .213, (1974).
&NEUTRON DIFFRACTION STUDIES OF THE MORPHOLOGY OF FLUX LINE CRYSTALS IN TYPE II SUPERCONDUCTORS.
&SCHELTEN(J.), LIPPMANN(G.), ULLMAIER(H.).

2267 PB*CR*03,. . . . . .A1,2. . . . . .J. APP. PHYS. 38,. . . . . 951, (1967).
&CRYSTAL AND MAGNETIC STRUCTURE OF PB*CR*03.
&ROTH(W.L.), DEVRIES(R.C.).

2268 PB*F2, ...A1. . . . . . .ACTA CRYST. 22,. . . . . .744, (1967).
&NEUTRON DIFFRACTION INVESTIGATION OF ORTHORHOMBIC LEAD (II) FLUORIDE.
&BOLDRINI(P.), LOOPSTRA(B.O.).

2269 PB*F2, ...A1. . . . . . .SOV. PHYS. DOKLADY VOL.6,. . . . 370, (1961).
&NEUTRON INVESTIGATION OF THE CUBIC PHASE OF PB*F2.
&YAMZIN(I.I.), NOZIK(YU.Z.), BELOV(N.V.).

2270 PB*(FE-NB)*03, ...A2. . . . . .SOV. PHYS. SOL. STATE VOL.7,. . .997, (1965).
&INVESTIGATION OF THE MAGNETIC ORDERING IN FERROELECTRIC PB*(FE-NB)*03.
&DRABKIN(G.M.), MAL≠TSEV(E.I.), PLAKHTII(V.P.).

2271 PB*(FE-W)*03,. . . . . .A2. . . . . .BULL. ACAD. SCI. USSR VOL.28,. . .350, (1964).
&NEUTRON DIFFRACTION STUDIES OF SOME COMPOUNDS WITH THE PEROVSKITE STRUCTURE.
&PLAKHTII(V.P.), MAL≠TSEV(E.I.), KAMINKER(D.M.).

2272 PB-GLASS,. . . . . .A4. . . . . .CAN. J. PHYS. VOL.52,. . . . . . . 748, (1974).
&NEUTRON SCATTERING FROM SMALL LEAD PARTICLES.
&NOVOTNY(V.), HOLDEN(T.M.), DOLLING(G.).

2273 PB*MO*04,. . . . . .A1. . . . . . .Z. KRIST. VOL.121,. . . . . . 158, (1965).
&A NEUTRON CRYSTALLOGRAPHIC INVESTIGATION OF LEAD MOLYBDENUM OXIDE.
&LECIEJEWICZ(J.).

2274 PB*(N3)2,. . . . . .A1. . . . . . .J. AMER. CHEM. SOC. VOL.85,. . . 3892, (1963).
&A NEUTRON DIFFRACTION STUDY OF ALPHA-LEAD AZIDE.
&GLEN(G.L.).

2275 PB*(N3)2,. . . . . .A1. . . . . . .J. CHEM. PHYS. VOL.48,. . . . . .1397, (1968).
&NEUTRON DIFFRACTION STUDY OF ALPHA-LEAD AZIDE.
&CHOI(C.S.), BOUTIN(H.).

2276 PB*(N3)2,. . . . . .A1. . . . . . .ACTA CRYST. VOL.B25,. . . . . . . 982, (1969).
&NEUTRON DIFFRACTION STUDY OF ALPHA-PB*(N3)2.
&CHOI(C.S.), BOUTIN(H.P.).

2277 PB*(N*03)2,. . . . . .A1. . . . . . .ACTA CRYST. VOL.10,. . . . . . . 1C3, (1957).
&A NEUTRON CRYSTALLOGRAPHIC STUDY OF LEAD NITRATE.
&HAMILTON(W.C.).

2278 PB2*03,. . . . . .A1. . . . . .ACTA CRYST. VOL.A26,. . . . . . . .501, (1970).
&SESQUIOXYDE DE PLOMB, PB2*03 .I. DETERMINATION DE LA STRUCTURE.
&BOUVAIST(J.), WEIGEL(D.).

2279 PB*02, ...NATURWISSENSCHAFTEN VOL.49,. . . 373, (1962).
&NOTE ON THE OXYGEN PARAMETER IN TETRAGONAL PB*02.
&LECIEJEWICZ(J.), PADLO(I.).

2280 PB*0,. . . . . .A1. . . . . . .ACTA CRYST. VOL.14,. . . . . . . 1304, (1961).
&ON THE CRYSTAL STRUCTURE OF TETRAGONAL (RED) PB*0.
&LECIEJEWICZ(J.).

2281 PB*0,. . . . . .A1,A2. . . . .NUKLEONIKA VOL.5,. . . . . 414, (1960).
&NEUTRON DIFFRACTION STUDIES IN SOLID STATE PHYSICS.
&BLINOWSKI(K.).

2282 PB*0,. . . . . .A1. . . . . . .ACTA CRYST. VOL.14,. . . . . 8C, (1961).
&A NEUTRON DIFFRACTION STUDY OF ORTHORHOMBIC PB*0.
&KAY(M.I.).

2283 PB*0,. . . . . .A1. . . . . . .ACTA CRYST. VOL.14,. . . . . . . 66, (1961).
&NEUTRON-DIFFRACTION STUDY OF ORTHORHOMBIC LEAD MONOXIDE.
&LECIEJEWICZ(J.).

SECTION   5 -STRUCTURE AND CROSS-SECTION DETERMINATIONS . . . .

2284 (PB*O)5-(GE*O2)3,. . . . . . .A1. . . . . .J. PHYS. SOC. JAPAN VOL.35, ... 314, (1973).
 ↦CRYSTAL STRUCTURE DETERMINATION OF FERROELECTRIC PHASE 5PB*0.3GE*02.
 ↦IWATA(Y.), KOIZUMI(H.), KOYANO(N.), SHIBUYA(I.), NIIZEKI(N.).

2285 (PB*O)5-(GE*O2)3,. . . . . . .A1. . . . . .PHYS. SOC. JAPAN VOL.35, ... 1269, (1973).
 ↦NEUTRON DIFFRACTION STUDIES OF FERROELECTRIC 5PB*0.3GE*02 ABOVE THE
 CURIE POINT.
 ↦IWATA(Y.), KOYANO(N.), SHIBUYA(I.).

2286 PB*TI*03,. . . . . .A1. . . . . . .PHYS. REV. VOL.97, ... 1179, (1955).
 ↦X-RAY AND NEUTRON DIFFRACTION STUDY OF FERROELECTRIC PB*TI*03.
 ↦SHIRANE(G.), PEPINSKY(R.).

2287 PB*TI*03,. . . . . .A1. . . . . .ACTA CRYST. VOL.9, ... 131, (1956).
 ↦X-RAY AND NEUTRON DIFFRACTION STUDY OF FERROELECTRIC PB*TI*03.
 ↦SHIRANE(G.), PEPINSKY(R.), FRAZER(B.C.).

2288 PB*(X-Y)*03, ...A1,2 . . . .SOV. PHYS. SOL. STATE, 11, ... 1133, (1969).
 ↦NEUTRON DIFFRACTION INVESTIGATION OF ATOMIC AND MAGNETIC ORDERING
 IN CERTAIN PEROVSKITE COMPOUNDS.
 ↦KISELEV(S.V.), OZEROV(R.P.).

2289 PB*ZR*03,. . . . . .A1. . . . . . .PHYS. REV. VOL.105, ... 849, (1957).
 ↦X-RAY AND NEUTRON DIFFRACTION STUDY OF ANTIFERROELECTRIC LEAD ZIRCONATE,
 PB*ZR*03.
 ↦JONA(F.), SHIRANE(G.), MAZZI(F.), PEPINSKY(R.).

2290 PD,. . . . . .A2. . . . . . .J. APP. PHYS. 36, ... 1093, (1965).
 ↦NEUTRON DIFFRACTION STUDIES OF PD, NI, FE*MN, AND CU(MN) SINGLE CRYSTALS
 ↦ARROTT(A.).

2291 PD,. . . . . .A2. . . . . . .J. PHYS. CHEM. SOLIDS VOL.24,. . 589, (1963).
 ↦ABSENCE OF ANTIFERROMAGNETISM IN PALLADIUM.
 ↦ABRAHAMS(S.C.).

2292 PD-AU-(H,D), ...A1. . . . . .CAN. J. PHYS. 46, ... 121, (1968).
 ↦A NEUTRON-DIFFRACTION STUDY OF THE ALPHA-PHASE IN THE PALLADIUM-GOLD-
 HYDROGEN AND PALLADIUM-GOLD-DEUTERIUM SYSTEMS.
 ↦MAELAND(A.J.).

2293 PD-CO, ...A2. . . . . .REF. SECTION 3, 65/CABLE,. . . . .

2294 PD-FE, ...A4. . . . . .PROC. INT. CONF. (NOTTINGHAM), . 288, (1964).
 ↦POLARIZED NEUTRON INVESTIGATION OF THE CO*PT3 AND FE*PD ALLOYS.
 ↦ANTONINI(B.), FELCHER(G.P.), MAZZONE(G.), MENZINGER(F.), PAOLETTI(A.).

2295 PD-FE, ...A4. . . . . .PROC. INT. CONF. (NOTTINGHAM), . 133, (1964).
 ↦NEUTRON DIFFRACTION STUDIES OF PD RICH FERROMAGNETIC ALLOYS.
 ↦LOW(G.G.).

2296 PD-FE, ...A2. . . . . .REF. SECTION 3, 66/SIDOROV,. . . . .
2297 PD-FE, ...A2. . . . . .REF. SECTION 3, 65/CABLE,. . . . .

2298 PD-FE, ...A2. . . . . . .J. PHYS. SOC. JAPAN VOL.17 B-III   38, (1962).
 ↦MAGNETIC MOMENT DISTRIBUTION IN PALLADIUM AND IRON GROUP ALLOYS.
 ↦WOLLAN(E.O.), CABLE(J.W.), KOEHLER(W.C.), WILKINSON(M.K.).

2299 PD-FE, ...A2. . . . . . .J. OF PHYS. C VOL.1, ... 528, (1968).
 ↦DISTRIBUTION OF THE FERROMAGNETIC POLARIZATION IN/PD-FE SINGLE CRYSTAL.
 ↦HICKS(T.J.), HOLDEN(T.M.), LOW(G.G.).

2300 PD3-FE,. . . . . .A4. . . . . .PROC. INT. CONF. (NOTTINGHAM), . 223, (1964).
 ↦SPIN DENSITIES OF 4D ATOMS IN FERROMAGNETIC ALLOYS.
 ↦SHIRANE(G.), NATHANS(R.), PICKART(S.J.), ALPERIN(H.A.).

2301 PD-FE,PD-CO, ...A2. . . . . . .J. APPL. PHYS. VOL.33, SUPPL., . 1340, (1962).
 ↦NEUTRON DIFFRACTION INVESTIGATIONS OF FERROMAGNETIC PALLADIUM AND IRON
 GROUP ALLOYS.
 ↦CABLE(J.W.), WOLLAN(E.O.), KOEHLER(W.C.), WILKINSON(M.K.).

2302 PD-(H), ...A4. . . . . . .PHYS. REV. 137, ...A483, (1965).
 ↦NEUTRON DIFFRACTION STUDY OF TEMPERATURE-DEPENDENT PROPERTIES OF
 PALLADIUM CONTAINING ABSORBED HYDROGEN.
 ↦FERGUSON(G.A.), SCHINDLER(A.I.), TANAKA(T.), MORITA(T.).

2303 PD-(H), ...A4. . . . . .SOL. STATE COMM. 4, ... 303, (1966).
 ↦NEUTRON STUDY OF THE DIFFUSION OF HYDROGEN IN PALLADIUM.
 ↦SKOLD(K.), NELIN(G.).

2304 PD-H,. . . . . .A1,P. . . . .PHYSICA VOL.26, ... 744, (1960).
 ↦THERMAL MOTION IN PALLADIUM HYDRIDE STUDIED BY MEANS OF ELASTIC AND
 INELASTIC SCATTERING OF NEUTRONS.
 ↦BERGSMA(J.), GOEDKOOP(J.A.).

2305 PD-H,. . . . . .A1. . . . . .PHYS. STAT. SOL. B VOL.45, . . . 527, (1971).
 ↦A NEUTRON DIFFRACTION STUDY OF PALLADIUM HYDRIDE.
 ↦NELIN(G.).

2306 PD-H,. . . . . .A1. . . . . .PLATINUM METALS REV. 10, ... 20, (1966).
 ↦THE HYDROGEN-PALLADIUM SYSTEM- THE ROLE OF X-RAY AND NEUTRON DIFFRACTION
 STUDIES.
 ↦MAELAND(A.), FLANAGAN(T.B.).

2307 PD-H,. . . . . .A1. . . . . .ENGELHARD IND. TECH. BULL. 7, . . 21, (1966).
 ↦LOW TEMPERATURE ELECTRONIC EFFECTS AND NEUTRON DIFFRACTION STUDIES OF
 PALLADIUM.
 ↦SCHINDLER(A.I.).

2308 PD-H,. . . . . .A1. . . . . .DISS. ABSTR. VOL.25, ... 3068, (1964).
 ↦A NEUTRON DIFFRACTION STUDY OF PALLADIUM CONTAINING ABSORBED HYDROGEN.
 ↦FERGUSON(G.A.).

2309 PD-H,PD-D, ...A1. . . . . . .J. PHYS. CHEM. SOLIDS VOL.3, . . 303, (1957).
 ↦NEUTRON-DIFFRACTION OBSERVATIONS ON THE PALLADIUM-HYDROGEN AND
 PALLADIUM-DEUTERIUM SYSTEMS.
 ↦WORSHAM-JR(J.E.), WILKINSON(M.K.), SHULL(C.G.).

2310 PD3-MN,. . . . . .A2. . . . . . .PHYS. REV. VOL.128, ...2118, (1962).
 ↦ANTIFERROMAGNETISM OF THE ANTIPHASE DOMAIN STRUCTURE OF PD3-MN.
 ↦CABLE(J.W.), WOLLAN(E.O.), KOEHLER(W.C.), CHILD(H.R.).

2311 PD3-MN2, ...A2. . . . . . .J. PHYS. SOC. JAPAN VOL.21, . . . 1626, (1966).
 ↦MAGNETIC STRUCTURE OF THE ZETA-PHASE OF PD3-MN2 ALLOY.
 ↦GONZALO(J.A.), KAY(M.I.).

SECTION   5 -STRUCTURE AND CROSS-SECTION DETERMINATIONS . . . .

2312 PO-MN, ...   A1,A2. . . .PHIL. MAG. VOL.16, ... 1063, (1967),
      *EQUIATOMIC TRANSITION METAL ALLOYS OF MANGANESE VI. STRUCTURAL AND
      MAGNETIC PROPERTIES OF PO-MN PHASES.
      *KJEKSHUS(A.), MOLLERUD(R.), ANDRESEN(A.F.), PEARSON(W.B.).

2313 PD2-MN-AL, . . .   A1,A2. . . .J. APP. PHYS. VOL.39, ... 471, (1968).
      *MAGNETIC AND CHEMICAL ORDER IN PD2-MN-AL IN RELATION TO ORDER IN THE
      HEUSLER ALLOYS PD2-MN-IN, PD2-MN-SN AND PD2-MN-SB.
      *WEBSTER(P.J.), TEBBLE(R.S.).

2314 PD2-MN-(GE,SN),.   A1,A2. . . .PHYS. STAT. SOLIDI A VOL.3, . . . 959, (1970).
      *ATOMIC AND MAGNETIC STRUCTURE OF THE HEUSLER ALLOYS PD2-MN-GE, PD2-MN-SN
      CU2-MN-IN, AND CO-MN-SB.
      *NATERA(M.G.), MURTHY(M.R.L.), BEGUM(R.J.), MURTHY(N.S.S.).

2315 PD2-MN-(IN,SN,SB), . . . A2. . . . . .PHIL. MAG. VOL.16, ... 347, (1967).
      *THE MAGNETIC AND CHEMICAL ORDERING OF THE HEUSLER ALLOYS PD2-MN-IN,
      PD2-MN-SN AND PD2-MN-SB.
      *WEBSTER(P.J.), TEBBLE(R.S.).

2316 PD(NI), ...   A4. . . . . .PHYS. REV. LETT. VOL.24, ... 897, (1970).
      *GIANT MOMENTS IN PD(NI) ALLOYS NEAR THE CRITICAL COMPOSITION.
      *ALDRED(A.T.), RAINFORD(B.D.), STRINGFELLOW(M.W.).

2317 PD-NI-(H), ...PHYS. REV. VOL.125, ... 1141, (1962).
      *SINGLE TRANSMISSION EFFECT FOR SLOW NEUTRONS PASSING THROUGH MAGNETIZED
      PD-NI ALLOY, CONTAINING ABSORBED HYDROGEN.
      *MITROFANOV(N.).

2318 PD*O, ...   A1. . . . . .ACTA CRYST. VOL.6, ... 661, (1953).
      *THE STRUCTURE OF PD*O.
      *WASER(J.), LEVY(H.A.), PETERSON(S.W.).

2319 PENTANOL, . . .   A3. . . . . .REV. ROUMAINE PHYS. VOL.13, . . . 287, (1968).
      *ANGULAR AND ENERGY DISTRIBUTIONS OF SLOW NEUTRONS IN PENTANOL.
      *RAPEANU(S.), ILIESCU(N.).

2320 PERDEUTERO-ALPHA-GLYCYLGLYCINE, ...A1. . . . . .ACTA CRYST. VOL.B26, ... 925, (1970).
      *A NEUTRON DIFFRACTION STUDY OF PERDEUTERO-ALPHA-GLYCYLGLYCINE.
      *FREEMAN(H.C.), PAUL(G.L.), SABINE(T.M.).

2321 PEROVSKITES, . .   A1,2. . . . .SOV. PHYS. SOL. STATE, 11, . . 1133, (1969).
      *NEUTRON DIFFRACTION INVESTIGATION OF ATOMIC AND MAGNETIC ORDERING
      IN CERTAIN PEROVSKITE COMPOUNDS.
      *KISELEV(S.V.), OZEROV(R.P.).

2322 PEROVSKITES, . .   A2. . . . . .SOV. PHYS. SOL. STATE 5, . . . 2425, (1963).
      *NEUTRON DIFFRACTION STUDY OF MAGNETIC ORDERING/ATOMIC DISPLACEMENTS IN
      PEROVSKYTELIKE SUBS, CONTAINING FE AND EXHIBITING PECULIAR DIELECTRIC...
      *KISELEV(S.V.), KSHNYAKINA(A.N.), OZEROV(R.P.), ZHDANOV(G.S.).

2323 PLUMBITE, ...   A1. . . . . .SOV. PHYS. CRYST. VOL.14, 3. . . 447, (1969).
      *NEUTRON DIFFRACTION INVESTIGATION OF THE STRUCTURE OF MAGNETOPLUMBITE.
      *ALESHKO-OZHEVSKII(O.P.), FAEK(M.K.), YAMZIN(I.I.).

2324 PM (ISO. A=147), ...P. . . . . . .PROC/CONF/NEUTRON C. S. + TECH., 687, (1968).
      *TOTAL NEUTRON CROSS SECTION AND RESONANCE PARAMETERS FOR PM (ISO. A=147)
      *KIROUAC(G.J.), EILAND(H.M.), SLOVACEK(R.E.), CONRAD(C.A.), SEEMANN(K.W.)

2325 PM (ISOTOPE A=147), . . . A4. . . . . .NUCL. SCI. ENG. VOL.43, ... 58, (1971).
      *TOTAL NEUTRON CROSS SECTION AND RESONANCE PARAMETERS OF PROMETHIUM-147.
      *CODDING-JR(J.W.), TROMP(R.L.), SIMPSON(F.B.).

2326 PM (A=147,148), .   P. . . . . . .NUCL. SCI. ENG. VOL.52, ... 310, (1973).
      *RESONANCE AND THERMAL-NEUTRON TOTAL CROSS SECTIONS FOR PROMETHIUM-147
      AND PROMETHIUM-148 M.
      *KIROUAC(G.J.), EILAND(H.M.), CONRAD(C.A.), SLOVACEK(R.E.), SEEMANN(K.W.)

2327 PM (A=147), . . .   A2. . . . . .A.I.P. CONF. PROC. NO.10 PART 2, 1319, (1973).
      *NEUTRON SCATTERING STUDIES OF PM(A=147).
      *KOEHLER(W.C.), MOON(R.M.), CHILD(H.R.).

2328 POLYETHYLENE, . .   P. . . . . . .NUCL. SCI. ENG. VOL.23, ... 192, (1965).
      *THE ENERGY-DEPENDENT TOTAL NEUTRON CROSS SECTION OF POLYETHYLENE.
      *ARMSTRONG(S.B.).

2329 POLYMER SOLUTIONS,   A3. . . . . .J. CHEM. PHYS. VOL.57, ... 290, (1972).
      *NEUTRON DIFFRACTION IN DILUTE AND SEMIDILUTE POLYMER SOLUTIONS.
      *COTTON(J.P.), FARNOUX(B.), JANNINK(G.).

2330 POLYMERS, ...   T. . . . . . .J. APPL. CRYST. VOL.7, ... 189, (1974).
      *STUDY OF POLYMER SOLUTION BY SMALL-ANGLE NEUTRON SCATTERING IN THE
      INTERMEDIATE MOMENTUM RANGE.
      *COTTON(J.P.), FARNOUX(B.), JANNINK(G.), OBER(R.).

2331 POLYMERS, ...   T,E. . . . . .SPECTROSCOPY BIOL. CHEM., ... 323, (1974).
      *SMALL ANGLE NEUTRON SCATTERING FROM POLYMERS.
      *SUMMERFIELD(G.C.). (REFER TO SECTION 4).

2332 POLYMETHYLMETHACRYLATE, ...A4. . . . . .J. APPL. CRYST. VOL.7, ... 188, (1974).
      *THE DETERMINATION OF THE CONFORMATION OF POLYMERS IN BULK BY NEUTRON
      DIFFRACTION.
      *KIRSTE(R.G.), KRUSE(W.A.), SCHELTEN(J.).

2333 PR, ...   A2. . . . . .J. PHYSIQUE TOME 32 COL.1 VOL.I, 370, (1971).
      *THE MAGNETIC STRUCTURES OF PRASEODYMIUM AND NEODYMIUM.
      *LEBECH(B.), RAINFORD(B.D.).

2334 PR, ...   A2. . . . . .REF. SECTION 3, 65/KOEHLER, ...

2335 PR, PR3*TL, . .   A2. . . . . .PHYS. REV. B VOL.6, ... 2724, (1972).
      *NEUTRON SCATTERING FROM FCC PR AND PR3*TL.
      *BIRGENEAU(R.J.), ALS-NIELSEN(J.), BUCHER(E.).

2336 PR, ...   A2. . . . . .PHYS. REV. LETT. VOL.25, ... 524, (1970).
      *CRYSTAL FIELDS AND THE MAGNETIC PROPERTIES OF PRASEODYMIUM AND NEODYMIUM
      *JOHANSSON(T.), LEBECH(B.), NIELSEN(M.), MOLLER(H.B.), MACKINTOSH(A.R.).

2337 PR, ...   A2. . . . . .PHYS. REV. LETT. VOL.12, ... 553, (1964).
      *ANTIFERROMAGNETISM OF PRASEODYMIUM.
      *CABLE(J.W.), MOON(R.M.), KOEHLER(W.C.), WOLLAN(E.O.).

2338 PR*AG, ...   A2,A4. . . . .PHYS. REV. B VOL.9, ... 248, (1974).
      *MAGNETIC STRUCTURE AND CRYSTAL-FIELD LEVELS IN PR*AG.
      *BRUN(T.O.), LANDER(G.H.), PRICE(D.L.), FELCHER(G.P.), REDDY(J.F.).

SECTION   5 -STRUCTURE AND CROSS-SECTION DETERMINATIONS . . . .

2339   PR*AL2,. . . . . .   .MAGNETIC PROPERTIES OF PR*AL2.......A2...........J. APP. PHYS. 38,. . . . . 1395, (1967).
                            .OLSEN(C.E.), ARNOLD(G.P.), NERESON(N.G.).

2340   PR*C2, ...CRYSTALS AND MAGNETIC STRUCTURES OF CE*C2, PR*C2 AND ND*C2.......A1,2.....PHYS. LETT. 22,.   21, (1966).
                            .ATOJI(M.).

2341   PR*C2, ...MAGNETIC AND   CRYSTAL   STRUCTURES OF CE*C2,PR*C2,ND*C2,TB*C2, AND HO*C2.......A1,A2....J. CHEM. PHYS. VOL.46,. . . . . . 1891, (1967).
                            AT LOW TEMPERATURES.
                            .ATOJI(M.).

2342   PR2*C3,. . . . .   .NEUTRON-DIFFRACTION STUDIES OF LA2*C3, CE2*C3, PR2*C3, AND TB2*C3.......A1,A4.....J. CHEM. PHYS. VOL.35,. . . 1960, (1961).
                            .ATOJI(M.), WILLIAMS(D.E.).

2343   PR*C*03, ...A2,A4....REF. SECTION 3, 66/BERTAUT,. . . . .

2344   PR*CO2,. . . . .   .MAGNETIC STRUCTURE OF INTERMETALLIC COMPOUND PR*CO2.......A2.......PHYS. LETT. VOL.24A,. . . . . . 739, (1967).
                            .SCHWEIZER(J.). (IN FRENCH).

2345   PR*CO5*O4, ...MAGNETIC STRUCTURE OF PR*CO5*O4.......A2........J. PHYSIQUE TOME 32 COL.1 VOL.II 657, (1971).
                            .KUIJPERS(F.A.), LOOPSTRA(B.O.).

2346   PR*CO5-D,. . . .   .A NEUTRON-DIFFRACTION STUDY ON THE STRUCTURAL RELATIONSHIPS OF R*CO5.......A1.......J. PHYS. CHEM. SOLIDS VOL.35,. . 301, (1974).
                            HYDRIDES.
                            .KUIJPERS(F.A.), LOOPSTRA(B.O.).

2347   PR-CU-AL,. . . .   .MAGNETIC AND NEUTRON DIFFRACTION STUDIES ON PR(.333)-CU(.1)-AL(.566).......A2........J. LESS-COMMON MET.(SWIZ.) 26, . 165, (1972).
                            .OESTERREICHER(H.).

2348   PR*FE*O3, ...LONG WAVELENGTH NEUTRON DIFFRACTION STUDY OF THE MAGNETIC STRUCTURES OF.......A2.......SOL. STATE COMM. VOL.10,. . . 663, (1972).
                            PR*FE*O3 AND ND*FE*O3.
                            .PINTO(H.), SHAKED(H.).

2349   PR2*O3,. . . . .   .NEUTRON-DIFFRACTION STUDY OF THE STRUCTURE OF THE A-FORM OF THE RARE.......A1........ACTA CRYST. VOL.6,. . . . . . 741, (1953).
                            EARTH SESQUIOXIDES.
                            .KOEHLER(W.C.), WOLLAN(E.O.).

2350   PROLINE(L-MONOHYDRATE).. . . . . . .A1........NATURE VOL.235,. . . . . 328, (1972).
                            .DIRECT METHODS IN NEUTRON CRYSTALLOGRAPHY: STRUCTURE OF L-PROLINE MONO-
                            HYDRATE BY SYMBOLIC ADDITION AND TANGENT REFINEMENT.
                            .VERBIST(J.J.), LEHMANN(M.S.), KOETZLE(T.F.), HAMILTON(W.C.).

2351   PR-TH, ...MAGNETIC PROPERTIES OF FCC RARE EARTH-THORIUM ALLOYS.......A2........J. APP. PHYS. 39,. . . . . 1329, (1968).
                            .CHILD(H.R.), KOEHLER(W.C.), MILLHOUSE(A.H.).

2352   PT3-CO,. . . . .   .POLARIZED NEUTRON INVESTIGATION OF THE CO-PT3 AND FE-PD ALLOYS.......A4.......PROC. INT. CONF. (NOTTINGHAM), . 288, (1964).
                            .ANTONINI(B.), FELCHER(G.P.), MAZZONE(G.), MENZINGER(F.), PAOLETTI(A.).

2353   PT-CO, ...THE MAGNETIC STRUCTURE OF CO-PT.......A2.......COLL. INTER. N.126 (GRENOBLE), . 176, (1963).
                            .VAN-LAAR(B.).

2354   PT3-FE,. . . . .   .EFFECTS OF ATOMIC ORDERING ON THE MAGNETIC STRUCTURE OF PT3-FE.......A2.......SOL. STATE COMM. 4,. . . . . . 31, (1966).
                            .KREN(E.), SZABO(P.), TARNOCZI(T.).

2355   PT-FE, ...CHEMICAL AND MAGNETIC ORDER IN PLATINUM-RICH PT-FE ALLOYS.......A2.......PROC. ROY. SOC. A VOL.272,. . . 387, (1963).
                            .BACON(G.E.), CRANGLE(J.).

2356   PT-FE-MN,. . . .   .MAGNETIC ORDER IN TERNARY PT-FE-MN ALLOYS.......A2.......PROC. PHYS. SOC. 88,. . . . . . 929, (1966).
                            .BACON(G.E.), MASON(E.W.).

2357   PT-MN3,. . . . .   .NEUTRON DIFFRACTION OF ORDERED PT-MN3.......A2.......Z. ANGEW. MATH. PHYS. VOL.16,. . 535, (1965).
                            .WUTTIG(M.).

2358   PT-MN, ...EQUIATOMIC TRANSITION METAL ALLOYS OF MANGANESE. IV. A NEUTRON.......A2.......PHIL. MAG. VOL.11,. . . . . . 1245, (1965).
                            DIFFRACTION STUDY OF MAGNETIC ORDERING IN THE PT-MN PHASE.
                            .ANDRESEN(A.F.), KJEKSHUS(A.), MOLLERUD(R.), PEARSON(W.B.).

2359   PT-MN3,. . . . .   .A NEUTRON DIFFRACTION STUDY OF CHEMICAL ORDER-DISORDER IN PT-MN3.......A4.......ACTA CRYST. 19,. . . . . . . 413, (1965).
                            .SIDHU(S.S.), ANDERSON(K.D.), ZAUBERIS(D.D.).

2360   PT-MN, ...EQUIATOMIC TRANSITION METAL ALLOYS OF MANGANESE V. ON THE MAGNETIC.......A2.......ACTA CHEM. SCAND. VOL.20,. . . 2529, (1966).
                            PROPERTIES OF THE PT-MN PHASE.
                            .ANDRESEN(A.F.), KJEKSHUS(A.), MOLLERUD(R.), PEARSON(W.B.).

2361   PU (A=239),. . .   .DELAYED GAMMA-RAYS FROM THERMAL NEUTRON FISSION OF U (A=235) AND.......P.......PHYS. REV. VOL.187,. . . . . . 1506, (1969).
                            PU (A=239).
                            .BERICK(A.C.), EVANS(A.E.), MEISSNER(J.A.).

2362   PU (ISOTOPE A=239),. . .   .NEUTRON SCATTERING CROSS SECTIONS OF U(ISOTOPES A=233, 235) AND PU(ISO-.......A4.......NUCL. PHYS. A VOL.A164,. . . . 34, (1971).
                            TOPE A=239).
                            .SIMPSON(F.B.), MILLER(L.G.), MOORE(M.S.), HOCKENBURG(R.W.), KING(T.J.).

2363   PU CARBIDES, . .   .CRYSTALLOGRAPHIC AND MAGNETIC ORDERING STUDIES OF PLUTONIUM CARBIDES.......A1,A2....J. NUCL. MATERIALS VOL.34,. . . 281, (1970).
                            USING NEUTRON DIFFRACTION.
                            .GREEN(J.L.), ARNOLD(G.P.), LEARY(J.A.), NERESON(N.G.).

2364   PU (ISOTOPE A=239),. . .   .COHERENT NEUTRON SCATTERING AMPLITUDES OF FISSIONABLE NUCLEI, U (A=233),.......P.......ACTA CRYST. 20,. . . . . . . 587, (1966).
                            U (A=235), AND PU (A=239).
                            .ATOJI(M.).

2365   PU,. . . . . . .   .NUCLEAR COHERENT SCATTERING AMPLITUDES FOR THORIUM, URANIUM AND.......P.......ACTA CRYST. VOL.15,. . . . . . 351, (1962).
                            PLUTONIUM.
                            .ROOF-JR(R.B.), ARNOLD(G.P.), GSCHNEIDNER-JR(K.A.).

SECTION   5 -STRUCTURE AND CROSS-SECTION DETERMINATIONS . . . .

2366   PU (ISOTOPE A=241), ... P . . . . CAN. J. PHYS. 42, ... 2384, (1964).
       ↪THE TOTAL NEUTRON CROSS SECTION OF 241PU BELOW 1000 EV.
       ↪CRAIG(D.S.), WESTCOTT(C.H.).

2367   PU (ISOTOPES A=240, 242), ... P . . . . ACTA CRYST. VOL.B27, ... 2284, (1971).
       ↪COHERENT SCATTERING AMPLITUDE OF PU(ISOTOPES A=240, 242).
       ↪LANDER(G.H.), MUELLER(M.H.).

2368   PU2*C3 TYPE, ... A4 . . . . J. PHYS. SOC. JAPAN VOL.17 B-II, 395, (1962).
       ↪NEUTRON DIFFRACTION STUDIES OF HIGHER CARBIDES OF HEAVY METALS.
       ↪ATOJI(M.).

2369   PU*PD3, ... A2 . . . . PHYS. REV. B VOL.9, ... 1041, (1974).
       ↪MAGNETIC PROPERTIES OF NP*PD3 AND PU*PD3 INTERMETALLIC COMPOUNDS.
       ↪NELLIS(W.J.), HARVEY(A.R.), LANDER(G.H.), DUNLAP(B.D.), BRODSKY(M.B.),
       MUELLER(M.H.), REDDY(J.F.), DAVIDSON(G.R.).

2370   PYRENE, ... A4 . . . . ADV. STRUC. RES. DIFFR. METHOD 4   1, (1972).
       ↪CONSTRAINED REFINEMENTS IN CRYSTALLOGRAPHY.
       ↪PAWLEY(G.S.).

2371   QUARTZ, ... A4 . . . . PHYS. STAT. SOL. 29, ... K51, (1968).
       ↪DIFFRACTION OF NEUTRONS ON A VIBRATING QUARTZ CRYSTAL.
       ↪CHALUPA(B.), MICHALEC(R.), PETRZILKA(V.), TICHY(J.), ZELENKA(J.).

2372   QUARTZ, ... A4 . . . . ACTA CRYST. VOL.A27, ... 410, (1971).
       ↪NEUTRON DIFFRACTION EFFECTS DUE TO THE LATTICE DISPLACEMENT OF A
       VIBRATING QUARTZ SINGLE CRYSTAL.
       ↪MICHALEC(R.), SEDLAKOVA(L.), CHALUPA(B.), GALOCIOVA(D.), PETRZILKA(V.).

2373   QUARTZ, ... A4 . . . . PHYS. STAT. SOL. VOL.42, ... 895, (1970).
       ↪NEUTRON DIFFRACTION ON LATTICE PLANES DEFORMED BY FLEXURAL VIBRATIONS
       OF QUARTZ SINGLE CRYSTALS.
       ↪PETRZILKA(V.), VRZAL(J.), MICHALEC(R.), CHALUPA(B.), MIKULA(P.),
       ZELENKA(J.).

2374   QUARTZ, ... A3 . . . . PHYS. CHEM. GLASSES VOL.10, ... 185, (1969).
       ↪NEUTRON DIFFRACTION FROM CONDENSED SILICA VAPOUR PARTICLES, SYNTHETIC
       VITREOUS SILICA AND FUSED QUARTZ GLASS.
       ↪LORCH(E.).

2375   QUARTZ, ... A4 . . . . BRIT. J. APPL. PHYS. VOL.2, ... 1041, (1969).
       ↪THE APPLICATION OF NEUTRON DIFFRACTION TO THE DETERMINATION OF THE
       QUALITY OF SYNTHETIC QUARTZ CRYSTALS.
       ↪ZELENKA(J.), TICHY(J.), CHALUPA(B.), MICHALEC(R.), PETRZILKA(V.).

2376   QUARTZ, ... A4 . . . . PHYS. LETT. 28A, ... 546, (1968).
       ↪TIME MODULATIONS OF A NEUTRON BEAM DIFFRACTED ON A SINGLE QUARTZ CRYSTAL
       EXCITED INTO VIBRATIONS BY HIGH FREQUENCY PULSES.
       ↪MICHALEC(R.), CHALUPA(B.), GALOCIOVA(D.), MIKULA(P.).

2377   QUARTZ, ... P . . . . PHIL. MAG. VOL.3, ... 1280, (1958).
       ↪THE SCATTERING OF LONG WAVELENGTH NEUTRONS BY IRRADIATED AND
       UNIRRADIATED QUARTZ.
       ↪MITCHELL(E.W.J.), WEDEPOHL(P.T.).

2378   QUARTZ, ... A4 . . . . APPL. PHYS. LETT. VOL.10, ... 293, (1967).
       ↪DIFFRACTION OF NEUTRONS AND X RAYS BY A VIBRATING QUARTZ CRYSTAL.
       ↪KLEIN(A.G.), PRAGER(P.), WAGENFELD(H.), ELLIS(P.J.), SABINE(T.M.).

2379   QUARTZ, ... P . . . . NUCL. INST. MET. 67, ... 357, (1969).
       ↪NEUTRON DIFFRACTION ON QUARTZ CRYSTALS AS A FUNCTION OF VIBRATING
       FREQUENCIES.
       ↪CHALUPA(B.), MICHALEC(R.), GALOCIOVA(D.).

2380   RARE EARTHS, ... A2 . . . . Z. ANGEW. PHYS. VOL.26, ... 67, (1968).
       ↪NEUTRON DIFFRACTION MEASUREMENTS OF THE MAGNETIC STRUCTURES AND FORM
       FACTORS OF THE RARE EARTH ELEMENTS.
       ↪WILL(G.).

2381   RARE-EARTHS, ... A2 . . . . CONF./RARE EARTHS/ACTINIDES, . . . 25, (1971).
       ↪RECENT DEVELOPMENTS IN NEUTRON SCATTERING FROM RARE EARTHS.
       ↪KOEHLER(W.C.).

2382   (RARE-EARTH)*IN3,(--)*SN3, . . . A2 . . . . PHYS. STAT. SOLIDI A VOL.15 NO.2 613, (1973).
       ↪PROPRIETES MAGNETIQUES ET STRUCTURE ORDONNEE DES COMPOSES T*M3 DES
       TERRES RARES AVEC L≠INDIUM ET L≠ETAIN.
       ↪LETHUILLIER(P.), PIERRE(J.), FILLION(G.), BARBARA(B.).

2383   (RARE-EARTH)*MN2*05, ... A2 . . . . PHYS. STAT. SOLIDI A VOL.16 NO.2 533, (1973).
       ↪ORDRE HELIMAGNETIQUE DU MANGANESE DANS LA SERIE T*MN2*05 (T=LU,ER,HO,TB,
       ND).
       ↪BUISSON(G.).

2384   RARE-EARTH METALS, ... A2 . . . . J. PHYS. SOC. JAPAN VOL.17 B-III   27, (1962).
       ↪RECENT MAGNETIC NEUTRON SCATTERING INVESTIGATIONS AT OAK RIDGE NATIONAL
       LABORATORY.
       ↪WILKINSON(M.K.), CHILD(H.R.), KOEHLER(W.C.), CABLE(J.W.), WOLLAN(E.O.).

2385   RARE-EARTH-ORTHOFERRITES, ... A2 . . . . J. PHYSIQUE TOME 30, ... 967, (1969).
       ↪MAGNETIC STRUCTURES OF RARE EARTH ORTHOFERRITES.
       ↪MARESCHAL(J.), SIVARDIERE(J.).

2386   RARE-EARTH METALS/COMPOUNDS, . . . A2 . . . . IAEA SYMP. (PILE RESEARCH)VIENNA 379, (1960).
       ↪NEUTRON-DIFFRACTION STUDIES OF RARE-EARTH METALS AND COMPOUNDS.
       ↪CABLE(J.W.), CHILD(H.R.), KOEHLER(W.C.), WILKINSON(M.K.), WOLLAN(E.O.).

2387   RB, ... P . . . . ACTA CRYST. VOL.A26, ... 377, (1970).
       ↪THE COHERENT NEUTRON SCATTERING AMPLITUDE OF RB: A NEUTRON DIFFRACTION
       STUDY OF RB*CL.
       ↪WANG(F.F.Y.), COX(D.E.).

2388   RB, ... P . . . . ACTA CRYST. VOL.A26, ... 376, (1970).
       ↪A REDETERMINATION OF THE COHERENT NEUTRON SCATTERING AMPLITUDE OF RB.
       ↪COPLEY(J.R.D.).

2389   RB, ... P . . . . C. R. ACAD. SCI. B TOME 270, ... 560, (1970).
       ↪NOUVELLE DETERMINATION DE L≠AMPLITUDE DE DIFFUSION COHERENTE DU RUBIDIUM
       POUR LES NEUTRONS.
       ↪MERIEL(P.).

2390   RB, ... A3 . . . . PHYS. LETT. VOL.29A, ... 296, (1969).
       ↪ATOMIC OVERLAP IN THE LIQUID STATE.
       ↪PAGE(D.I.), EGELSTAFF(P.A.), ENDERBY(J.E.), WINGFIELD(B.R.).

SECTION   5 -STRUCTURE AND CROSS-SECTION DETERMINATIONS . . . .

2391 RB*CO*BR3, ...A2. . . . . . .J. PHYSIQUE TOME 32 COL.1 VOL.II 892, (1971).
                       *NEUTRON SCATTERING FROM RB*CO*BR3 AND RB*NI*CL3.
                       *MINKIEWICZ(V.J.), COX(D.E.), SHIRANE(G.).

2392 RB*CO*F3,. . . . . .A2. . . . . . .J. PHYSIQUE TOME 32 COL.1 VOL.II 611, (1971).
                       *ETUDE DES PROPRIETES CRISTALLOGRAPHIQUES, MAGNETIQUES ET ELASTIQUES DE
                       RB*CO*F3.
                       *ALLAIN(Y.), DENIS(J.), HERPIN(A.), LECOMTE(J.), MERIEL(P.), NOUET(J.),
                       PLICQUE(F.), ZAREMBOVITCH(A.).

2393 RB2*CO*F4, ...A2. . . . . . .J. PHYS. CHEM. SOLIDS VOL.35,. . . 785, (1974).
                       *CRITICAL BEHAVIOR OF THE TWO-DIMENSIONAL ISING ANTIFERROMAGNETS K2*CO*F4
                       AND RB2*CO*F4.
                       *SAMUELSEN(E.J.).

2394 RB2*CU*H12*O14*S2, ...A1. . . . . . . .CRYST. STRUCT. COMM. VOL.1 N.4,. 367, (1972).
                       *RUBIDIUM HEXA-AQUACOPPER(II) SULPHATE.
                       *VAN-DER-ZEE(J.J.), SHIELDS(K.G.), GRAHAM(A.J.), KENNARD(C.H.L.).

2395 RB*FE*CL3, ...A2. . . . . . .MAGN. AND MAG. MATERIALS-1971, . 436, (1972).
                       *MAGNETIC ORDERING IN A LINEAR CHAIN COMPOUND RB*FE*CL3.
                       *DAVIDSON(G.R.), EIBSCHUTZ(M.), COX(D.E.), MINKIEWICZ(V.J.). (AIP CONF.
                       PROC. NO.5 PART I).

2396 RB*FE*F3,. . . . . .A2. . . . . . .PHYS. REV. B VOL.3,. . . . . . . 3946, (1971).
                       *NEUTRON DIFFRACTION STUDY OF RB*FE*F3.
                       *WANG(F.F.Y.), COX(D.E.), KESTIGIAN(M.).

2397 RB*FE*F4,. . . . . .A2. . . . . . .MAGN. AND MAG. MATERIALS-1971, 670, (1972).
                       *MAGNETIC STRUCTURE AND MOSSBAUER STUDIES OF THE LAYER ANTIFERROMAGNETS
                       ALPHA-RB*FE*F4 AND K*FE*F4.
                       *EIBSCHUTZ(M.), DAVIDSON(G.R.), GUGGENHEIM(H.J.), COX(D.E.). (AIP CONF.
                       PROC. NO.5 PART I).

2398 RB*FE*F4,. . . . . .A2. . . . . . .SOL. STATE COMM. VOL.10,. . . . 1299, (1972).
                       *MAGNETIC BEHAVIOUR OF THE TWO-DIMENSIONAL ANTIFERROMAGNET RB*FE*F4.
                       *HEGER(G.), DACHS(H.).

2399 RB*H*O*(C*H2*C*O*O)2,. . . . . . . .A1. . . . . . .ACTA CRYST. VOL.B29,. . . . . . . 2751, (1973).
                       *A NEUTRON DIFFRACTION STUDY OF POTASSIUM AND RUBIDIUM HYDROGEN OXYDI-
                       ACETATE. THE DYNAMICS OF THEIR HYDROGEN BONDS.
                       *ALBERTSSON(J.), GRENTHE(I.).

2400 RB*MN*BR3, ...A2. . . . . . .A.I.P. CONF. PROC. NO.10 PART 1, 659, (1973).
                       *THE MAGNETIC STRUCTURE OF RB*MN*BR3.
                       *GLINKA(C.J.), MINKIEWICZ(V.J.), COX(D.E.), KHATTAK(C.P.).

2401 RB2*MN*CL4,. . . . . .A2. . . . . . .PHYS. REV. LETT. VOL.25, ... 1713, (1970).
                       *MAGNETIC POLYTYPES IN RB2*MN*CL4.
                       *GUREWITZ(E.), EPSTEIN(A.), MAKOVSKY(J.), SHAKED(H.).

2402 RB*MN*CL3, ...A2. . . . . . .PHYS. REV. B VOL.3,. . . . . . . 821, (1971).
                       *MAGNETIC STRUCTURE OF RB*MN*CL3.
                       *MELAMUD(M.), MAKOVSKY(J.), SHAKED(H.), SHTRIKMAN(S.).

2403 RB2*MN*CL4,. . . . . .A2. . . . . . .PHYS. REV. B VOL.2,. . . . . . . 3703, (1970).
                       *MAGNETIC STRUCTURE AND TWO-DIMENSIONAL BEHAVIOUR OF RB2*MN*CL4 AND
                       CS2*MN*CL4.
                       *EPSTEIN(A.), GUREWITZ(E.), MAKOVSKY(J.), SHAKED(H.).

2404 RB*MN*CL3.2H2*O, ...A1. . . . . . .ACTA CHEM. SCAND. VOL.24,. . . . 3422, (1970).
                       *NEUTRON DIFFRACTION STUDY OF BETA-RB*MN*CL3.2H2*O.
                       *JENSEN(S.J.), LEHMANN(M.S.).

2405 RB*MN*F3,. . . . . .A2. . . . . . .COLL. INTER. N.126 (GRENOBLE), . 141, (1963).
                       *MAGNETIC STRUCTURE OF BINARY FLUORIDES CONTAINING MN(+2).
                       *PICKART(S.J.), ALPERIN(H.A.).

2406 RB*NI*CL3, ...A2. . . . . . .SOL. STATE COMM. VOL.8,. . . . . 1001, (1970).
                       *THE MAGNETIC STRUCTURES OF RB*NI*CL3 AND CS*NI*CL3.
                       *MINKIEWICZ(V.J.), COX(D.E.), SHIRANE(G.).

2407 RB*NI*CL3, ...A2. . . . . . .PHYS. REV. B VOL.6,. . . . . . . 204, (1972).
                       *MAGNETIC ORDERING IN RB*NI*CL3.
                       *YELON(W.B.), COX(D.E.).

2408 RB*NI*CL3, ...A2. . . . . . .J. PHYSIQUE TOME 32 COL.1 VOL.II 892, (1971).
                       *NEUTRON SCATTERING FROM RB*CO*BR3 AND RB*NI*CL3.
                       *MINKIEWICZ(V.J.), COX(D.E.), SHIRANE(G.).

2409 RB*NI*F3,. . . . . .A2. . . . . . .J. APPL. PHYS. VOL.42, ... 1617, (1971).
                       *MAGNETIC STRUCTURES OF FERRIMAGNETIC RB*NI*F3 AND CS*FE*F3.
                       *PICKART(S.J.), ALPERIN(H.A.).

2410 RB*U*O2*(N*O3)3, ...A1. . . . . . .ACTA CRYST. 19,. . . . . 205, (1965).
                       *THE CRYSTAL STRUCTURE OF RUBIDIUM URANYL NITRATE: A NEUTRON-DIFFRACTION
                       STUDY.
                       *BARCLAY(G.A.), SABINE(T.M.), TAYLOR(J.C.).

2411 RE (A=185,187),. . . . . .P. . . . . . .SOV. AT. ENERGY VOL.19,. . . . . 1162, (1965).
                       *TOTAL NEUTRON CROSS-SECTIONS OF RE (A=185,187).
                       *VERTEBNYI(V.P.), VLASOV(M.F.), KIRILYUK(A.L.), KOLOTYI(V.V.),
                       PISANKO(ZH.I.), TROFIMOVA(N.A.).

2412 RE*K2*H9,. . . . . .A4. . . . . . .COLL. INTER. N.126 (GRENOBLE), .   37, (1963).
                       *THE HYDROGEN ATOM ARRANGEMENT IN RE*H9(2-).
                       *ABRAHAMS(S.C.), KNOX(K.).

2413 RESORCINOL(ALPHA). . . . . .A1. . . . . . .ACTA CRYST. VOL.A28 PART S-4,. . 193, (1972).
                       *NEUTRON DIFFRACTION STUDIES OF SALICYLIC ACID AND ALPHA-RESORCINOL.
                       *BACON(G.E.), JUDE(R.J.). (ABSTRACT ONLY).

2414 RESORCINOL (C6*H6*O2). . . . . .A1. . . . . . .PROC. ROY. SOC. A VOL.235, . . . 552, (1956).
                       *A STUDY OF ALPHA-RESORCINOL BY NEUTRON DIFFRACTION.
                       *BACON(G.E.), CURRY(N.A.).

2415 RH,. . . . . .P. . . . . . .Z. PHYSIK VOL.266, ... 157, (1974).
                       *NEUTRON TOTAL CROSS SECTION OF SC, V, CU AND RH AT MICRO-EV ENERGIES.
                       *DILG(W.), MANNHART(W.).

2416 RH*(DY,HO,ER), ...A2. . . . . . .SOL. STATE COMM. VOL.10,. . . . . 685, (1972).
                       *STRUCTURES MAGNETIQUES DES COMPOSES EQUIATOMIQUES TERRES RARES-RHODIUM
                       T*RH (T=DY, HO, ER).
                       *CHAMARD-BOIS(R.), NGUYEN-VAN-NHUNG, YAKINTHOS(J.), WINTENBERGER(M.).

SECTION   5 -STRUCTURE AND CROSS-SECTION DETERMINATIONS . . . .

2417 RH-FE, ...SPIN DENSITIES OF 4D ATOMS IN FERROMAGNETIC ALLOYS. 223, (1964).
                           →SHIRANE(G.), NATHANS(R.), PICKART(S.J.), ALPERIN(H.A.).

2418 RH-FE, ... →INVESTIGATION   OF   THE   ANTIFERROMAGNETIC-FERROMAGNETIC TRANSFORMATION
                           IN IRON-RHODIUM ALLOYS.
                           →PAL(L.), TARNOCZI(T.), SZABO(P.), KREN(E.), TOTH(J.).

2419 S, ...A3. . . . . .PHYS. REV. VOL.77, ... 305, (1950).
                           →NEUTRON DIFFRACTION IN LIQUID SULFUR, LEAD, AND BISMUTH.
                           →CHAMBERLAIN(O.).

2420 S, ...P, A3 . . . . .Z. NATURFORSCH. VOL.26A, ... 400, (1971).
                           →ABSOLUTE DETERMINATION OF THE COHERENT SCATTERING AMPLITUDE OF SULPHUR
                           AND ITS TOTAL EFFECTIVE CROSS SECTION TO SUBTHERMAL NEUTRONS.
                           →TRUSTEDT(W.D.). (IN GERMAN).

2421 S4*N4*H4, ...A1. . . . . . ACTA CRYST. 23, ... 574, (1967).
                           →A NEUTRON STRUCTURE ANALYSIS OF S4*N4*H4.
                           →SABINE(T.M.), COX(G.W.).

2422 SB, ...A3. . . . . .PHYS. LETT. VOL.35A, ... 315, (1971).
                           →STRUCTURE FACTOR OF LIQUID ANTIMONY AT TWO TEMPERATURES.
                           →WASEDA(Y.), SUZUKI(K.).

2423 SB, ...A3. . . . . .PHYS. STAT. SOL. VOL.47, . . . 581, (1971).
                           →STRUCTURE OF LIQUID ANTIMONY BY NEUTRON DIFFRACTION.
                           →WASEDA(Y.), SUZUKI(K.).

2424 SB, ...P. . . . . . .J. NUCL. ENERGY VOL.22, ... 389, (1968).
                           →MEASUREMENT OF THE TOTAL NEUTRON CROSS-SECTION OF ZINC AND ANTIMONY IN
                           THE ENERGY RANGE 0.002-0.4 EV USING A TIME-OF-FLIGHT SPECTROMETER.
                           →EL-ELA(M.A.),   SALAMA(M.),   ABDEL-KAWY(A.),   ADIB(M.),   HAMOUDA(I.).

2425 SB*(CE,DY,ND), . →NEUTRON DIFFRACTION STUDIES ON DY*SB, ND*SB, AND CE*SB. 1385, (1971).
                           →NERESON(N.), STRUEBING(V.).

2426 SC, ...P. . . . . . .Z. PHYSIK VOL.266, ... 157, (1974).
                           →NEUTRON TOTAL CROSS SECTION OF SC, V, CU AND RH AT MICRO-EV ENERGIES.
                           →DILG(W.), MANNHART(W.).

2427 SC*H3, ...A1. . . . . . .SOV. PHYS. CRYST. VOL.17, . . . 822, (1971).
                           →NEUTRON DIFFRACTION INVESTIGATION OF SCANDIUM HYDRIDE.
                           →MIRON(N.F.), SHCHERBAK(V.I.), BYKOV(V.N.), LEVDIK(V.A.).

2428 SC*MN*O3, ...A2. . . . . . .REF. SECTION 3, 64/KOEHLER(1), ...

2429 SC*(N*D2)2, SC*(N*H2)2, ...A1. . . . . .ACTA CRYST. VOL.A24, ... 410, (1968).
                           →NEUTRON DIFFRACTION DETERMINATION OF THE CRYSTAL STRUCTURES OF THIOUREA
                           AND DEUTERATED THIOUREA ABOVE AND BELOW THE FERROELECTRIC TRANSITION.
                           →ELCOMBE(M.M.), TAYLOR(J.C.).

2430 SC*V*O4,SC2*O3, ...A1. . . . . .J. PHYS. CHEM. VOL.57, ... 535, (1953).
                           →NEUTRON DIFFRACTION STUDIES ON SCANDIUM ORTHOVANADATE AND SCANDIUM
                           OXIDE.
                           →MILLIGAN(W.O.), VERNON(L.W.), LEVY(H.A.), PETERSON(S.W.).

2431 SE, ...A4. . . . . .J. CHEM. PHYSICS VOL.46, ... 586, (1967).
                           →ATOMIC RADIAL DISTRIBUTION IN AMORPHOUS SELENIUM BY X-RAY AND NEUTRON
                           DIFFRACTION.
                           →HENNINGER(E.H.), BUSCHERT(R.C.) HEATON(L.).

2432 SE, ...A1. . . . . . .THESIS, ... (1971).
                           →HIGH PRESSURE X-RAY AND NEUTRON DIFFRACTION INVESTIGATION OF HEXAGONAL
                           SELENIUM.
                           →MCCANN(D.R.). (ED. NOTE: AVAIL. UNIV. MICROFILMS ORDER NO.72-20,393 (226
                           PP.) SEE: DISS. ABSTRACTS B VOL.33 NO.1, PG.312).

2433 SE, ...P. . . . . . .SOV. J. NUCL. PHYS. VOL.9, 6, . 1119, (1969).
                           →RADIATIVE CAPTURE AND TOTAL CROSS SECTIONS OF NEUTRON INTERACTION WITH
                           SE ISOTOPES.
                           →MALETSKI(K.H.),   PIKEL#NER(L.B.),   SALAMATIN(I.M.),   SHARAPOV(E.I.).

2434 SE, ...A3. . . . . . .J. PHYSIQUE TOME 34, ... 937, (1973).
                           →ETUDE DE LA STRUCTURE DU SELENIUM LIQUIDE PAR DIFFRACTION DE NEUTRONS.
                           →TOURAND(G.).

2435 SI, ...A4. . . . . .PHYS. LETT. VOL.37A, ... 434, (1971).
                           →ANHARMONICITY AND THE TEMPERATURE DEPENDENCE OF THE FORBIDDEN (222)
                           REFLECTION IN SILICON.
                           →PHILLIPS(J.C.).

2436 SI, ...A4. . . . . .ACTA CRYST. VOL.A27, ... 219, (1971).
                           →A NEUTRON DIFFRACTION SEARCH FOR NON-CENTROSYMMETRIC THERMAL
                           OSCILLATIONS IN GERMANIUM AND SILICON.
                           →NUNES(A.C.).

2437 SI, ...GENERA. . .PHYS. REV. LETT. 21, ... 1585, (1968).
                           →OBSERVATION OF PENDELLOSUNG FRINGE STRUCTURE IN NEUTRON DIFFRACTION.
                           →SHULL(C.G.).

2438 SI, ...A4. . . . . . .THESIS (132PP.), ... , (1969).
                           →FLUCTUATIONS IN THE NEUTRON CROSS SECTIONS OF SILICON.
                           →GRIMES(S.M.).   (ED. NOTE: AVAIL. UNIV. MICROFILMS, ANN ARBOR, MICH. USA
                           ORDER NO.68-17897).

2439 SI, ...P. . . . . . .J. PHYS. SOC. JAPAN VOL.31, . . 954, (1971).
                           →AN OBSERVATION OF NEUTRON PENDELLOSUNG FRINGES IN A WEDGE-SHAPED SILICON
                           SINGLE CRYSTAL.
                           →KIKUTA(S.), KOHRA(K.), MINAKAWA(N.), DOI(K.).

2440 SI, ...A4. . . . . .PHYS. REV. LETT. VOL.29, ... 871, (1972).
                           →SPHERICAL-WAVE NEUTRON PROPAGATION AND PENDELLOSUNG FRINGE STRUCTURE IN
                           SILICON.
                           →SHULL(C.G.), OBERTEUFFER(J.A.).

2441 SI, ...A4, T. . . . .PHYS. REV. B VOL.4, ... 2472, (1971).
                           →FORBIDDEN (222) NEUTRON REFLECTION IN SILICON: ANHARMONICITY AND THE
                           BONDING ELECTRONS.
                           →KEATING(D.), NUNES(A.), BATTERMAN(B.), HASTINGS(J.M.).

2442 SI, ...A4. . . . . .PHYS. REV. LETT. VOL.27, ... 320, (1971).
                           →ANHARMONICITY AND THE TEMPERATURE DEPENDENCE OF THE FORBIDDEN (222)
                           REFLECTION IN SILICON.
                           →KEATING(D.), NUNES(A.), BATTERMAN(B.), HASTINGS(J.M.).

SECTION 5 -STRUCTURE AND CROSS-SECTION DETERMINATIONS . . . .

2443 SI,. . . . . *A NEW METHOD FOR NEUTRON DIFFRACTION CRYSTAL STRUCTURE INVESTIGATION. A2 PHYS. STAT. SOL. 4, 349, (1964).
*BURAS(B.), LECIEJEWICZ(J.).

2444 SI,. . . . . *ANHARMONIC VIBRATION AND FORBIDDEN REFLECTIONS IN SILICON AND GERMANIUM. P PROC. ROY. SOC. 298, 307, (1967).
*DAWSON(B.), WILLIS(B.T.M.).

2445 SI,. . . . . *PENDELLÖSUNG INTERFERENCE WITH THERMAL NEUTRONS IN SI SINGLE CRYSTALS. A4 PHYS. LETT. VOL.14, 174, (1965).
*SIPPEL(D.), KLEINSTUCK(K.), SCHULZE(G.E.R.). (IN GERMAN).

2446 SI,. . . . . *THE APPLICATION OF THE NEUTRON TIME-OF-FLIGHT SPECTROMETRY TO SEARCHING THE FORBIDDEN (666) REFLECTION IN SILICON. A1 J. PHYS. SOC. JAPAN VOL.33, 1493, (1972).
*NIIMURA(N.), KIMURA(M.).

2447 SI,. . . . . *BACKSCATTERING OF NEUTRONS FROM PERFECT SILICON SINGLE CRYSTALS. E Z. PHYSIK VOL.263, 291, (1973).
*HEIDEMANN(A.), SCHOLZ(J.). (IN GERMAN).

2448 SI,. . . . . *PERFECT CRYSTALS AND IMPERFECT NEUTRONS. J. APPL. CRYST. VOL.6, 257, (1973).
*SHULL(C.G.).

2449 SI,. . . . . *NEUTRON DIFFRACTION EFFECT ON THE ROCKING CURVE OF A VIBRATING SINGLE CRYSTAL. A4 PHYS. STAT. SOLIDI A VOL.17 NO.1 163, (1973).
*MIKULA(P.), MICHALEC(R.), SEDLAKOVA(L.), CECH(J.), CHALUPA(B.), PETRZILKA(V.).

2450 SI,. . . . . *DIFFRACTION STUDIES OF THE (222) REFLECTION IN GE AND SI: ANHARMONICITY AND THE BONDING ELECTRONS. A4 PHYS. REV. B VOL.9, 2590, (1974).
*ROBERTO(J.B.), BATTERMAN(B.W.), KEATING(D.T.).

2451 SI,. . . . . *INFLUENCE OF OXYGEN SEGREGATIONS IN SILICON SINGLE CRYSTALS ON THE HALF WIDTH OF THE DOUBLE-CRYSTAL ROCKING CURVE OF THERMAL NEUTRONS. A4 PHYS. STAT. SOL. VOL.23, 237, (1967).
*EICHHORN(F.), SIPPEL(D.), KLEINSTUCK(K.).

2452 SI-FE, ... *MAGNETISM RESEARCH WITH POLARIZED NEUTRONS AT THE CENTRO DI STUDI NUCLEARI DELLA CASACCIA DEL C.N.E.N., ROMA, ITALY. A2 COLL. INTER. N.126 (GRENOBLE), 180, (1963).
*ANTONINI(B.), FELCHER(G.P.), MENZINGER(F.), PAOLETTI(A.), RICCI(F.P.), PASSARI(L.).

2453 SI*(HO,ER,TM), . *PROPRIETES MAGNETIQUES ET STRUCTURES MAGNETIQUES DES COMPOSES TB*(GE-SI), HO*SI, ER*SI ET TM*SI DU TYPE CR*B. A2 J. PHYSIQUE TOME 32 COL.1 VOL.II 1133, (1971).
*NGUYEN-VAN-NHUNG, BARLET(A.), LAFOREST(J.).

2454 SILICA (VITREOUS), *NEUTRON DIFFRACTION BY GERMANIA, SILICA AND RADIATION-DAMAGED SILICA GLASSES. A4 J. OF PHYS.-C- SER. 2, VOL.2, 229, (1969).
*LORCH(E.).

2455 SILICA,. . . . . *A STUDY OF THE STRUCTURE AND ATOMIC MOTIONS OF THE VITREOUS SILICA BY THERMAL NEUTRON SCATTERING TECHNIQUES. A1 DISS. ABS. VOL.32, 4156B, (1972).
*SUTTON(J.D.).

2456 SILICA,. . . . . *NEUTRON DIFFRACTION FROM CONDENSED SILICA VAPOUR PARTICLES, SYNTHETIC VITREOUS SILICA AND FUSED QUARTZ GLASS. A3,A4 PHYS. CHEM. GLASSES VOL.10, 185, (1969).
*LORCH(E.).

2457 SILICA,. . . . . *ATOMIC STRUCTURE AND CORRELATION IN VITREOUS SILICA BY X-RAY AND NEUTRON DIFFRACTION. A4 J. PHYS. CHEM. SOL. 28, 423, (1967).
*HENNINGER(E.H.), BUSCHERT(R.C.), HEATON(L.).

2458 SI2*O3,. . . . . *CONVENTIONAL AND ELASTIC NEUTRON DIFFRACTION FROM VITREOUS SILICA. A4 J. OF PHYS.-C-, SER.2, VOL.3, 1314, (1970).
*LORCH(E.).

2459 SI*O2, ... *NEUTRON DIFFRACTION RESEARCH ON VITREOUS SILICA. PHYSICS/NON-CRYSTALLINE SOLIDS, 152, (1965).
*CARRARO(G.), DOMENICI(M.), ZUCCA(T.). (ED. NOTE: PHYSICS OF NON-CRYSTAL-LINE SOLIDS PUBL. BY NORTH-HOLLAND: AMSTERDAM).

2460 SI*O2, ... *NEUTRON DIFFRACTION IN NA2*SI2*O5-, NA2*SI*O3- AND SI*O2-GLASSES AND THE ADDITIVE PROPERTIES OF THE INTENSITY DATA. A4 NATURWISSENSCHAFTEN VOL.53, 16, (1966).
*HOFFMAN(W.), FISCHER(P.), MAIER(G.). (IN GERMAN).

2461 SI*O2, ... *ON THE SCATTERING OF THERMAL NEUTRONS BY SOLIDS; III.THE EFFECT OF SURFACE TREATMENT FOR QUARTZ GRAIN SCATTERERS. E PROC. JAPAN ACAD. VOL.19, 152, (1943).
*KIMURA(M.).

2462 SI2*O3,. . . . . *TOTAL NEUTRON SCATTERING IN VITREOUS SILICA. P PHYS. REV. VOL.105, 517, (1957).
*BREEN(R.J.), DELANEY(R.M.), PERSIANI(P.J.), WEBER(A.H.).

2463 SI*O2, ... *SCATTERING OF THERMAL NEUTRONS BY SOLIDS, II.EFFECT OF THERMAL STRAIN IN THE DIFFRACTIVE SCATTERING OF NEUTRONS BY QUARTZ. P PROC. PHYS. MATH. SOC. JAPAN 25, 530, (1943).
*KIMURA(M.), HASHIGUCHI(R.).

2464 SI*O2, ... *TRANSMISSION OF SLOW NEUTRONS THROUGH CRYSTALS. P PHYS. REV. VOL.55, 1101, (1939).
*WHITAKER(M.D.), BEYER(H.G.).

2465 SI*O2, ... *ON THE SCATTERING OF THERMAL NEUTRONS BY SOLIDS, II.THE EFFECT OF THERMAL STRAIN IN THE DIFFRACTIVE SCATTERING OF NEUTRONS BY QUARTZ. E PROC. JAPAN ACAD. VOL.19, 26, (1943).
*KIMURA(M.), HASHIGUCHI(R.).

2466 SI*O2-TI*O2, . . *SMALL-ANGLE NEUTRON SCATTERING ON SILICA GLASSES CONTAINING TITANIA. A1 PHYS. STAT. SOLIDI A VOL.18 NO.2, K91, (1973).
*LOSHMANOV(A.A.), SIGAEV(V.N.), KHODAKOVSKAYA(R.YA.), PAVLUSHKIN(N.M.), YAMZIN(I.I.).

SECTION   5 -STRUCTURE AND CROSS-SECTION DETERMINATIONS . . . .

2467 SI*02-TI*02, ...A1. . . . . . .J. APPL. CRYST. VOL.7, ... 207, (1974).
                       ↪SMALL-ANGLE NEUTRON SCATTERING ON SILICA GLASSES CONTAINING TITANIA.
                       ↪LOSHMANOV(A.A.), SIGAEV(V.N.), KHODAKOVSKAYA(R.YA.), PAVLUSHKIN(N.M.),
                       YAMZIN(I.I.).

2468 SM (A=154),. . . . . .A2. . . . . . .PHYS. REV. LETT. VOL.29, ... 1468, (1972).
                       ↪MAGNETIC STRUCTURES OF SAMARIUM.
                       ↪KOEHLER(W.C.), MOON(R.M.).

2469 SM,. . . . . .A2. . . . . . .REF. SECTION 3, 65/KOEHLER,. . . . .

2470 SM,. . . . . .A2. . . . . . .ACTA CRYST. VOL.A28 PART S-4,. . . 197, (1972).
                       ↪THE MAGNETIC PROPERTIES OF METALLIC SAMARIUM.
                       ↪KOEHLER(W.C.), MOON(R.M.), CABLE(J.W.), CHILD(H.R.). (ABSTRACT ONLY).

2471 SM,. . . . . .P. . . . . . .PHYS. REV. VOL.91, ... 597, (1953).
                       ↪SLOW NEUTRON SCATTERING CROSS SECTIONS FOR RARE EARTH NUCLIDES.
                       ↪KOEHLER(W.C.), WOLLAN(E.O.).

2472 SM,. . . . . .T,P. . . . . .A.I.P. CONF. PROC. NO.10 PART 2, 1314, (1973).
                       ↪MAGNETIC SCATTERING AMPLITUDES OF SAMARIUM.
                       ↪MOON(R.M.), KOEHLER(W.C.).

2473 SM*(BR*O3)3-9H2*O, ...A1,P. . . . . .ACTA CRYST. A25 . . . . 621, (1969).
                       ↪THE NEUTRON RESONANCE SCATTERING IN THE STRUCTURE DETERMINATION OF
                       SM*(BR*O3)3-9H2*O.
                       ↪SIKKA(S.K.).

2474 SM*C*03, ...A2,4. . . . . .REF. SECTION 3, 66/BERTAUT,. . . . .

2475 SN,. . . . . .A3. . . . . . .IAEA SYMPOSIUM VIENNA . . . . . . 265, (1966).
                       ↪A STUDY OF THE DIFFUSIVE MOTIONS OF LIQUIDS BY MEANS OF COLD-NEUTRON
                       SCATTERING EXPERIMENT.
                       ↪PALEVSKY(H.).

2476 SN,. . . . . .P. . . . . . .J. OF PHYS. C VOL.1, ... 1075, (1968).
                       ↪THE STRUCTURE FACTOR FOR LIQUID METALS. II. RESULTS FOR LIQUID ZN, TL,
                       PB, SN AND BI.
                       ↪NORTH(D.M.), ENDERBY(J.E.), EGELSTAFF(P.A.).

2477 SN,. . . . . .A3. . . . . .PHYS. MET. METALLOGR. VOL.29,. . . 188, (1970).
                       ↪TRANSVERSE SCATTERING CROSS SECTION OF SLOW NEUTRONS IN LIQUID TIN.
                       ↪KHRUSHCHEV(B.I.), SHARIPOVA(L.S.).

2478 SN (ISO. A=117), ...P. . . . . . .SOV. J. NUCL. PHYS. VOL.10, 1, . 18, (1969).
                       ↪STUDY OF SN (ISO. A=117) NEUTRON CROSS SECTIONS.
                       ↪ADAMCHUK(YU.V.), ZENKEVICH(V.S.), MOSKALEV(S.S.), MURADYAN(G.V.),
                       SHCHEPKIN(YU.G.).

2479 SN,. . . . . .P. . . . . . .PHYS. LETT. 21, ... 286, (1966).
                       ↪THE STRUCTURE FACTOR FOR LIQUID METALS AT LOW ANGLES. . . .
                       ↪EGELSTAFF(P.A.), DUFFILL(C.), RAINEY(V.S.), ENDERBY(J.E.), NORTH(D.M.).

2480 SN,. . . . . .A3. . . . . .PHYS. MET. METALLOGR. VOL.18 N.3 134,(1964).
                       ↪NOTES ON THE INTERPRETATION OF INTENSITY CURVES OF RAYS DIFFRACTED BY
                       LIQUIDS.
                       ↪SPEKTOR(E.Z.), KHOKHLOV(S.F.).

2481 SN,. . . . . .P. . . . . . .ACTA CRYST. 23. . . . . 868, (1967).
                       ↪THE COHERENT NEUTRON SCATTERING AMPLITUDES FOR SEVEN ISOTOPES OF TIN.
                       ↪KAY(M.I.), RITTER(H.L.).

2482 SN,. . . . . .A3. . . . . .JAP. J. APPL. PHYS. VOL.12,. . . 1119, (1973).
                       ↪A FEASIBILITY STUDY OF APPLYING NEUTRON T-O-F METHOD FOR THE LIQUID
                       STRUCTURE ANALYSIS.
                       ↪TOMIYOSHI(S.), WATANABE(N.), MISAWA(M.), KAI(K.), KIMURA(M.).

2483 SPINELS, ...A4. . . . . . .REPORT CEA-R-3756 (148PP.). . . . . (1969).
                       ↪MAGNETIC INTERACTIONS IN SOME FERRI- AND ANTIFERROMAGNETIC NORMAL
                       SPINELS STUDIED BY NEUTRON DIFFRACTION.
                       ↪PLUMIER(R.). (ED. NOTE: AVAIL. CFSTI, SPRINGFIELD, VA. 22151, USA - IN
                       FRENCH).

2484 SR,. . . . . .P. . . . . . .NUCL. PHYS. VOL.A139, ... 625, (1969).
                       ↪STUDY OF THERMAL NEUTRON CAPTURE IN STRONTIUM.
                       ↪LYCKLAMA(H.), KENNETT(T.J.).

2485 SR,. . . . . .P. . . . . . .Z. KRISTALLOGR. VOL.135, ... 316, (1972).
                       ↪THE NEUTRON SCATTERING AMPLITUDES OF POTASSIUM, STRONTIUM AND BARIUM.
                       ↪COOPER(M.J.), ROUSE(K.D.).

2486 (SR-BA)*BI2*TA2*09, ...A1. . . . . .MATER. RES. BULL. VOL.8, ... 1183, (1973).
                       ↪CRYSTAL STRUCTURE OF (SR,BA)*BI2*TA2*09.
                       ↪NEWNHAM(R.E.), WOLFE(R.W.), HORSEY(R.S.), DIAZ-COLON(F.A.), KAY(M.I.).

2487 (SR-BA)2*ZN2*FE12*022, ...A2. . . . . .SOV. PHYS. J.E.T.P. VOL.25,. . . . 266, (1967).
                       ↪HELICOIDAL SPIN ORDERING IN THE HEXAGONAL FERRITE (SR-BA)2*ZN2*FE12*022.
                       ↪PEREKALINA(T.M.), SIZOV(V.A.), SIZOV(R.A.), YAMZIN(I.I.), VOSKANYAN(R.).

2488 (SR-BA)2*ZN2*FE12*022(Y), ...A2. . . . . .SOV. PHYS. JETP LETT. VOL.6,. . . . 176, (1967).
                       ↪A NEW TYPE OF SPIN ORDERING IN THE HEXAGONAL FERRITE ...(Y).
                       ↪SIZOV(V.A.), SIZOV(R.A.), YAMZIN(I.I.).

2489 SR*CO*0(2.5),. . . . . .A2. . . . . . .J. PHYS. SOC. JAPAN VOL.33,. . . . 970, (1972).
                       ↪MAGNETIC STRUCTURE OF SR*CO*0(2.5).
                       ↪TAKEDA(T.), YAMAGUCHI(Y.), WATANABE(H.).

2490 SR*F2, ...A4. . . . . . .ACTA CRYST. VOL.A27 . . . . 622, (1971).
                       ↪A NEUTRON DIFFRACTION STUDY OF SR*F2 AND CA*F2.
                       ↪COOPER(M.J.), ROUSE(K.D.).

2491 SR2*(FE-MO)*06,. . . . . .A2. . . . . . .J. PHYS. SOC. JAPAN VOL.24,. . . . 219, (1968).
                       ↪NEUTRON DIFFRACTION STUDY OF SR2*(FE-MO)*06.
                       ↪NAKAYAMA(S.), NAKAGAWA(T.), NOMURA(S.).

2492 SR2*FE2*05, ...A1,A2. . . . . .J. PHYS. SOC. JAPAN VOL.26,. . . . 1320, (1969).
                       ↪CRYSTAL AND MAGNETIC STRUCTURES OF SR2*FE2*05.
                       ↪TAKEDA(T.), YAMAGUCHI(Y.), WATANABE(H.), TOMIYOSHI(S.), YAMAMOTO(H.).

2493 SR*FE*03,. . . . . .A2. . . . . . .J. PHYS. SOC. JAPAN VOL.33,. . . . 967, (1972).
                       ↪MAGNETIC STRUCTURE OF SR*FE*03.
                       ↪TAKEDA(T.), YAMAGUCHI(Y.), WATANABE(H.).

2494 SR3*FE2*W*09,. . . . . .A4. . . . . . .2ND INTERNAT. MEET./FERROELECTR. 219, (1969).
                       ↪NEUTRON DIFFRACTION STUDY OF SR3*FE2*W*09.
                       ↪SOLOV≠EV(S.P.), SMIRNOV(V.F.), FADEEVA(N.V.), IVANOVA(V.V.),
                       KAPISHEV(A.G.). (ED. NOTE: PUBL. PHYS. SOC. JAP., TOKYO, JAPAN 1969).

SECTION   5 -STRUCTURE AND CROSS-SECTION DETERMINATIONS . . . .

2495   SR*MO*O4,. . . . . .A1. . . . . . .J. CHEM. PHYSICS VOL.55,. . . . . 1093, (1971).
       *CRYSTAL STRUCTURE REFINEMENT OF SR*MO*O4, SR*W*O4, CA*MO*O4, AND BA*W*O4
       BY NEUTRON DIFFRACTION.
       *GURMEN(E.), DANIELS(E.), KING(J.S.).

2496   SR*TB2*FE2*O7,. . . . . .A1. . . . . . .J. SOLID STATE CHEM. VOL.7,. . . . . 337, (1973).
       *DETERMINATION DES STRUCTURES DE DEUX FERRITES MIXTES NOUVEAUX DE
       FORMULE BA*LA2*FE2*O7 ET SR*TB2*FE2*O7.
       *SAMARAS(D.), COLLOMB(A.), JOUBERT(J.C.).

2497   SR*TI*O3,. . . . . .A4. . . . . . .CONF. INTERN., RENNES,. . . . . . . 155, (1971).
       *NEW OBSERVATIONS FROM NEUTRON SCATTERING STUDIES OF THE STRUCTURAL PHASE
       TRANSITION IN SR*TI*O3.
       *SHAPIRO(S.M.), AXE(J.D.), SHIRANE(G.).

2498   SR*TI*O3,. . . . . .A1. . . . . . .Z. PHYSIK VOL.258,. . . . . 429, (1973).
       *THE CHANGE OF LATTICE PARAMETERS OF SR*TI*O3.
       *HEIDEMANN(A.), WETTENGEL(H.). (IN GERMAN).

2499   SR3*U*O6,. . . . . .A1. . . . . . .ACTA CRYST. 20,. . . . . 508, (1966).
       *THE CRYSTAL STRUCTURE OF SOME ALKALINE EARTH METAL URANATES OF THE TYPE
       M3*U*O6.
       *RIETVELD(H.M.).

2500   SR3*U*O6,. . . . . .A1. . . . . . .ACTA CRYST. B25,. . . . . . . 787, (1969).
       *THE STRUCTURE OF SOME ALKALINE-EARTH METALS URANATES.
       *LOOPSTRA(B.O.), RIETVELD(H.M.).

2501   SR*U*O4, SR2*U*O5,. . . . . .A1. . . . . . .ACTA CRYST. B25,. . . . . . . 787, (1969).
       *THE STRUCTURE OF SOME ALKALINE-EARTH METAL URANATES.
       *LOOPSTRA(B.O.), RIETVELD(H.M.).

2502   SR*W*O4,. . . . . .A1. . . . . . .J. CHEM. PHYSICS VOL.55,. . . . . 1093, (1971).
       *CRYSTAL STRUCTURE REFINEMENT OF SR*MO*O4, SR*W*O4, CA*MO*O4, AND BA*W*O4
       BY NEUTRON DIFFRACTION.
       *GURMEN(E.), DANIELS(E.), KING(J.S.).

2503   STEELS (ROLLED ELECTRICAL),. . . . . .A4. . . . . . .DISS. ABS. 24,. . . . . . . 5316, (1964).
       *THE DETERMINATION OF PREFERRED ORIENTATION IN ROLLED ELECTRICAL STEELS
       USING SINGLE DIFFRACTION OF NEUTRONS.
       *EUGENIO(M.R.).

2504   SUCROSE (C12*H22*O11),. . . . . .A1. . . . . . .ACTA CRYST. VOL.B29,. . . . . . 797, (1973).
       *SUCROSE: X-RAY REFINEMENT AND COMPARISON WITH NEUTRON REFINEMENT.
       *HANSON(J.C.), SIEKER(L.C.), JENSEN(L.H.).

2505   SUCROSE (C12*H22*O11),. . . . . .A1. . . . . . .ACTA CRYST. VOL.B29,. . . . . . 790, (1973).
       *FURTHER REFINEMENT OF THE STRUCTURE OF SUCROSE BASED ON NEUTRON-
       DIFFRACTION DATA.
       *BROWN(G.M.), LEVY(H.A.).

2506   SUCROSE,. . . . . .A1. . . . . . .SCIENCE VOL.141,. . . . . 921, (1963).
       *SUCROSE: PRECISE DETERMINATION OF CRYSTAL AND MOLECULAR STRUCTURE BY
       NEUTRON DIFFRACTION.
       *BROWN(G.M.), LEVY(H.A.).

2507   SULPHONE(DICHLORO-DIPHENYL),. . . . . .A1. . . . . . .ACTA CRYST. VOL.13,. . . . . . . 10, (1960).
       *A STUDY OF 4,4≠-DICHLORO DIPHENYL SULPHONE BY NEUTRON DIFFRACTION.
       *BACON(G.E.), CURRY(N.A.).

2508   TA (ISO. A=182),. . . . . .P. . . . . . . .NUCL. SCI. ENGNG. VOL.33,. . . . . . 16, (1968).
       *THE TOTAL NEUTRON CROSS SECTION OF 115-DAY TANTALUM-182 FROM 0.01 TO
       1000 EV.
       *STOKES(G.E.), SCHUMAN(R.P.), SIMPSON(O.D.).

2509   TA,. . . . . .A1. . . . . . .PHYS. LETT. 20,. . . . . 470, (1966).
       *ERHOLUNG VON STRAHLENSCHADEN IN KUBISCH-RAUMZENTRIERTEN METALLEN NACH
       NEUTRONEN BESTRAHLUNG BEI 4.5 DEG. K.
       *BURGER(G.), ISEBECK(K.), KERLER(R.), VOLKL(J.), WENZL(H.).

2510   TA,. . . . . .P. . . . . . . .J. NUCL. ENERGY VOL.21,. . . . . . 425, (1967).
       *MEASUREMENT OF THE TOTAL NEUTRON CROSS SECTION OF TANTALUM IN THE
       ENERGY RANGE 0.002-0.2 EV USING A TIME-OF-FLIGHT SPECTROMETER.
       *ADIB(M.), ABU-EL-ELA(M.), SALAMA(M.), ABDEL-KAWY(A.), HAMOUDA(I.).

2511   TA2*C,. . . . . .A1. . . . . . .ACTA CRYST. 19,. . . . . 6, (1965).
       *THE CRYSTAL STRUCTURES OF V2*C AND TA2*C.
       *BOWMAN(A.L.), WALLACE(T.C.), YARNELL(J.L.), WENZEL(R.G.), STORMS(E.K.).

2512   TA2*D,. . . . . .A1. . . . . . .J. CHEM. PHYSICS VOL.41,. . . . . . 3261, (1964).
       *REPLY TO COMMENTS BY PALENIK ON RE-EXAMINATION OF THE NEUTRON DIFFRAC-
       TION RESULTS FOR TANTALUM DEUTERIDE TA2*D.
       *WALLACE(W.E.).

2513   TA2*D,. . . . . .T. . . . . . . .J. CHEM. PHYS. VOL.41,. . . . . . 3260, (1964).
       *RE-EXAMINATION OF THE NEUTRON DIFFRACTION RESULTS FOR TANTALUM
       DEUTERIDE TA2*D.
       *PALENIK(G.J.).

2514   TA2*D,. . . . . .A1. . . . . . .J. CHEM. PHYS. VOL.35,. . . . . . 2156, (1961).
       *NEUTRON DIFFRACTION DATA FOR TA2*D AND THE PROBABLE ARRANGEMENT OF THE
       DEUTERIUMS IN TWO OF ITS POLYMORPHIC VARIETIES.
       *WALLACE(W.E.).

2515   TA-D, TA-H,. . . . . .A1. . . . . . .SOV. PHYS. SOL. STATE VOL.10,. . . . . . 212, (1968).
       *NEUTRON DIFFRACTION STUDY OF THE STRUCTURE AND PHASE TRANSFORMATIONS OF
       TANTALUM HYDRIDES AND DEUTERIDES.
       *SOMENKOV(V.A.), GURSKAYA(A.V.), ZEMLYANOV(M.G.), KOST(M.E.),
       CHERNOPLEKOV(N.A.), CHERTKOV(A.A.).

2516   TA2*M4*O9 (M=MG,MN,FE,CO,NI),. . . . . . . .A2. . . . . . .C.R. ACAD. SCI. VOL.251,. . . . . . 1733, (1960).
       *THE CRYSTAL STRUCTURE AND MAGNETIC STRUCTURE OF NIOBATES AND TANTALATES
       OF BIVALENT TRANSITION METALS.
       *BERTAUT(E.F.), CORLISS(L.), FORRAT(F.). (IN FRENCH).

2517   TANTALATES (RUTILE-TYPE),. . . . . .A4. . . . . . .J. PHYSIQUE TOME 32 COL.1 VOL.I, 503, (1971).
       *NEUTRON DIFFRACTION, MOSSBAUER AND MAGNETIC INVESTIGATION OF RUTILE-TYPE
       TANTALATES M*TA*O4 (M=TI,V,CR,FE).
       *FADEEVA(N.A.), OZEROV(R.P.), SMIRNOV(V.P.),...ET AL.

2518   TA*O*N,. . . . . .A1. . . . . . .ACTA CRYST. VOL.B30,. . . . . . 809, (1974).
       *ANION ORDERING IN TA*O*N: A POWDER NEUTRON-DIFFRACTION INVESTIGATION.
       *ARMYTAGE(D.), FENDER(B.E.F.).

SECTION   5 -STRUCTURE AND CROSS-SECTION DETERMINATIONS . . . .

2519 TA,TI, ... P . . . . NUCL. SCI. ENGNG. VOL.7, ... 193, (1960).
      *TOTAL CROSS SECTIONS OF TI, V, Y, TA, AND W.
      *SCHMUNK(R.E.), RANDOLPH(P.D.), BRUGGER(R.M.).

2520 TB,. . . . . A5 . . . . J. PHYS. CHEM. SOLIDS VOL. 30, . 2175, (1969).
      *A NEUTRON DEPOLARIZATION STUDY ON DYSPROSIUM AND TERBIUM.
      *LOFFLER(E.), RAUCH(H.).

2521 TB,. P,A2 . . . . . DISS. ABS. VOL.31, ...5560B, (1971).
      *MAGNETIC FORM FACTOR OF THULIUM AND TERBIUM METALS.
      *BRUN(T.O.).

2522 TB,. . . . . P . . . . J. PHYSIQUE TOME 32 COL.1 VOL.I,   571, (1971).
      *THE MAGNETIC FORM FACTOR OF TERBIUM.
      *BRUN(T.O.), LANDER(G.H.).

2523 TB,. . . . . A2 . . . . REF. SECTION 3, 65/KOEHLER,. . . . .

2524 TB,. . . . . A2 . . . . PHYS. REV. 162,. . . . . 315, (1967).
      *NEUTRON DIFFRACTION STUDY OF THE MAGNETIC LONG-RANGE ORDER IN TB.
      *DIETRICH(O.W.), ALS-NIELSEN(J.).

2525 TB,. . . . . A2 . . . . J. APP. PHYS. 34,. . . . . 1335, (1963).
      *SOME MAGNETIC STRUCTURE PROPERTIES OF TERBIUM AND OF TERBIUM-YTTRIUM
      ALLOYS.
      *KOEHLER(W.C.), CHILD(H.R.), WOLLAN(E.O.), CABLE(J.W.).

2526 TB,. . . . . A2 . . . . PHYS. REV. VOL.165, ... 688, (1968).
      *NEUTRON DIFFRACTION STUDY OF TB AND HO UNDER PRESSURE.
      *UMEBAYASHI(H.), SHIRANE(G.), FRAZER(B.C.), DANIELS(W.B.).

2527 TB,. . . . . P . . . . PHYS. REV. VOL.121,. . . . . . . 610, (1961).
      *SLOW-NEUTRON SCATTERING CROSS SECTIONS OF TERBIUM, YTTERBIUM, AND
      LUTETIUM.
      *ATOJI(M.).

2528 TB,. . . . . P . . . . PHYS. REV. B VOL.8,. . . . . . . 2595, (1973).
      *ATOMIC AND NUCLEAR EFFECTS IN THE SLOW-NEUTRON TOTAL CROSS SECTION OF
      TERBIUM.
      *MALIK(S.S.), KAMAL(M.), TURANO(T.), DESJARDINS(J.S.).

2529 TB,. . . . . A4 . . . . SOV. PHYS. SOL. STATE VOL.15,. . 1295, (1973).
      *DEPOLARIZATION OF NEUTRONS IN TERBIUM IN THE MAGNETIC PHASE TRANSITION
      REGION.
      *BAAZOV(N.G.), MANDZHAVIDZE(A.G.).

2530 TB,. . . . . A2 . . . . J. APPL. PHYS. VOL.38, ... 969, (1967).
      *NEUTRON DIFFRACTION STUDIES OF TERBIUM.
      *ARROTT(A.), WERNER(S.A.), COOPER(M.J.), NATHANS(R.), SHIRANE(G.).

2531 TB*AG, ... A2 . . . . REF. SECTION 3, 64/CABLE,. . . . .

2532 TB*AG2,. . . . . A2 . . . . J. CHEM. PHYS. VOL.48, ... 3380, (1968).
      *MAGNETIC STRUCTURE OF TB*AG2.
      *ATOJI(M.).

2533 TB(AG,IN), ... A2 . . . . J. APP. PHYS. 36,. . . . . 1096, (1965).
      *MAGNETIC STRUCTURE VERSUS ELECTRON NUMBER FOR SOME RARE-EARTH
      INTERMETALLIC COMPOUNDS.
      *CABLE(J.W.), KOEHLER(W.C.), CHILD(H.R.).

2534 TB-AL GARNETS, ... A1 . . . . ACTA CRYST. B25, ... 1853, (1969).
      *ETUDE PAR DIFFRACTION DE NEUTRONS A 0.31 DEG. K DE LA STRUCTURE
      ANTIFERROMAGNETIQUE DES GRENATS D≠ALUMINUM-TERBIUM ET D≠ALUMINUM-HOLMIUM
      *HAMMANN(J.).

2535 TB3*AL2, ... A2 . . . . Z. ANGEW PHYS. VOL.32, ... 113, (1971).
      *MAGNETIC PROPERTIES AND MAGNETIC STRUCTURES OF TB3*AL2 AND DY3*AL2.
      *BARBARA(B.), BECLE(C.), LEMAIRE(R.). (ED. NOTE: IN FRENCH).

2536 TB*AL2,. . . . . A2 . . . . SOL. STATE COMM. VOL.11 NO.5,. . 707, (1972).
      *ELASTIC NEUTRON DIFFRACTION STUDY OF TB*AL2 AND HO*AL2.
      *MILLHOUSE(A.H.), PURWINS(H.G.), WALKER(E.).

2537 TB*AL, ... A2 . . . . SOL. STATE COMM. VOL.6 NO.2,. . 115, (1968).
      *ETUDE DE LA STRUCTURE MAGNETIQUE DU COMPOSE TB*AL PAR DIFFRACTION
      NEUTRONIQUE.
      *BECLE(C.), LEMAIRE(R.), PARTHE(E.).

2538 TB*AL*03,. . . . . A2 . . . . REF. SECTION 3, 68/MARESCHAL,. . . . .

2539 TB*AL*03,. . . . . A2 . . . . Z. ANGEW. PHYS. VOL.23,. . . . . 243, (1967).
      *MAGNETIC STRUCTURE OF THE PEROVSKITE-TYPE COMPOUND TB*AL*03.
      *BIELEN(J.), MARESCHAL(J.), SIVARDIERE(J.). (IN GERMAN).

2540 TB*AS, ... A2 . . . . PHYS. REV. VOL.131,. . . . . . . 922, (1963).
      *NEUTRON DIFFRACTION INVESTIGATION OF THE MAGNETIC PROPERTIES OF COM-
      POUNDS OF RARE-EARTH METALS WITH GROUP V ANIONS.
      *CHILD(H.R.), WILKINSON(M.K.), CABLE(J.W.), KOEHLER(W.C.), WOLLAN(E.O.).
      (ED. NOTE: REF. SECTION 3, 63/CHILD).

2541 TB*AU2,. . . . . A2 . . . . PHYS. LETT. 25A, ... 528, (1967).
      *THE MAGNETIC STRUCTURE OF TB*AU2.
      *ATOJI(M.).

2542 TB*AU2,. . . . . A2 . . . . AIP CONF. PROC. VOL.5,. . . . . 1390, (1971).
      *ON THE MAGNETIC STRUCTURE OF R*AU2 COMPOUNDS (R:TB-TM).
      *FEDRO(A.J.), SHAFFER(J.C.).

2543 TB*AU2,. . . . . A2 . . . . J. CHEM. PHYS. VOL.48, ... 560, (1968).
      *MAGNETIC STRUCTURES OF TB*AU2.
      *ATOJI(M.).

2544 TB*BI, ... A2 . . . . J. APPL. PHYS. VOL.42, ... 1625, (1971).
      *MAGNETIC PROPERTIES OF CE*BI, ND*BI, TB*BI, AND DY*BI.
      *NERESON(N.), ARNOLD(G.).

2545 TB2*C, ... A1,A2 . . . . J. CHEM. PHYSICS VOL.51,. . . . 3872, (1969).
      *MAGNETIC AND CRYSTAL STRUCTURES OF THE TRIGONAL TB2*C.
      *ATOJI(M.).

2546 TB2*C3 (PHASE: ALPHA),. . . . . A2 . . . . J. CHEM. PHYSICS VOL.54, ... 3504, (1971).
      *MAGNETIC STRUCTURE OF ALPHA-TB2*C3.
      *ATOJI(M.).

SECTION   5 -STRUCTURE AND CROSS-SECTION DETERMINATIONS . . . .

2547   TB2*C3,. . . . . .A2. . . . . . .J. APPL. PHYS. 42, ... 1630, (1971).
       #MAGNETIC STRUCTURES OF TB2*C3 AND HO2*C3.
       #ATOJI(M.).

2548   TB*C2, ...A2. . . . .PHYS. LETT. 23,. . . . .   208, (1966).
       #MAGNETIC STRUCTURES OF TB*C2 AND HO*C2.
       #ATOJI(M.).

2549   TB*C2, ...A1,A2. . . .J. CHEM. PHYS. VOL.46,. . . . . . 1891, (1967).
       #MAGNETIC AND CRYSTAL STRUCTURES OF CE*C2,PR*C2,ND*C2,TB*C2, AND HO*C2
       AT LOW TEMPERATURES.
       #ATOJI(M.).

2550   TB2*C3,. . . . . .A1,A4. . . J. CHEM. PHYS. VOL.35,. . . . 1960, (1961).
       #NEUTRON-DIFFRACTION STUDIES OF LA2*C3, CE2*C3, PR2*C3, AND TB2*C3.
       #ATOJI(M.), WILLIAMS(D.E.).

2551   TB*C2, ...A1,A4. . . .REF. SECTION 3, 61/ATOJI,. . . . .

2552   TB*C*O3, ...A2,A4. . . .REF. SECTION 3, 66/BERTAUT,. . . . .

2553   TB*CO2,. . . . . .A2. . . . .J. APP. PHYS. 36,. . . . . 978, (1965).
       #MAGNETIC STRUCTURE OF RARE-EARTH-COBALT R(CO2) INTERMETALLIC COMPOUNDS.
       #MOON(R.M.), KOEHLER(W.C.), FARRELL(J.).

2554   TB*CO5,. . . . . .A2. . . . .J. PHYSIQUE TOME 28. . . . . . 216, (1967).
       #MAGNETIC STRUCTURES OF THE INTERMETALLIC COMPOUNDS CE*CO5 AND TB*CO5.
       #LEMAIRE(R.), SCHWEIZER(J.). (IN FRENCH).

2555   TB*(CO2,FE2)-TB*AL2,. . . . .A2. . . . .J. PHYS. CHEM. SOLIDS VOL.34,. . . 1267, (1973).
       #STRUCTURAL, MAGNETIC AND NEUTRON DIFFRACTION STUDIES ON TB*FE2-TB*AL2,
       TB*CO2-TB*AL2 AND HO*CO2-HO*AL2.
       #OESTERREICHER(H.).

2556   TB*CO*O3,. . . . . .A2. . . . . .REF. SECTION 3, 68/MARESCHAL,. . . . .

2557   TB*CR*O3,. . . .   . .A2,T. . . . .ACTA CRYST A24,. . . . .   217, (1968).
       #REPRESENTATION ANALYSIS OF MAGNETIC STRUCTURES.
       #BERTAUT(E.F.).

2558   TB*CR*O3,. . . . . .A2. . . . . .REF. SECTION 3, 68/MARESCHAL,. . . . .

2559   TB*CR*O3,. . . .   . .A2. . . . .J. PHYS. CHEM. SOL. 28,. . . . . 2143, (1967).
       #ETUDE PAR DIFFRACTION NEUTRONIQUE DE LA STRUCTURE MAGNETIQUE DU
       CHROMITE DE TERBIUM.
       #BERTAUT(E.F.), MARESCHAL(J.), DE-VRIES(G.F.).

2560   TB*CR*O3,. . . .   . .A2. . . . .IEEE TRANS. MAGNETICS MAG-2, . . 453, (1966).
       #ETUDES DES PROPRIETES MAGNETOSTATIQUES ET DES STRUCTURES MAGNETIQUES
       DES CHROMITES DES TERRES RARES ET D#YTTRIUM.
       #BERTAUT(E.F.), MARESCHAL(J.). . .ET AL.

2561   TB*CU2,. . . . . .A2. . . . .AIP CONF. PROC. VOL.5,. . . . . .1376, (1971).
       #MAGNETIC STRUCTURE AND PROPERTIES OF THE METAMAGNET TB*CU2.
       #BRUN(T.O.), FELCHER(G.P.), KOUVEL(J.S.).

2562   TB*CU, ...A2. . . . .PHYS. REV. VOL.136,. . . . . . . A240, (1964).
       #MAGNETIC ORDER IN RARE-EARTH INTERMETALLIC COMPOUNDS.
       #CABLE(J.W.), KOEHLER(W.C.), WOLLAN(E.O.). (ED. NOTE: REF. SECT. 3,
       64/CABLE).

2563   TB*(CU-ZN),. . .   . . .A2. . . . .COMPTES RENDUS, SERIE B, 265,. . 1169, (1967).
       #PROPRIETES MAGNETIQUES DES ALLIAGES TB*CU(1-X)*ZN(X).
       #PIERRE(J.).

2564   TB*D2, ...A2. . . . .J. APP. PHYS. 34,. . . . . 1352, (1963).
       #MAGNETIC STRUCTURES OF TB*D2 AND HO*D2.
       #COX(D.E.), SHIRANE(G.), TAKEI(W.J.), WALLACE(W.E.).

2565   TB-FE, ...A4. . . . .PHYS. LETT. VOL.47A NO.1,. . . . 73, (1974).
       #NEUTRON DIFFRACTION STUDY OF SPUTTERED AND ANNEALED TB-FE ALLOYS.
       #PICKART(S.J.), RHYNE(J.J.), ALPERIN(H.A.), SAVAGE(H.).

2566   TB-FE2,. . . . . .A2. . . . .PHYS. REV. LETT. VOL.29,. . . . .1562, (1972).
       #DIRECT OBSERVATION OF AN AMORPHOUS SPIN-POLARIZATION DISTRIBUTION.
       #RHYNE(J.J.), PICKART(S.J.), ALPERIN(H.A.).

2567   TB*(FE,CO)*O3, ...A2. . . . .SOL. STATE COMM. VOL. 9,. . . . .435, (1971).
       #PROPRIETES MAGNETIQUES DES SOLUTIONS SOLIDES TB*(FE,CO)*O3 A STRUCTURE
       PEROVSKITE DEFORMEE.
       #KAPPATSCH(A.), BURLET(P.), SIVARDIERE(J.).

2568   TB3*FE5*O12, ...A2. . . . .SOL. STATE COMM. VOL.8,. . . . . 239, (1970).
       #ORDRE MAGNETIQUE DE TB+++ DANS TB3*FE5*O12 A BASSE TEMPERATURE.
       #BERTAUT(E.F.), SAYETAT(F.), TCHEOU(F.).

2569   TB*FE*O3,. . . .   . .A2,T. . . . .ACTA CRYST A24,. . . . . 217, (1968).
       #REPRESENTATION ANALYSIS OF MAGNETIC STRUCTURES.
       #BERTAUT(E.F.).

2570   TB*FE*O3,. . . .   . .A2. . . . .J. APPL. PHYS. VOL.39,. . . . . .1364, (1968).
       #MAGNETIC ORDERING OF TERBIUM IN SOME PEROVSKITE COMPOUNDS.
       #MARESCHAL(J.), SIVARDIERE(J.), DE-VRIES(G.F.), BERTAUT(E.F.). (ED. NOTE:
       REF. SECTION 3, 68/MARESCHAL).

2571   TB*FE*O3,. . . .   . .A2,D4. . . .SOL. STATE COMM. 5,. . . . . 293, (1967).
       #STRUCTURES MAGNETIQUES DE TB*FE*O3.
       #BERTAUT(E.F.), CHAPPERT(J.), MARESCHAL(J.), REBOUILLAT(J.P.),
       SIVARDIERE(J.).

2572   TB*GA, ...A2. . . . . .REF. SECTION 3, 64/CABLE,. . . . .

2573   TB*(GE-SI),. . .   . . .A2. . . . .J. PHYSIQUE TOME 32 COL.1 VOL.II 1133, (1971).
       #PROPRIETES MAGNETIQUES ET STRUCTURES MAGNETIQUES DES COMPOSES
       TB*(GE-SI), HO*SI, ER*SI ET TM*SI DU TYPE CR*B.
       #NGUYEN-VAN-NHUNG, BARLET(A.), LAFOREST(J.).

2574   TB*HG, ...A2. . . . . .REF. SECTION 3, 64/CABLE,. . . . .

2575   TB-HO, ...A2. . . . .SOL. STATE COMM. VOL.6,. . . . . 761, (1968).
       #ANTIFERROMAGNETISM IN A 20 PCT. HO- 80 PCT. TB ALLOY SINGLE CRYSTAL.
       #LEBECH(B.).

2576   TB*IN, ...A2. . . . . .REF. SECTION 3, 64/CABLE,. . . . .

SECTION   5 -STRUCTURE AND CROSS-SECTION DETERMINATIONS . . . .

2577 TB*IR2,. . . . . .A2. . . . . .PHYS. REV. 131, ... 1518, (1963).
&MAGNETIC STRUCTURES OF RARE-EARTH IRIDIUM COMPOUNDS R*IR2.
&FELCHER(G.P.), KOEHLER(W.C.).

2578 TB-IR2,. . . . . .A2. . . . . .COLL. INTER. N.126 (GRENOBLE), . 190, (1963).
&STRUCTURE MAGNETIQUE DES COMPOSES R*IR2 TERRES RARES-IRIDIUM.
&FELCHER(G.P.), KOEHLER(W.C.).

2579 TB-(LA,PR),. . . . . .A2. . . . . .ACTA CRYST. VOL.A28 PART S-4, . . 99, (1972).
&MAGNETIC STRUCTURE OF SAMARIUM TYPE TB8-LA2 AND TB8-PR2 ALLOYS.
&ACHIWA(N.), KAWANO(S.). (ABSTRACT ONLY).

2580 TB-(LA,PR,ND), ...A1,A2. . . .J. PHYS. SOC. JAPAN VOL.35, . . . 303, (1973).
&MAGNETIC AND CRYSTAL STRUCTURE OF TB-LIGHT RARE EARTH ALLOYS.
&ACHIWA(N.), KAWANO(S.).

2581 TB-LU, ...A2. . . . . .REF. SECTION 3, 65/CHILD,. . . . .

2582 TB-MN2,. . . . . .A2. . . . . .J. APP. PHYS. 35, ... 1051, (1964).
&A NEUTRON DIFFRACTION INVESTIGATION OF THE MAGNETIC STRUCTURE OF TB-MN2.
&CORLISS(L.M.), HASTINGS(J.M.).

2583 TB-MN-AL,. . . . . .A2,A1. . . .J. PHYS. CHEM. SOLIDS VOL.33, . . 1031, (1972).
&STRUCTURAL, MAGNETIC AND NEUTRON DIFFRACTION STUDIES ON ER-MN2, ER-AL2
AND TB-MN-AL.
&OESTERREICHER(H.).

2584 TB*MN2*O5, ...A2. . . . . .PHYS. STAT. SOLIDI A VOL.17 NO.1 191, (1973).
&STRUCTURES MAGNETIQUES SINUSOIDALES ET HELICOIDALES DE LA TERRE RARE
DANS T*MN2*O5 (T=ND,TB,ER).
&BUISSON(G.).

2585 TB*N,. . . . . .A2. . . . . .REF. SECTION 3, 63/CHILD,. . . . .

2586 TB*N,. . . . . .A2. . . . . .J. APPL. PHYS. VOL.31, SUPPL., . . 358, (1960).
&NEUTRON DIFFRACTION INVESTIGATIONS OF THE MAGNETIC ORDERING IN RARE
EARTH NITRIDES.
&WILKINSON(M.K.),   CHILD(H.R.), CABLE(J.W.), WOLLAN(E.O.), KOEHLER(W.C.).

2587 TB*NI5,. . . . . .A2. . . . . .C. R. ACAD. SCI. B TOME 270, . . 1131, (1970).
&STRUCTURE MAGNETIQUE DU COMPOSE INTERMETALLIQUE TB*NI5.
&LEMAIRE(R.), PACCARD(D.).

2588 TB*NI2,. . . . . .A2. . . . . .J. APP. PHYS. 36, ... 1001, (1965).
&INVESTIGATION OF THE MAGNETIC STRUCTURE OF ER*MN2; TM*MN2, TB*NI2 BY
NEUTRON DIFFRACTION.
&FELCHER(G.P.), CORLISS(L.M.), HASTINGS(J.M.).

2589 TB*O2, ...A2. . . . . .SOL. STATE COMM. VOL.11 NO.5,. . 605, (1972).
&STRUCTURE MAGNETIQUE DE L#OXYDE DE TERBIUM TB*O2.
&QUEZEL-AMBRUNAZ(S.), BERTAUT(E.F.).

2590 TB*O*CL, ...A2. . . . . .J. PHYSIQUE TOME 32 COL.1 VOL.II 741, (1971).
&ETUDE MAGNETIQUE ET DIFFRACTION NEUTRONIQUE DANS LES COMPOSES DY*O*CL ET
TB*O*CL.
&ELMALEH(D.), FRUCHART(D.), JOUBERT(J.C.).

2591 (TB2*O3)3-(FE2*O3)5, ...A2. . . . . .SOV. PHYS. CRYST. VOL.9, ... 159, (1964).
&A NEUTRON-DIFFRACTION STUDY OF GARNET FERRITES.
&KUZ#MINOV(YU.S.).

2592 TB*(O*H)3, ...A4. . . . . .ACTA CRYST. VOL.A29, ... 684, (1973).
&NEUTRON-DIFFRACTION STUDY OF TB*(O*H)3 - A CASE OF SEVERE EXTINCTION.
&LANDER(G.H.), BRUN(T.O.).

2593 TB*(O*H)3, ...A2. . . . . .PHYS. REV. B VOL.8, ... 3237, (1973).
&NEUTRON MAGNETIC FORM FACTOR AND RADIAL DENSITY OF 4F ELECTRONS IN
TB(3+)! EXPERIMENT AND THEORY.
&LANDER(G.H.), BRUN(T.O.), DESCLAUX(J.P.), FREEMAN(A.J.).

2594 TB*(O*H)3, ...A2. . . . . .PHYS. REV. B VOL.9,. . . . . . . 3003, (1974).
&MAGNETIC FORM FACTOR OF TERBIUM IN TB*(O*H)3.
&BRUN(T.O.), LANDER(G.H.).

2595 TB*O*O*H,. . . . . .A1,A2. . . .J. SOLID STATE CHEM. VOL.9,. . . 234, (1974).
&CRYSTAL STRUCTURE AND MAGNETIC STRUCTURE OF TB*O*O*H.
&CHRISTENSEN(A.N.), QUEZEL(S.).

2596 TB2*O2*S,. . . . . .A2. . . . . .J. PHYS. CHEM. SOLIDS 29, . . . 1001, (1968).
&ETUDE PAR DIFFRACTION NEUTRONIQUE ET MESURES MAGNETIQUES DES OXYSULFURES
DE TERRES RARES T2*O2*S.
&BALLESTRACCI(R.), BERTAUT(E.F.), QUEREL(G.).

2597 TB2*O2*(S,SE), ...A2. . . . . .SOL. STATE COMM. VOL.12, ... 985, (1973).
&MAGNETIC STRUCTURES AND MAGNETIC SUSCEPTIBILITIES OF TERBIUM OXYSULFIDE
AND OXYSELENIDE.
&ABBAS(Y.), ROSSAT-MIGNOD(J.), QUEZEL(G.).

2598 TB*P, ...A2. . . . . .REF. SECTION 3, 63/CHILD,. . . . .

2599 TB*(PD,AG),. . . . . .A2. . . . . .J. APP. PHYS. 36, ... 1096, (1965).
&MAGNETIC STRUCTURE OF RARE-EARTH-COBALT INTERMETALLIC COMPOUNDS.
&CABLE(J.W.), KOEHLER(W.C.), CHILD(H.R.).

2600 TB*RH*O3,. . . . . .A4. . . . . .C. R. ACAD. SCI. B TOME 273, . . 619, (1971).
&ETUDE PAR DIFFRACTION NEUTRONIQUE DES PEROVSKITES ER*AL*O3, HO*AL*O3;
TB*RH*O3, ETC.
&SIVARDIERE(J.), QUEZEL-AMBRUNAZ(S.).

2601 TB*SB, ...A2. . . . . .REF. SECTION 3, 63/CHILD,. . . . .

2602 TB-SC, ...A2. . . . . .J. APP. PHYS. 37, ... 1353, (1966).
&MAGNETIC STRUCTURE PROPERTIES OF TB-SC ALLOYS.
&CHILD(H.R.), KOEHLER(W.C.).

2603 TB-SC, ...A2. . . . . .PHYS. REV. 174, ... 562, (1968).
&MAGNETIC STRUCTURE PROPERTIES OF ALLOYS OF TB, HO AND ER WITH SC.
&CHILD(H.R.), KOEHLER(W.C.).

2604 TB*SI, ...A2. . . . . .SOL. STATE COMM. VOL.8 NO.1, . . 23, (1970).
&STRUCTURE ET PROPRIETES MAGNETIQUES DU MONOSILICIURE DE TERBIUM TB*SI.
&VAN-NHUNG(N.), LAFOREST(J.), SIVARDIERE(J.).

2605 TB-TH, ...A2. . . . . .J. APP. PHYS. 39, ... 1329, (1968).
&MAGNETIC PROPERTIES OF FCC RARE EARTH-THORIUM ALLOYS.
&CHILD(H.R.), KOEHLER(W.C.), MILLHOUSE(A.H.).

SECTION   5 -STRUCTURE AND CROSS-SECTION DETERMINATIONS . . . .

2606   TB,TM, ... A2. . . . . . .J. PHYS. SOC. JAPAN VOL.17 B-III   32, (1962).
       ~RECENT PROGRESS IN MAGNETIC STRUCTURE DETERMINATIONS OF RARE EARTH
       METALS.
       ~KOEHLER(W.C.), CABLE(J.W.), WOLLAN(E.O.), WILKINSON(M.K.).

2607   TB*V*03, ... A2. . . . . . .REF. SECTION 3, 68/MARESCHAL,. . . . .

2608   TB*V*04, ... A1. . . . . . .J. SOLID STATE CHEM. VOL.5,. . .   11, (1972).
       ~REFINEMENT OF THE CRYSTAL STRUCTURE OF SOME RARE EARTH VANADATES R*V*04
       (R=DY,TB,HO,YB).
       ~FUESS(H.), KALLEL(A.).

2609   TB-Y,. . . . . .A1. . . . . . .SOV. PHYS. J.E.T.P. 26,. . . . . 1086, (1968).
       ~CRYSTAL STRUCTURE OF TERBIUM-YTTRIUM ALLOYS AT 77-330 DEG. K.
       ~FINKEL(V.A.), VOROBEV(V.V.).

2610   TB-Y,. . . . . .A2. . . . . . .J. APP. PHYS. 34,. . . . . . . 1335, (1963).
       ~SOME MAGNETIC STRUCTURE PROPERTIES OF TERBIUM AND OF TERBIUM-YTTRIUM
       ALLOYS.
       ~KOEHLER(W.C.), CHILD(H.R.), WOLLAN(E.O.), CABLE(J.W.).

2611   (TB-Y)-NI, ...A2. . . . . . .SOL. STATE COMM. VOL.14 NO.9,. .   877, (1974).
       ~NARROW DOMAIN WALL PROPAGATION AND METAMAGNETIC PROPERTIES OF A
       TB(.5)-Y(.5)-NI SINGLE CRYSTAL.
       ~GIGNOUX(D.), LEMAIRE(R.).

2612   (TB-Y)*SB, ...A4. . . . . . .A.I.P. CONF. PROC. NO.10 PART 2, 1554, (1973).
       ~SUBLATTICE MAGNETIZATION AND THE EXCHANGE INTERACTION IN (TB-Y)*SB.
       ~CABLE(J.W.), COMLY(J.B.), COOPER(B.R.), JACOBS(I.S.), KOEHLER(W.C.),
       VOGT(O.).

2613   TB*ZN2,. . . . . .A2. . . . . . .CONF./RARE EARTHS/ACTINIDES, . .   218, (1971).
       ~MAGNETIC STRUCTURES OF CE*ZN2 AND TB*ZN2.
       ~DEBRAY(D.), SOUGI(M.), MERIEL(P.).

2614   TB*ZN, ...A2. . . . . . .REF. SECTION 3, 64/CABLE,. . . . .

2615   TB*ZN2,. . . . . .A2. . . . . . .J. CHEM. PHYS. VOL.56, ... 4325, (1972).
       ~MAGNETIC STRUCTURES OF CE*ZN2 AND TB*ZN2.
       ~DEBRAY(D.), SOUGI(M.), MERIEL(P.).

2616   TB*ZN2,. . . . . .A2. . . . . . .PHYS. STAT. SOLIDI A VOL.18 NO.1 227, (1973).
       ~MAGNETOCRYSTALLINE ANISOTROPY AND EFFECT OF APPLIED FIELD ON THE
       SINUSOIDAL MAGNETIC STRUCTURES OF TB*ZN2 AND HO*ZN2.
       ~DEBRAY(D.).

2617   TC (ISOTOPE A=99),. . . . . A4. . . . . . .NUCL. SCI. ENGNG. VOL.41,. . . . 188, (1970).
       ~TOTAL NEUTRON CROSS SECTION OF TECHNETIUM-99 FROM 0.01 TO 1000 EV.
       ~WATANABE(T.), REEDER(S.D.).

2618   TC,. . . . . .A4. . . . . . .J. LOW TEMP. PHYS. VOL.14,. . . . 213, (1974).
       ~NEUTRON DIFFRACTION STUDIES OF THE MORPHOLOGY OF FLUX LINE CRYSTALS IN
       TYPE II SUPERCONDUCTORS.
       ~SCHELTEN(J.), LIPPMANN(G.), ULLMAIER(H.).

2619   TE,. . . . . .A3. . . . . . .C. R. ACAD. SCI. B (FRANCE) 270, 109, (1970).
       ~STRUCTURE DU TELLURE LIQUIDE.
       ~TOURAND(G.), BREUIL(M.).

2620   TE,. . . . . .A3. . . . . . .J. PHYSIQUE (FRA.) VOL.32, . . .   813, (1971).
       ~NEUTRON DIFFRACTION STUDY OF THE STRUCTURE OF LIQUID TE AS A FUNCTION OF
       TEMPERATURE.
       ~TOURAND(G.), BREUIL(M.).

2621   TE4*CR3, ...A2. . . . . . .COLL. INTER. N.126 (GRENOBLE), . 158, (1963).
       ~STRUCTURES MAGNETIQUES DE CR3*X4 (X=S, SE, TE).
       ~BERTAUT(E.F.), ROULT(G.), ALEONARD(R.), PAUTHENET(R.), CHEVRETON(M.),
       JANSEN(R.).

2622   (TE*CR)-(SB*CR),. . . . . A2. . . . . . .PROC. INT. CONF. (NOTTINGHAM),   291, (1964).
       ~MAGNETIC STRUCTURES IN THE MN*SB-CR*SB AND CR*TE-CR*SB SYSTEMS.
       ~COX(D.E.), SHIRANE(G.), TAKEI(W.J.).

2623   TE*MN, ...A2. . . . . . .COLL. INTER. N.126 (GRENOBLE), . 144, (1963).
       ~NEUTRON DIFFRACTION STUDY ON MANGANESE.
       ~KUNITOMI(N.), HAMAGUCHI(Y.), ANZAI(S.).

2624   TE*02, ...A1. . . . . . .Z. KRIST. VOL.116, ... 345, (1961).
       ~THE CRYSTAL STRUCTURE OF TELLURIUM DIOXIDE. 'A' REDETERMINATION BY
       NEUTRON DIFFRACTION.
       ~LECIEJEWICZ(J.).

2625   TE*(O*H)6, ...A1. . . . . . .BU. SOC. FR. MIN. CRYSTALLOG. 94 172, (1971).
       ~NEUTRON DIFFRACTION STUDY OF TE*(O*H)6.
       ~COHEN-ADDAD(C.).

2626   TE*(O*H)6 (MON),. . . . . A1. . . . . . .ACTA CHEM. SCAND. VOL.27,. . . .   85, (1973).
       ~A NEUTRON DIFFRACTION REFINEMENT OF THE CRYSTAL STRUCTURE OF TELLURIC
       ACID, TE*(O*H)6 (MON).
       ~LINDQVIST(O.), LEHMANN(M.S.).

2627   TERPHENYL-(P), ...A1. . . . . . .ACTA CRYST. VOL.B26, ... 693, (1970).
       ~AN X-RAY AND NEUTRON DIFFRACTION REFINEMENT OF THE STRUCTURE OF
       P-TERPHENYL.
       ~RIETVELD(H.M.), MASLEN(E.N.), CLEWS(C.J.B.).

2628   TH (ISO. A=230), ...P. . . . . . .ATOMNAYA ENERGIYA (USSR) VOL.24, 243, (1968).
       ~TOTAL NEUTRON CROSS SECTION OF TH (ISO. A=230) BELOW 1 EV.
       ~KALEBIN(S.M.), PLAEI(P.N.), IVANOV(R.N.), ZARALOVA(Z.K.),
       KUKAVADZE(G.M.), PYZHOVA(Z.I.), RUKOLAINE(G.V.).

2629   TH,. . . . . .P. . . . . . .ACTA CRYST. VOL.15,. . . . . . .   351, (1962).
       ~NUCLEAR COHERENT SCATTERING AMPLITUDES FOR THORIUM, URANIUM AND
       PLUTONIUM.
       ~ROOF-JR(R.B.), ARNOLD(G.P.), GSCHNEIDNER-JR(K.A.).

2630   TH2*AL-(D2,H2),. . . . . .A1. . . . . . .PHILIPS RES. REP. VOL.18,. . . .   35, (1963).
       ~NEUTRON DIFFRACTION AND PROTON MAGNETIC RESONANCE OF DEUTERIUM AND
       HYDROGEN SOLUTIONS IN TH2*AL.
       ~VAN-VUCHT(J.H.).

2631   TH*C2, ...A1. . . . . . .ACTA CRYST. VOL.B24, ... 1121, (1968).
       ~THE CRYSTAL STRUCTURES OF TH*C2.
       ~BOWMAN(A.L.), KRIKORIAN(N.H.), ARNOLD(G.P.), WALLACE(T.C.),
       NERESON(N.G.).

SECTION   5 -STRUCTURE AND CROSS-SECTION DETERMINATIONS . . . .

2632 TH*C*N, ...A1. . . . . . .ACTA CRYST. VOL.B28, ... 1724, (1972).
      ↝TH*C*N CRYSTAL STRUCTURE.
      ↝BENZ(R.), ARNOLD(G.P.), ZACHARIASEN(W.H.).

2633 TH-(CO-FE)5, ...A1. . . . . . .IEEE TRANS. MAGN. VOL.9, . . . 217, (1973).
      ↝NEUTRON DIFFRACTION STUDY OF THE TH-(CO-FE)5 ALLOYS.
      ↝LAFOREST(J.), SHAH(J.S.).

2634 TH-(ER,HO), ...A4. . . . . . .J. APP. PHYS. 38, ... 1384, (1967).
      ↝MAGNETIC PROPERTIES OF SOME ALLOYS OF THORIUM WITH HOLMIUM AND ERBIUM.
      ↝KOEHLER(W.C.), CHILD(H.R.), CABLE(J.W.), MOON(R.M.).

2635 TH*FE5,TH2*FE17,TH*(FE-CO)5, ...A2. . . . . . .PHYS. STAT. SOLIDI B VOL.57 NO.2 K155, (1973).
      ↝MAGNETIC STRUCTURES OF TH*(FE-CO)5 AND TH2*FE17.
      ↝ELEMANS(J.B.A.), BUSCHOW(K.H.J.).

2636 TH*H2,TH*D2, ...A1. . . . . . .ACTA CRYST. VOL.5, ... 22, (1952).
      ↝THE CRYSTAL STRUCTURE OF THORIUM AND ZIRCONIUM DIHYDRIDES BY X-RAY
      AND NEUTRON DIFFRACTION.
      ↝RUNDLE(R.E.), SHULL(C.G.), WOLLAN(E.O.).

2637 TH-(HO,ER), ...A2. . . . . . .J. PHYSIQUE TOME 32 COL.1 VOL.II 1128, (1971).
      ↝SINGLE CRYSTAL NEUTRON DIFFRACTION STUDIES OF HCP RARE EARTH THORIUM
      ALLOYS.
      ↝CHILD(H.R.), KOEHLER(W.C.).

2638 TH3*N4, ...A1. . . . . . .ACTA CRYST. VOL.B27, ... 243, (1971).
      ↝THE CRYSTAL STRUCTURE OF TH3*N4.
      ↝BOWMAN(A.L.), ARNOLD(G.P.).

2639 TH*(N*O3)4.5(H2*O) . . . . .A1. . . . . . .ACTA CRYST. 20, ... 842, (1966).
      ↝CRYSTAL STRUCTURE OF THORIUM NITRATE PENTAHYDRATE BY NEUTRON DIFFRACTION
      ↝TAYLOR(J.C.), MUELLER(M.H.), HITTERMAN(R.L.).

2640 TH*NI2*CO3,TH*NI3*CO2, ...A2. . . . . . .J. APPL. PHYS. VOL.44, ... 5096, (1973).
      ↝NEUTRON DIFFRACTION STUDIES OF TH*NI2*CO3 AND TH*NI3*CO2 AT 5-297 K.
      ↝ATOJI(M.), ATOJI(I.), DO-DINH(C.), WALLACE(W.E.).

2641 TH*O2, ...P. . . . . . . .PROC. ROY. SOC. 274, ... 134, (1963).
      ↝NEUTRON DIFFRACTION STUDIES OF THE ACTINIDE OXIDES II. THERMAL MOTIONS
      OF THE ATOMS IN U*O2 AND TH*O2 BETWEEN ROOM TEMPERATURE AND 1110 DEG. C.
      ↝WILLIS(B.T.M.).

2642 TH*O2, ...P. . . . . . . .PROC. ROY. SOC. 274, ... 122, (1963).
      ↝NEUTRON DIFFRACTION STUDIES OF THE ACTINIDE OXIDES I. URANIUM DIOXIDE
      AND THORIUM DIOXIDE AT ROOM TEMPERATURE.
      ↝WILLIS(B.T.M.).

2643 TI (ISO. A=50), ...P. . . . . . . .PHYS. SCRIPTA(SWEDEN) VOL.4, . . 89, (1971).
      ↝THERMAL NEUTRON CAPTURE IN TI(ISOTOPE A=50) AND NI(ISOTOPE A=64).
      ↝ARNELL(S.E.), HARDELL(R.), HASSELGREN(A.), MATTSSON(C.G.),
      SKEPPSTEDT(O.).

2644 TI-C, ...A1. . . . . . .PHYS. STAT. SOL. 20, ... K141, (1967).
      ↝NEUTRON DIFFRACTION STUDIES ON TITANIUM-CARBON AND ZIRCONIUM-CARBON
      ALLOYS.
      ↝GORETZKI(H.).

2645 TI*C, ...P. . . . . . . .PHYS. REV. VOL.75, ... 975, (1949).
      ↝TOTAL NEUTRON CROSS SECTIONS OF COMPOUNDS WITH DIFFERENT CRYSTALLINE
      STRUCTURES.
      ↝WINSBERG(L.), MENEGHETTI(D.), SIDHU(S.S.).

2646 TI*(C-O), ...A1. . . . . . .SOV. PHYS. DOKLADY VOL.15, . . . 276, (1970).
      ↝NEUTRON-DIFFRACTION STUDY OF THE STRUCTURE OF TITANIUM OXICARBIDES.
      ↝ZUBKOV(V.G.), MATVEENKO(I.I.), DUBROVSKAYA(L.B.), BOGOMOLOV(G.D.),
      GEL*D(P.V.).

2647 TI*C-W*C, ...P. . . . . . . .J. APPL. PHYS. VOL.19, ... 639, (1948).
      ↝TRANSMISSION OF MONOENERGETIC SLOW NEUTRONS THROUGH SOLID SOLUTIONS
      AND MECHANICAL MIXTURES OF TI*C AND W*C.
      ↝SIDHU(S.S.).

2648 TICONAL X, ...A2. . . . . . .Z. ANGEW. PHYS. VOL.27, ... 73, (1969).
      ↝A NEUTRON DIFFRACTION STUDY OF THE MAGNETIC MICROSTRUCTURE OF TICONAL X.
      ↝MAIER(G.).

2649 TI-CO,TI-FE, ...A1. . . . . . .PHYS. MET. METALLOGR. VOL.23 N.3 168, (1967).
      ↝NEUTRON DIFFRACTION DETERMINATION OF THE STRUCTURE OF TI-FE AND TI-CO
      ALLOYS.
      ↝DOROSHENKO(A.V.), NEMNONOV(S.A.), SIDOROV(S.K.).

2650 TI*CR2*(SE4,S4), ...A1,A2. . . . .BULL. SOC. FRANC. MIN. CRIST. 91 88, (1968).
      ↝NEUTRON DIFFRACTION STUDY OF TI*CR2*SE4 AND TI*CR2*S4.
      ↝LAMBERT-ANDRON(B.), BERODIAS(G.), CHEVRETON(M.). (IN FRENCH).

2651 TI*CR2*TE4, ...A1. . . . . . .COMP. REND. 263, ... 621B, (1966).
      ↝ETUDE PAR DIFFRACTION NEUTRONIQUE DU COMPOSE FERROMAGNETIQUE TI*CR2*TE4.
      ↝ANDRON(B.), BERODIAS(G.), CHEVRETON(M.), MOLLARD(P.).

2652 TI*(FE-CO), ...A2,A4. . . . .J. APP. PHYS. 39, ... 2221, (1968).
      ↝NEUTRON DIFFRACTION STUDY OF TI*(FE-CO) ALLOYS.
      ↝PICKART(S.J.), NATHANS(R.), MENZINGER(F.).

2653 TI*H2, ...A1,2. . . . . .COLL. INTER. N.126 (GRENOBLE), . 38, (1963).
      ↝NEUTRON DIFFRACTION AND SCATTERING STUDIES AT J.A.E.R.I.
      ↝KUNITOMI(N.), HAMAGUCHI(Y.), SAKAMOTO(M.), DOI(K.), KOMURA(S.).

2654 TI-H,TI-D, ...A1. . . . . . .ACTA CRYST. VOL.9, ... 607, (1956).
      ↝NEUTRON DIFFRACTION STUDIES OF HAFNIUM-HYDROGEN AND TITANIUM-HYDROGEN
      SYSTEMS.
      ↝SIDHU(S.S.), HEATON(L.), ZAUBERIS(D.D.).

2655 TI*NB2*O7,TI2*NB10*O29, ...A1. . . . . . .NATURE (PHYS. SCI.) VOL.244, . . 139, (1973).
      ↝CATION DISTRIBUTIONS IN NIOBIUM OXIDE BLOCK STRUCTURES.
      ↝CHEETHAM(A.K.), VON-DREELE(R.B.).

2656 TI2*O3, ...A2. . . . . . .J. APP. PHYS. 39, ... 585, (1968).
      ↝NEUTRON DIFFRACTION DATA ON TI2*O3 AND V2*O3.
      ↝KENDRICK(H.), ARROTT(A.), WERNER(S.A.).

2657 TI2*O3, ...A1,2. . . . . .PHYS. REV. 130, ... 2230, (1963).
      ↝MAGNETIC AND CRYSTAL STRUCTURE OF TITANIUM SESQUIOXIDE.
      ↝ABRAHAMS(S.C.).

SECTION 5 -STRUCTURE AND CROSS-SECTION DETERMINATIONS . . . .

2658 TI-O,. . . . . .A4.. . .. . .NATURWISSENSCHAFTEN VOL.54,. . . . 163, (1967).
⇌NEUTRON DIFFRACTION BY HYDROGENATED AND NON-HYDROGENATED MIXED CRYSTALS
OF ALPHA-TITANIUM AND OXYGEN.
⇌GORETZKI(H.). (IN GERMAN).

2659 TI2*O3,. . . . . . .A2. . . . . . .J. SOLID STATE CHEM. VOL.4,. . . . 400, (1972).
⇌NEUTRON DIFFRACTION AND OTHER STUDIES OF MAGNETIC ORDERING IN PHASES
BASED ON CR2*O3, V2*O3, AND TI2*O3.
⇌REID(A.F.), SABINE(T.M.), WHEELER(D.A.).

2660 TI-O,. . . . . . .A1. . . . . . .J. PHYS. SOC. JAPAN VOL.28,. . . 1014, (1970).
⇌INTERSTITIAL ORDER-DISORDER TRANSFORMATION IN THE TI-O SOLID SOLUTION.
IV. A NEUTRON DIFFRACTION STUDY.
⇌YAMAGUCHI(S.), HIRAGA(K.), HIRABAYASHI(M.).

2661 TI2*O3,. . . . . .A2. . . . . . .J. PHYS. CHEM. SOLIDS VOL.13,. . 166, (1960).
⇌NEUTRON-DIFFRACTION STUDY OF TI2*O3.
⇌SHIRANE(G.), PICKART(S.J.), NEWNHAM(R.).

2662 TI*O,. . . . . . .P. . . . . . .PHYS. REV. VOL.118,. . . . . . . 797, (1960).
⇌SLOW NEUTRON SCATTERING BY THE TITANIUM ISOTOPES,
⇌SHULL(C.G.), WILKINSON(M.K.), MUELLER(M.H.).

2663 TI*TA*O4,. . . . .A1. . . . . . .SOV. PHYS. CRYST. VOL.17,. . . . 1017, (1973).
⇌ATOMIC AND MOLECULAR ORDERING IN ME*TA*O4 (ME=TI,V,CR,FE) WITH A RUTILE
STRUCTURE.
⇌ASTROV(D.N.), KRYUKOVA(N.A.), ZORIN(R.B.), MAKAROV(V.A.), OZEROV(R.P.),
ROZHDESTVENSKII(F.A.), SMIRNOV(V.P.), TURCHANINOV(A.M.), FADEEVA(N.V.).

2664 TI-ZR, ...A4. . . . . . .J. APPL. PHYS. VOL.27,. . . . . 1040, (1956).
⇌NEUTRON DIFFRACTION STUDY OF TITANIUM-ZIRCONIUM SYSTEM.
⇌SIDHU(S.S.), HEATON(L.), ZAUBERIS(D.D.), CAMPOS(F.P.).

2665 TI-ZR-O, ...A4. . . . . . .DOKL. AKAD. NAUK SSSR VOL.194,. 1374, (1970).
⇌X-RAY AND NEUTRON DIFFRACTION STUDY OF ORDERING IN SOME ALLOYS OF THE
TITANIUM-ZIRCONIUM-OXYGEN SYSTEM.
⇌FYKIN(L.E.), VAVILOVA(V.V.), KORNILOV(I.I.), OZEROV(R.P.),
SOLOV≠EV(S.P.).

2666 TL,. . . . . . .P. . . . . . .J. OF PHYS. C VOL.1,. . . . . . .1075, (1968).
⇌THE STRUCTURE FACTOR FOR LIQUID METALS II. RESULTS FOR LIQUID ZN, TL,
PB, SN AND BI.
⇌NORTH(D.M.), ENDERBY(J.E.), EGELSTAFF(P.A.).

2667 TL (ISO. A=203), ...P. . . . . . .J. INORG. NUCL. CHEM. VOL.32,. . 2839, (1970).
⇌THE THERMAL NEUTRON CAPTURE CROSS-SEC. AND RES. CAPTURE INTEG. OF CA(IS.
A=44), NI(IS. A=62), YB(IS. A=168, 174), TM(IS. A=169) AND TL(IS. A=203)
⇌SIMS(G.H.E.), JUHNKE(D.G.).

2668 TL,. . . . . . .P. . . . . . .PHYS. LETT. 21,. . . . . . . 286, (1966).
⇌THE STRUCTURE FACTOR FOR LIQUID METALS AT LOW ANGLES.
⇌EGELSTAFF(P.A.), DUFFILL(C.), RAINEY(V.S.), ENDERBY(J.E.), NORTH(D.M.).

2669 TL,. . . . . . .P. . . . . . .PHYS. REV. VOL.105,. . . . . 216, (1957).
⇌THERMAL-NEUTRON COHERENT SCATTERING AMPLITUDES OF THALLIUM AND OSMIUM.
⇌HEATON(L.), SIDHU(S.S.).

2670 TL-(BI,PB),. . . .A4. . . . . . .MATERIALS RES. BULL. VOL.2,. . . 69, (1967).
⇌CHARACTERIZATION OF LARGE ALLOY SINGLE CRYSTALS BY NEUTRON DIFFRACTION.
⇌NG(S.C.), BROCKHOUSE(B.N.), HALLMAN(E.D.).

2671 TL*BR, ...P. . . . . . .PHYS. REV. VOL.75,. . . . . . 975, (1949).
⇌TOTAL NEUTRON CROSS SECTIONS OF COMPOUNDS WITH DIFFERENT CRYSTALLINE
STRUCTURES.
⇌WINSBERG(L.), MENEGHETTI(D.), SIDHU(S.S.).

2672 TL*BR, ...T.E. . . . . . .PHYS. STAT. SOLIDI B VOL.61 NO.1 241, (1974).
⇌MEAN-SQUARE IONIC DISPLACEMENTS AND FORCE CONSTANTS OF CRYSTALS WITH THE
CS*CL TYPE STRUCTURE.
⇌JEX(H.), MULLNER(M.), DYCK(W.).

2673 TL*CO*F3,. . . .A2. . . . . . .C.R. ACAD. SCI. B TOME 276 NO.14 579, (1973).
⇌ETUDE DE LA STRUCTURE MAGNETIQUE DE TL*CO*F3.
⇌LECOMTE(M.), NOUET(J.), PORTES(M.).

2674 TL2*CU*H12*O14*S2, ...A1. . . . . . .CRYST. STRUCT. COMM. VOL.1 N.4,. 371, (1972).
⇌THALLIUM HEXA-AQUACOPPER(II) SULPHATE,...
⇌SHIELDS(K.G.), VAN-DER-ZEE(J.J.), KENNARD(C.H.L.).

2675 TL(ISO. A=204),. . . . . . .P. . . . . . .PROC/CONF/NEUTRON C. S. + TECH., 693, (1968).
⇌TOTAL NEUTRON CROSS SECTION OF TL (ISO. A=204) FROM 0.2 EV TO 1000 EV.
⇌WATANABE(T.), STOKES(G.E.), SCHUMAN(R.P.).

2676 TL*MN*CL3, . . .A2. . . . . . .PHYS. REV. B VOL.3,. . . . . . 2344, (1971).
⇌NEUTRON DIFFRACTION AND THE MAGNETIC STRUCTURE OF TL*MN*CL3.
⇌MELAMUD(M.), PINTO(H.), SHACHAR(G.), MAKOVSKY(J.), SHAKED(H.).

2677 TL*MN*F3,. . . .A2. . . . . . .J. PHYSIQUE TOME 32 COL.1 VOL.II 617, (1971).
⇌MAGNETIC STRUCTURE AND EXCHANGE INTEGRAL MEASUREMENT IN TL*MN*F3.
⇌MADHAV-RAO(L.), MURTHY(N.S.S.).

2678 TM,. . . . . . .P,A2. . . . . . .DISS. ABS. VOL.31,. . . . . . .5560B, (1971).
⇌MAGNETIC FORM FACTOR OF THULIUM AND TERBIUM METALS.
⇌BRUN(T.O.).

2679 TM (ISO. A=171), ...P. . . . . . .J. PHYS. SOC. JAP. VOL.31,. . . 1304, (1971).
⇌THE SLOW NEUTRON CAPTURE CROSS SECTIONS FOR THE LONG-LIVED ACTIVITIES,
ER(ISOTOPE A=171) AND TM(ISOTOPE A=171).
⇌MIYANO(K.).

2680 TM (ISO. A=169), ...P. . . . . . .J. INORG. NUCL. CHEM. VOL.32,. . 2839, (1970).
⇌THE THERMAL NEUTRON CAPTURE CROSS-SEC. AND RES. CAPTURE INTEG. OF CA(IS.
A=44), NI(IS. A=62), YB(IS. A=168, 174), TM(IS. A=169) AND TL(IS. A=203)
⇌SIMS(G.H.E.), JUHNKE(D.G.).

2681 TM,. . . . . . .A2. . . . . . .REF. SECTION 3, 65/KOEHLER,. . . . .

2682 TM,. . . . . . .A2. . . . . . .J. APPL. PHYS. VOL.33, SUPPL.,. . 1124, (1962).
⇌NEUTRON DIFFRACTION STUDY OF MAGNETIC ORDERING IN THULIUM.
⇌KOEHLER(W.C.), CABLE(J.W.), WOLLAN(E.O.), WILKINSON(M.K.).

2683 TM,. . . . . . .A2. . . . . . .PHYS. REV. VOL.126,. . . . . . 1672, (1962).
⇌MAGNETIC STRUCTURES OF THULIUM.
⇌KOEHLER(W.C.), CABLE(J.W.), WOLLAN(E.O.), WILKINSON(M.K.).

SECTION 5 -STRUCTURE AND CROSS-SECTION DETERMINATIONS . . . .

2684 TM*AG2,. . . . . .
.NEUTRON DIFFRACTION STUDIES OF TM*AU2, YB*AU2, AND TM*AG2 AT 300-1.7 K.
.ATOJI(M.).
... A1 ... J. CHEM. PHYSICS VOL.52, 6433, (1970).

2685 TM*AU2,. . . . . .
.NEUTRON DIFFRACTION STUDIES OF TM*AU2, YB*AU2, AND TM*AG2 AT 300-1.7 K.
.ATOJI(M.).
... A1 ... J. CHEM. PHYSICS VOL.52, 6433, (1970).

2686 TM*AU2,. . . . . .
.ON THE MAGNETIC STRUCTURE OF R*AU2 COMPOUNDS (R=TB-TM).
.FEDRO(A.J.), SHAFFER(J.C.).
... A2 ... AIP CONF. PROC. VOL.5, 1390, (1971).

2687 TM*C2, ...
.NEUTRON DIFFRACTION STUDY OF TM*C2 AT 300-1.7 K.
.ATOJI(M.).
... A1 ... J. CHEM. PHYSICS VOL.52, 6431, (1970).

2688 TM*C*O3,. . . . . .A2,A4....REF. SECTION 3, 66/BERTAUT,. . . . .

2689 TM*FE*O3,. . .
.INVESTIGATION OF IRON SUBLATTICES IN A MONOCRYSTAL OF THE ORTHOFERRITE
TM*FE*O3.
.KOCHAROV(A.G.), LOSHMANOV(A.A.), YAMZIN(I.I.), CHERVONENKIS(A.YA.).
... A2 ... SOV. PHYS. J.E.T.P. VOL.31, 808, (1970).

2690 TM*FE*O3,. . . .
.THE MAGNETIC STRUCTURE OF THULIUM ORTHOFERRITE, TM*FE*O3.
.LEAKE(J.A.), SHIRANE(G.), REMEIKA(J.P.).
... A2 ... SOL. STATE COMM. VOL.6, 15, (1968).

2691 TM*MN2,. . . . .
.INVESTIGATION OF THE MAGNETIC STRUCTURE OF ER*MN2, TM*MN2, TB*NI2 BY
NEUTRON DIFFRACTION.
.FELCHER(G.P.), CORLISS(L.M.), HASTINGS(J.M.).
... A2 ... J. APP. PHYS. 36, 1001, (1965).

2692 TM*MN*O3,. . . . . .A2 ... REF. SECTION 3, 64/KOEHLER(1),. . . . .

2693 TM*N,. . . . . .A2 ... REF. SECTION 3, 63/CHILD,. . . . .

2694 TM*NI, ...
.MAGNETIC STRUCTURE OF THE TM*NI COMPOUND CRYSTAL FIELD EFFECT.
.GIGNOUX(D.), ROSSAT-MIGNOD(J.), TCHEOU(F.).
... A2 ... PHYS. STAT. SOLIDI A VOL.14 NO.2 483, (1972).

2695 TM2*O3,. . . . .
.NEUTRONOGRAPHISCHE BESTIMMUNG DER KRISTALLSTRUKTURPARAMETER VON DY2*O3
TM2*O3 UND (ALPHA)-MN2*O3.
.HASE(W.).
... A1 ... PHYS. STAT. SOL. 3, K446, (1963).

2696 TM2*O3,. . . . .
.MESSUNG DER KOHARENTEN STREUAMPLITUDEN VON DYSPROSIUM UND THULIUM FUR
THERMISCHE NEUTRONEN.
.BETZL(M.), HASE(W.), KLEINSTUCK(K.), TOBISCH(J.).
... P ... PHYS. STAT. SOL. 2, K164, (1962).

2697 TM*SB, ...
.POLARIZED-NEUTRON STUDY OF THE INDUCED MAGNETIC MOMENT IN TM*SB.
.LANDER(G.H.), BRUN(T.O.), VOGT(O.).
... A2 ... PHYS. REV. B VOL.7, 1988, (1973).

2698 TM*SB, ...
.CRYSTAL-FIELD EFFECTS IN THE MAGNETIC FORM FACTOR OF TM*SB.
.BRUN(T.O.), LANDER(G.H.).
... A2 ... PHYS. REV. LETT. VOL.29, 1172, (1972).

2699 U, ...
.NEUTRON DIFFRACTION STUDY OF ALPHA-URANIUM AT LOW TEMPERATURES.
.LANDER(G.H.), MUELLER(M.H.).
... A1 ... ACTA CRYST. VOL.B26, 129, (1970).

2700 U, ...
.ANALYSIS OF ANGULAR DISTRIBUTIONS OF NEUTRONS SCATTERED BY URANIUM AT
SMALL ANGLES AND NEUTRON POLARIZABILITY.
.ANIKIN(G.V.), KOTUKHOV(I.I.).
... A4 ... SOV. J. NUCL. PHYS. VOL.14, 269, (1971).

2701 U (ISOTOPES A=233, 235),
.NEUTRON SCATTERING CROSS SECTIONS OF U(ISOTOPES A=233, 235) AND PU(ISO-
TOPE A=239).
.SIMPSON(F.B.), MILLER(L.G.), MOORE(M.S.), HOCKENBURG(R.W.), KING(T.J.).
... A4 ... NUCL. PHYS. A VOL.A164, 34, (1971).

2702 U (ISOTOPE A=235),
.THE SCATTERING CROSS SECTION OF U(ISOTOPE A=235) BETWEEN 0.025 EV AND
1 EV.
.CEULEMANS(H.), POORTMANS(F.).
... A4 ... 2ND INTERNAT. CONF./NUCL. DATA, (1970).

2703 U (ISO. A=238),.
.EVALUATION OF URANIUM 238 NEUTRON DATA IN THE ENERGY RANGE .0001 EV - 15
MEV.
.VASTEL(M.), RAVIER(J.).
... P ... PROC/CONF/NEUTRON C. S. + TECH., 1129, (1968).

2704 U (A=235), . . .
.DELAYED GAMMA-RAYS FROM THERMAL NEUTRON FISSION OF U (A=235) AND
PU (A=239).
.BERICK(A.C.), EVANS(A.E.), MEISSNER(J.A.).
... P ... PHYS. REV. VOL.187, 1506, (1969).

2705 U (ALPHA), . . .
.CRYSTAL STRUCTURE VARIATIONS IN ALPHA URANIUM AT LOW TEMPERATURES.
.BARRETT(C.S.), MUELLER(M.H.), HITTERMAN(R.L.).
... A1 ... PHYS. REV. VOL.129, 625, (1963).

2706 U (ISOTOPES A=233, 235),
.COHERENT NEUTRON SCATTERING AMPLITUDES OF FISSIONABLE NUCLEI, U(A=233),
U(A=235), AND PU(A=239).
.ATOJI(M.).
... P ... ACTA CRYST. 20, 587, (1966).

2707 U,. . . . . . .
.NUCLEAR COHERENT SCATTERING AMPLITUDES FOR THORIUM, URANIUM AND
PLUTONIUM.
.ROOF-JR(R.B.), ARNOLD(G.P.), GSCHNEIDNER-JR(K.A.).
... P ... ACTA CRYST. VOL.15, 351, (1962).

2708 U,. . . . . . .
.THE CROSS SECTIONS OF METALLIC URANIUM FOR SLOW NEUTRONS.
.WHITAKER(M.D.), BARTON(C.A.), BRIGHT(W.C.), MURPHY(E.J.).
... P ... PHYS. REV. VOL.55, 793, (1939).

2709 U (A=235), . . .
.A PRECISE DETERMINATION OF THE TOTAL CROSS SECTION OF URANIUM-235 FROM
0.000818 TO 0.0818 EV.
.SAFFORD(G.J.), HAVENS-JR(W.H.), RUSTAD(B.M.).
... P ... NUCL. SCI. ENGNG. VOL.6, 433, (1959).

2710 U,. . . . . . .
.STUDY OF NUCLEAR MATERIALS BY NEUTRON CRYSTALLOGRAPHY (REVIEW).
.LANIESSE(J.), ENGLANDER(M.), MERIEL(P.).
... BULL. INFORM. SCI. TECH. NO.49, 69, (1961).

SECTION   5 -STRUCTURE AND CROSS-SECTION DETERMINATIONS . . . .

2711 U (A=233), ...NEUTRON SCATTERING P. CROSS SECTION OF U (A=233). PHYS. REV. VOL.109, ... 1645, (1958).
      ..OLEKSA(S.).

2712 U (A=235), ...NEUTRON SCATTERING P. CROSS SECTION OF U (A=235). PHYS. REV. VOL.109, ... 1641, (1958).
      ..FOOTE-JR(H.L.).

2713 U (A=233), ...PRECISION MEASUREMENT P. OF THE TOTAL NEUTRON CROSS SECTION OF U(A=233) PHYS. REV. VOL.118, 799, (1960)
      BETWEEN 0.000818 AND 0.0818 EV.
      ..SAFFORD(G.J.), HAVENS-JR(W.W.), RUSTAD(B.M.).

2714 U, ...ETUDE AUX NEUTRONS DE LA TEXTURE CRISTALLINE DE BARREAUX D≠URANIUM. J. NUCL. MATERIALS VOL.2, ... 69, (1960).
      ..LANIESSE(J.), MERIEL(P.), ENGLANDER(M.).

2715 U (A=233,235), ...TOTAL NEUTRON CROSS SECTIONS OF U(A=233) AND U(A=235) FROM .02 TO .08 EV NUCL. SCI. ENG. VOL.7, ... 187, (1960).
      ..SIMPSON(O.D.), MOORE(M.S.), SIMPSON(F.B.).

2716 U*AS, ...THE MAGNETIC ORDERING IN URANIUM MONOARSENIDE. A2. PHYS. STAT. SOL. 23, ... K123, (1967).
      ..TROC(R.), MURASIK(A.), ZYGMUNT(A.), LECIEJEWICZ(J.).

2717 U*AS, ...THE ANTIFERROMAGNETIC ORDERING IN URANIUM MONOARSENIDE AND A1,2. PHYS. STAT. SOL. 30, . . . 157, (1968).
      MONOANTIMONIDE.
      ..LECIEJEWICZ(J.), MURASIK(A.), TROC(R.).

2718 U*AS2, ...NEUTRON DIFFRACTION STUDY OF U*AS2. A2. J. PHYSIQUE TOME 26, ... 561, (1965).
      ..OLES(A.). (IN FRENCH).

2719 U*AS, ...NEUTRON DIFFRACTION STUDY OF THE ANTIFERROMAGNETISM OF URANIUM A2. J. PHYS. CHEM. SOLIDS 29, . . 1702, (1968).
      MONOARSENIDE.
      ..WILLIAMS(J.M.), HEATON(L.), CAMPOS(F.P.).

2720 U*(AS-S), ...MAGNETIC PHASE DIAGRAM OF THE U*(AS-S) SYSTEM. A2,A4. MAGN. AND MAG. MATERIALS-1971, . 1371, (1972).
      ..LANDER(G.H.), MUELLER(M.H.), REDDY(J.F.). (AIP CONF. PROC. NO.5 PART II)

2721 U*(AS,SB)*SE, ...NEUTRON DIFFRACTION STUDY OF U*AS*SE AND U*SB*SE. A2. PHYS. STAT. SOLIDI A VOL.13, . . 657, (1972).
      ..LECIEJEWICZ(J.), ZYGMUNT(A.).

2722 U*(AS-SE), ...A NOTE ON ANTIFERROMAGNETISM IN U*(AS-SE). A4. PHYS. STAT. SOL. VOL.38, ... K89, (1970).
      ..LECIEJEWICZ(J.), MURASIK(A.), PALEWSKI(T.), TROC(R.).

2723 U*(AS-SE), ...THE MAGNETIC PHASE DIAGRAM OF THE U*(AS(2-X)-SE(X)) SOLID SOLUTION. PHYS. STAT. SOLIDI A VOL.16 NO.2 K171, (1973).
      ..LIGENZA(S.), MURASIK(A.), ZYGMUNT(A.), LECIEJEWICZ(J.).

2724 U*AS-U*S, ...MAGNETIC PHASE DIAGRAM OF THE U*AS-U*S SYSTEM. A4. PHYS. REV. B VOL.6, ... 1880, (1972).
      ..LANDER(G.H.), MUELLER(M.H.), REDDY(J.F.).

2725 U*(B*H4)4, ...14-COORDINATE URANIUM(IV). THE STRUCTURE OF URANIUM BOROHYDRIDE BY A1. INORG. CHEM. VOL.11, . . . 3009, (1972).
      SINGLE-CRYSTAL NEUTRON DIFFRACTION.
      ..BERNSTEIN(E.R.), HAMILTON(W.C.), KEIDERLING(T.A.), LA-PLACA(S.J.),
      LIPPARD(S.J.), MAYERLE(J.J.).

2726 U-BE, ...X-RAY AND NEUTRON DIFFRACTION STUDIES OF THE M*BE13 INTERMETALLIC A1. ACTA CRYST. VOL.5, . . . 394, (1952).
      COMPOUNDS.
      ..KOEHLER(W.C.), SINGER(J.), COFFINBERRY(A.S.).

2727 U*BI, ...NEUTRON-DIFFRACTION INVESTIGATIONS ON U*SB, U*BI, AND U3*SB4. A1. J. APP. PHYS. VOL.40, . . 1135, (1969).
      ..OLSEN(C.E.), KOEHLER(W.C.).

2728 U*BI, ...ANTIFERROMAGNETIC STRUCTURES OF U*SB AND U*BI. A2. J. PHYS. CHEM. SOLIDS 30, ... 1642, (1969).
      ..KUZNIETZ(M.), LANDER(G.H.), CAMPOS(F.P.).

2729 U*BI2, ...NEUTRON-DIFFRACTION STUDY OF ANTIFERROMAGNETISM IN U*SB2 AND U*BI2. A2. PHYS. STAT. SOL. 22, . . 517, (1967).
      ..LECIEJEWICZ(J.), TROC(R.), MURASIK(A.), ZYGMUNT(A.).

2730 U2*C3, ...ETUDE DE U2*C3 PAR DIFFRACTION DE NEUTRONS. A1,2. COMP. REND. 263, ... 457B, (1966).
      ..DE-NOVION(C.), KREBS(K.H.), MERIEL(P.).

2731 U*C2, ...THE CRYSTAL STRUCTURES OF U*C2. A1. ACTA CRYST. 21, ... 670, (1966).
      ..BOWMAN(A.L.), ARNOLD(G.P.), WITTEMAN(W.G.), WALLACE(T.C.),
      NERESON(N.G.).

2732 U*C, U*C2, U2*C3, . ..INTERATOMIC DISTANCES OF URANIUM CARBIDES. A1. ACTA CRYST. 16, ... 322, (1963).
      ..SHARMA(P.K.).

2733 U*C2, ...NEUTRON-DIFFRACTION STUDY OF U*C2 AT 300 -5 K. A1. J. CHEM. PHYS. VOL.47, ... 1188, (1967).
      ..ATOJI(M.).

2734 U*C2, ...STRUCTURES OF CALCIUM DICARBIDE AND URANIUM DICARBIDE BY NEUTRON A1. J. CHEM. PHYS. VOL.31, . . . 332, (1959).
      DIFFRACTION.
      ..ATOJI(M.), MEDRUD(R.C.).

2735 U*C2, ... A1,A4....REF. SECTION 3, 61/ATOJI, ...

2736 U*C,U*C2,U2*C3,. ..CARBON POSITIONS IN URANIUM CARBIDES. A1. ACTA CRYST. VOL.12, ... 159, (1959).
      ..AUSTIN(A.E.).

2737 U*CL4, ...A NEUTRON-DIFFRACTION STUDY OF ANHYDROUS URANIUM TETRACHLORIDE. A1. ACTA CRYST. VOL.B29, . . . 1942, (1973).
      ..TAYLOR(J.C.), WILSON(P.W.).

SECTION   5 -STRUCTURE AND CROSS-SECTION DETERMINATIONS . . . .

2738 U*CO*O4, ...A2. . . . . .J. APPL. PHYS. VOL.33. SUPPL. . : 1123, (1962).
                        *MAGNETIC STRUCTURE WORK AT THE NUCLEAR CENTER OF GRENOBLE.
                        *BERTAUT(E.F.), DELAPALME(A.), FORRAT(F.), ROULT(G.), DE-BERGEVIN(F.),
                        PAUTHENET(R.).

2739 U*CR*O4, ...A2. . . . . .BULL. SOC. FRANC. MIN. CRIST. 88 214, (1965).
                        *ATOMIC PARAMETERS AND MAGNETIC STRUCTURE OF U*CR*O4.
                        *BACHMANN(M.), BERTAUT(E.F.), BASSI(G.). (IN FRENCH).

2740 U*D3,. . . . . .A1. . . . . .J. AMER. CHEM. SOC. VOL.73. . . . 4172, (1951).
                        *HYDROGEN POSITIONS IN URANIUM HYDRIDE BY NEUTRON DIFFRACTION.
                        *RUNDLE(R.E.).

2741 U*F4,. . . . . .A1,2. . . . .COLL. INTER. N.126 (GRENOBLE), .   38, (1963).
                        *NEUTRON DIFFRACTION AND SCATTERING STUDIES AT J.A.E.R.I.
                        *KUNITOMI(N.), HAMAGUCHI(Y.), SAKAMOTO(M.), DOI(K.),   KOMURA(S.).

2742 U2*F9, ...A1. . . . . .BULL. SOC. FRANC. MIN. CRIST. 90 3C8, (1967).
                        *A NEUTRON DIFFRACTION STUDY OF THE CRYSTAL STRUCTURE OF U2*F9.
                        *LAVEISSIERE(J.). (IN FRENCH).

2743 U*F3,. . . . . .A1. . . . . .BULL. SOC. FRANC. MIN. CRIST. 90 3C4, (1967).
                        *A NEUTRON DIFFRACTION STUDY OF THE CRYSTAL STRUCTURE OF U*F3.
                        *LAVEISSIERE(J.). (IN FRENCH).

2744 U*F6,. . . . . .A1. . . . . .ACTA CRYST. VOL.B29 PART 1. . . .   7, (1973).
                        *THE STRUCTURES OF FLUORIDES.I.DEVIATIONS FROM IDEAL SYMMETRY IN THE
                        STRUCTURE OF CRYSTALLINE U*F6: A NEUTRON DIFFRACTION ANALYSIS.
                        *TAYLOR(J.C.), WILSON(P.W.), KELLY(J.W.).

2745 U*FE2, ...A2. . . .J. APP. PHYS. VOL.40. . . . . . 1133, (1969).
                        *MAGNETIC STRUCTURE AND SPIN-DENSITY DISTRIBUTION IN U-FE.
                        *YESSIK(M.).

2746 U*FE2, ...MAGNETIC PROPERTIES. . . A2. . . .J. PHYS. SOC. JAPAN VOL.17 B-III 46, (1962).
                        *MAGNETIC PROPERTIES AND STRUCTURES OF URANIUM-3D TRANSITION ELEMENT
                        ALLOYS.
                        *HAMAGUCHI(Y.), KOMURA(S.), KUNITOMI(N.), SAKAMOTO(M.).

2747 U*(FE-CR)*O4,. . . . . .A2. . . . . .J. PHYSIQUE TOME 32 COL.1 VOL.II 859, (1971).
                        STRUCTURE MAGNETIQUE DE LA SOLUTION SOLIDE U(FE-CR)*O4.
                        *WOLFERS(P.), BACHMANN(M.), BERTAUT(E.F.).

2748 U*FE*O4, ...A1. . . . . .COMPTES RENDUS 267, SERIE B, . . . 518, (1968).
                        *ETUDE PAR EFFET MOSSBAUER ET PAR DIFFRACTION NEUTRONIQUE DE U*FE*O4.
                        *BACHMANN(M.), CHEVALIER(R.), BERTAUT(E.F.), ROULT(G.), BELAKHOVSKY(M.).

2749 U*FE*O4, ...A2. . . . . .COMPTES RENDUS 266 SERIE B. . . . 45, (1968).
                        *ETUDE PAR DIFFRACTION NEUTRONIQUE ET MESURES MAGNETIQUES DE U*FE*O4
                        FERROMAGNETIQUE.
                        *BACHMANN(M.), BERTAUT(E.F.), BLAISE(A.).

2750 U*FE*O4, ...A2. . . . . .J. APP. PHYS. VOL.40. . . . . . 1131, (1969).
                        *MAGNETIC STRUCTURES AND PROPERTIES OF U*FE*O4.
                        *BACHMANN(M.), BERTAUT(E.F.), BLAISE(A.), CHEVALIER(R.), ROULT(G.).

2751 U*IN2, ...A2. . . . . .PHYS. STAT. SOLIDI A VOL.20, . . 395, (1973).
                        *MAGNETIC ORDERING IN URANIUM COMPOUNDS WITH AU*CU3-TYPE LATTICE.
                        *MURASIK(A.), LECIEJEWICZ(J.), LIGENZA(S.), MISIUK(A.).

2752 U2*IR*C2,. . . . . .A1. . . . . .ACTA CRYST. VOL.B27. . . . . . 1067, (1971).
                        *THE CRYSTAL STRUCTURE OF U2*IR*C2.
                        *BOWMAN(A.L.), ARNOLD(G.P.), KRIKORIAN(N.H.), ZACHARIASEN(W.H.).

2753 U*MN2, ...A1. . . . . .BULL. AMER. PHYS. SOC. SER.II 4, 184, (1959).
                        *NEUTRON DIFFRACTION STUDY OF U*MN2.
                        *IYENGAR(P.K.), BROCKHOUSE(B.N.). (ABSTRACT ONLY).

2754 U*MO*C2, ...A4. . . . . .J. APPL. PHYS. 41, ... 5080, (1970).
                        *ANISOTROPIC THERMAL EXPANSION OF REFRACTORY CARBIDES BY HIGH-TEMPERATURE
                        NEUTRON DIFFRACTION.
                        *BOWMAN(A.L.), ARNOLD(G.P.), KRIKORIAN(N.H.).

2755 U*(MONOCHALCOGENIDES), ...A2. . . . . .CONF./RARE EARTHS/ACTINIDES. . . 168, (1971).
                        *STUDY OF MAGNETIC ORDERING IN URANIUM MONOCHALCOGENIDES BY NEUTRON DIFF.
                        *WEDGWOOD(F.A.), KUZNIETZ(M.).

2756 U*N, ...A2. . . . . .PROC. PHYS. SOC. 86, ... 1193, (1965).
                        *AN INVESTIGATION OF THE MAGNETIC STRUCTURE OF URANIUM NITRIDE BY NEUTRON
                        DIFFRACTION.
                        *CURRY(N.A.).

2757 U*N, ...A1. . . . . .ACTA CRYST. VOL.11. . . . . . . 751, (1958).
                        *THE CRYSTAL STRUCTURE OF U*N BY NEUTRON DIFFRACTION.
                        *MUELLER(M.H.), KNOTT(H.W.).

2758 U2*N(3+X), ...A1. . . . . .PHYS. STAT. SOL. 21. . . . . . K11, (1967).
                        *STRUCTURE INVESTIGATIONS OF TWO URANIUM NITRIDES: U2*N(3+X).
                        *TOBISCH(J.), HASE(W.).

2759 U4*O9, ...A4. . . . . .J. PHYS. CHEM. SOLIDS VOL.32, . . 235, (1971).
                        *AN X-RAY AND NEUTRON DIFFRACTION STUDY ON A PHASE TRANSITION IN THE
                        U4*O9 PHASE.
                        *ISHII(T.), NAITO(K.), OSHIMA(K.).

2760 U*O2,. . . . . .A2. . . . . .PHYS. REV. 140. . . . . . . .A1448, (1965).
                        *NEUTRON-DIFFRACTION STUDY OF ANTIFERROMAGNETISM IN U*O2.
                        *FRAZER(B.C.), SHIRANE(G.), COX(D.E.), OLSEN(C.E.).

2761 U3*O8, ...A1. . . . . .ACTA CRYST. VOL.B25. . . . . . 2505, (1969).
                        *THE CRYSTAL STRUCTURE OF THE HIGH TEMPERATURE MODIFICATION OF U3*O8.
                        *HERAK(R.).

2762 U3*O8, ...A1. . . . . .ACTA CRYST. VOL.B26, ... 656, (1970).
                        *THE STRUCTURE OF BETA-U3*O8.
                        *LOOPSTRA(B.O.).

2763 U4*O9, ...A3. . . . . .ACTA CRYST B24, ... 1393, (1968).
                        *NEW SUPER LATTICE REFLECTIONS OF U4*O9.
                        *MASAKI(N.), DOI(K.).

2764 U3*O8, ...A1. . . . . .COLL. INTER. NO.126 (GRENOBLE),.   5, (1963).
                        *NEUTRON DIFFRACTION INVESTIGATION OF U3*O8.
                        *LOOPSTRA(B.O.).

SECTION   5 -STRUCTURE AND CROSS-SECTION DETERMINATIONS . . . .

2765 U4*09, ...ANALYSIS OF THE SUPERSTRUCTURE OF U4*09 BY NEUTRON DIFFRACTION.
A1. . . . . .ACTA CRYST. VOL.B28, ... 785, (1972).
MASAKI(N.), DOI(K.).

2766 U*03 (ALPHA),. . .THE STRUCTURE OF ALPHA-U*03 BY NEUTRON AND ELECTRON DIFFRACTION.
A1. . . . . .ACTA CRYST. VOL.B28, ... 3609, (1972).
GREAVES(C.), FENDER(B.E.F.).

2767 U*02,. . . . . . .NEUTRON DIFFRACTION STUDIES OF THE ACTINIDE OXIDES II. THERMAL MOTIONS
P. . . . . .PROC. ROY. SOC. 274, ... 134, (1963).
OF THE ATOMS IN U*02 AND TH*02 BETWEEN ROOM TEMPERATURE AND 1110 DEG. C.
WILLIS(B.T.M.).

2768 U*02,. . . . . . .NEUTRON DIFFRACTION STUDIES OF THE ACTINIDE OXIDES I. URANIUM DIOXIDE
P. . . . . .PROC. ROY. SOC. 274, ... 122, (1963).
AND THORIUM DIOXIDE AT ROOM TEMPERATURE.
WILLIS(B.T.M.).

2769 U3*08, ...NEUTRON DIFFRACTION INVESTIGATION OF U3*08.
A1. . . . . .ACTA CRYST. 17, ... 651, (1964).
LOOPSTRA(B.O.).

2770 U*03,. . . . . . .THE STRUCTURE OF BETA-U*03.
A1. . . . . .ACTA CRYST. 21, ... 589, (1966).
DEBETS(P.C.).

2771 U*02,. . . . . . .ANHARMONIC VIBRATION IN FLUORITE STRUCTURES.
T,P. . . . .PROC. ROY. SOC. 298, ... 289, (1967).
DAWSON(B.), HURLEY(A.C.), MASLEN(V.W.).

2772 U*02,. . . . . . .NEUTRON DIFFRACTION STUDY OF ANTIFERROMAGNETISM IN U*02.
A2. . . . . .PHYS. LETT. 17, ... 188, (1965).
WILLIS(B.T.M.), TAYLOR(R.I.).

2773 U*02,. . . . . . .A CRYSTAL FIELD CALCULATION IN URANIUM DIOXIDE.
T. . . . . . .J. PHYS. CHEM. SOL. 27, ... 1833, (1966).
RAHMAN(H.U.), RUNCIMAN(W.A.).

2774 U*02,. . . . . . .NEUTRON DIFFRACTION STUDIES OF NON-STOICHIOMETRIC COMPOUNDS.
A4. . . . . .PAPER NO.IIE-1 (2 PP.) . . , , (1966).
WILLIS(B.T.M.). (ED.NOTE* INTERNATIONAL CONFERENCE ON ELECTRON
DIFFRACTION AND/DEFECTS IN CRYSTALS, MELBOURNE, 1965; PERGAMON PRESS).

2775 U3*08, ...THE PHASE TRANSITION IN ALPHA-U3*08 AT 210 DEGREES C.
A1. . . . . .J. APPL. CRYST. VOL.3, ... 94, (1970).
LOOPSTRA(B.O.).

2776 U*02,. . . . . . .ANHARMONIC CONTRIBUTIONS TO THE DEBYE-WALLER FACTORS IN U*02.
A3. . . . . .ACTA CRYST. VOL.B24, ... 117, (1968).
ROUSE(K.O.), WILLIS(B.T.M.), PRYOR(A.W.).

2777 U3*08, ...THE STRUCTURE OF U3*08 DETERMINED BY NEUTRON DIFFRACTION.
A1. . . . . .ACTA CRYST. VOL.11, ... 612, (1958).
ANDRESEN(A.F.).

2778 U*0(2.13), . . .POSITIONS OF THE OXYGEN ATOMS IN U*0(2.13).
A1. . . . . .NATURE VOL.197, ... 755, (1963).
WILLIS(B.T.M.).

2779 U*02,. . . . . .ANTIFERROMAGNETIC ORDERING IN U*02.
A2. . . . . .BULL. AMER. PHYS. SOC. SER.II 2, 9, (1957).
HENSHAW(D.G.), BROCKHOUSE(B.N.). (ABSTRACT ONLY).

2780 U4*09, ...THE TRANSITION OF U4*09 AT ABOUT 70 DEGREES C.
A1. . . . . .J. APPL. CRYST. VOL.7, ... 247, (1974).
MASAKI(N.).

2781 U*02-(CA,TH,ZR)*02,. . .NEUTRON DIFFRACTION STUDIES OF DILUTE MAGNETIC SYSTEMS.
A2,A4. . . .ACTA CRYST. VOL.A28 PART S-4,. . 176, (1972).
SABINE(T.M.). (ABSTRACT ONLY).

2782 U*02*CL2,. . . . .THE STRUCTURE OF ANHYDROUS URANYL CHLORIDE BY POWDER NEUTRON DIFFRACTION
A1. . . . . .ACTA CRYST. VOL.B29, ... 1073, (1973).
TAYLOR(J.C.), WILSON(P.W.).

2783 U*0*CL2, ...THE STRUCTURE OF U*0*CL2 BY NEUTRON DIFFRACTION.
A1. . . . . .ACTA CRYST. VOL.B30, ... 175, (1974).
TAYLOR(J.C.), WILSON(P.W.).

2784 U*02*CL2.(H2*0,D2*0),. . .THE STRUCTURE OF URANYL CHLORIDE MONOHYDRATE BY NEUTRON DIFFRACTION
A1. . . . . .ACTA CRYST. VOL.B30, ... 169, (1974).
AND THE DISORDER OF THE WATER MOLECULE.
TAYLOR(J.C.), WILSON(P.W.).

2785 U*02*F2, ...THE CRYSTAL STRUCTURE OF ANHYDROUS U*02*F2.
A1. . . . . .ACTA CRYST. VOL.B26, ... 1540, (1970).
ATOJI(M.), MCDERMOTT(M.J.).

2786 U*03*(H2*0), . .THE HYDROGEN ATOM LOCATIONS IN THE ALPHA AND BETA FORMS OF URANYL
A1. . . . . .ACTA CRYST. VOL.B27, ... 2018, (1971).
HYDROXIDE.
TAYLOR(J.C.), HURST(H.J.).

2787 (U*02)*(N*03)2*6(H2*0),. . .A NEUTRON DIFFRACTION STUDY OF URANYL NITRATE HEXAHYDRATE.
A1. . . . . .ACTA CRYST. 19, ... 536, (1965).
TAYLOR(J.C.), MUELLER(M.H.).

2788 U*02*(0*H)2 (BETA),. . .A NEUTRON DIFFRACTION STUDY OF THE ANISOTROPIC THERMAL EXPANSION OF
A1,A4. . . .ACTA CRYST. VOL.B28, ... 2995, (1972).
BETA-URANYL DIHYDROXIDE.
TAYLOR(J.C.), BANNISTER(M.J.).

2789 U*0*S, ...MAGNETIC STRUCTURE INVESTIGATIONS AT THE NUCLEAR CENTER.
A2. . . . . .J. APP. PHYS. 34, ... 1333, (1963).
BALLESTRACCI(R.), BERTAUT(E.F.), COING-BOYAT(J.), DELAPALME(A.),
JAMES(W.J.), LEMAIRE(R.).

2790 U*0*S, ...STRUCTURE ET PROPRIETES MAGNETIQUES DE L*OXYSULFURE D*URANIUM U*0*S.
A1,2. . . . .J. PHYS. CHEM. SOL. 24, ... 487, (1963).
BALLESTRACCI(R.), BERTAUT(E.F.), PAUTHENET(R.).

2791 U*0*SE,. . . . .THE MAGNETIC PROPERTIES OF URANIUM OXYSELENIDE.
A1,2. . . . .PHYS. STAT. SOL. 30, ... 61, (1968).
MURASIK(A.), SUSKI(W.), TROC(R.), LECIEJEWICZ(J.).

SECTION 5 -STRUCTURE AND CROSS-SECTION DETERMINATIONS . . . .

2792 U*O*TE,. . . . .A2 MEAS.....PHYS. STAT. SOL. 34,...... K157, (1969).
*NEUTRON DIFFRACTION MEASUREMENTS ON ANTIFERROMAGNETIC U*O*TE AT
4.2 DEG. K.
*MURASIK(A.), SUSKI(W.), LECIEJEWICZ(J.).

2793 U*O2-TH*O2-BE*O, ...A4......PROC. AUSTRAL. AT. ENERGY SYMP., 168, (1958).
*NEUTRON-DIFFRACTION STUDIES IN SOME SINTERED U*O2-TH*O2-BE*O COMPACTS.
*REEVE(K.D.), SABINE(T.M.).

2794 U*O2,U4*O9,. . . ..A1.....COLL. INTER. N.126 (GRENOBLE), . . 7, (1963).
*STRUCTURES OF U*O2, U*O(2+X) AND U4*O9 BY NEUTRON DIFFRACTION.
*WILLIS(B.T.M.).

2795 U*O2,U*O(2+X), ...A1,A4....PROC. BR. CERAMIC SOC. NO.1, .. 9, (1964).
*POINT DEFECTS IN URANIUM OXIDES.
*WILLIS(B.T.M.).

2796 U3*P4, ...A2.......PHYS. STAT. SOL. 10,. . . . . . K85, (1965).
*NEUTRON DIFFRACTION STUDY OF FERROMAGNETIC U3*P4.
*CISZEWSKI(R.), MURASIK(A.), TROC(R.).

2797 U*P, ...A2.......DISS. ABS. VOL.32,...... 507B, (1971).
*TEMPERATURE DEPENDENT ELECTROSTATIC INTERACTIONS BETWEEN MAGNETIC IONS
IN URANIUM MONOPHOSPHIDE.
*LONG(C.E.).

2798 U3*P4, ...T,A2.....PHYS. STAT. SOL. VOL.42, ... K15, (1970).
*NOTE ON THE MAGNETIC STRUCTURE OF U3*P4.
*DOBRZYNSKI(L.), PRZYSTAWA(J.).

2799 U*P, ...A2.......J. PHYS. CHEM. SOLIDS 30,. . . . . 453, (1969).
*NEUTRON DIFFRACTION STUDY OF THE LOW TEMPERATURE TRANSITION IN U*P.
*HEATON(L.), MUELLER(M.H.), ANDERSON(K.D.), ZANBERIS(D.D.).

280U U*P2,. . . . .A2.......PHYS. STAT. SOL. 15,. . . . . 515, (1966).
*ANTIFERROMAGNETIC STRUCTURE OF URANIUM DIPHOSPHIDE.
*TROC(R.), LECIEJEWICZ(J.), CISZEWSKI(R.).

2801 U*P, ...A2......PROC. PHYS. SOC. 89,. . . . . 427, (1966).
*THE MAGNETIC STRUCTURE OF URANIUM MONOPHOSPHIDE.
*CURRY(N.A.).

2802 U*P, ...A2.......J. PHYS. CHEM. SOLIDS VOL.27,. . 1197, (1966).
*THE ANTIFERROMAGNETISM OF URANIUM MONOPHOSPHIDE.
*SIDHU(S.S.), VOGELSANG(W.), ANDERSON(K.D.).

2803 U3*P4, ...A2.......PHYS. STAT. SOLIDI A VOL.19 NO.1 K89, (1973).
*A NOTE ON THE MAGNETIC STRUCTURE OF U3*P4.
*MURASIK(A.), LIGENZA(S.), LECIEJEWICZ(J.), TROC(R.).

2804 U*(P-S), ...A2.......PHYS. REV. VOL.188,. . . . . 963, (1969).
NEUTRON-DIFFRACTION STUDY OF MAGNETIC ORDERING IN U(P-S).
*LANDER(G.H.), KUZNIETZ(M.), COX(D.E.).

2805 U*(P-S), ...A2.......SOL. STATE COMM. VOL.6,. . . . . 877, (1968).
MAGNETIC STRUCTURES IN U(P(.95)-S(.05)).
*LANDER(G.H.), KUZNIETZ(M.), BASKIN(Y.).

2806 U*(P-S), ...A2.......J. APP. PHYS. VOL.40,. . . . . . 1130, (1969).
MAGNETIC PHASE DIAGRAM OF THE U(P-S) SYSTEM.
*KUZNIETZ(M.), LANDER(G.H.), BASKIN(Y.).

2807 U*PB3, ...A2.......PHYS. STAT. SOLIDI A VOL.13 NO.1 K79, (1972).
*MAGNETIC STRUCTURE OF U*PB3.
*LECIEJEWICZ(J.), MISIUK(A.).

2808 U2*PT*C2,. . . . .A4.......J. APPL. PHYS. 41,. . . . . . 5080, (1970).
*ANISOTROPIC THERMAL EXPANSION OF REFRACTORY CARBIDES BY HIGH-TEMPERATURE
NEUTRON DIFFRACTION.
*BOWMAN(A.L.), ARNOLD(G.P.), KRIKORIAN(N.H.).

2809 U*S, ...A2.......J. PHYS. C VOL.5,. . . . . 2427, (1972).
*MAGNETIC MOMENT DISTRIBUTION IN URANIUM MONOSULPHIDE.
*WEDGWOOD(F.A.).

2810 U*(S,SE,TE), ...A2.......J. PHYS. C VOL.5,. . . . . 3012, (1972).
*ACTINIDE PNICTIDES AND CHALCOGENIDES! I. STUDY OF MAGNETIC ORDERING AND
ORDERED MOMENTS IN URANIUM MONOCHALCOGENIDES BY NEUTRON DIFFRACTION.
*WEDGWOOD(F.A.), KUZNIETZ(M.).

2811 U*SB,. . . . .A1,2.....PHYS. STAT. SOL. 30,. . . . . 157, (1968).
*THE ANTIFERROMAGNETIC ORDERING OF URANIUM MONOARSENIDE AND URANIUM
MONOANTIMONIDE.
*LECIEJEWICZ(J.), MURASIK(A.), TROC(R.).

2812 U*SB,. . . . .A2.......J. PHYS. CHEM. SOLIDS 30,. . . . 1642, (1969).
*ANTIFERROMAGNETIC STRUCTURES OF U*SB AND U*BI.
*KUZNIETZ(M.), LANDER(G.H.), CAMPOS(F.P.).

2813 U*SB2, ...A2.......PHYS. STAT. SOL. 22,. . . . 517, (1967).
*NEUTRON-DIFFRACTION STUDY OF ANTIFERROMAGNETISM IN U*SB2 AND U*BI2.
*LECIEJEWICZ(J.), TROC(R.), MURASIK(A.), ZYGMUNT(A.).

2814 U*SB,U3*SB4, ...A1.......J. APP. PHYS. VOL.40,. . . . . 1135, (1969).
*NEUTRON-DIFFRACTION INVESTIGATIONS ON U*SB, U*BI AND U3*SB4.
*OLSEN(C.E.), KOEHLER(W.C.).

2815 U3*SE4,. . . . .A1.......ACTA PHYS. POLON. VOL.A43, . . . 631, (1973).
*NEUTRON DIFFRACTION STUDY OF U3*SE4.
*SZYTULA(A.), SUSKI(W.).

2816 U*TE,. . . . .A2.......J. PHYS. C VOL.6,. . . . . 1652, (1973).
*ACTINIDE PNICTIDES AND CHALCOGENIDES! II.TELLURIUM HYPERFINE FIELD IN
U*TE FROM MOSSBAUER SPECTROSCOPY AND ITS RELATION TO U ORDERED MOMENT.
*LONGWORTH(G.), WEDGWOOD(F.A.), KUZNIETZ(M.).

2817 U*TL3, ...A2.......PHYS. STAT. SOLIDI A VOL.20,. . 395, (1973).
*MAGNETIC ORDERING IN URANIUM COMPOUNDS WITH AU*CU3-TYPE LATTICE.
*MURASIK(A.), LECIEJEWICZ(J.), LIGENZA(S.), MISIUK(A.).

2818 UREA,. . . . .A4.......ACTA CRYST. VOL.A26,. . . . . 543, (1970).
*COLLECTION AND INTERPRETATION OF NEUTRON DIFFRACTION MEASUREMENTS
ON UREA.
*PRYOR(A.W.), SANGER(P.L.).

SECTION 5 -STRUCTURE AND CROSS-SECTION DETERMINATIONS . . . .

2819 UREA,. . . . . *THE POSITIONS OF HYDROGEN ATOMS IN UREA BY NEUTRON DIFFRACTION.
*A1. . . . . . ACTA CRYST. VOL.10, ... 319, (1957).
*WORSHAM-JR(J.E.), LEVY(H.A.), PETERSON(S.W.).

2820 V,. . . . . *SPIN DEPENDENCE OF THE NEUTRON-CAPTURE CROSS SECTION OF VANADIUM AND
*P. . . . . . . PHYS. REV. B-3, ... 128, (1971).
THE INTERNAL MAGNETIC FIELD AT VANADIUM NUCLEI.
*POSTMA(H.), VANNESTE(L.), SAILOR(V.L.).

2821 V,. . . . . *KEY NEUTRON CAPTURE AND TRANSMISSION MEASUREMENTS ON CR(ISOTOPES A=50,
*P. . . . . . . THESIS (231PP.) . . . . (1970).
52, 53, 54), NI(ISOTOPE A=60), AND V.
*STIEGLITZ(R.G.). (ED. NOTE: AVAIL. UNIV. MICROFILMS, ANN ARBOR, MICH.
ORDER NO. 71-12829).

2822 V (ISO. A=51),. . *THERMAL NEUTRON CAPTURE CROSS-SECTION MEASUREMENTS FOR NA(ISOTOPE A=23),
*P. . . . . . . J. NUCL. ENERGY VOL.24,. . . . 419, (1970).
AL(ISOTOPE A=27), CL(ISOTOPE A=37) AND V(ISOTOPE A=51).
*RYVES(T.B.), PERKINS(D.R.).

2823 V,. . . . . *NEUTRON ELASTIC SCATTERING CROSS SECTIONS OF VANADIUM, CHROMIUM, IRON
*A4. . . . . 2ND INTERNAT. CONF./NUCL. DATA/. . . . (1970).
AND NICKEL.
*HOLMQVIST(B.), WIEDLING(T.).

2824 V,. . . . . *NEUTRON-CAPTURE CROSS-SECTIONS OF V, MN, CS, DY AND LU IN THE ENERGY
*P. . . . . . . HELV. PHYS. ACTA VOL.45, . . . 46, (1972).
RANGE FROM 0.01 TO 100 EV.
*WIDDER(F.).

2825 V,. . . . . *ANALYSIS AND INTERPRETATION OF THE NEUTRON CROSS-SECTION OF VANADIUM
*P. . . . . . . PROC. PHYS. SOC. 82, ... 477, (1963).
BELOW 25 KEV.
*FIRK(F.W.K.), LYNN(J.W.), MOXON(M.C.).

2826 V,. . . . . *NEUTRON SPIN-NEUTRON ORBIT INTERACTION WITH SLOW NEUTRONS.
*A4. . . . . . PHYS. REV. LETT. VOL.10, . . . 297, (1963).
*SHULL(C.G.).

2827 V,. . . . . *TOTAL CROSS SECTIONS OF TI, V, Y, TA, AND W.
*P. . . . . . . NUCL. SCI. ENGNG. VOL.7, . . . 193, (1960).
*SCHMUNK(R.E.), RANDOLPH(P.D.), BRUGGER(R.M.).

2828 V,. . . . . *NEUTRON DIFFRACTION STUDIES OF VARIOUS TRANSITION ELEMENTS.
*A2. . . . . . REV. MOD. PHYS. VOL.25, . . . 100, (1953).
*SHULL(C.G.), WILKINSON(M.K.).

2829 V,. . . . . *COHERENT NEUTRON SCATTERING CROSS SECTION OF V(A=51).
*P. . . . . . . PHYS. REV. VOL.83, ... 171, (1951).
*MCREYNOLDS(A.W.), WEISS(R.J.).

2830 V (V*N,V*C),. . *THE COHERENT NEUTRON SCATTERING CROSS SECTIONS OF NITROGEN AND VANADIUM.
*P. . . . . . . PHYS. REV. VOL.87, ... 462, (1952).
*PETERSON(S.W.), LEVY(H.A.).

2831 V,. . . . . *NEUTRON TOTAL CROSS SECTION OF SC, V, CU AND RH AT MICRO-EV ENERGIES.
*P. . . . . . . Z. PHYSIK VOL.266, ... 157, (1974).
*DILG(W.), MANNHART(W.).

2832 V8*C7,. . . . . *A NEUTRON DIFFRACTION INVESTIGATION OF V8*C7.
*A1. . . . . . ACTA CRYST. VOL.B26, ... 1882, (1970).
*HENFREY(A.W.), FENDER(B.E.F.).

2833 V2*C,. . . . . . *THE CRYSTAL STRUCTURES OF V2*C AND TA2*C.
*A1. . . . . . ACTA CRYST. 19,. . . . . . . 6, (1965).
*BOWMAN(A.L.), WALLACE(T.C.), YARNELL(J.L.), WENZEL(R.G.), STORMS(E.K.).

2834 V2*CA*O4,. . . . *REPRESENTATION ANALYSIS OF MAGNETIC STRUCTURES.
*A2,T. . . . . ACTA CRYST A24,. . . . . . 217, (1968).
*BERTAUT(E.F.).

2835 V2*D,. . . . . . *ABNORMAL PHASE TRANSITION IN THE VANADIUM DEUTERIDE, V2*D.
*A4. . . . . . SOV. PHYS. SOL. STATE VOL. 13, . 2178, (1972).
*SOMENKOV(V.A.), ENTIN(I.R.), CHERVYAKOV(A.YU.), SHIL≠SHTEIN(S.SH.),
CHERTKOV(A.A.).

2836 V-D,. . . . . . *INTERSTITIAL SUPERSTRUCTURES OF VANADIUM DEUTERIDES.
*A1. . . . . . PHYS. STAT. SOLIDI A VOL.15 NO.1 267, (1973).
*ASANO(H.), HIRABAYASHI(M.).

2837 V-D(0.8),. . . . *ORDER-DISORDER TRANSITION IN V-D(0.8).
*A4. . . . . . SOV. PHYS. SOL. STATE VOL. 13, . 2172, (1972).
*CHERVYAKOV(A.YU.), ENTIN(I.R.), SOMENKOV(V.A.), SHIL≠SHTEIN(S.SH.),
CHERTKOV(A.A.).

2838 V2*D,V4*D3,. . *STRUCTURAL TRANSITIONS AT LOW TEMPERATURES IN VANADIUM DEUTERIDES.
*A1. . . . . . J. APPL. CRYST. VOL.6, . . . 206, (1973).
*WESTLAKE(D.G.), MUELLER(M.H.), KNOTT(H.W.).

2839 V*F3,. . . . . . *ANTIFERROMAGNETIC PROPERTIES OF THE IRON GROUP TRIFLUORIDES.
*A2. . . . . . PHYS. REV. VOL.112, ... 1132, (1958).
*WOLLAN(E.O.), CHILD(H.R.), KOEHLER(W.C.), WILKINSON(M.K.).

2840 V*F2,. . . . . . *NEUTRON DIFFRACTION STUDY OF V*F2.
*A2. . . . . . J. APPL. PHYS. VOL.40, . . . 1136, (1969).
*LAU(H.Y.), STOUT(J.W.), KOEHLER(W.C.), CHILD(H.R.).

2841 V-FE,. . . . . . *ATOMIC MAGNETIC MOMENTS IN B2 TRANSITION ALLOYS I.ALLOY V-FE.
*A2. . . . . . J. PHYS. SOC. JAPAN VOL.17 B-III 16, (1962).
*CHANDROSS(R.J.), SHOEMAKER(D.P.).

2842 V-FE,. . . . . . *POLARIZED NEUTRON DIFFRACTION STUDY OF ORDERED V-FE.
*A2. . . . . . J. PHYS. CHEM. SOLIDS 29, . . 184, (1968).
*FEINSTEIN(L.G.), SHOEMAKER(D.P.).

2843 V-FE,V-NI,. . . *ORDERING OF ATOMS IN THE SIGMA PHASE.
*A1. . . . . . ACTA CRYST. VOL.9, ... 289, (1956).
*KASPER(J.S.), WATERSTRAT(R.M.).

2844 (V-MN)2-GA5,. . *NEUTRON DIFFRACTION, MAGNETIC AND SUPERCONDUCTING MEASUREMENTS ON
*A2. . . . . . J. APPL. CRYST. VOL.6, . . . 240, (1973).
VANADIUM-MANGANESE-GALLIUM ALLOYS.
*KITCHINGMAN(W.J.), NORMAN(P.L.).

SECTION   5 -STRUCTURE AND CROSS-SECTION DETERMINATIONS . . . .

2845 V2*MN*O4,. . . . .
  *A NEUTRON DIFFRACTION STUDY OF THE INFLUENCE OF TEMPERATURE ON SPIN
  CONFIGURATIONS IN SPINEL MN*V2*O4.
  *PLUMIER(R.).
  A4 . . . . .PROC. INT. CONF. (NOTTINGHAM), . . 295, (1964).

2846 V*N, ...
  *NEUTRON-DIFFRACTION STUDY OF VANADIUM NITRIDE.
  *VINTAIKIN(E.Z.), DMITRIEV(V.B.), TOMILIN(I.A.), SHCHURIK(A.G.).
  A1 . . . . . . .SOV. PHYS. DOKLADY VOL.15, . . . 776, (1971).

2847 V2*O3, ...
  *ANTIFERROMAGNETISM IN V2*O3.
  *MOON(R.M.).
  A2 . . . . . . .PHYS. REV. LETT. VOL.25, ... 527, (1970).

2848 V2*O3, ...
  *NEUTRON-DIFFRACTION DATA ON TI2*O3 AND V2*O3.
  *KENDRICK(H.), ARROTT(A.), WERNER(S.A.).
  A2 . . . . . .J. APP. PHYS. 39, ... 585, (1968).

2849 V2*O3, ...
  *NEUTRON DIFFRACTION AND OTHER STUDIES OF MAGNETIC ORDERING IN PHASES
  BASED ON CR2*O3, V2*O3, AND TI2*O3.
  *REID(A.F.), SABINE(T.M.), WHEELER(D.A.).
  A2 . . . . . .J. SOLID STATE CHEM. VOL.4, . . 400, (1972).

2850 V-O, ...
  *THE LOCATION OF OXYGEN ATOMS IN VANADIUM-OXYGEN ALLOYS BY MEANS OF
  NEUTRON DIFFRACTION.
  *TUCKER-JR(C.W.), SEYBOLT(A.U.), SUMSION(H.T.).
  A1 . . . . . . .ACTA MET. VOL.1, ... 390, (1953).

2851 V2*O3, ...
  *THE ELASTIC MAGNETIC NEUTRON CROSS-SECTION FOR 3D TRANSITION COMPOUNDS:
  CALCULATIONS FOR V2*O3, FE*F2 AND FE*C*O3.
  *BALCAR(E.), LOVESEY(S.W.), WEDGWOOD(F.A.).
  P,T. . . . . . .J. PHYS. C VOL.6, ... 3746, (1973).

2852 V2*O3, ...
  *STUDY OF RHOMBOHEDRAL V2*O3 BY NEUTRON DIFFRACTION.
  *PAOLETTI(A.), PICKART(S.J.).
  A1 . . . . . .J. CHEM. PHYS. VOL.32, ... 308, (1960).

2853 V2*O4*(MN,CO), .
  *MAGNETIC PROPERTIES OF V+++ IONS IN CUBIC SPINELS.
  *DWIGHT(K.), MENYUK(N.), ROGERS(D.B.), WOLD(A.).
  A4 . . . . . .PROC. INT. CONF. (NOTTINGHAM), . 538, (1964).

2854 V3*SI, ...
  *SEARCH FOR ANTIFERROMAGNETISM IN THE SILICIDES V3*SI, CR3*SI, AND MO3*SI
  *KOEHLER(W.C.), WOLLAN(E.O.).
  A2 . . . . . .PHYS. REV. VOL.95, ... 280, (1954).

2855 V3*SI, ...
  *NEUTRON-DIFFRACTION STUDIES OF ELECTRON-SPIN PAIRING IN SUPERCONDUCTING
  V3*SI.
  *SHULL(C.G.), WEDGWOOD(F.A.).
  D4 . . . . . .PHYS. REV. LETT. VOL.16, . . . 513, (1966).

2856 V*TA*O4, ...
  *ATOMIC AND MOLECULAR ORDERING IN ME*TA*O4 (ME=TI,V,CR,FE) WITH A RUTILE
  STRUCTURE.
  *ASTROV(D.N.), KRYUKOVA(N.A.), ZORIN(R.B.) MAKAROV(V.A.), OZEROV(R.P.),
  ROZHDESTVENSKII(F.A.), SMIRNOV(V.P.), TURCHANINOV(A.M.), FADEEVA(N.V.).
  A1 . . . . . .SOV. PHYS. CRYST. VOL.17, . . 1017, (1973).

2857 VITAMIN C, . . .
  *THE CRYSTAL STRUCTURE OF L-ASCORBIC ACID, ≠VITAMIN C≠. II. THE NEUTRON
  DIFFRACTION ANALYSIS.
  *HVOSLEF(J.).
  A1 . . . . . .ACTA CRYST. B24, ... 1431, (1968).

2858 VITAMIN B12(ACID DERIVATIVE),. . . . .
  *CRYSTAL AND MOLECULAR STRUCTURE FROM NEUTRON DIFFRACTION ANALYSIS.
  *MOORE(F.M.), WILLIS(B.T.M.), HODGKIN(D.C.).
  A1 . . . . . .NATURE VOL.214, ... 130, (1967).

2859 W. . . . . . .
  *ERHOLUNG VON STRAHLENSCHADEN IN KUBISCH-RAUMZENTRIERTEN METALLEN NACH
  NEUTRONEN BESTRAHLUNG BEI 4.5 DEG. K.
  *BURGER(G.), ISEBECK(K.), KERLER(R.), VOLKL(J.), WENZL(H.).
  A1 . . . . . .PHYS. LETT. 20, ... 470, (1966).

2860 W, ...
  *MESSUNG KOHARENTER STREULANGEN DURCH KLEINWINKELSTREUUNG VON NEUTRONEN
  IM SCHWERKRAFT-REFRAKTOMETER.
  *KOESTER(L.), UNGERER(H.).
  P. . . . . . .Z. PHYSIK VOL.219, ... 300, (1969).

2861 W2*C,. . . . .
  *NEUTRON-DIFFRACTION STUDY OF W2*C RHOMBIC MODIFICATION.
  *NOZIK(YU.Z.), LIPIN(YU.V.), KUVALDIN(B.V.).
  A4 . . . . . .LATV. PSR. FIZ. TEHN. SER. 6, . 30, (1968).

2862 W*C, ...
  *A NOTE ON THE STRUCTURE OF TUNGSTEN CARBIDE.
  *LECIEJEWICZ(J.).
  A1 . . . . . .ACTA CRYST. VOL.14, ... 200, (1961).

2863 W*C, ...
  *NEUTRON DIFFRACTION STUDIES IN SOLID STATE PHYSICS.
  *BLINOWSKI(K.).
  A1,A2 . . . . .NUKLEONIKA VOL.5, ... 414, (1960).

2864 W*C-TI*C,. . . .
  *TRANSMISSION OF MONOENERGETIC SLOW NEUTRONS THROUGH SOLID SOLUTIONS
  AND MECHANICAL MIXTURES OF TI*C AND W*C.
  *SIDHU(S.S.).
  P. . . . . . .J. APPL. PHYS. VOL.19, ... 639, (1948).

2865 W*CA*O4, ...
  *NEUTRON DIFFRACTION STUDIES AT THE PUERTO RICO NUCLEAR CENTER.
  *ALMODOVAR(I.), BIELEN(H.J.), FRAZER(B.C.), KAY(M.I.).
  A1,2 . . . . .COLL. INTER. N.126 (GRENOBLE), . 18, (1963).

2866 W*CL6, ...
  *THE STRUCTURE OF BETA-TUNGSTEN HEXACHLORIDE BY POWDER NEUTRON AND X-RAY
  DIFFRACTION.
  *TAYLOR(J.C.), WILSON(P.W.).
  A1 . . . . . .ACTA CRYST. VOL.B30, ... 1216, (1974).

2867 W*O3,. . . . .
  *FURTHER REFINEMENT OF THE STRUCTURE OF W*O3.
  *LOOPSTRA(B.O.), RIETVELD(H.M.).
  A1 . . . . . .ACTA CRYST. B25, ... 1420, (1969).

2868 W*O3,. . . . .
  *NEUTRON DIFFRACTION INVESTIGATION OF W*O3.
  *LOOPSTRA(B.O.), BOLDRINI(P.).
  A1 . . . . . .ACTA CRYST. 21, ... 158, (1966).

2869 W,Y, ...
  *TOTAL CROSS SECTIONS OF TI, V, Y, TA, AND W.
  *SCHMUNK(R.E.), RANDOLPH(P.D.), BRUGGER(R.M.).
  P. . . . . . .NUCL. SCI. ENGNG. VOL.7, . . . 193, (1960).

2870 XE,. . . . . .
  *COHERENT NEUTRON SCATTERING AMPLITUDES OF KRYPTON AND XENON, AND THE
  ELECTRON-NEUTRON INTERACTION.
  *CROUCH(M.F.), KROHN(V.E.), RINGO(G.R.).
  P. . . . . . .PHYS. REV. VOL.102, . . . 1321, (1956).

SECTION   5 -STRUCTURE AND CROSS-SECTION DETERMINATIONS . . . .

2871 XE*F4, ...A1. . . . . .SCIENCE VOL.139, ...1208, (1963).
      *XENON TETRAFLUORIDE MOLECULE AND ITS THERMAL MOTION: A NEUTRON
      DIFFRACTION STUDY.
      *BURNS(J.H.), AGRON(P.A.), LEVY(H.A.).

2872 XE*F2,XE*F4, ...A1. . . . . .COLL. INTER. N.126 (GRENOBLE), . . 45, (1963).
      *RECENT CRYSTAL STRUCTURE DETERMINATIONS BY NEUTRON DIFFRACTION AT
      OAK RIDGE.
      *BROWN(G.M.), LEVY(H.A.).

2873 XE*O2*F2,. . . . . .A1. . . . . .J. CHEM. PHYS. VOL.59, ... 453, (1973).
      *SYMMETRY AND STRUCTURE OF XE*O2*F2 BY NEUTRON DIFFRACTION.
      *PETERSON(S.W.), WILLETT(R.D.), HUSTON(J.L.).

2874 Y, ...A4. . . . . .BU. ACAD. SCI. USSR PHYS. SR. 32 579, (1968).
      *ELASTIC SCATTERING OF NEUTRONS ON YTTRIUM.
      *POPOV(V.I.), SLUCHEVSKAYA(V.M.), TRYKOVA(V.I.).

2875 Y. . . . . .P. . . . . . .ACTA CRYST., 19, ...679, (1965).
      *THE SCATTERING LENGTH OF YTTRIUM FOR THERMAL NEUTRONS.
      *PATON(M.G.), MASLEN(E.N.).

2876 Y, ...P. . . . . . .SOV. PHYS. CRYST. VOL.7, ...767, (1963).
      *COHERENT-SCATTERING AMPLITUDE FOR THERMAL NEUTRONS AND YTTRIUM NUCLEI.
      *KUZ#MINOV(YU.S.), YAMZIN(I.I.), MAL#TSEV(E.I.), BELOV(N.V.).

2877 Y*C2, ...A1,A4. . .REF. SECTION 3, 61/ATOJI, ...

2878 Y*C*O3,. . . . . .A2,A4. . .REF. SECTION 3, 66/BERTAUT, ...

2879 (Y-CA)3-(FE-ZR)2-FE3*O12, ...A2. . . . . .SOV. PHYS. JETP LETT. VOL.18,. . 47, (1973).
      *PHASE TRANSITIONS IN DILUTE FERRIMAGNETS AND THE PROBLEM OF LIQUID
      INFILTRATION.
      *PLAKHTII(V.P.), GOLOSOVSKII(I.V.), KUDRYASHEV(V.A.), PARFENOVA(N.N.),
      SMIRNOV(O.P.).

2880 (Y*CA2)*(ZR2)*(FE-GA)*O12, ...A2. . . . . .PHYS. STAT. SOL. B VOL.53 NO.1, . K37, (1972).
      *NEUTRON DIFFRACTION STUDY OF A GARNET WITH ONLY TETRAHEDRAL MAGNETIC
      SUBLATTICE.
      *PLAKHTII(V.P.), GOLOSOVSKII(I.V.).

2881 Y-CO,. . . . . .A2. . . . . .PHYS. REV. VOL.186,. . . . . . . .479, (1969).
      *POLARIZED-NEUTRON-DIFFRACTION STUDY OF MAGNETIC MOMENTS IN YTTRIUM-
      COBALT ALLOYS.
      *KREN(E.), SCHWEIZER(J.), TASSET(F.).

2882 Y-CO5, ...A2. . . . . .C.R. ACAD. SCI. VOL.255, ...896, (1962).
      *MAGNETIC STRUCTURE OF THE ALLOYS Y-CO5 AND HO-CO5.
      *JAMES(W.), LEMAIRE(R.), BERTAUT(E.F.). (IN FRENCH).

2883 Y-CO,. . . . . .A1. . . . . .J. APP. PHYS. 40, ...1454, (1969).
      *STUDY OF YTTRIUM-COBALT ALLOYS BY POLARIZED-NEUTRON DIFFRACTION.
      *KREN(E.), SCHWEIZER(J.), TASSET(F.).

2884 Y-CO5, ...A2. . . . . .J. APP. PHYS. 34,. . . . . 1333, (1963).
      *MAGNETIC STRUCTURE INVESTIGATIONS AT THE NUCLEAR CENTRE.
      *BALLESTRACCI(R.), BERTAUT(E.F.), COING-BOYAT(J.), DELAPALME(A.),
      JAMES(W.J.), LEMAIRE(R.).

2885 Y3-CO(2.5)-GE(2.5)*O12, ...A4. . . . . .LATV. PSR..FIZ. TEHN. SER. NO.3 28, (1970).
      *NEUTRON-DIFFRACTION STUDY OF Y3-CO(2.5)-GE(2.5)*O12 GARNET.
      *LIPIN(YU.), NOZIK(YU.).

2886 Y-DY,. . . . . .A2. . . . . .REF. SECTION 3, 65/CHILD, ...

2887 Y-ER,. . . . . .A2. . . . . .REF. SECTION 3, 65/CHILD, ...

2888 Y3*(FE-AL)5*O12, ...A1,A2. . . .HELV. PHYS. ACTA VOL.46, ...429, (1973).
      *STUDY OF THE NEUTRON DIFFRACTION OF THE NEEL FERRIMAGNET SYSTEM ...
      *FISCHER(P.), HALG(W.), CZERLINSKY(E.R.).

2889 Y*(FE-CO)*O3,Y*(FE-CR)*O3, ...A2. . . . . .A.I.P. CONF. PROC. NO.10 PART 2, 1603, (1973).
      *STUDY OF THE SPIN REORIENTATION IN CO- AND CR-SUBSTITUTED Y*FE*O3.
      *KREN(E.), PARDAVI(M.), POKO(Z.), SVAB(E.), ZSOLDOS(E.).

2890 Y-FE-GA GARNET,. . . . . .A4. . . . . .ACTA. CRYST. 21, ...765, (1966).
      *X-RAY AND NEUTRON DIFFRACTION STUDY OF THE SUBSTITUTIONAL DISORDER OF
      THE YTTRIUM-IRON-GALLIUM GARNETS.
      *FISCHER(P.), HALG(W.), STOLL(E.).

2891 Y-FE/GARNET, ...A2. . . . . .J. APPL. PHYS. VOL.36, ...1845, (1965).
      *SUBLATTICE MAGNETIZATIONS OF YTTRIUM IRON GARNET AS A FUNCTION OF
      TEMPERATURE.
      *PRINCE(E.).

2892 Y*(FE-MN)*O3,. . . . . .A2. . . . . .SOV. PHYS. SOL. STATE VOL.14,. . .3037, (1973).
      *MOSSBAUER AND NEUTRON-DIFFRACTION INVESTIGATIONS OF SPIN REORIENTATION
      IN Y*(FE-MN)*O3.
      *BYSTROV(M.V.), BOKOV(V.A.), POPOV(G.V.), KOCHAROV(A.G.), LIGENZA(S.),
      FAEK(M.K.), KOVALEVA(N.P.).

2893 Y*FE*O3, ...A2. . . . . .SOL. STATE COMM. VOL.8,. . . . . .597, (1970).
      *SUBLATTICE MAGNETIZATION IN YB*FE*O3 AND Y*FE*O3 AS OBTAINED BY NEUTRON
      DIFFRACTION AND ITS RELATION TO THE HYPERFINE FIELD.
      *PINTO(H.), SHACHAR(G.), SHAKED(H.).

2894 Y*FE*O3, ...A2,T. . . . .ACTA CRYST A24,. . . . . .217, (1968).
      *REPRESENTATION ANALYSIS OF MAGNETIC STRUCTURES.
      *BERTAUT(E.F.).

2895 Y3*FE5*O12,. . . . . .A2. . . . . .SOV. PHYS. CRYST. VOL.7, ...765, (1963).
      *MAGNETIC STRUCTURE OF YTTRIUM FERRITE.
      *KUZ#MINOV(YU.S.), YAMZIN(I.I.), BELOV(N.V.).

2896 Y*FE*O3, ...A2. . . . . .J. PHYS. CHEM. SOLIDS VOL.31,. . .2741, (1970).
      *COVALENCY PARAMETERS FOR CR(3+), FE(3+), AND MN(4+) IN AN OXIDE
      ENVIRONMENT.
      *TOFIELD(B.C.), FENDER(B.E.F.).

2897 Y3*FE5*O12,. . . . . .A2. . . . . .COMP. REND. TOME 243, ...898, (1956).
      *ETUDE PAR DIFFRACTION DE NEUTRONS DU GRENAT FERRIMAGNETIQUE Y3*FE5*O12.
      *BERTAUT(E.F.), FORRAT(F.), HERPIN(A.), MERIEL(P.).

2898 Y3*GA*FE4*O12, ...A4. . . . . .LATV. PSR...FIZ. TEHN. SER. 2, . . 49, (1969).
      *NEUTRON DIFFRACTION STUDY OF YTTRIUM-GALLIUM-FERRITE GARNET
      Y3*GA*FE4*O12.
      *NOZIK(YU.E.), KUVALDIN(B.V.), LIPIN(YU.V.).

SECTION   5 -STRUCTURE AND CROSS-SECTION DETERMINATIONS . . . .

2899 Y*H3,. . . . . .A1.. . . . . .SOV. PHYS. CRYST. VOL.17,. . . . .   342, (1972).
      ⊷NEUTRON-DIFFRACTION STUDY OF YTTRIUM TRIHYDRIDE.
      ⊷MIRON(N.F.), SHCHERBAK(V.I.), BYKOV(V.N.), LEVDIK(V.A.).

2900 Y-HO,. . . . . .A2. . . . . . .REF. SECTION 3, 65/CHILD,. . . . .

2901 Y*MN2*05,. . . . . .A2,T. . . . .ACTA CRYST A24. . . . .   217, (1968).
      ⊷REPRESENTATION ANALYSIS OF MAGNETIC STRUCTURES.
      ⊷BERTAUT(E.F.).

2902 Y*MN*03, ...A2. . . . . . .COLL. INTER. N.126 (GRENOBLE), . 126, (1963).
      ⊷ORDRE MAGNETIQUE ET PROPRIETES MAGNETIQUES DE MN*Y*03.
      ⊷BERTAUT(E.F.), MERCIER(M.), PAUTHENET(R.).

2903 Y*MN2*05,. . . . . .A2. . . . . .PHYS. STAT. SOLIDI A VOL.16 NO.2   533, (1973).
      ⊷ORDRE HELIMAGNETIQUE DU MANGANESE DANS LA SERIE T*MN2*05 (T=LU,ER,HO,TB,
      ND).
      ⊷BUISSON(G.).

2904 Y*MN*03, ...A4. . . . . . .SOV. PHYS. SOL. STATE 8, ...   215, (1966).
      ⊷MAGNETIC PROPERTIES OF Y*MN*03.
      ⊷KIZHAEV(S.A.), BOKOV(V.A.), KOCHALOV(O.V.).

2905 Y2*03, ...A1. . . . . . .ACTA CRYST. B25. . . . . 2140, (1969).
      ⊷A NEUTRON DIFFRACTION STUDY OF THE CRYSTAL STRUCTURE OF THE C-FORM
      YTTRIUM SESQUIOXIDE.
      ⊷O≠CONNOR(B.H.), VALENTINE(T.M.).

2906 (Y2*03)5-(ND2*03)1.5-(FE2*03)5. . . .A2. . . . . . .SOV. PHYS. CRYST. VOL.9, ...   159, (1964).
      ⊷A NEUTRON-DIFFRACTION STUDY OF GARNET FERRITES.
      ⊷KUZ≠MINOV(YU.S.).

2907 (Y2*03)1.5-(ND2*03)1.5-(FE2*03)5. . .A2. . . . . . .SOV. PHYS. CRYST. VOL.8, ...   15, (1963).
      ⊷NEUTRON DIFFRACTION INVESTIGATION OF AN YTTRIUM-NEODYMIUM FERRITE WITH
      A GARNET STRUCTURE.
      ⊷KUZ≠MINOV(YU.S.), YAMZIN(I.I.), BELOV(N.V.).

2908 Y2*SI*05,. . . . . .A1. . . . . . .COMP. REND. 264, ... 397B, (1967).
      ⊷STRUCTURE DE Y2*SI*05.
      ⊷MICHEL(C.), BUISSON(G.), BERTAUT(E.F.).

2909 Y-TB,. . . . . .A2. . . . . . .PHYS. REV. 138, ... A1655, (1965).
      ⊷MAGNETIC PROPERTIES OF HEAVY RARE-EARTHS DILUTED BY YTTRIUM AND LUTETIUM
      ⊷CHILD(H.R.), KOEHLER(W.C.), WOLLAN(E.O.), CABLE(J.W.). (ED. NOTE: REF.
      SECTION 3, 65/CHILD).

2910 Y2*TI2*07, ...A1. . . . . . .Z. KRIST. VOL.131, ...   278, (1970).
      ⊷ROENTGEN- UND NEUTRONENBEUGUNGSUNTERSUCHUNGEN AN Y2*TI2*07.
      ⊷BECKER(W.J.), WILL(G.).

2911 Y-TM,. . . . . .A2. . . . . . .REF. SECTION 3, 65/CHILD,. . . . .

2912 YB (ISO. A=168, 174),. . . . . .P. . . . . . .J. INORG. NUCL. CHEM. VOL.32,. . 2839, (1970).
      ⊷THE THERMAL NEUTRON CAPTURE CROSS-SEC. AND RES. CAPTURE INTEG. OF CA(IS.
      A=44), NI(IS. A=62), YB(IS. A=168, 174), TM(IS. A=169) AND TL(IS. A=203)
      ⊷SIMS(G.H.E.), JUHNKE(D.G.).

2913 YB,. . . . . .P. . . . . . .PROC/CONF/NEUTRON C. S. + TECH.,   875, (1968).
      ⊷THE THERMAL CROSS SECTIONS AND PARAMAGNETIC SCATTERING CROSS SECTIONS
      OF THE YB ISOTOPES.
      ⊷MUGHABGHAB(S.F.), CHRIEN(R.E.).

2914 YB,. . . . . .P. . . . . . .PHYS. REV. VOL.121, ...   610, (1961).
      ⊷SLOW-NEUTRON SCATTERING CROSS SECTIONS OF TERBIUM, YTTERBIUM, AND
      LUTETIUM.
      ⊷ATOJI(M.).

2915 YB*AU2,. . . . . .A1. . . . . . .J. CHEM. PHYSICS VOL.52, ... 6433, (1970).
      ⊷NEUTRON DIFFRACTION STUDIES OF TM*AU2, YB*AU2, AND TM*AG2 AT 300-1.7 K.
      ⊷ATOJI(M.).

2916 YB*C2, ...A1. . . . . . .J. CHEM. PHYSICS VOL.52, ... 6430, (1970).
      ⊷NEUTRON DIFFRACTION STUDY OF YB*C2 AT 300-2 K.
      ⊷ATOJI(M.), FLOWERS(R.H.).

2917 YB*C2, ...A1,A4. . . .REF. SECTION 3, 61/ATOJI,. . . . .

2918 YB*C*03, ...A2,A4. . . .REF. SECTION 3, 66/BERTAUT,. . . . .

2919 YB3-CO(2.5)-GE(2.5)*012, ...A4. . . . . . .LATV. PSR. FIZ. TEHN. SER. 6, .   46, (1970).
      ⊷A NEUTRON DIFFRACTION STUDY OF THE CATIONIC DISTRIBUTION IN THE GARNET
      YB3-CO(2.5)-GE(2.5)*012.
      ⊷LIPIN(YU.), NOZIK(YU.).

2920 YB*FE-GARNET,. . . . . .A2. . . . . . .J. APPL. PHYS. VOL.41, ... 1192, (1970).
      ⊷RARE-EARTH SUBLATTICE CANTING IN DY*IG, ER*IG, AND YB*IG.
      ⊷PICKART(S.J.), ALPERIN(H.A.), CLARK(A.E.).

2921 YB*FE*03, ...A2. . . . . . .SOL. STATE COMM. VOL.8, ...   597, (1970).
      ⊷SUBLATTICE MAGNETIZATION IN YB*FE*03 AND Y*FE*03 AS OBTAINED BY NEUTRON
      DIFFRACTION AND ITS RELATION TO THE HYPERFINE FIELD.
      ⊷PINTO(H.), SHACHAR(G.), SHAKED(H.).

2922 YB2*03,. . . . . .A2. . . . . . .PHYS. REV. 176,. . . . .   722, (1968).
      ⊷MAGNETIC STRUCTURES OF ER2*03 AND YB2*03,
      ⊷MOON(R.M.), KOEHLER(W.C.), CHILD(H.R.), RAUBENHEIMER(L.J.).

2923 YB2*02*S,. . . . . .A2. . . . . . .SOL. STATE COMM. VOL. 7 NO.14, . 1011, (1969).
      ⊷STRUCTURE MAGNETIQUE DE L≠OXYSULFURE D≠YTTERBIUM.
      ⊷BALLESTRACCI(R.), ROSSAT-MIGNOD(J.).

2924 YB2*02*SE, ...A2. . . . . . .SOL. STATE COMM. VOL.10, ...   735, (1972).
      ⊷PROPRIETES MAGNETIQUES DES OXYSELENIURES DE GD, TB, DY, HO, ER, TM ET
      STRUCTURE MAGNETIQUE DE HO2*02*SE ET DE YB2*02*SE.
      ⊷QUEZEL(G.), ROSSAT-MIGNOD(J.), LANG(H.Y.).

2925 YB*V*04, ...A1. . . . . . .J. SOLID STATE CHEM. VOL.5, . . .   11, (1972).
      ⊷REFINEMENT OF THE CRYSTAL STRUCTURE OF SOME RARE EARTH VANADATES R*V*04
      (R=DY,TB,HO,YB).
      ⊷FUESS(H.), KALLEL(A.).

2926 ZN,. . . . . .P. . . . . . .J. OF PHYS. C VOL.1, ... 1075, (1968).
      ⊷THE STRUCTURE FACTOR FOR LIQUID METALS. II. RESULTS FOR LIQUID ZN, TL,
      PB, SN AND BI.
      ⊷NORTH(D.M.), ENDERBY(J.E.), EGELSTAFF(P.A.).

SECTION   5 -STRUCTURE AND CROSS-SECTION DETERMINATIONS . . . .

2927   ZN,. . . . . .A1. . . . . .PHYS. MET. METALLOGR. VOL.31,. .   79, (1971).
. NEUTRON DIFFRACTION OF ZINC.
. KHRUSHCHEV(B.I.), BOGOMOLOV(A.M.).

2928   ZN,. . . . . .A3,T. . .NUOVO CIMENTO B VOL.60,. . . . . .   48, (1969).
. NEUTRON DIFFRACTION ANALYSIS OF LIQUID ZINC. II.
. CAGLIOTI(G.), CORCHIA(M.), RIZZI(G.).

2929   ZN,. . . . . .P. . . . . .PHYS. LETT. 21. . . . .   286, (1966).
. THE STRUCTURE FACTOR FOR LIQUID METALS AT LOW ANGLES.
. EGELSTAFF(P.A.), DUFFILL(C.), RAINEY(V.S.), ENDERBY(J.E.), NORTH(D.M.).

2930   ZN,. . . . . .A1. . . . .SOV. PHYS. SOL. STATE 6,. .   1070, (1964).
. STRUCTURE INVESTIGATION BY NEUTRON DIFFRACTION USING A PULSED FAST
REACTOR (IBR).
. NITTS(V.V.), PAPULOVA(Z.G.), SOSNOVSKAYA(I.), SOSNOVSKII(E.).

2931   ZN,. . . . . .A3. . . . . .NUOVO CIMENTO B VOL.49,. . . . . .   222, (1967).
. NEUTRON DIFFRACTION ANALYSIS OF LIQUID ZINC. I.
. CAGLIOTI(G.), CORCHIA(M.), RIZZI(G.).

2932   ZN,. . . . . .A4. . . . . .ACTA CRYST. 22,. . . . .   170, (1967).
. MEAN-SQUARE ATOMIC DISPLACEMENTS IN ZINC.
. BARRON(T.H.K.), MUNN(R.W.).

2933   ZN,. . . . . .P. . . . . .J. NUCL. ENERGY VOL.22,. . . . . .   389, (1968).
. MEASUREMENT OF THE TOTAL NEUTRON CROSS-SECTION OF ZINC AND ANTIMONY IN
THE ENERGY RANGE 0.002-0.4 EV USING A TIME-OF-FLIGHT SPECTROMETER.
. EL-ELA(M.A.),   SALAMA(M.),   ABDEL-KAWY(A.),   ADIB(M.),   HAMOUDA(I.).

2934   ZN*AL2*O4,. . .   . . .A1. . . . .Z. KRIST. VOL.124,. . . . . .   275, (1967).
. NEUTRONENBEUGUNGSUNTERSUCHUNG DER STRUKTUREN VON MG*AL2*O4- UND
ZN*AL2*O4- SPINELLEN, IN ABHANGIGKEIT VON DER VORGESCHICHTE.
. FISCHER(P.).

2935   ZN*CR2*SE4,. .   . . . .A2. . . . .J. APP. PHYS. 37,. . . . . . .   964, (1966).
. NEUTRON DIFFRACTION STUDY OF HELIMAGNETIC SPINEL ZN*CR2*SE4.
. PLUMIER(R.).

2936   ZN*CR2*SE4,. .   . . . .A2. . . . .COMP. REND. 260,. . . . . . .   3348, (1965).
. ETUDE PAR DIFFRACTION DE NEUTRONS DU COMPOSE SPINELLE
ANTIFERROMAGNETIQUE.
. PLUMIER(R.).

2937   ZN*CR2*SE4,. .   . . . .A2. . . . . . .J. PHYSIQUE TOME 27,. . . . .   213, (1966).
. NEUTRON DIFFRACTION STUDY OF HELICOIDAL ANTIFERROMAGNETISM OF THE
SPINEL ZN*CR2*SE4 IN A MAGNETIC FIELD.
. PLUMIER(R.). (IN FRENCH).

2938   (ZN-FE-CR)*O4, ...O4. . . . . .PHYS. STAT. SOL. A VOL.3,. .   569, (1970).
. NEUTRON DIFFRACTION STUDY OF THE (ZN-FE(2-X)-CR(X))*O4 SERIES.
. OLES(A.).

2939   ZN*(FE-CR)*O4, ...A2. . . . .J. PHYSIQUE TOME 32 COL.1 VOL.I,   328, (1971).
. ETUDE PAR DIFFRACTION DE NEUTRONS DES SPINELLES ZN*(FE(2-X)-CR(X))*O4.
. OLES(A.).

2940   ZN*FE2*O4, ...A1. . . . .PHYS. STAT. SOL. 32,. . . . . .   K91, (1969).
. DETERMINATION OF THE OXYGEN PARAMETER IN ZN*FE2*O4 BY NEUT. DIFFRACTION.
. BALANDA(M.), SZYTULA(A.), DIMITRIJEVIC(Z.), TODOROVIC(J.).

2941   ZN*FE2*O4, ...A2. . . . .PHYS. STAT. SOL. VOL.40,. . . . . .   171, (1970).
. STRUCTURE MAGNETIQUE DU SPINELLE ANTIFERROMAGNETIQUE ZN*FE2*O4.
. BOUCHER(B.), BUHL(R.), PERRIN(M.).

2942   ZN*FE*O3,. . . .   . .A2. . . . .J. PHYSIQUE TOME 32 COL.1 VOL.I,   320, (1971).
. NEUTRON DIFFRACTION AND MOSSBAUER STUDIES OF ZINC FERRITE.
. KONIG(U.), BERTAUT(E.F.), GROS(Y.), CHOL(G.).

2943   ZN*FE2*O4, ...T,A2. . . . .J. PHYS. SOC. JAPAN VOL.36,. . .   84, (1974).
. THEORY OF MAGNETIC STRUCTURE OF ZINC FERRITE.
. AKINO(T.).

2944   ZN*FE2*O4, ...A1,A2. . . . .REV. MOD. PHYS. VOL.25,. . . . . .   114, (1953).
. NEUTRON DIFFRACTION STUDIES OF ZINC FERRITE AND NICKEL FERRITE.
. HASTINGS(J.M.), CORLISS(L.M.).

2945   ZN*FE2*O4, ...A2. . . . .PHYS. REV. VOL.102,. . . . . . .   1460, (1956).
. AN ANTIFERROMAGNETIC TRANSITION IN ZINC FERRITE.
. HASTINGS(J.M.), CORLISS(L.M.).

2946   (ZN-FE)*O4,(ZN-(NI,CO,MN)-FE)*O4. . .A2. . . . . .PHYS. STAT. SOLIDI A VOL.4 NO.1,   53, (1971).
. THE INFLUENCE OF CATION SUBSTITUTION ON THE MAGNETIC BEHAVIOUR OF
ZINC FERRITE.
. KOCHAROV(A.G.), LECIEJEWICZ(J.), FAYEK(M.K.), MURASIK(A.).

2947   ZN*GE*N2,. . . .   . .A1. . . . .MATER. RES. BULL. VOL.8,. . .   1049, (1973).
. GROUPE SPATIAL ET ORDRE DES ATOMES DE ZINC ET DE GERMANIUM DANS ZN*GE*N2.
. WINTENBERGER(M.), MAUNAYE(M.), LAURENT(Y.).

2948   ZN*MN3*C,. . . .   . .A2. . . . .J. SOLID STATE CHEM. VOL.8,. . .   182, (1973).
. ETUDES PAR DIFFRACTION NEUTRONIQUE ET RMN DE ZN*MN3*C ET GA*MN3*C(.935).
. FRUCHART(D.), BERTAUT(E.F.), LE-CLERC(B.), LE-DANG-KHOI, VEILLET(P.)...

2949   ZN-MN-FE,. . . .   . .A2. . . . .SOV. PHYS. CRYST. VOL.8,. . .   18, (1963).
. A NEUTRON DIFFRACTION STUDY OF A MANGANESE-ZINC FERRITE.
. KONAKHOVICH(YU.YA.), SAKSONOV(YU.G.).

2950   (ZN-MN)*FE2*O4,.   . . . . .A2. . . . .PHYS. STAT. SOL. VOL.37,. . . . .   843, (1970).
. ANTIFERROMAGNETISM OF (ZN-MN)*FE2*O4.
. FAYEK(M.K.), LECIEJEWICZ(J.), MURASIK(A.), YAMZIN(I.I.).

2951   (ZN-MN)*FE2*O4,.   . . . . .A2. . . . .PHYS. STAT. SOLIDI A VOL.7 NO.2,   K71, (1971).
. ON THE MAGNETIC TRANSITION IN (ZN(X)-MN(1-X))*FE2*O4.
. LOSHMANOV(A.), LIGENZA(S.), FAYEK(M.K.), KOCHAROV(A.G.).

2952   ZN*MN2*O4, ...A1,2. . . . .J. PHYS. CHEM. SOLIDS 30,. . . . .   805, (1969).
. MANGANITES SPINELLES PURES D*ELEMENTS DE TRANSITION: PREPARATIONS ET
STRUCTURES CRISTALLOGRAPHIQUES.
. BUHL(R.).

2953   ZN-NI, ...P. . . . . .Z. ANGEW. MATH. PHYS. VOL.16,. .   820, (1965).
. REDETERMINATION OF THE NEUTRON SCATTERING LENGTH OF ZINC.
. FISCHER(P.), HALG(W.), STOLL(E.).

SECTION   5 -STRUCTURE AND CROSS-SECTION DETERMINATIONS . . . .

2954 (ZN-NI-FE)*O4, .   .A4.......SOV. PHYS. SOL. STATE VOL.13, . 321, (1971).
      *MAGNETIC HYPERFINE STRUCTURE OF THE NUCLEAR LEVELS OF FE(ISOTOPE A=57)
      ZN-NI FERRITE ABOVE T-CURIE, OBSERVED USING MOSSBAUER SPECTROSCOPY.
      *PEKOSHEVSKII(E.), SUVAL≠SKI(YA.), DOMBROVSKI(L.).

2955 (ZN-NI)*FE2*O4,.   .A2.......PHYS. REV. VOL.181, ... 969, (1969).
      *YAFET-KITTEL ANGLES IN ZINC-NICKEL FERRITES.
      *MURTHY(N.S.S.), NATERA(M.G.), YOUSSEF(S.I.), BEGUM(R.J.),
      SRIVASTAVA(C.M.).

2956 (ZN-NI)*FE2*O4,.   .A1,A2.....PHYS. REV. VOL.95, . . . 1408, (1954).
      *NEUTRON DIFFRACTION STUDIES OF A NICKEL ZINC FERRITE.
      *WILSON(V.C.), KASPER(J.S.).

2957 ZN*O,. . . . . .   .A1.......SOV. PHYS. SOL. STATE 6, . 1070, (1964).
      *STRUCTURE INVESTIGATION BY NEUTRON DIFFRACTION USING PULSED FAST
      REACTOR (IBR).
      *NITTS(V.V.), PAPULOVA(Z.G.), SOSNOVSKAYA(I.), SOSNOVSKII(E.).

2958 ZN*O,. . . . . .   .P.......PHYS. REV. VOL.75, ... 975, (1949).
      *TOTAL NEUTRON CROSS SECTIONS OF COMPOUNDS WITH DIFFERENT CRYSTALLINE
      STRUCTURES.
      *WINSBERG(L.), MENEGHETTI(D.), SIDHU(S.S.).

2959 ZN*O,. . . . . .   .A4.......ACTA CRYST. VOL.B25, . . . 2254, (1969).
      *THE WURTZITE Z PARAMETER FOR BERYLLIUM OXIDE AND ZINC OXIDE.
      *SABINE(T.M.), HOGG(S.).

2960 ZN*S, ZN*TE, . .   .A1.......ACTA CRYST. VOL.A29, ... 49, (1973).
      *A NEUTRON-DIFFRACTION STUDY OF ZN*S AND ZN*TE.
      *COOPER(M.J.), ROUSE(K.D.), FUESS(H.).

2961 ZN*SN*(O*H)6,. .   .A1.......ACTA CRYST B24, ... 1358, (1968).
      *APPLICATION DE LA METHODE DU SIMPLEX A LA DETERMINATION DIRECTE DES
      STRUCTURES CRISTALLINES.
      *BASSI(G.).

2962 ZN*V2*O4,. . . .   .A2.......PHYS. STAT. SOLIDI A VOL.18 NO.1 K11, (1973).
      *INVESTIGATION OF MAGNETIC PROPERTIES OF ZN*V2*O4 SPINEL.
      *NIZIOL(S.).

2963 ZN2*ZR,. . . . .   .A4.......PROC. INT. CONF. (NOTTINGHAM), . 223, (1964).
      *SPIN DENSITIES OF 4D ATOMS IN FERROMAGNETIC ALLOYS.
      *SHIRANE(G.), NATHANS(R.), PICKART(S.J.), ALPERIN(H.A.).

2964 ZR (ISO. A=94, 96), .P.......NUCL. SCI. ENG. VOL.46, . 314, (1971).
      *NEUTRON ABSORPTION CROSS SECTIONS FOR ZIRCONIUM-94 AND ZIRCONIUM-96.
      *FULMER(R.H.), STRICOS(D.P.), RUANE(T.F.).

2965 ZR,. . . . . . .   .P.......ATOMKERNENERGIE VOL.22, . 87, (1973).
      *MEASUREMENTS OF THE TOTAL NEUTRON CROSS-SECTION OF CHROMIUM, COBALT
      AND ZIRCONIUM IN THE ENERGY RANGE .003 EV TO 1 EV.
      *SALAMA(M.), ADIB(M.), ABDEL-KAWI(A.), EL-MINIAWY(S.), EL-KHOSHT(M.),
      HAMOUDA(I.).

2966 ZR-C,. . . . . .   .A1.......PHYS. STAT. SOL. 20, . . K141, (1967).
      *NEUTRON DIFFRACTION STUDIES ON TITANIUM-CARBON AND ZIRCONIUM-CARBON
      ALLOYS.
      *GORETZKI(H.).

2967 ZR-CARBOHYDRIDE,   .A4.......SOV. PHYS. CRYST. VOL.14, . . . 913, (1969).
      *NEUTRON DIFFRACTION STUDY OF ZIRCONIUM CARBOHYDRIDE.
      *BYKOV(V.N.), GOLOVKIN(V.S.), KALININ(V.P.), LEVDIK(V.A.),
      SHCHERBAK(V.I.).

2968 ZR*D2, ...A1.......ACTA CRYST. VOL.5, . . . 22, (1952).
      *THE CRYSTAL STRUCTURE OF THORIUM AND ZIRCONIUM DIHYDRIDES BY X-RAY
      AND NEUTRON DIFFRACTION.
      *RUNDLE(R.E.), SHULL(C.G.), WOLLAN(E.O.).

2969 ZR*H2, ...A1.......SOV. PHYS. CRYST. VOL.15, . . . 376, (1970).
      *NEUTRON DIFFRACTION INVESTIGATION OF ZIRCONIUM NITRIDE-HYDRIDE.
      *BYKOV(V.N.), GOLOVKIN(V.S.), LEVDIK(V.A.), KALININ(V.P.), MIRON(N.F.).

2970 ZR*H,. . . . . .   .A4.......BROOKHAVEN SYMPOSIUM, . . 96, (1965).
      *A MULTIPHONON NEUTRON SCATTERING STUDY OF HYDROGEN BONDING IN ZIRCONIUM
      HYDRIDE.
      *HARLING(O.K.), LEONARD-JR(B.R.).

2971 ZR*H2, ...P.......J. PHYS. RADIUM VOL.22, . . . 648, (1961).
      *SCATTERING CROSS-SECTION FOR SLOW NEUTRONS FROM HYDROGEN IN METALLIC
      HYDRIDES.
      *GENIN(R.), RIBRAG(M.). (IN FRENCH).

2972 ZR-H(X). . . . .   .P.......BULL. AMER. PHYS. SOC. VOL.17, . 125, (1972).
      *HYDROGEN BONDING EFFECTS IN PRECISION TOTAL NEUTRON CROSS SECTION
      MEASUREMENTS OF ZR*H(X).
      *RORER(D.C.), BRUNHART(G.), MALIK(S.S.). (ED. NOTE: PAPER PRESENTED AT
      ANNUAL MEETING OF AMER. PHYS. SOC. IN SAN FRANCISCO).

2973 ZR*N,. . . . . .   .A1.......SOV. PHYS. CRYST. VOL.15, . . . 376, (1970).
      *NEUTRON DIFFRACTION INVESTIGATION OF ZIRCONIUM NITRIDE-HYDRIDE.
      *BYKOV(V.N.), GOLOVKIN(V.S.), LEVDIK(V.A.), KALININ(V.P.), MIRON(N.F.).

2974 ZR-NB, ...T.......ACTA CRYST. VOL.A29, . . . 594, (1973).
      *THE SHORT-RANGE STRUCTURE OF TI AND ZR B.C.C. SOLID SOLUTIONS CONTAINING
      THE OMEGA PHASE. II. SOLUTION OF THE STRUCTURE DETERMINATION.
      *BORIE(B.), SASS(S.L.), ANDREASSEN(A.).

2975 ZR-NB, ...A1.......J. PHYS. CHEM. SOLIDS VOL.35, . 879, (1974).
      *NEUTRON DIFFRACTION DETERMINATION OF THE NUMBER AND DISPLACEMENT OF
      THE ATOMS IN THE DIFFUSE OMEGA-PHASE OF ZR(.8)-NB(.2).
      *KEATING(D.T.), LA-PLACA(S.J.).

2976 ZR*NI*H3,. . . .   .A4.......COLL. INTER. N.126 (GRENOBLE), . 27, (1963).
      *NEUTRON DIFFRACTION STUDY OF NICKEL ZIRCONIUM HYDRIDE.
      *PETERSON(S.W.), SADANA(V.N.), KORST(W.L.).

2977 ZR-O,. . . . . .   .A1.......ACTA CRYST. VOL.A28 PART S-4, . 99, (1972).
      *INTERSTITIAL ORDER-DISORDER TRANSFORMATION IN THE ZIRCONIUM-OXYGEN SOLID
      SOLUTION.
      *YAMAGUCHI(S.), HIRABAYASHI(M.), ASANO(H.), ARAI(T.), HASHIMOTO(S.).
      (ABSTRACT ONLY).

SECTION   5 -STRUCTURE AND CROSS-SECTION DETERMINATIONS . . . .

2978   ZR-O,. . . . . .A1. . . . . . .J. APPL. CRYST. VOL.7, ...   67, (1974).
                         ⊢ATOMIC ORDERING AND LATTICE DISTORTION IN THE ZIRCONIUM-OXYGEN ALLOYS
                          WITH 28.2 AND 29.2 AT. PCT. OXYGEN.
                         ⊢HASHIMOTO(S.), IWASAKI(M.), OGAWA(S.), YAMAGUCHI(S.), HIRABAYASHI(M.).

2979   ZR*O2-Y*O(1.5),. . . . . .A1. . . . . . .J. PHYS. C VOL.7, ...   1, (1974).
                         ⊢THE STRUCTURE OF CUBIC ZR*O2-Y*O(1.5) SOLID SOLUTIONS BY NEUTRON
                          SCATTERING.
                         ⊢STEELE(D.), FENDER(B.E.F.).

2980   ZR*O2,ZR*O2-SC2*O3. . . . . .A4. . . . . . .J. SOLID STATE CHEM. VOL.7,. . . .   448, (1973).
                         ⊢A HIGH-TEMPERATURE NEUTRON DIFFRACTION STUDY OF PURE AND SCANDIA-
                          STABILIZED ZIRCONIA.
                         ⊢BARKER(W.W.), BAILEY(F.P.), GARRETT(W.).

2981   ZR-V2, ...A4. . . . . . .SOL. STATE COMM. VOL.13, ...   1779, (1973).
                         ⊢LATTICE TRANSFORMATION IN THE SUPERCONDUCTOR ZR-V2 BY NEUTRON
                          DIFFRACTION.
                         ⊢MONCTON(D.E.).

2982   ZR*ZN2,. . . . . .A4. . . . . . .PROC. INT. CONF. (NOTTINGHAM), .   223, (1964).
                         ⊢SPIN DENSITIES OF 4D ATOMS IN FERROMAGNETIC ALLOYS.
                         ⊢SHIRANE(G.), NATHANS(R.), PICKART(S.J.), ALPERIN(H.A.).

2983   ZR*ZN2,. . . . . .A2. . . . . . .PHYS. REV. LETT. VOL.12, ...   444, (1964).
                         ⊢ITINERANT ELECTRON FERROMAGNETISM IN ZR*ZN2.
                         ⊢PICKART(S.J.), ALPERIN(H.A.), SHIRANE(G.), NATHANS(R.).

SECTION   6 -QUASI-ELASTIC AND INELASTIC SCATTERING STUDIES . . . .

1 ACETONITRILE (C*H3*C*N*), ...P. . . . . . . .BER. BUNSENGES. PHYS. CHEM. 75, . 769, (1971).
                  *A COMPARATIVE STUDY OF QUASIELASTIC NEUTRON SCATTERING AND NMR RELAXA-
                  TION IN LIQUID ACETONITRILE.
                  *ZEIDLER(M.D.).

2 ACID/ACETIC, ...C3. . . . . . .DISC. FARADAY SOC. NO.43,. . . . 169, (1967).
                  *NEUTRON SCATTERING SPECTROSCOPY OF LIQUIDS.
                  *ALDRED(B.K.), EDEN(R.C.), WHITE(J.W.).

3 ACID/ALPHA-AMINO-ISO-BUTYRIC, ...C1. . . .CHEM. PHYS. LETT. VOL.17, . . 53, (1972).
                  *INELASTIC INCOHERENT NEUTRON SCATTERING ON POLYCRYSTALLINE TAURINE AND
                  ALPHA-AMINO-ISO-BUTYRIC ACID.
                  *DAVIDOVIC(M.), RATKOVIC(S.), JOVIC(D.), ZIVANOVIC(M.).

4 ACID/CARBOXYLIC, ...B1,B4. . .J. CHEM. PHYS. VOL.52, ... 5740, (1970).
                  *VIBRATION SPECTRA OF CARBOXYLIC ACIDS BY NEUTRON SPECTROSCOPY.
                  *COLLINS(M.F.), HAYWOOD(B.C.).

5 ACID/CARBOXYLIC, ...B1,B4. . .J. CHEM. PHYS. VOL.55, ... 4156, (1971).
                  *ON THE VIBRATIONAL SPECTRA OF CARBOXYLIC ACIDS BY NEUTRON SPECTROSCOPY.
                  *LURIE(N.A.), DANNER(H.R.).

6 ACID/IODIC, ...C1. . . . . . .NATURE (PHYS. SCI.) VOL.241, . . 40, (1973).
                  *INCOHERENT NEUTRON SCATTERING STUDY OF HYDROGEN BONDING IN PARAPERIODIC
                  AND IODIC ACIDS.
                  *TEMME(F.P.), WADDINGTON(T.C.), JONES(L.V.), SMITH(J.A.S.).

7 ACID/NA-POLY-L-GLUTAMIC, ...C3. . . .IAEA SYMP. COPENHAGEN, VOL.2, . 175, (1968).
                  *AN INVESTIGATION OF THE POLYPEPTIDE; POLY-L-GLUTAMIC ACID; USING NEUTRON
                  INELASTIC SCATTERING.
                  *WHITTEMORE(W.L.).

8 ACID/PARAPERIODIC, ...C1. . . . . . .NATURE (PHYS. SCI.) VOL.241, . . 40, (1973).
                  *INCOHERENT NEUTRON SCATTERING STUDY OF HYDROGEN BONDING IN PARAPERIODIC
                  AND IODIC ACIDS.
                  *TEMME(F.P.), WADDINGTON(T.C.), JONES(L.V.), SMITH(J.A.S.).

9 ACID/POLY-L-GLUTAMIC, ...C3. . . .IAEA SYMP. COPENHAGEN, VOL.2, . 175, (1968).
                  *AN INVESTIGATION OF THE POLYPEPTIDE; POLY-L-GLUTAMIC ACID; USING NEUTRON
                  INELASTIC SCATTERING.
                  *WHITTEMORE(W.L.).

10 ACID/POLYGLUTAMIC, ...C2. . . . . . .J. CHEM. PHYSICS VOL.44, ... 3127, (1966).
                  *COMMUNICATION; INVESTIGATION OF THE ALPHA-HELIX-RANDOM-COIL TRANSITION
                  IN POLYGLUTAMIC ACID BY SLOW NEUTRON SCATTERING (LETTER TO THE EDITOR).
                  *BOUTIN(H.), WHITTEMORE(W.L.).

11 ADAMANTANE (C10*H16), ...C1. . . . . . .DISC. FARADAY SOC. NO.48, . . . 156, (1969).
                  *DYNAMICS OF ORIENTATIONAL DEFECTS IN SOLID ADAMANTANE.
                  *STOCKMEYER(R.).

12 AG, ...B1. . . . . . .PHYS. LETT. VOL.28A, ... 531, (1969).
                  *PHONON DISPERSION IN SILVER.
                  *DREXEL(W.), GLASER(W.), GOMPF(F.).

13 AG, ...B1. . . . . .Z. PHYS. VOL.255, ... 281, (1972).
                  *LATTICE DYNAMICS OF SILVER.
                  *DREXEL(W.).

14 AG, ...T. . . . . . .PHYS. LETT. VOL.36A, ... 337, (1971).
                  *AB INITIO PSEUDOPOTENTIAL CALCULATION OF PHONON FREQUENCIES IN SILVER.
                  *NIKULIN(Y.K.), TSAREV(YU.N.).

15 AG, ...B1. . . . . . .PHYS. LETT. 29A, ... 639, (1969).
                  *CRYSTAL DYNAMICS OF SILVER.
                  *KAMITAKAHARA(W.A.), BROCKHOUSE(B.N.).

16 AG*CL, ...B1. . . . . . .PHYS. REV. B-1, ... 4819, (1970).
                  *LATTICE DYNAMICS OF SILVER CHLORIDE.
                  *VIJAYARAGHAVAN(P.R.), NICKLOW(R.M.), SMITH(H.G.), WILKINSON(M.K.).

17 AG*CL, ...T. . . . . . .PHYS. STAT. SOL. B VOL.51 NO.1, . 389, (1972).
                  *THE DYNAMICAL BEHAVIOUR OF SILVER CHLORIDE WITH MODIFIED SHELL MODEL.
                  *SINGH(R.K.), UPADHYAYA(K.S.).

18 AG*CL, ...T,B1. . . . .PHYS. REV. VOL.126, ... 933, (1962).
                  *DISPERSION OF PHONONS IN SILVER CHLORIDE.
                  *JOSHI(S.K.), GUPTA(R.).

19 AG2*H3*I*O6, ...C1. . . . . . .J.C.S. FARADAY TRANS. II VOL.68, . 350, (1972).
                  *PROTON MOTIONS IN H-BONDED FERROELECTRIC AND ANTIFERROELECTRIC SOLIDS.
                  1.ENERGY-LOSS NEUTRON INCOHERENT SCATT/K*H2*P*O4/AG2*H3*I*O6/T<T(CRIT.);
                  *TEMME(F.P.), WADDINGTON(T.C.).

20 AG2*H3*I*O6,AG4*H2*I2*O10, ...C1. . . . . . .J.C.S. FARADAY TRANS. II VOL.69, . 1, (1973).
                  *PROTON MOTIONS IN HYDROGEN-BONDED FERROELECTRIC AND ANTIFERROELECTRIC
                  SOLIDS.2. NEUTRON INCOHERENT SCATTERING IN POLYCRYSTALLINE...
                  *TEMME(F.P.), SMITH(J.A.S.), WADDINGTON(T.C.).

21 AIR, ...SOV. AT. ENERGY VOL.28, ... 214, (1970).
                  *SCATTERING OF NEUTRONS IN AIR.
                  *MORDASHEV(V.M.).

22 AL, ...B1,B7. . . .PHYS. REV. B-2, ... 4743, (1970).
                  *FERMI SURFACE OF ALUMINUM FROM KOHN ANOMALIES.
                  *WEYMOUTH(J.W.), STEDMAN(R.).

23 AL, ...B1. . . . . . .IAEA SYMP. BOMBAY, VOL.1, ... 211, (1964).
                  *PHONONS IN ALUMINUM AT 80 DEGREES KELVIN.
                  *STEDMAN(R.), NILSSON(G.).

24 AL, ...B1. . . . . . .IAEA SYMP. COPENHAGEN, VOL.1, . 101, (1968).
                  *LATTICE DYNAMICS AND ELECTRONIC STRUCTURE OF SODIUM, MAGNESIUM AND
                  ALUMINIUM.
                  *SCHNEIDER(T.), STOLL(E.).

25 AL, ...T. . . . . . .IAEA SYMP. CHALK RIVER, VOL.2, . 55, (1962).
                  *THE FREQUENCIES OF THE NORMAL MODES OF ALUMINIUM.
                  *SQUIRES(G.L.).

SECTION 6 -QUASI-ELASTIC AND INELASTIC SCATTERING STUDIES . . . .

26 AL,. . . . . *LIQUID DYNAMICS FROM NEUTRON SCATTERING (REVIEW PAPER). 397, (1968).
        *LARSSON(K.E.).

27 AL,. . . . . *EXPERIMENTAL DISPERSION CURVES FOR PHONONS IN ALUMINUM. 57, (1963).
        *YARNELL(J.L.), WARREN(J.L.), KOENIG(S.H.).

28 AL,. . . . . *LATTICE VIBRATIONS OF ALUMINUM: A MODEL POTENTIAL APPROACH. 527, (1972).
        *PRASAD(B.), SRIVASTAVA(R.S.).

29 AL,. . . . . *INELASTIC NEUTRON SCATTERING FROM FE, AL, AND MN. . . . . .2171B, (1970).
        *LAMB(R.C.).

30 AL,. . . . . *DEPENDENCE OF PHONON FREQUENCIES ON THE PSEUDOPOTENTIAL FORM FACTOR FOR
        ALUMINUM.   991, (1969).
        *WALLACE(D.C.).

31 AL,. . . . . *ANHARMONIC INTERACTIONS IN ALUMINUM. II.   PHYS. REV. B VOL.3,. . . . 3568, (1971).
        *GILLIS(N.S.), KOEHLER(T.R.).

32 AL,. . . . . *TOTAL COHERENT CROSS SECTIONS FOR THE SCATTERING OF NEUTRONS FROM
        CRYSTALS.   PHYS. STAT. SOL. VOL.41. . . . 767, (1970).
        *BINDER(K.).

33 AL,. . . . . *STUDIES ON THE DETERMINATION OF THE PHONON DENSITY OF STATES BY COHERENT
        INELASTIC SCATTERING OF SLOW NEUTRONS FROM POLYCRYSTALLINE SAMPLES. 137, (1972).
        *GOMPF(F.), LAU(H.), REICHARDT(W.), SALGADO(J.).

34 AL,. . . . . *CALCULATIONS OF THE SUPERCONDUCTING TRANSITION TEMPERATURE IN ALUMINIUM.
        *CARBOTTE(J.B.), DYNES(R.C.).   PHYS. LETT. 25A. . . . . . . 685, (1967).

35 AL,. . . . . *PHONON FREQUENCY DISTRIBUTIONS AND HEAT CAPACITIES OF ALUMINUM AND LEAD.
        *STEDMAN(R.), ALMQVIST(L.), NILSSON(G.).   PHYS. REV. 162. . . . . 549, (1967).

36 AL,. . . . . *THE MEASUREMENT OF VERY SMALL DISORDERED SCATTERING CROSS-SECTIONS
        USING COLD NEUTRONS.   Z. NATURFORSCH. VOL.19A. . . . 354, (1964).
        *SCHERM(R.), SCHMATZ(W.). (IN GERMAN).

37 AL,. . . . . *SCATTERING OF SUB-THERMAL NEUTRONS BY DEFECTS IN NEUTRON IRRADIATED AL.
        *NIKLAUS(J.P.), SCHMATZ(W.), SIMSON(R.), WENZL(H.). (IN GERMAN). Z. ANGEW. PHYS. VOL.24,NO.6. . . . 313, (1968).

38 AL,. . . . . *COLLECTIVE ATOMIC MOTIONS IN LIQUID ALUMINUM STUDIED BY COLD NEUTRON
        SCATTERING.   ARK. FYS. VOL.33, PAPER 16. . . 271, (1966).
        *DAHLBORG(U.), LARSSON(K.E.).

39 AL,. . . . . *THE FREQUENCIES OF THE NORMAL MODES OF ALUMINUM. ARK. FYS. VOL.26 PAPER 17. . . . 223, (1964).
        *SQUIRES(G.L.).

40 AL,. . . . . *HIGH FREQUENCY WAVES IN LIQUID METALS.   ADV. IN PHYSICS VOL.16,. . . . . 189, (1967).
        *COCKING(S.J.).

41 AL,. . . . . *DIELECTRIC SCREENING AND PHONON DISPERSION IN AL. PHYSICA VOL.65,. . . . . . 63, (1973).
        *KACHHAVA(C.M.).

42 AL,. . . . . *A STUDY OF SOME TEMPERATURE EFFECTS ON THE PHONONS IN ALUMINUM BY USE
        OF COLD NEUTRONS.   ARK. FYS. VOL.17, PAPER 21. . . . 369, (1960).
        *LARSSON(K.E.), DAHLBORG(U.), HOLMRYD(S.).

43 AL,. . . . . *SCATTERING OF NEUTRONS BY PHONONS IN AN ALUMINUM SINGLE CRYSTAL.
        *BROCKHOUSE(B.N.), STEWART(A.T.).   PHYS. REV. VOL.100,. . . . . . . 756, (1955).

44 AL,. . . . . *ENERGY DISTRIBUTION OF SLOW NEUTRONS SCATTERED FROM SOLIDS. PHYS. REV. VOL.88,. . . . 542, (1952).
        *BROCKHOUSE(B.N.), HURST(D.G.).

45 AL,. . . . . *THE INELASTIC SCATTERING OF VERY SLOW NEUTRONS BY ALUMINIUM. PROC. ROY. SOC. A VOL.208,. . . . 527, (1951).
        *CASSELS(J.M.).

46 AL,. . . . . *ONE-PHONON COHERENT SCATTERING OF SLOW NEUTRONS FROM POLYCRYSTALLINE
        ALUMINUM.   J. NUCL. SCI. TECHNOL. VOL.3,. . . . 160, (1966).
        *IIJIMA(S.).

47 AL,. . . . . *OBSERVATIONS ON THE FERMI SURFACE OF ALUMINUM BY NEUTRON SPECTROMETRY. PHYS. REV. LETT. 15,. . . . 634, (1965).
        *STEDMAN(R.), NILSSON(G.).

48 AL,. . . . . *INVESTIGATION OF THE PHONON SPECTRUM OF ALUMINUM. SOV. PHYS. SOL. STATE 7, ... 1138, (1965).
        *BREDOV(M.M.), KOTOV(B.A.), OKUNEVA(N.M.), SHAKH-BUDAGOV(A.L.).

49 AL,. . . . . *NORMAL VIBRATIONS IN ALUMINUM AND DERIVED THERMODYNAMIC PROPERTIES. PHYS. REV. 143,. . . . 487, (1966).
        *GILAT(G.), NICKLOW(R.M.).

50 AL,. . . . . *DISPERSION RELATIONS FOR PHONONS IN ALUMINUM AT 80 AND 300 DEG. K. PHYS. REV. 145,. . . . 492, (1966).
        *STEDMAN(R.), NILSSON(G.).

51 AL,. . . . . *LATTICE VIBRATIONS OF ALUMINUM ON THE BASIS OF KREBS≠S MODEL. PHYS. STAT. SOL. 16,. . . . 513, (1966).
        *SHUKLA(M.M.), DAYAL(B.).

52 AL,. . . . . *MEASUREMENTS OF THE TEMPERATURE DEPENDENCE OF THE PHONONS IN ALUMINUM. IAEA SYMPOSIUM VIENNA,. . . . 587, (1960).
        *LARSSON(K.E.), HOLMRYD(S.), DAHLBORG(U.).

SECTION   6 -QUASI-ELASTIC AND INELASTIC SCATTERING STUDIES . . . .

53 AL,. . . . .   B1   IAEA SYMP. BOMBAY, VOL.1,   379, (1964).
    ~INTERFERENCE EFFECTS IN SLOW NEUTRON INELASTIC SCATTERING FROM POLY-
    CRYSTALLINE SOLIDS.
    ~SCHMUNK(R.E.), BRUGGER(R.M.), RANDOLPH(P.D.).

54 AL,. . . . .   B4   IAEA SYMP, BOMBAY, VOL.2,   117, (1964).
    ~COLLECTIVE ATOMIC MOTIONS IN LIQUID ALUMINIUM STUDIED BY COLD NEUTRON
    SCATTERING.
    ~LARSSON(K.E.), DAHLBORG(U.), JOVIC(D.).

55 AL,. . . . .   T   PHIL. MAG. VOL.4,   1325, (1959).
    ~SCATTERING OF THERMAL NEUTRONS IN ALUMINIUM.
    ~KOTHARI(L.S.).

56 AL,. . . . .   B1   PHYS. REV. VOL.106,   1168, (1957).
    ~INELASTIC SCATTERING OF SLOW NEUTRONS BY LATTICE VIBRATIONS IN ALUMINUM.
    ~CARTER(R.S.), PALEVSKY(H.), HUGHES(D.J.).

57 AL,. . . . .   B1   REV. MOD. PHYS. VOL.30,   236, (1958).
    ~NORMAL MODES OF ALUMINUM BY NEUTRON SPECTROMETRY.
    ~BROCKHOUSE(B.N.), STEWART(A.T.).

58 ALCOHOLS(N-AMYL),.   C3   IAEA SYMP. CHALK RIVER, VOL.1,   413, (1962).
    ~INELASTIC SCATTERING OF NEUTRONS BY METHYL-, ETHYL- AND N-AMYL ALCOHOLS.
    ~SAUNDERSON(D.H.), RAINEY(V.S.).

59 ALCOHOLS,. . . .   P,C3   J. NUCL. ENERGY VOL.26,   379, (1972).
    ~SLOW-NEUTRON SCATTERING CROSS-SECTION FOR METHANOL, ETHANOL, PROPANOL,
    ISO-PROPANOL, BUTANOL, ETHANEDIOL AND PROPANETRIOL.
    ~RODRIGUES(C.), VINHAS(L.A.), HERDADE(S.B.), DO-AMARAL(L.Q.).

60 AL-CU, ...   T   PHYS. REV. B VOL.9,   353, (1974).
    ~LOCAL MODES IN AL(.1)-CU(.9) AND (N*H4(.1)-K(.9))*CL IN THE COHERENT-
    POTENTIAL APPROXIMATION.
    ~KAPLAN(T.), MOSTOLLER(M.).

61 ALKALI METALS, .   B1   PHYSICA VOL.50,   10, (1970).
    ~PHONON DISPERSION IN ALKALI METALS.
    ~SINGH(S.N.), PRAKASH(S.).

62 AL(MN),. . . . .   D3   PHYS. LETT. VOL.40A,   173, (1972).
    ~NEUTRON SCATTERING BY SPIN FLUCTUATIONS IN DILUTE AL(MN) ALLOY.
    ~KROO(N.), SZENTIRMAY(Z.).

63 AL2*O3 (SAPPHIRE),   B1   PHYS. LETT. VOL.43A NO.2,   97, (1973).
    ~DISPERSION OF ACOUSTIC PHONONS IN SAPPHIRE.
    ~BIALAS(H.), WEIS(O.), WENDEL(H.).

64 AL2*O3-TYPE STRUCTURE,   D1,T   PHYSICA VOL.43,   353, (1969).
    ~SPIN WAVES IN ANTIFERROMAGNETS WITH CORUNDUM STRUCTURE.
    ~SAMUELSEN(E.J.).

65 AL*TH2-H,. . . .   C1   IAEA SYMPOSIUM VIENNA,   501, (1960).
    ~INELASTIC NEUTRON SCATTERING EXPERIMENTS ON A FEW METAL HYDRIDES.
    ~BERGSMA(J.), GOEDKOOP(J.A.).

66 AMMONIA, ...   C2   PHYS. STAT. SOL. VOL.50,   701, (1972).
    ~FREQUENCY DISTRIBUTION FUNCTION OF SOLID AMMONIA.
    ~GOYAL(P.S.), DASANNACHARYA(B.A.), THAPER(C.L.), IYENGAR(P.K.).

67 AMMONIUM SALTS,.   C1   J. CHEM. PHYSICS VOL.37,   234, (1962).
    ~ROTATIONAL FREEDOM OF AMMONIUM IONS AND METHYL GROUPS BY CROSS-SECTION
    MEASUREMENTS WITH SLOW NEUTRONS.
    ~RUSH(J.J.), TAYLOR(T.I.), HAVENS-JR(W.W.).

68 ANTHRACENE,. . .   B1   SOL. STATE COMM. VOL.8,   165, (1970).
    ~LATTICE DYNAMICS OF DEUTERATED ANTHRACENE.
    ~LUTZ(U.A.), HALG(W.).

69 AR,. . . . . . .   B4   IAEA SYMP. CHALK RIVER, VOL.1,   189, (1962).
    ~LIQUID DYNAMICS FROM NEUTRON SPECTROMETRY.
    ~BROCKHOUSE(B.N.), BERGSMA(J.), DASANNACHARYA(B.A.), POPE(N.K.).

70 AR,. . . . . . .   B4   IAEA SYMP. BOMBAY, VOL.2,   101, (1964).
    ~INELASTIC SCATTERING OF COLD NEUTRONS BY CONDENSED ARGON.
    ~KROO(N.), BORGONOVI(G.), SKOLD(K.), LARSSON(K.E.).

71 AR,. . . . . . .   C3   IAEA SYMP. COPENHAGEN, VOL.1,   545, (1968).
    ~ANALYSIS OF NEUTRON SCATTERING EXPERIMENTS ON LIQUIDS.
    ~AGRAWAL(A.K.), DESAI(R.C.), YIP(S.).

72 AR,. . . . . . .   C3   IAEA SYMP. COPENHAGEN, VOL.1,   457, (1968).
    ~TEMPERATURE DEPENDENCE OF QUASI-ELASTIC SCATTERING OF COLD NEUTRONS FROM
    LIQUID ARGON.
    ~ANDRUS(W.S.), MUETHER(H.R.), PALEVSKY(H.).

73 AR,. . . . . . .   B4   IAEA SYMP. COPENHAGEN, VOL.1,   397, (1968).
    ~LIQUID DYNAMICS FROM NEUTRON SCATTERING (REVIEW PAPER).
    ~LARSSON(K.E.).

74 AR,. . . . . . .   B5   PHYS. LETT. VOL.33A   338, (1970).
    ~ON THE COHERENT INTERMEDIATE SCATTERING FUNCTION OF GASEOUS ARGON AS DE-
    TERMINED BY INELASTIC NEUTRON SCATTERING/MOLECULAR DYNAMICS CALCULATIONS
    ~BRUIN(C.), HASMAN(A.).

75 AR,. . . . . . .   C4   PHYS. LETT. VOL.35A,   169, (1971).
    ~THE SELF-DIFFUSION COEFFICIENT IN ARGON AT LIQUID DENSITIES.
    ~VAN-LOEF(J.J.).

76 AR,. . . . . . .   B4   DISS. ABS. VOL.30,   4293B, (1970).
    ~INELASTIC SCATTERING OF COLD NEUTRONS FROM LIQUID ARGON AT ELEVATED
    TEMPERATURES.
    ~ANDRUS(W.S.).

77 AR,. . . . . . .   B4   DISS. ABS. VOL.30,   3807B, (1970).
    ~NEUTRON SCATTERING STUDY OF THE DYNAMICS IN DENSE ARGON GAS.
    ~ANDRIESSE(C.D.).

78 AR (ISO. A=36),.   B5   PHYS. LETT. VOL.28A,   642, (1969).
    ~INELASTIC NEUTRON SCATTERING BY AR (ISO. A=36) CLOSE TO CONDENSATION.
    ~ANDRIESSE(C.D.), COMPAGNER(A.), HASMAN(A.), VAN-LOEF(J.J.),
    VAN-ZEVENBERGEN(F.).

79 AR,. . . . . . .   B4   PHYS. REV. VOL.161,   102, (1967).
    ~ATOMIC MOTION IN LIQUID ARGON.
    ~SKOLD(K.), LARSSON(K.E.).

SECTION   6 -QUASI-ELASTIC AND INELASTIC SCATTERING STUDIES . . . .

80 AR,. . . . . .B7,C4,. . .IAEA SYMP. GRENOBLE. . . . . . 413, (1972).
    *THE COHERENT SCATTERING FUNCTION OF LIQUID ARGON AT 85.2 DEGREES K.
    *ROWE(J.M.), SKOLD(K.).

81 AR,. . . . . .B4. . . .PHYS. LETT. VOL.19,. . . . . . . 269, (1965).
    *CO-OPERATIVE MODES OF MOTION IN SIMPLE LIQUIDS.
    *CHEN(S.H.),   EDER(O.J.), EGELSTAFF(P.A.), HAYWOOD(B.C.), WEBB(F.J.).

82 AR,. . . . . .C4. . . . . .J. PHYS. C VOL.5,. . . . . .3279, (1972).
    *DENSITY FLUCTUATIONS IN SIMPLE LIQUIDS: A GENERALIZED HYDRODYNAMIC
    APPROACH.
    *BARKER(M.I.), GASKELL(T.).

83 AR,. . . . . .T. . . . . .PHYS. REV. LETT. VOL.25, ... 1423, (1970).
    *DYNAMICS OF THE LIQUID-SOLID TRANSITION.
    *SCHNEIDER(T.), BROUT(R.), THOMAS(H.), FEDER(J.).

84 AR,. . . . . .B4,T. . . . .PHYSICA VOL.31,. . . . . . . 1257, (1965).
    *COHERENT SCATTERING OF SLOW NEUTRONS BY LIQUID ARGON. II.
    *SINGWI(K.S.).

85 AR,. . . . . .B4,C3. . .PHYS. REV. A VOL.6,. . . . . . 1107, (1972).
    *COHERENT- AND INCOHERENT-SCATTERING LAWS OF LIQUID ARGON.
    *SKOLD(K.), ROWE(J.M.), OSTROWSKI(G.), RANDOLPH(P.D.).

86 AR,. . . . . .C3. . . . . .PHYS. REV. VOL.137,. . . . . . . A417, (1965).
    *NEUTRON SCATTERING FROM LIQUID ARGON.
    *DASANNACHARYA(B.A.), RAO(K.R.).

87 AR,. . . . . .B1. . . . . .J. PHYS. C, SER.2, VOL.3,. . . . 249, (1970).
    *LATTICE DYNAMICS OF ARGON AT 4 DEGREES KELVIN.
    *BATCHELDER(D.N.), COLLINS(M.F.), HAYWOOD(B.C.G.), SIDEY(G.R.).

88 AR,. . . . . .C3. . . . . .PHYS. REV. LETT. 12,. . . . . 721, (1964).
    *INELASTIC SCATTERING OF COLD NEUTRONS BY CONDENSED ARGON.
    *KROO(N.), BORGONOVI(G.M.), SKOLD(K.), LARSSON(K.E.).

89 AR,. . . . . .T,B4. . . . . .PHYS. REV. LETT. VOL.16, ... 839, (1966).
    *COLLECTIVE MOTION IN LIQUID ARGON.
    *DESAI(R.C.), NELKIN(M.).

90 AR,. . . . . .B5. . . . . .PHYSICA VOL.63 NO.3,. . . . . 499, (1973).
    *INTERMEDIATE SCATTERING FUNCTIONS OF HIGH-DENSITY GASEOUS ARGON OBTAINED
    BY QUASIELASTIC NEUTRON-SCATTERING STUDIES.
    *HASMAN(A.).

91 AR,. . . . . .T,B1. . . . .J. SOLID STATE CHEM. VOL.5,. . . 477, (1972).
    *INTERPRETATION OF PHONON DISPERSION IN SOLID ARGON AND NEON.
    *GUPTA(N.P.).

92 AR,. . . . . .B4. . . . . .PHYS. LETT. VOL.26A, ... 152, (1968).
    *INELASTIC SCATTERING OF COLD NEUTRONS FROM LIQUID ARGON AT ELEVATED
    TEMPERATURES.
    *ANDRUS(W.S.), MUETHER(H.R.), PALEVSKY(H.).

93 AR,. . . . . .T,B4. . . . . .PHYS. REV. VOL.166,. . . . . . 129, (1968).
    *DYNAMICAL CORRELATIONS IN SIMPLE LIQUIDS AND COLD-NEUTRON SCATTERING BY
    ARGON.
    *DESAI(R.C.), YIP(S.).

94 AR,. . . . . .P. . . . . . . .PHYS. REV. VOL.125,. . . . . . . 275, (1962).
    *ANOMALOUS NEUTRON SCATTERING OF AR (A=36).
    *CHRIEN(R.E.), JAIN(A.P.), PALEVSKY(H.).

95 AR,. . . . . .P,B1. . . . . .PHYS. STAT. SOL. B VOL.43,. . . 611, (1971).
    *ELASTIC CONSTANTS OF ARGON AT 4.2 DEGREES K FROM PHONON MEASUREMENTS.
    *DORNER(B.), EGGER(H.).

96 AR,. . . . . .B1,T. . . . .PHYS. REV. LETT. 21, ... 881, (1968).
    *ZERO SOUND IN CLASSICAL LIQUIDS.
    *SINGWI(K.S.), SKOLD(K.), TOSI(M.P.).

97 AR,. . . . . .B1. . . . . . .PHYS. LETT. 28A, ... 433, (1968).
    *PHONON DISPERSION MEASUREMENTS ON AN ARGON SINGLE CRYSTAL AT 4.2 DEG. K.
    *EGGER(H.), GSANGER(M.), LUSCHER(E.), DORNER(B.).

98 AR,. . . . . .B2. . . . . . .J. PHYS. C, SER.2, VOL.4,. . . . 910, (1971).
    *TEMPERATURE DEPENDENCE OF PHONON FREQUENCIES IN ARGON BY INELASTIC
    NEUTRON SCATTERING.
    *BATCHELDER(D.N.), HAYWOOD(B.C.G.), SAUNDERSON(D.H.).

99 AR,. . . . . .B4. . . . . .PHYSICA VOL.50,. . . . . . . 511, (1970).
    *TEMPERATURE DEPENDENCE OF THE ATOMIC SELF-MOTION IN LIQUID ARGON.
    *ZANDVELD(P.), ANDRIESSE(C.D.), BREGMAN(J.D.), HASMAN(A.), VAN-LOEF(J.J.)

100 AR,. . . . . .B5. . . . . .PHYSICA VOL.48,. . . . . . . 61, (1970).
    *ATOMIC MOTION IN GASEOUS ARGON. I.DETERMINATION OF THE INTERMEDIATE
    SCATTERING FUNCTION OF GASEOUS ARGON USING SUBTHERMAL NEUTRONS.
    *ANDRIESSE(C.D.).

101 AR,. . . . . .B4. . . . . .THESIS. . . . . , (1971).
    *NEUTRON QUASI-ELASTIC SCATTERING STUDIES ON FLUID ARGON.
    *HASMAN(A.). (PH.D. DISSERTATION - DELFT, THE NETHERLANDS).

102 AR,. . . . . .B4. . . . . .PHYSICA VOL.60,. . . . . . . 27, (1972).
    *LIQUID-ARGON DYNAMICS FROM...FIRST MOMENTS OF...COHERENT SCATT. FCN.:THE
    DENSITY RESPONSE FCN./DISPERSION CURVE/INTERMEDIATE SCATT. FCN.
    *DUBOIS(D.M.).

103 AR-C*H4, ...B4,C3. . .PHYSICA VOL.72,. . . . . . . 300, (1974).
    *A COMPARATIVE STUDY OF THE MOTIONS OF METHANE MOLECULES AND ARGON ATOMS
    IN LIQUID AND HIGH-PRESSURE GAS STATES BY NEUTRON SCATTERING.
    *OLSSON(L.G.), LARSSON(K.E.).

104 ARGON-LIKE LIQUID,. . . . .C3. . . . . .IAEA SYMP. COPENHAGEN, VOL.1,. . 561, (1968).
    *CURRENT FLUCTUATIONS IN CLASSICAL LIQUIDS.
    *RAHMAN(A.).

105 AR(H), ...C3. . . . . .IAEA SYMP. COPENHAGEN, VOL.1,. . 545, (1968).
    *ANALYSIS OF NEUTRON SCATTERING EXPERIMENTS ON LIQUIDS.
    *AGRAWAL(A.K.), DESAI(R.C.), YIP(S.).

106 AR(H), ...C3. . . . . .PHYS. REV. LETT. 19, ... 1023, (1967).
    *SMALL ENERGY TRANSFER SCATTERING OF COLD NEUTRONS FROM LIQUID ARGON.
    *SKOLD(K.).

SECTION   6 -QUASI-ELASTIC AND INELASTIC SCATTERING STUDIES . . . .

107   AR(H), ... C3 . . . . . . PROC. PHYS. SOC. 89, ... 833, (1966).
      *MOLECULAR MOTION OF HYDROGEN IN LIQUID ARGON AND NEON.
      *EDER(O.J.), CHEN(S.H.), EGELSTAFF(P.A.).

108   AR-(H), ... T . . . . . . PROC. PHYS. SOC. 86, ... 965, (1965).
      *THEORY OF COLD NEUTRON SCATTERING BY HYDROGEN IN LIQUID ARGON.
      *SEARS(V.F.).

109   AROMATIC CRYSTALS, ... B7 . . . . . DISC. FARADAY SOC. NO.48, ... 131, (1969).
      *INELASTIC NEUTRON SCATTERING FROM MOLECULAR AND CRYSTAL EXCITATIONS IN
      AROMATIC MOLECULAR CRYSTALS.
      *REYNOLDS(P.A.), WHITE(J.W.).

110   AU, ... P . . . . . . DISS. ABS. VOL.31, ... 867B, (1970).
      *FAST NEUTRON SPUTTERING FROM POLYCRYSTALLINE AND MONOCRYSTALLINE GOLD
      CRYSTALS.
      *FAIRAND(B.P.).

111   AU, ... B1 . . . . . . PHYS. REV. B VOL.8, ... 3493, (1973).
      *LATTICE DYNAMICS OF GOLD.
      *LYNN(J.W.), SMITH(H.G.), NICKLOW(R.M.).

112   AU-CR, ... D2 . . . . . SOL. STATE COMM. VOL. 9, ... 921, (1971).
      *DIFFUSE SCATTERING OF NEUTRONS NEAR THE CRITICAL COMPOSITION IN AU-CR
      ALLOYS.
      *NAKAI(Y.), KUNITOMI(N.), ENDOH(Y.), ISHIKAWA(Y.).

113   AU-CR, ... D2 . . . . . J. PHYS. SOC. JAPAN VOL.34, . . . 1197, (1973).
      *ANISOTROPIC DIFFUSE SCATTERING OF NEUTRONS IN AU-CR ALLOYS.
      *NAKAI(Y.), URANO(T.), KUNITOMI(N.).

114   AZOXYANISOLE, ... C3 . . . . . IAEA SYMP. GRENOBLE, ... 515, (1972).
      *A STUDY OF THE ANISOTROPY OF SELF-DIFFUSION IN MAGNETICALLY
      ORIENTED PARA-AZOXYANISOLE.
      *JANIK(J.A.), JANIK(J.M.), OTNES(K.), PYNN(R.).

115   AZOXYANISOLE, ... C3,C1 . . . . ACTA PHYS. POLON. VOL.25, ... 845, (1964).
      *TOTAL NEUTRON SCATTERING CROSS-SECTION STUDY OF MOLECULAR MOTIONS IN
      DIMETHOXYAZOXYBENZENE (P-AZOXYANISOLE).
      *JANIK(J.A.), JANIK(J.M.).

116   AZOXYANISOLE(PARA) . . . . . . C1 . . . . . FARADAY SYMP. CHEM. SOC. NO.6, . 48, (1972).
      *ROTATIONAL DIFFUSION IN PARA-AZOXYANISOLE.
      *JANIK(J.A.), WROBEL(S.), JANIK(J.M.), MIGDAL(A.), URBAN(S.).

117   (BA*CL2)*2(H2*O), ... C1 . . . . . IAEA SYMP. BOMBAY, VOL.2, ... 355, (1964).
      *INVESTIGATION OF THE DYNAMICS OF N*H4+ AND H2*O MOLECULAR GROUPS IN
      CRYSTALS.
      *.C.C.C.P.

118   BA*(CL*O3)2*(H2*O) . . . . C2 . . . . . SOL. STATE COMM. VOL.8, ... 497, (1970).
      *OBSERVATION OF LIBRATIONAL MODES OF WATER MOLECULES IN SINGLE CRYSTAL
      HYDRATES BY NEUTRON SCATTERING.
      *THAPER(C.L.),   DASANNACHARYA(B.A.),   SEQUEIRA(A.),   IYENGAR(P.K.).

119   BA*(CL*O3)2.H2*O, ... C1 . . . . . J. CHEM. PHYSICS VOL.45, . . . 699, (1966).
      *LOW-FREQUENCY MOTIONS OF H2*O MOLECULES IN CRYSTALS. II.
      *PRASK(H.J.), BOUTIN(H.).

120   BA*F2, ... B1 . . . . . . SOL. STATE COMM. VOL.8, ... 463, (1970).
      *THE CRYSTAL DYNAMICS OF BARIUM FLUORIDE.
      *HURRELL(J.P.), MINKIEWICZ(V.J.).

121   BA*F2(ER3+), ... D1 . . . . . SOV. PHYS. SOL. STATE 6, ... 76, (1964).
      *SPIN-LATTICE RELAXATION OF THE ER +3 ION IN CD*F2, BA*F2, AND CA*F2
      SINGLE CRYSTALS.
      *ZVEREV(G.M.), SMIRNOV(A.I.).

122   BA*H2, ... C1 . . . . . J. CHEM. PHYS. VOL.52, ... 3952, (1970).
      *VIBRATION SPECTRA OF THE ORTHORHOMBIC ALKALINE-EARTH HYDRIDES BY THE
      INELASTIC SCATTERING OF COLD NEUTRONS AND BY INFRARED TRANSM. MEASUR.
      *MAELAND(A.J.).

123   BA*LI*H3, ... C1 . . . . . J. CHEM. PHYSICS VOL.51, ... 2915, (1969).
      *TERNARY HYDRIDES POSSESSING THE CUBIC PEROVSKITE STRUCTURE. II.
      VIBRATION SPECTRA BY THE INELASTIC SCATTERING OF COLD NEUTRONS.
      *MAELAND(A.J.).

124   BA*TI*O3, ... B1,B6 . . . . PHYS. REV. B-2, ... 3651, (1970).
      *INELASTIC NEUTRON SCATTERING FROM SINGLE-DOMAIN BA*TI*O3.
      *SHIRANE(G.), AXE(J.D.), HARADA(J.), LINZ(A.).

125   BA*TI*O3, ... C2 . . . . . IAEA SYMPOSIUM VIENNA, ... 601, (1960).
      *INELASTIC NEUTRON SPECTRA FROM FERRO-ELECTRIC AND PARA-ELECTRIC
      BA*TI*O3.
      *PELAH(I.), LEFKOWITZ(I.).

126   BA*TI*O3, ... B1 . . . . . Z. PHYS. VOL.220, ... 145, (1969).
      *SOFT PHONON DISPERSION IN BA*TI*O3.
      *HULLER(A.). (ED. NOTE: IN ENGLISH).

127   BA*TI*O3, ... C1 . . . . . SOV. PHYS. SOL. STATE 8, ... 2156, (1966).
      *SCATTERING OF COLD NEUTRONS BY POLYCRYSTALLINE BARIUM, STRONTIUM AND
      LEAD TITANATES.
      *SOLOV*EV(S.P.), KUKHTO(O.L.), CHERNOPLEKOV(N.A.), ZETLYANOV(M.G.).

128   BA*TI*O3, ... B1 . . . . . PHYS. REV. LETT. 19, ... 234, (1967).
      *SOFT OPTIC MODES IN BARIUM TITANATE.
      *SHIRANE(G.),   FRAZER(B.C.),   MINKIEWICZ(V.J.), LEAKE(J.A.), LINZ(A.).

129   BA*TI*O3, ... B6 . . . . . PHYS. REV. B VOL.4, ... 155, (1971).
      *NEUTRON-SCATTERING STUDY OF SOFT MODES IN CUBIC BA*TI*O3.
      *HARADA(J.), AXE(J.D.), SHIRANE(G.).

130   BA*TI*O3, ... D2 . . . . . PHYS. REV. VOL.177, ... 848, (1969).
      *STUDY OF CRITICAL FLUCTUATIONS IN BA*TI*O3 BY NEUTRON SCATTERING.
      *YAMADA(Y.), SHIRANE(G.), LINZ(A.).

131   BE, ... C2 . . . . . IAEA SYMP. CHALK RIVER, VOL.2, . 199, (1962).
      *THE NEUTRON SCATTERING LAW AND THE FREQUENCY DISTRIBUTION OF THE NORMAL
      MODES OF BERYLLIUM AND BERYLLIUM OXIDE.
      *SINCLAIR(R.N.).

132   BE, ... T . . . . . . IAEA SYMP. COPENHAGEN, VOL.1, . . 165, (1968).
      *THEORY OF VIBRATIONAL SPECTRUM IN HEXAGONAL METALS.
      *.C.C.C.P.

SECTION   6 -QUASI-ELASTIC AND INELASTIC SCATTERING STUDIES . . . .

133 BE,. . . . . .THE ROLE OF ELECTRONS IN PHONON SPECTRUM FORMATION IN METALS. IAEA SYMP. COPENHAGEN, VOL.1,.   3, (1968).
.C.C.C.P.

134 BE,. . . . . .PHONON DISPERSION CURVES IN BERYLLIUM. CONF. INTERN., RENNES, ...   140, (1971).
.THAPER(C.L.), RAO(K.R.), DASANNACHARYA(B.A.), ROY(A.P.), IYENGAR(P.K.).

135 BE,. . . . . .CALCULATION AND INTERPRETATION OF NEUTRON THERMALIZATION IN FINITE DISS. ABS. 27,.   3625B, (1966).
BLOCKS OF BERYLLIUM.
.LEE(R.R.).

136 BE,. . . . . .THE EFFECTS OF COHERENT SCATTERING ON THE THERMALIZATION OF NEUTRONS IN DISS. ABS. 27,. . . . . 573B, (1966).
BERYLLIUM.
.FULLWOOD(R.R.).

137 BE,. . . . . .LATTICE-DYNAMICAL CALCULATIONS FOR ZINC AND BERYLLIUM. PHYS. REV. B VOL.4,. . . 1390, (1971).
.BEZDEK(H.F.), FINEGOLD(L.).

138 BE,. . . . . .COHERENT INELASTIC SCATTERING FROM POLYCRYSTALLINE BERYLLIUM. PHYS. LETT. 16,.   235, (1965).
.YOUNG(J.A.), KOPPEL(J.U.).

139 BE,. . . . . .LATTICE VIBRATIONAL SPECTRA OF BERYLLIUM, MAGNESIUM AND ZINC. PHYS. REV. VOL.134,. . .A1476, (1964).
.YOUNG(J.A.), KOPPEL(J.U.).

140 BE,. . . . . .SLOW NEUTRON INELASTIC SCATTERING FROM BERYLLIUM POWDERS. PHYS. REV. VOL.136,. . . .A1303, (1964).
.SCHMUNK(R.E.).

141 BE,. . . . . .SCATTERING KERNEL FOR BERYLLIUM. NUCL. SCI. ENGNG. VOL.19,. . . . 367, (1964).
.YOUNG(J.A.), KOPPEL(J.U.).

142 BE,. . . . . .THE MEASUREMENT OF VERY SMALL DISORDERED SCATTERING CROSS-SECTIONS Z. NATURFORSCH. VOL.19A,. . . . . 354, (1964).
USING COLD NEUTRONS.
.SCHERM(R.), SCHMATZ(W.). (IN GERMAN).

143 BE,. . . . . .THE SCATTERING OF COLD NEUTRONS BY METALS. UKAEA AERE REP. R4101 (24 PP.),. . . . , (1962).
.EGELSTAFF(P.A.).

144 BE,. . . . . .UNTERSUCHUNG ÜBER DIE STREUUNG LANGSAMER NEUTRONEN AN BESTRAHLTEM QUARTZ PHYS. STAT. SOL. 7,. . . . . 415, (1964).
AND BERYLLIUM.
.BAIERLEIN(K.).

145 BE,. . . . . .MESURE DE L≠ENERGIE DE NEUTRONS TRES LENTS APRES UNE DIFFUSION C.R. ACAD. SCI. TOME 240,. . . . . 745, (1955).
INELASTIQUE PAR DES POLYCRISTAUX ET DES MONOCRISTAUX.
.JACROT(B.).

146 BE,. . . . . .LATTICE DYNAMICS OF BERYLLIUM. PHYS. REV. VOL.128,. . . . . 562, (1962).
.SCHMUNK(R.E.), BRUGGER(R.M.), RANDOLPH(P.D.), STRONG(K.A.).

147 BE,. . . . . .EXTENSION OF THE DISPERSION-RELATIONS MEASUREMENTS OF BERYLLIUM. PHYS. REV. 149,.   450, (1966).
.SCHMUNK(R.E.).

148 BE,. . . . . .INTERFERENCE EFFECTS IN SLOW NEUTRON INELASTIC SCATTERING FROM POLY- IAEA SYMP. BOMBAY, VOL.1,.   379, (1964).
CRYSTALLINE SOLIDS.
.SCHMUNK(R.E.), BRUGGER(R.M.), RANDOLPH(P.D.).

149 BE,. . . . . .SCATTERING OF THERMAL NEUTRONS IN BERYLLIUM. J. NUCL. ENERGY VOL.6,. . . . . 104, (1957).
.BHANDARI(R.C.).

150 BE-CU, ...LOCAL MODES IN LI-MG AND BE-CU ALLOYS. PHYS. LETT. 24A, ... 517, (1967).
.NATKANIEC(I.), PARLINSKI(K.), BAJOREK(A.),   SUDNIK-HRYNKIEWICZ(M.).

151 BE*F2, ...INELASTIC COLD NEUTRON SCATTERING FROM CRYSTALLINE AND VITREOUS J. NON-CRYST. SOLIDS VOL.3,. . . . 239, (1970).
BERYLLIUM FLUORIDE.
.LEADBETTER(A.J.), WRIGHT(A.C.).

152 BE*O,. . . . . . .THE NEUTRON SCATTERING LAW AND THE FREQUENCY DISTRIBUTION OF THE NORMAL IAEA SYMP. CHALK RIVER, VOL.2,. 199, (1962).
MODES OF BERYLLIUM AND BERYLLIUM OXIDE.
.SINCLAIR(R.N.).

153 BE*O,. . . . . . .PHONON DISPERSION RELATION OF BERYLLIUM OXIDE. IAEA SYMP. COPENHAGEN, VOL.1,. . 315, (1968).
.OSTHELLER(G.L.), SCHMUNK(R.E.), BRUGGER(R.M.), KEARNEY(R.J.).

154 BE*O,. . . . . . .MODES NORMAUX DE VIBRATION DE L≠OXYDE DE BERYLLIUM. COMP. REND. 268,. . . . . 755, (1969).
.NUSIMOVICI(M.A.).

155 BE*O,. . . . . . .STUDY OF THE NEUTRON SPECTRA EMERGING FROM MODERATING ASSEMBLIES. IAEA SYMPOSIUM (VIENNA),. . . . . 631, (1960).
.RAMANNA(R.), SARMA(N.).

156 BE*O,. . . . . . .MEASUREMENT OF THE THERMAL NEUTRON WAVE DISPERSION RELATIONS IN BE*O. J. NUCL. ENERGY(GB) VOL.26,. . . 27, (1972).
.RITCHIE(A.I.M.), WHITTLESTONE(S.).

157 BE*O,. . . . . . .INELASTIC SCATTERING OF COLD NEUTRONS BY POLYCRYSTALLINE BERYLLIUM J. PHYS. CHEM. SOLIDS VOL.23,. . 1747, (1962).
OXIDE IN THE TEMPERATURE RANGE FROM 100 TO 1000 DEGREES K.
.BEGUM(R.J.), MADHAV-RAO(L.), UMAKANTHA(N.).

158 BE*O,. . . . . . .THE OBSERVATION OF RADIATION-INDUCED DEFECTS IN BE*O BY MEANS OF LONG- PROC. BRIT. CERAM. SOC. NO.7,. 391, (1967).
WAVELENGTH NEUTRON-SCATTERING MEASUREMENTS.
.MARTIN(D.G.).

SECTION   6 -QUASI-ELASTIC AND INELASTIC SCATTERING STUDIES . . . .

159 BE*O,. . . . . . . ↱THERMAL VIBRATIONS IN BE*O. . . . .J. NUCLEAR MATERIALS VOL.14, . .   275, (1964).
     ↱PRYOR(A.W.), SABINE(T.M.).

160 BE*O,. . . . . . ↱THE SCATTERING OF LONG WAVELENGTH NEUTRONS BY IRRADIATED BERYLLIUM
     OXIDE.   ↱PHIL. MAG. VOL.8, ...   43, (1963).
     ↱SABINE(T.M.), PRYOR(A.W.), HICKMAN(B.S.).

161 BE*O,. . . . . . ↱INELASTIC SCATTERING OF THERMAL NEUTRONS PRODUCED BY AN ELECTRON
     ACCELERATOR.   ↱IAEA SYMPOSIUM VIENNA, ...   511, (1960).
     ↱WHITTEMORE(W.L.), MCREYNOLDS(A.W.).

162 BE*O,. . . . . . ↱LATTICE VIBRATIONS OF BE*O. . . . . .J. PHYS. CHEM. SOL. 28,. . . . .   249, (1967).
     ↱BRUGGER(R.M.), STRONG(K.A.), CARPENTER(J.M.).

163 BE*O,. . . . . . ↱SCATTERING OF THERMAL NEUTRONS IN BERYLLIUM OXIDE.   J. NUCL. ENERGY VOL.7, ...   45, (1958).
     ↱BHANDARI(R.C.), KOTHARI(L.S.), SINGWI(K.S.).

164 BI,. . . . . . . ↱ATOMIC VIBRATIONS IN FACE-CENTERED-CUBIC ALLOYS OF BI, PB AND TL.   IAEA SYMP. COPENHAGEN, VOL.1, . 253, (1968).
     ↱NG(S.C.), BROCKHOUSE(B.N.).

165 BI,. . . . . . . ↱PHONON DISPERSION RELATIONS IN BISMUTH.   IAEA SYMP. COPENHAGEN, VOL.1,. .   157, (1968).
     ↱SOSNOWSKI(J.), BEDNARSKI(S.), CZACHOR(A.).

166 BI,. . . . . . . ↱THE TRANSMISSION OF THERMAL NEUTRONS THROUGH A LARGE SINGLE CRYSTAL OF
     BISMUTH AT LIQUID NITROGEN TEMPERATURE.   IAEA SYMP. CHALK RIVER, VOL.1, . 139, (1962).
     ↱MENARDI(S.), HAAS(R.), KLEY(W.).

167 BI,. . . . . . . ↱MOLECULAR DYNAMICS IN LIQUID NITROGEN AND LIQUID BISMUTH INVESTIGATED BY
     COLD NEUTRON SCATTERING.   IAEA SYMP. COPENHAGEN, VOL.1,. .   439, (1968).
     ↱MATEESCU(N.), TEUTSCH(H.), DIACONESCU(A.), NAHORNIAK(V.).

168 BI,. . . . . . . ↱PHONONS IN LIQUID BISMUTH STUDIED BY NEUTRON INELASTIC SCATTERING.   IAEA SYMP. COPENHAGEN, VOL.1,. .   431, (1968).
     ↱TUNKELO(E.), KUOPPAMAKI(R.), PALMGREN(A.).

169 BI,. . . . . . . ↱TEMPERATURE DEPENDENCE OF THE COHERENT SCATTERING AMPLITUDE FOR FORWARD
     SCATTERED NEUTRONS.   DISS. ABS. VOL. 28,. . . . . . .   1107B, (1967).
     ↱EDWARDS(T.R.).

170 BI,. . . . . . . ↱LATTICE DYNAMICS OF BISMUTH.   DISS. ABS. 29, ...   736B, (1968).
     ↱SMITH(D.B.).

171 BI,. . . . . . . ↱THERMAL VIBRATION SPECTRUM OF THE BISMUTH LATTICE.   SOV. PHYS. SOL. STATE VOL.11,. . 1615, (1969).
     ↱KOTOV(B.A.), OKUNEVA(N.M.), PLACHENOVA(E.L.).

172 BI,. . . . . . . ↱TOTAL COHERENT CROSS SECTIONS FOR THE SCATTERING OF NEUTRONS FROM
     CRYSTALS.   PHYS. STAT. SOL. VOL.41,. . . . .   767, (1970).
     ↱BINDER(K.).

173 BI (ISO. A=209), ↱INELASTIC NEUTRON SCATTERING BY BI (ISO. A=209) WITH EXCITATION OF LOW-
     LYING LEVELS.   BU. ACAD. SCI. USSR PHYS. SR. 32, 651, (1968).
     ↱DEGTYAREV(YU.G.), PROTOPOPOV(V.N.).

174 BI,. . . . . . . ↱MEASUREMENT OF INCOHERENT SCATTERING CROSS-SECTION OF LEAD AND BISMUTH
     WITH SLOW NEUTRONS.   NUKLEONIC VOL.12,. . . . .   4, (1968).
     ↱SCHERM(R.).

175 BI,. . . . . . . ↱THE MEASUREMENT OF VERY SMALL DISORDERED SCATTERING CROSS-SECTIONS
     USING COLD NEUTRONS.   Z. NATURFORSCH. VOL.19A, ...   354, (1964).
     ↱SCHERM(R.), SCHMATZ(W.). (IN GERMAN).

176 BI,. . . . . . . ↱HIGH FREQUENCY WAVES IN LIQUID METALS.   ADV. IN PHYSICS VOL.16,. . . . . .   189, (1967).
     ↱COCKING(S.J.).

177 BI,. . . . . . . ↱MEASUREMENT OF COHERENT SCATTERING OF NEUTRONS AT MOLTEN SODIUM, CESIUM
     AND BISMUTH AT DIFFERENT TEMPERATURES.   NATURWISSENSCHAFTEN VOL.53,. . . .   16, (1966).
     ↱OEHME(H.), RICHTER(H.). (IN GERMAN).

178 BI,. . . . . . . ↱MODEL OF LATTICE DYNAMICS FOR BISMUTH.   ACTA PHYS. POLON. VOL.A43, . . .   37, (1973).
     ↱CZACHOR(A.), RAJCA(A.), SOSNOWSKI(J.), PINDOR(A.).

179 BI,. . . . . ↱REF. SECTION 3, 48/HAVENS, ...

180 BI,. . . . . . . ↱PHONON DISPERSION CURVES IN BISMUTH.   IBM J. RES. DEVELOP. VOL.8,. . .   234, (1964).
     ↱YARNELL(J.L.), WARREN(J.L.), WENZEL(R.G.), KOENIG(S.H.).

181 BI,. . . . . . . ↱COLD NEUTRON SCATTERING IN BISMUTH.   PHIL. MAG. VOL.3,. . . . . . .   798, (1958).
     ↱BHANDARI(R.C.), KHUBCHANDANI(P.G.).

182 BI,. . . . . . . ↱LATTICE VIBRATIONS OF BISMUTH.   BULL. AMER. PHYS. SOC. VOL.12, .   689, (1967).
     ↱SMITH(D.B.), WARREN(J.L.), YARNELL(J.L.). (ABSTRACT ONLY).

183 BI*(FE-MN)*O3,. ↱VIRTUAL SPIN-WAVE STATES IN ANTIFERROMAGNETIC INSULATORS.   IAEA SYMP. GRENOBLE, ...   595, (1972).
     ↱KROO(N.), PEPY(G.), SZENTIRMAY(Z.).

184 BI-GA,. . . . . ↱CRITICAL OPALESCENCE IN BINARY LIQUID METAL MIXTURES. I. TEMPERATURE
     DEPENDENCE.   J. OF PHYS. C VOL.1, ...   1088, (1968).
     ↱WIGNALL(G.D.), EGELSTAFF(P.A.).

185 BI-GA,. . . . . ↱CRITICAL OPALESCENCE IN BINARY LIQUID METAL MIXTURES II. CONCENTRATION
     DEPENDENCE.   J. PHYS. C. SER.2 VOL.3, . . 1673, (1970).
     ↱EGELSTAFF(P.A.), WIGNALL(G.D.).

SECTION 6 -QUASI-ELASTIC AND INELASTIC SCATTERING STUDIES . . . .

186 BI-PB-TL,. . . . . .B1 . . . . . . .THESIS (137 PP.). . . . . , (1967).
  *CRYSTAL DYNAMICS OF FACE CENTRED CUBIC ALLOYS OF BI, PB AND TL.
  *NG(S.C.). (ED. NOTE: PH.D. DISSERTATION, MCMASTER UNIVERSITY).

187 BI-PB-TL,. . . . . .B1,B2 . . . .THESIS (209 PP.). . . . . , (1970).
  *CRYSTAL DYNAMICS AND ANHARMONIC PROPERTIES OF BI-PB-TL ALLOYS.
  *ROY(A.P.). (ED. NOTE: PH.D. DISSERTATION, MCMASTER UNIVERSITY).

188 BI-PB-TL,. . . . . .B1 . . . . . . .SOL. STATE COMM. 5,. . . . . . . 79, (1967).
  *LATTICE DYNAMICS OF THE ALLOY SYSTEMS BI-PB-TL.
  *NG(S.C.), BROCKHOUSE(B.N.).

189 BI-ZN, ...D2 . . . . . . .J. OF PHYS. C VOL.1,. . . . . . .1088, (1968).
  *CRITICAL OPALESCENCE IN BINARY LIQUID METAL MIXTURES. I. TEMPERATURE
  DEPENDENCE.
  *WIGNALL(G.D.), EGELSTAFF(P.A.).

190 BR,. .C3 . . . . . .IAEA SYMP. CHALK RIVER, VOL.1, . 259, (1962).
  *NEUTRON SPECTROMETRY WORK AT THE CNEN.
  *CAGLIOTI(G.), ASCARELLI(P.).

191 BR,. .C3 . . . . . .IAEA SYMP. CHALK RIVER, VOL.1, . 249, (1962).
  *SCATTERING OF NEUTRONS BY LIQUID BROMINE.
  *COOTE(G.E.), HAYWOOD(B.C.).

192 C (GRAPHITE),. . . . . .P . . . . . .NUCL. SCI. ENG. VOL.44 . . . . . 444, (1971).
  *SLOWING DOWN AND HEATING UP OF NEUTRONS IN GRAPHITE AT LOW TEMPERATURES.
  *BHUSHAN(V.), TRIKHA(S.K.).

193 C (GRAPHITE),. . . . . .B7 . . . . . .NUCL. SCI. ENGNG. VOL.34 . . . . 224, (1968).
  *INELASTIC SCATTERING OF THERMAL NEUTRONS IN GRAPHITE.
  *CARVALHO(F.).

194 C (DIAMOND), ...T . . . . . .BROOKHAVEN SYMPOSIUM . . . . 88, (1965).
  *CHARGED BOND CORRECTION TO THE DISPERISON CURVES OF DIAMOND.
  *WARREN(J.L.).

195 C (GRAPHITE),. . . . . .B7 . . . . . .NUCL. SCI. ENGNG. VOL.33,. . . . . 31, (1968).
  *NEUTRON SCATTERING BY REACTOR-GRADE GRAPHITE.
  *WHITTEMORE(W.L.).

196 C (DIAMOND), ...B1 . . . . . .IAEA SYMP. BOMBAY, VOL.1,. . . . 361, (1964).
  *DISPERSION CURVES FOR PHONONS IN DIAMOND.
  *WARREN(J.L.), MENZEL(R.G.), YARNELL(J.L.).

197 C (GRAPHITE),. . . . . .B1 . . . . . .IAEA SYMP. COPENHAGEN, VOL.1,. . 325, (1968).
  *THE PHONON DISTRIBUTION OF GRAPHITE AT HIGH TEMPERATURES.
  *PAGE(D.I.).

198 C (GRAPHITE),. . . . . .B7 . . . . . .CAN. J. PHYS. 49, ... 277, (1971).
  *TEMPERATURE VARIATION OF THE FREQUENCY OF LONGITUDINAL INTER-PLANAR
  OSCILLATIONS IN PYROLITIC GRAPHITE.
  *ROY(A.P.).

199 C (GRAPHITE),. . . . . .C2 . . . . . .IAEA SYMP. CHALK RIVER, VOL.2, . 111, (1962).
  *FREQUENCY DISTRIBUTION OF NORMAL MODES IN GRAPHITE.
  *HAYWOOD(B.C.), THORSON(I.M.).

200 C (DIAMOND), ...C2 . . . . . .IAEA SYMP. CHALK RIVER, VOL.2, . 49, (1962).
  *THE DISTRIBUTION OF HIGHER ENERGY MODES IN DIAMOND.
  *MITCHELL(E.W.J.), HARDY(J.R.), SAUNDERSON(D.H.).

201 C (GRAPHITE),. . . . . .B7 . . . . . .BULL. AMER. PHYS. SOC. VOL.17, . 123, (1972).
  *TEMPERATURE DEPENDENCE OF PHONONS IN PYROLYTIC GRAPHITE.
  *BROCKHOUSE(B.N.), SHIRANE(G.). (ED. NOTE: PAPER PRESENTED AT ANNUAL
  MEETING OF AMER. PHYS. SOC. IN SAN FRANCISCO).

202 C (DIAMOND), ...B1 . . . . . .SOV. PHYS. SOL. STATE VOL.4, . .1747, (1962).
  *CONCERNING THE SIMILARITY BETWEEN THE CHARACTERISTIC FREQUENCY
  DISPERSION CURVES OF DIAMOND-TYPE CRYSTALS.
  *KUCHER(T.I.).

203 C (GRAPHITE),. . . . . .P . . . . . . .IAEA SYMPOSIUM (VIENNA), ... 487, (1960).
  *ENERGY DISTRIBUTIONS OF NEUTRONS SCATTERED FROM GRAPHITE, LIGHT AND
  HEAVY WATER, ZR*H, LI*H, NA*H, AND N*H4*CL BY THE BE DETECTOR METHOD.
  *WOODS(A.D.B.), BROCKHOUSE(B.N.), SAKAMOTO(M.), SINCLAIR(R.W.). (ED.
  NOTE: REF. SECTION 3, 60/WOODS).

204 C (PYROLYTIC GRAPHITE),. . . . . .B1 . . . . . .PHYS. REV. B VOL.5,. . . . . . .4951, (1972).
  *LATTICE DYNAMICS OF PYROLYTIC GRAPHITE.
  *NICKLOW(R.M.), WAKABAYASHI(N.), SMITH(H.G.).

205 C (GRAPHITE),. . . . . .T,B1 . . . . .C.R. ACAD. SCI. VOL.257, ... 1843, (1963).
  *DYNAMICS OF THE GRAPHITE LATTICE.
  *CHAMPIER(G.), GENIN(J.M.), JANOT(C.). (IN FRENCH).

206 C (GRAPHITE),. . . . . .B1 . . . . . .PHYS. REV. VOL.128,. . . . . . .1120, (1962).
  *LATTICE VIBRATIONS IN PYROLITIC GRAPHITE.
  *DOLLING(G.), BROCKHOUSE(B.N.).

207 C (GRAPHITE),. . . . . .B3 . . . . . .PHIL. MAG. VOL.9,. . . . . 659, (1964).
  *THE SCATTERING OF LONG WAVELENGTH NEUTRONS BY DEFECTS IN
  NEUTRON-IRRADIATED GRAPHITE.
  *MARTIN(D.G.), HENSON(R.W.).

208 C (GRAPHITE),. . . . . .B1 . . . . . .PHYS. LETT. VOL.7,. . . . . . . .220, (1963).
  *HIGH-ENERGY MODES IN THE FREQUENCY DISTRIBUTION OF GRAPHITE.
  *EGELSTAFF(P.A.), HARRIS(D.H.C.).

209 C (GRAPHITE),. . . . . .B1 . . . . . .UKAEA AERE MEM.1199 (5 PP.), ... , (1963).
  *PHONON FREQUENCY DISTRIBUTION IN GRAPHITE.
  *SAUNDERSON(D.H.).

210 C (GRAPHITE),. . . . . .T,B1 . . . . .PHYS. REV. B VOL.7,. . . . . . .4527, (1973).
  *DISPERSION CURVES AND ELASTIC CONSTANTS OF GRAPHITE.
  *AHMADIEH(A.A.), RAFIZADEH(H.A.).

211 C (GRAPHITE),. . . . . .B1 . . . . . .UKAEA AERE REP.5574 (24 PP.), ... , (1967).
  *SCATTERING LAW S(ALPHA;BETA)-VALUES FOR GRAPHITE AT 1300 DEGREES K AND
  1800 DEGREES K.
  *PAGE(D.I.).

212 C (GRAPHITE),. . . . . .B2 . . . . . .J. PHYS. C VOL.6,. . . . . . . .3525, (1973).
  *INELASTIC NEUTRON SCATTERING FROM POLYCRYSTALLINE GRAPHITE AT
  TEMPERATURES UP TO 1920 DEGREES C.
  *ROSS(D.K.).

SECTION 6 -QUASI-ELASTIC AND INELASTIC SCATTERING STUDIES . . . .

213 C (DIAMOND,GRAPHITE), ... B7 . . . . . PHYS. REV. VOL.88, ... 542, (1952).
ⴲENERGY DISTRIBUTION OF SLOW NEUTRONS SCATTERED FROM SOLIDS.
ⴲBROCKHOUSE(B.N.), HURST(D.G.).

214 C (DIAMOND), ... B1 . . . . . PROC. PHYS. SOC. 88, ... 463, (1966).
ⴲTHE THERMODYNAMIC AND OPTICAL PROPERTIES OF GERMANIUM, SILICON, DIAMOND
AND GALLIUM ARSENIDE.
ⴲDOLLING(G.), COWLEY(R.A.).

215 C (DIAMOND), ... B1 . . . . . PHYS. REV. LETT. 13, ... 13, (1964).
ⴲLATTICE VIBRATIONS IN DIAMOND.
ⴲYARNELL(J.L.), WARREN(J.L.), WENZEL(R.G.).

216 C (DIAMOND), ... B1 . . . . . PROC. PHYS. SOC. 91, ... 381, (1967).
ⴲLATTICE DYNAMICS OF DIAMOND.
ⴲAGGARWAL(K.G.).

217 C (DIAMOND), ... B1 . . . . . PHYS. REV. 158, ... 805, (1967).
ⴲLATTICE DYNAMICS OF DIAMOND.
ⴲWARREN(J.L.), YARNELL(J.L.), DOLLING(G.), COWLEY(R.A.).

218 C (DIAMOND), ... T . . . . . SOV. PHYS. SOL. STATE VOL.8, . . 261, (1966).
ⴲLATTICE VIBRATION FREQUENCIES OF DIAMOND.
ⴲKUCHER(T.I.), NECHIPORUK(V.V.).

219 C (GRAPHITE), ... B1 . . . . . NUCL. SCI. ENGNG. VOL.40, ... 17, (1970).
ⴲAPPROXIMATE DISPERSION RELATIONS AND THE TOTAL COHERENT INELASTIC
NEUTRON SCATTERING CROSS SECTION FOR GRAPHITE.
ⴲCONN(R.).

220 C. . . . . B1 . . . . . CARBON 7, ... 663, (1969).
ⴲATOMIC VIBRATIONS IN CARBON FIBERS.
ⴲCOLLINS(M.F.), HAYWOOD(B.C.G.).

221 C (GRAPHITE), ... T,P . . . . . PHYS. REV. VOL.106, ... 230, (1957).
ⴲTHERMAL INELASTIC SCATTERING OF COLD NEUTRONS IN POLYCRYSTALLINE
GRAPHITE.
ⴲKOTHARI(L.S.), SINGWI(K.S.).

222 C (GRAPHITE), ... P . . . . . J. NUCL. ENERGY VOL.5, ... 203, (1957).
ⴲTHE SLOW-NEUTRON CROSS-SECTION OF GRAPHITE.
ⴲEGELSTAFF(P.A.).

223 C (GRAPHITE), ... P,T . . . . . PHYS. REV. VOL.110, ... 70, (1958).
ⴲTHERMAL INELASTIC SCATTERING OF COLD NEUTRONS IN POLYCRYSTALLINE
GRAPHITE. II.
ⴲKHUBCHANDANI(P.G.), KOTHARI(L.S.), SINGWI(K.S.).

224 C (DIAMOND), ... D2 . . . . . PROC. ROY. SOC. 281, ... 274, (1964).
ⴲCRITICAL-POINT ANALYSIS OF THE PHONON SPECTRA OF DIAMOND, SILICON AND
GERMANIUM.
ⴲJOHNSON(F.A.), LOUDON(R.).

225 C (GRAPHITE), ... C1 . . . . . BROOKHAVEN SYMPOSIUM, ... 94, (1965).
ⴲINELASTIC SCATTERING OF NEUTRONS BY REACTOR TYPE GRAPHITE.
ⴲWHITTEMORE(W.L.).

226 C (DIAMOND), ... B1 . . . . . SOL. STATE COMM. 5, ... 311, (1967).
ⴲTHE PHONON DISPERSION RELATION FOR DIAMOND.
ⴲPECKHAM(G.).

227 C (GRAPHITE), ... C2 . . . . . IAEA SYMPOSIUM VIENNA, ... 569, (1960).
ⴲTHE PHONON FREQUENCY DISTRIBUTION IN GRAPHITE AT SEVERAL TEMPERATURES.
ⴲEGELSTAFF(P.A.), COCKING(S.J.).

228 C (GRAPHITE)-H2*O, ... P . . . . . SOV. AT. ENERGY VOL.31, ... 459, (1971).
ⴲANALYSIS OF EXPERIMENTS ON NEUTRON THERMALIZATION IN GRAPHITE-WATER
SYSTEMS.
ⴲMOSTOVO(V.I.), TRUKHANOV(G.YA.), SAFIN(YU.A.), MOSKOVKIN(V.N.).

229 C*CL4, ... B4 . . . . . MOL. PHYS. VOL.20, ... 881, (1971).
ⴲORIENTATIONAL CORRELATIONS IN MOLECULAR LIQUIDS BY NEUTRON SCATTERING.
CARBON TETRACHLORIDE AND GERMANIUM TETRABROMIDE.
ⴲEGELSTAFF(P.A.), PAGE(D.I.), POWLES(J.G.).

230 C2*CL2*H4, ... C1 . . . . . CHEM. PHYS. LETT. VOL.18, . . . 306, (1973).
ⴲNEUTRON INELASTIC SCATTERING SPECTRA AND PHASE TRANSITION OF 1,2-DI-
CHLOROETHANE CRYSTAL.
ⴲOZORA(A.), ITO(M.), NIIMURA(N.), WATANABE(N.).

231 C*D4, ... B5 . . . . . PHYS. REV. VOL.148, ... 163, (1966).
ⴲINTRAMOLECULAR COHERENT SCATTERING OF NEUTRONS BY C*D4 GAS.
ⴲWEST(R.E.), BRUGGER(R.M.), GRIFFING(G.W.).

232 C*D4, ... B5,T . . . . . IAEA SYMP. COPENHAGEN, VOL.2, . . 205, (1968).
ⴲCOHERENT INELASTIC NEUTRON SCATTERING BY MOLECULAR GASES: MEASUREMENTS
ON C2*F6 AND ROTATION-VIBRATION COUPLING CALCULATIONS FOR C*D4.
ⴲCARPENTER(J.M.), LURIE(N.A.).

233 C*D4, ... D2 . . . . . PHYS. REV. LETT. VOL.29, ... 266, (1972).
ⴲCRITICAL SCATTERING IN SOLID C*D4.
ⴲHULLER(A.), PRESS(W.).

234 C10*D8 (PDN), ... T,B1 . . . . . DISC. FARADAY SOC. NO.48, . . . 125, (1969).
ⴲMOLECULAR CRYSTAL PHONON DISPERSION CURVES AND MODEL FITTING.
ⴲPAWLEY(G.S.).

235 C6*(D4,C4)*CL2, ... B1 . . . . . IAEA SYMP. GRENOBLE, ... 195, (1972).
ⴲEXPERIMENTAL LATTICE VIBRATIONAL DISPERSION CURVES FOR BETA-PHASE
P-C6*D4*CL2...AND ACOUSTIC VEL. FOR/C6*H4*CL2...CRYSTAL POT./CHLOROBENZ.
ⴲREYNOLDS(P.A.), KJEMS(J.K.), WHITE(J.W.).

236 C6*D6,C6*H6, ... C1,C3 . . . . . MOL. PHYS. VOL.24, ... 753, (1972).
ⴲTHE QUASI-ELASTIC SCATTERING OF NEUTRONS FROM C6*H6 AND C6*D6.
ⴲWINFIELD(D.J.), ROSS(D.K.).

237 (C*D3)4*N*MN*CL3, ... A2,D . . . . . J. APPL. PHYS. 42, ... 1265, (1971).
ⴲMAGNETIC NEUTRON SCATTERING FROM A NEARLY IDEAL ONE-DIMENSIONAL
ANTIFERROMAGNET.
ⴲHUTCHINGS(M.T.), SHIRANE(G.), BIRGENEAU(R.J.), DINGLE(R.), HOLT(S.L.).

238 (C*D3)4*N*MN*CL3, ... D1 . . . . . PHYS. REV. B VOL.5, ... 1999, (1972).
ⴲSPIN DYNAMICS IN THE ONE-DIMENSIONAL ANTIFERROMAGNET (C*D3)4*N*MN*CL3.
ⴲHUTCHINGS(M.T.), SHIRANE(G.), BIRGENEAU(R.J.), HOLT(S.L.).

SECTION   6 -QUASI-ELASTIC AND INELASTIC SCATTERING STUDIES . . . .

239 C3*O5*(O*D)3 (GLYCEROL),. . . . . . .B4. . . . . . . .IAEA SYMP. BOMBAY, VOL.2,. . . . .   3, (1964).
     *LIQUID DYNAMICS.
     *LARSSON(K.E.).

240 C*D3*O*H,. . . . . .C3. . . . . .DISC. FARADAY SOC. NO.43,. . . . .   169, (1967).
     *NEUTRON SCATTERING SPECTROSCOPY OF LIQUIDS.
     *ALDRED(B.K.), EDEN(R.C.), WHITE(J.W.).

241 C*D3*O*H,. . . . .B4. . . . .IAEA SYMP. COPENHAGEN, VOL.1,. .   397, (1968).
     *LIQUID DYNAMICS FROM NEUTRON SCATTERING (REVIEW PAPER).
     *LARSSON(K.E.).

242 C*F4,. . . . . . .B4. . . . .PHYS. LETT. VOL.19,. . . . . .   269, (1965).
     *CO-OPERATIVE MODES OF MOTION IN SIMPLE LIQUIDS.
     *CHEN(S.H.),   EDER(O.J.), EGELSTAFF(P.A.), HAYWOOD(B.C.), WEBB(F.J.).

243 C2*F6,. . . . . .B5. . . . .IAEA SYMP. COPENHAGEN, VOL.2,. .   205, (1968).
     *COHERENT INELASTIC NEUTRON SCATTERING BY MOLECULAR GASES: MEASUREMENTS
     ON C2*F6 AND ROTATION-VIBRATION COUPLING CALCULATIONS FOR C*D4.
     *CARPENTER(J.M.), LURIE(N.A.).

244 C6*F12,C6*F11*H,C6*F9*H3,. . . . . .C1,C3. . . .IAEA SYMP. GRENOBLE,. . . . .   231, (1972).
     *ROTATIONAL MOTIONS IN THE PLASTIC CRYSTAL PHASES OF C6*F12, C6*F11*H AND
     C6*F9*H3.
     *LEADBETTER(A.J.), LITCHINSKY(D.), TURNBULL(A.).

245 C*F3*C*O*O*H,. . . . . .C1,C3. . . . .J. AMER. CHEM. SOC. VOL.95,. . . . .   708, (1973).
     *SPECTROSCOPY   OF C*F3*C*O*Z COMPOUNDS. V. VIBRATIONAL SPECTRA AND
     STRUCTURE OF SOLID TRIFLUOROACETIC ACID.
     *BERNEY(C.V.).

246 C*H2,. . . . . .C3. . . . . .IAEA SYMPOSIUM VIENNA,. . . . .   511, (1960).
     *INELASTIC SCATTERING OF THERMAL NEUTRONS PRODUCED BY AN ELECTRON
     ACCELERATOR.
     *WHITTEMORE(W.L.), MCREYNOLDS(A.W.).

247 C*H3 GROUPS,. . . . . .C3. . . . . . . .J. CHEM. PHYS. VOL.32,. . . . .   476, (1960).
     *TRANSMISSION MEASUREMENTS WITH COLD NEUTRONS IN HYDROGENOUS LIQUIDS.
     *NASUHOGLU(R.), RINGO(G.R.).

248 C*H4,. . . . . .P. . . . . . .PORTUGALIAE PHYS. VOL.2,. . . . .   243, (1947).
     *SCATTERING OF THERMAL NEUTRONS BY METHANE BETWEEN 20 AND 200 DEGREES K.
     *GIBERT(A.).

249 C*H4,. . . . . .T,P. . . . . .PHYS. REV. VOL.84,. . . . . . .   204, (1951).
     *SCATTERING OF SLOW NEUTRONS BY H2 AND C*H4.
     *MESSIAH(A.M.L.).

250 C*H4,. . . . . .A3. . . . . .IAEA SYMPOSIUM VIENNA,. . . . .   277, (1960).
     *SLOW-NEUTRON INELASTIC SCATTERING MEASUREMENTS AT THE MATERIALS TESTING
     REACTOR.
     *BRUGGER(R.M.), EVANS(J.E.).

251 C*H4,. . . . . .C3. . . . . .PHYS. REV. VOL.124,. . . . . . .   460, (1961).
     *INELASTIC SCATTERING OF SLOW NEUTRONS FROM METHANE.
     *RANDOLPH(P.D.), BRUGGER(R.M.), STRONG(K.A.), SCHMUNK(R.E.).

252 C*H4,. . . . . .C5. . . . . .IAEA SYMP. CHALK RIVER, VOL.2,.   281, (1962).
     *ENERGY DISTRIBUTION OF NEUTRONS SCATTERED FROM SOLID METHANE.
     *STILLER(H.), HAUTECLER(S.).

253 C*H4,. . . . . .C3. . . . . .IAEA SYMP. CHALK RIVER, VOL.1,.   405, (1962).
     *SYSTEMATIC STUDY ON TOTAL NEUTRON SCATTERING CROSS-SECTIONS OF HYDROGEN
     CONTAINING MOLECULES.
     *JANIK(J.A.), JANIK(J.), MANIAWSKI(F.), RZANY(H.), ROGALSKA(Z.),
     SCIESINSKI(J.), SAGAN(O.), TUBBS(N.).

254 C*H4,. . . . . .C3. . . . . .IAEA SYMP. CHALK RIVER, VOL.1, .   423, (1962).
     *MOUVEMENTS MOLECULAIRES DANS LE METHANE LIQUIDE.
     *HAUTECLER(S.), STILLER(H.).

255 C*H4,. . . . . .C3. . . . . .IAEA SYMP. CHALK RIVER, VOL.1, .   435, (1962).
     *SCATTERING OF SLOW NEUTRONS BY GASEOUS METHANE.
     *GRIFFING(G.W.).

256 C*H4,. . . . . .C3. . . . . .IAEA SYMP. CHALK RIVER, VOL.1,   457, (1962).
     *THE INELASTIC SCATTERING OF COLD NEUTRONS BY METHANE, AMMONIA AND
     HYDROGEN.
     *WEBB(F.J.).

257 C*H4,. . . . . .C3. . . . . .PHYS. REV. VOL.127,. . . . . . .   1179, (1962).
     *INFLUENCE OF INTERFERENCE SCATTERING ON THE SCATTERING OF SLOW NEUTRONS
     BY GASEOUS METHANE.
     *GRIFFING(G.W.).

258 C*H4,. . . . . .C3. . . . . .Z. PHYSIK VOL.166,. . . . . . .   393, (1962).
     *ENERGY DISTRIBUTION OF COLD NEUTRONS SCATTERED IN LIQUID METHANE.
     *STILLER(H.), HAUTECLER(S.).

259 C*H4,. . . . . .C3. . . . . .PHYSICA VOL.29,. . . . . . .   491, (1963).
     *SLOW NEUTRON SCATTERING BY MOLECULES OF LIQUID METHANE.
     *ROGALSKA(Z.).

260 C*H4,. . . . . .C3. . . . . .IAEA SYMP. BOMBAY, VOL.2,. . . . .   157, (1964).
     *COLD NEUTRON SCATTERING BY METHANE.
     *DASANNACHARYA(B.A.), VENKATARAMAN(G.), USHA-DENIZ(K.).

261 C*H4,. . . . . .B4. . . . . .IAEA SYMP. BOMBAY, VOL.2,. . . . .   243, (1964).
     *MOLECULAR DYNAMICS INVESTIGATED BY NEUTRON SCATTERING.
     *JANIK(J.A.).

262 C*H4,. . . . . .B1. . . . . .IAEA SYMP. BOMBAY, VOL.2,. . . . .   291, (1964).
     *LATTICE DYNAMICS OF SOLID METHANE.
     *DORNER(B.), STILLER(H.).

263 C*H4,. . . . . .C1,C3. . . . .J. PHYS. CHEM. SOL. VOL.25,. . . .   1091, (1964).
     *STUDY OF MOLECULAR ROTATIONS IN SOLIDS AND LIQUIDS BY THE INELASTIC
     SCATTERING OF COLD NEUTRONS.
     *JANIK(J.A.), JANIK(J.M.), MELLOR(J.), PALEVSKY(H.). (ED. NOTE: REF.
     SECTION 3, 64/JANIK).

264 C*H4,. . . . . .C3. . . . . .NUCL. SCI. ENGNG. VOL.18,. . . . .   182, (1964).
     *INELASTIC NEUTRON SCATTERING IN LIQUID METHANE AND LIQUID PARAHYDROGEN.
     *WHITTEMORE(W.L.).

SECTION   6 -QUASI-ELASTIC AND INELASTIC SCATTERING STUDIES . . . .

265 C*H4,. . . . . .C3. . . . . .PHYS. REV. VOL.136,. . . . . . . A106, (1964).
                        ~SCATTERING OF SLOW NEUTRONS FROM METHANE GAS.
                        ~BRUGGER(R.M.), RAINEY(V.S.), MCMURRY(H.L.).

266 C*H4,. . . . . .C2 in sol.PHYS. STAT. SOL. 5,. . . . . 511, (1964).
                        ~FREQUENCY SPECTRUM IN SOLID METHANE AT 6.5 DEG. K.
                        ~DORNER(B.), STILLER(H.).

267 C*H4,. . . . . .C1. . . . .ACTA PHYS. POLONICA VOL.27,. . . . 581, (1965).
                        ~ROTATIONAL DYNAMICS OF SOLID METHANE MOLECULES BY SLOW NEUTRON CROSS-
                        SECTION MEASUREMENTS.
                        ~ROGALSKA(G.).

268 C*H4,. . . . . .T,C3. . . . .BROOKHAVEN SYMPOSIUM. . . . . 149, (1965).
                        ~APPLICATION OF THE LANGEVIN EQUATION TO THE SCATTERING OF NEUTRONS FROM
                        LIQUID METHANE.
                        ~GRIFFING(G.W.).

269 C*H4,. . . . . .C4. . . . . .J. CHEM. PHYSICS VOL.42,. . . . 275, (1965).
                        ~NEUTRON INELASTIC SCATTERING FROM LOW-TEMPERATURE GASEOUS METHANE,
                        LIQUID METHANE, AND SOLID METHANE.
                        ~HARKER(Y.D.), BRUGGER(R.M.).

270 C*H4,. . . . . .C1. . . . . . .PHYSICA VOL.32,. . . . . . 1571, (1966).
                        ~SLOW NEUTRON SCATTERING BY SOLID C*H4 AND N*H4*I.
                        ~KOSALY(G.), SOLT(G.).

271 C*H4,. . . . . .B1. . . . . . .PHYS. STAT. SOL. 18,. . . . . 795, (1966).
                        ~DIE INNERE DYNAMIK DER TIEFTEMPERATURPHASEN DES MOLEKULKRISTALLS C*H4.
                        ~DORNER(B.), STILLER(H.).

272 C*H4,. . . . . .T,C3. . . . .PROC. PHYS. SOC. 89,. . . . . . 379, (1966).
                        ~NEUTRON SCATTERING FROM GASEOUS METHANE AND AMMONIA.
                        ~VENKATARAMAN(G.),   RAO(K.R.),   DASANNACHARYA(B.A.),   DAYANIDHI(P.K.).

273 C*H4,. . . . . .C3. . . . . .CAN. J. PHYS. 45,. . . . . 237, (1967).
                        ~COLD NEUTRON SCATTERING BY MOLECULAR LIQUIDS.
                        ~SEARS(V.F.).

274 C*H4,. . . . . .C3. . . . . .CAN. JOURN. PHYS. 45,. . . . . . 3185, (1967).
                        ~COLD NEUTRON SCATTERING BY LIQUID METHANE.
                        ~RAO(K.R.), VENKATARAMAN(G.), DASANNACHARYA(B.A.).

275 C*H4,. . . . . .C2. . . . .J. CHEM. PHYSICS VOL.46,. . . . . 2201, (1967).
                        ~INVESTIGATION OF THE LOW-TEMPERATURE PHASE TRANSITION IN SOLID METHANE
                        BY SLOW NEUTRON INELASTIC SCATTERING.
                        ~HARKER(Y.D.), BRUGGER(R.M.).

276 C*H4,. . . . . .C1. . . . . .PHYSICA VOL.37,. . . . . 253, (1967).
                        ~SLOW NEUTRON SCATTERING BY SOLID METHANE.
                        ~SOLT(G.).

277 C*H4, C2*H2, C2*H4,. . . . . . .C3. . . . . .PHYS. LETT. 25A,. . . . . . . 435, (1967).
                        ~STUDIES OF THE DYNAMICAL BEHAVIOUR OF ADSORBED MOLECULES BY SLOW NEUTRON
                        SPECTROSCOPY.
                        ~VERDAN(G.).

278 C*H4,. . . . . .C3. . . . . .PHYS. REV. 156,. . . . . . . 196, (1967).
                        ~DYNAMICS OF LIQUID C*H4 FORM COLD NEUTRON SCATTERING.
                        ~DASANNACHARYA(B.A.), VENKATARAMAN(G.).

279 C*H4,. . . . . .C3. . . . . .PROC. PHYS. SOC. 92,. . . . . . 912, (1967).
                        ~THE INELASTIC SCATTERING OF COLD NEUTRONS BY METHANE AND AMMONIA.
                        ~WEBB(F.J.).

280 C*H4,. . . . . .B4. . . . . .IAEA SYMP. COPENHAGEN, VOL.1,. . 397, (1968).
                        ~LIQUID DYNAMICS FROM NEUTRON SCATTERING (REVIEW PAPER).
                        ~LARSSON(K.E.).

281 C*H4,. . . . . .C3. . . . . .IAEA SYMP. COPENHAGEN, VOL.1,. . 545, (1968).
                        ~ANALYSIS OF NEUTRON SCATTERING EXPERIMENTS ON LIQUIDS.
                        ~AGRAWAL(A.K.), DESAI(R.C.), YIP(S.).

282 C*H4,. . . . . .C5. . . . . .J. CHEM. PHYSICS VOL.49,. . . . . 2443, (1968).
                        ~EFFECTS OF MOLECULAR REORIENTATION IN SOLID METHANE ON THE QUASIELASTIC
                        SCATTERING OF THERMAL NEUTRONS.
                        ~SKOLD(K.).

283 C*H4,. . . . . .C3. . . . . .PHYS LETT 27A,. . . . . 9, (1968).
                        ~SCATTERING SLOW NEUTRONS BY GASEOUS METHANE AND INTERMOLECULAR
                        INTERACTIONS.
                        ~FULINSKI(A.), ZGIERSKI(M.).

284 C*H4,. . . . . .C3. . . . . .PHYS. REV. VOL.171,. . . . . . 263, (1968).
                        ~ROTATIONAL CORRELATION FUNCTIONS IN NEUTRON SCATTERING BY MOLECUL. GASES
                        ~AGRAWAL(A.K.), YIP(S.).

285 C*H4,. . . . . .C1. . . . . .DISCUSSIONS FARADAY SOC. NO.48,. 87, (1969).
                        ~LATTICE VIBRATIONS AND MOLECULAR ROTATION IN SOLID METHANE NEAR THE
                        MELTING POINT.
                        ~JANIK(J.A.), OTNES(K.), SOLT(G.), KOSALY(G.).

286 C*H4,. . . . . .C3,C1. . . .PHYSICA VOL.41,. . . . . 397, (1969).
                        ~MOLECULAR DYNAMICS IN GASEOUS AND SOLID METHANE STUDIED BY THE INELASTIC
                        NEUTRON SCATTERING METHOD.
                        ~BAJOREK(A.),   NATKANIEC(I.),   PARLINSKI(K.),   SUDNIK-HRYNKIEWICZ(M.),
                        JANIK(J.A.), JANIK(J.M.), OTNES(K.), TUNKELO(E.).

287 C*H4,. . . . . .P. . . . . . .PHYS. LETT. 30A,. . . . . . . 367, (1969).
                        ~POSSIBLE PHASE TRANSITIONS IN A PHYSISORBED STATE DETECTED BY COLD
                        NEUTRON SCATTERING.
                        ~TODIREANU(S.).

288 C*H4,. . . . . .T. . . . . . .PHYSICA VOL.48,. . . . . . . 126, (1970).
                        ~MOLECULAR INTERACTION EFFECT IN NEUTRON SCATTERING BY GASEOUS METHANE.
                        ~FULINSKI(A.), JANIK(J.A.), BLOCKI(J.).

289 C*H4,. . . . . .C2. . . . . .PHYS. LETT. VOL.31A,. . . . . . 158, (1970).
                        ~SLOW NEUTRON SCATTERING IN THE LOW TEMPERATURE PHASE OF SOLID METHANE.
                        ~KAPULLA(H.), GLASER(W.).

290 C*H4,. . . . . .T. . . . . . .PHYS. LETT. VOL.33A,. . . . . . 209, (1970).
                        ~A MODEL FOR THE LATTICE VIBRATIONS OF SOLID METHANE.
                        ~BANSAL(R.M.), KOTHARI(L.S.), TEWARI(S.P.).

SECTION   6 -QUASI-ELASTIC AND INELASTIC SCATTERING STUDIES . . . .

291 C*H4,. . . . . T OF ACTA OF PHYS.-C-. SER.2, VOL.4,. . 2725, (1971).
     ~MULTIPLE SCATTERING OF NEUTRONS IN LIQUID METHANE.
     ~RAO(K.R.), DASANNACHARYA(B.A.), YIP(S.).

292 C*H4,. . . . . PHYSICA VOL.54,. . . . . 393, (1971).
     ~MULTIPLE SCATTERING IN AN INELASTIC NEUTRON SCATTERING WITH LIQUID
     METHANE.
     ~TIITTA(A.), TUNKELO(E.).

293 C*H4,. . . . . .C3........IAEA SYMP. GRENOBLE, ... 489, (1972).
     ~MOLECULAR MOTIONS IN SILANE.
     ~HAUTECLER(S.), VORDERWISCH(P.).

294 C*H4,. . . . . .C1.......IAEA SYMP. GRENOBLE, ... 841, (1972).
     ~MOLECULAR ROTATIONS IN THE LOW-TEMPERATURE PHASE OF SOLID METHANE.
     ~KAPULLA(H.), GLASER(W.).

295 C*H4,. . . . . .C1,C3....J. NUCL. SCI. TECHNOL. VOL.9,. . 374, (1972).
     ~SLOW NEUTRON SPECTRA IN THE LIQUID AND SOLID METHANE.
     ~INOUE(K.), OTOMO(N.), UTSURO(H.), FUJITA(Y.).

296 C*H4,. . . . . .C3.......NUCL. INST. MET. VOL.103,. . . . 575, (1972).
     ~CORRECTION FOR MULTIPLE SCATTERING IN COLD NEUTRON EXPERIMENTS.
     ~TIITTA(A.), TUNKELO(E.).

297 C*H4,. . . . . .C3.......PHYS. LETT. VOL.43A NO.2,. . . . 189, (1973).
     ~THE STUDY OF MOBILE ADSORPTION BY COLD NEUTRON SCATTERING.
     ~TODIREANU(S.), HAUTECLER(S.).

298 C*H4,. . . . . .T........PROGR. THEOR. PHYS. VOL.50,. . . 1142, (1973).
     ~SPIN CORRELATION EFFECT ON THE SLOW NEUTRON SCATTERING BY POLYATOMIC
     MOLECULES WITH PARTICULAR REFERENCE TO METHANE.
     ~HAMA(J.), MIYAGI(H.).

299 C*H4,. . . . . .B4,C3....PHYSICA VOL.72,. . . . . 300, (1974).
     ~A COMPARATIVE STUDY OF THE MOTIONS OF METHANE MOLECULES AND ARGON ATOMS
     IN LIQUID AND HIGH-PRESSURE GAS STATES BY NEUTRON SCATTERING.
     ~OLSSON(L.G.), LARSSON(K.E.).

300 (C*H4) -AR,. . . . . . . .C3.......IAEA SYMP. COPENHAGEN, VOL.2,. . 223, (1968).
     ~THE MOLECULAR DYNAMICS OF METHANE IN ARGON.
     ~EDER(O.J.), EGELSTAFF(P.A.).

301 C2*H4, ...T,C1.....ACTA PHYS. POLON. VOL.17,. . . . 489, (1958).
     ~THEORETICAL CALCULATIONS OF THE SLOW NEUTRON SCATTERING CROSS SECTION
     OF THE ETHYLENE MOLECULE.
     ~JANIK(J.A.), MANIAWSKI(F.), RZANY(H.).

302 C2*H4, ...C3.......IAEA SYMP. CHALK RIVER, VOL.1, . 451, (1962).
     ~THE SCATTERING OF SLOW NEUTRONS FROM HYDROGEN AND ETHYLENE.
     ~BALLY(D.), TARINA(V.), TODIREANU(S.).

303 C2*H4 (ETHYLENE),. . . . .C1.......IAEA SYMP. BOMBAY, VOL.2,. . . . 421, (1964).
     ~INELASTIC SCATTERING OF NEUTRONS IN ETHYLENE.
     ~BALLY(D.), TODIREANU(S.), TARINA(V.).

304 C2*H4 LIGANDS, ...C1.......J.C.S. FARADAY TRANS. II VOL.69, 275, (1973).
     ~CHARACTERISATION OF THE TORSION POTENTIAL FOR ETHYLENE LIGANDS USING
     INELASTIC NEUTRON SCATTERING.
     ~GHOSH(R.E.), WADDINGTON(T.C.), WRIGHT(C.J.).

305 C2*H6, ...C1,3.....BROOKHAVEN SYMPOSIUM,. . . . . . 142, (1965).
     ~INELASTIC SCATTERING OF THERMAL NEUTRONS FROM PRESSURIZED ETHANE.
     ~STRAKER(E.A.), KING(J.S.), CARPENTER(J.M.), VINCENT(D.H.).

306 C2*H6, ...C3.......J. CHEM. PHYS. VOL.43,. . . . . . 4134, (1965).
     ~INELASTIC NEUTRON SCATTERING FROM PRESSURIZED ETHANE.
     ~STRAKER(E.A.).

307 C2*H6, ...C3.......DISS. ABS. 27,. . . . . 5078, (1967).
     ~INELASTIC NEUTRON SCATTERING FROM PRESSURIZED ETHANE.
     ~STRAKER(E.A.).

308 C2*H6 (ETHANE),. .C1,C3....J. CHEM. PHYS. VOL.46,. . . . . 2285, (1967).
     ~COLD-NEUTRON STUDY OF HINDERED ROTATIONS IN SOLID AND LIQUID METHYL-
     CHLOROFORM, NEOPENTANE, AND ETHANE.
     ~RUSH(J.J.).

309 C2*H6 (ETHANE),. .C3.......J. CHEM. PHYSICS VOL.47,. . . . . 421, (1967).
     ~MEASUREMENT OF THE FREQUENCY OF TORSIONAL VIBRATION IN THE ETHANE MOL.
     ~STRONG(K.A.), BRUGGER(R.M.).

310 C2*H6, ...C3.......IAEA SYMP. COPENHAGEN, VOL.1,. . 525, (1968).
     ~SELF-DIFFUSION COEFFICIENT OF ETHANE NEAR THE LIQUID-GAS TRANSITION
     POINT.
     ~.C.C.C.P.

311 C3*H8 (PROPANE), ...C3.......PHYS. REV. VOL.125,. . . . . . . 933, (1962).
     ~SCATTERING OF SLOW NEUTRONS FROM PROPANE GAS.
     ~STRONG(K.A.), MARSHALL(G.D.), BRUGGER(R.M.), RANDOLPH(P.D.).

312 C4*H6 (DIMETHYL ACETYLENE),. . . .C1.......NUCL. SCI. ENGNG. VOL.14,. . 339, (1962).
     ~THE EFFECT OF ROTATIONAL FREEDOM IN SEVERAL AMMONIUM SALTS AND DIMETHYL
     ACETYLENE ON THE INELASTIC SCATTERING OF SLOW NEUTRONS.
     ~RUSH(J.J.), SAFFORD(G.J.), TAYLOR(T.I.), HAVENS-JR(W.W.).

313 C4*H8, C8*H10, C10*H14,. .C1,C3....J. CHEM. PHYS. VOL.58,. . . . . . 1438, (1973).
     ~TORSIONAL FREQUENCIES AND BARRIERS TO METHYL ROTATION IN ISOBUTYLENE,
     O-XYLENE, AND DURENE.
     ~LIVINGSTON(R.C.), GRANT(D.M.), PUGMIRE(R.J.), STRONG(K.A.),
     BRUGGER(R.M.).

314 C5*H12 (PENTANE),. . . . .C3.......REV. ROUMAINE PHYS. VOL.11,. . . 265, (1966).
     ~THE INELASTIC SCATTERING OF SLOW NEUTRONS IN PENTANE.
     ~RIPEANU(S.).

315 C5*H12 (NEOPENTANE),. . . .C1,C3....J. CHEM. PHYS. VOL.46,. . . . . . 2285, (1967).
     ~COLD-NEUTRON STUDY OF HINDERED ROTATIONS IN SOLID AND LIQUID METHYL-
     CHLOROFORM, NEOPENTANE, AND ETHANE.
     ~RUSH(J.J.).

316 C5*H12, ...,C*H3*O*H,. . .C3.......REV. ROUMAINE PHYS. VOL.12,. . . 943, (1967).
     ~PROTON MOTIONS IN SOME HYDROGENOUS LIQUIDS BY NEUTRON SCATTERING CROSS-
     SECTION MEASUREMENTS.
     ~RAPEANU(S.), ILIESCU(N.), PREDA(I.M.).

SECTION   6 -QUASI-ELASTIC AND INELASTIC SCATTERING STUDIES . . . .

317   C5*H12,. . . . .   B4   IAEA SYMP. COPENHAGEN, VOL.1,. .   397, (1968).
                          ⊷LIQUID DYNAMICS FROM NEUTRON SCATTERING (REVIEW PAPER).
                          ⊷LARSSON(K.E.).

318   C5*H12,. . . . .   C3   IAEA SYMP. COPENHAGEN, VOL.1,. .   581, (1968).
                          ⊷NEW RESULTS ON *QUASI-ELASTIC* SCATTERING FROM SOME HYDROGENOUS LIQUIDS.
                          ⊷DAHLBORG(U.), FRIBERG(B.), LARSSON(K.E.), PIRKMAJER(E.).

319   C5*H12 (NEOPENTANE)   C3   PHYS. REV. LETT. VOL.20,. .   983, (1968).
                          ⊷DIRECT OBSERVATION OF METHYL LIBRATIONS IN NEOPENTANE.
                          ⊷GRANT(D.M.), STRONG(K.A.), BRUGGER(R.M.).

32C   C5*H12 (PENTANE),. . . . . .   C3. . . . . .DISC. FARADAY SOC. NO.49,. . . .   193, (1970).
                          ⊷LOW FREQUENCY MOLECULAR MODES IN LIQUID HYDROCARBONS.
                          ⊷EGELSTAFF(P.A.), HARRIS(O.H.C.).

321   C5*H12 (NEOPENTANE),. . . . . .C1. . . . . .PHYSICA VOL.48,. . . . .   79, (1970).
                          ⊷STUDY OF MOLECULAR MOTIONS IN PLASTIC-CRYSTALLINE NEOPENTANE BY COLD-
                          NEUTRON SCATTERING.
                          ⊷DE-GRAAF(L.A.), SCIESINSKI(J.).

322   C5*H12 (NEOPENTANE),. . . . .C1,C3. . . .PHYSICA VOL.59,. . . . . .   672, (1972).
                          ⊷PHASE TRANSITIONS IN NEOPENTANE STUDIED BY COLD NEUTRONS.
                          ⊷DAHLBORG(U.), GRASLUND(C.), LARSSON(K.E.).

323   C5*H12 (NEOPENTANE),. . . . .T. . . . . . .SOL. STATE COMM. VOL.10,. . . 1247, (1972).
                          ⊷ROTATIONAL MOTION OF MOLECULES IN PLASTIC NEOPENTANE.
                          ⊷LECHNER(R.E.).

324   C5*H12, C6*H14, C7*H16,. . . . . .C1. . . . . .J. CHEM. PHYS. VOL.59,. . . . 2305, (1973).
                          ⊷LOW-FREQUENCY MOLECULAR VIBRATIONS IN SOLID N-PARAFFINS BY NEUTRON IN-
                          ELASTIC SCATTERING: N-PENTANE, N-HEXANE, N-HEPTANE, AND N-OCTANE.
                          ⊷LOGAN(K.W.), DANNER(H.R.), GAULT(J.D.).

325   C6*H6, ...C3. . . . . .NUCL. SCI. ENGNG. VOL.5,. . . .   99, (1960).
                          ⊷INELASTIC SCATTERING OF COLD NEUTRONS FROM SEVERAL HYDROGENOUS LIQUIDS.
                          ⊷BRUGGER(R.M.), MCCLELLAN(L.W.), STREETMAN(G.B.), EVANS(J.E.).

326   C6*H6, ...C3. . . . . .IAEA SYMP. CHALK RIVER, VOL.1, .   285, (1962).
                          ⊷SLOW NEUTRON SCATTERING BY BENZENE.
                          ⊷BOFFI(V.C.), MOLLINARI(V.G.), PARKS(D.E.).

327   C6*H6, ...C3. . . . .IAEA SYMP. CHALK RIVER, VOL.1,   297, (1962).
                          ⊷A STUDY OF COLD NEUTRON INELASTIC SCATTERING ON CERTAIN HYDROGENOUS
                          SUBSTANCES.
                          ⊷.C.C.C.P.

328   C6*H6, C12*H10(DIPHENYL),. . . . . .C3. . . . . .SOV. AT. ENERGY VOL.14,. . . . .   252, (1963).
                          ⊷INVESTIGATION OF COLD NEUTRON INELASTIC SCATTERING ON SOME HYDROGEN-
                          CONTAINING MATERIALS.
                          ⊷ZEMLYANOV(M.G.), CHERNOPLEKOV(N.A.).

329   C6*H6, ...C4. . . . . .IAEA SYMP. BOMBAY, VOL.2,. . . . .   167, (1964).
                          ⊷RECENT MEASUREMENTS OF THE SCATTERING LAWS OF SOME HYDROGENOUS
                          MODERATORS.
                          ⊷GLASER(W.), EHRET(G.), MERKEL(A.).

330   C6*H6, ...C3. . . . . .IAEA SYMP. BOMBAY, VOL.2,. . . . .   201, (1964).
                          ⊷SLOW NEUTRON SCATTERING IN WATER AND SOME ORGANIC SUBSTANCES.
                          ⊷.C.C.C.P.

331   C6*H6, ...C3. . . . . .IAEA SYMP. BOMBAY, VOL.2,. . . . .   221, (1964).
                          ⊷COLD NEUTRON SCATTERING BY POLYPHENYLS.
                          ⊷.C.C.C.P.

332   C6*H6, C12*H10(DIPHENYL),. . . . . .C3. . . . . .NUCL. SCI. ENGNG. VOL.20,. . . . . .   236, (1964).
                          ⊷INELASTIC NEUTRON SCATTERING BY SOME HYDROGENOUS MODERATORS.
                          ⊷GLASER(W.), BECKURTS(K.H.).

333   C6*H6, C12*H10(DIPHENYL),. . . . . .C1. . . . . .NUKLEONIK VOL.7,. . . . . .   64, (1965).
                          ⊷DETERMINATION OF THE LAWS OF SCATTERING OF SOME ORGANIC SUBSTANCES WITH
                          SLOW NEUTRONS.
                          ⊷GLASER(W.). (IN GERMAN).

334   C6*H6, ...C1. . . . . .J. CHEM. PHYS. VOL.46,. . . . 3273, (1967).
                          ⊷ON THE LOW-FREQUENCY MOTIONS IN SOLID BENZENE BY INELASTIC SCATTERING
                          OF COLD NEUTRONS.
                          ⊷TARINA(V.).

335   C6*H6, ...C3. . . . . .CONF. NEUTRON THERMALIZ. VOL.I,.   477, (1968).
                          ⊷INELASTIC SCATTERING OF NEUTRONS FROM BENZENE AND WATER.
                          ⊷ROSS(D.K.), SZABO(F.P.), SANALAN(Y.). (ED. NOTE: PROC. CONF.
                          NEUTRON THERMALIZATION AND REACTOR SPECTRA; PUBL. IAEA:VIENNA).

336   C6*H6, ...B4. . . . . .IAEA SYMP. COPENHAGEN, VOL.1,. .   475, (1968).
                          ⊷DIFFUSIVE MOTIONS IN BENZENE AND TOLUENE STUDIED WITH SLOW NEUTRONS.
                          ⊷HOLMRYD(S.), NELIN(G.).

337   C6*H6, ...C2. . . . . .J. CHEM. PHYSICS VOL.53, ... 3417, (1970).
                          ⊷INELASTIC NEUTRON SCATTERING FROM SOLID BENZENE.
                          ⊷LOGAN(K.W.), TREVINO(S.F.), PRASK(H.J.), GAULT(J.D.).

338   C6*H6, ...C3. . . . . .J. CHEM. PHYS. VOL.60,. . . . . . 2832, (1974).
                          ⊷STUDY OF MOLECULAR ROTATIONS IN SOME AROMATIC COMPOUNDS BY COLD-NEUTRON
                          SCATTERING.
                          ⊷TREPADUS(V.), RAPEANU(S.), PADUREANU(I.), PARFENOV(V.A.), NOVIKOV(A.G.).

339   C6*H12,. . . . .   . .C1,3. . . . . .BROOKHAVEN SYMPOSIUM. . . . . .   138, (1965).
                          ⊷INELASTIC NEUTRON SCATTERING BY SOLID AND LIQUID CYCLOHEXANE.
                          ⊷DE-GRAAF(L.A.).

340   C6*H12, C5*H10,. .   . . .C3. . . . . .PHYSICA VOL.40,. . . . . . .   497, (1969).
                          ⊷STUDY OF MOLECULAR MOTIONS IN CYCLOHEXANE AND CYCLOPENTANE BY COLD-
                          NEUTRON SCATTERING.
                          ⊷DE-GRAAF(L.A.).

341   C6*H14,. . . . .   . . .C3. . . . . .IAEA SYMP. COPENHAGEN, VOL.1,. .   501, (1968).
                          ⊷STUDY OF THE SOLID-LIQUID TRANSITION IN CYCLOHEXANE BY COLD NEUTRONS.
                          ⊷TARINA(V.).

342   C8*H1C,. . . . .   . .C1,3. . . . .COLL. INTER. N.126 (GRENOBLE),.   224, (1963).
                          ⊷DETERMINATION PAR DIFFUSION DE NEUTRONS FROIDS DU SPECTRE DES FREQUENCES
                          DES PROTONS DANS L'ORTHO-, PARA-, ET META-XYLENE.
                          ⊷VAN-DINGENEN(W.), HAUTECLER(S.).

SECTION 6 -QUASI-ELASTIC AND INELASTIC SCATTERING STUDIES . . . .

343 C10*H8,. . . . .
    *LOW FREQUENCY PHONONS IN NAPHTHALENE.   SOL. STATE COMM. 7,. . . . .   385, (1969).
    *PAWLEY(G.S.), YEATS(E.A.).

344 C10*H16, ...
    *LATTICE DYNAMICS OF SOLID ADAMANTANE.   PHYS. STAT. SOL. 19, ...   781, (1967).
    *STOCKMEYER(R.), STILLER(H.).

345 C10*H16, ...
    *PHONONS, TORSONS, AND ROTATIONAL DIFFUSION IN ADAMANTANE.   PHYS. STAT. SOL. 27, ...   269, (1968).
    *STOCKMEYER(R.), STILLER(H.).

346 C10*H16 (ADAMANTANE), ...
    *DYNAMICS OF ORIENTATIONAL DEFECTS IN SOLID ADAMANTANE.   DISC. FARADAY SOC. NO.48,. . .   156, (1969).
    *STOCKMEYER(R.).

347 C12*H10 (DIPHENYL), . . .
    *COLD NEUTRON SCATTERING BY POLYPHENYLS.   IAEA SYMP. BOMBAY, VOL.2,. . . . .   221, (1964).
    *.C.C.C.P.

348 C12*H10 (BIPHENYL), . . .
    *LOW-FREQUENCY VIBRATIONS OF CRYSTALLINE BIPHENYL.   J. CHEM. PHYS. VOL.40,. . . . . .   3502, (1964).
    *KREBS(K.), SANDRONI(S.), ZERBI(G.).

349 C12*H10 (DIPHENYL), . . .
    *SLOW-NEUTRON SCATTERING BY DIPHENYL.   NUCL. SCI. ENG. VOL.35,. . . . . .   80, (1969).
    *SPREVAK(D.), KOPPEL(J.U.).

350 C12*H18 (HEXAMETHYLBENZENE), . . .
    *ROTATIONAL MOTIONS IN HEXAMETHYLBENZENE AND AMMONIUM PERCHLORATE BY CROSS SECTION MEASUREMENTS WITH SLOW NEUTRONS.   J. PHYS. CHEM. VOL.68, ...   2534, (1964).
    *RUSH(J.J.), TAYLOR(T.I.).

351 C12*H18 (HEXAMETHYLBENZENE), . . .
    *NEUTRON-SCATTERING STUDY OF HINDERED ROTATIONAL MOTIONS AND PHASE TRANSITIONS IN HEXAMETHYLBENZENE.   J. CHEM. PHYSICS VOL.44, ...   2749, (1966).
    *RUSH(J.J.), TAYLOR(T.I.).

352 C15*H16 (MONOISOPROPYLDIPHENYL), . . .
    *COLD NEUTRON SCATTERING BY POLYPHENYLS.   IAEA SYMP. BOMBAY, VOL.2,. . . . .   221, (1964).
    *.C.C.C.P.

353 C18*H14 (P-TERPHENYL), . . .
    *INELASTIC NEUTRON SCATTERING SPECTRUM OF P-TERPHENYL.   J. CHEM. SOC. FARADAY II VOL.68, 1434, (1972).
    *REYNOLDS(P.A.), WHITE(J.W.).

354 C19*H40 (N-NONADECANE), . . .
    *NEUTRON INELASTIC SCATTERING STUDY OF THE ROTATOR PHASE TRANSITION IN N-ALKANES.   IAEA SYMP. GRENOBLE, ...   287, (1972).
    *BARNES(J.D.).

355 C19*H40 (N-NONADECANE), . . .
    *INELASTIC NEUTRON SCATTERING STUDY OF THE #ROTATOR# PHASE TRANSITION IN N-NONADECANE.   J. CHEM. PHYS. VOL.58, ...   5193, (1973).
    *BARNES(J.D.).

356 C*H4,. . . . . .
    *MULTIPLE SCATTERING OF NEUTRONS IN GASEOUS AND LIQUID METHANE.   PHYS. REV. A VOL.4,. . . . .   1560, (1971).
    *AGRAWAL(A.K.).

357 C*H2*BR2,C*H*BR3,. . . . .
    *MOTIONS OF PROTONS IN C*H-, C*H2-, C*H3- AND C2*H5-HALIDES BY COLD NEUTRON TRANSMISSION EXPERIMENTS.   BER. BUNSENGES. PHYS. CHEM. 75,. . .   361, (1971).
    *FISCHER(C.O.).

358 C*H2*BR-C*H*F*BR,. . . .
    *NEUTRON SCATTERING AND FAR INFRA-RED SPECTRA OF A HEAVY HINDERED ROTATOR: C*H2*BR-C*H*F*BR.   MOL. PHYS. VOL.17, ...   1, (1969)
    *LONGSTER(G.F.), WHITE(J.W.).

359 C*H3*C*CL3 (METHYLCHLORFORM), . . .
    *COLD-NEUTRON STUDY OF HINDERED ROTATIONS IN SOLID AND LIQUID METHYL-CHLOROFORM, NEOPENTANE, AND ETHANE.   J. CHEM. PHYS. VOL.46, ...   2285, (1967).
    *RUSH(J.J.).

360 C*H3*C*CL3,C*H3*C*F3,. . . .
    *THE BARRIERS TO INTERNAL ROTATION IN SOME CHLORO- AND FLUORO-SUBSTITUTED ETHANES.   J. MOLEC. STRUC. VOL.6,. . . . .   23, (1970).
    *BRIER(P.N.).

361 C*H3*C*CL3,C*H3*C*H*CL2,. . . .
    *NEUTRON INELASTIC SCATTERING STUDY OF THE CHLOROETHANES.   MOL. PHYS. VOL.21,. . . . .   721, (1971).
    *BRIER(P.N.), HIGGINS(J.S.), BRADLEY(R.H.).

362 C*H3*C*F3, . . .
    *NEUTRON INELASTIC SCATTERING MEASUREMENTS ON 1,1,1-TRIFLUOROETHANE.   MOLEC. PHYS. VOL.19, ...   645, (1970).
    *BRIER(P.N.), HIGGINS(J.S.).

363 (C*H2*C*H*CL) N,. . .
    *FREQUENCY DISTRIBUTIONS OF SYNDIOTACTIC POLYVINYLCHLORIDE.   IAEA SYMP. COPENHAGEN, VOL.2,. . .   167, (1968).
    *LYNCH-JR(J.E.), SUMMERFIELD(G.C.).

364 C*H3*(C*H,N*O2), . . .
    *SLOW-NEUTRON SCATTERING AND ROTATIONAL FREEDOM OF METHYL GROUPS IN SEVERAL ORGANIC COMPOUNDS.   IAEA SYMP. COPENHAGEN, VOL.2,. . . . .   197, (1968).
    *HERDADE(S.B.).

365 C6*H5*C*H3,. . .
    *DIFFUSIVE MOTIONS IN BENZENE AND TOLUENE STUDIED WITH SLOW NEUTRONS.   IAEA SYMP. COPENHAGEN, VOL.1,. . . .   475, (1968).
    *HOLMRYD(S.), NELIN(G.).

366 C6*H5*(C*H3),. . .
    *MEASUREMENT OF THE APPARENT DIFFUSION COEFFICIENT OF TOLUENE BY QUASI-ELASTIC NEUTRON SCATTERING.   J. CHEM. PHYS. VOL.54,. . . . .   3643, (1971).
    *WINFIELD(O.J.).

367 C*H3*C*N (ACETONITRILE), . . .
    *A COMPARATIVE STUDY OF QUASIELASTIC NEUTRON SCATTERING AND NMR RELAXATION IN LIQUID ACETONITRILE.   BER. BUNSENGES. PHYS. CHEM. 75,. . .   769, (1971).
    *ZEIDLER(M.O.).

368 C*H3*C*O*O*H (ACETIC ACID), . . .
    *SLOW NEUTRON SCATTERING IN WATER AND SOME ORGANIC SUBSTANCES.   IAEA SYMP. BOMBAY, VOL.2,. . . . .   201, (1964).
    *.C.C.C.P.

SECTION   6 -QUASI-ELASTIC AND INELASTIC SCATTERING STUDIES . . . .

369 (C*H3)2*(C*O2,SI*CL2,S*O), ...C3. . . . . .IAEA SYMP. COPENHAGEN, VOL.2, . . 197, (1968).
     ⌐SLOW-NEUTRON SCATTERING AND ROTATIONAL FREEDOM OF METHYL GROUPS IN
     SEVERAL ORGANIC COMPOUNDS.
     ⌐HERDADE(S.B.).

370 (C*H3*C*O)2*O, ...C3. . . . . .IAEA SYMP. COPENHAGEN, VOL.2, . . 197, (1968).
     ⌐SLOW-NEUTRON SCATTERING AND ROTATIONAL FREEDOM OF METHYL GROUPS IN
     SEVERAL ORGANIC COMPOUNDS.
     ⌐HERDADE(S.B.).

371 C*H2*CL2, ...T. . . . . .IAEA SYMP. GRENOBLE, ... 461, (1972).
     ⌐MODEL CALCULATIONS FOR THE LOW-ENERGY SCATTERING FROM C*H2*CL2.
     ⌐BRIER(P.N.), HIGGINS(J.S.).

372 C6*H4*CL2 (PARA-DICHLOROBENZENE). . .B2. . . . . .J. CHEM. PHYS. VOL.56, ... 2928, (1972).
     ⌐HIGH-RESOLUTION INELASTIC NEUTRON SCATTERING FROM SOME AROMATIC MOLECUL.
     POLYCRYSTALS FOR STUDY OF CRYSTAL EXCITATIONS AND ANHARMONIC EFFECTS.
     ⌐REYNOLDS(P.A.), KJEMS(J.K.), WHITE(J.W.).

373 C*H*CL3,C6*H6, ...C3. . . . . .THESIS (146 PP.), ... , (1972).
     ⌐COLD NEUTRON SCATTERING FROM LIQUID CHLOROFORM AND BENZENE.
     ⌐REICHERT(P.F.), (ED. NOTE:AVAIL. UNIV. MICROFILMS, ANN ARBOR, MICH. USA,
     ORDER NO.73-7075; SEE DISS. ABST. B VOL.33 NO.9, PG.4439).

374 C6*H4*CL2,C6*D4*CL2, ...B1,B2. . . .J. CHEM. PHYS. VOL.60, ... 824, (1974).
     ⌐LATTICE VIBRATIONS IN CHLOROBENZENES: EXPERIMENTAL DISPERSION CURVES FOR
     BETA-PARADICHLOROBENZENE BY NEUTRON SCATTERING.
     ⌐REYNOLDS(P.A.), KJEMS(J.K.), WHITE(J.W.).

375 C*(H-D)4, ...B5. . . . . .PHYS. REV. VOL.136, ... ,A988, (1964).
     ⌐SCATTERING OF SLOW NEUTRONS BY DEUTERATED METHANE.
     ⌐GRIFFING(G.W.).

376 C*(H-D)4, ...C3. . . . . .PHYS. LETT. 23, ... 226, (1966).
     ⌐COLD NEUTRON SCATTERING FROM LIQUID C*D4.
     ⌐VENKATARAMAN(G.), DASANNACHARYA(B.A.), RAO(K.R.).

377 C*(H-D)4, ...C3. . . . . .PHYS. REV. 161, ... 133, (1967).
     ⌐DYNAMICS OF LIQUID C*D4, FROM COLD NEUTRON SCATTERING.
     ⌐VENKATARAMAN(G.), DASANNACHARYA(B.A.), RAO(K.R.).

378 C*H3*F,C*H3*I, ...C3. . . . . .THESIS (173 PP.), ... , (1972).
     ⌐THERMAL NEUTRON SCATTERING FROM GASEOUS AND LIQUID C*H3*F AND C*H3*I.
     ⌐MALM(W.C.), (ED. NOTE: AVAIL. UNIV. MICROFILMS, ANN ARBOR, MICHIGAN USA,
     ORDER NO.73-7059; SEE:DISS. ABST. B VOL.33 NO.9, PG.4437).

379 C*H3*I, .C1,C3. . . .REF. SECTION 3, 64/JANIK, ...

380 C*H3*I, .C1,C3. . . .PHYSICA VOL.35, ... 451, (1967).
     ⌐MOLECULAR DYNAMICS BY THE NEUTRON INELASTIC SCATTERING METHOD.
     II.METHYL IODIDE.
     ⌐JANIK(J.A.), JANIK(J.M.), BAJOREK(A.), PARLINSKI(K.),
     SUDNIK-HRYNKIEWICZ(M.).

381 C*H3*I,C*H3*(O*H), ...P. . . . . .BULL. ACAD. POLON. SCI. III V.1, 45, (1953).
     ⌐SCATTERING OF SLOW NEUTRONS BY LIQUIDS. I.EXPERIMENTS WITH METHYL
     ALCOHOL AND METHYL IODIDE.
     ⌐JANIK(J.A.).

382 C*H3*I,C*H2*I2, ...C1. . . . . .ACTA PHYS. POLON. VOL.33, ... 419, (1968).
     ⌐NEUTRON INELASTIC SCATTERING DATA FOR CRYSTALLINE H3*O*N*O3, C*H3*I, AND
     C*H2*I2 AND THEIR COMPARISON WITH INFRA-RED AND RAMAN SPECTROSCOPY.
     ⌐JANIK(J.A.), BAJOREK(A.), JANIK(J.M.), NATKANIEC(I.), PARLINSKI(K.),
     SUDNIK-HRYNKIEWICZ(M.).

383 C*H3*I,C*H3*C*H2*I, ...C3,P. . . .BER. BUNSENGES. PHYS. CHEM. 75, 361, (1971).
     ⌐MOTIONS OF PROTONS IN C*H-, C*H2-, C*H3- AND C2*H5-HALIDES BY COLD
     NEUTRON TRANSMISSION EXPERIMENTS.
     ⌐FISCHER(C.O.).

384 C*H3*N*H2, ...C3. . . . . .ACTA PHYS. POLON. VOL.21, ... 529, (1962).
     ⌐SCATTERING OF THERMAL NEUTRONS ON C*H3*N*H2 MOLECULES IN THE LIQUID
     STATE.
     ⌐BAJOREK(A.), KRASNICKI(S.).

385 C6*H6*(N*H2)8, ...C1. . . . . .J. CHEM. PHYSICS VOL.37, ... 431, (1962).
     ⌐FREQUENCY DISTRIBUTION OF HEXAMETHYLENETETRAMINE CRYSTALS.
     ⌐BECKA(L.N.).

386 C6*H5*N*H2, ...C3. . . . . .J. CHEM. PHYS. VOL.55, ... 2807, (1971).
     ⌐STUDIES OF VIBRATION SPECTRA OF BONDED HYDROGEN ATOMS USING A PULSED
     NEUTRON SOURCE.
     ⌐DAY(D.H.), SINCLAIR(R.N.).

387 (C*H3)4*N*HN*CL3, ...T. . . . . .PHYS. REV. B VOL.5, ... 2014, (1972).
     ⌐SPIN WAVES IN FINITE SPIN-1/2 HEISENBERG CHAINS.
     ⌐RICHARDS(P.M.), CARBONI(F.).

388 C5*H9*N*O3, ...B7. . . . . .NUCL./S.S. PHYS. SYMP. ABSTR., . . . , (1972).
     ⌐DISPERSION CURVES OF POLY-L-PROLINE 1 AND 2.
     ⌐SINGH(R.D.), DWIVEDI(A.M.), GUPTA(V.D.).

389 C2*H6*O2, ...C3. . . . . .IAEA SYMP. BOMBAY, VOL.2, ... 201, (1964).
     ⌐SLOW NEUTRON SCATTERING IN WATER AND SOME ORGANIC SUBSTANCES.
     ⌐.C.C.C.P.

390 (C2*H5)2*O, ...C3. . . . . .PHYS. LETT. VOL.33A, ... 287, (1973).
     ⌐THE MOLECULAR DYNAMICS ALONG THE COEXISTENCE CURVE OF A LIQUID-GAS
     SYSTEM.
     ⌐BATA(L.), JOVIC(D.).

391 C*H3*O*D, ...B4. . . . . .IAEA SYMP. COPENHAGEN, VOL.1, . . 397, (1968).
     ⌐LIQUID DYNAMICS FROM NEUTRON SCATTERING (REVIEW PAPER).
     ⌐LARSSON(K.E.).

392 C3*H5*(O*D)3 (GLYCEROL), ...B4. . . . . .IAEA SYMP. BOMBAY, VOL.2, ... 3, (1964).
     ⌐LIQUID DYNAMICS.
     ⌐LARSSON(K.E.).

393 C3*H5*(O*D)3, ...C3. . . . . .PHYS. LETT. 28A, ... 31, (1968).
     ⌐QUASI-ELASTIC SCATTERING OF COLD NEUTRONS IN C3*H5*(O*D)3 AT VARIOUS
     TEMPERATURES.
     ⌐RAPEANU(S.).

SECTION   6 -QUASI-ELASTIC AND INELASTIC SCATTERING STUDIES . . . .

394 C3*H5*(O*D)3,. . . . . .C1. . . . REV. ROUMAINE PHYS. VOL.13, ... 913, (1968).
     *INELASTIC SCATTERING OF COLD NEUTRONS IN C3*H5*(O*D)3 AT VARIOUS
     TEMPERATURES.
     *RAPEANU(S.). (ED. NOTE: IN ENGLISH).

395 C*H3*O*H,. . . . . .C3. . . . . . .IAEA SYMP. CHALK RIVER, VOL.1, . 413, (1962).
     *INELASTIC SCATTERING OF NEUTRONS BY METHYL-, ETHYL- AND N-AMYL ALCOHOLS.
     *SAUNDERSON(D.H.), RAINEY(V.S.).

396 C*H3*O*H, C*H3*C*N,. . . . . .C1,C3. . .PHYS. REV. LETT. VOL.17, ... 533, (1966).
     *LOCALIZED VIBRATIONS OF TRAPPED MOLECULES.
     *DOWNES(J.S.), WHITE(J.W.), EGELSTAFF(P.A.), RAINEY(V.S.).

397 C*H3*O*H,. . . . . .C3. . . . . .IAEA SYMP. COPENHAGEN, VOL.1, . . 491, (1968).
     *THERMAL NEUTRON INELASTIC SCATTERING BY METHYL ALCOHOL AND METHYL
     MERCAPTAN.
     *SAMPSON(T.E.), CARPENTER(J.M.).

398 C*H3*O*H,. . . . . .C3. . . . . .J. CHEM. PHYS. VOL.51, ... 5543, (1969).
     *QUASIELASTIC SCATTERING OF THERMAL NEUTRONS BY C*H3*O*H AND C*H3*S*H.
     *SAMPSON(T.E.), CARPENTER(J.M.).

399 C*H3*O*H,. . . . . .NAT. BUR. STAND. PUBL. NO.301, . 463, (1969).
     *SPECTRAL ASSIGNMENT IN INELASTIC NEUTRON SCATTERING SPECTROSCOPY BY
     ATOMIC SUBSTITUTION TECHNIQUES.
     *WHITE(J.W.). (U.S. NBS SPEC. PUBL. 1967; PUBL. 1969).

400 C*H3*O*H,C*H3*C*H2*OH. . . . . .C3,P. . . .BER. BUNSENGES. PHYS. CHEM. 74, . . 696, (1970).
     *COLD NEUTRON SCATTERING AT C*H3-, C*H2- AND O*H-PROTONS IN METHANOL
     AND ETHANOL.
     *FISCHER(C.O.).

401 C*H3*O*H,. . . . . .A3. . . . . .REV. ROUMAINE PHYS. VOL.15, . . . 783, (1970).
     *COLD NEUTRON SCATTERING BY METHYL ALCOHOL, CYCLOHEXANE AND THEIR MIXTURE
     NEAR THE CONSOLUTION POINT.
     *DIACONESCU(A.), NAHORNIAK(V.), MATEESCU(N.), TEUTSCH(H.).

402 C*H3*O*H,. . . . . .C3. . . . . .FARADAY SYMP. CHEM. SOC. NO.6, . 135, (1972).
     *HIGH-FREQUENCY DYNAMICS OF LIQUID METHANOL AND TOLUENE, CONTRIBUTION OF
     MOLECULAR ROTATIONAL DIFFUSION TO INELASTIC NEUTRON SCATTERING.
     *ALDRED(B.K.), STIRLING(G.C.), WHITE(J.W.).

403 C*H3*O*H,. . . . . .C3. . . . . .J. CHEM. PHYS. VOL.56, . . . 3118, (1972).
     *PROTON MOTIONS IN METHANOL BY COLD NEUTRON SCATTERING.
     *RODRIGUEZ(C.), AMARAL(L.Q.), VINHAS(L.A.), HERDADE(S.B.).

404 C*H3*(O*H),C2*H5*(O*H),. . . . . . . .P,C3. . .J. NUCL. ENERGY VOL.26, . . 379, (1972).
     *SLOW-NEUTRON SCATTERING CROSS-SECTION FOR METHANOL, ETHANOL, PROPANOL,
     ISO-PROPANOL, BUTANOL, ETHANEDIOL AND PROPANETRIOL.
     *RODRIGUES(C.), VINHAS(L.A.), HERDADE(S.B.), DO-AMARAL(L.Q.).

405 C2*H5*O*H, ...C3. . . . . .NUCL. SCI. ENGNG. VOL.5, . . . 99, (1960).
     *INELASTIC SCATTERING OF COLD NEUTRONS FROM SEVERAL HYDROGENOUS LIQUIDS.
     *BRUGGER(R.M.), MCCLELLAN(L.W.), STREETMAN(G.B.), EVANS(J.E.).

406 C2*H5*O*H, ...C3. . . . . .IAEA SYMP. CHALK RIVER, VOL.1, . 413, (1962).
     *INELASTIC SCATTERING OF NEUTRONS BY METHYL-, ETHYL- AND N-AMYL ALCOHOLS.
     *SAUNDERSON(D.H.), RAINEY(V.S.).

407 C3*H5*(O*H)3 (GLYCEROL). . . . . .C1. . . . .Z. PHYSIK VOL.238, ... 221, (1970).
     *MEASUREMENT OF THE QUASI-ELASTIC LINE-BROADENING IN GLYCEROL WITH A
     NEUTRON CRYSTAL SPECTROMETER OF EXTREMELY HIGH ENERGY RESOLUTION.
     *BIRR(M.). (IN GERMAN).

408 C3*H7*O*H, C2*H4,. . . . . .C3. . . . .REV. ROUMAINE PHYS. VOL.18 NO.2, 135, (1973).
     *THE STUDY OF FREQUENCY SPECTRA IN SOME ORGANIC COMPOUNDS USING THE COLD
     NEUTRON TECHNIQUE.
     *RAPEANU(S.), PADUREANU(I.), TREPADUS(V.), DUMITRU(O.).

409 C5*H11*O*H,. . . . . .C3. . . . . .IAEA SYMP. COPENHAGEN, VOL.1, . 581, (1968).
     *NEW RESULTS ON #QUASI-ELASTIC# SCATTERING FROM SOME HYDROGENOUS LIQUIDS.
     *DAHLBORG(U.), FRIBERG(B.), LARSSON(K.E.), PIRKMAJER(E.).

410 C6*H5*(O*H), ...C3. . . . . .REV. ROUMAINE PHYS. VOL.18, . . 313, (1973).
     *THE STUDY OF THE MOLECULAR MOTION IN PHENOL BY NEUTRON SCATTERING AND
     N.M.R. METHODS.
     *TREPADUS(V.), RAPEANU(S.), GROSESCU(R.).

411 C*H3*O*H, C*H3*C*N,. . . . . .MOL. SIEVES. PAP. CONF. 1967. . 306, (1967).
     *MOLECULAR MOTION IN MOLECULAR SIEVES BY NEUTRON SCATTERING SPECTROSCOPY.
     *EGELSTAFF(P.A.), DOWNES(J.S.), WHITE(J.W.).

412 C*H3*S*H,. . . . . .C3. . . . . .IAEA SYMP. COPENHAGEN, VOL.1, . 491, (1968).
     *THERMAL NEUTRON INELASTIC SCATTERING BY METHYL ALCOHOL AND METHYL
     MERCAPTAN.
     *SAMPSON(T.E.), CARPENTER(J.M.).

413 C*H3*S*H,. . . . . .C3. . . . . .J. CHEM. PHYS. VOL.51, . . 5543, (1969).
     *QUASIELASTIC SCATTERING OF THERMAL NEUTRONS BY C*H3*O*H AND C*H3*S*H.
     *SAMPSON(T.E.), CARPENTER(J.M.).

414 C*H3*S*H,. . . . . .C3. . . . . .IAEA SYMP. CHALK RIVER, VOL.1, . 405, (1962).
     *SYSTEMATIC STUDY ON TOTAL NEUTRON SCATTERING CROSS-SECTIONS OF HYDROGEN
     CONTAINING MOLECULES.
     *JANIK(J.A.), JANIK(J.), MANIAWSKI(F.), RZANY(H.), ROGALSKA(Z.),
     SCIESINSKI(J.), SAGAN(U.), TUBBS(N.).

415 C*H3*S*H,. . . . . .P. . . . . .ACTA PHYS. POLON. VOL.16, . . . 335, (1957).
     *ESTIMATION OF HEIGHT OF/POTENTIAL BARRIER OF (INTERNAL) HINDERED ROTA-
     TION IN THE C*H3*S*H MOLECULE BY MEANS OF THERMAL NEUTRON SCATTERING.
     *BUDZANOWSKI(A.), GROTOWSKI(K.), JANIK(J.A.), KOLOS(W.), MANIAWSKI(F.),
     RZANY(H.), SZKATULA(A.), WANIC(A.).

416 (C*H3)2*SN*F2,. . . . . .P. . . . . . . .INORG. CHEM. VOL.5, ... 2238, (1966).
     *FREE ROTATION OF METHYL GROUPS IN DIMETHYLTIN DIFLUORIDE.
     *RUSH(J.J.), HAMILTON(W.C.).

417 C*(HF,NB,TA,U),. . . . . .T. . . . . . .PHYS. REV. B VOL.5, . . . 1260, (1972).
     *PHONON SPECTRA OF SOME TRANSITION METAL CARBIDES FROM A SIMPLE
     PSEUDOPOTENTIAL APPROACH.
     *MOSTOLLER(M.).

418 C-(HF,TA,NB),. . . . . .B7. . . . . .CONF. INTERN., RENNES, . . . 145, (1971).
     *PHONON SPECTRA AND SUPERCONDUCTIVITY IN SOME TRANSITION METAL CARBIDES.
     *SMITH(H.G.), GLASER(W.).

SECTION   6 -QUASI-ELASTIC AND INELASTIC SCATTERING STUDIES . . . .

419   C6*N4*D12 (DHMT),. . . . . .T,B1. . . . .DISC. FARADAY SOC. NO.48. . . . . 125, (1969).
       ₥MOLECULAR CRYSTAL PHONON DISPERSION CURVES AND MODEL FITTING.
       ₥PAWLEY(G.S.).

420   C4*N2*D4 (PYRAZINE). . . . . .B1. . . . . . .J. CHEM. PHYS. VOL.59. . . . . . 2777, (1973).
       ₥LATTICE DYNAMICS OF THE PYRAZINE CRYSTAL STUDIED BY COHERENT INELASTIC
       NEUTRON SCATTERING.
       ₥REYNOLDS(P.A.).

421   C6*(N*H2)3*(N*O2)3,. . . . . . .C1. . . . . . .MOL. PHYS. VOL.19. . . . . 589, (1970).
       ₥LOW FREQUENCY NEUTRON SPECTRUM OF 1,3,5-TRIAMINO-2,4,6-TRINITROBENZENE.
       ₥GUPTA(V.D.), DEOPURA(B.L.).

422   C*O, ...B7. . . . . . .IAEA SYMP. COPENHAGEN, VOL.1,. . 599, (1968).
       ₥BRILLOUIN SCATTERING OF NEUTRONS FROM LIQUID C*O.
       ₥MIKA(K.), DORNER(B.), STILLER(H.H.).

423   C*O2,. . . . . .B1,T. . . . . . IAEA SYMP. GRENOBLE, ... 207, (1972).
       ₥NORMAL MODES OF SOLID CARBON DIOXIDE.
       ₥POWELL(B.M.), DOLLING(G.), PISERI(L.), MARTEL(P.).

424   C*O2,. . . . . .D2. . . . . . . .PHYS. LETT. 24A. . . . . 57, (1967).
       ₥INELASTIC NEUTRON SCATTERING BY DENSITY FLUCTUATIONS IN C*O2 NEAR THE
       CRITICAL POINT.
       ₥BATA(L.), KROO(N.).

425   C*O2,. . . . . .B1. . . . . . .BULL. AMER. PHYS. SOC. VOL.17, . 291, (1972).
       ₥LATTICE DYNAMICS OF CARBON DIOXIDE.
       ₥DOLLING(G.), POWELL(B.M.), PISERI(L.), MARTEL(P.). (ED. NOTE: PAPER
       PRESENTED AT MARCH MEETING IN ATLANTIC CITY).

426   C*O2,. . . . . .T. . . . . .J. CHEM. PHYS. VOL.58. . . . . . 3647, (1973).
       ₥EMPIRICAL INTERMOLECULAR POTENTIALS FOR N2 AND C*O2 FROM CRYSTAL DATA.
       ₥JACOBI(N.), SCHNEPP(O.).

427   C*O2,. . . . . .B4. . . . . . . .IAEA SYMP. COPENHAGEN, VOL.1,. . 615, (1968).
       ₥INVESTIGATION OF THE DYNAMICAL PROPERTIES OF LIQUID-GAS SYSTEMS BY
       COHERENT COLD NEUTRON SCATTERING NEAR THE CRITICAL POINT.
       ₥BATA(L.), KROO(N.).

428   C*O2,. . . . . .T,E. . . . . .SPECTROSCOPY BIOL. CHEM.. . . . . . 297, (1974).
       ₥NEUTRON SCATTERING AND OPTICAL STUDIES OF MOLECULAR VIBRATIONS.
       ₥BERNEY(C.V.). (REFER TO SECTION 4).

429   C*O*(N*H2)2 (UREA),. . . . . . .C1. . . . . . .CHEM. COMMUN. 1967,. . . . . . 74, (1967).
       ₥LOW-FREQUENCY MOLECULAR MOTIONS OF UREA BY NEUTRON-SCATTERING SPEC-
       TROSCOPY.
       ₥BOOGER(E.O.), WHITE(J.W.).

430   C*S2,. . . . . .C3. . . . . . .IAEA SYMP. COPENHAGEN, VOL.1,. . 615, (1968).
       ₥INVESTIGATION OF THE DYNAMICAL PROPERTIES OF LIQUID-GAS SYSTEMS BY
       COHERENT COLD NEUTRON SCATTERING NEAR THE CRITICAL POINT.
       ₥BATA(L.), KROO(N.).

431   C*S2 SOLUTIONS,. . . . . . .C3. . . . . . .J. CHEM. PHYSICS VOL.55. . . . . . 2384, (1971).
       ₥NEUTRON SCATTERING BY SMALL CHAINLIKE MOLECULES DISPERSED IN C*S2.
       ₥USHA-DENIZ(K.), JANNINK(G.), TAUPIN(D.).

432   CA,. . . . . .B7. . . . . . .IAEA SYMP. GRENOBLE. . . . . . . 137, (1972).
       ₥STUDIES ON THE DETERMINATION OF THE PHONON DENSITY OF STATES BY COHERENT
       INELASTIC SCATTERING OF SLOW NEUTRONS FROM POLYCRYSTALLINE SAMPLES.
       ₥GOMPF(F.), LAU(H.), REICHARDT(W.), SALGADO(J.).

433   CA*C*O3,. . . . . .P. . . . . . . .REF. SECTION 3, 48/HAVENS,. . . . .

434   CA*C*O3,. . . . . .B1. . . . . . .PHYS. REV. B VOL.8,. . . . . 4795, (1973).
       ₥LATTICE DYNAMICS OF CALCITE.
       ₥COWLEY(E.R.), PANT(A.K.).

435   CA*C*O3,. . . . . .T. . . . . . . .CAN. J. PHYS. 47. . . . . 1381, (1969).
       ₥SYMMETRY PROPERTIES OF THE NORMAL MODES OF VIBRATIONS OF CALCITE AND
       (ALPHA)-CORUNDUM.
       ₥COWLEY(E.R.).

436   CA*F2, ...C2. . . . . . .IAEA SYMP. CHALK RIVER, VOL.2,. 225, (1962).
       ₥FREQUENCES DE VIBRATIONS OPTIQUES DANS LA FLUORINE CA*F2.
       ₥CRIBIER(D.), FARNOUX(B.), JACROT(B.).

437   CA*F2, ...B7. . . . . . .J. PHYS. SOC. JAPAN VOL.33. . . . . . 647, (1972).
       ₥A METHOD OF DYNAMICAL STRUCTURE ANALYSIS AND ITS APPLICATION TO CALCIUM
       FLUORIDE.
       ₥IIZUMI(M.).

438   CA*F2, ...B1. . . . . . .J. PHYS. C, SER.2, VOL.4,. . . . 492, (1971).
       ₥THE LATTICE DYNAMICS OF CALCIUM FLUORIDE.
       ₥ELCOMBE(M.M.), PRYOR(A.W.).

439   CA*F2, ...B2. . . . . . .J. PHYS. SOC. JAPAN VOL.35. . . . . 204, (1973).
       ₥NEUTRON SCATTERING STUDY OF THE ANHARMONIC LATTICE VIBRATIONS IN
       CALCIUM FLUORIDE.
       ₥IIZUMI(M.).

440   CA*F2(ER3+), ...D1. . . . . . .SOV. PHYS. SOL. STATE 6,. . . . . . 76, (1964).
       ₥SPIN-LATTICE RELAXATION OF THE ER+3 ION IN CD*F2, BA*F2, AND CA*F2
       SINGLE CRYSTALS.
       ₥ZVEREV(G.M.), SMIRNOV(A.I.).

441   CA*F2-(Y*F3),. . . . . . . .B7. . . . . . .J. PHYS. C VOL.5,. . . . . 2677, (1972).
       ₥DEFECT STRUCTURE OF CALCIUM FLUORIDE CONTAINING EXCESS ANIONS:
       II. DIFFUSE SCATTERING.
       ₥STEELE(D.), CHILDS(P.E.), FENDER(B.E.F.).

442   CA*H2, ...C1. . . . . . .J. CHEM. PHYS. VOL.52. . . . . . . 3952, (1970).
       ₥VIBRATION SPECTRA OF THE ORTHORHOMBIC ALKALINE-EARTH HYDRIDES BY THE
       INELASTIC SCATTERING OF COLD NEUTRONS AND BY INFRARED TRANSM. MEASUR.
       ₥MAELAND(A.J.).

443   CA*H2, ...C1. . . . . . .IAEA SYMPOSIUM VIENNA, ... 501, (1960).
       ₥INELASTIC NEUTRON SCATTERING EXPERIMENTS ON A FEW METAL HYDRIDES.
       ₥BERGSMA(J.), GOEDKOOP(J.A.).

444   CA2*MN*O4, ...D1. . . . . . .SOL. STATE COMM. VOL.13,. . . . . 919, (1973).
       ₥SPIN WAVE SPECTRUM OF CA2*MN*O4 AT 4.2 K.
       ₥HENNION(B.), MOUSSA(F.), PEPY(G.), OLLIVIER(G.).

SECTION   6 -QUASI-ELASTIC AND INELASTIC SCATTERING STUDIES . . . .

445 CA*O,. . . . . *DISPERSION RELATIONS IN CALCIUM OXIDE BY NEUTRON SPECTROSCOPY. . . . . . B1. IN . . CONF. INTERN. RENNES . . . . 171, (1971).
*SAUNDERSON(D.H.), PECKHAM(G.E.).

446 CA*O,. . . . . *LATTICE DYNAMICS OF CALCIUM OXIDE. . . . . B1. . . . . . . IAEA SYMP. GRENOBLE, ... 95, (1972).
*VIJAYARAGHAVAN(P.R.), MARSONGKOHADI, IYENGAR(P.K.).

447 CA*O,. . . . . *THREE-BODY FORCE SHELL MODEL-AN APPLICATION TO CA*O. . . . . T,B1. . . . . . SOL. STATE COMM. VOL.11 NO.4,. . 567, (1972).
*UPADHYAYA(K.S.), SINGH(R.K.).

448 CA*O,. . . . . *THE LATTICE DYNAMICS OF CALCIUM OXIDE. . . . . B1. . . . . J. PHYS. C, SER.2, VOL.4,. . . . . 2009, (1971).
*SAUNDERSON(D.H.), PECKHAM(G.E.).

449 CA*(O*D)2, . . . *INELASTIC NEUTRON SPECTRA AND THE VIBRATIONAL MODES OF THE HYDROGEN LAYER IN ALKALI AND ALKALINE-EARTH HYDROXIDES. . . . . . C2. . . . . J. CHEM. PHYSICS VOL.43,. . . . . . 1864, (1965).
*PELAH(I.), KREBS(K.), IMRY(Y.).

450 CA*(O*H)2, . . . *INELASTIC NEUTRON SPECTRA AND THE VIBRATIONAL MODES OF THE HYDROGEN LAYER IN ALKALI AND ALKALINE-EARTH HYDROXIDES. . . . . . C2. . . . . J. CHEM. PHYSICS VOL.43,. . . . . . 1864, (1965).
*PELAH(I.), KREBS(K.), IMRY(Y.).

451 CA*(O*H)2, . . . *AN INVESTIGATION OF THE ENERGY LEVELS IN ALKALINE EARTH HYDROXIDES BY INELASTIC SCATTERING OF SLOW NEUTRONS. . . . . . C1. . . . . J. PHYS. CHEM. SOL. 24,. . . . . . 771, (1963).
*SAFFORD(G.J.), BRAJOVIC(V.), BOUTIN(H.).

452 CA2*SR*(C2*H5*C*O*O)6, . . *A COLD NEUTRON SCATTERING STUDY OF DICALCIUM STRONTIUM PROPIONATE. . . . . . C1. . . . . . . CHEM. PHYS. LETT. VOL.22,. . . . . 476, (1973).
*VAN-TRICHT(J.B.), DE-MUL(F.F.M.), MEIJERMANS(J.P.).

453 CA*W*O4, ... *EXTERNAL MODES IN CA*W*O4. . . . . B1. . . . . . . IAEA SYMP. GRENOBLE, ... 219, (1972).
*STEINMAN(D.K.), KING(J.S.), SMITH(H.G.).

454 CB,. . . . . P,. . . . . REF. SECTION 3, 47/WU, ...

455 CD,. . . . . . . *STUDY OF THERMAL NEUTRON INTERACTION WITH CADMIUM ISOTOPES. . . . . . P. . . . . . . UKR. FIZ. ZH.(USSR) VOL.17,. . . . 38, (1972).
*VERTEBNYI(V.P.), VLASOV(M.F.), GNIDAK(N.L.), GRISHANIN(E.I.), ZATSERKOVSKII(R.A.), KIZILYUK(A.L.), LEPENDIN(V.I.), TROFIMOVA(N.A.).

456 CD,. . . . . . . *STATE DENSITY OF PHONONS IN CADMIUM. . . . . . B1,P. . . . . SOV. PHYS. JETP LETT. VOL.18,. . . 177, (1973).
*EREMEEV(I.P.), CHERNYSHOV(A.A.), SADIKOV(I.P.).

457 CD*F2(ER3+), . . *SPIN-LATTICE RELAXATION OF THE ER+3 ION IN CD*F2, BA*F2, AND CA*F2 SINGLE CRYSTALS. . . . . . D1. . . . . . SOV. PHYS. SOL. STATE 6,. . . . . . 76, (1964).
*ZVEREV(G.M.), SMIRNOV(A.I.).

458 CD*I2, ... *ANOMALOUS DISPERSION OF SLOW NEUTRONS IN CRYSTALS. . . . . . B. . . . . . . COLL. INTER. N.126 (GRENOBLE), . 191, (1963).
*SMITH(H.G.), PETERSON(S.W.).

459 CD*S,. . . . . . *ANOMALOUS NEUTRON SCATTERING BY CRYSTALS AND THE AMPLITUDES OF VIBRATION OF LATTICE WAVES. . . . . . P. . . . . . . ACTA CRYST. VOL.A26,. . . . . 364, (1970).
*RAMASESHAN(S.), VISWANATHAN(K.S.).

460 CD*S,. . . . . . *LATTICE DYNAMICS OF WURTZITE CD*S. . . . . . B1. . . . . . . PHYS. REV. 156,. . . . . 925, (1967).
*NUSIMOVICI(M.A.), BIRMAN(J.L.).

461 CD*TE, ... *LATTICE DYNAMICS OF CADMIUM TELLURIDE. . . . . . B1. . . . . . . PHYS. REV. B VOL.10 NO.2,. . . . . , (1974).
*ROWE(J.M.), NICKLOW(R.M.), PRICE(D.L.), ZANIO(K.). (REF. OBTAINED FROM PHYS. REV. ABSTRACTS).

462 CE,. . . . . . . *INELASTIC MAGNETIC SCATTERING OF SLOW NEUTRONS IN METALLIC CERIUM. . . . . . D4,T. . . . . FIZ. MET. METALLOV. VOL.10,. . . . 825, (1960).
*KHABIBULLIN(B.M.). (IN RUSSIAN).

463 CE*AS, ... *MAGNETIC PROPERTIES OF CE*AS AND CE*SB DEDUCED FROM NEUTRON SCATTERING DATA. . . . . . A2,D4. . . J. OF PHYS. C VOL.1,. . . . . 679, (1968).
*RAINFORD(B.D.), TURBERFIELD(K.C.), BUSCH(G.), VOGT(O.).

464 CE-(AS,Y), . . . *PARAMAGNETIC SCATTERING OF NEUTRONS BY TRIVALENT RARE-EARTH IONS IN AN OCTAHEDRAL CRYSTAL FIELD. . . . . . D3. . . . . . IAEA SYMP. COPENHAGEN, VOL.2,. . . 133, (1968).
*FURRER(A.), HALG(W.), SCHNEIDER(T.).

465 CE*H2,CE*H3, . . *A LATTICE-DYNAMICAL INVESTIGATION OF THE HYDRIDES CE*H2, CE*H3, TH*H2 AND U*H(2.7) USING INELASTIC SCATTERING OF COLD NEUTRONS. . . . . . C1. . . . . . . SOV. PHYS. SOL. STATE 9,. . . . . . 1366, (1967).
*KARIMOV(I.), ZEMLYANOV(M.G.), KOST(M.E.), SOMENKOV(V.A.), CHERNOPLEKOV(N.A.).

466 CE*(P,AS,SB,BI), . *CRYSTAL FIELD SPLITTING IN CERIUM MONOPNICTIDES. . . . . . D4. . . . . . IAEA SYMP. GRENOBLE, ... 563, (1972).
*FURRER(A.), BUHRER(W.), HEER(H.), HALG(W.), BENES(J.), VOGT(O.).

467 CE*SB, ... *MAGNETIC PROPERTIES OF CE*AS AND CE*SB DEDUCED FROM NEUTRON SCATTERING DATA. . . . . . A2,D4. . . J. OF PHYS. C VOL.1,. . . . . 679, (1968).
*RAINFORD(B.D.), TURBERFIELD(K.C.), BUSCH(G.), VOGT(O.).

468 CE-Y-AS, ... *PARAMAGNETIC SCATTERING OF NEUTRONS BY TRIVALENT RARE-EARTH IONS IN AN OCTAHEDRAL CRYSTAL FIELD. . . . . . D3. . . . . . IAEA SYMP. COPENHAGEN, VOL.2,. . . 133, (1968).
*FURRER(A.), HALG(W.), SCHNEIDER(T.).

469 (CE-(Y,LA))*SB,. *CRYSTAL-FIELD SPLITTING IN CE(X)-Y(1-X)*SB AND CE(X)-LA(1-X)*SB. . . . . . B7. . . . . . SOL. STATE COMM. VOL.11,. . . . . . 21, (1972).
*COOPER(B.R.), FURRER(A.), BUHRER(W.).

470 CO,. . . . . . . *SMALL-ANGLE CRITICAL MAGNETIC SCATTERING OF NEUTRONS IN CO. . . . . . D2. . . . . . IAEA SYMP. COPENHAGEN, VOL.2,. . . 75, (1968).
*BALLY(D.), POPOVICI(M.), TOTIA(M.), GRABCEV(B.), LUNGU(A.M.).

SECTION   6 -QUASI-ELASTIC AND INELASTIC SCATTERING STUDIES . . . .

471 CO,. . . . . P. . . . . PHYS. REV. VOL.185, . . .   961, (1969).
     ⊬COHERENT   NEUTRON   SCATTERING   BY   COBALT   WITH   NUCLEAR   POLARIZATION.
     ⊬ITO(Y.), SHULL(C.G.).

472 CO,. . . . . .D1. . . . . . .J. APP. PHYS. 39, ...   383, (1968).
     ⊬SPIN WAVES IN 3D METALS.
     ⊬SHIRANE(G.), MINKIEWICZ(V.J.), NATHANS(R.).

473 CO,. . . . . .D1. . . . . . .SOL. STATE COMM. 3, ...   339, (1965).
     ⊬EXPERIMENTELLER NACHWEIS DER S-D WECHSELWIRKUNG IN F.C.C. COBALT DURCH
     SPINWELLENSTREUUNG POLARISIERTER NEUTRONEN.
     ⊬FURRER(A.), SCHNEIDER(T.), HALG(W.).

474 CO,. . . . . .D4. . . . . . .PHYS. REV. LETT. 10, ...   290, (1963).
     ⊬COHERENT MAGNETIC SCATTERING OF THERMAL NEUTRONS BY COBALT AT HIGH
     TEMPERATURE.
     ⊬MENZINGER(F.), PAOLETTI(A.).

475 CO,. . . . . .D1. . . . . . .IAEA SYMP. COPENHAGEN, VOL.2, . . .   3, (1968).
     ⊬NEUTRON SPIN SCATTERING BY SPIN WAVES IN METALS (REVIEW PAPER).
     ⊬MOLLER(H.B.).

476 CO,. . . . . .P. . . . . . . .PHYS. REV. 130, ...   1907, (1963).
     ⊬SPIN DEPENDENCE OF THE THERMAL NEUTRON CROSS SECTION OF COBALT (A=59).
     ⊬SCHERMER(R.I.).

477 CO,. . . . . .D1. . . . . .J. APP. PHYS. 36, ...   1076, (1965).
     ⊬SPIN-WAVE SCATTERING OF POLARIZED NEUTRONS FROM NICKEL AND COBALT.
     ⊬RISTE(T.), SHIRANE(G.), ALPERIN(H.A.), PICKART(S.J.).

478 CO,. . . . . .D1. . . . . .J. APP. PHYS. 37, ...   1052, (1966).
     ⊬OBSERVATION OF THE DISPERSION RELATION FOR SPIN WAVES IN HEXAGONAL
     COBALT.
     ⊬ALPERIN(H.A.), STEINSVOLL(O.), SHIRANE(G.), NATHANS(R.).

479 CO,. . . . . .D1. . . . . .PHYS. REV. 156, ...   623, (1967).
     ⊬SPIN WAVE DISPERSION IN FERROMAGNETIC NI AND FCC CO.
     ⊬PICKART(S.J.), ALPERIN(H.A.), MINKIEWICZ(V.J.), NATHANS(R.),
       SHIRANE(G.), STEINSVOLL(O.).

480 CO,. . . . . .D2. . . . . .PHYS. LETT. 25A, ...   595, (1967).
     ⊬CRITICAL MAGNETIC SCATTERING OF NEUTRONS IN COBALT.
     ⊬BALLY(D.), POPOVICI(M.), TOTIA(M.), GRABCEV(B.), LUNGU(A.M.).

481 CO,. . . . . .D1. . . . . .PHYSICA VOL.32, ...   2149, (1966).
     ⊬INELASTIC SCATTERING OF NEUTRONS BY SPIN WAVES IN F.C.C. COBALT.
     ⊬FRIKKEE(E.).

482 CO (ALPHA),. . . . . .D1. . . . . .PHYSICA VOL.37, ...   501, (1967).
     ⊬INVESTIGATION OF MAGNONS IN HEXAGONAL COBALT.
     ⊬KRASNICKI(S.), WANIC(A.), DIMITRIJEVIC(A.), TODOROVIC(J.).

483 CO,. . . . . .D2. . . . . .J. APPL. PHYS. VOL.39, ...   459, (1968).
     ⊬CRITICAL SCATTERING OF THERMAL NEUTRONS IN SOME FERROMAGNETIC METALS.
     ⊬BALLY(D.), GRABCEV(B.), POPOVICI(M.), TOTIA(M.), LUNGU(A.M.).

484 CO,. . . . . T,D4. . . . .PHYS. REV. LETT. VOL.11, ...   264, (1963).
     ⊬SPIN DENSITY IN COBALT.
     ⊬WEISS(R.J.).

485 CO,. . . . . .D1. . . . . .PROC. PHYS. SOC. VOL.79, . . .   473, (1962).
     ⊬SPIN WAVES AND NEUTRON SCATTERING IN HEXAGONAL COBALT.
     ⊬LOW(G.G.E.).

486 CO,. . . . . .P. . . . . . . .REF. SECTION 3, 47/WU, ...

487 CO,. . . . . .D1. . . . . .PROC. INT. CONF. (NOTTINGHAM), .   299, (1964).
     ⊬IMAGE OF THE FERMI SURFACE IN THE MAGNON SPECTRUM OF FACE CENTERED CUBIC
     COBALT.
     ⊬FRIKKEE(E.), RISTE(T.).

488 CO*C*O3,. . . . .A2,D4. . . .J. PHYS. C VOL.6, ...   1405, (1973).
     ⊬MAGNETIZATION DENSITY AND THE MAGNETIC STRUCTURE OF COBALT CARBONATE.
     ⊬BROWN(P.J.), WELFORD(P.J.), FORSYTH(J.B.).

489 CO*CL2,. . . . . .D1. . . . . .J. PHYS. C VOL.6, ...   3143, (1973).
     ⊬NEUTRON SCATTERING INVESTIGATION OF MAGNETIC EXCITATIONS IN CO*CL2.
     ⊬HUTCHINGS(M.T.).

490 CO*CL2.2*H2*O,. . . . . . .T. . . . . .PHYS. REV. B VOL.5, ...   1941, (1972).
     ⊬INTERACTING MAGNONS IN THE LINEAR CHAIN.
     ⊬FOGEDBY(H.C.).

491 CO*F2,. . . . . .T. . . . . .PHYS. LETT. VOL.39A, ...   105, (1972).
     ⊬TWO MAGNON LIGHT AND NEUTRON SCATTERING IN CO*F2.
     ⊬NATOLI(C.R.), RANNINGER(J.).

492 CO*F2,. . . . . .D1,T. . . . .PHYS. REV. LETT. VOL.23, ...   86, (1969).
     ⊬TWO-MAGNON SCATTERING OF NEUTRONS.
     ⊬COWLEY(R.A.), SVENSSON(E.C.), BUYERS(W.J.L.), MARTEL(P.),
       STEVENSON(R.W.H.).

493 CO*F2,. . . . . .D1. . . . . .PHYS. REV. LETT. 18, ...   162, (1967).
     ⊬SPIN-WAVE AND EXCITATION DISPERSION RELATIONS OF COBALT FLUORIDE.
     ⊬COWLEY(R.A.), MARTEL(P.), STEVENSON(R.W.H.).

494 CO*F2,. . . . . .D4. . . . . .J. APP. PHYS. 39, ...   1116, (1968).
     ⊬EXCITATIONS OF COBALT FLUORIDE NEAR NEEL TEMPERATURE.
     ⊬MARTEL(P.), COWLEY(R.A.), STEVENSON(R.W.H.).

495 CO*F2,. . . . . .T. . . . . .J. PHYS. C VOL.6, ...   370, (1973).
     ⊬TWO-MAGNON NEUTRON SCATTERING IN CO*F2.
     ⊬NATOLI(C.R.), RANNINGER(J.).

496 CO*F2,. . . . . .B1. . . . . .CAN. J. PHYS. 46, ...   1355, (1968).
     ⊬EXPERIMENTAL STUDIES OF THE MAGNETIC AND PHONON EXCITATIONS IN COBALT
     FLUORIDE.
     ⊬MARTEL(P.), COWLEY(R.A.), STEVENSON(R.W.H.).

497 CO*F2,. . . . . .D1,D2. . . .J. PHYS. C VOL.6, ...   2997, (1973).
     ⊬MAGNETIC EXCITATIONS AND MAGNETIC CRITICAL SCATTERING IN COBALT FLUORIDE
     ⊬COWLEY(R.A.), BUYERS(W.J.L.), MARTEL(P.), STEVENSON(R.W.H.).

498 CO-FE,. . . . . .D1. . . . . .PHYS. REV. VOL.120, ...   1638, (1960).
     ⊬DISPERSION RELATION FOR SPIN WAVES IN A F.C.C. COBALT ALLOY.
     ⊬SINCLAIR(R.N.), BROCKHOUSE(B.N.).

SECTION   6 -QUASI-ELASTIC AND INELASTIC SCATTERING STUDIES . . . .

499   CO-FE, ... D1 . . . . . . J. PHYS. F VOL.3, ... L173, (1973).
          *TEMPERATURE DEPENDENCE OF THE MAGNETIC EXCITATIONS IN A CO-FE ALLOY.
          *MIKKE(K.), JANKOWSKA(J.), MODRZEJEWSKI(A.).

500   CO*FE2*O4, ... B1,D1 . . . . CAN. J. PHYS. VOL.52, ... 396, (1974).
          *MAGNONS AND PHONONS IN COBALT FERRITE.
          *TEH(H.C.), COLLINS(M.F.), MOOK(H.A.).

501   (CO,MN)*F2, ... C1 . . . . . . PHYS. REV. LETT. VOL.27, ... 1442, (1971).
          *CHARACTER OF EXCITATIONS IN SUBSTITUTIONALLY DISORDERED ANTIFERROMAGNETS
          *BUYERS(W.J.L.), HOLDEN(T.M.), SVENSSON(E.C.), COWLEY(R.A.),
          STEVENSON(R.W.H.).

502   CO*(N*H3)6*(I3,CL3), ... C1 . . . . PHYSICA VOL.35, ... 441, (1967).
          *MOLECULAR DYNAMICS STUDY BY THE NEUTRON INELASTIC SCATTERING METHOD I.
          COMPLEXES.
          *JAKOB(W.),   JANIK(J.M.),   JANIK(J.A.),   BAJOREK(A.),   PARLINSKI(K.),
          SUDNIK-HRYNKIEWICZ(M.).

503   CO*(N*O3)2.2H2*O, ... C1 . . . . . . INORG. CHEM. VOL.6, ... 346, (1967).
          *NEUTRON-SCATTERING STUDY OF THE MOTIONS OF WATER MOLECULES IN HYDRATED
          SALTS OF TRANSITION METALS.
          *RUSH(J.J.), FERRARO(J.R.), WALKER(A.).

504   CO-NI, ... D4 . . . . . . PROC. PHYS. SOC. 82, ... 633, (1963).
          *MAGNETIC MOMENTS AND THE DEGREE OF ORDER IN COBALT-NICKEL ALLOYS.
          *COLLINS(M.F.), WHEELER(D.A.).

505   CO*O, ... D1 . . . . . . IAEA SYMP. COPENHAGEN, VOL.2, . . 123, (1968).
          *MAGNETIC EXCITATIONS IN COBALTOUS OXIDE.
          *BUYERS(W.J.L.), DOLLING(G.), SAKURAI(J.), COWLEY(R.A.).

506   CO*O, ... D2 . . . . . . PHYS. REV. LETT. VOL.26, ... 1480, (1971).
          *CRITICAL MAGNETIC NEUTRON SCATTERING FROM CO*O.
          *RECHTIN(M.D.), AVERBACH(B.L.).

507   CO*O, ... T, D4 . . . . PHYS. REV. 177, ... 932, (1969).
          *MAGNETIC SYMMETRY AND ANTIFERROMAGNETIC RESONANCE IN CO*O.
          *DANIEL(M.R.), CRACKNELL(A.P.).

508   CO*O, ... B1,T . . . . PHYS. LETT. VOL.40A NO.4, ... 291, (1972).
          *PHONON SPECTRA OF PARAMAGNETIC COBALTOUS OXIDE.
          *UPADHYAYA(K.S.), SINGH(R.K.).

509   CO*O, ... B1,D1 . . . PHYS. REV. 167, ... 510, (1968).
          *CRYSTAL DYNAMICS AND MAGNETIC EXCITATION IN COBALTOUS OXIDE.
          *SAKURAI(J.), BUYERS(W.J.L.), COWLEY(R.A.), DOLLING(G.).

510   CO*O, ... D2 . . . . . . J. PHYS. RADIUM VOL.20, ... 175, (1959).
          *CRITICAL SCATTERING OF NEUTRONS FROM CO*O.
          *MCREYNOLDS(A.W.), RISTE(T.).

511   CO*O, ... D4 . . . . . . REV. MOD. PHYS. VOL.30, ... 89, (1958).
          *THEORY OF THE MAGNETIC SCATTERING OF NEUTRONS BY CO*O.
          *NAGAMIYA(T.), MOTIZUKI(K.).

512   CO*PT3, ... D4 . . . . . . PHYS. REV. 143, ... 365, (1966).
          *MAGNETIC MOMENTS AND UNPAIRED-ELECTRON DENSITIES IN CO*PT3.
          *MENZINGER(F.), PAOLETTI(A.).

513   CORUNDUM (ALPHA), ... T . . . . . . CAN. J. PHYS. 47, ... 1381, (1969).
          *SYMMETRY PROPERTIES OF THE NORMAL MODES OF VIBRATION OF CALCITE AND
          (ALPHA)-CORUNDUM.
          *COWLEY(E.R.).

514   CR, ... D1,D2 . . . NUOV. APPL. PHYS.42, ... 1666, (1971).
          *SPIN-WAVE AND CRITICAL NEUTRON SCATTERING FROM CHROMIUM.
          *ALS-NIELSEN(J.), AXE(J.D.), SHIRANE(G.).

515   CR, ... B1 . . . . . . PHYS. REV. B-4, ... 969, (1971).
          *INVESTIGATION OF THE PHONON DISPERSION RELATIONS OF CHROMIUM BY
          INELASTIC NEUTRON SCATTERING.
          *SHAW(W.M.), MUHLESTEIN(L.D.).

516   CR, ... B1 . . . . . . IAEA SYMP. BOMBAY, VOL.1, ... 95, (1964).
          *INELASTIC SCATTERING OF NEUTRONS IN CHROMIUM.
          *MOLLER(H.B.), MACKINTOSH(A.R.).

517   CR, ... B1 . . . . . . DISS. ABS. VOL.32, ... 1795B, (1971).
          *THE STUDY OF INSTRUMENTAL RESOLUTION AND THE LATTICE DYNAMICS OF
          CHROMIUM.
          *SHAW(W.M.).

518   CR, ... T . . . . . . PHYS. REV. VOL.187, ... 584, (1969).
          *THEORY OF LONGITUDINAL SPIN FLUCTUATIONS AND THE ANTIFERROMAGNETIC PHASE
          TRANSITION IN CHROMIUM METAL.
          *SOKOLOFF(J.B.).

519   CR, ... T . . . . . . PHYS. REV. B VOL.3, ... 2367, (1971).
          *EFFECTS OF PHONONS AND IMPURITIES ON SINGLE-PARTICLE-MODE NEUTRON
          SCATTERING IN CHROMIUM.
          *SOKOLOFF(J.B.).

520   CR, ... D1 . . . . . . J. APP. PHYS. VOL.40, ... 1447, (1969).
          *SPIN DIRECTIONS IN PURE CHROMIUM.
          *WERNER(S.A.), ARROTT(A.), ATOJI(M.).

521   CR, ... D1,3 . . . . PHYS. REV. LETT. 22, ... 290, (1969).
          *SPIN WAVES AND THE ORDER-DISORDER TRANSITION IN CHROMIUM.
          *ALS-NIELSEN(J.), DIETRICH(O.W.).

522   CR, ... D1 . . . . . . J. PHYS. SOC. JAPAN VOL.32, ... 394, (1972).
          *NEUTRON SCATTERING MEASUREMENTS IN CHROMIUM NEAR THE NEEL TEMPERATURE.
          *TSUNODA(Y.), HAMAGUCHI(Y.), KUNITOMI(N.).

523   CR, ... D2 . . . . . . SOL. STATE COMM. 2, ... 109, (1964).
          *MAGNETIC SCATTERING OF NEUTRONS IN CHROMIUM.
          *MOLLER(H.B.), BLINOWSKI(K.), MACKINTOSH(A.R.), BRUN(T.O.).

524   CR, ... D2 . . . . . . PROC. INT. CONF. (NOTTINGHAM), . . 101, (1964).
          *MAGNETIC SCATTERING OF NEUTRONS IN CHROMIUM.
          *MOLLER(H.B.), BLINOWSKI(K.), NIELSEN(P.), MACKINTOSH(A.R.), BRUN(T.O.).

525   CR, ... D2 . . . . . . J. APP. PHYS. 39, ... 1227, (1968).
          *CRITICAL MAGNETIC SCATTERING OF NEUTRONS IN CHROMIUM.
          *HAMAGUCHI(Y.), TSUNODA(Y.), KUNITOMI(N.).

SECTION 6 -QUASI-ELASTIC AND INELASTIC SCATTERING STUDIES . . . .

526 CR,. . . . . .T. . . . . .PHYS. REV. 157. . . . . 540, (1967).
EXISTENCE OF AN INFINITY IN THE FREQUENCY DISTRIBUTION G(NU) OF
MONOATOMIC BODY-CENTERED CUBIC CRYSTALS.
GILAT(G.).

527 CR,. . . . . .D3. . . . . .PHYS. REV. VOL.127,. . . . . 2080, (1962).
INVESTIGATIONS OF PARAMAGNETIC NEUTRON SCATTERING FROM CHROMIUM.
WILKINSON(M.K.), WOLLAN(E.O.), KOEHLER(W.C.), CABLE(J.W.).

528 CR,. . . . . .B1. . . . . .IAEA SYMP. GRENOBLE, ... 53, (1972).
INVESTIGATION OF THE PHONON DISPERSION RELATIONS OF PARAMAGNETIC AND
ANTIFERROMAGNETIC CHROMIUM.
MUHLESTEIN(L.D.), GURMEN(E.), CUNNINGHAM(R.M.).

529 CR,. . . . . .B1. . . . . .A.I.P. CONF. PROC. NO.10 PART 1, . 41, (1973).
ANOMALIES IN THE PHONON-DISPERSION RELATIONS OF CHROMIUM IN A
SINGLE-Q STATE.
MUHLESTEIN(L.D.), GURMEN(E.), CUNNINGHAM(R.M.).

530 CR*BR3,. . . . . .D1. . . . . .PHYS. REV. B VOL.4. . . . . 2280, (1971).
RENORMALIZATION OF LARGE-WAVE-VECTOR MAGNONS IN FERROMAGNETIC CR*BR3
STUDIED BY INELASTIC NEUTRON SCATTERING: SPIN-WAVE CORRELATION EFFECTS.
YELON(W.B.), SILBERGLITT(R.).

531 CR*BR3,. . . . . . . .D1. . . . . .PHYS. REV. B VOL.3. . . . . 157, (1971).
SPIN WAVES IN FERROMAGNETIC CR*BR3 STUDIED BY INELASTIC NEUTRON
SCATTERING.
SAMUELSEN(E.J.), SILBERGLITT(R.), SHIRANE(G.), REMEIKA(J.P.).

532 CR*BR3,CR*F3,. . . . .D3. . . . . .J. APPL. PHYS. VOL.39 NO.2,. . . 1113, (1968).
NEUTRON SCATTERING FROM PARAMAGNETIC CHROMIUM HALIDES.
RAO(L.M.), MURTHY(N.S.S.), VENKATARAMAN(G.), IYENGAR(P.K.).

533 CR*BR3,CR*F3,. . . . .D3. . . . . .PHYS. LETT. VOL.26A. . . . . . . 108, (1968).
NEUTRON SCATTERING FROM PARAMAGNETIC CHROMIUM HALIDES.
RAO(L.M.), MURTHY(N.S.S.), VENKATARAMAN(G.), IYENGAR(P.K.).

534 CR*F2, ...D3. . . . . .SOL. STATE COMM. 6, ... 593, (1968).
EXCHANGE INTEGRALS IN CR*F2 AND FE*F2 BY PARAMAGNETIC NEUTRON SCATTERING
MADHAV-RAO(L.), NATERA(M.G.), MURTHY(N.S.S.), DASANNACHARYA(B.A.),
IYENGAR(P.K.).

535 CR-FE, ...B1. . . . . .PHYS. STAT. SOLIDI B VOL.55 NO.1 K1, (1973).
ANOMALIES IN THE PHONON DISPERSION RELATION IN CR-FE ALLOY BY NEUTRON
SCATTERING.
MIKKE(K.), JANKOWSKA(J.).

536 CR-FE, ...D3. . . . . .PHYS. STAT. SOLIDI B VOL.57 NO.1 K1, (1973).
MAGNETIC EXCITATIONS IN THE PARAMAGNETIC PHASE OF A CR-FE ALLOY BY
INELASTIC NEUTRON SCATTERING.
MIKKE(K.), JANKOWSKA(J.), BEDNARSKI(S.).

537 CR-(MN), ...D4. . . . . .J. APPL. PHYS. 41. . . . . 1365, (1970).
MAGNETIC EXCITATIONS IN A CR-MN ALLOY INELASTIC NEUTRON SCATTERING.
SINHA(S.K.), LIU(S.H.), WAKABAYASHI(N.), MUHLESTEIN(L.D.).

538 CR-MN, ...D1,3. . . . . .PHYS. REV. LETT. 22. . . . . . . 290, (1969).
SPIN WAVES AND THE ORDER-DISORDER TRANSITION IN CHROMIUM.
ALS-NIELSEN(J.), DIETRICH(O.W.).

539 CR-MN, ...D3,D1. . . . . .PHYS. REV. LETT. VOL.23. . . . . 311, (1969).
NEUTRON-SCATTERING STUDY OF MAGNONS AND PARAMAGNONS IN A CHROMIUM-
MANGANESE ALLOY.
SINHA(S.K.), LIU(S.H.), MUHLESTEIN(L.D.), WAKABAYASHI(N.).

540 CR-MN, ...D4. . . . . .SOV. PHYS. CRYST. VOL.9, ... 301, (1964).
ANTIFERROMAGNETISM OF CHROMIUM-MANGANESE ALLOYS.
LOSHMANOV(A.A.).

541 CR-MN, ...D3. . . . . .J. PHYS. SOC. JAPAN VOL.33,. . . 1348, (1972).
PARAMAGNETIC SCATTERING OF NEUTRONS IN EQUIATOMIC CR-MN ALLOY.
NAKAI(Y.).

542 CR2*03,. . . . . . . .D4. . . . . .IAEA SYMP. CHALK RIVER, VOL.2, . 327, (1962).
SOME INVESTIGATIONS OF NEUTRON INELASTIC SCATTERING ON MAGNETICS.
KRASNICKI(SZ.), RUTA-WALA(K.), WANIC(A.), MURASIK(A.), RISTE(T.).

543 CR2*03,. . . . . . . .D1. . . . . .PHYSICA VOL.48. . . . . 13, (1970).
INELASTIC NEUTRON SCATTERING INVESTIGATION OF SPIN WAVES AND MAGNETIC
INTERACTIONS IN CR2*03.
SAMUELSEN(E.J.), HUTCHINGS(M.T.), SHIRANE(G.).

544 CR2*03,. . . . . . . .D1. . . . . .PHYSICA VOL.45. . . . . 12, (1969).
TEMPERATURE EFFECTS ON SPIN WAVES IN CR2*03 STUDIED BY MEANS OF
INELASTIC NEUTRON SCATTERING.
SAMUELSEN(E.J.).

545 CR2*03,. . . . . . . .D1. . . . . .PHYS. STAT. SOL. 32. . . . . . . 41, (1969).
NEUTRON INVESTIGATION OF THE SPIN SYSTEM DYNAMICS IN (ALPHA)-CR2*03.
ALIKHANOV(R.A.), DINITRIJEVIC(L.), KOWALSKA(A.), KRASNICKI(S.),
RZANY(H.), TODOROVIC(J.), WANIC(A.).

546 CR2*03,. . . . . . . .D1. . . . . .SOL. STATE COMM. 7, NO15. . . . . 1043, (1969).
INELASTIC NEUTRON SCATTERING INVESTIGATION OF SPIN WAVES AND MAGNETIC
INTERACTIONS IN CR2*03.
SAMUELSEN(E.J.), HUTCHINGS(M.T.), SHIRANE(G.).

547 CR2*03,. . . . . . . .D1. . . . . .PHYS. LETT. VOL.26A. . . . . . . 160, (1968).
SPIN WAVES IN CR2*03 STUDIED BY MEANS OF NEUTRON SCATTERING.
SAMUELSEN(E.J.).

548 CR*02, ...D2. . . . . .J. APP. PHYS. 38,. . . . . 979, (1967).
MAGNETIC CRITICAL-POINT BEHAVIOR OF CR*02.
KOUVEL(J.S.), RODBELL(D.S.).

549 (CR2*03)-(FE2*03),. . .T. . . . . .PROC. INT. CONF. (NOTTINGHAM), . 516, (1964).
THEORY OF SOLID SOLUTIONS (1-X)*CR2*03*(X)*FE2*03 WITH HELICAL SPIN
CONFIGURATION.
BERTAUT(E.F.).

550 CR-(W),. . . . . . . .B1,B3. . . . . .PHYS. REV. B-2. . . . . 4864, (1970).
INVESTIGATION OF IN-BAND RESONANT MODES IN CR-W ALLOYS BY INELASTIC
NEUTRON SCATTERING.
CUNNINGHAM(R.M.), MUHLESTEIN(L.D.), SHAW(W.M.), TOMPSON(C.W.).

SECTION 6 -QUASI-ELASTIC AND INELASTIC SCATTERING STUDIES . . . .

551 CR-W,. . . . . *THE STUDY OF 'IN-BAND' RESONANT 'MODES' IN CHROMIUM-TUNGSTEN ALLOYS BY
                          INELASTIC NEUTRON SCATTERING. THESIS (139PP.) (1970)
                          *CUNNINGHAM(R.M.). (ED. NOTE: AVAIL. UNIV. MICROFILMS, ANN ARBOR, MICH.,
                          USA ORDER NO.70-20774).

552 CR-W,. . . . . T. SOL. STATE COMM. VOL.11 NO.6,. . 771, (1972).
                          *PHONON FREQUENCIES IN CR-1.6AT.PCT. W ALLOY.
                          *KESHARWANI(K.M.), AGRAWAL(B.K.).

553 CR-W,. . . . . B3 PHYS. REV. B VOL.6,. . . . . 2178, (1972).
                          *FREQUENCIES AND WIDTHS OF PHONONS IN DILUTE CR-W ALLOYS.
                          *KESHARWANI(K.M.), AGRAWAL(B.K.).

554 CR-W,. . . . . B3,T. SOL. STATE COMM. VOL.11,. . . . 1269, (1972).
                          *PHONON FREQUENCY SHIFTS AND FORCE-CONSTANT CHANGES IN DILUTE ALLOYS.
                          *COHEN(S.S.), GILAT(G.).

555 CR-(W),. . . . . . . B. PHYS. REV. LETT. 15,. . . . . . 623, (1965).
                          *OBSERVATION OF RESONANT LATTICE MODES BY INELASTIC NEUTRON SCATTERING.
                          *MOLLER(H.B.), MACKINTOSH(A.R.).

556 CR-(W),. . . . . . . T,C2. PROC. ROY. SOC. 296,. . . . . . 161, (1967).
                          *VIBRATIONS OF RANDOM DILUTE ALLOYS.
                          *ELLIOTT(R.J.), TAYLOR(D.W.).

557 CS,. . . . . B7 NATURWISSENSCHAFTEN VOL.53,. . . 16, (1966).
                          *MEASUREMENT OF COHERENT SCATTERING OF NEUTRONS AT MOLTEN SODIUM, CESIUM
                          AND BISMUTH AT DIFFERENT TEMPERATURES.
                          *OEHME(H.), RICHTER(H.). (IN GERMAN).

558 CS*BR, ... B1 PHYS. REV. B-4,. . . . . 4617, (1971).
                          *LATTICE DYNAMICS OF CS*BR.
                          *ROLANDSON(S.), RAUNIO(G.).

559 CS*BR, ... B1 PHYS. LETT. VOL.32A, ... 437, (1970).
                          *PHONON DISPERSION IN CS*BR.
                          *DAUBERT(J.).

560 CS*BR, ... B1 IAEA SYMP. GRENOBLE, ... 85, (1972).
                          *PHONON DISPERSION IN CS*BR.
                          *DAUBERT(J.).

561 CS*CL, ... B1 PHYS. REV. B VOL.6,. . . . . . . 3956, (1972).
                          *LATTICE DYNAMICS OF CESIUM CHLORIDE.
                          *AHMAD(A.A.Z.), SMITH(H.G.), WAKABAYASHI(N.), WILKINSON(M.K.).

562 CS*(CL,BR,I), . . . T,B1. PHYS. REV. B VOL.7,. . . . . . 4001, (1973).
                          *LATTICE DYNAMICS OF CS*CL, CS*BR, AND CS*I.
                          *VETELINO(J.F.), NAMJOSHI(K.V.), MITRA(S.S.).

563 CS*F, ... B1 PHYS. STAT. SOL. VOL.52 NO.2,. . 643, (1972).
                          *LATTICE DYNAMICS OF CESIUM FLUORIDE.
                          *ROLANDSON(S.).

564 CS*F, ... B1 J. PHYS. C VOL.6,. . . . . . . 2931, (1973).
                          *CRYSTAL DYNAMICS OF CAESIUM FLUORIDE.
                          *BUHRER(W.).

565 CS*H*CL2,. . . . . . C1 IAEA SYMP. GRENOBLE, ... 345, (1972).
                          *ENERGY-LOSS NEUTRON SPECTROSCOPY APPLIED TO STRONGLY HYDROGEN
                          BONDED MOLECULAR SYSTEMS.
                          *GHOSH(R.E.), WADDINGTON(T.C.), TEMME(F.P.).

566 CS*H*CL2, CS*D*CL2,. C1 J. CHEM. PHYS. VOL.52,. . . . . 1828, (1970).
                          *VIBRATION SPECTRA OF HYDROGEN BONDS BY NEUTRON ENERGY-LOSS SPECTROMETRY.
                          *COLLINS(M.F.), HAYWOOD(B.C.), STIRLING(G.C.).

567 CS*H*CL2,CS*D*CL2,. C1 J. CHEM. PHYS. VOL.52,. . . . . 2730, (1970).
                          *INELASTIC NEUTRON SCATTERING SPECTRA AND RAMAN SPECTRA OF CS*H*CL2 AND
                          CS*D*CL2.
                          *STIRLING(G.C.), LUDMAN(C.J.), WADDINGTON(T.C.).

568 CS*I,. . . . . B1 UKR. FIZ. ZH. (USSR) VOL.18, . . 92, (1973).
                          *DISPERSION CURVES OF CS*I CRYSTAL.
                          *GORBACHEV(B.I.), IVANITSKII(P.G.), KROTENKO(V.T.), PASECHNIK(M.V.). (IN
                          RUSSIAN).

569 CS*I,. . . . . B1 PHYS. STAT. SOL. B VOL.46, . . . 679, (1971).
                          *CRYSTAL DYNAMICS OF CESIUM IODIDE.
                          *BUHRER(W.), HALG(W.).

570 CS*MN*CL3(2*D2*O), . D1 PHYS. REV. B-2,. . . . . 4632, (1970).
                          *MAGNONS IN THE LINEAR-CHAIN ANTIFERROMAGNET CS*MN*CL3(2D2*O).
                          *SKALYO-JR(J.), SHIRANE(G.), FRIEDBERG(S.A.), KOBAYASHI(H.).

571 CS*NI*CL3, ... T. PHYS. REV. B VOL.6,. . . . . . . 1053, (1972).
                          *SPIN-WAVE APPROACH TO ONE-DIMENSIONAL ANTIFERROMAGNETIC CS*NI*CL3 AND
                          RB*NI*CL3.
                          *MONTANO(P.A.), COHEN(E.), SHECHTER(H.).

572 CS*NI*F3,. . . . . . D1 SOL. STATE COMM. VOL.12 NO.6,. . 537, (1973).
                          *SPIN WAVE MEASUREMENTS IN THE ONE-DIMENSIONAL FERROMAGNET CS*NI*F3.
                          *STEINER(M.), DORNER(B.).

573 CS*NI*F3,. . . . . . D1 INT. J. MAGN. VOL.5,. . . . . . 95, (1973).
                          *PROOF OF SPIN WAVES IN THE ONE-DIMENSIONAL FERROMAGNET CS*NI*F3 BY
                          INELASTIC NEUTRON SCATTERING.
                          *STEINER(M.). (IN GERMAN).

574 CS*PB*CL3, ... B6 PHYS. REV. B VOL.9,. . . . . . . 4549, (1974).
                          *NEUTRON-SCATTERING STUDY ON PHASE TRANSITIONS OF CS*PB*CL3.
                          *FUJII(Y.), HOSHINO(S.), YAMADA(Y.), SHIRANE(G.).

575 CS*S*H,. . . . . . . C1 IAEA SYMP. GRENOBLE, ... 247, (1972).
                          *NEUTRON SCATTERING STUDY OF THE ROTATIONAL MOTIONS AND PHASE TRANSITIONS
                          IN SODIUM- AND CAESIUM-HYDROSULFIDES.
                          *DE-GRAAF(L.A.), RUSH(J.J.), LIVINGSTON(R.C.).

576 CS*S*H,. . . . . . . C1 J. CHEM. PHYS. VOL.58,. . . . . 5469, (1973).
                          *NEUTRON QUASIELASTIC SCATTERING STUDY OF S*H(-) REORIENTATION IN THE
                          CUBIC PHASES OF CESIUM AND RUBIDIUM HYDROSULFIDE.
                          *ROWE(J.M.), LIVINGSTON(R.C.), RUSH(J.J.).

SECTION   6 -QUASI-ELASTIC AND INELASTIC SCATTERING STUDIES . . . .

577 CS*S*H,. . . . . .   .NEUTRON SCATTERING...C1...INV.J. CHEM. PHYS. VOL.58,. . . . 3439, (1973).
                           .NEUTRON SCATTERING INVESTIGATION OF THE ROTATIONAL DYNAMICS AND PHASE
                           TRANSITIONS IN SODIUM AND CESIUM HYDROSULFIDES.
                           .RUSH(J.J.), DE-GRAAF(L.A.), LIVINGSTON(R.C.).

578 CU,. . . . .B2,B7....CAN. J. PHYS. 49,. . . . . . . 704, (1971).
                           .CRYSTAL DYNAMICS AND ELECTRONIC SPECIFIC HEATS OF PALLADIUM AND COPPER.
                           .MIILLER(A.P.), BROCKHOUSE(B.N.).

579 CU,. . . . .B1......IAEA SYMP. COPENHAGEN, VOL.1,. . 187, (1968).
                           .OBSERVATIONS OF THE KOHN EFFECT IN COPPER.
                           .NILSSON(G.).

580 CU,. . . . .C1......COPENHAGEN CONF.-LATT.DYNAMICS,. 53, (1963).
                           .THE FREQUENCIES OF THE NORMAL MODES OF COPPER.
                           .SINHA(S.K.), SQUIRES(G.L.).

581 CU,. . . . .T......PHYS. REV. VOL.137,. . . . . . .A1113, (1965).
                           .MULTIPLE SCATTERING OF NEUTRONS IN VANADIUM AND COPPER.
                           .BLECH(I.A.), AVERBACH(B.L.).

582 CU,. . . . .C1......IAEA SYMPOSIUM VIENNA,. . . . . 549, (1960).
                           .DIFFUSION DES NEUTRONS PAR LES PHONONS DANS UN MONOCRISTAL.
                           .CRIBIER(D.), JACROT(B.), SAINT-JAMES(D.).

583 CU,. . . . .B1......PHYSICA VOL.53,. . . . . 628, (1971).
                           .PHONON DISPERSION IN COPPER.
                           .BAJPAI(R.P.), NEELAKANDAN(K.).

584 CU,. . . . .P......ACTA PHYS. ACAD. SCI. HUNG. 30,. 231, (1971).
                           .PHONON FREQUENCY DISTRIBUTION FUNCTIONS OF COPPER, NICKEL AND VANADIUM.
                           .YAI-PRAKASH, SEMUEAL(B.S.), SHARMA(P.K.).

585 CU,. . . . .B1......HELV. PHYS. ACTA VOL.41, . . . 399, (1968).
                           .PHONON DISPERSION IN COPPER.
                           .BUHRER(W.), SCHNEIDER(T.), STOLL(E.). (ED. NOTE: IN GERMAN).

586 CU,. . . . .T,B1.....J. PHYS. F VOL.3,. . . . . . . L1, (1973).
                           .A MODEL FOR LATTICE DYNAMICS OF FCC METALS.
                           .SHUKLA(M.M.), CLOSS(H.).

587 CU, CU(AU),. . .   ..B1,B3....THESIS (190 PP.),. . . . . . . , (1967).
                           .THE CRYSTAL DYNAMICS OF COPPER AND COPPER-GOLD ALLOYS.
                           .SVENSSON(E.C.).   (ED. NOTE: PH.D. DISSERTATION, MCMASTER UNIVERSITY).

588 CU,. . . . .C1......PHYS. STAT. SOL. 4,. . . . . . 95, (1964).
                           .KLEINWINKELSTREUUNG AN NEUTRONENBESTRAHLTEN KUPFER.
                           .SCHILLING(W.), SCHMATZ(W.).

589 CU,. . . . .T,B......SOL. STATE COMM. 4,. . . . . . 79, (1966).
                           .BERECHNUNG DER PHONONEN-DISPERSIONSKURVEN VON NA UND CU.
                           .SCHNEIDER(T.), STOLL(E.).

590 CU,. . . . .C1......PHYS. STAT. SOL. 7,. . . . . . 557, (1964).
                           .KLEINWINKELSTREUUNG VON LANGWELLIGEN NEUTRONEN AN VERSETZUNGEN IN
                           KALTVERFORMTEM KUPFER.
                           .CHRIST(J.).

591 CU,. . . . .B7......BULL. SOC. FRANC. MIN. CRIST. 90 428, (1967).
                           .INVESTIGATIONS ON SMALL ANGLE SCATTERING WITH NEUTRONS.
                           .SPRINGER(T.), SCHMATZ(W.).

592 CU,. . . . .T,B1.....PHYS. LETT. VOL.43A NO.5,. . . 429, (1973).
                           .LATTICE DYNAMICS OF COPPER ON MODIFIED BHATIA*S MODEL.
                           .SHUKLA(M.M.), SALZBERG(J.B.).

593 CU,. . . . .B4......PROPERTIES LIQUID METALS,. . . . 111, (1973).
                           .INVESTIGATIONS OF COLLECTIVE EXCITATIONS IN LIQUID METALS BY INELASTIC
                           NEUTRON SCATTERING.
                           .GLASER(W.), HAGEN(S.), LOFFLER(U.), SUCK(J.B.), SCHOMMERS(W.). (ED.
                           NOTE: REFER SECT. 4).

594 CU,. . . . .B1......PHYS. REV. B VOL.7,. . . . . . . 2393, (1973).
                           .LATTICE DYNAMICS OF COPPER AT 80 K.
                           .NILSSON(G.), ROLANDSON(S.).

595 CU,. . . . .B1......J. PHYS. RADIUM VOL.21,. . . . . 67, (1960).
                           .SCATTERING OF NEUTRONS BY PHONONS IN A SINGLE CRYSTAL.
                           .CRIBIER(D.), JACROT(B.), SAINT-JAMES(D.). (IN FRENCH).

596 CU,. . . . .B1......J. PHYS. CHEM. SOLIDS VOL.23,. . 1021, (1962).
                           .PHONON DISPERSION RELATIONS FOR COPPER SINGLE CRYSTAL IN THE (100)
                           DIRECTION.
                           .SOSNOWSKI(J.J.), KOZUBOWSKI(J.).

597 CU,. . . . .T......IAEA SYMPOSIUM BOMBAY, VOL.1,. . . 49, (1965).
                           .A COMPARISON BETWEEN THE ELECTRON-PHONON INTERACTIONS OF TOYA AND BAILYN
                           .SRIVASTAVA(P.L.).

598 CU,. . . . .P......PHYS. REV. VOL.73,. . . . . 963, (1948).
                           .SLOW NEUTRON VELOCITY SPECTROMETER STUDIES OF CU, NI, BI, FE, SN, AND
                           CALCITE.
                           .HAVENS-JR(W.W.), RAINWATER(L.J.), WU(C.S.), DUNNING(J.R.).

599 CU,. . . . .T,B1.....PHYS. LETT. VOL.46A NO.3,. . . . 221, (1973).
                           .PHONON DISPERSION RELATIONS IN COPPER.
                           .GOEL(C.M.).

600 CU,. . . . .B1......PHYS. REV. 164,. . . . . 922, (1967).
                           .PHONON FREQUENCIES IN COPPER AT 49 AND 298 DEG. K.
                           .NICKLOW(R.M.), GILAT(G.), SMITH(H.G.), RAUBENHEIMER(L.J.),
                           WILKINSON(M.K.).

601 CU,. . . . .B1......SOV. PHYS. SOL. STATE 7,. . . . . 296, (1965).
                           .INVESTIGATION OF THE THERMAL VIBRATIONS OF THE ATOMS IN COPPER BY
                           NEUTRON SPECTROMETRY.
                           .VINTAIKIN(E.Z.), GORBACHEV(V.V.), GRUZIN(P.L.).

602 CU,. . . . .B1......PHYS. LETT. 23,. . . . . 309, (1966).
                           .LATTICE DYNAMICS OF COPPER WITH A MORSE POTENTIAL.
                           .DE-WETTE(F.W.), COTTERILL(R.M.J.), DOYAMA(M.).

603 CU,. . . . .B1......PHYS. REV. 143,. . . . . 422, (1966).
                           .LATTICE DYNAMICS OF COPPER.
                           .SINHA(S.K.).

SECTION   6 -QUASI-ELASTIC AND INELASTIC SCATTERING STUDIES . . . .

604 CU,. . . . . .B1. . . . . .SOL. STATE COMM. 4,. . . . .   443, (1966).
               PHONON DISPERSION IN COPPER.
               BUHRER(W.), SCHNEIDER(T.), GLASER(W.).

605 CU,. . . . . .B1, T. . . . .PHYS. LETT. VOL.29A, ...   313, (1969).
               PHONON DISPERSION IN COPPER.
               BEHARI(J.), TRIPATHI(B.B.).

606 CU,. . . . . .B1. . . . . . .PHYS. REV. 155,. . . . .   619, (1967).
               CRYSTAL DYNAMICS OF COPPER.
               SVENSSON(E.C.), BROCKHOUSE(B.N.), ROWE(J.M.).

607 CU,. . . . . .B1. . . . . . .IAEA SYMP. CHALK RIVER, VOL.2, .   87, (1962).
               NEUTRON-PHONON INTERACTION STUDIES IN COPPER, ZINC AND MAGNESIUM SINGLE
               CRYSTALS.
               MALISZEWSKI(E.), SOSNOWSKI(J.), BLINOWSKI(K.), KOZUBOWSKI(J.),
               PADLO(I.), SLEDZIEWSKA(D.).

608 CU,. . . . . .B7. . . . . . .UKR. FIZ. ZH. (USSR) VOL.18, . . 1390, (1973).
               PHONON SPECTRUM OF THE COPPER LATTICE.
               GORBACHEV(B.I.), IVANITSKII(P.G.), KROTENKO(V.T.), PASECHNIK(M.V.). (IN
               RUSSIAN).

609 CU,. . . . . .B2. . . . . . .BULL. AMER. PHYS. SOC. VOL.19, . 320, (1974).
               TEMPERATURE DEPENDENCE OF PHONON DISPERSION CURVES IN COPPER.
               LAROSE(A.), BROCKHOUSE(B.N.), JACKSON(H.E.). (ED. NOTE: ABSTRACT ONLY).

610 CU,. . . . . .B2. . . . . . .PHYS. REV. B VOL.9, ... 3278, (1974).
               KOHN EFFECT AND THE FERMI SURFACE IN COPPER STUDIED BY NEUTRON SPECTROS-
               COPY.
               NILSSON(G.), ROLANDSON(S.).

611 CU-AL, ...B3. . . . . . .IAEA SYMP. COPENHAGEN, VOL.1,. .   47, (1968).
               COHERENT INELASTIC SCATTERING STUDY OF LOCAL AND IN-BAND MODES IN
               CU(1-X)AL(X) CRYSTALS.
               NICKLOW(R.M.), VIJAYARAGHAVAN(P.R.), SMITH(H.G.), DOLLING(G.),
               WILKINSON(M.K.).

612 CU-AL, ...B1. . . . . . .PHYS. REV. LETT. VOL.20, ... 1245, (1968).
               OBSERVATION OF LOCALIZED VIBRATIONS IN CU 4/100, AL
               BY COHERENT INELASTIC NEUTRON SCATTERING.
               NICKLOW(R.M.),   VIJAYARAGHAVAN(P.R.),   SMITH(H.G.),   WILKINSON(M.K.).

613 CU97-AU03, ...T. . . . . . .CAN. J. PHYS. 49,. . . . . 2496, (1971).
               CALCULATION OF THE FREQUENCIES AND WIDTHS OF PHONONS IN CU97-AU03.
               BRUNO(R.), TAYLOR(D.W.).

614 CU-(AU), ...B3. . . . . . .CAN. J. PHYS. 49,. . . . . 2291, (1971).
               RESONANT PERTURBATION OF PHONONS IN CU(AU) ALLOYS STUDIED BY NEUTRON
               SCATTERING.
               SVENSSON(E.C.), KAMITAKAHARA(W.A.).

615 CU-AU, ...B1. . . . . . .NUCL./S.S. PHYS. SYMP. ABSTR., ... , (1972).
               DISPERSION CURVES OF DILUTE CU-AU ALLOYS.
               KESHARWANI(K.M.), AGRAWAL(B.K.).

616 CU-AU, ...B3. . . . . . .PHYS. REV. LETT. VOL.18, ... 858, (1967).
               IN-BAND MODES OF VIBRATION OF A COPPER-3 AT. PCT. GOLD ALLOY.
               SVENSSON(E.C.), BROCKHOUSE(B.N.).

617 CU-(AU), ...T, C2. . . . .PROC. ROY. SOC. 296, ... 161, (1967).
               VIBRATIONS OF RANDOM DILUTE ALLOYS.
               ELLIOTT(R.J.), TAYLOR(D.W.).

618 CU-(AU), ...T. . . . . . .PHYS. REV. B VOL.7,. . . . . . . 5153, (1973).
               PHONON FREQUENCIES AND WIDTHS IN DILUTE CU-AU ALLOYS.
               KESHARWANI(K.M.), AGRAWAL(B.K.).

619 CU-(AU), ...B1. . . . . . .SOL. STATE COMM. 3,. . . . . . . 245, (1965).
               ≠IN-BAND≠ MODES OF VIBRATION OF A DILUTE DISORDERED ALLOY CU(AU).
               SVENSSON(E.C.), BROCKHOUSE(B.N.), ROWE(J.M.).

620 CU-(AU), ...B1. . . . . . .PHYS. REV. B VOL.10 NO.2, ... , (1974).
               VIBRATIONS OF A MIXED CRYSTAL: NEUTRON SCATTERING FROM NI55-PD45.
               KAMITAKAHARA(W.A.), BROCKHOUSE(B.N.). (REF. OBTAINED FROM PHYS. REV.
               ABSTRACTS).

621 CU-(BE,MG),. . . . . .B3. . . . . . .IAEA SYMP. COPENHAGEN, VOL.1,. .   65, (1968).
               LOCAL VIBRATIONS OF IMPURITY ATOMS IN COPPER AND LEAD.
               NATKANIEC(I.), PARLINSKI(K.), JANIK(J.A.), BAJOREK(A.),
               SUDNIK-HRYNKIEWICZ(M.).

622 CU*BR, ...B1. . . . . . .SOL. STATE COMM. VOL.13, ... 1725, (1973).
               NORMAL MODES OF VIBRATIONS IN CU*BR.
               PREVOT(B.), CARABATOS(C.), SCHWAB(C.), HENNION(B.), MOUSSA(F.).

623 CU*CL, ...B1. . . . . . .PHYS. REV. LETT. VOL.26, ... 770, (1971).
               NORMAL MODES OF VIBRATIONS IN CU*CL.
               CARABATOS(C.), HENNION(B.), KUNC(K.), MOUSSA(F.), SCHWAB(C.).

624 CU*CL, ...T, B. . . . . .COMP. REND. 268, ... 1658, (1969).
               VIBRATIONS DU RESEAU DU CHLORURE CUIVREUX: MODELE DES ION RIGIDES AVEC
               INTERACTION COULOMBIENNE.
               CARABATOS(C.).

625 CU*CL2.2N*(C5*D5), ...D1. . . . . . .PHYS. REV. LETT. VOL.32 NO.4,. . 170, (1974).
               DYNAMICS OF AN S=1/2, ONE-DIMENSIONAL HEISENBERG ANTIFERROMAGNET.
               ENDOH(Y.), SHIRANE(G.), BIRGENEAU(R.J.), RICHARDS(P.M.), HOLT(S.L.).

626 CU*FE*S2,. . . . . .D1. . . . . . .ACTA PHYS. POLON. VOL.24,. . . . 249, (1963).
               DISPERSION RELATION FOR SPIN WAVES IN CU*FE*S2.
               MURASIK(A.).

627 CU*I,. . . . . .B1. . . . . . .PHYS. REV. LETT. VOL.28, ... 964, (1972).
               NORMAL MODES OF VIBRATIONS IN CU*I.
               HENNION(B.), MOUSSA(F.), PREVOT(B.), CARABATOS(C.), SCHAWB(C.).

628 CU*(N*O3)2.3H2*O, ...C1. . . . . . .INORG. CHEM. VOL.6,. . . . . . . 346, (1967).
               NEUTRON-SCATTERING STUDY OF THE MOTIONS OF WATER MOLECULES IN HYDRATED
               SALTS OF TRANSITION METALS.
               RUSH(J.J.), FERRARO(J.R.), WALKER(A.).

629 CU-NI, ...B1. . . . . . .IAEA SYMP. COPENHAGEN, VOL.1,. . 181, (1968).
               LATTICE DYNAMICS OF CU(1-X)NI(X).
               SAKAMOTO(M.), HAMAGUCHI(Y.).

SECTION   6 -QUASI-ELASTIC AND INELASTIC SCATTERING STUDIES . . . .

630 CU-NI, ... *GIANT MOMENTS IN PARAMAGNETIC NI-CU ALLOYS. . . . . P. . . . . . . PHYS. REV. LETT. VOL.24, ... 598, (1970).
     *KOUVEL(J.S.), COMLY(J.B.).

631 CU-NI, ... *NEUTRON SCATTERING FROM NEAR CRITICAL NICKEL-COPPER ALLOYS. . . T,D2. . . . . PHYS. LETT. VOL.32A, ... 410, (1970).
     *HICKS(T.J.).

632 CU-NI, ... *INCOHERENT NEUTRON SCATTERING FROM LIQUID AND SOLID CU-NI. . . C1,3. . . . . PHYS. LETT. 30A, ... 206, (1969).
     *MOZER(B.), PRICE(D.L.), KEATING(D.T.), MEISTER(H.).

633 CU-NI, ... *NEUTRON SCATTERING STUDIES OF THE DYNAMICS OF IMPERFECT CRYSTALS. . . B3. . . . . . THESIS (246 PP.). . . . . (1972).
     *KAMITAKAHARA(W.A.). (ED. NOTE: PH.D. DISSERTATION, MCMASTER UNIVERSITY).

634 CU-NI-ZN, ... *A STUDY OF THE RESOLUTION OF A TRIPLE-AXIS SPECTROMETER AND MEASUREMENT . . B1. . . . . . THESIS (166 PP.). . . . . (1970).
     OF THE PHONON SPECTRUM OF THE TERNARY ALLOY CU(.63)-NI(.21)-ZN(.16).
     *LAROSE(A.).   (ED. NOTE: M.SC. DISSERTATION, MCMASTER UNIVERSITY).

635 CU2*O, ... *RIGID ION MODEL LATTICE DYNAMICS OF CUPRITE (CU2*O). . . T. . . . . . PHYS. STAT. SOL. VOL.44, ... 701, (1971).
     *CARABATOS(C.), PREVOT(B.).

636 (CU*S*O4)*5(H2*O) *INVESTIGATION OF THE DYNAMICS OF N*H4+ AND H2*O MOLECULAR GROUPS IN . . C1. . . . . . IAEA SYMP. BOMBAY, VOL.2, ... 355, (1964).
     CRYSTALS.
     *. C.C.C.P.

637 CU*S*O4.H2*O,CU*S*O4.5H2*O, ... C1. . . . . . INORG. CHEM. VOL.6, ... 346, (1967).
     *NEUTRON-SCATTERING STUDY OF THE MOTIONS OF WATER MOLECULES IN HYDRATED
     SALTS OF TRANSITION METALS.
     *RUSH(J.J.), FERRARO(J.R.), WALKER(A.).

638 CU2*SB, ... *THE MAGNETIC STRUCTURE OF CU2*SB. . . A2. . . . . . SOV. PHYS. CRYST. VOL.15, ... 935, (1971).
     *GOLOVKIN(V.S.), BYKOV(V.N.), LEVDIK(V.A.).

639 CU-SI, ... *LOCAL MODE IN CU(4.5 PCT. SI) STUDIED BY NEUTRON SPECTROSCOPY. . . B3. . . . . . BULL. AMER. PHYS. SOC. VOL.18, . . 112, (1973).
     *BROCKHOUSE(B.N.), NICKLOW(R.M.). (ED. NOTE: PRESENTED AT ANNUAL MEETING
     OF AMER. PHYS. SOC. IN NEW YORK).

640 CU-ZN, ... *THERMAL VIBRATIONS OF BETA-BRASS AND THE ORDER-DISORDER TRANSITION. . . B1. . . . . . IAEA SYMP. BOMBAY, VOL.1, ... 343, (1964).
     *DOLLING(G.), GILAT(G.).

641 CU-ZN, ... *ON FLAT PHONONS IN BETA-BRASS. . . T. . . . . . . SOL. STATE COMM. VOL.8, ... 2053, (1970).
     *GILAT(G.).

642 CU-ZN, ... *NORMAL VIBRATIONS OF BETA-BRASS. . . B1. . . . . . PHYS. REV. VOL.138, ...A1053, (1965).
     *GILAT(G.), DOLLING(G.).

643 CU-ZN, ... *CRITICAL NEUTRON SCATTERING FROM BETA-BRASS. . . D2. . . . . . NBS MISC. PUB.273, ... 144, (1966).
     *DIETRICH(O.W.), ALS-NIELSEN(J.).   (ED. NOTE:PROCEEDINGS CONFERENCE ON
     PHENOMENA IN THE NEIGHBOURHOOD OF CRITICAL POINTS, WASHINGTON 1965).

644 CU-ZN, ... *LONG RANGE ORDER AND CRITICAL SCATTERING OF NEUTRONS BELOW THE . . D2. . . . . . PHYS. REV. 153, ... 717, (1967).
     TRANSITION TEMPERATURE IN BETA-BRASS.
     *ALS-NIELSEN(J.), DIETRICH(O.W.).

645 CU-ZN, ... *ON TEMPERATURE DEPENDENCE OF PHONONS IN BETA-BRASS. . . B1. . . . . . SOL. STATE COMM. 2, ... 79, (1964).
     *DOLLING(G.), GILAT(G.).

646 CU-ZN, ... *CRYSTAL DYNAMICS OF NICKEL-IRON AND COPPER-ZINC ALLOYS. . . B1. . . . . . CAN. J. PHYS. 47, ... 1117, (1969).
     *HALLMAN(E.D.), BROCKHOUSE(B.N.).

647 CYCLIC-HYDROCARBONS. . . . . C3. . . . . . THESIS . . . . (1967).
     *COLD-NEUTRON SCATTERING EXPERIMENTS ON CYCLIC HYDROCARBONS WITH A
     ROTATING-CRYSTAL SPECTROMETER.
     *DE-GRAAF(L.A.). (PH.D. DISSERTATION - DELFT, THE NETHERLANDS).

648 CYCLOHEXANE ETC. . . . . C2,C4. . . . J. CHEM. PHYSICS VOL.38, ... 1685, (1963).
     *NEUTRON INELASTIC SCATTERING STUDIES OF GLOBULAR COMPOUNDS.
     *BECKA(L.N.).

649 CYCLOHEXANOL, . . *QUASI-ELASTIC INCOHERENT COLD NEUTRON SCATTERING IN LIQUID CYCLOHEXANOL. . . C3. . . . . . PHYS. LETT. VOL.33A, ... 87, (1970).
     *DE-MUL(F.F.), BREGMAN(J.D.).

650 CYCLOHEXANE, . . *INELASTIC NEUTRON SCATTERING BY SOLID AND LIQUID CYCLOHEXANE. . . C1,C3. . . . BROOKHAVEN SYMPOSIUM, ... 138, (1965).
     *DE-GRAAF(L.A.).

651 CYCLOHEXANOL, . . *ON THE MODELS FOR MOLECULAR ROTATION IN PLASTIC CRYSTALS STUDIED BY . . C1,T. . . . . PHYS. LETT. VOL.41A NO.3, ... 272, (1972).
     COLD NEUTRON SCATTERING.
     *DE-MUL(F.F.).

652 CYCLOHEXANE, . . *OBSERVATION OF THE LEAST-ENERGETIC E(SUB U) VIBRATION OF CYCLOHEXANE. . . C1,C3. . . . J. CHEM. PHYS. VOL.50, ...1030, (1969).
     *BRUGGER(R.M.), STRONG(K.A.), PUGMIRE(R.J.), GRANT(D.M.).

653 CYCLOHEXANE, . . *COLD NEUTRON SCATTERING BY METHYL ALCOHOL, CYCLOHEXANE AND THEIR MIXTURE . . A3. . . . . . REV. ROUMAINE PHYS. VOL.15, . . . 783, (1970).
     NEAR THE CONSOLUTION POINT.
     *DIACONESCU(A.), NAHORNIAK(V.), MATEESCU(N.), TEUTSCH(H.).

654 D2, ... *LATTICE DYNAMICS OF SOLID DEUTERIUM BY INELASTIC NEUTRON SCATTERING. . . B1. . . . . . PHYS. REV. B-3, ... 4383, (1971).
     *NIELSEN(M.), MOLLER(H.B.).

655 D2, ... *LATTICE DYNAMICS OF HCP ORTHO-DEUTERIUM. . . T. . . . . . . PHYS. LETT. VOL.33A, ... 253, (1970).
     *KLEIN(M.L.), KOEHLER(T.R.).

656 D2, ... *CO-OPERATIVE MODES OF MOTION IN SIMPLE LIQUIDS. . . B4. . . . . . PHYS. LETT. VOL.19, ... 269, (1965).
     *CHEN(S.H.), EDER(O.J.), EGELSTAFF(P.A.), HAYWOOD(B.C.), WEBB(F.J.).

SECTION   6 -QUASI-ELASTIC AND INELASTIC SCATTERING STUDIES . . . .

657 D2,. . . . . .T. . . . . .PHYS. REV. VOL.135. . . . . . . .A603, (1964).
     *SLOW NEUTRON SCATTERING BY MOLECULAR HYDROGEN AND DEUTERIUM.
     *YOUNG(J.A.), KOPPEL(J.U.).

658 D2,. . . . . .C1,3. . . . .PROC. PHYS. SOC. 90,. . . . . . .681, (1967).
     *MOLECULAR MOTIONS IN LIQUID AND SOLID HYDROGEN AND DEUTERIUM.
     *EGELSTAFF(P.A.), HAYWOOD(B.C.), WEBB(F.J.).

659 D2,. . . . . .B1. . . . . .PHYS. REV. B VOL.7,. . . . . 1626, (1973).
     *PHONONS IN SOLID HYDROGEN AND DEUTERIUM STUDIED BY INELASTIC COHERENT
     NEUTRON SCATTERING.
     *NIELSEN(M.).

660 D2,. . . . . .T. . . . . .PHYS. REV. VOL.69. . . . . . . .145, (1946).
     *THE SCATTERING OF SLOW NEUTRONS BY ORTHO- AND PARADEUTERIUM.
     *HAMERMESH(M.), SCHWINGER(J.).

661 D*BR-H*BR, ...97. . . . . .IAEA SYMP. COPENHAGEN, VOL.1,. . .379, (1968).
     *NUCLEAR MAGNETIC RESONANCE IN LIQUIDS AND SOLIDS (REVIEW PAPER).
     *POWLES(J.G.).

662 D*CL,. . . . . .B7. . . . . .DISC. FARADAY SOC. NO.48,. . . . .78, (1969).
     *NEUTRON DIFFRACTION STUDY OF MOLECULAR MOTION IN SOLID DEUTERIUM
     CHLORIDE.
     *SANDOR(E.), FARROW(R.F.C.).

663 D*CR*02,. . . . . .J. CHEM. PHYS. VOL.44,. . . . .2496, (1966).
     *NEUTRON AND INFRARED SPECTRA OF H*CR*02 AND D*CR*02.
     *RUSH(J.J.), FERRARO(J.R.).

664 D2*0,. . . . . .C1,C3. . . .IAEA SYMP. COPENHAGEN, VOL.1,. . .507, (1968).
     *THE DYNAMICS OF LIQUID H2*0 AND D2*0 AND SOLID H2*0 FROM THE INELASTIC
     SCATTERING OF EPITHERMAL NEUTRONS.
     *HARLING(O.K.).

665 D2*0,. . . . . .B1. . . . . .CONF. INTERN., RENNES, ... 167, (1971).
     *LATTICE DYNAMICS OF D2*0 - ICE.
     *RENKER(B.).

666 D2*0,. . . . . .T. . . . . .PROC. PHYS. SOC. 81 . . . . . .276, (1963).
     *THE SCATTERING OF SLOW NEUTRONS BY HEAVY WATER.I. INTRAMOLECULAR
     SCATTERING.
     *BUTLER(D.).

667 D2*0,. . . . . .P. . . . . .IAEA SYMPOSIUM (VIENNA),. . . . . 631, (1960).
     *STUDY OF THE NEUTRON SPECTRA EMERGING FROM MODERATING ASSEMBLIES.
     *RAMANNA(R.), SARMA(N.).

668 D2*0,. . . . . .C3. . . . . .IAEA SYMPOSIUM (VIENNA),. . . . . 351, (1960).
     *THE SCATTERING OF SLOW NEUTRONS BY LIGHT AND HEAVY WATER.
     *MIKKE(K.).

669 D2*0,. . . . . .C3. . . . . .IAEA SYMPOSIUM (VIENNA),. . . . .329, (1960).
     *COLD-NEUTRON SCATTERING EXPERIMENTS ON LIGHT AND HEAVY WATER.
     *LARSSON(K.E.), HOLMRYD(S.), OTNES(K.).

670 D2*0,. . . . . .B1. . . . . .PHYS. LETT. 30A, ... 493, (1969).
     *PHONON DISPERSION IN D2*0.
     *RENKER(B.).

671 D2*0,. . . . . .B4. . . . . .NUCL. SCI. ENGNG. VOL.33,. . . . 195, (1968).
     *INELASTIC THERMAL-NEUTRON SCATTERING BY LIQUID D2*0.
     *WHITTEMORE(W.L.).

672 D2*0,. . . . . .B4. . . . . .NUCL. SCI. ENGNG. VOL.33,. . . . . 41, (1968).
     *SLOW-NEUTRON, INELASTIC SCATTERING AND THE DYNAMICS OF HEAVY WATER.
     *HARLING(O.K.).

673 D2*0,. . . . . .P. . . . . .REF. SECTION 3, 60/WOODS,. . . . .

674 D2*0,. . . . . .C1. . . . . .PHYS. LETT. 15,. . . . . 231, (1965).
     *SMALL-ANGLE SCATTERING OF SUBTHERMAL NEUTRONS ON HEAVY ICE.
     *DENGEL(O.), CHRIST(J.), SCHMATZ(W.).

675 D2*0,. . . . . .T,C3. . . . .IAEA SYMPOSIUM (VIENNA),. . . . . 309, (1960).
     *THE THERMAL NEUTRON SCATTERING LAW FOR LIGHT AND HEAVY WATER.
     *EGELSTAFF(P.A.), COCKING(S.J.), ROYSTON(R.), THORSON(I.M.).

676 D2*0,. . . . . .C3. . . . . .IAEA SYMP. CHALK RIVER, VOL.1, . 317, (1962).
     *A STUDY OF THE DIFFUSIVE ATOMIC MOTIONS IN GLYCEROL AND OF THE
     VIBRATORY MOTIONS IN GLYCEROL AND HEAVY WATER BY COLD NEUTRON SCATTERING
     *LARSSON(K.E.), DAHLBORG(U.).

677 D2*0,. . . . . .B7. . . . . .Z. NATURFORSCH VOL.19A. . . . .1422, (1964).
     *INVESTIGATION OF THE SMALL ANGLE SCATTERING OF NEUTRONS BY HEAVY WATER.
     *GISSLER(W.). (IN GERMAN).

678 D2*0,. . . . . .B1. . . . . .PHYS. LETT. VOL.42A NO.7,. . . . 509, (1973).
     *DISPERSION RELATIONS IN HEAVY ICE.
     *JOVIC(D.), DAVIDOVIC(M.), ZIVANOVIC(M.).

679 D2*0,. . . . . .P. . . . . .UKAEA AERE REP.4582 (13 PP.),. . .(1964).
     *A COMPILATION OF THE SCATTERING LAW FOR HEAVY WATER AT 22 DEGREES C AND
     150 DEGREES C.
     *HAYWOOD(B.C.).

680 D2*0,. . . . . .P. . . . . .UKAEA AERE REPORT 5408 (21 PP.),. .(1967).
     *SCATTERING LAW S(ALPHA,BETA)-VALUES FOR HEAVY WATER AT 540 DEGREES K.
     *PAGE(D.I.).

681 D2*0,. . . . . .P. . . . . .REF. SECTION 3, 48/RAINWATER,. . . . .

682 D2*0,. . . . . .P. . . . . .CONF. NEUTRON THERMALIZ. VOL.I,. .361, (1968).
     *THE SCATTERING LAW FOR HEAVY WATER AT 540 DEGREES K AND LIGHT WATER
     AT 500 DEGREES K.
     *HAYWOOD(B.C.), PAGE(D.I.). (IN BOOK: NEUTRON THERMALIZATION AND REACTOR
     SPECTRA. PUBL. IAEA: VIENNA, 1968).

683 D2*0,. . . . . .B4. . . . . .NUOVO CIMENTO SUPPL. VOL.9,. . . . 45, (1958).
     *STRUCTURAL DYNAMICS OF WATER BY NEUTRON SPECTROMETRY.
     *BROCKHOUSE(B.N.).

684 D2*0,. . . . . .T. . . . . .PROC. PHYS. SOC. 81 . . . . . .294, (1963).
     *THE SCATTERING OF SLOW NEUTRONS BY HEAVY WATER. II. INTERMOLECULAR
     SCATTERING.
     *BUTLER(D.).

SECTION   6 -QUASI-ELASTIC AND INELASTIC SCATTERING STUDIES . . . .

685   DEUTERIDES,. . . . . .ATOMKERNENERGIE VOL.12,. . . . . 385, (1967).
      *INVESTIGATION OF HYDROGEN AND DEUTERIUM VIBRATIONS IN METAL HYDRIDES,
      HYDRODEUTERIDES AND DEUTERIDES WITH TOTAL NEUTRON CROSS-SECTIONS.
      *SCHMIDT(U.).

686   DIMETHOXYAZOXYBENZENE, ...C1,C3....REF. SECTION 3, 64/JANIK,. . . . .

687   2-2-DIMETHYLBUTANE,. . . . . .C2,C4....J. CHEM. PHYSICS VOL.38, ... 1685, (1963).
      *NEUTRON INELASTIC SCATTERING STUDIES OF GLOBULAR COMPOUNDS.
      *BECKA(L.N.).

688   DIMETHYL POLYSILOXANE, ...C3......IAEA SYMP. COPENHAGEN, VOL.2,. . 197, (1968).
      *SLOW-NEUTRON SCATTERING AND ROTATIONAL FREEDOM OF METHYL GROUPS IN
      SEVERAL ORGANIC COMPOUNDS.
      *HERDADE(S.B.).

689   DIMETHYLACETYLENE, ...C1,C3....REF. SECTION 3, 64/JANIK,. . . . .

690   DIOXANE, ...C3........IAEA SYMP. BOMBAY, VOL.2,. . . . . 201, (1964).
      *SLOW NEUTRON SCATTERING IN WATER AND SOME ORGANIC SUBSTANCES.
      *.C.C.C.P.

691   DIPHENYL,. . . . . .C3......IAEA SYMP. CHALK RIVER, VOL.1,. 297, (1962).
      *A STUDY OF COLD NEUTRON INELASTIC SCATTERING ON CERTAIN HYDROGENOUS
      SUBSTANCES.
      *.C.C.C.P.

692   DIPHENYL,. . . . . .C3......IAEA SYMP. CHALK RIVER, VOL.1,. . 307, (1962).
      *INELASTIC SCATTERING OF THERMAL NEUTRONS FROM DOWTHERM #A#.
      *GLASER(W.).

693   DNA, ...T........BROOKHAVEN SYMPOSIUM,. . . . . . . 57, (1965).
      *ON THE THEORY OF COILING AND UNCOILING OF DNA MOLECULES.
      *MONTROLL(E.W.).

694   DY,. . . . . .D1........J. APPL. PHYS. 42, ... 1672, (1971).
      *SPIN-WAVE DISPERSION RELATION IN RARE-EARTH METALS.
      *NICKLOW(R.M.).

695   DY,. . . . . .D1........PHYS. REV. LETT. VOL.26, ... 140, (1971).
      *SPIN-WAVE DISPERSION RELATION IN DYSPROSIUM METAL.
      *NICKLOW(R.M.), WAKABAYASHI(N.), WILKINSON(M.K.), REED(R.E.).

696   DY,. . . . . .D1........IAEA SYMP. GRENOBLE, ... 611, (1972).
      *SPIN WAVES IN FERROMAGNETIC DYSPROSIUM METAL.
      *NICKLOW(R.M.), WAKABAYASHI(N.),.

697   DY,. . . . . .D1........MAGN. AND MAG. MATERIALS-1971, . 1446, (1972).
      *MAGNON ENERGY SPECTRUM IN FERROMAGNETIC DY.
      *NICKLOW(R.M.), WAKABAYASHI(N.). (AIP CONF. PROC. NO.5 PART II).

698   DY,. . . . . .P........PHYS. STAT. SOL. 2, ... K164, (1962).
      *MESSUNG DER KOHARENTEN STREUAMPLITUDEN VON DYSPROSIUM UND THULIUM FUR
      THERMISCHE NEUTRONEN.
      *BETZL(M.), HASE(W.), KLEINSTUCK(K.), TOBISCH(J.).

699   DY3*AL5*012, ...D2........J. APP. PHYS. 39,. . . . . . . 1232, (1968).
      *CRITICAL MAGNETIC SCATTERING FROM DYSPROSIUM ALUMINUM GARNET (DAG)
      *NORVELL(J.C.), WOLF(W.P.), CORLISS(L.M.), HASTINGS(J.M.), NATHANS(R.).

700   DY-LA, ...D4........REF. SECTION 3, 64/KOEHLER(2), ...

701   DY-Y, ...D4........REF. SECTION 3, 64/KOEHLER(2), ...

702   ER,. . . . . .D1........CAN. J. PHYS. 49,. . . . . 2875, (1971).
      *SPIN-WAVE EXCITATIONS IN ERBIUM.
      *HOLDEN(T.M.), POWELL(B.M.), STRINGFELLOW(M.W.), WOODS(A.D.B.).

703   ER,. . . . . .D1........PHYS. REV. LETT. VOL.27, ... 334, (1971).
      *SPIN-WAVE DISPERSION RELATION FOR ER METAL AT 4.5 DEG. K.
      *NICKLOW(R.M.), WAKABAYASHI(N.), WILKINSON(M.K.), REED(R.E.).

704   ER,. . . . . .D1........COLLOQUE INTERNAT. NO.180 TOME 2 283, (1970).
      *EXCHANGE AND CRYSTAL FIELD INTERACTIONS IN ERBIUM AND HOLMIUM FROM
      INELASTIC NEUTRON SCATTERING MEASUREMENTS.
      *WOODS(A.D.B.), STRINGFELLOW(M.W.), HOLDEN(T.M.), POWELL(B.M.). (COLL. ON
      RARE-EARTH ELEMENTS MAY, 1969. AVAIL. 15, QUAI ANATOLE-FRANCE-PARIS VII)

705   ER,. . . . . .D1........J. PHYSIQUE TOME 32 COL.1 VOL.II 1179, (1971).
      *TEMPERATURE DEPENDENCE OF SPIN WAVE ENERGIES IN ERBIUM.
      *WOODS(A.D.B.), HOLDEN(T.M.), POWELL(B.M.), STRINGFELLOW(M.W.).

706   ER,. . . . . .D3........J. APP. PHYS. 39,. . . . . 457, (1968).
      *NEUTRON PARAMAGNETIC SCATTERING FROM HOLMIUM AND ERBIUM.
      *HOLDEN(T.M.), POWELL(B.M.), WOODS(A.D.B.).

707   ER,. . . . . .D1........IAEA SYMP. COPENHAGEN, VOL.2,. . 3, (1968).
      *NEUTRON SPIN SCATTERING BY SPIN WAVES IN METALS (REVIEW PAPER).
      *MOLLER(H.B.).

708   ER,. . . . . .D1........PHYS. REV. LETT. 19, ... 908, (1967).
      *OBSERVATION OF SPIN WAVES IN ERBIUM.
      *WOODS(A.D.B.), HOLDEN(T.M.), POWELL(B.M.).

709   ER*(AG,CU,ZN), ...B7........PHYS. REV. B VOL.9 NO.11,. . . . . , (1974).
      *CRYSTAL FIELDS IN ER*CU, ER*AG, AND ER*ZN.
      *MORIN(P.), PIERRE(J.), ROSSAT-MIGNOD(J.), KNORR(K.), DREXEL(W.). (REF.
      OBTAINED FROM PHYS. REV. ABSTRACTS).

710   ER*CO2,. . . . .D4........REF. SECTION 3, 64/KOEHLER(2), ...

711   ER-LA, ...D4........REF. SECTION 3, 64/KOEHLER(2), ...

712   ER2*03,. . . . . .B7........J. PHYS. SOC. JAPAN VOL.17 B-III 63, (1962).
      *CRYSTAL FIELD SPECTRA IN RARE EARTH OXIDES.
      *BROCKHOUSE(B.N.), BECKA(L.N.), RAO(K.R.), SINCLAIR(R.N.), WOODS(A.D.B.).

713   ER2*03,. . . . . .D3........PHYS. REV. VOL.92,. . . . . . . 1380, (1953).
      *PARAMAGNETIC SCATTERING OF NEUTRONS BY RARE EARTH OXIDES.
      *KOEHLER(W.C.), WOLLAN(E.O.).

714   ER2*03,. . . . . .B7........C.R. ACAD. SCI. VOL.250,. . . . 2871, (1960).
      *NEUTRON SCATTERING AND STARK EFFECT IN CRYSTALLINE RARE EARTH OXIDES.
      *CRIBIER(D.), JACROT(B.). (IN FRENCH).

SECTION   6 -QUASI-ELASTIC AND INELASTIC SCATTERING STUDIES . . . .

715 ER-Y,. . . . . .D4. . . . . . .REF. SECTION 3, 64/KOEHLER(2), ...

716 (ER-Y)-AL2,. . . . . .B7. . . . .J. PHYS. C VOL.7 . . . . .1207, (1974).
                      *NEUTRON CRYSTAL-FIELD SPECTROSCOPY AND SUSCEPTIBILITY IN (ER-Y)-AL2.
                      *HEER(H.), FURRER(A.), WALKER(E.), TREYVAUD(A.), PURWINS(H.G.), KJEMS(J.)

717 ETHYL-ETHER, ...C3. . . . . PHYS. LETT. 19 . . . . .15, (1965).
                      *INELASTIC SCATTERING OF COLD NEUTRONS IN ETHYL ETHER NEAR THE
                      CRITICAL POINT.
                      *BATA(L.), KOSZO(E.), KROO(N.), PAL(L.).

718 EU*O,. . . . . .D2. . . . . PHYS. REV. LETT. VOL.27, ... 741, (1971).
                      *CRITICAL BEHAVIOR OF THE HEISENBERG FERROMAGNETS EU*O AND EU*S.
                      *ALS-NIELSEN(J.), DIETRICH(O.W.), KUNNMANN(W.), PASSELL(L.).

719 EU*O,. . . . . .D1. . . . . IAEA SYMP. GRENOBLE, ... 619, (1972).
                      *RELAXATION AND RENORMALIZATION OF SPIN WAVES IN EU*O.
                      *PASSELL(L.), ALS-NIELSEN(J.), DIETRICH(O.W.).

720 EU*O, EU*S,. . . . . .D1. . . . . MAGN. AND MAG. MATERIALS-1971, . 1251, (1972).
                      *EXCHANGE INTERACTIONS IN EU*O AND EU*S.
                      *PASSELL(L.), DIETRICH(O.W.), ALS-NIELSEN(J.). (AIP CONF. PROC. NO.5
                      PART 2; GRAHAM-JR., RHYNE-EDS.).

721 EU*S,. . . . . .D2. . . . . PHYS. REV. LETT. VOL.27, ... 741, (1971).
                      *CRITICAL BEHAVIOR OF THE HEISENBERG FERROMAGNETS EU*O AND EU*S.
                      *ALS-NIELSEN(J.), DIETRICH(O.W.), KUNNMANN(W.), PASSELL(L.).

722 F,. . . . . .P. . . . . .PHYS. REV. LETT. VOL.28 . . . . . 805, (1972).
                      *MEASUREMENT OF THE SPIN-DEPENDENT PART OF THE SCATTERING AMPLITUDE OF
                      SLOW NEUTRONS ON F(ISOTOPE: A=19) USING POLARIZED BEAM/POLARIZED TARGET.
                      *ABRAGAM(A.), BACCHELLA(G.L.), LONG(C.), MERIEL(P.), PEISVAUX(J.),
                      PINOT(M.).

723 F2,. . . . . .P. . . . . .REF. SECTION 3, 48/RAINWATER, ...

724 F2*MG, ...B1. . . . . C.R. ACAD. SC. (PARIS) 267-B, . . 61, (1968).
                      *DETERMINATION DES FREQUENCES DE VIBRATION DANS F2*MG PAR DIFFUSION
                      INELASTIQUE DE NEUTRONS.
                      *BENOIT(C.), KAHN(R.), TROTIN(J.P.), STEVENSON(R.).

725 FE (ALPHA),. . . . . .B1. . . . . IAEA SYMP. COPENHAGEN, VOL.1,. . 233, (1968).
                      *LATTICE DYNAMICS OF ALPHA-FE AND FE3*AL.
                      *VAN-DYJK(C.), BERGSMA(J.).

726 FE,. . . . . .D2. . . . . IAEA SYMP. COPENHAGEN, VOL.2,. . 55, (1968).
                      *NEUTRON SCATTERING INVESTIGATION OF THE DYNAMICS OF THE CRITICAL STATE
                      IN IRON.
                      *GORDON(J.), KISDI-KOSZO(E.), PAL(L.), VIZI(I.).

727 FE,. . . . . .D1. . . . . IAEA SYMP. COPENHAGEN, VOL.2,. . 45, (1968).
                      *INELASTIC SCATTERING OF POLARIZED NEUTRONS BY MAGNETO-VIBRATIONAL WAVES
                      IN A SINGLE CRYSTAL OF BCC IRON.
                      *STEINSVOLL(O.).

728 FE,. . . . . .D2. . . . . KYOTO CONF. STAT. MECHANICS, . . 169, (1968).
                      *HIGH RESOLUTION STUDIES OF THE CRITICAL SCATTERING OF NEUTRONS FROM IRON
                      *COLLINS(M.F.), MINKIEWICZ(V.J.), NATHANS(R.), PASSELL(L.), SHIRANE(G.).

729 FE,. . . . . .D2. . . . . PHYS. REV. LETT. VOL.25, ... 730, (1970).
                      *ANALYSIS OF CRITICAL NEUTRON-SCATTERING DATA FROM IRON AND DYNAMICAL
                      SCALING THEORY.
                      *ALS-NIELSEN(J.).

730 FE,. . . . . .D2. . . . . PHYS. LETT. VOL.31A, ... 561, (1970).
                      *DIFFUSION CRITIQUE DES NEUTRONS PAR LE FER.
                      *BOURDONNAY(R.), . . .ET AL.

731 FE,. . . . . .D2. . . . . PHYS. LETT. VOL.32A, ... 341, (1970).
                      *ON THE TEMPERATURE SHIFT OF CRITICAL NEUTRON SCATTERING MAXIMA.
                      *GORDON(J.), POPOVICI(M.).

732 FE,. . . . . .D2. . . . . PHYS. LETT. VOL.34A, ... 319, (1971).
                      *WAVELENGTH-DEPENDENT SUSCEPTIBILITY OF IRON IN THE CRITICAL REGION.
                      *POPOVICI(M.).

733 FE,. . . . . .P. . . . .DISS. ABS. VOL.31, ...2171B, (1970).
                      *INELASTIC NEUTRON SCATTERING FROM FE, AL, AND MN.
                      *LAMB(R.C.).

734 FE,. . . . . .D1. . . . . J. APP. PHYS. VOL.40,. . . . . . 1442, (1969).
                      *STUDY OF SPIN DYNAMICS IN IRON WITH SLOW NEUTRONS.
                      *COLLINS(M.F.), MINKIEWICZ(V.J.), NATHANS(R.), PASSELL(L.), SHIRANE(G.).

735 FE,. . . . . .D1. . . . . J. OF PHYS. C VOL.1, ... 950, (1968).
                      *OBSERVATION OF SPIN-WAVE RENORMALIZATION EFFECTS IN IRON AND NICKEL.
                      *STRINGFELLOW(M.W.).

736 FE,. . . . . .D2. . . . . PHYS. LETT. 29A, ... 513, (1969).
                      *NEUTRON CRITICAL SCATTERING IN IRON WITH VERY HIGH TEMPERATURE STABILITY
                      *CISZEWSKI(R.), BLINOWSKI(K.).

737 FE (ISOTOPE A=56),. . . . . .T. . . . . NUCL. SCI. ENG. VOL.46, ... 255, (1971).
                      *STATISTICAL MODEL CALCULATIONS OF IRON-56 NEUTRON CROSS SECTIONS.
                      *BLOOM(S.D.), GREEN(J.M.), HUBBARD(H.W.), MOSZKOWSKI(S.A.).

738 FE,. . . . . .D2. . . . . J. PHYS. SOC. JAPAN (SUPPL.) 26, 169, (1969).
                      *HIGH RESOLUTION STUDIES OF THE CRITICAL SCATTERING OF NEUTRONS FROM IRON
                      *COLLINS(M.F.), MINKIEWICZ(V.J.), NATHANS(R.), PASSELL(L.), SHIRANE(G.).

739 FE (ALPHA),. . . . . .B1. . . . . RCN-129 (122 PP.), ... (1970).
                      *INVESTIGATION OF THE LATTICE DYNAMICS OF ALPHA-FE AND FE3*AL BY NEUTRON
                      INELASTIC SCATTERING.
                      *VAN-DIJK(C.). (ED. NOTE: AVAIL. REACTOR CENTRUM NEDERLAND, PETTEN,
                      THE NETHERLANDS).

740 FE,. . . . . .D2. . . . . J. PHYSIQUE TOME 32 COL.1 VOL.I, 523, (1971).
                      *DIFFUSION CRITIQUE DES NEUTRONS LENTS PAR LE FER.
                      *KAHN(R.), PARETTE(G.).

741 FE,. . . . . .D2. . . . . PHYS. LETT. VOL.26A, ... 396, (1968).
                      *EVIDENCE FOR FISHER*S CORRELATION FUNCTION IN IRON FROM CRITICAL NEUTRON
                      SCATTERING.
                      *BALLY(D.), POPOVICI(M.), TOTIA(M.), GRABCEV(B.), LUNGU(A.M.).

SECTION   6 -QUASI-ELASTIC AND INELASTIC SCATTERING STUDIES . . . .

742 FE,. . . . . .D1. . .PHYS. REV. B VOL.7,. . . . . 336, (1973).
                            *NEUTRON SCATTERING INVESTIGATION OF THE MAGNETIC EXCITATIONS IN IRON.
                            *HOOK(H.A.), NICKLOW(R.M.).

743 FE,. . . . . .D1 .MAGN. AND MAG. MATERIALS-1971, . 1340, (1972).
                            *COLLAPSE OF THE SMALL-ANGLE MAGNON SCATTERING IN FE AS A FUNCTION OF
                            MAGNETIC FIELD.
                            *GURMEN(E.), WERNER(S.A.), ARROTT(A.). (AIP CONF. PROC. NO.5 PART II).

744 FE (ALPHA),. . . . . .T,B1. . . . .J. PHYS. SOC. JAP. VOL.33, . . . 1207, (1972).
                            *LATTICE DYNAMICS OF SOME BCC TRANSITION METALS.
                            *BEHARI(J.), TRIPATHI(B.B.).

745 FE,. . . . . .D1. . . . .J. PHYSIQUE TOME 32 COL.1 VOL.II 1177, (1971).
                            *SPIN WAVES AND STONER MODES IN IRON.
                            *HOOK(H.A.), NICKLOW(R.M.).

746 FE,. . . . . .D2. . . . . . . .J. APP. PHYS. 39,. . . . . 528, (1968).
                            *NEUTRON-SCATTERING STUDIES OF CRITICAL MAGNETIC PHENOMENA.
                            *RISTE(T.).

747 FE,. . . . . .T,P. . . . .PHYS REV. 162,. . . . . 486, (1967).
                            *CALCULATIONS OF NEUTRON AND X-RAY SCATTERING AMPLITUDE FOR BCC IRON.
                            *DE-CICCO(P.D.), KITZ(A.).

748 FE,. . . . . .D2. . . . . . . .SOL. STATE COMM. 4,. . . . . 425, (1966).
                            *VERIFICATION DE LA SUSCEPTIBILITE MAGNETIQUE EN FONCTION DE LA
                            TEMPERATURE PAR LA DIFFUSION CRITIQUE DES NEUTRONS PAR LE FER.
                            *KONSTANTINOVIC(J.).

749 FE,. . . . . .T,D1,. . . .J. APP. PHYS. 39,. . . . . 383, (1968).
                            *SPIN WAVES IN 3D METALS.
                            *SHIRANE(G.), MINKIEWICZ(V.J.), NATHANS(R.).

750 FE,. . . . . .D4. . . . . .PHYS. REV. 142,. . . . . 291, (1966).
                            *SPIN CORRELATION IN IRON.
                            *SPOONER(S.), AVERBACH(B.L.).

751 FE,. . . . . .D1. . . . . .PHYS. REV. LETT. 15, ... 146, (1965).
                            *MEASUREMENT OF THE MAGNON DISPERSION RELATION OF IRON.
                            *ALPERIN(H.A.), PICKART(S.J.), SHIRANE(G.), NATHANS(R.), STEINSVOLL(O.).

752 FE,. . . . . .D2. . . . . .PHYS. REV. LETT. 21,. . . . . . . 99, (1968).
                            *CRITICAL BEHAVIOR OF IRON ABOVE ITS CURIE TEMPERATURE.
                            *COLLINS(M.F.), NATHANS(R.), PASSELL(L.), SHIRANE(G.).

753 FE,. . . . . .D2. . . . . .PROC. INT. CONF. (NOTTINGHAM), . 99, (1964).
                            *CRITICAL MAGNETIC SCATTERING OF NEUTRONS IN IRON.
                            *PASSELL(L.), BLINOWSKI(K.), NIELSEN(P.), BRUN(T.O.).

754 FE,. . . . . .D1. . . . . .IAEA SYMP. COPENHAGEN, VOL.2, . 3, (1968).
                            *NEUTRON SPIN SCATTERING BY SPIN WAVES IN METALS (REVIEW PAPER).
                            *MOLLER(H.B.).

755 FE,. . . . . .D2. . . . . . . .J. PHYS. CHEM. SOL. 28,. . . . . 1947, (1967).
                            *SMALL-ANGLE CRITICAL MAGNETIC SCATTERING OF NEUTRONS IN IRON.
                            *BALLY(D.), GRABCEV(B.), LUNGU(A.M.), POPOVICI(M.), TOTIA(M.).

756 FE,. . . . . .D2. . . . . .PHYS. REV. VOL.139,. . . . . . .A1866, (1965).
                            *CRITICAL MAGNETIC SCATTERING OF NEUTRONS IN IRON.
                            *PASSELL(L.), BLINOWSKI(K.), BRUN(T.O.), NIELSEN(P.).

757 FE,. . . . . .T. . . . . .PHYS. REV. 157,. . . . . 540, (1967).
                            *EXISTENCE OF AN INFINITY IN THE FREQUENCY DISTRIBUTION OF G(NU) OF
                            MONOATOMIC BODY-CENTERED CUBIC CRYSTAL.
                            *GILAT(G.).

758 FE (ALPHA),. . . . . .B1,D1. . . .PHYSICA VOL.37,. . . . . 603, (1967).
                            *DISPERSION RELATIONS FOR PHONONS AND MAGNONS IN ALPHA-FE.
                            *VAN-DINGENEN(W.), HAUTECLER(S.).

759 FE,. . . . . .D2. . . . . .J. APPL. PHYS. VOL.39,. . . . . . 459, (1968).
                            *CRITICAL SCATTERING OF THERMAL NEUTRONS IN SOME FERROMAGNETIC METALS.
                            *BALLY(D.), GRABCEV(B.), POPOVICI(M.), TOTIA(M.), LUNGU(A.M.).

760 FE,. . . . . .D2. . . . . .BULL/KIDRICH INST/NUCL. SCI. 17, 329, (1966).
                            *CONTRIBUTION TO AN EXPERIMENTAL STUDY OF COLD NEUTRON CRITICAL SCAT-
                            TERING BY IRON.
                            *KONSTANTINOVIC(J.). (IN FRENCH).

761 FE,. . . . . .D4. . . . . .ATOMKERNENERGIE VOL.13,. . . . . 50, (1968).
                            *MEASUREMENTS ON MAGNETIC SMALL ANGLE SCATTERING OF THERMAL NEUTRONS IN
                            PASSING THROUGH UNMAGNETIZED IRON.
                            *MATHUR(J.N.). (IN GERMAN).

762 FE,. . . . . .T. . . . . .ACTA PHYS. HUNGAR. VOL.15,. . . . 29, (1962).
                            *A CONTRIBUTION TO THE PROBLEM OF INELASTIC MAGNETIC SCATTERING OF
                            POLARIZED NEUTRONS IN FE AND NI.
                            *VALENTA(L.), ZAJAC(ST.).

763 FE,. . . . . .B1. . . . . .PROC. PHYS. SOC. VOL.79,. . . . . 479, (1962).
                            *SOME MEASUREMENTS OF PHONON DISPERSION RELATIONS IN IRON.
                            *LOW(G.G.E.).

764 FE,. . . . . .D2. . . . . .J. APPL. PHYS. VOL.35,. . . . . . 933, (1964).
                            *CRITICAL MAGNETIC SCATTERING OF NEUTRONS IN IRON.
                            *PASSELL(L.), BLINOWSKI(K.), BRUN(T.O.), NIELSEN(P.).

765 FE (ALPHA),. . . . . .B1,D1. . . .HELV. PHYS. ACTA VOL.40, ... 378, (1967).
                            *PHONON AND SPIN-WAVE DISPERSIONS IN IRON.
                            *SCHWEISS(P.), FURRER(A.), BUHRER(W.).

766 FE,. . . . . .P. . . . . . . .REF. SECTION 3, 48/HAVENS, ...

767 FE,. . . . . .E,D4. . . . .PHYS. REV. VOL.75,. . . . . 565, (1949).
                            *MAGNETIC REFRACTION OF NEUTRONS AT DOMAIN BOUNDARIES.
                            *HUGHES(D.J.), BURGY(M.T.), HELLER(R.B.), WALLACE(J.W.).

768 FE,. . . . . .D2. . . . . .PHYS. REV. VOL.103,. . . . . . . 525, (1956).
                            *CRITICAL MAGNETIC SCATTERING OF NEUTRONS BY IRON.
                            *GERSCH(H.A.), SHULL(C.G.), WILKINSON(M.K.).

769 FE,. . . . . .P. . . . . .PHYS. REV. VOL.80, ... 481, (1950).
                            *SCATTERING AND POLARIZATION OF NEUTRONS IN AN IRON SINGLE CRYSTAL.
                            *HUGHES(D.J.), BURGY(M.T.), WOOLF(W.E.).

SECTION   6 -QUASI-ELASTIC AND INELASTIC SCATTERING STUDIES . . . .

770 FE,. . . . .SPIN FLUCTUATION..D1.......PROC. ROY. SOC. A VOL.235,. . . . . 305, (1956).
     ..SPIN FLUCTUATION SCATTERING OF NEUTRONS AND THE FERROMAGNETIC STATE IN
     IRON.
     ..LOWDE(R.D.).

771 FE,. . . . .P.........PROC. ROY. SOC. A VOL.221,. . . . 206, (1954).
     ..ON THE DIFFUSE REFLEXION OF NEUTRONS BY A SINGLE CRYSTAL.
     ..LOWDE(R.D.).

772 FE,. . . . .D4.......PHYS. REV. VOL.92, ... 202, (1953).
     ..MAGNETIC INELASTIC SCATTERING OF SLOW NEUTRONS.
     ..PALEVSKY(H.), HUGHES(D.J.).

773 FE,. . . . .P,T.....PHYS. REV. VOL.74,. . . . . . . 103, (1948).
     ..INELASTIC SCATTERING OF NEUTRONS.
     ..CASSELS(J.M.), LATHAM(R.).

774 FE,. . . . .T,D1.....SOV. PHYS. JETP 20,. . . . . 1548, (1965).
     ..NEUTRON SCATTERING BY SPIN WAVES IN IRON.
     ..DRABKIN(G.M.), ZABIDAROV(E.I.)...ET AL.

775 FE,. . . . .B1.......PHYS. LETT. 24A, ... 270, (1967).
     ..NORMAL VIBRATIONS (ALPHA)-IRON.
     ..BERGSMA(J.), VAN-DIJK(C.), TOCCHETTI(D.).

776 FE,. . . . .T,B1.....PHYS. STAT. SOL. 12, ... 305, (1965).
     ..PHONON DISPERSION IN (ALPHA)-IRON.
     ..GUPTA(R.P.), SHARMA(P.K.).

777 FE,. . . . .B1,T.....PHYS. REV. 143,. . . . . 443, (1966).
     ..LATTICE DYNAMICS AND SPECIFIC HEATS OF SOME TRANSITION METALS IN
     KREBS#S MODEL.
     ..MAHESH(P.S.), DAYAL(B.).

778 FE,. . . . .B1.......PHYS. REV. 162,. . . . . 528, (1967).
     ..PHONON DISPERSION RELATIONS FOR IRON.
     ..MINKIEWICZ(V.J.), SHIRANE(G.), NATHANS(R.).

779 FE,. . . . .D2.......J. PHYSIQUE TOME 32,. . . . . . . 447, (1971).
     ..STUDY OF THE CRITICAL SCATTERING OF NEUTRONS BY IRON IN THE HYDRODYNAMIC
     AND QUASI HYDRODYNAMIC REGIONS.
     ..PARETTE(G.), KAHN(R.).

780 FE,. . . . .B1.......SOL. STATE COMM. 5,. . . . . . . 211, (1967).
     ..LATTICE VIBRATIONS IN IRON AT 296 DEG. K.
     ..BROCKHOUSE(B.N.), ABOU-HELAL(H.E.), HALLMAN(E.D.).

781 FE,. . . . .D2.......PHYS. LETT. 28A,. . . . . 389, (1968).
     ..THE TEMPERATURE DEPENDENCE OF THE #CURIE POINT# FOR IRON OBSERVED IN
     THE CRITICAL SCATTERING OF NEUTRONS.
     ..BLINOWSKI(K.), CISZEWSKI(R.).

782 FE,. . . . .B1.......IAEA SYMPOSIUM VIENNA,. . . . . . 555, (1960).
     ..INELASTIC SCATTERING OF NEUTRONS FROM IRON.
     ..IYENGAR(P.K.), MURTHY(N.S.S.), DASANNACHARYA(B.A.).

783 FE,. . . . .D4,P.....PHYS. REV. B VOL.3,. . . . . . . 830, (1971).
     ..FORWARD MAGNETIC SCATTERING AMPLITUDE OF IRON FOR THERMAL NEUTRONS.
     ..SCHNEIDER(C.S.), SHULL(C.G.).

784 FE,. . . . .B7.......UKR. FIZ. ZH. (USSR) VOL.18,. . 1528, (1973).
     ..STUDY OF SLOW NEUTRON INELASTIC SCATTERING IN AN IRON SPECIMEN WITHIN
     A TEMPERATURE RANGE FROM 24 TO 874 DEGREES C.
     ..GORBACHEV(B.I.),   IVANITSKII(P.G.),   KROTENKO(V.T.), PASECHNIK(M.V.),
     PASTUSHENKO(S.N.). (IN RUSSIAN).

785 FE,. . . . .B7.......UKR. FIZ. ZH. (USSR) VOL.18,. . . 1384, (1973).
     ..A STUDY OF SLOW NEUTRON INELASTIC SCATTERING IN POLYCRYSTALLINE SAMPLES
     OF IRON.
     ..GORBACHEV(B.I.),   IVANITSKII(P.G.), KROTENKO(V.T.), PASECHNIK(M.V.). (IN
     RUSSIAN).

786 FE,. . . . .D4.......C.R. ACAD. SCI. VOL.246,. . . . . 1018, (1958).
     ..EXPERIMENTAL STUDY OF THE KINETICS OF THE MAGNETIC MOMENTS OF IRON ABOVE
     THE CURIE POINT.
     ..ERICSON(M.), JACROT(B.). (IN FRENCH).

787 FE,. . . . .D1.......PHYS. REV. LETT. VOL.4,. . . . . 452, (1960).
     ..NEUTRON SMALL-ANGLE SCATTERING BY SPIN WAVES IN IRON.
     ..LOWDE(R.D.), UMAKANTHA(N.).

788 FE,. . . . .D2.......J. PHYS. CHEM. SOLIDS VOL.13,. . 235, (1960).
     ..EXPERIMENTAL STUDY OF THE CRITICAL SCATTERING OF NEUTRONS IN IRON.
     ..ERICSON(M.), JACROT(B.).

789 FE,. . . . .T.......J. PHYS. RADIUM VOL.20,. . . . . . 178, (1959).
     ..A NEUTRON SCATTERING STUDY OF THE KINETICS OF MAGNETIC MOMENTS IN IRON
     IN THE REGION OF THE CURIE POINT.
     ..ERICSON(M.), JACROT(B.). (IN FRENCH).

790 FE,. . . . .D2.......REV. MOD. PHYS. VOL.30,. . . . . 69, (1958).
     ..CRITICAL MAGNETIC SCATTERING OF NEUTRONS BY IRON.
     ..LOWDE(R.D.).

791 FE,. . . . .D2.......IAEA SYMP. CHALK RIVER, VOL.2, . 317, (1962).
     ..DIFFUSION AUX PETITS ANGLES DES NEUTRONS PAR LE FER ET LE NICKEL AU
     VOISINAGE DU POINT DE CURIE.
     ..JACROT(B.), KONSTANTINOVIC(J.), PARETTE(G.), CRIBIER(D.).

792 FE,. . . . .D2.......INT. J. MAGN. VOL.4, ... 205, (1973).
     ..CRITICAL SCATTERING OF NEUTRONS FROM IRON.
     ..BORONKAY(S.), COLLINS(M.F.).

793 FE,. . . . .T.......PHYS. REV. VOL.93,. . . . . 268, (1954).
     ..TEMPERATURE VARIATION OF THE MAGNETIC INELASTIC SCATTERING OF SLOW
     NEUTRONS.
     ..VAN-HOVE(L.).

794 FE,. . . . .D1,D2....PHYS. REV. VOL.179,. . . . . . . 417, (1969).
     ..CRITICAL AND SPIN-WAVE SCATTERING OF NEUTRONS FROM IRON.
     ..COLLINS(M.F.), MINKIEWICZ(V.J.), NATHANS(R.), PASSELL(L.), SHIRANE(G.).

795 FE3*AL,. . . . .B1.......IAEA SYMP. COPENHAGEN, VOL.1,. . 233, (1968).
     ..LATTICE DYNAMICS OF ALPHA-FE AND FE3*AL.
     ..VAN-DYJK(C.), BERGSMA(J.).

SECTION   6 -QUASI-ELASTIC AND INELASTIC SCATTERING STUDIES . . . .

796 FE3*AL,. . . . . LATTICE DYNAMICS OF FE3*AL.   B1   PHYS. LETT. VOL.32A, ...   255, (1970).
                           VAN-DIJK(C.).

797 FE3*AL,. . . . . AN ANALYSIS OF LATTICE VIBRATIONS OF ORDERED FE3*AL.   B1   J. PHYS. CHEM. SOL. 28,. . .   467, (1967).
                           BORGONOVI(G.), LOGIUOICE(G.), TOCCHETTI(D.).

798 FE3*AL,. . . . . INVESTIGATION OF THE LATTICE DYNAMICS OF ALPHA-FE AND FE3*AL BY NEUTRON   B1   RCN-129 (122 PP.)   (1970).
                           INELASTIC SCATTERING.
                           VAN-DIJK(C.). (ED. NOTE: AVAIL. REACTOR CENTRUM NEDERLAND, PETTEN,
                           THE NETHERLANDS).

799 FE3*AL,. . . . . SPIN WAVE DISPERSION RELATION IN ORDERED FE3*AL.   D1   J. PHYSIQUE TOME 32 COL.1 VOL.II 1188, (1971).
                           ANTONINI(B.), MENZINGER(F.), PAOLETTI(A.).

800 FE-AL, ... SPIN-WAVES STIFFNESS AND ELECTRONIC STRUCTURE OF IRON-ALUMINUM AND   D1   PROC. PHYS. SOC. 89,   419, (1966).
                           IRON-GALLIUM ALLOYS.
                           ANTONINI(B.), STRINGFELLOW(M.W.).

801 FE*C*03, ... NEUTRON INELASTIC SCATTERING STUDIES OF THE ISING SYSTEM FE*C*03.   D4   AIP CONF. PROC. VOL.5, ... 1334, (1971).
                           WREGE(D.E.), SPOONER(S.), GERSCH(H.A.).

802 FE*C*03, ... NEUTRON INELASTIC SCATTERING STUDIES OF THE ISING SYSTEM FE*C*03.   D4   THESIS,   (1971).
                           WREGE(D.E.). (ED. NOTE: AVAIL. UNIV. MICROFILMS ORDER NO.72-19,714 (108
                           PP.) SEE: DISS. ABSTRACTS B VOL.33 NO.1, PG.404).

803 FE*CL2,. . . . . MAGNETIC PROPERTIES OF FE*CL2 IN ZERO FIELD. I. EXCITATIONS.   D1   PHYS. REV. B VOL.5, ... 2607, (1972).
                           BIRGENEAU(R.J.), YELON(W.B.), COHEN(E.), MAKOVSKY(J.).

804 FE*CL2,. . . . . SMALL-ANGLE SCATTERING OF NEUTRONS BY FE*CL2 NEAR THE NEEL TEMPERATURE.   D4,T   J. PHYS. SOC. JAPAN VOL.18,. . .   74, (1963).
                           NAGAI(O.).

805 FE(CR),. . . . . CRITICAL NEUTRON SCATTERING IN FE(CR) DILUTE ALLOYS.   D2   ACTA CRYST. (INTERACT.) VOL.A25, S262, (1969).
                           BALLY(D.), GRABCEV(B.), LUNGU(A.M.), POPOVICI(M.), TOTIA(M.).

806 FE(CR),. . . . . VIRTUAL SPIN WAVE STATE BELOW AND ABOVE THE CURIE-TEMPERATURE IN DILUTE   D2   PHYS. LETT. VOL.28A, ...   213, (1968).
                           FE(CR) ALLOY.
                           KROO(N.), PAL(L.), ARIC(M.), JOVIC(D.).

807 FE(CR,ER,MN,V),. . . . VIRTUAL MAGNON STATES IN DILUTE ALLOYS.   D1   IAEA SYMP. COPENHAGEN, VOL.2,. .   37, (1968).
                           KROO(N.), PAL(L.), JOVIC(D.).

808 FE(CR,MO), ... IMPURITY INFLUENCE ON NEUTRON CRITICAL SCATTERING IN FERROMAGNETICS.   D2   SOL. STATE COMM. VOL. 9, . . .   353, (1971).
                           BALLY(D.), TOTIA(M.), LUNGU(A.M.), POPOVICI(M.), HASE(W.).

809 FE-CR-NI,. . . . . . . NEUTRON MAGNETIC SCATTERING FROM F.C.C. IRON ALLOYS.   D3   J. PHYS. CHEM. SOLIDS VOL.25,. .   183, (1964).
                           NATHANS(R.), PICKART(S.J.).

810 FE-ER, ... INELASTIC SCATTERING OF NEUTRONS BY VIRTUAL MAGNON STATES IN DILUTE   D1   J. APP. PHYS. 39,   453, (1968).
                           ALLOYS.
                           KROO(N.), PAL(L.).

811 FE*F2, ... SPIN WAVES IN ANTIFERROMAGNETIC FE*F2.   D1   J. OF PHYS.-C-, SER.2, VOL.4,. .   307, (1971).
                           HUTCHINGS(M.T.), RAINFORD(B.D.), GUGGENHEIM(H.J.).

812 FE*F2, ... MAGNONS AND THE MAGNON-PHONON INTERACTION IN FE*F2.   D1   IAEA SYMP. GRENOBLE, ...   655, (1972).
                           RAINFORD(B.D.), HOUMANN(J.G.), GUGGENHEIM(H.J.).

813 FE*F2, ... THEORY OF THE MAGNON AND PHONON INTERACTION IN FE*F2.   T   J. PHYS. C VOL.5, . . .   2769, (1972).
                           LOVESEY(S.W.).

814 FE*F2, ... NEUTRON-SCATTERING DETERMINATION OF SPIN-WAVE DISPERSION RELATIONS   D1   J. APP. PHYS. 39,   1120, (1968).
                           IN FE*F2.
                           GUGGENHEIM(H.J.), HUTCHINGS(M.T.), RAINFORD(B.D.).

815 FE*F2, ... MEASUREMENT OF DYNAMICAL SCALING IN FE*F2.   D4   J. APPL. PHYS. VOL.42, ... 1376, (1971).
                           SCHULHOF(M.P.), HUTCHINGS(M.T.), GUGGENHEIM(H.J.).

816 FE*F2, ... EXCHANGE INTEGRALS IN CR*F2 AND FE*F2 BY PARAMAGNETIC NEUTRON   D3   SOL. STATE COMM. 6,   593, (1968).
                           SCATTERING.
                           MADHAV-RAO(L.),   NATERA(M.G.),   MURTHY(N.S.S.),   DASANNACHARYA(B.A.),
                           IYENGAR(P.K.).

817 FE*F2, ... CRITICAL MAGNETIC NEUTRON SCATTERING FROM FERROUS FLUORIDE.   D2   PHYS. REV. B VOL.5,   154, (1972).
                           HUTCHINGS(M.T.), SCHULHOF(M.P.), GUGGENHEIM(H.J.).

818 FE-GA, ... SPIN-WAVES STIFFNESS AND ELECTRONIC STRUCTURE OF IRON-ALUMINUM AND IRON-   D1   PROC. PHYS. SOC. 89,   419, (1966).
                           GALLIUM ALLOYS.
                           ANTONINI(B.), STRINGFELLOW(M.W.).

819 (FE-GA)*03,. . . . . . STUDY OF THE ACENTRIC COMPOUND (FE-GA)*03 WITH POLARIZED NEUTRONS.   P   J. APP. PHYS. 39,   1332, (1968).
                           DELAPALME(A.).

820 FE-MN, ... PARAMAGNETIC NEUTRON SCATTERING IN AN EQUIATOMIC IRON-MANGANESE ALLOY.   D3   J. PHYS. SOC. JAPAN VOL.25,   367, (1968).
                           VANCE(E.R.), DAVIS(R.L.).

821 FE-MN, ... NEUTRON MAGNETIC SCATTERING FROM F.C.C. IRON ALLOYS.   D3   J. PHYS. CHEM. SOLIDS VOL.25,. .   183, (1964).
                           NATHANS(R.), PICKART(S.J.).

SECTION   6 -QUASI-ELASTIC AND INELASTIC SCATTERING STUDIES . . . .

822 FE-MN, ...*MAGNETIC CRITICAL D2 SCATTERING J. PHYSIQUE TOME 32 COL.1 VOL.II 1017, (1971). FROM AN ITINERANT ANTIFERROMAGNET OF GAMMA-
FE(0.5)-MN(0.5) ALLOY.
*ISHIKAWA(Y.), ENDOH(Y.).

823 FE-MN, ...*INELASTIC D1 J. APP. PHYS. 39, 453, (1968). SCATTERING OF NEUTRONS BY VIRTUAL MAGNON STATES IN DILUTE
ALLOYS.
*KROO(N.), PAL(L.).

824 FE-(MN), ...*NEUTRON SCATTERING D1 PHYS. LETT. 24A, 22, (1967). BY VIRTUAL MAGNON STATE IN FE WITH MN IMPURITY.
*KROO(N.), BATA(L.).

825 FE-MN, ...*MAGNETIC CRITICAL D2 SCATTER J. PHYS. SOC. JAPAN VOL.35, 1616, (1973). FROM AN ITINERANT ANTIFERROMAGNET OF
GAMMA FE-MN ALLOY I.QUASI ELASTIC SCATTERING.
*ISHIKAWA(Y.), ENDOH(Y.), IKEDA(S.).

826 FE-MN, ...*SPIN-WAVE SCATTERING D1 SOL. STATE COMM. VOL.13, 1179, (1973). FROM A GAMMA-FE(.47)-MN(.53) ALLOY.
*ENDOH(Y.), SHIRANE(G.), ISHIKAWA(Y.), TAJIMA(K.).

827 FE-NI, ...*DIRECT OBSERVATION D1 J. APPL. PHYS. 41 1363, (1970). OF THE ANGULAR DISTRIBUTION OF NEUTRONS SCATTERED
AT SMALL ANGLES BY SPIN WAVES IN FE-NI ALLOYS.
*WERNER(S.A.), WIENER(E.), GURMEN(E.), ARROTT(A.).

828 FE-(NI), ...*THE COMPOSITIONAL D1 DISS. ABS. VOL.32, 4147B, (1972). DEPENDENCE OF THE SPINWAVE DISPERSION COEFFICIENT D IN
THE INVAR REGION.
*BAUER(C.A.).

829 FE-NI, ...*SPIN-WAVE DISPERSION D1 J. APP. PHYS. 39, 455, (1968). RELATION IN FE-NI ALLOY.
*MENZINGER(F.), CAGLIOTI(G.), SHIRANE(G.), NATHANS(R.), PICKART(S.J.),
ALPERIN(H.A.).

830 FE-NI, ...*SPIN WAVE ENERGIES D1 PROC. PHYS. SOC. 84, 55, (1964). AND EXCHANGE PARAMETERS IN IRON-NICKEL ALLOYS.
*HATHERLY(M.), HIRAKAWA(K.), LOWDE(R.D.), MALLETT(J.F.),
STRINGFELLOW(M.W.), TORRIE(B.H.).

831 FE-NI, ...*A MEASUREMENT BY NEUTRON D3 PROC. PHYS. SOC. 86, 973, (1965). SCATTERING OF MAGNETIC MOMENTS IN A PARA-
MAGNETIC IRON-NICKEL ALLOY.
*COLLINS(M.F.).

832 FE-NI, ...*MAGNETIC DIFFUSE SCATTERING D4 J. PHYS. SOC. JAPAN VOL.35, 706, (1973). OF NEUTRONS FROM FE-NI INVAR ALLOY.
*KOMURA(S.), TAKEDA(T.), OHARA(S.).

833 FE-(NI,CO,CR,V),*DENSITY-OF-STATES EFFECTS D1 PHYS. REV. LETT. VOL.14, 698, (1965). IN THE MAGNETIC STIFFNESS OF 3D-3D TRANSITION-
METAL ALLOYS.
*LOWDE(R.D.), SHIMIZU(M.), STRINGFELLOW(M.W.), TORRIE(B.H.).

834 FE3*04,. . . . .*SPIN WAVE AND CRITICAL D2 J. APPL. PHYS. 41, 1433, (1970). FLUCTUATIONS IN MAGNETITE.
*COLLINS(M.F.), SAUNDERSON(D.H.).

835 FE2*03 (ALPHA),.*NEUTRON INVESTIGATION D1 IAEA SYMP. BOMBAY, VOL.1, 443, (1964). OF MAGNON SPECTRUM IN HAEMATITE.
*DIMITRIJEVIC(Z.), RZANY(H.), TODOROVIC(J.), WANIC(A.).

836 FE3*04,. . . . .*SYMETRIES ET INTENSITES D1 SOL. STATE COMM. VOL.8, 2141, (1970). DIFFUSEES DANS LE SPECTRE D*ONDES DE SPIN DE LA
MAGNETITE.
*NAUCIEL-BLOCH(M.), HENNION(B.), SARMA(G.).

837 FE2*03 (PHASE: ALPHA),*INELASTIC NEUTRON D1 PHYS. STAT. SOL. VOL.42, 241, (1970). SCATTERING INVESTIGATION OF SPIN WAVES AND MAGNETIC
INTERACTIONS IN ALPHA-FE2*03.
*SAMUELSEN(E.J.), SHIRANE(G.).

838 FE3*04,. . . . .*SPIN WAVES IN MAGNETITE D1 SOL. STATE COMM. VOL.5, 715, (1967). AT A TEMPERATURE BELOW THE ELECTRONIC ORDERING
TRANSITION.
*TORRIE(B.H.).

839 FE3*04,. . . . .*THE SCATTERING FUNCTION D1,D4 IAEA SYMP. GRENOBLE, 649, (1972). OF MAGNETITE.
*EVANS(M.T.), WARMING(E.), SQUIRES(G.L.).

840 FE3*04,. . . . .*EXPERIMENTAL DETERMINATION D4 J. PHYSIQUE TOME 32 COL.1 VOL.II 1182, (1971). OF EXCHANGE INTEGRALS IN MAGNETITE.
*BOURDONNAY(H.)...ET AL. (GROUPE DE DIFFUSION INELASTIQUE DES NEUTRONS).

841 FE3*04,. . . . .*MAGNON SCATTERING D1 PHYS. REV. 154, 508, (1967). OF POLARIZED NEUTRONS BY THE DIFFRACTION METHOD:
MEASUREMENTS ON MAGNETITE.
*ALPERIN(H.A.), STEINSVOLL(O.), NATHANS(R.), SHIRANE(G.).

842 FE3*04,. . . . .*OBSERVATION OF OPTICAL D1 PHYS. LETT. 1, 189, (1962). AND ACOUSTICAL MAGNONS IN MAGNETITE.
*WATANABE(H.), BROCKHOUSE(B.N.).

843 FE3*04,. . . . .*NEUTRON INVESTIGATION D1 PHYS. STAT. SOL. 15, 119, (1966). OF TEMPERATURE EFFECTS IN THE MAGNON SPECTRUM OF
MAGNETITE.
*DIMITRIJEVIC(Z.), KRASNICKI(S.), RZANY(H.), TODOROVIC(J.), WANIC(A.).

844 FE3*04,. . . . .*EXCHANGE INTEGRALS IN MAGNETITE. D1 IAEA SYMP. COPENHAGEN, VOL.2, 117, (1968).
*MOGLESTUE(K.T.).

845 FE3*04,. . . . .*SPIN WAVES SPECTRA T,D1 PHYS. REV. 130, 1783, (1963). OF MAGNETITE.
*GLASSER(M.L.), MILFORD(F.J.).

846 FE2*03,. . . . .*POLARIZED NEUTRONS D4 PHYS. REV. VOL.136, A1641, (1964). BY STUDY OF HEMATITE.
*NATHANS(R.), PICKART(S.J.), ALPERIN(H.A.), BROWN(P.J.).

SECTION   6 -QUASI-ELASTIC AND INELASTIC SCATTERING STUDIES . . . .

847 FE3*04,. . . . . .D1. . . . . . .J. APP. PHYS. 39,. . . . . 1114, (1968).
     *NEUTRON SCATTERING FROM MAGNETITE BELOW 119 DEG. K.
     *SAMUELSEN(E.J.), BLEEKER(E.J.), DOBRZYNSKI(L.), RISTE(T.).

848 FE3*04,. . . . . .D1. . . . . . .PHYS. REV. 156,. . . . . 632, (1967).
     *SCATTERING OF POLARIZED NEUTRONS BY SPIN WAVES IN MAGNETITE AND YTTRIUM
     IRON GARNET.
     *FERGUSON(G.A.), SAENZ(A.W.).

849 FE3*04,. . . . . .D1. . . . . . .SOL. STATE COMM. VOL.12 NO.8,. . . 795, (1973).
     *OPTIC SPIN-WAVES IN MAGNETITE NEAR THE VERWEY TRANSITION.
     *EVANS(M.T.), WARMING(E.), HUTCHINGS(M.T.), STRINGFELLOW(M.W.).

850 FE3*04,. . . . . .D1,D2. . . .J. PHYS. RADIUM VOL.23,. . . . . 494, (1962).
     *SOME APPLICATIONS OF INELASTIC NEUTRON SCATTERING TO THE STUDY OF
     MAGNETIC COUPLINGS.
     *JACROT(B.), CRIBIER(D.). (IN FRENCH).

851 FE2*03,. . . . . .D1. . . . . . .PHYSICA VOL.34,. . . . . 241, (1967).
     *A STUDY OF THE MAGNETIC INTERACTIONS IN ALPHA-FE2*03 THROUGH SCATTERING
     OF NEUTRONS BY SPIN WAVES.
     *SAMUELSEN(E.J.).

852 FE2*03,. . . . . .D1. . . . . . .J. PHYS. CHEM. SOLIDS VOL.17,. . 318, (1961).
     *A NEUTRON DIFFRACTION STUDY OF SPIN FLUCTUATIONS IN ALPHA-FE2*03.
     *RISTE(T.), WANIC(A.).

853 FE3*04,. . . . . .D2. . . . . . .J. PHYS. CHEM. SOLIDS VOL.17,. . 308, (1961).
     *CRITICAL MAGNETIC SCATTERING OF NEUTRONS IN MAGNETITE.
     *RISTE(T.).

854 FE3*04,. . . . . .D1. . . . . . .J. PHYS. CHEM. SOLIDS VOL.23,. . 117, (1962).
     *SCATTERING OF POLARIZED NEUTRONS BY SPIN WAVES IN MAGNETITE.
     *FERGUSON(G.A.), SAENZ(A.W.).

855 FE3*04,. . . . . .D1. . . . . . .PHYS. SOC. JAPAN VOL.17 B-III 6C, (1962).
     *SOME EXPERIMENTS ON MAGNETIC INELASTIC SCATTERING OF NEUTRONS.
     *RISTE(T.).

856 FE2*03,. . . . . .D3. . . . . . .PHYS. REV. VOL.76,. . . . . . . 1572, (1949).
     *THE SCATTERING OF SLOW NEUTRONS BY PARAMAGNETIC CRYSTALS.
     *RUDERMAN(I.W.).

857 FE3*04,. . . . . .D1. . . . . . .PHYS. REV. VOL.111,. . . . . . 1273, (1958).
     *FIELD DEPENDENCE OF NEUTRON SCATTERING BY SPIN WAVES.
     *BROCKHOUSE(B.N.).

858 FE3*04,. . . . . .D1. . . . . . .PHYS. REV. VOL.106,. . . . . . 859, (1957).
     *SCATTERING OF NEUTRONS BY SPIN WAVES IN MAGNETITE.
     *BROCKHOUSE(B.N.).

859 FE*0,. . . . . .D1. . . . . . .NATURE VOL.185,. . . . . 450, (1960).
     *NEUTRON DIFFRACTION STUDY OF ANTIFERROMAGNETIC SPIN WAVES IN ALPHA-
     FERRIC OXIDE.
     *GOEDKOOP(J.A.), RISTE(T.).

860 FE3*04,. . . . . .D1. . . . . . .J. PHYS. CHEM. SOLIDS VOL.9,. . 153, (1959).
     *SPIN FLUCTUATION SCATTERING OF NEUTRONS IN MAGNETITE.
     *RISTE(T.), BLINOWSKI(K.), JANIK(J.).

861 FE3*04,. . . . . .D1. . . . . . .IAEA SYMP. CHALK RIVER, VOL.2, . 297, (1962).
     *SPIN WAVES IN MAGNETITE FROM NEUTRON SCATTERING.
     *BROCKHOUSE(B.N.), WATANABE(H.).

862 FE2*03,. . . . . .D1. . . . . . .PHYS. STAT. SOL. 21,. . . . . . K163, (1967).
     *ANISOTROPY OF MAGNON DISPERSION RELATION IN HAEMATITE.
     *DIMITRIJEVIC(Z.), KRASNICKI(S.), RZANY(H.), TODOROVIC(J.), WANIC(A.),
     CURIEN(H.), MILOJEVIC(A.).

863 FE2*03,. . . . . .D1. . . . . . .PHYS. STAT. SOL. 41,. . . . . . K103, (1970).
     *INVESTIGATION OF MAGNON DISPERSION RELATION IN ALPHA-FE2*03 - ADDITIONAL
     DATA.
     *ALIKHANOV(R.A.), DIMITRIJEVIC(Z.), KRASNICKI(S.), RZANY(H.),
     TODOROVIC(J.), WANIC(A.).

864 FE3*04,. . . . . .D1. . . . . . .PHYS. STAT. SOL. 22, ... K55, (1967).
     *NEUTRON INVESTIGATION OF MAGNONS IN MAGNETITE.
     *DIMITRIJEVIC(Z.), KRASNICKI(S.), TODOROVIC(J.), WANIC(A.).

865 (FE2*03)-(CR2*03),. . . . . . . .T. . . . . . .PROC. INT. CONF. (NOTTINGHAM), . 516, (1964).
     *THEORY OF SOLID SOLUTIONS (1-X)*CR2*03(X)*FE2*03.
     *BERTAUT(E.F.).

866 FE2*03,FE3*04,. . . . . .D2. . . . . . .J. APP. PHYS. 33,. . . . . 528, (1968).
     *NEUTRON-SCATTERING STUDIES OF CRITICAL MAGNETIC PHENOMENA.
     *RISTE(T.).

867 FE2*03-(GA),. . . . . .D4. . . . . . .PHYS. LETT. 7,. . . . . 177, (1963).
     *MAGNETIC TRANSITION IN PURE AND GA DOPED (ALPHA)-FE2*03.
     *MORRISH(A.H.), JOHNSTON(G.B.), CURRY(N.A.).

868 FE-P,. . . . . .D1. . . . . . .SOV. PHYS. JETP LETT. VOL.16,. . 6, (1972).
     *MIXED EXCHANGE INTERACTION IN F.C.C. IRON-PALLADIUM ALLOYS (NEUTRON
     SCATTERING BY SPIN WAVES).
     *MEN≠SHIKOV(A.Z.), SIDOROV(S.K.), KUZ≠MIN(N.N.).

869 FE-PD,. . . . . .D1,T. . . . . .J. OF PHYS. C VOL.1,. . . . . . 1699, (1968).
     *THE SPIN-WAVE STIFFNESS OF DILUTE IRON-PALLADIUM ALLOYS.
     *STRINGFELLOW(M.W.).

870 FE-PD3,. . . . . .D1. . . . . . .SOL. STATE COMM. VOL. 9,. . . . 1579, (1971).
     *EXCHANGE INTERACTIONS IN PARTIALLY ORDERED FE-PD3 FERROMAGNETIC ALLOY.
     *MENZINGER(F.), SACCHETTI(F.), TEICHNER(R.).

871 FE-PD3,. . . . . .D1. . . . . . .SOL. STATE COMM. VOL. 9,. . . . 257, (1971).
     *SPIN WAVES AND EXCHANGE INTERACTIONS IN ORDERED FE-PD3 ALLOY.
     *ANTONINI(B.), MEDINA(R.), MENZINGER(F.).

872 FE*RH,. . . . . .D4. . . . . . .PHYS. LETT. VOL.37A,. . . . . . 333, (1971).
     *DIFFUSE SCATTERING OF NEUTRONS IN THE ANTIFERROMAGNETIC PHASE OF FE*RH.
     *KUNITOMI(N.), KOHGI(M.), NAKAI(Y.).

873 FERRITES,. . . . . .T,D4. . . . . .SOV. PHYS. SOL. STATE 5,. . . . 668, (1963).
     *EFFECT OF THE SPIN-PHONON INTERACTION ON ONE-QUANTUM SCATTERING
     OF SLOW NEUTRONS IN FERRITES.
     *KASHCHEEV(V.N.).

SECTION   6 -QUASI-ELASTIC AND INELASTIC SCATTERING STUDIES . . . .

874 FERRITES,. . . . . .T.D4..........SOV. PHYS. SOL. STATE. 5,. . . . . . 635, (1963).
&THE EFFECT OF SPIN-SPIN INTERACTION ON THE SCATTERING OF SLOW NEUTRON
IN FERRITES.
&KASHCHEEV(V.N.).

875 FERROELECTRICS,. . . .   . .B7.........PHYSICA VOL.44,. . . . . 69, (1969).
&INELASTIC SCATTERING CROSS SECTION OF NEUTRONS IN FERROELECTRICS.
&JAISWAL(V.K.), SHARMA(P.K.).

876 FERROELECTRICS,. . . .   . .T.B7......J. LOW TEMP. PHYS. VOL.9,. . . . . . 485, (1972).
&THEORY OF COHERENT NEUTRON SCATTERING BY H-BONDED FERROELECTRICS AT LOW
TEMP. II.SCATT. CHARACTERISTICS/CONCEPTIONS OF/TUNNELING QUASISPIN MODEL
&STAMENKOVIC(S.).

877 FERROELECTRICS,. . . .   . .T.B7......J. LOW TEMP. PHYS. VOL.9,. . . . . . 475, (1972).
&THEORY OF COHERENT NEUTRON SCATTERING BY HYDROGEN-BONDED FERROELECTRICS
AT LOW TEMPERATURES. I.GENERAL EXPRESSION FOR/SCATTERING/THERMAL FACTORS
&STAMENKOVIC(S.).

878 FERROMAGNETS,. .   . . . .D4.......UKR. FIZ. ZH. (USSR) VOL.13, . . .1682, (1968).
&TEMPERATURE DEPENDENCE OF SLOW NEUTRON INELASTIC SCATTERING BY IMPERFECT
FERROMAGNETS.
&DZYUB(I.P.).

879 FE-S,. . . . . .D1,4.....COLL. INTER. N.126 (GRENOBLE),. . 210, (1963).
&TEMPERATURE DEPENDENCE OF SPIN FLUCTUATION SCATTERING OF NEUTRONS ON
PYRRHOTITE.
&KRASNICKI(SZ.), WANIC(A.), DIMITRIJEVIC(Z.), MAGLIC(R.), MARKOVIC(V.),
TODOROVIC(J.).

880 FE-S,. . . . . .D1........COLL. INTER. N.126 (GRENOBLE),. . 203, (1963).
&MAGNON SCATTERING OF SLOW NEUTRONS ON A PYRRHOTITE SINGLE CRYSTAL.
&WANIC(A.).

881 FE-SI, ...A2,D4.....J. PHYS.F: METAL PHYS. VOL.2,. . 358, (1972).
&SPIN DENSITY DISTRIBUTION IN IRON-SILICON ALLOYS.
&MOSS(J.), BROWN(P.J.).

882 FE-(SI), ...D1.......IAEA SYMP. COPENHAGEN, VOL.2,. . . 3, (1968).
&NEUTRON SPIN SCATTERING BY SPIN WAVES IN METALS (REVIEW PAPER).
&MOLLER(H.B.).

883 FE-SI, ...D1.......PHYS. REV. VOL.178,. . . . . 833, (1969).
&SPIN WAVES DISPERSION RELATIONS IN FERROMAGNETIC FE-SI ALLOYS.
&ANTONINI(B.), MENZINGER(F.), PAOLETTI(A.), TUCCIARONE(A.).

884 FE-(X),. . . . . .D4.......J. APP. PHYS. VOL. 39,. . . . . . 1174, (1968).
&MAGNETIC NEUTRON SCATTERING FROM ATOMS DISSOLVED IN FERROMAGNETIC
IRON AND NICKEL.
&LOW(G.G.).

885 FRANKLINITE, ...D1........J. PHYS. SOC. JAPAN VOL.17 B-III 69, (1962).
&AN INVESTIGATION OF MAGNONS IN FRANKLINITE BY THE NEUTRON SCATTERING
METHOD.
&MURASIK(A.), RUTA-WALA(K.), WANIC(A.).

886 FRANKLINITE, ...D1.......PHYSICS VOL.27,. . . . . 883, (1961).
&AN EXAMINATION OF THE SPIN WAVES DISPERSION RELATION IN A FRANKLINITE
MONOCRYSTAL BY THE NEUTRON SCATTERING METHOD.
&MURASIK(A.), RUTA-WALA(K.), WANIC(A.).

887 GA,. . . . . .B4.......NUOVO CIMENTO B VOL.10B SER.11,. 117, (1972).
&MULTIPLE SCATTERING OF NEUTRONS IN LIQUID GALLIUM.
&ANTONINI(M.), CORCHIA(N.).

888 GA,. . . . . .B4.......J. PHYS. C VOL.6A,. . . . . 212, (1973).
&NEUTRON SCATTERING FROM SUPERCOOLED GALLIUM.
&PAGE(D.I.), SAUNDERSON(D.H.), WINDSOR(C.G.).

889 GA,. . . . . .B4.......PROPERTIES LIQUID METALS,. . . . 119, (1973).
&DYNAMICS OF LIQUID GALLIUM IN THE SUPERCOOLED STATE.
&CHEN(S.H.), LEFEVRE(Y.), YIP(S.). (ED. NOTE: REFER SECT.4).

890 GA,. . . . . .B4.......PROPERTIES LIQUID METALS,. . . . 111, (1973).
&INVESTIGATIONS OF COLLECTIVE EXCITATIONS IN LIQUID METALS BY INELASTIC
NEUTRON SCATTERING.
&GLASER(W.), HAGEN(S.), LOFFLER(U.), SUCK(J.B.), SCHOMMERS(W.). (ED.
NOTE: REFER SECT. 4).

891 GA,. . . . . .B4,C3.....PROPERTIES LIQUID METALS,. . . . 99, (1973).
&RELATION BETWEEN INCOHERENT AND COHERENT NEUTRON SCATTERING FROM LIQUID
METALS, WITH PARTICULAR REFERENCE TO MEASUREMENTS ON RB AND GA.
&BARKER(M.I.), JOHNSON(M.W.), MARCH(N.H.), PAGE(D.I.). (ED. NOTE: REFER
SECT. 4).

892 GA,. . . . . .B1......J. PHYS. C1 SER.2 VOL.2,. . . . . 903, (1969).
&LATTICE VIBRATIONS OF GALLIUM METAL.II. EXPERIMENTAL DETERMINATION OF
THE PHONON DISPERSION RELATION.
&WAEBER(W.B.).

893 GA,. . . . . .B7.......SOV. PHYS. SOL. STATE VOL.13,. . 1256, (1971).
&PHONON SPECTRUM OF GALLIUM IN POROUS GLASS.
&BOGOMOLOV(V.N.), KLUSHIN(N.A.), OKUNEVA(N.M.), PLACHENOVA(E.L.),
POGREBNOI(V.I.), CHUDNOVSKII(F.A.).

894 GA,. . . . . .T.......J. PHYS. C, SER.2, VOL.2,. . . . . 882, (1969).
&LATTICE VIBRATIONS OF GALLIUM METAL I. GROUP-THEORETICAL ANALYSIS.
&WAEBER(W.B.).

895 GA*AS, ...B1.......COPENHAGEN CONF.-LATT.DYNAMICS,. 19, (1963).
&NORMAL VIBRATIONS IN GALLIUM ARSENIDE.
&DOLLING(G.), WAUGH(J.L.T.).

896 GA*AS, ...T.......SOV. PHYS. SOL. STATE VOL.12,. . 497, (1970).
&PHONON DISPERSION IN GA*AS.
&KOROL(E.N.).

897 GA*AS, ...B1.......PHYS. REV. 132,. . . . . 2410, (1963).
&CRYSTAL DYNAMICS OF GALLIUM ARSENIDE.
&WAUGH(J.L.T.), DOLLING(G.).

898 GA*AS, ...B1.......PROC. PHYS. SOC. 88,. . . . . . . 463, (1966).
&THE THERMODYNAMIC AND OPTICAL PROPERTIES OF GERMANIUM, SILICON, DIAMOND
AND GALLIUM ARSENIDE.
&DOLLING(G.), COWLEY(R.A.).

SECTION   6 -QUASI-ELASTIC AND INELASTIC SCATTERING STUDIES . . . .

899 GA-BI, ...B6. . . . .J. PHYS. C. SER.2, VOL.3,. . . 1673, (1970).
                          ↦CRITICAL OPALESCENCE IN BINARY LIQUID METAL MIXTURES II: CONCENTRATION
                          DEPENDENCE.
                          ↦EGELSTAFF(P.A.), WIGNALL(G.D.).

900 GA-FE*O4,. . . . . .D1. . . . . . .PHYS REV 178,. . . . . 781, (1969).
                          ↦SPIN WAVES IN GA-FE-O4.
                          ↦MOGLESTUE(K.T.).

901 GA*P,. . . . . .B1. . . . . . .IAEA SYMP. COPENHAGEN, VOL.1,. . 301, (1968).
                          ↦LATTICE DYNAMICS OF GALLIUM PHOSPHIDE.
                          ↦YARNELL(J.L.), WARREN(J.L.), WENZEL(R.G.), DEAN(P.J.).

902 GARNETS, ...T. . . . . . .J. PHYS. CHEM. SOL. 28,. . . . . 2225, (1967).
                          ↦CATION DISTRIBUTIONS IN MULTISUBLATTICE IONIC CRYSTALS AND APPLICATIONS
                          TO SOLID SOLUTIONS OF FERROMAGNETIC GARNETS AND SPINELS.
                          ↦BORGHESE(C.).

903 GA*SE, ...B1. . . . . . .SOL. STATE COMM. VOL.13, ...1555, (1973).
                          ↦NEUTRON SCATTERING MEASUREMENTS OF THE INTERLAYER INTERACTION IN GA*SE.
                          ↦BREBNER(J.L.), JANDL(S.), POWELL(B.M.).

904 GD,. . . . . .T. . . . . . .J. OF PHYS.-C-, SER.2, VOL.4,. . 3215, (1971).
                          ↦THE SPIN WAVE CONTRIBUTION TO THE SPECIFIC HEAT OF GD.
                          ↦SEDAGHAT(A.K.), CRACKNELL(A.P.).

905 GD,. . . . . .D1. . . . . . .J. APPL. PHYS. 42,. . . . . . 1672, (1971).
                          ↦SPIN-WAVE DISPERSION RELATION IN RARE-EARTH METALS.
                          ↦NICKLOW(R.M.).

906 GD,. . . . . .D1. . . . . . .J. APPL. PHYS. 41,. . . . . . . 1182, (1970).
                          ↦SPIN-WAVE DISPERSION RELATION FOR GADOLINIUM.
                          ↦KOEHLER(W.C.),   CHILD(H.R.),   NICKLOW(R.M.),   SMITH(H.G.),   MOON(R.M.),
                          CABLE(J.W.).

907 GD,. . . . . .D1. . . . . . .PHYS. REV. LETT. VOL.24, ... 16, (1970).
                          ↦SPIN-WAVE DISPERSION RELATIONS IN GADOLINIUM.
                          ↦KOEHLER(W.C.),   CHILD(H.R.),   NICKLOW(R.M.),   SMITH(H.G.),   MOON(R.M.),
                          CABLE(J.W.).

908 GD,. . . . . .P. . . . . . .PHYS. REV. VOL.83,. . . . . . . 841, (1951).
                          ↦RESONANT SCATTERING IN SAMARIUM AND GADOLINIUM.
                          ↦BROCKHOUSE(B.N.), HURST(D.G.).

909 GD,. . . . . .D1. . . . . . .J. PHYSIQUE TOME 32 COL.1 VOL.I, 296, (1971).
                          ↦NEUTRON SCATTERING EXPERIMENTS ON GADOLINIUM.
                          ↦KOEHLER(W.C.), MOON(R.M.), CABLE(J.W.), CHILD(H.R.).

910 GD,. . . . . .T,D4. . . . .J. PHYS. -C-, VOL.1,. . . . . . 1279, (1968).
                          ↦SPIN-WAVE THEORY OF THE MAGNETOCRYSTALLINE ANISOTROPY IN GADOLINIUM
                          METAL.
                          ↦BROOKS(M.S.S.), GOODINGS(D.A.).

911 GD-LA, ...D4. . . . . . .REF. SECTION 3, 64/KOEHLER(2), ...

912 GD*N,. . . . . .A2,D3. . . .J. APPL. PHYS. 41,. . . . . . . 933, (1970).
                          ↦MAGNETIC PROPERTIES AND STRUCTURE OF GD*N AND GD*(N-O).
                          ↦GAMBINO(R.J.), MCGUIRE(T.R.), ALPERIN(H.A.), PICKART(S.J.).

913 GD*(N-O),. . . . . . . .A2,D3. . . .J. APPL. PHYS. 41,. . . . . . . 933, (1970).
                          ↦MAGNETIC PROPERTIES AND STRUCTURE OF GD*N AND GD*(N-O).
                          ↦GAMBINO(R.J.), MCGUIRE(T.R.), ALPERIN(H.A.), PICKART(S.J.).

914 GD-Y,. . . . . .D4. . . . . . .REF. SECTION 3, 64/KOEHLER(2), ...

915 GE,. . . . . .B1. . . . . . .PHYS. REV. B-3,. . . . . 364, (1971).
                          ↦PHONON DISPERSION RELATIONS IN GE AT 80 K.
                          ↦NILSSON(G.), NELIN(G.).

916 GE,. . . . . .B1. . . . . . .IAEA SYMP. BOMBAY, VOL.1,. . . . 249, (1964).
                          ↦INELASTIC NEUTRON SCATTERING FROM DOPED GERMANIUM AND SILICON.
                          ↦DOLLING(G.).

917 GE,. . . . . .T. . . . . . .IAEA SYMP. BOMBAY, VOL.1,. . . . 285, (1964).
                          ↦THE NATURE OF THE PHONON SPECTRUM AND THE ANALYSIS OF LATTICE THERMAL
                          CONDUCTIVITY.
                          ↦JOSHI(S.K.), SHARMA(K.C.).

918 GE,. . . . . .B1. . . . . . .PHYS. REV. B VOL.5,. . . . . 3151, (1972).
                          ↦PHONON DENSITY OF STATES IN GERMANIUM AT 80 K MEASURED BY NEUTRON
                          SPECTROMETRY.
                          ↦NELIN(G.), NILSSON(G.).

919 GE,. . . . . .T,P. . . . .PHYS. STAT. SOL. A VOL.4,. . . . 445, (1971).
                          ↦EXPERIMENTAL AND THEORETICAL INVESTIGATIONS OF DYNAMIC NEUTRON DIF-
                          FRACTION BY USING (GE) CRYSTALS WITH A LOW DISLOCATION DENSITY.
                          ↦EICHHORN(F.), KOSMOWSKI(M.), SCHOPF(H.G.), SCHULZE(G.E.R.).

920 GE,. . . . . .B7. . . . . . .BU. ACAD. SCI. USSR PHYS. SR. 33 1754, (1969).
                          ↦INELASTIC SCATTERING OF NEUTRONS ON GERMANIUM WITH EVEN NUMBER OF N IN
                          THE VICINITY OF FIRST EXCITED LEVEL.
                          ↦KONOBEEVSKII(E.S.),   MUSAELYAN(R.M.),   POPOV(V.I.),   SURKOVA(I.V.),
                          SHTRANIKH(I.V.).

921 GE,. . . . . .T,B2. . . . .IAEA SYMP. GRENOBLE, ... 29, (1972).
                          ↦ANHARMONIC INTERACTIONS IN GE AND SI.
                          ↦JEX(H.).

922 GE,. . . . . .T. . . . . . .SOL. STATE COMM. VOL.11 NO.6,. . 775, (1972).
                          ↦LATTICE DYNAMICS OF GERMANIUM BY THE C.G.W. TYPE OF ANGULAR FORCE MODEL.
                          ↦PANDEY(B.P.), DAYAL(B.).

923 GE,. . . . . .B1. . . . . . .SOV. PHYS. SOL. STATE VOL.4,. . 1747, (1962).
                          ↦CONCERNING THE SIMILARITY BETWEEN THE CHARACTERISTIC FREQUENCY
                          DISPERSION CURVES OF DIAMOND-TYPE CRYSTALS.
                          ↦KUCHER(T.I.).

924 GE,. . . . . .B1. . . . . . .PHYS. REV. B VOL.6,. . . . . . 3777, (1972).
                          ↦STUDY OF THE HOMOLOGY BETWEEN SILICON AND GERMANIUM BY THERMAL-NEUTRON
                          SPECTROMETRY.
                          ↦NILSSON(G.), NELIN(G.).

925 GE,. . . . . .D2. . . . . . .PROC. ROY. SOC. 281,. . . . . . 274, (1964).
                          ↦CRITICAL-POINT ANALYSIS OF THE PHONON SPECTRA OF DIAMOND, SILICON
                          AND GERMANIUM.
                          ↦JOHNSON(F.A.), LOUDON(R.).

SECTION   6 -QUASI-ELASTIC AND INELASTIC SCATTERING STUDIES . . . .

926 GE,. . . . . .P. . . . . . . .REF. SECTION 3, 47/WU, ...

927 GE,. . . . . .B1. . . . . . .SOL. STATE COMM. 1, ... 205, (1963).
     *TEMPERATURE EFFECTS ON LATTICE VIBRATIONS IN GERMANIUM.
     *BROCKHOUSE(B.N.), DASANNACHARYA(B.A.).

928 GE,. . . . . .B1. . . . . . .PROC. PHYS. SOC. 88, ... 463, (1966).
     *THE THERMODYNAMIC AND OPTICAL PROPERTIES OF GERMANIUM, SILICON, DIAMOND
     AND GALLIUM ARSENIDE.
     *DOLLING(G.), COWLEY(R.A.).

929 GE,. . . . . .B1. . . . . .IAEA SYMP. CHALK RIVER, VOL.2, . 23, (1962).
     *DETERMINATION OF POLARIZATION VECTORS FROM NEUTRON GROUP INTENSITIES.
     *BROCKHOUSE(B.N.), BECKA(L.N.), RAO(K.R.), WOODS(A.D.B.).

930 GE,. . . . . .PHYS. REV. VOL.108, ... 1091, (1957).
     *DETECTION OF OPTICAL LATTICE VIBRATIONS IN GE AND ZR*H BY SCATTERING OF
     COLD NEUTRONS.
     *PELAH(I.), EISENHAUER(C.M.), HUGHES(D.J.), PALEVSKY(H.).

931 GE,. . . . . .B1. . . . . . .PHYS. REV. VOL.108, ... 894, (1957).
     *NORMAL VIBRATIONS OF GERMANIUM BY NEUTRON SPECTROMETRY.
     *BROCKHOUSE(B.N.), IYENGAR(P.K.).

932 GE,. . . . . .B1. . . . . . .PHYS. REV. VOL.111, ... 747, (1958).
     *NORMAL MODES OF GERMANIUM BY NEUTRON SPECTROMETRY.
     *BROCKHOUSE(B.N.), IYENGAR(P.K.).

933 GE,. . . . . .T,B1. . . . .PHYS. REV. LETT. VOL.2, ... 495, (1959).
     *THEORY OF THE LATTICE VIBRATIONS OF GERMANIUM.
     *COCHRAN(W.).

934 GE,. . . . . .T,B1. . . . .PROC. ROY. SOC. A VOL.253, ... 260, (1959).
     *THEORY OF THE LATTICE VIBRATIONS OF GERMANIUM.
     *COCHRAN(W.).

935 GE,. . . . . .B1. . . . . .PHYS. REV. LETT. VOL.2, ... 256, (1959).
     *LATTICE VIBRATIONS IN SILICON AND GERMANIUM.
     *BROCKHOUSE(B.N.).

936 GE,. . . . . .SOV. PHYS. USPEKHI VOL.1, ... 165, (1958).
     *DIRECT OBSERVATION OF OPTICAL VIBRATIONS OF A CRYSTAL LATTICE OF A
     SOLID BY MEANS OF NEUTRON SCATTERING.
     *CHENTSOV(P.).

937 GE,. . . . . .B1. . . . . .PHYS. REV. VOL.113, ... 49, (1959).
     *LATTICE VIBRATIONS IN GERMANIUM BY SCATTERING OF COLD NEUTRONS.
     *GHOSE(A.), PALEVSKY(H.), HUGHES(D.J.), PELAH(I.), EISENHAUER(C.M.).

938 GE,. . . . . .B2. . . . . .PHYS. REV. B VOL.9 NO.12, ... (1974).
     *PHONON ANHARMONICITY OF GERMANIUM IN THE TEMPERATURE RANGE 80-880 K.
     *NELIN(G.), NILSSON(G.). (REF. OBTAINED FROM PHYS. REV. ABSTRACTS).

939 GE*BR4,. . . . . .B4. . . . . .MOL. PHYS. VOL.20, ... 881, (1971).
     *ORIENTATIONAL CORRELATIONS IN MOLECULAR LIQUIDS BY NEUTRON SCATTERING.
     CARBON TETRACHLORIDE AND GERMANIUM TETRABROMIDE.
     *EGELSTAFF(P.A.), PAGE(D.I.), POWLES(J.G.).

940 GE*O2,. . . . . .B1. . . . . .DISC. FARADAY SOC. NO.50, ... 62, (1970).
     *VIBRATIONAL PROPERTIES OF VITREOUS GERMANIA BY INELASTIC COLD NEUTRON
     SCATTERING.
     *LEADBETTER(A.J.), LITCHINSKY(D.).

941 GE(SI),. . . . .B3. . . . . .PHYS. REV. B-4, ... 2558, (1971).
     *NEUTRON SCATTERING INVESTIGATION OF IMPURITY PHONON MODES IN GE(SI).
     *WAKABAYASHI(N.), NICKLOW(R.M.), SMITH(H.G.).

942 GE-SI, ...T,B3. . . .PHYS. REV. VOL.156, ... 1017, (1967).
     *VIBRATIONAL PROPERTIES OF IMPERFECT CRYSTALS WITH LARGE DEFECT
     CONCENTRATIONS.
     *TAYLOR(D.W.).

943 GE-SI, ...T. . . . . . .PHYS. REV. B VOL.8, ... 6015, (1973).
     *IMPURITY PHONON MODES IN GE-SI.
     *WAKABAYASHI(N.).

944 GLASSES, ...CHEM. APPL. THERMAL NEUTRON SCAT 146, (1973).
     *STRUCTURE AND ATOMIC MOTION IN GLASSES.
     *LEADBETTER(A.J.). (IN BOOK: CHEMICAL APPLICATIONS OF THERMAL NEUTRON
     SCATTERING. ED. BY B.T.M. WILLIS: OXFORD UNIV. PRESS: LONDON).

945 GLYCERINE, . . .C3. . . . . .REV. ROUMAINE PHYS. VOL.18 NO.2, 135, (1973).
     *THE STUDY OF FREQUENCY SPECTRA IN SOME ORGANIC COMPOUNDS USING THE COLD
     NEUTRON TECHNIQUE.
     *RAPEANU(S.), PADUREANU(I.), TREPADUS(V.), DUMITRU(O.).

946 GLYCEROL,. . . .C3. . . . . .IAEA SYMP. CHALK RIVER, VOL.1, . 317, (1962).
     *A STUDY OF THE DIFFUSIVE ATOMIC MOTIONS IN GLYCEROL AND OF THE VIBRATORY
     MOTIONS IN GLYCEROL AND HEAVY WATER BY COLD NEUTRON SCATTERING.
     *LARSSON(K.E.), DAHLBORG(U.).

947 GLYCEROL,. . . .C1. . . . . .Z. PHYSIK VOL.238, ... 221, (1970).
     *MEASUREMENT OF THE QUASI-ELASTIC LINE-BROADENING IN GLYCEROL WITH A
     NEUTRON CRYSTAL SPECTROMETER OF EXTREMELY HIGH ENERGY RESOLUTION.
     *BIRR(M.). (IN GERMAN).

948 H2,. . . . . . .C3. . . . . .IAEA SYMP. CHALK RIVER, VOL.1, . 273, (1962).
     *NEUTRON INTERACTIONS IN LIQUID PARA- AND ORTHO-HYDROGEN.
     *WHITTEMORE(W.L.), DANNER(H.R.).

949 H2,. . . . . . .C3. . . . . .IAEA SYMP. CHALK RIVER, VOL.1, . 263, (1962).
     *DYNAMICS OF LIQUID HYDROGEN BY NEUTRON SCATTERING.
     *MCREYNOLDS(A.W.), WHITTEMORE(W.L.).

950 H2,. . . . . . .C3. . . . . .IAEA SYMP. CHALK RIVER, VOL.1, . 457, (1962).
     *THE INELASTIC SCATTERING OF COLD NEUTRONS BY METHANE, AMMONIA AND
     HYDROGEN.
     *WEBB(F.J.).

951 H2,. . . . . . .C3. . . . . .IAEA SYMP. CHALK RIVER, VOL.1, . 451, (1962).
     *THE SCATTERING OF SLOW NEUTRONS FROM HYDROGEN AND ETHYLENE.
     *BALLY(D.), TARINA(V.), TODIREANU(S.).

952 H2,. . . . . . .T. . . . . . .SOL. STATE COMM. VOL. 9, ... 1809, (1971).
     *DETERMINATION OF QUADRUPOLAR COUPLING CONSTANT BY NEUTRON SCATTERING IN
     SOLID H2.
     *NOOLANDI(J.).

SECTION   6 -QUASI-ELASTIC AND INELASTIC SCATTERING STUDIES . . . .

953 H2,. . . . . .T. . . . . . .SOL. STATE COMM. VOL.8,. . . . .373, (1970).
     NEUTRON SCATTERING BY SOLID ORTHO-HYDROGEN BELOW THE LAMBDA-TEMPERATURE.
     MISENTA(R.), OLIVI(L.).

954 H2,. . . . . .C1. . . . . .CONF. INTERN., RENNES,. . . . . .156, (1971).
     SLOW NEUTRON SCATTERING FROM H.C.P. AND F.C.C. SOLID HYDROGEN.
     STEIN(H.), STOCKMEYER(R.), STILLER(H.).

955 H2,. . . . . .T. . . . . . .PROC. PHYS. SOC. 90,. . . . . . .671, (1967).
     THEORY OF NEUTRON SCATTERING FROM LIQUID AND SOLID HYDROGEN.
     ELLIOTT(R.J.), HARTMANN(W.M.).

956 H2,. . . . . .P. . . . . .IAEA SYMPOSIUM (VIENNA),. . . . . 397, (1960).
     DIFFUSION DES NEUTRONS LENTS PAR L#HYDROGENE LIQUIDE.
     SARMA(G.).

957 H2 (FCC, HCP),. . . . . .J. CHEM. PHYS. VOL.57,. . .1726, (1972).
     PHONONS, LIBRONS, AND THE ROTATIONAL STATE J=1 IN HCP AND FCC SOLID
     HYDROGEN BY NEUTRON SPECTROSCOPY.
     STEIN(H.), STILLER(H.), STOCKMEYER(R.).

958 H,. . . . . .C1,T. . . . .Z. NATURFORSCH. A VOL.26A NO.3,. 575, (1971).
     QUASI-ELASTIC SCATTERING OF NEUTRONS FOR THE STUDY OF RANDOM MOTIONS IN
     SOLIDS.
     STILLER(H.), SPRINGER(T.).

959 H2,. . . . . .C1,C3. . . .Z. PHYS. VOL.231,. . . . . . . .243, (1970).
     INELASTIC SCATTERING OF SLOW NEUTRONS BY SOLID AND LIQUID HYDROGEN.
     SCHOTT(W.). (ED. NOTE: IN GERMAN).

960 H2 (PARAHYDROGEN),. . . . . . . .B1. . . . . .IAEA SYMP. GRENOBLE,. . . . . . .111, (1972).
     LATTICE DYNAMICS OF SOLID HYDROGEN.
     NIELSEN(M.), CARNEIRO(K.).

961 H2,. . . . . .B5. . . . . .IAEA SYMP. GRENOBLE,. . . . . . .445, (1972).
     TRANSLATIONAL LINE NARROWING IN PRESSURIZED HYDROGEN GAS.
     LEFEVRE(Y.), CHEN(S.H.), YIP(S.).

962 H2,. . . . . .C3. . . . . .PHYS. LETT. 25A. . . . . .435, (1967).
     STUDIES OF THE DYNAMICAL BEHAVIOUR OF ABSORBED MOLECULES BY SLOW NEUTRON
     SPECTROSCOPY.
     VERDAN(G.).

963 H2,. . . . . .P. . . . . .IAEA SYMPOSIUM (VIENNA),. . . . . 411, (1960).
     EXPERIENCE DE THERMALISATION DE NEUTRONS LENTS PAR L#HYDROGENE LIQUIDE.
     CRIBIER(D.), JACROT(B.), LACAZE(A.), ROUBEAU(P.).

964 H2,. . . . . .T. . . . . .PHYS. REV. VOL.135,. . . . . . . .A603, (1964).
     SLOW NEUTRON SCATTERING BY MOLECULAR HYDROGEN AND DEUTERIUM.
     YOUNG(J.A.), KOPPEL(J.U.).

965 H2,. . . . . .C1,3. . . . .PROC. PHYS. SOC. 90,. . . . . . .681, (1967).
     MOLECULAR MOTIONS IN LIQUID AND SOLID HYDROGEN AND DEUTERIUM.
     EGELSTAFF(P.A.), HAYWOOD(B.C.), WEBB(F.J.).

966 H2 (PARA),. . . . . .B1. . . . . .PHYS. REV. B VOL.7,. . . . . . .1626, (1973).
     PHONONS IN SOLID HYDROGEN AND DEUTERIUM STUDIED BY INELASTIC COHERENT
     NEUTRON SCATTERING.
     NIELSEN(M.).

967 H2,. . . . . .B4. . . . . .PHYS. REV. LETT. VOL.30,. . . . .481, (1973).
     COLLECTIVE EXCITATIONS IN LIQUID HYDROGEN OBSERVED BY COHERENT NEUTRON
     SCATTERING.
     CARNEIRO(K.), NIELSEN(M.), MCTAGUE(J.P.).

968 H2,. . . . . .T. . . . . .ACTA PHYS. POLON. VOL.30,. . . . .323, (1966).
     ON THE SCATTERING OF SLOW NEUTRONS BY ORTHO AND PARAHYDROGEN.
     ATANASOV(A.A.), IVANOV(K.I.).

969 H2,. . . . . .C1,C3. . . .PHYS LETT. VOL.12,. . . . . . . .188, (1964).
     MOLECULAR MOTIONS IN LIQUID AND SOLID ORTHO-HYDROGEN.
     EGELSTAFF(P.A.), HAYWOOD(B.C.), WEBB(F.J.), BASTON(A.H.).

970 H2,. . . . . .C3. . . . . .NUCL. SCI. ENGNG. VOL.18,. . . .182, (1964).
     INELASTIC NEUTRON SCATTERING IN LIQUID METHANE AND LIQUID PARAHYDROGEN.
     WHITTEMORE(W.L.).

971 H2,. . . . . .T. . . . . .ASEA RES. (SWEDEN) NO.7,. . . . .147, (1962).
     ON THE SCATTERING OF SLOW NEUTRONS BY HYDROGEN MOLECULES.
     BRIMBERG(S.).

972 H2,. . . . . .T,C3. . . .NUKLEONIK VOL.8,. . . . . 40, (1966).
     SLOW NEUTRON SPECTRA IN MOLECULAR HYDROGEN AT LOW TEMPERATURES.
     KOPPEL(J.U.), YOUNG(J.A.). (IN GERMAN).

973 H2,. . . . . .C3. . . . . .J. PHYS. RADIUM VOL.21,. . . . . 783, (1960).
     SCATTERING OF SLOW NEUTRONS BY LIQUID HYDROGEN.
     SARMA(G.). (IN FRENCH).

974 H2,. . . . . .C3. . . . . .NUKLEONIKA VOL.5,. . . . . . . .495, (1960).
     SCATTERING OF THERMAL NEUTRONS IN THE MOLECULES (OF GASES) CONTAINING
     HYDROGEN (MOLECULES).
     JANIK(J.A.), JANIK(J.), KRASNICKI(S.), MANIAWSKI(F.), MURASIK(A.),
     RZANY(H.), SZATULA(A.), SCIESINSKI(J.), WANIC(A.). (IN GERMAN).

975 H2,. . . . . .P. . . . . .PHYS. REV. VOL.73,. . . . . . . .733, (1948).
     SLOW NEUTRON VELOCITY SPECTROMETER STUDIES OF H, D, F, MG, S, SI, AND
     QUARTZ.
     RAINWATER(L.J.), HAVENS-JR(W.W.), DUNNING(J.R.), WU(C.S.).

976 H2 (PARA),. . . . . .T. . . . . .PHYS. REV. VOL.79,. . . . . . . .481, (1950).
     VARIATIONAL PRINCIPLES FOR SCATTERING PROCESSES. II. SCATTERING OF
     SLOW NEUTRONS BY PARA-HYDROGEN.
     LIPPMANN(B.A.).

977 H,. . . . . .C1. . . . . .NATURE VOL.168,. . . . . 290, (1951).
     INELASTIC SCATTERING OF COLD NEUTRONS.
     EGELSTAFF(P.A.).

978 H2,. . . . . .T. . . . . .PHYS. REV. VOL.52,. . . . . . . .286, (1937).
     THE SCATTERING OF NEUTRONS BY ORTHO- AND PARAHYDROGEN.
     SCHWINGER(J.), TELLER(E.).

979 H2,. . . . . .T,P. . . . .PHYS. REV. VOL.84,. . . . . . . .204, (1951).
     SCATTERING OF SLOW NEUTRONS BY H2 AND C#H4.
     MESSIAH(A.M.L.).

SECTION   6 -QUASI-ELASTIC AND INELASTIC SCATTERING STUDIES . . . .

980 H2,. . . . . *THE SCATTERING OF NEUTRONS BY PARA- AND ORTHO-HYDROGEN. . 115, (1957).
     *THE SCATTERING OF NEUTRONS BY PARA- AND ORTHO-HYDROGEN.
     *DROZDOV(S.I.).

981 H2,. . . . . *THE SCATTERING OF SLOW NEUTRONS BY ORTHO- AND PARA-HYDROGEN. 19, (1955).
     *THE SCATTERING OF SLOW NEUTRONS BY ORTHO- AND PARA-HYDROGEN.
     *SQUIRES(G.L.), STEWART(A.T.).

982 H2,. . . . . *ON THE SCATTERING OF SLOW NEUTRONS BY HYDROGEN MOLECULES. 451, (1965).
     *ON THE SCATTERING OF SLOW NEUTRONS BY HYDROGEN MOLECULES.
     *ATANASOV(A.A.), IVANOV(K.I.).

983 H2,. . . . . *KINETIC THEORY OF COLLISIONAL LINE NARROWING IN PRESSURIZED HYDROGEN GAS 3163, (1973).
     *KINETIC THEORY OF COLLISIONAL LINE NARROWING IN PRESSURIZED HYDROGEN GAS
     *CHEN(S.H.), LEFEVRE(Y.), YIP(S.).

984 H2,. . . . . *NEUTRON SCATTERING- A METHOD FOR INVESTIGATING THE LIQUID AND GASEOUS 305, (1973).
     STATES OF MATTER.
     *EDER(O.J.). (IN GERMAN).

985 H2,. . . . . *THE SCATTERING OF NEUTRONS BY PARA- AND ORTHO-HYDROGEN. . 115, (1957).
     *THE SCATTERING OF NEUTRONS BY PARA- AND ORTHO-HYDROGEN.
     *DROZDOV(S.I.).

986 H2,. . . . . *ON THE SCATTERING OF SLOW NEUTRONS BY HYDROGEN MOLECULES. . . , (1956).
     *ON THE SCATTERING OF SLOW NEUTRONS BY HYDROGEN MOLECULES.
     *BRIMBERG(S.).

987 (H)AR, ... *ANALYSIS OF NEUTRON SCATTERING EXPERIMENTS ON LIQUIDS. 545, (1968).
     *ANALYSIS OF NEUTRON SCATTERING EXPERIMENTS ON LIQUIDS.
     *AGRAWAL(A.K.), DESAI(R.C.), YIP(S.).

988 H2*(BA,CA,SR,YB),. . *VIBRATION SPECTRA OF THE ORTHORHOMBIC ALKALINE-EARTH HYDRIDES BY THE 3952, (1970).
     INELASTIC SCATTERING OF COLD NEUTRONS AND BY INFRARED TRANSM. MEASUR.
     *MAELAND(A.J.).

989 H-BONDS, ... *NEUTRON SCATTERING FROM HYDROGEN BONDS. 94, (1968).
     *NEUTRON SCATTERING FROM HYDROGEN BONDS.
     *STILLER(H.).

990 H-BONDS, ... *ON HYDROGEN BONDING IN INORGANIC CRYSTALS: SOME GENERALIZATIONS, SOME 466, (1968).
     RECENT RESULTS, AND SOME NEW TECHNIQUES (REVIEW).
     *HAMILTON(W.C.). (IN BOOK: STRUCTURAL CHEMISTRY AND MOLECULAR BIOLOGY;
     ED. BY A. RICH AND N. DAVIDSON.).

991 H-BONDS, ... *STRUCTRUE AND DYNAMICS IN MOLECULAR CRYSTALS-SOME COMPARISONS. 193, (1969).
     *STRUCTRUE AND DYNAMICS IN MOLECULAR CRYSTALS-SOME COMPARISONS.
     *HAMILTON(W.C.).

992 H*BR,. . . . . *STUDIES OF THE SOLID AND LIQUID PHASES OF H*F, H*CL AND H*BR BY SLOW 393, (1964).
     NEUTRON INELASTIC SCATTERING.
     *BOUTIN(H.), SAFFORD(G.J.).

993 H*BR,H*BR-D*BR,. . . *NUCLEAR MAGNETIC RESONANCE IN LIQUIDS AND SOLIDS (REVIEW PAPER). 379, (1968).
     *NUCLEAR MAGNETIC RESONANCE IN LIQUIDS AND SOLIDS (REVIEW PAPER).
     *POWLES(J.G.).

994 H-C BONDS (C6*H6),. . *RIDING MOTION AND HIGHER CUMULANTS IN CRYSTALLOGRAPHIC MODELS. 225, (1972).
     *RIDING MOTION AND HIGHER CUMULANTS IN CRYSTALLOGRAPHIC MODELS.
     *PAWLEY(G.S.).

995 H*CL,. . . . . *LOW-ENERGY NEUTRON SCATTERING FROM HYDROGEN CHLORIDE. 5193, (1971).
     *LOW-ENERGY NEUTRON SCATTERING FROM HYDROGEN CHLORIDE.
     *LURIE(N.A.), CARPENTER(J.M.).

996 H*CL,. . . . . *STUDIES OF THE SOLID AND LIQUID PHASES OF H*F, H*CL AND H*BR BY SLOW 393, (1964).
     NEUTRON INELASTIC SCATTERING.
     *BOUTIN(H.), SAFFORD(G.J.).

997 H*CL,. . . . . *NEUTRON SCATTERING BY LIQUID H*CL. 584, (1968).
     *NEUTRON SCATTERING BY LIQUID H*CL.
     *AGRAWAL(A.K.), YIP(S.), GORDON(R.G.).

998 H*CL,. . . . . *ROTATIONAL CORRELATION FUNCTIONS IN NEUTRON SCATTERING BY MOLECUL. GASES 263, (1968).
     *ROTATIONAL CORRELATION FUNCTIONS IN NEUTRON SCATTERING BY MOLECUL. GASES
     *AGRAWAL(A.K.), YIP(S.).

999 H*CL,. . . . . *LATTICE DYNAMICS OF HYDROGEN-BONDED CRYSTALS. 345, (1968).
     *LATTICE DYNAMICS OF HYDROGEN-BONDED CRYSTALS.
     *TREVINO(S.F.), PRASK(H.), WALL(T.), YIP(S.).

1000 H*CL2(-),H*F2(-),H*BR2(-),. *NEUTRON INELASTIC SCATTERING STUDIES ON THE HYDROGEN DIHALIDES. 1477, (1973).
     *NEUTRON INELASTIC SCATTERING STUDIES ON THE HYDROGEN DIHALIDES.
     *SMITH(J.A.S.), TEMME(F.P.), LUDMAN(C.J.), WADDINGTON(T.C.).

1001 H*CO*(C*O)4, . . *INCOHERENT INELASTIC NEUTRON SPECTRA OF HYDRIDOCARBONYLS. MID-INFRARED 970, (1970).
     VIBRATIONAL FREQUENCIES FOR H*CO*(C*O)4.
     *WHITE(J.W.), WRIGHT(C.J.).

1002 H*CO*O2,H*CR*O2, . . *ENERGY LOSS NEUTRON INELASTIC SCATTERING SPECTRA OF H*CR*O2 AND H*CO*O2. 817, (1973).
     *ENERGY LOSS NEUTRON INELASTIC SCATTERING SPECTRA OF H*CR*O2 AND H*CO*O2.
     *TEMME(F.P.), WADDINGTON(T.C.).

1003 H*CR*O2, ... *ENERGY-LOSS NEUTRON SPECTROSCOPY APPLIED TO STRONGLY HYDROGEN 345, (1972).
     BONDED MOLECULAR SYSTEMS.
     *GHOSH(R.E.), WADDINGTON(T.C.), TEMME(F.P.).

1004 H*CR*O2, ... *NEUTRON AND INFRARED SPECTRA OF H*CR*O2 AND D*CR*O2. 2496, (1966).
     *NEUTRON AND INFRARED SPECTRA OF H*CR*O2 AND D*CR*O2.
     *RUSH(J.J.), FERRARO(J.R.).

1005 H*F, ... *PROTON MOTIONS IN ACID AQUEOUS SOLUTIONS. 179, (1964).
     *PROTON MOTIONS IN ACID AQUEOUS SOLUTIONS.
     *STILLER(H.).

1006 H*F, ... *STUDIES OF THE SOLID AND LIQUID PHASES OF H*F, H*CL AND H*BR BY SLOW 393, (1964).
     NEUTRON INELASTIC SCATTERING.
     *BOUTIN(H.), SAFFORD(G.J.).

SECTION   6 -QUASI-ELASTIC AND INELASTIC SCATTERING STUDIES . . . .

1007 H*F, ...C1. . . . . . .J. CHEM. PHYS. VOL.39, H*F, ... 3135, (1963).
      ↦STUDY OF LOW-FREQUENCY MOLECULAR MOTIONS IN H*F, K*H*F2, K*H2*F3 AND
      NA*H2*F3.
      ↦BOUTIN(H.), SAFFORD(G.J.), BRAJOVIC(V.).

1008 H*F, ...C1,C3. . . . . .J. CHEM. PHYS. VOL.51, ... 762, (1969).
      ↦HYDROGEN MOTIONS IN LIQUID HYDROGEN FLUORIDE.
      ↦RING(J.W.), EGELSTAFF(P.A.).

1009 H*F, ...E,T. . . . . .DISC. FARADAY SOC. NO.48,. . . . . 69, (1969).
      ↦LATTICE DYNAMICS OF SOLID HYDROFLUORIC ACID.
      ↦AXMANN(A.), BIEM(W.), BORSCH(P.), HOSSFELD(F.), STILLER(H.).

1010 H*F-H2*O,. . . . . .C3. . . . . . .PHYSICA VOL.30, ... 931, (1964).
      ↦SPECTRUM OF LOW FREQUENCIES IN H2*O, D2*O AND SOLUTIONS OF H*F IN H2*O.
      ↦STILLER(H.).

1011 H*FE*CO3*(C*O)12,. . . . . .C1. . . . . . .J. CHEM. SOC. SECTION A,. . . . . 2843, (1971).
      ↦METAL ATOM VIBRATIONS IN TRANSITION METAL CLUSTERS AND COMPLEXES BY
      INELASTIC NEUTRON SCATTERING SPECTROSCOPY.
      ↦WHITE(J.W.), WRIGHT(C.J.).

1012 (H)-METALS,. . . . . .T. . . . . . .PHYSICA VOL.65, ... 109, (1973).
      ↦THEORY OF QUASIELASTIC NEUTRON SCATTERING BY HYDROGEN IN METALS
      CONSIDERING FINITE JUMP TIMES.
      ↦GISSLER(W.), STUMP(N.).

1013 H-METALS,. . . . . .T. . . . . . .PHYSICA VOL.50, ... 380, (1970).
      ↦THEORY OF THE QUASIELASTIC NEUTRON SCATTERING BY HYDROGEN IN BCC METALS
      APPLYING A RANDOM FLIGHT METHOD.
      ↦GISSLER(W.), ROTHER(H.).

1014 H3*MN*(C*O)12, ...C1. . . . . . .J. CHEM. SOC. D 1970 . . . . . . 971, (1970).
      ↦INCOHERENT INELASTIC NEUTRON SPECTRA OF HYDRIDOCARBONYLS. MID-INFRARED
      HYDROGEN VIBRATIONS OF H3*MN*(C*O)12.
      ↦WHITE(J.W.), WRIGHT(C.J.).

1015 H3*(MN3,RE3)*(C*O)12, ...C1. . . . . . .J. CHEM. SOC. SECTION A,. . . . . 2843, (1971).
      ↦METAL ATOM VIBRATIONS IN TRANSITION METAL CLUSTERS AND COMPLEXES BY
      INELASTIC NEUTRON SCATTERING SPECTROSCOPY.
      ↦WHITE(J.W.), WRIGHT(C.J.).

1016 H*N*O3.H2*O, ...C1,C3. . . . . .PHYSICA VOL.35, ... 457, (1967).
      ↦MOLECULAR DYNAMICS STUDY BY THE NEUTRON INELASTIC SCATTERING METHOD III:
      NITRIC ACID MONOHYDRATE.
      ↦JANIK(J.M.), JANIK(J.A.), BAJOREK(A.), PARLINSKI(K.),
      SUDNIK-HRYNKIEWICZ(M.).

1017 H*N*O3.H2*O, ...C1. . . . . . .ACTA PHYS. POLON. VOL.33,. . . . . 419, (1968).
      ↦NEUTRON INELASTIC SCATTERING DATA FOR CRYSTALLINE H3*O*N*O3, C*H3*I AND
      C*H2*I2 AND THEIR COMPARISON WITH INFRA-RED AND RAMAN SPECTROSCOPY.
      ↦JANIK(J.A.), BAJOREK(A.), JANIK(J.M.), NATKANIEC(I.), PARLINSKI(K.),
      SUDNIK-HRYNKIEWICZ(M.).

1018 (H)NB, ...B3. . . . . . .PHYS. REV. LETT. VOL.27, ... 1576, (1971).
      ↦ANISOTROPIC DIFFUSION OF HYDROGEN IN NIOBIUM SINGLE CRYSTALS.
      ↦KISTNER(G.), RUBIN(R.), SOSNOWSKA(I.).

1019 H2*O,. . . . . .C4. . . . . . .J. CHEM. PHYSICS VOL.50, ... 5279, (1969).
      ↦SLOW NEUTRON INELASTIC SCATTERING STUDY OF LIGHT WATER AND ICE.
      ↦HARLING(O.K.).

1020 H2*O,. . . . . .C2. . . . . . .J. CHEM. PHYSICS VOL.48, ... 3367, (1968).
      ↦FREQUENCY SPECTRUM OF HYDROGENOUS MOLECULAR SOLIDS BY INELASTIC NEUTRON
      SCATTERING. HEXAGONAL H2*O ICE.
      ↦PRASK(H.), BOUTIN(H.), YIP(S.).

1021 H2*O,. . . . . .C1. . . . . . .J. CHEM. PHYSICS VOL.45, ... 1312, (1966).
      ↦MOTIONS OF WATER MOLECULES IN POTASSIUM FERROCYANIDE TRIHYDRATE, WATER,
      AND ICE: A NEUTRON SCATTERING STUDY.
      ↦RUSH(J.J.), LEUNG(P.S.), TAYLOR(T.I.).

1022 H2*O,. . . . . .C3. . . . . . .IAEA SYMP. CHALK RIVER, VOL.1, . . 215, (1962).
      ↦FREQUENCY SPECTRUM OF LIQUIDS AND COLD NEUTRON SPECTROMETRY.
      ↦SINGWI(K.S.), SJOLANDER(A.), RAHMAN(A.).

1023 H2*O,. . . . . .C3. . . . . . .IAEA SYMP. BOMBAY, VOL.2, ... 201, (1964).
      ↦SLOW NEUTRON SCATTERING IN WATER AND SOME ORGANIC SUBSTANCES.
      ↦.C.C.C.P.

1024 H2*O,. . . . . .B4. . . . . . .IAEA SYMP. CHALK RIVER, VOL.1, . 189, (1962).
      ↦LIQUID DYNAMICS FROM NEUTRON SPECTROMETRY.
      ↦BROCKHOUSE(B.N.), BERGSMA(J.), DASANNACHARYA(B.A.), POPE(N.K.).

1025 H2*O,. . . . . .C4. . . . . . .IAEA SYMP. BOMBAY, VOL.2, ... 167, (1964).
      ↦RECENT MEASUREMENTS OF THE SCATTERING LAWS OF SOME HYDROGENOUS
      MODERATORS.
      ↦GLASER(W.), EHRET(G.), MERKEL(A.).

1026 H2*O,. . . . . .C4. . . . . . .IAEA SYMP. BOMBAY, VOL.2,. . . . . 141, (1964).
      ↦THE VAN HOVE SCATTERING FUNCTION FOR WATER.
      ↦POPE(N.K.), NATION(R.).

1027 H2*O,. . . . . .C1,C3. . . . . .IAEA SYMP. COPENHAGEN, VOL.1, . . 507, (1968).
      ↦THE DYNAMICS OF LIQUID H2*O AND D2*O AND SOLID H2*O FROM THE INELASTIC
      SCATTERING OF EPITHERMAL NEUTRONS.
      ↦HARLING(O.K.).

1028 H2*O,. . . . . .C3. . . . . . .IAEA SYMP. CHALK RIVER, VOL.1, . 389, (1962).
      ↦CALCULATION OF THE INELASTIC SCATTERING OF NEUTRONS FROM POLYETHYLENE
      AND WATER.
      ↦GOLDMAN(D.T.), FEDERIGHI(F.D.).

1029 H2*O,. . . . . .C3. . . . . . .IAEA SYMP. CHALK RIVER, VOL.1, . 383, (1962).
      ↦QUASI-ELASTIC SCATTERING OF COLD NEUTRONS IN WATER.
      ↦.C.C.C.P.

1030 H2*O,. . . . . .C3. . . . . . .IAEA SYMPOSIUM VIENNA, ... 511, (1960).
      ↦INELASTIC SCATTERING OF THERMAL NEUTRONS PRODUCED BY AN ELECTRON
      ACCELERATOR.
      ↦WHITTEMORE(W.L.), MCREYNOLDS(A.W.).

1031 H2*O,. . . . . .P. . . . . . .IAEA SYMPOSIUM (VIENNA), ... 631, (1960).
      ↦STUDY OF THE NEUTRON SPECTRA EMERGING FROM MODERATING ASSEMBLIES.
      ↦RAMANNA(R.), SARMA(N.).

SECTION   6 -QUASI-ELASTIC AND INELASTIC SCATTERING STUDIES . . . .

1032 H2*O,. . . . . . .   QUASI-ELASTIC AND INELASTIC SCATTERING OF COLD NEUTRONS FROM WATER.
      C3   IAEA SYMPOSIUM (VIENNA) 363, (1960).
      STILLER(H.H.), DANNER(H.R.).

1033 H2*O,. . . . . .   THE SCATTERING OF SLOW NEUTRONS BY LIGHT AND HEAVY WATER.
      C3   IAEA SYMPOSIUM (VIENNA) 351, (1960).
      MIKKE(K.).

1034 H2*O,. . . . . .   DIFFUSION QUASI-ELASTIQUE DES NEUTRONS FROIDS PAR L'EAU ET COEFFICIENT
      C3   IAEA SYMPOSIUM (VIENNA) 347, (1960).
      D'AUTODIFFUSION DU LIQUIDE.
      CRIBIER(D.), JACROT (B.).

1035 H2*O,. . . . . .   COLD-NEUTRON SCATTERING EXPERIMENTS ON LIGHT AND HEAVY WATER.
      C3   IAEA SYMPOSIUM (VIENNA) 329, (1960).
      LARSSON(K.E.), HOLMRYD(S.), OTNES(K.).

1036 H2*O,. . . . . .   STUDY OF THE DIFFUSION AND THERMALIZATION OF NEUTRONS IN WATER AND ICE
      T,P   IAEA SYMPOSIUM (VIENNA) 377, (1960).
      WITHIN A WIDE RANGE OF TEMPERATURES, USING THE IMPULSE METHOD.
      .C.C.C.P.

1037 H2*O,. . . . . .   THE DIFFUSION LENGTH OF THERMAL NEUTRONS IN WATER BETWEEN 18 AND 255
      P   PROC. ROY. SOC. 289 342, (1966).
      DEG. C.
      BESANT(C.B.), GRANT(P.J.).

1038 H2*O,. . . . . .   THE TEMPERATURE DEPENDENCE OF NEUTRON INELASTIC SCATTERING FROM WATER.
      B7   NUCL. SCI. ENG. VOL.46 223, (1971).
      ESCH(L.J.), YEATER(M.L.), MOORE(W.E.), SEEMANN(K.W.).

1039 H2*O,. . . . . .   MEASUREMENTS OF NEUTRON SPECTRA IN LIGHT WATER.
      ANNU./REACT. INST. KYOTO UNIV. 3 121, (1970).
      FUJITA(Y.), AIZAWA(O.), FUJINO(M.).

1040 H2*O,. . . . .   A NEUTRON INELASTIC SCATTERING INVESTIGATION OF THE H2*O MOLECULES IN
      B4   J. PHYS. CHEM. VOL.74, 3710, (1970).
      AQUEOUS SOLNS. AND SOLID GLASSES OF LATHANUM NITRATE AND CHROMIC CHLOR.
      LEUNG(P.S.), SANBORN(S.M.), SAFFORD(G.J.).

1041 H2*O,. . . . .   A NEUTRON INELASTIC SCATTERING INVESTIGATION OF THE CONCENTRATION AND
      B4   J. PHYS. CHEM. VOL.74, 3696, (1970).
      ANION DEPENDENCE OF LOW FREQ. MOTIONS OF H2*O MOLECULES IN IONIC SOLNS.
      LEUNG(P.S.), SANBORN(S.M.), SAFFORD(G.J.).

1042 H2*O,. . . . .   MEASUREMENT OF NEUTRON SLOWING DOWN TIME IN LIGHT WATER, ICE, PARAFFIN
      P   J. NUCL. SCI. TECHNOL. VOL.6, 671, (1969).
      AND SANTOWAX.
      SAKAMOTO(S.), KANEKO(Y.), AKINO(F.).

1043 H2*O,. . . . .   PHONON SPECTRUM AND THERMAL NEUTRON SCATTERING IN LIGHT WATER ICE.
      C1   J. NUCL. SCI. TECHNOL. VOL.5, 635, (1968).
      NAKAHARA(Y.).

1044 H2*O,. . . . .   NEUTRON SCATTERING STUDIES OF MOLECULAR VIBRATIONS IN ICE.
      B7   MOL. DYN. AND STRUCT. OF SOLIDS, 335, (1969).
      PRASK(H.), BOUTIN(H.), YIP(S.).

1045 H2*O,. . . . .   SLOW NEUTRON SCATTERING FROM WATER.
      C1   NUCL. SCI. ENGNG. VOL.33, 187, (1968).
      BRUGGER(R.M.).

1046 H2*O,. . . . .   STUDIES OF VIBRATION SPECTRA OF BONDED HYDROGEN ATOMS USING A PULSED
      C3   J. CHEM. PHYS. VOL.55, 2807, (1971).
      NEUTRON SOURCE.
      DAY(D.H.), SINCLAIR(R.N.).

1047 H2*O,. . . . .   P   REF. SECTION 3, 60/WOODS,. . . . .

1048 H2*O,. . . . .   A SIMPLE INTERPRETATION OF THE QUASIELASTIC SCATTERING OF SLOW NEUTRONS
      T   PHYS. LETT. 21 248, (1966).
      IN WATER.
      PELAH(I.), IMRY(Y.).

1049 H2*O,. . . . .   THE THERMAL NEUTRON SCATTERING LAW FOR LIGHT AND HEAVY WATER.
      T,C3   IAEA SYMPOSIUM (VIENNA), 309, (1960).
      EGELSTAFF(P.A.), COCKING(S.J.), ROYSTON(R.), THORSON(I.M.).

1050 H2*O,. . . . .   LATTICE DYNAMICS OF HYDROGEN-BONDED CRYSTALS.
      B1   IAEA SYMP. COPENHAGEN, VOL.1, 345, (1968).
      TREVINO(S.F.), PRASK(H.), WALL(T.), YIP(S.).

1051 H2*O,. . . . .   THE SCATTERING LAW FOR ROOM TEMPERATURE LIGHT WATER.
      C3   IAEA SYMP. CHALK RIVER, VOL.1, 359, (1962).
      KOTTWITZ(D.A.), LEONARD-JR(B.R.).

1052 H2*O,. . . . .   QUASI-ELASTIC SCATTERING BY ROOM TEMPERATURE LIGHT WATER.
      C3   IAEA SYMP. CHALK RIVER, VOL.1, 373, (1962).
      KOTTWITZ(D.A.), LEONARD-JR(B.R.), SMITH(R.B.).

1053 H2*O,. . . . .   SMALL-ENERGY ROTATIONAL TRANSITION IN SLOW-NEUTRON SCATTERING BY WATER.
      C3   PHYS. REV. 131, 2547, (1963).
      YIP(S.), OSBORN(R.K.).

1054 H2*O,. . . . .   MEASUREMENTS OF NEUTRON SPECTRA IN WATER, POLYETHYLENE AND ZIRCONIUM
      P   IAEA SYMPOSIUM (VIENNA), 613, (1960).
      HYDRIDE.
      WALTON(R.B.), BEYSTER(J.R.), WOOD(J.L.), LOPEZ(W.M.).

1055 H2*O,. . . . .   INELASTIC SCATTERING OF NEUTRONS FROM BENZENE AND WATER.
      C3   CONF. NEUTRON THERMALIZ. VOL.I, 477, (1968).
      ROSS(D.K.), SZABO(F.P.), SANALAN(Y.), (ED. NOTE: PROC. CONF.
      NEUTRON THERMALIZATION AND REACTOR SPECTRA; PUBL. IAEA:VIENNA).

1056 H2*O,. . . . .   NEUTRON SCATTERING BY WATER TAKING INTO ACCOUNT THE ANISOTROPY OF THE
      T   NUCL. SCI. ENGNG. VOL.19, 412, (1964).
      MOLECULAR VIBRATIONS.
      KOPPEL(J.U.), YOUNG(J.A.).

1057 H2*O,. . . . .   INELASTIC SCATTERING CROSS SECTIONS FOR POLYETHYLENE AND WATER IN THE
      C3   NUCL. SCI. ENGNG. VOL.25, 300, (1966).
      GAS TRANSITION REGION.
      KIROUAC(G.J.), MOORE(W.E.), SEEMANN(K.W.).

SECTION 6 -QUASI-ELASTIC AND INELASTIC SCATTERING STUDIES . . . .

1058 H2*O,. . . . . .SLOW-NEUTRON SCATTERING BY WATER.....C3.......NUCL. SCI. ENGNG. VOL.25,. . . . . 248, (1966).
      .MCMURRY(H.L.), RUSSELL(G.J.), BRUGGER(R.M.).

1059 H2*O,. . . . . .T........NUCL. SCI. ENGNG. VOL.30,. . . . . 199, (1967).
      .MULTIPLE SCATTERING IN SLOW-NEUTRON DOUBLE-DIFFERENTIAL MEASUREMENTS.
      .SLAGGIE(E.L.).

1060 H2*O,. . . . . .C3........PHYSICA VOL.36,. . . . . . . 35, (1967).
      .DYNAMICAL AND THERMODYNAMICAL STRUCTURES OF WATER FROM THE SCATTERING
      OF COLD NEUTRONS.
      .SZKATULA(A.), FULINSKI(A.).

1061 H2*O,. . . . . .C3........PHYS. LETT. VOL.22,. . . . . . 15, (1966).
      .SLOW NEUTRON WIDTH OF 200 MEV VIBRATION LEVEL IN LIQUID H2*O.
      .HARLING(O.K.).

1062 H2*O,. . . . . .P........UKAEA AERE REP. 4484 (16 PP.) . . (1964).
      .A COMPILATION OF THE SCATTERING LAW FOR WATER AT 22 DEGREES C AND 150
      DEGREES C.
      .HAYWOOD(B.C.).

1063 H2*O,. . . . . .C3.......WATER-A COMPREHENSIVE TREATISE,. . . , (1973).
      .THE SCATTERING OF NEUTRONS BY LIQUID WATER.
      .PAGE(D.I.). (VOL.1-THE PHYSICS AND PHYSICAL CHEMISTRY OF WATER; PUBL. BY
      PLENUM CORP.:NEW YORK. ED. BY F. FRANKS (596 PP.)).

1064 H2*O,. . . . . .C3.......COMMENTAT. PHYS.-MATH.(FIN.) 42, 277, (1972).
      .ON THE QUASI-ELASTIC SCATTERING OF SLOW NEUTRONS FROM WATER NEAR ITS
      DENSITY MAXIMUM.
      .POVRY(H.O.).

1065 H2*O,. . . . . .C1,C3....DISS. ABS. VOL.27,. . . . . . . 2668, (1967).
      .EPITHERMAL NEUTRON INELASTIC SCATTERING BY ROOM TEMPERATURE WATER AND
      POLYETHYLENE.
      .KIROUAC(G.J.).

1066 H2*O,. . . . . .T........DISS. ABS. VOL.25,. . . . . . 6014, (1965).
      .A SIMPLIFIED CELL THEORY APPLIED TO THE CALCULATION OF THERMAL NEUTRON
      SPECTRA IN LIGHT WATER LATTICES.
      .MACVEAN(C.R.).

1067 H2*O,. . . . . .C3.......J. PHYS. RADIUM VOL.21,. . . . . 69, (1960).
      .QUASI-ELASTIC SCATTERING OF COLD NEUTRONS BY WATER AND COEFFICIENT
      OF SELF DIFFUSION OF THE LIQUID.
      .CRIBIER(D.), JACROT(B.). (IN FRENCH).

1068 H2*O,. . . . . .C1.......PHYSICA VOL.27,. . . . . 373, (1961).
      .LOW QUANTUM TRANSITIONS IN WATER AND ICE.
      .DANNER(H.R.), STILLER(H.).

1069 H2*O,. . . . . .Z. NATURFORSCH. VOL.16A. . . . . 112, (1961).
      .MEASUREMENT OF THE DIFFERENTIAL SCATTERING CROSS-SECTION AND THE MEAN
      ENERGY VARIATION IN THE SCATTERING OF SLOW NEUTRONS BY WATER AND ICE.
      .REINSCH(C.), SPRINGER(T.). (IN GERMAN).

1070 H2*O,. . . . . .C3........J. PHYS. SOC. JAPAN VOL.17 B-II, 370, (1962).
      .NEUTRON INELASTIC SCATTERING STUDY OF WATER.
      .SAKAMOTO(M.), BROCKHOUSE(B.N.), JOHNSON(R.G.), POPE(N.K.).

1071 H2*O,. . . . . .T,P......J. NUCL. ENERGY VOL.4,. . . . 109, (1957).
      .THE SCATTERING OF SLOW NEUTRONS BY WATER MOLECULES.
      .GORYUNOV(A.F.).

1072 H2*O,. . . . . .T,E......HELV. PHYS. ACTA VOL.20,. . . . 105, (1947).
      .ETUDE DES FORCES INTERMOLECULAIRES PAR DIFFUSION DES NEUTRONS LENTS.
      APPLICATION A N2, H2*O ET K*H2*P*O4.
      .ROSSEL(J.).

1073 H2*O,. . . . . .T,E......NUKLEONIK VOL.3,. . . . . 110, (1961).
      .SCATTERING OF SLOW NEUTRONS IN WATER, ICE AND WATER VAPOUR.
      .SPRINGER(T.). (IN GERMAN).

1074 H2*O,. . . . . .T,E......NUKLEONIK VOL.3,. . . . . 110, (1961).
      .DIE STREUUNG VON LANGSAMEN NEUTRONEN AN WASSER, EIS UND WASSERDAMPF.
      .SPRINGER(T.).

1075 H2*O,. . . . . .T........NUKLEONIK VOL.12,. . . . . 237, (1969).
      .REMARKS ON THE CALCULATION OF SLOW-NEUTRON CROSS SECTION AND THERMAL
      SPECTRA IN WATER.
      .KOSALY(G.), VALKO(J.).

1076 H2*O,. . . . . .T........J. CHEM. PHYS. VOL.55,. . . . . 3336, (1971).
      .MOLECULAR DYNAMICS STUDY OF LIQUID WATER.
      .RAHMAN(A.), STILLINGER(F.H.).

1077 H2*O,. . . . . .P........CONF. NEUTRON THERMALIZ. VOL.I,. 361, (1968).
      .THE SCATTERING LAW FOR HEAVY WATER AT 540 DEGREES K AND LIGHT WATER
      AT 500 DEGREES K.
      .HAYWOOD(B.C.), PAGE(D.I.). (IN BOOK: NEUTRON THERMALIZATION AND REACTOR
      SPECTRA. PUBL. IAEA: VIENNA, 1968).

1078 H2*O,. . . . . .E........ADV. COLLOID INTERFACE SCI. 2, 1, (1968).
      .APPLICATIONS OF SLOW NEUTRON SCATTERING TO STUDIES IN COLLOID AND
      SURFACE CHEMISTRY (REVIEW).
      .BOUTIN(H.), PRASK(H.), IYENGAR(R.D.).

1079 H2*O, H2*O-D2*O, ...B7,P......MOL. PHYS. VOL.24 NO.5,. . . . 1025, (1972).
      .COHERENT NEUTRON SCATTERING BY LIGHT WATER(H2*O) AND A LIGHT-HEAVY
      MIXTURE (64 PCT. H2*O/36 PCT. D2*O).
      .POWLES(J.G.), DORE(J.C.), PAGE(D.I.).

1080 H2*O (SOLID),. . . . . . . .C1........J. CHEM. PHYS. VOL.56,. . . . . 3217, (1972).
      .ICE I-LATTICE DYNAMICS AND INCOHERENT NEUTRON SCATTERING.
      .PRASK(H.), TREVINO(S.F.), GAULT(J.D.), LOGAN(K.W.).

1081 H2*O,. . . . . .34........NATURE (PHYS. SCI.) VOL.230,. . 192, (1971).
      .WATER DYNAMICS IN CLAYS BY NEUTRON SPECTROSCOPY.
      .HUNTER(R.J.), STIRLING(G.C.), WHITE(J.W.).

1082 H2*O,. . . . . .C1,3.....PHYS. REV. 170,. . . . . . . 808, (1968).
      .PROTON DYNAMICS IN WATER AND ICE STUDIED BY INELASTIC SCATTERING OF
      SLOW NEUTRONS.
      .BURGMAN(J.O.), SCIESINSKI(J.), SKOLD(K.).

SECTION   6 -QUASI-ELASTIC AND INELASTIC SCATTERING STUDIES . . . .

1083   H2*O IN CRYSTALS, ... C1 . . . . J. CHEM. PHYSICS VOL.45, . 3284, (1966).
       *LOW-FREQUENCY MOTIONS OF H2*O MOLECULES IN CRYSTALS. III.
       *PRASK(H.), BOUTIN(H.).

1084   H2*O, ... C3 . . . . PHYS. REV. VOL.119, ... 872, (1960).
       *ATOMIC MOTIONS IN WATER BY SCATTERING OF COLD NEUTRONS.
       *HUGHES(D.J.), PALEVSKY(H.), KLEY(W.), TUNKELO(E.).

1085   H2*O, ... T C3 . . . . PHYS. REV. VOL.119, ... 863, (1960).
       *DIFFUSIVE MOTIONS IN WATER AND COLD NEUTRON SCATTERING.
       *SINGWI(K.S.), SJOLANDER(A.).

1086   H2*O, ... T . . . . . PHYS. REV. VOL.119, ... 741, (1960).
       *SCATTERING OF SLOW NEUTRONS BY WATER.
       *NELKIN(M.).

1087   H2*O, ... C3 . . . . BER. BUNSENGES. PHYS. CHEM. 75, . 379, (1971).
       *THE DYNAMICS AND STRUCTURE OF WATER AND IONIC SOLUTIONS BY NEUTRON
       INELASTIC SCATTERING SPECTROSCOPY.
       *WHITE(J.W.).

1088   H2*O (HYDRATION COMPLEXES), ... C3 . . . . BER. BUNSENGES. PHYS. CHEM. 75, . 366, (1971).
       *NEUTRON INELASTIC SCATTERING INVESTIGATIONS OF HYDRATION COMPLEXES AND
       ASSOCIATED DIFFUSIVE KINETICS IN IONIC SOLUTIONS.
       *SAFFORD(G.J.), LEUNG(P.S.).

1089   H2*O, ... C3 . . . . NUCL. SCI. ENGNG. VOL.5, ... 99, (1960).
       *INELASTIC SCATTERING OF COLD NEUTRONS FROM SEVERAL HYDROGENOUS LIQUIDS.
       *BRUGGER(R.M.), MCCLELLAN(L.W.), STREETMAN(G.B.), EVANS(J.E.).

1090   H2*O, ... C3 . . . . PHYS. REV. LETT. VOL.3, ... 91, (1959).
       *ATOMIC MOTIONS IN WATER BY SCATTERING OF COLD NEUTRONS.
       *HUGHES(D.J.), PALEVSKY(H.), KLEY(W.), TUNKELO(E.).

1091   H2*O, ... B4 . . . . NUOVO CIMENTO SUPPL. VOL.9, . . . 45, (1958).
       *STRUCTURAL DYNAMICS OF WATER BY NEUTRON SPECTROMETRY.
       *BROCKHOUSE(B.N.).

1092   H2*O IN CRYSTALS, ... C1 . . . . J. CHEM. PHYSICS VOL.45, . II. 699, (1966).
       *LOW-FREQUENCY MOTIONS OF H2*O MOLECULES IN CRYSTALS. II.
       *PRASK(H.J.), BOUTIN(H.).

1093   H2*O, ... P . . . . . SOV. AT. ENERGY VOL.32, ... 33, (1972).
       *NEUTRON THERMALISATION IN H2*O AT 318 AND 77K.
       *ISHMAEV(I.P.), SADIKOV(I.P.), CHERNYSHOV(A.A.).

1094   H2*O IN CRYSTALS, ... C1 . . . . J. CHEM. PHYSICS VOL.40, . . . 2670, (1964).
       *LOW-FREQUENCY MOTIONS OF H2*O MOLECULES IN CRYSTALS.
       *BOUTIN(H.), SAFFORD(G.J.), DANNER(H.R.).

1095   H2*O-ALCOHOL, . . C3 . . . . PROC. ROY. SOC. A319, NO.1537, . 189, (1970).
       *MOTIONS OF WATER MOLECULES IN DILUTE AQUEOUS SOLUTIONS OF TERTIARY
       BUTYL ALCOHOL: A NEUTRON SCATTERING STUDY OF HYDROPHOBIC HYDRATION.
       *FRANKS(F.), RAVENHILL(J.), EGELSTAFF(P.A.), PAGE(D.I.).

1096   H2*O-ALUMINA,SILICA, ... C3 . . . . SURFACE SCI. VOL.2, ... 261, (1964).
       *STUDY OF WATER VAPOR ADSORBED ON GAMMA-ALUMINA AND SILICA BY SLOW
       NEUTRON INELASTIC SCATTERING.
       *BOUTIN(H.), PRASK(H.).

1097   H2*O-BERYL, ... C1 . . . . J. CHEM. PHYSICS VOL.42, ... 1469, (1965).
       *LOW-FREQUENCY MOTIONS OF H2*O MOLECULES IN BERYL FROM NEUTRON INELASTIC
       SCATTERING DATA.
       *BOUTIN(H.), PRASK(H.), SAFFORD(G.J.).

1098   H2*O-C, ... P SOV. AT. ENERGY(USA) VOL.31, . 459, (1971).
       *ANALYSIS OF EXPERIMENTS ON NEUTRON THERMALIZATION IN GRAPHITE-WATER
       SYSTEMS.
       *MOSTOVOI(V.I.), TRUKHANOV(G.YA.), SAFIN(YU.A.), MOSKOVKIN(V.N.).

1099   H3*O*CL*O4, ... B4 . . . . IAEA SYMP. BOMBAY, VOL.2, . . . 243, (1964).
       *MOLECULAR DYNAMICS INVESTIGATED BY NEUTRON SCATTERING.
       *JANIK(J.A.).

1100   H3*O*CL*O4, ... P . . . . . PHYS. STAT. SOL. 9, ... 905, (1965).
       *DETERMINATION OF PARAMETERS OF THE ROTATIONAL DYNAMICS OF THE GROUPS
       N*H4 IN N*H4*CL*O4 AND H3*O IN H3*O*CL*O4 BY INELASTIC SCATTERING OF...
       *JANIK(J.M.), JANIK(J.A.), BAJOREK(A.), PARLINSKI(K.).

1101   H3*O.CL*O4, ... P . . . . . ACTA PHYS. POLON. VOL.27, . 491, (1965).
       *DETERMINATION/BARRIER TO ROTATION HEIGHT OF THE H3*O GROUP IN CRYSTAL:
       PERCHLORIC ACID MONOHYDRATE BY TOTAL NEUTRON SCATT. C.S. MEASUREMENT.
       *JANIK(J.).

1102   H3*O*CL*O4, ... C1,C3 . . . . REF. SECTION 3, 64/JANIK, ...

1103   H3*O.CL*O4, ... C1 . . . . PHYSICA VOL.72, ... 168, (1974).
       *PROTON JUMPS IN CRYSTALLINE HYDRONIUM PERCHLORATE.
       *JANIK(J.M.), RACHWALSKA(M.), JANIK(J.A.).

1104   H2*O-CRYSTALLINE HYDRATES, ... C1 . . . . IAEA SYMP. COPENHAGEN VOL.II, . 143, (1968).
       *INVESTIGATION OF THE DYNAMICS OF WATER MOLECULES IN CRYSTALLO-HYDRATES
       BY NEUTRON INELASTIC SCATTERING.
       *BAJOREK(A.), JANIK(J.A.), JANIK(J.M.), NATKANIEC(I.), PARLINSKI(K.),
       POKOTILOVSKY(YU.N.). . . . ET AL. (ED. NOTE: REF. SECTION 3, 68/BAJOREK).

1105   H2*O,D2*O, ... C3 . . . . PHYSICA VOL.30, ... 931, (1964).
       *SPECTRUM OF LOW FREQUENCIES IN H2*O, D2*O AND SOLUTIONS OF H*F IN H2*O.
       *STILLER(H.).

1106   H2*O,D2*O, ... C3 . . . . REV. ROUMAINE PHYS. VOL.9, . . . 737, (1964).
       *COLD NEUTRON SCATTERING BY LIGHT AND HEAVY WATER.
       *TEUTSCH(H.), MATEESCU(N.), NAHORNIAK(V.), DIACONESCU(A.), TIMIS(P.).

1107   H2*O,D2*O, ... C3 . . . . NUKLEONIK VOL.10 NO.3, ... 129, (1967).
       *SCATTERING OF SLOW NEUTRONS BY WATER.
       *PUCHER(M.). (IN GERMAN).

1108   H2*O,D2*O, ... P . . . . . BNWL-436, ... (1967).
       *COMPILATION OF DOUBLY DIFFERENTIAL CROSS SECTIONS AND THE SCATTERING
       LAW FOR H2*O AND D2*O AT 299 DEGREES K AND FOR H2*O AT 268 DEGREES K.
       *HARLING(O.K.). (BATTELLE NORTHWEST REPORT; AVAIL.: BATTELLE LABORATORY,
       RICHLAND, WASHINGTON STATE, U.S.A.).

SECTION 6 -QUASI-ELASTIC AND INELASTIC SCATTERING STUDIES . . . .

1109 H2*O,D2*O, ... C3 . . . . J. NUCL. ENERGY VOL.16, ... 81, (1962).
*SOME VIBRATIONAL PROPERTIES OF SOLID AND LIQUID H2*O AND D2*O DERIVED
FROM DIFFERENTIAL CROSS-SECTION MEASUREMENTS.
*LARSSON(K.E.), DAHLBORG(U.).

1110 H2*O,D2*O, ... MOL. SIEVES, PAP. CONF. 1967, . . 306, (1967).
*MOLECULAR MOTION IN MOLECULAR SIEVES BY NEUTRON SCATTERING SPECTROSCOPY.
*EGELSTAFF(P.A.), DOWNES(J.S.), WHITE(J.W.).

1111 H2*O,D2*O, ... PHYS. REV. LETT. VOL.2, ... 287, (1959).
*DIFFUSIVE MOTIONS IN LIQUIDS AND NEUTRON SCATTERING.
*BROCKHOUSE(B.N.).

1112 H2*O,D2*O, ... B4,C3 . . . . ACTA CRYST. VOL.10, ... 827, (1957).
*STRUCTURAL DYNAMICS OF WATER BY NEUTRON SPECTROMETRY.
*BROCKHOUSE(B.N.).

1113 H2*O-(H), ... C3 . . . . . . IAEA SYMP. CHALK RIVER, VOL.1, . 343, (1962).
*THE MOTION OF HYDROGEN IN WATER.
*EGELSTAFF(P.A.), HAYWOOD(B.C.), THORSON(I.M.).

1114 H2*O,H2*S, ... T . . . . ACTA PHYS. POLON. VOL.18, ... 255, (1959).
*SCATTERING OF SLOW NEUTRONS BY MOLECULES OF LIQUID WATER, AMMONIA AND
HYDROGEN SULFIDE.
*WANIC(A.).

1115 H2*O-SOLUTIONS, ... C3 . . . . J. CHEM. PHYS. VOL.50, ... 4444, (1969).
*INVESTIGATION OF LOW-FREQUENCY MOTIONS OF H2*O MOLECULES IN IONIC
SOLUTIONS BY NEUTRON INELASTIC SCATTERING.
*SAFFORD(G.J.), LEUNG(P.S.), NAUMANN(A.W.), SCHAFFER(P.C.).

1116 H2*S, ... C4 . . . . J. CHEM. PHYSICS VOL.42, ... 1568, (1965).
*SCATTERING OF SLOW NEUTRONS FROM AMMONIA AND HYDROGEN SULFIDE.
*STRONG(K.A.), HARKER(Y.D.), BRUGGER(R.M.).

1117 H2*S, ... C3 . . . . PHYSICA VOL.29, ... 488, (1963).
*SCATTERING OF SLOW NEUTRONS BY LIQUID H2*S.
*RZANY(H.), SCIESINSKI(J.).

1118 H2*S, ... C3 . . . . IAEA SYMP. CHALK RIVER, VOL.1, . 405, (1962).
*SYSTEMATIC STUDY ON TOTAL NEUTRON SCATTERING CROSS-SECTIONS OF HYDROGEN
CONTAINING MOLECULES.
*JANIK(J.A.), JANIK(J.), MANIAWSKI(F.), RZANY(H.), ROGALSKA(Z.),
SCIESINSKI(J.), SAGAN(U.), TUBBS(N.).

1119 H2*S*O4-H2*O, ... ACTA PHYS. POLON. VOL.11, ... 146, (1951).
*INFLUENCE OF THE ELECTROLYTIC DISSOCIATION AND THE HYDRATION OF H2*S*O4-
MOLECULES ON THE SCATTERING OF SLOW NEUTRONS.
*JANIK(J.A.).

1120 H-TI,H-V,H-NB,H-TA, ... C1,P . . . . . J. PHYS. SOC. JAPAN VOL.19, . . . 1862, (1964).
*STUDIES OF HYDROGEN VIBRATIONS IN TRANSITION METAL HYDRIDES BY THERMAL
NEUTRON TRANSMISSIONS.
*SAKAMOTO(M.).

1121 HE, ... B4 . . . . CAN. J. PHYS. 49, ... 177, (1971).
*INELASTIC SCATTERING OF THERMAL NEUTRONS FROM LIQUID HELIUM.
*COWLEY(R.A.), WOODS(A.D.B.).

1122 HE (A=3), ... T . . . . CAN. J. PHYS. 49, ... 2997, (1971).
*LATTICE DYNAMICS AND SHORT RANGE CORRELATIONS IN B.C.C. HE(A=3).
*GLYDE(H.R.), KHANNA(F.C.).

1123 HE, ... B4 . . . . IAEA SYMP. BOMBAY, VOL.2, ... 191, (1964).
*LIQUID HELIUM DISPERSION CURVE AT LARGE MOMENTUM.
*WOODS(A.D.B.).

1124 HE, ... B4 . . . . IAEA SYMP. COPENHAGEN, VOL.1, . . 609, (1968).
*HIGH-ENERGY EXCITATIONS IN LIQUID HELIUM.
*WOODS(A.D.B.), COWLEY(R.A.).

1125 HE, ... B1 . . . . IAEA SYMP. COPENHAGEN, VOL.1, . . 339, (1968).
*LATTICE DYNAMICS OF HCP HE4 BY INELASTIC NEUTRON SCATTERING.
*BRUN(T.O.), SINHA(S.K.), SWENSON(C.A.), TILFORD(C.R.).

1126 HE (BCC,LIQUID), ... T . . . . PHYS. REV. LETT. VOL.28, ... 1102, (1972).
*INTERPRETATION OF INELASTIC NEUTRON-SCATTERING OBSERVATIONS IN BCC SOLID
AND SUPERFLUID HE (ISOTOPE: A=4).
*WERTHAMER(N.R.).

1127 HE, ... T . . . . PHYS. REV. LETT. VOL.24, ... 1424, (1970).
*PHONON ENERGIES AND LIFETIMES IN SOLID NE AND HE IN THE FIRST-ORDER
SELF-CONSISTENT APPROXIMATION.
*GOLDMAN(V.V.), HORTON(G.K.), KLEIN(M.L.).

1128 HE, ... B4 . . . . PHYS. REV. LETT. VOL.24, ... 1046, (1970).
*HIGH-MOMENTUM-TRANSFER NEUTRON-LIQUID-HELIUM SCATTERING BOSE
CONDENSATION.
*HARLING(O.K.).

1129 HE, ... T . . . . PHYS. REV. LETT. VOL.24, ... 1044, (1970).
*SUM RULES, STRUCTURE FACTORS, AND PHONON DISPERSION IN LIQUID HE (A=4)
AT LONG WAVELENGTHS AND LOW TEMPERATURES.
*PINES(D.), CHIA-WEI-WOO.

1130 HE, ... B4 . . . . PHYS. REV. LETT. VOL.24, ... 646, (1970).
*LONG-WAVELENGTH PHONONS IN LIQUID HELIUM.
*WOODS(A.D.B.), COWLEY(R.A.).

1131 HE, ... T . . . . PHYS. REV. LETT. VOL.25, ... 220, (1970).
*PHONON DISPERSION AND THE PROPAGATION OF SOUND IN LIQUID HELIUM-4 BELOW
0.6 DEG. K.
*MARIS(H.J.), MASSEY(W.E.).

1132 HE, ... T . . . . PHYS. REV. LETT. VOL.25, ... 147, (1970).
*ANHARMONIC PHONONS AND LATTICE SPECIFIC HEAT IN BCC HE.
*HORNER(H.).

1133 HE, ... B1 . . . . PHYS. STAT. SOL. VOL.23, ... K155, (1967).
*LATTICE DYNAMICS OF SOLID HELIUM AT 2.9 K AND 125 ATM BY NEUTRON
SCATTERING.
*RITTER(M.), GISSLER(W.), SPRINGER(T.).

1134 HE (ISOTOPE A=4), ... T . . . . DISS. ABS. VOL.31, ... 6173B, (1971).
*ELEMENTARY EXCITATIONS, BOSE-EINSTEIN CONDENSATION, AND NEUTRON
SCATTERING: A STUDY OF LIQUID HELIUM FOUR.
*TENN(J.S.).

SECTION   6 -QUASI-ELASTIC AND INELASTIC SCATTERING STUDIES . . . .

1135 HE (ISOTOPE A=4), ... T . . . . DISS. ABS. VOL.30, ... 2839B, (1969).
      *THE PAIR DISTRIBUTION FUNCTION IN THE THEORY OF THE GROUND STATE OF
      LIQUID HELIUM FOUR.
      *LEE(M.H.).

1136 HE, ... T . . . . PHYS. REV. A VOL.2, . . AT LARGE MOMENTUM TRANSFER 2416, (1970).
      *NEUTRON SCATTERING FROM LIQUID HELIUM II AT LARGE MOMENTUM TRANSFER
      AND THE CONDENSATE FRACTION.
      *KERR(W.C.), PATHAK(K.N.), SINGWI(K.S.).

1137 HE, ... T . . . . PHYS. REV. A VOL.4, ... 2413, (1971).
      *ADDENDUM TO *NEUTRON SCATTERING FROM LIQUID HELIUM II AT LARGE MOMENTUM
      TRANSFER AND THE CONDENSATE FRACTION*.
      *KERR(W.C.), PATHAK(K.N.), SINGWI(K.S.).

1138 HE, ... T . . . . PHYS. REV. A VOL.5, ... 2230, (1972).
      *COMPUTATION OF PHONON SPECTRAL FUNCTIONS AND GROUND-STATE ENERGY OF
      SOLID HELIUM. I. BCC PHASE.
      *KOEHLER(T.R.), WERTHAMER(N.R.).

1139 HE, ... T . . . . PHYS. REV. A VOL.2, ... 2050, (1970).
      *NEUTRON SCATTERING FROM PHONONS IN SOLID HELIUM.
      *WERTHAMER(N.R.).

1140 HE, ... T . . . . PHYS. REV. A VOL.1, ... 1699, (1970).
      *HIGH-ENERGY NEUTRON SCATTERING FROM LIQUID HE4 :II. INTERFERENCE AND
      TEMPERATURE EFFECTS.
      *SEARS(V.F.).

1141 HE, ... B2 . . . . . . PHYS. REV. A VOL.3, ... 1688, (1971).
      *PHONON DISPERSION RELATIONS FOR HCP HE4 AT A MOLAR VOLUME OF 16 CC.
      *REESE(R.A.), SINHA(S.K.), BRUN(T.O.), TILFORD(C.R.).

1142 HE, ... T . . . . PHYS. REV. A VOL.5, ... 1547, (1972).
      *CORRECTIONS TO THE IMPULSE APPROXIMATION FOR HIGH-ENERGY NEUTRON
      SCATTERING FROM LIQUID HELIUM.
      *GERSCH(H.A.), RODRIGUEZ(L.J.), SMITH(P.N.).

1143 HE, ... B1 . . . . . . PHYS. REV. A VOL.5, ... 1537, (1972).
      *INELASTIC-NEUTRON SCATTERING FROM 4HE.
      *OSGOOD(E.B.), MINKIEWICZ(V.J.), KITCHENS(T.A.), SHIRANE(G.).

1144 HE, ... T . . . . PHYS. REV. A VOL.1, ... 1536, (1970).
      *DISPERSION OF PHONONS IN LIQUID HE4.
      *HWA-WEN-LAI, HOCK-KEE-SIM, CHIA-WEI-WOO.

1145 HE, ... B4 . . . . . . PHYS. REV. A VOL.5, ... 1377, (1972).
      *NEUTRON SCATTERING BY ROTONS IN LIQUID HELIUM.
      *DIETRICH(O.W.), GRAF(E.H.), HUANG(C.H.), PASSELL(L.).

1146 HE, ... B4 . . . . . . PHYS. REV. A VOL.3, ... 1073, (1971).
      *HIGH-ENERGY NEUTRON SCATTERING MEASUREMENTS ON LIQUID HELIUM AND BOSE
      CONDENSATION IN HE II.
      *HARLING(O.K.).

1147 HE, ... T . . . . PHYS. REV. A VOL.3, ... 820, (1971).
      *HIGH-ENERGY NEUTRON SCATTERING AND THE MIHARA-PUFF THEORY OF THE
      STRUCTURE FACTOR FOR LIQUID HELIUM-FOUR.
      *SPOSITO(G.).

1148 HE, ... T . . . . PHYS. REV. A VOL.1, ... 125, (1970).
      *HIGH-ENERGY NEUTRON-LIQUID HE4-SCATTERING AND THE HE4 CONDENSATE DENSITY
      *PUFF(R.D.), TENN(J.S.).

1149 HE, ... C3 . . . . . . PHYS. REV. LETT. 21, ... 787, (1968).
      *NEUTRON SCATTERING FROM LIQUID HELIUM AT HIGH ENERGIES.
      *COWLEY(R.A.), WOODS(A.D.B.).

1150 HE, ... T . . . . PHYS. LETT. 30A, ... 261, (1969).
      *INELASTIC NEUTRON SCATTERING FROM LIQUID HELIUM IN ITS GROUND STATE.
      *FERNANDEZ(J.F.), GERSCH(H.A.).

1151 HE, ... T . . . . BROOKHAVEN SYMPOSIUM., ... 173, (1965).
      *THE DISPERSION RELATION IN LIQUID HELIUM.
      *BORST(L.B.).

1152 HE (ISO. A=4), ... T . . . . CAN. J. PHYS. VOL.50, ... 1152, (1972).
      *ANHARMONIC PHONONS IN B.C.C. HE (ISO. A=4).
      *GLYDE(H.R.), KHANNA(F.C.).

1153 HE, ... B4 . . . . . . SOV. PHYS. TECH. PHYS. VOL.17, . 180, (1972).
      *INVESTIGATION OF HE (ISO. A=3) DIFFUSION IN LIQUID HE (ISO. A=4) BY
      MEANS OF THERMAL NEUTRONS.
      *DRABKIN(G.M.), NOSKIN(V.A.), TRUNOV(V.A.), SHCHEBETOV(A.F.), YAGUD(A.Z.)

1154 HE, ... B4 . . . . . . HELV. PHYS. ACTA VOL.45, ... 34, (1972).
      *LONG WAVELENGTH PHONON DISPERSION IN HE II.
      *DROZ(M.). (ED. NOTE: IN FRENCH).

1155 HE (ISO. A=3), ... D4 . . . . . . NUCL. PHYS. VOL.A129, ... 666, (1969).
      *SCATTERING OF POLARIZED NEUTRONS BY HE (ISO. A=3).
      *BUSSER(F.W.), DUBENKROPP(H.), NIEBERGALL(F.), SINRAM(K.).

1156 HE (ISO. A=3), ... T . . . . BROOKHAVEN SYMPOSIUM., ... 69, (1965).
      *LOW TEMPERATURE PROPERTIES OF LIQUID HE (ISO. A=3).
      *GLASSGOLD(A.E.).

1157 HE, ... T . . . . PHYS. LETT. VOL.42A NO.2, ... 133, (1972).
      *PHONON SPECTRUM OF SUPERFLUID HELIUM.
      *HAVLIN(S.), LUBAN(M.).

1158 HE (FCC), ... B1 . . . . . . IAEA SYMP. GRENOBLE, ... 129, (1972).
      *PHONON SPECTRA OF FCC HE(A=4).
      *TRAYLOR(J.G.), STASSIS(C.), REESE(R.A.), SINHA(S.K.).

1159 HE (A=4), ... B4 . . . . . . BULL. AMER. PHYS. SOC. VOL.18, . 22, (1973).
      *INELASTIC NEUTRON SCATTERING FROM SUPERFLUID HE (A=4) WITH A FREE
      SURFACE.
      *SAAM(W.F.). (ED. NOTE: PAPER PRESENTED AT ANNUAL MEETING OF AMER. PHYS.
      SOC. IN NEW YORK).

1160 HE, ... B4 . . . . . . THESIS (150 PP.), ... (1972).
      *NEUTRON SCATTERING BY ROTONS IN LIQUID HELIUM.
      *CHIA-HO-HUANG. (ED. NOTE: AVAIL. UNIV. MICROFILMS ORDER NO.73-1133; SEE
      DISS. ABSTRACTS VOL.33 NO.7 PG.3256-B).

SECTION   6 -QUASI-ELASTIC AND INELASTIC SCATTERING STUDIES . . . .

1161 HE,. . . . . *SCATTERING FUNCTION S(Q,W) FOR SOLID HELIUM. C5 PHYS. REV. LETT. VOL.29, ... 556, (1972).
      *HORNER(H.).

1162 HE (A=4),. . . . *SINGLE-PARTICLE SCATTERING FROM SOLID HE (A=4). B7 PHYS. REV. LETT. VOL.29, ... 552, (1972).
      *KITCHENS(T.A.), SHIRANE(G.), MINKIEWICZ(V.J.), OSGOOD(E.B.).

1163 HE (A=4),. . . . *≠ANOMALOUS≠ DEBYE-WALLER FACTOR FOR BCC HE (ISO.A=4). B7 PHYS. REV. LETT. VOL.29, . . . 549, (1972).
      *SEARS(V.F.), KHANNA(F.C.).

1164 HE,. . . . . . . *LOW-MOMENTUM-TRANSFER NEUTRON SCATTERING IN LIQUID HELIUM. B4 IAEA SYMP. GRENOBLE, 359, (1972).
      *WOODS(A.D.B.), SVENSSON(E.C.), MARTEL(P.).

1165 HE,. . . . . . . *NEUTRON SCATTERING IN LIQUID HELIUM (ABSTRACT ONLY). B4 IAEA SYMP. GRENOBLE, . . . 357, (1972).
      *DIETRICH(O.W.), PASSELL(L.), GRAF(E.H.).

1166 HE (ISO. A=4), *NEUTRON SCATTERING IN SOLID HE (A=4) AT LARGE MOMENTUM TRANSFER. B2 SOL. STATE COMM. VOL.11 1307, (1972).
      *SEARS(V.F.).

1167 HE,. . . . . . . *PHONON DISPERSION IN LIQUID HELIUM UNDER PRESSURE. B4 PHYS. REV. LETT. VOL.29, ... 1148, (1972).
      *SVENSSON(E.C.), WOODS(A.D.B.), MARTEL(P.).

1168 HE,. . . . . . *SLOW NEUTRON CROSS SECTIONS FOR HE3, B AND AU. P PHYS. REV. VOL.133, B925, (1964).
      *ALS-NIELSEN(J.), DIETRICH(O.W.).

1169 HE (A=4),. . . *KINETIC ENERGIES OF ATOMIC MOTIONS IN LIQUID HE (A=4). B4 PHYS. REV. A VOL.3, 1713, (1971).
      *GIBBS(A.G.), HARLING(O.K.).

1170 HE,. . . . . . *EXPERIMENTAL STUDIES OF INELASTIC NEUTRON SCATTERING FROM LIQUID HELIUM. B4 SYMP. ON QUANTUM FLUIDS, 242, (1966).
      *WOODS(A.D.B.). (ED. NOTE: PROCEEDING INTERNATIONAL SYMPOSIUM ON QUANTUM
      FLUIDS, BRIGHTON, 1965; PUBL. NORTH-HOLLAND:AMSTERDAM).

1171 HE,. . . . . . *NEUTRON SCATTERING BY LIQUID HELIUM UNDER PRESSURE. B4 PHYS. LETT. VOL.43A NO.3, . . . 223, (1973).
      *WOODS(A.D.B.), SVENSSON(E.C.), MARTEL(P.).

1172 HE (A=4),. . . *ATOMIC KINETIC ENERGIES IN LIQUID HE (A=4). B4 PHYS. LETT. VOL.36A, 203, (1971).
      *GIBBS(A.G.), HARLING(O.K.).

1173 HE,. . . . . . *DISPERSION RELATION IN LIQUID HELIUM. T,B4 NATURE VOL. 209, 187, (1966).
      *BORST(L.B.).

1174 HE,. . . . . . . *PHONONS AND THERMAL PROPERTIES OF BCC AND FCC HELIUM FROM A SELF- T,B2 J. LOW TEMP. PHYS. VOL.8, 511, (1972).
      CONSISTENT ANHARMONIC THEORY.
      *HORNER(H.).

1175 HE (A=4),. . . *INFERRING THE DISTRIBUTION OF ATOMIC KINETIC ENERGIES FROM NEUTRON- B4 PHYS. REV. A VOL.7, 1748, (1973).
      SCATTERING EXPERIMENTS.
      *GIBBS(A.G.), HARLING(O.K.).

1176 HE,. . . . . . . *MODES OF ATOMIC MOTIONS IN LIQUID HELIUM BY INELASTIC SCATTERING OF B4 PHYS. REV. VOL.121 1266, (1961).
      NEUTRONS.
      *HENSHAW(D.G.), WOODS(A.D.B.).

1177 HE,. . . . . . *INELASTIC SCATTERING OF NEUTRONS FROM ROTATING LIQUID HELIUM II. B4 CAN. J. PHYS. VOL.39, 1082, (1961).
      *WOODS(A.D.B.).

1178 HE,. . . . . . *INVESTIGATION OF THE THERMAL STRUCTURE OF HELIUM II BY SCATTERING OF SOV. PHYS. USPEKHI VOL.3, 888, (1961).
      COLD NEUTRONS (REVIEW PAPER).
      *ANDRONIKASHVILI(E.L.).

1179 HE,. . . . . . *ELEMENTARY EXCITATIONS IN LIQUID HELIUM. B4 PHYS. REV. VOL.127, 1452, (1962).
      *MILLER(A.), PINES(D.), NOZIERES(P.).

1180 HE,. . . . . . *TEMPERATURE-DEPENDENT GREEN FUNCTIONS AND NEUTRON SCATTERING IN LIQUID T,P ANN. PHYS. VOL.17, 301, (1962).
      HELIUM II.
      *PARRY(W.E.), TURNER(R.E.).

1181 HE,. . . . . . *THE DISPERSION OF NEUTRONS WITH ENERGIES OF THE ORDER OF A FEW DEGREES T,B4 DOKL. AKAD. NAUK VOL.93, 799, (1953).
      IN LIQUID HELIUM II.
      *KHALATNIKOV(I.M.), ZHARKOV(V.N.). (IN RUSSIAN).

1182 HE,. . . . . . *ON THE THEORY OF SLOW NEUTRON SCATTERING BY LIQUID HELIUM. T PHYS. REV. VOL.77, 319, (1950).
      *GOLDSTEIN(L.), SWEENEY(D.), GOLDSTEIN(M.).

1183 HE,. . . . . . *ON THE THEORY OF SLOW NEUTRON SCATTERING BY SOLID AND LIQUID HELIUM. T,P PHYS. REV. VOL.80, 141, (1950).
      *GOLDSTEIN(L.), SWEENEY(O.W.).

1184 HE II, ... *ON THE SCATTERING OF LOW ENERGY NEUTRONS IN HELIUM II. T J. PHYS. (USSR) VOL.9, 461, (1945).
      *AKHIEZER(A.), POMERANCHUK(I.YA.).

1185 HE,. . . . . . *PHONON DISPERSION MEASUREMENTS ON A HCP HE4 SINGLE CRYSTAL AT 27 ATM. B1 PHYS. REV. LETT. 19, 1307, (1967).
      *LIPSCHULTZ(F.P.), MINKIEWICZ(V.J.), KITCHENS(T.A.), SHIRANE(G.),
      NATHANS(R.).

1186 HE,. . . . . . *NEUTRON LINE WIDTH OF LONGITUDINAL PHONONS IN SOLID HELIUM. B1,P PHYS. LETT. 30A, 127, (1969).
      *KLEMENS(P.G.), LIPSCHULTZ(F.P.).

1187 HE,. . . . . . *HIGH-MOMENTUM EXCITATIONS IN LIQUID HE(A=4). B4,T ACTA PHYS. POLON. VOL.A44, . . . 829, (1973).
      *PARLINSKI(K.).

SECTION   6 -QUASI-ELASTIC AND INELASTIC SCATTERING STUDIES . . . .

1188 HE II, ...NEUTRON. SCATTERING.AND..THE.DISPERSION.RELATION. IN. HE II AT SMALL
B1   PHYS. REV. A VOL.8,   3244, (1973).
       MOMENTUM TRANSFER.
       .HALLOCK(R.B.).

1189 HE,. . . . . .LATTICE EXCITATIONS OF THE HE (A=4) QUANTUM SOLIDS.
B1   PHYS. REV. A VOL.8,   1513, (1973).
       .MINKIEWICZ(V.J.), KITCHENS(T.A.), SHIRANE(G.), OSGOOD(E.B.).

1190 HE,. . . . . .EXCITATIONS IN LIQUID HELIUM: THERMODYNAMIC CALCULATIONS.
T,B4   PHYS. REV. VOL.113,   1386, (1959).
       .BENDT(P.J.), COWAN(R.D.), YARNELL(J.L.).

1191 HE,. . . . . .EXCITATIONS IN LIQUID HELIUM: NEUTRON SCATTERING MEASUREMENTS.
B4   PHYS. REV. VOL.113,   1379, (1959).
       .YARNELL(J.L.), ARNOLD(G.P.), BENDT(P.J.), KERR(E.C.).

1192 HE,. . . . . .EXCITATION OF ROTONS IN HELIUM II BY COLD NEUTRONS.
B4   PHYS. REV. VOL.108,   1346, (1957).
       .PALEVSKY(H.), OTNES(K.), LARSSON(K.E.), PAULI(R.), STEDMAN(R.).

1193 HE,. . . . . .STRUCTURE AND EXCITATIONS OF LIQUID HELIUM (REVIEW PAPER).
REP. PROGR. PHYS. VOL.36,   1135, (1973).
       .WOODS(A.D.B.), COWLEY(R.A.).

1194 HE,. . . . . .THEORY OF INELASTIC NEUTRON SCATTERING FROM SUPERFLUID HE (A=4) WITH
T,P   PHYS. REV. A VOL.8,   1048, (1973).
       A FREE SURFACE.
       .SAAM(W.F.).

1195 HE,. . . . . .INELASTIC NEUTRON SCATTERING FROM A LIQUID HE(A=3)-HE(A=4) MIXTURE.
B4   PHYS. REV. LETT. VOL.31,   510, (1973).
       .ROWE(J.M.), PRICE(D.L.), OSTROWSKI(G.E.).

1196 HE,. . . . . .THE SCATTERING OF COLD NEUTRONS BY LIQUID HELIUM.
P   PROC. ROY. SOC. A VOL.242,   374, (1957).
       .EGELSTAFF(P.A.), LONDON(H.).

1197 HE,. . . . . .POSSIBLE DETERMINATION OF THE HELICITY OF ELEMENTARY EXCITATIONS IN
B4   PHYS. REV. LETT. VOL.2,   284, (1959).
       LIQUID HELIUM II.
       .LEE(T.D.), MOHLING(F.).

1198 HE,. . . . . .A STUDY OF THE EXCITATIONS IN LIQUID HELIUM BY USE OF SLOW NEUTRONS.
B4   PHYSICA VOL.24, SUPPL.,   145, (1958).
       .LARSSON(K.E.), OTNES(K.).

1199 HE,. . . . . .ENERGY-MOMENTUM RELATION IN LIQUID HELIUM BY INELASTIC SCATTERING OF
B4   PHYS. REV. LETT. VOL.1,   127, (1958).
       NEUTRONS.
       .HENSHAW(D.G.).

1200 HE,. . . . . .RELATION BETWEEN INELASTIC NEUTRON SCATTERING AND THERMODYNAMIC
T   PHYS. REV. VOL.118,   27, (1960).
       FUNCTIONS OF LIQUID HELIUM.
       .COHEN(M.).

1201 HE,. . . . . .EXCITATION OF ROTONS IN HELIUM II BY COLD NEUTRONS.
B4   PHYS. REV. VOL.112,   11, (1958).
       .PALEVSKY(H.), OTNES(K.), LARSSON(K.E.).

1202 HE,. . . . . .ENERGY VS. MOMENTUM RELATIONS FOR EXCITATIONS IN LIQUID HELIUM.
B4   PHYS. REV. LETT. VOL.1,   9, (1958).
       .YARNELL(J.L.), ARNOLD(G.P.), BENDT(P.J.), KERR(E.C.).

1203 HE,. . . . . .HIGH-ENERGY NEUTRON SCATTERING FROM LIQUID HELIUM IN THE IMPULSE
T   PHYS. REV. A VOL.4,   281, (1971).
       APPROXIMATION.
       .GERSCH(H.A.), SMITH(P.N.).

1204 HE,. . . . . .NEUTRON INELASTIC SCATTERING FROM LIQUID HELIUM AT SMALL MOMENTUM
B4   PHYS. REV. LETT. 14,   355, (1965).
       TRANSFER.
       .WOODS(A.D.B.).

1205 HE,. . . . . .ANALYSIS OF EXPERIMENTAL DATA ON NEUTRON SCATTERING FROM SUPERFLUID
T,B4   PHYS. REV. A VOL.9,   2085, (1974).
       HELIUM AT LARGE MOMENTUM TRANSFERS.
       .RODRIGUEZ(L.J.), GERSCH(H.A.), MOOK(H.A.).

1206 HE,. . . . . .REEXAMINATION OF EVIDENCE FOR A BOSE-EINSTEIN CONDENSATE IN SUPERFLUID
E,T   PHYS. REV. A VOL.9 NO.6,   (1974).
       HE (A=4).
       .JACKSON(H.W.). (REF. OBTAINED FROM PHYS. REV. ABSTRACTS).

1207 HE,. . . . . .NEUTRON-SCATTERING STUDY OF THE MOMENTUM DISTRIBUTION OF HE (A=4).
P   PHYS. REV. LETT. VOL.32,   1167, (1974).
       .MOOK(H.A.).

1208 HE,. . . . . .SCATTERING AT LARGE MOMENTUM AND ENERGY TRANSFER: APPLICATION TO NEUTRON
B4   DISS. ABSTR. B VOL.34,   4560, (1974).
       SCATTERING ON LIQUID HELIUM.
       .RODRIGUEZ(L.J.).

1209 HEMOGLOBIN,. . . .NEUTRON SMALL-ANGLE SCATTERING FROM AQUEOUS SOLUTIONS OF OXY- AND
C3   J. MOL. BIOL. VOL.41,   231, (1969).
       DEOXYHAEMOGLOBIN.
       .SCHNEIDER(R.), MAYER(A.), SCHMATZ(W.), KAISER(B.), SCHERM(R.).

1210 HEMOGLOBIN,. . . .NEUTRON SMALL ANGLE SCATTERING OF HEMOGLOBIN.
C3   J. BIOL. CHEM. VOL.247,   5436, (1972).
       .SCHELTEN(J.), SCHLECHT(P.), SCHMATZ(W.), MAYER(A.).

1211 HEXAMETHYLENETETRAMINE,. . .INTERMOLECULAR DYNAMICS OF HEXAMETHYLENETETRAMINE.
B1   PROC. ROY. SOC. LOND. A.319,   209, (1970).
       .DOLLING(G.), POWELL(B.M.).

1212 HEXAMETHYLENETETRAMINE,. . .LATTICE DYNAMICS OF HEXAMETHYLENETETRAMINE.
B1   IAEA SYMP. COPENHAGEN, VOL.2,   185, (1968).
       .POWELL(B.M.).

1213 HEXANE,. . . . . .LOW-FREQUENCY MOLECULAR VIBRATIONS IN SOLID HEXANE BY NEUTRON INELASTIC
C1   J. CHEM. PHYSICS VOL.41,   3649, (1964).
       SCATTERING.
       .DANNER(H.R.), BOUTIN(H.), SAFFORD(G.J.).

SECTION   6 -QUASI-ELASTIC AND INELASTIC SCATTERING STUDIES . . . .

1214   HF,. . . . . .C5. . . . . .IAEA SYMP. CHALK RIVER, VOL.2, .   183, (1962).
       ↪SCATTERING LAW FOR METAL ATOMS IN CHEMICAL COMPOUNDS.
       ↪EGELSTAFF(P.A.), HOLT(G.).

1215   HF-C,. . . . . .B7. . . . . .CONF. INTERN. RENNES, ... 145, (1971).
       ↪PHONON SPECTRA AND SUPERCONDUCTIVITY IN SOME TRANSITION METAL CARBIDES.
       ↪SMITH(H.G.), GLASER(W.).

1216   HF*C,. . . . . .B1. . . . . .PHYS. REV. LETT. VOL.25, ... 1611, (1970).
       ↪PHONON SPECTRA IN TA*C AND HF*C.
       ↪SMITH(H.G.), GLASER(W.).

1217   HF*C,. . . . . .T. . . . . . .PHYS. REV. B VOL.5, ... 1260, (1972).
       ↪PHONON SPECTRA OF SOME TRANSITION METAL CARBIDES FROM A SIMPLE
        PSEUDOPOTENTIAL APPROACH.
       ↪MOSTOLLER(M.).

1218   HF*O2,. . . . . .C5. . . . . .IAEA SYMP. CHALK RIVER, VOL.2, .   183, (1962).
       ↪SCATTERING LAW FOR METAL ATOMS IN CHEMICAL COMPOUNDS.
       ↪EGELSTAFF(P.A.), HOLT(G.).

1219   HO,. . . . . .D1. . . . . .J. APPL. PHYS. 42, ... 1672, (1971).
       ↪SPIN-WAVE DISPERSION RELATION IN RARE-EARTH METALS.
       ↪NICKLOW(R.M.).

1220   HO,. . . . . .B1. . . . . .PHYS. REV. B-3,. . . . . 1229, (1971).
       ↪LATTICE DYNAMICS OF HOLMIUM.
       ↪NICKLOW(R.M.), WAKABAYASHI(N.), VIJAYARAGHAVAN(P.R.).

1221   HO,. . . . . .D1. . . . . .PHYS. REV. LETT. VOL.23, ... 81, (1969).
       ↪SOFT SPIN-WAVE MODES AND THE CONE-TO-SPIRAL TRANSITION IN HOLMIUM METAL.
       ↪WOODS(A.D.B.), HOLDEN(T.M.), POWELL(B.M.), STRINGFELLOW(M.W.).

1222   HO,. . . . . .B. . . . . . .SOL. STATE COMM. 7, ... 535, (1969).
       ↪PHONON DISPERSION RELATIONS FOR HOLMIUM.
       ↪LEAKE(J.A.), MINKIEWICZ(V.J.), SHIRANE(G.).

1223   HO,. . . . . .D1. . . . . .METAL PHYS. VOL.3, ...   189, (1970).
       ↪SPIN WAVES IN HOLMIUM.
       ↪STRINGFELLOW(M.W.), HOLDEN(T.M.), POWELL(B.M.), WOODS(A.D.B.).

1224   HO,. . . . . .D1. . . . . .COLLOQUE INTERNAT. NO.180 TOME 2 283, (1970).
       ↪EXCHANGE AND CRYSTAL FIELD INTERACTIONS IN ERBIUM AND HOLMIUM FROM
        INELASTIC NEUTRON SCATTERING MEASUREMENTS.
       ↪WOODS(A.D.B.), STRINGFELLOW(M.W.), HOLDEN(T.M.), POWELL(B.M.). (COLL. ON
        RARE-EARTH ELEMENTS MAY, 1969. AVAIL. 15, QUAI ANATOLE-FRANCE-PARIS VII)

1225   HO,. . . . . .D3. . . . . .J. APP. PHYS. 39,. . . . .   457, (1968).
       ↪NEUTRON PARAMAGNETIC SCATTERING FROM HOLMIUM AND ERBIUM.
       ↪HOLDEN(T.M.), POWELL(B.M.), WOODS(A.D.B.).

1226   HO,. . . . . .D1. . . . . .J. APP. PHYS. VOL.40,. . . . . . 1452, (1969).
       ↪SPIN-WAVE DISPERSION RELATION FOR HOLMIUM IN THE SPIRAL MAGNETIC PHASE.
       ↪NICKLOW(R.M.), MOOK(H.A.), SMITH(H.G.), REED(R.E.), WILKINSON(M.K.).

1227   HO,. . . . . .D1. . . . . .J. APP. PHYS. VOL.40. . . . . . . 1443, (1969).
       ↪SPIN-WAVE EXCITATIONS IN THE CONICAL AND SPIRAL MAGNETIC PHASES OF
        HOLMIUM METAL.
       ↪STRINGFELLOW(M.W.), HOLDEN(T.M.), POWELL(B.M.), WOODS(A.D.B.).

1228   HO-ER,. . . . . .D1. . . . . .J. APPL. PHYS. 41, ... 1176, (1970).
       ↪SPIN-WAVE EXCITATIONS IN HO-ER.
       ↪HOLDEN(T.M.), POWELL(B.M.), STRINGFELLOW(M.W.), WOODS(A.D.B.).

1229   HO-LA,. . . . . .D4. . . . . .REF. SECTION 3, 64/KOEHLER(2), ...

1230   HO2*O3,. . . . . .B7. . . . . .PHYS. SOC. JAPAN VOL.17 B-III   63, (1962).
       ↪CRYSTAL FIELD SPECTRA IN RARE EARTH OXIDES.
       ↪BROCKHOUSE(B.N.), BECKA(L.N.), RAO(K.R.), SINCLAIR(R.N.), WOODS(A.D.B.).

1231   HO2*O3,. . . . . .B7. . . . . .C.R. ACAD. SCI. VOL.250 . . . . 2871, (1960).
       ↪NEUTRON SCATTERING AND STARK EFFECT IN CRYSTALLINE RARE EARTH OXIDES.
       ↪CRIBIER(D.), JACROT(B.). (IN FRENCH).

1232   HO*RH,. . . . . .B7. . . . . .SOL. STATE COMM. VOL.13 . . . . 1543, (1973).
       ↪THE DETERMINATION OF CRYSTAL FIELD LEVELS IN THE INTERMETALLIC COMPOUND
        HO*RH BY INELASTIC NEUTRON SCATTERING.
       ↪CHAMARD-BOIS(R.), ROSSAT-MIGNOD(J.), KNORR(K.), DREXEL(W.).

1233   HYDRATES,. . . . . .C2. . . . . .SOL. STATE COMM. VOL.8,. . . . .   497, (1970).
       ↪OBSERVATION OF LIBRATIONAL MODES OF WATER MOLECULES IN SINGLE CRYSTAL
        HYDRATES BY NEUTRON SCATTERING.
       ↪THAPER(C.L.),   DASANNACHARYA(B.A.),   SEQUEIRA(A.),   IYENGAR(P.K.).

1234   HYDRATES,. . . . . .C2. . . . . .PHYS. STAT. SOL. 34 NO.1. . . . .   279, (1969).
       ↪NEUTRON INELASTIC SCATTERING STUDIES IN CRYSTAL HYDRATES.
       ↪THAPER(C.L.), SEQUEIRA(A.), DASANNACHARYA(B.A.), IYENGAR(P.K.).

1235   HYDRIDOCARBONYLS (H*CO*(C*O)4,. .C1. . . . .J. CHEM. SOC. FARADAY II VOL.68, 1423, (1972).
       ↪NEUTRON SCATTERING SPECTRA OF HYDRIDOCARBONYLS IN THE 900-200 CM(-1)
        ENERGY REGION.
       ↪WHITE(J.W.), WRIGHT(C.J.).

1236   HYDROCARBONS (CYCLIC). . . . . . .C3. . . . . .THESIS,. . . . . (1967).
       ↪COLD-NEUTRON SCATTERING EXPERIMENTS ON CYCLIC HYDROCARBONS WITH A
        ROTATING-CRYSTAL SPECTROMETER.
       ↪DE-GRAAF(L.A.). (PH.D. DISSERTATION - DELFT, THE NETHERLANDS).

1237   HYDRODEUTERIDES,. . . . . .ATOMKERNENERGIE VOL.12 . . . .   385, (1967).
       ↪INVESTIGATION OF HYDROGEN AND DEUTERIUM VIBRATIONS IN METAL HYDRIDES,
        HYDRODEUTERIDES AND DEUTERIDES WITH TOTAL NEUTRON CROSS-SECTIONS.
       ↪SCHMIDT(U.).

1238   HYDROQUINONE,. . . . . .C3. . . . . .IAEA SYMP. COPENHAGEN, VOL.1,. .   483, (1968).
       ↪INELASTIC SCATTERING OF THERMAL NEUTRONS BY SOME BENZENE DERIVATIVES.
       ↪TREPADUS(V.), BALLY(D.), PARFENOV(V.A.), LIFOROV(V.G.).

1239   IN,. . . . . .P. . . . . . .PHYS. REV. VOL.70, ...   832, (1946).
       ↪SLOW NEUTRON RESONANCES IN INDIUM.
       ↪MCDANIEL(B.D.).

1240   IN*SB,. . . . . .B1. . . . . .PHYS. REV. B-3,. . . . . 1268, (1971).
       ↪LATTICE DYNAMICS OF GREY TIN AND INDIUM ANTIMONIDE.
       ↪PRICE(D.L.), ROWE(J.M.), NICKLOW(R.M.).

SECTION 6 -QUASI-ELASTIC AND INELASTIC SCATTERING STUDIES . . . .

1241 IN*SB, ... LATTICE DYNAMICS OF II-VI, III-V COMPOUNDS. T,B1.. SOL. STATE COMM. VOL.11, ... 1691, (1972).
   TALWAR(D.N.), AGRAWAL(B.K.).

1242 IN*SB, ... FAR-INFRARED PHONON ABSORPTION IN IN*SB. B1... PHYS. REV. B VOL.9,. . . . . . . 572, (1974).
   KOTELES(E.S.), DATARS(W.R.), DOLLING(G.).

1243 K, ... NEUTRONS, FREE ELECTRONS AND HIGH MAGNETIC FIELDS. B1,T... IAEA SYMP. COPENHAGEN, VOL.1,. . 141, (1968).
   COWLEY(R.A.).

1244 K, ... CRYSTAL DYNAMICS OF POTASSIUM. II. THE ANHARMONIC EFFECTS. B2... PHYS. REV. VOL.180,. . . 755, (1969).
   BUYERS(W.J.L.), COWLEY(R.A.).

1245 K. . . . . . . EXISTENCE OF AN INFINITY IN THE FREQUENCY DISTRIBUTION G(NU) OF T... PHYS. REV. 157,. . . 540, (1967).
   MONOATOMIC BODY CENTERED CUBIC CRYSTALS.
   GILAT(G.).

1246 K. . . . . . . CRYSTAL DYNAMICS OF POTASSIUM I. PSEUDOPOTENTIAL ANALYSIS OF PHONON T,B1... PHYS. REV. 150,. . . 487, (1966).
   DISPERSION CURVES AT 9 DEG. K.
   COWLEY(R.A.), WOODS(A.D.B.), DOLLING(G.).

1247 K*BR,. . . . . . NEUTRON SCATTERING AND ELASTIC CONSTANTS. B1... IAEA SYMP. COPENHAGEN, VOL.1,. . 281, (1968).
   COWLEY(R.A.), SVENSSON(E.C.), BUYERS(W.J.L.).

1248 K*BR,. . . . . . LATTICE DYNAMICS OF ALKALI HALIDES CRYSTALS II. EXPERIMENTAL STUDIES OF B2... PHYS. REV. 131,. . . 1025, (1963).
   K*BR AND NA*I.
   WOODS(A.D.B.), BROCKHOUSE(B.N.), COWLEY(R.A.), COCHRAN(W.).

1249 K*BR,. . . . . . SCATTERING OF COLD NEUTRONS IN IRRADIATED K*BR AND NA*CL CRYSTALS. B3... JETP LETT. VOL.3,. . . 110, (1966).
   ANDRONIKASHVILI(E.L.), BEDBENOVA(D.S.), POLITOV(N.G.), TSAKADZE(D.S.).

1250 K*BR,. . . . . . LATTICE DYNAMICS OF ALKALI HALIDE CRYSTALS.III. THEORETICAL. T,B1... PHYS. REV. 131,. . . 1030, (1963).
   COCHRAN(W.), COWLEY(R.A.), BROCKHOUSE(B.N.), WOODS(A.D.B.).

1251 K*BR,. . . . . . ANHARMONIC INTERACTIONS IN ALKALI HALIDES. II. T,B1... PROC. ROY. SOC. 292,. . . . . . 209, (1966).
   COWLEY(E.R.), COWLEY(R.A.).

1252 K*BR,. . . . . : ANHARMONIC INTERACTION IN ALKALI HALIDES. T,B1... PROC. ROY. SOC. 287,. . . . . . 259, (1965).
   COWLEY(E.R.), COWLEY(R.A.).

1253 K*BR,. . . . . . EXPERIMENTAL DIFFERENCE BETWEEN FIRST AND ZERO SOUND IN K*BR. B1... PHYS. REV. 165,. . . 1063, (1968).
   SVENSSON(E.C.), BUYERS(W.J.L.).

1254 K*BR,. . . . . . CRYSTAL DYNAMICS FROM NEUTRON SPECTROMETRY. B1... IAEA SYMP. CHALK RIVER, VOL.2, . 3, (1962).
   WOODS(A.D.B.).

1255 K*BR,. . . . . . INTERFERENCE BETWEEN ONE- AND MULTIPHONON PROCESSES IN THE SCATTERING OF B2... PHYS. REV. LETT. VOL.23,. . . 525, (1969).
   NEUTRONS AND X-RAYS BY CRYSTALS.
   COWLEY(R.A.), SVENSSON(E.C.), BUYERS(W.J.L.).

1256 K*BR-RB*BR,. . . NORMAL MODES OF VIBRATION OF MIXED K*BR/RB*BR CRYSTALS. B3... IAEA SYMP. COPENHAGEN, VOL.1,. . 43, (1968).
   BUYERS(W.J.L.), COWLEY(R.A.).

1257 K2*C2*O4*(H2*O), OBSERVATION OF LIBRATIONAL MODES OF WATER MOLECULES IN SINGLE CRYSTAL C2... SOL. STATE COMM. VOL.8,. . . 497, (1970).
   HYDRATES BY NEUTRON SCATTERING.
   THAPER(C.L.), DASANNACHARYA(B.A.), SEQUEIRA(A.), IYENGAR(P.K.).

1258 K2*C2*O4.H2*O, . LOW-FREQUENCY MOTIONS OF H2*O MOLECULES IN CRYSTALS. II. C1... J. CHEM. PHYSICS VOL.45,. . . 699, (1966).
   PRASK(H.J.), BOUTIN(H.).

1259 K*CL,. . . . . . LATTICE DYNAMICS OF POTASSIUM CHLORIDE. ERRATA IN PHYS. REV. 182, P. 965 B1... PHYS. REV. B VOL.1,. . . 4193, (1970).
   COPLEY(J.R.D.), MACPHERSON(R.W.), TIMUSK(T.).

1260 K*CL,. . . . . . LATTICE DYNAMICS OF POTASSIUM CHLORIDE. B1... PHYS. REV. 182,. . . . . 965, (1969).
   COPLEY(J.R.D.), MACPHERSON(R.W.), TIMUSK(T.).

1261 K*CL,. . . . . . PHONON WIDTHS IN NA*CL, K*CL AND RB*CL FROM NEUTRON MEASUREMENTS. B1,T... PHYS. STAT. SOL. 35, NO1,. . . 299, (1969).
   RAUNIO(G.).

1262 K*CL,. . . . . . DISPERSION RELATIONS FOR PHONONS IN K*CL AT 80 AND 300 DEG. K. B1... PHYS. STAT. SOL. 33,. . . 209, (1969).
   RAUNIO(G.), ALMQVIST(L.).

1263 K*CL,. . . . . . ON THE SCATTERING OF SLOW NEUTRONS BY K*CL. T... ARK. FYS. VOL.13, PAPER 17,. . . 199, (1958).
   SJOLANDER(A.).

1264 K*CL(N*H4),. . . OBSERVATION OF LOCAL TORSIONAL AND TRANSLATIONAL MODES IN K*CL(N*H4) B7... CONF. INTERN., RENNES,. . . 144, (1971).
   BY INELASTIC NEUTRON SCATTERING.
   SMITH(H.G.), WAKABAYASHI(N.), NICKLOW(R.M.).

1265 K*CL*(N*H4), . . LOCALIZED TORSIONAL AND TRANSLATIONAL MODES IN K*CL*(N*H4). B7... IAEA SYMP. GRENOBLE,. . . 103, (1972).
   SMITH(H.G.), WAKABAYASHI(N.), NICKLOW(R.M.).

1266 K2*CO*F4,. . . . SPIN CORRELATIONS AND SUBLATTICE MAGNETIZATION OF K2*CO*F4. D2... J. PHYS. SOC. JAPAN VOL.35,. . . 1795, (1973).
   IKEDA(H.), HIRAKAWA(K.).

1267 K2*CO*F4,. . . . NEUTRON SCATTERING STUDY OF TWO-DIMENSIONAL ISING NATURE OF K2*CO*F4. D2... SOL. STATE COMM. VOL.14,. . . 529, (1974).
   IKEDA(H.), HIRAKAWA(K.).

SECTION   6 -QUASI-ELASTIC AND INELASTIC SCATTERING STUDIES . . . .

1268 K*CO*F3, ...EXCITATIONS IN K*CO*F3-II. THEORETICAL.J. PHYS. -C-, VOL. 4,. . . . . . 2139, (1971).
                       *BUYERS(W.J.L.), HOLDEN(T.M.), SVENSSON(E.C.), COWLEY(R.A.),
                       HUTCHINGS(M.T.).

1269 K*CO*F3, ...EXCITATIONS IN K*CO*F3-I. EXPERIMENTAL.J. PHYS. -C-, VOL. 4,. . . . . . 2127, (1971).
                       *HOLDEN(T.M.), BUYERS(W.J.L.), SVENSSON(E.C.), COWLEY(R.A.),
                       HUTCHINGS(M.T.), HUKIN(D.), STEVENSON(R.W.H.).

1270 K*(CO,MN)*F2,. . .CHARACTER OF EXCITATIONS IN SUBSTITUTIONALLY DISORDERED ANTIFERROMAGNETS PHYS. REV. LETT. VOL.27,. . . . 1442, (1971).
                       *BUYERS(W.J.L.), HOLDEN(T.M.), SVENSSON(E.C.), COWLEY(R.A.),
                       STEVENSON(R.W.H.).

1271 K*(CO-MN)*F3,. . .SPIN WAVES IN DISORDERED ANTIFERROMAGNETS.T.D1. . . . . PHYS. REV. B VOL.7,. . . . . 1128, (1973).
                       *MANOHAR(C.).

1272 K2*CU*F4,. . . . .OBSERVATION OF SPIN DENSITIES ON ANTIBONDING ORBITALS IN K2*CU*F4 BY J. PHYS. SOC. JAPAN VOL.35,. . . 1608, (1973).
                       NEUTRON DIFFUSE SCATTERING.
                       *HIRAKAWA(K.), IKEDA(H.).

1273 K2*CU*F4, ...INVESTIGATIONS OF TWO-DIMENSIONAL FERROMAGNET K2*CU*F4 BY NEUTRON J. PHYS. SOC. JAPAN VOL.35,. . . 1328, (1973).
                       SCATTERING.
                       *HIRAKAWA(K.), IKEDA(H.).

1274 K*D2*P*O4, ...FERROELECTRIC TRANSITION IN K*D2*P*O4.B6. . . . . . PHYS. REV. B-2,. . . . . 4603, (1970).
                       *PAUL(G.L.), COCHRAN(W.), BUYERS(W.J.L.), COWLEY(R.A.).

1275 K*D2*P*O4, ...LOW FREQUENCY HYDROGEN VIBRATIONS IN POTASSIUM DIHYDROGEN PHOSPHATE.C2. . . . IAEA SYMP. CHALK RIVER, VOL.2,. . 273, (1962).
                       *PALEVSKY(H.), OTNES(K.), WAKUTA(Y.).

1276 K*D2*P*O4, K*H2*P*O4,. . .ON THE PROPERTIES OF THE KDP TYPE FERROELECTRICS WITH IMPURITIES.T. . . . . . PHYS. STAT. SOL. VOL.49,. . . . . 277, (1972).
                       *STAMENKOVIC(S.), ZEKOVIC(S.).

1277 K*D2*P*O4, ...NEUTRON SCATTERING NEAR THE FERROELECTRIC TRANSITION IN K*D2*P*O4.B1. . . . IAEA SYMP. COPENHAGEN, VOL.1,. . 267, (1968).
                       *BUYERS(W.J.L.), COWLEY(R.A.), PAUL(G.L.), COCHRAN(W.).

1278 K*D2*P*O4, ...FERROELECTRIC-MODE MOTION IN K*D2*P*O4.B1. . . . . . PHYS. REV. B-1,. . . . . 278, (1970).
                       *SKALYO-JR(J.), FRAZER(B.C.), SHIRANE(G.).

1279 K*D2*P*O4-FERROELECTRICS,. . .THEORY OF COHERENT NEUTRON SCATTERING BY H-BONDED FERROELECTRICS AT LOW T,B7. . . . J. LOW TEMP. PHYS. VOL.9,. . . . . 485, (1972).
                       TEMP. II.SCATT. CHARACTERISTICS/CONCEPTIONS OF/TUNNELING QUASISPIN MODEL
                       *STAMENKOVIC(S.).

1280 K*D2*P*O4-FERROELECTRICS,. . .THEORY OF COHERENT NEUTRON SCATTERING BY HYDROGEN-BONDED FERROELECTRICS T,B7. . . . J. LOW TEMP. PHYS. VOL.9,. . . . . 475, (1972).
                       AT LOW TEMPERATURES. I.GENERAL EXPRESSION FOR/SCATTERING/THERMAL FACTORS
                       *STAMENKOVIC(S.).

1281 K*D2*P*O4, ...DETERMINATION OF THE EIGENVECTORS OF SOFT MODES IN K*D2*P*O4 AND IN T. . . . . J. PHYSIQUE TOME 33 COL.2,. . . . 59, (1972).
                       N*D4*D2*P*O4,
                       *WALLACE(E.A.), COCHRAN(W.), STRINGFELLOW(M.W.).

1282 K*F, ...LATTICE DYNAMICS OF POTASSIUM FLUORIDE.B1. . . . . . . PHYS. STAT. SOL. VOL.41,. . . . 789, (1970).
                       *BUHRER(W.).

1283 K*F.2H2*O, ...LOW-FREQUENCY MOTIONS OF H2*O MOLECULES IN CRYSTALS. II.C1. . . . . . J. CHEM. PHYSICS VOL.45,. . . . . 699, (1966).
                       *PRASK(H.J.), BOUTIN(H.).

1284 K4*FE*(C*N)6*3(H2*O),. . .MOTIONS OF WATER MOLECULES IN POTASSIUM FERROCYANIDE TRIHYDRATE, WATER,C1. . . . J. CHEM. PHYSICS VOL.45,. . . . 1312, (1966).
                       AND ICE: A NEUTRON SCATTERING STUDY.
                       *RUSH(J.J.), LEUNG(P.S.), TAYLOR(T.I.).

1285 (K4*FE*(C*N)6)*3(H2*O),. . .STUDY OF LOW-FREQUENCY MOTIONS IN SEVERAL FERROELECTRIC SALTS BY THE C1. . . . . IAEA SYMP. BOMBAY, VOL.2,. . . 333, (1964).
                       INELASTIC SCATTERING OF COLD NEUTRONS.
                       *RUSH(J.J.), TAYLOR(T.I.).

1286 K*FE*F4, ...NEUTRON AND MOSSBAUER STUDIES OF THE PLANAR ANTIFERROMAGNET K*FE*F4 IN D2. . . . . PHYS. STAT. SOL. B VOL.53 NO.1,. . 227, (1972).
                       THE CRITICAL REGION.
                       *HEGER(G.), GELLER(R.).

1287 K*H*(C*F3*C*O*O)2, . .VIBRATION SPECTRA OF HYDROGEN BONDS BY NEUTRON ENERGY-LOSS SPECTROMETRY.C1. . . . J. CHEM. PHYS. VOL.52,. . . . . 1828, (1970).
                       *COLLINS(M.F.), HAYWOOD(B.C.), STIRLING(G.C.).

1288 K*H*(C*O2*C*CL3)2, . .ENERGY-LOSS NEUTRON SPECTROSCOPY APPLIED TO STRONGLY HYDROGEN C1. . . . . IAEA SYMP. GRENOBLE,. . . . . . 345, (1972).
                       BONDED MOLECULAR SYSTEMS.
                       *GHOSH(R.E.), WADDINGTON(T.C.), TEMME(F.P.).

1289 K*H*F2,. . . . . .DYNAMICS OF SODIUM AND POTASSIUM BIFLUORIDE: INFRARED, RAMAN, AND C1. . . . . J. CHEM. PHYS. VOL.56,. . . . . 2793, (1972).
                       NEUTRON STUDIES.
                       *RUSH(J.J.), SCHROEDER(L.W.), MELVEGER(A.J.).

1290 K*H*F2,. . . . . .VIBRATION SPECTRA OF HYDROGEN BONDS BY NEUTRON ENERGY-LOSS SPECTROMETRY.C1. . . . J. CHEM. PHYS. VOL.52,. . . . . 1828, (1970).
                       *COLLINS(M.F.), HAYWOOD(B.C.), STIRLING(G.C.).

1291 K*H*F2,K*H2*F3,. . .STUDY OF LOW-FREQUENCY MOLECULAR MOTIONS IN H*F, K*H*F2, K*H2*F3 AND C1. . . . J. CHEM. PHYS. VOL.39,. . . . 3135, (1963).
                       NA*H2*F3.
                       *BOUTIN(H.), SAFFORD(G.J.), BRAJOVIC(V.).

1292 K*H2*P*O4, ...LOW FREQUENCY HYDROGEN VIBRATIONS IN POTASSIUM DIHYDROGEN PHOSPHATE.C2. . . . IAEA SYMP. CHALK RIVER, VOL.2,. . 273, (1962).
                       *PALEVSKY(H.), OTNES(K.), WAKUTA(Y.).

SECTION 6 -QUASI-ELASTIC AND INELASTIC SCATTERING STUDIES . . . .

1293 K*H2*P*O4, . . . B7 . . . IAEA SYMP. BOMBAY, VOL.2, 325, (1964).
INDICATION OF PROTON TUNNELLING IN K*H2*P*O4 BY COMPARING NEUTRON
INELASTIC SCATTERING, AND INFRARED MEASUREMENTS.
PELAH(I.), WIENER(E.), IMRY(J.).

1294 K*H2*P*O4, . . . C1 . . . PHYS LETT 27A, 582, (1968).
INTERFERENCE EFFECTS IN NEUTRON SCATTERING ON PROTONS IN DOUBLE
MINIMUM POTENTIAL.
SCHENK(C.), WECKERMANN(B.).

1295 K*H2*P*O4, . . . B7 . . . IAEA SYMP. GRENOBLE, 825, (1972).
NEUTRON MEASUREMENTS ON THE FERROELECTRIC PHASE TRANSFORMATION
IN K*H2*P*O4.
ARSIC-ESKINJA(M.), GRIMM(H.), STILLER(H.).

1296 K*H2*P*O4, . . . C1 . . . J.C.S. FARADAY TRANS. II VOL.68, 350, (1972).
PROTON MOTIONS IN H-BONDED FERROELECTRIC AND ANTIFERROELECTRIC SOLIDS:
1.ENERGY-LOSS NEUTRON INCOHERENT SCATT/K*H2*P*O4/AG2*H3*I*O6/T<T(CRIT.).
TEMME(F.P.), WADDINGTON(T.C.).

1297 K*H2*P*O4, . . . I . . . 2ND SUMMER SCHOOL S.S. PHYS., 120, (1968).
NEUTRON SCATTERING FROM HYDROGEN BONDS.
PLESSER(T.), STILLER(H.).

1298 K*H2*P*O4, . . . T,E . . . HELV. PHYS. ACTA VOL.20, 105, (1947).
ETUDE DES FORCES INTERMOLECULAIRES PAR DIFFUSION DES NEUTRONS LENTS.
APPLICATION A N2, H2*O ET K*H2*P*O4.
ROSSEL(J.).

1299 K*H2*P*O4, . . . C1 . . . PHYS. LETT. VOL.26A, 8, (1967).
STUDY OF PROTON DYNAMICS IN PARAELECTRIC K*H2*P*O4 BY QUASI-ELASTIC
COLD NEUTRON SCATTERING.
BLINC(R.), DIMIC(V.), PETKOVSEK(J.), PIRKMAJER(E.).

1300 K*H2*P*O4, . . . C1 . . . PHYS. REV. VOL.172, 576, (1968).
DISAPPEARANCE OF A VIBRATIONAL MODE IN THE FERROELECTRIC PHASE
TRANSITION OF K*H2*P*O4.
SCHENK(C.), WIENER(E.), WECKERMANN(B.), KLEY(W.).

1301 K*H2*P*O4, . . . C1 . . . J. PHYS. CHEM. SOLIDS VOL.30, 449, (1969).
NEUTRON INCOHERENT SCATTERING FROM KDP.
STEINMAN(D.K.), SUMMERFIELD(G.C.).

1302 K*H2*P*O4,K2*H*P*O4, . . . C1 . . . PHYS. REV. LETT. VOL.2, 94, (1959).
OBSERVATIONS OF HYDROGEN VIBRATION FREQUENCIES IN PHOSPHATES BY MEANS
OF INELASTIC SCATTERING OF COLD NEUTRONS.
PELAH(I.), LEFKOWITZ(I.), KLEY(W.), TUNKELO(E.).

1303 K*I, . . . B1 . . . INDIAN J. PURE APPL. PHYS. VOL.7 151, (1969).
DISPERSION RELATIONS IN POTASSIUM IODIDE.
SINGH(R.K.), VERMA(M.P.).

1304 K*I, . . . B1 . . . PHYS. REV. 147, 577, (1966).
NORMAL VIBRATIONS IN POTASSIUM IODIDE.
DOLLING(G.), COWLEY(R.A.), SCHITTENHELM(C.), THORSON(I.M.).

1305 K*I-(N*H4*I), . . . C1 . . . IAEA SYMP. BOMBAY, VOL.2, 383, (1964).
STUDIES OF AMMONIUM ION MOTIONS IN SOLID SOLUTIONS.
MIKKE(K.), DOBRZYNSKI(L.).

1306 K*(MN-CO)*F3, . . . D1,T . . . J. PHYS. C VOL.5, 2611, (1972).
THEORY OF SPIN WAVES IN DISORDERED ANTIFERROMAGNETS. I. APPLICATION TO
(MN,CO)*F2 AND K*(MN,CO)*F3.
BUYERS(W.J.L.), PEPPER(D.E.), ELLIOTT(R.J.).

1307 K*(MN-(CO,NI))*F3, . . . D1,D4 . . . IAEA SYMP. GRENOBLE, 581, (1972).
MAGNETIC MODES ASSOCIATED WITH IMPURITIES IN INSULATORS.
BUYERS(W.J.L.), SVENSSON(E.C.), HOLDEN(T.M.), COWLEY(R.A.),
STEVENSON(R.W.H.).

1308 K*MN*F3, . . . B2 . . . SOL. STATE COMM. VOL.8, 1941, (1970).
NEUTRON SCATTERING STUDY OF THE LATTICE DYNAMICAL PHASE TRANSITIONS IN
K*MN*F3.
SHIRANE(G.), MINKIEWICZ(V.J.), LINZ(A.).

1309 K2*MN*F4, . . . D2 . . . J. PHYS. SOC. JAPAN VOL.33, 393, (1972).
ELASTIC AND QUASI-ELASTIC NEUTRON SCATTERING FROM K2*MN*F4 NEAR THE
CRITICAL POINT.
IKEDA(H.), HIRAKAWA (K.).

1310 K*MN*F3, . . . B2 . . . J. PHYS. SOC. JAPAN VOL.26, 674, (1969).
SOFT PHONON MODES IN K*MN*F3.
MINKIEWICZ(V.J.), SHIRANE(G.).

1311 K*MN*F3, . . . D3 . . . J. PHYSIQUE TOME 32 COL.1 VOL.II 619, (1971).
PARAMAGNETIC SCATTERING OF NEUTRONS FROM K*MN*F3 IN THE SHORT RANGE
ORDERED REGION.
USHA-DENIZ(K.), GOYAL(P.S.).

1312 K*MN*F3, . . . B6 . . . PHYS. REV. B VOL.6, 4332, (1972).
CRITICAL NEUTRON SCATTERING IN SR*TI*O3 AND K*MN*F3.
SHAPIRO(S.M.), AXE(J.D.), SHIRANE(G.), RISTE(T.).

1313 K*MN*F3, . . . D3 . . . J. APP. PHYS. 36, 1092, (1965).
SOME MEASUREMENTS OF EXCHANGE ENERGIES BY PARAMAGNETIC NEUTRON INELASTIC
SCATTERING.
COLLINS(M.F.), NATHANS(R.).

1314 K*MN*F3, . . . D1 . . . J. APP. PHYS. 37, 1054, (1966).
SPIN-WAVE DISPERSION IN K*MN*F3.
PICKART(S.J.), COLLINS(M.F.), WINDSOR(C.G.).

1315 K*MN*F3, . . . D2 . . . J. APP. PHYS. 37, 1041, (1966).
CRITICAL MAGNETIC SCATTERING FROM K*MN*F3.
COOPER(M.J.), NATHANS(R.).

1316 K2*MN*F4, . . . A2,D1 . . . PHYS. REV. B VOL.8, 304, (1973).
SPIN WAVES AND MAGNETIC ORDERING IN K2*MN*F4.
BIRGENEAU(R.J.), GUGGENHEIM(H.J.), SHIRANE(G.).

1317 K*MN*F3, . . . B1,B6 . . . PHYS. REV. B VOL.5, 1933, (1972).
DISPERSION AND DAMPING OF SOFT ZONE-BOUNDARY PHONONS IN K*MN*F3.
GESI(K.), AXE(J.D.), SHIRANE(G.), LINZ(A.).

SECTION   6 -QUASI-ELASTIC AND INELASTIC SCATTERING STUDIES . . . .

1318   K*MN*F3, ...D3. . . . . . .IAEA SYMP. BOMBAY, VOL.1,. . . . .   433, (1964).
       ⇢NEUTRON SCATTERING BY PARAMAGNETS.
       ⇢MURTHY(N.S.S.),   VENKATARAMAN(G.),   USHA-DENIZ(K.), DASANNACHARYA(B.A.),
       IYENGAR(P.K.).

1319   K2*MN*F4,. . . . . .D2. . . . . . .J. PHYS. SOC. JAPAN VOL.35,. . . .   617, (1973).
       ⇢CRITICAL MAGNETIC SCATTERING IN K2*MN*F4.
       ⇢IKEDA(H.), HIRAKAWA(K.).

1320   K*MN*F3(CO), ...B1,B7. . . .J. PHYS. SOC. JAPAN VOL.28, SUP.   242, (1970).
       ⇢ANTIFERROELECTRICITY IN CO DOPED K*MN*F3.
       ⇢BUYERS(W.J.L.), COWLEY(R.A.), PAUL(G.L.).

1321   K*MN*F3(CO), K*MN*F3(NI),. . . . . . .D1. . . . . . .MAGN. AND MAG. MATERIALS-1971, . 1315, (1972).
       ⇢EFFECTS OF IMPURITIES IN ANTIFERROMAGNETS.
       ⇢SVENSSON(E.C.), BUYERS(W.J.L.), HOLDEN(T.M.), COWLEY(R.A.),
       STEVENSON(R.W.H.). (AIP CONF. PROC. NO.5 PART 2; GRAHAM-JR., RHYNE-EDS.)

1322   K*MN*F3(CO), ...D4. . . . . . .CAN. J. PHYS. VOL.47,. . . . . . 1983, (1969).
       ⇢OBSERVATION OF LOCALIZED MAGNETIC EXCITATION IN CO-DOPED K*MN*F3.
       ⇢SVENSSON(E.C.), BUYERS(W.J.L.), HOLDEN(T.M.), COWLEY(R.A.),
       STEVENSON(R.W.H.).

1323   K*MN*F3(NI), ...D1. . . . . . .J. PHYSIQUE TOME 32 COL.1 VOL.II 1184, (1971).
       ⇢MAGNETIC EXCITATIONS IN NI-DOPED K*MN*F3.
       ⇢HOLDEN(T.M.), COWLEY(R.A.), BUYERS(W.J.L.), SVENSSON(E.C.),
       STEVENSON(R.W.H.).

1324   K*N3,. . . . . .B1. . . . . . .DISS. ABS. VOL.31,. . . . . . . .36589, (1970).
       ⇢A STUDY OF THE LATTICE DYNAMICS OF POTASSIUM AZIDE USING COHERENT
       NEUTRON SCATTERING.
       ⇢MICAL(R.D.).

1325   K*N3,. . . . . .B1. . . . . . .PHYS. REV. B VOL.4,. . . . . . . 4551, (1971).
       ⇢LATTICE DYNAMICS OF K*N3.
       ⇢RAO(K.R.), TREVINO(S.F.), PRASK(H.), MICAL(R.D.).

1326   K*NB*03, ...B7. . . . . . .J. PHYS. C VOL.6 NO.6,. . . . . . 1074, (1973).
       ⇢SOFT MODES AND THE STRUCTURE, SPONTANEOUS POLARIZATION AND CURIE
       CONSTANTS OF PEROVSKITE FERROELECTRICS: TETRAGONAL POTASSIUM NIOBATE.
       ⇢HEWAT(A.W.).

1327   K2*NI*F4,. . . . . .D2. . . . . . .PHYS. REV. B-1,. . . . . 2211, (1970).
       ⇢NEUTRON SCATTERING INVESTIGATION OF PHASE TRANSITIONS AND MAGNETIC
       CORRELATIONS IN THE TWO-DIMENSIONAL ANTIFERROMAGNETS.
       ⇢BIRGENEAU(R.J.), GUGGENHEIM(H.J.), SHIRANE(G.).

1328   K2*NI*F4,. . . . . .D1. . . . . . .PHYS. REV. LETT. VOL.23, ... 1394, (1969).
       ⇢MAGNONS AT LOW AND HIGH TEMPERATURES IN THE PLANAR ANTIFERROMAGNET
       K2*NI*F4.
       ⇢SKALYO-JR(J.), SHIRANE(G.), BIRGENEAU(R.J.), GUGGENHEIM(H.J.).

1329   K2*NI*F4,. . . . . .D4. . . . . . .J. PHYSIQUE TOME 32 COL.1 VOL.II 882, (1971).
       ⇢MAGNETIC NEUTRON SCATTERING FROM K2*NI*F4.
       ⇢SKALYO-JR(J.), SHIRANE(G.), BIRGENEAU(R.J.).

1330   K2*NI*F4,. . . . . .D2. . . . . . .PHYS. REV. B VOL.3,. . . . . . . 1736, (1971).
       ⇢CRITICAL MAGNETIC SCATTERING IN K2*NI*F4.
       ⇢BIRGENEAU(R.J.), SKALYO-JR(J.), SHIRANE(G.).

1331   K2*NI*F4,. . . . . .D2,D4. . . .J. APPL. PHYS. VOL.41,. . . . . . 1303, (1970).
       ⇢PHASE TRANSITIONS AND MAGNETIC CORRELATIONS IN TWO-DIMENSIONAL ANTI-
       FERROMAGNETS.
       ⇢BIRGENEAU(R.J.), SKALYO-JR(J.), SHIRANE(G.).

1332   K*(NI-MN)*F3,. . . . . .D1. . . . . . .IAEA SYMP. GRENOBLE,. . . . . . . 595, (1972).
       ⇢VIRTUAL SPIN-WAVE STATES IN ANTIFERROMAGNETIC INSULATORS.
       ⇢KROO(N.), PEPY(G.), SZENTIRMAY(Z.).

1333   K2*PT*(C*N)4*BR(0.3).3H2*O,. . . . . .B7. . . . . . .PHYS. REV. LETT. VOL.30,. . . . . 1144, (1973).
       ⇢OBSERVATION OF GIANT KOHN ANOMALY IN THE ONE-DIMENSIONAL CONDUCTOR
       K2*PT*(C*N)4*BR(0.3).3H2*O.
       ⇢RENKER(B.), RIETSCHEL(H.), PINTSCHOVIUS(L.), GLASER(W.), BRUESCH(P.),
       KUSE(D.), RICE(M.J.).

1334   K2*PT*(C*N)4*BR(.3).302*O, ...B2. . . . . . .PHYS. REV. LETT. VOL.32, ... 836, (1974).
       ⇢NEUTRON-SCATTERING STUDY OF THE STRUCTURAL PHASE TRANSITION IN THE ONE-
       DIMENSIONAL CONDUCTOR K2*PT*(C*N)4*BR(.3).302*O.
       ⇢RENKER(B.), PINTSCHOVIUS(L.), GLASER(W.), RIETSCHEL(H.), COMES(R.),
       LIEBERT(L.), DREXEL(W.).

1335   K2*RE*H9,. . . . . .C1. . . . . . .J. CHEM. SOC. FARADAY II VOL.68, 1414, (1972).
       ⇢INTERNAL VIBRATIONS AND HINDERED ROTATION OF RE*H9 (2-) BY NEUTRON
       SCATTERING SPECTROSCOPY.
       ⇢WHITE(J.W.), WRIGHT(C.J.).

1336   K*(TA-NB)*03,. . . . . .B7. . . . . . .FERROELECTRICS(GB) VOL.2,. . . . . 261, (1971).
       ⇢NEUTRON SCATTERING STUDY OF THE SOFT MODES IN CUBIC POTASSIUM TANTALATE-
       NIOBATE.
       ⇢YELON(W.B.), COCHRAN(W.), SHIRANE(G.), LINZ(A.).

1337   K*(TA-NB)*03,. . . . . .B7. . . . . . .FERROELECT. VOL.4,. . . . . 153, (1972).
       ⇢SOFT MODES AND THE STRUCTURE OF FERROELECTRIC TETRAGONAL POTASSIUM
       TANTALATE NIOBATE.
       ⇢HEWAT(A.W.), ROUSE(K.D.), ZACCAI(G.).

1338   K*(TA-NB)*03,. . . . . .B1. . . . . . .J. PHYSIQUE TOME 33 COL.2, . . . 133, (1972).
       ⇢KTN, STRUCTURAL AND DYNAMICAL STUDIES.
       ⇢ZACCAI(G.), HEWAT(A.W.), ROUSE(K.D.).

1339   K*(TA-NB)*03,. . . . . .B2. . . . . . .J. PHYS. C VOL.7,. . . . . 15, (1974).
       ⇢SOFT MODES AND THE STRUCTURE OF FERROELECTRIC TETRAGONAL KTN: II. THE
       LATTICE DYNAMICS OF THE CUBIC PHASE.
       ⇢ZACCAI(G.), HEWAT(A.W.).

1340   K*TA*03, ...B1,B6. . . .PHYS. REV. B-1,. . . . . 1227, (1970).
       ⇢ANOMALOUS ACOUSTIC DISPERSION IN CENTROSYMMETRIC CRYSTALS WITH SOFT
       OPTIC PHONONS.
       ⇢AXE(J.D.), HARADA(J.), SHIRANE(G.).

1341   K*TA*03, ...B7. . . . . . .ACTA CRYST. VOL.A26,. . . . . . . 608, (1970).
       ⇢DETERMINATION OF THE NORMAL VIBRATIONAL DISPLACEMENTS IN SEVERAL
       PEROVSKITES BY INELASTIC NEUTRON SCATTERING.
       ⇢HARADA(J.), AXE(J.D.), SHIRANE(G.).

SECTION 6 -QUASI-ELASTIC AND INELASTIC SCATTERING STUDIES . . . .

1342 K*TA*03, ... &TEMPERATURE DEPENDENCE OF THE SOFT FERROELECTRIC MODE IN K*TA*03.
. . . . . B1 . . . . PHYS. REV. 157 . . . . . 396, (1967).
&SHIRANE(G.), NATHANS(R.), MINKIEWICZ(V.J.).

1343 K*TA*03, ... &NEUTRON-SCATTERING ANALYSIS OF THE LINEAR-DISPLACEMENT CORRELATIONS IN
. . . . . B1 . . . . PHYS. REV. B VOL.5 . . . . . 1866, (1972).
K*TA*03.
&COMES(R.), SHIRANE(G.).

1344 KR, ... T . . . . PHYS. LETT. VOL.34A, ... 415, (1971).
&LATTICE DYNAMICS OF SOLID KR.
&BARKER(J.A.), BOBETIC(M.V.), KLEIN(M.L.).

1345 KR, ... B7 . . . . . . BULL. AMER. PHYS. SOC. VOL.18, . 111, (1973).
&ZERO SOUND ELASTIC CONSTANTS OF SOLID KRYPTON AT T=114 DEGREES K.
&SKALYO-JR(J.), ENDOH(Y.). (ED. NOTE: PRESENTED AT ANNUAL MEETING OF
AMER. PHYS. SOC. IN NEW YORK).

1346 KR, ... &ELASTIC CONSTANTS OF SOLID KRYPTON AT T=77 K DETERMINED BY INELASTIC
. . . . . B1 . . . . J. PHYS. CHEM. SOLIDS VOL.34 . . . . 255, (1973).
NEUTRON SCATTERING.
&PETER(H.), SKALYO-JR(J.), GRIMM(H.), LUSCHER(E.), KORPIUN(P.).

1347 KR, ... &ZERO-SOUND ELASTIC CONSTANTS OF SOLID KRYPTON AT T=114 DEGREES K.
. . . . . B1 . . . . . PHYS. REV. B VOL.7, ... 4670, (1973).
&SKALYO-JR(J.), ENDOH(Y.).

1348 KR, ... &OBSERVED DIFFERENCES IN ZERO- AND FIRST-SOUND PROPAGATION IN SOLID
. . . . . B7 . . . . . PHYS. REV. LETT. VOL.31, ... 296, (1973).
KRYPTON.
&JACKSON(H.E.), LANDHEER(D.), STOICHEFF(B.P.).

1349 KR, ... &PHONON-DISPERSION MEASUREMENTS ON KRYPTON SINGLE CRYSTAL.
. . . . . B1 . . . . . PHYS. REV. LETT. 18, ... 548, (1967).
&DANIELS(W.B.), SHIRANE(G.), FRAZER(B.C.), UMEBAYASHI(H.), LEAKE(J.A.).

1350 KR, ... &INELASTIC NEUTRON SCATTERING FROM SOLID KRYPTON AT 10 DEGREES K.
. . . . . B1 . . . . . PHYS. REV. B VOL.9, ... 1797, (1974).
&SKALYO-JR(J.), ENDOH(Y.), SHIRANE(G.).

1351 LA*AL*03, ... &SOFT PHONON RESPONSE FUNCTION: INELASTIC NEUTRON SCATTERING FROM
. . . . . B7 . . . . . BULL. AMER. PHYS. SOC. VOL.18, . 111, (1973).
LA*AL*03.
&KJEMS(J.K.), SHIRANE(G.), SCHEEL(H.J.), MULLER(K.A.). (ED. NOTE: PAPER
PRESENTED AT ANNUAL MEETING OF AMER. PHYS. SOC. IN NEW YORK).

1352 LA*AL*03, ... &ZONE-BOUNDARY PHONON INSTABILITY IN CUBIC LA*AL*03.
. . . . . B7 . . . . . PHYS. REV. VOL.183, ... 820, (1969).
&AXE(J.D.), SHIRANE(G.), MULLER(K.A.).

1353 LA*AL*03, ... &SOFT-PHONON RESPONSE FUNCTION: INELASTIC NEUTRON SCATTERING FROM
. . . . . B7 . . . . . PHYS. REV. B VOL.8, ... 1119, (1973).
LA*AL*03.
&KJEMS(J.K.), SHIRANE(G.), MULLER(K.A.), SCHEEL(H.J.).

1354 LA*CR*03, ... &SOME MEASUREMENTS OF EXCHANGE ENERGIES BY PARAMAGNETIC NEUTRON
. . . . . D3 . . . . . J. APP. PHYS. 36 . . . . 1092, (1965).
INELASTIC SCATTERING.
&COLLINS(M.F.), NATHANS(R.).

1355 LA-(DIHYDRIDE,TRIHYDRIDE), . . . P . . . . . . J. CHEM. PHYS. VOL.54, ... 3979, (1971).
&INELASTIC NEUTRON SCATTERING SPECTRA FROM LANTHANUM DIHYDRIDE AND
LANTHANUM TRIHYDRIDE.
&MAELAND(A.J.), HOLMES(D.E.).

1356 LA-(DY,ER,GD,HO,TB,TM), ... D4 . . . . . REF. SECTION 3, 64/KOEHLER(2), ...

1357 LA*FE3*04, . . . &SOME MEASUREMENTS OF EXCHANGE ENERGIES BY PARAMAGNETIC NEUTRON
. . . . . D3 . . . . . J. APP. PHYS. 36 . . . . 1092, (1965).
INELASTIC SCATTERING.
&COLLINS(M.F.), NATHANS(R.).

1358 LA2*MG3*(N*03)12.(H2*0)24, ... C1 . . . . . REF. SECTION 3, 68/BAJOREK, ...

1359 LA*MN*03, ... &SOME MEASUREMENTS OF EXCHANGE ENERGIES BY PARAMAGNETIC NEUTRON
. . . . . D3 . . . . . J. APP. PHYS. 36, ... 1092, (1965).
INELASTIC SCATTERING.
&COLLINS(M.F.), NATHANS(R.).

1360 LI, ... &PHONON DISPERSION CURVES IN LITHIUM.
. . . . . B1 . . . . . IAEA SYMP. COPENHAGEN, VOL.1, . . 149, (1968).
&SMITH(H.G.), DOLLING(G.), NICKLOW(R.M.), VIJAYARAGHAVAN(P.R.),
WILKINSON(M.K.).

1361 LI, ... &PHONON FREQUENCIES OF METALLIC LITHIUM.
. . . . . T,B1 . . . . PHYS. STAT. SOL. VOL.B-54, . . . K29, (1972).
&KACHHAVA(C.M.).

1362 LI, ... &CRYSTAL DYNAMICS OF LITHIUM BASED ON THE PSEUDOPOTENTIAL TECHNIQUE.
. . . . . T . . . . . . PHYS. REV. B VOL.6, ... 2192, (1972).
&PRASAD(B.), SRIVASTAVA(R.S.).

1363 LI, ... &LATTICE DYNAMICS OF METALLIC LITHIUM.
. . . . . B1,T . . . . J. PHYS. SOC. JAPAN VOL.34, . . . 26, (1973).
&ONO(M.).

1364 LI, ... &PHONON SPECTRUM OF LITHIUM.
. . . . . T,B1 . . . . J. PHYS. F VOL.3, ... 1388, (1973).
&SRIVASTAVA(P.L.), MITRA(N.R.), MISHRA(N.).

1365 LI, ... &LATTICE DYNAMICS OF LITHIUM.
. . . . . T,B1 . . . . J. PHYS. F VOL.3, ... 1308, (1973).
&SHARAN(B.), KUMAR(A.), NEELAKANDAN(K.).

1366 LI, ... &PSEUDOPOTENTIAL CALCULATION OF PHONON FREQUENCIES AND GRUNEISEN
. . . . . T . . . . . . PHYS. REV. 178, ... 900, (1969).
PARAMETERS FOR LITHIUM.
&WALLACE(D.C.).

1367 LI*AL*H4, ... &LIBRATIONAL MOTION IN SODIUM AND LITHIUM ALUMINIUM HYDRIDES, STUDIED BY
. . . . . B7 . . . . . J.C.S. FARADAY TRANS. II VOL.69, 783, (1973).
INELASTIC NEUTRON SCATTERING.
&TEMME(F.P.), WADDINGTON(T.C.).

SECTION   6 -QUASI-ELASTIC AND INELASTIC SCATTERING STUDIES . . . .

1368 LI*CL.H2*O,. . . . . .C1. . . . . . .REF. SECTION 3, 68/BAJOREK,. . . . .

1369 LI*CL.H2*O,LI*CL*O4.3H2*O,. . . . . .C1. . . . . .SOV. PHYS. CRYST. VOL.15,. . . 1010, (1971).
      *NEUTRON-DIFFRACTION STUDY OF THE CRYSTALLINE HYDRATES OF LITHIUM SALTS.
      III.STUDY OF THE DYNAMICS OF WATER MOLECULES BY I.I.S. OF NEUTRONS.
      *BAJOREK(A.), KOMAROV(V.E.), NATKANIEC(I.), OZEROV(R.P.), PARLINSKI(K.),
      SOLOV≠EV(S.P.),   SUDNIK-HRYNKIEWICZ(M.),   JANIK(J.M.),   JANIK(J.A.).

1370 LI*CL*O4.(H2*O)3,. . . . . .C1. . . . . .REF. SECTION 3, 68/BAJOREK,. . . . .

1371 LI*D,. . . . . .T.B1. . . . . .SOL. STATE COMM. VOL.11 NO.4,. . . 559, (1972).
      *DISPERSION OF PHONONS IN LITHIUM DEUTERIDE.
      *SINGH(R.K.).

1372 LI*(D,H),. . . . . .B. . . . . . .IAEA SYMP. BOMBAY, VOL.2,. . . . . . 431, (1964).
      *STUDY OF THE DYNAMICS OF LITHIUM HYDRIDE AND DEUTERIDE IN THE INELASTIC
      SCATTERING OF COLD NEUTRONS.
      *.C.C.C.P.

1373 LI*F,. . . . . .B1. . . . . .PHYS. REV. 168,. . . . . 970, (1968).
      *LATTICE DYNAMICS OF LITHIUM FLUORIDE.
      *DOLLING(G.), SMITH(H.G.), NICKLOW(R.M.), VIJAYARAGHAVAN(P.R.),
      WILKINSON(M.K.).

1374 (LI-FE)*O4,. . . . . .D1. . . . . .PHYS. REV. LETT. 20,. . . . . 997, (1968).
      *EXPERIMENTAL STUDIES OF POLARIZATION EFFECTS IN SPIN-WAVE SCATTERING
      OF NEUTRONS.
      *RISTE(T.), MOON(R.M.), KOEHLER(W.C.).

1375 (LI-FE)*O4,. . . . . .D1. . . . . .INT. J. MAGN. VOL.3 NO.4,. . . . . 349, (1972).
      *NEUTRON SCATTERING INVESTIGATION OF SPIN WAVES AND EXCHANGE INTERACTIONS
      IN LITHIUM FERRITE.
      *WANIC(A.), RISTE(T.), STEINSVOLL(O.).

1376 LI*H,. . . . . .T. . . . . . .PHYS. REV. B-1,. . . . . . 3510, (1970).
      *NON-CENTRAL-FORCE MODEL OF LI*H PHONON DISPERSION CURVES AND HE
      MIGRATION.
      *WILSON(W.D.), JOHNSON(R.A.).

1377 LI*H,. . . . . .P. . . . . . .REF. SECTION 3, 60/WOODS,. . . . .

1378 LI*H,. . . . . .B1. . . . . .PHYS REV 168,. . . . . 980, (1968).
      *LATTICE DYNAMICS OF LITHIUM HYDRIDE.
      *VERBLE(J.L.), WARREN(J.L.), YARNELL(J.L.).

1379 LI-MG,. . . . . .B1. . . . . .PHYS. LETT. 24A,. . . . . 517, (1967).
      *LOCAL MODES IN LI-MG AND BE-CU ALLOYS.
      *NATKANIEC(I.),   PARLINSKI(K.),   BAJOREK(A.),   SUDNIK-HRYNKIEWICZ(M.).

1380 LI*NB*O3,. . . . . . . .T. . . . . . .J. PHYS. C. SER.2 VOL.4,. . . . . 1091, (1971).
      *GROUP THEORETICAL SELECTION RULES FOR INELASTIC NEUTRON SCATTERING WITH
      APPLICATIONS TO LI*NB*O3.
      *DEVINE(S.), PECKHAM(G.E.).

1381 LI*NB*O3,. . . . . .B1. . . . . .J. PHYS. C VOL.7,. . . . . L99, (1974).
      *LATTICE DYNAMICS OF LITHIUM NIOBATE.
      *CHOWDHURY(M.R.), PECKHAM(G.E.), ROSS(R.T.), SAUNDERSON(D.H.).

1382 LI*O*H,. . . . . .C1. . . . . .J. CHEM. PHYSICS VOL.44,. . . . . 345, (1966).
      *STUDY OF THE LOW-FREQUENCY OH- MOTIONS OF LI*O*H BY NEUTRON INELASTIC
      SCATTERING.
      *SAFFORD(G.J.), LOSACCO(F.J.).

1383 LIQUID CRYSTAL,. . . . . .C3. . . . . .PHYS. LETT. VOL.31A,. . . . . . 531, (1970).
      *NEUTRON SCATTERING STUDY OF SELF DIFFUSION IN LIQUID CRYSTALS.
      *BLINC(R.), DIMIC(V.).

1384 LI2*S*O4*(H2*O),. . . . . . . .C1. . . . . .PHYS. STAT. SOL. VOL.44,. . . . 497, (1971).
      *TORSIONAL VIBRATIONS OF H2*O MOLECULES IN CRYSTALLINE LI2*S*O4*(H2*O).
      *JANIK(J.M.), JANIK(J.A.), PYTASZ(G.), WASIUTYNSKI(T.).

1385 LI2*S*O4.H2*O,. . . . . .C1. . . . . .REF. SECTION 3, 68/BAJOREK,. . . . .

1386 LI2*S*O4.H2*O,. . . . . .C1. . . . . .J. CHEM. PHYSICS VOL.45,. . . 699, (1966).
      *LOW-FREQUENCY MOTIONS OF H2*O MOLECULES IN CRYSTALS: II.
      *PRASK(H.J.), BOUTIN(H.).

1387 LI2*S*O4.H2*O,. . . . . .C1. . . . . .ACTA PHYS. POLON. VOL.A44,. . . . 731, (1973).
      *LOW FREQUENCY MOTIONS OF WATER MOLECULES IN CRYSTALLINE LI2*S*O4.H2*O.
      *MIKULI(E.),   JANIK(J.M.),   PYTASZ(G.),   JANIK(J.A.),   SCIESINSKI(J.),
      SCIESINSKA(E.), MAZURKIEWICZ(A.).

1388 LI2*S*O4.H2*O,. . . . . .C1. . . . . .SOV. PHYS. CRYST. VOL.15,. . . 1010, (1971).
      *NEUTRON-DIFFRACTION STUDY OF THE CRYSTALLINE HYDRATES OF LITHIUM SALTS.
      III.STUDY OF THE DYNAMICS OF WATER MOLECULES BY I.I.S. OF NEUTRONS.
      *BAJOREK(A.), KOMAROV(V.E.), NATKANIEC(I.), OZEROV(R.P.), PARLINSKI(K.),
      SOLOV≠EV(S.P.),   SUDNIK-HRYNKIEWICZ(M.),   JANIK(J.M.),   JANIK(J.A.).

1389 MESITYLENE,. . . . . .C1. . . . . .J. CHEM. PHYSICS VOL.37,. . . . 234, (1962).
      *ROTATIONAL FREEDOM OF AMMONIUM IONS AND METHYL GROUPS BY CROSS-SECTION
      MEASUREMENTS WITH SLOW NEUTRONS.
      *RUSH(J.J.), TAYLOR(T.I.), HAVENS-JR(W.W.).

1390 METAL HYDRIDES,. . . . . .ATOMKERNENERGIE VOL.12,. . . . . 385, (1967).
      *INVESTIGATION OF HYDROGEN AND DEUTERIUM VIBRATIONS IN METAL HYDRIDES,
      HYDRODEUTERIDES AND DEUTERIDES WITH TOTAL NEUTRON CROSS-SECTIONS.
      *SCHMIDT(U.).

1391 METAL HYDRIDES,. . . . . .B7. . . . . .MOL. DYN. AND STRUCT. OF SOLIDS, 315, (1969).
      *EPITHERMAL NEUTRON INELASTIC SCATTERING FROM METAL HYDRIDES.
      *PAN(S.S.), YEATER(M.L.), MOORE(W.E.).

1392 METHOXYBENZYLIDENE-BUTYLANILINE,. . . .C3. . . . . .PHYS. STAT. SOL. VOL.B-54,. . . 121, (1972).
      *STUDY OF MOLECULAR MOTIONS IN LIQUID CRYSTALLINE MBBA BY COLD NEUTRON
      SCATTERING.
      *DIMIC(V.), BARBIC(L.), BLINC(R.).

1393 METHYL BENZENES,. . . . . . . .C1. . . . . .J. CHEM. PHYSICS VOL.37,. . . . 234, (1962).
      *ROTATIONAL FREEDOM OF AMMONIUM IONS AND METHYL GROUPS BY CROSS-SECTION
      MEASUREMENTS WITH SLOW NEUTRONS.
      *RUSH(J.J.), TAYLOR(T.I.), HAVENS-JR(W.W.).

1394 METHYL ALCOHOL,. . . . . . . .C3. . . . . .DISS. ABS.INT. 30,. . . . . . .2380B, (1969).
      *THERMAL NEUTRON INELASTIC SCATTERING BY METHYL ALCOHOL AND METHYL
      MERCAPTAN.
      *SAMPSON(T.E.).

SECTION 6 -QUASI-ELASTIC AND INELASTIC SCATTERING STUDIES . . . .

1395 METHYL MERCAPTAN,. . . . . .C3. . . . . . . .DISS. ABS.INT.,30, ... 2380B, (1969).
&THERMAL NEUTRON INELASTIC SCATTERING BY METHYL ALCOHOL AND METHYL
MERCAPTAN.
&SAMPSON(T.E.).

1396 METHYLBENZENES,. . . . . .C1,C3. . . .J. CHEM. PHYS. VOL.47, ... 3936, (1967).
&NEUTRON-SCATTERING STUDY OF HINDERED ROTATIONS IN METHYLBENZENES.
&RUSH(J.J.).

1397 MG,. . . . . .B1. . . . . . .IAEA SYMP. COPENHAGEN, VOL.1,. . 215, (1968).
&MEASUREMENTS OF FREQUENCIES OF NORMAL MODES OF MAGNESIUM.
&PYNN(R.), SQUIRES(G.L.).

1398 MG,. . . . . .T. . . . . . .IAEA SYMP. COPENHAGEN, VOL.1,. . 165, (1968).
&THEORY OF VIBRATIONAL SPECTRUM IN HEXAGONAL METALS.
&.C.C.C.P.

1399 MG,. . . . . .B1. . . . . . .IAEA SYMP. BOMBAY, VOL.1,. . . . 153, (1964).
&LATTICE DYNAMICS OF MAGNESIUM.
&IYENGAR(P.K.), VENKATARAMAN(G.), VIJAYARAGHAVAN(P.R.), ROY(A.P.).

1400 MG,. . . . . .B1. . . . . . .IAEA SYMP. COPENHAGEN, VOL.1,. . 101, (1968).
&LATTICE DYNAMICS AND ELECTRONIC STRUCTURE OF SODIUM, MAGNESIUM AND
ALUMINIUM.
&SCHNEIDER(T.), STOLL(E.).

1401 MG,. . . . . .T. . . . . . .IAEA SYMP. COPENHAGEN, VOL.1,. . . 3, (1968).
&THE ROLE OF ELECTRONS IN PHONON SPECTRUM FORMATION IN METALS.
&.C.C.C.P.

1402 MG,. . . . . .B1. . . . . .COPENHAGEN CONF.-LATT.DYNAMICS,. 223, (1963).
&DISPERSION RELATIONS FOR PHONONS IN MAGNESIUM.
&IYENGAR(P.K.), VENKATARAMAN(G.), VIJAYARAGHAVAN(P.R.), ROY(A.P.).

1403 MG,. . . . . .B1. . . . . .PROC. ROY. SOC. A 326 NO.1566, . 347, (1972).
&MEASUREMENTS OF THE NORMAL-MODE FREQUENCIES OF MAGNESIUM.
&PYNN(R.), SQUIRES(G.L.).

1404 MG,. . . . . .T. . . . . . .PHYS. LETT. VOL.41A NO.2,. . . . 141, (1972).
&LATTICE VIBRATIONS OF MAGNESIUM: A MODEL POTENTIAL APPROACH.
&BAJPAI(R.P.), ONO(M.).

1405 MG,. . . . . .B1. . . . . .PROC. PHYS. SOC. VOL.80, ... 362, (1962).
&LATTICE DYNAMICS OF MAGNESIUM.
&COLLINS(M.F.).

1406 MG,. . . . . .B1,T. . . . .PHYS. REV. VOL.134,. . . . . . . A1476, (1964).
&LATTICE VIBRATIONAL SPECTRA OF BERYLLIUM, MAGNESIUM AND ZINC.
&YOUNG(J.A.), KOPPEL(J.U.).

1407 MG,. . . . . .P. . . . . . .REF. SECTION 3, 48/RAINWATER,. . . . .

1408 MG,. . . . . .E,T. . . . . .SOV. PHYS. SOLID STATE VOL.15,. . 1309, (1974).
&DETERMINATION OF THE DENSITY OF PHONON STATES FOR COHERENTLY SCATTERING
CRYSTALS.
&EREMEEV(I.P.), SADIKOV(I.P.), CHERNYSHOV(A.A.).

1409 MG,. . . . . .E,T. . . . . .PROC. ROY. SOC. A VOL.212, . . . 192, (1952).
&MULTI-OSCILLATOR PROCESSES IN THE SCATTERING OF NEUTRONS BY CRYSTALS.
&SQUIRES(G.L.).

1410 MG,. . . . . .B1. . . . . .PROC. PHYS. SOC. 88, ... 919, (1966).
&MEASUREMENTS OF THE FREQUENCIES OF THE NORMAL MODES IN MAGNESIUM.
&SQUIRES(G.L.).

1411 MG,. . . . . .B1. . . . . .BROOKHAVEN SYMPOSIUM,. . . . . . 78, (1965).
&MEASUREMENTS OF THE FREQUENCIES OF THE NORMAL MODES OF MAGNESIUM.
&SQUIRES(G.L.).

1412 MG,. . . . . .B1. . . . . .IAEA SYMP. CHALK RIVER, VOL.2, . 87, (1962).
&NEUTRON-PHONON INTERACTION STUDIES IN COPPER, ZINC AND MAGNESIUM SINGLE
CRYSTALS.
&MALISZEWSKI(E.), SOSNOWSKI(J.), BLINOWSKI(K.), KOZUBOWSKI(J.),
PADLO(I.), SLEDZIEWSKA(D.).

1413 MG,. . . . . .B1. . . . . .IAEA SYMP. CHALK RIVER, VOL.2, . 99, (1962).
&DISPERSION RELATIONS FOR PHONONS IN MAGNESIUM.
&IYENGAR(P.K.), VENKATARAMAN(G.), RAO(K.R.), VIJAYARAGHAVAN(P.R.),
ROY(A.P.).

1414 MG*F2, ...B1. . . . . .IAEA SYMP. COPENHAGEN, VOL.1,. . 289, (1968).
&ETUDE DE LA DYNAMIQUE DU RESEAU DE MG*F2 PAR DIFFUSION INELASTIQUE DE
NEUTRONS.
&KAHN(R.), TROTIN(J.P.), CRIBIER(D.), BENOIT(C.).

1415 MG*F2, ...T. . . . . . .J. OF PHYS.-C-, SER.2, VOL.3,. . 1693, (1970).
&DYNAMICS OF THE RUTILE STRUCTURE II. LATTICE DYNAMICS OF MAGNESIUM
FLUORIDE.
&KATIYAR(R.S.).

1416 MG*O,. . . . . .C1. . . . . .COPENHAGEN CONF.-LATT.DYNAMICS,. 49, (1963).
&PHONON DISPERSION RELATIONS IN MAGNESIUM OXIDE.
&PECKHAM(G.).

1417 MG*O,. . . . . .T,B1. . . . .PHYS. REV. B VOL.6,. . . . . . . 1589, (1972).
&CRYSTAL DYNAMICS OF MAGNESIUM OXIDE.
&SINGH(R.K.), UPADHYAYA(K.S.).

1418 MG*O,. . . . . .T. . . . . . .J. PHYS. C, VOL.3, ... 86, (1970).
&THE SCATTERING OF LONG WAVELENGTH NEUTRONS BY NEUTRON IRRADIATED MG*O.
&SABINE(T.M.), SVENSON(A.C.).

1419 MG*O,. . . . . .B. . . . . . .PHYS. REV. LETT. 20,. . . . . . 209, (1968).
&SEARCH FOR SURFACE MODES OF LATTICE VIBRATIONS IN MAGNESIUM OXIDE.
&RIEDER(K.H.), HORL(E.M.).

1420 MG*O,. . . . . .B1. . . . . .J. PHYS. C, SER.2, VOL.3,. . . . 1026, (1970).
&LATTICE DYNAMICS OF MAGNESIUM OXIDE.
&SANGSTER(M.J.L.), PECKHAM(G.E.), SAUNDERSON(D.H.).

1421 MG*O,. . . . . .B1. . . . . .PROC. PHYS. SOC. 90, ... 657, (1967).
&THE PHONON DISPERSION RELATION FOR MAGNESIUM OXIDE.
&PECKHAM(G.E.).

SECTION   6 -QUASI-ELASTIC AND INELASTIC SCATTERING STUDIES . . . .

1422 MG*O,. . . . . .B1. . . . . . .PROC. PHYS. SOC. 89,. . . . . . . 153, (1966).
      *ZERO-PHONON LINES UNDER UNIAXIAL STRESS IN MAGNESIUM OXIDE.
      *KING(R.D.), HENDERSON(B.).

1423 MG*O,. . . . . .B1. . . . . . .PHYS REV 174,. . . . . 953, (1968).
      *NEUTRON SCATTERING FROM MG*O.
      *BORGONOVI(G.M.), CARRIVEAU(G.W.).

1424 MG*(O*H)2, ...C2. . . . . . .J. CHEM. PHYSICS VOL.43, ...1864, (1965).
      *INELASTIC NEUTRON SPECTRA AND THE VIBRATIONAL MODES OF THE HYDROGEN
      LAYER IN ALKALI AND ALKALINE-EARTH HYDROXIDES.
      *PELAH(I.), KREBS(K.), IMRY(Y.).

1425 MG*(O*H)2, ...C1. . . . . . .J. PHYS. CHEM. SOL 24,. . . . . . 771, (1963).
      *AN INVESTIGATION OF THE ENERGY LEVELS IN ALKALINE EARTH HYDROXIDES BY
      INELASTIC SCATTERING OF SLOW NEUTRONS.
      *SAFFORD(G.J.), BRAJOVIC(V.), BOUTIN(H.).

1426 MG2*PB,. . . . . .B1. . . . . . .PHYS. REV. B VOL.5,. . . . . . . 2103, (1972).
      *LATTICE DYNAMICS OF MG2*PB AT ROOM TEMPERATURE.
      *WAKABAYASHI(N.), AHMAD(A.A.Z.), SHANKS(H.R.), DANIELSON(G.C.).

1427 MG-(PB), ...B3. . . . . . .SOV. PHYS. J.E.T.P. VOL.22,. . . . 315, (1966).
      *INVESTIGATION OF THE QUASILOCAL LEVEL IN THE VIBRATION SPECTRUM OF A
      LATTICE WITH HEAVY IMPURITY IONS.
      *CHERNOPLEKOV(N.A.), ZEMLYANOV(M.G.).

1428 MG2*SN,. . . . . .B1. . . . . . .J. PHYS. CHEM. SOLIDS VOL.31,. . 1085, (1970).
      *LATTICE DYNAMICS OF MAGNESIUM STANNIDE AT ROOM TEMPERATURE.
      *KEARNEY(R.J.), WORLTON(T.G.), SCHMUNK(R.E.).

1429 MG2*SN,. . . . . .B1. . . . . . .RCN-121 (119 PP.),. . . . . , (1970).
      *LATTICE DYNAMICS OF MAGNESIUM STANNIDE AND ZINC BLENDE.
      *BERGSMA(J.). (ED. NOTE: AVAIL. REACTOR CENTRUM NEDERLAND, PETTEN, THE
      NETHERLANDS).

1430 MG*ZN2,. . . . . .T. . . . . . .PHYS. STAT. SOL. VOL.49,. . . . . 807, (1972).
      *CALCULATED PHONON SPECTRA AND INS CROSS SECTIONS OF MG*ZN2.
      *ESCHRIG(H.), URWANK(P.), WONN(H.).

1431 MG*ZN2,. . . . . .B1,T. . . . . . .IAEA SYMP. GRENOBLE,. . . . . . . 157, (1972).
      *PHONONS IN MG*ZN2: MODEL POTENTIAL CALCULATIONS AND NEUTRON SCATTERING
      EXPERIMENTS.
      *ESCHRIG(H.), FELDMANN(K.), HENNIG(K.), WEISS(L.).

1432 MG*ZN2,. . . . . .T. . . . . . .PHYS. STAT. SOLIDI B VOL.58 NO.2 K159, (1973).
      *PHONON DENSITY OF STATES AND COHERENT INELASTIC NEUTRON SCATTERING
      ON COMPLEX STRUCTURES.
      *ESCHRIG(H.), FELDMANN(K.), HENNIG(K.), JOHN(W.), VON-LOYEN(L.).

1433 MN,. . . . . .P. . . . . . .DISS. ABS. VOL.31, ...21718, (1970).
      *INELASTIC NEUTRON SCATTERING FROM FE, AL, AND MN.
      *LAMB(R.C.).

1434 MN (PHASE: BETA),. . . . . .D1. . . . . . .PHYS. STAT. SOL. VOL.47,. . . . . K11, (1971).
      *PARAMAGNETIC NEUTRON SCATTERING IN BETA-MANGANESE.
      *VANCE(E.R.), SMITH(J.H.).

1435 MN (GAMMA),. . . . . .D1. . . . . . .J. PHYSIQUE TOME 32 COL.1 VOL.II 1186, (1971).
      *SPIN CORRELATIONS IN GAMMA MANGANESE.
      *HAYWOOD(B.C.G.), LOWDE(R.D.), STRINGFELLOW(M.W.), WAEBER(W.B.).

1436 MN*AS, ...D1. . . . . . .PHYS. STAT. SOL. VOL.43, ... K165, (1971).
      *DIFFUSE NEUTRON SCATTERING IN MANGANESE ARSENIDE.
      *SIROTA(N.N.), GOVOR(G.G.).

1437 MN*C*O3, ...D1. . . . . . .CAN. J. PHYS. 50,. . . . . 687, (1972).
      *SPIN-WAVE EXCITATIONS IN MANGANESE CARBONATE.
      *HOLDEN(T.M.), SVENSSON(E.C.), MARTEL(P.).

1438 MN-CO, ...C1. . . . . . .REV. MOD. PHYS. VOL.30,. . . . . . 250, (1958).
      *VIBRATION SPECTRA OF VANADIUM AND A MN-CO ALLOY BY NEUTRON SPECTROMETRY.
      *STEWART(A.T.), BROCKHOUSE(B.N.).

1439 (MN-CO)*F2,. . . . . .D1,T. . . . . . .J. PHYS. C VOL.5,. . . . . . . .2611, (1972).
      *THEORY OF SPIN WAVES IN DISORDERED ANTIFERROMAGNETS: I. APPLICATION TO
      (MN,CO)*F2 AND K*(MN,CO)*F3.
      *BUYERS(W.J.L.), PEPPER(D.E.), ELLIOTT(R.J.).

1440 MN*CR2*O4, ...D3. . . . . . .J. APP. PHYS. 36,. . . . . . . .1090, (1965).
      *DIFFUSE PARAMAGNETIC NEUTRON SCATTERING IN CHROMIUM SPINELS.
      *DWIGHT(K.), MENYUK(N.), KAPLAN(S.F.).

1441 MN*CR2*S4, ...D3. . . . . . .J. APP. PHYS. 36,. . . . . . . .1090, (1965).
      *DIFFUSE PARAMAGNETIC NEUTRON SCATTERING IN CHROMIUM SPINELS.
      *DWIGHT(K.), MENYUK(N.), KAPLAN(S.F.).

1442 (MN-CR)2*SB, ...D1. . . . . . .PHYSICA VOL.57,. . . . . 628, (1972).
      *DISCOVERY OF A MAGNON SOFT MODE IN CHROMIUM-MODIFIED MN2*SB.
      *ALIKHANOV(R.A.), DIMITRIJEVIC(Z.), RZANY(H.), TODOROVIC(J.), WANIC(A.).

1443 MN90-CU10, ...D3. . . . . . .J. PHYS. CHEM. SOLIDS VOL.31,. . . 485, (1970).
      *PARAMAGNETIC DIFFUSE SCATTERING OF NEUTRONS BY A 90 AT PCT MANGANESE
      COPPER ALLOY.
      *VANCE(E.R.), SMITH(J.H.).

1444 MN-CU, ...D4. . . . . . .J. APPL. PHYS. 41,. . . . . . . .1857, (1970).
      *NEUTRON SCATTERING AT SMALL ANGLES FROM MN-CU ALLOYS.
      *VANCE(E.R.), SMITH(J.H.).

1445 MN-CU, ...B7,D3. . . . . . .J. PHYS.F: METAL PHYS. VOL.1,. . . 763, (1971).
      *NEUTRON DIFFRACTION STUDY OF SHORT RANGE ORDER IN GAMMA MN-CU ALLOYS.
      *WELLS(P.), SMITH(J.H.).

1446 MN*F2, ...D2. . . . . . .PHYS. REV. B-1,. . . . . .2304, (1970).
      *CRITICAL MAGNETIC SCATTERING IN MANGANESE FLUORIDE.
      *SCHULHOF(M.P.), HELLER(P.), NATHANS(R.), LINZ(A.).

1447 MN*F2, ...D2. . . . . . .PHYS. REV. B-4,. . . . . .2254, (1971).
      *INELASTIC NEUTRON SCATTERING FROM MN*F2 IN THE CRITICAL REGION.
      *SCHULHOF(M.P.), NATHANS(R.), HELLER(P.), LINZ(A.).

1448 MN*F2, ...D2. . . . . . .J. APPL. PHYS. 42,. . . . . . . .1258, (1971).
      *INELASTIC NEUTRON SCATTERING STUDIES OF CRITICAL FLUCTUATIONS IN MN*F2
      ABOVE AND BELOW T-NEEL.
      *HELLER(P.), SCHULHOF(M.P.), NATHANS(R.), LINZ(A.).

SECTION   6 -QUASI-ELASTIC AND INELASTIC SCATTERING STUDIES . . . .

1449 MN*F2, ... D4. . . . . . . .J. APPL. PHYS. 41, ... 896, (1970).
      *TWO-MAGNON SCATTERING OF NEUTRONS BY MN*F2.
      *HOLDEN(T.M.), SVENSSON(E.C.), BUYERS(W.J.L.), COWLEY(R.A.),
      STEVENSON(R.W.H.).

1450 MN*F2, ... D3. . . . . . . .IAEA SYMP. CHALK RIVER, VOL.2, . 309, (1962).
      *DIFFUSION PAR MN*F2 A L#ETAT PARAMAGNETIQUE.
      *CRIBIER(D.), JACROT(B.).

1451 MN*F2, ... D2. . . . . . . .PHYS. REV. LETT. VOL.24, ...1184, (1970).
      *INELASTIC NEUTRON SCATTERING FROM MN*F2 IN THE CRITICAL REGION.
      *SCHULHOF(M.P.), HELLER(P.), NATHANS(R.), LINZ(A.).

1452 MN*F2, ... D1. . . . . . . .DISS. ABS. VOL.31, ...36599, (1970).
      *CRITICAL MAGNETIC SCATTERING OF NEUTRONS FROM MANGANESE FLUORIDE.
      *SCHULHOF(M.P.).

1453 MN*F2, ... D2. . . . . . . .PROC. INT. CONF. (NOTTINGHAM), .   92, (1964).
      *CRITICAL MAGNETIC SCATTERING OF NEUTRONS IN MN*F2.
      *OKAZAKI(A.), STEVENSON(R.W.H.), TURBERFIELD(K.C.).

1454 MN*F2, ... D4. . . . . . . .J. PHYS. C VOL.4, NO.13, ...L258, (1971).
      *DIFFUSE MAGNETIC SCATTERING OF NEUTRONS FROM POLYCRYSTALLINE MN*F2
      AT 4.2K.
      *SMITH(J.H.), VANCE(E.R.).

1455 MN*F2, ... D4. . . . . . . .SOL. STATE COMM. 6, ... 145, (1968).
      *OBSERVATION OF LOCALIZED MAGNON BY NEUTRON SCATTERING.
      *HOLDEN(T.M.), COWLEY(R.A.), BUYERS(W.J.L.), STEVENSON(R.W.H.).

1456 MN*F2, ... T,D4. . . . .PROC. PHYS. SOC. 89, ... 77, (1966).
      *TRANVERSE MAGNETIC FLUCTUATIONS IN MN*F2 NEAR THE CRITICAL POINT.
      *TORRIE(B.H.).

1457 MN*F2, ... D2. . . . . . .J. PHYS. CHEM. SOL. 28, ... 11, (1967).
      *DIFFUSION MAGNETIQUE DES NEUTRONS PAR MN*F2 PRES DU POINT DE NEEL.
      *ANTONINI(M.).

1458 MN*F2, ... D4. . . . . . .PHYS. LETT. 8, ... 9, (1964).
      *NEUTRON INELASTIC SCATTERING MEASUREMENTS OF ANTIFERROMAGNETIC EXCITA-
      TIONS IN MN*F2 AT 4.2 DEG. K AND AT TEMPERATURES UP TO THE NEEL POINT.
      *OKAZAKI(A.), TURBERFIELD(K.C.), STEVENSON(R.W.H.).

1459 MN*F2, ... D2. . . . . . .J. APP. PHYS. 39, ...1232, (1968).
      *MAGNETIC SCATTERING OF NEUTRONS IN MN*F2 NEAR THE CRITICAL POINT.
      *PARETTE(G.), USHA-DENIZ(K.).

1460 MN*F2, ... D1. . . . . . .J. APP. PHYS. 34, ...1182, (1963).
      *MEASUREMENT OF THE COVALENT SPIN DISTRIBUTION IN MANGANOUS FLUORIDE
      USING POLARIZED NEUTRONS.
      *NATHANS(R.), ALPERIN(H.A.), PICKART(S.J.), BROWN(P.J.).

1461 MN*F2, ... D1. . . . . . .PROC. PHYS. SOC. 85, ... 743, (1965).
      *THE DEVELOPMENT OF THE MAGNETIC EXCITATION SPECTRA OF MN*F2 WITH
      INCREASING TEMPERATURE.
      *TURBERFIELD(K.C.), OKAZAKI(A.), STEVENSON(R.W.H.).

1462 MN*F2, ... D1. . . . . . .J. APPL. PHYS. VOL.35, ... 998, (1964).
      *A MEASUREMENT OF SPIN-WAVE DISPERSION IN MN*F2 AT 4.2 DEGREES K.
      *LOW(G.G.), OKAZAKI(A.), STEVENSON(R.W.H.), TURBERFIELD(K.C.).

1463 MN*F2, ... D1,D2. . . .J. PHYS. RADIUM VOL.23, ... 494, (1962).
      *SOME APPLICATIONS OF INELASTIC NEUTRON SCATTERING TO THE STUDY OF
      MAGNETIC COUPLINGS.
      *JACROT(B.), CRIBIER(D.). (IN FRENCH).

1464 MN*F2, ... D1,T. . . .PROGR. THEOR. PHYS. VOL.25, ... 595, (1961).
      *SPIN WAVES IN MN*F2 AND THE INELASTIC SCATTERING OF NEUTRONS.
      *NAGAI(O.), YOSHIMORI(A.).

1465 MN*F2, ... D3. . . . . . .PHYS. REV. VOL.75, ... 895, (1949).
      *THE SCATTERING OF SLOW NEUTRONS BY PARAMAGNETIC CRYSTALS.
      *RUDERMAN(I.W.), HAVENS-JR(W.W.), TAYLOR(T.I.), RAINWATER(L.J.).

1466 MN*F2, ... D1. . . . . . .J. PHYS. C, SER.2, VOL.2, ...1168, (1969).
      *MAGNON DISPERSION RELATION AND EXCHANGE INTERACTIONS IN MN*F2.
      *NIKOTIN(O.P.), LINDGARD(P.A.), DIETRICH(O.W.).

1467 MN*F2, ... D2. . . . . . .J. PHYS. C, SER.2, VOL.2, ... 2022, (1969).
      *CRITICAL MAGNETIC FLUCTUATIONS IN MN*F2.
      *DIETRICH(O.W.).

1468 MN*F2, ... C1. . . . . . .C.R. ACAD. SCI. VOL.248, ...1631, (1959).
      *SCATTERING OF COLD NEUTRONS BY MANGANESE FLUORIDE.
      *CRIBIER(D.), ERICSON(M.), JACROT(B.), SARMA(G.). (IN FRENCH).

1469 MN*F2, ... D2. . . . . . .A.I.P. CONF. PROC. NO.10 PART 1, . 93, (1973).
      *ABSOLUTE MEASUREMENT OF THE TRANSVERSE K-DEPENDENT SUSCEPTIBILITY IN
      ANTIFERROMAGNETIC MN*F2.
      *LURIE(N.A.), SHIRANE(G.), HELLER(P.), LINZ(A.).

1470 MN*F2, ... T,D1. . . .J. APPL. PHYS. 36, ... 884, (1965).
      *SPIN WAVES AND NEUTRON SCATTERING.
      *LOWDE(R.D.).

1471 MN*F2, ... T,D4. . . .PHYS. REV. B VOL.5, ...1993, (1972).
      *MAGNON-PHOTON COUPLING IN ANTIFERROMAGNETS.
      *MANOHAR(C.), VENKATARAMAN(G.).

1472 MN*F2, ... D3. . . . . . .PHYS. REV. VOL.89, ... 561, (1953).
      *NEUTRON SCATTERING BY COUPLED PARAMAGNETIC IONS.
      *BENDT(P.J.).

1473 MN*F2, ... D1. . . . . . .IAEA SYMPOSIUM BOMBAY, VOL.1, . . . 453, (1964).
      *COMPARISON OF SPIN WAVE THEORY WITH NEUTRON SCATTERING RESULTS FOR MN*F2
      *LOW(G.G.).

1474 MN*F2(CO), MN*F2(ZN), ...D1. . . . .MAGN. AND MAG. MATERIALS-1971, .1315, (1972).
      *EFFECTS OF IMPURITIES IN ANTIFERROMAGNETS.
      *SVENSSON(E.C.), BUYERS(W.J.L.), HOLDEN(T.M.), COWLEY(R.A.),
      STEVENSON(R.W.H.). (AIP CONF. PROC. NO.5 PART 2; GRAHAM-JR., RHYNE-EDS.)

1475 MN*F2(CO), ... D4. . . . . . .J. APPL. PHYS. VOL.40, ... 991, (1969).
      *TEMPERATURE DEPENDENCE OF THE LOCAL MODE IN COBALT-DOPED MN*F2.
      *HOLDEN(T.M.), BUYERS(W.J.L.), STEVENSON(R.W.H.).

SECTION 6 -QUASI-ELASTIC AND INELASTIC SCATTERING STUDIES . . . .

1476 MN*F2(CO), ...D1. . . . . . . .J. APP. PHYS. 39, ... 1118, (1968).
⇢OBSERVATION OF A LOCALIZED MAGNON IN CO-DOPPED MN*F2.
⇢BUYERS(W.J.L.), COWLEY(R.A.), HOLDEN(T.M.), STEVENSON(R.W.H.).

1477 MN*F2(CO,NI,ZN), ...D1,D4. . . . .IAEA SYMP. GRENOBLE, ... 581, (1972).
⇢MAGNETIC MODES ASSOCIATED WITH IMPURITIES IN INSULATORS.
⇢BUYERS(W.J.L.), SVENSSON(E.C.), HOLDEN(T.M.), COWLEY(R.A.),
STEVENSON(R.W.H.).

1478 MN*F2(FE,NI,CO), ...T. . . . . . . .PROC. ROYAL SOC. EDINBURGH A-70, 75, (1972).
⇢THE TEMPERATURE DEPENDENCE OF LOCALISED MAGNONS.
⇢COWLEY(R.A.).

1479 MN*F2,MN*O,. . . . . .D3. . . . . . .PHYS. REV. VOL.76,. . . . . 1572, (1949).
⇢THE SCATTERING OF SLOW NEUTRONS BY PARAMAGNETIC CRYSTALS.
⇢RUDERMAN(I.W.).

1480 MN*F2,MN*O,. . . . . .D3. . . . . . .DISS. ABSTR. VOL.11,. . . . . 1080, (1951).
⇢THE DEPENDENCE OF PARAMAGNETIC NEUTRON SCATTERING ON COUPLING BETWEEN
MN(++) IONS.
⇢BENDT(P.J.).

1481 MN*F2(ZN), ...D1. . . . . . . .SOL. STATE COMM. VOL.7,. . . . . . 1693, (1969).
⇢ON THE RESONANT PERTURBATION OF SPIN WAVES BY IMPURITIES.
⇢SVENSSON(E.C.), HOLDEN(T.M.), BUYERS(W.J.L.), COWLEY(R.A.).

1482 MN*FE2*O4, ...D1. . . . . . . .PHYS. LETT. VOL.40A, ... 101, (1972).
⇢MAGNON DISPERSION MEASUREMENT IN MN*FE2*O4.
⇢RAKHECHA(V.C.), MADHAV-RAO(L.), MURTHY(N.S.S.), SRINIVASAN(B.S.).

1483 MN*O,. . . . . .D1. . . . . . . .J. OF PHYS.-C; SER. 2, VOL.2, . 874, (1969).
⇢MAGNONS IN ANTIFERROMAGNETIC NI*O, MN*O AND MN*S.
⇢REISSLAND(J.A.), BEGUM(N.A.).

1484 MN*O,. . . . . .D1. . . . . . . .IAEA SYMP. COPENHAGEN, VOL.2,. . 111, (1968).
⇢SPIN WAVES AROUND THE NEEL TEMPERATURE IN MN*O.
⇢KROO(N.), BATA(L.).

1485 MN*O,. . . . . .B1. . . . . . . .J. PHYS. C; SOLID ST. PHYS. V-4, 1299, (1971).
⇢OPTICAL PHONONS IN MN*O.
⇢HAYWOOD(B.C.), COLLINS(M.F.).

1486 MN*O,. . . . . .D1. . . . . . . .SOL. STATE COMM. VOL.10, ... 553, (1972).
⇢SPIN WAVES IN MN*O AT 4.2 DEG. K.
⇢BONFANTE(M.), HENNION(B.), MOUSSA(F.), PEPY(G.).

1487 MN*O2, ...T. . . . . . . .PHYS. LETT. VOL.36A, ... 439, (1971).
⇢MAGNON DISPERSION RELATIONS IN ANTIFERROMAGNETIC BETA-MN*O2.
⇢MORGAN(D.), JOSHUA(S.J.).

1488 MN*O,. . . . . .D1. . . . . . . .PROC. INT. CONF. (NOTTINGHAM), 319, (1964).
⇢SPIN WAVES IN MN*O.
⇢COLLINS(M.F.).

1489 MN*O,. . . . . .T,B1. . . . .SOL. STATE COMM. VOL.11, ... 109, (1972).
⇢DISPERSION PROPERTIES OF MANGANOUS OXIDE.
⇢UPADHYAYA(K.S.), SINGH(R.K.).

1490 MN*O,. . . . . .D1. . . . . . . .IAEA SYMP. GRENOBLE, ... 631, (1972).
⇢TEMPERATURE DEPENDENCE OF THE SPIN-WAVE SPECTRUM IN MN*O.
⇢BONFANTE(M.), HENNION(B.), MOUSSA(F.), PEPY(G.).

1491 MN*O,. . . . . .D1. . . . . . . .PHYSICS, ... 31, (1964).
⇢SPIN CORRELATIONS IN MN*O.
⇢BLECH(I.A.), AVERBACH(B.L.).

1492 MN*O,. . . . . .D1. . . . . . . .ACTA CRYST. VOL.A28 PART S-4,. 208, (1972).
⇢INELASTIC NEUTRON SCATTERING STUDY OF SPIN WAVES IN MN*O.
⇢ISHIKAWA(Y.), KOHGI(M.), ENDOH(Y.), TAJIMA(K.). (ABSTRACT ONLY).

1493 MN2*O3,. . . . . .D3. . . . . . .PHYS. REV. VOL.99,. . . . . 601, (1955).
⇢ENERGY DISTRIBUTION OF NEUTRONS SCATTERED BY PARAMAGNETIC SUBSTANCES.
⇢BROCKHOUSE(B.N.).

1494 MN*O,. . . . . .D1. . . . . . . .J. PHYS. SOC. JAPAN VOL.36,. . . 112, (1974).
⇢SPIN WAVES IN MANGANESE MONOXIDE.
⇢KOHGI(M.), ISHIKAWA(Y.), HARADA(I.), MOTIZUKI(K.).

1495 MN*O,. . . . . .D1. . . . . . . .J. PHYS. CHEM. SOLIDS VOL.35,. . 433, (1974).
⇢SPIN WAVES IN MN*O; FROM 4 DEGREES K TO TEMPERATURES CLOSE TO T(NEEL).
⇢PEPY(G.).

1496 MN*O,. . . . . .D3. . . . . . .SOL. STATE COMM. 7,. . . . . 123, (1969).
⇢NEUTRON SCATTERING FROM SHORT RANGE ORDERED PARAMAGNETS.
⇢MADHAV-RAO(L.), DASANNACHARYA(B.A.), MURTHY(N.S.S.), IYENGAR(P.K.).

1497 MN*O,. . . . . .D1. . . . . . . .PROC. INT. CONF. (NOTTINGHAM), 322, (1964).
⇢ENERGY SPECTRA OF NEUTRONS SCATTERED FROM PARAMAGNETIC MN*O.
⇢USHA-DENIZ(K.), VENKATARAMAN(G.), MURTHY(N.S.S.), DASANNACHARYA(B.A.),
IYENGAR(P.K.).

1498 MN*O,. . . . . .D1. . . . . . . .SOL. STATE COMM. VOL.11, ... 391, (1972).
⇢INELASTIC NEUTRON SCATTERING STUDY OF SPIN WAVES IN MN*O.
⇢KOHGI(M.), ISHIKAWA(Y.), ENDOH(Y.).

1499 MN*O,. . . . . .D3. . . . . . .IAEA SYMP. BOMBAY, VOL.1,. . . . 433, (1964).
⇢NEUTRON SCATTERING BY PARAMAGNETICS.
⇢MURTHY(N.S.S.), VENKATARAMAN(G.), USHA-DENIZ(K.), DASANNACHARYA(B.A.),
IYENGAR(P.K.).

1500 MN*O,. . . . . .B1. . . . . . . .J. PHYS. C, SER.2, VOL.2,. . . . 46, (1969).
⇢LATTICE DYNAMICS OF MN*O.
⇢HAYWOOD(B.C.G.), COLLINS(M.F.).

1501 MN*O,. . . . . .D3. . . . . . .BULL. AMER. PHYS. SOC. VOL.3,. . 195, (1958).
⇢NEUTRON SCATTERING BY PARAMAGNETIC MN*O.
⇢IYENGAR(P.K.), BROCKHOUSE(B.N.). (ABSTRACT ONLY).

1502 MN*O,. . . . . .D1. . . . . . . .INT. J. MAGN. VOL.4 NO.1,. . . . 17, (1973).
⇢SPIN WAVES IN MN*O AT LOW TEMPERATURES.
⇢COLLINS(M.F.), TONDON(V.K.), BUYERS(W.J.L.).

1503 MN*O,. . . . . .D3. . . . . . .PHYS. REV. VOL.89,. . . . . 561, (1953).
⇢NEUTRON SCATTERING BY COUPLED PARAMAGNETIC IONS.
⇢BENDT(P.J.).

SECTION 6 -QUASI-ELASTIC AND INELASTIC SCATTERING STUDIES . . . .

1504 MN*O,MN*S, ... D3 . . . . PHYS. REV. VOL.54 . . . . 771, (1938).
SCATTERING OF SLOW NEUTRONS BY PARAMAGNETIC SALTS.
WHITAKER(M.D.), BEYER(H.G.), DUNNING(J.R.).

1505 MN*P, ... D2 . . . . . J. APPL. PHYS. 42 . . . . . . NEUTRON SCATTERING. 1374, (1971).
MAGNETIC CRITICAL PHENOMENA IN MN*P BY NEUTRON SCATTERING.
MINKIEWICZ(V.J.), GESI(K.), HIRAHARA(E.).

1506 MN*P, ... D4 . . . . . . PROC. PHYS. SOC. 88 . . . . . . 333, (1966).
THE STRUCTURE OF THE METAMAGNETIC PHASE OF MN*P.
FORSYTH(J.B.), PICKART(S.J.), BROWN(P.J.).

1507 MN-PD, ... D4 . . . . . PROC. PHYS. SOC. 86 . . . . . 139, (1965).
DIFFUSE MAGNETIC SCATTERING OF NEUTRONS FROM A METALLIC ANTIFERROMAGNET.
HICKS(T.J.), BROWNE(J.D.).

1508 MN-PT3, ... D1 . . . . . PHYS. LETT. 30A . . . . IN MN-PT3 FERROMAGNETIC ALLOY. 310, (1969).
SPIN WAVES IN EXCHANGE INTERACTIONS IN MN-PT3 FERROMAGNETIC ALLOY.
ANTONINI(B.), FELICI(M.), MENZINGER(F.).

1509 MN-PT3, ... D1 . . . . . SOL. STATE COMM. VOL.10 . . . . 203, (1972).
NEUTRON SCATTERING STUDY OF SPIN WAVES AND PHONONS IN MN-PT3.
ANTONINI(B.), MINKIEWICZ(V.J.).

1510 MN*S (ALPHA), . . D1 . . . . . J. OF PHYS.-C-, SER. 2, VOL.2, . 874, (1969).
MAGNONS IN ANTIFERROMAGNETIC NI*O, MN*O AND MN*S.
REISSLAND(J.A.), BEGUM(N.A.).

1511 MN*S*O4, ... D1 . . . . . J. APP. PHYS. 36 . . . MN*S*O4. . . . 1095, (1965).
CYCLOIDAL SPIN CONFIGURATION OF ORTHORHOMBIC MN*S*O4.
WILL(G.), FRAZER(B.C.), SHIRANE(G.), COX(D.E.), BROWN(P.J.).

1512 MN*S*O4, ... D3 . . . . . PHYS. REV. VOL.99 . . . . . 601, (1955).
ENERGY DISTRIBUTION OF NEUTRONS SCATTERED BY PARAMAGNETIC SUBSTANCES.
BROCKHOUSE(B.N.).

1513 MN2*SB, ... D1 . . . . . J. APP. PHYS. 34 . . . . . 1201, (1963).
ASPHERICAL SPIN DENSITY IN THE FERROMAGNETIC COMPOUND MN2*SB.
ALPERIN(H.A.), BROWN(P.J.), NATHANS(R.).

1514 (MN-ZN)*F2, . . T,D1 . . . . . J. PHYS. C VOL.6, ... 1933, (1973).
SPIN WAVES IN DISORDERED SYSTEMS: II. THE DILUTE ANTIFERROMAGNET (MN,ZN)*F2.
BUYERS(W.J.L.), PEPPER(D.E.), ELLIOTT(R.J.).

1515 MO, ... B1 . . . . . . IAEA SYMP. BOMBAY, VOL.1, ... 87, (1964).
LATTICE DYNAMICS OF TRANSITION METALS.
WOODS(A.D.B.).

1516 MO, ... B1 . . . . . . THESIS (208 PP.) . . . . . . , (1964).
NEUTRON SCATTERING STUDIES OF LATTICE VIBRATIONS IN METALS.
CHEN(S.H.). (ED. NOTE: PH.D. DISSERTATION, MCMASTER UNIVERSITY).

1517 MO, ... B7 . . . . . . CAN. J. PHYS. VOL.50 . . . . . 3069, (1972).
TEMPERATURE DEPENDENCE OF A KOHN ANOMALY IN MOLYBDENUM.
BUYERS(W.J.L.), POWELL(B.M.), WOODS(A.D.B.).

1518 MO, ... T,B1 . . . . . J. PHYS. SOC. JAP. VOL.33, . . . 1207, (1972).
LATTICE DYNAMICS OF SOME BCC TRANSITION METALS.
BEHARI(J.), TRIPATHI(B.B.).

1519 MO, ... B1 . . . . . . SOL. STATE COMM. 2, ... 233, (1964).
LATTICE DYNAMICS OF MOLYBDENUM.
WOODS(A.D.B.), CHEN(S.H.).

1520 MO, ... T,B1 . . . . . PHYS. REV. 143, ... 443, (1966).
LATTICE DYNAMICS AND SPECIFIC HEATS OF SOME TRANSITION METALS ON KREBS#S MODEL.
MAHESH(P.S.), DAYAL(B.).

1521 MO, ... B1 . . . . . . PHYS. REV. 177, ... 1111, (1969).
LATTICE VIBRATIONS IN MOLYBDENUM.
WALKER(C.B.), EGELSTAFF(P.A.).

1522 MO*S2, ... B1 . . . . . . BULL. AMER. PHYS. SOC. VOL.17, . 292, (1972).
PHONON DISPERSION CURVES FOR HEXAGONAL MO*S2.
WAKABAYASHI(N.), SMITH(H.G.), NICKLOW(R.M.). (ED. NOTE: PAPER PRESENTED AT MARCH MEETING IN ATLANTIC CITY).

1523 N2, ... C3 . . . . . . IAEA SYMP. COPENHAGEN, VOL.1, . 439, (1968).
MOLECULAR DYNAMICS IN LIQUID NITROGEN AND LIQUID BISMUTH INVESTIGATED BY COLD NEUTRON SCATTERING.
MATEESCU(N.), TEUTSCH(H.), DIACONESCU(A.), NAHORNIAK(V.).

1524 N2, ... C3 . . . . . . PHYS. LETT. 22 . . . . . . 558, (1966).
COLD NEUTRON SCATTERING BY LIQUID NITROGEN.
TEUTSCH(H.), MATEESCU(N.), NAHORNIAK(V.), DIACONESCU(A.).

1525 N2, ... T,E . . . . . HELV. PHYS. ACTA VOL.20, . . . 105, (1947).
ETUDE DES FORCES INTERMOLECULAIRES PAR DIFFUSION DES NEUTRONS LENTS. APPLICATION A N2, H2*O ET K*H2*P*O4.
ROSSEL(J.).

1526 N2, ... T . . . . CONF. NEUTRON THERMALIZ. VOL.I, 303, (1968).
NEUTRON SCATTERING FROM MOLECULAR NITROGEN.
YOUNG(J.A.). (IN BOOK: NEUTRON THERMALIZATION AND REACTOR SPECTRA. PUBL. IAEA: VIENNA, 1968).

1527 N*(C*H3)4*MN*CL3, . . P . . . . . J. PHYSIQUE TOME 34 . . . . 473, (1973).
NEUTRON SCATTERING AND CATION ROTATIONAL MOTION IN TETRAMETHYLAMMONIUM MANGANESE CHLORIDE.
LASSIER(B.), BROT(C.), WHITE(J.W.).

1528 N*D4*BR, ... C1 . . . . . REF. SECTION 3, 66/VENKATARAMAN, ...

1529 N*D4*BR, ... B1,B6 . . PHYS. REV. B VOL.9, ... 4429, (1974).
DYNAMICAL CRITICAL PHENOMENA IN N*D4*BR.
YAMADA(Y.), NODA(Y.), AXE(J.D.), SHIRANE(G.).

1530 N*D4*(BR,CL), . . T,B7 . . . . PHYS. REV. B VOL.7 . . . . . 1644, (1973).
THEORY OF INELASTIC NEUTRON SCATTERING FROM ORIENTATIONALLY DISORDERED MOLECULAR CRYSTALS, WITH PARTICULAR APPLICATION TO N*D4*BR AND N*D4*CL.
SOKOLOFF(J.B.), LOVELUCK(J.M.).

SECTION   6 -QUASI-ELASTIC AND INELASTIC SCATTERING STUDIES . . . .

1531 N*D4*CL, ...B1 . . . . . . .PHYS. REV. B-3, ... 2733, (1971).
      *LATTICE VIBRATIONS IN DEUTERATED AMMONIUM CHLORIDE AT 85 K.
      I. EXPERIMENTAL.
      *TEH(H.C.), BROCKHOUSE(B.N.).

1532 N*D4*CL, ...T. . . . . . .PHYS. LETT. VOL.34A, ...   118, (1971).
      *LATTICE DYNAMICS OF N*D4*CL.
      *JEX(H.).

1533 N*D4*CL, ...B1. . . . . . .THESIS (183 PP.), ... , (1971).
      *CRYSTAL DYNAMICS AND PHASE TRANSITION IN DEUTERATED AMMONIUM CHLORIDE.
      *TEH(H.C.). (ED. NOTE: PH.D. DISSERTATION, MCMASTER UNIVERSITY).

1534 N*D4*CL, ...C1. . . . . . .REF. SECTION 3, 66/VENKATARAMAN, ...

1535 N*D4*CL, ...B2,B6. . .CONF. NATO SCHOOL, ... , (1973).
      *TRICRITICAL STUDIES OF N*D4*CL.
      *YELON(W.B.). (PROC. CONF. NATO SCHOOL ON ANHARMONIC LATTICE STRUCTURAL
      TRANSITIONS AND MELTING, GEILO, NORWAY, APRIL 25-MAY 1, 1973).

1536 N*D4*CL, ...B2. . .PHYS. REV. B VOL.8, ... 3928, (1973).
      *TEMPERATURE DEPENDENCE OF THE LATTICE VIBRATIONS IN N*D4*CL.
      *TEH(H.C.), BROCKHOUSE(B.N.).

1537 N*D4*CL, ...T. . . . . .CAN. J. PHYS. VOL.50, ... 2807, (1972).
      *MODEL CALCULATIONS FOR LATTICE VIBRATIONS IN N*D4*CL AT 85 DEGREES K.
      *TEH(H.C.).

1538 N*D4*CL, ...B1. . . . . . .IAEA SYMP. CHALK RIVER, VOL.2, .   253, (1962).
      *STUDY OF AMMONIUM HALIDES BY NEUTRON SPECTROMETRY.
      *VENKATARAMAN(G.),   USHA(K.),   IYENGAR(P.K.),   VIJAYARAGHAVAN(P.R.),
      ROY(A.P.).

1539 N*D4*O2*P*O4, ...B1. . . . . . .PHYS. REV. VOL.184, ... 550, (1969).
      *LATTICE-DYNAMICAL ASPECTS OF THE ANTIFERROELECTRIC PHASE TRANSITION IN
      N*D4*O2*P*O4.
      *MEISTER(H.), SKALYO-JR(J.), FRAZER(B.C.), SHIRANE(G.).

1540 N*D4*O2*P*O4, ...T. . . . . . .J. PHYSIQUE TOME 33 COL.2, ... 59, (1972).
      *DETERMINATION OF THE EIGENVECTORS OF SOFT MODES IN K*D2*P*O4 AND IN
      N*D2*P*O4.
      *WALLACE(E.A.), COCHRAN(W.), STRINGFELLOW(M.W.).

1541 N*(D-H)4*CL, ...B1. . . . . . .PHYS. LETT. 29A, ... 694, (1969).
      *LATTICE VIBRATIONS IN DEUTERATED AMMONIUM CHLORIDE.
      *TEH(H.C.), BROCKHOUSE(B.N.), DEWIT(G.A.).

1542 N*H3, ...C4. . . . . . .J. CHEM. PHYSICS VOL.42, ... 1568, (1965).
      *SCATTERING OF SLOW NEUTRONS FROM AMMONIA AND HYDROGEN SULFIDE.
      *STRONG(K.A.), HARKER(Y.D.), BRUGGER(R.M.).

1543 N*H3, ...C3. . . . . . .IAEA SYMP. CHALK RIVER, VOL.1, .   457, (1962).
      *THE INELASTIC SCATTERING OF COLD NEUTRONS BY METHANE, AMMONIA AND
      HYDROGEN.
      *WEBB(F.J.).

1544 N*H3, ...C3. . . . . . .IAEA SYMP. GRENOBLE, ... 477, (1972).
      *NEUTRON SCATTERING FROM LIQUID AMMONIA.
      *DASANNACHARYA(B.A.), THAPER(C.L.), GOYAL(P.S.).

1545 N*H3, ...C3. . . . . . .PHYS. REV. VOL.171, ... 263, (1968).
      *ROTATIONAL CORRELATION FUNCTIONS IN NEUTRON SCATTERING BY MOLECUL. GASES
      *AGRAWAL(A.K.), YIP(S.).

1546 N*H3, ...C3. . . . . . .PROC. PHYS. SOC. 89, ... 379, (1966).
      *NEUTRON SCATTERING FROM GASEOUS METHANE AND AMMONIA.
      *VENKATARAMAN(G.), RAO(K.R.), DASANNACHARYA(B.A.), DAYANIDHI(P.K.).

1547 N*H3, ...C1,C3. . . . . . .PHYS. REV. LETT. VOL.17, ... 533, (1966).
      *LOCALIZED VIBRATIONS OF TRAPPED MOLECULES.
      *DOWNES(J.S.), WHITE(J.W.), EGELSTAFF(P.A.), RAINEY(V.S.).

1548 N*H3, ...C3. . . . . . .PROC. PHYS. SOC. 92, ... 912, (1967).
      *THE INELASTIC SCATTERING OF COLD NEUTRONS BY METHANE AND AMMONIA.
      *WEBB(F.J.).

1549 N*H3, ...C3,P. . . . .PHYSICA VOL.29, ... 485, (1963).
      *SCATTERING OF SLOW NEUTRONS BY N*H3 MOLECULES.II.
      SCATTERING BY LIQUID N*H3.
      *JANIK(J.A.), RZANY(H.), SCIESINSKI(J.).

1550 N*H4 SALTS, ...C1. . . . . . .NUCL. SCI. ENGNG. VOL.14, ... 339, (1962).
      *THE EFFECT OF ROTATIONAL FREEDOM IN SEVERAL AMMONIUM SALTS AND DIMETHYL
      ACETYLENE ON THE INELASTIC SCATTERING OF SLOW NEUTRONS.
      *RUSH(J.J.), SAFFORD(G.J.), TAYLOR(T.I.), HAVENS-JR(W.W.).

1551 N*H3, ...T,C3. . . . . . .ACTA PHYS. POLON. VOL.25, ... 141, (1964).
      *DOUBLE DIFFERENTIAL CROSS-SECTION FOR SLOW NEUTRONS SCATTERING ON
      GASEOUS AMMONIA MOLECULES.
      *CZERLUNCZARKIEWICZ(B.), KOWALSKA(B.).

1552 N*H3, ...C3. . . . . . .IAEA SYMP. CHALK RIVER, VOL.1, .   405, (1962).
      *SYSTEMATIC STUDY ON TOTAL NEUTRON SCATTERING CROSS-SECTIONS OF HYDROGEN
      CONTAINING MOLECULES.
      *JANIK(J.A.), JANIK(J.), MANIAWSKI(F.), RZANY(H.), ROGALSKA(Z.),
      SCIESINSKI(J.), SAGAN(U.), TUBBS(N.).

1553 N*H3, ...MOL. SIEVES, PAP. CONF. 1967, . . . 306, (1967).
      *MOLECULAR MOTION IN MOLECULAR SIEVES BY NEUTRON SCATTERING SPECTROSCOPY.
      *EGELSTAFF(P.A.), DOWNES(J.S.), WHITE(J.W.).

1554 N*H3, ...T. . . . . . .ACTA PHYS. POLON. VOL.18, ... 255, (1959).
      *SCATTERING OF SLOW NEUTRONS BY MOLECULES OF LIQUID WATER, AMMONIA AND
      HYDROGEN SULFIDE.
      *WANIC(A.).

1555 (N*H4)2*BE*F4, ...C1. . . . . . .IAEA SYMP. BOMBAY, VOL.2, ... 333, (1964).
      *STUDY OF LOW-FREQUENCY MOTIONS IN SEVERAL FERROELECTRIC SALTS BY THE
      INELASTIC SCATTERING OF COLD NEUTRONS.
      *RUSH(J.J.), TAYLOR(T.I.).

1556 N*H4*BR, ...C1. . . . . . .REF. SECTION 3, 66/VENKATARAMAN, ...

1557 N*H4*BR, ...B1. . . . . . .IAEA SYMP. CHALK RIVER, VOL.2, .   253, (1962).
      *STUDY OF AMMONIUM HALIDES BY NEUTRON SPECTROMETRY.
      *VENKATARAMAN(G.),   USHA(K.),   IYENGAR(P.K.),   VIJAYARAGHAVAN(P.R.),
      ROY(A.P.).

SECTION 6 -QUASI-ELASTIC AND INELASTIC SCATTERING STUDIES . . . .

1558 N*H4*(BR,CL,F,I,N*03), ... B4 . . . . . IAEA SYMP. CHALK RIVER, VOL.2, . 237, (1962).
&LATTICE DYNAMICS OF SOME AMMONIUM SALTS BY INVERTED FILTER METHOD.
&MIKKE(K.), KROH(A.).

1559 (N*H4)2*C*03,(N*H4)2*H*C6*H5*07, . . . C1 . . . . J. CHEM. PHYS. VOL.46, ... 4034, (1967).
&SCATTERING OF COLD NEUTRONS IN AMMONIUM CARBONATE, AMMONIUM CITRATE,
AND AMMONIUM ACETATE.
&MYERS(V.W.).

1560 N*H2*C*O*N*H2 (UREA), ... C1 . . . . . J. CHEM. PHYS. VOL.47, ... 4278, (1967).
&NEUTRON-SCATTERING STUDY OF LOW FREQUENCY MODES IN UREA AND AND FERRO-
ELECTRIC THIOUREA.
&RUSH(J.J.).

1561 N*H2*C*S*N*H2 (THIOUREA), ... C1 . . . . . J. CHEM. PHYS. VOL.47, ... 4278, (1967).
&NEUTRON-SCATTERING STUDY OF LOW FREQUENCY MODES IN UREA AND AND FERRO-
ELECTRIC THIOUREA.
&RUSH(J.J.).

1562 N*H4*CL, ... B1 . . . . . . PHYS. REV. VOL.181, ... 1218, (1969).
&COHERENT INELASTIC NEUTRON SCATTERING IN N*H4*CL.
&SMITH(H.G.), TRAYLOR(J.G.), REICHARDT(W.).

1563 N*H4*CL, ... C1 . . . . . . J. CHEM. PHYS. VOL.57, ... 2291, (1972).
&LATTICE VIBRATIONS IN AMMONIUM CHLORIDE IN THE LOW-TEMPERATURE ORDERED
PHASE.
&KIM(C.H.), RAFIZADEH(H.A.), YIP(S.).

1564 N*H4*CL, ... C1 . . . . . . REF. SECTION 3, 66/VENKATARAMAN, ...

1565 N*H4*CL, ... P . . . . REF. SECTION 3, 60/WOODS, ...

1566 N*H4*CL, ... C3 . . . . . . SOL. STATE COMM. 2, ... 17, (1964).
&ANHARMONICITY OF THE TORSIONAL OSCILLATIONS OF THE AMMONIUM ION IN
N*H4*CL.
&VENKATARAMAN(G.), USHA-DENIZ(K.), IYENGAR(P.K.), VIJAYARAGHAVAN(P.R.).

1567 N*H4*CL, ... B1 . . . . . IAEA SYMP. CHALK RIVER, VOL.2, . 253, (1962).
&STUDY OF AMMONIUM HALIDES BY NEUTRON SPECTROMETRY.
&VENKATARAMAN(G.), USHA(K.), IYENGAR(P.K.), VIJAYARAGHAVAN(P.R.),
ROY(A.P.).

1568 N*H4*CL, ... C1 . . . . . SOL. STATE COMM. VOL.13 NO.5, . 543, (1973).
&REORIENTATION OF THE N*H4(+) ION IN N*H4*CL IN PHASE II.
&SKOLD(K.), DAHLBORG(U.).

1569 N*H4*CL, ... T . . . . . . ACTA PHYS. POLON. VOL.35, ... 223, (1969).
&LATTICE DYNAMICS OF AMMONIUM CHLORIDE.
&PARLINSKI(K.). (ED. NOTE: IN RUSSIAN).

1570 N*H4*(CL,BR,I,F,N*03,C*N*S), . . . C1 . . . . . IAEA SYMP. BOMBAY, VOL.2, . . . 355, (1964).
&INVESTIGATION OF THE DYNAMICS OF N*H4+ AND H2*O MOLECULAR GROUPS IN
CRYSTALS.
&.C.C.C.P.

1571 N*H4*CL*04, ... B4 . . . . . IAEA SYMP. BOMBAY, VOL.2, . . . 243, (1964).
&MOLECULAR DYNAMICS INVESTIGATED BY NEUTRON SCATTERING.
&JANIK(J.A.).

1572 N*H4*CL*04, ... C1 . . . . . J. PHYS. CHEM. VOL.68, ... 2534, (1964).
&ROTATIONAL MOTIONS IN HEXAMETHYLBENZENE AND AMMONIUM PERCHLORATE BY
CROSS SECTION MEASUREMENTS WITH SLOW NEUTRONS.
&RUSH(J.J.), TAYLOR(T.I.).

1573 N*H4*CL*04, ... P . . . . . . PHYS. STAT. SOL. 9, ... 905, (1965).
&DETERMINATION OF PARAMETERS OF THE ROTATIONAL DYNAMICS OF THE GROUPS
N*H4 IN N*H4*CL*04 AND H3*O IN H3*O*CL*04 BY INELASTIC SCATTERING OF...
&JANIK(J.M.), JANIK(J.A.), BAJOREK(A.), PARLINSKI(K.).

1574 N*H4*CL*04, ... C1,C3 . . . . REF. SECTION 3, 64/JANIK, ...

1575 N*H4-COMPOUNDS, ... B4 . . . . . IAEA SYMP. BOMBAY, VOL.2, ... 243, (1964).
&MOLECULAR DYNAMICS INVESTIGATED BY NEUTRON SCATTERING.
&JANIK(J.A.).

1576 N*H4*F, ... C3 . . . . . IAEA SYMP. BOMBAY, VOL.2, ... 179, (1964).
&PROTON MOTIONS IN ACID AQUEOUS SOLUTIONS.
&STILLER(H.).

1577 N*H4*(F,CL,...),N*H4*CL*04, . . . C1 . . . . . J. CHEM. PHYSICS VOL.48, ... 4912, (1968).
&STUDIES OF PHASE TRANSITIONS IN AMMONIUM SALTS AND BARRIERS TO ROTATION
OF AMMONIUM IONS BY NEUTRON-SCATT. CROSS-SECTIONS AS A FUNCTION OF TEMP.
&LEUNG(P.S.), TAYLOR(T.I.), HAVENS-JR(W.W.).

1578 N*H4*(F,CL,...), ... C1 . . . . . PHYS. REV. LETT. VOL.5, ... 507, (1960).
&PROTON MOTIONS IN AMMONIUM HALIDES BY SLOW NEUTRON CROSS-SECTION
MEASUREMENTS.
&RUSH(J.J.), TAYLOR(T.I.), HAVENS-JR(W.W.).

1579 N*H4*(H2*P*O4),N*H4*(H2*AS*04), . . . C1 . . . . . PHYS. STAT. SOLIDI B VOL.59 NO.2 471, (1973).
&STUDY OF N*H4 MOTION IN SOME ANTIFERROELECTRIC CRYSTALS BY INELASTIC
SCATTERING OF COLD NEUTRONS.
&DIMIC(V.), OSREDKAR(M.), SLAK(J.), KANDUSAR(A.).

1580 N*H4*H*S*04, ... C1 . . . . . IAEA SYMP. BOMBAY, VOL.2, ... 333, (1964).
&STUDY OF LOW-FREQUENCY MOTIONS IN SEVERAL FERROELECTRIC SALTS BY THE
INELASTIC SCATTERING OF COLD NEUTRONS.
&RUSH(J.J.), TAYLOR(T.I.).

1581 N*H4*I, ... C1 . . . . . PHYSICA VOL.32, ... 1571, (1966).
&SLOW NEUTRON SCATTERING BY SOLID C*H4 AND N*H4*I.
&KOSALY(G.), SOLT(G.).

1582 N*H4*I, ... C1 . . . . . J. PHYS. CHEM. SOL. VOL.27, . . . 1103, (1966).
&STUDY OF THE ROTATIONAL BEHAVIOUR OF THE AMMONIUM ION IN SEVERAL SALTS
BY NEUTRON SPECTROMETRY.
&VENKATARAMAN(G.), USHA-DENIZ(K.), IYENGAR(P.K.), ROY(A.P.),
VIJAYARAGHAVAN(P.R.). (ED. NOTE: REF. SECTION 3, 66/VENKATARAMAN).

1583 N*H4*I, ... C1 . . . . . J. PHYS. SOC. JAPAN VOL.17 B-II, 367, (1962).
&LATTICE DYNAMICS OF THE AMMONIUM HALIDES.
&PALEVSKY(H.).

1584 (N*H4*I)-K*I, ... C1 . . . . . IAEA SYMP. BOMBAY, VOL.2, ... 383, (1964).
&STUDIES OF AMMONIUM ION MOTIONS IN SOLID SOLUTIONS.
&MIKKE(K.), DOBRZYNSKI(L.).

SECTION   6 -QUASI-ELASTIC AND INELASTIC SCATTERING STUDIES . . . .

1585 (N*H4-K)*CL, ...LOCAL MODES IN AL(.1)-CU(.9) AND (N*H4(.1)-K(.9))*CL IN THE COHERENT-
      POTENTIAL APPROXIMATION.
      .KAPLAN(T.), MOSTOLLER(M.).

1586 (N*H4)*N3, ...C1. . . . . . J. CHEM. PHYSICS VOL.45, ... 401, (1966).
      .LOW-FREQUENCY MOLECULAR MOTIONS IN AMMONIUM AZIDE.
      .BOUTIN(H.), TREVINO(S.F.), PRASK(H.).

1587 N*H4*P*F6, ...C1. . . . . . J. PHYS. CHEM. SOL. 24, ... 617, (1963).
      .A STUDY OF THE ROTATIONAL FREEDOM IN SEVERAL AMMONIUM SALTS BY SLOW
      NEUTRON INELASTIC SCATTERING.
      .BRAJOVIC(V.), BOUTIN(H.), SAFFORD(G.J.), PALEVSKY(H.). (ED. NOTE: REF.
      SECTION 3, 63/BRAJOVIC).

1588 N*H4*P*F6, ...C1,C3. . . .REF. SECTION 3, 64/JANIK, ...

1589 (N*H4)2*(S*O4,S2*O8,CR2*O7,C*O3). . .C1. . . . . .IAEA SYMP. BOMBAY, VOL.2, ... 355, (1964).
      .INVESTIGATION OF THE DYNAMICS OF N*H4+ AND H2*O MOLECULAR GROUPS IN
      CRYSTALS.
      .C.C.C.P.

1590 (N*H4)2*S*O4, ...C1. . . . .IAEA SYMP. BOMBAY, VOL.2, ... 333, (1964).
      .STUDY OF LOW-FREQUENCY MOTIONS IN SEVERAL FERROELECTRIC SALTS BY THE
      INELASTIC SCATTERING OF COLD NEUTRONS.
      .RUSH(J.J.), TAYLOR(T.I.).

1591 (N*H4)2*S*O4, ...C4. . . . . . .SOL. STATE COMM. VOL.8, ... 889, (1970).
      .STUDY OF MOLECULAR REORIENTATION IN SOLIDS BY SLOW-NEUTRON SCATTERING.
      .KIM(H.J.), GOYAL(P.S.), VENKATARAMAN(G.), DASANNACHARYA(B.A.),
      THAPER(C.L.).

1592 (N*H4)2*S*O4, ...C1. . . . . . .PHYSICA VOL.49, ... 1, (1970).
      .ROTATIONAL MOTIONS IN SOLIDS. THE FERROELECTRIC TRANSITION IN
      (N*H4)2*S*O4 STUDIED BY COLD NEUTRONS.
      .DAHLBORG(U.), LARSSON(K.E.), PIRKMAJER(E.).

1593 (N*H4)2*SI*F6, ...A1,C1. . . .J. CHEM. PHYSICS VOL.44, ... 2499, (1966).
      .STRUCTURE OF CUBIC AMMONIUM FLUOSILICATE: NEUTRON-DIFFRACTION AND
      NEUTRON-INELASTIC-SCATTERING STUDIES.
      .SCHLEMPER(E.O.), HAMILTON(W.C.), RUSH(J.J.).

1594 (N*H4)2*SN*BR6, ...C1. . . . . .REF. SECTION 3, 66/VENKATARAMAN, ...

1595 (N*H4)2*SN*CL6, ...C1. . . . . .REF. SECTION 3, 66/VENKATARAMAN, ...

1596 N*H4*V*O3, ...C1. . . . . .PHYSICA VOL.35, ... 465, (1967).
      .MOLECULAR DYNAMICS STUDY BY THE NEUTRON INELASTIC SCATTERING METHOD. IV.
      AMMONIUM METAVANADATE.
      .BAJOREK(A.), PARLINSKI(K.), SUDNIK-HRYNKIEWICZ(M.).

1597 NA, ...B1. . . .J. APPL. PHYS. 42, ... 4736, (1971).
      .RESOLUTION EFFECTS IN THE MEASUREMENT OF PHONONS IN SODIUM METAL.
      .WERNER(S.A.), PYNN(P.).

1598 NA, ...C3. . . . . .IAEA SYMP. CHALK RIVER, VOL.1, . 227, (1962).
      .STUDIES OF LIQUID SODIUM BY INELASTIC SCATTERING OF SLOW NEUTRONS.
      .COCKING(S.J.).

1599 NA, ...B1. . . . . .IAEA SYMP. COPENHAGEN, VOL.1, . . 101, (1968).
      .LATTICE DYNAMICS AND ELECTRONIC STRUCTURE OF SODIUM, MAGNESIUM AND
      ALUMINIUM.
      .SCHNEIDER(T.), STOLL(E.).

1600 NA, ...B4. . . . . .IAEA SYMP. BOMBAY, VOL.2, ... 85, (1964).
      .COHERENT SCATTERING OF SLOW NEUTRONS BY LIQUID SODIUM.
      .SINGWI(K.S.), FELDMANN(G.).

1601 NA, ...B1. . . . . .IAEA SYMP. BOMBAY, VOL.1, ... 77, (1964).
      .PHONON DISPERSION RELATIONS IN ALKALI METALS.
      .SINGH(R.P.), MANI(K.K.), KELKAR(V.H.).

1602 NA, ...C3. . . . . .IAEA SYMP. COPENHAGEN, VOL.1, . . 545, (1968).
      .ANALYSIS OF NEUTRON SCATTERING EXPERIMENTS ON LIQUIDS.
      .AGRAWAL(A.K.), DESAI(R.C.), YIP(S.).

1603 NA, ...B4. . . . . .IAEA SYMP. COPENHAGEN, VOL.1, . . 449, (1968).
      .INELASTIC SCATTERING IN LIQUID SODIUM AT SMALL MOMENTUM TRANSFER.
      .RANDOLPH(P.D.).

1604 NA, ...B4. . . . . .IAEA SYMP. COPENHAGEN, VOL.1, . . 397, (1968).
      .LIQUID DYNAMICS FROM NEUTRON SCATTERING (REVIEW PAPER).
      .LARSSON(K.E.).

1605 NA, ...B1. . . . . .J. PHYS.F: METAL PHYS. VOL.1, . . 244, (1971).
      .MEASUREMENTS OF THE FREQUENCIES OF THE NORMAL MODES IN SODIUM.
      .MILLINGTON(A.J.), SQUIRES(G.L.).

1606 NA, ...T. . . . . .PHYS. LETT. 3, ... 162, (1962).
      .SLOW NEUTRON SCATTERING EVIDENCE FOR VIOLATING OF THE F-SUM RULE IN
      LIQUID SODIUM.
      .RANDOLPH(P.D.).

1607 NA, ...B4. . . . . .PHYS. REV. VOL.180, ... 308, (1969).
      .INELASTIC SCATTERING OF THERMAL NEUTRONS BY LIQUID SODIUM AND LEAD.
      .DESAI(R.C.), YIP(S.).

1608 NA, ...B2. . . . . .ACTA PHYS. AUSTRIACA VOL.33, . . . 27, (1971).
      .PRESSURE DEPENDENT FREQUENCY-SHIFT OF PHONONS IN SODIUM MEASURED BY
      INELASTIC NEUTRON SCATTERING.
      .ERNST(G.). (ED. NOTE: IN GERMAN).

1609 NA, ...C3. . . . . .J. PHYS. C SER.2 VOL.2, ... 2047, (1969).
      .ATOMIC MOTION IN LIQUID SODIUM I.DIFFUSIVE MOTION.
      .COCKING(S.J.).

1610 NA, ...B1,T. . . . . .SOL. STATE COMM. VOL.11, ... 1431, (1972).
      .A MODEL FOR LATTICE DYNAMICS OF B.C.C. METALS.
      .SHUKLA(M.M.), DA-CUNHA-LIMA(I.C.), BRESCANSIN(L.M.).

1611 NA, ...C2. . . . . .PROC. PHYS. SOC. 86, ... 181, (1965).
      .THE FREQUENCY DISTRIBUTION OF THE LATTICE VIBRATIONS OF SODIUM.
      .NUTKINS(M.A.E.).

SECTION 6 -QUASI-ELASTIC AND INELASTIC SCATTERING STUDIES . . . .

1612 NA,. . . . .T,C3.. ..PHYS. REV. 158 97, (1967).
*COLLECTIVE EFFECTS AND HYDRODYNAMIC APPROXIMATION IN NEUTRON SCATTERING
FROM FLUIDS.
*FERZIGER(J.H.), FEINSTEIN(D.L.).

1613 NA,. . . . .B4.. ......PHYS. REV. VOL.134, ...A1238, (1967).
*SLOW NEUTRON INELASTIC SCATTERING FROM LIQUID SODIUM.
*RANDOLPH(P.D.).

1614 NA,. . . . .C2.........PROC. PHYS. SOC. 81, 973, (1963).
*FREQUENCY DISTRIBUTION OF THE LATTICE VIBRATIONS IN SODIUM.
*DIXON(D.R.), WOODS(A.D.B.), BROCKHOUSE(B.N.).

1615 NA,. . . . .T........PHYS. REV. 157, 540, (1967).
*EXISTENCE OF AN INFINITY IN THE FREQUENCY DISTRIBUTION G(NU) OF
MONOATOMIC BODY-CENTERED CUBIC CRYSTALS.
*GILAT(G.).

1616 NA,. . . . .B4........PROC. PHYS. SOC. VOL.80, . . . 1201, (1962).
*THE DYNAMICS OF LIQUID SODIUM BY NEUTRON SCATTERING.
*COCKING(S.J.), TURBERFIELD(K.C.).

1617 NA,. . . . .T,B4.......NUOVO CIMENTO B VOL.13B NO.1.. . 185, (1973).
*COLLECTIVE MOTIONS IN CLASSICAL LIQUIDS, III.LIQUID SODIUM.
*PATHAK(K.N.), SINGWI(K.S.), CUBIOTTI(G.), TOSI(M.P.).

1618 NA,. . . . .B7.........NATURWISSENSCHAFTEN VOL.53, . . . 16, (1966).
*MEASUREMENT OF COHERENT SCATTERING OF NEUTRONS AT MOLTEN SODIUM, CESIUM
AND BISMUTH AT DIFFERENT TEMPERATURES.
*OEHME(H.), RICHTER(H.). (IN GERMAN).

1619 NA,. . . . .T,B1......J. PHYS. F VOL.3, . . L99, (1973).
*LATTICE DYNAMICS OF SODIUM ON MODIFIED BHATIA#S MODEL.
*SHUKLA(M.M.), SALZBERG(J.B.).

1620 NA,. . . . .T........PROPERTIES LIQUID METALS, ... 125, (1973).
*DYNAMIC STRUCTURE FACTOR CALCULATIONS AND NEUTRON INELASTIC SCATTERING.
*CHUNG(C.H.), YIP(S.), EGELSTAFF(P.A.). (ED. NOTE: REFER SECT. 4).

1621 NA,. . . . .B1.........PROC. PHYS. SOC. VOL.79, ... 440, (1962).
*NORMAL VIBRATIONS OF SODIUM.
*WOODS(A.D.B.), BROCKHOUSE(B.N.), MARCH(R.H.), BOWERS(R.).

1622 NA,. . . . .T........IAEA SYMPOSIUM BOMBAY, VOL.1.. . 49, (1964).
*A COMPARISON BETWEEN THE ELECTRON-PHONON INTERACTIONS OF TOYA AND BAILYN
*SRIVASTAVA(P.L.).

1623 NA,. . . . .T,B1......PROC. ROY. SOC. 283, ... 33, (1965).
*A CALCULATION OF THE PHONON FREQUENCIES IN SODIUM.
*SHAM(L.J.).

1624 NA,. . . . .T,B1......SOL. STATE COMM. 4, ... 79, (1966).
*BERECHNUNG DER PHONONEN-DISPERSIONSKURVEN VON NA UND CU.
*SCHNEIDER(T.), STOLL(E.).

1625 NA,. . . . .T,B1......PROC. ROY. SOC. 276, ... 308, (1963).
*LATTICE DYNAMICS OF SODIUM.
*COCHRAN(W.).

1626 NA,. . . . .T,B1......PHYS. STAT. SOL. 23, ... 489, (1967).
*A QUASI-HARMONIC CALCULATION OF LATTICE DYNAMICS FOR NA.
*HO(P.S.), RUOFF(A.L.).

1627 NA,. . . . .B1.........IAEA SYMP. CHALK RIVER, VOL.2, . 3, (1962).
*CRYSTAL DYNAMICS FROM NEUTRON SPECTROMETRY.
*WOODS(A.D.B.).

1628 NA,. . . . .B1.........IAEA SYMP. CHALK RIVER, VOL.2, . 23, (1962).
*DETERMINATION OF POLARIZATION VECTORS FROM NEUTRON GROUP INTENSITIES.
*BROCKHOUSE(B.N.), BECKA(L.N.), RAO(K.R.), WOODS(A.D.B.).

1629 NA,. . . . .B1.........PHYS. REV. VOL.128, ... 1112, (1962).
*CRYSTAL DYNAMICS OF SODIUM AT 90 DEG. K.
*WOODS(A.D.B.), BROCKHOUSE(B.N.), MARCH(R.H.), STEWART(A.T.), BOWERS(R.).

1630 NA,. . . . .T,B4......PHYS. REV. A VOL.9, ... 2128, (1974).
*SELF-MOTION IN LIQUID SODIUM.
*CHATURVEDI(D.K.), BAIJAL(J.S.), PATHAK(K.N.).

1631 NA*AL*H4,. . . . .B7.........J.C.S. FARADAY TRANS. II VOL.69, 783, (1973).
*LIBRATIONAL MOTION IN SODIUM AND LITHIUM ALUMINIUM HYDRIDES, STUDIED BY
INELASTIC NEUTRON SCATTERING.
*TEMME(F.P.), WADDINGTON(T.C.).

1632 NA*BR, ...B1........PHYS. REV. B-1, ... 1833, (1970).
*PHONON FREQUENCIES IN NA*BR.
*REID(J.S.), SMITH(T.), BUYERS(W.J.L.).

1633 NA*CL, ...B1........J. PHYS. CHEM. SOLIDS VOL.31, . 131, (1970).
*LATTICE DYNAMICS OF SODIUM CHLORIDE AT ROOM TEMPERATURE.
*SCHMUNK(R.E.), MINDER(D.R.).

1634 NA*CL, ...B1........IAEA SYMP. COPENHAGEN, VOL.1,. . 295, (1968).
*PHONON DISPERSION RELATIONS IN NA*CL AT 80 DEGREES KELVIN.
*ALMQVIST(L.), RAUNIO(G.), STEDMAN(R.).

1635 NA*CL, ...B1........UKR. FIZ. ZH.(USSR) VOL.15,. . . 1409, (1970).
*PHONON DISPERSION IN NA*CL CRYSTALS.
*KUCHER(T.I.), SHCHUR(I.A.).

1636 NA*CL, ...B3........JETP LETT. VOL.3,. . . . . . . 110, (1966).
*SCATTERING OF COLD NEUTRONS IN IRRADIATED K*BR AND NA*CL CRYSTALS.
*ANDRONIKASHVILI(E.L.), BEDBENOVA(D.S.), POLITOV(N.G.), TSAKADZE(D.S.).

1637 NA*CL, ...B1........PHYS. REV. 178,. . . . . 1496, (1969).
*PHONON DISPERSION RELATION IN NA*CL.
*RAUNIO(G.), ALMQVIST(L.), STEDMAN(R.).

1638 NA*CL, ...B1,T......PHYS. STAT. SOL. 35, NO1,. . . . 299, (1969).
*PHONON WIDTHS IN NA*CL, K*CL AND RB*CL FROM NEUTRON MEASUREMENTS.
*RAUNIO(G.).

1639 NA*CL, ...B2........PHYS. REV. LETT. VOL.23,. . . . 525, (1969).
*INTERFERENCE BETWEEN ONE- AND MULTIPHONON PROCESSES IN THE SCATTERING OF
NEUTRONS AND X-RAYS BY CRYSTALS.
*COWLEY(R.A.), SVENSSON(E.C.), BUYERS(W.J.L.).

SECTION   6 -QUASI-ELASTIC AND INELASTIC SCATTERING STUDIES . . . .

1640 NA*CL-TYPE STRUCTURE,. . . . . .T,P. . . . . .ACTA CRYST. 19,. . . . . 224, (1965).
      *COMPARISON OF SCATTERING FACTORS COMPUTED FROM FOUR DIFFERENT ATOMIC
      MODELS.
      *CROMER(D.T.).

1641 NA*CL-TYPE STRUCTURE,. . . . . .T. . . . . . . .SOV. PHYS. SOL. STATE, 4,. . . . . 2459, (1962).
      *CALCULATION OF STRUCTURE FACTORS FOR THE INELASTIC SCATTERING OF SLOW
      NEUTRONS IN NA*CL-TYPE CRYSTALS.
      *DEMIDENKO(Z.A.).

1642 NA*F,. . . . . .B1. . . . . .PHYS. REV. 153,. . . . . 923, (1967).
      *LATTICE DYNAMICS OF SODIUM FLUORIDE.
      *BUYERS(W.J.L.).

1643 NA*H,. . . . . .P. . . . . . .REF. SECTION 3, 60/WOODS,. . . . .

1644 NA*H*C*O3,. . . . . .C1. . . . . . .J. CHEM. PHYS. VOL.52, ... 1828, (1970).
      *VIBRATION SPECTRA OF HYDROGEN BONDS BY NEUTRON ENERGY-LOSS SPECTROMETRY.
      *COLLINS(M.F.), HAYWOOD(B.C.), STIRLING(G.C.).

1645 NA*H*F2, ...C1. . . . . . .J. CHEM. PHYS. VOL.56, ... 2793, (1972).
      *DYNAMICS OF SODIUM AND POTASSIUM BIFLUORIDE: INFRARED, RAMAN, AND
      NEUTRON STUDIES.
      *RUSH(J.J.), SCHROEDER(L.W.), MELVEGER(A.J.).

1646 NA*H2*F3,. . . . . .C1. . . . . . .J. CHEM. PHYS. VOL.39, ... 3135, (1963).
      *STUDY OF LOW-FREQUENCY MOLECULAR MOTIONS IN H*F, K*H*F2, K*H2*F3 AND
      NA*H2*F3.
      *BOUTIN(H.), SAFFORD(G.J.), BRAJOVIC(V.).

1647 NA*I,. . . . . .B1. . . . . . .IAEA SYMP. COPENHAGEN, VOL.1,. . . 367, (1968).
      *PRESSURE DEPENDENCE OF THE FREQUENCY OF SOME PHONONS IN SODIUM IODIDE.
      *QUITTNER(G.), VUKOVICH(S.), ERNST(G.).

1648 NA*I,. . . . . .T. . . . . . .SOV. PHYS. SOL. STATE VOL.12,. . . 423, (1970).
      *LATTICE DYNAMICS OF NA*I AND THE FOURIER COMPONENTS OF THE LAGLESS
      DIELECTRIC CONSTANT.
      *KUCHER(T.I.), TOMASEVICH(O.F.).

1649 NA*I,. . . . . .T,B2. . . . .PHYS. REV. VOL 119,. . . . . 980, (1960).
      *LATTICE DYNAMICS OF ALKALI HALIDE CRYSTALS.
      *WOODS(A.D.B.), COCHRAN(W.), BROCKHOUSE(B.N.).

1650 NA*I,. . . . . .B2. . . . . .PHYS. REV. 131,. . . . . 1025, (1963).
      *LATTICE DYNAMICS OF ALKALI HALIDE CRYSTALS.II. EXPERIMENTAL STUDIES
      OF K*BR AND NA*I.
      *WOODS(A.D.B.), BROCKHOUSE(B.N.), COWLEY(R.A.), COCHRAN(W.).

1651 NA*I,. . . . . .T,B1. . . . .PHYS. REV. 131,. . . . . 1030, (1963).
      *LATTICE DYNAMICS OF ALKALI HALIDE CRYSTALS, III. THEORETICAL.
      *COCHRAN(W.), COWLEY(R.A.), BROCKHOUSE(B.N.), WOODS(A.D.B.).

1652 NA*I,. . . . . .T,B1. . . . .PROC. ROY. SOC. 292,. . . . . . 209, (1966).
      *ANHARMONIC INTERACTIONS IN ALKALI HALIDES. II.
      *COWLEY(E.R.), COWLEY(R.A.).

1653 NA*I,. . . . . .T,B1. . . . .PROC. ROY. SOC. 287,. . . . . . 259, (1965).
      *ANHARMONIC INTERACTIONS IN ALKALI HALIDES.
      *COWLEY(E.R.), COWLEY(R.A.).

1654 NA*I,. . . . . .B1. . . . . .IAEA SYMP. CHALK RIVER, VOL.2, .   3, (1962).
      *CRYSTAL DYNAMICS FROM NEUTRON SPECTROMETRY.
      *WOODS(A.D.B.).

1655 NA*MN*F3,. . . . . .D3. . . . . .J. APP. PHYS. 36,. . . . . . . 1092, (1965).
      *SOME MEASUREMENTS OF EXCHANGE ENERGIES BY PARAMAGNETIC NEUTRON INELASTIC
      SCATTERING.
      *COLLINS(M.F.), NATHANS(R.).

1656 NA*N3 (RHOMBOHEDRAL). . . . . .B1. . . . . .J. CHEM. PHYS. VOL.56 NO.11, . . 5377, (1972).
      *LATTICE DYNAMICS OF RHOMBOHEDRAL SODIUM AZIDE.
      *RAFIZADEH(H.A.), YIP(S.), PRASK(H.).

1657 NA*N*H2, ...C3. . . . . .J. CHEM. PHYS. VOL.55, ... 2807, (1971).
      *STUDIES OF VIBRATION SPECTRA OF BONDED HYDROGEN ATOMS USING A PULSED
      NEUTRON SOURCE.
      *DAY(D.H.), SINCLAIR(R.N.).

1658 NA*N*O3, ...T,B1. . . . .CONF. INTERN., RENNES, ... 104, (1971).
      *THE LATTICE DYNAMICS OF NA*N*O3.
      *LOGAN(K.W.), TREVINO(S.F.), CASELLA(R.C.), SHAW(W.M.), MUHLESTEIN(L.D.),
      MICAL(R.D.).

1659 NA*N*O2, ...B1. . . . . .J. PHYS. SOC. JAPAN VOL.28, SUP. 258, (1970).
      *CRYSTAL DYNAMICS OF SODIUM NITRITE.
      *DOLLING(G.), SAKURAI(J.), COWLEY(R.A.).

1660 NA*N*O2, ...B1,B6. . . . .J. PHYS. SOC. JAPAN VOL.28,. . . 1426, (1970).
      *CRYSTAL DYNAMICS AND THE FERROELECTRIC PHASE TRANSITION OF SODIUM
      NITRITE.
      *SAKURAI(J.), COWLEY(R.A.), DOLLING(G.).

1661 NA*N*O3, ...E,T. . . . . .PHYS. REV. B VOL.10 NO.2,. . . . . , (1974).
      *GROUP-THEORETICAL SELECTION RULES AND EXPERIMENTAL DETERMINATION OF
      LATTICE MODES IN NA*N*O3 VIA NEUTRON SCATTERING.
      *TREVINO(S.F.), PRASK(H.), CASELLA(R.C.). (REF. OBTAINED FROM PHYS. REV.
      ABSTRACTS).

1662 NAPHTHALENE, ...C3. . . . . .IAEA SYMP. BOMBAY, VOL.2,. . . . . 201, (1964).
      *SLOW NEUTRON SCATTERING IN WATER AND SOME ORGANIC SUBSTANCES.
      *.C.C.C.P.

1663 NA2*RE*H9, ...C1. . . . . .J. CHEM. SOC. FARADAY II VOL.68, 1414, (1972).
      *INTERNAL VIBRATIONS AND HINDERED ROTATION OF RE*H9 (2-) BY NEUTRON
      SCATTERING SPECTROSCOPY.
      *WHITE(J.W.), WRIGHT(C.J.).

1664 NA*S*H,. . . . . .C1. . . . . .IAEA SYMP. GRENOBLE,. . . . . . 247, (1972).
      *NEUTRON SCATTERING STUDY OF THE ROTATIONAL MOTIONS AND PHASE TRANSITIONS
      IN SODIUM- AND CAESIUM-HYDROSULFIDES.
      *DE-GRAAF(L.A.), RUSH(J.J.), LIVINGSTON(R.C.).

1665 NA*S*H,. . . . . .C1. . . . . .J. CHEM. PHYS. VOL.58,. . . . . 3439, (1973).
      *NEUTRON SCATTERING INVESTIGATION OF THE ROTATIONAL DYNAMICS AND PHASE
      TRANSITIONS IN SODIUM AND CESIUM HYDROSULFIDES.
      *RUSH(J.J.), DE-GRAAF(L.A.), LIVINGSTON(R.C.).

SECTION 6 -QUASI-ELASTIC AND INELASTIC SCATTERING STUDIES . . . .

1666 NB,. . . . . *THE LATTICE DYNAMICS OF NIOBIUM II. KOHN ANOMALIES IN NIOBIUM.   B7 . . . J. OF PHYS.-C-, SER.2, VOL.2,   432, (1969).
                           *SHARP(R.I.).

1667 NB,. . . . . *THE LATTICE DYNAMICS OF NIOBIUM I. MEASUREMENTS OF THE PHONON FREQUEN-   B1 . . . J. OF PHYS.-C-, SER.2, VOL.2,   421, (1969).
                           CIES.
                           *SHARP(R.I.).

1668 NB,. . . . . *LATTICE DYNAMICS OF TRANSITION METALS.   B1 . . . . . IAEA SYMP. BOMBAY, VOL.1,. . . . .   87, (1964).
                           *WOODS(A.D.B.).

1669 NB,. . . . . *LATTICE DYNAMICS OF NIOBIUM.   B1 . . . . . COPENHAGEN CONF.-LATT.DYNAMICS,.   39, (1963).
                           *NAKAGAWA(Y.), WOODS(A.D.B.).

1670 NB,. . . . . *INVESTIGATION OF THE MIXED STATE USING POLARIZED NEUTRONS.   04 . . . . PHYS. LETT. VOL.34A, ...   35, (1971).
                           *PFEIFFER(K.), RAUCH(H.), WEBER(H.W.).

1671 NB,. . . . . *DISPERSION RELATIONSHIP FOR THE PHONON SPECTRUM OF NIOBIUM IN THREE   T . . . . . SOV. PHYS. SOL. STATE VOL. 13,   1793, (1972).
                           HIGH SYMMETRY DIRECTIONS.
                           *KORSUNSKII(M.I.), GENKIN(YA.E.), MARKOVNIN(V.I.), ZAVODINSKII(V.G.),
                           SATSUK(V.V.).

1672 NB,. . . . . *TEMPERATURE DEPENDENCE OF THE NORMAL MODES OF NIOBIUM.   B7 . . . . . IAEA SYMP. GRENOBLE, ...   43, (1972).
                           *POWELL(B.M.), WOODS(A.D.B.), MARTEL(P.).

1673 NB,. . . . . *LATTICE DYNAMICS OF NIOBIUM.   B1 . . . . . PHYS. REV. LETT. 11, ...   271, (1963).
                           *NAKAGAWA(Y.), WOODS(A.D.B.).

1674 NB,. . . . . *ELECTRON-PHONON INTERACTION IN SUPERCONDUCTING NB.   B7 . . . . . SOL. STATE COMM. VOL.13, ...   1893, (1973).
                           *SHIRANE(G.), AXE(J.D.), SHAPIRO(S.M.).

1675 NB-C,. . . . . .   *PHONON SPECTRA AND SUPERCONDUCTIVITY IN SOME TRANSITION METAL CARBIDES.   B7 . . . . . CONF. INTERN., RENNES, ...   145, (1971).
                           *SMITH(H.G.), GLASER(W.).

1676 NB*C,. . . . . .   *PHONON SPECTRA OF SOME TRANSITION METAL CARBIDES FROM A SIMPLE   T . . . . . PHYS. REV. B VOL.5, ...   1260, (1972).
                           PSEUDOPOTENTIAL APPROACH.
                           *MOSTOLLER(M.).

1677 NB(H),. . . . .   *THE DYNAMICS OF HYDROGEN IMPURITIES IN NIOBIUM AND VANADIUM.   B1 . . . . . IAEA SYMP. COPENHAGEN, VOL.1,. .   223, (1968).
                           *VERDAN(G.), RUBIN(R.), KLEY(W.).

1678 NB(H),. . . . .   *ANISOTROPIC DIFFUSION OF HYDROGEN IN NIOBIUM SINGLE CRYSTALS.   B3 . . . . . PHYS. REV. LETT. VOL.27, ...   1576, (1971).
                           *KISTNER(G.), RUBIN(R.), SOSNOWSKA(I.).

1679 NB*H,. . . . .   *INELASTIC AND QUASI-ELASTIC SCATTERING OF SUBTHERMAL NEUTRONS IN TH*H   C1 . . . . . SOL. STATE COMM. VOL.8, ...   1321, (1970).
                           AND NB*H.
                           *RUBIN(R.), CLAESSEN(Y.).

1680 NB-H,. . . . .   *STUDY OF THE DYNAMICS OF GROUP V TRANSITION-METAL HYDRIDES BY MEANS OF   C1 . . . . . SOV. PHYS. SOL. STATE VOL.11, .   2343, (1969).
                           INELASTIC NEUTRON SCATTERING.
                           *CHERNOPLEKOV(N.A.), ZEMLYANOV(M.G.), SOMENKOV(V.A.), CHERTKOV(A.A.).

1681 NB-H,. . . . .   *HYDROGEN MOTION IN PRIMARY SOLUTIONS OF HYDROGEN IN SOME TRANSITION   T,C3 . . . . BROOKHAVEN SYMPOSIUM, ...   105, (1965).
                           ELEMENTS.
                           *KLEY(W.), PERETTI(J.), RUBIN(R.), VERDAN(G.).

1682 NB(H),. . . . .   *A NEUTRON SCATTERING STUDY OF HYDROGEN MOTIONS IN NIOBIUM.   C1 . . . . . BER. BUNSENGES. PHYS. CHEM. 76,   782, (1972).
                           *BIRCHALL(J.H.L.), ROSS(D.K.).

1683 NB(H),. . . . .   *QUASI-ELASTIC NEUTRON SCATTERING ON HYDROGEN IN NIOBIUM SINGLE CRYSTALS.   C1 . . . . . PHYS. STAT. SOL. VOL.B-54,   295, (1972).
                           *STUMP(N.), GISSLER(W.), RUBIN(R.).

1684 NB-H,. . . . .   *INTEGRATED QUASI-ELASTIC NEUTRON SCATTERING INTENSITY OF HYDROGEN IN   PHYS. LETT. VOL.43A NO.3,   279, (1973).
                           NIOBIUM.
                           *GISSLER(W.), JAY(B.), RUBIN(R.), VINHAS(L.A.).

1685 NB-MO,. . . . .   *PHONONS IN DISORDERED NIOBIUM-MOLYBDENUM ALLOYS.   B1 . . . . . PHYS. REV. LETT. 15, ...   778, (1965).
                           *WOODS(A.D.B.), POWELL(B.M.).

1686 NB-MO,. . . . .   *LATTICE DYNAMICS OF NIOBIUM-MOLYBDENUM ALLOYS.   B1 . . . . . PHYS. REV. 171, ...   727, (1968).
                           *POWELL(B.M.), MARTEL(P.), WOODS(A.D.B.).

1687 NB3*SN,. . . . .   *EXPLANATION OF THE TEMPERATURE DEPENDENCE OF THE DEBYE-WALLER FACTOR IN   T . . . . . PHYS. LETT. VOL.35A,   31, (1971).
                           NB3*SN.
                           *SCHUSTER(H.), BOSTOCK(J.).

1688 NB3*SN,. . . . .   *INFLUENCE OF THE SUPERCONDUCTING ENERGY GAP ON PHONON LINEWIDTHS IN   B7 . . . . . PHYS. REV. LETT. VOL.30, ...   214, (1973).
                           NB3*SN.
                           *AXE(J.D.), SHIRANE(G.).

1689 NB3*SN,. . . . .   *ACOUSTIC-PHONON INSTABILITY AND CRITICAL SCATTERING IN NB3*SN.   B6 . . . . . PHYS. REV. LETT. VOL.27, . . .   1803, (1971).
                           *SHIRANE(G.), AXE(J.D.).

1690 NB3*SN,. . . . .   *THEORY OF LATTICE DYNAMICS OF NB3*SN-TYPE COMPOUNDS.   B7,T . . . . PHYS. REV. B VOL.6, ...   3584, (1972).
                           *SHAM(L.J.).

1691 NB3*SN,. . . . .   *INELASTIC-NEUTRON-SCATTERING STUDY OF ACOUSTIC PHONONS IN NB3*SN.   B2 . . . . . PHYS. REV. B VOL.8, ...   1965, (1973).
                           *AXE(J.D.), SHIRANE(G.).

SECTION   6 -QUASI-ELASTIC AND INELASTIC SCATTERING STUDIES . . . .

1692 NB3*SN,. . . . . B2 . . . . . . PHYSICS TODAY (MARCH), ...   17, (1973).
                        *PHONON LINEWIDTHS MEASURED BY NEUTRON SCATTERING.
                        *LUBKIN(G.B.).

1693 ND*(AS,P,SB),. . *. . . . . B7 . . . . . . J. PHYS. C VOL.5, ... 2246, (1972).
                        *CRYSTALLINE ELECTRIC FIELD LEVELS IN THE NEODYMIUM MONOPNICTIDES
                        DETERMINED BY NEUTRON SPECTROSCOPY.
                        *FURRER(A.), KJEMS(J.), VOGT(O.).

1694 ND*CO2,. . . . . D4 . . . . . . REF. SECTION 3, 64/KOEHLER(2), ...

1695 ND2*03,. . . . . D3 . . . . . . PHYS. REV. VOL.92, ... 1380, (1953).
                        *PARAMAGNETIC SCATTERING OF NEUTRONS BY RARE EARTH OXIDES.
                        *KOEHLER(W.C.), WOLLAN(E.O.).

1696 NE,. . . . . . . *. . . . . T. . . . . . . SOL. STATE COMM. VOL.10, ... 1219, (1972).
                        *INTERFERENCE EFFECTS IN THE INELASTIC SCATTERING OF NEUTRONS BY AN
                        ANHARMONIC NEON CRYSTAL.
                        *BOHLIN(L.).

1697 NE,. . . . . . . *. . . . . T. . . . . . . PHYS. REV. LETT. VOL.24, ... 1424, (1970).
                        *PHONON ENERGIES AND LIFETIMES IN SOLID NE AND HE IN THE FIRST-ORDER
                        SELF-CONSISTENT APPROXIMATION.
                        *GOLDMAN(V.V.), HORTON(G.K.), KLEIN(M.L.).

1698 NE,. . . . . . . *. . . . . T. . . . . . . DISS. ABS. VOL.31, ...4905B, (1971).
                        *LATTICE DYNAMICS OF SOLID NEON IN THE SELF CONSISTENT PHONON
                        APPROXIMATION.
                        *GOLDMAN(V.V.).

1699 NE,. . . . . . . *. . . . . B4 . . . . . . IAEA SYMP. GRENOBLE, ...   399, (1972).
                        *COLLECTIVE AND SINGLE-PARTICLE EXCITATIONS IN LIQUID NEON.
                        *BUYERS(W.J.L.), SEARS(V.F.), LONNGI(P.A.), LONNGI(D.A.).

1700 NE,. . . . . . . *. . . . . B4 . . . . . . PHYS. LETT. VOL.19, ...   269, (1965).
                        *CO-OPERATIVE MODES OF MOTION IN SIMPLE LIQUIDS.
                        *CHEN(S.H.),   EDER(O.J.), EGELSTAFF(P.A.), HAYWOOD(B.C.),   WEBB(F.J.).

1701 NE,. . . . . . . *. . . . . B1 . . . . . . PHYS. REV. B VOL.6, ... 4766, (1972).
                        *INELASTIC NEUTRON SCATTERING FROM SOLID NEON.
                        *SKALYO-JR(J.), MINKIEWICZ(V.J.), SHIRANE(G.), DANIELS(W.B.).

1702 NE,. . . . . . . *. . . . . B1 . . . . . . PHYS. REV. VOL.181, ... 1251, (1969).
                        *LATTICE DYNAMICS OF NEON AT TWO DENSITIES FROM COHERENT INELASTIC
                        NEUTRON SCATTERING.
                        *LEAKE(J.A.),   DANIELS(W.B.), SKALYO-JR(J.), FRAZER(B.C.),   SHIRANE(G.).

1703 NE,. . . . . . . *. . . . . T, B1 . . . . J. SOLID STATE CHEM. VOL.5, ... 477, (1972).
                        *INTERPRETATION OF PHONON DISPERSION IN SOLID ARGON AND NEON.
                        *GUPTA(N.P.).

1704 NE,. . . . . . . *. . . . . T, B4 . . . . PHYS. REV. A VOL.7, ... 1043, (1973).
                        *NEUTRON SCATTERING BY LIQUID NEON.
                        *KERR(W.C.), SINGWI(K.S.).

1705 NE,. . . . . . . *. . . . . B4 . . . . . . PHYS. LETT. VOL.45A, ... 479, (1973).
                        *INVESTIGATION OF COLLECTIVE EXCITATIONS IN LIQUID NEON BY MEANS OF
                        NEUTRON SCATTERING AT SMALL SCATTERING VECTORS.
                        *BELL(H.G.), KOLLMAR(A.), ALEFELD(B.), SPRINGER(T.).

1706 NE(H), ... *. . . . . C3 . . . . . . PROC. PHYS. SOC. 89, ... 833, (1966).
                        *MOLECULAR MOTION OF HYDROGEN IN LIQUID ARGON AND NEON.
                        *EDER(O.J.), CHEN(S.H.), EGELSTAFF(P.A.).

1707 NI,. . . . . . . *. . . . . C2 . . . . . . IAEA SYMP. CHALK RIVER, VOL.2, . 159, (1962).
                        *INVESTIGATION OF THE PHONON SPECTRUM OF NICKEL.
                        *.C.C.C.P.

1708 NI,. . . . . . . *. . . . . D1 . . . . . . IAEA SYMP. COPENHAGEN, VOL.2,. . 101, (1968).
                        *MAGNETIC EXCITATIONS IN NICKEL.
                        *KOMURA(S.), LOWDE(R.D.), WINDSOR(C.G.).

1709 NI,. . . . . . . *. . . . . C1 . . . . . . COPENHAGEN CONF.-LATT.DYNAMICS,. 63, (1963).
                        *MEASURED VIBRATIONAL FREQUENCY DISTRIBUTIONS OF NI, V, TI AND TI-ZR.
                        *MOZER(B.), OTNES(K.), PALEVSKY(H.).

1710 NI,. . . . . . . *. . . . . T. . . . . . PHYS. REV. B VOL.4, ... 3048, (1971).
                        *NEUTRON ORBITAL CROSS SECTION FOR A TIGHT-BINDING MODEL OF PARAMAGNETIC
                        NICKEL.
                        *LOVESEY(S.W.), WINDSOR(C.G.).

1711 NI,. . . . . . . *. . . . . D1 . . . . . J. APP. PHYS. VOL.40,. . . . . . . 1142, (1969).
                        *MAGNETIC EXCITATIONS IN NICKEL.
                        *LOWDE(R.D.), WINDSOR(C.G.).

1712 NI,. . . . . . . *. . . . . D1 . . . . . . J. APP. PHYS. VOL.40,. . . . . . . 1450, (1969).
                        *SPIN-WAVE SPECTRUM OF NICKEL METAL.
                        *MOOK(H.A.), NICKLOW(R.M.), THOMPSON(E.D.), WILKINSON(M.K.).

1713 NI,. . . . . . . *. . . . . D1 . . . . . . J. OF PHYS. C VOL.1,. . . . . . . 950, (1968).
                        *OBSERVATION OF SPIN-WAVE RENORMALIZATION EFFECTS IN IRON AND NICKEL.
                        *STRINGFELLOW(M.W.).

1714 NI,. . . . . . . *. . . . . T. . . . . . PHYS. LETT. 29A, ... 75, (1969).
                        *TEMPERATURE SHIFT AND CRYSTALLINE PERFECTION IN CRITICAL MAGNETIC
                        SCATTERING.
                        *STUMP(N.), MAIER(G.).

1715 NI +(IMPURITIES), *. . . . . D4 . . . . . J. OF PHYS. C VOL.1,. . . . . . . 458, (1968).
                        *THE MAGNETIC DISTURBANCES ASSOCIATED WITH SUBSTITUTIONAL IMPURITIES IN
                        NICKEL.
                        *COMLY(J.B.), HOLDEN(T.M.), LOW(G.G.).

1716 NI,. . . . . . . *. . . . . D1,D2 . . . . PHYS. REV. VOL.182, ... 624, (1969).
                        *CRITICAL AND SPIN-WAVE FLUCTUATIONS IN NICKEL BY NEUTRON SCATTERING.
                        *MINKIEWICZ(V.J.), COLLINS(M.F.), NATHANS(R.), SHIRANE(G.).

1717 NI,. . . . . . . *. . . . . P. . . . . . ACTA PHYS. ACAD. SCI. HUNG. 30, . 231, (1971).
                        *PHONON FREQUENCY DISTRIBUTION FUNCTIONS OF COPPER, NICKEL AND VANADIUM.
                        *YAI-PRAKASH, SEMUEAL(B.S.), SHARMA(P.K.).

1718 NI,. . . . . . . *. . . . . 1ST EUR. CONF./CONDENSED MATTER, 41, (1971).
                        *NEUTRON DEPOLARIZATION EXPERIMENTS IN THE IMMEDIATE VICINITY OF THE
                        CURIE TEMPERATURE OF NICKEL.
                        *REKVELDT(M.TH.). (ED. NOTE: PUBL. EUROPEAN PHYS. SOC. 1971, GENEVA,
                        SWITZ.).

SECTION 6 -QUASI-ELASTIC AND INELASTIC SCATTERING STUDIES . . . .

1719 NI,. . . . . .D2. . . . . . .Z. PHYS. VOL.238,. . . . . 389, (1970).
*SMALL ANGLE CRITICAL SCATTERING OF NEUTRONS BY NI.
*MAIER(G.), STUMP(N.).

1720 NI,. . . . . .D1. . . . . .SOL. STATE COMM. VOL.6,. . . .189, (1968).
*SPIN WAVES, STONER MODES AND CRITICAL FLUCTUATIONS IN NICKEL.
*LOWDE(R.D.), WINDSOR(C.G.).

1721 NI,. . . . . .B7,D4. . . . .PHYS. LETT. VOL.34A,. . . . . . 23, (1971).
*OBSERVATION OF ELECTRON SPIN-ORBIT COUPLING; ELECTRON-PHONON AND
ELECTRON-MAGNON INTERACTION IN NICKEL.
*FRIKKEE(E.).

1722 NI,. . . . . .B7,D4. . . . .RCN REPORT 140,. . . . . , (1971).
*INTERACTION BETWEEN ELECTRONS, MAGNONS AND PHONONS IN NICKEL.
*FRIKKEE(E.). (ED. NOTE: AVAIL. REACTOR CENTRUM NEDERLAND, PETTEN, THE
NETHERLANDS).

1723 NI,. . . . . .D4. . . . . .J. PHYSIQUE TOME 32 COL.1 VOL.II 679, (1971).
*INVESTIGATION OF MICROSTRAIN FIELDS BY NEUTRON SCATTERING AND
DEPOLARIZATION.
*SCHMATZ(W.), BERNDORFER(K.), DURCANSKY(G.).

1724 NI,. . . . . .D3. . . . . .PHYS. REV. LETT. VOL.18, ... 1136, (1967).
*GENERALIZED SUSCEPTIBILITY OF NICKEL.
*LOWDE(R.D.), WINDSOR(C.G.).

1725 NI,. . . . . .ADVANCES IN PHYS. VOL.19 NO.82,. 813, (1970).
*ON THE MAGNETIC EXCITATIONS IN NICKEL (REVIEW PAPER).
*LOWDE(R.D.), WINDSOR(C.G.).

1726 NI,. . . . . .D4. . . . . .PROC. PHYS. SOC. 86, ... 535, (1965).
*THE MAGNETIC MOMENT DISTRIBUTION AROUND TRANSITION ELEMENT IMPURITIES
IN IRON AND NICKEL.
*COLLINS(M.F.), LOW(G.G.E.).

1727 NI,. . . . . .D2. . . . . .J. APP. PHYS 39,. . . . . 528, (1968).
*NEUTRON SCATTERING STUDIES OF CRITICAL MAGNETIC PHENOMENA.
*RISTE(T.).

1728 NI,. . . . . .D1. . . . . .J. APP. PHYS. 39,. . . . . 383, (1968).
*SPIN WAVES IN 3D METALS.
*SHIRANE(G.), MINKIEWICZ(V.J.), NATHANS(R.).

1729 NI,. . . . . .D1. . . . . .SOL. STATE COMM. 4,. . . . . . 99, (1966).
*S-D AUSTAUSCHWECHSELWIRKUNG UND FERMIFLAECHE IN NICKEL.
*FURRER(A.), SCHNEIDER(T.), HALG(W.).

1730 NI,. . . . . .D1. . . . . .IAEA SYMP. COPENHAGEN, VOL.2,. . 3, (1968).
*NEUTRON SPIN SCATTERING BY SPIN WAVES IN METALS (REVIEW PAPER).
*MOLLER(H.B.).

1731 NI,. . . . . .B1. . . . . .PHYS. REV. VOL.136,. . . . . . .A1359, (1964).
*NORMAL MODES OF VIBRATION IN NICKEL.
*BIRGENEAU(R.J.), CORDES(J.), DOLLING(G.), WOODS(A.D.B.).

1732 NI,. . . . . .D3. . . . . .J. APP. PHYS. 38,. . . . . 1247, (1967).
*SPIN CORRELATIONS IN PARAMAGNETIC NICKEL.
*CABLE(J.W.), LOWDE(R.D.), WINDSOR(C.G.), WOODS(A.D.B.).

1733 NI,. . . . . .D2. . . . . .J. APP. PHYS. 38,. . . . . 1245, (1967).
*MAGNETIC FORM FACTOR OF NI IN THE PARAMAGNETIC STATE.
*CAGLIOTI(G.), COOPER(M.F.), MINKIEWICZ(V.J.), PICKART(S.J.).

1734 NI,. . . . . .D4. . . . . .J. APP. PHYS. 38,. . . . . . . .1240, (1967).
*SIMILARITY OF THE MAGNETIZATION DISTRIBUTION AROUND TRANSITION AND
NON-TRANSITION ELEMENT IMPURITIES IN NICKEL.
*COMLY(J.B.), HOLDEN(T.M.), LOW(G.G.E.).

1735 NI,. . . . . .D1. . . . . .J. APP. PHYS. 36,. . . . . 1076, (1965).
*SPIN WAVE SCATTERING OF POLARIZED NEUTRONS FROM NICKEL AND COBALT.
*RISTE(T.), SHIRANE(G.), ALPERIN(H.A.), PICKART(S.J.).

1736 NI,. . . . . .D1. . . . . .J. APP. PHYS. 37,. . . . . 1034, (1966).
*MAGNETIC-MOMENT DISTRIBUTION IN NICKEL METALS.
*MOOK(H.A.), SHULL(C.G.).

1737 NI,. . . . . .D2. . . . . .PHYS. LETT. 24A,. . . . . 625, (1967).
*EVIDENCE OF KOCINSKIS SPIN CORRELATION FUNCTION FROM CRITICAL MAGNETIC
SCATTERING BY NICKEL.
*STUMP(N.), MAIER(G.).

1738 NI,. . . . . .D1. . . . . .PHYS. REV. 156,. . . . . 623, (1967).
*SPIN-WAVE DISPERSION IN FERROMAGNETIC NI AND FCC CO.
*PICKART(S.J.), ALPERIN(H.A.), MINKIEWICZ(V.J.), NATHANS(R.),
SHIRANE(G.), STEINSVOLL(O.).

1739 NI,. . . . . .D1. . . . . .PHYS. REV. LETT. VOL.30,. . . . 556, (1973).
*TEMPERATURE DEPENDENCE OF THE MAGNETIC EXCITATIONS IN NICKEL.
*MOOK(H.A.), LYNN(J.W.), NICKLOW(R.M.).

1740 NI,. . . . . .B7. . . . . .BULL. SOC. FRANC. MIN. CRIST. 90 428, (1967).
*INVESTIGATIONS ON SMALL ANGLE SCATTERING WITH NEUTRONS.
*SPRINGER(T.), SCHMATZ(W.).

1741 NI,. . . . . .D4. . . . . .J. PHYS. CHEM. SOLIDS VOL.24,. . 387, (1963).
*ELECTRONIC STRUCTURE OF IMPURITIES IN NICKEL. APPLICATION TO MAGNETIC
NEUTRON SCATTERING BY DISORDERED NICKEL-BASE ALLOYS.
*GAUTIER(F.).

1742 NI,. . . . . .B1. . . . . .PHYSICA VOL.34,. . . . . 257, (1967).
*LATTICE DYNAMICS OF NICKEL.
*HAUTECLER(S.), VAN-DINGENEN(W.).

1743 NI,. . . . . .T. . . . . . .ACTA PHYS. HUNGAR. VOL.15,. . . . 29, (1962).
*A CONTRIBUTION TO THE PROBLEM OF INELASTIC MAGNETIC SCATTERING OF
POLARIZED NEUTRONS IN FE AND NI.
*VALENTA(L.), ZAJAC(ST.).

1744 NI,. . . . . .D2. . . . . .J. PHYS. SOC. JAPAN VOL.17 B-III 67, (1962).
*CRITICAL SCATTERING OF NEUTRONS BY NICKEL.
*CRIBIER(D.), JACROT(B.), PARETTE(G.).

1745 NI,. . . . . .P. . . . . .REF. SECTION 3, 48/HAVENS,. . . . .

SECTION   6 -QUASI-ELASTIC AND INELASTIC SCATTERING STUDIES . . . .

1746 NI,. . . . .   ↝MULTI-OSCILLATOR PROCESSES IN THE SCATTERING OF NEUTRONS BY CRYSTALS.
                           ↝SQUIRES(G.L.).   E.T.......PROC. ROY. SOC. A VOL.212,.....192, (1952).

1747 NI,. . . . . . .   ↝TEMPERATURE DEPENDENCE OF THE GENERALIZED SUSCEPTIBILITY OF NICKEL.
                          ↝LOWDE(R.D.), WINDSOR(C.G.).   D4.........J. APP. PHYS. VOL.39,.........449, (1968).

1748 NI,. . . . . . .   ↝STUDY OF INELASTIC SCATTERING OF SLOW NEUTRONS BY NICKEL SPECIMENS OF
                          DIFFERENT ISOTOPIC COMPOSITION.
                          ↝GORBACHEV(B.I.), IVANITSKII(P.G.), KROTENKO(V.T.), PASECHNIK(M.V.). (IN
                          RUSSIAN).   P,T......UKR. FIZ. ZH. (USSR) VOL.18,.....558, (1973).

1749 NI,. . . . . . .   ↝CRITICAL SCATTERING OF POLARIZED NEUTRONS IN NICKEL (NEAR THE CURIE
                          POINT).
                          ↝DRABKIN(G.M.), ZABIDAROV(E.I.), KASMAN(YA.A.), OKOROKOV(A.I.).   D2.....SOV. PHYS. JETP LETT. VOL.2,.....336, (1965).

1750 NI,. . . . . . .   ↝INVESTIGATION OF PHONON SPECTRUM OF NICKEL.
                          ↝CHERNOPLEKOV(N.A.), ZEMLYANOV(M.G.), CHICHERIN(A.G.), LYASHCHENKO(B.G.).   B1.......SOV. PHYS. J.E.T.P. VOL.17,... 584, (1963).

1751 NI,. . . . . . .   ↝THE LATTICE VIBRATIONS OF NICKEL IN KREBS*S MODEL.
                          ↝SHUKLA(M.M.), DAYAL(B.).   T,B1.....PHYS. STAT. SOL. 19,........729, (1967).

1752 NI,. . . . . . .   ↝PHONON FREQUENCIES ALONG THE (100) DIRECTION IN THE NICKEL LATTICE.
                          ↝VINTAIKIN(E.Z.), GORBACHEV(B.I.).   B1......SOV. PHYS. SOL. STATE 7, . . . 1547, (1965).

1753 NI,. . . . . . .   ↝PHONON DISPERSION RELATIONS IN NICKEL.
                          ↝PAL(S.), GUPTA(R.P.).   T,B1......SOL. STATE COMM. 4,........83, (1966).

1754 NI,. . . . . . .   ↝THE LATTICE DYNAMICS OF FERROMAGNETIC AND PARAMAGNETIC NICKEL.
                          ↝DEWIT(G.A.), BROCKHOUSE(B.N.).   B1.......J. APP. PHYS. 39,.........451, (1968).

1755 NI,. . . . . . .   ↝DIFFUSION DES NEUTRONS FROIDS PAR LES PHONONS DANS LE NICKEL.
                          ↝HAUTECLER(S.), VAN-DINGENEN(W.).   B1......COLL. INTER. N.126 (GRENOBLE), . 229, (1963).

1756 NI,. . . . . . .   ↝LATTICE DYNAMICS OF NICKEL AND KREBS*S MODEL.
                          ↝HAUTECLER(S.), VAN-DINGENEN(W.).   B1......BROOKHAVEN SYMPOSIUM.,.......83, (1965).

1757 NI,. . . . . . .   ↝DIFFUSION AUX PETITS ANGLES DES NEUTRONS PAR LE FER ET LE NICKEL AU
                          VOISINAGE DU POINT DE CURIE.
                          ↝JACROT(B.), KONSTANTINOVIC(J.), PARETTE(G.), CRIBIER(D.).   D2.......IAEA SYMP. CHALK RIVER, VOL.2,. 317, (1962).

1758 NI,. . . . . . .   ↝CRITICAL FLUCTUATIONS IN NI ABOVE THE CURIE POINT.
                          ↝GOTTLIER(A.M.), HOHENEMSER(C.).   T,D2.....PHYS. REV. LETT. VOL.31,. . . . 1222, (1973).

1759 NI-(BE), ...   ↝AN INVESTIGATION OF INELASTIC NEUTRON SCATTERING BY A CRYSTAL DOPED WITH
                          LIGHT IMPURITY ATOMS.
                          ↝ZEMLYANOV(M.G.), SOMENKOV(V.A.), CHERNOPLEKOV(N.A.).   B3.......SOV. PHYS. J.E.T.P. VOL. 25,. . . 436, (1967).

1760 NI(CR,V),. . . .   ↝VIRTUAL MAGNON STATES IN DILUTE ALLOYS.
                          ↝KROO(N.), PAL(L.), JOVIC(D.).   D1.......IAEA SYMP. COPENHAGEN, VOL.2,. . 37, (1968).

1761 NI*F2, ...   ↝NEUTRON AND OPTICAL INVESTIGATION OF MAGNONS AND MAGNON-MAGNON
                          INTERACTION EFFECTS IN NI*F2.
                          ↝HUTCHINGS(M.T.), THORPE(M.F.), BIRGENEAU(R.J.), FLEURY(P.A.),
                          GUGGENHEIM(H.J.).   D1,D4....PHYS. REV. B-2,.........1362, (1970).

1762 NI*F2, ...   ↝SPACE-GROUP COREPRESENTATIONS AND MAGNON DISPERSION RELATIONS IN NI*F2.
                          ↝JOSHUA(S.J.), CRACKNELL(A.P.).   D1,T......J. OF PHYS. C VOL.2,.........24, (1969).

1763 NI-FE, ...   ↝MAGNETIC EXCITATIONS IN A NI-FE ALLOY AT ELEVATED TEMPERATURES.
                          ↝MIKKE(K.), JANKOWSKA(J.), MODRZEJEWSKI(A.).   D1,D2....PHYS. STAT. SOLIDI B VOL.59 NO.2 K97, (1973).

1764 NI-FE, ...   ↝CRYSTAL DYNAMICS OF NICKEL-IRON AND COPPER-ZINC ALLOYS.
                          ↝HALLMAN(E.D.), BROCKHOUSE(B.N.).   B1.......CAN. J. PHYS. 47,.. . . . . . . 1117, (1969).

1765 NI-(FE,CU),. . .   ↝THE CRYSTAL DYNAMICS OF FACE CENTRED CUBIC IRON GROUP ALLOYS. . , (1969).
                          ↝HALLMAN(E.D.). (ED. NOTE: PH.D. DISSERTATION, MCMASTER UNIVERSITY).   B1.......THESIS (183 PP.),

1766 NI3*MN,. . . . .   ↝NEUTRON-SCATTERING OBSERVATIONS OF CRITICAL SLOWING DOWN OF AN ISING
                          SYSTEM.
                          ↝COLLINS(M.F.), TEH(H.C.).   B6.......PHYS. REV. LETT. VOL.30 NO.17,. 781, (1973).

1767 NI-MN, ...   ↝ATOMIC MAGNETIC MOMENTS IN NICKEL-MANGANESE ALLOYS.
                          ↝LOSHMANOV(A.A.).   D4.......PHYS. MET. METALLOGR. VOL.18 N.2 22, (1964).

1768 NI*O,. . . . . .   ↝MAGNONS IN ANTIFERROMAGNETIC NI*O, MN*O AND MN*S.
                          ↝REISSLAND(J.A.), BEGUM(N.A.).   D1.......J. OF PHYS.-C-, SER. 2, VOL.2, . 874, (1969).

1769 NI*O,. . . . . .   ↝INELASTIC NEUTRON SCATTERING MEASUREMENT OF SPIN WAVES AND MAGNETIC
                          INTERACTIONS IN NI*O.
                          ↝HUTCHINGS(M.T.), SAMUELSEN(E.J.).   D1.......SOL. STATE COMM. VOL. 9,......1011, (1971).

1770 NI*O,. . . . . .   ↝MEASUREMENT OF SPIN-WAVE DISPERSION IN NI*O BY INELASTIC NEUTRON SCAT-
                          TERING AND ITS RELATION TO MAGNETIC PROPERTIES.
                          ↝HUTCHINGS(M.T.), SAMUELSEN(E.J.).   D1.......PHYS. REV. B VOL.6,.. . . . . . 3447, (1972).

1771 NI-PD, ...   ↝MEASUREMENTS OF THE VIBRATIONAL FREQUENCY SPECTRUM OF NICKEL-PALLADIUM
                          ALLOYS.
                          ↝MOZER(B.), OTNES(K.).   C2.......IAEA SYMP. CHALK RIVER, VOL.2,. 167, (1962).

SECTION   6 -QUASI-ELASTIC AND INELASTIC SCATTERING STUDIES . . . .

1772 NI-PD, ...B3. . . . . .IAEA SYMP. GRENOBLE, ... 73, (1972):
      *NEUTRON SCATTERING BY VIBRATIONS IN A MASS-DISORDERED LATTICE: NI1.55-
      PD(.45)
      *KAMITAKAHARA(W.A.), BROCKHOUSE(B.N.).

1773 NI-PD, ...T. . . . . . .PHYS. REV. B VOL.10 NO.2, ... (1974).
      *COMPARISON OF SINGLE-SITE APPROXIMATIONS FOR THE LATTICE DYNAMICS OF
      MASS-DISORDERED ALLOYS.
      *KAMITAKAHARA(W.A.), TAYLOR(D.W.). (REF. OBTAINED FROM PHYS. REV.
      ABSTRACTS).

1774 NI-PD, ...B1. . . . . .PHYS. REV. B VOL.10 NO. 2, ... , (1974).
      *VIBRATIONS OF A MIXED CRYSTAL: NEUTRON SCATTERING FROM NI55-PD45.
      *KAMITAKAHARA(W.A.), BROCKHOUSE(B.N.). (REF. OBTAINED FROM PHYS. REV.
      ABSTRACTS).

1775 NI-PD-(H), ...D4. . . . . .PHYSICA VOL.28, ... 917, (1962):
      *THE INFLUENCE OF HYDROGEN ON THE MAGNETIC SCATTERING OF NEUTRONS BY A
      NICKEL-PALLADIUM ALLOY.
      *VAN-LOEF(J.J.), VAN-DINGENEN(W.).

1776 NI-PT, ...B3. . . . . .SOL. STATE COMM. VOL.13 NO.4, . . 495, (1973).
      *PHONON MODES IN MASS-DEFECT DISORDERED ALLOY NI-PT.
      *KUNITOMI(N.), TSUNODA(Y.), HIRAI(Y.).

1777 NI*S, ...B1,D1. . . . .IAEA SYMP. GRENOBLE, ... 669, (1972).
      *PHONON AND MAGNON EXCITATIONS IN NI*S.
      *BRIGGS(G.A.), DUFFILL(C.), HUTCHINGS(M.T.), LOWDE(R.D.),
      MURTHY(N.S.S.), SAUNDERSON(D.H.), STRINGFELLOW(M.W.)... ET AL.

1778 NI*S, ...D1. . . . . .A.I.P. CONF. PROC. NO.10 PART 2, 14C3, (1973).
      *MAGNONS IN NICKEL SULPHIDE.
      *HUTCHINGS(M.T.), LOWDE(R.D.), SAUNDERSON(D.H.), STRINGFELLOW(M.W.),
      WINDSOR(C.G.).

1779 NI*S*O4.(H2*O)6, NI*S*O4.(H2*O)7...C1. . . . . .REF. SECTION 3, 68/BAJOREK, ...

1780 NI-V, ...D1. . . . . .PHYS. STAT. SOL. B VOL.54 NO.2, . K99, (1972).
      *SPIN WAVES IN NI-V ALLOY BY NEUTRON SCATTERING.
      *MIKKE(K.), DOBRZYNSKI(L.), JANKOWSKA(J.).

1781 NI-(X), ...D4. . . . . .J. APP. PHYS. VOL. 39, ... 1174, (1968).
      *MAGNETIC NEUTRON SCATTERING FROM ATOMS DISSOLVED IN FERROMAGNETIC IRON
      AND NICKEL.
      *LOW(G.G.).

1782 NP, ...P. . . . . . .J. PHYS. CHEM. SOL. 28, ... 1651, (1967).
      *A NEUTRON DIFFRACTION DETERMINATION OF THE COHERENT SCATTERING AMPLITUDE
      OF NP AND THE POSSIBLE ANTIFERROMAGNETISM OF NEPTUNIUM DIOXIDE.
      *HEATON(L.), MUELLER(M.H.), WILLIAMS(J.M.).

1783 NYLON, ...C1. . . . . .J. CHEM. PHYSICS VOL.43, ... 34C4, (1965).
      *STUDY OF LOW-FREQUENCY MOTIONS IN NYLON-6.
      *SAFFORD(G.J.), LOSACCO(F.J.).

1784 O2, ...T. . . . . . .PHYS. REV. VOL.90, ... 869, (1953).
      *THE SCATTERING OF SLOW NEUTRONS BY O2 MOLECULES.
      *HALPERN(O.), APPLETON(G.L.).

1785 O2, ...D3. . . . . .PHYS. REV. VOL.98, ... 492, (1955).
      *MAGNETIC SCATTERING OF SLOW NEUTRONS BY GASEOUS OXYGEN.
      *PALEVSKY(H.), EISBERG(R.M.).

1786 O2, ...T,D3. . . . .PHYS. REV. VOL.97, ... 411, (1955).
      *MAGNETIC SCATTERING OF SLOW NEUTRONS FROM O2 GAS.
      *KLEINER(W.H.).

1787 OS, ...P. . . . . . .REF. SECTION 3, 47/WU, ...

1788 P*H4*I, ...C1. . . . .J. CHEM. PHYS. VOL.44, ... 1722, (1966).
      *LOW-FREQUENCY MOTIONS AND BARRIER TO ROTATION IN PHOSPHONIUM IODIDE.
      *RUSH(J.J.).

1789 PARAFFIN, ...C1. . . . .J. CHEM. PHYSICS VOL.40, ... 1417, (1964).
      *STUDY OF LOW-FREQUENCY MOTIONS IN POLYETHYLENE AND THE PARAFFIN
      HYDROCARBONS BY NEUTRON INELASTIC SCATTERING.
      *DANNER(H.R.), SAFFORD(G.J.), BOUTIN(H.), BERGER(M.).

1790 PARAFFINS, ...C1,C3. . . .DISS. ABS. VOL.30, ...4740B, (1970).
      *THERMAL NEUTRON SCATTERING FROM THE SOLID AND LIQUID PHASES OF SOME
      N-PARAFFINS.
      *LOGAN(K.W.).

1791 PARAFFIN(N), . . .C1. . . . .IAEA SYMP. BOMBAY, VOL.2, ... 4C7, (1964).
      *STUDY OF THE LOW-FREQUENCY MOLECULAR MOTIONS IN POLYETHYLENE AND THE
      N-PARAFFINS BY SLOW NEUTRON INELASTIC SCATTERING.
      *BOUTIN(H.), PRASK(H.), TREVINO(S.F.), TREVINO(H.R.).

1792 PARRAFIN, ...P. . . . . . .J. NUCL. SCI. TECHNOL. VOL.6, . . 671, (1969).
      *MEASUREMENT OF NEUTRON SLOWING DOWN TIME IN LIGHT WATER, ICE, PARAFFIN
      AND SANTOWAX.
      *SAKAMOTO(S.), KANEKO(Y.), AKINO(F.).

1793 PB, ...B1. . . . .CAN. J. PHYS. 48, ... 1781, (1970).
      *LATTICE FREQUENCY SPECTRA OF PB AND PB-TL BY NEUTRON SPECTROMETRY.
      *ROY(A.P.), BROCKHOUSE(B.N.).

1794 PB, ...B1. . . . .IAEA SYMP. COPENHAGEN, VOL.1, . 253, (1968).
      *ATOMIC VIBRATIONS IN FACE-CENTERED-CUBIC ALLOYS OF BI, PB AND TL.
      *NG(S.C.), BROCKHOUSE(B.N.).

1795 PB, ...C3. . . . .IAEA SYMP. CHALK RIVER, VOL.1, 215, (1962).
      *FREQUENCY SPECTRUM OF LIQUIDS AND COLD NEUTRON SPECTROMETRY.
      *SINGWI(K.S.), SJOLANDER(A.), RAHMAN(A.).

1796 PB, ...C3. . . . .IAEA SYMP. CHALK RIVER, VOL.1, 2C3, (1962).
      *INTERPRETATION OF COHERENT NEUTRON SCATTERING BY LIQUIDS.
      *EGELSTAFF(P.A.).

1797 PB, ...C3. . . . .IAEA SYMP. COPENHAGEN, VOL.1, . 545, (1968).
      *ANALYSIS OF NEUTRON SCATTERING EXPERIMENTS ON LIQUIDS.
      *AGRAWAL(A.K.), DESAI(R.C.), YIP(S.).

1798 PB, ...B4. . . . .IAEA SYMP. COPENHAGEN, VOL.1, . 463, (1968).
      *A QUASI-PHONON TREATMENT OF COHERENT NEUTRON SCATTERING BY LIQUID LEAD.
      *COCKING(S.J.).

SECTION   6 -QUASI-ELASTIC AND INELASTIC SCATTERING STUDIES . . . .

1799 PB,. . . . .   *LIQUID DYNAMICS FROM NEUTRON SCATTERING (REVIEW PAPER). IAEA SYMP. COPENHAGEN, VOL.1,   397, (1968).
                            *LARSSON(K.E.).

1800 PB,. . . . .   *EXPERIMENTAL PHONON FREQUENCIES AND WIDTHS OF LEAD AT 5, 80, AND 293 DEG. K. PHYS. STAT. SOL. VOL.42, 821, (1971).
                            *FURRER(A.), HALG(W.).

1801 PB,. . . . .   *CRYSTAL DYNAMICS OF LEAD. IAEA SYMPOSIUM VIENNA,   531, (1960).
                            *BROCKHOUSE(B.N.), ARASE(T.), CAGLIOTI(G.), SAKAMOTO(M.), SINCLAIR(R.N.),
                            WOODS(A.D.B.).

1802 PB,. . . . .   *CO-OPERATIVE MODES OF MOTION IN LIQUID LEAD. J. OF PHYS. C VOL.1,   507, (1968).
                            *COCKING(S.J.), EGELSTAFF(P.A.).

1803 PB,. . . . .   *INELASTIC SCATTERING OF THERMAL NEUTRONS BY LIQUID SODIUM AND LEAD. PHYS. REV. VOL.180,   308, (1969).
                            *DESAI(R.C.), YIP(S.).

1804 PB (ISO. A=206, 207) *NEUTRON INELASTIC SCATTERING IN PB 206 AND PB 207. INDIAN J. PHYS. VOL.42,   408, (1968).
                            *GUPTA(J.B.), NATH(N.).

1805 PB,. . . . .   *MEASUREMENT OF INCOHERENT SCATTERING CROSS-SECTION OF LEAD AND BISMUTH WITH SLOW NEUTRONS. NUKLEONIC VOL.12,   4, (1968).
                            *SCHERM(R.).

1806 PB,. . . . .   *STUDIES ON THE DETERMINATION OF THE PHONON DENSITY OF STATES BY COHERENT INELASTIC SCATTERING OF SLOW NEUTRONS FROM POLYCRYSTALLINE SAMPLES. IAEA SYMP. GRENOBLE,   137, (1972).
                            *GOMPF(F.), LAU(H.), REICHARDT(W.), SALGADO(J.).

1807 PB,. . . . .   *SLOW-NEUTRON SCATTERING AND COLLECTIVE MOTIONS IN LIQUID LEAD. PHYS. REV. VOL.152,   99, (1966).
                            *RANDOLPH(P.D.), SINGWI(K.S.).

1808 PB,. . . . .   *PHONON DISPERSION IN PB. J. PHYS. F VOL.3,   24, (1973).
                            *KACHHAVA(C.M.).

1809 PB,. . . . .   *DISPERSION RELATIONS AND ATOMIC FORCE CONSTANTS OF LEAD BY NEUTRON SPEC. THESIS,   (1959).
                            *ARASE(T.).   (ED. NOTE: AVAIL. UNIV. MICROFILMS ORDER NO.72-24,465 (84
                            PP.) SEE: DISS. ABSTRACTS B VOL.33 NO.3, PG.1248).

1810 PB,. . . . .   *CRYSTAL DYNAMICS OF LEAD. I, DISPERSION CURVES AT 100 DEGREES K. PHYS. REV. VOL.128,   1099, (1962).
                            *BROCKHOUSE(B.N.), ARASE(T.), CAGLIOTI(G.), RAO(K.R.), WOODS(A.D.B.).

1811 PB,. . . . .   *MASS-CURRENT FLUCTUATIONS IN LIQUID LEAD. PHYS. REV. LETT. 20,   531, (1968).
                            *RANDOLPH(P.D.).

1812 PB,. . . . .   *PHONON-FREQUENCY DISTRIBUTION AND HEAT CAPACITIES OF ALUMINUM AND LEAD. PHYS. REV. 162,   549, (1967).
                            *STEDMAN(R.), ALMQVIST(L.), NILSSON(G.).

1813 PB,. . . . .   *MESSUNGEN DES FREIEN UND INKOHARENTEN NEUTRONENWIRKUNGSQUERSCHNITTS AN SILIZIUM UND BEI. Z. PHYSIK VOL.190   295, (1966).
                            *NIKLAUS(J.P.), SIMSON(R.), TRIFTSHAUSER(W.), SCHMATZ(W.).

1814 PB,. . . . .   *HIGH FREQUENCY WAVES IN LIQUID METALS. ADV. IN PHYSICS VOL.16,   189, (1967).
                            *COCKING(S.J.).

1815 PB,. . . . .   *BRILLOUIN SCATTERING OF NEUTRONS FROM LIQUIDS. DISC. FARADAY SOC. NO.43,   160, (1967).
                            *DORNER(B.), PLESSER(T.), STILLER(H.).

1816 PB,. . . . .   *IMAGE OF THE FERMI SURFACE IN THE LATTICE VIBRATIONS OF LEAD. PHYS. REV. LETT. VOL.7,   93, (1961).
                            *BROCKHOUSE(B.N.), RAO(K.R.), WOODS(A.D.B.).

1817 PB,. . . . .   *THE SCATTERING OF COLD NEUTRONS BY METALS. UKAEA AERE REP. R4101 (24 PP.),   (1962).
                            *EGELSTAFF(P.A.).

1818 PB,. . . . .   *THE LATTICE VIBRATIONS IN LEAD. J. PHYS. SOC. JAPAN VOL.34,   1002, (1973).
                            *CAVALHEIRO(R.), SHUKLA(M.M.).

1819 PB,. . . . .   *DYNAMIC STRUCTURE FACTOR CALCULATIONS AND NEUTRON INELASTIC SCATTERING. PROPERTIES LIQUID METALS,   125, (1973).
                            *CHUNG(C.H.), YIP(S.), EGELSTAFF(P.A.). (ED. NOTE: REFER SECT. 4).

1820 PB,. . . . .   *TEMPERATURE DEPENDENCE OF THE COHERENT SCATTERING AMPLITUDE FOR FORWARD SCATTERED NEUTRONS. DISS. ABS. VOL. 28,   1107B, (1967).
                            *EDWARDS(T.R.).

1821 PB,. . . . .   *THE ATOMIC MOTIONS IN LIQUID LEAD. PROC. PHYS. SOC. VOL.80,   395, (1962).
                            *TURBERFIELD(K.C.).

1822 PB,. . . . .   *FERMI SURFACE OF LEAD FROM KOHN ANOMALIES. PHYS. REV. 163,   567, (1967).
                            *STEDMAN(R.), ALMQVIST(L.), NILSSON(G.), RAUNIO(G.).

1823 PB,. . . . .   *INELASTIC SCATTERING OF THERMAL NEUTRONS IN LEAD. PHYS. REV. VOL.91,   1368, (1953).
                            *MCREYNOLDS(A.W.).

1824 PB,. . . . .   *ENERGY DISTRIBUTION OF SLOW NEUTRONS SCATTERED FROM SOLIDS. PHYS. REV. VOL.88,   542, (1952).
                            *BROCKHOUSE(B.N.), HURST(D.G.).

1825 PB,. . . . .   *THREE-BODY FORCES IN THE LATTICE DYNAMICS OF LEAD. PHYS. LETT. 28A,   226, (1968).
                            *SCHMUCK(PH.), QUITTNER(G.).

SECTION   6 -QUASI-ELASTIC AND INELASTIC SCATTERING STUDIES . . . .

1826 PB,. . . . . *FREQUENCY-WAVE NUMBER RELATIONSHIP FOR CO-OPERATIVE MODES OF MOTION IN
                          LIQUID LEAD AND TIN. PHYS. LETT. 16,   130, (1965).
                          *COCKING(S.J.), EGELSTAFF(P.A.).

1827 PB,. . . . . *DISPERSION RELATION FOR PHONONS IN LEAD AT 80 AND 300 DEG. K. 545, (1967).
                          *STEDMAN(R.), ALMQVIST(L.), NILSSON(G.), RAUNIO(G.).   PHYS. REV. 162,

1828 PB,. . . . . *PRESSURE-INDUCED PHONON FREQUENCY SHIFTS IN LEAD MEASURED BY INELASTIC
                          NEUTRON SCATTERING. PHYS. REV. LETT. 17,   1259, (1966).
                          *LECHNER(R.E.), QUITTNER(G.).

1829 PB,. . . . . *PHONON DENSITY OF STATES IN LEAD. SOL. STATE COMM. 3,   101, (1965).
                          *GILAT(G.).

1830 PB,. . . . . *THE TEMPERATURE DEPENDENCE OF CO-OPERATIVE MODES OF MOTION IN LIQUID PB.
                          *WIGNALL(G.D.), EGELSTAFF(P.A.).   J. OF PHYS. C VOL.1,   519, (1968).

1831 PB,. . . . . *CRYSTAL DYNAMICS FROM NEUTRON SPECTROMETRY. IAEA SYMP. CHALK RIVER, VOL.2,   3, (1962).
                          *WOODS(A.D.B.).

1832 PB,. . . . . *A BORN-VON KARMAN MODEL FOR LEAD. SOL. STATE COMM. VOL.14,   587, (1974).
                          *COWLEY(E.R.).

1833 PB,. . . . . *HIGH FREQUENCY WAVES IN LIQUID LEAD. PHYSICA VOL.31,   1537, (1965).
                          *DORNER(B.), PLESSER(T.), STILLER(H.).

1834 PB,. . . . . *ENERGY DISTRIBUTION OF NEUTRONS SCATTERED BY LIQUID LEAD. 767, (1959).
                          *PELAH(I.), WHITTEMORE(W.L.), MCREYNOLDS(A.W.).   PHYS. REV. VOL.113,

1835 PB,. . . . . *MEAN FREE PATH OF PHONONS IN A LEAD CRYSTAL. NATURE VOL.181,   643, (1958).
                          *MCCALLUM(G.J.), EGELSTAFF(P.A.).

1836 PB,. . . . . *TIME-DEPENDENT PAIR CORRELATIONS IN LIQUID LEAD. PHYS. REV. LETT. VOL.3,   259, (1959).
                          *BROCKHOUSE(B.N.), POPE(N.K.).

1837 PB,. . . . . *INELASTIC SCATTERING OF COLD NEUTRONS BY SOLID AND LIQUID LEAD. 3944, (1963).
                          *COTTER(M.J.).   DISS. ABSTR. VOL.23,

1838 PB-(IN),. . . . . *VIBRATIONS OF RANDOM DILUTE ALLOYS. PROC. ROY. SOC. 296,   161, (1967).
                          *ELLIOTT(R.J.), TAYLOR(D.W.).

1839 PB-(NA,MG),. . . *LOCAL VIBRATIONS OF IMPURITY ATOMS IN COPPER AND LEAD.   65, (1968).
                          *NATKANIEC(I.), PARLINSKI(K.), JANIK(J.A.), BAJOREK(A.),   IAEA SYMP. COPENHAGEN, VOL.1,
                          SUDNIK-HRYNKIEWICZ(M.).

1840 PB*S,. . . . . . *THE CRYSTAL DYNAMICS OF LEAD SULPHIDE. PROC. ROY. SOC. 300,   210, (1967).
                          *ELCOMBE(M.M.).

1841 (PB-SN)-TE,. . . *SOFT MODES AND LANDAU TRANSITIONS IN (PB(1-X)-SN(X))-TE ALLOYS. 159, (1973).
                          *DOLLING(G.), BUYERS(W.J.L.).   J. NONMET. VOL.1,

1842 PB*TE,. . . . . *THE CRYSTAL DYNAMICS OF LEAD TELLURIDE. PROC. ROY. SOC. 293,   433, (1966).
                          *COCHRAN(W.), COWLEY(R.A.), DOLLING(G.), ELCOMBE(M.M.).

1843 PB*TE,. . . . . *ELASTIC CONSTANTS AND FUNDAMENTAL LATTICE FREQUENCY OF LEAD TELLURIDE. 51, (1966).
                          *BYLANDER(E.G.), HAAS(M.).   SOL. STATE COMM. 4,

1844 PB*TE,. . . . . *CONDUCTION ELECTRONS AND OPTIC MODES OF IONIC CRYSTALS. 549, (1965).
                          *COWLEY(R.A.), DOLLING(G.).   PHYS. REV. LETT. 14,

1845 PB*TI*O3,. . . . *SCATTERING OF COLD NEUTRONS BY POLYCRYSTALLINE BARIUM, STRONTIUM AND
                          LEAD TITANATES. SOV. PHYS. SOL. STATE 8,   2156, (1966).
                          *SOLOV≠EV(S.P.), KUKHTO(O.L.), CHERNOPLEKOV(N.A.), ZETLYANOV(M.G.).

1846 PB*TI*O3,. . . . *SOFT FERROELECTRIC MODES IN LEAD TITANATE. PHYS. REV. B-2,   155, (1970).
                          *SHIRANE(G.), AXE(J.D.), HARADA(J.), REMEIKA(J.P.).

1847 PB-TL,. . . . . *LATTICE FREQUENCY SPECTRA OF PB AND PB-TL BY NEUTRON SPECTROMETRY. 1781, (1970).
                          *ROY(A.P.), BROCKHOUSE(B.N.).   CAN. J. PHYS. 48,

1848 PB-TL,. . . . . *PHONON SPECTRA IN PB AND PB(40)TL(60) DETERMINED BY TUNNELING AND
                          NEUTRON SCATTERING. PHYS. REV. 178,   897, (1969).
                          *ROWELL(J.M.), MCMILLAN(W.L.), FELDMANN(W.L.).

1849 PB-TL-BI,. . . . *LATTICE DYNAMICS OF THE ALLOY SYSTEMS BI-PB-TL. SOL. STATE COMM. 5,   79, (1967).
                          *NG(S.C.), BROCKHOUSE(B.N.).

1850 PD,. . . . . . . *CRYSTAL DYNAMICS AND ELECTRONIC SPECIFIC HEATS OF PALLADIUM AND COPPER. 704, (1971).
                          *MIILLER(A.P.), BROCKHOUSE(B.N.).   CAN. J. PHYS. 49,

1851 PD,. . . . . . . *THE CRYSTAL DYNAMICS OF PALLADIUM. THESIS (181 PP.),   , (1969).
                          *MIILLER(A.P.). (ED. NOTE: PH.D. DISSERTATION, MCMASTER UNIVERSITY).

1852 PD,. . . . . . . *ANOMALOUS BEHAVIOR OF THE LATTICE VIBRATIONS AND THE ELECTRONIC SPECIFIC
                          HEAT OF PALLADIUM. PHYS. REV. LETT. 20,   798, (1968).
                          *MIILLER(A.P.), BROCKHOUSE(B.N.).

SECTION   6 -QUASI-ELASTIC AND INELASTIC SCATTERING STUDIES . . . .

1853 PD,. . . . . . .   .INVESTIGATION OF THE PARAMAGNETIC NEUTRON SCATTERING FROM METALLIC
      ....D3......PHYS. REV. 140. . . . . . .A2003, (1965).
      PALLADIUM.
      .CABLE(J.W.), WOLLAN(E.O.).

1854 PD(FE),. . . . .   .LATTICE DYNAMICS OF THE PD-(1 PERCENT FE) ALLOY.
      ....B1.......J. PHYS.F: METAL PHYS. VOL.1,. .   339, (1971).
      .MALISZEWSKI(E.F.), SOSNOWSKI(J.), CZACHOR(A.).

1855 PD-FE, ...ALIGNMENT OF RARE EARTH MOMENTS IN DILUTE PD-FE ALLOYS.
      ....D4......J. PHYS. CHEM. SOL. 25. . . . . 1453, (1964).
      .CABLE(J.W.), WOLLAN(E.O.), CHILD(H.R.), KOEHLER(W.C.).

1856 PD-FE, ...DISTRIBUTION OF THE FERROMAGNETIC POLARIZATION IN PD-FE SINGLE CRYSTAL.
      .....A2.D4......J. OF PHYS. C VOL.1. . . . . . 528, (1968).
      .HICKS(T.J.), HOLDEN(T.M.), LOW(G.G.).

1857 PD3-FE,. . . . .   .CRYSTAL DYNAMICS OF PD3-FE AT 80 DEG. K.
      ....B1.......J. PHYS.F: METAL PHYS. VOL.2,. .   421, (1972).
      .STIRLING(W.G.), COWLEY(R.A.), STRINGFELLOW(M.W.).

1858 PD3-FE,. . . . .   .SPIN WAVES IN PD3-FE.
      ....D1.......SOL. STATE COMM. VOL.11 NO.1,. .   271, (1972).
      .STIRLING(W.G.), COWLEY(R.A.).

1859 PD-FE-CO,. . . .   .DISTRIBUTION OF THE FERROMAGNETIC POLARIZATION INDUCED BY IRON AND
      ....D4......PROC. PHYS. SOC. 89. . . . . . 119, (1966).
      COBALT ATOMS IN PALLADIUM.
      .LOW(G.G.E.), HOLDEN(T.M.).

1860 PD-H,. . . . . .   .HYDROGEN MOTION IN PRIMARY SOLUTIONS OF HYDROGEN IN SOME TRANSITION
      ....T.C3.....BROOKHAVEN SYMPOSIUM. . . . . . 105, (1965).
      ELEMENTS.
      .KLEY(W.), PERETTI(J.), RUBIN(R.), VERDAN(G.).

1861 PD-H,. . . . . .   .DIFFUSION OF HYDROGEN IN THE ALPHA-PHASE OF PD-H STUDIED BY SMALL ENERGY
      ....C1......J. PHYS. CHEM. SOLIDS VOL.28,. .2369, (1967).
      TRANSFER NEUTRON SCATTERING.
      .SKOLD(K.), NELIN(G.).

1862 PD(H), ...NEUTRON QUASIELASTIC SCATTERING STUDY OF HYDROGEN DIFFUSION IN A SINGLE
      ....C1......PHYS. REV. LETT. VOL.29, ... 1250, (1972).
      CRYSTAL OF PALLADIUM.
      .ROWE(J.M.), RUSH(J.J.), DE-GRAAF(L.A.), FERGUSON(G.A.).

1863 PD-H,. . . . . .   .THERMAL MOTION IN PALLADIUM HYDRIDE STUDIED BY MEANS OF ELASTIC AND
      ....C1......PHYSICA VOL.26. . . . . 744, (1960).
      INELASTIC SCATTERING OF NEUTRONS.
      .BERGSMA(J.), GOEDKOOP(J.A.).

1864 PD-H,. . . . . .   .THE QUASIELASTIC SCATTERING OF COLD NEUTRONS FROM THE BETA-PHASE OF
      ....C1......J. PHYS. C. VOL.3. . . . . . . 2487, (1970).
      PALLADIUM HYDRIDE.
      .BEG(M.M.), ROSS(D.K.).

1865 PD-H,. . . . . .   .≠IN-BAND≠ MODES OF VIBRATION OF PD-H(0.03).
      ....B3......PHYS. REV. B VOL.8,. . . . . . . 6013, (1973).
      .ROWE(J.M.), RUSH(J.J.), SMITH(H.G.).

1866 PD-H,. . . . . .   .INELASTIC NEUTRON SCATTERING EXPERIMENTS ON A FEW METAL HYDRIDES.
      ....C1......IAEA SYMPOSIUM VIENNA, ... 501, (1960).
      .BERGSMA(J.), GOEDKOOP(J.A.).

1867 PD(H), ...NEUTRON QUASIELASTIC SCATTERING STUDY OF HYDROGEN DIFFUSION IN A SINGLE
      ....PHYS. REV. LETT. VOL.32, ... 1087, (1974).
      CRYSTAL OF PALLADIUM (ERRATUM).
      .ROWE(J.M.), RUSH(J.J.), DE-GRAAF(L.A.), FERGUSON(G.A.).

1868 PD-H,PD-AG-H,. .   .A NEUTRON SCATTERING STUDY OF THE VIBRATIONAL MODES OF HYDROGEN IN THE
      ....C1......SOL. STATE COMM. VOL.13 NO.2,. . 229, (1973).
      BETA-PHASES OF PD-H, PD-10AG-H AND PD-20AG-H.
      .CHOWDHURY(M.R.), ROSS(D.K.).

1869 PD-(NI), ...MEASUREMENT OF A SIMPLE DEFECT MODE OF VIBRATION.
      ....B3......PHYS. REV. LETT. VOL.8,. . . . . 278, (1962).
      .MOZER(B.), OTNES(K.), MYERS(V.W.).

1870 PEROVSKITES, . .   .DETERMINATION OF THE NORMAL VIBRATIONAL DISPLACEMENTS IN SEVERAL
      ....B7......ACTA CRYST. VOL.A26, ... 608, (1970).
      PEROVSKITES BY INELASTIC NEUTRON SCATTERING.
      .HARADA(J.), AXE(J.D.), SHIRANE(G.).

1871 PHENANTHRENE,. .   .DIELECTRIC AND NEUTRON INELASTIC SCATTERING MEASUREMENTS ON PHENANTHRENE
      ....C1......J. CHEM. PHYSICS VOL.54. . . . 2597, (1971).
      .SPIELBERG(D.H.), ARNDT(R.A.), DAMASK(A.C.), LEFKOWITZ(I.).

1872 PHENOL,. . . . .   .INELASTIC SCATTERING OF THERMAL NEUTRONS BY SOME BENZENE DERIVATIVES.
      ....C3......IAEA SYMP. COPENHAGEN, ... 483, (1968).
      .TREPADUS(V.), BALLY(D.), PARFENOV(V.A.), LIFOROV(V.G.).

1873 PHENOL,. . . . .   .STUDY OF MOLECULAR MOTION IN PHENOL BY NEUTRON SCATTERING AND NUCLEAR
      ....C3......REV. ROUMAINE PHYS. VOL.18. . . 313, (1973).
      MAGNETIC RESONANCE METHODS.
      .TREPADUS(V.), RAPEANU(S.), GROSESCU(R.).

1874 PHENOL,. . . . .   .THE STUDY OF THE MOLECULAR MOTION IN PHENOL BY NEUTRON SCATTERING AND
      ....C3......REV. ROUMAINE PHYS. VOL.18,. . . 313, (1973).
      N.M.R. METHODS.
      .TREPADUS(V.), RAPEANU(S.), GROSESCU(R.).

1875 PHENOL,. . . . .   .STUDY OF MOLECULAR ROTATIONS IN SOME AROMATIC COMPOUNDS BY COLD-NEUTRON
      ....C3......J. CHEM. PHYS. VOL.60. . . . . 2832, (1974).
      SCATTERING.
      .TREPADUS(V.), RAPEANU(S.), PADUREANU(I.), PARFENOV(V.A.), NOVIKOV(A.G.).

1876 POLYDIMETHYLSILOXANE,.   .STUDY OF LOW-FREQUENCY MOLECULAR MOTIONS IN POLYDIMETHYLSILOXANE
      ....C1......J. POLYMER SCI. A-2 VOL.7,. . . 433, (1969).
      POLYMERS BY NEUTRON INELASTIC SCATTERING.
      .HENRY(A.W.), SAFFORD(G.J.).

1877 POLYETHYLENE,. .   .NEUTRON SCATTERING FROM ORIENTED POLYETHYLENE.
      ....C1......J. CHEM. PHYSICS VOL.45, ... 757, (1966).
      .TREVINO(S.F.).

SECTION   6 -QUASI-ELASTIC AND INELASTIC SCATTERING STUDIES . . . .

1878 POLYETHYLENE,. . . . . .C1. . . . . . .J. CHEM. PHYSICS VOL.40, ... 1417, (1964).
      *STUDY OF LOW-FREQUENCY MOTIONS IN POLYETHYLENE AND THE PARAFFIN
      HYDROCARBONS BY NEUTRON INELASTIC SCATTERING.
      *DANNER(H.R.), SAFFORD(G.J.), BOUTIN(H.), BERGER(M.).

1879 POLYETHYLENE GLYCOL,. . . . . . . .C2. . . . . . .J. CHEM. PHYSICS VOL.50, ... 915, (1969).
      *FREQUENCY DISTRIBUTION AND NEUTRON SCATTERING OF POLYETHYLENE GLYCOL
      CHAIN.
      *MATSUURA(H.), MIYAZAWA(T.).

1880 POLYETHYLENE (DEUTERATED). . . . .B1. . . . . .IAEA SYMP. COPENHAGEN, VOL.2, . . 159, (1968).
      *DISPERSION RELATION FOR SKELETAL VIBRATIONS IN DEUTERATED POLYETHYLENE.
      *FELDKAMP(L.A.), VENKATARAMAN(G.), KING(J.S.).

1881 POLYETHYLENE,. . . . . .C3. . . . . .IAEA SYMP. CHALK RIVER, VOL.1, . . 389, (1962).
      *CALCULATION OF THE INELASTIC SCATTERING OF NEUTRONS FROM POLYETHYLENE
      AND WATER.
      *GOLDMAN(D.T.), FEDERIGHI(F.D.).

1882 POLYETHYLENE,. . . . . .C3. . . . . .IAEA SYMP. CHALK RIVER, VOL.1, . . 297, (1962).
      *A STUDY OF COLD NEUTRON INELASTIC SCATTERING ON CERTAIN HYDROGENOUS
      SUBSTANCES.
      *.C.C.C.P.

1883 POLYETHYLENE,. . . . . .B1. . . . . .DISS. ABS. VOL.30, ... 2369B, (1969).
      *STUDY OF THE LATTICE DYNAMICS OF DEUTERATED POLYETHYLENE AND ZINC
      SULPHIDE USING COHERENT NEUTRON SCATTERING.
      *FELDKAMP(L.A.).

1884 POLYETHYLENE,. . . . . .C2. . . . . .DISS. ABS. 25, ... 7333, (1965).
      *MEASUREMENT OF THE PHONON FREQUENCY DISTRIBUTION FOR POLYETHYLENE BY
      NEUTRON SCATTERING.
      *DONOVAN(J.L.).

1885 POLYETHYLENE,. . . . . .C1. . . . . .DISS. ABS. VOL.27, ... 5868, (1966).
      *NEUTRON SCATTERING IN STRETCH-ORIENTED POLYETHYLENE.
      *MYERS(W.R.).

1886 POLYETHYLENE,. . . . . .C1,C3. . . . .DISS. ABS. VOL.27, ... 2668, (1967).
      *EPITHERMAL NEUTRON INELASTIC SCATTERING BY ROOM TEMPERATURE WATER AND
      POLYETHYLENE.
      *KIROUAC(G.J.).

1887 POLYETHYLENE,. . . . . .C1,3. . . . .BROOKHAVEN SYMPOSIUM, ... 131, (1965).
      *EXPERIMENTAL DETERMINATION OF NEUTRON INELASTIC SCATTERING IN
      POLYETHYLENE.
      *WHITTEMORE(W.L.).

1888 POLYETHYLENE,. . . . . .C1,3. . . . .BROOKHAVEN SYMPOSIUM, ... 126, (1965).
      *NEUTRON SCATTERING IN STRETCH-ORIENTED POLYETHYLENE.
      *MYERS(W.), SUMMERFIELD(G.C.), KING(J.S.).

1889 POLYETHYLENE,. . . . . .C1. . . . . .UKRAINIAN PHYS. J. VOL.14, 9, . .1525, (1969).
      *STUDY OF INELASTIC SCATTERING OF SLOW NEUTRONS BY POLYETHYLENE AT
      TEMPERATURES 290 AND 100 K.
      *IVANITSKII(P.G.), KROTENKO(V.T.), GORBACHEV(B.I.), TSIBULNIK(A.N.).

1890 POLYETHYLENE,. . . . . .C1. . . . . .MOL. DYN. AND STRUCT. OF SOLIDS, 547, (1969).
      *EPITHERMAL NEUTRON INELASTIC SCATTERING BY POLYETHYLENE.
      *MOORE(W.E.), KIROUAC(G.J.), ESCH(L.J.), SEEMANN(K.W.), YEATER(M.L.).

1891 POLYETHYLENE (DEUTERATED),. . . . .B7. . . . . .MOL. DYN. AND STRUCT. OF SOLIDS, 543, (1969).
      *NEUTRON SCATTERING IN DEUTERATED POLYETHYLENE.
      *FELDKAMP(L.A.), KING(J.S.).

1892 POLYETHYLENE,. . . . . .B7. . . . . .POLYM. SCI(A-2)POLYM. PHYS.7, . .465, (1969).
      *OBSERVATION OF OPTICAL MODES IN POLYETHYLENE BY NEUTRON SCATTERING.
      *CHANG(Y.I.), SUMMERFIELD(G.C.).

1893 POLYETHYLENE (DEUTERO),. . . . . .B1. . . . . .IAEA SYMP. GRENOBLE, ... 301, (1972).
      *INTERCHAIN FORCE FIELD OF POLYETHYLENE BY NEUTRON SCATTERING.
      *TWISLETON(J.F.), WHITE(J.W.).

1894 POLYETHYLENE,. . . . . .C1. . . . . .IAEA SYMP. BOMBAY, VOL.2, . . . 407, (1964).
      *STUDY OF THE LOW-FREQUENCY MOLECULAR MOTIONS IN POLYETHYLENE AND THE
      N-PARAFFINS BY SLOW NEUTRON INELASTIC SCATTERING.
      *BOUTIN(H.), PRASK(H.), TREVINO(S.F.), DANNER(H.R.).

1895 POLYETHYLENE,. . . . . .P. . . . . . .IAEA SYMPOSIUM (VIENNA), ... 613, (1960).
      *MEASUREMENTS OF NEUTRON SPECTRA IN WATER, POLYETHYLENE AND ZIRCONIUM
      HYDRIDE.
      *WALTON(R.B.), BEYSTER(J.R.), WOOD(J.L.), LOPEZ(W.M.).

1896 POLYETHYLENE,. . . . . .C1. . . . . .J. CHEM. PHYS. VOL.42, ...4299, (1965).
      *POLYETHYLENE FREQUENCY SPECTRUM FROM #WARM#-NEUTRON SCATTERING.
      *MYERS(W.), DONOVAN(J.L.), KING(J.S.).

1897 POLYETHYLENE,. . . . . .C1. . . . . .J. CHEM. PHYS. VOL.43, ...1079, (1965).
      *DETERMINATION OF THE PHONON SPECTRUM OF POLYETHLENE BY NEUTRON
      SCATTERING.
      *SUMMERFIELD(G.C.).

1898 POLYETHYLENE,. . . . . .C1. . . . . .NUCL. SCI. ENGNG. VOL.24, . . . 394, (1966).
      *SCATTERING OF NEUTRONS BY POLYETHYLENE.
      *WHITTEMORE(W.L.).

1899 POLYETHYLENE,. . . . . .C3. . . . . .NUCL. SCI. ENGNG. VOL.25, . . . 300, (1966).
      *INELASTIC SCATTERING CROSS SECTIONS FOR POLYETHYLENE AND WATER IN THE
      GAS TRANSITION REGION.
      *KIROUAC(G.J.), MOORE(W.E.), SEEMANN(K.W.).

1900 POLYETHYLENE,. . . . . .C1. . . . . .J. CHEM. PHYS. VOL.44, ...184, (1966).
      *NEUTRON SCATTERING IN STRETCH-ORIENTED POLYETHYLENE.
      *MYERS(W.), SUMMERFIELD(G.C.), KING(J.S.).

1901 POLYETHYLENE,. . . . . .C5. . . . . .J. CHEM. PHYSICS VOL.49, ...1043, (1968).
      *LOW-FREQUENCY VIBRATIONS IN POLYETHYLENE AT 300, 77 AND 4.2 DEGREES K.
      *MYERS(W.R.), RANDOLPH(P.D.).

1902 POLYETHYLENE,. . . . . .P. . . . . . .REP. PROGR. POLYMER PHYS/JAP. 10 185, (1967).
      *INELASTIC SCATTERING CROSS SECTION OF NEUTRONS BY CRYSTAL VIBRATIONS OF
      POLYETHYLENE.
      *KITAGAWA(T.), MIYAZAWA(T.).

SECTION 6 -QUASI-ELASTIC AND INELASTIC SCATTERING STUDIES . . . .

1903 POLYETHYLENE,. . . . . .T. . . . . . .NUKLEONIK VOL.12,. . . . . 87, (1969).
        *NEUTRON SCATTERING BY POLYETHYLENE.
        *SPREVAK(D.), KOPPEL(J.U.).

1904 POLYETHYLENE,. . . . . .C1. . . . . . .SOV. AT. ENERGY VOL.20,. . . . . . 36, (1966).
        *INVESTIGATION OF INELASTIC SCATTERING OF SLOW NEUTRONS BY POLYETHYLENE.
        *IVANITSKII(P.G.), KROTENKO(V.T.).

1905 POLYETHYLENE,. . . . . .C1. . . . . . .J. CHEM. PHYSICS VOL.45,. . . . . 3787, (1966).
        *NEUTRON-SCATTERING STUDY OF THE INTRAMOLECULAR AND CRYSTALLINE MODES
        OF POLYETHYLENE.
        *SAFFORD(G.J.), NAUMANN(A.W.), SIMON(F.T.).

1906 POLYETHYLENE,. . . . . .C1,C3. . . . .J. POLYMER SCI. A-2 VOL.9,. . . . 1219, (1971).
        *STUDY OF LOW-FREQUENCY MOTIONS OF EXTENDED CHAINS IN POLYETHYLENE BY
        NEUTRON INELASTIC SCATTERING.
        *BERGHMANS(H.), SAFFORD(G.J.), LEUNG(P.S.).

1907 POLYETHYLENE,. . . . . .T,C1. . . . . .J. POLYMER SCI. B VOL.6,. . . . . . 83, (1968).
        *CROSS-SECTION FOR MULTI-PHONON SCATTERING OF NEUTRONS BY CRYSTALLINE
        POLYETHYLENE.
        *KITAGAWA(T.), MIYAZAWA(T.).

1908 POLYETHYLENE, POLYOXYMETHYLENE,. . . .C1. . . . . . .J. MACROMOL. SCI. VOL.A1,. . . . . 723, (1967).
        *STUDIES OF LOW-FREQUENCY MOLECULAR MOTIONS IN POLYMERS BY NEUTRON
        INELASTIC SCATTERING.
        *TREVINO(S.F.), BOUTIN(H.).

1909 POLYETHYLENE,. . . . . .SPECTROSCOPY BIOL. CHEM.. . . . . . 235, (1974).
        *NEUTRON SPECTROSCOPY OF CHAIN POLYMERS: A CRITICAL REVIEW OF POLY-
        ETHYLENE.
        *KING(J.S.). (REFER TO SECTION 4).

1910 POLYETHYLENE,. . . . . .T.E. . . . .MOL. PHYS. VOL.22,. . . . . 241, (1971).
        *VIBRATIONAL SPECTRUM OF CHAIN MOLECULES WITH CONFORMATIONAL DISORDER:
        POLYETHYLENE.
        *ZERBI(G.), PISERI(L.), CABASSI(F.).

1911 POLYETHYLENE,. . . . . .C1. . . . . . .J. CHEM. PHYSICS VOL.48,. . . . . 912, (1968).
        *NEUTRON SCATTERING IN NORMAL AND DEUTERATED POLYETHYLENE.
        *LYNCH-JR(J.E.), SUMMERFIELD(G.C.), FELDKAMP(L.A.), KING(J.S.).

1912 POLY-L-PROLINE,. . . . . .B7. . . . . . .NUCL./S.S. PHYS. SYMP. ABSTR., ... , (1972).
        *DISPERSION CURVES OF POLY-L-PROLINE 1 AND 2.
        *SINGH(R.D.), DWIVEDI(A.M.), GUPTA(V.D.).

1913 POLYMERS,. . . . . .C1. . . . . . .J. CHEM. PHYS. VOL.48,. . . . . 5271, (1968).
        *VIBRATIONAL ASSIGNMENTS IN POLYMER NEUTRON-SCATTERING SPECTRA BY ATOMIC
        SUBSTITUTION.
        *LONGSTER(G.F.), WHITE(J.W.).

1914 POLYMERS,. . . . . .J. MACROMOL. SCI. VOL.A4,. . . . . 1275, (1970).
        **HYDROPHOBIC BONDS: A STUDY OF LONG-RANGE INTERACTIONS IN POLYMERS.
        *WHITE(J.W.).

1915 POLYMERS,. . . . . .C1,B1. . . . .CHEM. APPL. THERMAL NEUTRON SCAT 97, (1973).
        *NEUTRON SCATTERING STUDIES OF THE DYNAMICS OF POLYMER CHAINS.
        *ALLEN(G.). (IN BOOK: CHEMICAL APPLICATIONS OF THERMAL NEUTRON
        SCATTERING. ED. BY B.T.M. WILLIS; OXFORD UNIV. PRESS: LONDON).

1916 POLYMERS,. . . . . .C1. . . . . . .ADV. POLYMER SCI. VOL.5,. . . . . 1, (1967).
        *LOW FREQUENCY MOTIONS IN POLYMERS AS MEASURED BY NEUTRON INELASTIC
        SCATTERING (REVIEW PAPER).
        *SAFFORD(G.J.), NAUMANN(A.W.).

1917 POLY(METHYL METHACRYLATE),. . . . . .C1. . . . . .POLYMER VOL.13,. . . . . 157, (1972).
        *METHYL GROUP MOTION IN POLY(PROPYLENE OXIDE), POLYPROPYLENE AND
        POLY(METHYL METHACRYLATE).
        *HIGGINS(J.S.), ALLEN(G.), BRIER(P.N.).

1918 POLYOXYMETHYLENE,. . . . . .C1. . . . . .J. CHEM. PHYSICS VOL.45,. . . . . 2700, (1966).
        *LOW-ENERGY VIBRATIONAL MODES OF POLYOXYMETHYLENE BY NEUTRON SCATTERING.
        *TREVINO(S.), BOUTIN(H.).

1919 POLYPROPYLENE, ...C1. . . . . .J. CHEM. PHYSICS VOL.40,. . . . . 1426, (1964).
        *INVESTIGATION OF THE LOW-FREQUENCY MOTIONS IN ISOTATIC AND ATACTIC
        POLYPROPYLENE BY NEUTRON INELASTIC SCATTERING.
        *SAFFORD(G.J.), DANNER(H.R.), BERGER(M.).

1920 POLYPROPYLENE, ...C1. . . . . .J. CHEM. PHYSICS VOL.55,. . . . . 983, (1971).
        *NEUTRON DOWN-SCATTERING SPECTRUM OF ISOTATIC POLYPROPYLENE.
        *YASUKAWA(T.), KIMURA(M.), WATANABE(N.), YAMADA(Y.).

1921 POLYPROPYLENE,. . . .OXIDE,. . . . . .C1. . . . . .POLYMER VOL.13,. . . . . 157, (1972).
        *METHYL GROUP MOTION IN POLY(PROPYLENE OXIDE), POLYPROPYLENE AND
        POLY(METHYL METHACRYLATE).
        *HIGGINS(J.S.), ALLEN(G.), BRIER(P.N.).

1922 POLYSTYRENE, ...B7,C3. . . . .PHYS. REV. LETT. VOL.32,. . . . . 1170, (1974).
        *EXPERIMENTAL DETERMINATIONS OF THE EXCLUDED-VOLUME EXPONENT IN DIFFERENT
        ENVIRONMENTS.
        *COTTON(J.P.), DECKER(D.), FARNOUX(B.), JANNINK(G.), OBER(R.), PICOT(C.).

1923 POLYTETRAFLUOROETHYLENE,. . . . . .B1. . . . . . .DISC. FARADAY SOC. NO.48,. . . . . 15, (1969).
        *VIBRATIONS IN TEFLON.
        *LAGARDE(V.), PRASK(H.), TREVINO(S.).

1924 POLYTETRAFLUOROETHYLENE,. . . . . .B1. . . . . .J. POLYM. SCI./POLYM. LETT. 11,. . . . 377, (1973).
        *MEASUREMENTS OF LONGITUDINAL ACOUSTIC PHONONS IN POLYTETRAFLUOROETHYLENE
        *SAKAMOTO(M.), IIZUMI(M.), MASAKI(N.), MOTOHASHI(H.), MINAKAWA(N.),
        DOI(K.), KURIYAMA(I.), YODA(O.), TAMURA(N.), ODAJIMA(A.).

1925 POLYTETRAFLUOROETHYLENE,. . . . . .C1. . . . . .POLYMER VOL.13 NO.1,. . . . . . 40, (1972).
        *INTERCHAIN FORCE FIELD AND ELASTIC CONSTANTS OF POLYTETRAFLUOROETHYLENE.
        *TWISLETON(J.F.), WHITE(J.W.).

1926 POLYTETRAFLUOROETHYLENE,. . . . . .B1. . . . . .J. CHEM. PHYS. VOL.58,. . . . . 158, (1973).
        *LATTICE DYNAMICS OF POLYTETRAFLUOROETHYLENE.
        *PISERI(L.), POWELL(B.M.), DOLLING(G.).

1927 PR,. . . . . .C1. . . . . .PHYS. REV. LETT. VOL.27,. . . . . 1530, (1971).
        *MAGNETIC EXCITONS IN SINGLET-GROUND-STATE FERROMAGNETS.
        *BIRGENEAU(R.J.), ALS-NIELSEN(J.), BUCHER(E.).

SECTION 6 -QUASI-ELASTIC AND INELASTIC SCATTERING STUDIES . . . .

1928 PR,. . . . .D1.......PHYS. REV. LETT. VOL.27, ... 223, (1971).
*ERRATA: MAGNETIC EXCITON DISPERSION IN PRASEODYMIUM, PHYS. REV. LETT. VOL.26, 1254 (1971).
*RAINFORD(B.D.), HOUMANN(J.G.).

1929 PR,. . . . .D1.......PHYS. REV. LETT. VOL.26, ... 1254, (1971).
*MAGNETIC EXCITON DISPERSION IN PRASEODYMIUM.
*RAINFORD(B.D.), HOUMANN(J.G.).

1930 PR COMPOUNDS,. . . . .D4.......PHYS. REV. LETT. VOL.25, ... 752, (1970).
*CRYSTAL FIELDS IN RARE-EARTH METALLIC COMPOUNDS.
*TURBERFIELD(K.C.), PASSELL(L.), BIRGENEAU(R.J.), BUCHER(E.).

1931 PR,. . . . .D1.......CONF./RARE EARTHS/ACTINIDES, . . 40, (1971).
*MAGNETIC EXCITATIONS IN PRASEODYMIUM.
*RAINFORD(B.D.), HOUMANN(J.G.).

1932 PR, PR3*TL, (PR-LA)3*TL, ...D4.......IAEA SYMP. GRENOBLE, ... 543, (1972).
*MAGNETIC EXCITONS IN GAMMA (SUB 1)-GAMMA (SUB 2) SINGLET GROUND STATE SYSTEMS.
*BIRGENEAU(R.J.), ALS-NIELSEN(J.), BUCHER(E.).

1933 PR3+,. . . . .T.......PHYS. LETT. VOL.31A, ... 67, (1970).
*CRYSTAL FIELD EFFECTS IN INELASTIC NEUTRON SCATTERING FROM PR+++.
*BALCAR(E.), LOVESEY(S.W.).

1934 PR*AL*O3,. . . . .B1,B7....PHYS. REV. LETT. VOL.31, ... 1300, (1973).
*QUADRUPOLE EXCITON-PHONON DYNAMICS AT THE 151-K PHASE TRANSITION IN PR*AL*O3.
*KJEMS(J.K.), SHIRANE(G.), BIRGENEAU(R.J.), VAN-UITERT(L.G.).

1935 PR-(BI,SB,AS,P,TE,SE,S), ...D4.......J. APPL. PHYS. 42, ... 1746, (1971).
*NEUTRON CRYSTAL-FIELD SPECTROSCOPY IN RARE-EARTH METALLIC COMPOUNDS.
*TURBERFIELD(K.C.), PASSELL(L.), BIRGENEAU(R.J.), BUCHER(E.).

1936 PR*F3, ...D3.......PHYS. STAT. SOL.45,. . . . . . . K105, (1971).
*CRYSTAL FIELD SPLITTING OF PR+++ IN PR*F3 STUDIED WITH INELASTIC PARAMAGNETIC NEUTRON SCATTERING.
*FELDMANN(K.), HENNIG(K.), NATKANIEC(I.), SAWENKO(B.N.), TEMPELHOFF(K.).

1937 PR*N,. . . . .D4.......A.I.P. CONF. PROC. NO.10 PART 2, 1548, (1973).
*DIRECT MEASUREMENT OF CRYSTAL-FIELD SPLITTINGS IN PR*N.
*DAVIS(H.L.), MOOK(H.A.).

1938 PR2*O3,. . . . .D3.......PHYS. REV. VOL.92, ... 1380, (1953).
*PARAMAGNETIC SCATTERING OF NEUTRONS BY RARE EARTH OXIDES.
*KOEHLER(W.C.), WOLLAN(E.O.).

1939 PROPANE, ...C4.......J. CHEM. PHYS. VOL.52, ... 4424, (1970).
*METHYL LIBRATION IN PROPANE MEASURED WITH NEUTRON INELASTIC SCATTERING.
*GRANT(D.M.), PUGMIRE(R.J.), LIVINGSTON(R.C.), STRONG(K.A.), MCMURRY(H.L.), BRUGGER(R.M.).

1940 PR3*TL,. . . . .C1.......PHYS. REV. LETT. VOL.27, ... 1530, (1971).
*MAGNETIC EXCITONS IN SINGLET-GROUND-STATE FERROMAGNETS.
*BIRGENEAU(R.J.), ALS-NIELSEN(J.), BUCHER(E.).

1941 PR3*TL,(PR-LA)3*TL. . . . .D1.......PHYS. REV. B VOL.9, ... 3797, (1974).
*TEMPERATURE DEPENDENCE OF THE MAGNETIC EXCITATIONS IN SINGLET-GROUND-STATE SYSTEMS: PARAMAGNETIC AND ZERO-TEMPERATURE BEHAVIOR OF ...
*HOLDEN(T.M.), BUYERS(W.J.L.).

1942 PT,. . . . .B1.......IAEA SYMP. COPENHAGEN, VOL.1,. . 203, (1968).
*LATTICE DYNAMICS OF PLATINUM.
*OHRLICH(R.), DREXEL(W.).

1943 PT,. . . . .B1.......THESIS (68 PP.), ... , (1970).
*CRYSTAL DYNAMICS OF PLATINUM.
*DUTTON(D.H.). (ED. NOTE: M.SC. DISSERTATION, MCMASTER UNIVERSITY).

1944 PT,. . . . .B1.......CAN. J. PHYS. VOL.50, ... 2915, (1972).
*CRYSTAL DYNAMICS OF PLATINUM BY INELASTIC NEUTRON SCATTERING.
*DUTTON(D.H.), BROCKHOUSE(B.N.), MIILLER(A.P.).

1945 PT,. . . . .P........REF. SECTION 3, 47/HAVENS, ...

1946 PU*(N*O3)3,. . . . .C3.......NUCL. SCI. ENG. VOL.46, ... 244, (1971).
*MEASUREMENTS OF THERMAL-NEUTRON SPECTRA IN PLUTONIUM NITRATE SOLUTIONS.
*NEILL(J.M.), YOUNG(J.C.), PRESKITT(C.A.), TRIMBLE(G.D.), LLOYD(R.C.), BROWN(C.L.).

1947 PYROGALLOL,. . . . .C3.......IAEA SYMP. COPENHAGEN, VOL.1,. . 483, (1968).
*INELASTIC SCATTERING OF THERMAL NEUTRONS BY SOME BENZENE DERIVATIVES.
*TREPADUS(V.), BALLY(D.), PARFENOV(V.A.), LIFOROV(V.G.).

1948 QUARTZ (ALPHA),. . . . .B7.......IAEA SYMP. GRENOBLE, ... 501, (1972).
*THE FREQUENCY SPECTRUM OF THE NORMAL MODES OF VIBRATION OF VITREOUS SILICA AND ALPHA QUARTZ.
*LEADBETTER(A.J.), STRINGFELLOW(M.W.).

1949 QUARTZ,. . . . .P........REF. SECTION 3, 48/RAINWATER,. . . . .

1950 RARE-EARTH HYDRIDES, ...C1.......CONF./RARE EARTHS/ACTINIDES, . . 46, (1971).
*SLOW NEUTRON SCATTERING STUDIES OF RARE EARTH HYDRIDES.
*HUNT(D.G.), ROSS(D.K.).

1951 RARE-EARTHS, ...D3.......C.R. ACAD. SCI. VOL.253, ... 2884, (1961).
*THE LINE WIDTH IN THE PARAMAGNETIC SCATTERING OF SLOW NEUTRONS BY RARE EARTHS.
*SAINT-JAMES(D.). (IN FRENCH).

1952 RARE-EARTH-METALS, ...T,D3....CAN. J. PHYS. 46, ... 1499, (1968).
*CRYSTAL FIELD EFFECTS ON THE INELASTIC NEUTRON SCATTERING FROM HEAVY RARE EARTH METALS IN THE PARAMAGNEITC PHASE.
*WOODS(A.D.B.).

1953 (RARE-EARTH)*P,. . . . .B7.......PHYS. REV. B VOL.8, ... 5345, (1973).
*CRYSTAL FIELDS AND THE EFFECTIVE-POINT-CHARGE MODEL IN THE RARE-EARTH PNICTIDES.
*BIRGENEAU(R.J.), BUCHER(E.), MAITA(J.P.), PASSELL(L.), TURBERFIELD(K.C.)

1954 RB,. . . . .B1.......IAEA SYMP. COPENHAGEN, VOL.1,. . 209, (1968).
*LATTICE DYNAMICS OF RUBIDIUM.
*COPLEY(J.R.D.), BROCKHOUSE(B.N.), CHEN(S.H.).

SECTION 6 -QUASI-ELASTIC AND INELASTIC SCATTERING STUDIES . . . .

1955 RB,. . . . . .91,B2. . . .THESIS (274 PP.), ... , (1970).
→HARMONIC AND ANHARMONIC VIBRATIONS IN RUBIDIUM METAL.
→COPLEY(J.R.D.). (ED. NOTE: PH.D. DISSERTATION, MCMASTER UNIVERSITY).

1956 RB,. . . . . . . .B4,B7. . . .IAEA SYMP. GRENOBLE, ... 435, (1972).
→COLLECTIVE EXCITATIONS IN LIQUID RUBIDIUM AND THE LIQUID-SOLID PHASE
TRANSITION.
→SUCK(J.B.), GLASER(W.).

1957 RB,. . . . . .B7. . . . . . .CAN. J. PHYS. VOL.50. . . . . . 3062, (1972).
→THE MEAN SQUARE FORCE ON IONS IN LIQUID RUBIDIUM.
→EGELSTAFF(P.A.), PAGE(D.I.), DUFFILL(C.).

1958 RB,. . . . . .B1. . . . . . .CAN. J. PHYS. VOL.51. . . . . . . 657, (1973).
→CRYSTAL DYNAMICS OF RUBIDIUM. I.MEASUREMENTS AND HARMONIC ANALYSIS.
→COPLEY(J.R.D.), BROCKHOUSE(B.N.).

1959 RB,. . . . . .B4. . . . . . .ADV. IN PHYSICS VOL.16,. . . . . . 189, (1967).
→HIGH FREQUENCY WAVES IN LIQUID METALS.
→COCKING(S.J.).

1960 RB,. . . . . .B4. . . . . . .PROPERTIES LIQUID METALS,. . . . 111, (1973).
→INVESTIGATIONS OF COLLECTIVE EXCITATIONS IN LIQUID METALS BY INELASTIC
NEUTRON SCATTERING.
→GLASER(W.), HAGEN(S.), LOFFLER(U.), SUCK(J.B.), SCHOMMERS(W.). (ED.
NOTE: REFER SECT. 4).

1961 RB,. . . . . .B4,C3. . . .PROPERTIES LIQUID METALS,. . . . 99, (1973).
→RELATION BETWEEN INCOHERENT AND COHERENT NEUTRON SCATTERING FROM LIQUID
METALS, WITH PARTICULAR REFERENCE TO MEASUREMENTS ON RB AND GA.
→BARKER(M.I.), JOHNSON(M.W.), MARCH(N.H.), PAGE(D.I.). (ED. NOTE: REFER
SECT. 4).

1962 RB,. . . . . .B2,T. . . .CAN. J. PHYS. VOL.51. . . . . . 2564, (1973).
→CRYSTAL DYNAMICS OF RUBIDIUM. II.ANHARMONIC CALCULATIONS OF THE PHONON
SELF-ENERGY AND THE HEAT CAPACITY.
→COPLEY(J.R.D.).

1963 RB,. . . . . .T,B4. . . .PHYS. REV. LETT. VOL.32 NO.2,. . 52, (1974).
→PROPAGATION OF DENSITY FLUCTUATIONS IN LIQUID RUBIDIUM: A MOLECULAR-
DYNAMICS STUDY.
→RAHMAN(A.).

1964 RB,. . . . . .B4. . . . . . .PHYS. REV. LETT. VOL.32 NO.2,. . 49, (1974).
→SHORT-WAVELENGTH COLLECTIVE EXCITATIONS IN LIQUID RUBIDIUM OBSERVED
BY COHERENT NEUTRON SCATTERING.
→COPLEY(J.R.D.), ROWE(J.M.).

1965 RB,. . . . . .T,B4. . . .PHYS. REV. A VOL.9,. . . . . . 1667, (1974).
→DENSITY FLUCTUATIONS IN LIQUID RUBIDIUM. II. MOLECULAR-DYNAMICS CALCU-
LATIONS.
→RAHMAN(A.).

1966 RB,. . . . . .B4. . . . . . .PHYS. REV. A VOL.9,. . . . . . . 1656, (1974).
→DENSITY FLUCTUATIONS IN LIQUID RUBIDIUM. I. NEUTRON-SCATTERING MEASURE-
MENTS.
→COPLEY(J.R.D.), ROWE(J.M.).

1967 RB*BR,. . . . . .B1. . . . . . .J. OF PHYS.-C-, SER.2, VOL.4,. . 958, (1971).
→LATTICE DYNAMICS OF RB*BR.
→ROLANDSON(S.), RAUNIO(G.).

1968 RB*BR-K*BR,. . .B3. . . . . . .IAEA SYMP. COPENHAGEN, VOL.1,. . 43, (1968).
→NORMAL MODES OF VIBRATION OF MIXED K*BR/RB*BR CRYSTALS.
→BUYERS(W.J.L.), COWLEY(R.A.).

1969 RB*CL,. . . . . .B1. . . . . . .J. OF PHYS.-C-, SER.2, VOL.3,. . 1013, (1970).
→PHONON DISPERSION RELATIONS IN RB*CL AND RB*F AT 80 K.
→RAUNIO(G.), ROLANDSON(S.).

1970 RB*CL,. . . . . .T. . . . . . .SOL. STATE COMM. VOL.11,. . . . 185, (1972).
→PHONON DISPERSION-RELATIONS IN RUBIDIUM CHLORIDE ON THE BASIS OF SHELL
MODEL.
→PANDEY(R.N.), DAYAL(B.).

1971 RB*CL,. . . . . .B1,T. . . .PHYS. STAT. SOL. 35, NO1,. . . . 299, (1969).
→PHONON WIDTHS IN NA*CL, K*CL AND RB*CL FROM NEUTRON MEASUREMENTS.
→RAUNIO(G.).

1972 RB2*CO*F4,. . .D2. . . . . . .PHYS. REV. LETT. VOL.31,. . . . 936, (1973).
→EXPERIMENTAL STUDY OF THE TWO-DIMENSIONAL ISING ANTIFERROMAGNET
RB2*CO*F4.
→SAMUELSEN(E.J.).

1973 RB*F,. . . . . .B1. . . . . . .J. OF PHYS.-C-, SER.2, VOL.3,. . 1013, (1970).
→PHONON DISPERSION RELATIONS IN RB*CL AND RB*F AT 80 K.
→RAUNIO(G.), ROLANDSON(S.).

1974 RB2*FE*F4,. . .D2. . . . . . .PHYS. REV. B-1,. . . . . . 2211, (1970).
→NEUTRON SCATTERING INVESTIGATION OF PHASE TRANSITIONS AND MAGNETIC
CORRELATIONS IN THE TWO-DIMENSIONAL ANTIFERROMAGNETS.
→BIRGENEAU(R.J.), GUGGENHEIM(H.J.), SHIRANE(G.).

1975 RB-HALIDES,. . .T. . . . . . .J. PHYS. C: SOLID ST. PHYS. V-5, 543, (1972).
→LATTICE DYNAMICS OF RUBIDIUM HALIDES.
→LAL(H.H.), VERMA(M.P.).

1976 RB*I,. . . . . .B1. . . . . . .PHYS. STAT. SOL. VOL.40,. . . . 749, (1970).
→LATTICE DYNAMICS OF RB*I AT 80 DEG. K.
→RAUNIO(G.), ROLANDSON(S.).

1977 RB*I,. . . . . .B4. . . . . . .PHYS. REV. LETT. 17,. . . . . . 530, (1966).
→PRESSURE DEPENDENCE OF THE TA(100) ZONE-BOUNDRY PHONON FREQUENCY IN
RUBIDIUM IODIDE.
→SAUNDERSON(D.H.).

1978 RB*MN*F3,. . .D2. . . . . . .PHYS. REV. B-4,. . . . . . 3206, (1971).
→QUANTITATIVE ANALYSIS OF INELASTIC SCATTERING IN TWO-CRYSTAL AND
THREE-CRYSTAL NEUTRON SPECTROMETRY: CRITICAL SCATTERING FROM RB*MN*F3.
→TUCCIARONE(A.), LAU(H.Y.), CORLISS(L.M.), DELAPALME(A.), HASTINGS(J.M.).

1979 RB*MN*F3,. . .T. . . . . . .PHYS. REV. B-1,. . . . . . 3178, (1970).
→SHORT-RANGE-ORDER EFFECTS IN NEUTRON SCATTERING FROM HEISENBERG
PARAMAGNETS: APPLICATIONS TO RB*MN*F3.
→TAHIR-KHELI(R.A.), MCFADDEN(D.G.).

SECTION   6 -QUASI-ELASTIC AND INELASTIC SCATTERING STUDIES . . . .

1980 RB2*MN*F4, . . .   D2   PHYS. REV. B-1 PHASE TRANSITIONS 2211, (1970).
    *NEUTRON SCATTERING INVESTIGATION OF PHASE TRANSITIONS AND MAGNETIC
    CORRELATIONS IN THE TWO-DIMENSIONAL ANTIFERROMAGNETS.
    *BIRGENEAU(R.J.), GUGGENHEIM(H.J.), SHIRANE(G.).

1981 RB*MN*F3,. . . .   D2,D3....J. APPL. PHYS. 42 . . . . 1378, (1971).
    *NEUTRON INVESTIGATION OF THE SPIN DYNAMICS IN PARAMAGNETIC RB*MN*F3.
    *TUCCIARONE(A.), CORLISS(L.M.), HASTINGS(J.M.).

1982 RB*MN*F3,. . . .   D1.......IAEA SYMP. COPENHAGEN, VOL.2,. . 83, (1968).
    *SPIN CORRELATIONS IN ONE, TWO AND THREE-DIMENSIONAL HEISENBERG
    PARAMAGNETS.
    *WINDSOR(C.G.).

1983 RB*MN*F3,. . . .   D3.......PHYS. REV. LETT. VOL.26, ... 257, (1971).
    *THEORETICAL AND EXPERIMENTAL SPIN-RELAXATION FUNCTIONS IN PARAMAGNETIC
    RB*MN*F3.
    *TUCCIARONE(A.), HASTINGS(J.M.), CORLISS(L.M.).

1984 RB*MN*F3,. . . .   T........PHYS. REV. LETT. VOL.24, ... 111, (1970).
    *THEORETICAL ESTIMATES FOR THE DECAY RATES OF THE SPIN FLUCTUATIONS IN
    RB*MN*F3.
    *HUBER(D.L.), KRUEGER(D.A.).

1985 RB*MN*F3,. . . .   D2.......PHYS. REV. LETT. VOL.23, ...1225, (1969).
    *TEST OF DYNAMIC SCALING BY NEUTRON SCATTERING FROM RB*MN*F3.
    *LAU(H.Y.), CORLISS(L.M.), DELAPALME(A.), HASTINGS(J.M.), NATHANS(R.),
    TUCCIARONE(A.).

1986 RB*MN*F3,. . . .   T,D1.....PHYS. REV. B VOL.5 . . . . . . .4561, (1972).
    *DYNAMICAL SPIN CORRELATIONS IN A HEISENBERG SPIN CLUSTER.
    *KWON(T.H.).

1987 RB*MN*F3,. . . .   B7.......ACTA CRYST. VOL.A26, ... 608, (1970).
    *DETERMINATION OF THE NORMAL VIBRATIONAL DISPLACEMENTS IN SEVERAL
    PEROVSKITES BY INELASTIC NEUTRON SCATTERING.
    *HARADA(J.), AXE(J.D.), SHIRANE(G.).

1988 RB*MN*F3,. . . .   D3.......J. OF PHYS. C VOL.1, ... 940, (1968).
    *SPIN CORRELATIONS IN PARAMAGNETIC RB*MN*F3.
    *WINDSOR(C.G.), BRIGGS(G.A.), KESTIGIAN(M.).

1989 RB*MN*F3,. . . .   D1.......IAEA SYMP. GRENOBLE, ... 639, (1972).
    *SPIN-WAVE DAMPING IN RB*MN*F3.
    *SAUNDERSON(D.H.), WINDSOR(C.G.), BRIGGS(G.A.), EVANS(M.T.),
    HUTCHISON(E.A.).

1990 RB*MN*F3,. . . .   C5.......J. PHYSIQUE TOME 32 COL.1 VOL.II 614, (1971).
    *THE SCATTERING FUNCTION S(Q,W) OF RB*MN*F3.
    *EVANS(M.T.), WINDSOR(C.G.).

1991 RB*MN*F3,. . . .   D1.......PROC. PHYS. SOC. 87, ... 501, (1966).
    *SPIN WAVES IN RB*MN*F3.
    *WINDSOR(C.G.), STEVENSON(R.W.H.).

1992 RB*MN*F3,. . . .   D2.......J. APP. PHYS. 39 . . . . .1237, (1968).
    *INELASTIC MAGNETIC SCATTERING FROM RB*MN*F3 IN THE NEIGHBOURHOOD OF ITS
    NEEL POINT.
    *NATHANS(R.), MENZINGER(F.), PICKART(S.J.).

1993 RB*MN*F3,. . . .   D3.......PROC. PHYS. SOC. 89, ... 825, (1966).
    *NEUTRON SCATTERING FROM PARAMAGNETIC RB*MN*F3.
    *WINDSOR(C.G.).

1994 RB*MN*F3,. . . .   D4.......J. PHYS. C VOL.6, ... 495, (1973).
    *SPIN CORRELATIONS IN RB*MN*F3 AT 1.17 T-NEEL.
    *EVANS(M.T.), WINDSOR(C.G.).

1995 RB*MN*F3,. . . .   D2.......J. APPL. PHYS. VOL.40, ...1278, (1969).
    *CRITICAL MAGNETIC SCATTERING FROM RB*MN*F3.
    *CORLISS(L.M.), DELAPALME(A.), HASTINGS(J.M.), LAU(H.Y.), NATHANS(R.).

1996 RB*MN*F3,. . . .   D2.......J. APPL. PHYS. VOL.41, ...1384, (1970).
    *CRITICAL SCATTERING OF NEUTRONS FROM RB*MN*F3.
    *LAU(H.Y.), CORLISS(L.M.), DELAPALME(A.), HASTINGS(J.M.), NATHANS(R.),
    TUCCIARONE(A.).

1997 RB*MN*F3,. . . .   T........J. APPL. PHYS. VOL.41, ...1370, (1970).
    *DYNAMICAL CHARACTERISTICS OF THE OGUCHI MODEL.
    *KWON(T.H.), GERSCH(H.A.).

1998 RB*MN*F3,. . . .   D4.......PHYS. REV. B VOL.8, ... 1103, (1973).
    *SPIN DIFFUSION IN RB*MN*F3.
    *TUCCIARONE(A.), HASTINGS(J.M.), CORLISS(L.M.).

1999 RB*NI*CL3, . . .   T........PHYS. REV. B VOL.6 . . . . . . .1053, (1972).
    *SPIN-WAVE APPROACH TO ONE-DIMENSIONAL ANTIFERROMAGNETIC CS*NI*CL3 AND
    RB*NI*CL3.
    *MONTANO(P.A.), COHEN(E.), SHECHTER(H.).

2000 RB*NI*F3,. . . .   D1.......PHYS. REV. B VOL.6 . . . . . . .2030, (1972).
    *NEUTRON-SCATTERING STUDY OF SPIN WAVES IN THE FERRIMAGNET RB*NI*F3.
    *ALS-NIELSEN(J.), BIRGENEAU(R.J.), GUGGENHEIM(H.J.).

2001 RB*S*H,. . . . .   C1.......J. CHEM. PHYS. VOL.58, ...5469, (1973).
    *NEUTRON QUASIELASTIC SCATTERING STUDY OF S*H(-) REORIENTATION IN THE
    CUBIC PHASES OF CESIUM AND RUBIDIUM HYDROSULFIDE.
    *ROWE(J.M.), LIVINGSTON(R.C.), RUSH(J.J.).

2002 RB*S*H,. . . . .   C1.......J. CHEM. PHYS. VOL.59, ...6652, (1973).
    *NEUTRON QUASIELASTIC SCATTERING STUDY OF S*H(-) REORIENTATION IN
    RUBIDIUM HYDROSULFIDE IN THE INTERMEDIATE TEMPERATURE TRIGONAL PHASE.
    *ROWE(J.M.), LIVINGSTON(R.C.), RUSH(J.J.).

2003 RUBBER,. . . . .   C2.......IAEA SYMP. COPENHAGEN, VOL.2,. . 237, (1968).
    *STUDY OF THE FREQUENCY DISTRIBUTION IN UNSTRETCHED AND STRETCHED RUBBER
    BY NEUTRON INELASTIC SCATTERING.
    *TUNKELO(E.), BAJOREK(A.), NATKANIEC(I.), PARLINSKI(K.),
    SUDNIK-HRYNKIEWICZ(M.).

2004 RUBBERS, ...   C1.......J.C.S. FARADAY TRANS. II VOL.70, 348, (1974).
    *SEGMENTAL DIFFUSION IN RUBBERS STUDIED BY NEUTRON QUASIELASTIC INCO-
    HERENT SCATTERING.
    *ALLEN(G.), HIGGINS(J.S.), WRIGHT(C.J.).

SECTION   6 -QUASI-ELASTIC AND INELASTIC SCATTERING STUDIES . . . .

2005 S, .P. . . . . . . .REF. SECTION 3, 48/RAINWATER,. . . . .

2006 S, ...B1. . . . . . .CONF. INTERN., RENNES, ...   223, (1971).
      ⇒THE LATTICE DYNAMICS OF ORTHORHOMBIC SULPHUR.
      ⇒PAWLEY(G.S.), RINALDI(R.P.), WINDSOR(C.G.).

2007 S*C*(N*H2)2, ...C1. . . . . . .PHYSICA VOL.35, ...469, (1967).
      ⇒MOLECULAR DYNAMICS STUDY BY THE NEUTRON INELASTIC SCATTERING METHOD V.
      S*C*(N*H2)2.
      ⇒9AJOREK(A.),   PARLINSKI(K.),   SUDNIK-HRYNKIEWICZ(M.),   JANIK(J.A.).

2008 S-FE,. . . . . .D1,4. . . . .COLL. INTER. N.126 (GRENOBLE), . 210, (1963).
      ⇒TEMPERATURE DEPENDENCE OF SPIN FLUCTUATION SCATTERING OF NEUTRONS ON
      PYRRHOTITE.
      ⇒KRASNICKI(SZ.), WANIC(A.), DIMITRIJEVIC(Z.), MAGLIC(R.), MARKOVIC(V.),
      TODOROVIC(J.).

2009 S-FE,. . . . . .D1. . . . . . .COLL. INTER. N.126 (GRENOBLE), . 203, (1963).
      ⇒MAGNON SCATTERING OF SLOW NEUTRONS ON A PYRRHOTITE SINGLE CRYSTAL.
      ⇒WANIC(A.).

2010 S*O2*(C*H3)2,. . . . . .C2. . . . . . .J. CHEM. PHYS. VOL.50, ...2140, (1969).
      ⇒NEUTRON INELASTIC SCATTERING AND X-RAY STUDIES OF AQUEOUS SOLUTIONS OF
      DIMETHYLSULPHOXIDE AND DIMETHYLSULPHONE.
      ⇒SAFFORD(G.J.), SCHAFFER(P.C.), LEUNG(P.S.), DOEBBLER(G.F.), BRADY(G.W.),
      LYDEN(E.F.X.).

2011 SB,. . . . . .B1. . . . . . .J. PHYS.F: METAL PHYS. VOL.1,. . 570, (1971).
      ⇒THE LATTICE DYNAMICS OF ANTIMONY.
      ⇒SHARP(R.I.), WARMING(E.).

2012 SB,. . . . . .B1. . . . . . .PHYS. STAT. SOL. VOL.44, ... K65, (1971).
      ⇒PHONON DISPERSION RELATIONS IN ANTIMONY.
      ⇒SOSNOWSKI(J.), MALISZEWSKI(E.F.), BEDNARSKI(S.), CZACHOR(A.).

2013 SB,. . . . . .B1. . . . . . .IAEA SYMP. GRENOBLE, ...   61, (1972).
      ⇒PHONON DISPERSION CURVES IN ANTIMONY.
      ⇒SOSNOWSKI(J.), CZACHOR(A.), MALISZEWSKI(E.).

2014 SB,. . . . . .P. . . . . . . .PHYS. REV. 131, ...2098, (1963).
      ⇒NEUTRON COHERENT-SCATTERING AMPLITUDES OF GA, IN, AS AND SB.
      ⇒ARNOLD(G.P.), NERESON(N.G.).

2015 SC,. . . . . .B1,D3. . . . .PHYS. REV. B-4,. . . . . .2398, (1971).
      ⇒PHONON SPECTRUM OF SCANDIUM METAL BY INELASTIC SCATTERING OF NEUTRONS.
      ⇒WAKABAYASHI(N.), SINHA(S.K.), SPEDDING(F.H.).

2016 SC,. . . . . .B1. . . . . . .DISS. ABS. VOL.30, ...5202B, (1970).
      ⇒LATTICE DYNAMICS OF SCANDIUM.
      ⇒WAKABAYASHI(N.).

2017 SC,. . . . . .T. . . . . . . .SOL. STATE COMM. VOL.12 NO.6,. . 527, (1973).
      ⇒LATTICE DYNAMICS AND ELASTIC CONSTANTS OF SCANDIUM.
      ⇒RAO(R.R.), MENON(C.S.).

2018 SC,. . . . . .T,B1. . . . .PHYS. LETT. VOL.43A NO.4,. . . . 365, (1973).
      ⇒PHONON DISPERSION IN SCANDIUM.
      ⇒BOSE(G.), GUPTA(H.C.), TRIPATHI(B.B.).

2019 SE,. . . . . .C1. . . . . . .PHYS. STAT. SOL. 19,. . . . . . 721, (1967).
      ⇒STREUUNG LANGSAMER NEUTRONEN AN POLYKRISTALLINEN SELEN UND TELLUR.
      ⇒AXMANN(A.), GISSLER(W.).

2020 SE,. . . . . .B4,B7. . . . .PROC/INTERNAT. SYMP/PHYS/SE/TE,. 299, (1969).
      ⇒INVESTIGATIONS ON CRYSTALLINE TELLURIUM AND SOLID AMORPHOUS AND LIQUID
      SELENIUM WITH INELASTIC NEUTRON SCATTERING.
      ⇒AXMANN(A.), GISSLER(W.), SPRINGER(T.). (ED. NOTE: PUBL. PERGAMON,
      OXFORD, ENGLAND 1969).

2021 SE,. . . . . .B1,B4. . . . .IAEA SYMP. COPENHAGEN, VOL.1,. . 245, (1968).
      ⇒INELASTIC NEUTRON SCATTERING ON SOLID AND LIQUID TELLURIUM AND SELENIUM.
      ⇒GISSLER(W.), AXMANN(A.), SPRINGER(T.).

2022 SE,. . . . . .B1. . . . . . .AMER. CRYSTALL. ASSOC.,. . . . .   27, (1971).
      ⇒PHONON DISPERSION CURVES IN TRIGONAL SELENIUM: A NEUTRON INELASTIC
      SCATTERING STUDY OF NON-BONDED FORCES IN A SEMI-MOLECULAR CRYSTAL.
      ⇒HAMILTON(W.C.), KAY(M.I.), LASSIER(B.). (ED. NOTE: PUBL. AMER. CRYSTALL.
      ASSOC. 1971, NEW YORK, USA).

2023 SE,. . . . . .C1. . . . . . .SOV. PHYS. SOL. STATE 9,. . . . 955, (1967).
      ⇒COMPARISON OF SPECTRA OF COLD NEUTRONS SCATTERED BY AMORPHOUS AND
      CRYSTALLINE SELENIUM.
      ⇒KOTOV(B.A.), OKUNEVA(N.M.), REGEL(A.R.), SHAKH-BUDAGOV(A.L.).

2024 SE,. . . . . .B4,B7. . . . .DISC. FARADAY SOC. NO.50,. . . .   74, (1970).
      ⇒THE INVESTIGATION OF ATOMIC MOTIONS IN CRYSTALLINE, AMORPHOUS AND LIQUID
      SELENIUM, AND IN CRYSTALLINE AND LIQUID TELLURIUM BY NEUTRON SPECTROS.
      ⇒AXMANN(A.), GISSLER(W.), KOLLMAR(A.), SPRINGER(T.).

2025 SI,. . . . . .B1. . . . . . .IAEA SYMP. BOMBAY, VOL.1,. . . . 249, (1964).
      ⇒INELASTIC NEUTRON SCATTERING FROM DOPED GERMANIUM AND SILICON.
      ⇒DOLLING(G.).

2026 SI,. . . . . .T,B2. . . . . .IAEA SYMP. GRENOBLE, ...   29, (1972).
      ⇒ANHARMONIC INTERACTIONS IN GE AND SI.
      ⇒JEX(H.).

2027 SI,. . . . . .B1. . . . . . .SOV. PHYS. SOL. STATE VOL.4,. . 1747, (1962).
      ⇒CONCERNING THE SIMILARITY BETWEEN THE CHARACTERISTIC FREQUENCY
      DISPERSION CURVES OF DIAMOND-TYPE CRYSTALS.
      ⇒KUCHER(T.I.).

2028 SI,. . . . . .B1. . . . . . .PHYS. REV. B VOL.6,. . . . . . .3777, (1972).
      ⇒STUDY OF THE HOMOLOGY BETWEEN SILICON AND GERMANIUM BY THERMAL-NEUTRON
      SPECTROMETRY.
      ⇒NILSSON(G.), NELIN(G.).

2029 SI,. . . . . .D2. . . . . . .PROC. ROY. SOC. 281,. . . . . . 274, (1964).
      ⇒CRITICAL-POINT ANALYSIS OF THE PHONON SPECTRA OF DIAMOND, SILICON AND
      GERMANIUM.
      ⇒JOHNSON(F.A.), LOUDON(R.).

2030 SI,. . . . . .T,B. . . . . .PHYS. REV LETT. 21,. . . . . . . 536, (1968).
      ⇒LATTICE VIBRATIONS IN SILICON: MICROSCOPIC DIELECTRIC MODEL.
      ⇒MARTIN(R.M.).

SECTION   6 -QUASI-ELASTIC AND INELASTIC SCATTERING STUDIES . . . .

2031 SI,. . . . . P.....Z PHYSIK VOL.190. . . . . . 295, (1966).
MESSUNGEN DES FREIEN UND INKOHARENTEN NEUTRONENWIRKUNGSQUERSCHNITTS AN
SILIZIUM UND BEI
NIKLAUS(J.P.), SIMSON(R.), TRIFTSHAUSER(W.), SCHMATZ(W.).

2032 SI,. . . . . P.......REF. SECTION 3, 48/RAINWATER,. . . . .

2033 SI,. . . . . . . B1.......PROC. PHYS. SOC. 88, . . . 463, (1966).
THE THERMODYNAMIC AND OPTICAL PROPERTIES OF GERMANIUM, SILICON, DIAMOND
AND GALLIUM ARSENIDE.
DOLLING(G.), COWLEY(R.A.).

2034 SI,. . . . . . . B1.....IAEA SYMPOSIUM VIENNA . . . . . 563, (1960).
PHONON DISPERSION RELATIONS FOR A SILICON SINGLE CRYSTAL.
DOLLING(G.).

2035 SI,. . . . . . . B1.....IAEA SYMP. CHALK RIVER, VOL.2, 37, (1962).
LATTICE VIBRATIONS IN CRYSTALS WITH THE DIAMOND STRUCTURE.
DOLLING(G.).

2036 SI,. . . . . . . B1.....PHYS. REV. LETT. VOL.2, 258, (1959).
LATTICE VIBRATIONS IN SILICON BY SCATTERING OF COLD NEUTRONS.
PALEVSKY(H.), HUGHES(D.J.), KLEY(W.), TUNKELO(E.).

2037 SI,. . . . . . . B1.....PHYS. REV. LETT. VOL.2,. . . . 256, (1959).
LATTICE VIBRATIONS IN SILICON AND GERMANIUM.
BROCKHOUSE(B.N.).

2038 SI*C,. . . . . . T.......PHYS. REV. 178,. . . . . 1349, (1969).
LATTICE DYNAMICS OF CUBIC SI*C.
VETELINO(J.F.), MITRA(S.S.).

2039 (SI)GE,. . . . . B3.......PHYS. REV. B-4, ... 2558, (1971).
NEUTRON SCATTERING INVESTIGATION OF IMPURITY PHONON MODES IN GE(SI).
WAKABAYASHI(N.), NICKLOW(R.M.), SMITH(H.G.).

2040 SI*H4, ... C3.......IAEA SYMP. GRENOBLE, ... 489, (1972).
MOLECULAR MOTIONS IN SILANE.
HAUTECLER(S.), VORDERWISCH(P.).

2041 SI*H4, ... C3,T.....PHIL. MAG. VOL.28, ... 1353, (1973).
ROTATIONAL DIFFUSION MODEL   CALCULATION OF INCOHERENT COLD NEUTRON
SCATTERING BY LIQUID SILANE.
CVIKL(B.).

2042 SILICA-H2*0, . . E........ADV. COLLOID INTERFACE SCI. 2, 1, (1968).
APPLICATIONS OF SLOW NEUTRON SCATTERING TO STUDIES IN COLLOID AND
SURFACE CHEMISTRY (REVIEW).
BOUTIN(H.), PRASK(H.), IYENGAR(R.D.).

2043 SILICATES, . . . C1.......SPEC. DISC. FARADAY SOC. NO.1, 194, (1970).
NEUTRON SCATTERING STUDIES OF HYDRATED LAYER SILICATES.
OLEJNIK(S.), STIRLING(G.C.), WHITE(J.W.).

2044 SILOXANE (POLY-DIMETHYL),. . . . . . . .FARADAY SYMP. CHEM. SOC. NO.6, 169, (1972).
MOTIONAL BROADENING OF THE QUASI-ELASTIC PEAK IN NEUTRONS SCATTERED FROM
POLYMERIC MATERIALS.
ALLEN(G.), BRIER(P.N.), GOODYEAR(G.), HIGGINS(J.S.).

2045 SILOXANE (POLY-DIMETHYL),. . . C1.......J. POLYMER SCI. A-2 VOL.7, 433, (1969).
STUDY OF LOW-FREQUENCY MOLECULAR MOTIONS IN POLYDIMETHYLSILOXANE
POLYMERS BY NEUTRON INELASTIC SCATTERING.
HENRY(A.W.), SAFFORD(G.J.).

2046 SILOXANES (DIMETHYL),. . . . . . .C1,B1.....CHEM. APPL. THERMAL NEUTRON SCAT 97, (1973).
NEUTRON SCATTERING STUDIES OF THE DYNAMICS OF POLYMER CHAINS.
ALLEN(G.). (IN BOOK: CHEMICAL APPLICATIONS OF THERMAL NEUTRON
SCATTERING. ED. BY B.T.M. WILLIS; OXFORD UNIV. PRESS; LONDON).

2047 SI*02, ... B7.......J. CHEM. PHYSICS VOL.51, 779, (1969).
INELASTIC COLD NEUTRON SCATTERING FROM DIFFERENT FORMS OF SILICA.
LEADBETTER(A.J.).

2048 SI*02, ... B1.......DISS. ABS. VOL.31, 6839B, (1971).
EXPERIMENTAL PHONON DISPERSION RELATION IN VITREOUS SILICA AT 47 DEG. K.
SAUER(G.E.).

2049 SI*02, ... B7.......IAEA SYMP. GRENOBLE, 501, (1972).
THE FREQUENCY SPECTRUM OF THE NORMAL MODES OF VIBRATION OF VITREOUS
SILICA AND ALPHA QUARTZ.
LEADBETTER(A.J.), STRINGFELLOW(M.W.).

2050 SI*02, ... B7.......CONF./PHONON SCATT. IN SOLIDS, . 338, (1972).
VIBRATIONAL EXCITATIONS IN GLASSES.
LEADBETTER(A.J.).   (ED. NOTE: INTERNAT. CONF. ON PHONON SCATT. IN SOLIDS
AT PARIS, JULY, 1972. PUBL. DOCUMENTATION SERVICE OF CENT SACLAY, FRA.).

2051 SI*02, ... P.......PHYS. STAT. SOL. 7, 415, (1964).
UNTERSUCHUNG UBER DIE STREUUNG LANGSAMER NEUTRONEN AN BESTRAHLTEN QUARTZ
UND BERYLLIUM.
BAIERLEIN(K.).

2052 SI*02, ... B7.......NATURE (PHYS. SCI.) VOL.242, 109, (1973).
NEUTRON TOPOGRAPHY OF VIBRATING SINGLE CRYSTALS OF ALPHA-SI*02.
SEDLAKOVA(L.),   MICHALEC(R.), MIKULA(P.), HRDLICKA(Z.), ZELENKA(J.),
PETRZILKA(V.).

2053 SI*02, ... B1.......PROC. PHYS. SOC. 91, 946, (1967).
SOME ASPECTS OF THE LATTICE DYNAMICS OF QUARTZ.
ELCOMBE(M.M.).

2054 SI*02, ... D2.......PHYS. REV. B-1, 342, (1970).
STUDY OF THE ALPHA-BETA QUARTZ PHASE TRANSFORMATION BY INELASTIC NEUTRON
SCATTERING.
AXE(J.D.), SHIRANE(G.).

2055 SM,. . . . . . . P.......PHYS. REV. VOL.83, 841, (1951).
RESONANT SCATTERING IN SAMARIUM AND GADOLINIUM.
BROCKHOUSE(B.N.), HURST(D.G.).

2056 SM2*03,. . . . . E.......COLL. INTER. N.126 (GRENOBLE), 69, (1963).
UN EMPLOI POSSIBLE DE FILTRES D'ABSORPTION EN DIFFRACTION DE NEUTRONS.
DOMENICI(M.).

SECTION   6 -QUASI-ELASTIC AND INELASTIC SCATTERING STUDIES . . . .

2057 SN (ALPHA),. . . . . .LATTICE DYNAMICS OF GREY TIN AND INDIUM ANTIMONIDE.. . . . . 1268, (1971).
      ↦PRICE(D.L.), ROWE(J.M.), NICKLOW(R.M.).

2058 SN,. . . . . .COMPARATIVE STUDIES OF SLOW NEUTRON SCATTERING BY SOLID AND LIQUID TIN.
      ↦COCKING(S.J.), GUNER(Z.).

2059 SN,. . . . . . .INTERPRETATION OF COHERENT NEUTRON SCATTERING BY LIQUIDS. 203, (1962).
      ↦EGELSTAFF(P.A.).

2060 SN,. . . . . . .THE VIBRATION SPECTRUM OF THE WHITE TIN LATTICE. 131, (1964).
      ↦.C.C.C.P.

2061 SN,. . . . . . .RECENT RESULTS ON THE CRYSTAL PHYSICS OF WHITE TIN. 117, (1964).
      ↦BORGONOVI(G.), CAGLIOTI(G.), ANTONINI(M.).

2062 SN,. . . . . . .SUPERCONDUCTIVITY AND LATTICE DYNAMICS OF WHITE TIN. 4065, (1971).
      ↦ROWELL(J.M.), MCMILLAN(W.L.), FELDMANN(W.L.).

2063 SN (PHASE: ALPHA),. . . . .THE CRYSTAL DYNAMICS OF GREY (ALPHA) TIN AT 90 DEG. K. 1433, (1969).
      ↦PRICE(D.L.), ROWE(J.M.).

2064 SN,. . . . . .NEUTRON SPECTROSCOPY: THE DESIGN AND CONSTRUCTION OF A TRIPLE-AXIS
      CRYSTAL SPECTROMETER, AND A STUDY OF THE LATTICE DYNAMICS OF METALS.
      ↦ROWE(J.M.). (ED. NOTE: PH.D. DISSERTATION, MCMASTER UNIVERSITY).

2065 SN,. . . . . .CONCERNING THE SIMILARITY BETWEEN THE CHARACTERISTIC FREQUENCY
      DISPERSION CURVES OF DIAMOND-TYPE CRYSTALS.
      ↦KUCHER(T.I.).

2066 SN,. . . . . .PHONONS IN WHITE TIN. 109, (1964).
      ↦PRICE(D.L.).

2067 SN,. . . . . .INELASTIC SCATTERING OF COLD NEUTRONS IN POLYCRYSTALLINE GRAY TIN.
      ↦MYERS(V.W.).

2068 SN,. . . . . .DISPERSION RELATIONS OF WHITE TIN. 108, (1965).
      ↦NICHOLS(H.), ROSTOKER(W.).

2069 SN,. . . . . . . .REF. SECTION 3, 48/HAVENS, ...

2070 SN,. . . . . .CRYSTAL DYNAMICS OF METALLIC BETA-SN AT 110 DEG. K. 547, (1967).
      ↦ROWE(J.M.).

2071 SN,. . . . . .LATTICE DYNAMICS OF WHITE TIN. 554, (1965).
      ↦ROWE(J.M.), BROCKHOUSE(B.N.), SVENSSON(E.C.).

2072 SN,. . . . . .PHONON SPECTRUM OF WHITE TIN. 1120, (1966).
      ↦BROVMAN(E.G.), KAGAN(YU.).

2073 SN,. . . . . .LATTICE DYNAMICS OF WHITE TIN. 25, (1967).
      ↦PRICE(D.L.).

2074 SN,. . . . . .DISPERSION RELATIONS OF WHITE TIN. 44, (1965).
      ↦SCHMUNK(R.E.), GAVIN(W.R.).

2075 SN,. . . . . .FREQUENCY WAVE NUMBER RELATIONSHIP FOR CO-OPERATIVE MODES OF MOTION IN
      LIQUID LEAD AND TIN.
      ↦COCKING(S.J.), EGELSTAFF(P.A.).

2076 SN,. . . . . .THE PHONON SPECTRUM OF METALS. 365, (1967).
      ↦BROVMAN(E.G.), KAGAN(YU.).

2077 SN,. . . . . .EXPERIMENTAL DETERMINATION OF THE THERMAL VIBRATION SPECTRUM OF WHITE
      TIN.
      ↦KOTOV(B.A.), OKUNEVA(N.M.), PLACHENOVA(E.L.).

2078 SN*TE,. . . . .DIATOMIC FERROELECTRICS. 753, (1966).
      ↦PAWLEY(G.S.), COCHRAN(W.), COWLEY(R.A.), DOLLING(G.).

2079 SN*TE,. . . . .THE LATTICE DYNAMICS OF TIN TELLURIDE. 1916, (1969).
      ↦COWLEY(E.R.), DARBY(J.K.), PAWLEY(G.S.).

2080 SN*TE-GE*TE,. . .THE TRANSITION IN SN*TE-GE*TE ALLOYS. 249, (1970).
      ↦LEFKOWITZ(I.), SHIELDS(M.), DOLLING(G.), BUYERS(W.J.L.), COWLEY(R.A.).

2081 SPINELS,. . . . .CATION DISTRIBUTIONS IN MULTISUBLATTICE IONIC CRYSTALS AND APPLICATIONS
      TO SOLID SOLUTIONS OF FERRIMAGNETIC GARNETS AND SPINELS.
      ↦BORGHESE(C.).

2082 SPINELS,. . . . .NEUTRON SCATTERING STUDIES OF SOME SPINELS. 318, (1971).
      ↦MURTHY(N.S.S.), MADHAV-RAO(L.), BEGUM(R.J.), NATERA(M.G.), YOUSSEF(S.I.)

2083 (SR*CL2)*6(H2*O),. . . . .INVESTIGATION OF THE DYNAMICS OF N*H4+ AND H2*O MOLECULAR GROUPS IN
      CRYSTALS.
      ↦.C.C.C.P.

2084 SR*F2,. . . . .NEUTRON DIFFRACTION STUDY OF ANHARMONIC VIBRATIONAL EFFECTS IN STRONTIUM
      FLUORIDE.
      ↦MAIR(S.L.), BARNEA(Z.).

SECTION   6 -QUASI-ELASTIC AND INELASTIC SCATTERING STUDIES . . . .

2085 SR*F2, ... B1 . . . . . . J. PHYS. C VOL.5, ... 2702, (1972).
      *THE LATTICE DYNAMICS OF STRONTIUM FLUORIDE.
      *ELCOMBE(M.M.).

2086 SR*H2, ... C1 . . . . J. CHEM. PHYS. VOL.52, ... 3952, (1970).
      *VIBRATION SPECTRA OF THE ORTHORHOMBIC ALKALINE-EARTH HYDRIDES BY THE
      INELASTIC SCATTERING OF COLD NEUTRONS AND BY INFRARED TRANSM. MEASUR.
      *MAELAND(A.J.).

2087 SR*LI*H3, ... C1 . . . . . . J. CHEM. PHYSICS VOL.51, ... 2915, (1969).
      *TERNARY HYDRIDES POSSESSING THE CUBIC PEROVSKITE STRUCTURE. II.
      VIBRATION SPECTRA BY THE INELASTIC SCATTERING OF COLD NEUTRONS.
      *MAELAND(A.J.).

2088 SR*TI*O3, ... B1 . . . . . . IAEA SYMP. COPENHAGEN, VOL.1, . . 281, (1968).
      *NEUTRON SCATTERING AND ELASTIC CONSTANTS.
      *COWLEY(R.A.), SVENSSON(E.C.), BUYERS(W.J.L.).

2089 SR*TI*O3, ... B1,T . . . . . IAEA SYMP. BOMBAY, VOL.1, ... 297, (1964).
      *ANHARMONIC EFFECTS AND THE SCATTERING OF NEUTRONS FROM A CRYSTAL.
      *COWLEY(R.A.).

2090 SR*TI*O3, ... B6 . . . . . . SOL. STATE COMM. VOL. 9, ... 1455, (1971).
      *CRITICAL BEHAVIOUR OF SR*TI*O3 NEAR THE 105 DEG. K PHASE TRANSITION.
      *RISTE(T.), SAMUELSEN(E.J.), OTNES(K.), FEDER(J.).

2091 SR*TI*O3, ... B1 . . . . . . SOL. STATE COMM. VOL. 9, ... 1103, (1971).
      *TEMPERATURE DEPENDENCE OF THE SOFT MODE IN SR*TI*O3 ABOVE THE 105 DEG. K
      TRANSITION.
      *OTNES(K.), RISTE(T.), SHIRANE(G.), FEDER(J.).

2092 SR*TI*O3, ... B1,T . . . . . PHYS. REV. VOL.134, ... A981, (1964).
      *LATTICE DYNAMICS AND PHASE TRANSITIONS OF STRONTIUM TITANATE.
      *COWLEY(R.A.).

2093 SR*TI*O3, ... B7 . . . . . . ACTA CRYST. VOL.A26, ... 608, (1970).
      *DETERMINATION OF THE NORMAL VIBRATIONAL DISPLACEMENTS IN SEVERAL
      PEROVSKITES BY INELASTIC NEUTRON SCATTERING.
      *HARADA(J.), AXE(J.D.), SHIRANE(G.).

2094 SR*TI*O3, ... D4 . . . . . . SOL. STATE COMM. 7, ... 181, (1969).
      *RELATIONSHIP OF NORMAL MODES OF VIBRATION OF STRONTIUM TITANATE AND ITS
      ANTIFERROELECTRIC PHASE TRANSITION AT 110 DEG. K.
      *COWLEY(R.A.), BUYERS(W.J.L.), DOLLING(G.).

2095 SR*TI*O3, ... B7 . . . . . . J. PHYS. SOC. JAPAN VOL.26, . . . 396, (1969).
      *NEUTRON SCATTERING AND NATURE OF THE SOFT OPTICAL PHONON IN SR*TI*O3.
      *YAMADA(Y.), SHIRANE(G.).

2096 SR*TI*O3, ... T . . . . . . SOL. STATE COMM. VOL.11 NO.1, . . 247, (1972).
      *MICROSCOPIC CALCULATION OF THE RESPONSE FUNCTION OF THE SOFT ZONE BOUND-
      ARY PHONON IN SR*TI*O3.
      *SILBERGLITT(R.).

2097 SR*TI*O3, ... T,B7 . . . . . PHYS. REV. B VOL.6, ... 4695, (1972).
      *LATTICE DYNAMICS ABOVE STRUCTURAL PHASE TRANSITIONS: SR*TI*O3.
      *ENZ(C.P.).

2098 SR*TI*O3, ... B6 . . . . . . PHYS. REV. B VOL.6, ... 4332, (1972).
      *CRITICAL NEUTRON SCATTERING IN SR*TI*O3 AND K*MN*F3.
      *SHAPIRO(S.M.), AXE(J.D.), SHIRANE(G.), RISTE(T.).

2099 SR*TI*O3, ... B1 . . . . . . J. PHYS. C VOL.5, ... 2711, (1972).
      *NEUTRON INELASTIC SCATTERING STUDY OF THE LATTICE DYNAMICS OF STRONTIUM
      TITANATE: HARMONIC MODELS.
      *STIRLING(W.G.).

2100 SR*TI*O3, ... B . . . . . . SOL. STAT. COMM. 5, ... 387, (1967).
      *EVIDENCE FOR THE PRESENCE OF A FIRST ORDER RAMAN SPECTRUM IN THE HIGH
      TEMPERATURE PHASE OF SR*TI*O3.
      *RIMAI(L.), PARSONS(J.L.).

2101 SR*TI*O3, ... C1 . . . . . . SOV. PHYS. SOL. STATE 8, ... 2156, (1966).
      *SCATTERING OF COLD NEUTRONS BY POLYCRYSTALLINE BARIUM, STRONTIUM AND
      LEAD TITANATES.
      *SOLOV≠EV(S.P.), KUKHTO(O.L.), CHERNOPLEKOV(N.A.), ZETLYANOV(M.G.).

2102 SR*TI*O3, ... B2 . . . . . . PHYS. REV. LETT. VOL.9, ... 159, (1962).
      *TEMPERATURE DEPENDENCE OF A TRANSVERSE OPTIC MODE IN STRONTIUM TITANATE.
      *COWLEY(R.A.).

2103 SR*TI*O3, ... B1 . . . . . . IAEA SYMP. CHALK RIVER, VOL.2, . . 229, (1962).
      *THE TEMPERATURE DEPENDENCE OF SOME NORMAL MODES IN STRONTIUM TITANATE.
      *COWLEY(R.A.).

2104 SR*TI*O3, ... B1 . . . . . . J. PHYS. C VOL.6, ... 3021, (1973).
      *DISPERSION RELATIONS OF THE NORMAL VIBRATIONS IN STRONTIUM TITANATE.
      *IIZUMI(M.), GESI(K.), HARADA(J.).

2105 SR*TI*O3, ... T . . . . . . J. PHYS. C VOL.6, ... 2422, (1973).
      *LATTICE DYNAMICS OF STRONTIUM TITANATE: ANHARMONIC INTERACTIONS AND
      STRUCTURAL PHASE TRANSITIONS.
      *BRUCE(A.D.), COWLEY(R.A.).

2106 SR*TI*O3, ... B1 . . . . . . J. PHYSIQUE TOME 33 COL.2, . . . 135, (1972).
      *THE LATTICE DYNAMICS OF STRONTIUM TITANATE.
      *STIRLING(W.G.), COWLEY(R.A.).

2107 SR*TI*O3, ... B1 . . . . . . PHYS. REV. VOL.177, ... 858, (1969).
      *LATTICE-DYNAMICAL STUDY OF THE 110 DEG. K PHASE TRANSITION IN SR*TI*O3.
      *SHIRANE(G.), YAMADA(Y.).

2108 SUCCINOTRILE, ... C1 . . . . . . SYMP. CHEM. ORGANIC SOL. STATE, . . . , (1972).
      *NEUTRON SCATTERING STUDY OF MOLECULAR ROTATION IN SUCCINOTRILE.
      *LEADBETTER(A.J.), TURNBULL(A.). (THIRD INTERNATIONAL SYMPOSIUM ON
      CHEMISTRY OF THE ORGANIC SOLID STATE, GLASGOW, SCOTLAND, SEPT. 18-22).

2109 TA, ... C5 . . . . . . IAEA SYMP. CHALK RIVER, VOL.2, . . 183, (1962).
      *SCATTERING LAW FOR METAL ATOMS IN CHEMICAL COMPOUNDS.
      *EGELSTAFF(P.A.), HOLT(G.).

2110 TA, ... B1 . . . . . . IAEA SYMP. BOMBAY, VOL.1, ... 87, (1964).
      *LATTICE DYNAMICS OF TRANSITION METALS.
      *WOODS(A.D.B.).

SECTION   6 -QUASI-ELASTIC AND INELASTIC SCATTERING STUDIES . . . .

2111 TA,. . . . . .LATTICE DYNAMICS OF TANTALUM. B1 PHYS. REV. VOL.136,. . . . . . . A781, (1964).
      .WOODS(A.D.B.).

2112 TA,. . . . . .SLOW NEUTRON VELOCITY SPECTROMETER STUDIES. II. AU, IN, TA, W, PT, ZR. P PHYS. REV. VOL.71, 165, (1947).
      .HAVENS-JR(W.W.), WU(C.S.), RAINWATER(L.J.), MEAKER(C.L.).

2113 TA*C,. . . . . .PHONON SPECTRA IN TA*C AND HF*C. B1 PHYS. REV. LETT. VOL.25, ... 1611, (1970).
      .SMITH(H.G.), GLASER(W.).

2114 TA*C,. . . . . .PHONON ANOMALIES IN TRANSITION-METAL CARBIDES. B1 PHYS. REV. LETT. VOL.29, ... 353, (1972).
      .SMITH(H.G.).

2115 TA*C,. . . . . .PHONON SPECTRA AND SUPERCONDUCTIVITY IN SOME TRANSITION METAL CARBIDES. B7 CONF. INTERN., RENNES, ... 145, (1971).
      .SMITH(H.G.), GLASER(W.).

2116 TA*C,. . . . . .PHONON SPECTRA OF SOME TRANSITION METAL CARBIDES FROM A SIMPLE T PHYS. REV. B VOL.5, . . . 1260, (1972).
      PSEUDOPOTENTIAL APPROACH.
      .MOSTOLLER(M.).

2117 TA-H,. . . . . .STUDY OF HYDROGEN DIFFUSION IN TANTALUM HYDRIDES BY INELASTIC NEUTRON C1 J. CHEM. PHYS. VOL.59 . . . . . . 6570, (1973).
      SCATTERING.
      .RUSH(J.J.), LIVINGSTON(R.C.), DE-GRAAF(L.A.), FLOTOW(H.E.), ROWE(J.M.).

2118 TA*H5,. . . . . . .STUDY OF HYDROGEN DIFFUSION IN VANADIUM AND TANTALUM HYDRIDE BY QUASI- C1 JULICH CONF/H IN METALS VOL.I, . . . 301, (1972).
      ELASTIC NEUTRON SCATTERING.
      .DE-GRAAF(L.A.), RUSH(J.J.), LIVINGSTON(R.C.), FLOTOW(H.E.), ROWE(J.M.).

2119 TA-(H),. . . . . . .NEUTRON-QUASIELASTIC-SCATTERING STUDY OF HYDROGEN DIFFUSION IN A SINGLE C1 PHYS. REV. B VOL.9 NO.12, ... (1974).
      CRYSTAL OF TANTALUM.
      .ROWE(J.M.), RUSH(J.J.), FLOTOW(H.E.). (REF. OBTAINED FROM PHYS. REV.
      ABSTRACTS).

2120 TA88-NB12,. . . .LOCALIZED AND BAND PHONONS IN TA88-NB12 STUDIED BY COHERENT INELASTIC B3 IAEA SYMP. COPENHAGEN, VOL.1,. . . 35, (1968).
      NEUTRON SCATTERING.
      .ALS-NIELSEN(J.).

2121 TA2*O5,. . . . . .SCATTERING LAW FOR METAL ATOMS IN CHEMICAL COMPOUNDS. C5 IAEA SYMP. CHALK RIVER, VOL.2, . 183, (1962).
      .EGELSTAFF(P.A.), HOLT(G.).

2122 TAURINE,. . . . .INELASTIC INCOHERENT NEUTRON SCATTERING ON POLYCRYSTALLINE TAURINE AND C1 CHEM. PHYS. LETT. VOL.17,. . . . . 53, (1972).
      ALPHA-AMINO-ISO-BUTYRIC ACID.
      .DAVIDOVIC(M.), RATKOVIC(S.), JOVIC(D.), ZIVANOVIC(M.).

2123 TB,. . . . . . . .LATTICE DYNAMICS OF TERBIUM. B1 PHYS. REV. B-1,. . . . . 3943, (1970).
      .HOUMANN(J.C.G.), NICKLOW(R.M.).

2124 TB,. . . . . . . .MAGNON INTERACTIONS IN TERBIUM. D1 J. APPL. PHYS. 41, ... 1174, (1970).
      .NIELSEN(M.), MOLLER(H.B.), MACKINTOSH(A.R.).

2125 TB,. . . . . . . .LINE SHAPE OF THE MAGNETIC SCATTERING FROM ANISOTROPIC PARAMAGNETS. T IAEA SYMP. COPENHAGEN, VOL.2, 93, (1968).
      .LINDGARD(P.A.).

2126 TB,. . . . . . . .INELASTIC CRITICAL NEUTRON SCATTERING IN TERBIUM. D2 IAEA SYMP. COPENHAGEN, VOL.2,. . 63, (1968).
      .DIETRICH(O.W.), ALS-NIELSEN(J.).

2127 TB,. . . . . . . .MAGNON ENERGIES AND EXCHANGE INTERACTIONS IN TERBIUM. T,D1 IAEA SYMP. COPENHAGEN, VOL.2,. . 29, (1968).
      .GYLDEN-HOUMANN(J.C.).

2128 TB,. . . . . . . .INELASTIC SCATTERING OF NEUTRONS BY SPIN WAVES IN TERBIUM. D1 BROOKHAVEN SYMPOSIUM,. . . . . . . 159, (1965).
      .MOLLER(H.B.), MOGENSEN(P.A.), HOUMANN(J.C.G.), KOWALSKA(A.).

2129 TB,. . . . . . . .INELASTIC CRITICAL NEUTRON SCATTERING FROM TERBIUM. D2 J. PHYS. SOC. JAPAN VOL.26,. . . 183, (1969).
      .ALS-NIELSEN(J.), DIETRICH(O.W.).

2130 TB,. . . . . . . .MAGNETIC FORM FACTORS OF TERBIUM. P PHYS. REV. 161,. . . . . 499, (1967).
      .STEINSVOLL(O.), SHIRANE(G.), NATHANS(R.), BLUME(M.), ALPERIN(H.A.),
      PICKART(S.J.).

2131 TB,. . . . . . . .MAGNETIC INTERACTIONS IN RARE-EARTH METALS FROM INELASTIC NEUTRON D1 PHYS. REV. LETT. 19, . . . 312, (1967).
      SCATTERING.
      .MOLLER(H.B.), HOUMANN(J.G.), MACKINTOSH(A.R.).

2132 TB,. . . . . . . .ANISOTROPIC COUPLING BETWEEN MAGNETIC IONS IN TERBIUM. D4 IAEA SYMP. GRENOBLE, ... 603, (1972).
      .MOLLER(H.B.), HOUMANN(J.G.), JENSEN(J.), MACKINTOSH(A.R.).

2133 TB,. . . . . . . .NEUTRON SPIN SCATTERING BY SPIN WAVES IN METALS (REVIEW PAPER). D1 IAEA SYMP. COPENHAGEN, VOL.2, 3, (1968).
      .MOLLER(H.B.).

2134 TB,. . . . . . . .CRITICAL SCATTERING OF NEUTRONS FROM TERBIUM. D2 J. APP. PHYS. 39,. . . . . . . 1229, (1968).
      .ALS-NIELSEN(J.), DIETRICH(O.W.), MARSHALL(W.), LINDGARD(P.A.).

2135 TB,. . . . . . . .MAGNETIC INTERACTIONS IN TB AND TB-10PERCENT,HO FROM INELASTIC NEUTRON D1 J. APP. PHYS. 39,. . . . . . 807, (1968).
      SCATTERING.
      .MOLLER(H.B.), HOUMANN(J.G.), MACKINTOSH(A.R.).

2136 TB,. . . . . . . .INELASTIC SCATTERING OF NEUTRONS BY SPIN WAVES IN TERBIUM. D1 PHYS. REV. LETT. 16, . . . 737, (1966).
      .MOLLER(H.B.), HOUMANN(J.G.).

SECTION   6 -QUASI-ELASTIC AND INELASTIC SCATTERING STUDIES . . . .

2137 TB,. . . . .*INELASTIC CRITICAL SCATTERING OF NEUTRONS FROM TERBIUM.   607, (1967).
                            D2   SOL. STATE COMM. 5,
                            *ALS-NIELSEN(J.), DIETRICH(O.W.), MARSHALL(W.), LINDGARD(P.A.).

2138 TB,. . . . . .T,D1. . . . .J. APP. PHYS. VOL.39,   887, (1968).
                            *EXCHANGE INTERACTIONS AND THE SPIN-WAVE SPECTRUM OF TERBIUM.
                            *GOODINGS(D.A.).

2139 TB,. . . . . .D4,T. . . . .J. OF PHYS. C VOL.1,   1596, (1968).
                            *MAGNETIZATION AND MAGNETIC SPECIFIC HEATS OF TERBIUM.
                            *BROOKS(M.S.S.), GOODINGS(D.A.), RALPH(H.I.).

2140 TB,. . . . . .T,D1. . . . .J. OF PHYS. C VOL.1,   132, (1968).
                            *TEMPERATURE-DEPENDENT CRYSTAL FIELD EFFECTS IN THE SPIN WAVE SPECTRUM OF
                            TERBIUM.
                            *BROOKS(M.S.S.), GOODINGS(D.A.), RALPH(H.I.).

2141 TB,. . . . . .T,D1. . . . .J. OF PHYS. C VOL.1,   125, (1968).
                            *EXCHANGE INTERACTIONS AND THE SPIN-WAVE SPECTRUM OF TERBIUM.
                            *GOODINGS(D.A.).

2142 TB,. . . . . .D2. . . . . .J. PHYS. C, SER.2, VOL.4,   71, (1971).
                            *SPIN DYNAMICS IN TB STUDIED BY CRITICAL NEUTRON SCATTERING.
                            *DIETRICH(O.W.), ALS-NIELSEN(J.).

2143 TB,. . . . . .P. . . . . . .PHYS. REV. B VOL.8,   2595, (1973).
                            *ATOMIC AND NUCLEAR EFFECTS IN THE SLOW-NEUTRON TOTAL CROSS SECTION OF
                            TERBIUM.
                            *MALIK(S.S.), KAMAL(M.), TURANO(T.), DESJARDINS(J.S.).

2144 TB*AG, ...D4. . . . . .REF. SECTION 3, 64/KOEHLER(2), ...

2145 TB-AL2,. . . . . .D1. . . . . .PHYS. REV. LETT. VOL.31,   1585, (1973).
                            *INTERACTION BETWEEN MAGNONS AND MAGNETIC EXCITONS IN TB-AL2.
                            *PURWINS(H.G.), HOUMANN(J.G.), BAK(P.), WALKER(E.).

2146 TB-AL2,. . . . . .D1. . . . . .SOL. STATE COMM. VOL.13,   881, (1973).
                            *MAGNON DISPERSION IN TB-AL2.
                            *BUHRER(W.), GODET(M.), PURWINS(H.G.), WALKER(E.).

2147 TB*CU, ...D4. . . . . .PROC. INT. CONF. (NOTTINGHAM),   271, (1964).
                            *NEUTRON MAGNETIC SCATTERING STUDIES AT THE OAK RIDGE NATIONAL LABORATORY
                            *KOEHLER(W.C.), CABLE(J.W.), CHILD(H.R.), MOON(R.M.), WOLLAN(E.O.). (ED.
                            NOTE: REF. SECTION 3, 64/KOEHLER(2)).

2148 TB*HG, ...D4. . . . . .REF. SECTION 3, 64/KOEHLER(2), ...

2149 TB-(HO), ...D1. . . . . .J. APPL. PHYS. 41,   1174, (1970).
                            *MAGNON INTERACTIONS IN TERBIUM.
                            *NIELSEN(M.), MOLLER(H.B.), MACKINTOSH(A.R.).

2150 TB-HO, ...D1. . . . . .PHYS. REV. LETT. 19,   312, (1967).
                            *MAGNETIC INTERACTIONS IN RARE-EARTH METALS FROM INELASTIC NEUTRON
                            SCATTERING.
                            *MOLLER(H.B.), HOUMANN(J.G.), MACKINTOSH(A.R.).

2151 TB(HO),. . . . . .D1. . . . . .PHYS. REV. LETT. VOL.25,   1451, (1970).
                            *MAGNETIC ANISOTROPY IN RARE-EARTH METALS.
                            *NIELSEN(M.), MOLLER(H.B.), LINDGARD(P.A.), MACKINTOSH(A.R.).

2152 TB-HO, ...D1. . . . . .IAEA SYMP. COPENHAGEN, VOL.2,   3, (1968).
                            *NEUTRON SPIN SCATTERING BY SPIN WAVES IN METALS (REVIEW PAPER).
                            *MOLLER(H.B.).

2153 TB-HO, ...D1. . . . . .J. APP. PHYS. 39,   807, (1968).
                            *MAGNETIC INTERACTIONS IN TB AND TB-10PERCENT,HO FROM INELASTIC NEUTRON
                            SCATTERING.
                            *MOLLER(H.B.), HOUMANN(J.G.), MACKINTOSH(A.R.).

2154 TB-LA, ...D4. . . . . .REF. SECTION 3, 64/KOEHLER(2), ...

2155 TB2*(MO*O4)3,. .   . . . .B2. . . . . .PHYS. REV. LETT. VOL.26,   519, (1971).
                            *MECHANISM OF THE FERROELECTRIC PHASE TRANSFORMATION IN RARE-EARTH
                            MOLYBDATES.
                            *AXE(J.O.), DORNER(B.), SHIRANE(G.).

2156 TB2*(MO*O4)3,. .   . . . .B2. . . . . .PHYS. REV. B VOL.6,   1950, (1972).
                            *NEUTRON-SCATTERING STUDY OF THE FERROELECTRIC PHASE TRANSFORMATION IN
                            TB2*(MO*O4)3.
                            *DORNER(B.), AXE(J.D.), SHIRANE(G.).

2157 TB2*O3,. . . . . .B7. . . . . .J. PHYS. SOC. JAPAN VOL.17 B-III   63, (1962).
                            *CRYSTAL FIELD SPECTRA IN RARE EARTH OXIDES.
                            *BROCKHOUSE(B.N.), BECKA(L.N.), RAO(K.R.), SINCLAIR(R.N.), WOODS(A.D.B.).

2158 TB*SB, ...D1. . . . . .IAEA SYMP. GRENOBLE,   553, (1972).
                            *DISPERSION RELATIONS FOR MAGNETIC EXCITONS IN TERBIUM ANTIMONIDE.
                            *HOLDEN(T.M.), SVENSSON(E.C.), BUYERS(W.J.L.), VOGT(O.).

2159 TB-Y,. . . . . .D4. . . . . .REF. SECTION 3, 64/KOEHLER(2), ...

2160 TB*ZN, ...D4. . . . . .REF. SECTION 3, 64/KOEHLER(2), ...

2161 TE,. . . . . .B7. . . . . .PHYS. REV. B VOL.4,   356, (1971).
                            *RAMAN SPECTRA AND LATTICE DYNAMICS OF TELLURIUM.
                            *PINE(A.S.), DRESSELHAUS(G.).

2162 TE,. . . . . .C1. . . . . .PHYS. STAT. SOL. 19,   721, (1967).
                            *STREUUNG LANGSAMER NEUTRONEN AN POLYKRISTALLINEN SELEN UND TELLUR.
                            *AXMANN(A.), GISSLER(W.).

2163 TE,. . . . . .B1. . . . . .PROC/10TH INT. CONF/PHYS/SEMIC.,   338, (1970).
                            *ELECTRON AND PHONON DISPERSION RELATIONS IN TELLURIUM.
                            *DRESSELHAUS(G.), DRESSELHAUS(M.S.). (ED. NOTE: PUBL. US ATOMIC ENERGY
                            COMMISION, WASHINGTON, D.C.).

2164 TE,. . . . . .B4,B7. . . . .PROC/INTERNAT. SYMP/PHYS/SE/TE,   299, (1969).
                            *INVESTIGATIONS ON CRYSTALLINE TELLURIUM AND SOLID AMORPHOUS AND LIQUID
                            SELENIUM WITH INELASTIC NEUTRON SCATTERING.
                            *AXMANN(A.), GISSLER(W.), SPRINGER(T.). (ED. NOTE: PUBL. PERGAMON,
                            OXFORD, ENGLAND 1969).

2165 TE,. . . . . .B1,B4. . . . .IAEA SYMP. COPENHAGEN, VOL.1,   245, (1968).
                            *INELASTIC NEUTRON SCATTERING ON SOLID AND LIQUID TELLURIUM AND SELENIUM.
                            *GISSLER(W.), AXMANN(A.), SPRINGER(T.).

SECTION   6 -QUASI-ELASTIC AND INELASTIC SCATTERING STUDIES . . . .

2166 TE,. . . . . .T,B1. . . . .CAN. J. PHYS. VOL.51, ... 843, (1973).
      ~A SIMPLE BOND CHARGE MODEL FOR THE PHONON DISPERSION RELATIONS IN
      TELLURIUM.
      ~COWLEY(E.R.).

2167 TE,. . . . . .B4,B7. . . . .DISC. FARADAY SOC. NO.50, ... 74, (1970).
      ~THE INVESTIGATION OF ATOMIC MOTIONS IN CRYSTALLINE, AMORPHOUS AND LIQUID
      SELENIUM, AND IN CRYSTALLINE AND LIQUID TELLURIUM BY NEUTRON SPECTROS.
      ~AXMANN(A.), GISSLER(W.), KOLLMAR(A.), SPRINGER(T.).

2168 TERPHENYLS,. . . . . .P. . . . . . .PHYS. REV. VOL.126, ... 29, (1962).
      ~SCATTERING OF SLOW NEUTRONS BY SOLID AND LIQUID TERPHENYLS.
      ~BRUGGER(R.M.).

2169 TH (ISO. A=232),. . . . . . .B7. . . . . . .BU. ACAD. SCI. USSR PHYS. SR. 32 600, (1968).
      ~SPECTRA OF NEUTRONS INELASTICALLY SCATTERED FROM TH (ISO. A=232);
      U (ISO. A=235) AND U (ISO. A=238).
      ~SAL-NIKOV(O.A.), FETISOV(N.I.), LOVCHIKOVA(G.N.), KOTEL-NIKOVA(G.V.),
      ANUFRIENKO(V.B.), DEVKIN(B.V.).

2170 TH,. . . . . .B1. . . . . .PHYS. REV. B VOL.8, ... 1332, (1973).
      ~PHONON SPECTRUM OF THORIUM.
      ~REESE(R.A.), SINHA(S.K.), PETERSON(D.T.).

2171 TH*H,. . . . . .C1. . . . . .SOL. STATE COMM. VOL.8, ... 1321, (1970).
      ~INELASTIC AND QUASI-ELASTIC SCATTERING OF SUBTHERMAL NEUTRONS IN TH*H
      AND NB*H.
      ~RUBIN(R.), CLAESSEN(Y.).

2172 TH*H2,. . . . . .C1. . . . . .SOV. PHYS. SOL. STATE 9, ... 1366, (1967).
      ~A LATTICE-DYNAMICAL INVESTIGATION OF THE HYDRIDES CE*H2, CE*H3, TH*H2
      AND U*H(2.7) USING INELASTIC SCATTERING OF COLD NEUTRONS.
      ~KARIMOV(I.), ZEMLYANOV(M.G.), KOST(M.E.), SOMENKOV(V.A.),
      CHERNOPLEKOV(N.A.).

2173 TH-H,TI-H, ...P. . . . . . .CONF. NEUTRON THERMALIZ. VOL.I, . . 407, (1968).
      ~INELASTIC NEUTRON SCATTERING IN METAL HYDRIDES, U*C AND U*O2 AND
      APPLICATIONS OF THE SCATTERING LAW.
      ~PUROHIT(S.N.), PAN(S.S.), BISCHOFF(F.), BRYANT(W.A.), LAJEUNESSE(C.).
      (IN: NEUTRON THERMALIZATION AND REACTOR SPECTRA. PUBL. IAEA: VIENNA).

2174 TI,. . . . . .C1. . . . . .COPENHAGEN CONF.-LATT.DYNAMICS,. . 63, (1963).
      ~MEASURED VIBRATIONAL FREQUENCY DISTRIBUTIONS OF NI, V, TI, AND TI-ZR.
      ~MOZER(B.), OTNES(K.), PALEVSKY(H.).

2175 TI-H (GAMMA),. . . . . .C2. . . . . .IAEA SYMP. CHALK RIVER, VOL.2, . 265, (1962).
      ~STUDIES OF PROTON VIBRATIONS IN GAMMA-TITANIUM HYDRIDE.
      ~SAUNDERSON(D.H.), COCKING(S.J.).

2176 TI*H2,. . . . . .C1. . . . . .J. PHYS. CHEM. SOLIDS VOL.34, . 725, (1973).
      ~ON THE MODEL FOR THE DIFFUSION OF HYDROGEN IN TITANIUM HYDRIDE.
      ~KORN(C.), ZAMIR(D.).

2177 TI*H2,. . . . . .C1. . . . . .NUCL. SCI. ENGNG. VOL.23, . . . 194, (1965).
      ~INELASTIC SCATTERING OF SLOW NEUTRONS BY ZIRCONIUM AND TITANIUM HYDRIDES
      ~PAN(S.S.), WEBB(F.J.).

2178 TI*02, ...B1. . . . . .PHYS. REV. B-3, ... 3457, (1971).
      ~LATTICE DYNAMICS OF RUTILE.
      ~TRAYLOR(J.G.), SMITH(H.G.), NICKLOW(R.M.), WILKINSON(M.K.).

2179 TI2*03,. . . . . .D1. . . . . .J. APP. PHYS. VOL.40, ... 1445, (1969).
      ~ABSENCE OF ANTIFERROMAGNETISM IN TI2*03.
      ~MOON(R.M.), RISTE(T.), KOEHLER(W.C.), ABRAHAMS(S.C.).

2180 TI*02, ...B1. . . . . .DISS. ABS. VOL.32, ...11648, (1971).
      ~THE LATTICE DYNAMICS OF RUTILE.
      ~TRAYLOR(J.G.).

2181 TI*02, ...B1,T. . . . . .CONF. INTERN., RENNES, ... 84, (1971).
      ~LATTICE DYNAMICS OF RUTILE USING SHELL MODEL.
      ~KATIYAR(R.S.).

2182 TI*02-H2*0,. . . . . .E. . . . . . .ADV. COLLOID INTERFACE SCI. 2, . 1, (1968).
      ~APPLICATIONS OF SLOW NEUTRON SCATTERING TO STUDIES IN COLLOID AND
      SURFACE CHEMISTRY (REVIEW).
      ~BOUTIN(H.), PRASK(H.), IYENGAR(R.D.).

2183 TI-(U),. . . . . .B1. . . . . .PHYS. STAT. SOL. 20, ... 767, (1967).
      ~INVESTIGATION OF THE QUASI-LOCAL LEVEL IN THE VIBRATIONAL SPECTRUM OF
      THE TI LATTICE WITH HEAVY IMPURITY ATOMS.
      ~CHERNOPLEKOV(N.A.),   PANOVA(G.KH.),   ZEMLYANOV(M.G.),   SAMOILOV(B.N.),
      KUTAITSEV(V.I.).

2184 TI-ZR, ...B. . . . . . .IAEA SYMP. CHALK RIVER, VOL.2, . 173, (1962).
      ~A STUDY OF THE INELASTIC SCATTERING OF NEUTRONS IN A TI-ZR ALLOY.
      ~.C.C.C.P.

2185 TI-ZR, ...C1. . . . . .COPENHAGEN CONF.-LATT.DYNAMICS,. . 63, (1963).
      ~MEASURED VIBRATION FREQUENCY DISTRIBUTIONS OF NI, V, TI, AND TI-ZR.
      ~MOZER(B.), OTNES(K.), PALEVSKY(H.).

2186 TI-ZR, ...B1. . . . . .SOV. PHYS. SOL. STATE 5, ... 78, (1963).
      ~INVESTIGATION OF THE INELASTIC SCATTERING OF NEUTRONS IN AN ALLOY OF
      TI AND ZR.
      ~CHERNOPLEKOV(N.A.),   ZEMLYANOV(M.G.),   BROVMAN(E.G.),   CHICHERIN(A.G.).

2187 TL,. . . . . .B1. . . . . .PHYS. REV. B-3, ... 4115, (1971).
      ~LATTICE VIBRATIONS OF THALLIUM AT 77 AND 296 K.
      ~WORLTON(T.G.), SCHMUNK(R.E.).

2188 TL,. . . . . .B1. . . . . .IAEA SYMP. COPENHAGEN, VOL.1, . 253, (1968).
      ~ATOMIC VIBRATIONS IN FACE-CENTERED-CUBIC ALLOYS OF BI, PB AND TL.
      ~NG(S.C.), BROCKHOUSE(B.N.).

2189 TL,. . . . . .P. . . . . . .REF. SECTION 3, 47/WU, ...
2190 TL-BI-PB,. . . . . .B1. . . . . .SOL. STATE COMM. 5, ... 79, (1967).
      ~LATTICE DYNAMICS OF THE ALLOY SYSTEMS BI-PB-TL.
      ~NG(S.C.), BROCKHOUSE(B.N.).

2191 TL*BR, ...B1. . . . . .PROC. ROY. SOC. 300, ... 45, (1967).
      ~THE LATTICE DYNAMICS OF THALLOUS BROMIDE.
      ~COWLEY(E.R.), OKAZAKI(A.).

SECTION 6 -QUASI-ELASTIC AND INELASTIC SCATTERING STUDIES . . . .

2192 TL-PB, ...B1..PB....PHYS. REV. 178, ... 897, (1969).
*PHONON SPECTRA IN PB AND PB(40)-TL(60) DETERMINED BY TUNNELING AND
NEUTRON SCATTERING.
*ROWELL(J.M.), MCMILLAN(W.L.), FELDMANN(W.L.).

2193 TM, ... P . . . . PHYS. STAT. SOL. 2, ... K164, (1962).
*MESSUNG DER KOHARENTEN STREUAMPLITUDEN VON DYSPROSIUM UND THULIUM FUR
THERMISCHE NEUTRONEN.
*BETZL(M.), HASE(W.), KLEINSTUCK(K.), TOBISCH(J.).

2194 TM-LA, ...D4.......REF. SECTION 3, 64/KOEHLER(2), ...

2195 TM*SB, ...D3.......PHYS. REV. B VOL.4, ... 718, (1971).
*NEUTRON-SCATTERING STUDY OF TM*SB: A MODEL CRYSTAL-FIELD-ONLY METALLIC
PARAMAGNET.
*BIRGENEAU(R.J.), BUCHER(E.), PASSELL(L.), TURBERFIELD(K.C.).

2196 (TM-Y)*AL2, ...B7.......SOL. STATE COMM. VOL.12 NO.2, . . 117, (1973).
*NEUTRON CRYSTAL-FIELD SPECTROSCOPY AND SUSCEPTIBILITY IN (TM-Y)*AL2.
*PURWINS(H.G.), WALKER(E.), DONZE(P.), TREYVAUD(A.), FURRER(A.),
BUHRER(W.), HEER(H.).

2197 (TM-Y)-AL2, ...D3.......INT. J. MAGN. VOL.4 NO.1, . . 63, (1973).
*NEUTRON CRYSTAL FIELD SPECTROSCOPY IN TM(.25)-Y(.75)-AL2.
*FURRER(A.), BUHRER(W.), HEER(H.), PURWINS(H.G.), WALKER(E.).

2198 TOLUENE, ...C1.......J. CHEM. PHYSICS VOL.37, ... 234, (1962).
*ROTATIONAL FREEDOM OF AMMONIUM IONS AND METHYL GROUPS BY CROSS-SECTION
MEASUREMENTS WITH SLOW NEUTRONS.
*RUSH(J.J.), TAYLOR(T.I.), HAVENS-JR(W.W.).

2199 TOLUENE, ...C3.......FARADAY SYMP. CHEM. SOC. NO.6, . . 135, (1972).
*HIGH-FREQUENCY DYNAMICS OF LIQUID METHANOL AND TOLUENE, CONTRIBUTION OF
MOLECULAR ROTATIONAL DIFFUSION TO INELASTIC NEUTRON SCATTERING.
*ALDRED(B.K.), STIRLING(G.C.), WHITE(J.W.).

2200 TOLUENE, ...C3.......J. CHEM. PHYS. VOL.60, ... 2832, (1974).
*STUDY OF MOLECULAR ROTATIONS IN SOME AROMATIC COMPOUNDS BY COLD-NEUTRON
SCATTERING.
*TREPADUS(V.), RAPEANU(S.), PADUREANU(I.), PARFENOV(V.A.), NOVIKOV(A.G.).

2201 TRANSITION METALS, ...E.......BROOKHAVEN SYMPOSIUM, ... 8, (1965).
*REVIEW OF EXPERIMENTAL MEASUREMENTS OF LATTICE MODES IN TRANSITION
METALS.
*WOODS(A.D.B.).

2202 TRANSITION METALS, ...B1.......REV. ROUMAINE PHYS. VOL.14, . . . 247, (1969).
*PHONON DISPERSION IN TRANSITION METALS.
*SHARMA(P.K.), PAL(S.), GUPTA(R.P.). (ED. NOTE: IN ENGLISH).

2203 U, ...C5.......IAEA SYMP. CHALK RIVER, VOL.2, . 183, (1962).
*SCATTERING LAW FOR METAL ATOMS IN CHEMICAL COMPOUNDS.
*EGELSTAFF(P.A.), HOLT(G.).

2204 U (ISO. A=235, 238), ...B7.......BU. ACAD. SCI. USSR PHYS. SR. 32 600, (1968).
*SPECTRA OF NEUTRONS INELASTICALLY SCATTERED FROM TH (ISO. A=232),
U (ISO. A=235) AND U (ISO. A=238).
*SAL-NIKOV(O.A.), FETISOV(N.I.), LOVCHIKOVA(G.N.), KOTEL-NIKOVA(G.V.),
ANUFRIENKO(V.B.), DEVKIN(B.V.).

2205 U (ISO. A=235), ...B7.......PROC/CONF/FAST CRIT. EXP./ANALY. 524, (1967).
*TIME-OF-FLIGHT MEASUREMENTS OF NEUTRON SPECTRA IN U (ISO. A=235) AND
TUNGSTEN.
*PROFIO(A.E.), YOUNG(J.C.), KROSBIE(K.L.), HACKNEY(R.), ANTUNEZ(H.M.),
RUSSELL-JR(J.L.). (ED. NOTE: PUBL. CFSTI, NBS, SPRINGFIELD, VA.).

2206 U (ISO. A=238), ...P.......CONF/NUCL. STRUC. STUDY/NEUTRONS 172, (1972).
*INELASTIC NEUTRON SCATTERING FROM U (ISO. A=238).
*ARMITAGE(B.H.), ROSE(J.L.). (ED. NOTE: AVAIL. CENTRAL RES. INST. PHYS.,
BUDAPEST, HUNGARY).

2207 U (ISO. A=238), ...B7.......2ND INTERNAT. CONF./NUCL. DATA, . . , (1970).
*ANALYSIS OF NEUTRON INELASTIC SCATTERING BY U(ISOTOPE A=238).
*IGARASI(S.), NAKAMURA(H.), MURATA(T.), NISHIMURA(K.).

2208 U*C, ...P.......NUCL. SCI. ENG. VOL.47, ... 349, (1972).
*LOW ENERGY NEUTRON INTERACTIONS WITH URANIUM CARBIDE.
*LAJEUNESSE(C.), MOORE(W.E.), YEATER(M.L.).

2209 U*C, ...T.......PHYS. REV. B VOL.5, ... 1260, (1972).
*PHONON SPECTRA OF SOME TRANSITION METAL CARBIDES FROM A SIMPLE
PSEUDOPOTENTIAL APPROACH.
*MOSTOLLER(M.).

2210 U*C,U*O2, ...P.......CONF. NEUTRON THERMALIZ. VOL.1, . 407, (1968).
*INELASTIC NEUTRON SCATTERING IN METAL HYDRIDES, U*C AND U*O2 AND
APPLICATIONS OF THE SCATTERING LAW.
*PUROHIT(S.N.), PAN(S.S.), BISCHOFF(F.), BRYANT(W.A.), LAJEUNESSE(C.).
(IN: NEUTRON THERMALIZATION AND REACTOR SPECTRA. PUBL. IAEA: VIENNA).

2211 U-H, ...C1.......SOV. PHYS. SOL. STATE 9, . . . 1366, (1967).
*A LATTICE DYNAMICAL INVESTIGATION OF THE HYDRIDES CE*H2, CE*H3, TH*H2
AND U*H(2.7) USING INELASTIC SCATTERING OF COLD NEUTRONS.
*KARIMOV(I.), ZEMLYANOV(M.G.), KOST(M.E.), SOMENKOV(V.A.),
CHERNOPLEKOV(N.A.).

2212 U*H3, ...C1.......J. CHEM. PHYS. VOL.45, ... 3817, (1966).
*VIBRATION SPECTRA OF YTTRIUM AND URANIUM HYDRIDES BY THE INELASTIC
SCATTERING OF COLD NEUTRONS.
*RUSH(J.J.), FLOTOW(H.E.), CONNOR(D.W.), THAPER(C.L.).

2213 U*O2, ...C5.......IAEA SYMP. CHALK RIVER, VOL.2, . 213, (1962).
*SCATTERING LAW FOR U*O2.
*THORSON(I.M.), HAYWOOD(B.C.).

2214 U*O2, ...C5.......IAEA SYMP. CHALK RIVER, VOL.2, . 183, (1962).
*SCATTERING LAW FOR METAL ATOMS IN CHEMICAL COMPOUNDS.
*EGELSTAFF(P.A.), HOLT(G.).

2215 U*O2, ...B1.......IAEA SYMP. BOMBAY, VOL.1, . . . 373, (1964).
*THE CRYSTAL DYNAMICS OF URANIUM DIOXIDE.
*WOODS(A.D.B.), DOLLING(G.), COWLEY(R.A.).

2216 U*O2, ...D1.......PHYS. STAT. SOL. VOL.36, ... 737, (1969).
*THE SPIN-WAVE DISPERSION RELATIONS AND THE SPIN-WAVE CONTRIBUTION TO THE
SPECIFIC HEAT OF ANTIFERROMAGNETIC U*O2.
*CRACKNELL(A.P.), JOSHUA(S.J.).

SECTION   6 -QUASI-ELASTIC AND INELASTIC SCATTERING STUDIES . . . .

2217 U*02,. . . . .,. . . . . .B7. . . . . .NUKLEONIK VOL.12,. . . . .   205, (1969).
      *NEUTRON SCATTERING FROM URANIUM DIOXIDE.
      *YOUNG(J.A.). (ED. NOTE: IN ENGLISH).

2218 U*02,. . . . .,. . . . . .T. . . . . . . .J. PHYS. CHEM. SOL. 27,. . . . . 1833, (1966).
      *A CRYSTAL FIELD CALCULATION IN URANIUM DIOXIDE.
      *RAHMAN(H.U.), RUNCIMAN(W.A.).

2219 U*02,. . . . . . .,. . . . . .D1. . . . . . .J. APP. PHYS. 39,. . . . . 1111, (1968).
      *TEMPERATURE DEPENDENCE OF THE MAGNETIC EXCITATIONS OF URANIUM DIOXIDE.
      *DOLLING(G.), COWLEY(R.A.).

2220 U*02,. . . . . . .,. . . . . .D1. . . . . . .PHYS. REV. LETT. 16,. . . . . 683, (1966).
      *OBSERVATION OF MAGNON-PHONON INTERACTION AT SHORT WAVELENGTHS.
      *DOLLING(G.), COWLEY(R.A.).

2221 U*02,. . . . . . .,. . . . . .B1. . . . . . .CAN. JOURN. PHYS. 43,. . . . . . . 1397, (1965).
      *THE CRYSTAL DYNAMICS OF URANIUM DIOXIDE.
      *DOLLING(G.), COWLEY(R.A.), WOODS(A.D.B.).

2222 U*02,. . . . . . .,. . . . . .D1. . . . . . .PHYS. REV. 167,. . . . .   464, (1968).
      *MAGNETIC EXCITATION IN URANIUM DIOXIDE.
      *COWLEY(R.A.), DOLLING(G.).

2223 U*02*(N*03)2.6H2*0,. . . . . . . .C1. . . . . .INORG. CHEM. VOL.6,. . . . . 346, (1967).
      *NEUTRON-SCATTERING STUDY OF THE MOTIONS OF WATER MOLECULES IN HYDRATED
      SALTS OF TRANSITION METALS.
      *RUSH(J.J.), FERRARO(J.R.), WALKER(A.).

2224 UREA (C*0*(N*H2)2),. . . . . . . .C1. . . . . .CHEM. COMMUN. 1967,. . . . . 74, (1967).
      *LOW-FREQUENCY MOLECULAR MOTIONS OF UREA BY NEUTRON-SCATTERING SPEC-
      TROSCOPY.
      *BODGER(E.O.), WHITE(J.W.).

2225 V,. . . . .,. . . . . .C2. . . . . .IAEA SYMP. CHALK RIVER, VOL.2, . 155, (1962).
      *OBSERVATION OF A LOW ENERGY PEAK IN THE PHONON FREQUENCY DISTRIBUTION OF
      VANADIUM.
      *PELAH(I.), HAAS(L.), KLEY(W.), KREBS(K.H.), PELETTI(J.), RUBIN(R.).

2226 V,. . . . .,. . . . . .C2. . . . . .IAEA SYMP. CHALK RIVER, VOL.2, . 145, (1962).
      *THE PHONON FREQUENCY DISTRIBUTION OF VANADIUM.
      *HAAS(R.), KLEY(W.), KREBS(K.H.), RUBIN(R.).

2227 V,. . . . .,. . . . . .C2. . . . . .IAEA SYMP. CHALK RIVER, VOL.2, . 125, (1962).
      *A STUDY OF THE PHONON SPECTRUM AND DISPERSION CURVES FOR VANADIUM.
      *.C.C.C.P.

2228 V,. . . . .,. . . . . .C2. . . . . .IAEA SYMP. BOMBAY, VOL.1,. . . . . 99, (1964).
      *PHONON FREQUENCY DISTRIBUTION OF VANADIUM.
      *GLASER(W.), CARVALHO(F.), EHRET(G.).

2229 V,. . . . .,. . . . . .C1. . . . . .COPENHAGEN CONF.-LATT.DYNAMICS,. . 63, (1963).
      *MEASURED VIBRATIONAL FREQUENCY DISTRIBUTIONS OF NI, V, TI, AND TI-ZR.
      *MOZER(B.), OTNES(K.), PALEVSKY(H.).

2230 V,. . . . .,. . . . . . . .T,C2. . . . . .PHYS. REV. 134,. . . . . . . .A1486, (1964).
      *FREQUENCY SPECTRA OF BODY-CENTERED CUBIC LATTICES.
      *CLARK(B.C.), GAZIS(D.C.), WALLIS(R.F.).

2231 V,. . . . .,. . . . . . . .T,. . . . . . .PHYS. REV. VOL.137,. . . . . . . .A1113, (1965).
      *MULTIPLE SCATTERING OF NEUTRONS IN VANADIUM AND COPPER.
      *BLECH(I.A.), AVERBACH(B.L.).

2232 V,. . . . .,. . . . . . . .D. . . . . . .ACTA PHYS. ACAD. SCI. HUNG. 30,. 231, (1971).
      *PHONON FREQUENCY DISTRIBUTION FUNCTIONS OF COPPER, NICKEL AND VANADIUM.
      *YAI-PRAKASH, SEMUEAL(B.S.), SHARMA(P.K.).

2233 V,. . . . .,. . . . . . . .B3. . . . . .REPORT EUR-4216E (41PP.),. . . . . . . (1969).
      *INELASTIC SCATTERING OF NEUTRONS BY LOCALIZED VIBRATIONS OF INTERSTITIAL
      HYDROGEN IN METAL LATTICES, CASE OF A VANADIUM LATTICE.
      *BLAESSER(G.), PERETTI(J.), TOTH(G.). (ED. NOTE: AVAIL. CFSTI,
      SPRINGFIELD, VA. 22151, USA).

2234 V,. . . . .,. . . . . . . .D4. . . . . .PHYS. REV. LETT. 10,. . . . . 295, (1963).
      *ELECTRONIC AND NUCLEAR POLARIZATION IN VANADIUM BY SLOW NEUTRON
      SCATTERING.
      *SHULL(C.G.), FERRIER(R.P.).

2235 V,. . . . .,. . . . . . . .C2. . . . . .PROC. PHYS. SOC. 91,. . . . . . .   76, (1967).
      *THE PHONON FREQUENCY DISTRIBUTION OF VANADIUM.
      *PAGE(D.I.).

2236 V,. . . . .,. . . . . . . .C1. . . . . .PHYSICA VOL.34,. . . . . 384, (1967).
      *LATTICE VIBRATIONS OF VANADIUM BY INELASTIC SCATTERING OF SLOW NEUTRONS.
      *ROY(A.P.), THAPER(C.L.), IYENGAR(P.K.).

2237 V,. . . . .,. . . . . . . .C1. . . . . .PHYS. REV. VOL.127,. . . . . . 1017, (1962).
      *PHONON FREQUENCY DISTRIBUTION IN VANADIUM AT SEVERAL TEMPERATURES.
      *TURBERFIELD(K.C.), EGELSTAFF(P.A.).

2238 V,. . . . .,. . . . . . . .C1. . . . . .NATURE VOL.168,. . . . . 290, (1951).
      *INELASTIC SCATTERING OF COLD NEUTRONS.
      *EGELSTAFF(P.A.).

2239 V,. . . . .,. . . . . . . .B1. . . . . .SOV. PHYS. J.E.T.P. VOL.16,. . . 1472, (1963).
      *INVESTIGATION OF THE PHONON SPECTRUM OF VANADIUM.
      *CHERNOPLEKOV(N.A.), ZEMLYANOV(M.G.), CHICHERIN(A.G.).

2240 V,. . . . .,. . . . . . . .C1. . . . . .PHYS. REV. VOL.109,. . . . . . 1046, (1958).
      *MEASUREMENT OF LATTICE VIBRATIONS IN VANADIUM BY NEUTRON SCATTERING.
      *EISENHAUER(C.M.), PELAH(I.), HUGHES(D.J.), PALEVSKY(H.).

2241 V,. . . . .,. . . . . . .C1,P. . . . .DISS. ABSTR. B VOL.34,. . . . . . 354, (1973).
      *INTEGRAL PHONON EFFECTS IN THE THERMAL NEUTRON SCATTERING FROM VANADIUM
      LATTICE.
      *KAMAL(M.).

2242 V,. . . . .,. . . . . . .T,C1. . . . .PHYS. REV. VOL.116,. . . . . . . 297, (1959).
      *VIBRATIONAL SPECTRUM OF VANADIUM.
      *SINGH(D.N.), BOWERS(W.A.).

2243 V,. . . . .,. . . . . . . .C1. . . . . .REV. MOD. PHYS. VOL.30,. . . . . . 250, (1958).
      *VIBRATION SPECTRA OF VANADIUM AND A MN-CO ALLOY BY NEUTRON SPECTROMETRY.
      *STEWART(A.T.), BROCKHOUSE(B.N.).

SECTION   6 -QUASI-ELASTIC AND INELASTIC SCATTERING STUDIES . . . .

2244 V. . . . . C2 . . . . IAEA SYMPOSIUM VIENNA at SEVERAL TEMPERATURES.   581, (1960).
THE PHONON FREQUENCY DISTRIBUTION IN VANADIUM AT SEVERAL TEMPERATURES.
TURBERFIELD(K.C.), EGELSTAFF(P.A.).

2245 V, ... C1 . . . . CAN. J. PHYS. VOL.33 . . . . 889, (1955).
NEUTRON SCATTERING AND THE FREQUENCY DISTRIBUTION OF THE NORMAL MODES OF
VANADIUM METAL.
BROCKHOUSE(B.N.).

2246 V, ... C1 . . . . PHYS. REV. VOL.104 . . . . 271, (1956).
INELASTIC SCATTERING OF LOW-ENERGY NEUTRONS BY LATTICE VIBRATIONS OF
VANADIUM.
CARTER(R.S.), HUGHES(D.J.), PALEVSKY(H.).

2247 V-BE, ... C1 . . . . PHYS. REV. VOL.152 . . . . 535, (1966).
EXPERIMENTAL STUDY OF THE LATTICE DYNAMICS OF DISORDERED VANADIUM ALLOYS
MOZER(B.), OTNES(K.), THAPER(C.L.).

2248 V-(BE,TA,W,PT,CR,NI), . . . B3 . . . . IAEA SYMP. COPENHAGEN, VOL.1, . . . 55, (1968).
LOCALIZED MODES, RESONANT MODES AND IMPURITY VIBRATIONAL BANDS IN
VANADIUM ALLOYS.
MOZER(B.).

2249 V-D, ... C1 . . . . SOV. PHYS. SOL. STATE VOL.11, . . 2343, (1969).
STUDY OF THE DYNAMICS OF GROUP V TRANSITION-METAL HYDRIDES BY MEANS OF
INELASTIC NEUTRON SCATTERING.
CHERNOPLEKOV(N.A.), ZEMLYANOV(M.G.), SOMENKOV(V.A.), CHERTKOV(A.A.).

2250 V*D5, ... B7 . . . . SOL. STATE COMM. VOL.11 . . . . . . 1299 (1972).
A NEUTRON SCATTERING STUDY OF THE VIBRATIONAL AND DIFFUSIONAL MOTIONS
OF DEUTERIUM IN THE ALPHA AND BETA PHASES OF V*D5.
ROWE(J.M.).

2251 V(H), ... B1 . . . . IAEA SYMP. COPENHAGEN, VOL.1, . . 223, (1968).
THE DYNAMICS OF HYDROGEN IMPURITIES IN NIOBIUM AND VANADIUM.
VERDAN(G.), RUBIN(R.), KLEY(W.).

2252 V-H, ... T, C3 . . . . BROOKHAVEN SYMPOSIUM . . . . 105, (1965).
HYDROGEN MOTION IN PRIMARY SOLUTIONS OF HYDROGEN IN SOME TRANSITION
ELEMENTS.
KLEY(W.), PERETTI(J.), RUBIN(R.), VERDAN(G.).

2253 V*H2, ... C1 . . . . J. PHYSIQUE TOME 28 COL.1, ... 26, (1967).
ETUDES DES MODES DE VIBRATIONS LOCALISEES DE L#HYDROGENE DANS L#HYDRURE
DE VANADIUM PAR DIFFUSION INELASTIQUE DES NEUTRONS FROIDS.
KLEY(W.), PERETTI(J.), RUBIN(R.), VERDAN(G.).

2254 V*H2, ... C1 . . . . . . PHYS. LETT. 14, ... 100, (1965).
INELASTIC SCATTERING OF COLD NEUTRONS BY LOCALISED MODES IN VANADIUM
HYDRIDE SYSTEMS.
RUBIN(R.), PERETTI(J.), VERDAN(G.), KLEY(W.).

2255 V*H2, ... C1 . . . . . . SOV. PHYS. SOL. STATE VOL.11, . . 2343, (1969).
STUDY OF THE DYNAMICS OF GROUP V TRANSITION-METAL HYDRIDES BY MEANS OF
INELASTIC NEUTRON SCATTERING.
CHERNOPLEKOV(N.A.), ZEMLYANOV(M.G.), SOMENKOV(V.A.), CHERTKOV(A.A.).

2256 V*H2, ... C1 . . . . . . J. PHYS. CHEM. SOLIDS VOL.32, . . 41, (1971).
QUASIELASTIC NEUTRON SCATTERING BY HYDROGEN IN THE ALPHA AND BETA PHASES
OF VANADIUM HYDRIDE.
ROWE(J.M.), SKOLD(K.), FLOTOW(H.E.), RUSH(J.J.).

2257 V*H2, ... C1 . . . . . . J. CHEM. PHYS. VOL.56, ... 4574, (1972).
QUASIELASTIC THERMAL NEUTRON SCATTERING BY HYDROGEN IN ALPHA-VANADIUM
HYDRIDE.
DE-GRAAF(L.A.), RUSH(J.J.), FLOTOW(H.E.), ROWE(J.M.).

2258 V-H, ... C1 . . . . . . J. CHEM. PHYS. VOL.48, ... 3795, (1968).
VIBRATION SPECTRA OF VANADIUM HYDRIDE IN THREE CRYSTAL PHASES BY
INELASTIC NEUTRON SCATTERING.
RUSH(J.J.), FLOTOW(H.E.).

2259 V*H2, ... C1 . . . . . . JULICH CONF/H IN METALS VOL.I, . 301, (1972).
STUDY OF HYDROGEN DIFFUSION IN VANADIUM AND TANTALUM HYDRIDE BY QUASI-
ELASTIC NEUTRON SCATTERING.
DE-GRAAF(L.A.), RUSH(J.J.), LIVINGSTON(R.C.), FLOTOW(H.E.), ROWE(J.M.).

2260 V-NI, ... C1 . . . . . . PHYS. REV. VOL.152 . . . . 535, (1966).
EXPERIMENTAL STUDY OF THE LATTICE DYNAMICS OF DISORDERED VANADIUM ALLOYS
MOZER(B.), OTNES(K.), THAPER(C.L.).

2261 V2*O3, ... D4 . . . . . . Z. PHYS. VOL.238 . . . . 208, (1970).
MEASUREMENT OF THE INTERNAL MAGNETIC FIELD IN V2*O3 USING THE INELASTIC
SPIN-FLIP-SCATTERING OF NEUTRONS.
HEIDEMANN(A.).

2262 V2*O3, ... B7 . . . . . . IAEA SYMP. GRENOBLE, ... 851, (1972).
HYPERFINE SPLITTING MEASUREMENTS IN V2*O3 BY INELASTIC NEUTRON
SCATTERING.
HEIDEMANN(A.), ALEFELD(B.).

2263 V3*O5, ... D4 . . . . . . PHYS. STAT. SOLIDI A VOL.16 NO.2 K129, (1973).
HYPERFINE SPLITTING IN V3*O5 MEASURED BY INELASTIC NEUTRON SCATTERING.
HEIDEMANN(A.).

2264 V-PT, ... C1 . . . . . . PHYS. REV. VOL.152, ... 535, (1966).
EXPERIMENTAL STUDY OF THE LATTICE DYNAMICS OF DISORDERED VANADIUM ALLOYS
MOZER(B.), OTNES(K.), THAPER(C.L.).

2265 V3*SI, ... B1 . . . . . . SOL. STATE COMM. VOL. 9, ... 397, (1971).
NEUTRON SCATTERING STUDY OF THE LATTICE DYNAMICAL PHASE TRANSITION
IN V3*SI.
SHIRANE(G.), AXE(J.D.), BIRGENEAU(R.J.).

2266 V3*SI, ... T . . . . . . PHYS. LETT. VOL.35A, ... 48, (1971).
THEORETICAL INTERPRETATION OF NEUTRON SCATTERING EXPERIMENTS IN V3*SI.
DIETERICH(W.), SCHUSTER(H.).

2267 V3*SI, ... D3 . . . . . . J. APPL. PHYS. VOL.39, ... 3501, (1968).
ABSENCE OF PARAMAGNETIC NEUTRON SCATTERING IN V3*SI.
VANCE(E.R.), FINLAYSON(T.R.).

2268 V-(TA,W), ... C2 . . . . . . IAEA SYMP. COPENHAGEN, VOL.1, . . 79, (1968).
INVESTIGATION OF ADMIXED STATES IN VANADIUM ALLOYS.
.C.C.C.P.

SECTION   6 -QUASI-ELASTIC AND INELASTIC SCATTERING STUDIES . . . .

2269 W, ...T,B1. . . . . .PHYS. REV. B-1, ... 509, (1970).
     ↪KOHN ANOMALIES IN TUNGSTEN AND OTHER CR-GROUP METALS.
     ↪RICE(T.M.), HALPERIN(B.I.).

2270 W, ...B7. . . . . .PROC/CONF/FAST CRIT. EXP./ANALY. 524, (1967).
     ↪TIME-OF-FLIGHT MEASUREMENTS OF NEUTRON SPECTRA IN U (ISO. A=235) AND
     TUNGSTEN.
     ↪PROFIO(A.E.), YOUNG(J.C.), KROSBIE(K.L.), HACKNEY(R.), ANTUNEZ(H.M.),
     RUSSELL-JR(J.L.). (ED. NOTE! PUBL. CFSTI, NBS, SPRINGFIELD, VA.).

2271 W, ...91. . . . . .THESIS (208 PP.) . . . . . . , (1964).
     ↪NEUTRON SCATTERING STUDIES OF LATTICE VIBRATIONS IN METALS.
     ↪CHEN(S.H.). (ED. NOTE! PH.D. DISSERTATION, MCMASTER UNIVERSITY).

2272 W, ...T,B1. . . . . .J. PHYS. SOC. JAP. VOL.33, . . . 1207, (1972).
     ↪LATTICE DYNAMICS OF SOME BCC TRANSITION METALS.
     ↪BEHARI(J.), TRIPATHI(B.B.).

2273 W, ...T. . . . . .PHYS. LETT. 19, ... 105, (1965).
     ↪DISPERSION OF LATTICE WAVES IN TUNGSTEN.
     ↪PAL(S.), SHARMA(P.K.).

2274 W, ...T. . . . . .PHYS. REV. 157, ... 540, (1967).
     ↪EXISTENCE OF AN INFINITY IN THE FREQUENCY DISTRIBUTION G(NU) OF
     MONOATOMIC BODY-CENTERED CUBIC CRYSTALS.
     ↪GILAT(G.).

2275 W, ...P. . . . . .REF. SECTION 3, 47/HAVENS, ...

2276 W. . . . . .B1. . . . . .SOL. STATE COMM. 2, ...   73, (1964).
     ↪LATTICE VIBRATIONS OF TUNGSTEN.
     ↪CHEN(S.H.), BROCKHOUSE(B.N.).

2277 W. . . . . .T,B1. . . . . .PHYS. REV. 143, ... 443, (1966)!
     ↪LATTICE DYNAMICS AND SPECIFIC HEATS OF SOME TRANSITION METALS ON KREBS≠S
     MODEL.
     ↪MAHESH(P.S.), DAYAL(B.).

2278 (W)-CR, ...B1,B3. . . . . .PHYS. REV. B-2, ... 4864, (1970).
     ↪INVESTIGATION OF IN-BAND RESONANT MODES IN CR-W ALLOYS BY INELASTIC
     NEUTRON SCATTERING.
     ↪CUNNINGHAM(R.M.), MUHLESTEIN(L.D.), SHAW(W.M.), TOMPSON(C.W.).

2279 XE, ...91. . . . . .PHYS. LETT. VOL.46A NO.5, . . . 357, (1974).
     ↪ON THE IMPORTANCE OF MANY-BODY FORCES IN SOLID XENON.
     ↪LURIE(N.A.), SKALYO-JR(J.).

2280 XE, ...B1. . . . . .J. PHYS. C VOL.6, ...L313, (1973).
     ↪PHONON FREQUENCIES IN XENON BY INELASTIC NEUTRON SCATTERING.
     ↪PALMER(B.J.), SAUNDERSON(D.H.), BATCHELDER(D.N.).

2281 XE, ...B1. . . . . .PHYS. REV. B VOL.9 NO.12, ... , (1974).
     ↪PHONON DISPERSION RELATIONS IN XENON AT 15 K.
     ↪LURIE(N.A.), SHIRANE(G.), SKALYO-JR(J.). (REF. OBTAINED FROM PHYS. REV.
     ABSTRACTS).

2282 XE, ...B1. . . . . .PHYS. REV. B VOL.9, ... 2661, (1974).
     ↪TEMPERATURE DEPENDENCE OF THE ZERO-SOUND ELASTIC CONSTANTS OF CRYSTAL-
     LINE XENON.
     ↪LURIE(N.A.), SHIRANE(G.), SKALYO-JR(J.).

2283 XYLENE, ...C1. . . . . .J. CHEM. PHYSICS VOL.37, ... 234, (1962).
     ↪ROTATIONAL FREEDOM OF AMMONIUM IONS AND METHYL GROUPS BY CROSS-SECTION
     MEASUREMENTS WITH SLOW NEUTRONS.
     ↪RUSH(J.J.), TAYLOR(T.I.), HAVENS-JR(W.W.).

2284 Y, ...B1. . . . . .PHYS. REV. B-1, ... 2430, (1970).
     ↪LATTICE DYNAMICS OF YTTRIUM AT 295K.
     ↪SINHA(S.K.), BRUN(T.O.), MUHLESTEIN(L.D.), SAKURAI(J.).

2285 Y (ISOTOPE A=89), ...B7. . . . . .NUCL. PHYS. VOL.A131, ... 561, (1969).
     ↪THE INELASTIC SCATTERING OF NEUTRONS FROM Y(ISOTOPE A=89).
     ↪TOWLE(J.H.).

2286 Y-CO, ...D4. . . . . .PHYS. REV. VOL.186, ... 479, (1969).
     ↪POLARIZED-NEUTRON-DIFFRACTION STUDY OF MAGNETIC MOMENTS IN YTTRIUM-
     COBALT ALLOYS.
     ↪KREN(E.), SCHWEIZER(J.), TASSET(F.).

2287 Y-(DY,GD,TB), ...D4. . . . . .REF. SECTION 3, 64/KOEHLER(2), ...

2288 Y-FE/GARNET, ...D1. . . . . .J. APP. PHYS. 37, . . . 1050, (1966).
     ↪SCATTERING OF POLARIZED NEUTRONS BY SPIN WAVES IN YIG.
     ↪FERGUSON(G.A.), SAENZ(A.W.), PODGOR(S.).

2289 Y-FE/GARNET, ...D1. . . . . .PHYS. REV. 156, ... 632, (1967).
     ↪SCATTERING OF POLARIZED NEUTRONS BY SPIN WAVES IN MAGNETITE AND YTTRIUM
     IRON GARNET.
     ↪FERGUSON(G.A.), SAENZ(A.W.).

2290 Y*FE-GARNET/RODS, ...T,D4. . . . . .J. APPL. PHYS. 40, ... 2359, (1969).
     ↪MAGNON LONGITUDINAL PHONON INTERACTION IN OBLIQUELY MAGNETIZED YIGS RODS
     ↪DE-SANTIS(P.).

2291 Y3*FE5*O12, ...D1. . . . . .SOV. PHYS. SOL. STATE VOL.10, . . 511, (1968).
     ↪SCATTERING OF NEUTRONS BY PARAMETRICALLY EXCITED SPIN WAVES.
     ↪GUREVICH(A.G.), DRABKIN(G.M.), LAZEBNIK(I.M.), MAL≠TSEV(E.I.),
     MARCHIK(I.I.), STARSBINETS(S.S.).

2292 Y*H2, ...C1. . . . . .J. CHEM. PHYS. VOL.45, ... 3817, (1966).
     ↪VIBRATION SPECTRA OF YTTRIUM AND URANIUM HYDRIDES BY THE INELASTIC
     SCATTERING OF COLD NEUTRONS.
     ↪RUSH(J.J.), FLOTOW(H.E.), CONNOR(D.W.), THAPER(C.L.).

2293 Y*MN*O3, ...T. . . . . .SOV. PHYS. SOL. STATE 8, ... 215, (1966).
     ↪MAGNETIC PROPERTIES OF Y*MN*O3.
     ↪KIZHAEV(S.A.), BOKOV(V.A.), KOCHALOV(O.V.).

2294 Y-TB, ...D1. . . . . .AIP CONF. PROC. VOL.5, ... 1450, (1971).
     ↪SPIN WAVE DISPERSION RELATION FOR Y(.1TB).
     ↪WAKABAYASHI(N.), NICKLOW(R.M.).

2295 Y*ZN, ...B1. . . . . .PHYS. REV. B VOL.6, ... 4438, (1972).
     ↪LATTICE DYNAMICS OF Y*ZN.
     ↪PREVENDER(T.S.), SINHA(S.K.), SMITH(J.F.).

SECTION   6 -QUASI-ELASTIC AND INELASTIC SCATTERING STUDIES . . . .

2296 YB*FE-GARNET,. . . . . .D4. . . . . . .PHYS. REV. VOL.128, ...   67, (1962).
        *EXCHANGE FIELD SPLITTING IN YTTERBIUM IRON GARNET.
        *WATANABE(H.), BROCKHOUSE(B.N.).

2297 YB*H2, ...C1. . . . . . .J. CHEM. PHYS. VOL.52, ... 3952, (1970).
        *VIBRATION SPECTRA OF THE ORTHORHOMBIC ALKALINE-EARTH HYDRIDES BY THE
        INELASTIC SCATTERING OF COLD NEUTRONS AND BY INFRARED TRANSM. MEASUR.
        *MAELAND(A.J.).

2298 ZN,. . . . . .B1. . . . . . .J. OF PHYS.-C-, SER. 2, VOL.2, . .2366, (1969).
        *MEASUREMENTS OF THE FREQUENCIES OF THE NORMAL MODES OF ZINC.
        *MILLINGTON(A.J.), SQUIRES(G.L.).

2299 ZN,. . . . . .B1. . . . . . .J. OF PHYS.-C-, SER. 2, VOL.2, . 1857, (1969).
        *PHONON DISPERSION CURVES FOR ZINC.
        *MCDONALD(D.L.), ELCOMBE(M.M.), PRYOR(A.W.).

2300 ZN,. . . . . .T. . . . . . .IAEA SYMP. BOMBAY, VOL.1,. . . . 205, (1964).
        *NOTE ON DISPERSION CURVE CALCULATIONS IN ZINC.
        *HOLAS(A.).

2301 ZN,. . . . . .B1. . . . . .IAEA SYMP. COPENHAGEN, VOL.1,. . 195, (1968).
        *KOHN ANOMALIES IN ZINC.
        *IYENGAR(P.K.), VENKATARAMAN(G.), GAMEEL(Y.H.), RAO(K.R.).

2302 ZN,. . . . . .T. . . . . . .IAEA SYMP. COPENHAGEN, VOL.1,. . 165, (1968).
        *THEORY OF VIBRATIONAL SPECTRUM IN HEXAGONAL METALS.
        *.C.C.C.P.

2303 ZN,. . . . . .T. . . . . . .IAEA SYMP. COPENHAGEN, VOL.1,. .   3, (1968).
        *THE ROLE OF ELECTRONS IN PHONON SPECTRUM FORMATION IN METALS.
        *.C.C.C.P.

2304 ZN,. . . . . .B2. . . . . . .IAEA SYMP. COPENHAGEN, VOL.1,. . 373, (1968).
        *PHONON-PHONON INTERACTIONS IN ZINC.
        *CAGLIOTI(G.), RIZZI(G.), CUBIOTTI(G.).

2305 ZN,. . . . . .B1. . . . . . .J. PHYS.F: METAL PHYS. VOL.1,. . 785, (1971).
        *PHONONS IN ZINC AT 80 DEG. K.
        *ALMQVIST(L.), STEDMAN(R.).

2306 ZN,. . . . . .B2. . . . . . .SOL. STATE COMM. VOL.8,. . . . . 367, (1970).
        *TEMPERATURE DEPENDENCE OF NEUTRON GROUPS FROM PHONONS IN ZINC. I.
        *CAGLIOTI(G.), RIZZI(G.), CUBIOTTI(G.).

2307 ZN,. . . . . .T. . . . . . .PHYS. REV. B VOL.4,. . . . . . .1390, (1971).
        *LATTICE-DYNAMICAL CALCULATIONS FOR ZINC AND BERYLLIUM.
        *BEZDEK(H.F.), FINEGOLD(L.).

2308 ZN,. . . . . .T,B1. . . . . .PHYS. LETT. VOL.38A, ... 497, (1972).
        *PHONON DISPERSION RELATIONS FOR ZINC.
        *RAJPUT(J.S.), KUSHWAHA(S.S.).

2309 ZN,. . . . . .B1. . . . . . .COPENHAGEN CONF.-LATT.DYNAMICS,.   33, (1963).
        *LATTICE DYNAMICS OF ZINC.
        *MALISZEWSKI(E.),   ROSOLOWSKI(J.),   SLEDZIEWSKA(D.),   CZACHOR(A.).

2310 ZN,. . . . . .B1. . . . . . .COPENHAGEN CONF.-LATT.DYNAMICS,.   17, (1963).
        *A STUDY OF THE CRYSTAL DYNAMICS OF ZINC. (ABSTRACT).
        *BORGONOVI(G.), CAGLIOTI(G.), ANTAL(J.J.).

2311 ZN,. . . . . .B7. . . . . . .THESIS (121 PP.), ...   (1972).
        *INVESTIGATION OF THE LATTICE DYNAMICS OF ZINC THROUGH INELASTIC NEUTRON
        SCATTERING INTENSITIES.
        *CHESSER(N.J.). (ED. NOTE: AVAIL. UNIV. MICROFILMS ORDER NO.73-585; SEE
        DISS. ABSTRACTS VOL.33 NO.7 PG.3253-B).

2312 ZN,. . . . . .C1. . . . . . .PHYS. LETT. 1, ... 338, (1962).
        *COLD NEUTRON MEASUREMENT OF THE PHONON DISPERSION-RELATION FOR ZINC
        SINGLE-CRYSTAL.
        *MALISZEWSKI(E.F.).

2313 ZN,. . . . . .B1,T. . . . . .PHYS. REV. VOL.134, ...A1476, (1964).
        *LATTICE VIBRATIONAL SPECTRA OF BERYLLIUM, MAGNESIUM AND ZINC.
        *YOUNG(J.A.), KOPPEL(J.U.).

2314 ZN,. . . . . .T,B1. . . . . .J. PHYS. F VOL.3,. . . . . 709, (1973).
        *DIELECTRIC SCREENING AND PHONON DISPERSION IN HCP ZINC: A PSEUDO-
        POTENTIAL APPROACH.
        *BAJPAI(R.P.).

2315 ZN,. . . . . .B1. . . . . . .PHYS. REV. 132,. . . . . 683, (1963).
        *A STUDY OF THE CRYSTAL DYNAMICS OF ZINC.
        *BORGONOVI(G.M.), CAGLIOTI(G.), ANTAL(J.J.).

2316 ZN,. . . . . .T,B1. . . . . .PHYS. STAT. SOL. 13, ... 519, (1966).
        *LATTICE DYNAMICS OF ZINC.
        *GUPTA(R.P.), DAYAL(B.).

2317 ZN,. . . . . .B1. . . . . . .IAEA SYMP. CHALK RIVER, VOL.2, .   87, (1962).
        *NEUTRON-PHONON INTERACTION STUDIES IN COPPER, ZINC AND MAGNESIUM SINGLE
        CRYSTALS.
        *MALISZEWSKI(E.), SOSNOWSKI(J.), BLINOWSKI(K.), KOZUBOWSKI(J.),
        PADLO(I.), SLEDZIEWSKA(D.).

2318 ZN,. . . . . .B1. . . . . . .PHYS. REV. B VOL.9,. . . . . . .4060, (1974).
        *LATTICE DYNAMICS OF ZINC: PHONON STRUCTURE FACTORS.
        *CHESSER(N.J.), AXE(J.D.).

2319 ZN*CR2*O4, ...D3. . . . . . .J. PHYS. CHEM. SOLIDS VOL.33,. . 759, (1972).
        *NEUTRON SCATTERING FROM PARAMAGNETIC ZN*CR2*O4.
        *BEGUM(R.J.), MURTHY(N.S.S.).

2320 ZN*FE2*O4, ...D3. . . . . . .PHYS. REV. VOL.98, ...1721, (1955).
        *MULTIPLE SCATTERING OF SLOW NEUTRONS BY FLAT SPECIMENS AND MAGNETIC
        SCATTERING BY ZINC FERRITE.
        *BROCKHOUSE(B.N.), CORLISS(L.M.), HASTINGS(J.M.).

2321 (ZN-MN-FE)*O4, ...D4. . . . . . .IAEA SYMP. CHALK RIVER, VOL.2, . 327, (1962).
        *SOME INVESTIGATIONS OF NEUTRON INELASTIC SCATTERING ON MAGNETICS.
        *KRASNICKI(SZ.), RUTA-WALA(K.), WANIC(A.), MURASIK(A.), RISTE(T.).

2322 ZN*O,. . . . . .B1. . . . . . .SOL. STATE COMM. VOL.8,. . . . . 187, (1970).
        *LATTICE DYNAMICS OF ZN*O AND BE*O.
        *HEWAT(A.W.).

SECTION   6 -QUASI-ELASTIC AND INELASTIC SCATTERING STUDIES . . . .

2323 ZN*O,. . . . . . . ~PHONON DISPERSION IN ZN*O.....B1.....PHYS. LETT. VOL.31A, ...   2, (1970).
~WEGENER(W.), HAUTECLER(S.).

2324 ZN*S,. . . . . . ~LATTICE DYNAMICS OF CUBIC ZINC SULFIDE BY NEUTRON SCATTERING..B1.....J. PHYS. CHEM. SOLIDS VOL.32,... 1573, (1971).
~FELDKAMP(L.A.), STEINMAN(D.K.), VAGELATOS(N.), KING(J.S.),
VENKATARAMAN(G.).

2325 ZN*S,. . . . . . ~STUDY OF THE LATTICE DYNAMICS OF DEUTERATED POLYETHYLENE AND ZINC.B1.....DISS. ABS. VOL.30,.....23698, (1969).
SULPHIDE USING COHERENT NEUTRON SCATTERING.
~FELDKAMP(L.A.).

2326 ZN*S,. . . . . . ~IONIC CHARGE AND LATTICE DYNAMICS OF CUBIC ZINC SULPHIDE..T.....PHYS. STAT. SOL. VOL.41,   491, (1970).
~KUNC(K.), BALKANSKI(M.), NUSIMOVICI(M.A.).

2327 ZN*S (CUBIC),. . ~LATTICE DYNAMICS OF CUBIC ZINC SULPHIDE..B1.....SOL. STATE COMM. 7, NO.21, . . . 1571, (1969).
~FELDKAMP(L.A.), VENKATARAMAN(G.), KING(J.S.).

2328 ZN*S (BETA), . . ~LATTICE DYNAMICS OF MAGNESIUM STANNIDE AND ZINC BLENDE..B1.....RCN-121 (119 PP.), ... , (1970).
~BERGSMA(J.). (ED. NOTE! AVAIL. REACTOR CENTRUM NEDERLAND, PETTEN, THE
NETHERLANDS).

2329 ZN*S (BETA), . . ~LATTICE DYNAMICS OF ZINC BLENDE..B1.....PHYS. LETT. VOL.32A, ...   324, (1970).
~BERGSMA(J.).

2330 ZN*(S,TE), . . . ~PHONON DISPERSION AND PHONON DENSITIES OF STATES FOR ZN*S AND ZN*TE..B1.....J. CHEM. PHYS. VOL.60, . . . 3613, (1974).
~VAGELATOS(N.), WEHE(D.), KING(J.S.).

2331 ZN*SE,. . . . . ~NORMAL MODES OF VIBRATIONS IN ZN*SE..B1.....PHYS. LETT. VOL.36A, ...   376, (1971).
~HENNION(B.), MOUSSA(F.), PEPY(G.), KUNC(K.).

2332 ZN*SE,. . . . . ~LATTICE DYNAMICS OF II-VI, III-V COMPOUNDS..T,B1..VI..SOL. STATE COMM. VOL.11, ... 1691, (1972).
~TALWAR(D.N.), AGRAWAL(B.K.).

2333 ZR,. . . . . . . ~A STUDY OF THE LATTICE DYNAMICS OF ZIRCONIUM USING THE TECHNIQUE OF.B1.....DISS. ABS. VOL.31,.....3649B, (1970).
THERMAL NEUTRON SCATTERING.
~BEZDEK(H.F.).

2334 ZR,. . . . . . . ~LATTICE DYNAMICS OF ZIRCONIUM..B1.....PHYS. STAT. SOL. VOL.42, ...   275, (1970).
~BEZDEK(H.F.), SCHMUNK(R.E.), FINEGOLD(L.).

2335 ZR (ISO. A=92, 94),. . . ~NEUTRON INELASTIC SCATTERING FROM ZR (ISO. A=92) AND ZR (ISO. A=94)..P.....PHYS. REV. C VOL.2, . . . 2390, (1970).
~TESSLER(G.), GLICKSTEIN(S.S.), CARROLL-JR(E.E.).

2336 ZR (ISO. A=90),. ~NEUTRON INELASTIC SCATTERING FROM ZR (ISO. A=90)..P.....PHYS. REV. C VOL.4, ... 1818, (1971).
~GLICKSTEIN(S.S.), TESSLER(G.), GOLDSMITH(M.).

2337 ZR (ISO. A=94, 96),. . . ~NEUTRON ABSORPTION CROSS SECTIONS FOR ZIRCONIUM-94 AND ZIRCONIUM-96..P.....NUCL. SCI. ENG. VOL.46, ...   314, (1971).
~FULMER(R.H.), STRICOS(D.P.), RUANE(T.F.).

2338 ZR,. . . . . .P.......REF. SECTION 3, 47/HAVENS, ...

2339 ZR,. . . . . . . ~TEMPERATURE DEPENDENCE OF THE SOFT ACOUSTIC SHEAR MODES IN H.C.P..B2.....SOL. STATE COMM. VOL.13, . . . 1465, (1973).
ZIRCONIUM.
~MOSS(S.C.), KEATING(D.T.), AXE(J.D.).

2340 ZR*C,. . . . . . ~PHONON SPECTRA IN ZR*C(X)..B1.....BULL. AMER. PHYS. SOC. VOL.17, . 292, (1972).
~SMITH(H.G.). (ED. NOTE! PAPER PRESENTED AT MARCH MEETING IN ATLANTIC
CITY).

2341 ZR*H2, ... ~TIME DEPENDENT THERMALIZATION OF NEUTRONS IN AN EINSTEIN SOLID WITH..P.....ATOMKERNENERGIE VOL.16, ... 201, (1970).
APPLICATION TO ZIRCONIUM HYDRIDE.
~VON-BALTZ(R.).

2342 ZR*H2, ... ~A MULTIPHONON NEUTRON SCATTERING STUDY OF HYDROGEN BONDING IN ZIRCONIUM.C1.....BROOKHAVEN SYMPOSIUM, ...   96, (1965).
HYDRIDE.
~HARLING(O.K.), LEONARD-JR(B.R.).

2343 ZR*H2, ... ~STUDIES OF VIBRATION SPECTRA OF BONDED HYDROGEN ATOMS USING A PULSED.C3.....J. CHEM. PHYS. VOL.55, ... 2807, (1971).
NEUTRON SOURCE.
~DAY(D.H.), SINCLAIR(R.N.).

2344 ZR*H2, ... ~INELASTIC SCATTERING OF SLOW NEUTRONS BY ZIRCONIUM AND TITANIUM HYDRIDES.C1.....NUCL. SCI. ENGNG. VOL.23, . . . 194, (1965).
~PAN(S.S.), WEBB(F.J.).

2345 ZR*H2, ... ~STRUCTURE IN THE NEUTRON SCATTERING SPECTRA OF ZIRCONIUM HYDRIDE.B3.....PHYS. REV. B-4, . . . 2675, (1971).
~COUCH(J.G.), HARLING(O.K.), CLUNE(L.C.).

2346 ZR*H2, ... ~STUDY OF INELASTIC SCATTERING OF SLOW NEUTRONS IN ZIRCONIUM HYDRIDE.C1.....IAEA SYMP. BOMBAY, VOL.2, . . . 317, (1964).
~.C.C.C.P.

2347 ZR*H2, ... ~THE NATURE OF HYDROGEN MOTION IN ZR*H2 DETERMINED FROM AN EXPERIMENTAL.C1.....IAEA SYMP. BOMBAY, VOL.2, . . . 305, (1964).
NEUTRON STUDY OF LARGE, BOUND ENERGY LEVELS.
~WHITTEMORE(W.L.).

2348 ZR*H2, ... ~RECENT MEASUREMENTS OF THE SCATTERING LAWS OF SOME HYDROGENOUS.C4.....IAEA SYMP. BOMBAY, VOL.2, . . . 167, (1964).
MODERATORS.
~GLASER(W.), EHRET(G.), MERKEL(A.).

2349 ZR*H2, ... ~SCATTERING CROSS SECTION AT SUBTHERMAL ENERGIES AND GENERALIZED PHONON.P,C1....J. NUCL. ENERGY VOL.25, . . . 189, (1971).
FREQUENCY SPECTRUM OF ZIRCONIUM HYDRIDE.
~SAASTAMOINEN(J.), PALMGREN(A.).

SECTION   6 -QUASI-ELASTIC AND INELASTIC SCATTERING STUDIES . . . .

2350   ZR*H2. . . . . .P. . . . . . .IAEA SYMPOSIUM (VIENNA),   613, (1960).
        ⌐MEASUREMENTS OF NEUTRON SPECTRA IN WATER, POLYETHYLENE AND ZIRCONIUM
        HYDRIDE.
        ⌐WALTON(R.B.), BEYSTER(J.R.), WOOD(J.L.), LOPEZ(W.M.).

2351   ZR*H2. . . . . .P. . . . . . .REF. SECTION 3, 60/WOODS,. . . . .

2352   ZR-H,. . . . . .P. . . . . . . .CONF. NEUTRON THERMALIZ. VOL.I,   407, (1968).
        ⌐INELASTIC NEUTRON SCATTERING IN METAL HYDRIDES, U*C AND U*O2 AND
        APPLICATIONS OF THE SCATTERING LAW.
        ⌐PUROHIT(S.N.), PAN(S.S.), BISCHOFF(F.), BRYANT(W.A.), LAJEUNESSE(C.);
        (IN: NEUTRON THERMALIZATION AND REACTOR SPECTRA. PUBL. IAEA: VIENNA).

2353   ZR*H2, ...C3. . . . . . .NUCL. SCI. ENGNG. VOL.5,   99, (1960).
        ⌐INELASTIC SCATTERING OF COLD NEUTRONS FROM SEVERAL HYDROGENOUS LIQUIDS.
        ⌐BRUGGER(R.M.), MCCLELLAN(L.W.), STREETMAN(G.B.), EVANS(J.E.).

2354   ZR*H2, ...C1. . . . . . .SOV. PHYS. USPEKHI VOL.1,   165, (1958).
        ⌐DIRECT OBSERVATION OF OPTICAL VIBRATIONS OF A CRYSTAL LATTICE OF A
        SOLID BY MEANS OF NEUTRON SCATTERING.
        ⌐CHENTSOV(P.).

2355   ZR*H2, ...C1. . . . . . .PHYS. REV. VOL.108,   1091, (1957).
        ⌐DETECTION OF OPTICAL LATTICE VIBRATIONS IN GE AND ZR*H BY SCATTERING OF
        COLD NEUTRONS.
        ⌐PELAH(I.), EISENHAUER(C.M.), HUGHES(D.J.), PALEVSKY(H.).

2356   ZR*H2, ...C1. . . . . . .PHYS. REV. VOL.108,. . . . . 1092, (1957).
        ⌐NEUTRON INVESTIGATION OF OPTICAL VIBRATION LEVELS IN ZIRCONIUM HYDRIDE.
        ⌐ANDRESEN(A.), MCREYNOLDS(A.W.), NELKIN(M.), ROSENBLUTH(M.),
        WHITTEMORE(W.L.).

SECTION   7A-CONCERNING NEUTRON-NUCLEON INTERACTIONS. . . . .

1   ACTA CRYST. VOL.A25, ... 396, (1969) . . . .
    *LOCATION OF THE ANOMALOUS SCATTERER IN NEUTRON ANOMALOUS SCATTERING
    STUDIES.
    *SIKKA(S.K.).

2   ACTA CRYST. VOL.A25, ... 391, (1969) . . . .P. . . . .
    *COHERENT NEUTRON-SCATTERING AMPLITUDES.
    *BACON(G.E.). (CHAIRMAN OF NEUTRON DIFFRACTION COMMISSION).

3   ACTA CRYST. VOL.A27, ... 148, (1971) . . . .
    *THE EVALUATION OF THERMAL DIFFUSE SCATTERING OF NEUTRONS FOR A
    ONE-VELOCITY MODEL.
    *COOPER(M.J.).

4   ACTA CRYST. VOL.A28, ... 357, (1972) . . . .P. . . . .
    *COHERENT NEUTRON SCATTERING AMPLITUDES.
    *BACON(G.E.).

5   ACTA CRYST. VOL.A29, ... 211, (1973) . . . .
    *USE OF THE TANGENT FORMULA TO RESOLVE THE PHASE AMBIGUITY IN THE NEUTRON
    ANOMALOUS-DISPERSION METHOD.
    *SIKKA(S.K.).

6   ACTA PHYS. POLON. VOL.13, ... 67, (1954) . . . .T. . . . .
    *THE INFLUENCE OF HINDERED ROTATION ON THE SCATTERING OF SLOW NEUTRONS
    BY BOUND PROTONS.
    *KOLOS(W.).

7   ACTA PHYS. POLON. VOL.22, ... 399, (1962) . . . .T. . . . .
    *INELASTIC SCATTERING OF SLOW NEUTRONS AND THE BEHAVIOUR OF A SINGLE
    PARTICLE SCATTERING SYSTEM.
    *FULINSKI(A.).

8   AM. J. PHYS. VOL.39, ... 324, (1971) . . . .T. . . . .
    *ON THE REPULSION OF SLOW NEUTRONS BY ATTRACTIVE POTENTIALS.
    *PESHKIN(M.), RINGO(G.R.).

9   ANN. DE PHYSIQUE TOME 2, ... 101, (1967) . . . .
    *ANALYSE DES RESONNANCES INDUITES PAR LES NEUTRONS S DANS LES EXPERIENCES
    PAR TEMPS DE VOL.
    *CORGE(C.R.).

10 ANN. OF PHYS. VOL.16, ... 387, (1961) . . . .T. . . . .
    *COMPLEX POTENTIAL MODEL FOR LOW-ENERGY NEUTRON SCATTERING.
    *FIEDELDEY(H.), FRAHN(W.E.).

11 ANN. OF PHYS. VOL.26, ... 72, (1964) . . . .
    *ON THE FERMI APPROXIMATION IN THERMAL NEUTRON SCATTERING.
    *SUMMERFIELD(G.C.).

12 ANN. OF PHYS. VOL.33, ... 15, (1965) . . . .
    *CALCULATIONS OF INELASTIC SCATTERING IN TERMS OF ELASTIC SCATTERING.
    *AUSTERN(N.), BLAIR(J.S.).

13 ANN. OF PHYS. VOL.73, ... 372, (1972) . . . .T. . . . .
    *A LOW-ENERGY S-MATRIX THEORY OF NEUTRON-DEUTERON SCATTERING.
    *BOWER(R.H.J.).

14 ANN. OF PHYS. VOL.75, ... 132, (1973) . . . .T,P. . . . .C (A=12) . . . . .
    *A STUDY OF THE NEUTRON PLUS C (A=12) SYSTEM USING THE UNIFIED-REACTION
    FORMALISM. I.ELASTIC SCATTERING.
    *LEUNG(T.T.), KOSHEL(R.D.).

15 ATLANTA SYMP/GEORGIA INST. TECH. . . 1, (1967) . . . .
    *NEUTRON INTERACTIONS WITH ATOMS.
    *SHULL(C.G.).

16 ATOMKERNENERGIE VOL.18, ... 219, (1971) . . . .P. . . . .
    *CALCULATION AND USE OF SYNTHETIC THERMAL SCATTERING KERNELS.
    *ROYL(P.).

17 AUSTRAL. J. PHYS. VOL.22, ... 145, (1969) . . . .T. . . . .
    *INELASTIC SCATTERING OF NEUTRONS. II. COMPARISON OF QCN THEORY WITH
    EXPERIMENT.
    *BERTRAM(W.K.).

18 AUSTRAL. J. PHYS. VOL.22, ... 135, (1969) . . . .T. . . . .
    *INELASTIC SCATTERING OF NEUTRONS. I. QUASI-COMPOUND NUCLEUS THEORY.
    *BERTRAM(W.K.).

19 AUSTRAL. J. PHYS. VOL.23, ... 823, (1970) . . . .A4. . . . .
    *A MULTILEVEL FORMALISM FOR NEUTRON ELASTIC SCATTERING CROSS SECTIONS.
    *CLAYTON(E.).

20 BU. ACAD. SCI. USSR PHYS. SER.34 2183, (1970) . . . .A4. . . . .
    *THIN-STRUCTURE RESONANCES DURING ELASTIC SCATTERING OF NEUTRONS BY
    ALMOST MAGICAL NUCLEI.
    *KOLOMIETS(V.M.).

21 BU. ACAD. SCI. USSR PHYS. SER.35 2349, (1971) . . . .P. . . . . . .ZN, GE, SE. . .
    *EXCITATION OF INITIAL STATES IN THE CAPTURE OF THERMAL NEUTRONS BY ZN
    (ISOTOPE A=68), GE(ISOTOPE A=70) AND SE(ISOTOPE A=80) NUCLEI.
    *MURZIN(A.V.), KOLOMIETS(V.M.).

22 CAN. J. PHYS. VOL.31, ... 432, (1953) . . . .E,P. . . . .
    *RESONANT SCATTERING OF SLOW NEUTRONS.
    *BROCKHOUSE(B.N.).

23 CAN. J. PHYS. 42, ... 1017, (1964) . . . .
    *PLANE WAVE APPROXIMATION FOR THE SCATTERING AMPLITUDE.
    *CHAN(H.H.), RAZAVY(M.).

24 CAN. J. PHYS. 48, ... 616, (1970) . . . .
    *CONTINUED FRACTION REPRESENTATION FOR SLOW NEUTRON SCATTERING. II.
    *SEARS(V.F.).

25 CONF/NUCL. STRUC. STUDY/NEUTRONS 144, (1972) . . . .D4. . . . .
    *THE NEUTRON-NUCLEUS SPIN-SPIN INTERACTION AND NEUTRON STRENGTH FUNCTIONS
    *NEWSTEAD(C.M.), DELAROCHE(J.P.) CAUVIN(B.). (ED. NOTE: AVAIL. CENTRAL
    RES. INST. PHYS., BUDAPEST, HUNGARY).

SECTION   7A-CONCERNING NEUTRON-NUCLEON INTERACTIONS. . . . .

26 DISS. ABS. VOL.24, 2410, (1964)
    *APPLICATION OF THE WIGNER REPRESENTATION TO THE THEORY OF SLOW NEUTRON
    SCATTERING.
    *ROSENBAUM(M.).

27 DISS. ABS. VOL.26, 298, (1966)
    *ON CERTAIN COMMON APPROXIMATIONS IN THERMAL NEUTRON SCATTERING.
    *PLUMMER(J.P.).

28 DISS. ABS. VOL.27, 2825B, (1967)
    *AN OPTICAL MODEL INVESTIGATION OF ELASTIC NEUTRON SCATTERING.
    *CASSOLA(R.L.).

29 DISS. ABS. VOL.32, 1746B, (1971)
    *S-MATRIX THEORY OF LOW ENERGY NEUTRON-DEUTERON SCATTERING.
    *BOWER(R.H.J.).

30 FIZIKA (YUGOSLAVIA) VOL.4, SUPL. 11, (1971) T
    *THE UNITARY OPERATOR METHOD IN THE THEORY OF SLOW NEUTRON SCATTERING
    BY A BOUND CENTRE SYSTEM.
    *STANCIC(V.).

31 IAEA SYMPOSIUM VIENNA, 97, (1960)
    *ELASTIC RESONANCE SCATTERING OF SLOW NEUTRONS IN CRYSTALS.
    *.C.C.C.P.

32 J. NUCL. ENERGY VOL.24, 35, (1970) P
    *ACTIVATION MEASUREMENTS OF THERMAL NEUTRON CAPTURE CROSS-SECTIONS AND
    RESONANCE INTEGRALS.
    *RYVES(T.B.).

33 J. NUCL. ENERGY VOL.25, 489, (1971) B7
    *A GENERALIZED ENERGY EXCHANGE KERNEL FOR INELASTIC NEUTRON SCATTERING
    AND THERMONUCLEAR REACTIONS.
    *WILLIAMS(M.M.R.).

34 J. PHYS. SOC. JAPAN VOL.28, 644, (1970) C1
    *INCOHERENT NEUTRON SCATTERING AND NUCLEAR MAGNETIC RESONANCE.
    *SUMMERFIELD(G.C.), KLEINBERG(R.).

35 NED. TIJDSCHR. NATUURK. VOL.38, 170, (1972) P
    *NEUTRON CAPTURE IN ORIENTED TARGET NUCLEI.
    *BOSMAN(J.J.).

36 NUCL. PHYS. A VOL.A177, 559, (1971) T
    *LOW-ENERGY NEUTRON SCATTERING BY A HARTREE-FOCK FIELD.
    *DOVER(C.B.), NGUYEN-VAN-GIAI.

37 NUCL. PHYS. A VOL.A169, 385, (1971) T
    *VELOCITY DEPENDENCE IN THE POTENTIALS FOR NEUTRON SCATTERING MODELS.
    *CANFIELD(E.H.), AMSTER(H.J.), KASPER(R.G.), MARK(H.).

38 NUCL. PHYS. A VOL.178, 249, (1971) T
    *A COMPARISON OF HARTREE-FOCK AND PEREY-BUCK NON-LOCAL POTENTIALS FOR
    LOW-ENERGY SCATTERING.
    *MACKELLAR(A.D.), SCHENTER(R.E.).

39 NUCL. PHYS. A VOL.182, 541, (1972) A4 D2
    *OBSERVABLES OF THE NEUTRONS ELASTICALLY SCATTERED BY DEUTERONS AT LOW
    ENERGY.
    *JACCARD(S.), VIENNET(R.).

40 NUCL. SCI. ENGNG. VOL.3, 29, (1958) T
    *SOME CHARACTERISTICS OF THE THERMAL NEUTRON SCATTERING PROBABILITY.
    *OSBORN(R.K.).

41 NUCL. SCI. ENGNG. VOL.34, 93, (1968) B7
    *ON THE CALCULATION OF NEUTRON INELASTIC SCATTERING CROSS SECTIONS.
    *GOLDSMITH(M.).

42 NUCL. SCI. ENG. VOL.43, 235, (1971) A4
    *MULTIGROUP ELASTIC SCATTERING CROSS SECTIONS FOR HEAVY ELEMENTS.
    *HENRYSON(H.).

43 NUCL. SCI. ENG. VOL.45, 167, (1971) T
    *APPLICATIONS OF THE DEGENERATE KERNEL TECHNIQUE TO THERMAL NEUTRON
    SPECTRA CALCULATIONS.
    *TURINSKY(P.J.), DUDERSTADT(J.J.).

44 NUCL. SCI. ENG. VOL.48, 119, (1972) T
    *ON THE SCATTERING MATRIX CONSTRUCTION FOR THE DISCRETE ENERGY REPRESEN-
    TATION OF THERMAL-NEUTRON SPECTRA.
    *MATAUSEK(M.V.).

45 NUKLEONIK VOL.6, 87, (1964) P
    *TRANSMISSION AND SCATTERING EXPERIMENTS WITH SLOW NEUTRONS IN THE
    FRM-REACTOR.
    *SPRINGER(T.). (IN GERMAN).

46 PHYS. CAN. VOL.27, 70, (1971) A4
    *DWBA CALCULATIONS OF THE MOTT-SCHWINGER EFFECT ON NEUTRON ELASTIC
    SCATTERING.
    *SHERIF(H.).

47 PHYSICA VOL.27, 260, (1961) T
    *THE QUASI-CLASSICAL APPROXIMATION FOR NEUTRON SCATTERING.
    *TURNER(R.E.).

48 PHYSICA VOL.32, 16, (1966) T
    *ON THE MASS TENSOR APPROXIMATION OF SLOW NEUTRON SCATTERING.
    *KOSALY(G.), SOLT(G.).

49 PHYS. LETT. 2, 266, (1962)
    *QUASI-CLASSICAL APPROXIMATION IN NEUTRON SCATTERING.
    *KOSALY(G.), TURNER(R.E.).

50 PHYS. LETT. 6, 51, (1963)
    *ON THE MASS TENSOR APPROXIMATION OF SLOW NEUTRONS SCATTERING.
    *KOSALY(G.), SOLT(G.).

51 PHYS. LETT. VOL.13, 223, (1964) T NELKIN
    *A MODIFICATION OF THE KRIEGER-NELKIN APPROXIMATION IN THE THEORY OF
    SLOW NEUTRON SCATTERING.
    *KOSALY(G.), SOLT(G.).

SECTION   7A-CONCERNING NEUTRON-NUCLEON INTERACTIONS. . . . .

52 PHYS. LETT. 28A,   376, (1968)
    *NEW MODEL FOR THE CALCULATION OF NEUTRON SCATTERING.
    *RICHTER(J.), VOSS(K.).

53 PHYS. LETT. VOL.29B,   33, (1969)....P...
    *MEASUREMENTS OF TOTAL CROSS-SECTIONS FOR VERY SLOW NEUTRONS WITH
    VELOCITIES FROM 100 M/SEC. TO 5 M/SEC.
    *STEYERL(A.).

54 PHYS. LETT. B VOL.36B,   560, (1971)....P...
    *NEUTRON DIFFRACTION DISSOCIATION AND COULOMB DISSOCIATION FROM NUCLEI.
    *LONGO(M.J.), JONES(L.W.), O*BRIEN(D.D.), VANDERVELDE(J.C.), DAVIS(M.B.),
    GIBBARD(B.G.), KREISLER(M.N.).

55 PHYS. LETT. B VOL.35B,   477, (1971)
    *SMALL-ANGLE SCATTERING OF NEUTRONS BY DEFORMED NUCLEI.
    *PALLA(G.).

56 PHYS. MET. METALLOG. VOL.11 NO.5 139, (1961)....T...
    *TIME FORMALISM IN THE BORN APPROXIMATION IN THE THEORY OF THE SCATTERING
    OF NEUTRONS BY A SUBSTANCE.
    *IZYUMOV(YU.A.).

57 PHYS. REV. VOL.48,   367, (1935)....T...
    *ON THE CROSS SECTION OF HEAVY NUCLEI FOR SLOW NEUTRONS.
    *VAN-VLECK(J.H.).

58 PHYS. REV. VOL.50,   899, (1936)....E...
    *ON THE ABSORPTION AND THE DIFFUSION OF SLOW NEUTRONS.
    *AMALDI(E.), FERMI(E.).

59 PHYS. REV. VOL.49,   519, (1936)....T...
    *CAPTURE OF SLOW NEUTRONS.
    *BREIT(G.), WIGNER(E.).

60 PHYS. REV. VOL.50,   133, (1936)....E...
    *SCATTERING OF SLOW NEUTRONS. II.
    *MITCHELL(A.C.G.), MURPHY(E.J.), WHITAKER(M.D.).

61 PHYS. REV. VOL.58,   26, (1940)....T....DEUTERONS,
    *THE SCATTERING OF THERMAL NEUTRONS BY DEUTERONS.
    *MOTZ(L.), SCHWINGER(J.).

62 PHYS. REV. VOL.60,   742, (1941)....P......H (PROTON),
    *THE INTERACTION OF SLOW NEUTRONS WITH NUCLEI.
    *CARROLL(H.).

63 PHYS. REV. VOL.77,   575, (1950)....T,E......CU,NI,MN,
    *A NEW METHOD FOR DETERMINING THE RELATIVE PHASE WITH WHICH SLOW
    NEUTRONS ARE SCATTERED BY NUCLEI.
    *BENDT(P.J.), RUDERMAN(I.W.).

64 PHYS. REV. VOL.79,   481, (1950)....T........H2 (PARA)
    *VARIATIONAL PRINCIPLES FOR SCATTERING PROCESSES. II. SCATTERING OF
    SLOW NEUTRONS BY PARA-HYDROGEN.
    *LIPPMANN(B.A.).

65 PHYS. REV. VOL.79,   469, (1950)....T...
    *VARIATIONAL PRINCIPLES FOR SCATTERING PROCESSES. I.
    *LIPPMANN(B.A.), SCHWINGER(J.).

66 PHYS. REV. VOL.81,   527, (1951)....E,P...
    *COHERENT SCATTERING AMPLITUDES AS DETERMINED BY NEUTRON DIFFRACTION.
    *SHULL(C.G.), WOLLAN(E.O.).

67 PHYS. REV. VOL.126,   632, (1962)....E...
    *STUDIES OF THE OPTICS OF NEUTRONS. I. MEASUREMENT OF THE NEUTRON-PROTON
    COHERENT SCATTERING AMPLITUDE BY MIRROR REFLECTION.
    *DICKINSON(W.C.), PASSELL(L.), HALPERN(O.).

68 PHYS. REV. 129,   1396, (1963)
    *HYPERVIRIAL THEOREMS FOR VARIATIONAL WAVE FUNCTIONS IN SCATTERING
    THEORY.
    *EPSTEIN(S.T.), ROBINSON(P.D.).

69 PHYS. REV. 131,   1153, (1963)
    *VALIDITY OF THE FERMI APPROXIMATION IN SLOW NEUTRON SCATTERING.
    *PLUMMER(J.P.), SUMMERFIELD(G.C.).

70 PHYS. REV. 129,   1391, (1963)
    *VIRIAL THEOREM AND ITS GENERALIZATIONS IN SCATTERING THEORY.
    *ROBINSON(P.D.), HIRSCHFELDER(J.O.).

71 PHYS. REV. VOL.135,   B895, (1964)
    *CALCULATIONS OF INELASTIC SCATTERING OF NEUTRONS BY HEAVY NUCLEI.
    *AUERBACH(E.H.), MOORE(S.O.).

72 PHYS. REV. LETT. 16,   495, (1966)
    *IMPORTANCE OF RELATIVISTIC EFFECTS IN THE SCATTERING OF SLOW NEUTRON.
    *BROWNE(H.N.), BAUER(F.).

73 PHYS. REV. VOL.177,   1706, (1969)....T...
    *INFLUENCE OF ELECTROMAGN. INTERACTION ON NEUTRON SCATTERING FROM NUCLEI.
    *HOGAN(W.S.).

74 PHYS. REV. VOL.178,   1647, (1969)....P...
    *ACCOUNT OF THE MAIN NUCLEAR-STRUCTURE PROPERTIES IN THE OPTICAL
    POTENTIAL FOR NEUTRON ELASTIC SCATTERING.
    *DOTSENKO(B.B.).

75 PHYS. STAT. SOL. 4,   31, (1964)
    *THEORY OF SCATTERING.
    *KAWEB(B.H.).

76 PROC. PHYS. SOC. VOL.71,   910, (1958)....P,T...
    *FLUCTUATIONS IN SLOW NEUTRON AVERAGE CROSS SECTIONS.
    *EGELSTAFF(P.A.).

77 PROC. PHYS. SOC. 81,   35, (1963)
    *QUANTUM DEFECTS AND SCATTERING LENGTHS.
    *MOISEIWITSCH(B.L.).

78 PROC. PHYS. SOC. 86,   363, (1965)
    *LOWER BOUND TO THE RECIPROCAL OF THE SCATTERING LENGTH.
    *MOISEIWITSCH(B.L.).

SECTION 7A-CONCERNING NEUTRON-NUCLEON INTERACTIONS. . . . .

79 PROC. PHYS. SOC. 89, 341, (1966)
*LOWER BOUND TO THE RECIPROCAL OF THE SCATTERING LENGTH.
*HOUSTON(S.K.), MOISEIWITSCH(B.L.).

80 PROC. PHYS. SOC. 91, 678, (1967)
*RELATIONS BETWEEN THE DIFFERENT SCATTERING THEORIES.
*LLOYD(P.), BERRY(M.V.).

81 RENSSELAER POL. INST. SYMP. 3, (1961)
*THE LOW ENERGY CROSS SECTIONS OF FISSILE NUCLIDES.
*LEONARD-JR(B.R.).

82 REV. GEN. SCI. TOME 50, 89, (1939)
*TRANSMUTATIONS PRODUCED BY SLOW NEUTRONS AND BOHR≠S NUCLEAR DYNAMICS.
*KAHAN(T.).

83 REV. MOD. PHYS. VOL.9, 113, (1937) T
*NUCLEAR PHYSICS B.NUCLEAR DYNAMICS, THEORETICAL.
*BETHE(H.A.).

84 REV. ROUMAINE PHYS. VOL.17, 57, (1972) P,C1
*THE USE OF INCOHERENT SCATTERING OF SLOW NEUTRONS FOR THE DETERMINATION
OF NONBONDED POTENTIAL PARAMETERS.
*TARINA(V.).

85 SOV. J. NUCL. PHYS. VOL.10,1, 47, (1969) P D2, N2, LA
*ON SPIN DEPENDENCE OF THE INTERACTION OF SLOW NEUTRONS WITH NUCLEI OF
DEUTERIUM, NITROGEN, AND LANTHANUM.
*IVANENKO(A.I.), LUSHCHIKOV(V.I.).

86 SOV. PHYS. JETP. 16, 1531, (1963)
*POLARIZATION OF INELASTICALLY SCATTERED NUCLEONS.
*ELAZIN(YU.P.).

87 SOV. PHYS. JETP. 16, 1321, (1963)
*THEORY OF NEUTRON SCATTERING IN THE COULOMB FIELD OF A NUCLEUS.
*GERASIMOV(S.B.), LEBEDEV(A.I.), PETRUN≠KIN(V.A.).

88 SOV. PHYS. JETP. 24, 946, (1967)
*EXTREMAL VALUE OF THE DIFFERENTIAL CROSS-SECTION OF ELASTIC SCATTERING.
*ARUSHANOV(G.G.).

89 SOV. PHYS. SOL. STATE, 10, 675, (1968) T
*SCATTERING OF SLOW NEUTRONS BY THE PROTONS IN HYDROGEN BONDS USING THE
DOUBLE-POTENTIAL-WELL MODEL.
*STAMENKOVICH(S.S.).

90 SOV. PHYS. CRYST. VOL.15, 252, (1970) C1
*INCOHERENT MULTIPLE SCATTERING OF NEUTRONS.
*MEN≠SHIKOV(A.Z.), BOGDANOV(S.G.).

91 SOV. PHYS. CRYST. VOL.15, 383, (1970) 97
*COHERENT DOUBLE SCATTERING OF NEUTRONS.
*MEN≠SHIKOV(A.Z.), BOGDANOV(S.G.).

92 SOV. PHYS. JETP LETT. VOL.14, 91, (1971) P
*SCATTERING OF NEUTRONS BY NUCLEI IN THE REGION OF NON-OVERLAPPING
RESONANCES OF THE NUCLEUS.
*MOROZOV(V.M.), ZUBOV(YU.G.), LEBEDEVA(N.S.).

93 SPECTROSCOPY BIOL. CHEM. 1, (1974) T
*INTRODUCTION TO NEUTRON, X-RAY, AND LASER SPECTROSCOPY.
*CHEN(S.H.). (REFER TO SECTION 4).

94 TH. NEUTRON SCATT./EGELSTAFF, 1, (1965)
*INTRODUCTORY THEORY (OF THERMAL NEUTRON SCATTERING).
*LOMER(W.M.), LOW(G.G.).

95 TRANS. AM. NUCL. SOC. VOL.13, 728, (1970) T C (ISOTOPE A=12)
*THE APPLICATION OF EXTENDED R-MATRIX THEORY TO ELASTIC SCATTERING OF
NEUTRONS BY C (ISOTOPE A=12).
*YOST(K.J.), PITKANEN(P.H.).

96 UKR. FIZ. ZH. (USSR) VOL.10, 1168, (1965) T
*QUASICLASSICAL APPROXIMATION IN THE THEORY OF SLOW NEUTRON SCATTERING.
*YUL≠MET≠EV(R.M.). (IN UKRAINIAN).

97 UKR. FIZ. ZH. (USSR) VOL.16, 280, (1971)
*ON DEPENDENCE OF ANGULAR DISTRIBUTIONS OF INELASTIC SCATTERED NEUTRONS
ON THE COMPOUND NUCLEUS INERTIA MOMENT.
*FEDOROV(M.B.).

98 Z. NATURFORSCH. A VOL.27A, 901, (1972) A4
*COHERENT NEUTRON SCATTERING AMPLITUDES OBTAINED FROM CHEMICAL COMPOUNDS.
*KOESTER(L.), KNOPF(K.).

99 Z. PHYS. VOL.210 NO.5, 434, (1968)
*THERMAL AND SUBTHERMAL NEUTRON SCATTERING.
*MEHRINGER(W.).

SECTION   7B-THEORY OF NEUTRON NUCLEAR SCATTERING IN CONDENSED MATTER . .

1   ANN. DE PHYSIQUE TOME 7 NO.5,   349, (1972)...............................
    ≠FUNDAMENTALS OF NEUTRON SCATTERING BY CONDENSED MATTER. . . . .
    ≠SCHERM(R.).

2   ANN. OF PHYS. VOL.45,   464, (1967) . . . .
    ≠SINGULAR SCATTERING EQUATIONS IN MOMENTUM SPACE.
    ≠PFAFFELHUBER(E.), BLOMER(R.).

3   ARK. FYSIK VOL.5,   53, (1952)...T.................................
    ≠ON THE SCATTERING OF SLOW NEUTRONS BY POLYCRYSTALS.
    ≠FROMAN(P.O.).

4   ARK. FYSIK VOL.32, PAPER 31,   537, (1966)...T.........................
    ≠ON THE DERIVATION OF THE VAN HOVE-GLAUBER FORMULA FOR THE SCATTERING OF
    THERMAL NEUTRONS BY A SYSTEM OF ATOMIC NUCLEI.
    ≠GOODMAN(B.), WALLER(I.).

5   BROOKHAVEN SYMPOSIUM,   1, (1965)...........................
    ≠COMPARISON OF ELECTROMAGNETIC AND NEUTRON STUDIES OF SOLIDS.
    ≠BLUME(M.).

6   BROOKHAVEN SYMPOSIUM,   175, (1965) . . . .
    ≠DENSITY EXPANSION OF THE SELF-CORRELATION FUNCTION FOR THERMAL NEUTRON
    SCATTERING IN DILUTE SYSTEMS.
    ≠SIGMAR(D.J.).

7   BROOKHAVEN SYMPOSIUM,   169, (1965)...........................
    ≠EFFECT OF NUCLEAR SPIN CORRELATION ON THE SCATTERING OF NEUTRONS BY
    MOLECULES.
    ≠SINHA(S.K.), VENKATARAMAN(G.).

8   BULL. ACAD. SCI. URSS, SER. PHYS 189, (1938) . . . .
    ≠SCATTERING OF SLOW NEUTRONS IN THE CRYSTALLINE LATTICE.
    ≠POMERANCHUK(I.YA.). (IN ENGLISH).

9   CHEM. APPL. THERMAL NEUTRON SCAT   1, (1973)...T.......................
    ≠BASIC THEORY OF THERMAL NEUTRON SCATTERING BY CONDENSED MATTER.
    ≠WINDSOR(C.G.). (IN BOOK: CHEMICAL APPLICATIONS OF THERMAL NEUTRON
    SCATTERING. ED. BY B.T.M. WILLIS; OXFORD UNIV. PRESS: LONDON).

10 COLL. INTER. N.126 (GRENOBLE),   217, (1963)...................
    ≠REMARKS ON THE SLOW NEUTRON SCATTERING BY ORGANIC MOLECULES.
    ≠ARDENTE(V.).

11 CONTEMPORARY PHYS. VOL.7,   278, (1966)...T...................
    ≠NEUTRON SPECTROSCOPY OF SOLIDS. I.
    ≠LOMER(W.M.).

12 DISS. ABS. VOL.25,   1131, (1965)...........................
    ≠DAMPING THEORY AND ITS APPLICATION TO THE INTERPRETATION OF SLOW NEUTRON
    SCATTERING EXPERIMENTS.
    ≠AKCASU(Z.A.).

13 IAEA SYMPOSIUM VIENNA,   25, (1960)...........................
    ≠THE THEORY OF THE THERMAL-NEUTRON SCATTERING LAW.
    ≠EGELSTAFF(P.A.).

14 IAEA SYMPOSIUM VIENNA,   39, (1960)...........................
    ≠SOME PROPERTIES OF THE SPACE-TIME CORRELATION FUNCTION.
    ≠SCHOFIELD(P.).

15 IAEA SYMPOSIUM VIENNA,   53, (1960)...........................
    ≠THERMODYNAMIC GREEN≠S FUNCTION METHODS IN NEUTRON SCATTERING BY
    CRYSTALS.
    ≠BAYM(G.).

16 IAEA SYMPOSIUM VIENNA,   75, (1960)...........................
    ≠THE SCATTERING OF NEUTRONS FROM POLYCRYSTALLINE MATERIALS.
    ≠MARSHALL(W.), STUART(R.).

17 IAEA SYMPOSIUM VIENNA,   87, (1960)...........................
    ≠NEUTRON THERMALIZATION AND THE PROPERTIES OF THE VAN HOVE PAIR
    CORRELATION FUNCTION.
    ≠.C.C.C.P.

18 IAEA SYMP. CHALK RIVER, VOL.1,   65, (1962)...................
    ≠PRACTICAL ANALYSIS OF NEUTRON SCATTERING DATA INTO SELF AND INTERFERENCE
    TERMS.
    ≠EGELSTAFF(P.A.).

19 IAEA SYMPOSIUM BOMBAY, VOL.2,   59, (1964)...................
    ≠THE SELF-CORRELATION FUNCTION OF REAL GASES.
    ≠SIGMAR(D.J.).

20 IAEA SYMPOSIUM BOMBAY, VOL.2,   35, (1964)...................
    ≠A KINETIC DESCRIPTION OF THE VAN HOVE CORRELATION FUNCTIONS.
    ≠NELKIN(M.), VAN-LEEUWEN(J.M.J.), YIP(S.).

21 IAEA SYMPOSIUM BOMBAY, VOL.1,   413, (1964)...................
    ≠DIFFUSE SCATTERING FROM ALLOYS AND DISORDERED SYSTEMS, EXPERIMENTAL
    TECHNIQUES AND POTENTIALITIES.
    ≠LOW(G.G.).

22 J. APPL. CRYST. VOL.4,   410, (1971)...P.....................
    ≠PRINCIPLES OF THE CORRELATION METHOD FOR NEUTRON SMALL-ANGLE SCATTERING
    RESEARCH.
    ≠HOSSFELD(F.), AMADORI(R.).

23 J. APPL. PHYS. 41,   5138, (1970)...AR...................
    ≠NEUTRON AND BRILLOUIN SCATTERING IN QUANTUM CRYSTALS.
    ≠HORTON(G.K.), GOLDMAN(V.V.), KLEIN(M.L.).

24 J. CHEM. PHYSICS VOL.47,   4923, (1967)...................
    ≠QUANTUM STATISTICS AND SLOW NEUTRON SCATTERING BY GASES.
    ≠PLUMMER(J.P.), SUMMERFIELD(G.C.), ZWEIFEL(P.F.).

25 J. CHEM. PHYSICS VOL.46,   465, (1967)...................
    ≠ORIENTATION-AVERAGED AMPLITUDE OF THE ONE-QUANTUM TERM IN THE NEUTRON
    SCATTERING LAW FOR MOLECULAR GASES.
    ≠CARPENTER(J.M.).

SECTION   7B-THEORY OF NEUTRON NUCLEAR SCATTERING IN CONDENSED MATTER . .

26   J. CHEM. PHYSICS VOL.46,   352, (1967)............N2, H*CL, H*F, H*C*N,.
     *INFLUENCE OF ROTATIONAL LEVELS ON SLOW-NEUTRON SCATTERING BY LINEAR
     GASES.
     *LURIE(N.A.).

27   J. CHEM. PHYSICS VOL.47,   337, (1967)............POLYETHELENE,.
     *INELASTIC-SCATTERING CROSS SECTION OF NEUTRON BY CRYSTAL VIBRATIONS OF
     POLYETHYLENE.
     *KITAGAWA(T.), MIYAZAWA(T.).

28   J. CHEM. PHYSICS VOL.48,   3016, (1968).FOR MIXTURES OF AMORPHOUS SOLIDS OR LIQUIDS.
     *ANALYSIS OF SCATTERING DATA FOR MIXTURES OF AMORPHOUS SOLIDS OR LIQUIDS.
     *PINGS(C.J.), WASER(J.).

29   J. CHEM. PHYSICS VOL.49,   890, (1968).BY GASEOUS O2.
     *INELASTIC NEUTRON SCATTERING BY GASEOUS O2.
     *SUMMERFIELD(G.C.), LURIE(N.A.).

30   J. CHEM. PHYS. VOL.58,   1143, (1973).T C1.
     *INCOHERENT NEUTRON SCATTERING AND MOLECULAR REORIENTATIONS IN CRYSTALS
     NEAR T CRITICAL.
     *MICHEL(K.H.).

31   J. EXPTL. THEORET. PHYS. VOL.17,   769, (1947)....T.
     *SCATTERING OF SLOW NEUTRONS IN CRYSTALS.
     *AKHIEZER(A.I.), POMERANCHUK(I.YA.). (IN RUSSIAN).

32   J. LOW TEMP. PHYS. VOL.9,   485, (1972).....T,B7.....K*D2*P*O4-FERROELECTRICS,.
     *THEORY OF COHERENT NEUTRON SCATTERING BY H-BONDED FERROELECTRICS AT LOW
     TEMP. II.SCATT. CHARACTERISTICS/CONCEPTIONS OF/TUNNELING QUASISPIN MODEL
     *STAMENKOVIC(S.).

33   J. LOW TEMP. PHYS. VOL.9,   475, (1972).....T,B7.....K*D2*P*O4-FERROELECTRICS,.
     *THEORY OF COHERENT NEUTRON SCATTERING BY HYDROGEN-BONDED FERROELECTRICS
     AT LOW TEMPERATURES. I.GENERAL EXPRESSION FOR/SCATTERING/THERMAL FACTORS
     *STAMENKOVIC(S.).

34   J. NUCL. SCI. TECHNOL. VOL.8,   406, (1971)....T.
     *SLOW NEUTRON SCATTERING AND CLASSICAL MEAN-SQUARE DISPLACEMENT FUNCTIONS
     OF ATOMS.
     *NISHIGORI(T.), SEKIYA(T.).

35   J. PHYS. -C-, VOL.2,   981, (1969).
     *THE THEORY OF THE ELECTROSTATIC SCATTERING OF THERMAL NEUTRONS BY SOLIDS
     *LOVESEY(S.W.).

36   J. PHYS.F: METAL PHYS. VOL.2,   209, (1972).
     *NEUTRON SCATTERING DUE TO LATTICE DISTORTION AROUND POINT DEFECTS.
     *MACDONALD(R.A.).

37   J. PHYS. RADIUM SER.7 VOL.8,   29, (1937)....E.
     *RECHERCHES SUR LES NEUTRONS LENTS.
     *VON-HALBAN(H.), PREISWERK(P.).

38   KERNTECHNIK VOL.13,   525, (1971)....P.
     *THE SCATTERING OF SLOW NEUTRONS.
     *SCHMATZ(W.), STILLER(H.).

39   METAL. SOC. CONF. LOS ANGELES,   1, (1967).
     *INTRODUCTION TO THE THEORY OF SLOW NEUTRON SCATTERING.
     *BLUME(M.).

40   NUCLEONICS VOL.6 NO.2,   66, (1950)....T,E.
     *NEUTRON SPECTROSCOPY FOR CHEMICAL ANALYSIS-II. RESONANCE EFFECTS;
     TRANSMISSION IN THE 1/V REGION.
     *HAVENS-JR(W.W.), TAYLOR(T.I.).

41   NUCL. INST. MET. VOL.80,   187, (1970)....P.
     *MULTIPLE SCATTERING CORRECTION FOR INELASTIC SCATTERING FROM CYLINDRICAL
     TARGETS.
     *ENGELBRECHT(C.A.).

42   NUCL. SCI. ENGNG. VOL.12,   260, (1962)....T.
     *ON THE EVALUATION OF THE THERMAL NEUTRON SCATTERING LAW.
     *EGELSTAFF(P.A.), SCHOFIELD(P.).

43   NUCL. SCI. ENG. VOL.48,   266, (1972)....T.
     *MONTE CARLO EVALUATION OF MULTIPLE SCATTERING AND RESOLUTION EFFECTS IN
     DOUBLE-DIFFERENTIAL NEUTRON SCATTERING CROSS-SECTION MEASUREMENTS.
     *BISCHOFF(F.G.), YEATER(M.L.), MOORE(W.E.).

44   NUOVO CIMENTO VOL.1,   233, (1955)....T.
     *INTERFERENCE EFFECTS IN THE TOTAL NEUTRON SCATTERING CROSS-SECTION OF
     CRYSTALS.
     *PLACZEK(G.), VAN-HOVE(L.).

45   PHIL. MAG. VOL.3,   213, (1958)....T.
     *THE THEORY OF SMALL ANGLE SCATTERING FROM DISLOCATIONS.
     *ATKINSON(H.H.), HIRSCH(P.B.).

46   PHYSIK. Z. SOWJETUNION VOL.13,   65, (1938)....T.
     *SCATTERING OF SLOW NEUTRONS IN A CRYSTAL LATTICE.
     *POMERANCHUK(I.YA.).

47   PHYS. KOND. MATER. (GER.) VOL.14, 336, (1972)....T.
     *NEUTRON SCATTERING FROM QUANTUM CRYSTALS.
     *BECK(H.), MEIER(P.F.).

48   PHYS. LETT. 3,   162, (1962)....NA.
     *SLOW NEUTRON SCATTERING EVIDENCE FOR VIOLATING OF THE F-SUM RULE IN
     LIQUID NA.
     *RANDOLPH(P.D.).

49   PHYS. LETT. 25A,   211, (1967).
     *SUM RULE CRITERION IN COHERENT SLOW NEUTRON SCATTERING IN LIQUIDS.
     *DESAI(R.C.), YIP(S.).

50   PHYS. LETT. VOL.27A,   93, (1968).
     *ON THE CONVOLUTION APPROXIMATION IN THE THEORY OF COHERENT NEUTRON
     SCATTERING.
     *ANDRIESSE(C.D.).

51   PHYS. REV. VOL.49,   229, (1936)....T.
     *THE ENERGY DISTRIBUTION OF NEUTRONS SLOWED BY ELASTIC IMPACTS.
     *CONDON(E.U.), BREIT(G.).

SECTION   78-THEORY OF NEUTRON NUCLEAR SCATTERING IN CONDENSED MATTER . .

52 PHYS. REV. VOL.55, 190, (1939). IN T A CRYSTAL.
       *CAPTURE OF NEUTRONS BY ATOMS IN A CRYSTAL.
       *LAMB-JR(W.R.).

53 PHYS. REV. VOL.57, 976, (1940). P E
       *INTERFERENCE PHENOMENA IN THE SCATTERING OF SLOW NEUTRONS.
       *BEYER(H.G.), WHITAKER(M.D.).

54 PHYS. REV. VOL.59, 981, (1941). T CRYSTALS AND POLYCRYSTALS.
       *THE PASSAGE OF NEUTRONS THROUGH CRYSTALS AND POLYCRYSTALS.
       *HALPERN(O.), HAMERMESH(M.), JOHNSON(M.H.).

55 PHYS. REV. VOL.62, 37, (1942). T
       *ON THE INELASTIC SCATTERING OF NEUTRONS BY CRYSTAL LATTICES.
       *SEEGER(R.J.), TELLER(E.).

56 PHYS. REV. VOL.65, 1, (1944). T
       *INELASTIC SCATTERING OF SLOW NEUTRONS.
       *WEINSTOCK(R.).

57 PHYS. REV. VOL.71, 666, (1947). E
       *INTERFERENCE PHENOMENA OF SLOW NEUTRONS.
       *FERMI(E.), MARSHALL(L.).

58 PHYS. REV. VOL.71, 232, (1947). T H2*O
       *THE SCATTERING OF SLOW NEUTRONS BY BOUND PROTONS II. HARMONIC BINDING-
       NEUTRONS OF ZERO ENERGY.
       *BREIT(G.), ZILSEL(P.R.).

59 PHYS. REV. VOL.71, 215, (1947). T
       *THE SCATTERING OF SLOW NEUTRONS BY BOUND PROTONS I. METHODS OF
       CALCULATION.
       *BREIT(G.).

60 PHYS. REV. VOL.76, 1811, (1949). T
       *MULTIPLE SCATTERING OF NEUTRONS. II. DIFFUSION IN A PLATE OF FINITE
       THICKNESS.
       *HALPERN(O.), LUNEBURG(R.K.).

61 PHYS. REV. VOL.82, 392, (1951). T
       *EFFECT OF SHORT WAVELENGTH INTERFERENCE ON NEUTRON SCATTERING BY DENSE
       SYSTEMS OF HEAVY NUCLEI.
       *PLACZEK(G.), NIJBOER(B.R.A.), VAN-HOVE(L.).

62 PHYS. REV. VOL.83, 379, (1951). E T
       *SMALL ANGLE SCATTERING OF NEUTRONS.
       *WEISS(R.J.).

63 PHYS. REV. VOL.86, 377, (1952). T
       *THE SCATTERING OF NEUTRONS BY SYSTEMS OF HEAVY NUCLEI.
       *PLACZEK(G.).

64 PHYS. REV. VOL.88, 1003, (1952). T
       *REMARKS ON SOME QUESTIONS OF NEUTRON OPTICS.
       *HALPERN(O.).

65 PHYS. REV. VOL.85, 633, (1952). T
       *GEOMETRICAL OPTICS AND THE THEORY OF MULTIPLE SMALL ANGLE SCATTERING.
       *VINEYARD(G.H.).

66 PHYS. REV. VOL.95, 249, (1954). T
       *CORRELATIONS IN SPACE AND TIME AND BORN APPROXIMATION SCATTERING IN
       SYSTEMS OF INTERACTING PARTICLES.
       *VAN-HOVE(L.).

67 PHYS. REV. VOL.94, 1228, (1954). T
       *THE SCATTERING OF NEUTRONS BY SYSTEMS CONTAINING LIGHT NUCLEI.
       *WICK(G.C.).

68 PHYS. REV. VOL.93, 895, (1954). T
       *INCOHERENT NEUTRON SCATTERING BY POLYCRYSTALS.
       *PLACZEK(G.).

69 PHYS. REV. VOL.105, 1240, (1957). T
       *INCOHERENT NEUTRON SCATTERING BY POLYCRYSTALS.
       *PLACZEK(G.).

70 PHYS. REV. VOL.117, 1029, (1960). T II.
       *SLOW-NEUTRON SCATTERING BY ROTATORS. II.
       *VOLKIN(H.C.).

71 PHYS. REV. LETT. VOL.4, 239, (1960). T
       *SPACE-TIME CORRELATION FUNCTION FORMALISM FOR SLOW NEUTRON SCATTERING.
       *SCHOFIELD(P.).

72 PHYS. REV. VOL.126, 1165, (1962). T
       *QUASI-CLASSICAL TREATMENT OF NEUTRON SCATTERING.
       *AAMODT(R.), CASE(K.M.), ROSENBAUM(M.), ZWEIFEL(P.F.).

73 PHYS. REV. 130,. 1334, (1963). T
       *INTERMEDIATE SCATTERING FUNCTION IN SLOW NEUTRON SCATTERING.
       *RAHMAN(A.).

74 PHYS. REV. 131,. 1149, (1963). T
       *SLOW NEUTRON SCATTERING, THE *SCATTERING LAW* AND G(R,T).
       *SUMMERFIELD(G.C.), ZWEIFEL(P.F.).

75 PHYS. REV. VOL.135, A1691, (1964). T
       *DIRECT CALCULATION OF ELECTRONIC PROPERTIES OF METALS FROM NEUTRON
       SCATTERING DATA.
       *BAYM(G.).

76 PHYS. REV. VOL.136, B112, (1964). T
       *NEUTRON POLARIZATION AT SMALL SCATTERING ANGLES NEAR AN S-WAVE RESONANCE
       *REDMOND(R.F.).

77 PHYS. REV. VOL.135, A603, (1964). T
       *SLOW NEUTRON SCATTERING BY MOLECULAR HYDROGEN AND DEUTERIUM.
       *YOUNG(J.A.), KOPPEL(J.U.).

78 PHYS. REV. VOL.133, A50, (1964). T
       *INCOHERENT INELASTIC NEUTRON SCATTERING AND SELF-DIFFUSION.
       *ZWANZIG(R.).

SECTION 7B-THEORY OF NEUTRON NUCLEAR SCATTERING IN CONDENSED MATTER . .

79 PHYS. REV. VOL.137, B271, (1965)
    *QUASI-CLASSICAL THEORY OF NEUTRON SCATTERING.
    *ROSENBAUM(M.), ZWEIFEL(P.F.).

80 PHYS. REV. VOL.139, A1138, (1965)
    *DERIVATION OF KINETIC EQUATIONS FOR SLOW-NEUTRON SCATTERING.
    *VAN-LEEUWEN(J.M.J.), YIP(S.).

81 PHYS. REV. 151,. 464, (1966)
    *SELECTION RULES FOR SCATTERING PROCESSES IN CRYSTALS.
    *ZAK(J.).

82 PHYS. REV. VOL.149,. 1, (1966) T
    *EFFECT OF NUCLEAR SPIN CORRELATIONS ON THE SCATTERING OF NEUTRONS BY
    MOLECULES.
    *SINHA(S.K.), VENKATARAMAN(G.).

83 PHYS. REV. 159,. 1, (1967)
    *APPLICATION OF THE QUASICLASSICAL METHOD TO INCOHERENT NEUTRON
    SCATTERING BY MOLECULES.
    *SUMMERFIELD(G.C.), ZWEIFEL(P.F.).

84 PHYS. REV. 168,. 752, (1968)
    *NEUTRON SCATTERING FROM A N-DIMENSIONAL LATTICE(N=1,2,3).
    *KOTHARI(L.S.), GOYAL(I.C.).

85 PHYS. REV. 168,. 725, (1968)
    *RANDOM IMPURITY PROBLEM.
    *PERSHAN(P.S.), LACINA(W.B.).

86 PHYS. REV. VOL.188,. 1445, (1969)
    *SYMMETRIES IN SCATTERING OF SLOW NEUTRONS.
    *KUSCER(I.), SUMMERFIELD(G.C.).

87 PHYS. REV. B VOL.6,. 4533, (1972) T
    *GROUP-THEORETICAL SELECTION RULES IN INELASTIC NEUTRON SCATTERING
    WITHIN THE RIGID-MOLECULE MODEL.
    *CASELLA(R.C.), TREVINO(S.F.).

88 PHYS. REV. A VOL.8,. 905, (1973) T
    *FINAL-STATE EFFECTS ON THERMAL-NEUTRON SCATTERING AT HIGH-ENERGY
    TRANSFER.
    *GERSCH(H.A.), RODRIGUEZ(L.J.).

89 PHYS. STAT. SOL. VOL.41,. 767, (1970) P AL,BI
    *TOTAL COHERENT CROSS SECTIONS FOR THE SCATTERING OF NEUTRONS FROM
    CRYSTALS.
    *BINDER(K.).

90 PHYS. STAT. SOL. VOL.42,. 757, (1970)
    *THE SEPARATION OF COHERENT AND INCOHERENT CROSS-SECTIONS IN INELASTIC
    NEUTRON SCATTERING USING POLARIZED NUCLEAR TARGETS.
    *JAUHO(P.), PIRILA(P.).

91 PHYS. STAT. SOL. VOL.47,. 143, (1971)
    *MULTIPLE SCATTERING OF THERMAL NEUTRONS BY A PERFECT CRYSTAL
    II. CALCULATION OF APPROXIMATE VALUES FOR THE INFORMATION QUANTITIES.
    *SOLBRIG(H.).

92 PHYS. STAT. SOL. VOL.46,. 273, (1971)
    *MULTIPLE SCATTERING OF THERMAL NEUTRONS BY A PERFECT CRYSTAL I. GENERAL
    FOUNDATIONS.
    *LENK(R.), SOLBRIG(H.).

93 PHYS. STAT. SOL. VOL.49,. K75, (1972)
    *NEUTRON SCATTERING FROM A FINITE LINEAR CHAIN.
    *SOBHANA(S.), GHATAK(A.K.).

94 PHYS. STAT. SOL. VOL.51,. 555, (1972) P
    *MULTIPLE SCATTERING OF THERMAL NEUTRONS BY A PERFECT CRYSTAL.
    III. CALCULATION OF THE NEUTRON WAVE FIELD.
    *SOLBRIG(H.).

95 PROC. ACAD. SCI. AMSTERDAM 39,. 1049, (1936) T
    *SCATTERING OF NEUTRONS IN MATTER III.
    *ORNSTEIN(L.S.).

96 PROC. ACAD. SCI. AMSTERDAM 39,. 914, (1936) T
    *SCATTERING OF NEUTRONS IN MATTER II.
    *ORNSTEIN(L.S.).

97 PROC. ACAD. SCI. AMSTERDAM 39,. 810, (1936) T
    *SCATTERING OF NEUTRONS IN MATTER.
    *ORNSTEIN(L.S.).

98 PROC. IMP. ACAD. JAPAN VOL.15,. 214, (1939) P HG*S
    *NON-ADDITIVITY IN SCATTERING CROSS-SECTIONS OF SLOW NEUTRONS.
    *KIMURA(M.).

99 PROC. PHYS. SOC. 91,. 903, (1967)
    *GROUP THEORETICAL SELECTION RULES IN INELASTIC NEUTRON SCATTERING.
    *ELLIOTT(R.J.), THORPE(M.F.).

100 PROC. PHYS. SOC. 92,. 551, (1967)
    *CLASSICAL DIFFERENTIAL SCATTERING CROSS-SECTIONS.
    *FLANNERY(M.R.).

101 PROC. PHYS. SOC. 92,. 889, (1967)
    *INFLUENCE OF ELASTIC COLLISIONS ON THE BROADENING OF ION LINES.
    *DAVIS(J.), ROBERTS(D.E.).

102 PROC. ROY. SOC. A VOL.138,. 460, (1932) T
    *PASSAGE OF NEUTRONS THROUGH MATTER.
    *MASSEY(H.S.W.).

103 PROC. ROY. SOC. A VOL.162,. 127, (1937) E,P
    *SCATTERING OF SLOW NEUTRONS.
    *GOLDHABER(M.), BRIGGS(G.H.).

104 PROC. ROY. SOC. 301,. 363, (1967)
    *DIFFRACTION BY A LATTICE OF S-WAVE SCATTERERS.
    *PLASKETT(J.S.).

105 PROGR. NUCL. PHYS. VOL.1,. 185, (1950) T,E
    *THE SCATTERING OF NEUTRONS BY CRYSTALS.
    *CASSELS(J.H.).

SECTION   7B-THEORY OF NEUTRON NUCLEAR SCATTERING IN CONDENSED MATTER . .

106   PROGR. THEOR. PHYS. VOL.37, . . . 1051, (1967)....T...........
      *SLOW NEUTRON SCATTERING AND SPACE-TIME CORRELATION FUNCTIONS.
      *SUNAKAWA(S.), YAMASAKI(S.), NIGHIGORI(T.).

107   REPORT AEEW-M730, ... (1969). . . .P............
      *AN INVESTIGATION OF A MONTE CARLO TREATMENT OF THERMAL NEUTRON SCATT.
      *BASHER(J.C.), PULL(I.C.). (ED. NOTE: AVAIL. HMSO, LONDON, ENGLAND).

108   REPORT FN-37/70 (22PP.), ... (1970)....P...........
      *SCATTERING OF NEUTRONS BY MOLECULAR SYSTEMS.
      *BALLY(D.). (ED. NOTE: AVAIL. COMITETUL PENTRU ENERGIA NUCLEARE INST.
      FIZICA ATOMICA, BUCHAREST).

109   REP. PROGR. PHYS. VOL.36, . . . 1073, (1973)............
      *PHYSICO-CHEMICAL ASPECTS OF NEUTRON STUDIES OF MOLECULAR MOTION (REVIEW)
      *ALLEN(G.), HIGGINS(J.S.).

110   RICERCA SCI. VOL.7, II, . . . 13, (1936)....T..........
      *MOTION OF NEUTRONS IN HYDROGENOUS SUBSTANCES.
      *FERMI(E.).

111   RICERCA SCI. VOL.8, I, . . . 400, (1937)....C1,T.....
      *DIFFUSION OF NEUTRONS IN CRYSTALS.
      *WICK(G.C.).

112   SOL. STATE PHYS. VOL.8, . . . 109, (1959)....T.....
      *INTERACTION OF THERMAL NEUTRONS WITH SOLIDS (REVIEW PAPER).
      *KOTHARI(L.S.), SINGWI(K.S.).

113   SOL. STATE COMM. VOL.12, . . . 1045, (1973)....T...
      *DETERMINATION BY POLARIZATION ANALYSIS OF/SCATTERING AMPLITUDES A(+),
      A(-) AND/DEGREE OF ORIENTATION OF NUCLEI ...IN MAGNETICALLY ORDERED SUB.
      *BERTAUT(E.F.).

114   SOV. AT. ENERGY VOL.28, ... 398, (1970).............
      *ANOMALOUS SCATTER OF NEUTRONS.
      *LEBEDEVA(N.S.), MOROZOV(V.M.).

115   SOV. PHYS. JETP VOL.5, . . . 1115, (1957)....T....
      *THEORY OF DIFFUSE SCATTERING OF X-RAYS AND THERMAL NEUTRONS IN SOLID
      SOLUTIONS. II. MICROSCOPIC THEORY.
      *KRIVOGLAZ(M.A.).

116   SOV. PHYS. JETP VOL.7, . . . 139, (1958)....T....
      *THEORY OF DIFFUSE SCATTERING OF X-RAYS AND THERMAL NEUTRONS. III.ACCOUNT
      OF GEOMETRICAL DISTORTIONS OF THE LATTICE.
      *KPIVOGLAZ(M.A.).

117   SOV. PHYS. JETP VOL.7, . . . 89, (1958)....T....
      *ON POLARIZATION OF SLOW NEUTRONS SCATTERED IN CRYSTALS.
      *MALEEV(S.V.).

118   SOV. PHYS. JETP. 16, . . . 1624, (1963).....
      *:OBSERVED: PROBABILITIES OF ELASTIC SCATTERING OF NEUTRONS, MOSSBAUER
      EFFECT IN DEGENERATE SYSTEMS, NEW POSSIBILITIES.
      *KAZARNOVSKII(M.V.), STEPANOV(A.V.).

119   SOV. PHYS. JETP. 17, . . . 390, (1963)............
      *SCATTERING OF NEUTRONS BY MOLECULES WITH LARGE ENERGY TRANSFER.
      *IVANOV(G.K.).

120   SOV. PHYS. JETP. 17, . . . 445, (1963)............
      *EXCLUSION OF COHERENT SCATTERING FROM CROSS SECTION FOR SCATTERING OF
      SLOW NEUTRONS BY SIMPLE CRYSTAL LATTICES.
      *OSKOTSKII(V.S.).

121   SOV. PHYS. USPEKHI VOL.80, . . . 359, (1963)....T.....
      *THEORY OF SCATTERING OF SLOW NEUTRONS IN MAGNETIC CRYSTALS.
      *IZYUMOV(YU.A.).

122   SOV. PHYS. SOL. STATE VOL.5, . . . 2526, (1963)...........
      *EFFECT OF LONG-RANGE FORCES ON FLUCTUATIONS AND THE SCATTERING OF WAVES
      IN CRYSTALS.
      *KRIVOGLAZ(M.A.).

123   SOV. PHYS. JETP. 16, . . . 109, (1963).............
      *CERENKOV RADIATION AND INELASTIC SCATTERING OF PARTICLES.
      *KAGANOV(M.I.).

124   SOV. PHYS. SOL. STATE, 6, . . . 2465, (1964)..............
      *SPIN-ELECTRON INTERACTION IN SEMICONDUCTORS, BROADENING, INELASTIC
      MAGNETIC SCATTERING OF NEUTRONS.
      *IVANOV(V.A.).

125   SOV. PHYS. JETP. 18, . . . 1006, (1964)............
      *SCATTERING THEORY IN THE IMPULSE APPROXIMATION.
      *IVANOV(G.K.), SAYASOV(YU.S.).

126   SOV. PHYS. JETP. 19, . . . 986, (1964)............
      *SCATTERING OF THERMAL NEUTRONS BY POLARIZED NUCLEI OF A FERROMAGNET.
      *KHARADZE(G.A.).

127   SOV. PHYS. JETP. 20, . . . 1037, (1965)............
      *THEORY OF SCATTERING OF NEUTRONS IN DISORDERED CRYSTALS.
      *PRIVOROTSKII(I.A.).

128   SOV. PHYS. JETP. 21, . . . 765, (1965)............
      *EFFECT OF THE INTERACTION BETWEEN NEUTRONS AND NUCLEI ON THE WIDTH OF
      PARAMAGNETIC RESONANCE IN A NEUTRON BEAM.
      *BARYSHEVSKII(V.G.), LYUBOSHITZ(V.L.), PODGORETSKII(M.I.).

129   SOV. PHYS. JETP. 23, . . . 1061, (1966)............
      *INELASTIC DIFFRACTION SCATTERING.
      *INOPIN(E.V.).

130   SOV. PHYS. JETP. 23, . . . 481, (1966)............
      *ELASTIC AND QUASIELASTIC SCATTERING OF NEUTRONS BY MOLECULES.
      *IVANOV(G.K.).

131   SOV. PHYS. SOL. STATE VOL.9, . . . 2040, (1968).............
      *THEORY OF DIFFUSION IN CRYSTALLINE SOLID SOLUTIONS, TIME EVOLUTION OF
      DIFFUSE SCATT. OF X-RAYS, THERMAL NEUTRON.
      *KHACHATURYAN(A.G.).

SECTION   7B-THEORY OF NEUTRON NUCLEAR SCATTERING IN CONDENSED MATTER . .

132   SOV. PHYS. SOL. STATE VOL.10, . 1243, (1968)
      *SINGULARITIES IN ENERGY DISTRIBUTION OF SLOW NEUTRONS INELASTICALLY
      SCATTERED IN CRYSTALS HAVING EXTENDED DEFECTS.
      *IOSILEVSKII(YA.A.).

133   SOV. PHYS. JETP. 26, ... 1139, (1968)................
      *A CRITERION IDENTIFYING WEAKLY PERTURBING SYSTEMS BASED ON MEASURED
      ELASTIC SCATTERING.
      *IWAVOV(V.V.), FEDOROVICH(G.V.).

134   SOV. PHYS. J. VOL.34, ... 138, (1970).....A4.......
      *ANOMALIES IN THE ELASTIC SCATTERING OF NEUTRONS AT SMALL ANGLES OF
      INCIDENCE.
      *GORLOV(G.V.), LEBEDEVA(N.S.), MOROZOV(V.M.).

135   SOV. PHYS. SOLID STATE VOL.14, . 1761, (1973)....T.........
      *FLUCTUATIONS AND SCATTERING OF SLOW NEUTRONS BY COLLECTIVE EXCITATIONS
      OF THE ELECTRON FERMI LIQUID IN METALS.
      *AKHIEZER(I.A.), CHUDNOVSKII(E.M.).

136   TECH. METALS RES. VOL.11, ... 641, (1969)................
      *NEUTRON SCATTERING TECHNIQUES (REVIEW PAPER).
      *BACON(G.E.).

137   UKAEA AERE REP.R 5479 (35 PP.). . . . , (1967)....T,B3....
      *THE MATHEMATICAL FORMULATION OF THE SCATTERING OF LONG WAVELENGTH
      NEUTRONS BY DEFECTS IN SOLIDS.
      *MARTIN(D.G.).

138   UKR. FIZ. ZH. (USSR) VOL.3, . . . 743, (1958)....T.........
      *THE THEORY OF NUCLEAR SCATTERING OF SLOW NEUTRONS IN ALLOYS.
      *DANYLENKO(V.M.), MATYSINA(Z.A.). (IN UKRAINIAN).

139   UKR. FIZ. ZH. (USSR) VOL.17, . . . 45, (1972)....T.........
      *ON THE PROBLEM OF THE G(DELTA-E) FUNCTION DERIVATION BY THE METHOD OF
      SLOW NEUTRON INELASTIC COHERENT SCATTERING.
      *GORBACHEV(B.I.), IVANITSKII(P.G.), KROTENKO(V.T.).

140   Z. NATURFORSCH. A VOL.26A, . . . 391, (1971)....E.........
      *MEASUREMENTS OF NEUTRON-SCATTERING-AMPLITUDES USING THE CHRISTIANSEN-
      FILTER-TECHNIQUE.
      *KOESTER(L.), KNOPF(K.).

SECTION   8 -THEORY OF NEUTRON MAGNETIC SCATTERING. . . . .

1 ACTA CRYST. VOL.10,. . . . .   598, (1957). . . . .
   *NEUTRON MAGNETIC SCATTERING FACTORS IN THE PRESENCE OF EXTINCTION.
   *CHANDRASEKHAR(S.), WEISS(R.J.).

2 ACTA PHYS. POLON. VOL.19,. . . .   691, (1960). . . .D1,T. . . . .
   *SCATTERING OF SLOW NEUTRONS IN FERRIMAGNETIC CRYSTALS.
   *KOCINSKI(J.).

3 ACTA PHYS. SINICA VOL.19,. . . .   673, (1963). . . .T. . . . .
   *INELASTIC SCATTERING OF NEUTRONS IN FERROMAGNETICS WITH IMPERFECTIONS.
   *LI-ZHI(F.), SHI-JIE(GU.). (IN CHINESE).

4 ACTA PHYS. POLON. VOL.26,. . . .   11, (1964). . . .T. . . . .
   *ON THE THEORY OF FERROMAGNETICS WITH HELICOIDAL STRUCTURES. V. EFFECT OF
   A MAGNETIC FIELD ON THE SCATTERING OF NEUTRONS.
   *KASHCHEEV(V.N.). (IN RUSSIAN).

5 ACTA PHYS. POLON. VOL.30,. . . .   591, (1966). . . .D2. . . . .
   *ON THE THERMODYNAMIC FLUCTUATIONS OF MAGNETIZATION IN FERROMAGNETS.
   *KOCINSKI(J.).

6 ACTA PHYS. POLON. VOL.33,. . . .   13, (1968). . . .T,D2. . . . .
   *ON THE CRITICAL MAGNETIC SCATTERING OF NEUTRONS.
   *KOCINSKI(J.).

7 A.I.P. CONF. PROC. NO.10 PART 2, 1664, (1973). . . .T. . . . . . . .TM*SB,TB*SB,(PR-LA)3*TL, ...
   *SINGLET-GROUND-STATE DYNAMICS.
   *BIRGENEAU(R.J.).

8 ANN. DE PHYSIQUE TOME 7 NO.5,. . .   371, (1972). . . . .
   *THERMAL NEUTRON SCATTERING BY MAGNETIC INTERACTION.
   *WILL(G.).

9 ANN. DE PHYSIQUE TOME 7 NO.5,. . .   329, (1972). . . .E. . . . . . . .PR*BI,(TM-Y)*AL2,. . . . . .
   *PHYSICAL ASPECTS OF THE CRYSTALLINE ELECTRIC FIELD (CEF).
   *PURWINS(H.G.).

10 ANN. DE PHYSIQUE TOME 7 NO.4,. . .   299, (1972). . . . .
   *CRITICAL PHENOMENA AND NEUTRON SCATTERING (REVIEW PAPER).
   *PARETTE(G.). (PRESENTED AT AUTRANS SUMMER SCHOOL 1972).

11 ANN. DE PHYSIQUE TOME 7 NO.4,. . .   287, (1972). . . . .
   *MAGNETIC NEUTRON SCATTERING IN RELATION TO SOME OTHER EXPERIMENTAL
   METHODS.
   *MOSSBAUER(R.L.). (PRESENTED AT AUTRANS SUMMER SCHOOL 1972).

12 ANN. DE PHYSIQUE TOME 7 NO.4,. . .   269, (1972). . . . .
   *DILUTE ALLOYS (REVIEW PAPER).
   *MEZEI(F.), RADHAKRISHNA(P.). (PRESENTED AT AUTRANS SUMMER SCHOOL 1972).

13 ANN. DE PHYSIQUE TOME 7 NO.4,. . .   233, (1972). . . . .
   *SPIN WAVES (REVIEW PAPER).
   *HENNION(B.), MOUSSA(F.). (PRESENTED AT AUTRANS SUMMER SCHOOL 1972).

14 ATLANTA SYMP/GEORGIA INST. TECH.   1, (1967). . . . .
   *NEUTRON INTERACTIONS WITH ATOMS.
   *SHULL(C.G.).

15 ATLANTA SYMP/GEORGIA INST. TECH.   53, (1967). . . . .
   *MAGNETIC SCATTERING OF NEUTRONS AND MICROSCOPIC MAGNETIC PROPERTIES OF
   SOLIDS.
   *KOEHLER(W.C.).

16 CONTEMPORARY PHYS. VOL.7,. . . . .   278, (1966). . . . .
   *NEUTRON SPECTROSCOPY OF SOLIDS. I.
   *LOMER(W.M.).

17 C.R. ACAD. SCI. URSS VOL.20,. . .   551, (1938). . . . .
   *SCATTERING OF NEUTRONS IN FERROMAGNETICS.
   *MIGDAL(A.). (IN ENGLISH).

18 C.R. ACAD. SCI. B VOL.274,. . .   423, (1972). . . .T,E. . . . .
   *RESONANCE NUCLEAIRE #PSEUDO-MAGNETIQUE# DU NEUTRON INDUITE PAR UN CHAMP
   NUCLEAIRE DE RADIOFREQUENCE.
   *ABRAGAM(A.), BACCHELLA(G.L.), GLATTLI(H.), MERIEL(P.), PIESVAUX(J.),
   PINOT(M.).

19 DISS. ABS. VOL.30,. . . . . . .4755B, (1970). . . . .
   *THE SPIN-DEPENDENT SCATTERING OF THERMAL NEUTRONS FROM POLARIZED NUCLEI
   IN MAGNETICALLY ORDERED SOLIDS.
   *MARKWORTH(A.J.).

20 INT. J. MAGN. VOL.1,. . . . . . .   45, (1970). . . .B7. . . . . .ANTIFERROMAGNETS,. . . . .
   *INELASTIC SCATTERING OF NEUTRONS FROM ANTIFERROMAGNETS.
   *SCHULHOF(M.P.).

21 J. APPL. PHYS. VOL.39,. . . . .   533, (1968). . . .D3. . . . .
   *TEMPORAL FLUCTUATIONS OF MOMENT IN PARAMAGNETS BY NEUTRON DIFFRACTION.
   *COLLINS(M.F.).

22 J. APPL. PHYS. VOL.40,. . . . . 1226, (1969). . . .P. . . . .
   *NEUTRON SCATTERING STUDIES OF THIN-FILM MAGNETIZATION.
   *SPOONER(S.), WREGE(D.E.), LIVESAY(B.R.).

23 J. APPL. PHYS. 40,. . . . . . 1552, (1969). . . . .
   *MAGNETIC SCATTERING OF NEUTRONS FROM HEISENBERG ANTIFERROMAGNETS.
   *BENNETT(H.S.).

24 J. APPL. PHYS. 41,. . . . . . 1365, (1970). . . . .
   *SHORT-RANGE ORDER EFFECTS IN PARAMAGNETIC NEUTRON SCATTERING.
   *TAHIR-KHELI(R.), MALINOSKI(F.).

25 J. APPL. PHYS. 42,. . . . . . 1610, (1971). . . . .
   *ANALYSIS OF DIFFUSE ELASTIC NEUTRON SCATTERING FROM MAGNETIC ALLOYS.
   *GARLAND(J.W.), BENNEMANN(K.H.).

26 J. APPL. PHYS. 42,. . . . . . 1390, (1971). . . . .
   *SPIN CORRELATIONS IN THE PARAMAGNETIC PHASE.
   *HUBBARD(J.).

SECTION   8 -THEORY OF NEUTRON MAGNETIC SCATTERING. . . . .

27 J. CHEM. PHYS. VOL.37, . . . 1245, (1962). . . . T. AND X-RARE-EARTH IONS.
    *NEUTRON MAGNETIC FORM FACTORS AND X-RAY ATOMIC SCATTERING FACTORS FOR
    RARE-EARTH IONS.
    *BLUME(M.), FREEMAN(A.J.), WATSON(R.E.).

28 J. CHEM. PHYSICS VOL.49, ... 890, (1968). . . . . 02; . . . .
    *INELASTIC NEUTRON SCATTERING BY GASEOUS O2.
    *SUMMERFIELD(G.C.), LURIE(N.A.).

29 J. PHYS. CHEM. SOLIDS VOL.4, . . . 223, (1958). . . . T. OF NEUTRONS AT HIGH TEMPERATURES.
    *INELASTIC MAGNETIC SCATTERING OF NEUTRONS AT HIGH TEMPERATURES.
    *DE-GENNES(P.G.).

30 J. PHYS. CHEM. SOLIDS VOL.6, ... 43, (1958). . . . T. ABOVE THE NEEL POINT.
    *NEUTRON SCATTERING BY ANTIFERROMAGNETS ABOVE THE NEEL POINT.
    *DE-GENNES(P.G.).

31 J. PHYS. C VOL.2, ... 317, (1969). . . . T. SCATTERING CROSS SECTION OF
    *TIME-DEPENDENT SUSCEPTIBILITY AND NEUTRON SCATTERING CROSS SECTION OF
    THE ISING MODEL WITH GENERAL SPIN.
    *ESSAM(J.W.), GARELICK(H.).

32 J. PHYS. -C-, VOL. 2, ... 470, (1969). . . . T. OF NEUTRONS BY MAGNETIC
    *SOME ASPECTS OF THE THEORY OF THE SCATTERING OF NEUTRONS BY MAGNETIC
    IONS.
    *LOVESEY(S.W.).

33 J. PHYS. CHEM. SOLIDS VOL.31, . . . 255, (1970). . . . T,D1. AS APPLIED TO
    *DYNAMICAL SPIN CORRELATIONS IN THE B-P-W CLUSTER MODEL AS APPLIED TO
    INELASTIC NEUTRON SCATTERING.
    *KWON(T.H.), GERSCH(H.A.).

34 J. PHYS. C VOL.3, ... 1292, (1970). . . . T. CONTAINING IONS WITH
    *THEORY OF THERMAL NEUTRON SCATTERING FROM CRYSTALS CONTAINING IONS WITH
    UNPAIRED F ELECTRONS.
    *BALCAR(E.), LOVESEY(S.W.), WEDGWOOD(F.A.).

35 J. PHYS. C VOL.4, ... 1168, (1971). . . . T. CROSS SECTION FOR
    *TRANSVERSE COLLECTIVE EXCITATIONS IN CHARGED SYSTEMS. CROSS SECTION FOR
    MAGNETIC NEUTRON SCATTERING AND DIAMAGNETIC SUSCEPTIBILITY.
    *SCHNEIDER(T.).

36 J. PHYS. C VOL.4, ... 3307, (1971). . . . T. UNFILLED SHELLS CON-
    *MAGNETIC SCATTERING OF NEUTRONS BY SEVERAL UNFILLED SHELLS CON-
    FIGURATIONS.
    *MAHENDRA(A.), KHAN(D.C.).

37 J. PHYS. C VOL.4, ... 53, (1971). . . . T. PHASE OF A HEISENBERG
    *SPIN-CORRELATION FUNCTIONS IN THE PARAMAGNETIC PHASE OF A HEISENBERG
    FERROMAGNET.
    *HUBBARD(J.).

38 J. PHYSIQUE TOME 32 COL.1 VOL.II 818, (1971). . . . T. FERROMAGNETIC METALS.
    *A NEW ANALYSIS OF NEUTRON DIFFRACTION IN FERROMAGNETIC METALS.
    *KIM(D.J.), SCHWARTZ(B.B.), PRADDAUDE(H.C.).

39 J. PHYSIQUE TOME 32 COL.1 VOL.II 812, (1971). . . . D4,T. AND MAGNETIC
    *ORBITAL CONTRIBUTION TO GENERALIZED SUSCEPTIBILITY AND MAGNETIC
    SCATTERING OF NEUTRONS IN TRANSITION METALS.
    *OBATA(Y.), SASAKI(K.), MORI(N.).

40 J. PHYSIQUE TOME 32 COL.1, ... 573, (1971). . . . T. HUBBARD HAMILTONIAN.
    *ORBITAL NEUTRON CROSS SECTION FOR HUBBARD HAMILTONIAN.
    *LOVESEY(S.W.), WINDSOR(C.G.).

41 J. PHYS. SOC. JAPAN VOL.17 B-III   60, (1962). . . . D1. FE3*O4.
    *SOME EXPERIMENTS ON MAGNETIC INELASTIC SCATTERING OF NEUTRONS.
    *RISTE(T.).

42 J. PHYS. SOC. JAPAN VOL.18, . . . 1025, (1963). . . . D2,D3. PHENOMENA IN
    *BAND THEORETICAL INTERPRETATION OF NEUTRON DIFFRACTION PHENOMENA IN
    FERROMAGNETIC METALS.
    *I7UYAMA(T.), DUK-JOO-KIM, KUBO(R.).

43 J. PHYS. SOC. JAPAN VOL.21, . . . 2178, (1966). . . . T,D2. BY FERROMAGNET AND
    *THEORY OF CRITICAL MAGNETIC SCATTERING OF NEUTRONS BY FERROMAGNET AND
    ANTIFERROMAGNET.
    *OGUCHI(T.), ONO(I.).

44 J. PHYS. SOC. JAPAN VOL.25, . . . 1001, (1968). . . . D1,T. WITH MULTIPLE BANDS. II.
    *THEORY OF SPIN WAVES IN FERROMAGNETIC METALS WITH MULTIPLE BANDS. II.
    INELASTIC SCATTERING OF NEUTRONS BY SPIN WAVES.
    *YAMADA(H.), SHIMIZU(M.).

45 LATV. PSR ZIN. AKAD. VESTIS NO.9   65, (1962). . . . T. IN CRYSTALS.
    *THEORY OF MAGNETIC NEUTRON SCATTERING IN CRYSTALS.
    *KASHCHEEV(V.N.). (IN RUSSIAN).

46 LATV. PSR ZIN. AKAD. VESTIS NO.7   53, (1962). . . . T. AN ADDITION CENTRE IN A
    *THEORY OF MAGNETIC NEUTRON SCATTERING FROM AN ADDITION CENTRE IN A
    CRYSTAL.
    *KASHCHEEV(V.N.). (IN RUSSIAN).

47 LATV. PSR ZIN. AKAD. VESTIS NO.3   39, (1962). . . . T. BY A SINGLE-AXIS
    *EFFECT OF DOMAIN WALLS ON SLOW NEUTRON SCATTERING BY A SINGLE-AXIS
    FERROMAGNETIC CRYSTAL.
    *BILENSKII(V.), KASHCHEEV(V.N.). (IN RUSSIAN).

48 LATV. PSR ZIN. AKAD. VESTIS NO.2   6, (1973). . . . T. THE ENERGY DISTRIBUTION OF
    *THE EFFECT OF SPIN-PHONON INTERACTION ON THE ENERGY DISTRIBUTION OF
    ONE-PHONON NEUTRON SCATTERING IN FERROMAGNETS.
    *KASHCHEEV(V.N.). (IN RUSSIAN).

49 LECT/ELEM. EXCIT./INTER. IN SOL. . . . . , (1968). . . . T. LOCALIZED SPIN WAVES.
    *THEORY OF SCATTERING IN SOLIDS AND LOCALIZED SPIN WAVES.
    *CALLAWAY(J.), (ED. NOTE: PUBL. NATO ADVANCED STUDY INSTITUTE, CORTINA
    D*AMPEYYO 1968, 29PP.).

50 METAL. SOC. CONF. LOS ANGELES, . . . 1, (1967). . . . T. SLOW NEUTRON SCATTERING.
    *INTRODUCTION TO THE THEORY OF SLOW NEUTRON SCATTERING.
    *BLUME(M.).

51 2ND SUMMER SCHOOL S.S. PHYS., . . . 321, (1968). . . . D1,T. SPIN-WAVE STATES.
    *NEUTRON SCATTERING ON MAGNETIC IMPURITY SPIN-WAVE STATES.
    *KROO(N.).

SECTION 8 -THEORY OF NEUTRON MAGNETIC SCATTERING. . . . .

52 NUCL./S.S. PHYS. SYMP. ABSTR. , . . . , (1972). . . . IN RARE EARTH CRYSTALS.
≈MAGNETIC SCATTERING OF NEUTRONS IN RARE EARTH CRYSTALS.
≈MAHENDRA(A.), KHAN(D.C.).

53 NUKLEONIK VOL.11, ... 113, (1968). . . . D4
≈SPIN FLIP FOR SCATTERING OF COLD NEUTRONS ON ORIENTED ATOMIC NUCLEI.
≈BINDER(K.), RAUCH(H.).

54 NUOVO CIMENTO SUPPL. VOL.6, . . 1183, (1957)
≈CRITICAL SCATTERING OF NEUTRONS FROM FERROMAGNETS.
≈MARSHALL(W.).

55 PHYSICA VOL.29, ... 617, (1963). . . . T
≈ON THE QUASI-ELASTIC MAGNETIC SCATTERING OF SLOW NEUTRONS.
≈RUIJGROK(T.W.).

56 PHYS. LETT. VOL.31A, ... 67, (1970). . . . T. . . . . PR3+
≈CRYSTAL FIELD EFFECTS IN INELASTIC NEUTRON SCATTERING FROM PR+++.
≈BALCAR(E.), LOVESEY(S.W.).

57 PHYS. LETT. VOL.34A, ... 306, (1971)
≈CRITICAL SCATTERING CROSS SECTION IN FERROMAGNETIC POLYCRYSTALS.
≈WOJTCZAK(L.), KOCINSKI(J.).

58 PHYS. LETT. VOL.37A, ... 63, (1971)
≈CRITICAL FLUCTUATIONS NEAR TRANSITION POINTS BETWEEN FERROMAGNETIC AND
ANTIFERROMAGNETIC PHASES.
≈AKHIEZER(I.A.), GINZBURG(A.E.).

59 PHYS. LETT. VOL.41A NO.2, . . . 175, (1972). . . . D2
≈THE BRILLOUIN-MANDELSTAM DOUBLET IN THE CRITICAL MAGNETIC SCATTERING
OF NEUTRONS.
≈MALINOWSKI(K.), KOCINSKI(J.).

60 PHYS. MET. METALLOGR. VOL.20 N.2 27, (1965). . . . T
≈MAGNETOELASTIC SCATTERING OF SLOW NEUTRONS IN ALLOYS.
≈RIZOWYANETSKII(D.R.), SMIRNOV(A.A.).

61 PHYS. MET. METALLOGR. VOL.20 N.2 11, (1965). . . . T
≈THEORY OF INELASTIC MAGNETIC SCATTERING OF NEUTRONS BY LOCALIZED SPIN
EXCITATIONS IN FERROMAGNETS.
≈IVANOV(M.A.), KRIVOGLAZ(M.A.), MASYUKEVICH(A.M.).

62 PHYS. MET. METALLOG. VOL.22 NO.6 1, (1966). . . . T
≈NEUTRON SCATTERING IN FERROMAGNETIC CRYSTALS CONTAINING IMPURITIES WITH
A NEGATIVE EXCHANGE BOND.
≈IZYUMOV(YU.A.), MEDVEDEV(M.V.).

63 PHYS. MET. METALLOG. VOL.22 NO.4 7, (1967). . . . T,P
≈SCATTERING CROSS-SECTION OF SLOW POLARIZED NEUTRONS IN A UNIAXIAL
ANTIFERROMAGNET.
≈ZAROCHENTSEV(E.V.), POPOV(V.A.).

64 PHYS. MET. METALLOGR. VOL.34 N.2 13, (1972). . . . T
≈MAGNETIC NEUTRON SCATTERING IN POLYDOMAIN FERROMAGNETS.
≈KURKIN(M.I.), TANKEEV(A.P.), TUROV(E.A.).

65 PHYS. REV. VOL.50, ... 259, (1936). . . . T
≈ON THE MAGNETIC SCATTERING OF NEUTRONS.
≈BLOCH(F.).

66 PHYS. REV. VOL.51, ... 994, (1937). . . . T
≈ON THE MAGNETIC SCATTERING OF NEUTRONS. II.
≈BLOCH(F.).

67 PHYS. REV. VOL.51, ... 544, (1937). . . . T
≈ON THE MAGNETIC SCATTERING OF NEUTRONS.
≈SCHWINGER(J.).

68 PHYS. REV. VOL.55, ... 898, (1939). . . . T
≈ON THE MAGNETIC SCATTERING OF NEUTRONS.
≈HALPERN(O.), JOHNSON(M.H.).

69 PHYS. REV. VOL.61, ... 17, (1942). . . . T,E
≈MAGNETIC SCATTERING OF NEUTRONS.
≈HAMERMESH(M.).

70 PHYS. REV. VOL.72, ... 1139, (1947). . . . E
≈ON THE INTERACTION BETWEEN NEUTRONS AND ELECTRONS.
≈FERMI(E.), MARSHALL(L.).

71 PHYS. REV. VOL.76, ... 1130, (1949). . . . T
≈ON MAGNETO-OPTICS OF NEUTRONS AND SOME RELATED PHENOMENA.
≈HALPERN(O.).

72 PHYS. REV. VOL.76, ... 994, (1949). . . . E
≈ON THE POLARIZATION OF SLOW NEUTRONS.
≈STEINBERGER(J.), WICK(G.C.).

73 PHYS. REV. VOL.76, ... 1328, (1949). . . . T
≈MAGNETIC INTERACTION BETWEEN NEUTRONS AND ELECTRONS.
≈EKSTEIN(H.).

74 PHYS. REV. VOL.78, ... 731, (1950). . . . T
≈REMARKS ON THE MAGNETIC SCATTERING OF NEUTRONS.
≈EKSTEIN(H.).

75 PHYS. REV. VOL.83, ... 1226, (1951). . . . T
≈MAGNETIC NEUTRON DIFFRACTION FROM EXCHANGE-COUPLED LATTICES AT HIGH
TEMPERATURES.
≈SLOTNICK(M.).

76 PHYS. REV. VOL.81, ... 498, (1951). . . . E,T. . . . . BE,FE,CO,
≈REFLECTION OF NEUTRONS FROM MAGNETIZED MIRRORS.
≈HUGHES(D.J.), BURGY(M.T.).

77 PHYS. REV. VOL.88, ... 1003, (1952). . . . T
≈REMARKS ON SOME QUESTIONS OF NEUTRON OPTICS.
≈HALPERN(O.).

78 PHYS. REV. VOL.88, ... 232, (1952). . . . T
≈MULTIPLE SCATTERING OF NEUTRONS. III. SCATTERING BY SPIN-DEPENDENT
FORCES AND POLARIZATION PHENOMENA.
≈HALPERN(O.).

SECTION   8 -THEORY OF NEUTRON MAGNETIC SCATTERING. . . . .

79   PHYS. REV. VOL.92,   1387, (1953)........T.....FROM.RARE.ER2*O3,ND2*O3, ...
     *MAGNETIC SCATTERING OF NEUTRONS FROM RARE EARTH IONS.
     *TRAMMELL(G.T.).

80   PHYS. REV. VOL.95,   1374, (1954)...T.............
     *TIME-DEPENDENT CORRELATIONS BETWEEN SPINS AND NEUTRON SCATTERING IN
     FERROMAGNETIC CRYSTALS.
     *VAN-HOVE(L.).

81   PHYS. REV. VOL.93, ... 268, (1954)....T.........FE,
     *TEMPERATURE VARIATION OF THE MAGNETIC INELASTIC SCATTERING OF SLOW
     NEUTRONS.
     *VAN-HOVE(L.).

82   PHYS. REV. VOL.119,.   1542, (1960)....T.......
     *MAGNETIC SCATTERING OF NEUTRONS BY EXCHANGE-COUPLED LATTICES. . . . . .
     *SAENZ(A.W.).

83   PHYS. REV. LETT. 10,. . . . . . 489, (1963) ....
     *MAGNETIC SCATTERING OF NEUTRONS BY NONCOLLINEAR SPIN DENSITIES.
     *BLUME(M.).

84   PHYS. REV. 130,.   .POLARIZATION 1670, (1963)....MAGNETIC ELASTIC SCATTERING OF SLOW
     EFFECTS IN THE
     NEUTRONS.
     *BLUME(M.).

85   PHYS. REV. VOL.133,. . . . . . . B581, (1964)........:::::
     *STUDIES OF THE OPTICS OF NEUTRONS. III. PROBLEMS OF POLARIZATION OF SLOW
     NEUTRONS.
     *HALPERN(O.).

86   PHYS. REV. VOL.136,.   B112, (1964)....SCATTERING ANGLES NEAR AN S-WAVE RESONANCE
     *NEUTRON POLARIZATION AT SMALL
     *REDMOND(R.F.).

87   PHYS. REV. VOL.134,.   A126, (1964).............
     *STUDIES OF THE OPTICS OF NEUTRONS. IV. MAGNETIC SATURATION AND NEUTRON
     DEPOLARIZATION.
     *HALPERN(O.).

88   PHYS. REV. VOL.133,.   B579, (1964)...........
     *STUDIES OF THE OPTICS OF NEUTRONS. II. SPIN-INDEPENDENT INTERACTION
     BETWEEN NEUTRONS AND ELECTRONS.
     *HALPERN(O.).

89   PHYS. REV. VOL.133,.   A1366, (1964)........
     *POLARIZATION EFFECTS IN SLOW NEUTRON SCATTERING. II. SPIN-ORBIT . . . .
     SCATTERING AND INTERFERENCE.
     *BLUME(M.).

90   PHYS. REV. LETT. 21,. . . . . . 1744, (1968)............
     *NEUTRON SCATTERING IN FERROMAGNETIC DILUTE ALLOYS. . . . . . .
     *KIM(D.J.), SCHWARTZ(B.B.).

91   PHYS. REV. 166,. . . . . 554, (1968)............. . . . . . .
     *POLARIZATION EFFECTS IN SLOW-NEUTRON SCATTERING III. NUCLEAR
     POLARIZATION.
     *SCHERMER(R.I.), BLUME(M.).

92   PHYS. REV. VOL.185,. . . . . . . 783, (1969)..T.........
     *THEORY OF NEUTRON SCATTERING IN THE ITINERANT MODEL OF ANTIFERROMAGNETIC
     METALS. II.
     *SOKOLOFF(J.B.).

93   PHYS. REV. VOL.185,.THEORY OF INELASTIC 770, (1969)NEUTRON T SCATTERING IN THE ITINERANT MODEL ANTI-
     FERROMAGNETIC METALS. I.
     *SOKOLOFF(J.B.).

94   PHYS. REV. 180,. . . . . 613, (1969)..............
     *MULTIBAND THEORY OF INELASTIC NEUTRON SCATTERING BY FERROMAGNETIC METALS
     AT LOW TEMPERATURES.
     *SOKOLOFF(J.B.).

95   PHYS. REV. LETT. VOL.24,.   1415, (1970)...............
     *DIAMAGNETIC SCATTERING OF SLOW NEUTRONS. . . . . . .
     *STASSIS(C.).

96   PHYS. REV. LETT. VOL.25,. . . . 110, (1970)..T.......
     *ELECTRON DELOCALIZATION AND THE NEUTRON MAGNETIC FORM FACTOR OF MN++ AND
     NI++ ANTIFERROMAGNETIC SALTS.
     *SOULES(T.F.), RICHARDSON(J.W.).

97   PHYS. REV. B VOL.2,. . . . . . . 2664, (1970)..............
     *INELASTIC NEUTRON SCATTERING FROM ITINERANT ELECTRON ANTIFERROMAGNETS:
     THEORY.
     *LIU(S.H.).

98   PHYS. REV. LETT. VOL.26,.   1568, (1971)SECOND MAGNON IN ANISOTROPIC ANTIFERROMAGNETS.
     *NEUTRON SCATTERING AND
     *MICHEL(K.H.), SCHWABL(F.).

99   PHYS. REV. B VOL.4,.THEORETICAL 3641, (1971)FACTORS OF CO*O, FE*O.
     MAGNETIC FORM FACTORS OF CO++ AND FE++ IN THEIR MONOXIDES.
     *MAHENDRA(A.), KHAN(D.C.).

100 PHYS. REV. LETT. VOL.28,.   596, (1972)..............
     *SPIN-ORBIT EFFECTS IN THE INDEX-OF-REFRACTION FORMALISM FOR POLARIZED
     NEUTRONS.
     *HANDEL(P.H.).

101 PHYS. REV. B VOL.5,. . . . . . . 2719, (1972)..............T9, DY,
     *STATIC AND DYNAMIC EFFECTS OF THE MAGNETOELASTIC INTERACTION IN TERBIUM
     AND DYSPROSIUM METALS.
     *VIGREN(D.T.), LIU(S.H.).

102 PHYS. REV. B VOL.5,.   1915, (1972)............
     *RAMAN AND NEUTRON SCATTERING FROM IMPURITY PAIRS IN FERROMAGNETS.
     *SHILES(E.), HONE(D.).

103 PHYS. REV. B VOL.7,.   5017, (1973)...........
     *ERRATUM: SPIN DYNAMICS OF LINEAR HEISENBERG MAGNETIC CHAINS (VOL.7,1149)
     *MCLEAN(F.B.), BLUME(M.).

SECTION   8 -THEORY OF NEUTRON MAGNETIC SCATTERING. . . . .

104 PHYS. REV. B VOL.7, ... 4142, (1973)....E. . . . .
     *REFRACTION OF THERMAL NEUTRONS BY SHAPED MAGNETIC FIELDS.
     *JUST(W.), SCHNEIDER(C.S.), CISZEWSKI(R.), SHULL(C.G.).

105 PHYS. REV. B VOL.7, ... 1149, (1973)....T,01.....TMMC. . . . .
     *SPIN DYNAMICS OF LINEAR HEISENBERG MAGNETIC CHAINS.
     *MCLEAN(F.B.), BLUME(M.).

106 PHYS. REV. B VOL.7, ... 1108, (1973)....T. . . . .
     *NEUTRON SCATTERING FROM ITINERANT-ELECTRON FERROMAGNETS.
     *COOKE(J.F.).

107 PHYS. REV. LETT. VOL.31, ... 1500, (1973)....T. . . . .
     *NEUTRON SCATTERING AND THE CORRELATION FUNCTIONS OF THE ISING MODEL
     NEAR T CRITICAL.
     *TRACY(C.A.), MCCOY(B.M.).

108 PHYS. STAT. SOL. VOL.49, ... 453, (1972)....NEUTRAL PARTICLES IN POLAR MEDIA AND THE
     *THE MAGNETIC SCATTERING OF NEUTRAL PARTICLES IN POLAR MEDIA AND THE
     STUDY OF THE PERMITTIVITY DISPERSION IN SPACE AND TIME.
     *DOGONADZE(R.R.), KORNYSHEV(A.A.).

109 PROC....CONF. ON POL. TARGETS, . 235, (1971). . . . .
     *NEUTRON PHYSICS EXPERIMENTS WITH ORIENTED NUCLEI.
     *DABBS(J.W.T.). (ED. NOTE: AVAIL. FROM NAT. TECH. INFORMATION SERVICE,
     SPRINGFIELD, VA., USA).

110 PROC. INDIAN ACAD. SCI. VOL.61, . 242, (1965)....T. . . . .
     *MAGNETIC INELASTIC SCATTERING OF NEUTRONS BY A FERROMAGNETIC CRYSTAL.
     *SUBRAMANIAM(R.), KHARADZE(G.A.).

111 PROC. PHYS. SOC. A VOL.64, . . . 1097, (1951)....T,01. . . . .
     *SLOW NEUTRON SCATTERING BY FERROMAGNETIC CRYSTALS.
     *MOORHOUSE(R.G.).

112 PROC. PHYS. SOC. A VOL.67, . . . 85, (1954)....T. . . . .
     *INELASTIC MAGNETIC SCATTERING OF NEUTRONS FROM A FERROMAGNETIC CRYSTAL.
     *MARSHALL(W.).

113 PROC. PHYS. SOC. 87, ... 521, (1966)....T. . . . .
     *THE THEORY OF INELASTIC SCATTERING OF NEUTRONS IN FERROMAGNETIC CRYSTALS
     CONTAINING IMPURITIES.
     *IZYUMOV(YU.A.).

114 PROC. PHYS. SOC. VOL.89, ... 613, (1966)....T. . . . .
     *DISTRIBUTION OF MAGNETIZATION IN MIXED MAGNETIC SYSTEMS I. NON-MAGNETIC
     IMPURITY.
     *LOVESEY(S.W.), MARSHALL(W.).

115 PROC. ROY. SOC. A VOL.230, . . . 46, (1955)....T. . . . .
     *THE INELASTIC SCATTERING OF NEUTRONS BY MAGNETIC SPIN WAVES.
     *ELLIOTT(R.J.), LOWDE(R.D.).

116 PROC. ROY. SOC. A VOL.235, . . . 289, (1956)....T. . . . .
     *THEORY OF NEUTRON SCATTERING BY CONDUCTION ELECTRONS IN A METAL AND ON
     THE COLLECTIVE-ELECTRON MODEL OF A FERROMAGNET.
     *ELLIOTT(R.J.).

117 PROG. LOW TEMP. PHYS. VOL.5, . . 161, (1967)....T. . . . . . N3, P8-91. . . . . .
     *STUDY OF THE SUPERCONDUCTIVE MIXED STATE BY NEUTRON-DIFFRACTION.
     *CRIBIER(D.), JACROT(B.), MADHAV-RAO(L.), FARNOUX(B.).

118 PROGR. THEOR. PHYS. VOL.25, . . . 723, (1961)....T. . . . .
     *ON THE CRITICAL SCATTERING OF NEUTRONS.
     *MORI(H.), KAWASAKI(K.).

119 REP. PROGR. PHYS. VOL.32, ... 333, (1969)....T. . . . .
     *THE THEORY OF ELASTIC SCATTERING OF NEUTRONS BY MAGNETIC SALTS.
     *LOVESEY(S.W.), RIMMER(D.E.).

120 REV. MOD. PHYS. VOL.30, ... 75, (1958)....T. . . . .
     *THEORY OF CRITICAL SCATTERING.
     *ELLIOTT(R.J.), MARSHALL(W.).

121 REV. MOD. PHYS. VOL.30, ... 1, (1958)....T. . . . .
     *SPIN WAVES.
     *VAN-KRANENDONK(J.), VAN-VLECK(J.H.).

122 REV. MOD. PHYS. VOL.44, ... 406, (1972)....T,E. . . . .
     *THE PROPERTIES OF DEFECTS IN MAGNETIC INSULATORS.
     *COWLEY(R.A.), BUYERS(W.J.L.).

123 REV. ROUMAINE PHYS. VOL.13, . . . 637, (1968)....T. . . . .
     *MAGNETIC NEUTRON SCATTERING IN THIN FILMS.
     *WOJTCZAK(L.). (ED. NOTE: IN ENGLISH).

124 SOLID/LIQUID STATE PHYSICS, . . . 1, (1965)....E,T. . . . .
     *ELASTIC DIFFUSE SCATTERING OF NEUTRONS BY DILUTE FERROMAGNETIC ALLOYS.
     *LOW(G.G.). (LECTURES ON SOLID AND LIQUID STATE PHYSICS. PUBL. ATOMIC
     ENERGY ESTABLISHMENT TROMBAY, BOMBAY, INDIA).

125 SOL. STATE COMM. VOL.8, ... 279, (1970). . . . .
     *RELATION BETWEEN THE PAULI- AND LANDAU-SUSCEPTIBILITY AND THE NEUTRON
     SCATTERING CROSS SECTION.
     *SCHNEIDER(T.).

126 SOL. STATE COMM. VOL.9, ... 1003, (1971). . . . .
     *ROTATION OF NEUTRON POLARISATION IN HELICAL MAGNETIC STRUCTURES.
     *NITYANANDA(R.), RAMASESHAN(S.).

127 SOL. STATE COMM. VOL.12, ... 1045, (1973)....T. . . . .
     *DETERMINATION BY POLARIZATION ANALYSIS OF/SCATTERING AMPLITUDES A(+),
     A(-) AND/DEGREE OF ORIENTATION OF NUCLEI ...IN MAGNETICALLY ORDERED SUR.
     *BERTAUT(E.F.).

128 SOV. PHYS. JETP VOL.6, ... 776, (1958)....T. . . . .
     *SCATTERING OF SLOW NEUTRONS IN FERROMAGNETS.
     *MALEEV(S.V.).

129 SOV. PHYS. DOKLADY VOL.3, ... 61, (1958)....T. . . . .
     *ON THE MAGNETIC SCATTERING OF THERMAL NEUTRONS NEAR THE CURIE POINT
     IN FERROMAGNETS AND ANTIFERROMAGNETS.
     *KRIVOGLAZ(M.A.).

SECTION   8 -THEORY OF NEUTRON MAGNETIC SCATTERING. . . . .

130   SOV. PHYS. JETP VOL.13,. . . . . 860, (1961)...A.. THE. SCATTERING OF (SLOW NEUTRONS BY
      *POLARIZATION RESULTING FROM THE SCATTERING OF (SLOW NEUTRONS BY
      FERROMAGNETIC SUBSTANCES.
      *MALEEV(S.V.).

131   SOV. PHYS. JETP VOL.12,. . . . . 995, (1961)....D1.......
      *THE SCATTERING OF SLOW NEUTRONS IN FERRITES AND ANTIFERROMAGNETS.
      *BAR≠YAKHTAR(V.G.), MALEEV(S.V.).

132   SOV. PHYS. SOL. STATE, 4,. . . 1054, (1962)................
      *THEORY OF MAGNETO-VIBRATIONAL SCATTERING OF SLOW NEUTRONS IN A UNIAXIAL
      ANTIFERROMAGNET.
      *KASHCHEEV(V.N.).

133   SOV. PHYS. CRYST. VOL.7,. . 286, (1962)...A.. T.........
      *THE THEORY OF SLOW NEUTRON SCATTERING IN SOLID SOLUTIONS WITH SPINEL
      TYPE STRUCTURE AS A FUNCTION OF THE COMPOSITION AND DEGREE OF INVERSION.
      *IZYUMOV(YU.A.), MEN≠(A.N.).

134   SOV. PHYS. SOLID STATE VOL.3,. . 1883, (1962)....T..........
      *SCATTERING OF SLOW NEUTRONS IN ANTIFERROMAGNETS.
      *POPOV(V.A.).

135   SOV. PHYS. USPEKHI VOL.5,. . . . 104, (1962)....D1,D2.......
      *NEUTRON DIFFRACTION STUDY OF MAGNETIC MATERIALS.
      *ZHDANOV(G.S.), OZEROV(R.P.).

136   SOV. PHYS. SOL. STATE, 5,. . . 635, (1963)................
      *EFFECT OF SPIN-SPIN INTERACTION ON THE SCATTERING OF SLOW NEUTRONS IN
      FERRITES.
      *KASHCHEEV(V.N.).

137   SOV. PHYS. SOL. STATE, 5,. . 668, (1963)................
      *EFFECT OF THE SPIN-PHONON INTERACTION ON ONE QUANTUM SCATTERING OF SLOW
      NEUTRONS IN FERRITES.
      *KASHCHEEV(V.N.).

138   SOV. PHYS. USPEKHI VOL.80,. . . 359, (1963)....T.........
      *THEORY OF SCATTERING OF SLOW NEUTRONS IN MAGNETIC CRYSTALS.
      *IZYUMOV(YU.A.).

139   SOV. PHYS. SOL. STATE, 6,. . . 339, (1964)................
      *SCATTERING OF SLOW NEUTRONS IN WEAKLY FERROMAGNETIC ANTIFERROMAGNETIC
      MATERIALS.
      *BAR≠YAKHTAR(V.G.), SINEPOL≠SKII(O.I.).

140   SOV. PHYS. SOL. STATE, 6,. . 2465, (1964)................
      *SPIN-ELECTRON INTERACTION IN SEMICONDUCTORS, BROADENING, INELASTIC
      MAGNETIC SCATTERING OF NEUTRONS.
      *IVANOV(V.A.).

141   SOV. PHYS. JETP. 19,. . . . . 986, (1964)................
      *SCATTERING OF THERMAL NEUTRONS BY POLARIZED NUCLEI OF A FERROMAGNET.
      *KHARADZE(G.A.).

142   SOV. PHYS. JETP LETT. VOL.2, . 338, (1965)................
      *SCATTERING OF POLARIZED NEUTRONS IN MAGNETS NEAR THE PHASE-TRANSITION
      POINT.
      *MALEEV(S.V.).

143   SOV. PHYS. CRYST. VOL.12,. . . . 199, (1967)....T.........
      *SYSTEMATIC EXTINCTION OF REFLECTIONS DURING COHERENT MAGNETIC SCATTERING
      OF SLOW NEUTRONS, CAUSED BY/SYMMETRY AND ANTISYMMETRY ELEMENTS...
      *OZEROV(R.P.).

144   SOV. PHYS. JETP. 25,. . 517, (1967)................
      *CONTRIBUTION TO THE THEORY OF FLUCTUATIONS AND THE SCATTERING OF SLOW
      NEUTRONS IN FERROMAGNETS.
      *AKHIEZER(I.A.), BOLOTIN(YU.L.).

145   SOV. PHYS. JETP. 24,. . . . . . 960, (1967)................
      *INCOHERENT SCATTERING OF NEUTRONS AND THE PROBLEM OF RECONSTRUCTING THE
      MAGNON SPECTRUM.
      *IZYUMOV(YU.A.), MEDVEDEV(M.V.).

146   SOV. PHYS. SOLID STATE VOL.10,. . 3142, (1968)....T.....
      *THEORY OF SLOW-NEUTRON SCATTERING IN PARAMAGNETS SUBJECTED TO STRONG
      MAGNETIC FIELDS.
      *AKHIEZER(I.A.), BARTS(B.I.).

147   SOV. PHYS. SOLID STATE VOL.9,. . 2016, (1968)................
      *FLUCTUATIONS AND SCATTERING PROCESSES OF SLOW NEUTRONS AND LIGHT IN
      ANTIFERROMAGNETS IN STRONG MAGNETIC FIELD.
      *AKHIEZER(I.A.), BOLOTIN(YU.L.).

148   SOV. PHYS. J.E.T.P. VOL.29,. . . 685, (1969)................
      *MAGNETIC SCATTERING OF NEUTRONS.
      *BARYSHEVSKII(V.G.), KORENNAYA(L.N.).

149   SOV. PHYS. J.E.T.P. VOL.31,. . . 583, (1970)................
      *INELASTIC MAGNETIC NEUTRON SCATTERING AT SMALL ANGLES.
      *BARYSHEVSKII(V.G.), KORENNAYA(L.N.).

150   SOV. PHYS. SOL. STATE VOL.12,. . . 1596, (1971)................
      *ASSOCIATED-WAVE SCATTERING OF LIGHT AND SLOW NEUTRONS IN SEMICONDUCTOR
      ANTIFERROMAGNETS.
      *BAR≠YAKHTAR(V.G.), BASS(F.G.), MAKHMUDOV(Z.Z.).

151   SOV. PHYS. SOLID STATE VOL.14,. . 142, (1972)....T,D2.......
      *FLUCTUATIONS AND SCATTERING OF SLOW NEUTRONS AND ELECTROMAGNETIC WAVES
      NEAR THE CRITICAL ANTIFERROMAGNETIC-FERROMAGNETIC TRANSITION POINT.
      *AKHIEZER(I.A.), GINZBURG(A.E.).

152   SOV. PHYS. SOLID STATE VOL.14,. . 1761, (1973)....T.........
      *FLUCTUATIONS AND SCATTERING OF SLOW NEUTRONS BY COLLECTIVE EXCITATIONS
      OF THE ELECTRON FERMI LIQUID IN METALS.
      *AKHIEZER(I.A.), CHUDNOVSKII(E.M.).

153   SOV. PHYS. SOLID STATE VOL.14,. . 2006, (1973)....T,D2.......
      *CRITICAL SCATTERING OF NEUTRONS IN FERRITES AND ANTIFERROMAGNETS IN A
      MAGNETIC FIELD.
      *BAR≠YAKHTAR(V.G.), TARASENKO(V.V.), YABLONSKII(D.A.).

154   SOV. PHYS. SOLID STATE VOL.14,. . 1774, (1973)....T,D2.......
      *INTERACTION OF NEUTRONS AND LIGHT WITH MAGNETOACOUSTIC WAVES IN CRYSTALS
      HAVING A FERROMAGNETIC-ANTIFERROMAGNETIC TRANSITION.
      *GINZBURG(A.E.), DAVYDOV(L.N.), SPOL≠NIK(Z.A.).

SECTION   8 -THEORY OF NEUTRON MAGNETIC SCATTERING. . . . .

155 SOV. PHYS. SOLID STATE VOL.15, . 656, (1973)....T........ . . . . . .
              ~INTERACTION OF A FERROMAGNET IN ITS METASTABLE STATE WITH A NEUTRON BEAM
              ~AKHIEZER(I.A.), KALINICHENKO(A.I.).

156 TECH. METALS RES. VOL.11,. . . . . 641, (1969)........... . . . .
              ~NEUTRON SCATTERING TECHNIQUES (REVIEW PAPER).
              ~BACON(G.E.).

157 TH. NEUTRON SCATT./EGELSTAFF,. . 251, (1965)............. . . . .
              ~MAGNETIC INELASTIC SCATTERING OF NEUTRONS.
              ~JACROT(B.), RISTE(T.).

158 TH. NEUTRON SCATT./EGELSTAFF,. . 1, (1965)............. . . . .
              ~INTRODUCTORY THEORY (OF THERMAL NEUTRON SCATTERING).
              ~LOMER(W.M.), LOW(G.G.).

159 UKR. FIZ. ZH. VOL.17,. . . . . . 1339, (1972)....D1....... . . . .
              ~SPIN WAVES AND SCATTERING PROCESSES OF SLOW NEUTRONS AND LIGHT IN
              ANTIFERROMAGNETICS WITH A SMALL EXCHANGE CONSTANT.
              ~AKHIEZER(I.A.), GINZBURG(A.E.). (ED. NOTE: IN RUSSIAN).

160 VESTN. MOSK. UNIV. FIZ. ASTRON. 149, (1972)........ . . . .
              ~DIFFRACTION OF COLD NEUTRONS BY FERROMAGNETIC BLOCH WALLS. NONELASTIC
              SCATTERING.
              ~BUSHEV(M.K.).

161 Z. ANGEW. PHYS. VOL.20,. . . . . 311, (1966)....T....... . . . .
              ~THE WAVE-MECHANICAL CALCULATION OF NEUTRON SCATTERING IN MAGNETIZED
              CRYSTALS.
              ~DORING(W.). (IN GERMAN).

162 Z. ANGEW. PHYS. VOL.32,. . . . . 178, (1971)....D1....... . . . .
              ~DISCUSSION OF METHODS TO DETERMINE THE SPIN DIFFUSION PARAMETER BY
              NEUTRON SCATTERING.
              ~BINDER(K.).

163 Z. NATURFORSCH VOL.24A,. . . . . 1646, (1969)....D4....... . . . .
              ~RECOILLESS SPIN-FLIP IN THE MAGNETIC SCATTERING OF COLD NEUTRONS.
              ~HANDEL(P.H.). (ED. NOTE: IN ENGLISH).

164 Z. NATURFORSCH. A VOL.26A,. . . . 432, (1971)....D4....... . . . .
              ~CROSS SECTION FOR ULTRACOLD NEUTRONS. 1.THEORY OF MAGNETIC SCATTERING.
              ~BINDER(K.).

SECTION   9A-DIFFRACTION: REVIEW PAPERS . . . .

1   ADV. INORG. CHEM. RADIOCHEM. 8,   225, (1966). APPLICATIONS IN INORGANIC CHEMISTRY (REVIEW)
    ₼NEUTRON DIFFRACTION AND ITS APPLICATIONS IN INORGANIC CHEMISTRY (REVIEW)
    ₼BACON(G.E.).

2   ADV. STRUC. RES. DIFFR. METHOD 1   1, (1964), ...A1...,...H.ATOM
    ₼THE DETERMINATION OF CRYSTAL STRUCTURES BY NEUTRON-DIFFRACTION
    MEASUREMENTS (REVIEW).
    ₼BACON(G.E.).

3   ADV. STRUC. RES. DIFFR. METHOD 2   1, (1966)..............
    ₼THE INVESTIGATION OF MAGNETIC STRUCTURES BY NEUTRON DIFFRACTION (REVIEW)
    ₼BACON(G.E.).

4   ADV. STRUC. RES. DIFFR. METHOD 2   1, (1966)..............
    ₼THE INVESTIGATION OF MAGNETIC STRUCTURES BY NEUTRON DIFFRACTION (REVIEW)
    ₼BACON(G.E.).

5   AMER. J. PHYS. VOL.22, ...   263, (1954)...............
    ₼TECHNIQUES AND APPLICATIONS OF NEUTRON DIFFRACTION (REVIEW).
    ₼WILKINSON(M.K.).

6   ANN. DE PHYSIQUE TOME 7 NO.4, . .   2C3, (1972) . . . .
    ₼MAGNETIC STRUCTURES (REVIEW PAPER).
    ₼BERTAUT(E.F.). (PRESENTED AT AUTRANS SUMMER SCHOOL 1972).

7   ANN. REV. BIOPHYS. BIOENGNG. 1,   529, (1972)............ . . .
    ₼NEUTRON SCATTERING (REVIEW).
    ₼SCHOENBORN(B.P.), NUNES(A.C.).

8   AT. ENERGY AUST. VOL.12, ...   2, (1969).................
    ₼NEUTRON DIFFRACTION IN STRUCTURAL AND QUANTUM CHEMISTRY (REVIEW PAPER).
    ₼MASLEN(E.N.).

9   CAHIERS DE PHYS. VOL.113, ...   29, (1960).................
    ₼THE APPLICATION OF NEUTRON DIFFRACTION TO THE STUDY OF MAGNETIC
    SUBSTANCES (REVIEW PAPER).
    ₼MERIEL(P.). (IN FRENCH).

10 CHEM. APPL. THERMAL NEUTRON SCAT 270, (1973)....T,E........ . . . .
    ₼NEUTRON DIFFRACTION AND COVALENCY (REVIEW).
    ₼JACOBSON(A.J.). (IN BOOK: CHEMICAL APPLICATIONS OF THERMAL NEUTRON
    SCATTERING. ED. BY B.T.M. WILLIS; OXFORD UNIV. PRESS: LONDON).

11 CHEM. APPL. THERMAL NEUTRON SCAT 270, (1973)....T,E........
    ₼NEUTRON DIFFRACTION AND COVALENCY (REVIEW).
    ₼JACOBSON(A.J.). (IN BOOK: CHEMICAL APPLICATIONS OF THERMAL NEUTRON
    SCATTERING. ED. BY B.T.M. WILLIS; OXFORD UNIV. PRESS: LONDON).

12 ENDEAVOUR VOL.25,. . . . .   129, (1966)................ . . . .
    ₼THE APPLICATIONS OF NEUTRON DIFFRACTION (REVIEW PAPER).
    ₼BACON(G.E.).

13 IAEA SYMP. GRENOBLE, ...   529, (1972)................ . . . .
    ₼MAGNETISM IN ONE AND TWO DIMENSIONS (REVIEW PAPER).
    ₼BLUME(M.).

14 INSTRUM. EXP. TECH. NO.2,. . . .   183, (1960)....T............ . . . . . .
    ₼THEORY OF NEUTRON CRYSTAL MONOCHROMATORS (REVIEW).
    ₼ABOV(YU.G.).

15 J. PHYS. SOC. JAPAN VOL.17 B-II, 324, (1962)....E........ . . . .
    ₼RECENT PROGRESS IN NEUTRON DIFFRACTION.
    ₼BACON(G.E.).

16 NAT. BUR. STAND. PUBL. NO.301,   57, (1969)................ . . . .
    ₼NEUTRON AND X-RAY DIFFRACTION TECHNIQUES (REVIEW).
    ₼BACON(G.E.). (U.S. NBS SPEC. PUBL. 1967; PUBL. 1969).

17 NATURE VOL.164,. . . . .   205, (1949)................ . . . .
    ₼NEUTRON DIFFRACTION BY CRYSTALS (REVIEW).
    ₼LONSDALE(K.).

18 NUCLEONICS VOL.3 NO.2,. . . . . . 17, (1948)................ . . . . . .
    ₼NEUTRON DIFFRACTION AND ASSOCIATED STUDIES- II. (REVIEW).
    ₼WOLLAN(E.O.), SHULL(C.G.).

19 NUCLEONICS VOL.3 NO.1,. . . . . . 8, (1948)................ . . . . .
    ₼NEUTRON DIFFRACTION AND ASSOCIATED STUDIES-I. (REVIEW).
    ₼WOLLAN(E.O.), SHULL(C.G.).

20 NUCLEONICS VOL.7 NO.6,. . . . . . 31, (1950)....E............
    ₼RECENT APPLICATIONS OF NEUTRON DIFFRACTION.
    ₼WEBER(A.H.).

21 PROC. ROY. SOC. A VOL.196,. . . . 50, (1949)....T............
    ₼NEUTRON DIFFRACTION.
    ₼BACON(G.E.), THEWLIS(J.).

22 PURE APPL. CHEM. VOL.18, ...   517, (1969).................
    ₼NEUTRON DIFFRACTION STUDIES OF ORGANIC MOLECULES (REVIEW PAPER).
    ₼BACON(G.E.).

23 REP. PROGR. PHYS. VOL.16,. . . .   1, (1953)....E............
    ₼NEUTRON DIFFRACTION (REVIEW).
    ₼BACON(G.E.), LONSDALE(K.).

24 SCIENCE VOL.108, ...   69, (1948)................
    ₼X-RAY, ELECTRON, AND NEUTRON DIFFRACTION (REVIEW).
    ₼SHULL(C.G.), WOLLAN(E.O.).

25 SOL. STATE PHYS. VOL.2,. . . . .   137, (1956)................
    ₼APPLICATIONS OF NEUTRON DIFFRACTION TO SOLID STATE PROBLEMS (REVIEW).
    ₼SHULL(C.G.), WOLLAN(E.O.). (REFER TO SECTION 4).

26 SOV. PHYS. USPEKHI VOL.7,. . . .   855, (1965).................
    ₼STRUCTURE OF ANTIFERROMAGNETS (REVIEW PAPER).
    ₼FARZTDINOV(M.M.).

27 USPEKHI FIZ. NAUK VOL.45,. . . .   481, (1951)....T,E........
    ₼STRUCTURAL NEUTRONOGRAPHY (REVIEW PAPER).
    ₼OZEROV(R.P.). (IN RUSSIAN).

SECTION   98-DIFFRACTION THEORY AND TECHNIQUE . . . .

1   ACTA CRYST. VOL.1,   303, (1948)....T..........
    ↝SECONDARY EXTINCTION AND NEUTRON CRYSTALLOGRAPHY.
    ↝BACON(G.E.), LOWDE(R.D.).

2   ACTA CRYST. VOL.9,   151, (1956)
    ↝A NEW RATIONALE OF STRUCTURE-FACTOR MEASUREMENT IN NEUTRON-DIFFRACTION
    ANALYSIS.
    ↝LOWDE(R.D.).

3   ACTA CRYST. VOL.10,   629, (1957)....T..........
    ↝THE EFFECT OF CRYSTAL SHAPE AND SETTING ON SECONDARY EXTINCTION.
    ↝HAMILTON(W.C.).

4   ACTA CRYST. 16,.   276, (1963)....
    ↝DYNAMICAL DIFFRACTION THEORY OF WAVES IN DISTORTED CRYSTALS I.GENERAL
    FORMULATION AND TREATMENT FOR PERFECT CRYSTALS.
    ↝KATO(N.).

5   ACTA CRYST. VOL.16,   174, (1963)....
    ↝FOURIER PROJECTIONS OF UNPAIRED ELECTRON DENSITIES.
    ↝PICKART(S.J.).

6   ACTA CRYST. VOL.A25,   391, (1969)....P.........
    ↝COHERENT NEUTRON-SCATTERING AMPLITUDES.
    ↝BACON(G.E.). (CHAIRMAN OF NEUTRON DIFFRACTION COMMISSION).

7   ACTA CRYST. VOL.A25,   396, (1969)....
    ↝LOCATION OF THE ANOMALOUS SCATTERER IN NEUTRON ANOMALOUS SCATTERING
    STUDIES.
    ↝SIKKA(S.K.).

8   ACTA CRYST. VOL.A25,   378, (1969)....
    ↝A CONTRIBUTION TO THE DYNAMICAL DIFFRACTION THEORY OF SCALAR WAVES IN
    CRYSTALS.
    ↝HIISMAKI(P.).

9   ACTA CRYST. VOL.A26,   457, (1970)
    ↝THE CORRECTION OF MEASURED INTEGRATED BRAGG INTENSITIES FOR ANISOTROPIC
    THERMAL DIFFUSE SCATTERING. A CORRECTION.
    ↝ROUSE(K.D.), COOPER(M.J.).

10 ACTA CRYST. VOL.A26,   447, (1970)
    ↝THERMAL DIFFUSE SCATTERING IN INTEGRATED INTENSITIES OF BRAGG
    REFLECTIONS.
    ↝WALKER(C.B.), CHIPMAN(D.R.).

11 ACTA CRYST. VOL.A26,   682, (1970)
    ↝ABSORPTION CORRECTIONS FOR NEUTRON DIFFRACTION.
    ↝ROUSE(K.D.), COOPER(M.J.), YORK(E.J.), CHAKERA(A.).

12 ACTA CRYST. VOL.A26,   214, (1970)....
    ↝EXTINCTION IN X-RAY AND NEUTRON DIFFRACTION.
    ↝COOPER(M.J.), ROUSE(K.D.).

13 ACTA CRYST. VOL.A27,   665, (1971)....
    ↝CHOICE OF SCANS IN NEUTRON DIFFRACTION.
    ↝WERNER(S.A.).

14 ACTA CRYST. VOL.A27,   563, (1971)....
    ↝ESTIMATE OF THE CONTRIBUTION OF MULTIPLE DIFFRACTION TO OBSERVED
    INTENSITIES IN SINGLE-CRYSTAL X-RAY AND NEUTRON DIFFRACTION.
    ↝PRAGER(P.R.).

15 ACTA CRYST. VOL.A28,   236, (1972)
    ↝A HIGHLY EFFICIENT NEUTRON DIFFRACTION TECHNIQUE: USE OF WHITE RADIATION
    ↝HUBBARD(C.R.), QUICKSALL(C.O.), JACOBSON(R.A.).

16 ACTA CRYST. VOL.A28,   218, (1972)....
    ↝MULTIPLE DIFFRACTION EFFECTS IN NEUTRON SINGLE-CRYSTAL DIFFRACTOMETRY.
    ↝COLELLA(R.).

17 ACTA CRYST. VOL.A28,   151, (1972)....
    ↝MODIFIED EWALD CONSTRUCTION FOR NEUTRONS REFLECTED BY MOVING LATTICES.
    ↝BURAS(B.), GIEBULTOWICZ(T.).

18 ACTA CRYST. VOL.A28 PART S-4, 215, (1972)....T..........
    ↝SOME CONTRIBUTIONS TO THE DYNAMICAL THEORY OF NEUTRON DIFFRACTION.
    ↝RUSTICHELLI(F.). (ABSTRACT ONLY).

19 ACTA CRYST. VOL.A29,   90, (1973)....E.........
    ↝WHITE RADIATION NEUTRON-DIFFRACTION TECHNIQUES.
    ↝WILSON(S.A.), COOPER(M.J.).

20 ACTA CRYST. VOL.A29,.   372, (1973)....T.........
    ↝THE DEPENDENCE OF THE INTENSITIES OF DIFFUSE PEAKS ON SCATTERING ANGLE
    IN NEUTRON DIFFRACTION.
    ↝YESSIK(M.), WERNER(S.A.), SATO(H.).

21 ACTA CRYST. VOL.A29,   577, (1973)....E.........
    ↝OPTIMUM SCANNING RATIO IN NEUTRON DIFFRACTION.
    ↝PANTAZATOS(P.), WERNER(S.A.).

22 ACTA CRYST. VOL.B30,   255, (1974)....E,T.......
    ↝SOME IMPLICATIONS OF COMBINED X-RAY AND NEUTRON DIFFRACTION STUDIES.
    ↝COPPENS(P.).

23 ACTA CRYST. VOL.A30,   207, (1974)....T.........
    ↝INSTRUMENTAL WIDTHS AND INTENSITIES IN NEUTRON CRYSTAL DIFFRACTOMETRY.
    ↝GRABCEV(B.).

24 ANN. ACAD. SCI. FENNICAE A 267, 3, (1967)....E.........
    ↝SOME SUGGESTIONS FOR THE APPLICATION OF CONSTRICTED MONOCHROMATOR CRYS-
    TALS FOR NEUTRON DIFFRACTION AND SCATTERING MEASUREMENTS.
    ↝MAIER-LEIBNITZ(H.). (ED. NOTE: IN GERMAN).

25 ANN. OF PHYS. VOL.33,   400, (1965)....
    ↝THE SCATTERING OF PLANE WAVES FROM PERIODIC SURFACES.
    ↝URETSKY(J.L.).

SECTION   9B-DIFFRACTION THEORY AND TECHNIQUE . . . .

26 ANN. OF PHYS. VOL.45,   404, (1967)....
    *SINGULAR SCATTERING EQUATIONS IN MOMENTUM SPACE.
    *PFAFFELHUBER(E.), BLOMER(R.).

27 ARCH. HUTNICTWA VOL.4,   81, (1959)....E....
    *PREPARATION OF LARGE METAL MONOCRYSTALS FOR NEUTRON DIFFRACTION.
    *MODRZEJEWSKI(A.), BURAS(B.), CZARNECKI(R.). (SEE: KERNTECHNIK 2, PG.153)

28 ARK. FYSIK VOL.4,   191, (1952)....T....
    *ON NEUTRON DIFFRACTION PHENOMENA ACCORDING TO THE KINEMATICAL THEORY. II
    *FROMAN(P.O.).

29 ARK. FYSIK VOL.4,   183, (1952)....T....
    *ON NEUTRON DIFFRACTION PHENOMENA ACCORDING TO THE KINEMATICAL THEORY. I.
    *WALLER(I.), FROMAN(P.O.).

30 ARK. FYSIK VOL.6,   113, (1953)....T....
    *ON THE INFLUENCE OF SPIN AND ISOTOPES IN THE KINEMATICAL THEORY OF
    NEUTRON DIFFRACTION.
    *FROMAN(P.O.).

31 ATOMKERNENERGIE VOL.13,   183, (1968)....A4....
    *TEMPERATURE DEPENDENCE OF THE COHERENT EXTINCTION FOR NEUTRON SCATTERING
    ON SINGLE CRYSTALS.
    *BROSCH(R.), RAUCH(H.).

32 BU. ACAD. POLON. SCI. TECH. 16,   330, (1968)....P....
    *X-RAY AND NEUTRON DIFFRACTION METHODS OF TESTING THE FIBRE TEXTURE OF
    ALUMINUM, COPPER AND ALPHA-BRASS.
    *SZPUNAR(J.), DUTKIEWICZ(J.).

33 BU. ACAD. SCI. USSR PHYS. SER.35 2406, (1971)....P....
    *INFLUENCE OF NUCLEAR SURFACE AND COULOMB INTERACTION ON DIFFRACTION OF
    CHOPPED NEUTRONS.
    *BEREZHNOI(YU.A.), ERLANOV(M.V.).

34 CAN. J. PHYS. VOL.30,   597, (1952)....T....
    *THE THEORY OF NEUTRON DIFFRACTION BY GASES. I.
    *POPE(N.K.).

35 C.N.E.N. SYMP. CASACCIA,   197, (1968)....
    *NEUTRON DIFFRACTION: A GENERAL TOOL IN PHYSICS.
    *SHULL(C.G.).

36 CZECH. J. PHYS. VOL.18,   1111, (1968)....P....
    *NEUTRON DIFFRACTION ON PIEZOELECTRIC VIBRATING RESONATORS.
    *PETRZILKA(V.).

37 CZECH. J. PHYS. B VOL.19,   1608, (1969)....A4....
    *DIFFRACTION EFFECT CONNECTED WITH SCATTERING AND ABSORPTION OF SLOW
    NEUTRONS IN SINGLE CRYSTALS.
    *CHALUPA(B.), MICHALEC(R.), GALOCIOVA(D.), BISCHOF(J.).

38 DISS. ABSTR. B VOL.34,   1125, (1973)....E....
    *INVESTIGATION OF THERMAL NEUTRON BEAM MANIPULATION BY VIBRATING
    CRYSTALLINE MEDIA.
    *JEFFRIES(J.D.).

39 HARWELL SUMMER SCHOOL,   1, (1968)....
    *THE COLLECTION OF HIGH-ACCURACY DIFFRACTOMETER DATA.
    *COOPER(M.J.), ROUSE(K.D.).

40 HARWELL SUMMER SCHOOL,   51, (1968)....
    *THE CORRECTION OF MEASURED BRAGG INTENSITIES FOR THERMAL DIFFUSE
    SCATTERING.
    *COOPER(M.J.).

41 INSTRUM. EXP. TECH. NO.3,   359, (1960)....E....
    *EXPERIMENTAL METHODS OF NEUTRON DIFFRACTION.
    *ABOV(YU.G.), LITVIN(O.F.).

42 INSTRUM. EXP. TECH. NO.2,   306, (1964)....E....
    *A UNIVERSAL APPARATUS FOR STRUCTURE STUDY BY NEUTRON DIFFRACTION.
    *ABESADZE(P.O.), DOIDZHASHVILI(G.I.), LITVIN(O.F.), LYASHCHENKO(B.G.),
    PROTOPOPOV(N.N.), CHIKOBAVA(V.S.),.

43 JAP. J. APPL. PHYS. VOL.12,   167, (1973)....
    *LORENTZ FACTOR IN SINGLE-CRYSTAL NEUTRON DIFFRACTION.
    *IIZUMI(M.).

44 JAP. J. APPL. PHYS. VOL.12,   167, (1973)....
    *LORENTZ FACTOR IN SINGLE-CRYSTAL NEUTRON DIFFRACTION.
    *IIZUMI(M.).

45 J. APPL. CRYST. VOL.2,   65, (1969)....E....
    *A PROFILE REFINEMENT METHOD FOR NUCLEAR AND MAGNETIC STRUCTURES.
    *RIETVELD(H.M.).

46 J. APPL. CRYST. VOL.4,   528, (1971)....T....
    *A TRIAL OF NEUTRON DIFFRACTION TOPOGRAPHY.
    *DOI(K.), MINAKAWA(N.), MOTOHASHI(H.), MASAKI(N.).

47 J. APPL. CRYST. VOL.5,   83, (1972)....E....
    *MODIFIED LAUE METHOD FOR THERMAL NEUTRONS.
    *THOMAS(P.).

48 J. APPL. CRYST. VOL.6,   257, (1973)....SI,..
    *PERFECT CRYSTALS AND IMPERFECT NEUTRONS.
    *SHULL(C.G.).

49 J. APPL. CRYST. VOL.7,   96, (1974)....E,T....
    *NEUTRON SMALL-ANGLE SCATTERING: EXPERIMENTAL TECHNIQUES AND APPLICATIONS
    *SCHMATZ(W.), SPRINGER(T.), SCHELTEN(J.), IBEL(K.).

50 J. APPL. PHYS. VOL.26,   1041, (1955)....T,E....MG*O.
    *CORRELATION OF DIFFRACTION AND TRANSMISSION EXPERIMENTS FOR X-RAY AND
    NEUTRON ELASTIC SCATTERING.
    *KEATING(D.T.), ANTAL(J.J.).

51 J. APPL. PHYS. VOL.30,   1323, (1959)....E....
    *NEUTRON DIFFRACTION TECHNIQUES AND THEIR APPLICATIONS TO SOME PROBLEMS
    IN PHYSICS.
    *SIDHU(S.S.), HEATON(L.), MUELLER(M.H.).

SECTION 99-DIFFRACTION THEORY AND TECHNIQUE . . . .

52 J. APPL. PHYS. VOL.40, ... 1697, (1969). . . . .
    ↵THEORY OF X-RAY DIFFRACTION BY VIBRATING CRYSTAL.
    ↵KURIYAMA(M.), MIYAKAWA(T.).

53 J. APP. PHYS. 37, ... 2343, (1966). . . . .
    ↵PROPAGATION OF BRAGG REFLECTED NEUTRONS IN BOUNDED MOSAIC CRYSTALS.
    ↵WERNER(S.A.), ARROTT(A.), KING(J.S.), KENDRICK(H.).

54 J. PHYS. CHEM. SOL. 25,. . . 1005, (1964). . . . .
    ↵NUMBER OF NEIGHBOURING ATOMS IN CALCULATIONS OF THE SLOW NEUTRONS
    SCATTERING CROSS SECTION OF DEFECTS IN SOLIDS.
    ↵MARTIN(O.G.).

55 J. PHYS. RADIUM SER.7 VOL.8, . . 29, (1937). . . . E. . . . .
    ↵RECHERCHES SUR LES NEUTRONS LENTS.
    ↵VON-HALBAN(H.), PREISWERK(P.).

56 J. PHYS. RADIUM VOL.18,. . . . . 649, (1957). . . . T. . . . .
    ↵NEUTRON PROPAGATION IN PERFECT CRYSTALS OF FINITE SIZES.
    ↵HERPIN(A.).

57 J. PHYS. SOC. JAPAN VOL.23,. . . 460, (1967). . . . E. . . . .
    ↵A NEW NEUTRON DIFFRACTION METHOD DIRECTLY OBSERVING THERMAL MOTIONS OF
    RADICALS.
    ↵SHIBUYA(I.), MITANI(S.), IWATA(Y.), TOKUNAGA(M.).

58 LA RECHERCHE VOL.3,. . . . . . 325, (1972). . . . .
    ↵LA DIFFUSION DES NEUTRONS.
    ↵JACROT(B.).

59 MOD. DIFF./IMAG. TECH./MAT. SCI. 521, (1970). . . . E. . . . .
    ↵ADVANCES IN X-RAY AND NEUTRON DIFFRACTION TECHNIQUES.
    ↵GUINIER(A.). (ED. NOTE: PUBL. NORTH-HOLLAND, AMSTERDAM, NETH. 1970).

60 MOL. CRYST. LIQUID CRYST. VOL.9, 11, (1969). . . . . .ORGANIC MOLECULES, ...
    ↵DIFFRACTION STUDIES OF MOLECULAR MOTION.
    ↵HAMILTON(W.C.).

61 MOL. DYN. AND STRUCT. OF SOLIDS, 355, (1969). . . . 87. . . . .
    ↵DIFFRACTION STUDIES OF MOLECULAR DYNAMICS IN ORGANIC CRYSTALS.
    ↵TRUEBLOOD(K.N.).

62 NATURE VOL.218,. . 80, (1968). . . . T. . . . .
    ↵NEUTRON DIFFRACTION ON PIEZOELECTRIC VIBRATING RESONATORS.
    ↵PETRZILKA(V.).

63 NATURE VOL.229,. . 111, (1971). . . . A4. . . . .
    ↵NEUTRON DIFFRACTION BY A PIEZOELECTRIC RESONATOR.
    ↵ENGLEHART(R.W.), JACOBS(A.M.).

64 NUCL. INST. MET. VOL.84,. . 153, (1970). . . . MEASUREMENT OF SECOND-ORDER CONTAMINATION IN DIFFRACTED NEUTRON BEAMS
    BY AN ABSORBING FOIL TECHNIQUE.
    ↵ALSTON(W.C.H.).

65 NUCL. INST. MET. VOL.94,. . 185, (1971). . . . P. . . . .
    ↵A NEW TECHNIQUE FOR NEUTRON DIFFRACTION. USE OF WHITE RADIATION.
    ↵HUBBARD(C.R.), QUICKSALL(C.O.), JACOBSON(R.A.),.

66 NUCL. INST. MET. 99,. . . . 453, (1972). . . . .
    ↵EQUAL-PATH TIME-OF-FLIGHT NEUTRON DIFFRACTION.
    ↵CARPENTER(J.M.), SUTTON(J.D.).

67 NUKLEONIKA VOL.13,. . . . . 171, (1968). . . . E. . . . .
    ↵NEUTRON DIFFRACTION METHOD FOR SHEET TEXTURE DETERMINATION.
    ↵OLES(A.), SZPUNAR(J.), SOSNOWSKA(I.).

68 NUKLEONIKA VOL.13,. . . . . 1111, (1968). . . . A4. . . . .
    ↵TEXTURE STUDIES OF COARSE-GRAIN METAL SHEETS BY MEANS OF NEUTRON
    DIFFRACTION.
    ↵SZPUNAR(J.), OLES(A.), BURAS(B.), SOSNOWSKA(I.), PIETRAS(E.). (ED. NOTE:
    IN ENGLISH).

69 NUOVO CIMENTO VOL.24,. . . . 1174, (1962). . . . E. . . . . .NI,AL,FE*S2,LI*F,NA*CL, ...
    ↵EXPERIMENTAL OBSERVATIONS ON THE MULTIPLE BRAGG REFLECTION OF NEUTRONS.
    ↵BORGONOVI(G.), CAGLIOTI(G.). (IN ITALIAN).

70 NUOVO CIMENTO VOL.13B NO.2,. . 249, (1973). . . . T. . . . .
    ↵DYNAMICAL NEUTRON DIFFRACTION BY IDEALLY CURVED CRYSTALS.
    ↵KLAR(B.), RUSTICHELLI(F.).

71 PHYS. LETT. 8, ... 241, (1964). . . . .
    ↵NEUTRON DIFFRACTION OF IDEAL CRYSTALS USING A DOUBLE CRYSTAL
    SPECTROMETER.
    ↵SIPPEL(D.), KLEINSTUCK(K.), SCHULZE(G.E.R.).

72 PHYS. LETT. VOL.37A,. . . 403, (1971). . . . PHASE SHIFT IN THE TIME SPECTRUM OF NEUTRONS DIFFRACTED BY A VIBRATING
    SINGLE CRYSTAL.
    ↵MICHALEC(R.), SEDLAKOVA(L.), CECH(J.), PETRZILKA(V.).

73 PHYS. REV. VOL.50,. . . . . 486, (1936). . . . E. . . . . . . .MG*O,AL, ...
    ↵BRAGG REFLECTION OF SLOW NEUTRONS.
    ↵MITCHELL(D.P.), POWERS(P.N.).

74 PHYS. REV. VOL.71,. . . . . 294, (1947). . . . T. . . . .
    ↵THEORY OF THE REFRACTION AND THE DIFFRACTION OF NEUTRONS BY CRYSTALS.
    ↵GOLDBERGER(M.L.), SEITZ(F.).

75 PHYS. REV. VOL.73,. . . . . 527, (1948). . . . E. . . . . . . .NA*CL, ...
    ↵LAUE PHOTOGRAPHY OF NEUTRON DIFFRACTION.
    ↵WOLLAN(E.O.), SHULL(C.G.), MARNEY(M.C.).

76 PHYS. REV. VOL.76,. . . . . 1117, (1949). . . . T. . . . .
    ↵SMALL ANGLE DIFFRACTION OF NEUTRONS AND SIMILAR WAVE PHENOMENA.
    ↵HALPERN(O.), GERJUOY(E.).

77 PHYS. REV. VOL.80,. . . . . 507, (1950). . . . E,T. . . . .
    ↵SMALL ANGLE SCATTERING OF THERMAL NEUTRONS.
    ↵KRUEGER(H.H.A.), MENEGHETTI(D.), RINGO(G.R.), WINSBERG(L.).

78 PHYS. REV. VOL.85,. . . . . 633, (1952). . . . T. . . . .
    ↵GEOMETRICAL OPTICS AND THE THEORY OF MULTIPLE SMALL ANGLE SCATTERING.
    ↵VINEYARD(G.H.).

SECTION 9B-DIFFRACTION THEORY AND TECHNIQUE . . . .

79 PHYS. REV. VOL.86,. . . . . 271, (1952). . . . .T,E. . . . . .AL,FE,NA*F,. . . .
        *EXTINCTION 'EFFECTS' IN 'NEUTRON' TRANSMISSION OF POLYCRYSTALLINE MEDIA.
        *WEISS(R.J.).

80 PHYS. REV. VOL.96,. . . . . 93, (1954). . . . .T. . . . .
        *MULTIPLE SCATTERING OF NEUTRONS.
        *VINEYARD(G.H.).

81 PHYS. REV. VOL.97,. . . . . 889, (1955). . . . .E. . . . . . . .V,PB,NB,. . . . . .
        *NEUTRON DIFFRACTION OBSERVATIONS ON THE SUPERCONDUCTING STATE.
        *WILKINSON(M.K.), SHULL(C.G.), ROBERTS(L.D.), BERNSTEIN(S.).

82 PHYS. REV. VOL.128,. . . . 2188, (1962). . . . .T. . . . .
        *MULTIPLE SCATTERING OF NEUTRONS IN THE STATIC APPROXIMATION.
        *FERZIGER(J.H.), LEONARD(A.).

83 PHYS. REV. LETT. 21,. . . . 1585, (1968). . . . .P. . . . .
        *OBSERVATION OF PENDELLOSUNG FRINGE STRUCTURE IN NEUTRON DIFFRACTION.
        *SHULL(C.G.).

84 PHYS. REV. 179,. . . . . . 752, (1969). . . . .
        *SINGLE SLIT DIFFRACTION OF NEUTRONS.
        *SHULL(C.G.).

85 PHYS. REV. B-3,. . . . . . 3173, (1971). . . . .
        *OBSERVABILITY OF CHARGE-DENSITY WAVES BY NEUTRON DIFFRACTION.
        *OVERHAUSER(A.W.).

86 PHYS. STAT. SOL. 2,. . . . K104, (1962). . . . .
        *NACHWEIS DER ANOMALEN ABSORPTION THERMISCHER NEUTRONEN BEI INTERFERENZ
        AM IDEALKRISTAL
        *SIPPEL(D.), KLEINSTUCK(K.), SCHULZE(G.E.R.).

87 PHYS. STAT. SOL. VOL.5,. . . K9, (1964). . . . .
        *SPLITTING OF DIFFRACTION LINES.
        *GOLD(L.).

88 PHYS. STAT. SOL. 4,. . . . 349, (1964). . . . . .AL,SI. . . . . . .
        *A NEW METHOD FOR NEUTRON DIFFRACTION CRYSTAL STRUCTURE INVESTIGATION.
        *BURAS(B.), LECIEJEWICZ(J.).

89 PHYS. STAT. SOL. A VOL.2,. . 211, (1970). . . . .P. . . . .
        *NEUTRON DIFFRACTION ON PIEZOELECTRIC VIBRATING BARS.
        *GALOCIOVA(D.), TICHY(J.), ZELENKA(J.), MICHALEC(R.), CHALUPA(B.).

90 PHYS. STAT. SOL. A VOL.4,. . 445, (1971). . . . .T,P. . . . . .GE. . . . .
        *EXPERIMENTAL AND THEORETICAL INVESTIGATIONS OF DYNAMIC NEUTRON DIF-
        FRACTION BY USING (GE) CRYSTALS WITH A LOW DISLOCATION DENSITY.
        *EICHHORN(F.), KOSMOWSKI(M.), SCHOPF(H.G.), SCHULZE(G.E.R.).

91 PHYS. STAT. SOLIDI A VOL.5 NO.2, 397, (1971). . . . . .PB. . . . .
        *NEUTRON REFLECTION PROPERTIES OF BENT SINGLE CRYSTALS.
        *KARAS(W.), RAUCH(H.), SEIDL(E.).

92 PHYS. STAT. SOLIDI A VOL.9,. . 423, (1972). . . . .P. . . . .
        *EXPLANATION OF NEUTRON DIFFRACTION PHENOMENA OBSERVED IN VIBRATING
        PIEZOELECTRIC CRYSTALS.
        *BURAS(B.), GIEBULTOWICZ(T.), MINOR(W.), RAJCA(A.).

93 PHYS. STAT. SOLIDI A VOL.17 NO.1 163, (1973). . . . .A4. . . . . .SI. . . . .
        *NEUTRON DIFFRACTION EFFECT ON THE ROCKING CURVE OF A VIBRATING SINGLE
        CRYSTAL.
        *MIKULA(P.), MICHALEC(R.), SEDLAKOVA(L.), CECH(J.), CHALUPA(B.),
        PETRZILKA(V.).

94 PROC. INT. CONF. (NOTTINGHAM). . 327, (1964). . . . . .NI*O,LA*(CR,FE,MN)*O3,MN*O,MN*F2
        *NEUTRON DIFFRACTION DATA AND COVALENCY EFFECTS.
        *NATHANS(R.), WILL(G.), COX(D.E.).

95 REPORT AERE-R5647 (33 PP.),. . . . (1968). . . . .E. . . . . . .
        *NEUTRON MONOCHROMATOR STUDIES AT A.E.R.E.
        *TURBERFIELD(K.C.). (ED. NOTE: AVAIL. H.M.S.O., 49 HIGH HOLBORN, LONDON,
        WC1, ENG.).

96 REPORT FMRB-37-71,. . . . . . (1971). . . . .T. . . . .
        *THEORY OF SLOW NEUTRON DIFFRACTION.
        *HEIMTZ(W.). (ED. NOTE: PUBL. PHYS.-TECH. BUNDESANSTALT, BRAUNSCHWEIG,
        GERMANY).

97 REPORT FN-43-1973 (31 PP.),. . . . (1973). . . . .T. . . . .
        *INSTRUMENTAL WIDTHS AND INTENSITIES IN NEUTRON CRYSTAL DIFFRACTOMETRY.
        *GRABCEV(B.). (AVAIL. INST. ATOMIC PHYS., BUCHAREST).

98 REP. PROGR. PHYS. VOL.25,. . . 395, (1962). . . . .T,E. . . . . . .
        *THE RADIAL DISTRIBUTION CURVES OF LIQUIDS BY DIFFRACTION METHODS.
        *FURUKAWA(K.).

99 RESEARCH VOL.7,. . . . . . 312, (1954). . . . .
        *NEUTRON OPTICS AND NEUTRON DIFFRACTION 2. APPLICATIONS.
        *BACON(G.E.).

100 RESEARCH VOL.7,. . . . . . 257, (1954). . . . .
        *NEUTRON OPTICS AND NEUTRON DIFFRACTION 1. AN OUTLINE OF THE PHYSICAL
        PRINCIPLES.
        *BACON(G.E.).

101 RESEARCH VOL.10,. . . . . 241, (1957). . . . .
        *NEUTRON DIFFRACTION 3. THE STUDY OF MOLECULAR STRUCTURE (REVIEW).
        *BACON(G.E.).

102 REV. SCI. INSTRUM. VOL.33,. . . 126, (1962). . . . .E. . . . .
        *PHOTOGRAPHY OF NEUTRON DIFFRACTION PATTERNS.
        *WANG(S.P.), SHULL(C.G.), PHILLIPS(W.C.).

103 SOV. PHYS. CRYST. VOL.6,. . . . 374, (1961). . . . .
        *ADJUSTMENT OF SINGLE-CRYSTAL SPECIMENS FOR NEUTRON DIFFRACTION STUDIES.
        *YAMZIN(I.I.), NOZIK(YU.Z.).

104 SOV. PHYS. CRYST. VOL.6,. . . . 404, (1962). . . . .
        *#NULL MATRIX# IN NEUTRON DIFFRACTION.
        *LYASHCHENKO(B.G.).

105 SOV. PHYS. DOKLADY VOL.3,. . . . 2272, (1962). . . . .T. . . . .
        *THE THEORY OF THE SCATTERING OF SLOW NEUTRONS BY A PARALLEL-SIDED
        CRYSTAL SPECIMEN.
        *ATANASOV(A.A.).

SECTION   9B-DIFFRACTION THEORY AND TECHNIQUE . . . .

106 SOV. PHYS. CRYST. VOL.10, ... 346, (1965). . . . .
                    +OBSERVATION OF THE EFFECTS OF PARTICLE SIZE ON THE EXTINCTION OF
                     NEUTRON DIFFRACTION.
                    +SIZOV(R.A.), YAMZIN(I.I.).

107 SOV. PHYS. JETP. 20, ... 94, (1965). . . . .
                    +ELASTIC SCATTERING OF NEUTRONS AND THE MOSSBAUER EFFECT IN SYSTEMS WITH
                     LOCAL DEGREES OF FREEDOM.
                    +KAZARNOVSKII(M.V.), STEPANOV(A.V.).

108 SOV. PHYS. JETP. 22, ... 1069, (1966). . . . .
                    +RESTRICTION OF THE ELASTIC SCATTERING CROSS SECTION IN THE DIFFRACTION
                     PEAK REGION.
                    +MALKOV(E.I.).

109 TECHNIQUE ORG. CHEM. 1 PT.1, . . 2361, (1949). . . . .
                    +NEUTRON DIFFRACTION.
                    +HASTINGS(J.M.), CORLISS(L.M.).

110 VESTN. MOSK. UNIV. FIZ. ASTR. 12 710, (1971). . . .A4. . . . .
                    +DIFFRACTION OF COLD NEUTRONS ON FERROMAGNETIC BLOCH WALLS. ELASTIC
                     SCATTERING.
                    +BUSHEV(M.K.).

111 Z. NATURFORSCH. VOL.28A, ... 657, (1973). . . .E. . . . . . . .GE. . . . .
                    +NEUTRON PENDELLOSUNG FRINGE STRUCTURE IN THE LAUE DIFFRACTION BY GE.
                    +SHULL(C.G.), SHAW(W.M.).

112 Z. PHYSIK BAND 252, ... 7, (1972). . . .T. . . . .
                    +ASYMMETRIC TOTAL REFLECTION OF POLARIZED NEUTRONS AND THE NEUTRON AS A
                     SURFACE PROBE.
                    +HANDEL(P.H.).

113 Z. PHYS. VOL.220, ... 419, (1969). . . .A4. . . . .
                    +DIFFRACTION OF THERMAL NEUTRONS BY A RULED GRATING.
                    +KURZ(H.), RAUCH(H.).

SECTION 9C-DIFFRACTION APPLIED TO CRYSTALLINE STRUCTURE DETERMINATION .

1 ACTA CRYST. 19,. .... 137, (1965)................
*DIFFRACTION BY A ONE DIMENSIONALLY DISORDERED CRYSTAL. I. THE INTENSITY
EQUATION.
*KAKINOKI(J.), KOMURA(Y.).

2 ACTA CRYST. 19,. .... 224, (1965)................
*COMPARISON OF SCATTERING FACTORS COMPUTED FROM FOUR DIFFERENT ATOMIC
MODELS.
*CROMER(D.T.).

3 ACTA CRYST. VOL.A26,. ...... 396, (1970)................
*THE CORRECTION OF MEASURED NEUTRON STRUCTURE FACTORS FOR THERMAL DIFFUSE
SCATTERING.
*WILLIS(B.T.M.).

4 ACTA CRYST. VOL.B27,. .... 2289, (1971)................
*A SIMPLE ≠DIRECT≠ SOLUTION OF THE CRYSTALLOGRAPHIC PHASE PROBLEM.
*MACDONALD(A.L.), ROBERTSON(J.M.), SPEAKMAN(J.C.).

5 ACTA CRYST. VOL.A27,. ...... 569, (1971)................
*DETERMINATION OF THE STATIC DISPLACEMENT OF ATOMS IN A BINARY ALLOY
SYSTEM USING ANOMALOUS SCATTERING.
*RAMESH(T.G.), RAMASESHAN(S.).

6 ACTA POLYTECH. SCANDINAVIA NO.70 6, (1970)............
*BACK-REFLECTION OF NEUTRONS FROM MOSAIC CRYSTALS.
*HIISMAKI(P.).

7 APPL. MATERIALS RES. VOL.1,. .. 160, (1962).....A4......
*THE INVESTIGATION OF STRUCTURAL DISORDER IN MATERIALS BY MEANS OF COLD
NEUTRON SCATTERING MEASUREMENTS.
*MARTIN(D.G.).

8 BER. BUNSENGES. PHYS. CHEM. 74,. 1202, (1970).....A1......MYOGLOBIN,
*NEUTRON DIFFRACTION ANALYSIS OF BIOLOGICAL STRUCTURES.
*SCHOENBORN(B.P.), NUNES(A.C.), NATHANS(R.).

9 CHEM. APPL. THERMAL NEUTRON SCAT 250, (1973).....T,E......
*DIFFUSE SCATTERING AND THE STUDY OF DEFECT SOLIDS.
*FENDER(B.E.F.). (IN BOOK: CHEMICAL APPLICATIONS OF THERMAL NEUTRON
SCATTERING. ED. BY B.T.M. WILLIS; OXFORD UNIV. PRESS: LONDON).

10 CHEM. APPL. THERMAL NEUTRON SCAT 225, (1973)................
*STRUCTURAL STUDIES ON NON-STOICHIOMETRIC COMPOUNDS BY THE BRAGG
SCATTERING OF NEUTRONS.
*CHEETHAM(A.K.). (IN BOOK: CHEMICAL APPLICATIONS OF THERMAL NEUTRON
SCATTERING. ED. BY B.T.M. WILLIS; OXFORD UNIV. PRESS: LONDON).

11 CHEM.-ING.-TECH. VOL.32,. .... 651, (1960).....T,E......
*STRUCTURE RESEARCH BY MEANS OF NEUTRON DIFFRACTION.
*BALKE(S.), LUTZ(G.).

12 COMP. REND. TOME 202,. .. 1029, (1936).....T......
*SUR LA DIFFRACTION DES NEUTRONS LENTS PAR LES SUBSTANCES CRISTALLINES.
*ELSASSER(W.M.).

13 CRYSTALLOGR./CRYSTAL PERFECTION, 269, (1963).....E......
*ANALYSIS OF HYDROGEN-BONDED CRYSTALS BY NEUTRON DIFFRACTION TECHNIQUES.
*SHANKAR(J.), PADMANABHAN(V.M.). (ED. NOTE: CRYSTALLOGRAPHY AND CRYSTAL
PERFECTION; PUBL. ACADEMIC PRESS: NEW YORK AND LONDON, 1963).

14 CURRENT SCI. (INDIA) VOL.35,. . 87, (1966)................
*THE USE OF ANOMALOUS SCATTERING OF NEUTRONS IN THE SOLUTION OF CRYSTAL
STRUCTURES CONTAINING LARGE MOLECULES.
*RAMASESHAN(S.).

15 DISS. ABS. VOL.27,. ...... 2487B, (1966)................
*DYNAMIC SCATTERING OF PARTICLES BY PERFECT CRYSTALS.
*HOWSMAN(A.J.).

16 HARWELL SUMMER SCHOOL,. ...... 161, (1968)................
*THE USE OF CONSTRAINTS IN CRYSTAL STRUCTURE REFINEMENT.
*PAWLEY(G.S.).

17 INTERN. SUMMER SCHOOL, MOL,. .. 643, (1963)................
*THE SCATTERING OF COLD NEUTRONS BY DEFECTS AS A MEANS OF STUDYING RADIA-
TION DAMAGE IN SOLIDS.
*MARTIN(D.G.).

18 INTERN. SUMMER SCHOOL, MOL,. .. 534, (1963)................
*ELASTIC SCATTERING OF NEUTRONS IN A CRYSTAL.
*GOEDKOOP(J.A.).

19 IZV. AKAD. NAUK...NEORG. MATER 8 1, (1972).....P......
*THE NEUTRON DIFFRACTION INVESTIGATION OF THE ATOMIC STRUCTURE OF
INORGANIC MATERIALS.
*YAMZIN(I.I.), LOSHMANOV(A.A.).

20 J. CHEM. EDUC. VOL.45,. ...... 296, (1968)................
*STRUCTURAL CHEMISTRY IN THE NUCLEAR AGE.
*HAMILTON(W.C.).

21 J. CHEM. PHYS. VOL.19,. .... 1416, (1951).....E,A1......NA*CL,K*CL,K*BR,K*H*F,
*THE USE OF SINGLE-CRYSTAL NEUTRON DIFFRACTION DATA FOR CRYSTAL STRUCTURE
DETERMINATION.
*PETERSON(S.W.), LEVY(H.A.).

22 J. NON-CRYST. SOLIDS VOL.7,. . 23, (1972).....T......GLASSES,
*DIFFRACTION STUDIES OF GLASS STRUCTURE I.THEORY AND QUASI-CRYSTALLINE
MODEL.
*LEADBETTER(A.J.), WRIGHT(A.C.).

23 J. PHYS. (USSR) VOL.5,. ...... 263, (1941).....T......
*THE EFFECT OF LONG RANGE ORDER IN ALLOYS UPON THE SCATTERING OF SLOW
NEUTRONS.
*SMIRNOV(A.A.), VONSOVSKY(S.V.).

24 KRISTALL TECH. VOL.4,. ...... 135, (1969).....A1......
*NEUTRON DIFFRACTION PROPERTIES OF MACHINED LARGE METALLIC SINGLE
CRYSTALS.
*MODRZEJEWSKI(A.), KOBLA(J.). (ED. NOTE: IN ENGLISH).

SECTION   9C-DIFFRACTION APPLIED TO CRYSTALLINE STRUCTURE DETERMINATION .

25 MATERIALS RES. BULL. VOL.2,. . . . . 69, (1967)....A4........TL-(BI,PB), CU-(AU,NI,ZN), NI-FE
      ⌐CHARACTERIZATION OF LARGE ALLOY SINGLE CRYSTALS BY NEUTRON DIFFRACTION.
      ⌐NG(S.C.), BROCKHOUSE(B.N.), HALLMAN(E.D.).

26 NAT. BUR. STAND., SPEC. PUB. 301 193, (1969)...........H-BONDS
      ⌐STRUCTURE AND DYNAMICS IN MOLECULAR CRYSTALS-SOME COMPARISONS.
      ⌐HAMILTON(W.C.).

27 NATURE VOL.167,. . . . . . . 243, (1951).....E.........NA*CL,NAPHTHALENE, ...
      ⌐SINGLE-CRYSTAL NEUTRON DIFFRACTION ANALYSIS.
      ⌐LOWDE(R.D.).

28 NATURWISSENSCHAFTEN VOL.58,. . . 444, (1971)....P.........
      ⌐INVESTIGATION OF CHEMICAL BONDING USING X-RAY AND NEUTRON DIFFRACTION.
      ⌐WILL(G.).

29 NED. TIJDSCHR. NATUURK. VOL.18,. 201, (1952)................
      ⌐DIFFRACTION OF NEUTRONS IN GASES AND CRYSTALS.
      ⌐JONKER(C.C.), BLOK(J.). (IN DUTCH).

30 PHYSICS TODAY (MARCH),. . . . . 19, (1973).............PROTEINS,. . . . .
      ⌐NEUTRONS PROBE PROTEIN STRUCTURE.
      ⌐ROTHENBERG(M.S.).

31 PHYS. LETT. VOL.14,. . . . . . . 174, (1965)....A4........SI.........
      ⌐PENDELLOSUNG INTERFERENCE WITH THERMAL NEUTRONS IN SI SINGLE CRYSTALS.
      ⌐SIPPEL(D.), KLEINSTUCK(K.), SCHULZE(G.E.R.). (IN GERMAN).

32 PHYS. LETT. 29A, ... 483, (1969)................
      ⌐IS THERE BRAGG SCATTERING OFF A TWO DIMENSIONAL CRYSTAL.
      ⌐IMRY(Y.), GUNTHER(L.).

33 PHYS. LETT. VOL.37A,. . . . . 29, (1971)................
      ⌐DETERMINATION OF X-RAY STRUCTURE FACTORS FROM HALF-VALUE WIDTHS OF
       DIFFRACTION   CURVES   OBTAINED WITH THE TRIPLE-CRYSTAL ARRANGEMENT.
      ⌐NAKAYAMA(K.), KIKUTA(S.), KOHRA(K.).

34 PHYS. STAT. SOL. 11,. . . . . 567, (1965)............AL,CR,
      ⌐THE TIME-OF-FLIGHT METHOD FOR INVESTIGATIONS OF SINGLE-CRYSTAL
       STRUCTURES.
      ⌐BURAS(B.), MIKKE(K.), LEBECH(B.), LECIEJEWICZ(J.).

35 PROC. ROY. SOC. A VOL.209,. . . 397, (1951)....E.........
      ⌐THE DIFFRACTION OF THERMAL NEUTRONS BY SINGLE CRYSTALS.
      ⌐BACON(G.E.).

36 PROC/SYMP/CRYS. STRUC/HIGH PRESS 141, (1969)....A4........ . . . .
      ⌐NEUTRON DIFFRACTION AT HIGH PRESSURES.
      ⌐BRUGGER(R.M.), BENNION(R.B.), WORLTON(T.G.), MYERS(W.R.).

37 REVUE CHIM. MINER. (FRANCE) VOL.8 185, (1971)....A1,A2........
      ⌐STUDY OF CRYSTALLINE AND MAGNETIC STRUCTURES BY NEUTRON DIFFRACTION.
      ⌐GUYADER(J.).

38 SCIENCE VOL.117,. . . . . . 1, (1953)................
      ⌐SOME ASPECTS OF NEUTRON SINGLE CRYSTAL ANALYSIS.
      ⌐PEPINSKY(R.).

39 SOL. STATE COMM. VOL. 9,. . . 2239, (1971)............HE (H.C.P.),
      ⌐FORBIDDEN BRAGG REFLECTIONS IN NEUTRON AND X-RAY SCATTERING FROM H.C.P.
       HELIUM.
      ⌐WERTHAMER(N.R.).

40 SOV. PHYS. JETP. 24,. . . . . 1068, (1967)................
      ⌐NEUTRON DIFFRACTION IN A POLARIZED CRYSTAL.
      ⌐BARYSHEVSKII(V.G.).

41 SOV. PHYS. J.E.T.P. VOL.29,. . . 655, (1969)................
      ⌐THE ANOMALOUS INTENSITY DISTRIBUTION IN SATELLITES IN NEUTRON
       DIFFRACTION INVESTIGATIONS OF BLOCK HELICOIDAL STRUCTURES.
      ⌐ALESHKO-OZHEVSKII(O.P.), YAMZIN(I.I.).

42 SPECTROS. INORG. CHEM,. . . . 1, (1971)....A1........
      ⌐MOLECULAR STRUCTURE DETERMINATION BY NEUTRON AND X-RAY DIFFRACTION.
      ⌐WILLIAMS(J.M.), PETERSON(S.W.), (IN BOOK: SPECTROSCOPY IN INORGANIC
       CHEMISTRY VOL.II, RAO, FERRARO (EDS.); PUBL. ACADEMIC: NEW YORK).

43 TRANS. AMER. CRYST. ASSOC. VOL.2 53, (1966)................
      ⌐THE APPLICATION OF DIRECT METHODS TO NEUTRON CRYSTALLOGRAPHY AND VICE
       VERSA.
      ⌐HAMILTON(W.C.).

44 Z. NATURFORSCH. VOL.28A,. . . 980, (1973)....T........
      ⌐ON THE SCATTERING OF X-RAYS FROM CRYSTALS CONTAINING A RANDOM
       DISTRIBUTION OF DEFECTS.
      ⌐TRINKAUS(H.). (IN GERMAN).

SECTION 9D-DIFFRACTION APPLIED TO MAGNETIC STRUCTURE DETERMINATION. . .

1 ACTA CRYST. VOL.9, ... 738, (1956)....T.........ZN*CR2*O4, ...
*ON THE SYSTEMATIC ABSENCE OF MAGNETIC REFLECTIONS OF NEUTRON DIFFRACTION
*YIN-YUAN-LI.

2 ACTA CRYST. VOL.10, ... 598, (1957)....T................. . . . .
*NEUTRON MAGNETIC SCATTERING FACTORS IN THE PRESENCE OF EXTINCTION.
*CHANDRASEKHAR(S.), WEISS(R.J.).

3 ACTA CRYST. VOL.11, ... 585, (1958)......................
*EXTINCTION EFFECTS IN NEUTRON SCATTERING FROM SINGLE MAGNETIC CRYSTALS.
*HAMILTON(W.C.).

4 ACTA CRYST. VOL.12, ... 282, (1959).....................
*A NOTE ON THE MAGNETIC INTENSITIES OF POWDER NEUTRON DIFFRACTION.
*SHIRANE(G.).

5 ACTA CRYST. VOL.14, ... 535, (1961).....................
*NEUTRON DIFFRACTION BY HELICAL SPIN (MAGNETIC) STRUCTURES.
*KOEHLER(W.C.).

6 ACTA CRYST. VOL.A28, ... 341, (1972)....A2.................
*AMBIGUITY IN THE MAGNETIC LATTICE DIMENSIONS OF CUBIC AND UNIAXIAL
COMPOUNDS FROM POWDER NEUTRON DIFFRACTION STUDIES.
*WINTENBERGER(M.), CHAMARD-BOIS(R.).

7 ACTA CRYST. VOL.A29, ... 453, (1973)....T................
*THE INFORMATION OF ORDERED MAGNETIC STRUCTURES WHICH CAN BE GAINED
FROM UNPOLARIZED-NEUTRON-DIFFRACTION DATA.
*WILKINSON(C.), LISHER(E.J.).

8 ACTA CRYST. VOL.A29, ... 449, (1973)....T................
*THE THEORY OF THE SPIN-DENSITY PATTERSON FUNCTION.
*WILKINSON(C.).

9 ACTA CRYST. VOL.A29, ... 651, (1973)....T................
*SPIN TRANSLATION GROUPS AND NEUTRON DIFFRACTION ANALYSIS.
*LITVIN(D.B.).

10 ACTA PHYS. POLON. VOL.35, ... 61, (1969)....D2.........
*CRITICAL MAGNETIC SCATTERING OF NEUTRONS IN ANTIFERROMAGNETS.
*KOCINSKI(J.).

11 ADV. STRUC. RES. DIFFR. METHOD 2 1, (1966)................
*THE INVESTIGATION OF MAGNETIC STRUCTURES BY NEUTRON DIFFRACTION (REVIEW)
*BACON(G.E.).

12 AIP CONF. PROC.(USA) VOL.5, . . . 1355, (1971)....D4..............
*GROUP REPORT ON MAGNETIC STRUCTURE WORK AT THE GRENOBLE NUCLEAR CENTER.
*BERTAUT(E.F.), BOLLER(H.), BURLET(P.), CHEVALIER(R.), DO-DINH(C.),
KALLEL(A.), ROSSAT-MIGNOD(J.), TCHEOU(F.), FRUCHART(D.), FRUCHART(R.).

13 ANN. DE PHYSIQUE TOME 7 NO.5, . . 371, (1972) BY MAGNETIC INTERACTION.
*THERMAL NEUTRON SCATTERING BY MAGNETIC INTERACTION.
*WILL(G.).

14 ANN. DE PHYSIQUE TOME 7 NO.5, . . 329, (1972)....E......HO*AL2*YB(3+).
*PHYSICAL ASPECTS OF THE CRYSTALLINE ELECTRIC FIELD (CEF).
*PURWINS(H.G.).

15 ANN. DE PHYSIQUE TOME 7 NO.4, . . 287, (1972)
*MAGNETIC NEUTRON SCATTERING IN RELATION TO SOME OTHER EXPERIMENTAL
METHODS.
*MOSSBAUER(R.L.). (PRESENTED AT AUTRANS SUMMER SCHOOL 1972).

16 ANN. DE PHYSIQUE TOME 7 NO.4, . . 203, (1972) . . . .
*MAGNETIC STRUCTURES (REVIEW PAPER).
*BERTAUT(E.F.). (PRESENTED AT AUTRANS SUMMER SCHOOL 1972).

17 ATLANTA SYMP/GEORGIA INST. TECH. 96, (1967).............
*SPIN DENSITIES.
*NATHANS(R.).

18 ATLANTA SYMP/GEORGIA INST. TECH. 74, (1967)............
*GENERALIZED SYMMETRY AND MAGNETIC SPACE GROUPS.
*DONNAY(J.D.H.).

19 CAHIERS DE PHYS. VOL.113, ... 29, (1960).............
*THE APPLICATION OF NEUTRON DIFFRACTION TO THE STUDY OF MAGNETIC
SUBSTANCES (REVIEW PAPER).
*MERIEL(P.). (IN FRENCH).

20 CAN. J. PHYS. VOL.50, ... 2991, (1972)....P...
*ZERO-POINT EFFECTS IN HEISENBERG ANTIFERROMAGNETS WITH ARBITRARY RANGE
OF INTERACTION.
*COLLINS(M.F.), TONDON(V.K.).

21 CHEM. APPL. THERMAL NEUTRON SCAT 270, (1973)....T,E........
*NEUTRON DIFFRACTION AND COVALENCY (REVIEW).
*JACOBSON(A.J.). (IN BOOK: CHEMICAL APPLICATIONS OF THERMAL NEUTRON
SCATTERING. ED. BY B.T.M. WILLIS; OXFORD UNIV. PRESS: LONDON).

22 C.N.E.N. SYMP. CASACCIA, ... 149, (1968)
*MAGNETIZATION DENSITY BY NEUTRONS SCATTERING IN MAGNETICALLY ORDERED
SYSTEMS.
*PAOLETTI(A.).

23 ELECTRON TECHNOL. (POL.) VOL.1, . 5, (1968)....A2.........
*THE MAGNETIC STRUCTURE OF FERROMAGNETIC MATERIALS OF UNIAXIAL SYMMETRY.
*SZYMCZAK(R.).

24 HARWELL SUMMER SCHOOL, ... 211, (1968)............
*COVALENCY IN MAGNETIC SALTS.
*RIMMER(D.E.).

25 HARWELL SUMMER SCHOOL, ... 190, (1968)................
*MAGNETIC MOMENT DISTRIBUTIONS IN METALS AND IONIC SOLIDS.
*FORSYTH(J.B.).

26 HARWELL SUMMER SCHOOL, ... 176, (1968)................
*THE INTERPRETATION OF MAGNETIC SCATTERING EXPERIMENTS.
*BROWN(P.J.).

SECTION   90-DIFFRACTION APPLIED TO MAGNETIC STRUCTURE DETERMINATION. . .

27 HARWELL SUMMER SCHOOL, ... 171, (1968). . . . .
                         *THE SIGNIFICANCE OF ACCURATE INTENSITY MEASUREMENTS IN CRYSTALLOGRAPHIC
                          MAGNETIC SCATTERING.
                         *LOMER(W.M.).

28 IAEA SYMP. (PILE RESEARCH)VIENNA 415, (1960). . . . .
                         *NEUTRON-DIFFRACTION EXPERIMENTS ON MAGNETIC INELASTIC SCATTERING OF
                          NEUTRONS.
                         *RISTE(T.), BLINOWSKI(K.), JANIK(J.), WANIC(A.).

29 IAEA SYMP. GRENOBLE . . . . 529, (1972). . . . .
                         *MAGNETISM IN ONE AND TWO DIMENSIONS (REVIEW PAPER).
                         *BLUME(M.).

30 IEEE TRANS. MAGN.(USA) VOL.MAG-8 161, (1972). . . .04. . . . .
                         *NEUTRON-DIFFRACTION DETERMINATION OF MAGNETIC STRUCTURES.
                         *COX(D.E.).

31 INT. J. MAGN. VOL.1 NO.3 . . . . . 219, (1971). . . . .
                         *FOURIER TECHNIQUES FOR ANALYSIS OF MAGNETIC STRUCTURE FACTORS:
                          APPLICATION TO 3D METALS.
                         *MOON(R.M.).

32 J. APPL. PHYS. VOL.31, SUPPL., . . 350, (1960). . . .E. . . . .
                         *PRECISE MEASUREMENT OF MAGNETIC-FORM FACTORS.
                         *NATHANS(R.).

33 J. APPL. PHYS. VOL.32, SUPPL., . . 20S, (1961). . . . .
                         *NEUTRON DIFFRACTION BY HELICAL SPIN STRUCTURES.
                         *KOEHLER(W.C.).

34 J. APPL. PHYS. VOL.33, SUPPL., . 1029, (1962). . . .T. . . . . . .MN*AU2,RARE-EARTH METALS. . .
                         *MODIFICATION OF SPIN SCREW STRUCTURE DUE TO ANISOTROPY ENERGY AND
                          APPLIED MAGNETIC FIELD.
                         *NAGAMIYA(T.).

35 J. APPL. PHYS. 41, ... 937, (1970). . . . .
                         *COVALENCY, SPIN DENSITIES, AND NEUTRON MAGNETIC SCATTERING.
                         *ELLIS(D.E.), FREEMAN(A.J.).

36 J. APPL. PHYS. VOL.41 . . . . 919, (1970). . . . .
                         *COMMENTS ON THE MAGNETIZATION AND NEUTRON SCATTERING IN LOCAL MOMENT
                          FERROMAGNETIC ALLOYS.
                         *SILVERSTEIN(S.D.), RICE(M.J.).

37 J. CHEM. PHYSICS VOL.53, ... 1387, (1970). . . . .
                         *CALCULATION OF NEUTRON MAGNETIC FORM FACTORS FOR RARE-EARTH IONS.
                         *LANDER(G.H.), BRUN(T.O.).

38 J. PHYS. CHEM. SOLIDS VOL.5 . . . . 180, (1958). . . . .
                         *TYPES OF MAGNETICALLY ORDERED CONFIGURATIONS ON SIMPLE LATTICES.
                         *GERSCH(H.A.), KOEHLER(W.C.).

39 J. PHYS. -C-, VOL. 1, ... 88, (1968). . . . .
                         *NEUTRON ELASTIC DIFFUSE SCATTERING FROM MIXED MAGNETIC SYSTEMS.
                         *MARSHALL(W.).

40 J. PHYS. -C-, VOL. 1, ... 966, (1968). . . . .
                         *SECOND ORDER EFFECTS IN DIFFUSE ELASTIC NEUTRON SCATTERING FROM
                          FERROMAGNETIC ALLOYS.
                         *BALCAR(E.), MARSHALL(W.).

41 J. PHYS. -C-, VOL. 2, ... 470, (1969). . . . .
                         *SOME ASPECTS OF THE THEORY OF SCATTERING OF NEUTRONS BY MAGNETIC IONS.
                         *LOVESEY(S.W.).

42 J. PHYS. C, SER.2, VOL.2, ... 1151, (1969). . . . .
                         *THEORY OF SCATTERING OF NEUTRONS BY MAGNETIC IONS IN CRYSTALS.
                         *JOHNSTON(D.F.), RIMMER(D.E.).

43 J. PHYSIQUE TOME 31 . . . . 369, (1970). . . .A2. . . . . .ORTHOCOBALTITES. . . . .
                         *MAGNETIC STRUCTURE AND PROPERTIES OF RARE EARTH ORTHOCOBALTITES.
                         *KAPPATSCH(A.), QUEZEL-AMBRUNAZ(S.), SIVARDIERE(J.).

44 J. PHYSIQUE TOME 32 COL.1 VOL.I, 462, (1971). . . .T. . . . .
                         *MAGNETIC STRUCTURE ANALYSIS AND GROUP THEORY.
                         *BERTAUT(E.F.).

45 J. PHYSIQUE TOME 32 COL.1 VOL.I, 579, (1971). . . .A2. . . . . .FERROMAGNETS. . . . .
                         *NEUTRON DEPOLARIZATION AS A METHOD TO DETERMINE THE MAGNETIZATION, THE
                          MEAN DOMAIN SIZE ... OF THE INNER MAGNETIZATION OF FERROMAGNETS.
                         *REKVELDT(M.TH.).

46 J. PHYS. RADIUM TOME 17, ... 72, (1956). . . . .
                         *APPLICATION DE LA DIFFRACTION DES NEUTRONS À L#ETUDE DE L#ANTI-
                          FERROMAGNETISME.
                         *JOHANNIN-GILLES(A.).

47 J. PHYS. RADIUM VOL.20, ... 169, (1959). . . .A2. . . . . .FE3*04,MN*O,NI*O. . . . .
                         *RECENT MAGNETIC STRUCTURE STUDIES BY NEUTRON DIFFRACTION.
                         *SHULL(C.G.).

48 J. PHYS. SOC. JAPAN VOL.17 B-II, 342, (1962). . . .E. . . . .
                         *A TECHNIQUE FOR MEASURING THE MAGNETIC DISORDER SCATTERING OF NEUTRONS.
                         *LOWDE(R.D.), WHEELER(D.A.).

49 J. PHYS. SOC. JAPAN VOL.17 B-II, 332, (1962). . . .P. . . . .
                         *TABLES OF MAGNETIC SPACE GROUPS. II,SPECIAL POSITIONS.
                         *BELOV(N.V.), NERONOVA(N.N.), DONNAY(J.D.H.), DONNAY(G.).

50 J. PHYS. SOC. JAPAN VOL.17 B-II, 330, (1962). . . .P. . . . .
                         *FERROMAGNETIC GROUPS.
                         *BELOV(N.V.), NERONOVA(N.N.).

51 MAGNETIC MATERIALS DIGEST 1964, . . 37, (1964). . . .A2. . . . .
                         *MAGNETIC STRUCTURES AND NEUTRON DIFFRACTION.
                         *PROSEN(R.J.), ANDERSON(F.B.). (PUBL. M.W. LADS, PHILADELPHIA).

52 METAL. SOC. CONF. LOS ANGELES, . . 99, (1967). . . . .
                         *THE MAGNETIC STRUCTURE OF RARE EARTH ALLOYS.
                         *KOEHLER(W.C.).

53 METAL. SOC. CONF. LOS ANGELES, . . 31, (1967). . . . .
                         *THE DISTRIBUTION OF MAGNETIC MOMENTS IN FERROMAGNETIC ALLOYS.
                         *CABLE(J.W.).

SECTION   90-DIFFRACTION APPLIED TO MAGNETIC STRUCTURE DETERMINATION. . .

54   2ND SUMMER SCHOOL S.S. PHYS,,   171, (1968)....T.......:
     ⌐FIRST-ORDER MAGNETIC PHASE TRANSITIONS:
     ⌐PAL(L.).

55   PHIL. MAG. VOL.17,. . . . . 609, (1968)..........OF MAGNETIC STRUCTURES FROM NEUTRON
     ⌐A METHOD FOR THE DETERMINATION OF MAGNETIC STRUCTURES FROM NEUTRON
     DIFFRACTION DATA BY THE USE OF A SPIN DENSITY PATTERSON FUNCTION.
     ⌐WILKINSON(C.).

56   PHYS. LETT. VOL.39A. . . . . . . 141, (1972)..............NI*O.
     ⌐OBSERVATION OF MAGNETIC SUPERLATTICE PEAKS BY X-RAY DIFFRACTION ON AN
     ANTIFERROMAGNETIC NI*O CRYSTAL.
     ⌐DE-BERGEVIN(F.), BRUNEL(M.).

57   PHYS. REV. VOL.76,. . . . 1256, (1949)....E.......MN*O.
     ⌐DETECTION OF ANTIFERROMAGNETISM BY NEUTRON DIFFRACTION.
     ⌐SHULL(C.G.), SMART(J.S.).

58   PHYS. REV. VOL.100,. . . . 627, (1955).........MN*O,FE*O,CO*O, ETC.
     ⌐MAGNETIC MOMENT ARRANGEMENTS AND MAGNETOCRYSTALLINE DEFORMATIONS IN
     ANTIFERROMAGNETIC COMPOUNDS.
     ⌐YIN-YUAN-LI.

59   PHYS. REV. VOL.112,. . . . 1917, (1958)....T.......CU*FE*S2.
     ⌐SYMMETRY OF MAGNETIC STRUCTURES: MAGNETIC STRUCTURE OF CHALCOPYRITE.
     ⌐DONNAY(G.), CORLISS(L.M.), DONNAY(J.D.H.), ELLIOTT(N.), HASTINGS(J.M.).

60   PHYS. REV. LETT. 10,. . . . 489, (1963).................
     ⌐MAGNETIC SCATTERING OF NEUTRONS BY NONCOLLINEAR SPIN DENSITIES.
     ⌐BLUME(M.).

61   PHYS. REV. 178,,. . . . . 783, (1969)................
     ⌐THEORY OF MAGNETIC ORDERING IN THE HEAVY RARE EARTHS.
     ⌐EVENSON(W.E.), LIU(S.H.).

62   PHYS. REV. LETT. VOL.26,. . . 718, (1971)....A2.......(C*D3)4*N*MN*CL3.
     ⌐SPIN CORRELATIONS IN A ONE-DIMENSIONAL HEISENBERG ANTIFERROMAGNET.
     ⌐BIRGENEAU(R.J.), DINGLE(R.), HUTCHINGS(M.T.), SHIRANE(G.), HOLT(S.L.).

63   PHYS. STAT. SOL. VOL.40,. . . . 59, (1970).................
     ⌐SPATIAL DISTRIBUTION OF MAGNETIC MOMENTS.
     ⌐GALPERIN(F.M.), ALPAEV(YU.A.), KIZHAEV(F.G.).

64   PROC. PHYS. SOC. 86,. . . . 561, (1965).................
     ⌐COVALENCY EFFECTS IN NEUTRON DIFFRACTION FROM FERROMAGNETIC AND
     ANTIFERROMAGNETIC SALTS.
     ⌐HUBBARD(J.), MARSHALL(W.).

65   PROC. PHYS. SOC. 88,. . . . 37, (1966).................
     ⌐ON THE THEORY OF THE ELECTRON ORBITAL CONTRIBUTION TO THE SCATTERING OF
     NEUTRONS BY MAGNETIC IONS IN CRYSTALS.
     ⌐JOHNSTON(D.F.).

66   PROGRESS/LOW TEMP. PHYS. VOL.4,. 265, (1964).....A2.......RARE-EARTH METALS,
     ⌐MAGNETIC STRUCTURES OF HEAVY RARE-EARTH METALS.
     ⌐YOSIDA(K.). (ED. NOTE: PROGRESS IN LOW TEMPERATURE PHYSICS VOL.IV; PUBL.
     NORTH-HOLLAND: AMSTERDAM).

67   REVUE CHIM. MINER.(FRANCE) VOL.8 185, (1971)....A1,A2.
     ⌐STUDY OF CRYSTALLINE AND MAGNETIC STRUCTURES BY NEUTRON DIFFRACTION.
     ⌐GUYADER(J.).

68   SOL. STATE COMM. 5,. . . . 289, (1967)...............
     ⌐DIFFRACTION DES NEUTRONS POLARISES SUR LES COMPOSES MAGNETIQUES À
     STRUCTURE NON-COLINEAIRE.
     ⌐SIVARDIERE(J.).

69   SOL. STATE COMM. VOL.12,. . . 1167, (1973)....A2,T.....HO.
     ⌐ON THE DIFFRACTION OF NEUTRONS BY MAGNETIC SPIRAL STRUCTURES.
     ⌐FELCHER(G.P.).

70   SOV. PHYS. SOL. STATE, 4,. . . 2533, (1962)...............
     ⌐THE SCATTERING OF SLOW NEUTRONS BY COMPLEX MAGNETIC STRUCTURES.
     ⌐MALEEV(S.V.), BAR≠YAKHTAR(V.G.), SURIS(R.A.).

71   SOV. PHYS. JETP VOL.15,. . . 1162, (1962)....T.
     ⌐THE SCATTERING OF POLARIZED NEUTRONS BY A HELICAL MAGNETIC STRUCTURE.
     ⌐IZYUMOV(YU.A.).

72   SOV. PHYS. SOL. STATE, 4,. . . 158, (1962)...............
     ⌐MAGNETIC SCATTERING OF SLOW NEUTRONS IN DILUTE ALLOYS OF A TRANSITION
     METAL WITH A SIMPLE METAL.
     ⌐IZYUMOV(YU.A.).

73   SOV. PHYS. USPEKHI VOL.5,. . . 104, (1962)....D1,D2.
     ⌐NEUTRON DIFFRACTION STUDY OF MAGNETIC MATERIALS.
     ⌐ZHDANOV(G.S.), OZEROV(R.P.).

74   SOV. PHYS. JETP. 20,. . . . 1505, (1965)...............
     ⌐SCATTERING OF NEUTRONS BY QUANTIZED MAGNETIC FLUX LINES IN TYPE II
     SUPERCONDUCTORS.
     ⌐KEMOKLIDZE(M.P.).

75   SOV. PHYS. CRYST. VOL.9,. . . 535, (1965)....T.......CR.
     ⌐DIFFRACTION OF NEUTRONS BY MAGNETIC SUPERLATTICES.
     ⌐LEVDIK(V.A.), BYKOV(V.N.), GOLOVKIN(V.S.).

76   SOV. PHYS. SOL. STATE, 8,. . . 1852, (1966)...............
     ⌐SLOW NEUTRON SCATTERING IN A SUPERCONDUCTOR.
     ⌐GINZBURG(S.L.), MALEEV(S.V.).

77   SOV. PHYS. SOLID STATE VOL.9,. 2016, (1968)...............
     ⌐FLUCTUATIONS AND SCATTERING PROCESSES OF SLOW NEUTRONS AND LIGHT IN
     ANTIFERROMAGNETS IN STRONG MAGNETIC FIELD.
     ⌐AKHIEZER(I.A.), BOLOTIN(YU.L.).

78   TRANS. AMER. CRYST. ASSOC. VOL.8 59, (1972)....A2.
     ⌐EXPERIMENTAL SPIN DENSITIES.
     ⌐MOON(R.M.).

79   TRANSITION METAL COMPOUNDS,. . . 29, (1964)............
     ⌐NEUTRON DIFFRACTION AND COVALENT BONDING IN MAGNETIC SALTS.
     ⌐MARSHALL(W.). (ED. NOTE: PUBL. BY GORDON AND BREACH:NEW YORK).

SECTION   90-DIFFRACTION APPLIED TO MAGNETIC STRUCTURE DETERMINATION. . .

80   TRIESTE/INTERN. COURSE. . . . . . . 561, (1967). . . . .
     ⇌MAGNETIZATION DENSITY IN FERROMAGNETIC METALS. (14 PAGES).
     ⇌PAOLETTI(A.).

81   TRIESTE/INTERN. COURSE. . . . . . . 501, (1967). . . . .
     ⇌INVESTIGATIONS OF MAGNETIC MATERIALS USING NEUTRON SCATTERING.
     (38 PAGES).
     ⇌MARSHALL(W.), LOW(G.G.).

82   USPEKHI FIZ. NAUK VOL.47,. . . . . 445, (1952)....E.........CR2*O3,MN*O,NI*O,CO*O,FE*O,ETC.,
     ⇌NEUTRONOGRAPHIC STUDY OF THE MAGNETIC STRUCTURE OF ANTIFERROMAGNETS.
     ⇌OZEROV(R.P.). (IN RUSSIAN).

83   WISS. Z. TECH. UNIV/DRESDEN 12,. 1159, (1963)....E,A2..... . . . .
     ⇌RESULTS OF MAGNETIC NEUTRON DIFFRACTION.
     ⇌SCHULZE(G.E.R.). (IN GERMAN).

84   Z. ANGEW. PHYS. VOL.21,. . . . . 259, (1966)....T......... . . . .
     ⇌MAGNETIC CONFIGURATIONS ACCORDING TO THE RESULTS OF NEUTRON DIFFRACTION.
     ⇌BERTAUT(E.F.). (IN GERMAN).

85   Z. ANGEW. PHYS. VOL.24,. . . . . 260, (1968)....T,E......FE,NI,CO,. . . . .
     ⇌LOCAL DISTRIBUTION OF MAGNETIC MOMENTS IN CRYSTALS BY NEUTRON
     DIFFRACTION.
     ⇌WILL(G.). (IN GERMAN).

86   Z. ANGEW. PHYS. VOL.24,. . . . . 254, (1968)....A2. . . . .
     ⇌MEASUREMENTS OF DIFFUSE MAGNETIC SCATTERING OF NEUTRONS IN FERROMAGNETIC
     ALLOYS.
     ⇌LOW(G.G.).

87   Z. ANGEW. PHYS. VOL.32,. . . . . 1, (1971)....A2......... . . . .
     ⇌NEUTRON DIFFRACTION AND MAGNETIC STRUCTURES OF THE RARE EARTH ELEMENTS
     AND OF 3D-4F-INTERMETALLIC COMPOUNDS.
     ⇌WILL(G.).

SECTION 10A-PHONON SCATTERING: REVIEW PAPERS . . . .

1 BROOKHAVEN SYMPOSIUM. . . . . . . 8, (1965). . . . .
   ⋆REVIEW OF EXPERIMENTAL MEASUREMENTS OF LATTICE MODES IN TRANSITION
   METALS.
   ⋆WOODS(A.D.B.).

2 IAEA SYMP. CHALK RIVER, VOL.2, . . 3, (1962). . . .B1. . . . . .K⋆BR,NA,NA⋆I,PB, ...
   ⋆CRYSTAL DYNAMICS FROM NEUTRON SPECTROMETRY.
   ⋆WOODS(A.D.B.).

3 IAEA SYMP. GRENOBLE. . . . . . . 3, (1972). . . .T. . . . .
   ⋆PHONONS IN METALS (REVIEW PAPER).
   ⋆HANKE(W.), BILZ(H.).

4 J. APPL. PHYS. VOL.33, SUPPL., . . 307, (1962). . . . .
   ⋆LATTICE VIBRATIONS IN SOLIDS (REVIEW PAPER).
   ⋆KRUMHANSL(J.A.).

5 J. PHYSIQUE TOME 33, COL.2, . . . 7, (1972). . . . . . .SR⋆TI⋆O3,K⋆TA⋆O3,NA⋆N⋆O2,DKDP. . .
   ⋆NEUTRON INELASTIC SCATTERING AT STRUCTURAL PHASE TRANSITIONS (REVIEW
   PAPER).
   ⋆COWLEY(R.A.).

6 2ND SUMMER SCHOOL S.S. PHYS., . . 241, (1968). . . .T,B3. . . . .
   ⋆VIBRATIONS OF DEFECTS IN LATTICES.
   ⋆ELLIOTT(R.).

7 2ND SUMMER SCHOOL S.S. PHYS., . . 137, (1968). . . . .
   ⋆LATTICE DYNAMICS AND NEUTRON SCATTERING (REVIEW PAPER).
   ⋆CAGLIOTI(G.).

8 NED. TIJDSCHRIFT NATUURKUNDE 29, 93, (1963). . . . .
   ⋆SCATTERING OF SLOW NEUTRONS IN CRYSTALS AND LIQUIDS. II.
   ⋆KOKKEDEE(J.J.), NIJBOER(B.R.). (IN DUTCH).

9 NED. TIJDSCHRIFT NATUURKUNDE 29, 61, (1963). . .T. . . . .
   ⋆SCATTERING OF SLOW NEUTRONS BY CRYSTALS AND FLUIDS. I.
   ⋆NIJBOER(B.R.), KOKKEDEE(J.J.). (IN DUTCH).

10 PHYSICS OF SEMICONDUCTORS, . . . 467, (1962). . . . . .GE,SI, ...
   ⋆REVIEW PAPER ON LATTICE VIBRATIONS.
   ⋆COCHRAN(W.). (ED. NOTE: PROCEEDINGS INTERNATIONAL CONFERENCE ON PHYSICS
   OF SEMICONDUCTORS,EXETER 1962; PUBL. INST./PHYS. AND PHYSICAL SOCIETY).

11 PROGR. NUCL. PHYS. VOL.1, ... 185, (1950). . . .T,E. . . . .
   ⋆THE SCATTERING OF NEUTRONS BY CRYSTALS.
   ⋆CASSELS(J.M.).

12 RENSSELAER POL. INST. SYMP., . . . 129, (1961). . . . .
   ⋆CRYSTAL AND LIQUID DYNAMICS FROM NEUTRON ENERGY DISTRIBUTIONS.
   ⋆BROCKHOUSE(B.N.).

13 REP. PROGR. PHYS. VOL.26, ... 1, (1963). . . .T. . . . . . .
   ⋆LATTICE VIBRATIONS.
   ⋆COCHRAN(W.).

14 REV. MOD. PHYS. VOL.42, ... 409, (1970). . . .T,B1. . . . .
   ⋆EXTERNAL VIBRATIONS IN COMPLEX CRYSTALS.
   ⋆VENKATARAMAN(G.), SAHNI(V.C.).

15 SCI. PROGR. VOL.51, ... 424, (1963). . . .E. . . .
   ⋆SPECIFIC HEAT, NEUTRONS AND PHONONS.
   ⋆COCHRAN(W.).

16 SCOTTISH UNIV. SUMMER SCHOOL, . . 377, (1965). . . . . . .
   ⋆VIBRATIONS OF DEFECTS IN LATTICES.
   ⋆ELLIOTT(R.J.).

17 SCOTTISH UNIV. SUMMER SCHOOL, . . 110, (1965). . . . .
   ⋆NEUTRON SCATTERING BY PHONONS.
   ⋆BROCKHOUSE(B.N.).

18 SOLID/LIQUID STATE PHYSICS, . . . 43, (1965). . . .T,E. . . .
   ⋆DYNAMICS OF CRYSTAL LATTICES.
   ⋆COCHRAN(W.). (LECTURES ON SOLID AND LIQUID STATE PHYSICS. PUBL. ATOMIC
   ENERGY ESTABLISHMENT TROMBAY, BOMBAY, INDIA).

19 SOL. STATE PHYS. VOL.8, ... 109, (1959). . . .T. . . . .
   ⋆INTERACTION OF THERMAL NEUTRONS WITH SOLIDS (REVIEW PAPER).
   ⋆KOTHARI(L.S.), SINGWI(K.S.).

20 SPRINGER TRACTS IN MOD. PHYS. 43 232, (1967). . . .T. . . . .
   ⋆INTERACTION OF PHONONS WITH THERMAL NEUTRONS.
   ⋆LUDWIG(W.).

21 TH. NEUTRON SCATT./EGELSTAFF, . . 193, (1965). . . . .
   ⋆THERMAL VIBRATIONS OF CRYSTAL LATTICES.
   ⋆DOLLING(G.), WOODS(A.D.B.).

SECTION   109-CRYSTAL DYNAMICS IN GENERAL (INCLUDING THE HARMONIC THEORY).

1   AARHUS SUMMER SCHOOL,. . . . . . 424, (1963)..........
        *PHONONS AND LATTICE IMPERFECTION.
        *MARADUDIN(A.A.).

2   AARHUS SUMMER SCHOOL,. . . . . .   76, (1963).........
        *LECTURE ON PHONONS AND EXTERNAL RADIATION.
        *SJOLANDER(A.).

3   ACTA CRYST. VOL.A27,. . . . . . . 556, (1971)........
        *THE RELATION BETWEEN PHONON FREQUENCIES AND INTERATOMIC FORCE CONSTANTS.
        *COCHRAN(W.).

4   ACTA CRYST. VOL.A28,. . . . . . . 170, (1972)........
        *IRRELEVANCE OF ATOMIC MASSES FOR DEBYE-WALLER B VALUES IN THE LIMIT
        OF HIGH TEMPERATURES.
        *HUISZOON(C.), GROENEWEGEN(P.P.M.).

5   ACTA CRYST. VOL.A28,. . . . . . . 166, (1972)....NA*CL, K*CL, MG*O, AG*CL.
        *DEBYE-WALLER B VALUES OF SOME NA*CL-TYPE STRUCTURES AND INTERIONIC
        INTERACTION.
        *GROENEWEGEN(P.P.M.), HUISZOON(C.).

6   ACTA PHYS. POLON. VOL.A43,. . . 247, (1973)....T........
        *AN ASPECT OF NORMAL MODES OF VIBRATIONS IN DISORDERED CRYSTAL.
        *PARLINSKI(K.).

7   ANN. DE PHYSIQUE TOME 5,. . . . . 77, (1970)........
        *ETAT ACTUEL DES ETUDES SUR LA DYNAMIQUE DES CRISTAUX NON CONDUCTEURS.
        I. CRISTAL DANS L≠APPROXIMATION HARMONIQUE.
        *LAPLAZE(D.), VERGNOUX(A.M.).

8   ANN. DE PHYSIQUE TOME 7 NO.5,. . 329, (1972)....E.........
        *PHYSICAL ASPECTS OF THE CRYSTALLINE ELECTRIC FIELD (CEF).
        *PURWINS(H.G.).

9   ARK. FYS. VOL.25, PAPER 3,. . .   21, (1963)..........
        *THE RELATION BETWEEN THE INTERATOMIC FORCES AND THE FREQUENCIES OF
        SYMMETRY PHONONS IN CUBIC CRYSTALS.
        *SQUIRES(G.L.).

10 CAN. J. PHYS. 47,.. . . . . 617, (1969).........
        *LATTICE DYNAMICS OF HEAVY RARE-GAS SOLIDS.
        *GUPTA(N.P.), GUPTA(R.K.).

11 CAN. J. PHYS. 47,.. . . . . 451, (1969).........
        *LATTICE DYNAMICS OF III-V COMPOUNDS.
        *BANERJEE(R.), VARSHNI(Y.P.).

12 CAN. J. PHYS. 47,.. . . . . 1381, (1969).......CALCITE, ALPHA-CORUNDUM,
        *SYMMETRY PROPERTIES OF THE NORMAL MODES OF VIBRATION OF CALCITE AND
        ALPHA-CORUNDUM.
        *COWLEY(E.R.).

13 CAN. J. PHYS. 48,.. . . . . 183, (1970)........
        *INFLUENCE OF THE STATIC ELECTRON GAS SCREENING FUNCTION ON THE LATTICE
        DYNAMICS OF SODIUM.
        *GELDART(D.J.W.), TAYLOR(R.)...ET AL.

14 CAN. J. PHYS. 49,.. . . . . 2496, (1971)....CU97-AU03.
        *CALCULATION OF THE FREQUENCIES AND WIDTHS OF PHONONS IN CU97-AU03.
        *BRUNO(R.), TAYLOR(D.W.).

15 CAN. J. PHYS. 50,.. . . . . 122, (1972)....CS*CL, CS*BR, CS*I,
        *SHELL-MODEL LATTICE DYNAMICS OF CS*CL, CS*BR AND CS*I.
        *CARABATOS(C.), PREVOT(B.).

16 CHEM. APPL. THERMAL NEUTRON SCAT   78, (1973)....T........
        *MODELS FOR CALCULATING THE PROPERTIES OF PHONONS IN MOLECULAR CRYSTALS.
        *PAWLEY(G.S.).   (IN BOOK: CHEMICAL APPLICATIONS OF THERMAL NEUTRON
        SCATTERING. ED. BY B.T.M. WILLIS; OXFORD UNIV. PRESS: LONDON).

17 CONF. INTERN., RENNES,. . . . . . 209, (1971)........
        *DENSITY OF STATE MEASUREMENTS FOR PHONONS IN AROMATIC CRYSTALS.
        *HARRYMAN(M.B.M.), REYNOLDS(P.A.), WHITE(J.W.), KJEMS(J.K.).

18 DISS. ABS. VOL.27,. . . . . . . .3247B, (1966)........
        *THE LATTICE DYNAMICS OF MASS DISORDERED ALLOYS.
        *LEATH(P.L.).

19 DISS. ABS. VOL.27,. . . . . . . .2492B, (1966)........
        *ONE DIMENSIONAL ELECTRON-PHONON MODELS.
        *VARGA(B.B.).

20 DISS. ABS. VOL.27,. . . . . . . .2487B, (1966)........
        *DYNAMIC SCATTERING OF PARTICLES BY PERFECT CRYSTALS.
        *HOWSMAN(A.J.).

21 DISS. ABS. VOL.28,. . . . . . . .4724B, (1967)........
        *LATTICE DYNAMICS OF SOLID HELIUM.
        *HORLEY(G.L.).

22 DISS. ABS. VOL.28,. . . . . . . .2586B, (1967)........
        *EFFECTS OF IMPURITIES ON LATTICE VIBRATIONS- A MODEL CALCULATION
        INCLUDING MASS AND FORCE CONSTANT CHANGES- NEUTRON SCATTERING.
        *LAKATOS(K.S.).

23 DISS. ABS. VOL.29,. . . . . . . .1815B, (1968)........
        *AB INITIO CALCULATION OF PHONON DISPERSION CURVES FOR BE.
        *KOPPEL(J.U.).

24 DISS. ABS. VOL.29,. . . . . 730B, (1968)........
        *A CALCULATION OF PHONON LIFETIMES IN SOLID ARGON.
        *JONES(H.D.).

25 ERGEB. EXAKT. NATURWISS. VOL.35,   1, (1964)....93......
        *DYNAMICS OF CRYSTALS WITH POINT DEFECTS.
        *LUDWIG(W.). (IN GERMAN).

26 HARWELL SUMMER SCHOOL,. . . . . . 132, (1968)........
        *GENERALIZED TREATMENTS FOR THERMAL MOTION.
        *JOHNSON(C.K.).

SECTION 10B-CRYSTAL DYNAMICS IN GENERAL (INCLUDING THE HARMONIC THEORY).

27 IAEA SYMP. CHALK RIVER, VOL.2, . 71, (1962)..INTERATOMIC FORCES AND THE FREQUENCIES OF
    *THE RELATION BETWEEN THE INTERATOMIC FORCES AND THE FREQUENCIES OF
    SYMMETRY PHONONS IN CUBIC CRYSTALS.
    *SQUIRES(G.L.).

28 IAEA SYMP. CHALK RIVER, VOL.1, . 59, (1962)..........................
    *DIELECTRIC CONSTANTS AND LATTICE VIBRATIONS IN IONIC CRYSTALS.
    *COCHRAN(W.), COWLEY(R.A.).

29 IAEA SYMP. CHALK RIVER, VOL.1, . 49, (1962)..........................
    *THE CALCULATION OF DEBYE-WALLER FACTORS FROM THERMODYNAMIC DATA.
    *BARRON(T.H.K.),   LEADBETTER(A.J.),   MORRISON(J.A.),   SALTER(L.S.).

30 IAEA SYMP. CHALK RIVER, VOL.1, . 3, (1962)..........................
    *THE MOSSBAUER EFFECT AND DYNAMICS OF ATOMIC MOTIONS IN CONDENSED SYSTEMS
    *SINGWI(K.S.).

31 IAEA SYMP. BOMBAY, VOL.1,. . . 25, (1964)....T...............
    *ELECTRON-PHONON INTERACTIONS AND LATTICE DYNAMICS.
    *TOYA(T.).

32 IAEA SYMPOSIUM BOMBAY, VOL.1,. . 313, (1964)..........................
    *THEORIE MICROSCOPIQUE DES TRANSITIONS S≠ACCOMPAGNANT D≠UNE MODIFICATION
    DE LA STRUCTURE CRISTALLINE.
    *BOCCARA(N.), SARMA(G.).

33 IAEA SYMPOSIUM BOMBAY, VOL.1,. . 181, (1964)
    *LATTICE DYNAMICS OF THE HEXAGONAL CLOSE-PACKED STRUCTURE.
    *CZACHOR(A.).

34 IAEA SYMPOSIUM BOMBAY, VOL.1,. . 49, (1964).............NA,CU,
    *A COMPARISON BETWEEN THE ELECTRON-PHONON INTERACTIONS OF TOYA AND BAILYN
    *SRIVASTAVA(P.L.).

35 IAEA SYMPOSIUM BOMBAY, VOL.1,. . 3, (1964)
    *THEORETICAL ASPECTS OF PHONON DISPERSION CURVES FOR METALS.
    *COCHRAN(W.).

36 IAEA SYMP. COPENHAGEN, VOL.1, . 165, (1968)...T........MG,ZN,BE,
    *THEORY OF VIBRATIONAL SPECTRUM IN HEXAGONAL METALS.
    *.C.C.C.P.

37 IAEA SYMP. COPENHAGEN, VOL.1, . 275, (1968)
    *A CORRELATION FUNCTION FOR PHONON EIGENVECTORS.
    *COCHRAN(W.).

38 IAEA SYMP. COPENHAGEN, VOL.1, . 119, (1968)
    *MICROSCOPIC THEORY OF FORCE CONSTANTS IN SOLIDS.
    *COHEN(M.H.), MARTIN(R.M.), PICK(R.M.).

39 IAEA SYMP. COPENHAGEN, VOL.1, . 91, (1968)
    *APPLICATION OF THE KREBS MODEL TO THE STUDY OF THE LATTICE DYNAMICS OF
    HEXAGONAL CLOSE-PACKED METALS.
    *HAUTECLER(S.), NEVE-DE-MEVERGNIES(M.).

40 IAEA SYMP. GRENOBLE, ... 813, (1972).....B3,T...........
    *NORMAL MODES OF VIBRATION IN CRYSTALS HAVING AN ORDER-DISORDER
    TRANSITION.
    *COWLEY(R.A.).

41 IAEA SYMP. GRENOBLE, ... 149, (1972)....T..OF MATERIALS.
    *PHONONS AND MECHANICAL PROPERTIES OF MATERIALS.
    *BOFFI(S.), CAGLIOTI(G.), RIZZI(G.), ROSSITTO(F.).

42 J. APPL. PHYS. 40, ... 2359, (1969)
    *MAGNON LONGITUDINAL PHONON INTERACTION IN OBLIQUELY MAGNETIZED YIG RODS.
    *DE-SANTIS(P.).

43 J. APPL. PHYS. 40, ... 4696, (1969)
    *PHONON SCATTERING BY COTTRELL ATMOSPHERES. II. TIME-DEPENDENT EFFECT.
    *KLEMENS(P.G.).

44 J. CHEM. PHYSICS VOL.39, . . . 2633, (1963)...........NA,
    *MODEL FOR THE LATTICE DYNAMICS OF METALS AND ITS APPLICATION TO SODIUM.
    *SHARMA(P.K.), JOSHI(S.K.).

45 J. CHEM. PHYSICS VOL.41, . . . 3158, (1964)
    *TRENDS IN THE CHARACTERISTIC PHONON FREQUENCIES OF THE NA*CL, DIAMOND-,
    ZINC-BLENDE-, AND WURTZITE-TYPE CRYSTALS.
    *MITRA(S.S.), MARSHALL(R.).

46 J. CHEM. PHYSICS VOL.40, . . . 662, (1964)..........CU,
    *MODEL FOR THE LATTICE DYNAMICS OF METALS. II. APPLICATION TO FACE-
    CENTERED CUBIC METAL COPPER.
    *SHARMA(P.K.), JOSHI(S.K.).

47 J. CHEM. PHYSICS VOL.40, . . . 531, (1964)........HG,
    *LATTICE DYNAMICS OF ALPHA-MERCURY.
    *SLUTSKY(L.J.), JELINEK(G.E.).

48 J. CHEM. PHYSICS VOL.48, . . . 4060, (1968)
    *GROUP-THEORETICAL ANALYSIS OF LATTICE VIBRATIONS IN MOLECULAR CRYSTALS.
    *CHEN(S.H.), DVORAK(V.).

49 J. CHEM. PHYSICS VOL.49, . . . 3840, (1968)......POLYPROPYLENE,
    *DISPERSION CURVES AND FREQUENCY DISTRIBUTIONS OF ISOTACTIC POLYPROPYLENE
    *ZERBI(G.), PISERI(L.).

50 J. CHEM. PHYSICS VOL.48, . . . 3561, (1968)
    *DISPERSION CURVES AND FREQUENCY DISTRIBUTION OF POLYMERS: SINGLE CHAIN
    MODEL.
    *PISERI(L.), ZERBI(G.).

51 J. CHEM. PHYSICS VOL.48, . . . 3173, (1968)........CS*CL, CS*BR, CS*I,
    *LATTICE DYNAMICS AND SPECIFIC-HEAT DATA OF CS*CL, CS*BR AND CS*I.
    *KARO(A.M.), HARDY(J.R.).

52 J. CHEM. PHYSICS VOL.48, . . . 5242, (1969)
    *LATTICE DYNAMICS OF IMPERFECT ALKALI HALIDES.
    *BENEDEK(G.), NARDELLI(G.F.).

53 J. CHEM. PHYSICS VOL.53, . . . 4661, (1970)..........K*N3,
    *LATTICE DYNAMICS OF POTASSIUM AZIDE. A GROUP-THEORETICAL ANALYSIS.
    *RAO(K.R.), TREVINO(S.F.).

SECTION 10B-CRYSTAL DYNAMICS IN GENERAL (INCLUDING THE HARMONIC THEORY).

54 J. CHEM. PHYSICS VOL.53, ... 4645, (1973). . . . . .NA*(N*O3). . . . .
    *LATTICE DYNAMICS OF SODIUM NITRATE: A GROUP-THEORETICAL ANALYSIS.
    *RAO(K.R.), TREVINO(S.F.), LOGAN(K.W.).

55 J. CHEM. PHYSICS VOL.53, ... 4624, (1970). . . . . .NA*N3. . . . .
    *LATTICE DYNAMICS OF SODIUM AZIDE: A GROUP-THEORETICAL ANALYSIS.
    *RAO(K.R.), TREVINO(S.F.).

56 J. CHEM. PHYSICS VOL.53, ... 1428, (1970). . . . . .H*C*O*O*H, H*C*O*O*D, D*C*O*O*H,
    *PHONON CURVES AND FREQUENCY DISTRIBUTION FOR H-BONDED SYSTEMS: BETA MO-
    DIFICATION OF SOLID H*C*O*O*H, H*C*O*O*D AND D*C*O*O*H (FORMIC ACID).
    *TUBINO(R.), ZERBI(G.).

57 J. CHEM. PHYSICS VOL.55, ... 3997, (1971). . . . . .PT. . . . .
    *LATTICE DYNAMICS AND THERMODYNAMIC PROPERTIES OF PLATINUM.
    *KONTI(A.).

58 J. CHEM. PHYSICS VOL.54, ... 3600, (1971). . . . . . . .C*H3*O*H, C*H3*O*D (METHANOL), . .
    *LATTICE DYNAMICS OF METHANOL: HYDROGEN BONDING AND INFRARED ABSORPTION.
    *DEMPSTER(A.B.), ZERBI(G.).

59 J. CHEM. PHYS. VOL.56, ... 1022, (1972). . . . . .C,SI,GE,SN, ...
    *LATTICE DYNAMICS AND SPECTROSCOPIC PROPERTIES BY A VALENCE FORCE
    POTENTIAL OF DIAMONDLIKE CRYSTALS: C, SI, GE, AND SN.
    *TUBINO(R.), PISERI(L.), ZERBI(G.).

60 J. CHEM. PHYS. VOL.59, ... 4578, (1973). . . . .T. H2*O,D2*O,
    *ON THE PROBLEM OF THE VIBRATIONAL SPECTRUM AND STRUCTURE OF ICE IH:
    LATTICE DYNAMICAL CALCULATIONS.
    *BOSI(P.), TUBINO(R.), ZERBI(G.).

61 J. PHYS. C, SER.2, VOL.2, ... 882, (1969). . . .T. . . . . .GA. GROUP-THEORETICAL ANALYSIS.
    *LATTICE VIBRATIONS OF GALLIUM METAL I. GROUP-THEORETICAL ANALYSIS.
    *WAEBER(W.B.).

62 J. PHYS. C: SOLID ST. PHYS. V-4, 2749, (1971). . . . . .LI*H, LI*F, LI*CL, ...
    *LATTICE DYNAMICS OF LI*H, LI*F AND LI*CL.
    *VERMA(M.P.), SINGH(R.K.).

63 J. PHYS. C: SOLID ST. PHYS. V-4, 1674, (1971). . . . . .LANTHANIDES, ...
    *AN ELECTRON GAS MODEL FOR THE LATTICE DYNAMICS OF LANTHANIDES.
    *KUSHWAHA(S.S.), KUMAR(A.).

64 J. PHYS. C: SOLID ST. PHYS. V-4, 20, (1971). . . . .
    *FORMULATION OF THE LATTICE DYNAMICS PROBLEM IN TERMS OF RIGIDLY MOVING
    WANNIER FUNCTIONS II: ELECTRON-PHONON INTERACTION.
    *FERREIRA(L.G.), PRATT-JR(G.W.).

65 J. PHYS. C: SOLID ST. PHYS. V-4,   3, (1971). . . . .
    *FORMULATION OF THE LATTICE DYNAMICS PROBLEM IN TERMS OF RIGIDLY MOVING
    WANNIER FUNCTIONS.
    *FERREIRA(L.G.).

66 J. PHYS. C, SER.2, VOL.4, ... 2304, (1971). . . .B1. . . . . .UREA, ...
    *LATTICE DYNAMICS OF UREA.
    *MCKENZIE(D.R.), PRYOR(A.W.).

67 J. PHYS. C, SER.2, VOL.4, ... 820, (1971). . . . . .NA, AL, PB, ...
    *PRESSURE DEPENDENCE OF PHONON DISPERSION CURVES IN SIMPLE METALS
    (NA, AL, PB)
    *COULTHARD(M.A.).

68 J. PHYS. C: SOLID ST. PHYS. V-5, 1038, (1972). . . . . .CS*CL, CS*BR, CS*I, ...
    *LATTICE DYNAMICS OF CS*CL, CS*BR AND CS*I.
    *LAL(H.H.), VERMA(M.P.).

69 J. PHYS. C: SOLID ST. PHYS. V-5, 543, (1972). . . . . .RB-HALIDES, ...
    *LATTICE DYNAMICS OF RUBIDIUM HALIDES.
    *LAL(H.H.), VERMA(M.P.).

70 J. PHYS. C: SOLID ST. PHYS. V-5, 293, (1972). . . . . .NA. . . . .
    *LATTICE DYNAMICS OF SODIUM-COMPARISON OF DE LAUNAY AND CGW MODELS.
    *KOTHARI(L.S.), SINGHAL(U.).

71 J. PHYS. C: SOLID ST. PHYS. V-5, 287, (1972). . . . .
    *LATTICE DYNAMICS OF A CRYSTAL CONTAINING A POINT DEFECT WITH A LONG
    RANGE INTERACTION.
    *LITZMAN(O.), BARTUSEK(M.), ZAVADIL(V.).

72 J. PHYS. C VOL.5, ... 3168, (1972). . . .T. . . . .
    *THEORY OF HYDROGEN-BONDED FERROELECTRICS: I.
    *MOOR(M.A.), WILLIAMS(H.C.W.).

73 J. PHYS. C VOL.6, ... 1521, (1973). . . .T. . . . .
    *ON THE SEPARATION OF INTERNAL AND EXTERNAL VIBRATIONS IN MOLECULAR
    CRYSTALS.
    *PISERI(L.).

74 J. PHYS. C VOL.6, ... 1149, (1973). . . .T. . . . . .CA*F2,SR*F2,BA*F2, ...
    *CALCULATION OF HYDROGEN VIBRATIONS IN ALKALINE EARTH FLUORIDES.
    *HAYES(W.), WILTSHIRE(M.C.K.).

75 J. PHYS. C VOL.6, ... 2943, (1973). . . .T,B1. . . . .SI,C(DIAMOND), ...
    *LATTICE DYNAMICS OF SILICON AND DIAMOND.
    *PANDEY(B.P.), DAYAL(B.).

76 J. PHYS. CHEM. SOLIDS VOL.34, . . 1867, (1973). . . .B1,T. . . . .SI,GE, ...
    *ANGULAR FORCES IN THE LATTICE DYNAMICS OF DIAMOND CRYSTAL LATTICES.
    *BOSE(G.), TRIPATHI(B.B.), GUPTA(H.C.).

77 J. PHYS. CHEM. SOLIDS VOL.35, . . 123, (1974). . . .T,B1. . . . .CU. . . . .
    *AN ANGULAR MODEL FOR LATTICE DYNAMICS OF F.C.C. METALS.
    *SHUKLA(M.M.), CLOSS(H.).

78 J. PHYS. CHEM. SOLIDS VOL.35, . . 669, (1974). . . .T,B1. . . . .AL. . . . .
    *PHONON DISPERSION RELATIONS, EFFECTIVE INTERIONIC POTENTIAL AND LIQUID
    RESISTIVITY OF AL.
    *RAO(P.V.S.).

79 J. PHYS. C VOL.7, ... 1443, (1974). . . .T. . . . . .NA. . . . .
    *THE DERIVATION OF FORCE CONSTANTS FROM PHONON FREQUENCIES.
    *NEWMAN(D.J.).

80 J. PHYS.F: METAL PHYS. VOL.1, . . 588, (1971). . . . . .PD, ...
    *LATTICE VIBRATIONS IN PALLADIUM.
    *PAL(S.).

SECTION 10B-CRYSTAL DYNAMICS IN GENERAL (INCLUDING THE HARMONIC THEORY).

81 J. PHYS.F: METAL PHYS. VOL.1,. . . . 554, (1971).....FUNCTION....CU..
    *ON A SEMI-CONTINUUM GREEN FUNCTION METHOD FOR LATTICE DYNAMICS WITH
    APPLICATION TO COPPER.
    *TEWARY(V.K.), BULLOUGH(R.).

82 J. PHYS.F: METAL PHYS. VOL.1,. . . . 377, (1971)..............TL,. . . . .
    *PHONON DISPERSION RELATIONS FOR THALLIUM.
    *KUSHWAHA(S.S.), RAJPUT(J.S.).

83 J. PHYS.F: METAL PHYS. VOL.1,. . . L46, (1971)..............NA,. . . . .
    *CALCULATION OF THE ROOM TEMPERATURE PHONON FREQUENCIES OF SODIUM.

84 J. PHYS.F: METAL PHYS. VOL.1,. . . 19, (1971)....T,B1..ALKALI,. . . . .
    *CRYSTAL DYNAMICS OF LIGHT WEIGHT ALKALI METALS.
    *TRIPATHI(B.B.), BEHARI(J.).

85 J. PHYS.F: METAL PHYS. VOL.1,. . . 12, (1971)....T,B1.....LI,. . . . .
    *LATTICE DYNAMICS OF LITHIUM USING A PSEUDOPOTENTIAL APPROACH.
    *GUPTA(H.C.), TRIPATHI(B.B.).

86 J. PHYS.F: METAL PHYS. VOL.2,. . . 247, (1972)....T,B1.....CU,. . . . .
    *LATTICE DYNAMICS OF NOBLE METALS.
    *PRASAD(B.), SRIVASTAVA(R.S.).

87 J. PHYS. F VOL.3,. . . . . . . L83, (1973)....T,B1.....CU,. . . . .
    *AN ANGULAR FORCE MODEL FOR LATTICE DYNAMICS OF FACE CENTRED CUBIC METALS
    *SHUKLA(M.M.), CLOSS(H.).

88 J. PHYS. F VOL.3,. . . . . . . 772, (1973)....T,B1.....NA,MG,AL,. . . . .
    *APPLICATIONS OF THE OPTIMIZED MODEL POTENTIAL.
    *WILLIAMS(A.R.), APPAPILLAI(M.).

89 J. PHYS. F VOL.3,. . . . . . . 640, (1973)....T,B1....BE,MG,HO,TL,. . . . .
    *DISPERSION RELATIONS IN SOME HEXAGONAL METALS.
    *UPADHYAYA(J.C.), VERMA(M.P.).

90 J. PHYS. F VOL.3,. . . . . . . 1672, (1973)....T,B1.....Y,TB,. . . . .
    *PHONON DISPERSION IN YTTRIUM AND TERBIUM.
    *UPADHYAYA(J.C.), VERMA(M.P.).

91 J. PHYS. F VOL.3,. . . . . . . 1531, (1973)....T,B1.....BE,MG,. . . . .
    *PHONON DISPERSION RELATIONS FOR BERYLLIUM AND MAGNESIUM.
    *RAJPUT(J.S.), KUSHWAHA(S.S.).

92 J. PHYS. F VOL.4,. . . . . . . 19, (1974)....T,B1.....LI,NA,AL,PB,. . . . .
    *THIRD ORDER PERTURBATION THEORY AND LATTICE DYNAMICS OF SIMPLE METALS.
    *BERTONI(C.M.), BORTOLANI(V.), CALANDRA(C.), NIZZOLI(F.).

93 J. PHYS. F VOL.4,. . . . . . . 11, (1974)....T,B1.....RB,CS,PT,. . . . .
    *CRYSTAL DYNAMICS OF SOME CUBIC METALS ON AN ANGULAR FORCE MODEL.
    *JOGI(S.).

94 J. PHYS. F VOL.4,. . . . . . . 466, (1974)....T........GD,TB,DY,. . . . .
    *A GROUP THEORETICAL STUDY OF SELECTION RULES FOR MAGNON-PHONON INTER-
    ACTIONS IN FERROMAGNETIC HCP METALS.
    *CRACKNELL(A.P.).

95 J. PHYSIQUE TOME 35,. . . . . 263, (1974)....T,B1.....NA*CL,K*CL,K*BR,K*I,. .
    *ROLE OF DIFFERENT POLARISATION BRANCHES IN THE PHONON CONDUCTIVITY OF
    NA*CL, K*CL, K*BR AND K*I IN THE TEMPERATURE RANGE 1-100 K.
    *SINGH(M.P.), VERMA(G.S.).

96 J. PHYS. SOC. JAPAN VOL.21,. . . 2208, (1966)....T,B1.....MO,TA,NB,. . . . .
    *PHONON DISPERSION IN TRANSITION METALS.
    *PAL(S.), GUPTA(R.P.).

97 J. PHYS. SOC. JAPAN VOL.31,. . . 1639, (1971)....T,B1.....NE,. . . . .
    *LATTICE DYNAMICS OF SOLIDIFIED NEON.
    *BEHARI(J.), TRIPATHI(B.B.).

98 J. PHYS. SOC. JAPAN VOL.34,. . . 1006, (1973)....T........BE,MG,ZN,. . . . .
    *THE LATTICE DYNAMICS OF HEXAGONAL CLOSE PACKED METALS.
    *BOSE(G.), TRIPATHI(B.B.), GUPTA(H.C.).

99 J. PHYS. SOC. JAPAN VOL.35,. . . 1487, (1973)....T,B1.....PD,. . . . .
    *PHONON DISPERSION, FREQUENCY SPECTRUM AND SPECIFIC HEAT OF PALLADIUM.
    *PAL(S.), SINGH(R.B.).

100 MOL. DYN. AND STRUCT. OF SOLIDS, 289, (1969)....B7........ . . . . .
    *LATTICE DYNAMICS OF MOLECULAR SOLIDS.
    *DOLLING(G.).

101 NUCL./S.S. PHYS. SYMP. ABSTR.,. . . . , (1972)............CS*CL-STRUCTURE,. .
    *LATTICE DYNAMICS OF SOLIDS CRYSTALLISING IN THE CS*CL STRUCTURE.
    *LAL(H.H.), VERMA(M.P.).

102 NUOVO CIM. VOL.17B NO.1,. . . . . 166, (1973)....T,B1.....KR,. . . . .
    *DISPERSION OF PHONON WAVES IN SOLID KRYPTON.
    *SALZBERG(J.B.), SHUKLA(M.M.).

103 PHYSICA VOL.30,. . . . . 2105, (1964)....T........NA,. . . . .
    *PHONON DISPERSION IN METALS.
    *SINGH(N.), JOSHI(S.K.).

104 PHYSICA VOL.62,. . . . . 587, (1972)....T,B1.....LI,NA,K,. . . . .
    *ON THE PHONON DISPERSION IN ALKALI METALS.
    *BAJPAI(R.P.), NEELAKANDAN(K.).

105 PHYSICA VOL.62,. . . . . 574, (1972)....T,B1.....MG,ZR,Y,. . . . .
    *ON PHONON DISPERSION IN HCP METALS.
    *BAJPAI(R.P.).

106 PHYSICA VOL.58,. . . . . 71, (1972)....T,B1.....
    *PHONON DISPERSION IN ALKALI METALS.2.
    *SINGH(S.N.), PRAKASH(S.).

107 PHYSICA VOL.59,. . . . . 109, (1972)....T........LI,NA,K,RB,. . . . .
    *LATTICE VIBRATIONS AND DEBYE TEMPERATURES OF ALKALI METALS.
    *SHARMA(P.K.), SINGH(N.).

108 PHYSICA VOL.64,. . . . . 625, (1973)....T,B1.....ZN,. . . . .
    *LATTICE DYNAMICS OF ANISOTROPIC HCP METALS.
    *KUSHWAHA(S.S.).

SECTION 10B-CRYSTAL DYNAMICS IN GENERAL (INCLUDING THE HARMONIC THEORY).

109 PHYSICA VOL.72,.   179, (1974)....T,B1.....K,RB,LI, ...
     *LATTICE DYNAMICS OF ALKALI METALS.
     *DA-CUNHA-LIMA(I.C.), BRESCANSIN(L.M.), SHUKLA(M.M.).

110 PHYS. KONDENS. MATERIE VOL.5,.   364, (1966)....T,B1.....R3,CS,PB ELECTRICAL PROPERTIES OF
     *LATTICE DYNAMICS, ELECTRONIC STRUCTURE AND
     SIMPLE METALS; II.RUBIDIUM,CESIUM AND LEAD.
     *SCHNEIDER(T.), STOLL(E.).

111 PHYS. KONDENS. MATERIE VOL.5,. .   331, (1966)....T,B1.....NA,K,AL, ...
     *LATTICE DYNAMICS, ELECTRONIC STRUCTURE AND ELECTRICAL PROPERTIES OF
     SIMPLE METALS I.SODIUM, POTASSIUM AND ALUMINUM.
     *SCHNEIDER(T.), STOLL(E.).

112 PHYS. LETT. VOL.10,.   12, (1964)....T,B1.....NA, ...
     *DYNAMICAL MODEL FOR LATTICE VIBRATIONS IN METALS.
     *KREBS(K.).

113 PHYS. LETT. VOL.8,.   304, (1964)....E, ...
     *A NEW SAMPLING METHOD FOR CALCULATING THE FREQUENCY DISTRIBUTION
     FUNCTION OF SOLIDS.
     *GILAT(G.), DOLLING(G.).

114 PHYS. LETT. 25A,. . . . . .   400, (1967) . . . .
     *LIFETIME OF SOFT MODES IN DISPLACIVE TYPE FERROELECTRICS.
     *TANI(K.).

115 PHYS. LETT. 27A,. . . . .   523, (1968) . . . .
     *BOUNDARY SCATTERING OF PHONONS IN SOLID SOLUTIONS.
     *GOLDSMITH(H.J.), PENN(A.W.).

116 PHYS. LETT. VOL.37A,. . . . . .   193, (1971) . . . .
     *THEORETICAL FORMULATION FOR THE LATTICE DYNAMICS OF H.C.P. METALS.
     *KUSHWAHA(S.S.).

117 PHYS. LETT. VOL.46A NO.3,.   163, (1973)....T.........NB3*SN,. . . . .
     *PHONON LINE WIDTH IN SUPERCONDUCTORS.
     *SCHUSTER(H.G.).

118 PHYS. MET. METALLOGR. VOL.33,. . 939, (1972)....T.. . . . .
     *PHONON SPECTRA OF NOBLE METALS. CALCULATIONS FROM THE FIRST PRINCIPLES
     ON THE METHOD OF PSEUDO-POTENTIAL.
     *TSAREV(YU.N.), NIKULIN(Y.K.), CHEVYCHELOV(A.D.).

119 PHYS. REV. VOL.113,.   147, (1959)....T.........GE, ...
     *VIBRATION SPECTRA AND SPECIFIC HEATS OF DIAMOND-TYPE LATTICES.
     *PHILLIPS(J.C.).

120 PHYS. REV. VOL 119,.   980, (1960)....T,B2.....NA*I,. . . . .
     *LATTICE DYNAMICS OF ALKALI HALIDE CRYSTALS.
     *WOODS(A.D.B.), COCHRAN(W.), BROCKHOUSE(B.N.).

121 PHYS. REV. 131,.   1995, (1963) . . . .
     *THEORY OF KOHN ANOMALIES IN THE PHONON SPECTRA OF METALS.
     *TAYLOR(P.L.).

122 PHYS. REV. 171,.   665, (1963) . . . .
     *THEORY OF LOCALIZED VIBRATIONS OF INTERSTITIAL ATOMS IN BCC LATTICES.
     *BLAESSER(G.), PERETTI(J.), TOTH(G.).

123 PHYS. REV. VOL.138,.   A143, (1965)....T,B1.....LI,NA,K, ...
     *DISPERSION CURVES AND LATTICE FREQUENCY DISTRIBUTION OF METALS.
     *KREBS(K.).

124 PHYS. REV. VOL.156,. . . . . .   1017, (1967)....T.. . . . .
     *VIBRATIONAL PROPERTIES OF IMPERFECT CRYSTALS WITH LARGE DEFECT CONCEN-
     TRATIONS.
     *TAYLOR(D.W.).

125 PHYS. REV. 175,. .   1083, (1968) . . . .
     *LATTICE DYNAMICS AND IONIC DEFORMATION IN SOME ALKALI HALIDES.
     *MELVIN(J.S.), PIRIE(J.D.), SMITH(T.).

126 PHYS. REV. 175,. .   1110, (1968) . . . .
     *CALCULATION OF THE SELF-CONSISTENT PHONON SPECTRUM OF HCP HE4.
     *GILLIS(N.S.), KOEHLER(W.C.), WERTHAMER(N.R.).

127 PHYS. REV. 175,. .   1156, (1968) . . . .
     *NUMERICAL METHOD FOR CALCULATING FREQUENCY DISTRIBUTION FUNCTIONS IN
     SOLIDS III. EXTENSION TO TETRAGONAL CRYSTALS.
     *KAM(Z.), GILAT(G.).

128 PHYS. REV. 167,. .   607, (1968) . . . .
     *TOWARDS A QUANTUM MANY-BODY THEORY OF LATTICE DYNAMICS II. COLLECTIVE
     FLUCTUATION APPROXIMATION.
     *GILLIS(N.S.), WERTHAMER(N.R.).

129 PHYS. REV. 166,. .   856, (1968) . . . .
     *ASYMPTOTIC DESCRIPTION OF LOCALIZED LATTICE MODES AND LOW-FREQUENCY
     RESONANCES.
     *KRUMHANSL(J.A.), MATTHEW(J.A.D.).

130 PHYS. REV. 165,. .   942, (1968) . . . .
     *NEW THEORY OF LATTICE DYNAMICS AT 0 DEG. K.
     *KOEHLER(T.R.).

131 PHYS. REV. 174,. .   766, (1968) . . . .
     *ANGULAR FORCES IN THE LATTICE DYNAMICS OF FACE-CENTERED CUBIC METALS.
     *VARSHNI(Y.P.), YUEN(P.S.).

132 PHYS. REV. 176,. .   1004, (1968) . . . .
     *VIBRATIONS OF DISORDERED SOLIDS.
     *ROSENSTOCK(H.B.), MCGILL(R.E.).

133 PHYS. REV. 175,. .   1201, (1968) . . . .
     *DYNAMICS OF DISORDERED HARMONIC LATTICES III. NORMAL-MODE SPECTRA FOR
     ABNORMAL ARRAYS.
     *PAYTON(D.N.), VISSCHER(W.M.).

134 PHYS. REV. 175,. .   1171, (1968) . . . .
     *DIELECTRIC SCREENING AND THE PHONON SPECTRA OF METALLIC AND NONMETALLIC
     CRYSTALS.
     *KEATING(P.N.).

SECTION 10B-CRYSTAL DYNAMICS IN GENERAL (INCLUDING THE HARMONIC THEORY).

135 PHYS. REV. 175,. DISPLACEMENT CORRELATIONS AND FREQUENCY SPECTRA FOR MASS-DISORDERED LATTICES II.
363, (1968)
LEATH(P.L.); GOODMAN(B.).

136 PHYS. REV. 172,. SCREENING AND ELECTRON CORRELATION EFFECTS ON THE ELECTRON-PHONON INTERACTION IN MONOVALENT METALS.
747, (1968)
TAKAHASHI(H.).

137 PHYS. REV. 172,. THEORY OF THE VIBRATIONS OF DILUTE ALLOYS WITH SHORT-RANGE ORDER.
677, (1968)
HARTMANN(W.M.).

138 PHYS. REV. 171,. SELF-CONSISTENT-FIELD APPROXIMATIONS IN DISORDERED ALLOYS.
725, (1968)
LEATH(P.L.).

139 PHYS. REV. 169,. LATTICE-DYNAMIC CALCULATION FOR ALKALI METALS.
523, (1968)
HO(P.S.).

140 PHYS. REV. 169,. MODEL FOR LATTICE DYNAMICS IN METALS.
496, (1968)
CHEVEAU(L.).

141 PHYS. REV. 165,. INFLUENCE OF FORCE-CONSTANT CHANGES ON THE LATTICE DYNAMICS OF CUBIC CRYSTALS WITH POINT DEFECTS.
1011, (1968)
MANNHEIM(P.P.).

142 PHYS. REV. VOL.188,. ELECTRONIC CONTRIBUTION TO LATTICE DYNAMICS IN INSULATING CRYSTALS.
1431, (1969)
SHAM(L.J.).

143 PHYS. REV. VOL.188,. COMPUTER SIMULATION OF THE LATTICE DYNAMICS OF SOLIDS.
1407, (1969)
DICKEY(J.M.), PASKIN(A.).

144 PHYS. REV. VOL.187,. TUNNELING DETERMINATION OF PHONON-ENERGY UNCERTAINTIES DUE TO FORCE-CONSTANT DISORDERS IN THE ALLOY PB-TL-BI.
821, (1969) PB-TL-BI,
DYNES(R.C.), ROWELL (J.M.).

145 PHYS. REV. VOL.187,. PHONON FREQUENCIES OF ALKALI METALS.
808, (1969)
PRAKASH(S.), JOSHI(S.K.).

146 PHYS. REV. LETT. 22,. THEORY OF SURFACE SCATTERING AND DETECTION OF SURFACE PHONONS.
346, (1969)
CABRERA(T.N.), CELLI(V.), MANSON(R.).

147 PHYS. REV. 178,. PSEUDOPOTENTIAL CALCULATION OF PHONON FREQUENCIES AND GRUNEISEN PARAMETERS FOR LITHIUM.
900, (1969) LI,
WALLACE(D.C.).

148 PHYS. REV. 177,. DERIVATION OF THE SHELL MODEL OF LATTICE DYNAMICS AND ITS RELATION TO THE THEORY OF THE DIELECTRIC CONSTANT.
1256, (1969)
SINHA(S.K.).

149 PHYS. REV. 177,. DISPERSION RELATIONS FOR HEXAGONAL CLOSE-PACKED LATTICES.
1139, (1969)
METZBOWER(E.A.).

150 PHYS. REV. B-2,. PHONON DISPERSION IN LITHIUM.
4741, (1970) LI,
PAL(S.).

151 PHYS. REV. B-1,. LINDEMANN LAW AND LATTICE DYNAMICS.
3982, (1970)
SHAPIRO(J.N.).

152 PHYS. REV. B-2,. EFFECT OF THE DIELECTRIC FUNCTION ON THE PHONON SPECTRUM OF MAGNESIUM.
3947, (1970) MG,
FLOYD(E.R.), KLEINMAN(L.).

153 PHYS. REV. B-2,. PHONON DISPERSION RELATIONS OF BODY-CENTERED-CUBIC METALS.
3943, (1970)
KUSHWAHA(S.S.), RAJPUT(J.S.).

154 PHYS. REV. B-1,. NON-CENTRAL-FORCE MODEL OF LI*H PHONON DISPERSION CURVES AND HE MIGRATION.
3510, (1970) LI*H,
WILSON(W.D.), JOHNSON(R.A.).

155 PHYS. REV. B-2,. LATTICE DYNAMICS OF ALKALI METALS IN THE SELF-CONSISTENT SCREENING THEORY.
2983, (1970)
PRICE(D.L.), SINGWI(K.S.), TOSI(M.P.).

156 PHYS. REV. B-2,. LATTICE DYNAMICS, MODE GRUNEISEN PARAMETERS AND COEFFICIENT OF THERMAL EXPANSION OF CS*CL, CS*BR, CS*I.
2167, (1970) CS*CL, CS*BR, CS*I,
VETELINO(J.F.), MITRA(S.S.), NAMJOSHI(K.V.).

157 PHYS. REV. B-2,. LATTICE DYNAMICS OF NA*CL, K*CL, RB*CL AND RB*F.
2098, (1970) NA*CL, K*CL, RB*CL, RB*F,
RAUNIO(G.), ROLANDSON(S.).

158 PHYS. REV. B-2,. LATTICE DYNAMICS OF ZN-TE: PHONON DISPERSION, MULTIPHONON INFRARED SPECTRUM, MODE GRUNEISEN PARAMETERS AND THERMAL EXPANSION.
967, (1970)
VETELINO(J.F.), MITRA(S.S.), NAMJOSHI(K.V.).

159 PHYS. REV. B-1,. KOHN ANOMALIES IN TUNGSTEN AND OTHER CR-GROUP METALS.
509, (1970) T,B1,,W,
RICE(T.M.), HALPERIN(B.I.).

160 PHYS. REV. B-2,. PSEUDOPOTENTIAL AND THE PHONON DISPERSION IN ALUMINUM.
248, (1970) AL,
GUPTA(H.C.), TRIPATHI(B.B.).

SECTION   108-CRYSTAL DYNAMICS IN GENERAL (INCLUDING THE HARMONIC THEORY).

161 PHYS. REV. LETT. VOL.25, . of . . . 1014, (1970)AES AND PHASE-W ALLOYS (PHASE:BETA)
        ~THEORY OF SOFT OPTIC MODES AND PHASE TRANSITIONS IN BETA-W STRUCTURE
        TRANSITION-METAL ALLOYS.
        ~KLEIN(B.M.), BIRMAN(J.L.).

162 PHYS. REV. B-4,. . . . . 4636, (1971).................
        ~PHONON DISPERSION IN NOBLE METALS.
        ~SHARMA(P.K.), SINGH(N.).

163 PHYS. REV. B-4,. . . . . . . 2774, (1971)................CS*I,.
        ~LATTICE DYNAMICS OF CS*I CONTAINING IMPURITY IONS.
        ~AGRAWAL(B.K.), RAM(P.N.).

164 PHYS. REV. B-3,. . . . . . . 2485, (1971)............MG..
        ~LATTICE DYNAMICS OF MAGNESIUM FROM A FIRST-PRINCIPLES NONLOCAL
        PSEUDOPOTENTIAL APPROACH.
        ~KING(W.F.), CUTLER(P.H.).

165 PHYS. REV. B-4,. . . . . . 1770, (1971).............NI.
        ~LATTICE DYNAMICS OF TRANSITION METALS-APPLICATION TO PARAMAGNETIC NICKEL
        ~PRAKASH(S.), JOSHI(S.K.).

166 PHYS. REV. B-4,. . . . . . . 567, (1971).....
        ~RELATIONSHIP BETWEEN ADIABATIC ELASTIC CONSTANTS AND THE SLOPES OF
        PHONON DISPERSION CURVES FOR RARE-GAS SOLIDS.
        ~GOLDMAN(V.V.), HORTON(G.K.), KLEIN(M.L.).

167 PHYS. REV. LETT. VOL.26, ... 1324, (1971).......T.
        ~GENERALIZED SCREENING MODEL FOR LATTICE DYNAMICS.
        ~SINHA(S.K.), GUPTA(R.P.), PRICE(D.L.).

168 PHYS. REV. A VOL.3,. . . . . . 1453, (1971)......
        ~PHONON DISPERSION IN A LOW-TEMPERATURE WEAKLY INTERACTING BOSE GAS.
        ~SHANG-KENG-MA, GOULD(H.), WONG(V.K.).

169 PHYS. REV. B VOL.3,. . . . . . 1482, (1971)........MG*O,
        ~SOME ASPECTS OF COVALENT BONDING IN NA*CL STRUCTURE CRYSTALS:
        APPLICATION TO THE LATTICE DYNAMICS OF MG*O.
        ~GILLIS(N.S.).

170 PHYS. REV. B-3,. 2743, (1971)
        ~LATTICE VIBRATIONS IN DEUTERATED AMMONIUM CHLORIDE AT 85 K.
        II. THEORETICAL.
        ~COWLEY(E.R.).

171 PHYS. REV. LETT. VOL.28, ... 1578, (1972)................
        ~DIELECTRIC MATRIX AND PHONON FREQUENCIES IN SILICON.
        ~BERTONI(C.M.), BORTOLANI(V.), CALANDRA(C.), TOSATTI(E.).

172 PHYS. REV. LETT. VOL.28, ... 600, (1972)............CARBIDES,
        ~RESONANT ELECTRONIC POLARIZATION IN THE LATTICE DYNAMICS OF TRANSITION-
        METAL COMPOUNDS.
        ~WEBER(W.), BILZ(H.), SCHRODER(U.).

173 PHYS. REV. B VOL.5,. . . . . . 2887, (1972)............SN.
        ~OPTIMUM-MODEL-POTENTIAL LATTICE DYNAMICS OF BETA-SN.
        ~KAM(Z.), GILAT(G.).

174 PHYS. REV. B VOL.5,. 2880, (1972)
        ~LATTICE DYNAMICS OF NOBLE METALS-APPLICATION TO COPPER.
        ~PRAKASH(S.), JOSHI(S.K.).

175 PHYS. REV. B VOL.6,. . . . . . 3581, (1972)......T.
        ~MICROSCOPIC THEORY OF LATTICE DYNAMICS IN CONDUCTING CRYSTALS.
        ~SHAH(L.J.).

176 PHYS. REV. LETT. VOL.29, ... 1593, (1972)....T.
        ~SUPERCONDUCTIVITY AND PHONON SOFTENING.
        ~ALLEN(P.B.), COHEN(M.L.).

177 PHYS. REV. LETT. VOL.29, ... 793, (1972)....T.
        ~NEW MAGNETOELASTIC INTERACTION.
        ~LIU(S.H.).

178 PHYS. REV. B VOL.5,. . . . . . 1260, (1972)............C*(HF,NB,TA,U),
        ~PHONON SPECTRA OF SOME TRANSITION METAL CARBIDES FROM A SIMPLE
        PSEUDOPOTENTIAL APPROACH.
        ~MOSTOLLER(M.).

179 PHYS. REV. B VOL.8,. . . . . . 593, (1973)....T,B1....MG,SC,ZR,HO,
        ~STUDY OF PHONON DISPERSION IN HCP METALS WITH CENTRAL PAIR POTENTIALS
        REPRESENTING ION-ION INTERACTIONS.
        ~UPADHYAYA(J.C.), VERMA(M.P.).

180 PHYS. REV. B VOL.7,. . . . . . 4338, (1973)....T.....AL,K,PB,
        ~DAMPING OF PHONONS IN SIMPLE METALS.
        ~JAIN(H.), JAIN(S.C.).

181 PHYS. REV. LETT. VOL.30, ... 1196, (1973)....T.......HE,
        ~PHONONS AND PHASE TRANSITIONS OF HELIUM.
        ~MATTIS(D.C.), LANDOVITZ(L.F.).

182 PHYS. REV. B VOL.7,. . . . . . 2285, (1973)....T,B3.....
        ~LOCALIZED BASIS FOR LATTICE VIBRATIONS.
        ~KOHN(W.).

183 PHYS. REV. B VOL.8,. . . . . . 5082, (1973)....T,B1.....CARBIDES,
        ~LATTICE DYNAMICS OF TRANSITION-METAL CARBIDES.
        ~WEBER(W.).

184 PHYS. REV. B VOL.8,. . . . . . 4880, (1973)....T,B1....MG*O,
        ~THREE-BODY-FORCE SHELL MODEL AND THE LATTICE DYNAMICS OF MAGNESIUM OXIDE
        ~VERMA(M.P.), AGARWAL(S.K.).

185 PHYS. REV. B VOL.8,. . . . . . 2982, (1973)....T........NA*(CL,BR),K*(I,CL,BR),.
        ~LATTICE DYNAMICS OF ALKALI HALIDES.
        ~BASU(A.N.), SENGUPTA(S.).

186 PHYS. REV. LETT. VOL.31, ... 1466, (1973)....T,B1....BE,.
        ~THREE-BODY FORCES IN THE LATTICE DYNAMICS OF BERYLLIUM.
        ~BERTONI(C.M.), BORTOLANI(V.), CALANDRA(C.), NIZZOLI(F.).

187 PHYS. REV. B VOL.10 NO.2,. . . . . (1974)....T.........NI-PD,
        ~COMPARISON OF SINGLE-SITE APPROXIMATIONS FOR THE LATTICE DYNAMICS OF
        MASS-DISORDERED ALLOYS.
        ~KAMITAKAHARA(W.A.), TAYLOR(D.W.).   (REF. OBTAINED FROM PHYS. REV.
        ABSTRACTS).

SECTION 10B-CRYSTAL DYNAMICS IN GENERAL (INCLUDING THE HARMONIC THEORY).

188 PHYS. REV. B VOL.9, 1710, (1974)....T.....SI,
  *DIELECTRIC-SCREENING MATRIX AND LATTICE DYNAMICS OF SI.
  *BERTONI(C.M.), BORTOLANI(V.), CALANDRA(C.), TOSATTI(E.).

189 PHYS. REV. B VOL.9, 2573, (1974)....T,B1.....SI,GE,
  *MICROSCOPIC THEORY OF DIELECTRIC SCREENING AND LATTICE DYNAMICS. II.
  PHONON SPECTRA AND EFFECTIVE CHARGES.
  *PRICE(D.L.), SINHA(S.K.), GUPTA(R.P.).

190 PHYS. STAT. SOL. 7, K11, (1964)....T,B1.....CU,
  *THE VIBRATIONAL FREQUENCIES OF COPPER ON KREBS' MODEL.
  *SHUKLA(M.M.).

191 PHYS. STAT. SOL. VOL.38, 141, (1970).....
  *CRYSTAL VIBRATIONS OF SILICON BY THE USE OF VALENCE FORCE POTENTIALS.
  *SINGH(B.D.), DAYAL(B.).

192 PHYS. STAT. SOL. VOL.39, K75, (1970).....PD,
  *KREBS MODEL CALCULATION OF PHONON DISPERSION IN PALLADIUM.
  *BROWN(J.S.).

193 PHYS. STAT. SOL.46, 361, (1971)....
  *THE LATTICE DYNAMICS OF THE HEXAGONAL CLOSE-PACKED METALS.
  *TROTT(A.J.), HEALD(P.T.).

194 PHYS. STAT. SOL. B VOL.53 NO.1, 279, (1972)....T.....SR*TI*O3, LA*AL*O3, K*MN*F3,
  *SELF-CONSISTENT LATTICE-DYNAMICAL THEORY OF STRUCTURAL PHASE TRANSITIONS
  IN PEROVSKITE-TYPE CRYSTALS.
  *PIETRASS(B.).

195 PHYS. STAT. SOL. B VOL.53, K33, (1972)....B7.....K*(TA-NB)*O3,
  *LOW FREQUENCY ZONE BOUNDARY MODES IN PEROVSKITE FERROELECTRICS INDICATED
  BY ANISOTROPIC DEBYE-WALLER FACTORS.
  *HEWAT(A.W.).

196 PHYS. STAT. SOLIDI B VOL.56 NO.2 591, (1973)....T.....
  *EXCHANGE-QUADRUPOLE FORCES AND PHONON DISPERSION FOR THE SIMPLEST CUBIC
  CRYSTALS.
  *TOLPYGO(K.B.).

197 PHYS. STAT. SOLIDI B VOL.56 NO.1 327, (1973)....T,B1.....R9,CS,BA,
  *PSEUDOPOTENTIAL AND LATTICE VIBRATIONS OF SIMPLE METALS.
  *PRASAD(B.), SRIVASTAVA(R.S.).

198 PHYS. STAT. SOLIDI B VOL.59, 279, (1973)....T,B1.....AG*CL,
  *DISPERSION PROPERTIES OF SILVER CHLORIDE.
  *UPADHYAYA(K.S.), MAHESH(P.S.).

199 PHYS. STAT. SOLIDI B VOL.59 NO.2 517, (1973)....T.....
  *ON THE THEORY OF PHONON-LIKE EXCITATIONS IN NONCRYSTALLINE SOLIDS.
  *BOTTGER(H.).

200 PHYS. STAT. SOLIDI B VOL.58 NO.1 315, (1973)....T,B1.....NA*N*O3,
  *PHONON DISPERSION AND PHONON DENSITIES OF NA*N*O3.
  *PLIHAL(M.).

201 PHYS. STAT. SOLIDI B VOL.61 NO.1 241, (1974)....T,E.....TL*BR,CS*(CL,BR,I),
  *MEAN-SQUARE IONIC DISPLACEMENTS AND FORCE CONSTANTS OF CRYSTALS WITH THE
  CS*CL TYPE STRUCTURE.
  *JEX(H.), MULLNER(M.), DYCK(W.).

202 PROC. PHYS. SOC. B. VOL.70, 1143, (1957)....T.....
  *LATTICE VIBRATIONS AND HARMONIC FORCES IN SOLIDS.
  *FOREMAN(A.J.E.), LOMER(W.M.).

203 PROC. ROYAL SOC. A VOL.268, 109, (1962)....T.....
  *THE LATTICE DYNAMICS OF IONIC AND COVALENT CRYSTALS.
  *COWLEY(R.A.).

204 PROC. ROY. SOC. A VOL.317, 279, (1970).....
  *A CLASSICAL THEORY OF LATTICE DYNAMICS.
  *JOHNSON(F.A.).

205 PROC. ROY. SOC. A VOL.317, 55, (1970).....CA*F2,
  *OPTICAL AND DIELECTRIC PROPERTIES AND LATTICE DYNAMICS OF SOME FLUORITE
  STRUCTURE IONIC CRYSTALS.
  *DENHAM(P.), FIELD(G.R.), MORSE(P.L.R.), WILKINSON(G.R.).

206 PROC. ROY. SOC. A VOL.320, 505, (1971)....T.....
  *FORCE CONSTANTS AND LATTICE FREQUENCIES.
  *LEIGH(R.S.), SZIGETI(B.), TEWARY(V.K.).

207 PROC. ROY. SOC. A VOL.333, 363, (1973)....T,B1.....HEXAMETHYLENETETRAMINE (DEUTER.)
  *INTERATOMIC FORCES IN HEXAMETHYLENETETRAMINE.
  *DOLLING(G.), PAWLEY(G.S.), POWELL(B.M.).

208 SEMICONDUCTORS, 50, (1963)....T.....
  *LATTICE DYNAMICS.
  *COCHRAN(W.). (ED. NOTE: PUBL. ACADEMIC PRESS: NEW YORK AND LONDON, 1963)

209 SOL. STATE COMM. 1, 132, (1963).....
  *COLLECTIVE MOTION OF HYDROGEN BONDS.
  *DE-GENNES(P.G.).

210 SOL. STATE COMM. 5, 159, (1967)....FOR THE LATTICE DYNAMICS OF METALS.
  *PSEUDO-ATOM POTENTIAL MODEL FOR THE LATTICE DYNAMICS OF METALS.
  *KREBS(K.), HOLZL(K.).

211 SOL. STATE COMM. 7, 295, (1969).....
  *PHONON DISPERSION-RELATIONS IN BERYLLIUM CALCULATED FROM A
  PSEUDOPOTENTIAL APPROACH.
  *CUTLER(P.H.).

212 SOL. STATE COMM. VOL.8, 991, (1970).....LI,
  *AB INITIO PSEUDOPOTENTIAL CALCULATION OF PHONON FREQUENCIES IN LITHIUM.
  *GUPTA(R.P.).

213 SOL. STATE COMM. VOL.9, 2057, (1971).....N*H4*CL, N*D4*CL,
  *LATTICE DYNAMICS OF N*H4*CL AND N*D4*CL.
  *JEX(H.).

214 SOL. STATE COMM. VOL.9, 1317, (1971).....
  *LOCALIZED EIGENSTATES IN DISORDERED SYSTEMS: APPLICATION TO PHONONS.
  *ECONOMOU(E.N.).

SECTION 10B-CRYSTAL DYNAMICS IN GENERAL (INCLUDING THE HARMONIC THEORY).

215   SOL. STATE COMM.   VOL. 9, ...   185, (1971). . . . .
                         *RIGID ION MODEL OF LATTICE DYNAMICS - A RE-EVALUATION.
                         *NEMJOSHI(K.V.), MITRA(S.S.), VETELINO(J.F.).

216   SOL. STATE COMM.   VOL. 9, ...   79, (1971). . . . . .HE, NE, ...
                         *INFLUENCE OF SHORT-RANGE CORRELATIONS ON THE PHONON SPECTRA OF F.C.C.
                         HELIUM AND NEON.
                         *HORNER(H.).

217   SOL. STATE COMM.   VOL. 9, ...   167, (1971). . . . . .NA, ...
                         *PHONON DISPERSION IN SODIUM.
                         *BAJPAI(R.P.), NEELAKANDAN(K.).

218   SOL. STATE COMM.   VOL. 10, ...   179, (1972). . . . . .TB, ...
                         *LATTICE DYNAMICS AND SOE CONSTANTS OF TERBIUM USING KEATING#S APPROACH.
                         *MENON(C.S.), RAO(R.R.).

219   SOL. STATE COMM.   VOL. 11, ...   1431, (1972). . . . . . .B, T, . . .NA, ...
                         *A MODEL FOR LATTICE DYNAMICS OF B.C.C. METALS.
                         *SHUKLA(M.M.), DA-CUNHA-LIMA(I.C.), BRESCANSIN(L.M.).

220   SOL. STATE COMM.   VOL. 13, ...   779, (1973). . . . . .T, ...Y, ...
                         *THE USE OF PAIR POTENTIALS IN THE LATTICE DYNAMICS OF YTTRIUM.
                         *VERMA(M.P.), UPADHYAYA(J.C.).

221   SOL. STATE COMM.   VOL. 14 NO.3, . .   239, (1974). . . .T,B1, . . .CS*(CL,BR,I), ...
                         *THE LATTICE DYNAMICS OF CS*CL, CS*BR AND CS*I.
                         *AGRAWAL(B.S.), HARDY(J.R.).

222   SOV. PHYS. SOL.   STATE, 4, . . .   556, (1962). . . . .
                         *EFFECTS OF SPIN-SPIN AND SPIN-PHONON INTERACTIONS IN ANTIFERROMAGNETS ON
                         THE ENERGY.
                         *KASHCHEEV(V.N.).

223   SOV. PHYS. SOL.   STATE, 8, ...   850, (1966). . . . .
                         *TEMPERATURE-DEPENDENCE OF THE WIDTH OF THE FUNDAMENTAL LATTICE VIBRATION
                         ABSORPTION PEAK IN IONIC CRYSTALS.
                         *WALLIS(R.F.), IPATOVA(I.P.), MARADUDIN(A.A.).

224   SOV. PHYS. SOL.   STATE 9, ...   214, (1967). . . . .
                         *MEASURING THE   THERMAL VIBRATION SPECTRUM G(W) FROM A POLYCRYSTALLINE
                         SAMPLE.
                         *BREDOV(M.M.), KOTOV(B.A.). . .ET AL.

225   SOV. PHYS. SOL.   STATE VOL.11, . .   733, (1969). . . . . .BE,MG,ZN, ...
                         *THEORY OF THE VIBRATIONAL SPECTRA OF HEXAGONAL METALS.
                         *BROVMAN(E.G.), KAGAN(YU.), KHOLAS(A.).

226   SOV. PHYS. SOL.   STATE VOL.12, . .   1508, (1971). . . . .
                         *DYNAMICS OF QUASILOCALIZED VIBRATIONS WITH A HIGH CONCENTRATION OF
                         IMPURITY CENTERS.
                         *IVANOV(M.A.).

227   SOV. PHYS. SOLID   STATE VOL.14, . .   2480, (1973). . . .T,B1, ...NE,AR,KR, ...
                         *FORCE CONSTANTS AND PHONON DISPERSION IN RARE-GAS CRYSTALS.
                         *TOLPYGO(K.B.), TROITSKAYA(E.P.).

228   TRIESTE-INTERN.   COURSE, ...   665, (1967). . . . .
                         *INTERATOMIC FORCES IN SOLIDS. (24 PAGES).
                         *IYENGAR(P.K.).

SECTION   10C-NEUTRON-PHONON INTERACTION IN THE HARMONIC THEORY. . . . . .

1   AARHUS SUMMER SCHOOL,   221, (1963)...............................
                          *PHONONS' AND' NEUTRON'SCATTERING.
                          *BROCKHOUSE(B.N.).

2   ACTA CRYST. VOL.A26,   364, (1970). BY CRYSTALS CO*S THE AMPLITUDES OF VIBRATION
                          *ANOMALOUS NEUTRON SCATTERING BY CRYSTALS AND THE AMPLITUDES OF VIBRATION
                          OF LATTICE WAVES.
                          *RAMASESHAN(S.), VISWANATHAN(K.S.).

3   ACTA CRYST. VOL.A27,   332, (1971). P
                          *DETERMINATION OF THE POLARIZATION VECTORS OF LATTICE WAVES BY ANOMALOUS
                          NEUTRON SCATTERING.
                          *RAMESH(T.G.), RAMASESHAN(S.).

4   ACTA PHYS. POLON. VOL.25,   337, (1964). T
                          *ON THE THEORY OF FERROMAGNETICS WITH HELICOIDAL STRUCTURES. III.
                          MAGNETO-OSCILLATORY SCATTERING OF NEUTRONS.
                          *KASHCHEEV(V.N.). (IN RUSSIAN).

5   ADV. IN PHYS. VOL.18,   157, (1969). T
                          *DYNAMICAL,   SCATTERING   AND DIELECTRIC PROPERTIES OF FERROELECTRIC
                          CRYSTALS.
                          *COCHRAN(W.).

6   ANN. DE PHYSIQUE TOME 7 NO.4, 247, (1972). T
                          *INFLUENCE OF THE THERMAL MOTION OF ATOMS IN A SOLID ON THE NEUTRON
                          SCATTERING.
                          *KALUS(J.).

7   ANN. OF PHYS. VOL.33,   15, (1965).....................
                          *CALCULATIONS OF INELASTIC SCATTERING IN TERMS OF ELASTIC SCATTERING.
                          *AUSTERN(N.), BLAIR(J.S.).

8   APPL. SPECTOSC. REV. VOL.6,   79, (1972).....................
                          *NEUTRON SCATTERING AND VIBRATIONAL SPECTRA OF MOLECULAR CRYSTALS.
                          *KITAGAWA(T.).

9   ARK. FYSIK VOL.4,   191, (1952). T
                          *ON NEUTRON DIFFRACTION PHENOMENA ACCORDING TO THE KINEMATICAL THEORY. II
                          *FROMAN(P.O.).

10 ARK. FYSIK VOL.4,   183, (1952). T
                          *ON NEUTRON DIFFRACTION PHENOMENA ACCORDING TO THE KINEMATICAL THEORY. I.
                          *WALLER(I.), FROMAN(P.O.).

11 ARK. FYS. VOL.13, PAPER 17,   199, (1958). T. K*CL.
                          *ON THE SCATTERING OF SLOW NEUTRONS BY K*CL.
                          *SJOLANDER(A.).

12 ATLANTA SYMP. GEORGIA INST. TECH   17, (1967).....................
                          *INVESTIGATION OF INTERATOMIC FORCES BY NEUTRON SPECTROSCOPY.
                          *COCHRAN(W.).

13 BRIT. J. APPL. PHYS. VOL.10,   1, (1959).....................
                          *SOLID AND LIQUID STATE RESEARCH WITH COLD NEUTRONS.
                          *EGELSTAFF(P.A.).

14 CHEM. APPL. THERMAL NEUTRON SCAT   49, (1973). C1,C2
                          *NEUTRON INELASTIC SCATTERING AND MOLECULAR SPECTROSCOPY.
                          *WHITE(J.W.). (IN BOOK: CHEMICAL APPLICATIONS OF THERMAL NEUTRON
                          SCATTERING. ED. BY B.T.M. WILLIS; OXFORD UNIV. PRESS: LONDON).

15 C.N.E.N. SYMP. CASACCIA,   13, (1968).....................
                          *LATTICE DYNAMICS BY NEUTRON SCATTERING.
                          *CAGLIOTI(G.).

16 CONF. INTERN., RENNES,   150, (1971).....................
                          *COMPARISON OF PHONON SPECTRA DETERMINED FROM TUNNELING EXPERIMENTS AND
                          NEUTRON SCATTERING.
                          *ROWELL(J.M.), DYNES(R.C.).

17 CONF./PHONON SCATT. IN SOLIDS, 314, (1972). T
                          *THE EFFECT OF SPIN-PHONON INTERACTION OF NEUTRON SCATTERING FROM LATTICE
                          VIBRATIONS.
                          *TANI(K.). (ED. NOTE: INTERNATIONAL CONF. ON PHONON SCATTERING IN SOLIDS
                          AT PARIS, JULY, 1972. PUBL. DOCUMENTATION SERVICE OF CEN: SACLAY, FRA.).

18 CONTEMPORARY PHYS. VOL.7,   401, (1966).....................
                          *NEUTRON SPECTROSCOPY OF SOLIDS. II.
                          *LOMER(W.M.).

19 COPENHAGEN CONF.-LATT.DYNAMICS, 427, (1963).
                          *QUANTUM RELAXATION, THE SHAPE OF LATTICE ABSORPTION AND INELASTIC
                          NEUTRON SCATTERING LINES.
                          *LAX(M.).

20 COPENHAGEN CONF.-LATT.DYNAMICS, 71, (1963).....................
                          *DETERMINATION OF DISPERSION CURVES OF PHONONS WITH THE AID OF DIFFUSE
                          SCATTERING OF X-RAYS AND INELASTIC SCATTERING OF NEUTRONS.
                          *CRIBIER(D.).

21 CRYSTALLOGR./CRYST. PERFECTION, 189, (1963). T,B3
                          *ON THE EFFECT OF THE IMPURITIES ON NEUTRON SCATTERING BY CRYSTALS.
                          *WALLER(I.). (ED. NOTE: CRYSTALLOGRAPHY AND CRYSTAL PERFECTION; PUBL. BY
                          ACADEMIC PRESS: NEW YORK AND LONDON, 1963).

22 CRYSTAL LATTICE DEFECTS VOL.2, 181, (1971). B3,T
                          *X-RAY AND NEUTRON SCATTERING BY DUMBBELL INTERSTITIALS IN F.C.C. METALS.
                          *EISENRIEGLER(E.).

23 CZECH. J. PHYS. B VOL.20,   950, (1970). A4
                          *ON SLOW-NEUTRON SCATTERING CROSS-SECTIONS OF DEBYE-SPECTRUM SPACE LAT-
                          TICES.
                          *WEIL(J.W.).

24 DISS. ABSTR. VOL.19,   2991, (1959). T. BE,AL,MG,
                          *EINSTEIN MODEL THERMAL NEUTRON SCATTERING.
                          *DELANEY(R.M.).

SECTION 10C-NEUTRON-PHONON INTERACTION IN THE HARMONIC THEORY. . . . . .

25 DISS. ABS. VOL.27,   3224B, (1966)
    *SLOW NEUTRON SCATTERING AND THE ROTATIONAL DYNAMICS OF MOLECULES.
    *ERICKSON(J.D.).

26 ENERGIA NUCLEARE VOL.7,   OF THE 772, (1960)
    *STUDY OF THE (LATTICE) DYNAMICS OF SOLID AND OF LIQUID MEDIA BY SLOW
    NEUTRON SPECTROMETRY.
    *FELCHER(G.P.), MOSCI(M.).

27 ENERGIA NUCLEARE (ITALY) VOL.16,   425, (1969)....B7........
    *LATTICE DYNAMICS BY NEUTRON SCATTERING.2.
    *BOFFI(S.), CAGLIOTI(G.). (ED. NOTE: IN ENGLISH).

28 ENERGIA NUCLEARE VOL.16, ... 369, (1969)....B7........
    *LATTICE DYNAMICS BY NEUTRON SCATTERING.1.
    *BOFFI(S.), CAGLIOTI(G.).

29 EXCIT./MAGN./PHONONS/MOL. CRYST.   43, (1968)....C2.....
    *DENSITY OF STATES MEASUREMENTS FOR OPTICAL AND ACOUSTIC PHONONS BY
    NEUTRON-SCATTERING SPECTROSCOPY.
    *WHITE(J.W.). (ED. NOTE: EXCITONS, MAGNONS AND PHONONS IN MOLECULAR
    CRYSTALS PUBL. BY CAMBRIDGE PRESS, LONDON, ENG. 1968).

30 IAEA SYMPOSIUM VIENNA, ... 75, (1960)
    *THE SCATTERING OF NEUTRONS FROM POLYCRYSTALLINE MATERIALS.
    *MARSHALL(W.), STUART(R.).

31 IAEA SYMPOSIUM VIENNA, ... 61, (1960)..............
    *LINE WIDTHS IN NEUTRON-PHONON AND NEUTRON-MAGNON SCATTERING.
    *ELLIOTT(R.J.), STERN(H.).

32 IAEA SYMPOSIUM VIENNA, ... 479, (1960)............
    *SPECTRUM FOR THERMAL OSCILLATION OF CRYSTALS OF ARBITRARY SYMMETRY
    SINGLE-PHONON SCATTERING OF COLD NEUTRONS.
    *.C.C.C.P.

33 IAEA SYMPOSIUM BOMBAY, VOL.1, . . 399, (1964).....
    *DYNAMICS OF MAGNETIC SYSTEMS INVESTIGATED BY MOSSBAUER AND NEUTRON
    SCATTERING EXPERIMENTS.
    *MARSHALL(W.).

34 IAEA SYMPOSIUM BOMBAY, VOL.2, . . 279, (1964)...........
    *LATTICE VIBRATIONS IN MOLECULAR CRYSTALS AND NEUTRON SCATTERING.
    *HAHN(H.).

35 IAEA SYMPOSIUM BOMBAY, VOL.1, . . 231, (1964)..........
    *THEORY OF NEUTRON SCATTERING BY LATTICE VIBRATIONS IN IMPERFECT CRYSTALS
    *ELLIOTT(R.J.), MARADUDIN(A.A.).

36 IAEA SYMPOSIUM BOMBAY, VOL.1, . . 61, (1964)............
    *INTERPRETATION OF SLOW-NEUTRON SCATTERING DATA IN METALS.
    *SJOLANDER(A.), JOHNSON(R.).

37 IAEA SYMP. COPENHAGEN, VOL.1, . . 281, (1968)....B1.......K*BR,SR*TI*03, . . .
    *NEUTRON SCATTERING AND ELASTIC CONSTANTS.
    *COWLEY(R.A.), SVENSSON(E.C.), BUYERS(W.J.L.).

38 IAEA SYMP. COPENHAGEN, VOL.1, . . 141, (1968)....B1,T.......K, . . .
    *NEUTRONS, FREE ELECTRONS AND HIGH MAGNETIC FIELDS.
    *COWLEY(R.A.).

39 J. CHEM. PHYS. VOL.57, ... 1726, (1972)..............H2 (FCC, HCP), . .
    *PHONONS, LIBRONS, AND THE ROTATIONAL STATE J=1 IN HCP AND FCC SOLID
    HYDROGEN BY NEUTRON SPECTROSCOPY.
    *STEIN(H.), STILLER(H.), STOCKMEYER(R.).

40 J. NUCL. SCI. TECHNOL. VOL.8, . . 319, (1971)....P........
    *PHONON TYPE EXPANSION FOR SLOW NEUTRON SCATTERING.
    *NISHIGORI(T.), SEKIYA(T.).

41 J. PHYS. CHEM. SOLIDS VOL.8, . . 460, (1959)....B1.....GE,SI.
    *LATTICE VIBRATIONS OF SEMICONDUCTORS BY NEUTRON SPECTROMETRY.
    *BROCKHOUSE(B.N.).

42 J. PHYS. CHEM. SOLIDS VOL.19, . . 173, (1961)....T.......
    *NEUTRON SCATTERING FROM ISOTROPIC LATTICES.
    *SUBRAMANIAN(R.).

43 J. PHYS. CHEM. SOL. 25, ... 487, (1964).....
    *QUANTUM RELAXATION, THE SHAPE OF LATTICE ABSORPTION AND INELASTIC
    NEUTRON SCATTERING LINES.
    *LAX(M.).

44 J. PHYS. CHEM. SOLIDS VOL.30, . . 449, (1969)..........
    *NEUTRON INCOHERENT SCATTERING FROM KDP.
    *STEINMAN(O.K.), SUMMERFIELD(G.C.).

45 J. PHYS. C, SER.2, VOL.2, . . . 2262, (1969).....
    *ASYMMETRIES IN THE SCATTERING OF X-RAYS AND NEUTRONS BY CRYSTALS.
    *COWLEY(R.A.), BUYERS(W.J.L.).

46 J. PHYS. CHEM. SOLIDS VOL.31, . . 1317, (1970).....
    *A NOTE ON THE DETERMINATION OF THE DENSITY OF PHONON EIGENSTATES FROM
    THE INELASTIC INCOHERENT NEUTRON SCATTERING.
    *KEATING(D.T.).

47 J. PHYS. C, SER.2, VOL.4, . . . 1091, (1971)..............LI*NB*03,.
    *GROUP THEORETICAL SELECTION RULES FOR INELASTIC NEUTRON SCATTERING WITH
    APPLICATIONS TO LI*NB*03.
    *DEVINE(S.), PECKHAM(G.E.).

48 J. PHYS. SOC. JAPAN VOL.17 B-II, 363, (1962).....
    *INTERATOMIC FORCES IN CRYSTALS FROM NEUTRON SCATTERING.
    *BROCKHOUSE(B.N.).

49 J. PHYS. SOC. JAPAN VOL.21, . . 1263, (1966)....T.....
    *ATTENUATION OF THERMAL NEUTRONS BY PHONONS IN A SINGLE CRYSTAL. I.
    *KASHIWASE(Y.).

50 J. PHYS. SOC. JAPAN VOL.29, . . 594, (1970)....T.....
    *THEORY OF NEUTRON SCATTERING FROM LATTICE VIBRATIONS.
    *TANI(K.).

51 LECT/ELEM. EXCIT./INTER. IN SOL. . (1968)....T.....
    *DETERMINATION OF PHONON AND MAGNON DISPERSION CURVES BY NEUTRON SPECTRO-
    SCOPY.
    *COLLINS(M.F.). (ED. NOTE: PUBL. NATO ADVANCED STUDY INSTITUTE, CORTINA
    D#AMPEYYO 1968, 43PP.).

SECTION 10C-NEUTRON-PHONON INTERACTION IN THE HARMONIC THEORY. . . . . .

52 METAL. SOC. CONF. LOS ANGELES, 161, (1967).AND ALLOYS STUDIED BY NEUTRON SPECTROSCOPY.
⊢ATOMIC VIBRATIONS IN METALS AND ALLOYS STUDIED BY NEUTRON SPECTROSCOPY.
⊢BROCKHOUSE(B.N.), HALLMAN(E.D.), NG(S.C.).

53 NATURE VOL.218,. 80, (1968).T
⊢NEUTRON DIFFRACTION ON PIEZOELECTRIC VIBRATING RESONATORS.
⊢PETRZILKA(V.).

54 2ND INTERNAT. MEET./FERROELECTR. 5, (1969)....B7
⊢NEUTRON INELASTIC SCATTERING STUDY ON SOFT MODES.
⊢SHIRANE(G.). (ED. NOTE: PUBL. PHYS. SOC. JAPAN, TOKYO, JAPAN 1969).

55 NUCL. INST. MET. VOL.112,. 571, (1973).E
⊢MEASUREMENT OF INTEGRATED INTENSITIES IN INELASTIC NEUTRON SCATTERING.
⊢SKALYO-JR(J.), LURIE(N.A.).

56 NUCL. INST. MET. VOL.113,. 15, (1973).E,T.....SI,BE
⊢PHONON MEASUREMENTS ON A FILTER-DETECTOR SPECTROMETER AND ITS RESOLUTION FUNCTION.
⊢THAPER(C.L.), DASANNACHARYA(B.A.), IYENGAR(P.K.).

57 NUCL. SCI. ENGNG. VOL.13,. 40, (1962)....T
⊢INELASTIC SCATTERING OF COLD NEUTRONS BY POLYCRYSTALS.
⊢KHUBCHANDANI(P.G.), SHARMA(R.R.).

58 NUKLEONIKA VOL.4,. 119, (1959)....E
⊢THE NEUTRON-PHONON INTERACTION IN SOLIDS.
⊢BURAS(B.), O≠CONNOR(D.).

59 NUOVO CIMENTO VOL.34,. 293, (1964)....T,83
⊢SCATTERING OF SLOW NEUTRONS BY A CRYSTAL WITH POINT DEFECTS.
⊢PERETTI(J.), JOUANIN(C.).

60 PHIL. MAG. VOL.5,. 1235, (1960)....T
⊢THE EFFECT OF LATTICE DISTORTION AROUND POINT DEFECTS ON THE SCATTERING OF LONG-WAVELENGTH NEUTRONS.
⊢MARTIN(D.G.).

61 PHYS. LETT. 16,. 225, (1965)....T
⊢ONE-PHONON COHERENT NEUTRON SCATTERING FROM CERTAIN POLYCRYSTALLINE MATERIALS.
⊢MYERS(W.), SUMMERFIELD(G.C.), KING(J.S.).

62 PHYS. LETT. VOL.44A NO.7,. 519, (1973)....T
⊢INTERFERENCE OF ZERO-POINT AND THERMAL OSCILLATIONS IN NEUTRON INELASTIC SCATTERING.
⊢SARMA(B.K.), TEWARI(S.P.).

63 PHYS. LETT. VOL.45A,. 481, (1973)....T
⊢SCATTERING OF NEUTRONS FROM THE ZERO-POINT VIBRATIONS OF A LATTICE AT ABSOLUTE ZERO.
⊢SARMA(B.K.), TEWARI(S.P.).

64 PHYS. LETT. VOL.46A NO.7,. 471, (1974)....T,C1
⊢MULTIQUANTUM INCOHERENT NEUTRON SCATTERING FROM MOLECULAR SOLIDS.
⊢SARMA(B.K.), TEWARI(S.P.).

65 PHYS. MET. METALLOGR. VOL.10 N.2 9, (1960)....T
⊢STATIC DISTORTIONS AND WEAKENING OF LINE INTENSITIES IN X-RAY OR NEUTRON DIFFRACTION PATTERNS OF SOLID SOLUTIONS WITH F.C.C. LATTICES.
⊢KRIVOGLAZ(M.A.).

66 PHYS. REV. VOL.72,. 907, (1947)....T........BE
⊢SCATTERING OF NEUTRONS IN POLYCRYSTALS.
⊢FINKELSTEIN(R.J.).

67 PHYS. REV. VOL.71,. 232, (1947)....T........H2*O
⊢THE SCATTERING OF SLOW NEUTRONS BY BOUND PROTONS II. HARMONIC BINDING-NEUTRONS OF ZERO ENERGY.
⊢BREIT(G.), ZILSEL(P.R.).

68 PHYS. REV. VOL.93,. 1207, (1954)....T
⊢CRYSTAL DYNAMICS AND INELASTIC SCATTERING OF NEUTRONS.
⊢PLACZEK(G.), VAN-HOVE(L.).

69 PHYS. REV. VOL.98,. 1692, (1955)....T
⊢TIME-DEPENDENT DISPLACEMENT CORRELATIONS AND INELASTIC SCATTERING BY CRYSTALS.
⊢GLAUBER(R.J.).

70 PHYS. REV. VOL.103,. 304, (1956)....T,B1......AL
⊢RELATION BETWEEN THE VIBRATION FREQUENCIES OF A CRYSTAL AND THE SCATTERING OF SLOW NEUTRONS.
⊢SQUIRES(G.L.).

71 PHYS. REV. VOL.121,. 1388, (1961)....T
⊢INFLUENCE OF ERGODIC BEHAVIOUR ON THE SCATTERING OF SLOW NEUTRONS BY A HARMONIC OSCILLATOR.
⊢KRIEGER(T.J.).

72 PHYS. REV. VOL.121,. 741, (1961)....T
⊢THERMODYNAMIC GREEN≠S FUNCTION METHODS IN NEUTRON SCATTERING BY CRYSTALS
⊢BAYM(G.).

73 PHYS. REV. VOL.135,. A1691, (1964)....T
⊢DIRECT CALCULATION OF ELECTRONIC PROPERTIES OF METALS FROM NEUTRON SCATTERING DATA.
⊢BAYM(G.).

74 PHYS. REV. VOL.135,. B895, (1964)....T
⊢CALCULATIONS OF INELASTIC SCATTERING OF NEUTRONS BY HEAVY NUCLEI.
⊢AUERBACH(E.H.), MOORE(S.O.).

75 PHYS. REV. VOL.136,. A1280, (1964)....T
⊢COHERENT INELASTIC SCATTERING OF SLOW NEUTRONS BY A POLYATOMIC LIQUID ELEMENT.
⊢ANTONINI(M.), ASCARELLI(P.), CAGLIOTI(G.).

76 PHYS. REV. LETT. 15,. 693, (1965)....T
⊢KOHN ANOMALY AND COHERENT SCATTERING OF SLOW NEUTRONS IN LIQUID METALS.
⊢SINGWI(K.S.), ANDERSON(L.E.).

77 PHYS. REV. 176,. 784, (1968)....T
⊢INELASTIC SCATTERING OF NEUTRONS BY POLYCRYSTALS.
⊢DE-WETTE(F.W.), RAHMAN(A.).

SECTION 10C-NEUTRON-PHONON INTERACTION IN THE HARMONIC THEORY. . . . . .

78   PHYS. REV. 175,.   841, (1968).
     ~EFFECT OF FORCE-CONSTANT CHANGES ON THE INCOHERENT NEUTRON SCATTERING
      FROM CUBIC CRYSTALS WITH POINT DEFECTS.
     ~LAKATOS(K.), KRUMHANSL(J.A.).

79   PHYS. REV. VOL.180   729, (1969)   B3
     ~EFFECT OF FORCE-CONSTANT CHANGES ON THE COHERENT NEUTRON SCATTERING
      FROM FACE-CENTERED CUBIC CRYSTALS WITH POINT DEFECTS.
     ~LAKATOS(K.), KRUMHANSL(J.A.).

80   PHYS. REV. B VOL.6,   2577, (1972)
     ~NEUTRON SPECTROSCOPY OF SUPERCONDUCTORS.
     ~ALLEN(P.B.).

81   PHYS. STAT. SOL. 13,   233, (1966)
     ~EFFECTS OF EXCITONS ON THE SCATTERING OF NEUTRONS BY CRYSTALS.
     ~GLAUBERMAN(A.E.), RUVINSKII(M.A.).

82   PHYS. STAT. SOL. VOL.50,   673, (1972)
     ~INELASTIC INCOHERENT NEUTRON SCATTERING FROM IMPURITIES IN CRYSTALS.
     ~COHEN(S.S.).

83   PHYS. STAT. SOL. B VOL.51 NO.2,   853, (1972)   B3,T
     ~INELASTIC SCATTERING OF NEUTRONS IN CRYSTALS CONTAINING LINEAR DEFECTS.
     ~IVANOV(M.A.), FISHMAN(A.YA.).

84   PHYS. STAT. SOLIDI B VOL.56 NO.1   313, (1973)   E
     ~BROAD NEUTRON GROUPS IN INELASTIC NEUTRON SCATTERING AND ANOMALOUS
      ANISOTROPY OF THE ENERGY OF THE DISPERSION BRANCHES.
     ~COMES(R.).

85   PHYS. STAT. SOLIDI B VOL.58 NO.2 K159, (1973)   T   MG*ZN2
     ~PHONON DENSITY OF STATES AND COHERENT INELASTIC NEUTRON SCATTERING
      ON COMPLEX STRUCTURES.
     ~ESCHRIG(H.), FELDMANN(K.), HENNIG(K.), JOHN(W.), VON-LOYEN(L.).

86   PHYS. ZEITS. VOL.38,   689, (1937)   T
     ~SCATTERING OF SLOW NEUTRONS BY ATOMIC LATTICES OF CRYSTALS. PART II.
     ~WICK(G.C.).

87   PHYS. ZEITS. VOL.38,   403, (1937)   T
     ~SCATTERING OF SLOW NEUTRONS BY ATOMIC LATTICES OF CRYSTALS.
     ~WICK(G.C.).

88   PROC. PHYS. SOC. 91,   903, (1967)
     ~GROUP THEORETICAL SELECTION RULES IN INELASTIC NEUTRON SCATTERING.
     ~ELLIOTT(R.J.), THORPE(M.F.).

89   PROC. PHYS. SOC. VOL.91,   917, (1967)   T
     ~THE DYNAMICAL THEORY OF THE INTERACTION OF X-RAYS AND NEUTRONS WITH
      PHONONS.
     ~O*CONNOR(D.A.).

90   PROC. ROY. SOC. A VOL.231,   293, (1955)   T
     ~THERMAL INELASTIC SCATTERING OF COLD NEUTRONS IN POLYCRYSTALLINE SOLIDS.
     ~KOTHARI(L.S.), SINGWI(K.S.).

91   PROC/1ST INT. CONF/LOC. EXC/SOL.   721, (1968)   B7
     ~INVESTIGATION OF LOCALIZED EXCITATIONS BY INELASTIC NEUTRON SCATTERING.
     ~MACKINTOSH(A.R.), MOLLER(H.B.). (ED. NOTE: PUBL. PLENUM: LONDON, ENGL.).

92   REPORT A.E.R.E. N/R 1165   (1953)   E   BE,MG,PB
     ~THE INELASTIC SCATTERING OF COLD NEUTRONS BY CRYSTALS II. EXPERIMENT.
     ~EGELSTAFF(P.A.).

93   REPORT A.E.R.E. N/R 1164   (1953)   T   BE,MG,PB
     ~THE INELASTIC SCATTERING OF COLD NEUTRONS BY CRYSTALS I. THEORY.
     ~EGELSTAFF(P.A.).

94   REPORT NAL-TN-21 (14PP.)   (1970)   B1
     ~ANOMALOUS SCATTERING OF NEUTRONS BY CRYSTALS AND THE AMPLITUDE OF
      VIBRATION OF LATTICE WAVES.
     ~RAMASESHAN(S.), VISWANATHAN(K.S.). (ED. NOTE: AVAIL. NTIS, SPRINGFIELD,
      VA. 22151, USA).

95   REV. MEX. FIS. VOL.2,   137, (1953)   T
     ~DIFFRACTION OF NEUTRONS BY VIBRATING CRYSTALS.
     ~MOSHINSKY(M.).

96   REV. MEX. FIS. VOL.3,   1, (1954)   T
     ~THERMAL VIBRATIONS OF CRYSTALS AND THE DIFFRACTION OF NEUTRONS.
     ~MOSHINSKY(M.).

97   REV. ROUMAINE PHYS. VOL.17,   757, (1972)   E
     ~REAL STRUCTURE AND DYNAMICS OF CRYSTALS BY NEUTRON AND X-RAY SCATTERING.
      SOME NEW EXPERIMENTAL METHODS.
     ~BALLY(D.).

98   SOL. STATE COMM. VOL.12,   1191, (1973)   T
     ~NEUTRON SCATTERING AS A METHOD FOR FERMI SURFACE INVESTIGATIONS.
     ~DIETERICH(W.).

99   SOV. PHYS. JETP VOL.7,   281, (1958)   T
     ~SCATTERING OF X-RAYS AND THERMAL NEUTRONS BY SINGLE-COMPONENT CRYSTALS
      NEAR PHASE-TRANSITION POINTS OF THE SECOND KIND.
     ~KRIVOGLAZ(M.A.).

100 SOV. PHYS. JETP VOL.7,   151, (1958)   T
     ~THE EXCITATION BY SLOW NEUTRONS OF OPTICAL OSCILLATIONS IN A CRYSTAL.
     ~TURCHIN(V.F.).

101 SOV. PHYS. JETP VOL.7,   139, (1958)   T
     ~THEORY OF DIFFUSE SCATTERING OF X-RAYS AND THERMAL NEUTRONS. III.ACCOUNT
      OF GEOMETRICAL DISTORTIONS OF THE LATTICE.
     ~KRIVOGLAZ(M.A.).

102 SOV. PHYS. SOL. STATE VOL.1,   1277, (1959)   T
     ~THE THEORY OF THE SCATTERING OF X-RAYS AND THERMAL NEUTRONS BY MULTI-
      COMPONENT SUBSTITUTION ALLOYS.
     ~SMIRNOV(A.A.), TIKHONOVA(E.A.).

103 SOV. PHYS. SOL. STATE VOL.2,   647, (1960)   T,C2
     ~THE POSSIBILITY OF INVESTIGATING THE DENSITY OF DISTRIBUTION OF PHONONS
      IN NON-CUBIC CRYSTALS WITH THE AID OF INCOHERENT NEUTRON SCATTERING.
     ~OSKOTSKII(V.S.).

SECTION 10C-NEUTRON-PHONON INTERACTION IN THE HARMONIC THEORY. . . . . .

104 SOV. PHYS. SOLID STATE VOL.3, . . 1039, (1961). . .E . . . .
     *THE EXPER. DETERMINATION OF THE SPECTRUM OF THERMAL VIBRATIONS/CRYSTALS
     WITH ARBITRARY SYMMETRY USING INCOH. ONE-PHONON SCAT. OF COLD NEUTRONS.
     *TARASOV(L.V.).

105 SOV. PHYS. JETP VOL.13, . . . 397, (1961)...B3,T. . . . .
     *THEORY OF INELASTIC SCATTERING OF NEUTRONS BY IMPERFECT CRYSTALS.
     *KRIVOGLAZ(M.A.).

106 SOV. PHYS. JETP VOL.12, ... 714, (1961). . . . .
     *INELASTIC MAGNETIC SCATTERING OF SLOW NEUTRONS BY PHONONS.
     *KHABIBULLIN(B.M.).

107 SOV. PHYS. SOL. STATE, 4, . . . 2459, (1962). . . . .
     *CALCULATION OF STRUCTURE FACTORS FOR THE INELASTIC SCATTERING OF SLOW
     NEUTRONS IN NA*CL-TYPE CRYSTALS.
     *DEMIDENKO(Z.A.).

108 SOV. PHYS. JETP VOL.13, . . . 1273, (1962). . .B3 . . . .
     *EFFECT OF DIFFUSION ON THE SCATTERING OF NEUTRONS AND PHOTONS BY
     CRYSTAL IMPERFECTIONS AND ON THE MOSSBAUER EFFECT.
     *KRIVOGLAZ(M.A.).

109 SOV. PHYS. SOL. STATE VOL.3, . . 2301, (1962)...T. . . . .
     *THE THEORY OF INELASTIC SCATTERING OF NEUTRONS BY IMPURITY CENTRES
     IN CRYSTALS.
     *KASHCHEEV(V.N.), KRIVOGLAZ(M.A.).

110 SOV. PHYS. DOKLADY VOL.3, . . . 2015, (1962). . . .T. . . . .
     *THE INFLUENCE OF THE CONDUCTION ELECTRONS ON THE SCATTERING OF NEUTRONS
     BY CRYSTALS.
     *KRIVOGLAZ(M.A.).

111 SOV. PHYS. JETP 15, ... 954, (1962). . . .T. . . . .
     *INELASTIC SCATTERING OF SLOW NEUTRONS ON ARBITRARY CRYSTALS AND THE
     GENERAL PROBLEM OF RECONSTRUCTION OF THE PHONON SPECTRUM.
     *KAGAN(YU.).

112 SOV. PHYS. SOL. STATE, 5, . . . 2222, (1963). . . . .
     *DETERMINATION OF THE PHONON SPECTRUM FROM THE CROSS SECTION OF . . . . .
     INCOHERENT SCATTERING OF NEUTRONS IN A CRYSTAL.
     *LAIKHTMAN(B.D.).

113 SOV. PHYS. JETP. 17, ... 925, (1963). . . . .
     *SCATTERING OF NEUTRONS BY CRYSTALS WITH IMPURITY NUCLEII AND THE PROBLEM
     OF RECONSTRUCTING THE VIBRATION SPECTRUM.
     *KAGAN(YU.) IOSILEVSKII(YA.).

114 SOV. PHYS. SOL. STATE, 6, . . 1469, (1964). . . . .
     *INELASTIC INCOHERENT SCATTERING OF SLOW NEUTRONS BY DISORDERED SOLID
     SOLUTIONS.
     *DZYUB(I.P.).

115 SOV. PHYS. SOL. STATE, 6, . . . 2162, (1964). . . . .
     *USE OF RESONANCE CAPTURE OF NEUTRONS FOR INVESTIGATING THE VIBRATIONAL
     SPECTRUM OF IMPURITY ATOMS IN CRYSTALS.
     *MALEEV(S.V.).

116 SOV. PHYS. SOL. STATE, 6, . . . 2955, (1964). . . . .
     *RESONANT SCATTERING OF PHONONS BY IMPURITY ATOMS AND ONE-PHONON COHERENT
     SCATTERING OF SLOW NEUTRONS.
     *DZYUB(I.P.).

117 SOV. PHYS. JETP. 20, ... 1422, (1965). . . . .
     *AN APPROXIMATE DETERMINATION OF FREQUENCY SPECTRUM OF PHONONS IN
     CRYSTALS BY NEUTRON SCATTERING.
     *KOTHARI(L.S.).

118 SOV. PHYS. JETP. 20, ... 1340, (1965). . . . .
     *ON THE NATURE OF THE *TAIL* IN THE INCOHERENT INELASTIC SCATTERING CROSS
     SECTION OF SLOW NEUTRONS IN CRYSTALS.
     *KAGAN(YU.), ZHERNOV(A.P.).

119 SOV. PHYS. SOL. STATE, 6,.. . 2618, (1965). . . . .
     *PHONON CORRELATION FUNCTION AND COHERENT INELASTIC SCATTERING OF
     NEUTRONS IN CRYSTALS CONTAINING SHALLOW ELECTRON IMPURITY CENTERS.
     *KRIVOGLAZ(M.A.), SHALDERVAN(P.I.).

120 SOV. PHYS. SOL. STATE, 7,. . . 1198, (1965). . . . .
     *EFFECT OF A TWO-LEVEL IMPURITY SYSTEM ON THE PHONON SPECTRUM OF A
     CRYSTAL AND ON ONE-QUANTUM INELASTIC NEUTRON SCATTERING.
     *IOLIN(E.M.).

121 SOV. PHYS. JETP. 23, ... 1061, (1966). . . . .
     *INELASTIC DIFFRACTION SCATTERING.
     *INOPIN(E.V.).

122 SOV. PHYS. SOL. STATE, 9,. . . 420, (1967). . . . .
     *MEASUREMENT OF THE PHONON DISTRIBUTION FUNCTION IN POLYCRYSTALLINE
     MATERIALS USING COHERENT SCATTERING.
     *OSKOTSKII(V.S.).

123 SOV. PHYS. J. NO.9,. . . . . 31, (1968). . . .B7. . . . .
     *PEAK WIDTH IN SINGLE PHONON SCATTERING OF NEUTRONS BY MOLECULAR CRYSTALS
     *ZHIDKOV(L.G.), KORSHVNOV(A.V.), TRET-YAKOV(A.G.).

124 SOV. PHYS. SOL. STATE VOL.12,. . 1031, (1970). . . . .
     *THEORY OF NEUTRON SCATTERING IN DIATOMIC CRYSTALS CONTAINING IMPURITIES.
     *AUGST(G.R.).

125 SOV. PHYS. SOL. STATE VOL.13,. . 215, (1971). . . . .
     *NEUTRON COUNTER OF *HOT* AND COHERENT PHONONS.
     *MITIN(A.V.), KHABIBULLIN(B.M.).

126 SOV. PHYS. SOLID STATE VOL.13, . 3012, (1972)....P. . . . .
     *POSSIBLE USE OF NEUTRONS TO STUDY THE VIBRATIONAL SPECTRA OF SOLIDS.
     *SADIKOV(I.P.).

127 SOV. PHYS. SOLID STATE VOL.14, . . 1, (1972). . . .T. . . . .
     *INELASTIC SCATTERING OF SLOW NEUTRONS BY LOCALIZED VIBRATIONS OF INTER-
     STITIAL ATOMS.
     *DZYUB(I.P.), KOCHMARSKII(V.Z.).

128 SOV. PHYS. JETP VOL.35,. . . . 399, (1972). . .T. . . . .
     *HARD PHOTON AND SLOW NEUTRON SCATTERING IN SEMICONDUCTOR CRYSTALS UNDER
     CONDITIONS OF SOUND INSTABILITY.
     *PUSTOVOIT(V.I.).

SECTION 10C-NEUTRON-PHONON INTERACTION IN THE HARMONIC THEORY. . . . . .

129 SOV. PHYS. SOLID STATE VOL.15, . 1309, (1974)....E,T......MG
⇔DETERMINATION OF THE DENSITY OF PHONON STATES FOR COHERENTLY SCATTERING
CRYSTALS.
⇔EREMEEV(I.P.), SADIKOV(I.P.), CHERNYSHOV(A.A.).

130 THESIS (246 PP.), ... , (1972)....B3. . . . .
⇔NEUTRON SCATTERING STUDIES OF THE DYNAMICS OF IMPERFECT CRYSTALS.
⇔KAMITAKAHARA(W.A.). (ED. NOTE: PH.D. DISSERTATION, MCMASTER UNIVERSITY).

131 UKAEA AERE REP. R4101 (24 PP.) . . . (1962)....T........BE,PB, ...
⇔THE SCATTERING OF COLD NEUTRONS BY METALS.
⇔EGELSTAFF(P.A.).

132 UKR. FIZ. ZH. (USSR) VOL.8,. . . 162, (1963)....T........
⇔THEORY OF THE SCATTERING OF X-RAYS AND NEUTRONS BY ORDERED ALLOYS.
⇔KRIVOGLAZ(M.A.). (IN RUSSIAN).

133 UKR. FIZ. ZH. (USSR) VOL.8,. . . 256, (1963)....B3.
⇔EFFECT OF LATTICE DISTORTIONS ON THE SCATTERING OF SLOW NEUTRONS IN
ALLOYS.
⇔ZYUGANOV(A.N.), MOLODKIN(V.B.), SMIRNOV(A.A.), TIKHONOVA(E.A.). (IN
RUSSIAN).

134 UKR. FIZ. ZH. (USSR) VOL.9,. . . 1331, (1964)....T........
⇔PHONON GREEN≠S FUNCTION, CORRELATION FCN. OF/PHONON AND/INELASTIC COHER-
ENT SCATT. OF NEUTRONS BY CRYSTALS CONTAINING SHALLOW ELECTRON IMP. CEN.
⇔KRYVOHLAZ(M.O.), SHALDERVAN(P.I.). (IN UKRAINIAN).

135 UKR. FIZ. ZH. (USSR) VOL.13,. . . 294, (1968)....T,B3.
⇔MONOPHONON SCATTERING OF SLOW NEUTRONS IN DISORDERED SOLID SOLUTIONS.
⇔KOVAL≠CHUK(V.G.). (IN RUSSIAN).

136 UKR. FIZ. ZH. (USSR) VOL.17,. . . 789, (1972)....T........
⇔ON DERIVATION OF THE G(NU)-FUNCTION FROM THE DATA ON INELASTIC COHERENT
SCATTERING OF SLOW NEUTRONS.
⇔GORBACHEV(B.I.), IVANITSKII(P.G.), KROTENKO(V.T.).

137 Z. ANGEW. PHYS. VOL.18,. . . . 295, (1965)....B3,E. . .
⇔SMALL-ANGLE SCATTERING OF NEUTRONS IN INVESTIGATION OF LATTICE DEFECTS.
⇔CHRIST(J.), SCHILLING(W.), SCHMATZ(W.), SPRINGER(T.).

138 Z. NATURFORSCH. VOL.14A,. . . 74, (1959)....T,B3.....
⇔THE THEORY OF SMALL ANGLE SCATTERING OF X-RAYS AND NEUTRONS FROM
EXTENDED DISLOCATIONS IN SOLIDS.
⇔SEEGER(A.), KRONER(E.).

139 Z. PHYS. VOL.171,. . . . . . 291, (1963)....T,B3.......
⇔DEPOLARIZATION AND SMALL-ANGLE SCATTERING OF NEUTRONS BY LATTICE
DEFECTS IN FERROMAGNETIC CRYSTALS.
⇔KRONMUELLER(H.), SEEGER(A.), WILKENS(M.). (IN GERMAN).

140 Z. PHYS. VOL.210 NO.5,. . . . . 434, (1968)....... . . . .
⇔THERMAL AND SUBTHERMAL NEUTRON SCATTERING.
⇔MEHRINGER(W.).

SECTION 100-CRYSTAL DYNAMICS IN THE ANHARMONIC THEORY. . . . .

1 ACTA CRYST. VOL.12, ... 251, (1959). . . . .
   *ANHARMONIC OSCILLATIONS OF NUCLEI.
   *IBERS(J.A.).

2 ADVANCES IN PHYS. VOL.12, ... 421, (1963). . . . T . . . .
   *THE LATTICE DYNAMICS OF AN ANHARMONIC CRYSTAL.
   *COWLEY(R.A.).

3 BROOKHAVEN SYMPOSIUM, ... 44, (1965). . . . P . . . .
   *LATTICE DYNAMICS AND PHASE TRANSITIONS.
   *BOCCARA(N.).

4 BROOKHAVEN SYMPOSIUM, ... 29, (1965). . . P . . . .
   *ELECTRONIC EFFECTS ON LATTICE VIBRATIONS.
   *SJOLANDER(A.).

5 CAN. J. PHYS. 49, ... 2997, (1971). . . . . HE (A=3) . . IN . B.C.C. . HE(A=3).
   *LATTICE DYNAMICS AND SHORT RANGE CORRELATIONS IN B.C.C. HE(A=3).
   *GLYDE(H.R.), KHANNA(F.C.).

6 CAN. J. PHYS. VOL.50, ... 1152, (1972). . . . T . . . . HE (ISO. A=4), ...
   *ANHARMONIC PHONONS IN B.C.C. HE (ISO. A=4).
   *GLYDE(H.R.), KHANNA(F.C.).

7 CAN. J. PHYS. VOL.51, ... 2564, (1973). . . . 92,T. . . . . . RB.
   *CRYSTAL DYNAMICS OF RUBIDIUM. II. ANHARMONIC CALCULATIONS OF THE PHONON
   SELF-ENERGY AND THE HEAT CAPACITY.
   *COPLEY(J.R.D.).

8 CRIT. REV. SOLID STATE SCI. 2, . . 181, (1971). . . . B2 . . . .
   *ANHARMONIC EFFECTS AND THE LATTICE DYNAMICS OF INSULATORS.
   *GLYDE(H.R.), KLEIN(M.L.).

9 HARWELL SUMMER SCHOOL, ... 124, (1968). . . . .
   *ANHARMONIC CONTRIBUTIONS TO THE DEBYE-WALLER FACTORS OF CUBIC CRYSTALS.
   *WILLIS(B.T.M.).

10 IAEA SYMPOSIUM BOMBAY, VOL.1, . . 391, (1964). . . . .
   *EFFECT OF ANHARMONICITY ON THE PHONON SPECTRUM NEAR ITS DISCONTINUITY.
   *.C.C.C.P.

11 IAEA SYMPOSIUM BOMBAY, VOL.1, . . 49, (1964). . . . .
   *A COMPARISON BETWEEN THE ELECTRON-PHONON INTERACTIONS OF TOYA AND BAILYN
   *SRIVASTAVA(P.L.).

12 IAEA SYMPOSIUM BOMBAY, VOL.1, . . 325, (1964). . . . .
   *STRAIN DEPENDENCE OF THE FREQUENCIES AND THERMAL EXPANSION OF THE
   HEXAGONAL CLOSE-PACKED LATTICE.
   *SRINIVASAN(R.), RAO(R.R.).

13 IAEA SYMP. GRENOBLE, ... 119, (1972). . . . T . . . .
   *THE EFFECT OF HARD-CORE INTERACTIONS ON CALCULATIONS OF PHONON FREQUEN-
   CIES AND DAMPING IN RARE GAS CRYSTALS.
   *HORNER(H.).

14 J. LOW TEMP. PHYS. VOL.8, . . . 511, (1972). . . . T,B2 . . . . . HE, ...
   *PHONONS AND THERMAL PROPERTIES OF BCC AND FCC HELIUM FROM A SELF-
   CONSISTENT ANHARMONIC THEORY.
   *HORNER(H.).

15 JOUR. APP. PHYS. 40, ... 2766, (1969). . . . IN ALKALI HALIDES I. TRANSPORT OF SECONDARY
   *ELECTRON PHONON INTERACTION IN ALKALI HALIDES I. TRANSPORT OF SECONDARY
   ELECTRONS WITH ENERGIES BETWEEN 0.25 + 7.5 EV.
   *LEACER(J.), GARWIN(E.L.).

16 JOUR. APP. PHYS. 40, ... 2776, (1969). . . . .
   *ELECTRON PHONON INTERACTION IN ALKALI HALIDES II. TRANSMISSION SECONDARY
   EMISSION FROM ALKALI HALIDES.
   *LEACER(J.), GARWIN(E.L.).

17 J. PHYS. C: SOLID ST. PHYS. V-4, 988, (1971). . . . . NA*CL . . . .
   *ANHARMONIC CONTRIBUTIONS TO THE THERMODYNAMIC PROPERTIES OF SODIUM
   CHLORIDE.
   *COWLEY(E.R.).

18 J. PHYS. C VOL.6, ... 2422, (1973). . . . T . . . . SR*TI*03 . . . .
   *LATTICE DYNAMICS OF STRONTIUM TITANATE: ANHARMONIC INTERACTIONS AND
   STRUCTURAL PHASE TRANSITIONS.
   *BRUCE(A.D.), COWLEY(R.A.).

19 KYOTO CONF. STAT. MECHANICS, . . 196, (1968). . . . .
   *NONLINEAR LATTICE DYNAMICS AND ENERGY SHARING.
   *ZABUSKY(N.J.).

20 2ND SUMMER SCHOOL S.S. PHYS., . . 7, (1968). . . . T . . . .
   *LIQUID DYNAMICS AND TRANSPORT PROPERTIES OF ANHARMONIC CRYSTALS.
   *SJOLANDER(A.).

21 NUCL./S.S. PHYS. SYMP. ABSTR., . . (1972). . . . T . . . .
   *PHONON DISPERSION RELATIONS AND DEBYE TEMPERATURE FOR ANISOTROPIC HCP
   METALS USING MODIFIED AXIALLY SYMMETRIC MODEL.
   *SINGH(O.N.), KUSHWAHA(S.S.).

22 PHYS. LETT. VOL.31A, ... 563, (1970). . . . .
   *FOURTH SOUND ATTENUATION BY A SUPERFLOW IN A POROUS MEDIUM.
   *FRANCOIS(M.), LE-RAY(M.), BATAILLE(J.).

23 PHYS. LETT. VOL.32A, ... 134, (1970). . . . .
   *SELF-CONSISTENT PHONONS IN FERROMAGNETIC CRYSTALS.
   *PLAKIDA(N.M.).

24 PHYS. LETT. VOL.37A, ... 287, (1971). . . . .
   *EXPERIMENTAL EVIDENCE FOR A DIFFERENCE BETWEEN THERMAL AND ELASTIC DEBYE
   THETAS OF A CRYSTAL NEAR 0 DEG. K.
   *OVERTON-JR(W.C.).

25 PHYS. LETT. VOL.39A, ... 327, (1972). . . . . HE, ...
   *LINE WIDTH OF ROTONS IN HE II ABOVE THE THRESHOLD FOR PHONON EMISSION.
   *JACKLE(J.), KEHR(K.W.).

SECTION 100-CRYSTAL DYNAMICS IN THE ANHARMONIC THEORY. . . . .

26 PHYS. REV. 131,. . . . . 1030, (1963)....T,B2......K*BR, NA*I,. . . . .
    *LATTICE DYNAMICS OF ALKALI HALIDE CRYSTALS, III. THEORETICAL.
    *COCHRAN(W.), COWLEY(R.A.), BROCKHOUSE(B.N.), WOODS(A.D.B.).

27 PHYS. REV. 169,. . . . . 477, (1968). . . . .
    *ELECTRON-PHONON INTERACTION AND THE PHONON DISPERSION RELATIONS USING
    THE AUGMENTED-PLANE-WAVE METHOD.
    *SINHA(S.K.).

28 PHYS. REV. 172,. . . . . 747, (1968). . . . .
    *SCREENING AND ELECTRON CORRELATION EFFECTS ON THE ELECTRON-PHONON
    INTERACTION IN MONOVALENT METALS.
    *TAKAHASHI(H.).

29 PHYS. REV. 178,. . . . . 1171, (1969). . . . .
    *ANHARMONIC CONTRIBUTIONS TO THE DEBYE-WALLER FACTOR. . . . .
    *WOLFE(G.A.), GOODMAN(B.).

30 PHYS. REV. B-1,. . . . . 4521, (1970). . . . . .AL,. . . . .
    *ANHARMONIC INTERACTION IN ALUMINUM I.
    *KOEHLER(T.R.), GILLIS(N.S.), WALLACE(D.C.).

31 PHYS. REV. B-2,. . . . . 4176, (1970). . . . . .AR,. . . . .
    *ELASTIC CONSTANTS AND PHONON DISPERSION CURVES FOR SOLID ARGON NEAR 0 K.
    *BARKER(J.A.), KLEIN(M.L.), BOBETIC(M.V.).

32 PHYS. REV. LETT. VOL.24,. . . . . 1101, (1970). . . . .
    *RENORMALIZED CALCULATION OF PHONON DAMPING IN QUANTUM CRYSTALS.
    *JACKLE(J.), KEHR(K.W.).

33 PHYS. REV. B VOL.1,. . . . . 572, (1970). . . . .
    *SELF-CONSISTENT PHONON FORMULATION OF ANHARMONIC LATTICE DYNAMICS.
    *WERTHAMER(N.R.).

34 PHYS. REV. B-4,. . . . . 1983, (1971). . . . . .AR,. . . . .
    *LATTICE DYNAMICS OF FCC ARGON WITH THREE-BODY FORCES.
    *KLEIN(M.L.), BARKER(J.A.), KOEHLER(T.R.).

35 PHYS. REV. LETT. VOL.27,. . . . . 1725, (1971). . . . . .NB3*SN,. . . . .
    *LATTICE DYNAMICS OF NB3*SN-TYPE COMPOUNDS.
    *SHAM(L.J.).

36 PHYS. REV. LETT. VOL.27,. . . . . 1134, (1971). . . . .
    *ANHARMONIC SELF-ENERGY OF A SOFT MODE.
    *LOWNDES(R.P.).

37 PHYS. REV. B VOL.3,. . . . . 4404, (1971). . . . .
    *TWO-PHONON PRODUCTION VIA SINGLE-PHONON SCATTERING.
    *LEATH(P.L.), WATSON(B.P.).

38 PHYS. REV. B VOL.4,. . . . . 3971, (1971). . . . .
    *SELF-CONSISTENT TREATMENT OF THE FREQUENCY SPECTRUM OF A MODEL
    PARAELECTRIC.
    *GILLIS(N.S.), KOEHLER(T.R.).

39 PHYS. REV. B VOL.3,. . . . . 3556, (1971). . . . .
    *COMMENTS ON TWO-PHONON RESONANCES IN QUANTUM CRYSTALS.
    *RUVALDS(J.).

40 PHYS. REV. LETT. VOL.28,. . . . . 1261, (1972). . . . .
    *THEORY OF THE VIBRATIONS OF DILUTE QUANTUM CRYSTAL ALLOYS.
    *NELSON(R.D.), HARTMANN(W.M.).

41 PHYS. REV. A VOL.5,. . . . . 1528, (1972). . . . .
    *PHASE TRANSITIONS AND SOFT MODES.
    *SCHNEIDER(T.), SRINIVASAN(G.), ENZ(C.P.).

42 PHYS. REV. B VOL.5,. . . . . 3758, (1972). . . . .
    *THEORY OF PEROVSKITE FERROELECTRICS.
    *PYTTE(E.).

43 PHYS. REV. B VOL.5,. . . . . 285, (1972). . . . .
    *TWO-PHONON RESPONSE OF ANHARMONIC LATTICES IN THE FIRST ORDER
    SELF-CONSISTENT PHONON APPROXIMATION.
    *WERTHAMER(N.R.).

44 PHYS. REV. B VOL.5,. . . . . 1236, (1972). . . . . .NA,. . . . .
    *ANHARMONIC LATTICE DYNAMICS IN NA.
    *GLYDE(H.R.), TAYLOR(R.).

45 PHYS. STAT. SOL. VOL.36,. . . . . 335, (1969). . . . . .K*BR, K*I,. . . . .
    *EFFECT OF THREE-BODY FORCES ON THE SHELL MODEL OF ALKALI HALIDES:
    APPLICATION TO K*BR AND K*I.
    *SINGH(R.K.), VERMA(M.P.).

46 PHYS. STAT. SOL. VOL.40,. . . . . K73, (1970). . . . . .AR,. . . . .
    *HARD-CORE POTENTIAL FOR SOLID ARGON.
    *AGARWAL(T.N.), GUPTA(R.K.).

47 PHYS. STAT. SOL. VOL.38,. . . . . K19, (1970). . . . . .CS*CL,. . . . .
    *CONTRIBUTION OF THE THREE-BODY OVERLAP FORCES TO THE DYNAMICAL MATRIX
    OF CESIUM CHLORIDE.
    *LAL(H.H.), VERMA(M.P.).

48 PHYS. STAT. SOL. VOL.45,. . . . . 537, (1971). . . . . .PB,. . . . .
    *LATTICE DYNAMICS OF LEAD: A PSEUDOPOTENTIAL APPROACH.
    *GUPTA(H.C.), TRIPATHI(B.B.).

49 PHYS. STAT. SOL. VOL.48,. . . . . 711, (1971). . . . .
    *MICROSCOPIC THEORY OF LATTICE DYNAMICS FOR IONIC CRYSTALS II. CONTRIBU-
    TION/LONG-RANGE THREE-CENTRE POTENTIALS/PHONON DISPERSION/ALKALI HALIDES
    *ZEYHER(R.).

50 PHYS. STAT. SOL. VOL.45,. . . . . 235, (1971). . . . . .PB,. . . . .
    *LATTICE DYNAMICS OF LEAD: AN ELASTIC FORCE MODEL APPROACH.
    *GUPTA(H.C.), TRIPATHI(B.B.).

51 PHYS. STAT. SOL. VOL.49,. . . . . 235, (1972). . . . .
    *LATTICE DYNAMICS OF HIGHLY POLARIZABLE HOMOPOLAR CRYSTALS WITH DIAMOND
    STRUCTURE.
    *KRESS(W.).

52 PROC. PHYS. SOC. VOL.90,. . . . . 1127, (1967). . . .T,. . . . .
    *ZERO SOUND, FIRST SOUND AND SECOND SOUND OF SOLIDS. . . . .
    *COWLEY(R.A.).

SECTION 100-CRYSTAL DYNAMICS IN THE ANHARMONIC THEORY. . . . .

53 PROC. ROY. SOC. 287,   259, (1965) ALKALI HB2 . . . .
         *ANHARMONIC INTERACTIONS IN ALKALI HALIDES.
         *COWLEY(E.R.), COWLEY(R.A.).

54 REPORT U.K.A.E.A. R7350 (62 PP.)   (1973) . . . .
         *THE RIETVELD COMPUTER PROGRAM FOR THE PROFILE ANALYSIS OF NEUTRON
         DIFFRACTION POWDER PATTERNS, MODIFIED FOR ANISOTROPIC THERMAL VIBRATION.
         *HEWAT(A.W.).

55 REV. MEXICANA FIS. VOL.17,   138, (1968)....T. . . . .
         *EXACT RESULTS FOR NEUTRON DISPERSION IN A SYSTEM OF ONE-DIMENSIONAL
         ANHARMONIC OSCILLATORS.
         *BRAUN(E.).

56 REV. MOD. PHYS. VOL.30,   197, (1958) . . . .
         *INTERACTION OF WAVES IN CRYSTALS.
         *SLATER(J.C.).

57 SOL. STATE COMM. VOL.8,   1415, (1970) SPECTRA OF S3*TI*O3, K*TA*O3. . . . . . .
         *COUPLED MODES IN THE PHONON SPECTRA OF SR*TI*O3 AND K*TA*O3.
         *MANTE(A.J.H.).

58 SOL. STATE COMM. VOL. 8,   923, (1970) . . . . T. . . . . . HE (A=3). . . . .
         *CUBIC ANHARMONIC CORRECTION TO SELF-CONSISTENT PHONONS IN B.C.C. 3HE.
         *GLYDE(H.R.), COWLEY(R.A.).

59 SOL. STATE COMM. VOL. 9,   141, (1971) ANHARMONIC NE. . . . .
         *CONVENTIONAL PERTURBATIONAL ANHARMONIC EFFECTS ON PHONONS IN SOLID NEON.
         *BOHLIN(L.).

60 SOL. STATE COMM. VOL. 9,   129, (1971) . . . .
         *RESONANCES OF TWO PHONONS FROM DIFFERENT DISPERSION BRANCHES.
         *RUVALDS(J.), ZAWADOWSKI(A.).

61 SOV. PHYS. JETP VOL.6,   96, (1958)....T. . . . .
         *MULTIPHONON PROCESSES IN CRYSTALS.
         *TURCHIN(V.F.).

62 SOV. PHYS. JETP 21,   646, (1965)....T. . . . .
         *EFFECT OF ANHARMONICITY ON THE PHONON SPECTRUM NEAR THE DEGENERACY
         POINT.
         *KAGAN(YU.), ZHERNOV(A.P.).

63 SOV. PHYS. SOL. STATE VOL.11,   69, (1969). . . . .
         *INFLUENCE OF ANHARMONICITY ON THE SPECTRAL DISTRIBUTION OF QUASILOCAL
         MODES.
         *KRIVOGLAZ(M.A.), PINKEVICH(I.P.).

64 SOV. PHYS. J.E.T.P. VOL.30,   170, (1970). . . . .
         *VARIATION OF THE NATURE OF THE ENERGY SPECTRUM THRESHOLD CHARACTERISTICS
         UNDER PRESSURE.
         *RECHESTER(A.B.).

65 SOV. PHYS. J.E.T.P. VOL.30,   147, (1970). . . . .
         *LOCALIZED LONG-LIVED VIBRATIONAL STATES IN MOLECULAR CRYSTALS. . . . . . .
         *OVCHINNIKOV(A.A.).

66 SOV. PHYS. SOL. STATE VOL.13,   112, (1971). . . . .
         *LATTICE VIBRATIONS WITH A RANDOM DISTRIBUTION OF INTERSTITIAL IMPURITIES
         *MURTAZIN(I.A.).

SECTION   10E-NEUTRON-PHONON INTERACTION IN THE ANHARMONIC THEORY. . . . .

1 ARK. FYS. VOL.13, PAPER 18,. . . . 215, (1958). IN T NEUTRON K*CL
   ↪ON TWO-PHONON PROCESSES IN NEUTRON DIFFRACTION AGAINST CRYSTALS.
   ↪SJOLANDER(A.).

2 ARK. FYS. VOL.34, PAPER 12,. . . . 121, (1967). . . . T . . .
   ↪FUNCTIONAL EXPANSION AND GREEN FUNCTION TECHNIQUES IN SCATTERING OF
   NEUTRONS BY AN ANHARMONIC CRYSTAL.
   ↪HOGBERG(T.).

3 BROOKHAVEN SYMPOSIUM,. . . . . . 23, (1965). . . . .
   ↪IMPURITY AND ANHARMONIC EFFECTS IN NEUTRON SCATTERING BY PHONONS.
   ↪ELLIOTT(R.J.).

4 COPENHAGEN CONF.-LATT.DYNAMICS,. . 261, (1963). . . . .
   ↪INELASTIC SCATTERING OF NEUTRONS BY ANHARMONIC CRYSTALS.
   ↪AMBEGAOKAR(V.), CONWAY(J.M.), BAYM(G.).

5 DISS. ABSTR. VOL.20,. . . . . . 3794, (1960). . . . .
   ↪SLOW NEUTRON SCATTERING BY ANISOTROPIC MATERIALS.
   ↪BOEDEKER(R.R.).

6 DISS. ABS. VOL.25,. . . . . . . 4781, (1965). . . . .
   ↪INELASTIC SCATTERING OF NEUTRONS BY ANHARMONIC CRYSTALS.
   ↪CONWAY(J.M.).

7 DISS. ABS. VOL.31,. . . . . . . 6173B, (1971). . . . T . . . . HE(ISOTOPE A=4), . . . NEUTRON
   ↪ELEMENTARY EXCITATIONS, BOSE-EINSTEIN CONDENSATION, AND NEUTRON
   SCATTERING: A STUDY OF LIQUID HELIUM FOUR.
   ↪TENN(J.S.).

8 IAEA SYMPOSIUM VIENNA,. . . . . 101, (1960). . . . .
   ↪THE EFFECT OF CRYSTAL ANHARMONICITIES ON NEUTRON SCATTERING.
   ↪SCALETTAR(R.).

9 IAEA SYMPOSIUM VIENNA,. . . . . . 61, (1960). . . . .
   ↪LINE WIDTHS IN NEUTRON-PHONON AND NEUTRON-MAGNON SCATTERING.
   ↪ELLIOTT(R.J.), STERN(H.).

10 IAEA SYMP. CHALK RIVER, VOL.1,. . . 37, (1962). . . . .
   ↪PERTURBATION THEORY OF ANHARMONICITY EFFECTS IN SLOW NEUTRON INELASTIC
   SCATTERING BY CRYSTALS.
   ↪HAHN(H.).

11 IAEA SYMP. CHALK RIVER, VOL.1,. . . 15, (1962). . . . .
   ↪THEORY OF THE INFLUENCE OF PHONON-PHONON AND ELECTRON-PHONON
   INTERACTIONS ON THE SCATTERING OF NEUTRONS BY CRYSTALS.
   ↪KOKKEDEE(J.J.).

12 IAEA SYMP. BOMBAY, VOL.1,. . . . 297, (1964). . . . B1,T . . . . . SR*TI*O3,. . . . . .
   ↪ANHARMONIC EFFECTS AND THE SCATTERING OF NEUTRONS FROM A CRYSTAL.
   ↪COWLEY(R.A.).

13 IAEA SYMPOSIUM BOMBAY, VOL.1,. . 261, (1964). . . . .
   ↪A STUDY OF INELASTIC NEUTRON SCATTERING BY ANHARMONIC CRYSTALS WITH THE
   DAMPING THEORY.
   ↪AKCASU(A.Z.), OSBORN(R.K.).

14 IAEA SYMPOSIUM BOMBAY, VOL.1,. . 225, (1964). . . . .
   ↪INFLUENCE OF ANHARMONICITY AND IMPURITIES ON NEUTRON SCATTERING BY
   CRYSTALS AND ON THE MOSSBAUER EFFECT.
   ↪WALLER(I.).

15 J. CHEM. PHYS. VOL.56,. . . . . 2928, (1972). . . . B2 . . . . C6*H4*CL2 (PARA-DICHLOROBENZENE)
   ↪HIGH-RESOLUTION INELASTIC NEUTRON SCATTERING FROM SOME AROMATIC MOLECUL.
   POLYCRYSTALS FOR STUDY OF CRYSTAL EXCITATIONS AND ANHARMONIC EFFECTS.
   ↪REYNOLDS(P.A.), KJEMS(J.K.), WHITE(J.W.).

16 J. PHYS. C, SER.2, VOL.2,. . . . 2262, (1969). . . . .
   ↪ASYMMETRIES IN THE SCATTERING OF X-RAYS AND NEUTRONS BY CRYSTALS.
   ↪COWLEY(R.A.), BUYERS(W.J.L.).

17 J. PHYS. SOC. JAPAN VOL.36,. . . 406, (1974). . . . T . . . .
   ↪THEORY OF NEUTRON SCATTERING FROM LATTICE VIBRATIONS. II. APPLICATION OF
   A PHONON THEORY.
   ↪TANI(K.).

18 KOLL. Z. Z. POLYM. BAND 250,. . . 993, (1972). . . . T,E . . . . . C*H4,C10*H16, ...
   ↪UNTERSUCHUNG VON MOLEKULBEWEGUNGEN IN KRISTALLEN MITTELS NEUTRONEN-
   SPEKTROSKOPIE.
   ↪SPRINGER(T.).

19 LATV. PSR ZIN. AKAD. VESTIS NO.5   67, (1962). . . . T . . . .
   ↪THE INFLUENCE OF ANHARMONICITY UPON ENERGETIC DISTRIBUTION OF INELASTIC
   SCATTERING OF NEUTRONS. II. FOUR-PHONON INTERACTIONS.
   ↪KASHCHEEV(V.N.). (IN RUSSIAN).

20 NUOVO CIMENTO VOL.38,. . . . . . 175, (1965). . . . T . . . .
   ↪DAMPING THEORY AND ITS APPLICATION TO NEUTRON SCATTERING BY ANHARMONIC
   CRYSTALS.
   ↪AKCASU(A.Z.), OSBORN(R.K.).

21 PHYSICA VOL.28,. . . . . 374, (1962). . . . T . . . .
   ↪ANHARMONIC EFFECTS IN THE COHERENT SCATTERING OF NEUTRONS BY CRYSTALS.
   A FORMAL TREATMENT OF SHIFT AND WIDTH OF THE PEAKS IN THE SCAT. SPECTRUM
   ↪KOKKEDEE(J.J.).

22 PHYSICA VOL.28,. . . . . 893, (1962). . . . T . . . .
   ↪INFLUENCE OF ELECTRON-PHONON INTERACTION ON THE SCATTERING OF NEUTRONS
   BY CONDUCTING CRYSTALS.
   ↪KOKKEDEE(J.J.).

23 PHYS. LETT. 17,. . . . . 228, (1965). . . . .
   ↪INELASTIC SCATTERING OF NEUTRONS BY AN ANHARMONIC CRYSTAL AT LOW
   TEMPERATURES.
   ↪MARIS(H.J.).

24 PHYS. REV. VOL.128,. . . . . . 2589, (1962). . . . T . . . .
   ↪SCATTERING OF NEUTRONS BY AN ANHARMONIC CRYSTAL.
   ↪MARADUDIN(A.A.), FEIN(A.E.).

SECTION 10E-NEUTRON-PHONON INTERACTION IN THE ANHARMONIC THEORY. . . . .

25 PHYS. REV. 131,. . . . . 1420, (1963). . . . .
    *NEUTRON SCATTERING BY AN ANHARMONIC CRYSTAL.
    *THOMPSON(B.V.).

26 PHYS. REV. VOL.135,. . . . . . . .A1071, (1964). . . . .
    *CALCULATION OF THE SCATTERING FUNCTION $S(K,W)$ FOR THE INELASTIC
    SCATTERING OF NEUTRONS BY ANHARMONIC CRYSTALS.
    *MARADUDIN(A.A.), AMBEGAOKAR(V.).

27 PHYS. STAT. SOL. VOL.4,. . . . . . 31, (1964). . . .T. . . . .
    *FORM OF THE ENERGY DISTRIBUTION OF ONE-PHONON COHERENT SCATTERING OF
    MONOCHROMATIC NEUTRONS.
    *KASHCHEEV(V.N.). (IN RUSSIAN).

28 PHYS. STAT. SOLIDI B VOL.56 NO.1 313, (1973). . . .E. . . . .
    *BROAD NEUTRON GROUPS IN INELASTIC NEUTRON SCATTERING AND ANOMALOUS
    ANISOTROPY OF THE ENERGY OF THE DISPERSION BRANCHES.
    *COMES(R.).

29 SOV. PHYS. SOL. STATE VOL.3, . . 1107, (1961). . . .T. . . . .
    *THE INFLUENCE OF ANHARMONICITY ON THE ENERGY DISTRIBUTION IN INELASTIC-
    ALLY SCATTERED NEUTRONS I.THE CASE OF WEAK INTERACTIONS.
    *KASHCHEEV(V.N.), KRIVOGLAZ(M.A.).

30 SOV. PHYS. SOL. STATE VOL.14,. . 2462, (1973). . . .T. . . . .
    *INTERFERENCE IN NEUTRON SCATTERING IN ANHARMONIC CRYSTALS.
    *PLAKIDA(N.M.).

31 UKR. FIZ. ZH.(USSR) VOL.16,. . . 1985, (1971). . . .T. . . . .
    *ON THE THEORY OF SLOW NEUTRON INELASTIC SCATTERING BY IMPURITY CRYSTALS.
    1.COHERENT EFFECTS OF LOCAL VIBRATIONS OF SUBSTITUTION ATOMS.
    *DZYUB(I.P.), KOCHMARSKII(V.Z.).

SECTION   11A-DYNAMICS OF MAGNETIC SYSTEMS: THEORY . . . .

1   ACTA PHYS. POLON. VOL.25,. . . . .   349, (1964)....T.........
    ~ON THE THEORY OF FERROMAGNETICS WITH HELICOIDAL STRUCTURES. IV.
    SPIN-PHONON INTERACTION AND RELAXATION OF MAGNONS.
    ~KASHCHEEV(V.N.). (IN RUSSIAN).

2   ACTA PHYS. POLON. VOL.25,. . . . .   337, (1964)....
    ~ON THE THEORY OF FERROMAGNETICS WITH HELICOIDAL STRUCTURES. III.
    MAGNETO-OSCILLATORY SCATTERING OF NEUTRONS.
    ~KASHCHEEV(V.N.). (IN RUSSIAN).

3   ACTA PHYS. POLON. VOL.26,. . . . .   257, (1964)....T. . . . .
    ~ON THE THEORY OF IMPURE FERROMAGNETICS.
    I. THE CASE OF LOW TEMPERATURES.
    ~KASHCHEEV(V.N.). (IN RUSSIAN).

4   ANN. DE PHYSIQUE TOME 5,. . . . .   139, (1970). . . . P. . . . .
    ~UNE METHODE DE RENORMALISATION DES ONDES DE SPIN.
    ~NAUCIEL-BLOCH(M.).

5   ANN. DE PHYSIQUE TOME 7 NO.4,. .   233, (1972)..............
    ~SPIN WAVES (REVIEW PAPER).
    ~HENNION(B.), MOUSSA(F.).   (PRESENTED   AT   AUTRANS   SUMMER SCHOOL 1972).

6   C.N.E.N. SYMP. CASACCIA,. . . .   235, (1968). . . . .
    ~CONSIDERATIONS   REGARDING THE MAGNETIC COUPLING IN RARE EARTH METALS AND
    ALLOYS.
    ~WOLLAN(E.O.).

7   C.N.E.N. SYMP. CASACCIA,. . . .   101, (1968). . . . .
    ~RECENT DEVELOPMENTS IN SPIN WAVE THEORIES.
    ~MARSHALL(W.).

8   IAEA SYMP. COPENHAGEN, VOL.2,. .   117, (1968)....D1.......FE2*03,. . . . .
    ~EXCHANGE INTEGRALS IN MAGNETITE.
    ~MOGLESTUE(K.T.).

9   IAEA SYMP. GRENOBLE,. . . . . .   529, (1972).................
    ~MAGNETISM IN ONE AND TWO DIMENSIONS (REVIEW PAPER).
    ~BLUME(M.).

10 J. APPL. PHYS. 40,. . . . . . . 2359, (1969).................
    ~MAGNON LONGITUDINAL PHONON INTERACTION IN OBLIQUELY MAGNETIZED YIG RODS.
    ~DE-SANTIS(P.).

11 J. PHYS. -C-, VOL. 1,. . . . . .   102, (1968).................
    ~SPIN   WAVE   THEORY OF IMPURITY STATES IN A HEISENBERG ANTIFERROMAGNET I.
    POSITIVE IMPURITY-HOST EXCHANGE COUPLING.
    ~LOVESEY(S.W.).

12 J. PHYS. -C-, VOL. 1,. . . . . .   118, (1968).................
    ~SPIN   WAVE   THEORY OF IMPURITY STATES IN A HEISENBERG ANTIFERROMAGNET II
    NEGATIVE IMPURITY-HOST EXCHANGE COUPLING.
    ~LOVESEY(S.W.).

13 J. PHYS. -C-, VOL. 1,. . . . . .   408, (1968).................
    ~THE   GROUND-STATE   SUBLATTICE MAGNETIZATION IN AN ANTIFERROMAGNET.
    ~LOVESEY(S.W.).

14 J. PHYS. -C-, VOL. 1,. . . . . . 1650, (1968).................
    ~GENERALIZED   SPIN   SUSCEPTIBILITY   IN   THE CORRELATED NARROW ENERGY-BAND
    MODEL.
    ~HUBBARD(J.), JAIN(K.P.).

15 J. PHYS. -C-, VOL. 1,. . . . . . 1279, (1968)....TD4.....GD
    ~SPIN   WAVE   THEORY   OF   THE   MAGNETOCRYSTALLINE   ANISOTROPY IN GD METAL.
    ~BROOKS(M.S.S.), GOODINGS(D.A.).

16 J. PHYS. -C-, VOL. 2,. . . . . .   84, (1969).................
    ~SPIN   WAVES   AND OTHER MAGNETIC EXCITATIONS IN NON FERROMAGNETIC METALS.
    ~EDWARDS(D.M.).

17 J. PHYS. C VOL.2,. . . . .   539, (1969).................
    ~SPIN-WAVE INTERACTIONS IN A HEISENBERG FERROMAGNET.
    ~MARSHALL(W.), MURRAY(G.).

18 J. PHYS. -C-, VOL. 4,. . . . . . 2139, (1971)....T.......K*CO*F3, ...
    ~EXCITATIONS IN K*CO*F3-II. THEORETICAL.
    ~BUYERS(W.J.L.), HOLDEN(T.M.), SVENSSON(E.C.), COWLEY(R.A.),
    HUTCHINGS(M.T.).

19 J. PHYS. C VOL.6,. . . . .   386, (1973)....T.........
    ~DYNAMICAL   PROPERTIES   OF   HEISENBERG   ANTIFERROMAGNETS: COMPARISON OF
    EQUATION OF MOTION AND DIAGRAMMATIC TECHNIQUES.
    ~NATOLI(C.R.), RANNINGER(J.).

20 J. PHYS. C VOL.6,. . . . .   323, (1973)....T.........
    ~TWO-MAGNON   CORRELATIONS   IN HEISENBERG ANTIFERROMAGNETS: AN EQUATION OF
    MOTION STUDY.
    ~NATOLI(C.R.), RANNINGER(J.).

21 J. PHYS. C VOL.6,. . . . .   L97, (1973)....T,D1........
    ~SPIN   WAVES   IN   THE ONE- OR TWO-DIMENSIONAL CLASSICAL HEISENBERG FERRO-
    MAGNET WITH A HARD MAGNETIZATION AXIS.
    ~VILLAIN(J.).

22 J. PHYS. C VOL.7,. . . . .   979, (1974)....T.......DY,TB,
    ~THEORY   OF   HIGHLY   ANISOTROPIC   FERROMAGNETS: III. RELATIONSHIP BETWEEN
    MICROSCOPIC   ANISOTROPY   AND   MAGNETOSTRICTION   IN   HCP   FERROMAGNETS.
    ~BROOKS(M.S.S.), EGAMI(T.).

23 J. PHYS. C VOL.7,. . . . .   947, (1974)....T.........
    ~HIGH-FREQUENCY MAGNON DAMPING IN HEISENBERG ANTIFERROMAGNETS.
    ~BOHNEN(K.P.), NATOLI(C.R.), RANNINGER(J.).

24 J. PHYS. F VOL.4,. . . . .   466, (1974)....T.......GD,TB,DY
    ~A   GROUP   THEORETICAL   STUDY OF SELECTION RULES FOR MAGNON-PHONON INTER-
    ACTIONS IN FERROMAGNETIC HCP METALS.
    ~CRACKNELL(A.P.).

SECTION 11A-DYNAMICS OF MAGNETIC SYSTEMS: THEORY . . . .

25 J. PHYSIQUE TOME 32 COL.1 VOL.II 585, (1971)....T.........
*TWO MAGNON PAIRING EFFECTS.
*ELLIOTT(R.J.), SMITH(A.J.).

26 J. PHYSIQUE TOME 35,   27, (1974).....T.........
*QUANTUM THEORY OF 1- AND 2-D FERRO- AND ANTIFERROMAGNETS WITH AN EASY
MAGNETIZATION PLANE I. IDEAL LATTICES WITHOUT IN-PLANE ANISOTROPY.
*VILLAIN(J.).

27 J. PHYS. SOC. JAPAN VOL.28, 327, (1970)....D1
*DYNAMICAL SPIN SUSCEPTIBILITY AND SPIN WAVES IN FERROMAGNETIC DILUTE
ALLOYS.
*YAMADA(H.), SHIMIZU(M.).

28 NUCL./S.S. PHYS. SYMP. ABSTR.   , (1972)....D3
*SPIN SUSCEPTIBILITY OF NEARLY FERROMAGNETIC FERMI SYSTMES.
*RAMAKRISHNAN(T.V.).

29 PHYSICA VOL.43,,   353, (1969)....D1,T.....AL2*03-TYPE STRUCTURE,
*SPIN WAVES IN ANTIFERROMAGNETS WITH CORUNDUM STRUCTURE.
*SAMUELSEN(E.J.).

30 PHYS. LETT. 24A,   117, (1967)........
*FLUCTUATION DRIVEN SPIN WAVES.
*BROUT(R.).

31 PHYS. LETT. 27A,   215, (1968)........
*SPIN WAVES IN FERROMAGNETIC METALS.
*CALLAWAY(J.).

32 PHYS. LETT. VOL.30A,   84, (1969)....T.........
*BOUND STATES IN HEISENBERG ANTIFERROMAGNETS.
*LOVESEY(S.W.), BALCAR(E.).

33 PHYS. LETT. VOL.31A,   508, (1970)........
*ON THE BEHAVIOUR OF THE MAGNON ENERGY AT THE CURIE TEMPERATURE.
*JOHANSSON(B.).

34 PHYS. MET. METALLOGR. VOL.30, 5, 1064, (1970)....T.........
*THEORY OF SPIN WAVES IN FERROMAGNETS WITH A PERIODIC DOMAIN STRUCTURE.
*TUROV(E.A.), FARZTDINOV(M.M.).

35 PHYS. MET. METALLOGR. VOL.33 N.1   22, (1972)....T.........
*THEORY OF SPIN WAVES IN METALS.
*DUNIN(S.Z.).

36 PHYS. REV. VOL.109,   782, (1958)....T.........
*APPROXIMATE THEORY OF FERRIMAGNETIC SPIN WAVES.
*KAPLAN(T.A.).

37 PHYS. REV. VOL.173,   603, (1968)....T.........
*ESTIMATE OF DAMPING OF SHORT-WAVE MAGNONS FOR A HEISENBERG FERROMAGNET
BELOW T CRITICAL.
*LOLY(P.D.), DONIACH(S.).

38 PHYS. REV. 173,,   617, (1968)........
*POLAR SPIN WAVES IN FERROMAGNETIC METALS.
*SOKOLOFF(J.B.).

39 PHYS. REV. VOL.188,   898, (1969)........
*HYDRODYNAMIC THEORY OF SPIN WAVES.
*HALPERIN(B.I.), HOHENBERG(P.C.).

40 PHYS. REV. VOL.188,   821, (1969)........FE*F2, MN*F2,
*TEMPERATURE-DEPENDENT MAGNON-ENERGY THEORY OF FE*F2 AND MN*F2.
*NAGAI(O.), TANAKA(T.).

41 PHYS. REV. 177,,   932, (1969)........
*MAGNETIC SYMMETRY AND ANTIFERROMAGNETIC RESONANCE IN CO*O.
*DANIEL(M.R.), CRACKNELL(A.P.).

42 PHYS. REV. LETT. VOL.25,   934, (1970)........CR2*03,
*TRANSLATIONAL SYMMETRY OF OPTICAL EXCITONS AND MAGNONS IN CR2*03.
*ALLEN(J.W.).

43 PHYS. REV. B VOL.3,   1025, (1971)........
*INDIRECT EXCHANGE COUPLING OF MAGNETIC MOMENTS IN RARE-EARTH METALS.
*ROBINSON(L.B.), FERGUSON(L.N.), MILSTEIN(F.).

44 PHYS. REV. B VOL.3,   961, (1971)........
*DYNAMICS OF AN ANTIFERROMAGNET AT LOW TEMPERATURES: SPIN-WAVE DAMPING
AND HYDRODYNAMICS.
*HARRIS(A.B.), KUMAR(D.), HALPERIN(B.I.), HOHENBERG(P.C.).

45 PHYS. REV. B VOL.4,   2236, (1971)........CR*BR3,
*INFLUENCE OF TRUE CRYSTALLOGRAPHIC STRUCTURE ON SPIN WAVES IN
FERROMAGNETIC CR*BR3.
*SIVARDIERE(J.), SILBERGLITT(R.).

46 PHYS. REV. LETT. VOL.28,   1206, (1972)........
*LOCALIZED MAGNETIC EXCITATIONS IN SUBSTITUTIONALLY DISORDERED ANTIFERRO-
MAGNETS.
*ECONOMOU(E.N.).

47 PHYS. REV. LETT. VOL.28,   1192, (1972)........
*MAGNON LOCALIZATION IN ANTIFERROMAGNETS.
*LYO(S.K.).

48 PHYS. REV. LETT. VOL.29,   793, (1972)....T.........
*NEW MAGNETOELASTIC INTERACTION.
*LIU(S.H.).

49 PHYS. REV. B VOL.5,   1941, (1972)........
*INTERACTING MAGNONS IN THE LINEAR CHAIN.
*FOGEDBY(H.C.).

50 PHYS. REV. B VOL.5,   1993, (1972)........MN*F2,
*MAGNON-PHOTON COUPLING IN ANTIFERROMAGNETS.
*MANOHAR(C.), VENKATARAMAN(G.).

51 PHYS. REV. LETT. VOL.28,   614, (1972)....T.........(C*03)4*N*MN*CL3,
*DYNAMIC PROPERTIES OF A ONE-DIMENSIONAL HEISENBERG MAGNET.
*LOVESEY(S.W.), MESERVE(R.A.).

SECTION 11A-DYNAMICS OF MAGNETIC SYSTEMS: THEORY . . . .

52 PHYS. STAT. SOL. VOL.49, . . .   199, (1972)....THE PARAMAGNETIC PHASE OF FERROMAGNETS.
    ~IMPURITY SPIN STATES IN THE PARAMAGNETIC PHASE OF FERROMAGNETS.
    ~TIMMESFELD(K.H.).

53 PHYS. STAT. SOLIDI B VOL.56 NO.1   157, (1973)....T.....MN-PT3,FE-PO3.
    ~THE SPIN-WAVE CONTRIBUTIONS TO THE LOW-TEMPERATURE SPECIFIC HEATS OF
    MN-PT3 AND FE-PO3.
    ~ANDERSON(D.A.), CRACKNELL(A.P.).

54 PROC. INT. CONF. (NOTTINGHAM), .   516, (1964)....X(X).....(1-X)CR2*O3.(X)FE2*O3 WITH HELICAL SPIN
    ~THEORY OF SOLID SOLUTIONS (1-X)CR2*O3.(X)FE2*O3 WITH HELICAL SPIN
    CONFIGURATION.
    ~BERTAUT(E.F.).

55 PROC. ROY. SOC. VOL.295, ...   182, (1966)....D1,T...... . . . .
    ~SPIN WAVES AND THEIR STABILITY IN METALS.
    ~KATSUKI(A.), WOHLFARTH(E.P.).

56 PROC. ROYAL SOC. EDINBURGH A-70,   75, (1972)....T.......MN*F2(FE,NI,CO), ...
    ~THE TEMPERATURE DEPENDENCE OF LOCALISED MAGNONS.
    ~COWLEY(R.A.).

57 REV. MOD. PHYS. VOL.30, ...   1, (1958)....T........ . . . .
    ~SPIN WAVES.
    ~VAN-KRANENDONK(J.), VAN-VLECK(J.H.).

58 REV. MOD. PHYS. VOL.44, ...   406, (1972)....T,E......... . . . .
    ~THE PROPERTIES OF DEFECTS IN MAGNETIC INSULATORS.
    ~COWLEY(R.A.), BUYERS(W.J.L.).

59 SOV. PHYS. SOL. STATE, 4, . . .   556, (1962)........... . . . .
    ~EFFECTS OF SPIN-SPIN AND SPIN-PHONON INTERACTIONS IN ANTIFERROMAGNETS ON
    THE ENERGY.
    ~KASHCHEEV(V.N.).

60 SOV. PHYS. J.E.T.P. 26,. . . . .   764, (1968)............ . . . .
    ~DISPERSION OF SPIN WAVES IN METALS IN A STRONG MAGNETIC FIELD.
    ~BLANK(A.Y.), KONDRATENKO(P.S.).

61 SOV. PHYS. J.E.T.P. 27,. . . . .   307, (1968)............ . . . .
    ~SPIN WAVES AND LOCALIZED SPIN STATES IN CRYSTALS WITH DISLOCATIONS.
    ~PUSHKAROV(K.I.), SAVCHENKO(M.A.), TARASENKO(V.V.).

62 SOV. PHYS. SOL. STATE VOL.11, . . 1566, (1969)............ . . . .
    ~SPIN WAVE SPECTRUM IN ANTIFERROMAGNETS HAVING A SPIRAL MAGNETIC
    STRUCTURE.
    ~BAR≠YAKHTAR(V.G.), STEFANOVSKII(E.P.).

63 SOV. PHYS. JETP LETT. VOL.9, . . .   204, (1969)....D2...... . . . .
    ~SPIN WAVES NEAR THE CURIE POINT.
    ~DRABKIN(G.M.), OKOROKOV(A.I.), ZABIDAROV(E.I.), KASMAN(YA.A.).

64 SOV. PHYS. SOLID STATE VOL.14, . . 1035, (1972)....T......... . . . .
    ~COUPLED ELECTROMAGNETIC-MAGNON WAVES AND SCATTERING OF PARTICLES IN
    ANTIFERROMAGNETIC METALS.
    ~CHUDNOVSKII(E.M.).

65 Z. ANGEW PHYS. VOL.32, ...   104, (1971)....D1........ . . . .
    ~THE SPIN WAVE SPECTRUM OF MAGNETIC THIN FILMS.
    ~HANSEN(P.), KREY(U.).

66 ZH. EKSPER. TEOR. FIZ.(USSR) 58,   918, (1970)....T......... . . . .
    ~THEORY OF SPIN WAVES IN ANTIFERROMAGNETIC SUBSTANCES WITH DOMAIN
    STRUCTURES.
    ~FARZTDINOV(M.M.), KHALFINA(A.A.).

67 Z. PHYSIK VOL.223, ...   277, (1969)....D1......... . . . .
    ~SPIN CORRELATION IN DILUTE MAGNETIC ALLOYS.
    ~MULLER-HARTMANN(E.).

SECTION   11B-DYNAMICS OF MAGNETIC SYSTEMS: NEUTRON-MAGNON INTERACTION . .

1   C.N.E.N. SYMP. CASACCIA, ...   185, (1968)...........•••• . . . .
        ⌐SCATTERING OF NEUTRONS BY SPIN WAVES.
        ⌐SHIRANE(G.).

2   CONF./PHONON SCATT. IN SOLIDS, . 314, (1972).....T.........
        ⌐THE EFFECT OF SPIN-PHONON INTERACTION OF NEUTRON SCATTERING FROM LATTICE
          VIBRATIONS.
        ⌐TANI(K.); (ED; NOTE: INTERNATIONAL CONF. ON PHONON SCATTERING IN SOLIDS
          AT PARIS; JULY; 1972. PUBL. DOCUMENTATION SERVICE OF CEN: SACLAY, FRA.).

3   CONTEMPORARY PHYS. VOL.7,. . . .   401, (1966)............. . . . .
        ⌐NEUTRON SPECTROSCOPY OF SOLIDS. II.
        ⌐LOMER(W.M.).

4   DISS. ABS. VOL.32, ...1786B, (1971)...........•••• . . . .
        ⌐SPIN DYNAMICS AND NEUTRON SCATTERING.
        ⌐YOON-IL-CHANG.

5   IAEA SYMPOSIUM VIENNA, ...   61, (1960).............. . . . .
        ⌐LINE WIDTHS IN NEUTRON-PHONON AND NEUTRON-MAGNON SCATTERING.
        ⌐ELLIOTT(R.J.), STERN(H.).

6   IAEA SYMPOSIUM BOMBAY, VOL.1,. .   399, (1964)............. . . . .
        ⌐DYNAMICS OF MAGNETIC SYSTEMS INVESTIGATED BY MOSSBAUER AND NEUTRON
          SCATTERING EXPERIMENTS.
        ⌐MARSHALL(W.).

7   J. APPL. PHYS. VOL.31, SUPPL., .   108, (1960)....T.........
        ⌐SPIN WAVES IN COMPLEX EXCHANGE-COUPLED LATTICES AND NEUTRON SCATTERING.
        ⌐SAENZ(A.W.).

8   J. APPL. PHYS. 36,. . . . .   884, (1965)........... . . . .
        ⌐SPIN WAVES AND NEUTRON SCATTERING.
        ⌐LOWDE(R.D.).

9   J. PHYS. CHEM. SOL. 28,. . . . . . 1357, (1967)........... . . . .
        ⌐MAGNON DISPERSION RELATIONS AND NEUTRON SCATTERING CROSS SECTIONS WITH
          SPECIAL ATTENTION TO ANISOTROPY EFFECTS.
        ⌐LINDGARD(P.A.), KOWALSKA(A.), LAUT(P.).

10 J. PHYS. -C-, VOL. 1,. . . . . .   673, (1968)...........
        ⌐TWO-MAGNON SCATTERING OF NEUTRONS INVOLVING ORBITAL TRANSITIONS IN
          ANTIFERROMAGNETS.
        ⌐BAKRE(R.V.), SINHA(K.P.).

11 J. PHYS. C VOL.6,. . . . . 2350, (1973)....T.........
        ⌐SELECTION RULES FOR INELASTIC NEUTRON SCATTERING BY MAGNETIC CRYSTALS:
          II.ONE-MAGNON INELASTIC SCATTERING,...FERROMAGNETIC FCC, BCC, HCP METALS
        ⌐CRACKNELL(A.P.), SEDAGHAT(A.K.).

12 J. PHYS. C VOL.6 NO.6,. . . . . . 1054, (1973).........FE*F2,
        ⌐SELECTION RULES FOR INELASTIC NEUTRON SCATTERING BY MAGNETIC CRYSTALS:
          I./SYMMETRIES OF MAGNETOELASTIC WAVES/SELEC. RULES/MAGNON-PHON. INTER...
        ⌐CRACKNELL(A.P.).

13 J. PHYS. SOC. JAPAN VOL.25,. . . 1001, (1968).....D1,T.....
        ⌐THEORY OF SPIN WAVES IN FERROMAGNETIC METALS WITH MULTIPLE BANDS. II.
          INELASTIC SCATTERING OF NEUTRONS BY SPIN WAVES.
        ⌐YAMADA(H.), SHIMIZU(M.).

14 LECT/ELEM. EXCIT./INTER. IN SOL. .   . (1968)....T.........
        ⌐DETERMINATION OF PHONON AND MAGNON DISPERSION CURVES BY NEUTRON SPECTRO-
          SCOPY.
        ⌐COLLINS(M.F.). (ED. NOTE: PUBL. NATO ADVANCED STUDY INSTITUTE, CORTINA
          D≠AMPEYYO 1968, 43PP.).

15 METAL. SOC. CONF. LOS ANGELES, .   143, (1967)........... . . . .
        ⌐SPIN-WAVE INELASTIC SCATTERING ON MAGNETIC METALS.
        ⌐NATHANS(R.).

16 2ND SUMMER SCHOOL S.S. PHYS.,. .   211, (1968)....T......... . . . .
        ⌐SOME METHODS OF INVESTIGATION OF MAGNONS AND THEIR RESULTS.
        ⌐WANIC(A.).

17 NUCL. INST. MET. 87,. . . . . . .   125, (1970)........... . . . .
        ⌐CORRECTIONS TO MAGNON DISPERSION RELATIONS AS DETERMINED BY THE NEUTRON
          DIFFRACTION TECHNIQUE.
        ⌐ANTONINI(B.), MEDINA(R.), MENZINGER(F.).

18 NUKLEONIKA VOL.11,. . . . .   839, (1964)....T......... . . . .
        ⌐SOME REMARKS ON THE INVESTIGATION OF NEUTRON MAGNON SCATTERING.
        ⌐WANIC(A.).

19 PHYS. LETT. 6, ...   47, (1963)............. . . . .
        ⌐SPIN WAVE SCATTERING OF POLARIZED NEUTRONS.
        ⌐SAMUELSEN(E.J.), RISTE(T.), STEINSVOLL(O.).

20 PHYS. LETT. 31A,. . . . .   53, (1970)........... . . . .
        ⌐A COMMENT ON THE INTENSITY OF THE SINGLE SPIN-WAVE NEUTRON
          CROSS-SECTION.
        ⌐LOVESEY(S.W.).

21 PHYS. REV. VOL.95,. . . . . 1374, (1954)....T......... . . . .
        ⌐TIME-DEPENDENT CORRELATIONS BETWEEN SPINS AND NEUTRON SCATTERING IN
          FERROMAGNETIC CRYSTALS.
        ⌐VAN-HOVE(L.).

22 PHYS. REV. VOL.125,. . . . . . . 1940, (1962).....D1.........
        ⌐SPIN WAVES IN EXCHANGE-COUPLED COMPLEX MAGNETIC STRUCTURES AND NEUTRON
          SCATTERING.
        ⌐SAENZ(A.W.).

23 PHYS. REV. LETT.19,. . . . . . .   635, (1967)............... . . . .
        ⌐INELASTIC NEUTRON SCATTERING IN FERROMAGNETIC METALS.
        ⌐THOMPSON(E.D.).

24 PHYS. REV. LETT. VOL.23,. . . . .   86, (1969)....D1,T.....CO*F2,
        ⌐TWO-MAGNON SCATTERING OF NEUTRONS.
        ⌐COWLEY(R.A.), SVENSSON(E.C.), BUYERS(W.J.L.), MARTEL(P.),
          STEVENSON(R.W.H.).

SECTION 11B-DYNAMICS OF MAGNETIC SYSTEMS: NEUTRON-MAGNON INTERACTION . .

25 PHYS. REV. 180,. . . . . 613, (1969). . . . .
       *MULTIBAND THEORY OF INELASTIC NEUTRON SCATTERING BY FERROMAGNETIC METALS
       AT LOW TEMPERATURES.
       *SOKOLOFF(J.B.).

26 PHYS. REV. B VOL.9,. . . . . . . 3786, (1974)....T........MN*F2,
       *SPIN-WAVE WIDTHS IN ANTIFERROMAGNETS AT INTERMEDIATE TEMPERATURES.
       *STINCHCOMBE(R.B.), REINECKE(T.L.).

27 PROC. PHYS. SOC. 87,. . . . . . . 521, (1966). . . . .
       *THEORY OF INELASTIC SCATTERING OF NEUTRONS IN FERROMAGNETIC CRYSTALS
       CONTAINING IMPURITIES.
       *IZYUMOV(YU.A.).

28 PROC. PHYS. SOC. 91,. . . . . . . 86, (1967). . . . .
       *INELASTIC NEUTRON SCATTERING IN NEARLY FERROMAGNETIC METALS.
       *DONIACH(S.).

29 PROC. PHYS. SOC. 91,. . . . . . . 97, (1967) INVOLVING ORBITAL TRANSITIONS.
       *NEUTRON MAGNON INTERACTION INVOLVING ORBITAL TRANSITIONS.
       *JOSHI(A.W.), SINHA(K.P.).

30 PROC. PHYS. SOC. 91,. . . . . . . 658, (1967). . . . .
       *INELASTIC NEUTRON CROSS SECTION FOR A MAGNETIC IMPURITY IN A HEISENBERG
       FERROMAGNET.
       *LOVESEY(S.W.).

31 PROC. PHYS. SOC. 91,. . . . . . . 903, (1967). . . . .
       *GROUP THEORETICAL SELECTION RULES IN INELASTIC NEUTRON SCATTERING.
       *ELLIOTT(R.J.), THORPE(M.F.).

32 PROC. PHYS. SOC. 92,. . . . . . . 845, (1967). . . . .
       *DIFFUSE INELASTIC NEUTRON CROSS-SECTIONS FOR IMPURITIES IN HEISENBERG
       FERROMAGNETS.
       *LOVESEY(S.W.).

33 PROC. ROY. SOC. A VOL.230,. . . . 46, (1955)....T........ . . . .
       *THE INELASTIC SCATTERING OF NEUTRONS BY MAGNETIC SPIN WAVES.
       *ELLIOTT(R.J.), LOWDE(R.D.).

34 REPORTS ON PROGRESS IN PHYS. 31,. 705, (1968)......04. . . . .
       *MAGNETIC CORRELATIONS AND NEUTRON SCATTERING.
       *MARSHALL(W.), LOWDE(R.D.). (ED. NOTE: PUBL. THE INSTITUTE OF PHYSICS AND
       THE PHYSICAL SOCIETY, LONDON, ENG., PEDERSON(C.I.)-ED.).

35 SOL. STATE COMM. VOL.9,. . . . . . 417, (1971). . . . .
       *SOME CONSIDERATIONS ON THE MAGNON DISPERSION RELATIONS AS OBTAINED BY
       DIFFERENT NEUTRON SCATTERING TECHNIQUES.
       *ANTONINI(B.), MENZINGER(F.).

36 SOV. PHYS. JETP VOL.7,. . . . . . 1048, (1958)....T. . . . .
       *MULTI-MAGNON PROCESSES IN THE SCATTERING OF SLOW NEUTRONS IN FERRO-
       MAGNETS.
       *MALEEV(S.V.).

37 SOV. PHYS. SOL. STATE, 4,. . . . . 1494, (1962). . . . .
       *THEORY OF INELASTIC MAGNETIC SCATTERING OF SLOW NEUTRONS IN DOPED
       FERROMAGNETS.
       *KASHCHEEV(V.N.).

38 SOV. PHYS. SOL. STATE, 5,. . . . . 668, (1963). . . . .
       *EFFECT OF THE SPIN-PHONON INTERACTION ON ONE QUANTUM SCATTERING OF SLOW
       NEUTRONS IN FERRITES.
       *KASHCHEEV(V.N.).

39 SOV. PHYS. SOL. STATE, 5,. . . . . 858, (1963). . . . .
       *INELASTIC SCATTERING OF SLOW NEUTRONS IN SUBSTANCES WITH A SPIRAL
       MAGNETIC STRUCTURE.
       *BAR≠YAKHTAR(V.G.), MALEEV(S.V.).

40 SOV. PHYS. JETP. 21,. . . . . . . 969, (1965). . . . .
       *INELASTIC SMALL ANGLE NEUTRON SCATTERING IN FERROMAGNETS.
       *MALEEV(S.V.).

41 SOV. PHYS. SOL. STATE, 7,. . . . . 2916, (1965). . . . .
       *CHARACTERISTICS OF THE SCATTERING OF NEUTRONS BY FERROMAGNETS IN THE
       REGION OF FERROACOUSTIC RESONANCE.
       *KLOCHIKHIN(A.A.), KORENBLIT(I.YA.).

42 SOV. PHYS. SOL. STATE, 7,. . . . . 891, (1965). . . . .
       *SLOW NEUTRON SCATTERING IN ANTIFERROMAGNETS HAVING WEAK FERROMAGNETISM.
       *BAR≠YAKHTAR(V.G.), KVIRIKADZE(A.G.).

43 SOV. PHYS. JETP 20,. . . . . 1548, (1965). . . . .
       *NEUTRON SCATTERING BY SPIN WAVES IN IRON.
       *DRABKIN(G.M.), ZABIDAROV(E.I.)...ET AL.

44 SOV. PHYS. SOL. STATE, 9,. . . . . 575, (1967). . . . .
       *SLOW NEUTRON SCATTERING BY SPIN WAVES EXCITED BY FERROMAGNETIC
       RESONANCE.
       *MALEEV(S.V.), STAMENKOVICH(S.S.).

45 SOV. PHYS. JETP. 24,. . . . . . . 960, (1967). . . . .
       *INCOHERENT SCATTERING OF NEUTRONS AND THE PROBLEM OF RECONSTRUCTING THE
       MAGNON SPECTRUM.
       *IZYUMOV(YU.A.), MEDVEDEV(M.V.).

46 SOV. PHYS. SOL. STATE VOL.13,. . . 462, (1971). . . . .
       *SURFACE FLUCTUATIONS AND SCATTERING OF SLOW NEUTRONS IN ANTIFERROMAGNETS
       *AKHIEZER(I.A.), SYSHCHENKO(V.G.).

47 SOV. PHYS. SOLID STATE VOL.14,. . . 387, (1972)....T. . . . .
       *SCATTERING OF ELECTRONS AND NEUTRONS BY COUPLED HELICON-MAGNON WAVES.
       *AKHIEZER(I.A.), CHUDNOVSKII(E.M.).

48 SOV. PHYS. JETP VOL.34,. . . . . . 913, (1972)......01. . . . .
       *SPIN WAVES AND SLOW NEUTRON AND LIGHT SCATTERING NEAR THE TRANSITION
       FROM THE FERROMAGNETIC TO ANTIFERROMAGNETIC PHASE.
       *AKHIEZER(I.A.), GINZBURG(A.E.).

49 1ST EUR. CONF./CONDENSED MATTER,. . 45, (1971). . . . .
       *THE EFFECTS OF MAGNON-MAGNON INTERACTION IN THE DYNAMICAL SUSCEPTIBILITY
       FOR NEUTRON AND LIGHT SCATTERING IN HEISENBERG ANTIFERROMAGNETS.
       *NATOLI(C.R.), RUNNINGER(J.). (ED. NOTE: PUBL. EUROPEAN PHYS. SOC. 1971,
       GENEVA, SWITZ.).

SECTION 11B-DYNAMICS OF MAGNETIC SYSTEMS: NEUTRON-MAGNON INTERACTION . .

50 UKAEA AERE REP. R 4299 (7 PP.),. . . (1963) SCATTERING FORM FACTORS ASSOCIATED WITH SPIN
                                      ↦ON MEASURING THE NEUTRON SCATTERING FORM FACTORS ASSOCIATED WITH SPIN
                                       WAVES.
                                      ↦LOW(G.G.), OKAZAKI(A.).

51 UKAEA AERE REP. R4535 (22 PP.),. . . . (1966)....T.........
                                      ↦THE ANGULAR DISTRIBUTION OF NEUTRONS SCATTERED AT SMALL ANGLES BY
                                       FERROMAGNETIC SPIN WAVES.
                                      ↦STRINGFELLOW(M.W.).

52 UKRAYIN FIZ. ZH. (USSR) VOL.13, 1553, (1968)....T........
                                      ↦ON SCATTERING OF ELECTROMAGNETIC WAVES AND SLOW NEUTRONS BY ANOMALOUS
                                       FLUCTUATIONS OF SPIN WAVES IN MAGNETIC SEMICONDUCTORS IN/EXT. ELECT. FD.
                                      ↦NGUYEN-VAN-TRONG.

53 Z. ANGEW. PHYS. VOL.28,. . . . . 325, (1970)....P.........
                                      ↦SPIN-CORRELATION EFFECTS ABOVE THE CURIE-POINT AND THEIR ACTION ON
                                       NEUTRONS.
                                      ↦BINDER(K.), RAUCH(H.).

54 Z. PHYS. VOL.243,. 382, (1971). . D4....
                                      ↦ON THE TWO-MAGNON SCATTERING OF LIGHT AND NEUTRONS.
                                      ↦SOLYOM(J.).

SECTION 11C-DYNAMICS OF MAGNETIC SYSTEMS: PARAMAGNETIC SCATTERING. . . .

1   CAN. J. PHYS. 45,.   2923, (1967).....T........TB.
    ↪TEMPERATURE DEPENDENCE OF PARAMAGNETIC NEUTRON SCATTERING.
    ↪SEARS(V.F.).

2   CAN. J. PHYS. 46,.   1499, (1968)....T,D3.
    ↪CRYSTAL FIELD EFFECTS ON THE INELASTIC NEUTRON SCATTERING FROM HEAVY
     RARE EARTH METALS IN THE PARAMAGNETIC PHASE.
    ↪WOODS(A.D.B.).

3   CAN. J. PHYS. 46,.   799, (1968)..................
    ↪THE TEMPERATURE AND FREQUENCY DEPENDENCE OF THE INELASTIC NEUTRON
     SCATTERING FROM AN ISING MAGNET.
    ↪ALLAN(G.A.T.), BETTS(D.D.).

4   C.N.E.N. SYMP. CASACCIA,   73, (1968)...........
    ↪NEUTRON SCATTERING BY PARAMAGNETS.
    ↪LOWDE(R.D.).

5   DISS. ABSTR. VOL.11,   1080, (1951).....E........MN*F2,MN*O,.
    ↪THE DEPENDENCE OF PARAMAGNETIC NEUTRON SCATTERING ON COUPLING BETWEEN
     MN(++) IONS.
    ↪BENDT(P.J.).

6   DISS. ABSTR. VOL.14,   162, (1954)....E........MN*F2,MN*O,.
    ↪SCATTERING OF SUB-THERMAL NEUTRONS BY PARAMAGNETIC IONS.
    ↪SMITH(R.R.).

7   IAEA SYMP. COPENHAGEN, VOL.2,   93, (1968).
    ↪LINE SHAPE OF THE MAGNETIC SCATTERING FROM ANISOTROPIC PARAMAGNETS.
    ↪LINDGARD(P.A.).

8   IAEA SYMP. COPENHAGEN, VOL.2,   83, (1968)....D1.......R3*MN*F3,.
    ↪SPIN CORRELATIONS IN ONE, TWO AND THREE-DIMENSIONAL HEISENBERG
     PARAMAGNETS.
    ↪WINDSOR(C.G.).

9   J. APPL. PHYS. VOL.39,   533, (1968)....D3.
    ↪TEMPORAL FLUCTUATIONS OF MOMENT IN PARAMAGNETS BY NEUTRON DIFFRACTION.
    ↪COLLINS(M.F.).

10 J. APPL. PHYS. 41,   1365, (1970).
    ↪SHORT-RANGE ORDER EFFECTS IN PARAMAGNETIC NEUTRON SCATTERING.
    ↪TAHIR-KHELI(R.), MALINOSKI(F.).

11 J. APPL. PHYS. 42,   1390, (1971).
    ↪SPIN CORRELATIONS IN THE PARAMAGNETIC PHASE.
    ↪HUBBARD(J.).

12 J. EXPER. THEOR. PHYS. (USSR) 10   5, (1940)....T.
    ↪SCATTERING OF NEUTRONS IN PARAMAGNETIC SUBSTANCES.
    ↪MIGDAL(A.B.).

13 J. PHYS. CHEM. SOLIDS VOL.4,   223, (1958)....T.
    ↪INELASTIC MAGNETIC SCATTERING OF NEUTRONS AT HIGH TEMPERATURES.
    ↪DE-GENNES(P.G.).

14 J. PHYS. CHEM. SOLIDS VOL.17, 117, (1960)..............RARE-EARTHS,.
    ↪PARAMAGNETIC DIFFRACTION OF NEUTRONS BY RARE-EARTH IONS.
    ↪ODIOT(S.), SAINT-JAMES(D.). (IN FRENCH).

15 J. PHYS. -C-, VOL. 1,   1563, (1968).
    ↪NEUTRON SCATTERING FROM PARAMAGNETS.
    ↪KWON(T.H.), GERSCH(H.A.).

16 J. PHYS. -C-, VOL. 1,   1650, (1968).
    ↪GENERALIZED SPIN SUSCEPTIBILITY IN THE CORRELATED NARROW ENERGY-BAND
     MODEL.
    ↪HUBBARD(J.), JAIN(K.P.).

17 J. PHYS. -C-, VOL. 1,   1706, (1968).
    ↪SPIN POLARIZATION IN DILUTE MAGNETIC ALLOYS.
    ↪MANOHAR(C.), KHUBCHANDANI(P.G.).

18 J. PHYS. C VOL.4,   53, (1971).
    ↪SPIN-CORRELATION FUNCTIONS IN THE PARAMAGNETIC PHASE OF A HEISENBERG
     FERROMAGNET.
    ↪HUBBARD(J.).

19 J. PHYS. C VOL.6,   79, (1973)....T.
    ↪DYNAMIC PROPERTIES OF HEISENBERG PARAMAGNETS.
    ↪LOVESEY(S.W.), MESERVE(R.A.).

20 J. PHYS. F VOL.2,   1145, (1972)....D3.
    ↪QUASIELASTIC NEUTRON SCATTERING FROM METALS WITH PARAMAGNETIC IMPURITIES
    ↪LLOYD(P.), OSBORNE(C.F.).

21 J. PHYS. SOC. JAPAN VOL.15,   429, (1960)....T.
    ↪ON THE PARAMAGNETIC INELASTIC SCATTERING OF NEUTRONS DUE TO IONS IN THE
     ANISOTROPIC CRYSTALLINE FIELD.
    ↪YAMADA(Y.).

22 J. PHYS. SOC. JAPAN VOL.16,   241, (1961)....T.
    ↪INELASTIC SCATTERING OF THERMAL NEUTRONS BY PARAMAGNETIC IONS.
    ↪TOSIMA(S.).

23 KYOTO CONF. STAT. MECHANICS,   157, (1968).
    ↪QUASI-COLLECTIVE MODES IN THE PARAMAGNETIC PHASE.
    ↪TOMITA(K.), KAWASAKI(T.).

24 NUCL./S.S. PHYS. SYMP. ABSTR.,   (1972)....D3.
    ↪SPIN SUSCEPTIBILITY OF NEARLY FERROMAGNETIC FERMI SYSTEMS.
    ↪RAMAKRISHNAN(T.V.).

25 PHYS. LETT. VOL.33A,   375, (1970).
    ↪EVIDENCE FOR YB+++ 4F RADIAL WAVEFUNCTION EXPANSION IN SCHEELITES.
    ↪BROWN(E.A.), NEMARICH(J.), KARAYIANIS(N.), MORRISON(C.A.).

26 PHYS. LETT. VOL.31A,   267, (1970).
    ↪CORRECTIONS TO THE MORI-KAWASAKI APPROXIMATION FOR THE HIGH TEMPERATURE
     LIMIT OF THE SPIN DIFFUSION CONSTANT FOR A MAGNETIC LATTICE.
    ↪HUBER(D.L.).

SECTION 11C-DYNAMICS OF MAGNETIC SYSTEMS: PARAMAGNETIC SCATTERING. . . .

27 PHYS. REV. VOL.55, 924, (1939)....T..........
ON 'THE' THEORY OF THE FORWARD SCATTERING OF NEUTRONS BY PARAMAGNETIC
MEDIA.
VAN-VLECK(J.H.).

28 PHYS. REV. VOL.60, 280, (1941)....E..........MN*S*O4,MN*F2, .
ON THE FORWARD SCATTERING OF NEUTRONS BY PARAMAGNETIC MEDIA.
WHITAKER(M.D.), BRIGHT(M.C.).

29 PHYS. REV. VOL.76, 1572, (1949)....E..........MN*F2,MN*O,FE2*O3,
THE SCATTERING OF SLOW NEUTRONS BY PARAMAGNETIC CRYSTALS.
RUDERMAN(I.W.).

30 PHYS. REV. VOL.83, 333, (1951)..............MN,
NEUTRON DIFFRACTION BY PARAMAGNETIC AND ANTIFERROMAGNETIC SUBSTANCES.
SHULL(C.G.), STRAUSER(W.A.), WOLLAN(E.O.).

31 PHYS. REV. VOL.98, 1721, (1955)..........O3...ZN*FE2*O4,
MULTIPLE SCATTERING OF SLOW NEUTRONS BY FLAT SPECIMENS AND MAGNETIC
SCATTERING BY ZINC FERRITE.
BROCKHOUSE(B.N.), CORLISS(L.M.), HASTINGS(J.M.).

32 PHYS. REV. VOL.99, 601, (1955)....E.........MN2*O3,MN*S*O4,
ENERGY DISTRIBUTION OF NEUTRONS SCATTERED BY PARAMAGNETIC SUBSTANCES.
BROCKHOUSE(B.N.).

33 PHYS. REV. VOL.98, 492, (1955)....D3....O2.
MAGNETIC SCATTERING OF SLOW NEUTRONS BY GASEOUS OXYGEN.
PALEVSKY(H.), EISBERG(R.M.).

34 PHYS. REV. VOL.97, 411, (1955)....T.D3....O2.
MAGNETIC SCATTERING OF SLOW NEUTRONS FROM O2 GAS.
KLEINER(W.H.).

35 PHYS. REV. VOL.110 37, (1958)....P....HO2*O3,
PARAMAGNETIC AND NUCLEAR SCATTERING CROSS SECTIONS OF HOLMIUM
SESQUIOXIDE.
KOEHLER(W.C.), WOLLAN(E.O.), WILKINSON(M.K.).

36 PHYS. REV. 147,. 457, (1966)
APPLICATION OF INVERTED-NEUTRON-FILTER TECHNIQUE FOR STUDY OF INELASTIC
SCATTERING BY PARAMAGNETIC SUBSTANCES.
FRIEDMAN(E.A), GOLAND(A.N.).

37 PHYS. REV. 167,. 458, (1968)
PARAMAGNETIC SCATTERING OF NEUTRONS FROM A HEISENBERG SYSTEM.
KWON(T.H.), GERSCH(H.A.).

38 PHYS. REV. 176,. 650, (1968)
PHENOMENOLOGY OF NEUTRON SCATTERING IN HEISENBERG SYSTEMS.
BENNETT(H.S.).

39 PHYS. REV. B-2,. 4552, (1970)
HIGH-TEMPERATURE WAVELENGTH-DEPENDENT PROPERTIES OF A HEISENBERG
PARAMAGNET.
COLLINS(M.F.).

40 PHYS. REV. B-1,. 3815, (1970)
SPIN CORRELATION FUNCTIONS AT HIGH TEMPERATURES.
BLUME(M.), HUBBARD(J.).

41 PHYS. REV. B-1,. 3178, (1970)....R8*MN*F3,
SHORT-RANGE-ORDER EFFECTS IN NEUTRON SCATTERING FROM HEISENBERG
PARAMAGNETS: APPLICATIONS TO RB*MN*F3.
TAHIR-KHELI(R.A.), MCFADDEN(D.G.).

42 PHYS. REV. B VOL.4, 3964, (1971)
SPIN-SPIN CORRELATIONS OF THE CLASSICAL HEISENBERG FERROMAGNET ON THE
FCC LATTICE, FOR TEMPERATURES ABOVE T(C).
FERER(M.).

43 PHYS. REV. B VOL.4, 1588, (1971)
SERIES EXPANSIONS FOR HIGH-TEMPERATURE DYNAMICS OF HEISENBERG
PARAMAGNETS.
COLLINS(M.F.).

44 PHYS. REV. B VOL.4, 3048, (1971)....T....NI,
NEUTRON ORBITAL CROSS SECTION FOR A TIGHT-BINDING MODEL OF PARAMAGNETIC
NICKEL.
LOVESEY(S.W.), WINDSOR(C.G.).

45 PHYS. STAT. SOL. VOL.50, 737, (1972)
THE MODIFIED CONVOLUTION APPROXIMATION MODEL FOR NEUTRON SCATTERING FROM
SHORT RANGE ORDERED PARAMAGNETS.
MADHAV-RAO(L.).

46 PROC. PHYS. SOC. 90, 1015, (1967)
PARAMAGNETIC SCATTERING OF NEUTRONS FROM AN ISING SYSTEM.
COLLINS(M.F.), WINDSOR(C.G.).

47 PROC. PHYS. SOC. 91, 86, (1967)
INELASTIC NEUTRON SCATTERING IN NEARLY FERROMAGNETIC METALS.
DONIACH(S.).

48 PROC. PHYS. SOC. 92, 390, (1967)
NEUTRON SCATTERING FROM PARAMAGNETS.
COLLINS(M.F.), MARSHALL(W.).

49 PROC. PHYS. SOC. 92, 837, (1967)
A NOTE ON THE PARAMAGNETIC SCATTERING OF NEUTRONS FROM AN ISING SYSTEM.
SEARS(V.F.).

50 SOL. STATE COMM. VOL.8, 1793, (1970)
DESCRIPTION OF NEARLY ANTIFERROMAGNETIC SYSTEMS.
JEROME(D.).

51 SOV. PHYS. JETP VOL.12, 714, (1961)
INELASTIC MAGNETIC SCATTERING OF SLOW NEUTRONS BY PHONONS.
KHABIBULLIN(B.M.).

52 SOV. PHYS. SOL. STATE, 4,. 428, (1962)
EFFECT OF SPIN-PHONON INTERACTION ON THE ENERGY DISTRIBUTION OF
SCATTERED NEUTRONS IN A PARAMAGNET.
KHABIBULLIN(B.M.).

SECTION 11C-DYNAMICS OF MAGNETIC SYSTEMS: PARAMAGNETIC SCATTERING. . . .

53   SOV. PHYS. SOL. STATE, 4, . . . 1339, (1962)
     ⋆INFLUENCE OF SPIN-SPIN INTERACTION ON NEUTRON SCATTERING IN PARAMAGNETIC
     MATERIALS.
     ⋆KHABIBULLIN(B.M.).

54   SOV. PHYS. SOL. STATE, 4, . . . 1449, (1962)
     ⋆SLOW NEUTRON SCATTERING BY THE SPIN SYSTEM OF A PARAMAGNETIC MATERIAL.
     ⋆KHABIBULLIN(B.M.).

55   SOV. PHYS. SOL. STATE VOL.4, . . . 1449, (1963)....T.
     ⋆SCATTERING OF SLOW NEUTRONS BY THE SPIN SYSTEM OF A PARAMAGNET.
     ⋆KHABIBULLIN(B.M.).

56   SOV. PHYS. SOL. STATE VOL.4, . . . 1339, (1963)....T.
     ⋆EFFECT OF SPIN-SPIN INTERACTION ON SCATTERING OF NEUTRONS IN A
     PARAMAGNET.
     ⋆KHABIBULLIN(B.M.).

57   SOV. PHYS. SOLID STATE VOL.10, . . . 3142, (1968)
     ⋆THEORY OF SLOW-NEUTRON SCATTERING IN PARAMAGNETS SUBJECTED TO STRONG
     MAGNETIC FIELDS.
     ⋆AKHIEZER(I.A.), BARTS(B.I.).

58   SOV. PHYS. J.E.T.P. VOL.29, . . . 1089, (1969)
     ⋆SPIN DENSITY FLUCTUATION SPECTRUM IN METALS WITH PARAMAGNETIC IMPURITIES
     ⋆GURGENISHVILI(G.E.), NERSESYAN(A.A.), KHARADZE(G.A.).

59   SOV. PHYS. SOL. STATE VOL.10, . . . 2483, (1969)
     ⋆THEORY OF SCATTERING OF SLOW NEUTRONS IN PARAMAGNETS IN A STRONG
     MAGNETIC FIELD.

60   TRIESTE/INTERN. COURSE, ... 501, (1967)
     ⋆INVESTIGATIONS OF MAGNETIC MATERIALS USING NEUTRON SCATTERING.
     (38 PAGES).
     ⋆MARSHALL(W.), LOW(G.G.).

61   UKRAINIAN PHYS. J. VOL.14, NO.2, 216, (1969).....03.
     ⋆FLUCTUATIONS AND SCATTERING OF SLOW NEUTRONS AND ELECTROMAGNETIC WAVES
     IN PARAMAGNETS IN A STRONG MAGNETIC FIELD.
     ⋆AKHIEZER(I.A.), BARTS(B.I.).

62   UKRAINIAN PHYS. J. VOL.14, . . . 1909, (1969)....03.......PARAMAGNETICS,
     ⋆THE MAGNETIC SCATTERING OF NEUTRONS BY LIQUID PARAMAGNETICS.
     ⋆ZATOVSKII(A.V.), KRASNYT(YU.P.).

SECTION 12A-POLARIZED NEUTRONS: THEORY AND APPLICATIONS. . . . .

1   ACTA PHYS. POLON. 36, ... 697, (1969), ... 86, ...
    ≈CRITICAL SCATTERING OF POLARIZED NEUTRONS IN FERROMAGNETS.
    ≈KOCINSKI(J.), WENTOWSKA(K.).

2   ANN. DE PHYSIQUE TOME 7 NO.5, . . . 371, (1972) . . . .
    ≈THERMAL NEUTRON SCATTERING BY MAGNETIC INTERACTION.
    ≈WILL(G.).

3   ANN. DE PHYSIQUE TOME 7 NO.4, . . . 269, (1972). . . . .
    ≈DILUTE ALLOYS (REVIEW PAPER).
    ≈MEZEI(F.), RADHAKRISHNA(P.). (PRESENTED AT AUTRANS SUMMER SCHOOL 1972).

4   ATOMKERNENERGIE VOL.7, ... 170, (1962) . . . .E. . . . .
    ≈STERN-GERLACH EXPERIMENT WITH NEUTRONS.
    ≈HASLER(H.G.), WEBER(G.).

5   ATOMKERNENERGIE VOL.10, ... 177, (1965) . . . .E. . . . .FE. . . . .
    ≈MEASUREMENTS OF POLARIZED NEUTRON TRANSMISSION THROUGH MAGNETIZED IRON
    BY A STERN-GERLACH EXPERIMENT.
    ≈WEBER(G.). (IN GERMAN).

6   BUTSURI (JAPAN) VOL.26, ... 464, (1971) . . . .T. . . . .
    ≈WHAT IS THE USE OF THE POLARIZED THERMAL NEUTRON.
    ≈ITO(Y.). (ED. NOTE: IN JAPANESE).

7   COMP. REND. 264, ... 840, (1967) . . . .
    ≈DIFFRACTION DES NEUTRONS POLARISES PAR UN HÉLIMAGNETIQUE.
    ≈HERPIN(A.).

8   CZECH. J. PHYS. B VOL.19, ... 278, (1969) . . . .P. . . . .
    ≈ANOMALOUS CAPTURE OF THERMAL POLARIZED NEUTRONS IN PERFECT FERROMAGNETIC
    SINGLE CRYSTALS.
    ≈MICHALEC(R.), STICH(V.).

9   CZECH. J. PHYS. B VOL.22, ... 38, (1972) . . . .E. . . . .
    ≈TARGET THICKNESS CORRECTION IN POLARIZED NEUTRON-AND-TARGET EXPERIMENTS.
    ≈HONZATKO(J.), KAJFOSZ(J.).

10 IAEA SYMP. (PILE RESEARCH)VIENNA 423, (1960) . . . .
    ≈INELASTIC SCATTERING OF POLARIZED NEUTRONS BY SPIN WAVES IN COMPLEX
    EXCHANGE-COUPLED LATTICES.
    ≈SAENZ(A.W.).

11 J. PHYS. CHEM. SOLIDS VOL.10, . . 138, (1959) . . . .E. . . . . . .FE,NI. . . .
    ≈THE USE OF POLARIZED NEUTRONS IN DETERMINING THE MAGNETIC SCATTERING
    BY IRON AND NICKEL.
    ≈NATHANS(R.), SHULL(C.G.), SHIRANE(G.), ANDRESEN(A.).

12 NED. TIJDSCHR. NATUURK. VOL.38, . 171, (1972) . . . .A4. . . . .
    ≈SCATTERING OF POLARIZED NEUTRONS.
    ≈ZIJP(E.), VAN-DER-WEY(R.), VAN-LEEUWEN(B.), ALONS(P.W.F.), JONKER(C.C.).

13 NUCL. INST. MET. 28, ... 125, (1964) . . . .
    ≈SOME CONSIDERATIONS ON THE PROPER USE OF THE POLARIZED NEUTRON
    SPECTROMETER.
    ≈PAOLETTI(A.), RICCI(F.P.).

14 NUCL. INST. MET. 63, ... 283, (1968) . . . .
    ≈ACCURACY OF POLARIZED SPECTROMETER OPERATING WITH SCATTERING VECTORS NOT
    NORMAL TO DIRECTION OF MAGNETIC FIELD.
    ≈DELAPALME(A.), GEORGES(R.), SCHWEIZER(J.).

15 NUCL. INST. MET. 68, ... 50, (1969) . . . .
    ≈THE DETERMINATION OF ERRORS IN POLARIZED NEUTRON DIFFRACTOMETRY.
    ≈KENDRICK(H.), WERNER(S.A.), ARROTT(A.).

16 NUCL. INST. MET. 75, ... 333, (1969) . . . .
    ≈A PROPOSED METHOD FOR THE ABSOLUTE MEASUREMENT OF NEUTRON POLARIZATION.
    ≈WHITE(R.E.), CHISHOLM(A.), GARRETT(R.).

17 NUCL. INST. MET. VOL.98, ... 385, (1972) . . . .P. . . . .
    ≈POLARIZATION STUDIES USING THE NEUTRON SPIN-PRECESSION METHOD WITH A
    CONTINUOUS ENERGY SPECTRUM OF NEUTRONS.
    ≈NATH(R.), FIRK(F.W.K.), HOLT(R.J.), SCHULTZ(H.L.).

18 NUCL. PHYS. A VOL.A176, ... 225, (1971) . . . .T. . . . . .F (ISO. A=19), . .
    ≈THEORY OF POLARIZED THERMAL NEUTRON SCATTERING ON F(ISOTOPE A=19).1.
    ≈GILLET(V.), NORMAND(J.M.).

19 NUCL. PHYS. A VOL.A181, ... 241, (1972) . . . .P. . . . .
    ≈A STUDY OF THE CO(ISOTOPE A=59)(N,GAMMA)CO(ISOTOPE A=60) REACTION WITH
    POLARIZED THERMAL NEUTRONS.
    ≈STECHER-RASMUSSEN(F.), ABRAHAMS(K.), KOPECKY(J.).

20 NUCL./S.S. PHYS. SYMP. ABSTR. . . . (1972) . . . .D4. . . . . .FERRITES. . . . .
    ≈MAGNETIC DIFFUSE SCATTERING STUDIES IN FERRITES BY POLARIZED NEUTRONS.
    ≈RAKHECHA(V.C.), MADHAV(L.), SRINIVASAN(B.S.), MURTHY(N.S.S.).

21 NUOVO CIMENTO A VOL.5A, ... 591, (1971) . . . .P. . . . .
    ≈ELECTROMAGNETIC EFFECTS IN THE DECAY OF POLARIZED NEUTRONS.
    ≈SHANN(R.T.).

22 PHYS. LETT. 6, ... 47, (1963) . . . .
    ≈SPIN WAVE SCATTERING OF POLARIZED NEUTRONS.
    ≈SAMUELSEN(E.J.), RISTE(T.), STEINSVOLL(O.).

23 PHYS. LETT. VOL.34A, ... 35, (1971) . . . .D4. . . . . .NB. . . . .
    ≈INVESTIGATION OF THE MIXED STATE USING POLARIZED NEUTRONS.
    ≈PFEIFFER(K.), RAUCH(H.), WEBER(H.W.).

24 PHYS. MET. METALLOG. VOL.22 NO.4 7, (1967) . . . .T,P. . . . .
    ≈SCATTERING CROSS-SECTION OF SLOW POLARIZED NEUTRONS IN A UNIAXIAL
    ANTIFERROMAGNET.
    ≈ZAROCHENTSEV(E.V.), POPOV(V.A.).

25 PHYS. REV. LETT. VOL.10, ... 297, (1963) . . . .A4. . . . . .V. . . . .
    ≈NEUTRON SPIN-NEUTRON ORBIT INTERACTION WITH SLOW NEUTRONS.
    ≈SHULL(C.G.).

SECTION 12A-POLARIZED NEUTRONS: THEORY AND APPLICATIONS. . . . .

26 PHYS. REV. LETT. VOL.25, ... 527, (1970)....A2........V2*O3, ...
    ~ANTIFERROMAGNETISM IN V2*O3.
    ~MOON(R.M.).

27 PHYS. REV. LETT. VOL.28 . . . 805, (1972)....P.....
    ~MEASUREMENT OF THE SPIN-DEPENDENT PART OF THE SCATTERING AMPLITUDE OF
    SLOW NEUTRONS ON F(ISOTOPE: A=19) USING POLARIZED BEAM/POLARIZED TARGET.
    ~ABRAGAM(A.), BACCHELLA(G.L.), LONG(C.), MERIEL(P.), PEISVAUX(J.),
    PINOT(M.).

28 PHYS. REV. LETT. VOL.28 . . . . 596, (1972).....
    ~SPIN-ORBIT EFFECTS IN THE INDEX-OF-REFRACTION FORMALISM FOR POLARIZED
    NEUTRONS.
    ~HANDEL(P.H.).

29 PHYS. REV. B VOL.7, ... 4142, (1973)....E. . . . .
    ~REFRACTION OF THERMAL NEUTRONS BY SHAPED MAGNETIC FIELDS.
    ~JUST(W.), SCHNEIDER(C.S.), CISZEWSKI(R.), SHULL(C.G.).

30 PHYS. STAT. SOL. VOL.40, ... 59, (1970)............... .
    ~SPATIAL DISTRIBUTION OF MAGNETIC MOMENTS.
    ~GALPERIN(F.M.), ALPAEV(YU.A.), KIZHAEV(F.G.).

31 PROC/CONF/POL. TARGETS/ION SOUR. 339, (1967)....P............ .
    ~POLARIZED NUCLEI AND NEUTRONS.
    ~SHAPIRO(F.L.). (ED. NOTE: PUBL. CENTRE D≠ETUDES NUCLEAIRES DE SACLAY,
    SACLAY 1967).

32 PROC....CONF. ON POL. TARGETS. . . 235, (1971)........... .
    ~NEUTRON PHYSICS EXPERIMENTS WITH ORIENTED NUCLEI.
    ~DABBS(J.W.T.). (ED. NOTE: AVAIL. FROM NAT. TECH. INFORMATION SERVICE,
    SPRINGFIELD, VA., USA).

33 REV. PHYS. APPL. (FRANCE) VOL.4, 115, (1969)....E.. . . . .
    ~APPLICATION OF POLARIZED-NEUTRON SCATTERING MEASUREMENTS.
    ~SENE(R.), KAHANE(J.), DELPIERRE(P.), HEYMAN(M.).

34 RIV. NUOVO CIMENTO VOL.2, . . . 451, (1970)....D4. . . . .
    ~MAGNETIC FORM FACTOR AND MAGNETIZATION DENSITIES BY MEANS OF POLARIZED
    NEUTRONS.
    ~PAOLETTI(A.).

35 SOL. STATE COMM. 5, ... 289, (1967)....
    ~DIFFRACTION DES NEUTRONS POLARISES SUR LES COMPOSES MAGNETIQUES A
    STRUCTURE NON-COLINEAIRE.
    ~SIVARDIERE(J.).

36 SOL. STATE COMM. VOL.8, ... 1, (1970)....A2.......CO-NI, ...
    ~MAGNETIC MOMENT DISTRIBUTION IN CO-NI ALLOY.
    ~ANTONINI(B.), LUCARI(F.), MENZINGER(F.).

37 SOL. STATE COMM. VOL.10, . . . 667, (1972)....A2.......CO, ...
    ~MAGNETIC FORM FACTOR OF PURE COBALT IN THE HIGH TEMPERATURE F.C.C. PHASE
    ~MENZINGER(F.), SACCHETTI(F.).

38 SOV. J. NUCL. PHYS. VOL.14, . . . 269, (1971)....A4.......U. . . . .
    ~ANALYSIS OF ANGULAR DISTRIBUTIONS OF NEUTRONS SCATTERED BY URANIUM AT
    SMALL ANGLES AND NEUTRON POLARIZABILITY.
    ~ANIKIN(G.V.), KOTUKHOV(I.I.).

39 SOV. PHYS. JETP VOL.14, ... 1168, (1962)....T. . . . .
    ~ON SCATTERING OF POLARIZED NEUTRONS IN FERROMAGNETIC AND ANTIFERRO-
    MAGNETIC SUBSTANCES.
    ~IZYUMOV(YU.A.), MALEEV(S.V.).

40 SOV. PHYS. SOL. STATE, 7, ... 2477, (1965). . . . .
    ~CERTAIN POLARIZATION EFFECTS IN THE SCATTERING OF NEUTRONS IN SOLIDS.
    ~GINZBURG(S.L.), MALEEV(S.V.).

41 SOV. PHYS. JETP LETT. VOL.2, . . 338, (1965). . . . .
    ~SCATTERING OF POLARIZED NEUTRONS IN MAGNETS NEAR THE PHASE-TRANSITION
    POINT.
    ~MALEEV(S.V.).

42 SOV. PHYS. JETP. 24, ... 1068, (1967)....... .. . .
    ~NEUTRON DIFFRACTION IN A POLARIZED CRYSTAL.
    ~BARYSHEVSKII(V.G.).

43 SOV. PHYS. SOL. STATE VOL.13, . . 258, (1971)..... . . . .
    ~COHERENT EXCITATION OF SPIN WAVES AND DOPPLER EFFECT EXPERIENCED BY
    POLARIZED NEUTRONS.
    ~DRABKIN(G.M.), MARCHIK(I.I.).

44 UKRAINIAN PHYS. J. VOL.14, . . . 1867, (1969)....D4. . . . .
    ~SCATTERING OF SLOW POLARIZED NEUTRONS IN ANTIFERROMAGNET PLACED IN A
    STRONG MAGNETIC FIELD.
    ~ZAROCHENTSEV(E.V.).

45 YADERNAYA FIZIKA VOL.4, ... 72, (1966)....T. . . . .
    ~NATURAL ROTATION OF THE NEUTRON SPIN DIRECTION.
    ~BARYSHEVSKII(V.G.). (IN RUSSIAN).

46 Z. PHYSIK BAND 252, ... 7, (1972)....T. . . . .
    ~ASYMMETRIC TOTAL REFLECTION OF POLARIZED NEUTRONS AND THE NEUTRON AS A
    SURFACE PROBE.
    ~HANDEL(P.H.).

SECTION 12B-POLARIZATION/DEPOLARIZATION EFFECTS AND MEASUREMENTS . . . .

1   ACTA CRYST. (INTERACT.) VOL.A25, S255, (1969).....E...................
      *A NEW MONOCHROMATOR FOR POLARIZED NEUTRON BEAM.
      *DELAPALME(A.), SCHWEIZER(J.).

2   ANN. DE PHYSIQUE TOME 7 NO.4,.   255, (1972).....................................
      *EXPERIMENTAL TECHNIQUES IN NEUTRON SCATTERING (REVIEW PAPER).
      *MERIEL(P.). (PRESENTED AT AUTRANS SUMMER SCHOOL 1972).

3   ANN. OF PHYSICS VOL.74,.   .   .   250, (1972).....E.........D2,T2,
      *ELASTIC SCATTERING AND POLARIZATION OF FAST NEUTRONS BY LIQUID DEUTERIUM
      AND TRITIUM.
      *SEAGRAVE(J.D.), HOPKINS(J.C.), DIXON(D.R.), KEATON-JR(P.W.), KERR(E.C.),
      NILLER(A.), SHERMAN(R.H.), WALTER(R.K.).

4   ATOMKERNENERGIE VOL.19,. . . . . 167, (1972)..............DY,. . . . .
      *NEUTRON DEPOLARIZATION MEASUREMENTS ON A DY-SINGLE CRYSTAL.
      *RAUCH(H.), ZEILINGER(A.).

5   BRIT. J. APPL. PHYS. VOL.15, . . 1529, (1964)........................
      *THE DETERMINATION OF BEAM POLARIZATION AND FLIPPING EFFICIENCY IN
      POLARIZED NEUTRON DIFFRACTOMETRY.
      *BROWN(P.J.), FORSYTH(J.B.).

6   CAN. J. PHYS. VOL.39,. . . . . . 1X, (1961).....E.....CO-FE,
      *A POLARIZED NEUTRON BEAM PRODUCED BY BRAGG REFLECTION FROM CO-FE ALLOY.
      *CLARK(M.A.), ROBSON(J.M.).

7   CZECH. J. PHYS. B VOL.13,. . . . 474, (1963).....E..................
      *THE BEAM OF POLARIZED NEUTRONS OBTAINED BY THE METHOD OF REFLECTION
      FROM A COBALT MIRROR.
      *KOPECKY(J.), CHALUPA(B.), MICHALEC(R.), KAJFOSZ(J.).

8   DISS. ABSTR. VOL.13,. . . . . . 248, (1953).....E.........FE,. . . . .
      *NEUTRON POLARIZATION.
      *FLEEMAN(J.).

9   DISS. ABS. VOL.26,. . . . . . . 1111, (1966).....................
      *THE ANGULAR THERMAL NEUTRON SPECTRUM IN THE VICINITY OF THE INTERFACE
      BETWEEN TWO MEDIA.
      *O#DELL(R.O.).

10   IAEA SYMP. COPENHAGEN, VOL.2,. . 429, (1968)....................
      *A NEW HIGH-EFFICIENCY TIME-OF-FLIGHT SYSTEM.
      *COLWELL(J.F.), MILLER(P.H.), WHITTEMORE(W.L.).

11   IAEA SYMP. COPENHAGEN, VOL.2,. . 407, (1968)....................
      *CORRELATION-TYPE TIME-OF-FLIGHT SPECTROMETER WITH MAGNETICALLY CHOPPED
      POLARIZED NEUTRON BEAM.
      *PAL(L.),   KROO(N.),   GORDON(G.),   PELLIONISZ(P.), SZLAVIK(F.), VIZI(I.).

12   IAEA SYMP. COPENHAGEN, VOL.2,. . 387, (1968)....................
      *A NEUTRON SPIN-FLIP CHOPPER FOR TIME-OF-FLIGHT MEASUREMENTS.
      *RAUCH(H.), HARMS(J.), MOLDASCHL(H.).

13   IAEA SYMP. COPENHAGEN, VOL.2,. . 395, (1968)....................
      *TIME-OF-FLIGHT SPECTROMETER USING AN ELECTRONIC CHOPPER FOR POLARIZED
      SLOW NEUTRONS.
      *STEINSVOLL(O.), VIRJO(A.).

14   IAEA SYMP. GRENOBLE,. . . . . . 797, (1972).....E................
      *PSEUDORANDOM MODULATION IN POLARIZED NEUTRON DIFFRACTOMETRY.
      *MEZEI(F.), PELLIONISZ(P.).

15   INDIAN J. PURE APPL. PHYS. VOL.7 546, (1969).....E................
      *A POLARIZED NEUTRON SPECTROMETER FOR MAGNETIC SCATTERING STUDIES.
      *MURTHY(N.S.S.),   SOMANATHAN(C.S.),   BEGUM(R.J.),   SRINIVASAN(B.S.),
      MURTHY(M.R.L.N.).

16   INSTRUM. EXP. TECH. NO.4,. . . . 571, (1960).....E.....CO,
      *PRODUCTION OF POLARIZED NEUTRONS BY REFLECTION FROM A COBALT MIRROR.
      *ABOV(YU.G.), BEKETOV(V.A.)...ET AL.

17   INSTRUM. EXP. TECH. 3,. . . . . 729, (1970).....E.........
      *DOUBLE THERMAL-NEUTRON POLARIZER.
      *OBINYAKOV(B.A.), MOSTOVOI(YU.A.).

18   INSTRUM. EXP. TECH. (USA) VOL.14 618, (1971).....E................
      *FABRICATION OF THIN FLAT COBALT MIRRORS FOR THE POLARIZATION OF THERMAL
      NEUTRONS.
      *SHATLOVSKAYA(N.S.), BEKETOV(V.A.).

19   J. APP. PHYS. 39,. . . . . . . 447, (1968)....................
      *A MAGNETICALLY PULSED NEUTRON BEAM FOR TIME OF FLIGHT MEASUREMENTS.
      *MOOK(H.A.), WILKINSON(M.K.).

20   J. PHYS. E VOL.6 NO.8,. . . . . 714, (1973).....E................
      *AN ASYMMETRIC SPLIT-PAIR SUPERCONDUCTING MAGNET FOR NUCLEAR POLARIZATION
      EXPERIMENTS.
      *GILBERT(E.), HANLEY(P.), HAYTER(J.B.), WHITE(J.W.).

21   J. PHYSIQUE TOME 24,. . . . . . 359, (1963).....E................
      *PRODUCTION AND STUDY OF MONOCHROMATIC POLARIZED BEAMS OF SLOW NEUTRONS.
      *BEIL(H.), CARLOS(P.), MATUSZEK(J.). (IN FRENCH).

22   J. PHYSIQUE TOME 24,. . . . . . 89A, (1963).....E................
      *CRYSTAL SPECTROMETER FOR POLARIZED NEUTRONS.
      *BEIL(H.), CARLOS(P.), GENIN(R.), SIGNARBIEUX(C.), JOLY(R.), MATUSZEK(J.)

23   J. PHYS. SOC. JAPAN VOL.28,. . . 1116, (1970).....A4,
      *NEUTRON DEPOLARIZATION IN ELASTIC SCATTERING.
      *KATORI(K.), NAGATA(T.), UCHIDA(A.), KOBAYASHI(S.).

24   NUCL. INST. MET. 30,. . . . . . 271, (1964)....................
      *ELECTROMAGNETIC FOCUSING AND POLARIZATION OF NEUTRON BEAMS.
      *FARAGO(P.S.).

25   NUCL. INST. MET. VOL.45,. . . . 293, (1966).....E................
      *ON A SYSTEM OF MAGNETIZED COBALT MIRRORS USED TO PRODUCE AN INTENSE
      BEAM OF POLARIZED THERMAL NEUTRONS.
      *ABRAHAMS(K.), RATYNSKI(W.), STECHER-RASMUSSEN(F.), WARMING(E.).

SECTION 128-POLARIZATION/DEPOLARIZATION EFFECTS AND MEASUREMENTS . . . .

26 NUCL. INST. MET. 48,   154, (1967)...........
    *IMPROVEMENTS IN NEUTRON POLARIMETER.
    *MILLER(T.G.).

27 NUCL. INST. MET. 64,   77, (1968)...........
    *A NEUTRON POLARIMETER WITH HIGH ENERGY RESOLUTION.
    *SIEMINSKI(M.), WILHELMI(Z.), ZYCH(W.), ZUPRANSKI(P.).

28 NUCL. INST. MET. VOL.95,   589, (1971)...........
    *STUDY OF A HEUSLER ALLOY (CU2-MN-AL) AS A MONOCHROMATOR FOR POLARIZED
    NEUTRONS.
    *DELAPALME(A.), SCHWEIZER(J.), COUDERCHON(G.), PERRIER-DE-LA-BATHIE(R.).
    (ED. NOTE: IN FRENCH).

29 NUCL. INST. MET. 96,   29, (1971)...........
    *A SPIN-FLIP-SELECTOR FOR POLARIZED NEUTRONS.
    *PAPP(R.), RAUCH(H.).

30 NUCL. INST. MET. VOL.99,   613, (1972)....E........
    *ENERGY SELECTION BY PSEUDORANDOM MODULATION IN POLARIZED NEUTRON DIFFR.
    *MEZEI(F.), PELLIONISZ(P.).

31 PHYS. LETT. 12,.   334, (1964)...........
    *NEUTRON POLARIZATION BY TRANSMISSION THROUGH A POLARIZED PROTON TARGET.
    *DRAGHICESCU(P.), LUSHCHIKOV(V.I.)...ET AL.

32 PHYS. REV. VOL.59,   981, (1941)....T........
    *THE PASSAGE OF NEUTRONS THROUGH CRYSTALS AND POLYCRYSTALS.
    *HALPERN(O.), HAMERMESH(M.), JOHNSON(M.H.).

33 PHYS. REV. VOL.59,   960, (1941)....T........
    *ON THE PASSAGE OF NEUTRONS THROUGH FERROMAGNETS.
    *HALPERN(O.), HOLSTEIN(T.).

34 PHYS. REV. VOL.64,   47, (1943)....E........FE,
    *NEUTRON POLARIZATION AND FERROMAGNETIC SATURATION.
    *BLOCH(F.), HAMERMESH(M.), STAUB(H.).

35 PHYS. REV. VOL.74,   1285, (1948)....T........
    *DEPOLARIZATION OF NEUTRONS DURING DIFFUSION.
    *BOROWITZ(S.), HAMERMESH(M.).

36 PHYS. REV. VOL.73,   1277, (1948)....E........FE,
    *NEUTRON POLARIZATION.
    *HUGHES(D.J.), WALLACE(J.R.), HOLTZMAN(R.H.).

37 PHYS. REV. VOL.76,   1413, (1949)....E........
    *REFLECTION AND POLARIZATION OF NEUTRONS BY MAGNETIZED MIRRORS.
    *HUGHES(D.J.), BURGY(M.T.).

38 PHYS. REV. VOL.76,   994, (1949)....E........
    *ON THE POLARIZATION OF SLOW NEUTRONS.
    *STEINBERGER(J.), WICK(G.C.).

39 PHYS. REV. VOL.80,   953, (1950)....E........
    *DOUBLE TRANSMISSION AND DEPOLARIZATION OF NEUTRONS.
    *BURGY(M.T.), HUGHES(D.J.), WALLACE(J.R.), HELLER(R.B.), WOOLF(W.E.).

40 PHYS. REV. VOL.81,   626, (1951)....E........FE3*O4,
    *PRODUCTION OF HIGHLY POLARIZED NEUTRON BEAMS BY BRAGG REFLECTION FROM
    FERROMAGNETIC CRYSTALS.
    *SHULL(C.G.).

41 PHYS. REV. VOL.94,   374, (1954)....E........FE,
    *NEUTRON POLARIZATION.
    *STANFORD(C.P.), STEPHENSON(T.E.), COCHRAN(L.W.), BERNSTEIN(S.).

42 PHYS. REV. 130,.   1670, (1963)...........
    *POLARIZATION EFFECTS IN THE MAGNETIC ELASTIC SCATTERING OF SLOW
    NEUTRONS.
    *BLUME(M.).

43 PHYS. REV. VOL.133,   B581, (1964)...........
    *STUDIES OF THE OPTICS OF NEUTRONS. III. PROBLEMS OF POLARIZATION OF SLOW
    NEUTRONS.
    *HALPERN(O.).

44 PHYS. REV. VOL.136,   B112, (1964)...........
    *NEUTRON POLARIZATION AT SMALL SCATTERING ANGLES NEAR AN S-WAVE RESONANCE
    *REDMOND(R.F.).

45 PHYS. REV. VOL.134,   A126, (1964)...........
    *STUDIES OF THE OPTICS OF NEUTRONS. IV. MAGNETIC SATURATION AND NEUTRON
    DEPOLARIZATION.
    *HALPERN(O.).

46 PHYS. REV. VOL.133,   A1366, (1964)...........
    *POLARIZATION EFFECTS IN SLOW NEUTRON SCATTERING. II. SPIN-ORBIT
    SCATTERING AND INTERFERENCE.
    *BLUME(M.).

47 PHYS. REV. 166,.   554, (1968)...........
    *POLARIZATION EFFECTS IN SLOW-NEUTRON SCATTERING III. NUCLEAR
    POLARIZATION.
    *SCHERMER(R.I.), BLUME(M.).

48 PHYS. REV. 181,.   920, (1969)....A4........NI, MN*F2,
    *POLARIZATION ANALYSIS OF THERMAL-NEUTRON SCATTERING.
    *MOON(R.M.), RISTE(T.), KOEHLER(W.C.).

49 REPORT RL-73-034 (68 PP.),   (1973)....E........
    *A DESIGN FOR THERMAL NEUTRON POLARIZATION ANALYSIS DIFFRACTOMETER WITH
    A FILTER CONTAINING SAMARIUM-149 NUCLEI AS THE SPIN ANALYSER.
    *WILLIAMS(W.G.). (AVAIL. HMSO, LONDON, ENGLAND).

50 REV. PHYS. APPL. (FRANCE) VOL.4, 254, (1969)....E........
    *MEASUREMENT OF NEUTRON POLARIZATION DURING EXPERIMENTS.
    *DELPIERRE(P.), HEYMAN(M.), KAHANE(J.), SENE(R.), SAGET(G.).

51 REV. ROUMAINE PHYS. VOL.8,   277, (1963)....E........
    *ON THE TOTAL REFLECTION OF NEUTRONS BY COBALT MIRRORS.
    *POPOVICI(M.), RIPEANU(S.), GRABCHEV(B.), VASILIU(V.). (IN RUSSIAN).

52 REV. SCI. INSTRUM. VOL.33,   524, (1962)....E........
    *REVERSAL OF THE SPIN OF POLARIZED THERMAL NEUTRONS WITHOUT DEPOLAR-
    IZATION.
    *ABRAHAMS(K.), STEINSVOLL(O.), BONGAARTS(P.J.M.), DE-LANGE(P.W.).

SECTION 12B-POLARIZATION/DEPOLARIZATION EFFECTS AND MEASUREMENTS . . . .

53 REV. SCI. INSTRUM. VOL.39, . . . 101, (1968).............
    *MEASUREMENT OF THE POLARIZATION OF THERMAL NEUTRON BEAMS OF MIXED
    VELOCITIES.
    *BARKAN(S.), BEIBER(E.), BURGY(M.T.), KETUDAT(S.)...ET AL.

54 REV. SCI. INSTRUM. VOL.42, . . . 1007, (1971).............
    *FAST SWITCHING PULSED NEUTRON POLARIZATION INVERTER.
    *FORGACS(R.L.).

55 SOV. J. NUCL. PHYS. VOL.10, . . . 1178, (1969).....P.......
    *POLARIZED PROTON TARGET AS A NEUTRON POLARIZER.
    *LUSHCHIKOV(V.I.), TARAN(YU.V.), SHAPIRO(F.L.).

56 SOV. J. NUCL. PHYS. VOL.12, . . . 1121, (1970).....A4.....
    *DIFFERENTIAL CROSS SECTIONS OF SMALL ANGLE NEUTRON SCATTERING AND
    NEUTRON POLARIZABILITY.
    *ANIKIN(G.V.), KOTUKHOV(I.I.).

57 SOV. PHYS. JETP VOL.13, ... 860, (1961)....T........
    *POLARIZATION RESULTING FROM THE SCATTERING OF (SLOW) NEUTRONS BY
    FERROMAGNETIC SUBSTANCES.
    *MALEEV(S.V.).

58 SOV. PHYS. JETP. 16, ... 1531, (1963)...............
    *POLARIZATION OF INELASTICALLY SCATTERED NUCLEONS.
    *ELAZIN(YU.P.).

59 SOV. PHYS. CRYST. VOL.10, ... 76, (1965)....E........
    *PRODUCING SINGLE-CRYSTAL PLATES OF COBALT-IRON ALLOY FOR THE
    POLARIZATION OF NEUTRONS.
    *OVCHAROV(V.P.), ALESHKO-OZHEVSKII(O.P.).

60 SOV. PHYS. CRYST. VOL.11, ... 597, (1967)....E........
    *PRODUCTION OF A MONOENERGETIC POLARIZED NEUTRON BEAM.
    *ABOV(YU.G.), ALESHKO-OZHEVSKII(O.P.), ERMAKOV(O.N.), YAMZIN(I.I.).

61 SOV. PHYS. JETP LETT. VOL.10, . . 345, (1969)....P........
    *RESONANT DEPOLARIZATION OF NEUTRONS BY THE DOMAIN STRUCTURE IN FERRO-
    MAGNETS.
    *MALEEV(S.V.), RUBAN(V.A.), TRUNOV(V.A.).

62 SOV. PHYS. J.E.T.P. VOL.31, . . . 378, (1970)...............
    *POLARIZATION OF PARTICLES AND QUANTA SCATTERED BY THICK LAYERS OF MATTER
    *GNEDIN(YU.N.), DOLGINOV(A.Z.), SILANT≠EV(N.A.).

63 SOV. PHYS. J.E.T.P. VOL.31, . . . 111, (1970)...............
    *DEPOLARIZATION OF NEUTRONS PASSING THROUGH A FERROMAGNET.
    *MALEEV(S.V.), RUBAN(V.A.).

64 SOV. PHYS. TECH. PHYS. VOL.40, . . 1317, (1970)...............
    *THE PASSAGE OF POLARIZED NEUTRONS THROUGH FERROMAGNETIC FILMS.
    *DRABKIN(G.M.), TRUNOV(V.A.), DMITRIEV(R.P.).

65 SOV. PHYS. SOL. STATE VOL.12, . . 2445, (1971)...............
    *DEPOLARIZATION OF NEUTRONS TRANSMITTED BY MAGNETIC MATERIALS.
    *TOPERVERG(B.P.).

66 SOV. PHYS. SOL. STATE VOL.13, . . 727, (1971)...............
    *DIFFRACTION POLARIZATION OF NEUTRONS IN A POLARIZED CRYSTAL.
    *KORENNAYA(L.N.).

67 SOV. PHYS. JETP VOL.35, ... 222, (1972)....T........
    *CRITICAL DEPOLARIZATION OF NEUTRONS TRAVERSING A FERROMAGNETIC BODY.
    *MALEEV(S.V.), RUBAN(V.A.).

68 SOV. PHYS. JETP LETT. VOL.15, . . 324, (1972)....P........
    *ANISOTROPY OF DEPOLARIZATION OF A NEUTRON BEAM. (FERROMAGNETIC MATER.).
    *DRABKIN(G.M.), OKOROKOV(A.I.), RUNOV(V.V.).

69 1ST EUR. CONF./CONDENSED MATTER, . 41, (1971)...............
    *NEUTRON DEPOLARIZATION EXPERIMENTS IN THE IMMEDIATE VICINITY OF THE
    CURIE TEMPERATURE OF NICKEL.
    *REKVELDT(M.TH.). (ED. NOTE: PUBL. EUROPEAN PHYS. SOC. 1971, GENEVA,
    SWITZ.).

70 THESIS, ... (1972)...............
    *NEUTRON DEPOLARIZATION STUDY OF FERROMAGNETIC DOMAIN STRUCTURES.
    *REKVELDT(M.TH.). (PH.D. DISSERTATION - DELFT, THE NETHERLANDS).

71 YADERNAYA FIZ. (USSR) VOL.12, . . 815, (1970)....P........
    *NEUTRON POLARIZATION IN THE SCATTERING BY THICK LAYERS OF MATTER.
    *GNEDIN(YU.N.), DOLGINOV(A.Z.), SILANT≠EV(N.A.), SHIBANOV(YU.A.).

72 Z. ANGEW PHYS. VOL.32, ... 109, (1971)....A4........DY.
    *A NEUTRON DEPOLARIZATION EXPERIMENT ON DY NEAR THE FERROMAGNETIC
    TRANSITION POINT.
    *RAUCH(H.), SEIDL(E.), ZEILINGER(A.).

73 Z. PHYSIK VOL.259 NO.5, . . . 391, (1973)....T........
    *STUDY OF FERROMAGNETIC BULK DOMAINS BY NEUTRON DEPOLARIZATION IN THREE
    DIMENSIONS.
    *REKVELDT(M.TH.).

74 Z. PHYS. VOL.171, ... 291, (1963)....T.B3........
    *DEPOLARIZATION AND SMALL-ANGLE SCATTERING OF NEUTRONS BY LATTICE
    DEFECTS IN FERROMAGNETIC CRYSTALS.
    *KRONMUELLER(H.), SEEGER(A.), WILKENS(M.). (IN GERMAN).

75 Z. PHYS. VOL.243, ... 188, (1971)....E........
    *PRODUCTION OF POLARIZED THERMAL NEUTRONS BY MEANS OF AN IRON-COBALT-
    NEUTRON GUIDE.
    *BERNDORFER(K.).

76 Z. PHYS. VOL.255, ... 146, (1972)....E........
    *NEUTRON SPIN ECHO: A NEW CONCEPT IN POLARIZED THERMAL NEUTRON TECHNIQUES
    *MEZEI(F.).

SECTION 13 -CRITICAL SCATTERING. . . . .

1 ACTA CRYST. VOL.A28 PART S-4,. . . 177, (1972)................
.NEUTRON SCATTERING STUDIES OF STRUCTURAL PHASE TRANSFORMATIONS IN SOLIDS
.AXE(J.D.). (ABSTRACT ONLY).

2 ACTA PHYS. POLON. VOL.33,. . . . . 13, (1968)....T,D2.........
.ON THE CRITICAL MAGNETIC SCATTERING OF NEUTRONS.
.KOCINSKI(J.).

3 ACTA PHYS. POLON. VOL.35,. . . . . 61, (1969)....D2.........
.CRITICAL MAGNETIC SCATTERING OF NEUTRONS IN ANTIFERROMAGNETS.
.KOCINSKI(J.).

4 ACTA PHYS. POLON. 36,. . . . . 697, (1969)....86.........
.CRITICAL SCATTERING OF POLARIZED NEUTRONS IN FERROMAGNETS.
.KOCINSKI(J.), WENTOWSKA(K.).

5 AIP CONF. PROC. VOL.5,. . . . . 125C, (1971)....D2.........
.CRITICAL SCATTERING IN THE HEISENBERG MODEL.
.RITCHIE(D.S.), FISHER(M.E.).

6 AIP CONF. PROC. NO.1C PART 1,. . 822, (1973)....T.........
.THE ROLE OF DIPOLE-DIPOLE INTERACTIONS IN THE CRITICAL BEHAVIOR OF
FERROMAGNETIC MATERIALS.
.ARROTT(A.), HEINRICH(B.), NOAKES(J.E.).

7 ANN. DE PHYSIQUE TOME 7 NO.5,. . 313, (1972)....E.........R3*MN*F3,FE,NI,EU*(S,O),MN*F2, .
.CRITICAL PHENOMENA AND NEUTRON SCATTERING (II).
.PARETTE(G.).

8 ANN. DE PHYSIQUE TOME 7 NO.4,. . 299, (1972)................
.CRITICAL PHENOMENA AND NEUTRON SCATTERING (REVIEW PAPER).
.PARETTE(G.). (PRESENTED AT AUTRANS SUMMER SCHOOL 1972).

9 BROOKHAVEN SYMPOSIUM,. . . . . 49, (1965)....P.........
.CRITICAL PHENOMENA.
.MORI(H.).

10 CAN. J. PHYS. 46,. . . . . . . 799, (1968)................
.THE TEMPERATURE AND FREQUENCY DEPENDENCE OF THE INELASTIC NEUTRON
SCATTERING FROM AN ISING MAGNET.
.ALLAN(G.A.T.), BETTS(D.D.).

11 C.N.E.N. SYMP. CASACCIA,. . . . 133, (1968)................
.CRITICAL SCATTERING OF NEUTRONS.
.NATHANS(R.).

12 COLL. INTER. NO.126 (GRENOBLE),. 194, (1963)................
.DIFFUSION CRITIQUE DES NEUTRONS.
.VILLAIN(J.).

13 C.R.ACAD.SC.(PARIS),TOME 260-B,. 457, (1965)................
.APPROXIMATION LANDAU (DIFFUSION MAGNETIQUE INELASTIQUE CRITIQUE PAR UN
FERROMAGNETIQUE) POINT DE CURIE.
.SEIDEN(J.).

14 C.R.ACAD.SC.(PARIS),TOME 260-B,. 833, (1965)................
.APPROXIMATION LANDAU (DIFFUSION MAGNETIQUE INELASTIQUE CRITIQUE PAR UN
FERROMAGNETIQUE) APPROX. DE DIFFUSION DE SPIN.
.SEIDEN(J.).

15 DISS. ABS. VOL.32,. . . . . 1166B, (1971)................
.A STUDY OF DYNAMIC CRITICAL PHENOMENA.
.YOUNG(P.L.C.).

16 DYN. ASPECTS OF CRITICAL PHEN.,. 69, (1972)................
.CRITICAL AND SPIN-WAVE FLUCTUATIONS IN FERROMAGNETS BY NEUTRON
SCATTERING.
.MINKIEWICZ(V.J.).

17 DYN. ASPECTS OF CRITICAL PHEN.,. 50, (1972)................
.AN INTERPRETATION OF THE BEHAVIOR OF THE LONGITUDINAL SCATTERING IN
MN*F2 BELOW THE CRITICAL TEMPERATURE.
.HELLER(P.).

18 DYN. ASPECTS OF CRITICAL PHEN.,. 32, (1972)................
.INELASTIC SCATTERING OF NEUTRON FROM ANTIFERROMAGNETS.
.SCHULHOF(M.P.).

19 DYN. ASPECTS OF CRITICAL PHEN.,. 19, (1972)................
.CRITICAL SPIN DYNAMICS, SYMMETRY AND CONSERVATION LAWS.
.RIEDEL(E.), WEGNER(F.).

20 INTERN. SUMMER SCHOOL, MOL.,. . 560, (1963)................
.SCATTERING OF NEUTRONS FROM MAGNETIC MATERIALS NEAR THEIR TRANSITION
TEMPERATURE.
.JACROT(B.).

21 INT. J. MAGN. VOL.1,. . . . . 149, (1971)....D1,D2.........
.CRITICAL AND SPIN-WAVE FLUCTUATIONS IN FERROMAGNETS BY NEUTRON
SCATTERING.
.MINKIEWICZ(V.J.).

22 J. CHEM. PHYS. VOL.58,. . . . 1143, (1973)....T,C1.........
.INCOHERENT NEUTRON SCATTERING AND MOLECULAR REORIENTATIONS IN CRYSTALS
NEAR T CRITICAL.
.MICHEL(K.H.).

23 J. PHYS. CHEM. SOLIDS VOL.13,. . 235, (1960)....D2.........FE
.EXPERIMENTAL STUDY OF THE CRITICAL SCATTERING OF NEUTRONS IN IRON.
.ERICSON(M.), JACROT(B.).

24 J. PHYS. C VOL.6,. . . . . 143, (1973)....T.........
.PARAELECTRIC, PIEZOELECTRIC AND PYROELECTRIC CRYSTALS:
II. PHASE TRANSITIONS.
.COWLEY(R.A.), COOMBS(G.J.).

25 J. PHYS. C VOL.7,. . . . . 573, (1974)....T.........
.A COMPARISON OF CALCULATED AND MEASURED CRITICAL TEMPERATURES IN
HEISENBERG MAGNETIC MATERIALS.
.COLLINS(M.F.).

SECTION 13 -CRITICAL SCATTERING. . . . .

26 J. PHYSIQUE TOME 24, ... 622, (1963)...T..on..scatt....
    ╓INFLUENCE OF A MAGNETIC FIELD ON NEUTRON SCATTERING IN FERROMAGNETIC
     SUBSTANCES AT THE CRITICAL POINT.
    ╓VILLAIN(J.). (IN FRENCH).

27 J. PHYSIQUE TOME 28 COL.1, ... 140, (1967)....D2........ . . . .
    ╓DIFFUSION MAGNETIQUE CRITIQUE DE PHONONS.
    ╓PAPOULAR(M.).

28 J. PHYSIQUE TOME 29, ... 687, (1968)....T........T9....
    ╓RELAXATION CRITIQUE AU-DESSUS DU POINT DE NEEL DANS LES ANTIFERRO-
     MAGNETIQUES ET LES HELIMAGNETIQUES.
    ╓VILLAIN(J.).

29 J. PHYSIQUE TOME 29, ... 488, (1968)....T........
    ╓ELARGISSEMENT PAR DIFFUSION D≠UN SPECTRE INCIDENT DANS LA LIMITE DES
     FAIBLES ELARGISSEMENTS. APPLICATION/DIFFUSION MAGNETIQUE CRIT./NEUTRONS.
    ╓VILLAIN(J.).

30 J. PHYSIQUE TOME 33, COL.2, . . . 7, (1972)............SR*TI*03,K*TA*03,NA*N*02,DKDP...
    ╓NEUTRON INELASTIC SCATTERING AT STRUCTURAL PHASE TRANSITIONS (REVIEW
     PAPER).
    ╓COWLEY(R.A.).

31 J. PHYS. RADIUM VOL.20, ... 175, (1959)....D2.......CO*0,, ...
    ╓CRITICAL SCATTERING OF NEUTRONS FROM CO*O.
    ╓MCREYNOLDS(A.W.), RISTE(T.).

32 J. PHYS. SOC. JAPAN VOL.21, . . . 2178, (1966)....T,D2.......
    ╓THEORY OF CRITICAL MAGNETIC SCATTERING OF NEUTRONS BY FERROMAGNET AND
     ANTIFERROMAGNET.
    ╓OGUCHI(T.), ONO(I.).

33 LATV. PSR ZIN. AKAD. VESTIS NO.2 22, (1966)....D2........
    ╓INFORMATION OBTAINED WITH THE HELP OF THE CRITICAL MAGNETIC SCATTERING
     OF NEUTRONS.
    ╓KASHCHEEV(V.N.). (IN RUSSIAN).

34 MAT. SCI. ENGNG., ... (1972)....E......
    ╓CRITICAL PHENOMENA IN ALLOYS, MAGNETS AND SUPERCONDUCTORS.
    ╓VILLAIN(J.). (ED. NOTE: IN MCGRAW-HILL BOOK CO.-SERIES IN MATERIALS
     SCIENCE AND ENGINEERING, 1972).

35 NBS MISC. PUB.273, ... 135, (1966)....D2.......
    ╓CRITICAL SCATTERING OF NEUTRONS BY FERROMAGNETS.
    ╓MARSHALL(W.). (ED. NOTE:PROCEEDINGS CONFERENCE ON PHENOMENA IN THE
     NEIGHBOURHOOD OF CRITICAL POINTS, WASHINGTON, 1965).

36 2ND SUMMER SCHOOL S.S. PHYS,,,,, 93, (1968)...T....
    ╓FERROELECTRICITY, NEUTRON SCATTERING AND CRITICAL SCATTERING.
    ╓COCHRAN(W.).

37 NUOVO CIMENTO SUPPL. VOL.6, . . . 1183, (1957)........
    ╓CRITICAL SCATTERING OF NEUTRONS FROM FERROMAGNETS.
    ╓MARSHALL(W.).

38 PHYS. LETT. 25A, ... 660, (1967)........
    ╓CRITICAL MAGNETIC SCATTERING OF NEUTRONS AND THE (SIN(KR)/R) SPIN
     CORRELATION.
    ╓KOCINSKI(J.), MRYGON(B.).

39 PHYS. LETT. 30A, ... 68, (1969)........
    ╓ANGULAR POSITION OF SIDE MAXIMUM IN THE CRITICAL SCATTERING FOR
     DIFFERENT NEUTRON WAVELENGTHS.
    ╓CISZEWSKI(R.), BLINOWSKI(K.).

40 PHYS. LETT. VOL.32A, ... 341, (1970)........FE.....
    ╓ON THE TEMPERATURE SHIFT OF CRITICAL NEUTRON SCATTERING MAXIMA.
    ╓GORDON(J.), POPOVICI(M.).

41 PHYS. LETT. VOL.33A, ... 287, (1970).....C3.......(C2*H5)2*0,..
    ╓THE MOLECULAR DYNAMICS ALONG THE COEXISTENCE CURVE OF A LIQUID-GAS
     SYSTEM.
    ╓BATA(L.), JOVIC(D.).

42 PHYS. LETT. VOL.32A, ... 273, (1970)............MN*F2,FE*F2.
    ╓CROSSOVER EFFECTS IN DYNAMIC CRITICAL PHENOMENA IN MN*F2 AND FE*F2.
    ╓RIEDEL(E.), WEGNER(F.).

43 PHYS. LETT. VOL.33A, ... 58, (1970)........
    ╓ON DYNAMIC SCALING FOR ISOTROPIC MAGNETS IN THE PARAMAGNETIC CRITICAL
     REGION.
    ╓WAGNER(H.).

44 PHYS. LETT. VOL.37A, ... 406, (1971)........
    ╓ON DYNAMIC SCALING IN THE HEISENBERG FERROMAGNET.
    ╓MALEEV(S.V.).

45 PHYS. LETT. VOL.34A, ... 306, (1971)........
    ╓CRITICAL SCATTERING CROSS SECTION IN FERROMAGNETIC POLYCRYSTALS.
    ╓WOJTCZAK(L.), KOCINSKI(J.).

46 PHYS. LETT. VOL.37A, ... 63, (1971)........
    ╓CRITICAL FLUCTUATIONS NEAR TRANSITION POINTS BETWEEN FERROMAGNETIC AND
     ANTIFERROMAGNETIC PHASES.
    ╓AKHIEZER(I.A.), GINZBURG(A.E.).

47 PHYS. LETT. VOL.41A NO.2,....... 175, (1972)....D2....
    ╓THE BRILLOUIN-MANDELSTAM DOUBLET IN THE CRITICAL MAGNETIC SCATTERING
     OF NEUTRONS.
    ╓MALINOWSKI(K.), KOCINSKI(J.).

48 PHYS. REV. VOL.156, ... 583, (1967)....T....
    ╓THEORY OF CRITICAL-POINT SCATTERING AND CORRELATIONS. I. THE ISING MODEL
    ╓FISHER(M.E.), BURFORD(R.J.).

49 PHYS. REV. VOL.177, ... 952, (1969)........
    ╓SCALING LAWS FOR DYNAMIC CRITICAL PHENOMENA.
    ╓HALPERIN(B.I.), HOHENBERG(P.C.).

50 PHYS. REV. B VOL.4,,,,,,,,,,,,, 3954, (1971)........
    ╓SOME CRITICAL PROPERTIES OF THE NEAREST-NEIGHBOR CLASSICAL HEISENBERG
     MODEL FOR THE FCC LATTICE IN FINITE FIELD FOR TEMPERATURES > T(C).
    ╓FERER(M.), MOORE(M.A.), WORTIS(M.).

SECTION 13 -CRITICAL SCATTERING. . . . .

51 PHYS. REV. B-4,. . 3206, (1971). INELASTIC SCATTERING IN TWO-CRYSTAL AND
    ⊹QUANTITATIVE ANALYSIS OF INELASTIC SCATTERING IN TWO-CRYSTAL AND
    THREE-CRYSTAL NEUTRON SPECTROMETRY: CRITICAL SCATTERING FROM RB*MN*F3.
    ⊹TUCCIARONE(A.), LAU(H.Y.), CORLISS(L.M.), DELAPALME(A.), HASTINGS(J.M.).

52 PHYS. REV. B VOL.5,. . . . . . 2668, (1972). . . . .
    ⊹THEORY OF CRITICAL-POINT SCATTERING AND CORRELATIONS. II. HEISENBERG
    MODELS.
    ⊹RITCHIE(D.S.), FISHER(M.E.).

53 PHYS. REV. LETT. VOL.28, . . 22, (1972). . . .E. . OF THE GA-HG
    ⊹SHAPE OF THE COEXISTENCE CURVE OF THE GA-HG SYSTEM NEAR T(CRITICAL).
    ⊹D≠ABRAMO(G.), RICCI(F.P.), MENZINGER(F.).

54 PHYS. REV. B VOL.7, ... 2611, (1973). . . .T. . . . .
    ⊹CRITICAL DYNAMICS AT STRUCTURAL TRANSITIONS FOR T≷T(CRITICAL). . . . . .
    ⊹SCHNEIDER(T.).

55 PHYS. REV. LETT. VOL.30 NO.17, . 781, (1973). . . .B6 OF NI3*MN.
    ⊹NEUTRON-SCATTERING OBSERVATIONS OF CRITICAL SLOWING DOWN OF AN ISING
    SYSTEM.
    ⊹COLLINS(M.F.), TEH(H.C.).

56 PHYS. REV. B VOL.8,. . . . . . . 4422, (1973). . .T. . . . .
    ⊹INELASTIC INCOHERENT NEUTRON SCATTERING AS AN ALTERNATIVE METHOD TO
    INVESTIGATE LOCAL CRITICAL BEHAVIOUR.
    ⊹SCHNEIDER(T.), MEIER(P.F.).

57 PHYS. REV. LETT. VOL.31, ... 1500, (1973). . .T. . . . .
    ⊹NEUTRON SCATTERING AND THE CORRELATION FUNCTIONS OF THE ISING MODEL
    NEAR T CRITICAL.
    ⊹TRACY(C.A.), MCCOY(B.M.).

58 PHYS. TODAY SEPT., 1973, ... 32, (1973). . . . . .SR*TI*O3,NB3*SN, ...
    ⊹EXPLORING PHASE TRANSFORMATIONS WITH NEUTRON SCATTERING.
    ⊹AXE(J.D.), SHIRANE(G.).

59 PROC. INT. CONF. (NOTTINGHAM), . 103, (1964). . . . .
    ⊹THEORY OF CRITICAL RELAXATION AND SCATTERING.
    ⊹TOMITA(K.).

60 PROGR. THEOR. PHYS. VOL.25,. . . 723, (1961). . . .T. . . . .
    ⊹ON THE CRITICAL SCATTERING OF NEUTRONS.
    ⊹MORI(H.), KAWASAKI(K.).

61 REPORT AERE-R5627 (15 PP.),. . . . . , (1967). . . .B6. . . . .
    ⊹CRITICAL INELASTIC SCATTERING OF SLOW NEUTRONS FROM A BINARY LIQUID.
    ⊹EGELSTAFF(P.A.), WIGNALL(G.D.). (ED. NOTE: AVAIL. H.M.S.O., 49 HIGH
    HOLBORN, LONDON, WC1, ENG.).

62 REPORT FN-40/70 (6PP.),. . . . . , (1970). . . .D2. . . . .
    ⊹IMPURITY INFLUENCE ON NEUTRON CRITICAL SCATTERING IN FERROMAGNETICS.
    ⊹BALLY(D.), TOTIA(M.), LUNGU(A.M.), POPOVICI(M.). (ED. NOTE: AVAIL.
    COMITETUL PENTRU ENERGIA NUCLEARĂ INST. FIZICA ATOMICA, BUCHAREST).

63 REV. MOD. PHYS. VOL.30,. . . . . 75, (1958). . .T. . . . .
    ⊹THEORY OF CRITICAL SCATTERING.
    ⊹ELLIOTT(R.J.), MARSHALL(W.).

64 SOL. STATE COMM. VOL. 9, . . . 2021, (1971). . . . .
    ⊹ON THE CRITICAL BEHAVIOUR AT SECOND ORDER STRUCTURAL PHASE TRANSITIONS.
    ⊹FEDER(J.).

65 SOV. PHYS. JETP VOL.7, ... 281, (1958). . . .T. . . . .
    ⊹SCATTERING OF X-RAYS AND THERMAL NEUTRONS BY SINGLE-COMPONENT CRYSTALS
    NEAR PHASE-TRANSITION POINTS OF THE SECOND KIND.
    ⊹KRIVOGLAZ(M.A.).

66 SOV. PHYS. CRYST. VOL.4, ... 290, (1960). . . .T. . . . .
    ⊹X-RAY AND THERMAL-NEUTRON SCATTERING NEAR A SECOND-ORDER PHASE . . . . .
    TRANSITION POINT.
    ⊹KRIVOGLAZ(M.A.).

67 SOV. PHYS. J.E.T.P. VOL.30,. . . 151, (1970). . . . .
    ⊹PROPERTIES OF LONG AND SHORT RANGE CORRELATIONS IN THE CRITICAL REGION.
    ⊹POLYAKOV(A.M.).

68 SOV. PHYS. SOLID STATE VOL.14, . 142, (1972). . .T.D2. . . . .
    ⊹FLUCTUATIONS AND SCATTERING OF SLOW NEUTRONS AND ELECTROMAGNETIC WAVES
    NEAR THE CRITICAL ANTIFERROMAGNETIC-FERROMAGNETIC TRANSITION POINT.
    ⊹AKHIEZER(I.A.), GINZBURG(A.E.).

69 SOV. PHYS. JETP VOL.34, ... 913, (1972). . . .D1. . . . .
    ⊹SPIN WAVES AND SLOW NEUTRON AND LIGHT SCATTERING NEAR THE TRANSITION
    FROM THE FERROMAGNETIC TO ANTIFERROMAGNETIC PHASE.
    ⊹AKHIEZER(I.A.), GINZBURG(A.E.).

70 SOV. PHYS. SOLID STATE VOL.14, . . 2006, (1973). . .T.D2. . . . .
    ⊹CRITICAL SCATTERING OF NEUTRONS IN FERRITES AND ANTIFERROMAGNETS IN A
    MAGNETIC FIELD.
    ⊹BAR≠YAKHTAR(V.G.), TARASENKO(V.V.), YABLONSKII(D.A.).

71 1ST EUR. CONF./CONDENSED MATTER, 34, (1971). . . . .
    ⊹STUDY OF PHASE CHANGES BY NEUTRON SCATTERING.
    ⊹NIELSON(J.A.). (ED. NOTE: PUBL. EUROPEAN PHYS. SOC., GENEVA, SWITZ.).

72 TRIESTE/INTERN. COURSE,. . . . 501, (1967). . . . .
    ⊹INVESTIGATIONS OF MAGNETIC MATERIALS USING NEUTRON SCATTERING. . . . . .
    (38 PAGES).
    ⊹MARSHALL(W.), LOW(G.G.).

73 TRIESTE/INTERN. COURSE,. . . . 483, (1967). . . . .
    ⊹NEUTRONS AND CRITICAL PHENOMENA. (18 PAGES).
    ⊹JACROT(B.).

SECTION 14A-STATIC AND DYNAMIC STRUCTURE OF FLUIDS . . . .

1   AARHUS SUMMER SCHOOL, ...   373, (1963). . . . .
    *PHONONS IN LIQUIDS.
    *DE-BOER(J.).

2   ADV. IN PHYSICS VOL.14, ... 453, (1965) STRUCTURE OF LIQUIDS.
    *INTERATOMIC FORCES AND THE STRUCTURE OF LIQUIDS.
    *ENDERBY(J.E.), MARCH(N.H.).

3   ADV. STRUC. RES. DIFFR. METHOD 4   65, (1972). . . . .
    *THE CORRELATION FUNCTIONS FOR SIMPLE LIQUIDS.
    *ENDERBY(J.E.).

4   ANN. OF PHYS. VOL.81, ... 414, (1973). . .T. . . . .
    *QUANTUM THEORY OF PURE LIQUID METALS AS TWO-COMPONENT SYSTEMS.
    *MARCH(N.H.), TOSI(M.P.).

5   ANN. REV. PHYS. CHEM. VOL.24, . 159, (1973). . . .T,E. . . . .
    *THE STRUCTURE OF SIMPLE LIQUIDS (REVIEW).
    *EGELSTAFF(P.A.).

6   BRIT. J. APPL. PHYS. VOL.16, . . 1219, (1965). . . .T. . . . .
    *THE THERMAL MOTION OF SIMPLE LIQUIDS.
    *EGELSTAFF(P.A.).

7   BROOKHAVEN SYMPOSIUM, ...   16, (1965). . . . .
    *LIQUIDS.
    *VINEYARD(G.H.).

8   BROOKHAVEN SYMPOSIUM, ... 155, (1965). . . .T. . . . .
    *A MODIFIED KINETIC MODEL FOR THE CALCULATION OF G-SUB S(R VECTOR, TIME)
    IN DENSE FLUIDS.
    *GIBBS(A.G.).

9   CHEM. PHYS. LETT. VOL.21 NO.1, . 109, (1973). . . .A3. . . . . . D2*O-(NI*CL2,NA*CL,BA*CL2), . . .
    *THE STRUCTURE OF AQUEOUS SOLUTIONS.
    *ENDERBY(J.E.), HOWELLS(W.S.), HOWE(R.A.).

10 C.N.E.N. SYMP. CASACCIA, ... 219, (1968). . . . .
    *ATOMIC MOTION IN SIMPLE LIQUIDS.
    *SJOLANDER(A.).

11 CONTEMP. PHYS. VOL.6, ... 274, (1965). . . .T,E. . . . .
    *THE STRUCTURE AND THERMAL MOTION OF SIMPLE LIQUIDS- I.
    *EGELSTAFF(P.A.), SCHOFIELD(P.).

12 CONTEMP. PHYS. VOL.6, ... 453, (1965). . . .T,E. . . . .
    *THE STRUCTURE AND THERMAL MOTION OF SIMPLE LIQUIDS- II.
    *EGELSTAFF(P.A.), SCHOFIELD(P.).

13 DISC. FARADAY SOC. NO.43, . . . 160, (1967). . . .B4. . . . . . .P3. . . . .
    *BRILLOUIN SCATTERING OF NEUTRONS FROM LIQUIDS.
    *DORNER(B.), PLESSER(T.), STILLER(H.).

14 DISC. FARADAY SOC. NO.43, . . . 149, (1967). . . .T. . . . .
    *RADIATION SCATTERING STUDIES OF THE STRUCTURE AND TRANSPORT PROPERTIES
    OF LIQUIDS.
    *EGELSTAFF(P.A.).

15 DISC. FARADAY SOC. NO.49, . . . 193, (1970). . . .C3. . . . . . C5*H12 (PENTANE), ...
    *LOW FREQUENCY MOLECULAR MODES IN LIQUID HYDROCARBONS.
    *EGELSTAFF(P.A.), HARRIS(D.H.C.).

16 IAEA SYMPOSIUM BOMBAY, VOL.2, . . 59, (1964). . . . .
    *THE SELF-CORRELATION FUNCTION OF REAL GASES.
    *SIGMAR(D.J.).

17 IAEA SYMP. COPENHAGEN, VOL.1, . . 561, (1968). . . .C3. . . . . ARGON-LIKE LIQUID, ...
    *CURRENT FLUCTUATIONS IN CLASSICAL LIQUIDS.
    *RAHMAN(A.).

18 IAEA SYMP. COPENHAGEN, VOL.1, . . 573, (1968). . . . .
    *THE THEORY OF HIGH-FREQUENCY DENSITY FLUCTUATION IN LIQUIDS.
    *SCHOFIELD(P.).

19 IAEA SYMP. GRENOBLE, ... 365, (1972). . . .T. . . . .
    *A MODELLED KINETIC EQUATION WITH TWO RELAXATION TIMES.
    *AKCASU(A.Z.), LINNEBUR(E.J.).

20 IAEA SYMP. GRENOBLE, ... 383, (1972). . . . .
    *TIME-DEPENDENT TRIPLET CORRELATION FUNCTIONS IN SIMPLE LIQUIDS.
    *EGELSTAFF(P.A.).

21 J. CHEM. PHYSICS VOL.46, . . . 1999, (1967). . . .C3. . . . . .C*H4, ...
    *CORRELATION OF INFRARED AND INELASTIC NEUTRON SCATTERING SPECTRA.
    *AGRAWAL(A.K.), YIP(S.).

22 J. CHEM. PHYSICS VOL.48, . . . 3016, (1968). . . . .
    *ANALYSIS OF SCATTERING DATA FOR MIXTURES OF AMORPHOUS SOLIDS OR LIQUIDS.
    *PINGS(C.J.), WASER(J.).

23 J. CHEM. PHYS. VOL.57, ... 4346, (1972). . . . .
    *EVALUATION OF LIQUID STRUCTURE DATA.
    *MOUNTAIN(R.D.).

24 J. PHYS. C VOL.2, ... 556, (1969). . . . .
    *COLLECTIVE MOTION IN LIQUIDS.
    *HUBBARD(J.), BEEBY(J.L.).

25 J. PHYS. C VOL.4, ... 1453, (1971). . . . .
    *EXPERIMENTAL STUDY OF THE TRIPLET CORRELATION FUNCTION FOR SIMPLE
    LIQUIDS.
    *EGELSTAFF(P.A.), PAGE(D.I.), HEARD(C.R.T.).

26 J. PHYS. C VOL.4, ... 3057, (1971). . . . .
    *DENSITY FLUCTUATIONS IN CLASSICAL MONATOMIC LIQUIDS.
    *LOVESEY(S.W.).

27 J. PHYS. C VOL.6, ... 1856, (1973). . . .T. . . . . .AR. . . . .
    *SINGLE-PARTICLE MOTION IN CLASSICAL MONATOMIC LIQUIDS.
    *LOVESEY(S.W.).

SECTION   14A-STATIC AND DYNAMIC STRUCTURE OF FLUIDS . . . .

28 J. PHYS. C VOL.6,..   L254, (1973)....T...IN LIQUID METALS. . . . .
                        ⇌ELECTRON-ION TRIPLET CORRELATIONS IN LIQUID METALS.
                        ⇌PARRINELLO(M.), TOSI(M.P.).

29 J. STAT. PHYS. VOL.8 NO.2,.   107, (1973)....T....AR
                        ⇌COLLECTIVE MODES, DAMPING, AND THE SCATTERING FUNCTION IN CLASSICAL
                          LIQUIDS.
                        ⇌KUGLER(A.A.).

30 2ND SUMMER SCHOOL S.S. PHYS.,...   7, (1968)....T
                        ⇌LIQUID   DYNAMICS   AND   TRANSPORT PROPERTIES OF ANHARMONIC CRYSTALS.
                        ⇌SJOLANDER(A.).

31 NUOVO CIMENTO B VOL.13B NO.1,.   185, (1973)....T,B4....NA
                        ⇌COLLECTIVE MOTIONS IN CLASSICAL LIQUIDS. III.LIQUID SODIUM.
                        ⇌PATHAK(K.N.), SINGWI(K.S.), CUBIOTTI(G.), TOSI(M.P.).

32 OPT. SPECTROSC. VOL.35 NO.3,.   342, (1973)....T
                        ⇌THE POSSIBILITY OF COMPARING NMR EXPERIMENTS WITH DATA OF OPTICAL AND
                          NEUTRON SPECTROSCOPY.
                        ⇌YULMETEV(R.M.).

33 PHYS. CHEM. LIQUIDS VOL.3,.   205, (1972)...T......AR
                        ⇌SELF-MOTION RESPONSE AND INCOHERENT SCATTERING FUNCTION IN CLASSICAL
                          LIQUIDS.
                        ⇌KUGLER(A.A.).

34 PHYSICA VOL.50,..   511, (1970)....B4.......AR
                        ⇌TEMPERATURE DEPENDENCE OF THE ATOMIC SELF-MOTION IN LIQUID ARGON.
                        ⇌ZANDVELD(P.), ANDRIESSE(C.D.), BREGMAN(J.D.), HASMAN(A.), VAN-LOEF(J.J.)

35 PHYSICA VOL.62,..   345, (1972).......NE,AR,KR
                        ⇌TEMPERATURE AND DENSITY DEPENDENCE OF THE SELF-DIFFUSION COEFFICIENT
                          IN SIMPLE LIQUIDS.
                        ⇌VAN-LOEF(J.J.).

36 PHYSICA VOL.60,..   27, (1972)...B4.....AR
                        ⇌LIQUID-ARGON DYNAMICS FROM...FIRST MOMENTS OF:..COHERENT SCATT. FCN.:THE
                          DENSITY RESPONSE FCN./DISPERSION CURVE/INTERMEDIATE SCATT. FCN.
                        ⇌DUBOIS(D.M.).

37 PHYS. LETT. 28A,   208, (1968) . . . .
                        ⇌CO-OPERATIVE MODES OF MOTION IN A SIMPLE MODEL OF A LIQUID.
                        ⇌SYKES(J.).

38 PHYS. LETT. VOL.31A,.   237, (1970).......AL
                        ⇌THE RADIAL DISTRIBUTION FUNCTION FOR LIQUID ALUMINIUM FROM A NEW
                          PSEUDOPOTENTIAL CALCULATION.
                        ⇌BRAUNECK(W.), GAHN(U.), SOMMER(F.), WERBER(K.), WILLEE(CH.).

39 PHYS. LETT. VOL.33A,.   419, (1970).......AR
                        ⇌EVIDENCE FOR A LONG TAIL OF THE VELOCITY AUTOCORRELATION FUNCTION IN
                          LIQUID ARGON.
                        ⇌ANDRIESSE(C.D.).

40 PHYS. LETT. VOL.37A,.   321, (1971)....T
                        ⇌DENSITY HIERARCHY FOR THE TIME-DEPENDENT CORRELATION FUNCTIONS.
                        ⇌EGELSTAFF(P.A.), GRAY(C.G.), GUBBINS(K.E.).

41 PHYS. REV. VOL.84,   466, (1951)....T
                        ⇌ON THE THEORY OF COHERENT SCATTERING PROCESSES IN LIQUIDS.
                        ⇌GOLDSTEIN(L.).

42 PHYS. REV. VOL.95,   249, (1954)....T
                        ⇌CORRELATIONS IN SPACE AND TIME AND BORN APPROXIMATION SCATTERING IN
                          SYSTEMS OF INTERACTING PARTICLES.
                        ⇌VAN-HOVE(L.).

43 PHYS. REV. VOL.164,.   222, (1967)....T.......PB,
                        ⇌COLLISIONLESS SOUND IN CLASSICAL FLUIDS.
                        ⇌NELKIN(M.), RANGANATHAN(S.).

44 PHYS. REV. LETT. 21,   881, (1968)
                        ⇌ZERO SOUND IN CLASSICAL LIQUIDS.
                        ⇌SINGWI(K.S.), SKOLD(K.), TOSI(M.P.).

45 PHYS. REV. VOL.167,.   171, (1968)....T
                        ⇌ROTATIONAL AND TRANSLATIONAL DIFFUSION IN COMPLEX LIQUIDS.
                        ⇌LARSSON(K.E.).

46 PHYS. REV. VOL.176,.   239, (1968)....T.......AR
                        ⇌UNIFIED APPROXIMATION FOR THE VELOCITY AUTOCORRELATION FUNCTION AND THE
                          STRUCTURE FUNCTION OF A SIMPLE LIQUID.
                        ⇌GLASS(L.), RICE(S.A.).

47 PHYS. REV. VOL.183,.   349, (1969)....T
                        ⇌EFFECTIVE-FIELD APPROXIMATIONS IN CLASSICAL LIQUIDS.
                        ⇌NELKIN(M.).

48 PHYS. REV. VOL.182,.   323, (1969)....T.......AR
                        ⇌GENERALIZED HYDRODYNAMICS AND TIME CORRELATION FUNCTIONS.
                        ⇌CHUNG(C.H.), YIP(S.).

49 PHYS. REV. A VOL.2,.   2427, (1970)....T
                        ⇌COLLECTIVE MOTIONS IN CLASSICAL LIQUIDS. II.
                        ⇌PATHAK(K.N.), SINGWI(K.S.).

50 PHYS. REV. A VOL.2,.   2158, (1970)....T
                        ⇌COMMENT ON DISPERSION OF PHONONS IN LIQUID HE-4.
                        ⇌FEENBERG(E.).

51 PHYS. REV. A VOL.1,.   454, (1970)....T
                        ⇌COLLECTIVE MOTIONS IN CLASSICAL LIQUIDS.
                        ⇌SINGWI(K.S.), SKOLD(K.), TOSI(M.P.).

52 PHYS. REV. A VOL.2,.   1095, (1970)....T
                        ⇌THEORIES OF COLLECTIVE MOTION IN LIQUIDS.
                        ⇌GLUCK(P.).

53 PHYS. REV. A VOL.3,.   2145, (1971)....T
                        ⇌THEORY OF THE LIQUID-SOLID PHASE TRANSITION.
                        ⇌SCHNEIDER(T.).

54 PHYS. REV. A VOL.3,.   1752, (1971)....T
                        ⇌SPACE-TIME PAIR CORRELATION FUNCTION OF A MAXWELL-BOLTZMANN FLUID IN
                          THE T-APPROXIMATION FOR TEMPERATURE GREEN≠S FUNCTIONS. I. THEORY.
                        ⇌LERIBAUX(H.R.), POPE(N.K.).

SECTION   14A-STATIC AND DYNAMIC STRUCTURE OF FLUIDS . . . .

55   PHYS. REV. A VOL.4, ... 1616, (1971). . . . .
     *GENERALIZED HYDRODYNAMICS AND ANALYSIS OF CURRENT CORRELATION FUNCTIONS.
     *AILAWADI(N.K.), RAHMAN(A.), ZWANZIG(R.).

56   PHYS. REV. A VOL.3, ... 717, (1971). . . . .
     *HYDRODYNAMIC EQUATIONS FOR THE CONDENSATE AND THE DEPLETION OF HELIUM II
     *HAUG(H.), WEISS(K.).

57   PHYS. REV. A VOL.5, ... 2629, (1972). . . . .
     *ASYMPTOTIC FORM OF THE PAIR-CORRELATION FUNCTION IN LIQUIDS.
     *ALBERS(J.), MOUNTAIN(R.D.).

58   PHYS. REV. A VOL.5, ... 2519, (1972). . . . .
     *DECAY OF PAIR CORRELATIONS.
     *THROOP(G.J.), FISK(S.).

59   PHYS. REV. A VOL.6, ... 2243, (1972). . . . .T. . . . .
     *PARACRYSTALLINE STRUCTURE OF MOLTEN METALS.
     *HOSEMANN(R.), WILLMANN(G.), ROESSLER(B.).

60   PHYS. STAT. SOLIDI B VOL.57 NO.1   351, (1973). . . .T,P. . . . . . .
     *INTERIONIC POTENTIALS IN LIQUID METALS INCLUDING LIQUID NOBLE AND
     TRANSITION METALS.
     *WASEDA(Y.), SUZUKI(K.).

61   PROC. ROY. SOC. A VOL.282, ... 283, (1964). . . .T. . . . .
     *ION-ION OSCILLATORY POTENTIALS IN LIQUID METALS.
     *JOHNSON(M.D.), HUTCHINSON(P.), MARCH(N.H.).

62   PROPERTIES LIQUID METALS, ... 43, (1973). . . .T. R8,NA,K,
     *EFFECTIVE PAIR-POTENTIALS FOR THE LIQUID ALKALI METALS.
     *HOWELLS(W.S.). (ED. NOTE: REFER SECT. 4).

63   PROPERTIES LIQUID METALS, ... 13, (1973). . . . . .R3.
     *LIQUID METALS: THEIR STRUCTURE AND DYNAMICS (REVIEW PAPER).
     *EGELSTAFF(P.A.). (ED. NOTE: REFER SECT. 4).

64   PROPERTIES LIQUID METALS, ... 3, (1973). R8,P8,
     *THE STRUCTURES OF LIQUID METALS AND ALLOYS (REVIEW PAPER).
     *ENDERBY(J.E.). (ED. NOTE: REFER SECT. 4).

65   REACTOR PHYSICS, ... 123, (1966). . . .T. . . . .
     *THE VELOCITY AUTOCORRELATION FUNCTION IN LIQUIDS FROM A NEW POINT OF
     VIEW.
     *RAHMAN(A.). (IN BOOK: REACTOR PHYSICS IN THE RESONANCE AND THERMAL
     REGIONS VOL.I, NEUTRON THERMALIZATION. PUBL. M.I.T. PRESS, 1966).

66   REPORT AF-SSP-27, ... , (1968). . . .T. . . . . . . .AR,
     *ZERO SOUND DISPERSION LAW IN CLASSICAL LIQUIDS.
     *SCHNEIDER(T.), STOLL(E.), SZABO(N.). (ED. NOTE: AVAIL. SWISS FED. INST.
     FOR REACTOR RESEARCH, 5303 WURENLINGEN, SWITZ.).

67   REP. PROGR. PHYS. VOL.29, ... 333, (1966). . . .T,E. . . . .
     *MICROSCOPIC TRANSPORT PHENOMENA IN LIQUIDS.
     *EGELSTAFF(P.A.).

68   SOV. PHYS. J.E.T.P. VOL.32, ... 1183, (1971). . . . .
     *CONTRIBUTION OF VAN DER WAALS FORCES TO THE FORM FACTOR OF A LIQUID.
     *KEMOKLIDZE(M.P.), PITAEVSKII(L.P.).

69   SOV. PHYS. J.E.T.P. VOL.32, ... 729, (1971). . . . .
     *THE BEHAVIOR OF THE STRUCTURE FACTOR OF A CLASSICAL LIQUID WITH SMALL
     WAVE NUMBERS.
     *FISHER(I.Z.).

SECTION 14B-NEUTRON SCATTERING BY FLUIDS . . . .

1 ACTA CRYST. VOL.A29, ... 692, (1973)....T,C3....
   *COLD-NEUTRON INCOHERENT SCATTERING BY HOMOGENEOUSLY ORIENTED NEMATIC
   LIQUID CRYSTALS.
   *OLIVEI(A.).

2 ACTA PHYS. POLON. VOL.26,. . . . . 19, (1964)....T
   *ON THE STATISTICAL MECHANICAL EVALUATION OF THE SLOW NEUTRON SCATTERING
   FUNCTION.
   *FULINSKI(A.).

3 ACTA PHYS. POLON. VOL.A41, . . . 549, (1972)....C3........
   *INCOHERENT CROSS-SECTION FOR NEUTRON QUASI-ELASTIC SCATTERING IN A
   LIQUID CRYSTAL.
   *ROSCISZEWSKI(K.).

4 ACTA PHYS. AUSTRIACA VOL.37, . . 365, (1973)....T,E....H2
   *NEUTRON SCATTERING- A METHOD FOR INVESTIGATING THE LIQUID AND GASEOUS
   STATES OF MATTER.
   *EDER(O.J.). (IN GERMAN).

5 ADV. IN PHYSICS VOL.11,. . . . . 203, (1962)....................
   *NEUTRON SCATTERING STUDIES OF LIQUID DIFFUSION.
   *EGELSTAFF(P.A.).

6 ADV. IN PHYSICS VOL.16,. . . . . 177, (1967)....NA-K,SI*O2....
   *STRUCTURE AND CORRELATION IN LIQUID ALLOYS BY X-RAY AND NEUTRON
   DIFFRACTION.
   *HENNINGER(E.H.), BUSCHERT(R.C.), HEATON(L.).

7 ADV. IN PHYSICS VOL.16,. . . . . 147, (1967)....................
   *RADIATION SCATTERING DATA ON LIQUID METALS.
   *EGELSTAFF(P.A.).

8 ADV. IN PHYSICS VOL.22,. . . . . 1, (1973)....E,T
   *THE STRUCTURE OF MOLECULAR LIQUIDS BY NEUTRON SCATTERING (REVIEW).
   *POWLES(J.G.).

9 ANN. PHYS. (GER.) VOL.23, ... 49, (1969)....T........AR
   *COLLECTIVE MOTIONS IN CLASSICAL SYSTEMS.3. THEORETICAL STUDY OF THERMAL
   NEUTRON SCATTERING IN LIQUID ARGON.
   *RICHTER(J.), VOSS(K.).

10 BER. BUNSENGES. PHYS. CHEM. 75,. 769, (1971)....P........C*H3*C*N (ACETONITRILE),
   *A COMPARATIVE STUDY OF QUASIELASTIC NEUTRON SCATTERING AND NMR RELAXA-
   TION IN LIQUID ACETONITRILE.
   *ZEIDLER(M.D.).

11 BER. BUNSENGES. PHYS. CHEM. 75,. 366, (1971)....C3........H2*O (HYDRATION COMPLEXES),.
   *NEUTRON INELASTIC SCATTERING INVESTIGATIONS OF HYDRATION COMPLEXES AND
   ASSOCIATED DIFFUSIVE KINETICS IN IONIC SOLUTIONS.
   *SAFFORD(G.J.), LEUNG(P.S.).

12 BRIT. J. APPL. PHYS. VOL.10,. . . 1, (1959)....................
   *SOLID AND LIQUID STATE RESEARCH WITH COLD NEUTRONS.
   *EGELSTAFF(P.A.).

13 CAN. J. PHYS. VOL.30,. . . . . . 597, (1952)....T
   *THE THEORY OF NEUTRON DIFFRACTION BY GASES. I.
   *POPE(N.K.).

14 CAN. J. PHYS. 44,. . . . . 1279, (1966)................
   *THEORY OF COLD NEUTRON SCATTERING BY HOMONUCLEAR DIATOMIC LIQUIDS. I.
   FREE ROTATION.
   *SEARS(V.F.).

15 CAN. J. PHYS. 44,. . . . . 1299, (1966)................
   *THEORY OF COLD NEUTRON SCATTERING BY HOMONUCLEAR DIATOMIC LIQUIDS. II.
   HINDERED ROTATION.
   *SEARS(V.F.).

16 CAN. J. PHYS. 44,. . . . . 867, (1966)................
   *THE LAW OF CORRESPONDING STATES AND COLD NEUTRON SCATTERING BY LIQUIDS.
   *SEARS(V.F.).

17 CHEM. APPL. THERMAL NEUTRON SCAT 118, (1973)....T
   *ATOMIC AND MOLECULAR MOTION IN LIQUIDS BY THERMAL NEUTRON SCATTERING.
   *POWLES(J.G.). (IN BOOK: CHEMICAL APPLICATIONS OF THERMAL NEUTRON
   SCATTERING. ED. BY B.T.M. WILLIS; OXFORD UNIV. PRESS: LONDON).

18 CHEM. APPL. THERMAL NEUTRON SCAT 173, (1973)................
   *THE STRUCTURE OF LIQUIDS BY NEUTRON SCATTERING.
   *PAGE(D.I.). (IN BOOK: CHEMICAL APPLICATIONS OF THERMAL NEUTRON
   SCATTERING. ED. BY B.T.M. WILLIS; OXFORD UNIV. PRESS: LONDON).

19 C.N.E.N. SYMP. CASSACCIA,. . . . 51, (1968)................
   *NEUTRON SCATTERING AND LIQUID STATE.
   *EGELSTAFF(P.A.).

20 DISS. ABSTR. VOL.21,. . . . . . . 3826, (1961)....T
   *SLOW-NEUTRON DIFFRACTION IN NORMAL LIQUIDS.
   *AMADO(A.M.).

21 DISS. ABS. VOL.24,. . . . . 4250, (1964)................
   *TEMPERATURE MEASUREMENTS IN HOT GASES BY THERMAL NEUTRON SCATTERING.
   *DOHERTY(P.D.).

22 DISS. ABS. VOL.24,. . . . . 685, (1964)................
   *THE SCATTERING OF SLOW NEUTRONS BY POLAR LIQUIDS.
   *YIP(S.).

23 DISS. ABS. VOL.26,. . . . . 6781, (1966)................
   *KINETIC MODELS FOR FLUIDS AND THEIR APPLICATION TO SLOW NEUTRON
   SCATTERING.
   *GIBBS(A.G.).

24 DISS. ABS. VOL.27,. . . . . 1964B, (1967)................
   *SCATTERING OF LOW-ENERGY NEUTRONS IN A MONATOMIC GAS MODEL OF A
   MULTIPLYING SYSTEM.
   *BLACKSHAW(G.L.).

SECTION 14B-NEUTRON SCATTERING BY FLUIDS . . . .

25 ENERGIA NUCLEARE VOL.7, ... 772, (1960) . . . .
&STUDY OF THE (LATTICE) DYNAMICS OF SOLID AND OF LIQUID MEDIA BY SLOW
NEUTRON SPECTROMETRY.
&FELCHER(G.P.), MUSCI(M.).

26 FORTSCHR. PHYS. VOL.14 NO.2, . . . 73, (1966) . . . . . A3 . . . .
&STRUCTURE OF MONATOMIC LIQUID METALS BY X-RAY, ELECTRON AND NEUTRON
DIFFRACTION.
&RICHTER(H.), BREITLING(G.). (IN GERMAN).

27 IAEA SYMPOSIUM VIENNA, ... 239, (1960) . . . .
&NEUTRON SCATTERING BY NORMAL LIQUIDS.
&DE-GENNES(P.G.).

28 IAEA SYMPOSIUM VIENNA, ... 251, (1960) . . . .
&SLOW-NEUTRON SCATTERING BY NORMAL LIQUIDS ACCORDING TO THE SMEARED
POTENTIAL MODEL.
&MORALES-AMADO(A.), OSBORN(R.K.).

29 IAEA SYMPOSIUM VIENNA, ... 259, (1960) . . . .
&SCATTERING OF NEUTRONS IN LIQUIDS.
&.C.C.C.P.

30 IAEA SYMP. CHALK RIVER, VOL.1, . . 215, (1962) . . . C3 . . . . . . H2*O,PB,
&FREQUENCY SPECTRUM OF LIQUIDS AND COLD NEUTRON SPECTROMETRY.
&SINGWI(K.S.), SJOLANDER(A.), RAHMAN(A.).

31 IAEA SYMP. CHALK RIVER, VOL.1, . . 203, (1962) . . . C3 . . . . . . PB,SN,
&INTERPRETATION OF COHERENT NEUTRON SCATTERING BY LIQUIDS.
&EGELSTAFF(P.A.).

32 IAEA SYMP. CHALK RIVER, VOL.1, . . 189, (1962) . . . B4 . . . . . . AR,H2*O,
&LIQUID DYNAMICS FROM NEUTRON SPECTROMETRY.
&BROCKHOUSE(B.N.), BERGSMA(J.), DASANNACHARYA(B.A.), POPE(N.K.).

33 IAEA SYMP. COPENHAGEN, VOL.1, . . 535, (1968) . . . .
&COLLECTIVE MODES IN LIQUIDS AND NEUTRON SCATTERING.
&NELKIN(M.), ORTOLEVA(P.J.).

34 IAEA SYMP. (INSTRUMENT.), VIENNA, 19, (1969) . . . .
&A COMPARISON OF DIFFERENT INSTRUMENTS WITH RESPECT TO MEASUREMENTS ON
LIQUIDS.
&DORNER(B.), STILLER(H.H.).

35 IAEA SYMP. GRENOBLE, ... 315, (1972) . . . .
&SOME APPLICATIONS OF INELASTIC NEUTRON SCATTERING SPECTROSCOPY TO
SURFACE CHEMISTRY AND CATALYSIS (REVIEW PAPER).
&WHITE(J.W.).

36 J. CHEM. PHYS. VOL.42, ... 3540, (1965) . . . A3 . . . . . . AR,
&NORMALIZATION OF DIFFRACTION DATA FROM LIQUIDS.
&RAHMAN(A.).

37 J. CHEM. PHYS. VOL.42, ... 1863, (1965) . . . T . . . .
&SCATTERING FROM FLUIDS OF NONSPHERICAL MOLECULES.
I. X RAYS AND NEUTRONS.
&STEELE(W.A.), PECORA(R.).

38 J. CHEM. PHYS. VOL.43, ... 3328, (1965) . . . . C*H4, ...
&ON THE USE OF THE LANGEVIN EQUATION OF BROWNIAN MOTION FOR THE ANALYSES
OF COLD AND SLOW NEUTRON SCATTERING BY LIQUID METHANE.
&GRIFFING(G.W.).

39 J. CHEM. PHYSICS VOL.47, ... 4923, (1967) . . . .
&QUANTUM STATISTICS AND SLOW NEUTRON SCATTERING BY GASES.
&PLUMMER(J.P.), SUMMERFIELD(G.C.), ZWEIFEL(P.F.).

40 J. CHEM. PHYSICS VOL.46, ... 465, (1967) . . . .
&ORIENTATION-AVERAGED AMPLITUDE OF THE ONE-QUANTUM TERM IN THE NEUTRON
SCATTERING LAW FOR MOLECULAR GASES.
&CARPENTER(J.M.).

41 J. CHEM. PHYS. VOL.56, ... 3118, (1972) . . . C3 . . . . . C*H3*O*H,
&PROTON MOTIONS IN METHANOL BY COLD NEUTRON SCATTERING.
&RODRIGUEZ(C.), AMARAL(L.Q.), VINHAS(L.A.), HERDADE(S.B.).

42 J. CHEM. PHYS. VOL.57, ... 290, (1972) . . . T . . . . . . POLYMER SOLUTIONS, . . .
&NEUTRON DIFFRACTION IN DILUTE AND SEMIDILUTE POLYMER SOLUTIONS.
&COTTON(J.P.), FARNOUX(B.), JANNINK(G.).

43 J. CHIM. PHYS. VOL.61, ... 97, (1964) . . . E . . . .
&COLD NEUTRON INVESTIGATION OF THE STRUCTURE OF LIQUIDS.
&JANIK(J.A.).

44 J. NUCL. SCI. TECHNOL. VOL.8, . . 1, (1971) . . . A3 . . . .
&THERMAL NEUTRON SCATTERING FROM SIMPLE LIQUID METALS.
&NAKAHARA(Y.).

45 J. OF PHYS. C VOL.1, ... 784, (1968) . . . .
&THE STRUCTURE FACTOR FOR LIQUID METALS. I. THE APPLICATION OF NEUTRON
DIFFRACTION TECHNIQUES.
&NORTH(D.M.), ENDERBY(J.E.), EGELSTAFF(P.A.).

46 J. OF PHYS.-C-, SER.2, VOL.4, . . 2725, (1971) . . . T C*H4,
&MULTIPLE SCATTERING OF NEUTRONS IN LIQUID METHANE.
&RAO(K.R.), DASANNACHARYA(B.A.), YIP(S.).

47 J. PHYS. C VOL.4, ... 3029, (1971) . . . .
&PARTIAL STRUCTURE FACTORS IN A LIQUID OR AMORPHOUS BINARY SYSTEM USING
ANOMALOUS SCATTERING.
&RAMESH(T.G.), RAMASESHAN(S.).

48 J. PHYS. C VOL.6, ... 2262, (1973) . . . T . . . .
&NEUTRON SCATTERING FROM AQUEOUS SOLUTIONS: THE LONG-WAVELENGTH LIMIT.
&BEEBY(J.L.).

49 LIQUIDS: STRUC.,PROP., SOL. INT. 201, (1965) . . . E,T H2*O,PB,
&INELASTIC NEUTRON SCATTERING BY LIQUIDS.
&PALEVSKY(H.). (ED. NOTE: LIQUIDS: STRUCTURE, PROPERTIES, SOLID
INTERACTIONS;PUBL. BY ELSEVIER: AMSTERDAM).

50 LIQUIDS: STRUC.,PROP., SOL. INT. 172, (1965) . . . E . . . .
&X-RAY AND NEUTRON DIFFRACTION STUDIES OF LIQUID STRUCTURE.
&GINGRICH(N.S.). (ED. NOTE: LIQUIDS: STRUCTURE, PROPERTIES, SOLID
INTERACTIONS; PUBL. BY ELSEVIER: AMSTERDAM).

SECTION 14B-NEUTRON SCATTERING BY FLUIDS . . . .

51 2ND SUMMER SCHOOL S.S. PHYS.,. . 43, (1968).............
   ~LIQUID DYNAMICS FROM NEUTRON SCATTERING (REVIEW PAPER).
   ~LARSSON(K.E.).

52 NED. TIJDSCHRIFT NATUURKUNDE 29, 93, (1963)..........
   ~SCATTERING OF SLOW NEUTRONS IN CRYSTALS AND LIQUIDS. II.
   ~KOKKEDEE(J.J.), NIJBOER(B.R.). (IN DUTCH).

53 NED. TIJDSCHRIFT NATUURKUNDE 29, 61, (1963).....T.......
   ~SCATTERING OF SLOW NEUTRONS BY CRYSTALS AND FLUIDS. I.
   ~NIJBOER(B.R.), KOKKEDEE(J.J.). (IN DUTCH).

54 NUCL. PHYS. VOL.40,. . . . 139, (1963).....T.......
   ~ON SCATTERING OF SLOW NEUTRONS IN A FERMI LIQUID.
   ~AKHIEZER(A.I.), AKHIEZER(I.A.), POMERANCHUK(I.YA.).

55 NUCL. SCI. ENGNG. VOL.12,. . . . 260, (1962).....T.......
   ~ON THE EVALUATION OF THE THERMAL NEUTRON SCATTERING LAW.
   ~EGELSTAFF(P.A.), SCHOFIELD(P.).

56 NUCL. SCI. ENGNG. VOL.15,. . . . 438, (1963)....T,P......C*H4,C3*H8..
   ~EVALUATION OF TECHNIQUES FOR COMPUTING PARTIAL DIFFERENTIAL SCATTERING
   CROSS-SECTIONS.
   ~MCMURRY(H.L.), GANNON(L.J.), HESTIR(W.A.).

57 NUCL. SCI. ENGNG. VOL.15,. . . . 429, (1963)....T,P......
   ~CALCULATION OF PARTIAL DIFFERENTIAL SCATTERING CROSS-SECTIONS FOR SLOW
   NEUTRONS.
   ~MCMURRY(H.L.).

58 NUCL. SCI. ENGNG. VOL.50,. . . . 164, (1973)....T.......
   ~AN INTEGRAL FORMULATION OF THE NEUTRON SCATTERING MOMENTS FOR MONATOMIC
   GASES.
   ~KUEHN(N.H.), MURRAY(R.L.).

59 NUOVO CIMENTO B VOL.43,. . . . 375, (1966).....E.......
   ~ACCURATE MEASUREMENTS OF STRUCTURE FACTORS OF LIQUIDS BY SLOW-NEUTRON
   SPECTROMETRY.
   ~ASCARELLI(P.), CAGLIOTI(G.).

60 PHIL. MAG. (8) VOL.1,. . . . 560, (1956)....T.......
   ~SCATTERING OF COLD NEUTRONS IN LIQUID METALS AND THE ENTROPY OF DISORDER
   ~KOTHARI(L.S.), SINGWI(K.S.), VISVANATHAN(S.).

61 PHYSICA VOL.25,. . . . 825, (1959)....T.......
   ~LIQUID DYNAMICS AND INELASTIC SCATTERING OF NEUTRONS.
   ~DE-GENNES(P.G.).

62 PHYSICA VOL.30,. . . . 1561, (1964).....C3......H2*O,GLYCEROL,OLEIC ACID,PENTANE
   ~PROTON MOTION IN SOME HYDROGENOUS LIQUIDS STUDIED BY COLD NEUTRON
   SCATTERING.
   ~LARSSON(K.E.), DAHLBORG(U.).

63 PHYSICA VOL.31,. . . . 1257, (1965)....B4,T......AR
   ~COHERENT SCATTERING OF SLOW NEUTRONS BY LIQUID ARGON. II.
   ~SINGWI(K.S.).

64 PHYSICA VOL.32,. . . . 415, (1966)....T.......
   ~TIME EXPANSION OF CORRELATION FUNCTIONS AND THE THEORY OF SLOW NEUTRON
   SCATTERING.
   ~NIJBOER(B.R.), RAHMAN(A.).

65 PHYSICS VOL.3 NO.1,. . . . 37, (1967)....T.......
   ~QUASI-ELASTIC SCATTERING OF NEUTRONS BY DILUTE POLYMER SOLUTIONS: I.
   FREE-DRAINING LIMIT.
   ~DE-GENNES(P.G.).

66 PHYSICA VOL.48,. . . . 61, (1970)....B5......AR
   ~ATOMIC MOTION IN GASEOUS ARGON. I.DETERMINATION OF THE INTERMEDIATE
   SCATTERING FUNCTION OF GASEOUS ARGON USING SUBTHERMAL NEUTRONS.
   ~ANDRIESSE(C.D.).

67 PHYSICA VOL.49,. . . . 502, (1970)....T......AR
   ~ATOMIC MOTION IN GASEOUS ARGON. II.COMPARISON OF MODELS ON THE ATOMIC
   MOTION IN FLUIDS WITH NEUTRON-SCATTERING DATA.
   ~ANDRIESSE(C.D.).

68 PHYSICA VOL.70,. . . . 105, (1973)....T......AR
   ~HIGH-FREQUENCY BEHAVIOUR OF THE NEUTRON-SCATTERING FUNCTION.
   ~KLEBAN(P.).

69 PHYS. LETT. 3, ... 162, (1962)....NA......
   ~SLOW NEUTRON SCATTERING EVIDENCE FOR VIOLATING OF THE F-SUM RULE IN
   LIQUID NA.
   ~RANDOLPH(P.D.).

70 PHYS. LETT. 3, ... 145, (1962)....T.......
   ~COLD NEUTRON SCATTERING AND DIFFUSIVE MOTIONS IN HYDROGEN BONDED LIQUIDS
   ~LARSSON(K.E.), SINGWI(K.S.).

71 PHYS. LETT. 25A, ... 211, (1967)....T.......
   ~SUM RULE CRITERION IN COHERENT SLOW NEUTRON SCATTERING IN LIQUIDS.
   ~DESAI(R.C.), YIP(S.).

72 PHYS. LETT. VOL.30A,. . . . 393, (1969).....E......HYDROGENOUS LIQUIDS,
   ~MICRODYNAMIC BEHAVIOUR OF HYDROGENOUS LIQUIDS BY CROSS SECTION MEASURE-
   MENTS WITH COLD NEUTRONS.
   ~FISCHER(C.O.).

73 PHYS/NON-CRYSTALLINE SOLIDS, . . 127, (1965).....E.......
   ~THE STUDY OF NON-CRYSTALLINE SOLIDS AND LIQUIDS BY INELASTIC NEUTRON
   SCATTERING.
   ~EGELSTAFF(P.A.). (ED. NOTE: PHYSICS OF NON-CRYSTALLINE SOLIDS PUBL. BY
   NORTH-HOLLAND: AMSTERDAM).

74 PHYS. REV. VOL.82,. . . . 392, (1951)....T.......
   ~EFFECT OF SHORT WAVELENGTH INTERFERENCE ON NEUTRON SCATTERING BY DENSE
   SYSTEMS OF HEAVY NUCLEI.
   ~PLACZEK(G.), NIJBOER(B.R.A.), VAN-HOVE(L.).

75 PHYS. REV. VOL.101,. . . . 129, (1956)....T......C*H4,.
   ~NEUTRON DIFFRACTION BY GASES.
   ~ZEMACH(A.C.), GLAUBER(R.J.).

SECTION 14B-NEUTRON SCATTERING BY FLUIDS . . . .

76 PHYS. REV. VOL.110, ... 999, (1958). . . .T. . . . .
*SCATTERING OF SLOW NEUTRONS BY A LIQUID.
*VINEYARD(G.H.).

77 PHYS. REV. VOL.109, ... 1564, (1958). . . .T. . . . .
*DIFFRACTION OF NEUTRONS BY IMPERFECT GASES.
*HAZO(R.M.), ZEMACH(A.C.).

78 PHYS. REV. VOL.119, ... 1150, (1960). . . . .
*NEUTRON SCATTERING BY FLUIDS AND THE LAW OF CORRESPONDING STATES.
*VINEYARD(G.H.).

79 PHYS. REV. VOL.122, ... 9, (1961). . . .T. . . . . . . .P8, ...
*DYNAMICS OF ATOMIC MOTIONS IN LIQUIDS AND COLD NEUTRON SCATTERING.
*RAHMAN(A.), SINGWI(K.S.), SJOLANDER(A.).

80 PHYS. REV. VOL.126, ... 997, (1962). . . .T. . . . .
*STOCHASTIC MODEL OF A LIQUID AND COLD NEUTRON SCATTERING.
*RAHMAN(A.), SINGWI(K.S.), SJOLANDER(A.).

81 PHYS. REV. VOL.126, ... 986, (1962). . . .T. . . . .
*THEORY OF SLOW NEUTRON SCATTERING BY LIQUIDS. I.
*RAHMAN(A.), SINGWI(K.S.), SJOLANDER(A.).

82 PHYS. REV. VOL.136, ... A969, (1964). . . . .
*COHERENT SCATTERING OF SLOW NEUTRONS BY A LIQUID.
*SINGWI(K.S.).

83 PHYS. REV. VOL.136, ... A1280, (1964). . . . .
*COHERENT 'INELASTIC' SCATTERING OF SLOW NEUTRONS BY A POLYATOMIC LIQUID
ELEMENT.
*ANTONINI(M.), ASCARELLI(P.), CAGLIOTI(G.).

84 PHYS. REV. VOL.133, ... A50, (1964). . . . .
*INCOHERENT INELASTIC NEUTRON SCATTERING AND SELF-DIFFUSION.
*ZWANZIG(R.).

85 PHYS. REV. LETT. 15, ... 693, (1965). . . . .
*KOHN ANOMALY AND COHERENT SCATTERING OF SLOW NEUTRONS IN LIQUID METALS.
*SINGWI(K.S.), ANDERSON(L.E.).

86 PHYS. REV. VOL.151, ... 126, (1966). . . .C3. . . . . .C5*H12, C3*H7*O*H, ...
*PROTON MOTIONS IN COMPLEX HYDROGENOUS LIQUIDS.
II. RESULTS GAINED FROM SOME NEUTRON SCATTERING EXPERIMENTS.
*LARSSON(K.E.), QUEROZ-DO-AMARAL(L.), IVANCHEV(N.), RIPEANU(S.),
BERGSTEDT(L.), DAHLBORG(U.).

87 PHYS. REV. VOL.148, ... 124, (1966). . . .T. . . . .
*SLOW-NEUTRON SCATTERING BY LIQUIDS: A HINDERED-TRANSLATOR MODEL.
*ARDENTE(V.), NARDELLI(G.F.), REATTO(L.).

88 PHYS. REV. VOL.151, ... 117, (1966). . . .T. . . . .
*PROTON MOTIONS IN COMPLEX HYDROGENOUS LIQUIDS.
I. A CROSS SECTION FOR QUASI-ELASTIC SCATTERING OF SLOW NEUTRONS.
*LARSSON(K.E.), BERGSTEDT(L.).

89 PHYS. REV. 158, ... 97, (1967). . . . .
*COLLECTIVE EFFECTS AND THE HYDRODYNAMIC APPROXIMATION IN NEUTRON
SCATTERING FROM FLUIDS.
*FERZIGER(J.H.), FEINSTEIN(D.L.).

90 PHYS. REV. VOL.165, ... 186, (1968). . . .T. . . . .
*NEW APPROXIMATION FOR THE CALCULATION OF NEUTRON SCATTERING
FROM A SIMPLE LIQUID.
*GLASS(L.), RICE(S.A.).

91 PHYS. REV. LETT. VOL.24, ... 1044, (1970). . . .T. . . . . . . .HE, ...
*SUM RULES, STRUCTURE FACTORS, AND PHONON DISPERSION IN LIQUID HE (A=4)
AT LONG WAVELENGTHS AND LOW TEMPERATURES.
*PINES(D.), CHIA-WEI-WOO.

92 PHYS. REV. A VOL.5, ... 452, (1972). . . . .
*INCOHERENT NEUTRON SCATTERING BY SIMPLE CLASSICAL LIQUIDS FOR LARGE
MOMENTUM TRANSFER.
*SEARS(V.F.).

93 PHYS. REV. A VOL.5, ... 320, (1972). . . . . .HE, ...
*X-RAY SCATTERING FROM LIQUID 4 HE.
*HALLOCK(R.B.).

94 PHYS. REV. A VOL.7 NO.1, ... 340, (1973). . . .T. . . . . . .NE,RB,H2, ...
*INCOHERENT NEUTRON SCATTERING FOR LARGE MOMENTUM TRANSFER.
II. QUANTUM EFFECTS AND APPLICATIONS.
*SEARS(V.F.).

95 PROC. PHYS. SOC. VOL.77, ... 353, (1961). . . .T. . . . .
*NEUTRON SCATTERING FROM A LIQUID ON A JUMP DIFFUSION MODEL.
*CHUDLEY(C.T.), ELLIOTT(R.J.).

96 PROC. PHYS. SOC. 81, ... 276, (1963). . . . .
*SCATTERING OF SLOW NEUTRONS BY HEAVY WATER.
*BUTLER(D.).

97 PROC. PHYS. SOC. 81, ... 294, (1963). . . . .
*SCATTERING OF SLOW NEUTRONS BY HEAVY WATER.
*BUTLER(D.).

98 PROC. PHYS. SOC. 86, ... 965, (1965). . . . .
*THEORY OF COLD NEUTRON SCATTERING BY HYDROGEN IN LIQUID ARGON.
*SEARS(V.F.).

99 PROC. PHYS. SOC. 89, ... 379, (1966). . . . .
*NEUTRON SCATTERING FROM GASEOUS METHANE AND AMMONIA.
*VENKATARAMAN(G.), RAO(K.R.). . .ET AL.

100 PROC. PHYS. SOC. 90, ... 671, (1967). . . . .
*THEORY OF NEUTRON SCATTERING FROM LIQUID AND SOLID HYDROGEN.
*ELLIOTT(R.J.), HARTMANN(W.M.).

101 PROGR. THEOR. PHYS. VOL.43, ... 1423, (1970). . . .T. . . . . . . .H*CL, ...
*QUASICLASSICAL APPROXIMATION FOR SLOW NEUTRON SCATTERING.
*NISHIGORI(T.).

102 PROPERTIES LIQUID METALS, ... 37, (1973). . . . .
*STRUCTURE AND PAIR-POTENTIAL OF LIQUID METALS BY X-RAY AND NEUTRON
DIFFRACTION.
*WASEDA(Y.), SUZUKI(K.). (ED. NOTE: REFER SECT.4).

SECTION 14B—NEUTRON SCATTERING BY FLUIDS . . . .

103 REACTOR PHYSICS, ... 73, (1966) . . . .T,E . . . .
    →INCOHERENT SCATTERING OF SLOW NEUTRONS BY A LIQUID: A HINDERED-
    TRANSLATOR MODEL.
    →ARDENTE(V.), NARDELLI(G.F.), REATTO(L.). (IN BOOK: REACTOR PHYSICS IN
    THE RESONANCE AND THERMAL REGIONS VOL.I. PUBL. M.I.T. PRESS, 1966).

104 REACTOR PHYSICS, ... 91, (1966) . . . .
    →INTERPRETATION OF SCATTERING-LAW DATA.
    →BRUGGER(R.M.). (IN BOOK: REACTOR PHYSICS IN THE RESONANCE AND THERMAL
    REGIONS VOL.I, NEUTRON THERMALIZATION. PUBL. M.I.T. PRESS, 1966).

105 RENSSELAER POL. INST. SYMP. . . . . 145, (1961) . . . .
    →DYNAMICS OF ATOMIC MOTIONS IN LIQUIDS AND COLD NEUTRON SCATTERING.
    →SINGWI(K.S.).

106 RENSSELAER POL. INST. SYMP. . . . . 129, (1961) . . . .
    →CRYSTAL AND LIQUID DYNAMICS FROM NEUTRON ENERGY DISTRIBUTIONS.
    →BROCKHOUSE(B.N.).

107 REPORT AERE-R5627 (15 PP.), ... (1967) . . . .96 . . . .
    →CRITICAL INELASTIC SCATTERING OF SLOW NEUTRONS FROM A BINARY LIQUID.
    →EGELSTAFF(P.A.), WIGNALL(G.D.). (ED. NOTE: AVAIL. H.M.S.O., 49 HIGH
    HOLBORN, LONDON, WC1, ENG.).

108 REPORT AECL-3189 (11 PP.), ... (1968) . . . .A3 . . . .
    →A CALCULATION OF THE NEUTRON SCATTERING BY LIQUIDS AS MEASURED WITH THE
    COLD NEUTRON TIME-OF-FLIGHT TECHNIQUE.
    →COWLEY(R.A.). (ED. NOTE: AVAIL. ATOMIC ENERGY OF CANADA LTD., CHALK
    RIVER, ONTARIO).

109 REPORT AERE-R5867 (137 PP.), ... (1968) . . . .B4 . . . .
    →STUDIES OF THE LIQUID STATE USING THE INELASTIC SCATTERING OF SLOW
    NEUTRONS.
    →COCKING(S.J.). (ED. NOTE: AVAIL. HMSO, 49 HIGH HOLBORN, LONDON, WC2,
    ENGLAND).

110 REPORT AERE-R-6277 (16PP.), ... (1969) . . . .B4 . . . .
    →THE ABC GUIDE TO THEORIES OF NEUTRON INELASTIC SCATTERING BY LIQUIDS.
    →NELKIN(M.). (ED. NOTE: AVAIL. HMSO, 49 HIGH HOLBORN, LONDON, WC1, ENG.).

111 REPORT AEEW-R 701 (31PP.), ... (1970) . . . .H2*O . . . .
    →THERMAL NEUTRON DIFFUSION DATA AND THE WIMS SCATTERING MODEL FOR LIGHT
    WATER.
    →BUTLAND(A.T.D.), CHUDLEY(C.T.). (ED. NOTE: AVAIL. HMSO, 49 HIGH HOLBORN,
    LONDON, WC1, ENG.).

112 REP. PROGR. PHYS. VOL.25, ... 395, (1962) . . . .T,E . . . .
    →THE RADIAL DISTRIBUTION CURVES OF LIQUIDS BY DIFFRACTION METHODS.
    →FURUKAWA(K.).

113 REP. PROGR. PHYS. VOL.36, ... 1135, (1973) . . . .HE . . . .
    →STRUCTURE AND EXCITATIONS OF LIQUID HELIUM (REVIEW PAPER).
    →WOODS(A.D.B.), COWLEY(R.A.).

114 REV. ROUMAINE PHYS. VOL.18, ... 997, (1973) . . . .T . . . .
    →CALCULATION OF THE THERMAL NEUTRON SCATTERING LAW.
    →RAPEANU(S.), PADUREANU(I.), DECIU(N.).

115 RIV. DEL NUOVO CIMENTO VOL.1, . . . 155, (1969) . . . .
    →THE USE OF NEUTRONS IN THE STUDY OF SOLIDS AND LIQUIDS.
    →MARSHALL(W.), LOVESEY(S.W.). (ED. NOTE: THIS REVIEW PAPER IS FOUND IN
    A SPECIAL ISSUE WITH SUBHEADING: NUMERO SPECIALE).

116 SCI. PROGR. VOL.56, ... 223, (1968) . . . .
    →THE STRUCTURE AND THERMAL MOTION OF SIMPLE LIQUIDS (REVIEW PAPER).
    →EGELSTAFF(P.A.).

117 SOLID/LIQUID STATE PHYSICS, ... 113, (1965) . . . .E,T . . . .
    →CO-OPERATIVE MODES OF MOTION IN LIQUID METALS.
    →EGELSTAFF(P.A.). (LECTURES ON SOLID AND LIQUID STATE PHYSICS. PUBL.
    ATOMIC ENERGY ESTABLISHMENT TROMBAY, BOMBAY, INDIA).

118 SOLID/LIQUID STATE PHYSICS, ... 140, (1965) . . . .T . . . .
    →THEORY OF NEUTRON SCATTERING FROM LIQUIDS.
    →SINGWI(K.S.). (LECTURES ON SOLID AND LIQUID STATE PHYSICS. PUBL. ATOMIC
    ENERGY ESTABLISHMENT TROMBAY, BOMBAY, INDIA).

119 SOV. AT. ENERGY VOL.18, ... 146, (1965) . . . .T . . . .
    →ASYMPTOTIC FORMULAE FOR THE SCATTERING OF SLOW NEUTRONS BY BOUND ATOMS.
    →TURCHIN(V.F.), TARASOV(V.A.).

120 SOV. PHYS. SOL. STATE, 5, ... 789, (1963) . . . .T . . . .
    →THEORY OF THE QUASI-ELASTIC SCATTERING OF COLD NEUTRONS IN LIQUIDS.
    →OSKOTSKII(V.S.).

121 SOV. PHYS. ACOUSTICS VOL.9, ... 349, (1964) . . . .E . . . .
    →HYPERSOUND AND THE SCATTERING OF SLOW NEUTRONS IN LIQUIDS.
    →KACHARSKAYA(L.V.), KOMAROV(L.I.), FISHER(I.Z.).

122 SOV. PHYS. J.E.T.P. VOL.21, ... 733, (1965) . . . .HE . . . .
    →SCATTERING OF SLOW NEUTRONS IN HE NEAR THE LAMBDA CURVE.
    →POKROVSKII(V.L.), SARDUTOVICH(G.I.).

123 SOV. PHYS. JETP. 24, ... 749, (1967) . . . .
    →ROLE OF DIFFUSION PROCESS IN THE SCATTERING OF SLOW NEUTRONS IN LIQUIDS.
    →IVANOV(G.K.).

124 SPECTROSCOPY BIOL. CHEM. . . . . . 53, (1974) . . . .T . . . .
    →QUASIELASTIC SCATTERING IN NEUTRON AND LASER SPECTROSCOPY.
    →YIP(S.). (REFER TO SECTION 4).

125 SPECTROSCOPY BIOL. CHEM. . . . . . 1, (1974) . . . .T . . . .
    →INTRODUCTION TO NEUTRON, X-RAY, AND LASER SPECTROSCOPY.
    →CHEN(S.H.). (REFER TO SECTION 4).

126 SPRINGER TRACTS IN MOD. PHYS. 47   1, (1968) . . . .A3 . . . .
    →EVALUATION OF ATOMIC DISTRIBUTION IN LIQUID METALS AND ALLOYS BY MEANS
    OF X-RAY-, NEUTRON-, AND ELECTRON DIFFRACTION.
    →STEEB(S.). (ED. NOTE: PUBL. SPRINGER-VERLAG, BERLIN, GER. 1968,
    HOHLER(G.)-ED. 225PP.).

127 TECH. ELECTROCHEM. VOL.2, ... 173, (1973) . . . .T,E . . . . .H2*O,IONIC SOLUTIONS,
    →STUDIES OF ELECTROLYTIC SOLUTIONS BY X-RAY DIFFRACTION AND NEUTRON
    INELASTIC SCATTERING.
    →SAFFORD(G.J.), LEUNG(P.S.). (IN BOOK: TECHNIQUES OF ELECTROCHEMISTRY.
    ED. BY YEAGER(E.) AND SALKIND(A.J.); PUBL. WILEY: SUSSEX, ENGLAND).

SECTION   14B-NEUTRON SCATTERING BY FLUIDS . . . .

128   THESIS,. . . . . (1967)....C3.......CYCLIC-HYDROCARBONS, ...
      *COLD-NEUTRON SCATTERING EXPERIMENTS ON CYCLIC HYDROCARBONS WITH A
       ROTATING-CRYSTAL SPECTROMETER.
      *DE-GRAAF(L.A.). (PH.D. DISSERTATION - DELFT, THE NETHERLANDS).

129   THESIS (154PP.), ... (1970)...A3 NEUTRON SCATTERING FROM SIMPLE LIQUIDS
      *QUANTUM CORRELATION FUNCTIONS AND NEUTRON SCATTERING FROM SIMPLE LIQUIDS
      *CHERIF(H.S.). (ED. NOTE: AVAIL. UNIV. MICROFILMS, ANN ARBOR, MICH. ORDER
       NO.71-15114).

130   THESIS,. . . . . (1971)....B4.......AR,. . . . .
      *NEUTRON QUASI-ELASTIC SCATTERING STUDIES ON FLUID ARGON.
      *HASMAN(A.). (PH.D. DISSERTATION - DELFT, THE NETHERLANDS).

131   TH. NEUTRON SCATT./EGELSTAFF,. . 347, (1965)............... . . . .
      *EXPERIMENTAL RESULTS ON LIQUIDS.
      *LARSSON(K.E.).

132   TH. NEUTRON SCATT./EGELSTAFF,. . 291, (1965)............. . . . .
      *THEORY OF NEUTRON SCATTERING BY LIQUIDS.
      *SJÖLANDER(A.).

133   TRIESTE-INTERN. COURSE,. . . . . . 603, (1967).............. . . . .
      *ATOMIC MOTIONS IN LIQUIDS AND NEUTRON SCATTERING. (36 PAGES).
      *SINGWI(K.S.).

134   UKAEA AERE REP. R4043 (28 PP.),. . . (1962)................ . . . .
      *NEUTRON SCATTERING STUDIES OF LIQUID DIFFUSION.
      *EGELSTAFF(P.A.).

135   UKRAINIAN PHYS. J. VOL.14,. . . . 1909, (1969)....D3......PARAMAGNETICS,. . . . . . .
      *THE MAGNETIC SCATTERING OF NEUTRONS BY LIQUID PARAMAGNETICS.
      *ZATOVSKII(A.V.), KRASNYT(YU.P.).

136   UKR. FIZ. ZH. (USSR) VOL.9,. . . 684, (1964).....P...............
      *THE EFFECT OF SHORT-RANGE ORDER IN LIQUIDS ON THE TOTAL CROSS-SECTION
       OF COLD NEUTRON INTERACTION.
      *VERTEBNYI(V.P.), DZYUB(I.P.), MAISTRENKO(A.N.), PASICHNYK(H.V.). (IN
       UKRAINIAN).

137   Z. NATURFORSCH. VOL.16A,. . . . . 187, (1961)....A3,T..............
      *THE STRUCTURE OF MONATOMIC MOLTEN METALS BY A NEW METHOD OF ANALYSIS.
      *RICHTER(H.), BREITLING(G.). (IN GERMAN).

SECTION 15 -NEUTRON SCATTERING BY MOLECULAR SYSTEMS. . . . .

1 ACTA PHYS. POLON. VOL.22, SUPP., 179, (1962)....T...........................
   -APPLICATION OF THE CLASSICAL SELFCORRELATION FUNCTION TO DETERMINE THE
   SLOW NEUTRON SCATTERING CROSS-SECTION OF FREE MOLECULES.
   -PARLINSKI(K.).

2 ASEA RES. (SWEDEN) NO.7, 147, (1962)....T......H2
   -ON THE SCATTERING OF SLOW NEUTRONS BY HYDROGEN MOLECULES.
   -BRIMBERG(S.).

3 ATLANTA SYMP/GEORGIA INST. TECH. 30, (1967).....................
   -MOLECULAR SPECTROSCOPY.
   -BOUTIN(H.P.).

4 BER. BUNSENGES. PHYS. CHEM. 75, 352, (1971)..........AR,NA,C*H4,
   -MOLECULAR MOTION AS SEEN BY NEUTRON SCATTERING (REVIEW PAPER).
   -LARSSON(K.E.).

5 BROOKHAVEN SYMPOSIUM, 169, (1965)....T..........
   -EFFECT OF NUCLEAR SPIN CORRELATION ON THE SCATTERING OF NEUTRONS BY
   MOLECULES.
   -SINHA(S.K.), VENKATARAMAN(G.).

6 COLL. INTER. N.126 (GRENOBLE). 217, (1963).....................
   -REMARKS ON THE SLOW NEUTRON SCATTERING BY ORGANIC MOLECULES.
   -ARDENTE(V.).

7 CONF. NEUTRON THERMALIZ. VOL.I. 283, (1968)....T.............
   -QUASI-CLASSICAL APPROXIMATION FOR SCATTERING FROM MOLECULES.
   -SUMMERFIELD(G.C.), ZWEIFEL(P.F.). (IN BOOK: NEUTRON THERMALIZATION AND
   REACTOR SPECTRA. PUBL. IAEA: VIENNA, 1968).

8 DISS. ABSTR. VOL.23, 3151, (1963)....E,P......N*H4 SALTS,N*H3,K*H2*P*O4,,,,
   -INVESTIGATION OF ROTATIONAL AND VIBRATIONAL FREEDOM IN MOLECULES BY
   CROSS-SECTION MEASUREMENTS WITH SLOW NEUTRONS.
   -RUSH(J.J.).

9 FARADAY SYMP. CHEM. SOC. NO.6. 122, (1972)....T,E......CU-NI,AR,AL
   -MOLECULAR DYNAMICS OF LIQUIDS IN RELATION TO THE SOLID AND GAS PHASES
   AS SEEN BY NEUTRON SCATTERING (REVIEW).
   -LARSSON(K.E.).

10 HARWELL SUMMER SCHOOL, 82, (1968)....................
   -NEW PERSPECTIVES IN NEUTRON DIFFRACTION ANALYSIS OF SMALL ORGANIC
   MOLECULES.
   -COPPENS(P.).

11 HARWELL SUMMER SCHOOL, 68, (1968)....................
   -MOLECULAR OBJECTIVES IN ELASTIC NEUTRON SCATTERING.
   -COULSON(C.A.).

12 HELV. PHYS. ACTA VOL.41, 533, (1968)....B7...........
   -NEUTRON SPECTROSCOPY AS A METHOD FOR THE INVESTIGATION OF THE DYNAMICS
   OF ADSORBED MOLECULES.
   -VERDAN(G.). (ED. NOTE: IN GERMAN).

13 IAEA SYMP. COPENHAGEN, VOL.2. 299, (1968).................
   -MEASUREMENT OF THE VIBRATIONAL SPECTRA OF MOLECULES BY MEANS OF THE DOWN
   SCATTERING OF NEUTRONS.
   -BEG(M.M.), ROSS(D.K.).

14 IAEA SYMP. GRENOBLE, 277, (1972)....C5,T.........
   -CONFORMATIONAL MOTIONS OF POLYMERS.
   -JANNINK(G.), SUMMERFIELD(G.C.).

15 IAEA SYMP. GRENOBLE, 261, (1972)...................
   -THE DYNAMICS OF POLYMER CHAINS (REVIEW PAPER).
   -ALLEN(G.).

16 IAEA SYMP. GRENOBLE, 175, (1972)...................
   -DYNAMICS OF MOLECULAR CRYSTALS (REVIEW PAPER).
   -PAWLEY(G.S.).

17 J. APPL. CRYST. VOL.7, 173, (1974)....E......FERRITIN,MYOGLOBIN
   -NEUTRON SMALL-ANGLE SCATTERING OF BIOLOGICAL MACROMOLECULES IN SOLUTION.
   -STUHRMANN(H.B.).

18 J. CHEM. PHYS. VOL.42, 1863, (1965)....T..............
   -SCATTERING FROM FLUIDS OF NONSPHERICAL MOLECULES.
   I. X RAYS AND NEUTRONS.
   -STEELE(W.A.), PECORA(R.).

19 J. CHEM. PHYS. VOL.53, 2590, (1970).........CYCLOHEXANE,
   -COOPERATIVE ROTATION OF SPHERICAL MOLECULES.
   -EGELSTAFF(P.A.).

20 J. CHEM. PHYS. VOL.59, 4612, (1973)....T...............
   -ROTATIONAL MOTION OF MOLECULES AND NEUTRON SCATTERING.
   -LARSSON(K.E.).

21 J. CHIM. PHYS. VOL.63, 157, (1966)....................
   -SLOW NEUTRON STUDIES OF MOLECULAR MOTIONS IN CONDENSED MATTER.
   -PALEVSKY(H.).

22 J. NUCL. ENERGY VOL.13, 128, (1961)....T.............
   -SCATTERING OF SLOW NEUTRONS BY MOLECULES.
   -RAHMAN(A.).

23 J. POLYMER SCI./B POLYM. LETT. 7, 865, (1969)....A4...........
   -COHERENT NEUTRON SCATTERING BY POLYMER SOLUTIONS AT INTERMEDIATE
   CONCENTRATIONS.
   -JANNINK(G.).

24 K. TEKN. HOGSK. AVHANDL. NO.116, (1956)....T,P......H2
   -ON THE SCATTERING OF SLOW NEUTRONS BY HYDROGEN MOLECULES.
   -BRIMBERG(S.).

25 MOL. CRYST. LIQUID CRYST. VOL.9, 11, (1969)...............ORGANIC MOLECULES,
   -DIFFRACTION STUDIES OF MOLECULAR MOTION.
   -HAMILTON(W.C.).

SECTION 15 -NEUTRON SCATTERING BY MOLECULAR SYSTEMS. . . . .

26 MOL. DYN. AND STRUCT. OF SOLIDS, 135, (1969)....B7
    *NEUTRON INELASTIC SCATTERING STUDIES OF MOLECULAR SOLIDS AND LIQUIDS.
    *EGELSTAFF(P.A.).

27 MOL. DYN. AND STRUCT. OF SOLIDS, 355, (1969)....B7
    *DIFFRACTION STUDIES OF MOLECULAR DYNAMICS IN ORGANIC CRYSTALS. . . . . .
    *TRUEBLOOD(K.N.).

28 MOL. PHYS. VOL.20, 881, (1971)....B4.......C*CL4, GE*BR4
    *ORIENTATIONAL CORRELATIONS IN MOLECULAR LIQUIDS BY NEUTRON SCATTERING.
    CARBON TETRACHLORIDE AND GERMANIUM TETRABROMIDE.
    *EGELSTAFF(P.A.), PAGE(D.I.), POWLES(J.G.).

29 MOL. PHYS. VOL.26 NO.6, 1325, (1973)....T,E.
    *THE ANALYSIS OF A TIME-OF-FLIGHT NEUTRON DIFFRACTOMETER FOR AMORPHOUS
    MATERIALS: THE STRUCTURE OF A MOLECULE IN A LIQUID.
    *POWLES(J.G.).

30 MOL. PHYS. VOL.25, 1353, (1973)..IN A3,T.......BR2,C*CL4.
    *ANGULAR CORRELATION EFFECTS IN NEUTRON DIFFRACTION FROM MOLECULAR FLUIDS
    *GUBBINS(K.E.), GRAY(C.G.), EGELSTAFF(P.A.), ANANTH(M.S.).

31 NUCL. SCI. ENGNG. VOL.37, 368, (1969)....A3.
    *SLOW-NEUTRON SCATTERING BY MOLECULAR LIQUIDS. . . . .
    *AGRAWAL(A.K.), YIP(S.).

32 PHYSICA VOL.48, 126, (1970)....T.......C*H4
    *MOLECULAR INTERACTION EFFECT IN NEUTRON SCATTERING BY GASEOUS METHANE.
    *FULINSKI(A.), JANIK(J.A.), BLOCKI(J.).

33 PHYS. LETT. 22, 288, (1966)
    *HINDRANCE EFFECTS IN COLD NEUTRON SCATTERING BY ROTATING MOLECULES.
    *SOLT(G.).

34 PHYS. NORVEG. VOL.1, 127, (1962)......T
    *THE SCATTERING OF SLOW NEUTRONS BY DIATOMIC MOLECULES. . . . . . .
    *LOVSETH(J.).

35 PHYS. REV. VOL.60, 18, (1941)....T.......H2,H2*O,C*H4,N*H3, ...
    *THE SCATTERING OF SLOW NEUTRONS BY MOLECULAR GASES.
    *SACHS(R.G.), TELLER(E.).

36 PHYS. REV. VOL.101, 118, (1956)....T.......
    *DYNAMICS OF NEUTRON SCATTERING BY MOLECULES. . . . .
    *ZEMACH(A.C.), GLAUBER(R.J.).

37 PHYS. REV. VOL.113, 866, (1959)....T.
    *SLOW-NEUTRON SCATTERING BY ROTATORS.
    *VOLKIN(H.C.).

38 PHYS. REV. VOL.117, 1029, (1960)....T.......II.
    *SLOW-NEUTRON SCATTERING BY ROTATORS. II.
    *VOLKIN(H.C.).

39 PHYS. REV. 130, 1860, (1963)
    *SLOW NEUTRON SCATTERING BY HINDERED ROTATORS. . . . .
    *YIP(S.), OSBORN(R.K.).

40 PHYS. REV. VOL.138A, 701, (1965)....C3.
    *MODELS FOR THE MOTION OF A MOLECULE IN A LIQUID AND THEIR APPLICATION
    TO SLOW-NEUTRON SCATTERING.
    *GIBBS(A.G.), FERZIGER(J.H.).

41 PHYS. REV. VOL.167, 97, (1968)....T.
    *ROTATION-VIBRATION INTERACTION IN SCATTERING OF SLOW NEUTRONS BY
    SPHERICAL-TOP MOLECULES.
    *BUZANO(C.), DEMICHELIS(F.), RASETTI(M.).

42 PHYS. REV. A VOL.2, 975, (1970)....T.......HE.
    *TIME-CORRELATION FUNCTIONS, MEMORY FUNCTIONS, AND MOLECULAR DYNAMICS.
    *HARP(G.D.), BERNE(B.J.).

43 PHYS. REV. A VOL.3, 1006, (1971)
    *ROTATIONAL MOLECULAR MOTION AND THE TRANSIENT NATURE OF LIQUIDS AS SEEN
    BY SLOW-NEUTRON SCATTERING.
    *LARSSON(K.E.).

44 PROGR. THEOR. PHYS. VOL.46, 1666, (1971)....T.
    *ROTATIONAL CORRELATION FUNCTION OF SPHERICAL ROTORS AND NEUTRON . . . .
    SCATTERING.
    *HAMA(J.), NAKAMURA(T.).

45 PROGR. THEOR. PHYS. VOL.50, 1142, (1973)....T.......C*H4.
    *SPIN CORRELATION EFFECT ON THE SLOW NEUTRON SCATTERING BY POLYATOMIC
    MOLECULES WITH PARTICULAR REFERENCE TO METHANE.
    *HAMA(J.), MIYAGI(H.).

46 PROPERTIES LIQUID METALS, 143, (1973)....T.......NA.
    *CHARACTERISTICS OF COLLECTIVE MOTIONS IN A LIQUID METAL: SIMULATION OF
    MOLECULAR DYNAMICS OF NA.
    *TAKEUCHI(S.), TANAKA(M.), FUKUI(Y.), WATABE(M.), HASEGAWA(M.). (ED.
    NOTE: REFER SECT. 4).

47 RAD. ZAVODA FIZ. (YUGOSLAVIA), 41, (1971)....T.......H2*O.
    *AN ANALYTICAL IMPROVEMENT OF NELKIN≠S FORMALISM FOR SLOW NEUTRON
    SCATTERING BY WATER MOLECULES. (IN ENGLISH).
    *STANCIC(V.).

48 REACTOR SCI. VOL.13, 128, (1961)....T. . . . .
    *SCATTERING OF SLOW NEUTRONS BY MOLECULES. . . . .
    *RAHMAN(A.).

49 REPORT EUR-4037.1 (50.PP.), (1968)....B4.
    *SLOW-NEUTRON SCATTERING AND MOLECULAR DYNAMICS OF THE LIQUID STATE.
    *ARDENTE(V.), NARDELLI(G.F.), REATTO(L.). (ED. NOTE: AVAIL. EUROPEAN
    ATOMIC ENERGY COMM., ISPRA, ITALY).

50 REPORT FN-37/70 (22PP.), (1970)....P.
    *SCATTERING OF NEUTRONS BY MOLECULAR SYSTEMS.
    *BALLY(D.). (ED. NOTE: AVAIL. COMITETUL PENTRU ENERGIA NUCLEARE INST.
    FIZICA ATOMICA, BUCHAREST).

51 REP. PROGR. PHYS. VOL.36, 1073, (1973)
    *PHYSICO-CHEMICAL ASPECTS OF NEUTRON STUDIES OF MOLECULAR MOTION (REVIEW)
    *ALLEN(G.), HIGGINS(J.S.).

SECTION 15 -NEUTRON SCATTERING BY MOLECULAR SYSTEMS. . . . .

52 SOV. AT. ENERGY VOL.12, ... 48 (1962). . . T. . . . .
    *THE PROBLEM OF THE SCATTERING OF NEUTRONS BY MOLECULES.
    *IVANOV(G.K.).

53 SOV. PHYS. JETP VOL.6, ... 480 (1958). . . T. . . . .
    *SCATTERING OF SLOW NEUTRONS BY DIATOMIC MOLECULES.
    *ANSEL#M(A.A.).

54 SOV. PHYS. JETP. 17, ... 390 (1963). . . . .
    *SCATTERING OF NEUTRONS BY MOLECULES WITH LARGE ENERGY TRANSFER.
    *IVANOV(G.K.).

55 SOV. PHYS. JETP. 22, ... 347 (1966). . . . .
    *ROLE OF ROTATIONAL STATES IN THE SCATTERING OF SLOW NEUTRONS BY GAS
    MOLECULES.
    *OZYUBLI.P.).

56 SOV. PHYS. JETP. 23, ... 481, (1966). . . . .
    *ELASTIC AND QUASIELASTIC SCATTERING OF NEUTRONS BY MOLECULES.
    *IVANOV(G.K.).

57 SOV. PHYS. USPEKHI VOL.9, . . . 670, (1967). . . . .
    *INTERACTION OF NEUTRONS WITH MOLECULES.
    *IVANOV(G.K.), SAYASOV(YU.S.).

58 SPECTROSCOPY BIOL. CHEM., ... 269, (1974). . . . E.T. . . . .
    *SELECTED CHEMICAL AND BIOLOGICAL APPLICATIONS OF NEUTRON INELASTIC
    SCATTERING.
    *EGELSTAFF(P.A.). (REFER TO SECTION 4).

59 STUD. CERCET. FIZ. (RUMANIA) 23, 699, (1971). . . . B4. . . . .
    *MOLECULAR DYNAMICS IN LIQUIDS FROM NEUTRON SPECTROMETRY.
    *TREPADUS(V.).

60 TH. NEUTRON SCATT./EGELSTAFF, . . 453, (1965). . . . .
    *NEUTRON SCATTERING EXPERIMENTS ON MOLECULES.
    *JANIK(J.A.), KOWALSKA(A.).

61 TH. NEUTRON SCATT./EGELSTAFF, . . 413, (1965). . . . .
    *THE THEORY OF NEUTRON SCATTERING BY MOLECULES.
    *JANIK(J.A.), KOWALSKA(A.).

62 UKR. FIZ. ZH. (USSR) VOL.9, . . 349, (1964). . . T. . . . .
    *NEUTRON AND OPTICAL SPECTRA AS SOURCES OF INFORMATION ON THE MOTION OF
    MOLECULES IN LIQUIDS.
    *KOMAROV(L.I.), FISHER(I.Z.). (IN RUSSIAN).

63 UKR. FIZ. ZH. (USSR) VOL.11, . . 581, (1966). . . T. . . . .
    *SCATTERING OF SLOW NEUTRONS BY LINEAR MOLECULES.
    *KOVAL#CHUK(V.G.). (IN UKRAINIAN).

64 Z. NATURFORSCH. VOL.20A, ... 380, (1965). . . . . T. . . . .
    *CALCULATION OF SCATTERING OF LOW-ENERGY NEUTRONS BY FREE MOLECULES.
    *ZECH(H.J.). (IN GERMAN).

SECTION 16 -NEUTRON TRANSPORT AND MODERATION (NON-TECHNICAL ASPECTS) . .

1   ARK. FYS. VOL.34, PAPER 40,,   481, (1967)
    ~A MONTE CARLO PROGRAM FOR CALCULATION OF NEUTRON ATTENUATION AND
        MULTIPLE SCATTERING CORRECTIONS.
    ~HOLMQVIST(B.), GUSTAVSSON(B.), WIEDLING(T.).

2   ATOMKERNENERGIE VOL.1C,. . . . . 243, (1965).....E........
    ~NEUTRON SPECTRA OF COOLED MODERATORS.
    ~RAUCH(H.), SCHMIDT(H.). (IN GERMAN).

3   ATOMKERNENERGIE VOL.16,,. . . 201, (1970).....P.....ZR*H2,
    ~TIME DEPENDENT THERMALIZATION OF NEUTRONS IN AN EINSTEIN SOLID WITH
        APPLICATION TO ZIRCONIUM HYDRIDE.
    ~VON-BALTZ(R.).

4   ATOMKERNENERGIE VOL.18,,. . 261, (1971).....P........
    ~NEUTRON THERMALIZATION IN FACE CENTERED CUBIC (F.C.C.) METAL HYDRIDES.
    ~KEINERT(J.).

5   ATOMKERNENERGIE VOL.17,.. . 229, (1971).....P........
    ~TIME DEPENDENT NEUTRON THERMALIZATION IN A METAL HYDRIDE MODERATOR.
    ~MEHNER(J.).

6   ATOMKERNENERGIE VOL.18,,. . . 158, (1971).....T........
    ~PARTICULAR SOLUTIONS IN MULTIGROUP TRANSPORT THEORY.
    ~LEUTHAUSER(K.D.).

7   CAN. J. PHYS. 43,. . . . 432, (1965)
    ~MODIFIED SPHERICAL HARMONIC METHOD AND NEUTRON TRANSPORT PROBLEM WITH
        FINITE SPHERICAL CORE.
    ~SEN(K.K.).

8   COMP. REND. TOME 203,. . . . 73, (1936).....E........
    ~PREUVE EXPERIMENTALE DE LA DIFFRACTION DES NEUTRONS.
    ~VON-HALBAN(H.), PREISWERK(P.).

9   CONF. NEUTRON THERMALIZ. VOL.I,. 235, (1968).....P........
    ~A REVIEW OF SCATTERING LAW STUDIES FOR MODERATORS.
    ~GLASER(W.). (IN BOOK: NEUTRON THERMALIZATION AND REACTOR SPECTRA. PUBL.
        IAEA: VIENNA, 1968).

10 CONF. NEUTRON THERMALIZ. VOL.II,. 283, (1968).....T........
    ~QUASI-CLASSICAL APPROXIMATION FOR SCATTERING FROM MOLECULES.
    ~SUMMERFIELD(G.C.), ZWEIFEL(P.F.). (IN BOOK: NEUTRON THERMALIZATION AND
        REACTOR SPECTRA. PUBL. IAEA: VIENNA, 1968).

11 C.R. ACAD. BULG. SCI. VOL.13,. 527, (1960).....T........
    ~ON THE PASSAGE OF SLOW NEUTRONS THROUGH A CRYSTAL PLATE.
    ~ATANASSOV(A.A.).

12 DISS. ABS. VOL.24, ... 5484, (1964).............
    ~MEASUREMENT OF NEUTRON DIFFUSION PARAMETERS USING NEUTRON WAVES.
    ~MORTENSEN(G.A.).

13 DISS. ABS. VOL.25, ... 1986, (1965).............
    ~ANALYTIC EXPRESSIONS FOR CALCULATING NEUTRON DISTRIBUTION FUNCTIONS BY
        MONTE-CARLO METHOD.
    ~TALLEY(W.K.).

14 DISS. ABS. VOL.25, ... 6014, (1965)............H2*O,
    ~A SIMPLIFIED CELL THEORY APPLIED TO THE CALCULATION OF THERMAL NEUTRON
        SPECTRA IN LIGHT WATER LATTICE.
    ~MACVEAN(C.R.).

15 DISS. ABS. 27, ...3625B, (1966).....T........BE,
    ~CALCULATION AND INTERPRETATION OF NEUTRON THERMALIZATION IN FINITE
        BLOCKS OF BERYLLIUM.
    ~LEE(R.R.).

16 DISS. ABS. VOL.26,. . . 2087, (1966) SPECTRA
    ~INVESTIGATION OF NEUTRON SPECTRA EMERGING FROM POSSIBLE COLD NEUTRON
        MODERATORS AT 80 AND 20 DEGREES KELVIN.
    ~ALTSCHULER(S.J.).

17 DISS. ABS. VOL.29,. . . .1120B, (1968).............
    ~THE THEORY OF NEUTRON WAVE PROPAGATION.
    ~DUDERSTADT(J.J.).

18 DISS. ABS. VOL.30,. . . .3819B, (1970).............
    ~THEORY OF COUPLED SPACE-TIME EFFECTS IN FAST PULSED NEUTRON KINETICS AND
        TRANSPORT THEORY.
    ~KNICKLE(H.N.).

19 ENERGIA NUCLEARE (ITALY) VOL.16, 700, (1969).....T........
    ~ANISOTROPIC DIFFUSION THEORY.
    ~MICHELINI(M.).

20 IAEA SYMPOSIUM VIENNA,. . . . 3, (1960).............
    ~SLOW NEUTRON INELASTIC SCATTERING AND NEUTRON THERMALIZATION.
    ~NELKIN(M.).

21 IAEA SYMPOSIUM VIENNA,. . . 87, (1960).............
    ~NEUTRON THERMALIZATION AND THE PROPERTIES OF THE VAN HOVE PAIR
        CORRELATION FUNCTION.
    ~.C.C.C.P.

22 IAEA SYMPOSIUM VIENNA,. . . 607, (1960).............
    ~CHEMICAL BINDING EFFECTS IN NEUTRON THERMALIZATION.
    ~CORNGOLD(N.).

23 IAEA SYMPOSIUM (VIENNA), . . . 613, (1960)............H2*O,ZR-H,POLYETHYLENE,
    ~MEASUREMENTS OF NEUTRON SPECTRA IN WATER, POLYETHYLENE AND ZIRCONIUM
        HYDRIDE.
    ~WALTON(R.B.), BEYSTER(J.R.), WOOD(J.L.), LOPEZ(W.M.).

24 IAEA SYMPOSIUM (VIENNA), . . . 631, (1960)........
    ~STUDY OF THE NEUTRON SPECTRA EMERGING FROM MODERATING ASSEMBLIES.
    ~RAMANNA(R.), SARMA(N.).

25 IAEA SYMPOSIUM VIENNA,. . . 437, (1960).............
    ~NEUTRON MODERATION TO LOW TEMPERATURES.
    ~BORST(L.B.).

SECTION 16 -NEUTRON TRANSPORT AND MODERATION (NON-TECHNICAL ASPECTS) . .

26 J. NUCL. ENERGY VOL.23, ... 505, (1969) . . . .
    ~NEUTRON DIFFUSION ACROSS A MEDIUM DISCONTINUITY: CRYSTALLINE MODERATORS.
    ~TRIKHA(S.K.), KOTHARI(L.S.), GOYAL(I.C.).

27 J. NUCL. ENERGY(GB) VOL.26, ... 61, (1972) . . . . P . . . .
    ~NEUTRON TRANSPORT IN A DOUBLY-PERIODIC ARRAY OF ABSORBING MULTIPOLES.
    ~AVIRAM(I.).

28 J. NUCL. ENERGY VOL.27, ... 171, (1973) . . . . T . . . .
    ~MULTIPLE SCATTERING AND TRANSPORT OF NEUTRONS.
    ~NISHIGORI(T.).

29 J. NUCL. SCI. TECHNOL. VOL.7, . . . 580, (1970) . . . . P . . . .
    ~SLOWING DOWN OF NEUTRONS TO VERY LOW TEMPERATURE BY COLD SOLID HYDRO-
    GENOUS MODERATORS.
    ~INOUE(K.).

30 J. NUCL. SCI. TECHNOL. VOL.8, . . 153, (1971) . . . . T . . . .
    ~NORMAL MODE EXPANSION IN NEUTRON THERMALIZATION THEORY WITH A SIMPLE
    SCATTERING KERNEL.
    ~YAMAGISHI(T.).

31 NUCLEAR SCI. ENGNG. VOL.24, . . . 417, (1966) . . . . E . . . . . GRAPHITE . . .
    ~VARIATION OF THE (THERMAL NEUTRON) TRANSPORT MEAN FREE PATH IN GRAPHITE
    WITH TEMPERATURE.
    ~DEJUREN(J.A.).

32 NUCL. INST. MET. 36, ... 88, (1965) . . . .
    ~AN ACCURATE DETERMINATION OF THE NEUTRON SPECTRUM IN THE THERMAL COLUMN
    OF THE TRIGA REACTOR FIR.1.
    ~PALMGREN(A.).

33 NUCL. INST. MET. 85, ... 163, (1970) . . . .
    ~PULSED MODERATOR STUDIES USING A TIME FOCUSSED CRYSTAL SPECTROMETER.
    ~GRAHAM(K.F.), CARPENTER(J.M.).

34 NUCL. INST. MET. VOL.101, ... 263, (1972) . . . . T . . . .
    ~A MONTE CARLO SOLUTION OF MULTIPLE SCATTERING IN COLD NEUTRON SCATTERING
    EXPERIMENTS.
    ~KALLI(H.).

35 NUCL. SCI. ENGNG. VOL.12, ... 250, (1962) . . . .
    ~THE SCATTERING OF THERMAL NEUTRONS BY MODERATORS.
    ~EGELSTAFF(P.A.).

36 NUCL. SCI. ENGNG. VOL.13, ... 250, (1962) . . . . T . . . .
    ~SCATTERING OF THERMAL NEUTRONS IN THE DOPPLER APPROXIMATION.
    ~PUROHIT(S.N.), RAJAGOPAL(A.K.).

37 NUCL. SCI. ENGNG. VOL.20, ... 236, (1964) . . . . C3 . . . . . . C6*H6,C12*H10(DIPHENYL). . . . .
    ~INELASTIC NEUTRON SCATTERING BY SOME HYDROGENOUS MODERATORS.
    ~GLASER(W.), BECKURTS(K.H.).

38 NUCL. SCI. ENGNG. VOL.24, ... 142, (1966) . . . . T . . . .
    ~ATOMIC MOTIONS IN A RIGID SPHERE GAS AS A PROBLEM IN NEUTRON TRANSPORT.
    ~DESAI(R.C.), NELKIN(M.).

39 NUCL. SCI. ENGNG. VOL.28, ... 494, (1967) . . . . E . . . .
    ~ON THE EXPERIMENTAL APPLICATION OF THE NEUTRON-WAVE TECHNIQUE TO THERMAL
    NEUTRON SYSTEMS.
    ~BOOTH(R.S.), HARTLEY(R.H.), PEREZ(R.B.).

40 NUCL. SCI. ENGNG. VOL.30, ... 199, (1967) . . . . T . . . . . . H2*O . . . .
    ~MULTIPLE SCATTERING IN SLOW-NEUTRON DOUBLE-DIFFERENTIAL MEASUREMENTS.
    ~SLAGGIE(E.L.).

41 NUCL. SCI. ENGNG. VOL.36, ... 351, (1969) . . . . P . . . .
    ~NEUTRON RETHERMALIZATION NEAR A TEMPERATURE DISCONTINUITY BETWEEN
    TERPHENYL AND LIGHT WATER.
    ~RASTAS(A.), SAASTAMOINEN(J.).

42 NUCL. SCI. ENGNG. VOL.37, ... 127, (1969) . . . . P . . . .
    ~THERMAL-NEUTRON DIFFUSION IN AQUEOUS ABSORBING SOLUTIONS.
    ~GODDARD(A.J.H.), JOHNSON(P.W.).

43 NUCL. SCI. ENGNG. VOL.40, ... 460, (1970) . . . . T . . . .
    ~SPACE-AND TIME-DEPENDENT NEUTRON SLOWING DOWN.
    ~DIAMOND(D.J.), YIP(S.).

44 NUCL. SCI. ENG. VOL.42, ... 97, (1970) . . . . P . . . .
    ~CALCULATIONS OF NEUTRON TIME-ENERGY DISTRIBUTIONS IN HEAVY MODERATORS.
    ~WILLIAMSON(T.J.), ALBRECHT(R.W.).

45 NUCL. SCI. ENG. VOL.43, ... 120, (1971) . . . . P . . . .
    ~THERMAL NEUTRON DIFFUSION AT THE ICE-WATER PHASE TRANSITION.
    ~WILLIAMS(P.M.), MUNNO(F.J.).

46 NUCL. SCI. ENG. VOL.44, ... 444, (1971) . . . . P . . . . . . C (GRAPHITE) . . .
    ~SLOWING DOWN AND HEATING UP OF NEUTRONS IN GRAPHITE AT LOW TEMPERATURES.
    ~BHUSHAN(V.), TRIKHA(S.K.).

47 NUCL. SCI. ENG. VOL.45, ... 308, (1971) . . . . P . . . . . . H2*O . . . .
    ~THERMAL-NEUTRON DIFFUSION PARAMETERS IN WATER BY THE POISONING METHOD.
    ~MARTINHO(E.), COSTA-PAIVA(M.M.).

48 NUCL. SCI. ENG. VOL.47, ... 349, (1972) . . . . P . . . . . . U*C . . . .
    ~LOW ENERGY NEUTRON INTERACTIONS WITH URANIUM CARBIDE.
    ~LAJEUNESSE(C.), MOORE(W.E.), YEATER(M.L.).

49 NUCL. SCI. ENG. VOL.47, ... 156, (1972) . . . . T . . . .
    ~TWO-GROUP NEUTRON TRANSPORT THEORY WITH ANISOTROPIC SCATTERING.
    ~REITH-JR(R.J.), SIEWERT(C.E.).

50 NUCL. SCI. ENG. VOL.47, ... 66, (1972) . . . . T . . . .
    ~IMPROVEMENTS TO NEUTRON SLOWING DOWN THEORY FOR FAST REACTORS.
    ~DUNN(F.E.), BECKER(M.).

51 NUCL. SCI. ENG. VOL.48, ... 10, (1972) . . . . T . . . .
    ~A GENERALIZED MODEL FOR THE ELASTIC AND INELASTIC SLOWING DOWN OF FAST
    NEUTRONS.
    ~ROCCA-VOLMERANGE(B.).

52 NUKLEONIK VOL.6, ... 87, (1964) . . . . P . . . .
    ~TRANSMISSION AND SCATTERING EXPERIMENTS WITH SLOW NEUTRONS IN THE
    FRM-REACTOR.
    ~SPRINGER(T.). (IN GERMAN).

SECTION 16 -NEUTRON TRANSPORT AND MODERATION (NON-TECHNICAL ASPECTS) . .

53 NUKLEONIK VOL.7,   408, (1965) SCATTERING IN D2*O
    *THE ROLE OF INTERFERENCE SCATTERING IN NEUTRON THERMALIZATION BY D2*O.
    *KOPPEL(J.U.), YOUNG(J.A.).

54 NUKLEONIK VOL.11,   282, (1968)   B7
    *INVESTIGATIONS OF INELASTIC SCATTERING OF SLOW NEUTRONS IN MODERATORS.
    *GLASER(W.).

55 NUOVO CIMENTO VOL.12 SER.8,   211, (1935)....E
    *THE PROPERTIES OF SLOW NEUTRONS.
    *PONTECORVO(B.). (IN ITALIAN).

56 NUOVO CIMENTO VOL.12 SER.8,   201, (1935)
    *SLOW NEUTRONS.
    *FERMI(E.), RASETTI(F.). (IN ITALIAN).

57 PHYS. LETT. VOL.39A,   327, (1972)   HE
    *LINE WIDTH OF ROTONS IN HE II ABOVE THE THRESHOLD FOR PHONON EMISSION.
    *JACKLE(J.), KEHR(K.W.).

58 PHYS. REV. VOL.50,   899, (1936)....E
    *ON THE ABSORPTION AND THE DIFFUSION OF SLOW NEUTRONS.
    *AMALDI(E.), FERMI(E.).

59 PHYS. REV. VOL.69,   423, (1946)....T
    *THE THEORY OF THE SLOWING DOWN OF NEUTRONS IN HEAVY SUBSTANCES.
    *PLACZEK(G.).

60 PHYS. REV. VOL.137,   A1686, (1965)
    *APPROACH TO EQUILIBRIUM OF A NEUTRON GAS.
    *SHAPIRO(C.S.), CORNGOLD(N.).

61 PHYS. REV. VOL.139,   A1138, (1965)
    *DERIVATION OF KINETIC EQUATIONS FOR SLOW-NEUTRON SCATTERING.
    *VAN-LEEUWEN(J.M.J.).

62 PHYS. REV. B VOL.7,   917, (1973)....T
    *POSSIBILITY OF GUIDED-NEUTRON-WAVE PROPAGATION IN THIN FILMS.
    *DE-WAMES(R.E.), SINHA(S.K.).

63 PROC. JAP. ACAD. VOL.46, 944, (1970)....P
    *NEUTRON TRANSPORT PROCESS ON BOUNDED HOMOGENEOUS DOMAIN.
    *MORI(T.).

64 PROC. PHYS. SOC. VOL.48, 642, (1936)....E
    *SOME EXPERIMENTS WITH NEUTRONS HAVING THERMAL ENERGIES.
    *TILLMAN(J.R.).

65 PROC. PHYS. SOC. 85,   413, (1965)
    *NEUTRON FLUX PERTURBATIONS DUE TO INFINITE PLANE ABSORBER. I. SPATIALLY
    CONSTANT SOURCE.
    *WILLIAMS(M.M.R.).

66 PROC. ROY. SOC. A VOL.149, 522, (1935)....E
    *ARTIFICIAL RADIOACTIVITY PRODUCED BY NEUTRON BOMBARDMENT.
    *AMALDI(E.),   D'AGOSTINO(O.), FERMI(E.), PONTECORVO(B.), RASETTI(F.),
    SEGRE(E.).

67 PROC. ROY. SOC. A VOL.153, 476, (1936)....E
    *NEUTRONS OF THERMAL ENERGIES.
    *MOON(P.B.), TILLMAN(J.R.).

68 PROC. ROY. SOC. 289,   342, (1966)
    *THE DIFFUSION LENGTH OF THERMAL NEUTRONS IN H2*O BETWEEN 18 AND 255
    DEGREES CENTIGRADE.
    *BESANT(C.B.), GRANT(P.J.).

69 REACTOR PHYSICS,   123, (1966)....T
    *THE VELOCITY AUTOCORRELATION FUNCTION IN LIQUIDS FROM A NEW POINT OF
    VIEW.
    *RAHMAN(A.). (IN BOOK: REACTOR PHYSICS IN THE RESONANCE AND THERMAL
    REGIONS VOL.I, NEUTRON THERMALIZATION. PUBL. M.I.T. PRESS, 1966).

70 REACTOR PHYSICS,   73, (1966)....T,E
    *INCOHERENT SCATTERING OF SLOW NEUTRONS BY A LIQUID: A HINDERED-
    TRANSLATOR MODEL.
    *ARDENTE(V.), NARDELLI(G.F.), REATTO(L.). (IN BOOK: REACTOR PHYSICS IN
    THE RESONANCE AND THERMAL REGIONS VOL.I. PUBL. M.I.T. PRESS, 1966).

71 REACTOR PHYSICS,   27, (1966)....E,T......H2*O,H2,ZR*H2,POLYETHYLENE,
    *NEUTRON SCATTERING BY HYDROGENOUS MODERATORS.
    *KOPPEL(J.U.). (IN BOOK: REACTOR PHYSICS IN THE RESONANCE AND THERMAL
    REGIONS VOL.I, NEUTRON THERMALIZATION. PUBL. M.I.T. PRESS, 1966).

72 REACTOR PHYSICS,   3, (1966)....E,T......BE,GRAPHITE,
    *ATOMIC MOTION IN MODERATORS.
    *YOUNG(J.A.). (IN BOOK: REACTOR PHYSICS IN THE RESONANCE AND THERMAL
    REGIONS VOL.I, NEUTRON THERMALIZATION. PUBL. BY M.I.T. PRESS, 1966).

73 RENSSELAER POL. INST. SYMP., 248, (1961)
    *MODERATION STUDIES BY NEUTRON TIME-OF-FLIGHT.
    *POOLE(M.J.), BARNARD(E.), COATES(M.S.), ROGERS(B.J.).

74 RENSSELAER POL. INST. SYMP., 215, (1961)
    *NEUTRON THERMALIZATION MEASUREMENTS USING AN ELECTRON LINEAR ACCELERATOR
    *BEYSTER(J.R.).

75 RENSSELAER POL. INST. SYMP., 187, (1961)
    *THEORETICAL INTERPRETATION OF PULSED NEUTRON EXPERIMENTS IN MODERATORS.
    *DAITCH(P.B.).

76 RENSSELAER POL. INST. SYMP., 162, (1961)
    *VARIATIONAL METHODS IN NEUTRON THERMALIZATION.
    *SELENGUT(D.S.).

77 REPORT AEEW-R 701 (31PP.)   (1970)   H2*O.
    *THERMAL NEUTRON DIFFUSION DATA AND THE WIMS SCATTERING MODEL FOR LIGHT
    WATER.
    *BUTLAND(A.T.D.), CHUDLEY(C.T.). (ED. NOTE: AVAIL. HMSO, 49 HIGH HOLBORN,
    LONDON, WC1, ENG.).

78 REPORT REPT-6-49 (69PP.)   (1970)....P
    *ANISOTROPIC NEUTRON DIFFUSION IN HETEROGENEOUS SYSTEMS.
    *SAUR(H.D.). (ED. NOTE: AVAIL. NTIS, SPRINGFIELD, VA. 22151 USA).

SECTION 16 -NEUTRON TRANSPORT AND MODERATION (NON-TECHNICAL ASPECTS) . .

79 REV. ROUMAINE PHYS. VOL.18, . . . 997, (1973)....T................
    ~CALCULATION OF THE THERMAL NEUTRON SCATTERING LAW.
    ~RAPEANU(S.), PADUREAND(I.), DECIU(N.).

80 RICERCA SCI. VOL.7, I, ... 393, (1936)....E........ . . . .
    ~DIFFUSION OF SLOW NEUTRONS.
    ~AMALDI(E.), FERMI(E.).

81 RICERCA SCI. VOL.7, II, ... 13, (1936)....T............. . . . .
    ~MOTION OF NEUTRONS IN HYDROGENOUS SUBSTANCES.
    ~FERMI(E.).

82 SOV. AT. ENERGY VOL.26, . . . 334, (1969)....B7..............
    ~MODERATING ABILITY OF NEUTRON INELASTIC SCATTERERS.
    ~KOZHEVNIKOV(D.A.), CHEKANOVA(S.S.).

83 SOV. AT. ENERGY VOL.32, . . . 33, (1972)....P........H2*0. . . . .
    ~NEUTRON THERMALISATION IN H2*O AT 318 AND 77K.
    ~ISHMAEV(I.P.), SADIKOV(I.P.), CHERNYSHOV(A.A.).

84 SOV. PHYS. SOL. STATE VOL.9, . . 2040, (1968).................
    ~THEORY OF DIFFUSION IN CRYSTALLINE SOLID SOLUTIONS; TIME EVOLUTION OF
    DIFFUSE SCATT. OF X-RAYS, THERMAL NEUTRON.
    ~KHACHATURYAN(A.G.).

85 SOV. PHYS. DOKLADY VOL.15, . . 724, (1971).....................
    ~THE FLUX OF SINGLY SCATTERED NEUTRONS FROM A POINT SOURCE IN AN
    INFINITE MEDIUM.
    ~DROZHZHINOV(YU.N.), MOSKALEV(O.B.).

86 SOV. PHYS. DOKLADY VOL.18, . . . 481, (1974)....T..................
    ~PHONONLESS COOLING OF NEUTRONS TO EXTREMELY LOW TEMPERATURES.
    ~NAMIOT(V.A.).

87 STUD. CERCETARI FIZ.(RUMAN.) 22, 365, (1970)....E.................
    ~NEW THEORETICAL AND EXPERIMENTAL METHODS FOR THE STUDY OF NEUTRON
    THERMALIZATION.
    ~PURICA(I.I.).

88 THESIS (88PP.), ... (1970)....P..................
    ~CALCULATION OF THERMAL NEUTRON PARAMETERS IN HYDROGENOUS LIQUID
    MODERATORS.
    ~BEN-AHMED(A.). (ED. NOTE: AVAIL. UNIV. MICROFILMS, ANN ARBOR, MICH., USA
    ORDER NO. 71-15477).

89 TRANS. AM. NUCL. SOC. VOL.13, . . 726, (1970)....P..................
    ~ANISOTROPIC ELASTIC MODERATION OF NEUTRONS.
    ~STACEY-JR(M.).

90 TRANS. AM. NUCL. SOC. VOL.13, . . 488, (1970)....P..................
    ~SLOW-NEUTRON TRANSMISSION ANALYSIS.
    ~MOTT(W.E.).

91 TRANSP. THEORY/STAT. PHYS. VOL.1 263, (1971)....T................
    ~MULTIGROUP TRANSPORT EQUATION WITH A DEGENERATE KERNEL.
    ~SILVENNOINEN(P.).

92 Z. PHYSIK BAND 252, ... 371, (1972)...E.T.....C.AU........
    ~INTERFERENCE IN THE ULTRACOLD NEUTRON TRANSMISSION THROUGH THIN LAYERS.
    ~STEYERL(A.).

SECTION 17A-THEORY AND TECHNIQUE OF MEASUREMENT. . . . .

1   ACTA CRYST. VOL.9, ... 151, (1956). . . . .
    *A NEW RATIONALE OF STRUCTURE-FACTOR MEASUREMENT IN NEUTRON-DIFFRACTION
    ANALYSIS.
    *LOWDE(R.D.).

2   ACTA CRYST. VOL.14, ... 90, (1961). . . E . . . .
    *USE OF A TWO-CIRCLE DEVICE TO OBTAIN THREE-DIMENSIONAL NEUTRON
    DIFFRACTION DATA.
    *WILLIS(B.T.M.).

3   ACTA CRYST. 17, ... 597, (1964). . . . .
    *GEOMETRICAL PROPERTIES OF A 4 CIRCLE NEUTRON DIFFRACTOMETER (INTENSITIES
    AT OPTIMUM AZIMUTH OF REFLECTING PLANE.).
    *SANTORO(A.), ZOCCHI(M.).

4   ACTA CRYST. 18, ... 184, (1965). . . . .
    *THE EFFECT OF EXPERIMENTAL RESOLUTION ON CRYSTAL REFLECTIVITY
    AND SECONDARY EXTINCTION IN NEUTRON DIFFRACTION.
    *DIETRICH(O.W.), ALS-NIELSEN(J.).

5   ACTA CRYST. 22, ... 331, (1967). . . . .
    *SIMULTANEOUS DIFFRACTION WITH THE THREE-CIRCLE DIFFRACTOMETER.
    *ZOCCHI(M.), SANTORO(A.).

6   ACTA CRYST. 22, ... 457, (1967). . . . .
    *ANGLE CALCULATIONS FOR 3 AND 4 CIRCLE X-RAY AND NEUTRON DIFFRACTOMETERS.
    *BUSING(W.R.), LEVY(H.A.).

7   ACTA CRYST. A24, ... 253, (1968). . . . .
    *ELIMINATION OF MULTIPLE REFLECTION ON THE 4-CIRCLE DIFFRACTOMETER.
    *COPPENS(P.).

8   ACTA CRYST. VOL.A25, . . 206, (1969). . . . .
    *INSTRUMENTS AND TECHNIQUES REQUIRED FOR ACCURATE RELATIVE INTENSITY
    MEASUREMENTS.
    *DEWOLFF(P.M.), SAS(W.H.).

9   ACTA CRYST. VOL.A26, . . 667, (1970). . . . .
    *PRIMARY AND SECONDARY EXTINCTIONS IN THE DYNAMICAL THEORY FOR AN
    IMPERFECT CRYSTAL.
    *KURIYAMA(M.), MIYAKAWA(T.).

10 ACTA CRYST. VOL.A26, . . 71, (1970). . . . .
    *ANISOTROPIC EXTINCTION CORRECTIONS IN THE ZACHARIASEN APPROXIMATION.
    *COPPENS(P.), HAMILTON(W.C.).

11 ACTA CRYST. VOL.A26, . . 18, (1970). . . . .
    *INTERN. UNION OF CRYST. COMM. ON CRYSTALLOGRAPHIC APPARATUS SINGLE CRYS-
    TAL. . . II. LEAST-SQUARES REFINEMENTS OF STRUCTURAL PARAMETERS.
    *HAMILTON(W.C.), ABRAHAMS(S.C.).

12 ACTA CRYST. VOL.A26, . . 1, (1970). . . . .
    *INTERN. UNION OF CRYST. COMM. ON CRYSTALLOGRAPHIC APPARATUS SINGLE CRYS-
    TAL INTENSITY MEASUREMENT PROJECT REPORT I. INTER-EXPERIMENTAL AGREEMENT
    *ABRAHAMS(S.C.), HAMILTON(W.C.), MATHIESON(A.MCL.).

13 ACTA CRYST. VOL.A29, . . 577, (1973). . . E . . . .
    *OPTIMUM SCANNING RATIO IN NEUTRON DIFFRACTION.
    *PANTAZATOS(P.), WERNER(S.A.).

14 ACTA CRYST. VOL.A30, . . 448, (1974). . . T . . . .
    *USE OF A TWO-AXIS NEUTRON DIFFRACTOMETER ON AN ANGLED REACTOR CHANNEL.
    *HAYWOOD(B.C.G.).

15 ACTA PHYS. AUSTRIACA VOL.20, . . 7, (1965). . . ET . . . .
    *COHERENT EXTINCTION AND SECOND-ORDER SCATTERING OF NEUTRONS BY SINGLE
    CRYSTALS.
    *KUICH(G.), RAUCH(H.). (IN GERMAN).

16 ACTA PHYS. POLON. VOL.A42, . . 259, (1972). . . E . . . .
    *STRUCTURE FACTOR DETERMINATION OF SINGLE CRYSTALS WITH HIGH EXTINCTION.
    *BURAS(B.), GIEBULTOWICZ(T.), MINOR(W.), RAJCA(A.), SOSNOWSKA(I.),
    SLEDZIEWSKA-BLOCKA(D.), WOJTCZAK(K.).

17 AMER. CRYSTALL. ASSOC. . . 65, (1971). . . E . . . .
    *CHOICE OF SCANS IN NEUTRON DIFFRACTION.
    *WERNER(S.A.). (ED. NOTE: PUBL. AMER. CRYSYALLOGRAPHIC ASSOC. 1971, NEW
    YORK, USA).

18 ANN. DE PHYSIQUE TOME 7 NO.4, . . 255, (1972). . . . .
    *EXPERIMENTAL TECHNIQUES IN NEUTRON SCATTERING (REVIEW PAPER).
    *MERIEL(P.). (PRESENTED AT AUTRANS SUMMER SCHOOL 1972).

19 ATOMKERNENERGIE VOL.12, . . 395, (1967). . . E . . . .
    *CRYSTAL MOSAIC SPREAD MEASUREMENT BY MEANS OF A NEUTRON TWO-AXIS SPEC.
    *RAUCH(H.). (ED. NOTE: IN GERMAN).

20 ATOMKERNENERGIE VOL.17, . . 15, (1971). . . T . . . .
    *THEORY OF A VELOCITY FOCUSING INSTRUMENT FOR NEUTRON SMALL ANGLE
    SCATTERING.
    *IBEL(K.), SCHMATZ(W.), SPRINGER(T.).

21 ATOMKERNENERGIE VOL.17, . . 93, (1971). . . T . . . .
    *BUNCHING OF NEUTRONS IN THE MILLI-EV REGION IN HIGH RESOLUTION NEUTRON
    SPECTROSCOPY.
    *PETERLIN(T.M.).

22 BU. ACAD. SCI. USSR PHYS. SR. 33 36, (1969). . . 87 . . . .
    *A METHOD OF MEASURING EXCITATION FUNCTIONS IN INELASTIC NEUTRON
    SCATTERING.
    *MUSAELYAN(R.M.), POPOV(V.I.), SURKOVA(I.V.), SHTRANIKH(I.V.).

23 CAN. J. PHYS. VOL.31, . . . 432, (1953). . . E,P . . . .
    *RESONANT SCATTERING OF SLOW NEUTRONS.
    *BROCKHOUSE(B.N.).

24 CHEM. APPL. THERMAL NEUTRON SCAT 31, (1973). . . E . . . .
    *EXPERIMENTAL TECHNIQUES.
    *STIRLING(G.C.). (IN BOOK: CHEMICAL APPLICATIONS OF THERMAL NEUTRON
    SCATTERING. ED. BY B.T.M. WILLIS; OXFORD UNIV. PRESS: LONDON).

SECTION 17A-THEORY AND TECHNIQUE OF MEASUREMENT. . . . .

25 C.R. ACAD. SCI. VOL.251, 230, (1960)....E...............................
*A METHOD OF INCREASING THE TRANSMISSION OF A TIME-OF-FLIGHT SPECTROMETER
FOR SLOW NEUTRONS.
*CRIBIER(D.). (IN FRENCH).

26 ENERGIA NUCLEARE VOL.7, 11, (1960)....E...........................
*DEVELOPMENT AND FUTURE AIMS OF NEUTRON DIFFRACTION.
*BACCHELLA(G.L.). (IN ITALIAN).

27 FORTSCHR. PHYS. VOL.14 NO.2, 73, (1966)....A3...............
*STRUCTURE OF MONATOMIC LIQUID METALS BY X-RAY, ELECTRON AND NEUTRON
DIFFRACTION.
*RICHTER(H.), BREITLING(G.). (IN GERMAN).

28 HARWELL SUMMER SCHOOL, 14, (1968)..........................
*CONVENTIONAL AND THREE-AXIS NEUTRON POWDER DIFFRACTION.
*CAGLIOTI(G.).

29 IAEA SYMP. (PILE RESEARCH) VIENNA 455, (1960)......................
*SECONDARY EXTINCTION AND NEUTRON DIFFRACTION.
*WILLIS(B.T.M.).

30 IAEA SYMPOSIUM BOMBAY, VOL.2, 455, (1964)..............................
*EXPERIMENTAL EQUIPMENT AND METHODS FOR INELASTIC NEUTRON SCATTERING
MEASUREMENTS.
*PALEVSKY(H.).

31 IAEA SYMPOSIUM BOMBAY, VOL.2, 513, (1964)..............................
*APPLICATION OF THE TIME-OF-FLIGHT METHOD TO NEUTRON-DIFFRACTION STUDIES.
*.C.C.C.P.

32 IAEA SYMPOSIUM BOMBAY, VOL.2, 5L5, (1964)..............................
*PERFORMANCE OF A THREE-AXIS SPECTROMETER IN EXPERIMENTS OF ELASTIC
DIFFRACTION OF NEUTRONS.
*CAGLIOTI(G.), TOCCHETTI(D.).

33 IAEA SYMP. BOMBAY, VOL.2, 483, (1964)..........................
*RECENT METHODS IN CRYSTAL SPECTROMETRY (ABSTRACT ONLY).
*IYENGAR(P.K.).

34 IAEA SYMP. COPENHAGEN, VOL.2, 323, (1968)......................
*THE #SMALL K# METHOD OF NEUTRON MOLECULAR SPECTROSCOPY.
*BRUGGER(R.M.), STRONG(K.A.), GRANT(D.M.).

35 IAEA SYMP. COPENHAGEN, VOL.2, 299, (1968)..............................
*MEASUREMENT OF THE VIBRATIONAL SPECTRA OF MOLECULES BY MEANS OF THE DOWN
SCATTERING OF NEUTRONS.
*BEG(M.M.), ROSS(D.K.).

36 IAEA SYMP. (INSTRUMENT.), VIENNA 211, (1969)......................
*SOME PROBLEMS OF NEUTRON SPECTROSCOPY AT PULSED SOURCES.
*KROO(N.).

37 IAEA SYMP. (INSTRUMENT.), VIENNA 105, (1969)......................
*THE REFLECTIVITY OF NEUTRONS BY DISTORTED COPPER CRYSTALS.
*DYMOND(R.R.), BROCKHOUSE(B.N.).

38 IAEA SYMP. (INSTRUMENT.), VIENNA 77, (1969)......................
*EFFICIENCY OF NEUTRON CRYSTAL SPECTROMETERS IN THE WAVELENGTH RANGE
4 TO 14A.
*BALLY(D.).

39 IAEA SYMP. (INSTRUMENT.), VIENNA 37, (1969)......................
*COMPARISON OF NEUTRON-SCATTERING TECHNIQUES WITH REGARD TO RESOLUTION
AND INTENSITY.
*STEDMAN(R.).

40 IAEA SYMP. (INSTRUMENT.), VIENNA 1, (1969)......................
*A CRITICAL COMPARISON OF TECHNIQUES FOR INELASTIC NEUTRON SCATTERING.
*WOODS(A.D.B.), DOLLING(G.), COWLEY(R.A.).

41 IAEA SYMP. GRENOBLE, 797, (1972)....E......................
*PSEUDORANDOM MODULATION IN POLARIZED NEUTRON DIFFRACTOMETRY.
*MEZEI(F.), PELLIONISZ(P.).

42 IAEA SYMP. GRENOBLE, 773, (1972)....T......................
*STATISTICAL CHOPPER FOR PERIODICALLY PULSED NEUTRON BEAMS.
*MATTHES(W.).

43 IAEA SYMP. GRENOBLE, 747, (1972)....T......................
*THEORY OF MULTIVARIABLE STATISTICAL CHOPPERS UTILIZING UNCORRELATED
PSEUDORANDOM SEQUENCES.
*AMADORI(R.), HOSSFELD(F.).

44 IAEA SYMP. GRENOBLE, 713, (1972)....E......................
*A HIGH RESOLUTION TIME FOCUSING SPECTROMETER FOR QUASI-ELASTIC
NEUTRON SCATTERING.
*MEISTER(H.), WECKERMANN(B.).

45 INSTRUM. EXP. TECH. NO.3, 359, (1960)....E......................
*EXPERIMENTAL METHODS OF NEUTRON DIFFRACTION.
*ABOV(YU.G.), LITVIN(D.F.).

46 INSTRUM. EXP. TECH.(USA) VOL.14, 733, (1971)....E......................
*TWO-CRYSTAL NEUTRON SPECTROMETER AND THE POSSIBILITIES OF ITS
APPLICATION.
*SHIL#SHTEIN(S.SH.), MARUKHIN(V.I.), KALANOV(M.), SOMENKOV(V.A.).

47 INSTRUM. EXP. TECH. VOL.15, 44, (1972)....E......................
*NEUTRON-SELECTOR PHASE STABILIZER.
*ORBAN(D.), SLAVIK(F.).

48 INTERN. SUMMER SCHOOL, MOL., 580, (1963)....NEUTRONS.............
*INELASTIC SCATTERING OF SLOW NEUTRONS.
*BROCKHOUSE(B.N.), HAUTECLER(S.), STILLER(H.).

49 JAP. J. APPL. PHYS. VOL.12, 1119, (1973)....A3....SN...............
*A FEASIBILITY STUDY OF APPLYING NEUTRON T-O-F METHOD FOR THE LIQUID
STRUCTURE ANALYSIS.
*TOMIYOSHI(S.), WATANABE(N.), MISAWA(M.), KAI(K.), KIMURA(M.).

50 J. APPL. CRYST. VOL.2, 65, (1969)....E......................
*A PROFILE REFINEMENT METHOD FOR NUCLEAR AND MAGNETIC STRUCTURES.
*RIETVELD(H.M.).

SECTION 17A-THEORY AND TECHNIQUE OF MEASUREMENT. . . . .

51 J. APPL. CRYST. VOL.5, ... 373, (1972). . . .E. . . . . . .
    *PRODUCTION OF SENSITIVE CONVERTER SCREENS FOR THERMAL NEUTRON
    DIFFRACTION PATTERNS.
    *THOMAS(P.).

52 J. APPL. CRYST. VOL.7, ... 96, (1974). . . .E,T. . . . . . .
    *NEUTRON SMALL-ANGLE SCATTERING: EXPERIMENTAL TECHNIQUES AND APPLICATIONS
    *SCHMATZ(W.), SPRINGER(T.), SCHELTEN(J.), IBEL(K.).

53 J. APPL. PHYS. VOL.30, ... 1323, (1959). . . .E. . . . . . .
    *NEUTRON DIFFRACTION TECHNIQUES AND THEIR APPLICATIONS TO SOME PROBLEMS
    IN PHYSICS.
    *SIDHU(S.S.), HEATON(L.), MUELLER(M.H.).

54 J. CRYST. GROWTH VOL.13, ... 247, (1972). . . .E. . . . . . .
    *TWO NEW EXPERIMENTAL DIFFRACTION METHODS FOR A PRECISE MEASUREMENT OF
    CRYSTAL PERFECTION.
    *FREUND(A.), SCHNEIDER(J.).

55 J. PHYS. SOC. JAPAN VOL.17 B-II, 342, (1962). . . .E. . . . . . .
    *A TECHNIQUE FOR MEASURING THE MAGNETIC DISORDER SCATTERING OF NEUTRONS.
    *LOWDE(R.D.), WHEELER(D.A.).

56 J. PHYS. SOC. JAPAN VOL.17 B-II, 363, (1962). . . .
    *INTERATOMIC FORCES IN CRYSTALS FROM NEUTRON SCATTERING.
    *BROCKHOUSE(B.N.).

57 J. PHYS. SOC. JAPAN VOL.23, ... 460, (1967). . . .E. . . . . . .
    *A NEW NEUTRON DIFFRACTION METHOD DIRECTLY OBSERVING THERMAL MOTIONS OF
    RADICALS.
    *SHIBUYA(I.), MITANI(S.), IWATA(Y.), TOKUNAGA(M.).

58 KERNTECHNIK VOL.14, ... 86, (1972). . . .
    *NEUTRON BACKSCATTERING AT HIGH DOPPLER VELOCITIES FOR MEASURING
    DIFFERENCES IN LATTICE PARAMETERS.
    *BAUER(G.).

59 MOL. PHYS. VOL.26 NO.6, ... 1325, (1973). . . .T,E. . . . . . .
    *THE ANALYSIS OF A TIME-OF-FLIGHT NEUTRON DIFFRACTOMETER FOR AMORPHOUS
    MATERIALS: THE STRUCTURE OF A MOLECULE IN A LIQUID.
    *POWLES(J.G.).

60 NUCL. INST. MET. VOL.12, ... 365, (1961). . . .E. . . . . . .
    *THE SCATTERING SURFACE METHOD OF MEASURING DISPERSION RELATIONS WITH A
    PHASED CHOPPER VELOCITY SELECTOR.
    *SCHMUNK(R.E.), BRUGGER(R.M.).

61 NUCL. INST. MET. VOL.25, ... 188, (1963). . . . .
    *PROPOSAL FOR EXPERIMENTAL DETERMINATION OF INCOHERENT INELASTIC
    NEUTRON SCATTERING.
    *GUTTMAN(L.).

62 NUCL. INST. MET. 42, ... 197, (1966). . . . .
    *A FILTER DIFFERENCE METHOD FOR MEASURING INELASTIC SCATTERING OF SLOW
    NEUTRONS.
    *PAN(S.S.), WEBB(F.J.), YEATER(M.L.).

63 NUCL. INST. MET. 47, ... 89, (1967). . . . .
    *A COMPARISON OF TWO DETECTING METHODS FOR TIME OF FLIGHT MEASUREMENTS OF
    THERMAL NEUTRON SPECTRA.
    *HEIBERG(E.).

64 NUCL. INST. MET. 70, ... 164, (1969). . . . .
    *THE USE OF VANADIUM SCATTERING TO INTERCALIBRATE AN ARRAY OF THERMAL
    NEUTRON DETECTORS.
    *DAY(D.H.), JOHNSON(D.A.G.), SINCLAIR(R.N.).

65 NUCL. INST. MET. VOL.84, ... 61, (1970). . . .E. . . . . . .
    *THE PSEUDO-SAMPLE METHOD: AN EXPERIMENTAL PROCEDURE FOR SAMPLE-HOLDER
    CORRECTIONS IN NEUTRON SCATTERING.
    *PRICE(D.L.), MEISTER(H.).

66 NUCL. INST. MET. 77, ... 13, (1970). . . . .
    *SPINNING SINGLE CRYSTAL T.O.F. METHOD FOR STRUCTURE ANALYSIS.
    *BURAS(B.), GIEBULTOWICZ(T.). . . .ET AL.

67 NUCL. INST. MET. VOL.91, ... 205, (1971). . . .P. . . . . . .
    *MAGNETIC NEUTRON SPECTROMETERS FOR NEUTRONS WITH WAVELENGTHS GREATER
    THAN 40 ANGSTROMS.
    *GOLUB(R.).

68 NUCL. INST. MET. VOL.94, ... 173, (1971). . . .E. . . . . . .
    *A THICK TARGET METHOD FOR THE MEASUREMENT OF NEUTRON DIFFERENTIAL
    SCATTERING CROSS SECTIONS.
    *ALBERT(T.E.), CARROLL-JR(E.E.).

69 NUCL. INST. MET. 99, ... 227, (1972). . . . .
    *DIFFRACTION METHOD FOR MAGNON STUDIES AT THE PULSED NEUTRON SOURCE.
    *DOBRZYNSKI(L.), KEPA(H.).

70 NUCL. INST. MET. VOL.108, ... 11, (1973). . . .T. . . . . . .
    *ON THE REDUCTION OF MULTIPLE SCATTERING BY A SANDWICH STRUCTURE OF THE
    SAMPLE IN NEUTRON SCATTERING EXPERIMENTS.
    *KALLI(H.).

71 NUCL. INST. MET. VOL.106, ... 461, (1973). . . .E. . . . . . .
    *MOVING CRYSTAL SLOW-NEUTRON WAVELENGTH ANALYSER.
    *BURAS(B.), KJEMS(J.K.).

72 NUCL. INST. MET. VOL.106, ... 419, (1973). . . .T,E. . . . . . .
    *DEPENDENCE OF THE INTEGRATED INTENSITY OF A SCATTERED NEUTRON GROUP ON
    THE EXPERIMENTAL CONDITIONS.
    *DOLLING(G.), SEARS(V.F.).

73 NUCL. INST. MET. VOL.112, ... 571, (1973). . . .E. . . . . . .
    *MEASUREMENT OF INTEGRATED INTENSITIES IN INELASTIC NEUTRON SCATTERING.
    *SKALYO-JR(J.), LURIE(N.A.).

74 NUCL. INST. MET. VOL.113, ... 15, (1973). . . .E,T. . . . .SI,BE.
    *PHONON MEASUREMENTS ON A FILTER-DETECTOR SPECTROMETER AND ITS RESOLUTION
    FUNCTION.
    *THAPER(C.L.), DASANNACHARYA(B.A.), IYENGAR(P.K.).

75 NUCL. SCI. ENG. VOL.48, ... 281, (1972). . . .E. . . . . . .
    *THE DOUBLE FILTER TECHNIQUE FOR THE INVESTIGATION OF THERMAL-NEUTRON
    SPECTRA.
    *ANTONINI(D.), OMICINI(E.), PISTELLA(F.).

SECTION 17A-THEORY AND TECHNIQUE OF MEASUREMENT. . . . .

76 NUKLEONIKA VOL.5, ... 414, (1961) . . . E. . . . .
    *NEUTRON DIFFRACTION STUDIES IN SOLID STATE PHYSICS.
    *BLINOWSKI(K.).

77 NUKLEONIKA VOL.9, ... 523, (1964) . . . E. . . . .
    *THE TIME-OF-FLIGHT METHOD FOR NEUTRON CRYSTAL STRUCTURE INVESTIGATIONS
    AND ITS POSSIBILITIES IN CONNECTION WITH VERY HIGH FLUX REACTORS.
    *BURAS(B.), LECIEJEWICZ(J.), NITC(W.), SOSNOWSKA(I.), SOSNOWSKI(J.),
    SHAPIRO(F.).

78 NUKLEONIK VOL.8, ... 61, (1966) . . . E. . . . .
    *BASIS FOR THE JUDGEMENT OF INTENSITY AND ACCURACY IN NEUTRON SCATTERING
    MEASUREMENTS.
    *MAIER-LEIBNITZ(H.). (IN GERMAN).

79 NUKLEONIK BAND 10, ... 101, (1967) . . . E. . . . .
    *DAS DEFOKUSSIERUNGSSPEKTROMETER, EIN GERAT HOHER AUFLOSUNG FUR DIE
    NEUTRONENSPEKTROSKOPIE IN MILLI EV-BEREICH.
    *MEISTER(H.).

80 NUKLEONIK BAND 10, ... 97, (1967) . . . E. . . . .
    *DER DOPPLER-EFFEKT BEI DER NEUTRONENREFLEXION AN ROTIERENDEN EINKRIST-
    ALLEN UND DESSEN AUSNUTZUNG/LAUFZEITFOKUSSIERUNG/REFLEK. NEUTRONENPAKETE
    *MEISTER(H.).

81 NUKLEONIK VOL.10 NO.6, ... 287, (1968) . . . E. . . . .
    *SUGGESTIONS FOR THE IMPROVEMENT OF COLD NEUTRON SPECTROSCOPY.
    *KLEY(W.).

82 NUOVO CIMENTO SUPPL. VOL.3, . . 187, (1965) . . . E. . . . .
    *A TECHNIQUE OF N-GAMMA DISCRIMINATION.
    *FABIANI(F.), LONGO(G.), SAPORETTI(F.), VENTURINI(G.).

83 NUOVO CIMENTO B VOL.43, ... 375, (1966) . . . E. . . . .
    *ACCURATE MEASUREMENTS OF STRUCTURE FACTORS OF LIQUIDS BY SLOW-NEUTRON
    SPECTROMETRY.
    *ASCARELLI(P.), CAGLIOTI(G.).

84 NUOVO CIMENTO B VOL.46, ... 248, (1966) . . . E.A1 . . . . N*H4*BR . . .
    *ENHANCEMENT OF THE SIGNAL-TO-BACKGROUND RATIO IN THE CRYSTAL STRUCTURE
    ANALYSIS OF HYDROGENOUS COMPOUNDS BY/ELASTIC DIFFRACTION OF SLOW NEUTRON
    *CAGLIOTI(G.), POMPA(F.).

85 PHYS. LETT. 31A, ... 78, (1970) . . . E. . . . .
    *MATHEMATICAL ANALYSIS OF DISTORTION EFFECTS OF SPECTROMETERS.
    *MOHOS(B.), TUDOS(F.).

86 PHYS. REV. VOL.50, ... 570, (1936) . . . E. . . . .
    *A METHOD FOR THE DETERMINATION OF ABSORPTION CROSS SECTIONS FOR THERMAL
    NEUTRONS.
    *ZAHN(C.T.), HARRINGTON(E.L.), GOUDSMIT(S.).

87 PHYS. REV. LETT. VOL.31, ... 776, (1973) . . . E,I . . H(A=1),V(A=51) . .
    *PSEUDO MAGNETIC MOMENTS OF H(A=1) AND V(A=51) MEASURED BY A NEW METHOD.
    *ABRAGAM(A.),   BACCHELLA(G.L.),   GLATTLI(H.),   MERIEL(P.),   PINOT(M.),
    PIESVAUX(J.).

88 PHYS. STAT. SOL. 4, ... 349, (1964) . . . . AL,SI . . . . .
    *A NEW METHOD FOR NEUTRON DIFFRACTION CRYSTAL STRUCTURE INVESTIGATION.
    *BURAS(B.), LECIEJEWICZ(J.).

89 PHYS. STAT. SOLIDI B VOL.58 NO.2 633, (1973) . . . .
    *MISALIGNMENT OF FLUX LINES IN A TYPE II SUPERCONDUCTOR STUDIED BY
    NEUTRON SMALL-ANGLE DIFFRACTION.
    *LIPPMANN(G.), SCHELTEN(J.), HENDRICKS(R.W.), SCHMATZ(W.).

90 REPORT NO. 702/II/PS, ... (1966) . . . E. . . . .
    *ON THE POSSIBILITIES OF SIMULTANEOUS STRUCTURE AND LATTICE DYNAMICS
    STUDIES OF SOLIDS USING THE TIME-OF-FLIGHT METHOD.
    *BURAS(B.). (AVAIL. INST. OF NUCLEAR RESEARCH, WARSAW, POLAND).

91 REV. ROUMAINE PHYS. VOL.17, . . 757, (1972) . . . E. . . . .
    *REAL STRUCTURE AND DYNAMICS OF CRYSTALS BY NEUTRON AND X-RAY SCATTERING.
    SOME NEW EXPERIMENTAL METHODS.
    *BALLY(D.).

92 REV. SCI. INSTRUM. VOL.45, . . . 526, (1974) . . . E,T . . . .
    *AUTOFRETTAGED HIGH PRESSURE CHAMBER FOR USE IN INELASTIC NEUTRON
    SCATTERING.
    *BLASCHKO(O.), ERNST(G.).

93 SOL. STATE COMM. VOL. 9, ... 531, (1971) . . . .
    *OBSERVATION OF #FORBIDDEN MODES# IN NEUTRON SCATTERING MEASUREMENTS.
    *COPLEY(J.R.D.).

94 SOV. AT. ENERGY VOL.19, ... 1387, (1965) . . . T. . . . .
    *CALCULATION OF SLOW NEUTRON DIFFERENTIAL SCATTERING CROSS-SECTIONS BY
    TIME INTEGRATION.
    *TURCHIN(V.F.).

95 SOV. AT. ENERGY(USA) VOL.32, . . 68, (1972) . . . E. . . . .
    *APPLICATION OF THRESHOLD DETECTORS FOR THE MEASUREMENT OF NEUTRON
    SPECTRA.
    *KOSHAEVA(K.K.), KRAYTOR(S.N.), PIKEL#NER(L.B.).

96 SOV. PHYS. CRYST. VOL.6, ... 404, (1962) . . . .
    *#NULL MATRIX# IN NEUTRON DIFFRACTION.
    *LYASHCHENKO(B.G.).

97 THESIS (120 PP.), ... (1971) . . . E. . . . .
    *THERMAL NEUTRON FOCUSING BY REFLECTIVE DISPERSION.
    *BANDOPADHYAY(P.K.). (ED. NOTE: AVAIL. UNIV. MICROFILMS, ANN ARBOR, MICH.
    USA ORDER NO.72-9787).

98 TH. NEUTRON SCATT./EGELSTAFF, . . 97, (1965) . . . .
    *CRYSTAL DIFFRACTION TECHNIQUES.
    *IYENGAR(P.K.).

99 TRIESTE/INTERN. COURSE, ... 443, (1967) . . . E. . . . .
    *METHODS OF NEUTRON SPECTROSCOPY. (40 PAGES).
    *BURAS(B.).

100 UKAEA AERE REP. R 4562 (9 PP.), . . . (1964) . . . E. . . . . . NI . . . . .
    *A NEW METHOD OF MEASURING FREQUENCY DISTRIBUTIONS OF CRYSTALS.
    *BRUGGER(R.M.).

SECTION 17A-THEORY AND TECHNIQUE OF MEASUREMENT. . . . .

101 Z. ANGEW. PHYS. VOL.12,   133, (1960). . .E. . . . . . .
     ⌐EXPERIMENTAL TECHNIQUES USING COLD NEUTRONS.
     ⌐NABAUER(M.), SCHMEISSNER(F.). (IN GERMAN).

102 Z. ANGEW. PHYS. VOL.18,   295, (1965). . . .83,E. . . .
     ⌐SMALL-ANGLE SCATTERING OF NEUTRONS IN INVESTIGATION OF LATTICE DEFECTS.
     ⌐CHRIST(J.), SCHILLING(W.), SCHMATZ(W.), SPRINGER(T.).

103 Z. NATURFORSCH. VOL.21 SUPPL.,   1770, (1966). . . .E. . . . . . .
     ⌐PROBLEM OF THE DETERMINATION OF THE FREQUENCY SPECTRUM IN SOLIDS BY
     MEANS OF THE INELASTIC SCATTERING OF COLD NEUTRONS.
     ⌐KLEY(W.). (IN GERMAN).

104 Z. PHYSIK VOL.189,   113, (1966). . . .E. . . . . . .
     ⌐DIFFRACTION EXPERIMENTS WITH SLOW NEUTRONS.
     ⌐LANDKAMMER(F.J.). (IN GERMAN).

105 Z. PHYSIK BAND 254,   162, (1972). . . .T. . . . . . .
     ⌐THE MEASUREMENT OF PHONON POLARIZATION VECTORS WITH THREE-AXIS
     SPECTROMETERS.
     ⌐KALUS(J.).

SECTION 17B-DATA COLLECTION AND ANALYSIS . . . .

1   ACTA CRYST. VOL.10,. . . . . 180, (1957). . . . .
      *HIGH-SPEED COMPUTATION OF THE ABSORPTION CORRECTION FOR SINGLE CRYSTAL
      DIFFRACTION MEASUREMENTS.
      *BUSING(W.R.), LEVY(H.A.).

2   ACTA CRYST. VOL.16,. . . . . 174, (1963). . . . .
      *FOURIER PROJECTIONS OF UNPAIRED ELECTRON DENSITIES.
      *PICKART(S.J.).

3   ACTA CRYST. 17,. . . . . 1202, (1964). . . . .
      *ACCURATE INTENSITY MEASUREMENTS IN SINGLE CRYSTAL ANALYSIS BY NEUTRON
      DIFFRACTION.
      *CAGLIOTI(G.).

4   ACTA CRYST. 18,. . . . . 705, (1965). . . . .
      *MULTIPLE DIFFRACTION IN IMPERFECT CRYSTALS.
      *ZACHARIASEN(W.H.).

5   ACTA CRYST. 19,. . . . . 68, (1965). . . . .
      *THE INTEGRATED INTENSITIES OF PERFECT CRYSTALS.
      *DEMARCO(J.J.), WEISS(R.J.).

6   ACTA CRYST. 19,. . . . . 224, (1965). . . . .
      *COMPARISON OF SCATTERING FACTORS COMPUTED FROM FOUR DIFFERENT ATOMIC
      MODELS.
      *CROMER(D.T.).

7   ACTA CRYST. 18,. . . . . 398, (1965). . . . .
      *DETERMINING A SET OF EXPERIMENTAL SPHERICALLY SYMMETRICAL FORM FACTORS
      FROM X-RAY AND NEUTRON DIFFRACTION MEASUREMENTS.
      *BROWN(P.J.), WILKINSON(C.).

8   ACTA CRYST. 20,. . . . . 138, (1966). . . . .
      *DEBYE-WALLER FACTORS IN CRYSTALS OF THE SODIUM CHLORIDE STRUCTURE.
      *PRYOR(A.W.).

9   ACTA CRYST. 20,. . . . . 881, (1966). . . . .
      *THE EVALUATION OF PHASES FOR STRUCTURE DETERMINATION BY NEUTRON
      DIFFRACTION.
      *KARLE(J.).

10 ACTA CRYST. 22,. . . . . 151, (1967). . . . .
      *LINE PROFILES OF NEUTRON POWDER-DIFFRACTION PEAKS FOR STRUCTURE
      REFINEMENT.
      *RIETVELD(H.M.).

11 ACTA CRYST. 23,. . . . . 54, (1967). . . . .
      *LEAST SQUARES WEIGHTING SCHEMES FOR DIFFRACTOMETER-COLLECTED DATA.
      *KILLEAN(R.C.G.).

12 ACTA CRYST. A24,. . . . . 347, (1968). . . . .
      *THE DERIVATION OF A SET OF SCATTERING FACTORS FROM X-RAY OR NEUTRON
      DIFFRACTION STRUCTURE-FACTOR MEASUREMENT.
      *WILKINSON(C.), BROWN(P.J.).

13 ACTA CRYST. VOL.B24,. . . . . 35, (1968). . . . .
      *THE USE OF NEUTRON ANOMALOUS SCATTERING IN CRYSTAL STRUCTURE ANALYSIS
      I. NON-CENTROSYMMETRIC STRUCTURES.
      *SINGH(A.K.), RAMASESHAN(S.).

14 ACTA CRYST. VOL.B24,. . . . . 1701, (1968). . . . .
      *THE USE OF NEUTRON ANOMALOUS SCATTERING IN CRYSTAL STRUCTURE ANALYSIS
      II. CENTROSYMMETRIC STRUCTURES.
      *SINGH(A.K.), RAMASESHAN(S.).

15 ACTA CRYST. VOL.A25,. . . . . 666, (1969). . . . .
      *SIMULTANEOUS REFLECTIONS AND THE MOSAIC SPREAD IN A CRYSTAL PLATE.
      *CATICHA-ELLIS(S.).

16 ACTA CRYST. VOL.A25,. . . . . 488, (1969). . . . .
      *THE IMPORTANCE OF THERMAL DIFFUSE SCATTERING IN THE COMPARISON OF X-RAY
      AND NEUTRON DIFFRACTION DATA.
      *COOPER(M.J.).

17 ACTA CRYST. VOL.A25,. . . . . 482, (1969). . . . .A4. . . . . . .HEXAMETHYLENETETRAMINE,. . . . . .
      *JOINT REFINEMENT OF NEUTRON AND X-RAY DIFFRACTION DATA.
      *DUCKWORTH(J.A.K.), WILLIS(B.T.M.), PAWLEY(G.S.).

18 ACTA CRYST. A25,. . . . . 539, (1969). . . . .
      *ON THE APPLICATION OF THE SYMBOLIC ADDITION PROCEDURE IN NEUTRON
      DIFFRACTION STRUCTURE DETERMINATION.
      *SIKKA(S.K.).

19 ACTA CRYST. A25,. . . . . 116, (1969). . . . . . .GE. . . . . . .
      *DEBYE WALLER FACTOR AND ANOMALOUS ABSORPTION(GE, 293-5 DEG. K).
      *LUDEWIG(J.), BORRMANN(G.).

20 ACTA CRYST. VOL.A25,. . . . . 194, (1969). . . . .
      *COMPARISION OF X-RAY AND NEUTRON DIFFRACTION STRUCTURAL RESULTS A STUDY
      IN METHODS OF ERROR ANALYSIS.
      *HAMILTON(W.C.).

21 ACTA CRYST. A25,. . . . . 12, (1969). . . . .
      *THE SIGNIFICANCE OF ACCURATE STRUCTURE FACTORS.
      *DAWSON(B.).

22 ACTA CRYST. A25,. . . . . 277, (1969). . . . .
      *LATTICE VIBRATIONS AND THE ACCURATE DETERMINATION OF STRUCTURE FACTORS
      FOR THE ELASTIC SCATTERING OF X-RAYS AND NEUTRONS.
      *WILLIS(B.T.M.).

23 ACTA CRYST. B25,. . . . . 977, (1969). . . . .
      *LEAST-SQUARES WEIGHTING SCHEMES FOR DIFFRACTOMETER-COLLECTED DATA
      III. OPTIMIZATION PROCESS.
      *KILLEAN(R.C.G.).

24 ACTA CRYST. B25,. . . . . 374, (1969). . . . .
      *LEAST-SQUARES WEIGHTING SCHEMES FOR DIFFRACTOMETER-COLLECTED DATA
      II. EFFECT OF RANDOM SETTING ERRORS.
      *GRANT(D.F.), KILLEAN(R.C.G.), LAURENCE(J.L.).

SECTION   178-DATA COLLECTION AND ANALYSIS . . . .

25 ACTA CRYST. VOL.A26,. . . . . . 662, (1971).. . . . .
    ~APPLICATION OF THE SYMBOLIC ADDITION PROCEDURE IN NEUTRON DIFFRACTION
    FOR NON-CENTROSYMMETRIC CRYSTALS.
    ~SIKKA(S.K.).

26 ACTA CRYST. VOL.A27, ... 280, (1971).. . . . .
    ~SOME PROPERTIES OF THE SINGLE-CRYSTAL ROCKING CURVE IN THE BRAGG CASE.
    ~FINGERLAND(A.).

27 ADV. IN PHYSICS VOL.22,. . . . . . 1, (1973). . . .E.T. . . . .
    ~THE STRUCTURE OF MOLECULAR LIQUIDS BY NEUTRON SCATTERING (REVIEW).
    ~POWLES(J.G.).

28 ARK. FYS. VOL.34, PAPER 40,. . . . 481, (1967). . . . .
    ~A MONTE CARLO PROGRAM FOR CALCULATION OF NEUTRON ATTENUATION AND
    MULTIPLE SCATTERING CORRECTIONS.
    ~HOLMQVIST(B.), GUSTAVSSON(B.), WIEDLING(T.).

29 BRIT. J. APPL. PHYS. VOL.15,. . . 1529, (1964). . . . .
    ~THE DETERMINATION OF BEAM POLARIZATION AND FLIPPING EFFICIENCY IN
    POLARIZED NEUTRON DIFFRACTOMETRY.
    ~BROWN(P.J.), FORSYTH(J.B.).

30 COLL. INTER. N.126 (GRENOBLE), . 73, (1963). . . . .
    ~PROBLEMS AND PROCEDURES IN COLLECTION OF THREE-DIMENSIONAL NEUTRON
    DIFFRACTION DATA FROM CRYSTAL STRUCTURE DETERMINATION.
    ~BROWN(G.M.), LEVY(H.A.).

31 CONF. INTERN., RENNES,. . . . . . 162, (1971). . . . .
    ~ON MULTIVARIATE CORRELATION TECHNIQUES FOR INELASTIC NEUTRON SCATTERING.
    ~AMADORI(R.), HOSSFELD(F.).

32 IAEA SYMP. CHALK RIVER, VOL.1,. . 65, (1962). . . . .
    ~PRACTICAL ANALYSIS OF NEUTRON SCATTERING DATA INTO SELF AND INTERFERENCE
    TERMS.
    ~EGELSTAFF(P.A.).

33 INSTRUM. EXP. TECH. NO.4,. . . . . 832, (1965). . . .E. . . . .
    ~MULTIDIMENSIONAL ANALYZER WITH PRELIMINARY DATA PROCESSING AND MIXED
    MEMORY.
    ~EKATOV(A.B.), IVCHENKO(V.E.), MATALIN(L.A.), MESHKOV(N.V.),
    SMIRNOV(V.I.), CHERNUKHIN(V.L.).

34 J. APP. PHYS. 37,. . . . . 2343, (1966). . . . .
    ~PROPAGATION OF BRAGG REFLECTED NEUTRONS IN BOUNDED MOSAIC CRYSTALS.
    ~WERNER(S.A.), ARROTT(A.), KING(J.S.), KENDRICK(H.).

35 J. CHEM. PHYSICS VOL.48,. . . . . 3016, (1968). . . . .
    ~ANALYSIS OF SCATTERING DATA FOR MIXTURES OF AMORPHOUS SOLIDS OR LIQUIDS.
    ~PINGS(C.J.), WASER(J.).

36 J. CHEM. PHYS. VOL.57,. . . . . . 4346, (1972). . . . .
    ~EVALUATION OF LIQUID STRUCTURE DATA.
    ~MOUNTAIN(R.D.).

37 J. PHYS. CHEM. SOL. 28,. . . . . . 2225, (1967). . . . .
    ~CATION DISTRIBUTIONS IN MULTISUBLATTICE IONIC CRYSTALS (SOLID SOLUTIONS
    OF FERRIMAGNETIC GARNETS AND SPINELS).
    ~BORGHESE(C.).

38 J. SCI. INSTRUM.(J. PHYS. E)S2,1 951, (1968). . . .E. . . . .
    ~COMPUTER INDEXING OF X-RAY AND NEUTRON POWDER DIFFRACTION PATTERNS OF
    CUBIC, TETRAGONAL, ORTHORHOMBIC AND HEXAGONAL STRUCTURES.
    ~HARDCASTLE(K.I.), STOCK(A.O.).

39 LATV. PSR...FIZ. TEHN. SER. NO.5 124, (1971). . . .P. . . . .
    ~LINE WIDTH FOR INTEGRAL INTENSITY MEASUREMENTS IN NEUTRON DIFFRACTOMETER
    WITH EQUATORIAL SURVEY CIRCUIT.
    ~NOZIK(YU.), HEIKER(D.), LIPIN(YU.).

40 NUCLEONICS VOL.20,. . . . . 157, (1962). . . . .
    ~NEUTRON-SPECTROMETRY UNITS.
    ~ANDERSON(C.A.), THOMPSON(T.J.).

41 NUCL. INST. MET. 26,. . . . . . . 346, (1964). . . . .
    ~FAST RECORDING OF NEUTRON DIFFRACTION INTENSITY DATA.
    ~CHIDAMBARAM(R.), SEQUEIRA(A.), SIKKA(S.K.).

42 NUCL. INST. MET. VOL.40,. . . . . 100, (1966). . . .E. . . . .
    ~A COMPUTER PROGRAM FOR THE ANALYSIS OF TIME-OF-FLIGHT SPECTRA.
    ~TEPEL(J.W.).

43 NUCL. INST. MET. 44,. . . . . . . 341, (1966). . . . .
    ~MANIAC, A COMPUTER CODE FOR CORRECTING NEUTRON SCATTERING ANGULAR
    DISTRIBUTIONS.
    ~BEVINGTON(P.R.), PETIT(G.A.).

44 NUCL. INST. MET. 62,. . . . . . . 29, (1968). . . . .
    ~CURVE FITTING AND STATISTICAL TECHNIQUES FOR USE IN THE MECHANISED
    EVALUATION OF NEUTRON SECTIONS.
    ~HORSLEY(A.), PARKER(J.B.), PARKER(K.), PRICE(J.A.).

45 NUCL. INST. MET. 72,. . . . . . . 205, (1969). . . . .
    ~GRAPHICAL ANALYSIS OF NEUTRON LINE SPECTRA.
    ~NISLE(R.G.).

46 NUCL. INST. MET. VOL.85,. . . . . 151, (1970). . . .P. . . . .
    ~ABSOLUTE NORMALIZATION OF NEUTRON SCATTERING CROSS SECTION DATA USING
    ORGANIC SCINTILLATORS AS SCATTERERS.
    ~LINDOW(J.T.), BOSCHUNG(P.), SHRADER(E.F.).

47 NUCL. INST. MET. VOL.93,. . . . . 109, (1971). . . .E.T. . . . .
    ~INTERMEDIATE SCATTERING FUNCTIONS OBTAINED BY FAST FOURIER TRANSFOR-
    MATION OF COLD NEUTRON TIME-OF-FLIGHT SPECTRA.
    ~BREGMAN(J.D.), DE-MUL(F.F.).

48 NUCL. INST. MET. VOL.105,. . . . . 533, (1972). . . .T. . . . .
    ~INTERPRETATION DES INTENSITES MESUREES EN DIFFRACTION NEUTRONIQUE DANS
    LE CAS DE LA DIFFRACTION ELASTIQUE A DEUX AXES AVEC BRAS LEVANT.
    ~DELAPALME(A.).

49 NUCL. INST. MET. VOL.98, ... 53, (1972). . . . .
    ~ON THE ANALYSIS OF COLD NEUTRON TIME-OF-FLIGHT SPECTRA USING DIRECT
    FAST FOURIER TRANSFORMATION.
    ~DE-MUL(F.F.), BREGMAN(J.D.).

SECTION 179-DATA COLLECTION AND ANALYSIS . . . .

50 NUCL. INST. MET. VOL.107, ...   501, (1973). . . . .
    +A SYSTEM OF PROGRAMS FOR THE REDUCTION OF DATA FROM A TIME-OF-FLIGHT
    SPECTROMETER.
    +COPLEY(J.R.D.), PRICE(D.L.), ROWE(J.M.).

51 NUCL. INST. MET. VOL.114, ...   411, (1974). . . . .
    +A SYSTEM OF PROGRAMS FOR THE REDUCTION OF DATA FROM A TIME-OF-FLIGHT
    SPECTROMETER (ERRATA).
    +COPLEY(J.R.D.), PRICE(D.L.), ROWE(J.M.). (PAPER PUBL. IN VOL.107: 501).

52 NUCL. SCI. ENGNG. VOL.12, ...   260, (1962). . . .E. . . . .
    +ON THE EVALUATION OF THE THERMAL NEUTRON SCATTERING LAW.
    +EGELSTAFF(P.A.), SCHOFIELD(P.).

53 NUCL. SCI. ENGNG. VOL.34, ...   93, (1968). . . .B7. . . . .
    +ON THE CALCULATION OF NEUTRON INELASTIC SCATTERING CROSS SECTIONS.
    +GOLDSMITH(M.).

54 NUCL. SCI. ENG. VOL.45, ... 167, (1971). . . .T. . . . .
    +APPLICATIONS OF THE DEGENERATE KERNEL TECHNIQUE TO THERMAL NEUTRON
    SPECTRA CALCULATIONS.
    +TURINSKY(P.J.), OUDERSTADT(J.J.).

55 NUCL. SCI. ENG. VOL.48, ...   119, (1972). . . .T. . . . .
    +ON THE SCATTERING MATRIX CONSTRUCTION FOR THE DISCRETE ENERGY REPRESEN-
    TATION OF THERMAL-NEUTRON SPECTRA.
    +MATAUSEK(M.V.).

56 PHYS. LETT. VOL.8, ...   304, (1964). . . .E. . . . .
    +A NEW SAMPLING METHOD FOR CALCULATING THE FREQUENCY DISTRIBUTION
    FUNCTION OF SOLIDS.
    +GILAT(G.), DOLLING(G.).

57 PHYS. REV. 132, ...   2764, (1963). . . . .
    +USE OF INTENSITY CORRELATIONS TO DETERMINE THE PHASE OF A SCATTERING
    AMPLITUDE.
    +GOLDBERGER(M.L.), LEWIS(H.W.), WATSON(B.P.).

58 PHYS. REV. A VOL.7, ...   1748, (1973). . . .HE (A=4). . . . .
    +INFERRING THE DISTRIBUTION OF ATOMIC KINETIC ENERGIES FROM NEUTRON-
    SCATTERING EXPERIMENTS.
    +GIBBS(A.G.), HARLING(O.K.).

59 PROC. PHYS. SOC. 85, ...   837, (1965). . . . .
    +ON VARIANCE AS A MEASURE OF LINE BROADENING IN DIFFRACTOMETRY EFFECT OF
    TRUNCATED INTEGRATED INTENSITY.
    +WILSON(A.J.C.).

60 PROC. ROY. SOC. 298, ...   255, (1967). . . . .
    +A GENERAL STRUCTURE FACTOR FORMALISM FOR INTERPRETING ACCURATE X-RAY AND
    NEUTRON DIFFRACTION DATA.
    +DAWSON(B.).

61 REACTOR PHYSICS, ...   91, (1966). . . . .
    +INTERPRETATION OF SCATTERING-LAW DATA.
    +BRUGGER(R.M.), (IN BOOK: REACTOR PHYSICS IN THE RESONANCE AND THERMAL
    REGIONS VOL.I, NEUTRON THERMALIZATION. PUBL. M.I.T. PRESS, 1966).

62 REP. NO. BNL-10238 (3 PP.), ...   (1966). . . .E. . . . .
    +AUTOMATIC ACQUISITION OF NEUTRON SPECTROMETER DATA.
    +HAMILTON(W.C.).

63 REPORT AERE-R5642 (8 PP.), ...   (1967). . . .E. . . . .
    +ASSESSMENT OF DATA FROM AN AUTOMATIC NEUTRON DIFFRACTOMETER INSTALLATION
    +HALL(J.W.), TURBERFIELD(K.C.). (ED. NOTE: AVAIL. H.M.S.O., 49 HIGH
    HOLBORN, LONDON WC1, ENG.).

64 REPORT U.K.A.E.A. R7350 (62 PP.) . . .   (1973). . . . .
    +THE RIETVELD COMPUTER PROGRAM FOR THE PROFILE ANALYSIS OF NEUTRON
    DIFFRACTION POWDER PATTERNS, MODIFIED FOR ANISOTROPIC THERMAL VIBRATION.
    +HEWAT(A.W.).

65 REPORT FN-43-1973 (31 PP.), ...   (1973). . . .T. . . . .
    +INSTRUMENTAL WIDTHS AND INTENSITIES IN NEUTRON CRYSTAL DIFFRACTOMETRY.
    +GRABCEV(B.). (AVAIL. INST. ATOMIC PHYS., BUCHAREST).

66 SOL. STATE COMM. VOL.8, ...   1799, (1970). . . . .
    +DECONVOLUTION DU FAISCEAU OBSERVE EN DIFFUSION ELASTIQUE DE NEUTRONS
    LENTS.
    +DELTOUR(J.), KONSTANTINOVIC(J.).

67 SOL. STATE COMM. VOL.9, ...   1353, (1971). . . .NAPHTHALENE. . . . .
    +A MODEL CALCULATION OF THE INELASTIC NEUTRON SCATTERING SPECTRA FROM
    POLYCRYSTALLINE NAPHTHALENE.
    +PAWLEY(G.S.), REYNOLDS(P.A.), KJEMS(J.K.), WHITE(J.W.).

68 SOL. STATE COMM. VOL.9, ...   417, (1971). . . . .
    +SOME CONSIDERATIONS ON THE MAGNON DISPERSION RELATIONS AS OBTAINED BY
    DIFFERENT NEUTRON SCATTERING TECHNIQUES.
    +ANTONINI(B.), MENZINGER(F.).

69 SOV. J. NUCL. PHYS. VOL.12, . .   960, (1970). . . .B7. . . . .
    +STATISTICAL ANALYSIS OF SPECTRA OF INELASTICALLY SCATTERED NEUTRONS.
    +STAVINSKII(V.S.).

70 UKAEA AERE REPORT 5237 (13 PP.), ...   (1966). . . .E. . . . .
    +A NEW TIME-OF-FLIGHT DATA PROCESSING SYSTEM.
    +HALL(J.W.).

71 UKAEA AERE REP.R5536 (41 PP.), ...   (1967). . . . .
    +SCAT AND SLAB: TWO COMPUTER CODES FOR THE COMPUTATION OF THERMAL NEUTRON
    SCATTERING CROSS-SECTIONS.
    +HUTCHINSON(P.), SCHOFIELD(P.).

72 UKAEA AERE REPORT M2570 (32 PP.) . . . .   (1972). . . .E. . . . .
    +THE COLLECTION AND PROCESSING OF DATA FROM THREE TIME-OF-FLIGHT
    NEUTRON SPECTROMETERS.
    +BASTON(A.H.). (AVAIL. HMSO, LONDON, ENGLAND).

73 UKR. FIZ. ZH. (USSR) VOL.17, . .   789, (1972). . . .T. . . . .
    +ON DERIVATION OF THE G(NU)-FUNCTION FROM THE DATA ON INELASTIC COHERENT
    SCATTERING OF SLOW NEUTRONS.
    +GORBACHEV(B.I.), IVANITSKII(P.G.), KROTENKO(V.T.).

SECTION 17B-DATA COLLECTION AND ANALYSIS . . . .

74 Z. ANGEW. MATH. PHYS. VOL.15, 481 (1964)
A PROGRAMMING AND AUTOMATIC DATA COLLECTING SYSTEM FOR A NEUTRON
DIFFRACTION SPECTROMETER.
HALG(W.), MEIER(F.), GASSER(F.).

SECTION 17C-DATA CORRECTION FACTORS. . . . .

1   ACTA CRYST. 16,. .SECONDARY EXTINCTION CORRECTIONS FOR CYLINDRICAL CRYSTALS. . . . .
                       619, (1963)
                       .HAMILTON(W.C.).

2   ACTA CRYST. 17,. .THE EFFECT OF THERMAL MOTION ON THE ESTIMATION OF BOND LENGTHS FROM
                       142, (1964)
                       DIFFRACTION MEASUREMENTS.
                       .BUSING(W.R.), LEVY(H.A.).

3   ACTA CRYST. 17,. .THE EFFECTS OF GRANULARITY ON THE DIFFRACTED INTENSITY IN POWDERS.
                       325, (1964)
                       .HARRISON(R.J.), PASKIN(A.).

4   ACTA CRYST. 17,. .THE EFFECTS OF SIMULTANEOUS REFLECTIONS ON SINGLE CRYSTAL NEUTRON
                       805, (1964)
                       DIFFRACTION INTENSITIES.
                       .MOON(R.M.), SHULL(C.G.).

5   ACTA CRYST. 17,. .CORRECTION OF INTENSITIES OF DIFFRACTION MAXIMA FOR ABSORPTION IN
                       1529, (1964)
                       CYLINDRICAL SAMPLES.
                       .BALBY(D.), GHEORGHIU(Z.), PASCULESCU(D.).

6   ACTA CRYST. 17,. .THE REPRESENTATION OF ABSORPTION CORRECTION FACTORS FOR SPHERICAL AND
                       1326, (1964)
                       CYLINDRICAL CRYSTALS BY GAUSSIAN FUNCTIONS.
                       .PALM(J.H.).

7   ACTA CRYST. 19,. .THE ABSORPTION CORRECTION IN CRYSTAL STRUCTURE ANALYSIS. . . . .
                       1014, (1965)
                       .DE-MEULENAER(J.), TOMPA(H.).

8   ACTA CRYST. 20,. .THE SECONDARY EXTINCTION CORRECTION APPLIED TO A CRYSTAL OF ARBITRARY
                       407, (1966)
                       SHAPE.
                       .ASBRINK(S.), WERNER(P.E.).

9   ACTA CRYST. 23,. .INCLUSION OF SECONDARY EXTINCTION IN LEAST SQUARES CALCULATIONS.
                       664, (1967)
                       .LARSON(A.C.).

10 ACTA CRYST. A24, .SPECIMEN MOTION EFFECTS IN NEUTRON DIFFRACTION. . . . .
                       160, (1968)
                       .SHULL(C.G.), MORASH(K.R.), ROGERS(J.G.).

11 ACTA CRYST. A24, .THE CORRECTION OF MEASURED INTEGRATED BRAGG INTENSITIES FOR FIRST ORDER
                       405, (1968)
                       THERMAL DIFFUSE SCATTERING.
                       .COOPER(M.J.), ROUSE(K.D.).

12 ACTA CRYST. A25, .THE CORRECTION OF MEASURED INTEGRATED BRAGG INTENSITIES FOR ANISOTROPIC
                       615, (1969)
                       THERMAL DIFFUSE SCATTERING.
                       .ROUSE(K.D.), COOPER(M.J.).

13 ACTA CRYST. A25, .THE CORRECTION OF MEASURED STRUCTURE FACTORS FOR THERMAL DIFFUSE
                       95, (1969)
                       SCATTERING.
                       .COCHRAN(W.).

14 ACTA CRYST. VOL.A26 .ARTIFICIAL SPLITTING OF ONE-PHONON NEUTRON GROUPS DUE TO RELAXED
                       539, (1970)
                       VERTICAL COLLIMATION.
                       .COWLEY(E.R.), PANT(A.K.).

15 ACTA CRYST. VOL.A30 .INSTRUMENTAL WIDTHS AND INTENSITIES IN NEUTRON CRYSTAL DIFFRACTOMETRY.
                       207, (1974)
                       .GRABCEV(B.).

16 BULL. SOC. FRANC. MIN. CRIST. 83 125, (1960) .CORRECTION OF THE SAMPLE-HOLDER ERROR IN NEUTRON DIFFRACTION. . . . . .
                       .BOUTRON(F.), MERIEL(P.). (IN FRENCH).

17 J. APPL. CRYST. VOL.3, .COLLIMATION CORRECTIONS IN SMALL ANGLE SCATTERING OF NEUTRONS. . . . . .
                       145, (1970)
                       .ANTONINI(M.), DANERI(A.), TOSELLI(G.).

18 J. APPL. CRYST. VOL.4, .APPLICATION OF SPLINE FUNCTIONS TO THE CORRECTION OF RESOLUTION ERRORS
                       210, (1971)
                       IN SMALL-ANGLE SCATTERING.
                       .SCHELTEN(J.), HOSSFELD(F.).

19 NUCL. INST. MET. 56, .MULTIPLE SCATTERING CORRECTION USING A COMBINATION OF MONTE CARLO AND
                       245, (1967)
                       ANALYTICAL METHODS.
                       .COX(S.A.).

20 NUCL. INST. MET. 83, .FINITE SAMPLE CORRECTIONS TO NEUTRON SCATTERING DATA. . . . .
                       15, (1970)
                       .KINNEY(W.E.).

21 NUCL. INST. MET. VOL.84, .THE PSEUDO-SAMPLE METHOD: AN EXPERIMENTAL PROCEDURE FOR SAMPLE-HOLDER
                       61, (1970)
                       CORRECTIONS IN NEUTRON SCATTERING.
                       .PRICE(D.L.), MEISTER(H.).

22 NUCL. INST. MET. 77, .A METHOD FOR CORRECTING THE TIME OF FLIGHT SPECTRA FOR THE NEUTRON
                       12, (1970)
                       EMISSION TIME AND THE DETECTOR EFFICIENCY.
                       .D#OULTREMONT(P.).

23 NUCL. INST. MET. 87, .CORRECTIONS TO MAGNON DISPERSION RELATIONS AS DETERMINED BY THE NEUTRON
                       125, (1970)
                       DIFFRACTION TECHNIQUE.
                       .ANTONINI(B.), MEDINA(R.), MENZINGER(F.).

24 REV. SCI. INSTRUM. VOL.40, .ATTENUATION CORRECTION FACTOR FOR SCATTERING FROM CYLINDRICAL TARGETS.
                       555, (1969)
                       .CARPENTER(J.M.).

SECTION 17C-DATA CORRECTION FACTORS. . . . .

25 REV. SCI. INSTRUM. VOL.45,   572 (1974).....T...............
    #GENERALIZED ATTENUATION CORRECTION FACTOR FOR SCATTERING FROM CYLIN-
    DRICAL TARGETS.
    #MILDNER(D.F.R.), CARPENTER(J.M.), PELIZZARI(C.A.).

26 Z. KRIST. VOL.130, ...   318, (1969)....T...............
    #INCOHERENT BACKGROUND IN THE NEUTRON DIFFRACTOMETRY OF POWDERED CRYSTALS
    #BURAS(B.).

SECTION   170-RESOLUTION EFFECTS . . . .

1   ACTA CRYST. 23,.   357, (1967).....T.E.
     *RESOLUTION FUNCTION IN NEUTRON DIFFRACTOMETRY. I. RESOLUTION FUNCTION
     OF A NEUTRON DIFFRACTOMETER AND ITS APPLICATION TO PHONON MEASUREMENTS.
     *COOPER(M.J.), NATHANS(R.).

2   ACTA CRYST. A24,   481, (1968).....................
     *RESOLUTION FUNCTION IN NEUTRON DIFFRACTOMETRY. II.RESOLUTION FUNCTION OF
     A CONVENTIONAL TWO CRYSTAL NEUTRON DIFFRACTOMETER FOR ELASTIC SCATTERING
     *COOPER(M.J.), NATHANS(R.).

3   ACTA CRYST. A24,   619, (1968).....................
     *RESOLUTION FUNCTION IN NEUTRON DIFFRACTOMETRY. III.
     DETERMINATION AND PROPERTIES OF ELASTIC TWO-CRYSTAL RESOLUTION FUNCTION.
     *COOPER(M.J.), NATHANS(R.).

4   ACTA CRYST. A24,   624, (1968)..................
     *RESOLUTION FUNCTION IN NEUTRON DIFFRACTOMETRY. IV. APPLICATION OF
     RESOLUTION FUNCTION TO MEASUREMENT OF BRAGG PEAKS.
     *COOPER(M.J.).

5   ACTA CRYST. VOL.A25,   547, (1969).SPECTROMETER.
     *RESOLUTION OF A TRIPLE AXIS SPECTROMETER.
     *NIELSEN(M.), MOLLER(H.B.).

6   ACTA CRYST. VOL.A28,   319, (1972)....................
     *THE NORMALIZATION OF THE RESOLUTION FUNCTION FOR INELASTIC NEUTRON
     SCATTERING AND ITS APPLICATION.
     *DORNER(B.).

7   ACTA CRYST. VOL.A29,   160, (1973)....T.E.............
     *DERIVATION AND EXPERIMENTAL VERIFICATION OF THE NORMALIZED RESOLUTION
     FUNCTION FOR INELASTIC NEUTRON SCATTERING.
     *CHESSER(N.J.), AXE(J.D.).

8   IAEA SYMPOSIUM VIENNA,   199, (1960).................
     *RESOLUTION OF VARIOUS DEVICES FOR NEUTRON SPECTROSCOPY.
     *ZWEIFEL(P.F.), CARPENTER(J.M.).

9   IAEA SYMP. (INSTRUMENT.), VIENNA   49, (1969)...............
     *RESOLUTION OF NEUTRON SPECTROMETERS FOR INELASTIC NEUTRON SCATTERING.
     *MOLLER(H.B.), NIELSEN(M.).

10 J. APPL. PHYS. 42,   4736, (1971).........NA........
     *RESOLUTION EFFECTS IN THE MEASUREMENT OF PHONONS IN SODIUM METAL.
     *WERNER(S.A.), PYNN(R.).

11 J. PHYS. SOC. JAPAN VOL.17 B-II, 347, (1962)....T............
     *RESOLUTION AND LUMINOSITY OF CRYSTAL SPECTROMETERS FOR NEUTRON
     DIFFRACTION.
     *CAGLIOTI(G.), RICCI(F.P.).

12 NUCL. INST. MET. VOL.9,   195, (1960)....T.............
     *ON RESOLUTION AND LUMINOSITY OF A NEUTRON DIFFRACTION SPECTROMETER FOR
     SINGLE CRYSTAL ANALYSIS.
     *CAGLIOTI(G.), PAOLETTI(A.), RICCI(F.P.).

13 NUCL. INST. MET. VOL.15,   155, (1962)....T.E...........
     *RESOLUTION AND LUMINOSITY OF CRYSTAL SPECTROMETERS FOR NEUTRON
     DIFFRACTION.
     *CAGLIOTI(G.), RICCI(F.P.).

14 NUCL. INST. MET. 42,   81, (1966)...................
     *INTENSITY AND RESOLUTION IN THE TIME OF FLIGHT METHOD FOR NEUTRON
     DIFFRACTION.
     *SCHWARTZ(L.H.).

15 REV. ROUMAINE PHYS. VOL.18,   373, (1973)....E.............
     *RESOLUTION IN NEUTRON CRYSTAL SPECTROSCOPY OF ISOTROPIC SYSTEMS.
     *GRABCHEV(B.).

SECTION 18A-TRIPLE-AXIS SPECTROMETRY: THEORY AND TECHNIQUE . . . .

1 ACTA CRYST. 18,. . . . . 184, (1965). . . . .
*THE EFFECT OF EXPERIMENTAL RESOLUTION ON CRYSTAL REFLECTIVITY
AND SECONDARY EXTINCTION.
*DIETRICH(O.W.), ALS-NIELSEN(J.).

2 ACTA CRYST. VOL.A25. . . . . 547, (1969). . . . .
*RESOLUTION OF A TRIPLE AXIS SPECTROMETER.
*NIELSEN(M.), MOLLER(H.B.).

3 ACTA CRYST. VOL.A27. . . . . 468, (1971). . . . .
*A METHOD OF PREDICTING LINE SHAPES IN THREE-AXIS SPECTROMETRY.
*HAYWOOD(B.C.).

4 ACTA CRYST. VOL.A27. . . . . 605, (1971). . . . .
*THE (Q,W) TRANSMISSION FUNCTION OF A TRIPLE-AXIS NEUTRON SPECTROMETER.
*QUITTNER(G.).

5 ACTA CRYST. VOL.A28. . . . . 474, (1972). . . . .
*THE (Q,W) TRANSMISSION FUNCTION OF A TRIPLE-AXIS NEUTRON SPECTROMETER.
ERRATUM.
*QUITTNER(G.).

6 ACTA CRYST. VOL.A28. . . . . 81, (1972). . . . .
*FOCUSING EFFECTS FOR THREE-CRYSTAL NEUTRON DIFFRACTOMETERS.
*COOPER(M.J.).

7 ACTA CRYST. VOL.A29. . . . . 160, (1973). . . . T,E. . . . .
*DERIVATION AND EXPERIMENTAL VERIFICATION OF THE NORMALIZED RESOLUTION
FUNCTION FOR INELASTIC NEUTRON SCATTERING.
*CHESSER(N.J.), AXE(J.D.).

8 ACTA PHYS. AUSTRIACA VOL.38. . . . 282, (1973). . . . T. . . . .
*CONFIGURATION CHOICES FOR SIMILAR THREE AXIS NEUTRON SPECTROMETERS.
*QUITTNER(G.). (IN GERMAN).

9 BRIT. J. APPL. PHYS. VOL.14, . . . 845, (1963). . . . E. . . . .
*FOCUSSING EFFECTS IN TRIPLE-AXIS NEUTRON SPECTROMETERS.
*COLLINS(M.F.).

10 BRIT. J. APPL. PHYS. VOL.18, . . . 473, (1967). . . . E. . . . .
*FOCUSING CONDITIONS FOR A TRIPLE-AXIS NEUTRON SPECTROMETER.
*PECKHAM(G.E.), SAUNDERSON(D.H.), SHARP(R.I.).

11 BROOKHAVEN SYMPOSIUM. . . . . 163, (1965). . . . .
*FOCUSING OF A TRIPLE-AXIS NEUTRON SPECTROMETER WITH THE AID OF A
GRAPHICAL METHOD.
*BERGSMA(J.), VAN-DIJK(C.).

12 BULL/KIDRICH INST. NUCL. SCI. 15 115, (1964). . . . E. . . . .
*SOME EXPERIENCE WITH THE THREE CRYSTAL NEUTRON SPECTROMETER.
*DIMITRIJEVIC(Z.), KRASNICKI(SZ.)... ET AL.

13 IAEA SYMPOSIUM VIENNA. . . . . . . 113, (1963). . . . .
*METHODS FOR NEUTRON SPECTROMETRY.
*BROCKHOUSE(B.N.).

14 IAEA SYMPOSIUM BOMBAY, VOL.2, . . . 505, (1964). . . . .
*PERFORMANCE OF A THREE-AXIS SPECTROMETER IN EXPERIMENTS OF ELASTIC
DIFFRACTION OF NEUTRONS.
*CAGLIOTI(G.), TOCCHETTI(D.).

15 IAEA SYMP. BOMBAY, VOL.2. . . . . . 483, (1964). . . . .
*RECENT METHODS IN CRYSTAL SPECTROMETRY.
*IYENGAR(P.K.).

16 KERNTECHNIK VOL.15. . . . . 228, (1973). . . . E. . . . .
*THE OPTIMALIZATION OF THE COUNTING RATE OF A THREE-AXIS-SPECTROMETER.
*KALUS(J.).

17 NUCL. INST. MET. 32. . . . . 181, (1965). . . . .
*RESOLUTION AND LUMINOSITY OF A TRIPLE AXIS SPECTROMETER IN EXPERIMENTS
OF ELASTIC NEUTRON DIFFRACTION.
*CAGLIOTI(G.), TOCCHETTI(D.).

18 NUCL. INST. MET. 61. . . . . 296, (1968). . . . .
*RESOLUTION AND INTENSITY IN NEUTRON SPECTROMETRY DETERMINED BY MONTE
CARLO SIMULATION.
*DIETRICH(O.W.).

19 NUCL. INST. MET. VOL.95. . . . . . 445, (1971). . . . E. . . . . . . C (GRAPHITE).
*VERTICALLY BENT PYROLYTIC GRAPHITE CRYSTALS APPLIED TO TRIPLE-AXIS
NEUTRON SPECTROMETRY.
*NUNES(A.C.), SHIRANE(G.).

20 NUCL. INST. MET. VOL.107. . . . . . 461, (1973). . . . T. . . . .
*GAUSSIAN APPROXIMATION TO THE TRANSMISSION FUNCTION OF A TRIPLE-AXIS
NEUTRON SPECTROMETER.
*QUITTNER(G.), ERNST(G.).

21 NUCL. INST. MET. VOL.106. . . . . . 419, (1973). . . . T,E. . . . .
*DEPENDENCE OF THE INTEGRATED INTENSITY OF A SCATTERED NEUTRON GROUP ON
THE EXPERIMENTAL CONDITIONS.
*DOLLING(G.), SEARS(V.F.).

22 NUCL. INST. MET. VOL.106. . . . . . 349, (1973). . . . T. . . . .
*THE RESOLUTION FUNCTION OF A TRIPLE-AXIS NEUTRON SPECTROMETER.
*GRABCEV(B.).

23 NUKLEONIK VOL.5. . . . . 121, (1963). . . . T. . . . .
*CALCULATION OF THE EXPERIMENTAL WIDTH OF PHONON LINES IN THE TRIAXIAL
NEUTRON SPECTROMETER.
*STOCKMEYER(R.). (IN GERMAN).

24 PHYS. REV. B-4,. . . . . 3206, (1971). . . . . RB*MN*F3.
*QUANTITATIVE ANALYSIS OF INELASTIC SCATTERING IN TWO-CRYSTAL AND
THREE-CRYSTAL NEUTRON SPECTROMETRY: CRITICAL SCATTERING FROM RB*MN*F3.
*TUCCIARONE(A.), LAU(H.Y.), CORLISS(L.M.), DELAPALME(A.), HASTINGS(J.M.).

25 REV. ROUMAINE PHYS. VOL.10. . . . . 1035, (1965). . . . E. . . . .
*SECONDARY EXTINCTION AND OPTICAL PROPERTIES OF NEUTRON CRYSTAL
SPECTROMETERS.
*POPOVICI(M.), GELBERG(D.). (IN FRENCH).

SECTION 18A-TRIPLE-AXIS SPECTROMETRY: THEORY AND TECHNIQUE . . . .

26 REV. SCI. INSTRUM. VOL.39,   878, (1968). . . . .
    ↔ENERGY RESOLUTION AND FOCUSSING IN INELASTIC SCATTERING EXPERIMENTS.
    ↔STEDMAN(R.).

27 UKAEA AERE REP. 4380 (6 PP.),   (1964). . . E. . . . .
    ↔A GRAPHICAL AID TO FOCUSSING A TRIPLE AXIS SPECTROMETER.
    ↔PECKHAM(G.).

28 Z. ANGEW. PHYS.(GER.) VOL.32,   296, (1971). . . . E. . . . .
    ↔FOCUSING CONDITIONS FOR A TRIPLE-AXIS SPECTROMETER.
    ↔MARX(D.). (ED. NOTE: IN ENGLISH).

29 Z. PHYSIK BAND 254,. . . . . 162, (1972). . . .T. . . . .
    ↔THE MEASUREMENT OF PHONON POLARIZATION VECTORS WITH THREE-AXIS
    SPECTROMETERS.
    ↔KALUS(J.).

30 Z. PHYSIK BAND 254,   148, (1972). . .T. . . . .
    ↔THE COUNTING RATE OF A THREE-AXIS SPECTROMETER.
    ↔KALUS(J.).

SECTION 18B-TRIPLE-AXIS SPECTROMETRY: INSTRUMENTS AND FACILITIES . . . .

1 ACTA PHYS. AUSTRIACA VOL.20, 166, (1965)....E........
  *H.A.N. NEUTRON DIFFRACTION INSTALLATION, SEIBERSDORF: AUTOMATIC CONTROL
   AND PROGRAMMING.
  *MAY(F.). (IN GERMAN).

2 DISS. ABS. VOL.32, ...1795B, (1971)....B1.......CR2...
  *THE STUDY OF INSTRUMENTAL RESOLUTION AND THE LATTICE DYNAMICS OF
   CHROMIUM.
  *SHAW(W.M.).

3 IAEA SYMPOSIUM VIENNA, ... 199, (1960)....E........
  *RESOLUTION OF VARIOUS DEVICES FOR NEUTRON SPECTROSCOPY.
  *ZWEIFEL(P.F.), CARPENTER(J.M.).

4 IAEA SYMPOSIUM VIENNA, ... 277, (1960)....E........
  *SLOW-NEUTRON INELASTIC SCATTERING MEASUREMENTS AT THE MATERIALS TESTING
   REACTOR.
  *BRUGGER(R.M.), EVANS(J.E.).

5 IAEA SYMP. GRENOBLE, ... 681, (1972)....E........
  *NEW INSTRUMENTATION (WITH SPECIAL REFERENCE TO THE HIGH FLUX
   REACTOR ILL).
  *MAIER-LEIBNITZ(H.).

6 JAD. ENERG. (CZECH.) VOL.18, . 367, (1972)....E........
  *TRIPLE AXIS CRYSTAL NEUTRON SPECTROMETER, TYPE TKSN-400.
  *PETRZILKA(V.). (ED. NOTE: IN CZECH.).

7 J. APPL. CRYST. VOL.2, ... 37, (1969)....E........
  *AN INEXPENSIVE TRIPLE-AXIS NEUTRON SPECTROMETER.
  *PAWLEY(G.S.).

8 J. SCI. INSTRUM. SER.2 VOL.1, . 528, (1968)....E........
  *A TRIPLE-AXIS SPECTROMETER FOR NEUTRON INELASTIC SCATTERING.
  *CHUMBLEY(L.), DYER(R.F.), WALLIS(D.E.).

9 REV. ROUMAINE PHYS. VOL.10, . . 445, (1965)....E........
  *THE THREE AXES SPECTROMETER OF THE INSTITUTE FOR ATOMIC PHYSICS
   (BUCHAREST).
  *BALLY(D.), TODIREANU(S.), BIRSAN(I.), PIRLOGEA(P.), TARINA(E.).

10 REV. SCI. INSTRUM. VOL.39, . . 637, (1968)....E........
   *THREE AXIS CRYSTAL SPECTROMETER FOR NEUTRONS.
   *STEDMAN(R.), NILSSON(G.).

11 THESIS (98 PP.), ... (1966)....E........
   *NEUTRON SPECTROSCOPY: THE DESIGN AND CONSTRUCTION OF A TRIPLE-AXIS
    CRYSTAL SPECTROMETER, AND A STUDY OF THE LATTICE DYNAMICS OF METALS.
   *ROWE(J.M.). (ED. NOTE: PH.D. DISSERTATION, MCMASTER UNIVERSITY).

12 UKAEA AERE REPORT R 4895 (22 PP) . . . (1965)....E........
   *THE DIDO TRIPLE AXIS CRYSTAL SPECTROMETER.
   *SAUNDERSON(D.H.), DUFFILL(C.), SHARP(R.I.).

13 UKAEA AERE REPORT 5259 (9PP.), ... (1966)....E........
   *A REVIEW OF THERMAL NEUTRON SCATTERING EQUIPMENT AT A.E.R.E. HARWELL.II.
    INELASTIC SCATTERING APPARATUS.
   *DYER(R.F.).

SECTION 19A-TIME-OF-FLIGHT SPECTROMETRY: THEORY AND TECHNIQUE. . . . . .

1 ACTA CRYST. (INTERACT.) VOL.A25, S250, (1969)....T.........
    *THE USE OF PSEUDORANDOM NEUTRON TIME-OF-FLIGHT TECHNIQUES FOR LATTICE
    DYNAMICS AND DIFFRACTION STUDIES.
    *GLASER(W.), GOMPF(F.), DREXEL(W.), BECKURTS(K.H.).

2 ACTA CRYST. VOL.A27,   461, (1971)
    *THE RESOLUTION FUNCTION OF A SLOW NEUTRON ROTATING CRYSTAL
    TIME-OF-FLIGHT SPECTROMETER I. APPLICATION TO PHONON MEASUREMENTS.
    *FURRER(A.).

3 ACTA CRYST. VOL.A28,   287, (1972)
    *THE RESOLUTION FUNCTION OF A SLOW NEUTRON ROTATING-CRYSTAL T.O.F.
    SPECTROMETER.II. APPLICATION TO THE MEASUREMENT OF GENERAL FREQUENCY...
    *FURRER(A.).

4 ACTA CRYST. VOL.A28 PART S-4,   197, (1972)....E.........
    *THE NEUTRON TIME-OF-FLIGHT SPECTROMETRY FOR SINGLE CRYSTALS.
    *NIIMURA(N.), TOMIYOSHI(S.), WATANABE(N.), KIMURA(M.). (ABSTRACT ONLY).

5 ARK. FYS. VOL.16 PAPER 19,   199, (1959)....T,E.
    *THE SLOW CHOPPER AND TIME OF FLIGHT SPECTROMETER IN THEORY AND
    EXPERIMENT.
    *LARSSON(K.E.), DAHLBORG(U.), HOLMRYD(S.), OTNES(K.), STEDMAN(R.).

6 C.R. ACAD. SCI. VOL.251,   230, (1960)....E.........
    *A METHOD OF INCREASING THE TRANSMISSION OF A TIME-OF-FLIGHT SPECTROMETER
    FOR SLOW NEUTRONS.
    *CRIBIER(D.). (IN FRENCH).

7 DISS. ABS. VOL.25,   1131, (1965)
    *PREDICTION AND MEASUREMENT OF NEUTRON CHOPPER BURST SHAPES.
    *CARPENTER(J.M.).

8 HARWELL SUMMER SCHOOL,   34, (1968)
    *TIME-OF-FLIGHT NEUTRON DIFFRACTOMETRY.
    *TURBERFIELD(K.C.).

9 HELV. PHYS. ACTA VOL.89,   569, (1966)....E.........
    *TIME OF FLIGHT SPECTROMETER FOR THE INVESTIGATION OF INELASTIC NEUTRON
    SCATTERING PROCESSES.
    *MEIER(F.), HALG(W.), BOSSEL(J.B.). (IN GERMAN).

10 IAEA SYMP. (PILE RESEARCH)VIENNA 559, (1960)
    *THE NATURAL FREQUENCIES OF ROTORS ON FLEXIBLE SHAFTS.
    *HAY(H.J.), EGELSTAFF(P.A.), RAFFLE(J.F.).

11 IAEA SYMPOSIUM VIENNA,   113, (1960)
    *METHODS FOR NEUTRON SPECTROMETRY.
    *BROCKHOUSE(B.N.).

12 IAEA SYMPOSIUM VIENNA,   159, (1960)
    *THE THEORY AND OPERATION OF THE ROTATING CRYSTAL SLOW-NEUTRON CHOPPER.
    *O#CONNOR(D.).

13 IAEA SYMPOSIUM VIENNA,   199, (1960)
    *RESOLUTION OF VARIOUS DEVICES FOR NEUTRON SPECTROSCOPY.
    *ZWEIFEL(P.F.), CARPENTER(J.M.).

14 IAEA SYMP. CHALK RIVER, VOL.1, 147, (1962)
    *PROBLEMS ASSOCIATED WITH THE MONOCHROMATIZATION OF SLOW NEUTRONS.
    *.C.C.C.P.

15 IAEA SYMPOSIUM BOMBAY, VOL.2,   519, (1964)
    *TIME-OF-FLIGHT SPECTROMETER WITH FILTER IN FRONT OF THE DETECTOR.
    *.C.C.C.P.

16 IAEA SYMPOSIUM BOMBAY, VOL.2,   513, (1964)
    *APPLICATION OF THE TIME-OF-FLIGHT METHOD TO NEUTRON-DIFFRACTION STUDIES.
    *.C.C.C.P.

17 IAEA SYMP. COPENHAGEN, VOL.2,   417, (1968)
    *THE USE OF PSEUDO-STATISTICAL CHOPPER FOR TIME-OF-FLIGHT MEASUREMENTS.
    *GOMPF(F.), REICHARDT(W.), GLASER(W.), BECKURTS(K.H.).

18 IAEA SYMP. COPENHAGEN, VOL.2,   407, (1968)
    *CORRELATION-TYPE TIME-OF-FLIGHT SPECTROMETER WITH MAGNETICALLY CHOPPED
    POLARIZED NEUTRON BEAM.
    *PAL(L.), KROO(N.), GORDON(G.), PELLIONISZ(P.), SZLAVIK(F.), VIZI(I.).

19 IAEA SYMP. COPENHAGEN, VOL.2,   387, (1968)
    *A NEUTRON SPIN-FLIP CHOPPER FOR TIME-OF-FLIGHT MEASUREMENTS.
    *RAUCH(H.), HARMS(J.), MOLDASCHL(H.).

20 IAEA SYMP. COPENHAGEN, VOL.2,   395, (1968)
    *TIME-OF-FLIGHT SPECTROMETER USING AN ELECTRONIC CHOPPER FOR POLARIZED
    SLOW NEUTRONS.
    *STEINSVOLL(O.), VIRJO(A.).

21 IAEA SYMP. (INSTRUMENT.) VIENNA 117, (1969)
    *ON THE THEORY AND OPTIMIZATION OF CORRELATION TIME-OF-FLIGHT EXPERIMENTS
    *HOSSFELD(F.), AMADORI(R.), SCHERM(R.).

22 IAEA SYMP. (INSTRUMENT.) VIENNA 173, (1969)
    *A MAGNETICALLY-PULSED TIME-OF-FLIGHT SPECTROMETER FOR INELASTIC NEUTRON
    SCATTERING.
    *MOOK(H.A.), WILKINSON(M.K.).

23 IAEA SYMP. (INSTRUMENT.) VIENNA 147, (1969)
    *INVESTIGATIONS ON THE CORRELATION METHOD IN TIME-OF-FLIGHT EXPERIMENTS.
    *REICHARDT(W.), GOMPF(F.), BECKURTS(K.H.), GLASER(W.), EHRET(G.),
    WILHELMI(G.).

24 IAEA SYMP. GRENOBLE,   803, (1972)....E,T......
    *INVERSE TIME-OF-FLIGHT METHOD.
    *HIISMAKI(P.).

25 IAEA SYMP. GRENOBLE,   787, (1972)....T,E......
    *STATISTICAL CHOPPER FOR TIME-OF-FLIGHT SPECTRUM FILTERING.
    *PELLIONISZ(P.), KROO(N.), MEZEI(F.).

SECTION 19A-TIME-OF-FLIGHT SPECTROMETRY: THEORY AND TECHNIQUE. . . . . .

26 IAEA SYMP. GRENOBLE, 773, (1972)
   *STATISTICAL CHOPPER FOR PERIODICALLY PULSED NEUTRON BEAMS.
   *MATTHES(W.).

27 IAEA SYMP. GRENOBLE, 763, (1972)
   *CORRELATION TIME-OF-FLIGHT SPECTROMETRY AT PULSED REACTORS.
   *KROO(N.), PELLIONISZ(P.), VIZI(I.), ZSIGMOND(G.), ZHUKOV(G.), NAGY(G.).

28 JAP. J. APPL. PHYS. VOL.9, 866, (1970)
   *THE RESOLUTION FUNCTION OF A TWIN-ROTOR NEUTRON TIME-OF-FLIGHT
   SPECTROMETER
   *KOMURA(S.), COOPER(M.J.).

29 JAP. J. APPL. PHYS. VOL.12, 1119, (1973)....A3........SN.
   *A FEASIBILITY STUDY OF APPLYING NEUTRON T-O-F METHOD FOR THE LIQUID
   STRUCTURE ANALYSIS.
   *TOMIYOSHI(S.), WATANABE(N.), MISAWA(M.), KAI(K.), KIMURA(M.).

30 J. APP. PHYS. 39, 447, (1968)
   *A MAGNETICALLY PULSED NEUTRON BEAM FOR TIME OF FLIGHT MEASUREMENTS.
   *MOOK(H.A.), WILKINSON(M.K.).

31 NUCL. INST. MET. VOL.7, 174, (1960)
   *A ≠TIME EXPANDER≠ FOR PRECISION NEUTRON TIME-OF-FLIGHT EXPERIMENTATION.
   *WATERS(J.R.).

32 NUCL. INST. MET. VOL.12, 365, (1961)
   *THE SCATTERING SURFACE METHOD OF MEASURING DISPERSION RELATIONS WITH A
   PHASED CHOPPER VELOCITY SELECTOR.
   *SCHMUNK(R.E.), BRUGGER(R.M.).

33 NUCL. INST. MET. VOL.15, 203, (1962)
   *A NOMOGRAM FOR NEUTRON TIME-OF-FLIGHT SPECTROMETRY.
   *TEUTSCH(H.). (IN GERMAN).

34 NUCL. INST. MET. 30, 293, (1964)
   *THE DETERMINATION OF BACKGROUNDS FOR NEUTRON TIME OF FLIGHT
   SPECTROMETERS.
   *SIMPSON(O.D.), FLUHARTY(R.G.), MOORE(M.S.)...ET AL.

35 NUCL. INST. MET. 42, 81, (1966)
   *INTENSITY AND RESOLUTION IN THE TIME OF FLIGHT METHOD FOR NEUTRON
   DIFFRACTION.
   *SCHWARTZ(L.H.).

36 NUCL. INST. MET. 49, 197, (1967)
   *THE ROTATING CRYSTAL TIME OF FLIGHT SPECTROMETER FOR SCATTERING LAW
   MEASUREMENTS WITH SLOW NEUTRONS.
   *CARVALHO(F.), EHRET(G.), GLASER(W.).

37 NUCL. INST. MET. 61, 296, (1968)
   *RESOLUTION AND INTENSITY IN NEUTRON SPECTROMETRY DETERMINED BY MONTE
   CARLO SIMULATION.
   *DIETRICH(O.W.).

38 NUCL. INST. MET. 63, 347, (1968)
   *ON A SPECIAL APPLICATION OF CORRELATION METHODS IN NEUTRON SPECTROSCOPY.
   *SKOLD(K.).

39 NUCL. INST. MET. 63, 351, (1968)
   *STATISTICAL ANALYSIS OF A CROSS CORRELATION CHOPPER FOR TIME OF FLIGHT
   MEASUREMENTS.
   *VIRJO(A.).

40 NUCL. INST. MET. 70, 262, (1969)
   *RANDOM BEAM PULSING AS A MEANS OF IMPROVING THE EFFICIENCY OF SOME TIME
   OF FLIGHT EXPERIMENTS.
   *GALLOWAY(R.B.).

41 NUCL. INST. MET. 73, 189, (1969)
   *THE FOURIER METHOD IN SLOW NEUTRON TIME OF FLIGHT SPECTROMETRY.
   *VIRJO(A.).

42 NUCL. INST. MET. 75, 77, (1969)
   *SLOW NEUTRON TIME OF FLIGHT SPECTROMETRY WITH A PSEUDO-RANDOM INPUT
   SIGNAL.
   *VIRJO(A.).

43 NUCL. INST. MET. 75, 163, (1969)
   *THE REDUCTION OF TIME OF FLIGHT ERRORS IN PULSED NEUTRON MEASUREMENTS.
   *MEADOWS(J.W.).

44 NUCL. INST. MET. 76, 135, (1969)
   *FOURIER ANALYSIS OF THERMAL NEUTRON TIME OF FLIGHT DATA; A HIGH
   EFFICIENCY NEUTRON CHOPPING SYSTEM,I.
   *COLWELL(J.F.), LEHINAN(S.R.), MILLER(P.H.)...ET AL.

45 NUCL. INST. MET. 69, 173, (1969)
   *FOCUSING OF A TIME-OF-FLIGHT DIFFRACTOMETER FOR STRUCTURE ANALYSIS.
   THE EXPERIMENTAL CHECK.
   *HOLAS(A.), HOLAS(J.), MALISZEWSKI(E.), SEDLAKOVA(L.).

46 NUCL. INST. MET. VOL.75, 32, (1969)
   *A SIMPLE METHOD FOR STABILIZING A TIME-OF-FLIGHT NEUTRON SPECTROMETER.
   *SONODA(M.), WAKUTA(Y.), TAMARA(H.), HYAKUTAKE(M.), IJIRI(H.).

47 NUCL. INST. MET. 77, 29, (1970)
   *FOURIER ANALYSIS OF THERMAL NEUTRON TIME OF FLIGHT DATA; A HIGH
   EFFICIENCY NEUTRON CHOPPING SYSTEM,II.
   *COLWELL(J.F.), LEHINAN(S.R.), MILLER(P.H.)...ET AL.

48 NUCL. INST. MET. VOL.82, 208, (1970)
   *A DETAILED EVALUATION OF THE MECHANICAL CORRELATION CHOPPER FOR NEUTRON
   TIME-OF-FLIGHT SPECTROMETRY.
   *PRICE(D.L.), SKOLD(K.).

49 NUCL. INST. MET. VOL.92, 125, (1971)
   *STATISTICAL CHOPPER METHOD FOR SEPARATION OF ELASTIC AND INELASTIC
   SCATTERED NEUTRONS IN TIME-OF-FLIGHT EXPERIMENTS.
   *PELLIONISZ(P.).

50 NUCL. INST. MET. VOL.93, 109, (1971)
   *INTERMEDIATE SCATTERING FUNCTIONS OBTAINED BY FAST FOURIER TRANSFOR-
   MATION OF COLD NEUTRON TIME-OF-FLIGHT SPECTRA.
   *BREGMAN(J.D.), DE-MUL(F.F.).

SECTION 19A-TIME-OF-FLIGHT SPECTROMETRY: THEORY AND TECHNIQUE. . . . . .

51 NUCL. INST. MET. VOL.98,. . . 53, (1972). . .NEUTRON TIME-OF-FLIGHT SPECTRA USING DIRECT
    *ON THE ANALYSIS OF COLD NEUTRON TIME-OF-FLIGHT SPECTRA USING DIRECT
    FAST FOURIER TRANSFORMATION.
    *DE-MUL(F.F.), BREGMAN(J.D.).

52 NUCL. INST. MET. 99, ... 453, (1972). . . . .
    *EQUAL-PATH TIME-OF-FLIGHT NEUTRON DIFFRACTION.
    *CARPENTER(J.M.), SUTTON(J.D.).

53 NUCL. INST. MET. VOL.107,. . . 501, (1973). . . . .
    *A SYSTEM OF PROGRAMS FOR THE REDUCTION OF DATA FROM A TIME-OF-FLIGHT
    SPECTROMETER.
    *COPLEY(J.R.D.), PRICE(D.L.), ROWE(J.M.).

54 NUCL. INST. MET. VOL.106,. . . 453, (1973). A . . TOF . . . .
    *THE RESOLUTION FUNCTION OF A HYBRID NEUTRON SPECTROMETER.
    *STEINSVOLL(O.).

55 NUCL. INST. MET. VOL.114,. . . 451, (1974). . . . T . . . .
    *STATIC APPROXIMATION DISTORTIONS AND NEUTRON TIME-OF-FLIGHT DIFFRACTION
    USING THE HARWELL LINAC.
    *SINCLAIR(R.N.), WRIGHT(A.C.).

56 NUCL. INST. MET. VOL.116,. . . 509, (1974). . . . E . . . .
    *A METHOD FOR THE ZERO-CHANNEL DETERMINATION IN NEUTRON TIME-OF-FLIGHT
    SPECTROMETRY.
    *ADIB(M.), SALAMA(M.), ABDEL-KAWI(A.), HAMOUDA(I.).

57 NUKLEONIKA VOL.8,. . . . . . 259, (1963). . . . . T . . . .
    *NOTE ON THE INTEGRATED INTENSITY IN THE TIME-OF-FLIGHT METHOD FOR
    CRYSTAL STRUCTURE INVESTIGATIONS.
    *BURAS(B.).

58 NUKLEONIKA VOL.8,. . . . . . 75, (1963). . . E . NEUTRON DIFFRACTION CRYSTAL STRUCTURE
    *A TIME-OF-FLIGHT METHOD FOR NEUTRON DIFFRACTION CRYSTAL STRUCTURE
    INVESTIGATIONS.
    *BURAS(B.), LECIEJEWICZ(J.).

59 NUKLEONIKA VOL.9,. . . . . . 523, (1964). . . . E . . . .
    *THE TIME-OF-FLIGHT METHOD FOR NEUTRON CRYSTAL STRUCTURE INVESTIGATIONS
    AND ITS POSSIBILITIES IN CONNECTION WITH VERY HIGH FLUX REACTORS.
    *BURAS(B.), LECIEJEWICZ(J.), NITC(W.), SOSNOWSKA(I.), SOSNOWSKI(J.),
    SHAPIRO(F.).

60 NUKLEONIKA (POL.) VOL.13,. . . 753, (1968). . . E . . . .
    *COLLIMATORS IN NEUTRON TIME-OF-FLIGHT POWDER DIFFRACTOMETRY.
    *HOLAS(A.).

61 NUKLEONIKA (POL.) VOL.13,. . . 591, (1968). . . E . . . .
    *INTENSITY AND RESOLUTION IN NEUTRON TIME-OF-FLIGHT POWDER DIFFRACTOMETRY
    *BURAS(B.), HOLAS(A.). (ED. NOTE: IN ENGLISH).

62 NUKLEONIKA VOL.14,. . . . . . 487, (1969). . . E . . . .
    *ROTATING SINGLE CRYSTAL NEUTRON TIME-OF-FLIGHT METHOD FOR STRUCTURE
    STUDIES.
    *BURAS(B.), GIEBULTOWICZ(T.).

63 NUKLEONIK (GER.) VOL.12,. . . 153, (1969). . . A1 . . . .
    *THE STATISTICAL TIME-OF-FLIGHT METHOD FOR STRUCTURE INVESTIGATIONS WITH
    SLOW NEUTRONS.
    *GLASER(W.), GOMPF(F.).

64 PHYS. STAT. SOL. 11, ... 567, (1965). . . . . . AL,CR, ...
    *THE TIME-OF-FLIGHT METHOD FOR INVESTIGATIONS OF SINGLE-CRYSTAL
    STRUCTURES.
    *BURAS(B.), MIKKE(K.), LEBECH(B.), LECIEJEWICZ(J.).

65 PROC./SEMINAR/LOW-EN. NUCL. PHYS 138, (1967). . . . .
    *TIME-OF-FLIGHT TECHNIQUES AND INELASTIC SCATTERING OF NEUTRONS.
    *STARFELT(N.). (ED. NOTE: INTERNATIONAL SEMINAR ON LOW-ENERGY NUCLEAR
    PHYSICS HELD AT DACCA,1967; PUBL. NORTH-HOLLAND:AMSTERDAM).

66 REPORT AECL-3189 (11 PP.). . . . (1968). . . A3 . . . .
    *A CALCULATION OF THE NEUTRON SCATTERING BY LIQUIDS AS MEASURED WITH THE
    COLD NEUTRON TIME-OF-FLIGHT TECHNIQUE.
    *COWLEY(R.A.). (ED. NOTE: AVAIL. ATOMIC ENERGY OF CANADA LTD., CHALK
    RIVER, ONTARIO).

67 REPORT NO.1108/II/PS. . . . . (1969). . . E . . . .
    *THE NEUTRON TIME-OF-FLIGHT METHOD FOR CRYSTAL STRUCTURE STUDIES.
    *BURAS(B.). (AVAIL. INST. OF NUCLEAR RESEARCH, WARSAW, POLAND).

68 REV. SCI. INSTRUM. VOL.35,. . . 1150, (1964). . . E . . . .
    *TIME-SHARED TIME-OF-FLIGHT ANALYSIS BY COMPUTER.
    *CHRIEN(R.E.), RANKOWITZ(S.), SPINRAD(R.J.).

69 RISO REPORT NO.229 (56 PP.). . . . (1970). . . E . . . .
    *THERMAL NEUTRON SPECTROSCOPY USING POSITION-SENSITIVE DETECTORS TOGETHER
    WITH FOURIER CORRELATION T.O.F. AND SINGLE-CRYSTAL TECHNIQUES.
    *KJEMS(J.). (ED. NOTE: AVAIL. DANISH ATOMIC ENERGY COMM. RESEARCH ESTAB-
    LISHMENT, RISO).

70 STUD. CERCETARI FIZ. VOL.14,. . . 353, (1963). . . E . . . .
    *USE OF THE MICROTRON AS A PULSED SOURCE FOR NEUTRON SPECTROSCOPY BY
    THE TIME-OF-FLIGHT METHOD.
    *APOSTOLESCU(S.), INDREAS(GR.). (IN RUMANIAN).

71 UKAEA AERE REPORT M2570 (32 PP.). . . (1972). . . . .
    *THE COLLECTION AND PROCESSING OF DATA FROM THREE TIME-OF-FLIGHT
    NEUTRON SPECTROMETERS.
    *BASTON(A.H.). (AVAIL. HMSO, LONDON, ENGLAND).

72 Z. ANGEW. MATH. PHYS. VOL.22,. . . 62, (1971). . . P . . . .
    *STATISTICS OF FLIGHT TIME MEASUREMENTS IN NEUTRON SCATTERING EXPERIMENTS
    *HALG(W.), STUDACH(T.), MEYR(H.).

SECTION 19B-TIME-OF-FLIGHT SPECTROMETRY: INSTRUMENTS AND FACILITIES...

1   ACTA CRYST. (INTERACT.) VOL.A25, S256, (1969).....E..........OF THE GRENOBLE NUCLEAR CENTER
        ~THE TIME OF FLIGHT NEUTRON SPECTROMETER OF THE GRENOBLE NUCLEAR CENTER
         FOR NEUTRON DIFFRACTION UNDER HIGH PRESSURE.
        ~ROULT(G.), BUEVOZ(J.L.).

2   ATOMKERNENERGIE VOL.9,. . . . .   307, (1964)....E............
        ~A NEUTRON CHOPPER AT THE FRG 1 FOR MEASUREMENTS OF TOTAL NEUTRON CROSS
         SECTIONS.
        ~RICHTER(G.), BAGGE(E.). (IN GERMAN).

3   ATOMKERNENERGIE VOL.15,. . . . .   91, (1970)....E............
        ~THE FAST CHOPPER TIME-OF-FLIGHT SPECTROMETER AT THE FRG-1 AT GEESTHACHT.
        ~BIEL(W.), JUNG(H.H.), RICHTER(G.).

4   IAEA SYMPOSIUM VIENNA,. . . . . .   277, (1960)............
        ~SLOW-NEUTRON INELASTIC SCATTERING MEASUREMENTS AT THE MATERIALS TESTING
         REACTOR.
        ~BRUGGER(R.M.), EVANS(J.E.).

5   IAEA SYMP. COPENHAGEN, VOL.2,. .   429, (1968)............
        ~A NEW HIGH-EFFICIENCY TIME-OF-FLIGHT SYSTEM.
        ~COLWELL(J.F.), MILLER(P.H.), WHITTEMORE(W.L.).

6   IAEA SYMP. (INSTRUMENT.), VIENNA 181, (1969)............
        ~DEVELOPMENT AND IMPROVEMENTS OF SPIN-FLIP-CHOPPER SPECTROMETERS.
        ~RAUCH(H.).

7   JAP. J. APPL. PHYS. VOL.10,. . . . 1090, (1971)....E............
        ~SOME IMPROVEMENTS IN A NEUTRON TIME-OF-FLIGHT SPECTROMETER FOR D+T
         NEUTRONS.
        ~HYAKUTAKE(M.), MATOBA(M.), TAWARA(H.), WAKUTA(Y.), KATASE(A.),
         SONODA(M.).

8   J. NUCL. ENERGY VOL.1,. . . . . .   57, (1954)....E............
        ~THE OPERATION OF A THERMAL NEUTRON TIME-OF-FLIGHT SPECTROMETER.
        ~EGELSTAFF(P.A.).

9   KERNTECHNIK VOL.14,. . . . . . .   13, (1972)....E............
        ~TIME-OF-FLIGHT SPECTROMETER FOR SLOW NEUTRONS.
        ~STOCKMEYER(R.).

10 NUCL. INST. MET. VOL.12,. . . .   355, (1961)....E............
        ~A NEUTRON TIME-OF-FLIGHT SPECTROMETER WITH A SEMI-MONOCHROMATIZING
         CHOPPER.
        ~HOLMRYD(S.), LARSSON(K.E.), OTNES(K.).

11 NUCL. INST. MET. VOL.29,. . . .   241, (1964)....E............
        ~AN ELECTRONIC DEVICE IMPROVING THE RESOLUTION OF THE TIME-OF-FLIGHT
         SPECTROMETER WITH MANY-LAYER TRAY OF COUNTERS.
        ~MOSZYNSKI(M.), TUROS(A.), WILHELMI(Z.).

12 NUCL. INST. MET. VOL.32,. . . .   300, (1965)....E............
        ~TIME-OF-FLIGHT MODULE FOR THE TRANSISTORIZED RCL 256 CHANNEL ANALYZER.
        ~BACHLI(A.), BEHRINGER(K.).

13 NUCL. INST. MET. 68,. . . . . .   290, (1969)............
        ~ON THE EFFICIENCY OF THE NOVEL HIGH DUTY CYCLE TIME OF FLIGHT METHODS OF
         NEUTRON SPECTROMETRY.
        ~QUITTNER(G.).

14 NUCL. INST. MET. VOL.72,. . . .   237, (1969)....E............
        ~NEUTRON MODERATOR ASSEMBLIES FOR PULSED THERMAL NEUTRON TIME-OF-FLIGHT
         EXPERIMENTS.
        ~DAY(D.H.), SINCLAIR(R.N.).

15 NUCL. INST. MET. VOL.81,. . . . . 301, (1970)....P............
        ~A SIX-FOOT LONG DETECTOR FOR NEUTRON TIME-OF-FLIGHT EXPERIMENTS.
        ~NEILSON(G.C.), GLAVINA(C.), DAWSON(W.K.), IYENGAR(K.V.), MCDONALD(W.J.).

16 NUCL. INST. MET. VOL.80,. . . . .   69, (1970)....P............
        ~THE STATISTICAL CHOPPER FOR NEUTRON TIME-OF-FLIGHT SPECTROSCOPY.
        ~VON-JAN(R.), SCHERM(R.).

17 NUCL. INST. MET. 92,. . . . . . .   51, (1971)............
        ~A NEW TIME-OF-FLIGHT ANALYSER FOR NEUTRON SCATTERING.
        ~HALL(J.W.), EGELSTAFF(P.A.).

18 NUCL. INST. MET. VOL.91,. . . .   541, (1971)....P............
        ~A 250 METER NEUTRON TIME-OF-FLIGHT FACILITY WITH MODULAR DETECTOR AND
         DIGITAL ELECTRONICS STABILIZATION.
        ~STOLER(P.), YERGIN(P.F.), CLEMENT(J.C.), MANN(D.), GOULDING(C.G.),
         FAIRCHILD(R.). (ED. NOTE: FACILITY AT RENSSELAER POLYTECH. INST.).

19 NUCL. INST. MET. VOL.101,. . . .   295, (1972)....E............
        ~A TIME-OF-FLIGHT SPECTROMETER FOR ULTRACOLD NEUTRONS.
        ~STEYERL(A.).

20 NUCL. INST. MET. VOL.106,. . . .   221, (1973)....E............
        ~A NEW #HYBRID# SPECTROMETER FOR INELASTIC THERMAL NEUTRON SCATTERING.
        ~KLEB(R.), OSTROWSKI(G.E.), PRICE(D.L.), ROWE(J.M.).

21 NUCL. INST. MET. VOL.114,. . . .   417, (1974)....E............
        ~A MECHANICAL TIME-OF-FLIGHT NEUTRON DIFFRACTOMETER.
        ~CARPENTER(J.M.), MILDNER(D.F.R.), PELIZZARI(C.A.), SUTTON(J.D.),
         GUNNING(J.E.).

22 NUCL. INST. MET. VOL.114,. . . .   198, (1974)....E............
        ~A TIME-OF-FLIGHT SPECTROMETER FOR ULTRACOLD NEUTRONS (ERRATA).
        ~STEYERL(A.). (PAPER PUBL. IN VOL.101, PG.295).

23 NUCL. INST. MET. VOL.116,. . . .   205, (1974)....E............
        ~A MAGNETICALLY PULSED NEUTRON TIME-OF-FLIGHT SPECTROMETER FOR INELASTIC
         SCATTERING.
        ~MOOK(H.A.), SNODGRASS(F.W.), BATES(D.D.).

24 NUKLEONIKA VOL.7,. . . . .   231, (1962)....E............
        ~A TIME OF FLIGHT SPECTROMETER FOR SLOW NEUTRONS.
        ~NIEWIADOMSKI(T.), SZKATULA(A.), SCIESINSKI(J.).

SECTION 19B-TIME-OF-FLIGHT SPECTROMETRY: INSTRUMENTS AND FACILITIES. . .

25 NUKLEONIKA VOL.8, , , 695, (1963) , , , , E, , , , , , , , , , , ,
*256-CHANNEL TIME-OF-FLIGHT ANALYSER WITH MAGNETIC MEMORY.
*HOFFMAN(Z.), KOMOR(Z.), SAWICKI(A.). (IN POLISH).

26 PHYS. LETT. VOL.44A NO.3, , , 165, (1973) , , , E, , , , , , , , ,
*A HIGH RESOLUTION NEUTRON TIME-OF-FLIGHT DIFFRACTOMETER.
*STEICHELE(E.), ARNOLD(P.).

27 RC ACCAD. NAZ. LINCEI VOL.45, , 163, (1968) , , , E, , , , , , , , ,
*PULSED VAN-DE-GRAAFF FOR TIME-OF-FLIGHT NEUTRON SPECTROMETRY.
*HEBERT(D.), LEROY(J.L.), ARNAUD(A.). (ED. NOTE: IN FRENCH).

28 RENSSELAER POL. INST. SYMP. , , 35, (1961) , , , , E, , , , , , , ,
*TIME-OF-FLIGHT METHODS IN LOW ENERGY NEUTRON CROSS SECTION MEASUREMENTS
AT THE MTR.
*FLUHARTY(R.G.), EVANS(J.E.).

29 REPORT KAPL-1657 (22 PP.) , , , , , , , , (1957) , , , , E, , , , , , , , , ,
*TWO-HUNDRED-CHANNEL TIME ANALYZER FOR NEUTRON SPECTROMETRY.
*BRYANT(W.A.), YEATER(M.L.). (AVAIL. KNOLLS ATOMIC POWER LAB,
SCHENECTADY, NEW YORK).

30 REPORT KFKI-07/1968 (14PP.) , , , , (1968) , , , E, , , , , , , , , ,
*CORRELATION TYPE TIME-OF-FLIGHT SPECTROMETER WITH MAGNETICALLY CHIPPED
POLARIZED NEUTRON BEAM.
*POIL(L.), KROO(N.), GORDON(J.) PILLIONISZ(P.), SZLAVIK(F.), VIZI(I.).
(ED. NOTE: AVAIL. HUNG. ACAD. SCI., CENTRAL RES. INST. PHYS., BUDAPEST).

31 REV. DE PHYSIQUE (BUCAREST) 4, , 327, (1959) , , , E, , , , , , , , , ,
*A NEUTRON TIME OF FLIGHT SPECTROMETER.
*TOTIA(H.), TIMIS(P.), LAZAROVICI(C.). (IN FRENCH).

32 REV. PHYS. APPL. (FRANCE) VOL.4, 111, (1969) , , , E, , , , , , , , , ,
*DOUBLE-TIME-OF-FLIGHT SPECTROMETER WITH MULTIPARAMETRIC ANALYSIS.
*PERRIN(C.), BOUCHEZ(R.), QUIVY(P.), GONDRAND(J.C.), GIORNI(A.),
POUXE(J.), CREMET(J.P.), PERRIN(P.).

33 REV. PHYS. APPL.(FRANCE) VOL.5, , 693, (1970) , , , E, , , , , , , , , ,
*HIGH EFFICIENCY NEUTRON TIME-OF-FLIGHT SPECTROMETER FOR RANDOM SOURCES.
*PICOT(A.).

34 REV. SCI. INSTRUM. VOL.37, , , , 697, (1966) , , , E, , , , , , , , , ,
*PHASED ROTATING CRYSTAL AND CHOPPER FOR TIME-OF-FLIGHT NEUTRON SPEC-
TROSCOPY.
*HARLING(O.K.).

35 SOV. AT. ENERGY VOL.28, , , , , 150, (1970) , , , E, , , , , , , , , ,
*TIME-OF-FLIGHT SPECTROMETER FOR MEASUREMENT OF SCATTERING NEUTRON CROSS-
SECTIONS.
*GERASIMOV(V.F.), ZENKEVICH(V.S.), MOSKALEV(S.S.).

36 STUD. CERCETARI FIZ. VOL.10, , , 89, (1959) , , , E, , , , , , , , ,
*NEUTRON TIME-OF-FLIGHT SPECTROMETER.
*TOTIA(H.), TIMIS(P.), LAZAROVICI(C.). (IN ROUMANIAN).

37 UKAEA AERE REPORT 5259 (9PP.) , , , , (1966) , , , E, , , , , , , , ,
*A REVIEW OF THERMAL NEUTRON SCATTERING EQUIPMENT AT A.E.R.E. HARWELL.II.
INELASTIC SCATTERING APPARATUS.
*DYER(R.F.).

38 UKAEA REP. AERE-R-6035 (8 PP.) , , , , (1969) , , , E, , , , , , , , ,
*TIME-OF-FLIGHT SPECTROMETER FOR COLD-NEUTRON STUDIES ON THE DIDO 6H HOLE
*STIRLING(G.C.).

SECTION   20A-ON NEUTRON BEAMS . . . .

1   ACTA CRYST. VOL.11,   228, (1958). . . E . . . . BE
       *THE SCATTERING OF 4 ANGSTROM NEUTRONS BY A BERYLLIUM CRYSTAL. . . . . .
       *HAY(H.J.), PATTENDEN(N.J.), EGELSTAFF(P.A.).

2   ACTA CRYST. VOL.14,   292, (1961). . . . E . . . . AL,CU
       *PARASITIC MULTIPLE BRAGG SCATTERING IN THE NEUTRON CRYSTAL SPECTROMETER.
       *O#CONNOR(D.A.), SOSNOWSKI(J.).

3   ACTA CRYST. 16,. .   435, (1963). . . . .
       *A GEOMETRICAL CONSIDERATION IN THE DESIGN OF NEUTRON SPECTROMETERS.
       *SABINE(T.M.), BROWNE(J.D.).

4   ACTA CRYST. 18,. .   184, (1965). . . . .
       *THE EFFECT OF EXPERIMENTAL RESOLUTION ON CRYSTAL REFLECTIVITY . . . . .
       AND SECONDARY EXTINCTION IN NEUTRON DIFFRACTION.
       *DIETRICH(O.W.), ALS-NIELSEN(J.).

5   ACTA CRYST. 20,. .   311, (1966). . . . .
       *EIN DOPPELT GEKRÜMMTER NEUTRONENLEITER ZUR UNTERDRÜCKUNG DER HÖHREN
       ORDNUNG BEI NEUTRONEN BEUGUNGS EXPERIMENTEN.
       *SCHMATZ(W.).

6   ACTA CRYST. VOL.A30,. . .   448, (1974). . . . T . . . .
       *USE OF A TWO-AXIS NEUTRON DIFFRACTOMETER ON AN ANGLED REACTOR CHANNEL.
       *HAYWOOD(B.C.G.).

7   ACTA PHYS. POLON. VOL.19,. . .   329, (1960). . . . E . . . .
       *MEASUREMENT OF THE SLOW NEUTRON SPECTRUM OF A NEUTRON BEAM FROM THE
       W.W.R.S. REACTOR BY MEANS OF A CRYSTAL NEUTRON SPECTROMETER.
       *O#CONNOR(D.A.), SOSNOWSKI(J.).

8   ACTA PHYS. AUSTRIACA VOL.20,. . .   7, (1965). . . . E,T . . . . . .
       *COHERENT EXTINCTION AND SECOND-ORDER SCATTERING OF NEUTRONS BY SINGLE
       CRYSTALS.
       *KUICH(G.), RAUCH(H.). (IN GERMAN).

9   ACTA PHYS. AUSTRIACA VOL.38,. . .   282, (1973). . . . T . . . . . .
       *CONFIGURATION CHOICES FOR SIMILAR THREE AXIS NEUTRON SPECTROMETERS.
       *QUITTNER(G.). (IN GERMAN).

10   ANN. DE PHYSIQUE TOME 7 NO.4,. . .   255, (1972). . . . .
       *EXPERIMENTAL TECHNIQUES IN NEUTRON SCATTERING (REVIEW PAPER). . . . . .
       *MERIEL(P.). (PRESENTED AT AUTRANS SUMMER SCHOOL 1972).

11   ANN. FAC. SCI. UNIV. TOULOUSE 30   9, (1966). . . . E . . . .
       *THERMAL NEUTRON FLUX MEASUREMENT BY MEANS OF ACTIVABLE-CATHODE COUNTERS.
       *BAYLE(P.), RAKOTONDRAFARA(H.), DARRES(M.).

12   ANN. REV. OF NUCLEAR SCI. VOL.16   207, (1966). . . . .
       *PRODUCTION AND USE OF THERMAL REACTOR NEUTRON BEAMS. . . . .
       *MAIER-LEIBNITZ(H.), SPRINGER(T.).

13   ATOMKERNENERGIE VOL.17,. . .   93, (1971). . . . T . . . .
       *BUNCHING OF NEUTRONS IN THE MILLI-EV REGION IN HIGH RESOLUTION NEUTRON
       SPECTROSCOPY.
       *PETERLIN(T.M.).

14   BULL. D#INFO. SCI. ET TECH. 166,. . .   (1972). . . E . . . . . .
       *LE REACTEUR FRANCO-ALLEMAND A HAUT FLUX. LES POSSIBILITES EXPERIMENTALES
       *AGERON(P.). . . ET AL.

15   ENDEAVOUR 31,. . .   67, (1972). . . . E . . . .
       *THE HIGH-FLUX REACTOR AT THE INSTITUT LAUE-LANGEVIN. . . . .
       *AGERON(P.).

16   IAEA SYMP. CHALK RIVER, VOL.1,   95, (1962). . . . .
       *A NEW APPARATUS FOR INELASTIC, QUASI-ELASTIC AND ELASTIC COLD NEUTRON
       MEASUREMENTS.
       *OTNES(K.), PALEVSKY(H.).

17   IAEA SYMP. GRENOBLE,   733, (1972). . . . E . . . .
       *A MULTI-ANGLE REFLECTING CRYSTAL SPECTROMETER. . . . .
       *KJEMS(J.K.), REYNOLDS(P.A.).

18   IAEA SYMP. GRENOBLE,. . . . .   681, (1972). . . . .
       *NEW INSTRUMENTATION (WITH SPECIAL REFERENCE TO THE HIGH FLUX . . . . .
       REACTOR ILL).
       *MAIER-LEIBNITZ(H.).

19   INSTRUM. EXP. TECH. NO.1,. . . . .   58, (1966). . . . E . . . .
       *DEVICE FOR COLD-NEUTRON MEASUREMENTS WITH A PULSE REACTOR.
       *GOLIKOV(V.V.), SHAPIRO(F.L.), SHKATULA(A.A.).

20   INT. J. APPL. RADIAT. ISOTOP. 22 529, (1971). . . . E . . . .
       *THE DETERMINATION OF THE THERMAL NEUTRON FLUENCE BY COBALT ACTIVATION
       MONITORS.
       *KOHLER(H.), VANINBROUKX(R.).

21   J. APPL. CRYST. VOL.4,. . . . . .   185, (1971). . . . E . . . . . . PB,CU
       *MEASUREMENTS OF ABSOLUTE REFLECTIVITIES OF MOSAIC CRYSTALS AND THEIR
       WAVELENGTH DEPENDENCE.
       *DORNER(B.).

22   J. APPL. CRYST. VOL.7,. . . . . .   38, (1974). . . . E . . . . C (GRAPHITE),. . . . . . .
       *IS PYROLYTIC GRAPHITE AN IDEAL MOSAIC CRYSTAL.
       *DORNER(B.), KOLLMAR(A.).

23   J. NUCL. ENERGY VOL.6,   222, (1958). . . . E . . . .
       *SLOW-NEUTRON SPECTRUM MEASUREMENTS FROM THE SWEDISH HEAVY-WATER . . . .
       REACTOR, R1.
       *LARSSON(K.E.), STEDMAN(R.), PALEVSKY(H.).

24   J. NUCL. ENERGY VOL.17,. . .   217, (1963). . . . T . . . .
       *THE USE OF NEUTRON OPTICAL DEVICES ON BEAM-HOLE EXPERIMENTS. . . . . .
       *MAIER-LEIBNITZ(H.), SPRINGER(T.).

25   J. PHYS. E VOL.6 NO.6,   568, (1973). . . . E . . . .
       *A SIMPLE NEUTRON GUIDE TUBE AND DIFFRACTOMETER FOR SMALL-ANGLE
       SCATTERING OF COLD NEUTRONS.
       *HAYWOOD(B.C.G.), WORCESTER(D.L.).

SECTION 20A-ON NEUTRON BEAMS . . . .

26 J. PHYS. SOC. JAPAN VOL.19,. . . 2280, (1964)....I.E........PB*LI*F,........
    *REFLECTIVITY OF COLLIMATED NEUTRONS BY A MOSAIC SINGLE CRYSTAL.
    *KUNITOMI(N.), SAKAMOTO(M.), HAMAGUCHI(Y.), BETSUYAKU(H.).

27 J. RES. NAT. BUR/STAND. VOL.67A, 215, (1963) OF...E...........................
    *AN ABSOLUTE CALIBRATION OF THE NATIONAL BUREAU OF STANDARDS THERMAL
    NEUTRON FLUX.
    *AXTON(E.J.).

28 KERNTECHNIK VOL.14,. . . . . . 86, (1972)....E.........................
    *THE SMALL ANGLE SCATTERING FACILITY AT THE BENT NEUTRON GUIDE TUBE OF
    THE COLD SOURCE AT FRJ-2.
    *SCHELTEN(J.).

29 KERNTECHNIK VOL.14,. . . . . . 9, (1972)....E.........................
    *NEUTRON BEAM GUIDE AND EXTERNAL NEUTRON LABORATORY OF A COLD NEUTRON
    SOURCE.
    *BAUER(G.), JOSWIG(G.), SCHELTEN(J.), SCHMATZ(W.).

30 NATURE VOL.142,. . . . . 829, (1938)....E.........................
    *VELOCITY DISTRIBUTION OF THERMAL NEUTRONS.
    *FERTEL(G.E.F.), MOON(P.B.), THOMSON(G.P.), WYNN-WILLIAMS(C.E.).

31 NATURE VOL.167,. . . . . 243, (1951)....E........NA*CL,NAPHTHALENE, ...
    *SINGLE-CRYSTAL NEUTRON DIFFRACTION ANALYSIS.
    *LOWDE(R.D.).

32 NATURE VOL.211,. . . . . 400, (1966)....E........QUARTZ,. . . . .
    *MODULATION OF DIFFRACTED NEUTRONS WITH A PIEZOELECTRIC CRYSTAL.
    *PARKINSON(T.F.), MOYER(M.W.).

33 NUCLEONICS VOL.6 NO.4, ... 54, (1950)....E.........................
    *NEUTRON SPECTROSCOPY FOR CHEMICAL ANALYSIS-III. THERMAL NEUTRONS,
    NEUTRON FLUX, ACTIVATION.
    *TAYLOR(T.I.), HAVENS-JR(W.W.).

34 NUCL. INST. MET. VOL.9,. . . . 341, (1960)....E........SLOW CHOPPER........
    *THE OPTICAL PROPERTIES OF SLOW CHOPPER AND TIME-OF-FLIGHT DEVICE.
    *MALISZEWSKI(E.).

35 NUCL. INST. MET. 40,. . . . . 77, (1966)....E.........................
    *ON OPTICAL PROPERTIES OF THE NEUTRON CRYSTAL SPECTROMETERS.
    *POPOVICI(M.), GELBERG(D.).

36 NUCL. INST. MET. 49,. . . . . 117, (1967)....E.........................
    *ELIMINATION OF SECOND ORDER EFFECTS IN TRIPLE-AXIS CRYSTAL
    SPECTROMETERS.
    *DOLLING(G.), NIEMAN(H.).

37 NUCL. INST. MET. 65,. . . . . 233, (1968)....E........GE,........
    *THE EFFECT OF TEMPERATURE AND PRESSURE ON THE MOSAIC SPREAD IN A
    GERMANIUM MONOCRYSTAL.
    *KONSTANTINOVIC(J.), ZIVANOVIC(M.), DAVIDOVIC(M.).

38 NUCL. INST. MET. 61,. . . . . 93, (1968)....E.........................
    *ON THE MEASUREMENT OF THE THERMAL NEUTRON FLUX WITH FOILS.
    *ILBERG(D.), SEGAL(Y.).

39 NUCL. INST. MET. 61,. . . . . 37, (1968)....E.........................
    *MEASUREMENT OF NEUTRON SPECTRA BY THE TRANSMISSION METHOD.
    *NAVALKAR(M.B.), RAMAKRISHNA(D.V.S.).

40 NUCL. INST. MET. 67,. . . . . 357, (1969)....E.........................
    *NEUTRON DIFFRACTION ON QUARTZ CRYSTALS AS A FUNCTION OF VIBRATING
    FREQUENCIES.
    *CHALUPA(B.), MICHALEC(R.), GALOCIOVA(D.).

41 NUCL. INST. MET. VOL.79,. . . 51, (1970)....E.........................
    *EXPERIMENTAL DETERMINATION OF THE SLOW-NEUTRON WAVELENGTH DISTRIBUTION.
    *LEBECH(B.), MIKKE(K.), SLEDZIEWSKA-BLOCKA(D.).

42 NUCL. INST. MET. VOL.95,. . . 445, (1971)....E.........C (GRAPHITE),........
    *VERTICALLY BENT PYROLYTIC GRAPHITE CRYSTALS APPLIED TO TRIPLE-AXIS
    NEUTRON SPECTROMETRY.
    *NUNES(A.C.), SHIRANE(G.).

43 NUCL. INST. MET. VOL.94,. . . 533, (1971)....E.........................
    *MICROGUIDES FOR NEUTRONS.
    *MARX(D.).

44 NUCL. INST. MET. VOL.101,. . . 183, (1972)....E........C (GRAPHITE),........
    *CHARACTERISTICS OF PYROLYTIC GRAPHITE AS AN ANALYZER AND HIGHER ORDER
    FILTER IN NEUTRON SCATTERING EXPERIMENTS.
    *SHAPIRO(S.M.), CHESSER(N.J.).

45 NUCL. INST. MET. VOL.102,. . . 193, (1972)....E.........................
    *FAST-NEUTRON FLUX MEASUREMENT WITH A LITHIUM-DRIFTED GERMANIUM DETECTOR.
    *SMITH(D.L.).

46 NUCL. INST. MET. VOL.102,. . . 87, (1972)....E.........................
    *A METHOD FOR OBTAINING A HOMOGENEOUS FLUX IN A REACTOR BEAM TUBE.
    *KEDEM(D.), MAHLAV(M.), PELAH(I.).

47 NUCL. INST. MET. VOL.107,. . . 13, (1973)....E........IR,........
    *THERMAL NEUTRON FLUX MEASUREMENT AT VERY HIGH TEMPERATURE USING IRIDIUM.
    *LAVI(N.).

48 NUCL. INST. MET. VOL.114,. . . 143, (1974)....E.........................
    *A COLD NEUTRON DIFFERENTIAL REFLECTION SPECTROMETER WHICH UTILIZES
    CYLINDRICAL REFLECTIONS.
    *ALI(S.A.), OLSON(N.T.).

49 NUCL. INST. MET. VOL.116,. . . 615, (1974)....E.........................
    *A THERMAL AND FAST NEUTRON MONITOR FOR USE IN A SUBCRITICAL FACILITY.
    *BIRSTEIN(L.), MARTINEZ(P.), FILEVICH(A.), LI(F.).

50 NUCL. SCI. ENGNG. VOL.14,. . . 397, (1962)....E.........................
    *NEUTRON DIFFRACTOMETER MEASUREMENT OF THE SLOW NEUTRON SPECTRUM FROM A
    MODIFIED SPLIT REACTOR CORE.
    *FERGUSON(G.A.), VOGT(R.H.).

51 NUCL. TECHNOL. VOL.14,. . . . 284, (1972)....E........LI*I(EU),........
    *MEASUREMENTS OF FAST- AND THERMAL-NEUTRON FLUXES USING A SMALL LI*I(EU)
    CRYSTAL DETECTOR.
    *NGUYEN(D.H.), BENNETT(R.G.).

SECTION 20A-ON NEUTRON BEAMS . . . .

52   NUKLEONIKA VOL.5,   414, (1960). . .IN. E. . . . .
     *NEUTRON DIFFRACTION STUDIES IN SOLID STATE PHYSICS.
     *BLINOWSKI(K.).

53   NUKLEONIK VOL.2,   41, (1960). . . . .E. . . . .
     *THE THERMAL NEUTRON SPECTRUM IN THE WWR-S REACTOR, BUCHAREST.
     *TEUTSCH(H.), APOSTOLESCU(S.), TIMIS(P.).

54   NUKLEONIK VOL.4,   23, (1962). . . . .E. . . . .
     *DEVELOPMENT OF A NEUTRON BEAM TUBE IN THE FRM REACTOR.
     *CHRIST(J.), SPRINGER(T.). (IN GERMAN).

55   NUOVO CIMENTO VOL.13B NO.2, 249, (1973). . . .T. . . . .
     *DYNAMICAL NEUTRON DIFFRACTION BY IDEALLY CURVED CRYSTALS.
     *KLAR(B.), RUSTICHELLI(F.).

56   PHYS. LETT. 28A,   546, (1968). . . . .
     *TIME MODULATION OF A NEUTRON BEAM DIFFRACTED ON A SINGLE QUARTZ CRYSTAL
     EXCITED INTO VIBRATIONS BY HIGH FREQUENCY PULSES.
     *MICHALEC(R.), CHALUPA(B.), GALOCIOVA(D.), MIKULA(P.).

57   PHYS. REV. VOL.48,   265, (1935). . . . .E. . . . .
     *INTERACTION OF NEUTRONS WITH MATTER.
     *DUNNING(J.R.), PEGRAM(G.B.), FINK(G.A.), MITCHELL(D.P.).

58   PHYS. REV. VOL.49,   103, (1936). . . . .
     *THE VELOCITIES OF SLOW NEUTRONS.
     *FINK(G.A.), DUNNING(J.R.), PEGRAM(G.B.), MITCHELL(D.P.).

59   PHYS. REV. VOL.59,   332, (1941). . . . .E. . . . .
     *EXPERIMENTS WITH A SLOW NEUTRON VELOCITY SPECTROMETER.
     *BAKER(C.P.), BACHER(R.F.).

60   PHYS. REV. VOL.70,   136, (1946). . . .E. . . . .B,CD,PARAFFIN. . . .
     *NEUTRON BEAM SPECTROMETER STUDIES OF BORON, CADMIUM, AND THE ENERGY
     DISTRIBUTION FROM PARAFFIN.
     *RAINWATER(L.J.), HAVENS-JR(W.W.).

61   PHYS. REV. VOL.72,   193, (1947). . . .E. . . . . . .B. . . . .
     *A THERMAL NEUTRON VELOCITY SELECTOR AND ITS APPLICATION TO THE MEASURE-
     MENT OF THE CROSS SECTION OF BORON.
     *FERMI(E.), MARSHALL(J.), MARSHALL(L.).

62   PHYS. REV. VOL.71,   752, (1947). . . . .E. . . . . . .CALCITE,CD,. .
     *DIFFRACTION OF NEUTRONS BY A SINGLE CRYSTAL.
     *ZINN(W.H.).

63   PHYS. REV. VOL.71,   757, (1947). . . . .E. . . . .
     *MEASUREMENT OF NEUTRON CROSS SECTIONS WITH A CRYSTAL SPECTROMETER.
     *STURM(W.J.).

64   PHYS. REV. VOL.72,   585, (1947). . . . .E. . . . .
     *NEUTRON CROSS-SECTION STUDIES WITH THE ROTATING SHUTTER MECHANISM.
     *BRILL(T.), LICHTENBERGER(H.V.).

65   PHYS. REV. VOL.77,   291, (1950). . . . .E. . . . .
     *COHERENT NEUTRON-PROTON SCATTERING BY LIQUID MIRROR REFLECTION.
     *HUGHES(D.J.), BURGY(M.T.), RINGO(G.R.).

66   PHYS. REV. VOL.81,   498, (1951). . . . .E,T. . . . . .BE,FE,CO,. . .
     *REFLECTION OF NEUTRONS FROM MAGNETIZED MIRRORS.
     *HUGHES(D.J.), BURGY(M.T.).

67   PHYS. STAT. SOLIDI A VOL.5 NO.2, 397, (1971). . . . . .PB,. . . .
     *NEUTRON REFLECTION PROPERTIES OF BENT SINGLE CRYSTALS.
     *KARAS(W.), RAUCH(H.), SEIDL(E.).

68   PROC. INDIAN ACAD. SCI. A VOL.53   59, (1961). . . .E. . . . .
     *ANOMALOUS REFLECTIONS IN A SINGLE-CRYSTAL NEUTRON SPECTROMETER.
     *DUGGAL(V.P.), RAO(K.R.), THAPER(C.L.), SINGH(V.).

69   PROC. ROY. SOC. A VOL.162, 127, (1937). . . .E,P. . . . . .
     *SCATTERING OF SLOW NEUTRONS.
     *GOLDHABER(M.), BRIGGS(G.H.).

70   PULSED NEUTRON SOURCES/UTILIZ.,   (1971). . . .E. . . . .
     *NEUTRON SPECTROSCOPY WITH LINAC-BOOSTER (PROJECTED IN JAPAN).
     *ISHIKAWA(Y.), WATANABE(N.), KIMURA(M.). (ED. NOTE: PRES. AT EURATOM-
     JAPAN JOINT MEETING ON PULS. NEUT. SOUR. AND UTILIZ., ISPRA, SEPT.17).

71   REACTOR SCI. TECHNOL. VOL.14,   25, (1961). . . . .
     *A RECALIBRATION OF THE N.B.S. STANDARD THERMAL NEUTRON FLUX.
     *MOSBURG-JR(E.R.), MURPHEY(W.M.).

72   REACTOR SCI. TECHNOL. VOL.17,   147, (1963). . . . .
     *FORMATION OF NEUTRON BEAMS WITH SPHERICAL PROPERTIES.
     *BECKURTS(K.H.), EGELSTAFF(P.A.), GOLDSTEIN(H.), SJOSTRAND(N.G.).

73   REPORT A.E.R.E. N/R 1165,   (1953). . . .E. . . . . .BE,MG,PB,. . . .
     *THE INELASTIC SCATTERING OF COLD NEUTRONS BY CRYSTALS II. EXPERIMENT.
     *EGELSTAFF(P.A.).

74   REV. ROUMAINE PHYS. VOL.8,   277, (1963). . . .E. . . . .
     *ON THE TOTAL REFLECTION OF NEUTRONS BY COBALT MIRRORS.
     *POPOVICI(M.), RIPEANU(S.), GRABCHEV(B.), VASILIU(V.). (IN RUSSIAN).

75   REV. ROUMAINE PHYS. VOL.10, 1035, (1965). . . .E. . . . .
     *SECONDARY EXTINCTION AND OPTICAL PROPERTIES OF NEUTRON CRYSTAL
     SPECTROMETERS.
     *POPOVICI(M.), GELBERG(D.). (IN FRENCH).

76   REV. SCI. INSTRUM. VOL.30,   17, (1959). . . .T,E. . . . .NA*CL,BE,. . . .
     *SECOND-ORDER CONTAMINATION IN A NEUTRON CRYSTAL SPECTROMETER.
     *HAAS(R.), SHORE(F.J.).

77   REV. SCI. INSTRUM. VOL.31,   640, (1960). . . .T. . . . .
     *A PLANE-CRYSTAL AND BENT-CRYSTAL HIGH RESOLVING-POWER NEUTRON
     SPECTROMETER.
     *BALLY(D.), TODIREANU(S.), TARINA(E.), OLTEANU(I.).

78   REV. SCI. INSTRUM. VOL.33,   49, (1962). . . .E. . . . .
     *REMOVAL OF HIGHER ORDERS IN THE THERMAL REGION FROM A NEUTRON CRYSTAL
     SPECTROMETER.
     *DUGGAL(V.P.), THAPER(C.L.).

SECTION   20A-ON NEUTRON BEAMS . . . .

79   REV. SCI. INSTRUM. VOL.33,   916, (1962).   #E.   CU,FE,CD.
        #TOTAL REFLECTION OF NEUTRONS FROM METALLIC MIRRORS.
        #BALLY(D.), TODIREANU(S.), RIPEANU(S.), BELLONI(M.G.).

80   REV. SCI. INSTRUM. VOL.42,   240, (1971).   #T.
        #A PROTON RECOIL MONITOR FOR NEUTRON FLUX MEASUREMENTS.
        #JASZCZAK(R.J.), MACKLIN(R.L.), TAYLOR(M.C.).

81   THESIS (86 PP.),   (1970).   #T.   CU.
        #THE REFLECTIVITY OF NEUTRONS BY DISTORTED COPPER CRYSTALS.
        #DYMOND(R.R.).   (ED. NOTE: M.SC. DISSERTATION, MCMASTER UNIVERSITY).

82   UKAEA AERE TRANS 1074 (20 PP.),   (1967).   #E.
        #NEUTRON GUIDES. A REPORT ON THE PRESENT STATE OF DEVELOPMENT.
        #ALEFELD(B.), CHRIST(J.), KUKLA(D.), SCHERM(R.), SCHMATZ(W.).

83   Z. KRIST. VOL.115,   80, (1961).   #T.
        #BEAM PATH IN A NEUTRON SPECTROMETER FOR SINGLE CRYSTAL INVESTIGATIONS.
        #DACHS(H.). (IN GERMAN).

84   Z. PHYSIK BAND 254,   169, (1972).   #T.
        #EFFECT OF SURFACE ROUGHNESS ON THE TOTAL REFLECTION AND TRANSMISSION
        OF SLOW NEUTRONS.
        #STEYERL(A.).

SECTION 208-ON NEUTRON SOURCES . . . .

1   ACTA CRYST. A24, ... 117, (1968). . . . .
    *HIGH-RESOLUTION MONOCHROMATIZATION OF NEUTRONS AND X-RAYS BY MULTIPLE
    BRAGG REFLECTION.
    *KOTTWITZ(D.A.).

2   ACTA CRYST. VOL.A25 . . . . . . 459, (1969). . . . .
    *HIGH-RESOLUTION MONOCHROMATIZATION OF NEUTRONS BY MULITIPLE BRAGG
    REFLECTION IN HEXAGONAL CLOSE-PACKED CRYSTALS.
    *KOTTWITZ(D.A.).

3   ACTA POLYTECH. SCANDIN. PH 38, ... (1966). . . . .E. . . . .
    *CONSTRUCTION AND PERFORMANCE OF A COLD NEUTRON SOURCE.
    *TUNKELO(E.).

4   ANN. ACAD. SCI. FENNICAE A 267, . . . 3, (1967). . . .E. . . . .
    *SOME SUGGESTIONS FOR THE APPLICATION OF CONSTRICTED MONOCHROMATOR CRYS-
    TALS FOR NEUTRON DIFFRACTION AND SCATTERING MEASUREMENTS.
    *MAIER-LEIBNITZ(H.). (ED. NOTE IN GERMAN).

5   ANN. PHYS. (GERMANY) VOL.7, . . . 50, (1961). . . .E. . . . .
    *A MECHANICAL VELOCITY SELECTOR FOR SUB-THERMAL NEUTRONS AND ITS USE IN
    THE MEASUREMENT OF THE TOTAL EFFECTIVE C.S. OF CD, SM, EU AND GD.
    *HOEHNE(P.). (IN GERMAN).

6   AN. REAL SOC. ESPAN/FIS. VOL.63A 211, (1967). . . .E. . . . . . . .CA*C*O3. . .
    *USE OF A CALCITE CRYSTAL AS NEUTRON MONOCHROMATOR.
    *DIAZ(L.M.), DE-SALAMANCA(M.E.), PAREDES(M.C.). (IN SPANISH).

7   ARK. FYS. VOL.16 PAPER 19, . . . 199, (1959). . .T. . . . .
    *THE SLOW CHOPPER AND TIME OF FLIGHT SPECTROMETER IN THEORY AND
    EXPERIMENT.
    *LARSSON(K.E.),   DAHLBORG(U.),   HOLMRYD(S.),   OTNES(K.),   STEDMAN(R.).

8   ATOMKERNENERGIE VOL.9, ... 307, (1964). . . .E. . . . .
    *A NEUTRON CHOPPER AT THE FRG 1 FOR MEASUREMENTS OF TOTAL NEUTRON CROSS
    SECTIONS.
    *RICHTER(G.), BAGGE(E.). (IN GERMAN).

9   ATOMKERNENERGIE VOL.11, ... 381, (1966). . . .E. . . . . . . .AL,PB,GE,CU. .
    *THE USE OF ALUMINUM, LEAD, GERMANIUM AND COPPER AS MONOCHROMATING
    CRYSTALS FOR NEUTRONS.
    *BROCKER(B.). (IN GERMAN).

10 ATOMKERNENERGIE VOL.17, ... 277, (1971). . .E. . . . .
    *DOUBLE STATISTICAL CHOPPER FOR THE ELIMINATION OF INELASTIC SCATTERED
    NEUTRONS IN TIME-OF-FLIGHT EXPERIMENTS.
    *PELLIONISZ(P.).

11 CAN. J. PHYS. VOL.39, ... 1, (1961). . . .E. . . . . . . .CO-FE. . . . . .
    *A POLARIZED NEUTRON BEAM PRODUCED BY BRAGG REFLECTION FROM CO-FE ALLOY.
    *CLARK(M.A.), ROBSON(J.M.).

12 C. R. ACAD. SCI. B TOME 270, . . 185, (1970). . . .E. . . . .
    *UTILISATION D'UNE CIBLE GAZEUSE CONSTITUEE DE VAPEUR D2*O COMME SOURCE
    DE NEUTRONS A PARTIR DES REACTIONS (DD).
    *DEBIESSE(J.), KLEIN(S.), NGO-VAN-HAI, ORIA(M.), LEPETIT(J.).

13 CRYOGENICS VOL.11, ... 107, (1971). . . .E. . . . .
    *THE COLD NEUTRON SOURCE IN JULICH.
    *DOOSE(C.), PREUSSNER(A.), STELZER(F.), STILLER(H.), THOLEN(A.).

14 CZECH. J. PHYS. VOL.8, ... 592, (1958). . . .E. . . . .
    *FUNDAMENTAL RELATIONS FOR THE CONSTRUCTION OF MECHANICAL NEUTRON
    SELECTORS.
    *JUNA(J.). (IN RUSSIAN).

15 CZECH. J. PHYS. B VOL.13, . . . 474, (1963). . . .E. . . . .
    *THE BEAM OF POLARIZED NEUTRONS OBTAINED BY THE METHOD OF REFLECTION
    FROM A COBALT MIRROR.
    *KOPECKY(J.), CHALUPA(B.), MICHALEC(R.), KAJFOSZ(J.).

16 DISS. ABS. VOL.25, ... 1131, (1965). . . . .
    *PREDICTION AND MEASUREMENT OF NEUTRON CHOPPER BURST SHAPES.
    *CARPENTER(J.M.).

17 ENERGIA NUCLEARE VOL.10, . . . 173, (1963). . . .E. . . . .
    *A CRYSTAL MONOCHROMATOR FOR SLOW NEUTRONS.
    *GIACCHETTI(G.), MUSCI(M.), POLETTI(G.).

18 ENERGIE NUCLEAIRE VOL.13, . . . 15, (1971). . . .E. . . . .
    *LA SOURCE DE NEUTRONS FROIDS POUR LE REACTEUR A HAUT FLUX FRANCO-
    ALLEMAND DE GRENOBLE.
    *AGERON(P.), EWALD(R.), HARIG(H.D.), VERDIER(J.).

19 ENG. J. (CANADA), ... 22, (1972). . . . . .CF. . . . . .
    *CALIFORNIUM-252, A NEW NEUTRON SOURCE FOR SCIENCE AND ENGINEERING.
    *WIGGINS(P.F.), DUFFEY(D.).

20 IAEA SYMP. (PILE RESEARCH)VIENNA 549, (1960). . . . .
    *A MULTI-ROTOR FAST CHOPPER SYSTEM.
    *EGELSTAFF(P.A.), HAY(H.J.), HAYWOOD(B.C.G.).

21 IAEA SYMP. (PILE RESEARCH)VIENNA 491, (1960). . . . .
    *FLUCTUATION OF THE NEUTRON YIELD IN PULSED FAST REACTOR.
    *..C.C.C.P.

22 IAEA SYMP. (PILE RESEARCH)VIENNA 469, (1960). . . . .
    *PULSED-REACTOR AND ACCELERATOR NEUTRON SOURCES FOR PHYSICS RESEARCH.
    *MCREYNOLDS(A.W.).

23 IAEA SYMP. (PILE RESEARCH)VIENNA 409, (1960). . . . .
    *COLD-NEUTRON FACILITY FOR STUDYING THE DYNAMICS OF CONDENSED MATTER
    UTILIZING INELASTIC SCATTERING OF COLD NEUTRONS.
    *..C.C.C.P.

24 IAEA SYMP. (PILE RESEARCH)VIENNA 393, (1960). . . . .
    *REFROIDISSEMENT DES NEUTRONS ET SOURCES DE NEUTRONS FROIDS.
    *JACROT(B.).

SECTION   20B-ON NEUTRON SOURCES . . . .

25   IAEA SYMPOSIUM VIENNA,   421, (1960).............................
     ↱INELASTIC SCATTERING OF NEUTRONS FROM VERY COLD MATERIALS.
     ↱MCREYNOLDS(A.W.), WHITTEMORE(W.L.).

26   IAEA SYMPOSIUM VIENNA,   453, (1960)............................
     ↱ETUDE SYSTEMATIQUE DE SOURCES DE NEUTRONS FROIDS.
     ↱HAUTECLER(S.), VAN-DINGENEN(W.).

27   IAEA SYMPOSIUM VIENNA,   447, (1960)...........................
     ↱THE PRODUCTION OF COLD NEUTRONS BY INELASTIC SCATTERING IN
     HYDROGEN-CONTAINING SUBSTANCES.
     ↱O#CONNOR(D.), MALISZEWSKI(E.), BEDELIK(W.).

28   IAEA SYMPOSIUM (VIENNA),   411, (1960).........................
     ↱EXPERIENCE   DE   THERMALISATION   DE   NEUTRONS   LENTS   PAR   DE L#HYDROGENE
     LIQUIDE.
     ↱CRIBIER(D.), JACROT(B.), LACAZE(A.), ROUBEAU(P.).

29   IAEA SYMPOSIUM (VIENNA),   397, (1960).........................
     ↱DIFFUSION DES NEUTRONS LENTS PAR L#HYDROGENE LIQUIDE.
     ↱SARMA(G.).

30   IAEA SYMP. (PILE RESEARCH)VIENNA 527, (1960)......................
     ↱THE   RELATIVE   MERITS   OF   CHOPPERS AND LINEAR ELECTRON ACCELERATORS FOR
     NEUTRON TIME-OF-FLIGHT MEASUREMENTS.
     ↱EGELSTAFF(P.A.).

31   IAEA SYMP. (PILE RESEARCH)VIENNA 509, (1960)......................
     ↱TIME-DEPENDENT   NEUTRON MEASUREMENTS UTILIZING A CHOPPED REACTOR-BEAM OR
     A PULSED ACCELERATOR.
     ↱SILVER(E.G.), DE-SAUSSURE(G.), PEREZ(R.B.).

32   IAEA SYMP. (PILE RESEARCH)VIENNA 445, (1960)......................
     ↱PARASITIC REFLECTIONS IN NEUTRON MONOCHROMATORS.
     ↱BLINOWSKI(K.), SOSNOWSKI(J.).

33   IAEA SYMP. (PILE RESEARCH)VIENNA 433, (1960)......................
     ↱COMPOSITION OF NEUTRON BEAMS FROM CRYSTAL MONOCHROMATORS.
     ↱SPENCER(R.R.), SMITH(J.R.), BRUGGER(R.M.).

34   IAEA SYMP. CHALK RIVER, VOL.1,   107, (1962)......................
     ↱A COLD NEUTRON MONOCHROMATOR AND SCATTERING APPARATUS.
     ↱HARRIS(D.), COCKING(S.J.), EGELSTAFF(P.A.), WEBB(F.J.).

35   IAEA SYMP. CHALK RIVER, VOL.1,   83, (1962)......................
     ↱THE   PERFORMANCE   AND   AUTOMATIC   OPERATION   OF THE LIQUID HYDROGEN COLD
     NEUTRON SOURCE IN DIDO.
     ↱WEBB(F.J.), PEARCE(D.G.).

36   IAEA SYMP. COPENHAGEN, VOL.2,   341, (1968).....................
     ↱THE DESIGN AND PERFORMANCE OF THE HERALD COLD SOURCE.
     ↱DAVIES(F.),   RODGERS(A.L.),   TODD(M.C.J.),   ROSS(D.K.),   SANALAN(Y.),
     WALKER(J.),   BELSON(J.),   CLARK(C.D.), MITCHELL(E.W.J.), TUCKEY(G.S.G.).

37   IAEA SYMP. COPENHAGEN, VOL.2,   331, (1968).....................
     ↱THE   PROPERTIES   AND   PERFORMANCE   OF   THE HOT NEUTRON SOURCE AT THE FR2
     REACTOR.
     ↱ABELN(O.), DREXEL(W.), GLASER(W.), GOMPF(F.), REICHARDT(W.), RIPFEL(H.).

38   IAEA SYMP. (INSTRUMENT.), VIENNA 269, (1969) . . . .
     ↱DISCUSSION ON NEUTRON SOURCES.

39   IAEA SYMP. (INSTRUMENT.), VIENNA 249, (1969).....................
     ↱REVIEW OF HOT NEUTRON SOURCES.
     ↱EGELSTAFF(P.A.).

40   IAEA SYMP. (INSTRUMENT.) VIENNA   225, (1969).....................
     ↱UTILIZATION   OF   NEUTRON   GUIDE TUBES FOR NEUTRON INELASTIC SCATTERING.
     ↱JACROT(B.).

41   I.A.E.A. SYMP., STUDSVIK,   93, (1969)...INVITED.TALK.......
     ↱NEUTRON CONDUCTING TUBES (INVITED TALK).
     ↱MAIER-LEIBNITZ(H.), (SYMPOSIUM   ON #NEUTRON CAPTURE GAMMA-RAY SPECTROS-
     COPY#, PUBL. IAEA#VIENNA).

42   IAEA SYMP. (INSTRUMENT.), VIENNA   91, (1969).....................
     ↱ORIENTED GRAPHITE AS A NEUTRON MONOCHROMATOR.
     ↱RISTE(T.).

43   IAEA SYMP. GRENOBLE,   697, (1972)....T.E......CU-GE.........
     ↱CRYSTALS WITH A GRADIENT IN THE LATTICE PARAMETER AS NEUTRON
     MONOCHROMATORS.
     ↱RUSTICHELLI(F.).

44   IEEE TRANS. NUCL. SCI. VOL.NS-13 311, (1966)....E................
     ↱SUBMICROSECOND SYNCHRONIZATION OF MULTIPLE CHOPPER SYSTEMS.
     ↱HAUMANN(J.), OSTROWSKI(G.), CONNOR(D.).

45   INSTRUM. EXP. TECH. NO.4,   571, (1960)....E................
     ↱PRODUCTION   OF   POLARIZED   NEUTRONS BY REFLECTION FROM A COBALT MIRROR.
     ↱ABOV(YU.G.), BEKETOV(V.A.)...ET AL.

46   INSTRUM. EXP. TECH. NO.2,   183, (1960)....T................
     ↱THEORY OF NEUTRON CRYSTAL MONOCHROMATORS (REVIEW).
     ↱ABOV(YU.G.).

47   INSTRUM. EXP. TECH. NO.3,   444, (1961)....E................
     ↱A MECHANICAL MONOCHROMATOR FOR NEUTRONS IN THE ENERGY RANGE .001 TO 2 EV
     ↱EFIMOV(B.V.), DANELYAN(L.S.)... ET AL.

48   INSTRUM. EXP. TECH. NO.4,   632, (1963)....E................
     ↱A   MECHANICAL   NEUTRON CHOPPER WITH A SYNCHRONOUSLY ROTATING COLLIMATOR.
     ↱KALEBIN(S.M.), RUKOLAINE(G.V.), SOKOLOVSKII(V.V.).

49   INSTRUM. EXP. TECH. NO.2,   347, (1963)....E.T..............
     ↱TRANSMISSION FUNCTION OF A TWO-ROTOR NEUTRON CHOPPER.
     ↱VERTEBNYI(V.P.), KOLOTYI(V.V.).

50   INSTRUM. EXP. TECH. 3,   729, (1970)....E................
     ↱DOUBLE THERMAL-NEUTRON POLARIZER.
     ↱OBINYAKOV(B.A.), MOSTOVOI(YU.A.).

51   INSTRUM. EXP. TECH. VOL.16 NO.2, 399, (1973)....E......K*MG3*(AL*SI3*O3*O10)*F2.
     ↱FLUORPHLOGOPITE   SINGLE   CRYSTAL   AS A LONG-WAVE NEUTRON MONOCHROMATOR.
     ↱ZELENYUK(F.M.), ZAITSEV(K.N.)... ET AL.

SECTION 20B-ON NEUTRON SOURCES . . . .

52 J. APPL. CRYST. VOL.3, ... 220, (1973)....T....
    ~GUINIER-ANORDNUNG AM NEUTRONENLEITER.
    ~DACHS(H.).

53 J. APPL. CRYST. VOL.3, ... 214, (1970)....T....
    ~VERWENDUNG GEBOGENER IDEALKRISTALLE ALS NEUTRONENMONOCHROMATOREN.
    ~EGERT(G.), DACHS(H.).

54 J. APPL. CRYST. VOL.3, ... 361, (1970)....T....
    ~NEUTRON MONOCHROMATOR RESPONSE BASED ON THE SECONDARY EXTINCTION
    GREEN FUNCTION FOR BRAGG GEOMETRY.
    ~DESJARDINS(J.S.).

55 J. APPL. CRYST. VOL.4, ... 324, (1971)....
    ~A MULTI-WAVELENGTH NEUTRON MONOCHROMATOR SYSTEM.
    ~COOPER(M.J.), FORSYTH(J.B.).

56 J. APPL. CRYST. VOL.5, ... 78, (1972)....E.........CU..
    ~PROPERTIES OF BENT COPPER MONOCHROMATORS FOR THERMAL NEUTRONS.
    ~THOMAS(P.).

57 J. NUCL. ENERGY A:REAC. SCI. 11, 69, (1960)....T....
    ~THE PRINCIPLES OF MECHANICAL NEUTRON-VELOCITY SELECTION.
    ~LOWDE(R.D.).

58 J. NUCL. ENERGY VOL.17, ... 187, (1963)....E....
    ~COLD NEUTRON SOURCES.
    ~WEBB(F.J.).

59 J. NUCL. SCI. TECHNOL. VOL.7, . . 381, (1970)....E....
    ~A SINUSOIDAL NEUTRON SOURCE BY PULSE FREQUENCY MODULATION OF A PULSED
    NEUTRON SOURCE.
    ~YASUDA(H.), SANDA(T.), SUMITA(K.).

60 J. PHYS. E VOL.4 NO.11, ... 965, (1971)....E....
    ~CONSTRUCTION AND PERFORMANCE OF THE LIQUID METHANE COLD SOURCE.
    ~DIMIC(V.), PETKOVSEK(J.).

61 J. PHYS. E VOL.6 NO.5, ... 488, (1973)....E....
    ~THERMALLY BENT IDEAL CRYSTALS AS MONOCHROMATORS FOR NEUTRON SCATTERING.
    ~KALUS(J.), GOBERT(G.), SCHEDLER(E.).

62 J. PHYSIQUE TOME 24, ... 359, (1963)....E....
    ~PRODUCTION AND STUDY OF MONOCHROMATIC POLARIZED BEAMS OF SLOW NEUTRONS.
    ~BEIL(H.), CARLOS(P.), MATUSZEK(J.). (IN FRENCH).

63 J. PHYS. RADIUM VOL.19 SUPPL.4, 51A, (1958)....E....
    ~A MECHANICAL SELECTOR FOR SLOW NEUTRONS.
    ~GOBERT(G.), JACROT(B.). (IN FRENCH).

64 J. PHYS. SOC. JAPAN VOL.15, . . . 630, (1960)....E........GE..
    ~GERMANIUM CRYSTAL AS A NEUTRON MONOCHROMATOR AND THE DETERMINATION OF
    ITS HIGHER ORDER CONTAMINATIONS.
    ~WAJIMA(J.T.), RUSTAD(B.M.), MELKONIAN(E.).

65 J. SCI. INSTRUM. VOL.43, ... 1, (1966)....E,T....
    ~THE DESIGN OF A VELOCITY SELECTOR FOR LONG WAVELENGTH NEUTRONS.
    ~CLARK(C.D.), MITCHELL(E.W.J.), PALMER(D.W.), WILSON(I.H.).

66 METROLOGIA (GER.) VOL.4, ... 8, (1968)....E....
    ~RECENT IMPROVEMENTS IN THE ABSOLUTE CALIBRATION OF NEUTRON SOURCES.
    ~GEIGER(K.W.).

67 NUCL. INSTRUM. VOL.1, ... 92, (1957)....E....
    ~MODIFICATION OF THE BROOKHAVEN FAST CHOPPER.
    ~SEIDL(F.G.P.), PALEVSKY(H.), HUGHES(D.J.), ZIMMERMAN(R.L.).

68 NUCL. INST. MET. VOL.8, ... 203, (1960)....
    ~CALCULATED SHAPES FOR NEUTRON BURSTS PRODUCED BY PHASED CHOPPER
    VELOCITY SELECTORS.
    ~TOLK(N.H.), BRUGGER(R.M.).

69 NUCL. INST. MET. VOL.8, ... 244, (1960)....T....
    ~THE THEORY AND OPERATION OF THE ROTATING CRYSTAL SLOW NEUTRON CHOPPER.
    ~O'CONNOR(D.A.).

70 NUCL. INST. MET. VOL.10, ... 289, (1961)....E........CU..
    ~PARASITIC REFLECTIONS OF NEUTRONS IN CRYSTAL MONOCHROMATORS.
    ~BLINOWSKI(K.), SOSNOWSKI(J.).

71 NUCL. INST. MET. VOL.13, ... 1, (1961)....E....
    ~THE U.S.-CANADIAN FAST CHOPPER INSTALLATION AT CHALK RIVER.
    ~ZIMMERMAN(R.L.), PALEVSKY(H.), CHRIEN(R.E.), OLSEN(W.C.), SINGH(P.P.),
    WESTCOTT(C.H.).

72 NUCL. INST. MET. VOL.11, ... 144, (1961)....
    ~REACTOR PHYSICS RESEARCH WITH PULSED NEUTRON SOURCES.
    ~BECKURTS(K.H.).

73 NUCL. INST. MET. VOL.12, ... 75, (1961)....E....
    ~M.T.R. PHASED CHOPPER VELOCITY SELECTOR.
    ~BRUGGER(R.M.), EVANS(J.E.).

74 NUCL. INST. MET. VOL.17, ... 129, (1962)....E....
    ~PHASING TWO CHOPPERS FROM A MOTOR GENERATOR SOURCE.
    ~BRUGGER(R.M.), STRONG(K.A.).

75 NUCL. INST. MET. VOL.16, ... 116, (1962)....E....
    ~SYSTEMATIC STUDY OF SOME COLD-NEUTRON SOURCES.
    ~VAN-DINGENEN(W.).

76 NUCL. INST. MET. 33, ... 229, (1965)....E....
    ~A SLOW NEUTRON CHOPPER FOR ENERGIES BETWEEN 0.1 AND 0.001 EV.
    ~ANTONINI(B.), MERZAGORA(N.), PAULI(G.).

77 NUCL. INST. MET. VOL.45, ... 293, (1966)....E....
    ~ON A SYSTEM OF MAGNETIZED COBALT MIRRORS USED TO PRODUCE AN INTENSE
    BEAM OF POLARIZED THERMAL NEUTRONS.
    ~ABRAHAMS(K.), RATYNSKI(W.), STECHER-RASMUSSEN(F.), WARMING(E.).

78 NUCL. INST. MET. 46, ... 70, (1966)....
    ~A NEW TYPE OF NEUTRON CHOPPER FOR MEASUREMENT WITH VERY SLOW NEUTRONS.
    ~TUNKELO(E.), PALMGREN(A.).

SECTION 20B-ON NEUTRON SOURCES . . . .

79   NUCL. INST. MET. 45,   255, (1966)
     *SLOW NEUTRON CHOPPER-MONOCHROMATOR WITH CURVED SLITS. . . . .
     *ZSIGMOND(G.), BATA(L.), KISDI-KOSZO(E.).

80   NUCL. INST. MET. 46,   141, (1966)
     *THE ENERGY CALIBRATION OF NEUTRON VELOCITY SPECTROMETERS. . . . .
     *WYNCHANK(S.A.R.).

81   NUCL. INST. MET. 46,   266, (1966)
     *THE COLD NEUTRON FACILITY AT THE FI RI. . . . .
     *TUNKELO(E.), PALMGREN(A.).

82   NUCL. INST. MET. 55,   141, (1967)
     *4 PI NEUTRON SOURCE TECHNIQUE FOR NEUTRON SCATTERING MEASUREMENTS.
     *SAUTER(G.D.), BOWMAN(C.D.).

83   NUCL. INST. MET. 55,   288, (1967)
     *SLOW NEUTRON SPECTRUM MEASUREMENTS WITH A LI-F SINGLE CRYSTAL AS
     MONOCHROMATOR.
     *ROSSITTO(F.), TERRANI(M.).

84   NUCL. INST. MET. 60,   182, (1968)
     *INVESTIGATION OF THE DOUBLE REFLECTION NEUTRON CRYSTAL MONOCHROMATOR.
     *CHOI(C.S.).

85   NUCL. INST. MET. 59,   136, (1968)
     *CHOPPER WITH TYPICAL CURVED SLITS AS MONOCHROMATOR FOR COLD NEUTRONS.
     *DEMICHELIS(F.), RASETTI(M.), VADACCHINO(M.).

86   NUCL. INST. MET. 64,   77, (1968)
     *A NEUTRON POLARIMETER WITH HIGH ENERGY RESOLUTION.
     *SIEMINSKI(M.), WILHELMI(Z.), ZYCH(W.), ZUPRANSKI(P.).

87   NUCL. INST. MET. 60,   349, (1968)
     *ON THE DEPENDENCE OF START PULSE POSITION ON ROTOR SPEED IN NEUTRON TIME
     OF FLIGHT SPECTROMETRY.
     *SALAMA(M.), HAMOUDA(I.).

88   NUCL. INST. MET. VOL.59,   245, (1968)
     *THE DIDO HOT SOURCE; ENHANCEMENT OF .1-.4 EV NEUTRON INTENSITIES BY
     RETHERMALIZATION.
     *EGELSTAFF(P.A.), MOFFITT(R.D.), SAUNDERSON(D.H.).

89   NUCL. INST. MET. 69,   125, (1969)
     *EFFICIENCY OF NEUTRON CRYSTAL MONOCHROMATORS.
     *POPOVICI(M.), GHEORGHIU(Z.), GELBERG(D.).

90   NUCL. INST. MET. 75,   197, (1969)
     *ORIENTED GRAPHITE AS A NEUTRON MONOCHROMATOR.
     *RISTE(T.), OTNES(K.).

91   NUCL. INST. MET. 74,   219, (1969)
     *ANALYSIS OF COMPOSITE NEUTRON MONOCHROMATOR SYSTEMS CONSISTING OF CURVED
     CRYSTALLINE LAMELLAS.
     *RUSTICHELLI(F.).

92   NUCL. INST. MET. 72,   237, (1969)
     *NEUTRON MODERATOR ASSEMBLIES FOR PULSED THERMAL NEUTRON TIME OF FLIGHT
     EXPERIMENTS.
     *DAY(D.H.), SINCLAIR(R.N.).

93   NUCL. INST. MET. 72,   77, (1969)
     *A SIMPLE METHOD OF INCREASING THE MOSAIC SPREAD OF GERMANIUM NEUTRON
     MONOCHROMATORS.
     *NIKOTIN(O.P.).

94   NUCL. INST. MET. 68,   353, (1969)
     *A LITHIUM GLASS DETECTOR BANK FOR NEUTRON TIME OF FLIGHT MEASUREMENTS.
     *JUNG(H.H.), PRIESMEYER(H.G.).

95   NUCL. INST. MET. VOL.86,   1, (1970) . . .E. . . . .C (GRAPHITE),. . . . .
     *SINGLY BENT GRAPHITE MONOCHROMATORS FOR NEUTRONS.
     *RISTE(T.).

96   NUCL. INST. MET. 83,   124, (1970)
     *STUDY OF A COMPOSITE NEUTRON MONOCHROMATOR SYSTEM CONSISTING OF CRYSTALS
     WITH A GRADIENT OF THE LATTICE SPACING.
     *RUSTICHELLI(F.).

97   NUCL. INST. MET. VOL.86,   55, (1970) . . .E. . . . .
     *NEUTRON FOCUSING BY A CURVED SOLLER COLLIMATOR SYSTEM.
     *FRIEDMANN(M.), RAUCH(H.).

98   NUCL. INST. MET. VOL.95,   589, (1971) . . . .P. CU2-MN-AL,
     *STUDY OF A HEUSLER ALLOY (CU2-MN-AL) AS A MONOCHROMATOR FOR POLARIZED
     NEUTRONS.
     *DELAPALME(A.), SCHWEIZER(J.), COUDERCHON(G.), PERRIER-DE-LA-BATHIE(R.).
     (ED. NOTE: IN FRENCH).

99   NUCL. INST. MET. VOL.91,   159, (1971) . . . .E. . . . .
     *THE DETERMINATION OF THE TRANSMISSION FUNCTION OF A NEUTRON CHOPPER WITH
     CIRCULAR SLITS, AND ITS USE IN MEASURING COLD SOURCE SPECTRA.
     *ROSS(D.K.), WINFIELD(D.J.).

100 NUCL. INST. MET. 98,   87, (1972)
     *A DOUBLE CHOPPER SPECTROMETER FOR COLD NEUTRONS.
     *WESTPHAL(G.P.), RAUCH(H.), BREITFUSS(G.).

101 NUCL. INST. MET. VOL.102,   501, (1972) . . . .E. . . . .
     *TRITIUM-CONTAINING TARGETS FOR HIGH-YIELD NEUTRON GENERATORS.
     *O#DONNELL(F.R.), ADAIR(H.L.).

102 NUCL. INST. MET. VOL.101,   221, (1972) . . . .T. . . . .
     *PARABOLISCH GEKRUMMTE SPIEGEL ALS KONDENSOREN THERMISCHER UND SUBTHERM-
     ISCHER NEUTRONENSTRAHLEN.
     *ROTHBAUER(R.).

103 NUCL. INST. MET. VOL.104,   147, (1972) . . . .E. . . . . . .SI,
     *CURVED SILICON CRYSTALS AS NEUTRON MONOCHROMATORS.
     *ANTONINI(M.), CORCHIA(M.), NICOTERA(E.), RUSTICHELLI(F.).

104 NUCL. INST. MET. VOL.98,   61, (1972) . . . .E. . . . .
     *NEUTRON SPIN-FLIP-CHOPPER WITH HIGH REPETITION RATES.
     *FREISLEBEN(H.), RAUCH(H.).

SECTION 20B-ON NEUTRON SOURCES . . . .

105 NUCL. INST. MET. VOL.109, 313, (1973)....T....
~AN EXPERIMENTAL DETERMINATION OF THE TRANSMISSION FUNCTION OF A NEUTRON
CHOPPER.
~BROWN(R.A.), ROSS(D.K.).

106 NUCL. INST. MET. VOL.107, 429, (1973)....E........
~COMPOSITE FOCUSING NEUTRON MONOCHROMATOR SYSTEM.
~BOEUF(A.), RUSTICHELLI(F.).

107 NUCL. INST. MET. VOL.107, 405, (1973)....E........C*04,
~AN IMPROVED VERSION OF THE METHANE COLD NEUTRON SOURCE.
~DIMIC(V.).

108 NUCL. INST. MET. VOL.107, 37, (1973)....T....
~A DETAILED EVALUATION OF NEUTRON MIRROR MONOCHROMATIZING SYSTEMS.
~FIALA(W.), DELSANTO(P.P.).

109 NUCL. INST. MET. VOL.107, 33, (1973)....E........
~COLD NEUTRON MONOCHROMATIZATION BY MEANS OF TOTAL REFLECTION.
~FIALA(W.), FOOTE-JR(H.L.).

110 NUCL. INST. MET. VOL.107, 21, (1973)....T....
~THE EFFICIENCY OF VERTICALLY BENT NEUTRON MONOCHROMATORS.
~CURRAT(R.).

111 NUCL. INST. MET. VOL.113, 461, (1973)....T....
~HYDROGEN COLD NEUTRON SOURCE CALCULATIONS FOR DIFFERENT ORTHO/PARA
MIXTURES.
~KALLI(H.).

112 NUCL. INST. MET. VOL.115, 277, (1974)....E........
~A PACKET OF IDEAL-CRYSTALLINE LAMELLAE AS NEUTRON MONOCHROMATOR.
~FREY(F.).

113 NUCL. SCI. ENGNG. VOL.8, 393, (1960)....E........
~COMPETITIVE EXTINCTION IN NEUTRON MONOCHROMATING CRYSTALS.
~SPENCER(R.R.), SMITH(J.R.).

114 NUCL. TECHNOL. VOL.15, 14, (1972)....T....
~THE POTENTIAL OF A LASER-INDUCED FUSION DEVICE AS A THERMAL-NEUTRON
SOURCE.
~BRUGGER(R.M.).

115 NUKLEONIK VOL.3, 15, (1961)....T....
~THE CURVED SLIT NEUTRON CHOPPER.
~TEUTSCH(H.). (IN GERMAN).

116 NUKLEONIK VOL.4, 266, (1962)....E........
~PRODUCTION OF MONOENERGETIC NEUTRON PULSES BY TOTAL REFLECTION.
~STOCKMEYER(R.). (IN GERMAN).

117 PATENT USA 3683190, (1969)....E........
~TRITIUM AND DEUTERIUM IMPREGNATED TARGETS FOR NEUTRON GENERATORS.
~STARK(D.S.). (ED. NOTE: PUBL. AUG.8, 1972).

118 PHIL. MAG. (8) VOL.2, 917, (1957)....E........
~THE PRODUCTION OF INTENSE COLD NEUTRON BEAMS.
~BUTTERWORTH(I.), EGELSTAFF(P.A.), LONDON(H.), WEBB(F.J.).

119 PHYS. LETT. VOL.26A, 469, (1968)....E........
~A FLIPPER-CHOPPER FOR POLARIZED SLOW NEUTRONS.
~STEINSVOLL(O.), VIRJO(A.).

120 PHYS. LETT. VOL.38A, 177, (1972)....T....
~THE PRODUCTION OF VERY COLD NEUTRONS.
~GOLUB(R.).

121 PHYS. NORVEG. VOL.3, 208, (1969)....E........
~PYROLYTIC GRAPHITE AS A MONOCHROMATOR FOR NEUTRONS AND X-RAYS.
~RISTE(T.).

122 PHYS. REV. VOL.45, 507, (1934)....E........
~ARTIFICIAL PRODUCTION OF NEUTRONS.
~CRANE(H.R.), LAURITSEN(C.C.), SOLTAN(A.).

123 PHYS. REV. VOL.48, 704, (1935)....E........
~VELOCITY OF SLOW NEUTRONS BY MECHANICAL VELOCITY SELECTOR.
~DUNNING(J.R.), PEGRAM(G.B.), FINK(G.A.), MITCHELL(D.P.), SEGRE(E.).

124 PHYS. REV. VOL.50, 738, (1936)....E........
~THE PRODUCTION AND ABSORPTION OF THERMAL ENERGY NEUTRONS.
~FINK(G.A.).

125 PHYS. REV. VOL.52, 592, (1937)....E........
~THE PRODUCTION AND PROPERTIES OF LOW TEMPERATURE NEUTRONS.
~LIBBY(W.F.), LONG(E.A.).

126 PHYS. REV. VOL.83, 863, (1951)....E........
~NEUTRON CRYSTAL MONOCHROMATORS.
~WEISS(R.J.), HASTINGS(J.M.), CORLISS(L.M.).

127 PHYS. REV. VOL.88, 958, (1952)....E........GE,FE3*04,
~NEUTRON MONOCHROMATOR CRYSTALS - FE3*04 AND GE.
~MCREYNOLDS(A.W.).

128 PHYS. REV. VOL.140, A675, (1965)....E........
~PROPAGATION OF BRAGG-REFLECTED NEUTRONS IN LARGE MOSAIC CRYSTALS AND THE
EFFICIENCY OF MONOCHROMATORS.
~WERNER(S.A.), ARROTT(A.).

129 PHYS. TODAY VOL.21, NO.12, 23, (1968)....P....
~WE NEED MORE INTENSE THERMAL-NEUTRON BEAMS.
~BRUGGER(R.M.).

130 PRIB. TEKH. EKSPER. NO.2, 3, (1956)....E........
~NEUTRON SPECTROMETER. I.MECHANICAL BEAM CHOPPER.
~SOKOLOVSKII(V.V.), VLADIMIRSKII(V.V.), RADKEVICH(I.A.). (IN RUSSIAN).

131 PRIB. TEKH. EKSPER. VOL.2, 36, (1964)....T....
~CALCULATION OF THE CHARACTERISTICS OF A MULTIROTOR MECHANICAL NEUTRON
CHOPPER.
~ZOLOTUKHIN(V.G.), KHAMYANOV(L.P.), BLYSKAVKA(A.A.). (IN RUSSIAN).

132 PROC. ROY. SOC. A VOL.175, 316, (1940)....E........
~EXPERIMENTS WITH A VELOCITY-SPECTROMETER FOR SLOW NEUTRONS.
~FERTEL(G.E.F.), GIBBS(D.F.), MOON(P.B.), THOMSON(G.P.),
WYNN-WILLIAMS(C.E.).

SECTION 20B-ON NEUTRON SOURCES . . . .

133 REPORT AERE-R5647 (33 PP.),. . . . . (1968). . . .E. . . . .
     ⌐NEUTRON MONOCHROMATOR STUDIES AT A.E.R.E.
     ⌐TURBERFIELD(K.C.). (ED. NOTE: AVAIL. H.M.S.O., 49 HIGH HOLBORN, LONDON,
     WC1, ENG.).

134 REPORT ANL-8032 (104 PP.),. . . . . . (1973). . . . .
     ⌐APPLICATIONS OF A PULSED SPALLATION NEUTRON SOURCE.
     ⌐CARPENTER(J.M.), PRICE(D.L.),-EDITORS (REPORT OF A WORKSHOP HELD AT
     ARGONNE NATIONAL LABORATORY, APRIL 29- MAY 4, 1973).

135 RESEARCH APPL./PULSED REACTORS,. . . . (1970). . . .E. . . . .
     ⌐NEUTRON SPECTROSCOPY BY REPETITIVELY PULSED NEUTRON SOURCES.
     ⌐KLEY(W.). (ED. NOTE: PROCEEDINGS OF A PANEL ON RESEARCH APPLICATIONS OF
     PULSED REACTORS, HELD AT VIENNA, 17-21 AUGUST).

136 REV. DE PHYSIQUE VOL.6,. . . . . 411, (1961). . . .E. . . . .
     ⌐FEATURES OF THE CURVED-SLIT NEUTRON CHOPPER OF THE INSTITUTE FOR ATOMIC
     PHYSICS AT BUCHAREST.
     ⌐TEUTSCH(H.),   MATEESCU(N.),   PIRLOGEA(P.), RADULESCU(C.), TIMIS(P.),
     VASILIU(V.). (IN GERMAN).

137 REV. FAC. SCI/ISTANBUL C VOL.25,   40, (1960). . . .E. . . . .
     ⌐CALIBRATION OF A NEUTRON SOURCE.
     ⌐ELBRUS(D.T.). (IN GERMAN).

138 REV. ROUMAINE PHYS. VOL.11,. . . . . 601, (1966). . . .E. . . . .
     ⌐MAGNETIC MIRROR THERMAL NEUTRON POLARIZER.
     ⌐PETRASCU(M.), APOSTOLESCU(S.), IONESCU(D.R.), MARINESCU(L.).

139 REV. SCI. INSTRUM. VOL.24,. . . . . 91, (1953). . . .E. . . . .
     ⌐A HIGH TRANSMISSION SLOW NEUTRON VELOCITY SELECTOR.
     ⌐DASH(J.G.), SOMMERS-JR(H.S.).

140 REV. SCI. INSTRUM. VOL.30,. . . . 269, (1959). . . .E. . . . .
     ⌐SYNTHETIC MICA AS A MONOCHROMATOR FOR LONG WAVELENGTH NEUTRONS.
     ⌐GOLAND(A.N.), SONDERICKER(J.H.), ANTAL(J.J.).

141 REV. SCI. INSTRUM. VOL.31,. . . . 214, (1960). . . .E. . . . . . .FE2*O3,. . . . . . .
     ⌐ALPHA-FE2*O3 AS A CRYSTAL MONOCHROMATOR.
     ⌐WANIC(A.), RISTE(T.).

142 REV. SCI. INSTRUM. VOL.32,. . . . 297, (1961). . . .T. . . . .
     ⌐OPTICAL CHARACTERISTICS OF A MECHANICAL NEUTRON MONOCHROMATOR WITH
     HELICAL SLOTS.
     ⌐BALLY(D.), TARINA(E.), PIRLOGEA(P.).

143 REV. SCI. INSTRUM. VOL.32,. . . . 870, (1961). . . .E. . . . .
     ⌐EFFECT OF REFLECTION ON THE PERFORMANCE OF MECHANICAL NEUTRON
     MONOCHROMATORS.
     ⌐PASSELL(L.), DICKINSON(W.C.), BARTOLINI(W.).

144 REV. SCI. INSTRUM. VOL.32,. . . . 654, (1961). . . .E. . . . .
     ⌐NEUTRON SPECTROMETER FOR PRODUCING PURE MONOCHROMATIC BEAMS IN THE
     THERMAL REGION.
     ⌐MOLLER(H.B.), SHORE(F.J.), SAILOR(V.L.).

145 REV. SCI. INSTRUM. VOL.32,. . . . 602, (1961). . . .E. . . . .
     ⌐CONICAL TWO-CRYSTAL MONOCHROMATOR FOR SCATTERING, DIFFRACTION, AND
     ABSORPTION CROSS-SECTION WORK WITH SLOW NEUTRONS.
     ⌐DAS-GUPTA(K.).

146 REV. SCI. INSTRUM. VOL.34,. . . . 847, (1963). . . .E. . . . . . .GE,. . . . .
     ⌐GERMANIUM AS A NEUTRON MONOCHROMATOR.
     ⌐BARRETT(C.S.), MUELLER(M.H.), HEATON(L.).

147 REV. SCI. INSTRUM. 36,. . . . . 122, (1965). . . . .
     ⌐NEUTRON GENERATOR (NEW INSTRUMENTS) PP. 122, 873, 1904.

148 REV. SCI. INSTRUM. VOL.37,. . . . 742, (1966). . . .E. . . . .
     ⌐SMALL NEUTRON BEAM CHOPPER.
     ⌐WHITTEMORE(W.L.).

149 REV. SCI. INSTRUM. 38,. . . . . 151, (1967). . . . .
     ⌐NEUTRON GENERATOR (NEW INSTRUMENTS).

150 SOV. PHYS. CRYST. VOL.4,. . . . . 623, (1960). . . .E. . . . . . .PB,. . . . .
     ⌐CONSTRUCTION OF A LEAD CRYSTAL MONOCHROMATOR FOR NEUTRON DIFFRACTION
     STUDIES.
     ⌐LITVIN(D.F.).

151 SOV. PHYS. CRYST. VOL.11,. . . . . 597, (1967). . . .E. . . . .
     ⌐PRODUCTION OF A MONOENERGETIC POLARIZED NEUTRON BEAM.
     ⌐ABOV(YU.G.), ALESHKO-OZHEVSKII(O.P.), ERMAKOV(O.N.), YAMZIN(I.I.).

152 SOV. PHYS. CRYST. VOL.17,. . . . . 826, (1973). . . .T. . . . .
     ⌐BEAM DIVERGENCE IN NEUTRON DIFFRACTION WITH A CRYSTAL MONOCHROMATOR.
     ⌐LEVDIK(V.A.), BYKOV(V.N.), GOLOVKIN(V.S.).

153 SOV. PHYS. JETP VOL.37,. . . . . 41, (1973). . . .T,E. . . . .
     ⌐PRODUCTION OF VERY COLD NEUTRONS.
     ⌐GOLIKOV(V.V.), LUSHCHIKOV(V.I.), SHAPIRO(F.L.).

154 STUD. CERCETARI FIZ. VOL.14,. . . 353, (1963). . . . .
     ⌐USE OF THE MICROTRON AS A PULSED SOURCE FOR NEUTRON SPECTROSCOPY BY
     THE TIME-OF-FLIGHT METHOD.
     ⌐APOSTOLESCU(S.), INDREAS(GR.). (IN RUMANIAN).

155 UKAEA AERE REP. R4263 (30 PP.),. . . . (1965). . . .E. . . . .
     ⌐LOW ENERGY NEUTRON INELASTIC SCATTERING EXPERIMENTS WITH A LINEAR
     ACCELERATOR.
     ⌐WEBB(F.J.).

156 UKAEA AERE REP. 0-98/65 (16 PP.),. . . (1965). . . .E. . . . .
     ⌐A CRYSTAL NEUTRON MONOCHROMATOR.
     ⌐REICHELT(J.M.A.).

157 Z. KRIST. VOL.117,. . . . . 135, (1962). . . .E. . . . .
     ⌐THE USE OF FLAT AND CURVED MONOCHROMATOR CRYSTALS IN NEUTRON SPECTRO-
     METERS FOR INVESTIGATING SINGLE CRYSTALS.
     ⌐OACHS(H.), STEHR(H.). (IN GERMAN).

158 Z. KRIST. VOL.118,. . . . . 263, (1963). . . .T,E. . . . .
     ⌐THE REFLECTION ABILITY OF MOSAIC CRYSTALS WITH REGARD TO THEIR USE AS
     MONOCHROMATORS IN NEUTRON DIFFRACTION SPECTROMETERS.
     ⌐STEHR(H.). (IN GERMAN).

SECTION 20C-ON NEUTRON DETECTORS . . . .

1 ACTA CRYST. VOL.A28 PART S-4, 250, (1972)....E..........
   *A NEW SOLID STATE DETECTOR FOR NEUTRON DIFFRACTION MEASUREMENTS.
   *HOSHINO(S.), TAKAHASHI(S.). (ABSTRACT ONLY).

2 ACTA CRYST. VOL.A28 PART S-4, 250, (1972)....E..........
   *SINGLE CRYSTAL NEUTRON DIFFRACTOMETER WITH 100 MOVABLE DETECTORS.
   *KLAR(B.). (ABSTRACT ONLY).

3 ACTA PHYS. HUNGAR. VOL.12, 333, (1960)....E..........
   *THE #RADIUS EFFECT# IN LARGE B#F3 COUNTER TUBES.
   *GORDON(J.), SZABO(P.).

4 AERE REPORT N/R-1639 (7 PP.), (1955)....E..........
   *DELAYS IN B#F3 PROPORTIONAL COUNTERS.
   *NICHOLSON(K.P.). (AVAIL. ATOMIC ENERGY RES. ESTABLISHMENT, HARWELL).

5 APPL. SCI. RES. B VOL.7, 87, (1958)....E..........
   *NEW SCINTILLATORS FOR THE DETECTION OF THERMAL NEUTRONS.
   *THIELENS(G.).

6 APPL. SCI. RES. B VOL.10, 247, (1963)....E..........
   *PREPARATION AND PROPERTIES OF A BORON CONTAINING SCINTILLATOR FOR THE
   DETECTION OF SLOW NEUTRONS.
   *VERHAEGHE(J.L.), THIELENS(G.), CRETEN(W.L.).

7 ATOMKERNENERGIE VOL.19 NO.4, 312, (1972)....E..........
   *SCATTERED NEUTRONS AND THEIR EFFECTS ON THE CALIBRATION OF BONNER SPHERE
   COUNTERS.
   *HEINZELMANN(M.), ROHLOFF(F.), SCHUREN(H.). (ED. NOTE: IN GERMAN).

8 BULL. D#INSTRUM. SCI. TECH. 172, 55, (1972)....E..........
   *MESURE DU RENDEMENT COMPARATIF DE DIFFERENTS CONVERTISSEURS UTILISES
   POUR LA DETECTION PHOTOGRAPHIQUE DE NEUTRONS THERMIQUES.
   *BREYNAT(G.), THOMAS(P.), VU-HONG-LAC.

9 ENERGIA NUCLEARE VOL.15, 311, (1968)....E..........
   *SELF-POWERED NEUTRON DETECTORS.
   *ACCINNI(F.), TONOLINI(F.).

10 ENERGIA NUCLEARE VOL.16, 400, (1969)....E..........
   *ON THE ENERGY RESPONSE OF A MODERATED NEUTRON DETECTOR.
   *CAVALLARI(F.), TERRANI(M.).

11 IAEA SYMP. CHALK RIVER, VOL.1, 171, (1962)....E..........
   *SCINTILLATION COUNTERS FOR NEUTRON SCATTERING EXPERIMENTS.
   *HARRIS(D.), WRAIGHT(L.A.), DUFFILL(C.).

12 IAEA SYMP. (INSTRUMENT.), VIENNA 275, (1969)..............
   *DISCUSSION ON DETECTORS.

13 IEEE TRANS. NUCL. SCI. VOL.S-16, 419, (1969)....E..........
   *MINIATURE NEUTRON DETECTOR.
   *LONG(W.R.).

14 INSTRUM. EXP. TECH. NO.2, 243, (1963)....E..........
   *HIGH-EFFICIENCY DETECTOR OF SLOW NEUTRONS BASED ON A ZN*S(AG)+B2#O3
   MIXTURE.
   *GOLIKOV(V.V.), SHIMCHAK(G.F.), SHKATULA(A.A.).

15 INSTRUM. EXP. TECH. NO.4, 808, (1965)....E..........
   *COMPARATIVE CHARACTERISTICS OF FLAT SLOW-NEUTRON SCINTILLATION DETECTORS
   *TSIRLIN(YU.A.)... ET AL.

16 INSTRUM. EXP. TECH. NO.4, 805, (1965)....E..........
   *A LIQUID SCINTILLATION NEUTRON DETECTOR WITH COOLED PHOTOMULTIPLIERS.
   *POPOV(A.B.), YAZVITSKII(YU.S.).

17 INSTRUM. EXP. TECH. NO.2, 305, (1965)....E..........
   *SLOW-NEUTRON DETECTOR WITH IMPROVED LIGHT YIELD.
   *VEDEKHIN(A.F.), KUCHERNYUK(V.D.).

18 INSTRUM. EXP. TECH. NO.1, 87, (1966)....E..........
   *HIGH-PRESSURE HELIUM-3 IONIZATION CHAMBER AS A DETECTOR OF SLOW
   NEUTRONS.
   *DMITRIEV(A.B.), ISTOMIN(I.V.), NOVIKOV(A.G.), MALYSHEV(E.K.).

19 INSTRUM. EXPER. TECH. NO.2, 342, (1967)....E..........
   *A FLAT SLOW-NEUTRON DETECTOR.
   *BEDEKHIN(A.F.).

20 INSTRUM. EXPER. TECH. NO.3, 529, (1968)....E..........
   *DIRECTIONAL FLAT-RESPONSE NEUTRON DETECTOR.
   *GORYACHEV(I.V.), GVOZDEV(M.M.), KUKHTEVICH(V.I.), TRYKOV(L.A.).

21 INSTRUM. EXP. TECH. NO.1, 77, (1970)....E..........
   *SENSITIVITY OF SPHERICAL NEUTRON DETECTORS IN THE ENERGY RANGE FROM
   THERMAL TO 10 MEV.
   *ANDREEVA(L.S.), KEIRIM-MARKUS(I.B.), USPENSKII(L.N.), FILYUSHKIN(I.V.),
   CHERNOV(E.N.).

22 INSTRUM. EXP. TECH. 4, 1013, (1970)....E..........
   *SCINTILLATOR FOR RECORDING SLOW NEUTRONS.
   *MARKOV(YU.YA.), RYAZHSKAYA(O.G.).

23 INSTRUM. EXP. TECH. NO.2, 399, (1970)....E..........
   *SCINTILLATING COUNTER WITH A LI#I(EU) CRYSTAL FOR REGISTERING THERMAL
   NEUTRONS.
   *KUZ#MINOV(YU.S.).

24 INSTRUM. EXP. TECH. VOL.14, NO.5 1343, (1971)....E..........
   *END-WINDOW MULTIFILAMENT NEUTRON DETECTOR.
   *VORONA(P.N.), MAISTRENKO(A.N.).

25 INSTRUM. EXP. TECH. VOL.15 PT.1, 351, (1972)....E..........
   *EFFICIENCY OF NEUTRON COUNTERS HAVING A GAS RADIATOR.
   *TOLCHENOV(YU.M.), CHAIKOVSKII(V.G.).

26 INSTRUM. EXP. TECH. VOL.16 NO.2, 345, (1973)....E..........
   *SLOW-NEUTRON DETECTORS THAT ARE SENSITIVE TO THE SPOT OF PARTICLE
   PASSAGE (REVIEW).
   *DEME(SH.), PEPELYSHEV(YU.N.).

SECTION 20C-ON NEUTRON DETECTORS . . . .

27 J. APPL. PHYS. VOL.33, ... 48, (1962)....E...
    ~COMPARISON OF SEVERAL METHODS FOR THE PHOTOGRAPHIC DETECTION OF THERMAL
    NEUTRON IMAGES.
    ~BERGER(H.).

28 J. NUCL. ENERGY VOL.27, . . . 677, (1973)....E...
    ~CADMIUM CORRECTION FACTORS OF SEVERAL THERMAL NEUTRON FOIL DETECTORS.
    ~MUECK(K.), BENSCH(F.).

29 J. NUCL. SCI. TECHNOL. JAP. 4 22, (1967)....E...
    ~A SCINTILLATION DETECTOR FOR THE MEASUREMENT OF THERMAL NEUTRONS IN
    SUBCRITICAL ASSEMBLY.
    ~TOJO(T.), NIWA(T.), NAKAJIMA(T.), KONDO(M.).

30 J. PHYS. SOC. JAPAN VOL.17 B-II, 340, (1962)....E...
    ~PHOTOGRAPHY OF NEUTRON DIFFRACTION PATTERNS.
    ~WANG(S.P.), SHULL(C.G.).

31 J. SCI. INSTRUM. VOL.24, . . . 331, (1947)....E...
    ~SMALL BORON CHAMBERS FOR SLOW NEUTRON MEASUREMENTS.
    ~VEALL(N.).

32 J. SCI. INSTRUM. VOL.39, . . . 124, (1962)....E...
    ~CHARACTERISTICS OF A BORON-COATED PROPORTIONAL NEUTRON COUNTER IN THE
    CORONA ZONE.
    ~TAVENDALE(A.J.).

33 MAGYAR FIZ. FOLYOIRAT (HUNG.) 16 279, (1968)....
    ~A DETECTOR-SYSTEM FOR NEUTRON SCATTERING EXPERIMENTS.
    ~KONCZOS(G.), MADARASZ(Z.).

34 2ND INTERNAT. CONF./NUCL. DATA, . . . (1970)....
    ~A NEUTRON DETECTOR WITH A FLAT ENERGY RESPONSE FOR USE IN TIME-OF-FLIGHT
    EXPERIMENTS.
    ~COATES(M.S.), HUNT(G.J.), RAE(E.R.).

35 NUCL. ENG. INT. VOL.17, ... 399, (1972)....E...
    ~SELF-POWERED NEUTRON DETECTORS.
    ~JOSLIN(C.W.).

36 NUCLEONICS VOL.17 APRIL, ... 116, (1959)....E...
    ~COATED SEMICONDUCTOR IS TINY NEUTRON DETECTOR.
    ~BABCOCK(R.V.),   DAVIS(R.E.),   RUBY(S.L.),   SUN(K.H.),   WOLLEY(E.D.).

37 NUCL. INST. MET. VOL.17, . . . 97, (1962)....E...
    ~NEUTRON DETECTION WITH GLASS SCINTILLATORS.
    ~BOLLINGER(L.M.), THOMAS(G.E.), GINTHER(R.J.).

38 NUCL. INST. MET. VOL.27, . . . 351, (1964)....E...
    ~HE (A=3)-NE-C*H4 MIXTURES IN PROPORTIONAL COUNTERS FOR THERMAL NEUTRONS.
    ~KROHN-JR(V.E.).

39 NUCL. INST. MET. 34, ... 21, (1965)....
    ~A LOW THRESHOLD NEUTRON DETECTOR FOR NANOSECOND TIME OF FLIGHT
    SPECTROMETRY.
    ~ADAMS(J.M.), BARNARD(E.), FERGUSON(A.T.G.)...ET AL.

40 NUCL. INST. MET. 33, . . . 181, (1965)....
    ~IMPROVEMENTS IN THERMAL NEUTRON SCINTILLATION DETECTORS FOR TIME OF
    FLIGHT STUDIES.
    ~WRAIGHT(L.A.), HARRIS(D.H.C.), EGELSTAFF(P.A.).

41 NUCL. INST. MET. VOL.33, ... 194, (1965)....
    ~N,N,N,TRIMETHYLBORAZINE AS A LIQUID SCINTILLATOR SOLVENT FOR THERMAL
    NEUTRON DETECTION.
    ~ROSS(H.H.), HOLSOPPLE(H.L.).

42 NUCL. INST. MET. VOL.34, . . . 141, (1965)....E...
    ~DETECTOR TIME JITTER IN HE (A=3) AND B*F3 PROPORTIONAL COUNTERS.
    ~HARLING(O.K.).

43 NUCL. INST. MET. 35, ... 203, (1965)....
    ~LARGE AREA LI6I(EU) DETECTOR BANDS FOR SLOW NEUTRON TIME OF FLIGHT
    MEASUREMENTS.
    ~STRUB(A.).

44 NUCL. INST. MET. VOL.45, . . . 151, (1966)....E...
    ~A SPHERICAL DOSE EQUIVALENT NEUTRON DETECTOR.
    ~LEAKE(J.W.).

45 NUCL. INST. MET. 46, . . . 153, (1966)....
    ~EIN NEUARTIGER DETEKTOR FUR LANGSAME NEUTRONEN.
    ~RAUCH(H.), GRASS(F.), FEIGL(B.).

46 NUCL. INST. MET. 57, . . . 237, (1967)....
    ~DETECTOR FOR NEUTRON TIME OF FLIGHT SPECTROMETRY WITH IMPROVED RESPONSE
    TO LOW ENERGY NEUTRONS.
    ~WISHART(L.P.), PLATTNER(R.), CRANBERG(L.).

47 NUCL. INST. MET. 52, . . . 175, (1967)....
    ~THE 3HE COUNTER AS A DETECTOR FOR TIME OF FLIGHT MEASUREMENTS OF THERMAL
    NEUTRON SPECTRA.
    ~HEIBERG(E.).

48 NUCL. INST. MET. 53, . . . 163, (1967)....
    ~PREPARATION OF THERMAL NEUTRON SCINTILLATORS BASED ON A MIXTURE OF
    ZN*S(AG), 6LI*F AND POLYETHYLENE.
    ~TOJO(T.), NAKAJIMA(T.).

49 NUCL. INST. MET. 50, . . . 181, (1967)....
    ~PRECISION MEASUREMENT OF THERMAL NEUTRON BEAM DENSITIES USING A 3HE
    PROPORTIONAL COUNTER.
    ~ALS-NIELSEN(J.), BAHNSEN(A.), BROWN(W.K.).

50 NUCL. INST. MET. 53, . . . 325, (1967)....
    ~DISCRIMINATION BY HALF WIDTH AND PULSE LENGTH IN THE 6LI*F-ZN*S
    SCINTILLATION DETECTOR FOR SLOW NEUTRONS.
    ~ALBOLD(E.), HUBER(U.).

51 NUCL. INST. MET. 55, . . . 349, (1967)....
    ~METHOD FOR NEUTRON DETECTION-EFFICIENCY MEASUREMENTS AND NEUTRON-
    CHARGED PARTICLE COINCIDENCE DETECTION.
    ~JACKSON(W.R.), DIVATIAIA.S.), BONNER(B.E.)...ET AL.

SECTION 20C-ON NEUTRON DETECTORS . . . .

52 NUCL. INST. MET. 65, 110, (1968).
≠THERMAL NEUTRONS GD-DETECTOR.
≠FOGLIO-PARA(A.), GOTTARDI(N.A.), MANDELLI-BETTONI(M.).

53 NUCL. INST. MET. 65, 113, (1968).
≠AN EFFECT OF SCATTERED NEUTRONS FROM PHOTOMULTIPLIERS ON THE NEUTRON
SPECTRUM MEASURED BY 6LI GLASS SCINTILLATORS.
≠OHKUBO(M.).

54 NUCL. INST. MET. 63, 185, (1968).
≠A HIGH EFFICIENCY GRAPHITE-MODERATED NEUTRON COUNTER.
≠JEWELL(R.W.), JOHN(W.), WHITE(D.H.).

55 NUCL. INST. MET. 62, 43, (1968).
≠RELATIVE CALIBRATION OF NEUTRON SOURCES BY A POINT DETECTOR.
≠NOTEA(A.), NIR-EL(Y.).

56 NUCL. INST. MET. VOL.73, 225, (1969).
≠A POSITION SENSITIVE NEUTRON DETECTOR.
≠BERKOWITZ(E.H.).

57 NUCL. INST. MET. 72, 307, (1969).
≠ETUDE D'UN DETECTEUR POUR DES MESURES DE DIFFUSION ELASTIQUE DE NEUTRONS
DE RESONANCES.
≠TROCHON(J.), ASGHAR(M.), LUCAS(B.), RIBON(P.).

58 NUCL. INST. MET. 68, 277, (1969).
≠INVESTIGATION OF THE FLATNESS OF RESPONSE OF A MULTI B-F3 COUNTER
NEUTRON DETECTION SYSTEM.
≠BARRET(R.F.), BIRKELUND(J.R.), THIES(H.H.).

59 NUCL. INST. MET. 74, 123, (1969).
≠COLLIMATED DETECTOR RESPONSE TO POINT, LINE AND PLANE SOURCES.
≠STEYN(J.J.), ANDREWS(D.G.), DIXMIER(M.).

60 NUCL. INST. MET. 74, 322, (1969).
≠NEUTRON SPECTROSCOPY WITH A HE-3 PROPORTIONAL COUNTER.
≠FUSE(T.), MIURA(T.), YAMAJI(A.), YOSHIMURA(T.).

61 NUCL. INST. MET. 74, 355, (1969).
≠RESPONSE OF B-F3 COUNTERS TO NEUTRONS IN MODERATE GAMMA FIELDS.
≠VERGHESE(K.), BOHANNON(J.R.), KOWALCZUK(A.D.).

62 NUCL. INST. MET. 75, 35, (1969).
≠MEASUREMENT OF THE ABSOLUTE SCINTILLATION EFFICIENCY OF GRANULAR AND
GLASS NEUTRON SCINTILLATORS.
≠SPOWART(A.R.).

63 NUCL. INST. MET. 72, 161, (1969).
≠POLYETHYLENE MODERATED HE-3 NEUTRON DETECTORS.
≠EAST(L.V.), WALTON(R.B.).

64 NUCL. INST. MET. 72, 120, (1969).
≠EXPERIMENTAL DETERMINATION OF THE EFFICIENCY OF THE GREY NEUTRON
DETECTOR.
≠POENITZ(W.P.).

65 NUCL. INST. MET. 71, 292, (1969).
≠THE WALL EFFECT IN HE-3 COUNTERS.
≠SHALEV(S.), FISHELSON(Z.)...ET AL.

66 NUCL. INST. MET. 67, 267, (1969).
≠A GAS RECOIL NEUTRON POLARIMETER.
≠MANDUCHI(C.), MOSCHINI(N.G.), TORNIELLI(G.), ZANNONI(G.).

67 NUCL. INST. MET. 74, 224, (1969).
≠RESPONSE OF 25 CUBIC CM. GE(LI) DETECTOR TO NEUTRONS- SHIELDING FACTORS.
≠RODDA(J.L.), MACKLIN(R.L.), GIBBONS(J.H.).

68 NUCL. INST. MET. 70, 164, (1969).
≠THE USE OF VANADIUM SCATTERING TO INTERCALIBRATE AN ARRAY OF THERMAL
NEUTRON DETECTORS.
≠DAY(D.H.), JOHNSON(D.A.G.), SINCLAIR(R.N.).

69 NUCL. INST. MET. VOL.81, 301, (1970).
≠A SIX-FOOT LONG DETECTOR FOR NEUTRON TIME-OF-FLIGHT EXPERIMENTS.
≠NEILSON(G.C.), GLAVINA(C.), DAWSON(W.K.), IYENGAR(K.V.), MCDONALD(W.J.).

70 NUCL. INST. MET. VOL.84, 67, (1970).
≠A HIGH EFFICIENCY, GADOLINIUM LOADED MINERAL OIL SCINTILLATOR FOR
NEUTRON DETECTION.
≠CLARK(G.E.), MORRISON(R.C.), O'LAUGHLIN(J.W.), BURKHOLDER(H.R.).

71 NUCL. INST. MET. VOL.87, 299, (1970).
≠TRANSIENT RESPONSE OF COBALT SELF-POWERED NEUTRON DETECTORS.
≠BOCK(H.), STIMLER(M.), STRINDEHAG(O.).

72 NUCL. INST. MET. VOL.83, 111, (1970).
≠A MULTIDETECTOR SYSTEM FOR NEUTRON SMALL-ANGLE SCATTERING EXPERIMENTS.
≠ABEND(K.), SCHMATZ(W.), SCHELTEN(J.), MULLER(K.D.).

73 NUCL. INST. MET. 78, 173, (1970).
≠DIFFERENTIAL SENSITIVITY OF LONG SELF-POWERED NEUTRON DETECTORS.
≠STRINDEHAG(O.), SODERLUND(B.).

74 NUCL. INST. MET. VOL.96, 551, (1971).
≠MEASUREMENTS AND CALCULATIONS OF NEUTRON DETECTOR EFFICIENCIES.
≠THORNTON(S.T.), SMITH(J.R.).

75 NUCL. INST. MET. VOL.94, 493, (1971).
≠A CONVENIENT CALIBRATION TECHNIQUE FOR NEUTRON DETECTORS.
≠CHASTEL(A.), DAVIS(M.B.), HOFFMAN(C.M.), KREISLER(M.N.), SMITH(A.J.S.).

76 NUCL. INST. MET. VOL.92, 221, (1971).
≠EFFECTS OF ELASTIC SCATTERING ON THE LOW ENERGY NEUTRON DETECTION
EFFICIENCY OF A LITHIUM GLASS AND BORON-LOADED LIQUID SCINTILLATOR.
≠DALTON(A.W.).

77 NUCL. INST. MET. VOL.106, 279, (1973).
≠ON THE RESOLUTION OF A POSITION SENSITIVE B-F3 PROPORTIONAL COUNTER.
≠WESTPHAL(G.P.).

78 NUCL. INST. MET. VOL.106, 461, (1973).
≠MOVING CRYSTAL SLOW-NEUTRON WAVELENGTH ANALYSER.
≠BURAS(B.), KJEMS(J.K.).

SECTION   20C-ON NEUTRON DETECTORS . . . .

79 NUCL. SCI. ENGNG. VOL.20,   23 (1964)....E....
    *A NEW LIQUID SCINTILLATOR FOR THERMAL NEUTRON DETECTION.
    *ROSS(H.H.), YERICK(R.E.).

80 NUCL. SCI. APPL. B(PAKIS.) VOL.5   43 (1969)....E....
    *BORON-10 SLAB DETECTOR FOR TOTAL NEUTRON CROSS-SECTION MEASUREMENTS.
    *HUSSAIN(M.), SHAHABUDDIN(A.M.), VUISTER(P.H.), ISLAM-MOLLA(N.).

81 NUKLEONIKA VOL.8,   489 (1963)....E....
    *SILICON DETECTORS OF THERMAL NEUTRONS.
    *KHVASHCHEVSKA(YA.), SHEKHTER(A.). (IN RUSSIAN).

82 NUOVO CIMENTO SUPPL. VOL.3,   187, (1965)....E....
    *A TECHNIQUE OF N-GAMMA DISCRIMINATION.
    *FABIANI(F.), LONGO(G.), SAPORETTI(F.), VENTURINI(G.).

83 PHYS. LETT. B VOL.40B,   537, (1972)....E....
    *THE DETECTION OF ULTRACOLD NEUTRONS BY THE ACTIVATION OF MANGANESE.
    *ROBSON(J.M.), WINFIELD(D.).

84 PRIB. TEKH. EKSPER NO.1,   61 (1959)....T....
    *INVESTIGATION INTO SLOW NEUTRON COUNTERS.
    *SAVEL*EV(V.YA.), KONONENKO(V.A.). (IN RUSSIAN).

85 REV. PHYS. APPL. VOL.1,   149, (1966)....E....
    *SILICON DIODES USED IN NEUTRON DIFFRACTION.
    *GILLY(L.), ROBERT(A.), ROULT(G.). (IN FRENCH).

86 REV. PHYS. APPL. (FRANCE) VOL.4,   259, (1969)....E....
    *PRACTICAL NEUTRON SPECTROMETRY WITH VISUAL IONOGRAPHIC DETECTORS.
    *DEBEAUVAIS(M.), STEIN(R.), REMY(G.), RALAROSY(J.), TRIPIER(J.).

87 REV. ROUMAINE PHYS. VOL.6,   207, (1961)....E....
    *GLASS PROPORTIONAL COUNTERS FOR THERMAL NEUTRON DETECTION.
    *DREGICHESKU(M.)...ET AL.

88 REV. SCI. INSTRUM. VOL.21,   835, (1950)....T,E....
    *THE DESIGN OF NEUTRON COUNTERS USING MULTIPLE DETECTING LAYERS.
    *LOWDE(R.D.).

89 REV. SCI. INSTRUM. VOL.21,   734, (1950)....E....B*F3,
    *BORON TRIFLUORIDE PROPORTIONAL COUNTERS.
    *FOWLER(I.L.), TUNNICLIFFE(P.R.).

90 REV. SCI. INSTRUM. VOL.23,   769, (1952)....E....B2*O3-ZN*S,
    *A SLOW NEUTRON DETECTOR.
    *ALBURGER(D.E.).

91 REV. SCI. INSTRUM. VOL.30,   442, (1959)....E....
    *PLATEAU SLOPES AND PULSE CHARACTERISTICS OF LARGE, HIGH-PRESSURE B*F3
    COUNTERS.
    *MENDELL(R.B.), KORFF(S.A.).

92 REV. SCI. INSTRUM. VOL.30,   1135, (1959)....E....
    *GLASS SCINTILLATORS FOR NEUTRON DETECTION.
    *BOLLINGER(L.M.), THOMAS(G.E.), GINTHER(R.G.).

93 REV. SCI. INSTRUM. VOL.31,   1156, (1960)....E....LI*F,ZN*S*(AG),
    *SCINTILLATOR FOR THERMAL NEUTRONS USING LI*F AND ZN*S*(AG).
    *STEDMAN(R.).

94 REV. SCI. INSTRUM. VOL.33,   866, (1962)....E....
    *LOW VOLTAGE HE (A=3)-FILLED PROPORTIONAL COUNTER FOR EFFICIENT DETECTION
    OF THERMAL AND EPITHERMAL NEUTRONS.
    *MILLS(W.R.), CALDWELL(R.L.), MORGAN(I.L.).

95 REV. SCI. INSTRUM. VOL.33,   844, (1962)....E....
    *PHOTOGRAPHIC DETECTORS FOR NEUTRON DIFFRACTION.
    *BERGER(H.).

96 REV. SCI. INSTRUM. VOL.33,   128, (1962)....E....
    *USE OF POLAROID FILM IN NEUTRON AND X-RAY DIFFRACTION.
    *SMITH(H.G.).

97 REV. SCI. INSTRUM. VOL.33,   126, (1962)....E....
    *PHOTOGRAPHY OF NEUTRON DIFFRACTION PATTERNS.
    *WANG(S.P.), SHULL(C.G.), PHILLIPS(W.C.).

98 REV. SCI. INSTRUM. 34,   731, (1963)....E....
    *VERY LARGE B*F3 PROPORTIONAL COUNTERS.

99 REV. SCI. INSTRUM. VOL.35,   853, (1964)....E....
    *HE(A=3)-NE-C*H4 MIXTURES IN A PROPORTIONAL COUNTER FOR THERMAL NEUTRONS.
    *KROHN-JR(V.E.).

100 REV. SCI. INSTRUM. 35,   469, (1964)....E....
    *DESIGN OF 3HE-FILLED FAST PROPORTIONAL COUNTERS.

101 REV. SCI. INSTRUM. VOL.36,   419, (1965)....E....
    *BORON-LOADED NEUTRON DETECTOR WITH VERY LOW GAMMA-RAY SENSITIVITY.
    *JACKSON(H.E.), THOMAS(G.E.).

102 REV. SCI. INSTRUM. 36,   1074, (1965)....
    *NEUTRON DETECTOR.

103 REV. SCI. INSTRUM. 37,   1763, (1966)....
    *NEUTRON DETECTOR. F*XLIUM.

104 REV. SCI. INSTRUM. 37,   525, (1966)....
    *SIMPLE DETECTORS FOR NEUTRONS OR HEAVY COSMIC RAY NUCLEI.

105 REV. SCI. INSTRUM. 38,   1196, (1967)....
    *FAST-RESPONSE NEUTRON FLUX DETECTOR (NEW MATERIAL).

106 REV. SCI. INSTRUM. 39,   1588, (1968)....
    *SILVER COUNTER FOR BURSTS OF NEUTRONS.
    *LANTER(R.J.), BANNERMAN(D.E.).

107 RISO REPORT NO.229 (56 PP.)   (1970)....E....
    *THERMAL NEUTRON SPECTROSCOPY USING POSITION-SENSITIVE DETECTORS TOGETHER
    WITH FOURIER CORRELATION T.O.F. AND SINGLE-CRYSTAL TECHNIQUES.
    *KJEMS(J.).   (ED. NOTE: AVAIL. DANISH ATOMIC ENERGY COMM. RESEARCH ESTAB-
    LISHMENT, RISO).

SECTION   20C-ON NEUTRON DETECTORS . . . .

108   SCI/INST. PHYS. CHEM. RES. 54,   271, (1960)....E........ . . . .
             ↱B-F3 NEUTRON COUNTER.
             ↱KOBAYASHI(H.).
109   SOV. AT. ENERGY VOL.32, NO.5,   416, (1972)....E....... . . . .
             ↱LI-SCINTILLATION COUNTERS FOR NEUTRONS.
             ↱ORLOV(N.F.), ANDREEVA(N.Z.), GERASIMOV(V.F.), ZENKEVICH(V.S.),
             MOSKALEV(S.S.).
110   SOV. PHYS. J.E.T.P. VOL.31,   59, (1970)........... . . . .
             ↱A CO-ORDINATE PROPORTIONAL COUNTER.
             ↱BIRYUKOV(V.A.), ZINOV(V.G.), KONIN(A.D.).
111   TH. NEUTRON SCATT./EGELSTAFF,   141, (1965).......... . . . .
             ↱NEUTRON SOURCES AND DETECTORS.
             ↱COCKING(S.J.), WEBB(F.J.).

SECTION 20D-ON NEUTRON COLLIMATORS AND FILTERS . . . .

1 ACTA CRYST. VOL.A27, ... 341, (1971)....T,E. . . . . I . ACCURATE TRANSMITTED
   *THE PERFORMANCES OF NEUTRON COLLIMATORS. I. ACCURATE TRANSMITTED
   INTENSITY EVALUATIONS FOR NEUTRON COLLIMATOR SYSTEMS.
   *ROSSITTO(F.), POLETTI(G.).

2 ACTA CRYST. VOL.A29, ... 440, (1973)....T,E.
   *THE PERFORMANCES OF NEUTRON COLLIMATORS. II.CHOICE OF THE PARAMETERS
   OF A PRIMARY COLLIMATOR.
   *POLETTI(G.), ROSSITTO(F.).

3 ACTA CRYST. VOL.A29, ... 526, (1973)....T. . . . .
   *ON THE USE OF IN-PILE COLLIMATORS IN INELASTIC NEUTRON SCATTERING.
   *KALUS(J.), DORNER(B.).

4 ARCH. SCI. VOL.12, ... 676, (1959)....E. . . . .
   *A NEW TYPE OF COLLIMATOR FOR SLOW NEUTRONS.
   *DENIS(P.), ROUX(D.).

5 COLL. INTER. N.126 (GRENOBLE), . . 69, (1963). . . . .
   *UN EMPLOI POSSIBLE DE FILTRES D≠ABSORPTION EN DIFFRACTION DE NEUTRONS.
   *DOMENICI(M.).

6 C.R. ACAD. SCI. TOME 240,. . . . . . 745, (1955)....B7. . . . . . . .BE,. . . . .
   *MESURE DE L≠ENERGIE DE NEUTRONS TRES LENTS APRES UNE DIFFUSION
   INELASTIQUE PAR DES POLYCRISTAUX ET DES MONOCRISTAUX.
   *JACROT(B.).

7 IAEA SYMP. CHALK RIVER, VOL.1 . 139, (1962). . . . .
   *TRANSMISSION OF THERMAL NEUTRONS THROUGH A LARGE SINGLE CRYSTAL OF
   BISMUTH AT LIQUID NITROGEN TEMPERATURE.
   *MENARDI(S.), HAAS(R.), KLEY(W.).

8 JAPAN. J. APPL. PHYS. VOL.4, . . . 911, (1965)....E. . . . .
   *CHOICE OF COLLIMATORS FOR NEUTRON DIFFRACTION.
   *SAKAMOTO(M.), KUNITOMI(N.), MOTOHASHI(H.), MINAKAWA(N.).

9 J. APPL. CRYST. VOL.2, ... 141, (1969)....E. . . . .
   *THE INFLUENCE OF COLLIMATOR GEOMETRY ON NEUTRON FLUX.
   *SABINE(T.M.), WEINSTOCK(E.V.).

10 J. NUCL. ENERGY VOL.17, ... 227, (1963)....E. . . . .
   *THE PERFORMANCE OF REFLECTING MULTI-CHANNEL COLLIMATORS AS A NEUTRON
   BEAM FILTER AND POLARIZER.
   *MOLLER(H.B.), PASSELL(L.), STECHER-RASMUSSEN(F.).

11 J. SCI. INSTRUM. VOL.31, ... 207, (1954)....T,E. . . . .
   *THE DESIGN OF COLD NEUTRON FILTERS.
   *EGELSTAFF(P.A.), PEASE(R.S.).

12 NUCL. INSTRUM. VOL.3 . . . . 223, (1958)....T. . . . .
   *CHOICE OF COLLIMATORS FOR A CRYSTAL SPECTROMETER FOR NEUTRON DIFFRACTION
   *CAGLIOTI(G.), PAOLETTI(A.), RICCI(F.P.).

13 NUCL. INST. MET. VOL.5, ... 184, (1959)....T. . . . .
   *ON THE OPTIMUM DIMENSIONS OF COLLIMATORS FOR NEUTRON DIFFRACTION.
   *SZABO(P.).

14 NUCL. INST. MET. VOL.6, ... 183, (1960)....T. . . . .
   *ON THE EFFECT OF TOTAL REFLECTION ON THE OPTIMUM DIMENSIONS OF COLLIMA-
   TORS FOR NEUTRON CRYSTAL SPECTROMETERS AND DIFFRACTOMETERS.
   *SZABO(P.).

15 NUCL. INST. MET. 25, ... 367, (1964) . . . .
   *A ≠WINDOW FILTER≠ FOR NEUTRON SPECTROMETRY.
   *IYENGAR(P.K.).

16 NUCL. INST. MET. 37, ... 121, (1965). . . . .
   *REMOVAL OF SECOND ORDER NEUTRONS BY ORIENTED SINGLE CRYSTAL FILTERS.
   *IYENGAR(P.K.), SONI(J.N.), NAVARRO(Q.O.), PINEDA(V.M.)...ET AL.

17 NUCL. INST. MET. 36, ... 179, (1965). . . . .
   *CHOICE OF COLLIMATORS FOR NEUTRON POWDER DIFFRACTOMETRY.
   *POPOVICI(M.).

18 NUCL. INST. MET. 51, ... 121, (1967). . . . .
   *PYROLYTIC GRAPHITE AS A SECOND ORDER NEUTRON FILTER.
   *BERGSMA(J.), VAN-DIJK(C.).

19 NUCL. INST. MET. 47, ... 320, (1967). . . . .
   *DOUBLY BENT TOTAL REFLECTING TUBE AS COLD NEUTRON FILTER.
   *BUZANO(C.), DEMICHELIS(F.), RASETTI(M.).

20 NUCL. INST. MET. 52, ... 15, (1967). . . . .
   *ABLENKUNG LANGSAMER, NEUTRONEN MIT EINEM GEKRUMMTEN,
   TOTALREFLEKTIERENDEN KOLLIMATORSYSTEM.
   *FIALA(W.), RAUCH(H.).

21 NUCL. INST. MET. 60, ... 246, (1968). . . . .
   *A NEUTRON SPECTROMETER EMPLOYING CHARGED PARTICLE COLLIMATION TO IMPROVE
   RESOLUTION.
   *WOLFE(R.A), STUBBINS(W.F.).

22 NUCL. INST. MET. 69, ... 325, (1969). . . . .
   *THE LEAKAGE OF THERMAL NEUTRON THROUGH BERYLLIUM FILTERS.
   *WEBB(F.J.).

23 NUCL. INST. MET. 68, ... 293, (1969). . . . .
   *A MONTE CARLO STUDY OF NEUTRON COLLIMATION BY RECTANGULAR TUBES.
   *CHOUDRY(A.), BANDOPADHYAY(P.K.).

24 NUCL. INST. MET. 69, ... 325, (1969). . . . .
   *THE LEAKAGE OF THERMAL NEUTRONS THROUGH BERYLLIUM FILTERS.
   *WEBB(F.J.).

25 NUCL. INST. MET. VOL.89, ... 109, (1970)....E. . . . . . . .C. (GRAPHITE)
   *PYROLYTIC GRAPHITE AS A HIGH EFFICIENCY FILTER FOR .013-.315 EV NEUTRONS
   *SHIRANE(G.), MINKIEWICZ(V.J.).

26 NUCL. INST. MET. VOL.108, ... 107, (1973)....E. . . . .
   *NEUTRON COLLIMATORS WITH PLATES OF SELF-CONTRACTING FOILS.
   *MEISTER(H.), WECKERMANN(B.).

SECTION 200-ON NEUTRON COLLIMATORS AND FILTERS . . . .

27 PHYS. REV. VOL.54, ... 609, (1938)....E......... . . . .
    *THE PRODUCTION OF COLLIMATED BEAMS OF MONOCHROMATIC NEUTRONS IN THE
    TEMPERATURE RANGE 300-310 DEGREES K.
    *ALVAREZ(L.W.).

28 PHYS. REV. VOL.70, ... 815, (1946)...E......... . . . .
    *PRODUCTION OF LOW ENERGY NEUTRONS BY FILTERING THROUGH GRAPHITE.
    *ANDERSON(H.L.), FERMI(E.), MARSHALL(L.).

29 REPORT RL-73-034 (68 PP.), ... , (1973)....E......... . . . .
    *A DESIGN FOR THERMAL NEUTRON POLARIZATION ANALYSIS DIFFRACTOMETER WITH
    A FILTER CONTAINING/SAMARIUM-149 NUCLEI AS THE SPIN ANALYSER.
    *WILLIAMS(W.G.). (AVAIL. HMSO, LONDON, ENGLAND).

30 REV. SCI. INSTRUM. VOL.30, ... 136, (1959)....E......... . . . .
    *CRYSTAL FILTER TO PRODUCE PURE THERMAL NEUTRON BEAMS FROM REACTORS.
    *BROCKHOUSE(B.N.).

31 REV. SCI. INSTRUM. VOL.33, ... 1399, (1962)....T......... . . . .
    *CALCULATION OF THE OPTIMUM DIMENSIONS OF COLLIMATORS FOR NEUTRON
    DIFFRACTION.
    *JONES(I.R.).

32 REV. SCI. INSTRUM. VOL.33, ... 1103, (1962)....E......... . . . .
    *RUBBER SOLLER SLIT COLLIMATORS FOR NEUTRON SPECTROMETRY.
    *CAGLIOTI(G.), FARFALETTI-CASALI(F.).

33 REV. SCI. INSTRUM. VOL.34, ... 28, (1963)....E......... . . . .
    *SUPPRESSION OF TOTAL REFLECTION OF NEUTRONS FROM COLLIMATOR SURFACES.
    *JONES(I.R.), BARTOLINI(W.).

34 REV. SCI. INSTRUM. VOL.36, ... 48, (1965)....E.........BI,QUARTZ, ...
    *SINGLE-CRYSTAL FILTERS FOR ATTENUATING EPITHERMAL NEUTRONS AND GAMMA
    RAYS IN REACTOR BEAMS.
    *RUSTAD(B.M.), ALS-NIELSEN(J.), BAHNSEN(A.), CHRISTENSEN(C.J.),
    NIELSEN(A.).

35 REV. SCI. INSTRUM. VOL.40, ... 49, (1968)....E......... . . . .
    *SI*O2, MG*O, PB*F2, AND BI AS LOW-PASS NEUTRON VELOCITY FILTERS.
    *HOLMRYD(S.), CONNOR(D.).

36 Z. KRIST. VOL.116, ... 48, (1961)....T......... . . . . . .
    *THE INFLUENCE OF TOTAL REFLECTION AT THE COLLIMATOR LAMELLAE ON NEUTRON
    POWDER DIAGRAMS.
    *HASE(W.), KLEINSTUCK(K.). (IN GERMAN).

SECTION   21 -MECHANICAL ASPECTS OF INSTRUMENTS. . . . .

1 ACTA PHYS. POLON. VOL.18, ... 265, (1959). . . .E. . . . .
            ~A UNIVERSAL DOUBLE-CRYSTAL NEUTRON SPECTROMETER. . . . .
            ~O±CONNOR(D.), BONKOWSKI(L.).

2 ACTA PHYS. HUNGAR. VOL.15, ... 203, (1963). . . .E. . . . .
            ~HIGH INTENSITY NEUTRON DIFFRACTOMETER.
            ~SZABO(P.), KREN(E.), GORDON(J.).

3 ACTA POLYTECH. SCANDIN. PH 38 . . . . (1966). . . .E. . . . .
            ~CONSTRUCTION AND PERFORMANCE OF A COLD NEUTRON SOURCE. . . . .
            ~TUNKELO(E.).

4 AN. REAL SOC. ESPAN/FIS. VOL.58A 119, (1962). . . .E. . . . .
            ~CONSTRUCTION AND SETTING-UP OF A CRYSTAL NEUTRON SPECTROMETER.
            ~PONCE-DE-LEON(J.M.), DIAZ(L.M.). (IN SPANISH).

5 ARCH. SCI. VOL.13, ... 375, (1960). . . .E. . . . .
            ~CONSTRUCTION OF A NEUTRON SPECTROMETER.
            ~COTTIER(J.M.), DENIS(P.), PHILIPPE(M.), ROUX(D.). (IN FRENCH).

6 IAEA SYMPOSIUM VIENNA . . . . 113, (1960). . . . .
            ~METHODS FOR NEUTRON SPECTROMETRY.
            ~BROCKHOUSE(B.N.).

7 IAEA SYMPOSIUM VIENNA . . . . 179, (1960). . . . .
            ~A TWIN-ROTOR NEUTRON SPECTROMETER FOR SOLID-STATE RESEARCH.
            ~DYER(R.F.), LOW(G.G.E.).

8 IAEA SYMPOSIUM VIENNA . . . . 165, (1960). . . . .
            ~A FOUR-ROTOR THERMAL-NEUTRON ANALYSER.
            ~EGELSTAFF(P.A.), COCKING(S.J.), ALEXANDER(T.K.).

9 IAEA SYMP. CHALK RIVER, VOL.1, . . 127, (1962). . . . .
            ~A DOUBLE SLOW NEUTRON SPECTROMETER.
            ~.C.C.C.P.

10 IAEA SYMP. CHALK RIVER, VOL.1, . . 119, (1962). . . . .
            ~INSTALLATION FOR STUDYING THE SCATTERING OF COLD NEUTRONS.
            ~.C.C.C.P.

11 IAEA SYMP. CHALK RIVER, VOL.1, . . 107, (1962). . . . .
            ~A COLD NEUTRON MONOCHROMATOR AND SCATTERING APPARATUS.
            ~HARRIS(D.), COCKING(S.J.), EGELSTAFF(P.A.), WEBB(F.J.).

12 IAEA SYMP. COPENHAGEN, VOL.2, . . 429, (1968). . . . .
            ~A NEW HIGH-EFFICIENCY TIME-OF-FLIGHT SYSTEM.
            ~COLWELL(J.F.), MILLER(P.H.), WHITTEMORE(W.L.).

13 IAEA SYMP. COPENHAGEN, VOL.2, . . 417, (1968). . . . .
            ~THE USE OF PSEUDO-STATISTICAL CHOPPER FOR TIME-OF-FLIGHT MEASUREMENTS.
            ~GOMPF(F.), REICHARDT(W.), GLASER(W.), BECKURTS(K.H.).

14 IAEA SYMP. COPENHAGEN, VOL.2, . . 407, (1968). . . . .
            ~CORRELATION-TYPE TIME-OF-FLIGHT SPECTROMETER WITH MAGNETICALLY CHOPPED
            POLARIZED NEUTRON BEAM.
            ~PAL(L.), KROO(N.), GORDON(G.), PELLIONISZ(P.), SZLAVIK(F.), VIZI(I.).

15 IAEA SYMP. COPENHAGEN, VOL.2, . . 313, (1968). . . . .
            ~THE SPECTROMETER FOR THERMAL-NEUTRON INELASTIC SCATTERING STUDIES AT THE
            IBR PULSED REACTOR.
            ~MALISZEWSKI(E.), NITC(V.V.), SOSNOWSKA(I.), SOSNOWSKI(J.).

16 IAEA SYMP. COPENHAGEN, VOL.2, . . 395, (1968). . . . .
            ~TIME-OF-FLIGHT SPECTROMETER USING AN ELECTRONIC CHOPPER FOR POLARIZED
            SLOW NEUTRONS.
            ~STEINSVOLL(O.), VIRJO(A.).

17 INSTRUM. EXP. TECH. NO.6, ... 1176, (1964). . . .E. . . . .
            ~CENTER FOR NEUTRON-SPECTROMETER MEASUREMENTS.
            ~ZHUKOV(G.P.), ZHURAVLEV(B.E.), ZABIYAKIN(G.I.), ZAMRII(V.N.).

18 JAP. J. APPL. PHYS. VOL.10, . . . 933, (1971). . . .E. . . . .
            ~A NEW CRYOSTAT FOR NEUTRON DIFFRACTION EXPERIMENTS.
            ~SHIMAOKA(K.), NIIMURA(N.), HOSHINO(S.).

19 JAP. J. APPL. PHYS. VOL.11, . . . 82, (1972). . . .E. . . . .
            ~A FACILE CRYOSTAT FOR NEUTRON DIFFRACTION EXPERIMENTS.
            ~HIRAKAWA(K.), IKEDA(H.).

20 J. APPL. CRYST. VOL.2, ... 109, (1969). . . .E. . . . .
            ~AN AUTOMATIC LOW-TEMPERATURE APPARATUS FOR SINGLE CRYSTAL DIFFRACTOMETRY
            ~RUDMAN(R.), GODEL(J.B.).

21 J. APPL. CRYST. VOL.3, ... 289, (1970). . . .P. . . . .
            ~NEUTRON DIFFRACTION CRYO-ORIENTER.
            ~HEATON(L.), MUELLER(M.H.), ADAM(M.F.), HITTERMAN(R.L.).

22 J. APPL. CRYST. VOL.4, ... 254, (1971). . . . .
            ~A SELF-REGULATING MINI-COOLER FOR THREE-DIMENSIONAL SINGLE-CRYSTAL
            NEUTRON-DIFFRACTION MEASUREMENTS.
            ~EBDON(F.R.), WHEELER(D.A.).

23 J. APPL. CRYST. VOL.6, ... 42, (1973). . . .E. . . . .
            ~THREE-DIMENSIONAL VACUUM FURNACE FOR NEUTRON DIFFRACTION FROM FERRO-
            ELECTRIC CRYSTALS.
            ~HEWAT(A.W.).

24 J. APP. PHYS. 39, ... 447, (1968). . . . .
            ~A MAGNETICALLY PULSED NEUTRON BEAM FOR TIME OF FLIGHT MEASUREMENTS.
            ~MOOK(H.A.), WILKINSON(M.K.).

25 J. PHYS. E VOL.4 NO.11, ... 905, (1971). . . .E. . . . .
            ~CONSTRUCTION AND PERFORMANCE OF THE LIQUID METHANE COLD SOURCE.
            ~DIMIC(V.), PETKOVSEK(J.).

26 J. PHYS. E VOL.6 NO.6, ... 576, (1973). . . .E. . . . .
            ~A FURNACE OF HIGH ACCURACY AND VERSATILITY FOR NEUTRON SCATTERING
            EXPERIMENTS ON SOLIDS AND LIQUIDS UP TO 1100 K.
            ~KUNSCH(B.), EDER(O.J.), SCHWARA(R.).

SECTION 21 -MECHANICAL ASPECTS OF INSTRUMENTS. . . . .

27 J. PHYS. E VOL.6 NO.1, 51, (1973) . . . . . E . . . .
    *A FACILITY FOR NEUTRON SCATTERING EXPERIMENTS.
    *ALSTON(W.C.H.), JONES(T.C.).

28 J. PHYSIQUE TOME 24, 15A, (1963) . . . E . . . . CO-FE. . . . . .
    *HIGH PRECISION CUTTING OF CO-FE SINGLE CRYSTALS FOR USE IN A NEUTRON
    SPECTROMETER.
    *BEIL(H.), CARLOS(P.), GENIN(R.).

29 J. SCI. INSTRUM. VOL.27, 330, (1950) . . . . . E . . . .
    *SOME MECHANICAL FEATURES OF A DOUBLE CRYSTAL NEUTRON SPECTROMETER.
    *BACON(G.E.), SMITH(J.A.G.), WHITEHEAD(C.D.).

30 J. SCI. INSTRUM. VOL.40, 14, (1963) . . . . . E . . . .
    *GEOMETRICAL FACTORS INFLUENCING THE DESIGN OF AUTOMATIC SINGLE-CRYSTAL
    X-RAY AND NEUTRON DIFFRACTOMETERS.
    *WOOSTER(W.A.).

31 KERNTECHNIK VOL.15, 450, (1973) . . . . E . . . .
    *CONSTRUCTION OF A NEUTRON SMALL ANGLE SCATTERING FACILITY FOR THE
    GRENOBLE HIGH FLUX REACTOR.
    *DEGENKOLBE(G.), GREISS(H.B.).

32 MITSUB. DENKI LAB. REP. VOL.3, 111, (1962) . . . . . E . . . . . .
    *THREE NEUTRON DIFFRACTOMETERS.
    *HAGIHARA(S.), MIYASHITA(K.), YOSHIE(T.), OHNO(E.), MOGI(M.).

33 NUCL. INST. MET. 35, 13, (1965) . . . .
    *MULTIPURPOSE NEUTRON DIFFRACTION INSTRUMENTATIONS.
    *ATOJI(M.).

34 NUCL. INST. MET. 56, 305, (1967) . . . .
    *A SPECIMEN CONTAINER FOR INELASTIC NEUTRON SCATTERING STUDIES OF LIQUIDS
    AT HIGH PRESSURES.
    *SKOLD(K.), KARLEN(L.), SCIESINSKI(J.).

35 NUCL. INST. MET. 50, 251, (1967) . . . .
    *ROTATING CRYSTAL SPECTROMETERS WITH AND WITHOUT AN AUXILIARY DEVICE FOR
    SUPPRESION OF UNWANTED REFLECTIONS.
    *QUITTNER(G.), LEGLER(E.).

36 NUCL. INST. MET. 63, 114, (1968) . . . .
    *A MECHANICAL CORRELATION CHOPPER FOR THERMAL NEUTRON SPECTROSCOPY.
    *SKOLD(K.).

37 NUCL. INST. MET. 64, 52, (1968) . . . .
    *A CONDITIONER WITH INTEGRATED CIRCUITS FOR NEUTRON TIME OF FLIGHT
    EXPERIMENT DATA SELECTION.
    *COLLING(F.), STUBER(W.).

38 NUCL. INST. MET. 65, 125, (1968) . . . .
    *BERYLLIUM FILTERED NEUTRON BEAM QUALITY IMPROVEMENT AT A POOL REACTOR
    THROUGH CORE ELEMENT REARRANGEMENTS.
    *ANTAL(J.J.), WARNAS(A.A,).

39 NUCL. INST. MET. 58, 261, (1968) . . . .
    *EIN NEUTRONEN-SPIN-FLIP-CHOPPER.
    *RAUCH(H.), HARMS(J.), MOLDASCHL(H.).

40 NUKLEONIKA VOL.12, 385, (1967) . . . . E . . . .
    *THERMOSTAT FOR NEUTRON CROSS-SECTION MEASUREMENTS WITHIN THE RANGE
    FROM -196 DEGREES C TO 200 DEGREES C.
    *SUDNIK-HRYNKIEWICZ(M.), OLEJARCZYK(W.), ZIELENIEWSKI(R.), SZKATULA(A.),
    JANIK(J.), BAJOREK(A.).

41 PHYS. REV. 147, 457, (1966) . . . .
    *APPLICATION OF INVERTED-NEUTRON-FILTER TECHNIQUE FOR STUDY OF INELASTIC
    SCATTERING BY PARAMAGNETIC SUBSTANCES.
    *FRIEDMAN(E.A), GOLAND(A.N.).

42 REPORT NP/R-2551 (26 PP.), (1957) . . . . E . . . .
    *CONSTRUCTION AND OPERATION OF A SINGLE-CRYSTAL NEUTRON SPECTROMETER.
    *PATTENDEN(N.J.), BASTON(A.H.). (AVAIL. AT. ENERGY RES. ESTABLISHMENT,
    HARWELL,ENGLAND).

43 REV. SCI. INSTRUM. VOL.32, 456, (1961) . . . . E . . . .
    *STEPPING MECHANISM FOR X-RAY AND NEUTRON DIFFRACTOMETERS AND SPEC-
    TROMETERS.
    *MUELLER(M.H.), HEATON(L.), JOHANSON(E.W.).

44 REV. SCI. INSTRUM. VOL.34, 113, (1963) . . . . E . . . .
    *GONIOMETER-MOUNTED EVACUATED FURNACE FOR SINGLE CRYSTAL NEUTRON
    DIFFRACTOMETRY.
    *ABRAHAMS(S.C.).

45 REV. SCI. INSTRUM. VOL.37, 1543, (1966) . . . .
    *HIGH PRESSURE VESSEL FOR MEASUREMENT OF PRESSURE INDUCED PHONON ENERGY
    SHIFTS BY INELASTIC SCATTERING OF NEUTRONS.

46 REV. SCI. INSTRUM. VOL.37, 697, (1966) . . . .
    *PHASE ROTATING CRYSTAL AND CHOPPER FOR TIME-OF-FLIGHT NEUTRON
    SPECTROSCOPY.

47 REV. SCI. INSTRUM. VOL.40, 249, (1968) . . . .
    *NEW CRYSTAL SPECTROMETER FOR NEUTRONS.
    *STEDMAN(R.), ALMQVIST(L.), RAUNIO(G.), NILSSON(G.).

48 REV. SCI. INSTRUM. 41, 61, (1970) . . . .
    *LOW TEMPERATURE CRYSTAL GROWTH CRYOSTAT FOR USE IN OPTICAL AND INELASTIC
    NEUTRON STUDIES OF RARE GAS SOLIDS.
    *DYER(R.H.), LEFKOWITZ(I.), MARTEL(P.).

49 REV. SCI. INSTRUM. VOL.45, 643, (1974) . . . . E . . . .
    *APPARATUS FOR NEUTRON DIFFRACTION AT HIGH PRESSURE.
    *MCWHAN(D.B.), BLOCH(D.), PARISOT(G.).

50 SOV. PHYS. CRYST. VOL.4, 623, (1960) . . . . E . . . . . . PB. . .
    *CONSTRUCTION OF A LEAD CRYSTAL MONOCHROMATOR FOR NEUTRON DIFFRACTION
    STUDIES.
    *LITVIN(D.F.).

51 SOV. PHYS. SOL. STATE 7, 1146, (1965) . . . .
    *HYBRID NEUTRON SPECTROMETER SYSTEM.
    *KOTOV(B.A.), OKUNEVA(N.M.), SHAKH-BUDAGOV(A.L.).

SECTION 21 -MECHANICAL ASPECTS OF INSTRUMENTS. . . . .

52 SOV. PHYS. CRYST. VOL.12, ... 477, (1967)....E....... . . . .
                *HELIUM CRYOSTAT FOR A NEUTRON DIFFRACTOMETER.
                *GOLOVKIN(V.S.), BYKOV(V.N.), KHOTEEV(N.V.), LEVDIK(V.A.).

53 SOV. PHYS. CRYST. VOL.12, ... 978, (1968)....E....... . . . .
                *A SMALL CRYSTAL NEUTRON SPECTROMETER FOR STUDIES ON CRYSTAL DYNAMICS.
                *GORBACHEV(V.V.), PLATONOV(V.I.).

54 THESIS (98 PP.), ... (1966)....E....... AND CONSTRUCTION OF A TRIPLE-AXIS
                *NEUTRON SPECTROSCOPY: THE DESIGN AND CONSTRUCTION OF A TRIPLE-AXIS
                *CRYSTAL SPECTROMETER; AND A STUDY OF THE LATTICE DYNAMICS OF METALS.
                *ROWE(J.M.).   (ED. NOTE: PH.D. DISSERTATION, MCMASTER UNIVERSITY).

55 TH. NEUTRON SCATT./EGELSTAFF . . . 53, (1965) . . . .
                *MECHANICAL AND TIME-OF-FLIGHT TECHNIQUE.
                *BRUGGER(R.M.).

SECTION 22A-GENERAL INSTRUMENTATION: DIFFRACTOMETERS/SPECTROMETERS . . .

1   ACTA CRYST. VOL.13,   763, (1960)....I.........
    *SOME THEORETICAL PROPERTIES OF THE DOUBLE-CRYSTAL SPECTROMETER USED IN
    NEUTRON DIFFRACTION.
    *WILLIS(B.T.M.).

2   ACTA CRYST. VOL.A28 PART S-4,.   250, (1972)....E.........
    *SINGLE CRYSTAL NEUTRON DIFFRACTOMETER WITH 100 MOVABLE DETECTORS.
    *KLAR(B.). (ABSTRACT ONLY).

3   ACTA PHYS. POLON. VOL.19,. . .   329, (1960)....E.........
    *MEASUREMENT OF THE SLOW NEUTRON SPECTRUM OF A NEUTRON BEAM FROM THE
    W.W.R.S. REACTOR BY MEANS OF A CRYSTAL NEUTRON SPECTROMETER.
    *O#CONNOR(D.A.), SOSNOWSKI(J.).

4   ACTA PHYS. SINICA VOL.17,.   222, (1961)....E.........
    *THE CONSTRUCTION AND CHARACTERISTICS OF A NEUTRON DIFFRACTOMETER.
    *WANG-SHOU(A.)...ET AL. (IN CHINESE).

5   ACTA PHYS. HUNGAR. VOL.15,.   203, (1963)....E.........
    *HIGH INTENSITY NEUTRON DIFFRACTOMETER.
    *SZABO(P.), KREN(E.), GORDON(J.).

6   BULL. INST. NUCL. SCI.(YUGO.) 11   59, (1961)....E.........
    *DOUBLE-CRYSTAL NEUTRON SPECTROMETER.
    *ZIVANOVIC(M.O.), JOVIC(D.M.), KONSTANTINOVIC(J.M.).

7   CAN. JOURN. PHYS. 42,. . .   1593, (1964)....E.........
    *A NEW THERMAL NEUTRON FLUX DENSITY STANDARD.
    *HARGROVE(C.K.), GEIGER(K.W.).

8   CAN. JOURN. PHYS. 42,.   2443, (1964)....E.........
    *EXPERIMENTAL INVESTIGATION OF THE NM-64 NEUTRON MONITOR.
    *HATTON (C.J.), CARMICHAEL(H.).

9   CHIN. J. PHYS. (TAIWAN) VOL.8,.   31, (1970)....E.........
    *A SIMPLE NEUTRON SPECTROMETER.
    *LIU(Y.C.), TZENG(H.S.), CHENG(V.K.C.), HSU(W.S.).

10 CZECH. J. PHYS. B VOL.16,.   942, (1966)....E.........
    *SECOND-ORDER CONTAMINATION IN A NEUTRON SPECTROMETER WITH CO-FE CRYSTAL.
    *CHALUPA(B.), MICHALEC(R.), BISCHOF(J.), GALOCIOVA(D.).

11 IAEA SYMP. (PILE RESEARCH)VIENNA   597, (1960)....E.........
    *A CURVED-CRYSTAL NEUTRON SPECTROMETER.
    *PATTENDEN(N.J.), SAUNDERSON(D.H.), HAY(H.J.).

12 IAEA SYMP. BOMBAY, VOL.2,. . .   487, (1964)....E.........
    *SLOW NEUTRON SPECTROMETERS AT THE SWEDISH REACTORS.
    *DAHLBORG(U.), SKOLD(K.), LARSSON(K.E.).

13 IAEA SYMP. COPENHAGEN, VOL.2,. .   381, (1968)....E.........
    *A NEUTRON CRYSTAL SPECTROMETER WITH EXTREMELY HIGH ENERGY RESOLUTION.
    *ALEFELD(B.), BIRR(H.), HEIDEMANN(A.).

14 IAEA SYMP. GRENOBLE,.   733, (1972)....E.........
    *A MULTI-ANGLE REFLECTING CRYSTAL SPECTROMETER.
    *KJEMS(J.K.), REYNOLDS(P.A.).

15 IAEA SYMP. GRENOBLE,.   713, (1972)....E.........
    *A HIGH RESOLUTION TIME FOCUSING SPECTROMETER FOR QUASI-ELASTIC
    NEUTRON SCATTERING.
    *MEISTER(H.), WECKERMANN(B.).

16 IEEE TRANS. ELECTRON DEVICES 17,   236, (1970)....E.........
    *A NEW ACOUSTIC PHONON SPECTROMETER.
    *ANDERSON(C.H.).

17 INSTRUM. EXP. TECH. NO.4,.   533, (1959)....E.........
    *A NEUTRON SPECTROMETER WITH A #PULSED# CYCLOTRON BEAM.
    *IGNAT#EV(K.G.)...ET AL.

18 INSTRUM. EXP. TECH. NO.6,. . .   1176, (1964)....E.........
    *CENTER FOR NEUTRON-SPECTROMETER MEASUREMENTS.
    *ZHUKOV(G.P.), ZHURAVLEV(B.E.), ZABIYAKIN(G.I.), ZAMRII(V.N.).

19 INSTRUM. EXP. TECH. NO.2,.   306, (1964)....E.........
    *A UNIVERSAL APPARATUS FOR STRUCTURE STUDY BY NEUTRON DIFFRACTION.
    *ABESADZE(P.D.), DOIDZHASHVILI(G.I.), LITVIN(D.F.), LYASHCHENKO(B.G.),
    PROTOPOPOV(N.N.), CHIKOBAVA(V.S.),.

20 INSTRUM. EXPER. TECH. NO.6,.   1294, (1967)....E.........
    *TWO-CRYSTAL NEUTRON DIFFRACTOMETER WITH A MOVABLE MONOCHROMATOR.
    *KHRUSHCHEV(B.I.), SHARIPOVA(L.S.).

21 INSTRUM. EXP. TECH. NO.4,.   808, (1967)....E.........
    *A TWO-AXIS CRYSTAL NEUTRON SPECTROMETER.
    *CHIKOBAVA(V.S.)...ET AL.

22 INSTRUM. EXP. TECH. 3,.   726, (1970)....E.........
    *TWO-CRYSTAL NEUTRON DIFFRACTOMETER.
    *KHRUSHCHEV(B.I.), IGAMBERDIEV(SH.G.).

23 J. KOREAN PHYS. SOC. VOL.4,.   47, (1971)....E.........
    *FIXED ANGLE NEUTRON DIFFRACTOMETER.
    *YUN-PEEL-LEE.

24 J. NUCL. ENERGY VOL.1,.   57, (1954)....E.........
    *THE OPERATION OF A THERMAL NEUTRON TIME-OF-FLIGHT SPECTROMETER.
    *EGELSTAFF(P.A.).

25 J. PHYS. E VOL.6 NO.6,.   568, (1973)....E.........
    *A SIMPLE NEUTRON GUIDE TUBE AND DIFFRACTOMETER FOR SMALL-ANGLE
    SCATTERING OF COLD NEUTRONS.
    *HAYWOOD(B.C.G.), WORCESTER(D.L.).

26 J. PHYS. E VOL.6 NO.1,.   51, (1973)....E.........
    *A FACILITY FOR NEUTRON SCATTERING EXPERIMENTS.
    *ALSTON(W.C.H.), JONES(T.C.).

SECTION 22A-GENERAL INSTRUMENTATION: DIFFRACTOMETERS/SPECTROMETERS . . .

27 J. PHYSIQUE TOME 24,   89A (1963).....E..............
    ~CRYSTAL SPECTROMETER FOR POLARIZED NEUTRONS.
    ~BEIL(H.), CARLOS(P.), GENIN(R.), SIGNARBIEUX(C.), JOLY(R.), MATUSZEK(J.)

28 J. SCI. INSTRUM. VOL.32,   256, (1955).....E..............
    ~A SPECTROMETER FOR SINGLE CRYSTAL NEUTRON DIFFRACTION.
    ~BACON(G.E.), DYER(R.F.).

29 KERNTECHNIK (GER.) VOL.8,   560, (1966).....E..............
    ~A MULTIPURPOSE NEUTRON DIFFRACTION APPARATUS.
    ~KONIG(U.).

30 KERNTECHNIK VOL.14,   20, (1972).....E..............
    ~SPECTROMETER FOR MEASURING DIFFUSE NEUTRON SCATTERING.
    ~BAUER(G.).

31 KERNTECHNIK VOL.14,   17, (1972).....E..............
    ~HIGH RESOLUTION NEUTRON SPECTROMETER.
    ~SCHMIDT(H.H.).

32 KERNTECHNIK VOL.14,   15, (1972).....E..............
    ~NEUTRON BACK-SCATTERING SPECTROMETER.
    ~ALEFELD(B.).

33 LATV. PSR...FIZ. TEHN. SER. NO.3   64, (1970)....E...............
    ~A CHOICE OF THE METHOD OF MEASUREMENT OF THE INTEGRAL INTENSITY IN
    NEUTRON DIFFRACTOMETERS WITH THE EQUATORIAL SURVEY SCHEME.
    ~NOZIK(YU.Z.), KHEIER(D.M.).

34 NUCL. INST. MET. VOL.9,   195, (1960).....T..............
    ~ON RESOLUTION AND LUMINOSITY OF A NEUTRON DIFFRACTION SPECTROMETER FOR
    SINGLE CRYSTAL ANALYSIS.
    ~CAGLIOTI(G.), PAOLETTI(A.), RICCI(F.P.).

35 NUCL. INST. MET. VOL.15,   155, (1962)....T.E.............
    ~RESOLUTION AND LUMINOSITY OF CRYSTAL SPECTROMETERS FOR NEUTRON
    DIFFRACTION.
    ~CAGLIOTI(G.), RICCI(F.P.).

36 NUCL. INST. MET. 26,   125, (1964)..............
    ~SOME CONSIDERATIONS ON THE PROPER USE OF THE POLARIZED NEUTRON
    SPECTROMETER.
    ~PAOLETTI(A.), RICCI(F.P.).

37 NUCL. INST. MET. 40,   153, (1966)..............
    ~MEASUREMENT OF THE SLOW NEUTRON SPECTRUM OF A NEUTRON BEAM FROM THE
    UA-RR-1 REACTOR BY TIME OF FLIGHT SPECTROMETRY.
    ~HAMOUDA(I.), MAAYOUF(R.), STUPAK(A.)...ET AL.

38 NUCL. INST. MET. 62,   19, (1968)..............
    ~A NEUTRON SPECTROMETER USING PULSE SHAPE DISCRIMINATION.
    ~JONES(D.W.).

39 NUCL. INST. MET. 63,   157, (1968)..............
    ~A DOUBLE SCATTER NEUTRON SPECTROMETER.
    ~LEGGE(G.J.F.), VAN-DER-MERWE(P.).

40 NUCL. INST. MET. 63,   13, (1968)..............
    ~CERTAIN ASPECTS OF THE CALIBRATION AND RESOLUTION OF SLOW NEUTRON
    SPECTROMETERS.
    ~AMARAL(L.Q.), VINHAS(L.A.), RODRIGUEZ(C.), HERDADE(S.B.).

41 NUCL. INST. MET. 81,   317, (1970)..............
    ~THE THEORY OF POLARIZATION SPECTROMETERS FOR THERMAL NEUTRONS.
    ~MISENTA(R.).

42 NUCL. INST. MET. 95,   435, (1971)..............
    ~A NEUTRON CRYSTAL SPECTROMETER WITH EXTREMELY HIGH ENERGY RESOLUTION.
    ~BIRR(M.), HEIDEMANN(A.), ALEFELD(B.).

43 NUCL. INST. MET. VOL.108,   401, (1973).....E..............
    ~TEMPERATURE COEFFICIENTS OF A NE 213 NEUTRON SPECTROMETER.
    ~WEINERT(M.).

44 NUCL. INST. MET. VOL.108,   189, (1973).....E..............
    ~A SIMPLE NEUTRON DIFFRACTOMETER FOR LOW ANGLE BIOLOGICAL STUDIES.
    ~NUNES(A.C.).

45 NUCL. INST. MET. VOL.106,   453, (1973)....T.E.............
    ~THE RESOLUTION FUNCTION OF A HYBRID NEUTRON SPECTROMETER.
    ~STEINSVOLL(O.).

46 NUCL. INST. MET. VOL.106,   221, (1973)....E..............
    ~A NEW *HYBRID* SPECTROMETER FOR INELASTIC THERMAL NEUTRON SCATTERING.
    ~KLEB(R.), OSTROWSKI(G.E.), PRICE(D.L.), ROWE(J.M.).

47 NUCL. INST. MET. VOL.113,   15, (1973)....E.T........SI..BE.
    ~PHONON MEASUREMENTS ON A FILTER-DETECTOR SPECTROMETER AND ITS RESOLUTION
    FUNCTION.
    ~THAPER(C.L.), DASANNACHARYA(B.A.), IYENGAR(P.K.).

48 NUCL. INST. MET. VOL.114,   417, (1974)....E..............
    ~A MECHANICAL TIME-OF-FLIGHT NEUTRON DIFFRACTOMETER.
    ~CARPENTER(J.M.),   MILDNER(D.F.R.),   PELIZZARI(C.A.),   SUTTON(J.D.),
    GUNNING(J.E.).

49 NUCL. INST. MET. VOL.114,   21, (1974)....E..............
    ~A COLD-NEUTRON SPECTROMETER FOR INELASTIC-SCATTERING STUDIES.
    ~CASTAGNO(G.),   COPPO(M.),   DEMICHELIS(F.),   SAITTA(L.),   TARTAGLIA(A.).

50 NUCL. SCI. ENGNG. VOL.8,   453, (1960)....P........AU..B.
    ~LONG WAVELENGTH CRYSTAL SPECTROMETER AND THE NEUTRON ABSORPTION CROSS
    SECTIONS OF GOLD AND BORON.
    ~GOULD(F.T.), TAYLOR(T.I.), HAVENS-JR(W.W.), RUSTAD(B.M.), MELKONIAN(E.).

51 NUCL. TECHNOL. VOL.10,   215, (1971)....E..............
    ~A SIMPLE TOTAL-REFLECTING LOW-ENERGY NEUTRON SPECTROMETER.
    ~FIALA(W.), WHEELER(C.V.).

52 PHYS. LETT. VOL.44A NO.3,   165, (1973)....E..............
    ~A HIGH RESOLUTION NEUTRON TIME OF FLIGHT DIFFRACTOMETER.
    ~STEICHELE(E.), ARNOLD(P.).

53 PHYS. LETT. VOL.47A NO.5,   369, (1974)....E..............
    ~TEST OF A SINGLE CRYSTAL NEUTRON INTERFEROMETER.
    ~RAUCH(H.), TREIMER(W.), BONSE(U.).

SECTION 22A-GENERAL INSTRUMENTATION: DIFFRACTOMETERS/SPECTROMETERS . . .

54 PHYS. REV. VOL.70, 136, (1946).....E......
 *NEUTRON BEAM SPECTROMETER STUDIES OF BORON, CADMIUM, AND THE ENERGY
 DISTRIBUTION FROM PARAFFIN.
 *RAINWATER(L.J.), HAVENS-JR(W.W.).

55 PHYS. REV. VOL.71, 65, (1947).....E......
 *SLOW NEUTRON VELOCITY SPECTROMETER STUDIES I. CO, AG, SB, IR, MN.
 *RAINWATER(L.J.), HAVENS-JR(W.W.), WU(C.S.), DUNNING(J.R.).

56 PHYS. REV. VOL.72, 109, (1947).....E......
 *A BENT CRYSTAL NEUTRON SPECTROMETER AND ITS APPLICATION TO NEUTRON
 CROSS-SECTION MEASUREMENTS.
 *SAWYER(R.B.), WOLLAN(E.O.), BERNSTEIN(S.), PETERSON(K.C.).

57 PRIBORY TEKH. EKSPER. NO.3, 26, (1959).....E......
 *A PLANE CRYSTAL NEUTRON SPECTROMETER.
 *KONAKHOVICH(YU.YA.), PANASYUK(I.S.). (IN RUSSIAN).

58 REPORT 132-67-02, (1967).....E......
 *THE ROTATING-CRYSTAL SPECTROMETER FOR INELASTIC SCATTERING EXPERIMENTS
 WITH COLD NEUTRONS AT THE HOR.
 *DE-GRAAF(L.A.). (ED. NOTE: AVAIL. REACTOR INSTITUUT, DELFT, NETHERLANDS)

59 REPORT NASA-TM-X-52828 (11PP.), (1970).....E......
 *A SMALL DIFFERENTIABLE LIQUID SCINTILLATOR NEUTRON SPECTROMETER.
 *SHOOK(D.). (ED. NOTE: AVAIL. CFSTI, SPRINGFIELD, VA. 22151, USA).

60 REV. ROUMAINE PHYS. VOL.10, 1035, (1965).....E......
 *SECONDARY EXTINCTION AND OPTICAL PROPERTIES OF NEUTRON CRYSTAL
 SPECTROMETERS.
 *POPOVICI(M.), GELBERG(D.). (IN FRENCH).

61 REV. SCI. INSTRUM. VOL.27, 26, (1956).....E......
 *HIGH RESOLUTION CRYSTAL SPECTROMETER FOR NEUTRONS.
 *SAILOR(V.L.), FOOTE-JR(H.L.), LANDON(H.H.), WOOD(R.E.).

62 REV. SCI. INSTRUM. VOL.31, 481, (1960).....E......
 *NEVIS SYNCHROCYCLOTRON SLOW NEUTRON VELOCITY SPECTROMETER.
 *RAINWATER(L.J.), HAVENS-JR(W.W.), DESJARDINS(J.S.), ROSEN(J.L.).

63 REV. SCI. INSTRUM. VOL.34, 224, (1963).....E......
 *AUTOMATIC NEUTRON DIFFRACTOMETER FOR THREE-DIMENSIONAL STRUCTURE-
 FACTOR DETERMINATION.
 *ARNDT(U.W.), WILLIS(B.T.M.).

64 REV. SCI. INSTRUM. VOL.36, 887, (1965).....E......
 *NEUTRON CRYSTAL SPECTROMETER WITH RANGE EXTENDED TO SUBTHERMAL ENERGIES.
 *RUSTAD(B.M.), MELKONIAN(E.), HAVENS-JR(W.W.), TAYLOR(T.I.), GOULD(F.T.),
 MOORE(J.A.).

65 REV. SCI. INSTRUM. VOL.44, 1594, (1973).....E......
 *A PRECISE REFRACTOMETER FOR THERMAL NEUTRONS.
 *SCHNEIDER(C.S.).

66 SOV. PHYS. CRYST. VOL.2, 626, (1957).....E......
 *A TWO-CRYSTAL NEUTRON SPECTROMETER.
 *BYKOV(V.N.), VINOGRADOV(S.I.), LEVDIK(V.A.), GOLOVKIN(V.S.).

67 SOV. PHYS. CRYST. VOL.5, 294, (1960).....E......
 *A REMOTELY CONTROLLED NEUTRON DIFFRACTOMETER BASED ON A GUR-3 UNIT.
 *OZEROV(R.P.), KISELEV(S.V.), KARPOVICH(I.R.), GOMAN≠KOV(V.I.),
 LOSHMANOV(A.A.).

68 SOV. PHYS. CRYST. VOL.7, 58, (1962).....E......
 *A SMALL NEUTRON DIFFRACTOMETER.
 *YAMZIN(I.I.), STARITSYN(V.E.), NOZIK(YU.Z.).

69 SOV. PHYS. CRYST. VOL.8, 234, (1963).....E......
 *A NEUTRON DIFFRACTOMETER.
 *YAMZIN(I.I.), KUZ≠MINOV(YU.S.), STARITSYN(V.E.), MAL≠TSEV(E.I.).

70 SOV. PHYS. CRYST. VOL.9, 797, (1965).....E......
 *A TWO-COORDINATE NEUTRON DIFFRACTOMETER.
 *YAMZIN(I.I.), SIZOV(R.A.).

71 UKAEA AERE TRANS 1080 (15 PP.), (1967).....E......
 *A HIGH-RESOLUTION SPECTROMETER FOR VERY SLOW NEUTRONS FOR USE WITH THE
 COLD SOURCE OF A HIGH FLUX REACTOR.
 *SCHERM(R.).

72 Z. ANGEW. PHYS. VOL.14, 738, (1962).....T,E......
 *THEORY OF A GRAVITY NEUTRON REFRACTOMETER FOR THE ABSOLUTE DETERMINATION
 OF COHERENT SCATTERING CROSS-SECTIONS.
 *MAIER-LEIBNITZ(H.).

73 Z. ANGEW. MATH. PHYS. VOL.16, 817, (1965).....E......
 *A DOUBLE AXIS NEUTRON DIFFRACTOMETER AND CRITERIA FOR OPTIMAL ADJUSTMENT
 FOR POWDER INVESTIGATIONS.
 *STOLL(E.), HALG(W.).

74 Z. PHYSIK VOL.167, 386, (1962).....E......
 *INTERFEROMETER FOR SLOW NEUTRONS.
 *MAIER-LEIBNITZ(H.), SPRINGER(T.). (IN GERMAN).

SECTION 228-GENERAL INSTRUMENTATION: AUTOMATION AND ELECTRONICS. . . . .

1 ACTA CRYST. 17,. . . . . 1190, (1964)............................
       *EVALUATION OF DIGITAL AUTOMATIC DIFFRACTOMETER SYSTEMS.
       *ABRAHAMS(S.C.).

2 ACTA CRYST. 17,. . . . . 1183, (1964)....................
       *ANALOGUE AND DIGITAL SINGLE CRYSTAL DIFFRACTOMETERS.
       *ARNDT(U.W.).

3 ACTA CRYST. A24,. . . . . 136, (1968)...................
       *OPTIMIZATION OF COUNTING TIMES IN COMPUTER CONTROLLED X-RAY AND NEUTRON
       SINGLE-CRYSTAL DIFFRACTOMETRY.
       *SHOEMAKER(D.P.).

4 ACTA PHYS. POLON. VOL.16,. . . 293, (1957)....E............
       *AN AUTOMATIC TIMING AND RECORDING CIRCUIT FOR USE WITH A NEUTRON
       CRYSTAL SPECTROMETER.
       *O#CONNOR(D.), BLINOWSKI(K.).

5 ACTA PHYS. AUSTRIACA VOL.20,. . 166, (1965).....E..........
       *M.A.N. NEUTRON DIFFRACTION INSTALLATION, SEIBERSDORF. AUTOMATIC CONTROL
       AND PROGRAMMING.
       *MAY(F.). (IN GERMAN).

6 AUTOMATIC/NUCLEAR DATA. . . . . . 233, (1964)....E.........
       *A PROGRAMMING AND AUTOMATIC DATA COLLECTING SYSTEM FOR A NEUTRON
       DIFFRACTION SPECTROMETER.
       *HALG(W.), MEIER(F.), GASSER(F.). (ED. NOTE: AUTOMATIC ACQUISITION AND
       REDUCTION OF NUCLEAR DATA: PUBL. GES. FUR KERNFORSCHUNG: KARLSRUHE).

7 AUTOMATIC/NUCLEAR DATA. . . . . . 83, (1964)....E.........
       *GENERAL SURVEY: MAGNETIC TAPE DATA RECORDING SYSTEMS.
       *EGELSTAFF(P.A.), RAE(E.R.). (ED. NOTE: AUTOMATIC ACQUISITION AND
       REDUCTION OF NUCLEAR DATA: PUBL. GES. FUER KERNFORSCHUNG: KARLSRUHE).

8 IAEA SYMP. COPENHAGEN, VOL.2,. . 253, (1968)...............
       *ON-LINE COMPUTER-CONTROLLED TRIPLE-AXIS NEUTRON SPECTROMETERS AT THE
       HFIR.
       *WILKINSON(M.K.), SMITH(H.G.), KOEHLER(W.C.), NICKLOW(R.M.), MOON(R.M.).

9 IAEA SYMP. (INSTRUMENT.) VIENNA 281, (1969)................
       *DISCUSSION ON COMPUTERS AND DATA HANDLING.

10 IEEE TRANS. NUCL. SCI. VOL.NS-13 311, (1966)....E......
       *SUBMICROSECOND SYNCHRONIZATION OF MULTIPLE CHOPPER SYSTEMS.
       *HAUMANN(J.), OSTROWSKI(G.), CONNOR(D.).

11 INSTRUM. EXP. TECH. NO.4,. . . . 832, (1965)....E.....
       *MULTIDIMENSIONAL ANALYZER WITH PRELIMINARY DATA PROCESSING AND MIXED
       MEMORY.
       *EKATOV(A.B.), IVCHENKO(V.E.), MATALIN(L.A.), MESHKOV(N.V.),
       SMIRNOV(V.I.), CHERNUKHIN(V.L.).

12 INSTRUM. EXP. TECH. NO.2,. . . . 415, (1970)....E......
       *CIRCUITS FOR THE IDENTIFICATION OF NEUTRONS.
       *BROVCHENKO(V.G.), KONDRAT#EV(L.G.).

13 INSTRUM. EXP. TECH. VOL.15,. . . 67, (1972)....E....
       *MULTICHANNEL SYSTEM FOR RECORDING NEUTRONS USING A LINE WITH AN
       ELECTRONIC COMPUTER.
       *BYSTRITSKII(V.M.), DZHELEPOV(V.P.), ERMOLOV(P.F.), OGANESYAN(K.O.),
       OMEL#YANENKO(M.N.), POROKHOVOI(S.YU.), FIH#CHENKOV(V.V.).

14 INTERNAT. COMPUTING SYMP.,. . . 291, (1974)............
       *CARINE- A MULTI-USER REAL-TIME SYSTEM FOR CONTROL AND DATA ACQUISITION
       OF NEUTRON BEAM EXPERIMENTS.
       *BARTHELEMY(A.), KAISER(W.), LE-SOURNE(M.), TAESCHNER(M.), DARIER(P.),
       VINIT(A.), GIROD(J.J.). (SYMPOSIUM AT DAVOS, SWITZERLAND, SEPT., 1973).

15 J. APPL. CRYST. VOL.1,. . . . . 272, (1968)....E..........
       *A COMPUTER-CONTROLLED NEUTRON DIFFRACTOMETER.
       *PRYOR(A.W.), ELLIS(P.J.), DULLOW(R.J.).

16 J. PHYS. SOC. JAPAN VOL.17 B-II, 358, (1962)....E........
       *SINGLE-CRYSTAL NEUTRON DIFFRACTOMETER WITH AUTOMATIC PROGRAMMING-
       CONTROL SYSTEM.
       *MIYAKE(S.), HOSHINO(S.), SUZUKI(K.), KATSURAGI(H.), HAGIWARA(S.),
       YOSHIE(T.), MIYASHITA(K.).

17 LATV. PSR...FIZ. TEHN. SER. NO.5 124, (1971)....P.......
       *LINE WIDTH FOR INTEGRAL INTENSITY MEASUREMENTS IN NEUTRON DIFFRACTOMETER
       WITH EQUATORIAL SURVEY CIRCUIT.
       *NOZIK(YU.), HEIKER(D.), LIPIN(YU.).

18 NUCL. INST. MET. 24,. . . . . . 255, (1963)...................
       *A PROGRAMMED MULTICHANNEL NEUTRON DIFFRACTOMETER INSTALLATION.
       *ARNDT(U.W.), WILLIS(B.T.M.).

19 NUCL. INST. MET. 23,. . . . . . 181, (1963)................
       *A PROGRAMMED CONTROL SYSTEM FOR AN AUTOMATIC NEUTRON SPECTROMETER.
       *ALLENDEN(D.), WINKWORTH(R.).

20 NUCL. INST. MET. 26,. . . . . . 340, (1964)................
       *FAST RECORDING OF NEUTRON DIFFRACTION INTENSITY DATA.
       *CHIDAMBARAM(R.), SEQUEIRA(A.), SIKKA(S.K.).

21 NUCL. INST. MET. VOL.29,. . . . 241, (1964)....E.......
       *AN ELECTRONIC DEVICE IMPROVING THE RESOLUTION OF THE TIME-OF-FLIGHT
       SPECTROMETER WITH MANY-LAYER TRAY OF COUNTERS.
       *MOSZYNSKI(M.), TUROS(A.), WILHELMI(Z.).

22 NUCL. INST. MET. 25,. . . . . . 288, (1964)....................
       *AUTOMATIC DATA RECORDING SYSTEM FOR THE NEUTRON SPECTROMETER ON THE
       MERLIN RESEARCH REACTOR.
       *WINKWORTH(R.).

23 NUCL. INST. MET. 35,. . . . . . 13, (1965)...................
       *MULTIPURPOSE NEUTRON DIFFRACTION INSTRUMENTATIONS.
       *ATOJI(M.).

SECTION 22B-GENERAL INSTRUMENTATION: AUTOMATION AND ELECTRONICS. . . . .

24 NUCL. INST. MET. 40, ... 125, (1966). . . . .
+A SMALL NEUTRON DIFFRACTOMETER.
+TUNKELO(E.), KAJAMA(J.).

25 NUCL. INST. MET. 41, ... 61, (1966). . . . .
+INSTRUMENTATION OF A MANUALLY PROGRAMMED NEUTRON DIFFRACTOMETER.
+HANSEN(K.B.), NEISIG(K.E.).

26 NUCL. INST. MET. 45, ... 245, (1966). . . . .
+NEUTRON DIFFRACTION BY TIME OF FLIGHT.
+REICHELT(J.M.A.), RODGERS(A.L.).

27 NUCL. INST. MET. 44, ... 181, (1966). . . . .
+NEUTRON POWDER DIFFRACTOMETRY USING A WAVELENGTH OF 2.6Å.
+LOOPSTRA(B.O.).

28 NUCL. INST. MET. 42, ... 29, (1966). . . . .
+A PROGRAMMED AUTOMATIC CONTROL SYSTEM FOR A TRIPLE AXIS NEUTRON
SPECTROMETER.
+BEVILACQUA(S.), DE-AGOSTINO(E.)...ET AL.

29 NUCL. INST. MET. 47, ... 179, (1967). . . . .
+EXTENDED DETECTORS IN NEUTRON TIME-OF-FLIGHT DIFFRACTION EXPERIMENTS.
+CARPENTER(J.M.).

30 NUCL. INST. MET. 60, ... 237, (1968). . . . .
+ON LINE COMPUTER CONTROL OF A TRIPLE AXIS NEUTRON DIFFRACTOMETER.
+RUSSELL(M.C.B.), MARTIN(R.), BROWN(L.H.), SMITH(D.B.J.).

31 NUCL. INST. MET. 69, ... 173, (1969). . . . .
+FOCUSING OF A TIME OF FLIGHT DIFFRACTOMETER FOR STRUCTURE ANALYSIS THE
EXPERIMENT CHECK.
+HOLAS(A.), HOLAS(J.)...ET AL.

32 NUCL. INST. MET. 75, ... 309, (1969). . . . .
+PULSE RISETIME DISCRIMINATION FOR HE-3 COUNTERS.
+CUTTLER(J.M.), GREENBERGER(S.), SHALEV(S.).

33 NUCL. INST. MET. 75, ... 1, (1969). . . . .
+ON LINE COMPUTER CONTROL OF A TRIPLE AXIS NEUTRON SPECTROMETER.
+SKAARUP(P.).

34 NUCL. INST. MET. VOL.96, ... 609, (1971). . . . E. . . . .
+A PHASE CONTROL CIRCUIT FOR HIGH SPEED ROTORS.
+HARLING(O.K.), SVOBODA(H.J.).

35 NUCL. INST. MET. VOL.91, ... 541, (1971). . . . P. . . . .
+A 250 METER NEUTRON TIME-OF-FLIGHT FACILITY WITH MODULAR DETECTOR AND
DIGITAL ELECTRONICS STABILIZATION.
+STOLER(P.), YERGIN(P.F.), CLEMENT(J.C.), MANN(D.), GOULDING(C.G.),
FAIRCHILD(R.). (ED. NOTE: FACILITY AT RENSSELAER POLYTECH. INST.).

36 NUCL. INST. MET. VOL.111, ... 375, (1973). . . . E. . . . .
+A COMPUTER CONTROLLED TRIPLE AXIS NEUTRON SPECTROMETER (TAS) WITH CAMAC
INSTRUMENTATION AND A HIGH LEVEL COMPUTER LANGUAGE CONTROL PROGRAM.
+MAY(F.), EDER(O.J.), SCHOITSCH(E.), SCHWARZER(J.).

37 NUCL. INST. MET. VOL.115, ... 393, (1974). . . . E. . . . .
+GENERAL INTERFACE CIRCUIT FOR TIME-OF-FLIGHT MEASUREMENTS WITH COLD
NEUTRONS.
+COPPO(M.), MAINO(G.), VILLA(S.).

38 NUCL. INST. MET. VOL.116, ... 561, (1974). . . . E. . . . .
+A COMPUTERIZED MEASURING SYSTEM FOR NEUTRON PHYSICS EXPERIMENTS.
+GROSSHOG(G.).

39 PHYS. REV. VOL.72, ... 585, (1947). . . . E. . . . .
+NEUTRON CROSS-SECTION STUDIES WITH THE ROTATING SHUTTER MECHANISM.
+BRILL(T.), LICHTENBERGER(H.V.).

40 PROC/NAT. ELECTRONICS CONF. 21, . 816, (1965). . . . .
+ARGONNE AUTOMATIC NEUTRON DIFFRACTOMETER.
+JOHANSON(E.W.). (ED. NOTE: PROCEEDINGS OF THE NATIONAL ELECTRONICS
CONFERENCE, CHICAGO,1965).

41 REP. NO. BNL-10238 (3 PP.), ... (1966). . . . E. . . . .
+AUTOMATIC ACQUISITION OF NEUTRON SPECTROMETER DATA.
+HAMILTON(W.C.).

42 REV. DE PHYSIQUE (BUCAREST) 4, . 327, (1959). . . . E. . . . .
+A NEUTRON TIME OF FLIGHT SPECTROMETER.
+TOTIA(H.), TIMIS(P.), LAZAROVICI(C.). (IN FRENCH).

43 REV. SCI. INSTRUM. VOL.30, ... 997, (1959). . . . E. . . . .
+AUTOMATIC CONTROL AND PROGRAMMING SYSTEM FOR SINGLE CRYSTAL
DIFFRACTOMETRY.
+LANGDON(F.), FRAZER(B.C.).

44 REV. SCI. INSTRUM. VOL.30, ... 581, (1959). . . . E. . . . .
+SINGLE CRYSTAL AUTOMATIC NEUTRON DIFFRACTOMETER.
+PRINCE(E.), ABRAHAMS(S.C.).

45 REV. SCI. INSTRUM. VOL.31, ... 490, (1960). . . . E. . . . .
+A 2000-CHANNEL ANALYZER FOR NEUTRON SPECTROSCOPY.
+HAHN(J.), HAVENS-JR(W.W.).

46 REV. SCI. INSTRUM. VOL.32, ... 456, (1961). . . . E. . . . .
+STEPPING MECHANISM FOR X-RAY AND NEUTRON DIFFRACTOMETERS AND SPEC-
TROMETERS.
+MUELLER(M.H.), HEATON(L.), JOHANSON(E.W.).

47 REV. SCI. INSTRUM. VOL.34, ... 224, (1963). . . . C. . . . .
+AUTOMATIC NEUTRON DIFFRACTOMETER FOR THREE-DIMENSIONAL STRUCTURE-
FACTOR DETERMINATION.
+ARNDT(U.W.), WILLIS(B.T.M.).

48 REV. SCI. INSTRUM. VOL.34, ... 224, (1963). . . . .
+AUTOMATIC NEUTRON DIFFRACTOMETER FOR THREE-DIMENSIONAL STRUCTURE-FACTOR
DETERMINATION.

49 REV. SCI. INSTRUM. VOL.34, ... 74, (1963). . . . .
+FULL CIRCLE GONIOSTAT FOR DIFFRACTION INTENSITY DATA.

50 REV. SCI. INSTRUM. VOL.35, ... 1150, (1964). . . . E. . . . .
+TIME-SHARED TIME-OF-FLIGHT ANALYSIS BY COMPUTER.
+CHRIEN(R.E.), RANKOWITZ(S.), SPINRAD(R.J.).

SECTION 22B-GENERAL INSTRUMENTATION: AUTOMATION AND ELECTRONICS. . . . .

51 REV. SCI. INSTRUM. VOL.36, . . . 887, (1965). . . .E. . . . .
    *NEUTRON CRYSTAL SPECTROMETER WITH RANGE EXTENDED TO SUBTHERMAL ENERGIES.
    *RUSTAD(R.M.), MELKONIAN(E.), HAVENS-JR(W.W.), TAYLOR(T.I.), GOULD(F.T.),
    MOORE(J.A.).

52 REV. SCI. INSTRUM. VOL.40, . . . 1144, (1969). . . . .
    *DISCRETE AVERAGING OF X-RAY DIFFRACTION DATA USING A MULTICHANNEL
    ANALYSER.
    *BERMAN(M.), ERGUN(S.).

53 REV. SCI. INSTRUM. 41, ... 117, (1970). . . . .
    *AN APPARATUS FOR NEUTRON DIFFRACTION FROM SINGLE CRYSTALS.
    *KHAN(D.C.), ERICKSON(R.A.).

54 REV. SCI. INSTRUM. VOL.41, ... 11, (1970). . . . .
    *NEW PULSE HEIGHT TO CLOCK PULSE CONVERTER AND ITS APPLICATION TO NEUTRON
    DOSIMETRY.
    *FURUTA(Y.), KINBARA(S.), MIYAKOSHI(J.).

55 REV. SCI. INSTRUM. VOL.41, . . . 1539, (1970). . . . .
    *FAST FISSION CHAMBER AMPLIFIER-DISCRIMINATOR SYSTEM.
    *INGLE(R.W.), GILLESPIE(F.E.), WESTON(L.W.).

56 SOV. PHYS. CRYST. VOL.5, ... 294, (1960). . . . .E. . . . .
    *A REMOTELY CONTROLLED NEUTRON DIFFRACTOMETER BASED ON A GUR-3 UNIT.
    *OZEROV(R.P.), KISELEV(S.V.), KARPOVICH(I.R.), GOMAN#KOV(V.I.),
    LOSHMANOV(A.A.).

57 Z. ANGEW. MATH. PHYS. VOL.15,. . . 481, (1964). . . . .E. . . . .
    *A PROGRAMMING AND AUTOMATIC DATA COLLECTING SYSTEM FOR A NEUTRON
    DIFFRACTION SPECTROMETER.
    *HALG(W.), MEIER(F.), GASSER(F.).

SECTION 22C-GENERAL INSTRUMENTATION: AUXILIARY EQUIPMENT . . . .

1   ACTA CRYST. VOL. A25, ... S-65, (1969). . . . .
      *THREE CRYOSTATS FOR NEUTRON DIFFRACTION.
      *HOSHINO(S.), ISHIKAWA(Y.), ENDOH(Y.), SHIMAOKA(K.).

2   ACTA CRYST. VOL.A27, ... 284, (1971). . . . .
      *A NEUTRON FOURIER CHOPPER FOR SINGLE CRYSTAL REFLECTIVITY MEASUREMENTS:
      SOME GENERAL DESIGN CONSIDERATIONS.
      *NUNES(A.C.), NATHANS(R.), SCHOENBORN(B.P.).

3   ACTA CRYST. VOL.A27, ... 391, (1971). . . . .
      *PARALLEL BEAMS OF NEUTRONS OR X-RAYS BY MULTIPLE BRAGG REFLECTION.
      *KOTTWITZ(D.A.).

4   ANN. ACAD. SCI. FENNICAE A 267, . . . 3, (1967). . . . .
      *SOME SUGGESTIONS FOR THE APPLICATION OF CONSTRICTED MONOCHROMATOR CRYS-
      TALS FOR NEUTRON DIFFRACTION AND SCATTERING MEASUREMENTS.
      *MAIER-LEIBNITZ(H.). (ED. NOTE: IN GERMAN).

5   ANN. OF PHYSICS VOL.74, ... 250, (1972). . . . E. . . . . .D2,T2, ...
      *ELASTIC SCATTERING AND POLARIZATION OF FAST NEUTRONS BY LIQUID DEUTERIUM
      AND TRITIUM.
      *SEAGRAVE(J.D.), HOPKINS(J.C.), DIXON(D.R.), KEATON-JR(P.W.), KERR(E.C.),
      NIILER(A.), SHERMAN(R.H.), WALTER(R.K.).

6   ANN. PHYS. (GERMANY) VOL.7, ... 50, (1961). . . . E. . . . .
      *A MECHANICAL VELOCITY SELECTOR FOR SUB-THERMAL NEUTRONS AND ITS USE IN
      THE MEASUREMENT OF THE TOTAL EFFECTIVE C.S. OF CD, SM, EU AND GD.
      *HOEHNE(P.). (IN GERMAN).

7   AUSTRAL. J. PHYS. VOL.16, ... 272, (1963). . . . T. . . . .
      *AN ANALYTICAL TREATMENT OF THE GEOMETRICAL PROPERTIES OF THE THREE-
      CIRCLE GONIOMETER.
      *SABINE(T.M.).

8   BRIT. J. APPL. PHYS. VOL.13, . . . 547, (1962). . . . E. . . . .
      *USE OF A THREE-CIRCLE GONIOMETER FOR DIFFRACTION MEASUREMENTS.
      *WILLIS(B.T.M.).

9   CAN. JOURN. PHYS. 41, ... 1519, (1963). . . . .
      *CORRECTION FACTORS FOR NEUTRON FLUX MEASUREMENTS WITH CADMIUM-COVERED
      COBALT WIRES.
      *EASTWOOD(T.A.), MATYAS(E.), HNATOWICH(D.J.).

10 CHEM. APPL. THERMAL NEUTRON SCAT . 31, (1973). . . . E. . . . .
      *EXPERIMENTAL TECHNIQUES.
      *STIRLING(G.C.). (IN BOOK: CHEMICAL APPLICATIONS OF THERMAL NEUTRON
      SCATTERING. ED. BY B.T.M. WILLIS; OXFORD UNIV. PRESS: LONDON).

11 CONF. NEUTRON BEAMS, ... (1972). . . . E. . . . .
      *HIGH RESOLUTION APPARATUS FOR NEUTRON SCATTERING EXPERIMENTS.
      *KALUS(J.). (PROC. CONF. ON NEUTRON BEAMS IN RESEARCH, HARWELL, ENGLAND,
      SEPT., 1972).

12 CRYOGENICS VOL.10, ... 440, (1970). . . . E. . . . .
      *A CRYOSTAT FOR A NEUTRON SCINTILLATION SPECTROMETER.
      *ADAMOVICH(N.I.), ZUEV(V.I.), PAVLOVSKAYA(T.F.).

13 DISS. ABSTR. B VOL.33 NO.11, . . . 5435, (1973). . . . E. . . . CRYSTAL NEUTRON DIFFRACTION
      *A SMALL HE(A=3) CRYOSTAT FOR SINGLE CRYSTAL NEUTRON DIFFRACTION
      APPLICATIONS WITH A NEW THERMOMETER CALIBRATION TECHNIQUE.
      *STARR-JR(E.F.).

14 ENERGIA NUCLEARE VOL.10, ... 173, (1963). . . . .
      *A CRYSTAL MONOCHROMATOR FOR SLOW NEUTRONS.
      *GIACCHETTI(G.), MUSCI(M.), POLETTI(G.).

15 EXPER. TECH. PHYS. VOL.2, ... 111, (1965). . . . .
      *LOW TEMPERATURE CHAMBERS FOR NEUTRON DIFFRACTION EXPERIMENTS.
      *WEISS(L.), KLEINSTUCK(K.). (IN GERMAN).

16 IAEA SYMP. CHALK RIVER, VOL.1, . . 95, (1962). . . . .
      *A NEW APPARATUS FOR INELASTIC, QUASI-ELASTIC AND ELASTIC COLD NEUTRON
      MEASUREMENTS.
      *OTNES(K.), PALEVSKY(H.).

17 INSTRUM. EXP. TECH. NO.3, ... 359, (1960). . . . E. . . . .
      *EXPERIMENTAL METHODS OF NEUTRON DIFFRACTION.
      *ABOV(YU.G.), LITVIN(D.F.).

18 INSTRUM. EXP. TECH. NO.2, ... 285, (1961). . . . E. . . . .
      *A GONIOMETER FOR THE ORIENTATION OF LARGE SINGLE CRYSTALS USED IN THE
      MONOCHROMATIZATION OF NEUTRONS.
      *SABO(P.), KREN(E.).

19 J. APPL. CRYST. VOL.5 PT.6, . . . 432, (1972). . . . E. . . . .
      *THERMOELECTRIC COOLING DEVICE FOR SINGLE-CRYSTAL NEUTRON DIFFRACTOMETER.
      *AGRON(P.A.), LEVY(H.A.).

20 J. APPL. CRYST. VOL.7, ... 96, (1974). . . . E,T. . . . .
      *NEUTRON SMALL-ANGLE SCATTERING: EXPERIMENTAL TECHNIQUES AND APPLICATIONS
      *SCHMATZ(W.), SPRINGER(T.), SCHELTEN(J.), IBEL(K.).

21 J. NUCL. ENERGY VOL.24, ... 385, (1970). . . . E. . . . .
      *A DETECTOR SHIELDING FOR THE MEASUREMENT OF TOTAL NEUTRON CROSS-SECTION
      AT BACKWARD ANGLES.
      *HUSSAIN(M.),   HUSSAIN(S.),   ENAYETULLAN(H.),   ISLAM(E.),   AMEEN(N.),
      ENAYETULLAN(M.), ISLAM(N.).

22 J. OPTIMIZ. THEORY APPLIC. VOL.5 301, (1970). . . . E. . . . .
      *MAXIMUM NEUTRON FLUX IN THERMAL RESEARCH REACTORS.
      *STRUGAR(P.V.).

23 J. PHYS. E VOL.6 NO.8, ... 714, (1973). . . . E. . . . .
      *AN ASYMMETRIC SPLIT-PAIR SUPERCONDUCTING MAGNET FOR NUCLEAR POLARIZATION
      EXPERIMENTS.
      *GILBERT(E.), HANLEY(P.), HAYTER(J.B.), WHITE(J.W.).

24 J. SCI. INSTRUM. VOL.36, ... 419, (1959). . . . E. . . . .
      *NEUTRON DIFFRACTION INSTRUMENTS AT A HIGH-FLUX NUCLEAR REACTOR.
      *BACON(G.E.), DYER(R.F.).

SECTION 22C-GENERAL INSTRUMENTATION: AUXILIARY EQUIPMENT . . . .

25 J. SCI. INSTRUM. VOL.41, ... 376, (1964)....E........
    *A THIN-WALLED SPECIMEN CONTAINER FOR NEUTRON SCATTERING STUDIES OF
    LIQUIDS AT HIGH TEMPERATURES AND PRESSURES.
    *COCKING(S.J.), BALL(A.R.).

26 J. SCI. INSTRUM. SER.2 VOL.1, ... 367, (1968)....E........
    *A SIMPLE HEATING ARRANGEMENT FOR USE IN NEUTRON SCATTERING EXPERIMENTS.
    *COCKING(S.J.), MCPHEE(R.C.), WIGNALL(G.O.).

27 KERNTECHNIK VOL.10 NO.7, ... 371, (1968)....E........
    *A CONTINUOUS FLOW CRYOSTAT FOR NEUTRON SCATTERING EXPERIMENTS.
    *KONIG(U.), TIPPE(A.), DALLUGGE(W.).

28 NUCL. INST. MET. 30, ... 271, (1964)....E........
    *ELECTROMAGNETIC FOCUSING AND POLARIZATION OF NEUTRON BEAMS.
    *FARAGO(P.S.).

29 NUCL. INST. MET. VOL.32, ... 300, (1965)....E........
    *TIME-OF-FLIGHT MODULE FOR THE TRANSISTORIZED RCL 256 CHANNEL ANALYZER.
    *BACHLI(A.), BEHRINGER(K.).

30 NUCL. INST. MET. 33, ... 155, (1965)....E........
    *FUSED LI26-CO3 FOR ABSORBING SLOW NEUTRONS IN COLLIMATING SYSTEMS.
    *RUSTAD(B.M.), CHRISTENSEN(C.J.), SKYTTE-JENSEN(B.).

31 NUCL. INST. MET. 40, ... 93, (1966)....E........
    *A LIQUID HELIUM TIME OF FLIGHT COINCIDENCE NEUTRON POLARIMETER.
    *MILLER(T.G.).

32 NUCL. INST. MET. 42, ... 309, (1966)....E........
    *A METHOD OF DETERMINATION OF THERMAL NEUTRON BEAM COLLIMATION.
    *INOUYE(T.), OGAWA(K.).

33 NUCL. INST. MET. 49, ... 161, (1967)....E........
    *MEASUREMENT OF ELASTIC NEUTRONS BY ORGANIC SCINTILLATORS.
    *CHATTERJEE(A.), GHOSE(A.M.).

34 NUCL. INST. MET. 50, ... 191, (1967)....E........
    *CORRECTIONS IN THE GOLD FOIL ACTIVATION METHOD FOR DETERMINATION OF
    NEUTRON BEAM DENSITY.
    *ALS-NIELSEN(J.).

35 NUCL. INST. MET. 53, ... 299, (1967)....E........
    *HIGH FREQUENCY MODULATION OF MONOENERGETIC NEUTRONS WITH A QUARTZ
    PIEZOELECTRIC CRYSTAL.
    *MOYER(M.W.), PARKINSON(T.F.).

36 NUCL. INST. MET. 52, ... 15, (1967)....E........
    *ABLENKUNG LANGSAMER NEUTRONEN MIT EINEM GEKRUMMTEN,
    TOTALREFLEKTIERENDEN KOLLIMATORSYSTEM.
    *FIALA(W.), RAUCH(H.).

37 NUCL. INST. MET. 59, ... 117, (1968)....E........
    *NEW NEUTRON DETECTOR FILMS FROM POLYVINYL ALCOHOL.
    *NAPOLI(C.), CORTELLESSA(G.).

38 NUCL. INST. MET. 62, ... 233, (1968)....E........
    *NEUTRON ENRICHMENT BY TRANSMISSION IN A TOTAL REFLECTING CONICAL TUBE.
    *CAMBIAGHI(M.), FOSSATI(F.), PINELLI(T.).

39 NUCL. INST. MET. 65, ... 119, (1968)....E........
    *WEIGHTED EFFICIENCY CALCULATIONS FOR NEUTRON DETECTORS.
    *REIER(M.).

40 NUCL. INST. MET. 65, ... 152, (1968)....E........
    *A DIGITAL CONTROL SYSTEM FOR NEUTRON SPECTROMETERS.
    *HANSEN(K.B.), SKAARUP(P.).

41 NUCL. INST. MET. 62, ... 57, (1968)....E........
    *ABSOLUTE NEUTRON FLUX DETERMINATION BY A BACK TO BACK FISSION CHAMBER
    AND ACTIVATION DETECTORS.
    *HARACCI(G.), RUSTICHELLI(F.), AIELLO(V.)...ET AL.

42 NUCL. INST. MET. 71, ... 13, (1969)....E........
    *USE OF A PDP-8 COMPUTER FOR OPERATION OF SEVERAL NEUTRON BEAM
    EXPERIMENTS.
    *MASLIN(E.E.), TERRY(S.H.), WRAIGHT(L.A.).

43 NUCL. INST. MET. 71, ... 102, (1969)....E........
    *NEUTRON DEBYE-SCHERRER DIFFRACTION WORKS USING A LINEAR ELECTRON
    ACCELERATOR.
    *KIMURA(M.), SUGAWARA(M.), OYAMADA(M.), YAMADA(Y.)...ET AL.

44 PHYS. LETT. 29A, ... 679, (1969)....E........
    *NEUTRON DIFFRACTION BY A SINGLE CRYSTAL OF IRON EXCITED BY MEANS OF A
    MAGNETOSTRICTIVE RESONNATOR.
    *MICHALEC(R.), CHALUPA(B.), CECH(J.)...ET AL.

45 PHYS. MET. METALLOGR. VOL.14, ... 25, (1962)....E........NI-CR.
    *NEUTRON DIFFRACTION STUDY OF NI-CR ALLOYS.
    *GOMAN#KOV(V.I.), LITVIN(D.F.), LOSHMANOV(A.A.), LYASHCHENKO(B.G.).

46 PHYS. REV. VOL.71, ... 752, (1947)....E........
    *DIFFRACTION OF NEUTRONS BY A SINGLE CRYSTAL.
    *ZINN(W.H.).

47 PHYS. REV. 175, ... 1056, (1968)....E........
    *INTENSITY OF FORBIDDEN NEUTRON REFLECTIONS STIMULATED BY MULTIPLE BRAGG
    REFLECTION.
    *KOTTWITZ(D.A.).

48 REV. SCI. INSTRUM. VOL.25, ... 699, (1954)....E........
    *NEW GONIOMETERS FOR SINGLE-CRYSTAL NEUTRON DIFFRACTION STUDIES.
    *PEPINSKY(R.), FRAZER(B.C.), MCKEOWN(M.L.).

49 REV. SCI. INSTRUM. VOL.28, ... 916, (1957)....E........
    *SINGLE CRYSTAL GONIOMETER FOR X-RAY AND NEUTRON DIFFRACTION.
    *BARSTAD(G.E.B.), ANDRESEN(A.F.).

50 REV. SCI. INSTRUM. VOL.31, ... 1355, (1960)....E........
    *SAMPLE ASSEMBLY FOR NEUTRON DIFFRACTION STUDIES OF LIQUEFIED GASES.
    *CLAYTON(G.T.), HEATON(L.).

51 REV. SCI. INSTRUM. VOL.31, ... 174, (1960)....E........
    *LIQUID-HELIUM GONIOMETER-MOUNTED CRYOSTAT FOR A SINGLE CRYSTAL
    AUTOMATIC NEUTRON DIFFRACTOMETER.
    *ABRAHAMS(S.C.).

SECTION 22C-GENERAL INSTRUMENTATION: AUXILIARY EQUIPMENT . . . .

52 REV. SCI. INSTRUM. VOL.34,   194, (1963).....E........
    *TEMPERATURE REGULATOR AND SAMPLE CHANGER FOR NEUTRON TOTAL CROSS-SECTION
    MEASUREMENTS AT LOW TEMPERATURES.
    *KISTNER(A.M.), KISTNER(O.C.).

53 REV. SCI. INSTRUM. VOL.34,   74, (1963).....E........INTENSITY DATA.
    *FULL CIRCLE GONIOSTAT FOR DIFFRACTION INTENSITY DATA.
    *MUELLER(M.H.), HEATON(L.), SIDHU(S.S.).

54 REV. SCI. INSTRUM. 34,   1441, (1963).....E........
    *MODIFIED POLAROID FILM HOLDER FOR X-RAY AND NEUTRON DIFFRACTION.

55 REV. SCI. INSTRUM. VOL.34,   194, (1963).....E........
    *TEMPERATURE REGULATOR AND SAMPLE CHANGER FOR NEUTRON TOTAL CROSS-SECTION
    MEASUREMENTS AT LOW TEMPERATURES.

56 REV. SCI. INSTRUM. 36,   1167, (1965).....E........
    *PULSE SHAPE DISCRIMINATION SYSTEM FOR $6LI*F(ZN*S)$ SCINTILLATION
    COUNTERS.

57 REV. SCI. INSTRUM. 36,   1161, (1965).....E........
    *4C96 CHANNEL TIME ANALYSER WITH VARIABLE CHANNEL WIDTH (TIME-OF-FLIGHT).

58 REV. SCI. INSTRUM. VOL.37,   1543, (1966).....E........
    *HIGH TEMPERATURE NEUTRON DIFFRACTION FURNACE.
    *BOWMAN(M.G.), HULL(O.E.), WITTEMAN(W.G.), ARNOLD(G.P.), BOWMAN(A.L.).

59 REV. SCI. INSTRUM. VOL.37,   435, (1966).....E........
    *CRYOSTAT FOR NEUTRON DIFFRACTION STUDIES BETWEEN LIQUID HELIUM
    TEMPERATURES AND ROOM TEMPERATURE.
    *SPARKS(J.T.), KOMOTO(T.), WADE(W.R.).

60 REV. SCI. INSTRUM. VOL.37,   1534, (1966).....E........
    *HIGH PRESSURE VESSEL FOR MEASUREMENT OF PRESSURE INDUCED PHONON ENERGY
    SHIFTS BY INELASTIC SCATTERING OF NEUTRONS.
    *LECHNER(R.E.).

61 REV. SCI. INSTRUM. VOL.38,   275, (1967).....E........
    *SIMPLE CRYOSTAT FOR NEUTRON DIFFRACTION.
    *SPARKS(J.T.), KOMOTO(T.), WADE(W.R.).

62 REV. SCI. INSTRUM. 38,   1333, (1967).....E........
    *USE OF A JOULE-THOMSON REFRIGERATOR FOR NEUTRON DIFFRACTION AT LIQUID
    NITROGEN TEMPERATURES.

63 REV. SCI. INSTRUM. 40,   1397, (1969).....E........
    *AN OVEN FOR A NEUTRON SPECTROMETER, TO 1000 K.
    *BEZDEK(H.F.), SCHMUNK(R.E.), FINEGOLD(L.).

64 REV. SCI. INSTRUM. 41,   107, (1970).....E........
    *AN APPARATUS FOR NEUTRON DIFFRACTION FROM SINGLE CRYSTALS.
    *KHAN(O.C.), ERICKSON(R.A.).

65 REV. SCI. INSTRUM. 41,   1877, (1970).....E........
    *THE DYNAMICAL BEHAVIOR OF A CHOPPER-ROTATING CRYSTAL SYSTEM IN A NEUTRON
    TIME-OF-FLIGHT EXPERIMENT.
    *EDER(O.J.).

66 REV. SCI. INSTRUM. VOL.45,   341, (1974).....E........
    *A 4.2*30J K CRYOSTAT FOR NEUTRON DIFFRACTION UNDER HIGH PRESSURE.
    *CLAUDET(G.), DISDIER(F.), SENET(L.), BUEVOZ(J.L.), ROULT(G.).

67 REV. SCI. INSTRUM. VOL.45,   643, (1974).....E........
    *APPARATUS FOR NEUTRON DIFFRACTION AT HIGH PRESSURE.
    *MCWHAN(D.B.), BLOCH(D.), PARISOT(G.).

68 REV. SCI. INSTRUM. VOL.45,   526, (1974).....E.T........
    *AUTOFRETTAGED HIGH PRESSURE CHAMBER FOR USE IN INELASTIC NEUTRON
    SCATTERING.
    *BLASCHKO(O.), ERNST(G.).

69 REVUE PHYS. APPL. VOL.7,   29, (1972).....E........
    *CRYOSTAT FOR NEUTRON DIFFRACTION AT VARIOUS TEMPERATURES BETWEEN 1 AND
    300 K.
    *SOUGI(M.). (ED. NOTE: IN FRENCH).

70 SCIENCE VOL.114,   341, (1951).....E........
    *CHEMICAL ANALYSIS BY NEUTRON SPECTROSCOPY.
    *TAYLOR(T.I.), ANDERSON(R.H.), HAVENS-JR(W.W.).

71 SOV. PHYS. CRYST. VOL.5,   297, (1960).....E........
    *A CRYOSTAT FOR USE AT LIQUID HYDROGEN AND HELIUM TEMPERATURES IN
    NEUTRON DIFFRACTION STUDIES.
    *KOGAN(V.S.), LAZAREV(B.G.), ZHDANOV(G.S.), OZEROV(R.P.).

72 SOV. PHYS. JETP VOL.11,   585, (1960).....E........
    *A CRYOSTAT FOR NEUTRON DIFFRACTION INVESTIGATIONS.
    *ALIKHANOV(R.A.).

73 SOV. PHYS. CRYST. VOL.4,   397, (1960).....E........
    *APPARATUS FOR NEUTRON DIFFRACTION STRUCTURAL ANALYSIS.
    *YAMZIN(I.I.).

74 SOV. PHYS. CRYST. VOL.11,   322, (1966).....E........
    *AN APPARATUS FOR NEUTRON DIFFRACTOMETRY OF SPECIMENS UNDER HIGH
    HYDROSTATIC PRESSURE.
    *LITVIN(D.F.), PONYATOVSKII(E.G.).

75 TECH. CEM (FRANCE) NO.84,   16, (1972).....E........
    *CRYOMAGNETIC ASSEMBLY FOR NEUTRON DIFFRACTION STUDIES.
    *GIRARD(B.), HELLEGOUARCH(J.).

76 Z. PHYSIK VOL.263,   291, (1973).....E........SI..
    *BACKSCATTERING OF NEUTRONS FROM PERFECT SILICON SINGLE CRYSTALS.
    *HEIDEMANN(A.), SCHOLZ(J.). (IN GERMAN).

SECTION 23 -RESEARCH PROGRAMMES AND FACILITIES AT VARIOUS LABORATORIES .

1 ACTA PHYS. POLON. VOL.19,. . . . 329, (1960)....E.........
  ↦MEASUREMENT OF THE SLOW NEUTRON SPECTRUM OF A NEUTRON BEAM FROM THE
  W.W.R.S. REACTOR BY MEANS OF A CRYSTAL NEUTRON SPECTROMETER.
  ↦O≠CONNOR(D.A.), SOSNOWSKI(J.).

2 ACTA PHYS. AUSTRIACA VOL.20,. . . 166, (1965)....E.........
  ↦M.A.N. NEUTRON DIFFRACTION INSTALLATION, SEIBERSDORF: AUTOMATIC CONTROL
  AND PROGRAMMING.
  ↦MAY(F.). (IN GERMAN).

3 AT. ENERGY AUST. VOL.15, ... 2, (1972)....P.........
  ↦NEUTRON DIFFRACTION FACILITIES.
  ↦MOORE(F.H.).

4 BROOKHAVEN SYMPOSIUM,. . . . . . 179, (1965)...............
  ↦TROMBAY MULTIARM SPECTROMETER.
  ↦IYENGAR(P.K.).

5 BULL. D≠INFO. SCI. ET TECH. 166, . . (1972)....E.........
  ↦LE REACTEUR FRANCO-ALLEMAND A HAUT FLUX. LES POSSIBILITES EXPERIMENTALES
  ↦AGERON(P.)...ET AL.

6 BULL. INF. SCI. TECH. NO.170,. . . 61, (1972)....E.........
  ↦MEASUREMENTS OF FUNDAMENTAL NEUTRON DATA WITH THE VAN DE GRAAFF AT
  CADARACHE.
  ↦ABRAMSON(D.), ARNAUD(A.), FILIPPI(G.), FORT(E.), HUET(J.L.),
  LE-RIGOLEUR(C.), LEROY(J.L.), SZABO(I.). (ED. NOTE: IN FRENCH).

7 COLL. INTER. N.126 (GRENOBLE),. . 71, (1963)...............
  ↦EXPERIENCE WITH THE OAK RIDGE AUTOMATIC THREE CIRCLE NEUTRON
  DIFFRACTOMETER.
  ↦BUSING(W.R.), SMITH(H.G.), PETERSON(S.W.), LEVY(H.A.).

8 COMMENTAT. PHYS.-MATH.(FIN.) 42, 276, (1972)....E.........
  ↦NEUTRON DIFFRACTION RESEARCH AT FIR 1.
  ↦SZPUNAR(J.), OJANEN(M.).

9 CRYOGENICS VOL.11,. . . . . . . 107, (1971)....E.........
  ↦THE COLD NEUTRON SOURCE IN JULICH.
  ↦DOOSE(C.), PREUSSNER(A.), STELZER(F.), STILLER(H.), THOLEN(A.).

10 CRYSTALLOGR./CRYSTAL PERFECTION, 279, (1963)....E.........
  ↦NEUTRON SPECTROMETRY AT TROMBAY.
  ↦IYENGAR(P.K.). (ED. NOTE: CRYSTALLOGRAPHY AND CRYSTAL PERFECTION; PUBL.
  ACADEMIC PRESS: NEW YORK AND LONDON, 1963).

11 DISS. ABS. VOL.31,. . . . . . .2897B, (1970)...............
  ↦THE THEORY, DESIGN, AND OPERATION OF THE MISSOURI UNIVERSITY
  TIME-OF-FLIGHT SPECTROMETER.
  ↦GAULT(J.D.).

12 ENDEAVOUR 31,. . . . . 67, (1972)....E.........
  ↦THE HIGH-FLUX REACTOR AT THE INSTITUT LAUE-LANGEVIN.
  ↦AGERON(P.).

13 ENERGIE NUCLEAIRE VOL.13,. . . . 15, (1971)....E.........
  ↦LA SOURCE DE NEUTRONS FROIDS POUR LE REACTEUR A HAUT FLUX FRANCO-
  ALLEMAND DE GRENOBLE.
  ↦AGERON(P.), EWALD(R.), HARIG(H.D.), VERDIER(J.).

14 EVOLUTION OF PARTICLE PHYSICS, . 204, (1970)....E.........
  ↦STRUCTURE OF MATTER INVESTIGATIONS BY THERMAL NEUTRONS IN ROME.
  ↦PAOLETTI(A.), SCIUTI(S.). (ED. NOTE: PUBL. ACADEMIC, LONDON, ENGLAND.
  ED.- CONVERSI(M.), 342PP.).

15 IAEA SYMPOSIUM VIENNA,. . . . . 277, (1960)...............
  ↦SLOW-NEUTRON INELASTIC SCATTERING MEASUREMENTS AT THE MATERIALS TESTING
  REACTOR.
  ↦BRUGGER(R.M.), EVANS(J.E.).

16 IAEA SYMP. (PILE RESEARCH)VIENNA 585, (1960)...............
  ↦THE JAERI NEUTRON CRYSTAL SPECTROMETER.
  ↦OHNO(Y.), ASAMI(T.), OKAMOTO(K.).

17 IAEA SYMP. (PILE RESEARCH)VIENNA 577, (1960)...............
  ↦METHODOLOGY WORK IN THE RESEARCH LABORATORIES ATTACHED TO THE REACTOR
  OF THE INSTITUTE FOR ATOMIC PHYSICS OF THE ROMANIAN PEOPLE≠S REPUBLIC.
  ↦BALLY(D.), CRISTU(M.), TEUTSCH(H.).

18 IAEA SYMP. (PILE RESEARCH)VIENNA 351, (1960)...............
  ↦NEUTRON-DIFFRACTION RESEARCH AT BROOKHAVEN.
  ↦HASTINGS(J.M.).

19 IAEA SYMP. (PILE RESEARCH)VIENNA 369, (1960)...............
  ↦NEUTRON-DIFFRACTION INVESTIGATIONS OF SOLIDS AT THE EWA REACTOR.
  ↦BLINOWSKI(K.).

20 IAEA SYMP. (PILE RESEARCH)VIENNA 117, (1960)...............
  ↦REVIEW OF PILE NEUTRON RESEARCH IN PHYSICS AT THE JAPAN ATOMIC
  ENERGY INSTITUTE.
  ↦SUGIMOTO(A.), KAKIHARA(K.).

21 IAEA SYMP. (PILE RESEARCH)VIENNA 123, (1960)...............
  ↦PILE NEUTRON RESEARCH PROGRAMMES AT THE CENTRO NUCLEARE DELLA CASACCIA;
  ROME.
  ↦CLEMENTEL(E.), GIANNINI(M.), MARSEGUERRA(M.), PAOLETTI(A.), PAULI(G.),
  PROSPERI(D.), SCIUTI(S.).

22 IAEA SYMP. (PILE RESEARCH)VIENNA 129, (1960)...............
  ↦REVIEW OF PILE NEUTRON PHYSICS AT AAEC RESEARCH ESTABLISHMENT.
  ↦NICHOLSON(K.P.), SYMONDS(J.L.).

23 IAEA SYMP. (PILE RESEARCH)VIENNA 337, (1960)...............
  ↦ETUDES DE PHYSIQUE DU SOLIDE REALISEES A LA PILE DE SACLAY.
  ↦HERPIN(A.).

24 IAEA SYMP. (PILE RESEARCH)VIENNA 49, (1960)...............
  ↦REVIEW OF PILE NEUTRON RESEARCH AND PHYSICS AT OAK RIDGE NATIONAL LAB.
  ↦SNELL(A.H.).

SECTION 23 -RESEARCH PROGRAMMES AND FACILITIES AT VARIOUS LABORATORIES .

25 IAEA SYMP. (PILE RESEARCH)VIENNA   63, (1960)............... . . . .
    ⌐NEUTRON RESEARCH WITH THE FRM REACTOR.
    ⌐MAIER-LEIBNITZ(H.).

26 IAEA SYMP. (PILE RESEARCH)VIENNA   83, (1960).............. . . . .
    ⌐RESEARCH PROGRAMME AT THE LIVERMORE POOL-TYPE REACTOR.
    ⌐KIRSCHBAUM(A.J.), JOHN(W.).

27 IAEA SYMP. (PILE RESEARCH)VIENNA   93, (1960)............ . . . .
    ⌐THE MTR NUCLEAR PHYSICS PROGRAMME.
    ⌐EVANS(J.E.), FLUHARTY(R.G.), MOORE(M.S.).

28 IAEA SYMP. (PILE RESEARCH)VIENNA   535, (1960)......... . . . .
    ⌐THE ORNL FAST-CHOPPER TIME-OF-FLIGHT NEUTRON SPECTROMETER.
    ⌐BLOCK(R.), SLAUGHTER(G.), PATTENDEN(N.J.), HARVEY(J.).

29 IAEA SYMP. (PILE RESEARCH)VIENNA   25, (1960)........ . . . .
    ⌐PHYSICS RESEARCH AT THE CHALK RIVER REACTORS.
    ⌐BROCKHOUSE(B.N.), CLARK(M.A.), KNOWLES(J.W.), MILTON(J.C.D.),
    SINCLAIR(R.N.).

30 IAEA SYMP. (PILE RESEARCH)VIENNA   361, (1960).......... . . . .
    ⌐REVIEW OF THE NEUTRON-DIFFRACTION INVESTIGATIONS PERFORMED AT THE
    KJELLER RESEARCH ESTABLISHMENT.
    ⌐GOEDKOOP(J.A.).

31 IAEA SYMP. COPENHAGEN, VOL.2,   353, (1968)........ . . . .
    ⌐DESCRIPTION ET CARACTÉRISTIQUES NEUTRONIQUES DU TUBE CONDUCTEUR DE
    NEUTRONS INSTALLE PRES DU REACTEUR EL3.
    ⌐FARNOUX(B.), HENNION(B.), FAGOT(J.).

32 IAEA SYMP. COPENHAGEN, VOL.2,   331, (1968)....... . . . .
    ⌐THE PROPERTIES AND PERFORMANCE OF THE HOT NEUTRON SOURCE AT THE FR2
    REACTOR.
    ⌐ABELN(O.), DREXEL(W.), GLASER(W.), GOMPF(F.), REICHARDT(W.), RIPFEL(H.).

33 IAEA SYMP. COPENHAGEN, VOL.2,   259, (1968)...... . . . .
    ⌐MCMASTER UNIVERSITY NEUTRON CRYSTAL SPECTROMETERS.
    ⌐BROCKHOUSE(B.N.), DEWIT(G.A.), HALLMAN(E.D.), ROWE(J.M.).

34 IAEA SYMP. COPENHAGEN, VOL.2,   271, (1968)....... . . . .
    ⌐THE BATTELLE NORTHWEST ROTATING CRYSTAL AND PHASED CHOPPER SLOW NEUTRON
    SPECTROMETER.
    ⌐HARLING(O.K.).

35 IAEA SYMP. COPENHAGEN, VOL.2,   281, (1968)...... . . . .
    ⌐THE CHALK RIVER ROTATING CRYSTAL SPECTROMETER.
    ⌐WOODS(A.D.B.), GLASER(E.A.), COWLEY(R.A.).

36 IAEA SYMP. COPENHAGEN, VOL.2,   289, (1968)...... . . . .
    ⌐THE DOUBLE-CHOPPER NEUTRON SPECTROMETER AT ISPRA.
    ⌐KREBS(K.).

37 IAEA SYMP. (INSTRUMENT.) VIENNA 197, (1969)..... . . . .
    ⌐DEVELOPMENTS IN NEUTRON SCATTERING AT THE MTR.
    ⌐BRUGGER(R.M.).

38 IAEA SYMP. GRENOBLE,   681, (1972)..... . . . .
    ⌐NEW INSTRUMENTATION (WITH SPECIAL REFERENCE TO THE HIGH FLUX
    REACTOR ILL)
    ⌐MAIER-LEIBNITZ(H.).

39 IEEE TRANS NUCL. SCI. NS-13 NO.2   69, (1966)....E......... . . . .
    ⌐ARGONNE AUTOMATIC NEUTRON DIFFRACTOMETER.
    ⌐JOHANSON(E.W.).

40 INDIAN J. PURE APPL. PHYS. VOL.6 550, (1968)....E...... . . . .
    ⌐TROMBAY AUTOMATIC DIFFRACTOMETERS: A SEMI AUTOMATIC NEUTRON DIFFRACT-
    OMETER (SAND)
    ⌐CHIDAMBARAM(R.), SEQUEIRA(A.), MOMIN(S.N.).

41 INSTRUM. EXP. TECH. NO.6,   902, (1961)....E...... . . . .
    ⌐NEUTRON DIFFRACTION APPARATUS ON THE IRT REACTOR.
    ⌐GOMAN≠KOV(V.I.),   KASATKIN(S.N.),   KISELEV(S.V.),   LOSHMANOV(A.A.),
    OZEROV(R.P.).

42 J. APPL. CRYST. VOL.4,   410, (1971)....E...... . . . .
    ⌐A NEUTRON SMALL-ANGLE SCATTERING APPARATUS WITH MULTIDETECTOR AT THE
    FRJ-2.
    ⌐SCHMATZ(W.), SCHELTEN(J.).

43 J. NUCL. ENERGY VOL.6,   222, (1958)....E...... . . . .
    ⌐SLOW-NEUTRON SPECTRUM MEASUREMENTS FROM THE SWEDISH HEAVY-WATER
    REACTOR, R1.
    ⌐LARSSON(K.E.), STEDMAN(R.), PALEVSKY(H.).

44 J. PHYS. SOC. JAPAN VOL.17 B-II, 354, (1962)....E...... . . . .
    ⌐NEUTRON DIFFRACTOMETER J.A.E.R.I.
    ⌐KUNITOMI(N.), HAMAGUCHI(Y.), SAKAMOTO(M.), KOMURA(S.).

45 J. PHYS. SOC. JAPAN VOL.17 B-II, 352, (1962)....E...... . . . .
    ⌐NEUTRON DIFFRACTION RESEARCH IN AUSTRALIA.
    ⌐SABINE(T.M.).

46 J. SCI. INSTRUM. VOL.39,   590, (1962)....E...... . . . .
    ⌐SINGLE-CRYSTAL NEUTRON DIFFRACTION EQUIPMENT AT THE DIDO REACTOR.
    ⌐WILLIS(B.T.M.).

47 KERNTECHNIK VOL.15,   450, (1973)....E...... . . . .
    ⌐CONSTRUCTION OF A NEUTRON SMALL ANGLE SCATTERING FACILITY FOR THE
    GRENOBLE HIGH FLUX REACTOR.
    ⌐DEGENKOLBE(G.), GREISS(H.B.).

48 MITSUBISHI DENHI ENG. NO.26,   10, (1970)....E...... . . . .
    ⌐NEUTRON DIFFRACTOMETER FOR THE ATOMIC ENERGY COMMISSION OF IRAQ.
    ⌐IMAI(H.), MITOMI(Y.), MATSUMIYA(M.), HIRONAKA(K.). (ED. NOTE: IN ENGL.).

49 MON. DEUTSCHEN AKAD. WISS. VOL.4 586, (1962)....E...... . . . .
    ⌐THE POSSIBILITIES OF STRUCTURAL INVESTIGATIONS WITH NEUTRONS IN THE
    CENTRAL INSTITUTE FOR NUCLEAR PHYSICS AT ROSSENDORF.
    ⌐KLEINSTUCK(K.). (IN GERMAN).

50 NATURE VOL.183,   35, (1959)....E...... . . . .
    ⌐NEUTRON DIFFRACTION AT A HIGH-FLUX REACTOR.
    ⌐BACON(G.E.), DYER(R.F.).

SECTION 23 -RESEARCH PROGRAMMES AND FACILITIES AT VARIOUS LABORATORIES .

51 NATURE VOL.228,.. . 324, (1970)...............................
    *BRITISH HIGH FLUX BEAM REACTOR.
    *EGELSTAFF(P.A.).

52 NUCL. INSTRUM. VOL.1,.. 92, (1957)...E. FAST CHOPPER. . . . .
    *MODIFICATION OF THE BROOKHAVEN FAST CHOPPER.
    *SEIDL(F.G.P.), PALEVSKY(H.), HUGHES(D.J.), ZIMMERMAN(R.L.).

53 NUCL. INST. MET. VOL.12, 75, (1961)....E. . . . .
    *M.T.R. PHASED CHOPPER VELOCITY SELECTOR.
    *BRUGGER(R.M.), EVANS(J.E.).

54 NUCL. INST. MET. VOL.13, 1, (1961). . . . .
    *THE U.S.-CANADIAN FAST CHOPPER INSTALLATION AT CHALK RIVER.
    *ZIMMERMAN(R.L.), PALEVSKY(H.), CHRIEN(R.E.), OLSEN(W.C.), SINGH(P.P.),
    WESTCOTT(C.H.).

55 NUCL. INST. MET. 26, 1, (1964)..................
    *A METHOD OF EXTENDING THE ENERGY RANGE OF THE HE3 NEUTRON SPECTROMETER.
    *BROWN(W.K.).

56 NUCL. INST. MET. 25, 205, (1964).................
    *DESIGN OF A THIRD AXIS FOR THE CHALK RIVER CRYSTAL NEUTRON SPECTROMETER.
    *MCALPIN(W.).

57 NUCL. INST. MET. 27, 61, (1964)..................
    *A SLOW NEUTRON CHOPPER TIME OF FLIGHT SPECTROMETER AT THE REACTOR R2 IN
    SWEDEN.
    *HOLMRYD(S.), SKOLD(K.), PILCHER(E.), LARSSON(K.E.).

58 NUCL. INST. MET. 46, 266, (1966). . . . .
    *THE COLD NEUTRON FACILITY AT THE FI R1.
    *TUNKELO(E.), PALMGREN(A.).

59 NUCL. INST. MET. 53, 93, (1967).................
    *THE BROOKHAVEN HIGH FLUX BEAM FAST CHOPPER FACILITY.
    *CHRIEN(R.E.), REICH(M.).

60 NUCL. INST. MET. 55, 151, (1967).................
    *A SINGLE CRYSTAL SLOW NEUTRON SPECTROMETER AT THE PULSED REACTOR IN
    DUBNA.
    *GORDON(J.), PELLIONISZ(P.), VIZI(I.), ZSIGMOND(G.), SZKATULA(A.).

61 NUCL. INST. MET. VOL.59, 245, (1968)....E.................
    *THE DIDO HOT SOURCE; ENHANCEMENT OF .1-.4 EV NEUTRON INTENSITIES BY
    RETHERMALIZATION.
    *EGELSTAFF(P.A.), MOFFITT(R.D.), SAUNDERSON(D.H.).

62 NUKLEONIK VOL.2, 41, (1963)....E.................
    *THE THERMAL NEUTRON SPECTRUM IN THE WWR-S REACTOR, BUCHAREST.
    *TEUTSCH(H.), APOSTOLESCU(S.), TIMIS(P.).

63 NUKLEONIKA VOL.7,.. 223, (1962) . . . .
    *THE CRACOW NEUTRON CRYSTAL SPECTROMETER.
    *KRASNICKI(S.), PAWELCZYK(J.), RAPACKI(H.).

64 NUKLEONIK VOL.6, 87, (1964)....P...............
    *TRANSMISSION AND SCATTERING EXPERIMENTS WITH SLOW NEUTRONS IN THE
    FRM-REACTOR.
    *SPRINGER(T.). (IN GERMAN).

65 NUOVO CIMENTO SUPPL. VOL.23, 17, (1962)....E.............
    *CRYSTAL SPECTROMETER FOR THE DIFFRACTION OF NEUTRONS AT THE CENTRE OF
    ISPRA.
    *CAGLIOTI(G.), DE-AGOSTINO(E.), MARSILI(F.), PAOLETTI(A.),
    PELLEGRINI(U.), RICCI(F.P.).

66 PROC. INT. CONF. (NOTTINGHAM), . 271, (1964)................
    *NEUTRON MAGNETIC SCATTERING STUDIES AT THE OAK RIDGE NATIONAL LABORATORY
    *KOEHLER(W.C.), CABLE(J.W.), CHILD(H.R.), MOON(R.M.), WOLLAN(E.O.).

67 PULSED NEUTRON SOURCES/UTILIZ.,.. . (1971)....E............
    *NEUTRON SPECTROSCOPY WITH LINAC-BOOSTER (PROJECTED IN JAPAN).
    *ISHIKAWA(Y.), WATANABE(N.), KIMURA(M.). (ED. NOTE: PRES. AT EURATOM-
    JAPAN JOINT MEETING ON PULS. NEUT. SOUR. AND UTILIZ., ISPRA, SEPT.17).

68 RENSSELAER POL. INST. SYMP.,.. 124, (1961). . . . .
    *BROOKHAVEN RESPONSIBILITY FOR CROSS SECTION COMPILATION AND REVIEW.
    *KOUTS(H.).

69 REPORT AERE-R5698 (14 PP.),.. (1968)....E............
    *THE DIDO WHITE BEAM SCATTERING APPARATUS.
    *BIERNE(T.). (ED.NOTE: AVAIL. H.M.S.O., 49 HIGH HOLBORN, LONDON WC1,
    ENG.).

70 REPORT AERE-R5647 (33 PP.),.. (1968)....E...........
    *NEUTRON MONOCHROMATOR STUDIES AT A.E.R.E.
    *TURBERFIELD(K.C.). (ED. NOTE: AVAIL. H.M.S.O., 49 HIGH HOLBORN, LONDON,
    WC1, ENG.).

71 REPORT AERE-R 6246 (25PP.),.. (1970)....E...........
    *THE DIDO (6H) LONG WAVELENGTH INELASTIC NEUTRON SPECTROMETER.
    *BUNCE(L.J.), HARRIS(D.H.C.), STIRLING(G.C.). (ED. NOTE: AVAIL. HMSO,
    49 HIGH HOLBORN, LONDON WC1, ENGLAND).

72 REV. DE PHYSIQUE (BUCAREST) 5, 83, (1960)....E............
    *THE NEUTRON CRYSTAL SPECTROMETER OF THE INSTITUTE OF ATOMIC PHYSICS-
    BUCHAREST.
    *BALLY(D.), TARINA(E.), TODIREANU(S.), OLTEANU(I.).

73 REV. DE PHYSIQUE VOL.6,.. 411, (1961)....E............
    *FEATURES OF THE CURVED-SLIT NEUTRON CHOPPER OF THE INSTITUTE FOR ATOMIC
    PHYSICS AT BUCHAREST.
    *TEUTSCH(H.), MATEESCU(N.), PIRLOGEA(P.), RADULESCU(C.), TIMIS(P.),
    VASILIU(V.). (IN GERMAN).

74 REV. ROUMAINE PHYS. VOL.10,.. 445, (1965)....E............
    *THE THREE AXES SPECTROMETER OF THE INSTITUTE FOR ATOMIC PHYSICS
    (BUCHAREST).
    *BALLY(D.), TODIREANU(S.), BIRSAN(I.), PIRLOGEA(P.), TARINA(E.).

75 REV. SCI. INSTRUM. VOL.21, 705, (1950)....E............
    *THE CHALK RIVER SINGLE CRYSTAL NEUTRON SPECTROMETER.
    *HURST(D.G.), PRESSESKY(A.J.), TUNNICLIFFE(P.R.).

SECTION 23 -RESEARCH PROGRAMMES AND FACILITIES AT VARIOUS LABORATORIES .

76   SOV. PHYS. JOURNAL VOL.10 NO.4, .   63, (1967)....E....
       *THE NEUTRON-DIFFRACTION SYSTEM ON THE THERMAL COLUMN OF THE VVR-S
       REACTOR.
       *KHRUSHCHEV(B.I.), BOGOMOLOV(A.M.), SHARIPOVA(L.S.).

77   STUD. CERCETARI FIZ. VOL.20, . , 1033, (1968)....E....
       *NEUTRON DIFFRACTOMETER AT THE BUCHAREST INSTITUTE OF ATOMIC PHYSICS.
       *GHEORGHIU(Z.), DUTESCU(N.), GELLBERG(D.), BALLY(D.). (ED. NOTE: IN RUM.)

78   THESIS (106 PP.), ... , (1969)....B7....
       *NEUTRON INELASTIC SCATTERING WITH THE WEST VIRGINIA UNIVERSITY ELECTRO-
       STATIC ACCELERATOR.
       *ERRINGTON(P.R.).   (ED.   NOTE: AVAIL. UNIV. MICROFILMS, ANN ARBOR, MICH.,
       ORDER NO. 67-10848).

79   UKAEA AERE REPORT R 4895 (22 PP) . , (1965)....E....
       *THE DIDO TRIPLE AXIS CRYSTAL SPECTROMETER.
       *SAUNDERSON(D.H.), DUFFILL(C.), SHARP(R.I.).

80   UKAEA AERE REPORT 5259 (9PP.), . . , (1966)....E....
       *A REVIEW OF THERMAL NEUTRON SCATTERING EQUIPMENT AT A.E.R.E. HARWELL.II.
       INELASTIC SCATTERING APPARATUS.
       *DYER(R.F.).

81   UKAEA AERE REPORT 5259 (13 PP.), .   , (1966)....E....
       *A REVIEW OF THERMAL NEUTRON SCATTERING EQUIPMENT AT A.E.R.E. HARWELL. I.
       ELASTIC SCATTERING APPARATUS.
       *DYER(R.F.).

SECTION   24 -MISCELLANEOUS. . . . .

1   ACTA CRYST. VOL.9, ... 61, (1956). . . . .E. . . . . . .CALCITE,CO*S*O4 IN NEARLY PERFECT
                      ↪ANOMALOUS ABSORPTION OF SLOW NEUTRONS AND X-RAYS IN NEARLY PERFECT
                      SINGLE CRYSTALS.
                      ↪KNOWLES(J.W.).

2   ACTA CRYST. VOL.A29, ... 211, (1973). . . . .
                      ↪USE OF THE TANGENT FORMULA TO RESOLVE THE PHASE AMBIGUITY IN THE NEUTRON
                      ANOMALOUS-DISPERSION METHOD.
                      ↪SIKKA(S.K.).

3   ADV. COLLOID INTERFACE SCI. 2, . . 1, (1968). . . . .E. . . . .
                      ↪APPLICATIONS OF SLOW NEUTRON SCATTERING TO STUDIES IN COLLOID AND
                      SURFACE CHEMISTRY (REVIEW).
                      ↪BOUTIN(H.), PRASK(H.), IYENGAR(R.D.).

4   ANN. REV. BIOPHYS. BIOENGNG. 1, . 529, (1972). . . . .
                      ↪NEUTRON SCATTERING (REVIEW).
                      ↪SCHOENBORN(B.P.), NUNES(A.C.).

5   ARCH. HUTNICTWA VOL.4, ... 81, (1959). . . . .E. . . . .
                      ↪PREPARATION OF LARGE METAL MONOCRYSTALS FOR NEUTRON DIFFRACTION.
                      ↪MODRZEJEWSKI(A.), BURAS(B.), CZARNECKI(R.). (SEE: KERNTECHNIK 2, PG.153)

6   ATOMKERNENERGIE VOL.7, ... 170, (1962). . . .E. . . . .
                      ↪STERN-GERLACH EXPERIMENT WITH NEUTRONS.
                      ↪HASLER(H.G.), WEBER(G.).

7   ATOMKERNENERGIE VOL.15, ... 275, (1970). . . . .
                      ↪INTERACTIONS OF NEUTRONS WITH BLOCH WALLS.
                      ↪RAUCH(H.), ROCKENBAUER(G.), RUPAR(H.).

8   ATOMKERNENERGIE VOL.18, ... 215, (1971). . . .P. . . . .
                      ↪TRANSMISSION AND REFLECTION COEFFICIENT WITH ANISOTROPIC SCATTERING.
                      ↪EL-WAKIL(S.A.).

9   BR. J. RADIOL. VOL.45, ... 529, (1972). . . .P. . . . .
                      ↪ASPECTS OF NEUTRON THERAPY BASED ON AN ANALYSIS OF RELATIONSHIPS BETWEEN
                      RBE AND DEOSE.
                      ↪ALPER(T.).

10 COMMENTS SOLID STATE PHYS. VOL.2   88, (1969). . .P. . . . .
                      ↪THE INTERACTIONS OF NEUTRONS WITH SOLIDS.
                      ↪MARSHALL(W.), LOVESEY(S.W.).

11 DISS. ABS. VOL.30, ... 3817B, (1970). . . . .
                      ↪THE LOCAL PILE OSCILLATOR AS A DEVICE FOR MEASURING TEMPERATURE
                      DEPENDENCE IN EPITHERMAL-NEUTRON ABSORPTION.
                      ↪DEKKER(A.L.).

12 ELECTRONIC STRUCTURES IN SOLIDS, 153, (1969). . . . .A4. . . . .
                      ↪METHOD AND APPLICATION OF NEUTRON SCATTERING FOR THE STUDY OF FLUCTUA-
                      TIONS IN CONDENSED MATTER.
                      ↪SPRINGER(T.), GLASER(W.).   (ED. NOTE: PUBL. PLENUM, LONDON, ENG. 1969).

13 ENERGIA NUCLEARE (ITALY) VOL.16, 65, (1969). . . .P. . . . .
                      ↪NEUTRON RADIOGRAPHY AND ITS APPLICATIONS.
                      ↪ACCINNI(F.), TONOLINI(F.).

14 HELV. PHYS. ACTA VOL.41, ... 868, (1968). . . .P. . . . .
                      ↪INVESTIGATION OF CONDENSED MATERIAL BY MEANS OF NEUTRON SCATTERING.
                      ↪HALG(W.). (ED. NOTE: IN GERMAN).

15 IAEA SYMP. (PILE RESEARCH)VIENNA 633, (1960). . . . .
                      ↪PAIRED-CHAMBER TYPE PILE OSCILLATOR.
                      ↪FUKETA(T.), ISHII(M.), OTOMO(S.).

16 IAEA SYMP. (PILE RESEARCH)VIENNA 615, (1960). . . . .
                      ↪AN APPARATUS FOR THE STUDY OF GAMMA-RAY SPECTRA FROM THE CAPTURE OF
                      MONOCHROMATIC NEUTRONS.
                      ↪MONAHAN(J.), RABOY(S.), RINGO(G.R.), TRAIL(C.C.).

17 IAEA SYMP. (PILE RESEARCH)VIENNA 609, (1960). . . . .
                      ↪MEASUREMENT OF CAPTURED GAMMA-RAYS USING A BENT CRYSTAL SPECTROMETER.
                      ↪RASMUSSEN(N.C.), MARK(H.), KAZI(A.H.).

18 IAEA SYMPOSIUM BOMBAY, VOL.2, . . 537, (1964). . . . .
                      ↪DYNAMICS OF LIQUIDS AND SOLIDS BY OPTICAL SPECTROSCOPY.
                      ↪KRISHNAN(R.S.).

19 IAEA SYMPOSIUM BOMBAY, VOL.2, . . 553, (1964). . . . .
                      ↪SUMMARY OF THE SYMPOSIUM ON INELASTIC SCATTERING OF NEUTRONS.
                      ↪EGELSTAFF(P.A.).

20 IAEA SYMP. COPENHAGEN, VOL.1, . . 623, (1968). . . . .
                      ↪RELAXATION SPECTROSCOPY IN LIQUIDS (REVIEW PAPER).
                      ↪MONTROSE(C.J.), LITOVITZ(T.A.).

21 IAEA SYMP. COPENHAGEN, VOL.1, . . 379, (1968). . . .87. . . .H*BR,H*BR-D*BR. . . . .
                      ↪NUCLEAR MAGNETIC RESONANCE IN LIQUIDS AND SOLIDS (REVIEW PAPER).
                      ↪POWLES(J.G.).

22 IEEE TRANS. NUCL. SCI. VOL.17, . 138, (1970). . . .E. . . . .
                      ↪INSTRUMENTATION FOR IN VIVO NEUTRON ACTIVATION ANALYSIS IN HUMANS.
                      ↪PALMER(H.E.), PAILTHORP(K.G.), SHEEN(E.M.).

23 J. CHEM. EDUC. VOL.45, ... 296, (1968). . . . .
                      ↪STRUCTURAL CHEMISTRY IN THE NUCLEAR AGE.
                      ↪HAMILTON(W.C.).

24 J. CHEM. PHYS. VOL.57, ... 175, (1972). . . .P. . . . .
                      ↪STUDY OF HINDERED ROTATION IN CRYSTALS BY NEUTRON TRANSMISSION MEASURE-
                      MENTS: COMPARISON OF CALCULATED AND MEASURED SCATTERING CROSS SECTIONS.
                      ↪LEUNG(P.S.), RUSH(J.J.), TAYLOR(T.I.).

25 J. NUCL. ENERGY VOL.25, ... 129, (1971). . . .P. . . . .
                      ↪FURTHER ACTIVATION THERMAL NEUTRON CAPTURE CROSS-SECTIONS AND RESONANCE
                      INTEGRALS.
                      ↪RYVES(T.B.).

SECTION 24 -MISCELLANEOUS. . . . .

26 J. NUCL. SCI. TECHNOL.(JAP.) 8, 470, (1971)....I........
    *APPLICATION OF NORMAL MODE EXPANSION FOR THE THERMAL-NEUTRONS MILNE
    PROBLEM.
    *YAMAGISHI(T.).

27 KERNTECHNIK VOL.2, 153, (1960)....E........AL,CU,PB,ZN,.......
    *GROWTH OF LARGE METAL SINGLE CRYSTALS FOR NEUTRON DIFFRACTION.
    *MODRZEJEWSKI(A.), BURAS(B.), CZARNECKI(R.). (IN GERMAN).

28 NATURE VOL.129,. 312, (1932).......................
    *POSSIBLE EXISTENCE OF A NEUTRON.
    *CHADWICK(J.).

29 NATURE VOL.130,. 57, (1932)....E........
    *NEW EVIDENCE FOR THE NEUTRON.
    *CURIE(I.), JOLIOT(F.).

30 NATURE VOL.135,. 904, (1935)....E........
    *EVIDENCE ON THE VELOCITIES OF ≠SLOW≠ NEUTRONS.
    *MOON(P.B.), TILLMAN(J.R.).

31 2ND INTERNAT. CONF./NUCL. DATA,. (1970)....P........
    *NEUTRON CROSS-SECTION EVALUATION IN THE THERMAL AND RESONANCE ENERGY
    RANGE FOR NUCLIDES OF MASS LESS THAN 220.
    *STORY(J.S.).

32 NED. TIJDSCHR. NATUURK. VOL.36,. 250, (1970)..............
    *NEUTRONS AND CRYSTALS.
    *GOEDKOOP(J.A.).

33 NUCLEONICS VOL.5 OCTOBER, 51, (1949)....P........
    *THERMAL NEUTRON CROSS SECTIONS AND RELATED DATA.
    *KROEGER(H.R.).

34 NUCLEONICS VOL.5 DECEMBER, 4, (1949)........
    *NEUTRON SPECTROSCOPY FOR CHEMICAL ANALYSIS- I. NEUTRON PROPERTIES,
    INTERACTIONS, VELOCITY SELECTION.
    *TAYLOR(T.I.), HAVENS-JR(W.W.).

35 NUCLEONICS VOL.6 NO.2,. 66, (1950)....T,E........
    *NEUTRON SPECTROSCOPY FOR CHEMICAL ANALYSIS-II. RESONANCE EFFECTS,
    TRANSMISSION IN THE 1/V REGION.
    *HAVENS-JR(W.W.), TAYLOR(T.I.).

36 NUCLEONICS VOL.6 NO.4,. 54, (1950)....E........
    *NEUTRON SPECTROSCOPY FOR CHEMICAL ANALYSIS-III. THERMAL NEUTRONS,
    NEUTRON FLUX, ACTIVATION.
    *TAYLOR(T.I.), HAVENS-JR(W.W.).

37 NUCLEONICS VOL.6 NO.5,. 38, (1950)....E........
    *PILE NEUTRON RESEARCH TECHNIQUES-II.
    *HUGHES(D.J.).

38 NUCL. INST. MET. 20,. 181, (1963)........
    *EFFECT OF TARGET THICKNESS AND BEAM STRAGGLING ON THE MAGNITUDE OF
    RESONANCES IN ELASTIC SCATTERING EXPERIMENTS.
    *WEINMAN(J.A.).

39 NUCL. INST. MET. 30,. 77, (1964)........
    *MULTIPLE SCATTER CORRECTIONS USING THE MONTE-CARLO PROGRAM MAGGIE.
    *PARKER(J.B.), TOWLE(J.H.), SAMS(D.), GILBOY(W.B.)...ET AL.

40 NUCL. INST. MET. 31,. 61, (1964)........
    *INTENSITY BACKGROUND AND RESOLUTION WIDTH IN NUCLEAR COUNTING EXPERI-
    MENTS.
    *QUITTNER(G.).

41 NUCL. INST. MET. 35,. 153, (1965)........
    *AIR SCATTERING IN NEUTRON CAPTURE CROSS SECTION MEASUREMENTS.
    *ROBERTSON(J.C.), RYVES(T.B.).

42 NUCL. INST. MET. 34,. 29, (1965)........
    *THICK TARGET METHOD FOR THE MEASUREMENT OF DIFFERENTIAL NEUTRON
    SCATTERING CROSS-SECTIONS.
    *BARNARD(E.), FERGUSON(A.T.G.), MCMURRAY(W.R.)...ET AL.

43 NUCL. INST. MET. 41,. 89, (1966)........
    *A CONVERSION FUNCTION FOR CALCULATING ELASTIC SCATTERING ANGULAR
    DISTRIBUTIONS.
    *NURZYNSKI(J.).

44 NUCL. INST. MET. 39,. 350, (1966)........
    *THERMAL NEUTRON ABSORPTION CROSS SECTION OF THE AERE BORON STANDARD.
    *WYNCHANK(S.A.R.), COX(A.E.), COLLIE(C.H.).

45 NUCL. INST. MET. 52,. 321, (1967)........
    *A NEW TECHNIQUE FOR MEASURING WEAK NEUTRON SCATTERING IN LOW ENERGY
    RESONANCES.
    *KING(T.J.), FULLWOOD(R.R.), BLOCK(R.C.).

46 NUCL. INST. MET. 57,. 82, (1967)........
    *SLOW NEUTRON SCATTERING BY THE BRIGHT LINE TECHNIQUE.
    *GARNSWORTHY(R.K), BROOMHALL(G.J.).

47 NUCL. INST. MET. 59,. 120, (1968)........
    *A METHOD OF TIME CALIBRATION FOR TIME OF FLIGHT EXPERIMENTS IN NUCLEAR
    PHYSICS.
    *LANGSFORD(A.), DOLLEY(P.E.).

48 NUCL. INST. MET. 61,. 349, (1968)..............
    *DER GD-NEUTRONENGAHLER.
    *FEIGL(B.), RAUCH(H.).

49 NUCL. INST. MET. 61,. 37, (1968)........
    *MEASUREMENT OF NEUTRON SPECTRA BY THE TRANSMISSION METHOD.
    *NAVALKAR(M.B.), RAMAKRISHNA(D.V.S.).

50 NUCL. INST. MET. 61,. 45, (1968)........
    *SELF INDICATION IN AREA ANALYSIS OF NEUTRON RESONANCES 1. METHOD.
    *FLECK(C.M.), WEISSHAUPL(H.A.).

51 NUCL. INST. MET. 61,. 53, (1968)........
    *SELF INDICATION IN AREA ANALYSIS OF NEUTRON RESONANCES 2. CORRECTIONS.
    *FLECK(C.M.), WEISSHAUPL(H.A.).

SECTION 24 -MISCELLANEOUS. . . . .

52 NUCL. INST. MET. 61, 245, (1968)
    *A TECHNIQUE TO MEASURE NEUTRON CROSS SECTIONS IN THE LOW KEV ENERGY
    REGION.
    *SIMPSON(O.D.), MILLER(L.G.).

53 NUCL. INST. MET. 62, 62, (1968)
    *RECURSION FORMULA FOR SEMI-ANALYTICAL MULTIPLE SCATTERING CALCULATIONS
    IN SLAB GEOMETRY.
    *FROHNER(F.H.).

54 NUCL. INST. MET. 62, 311, (1968)
    *A THERMAL NEUTRON TELEVISION SYSTEM USING A HIGH YIELD NEUTRON
    GENERATOR.
    *KAWASAKI(S.).

55 NUCL. INST. MET. 63, 307, (1968)
    *THE BEHAVIOUR OF IONIZATION CHAMBER-NEUTRON DETECTORS IN INTENSE
    TRANSIENT GAMMA FIELDS.
    *GOZANI(T.).

56 NUCL. INST. MET. 63, 329, (1968)
    *AN IMPROVED SPHERICAL DOSE EQUIVALENT NEUTRON DETECTOR.
    *LEAKE(J.W.).

57 NUCL. INST. MET. 66, 141, (1968)
    *NEUTRON RESONANCE AREA ANALYSIS WITH MUTUAL-INDICATION.
    *FLECK(C.M.), WEISSHAUPL(H.A.).

58 NUCL. INST. MET. 66, 149, (1968)
    *SELECTION OF THE FOREIGN NEUTRON RESONANCE IN MUTUAL-INDICATION.
    *FLECK(C.M.), WEISSHAUPL(H.A.).

59 NUCL. INST. MET. 66, 229, (1968)
    *A FLAT CRYSTAL SPECTROMETER FOR (MU,GAMMA) STUDIES.
    *NILSSON(S.), FALKSTROM(E.), BOREVING(S.).

60 NUCL. INST. MET. 61, 198, (1968)
    *DECOMPOSITION AUTOMATIQUE DES COURBES DE SECTION EFFICACE EN SOMME DE
    TERMES DE BREIT ET WIGNER.
    *NIFENECKER(H.), PERRIN(G.), AUDIAS(A.).

61 NUCL. INST. MET. 58, 1, (1968)
    *ABOUT THE DETERMINATION OF SLOW NEUTRON RESONANCE PARAMETERS.
    *COCEVA(C.), CORVI(F.), GIACOBBE(P.), STEFANON(M.).

62 NUCL. INST. MET. 60, 269, (1968)
    *A COMBINATION NEUTRON AND GAMMA RAY SPECTROMETER.
    *CIALELLA(C.M.), DEVANNEY(J.A).

63 NUCL. INST. MET. 70, 52, (1969)
    *COUNTING STATISTICS OF NUCLEAR REACTIONS.
    *FOGLIO-PARA(A.), MANDELLI-BETTONI(M.).

64 NUCL. INST. MET. 70, 221, (1969)
    *COMMENTS ON A COMBINATION NEUTRON AND GAMMA RAY SPECTROMETER.
    *MEYER(W.), SIMONS(G.G.), DONNER(H.J.).

65 NUCL. INST. MET. 70, 225, (1969)
    *COMMENTS ON COMMENTS ON A COMBINATION NEUTRON AND GAMMA RAY
    SPECTROMETER.
    *GIALELLA(C.M.).

66 NUCL. INST. MET. 71, 221, (1969)
    *MULTIPLE SCATTERING AND ABSORPTION OF THERMAL NEUTRON IN THIN TARGETS.
    *CHOUDRY(A.).

67 NUCL. INST. MET. 72, 301, (1969)
    *EIN UNTERGRUNDARMES SPEKTROMETER FUR SCHNELLE NEUTRONEN.
    *SCHERBER(W.), HOFMANN(A.).

68 NUCL. INST. MET. 74, 109, (1969)
    *DESIGN OF A NEUTRON SCATTERING CHAMBER USING MONTE CARLO CALCULATIONS.
    *GOLLNITZ(H.), HEIDBREDER(E.)...ET AL.

69 NUCL. INST. MET. 72, 213, (1969)
    *SHIELDING AGAINST NEUTRONS IN THE ENERGY RANGE 50 TO 400 MEV.
    *ALSMILLER(R.G.), MYNATT(F.R.), BARISH(J.), ENGLE(W.W.).

70 NUCL. INST. MET. 76, 103, (1969)
    *VERWENDUNG VON RECHENPROGRAMMEN BEI BESTIMMUNG WIRKUNGSQUERSCHNITTEN
    SCHNELLER NEUTRONEN NACH AKTIVIERUNGSVERFAHREN.
    *ABELS(C.), BISSEM(H.), BORMANN(M.), LAMMERS(B.)...ET AL.

71 NUCL. INST. MET. 77, 189, (1970)
    *SEMICONDUCTOR NEUTRON DETECTORS UTILIZING RADIOACTIVE DECAY.
    *JOHNSON(R.T.).

72 NUCL. INST. MET. VOL.79, 51, (1970)...E
    *EXPERIMENTAL DETERMINATION OF THE SLOW-NEUTRON WAVELENGTH DISTRIBUTION.
    *LEBECH(B.), MIKKE(K.), SLEDZIEWSKA-BLOCKA(D.).

73 NUCL. INST. MET. 77, 55, (1970)
    *PLURAL NEUTRON SCATTERING CORRECTION BY AN ANALYTICAL METHOD.
    *KUIJPER(P.), VEEKIND(J.C.), JONKER(C.C.).

74 NUCL. INST. MET. 77, 245, (1970)
    *NEUTRON INELASTIC SCATTERING MEASUREMENTS WITH A GATED PHOTOMULTIPLIER.
    *FULLWOOD(R.R.), HOCKENBURG(R.W.).

75 NUCL. INST. MET. VOL.91, 13, (1971)...P
    *TARGET ASSEMBLY FOR LOW YIELD NEUTRON AND GAMMA-RAY EXPERIMENTS.
    *ROBB(A.D.), SCHIER(W.A.).

76 NUCL. INST. MET. VOL.103, 229, (1972)...P
    *CONVERTER-THICKNESS FOR OPTIMUM INTENSITY IN NEUTRON-RADIOGRAPHY.
    *MULLNER(M.), JEX(H.).

77 NUCL. INST. MET. VOL.104, 217, (1972)...E
    *TEMPERATURE SENSING BY NEUTRON TRANSMISSION.
    *HARMS(A.A.), GARLAND(W.J.).

78 NUCL. INST. MET. VOL.106, 611, (1973)
    *NEUTRON MIRAGE FORMATION.
    *BATES(J.C.).

SECTION 24 -MISCELLANEOUS. . . . .

79 NUCL. SCI. ENGNG. VOL.6, ... 76, (1959) . . . . .E . . . .
  *THERMAL NEUTRON INELASTIC SCATTERING EFFECTS IN A SINGLE CRYSTAL
  NEUTRON SPECTROMETER.
  *DUGGAL(V.P.).

80 NUCL. SCI. ENGNG. VOL.34, ... 114, (1968) . . . . .E . . . .
  *EXPERIMENTAL EVALUATION OF MINIMA IN THE TOTAL NEUTRON CROSS SECTIONS OF
  SEVERAL SHIELDING MATERIALS.
  *STRAKER(E.A.).

81 NUCL. SCI. ENGNG. VOL.40, ... 25, (1970) . . . .
  *THE CONSTRUCTION OF NEUTRON CROSS SECTIONS IN THE UNRESOLVED RESONANCE
  REGION.
  *ISHIGURO(Y.), KATSURAGI(S.), NAKAGAWA(M.), TAKANO(H.).

82 NUCL. SCI. ENGNG. VOL.42, ... 104, (1970) . . . . .
  *THE AVERAGE ANGLE OF SCATTERING IN ENERGY-DEPENDENT PROBLEMS.
  *MCGIRT(F.), BECKER(M.).

83 NUCL. TECHNOL. VOL.15, ... 56, (1972) . . . .E . . . . . .
  *DEVELOPMENTS IN THE USE OF CALIFORNIUM-252 FOR NEUTRON RADIOGRAPHY.
  *BARTON(J.P.).

84 NUOVO CIMENTO A VOL.68, ... 657, (1970) . . . .P . . . .
  *AN ESTIMATE OF THE NEUTRON ELECTRIC-DIPOLE MOMENT.
  *MCCLIMENT(E.R.), TEETERS(W.D.).

85 PHYS. EDUC.(GB) VOL.6, ... 396, (1971) . . . .P . . . .
  *THE NEUTRON- SOME OF ITS PROPERTIES AND THEIR APPLICATION.
  *ALSTON(W.C.H.).

86 PHYS. MED. BIOL.(GB) VOL. 16, . . . 439, (1971) . . . .T . . . .
  *NEUTRON RADIOGRAPHY AND DOSIMETRY IN HUMAN BEINGS; THEORETICAL STUDIES.
  *BUDINGER(T.F.), HOWERTON(R.F.), PLECHATY(E.F.).

87 PHYS. REV. VOL.83, ... 641, (1951) . . . .E . . . .
  *THERMAL NEUTRON CAPTURE CROSS SECTIONS.
  *POMERANCE(H.).

88 PHYS. REV. VOL.88, ... 412, (1952) . . . .P . . . .
  *THERMAL NEUTRON CAPTURE CROSS SECTIONS.
  *POMERANCE(H.).

89 PHYS. REV. VOL.114, ... 830, (1959) . . . .E . . . .
  *ELECTRIC POLARIZABILITY OF THE NEUTRON.
  *BREIT(G.), RUSTGI(M.L.).

90 PHYS. REV. VOL.114, ... 827, (1959) . . . .E . . . .
  *POLARIZABILITY OF THE NEUTRON.
  *THALER(R.M.).

91 PHYS. REV. LETT. VOL.3, ... 105, (1959) . . . .E . . . .
  *ELECTRIC POLARIZABILITY OF THE NEUTRON.
  *FOLDY(L.L.).

92 PHYS. REV. 145, ... 1023, (1966) . . . .
  *COSMIC-RAY NEUTRONS NEAR THE EARTH.
  *HENDRICK(L.D.), EDGE(R.D.).

93 PHYS. REV. VOL.153, ... 1415, (1967) . . . .E . . . .
  *EXPERIMENTAL LIMIT FOR THE NEUTRON CHARGE.
  *SHULL(C.G.), BILLMAN(K.W.), WEDGWOOD(F.A.).

94 PHYS. REV. B VOL.3, ... 2929, (1971) . . . .
  *SEMICLASSICAL ESTIMATION OF NEUTRON STOPPING POWER.
  *VORA(R.B.), NEELAVATHI(V.N.), TURNER(J.E.), SUBRAMANIAN(T.S.),
  PRASAD(M.A.).

95 PROC/CONF/NEUTRON C. S. + TECH., 951, (1968) . . . .A4 . . . .
  *THERMAL NEUTRON CROSS SECTIONS AND RESONANCE INTEGRALS FOR TRANSURANIUM
  ISOTOPES.
  *PRINCE(A.).

96 PROC. ROY. SOC. A VOL.136, . . . 692, (1932) . . . .E . . . .
  *EXISTENCE OF A NEUTRON.
  *CHADWICK(J.).

97 PROC. ROY. SOC. A VOL.142, ... 1, (1933) . . . .
  *THE NEUTRON (BAKERIAN LECTURE).
  *CHADWICK(J.).

98 PROC/7TH/BIOMED/INSTRUM. SYMP., . 93, (1969) . . . .E . . . .
  *SELECTED BIOMEDICAL APPLICATIONS OF THERMAL NEUTRON RADIOGRAPHY.
  *ALLEN(J.J.), HARMER(D.S.). (ED. NOTE: PUBL. INSTRUMENT SOC. AMERICA,
  PITTSBURGH, PA., USA 1969).

99 RADIATION RESEARCH VOL.35, ... 1, (1968) . . . .
  *ENERGY TRANSFER TO MATTER BY NEUTRON.
  *BACH(R.L.), CASWELL(R.S.).

100 REPORT AERE NP/GEN/13, ... (1960) . . . .P . . . .
  *TABLES OF MOMENTUM AND ENERGY TRANSFERS IN THE SCATTERING OF THERMAL
  NEUTRONS.
  *EGELSTAFF(P.A.), HEARD(C.R.T.). (AVAIL. ATOMIC ENERGY RESEARCH ESTBL.,
  HARWELL).

101 RES. APPL./NUCL. PULSED SYSTEMS, 17, (1967) . . . .E . . . .
  *APPLICATION OF REPETITIVELY-PULSED REACTORS TO STRUCTURE AND DYNAMICS
  STUDIES OF SOLIDS.
  *BURAS(B.). (IN ≠RESEARCH APPLICATIONS OF NUCLEAR PULSED SYSTEMS≠. PUBL.
  IAEA: VIENNA).

102 REV. GEN. SCI. TOME 50, ... 89, (1939) . . . .
  *TRANSMUTATIONS PRODUCED BY SLOW NEUTRONS AND BOHR≠S NUCLEAR DYNAMICS.
  *KAHAN(T.).

103 REV. MEXICANA FIS. VOL.18, . . . 115, (1969) . . . .P . . . .
  *IMPROVEMENT IN THE MEASUREMENT OF NEUTRON TOTAL CROSS-SECTIONS.
  *VELAZQUEZ(R.), RICKARDS(J.).

104 REV. SCI. INSTRUM. VOL.41, . . . 1960, (1970) . . . .
  *A TECHNIQUE FOR OBTAINING NEUTRON RADIOGRAPHS IN THE RESONANCE REGION.
  *FORMAN(L.), BENTON(C.U.), GARRETT(D.A.), SCHELBERG(A.D.).

SECTION   24 -MISCELLANEOUS. . . . .

105 RIV. DEL NUOVO CIMENTO VOL.1, . . . 155, (1969) STUDY OF SOLIDS AND LIQUIDS. . . . .
            *THE USE OF NEUTRONS IN THE STUDY OF SOLIDS AND LIQUIDS. . . . .
            *MARSHALL(W.), LOVESEY(S.W.). (ED. NOTE: THIS REVIEW PAPER IS FOUND IN
            A SPECIAL ISSUE WITH SUBHEADING: NUMERO SPECIALE).

106 ROY. INST. GT. BRIT., PROC. 41, . 510, (1967). . . . .
            *LOOKING AT SOLIDS WITH NEUTRONS.
            *BACON(G.E.).

107 SCIENCE VOL.114, ... 341, (1951). . . .E. . . . .
            *CHEMICAL ANALYSIS BY NEUTRON SPECTROSCOPY.
            *TAYLOR(T.I.), ANDERSON (R.H.), HAVENS-JR(W.W.).

108 SOL. STATE COMM. VOL.12, ... 1191, (1973). . . .T. . . . .
            *NEUTRON SCATTERING AS A METHOD FOR FERMI SURFACE INVESTIGATIONS.
            *DIETERICH(W.).

109 SOV. AT. ENERGY VOL.25, ... 435, (1968). . . .P. . . . .
            *MULTILEVEL DESCRIPTION OF ENERGY STRUCTURE OF NEUTRON CROSS-SECTIONS.
            *LUK-YANOV(A.A.), EL-VAKIL(S.A.).

110 SOV. AT. ENERGY VOL.28, ... 214, (1970). . . . . .AIR, ...
            *SCATTERING OF NEUTRONS IN AIR.
            *MORDASHEV(V.M.).

111 SOV. PHYS. JETP VOL.14, ... 838, (1962). . . .T. . . . .
            METALLIC REFLECTION OF NEUTRONS.
            *GUREVICH(I.I.), NEMIROVSKII(P.E.).

112 SOV. PHYS. DOKLADY VOL.16, ... 154, (1971). . . . .
            *PRECISION DETERMINATION   OF THE DENSITY OF NATURAL DIAMONDS AFTER THEIR
            IRRADIATION BY SLOW NEUTRONS.
            *DERYAGIN(B.V.), BOCHKO(A.V.), KOCHERGIN(A.V.).

113 SOV. PHYS. DOKLADY VOL.16, ... 49, (1971). . . . .
            *A NEUTRON ACTIVATION METHOD FOR RAPID DETERMINATION OF CARBON CONTENT.
            *TUSTANOVSKII(V.T.), ANDRYUSHCHENKO(V.I.), VOL*GEMUT(A.A.), PRONMAN(I.M.)

114 1ST EUR. CONF./CONDENSED MATTER, . 34, (1971). . . . .
            *STUDY OF PHASE CHANGES BY NEUTRON SCATTERING.
            *NIELSON(J.A.). (ED. NOTE: PUBL. EUROPEAN PHYS. SOC., GENEVA, SWITZ.).

115 TEK. TIDSKR. (SWEDEN) VOL.103, . . 37, (1973). . . . .
            *NEUTRON BEAM METHODS REVEAL THE SECRETS OF MATERIALS.
            *SCHMIDT(W.F.). (IN SWEDISH).

116 TRIESTE/INTERN. COURSE, ... 577, (1967). . . . .
            *NEUTRONS AND MOLECULES. (26 PAGES).
            *JANIK(J.A.).

117 TRIESTE/INTERN. COURSE, ... 539, (1967). . . . .
            *SELECTED TOPICS IN NEUTRON SPECTROMETRY. (22 PAGES).
            *CAGLIOTI(G.).

118 Z. PHYS. VOL.233, ... 178, (1970). . . .B7. . . . .
            *NEUTRON SCATTERING IN A LEIBFRIED-BRENIG SPECTRUM SPACE LATTICE.
            *WEIL(J.W.). (ED. NOTE: IN ENGLISH).

SECTION 25 -PUBLICATIONS PRIOR TO 1945 . . . .

1 BULL. ACAD. SCI. URSS, SER. PHYS 189, (1938). . . . .
   *SCATTERING OF SLOW NEUTRONS IN THE CRYSTALLINE LATTICE.
   *POMERANCHUK(I.YA.). (IN ENGLISH).

2 COMP. REND. TOME 202,. . . . . 1029, (1936). . . .T. . . . .
   *SUR LA DIFFRACTION DES NEUTRONS LENTS PAR LES SUBSTANCES CRISTALLINES.
   *ELSASSER(W.M.).

3 COMP. REND. TOME 203,. . . . . . . 73, (1936). . . .A4. . . . . . . FE, ...
   *PREUVE EXPERIMENTALE DE LA DIFFRACTION DES NEUTRONS.
   *VON-HALBAN(H.), PREISWERK(P.).

4 C.R. ACAD. SCI. URSS VOL.20,. . . 551, (1938). . . . .
   *SCATTERING OF NEUTRONS IN FERROMAGNETICS.
   *MIGDAL(A.). (IN ENGLISH).

5 J. EXPER. THEOR. PHYS. (USSR) 10   5, (1940). . . .T. . . . .
   *SCATTERING OF NEUTRONS IN PARAMAGNETIC SUBSTANCES.
   *MIGDAL(A.B.).

6 J. PHYS. RADIUM SER.7 VOL.8,. . . 29, (1937). . . .E. . . . .
   *RECHERCHES SUR LES NEUTRONS LENTS.
   *VON-HALBAN(H.), PREISWERK(P.).

7 J. PHYS. (USSR) VOL.5,. . . . . . 263, (1941). . . .T. . . . .
   *THE EFFECT OF LONG RANGE ORDER IN ALLOYS UPON THE SCATTERING OF SLOW
   NEUTRONS.
   *SMIRNOV(A.A.), VONSOVSKY(S.V.).

8 NATURE VOL.129,. . . . . 312, (1932). . . . .
   *POSSIBLE EXISTENCE OF A NEUTRON.
   *CHADWICK(J.).

9 NATURE VOL.130,. . . . .   57, (1932). . . .E. . . . .
   *NEW EVIDENCE FOR THE NEUTRON.
   *CURIE(I.), JOLIOT(F.).

10 NATURE VOL.135,. . . . . 904, (1935). . . .E. . . . .
    *EVIDENCE ON THE VELOCITIES OF *SLOW* NEUTRONS.
    *MOON(P.B.), TILLMAN(J.R.).

11 NATURE VOL.142,. . . . . 829, (1938). . . .E. . . . .
    *VELOCITY DISTRIBUTION OF THERMAL NEUTRONS.
    *FERTEL(G.E.F.), MOON(P.B.), THOMSON(G.P.), WYNN-WILLIAMS(C.E.).

12 NUOVO CIMENTO VOL.12 SER.8,. . . 211, (1935). . . .E . . . .
    *THE PROPERTIES OF SLOW NEUTRONS.
    *PONTECORVO(B.). (IN ITALIAN).

13 NUOVO CIMENTO VOL.12 SER.8,. . . 201, (1935). . . . .
    *SLOW NEUTRONS.
    *FERMI(E.), RASETTI(F.). (IN ITALIAN).

14 PHYSIK. Z. SOWJETUNION VOL.13, .   65, (1938). . . .T. . . . .
    *SCATTERING OF SLOW NEUTRONS IN A CRYSTAL LATTICE.
    *POMERANCHUK(I.YA.).

15 PHYS. REV. VOL.45,. . . . . 507, (1934). . . .E. . . . .
    *ARTIFICIAL PRODUCTION OF NEUTRONS.
    *CRANE(H.R.), LAURITSEN(C.C.), SOLTAN(A.).

16 PHYS. REV. VOL.48,. . . . . 704, (1935). . . .E. . . . .
    *VELOCITY OF SLOW NEUTRONS BY MECHANICAL VELOCITY SELECTOR.
    *DUNNING(J.R.), PEGRAM(G.B.), FINK(G.A.), MITCHELL(D.P.), SEGRE(E.).

17 PHYS. REV. VOL.48,. . . . . 367, (1935). . . .T. . . . .
    *ON THE CROSS SECTION OF HEAVY NUCLEI FOR SLOW NEUTRONS.
    *VAN-VLECK(J.H.).

18 PHYS. REV. VOL.48,. . . . . 265, (1935). . . .E. . . . .
    *INTERACTION OF NEUTRONS WITH MATTER.
    *DUNNING(J.R.), PEGRAM(G.B.), FINK(G.A.), MITCHELL(D.P.).

19 PHYS. REV. VOL.50,. . . . . 899, (1936). . . . .
    *ON THE ABSORPTION AND THE DIFFUSION OF SLOW NEUTRONS.
    *AMALDI(E.), FERMI(E.).

20 PHYS. REV. VOL.50,. . . . . 738, (1936). . . .E. . . . .
    *THE PRODUCTION AND ABSORPTION OF THERMAL ENERGY NEUTRONS.
    *FINK(G.A.).

21 PHYS. REV. VOL.50,. . . . . 570, (1936). . . .E. . . . .
    *A METHOD FOR THE DETERMINATION OF ABSORPTION CROSS SECTIONS FOR THERMAL
    NEUTRONS.
    *ZAHN(C.T.), HARRINGTON(E.L.), GOUDSMIT(S.).

22 PHYS. REV. VOL.49,. . . . . 519, (1936). . . .T. . . . .
    *CAPTURE OF SLOW NEUTRONS.
    *BREIT(G.), WIGNER(E.).

23 PHYS. REV. VOL.50,. . . . . 486, (1936). . . .E. . . . . MG*O,AL, ...
    *BRAGG REFLECTION OF SLOW NEUTRONS.
    *MITCHELL(D.P.), POWERS(P.N.).

24 PHYS. REV. VOL.50,. . . . . 259, (1936). . . .T. . . . .
    *ON THE MAGNETIC SCATTERING OF NEUTRONS.
    *BLOCH(F.).

25 PHYS. REV. VOL.49,. . . . . 229, (1936). . . .T. . . . .
    *THE ENERGY DISTRIBUTION OF NEUTRONS SLOWED BY ELASTIC IMPACTS.
    *CONDON(E.U.), BREIT(G.).

26 PHYS. REV. VOL.50,. . . . . 133, (1936). . . .E. . . . .
    *SCATTERING OF SLOW NEUTRONS. II.
    *MITCHELL(A.C.G.), MURPHY(E.J.), WHITAKER(M.D.).

27 PHYS. REV. VOL.49,. . . . . 103, (1936). . . .E. . . . .
    *THE VELOCITIES OF SLOW NEUTRONS.
    *FINK(G.A.), DUNNING(J.R.), PEGRAM(G.B.), MITCHELL(D.P.).

SECTION   25 -PUBLICATIONS PRIOR TO 1945 . . . .

28   PHYS. REV. VOL.52, ... 1076, (1937)....P.......H2. . . . .
     *THE INTERACTION OF NEUTRONS WITH NORMAL AND PARAHYDROGEN.
     *DUNNING(J.R.), MANLEY(J.H.), HOGE(H.J.), BRICKWEDDE(F.G.).

29   PHYS. REV. VOL.51, ... 994, (1937)....T.......
     *ON THE MAGNETIC SCATTERING OF NEUTRONS. II.
     *BLOCH(F.).

30   PHYS. REV. VOL.52, ... 592, (1937)....E.......
     *THE PRODUCTION AND PROPERTIES OF LOW TEMPERATURE NEUTRONS. . . . .
     *LIBBY(W.F.), LONG(E.A.).

31   PHYS. REV. VOL.51, ... 544, (1937)....T.......
     *ON THE MAGNETIC SCATTERING OF NEUTRONS.
     *SCHWINGER(J.).

32   PHYS. REV. VOL.52, ... 286, (1937)....T.......H2.
     *THE SCATTERING OF NEUTRONS BY ORTHO- AND PARAHYDROGEN. . . . . . .
     *SCHWINGER(J.), TELLER(E.).

33   PHYS. REV. VOL.54, ... 827, (1938)....P.......FE,.
     *THE MAGNETIC SCATTERING OF NEUTRONS.
     *POWERS(P.N.).

34   PHYS. REV. VOL.54, ... 609, (1938)....E.......
     *THE PRODUCTION OF COLLIMATED BEAMS OF MONOCHROMATIC NEUTRONS IN THE
     TEMPERATURE RANGE 300-310 DEGREES K.
     *ALVAREZ(L.W.).

35   PHYS. REV. VOL.54, ... 266, (1938)....P,A3.....H2.
     *NEUTRON SCATTERING CROSS SECTIONS OF PARA- AND ORTHOHYDROGEN, AND OF
     N2, O2, AND H2*O.
     *BRICKWEDDE(F.G.), DUNNING(J.R.), HOGE(H.J.), MANLEY(J.H.).

36   PHYS. REV. VOL.54, ... 771, (1938)....D3.......MN*O,MN*S,
     *SCATTERING OF SLOW NEUTRONS BY PARAMAGNETIC SALTS. . . . .
     *WHITAKER(M.D.), BEYER(H.G.), DUNNING(J.R.).

37   PHYS. REV. VOL.55, ... 1101, (1939)....P.......SI*O2. . . . . .
     *TRANSMISSION OF SLOW NEUTRONS THROUGH CRYSTALS.
     *WHITAKER(M.D.), BEYER(H.G.).

38   PHYS. REV. VOL.55, ... 924, (1939)....T.......
     *ON THE THEORY OF THE FORWARD SCATTERING OF NEUTRONS BY PARAMAGNETIC
     MEDIA.
     *VAN-VLECK(J.H.).

39   PHYS. REV. VOL.55, ... 898, (1939)....T.......
     *ON THE MAGNETIC SCATTERING OF NEUTRONS.
     *HALPERN(O.), JOHNSON(M.H.).

40   PHYS. REV. VOL.55, ... 793, (1939)....P.......U, ...
     *THE CROSS SECTIONS OF METALLIC URANIUM FOR SLOW NEUTRONS.
     *WHITAKER(M.D.), BARTON(C.A.), BRIGHT(W.C.), MURPHY(E.J.).

41   PHYS. REV. VOL.55, ... 339, (1939)....P,A3.....H2,
     *THE SCATTERING OF SLOW NEUTRONS BY GASEOUS ORTHO- AND PARAHYDROGEN: SPIN
     DEPENDENCE OF THE NEUTRON-PROTON FORCE.
     *LIBBY(W.F.), LONG(E.A.).

42   PHYS. REV. VOL.55, ... 190, (1939)....T.......
     *CAPTURE OF NEUTRONS BY ATOMS IN A CRYSTAL.
     *LAMB-JR(W.R.).

43   PHYS. REV. VOL.58, ... 1031, (1940)....A1.......NI-FE, ...
     *NEUTRON STUDIES OF ORDER IN FE-NI ALLOYS.
     *NIX(F.C.), BEYER(H.G.), DUNNING(J.R.).

44   PHYS. REV. VOL.58, ... 1003, (1940)....P.......H2,
     *SCATTERING OF 20 DEGREE NEUTRONS IN ORTHO- AND PARAHYDROGEN.
     *ALVAREZ(L.W.), PITZER(K.S.).

45   PHYS. REV. VOL.57, ... 976, (1940)....P,E.......
     *INTERFERENCE PHENOMENA IN THE SCATTERING OF SLOW NEUTRONS.
     *BEYER(H.G.), WHITAKER(M.D.).

46   PHYS. REV. VOL.58, ... 321, (1940)....P.......CA*C*O3, ...
     *SCATTERING OF THERMAL NEUTRONS BY CRYSTALS.
     *RASETTI(F.).

47   PHYS. REV. VOL.58, ... 26, (1940)....T.......DEUTERONS,
     *THE SCATTERING OF THERMAL NEUTRONS BY DEUTERONS.
     *MOTZ(L.), SCHWINGER(J.).

48   PHYS. REV. VOL.58, ... 981, (1941)....T.......
     *THE PASSAGE OF NEUTRONS THROUGH CRYSTALS AND POLYCRYSTALS. . . . .
     *HALPERN(O.), HAMERMESH(M.), JOHNSON(M.H.).

49   PHYS. REV. VOL.59, ... 960, (1941)....T.......
     *ON THE PASSAGE OF NEUTRONS THROUGH FERROMAGNETS.
     *HALPERN(O.), HOLSTEIN(T.).

50   PHYS. REV. VOL.60, ... 702, (1941)....P.......H (PROTON),.
     *THE INTERACTION OF SLOW NEUTRONS WITH NUCLEI.
     *CARROLL(H.).

51   PHYS. REV. VOL.59, ... 332, (1941)....E.......
     *EXPERIMENTS WITH A SLOW NEUTRON VELOCITY SPECTROMETER.
     *BAKER(C.P.), BACHER(R.F.).

52   PHYS. REV. VOL.60, ... 280, (1941)....E.......MN*S*O4,MN*F2,
     *ON THE FORWARD SCATTERING OF NEUTRONS BY PARAMAGNETIC MEDIA.
     *WHITAKER(M.D.), BRIGHT(W.C.).

53   PHYS. REV. VOL.60, ... 18, (1941)....T.......H2,H2*O,C*H4,N*H3, ...
     *THE SCATTERING OF SLOW NEUTRONS BY MOLECULAR GASES.
     *SACHS(R.G.), TELLER(E.).

54   PHYS. REV. VOL.62, ... 37, (1942)....T.......
     *ON THE INELASTIC SCATTERING OF NEUTRONS BY CRYSTAL LATTICES. . . . .
     *SEEGER(R.J.), TELLER(E.).

55   PHYS. REV. VOL.61, ... 17, (1942)....T,E.......
     *MAGNETIC SCATTERING OF NEUTRONS.
     *HAMERMESH(M.).

SECTION 25 -PUBLICATIONS PRIOR TO 1945 . . . .

56   PHYS. REV. VOL.64, ...   47, (1943)....E.........FE,. . . . .
     *NEUTRON POLARIZATION AND FERROMAGNETIC SATURATION.
     *BLOCH(F.), HAMERMESH(M.), STAUB(H.).

57   PHYS. REV. VOL.65, ...   1, (1944)....T.........   . . . .
     *INELASTIC SCATTERING OF SLOW NEUTRONS.
     *WEINSTOCK(R.).

58   PHYS. ZEITS. VOL.38, ... 689, (1937)....T......   . . . .
     *SCATTERING OF SLOW NEUTRONS BY ATOMIC LATTICES OF CRYSTALS. PART II.
     *WICK(G.C.).

59   PHYS. ZEITS. VOL.38, ... 403, (1937)....T......   . . . .
     *SCATTERING OF SLOW NEUTRONS BY ATOMIC LATTICES OF CRYSTALS.
     *WICK(G.C.).

60   PROC. ACAD. SCI. AMSTERDAM 39, . 1049, (1936)....T....... . . . .
     *SCATTERING OF NEUTRONS IN MATTER III.
     *ORNSTEIN(L.S.).

61   PROC. ACAD. SCI. AMSTERDAM 39, . 904, (1936)....T....... . . . .
     *SCATTERING OF NEUTRONS IN MATTER II.
     *ORNSTEIN(L.S.).

62   PROC. ACAD. SCI. AMSTERDAM 39, . 810, (1936)....T....... . . . .
     *SCATTERING OF NEUTRONS IN MATTER.
     *ORNSTEIN(L.S.).

63   PROC. IMP. ACAD. JAPAN VOL.15, . 214, (1939)....P........HG*S,. . . . .
     *NON-ADDITIVITY IN SCATTERING CROSS-SECTIONS OF SLOW NEUTRONS.
     *KIMURA(M.).

64   PROC. JAPAN ACAD. VOL.19,. . . . 152, (1943)....E........SI*02,. . . . .
     *ON THE SCATTERING OF THERMAL NEUTRONS BY SOLIDS, III.THE EFFECT OF
     SURFACE TREATMENT FOR QUARTZ GRAIN SCATTERERS.
     *KIMURA(M.).

65   PROC. JAPAN ACAD. VOL.19,. . . .   26, (1943)....E.......SI*02, SOLIDS, II.THE EFFECT OF
     *ON THE SCATTERING OF THERMAL NEUTRONS BY
     THERMAL STRAIN IN THE DIFFRACTIVE SCATTERING OF NEUTRONS BY QUARTZ.
     *KIMURA(M.), HASHIGUCHI(R.).

66   PROC. PHYS. SOC. VOL.48, ... 642 (1936)....E........   . . . .
     *SOME EXPERIMENTS WITH NEUTRONS HAVING THERMAL ENERGIES.
     *TILLMAN(J.R.).

67   PROC. PHYS.-MATH. SOC. JAPAN 22, 391, (1940)....P........AG,HG, ...
     *SCATTERING OF SLOW NEUTRONS BY SOME ELEMENTS.
     *KIMURA(M.).

68   PROC. PHYS. MATH. SOC. JAPAN 25, 530, (1943)....P........SI*02,. . . . .
     *SCATTERING OF THERMAL NEUTRONS BY SOLIDS. II.EFFECT OF THERMAL STRAIN
     IN THE DIFFRACTIVE SCATTERING OF NEUTRONS BY QUARTZ.
     *KIMURA(M.), HASHIGUCHI(R.).

69   PROC. PHYS. MATH. SOC. JAPAN 25, 495, (1943)....P........AL,CU,FE,. . . . . . .
     *SCATTERING OF THERMAL NEUTRONS BY SOLIDS. I.VARIATION IN TOTAL CROSS
     SECTION OF METALS BY COLD-WORKING.
     *KIMURA(M.), HASHIGUCHI(R.).

70   PROC. ROY. SOC. A VOL.136, . . . 692, (1932)....E........   . . . .
     *EXISTENCE OF A NEUTRON.
     *CHADWICK(J.).

71   PROC. ROY. SOC. A VOL.138, . . . 460, (1932)....T....   . . . .
     *PASSAGE OF NEUTRONS THROUGH MATTER.
     *MASSEY(H.S.W.).

72   PROC. ROY. SOC. A VOL.142, . . .   1, (1933)........   . . . .
     *THE NEUTRON (BAKERIAN LECTURE).
     *CHADWICK(J.).

73   PROC. ROY. SOC. A VOL.149, . . . 522, (1935)....E........ . . . .
     *ARTIFICIAL RADIOACTIVITY PRODUCED BY NEUTRON BOMBARDMENT.
     *AMALDI(E.), D*AGOSTINO(O.), FERMI(E.), PONTECORVO(B.), RASETTI(F.),
     SEGRE(E.).

74   PROC. ROY. SOC. A VOL.153, . . . 476, (1936)....E........ . . . .
     *NEUTRONS OF THERMAL ENERGIES.
     *MOON(P.B.), TILLMAN(J.R.).

75   PROC. ROY. SOC. A VOL.162, . . . 127, (1937)....E,P...... . . . . . .
     *SCATTERING OF SLOW NEUTRONS.
     *GOLDHABER(M.), BRIGGS(G.H.).

76   PROC. ROY. SOC. A VOL.175, . . . 316, (1940)....E........ . . . .
     *EXPERIMENTS WITH A VELOCITY-SPECTROMETER FOR SLOW NEUTRONS.
     *FERTEL(G.E.F.), GIBBS(D.F.), MOON(P.B.), THOMSON(G.P.),
     WYNN-WILLIAMS(C.E.).

77   REV. GEN. SCI. TOME 50, ...   89, (1939) . . . .
     *TRANSMUTATIONS PRODUCED BY SLOW NEUTRONS AND BOHR*S NUCLEAR DYNAMICS.
     *KAHAN(T.).

78   REV. MOD. PHYS. VOL.9, ... 113, (1937)....T....... . . . . . .
     *NUCLEAR PHYSICS B.NUCLEAR DYNAMICS, THEORETICAL.
     *BETHE(H.A.).

79   RICERCA SCI. VOL.7, I,. . . . . . 393, (1936)....E........ . . . . . .
     *DIFFUSION OF SLOW NEUTRONS.
     *AMALDI(E.), FERMI(E.).

80   RICERCA SCI. VOL.7, II,. . . . .   13, (1936)....T......... . . . .
     *MOTION OF NEUTRONS IN HYDROGENOUS SUBSTANCES.
     *FERMI(E.).

81   RICERCA SCI. VOL.8, I,. . . . . . 400, (1937)....C1,T...... . . . .
     *DIFFUSION OF NEUTRONS IN CRYSTALS.
     *WICK(G.C.).

AUTHOR INDEX . . . .

37 AGARWAL.........(T.N.)..:   (10D,   46) PHYS. STAT. SOL. VOL.40, ...   K73, (1970).

38 AGEEV...........(N.V.)..:   (5 ,   553) SOV. PHYS. DOKLADY VOL.4,. . . .   1070, (1960).

39 AGERON..........(P.)....:   (23 ,   13) ENERGIE NUCLEAIRE VOL.13, ...   15, (1971).
                                (23 ,   5) BULL. D±INFO. SCI. ET TECH. 166, . ., (1972).
                                (23 ,   12) ENDEAVOUR 31,. . . . .   67, (1972).

40 AGGARWAL........(K.G.)..:   (6 ,   216) PROC. PHYS. SOC. 91, ...   381, (1967).

41 AGRAWAL.........(A.K.)..:   (14A,   21) J. CHEM. PHYSICS VOL.46,. . . .   1999, (1967).
                                (6 ,   997) CHEM. PHYS. LETTERS VOL.2, . . .   584, (1968).
                                (6 ,   71) IAEA SYMP. COPENHAGEN, VOL.1,. .   545, (1968).
                                (6 ,   998) PHYS. REV. VOL.171,. . . . . . .   263, (1968).
                                (15 ,   31) NUCL. SCI. ENGNG. VOL.37,. . . .   368, (1969).
                                (6 ,   356) PHYS. REV. A VOL.4,. . . . . . .   1560, (1971).

42 AGRAWAL.........(B.K.)..:   (10B,   163) PHYS. REV. B-4,. . . . .   2774, (1971).
                                (6 ,   615) NUCL./S.S. PHYS. SYMP. ABSTR.,. .   ..., (1972).
                                (6 ,   553) PHYS. REV. B VOL.6,. . . . . . .   2178, (1972).
                                (6 ,   552) SOL. STATE COMM. VOL.11 NO.6,. .   771, (1972).
                                (6 , 1241) SOL. STATE COMM. VOL.11,. . . .   1691, (1972).
                                (6 ,   618) PHYS. REV. B VOL.7,. . . . . . .   5153, (1973).

43 AGRAWAL.........(B.S.)..:   (10B,   221) SOL. STATE COMM. VOL.14 NO.3,. .   239, (1974).

44 AGRON...........(P.A.)..:   (5 , 1548) ANN. N.Y. ACAD. SCI. VOL.79, . .   762, (1960).
                                (5 , 2871) SCIENCE VOL.139, ...   1208, (1963).
                                (5 , 1599) ACTA CRYST. VOL.A25,. . . . . .   S118, (1969).
                                (22C,   19) J. APPL. CRYST. VOL.5 Pt.6,. . .   432, (1972).

45 AHMADIEH........(A.A.)..:   (6 ,   210) PHYS. REV. B VOL.7,. . . . . . .   4527, (1973).

46 AHMAD...........(A.A.Z.).: (6 , 1426) PHYS. REV. B VOL.5,. . . . . . .   2103, (1972).
                                (6 ,   561) PHYS. REV. B VOL.6,. . . . . . .   3956, (1972).

47 AIELLO..........(V.)....:   (22C,   41) NUCL. INST. MET. 62, ...   57, (1968).

48 AILAWADI........(N.K.)..:   (14A,   55) PHYS. REV. A VOL.4,. . . . . . .   1616, (1971).

49 AIZAWA..........(O.)....:   (6 , 1039) ANNU./REACT. INST. KYOTO UNIV. 3 121, (1970).

50 AKCASU..........(A.Z.)..:   (10E,   13) IAEA SYMPOSIUM BOMBAY, VOL.1,. .   261, (1964).
                                (10E,   20) NUOVO CIMENTO VOL.38,. . . . . .   175, (1965).
                                (14A,   19) IAEA SYMP. GRENOBLE,. . . . . .   365, (1972).

51 AKCASU..........(Z.A.)..:   (7B,   12) DISS. ABS. VOL.25,. . . . . . .   1131, (1965).

52 AKHIEZER........(A.I.)..:   (7B,   31) J. EXPTL. THEORET. PHYS. VOL.17,   769, (1947).
                                (14B,   54) NUCL. PHYS. VOL.40,. . . . . . .   139, (1963).

53 AKHIEZER........(A.)....:   (6 , 1184) J. PHYS. (USSR) VOL.9, ...   461, (1945).

54 AKHIEZER........(I.A.)..:   (14B,   54) NUCL. PHYS. VOL.40,. . . . . . .   139, (1963).
                                (8 ,   144) SOV. PHYS. JETP. 25,. . . . . .   517, (1967).
                                (5 , 1137) SOV. PHYS. JETP LETT. VOL.8,. .   128, (1968).
                                (8 ,   146) SOV. PHYS. SOLID STATE VOL.10, .   3142, (1968).
                                (9D,   77) SOV. PHYS. SOLID STATE VOL.9,. .   2016, (1968).
                                (11C,   61) UKRAINIAN PHYS. J. VOL.14, NO.2,   216, (1969).
                                (13 ,   46) PHYS. LETT. VOL.37A,. . . . . .   63, (1971).
                                (11B,   46) SOV. PHYS. SOL. STATE VOL.13,. .   462, (1971).
                                (13 ,   69) SOV. PHYS. JETP VOL.34,. . . . .   913, (1972).
                                (13 ,   68) SOV. PHYS. SOLID STATE VOL.14, .   142, (1972).
                                (11B,   47) SOV. PHYS. SOLID STATE VOL.14, .   387, (1972).
                                (8 ,   159) UKR. FIZ. ZH. VOL.17,. . . . . .   1339, (1972).
                                (8 ,   152) SOV. PHYS. SOLID STATE VOL.14, .   1761, (1973).
                                (8 ,   155) SOV. PHYS. SOLID STATE VOL.15, .   656, (1973).

55 AKIMITSU........(J.)....:   (5 , 1834) SOL. STATE COMM. VOL.8,. . . . .   87, (1970).
                                (5 , 2117) ACTA CRYST. VOL.A28 PART S-4,. .   196, (1972).
                                (5 , 2118) J. PHYS. SOC. JAPAN VOL.35,. . .   1000, (1973).
                                (5 , 2111) J. PHYS. SOC. JAPAN VOL.36,. . .   431, (1974).

56 AKINO...........(F.)....:   (6 , 1042) J. NUCL. SCI. TECHNOL. VOL.6,. .   671, (1969).

57 AKINO...........(T.)....:   (5 , 2943) J. PHYS. SOC. JAPAN VOL.36,. . .   84, (1974).

58 AKYUZ...........(R.O.)..:   (5 ,   95) NUCL. PHYS. VOL.A133,. . . . . .   410, (1969).

59 ALBERS..........(J.)....:   (14A,   57) PHYS. REV. A VOL.5,. . . . . . .   2629, (1972).

60 ALBERTSSON......(J.)....:   (5 , 2399) ACTA CRYST. VOL.B29,. . . . . .   2751, (1973).

61 ALBERT..........(T.E.)..:   (17A,   68) NUCL. INST. MET. VOL.94,. . . .   173, (1971).

62 ALBOLD..........(E.)....:   (20C,   50) NUCL. INST. MET. 53,. . . . . .   325, (1967).

63 ALBRECHT........(G.)....:   (5 ,   525) PHYS. STAT. SOL. 3,. . . . . . .   K249, (1963).

64 ALBRECHT........(R.W.)..:   (16 ,   44) NUCL. SCI. ENG. VOL.42,. . . . .   97, (1970).

65 ALBURGER........(D.E.)..:   (20C,   90) REV. SCI. INSTRUM. VOL.23, . . .   769, (1952).

66 ALCOCK..........(N.Z.)..:   (5 , 2234) PHYS. REV. VOL.75,. . . . . . .   1609, (1949).
                                (5 ,   772) CAN. J. PHYS. VOL.29,. . . . . .   36, (1951).
                                (5 , 1886) PHYS. REV. VOL.83,. . . . . . .   1100, (1951).

67 ALDRED..........(A.T.)..:   (5 , 2316) PHYS. REV. LETT. VOL.24,. . . .   897, (1970).
                                (5 , 2106) J. PHYSIQUE TOME 32 COL.1 VOL.I,   575, (1971).
                                (5 , 2216) A.I.P. CONF. PROC. NO.10 PART 1,   88, (1973).
                                (5 , 2115) PHYS. REV. B VOL.7,. . . . . . .   218, (1973).
                                (5 , 2212) PHYS. REV. B VOL.9,. . . . . . .   3766, (1974).

68 ALDRED..........(B.K.)..:   (6 ,   2) DISC. FARADAY SOC. NO.43,. . . .   169, (1967).
                                (6 , 2199) FARADAY SYMP. CHEM. SOC. NO.6,. .   135, (1972).

69 ALEFELD.........(B.)....:   (20A,   82) UKAEA AERE TRANS 1074 (20 PP.),.   ..., (1967).
                                (22A,   13) IAEA SYMP. COPENHAGEN, VOL.2,. .   381, (1968).
                                (22A,   42) NUCL. INST. MET. 95,. . . . . .   435, (1971).
                                (6 , 2262) IAEA SYMP. GRENOBLE,. . . . . .   851, (1972).
                                (22A,   32) KERNTECHNIK VOL.14,. . . . . . .   15, (1972).
                                (6 , 1705) PHYS. LETT. VOL.45A, ...   479, (1973).

70 ALEFOLD.........(G.)....:   (5 , 2013) J. PHYS. CHEM. SOLIDS VOL.31,. .   2361, (1970).

71 ALEONARD........(R.)....:   (5 , 2018) J. PHYS. CHEM. SOLIDS VOL.21,. .   234, (1961).
                                (5 ,   615) COLL. INTER. N.126 (GRENOBLE),. .   158, (1963).
                                (5 ,   591) J. APPL. PHYS. VOL.37,. . . . .   1038, (1966).

72 ALESHKO-OZHEVSKI(O.P.)..:   (12B,   59) SOV. PHYS. CRYST. VOL.10,. . . .   76, (1965).
                                (12B,   60) SOV. PHYS. CRYST. VOL.11,. . . .   597, (1967).
                                (5 ,   176) SOV. PHYS. JETP LETT. VOL.7,. . .   156, (1968).
                                (5 , 2323) SOV. PHYS. CRYST. VOL.14,. . . .   447, (1969).
                                (9C,   41) SOV. PHYS. J.E.T.P. VOL.29,. . .   655, (1969).
                                (5 , 1132) SOV. PHYS. CRYST. VOL.16,. . . .   935, (1971).
                                (5 ,   168) SOV. PHYS. CRYST. VOL.16,. . . .   711, (1972).

```
72  ALESHKO-OZHEVSKI(O.P.)..?   ( 5 ,  182) SOV. PHYS. JETP VOL.35,.     370, (1972).
                                ( 5 ,  169) SOV. PHYS. CRYST. VOL.18,.: : : : 393; (1973):

73  ALEXANDER........(T.K.)..?  (21 ,    8) IAEA SYMPOSIUM VIENNA, . . . . . 165, (1960).

74  ALIKHANOV.......(R.A.)..?   ( 5 , 1661) SOV. PHYS. JETP VOL.9,.     1204, (1959).
                                ( 5 , 2125) SOV. PHYS. JETP VOL.10,.: : : : 814, (1960).
                                (22C,   72) SOV. PHYS. JETP VOL.11,.     585, (1960).
                                ( 5 ,  447) SOV. PHYS. JETP VOL.12,.     1029, (1961).
                                ( 5 , 2094) J. PHYS. SOC. JAPAN VOL.17 B-III  58, (1962).
                                ( 5 , 2220) COLL. INTER. N.126 (GRENOBLE),    29, (1963).
                                ( 5 , 2223) SOV. PHYS. J.E.T.P. VOL.18,. .   556, (1964).
                                ( 5 , 2225) JETP LETT. VOL.5,.     349, (1967).
                                ( 6 ,  545) PHYS. STAT. SOL. 32, . . . . .    41, (1969).
                                ( 6 ,  863) PHYS. STAT. SOL. 41, . . . . . K103, (1970).
                                ( 5 , 1911) SOV. PHYS. DOKLADY VOL.16, . .     9, (1971).
                                ( 6 , 1442) PHYSICA VOL.57,.     628, (1972).
                                ( 5 , 2219) PHYS. STAT. SOL. VOL.50,. . . .   385, (1972).

75  ALI.............(S.A.)..?   (20A,   48) NUCL. INST. MET. VOL.114,. . .   143, (1974).

76  ALLAIN..........(Y.)....?   ( 5 ,  361) J. PHYSIQUE TOME 26,.     789, (1965).
                                ( 5 ,  343) COMP. REND. 263,.     98, (1966).
                                ( 5 , 2392) J. PHYSIQUE TOME 32 COL.1 VOL.II 611, (1971).
                                ( 5 , 1029) SOL. STATE COMM. VOL.14 NO.2,. . 187, (1974).

77  ALLAN...........(G.A.T.)?   (13 ,   10) CAN. J. PHYS. 46,.     799, (1968).

78  ALLENDEN........(D.)....?   (229,   19) NUCL. INST. MET. 23, . . . . .   181, (1963).

79  ALLEN...........(G.)....?   ( 6 , 2044) FARADAY SYMP. CHEM. SOC. NO.6, . 169, (1972).
                                (15 ,   15) IAEA SYMP. GRENOBLE, . . . . .   261, (1972).
                                ( 6 , 1921) POLYMER VOL.13,.     157, (1972).
                                ( 6 , 2046) CHEM. APPL. THERMAL NEUTRON SCAT  97, (1973).
                                (15 ,   51) REP. PROGR. PHYS. VOL.36, . .   1073, (1973).
                                ( 6 , 2004) J.C.S. FARADAY TRANS. II VOL.70, 348, (1974).

80  ALLEN...........(J.J.)..?   (24 ,   98) PROC/7TH/BIOMED/INSTRUM. SYMP...  93, (1969).

81  ALLEN...........(J.W.)..?   (11A,   42) PHYS. REV. LETT. VOL.25, . . . . 934, (1970).

82  ALLEN...........(P.B.)..?   (10C,   80) PHYS. REV. B VOL.6,. . . . . .  2577, (1972).
                                (10B,  176) PHYS. REV. LETT. VOL.29, . . . . 1593, (1972).

83  ALMLOF..........(J.)....?   ( 5 ,    9) J. CHEM. PHYS. VOL.59, . . . . 3901, (1973).

84  ALMODOVAR.......(I.)....?   ( 5 ,  742) COLL. INTER. N.126 (GRENOBLE), . 18, (1963).
                                ( 5 ,  376) J. CHEM. PHYSICS VOL.40,.     504, (1964).
                                ( 5 ,  744) PHYS. REV. 138,.     A153, (1965).
                                ( 5 ,  716) J. CHEM. PHYSICS VOL.44, . . .  1648, (1966).
                                ( 5 , 1715) ACTA CRYST B24,. . . . . . .   1312, (1968).

85  ALMQVIST........(L.)....?   ( 6 , 1827) PHYS. REV. 162,.     545, (1967).
                                ( 6 , 1812) PHYS. REV. 162,.     549, (1967).
                                ( 6 , 1822) PHYS. REV. 163,.     567, (1967).
                                ( 6 , 1634) IAEA SYMP. COPENHAGEN, VOL.1,.   295, (1968).
                                (21 ,   47) REV. SCI. INSTRUM. VOL.40,. . .  249, (1968).
                                ( 6 , 1637) PHYS. REV. 178,.     1496, (1969).
                                ( 6 , 1262) PHYS. STAT. SOL. 33,.     209, (1969).
                                ( 6 , 2305) J. PHYS.F: METAL PHYS. VOL.1,.   785, (1971).

86  ALONS...........(P.W.F.)?   (12A,   12) NED. TIJOSCHR. NATUURK. VOL.38,. 171, (1972).

87  ALPAEV..........(YU.A.).?   ( 9D,   63) PHYS. STAT. SOL. VOL.40,. . . .   59, (1970).

88  ALPERIN.........(H.A.)..?   ( 5 , 2172) J. APPL. PHYS. VOL.31, SUPPL.,.  354, (1960).
                                ( 5 , 2176) DISS. ABSTR. VOL.21, . . . . .  1979, (1961).
                                ( 5 , 2171) J. PHYS. SOC. JAPAN VOL.17 B-III  12, (1962).
                                ( 5 ,  933) J. PHYS. SOC. JAPAN VOL.17 B-III   7, (1962).
                                ( 5 , 1683) PHYS. REV. LETT. VOL.8,. . . .   237, (1962).
                                ( 5 , 1090) COLL. INTER. N.126 (GRENOBLE),   118, (1963).
                                ( 5 , 2405) COLL. INTER. N.126 (GRENOBLE),   141, (1963).
                                ( 6 , 1460) J. APP. PHYS. 34,.     1182, (1963).
                                ( 6 , 1513) J. APP. PHYS. 34,. . . . . .    1201, (1963).
                                ( 5 ,  915) J. PHYS. CHEM. SOL. 24,. . . .  1679, (1963).
                                ( 5 , 1206) J. APPL. PHYS. VOL.35,.     1045, (1964).
                                ( 5 , 2983) PHYS. REV. LETT. VOL.12, . . .   444, (1964).
                                ( 6 ,  846) PHYS. REV. VOL.136,.     A1641, (1964).
                                ( 5 , 1122) PROC. INT. CONF. (NOTTINGHAM),   223, (1964).
                                ( 6 , 1735) J. APP. PHYS. 36,.     1076, (1965).
                                ( 6 ,  751) PHYS. REV. LETT. 15,.     146, (1965).
                                ( 6 ,  478) J. APP. PHYS. 37,. . . . . .    1052, (1966).
                                ( 5 ,  917) J. APP. PHYS. 37,.     981, (1966).
                                ( 6 ,  841) PHYS. REV. 154,.     508, (1967).
                                ( 6 , 1738) PHYS. REV. 156,.     623, (1967).
                                ( 6 , 2130) PHYS. REV. 161,.     499, (1967).
                                ( 6 ,  829) J. APP. PHYS. 39,.     455, (1968).
                                ( 5 ,  913) J. PHYS. CHEM. SOLIDS 29,. . .   414, (1968).
                                ( 5 , 1213) J. APPL. PHYS. VOL.40,.     1009, (1969).
                                ( 5 , 2920) J. APPL. PHYS. VOL.41,.     1192, (1970).
                                ( 6 ,  912) J. APPL. PHYS. 41,.     933, (1970).
                                ( 5 , 2409) J. APPL. PHYS. VOL.42,.     1617, (1971).
                                ( 5 , 2566) PHYS. REV. LETT. VOL.29,.     1562, (1972).
                                ( 5 ,  911) A.I.P. CONF. PROC. NO.10 PART 2, 1569, (1973).
                                ( 5 , 2565) PHYS. LETT. VOL.47A NO.1,. . .    73, (1974).

89  ALPER...........(T.)....?   (24 ,    9) BR. J. RADIOL. VOL.45, . . . . . 529, (1972).

90  ALSMILLER.......(R.G.)..?   (24 ,   69) NUCL. INST. MET. 72, . . . . .   213, (1969).

91  ALSTON..........(W.C.H.)?   ( 9B,   64) NUCL. INST. MET. VOL.84, . . .   153, (1970).
                                (24 ,   85) PHYS. EDUC.(GB) VOL.6,. . . . .  396, (1971).
                                (22A,   26) J. PHYS. E VOL.6 NO.1, . . . . .  51, (1973).

92  ALS-NIELSEN.....(J.)....?   ( 6 , 1168) PHYS. REV. VOL.133,. . . . . . B925, (1964).
                                (20A,    4) ACTA CRYST. 18,.     184, (1965).
                                (20D,   34) REV. SCI. INSTRUM. VOL.36, . .    48, (1965).
                                ( 6 ,  643) NBS MISC. PUB.273,.     144, (1966).
                                (20C,   49) NUCL. INST. MET. 50,.     181, (1967).
                                (22C,   34) NUCL. INST. MET. 50,.     191, (1967).
                                ( 5 ,  753) PHYS. REV. 153,. . . . . .      706, (1967).
                                ( 5 ,  752) PHYS. REV. 153,.     711, (1967).
                                ( 6 ,  644) PHYS. REV. 153,.     717, (1967).
                                ( 6 , 2524) PHYS. REV. 162,.     315, (1967).
                                ( 6 , 2137) SOL. STATE COMM. 5,.     607, (1967).
                                ( 6 , 2120) IAEA SYMP. COPENHAGEN, VOL.1,.    35, (1968).
                                ( 6 , 2126) IAEA SYMP. COPENHAGEN, VOL.2,.    63, (1968).
                                ( 6 , 2134) J. APP. PHYS. 39,.     1229, (1968).
                                ( 6 , 2129) J. PHYS. SOC. JAPAN VOL.26,. . .  49, (1969).
                                ( 6 ,  521) PHYS. REV. LETT. 22,.     230, (1969).
                                ( 6 ,  751) PHYS. REV. B-2,.     277, (1970).
                                ( 6 ,  729) PHYS. REV. LETT. VOL.25,.     730, (1970).
                                ( 6 ,  514) J. APPL. PHYS. 42,.     1666, (1971).
                                ( 6 , 2142) J. PHYS. C, SER.2, VOL.4,.     71, (1971).
                                ( 6 , 1927) PHYS. REV. LETT. VOL.27,.     1530, (1971).
                                ( 6 ,  718) PHYS. REV. LETT. VOL.27,. . . .  741, (1971).
```

92 ALS-NIELSEN.....(J.)....: (6 , 1932) IAEA SYMP. GRENOBLE, ... 543, (1972).
                              (6 , 719) IAEA SYMP. GRENOBLE, ... 619, (1972).
                              (6 , 720) MAGN. AND MAG. MATERIALS-1971, : 1251, (1972).
                              (6 , 2000) PHYS. REV. B VOL.6, ... 2030, (1972).
                              (5 , 2335) PHYS. REV. B VOL.6, ... 2724, (1972).
                              (5 , 758) PHYS. REV. B VOL.9, ... 3921, (1974).
                              (5 , 1583) PHYS. REV. LETT. VOL.32, ... 610, (1974).

93 ALTMAN..........(R.F.)..: (5 , 974) A.I.P. CONF. PROC. NO.10 PART 2, 1163, (1973).

94 ALTSCHULER......(S.J.)..: (16 , 16) DISS. ABS. VOL.26, ... 2087, (1966).

95 ALVAREZ.........(L.W.)..: (20D, 27) PHYS. REV. VOL.54, ... 609, (1938).
                              (5 , 1254) PHYS. REV. VOL.58, ... 1003, (1940).

96 AL-NAIMI........(A.)....: (5 , 1595) ACTA CRYST. VOL.A28, ... 473, (1972).

97 AL-SAJI.........(A.)....: (5 , 1595) ACTA CRYST. VOL.A28, ... 473, (1972).

98 AL-SHAHERY......(G.Y.)..: (5 , 1595) ACTA CRYST. VOL.A28, ... 473, (1972).

99 AMADORI.........(R.)....: (19A, 21) IAEA SYMP. (INSTRUMENT.), VIENNA 117, (1969).
                              (17B, 31) CONF. INTERN., RENNES, ... 162, (1971).
                              (7B, 22) J. APPL. CRYST. VOL.4, ... 410, (1971).
                              (17A, 43) IAEA SYMP. GRENOBLE, ... 747, (1972).

100 AMADO..........(A.M.)..: (14B, 20) DISS. ABSTR. VOL.21, ... 3826, (1961).

101 AMALDI.........(E.)....: (16 , 66) PROC. ROY. SOC. A VOL.149, . . . 522, (1935).
                              (7A, 58) PHYS. REV. VOL.50, ... 899, (1936).
                              (16 , 80) RICERCA SCI. VOL.7, I, ... 393, (1936).

102 AMARAL.........(L.Q.)..: (22A, 40) NUCL. INST. MET. 63, ... 13, (1968).
                              (14B, 41) J. CHEM. PHYS. VOL.56, ... 3118, (1972).

103 AMBEGAOKAR.....(V.)....: (10E, 4) COPENHAGEN CONF.-LATT.DYNAMICS,. 261, (1963).
                              (13E, 26) PHYS. REV. VOL.135, ...A1071, (1964).

104 AMEEN..........(N.)....: (22C, 21) J. NUCL. ENERGY VOL.24,. . . . . 385, (1970).

105 AMIN...........(F.A.)..: (5 , 1595) ACTA CRYST. VOL.A28, ... 473, (1972).

106 AMSTER.........(H.J.)..: (7A, 37) NUCL. PHYS. A VOL.A169,. . . . . 385, (1971).

107 ANANTH.........(M.S.)..: (15 , 30) MOL. PHYS. VOL.25, ... 1353, (1973).

108 ANDERSON ......(R.H.)..: (24 , 107) SCIENCE VOL.114, ... 341, (1951).

109 ANDERSON.......(C.A.)..: (17B, 40) NUCLEONICS VOL.20, ... 157, (1962).

110 ANDERSON.......(C.H.)..: (22A, 16) IEEE TRANS. ELECTRON DEVICES 17, 236, (1970).

111 ANDERSON.......(D.A.)..: (11A, 53) PHYS. STAT. SOLIDI B VOL.56 NO.1 157, (1973).

112 ANDERSON.......(E.E.)..: (5 , 1256) PHYS. REV. VOL.72, ... 1147, (1947).

113 ANDERSON.......(F.B.)..: (9D, 51) MAGNETIC MATERIALS DIGEST 1964,. 37, (1964).

114 ANDERSON.......(H.L.)..: (20D, 28) PHYS. REV. VOL.70, ... 815, (1946).

115 ANDERSON.......(J.L.)..: (5 , 1554) J. PHYS. CHEM. SOLIDS VOL.31,. . 1193, (1970).

116 ANDERSON.......(K.D.)..: (5 , 1577) ACTA CRYST. 18,. . . . . 906, (1965).
                              (5 , 2359) ACTA CRYST. 19,. . . . . 413, (1965).
                              (5 , 2802) J. PHYS. CHEM. SOLIDS VOL.27,. . 1197, (1966).
                              (5 , 2677) PHYS. REV. VOL.156,. . . . . . . 1225, (1967).
                              (5 , 2799) J. PHYS. CHEM. SOLIDS 30,. . . . 453, (1969).
                              (5 , 2213) J. PHYS. CHEM. SOLIDS 30,. . . . 733, (1969).

117 ANDERSON.......(L.E.)..: (14B, 85) PHYS. REV. LETT. 15, ... 693, (1965).

118 ANDERSON.......(M.R.)..: (5 , 1570) ACTA CRYST. VOL.A28, ... 663, (1972).

119 ANDERSON.......(R.H.)..: (22C, 70) SCIENCE VOL.114, ... 341, (1951).

120 ANDO...........(M.)....: (5 , 524) PHYS. REV. LETT. VOL.29, ... 281, (1972).

121 ANDREASSEN.....(A.)....: (5 , 2974) ACTA CRYST. VOL.A29, ... 594, (1973).

122 ANDREEVA.......(L.S.)..: (20C, 21) INSTRUM. EXP. TECH. NO.1,. . . . 77, (1970).

123 ANDREEVA.......(N.Z.)..: (20C, 109) SOV. AT. ENERGY VOL.32, NO.5,. . 416, (1972).

124 ANDRESEN.......(A.F.)..: (5 , 718) ACTA CRYST. VOL.8, ... 118, (1955).
                              (5 , 308) ACTA CRYST. VOL.10, ... 107, (1957).
                              (22C, 49) REV. SCI. INSTRUM. VOL.28, . . . 916, (1957).
                              (5 , 2777) ACTA CRYST. VOL.11,. . . . . . . 612, (1958).
                              (5 , 1142) ACTA CHEM. SCAND. VOL.14,. . . . 919, (1960).
                              (5 , 1152) COLL. INTER. N.126 (GRENOBLE), . 150, (1963).
                              (5 , 2358) PHIL. MAG. VOL.11, ... 1245, (1965).
                              (5 , 2360) ACTA CHEM. SCAND. VOL.20,. . . . 2529, (1966).
                              (5 , 1654) ACTA CHEM. SCAND. VOL.21,. . . . 1543, (1967).
                              (5 , 1146) ACTA CHEM. SCAND. VOL.21,. . . . 2841, (1967).
                              (5 , 498) ACTA CHEM. SCAND. VOL.21,. . . . 833, (1967).
                              (5 , 2312) PHIL. MAG. VOL.16, ... 1063, (1967).
                              (5 , 1412) ACTA CHEM. SCAND. VOL.22,. . . . 3039, (1968).
                              (5 , 1158) ACTA CHEM. SCAND. VOL.22,. . . . 827, (1968).
                              (5 , 170) J. CHEM. PHYSICS VOL.48, ... 4660, (1968).
                              (5 , 2193) ACTA CHEM. SCAND. VOL.23,. . . . 2325, (1969).
                              (5 , 1812) ACTA CRYST. (INTERACT.) VOL.A25, S250, (1969).
                              (5 , 1811) PHYS. NORVEG. VOL.3, ... 203, (1969).
                              (5 , 1157) ACTA CHEM. SCAND. VOL.24,. . . . 2435, (1970).
                              (5 , 1149) ACTA CHEM. SCAND. VOL.24,. . . . 3309, (1970).
                              (5 , 561) ACTA CHEM. SCAND. VOL.25,. . . . 1703, (1971).
                              (5 , 966) ACTA CHEM. SCAND. VOL.26,. . . . 3101, (1972).
                              (5 , 1656) ACTA CHEM. SCAND. VOL.26,. . . . 175, (1972).
                              (5 , 1655) ACTA CRYST. VOL.A28 PART S-4,. . 196, (1972).

125 ANDRESEN.......(A.)....: (6 , 2356) PHYS. REV. VOL.108,. . . . . . . 1092, (1957).
                              (5 , 2083) J. PHYS. CHEM. SOLIDS VOL.10,. . 138, (1959).

126 ANDREWS........(D.G.)..: (20C, 59) NUCL. INST. MET. 74, ... 123, (1969).

127 ANDRIESSE......(C.D.)..: (7B, 50) PHYS. LETT. VOL.27A, ... 93, (1968).
                              (6 , 78) PHYS. LETT. VOL.28A, ... 642, (1969).
                              (6 , 77) DISS. ABS. VOL.30, ...3807B, (1970).
                              (14B, 66) PHYSICA VOL.48, ... 61, (1970).
                              (14B, 67) PHYSICA VOL.49, ... 502, (1970).
                              (14A, 34) PHYSICA VOL.50, ... 511, (1970).
                              (14A, 39) PHYS. LETT. VOL.33A, ... 419, (1970).
                              (5 , 101) PHYSICA VOL.57, ... 191, (1972).
                              (5 , 98) PHYSICA VOL.62,. . . . . 474, (1972).

```
128  ANDRONIKASHVILI.(E.L.)..:  ( 6 , 1178) SOV. PHYS. USPEKHI VOL.3,. . . .    888, (1961).
                                ( 6 , 1636) JETP LETT. VOL.3,. . . . . . . .    110, (1966).

129  ANDRON..........(B.)..:    ( 5 , 2651) COMP. REND. 263,. . . . . . . .   6218, (1966).
                                ( 5 ,  599) J. PHYSIQUE TOME 27,. . . . . .    619, (1966).
                                ( 5 , 1001) COMP. REND. 264,. . . . . . . .   3168, (1967).

130  ANDRUS..........(W.S.)..:  ( 6 ,   72) IAEA SYMP. COPENHAGEN, VOL.1,. .    457, (1968).
                                ( 6 ,   92) PHYS. LETT. VOL.26A,. . . . . .    152, (1968).
                                ( 6 ,   76) DISS. ABS. VOL.30,. . . . . . .  .4293B, (1970).

131  ANDRYUSHCHENKO..(V.I.)..:  (24 ,  113) SOV. PHYS. DOKLADY VOL.16, . . .     49, (1971).

132  ANIKIN..........(G.V.)..:  (12A,   56) SOV. J. NUCL. PHYS. VOL.12,. . .   1121, (1970).
                                ( 5 , 2700) SOV. J. NUCL. PHYS. VOL.14,. . .    269, (1971).

133  ANISIMOV........(I.S.)..:  ( 5 , 1955) SOV. AT. ENERGY VOL.18,. . . . .    350, (1965).

134  ANSEL≠M.........(A.A.)..:  (15 ,   53) SOV. PHYS. JETP VOL.6, . . . . .    480, (1958).

135  ANTAL..........(J.J.)..:   ( 9B,   50) J. APPL. PHYS. VOL.26, . . . . .   1041, (1955).
                                (20B,  140) REV. SCI. INSTRUM. VOL.30, . . .    269, (1959).
                                ( 6 , 2310) COPENHAGEN CONF.-LATT.DYNAMICS,.     17, (1963).
                                ( 6 , 2315) PHYS. REV. 132,. . . . . . . .     683, (1963).
                                (21 ,   38) NUCL. INST. MET. 65,. . . . . .    125, (1968).

136  ANTONINI........(B.)..:    ( 5 ,  951) COLL. INTER. N.126 (GRENOBLE),.    180, (1963).
                                ( 5 ,  271) PHYSICA VOL.30,. . . . . . . . .   1647, (1964).
                                ( 5 , 2294) PROC. INT. CONF. (NOTTINGHAM),.     288, (1964).
                                (20B,   76) NUCL. INST. MET. 33,. . . . . .    229, (1965).
                                ( 5 ,  273) PHYSICA VOL.32,. . . . . . . . .    119, (1966).
                                ( 6 ,  818) PROC. PHYS. SOC. 89,. . . . . .     419, (1966).
                                ( 5 ,  495) PHYS. LETT. 25A,. . . . . . . .     372, (1967).
                                ( 6 , 1508) PHYS. LETT. 30A,. . . . . . . .     310, (1969).
                                ( 6 ,  883) PHYS. REV. VOL.178,. . . . . . .    833, (1969).
                                ( 5 , 1790) PHYS. REV. VOL.187,. . . . . . .    611, (1969).
                                (11A,   17) NUCL. INST. MET. 87,. . . . . .     125, (1970).
                                ( 5 ,  478) SOL. STATE COMM. VOL.8,. . . . .      1, (1970).
                                ( 5 ,  479) INT. J. MAGN. VOL.1 NO.2,. . . .    183, (1971).
                                ( 6 ,  799) J. PHYSIQUE TOME 32 COL.1 VOL.II 1188, (1971).
                                ( 6 ,  871) SOL. STATE COMM. VOL. 9,. . . .     257, (1971).
                                (11A,   35) SOL. STATE COMM. VOL. 9,. . . .     417, (1971).
                                ( 6 , 1509) SOL. STATE COMM. VOL.10,. . . .     263, (1972).

137  ANTONINI........(D.)....:  (17A,   75) NUCL. SCI. ENG. VOL.48,. . . . .    281, (1972).

138  ANTONINI........(M.)....:  ( 6 , 2061) IAEA SYMP. BOMBAY, VOL.1,. . . .    117, (1964).
                                (10C,   75) PHYS. REV. VOL.136,. . . . . . . .A1280, (1964).
                                ( 5 , 1999) PROC. INT. CONF. (NOTTINGHAM),.     285, (1964).
                                ( 6 , 1457) J. PHYS. CHEM. SOL. 28,. . . . .     11, (1967).
                                (17C,   17) J. APPL. CRYST. VOL.3, . . . . .    145, (1970).
                                ( 5 , 1926) J. PHYSIQUE TOME 33 COL.2, . . .     83, (1972).
                                (20B,  103) NUCL. INST. MET. VOL.104,. . . .    147, (1972).
                                ( 6 ,  887) NUOVO CIMENTO B VOL.10B SER.11,.    117, (1972).
                                ( 5 , 1236) J. NON-CRYST. SOLIDS VOL.11 NO.5  417, (1973).

139  ANTSUPOV........(P.S.)..:  ( 5 , 1072) SOV. PHYS. SOL. STATE VOL.13,. .     44, (1971).

140  ANTUNEZ.........(H.M.)..:  ( 6 , 2205) PROC/CONF/FAST CRIT. EXP./ANALY.    524, (1967).

141  ANUFRIENKO......(V.B.)..:  ( 6 , 2169) BU. ACAD. SCI. USSR PHYS. SR. 32  600, (1968).

142  ANZAI...........(S.)....:  ( 5 , 1827) COLL. INTER. N.126 (GRENOBLE),.    144, (1963).

143  APLING..........(A.J.)..:  ( 5 , 1233) AMORPHOUS MATERIALS, . . . . . .    423, (1972).

144  APOSTOLESCU.....(S.)....:  (23 ,   62) NUKLEONIK VOL.2, . . . . . . . .     41, (1960).
                                (19A,   70) STUD. CERCETARI FIZ. VOL.14,. . .    353, (1963).
                                ( 5 ,  110) REV. ROUMAINE PHYS. VOL.9,. . . .    121, (1964).
                                (20A,  138) REV. ROUMAINE PHYS. VOL.11,. . .     601, (1966).

145  APOSTOLOV.......(A.)....:  ( 5 , 1002) ANNU. UNIV. SOFIA FAC. PHYS. 59,     47, (1966).
                                ( 5 , 1092) BULL. SOC. FRANC. MIN. CRIST. 89   206, (1966).
                                ( 5 ,  570) ANNU. UNIV. SOFIA FAC. PHYS. 63,    165, (1971).

146  APPAPILLAI......(M.)....:  (10B,   88) J. PHYS. F VOL.3,. . . . . . . .    772, (1973).

147  APPLETON........(G.L.)..:  ( 6 , 1784) PHYS. REV. VOL.90, . . . . . . .    869, (1953).

148  ARAI...........(T.)....:   ( 5 , 2977) ACTA CRYST. VOL.A28 PART S-4,. .     99, (1972).
                                ( 5 , 1345) J. PHYS. SOC. JAPAN VOL.35,. . .    473, (1973).

149  ARASE..........(T.)....:   ( 6 , 1809) THESIS,. . . . . . . . . . . . .          (1959).
                                ( 6 , 1801) IAEA SYMPOSIUM VIENNA, . . . . .    531, (1960).
                                ( 6 , 1810) PHYS. REV. VOL.128,. . . . . . .   1099, (1962).

150  ARDENTE.........(V.)....:  ( 7B,   10) COLL. INTER. N.126 (GRENOBLE),.    217, (1963).
                                (14B,   87) PHYS. REV. VOL.148,. . . . . . .    124, (1966).
                                (14A,  103) REACTOR PHYSICS,. . . . . . . .      73, (1966).
                                (15 ,   49) REPORT EUR-4037.1 (50.PP.),. . .         (1968).

151  ARGAY..........(G.)....:   ( 5 ,   47) ACTA CRYST. 19,. . . . . . . . .    180, (1965).

152  ARIC...........(M.)....:   ( 6 ,  806) PHYS. LETT. VOL.28A, . . . . . .    213, (1968).

153  ARKHIPOV........(V.E.)..:  ( 5 , 2133) SOV. PHYS. JETP VOL.34,. . . . .    163, (1972).
                                ( 5 , 1059) SOV. PHYS. JETP VOL.34,. . . . .    799, (1972).

154  ARMITAGE........(B.H.)..:  ( 6 , 2206) CONF/NUCL. STRUC. STUDY/NEUTRONS   172, (1972).

155  ARMSTRONG.......(J.R.)..:  ( 5 ,  661) INORG. NUCL. CHEM. LETT. VOL.9,.  1025, (1973).

156  ARMSTRONG.......(S.B.)..:  ( 5 , 2328) NUCL. SCI. ENG. VOL.23,. . . . .    192, (1965).

157  ARMYTAGE........(D.)....:  ( 5 , 2518) ACTA CRYST. VOL.B30, . . . . . .    809, (1974).

158  ARNAUD..........(A.)....:  (19B,   27) RC ACCAD. NAZ. LINCEI VOL.45,. .    163, (1968).
                                (23 ,    6) BULL. INF. SCI. TECH. NO.170,. .     61, (1972).

159  ARNDT..........(R.A.)..:   ( 6 , 1871) J. CHEM. PHYSICS VOL.54, . . . .   2597, (1971).

160  ARNDT..........(U.W.)..:   (22B,   18) NUCL. INST. MET. 24,. . . . . .     255, (1963).
                                (22A,   63) REV. SCI. INSTRUM. VOL.34, . . .    224, (1963).
                                (22B,    2) ACTA CRYST. 17,. . . . . . . . .   1183, (1964).

161  ARNELL..........(S.E.)..:  ( 5 , 2068) PHYS. SCRIPTA(SWEDEN) VOL.4, . .     89, (1971).

162  ARNOLD..........(G.P.)..:  ( 5 ,   63) PHYS. REV. VOL.73, . . . . . . .   1385, (1948).
                                ( 5 ,  239) PHYS. REV. VOL.75, . . . . . . .    217, (1949).
                                ( 6 , 1202) PHYS. REV. LETT. VOL.1,. . . . .      9, (1958).
                                ( 6 , 1191) PHYS. REV. VOL.113,. . . . . . .   1379, (1959).
                                ( 5 ,  907) PHYS. REV. VOL.124,. . . . . . .   1848, (1961).
                                ( 5 , 2707) ACTA CRYST. VOL.15,. . . . . . .    351, (1962).
                                ( 5 ,  908) J. APPL. PHYS. VOL.33, SUPPL.,.   1135, (1962).
                                ( 5 ,  912) PHYS. REV. VOL.127,. . . . . . .   2101, (1962).
                                ( 6 , 2014) PHYS. REV. 131,. . . . . . . . .   2098, (1963).
```

162 ARNOLD..........(G.P.)...:
(5 ,   906) PHYS. REV. 135,. . . . .   A176, (1964).
(5 ,   2731) ACTA CRYST. 21,. : . : . : . : . :   670, (1966).
(5 ,   2030) J. APP. PHYS. 37,. . . . .   4575, (1966).
(22C,   58) REV. SCI. INSTRUM. VOL.37;. . . . .   1543, (1966).
(5 ,   2339) J. APP. PHYS. 38,. . . . .   1395, (1967).
(5 ,   809) J. CHEM. PHYS. VOL.46,. . . . . . .   4041, (1967).
(5 ,   2631) ACTA CRYST. VOL.B24,. . . . .   1121, (1968).
(5 ,   1524) ACTA CRYST. VOL.R24,. . . . .   459, (1968).
(5 ,   796) J. CHEM. PHYSICS VOL.49,. . . . . .   2514, (1968).
(5 ,   782) J. CHEM. PHYSICS VOL.49,. . . . . .   4365, (1968).
(5 ,   2754) J. APPL. PHYS. 41,. . . . .   5080, (1970).
(5 ,   2363) J. NUCL. MATERIALS VOL.34,. . . . .   281, (1970).
(5 ,   1554) J. PHYS. CHEM. SOLIDS VOL.31,. . .   1193, (1970).
(5 ,   2772) ACTA CRYST. VOL.B27,. . . . .   1967, (1971).
(5 ,   2638) ACTA CRYST. VOL.B27,. . . . .   243, (1971).
(5 ,   784) J. CHEM. PHYSICS VOL.55,. . . . . .   589, (1971).
(5 ,   2632) ACTA CRYST. VOL.B28,. . . . .   1724, (1972).
(5 ,   565) ACTA CRYST. VOL.B28,. . . . .   3102, (1972).

163 ARNOLD..........(G.)....:
(5 ,   2544) J. APPL. PHYS. VOL.42,. . . . . . .   1625, (1971).
(5 ,   1652) A.I.P. CONF. PROC. NO.10 PART 1,   658, (1973).

164 ARNOLD..........(P.)....:
(19B,   26) PHYS. LETT. VOL.44A NO.3,. . . . .   165, (1973).
(5 ,   438) Z. PHYSIK VOL.264,. . . . .   427, (1973).

165 ARROTT..........(A.)....:
(5 ,   956) J. APPL. PHYS. VOL.29,. . . . . . .   515, (1958).
(5 ,   960) PHYS. REV. VOL.114,. . . . .   1420, (1959).
(5 ,   959) PHYS. REV. VOL.114,. . . . .   1427, (1959).
(5 ,   549) PHYS. REV. LETT. VOL.4,. . . . . .   226, (1960).
(5 ,   2290) J. APP. PHYS. 36,. . . . .   1093, (1965).
(5 ,   542) PHYS. REV. LETT. VOL.14,. . . . .   1022, (1965).
(20B,   128) PHYS. REV. VOL.140,. . . . .   A675, (1965).
(5 ,   537) J. APP. PHYS. 37,. . . . .   1260, (1966).
(9B,   53) J. APP. PHYS. 37,. . . . .   2343, (1966).
(5 ,   2530) J. APPL. PHYS. VOL.38,. . . . . .   969, (1967).
(5 ,   577) J. APP. PHYS. 38,. . . . .   1243, (1967).
(5 ,   556) METAL. SOC. CONF. LOS ANGELES,   59, (1967).
(5 ,   574) PHYS. REV. 153,. . . . .   624, (1967).
(5 ,   532) PHYS. REV. 155,. . . . .   528, (1967).
(5 ,   2848) J. APP. PHYS. 39,. . . . .   585, (1968).
(5 ,   531) J. APP. PHYS. 39,. . . . .   671, (1968).
(5 ,   133) J. APP. PHYS. VOL.40,. . . . . . .   1373, (1969).
(6 ,   520) J. APP. PHYS. VOL.40,. . . . . . .   1447, (1969).
(12A,   15) NUCL. INST. MET. 68,. . . . . . .   50, (1969).
(5 ,   1413) PHYS. REV. 186,. . . . .   705, (1969).
(6 ,   827) J. APPL. PHYS. 41,. . . . . 1363, (1970).
(6 ,   743) MAGN. AND MAG. MATERIALS-1971;   1340, (1972).
(13 ,   6) AIP CONF. PROC. NO.10 PART 1,.   822, (1973).

166 ARSIC-ESKINJA...(M.)....:
(5 ,   1458) 1ST EUR. CONF./CONDENSED MATTER,   60, (1971).
(5 ,   1461) IAEA SYMP. GRENOBLE,. . . . . . .   825, (1972).

167 ARUSHANOV.......(G.G.)..:
(7A,   88) SOV. PHYS. JETP. 24,. . . . . . .   946, (1967).

168 ARZI............(E.)....:
(5 ,   254) ACTA CRYST. VOL.A28 PART S-4,. .   188, (1972).

169 ASAMI...........(A.)....:
(5 ,   139) J. NUCL. ENERGY VOL.24,. . . . . .   85, (1970).
(5 ,   1544) 2ND INTERNAT. CONF./NUCL. DATA,. . . , (1970).

170 ASAMI...........(T.)....:
(23 ,   16) IAEA SYMP. (PILE RESEARCH)VIENNA   585, (1960).

171 ASANO...........(H.)....:
(5 ,   2977) ACTA CRYST. VOL.A28 PART S-4,. :   99, (1972).
(5 ,   2836) PHYS. STAT. SOLIDI A VOL.15 NO.1   267, (1973).

172 ASBRINK.........(S.)....:
(17C,   8) ACTA CRYST. 20,. . . . .   407, (1966).

173 ASCARELLI.......(P.)....:
(6 ,   190) IAEA SYMP. CHALK RIVER, VOL.1,   259, (1962).
(10C,   75) PHYS. REV. VOL.136,. . . . . . .   A1280, (1964).
(17A,   83) NUOVO CIMENTO B VOL.43,. . . . .   375, (1966).
(5 ,   1194) PHYS. REV. 143,. . . . .   36, (1966).

174 ASGHAR..........(M.)....:
(20C,   57) NUCL. INST. MET. 72, ...   307, (1969).

175 ASHMORE.........(J.P.)..:
(5 ,   1453) CAN. J. PHYS. 48,. . . . . 1091, (1970).

176 ASTROV..........(D.N.)..:
(5 ,   2856) SOV. PHYS. CRYST. VOL.17,. . . .   1017, (1973).

177 ATANASOV........(A.A.)..:
(9B,   105) SOV. PHYS. DOKLADY VOL.3,. . . .   2272, (1962).
(6 ,   982) ACTA PHYS. POLON. VOL.28,. . . .   451, (1965).
(6 ,   968) ACTA PHYS. POLON. VOL.30,. . . .   323, (1966).

178 ATANASSOV.......(A.A.)..:
(16 ,   11) C.R. ACAD. BULG. SCI. VOL.13,. .   527, (1960).

179 ATKINSON........(H.H.)..:
(7B,   45) PHIL. MAG. VOL.3,. . . . .   213, (1958).
(5 ,   689) PHIL. MAG. VOL.3,. : . : . : . : .   476, (1958).

180 ATOJI...........(I.)....:
(5 ,   2640) J. APPL. PHYS. VOL.44,. . . . . .   5096, (1973).

181 ATOJI...........(M.)....:
(5 ,   370) J. CHEM. PHYS. VOL.29,. . . . . .   1306, (1958).
(5 ,   1540) J. CHEM. PHYS. VOL.31,. . . . . .   329, (1959).
(5 ,   2734) J. CHEM. PHYS. VOL.31,. . . . . .   332, (1959).
(5 ,   1996) J. CHEM. PHYS. VOL.32,. . . . . .   627, (1960).
(5 ,   325) J. CHEM. PHYS. VOL.35,. . . . . .   1950, (1961).
(5 ,   2550) J. CHEM. PHYS. VOL.35,. . . . . .   1960, (1961).
(5 ,   2914) PHYS. REV. VOL.121,. . . . . . .   610, (1961).
(5 ,   2368) J. PHYS. SOC. JAPAN VOL.17 B-II,   395, (1962).
(5 ,   1343) ACTA CRYST. 17,. . . . . 1187, (1964).
(5 ,   69) J. CHEM. PHYS. VOL.43,. . . . . .   222, (1965).
(22B,   23) NUCL. INST. MET. 35,. . . . . . .   13, (1965).
(5 ,   2706) ACTA CRYST. 20,. . . . .   587, (1966).
(5 ,   397) PHYS. LETT. 22,. . . . .   21, (1966).
(5 ,   1372) PHYS. LETT. 23,. . . . .   208, (1966).
(5 ,   2549) J. CHEM. PHYS. VOL.46,. . . . . .   1891, (1967).
(5 ,   398) J. CHEM. PHYS. VOL.46,. . . . . .   4148, (1967).
(5 ,   2733) J. CHEM. PHYS. VOL.47,. . . . . .   1188, (1967).
(5 ,   2541) PHYS. LETT. 25A,. . . . .   528, (1967).
(5 ,   531) J. APP. PHYS. 39,. . . . .   671, (1968).
(5 ,   2532) J. CHEM. PHYS. VOL.48,. . . . . .   3380, (1968).
(5 ,   820) J. CHEM. PHYS. VOL.48,. . . . . .   3384, (1968).
(5 ,   2543) J. CHEM. PHYS. VOL.48,. . . . . .   560, (1968).
(6 ,   520) J. APP. PHYS. VOL.40,. . . . . . .   1447, (1969).
(5 ,   2545) J. CHEM. PHYSICS VOL.51,. . . . .   3872, (1969).
(5 ,   808) J. CHEM. PHYSICS VOL.51,. . . . .   3877, (1969).
(5 ,   1363) J. CHEM. PHYSICS VOL.51,. . . . .   3882, (1969).
(5 ,   1416) SOL. STATE COMM. VOL.7,. . . . .   1681, (1969).
(5 ,   2785) ACTA CRYST. VOL.B26,. . . . . . .   1540, (1970).
(5 ,   2916) J. CHEM. PHYSICS VOL.52,. . . . .   6430, (1970).
(5 ,   2687) J. CHEM. PHYSICS VOL.52,. . . . .   6431, (1970).
(5 ,   2684) J. CHEM. PHYSICS VOL.52,. . . . .   6433, (1970).
(5 ,   1371) J. APPL. PHYS. 42,. . . . .   1630, (1971).
(5 ,   407) J. CHEM. PHYSICS VOL.54,. . . . .   3226, (1971).
(5 ,   2546) J. CHEM. PHYSICS VOL.54,. . . . .   3510, (1971).
(5 ,   1370) J. CHEM. PHYSICS VOL.54,. . . . .   3514, (1971).
(5 ,   324) J. CHEM. PHYSICS VOL.54,. . . . .   3514, (1971).
(5 ,   851) ACTA CRYST. VOL.A28 PART S-4,. .   197, (1972).
(5 ,   225) ACTA CRYST. VOL.A28,. . . . . . .   663, (1972).

181 ATOJI.........(M.)...: (5 , 1954) J. CHEM. PHYS. VOL.56 NO.2, : 4947; (1972):
                           (5 , 854) J. CHEM. PHYS. VOL.57 NO.2, : : 851; (1972):
                           (5 , 1369) J. CHEM. PHYS. VOL.57, : : : 2402; (1972):
                           (5 , 864) J. CHEM. PHYS. VOL.57, : : : 2453; (1972):
                           (5 , 865) J. CHEM. PHYS. VOL.57, : : : 2410; (1972):
                           (5 , 2640) J. APPL. PHYS. VOL.44, : : : 5096; (1973):
                           (5 , 1953) J. CHEM. PHYS. VOL.60, : : : 3324; (1974):

182 AUDIAS.........(A.)...: (24 ,   60) NUCL. INST. MET. 61, ... 198, (1968).

183 AUERBACH.......(E.H.)..: (7A,   71) PHYS. REV. VOL.135, ... B895, (1964).

184 AUGST.........(G.R.)..: (10C, 124) SOV. PHYS. SOL. STATE VOL.12, . . 1031, (1970).

185 AURIVILLUS.....(K.)...: (5 , 1353) ARK. KEMI VOL.22, ... 517, (1964).

186 AUSTERN........(N.)...: (7A,   12) ANN. OF PHYS. VOL.33, ... 15, (1965).

187 AUSTIN.........(A.E.)..: (5 , 2736) ACTA CRYST. VOL.12, ... 159, (1959).
                           (5 , 1807) PHYS. REV. VOL.120, ... 1969, (1960).
                           (5 , 1640) J. APPL. PHYS. VOL.33, SUPPL., 1356, (1962).
                           (5 , 592) PHYS. REV. 131, ... 1511, (1963).
                           (5 , 1021) J. PHYS. CHEM. SOL. 26, ... 1795, (1965).
                           (5 , 1043) ACTA CRYST. A24, ... 513, (1968).

188 AVERBACH.......(B.L.)..: (6 , 1491) PHYSICS, ... 31, (1964).
                           (6 , 581) PHYS. REV. VOL.137, ...A1113, (1965).
                           (5 , 1750) PHYS. REV. 142, ... 287, (1966).
                           (6 , 750) PHYS. REV. 142, ... 291, (1966).
                           (5 , 1749) PHYS. REV. 147, ... 418, (1966).
                           (6 , 506) PHYS. REV. LETT. VOL.26, . . . 1480, (1971).
                           (5 , 483) PHYS. REV. B VOL.5, ... 2693, (1972).
                           (5 , 488) PHYS. REV. B VOL.6, ... 4294, (1972).

189 AVIRAM.........(I.)...: (16 ,   27) J. NUCL. ENERGY(GB) VOL.26, . . 61, (1972).

190 AXE...........(J.D.)..: (6 , 1352) PHYS. REV. VOL.183, ... 820, (1969).
                           (6 , 1341) ACTA CRYST. VOL.A26, ... 608, (1970).
                           (6 , 1340) PHYS. REV. B-1, ... 1227, (1970).
                           (6 , 2054) PHYS. REV. B-1, ... 342, (1970).
                           (6 , 124) PHYS. REV. B-2, ... 3651, (1970).
                           (6 , 1846) PHYS. REV. B-2, ... 155, (1970).
                           (5 , 2497) CONF. INTERN., RENNES, ... 155, (1971).
                           (6 , 514) J. APPL. PHYS. 42, ... 1666, (1971).
                           (5 , 2020) PHYS. REV. B VOL.4, ... 2957, (1971).
                           (6 , 129) PHYS. REV. B VOL.4, ... 155, (1971).
                           (6 , 2155) PHYS. REV. LETT. VOL.26, ... 519, (1971).
                           (6 , 1689) PHYS. REV. LETT. VOL.27, . . . 1803, (1971).
                           (5 , 2265) SOL. STATE COMM. VOL.9, ... 397, (1971).
                           (13 ,   1) ACTA CRYST. VOL.A28 PART S-4, . 177, (1972).
                           (6 , 1317) PHYS. REV. B VOL.5, ... 1933, (1972).
                           (6 , 2156) PHYS. REV. B VOL.6, ... 1950, (1972).
                           (6 , 1312) PHYS. REV. B VOL.6, ... 4332, (1972).
                           (18A,   7) ACTA CRYST. VOL.A29, ... 160, (1973).
                           (6 , 1691) PHYS. REV. B VOL.8, ... 1965, (1973).
                           (6 , 1688) PHYS. REV. LETT. VOL.30, ... 214, (1973).
                           (13 ,   58) PHYS. TODAY SEPT., 1973, ... 32, (1973).
                           (6 , 2339) SOL. STATE COMM. VOL.13, . . . 1465, (1973).
                           (6 , 1674) SOL. STATE COMM. VOL.13, . . . 1893, (1973).
                           (6 , 2318) PHYS. REV. B VOL.9, ... 4060, (1974).
                           (6 , 1529) PHYS. REV. B VOL.9, ... 4429, (1974).

191 AXMANN.........(A.)...: (6 , 2019) PHYS. STAT. SOL. 19, ... 721, (1967).
                           (6 , 2021) IAEA SYMP. COPENHAGEN, VOL.1, . 245, (1968).
                           (6 , 1009) DISC. FARADAY SOC. NO.48, . . . 69, (1969).
                           (6 , 2020) PROC/INTERNAT. SYMP/PHYS/SE/TE, 299, (1969).
                           (6 , 2167) DISC. FARADAY SOC. NO.50, . . . 74, (1970).

192 AXTON.........(E.J.)..: (20A,   27) J. RES. NAT. BUR/STAND. VOL.67A, 215, (1963).

193 AYAO..........(S.)...: (5 , 246) J. NUCL. SCI. TECHNOL. VOL.5, . 649, (1968).

194 AYASSE........(J.B.)..: (5 , 874) J. PHYSIQUE TOME 31, ... 607, (1970).

195 BAAZOV........(N.G.)..: (5 , 2529) SOV. PHYS. SOL. STATE VOL.15, . 1295, (1973).

196 BABCOCK.......(R.V.)..: (20C,   36) NUCLEONICS VOL.17 APRIL, ... 116, (1959).

197 BABEL.........(D.)...: (5 , 674) A.I.P. CONF. PROC. NO.10 PART 1, 664, (1973).

198 BABOT.........(O.)...: (5 , 1155) J. PHYSIQUE TOME 32 COL.1 VOL.II 985, (1971).
                           (5 , 1159) J. PHYS. CHEM. SOLIDS VOL.33, . 87, (1972).

199 BACCHELLA......(G.L.)..: (17A,   26) ENERGIA NUCLEARE VOL.7, ... 11, (1960).
                           (5 , 998) COLL. INTER. N.126 (GRENOBLE) , 113, (1963).
                           (5 , 1373) BULL. SOC. FRANC. MIN. CRIST. 89 216, (1966).
                           (8 ,   18) C.R. ACAD. SCI. B VOL.274, . . . 423, (1972).
                           (6 , 722) PHYS. REV. LETT. VOL.28, ... 805, (1972).
                           (17A,   87) PHYS. REV. LETT. VOL.31, ... 776, (1973).

200 BACHER........(R.F.)..: (20A,   59) PHYS. REV. VOL.59, ... 332, (1941).
                           (5 , 145) PHYS. REV. VOL.69, ... 443, (1946).

201 BACHLI........(A.)...: (19B,   12) NUCL. INST. MET. VOL.32, ... 300, (1965).

202 BACH..........(R.L.)..: (24 ,   99) RADIATION RESEARCH VOL.35, . . . 1, (1968).

203 BACMANN.......(M.)...: (5 , 2739) BULL. SOC. FRANC. MIN. CRIST. 88 214, (1965).
                           (5 , 1838) J. PHYSIQUE TOME 27, ... 726, (1966).
                           (5 , 2749) COMPTES RENDUS 266 SERIE B, . . 45, (1968).
                           (5 , 2748) COMPTES RENDUS 267, SERIE B, . 518, (1968).
                           (5 , 2750) J. APP. PHYS. VOL.40, ... 1131, (1969).
                           (5 , 2747) J. PHYSIQUE TOME 32 COL.1 VOL.II 859, (1971).
                           (5 , 514) ACTA CRYST. VOL.B29, ... 1570, (1973).

204 BACON.........(G.E.)..: (9B,   1) ACTA CRYST. VOL.1, ... 303, (1948).
                           (9A,   21) PROC. ROY. SOC. A VOL.196, . . . 50, (1949).
                           (21 ,   29) J. SCI. INSTRUM. VOL.27, ... 330, (1950).
                           (9C,   35) PROC. ROY. SOC. A VOL.209, . . . 397, (1951).
                           (5 , 238) ACTA CRYST. VOL.5, ... 492, (1952).
                           (5 , 1597) ACTA CRYST. VOL.5, ... 684, (1952).
                           (5 , 1605) ACTA CRYST. VOL.6, ... 57, (1953).
                           (5 , 473) PROC. ROY. SOC. A VOL.217, . . . 252, (1953).
                           (5 , 1467) PROC. ROY. SOC. A VOL.220, . . . 397, (1953).
                           (9A,   23) REP. PROGR. PHYS. VOL.16, . . . 1, (1953).
                           (9B, 100) RESEARCH VOL.7, ... 257, (1954).
                           (9B,   99) RESEARCH VOL.7, ... 312, (1954).
                           (22A,   28) J. SCI. INSTRUM. VOL.32, ... 256, (1955).
                           (5 , 1636) NATURE VOL.175, ... 518, (1955).
                           (5 , 1466) PROC. ROY. SOC. A VOL.230, . . . 359, (1955).
                           (5 , 1958) ACTA CRYST. VOL.9, ... 82, (1956).
                           (5 , 2414) PROC. ROY. SOC. A VOL.235, . . . 552, (1956).
                           (5 , 1444) ACTA CRYST. VOL.10, ... 524, (1957).
                           (5 , 1679) PROC. ROY. SOC. A VOL.241, . . . 223, (1957).
                           (9B, 101) RESEARCH VOL.10, ... 241, (1957).
                           (5 , 128) PROC. PHYS. SOC. VOL.72, ... 470, (1958).

```
204  BACON..........(G.E.)...‡  (  5 ,   580) PROC. ROY. SOC. A VOL.246, . . .        78, (1958).
                                (  5 ,  1259) REV. MOD. PHYS. VOL.30, . : . :         94, (1958).
                                ( 22C,     24) J. SCI. INSTRUM. VOL.36; . : . :       419, (1959).
                                ( 23 ,     50) NATURE VOL.183, . : . : . :             35, (1959).
                                (  5 ,  1443) ACTA CRYST. VOL.13, . : . : . :        717, (1960).
                                (  5 ,  2507) ACTA CRYST. VOL.13, . : . : . :         10, (1960).
                                (  5 ,   544) ACTA CRYST. VOL.14, . : . : . :        823, (1961).
                                (  9A,     15) J. PHYS. SOC. JAPAN VOL.17 B-II,       324, (1962).
                                (  5 ,   123) PROC. PHYS. SOC. VOL.79, . . .         938, (1962).
                                (  5 ,   745) PROC. ROY. SOC. A VOL.266, . .          95, (1962).
                                (  5 ,  2355) PROC. ROY. SOC. A VOL.272, . .         387, (1963).
                                (  5 ,  1267) ADV. STRUC. RES. DIFFR. METHOD 1         1, (1964).
                                (  5 ,   270) PROC. ROY. SOC. A VOL.279, . .          98, (1964).
                                (  9A,      1) ADV. INORG. CHEM. RADIOCHEM. 8,        225, (1965).
                                (  9A,      3) ADV. STRUC. RES. DIFFR. METHOD 2         1, (1966).
                                (  9A,     12) ENDEAVOUR VOL.25. . . . . . .          129, (1966).
                                (  5 ,  1506) J. APP. PHYS. 37, . . . . . .          979, (1966).
                                (  5 ,   126) J. PHYS. CHEM. SOL. 27, . . .          925, (1966).
                                (  5 ,  2355) PROC. PHYS. SOC. 88, . . . . :         923, (1966).
                                (  5 ,   129) PROC. PHYS. SOC. 92, . . . . :         713, (1967).
                                ( 24 ,   166) ROY. INST. GT. BRIT. PROC. 41,         510, (1967).
                                (  5 ,  1417) Z. KRIST. VOL.126, . . . . . :         460, (1968).
                                (  7A,      2) ACTA CRYST. VOL.A25, . . . . :         391, (1969).
                                (  5 ,   546) J. OF PHYS.-C- SER.2 VOL.2, .          239, (1969).
                                (  9A,     16) NAT. BUR. STAND. PUBL. NO.301,          57, (1969).
                                (  5 ,    32) PURE APPL. CHEM. VOL.18, . . .         517, (1969).
                                (  8 ,   156) TECH. METALS RES. VOL.11, . .          641, (1969).
                                (  5 ,  1630) J. OF PHYS.-C- SER.2 VOL.3, .          675, (1970).
                                (  5 ,  1339) J. PHYS.F: METAL PHYS. VOL.1,.         524, (1971).
                                (  5 ,  2413) ACTA CRYST. VOL.A28 PART S-4, .        193, (1972).
                                (  7A,      4) ACTA CRYST. VOL.A28, . . . . .         357, (1972).
                                (  5 ,   430) J. PHYS. F VOL.3, . : . : . :         2003, (1973).
                                (  5 ,  1732) J. PHYS. F VOL.3, . : . : . :            6, (1973).

205  BAFFIER.........(N.)....‡  (  5 ,  1757) J. PHYS. CHEM. SOLIDS VOL.33, .        737, (1972).
                                (  5 ,   469) J. SOLID STATE CHEM. VOL.8, .           50, (1973).

206  BAGGE...........(E.)....‡  ( 19B,      2) ATOMKERNENERGIE VOL.9, . . . .         307, (1964).

207  BAHNSEN.........(A.)....‡  ( 20D,     34) REV. SCI. INSTRUM. VOL.36, . .          48, (1965).
                                ( 20C,     49) NUCL. INST. MET. 50, . . . . :         181, (1967).

208  BAIERLEIN.......(K.)....‡  (  6 ,  2051) PHYS. STAT. SOL. 7, . . . . . .        415, (1964).

209  BAIJAL..........(J.S.)..‡  (  5 ,    94) J. PHYS. SOC. JAP. VOL.31, . .         1162, (1971).
                                (  6 ,  1630) PHYS. REV. A VOL.9, . . . . .          2128, (1974).

210  BAILEY..........(F.P.)..‡  (  5 ,  2980) J. SOLID STATE CHEM. VOL.7, . .        448, (1973).

211  BAILEY..........(R.L.)..‡  (  5 ,   144) NUCL. PHYS. VOL.17, . . . . .          109, (1960).

212  BAJOREK.........(A.)....‡  (  6 ,   384) ACTA PHYS. POLON. VOL.21, . .          529, (1962).
                                (  6 ,  1573) PHYS. STAT. SOL. 9, . . . . .          905, (1965).
                                ( 21 ,    40) NUKLEONIKA VOL.12, . : . : . :         385, (1967).
                                (  6 ,   502) PHYSICA VOL.35, . . . . . . :          441, (1967).
                                (  6 ,   380) PHYSICA VOL.35, . . . . . . :          451, (1967).
                                (  6 ,  1016) PHYSICA VOL.35, . . . . . . :          457, (1967).
                                (  6 ,  1596) PHYSICA VOL.35, . . . . . . :          465, (1967).
                                (  6 ,  2007) PHYSICA VOL.35, . . . . . . :          469, (1967).
                                (  6 ,  1379) PHYS. LETT. 24A, . . . . . . .         517, (1967).
                                (  6 ,  1017) ACTA PHYS. POLON. VOL.33, . . .        419, (1968).
                                (  6 ,  1104) IAEA SYMP. COPENHAGEN VOL.II, .         143, (1968).
                                (  6 ,   621) IAEA SYMP. COPENHAGEN, VOL.1,.          65, (1968).
                                (  6 ,  2003) IAEA SYMP. COPENHAGEN, VOL.2,.         237, (1968).
                                (  6 ,   286) PHYSICA VOL.41, . . . . . . .          397, (1969).
                                (  6 ,  1388) SOV. PHYS. CRYST. VOL.15, . . .       1010, (1971).

213  BAJPAI..........(R.P.)..‡  (  6 ,   583) PHYSICA VOL.53, . . . . . . :          628, (1971).
                                ( 10B,   217) SOL. STATE COMM. VOL. 9, . . .         167, (1971).
                                ( 10B,   105) PHYSICA VOL.62, . . . . . . :          574, (1972).
                                ( 10B,   104) PHYSICA VOL.62, . . . . . . :          587, (1972).
                                (  6 ,  1404) PHYS. LETT. VOL.41A NO.2, . .          141, (1972).
                                (  6 ,  2314) J. PHYS. F VOL.3, . . . . . .          709, (1973).

214  BAKER..........(C.P.)..‡  ( 20A,    59) PHYS. REV. VOL.59, . . . . . .         332, (1941).
                                (  5 ,   145) PHYS. REV. VOL.69, . . . . . .         443, (1946).

215  BAKKER.........(H.K.)..‡  (  5 ,  2078) PHYS. LETT. VOL.27A, . . . . .          69, (1968).

216  BAKRE..........(R.V.)..‡  ( 11B,    10) J. PHYS. -C-, VOL. 1, . . . . .        673, (1968).

217  BAK............(H.I.)..‡  (  5 ,  1956) J. KOREAN PHYS. SOC. VOL.1, . .        108, (1968).

218  BAK............(P.)....‡  (  6 ,  2145) PHYS. REV. LETT. VOL.31, . . .         1585, (1973).

219  BALANDA........(M.)....‡  (  5 ,  2943) PHYS. STAT. SOL. 32, . . . .           K91, (1969).
                                (  5 ,  1098) PHYS. STAT. SOL. A VOL.3, . .          1033, (1970).
                                (  5 ,  2181) PHYS. STAT. SOL. B VOL.43, . .          125, (1971).

220  BALASUBRAMANIAN.(R.)....‡  (  5 ,  1571) ACTA CRYST. 22, . . . . . . .          532, (1967).

221  BALBY..........(D.)....‡  ( 17C,     5) ACTA CRYST. 17, . . . . . . .          1529, (1964).

222  BALCAR.........(E.)....‡  (  9D,    40) J. PHYS. -C-, VOL. 1, . . . . .        966, (1968).
                                ( 11A,    32) PHYS. LETT. VOL.30A, . . . . .          84, (1969).
                                (  8 ,    34) J. PHYS. C VOL.3, . . . . . .          1292, (1970).
                                (  8 ,    56) PHYS. LETT. VOL.31A, . . . . .          67, (1970).
                                (  5 ,  2851) J. PHYS. C VOL.6, . . . . . .          3746, (1973).

223  BALKANSKI.......(M.)....‡  (  6 ,  2326) PHYS. STAT. SOL. VOL.41, . . .         491, (1970).

224  BALKE..........(S.)....‡  (  9C,    11) CHEM.-ING.-TECH. VOL.32, . . .         651, (1960).

225  BALLESTRACCI....(R.)....‡  (  5 ,   999) COLL. INTER. N.126 (GRENOBLE), .        102, (1963).
                                (  5 ,  2789) J. APP. PHYS. 34, . . . . . .          1333, (1963).
                                (  5 ,  2790) J. PHYS. CHEM. SOL. 24, . . .          487, (1963).
                                (  5 ,  1585) MATERIALS RES. BULL. VOL.2 NO.7,        473, (1967).
                                (  5 ,   836) J. PHYS. CHEM. SOLIDS 29, . .          1001, (1968).
                                (  5 ,  2923) SOL. STATE COMM. VOL. 7 NO.14,         1011, (1969).

226  BALLY..........(D.)....‡  ( 23 ,    17) IAEA SYMP. (PILE RESEARCH)VIENNA        577, (1960).
                                ( 23 ,    72) REV. DE PHYSIQUE (BUCAREST) 5, .         83, (1960).
                                ( 20A,    77) REV. SCI. INSTRUM. VOL.31, . .         640, (1960).
                                ( 20B,   142) REV. SCI. INSTRUM. VOL.32, . .         297, (1961).
                                (  6 ,   951) IAEA SYMP. CHALK RIVER, VOL.1,.        451, (1962).
                                (  5 ,    57) NUCL. SCI. ENGNG. VOL.12, . . .        157, (1962).
                                ( 23A,    79) REV. SCI. INSTRUM. VOL.33, . .         916, (1962).
                                (  6 ,   303) IAEA SYMP. BOMBAY, VOL.2, . .          421, (1964).
                                (  5 ,   437) REV. ROUMAINE PHYS. VOL.9, . .         245, (1964).
                                ( 18B,     9) REV. ROUMAINE PHYS. VOL.10, . .         445, (1965).
                                (  6 ,   755) J. PHYS. CHEM. SOL. 28, . . .          1947, (1967).
                                (  5 ,   263) PHYSICA VOL.33, . . . . . . .          523, (1967).
                                (  6 ,   480) PHYS. LETT. 25A, . . . . . . .         595, (1967).
                                (  6 ,  1238) IAEA SYMP. COPENHAGEN, VOL.1, .        483, (1968).
                                (  6 ,   470) IAEA SYMP. COPENHAGEN, VOL.2, .         75, (1968).
                                (  6 ,   759) J. APPL. PHYS. VOL.39, . . . .         459, (1968).
```

226 BALLY..........(D.)....‡ (6 ,   741) PHYS. LETT. VOL.26A, ...   396, (1968):
                               (23 ,   77) STUD. CERCETARI FIZ. VOL.20, . :   1033, (1968):
                               (6A,   805) ACTA CRYST. (INTERACT+),VOL.A25 S262, (1969):
                               (17A,   38) IAEA SYMP. (INSTRUMENT.), VIENNA   77, (1969):
                               (7B,   108) REPORT FN-37/70 (22PP.), ... :   . , (1970):
                               (13 ,   62) REPORT FN=4C/70 (6PP.), . . : . . . : . , (1970):
                               (6 ,   808) SOL. STATE COMM. VOL.9, . . .   353, (1971):
                               (17A,   91) REV. ROUMAINE PHYS. VOL.17, . :   757, (1972):

227 BALL...........(A.R.)..‡ (22C,   25) J. SCI. INSTRUM. VOL.41, ...   376, (1964).

228 BANDOPADHYAY....(P.K.)..‡ (20D,   23) NUCL. INST. MET. 68, ...   293, (1969).
                               (17A,   97) THESIS (120 PP.), ... , (1971)

229 BANERJEE........(R.)....‡ (10B,   11) CAN. J. PHYS. 47,. . . . .   451, (1969).

230 BANKS..........(E.)....‡ (5 , 1179) J. PHYS. CHEM. SOLIDS 29,. . . .   1359, (1968):
                               (5 ,   183) ACTA CRYST. VOL.B28, . . : . . .   3429, (1972).

231 BANNERMAN.......(D.E.)..‡ (20C,   106) REV. SCI. INSTRUM. 39, ...   1588, (1968).

232 BANNISTER.......(M.J.)..‡ (5 , 2788) ACTA CRYST. VOL.B28, ...   2995, (1972).

233 BANSAL.........(R.M.)..‡ (6 ,   290) PHYS. LETT. VOL.33A, ...   209, (1970):
                               (5 , 1905) NUCL./S.S. PHYS. SYMP. ABSTR., . , . , (1972).

234 BAN............(N.T.)..‡ (5 , 1510) ACTA CRYST. 23,. . . . .   578, (1967).

235 BARBARA........(B.)....‡ (5 ,   827) J. PHYSIQUE TOME 32 COL.1 VOL.II 1126, (1971).
                               (5 ,   810) Z. ANGEW PHYS. VOL.32, ...   113, (1971).
                               (5 ,   831) C. R. ACAD. SCI. B TOME 274, . .   319, (1972).
                               (5 , 2382) PHYS. STAT. SOLIDI A VOL.15 NO.2   613, (1973).

236 BARBERON.......(M.)....‡ (5 , 1791) INT. J. MAGN.(GB) VOL.1, ...   341, (1971).
                               (5 , 1770) J. PHYSIQUE TOME 32 COL.1 VOL.II   980, (1971).
                               (5 , 1793) SOL. STATE COMM. VOL. 9, ...   27, (1971).

237 BARBIC.........(L.)....‡ (6 , 1392) PHYS. STAT. SOL. VOL.B-54, . . .   121, (1972).

238 BARCLAY........(G.A.)..‡ (5 , 2410) ACTA CRYST. 19,. . . . .   205, (1965).

239 BARGOUTH.......(M.O.)..‡ (5 ,   859) CONF./RARE EARTHS/ACTINIDES, . ‡   196, (1971).
                               (5 ,   877) J. PHYSIQUE TOME 32 COL.1 VOL.II   675, (1971).
                               (5 ,   876) PHYS. KOND. MATER.(GER.) VOL.13,   137, (1971).
                               (5 ,   856) PHYS. LETT. VOL.36A, ...   50, (1971).
                               (5 ,   862) INT. J. MAGN. VOL.3, ...   87, (1972).
                               (5 ,   904) Z. NATURFORSCH. A VOL.27A, . . .   1581, (1972).

24C BARISH.........(J.)....‡ (24 ,   69) NUCL. INST. MET. 72, ...   213, (1969).

241 BARKAN.........(S.)....‡ (12B,   53) REV. SCI. INSTRUM. VOL.39, . . .   101, (1968).

242 BARKER.........(J.A.)..‡ (10D,   31) PHYS. REV. B-2,. . . . .   4176, (1970).
                               (6 , 1344) PHYS. LETT. VOL.34A, ...   415, (1971).
                               (10D,   34) PHYS. REV. B-4,. . . . .   1983, (1971).

243 BARKER.........(M.I.)..‡ (6 ,   82) J. PHYS. C VOL.5,. . . . .   3279, (1972).
                               (6 , 1961) PROPERTIES LIQUID METALS,. . . .   99, (1973).

244 BARKER.........(W.W.)..‡ (5 , 2980) J. SOLID STATE CHEM. VOL.7,. . .   448, (1973).

245 BARNARD........(E.)....‡ (16 ,   73) RENSSELAER POL. INST. SYMP., . .   248, (1961).
                               (20C,   39) NUCL. INST. MET. 34, ...   21, (1965).
                               (24 ,   42) NUCL. INST. MET. 34, ...   29, (1965).

246 BARNEA.........(Z.)....‡ (5 ,   184) ACTA CRYST. VOL.A26, ...   336, (1970).
                               (6 , 2084) PHYS. LETT. VOL.35A, ...   286, (1971).

247 BARNES.........(J.D.)..‡ (6 ,   354) IAEA SYMP. GRENOBLE, ...   287, (1972):
                               (6 ,   355) J. CHEM. PHYS. VOL.58, ...   5193, (1973).

248 BARNES.........(R.F.)..‡ (5 ,   422) PHYS. REV. VOL.134,. . . . . . . B1281, (1964).

249 BARRETT........(C.S.)..‡ (5 , 2705) PHYS. REV. VOL.129,. . . . . . .   625, (1963):
                               (20B,   146) REV. SCI. INSTRUM. VOL.34, . . .   847, (1963).

250 BARRET.........(R.F.)..‡ (20C,   58) NUCL. INST. MET. 68, ...   277, (1969).

251 BARRON.........(T.H.K.)‡ (10B,   29) IAEA SYMP. CHALK RIVER, VOL.1, .   49, (1962).
                               (5 , 2932) ACTA CRYST. 22,. . . . .   170, (1967).

252 BARSTAD........(G.E.B.)‡ (22C,   49) REV. SCI. INSTRUM. VOL.28, . . .   916, (1957).

253 BARTHELEMY.....(A.)....‡ (22B,   14) INTERNAT. COMPUTING SYMP., . . .   291, (1974).

254 BARTHOLIN......(H.)....‡ (5 , 2037) J. PHYS. CHEM. SOL. 27, ...   1287, (1966).

255 BARTL..........(H.)....‡ (5 ,   321) ACTA CRYST. (INTERACT.) VOL.A25 S119, (1969):
                               (5 , 1424) ACTA CRYST. VOL.A28 PART S-4,. .   93, (1972).

256 BARTOLINI......(W.)....‡ (20B,   143) REV. SCI. INSTRUM. VOL.32, . . .   870, (1961).
                               (20D,   33) REV. SCI. INSTRUM. VOL.34, . . .   28, (1963).
                               (5 , 2236) PHYS. REV. VOL.138,. . . . . . . B1116, (1965).
                               (5 ,   766) PHYS. REV. VOL.174,. . . . . . .   313, (1968).
                               (5 , 1244) PHYS. REV. C VOL.5,. . . . . . .   1952, (1972).

257 BARTON.........(C.A.)..‡ (5 , 2708) PHYS. REV. VOL.55, ...   793, (1939).

258 BARTON.........(J.P.)..‡ (24 ,   83) NUCL. TECHNOL. VOL.15, ...   56, (1972).

259 BARTS..........(B.I.)..‡ (8 ,   146) SOV. PHYS. SOLID STATE VOL.10, .   3142, (1968):
                               (11C,   61) UKRAINIAN PHYS. J. VOL.14, NO.2,   216, (1969).

260 BARTUSEK.......(M.)....‡ (10B,   71) J. PHYS. C‡ SOLID ST. PHYS. V-5,   287, (1972).

261 BARYSHEVSKII....(V.G.)..‡ (7B,   128) SOV. PHYS. JETP, 21, ...   765, (1965):
                               (12A,   45) YADERNAYA FIZIKA VOL.4,. . . . .   72, (1966):
                               (12A,   42) SOV. PHYS. JETP, 24, ...   1068, (1967):
                               (8 ,   148) SOV. PHYS. J.E.T.P. VOL.29,. . .   685, (1969):
                               (8 ,   149) SOV. PHYS. J.E.T.P. VOL.31,. . .   583, (1970).

262 BAR≠YAKHTAR.....(V.G.)..‡ (8 ,   131) SOV. PHYS. JETP VOL.12, ...   995, (1961):
                               (90 ,   70) SOV. PHYS. SOL. STATE, 4,. . . .   2533, (1962):
                               (11B,   39) SOV. PHYS. SOL. STATE, 5,. . . .   858, (1963):
                               (8 ,   139) SOV. PHYS. SOL. STATE, 6,. . . .   339, (1964):
                               (11B,   42) SOV. PHYS. SOL. STATE, 7,. . . .   891, (1965):
                               (11A,   62) SOV. PHYS. SOL. STATE VOL.11,. .   1566, (1969):
                               (8 ,   150) SOV. PHYS. SOL. STATE VOL.12,. .   1596, (1971):
                               (8 ,   153) SOV. PHYS. SOLID STATE VOL.14, .   2006, (1973).

263 BASHER.........(J.C.)..‡ (7B,   107) REPORT AEEW-M730,. . . . . , (1969).

264 BASKIN.........(Y.)....‡ (5 , 2805) SOL. STATE COMM. VOL.6,. . . . .   877, (1968):
                               (5 , 2806) J. APP. PHYS. VOL.40,. . . . . .   1130, (1969).

265 BASSI..........(G.)....! (5 , 1054) COLL. INTER. N.126 (GRENOBLE), .   121, (1963).
                              (5 , 1053) COMP. REND. 257, ...   421, (1963).
                              (5 , 2739) BULL. SOC. FRANC. MIN. CRIST. 88   214, (1965).
                              (5 , 2016) BULL. SOC. FRANC. MIN. CRIST. 88   345, (1965).
                              (5 , 591) J. APPL. PHYS. VOL.37, ... 1338, (1966).
                              (5 , 1011) J. PHYSIQUE TOME 27, ...   433, (1966).
                              (5 , 371) ACTA CRYST B24, ... 1358, (1968).

266 BASS...........(F.G.)..! (8 , 150) SOV. PHYS. SOL. STATE VOL.12, . 1596, (1971).

267 BASTON.........(A.H.)..! (21 ,   42) REPORT NP/R-2551 (26 PP.), ... , (1957).
                              (6 , 969) PHYS. LETT. VOL.12, ...   188, (1964).
                              (17B,   72) UKAEA AERE REPORT M2570 (32 PP.) . ., (1972).

268 BASTOW.........(T.J.)..! (5 , 540) PROC. PHYS. SOC. 86, ... 1143, (1965).
                              (5 , 533) PHYS. REV. 141, ...   510, (1966).
                              (5 , 585) PROC. PHYS. SOC. 88, ...   935, (1966).

269 BASU...........(A.N.)..! (10B, 185) PHYS. REV. B VOL.8, ... 2982, (1973).

270 BATAILLE.......(J.)....! (10D,   22) PHYS. LETT. VOL.31A, ... 563, (1970).

271 BATA...........(L.)....! (6 , 717) PHYS. LETT. 19, ...   15, (1965).
                              (2)B,   79) NUCL. INST. MET. 45, ...   255, (1966).
                              (6 , 824) PHYS. LETT. 24A, ...   22, (1967).
                              (6 , 424) PHYS. LETT. 24A, ...   57, (1967).
                              (6 , 430) IAEA SYMP. COPENHAGEN, VOL.1, .   615, (1968).
                              (6 , 1484) IAEA SYMP. COPENHAGEN, VOL.2, .   111, (1968).
                              (13 ,   41) PHYS. LETT. VOL.33A, ...   287, (1970).

272 BATCHELDER.....(D.N.)..! (6 ,   87) J. PHYS. C, SER.2, VOL.3, . . .   249, (1970).
                              (6 ,   98) J. PHYS. C, SER.2, VOL.4, . . .   910, (1971).
                              (6 , 2280) J. PHYS. C VOL.6, ... L313, (1973).

273 BATES..........(D.D.)..! (19B,   23) NUCL. INST. MET. VOL.116, . . .   205, (1974).

274 BATES..........(J.C.)..! (24 ,   78) NUCL. INST. MET. VOL.106, . . .   611, (1973).

275 BATTERMAN......(B.W.)..! (5 , 1222) PHYS. STAT. SOLIDI B VOL.59, .   K59, (1973).
                              (5 , 2450) PHYS. REV. B VOL.9, ... 2590, (1974).

276 BATTERMAN......(B.)....! (5 , 2441) PHYS. REV. B VOL.9, ... 2472, (1971).
                              (5 , 2442) PHYS. REV. LETT. VOL.27, ...   320, (1971).

277 BAUER..........(C.A.)..! (6 , 828) DISS. ABS. VOL.32, ...4147B, (1972).

278 BAUER..........(F.)....! (7A,   72) PHYS. REV. LETT. 16, ... 495, (1966).

279 BAUER..........(G.)....! (22A,   30) KERNTECHNIK VOL.14, ...   20, (1972).
                              (17A,   58) KERNTECHNIK VOL.14, ...   86, (1972).
                              (20A,   29) KERNTECHNIK VOL.14, ...   9, (1972).
                              (5 , 1626) SOLID STATE COMM. VOL.11, . . .   179, (1972).

280 BAUR...........(W.H.)..! (5 , 1614) ACTA CRYST. 17, ...   863, (1964).
                              (5 , 1284) ACTA CRYST. 19, ...   909, (1965).

281 BAYLE..........(P.)....! (20A,   11) ANN. FAC. SCI. UNIV. TOULOUSE 30   9, (1966).

282 BAYM...........(G.)....! (7B,   15) IAEA SYMPOSIUM VIENNA, ...   53, (1960).
                              (10C,   72) PHYS. REV. VOL.121, ...   741, (1961).
                              (10E,   4) COPENHAGEN CONF.-LATT.DYNAMICS,   261, (1963).
                              (7B,   75) PHYS. REV. VOL.135, ...A1691, (1964).

283 BAZUEV.........(G.V.)..! (5 , 1543) SOV. PHYS. SOL. STATE VOL.15, . 1079, (1973).

284 BECKA..........(L.N.)..! (6 , 1628) IAEA SYMP. CHALK RIVER, VOL.2, .   23, (1962).
                              (6 , 385) J. CHEM. PHYSICS VOL.37, ...   431, (1962).
                              (6 , 2157) J. PHYS. SOC. JAPAN VOL.17 B-III   63, (1962).
                              (6 , 648) J. CHEM. PHYSICS VOL.38, ... 1685, (1963).

285 BECKER.........(M.)....! (24 ,   82) NUCL. SCI. ENGNG. VOL.42, . . .   104, (1970).
                              (16 ,   50) NUCL. SCI. ENG. VOL.47, ...   66, (1972).

286 BECKER.........(W.J.)..! (5 , 2091) Z. KRIST. VOL.131, ... 139, (1970).
                              (5 , 2910) Z. KRIST. VOL.131, ... 278, (1970).

287 BECKURTS.......(K.H.)..! (20B,   72) NUCL. INST. MET. VOL.11, ... 144, (1961).
                              (20A,   72) REACTOR SCI. TECHNOL. VOL.17, .   147, (1963).
                              (16 ,   37) NUCL. SCI. ENGNG. VOL.20, . . .   236, (1964).
                              (19A,   17) IAEA SYMP. COPENHAGEN, VOL.2, .   417, (1968).
                              (19A,   1) ACTA CRYST. (INTERACT.) VOL.A25 S250, (1969).
                              (19A,   23) IAEA SYMP. (INSTRUMENT.) VIENNA   147, (1969).
                              (5 , 2222) J. PHYSIQUE TOME 32 COL.1 VOL.I, 582, (1971).
                              (5 , 2227) PHYS. REV. B VOL.7, ... 3102, (1973).

288 BECK...........(H.)....! (7B,   47) PHYS. KOND. MATER. (GER.) VOL.14 336, (1972).

289 BECLE..........(C.)....! (5 , 1365) COMPTES RENDUS, SERIE B, 266, .   994, (1968).
                              (5 , 857) PHYS LETT 27A, ... 541, (1968).
                              (5 , 2537) SOL. STATE COMM. VOL.6 NO.2, . .   115, (1968).
                              (5 , 827) J. PHYSIQUE TOME 32 COL.1 VOL.II 1126, (1971).
                              (5 , 810) Z. ANGEW PHYS. VOL.32, ...   113, (1971).

290 BEDBENOVA......(D.S.)..! (6 , 1636) JETP LETT. VOL.3, ...   110, (1966).

291 BEDEKHIN.......(A.F.)..! (20C,   19) INSTRUM. EXPER. TECH. NO.2, . .   342, (1967).

292 BEDELIK........(W.)....! (20B,   27) IAEA SYMPOSIUM VIENNA, ... 447, (1960).

293 BEDNARSKI......(S.)....! (6 , 165) IAEA SYMP. COPENHAGEN, VOL.1, .   157, (1968).
                              (6 , 2012) PHYS. STAT. SOL. VOL.44, ... K65, (1971).
                              (6 , 536) PHYS. STAT. SOLIDI B VOL.57 NO.1   K1, (1973).

294 BEEBY..........(J.L.)..! (14A,   24) J. PHYS. C VOL.2, ...   556, (1969).
                              (14B,   48) J. PHYS. C VOL.6, ... 2262, (1973).

295 BEER...........(C.)....! (5 ,   96) PHYS. SCRIPTA VOL.1, ...   85, (1970).

296 BEGUM..........(N.A.)..! (6 , 1768) J. OF PHYS.-C-, SER. 2, VOL.2, . 874, (1969).

297 BEGUM..........(R.J.)..! (6 , 157) J. PHYS. CHEM. SOLIDS VOL.23, . 1747, (1962).
                              (5 , 1824) PHYS. LETT. VOL.15, ...   225, (1965).
                              (5 , 1018) SOL. STATE COMM. 3, ...   113, (1965).
                              (12B,   15) INDIAN J. PURE APPL. PHYS. VOL.7 546, (1969).
                              (5 , 1628) J. APP. PHYS. VOL.40, ... 1870, (1969).
                              (5 , 2142) J. PHYS. CHEM. SOLIDS VOL.30, . 1941, (1969).
                              (5 , 1663) J. PHYS. CHEM. SOLIDS 30, . . .   939, (1969).
                              (5 , 2955) PHYS. REV. VOL.181, ...   969, (1969).
                              (5 , 2314) PHYS. STAT. SOLIDI A VOL.3, . .   959, (1970).
                              (5 , 2082) J. PHYSIQUE TOME 32 COL.1 VOL.I, 318, (1971).
                              (6 , 2319) J. PHYS. CHEM. SOLIDS VOL.33, .   759, (1972).
                              (5 , 522) NUCL./S.S. PHYS. SYMP. ABSTR., . . , (1972).

298   BEG............(M.M.)..：   (15 ,   13) IAEA SYMP. COPENHAGEN, VOL.2,. .   299, (1968).
                                 (6 , 1864) J. PHYS. C. VOL.3, ...   2487, (1970).

299   BEHARI.........(J.)....：   (6 ,   605) PHYS. LETT. VOL.29A,. . . . . . .   313, (1969).
                                 (10B,   84) J. PHYS.F: METAL PHYS. VOL.1,. .   19, (1971).
                                 (10B,   97) J. PHYS. SOC. JAPAN VOL.31,. . .   1639, (1971).
                                 (6 ,   744) J. PHYS. SOC. JAP. VOL.33,. . .   1207, (1972).

300   BEHRINGER......(K.)....：   (19B,   12) NUCL. INST. MET. VOL.32, ...   300, (1965).

301   BEIBER.........(E.)....：   (12B,   53) REV. SCI. INSTRUM. VOL.39, . . .   101, (1968).

302   BEIL...........(H.)....：   (21 ,   28) J. PHYSIQUE TOME 24, ...   15A, (1963).
                                 (20B,   62) J. PHYSIQUE TOME 24, ...   359, (1963).
                                 (12B,   22) J. PHYSIQUE TOME 24, ...   89A, (1963).
                                 (5 , 2060) J. PHYSIQUE TOME 24, ...   21, (1963).

303   BEKETOV........(V.A.)..：   (12B,   16) INSTRUM. EXP. TECH. NO.4,. . . .   571, (1969).
                                 (12B,   18) INSTRUM. EXP. TECH. (USA) VOL.14   618, (1971).

304   BELAKHOVSKY....(M.)....：   (5 , 2748) COMPTES RENDUS 267, SERIE B,. .   518, (1968).
                                 (5 , 1178) SOL. STATE COMM. VOL.6 NO.5,. .   317, (1968).
                                 (5 ,   844) PHYS. LETT. VOL.34A,. . . . . .   361, (1971).
                                 (5 ,   829) PHYS. STAT. SOL. VOL.47,. . . .   247, (1971).
                                 (5 ,   835) SOL. STATE COMM. VOL.9,. . . .   925, (1971).

305   BELLAVANCE.....(D.)....：   (5 , 1104) PHYS. REV. B-3,. . . . .   3046, (1971).

306   BELLONI........(M.G.)..：   (20A,   79) REV. SCI. INSTRUM. VOL.33, . . .   916, (1962).

307   BELL...........(H.G.)..：   (6 , 1705) PHYS. LETT. VOL.45A,. . . . . .   479, (1973).

308   BELOV..........(N.V.)..：   (5 , 2269) SOV. PHYS. DOKLADY VOL.6,. . . .   370, (1961).
                                 (5 , 1693) J. PHYS. SOC. JAPAN VOL.17 B-III   55, (1962).
                                 (9D,   50) J. PHYS. SOC. JAPAN VOL.17 B-II   330, (1962).
                                 (9D,   49) J. PHYS. SOC. JAPAN VOL.17 B-II   332, (1962).
                                 (5 , 2895) SOV. PHYS. CRYST. VOL.7, ...   765, (1963).
                                 (5 , 2875) SOV. PHYS. CRYST. VOL.7, ...   767, (1963).
                                 (5 , 2907) SOV. PHYS. CRYST. VOL.8, ...   15, (1963).
                                 (5 , 1195) SOV. PHYS. CRYST. VOL.8, ...   537, (1964).

309   BELOZOROV......(D.P.)..：   (5 , 1137) SOV. PHYS. JETP LETT. VOL.8, . .   128, (1968).

310   BELSON.........(J.)....：   (20B,   36) IAEA SYMP. COPENHAGEN, VOL.2,. .   341, (1968).

311   BEMIS-JR.......(C.E.)..：   (5 ,   414) NUCL. SCI. ENG. VOL.37,. . . . .   228, (1969).
                                 (5 , 1515) NUCL. SCI. ENG. VOL.47,. . . . .   371, (1972).

312   BENDT..........(P.J.)..：   (7A,   63) PHYS. REV. VOL.77,. . . . . . .   575, (1950).
                                 (6 , 1480) DISS. ABSTR. VOL.11,. . . . . .   1080, (1951).
                                 (6 , 1503) PHYS. REV. VOL.89, ...   561, (1953).
                                 (6 , 1202) PHYS. REV. LETT. VOL.1,. . . . .   9, (1958).
                                 (6 , 1191) PHYS. REV. VOL.113,. . . . . . .   1379, (1959).
                                 (6 , 1190) PHYS. REV. VOL.113,. . . . . . .   1386, (1959).

313   BENEDEK........(G.)....：   (10B,   52) J. CHEM. PHYSICS VOL.48, ...   5242, (1969).

314   BENES..........(J.)....：   (6 ,   466) IAEA SYMP. GRENOBLE, ...   563, (1972).

315   BENNEMANN......(K.H.)..：   (8 ,   25) J. APPL. PHYS. 42, ...   1610, (1971).

316   BENNETT........(H.S.)..：   (11C,   38) PHYS. REV. 176,. . . . .   650, (1968).
                                 (8 ,   23) J. APPL. PHYS. 40, ...   1552, (1969).

317   BENNETT........(R.A.)..：   (5 ,   685) NUCL. SCI. ENGNG. VOL.7, ...   184, (1960).

318   BENNETT........(R.G.)..：   (20A,   51) NUCL. TECHNOL. VOL.14, ...   284, (1972).

319   BENNION........(R.B.)..：   (5 ,   209) PHYS. LETT. 24A,. . . . .   714, (1967).
                                 (5 ,   603) J. PHYS. CHEM. SOLIDS VOL.29,. .   435, (1968).
                                 (9C,   36) PROC/SYMP/CRYS. STRUC/HIGH PRESS   141, (1969).

320   BENOIT.........(C.)....：   (6 ,   724) C.R. ACAD. SC. (PARIS), 267-B, .   61, (1968).
                                 (6 , 1414) IAEA SYMP. COPENHAGEN, VOL.1,. .   289, (1968).

321   BENSCH.........(F.)....：   (20C,   28) J. NUCL. ENERGY VOL.27,. . . . .   677, (1973).

322   BENTON.........(C.U.)..：   (24 ,   104) REV. SCI. INSTRUM. VOL.41, . .   1900, (1970).

323   BENZ...........(R.)....：   (5 , 2632) ACTA CRYST. VOL.B28, ...   1724, (1972).

324   BEN-AHMED......(A.)....：   (16 ,   88) THESIS (88PP.),. . . . . , (1970).

325   BEREZHNOI......(YU.A.)..： (9B,   33) BU. ACAD. SCI. USSR PHYS. SER.35 2456, (1971).

326   BERGER.........(H.)....：   (20C,   27) J. APPL. PHYS. VOL.33, ...   48, (1962).
                                 (20C,   95) REV. SCI. INSTRUM. VOL.33, . . .   844, (1962).

327   BERGER.........(M.)....：   (6 , 1789) J. CHEM. PHYSICS VOL.40, ...   1417, (1964).
                                 (6 , 1919) J. CHEM. PHYSICS VOL.40, ...   1426, (1964).

328   BERGHMANS......(H.)....：   (6 , 1906) J. POLYMER SCI. A-2 VOL.9, . . .   1219, (1971).

329   BERGSMA........(J.)....：   (6 ,   65) IAEA SYMPOSIUM VIENNA, ...   501, (1960).
                                 (5 , 2304) PHYSICA VOL.26,. . . . .   744, (1960).
                                 (5 ,   89) ACTA CRYST. VOL.14,. . . . . . .   223, (1961).
                                 (5 ,   353) ACTA CRYST. VOL.15,. . . . . . .   92, (1962).
                                 (6 ,   69) IAEA SYMP. CHALK RIVER, VOL.1, .   189, (1962).
                                 (18A,   11) BROOKHAVEN SYMPOSIUM,. . . . . .   163, (1965).
                                 (20D,   18) NUCL. INST. MET. 51, ...   121, (1967).
                                 (6 ,   775) PHYS. LETT. 24A, ...   270, (1967).
                                 (6 ,   725) IAEA SYMP. COPENHAGEN, VOL.1,. .   233, (1968).
                                 (6 , 2329) PHYS. LETT. VOL.32A,. . . . . .   324, (1970).
                                 (6 , 1429) RCN-121 (119 PP.),. . . . . , (1970).

330   BERGSTEDT......(L.)....：   (14B,   88) PHYS. REV. VOL.151,. . . . . . .   117, (1966).
                                 (14B,   86) PHYS. REV. VOL.151,. . . . . . .   126, (1966).

331   BERICK.........(A.C.)..：   (5 , 2361) PHYS. REV. VOL.187,. . . . . . .   1506, (1969).

332   BERKEBILE......(C.A.)..：   (5 , 1448) J. MOLEC. STRUC. VOL.1,. . . . .   283, (1968).

333   BERKOWITZ......(E.H.)..：   (20C,   56) NUCL. INST. MET. VOL.73, ...   225, (1969).

334   BERMAN.........(M.)....：   (22B,   52) REV. SCI. INSTRUM. VOL.40, . . .   1144, (1969).

335   BERNAL.........(I.)....：   (5 , 1551) MOL. DYN. AND STRUCT. OF SOLIDS,   249, (1969).
                                 (5 ,   295) ACTA CRYST. VOL.B29, ...   2278, (1973).
                                 (5 ,   17) ACTA CRYST. VOL.B30, ...   1220, (1974).

336   BERNARD........(J.L.)..：   (5 ,   469) J. SOLID STATE CHEM. VOL.8,. . .   50, (1973).

337   BERNDORFER.....(K.)....：   (6 , 1723) J. PHYSIQUE TOME 32 COL.1 VOL.II   679, (1971).
                                 (12B,   75) Z. PHYS. VOL.243,. . . . .   188, (1971).

```
338  BERNEY..........(C.V.)..:  ( 6 ,   245) J. AMER. CHEM. SOC. VOL.95,. . .   768, (1973).
                                ( 6 ,   428) SPECTROSCOPY BIOL. CHEM.,. . . .   297, (1974).

339  BERNE...........(B.J.)..:  (15 ,    42) PHYS. REV. A VOL.2,. . . . . . .   975, (1970).

340  BERNOLE.........(M.)....:  ( 5 ,    76) C.R. ACAD. SCI. B TOME 277,. . .   225, (1973).

341  BERNSTEIN.......(E.R.)..:  ( 5 ,  2725) INORG. CHEM. VOL.11, . . . . . . 3009, (1972).

342  BERNSTEIN.......(S.)....:  ( 5 ,  1411) PHYS. REV. VOL.72, . . . . . . .   109, (1947).
                                (12B,    41) PHYS. REV. VOL.94, . . . . . . .   374, (1954).
                                ( 9B,    81) PHYS. REV. VOL.97, . . . . . . .   889, (1955).

343  BERODIAS........(G.)....:  ( 5 ,  2651) COMP. REND. 263, . . . . . . .   621B, (1966).
                                ( 5 ,  2650) BULL. SOC. FRANC. MIN. CRIST. 91    88, (1968).
                                ( 5 ,  1154) SOL. STATE COMM. VOL.7,. . . . .   623, (1969).
                                ( 5 ,  1155) J. PHYSIQUE TOME 32 COL.1 VOL.II   985, (1971).
                                ( 5 ,  1159) J. PHYS. CHEM. SOLIDS VOL.33,. .    87, (1972).

344  BERRETH.........(J.R.)..:  ( 5 ,  2248) NUCL. SCI. ENG. VOL.20,. . . . .   235, (1964).

345  BERRY...........(M.V.)..:  ( 7A,    80) PROC. PHYS. SOC. 91, . . . . . .   678, (1967).

346  BERTAUT.........(E.F.)..:  ( 5 ,  2897) COMP. REND. TOME 243,. . . . . .   898, (1956).
                                ( 5 ,  2516) C.R. ACAD. SCI. VOL.251,. . . .  1733, (1960).
                                ( 5 ,  2018) J. PHYS. CHEM. SOLIDS VOL.21,. .   234, (1961).
                                ( 5 ,  2882) C.R. ACAD. SCI. VOL.255,. . . .   896, (1962).
                                ( 5 ,  2738) J. APPL. PHYS. VOL.33, SUPPL.,.  1123, (1962).
                                ( 5 ,  1837) J. PHYS. RADIUM VOL.23,. . . . .   477, (1962).
                                ( 5 ,  1128) J. PHYS. SOC. JAPAN VOL.17 B-III    53, (1962).
                                ( 5 ,   500) PHYS. LETT. 3,. . . . . . . . .   178, (1962).
                                ( 5 ,  1054) COLL. INTER. N.126 (GRENOBLE),.   121, (1963).
                                ( 5 ,  1849) COLL. INTER. N.126 (GRENOBLE),.   126, (1963).
                                ( 5 ,   615) COLL. INTER. N.126 (GRENOBLE),.   158, (1963).
                                ( 5 ,   427) COLL. INTER. N.126 (GRENOBLE),.   186, (1963).
                                ( 5 ,    84) COLL. INTER. N.126 (GRENOBLE),.    54, (1963).
                                ( 5 ,  1230) COLL. INTER. N.126 (GRENOBLE),.    92, (1963).
                                ( 5 ,  1126) COMP. REND. 256,. . . . . . . .  1688, (1963).
                                ( 5 ,  1053) COMP. REND. 257,. . . . . . . .   421, (1963).
                                ( 5 ,  2789) J. APP. PHYS. 34,. . . . . . . .  1333, (1963).
                                ( 5 ,  2790) J. PHYS. CHEM. SOL. 24,. . . . .   487, (1963).
                                ( 5 ,  1852) PHYS. LETT. 5,. . . . . . . . .    27, (1963).
                                ( 5 ,  1851) PHYS. LETT. 7,. . . . . . . . .   110, (1963).
                                ( 5 ,   435) SOL. STATE COMM. 1,. . . . . . .   81, (1963).
                                ( 5 ,    83) SOL. STATE COMM. 1,. . . . . . .   85, (1963).
                                ( 5 ,  1855) J. APPL. PHYS. VOL.35,. . . . . .  952, (1964).
                                ( 5 ,   579) PROC. INT. CONF. (NOTTINGHAM),.   275, (1964).
                                ( 6 ,   549) PROC. INT. CONF. (NOTTINGHAM),.   516, (1964).
                                ( 5 ,  2739) BULL. SOC. FRANC. MIN. CRIST. 88   214, (1965).
                                ( 5 ,   155) BULL. SOC. FRANC. MIN. CRIST. 88   413, (1965).
                                ( 5 ,  1853) PHYS. LETT. 18,. . . . . . . . .    13, (1965).
                                ( 5 ,  1017) SOL. STATE COMM. 3,. . . . . . .   117, (1965).
                                ( 5 ,  1139) SOL. STATE COMM. 3,. . . . . . .   335, (1965).
                                ( 5 ,  1092) BULL. SOC. FRANC. MIN. CRIST. 89   206, (1966).
                                ( 5 ,   893) COMP. REND. 262, . . . . . . . 1707B, (1966).
                                ( 5 ,  1542) COMP. REND. 262, . . . . . . .  962B, (1966).
                                ( 5 ,  2560) IEEE TRANS. MAGNETICS MAG-2,. .   453, (1966).
                                ( 5 ,   591) J. APPL. PHYS. VOL.37,. . . . .  1038, (1966).
                                ( 5 ,  1011) J. PHYSIQUE TOME 27,. . . . . .   433, (1966).
                                ( 5 ,   599) J. PHYSIQUE TOME 27,. . . . . .   619, (1966).
                                ( 5 ,  1838) J. PHYSIQUE TOME 27,. . . . . .   726, (1966).
                                ( 5 ,   646) SOL. STATE COMM. 4,. . . . . . .  249, (1966).
                                ( 9D,    84) Z. ANGEW. PHYS. VOL.21,. . . . .  259, (1966).
                                ( 5 ,   456) BULL. SOC. FRANC. MIN. CRIST. 90   109, (1967).
                                ( 5 ,  1674) COMP. REND. 264, . . . . . . .  323B, (1967).
                                ( 5 ,  2908) COMP. REND. 264, . . . . . . .  397B, (1967).
                                ( 5 ,  2559) J. PHYS. CHEM. SOL. 28,. . . . .  2143, (1967).
                                ( 5 ,   221) SOLID STATE COMM. 5,. . . . . .    25, (1967).
                                ( 5 ,  1675) SOL. STATE COMM. 5,. . . . . . .   279, (1967).
                                ( 5 ,  2571) SOL. STATE COMM. 5,. . . . . . .   293, (1967).
                                ( 5 ,  2039) SOL. STATE COMM. 5,. . . . . . .    93, (1967).
                                ( 5 ,  1747) SOL. STATE COMM. 5,. . . . . . .     7, (1967).
                                ( 5 ,   822) ACTA CRYST A24,. . . . . . . . .  217, (1968).
                                ( 5 ,  2749) COMPTES RENDUS 266 SERIE B,. .     45, (1968).
                                ( 5 ,  2748) COMPTES RENDUS 267, SERIE B,. .    518, (1968).
                                ( 5 ,  2570) J. APPL. PHYS. VOL.39,. . . . .  1364, (1968).
                                ( 5 ,   618) J. PHYSIQUE TOME 29,. . . . . .   813, (1968).
                                ( 5 ,   824) J. PHYSIQUE TOME 29,. . . . . .    67, (1968).
                                ( 5 ,   836) J. PHYS. CHEM. SOLIDS 29,. . .   1001, (1968).
                                ( 5 ,  1702) SOL. STATE COMM. VOL.6 NO.5,. .    251, (1968).
                                ( 5 ,  1832) SOL. STATE COMM. VOL.6 NO.5,. .    269, (1968).
                                ( 5 ,  1532) SOL. STATE COMM. VOL.6,. . . . .   751, (1968).
                                ( 5 ,   453) BULL. SOC. FRANC. MIN. CRIST. 92   264, (1969).
                                ( 5 ,   595) COMP. REND. 269, . . . . . . .   574, (1969).
                                ( 5 ,  2750) J. APP. PHYS. VOL.40,. . . . .   1131, (1969).
                                ( 5 ,   174) J. PHYSIQUE TOME 30,. . . . . .   566, (1969).
                                ( 5 ,  1681) SOL. STATE COMM. VOL.7 NO.19,.   1403, (1969).
                                ( 5 ,  1680) SOL. STATE COMM. VOL.7 NO.2,. .    343, (1969).
                                ( 5 ,  1713) SOL. STATE COMM. VOL.7,. . . . .   641, (1969).
                                ( 5 ,   521) ACTA CRYST. VOL.B26,. . . . . .  2036, (1970).
                                ( 5 ,   166) J. PHYSIQUE TOME 31,. . . . . .   401, (1970).
                                ( 5 ,  1203) SOL. STATE COMM. VOL.8,. . . . .  1755, (1970).
                                ( 5 ,  1202) SOL. STATE COMM. VOL.8,. . . . .  1751, (1970).
                                ( 5 ,  2568) SOL. STATE COMM. VOL.8,. . . . .   239, (1970).
                                ( 5 ,   736) SOL. STATE COMM. VOL.8,. . . . .   935, (1970).
                                ( 9D,    12) AIP CONF. PROC. (USA) VOL.5,. .   1355, (1971).
                                ( 5 ,  2747) J. PHYSIQUE TOME 32 COL.1 VOL.II   859, (1971).
                                ( 5 ,  1861) J. PHYSIQUE TOME 32 COL.1 VOL.II   876, (1971).
                                ( 5 ,  1015) J. PHYSIQUE TOME 32 COL.1 VOL.I,   202, (1971).
                                ( 5 ,  2942) J. PHYSIQUE TOME 32 COL.1 VOL.I,   320, (1971).
                                ( 9D,    44) J. PHYSIQUE TOME 32 COL.1 VOL.I,   462, (1971).
                                ( 5 ,  1028) SOL. STATE COMM. VOL. 9,. . . .   1633, (1971).
                                ( 5 ,  1741) SOL. STATE COMM. VOL. 9,. . . .   1793, (1971).
                                ( 5 ,   594) SOL. STATE COMM. VOL.9 S-4,. .     717, (1971).
                                ( 5 ,  2113) ACTA CRYST. VOL.A28 PART S-4,.    195, (1972).
                                ( 9D,    16) ANN. DE PHYSIQUE TOME 7 NO.4,.    263, (1972).
                                ( 5 ,  2589) SOL. STATE COMM. VOL.11 NO.5,.    605, (1972).
                                ( 5 ,  1027) J. PHYS. CHEM. SOLIDS VOL.34,.    151, (1973).
                                ( 5 ,  2948) J. SOLID STATE CHEM. VOL.8,. .    182, (1973).
                                ( 8 ,   127) SOL. STATE COMM. VOL.12,. . . .  1045, (1973).
                                ( 5 ,   642) SOL. STATE COMM. VOL.13,. . . .   905, (1973).
                                ( 5 ,  1659) SOL. STATE COMM. VOL.14,. . . .   665, (1974).

347  BERTINOTTI......(A.)....:  ( 5 ,  1418) ACTA CRYST. VOL.B26, . . . . . .  422, (1970).

348  BERTINOTTI......(C.)....:  ( 5 ,  1418) ACTA CRYST. VOL.B26, . . . . . .  422, (1970).

349  BERTONI.........(C.M.)..:  (10B,   171) PHYS. REV. LETT. VOL.28,. . . .  1578, (1972).
                                (10B,   186) PHYS. REV. LETT. VOL.31,. . . .  1466, (1973).
                                (10B,    92) J. PHYS. F VOL.4,. . . . . . . .    19, (1974).
                                (10B,   188) PHYS. REV. B VOL.9,. . . . . . .  1710, (1974).

350  BERTRAM.........(W.K.)..:  ( 7A,    18) AUSTRAL. J. PHYS. VOL.22,. . . .   135, (1969).
                                ( 7A,    17) AUSTRAL. J. PHYS. VOL.22,. . . .   145, (1969).
```

351 BESANT.........(C.B.)..: (6 , 1037) PROC. ROY. SOC. 289, ...   342, (1966).

352 BESSER.........(P.J.)..: (5 , 1094) PHIL. MAG. VOL.12, ...   221, (1965).

353 BETHE.........(H.A.)..: (7A,   83) REV. MOD. PHYS. VOL.9, ...   113, (1937).

354 BETSUYAKU......(H.)....: (20A, 26) J. PHYS. SOC. JAPAN VOL.19,. . 2280, (1964).

355 BETTS.........(D.D.)..: (13 ,   10) CAN. J. PHYS. 46,. . . . .   799, (1968).

356 BETZL.........(M.)....: (5 , 2696) PHYS. STAT. SOL. 2,. . . . . . .   K164, (1962).
                             (5 , 525) PHYS. STAT. SOL. 3,. . . . . . .   K249, (1963).
                             (5 , 1409) Z. KRIST. VOL.118,. : . . . . . .   473, (1963).

357 BEVILACQUA.....(S.)....: (22B,   28) NUCL. INST. MET. 42, ...   29, (1966).

358 BEVINGTON......(P.R.)..: (17B,   43) NUCL. INST. MET. 44, ...   341, (1966).

359 BEYERLEIN......(R.)....: (5 , 487) BULL. AMER. PHYS. SOC. VOL.17, .   667, (1972).

360 BEYER.........(H.G.)..: (25 ,   36) PHYS. REV. VOL.54, ...   771, (1938).
                             (5 , 2464) PHYS. REV. VOL.55, ...   1101, (1939).
                             (7B,   53) PHYS. REV. VOL.57, ...   976, (1940).
                             (5 , 2136) PHYS. REV. VOL.58, ...   1031, (1940).

361 BEYSTER........(J.R.)..: (6 , 1895) IAEA SYMPOSIUM (VIENNA), ...   613, (1960).
                             (16 ,   74) RENSSELAER POL. INST. SYMP., . .   215, (1961).

362 BEZDEK.........(H.F.)..: (22C,   63) REV. SCI. INSTRUM. 40, ...   1397, (1969).
                             (6 , 2333) DISS. ABS. VOL.31,. . . . .   36498, (1970).
                             (6 , 2334) PHYS. STAT. SOL. VOL.42, ...   275, (1970).
                             (6 , 137) PHYS. REV. B VOL.4,. . . . . . .   1390, (1971).

363 BHANDARI.......(R.C.)..: (5 , 199) J. NUCL. ENERGY VOL.6, ...   104, (1957).
                             (6 , 163) J. NUCL. ENERGY VOL.7, ...   45, (1958).
                             (6 , 181) PHIL. MAG. VOL.3,. . . . .   798, (1958).

364 BHUSHAN........(V.)....: (6 , 192) NUCL. SCI. ENG. VOL.44,. . . . .   444, (1971).

365 BIALAS.........(H.)....: (6 ,   63) PHYS. LETT. VOL.43A NO.2,. . . .   97, (1973).

366 BIDAUX.........(R.)....: (5 , 815) J. PHYSIQUE VOL.29 NO.2-3, . . .   220, (1968).

367 BIELEN.........(H.J.)..: (5 , 742) COLL. INTER. N.126 (GRENOBLE), .   18, (1963).

368 BIELEN.........(J.)....: (5 , 2539) Z. ANGEW. PHYS. VOL.23,. . . . .   243, (1967).

369 BIEL..........(W.)....: (19B,   3) ATOMKERNENERGIE VOL.15,. . . . .   91, (1970).

370 BIEM..........(W.)....: (6 , 1009) DISC. FARADAY SOC. NO.48,. . . .   69, (1969).

371 BIERNE.........(T.)....: (23 ,   69) REPORT AERE-R5698 (14 PP.),. . . . . , (1968).

372 BILENSKII......(V.)....: (8 ,   47) LATV. PSR ZIN. AKAD. VESTIS NO.3   39, (1962).

373 BILLMAN........(K.W.)..: (24 ,   93) PHYS. REV. VOL.153,. . . . . . .   1415, (1967).

374 BILZ..........(H.)....: (10A,   3) IAEA SYMP. GRENOBLE, ...   3, (1972).
                             (10B, 172) PHYS. REV. LETT. VOL.28, ...   600, (1972).

375 BINCZYCKA......(H.)....: (5 , 1697) PHYS. STAT. SOLIDI A VOL.19 NO.1 K13, (1973).

376 BINDER.........(F.)....: (5 , 669) SOL. STATE COMM. VOL.11, ...   1471, (1972).

377 BINDER.........(K.)....: (8 ,   53) NUKLEONIK VOL.11,. . . . .   113, (1968).
                             (6 ,   32) PHYS. STAT. SOL. VOL.41, ...   767, (1970).
                             (11B,   53) Z. ANGEW. PHYS. VOL.28,. . . . .   325, (1970).
                             (8 , 162) Z. ANGEW. PHYS. VOL.32,. . . . .   178, (1971).
                             (8 , 164) Z. NATURFORSCH. A VOL.26A,. . . .   432, (1971).

378 BIRCHALL.......(J.H.L.): (6 , 1682) BER. BUNSENGES. PHYS. CHEM. 76,.   782, (1972).

379 BIRGENEAU......(R.J.)..: (6 , 1731) PHYS. REV. VOL.136,. . . . . . .   A1359, (1964).
                             (6 , 1328) PHYS. REV. LETT. VOL.23, ...   1394, (1969).
                             (5 , 1493) PHYS. REV. LETT. 22, ...   720, (1969).
                             (5 , 360) PHYS. REV. VOL.188,. . . . . . .   930, (1969).
                             (6 , 1331) J. APPL. PHYS. VOL.41, ...   1303, (1970).
                             (6 , 1327) PHYS. REV. B-1, ...   2211, (1970).
                             (6 , 1761) PHYS. REV. B-2, ...   1362, (1970).
                             (6 , 1930) PHYS. REV. LETT. VOL.25, ...   752, (1970).
                             (6 , 237) J. APPL. PHYS. 42, ...   1265, (1971).
                             (6 , 1935) J. APPL. PHYS. 42,. . . . . . .   1746, (1971).
                             (6 , 1329) J. PHYSIQUE TOME 32 COL.1 VOL.II   882, (1971).
                             (6 , 1330) PHYS. REV. B VOL.3,. . . . . . .   1736, (1971).
                             (6 , 2195) PHYS. REV. B VOL.4,. . . . . . .   718, (1971).
                             (9D,   62) PHYS. REV. LETT. VOL.26, ...   718, (1971).
                             (6 , 1927) PHYS. REV. LETT. VOL.27, ...   1530, (1971).
                             (6 , 2265) SOL. STATE COMM. VOL. 9, ...   397, (1971).
                             (6 , 1932) IAEA SYMP. GRENOBLE, ...   543, (1972).
                             (6 , 238) PHYS. REV. B VOL.5,. . . . . . .   1999, (1972).
                             (6 , 803) PHYS. REV. B VOL.5,. . . . . . .   2607, (1972).
                             (5 , 976) PHYS. REV. B VOL.5,. . . . . . .   2615, (1972).
                             (6 , 2000) PHYS. REV. B VOL.6,. . . . . . .   2630, (1972).
                             (6 , 2335) PHYS. REV. B VOL.6,. . . . . . .   2724, (1972).
                             (5 , 838) PHYS. REV. LETT. VOL.28, ...   746, (1972).
                             (8 ,   7) A.I.P. CONF. PROC. NO.10 PART 2, 1664, (1973).
                             (6 , 1953) PHYS. REV. B VOL.8,. . . . . . .   5345, (1973).
                             (5 , 1483) PHYS. REV. B VOL.8,. . . . . . .   304, (1973).
                             (6 , 1934) PHYS. REV. LETT. VOL.31,. . . . .   1300, (1973).
                             (6 , 625) PHYS. REV. LETT. VOL.32 NO.4,. .   170, (1974).

380 BIRKELUND......(J.R.)..: (20C,   58) NUCL. INST. MET. 68, ...   277, (1969).

381 BIRMAN.........(J.L.)..: (6 , 460) PHYS. REV. 156,. . . . .   925, (1967).
                             (10B, 161) PHYS. REV. LETT. VOL.25, ...   1014, (1970).

382 BIRR..........(M.)....: (22A,   13) IAEA SYMP. COPENHAGEN, VOL.2,. .   381, (1968).
                             (6 , 947) Z. PHYSIK VOL.238, ...   221, (1970).
                             (22A,   42) NUCL. INST. MET. 95,. . . . . .   435, (1971).

383 BIRSAN.........(I.)....: (18B,   9) REV. ROUMAINE PHYS. VOL.10,. . .   445, (1965).

384 BIRSTEIN.......(L.)....: (20A,   49) NUCL. INST. MET. VOL.116,. . . .   615, (1974).

385 BIRYUKOV.......(V.A.)..: (20C, 110) SOV. PHYS. J.E.T.P. VOL.31,. . .   59, (1970).

386 BISCHOFF.......(F.G.)..: (7B,   43) NUCL. SCI. ENG. VOL.48,. . . . .   266, (1972).

387 BISCHOFF.......(F.)....: (6 , 2352) CONF. NEUTRON THERMALIZ. VOL.I,.   407, (1968).

388 BISCHOF........(J.)....: (22A,   10) CZECH. J. PHYS. B VOL.16,. . . .   942, (1966).
                             (9B,   37) CZECH. J. PHYS. B VOL.19,. . . .   1608, (1969).

389 BISSEM..........(H.)....: (24 ,   70) NUCL. INST. MET. 76, ...   103, (1969).

390 BITTERMANN......(K.)....: (5 , 1484) J. CRYST. GROWTH VOL.21, ...   82, (1974).

391 BITTER..........(M.)....: (6 , 1133) PHYS. STAT. SOL. VOL.23, ... K155, (1967).

392 BLACKSHAW.......(G.L.)..: (14B,   24) DISS. ABS. VOL.27, ...19648, (1967).

393 BLAESSER........(G.)....: (10B, 122) PHYS. REV. 171,. . . . . 665, (1963).
                             (6 , 2233) REPORT EUR-4216E (41PP.),. . . . . . ., (1969).

394 BLAIR...........(J.S.)..: (7A,   12) ANN. OF PHYS. VOL.33, ...   15, (1965).

395 BLAISE..........(A.)....: (5 , 2749) COMPTES RENDUS 266 SERIE B,. . . .   45, (1968).
                             (5 , 2750) J. APP. PHYS. VOL.40,. . . . . . 1131, (1969).

396 BLANK...........(A.Y.)..: (11A,   60) SOV. PHYS. J.E.T.P. 26,. . . . . 764, (1968).

397 BLASCHKO........(O.)....: (17A,   92) REV. SCI. INSTRUM. VOL.45, . . . 526, (1974).

398 BLECH...........(I.A.)..: (6 , 1491) PHYSICS, ...   31, (1964).
                             (6 , 581) PHYS. REV. VOL.137,. . . . . . .A1113, (1965).
                             (5 , 1750) PHYS. REV. 142,. . . . . 287, (1966).

399 BLEEKER.........(E.J.)..: (6 , 847) J. APP. PHYS. 39,. . . . . 1114, (1968).

400 BLINC...........(R.)....: (6 , 1299) PHYS. LETT. VOL.26A, ...   8, (1967).
                             (6 , 1383) PHYS. LETT. VOL.31A, ... 531, (1970).
                             (6 , 1392) PHYS. STAT. SOL. VOL.B-54, . . . 121, (1972).

401 BLINOWSKI.......(K.)....: (22B,   4) ACTA PHYS. POLON. VOL.16,. . . . 293, (1957).
                             (6 , 860) J. PHYS. CHEM. SOLIDS VOL.9,. . . 153, (1959).
                             (23 ,   19) IAEA SYMP. (PILE RESEARCH)VIENNA 369, (1960).
                             (9D,   28) IAEA SYMP. (PILE RESEARCH)VIENNA 415, (1960).
                             (20B,   32) IAEA SYMP. (PILE RESEARCH)VIENNA 445, (1960).
                             (20A,   52) NUKLEONIKA VOL.5,. . . . . 414, (1960).
                             (20B,   70) NUCL. INST. MET. VOL.10, ... 289, (1961).
                             (6 , 2317) IAEA SYMP. CHALK RIVER, VOL.2, .   87, (1962).
                             (6 , 764) J. APPL. PHYS. VOL.35, ... 933, (1964).
                             (6 , 524) PROC. INT. CONF. (NOTTINGHAM), . 101, (1964).
                             (6 , 753) PROC. INT. CONF. (NOTTINGHAM), .   99, (1964).
                             (6 , 523) SOL. STATE COMM. 2,. . . . . . . 109, (1964).
                             (6 , 756) PHYS. REV. VOL.139,. . . . . . .A1866, (1965).
                             (6 , 781) PHYS. LETT. 28A, ... 389, (1968).
                             (6 , 736) PHYS. LETT. 29A, ... 513, (1969).
                             (13 ,   39) PHYS. LETT. 30A, ...   68, (1969).

402 BLOCH...........(D.)....: (22C,   67) REV. SCI. INSTRUM. VOL.45, . . . 643, (1974).

403 BLOCH...........(F.)....: (8 ,   65) PHYS. REV. VOL.50, ... 259, (1936).
                             (8 ,   66) PHYS. REV. VOL.51, ... 994, (1937).
                             (12B,   34) PHYS. REV. VOL.64, ...   47, (1943).

404 BLOCKI..........(J.)....: (15 ,   32) PHYSICA VOL.48,. . . . . 126, (1970).

405 BLOCK...........(R.C.)..: (5 , 144) NUCL. PHYS. VOL.17,. . . . . . . 109, (1960).
                             (24 ,   45) NUCL. INST. MET. 52, ... 321, (1967).

406 BLOCK...........(R.)....: (23 ,   28) IAEA SYMP. (PILE RESEARCH)VIENNA 535, (1960).

407 BLOK............(J.)....: (9C,   29) NED. TIJDSCHR. NATUURK. VOL.18,. 201, (1952).

408 BLOMER..........(R.)....: (7B,   2) ANN. OF PHYS. VOL.45,. . . . . . 404, (1967).

409 BLOOM...........(M.)....: (5 , 384) PHYS. REV. VOL.83, ... 840, (1951).

410 BLOOM...........(S.D.)..: (5 , 924) NUCL. SCI. ENG. VOL.46,. . . . . 255, (1971).

411 BLUME...........(M.)....: (8 ,   27) J. CHEM. PHYS. VOL.37, ... 1245, (1962).
                             (8 ,   83) PHYS. REV. LETT. 11, ... 489, (1963).
                             (8 ,   84) PHYS. REV. 130,. . . . . 1670, (1963).
                             (12B,   46) PHYS. REV. VOL.133,. . . . . . .A1366, (1964).
                             (7B,   5) BROOKHAVEN SYMPOSIUM,. . . . . .   1, (1965).
                             (7B,   39) METAL. SOC. CONF. LOS ANGELES, .   1, (1967).
                             (6 , 2130) PHYS. REV. 161,. . . . . 499, (1967).
                             (8 ,   91) PHYS. REV. 166,. . . . . 554, (1968).
                             (11C,   40) PHYS. REV. B-1,. . . . . 3815, (1970).
                             (11A,   9) IAEA SYMP. GRENOBLE, ... 529, (1972).
                             (8 , 105) PHYS. REV. B VOL.7,. . . . . . . 1149, (1973).
                             (8 , 103) PHYS. REV. B VOL.7,. . . . . . . 5017, (1973).
                             (5 , 812) PHYS. REV. LETT. VOL.32, ... 544, (1974).

412 BLYSKAVKA.......(A.A.)..: (20B, 131) PRIB. TEKH. EKSPER. VOL.2, . . .   36, (1964).

413 BOBETIC.........(M.V.)..: (10D,   31) PHYS. REV. B-2,. . . . . 4176, (1970).
                             (6 , 1344) PHYS. LETT. VOL.34A, ... 415, (1971).

414 BOCCARA.........(N.)....: (10B,   32) IAEA SYMPOSIUM BOMBAY, VOL.1,. . 313, (1964).
                             (10D,   3) BROOKHAVEN SYMPOSIUM,. . . . . .   44, (1965).

415 BOCHKO..........(A.V.)..: (24 , 112) SOV. PHYS. DOKLADY VOL.16, . . . 154, (1971).

416 BOCK............(H.)....: (20C,   71) NUCL. INST. MET. VOL.87, ... 299, (1970).

417 BODGER..........(E.O.)..: (6 , 2224) CHEM. COMMUN. 1967,. . . . . . .   74, (1967).

418 BOEDEKER........(R.R.)..: (10E,   5) DISS. ABSTR. VOL.20, ... 3794, (1960).

419 BOEUF...........(A.)....: (20B, 106) NUCL. INST. MET. VOL.107,. . . . 429, (1973).

420 BOFFI...........(S.)....: (10C,   28) ENERGIA NUCLEARE VOL.16, ... 369, (1969).
                             (10C,   27) ENERGIA NUCLEARE (ITALY) VOL.16, 425, (1969).
                             (10B,   41) IAEA SYMP. GRENOBLE, ... 149, (1972).

421 BOFFI...........(V.C.)..: (6 , 326) IAEA SYMP. CHALK RIVER, VOL.1, . 285, (1962).

422 BOGDANOV........(S.G.)..: (7A,   90) SOV. PHYS. CRYST. VOL.15,. . . . 252, (1970).
                             (7A,   91) SOV. PHYS. CRYST. VOL.15,. . . . 383, (1970).

423 BOGOMOLOV.......(A.M.)..: (5 , 2256) PHYS. MET. METALLOGR. VOL.22 N.2 123, (1967).
                             (23 ,   76) SOV. PHYS. JOURNAL VOL.10 NO.4,.   63, (1967).
                             (5 , 208) PHYS. MET. METALLOGR. VOL.27, 6, 1011, (1969).
                             (5 , 2927) PHYS. MET. METALLOGR. VOL.31,. .   79, (1971).

424 BOGOMOLOV.......(G.D.)..: (5 , 2646) SOV. PHYS. DOKLADY VOL.15, . . . 276, (1970).

425 BOGOMOLOV.......(V.N.)..: (6 , 893) SOV. PHYS. SOL. STATE VOL.13,. . 1256, (1971).

426 BOHANNON........(J.R.)..: (20C,   61) NUCL. INST. MET. 74, ... 355, (1969).

427 BOHLIN..........(L.)....: (10D,   59) SOL. STATE COMM. VOL.9, ... 141, (1971).
                             (6 , 1696) SOL. STATE COMM. VOL.10, ... 1219, (1972).

428 BOHNEN..........(K.P.)..: (11A,   23) J. PHYS. C VOL.7,. . . . . . . 947, (1974).

429 BOKHENKOV.......(E.L.)..: (5 ,   181) SOV. PHYS. SOL. STATE VOL.10,.. 2537, (1969).

430 BOKOV...........(V.A.)..: (5 ,   218) PROC. INT. CONF. (NOTTINGHAM), . 354, (1964).
                            (5 , 1854) SOV. PHYS. SOL. STATE 8,. . . . 215, (1966).
                            (5 , 2892) SOV. PHYS. SOL. STATE VOL.14,. : 3037, (1973).

431 BOLDRINI........(P.)....: (5 , 2868) ACTA CRYST. 21,. . . . . 158, (1966).
                            (5 , 2268) ACTA CRYST. 22,. . . . . 744, (1967).

432 BOLLER..........(H.)....: (90,   12) AIP CONF. PROC.(USA) VOL.5,. . 1355, (1971).
                            (5 ,   557) SOL. STATE COMM. VOL. 9,. . . 1699, (1971).
                            (5 ,   562) SOL. STATE COMM. VOL.12 NO.7,. 665, (1973).

433 BOLLINGER.......(L.M.)..: (20C,   92) REV. SCI. INSTRUM. VOL.30, . . 1135, (1959).
                            (20C,   37) NUCL. INST. MET. VOL.17,. . .   97, (1962).
                            (5 , 1625) PHYS. REV. VOL.134,. . . . . .B1047, (1964).

434 BOLOTIN.........(YU.L.).: (8 ,   144) SOV. PHYS. JETP. 25,. . . . . 517, (1967).
                            (9D,   77) SOV. PHYS. SOLID STATE VOL.9,. 2016, (1968).

435 BONFANTE........(M.)....: (6 , 1490) IAEA SYMP. GRENOBLE, ... 631, (1972).
                            (6 , 1486) SOL. STATE COMM. VOL.10, . . . 553, (1972).

436 BONGAARTS.......(A.L.M.): (5 ,   655) ACTA CRYST. VOL.A28 PART S-4, . 194, (1972).
                            (5 , 2097) PHYS. LETT. VOL.41A NO.5,. . . 411, (1972).
                            (5 ,   656) PHYS. REV. B VOL.6,. . . . . . 2669, (1972).

437 BONGAARTS.......(P.J.M.): (12B,   52) REV. SCI. INSTRUM. VOL.33, . . 524, (1962).

438 BONKOWSKI.......(L.)....: (21 ,   1) ACTA PHYS. POLON. VOL.18,. . . 265, (1959).

439 BONNER..........(B.E.)..: (20C,   51) NUCL. INST. MET. 55, ... 349, (1967).

440 BONSE...........(U.)....: (22A,   53) PHYS. LETT. VOL.47A NO.5,. . . 369, (1974).

441 BOOTH...........(J.G.)..: (5 ,   607) PHYS. LETT. VOL.43A NO.4,. . . 381, (1973).

442 BOOTH...........(R.S.)..: (16 ,   39) NUCL. SCI. ENGNG. VOL.28,. . . 404, (1967).

443 BORELI..........(F.)....: (5 , 2221) FIZIKA (YUGOSLAVIA) VOL.2, . .   97, (1970).

444 BOREVING........(S.)....: (24 ,   59) NUCL. INST. MET. 66, ... 229, (1968).

445 BORGHESE........(C.)....: (6 ,   902) J. PHYS. CHEM. SOL. 28,. . . . 2225, (1967).

446 BORGONOVI.......(G.M.)..: (6 , 2315) PHYS. REV. 132,. . . . . 683, (1963).
                            (6 ,   88) PHYS. REV. LETT. 12,. . . . . 721, (1964).
                            (6 , 1423) PHYS REV 174,. . . . . 953, (1968).
                            (5 ,   196) NUCL. SCI. ENG. VOL.42,. . . . 137, (1970).

447 BORGONOVI.......(G.)....: (9B,   69) NUOVO CIMENTO VOL.24,. . . . . 1174, (1962).
                            (6 , 2310) COPENHAGEN CONF.-LATT.DYNAMICS,   17, (1963).
                            (6 , 2061) IAEA SYMP. BOMBAY, VOL.1,. . . 117, (1964).
                            (6 ,   70) IAEA SYMP. BOMBAY, VOL.2,. . . 101, (1964).
                            (6 ,   797) J. PHYS. CHEM. SOL. 28,. . . . 467, (1967).

448 BORIE...........(B.)....: (5 , 2974) ACTA CRYST. VOL.A29, ... 594, (1973).

449 BORMANN.........(M.)....: (24 ,   70) NUCL. INST. MET. 76, ... 103, (1969).

450 BORONKAY........(S.)....: (6 ,   792) INT. J. MAGN. VOL.4, ... 205, (1973).

451 BOROWITZ........(S.)....: (12B,   35) PHYS. REV. VOL.74,. . . . . . 1285, (1948).

452 BOROWSKI........(F.)....: (5 ,   303) NUKLEONIKA VOL.10, ... 201, (1965).

453 BORRMANN........(G.)....: (17B,   19) ACTA CRYST. A25, ... 116, (1969).

454 BORSCH..........(P.)....: (6 , 1009) DISC. FARADAY SOC. NO.48,. . .   69, (1969).

455 BORST...........(L.B.).: (16 ,   25) IAEA SYMPOSIUM VIENNA, ... 437, (1960).
                            (6 , 1151) BROOKHAVEN SYMPOSIUM,. . . . . 173, (1965).
                            (6 , 1173) NATURE VOL. 209, ... 187, (1966).

456 BORTOLANI.......(V.)....: (10B,   171) PHYS. REV. LETT. VOL.28,. . . 1578, (1972).
                            (10B,   186) PHYS. REV. LETT. VOL.31,. . . 1466, (1973).
                            (10B,   92) J. PHYS. F VOL.4,. . . . . . .   19, (1974).
                            (10B,   188) PHYS. REV. B VOL.9,. . . . . . 1710, (1974).

457 BOSCHUNG........(P.)....: (17B,   46) NUCL. INST. MET. VOL.85,. . . . 151, (1970).

458 BOSE............(G.)....: (10B,   76) J. PHYS. CHEM. SOLIDS VOL.34,. 1867, (1973).
                            (10B,   98) J. PHYS. SOC. JAPAN VOL.34,. . 1006, (1973).
                            (6 , 2018) PHYS. LETT. VOL.43A NO.4,. . . 365, (1973).

459 BOSI............(P.)....: (10B,   60) J. CHEM. PHYS. VOL.59, ... 4578, (1973).

460 BOSMAN..........(J.J.)..: (7A,   35) NED. TIJDSCHR. NATUURK. VOL.38,. 170, (1972).

461 BOSSEL..........(J.B.)..: (19A,   9) HELV. PHYS. ACTA VOL.89, . . . 569, (1966).

462 BOSTOCK.........(J.)....: (6 , 1687) PHYS. LETT. VOL.35A, ...   31, (1971).

463 BOTTERMAN.......(A.C.)..: (5 , 2097) PHYS. LETT. VOL.41A NO.5,. . . 411, (1972).

464 BOTTGER.........(H.)....: (10B,   199) PHYS. STAT. SOLIDI B VOL.59 NO.2 517, (1973).

465 BOUCHAUD........(J.P.)..: (5 , 1702) SOL. STATE COMM. VOL.6 NO.5, . . 251, (1968).

466 BOUCHERLE.......(J.X.)..: (5 ,   844) PHYS. LETT. VOL.34A, ... 361, (1971).

467 BOUCHER.........(B.)....: (5 , 2168) COMP. REND. 249, ... 514, (1959).
                            (5 ,   361) J. PHYSIQUE TOME 26, ... 789, (1965).
                            (5 , 2166) COMP. REND. 263, ... 3448, (1966).
                            (5 ,   343) COMP. REND. 263, ...   98, (1966).
                            (5 , 1703) J. APP. PHYS. 37,. . . . . . . 960, (1966).
                            (5 , 1704) J. PHYSIQUE TOME 27, ...   51, (1966).
                            (5 ,   472) J. APP. PHYS. 39,. . . . . . . 632, (1968).
                            (5 , 2167) ACTA CRYST. B25, ... 2326, (1969).
                            (5 , 1047) J. APP. PHYS. 40,. . . . . . . 1126, (1969).
                            (5 ,   471) J. PHYSIQUE TOME 31, ... 113, (1970).
                            (5 , 2941) PHYS. STAT. SOL. VOL.40, . . . 171, (1970).
                            (5 , 1748) J. APPL. PHYS. 42, ... 1615, (1971).
                            (5 , 1229) J. PHYSIQUE TOME 32 COL.1 VOL.II 853, (1971).
                            (5 , 1045) J. PHYSIQUE TOME 32 COL.1 VOL.I, 322, (1971).
                            (5 , 1742) J. PHYS. CHEM. SOLIDS VOL.32,. 2429, (1971).
                            (5 , 1712) PHYS. STAT. SOL. VOL.44, . . .   71, (1971).

468 BOUCHEZ.........(R.)....: (19B,   32) REV. PHYS. APPL. (FRANCE) VOL.4, 111, (1969).

469 BOURDONNAY......(H.)....: (6 ,   730) PHYS. LETT. VOL.31A,. . . . . 561, (1970).
                            (6 ,   840) J. PHYSIQUE TOME 32 COL.1 VOL.II 1182, (1971).

470   BOUTIN..........(H.P.)..:   (15 ,   3) ATLANTA SYMP/GEORGIA INST. TECH.   30, (1967).
                                    (5 , 2276) ACTA CRYST. VOL.B25, ...   982, (1969).
                                    (5 , 759) ACTA CRYST. VOL.B26, ...   1235, (1970).

471   BOUTIN..........(H.)....:   (6 , 1646) J. CHEM. PHYS. VOL.39, ...   3135, (1963).
                                    (5 , 1975) J. PHYS. CHEM. SOL. 24, ...   1341, (1963).
                                    (6 , 1587) J. PHYS. CHEM. SOL. 24, ...   617, (1963).
                                    (6 , 1425) J. PHYS. CHEM. SOL. 24, ...   771, (1963).
                                    (6 , 992) IAEA SYMP. BOMBAY, VOL.2, ...   393, (1964).
                                    (6 , 1894) IAEA SYMP. BOMBAY, VOL.2, ...   407, (1964).
                                    (6 , 1789) J. CHEM. PHYSICS VOL.40, ...   1417, (1964).
                                    (6 , 1919) J. CHEM. PHYSICS VOL.40, ...   1426, (1964).
                                    (6 , 1094) J. CHEM. PHYSICS VOL.40, ...   2670, (1964).
                                    (6 , 1213) J. CHEM. PHYSICS VOL.41, ...   3649, (1964).
                                    (6 , 1096) SURFACE SCI. VOL.2, ...   261, (1964).
                                    (6 , 1097) J. CHEM. PHYSICS VOL.42, ...   1469, (1965).
                                    (6 ,   10) J. CHEM. PHYSICS VOL.44, ...   3127, (1966).
                                    (6 , 1918) J. CHEM. PHYSICS VOL.45, ...   2700, (1966).
                                    (6 , 1083) J. CHEM. PHYSICS VOL.45, ...   3284, (1966).
                                    (6 , 1586) J. CHEM. PHYSICS VOL.45, ...   401, (1966).
                                    (6 , 1386) J. CHEM. PHYSICS VOL.45, ...   699, (1966).
                                    (6 , 1903) J. MACROMOL. SCI. VOL.A1, . . .   723, (1967).
                                    (6 , 2043) ADV. COLLOID INTERFACE SCI. 2, .   1, (1968).
                                    (6 , 1020) J. CHEM. PHYSICS VOL.48, ...   3367, (1968).
                                    (5 , 2275) J. CHEM. PHYS. VOL.48, ...   1397, (1968).
                                    (6 , 1044) MOL. DYN. AND STRUCT. OF SOLIDS,   335, (1969).

472   BOUTRON.........(F.)....:   (17C,   16) BULL. SOC. FRANC. MIN. CRIST. 83   125, (1960).

473   BOUVAIST........(J.)....:   (5 , 2278) ACTA CRYST. VOL.A26, ...   501, (1970).

474   BOUWMA..........(J.)....:   (5 , 1810) J. PHYSIQUE TOME 32 COL.1 VOL.I,   78, (1971).

475   BOWERS..........(R.)....:   (6 , 1629) PHYS. REV. VOL.128, ...   1112, (1962).
                                    (6 , 1621) PROC. PHYS. SOC. VOL.79, ...   446, (1962).

476   BOWERS..........(W.A.)..:   (6 , 2242) PHYS. REV. VOL.116, ...   297, (1959).

477   BOWER...........(R.H.J.):   (7A,   29) DISS. ABS. VOL.32, ... 17468, (1971).
                                    (7A,   13) ANN. OF PHYS. VOL.73, ...   372, (1972).

478   BOWMAN..........(A.L.)..:   (5 , 2833) ACTA CRYST. 19, ...   6, (1965).
                                    (5 , 2731) ACTA CRYST. 21, ...   670, (1966).
                                    (5 , 2019) ACTA CRYST. 21, ...   843, (1966).
                                    (22C,   58) REV. SCI. INSTRUM. VOL.37, . . .   1543, (1966).
                                    (5 , 2631) ACTA CRYST. VOL.B24, ...   1121, (1968).
                                    (5 , 1524) ACTA CRYST. VOL.B24, ...   459, (1968).
                                    (6 , 796) J. CHEM. PHYSICS VOL.49, ...   2514, (1968).
                                    (5 , 2754) J. APPL. PHYS. 41, ...   5080, (1970).
                                    (5 , 1554) J. PHYS. CHEM. SOLIDS VOL.31, . .   1193, (1970).
                                    (5 , 2752) ACTA CRYST. VOL.B27, ...   1067, (1971).
                                    (5 , 2638) ACTA CRYST. VOL.B27, ...   243, (1971).
                                    (5 , 784) J. CHEM. PHYSICS VOL.55, ...   589, (1971).
                                    (5 , 565) ACTA CRYST. VOL.B28, ...   3102, (1972).

479   BOWMAN..........(A.)....:   (5 , 1652) A.I.P. CONF. PROC. NO.10 PART 1,   658, (1973).

480   BOWMAN..........(C.D.)..:   (20B,   82) NUCL. INST. MET. 55, ...   141, (1967).

481   BOWMAN..........(M.G.)..:   (22C,   58) REV. SCI. INSTRUM. VOL.37, . . .   1543, (1966).

482   BRADLEY.........(J.G.)..:   (5 , 126) J. PHYS. CHEM. SOL. 27, ...   925, (1966).
                                    (5 , 1516) NUCL. SCI. ENG.(USA) VOL.47, . .   151, (1972).

483   BRADLEY.........(R.H.)..:   (6 , 361) MOL. PHYS. VOL.21, ...   721, (1971).

484   BRADY...........(G.W.)..:   (6 , 2010) J. CHEM. PHYS. VOL.50, ...   2140, (1969).

485   BRAJOVIC........(V.)....:   (6 , 1646) J. CHEM. PHYS. VOL.39, ...   3135, (1963).
                                    (6 , 1587) J. PHYS. CHEM. SOL. 24, ...   617, (1963).
                                    (6 , 1425) J. PHYS. CHEM. SOL. 24, ...   771, (1963).

486   BRAUNECK........(W.)....:   (14A,   38) PHYS. LETT. VOL.31A, ...   237, (1970).

487   BRAUN...........(E.)....:   (5 ,   79) ACTA CRYST. 16, ...   737, (1963).
                                    (10D,   55) REV. MEXICANA FIS. VOL.17, . . .   138, (1968).

488   BREBNER.........(J.L.)..:   (6 , 903) SOL. STATE COMM. VOL.13, ...   1555, (1973).

489   BREDIG..........(M.A.)..:   (5 , 1548) ANN. N.Y. ACAD. SCI. VOL.79, . .   762, (1960).

490   BREDOV..........(M.M.)..:   (6 ,   48) SOV. PHYS. SOL. STATE 7, ...   1138, (1965).
                                    (10B, 224) SOV. PHYS. SOL. STATE 9, ...   214, (1967).

491   BREED...........(D.J.)..:   (5 , 1481) PHYS. LETT. VOL.26A NO.11, . . .   526, (1968).

492   BREEN...........(R.J.)..:   (5 , 2462) PHYS. REV. VOL.105, ...   517, (1957).

493   BREGMAN.........(J.D.)..:   (14A,   34) PHYSICA VOL.50, ...   511, (1970).
                                    (6 , 649) PHYS. LETT. VOL.33A, ...   87, (1970).
                                    (19A,   50) NUCL. INST. MET. VOL.93, ...   109, (1971).
                                    (19A,   51) NUCL. INST. MET. VOL.98, ...   53, (1972).

494   BREITFUSS.......(G.)....:   (20B, 100) NUCL. INST. MET. 98, ...   87, (1972).

495   BREITLING.......(G.)....:   (14B, 137) Z. NATURFORSCH. VOL.16A, ...   187, (1961).
                                    (17A,   27) FORTSCHR. PHYS. VOL.14 NO.2, . .   73, (1966).

496   BREIT...........(G.)....:   (7B,   51) PHYS. REV. VOL.49, ...   229, (1936).
                                    (7A,   59) PHYS. REV. VOL.49, ...   519, (1936).
                                    (7B,   59) PHYS. REV. VOL.71, ...   215, (1947).
                                    (7B,   58) PHYS. REV. VOL.71, ...   232, (1947).
                                    (24 ,   89) PHYS. REV. VOL.114, ...   830, (1959).

497   BRENNAN.........(T.)....:   (5 , 415) J. PHARM. PHARMACOL. VOL.22, . .   724, (1970).

498   BRESCANSIN......(L.M.)..:   (6 , 1610) SOL. STATE COMM. VOL.11, ...   1431, (1972).
                                    (10B, 109) PHYSICA VOL.72, ...   179, (1974).

499   BREUIL..........(M.)....:   (5 , 1522) PHYS. LETT. 29A, ...   506, (1969).
                                    (5 , 2619) C. R. ACAD. SCI. B (FRANCE) 270,   109, (1970).
                                    (5 , 677) J. PHYS. CHEM. SOLIDS VOL.31, . .   549, (1970).
                                    (5 , 2620) J. PHYSIQUE (FRA.) VOL.32, . . .   813, (1971).

500   BREYNAT.........(G.)....:   (20C,   8) BULL. D#INSTRUM. SCI. TECH. 172,   55, (1972).

501   BRICKWEDDE......(F.G.)..:   (5 , 1255) PHYS. REV. VOL.52, ...   1076, (1937).
                                    (5 , 1250) PHYS. REV. VOL.54, ...   266, (1938).

502   BRIDGE..........(H.S.)..:   (5 , 1256) PHYS. REV. VOL.72, ...   1147, (1947).

503   BRIER...........(P.N.)..:   (6 , 360) J. MOLEC. STRUC. VOL.6, ...   23, (1970).
                                    (6 , 362) MOLEC. PHYS. VOL.19, ...   645, (1970).
                                    (6 , 361) MOL. PHYS. VOL.21, ...   721, (1971).
                                    (6 , 2044) FARADAY SYMP. CHEM. SOC. NO.6, .   169, (1972).
                                    (6 , 371) IAEA SYMP. GRENOBLE, ...   461, (1972).

```
503  BRIER..........(P.N.)..:  ( 6 , 1921) POLYMER VOL.13,. . . . . . . . .  157, (1972).

504  BRIGGS.........(G.A.)..:  ( 6 , 1988) J. OF PHYS. C VOL.1,. . . . . .  940, (1968).
                              ( 6 , 1989) IAEA SYMP. GRENOBLE,. . . . . .  639, (1972).
                              ( 6 , 1777) IAEA SYMP. GRENOBLE,. . . . . .  669, (1972).

505  BRIGGS.........(G.H.)..:  (20A,   69) PROC. ROY. SOC. A VOL.162,. . .  127, (1937).

506  BRIGHT.........(W.C.)..:  ( 5 , 2708) PHYS. REV. VOL.55,. . . . . . .  793, (1939).
                              (11C,   28) PHYS. REV. VOL.60,. . . . . . .  280, (1941).

507  BRILL..........(R.)....:  ( 5 ,  150) ACTA CRYST. VOL.B27,. . . . . . 2003, (1971).

508  BRILL..........(T.)....:  (20A,   64) PHYS. REV. VOL.72,. . . . . . .  585, (1947).

509  BRIMBERG.......(S.)....:  (15 ,   24) K. TEKN. HOGSK. AVHANDL. NO.116, . . , (1956).
                              (15 ,    2) ASEA RES. (SWEDEN) NO.7, . . . .  147, (1962).

510  BRISSE.........(F.)....:  ( 5 ,  902) CANAD. J. CHEM. VOL.43,. . . . . 2819, (1965).

511  BROCKER........(B.)....:  (20B,    9) ATOMKERNENERGIE VOL.11,. . . . .  381, (1966).

512  BROCKHOUSE.....(B.N.)..:  ( 5 ,  384) PHYS. REV. VOL.83,. . . . . . .  840, (1951).
                              ( 6 , 2055) PHYS. REV. VOL.83,. . . . . . .  841, (1951).
                              ( 6 , 1824) PHYS. REV. VOL.88,. . . . . . .  542, (1952).
                              ( 5 , 2086) CAN. J. PHYS. VOL.31,. . . . . .  339, (1953).
                              ( 7A,   22) CAN. J. PHYS. VOL.31,. . . . . .  432, (1953).
                              ( 5 ,  605) J. CHEM. PHYS. VOL.21,. . . . . .  961, (1953).
                              ( 5 ,  740) PHYS. REV. VOL.94,. . . . . . .  781, (1954).
                              ( 6 , 2245) CAN. J. PHYS. VOL.33,. . . . . .  889, (1955).
                              ( 6 ,   43) PHYS. REV. VOL.100,. . . . . . .  756, (1955).
                              (11C,   31) PHYS. REV. VOL.98,. . . . . . .  1721, (1955).
                              (11C,   32) PHYS. REV. VOL.99,. . . . . . .   601, (1955).
                              ( 6 , 1112) ACTA CRYST. VOL.10,. . . . . . .  827, (1957).
                              ( 5 , 2779) BULL. AMER. PHYS. SOC. SER.II 2, .  9, (1957).
                              ( 5 , 1859) CAN. J. PHYS. VOL.35,. . . . . .  313, (1957).
                              ( 6 ,  858) PHYS. REV. VOL.106,. . . . . . .  859, (1957).
                              ( 6 ,  931) PHYS. REV. VOL.108,. . . . . . .  894, (1957).
                              ( 6 , 1501) BULL. AMER. PHYS. SOC. VOL.3,. .  195, (1958).
                              ( 6 , 1091) NUOVO CIMENTO SUPPL. VOL.9,. . .   45, (1958).
                              ( 6 ,  857) PHYS. REV. VOL.111,. . . . . . . 1273, (1958).
                              ( 6 ,  932) PHYS. REV. VOL.111,. . . . . . .  747, (1958).
                              ( 6 ,   57) REV. MOD. PHYS. VOL.30,. . . . .  236, (1958).
                              ( 6 , 2243) REV. MOD. PHYS. VOL.30,. . . . .  250, (1958).
                              ( 5 , 2753) BULL. AMER. PHYS. SOC. SER.II 4, . 184, (1959).
                              (10C,   41) J. PHYS. CHEM. SOLIDS VOL.8,. . .  400, (1959).
                              ( 6 , 2037) PHYS. REV. LETT. VOL.2,. . . . .  256, (1959).
                              ( 6 , 1111) PHYS. REV. LETT. VOL.2,. . . . .  287, (1959).
                              ( 6 , 1836) PHYS. REV. LETT. VOL.3,. . . . .  259, (1959).
                              (20D,   30) REV. SCI. INSTRUM. VOL.30,. . . .  136, (1959).
                              (18A,   13) IAEA SYMPOSIUM VIENNA,. . . . .   113, (1960).
                              ( 6 , 1801) IAEA SYMPOSIUM VIENNA,. . . . .   531, (1960).
                              ( 6 ,  203) IAEA SYMPOSIUM (VIENNA),. . . .   487, (1960).
                              (23 ,   29) IAEA SYMP. (PILE RESEARCH) VIENNA  25, (1960).
                              ( 6 , 1649) PHYS. REV. VOL 119,. . . . . . .  980, (1960).
                              ( 6 ,  498) PHYS. REV. VOL.120,. . . . . . . 1638, (1960).
                              ( 6 , 1816) PHYS. REV. LETT. VOL.7,. . . . .   93, (1961).
                              (10A,   12) RENSSELAER POL. INST. SYMP.,. . .  129, (1961).
                              ( 6 ,   69) IAEA SYMP. CHALK RIVER, VOL.1,. .  189, (1962).
                              ( 6 ,  861) IAEA SYMP. CHALK RIVER, VOL.2,. .  297, (1962).
                              ( 6 , 1628) IAEA SYMP. CHALK RIVER, VOL.2,. .   23, (1962).
                              ( 6 , 2157) J. PHYS. SOC. JAPAN VOL.17 B-III,  63, (1962).
                              (10C,   48) J. PHYS. SOC. JAPAN VOL.17 B-II,  363, (1962).
                              ( 6 , 1070) J. PHYS. SOC. JAPAN VOL.17 B-II,  370, (1962).
                              ( 6 ,  842) PHYS. LETT. 1,. . . . . . . . .   189, (1962).
                              ( 6 , 1810) PHYS. REV. VOL.128,. . . . . . . 1099, (1962).
                              ( 6 , 1629) PHYS. REV. VOL.128,. . . . . . . 1112, (1962).
                              ( 6 ,  206) PHYS. REV. VOL.128,. . . . . . . 1120, (1962).
                              ( 6 , 2296) PHYS. REV. VOL.128,. . . . . . .   67, (1962).
                              ( 6 , 1621) PROC. PHYS. SOC. VOL.79,. . . . .  440, (1962).
                              (10C,    1) AARHUS SUMMER SCHOOL,. . . . . .  221, (1963).
                              (17A,   48) INTERN. SUMMER SCHOOL, MOL,. . .  580, (1963).
                              ( 6 , 1248) PHYS. REV. 131,. . . . . . . . . 1025, (1963).
                              (10D,   26) PHYS. REV. 131,. . . . . . . . . 1030, (1963).
                              ( 6 , 1614) PROC. PHYS. SOC, 81,. . . . . . .  973, (1963).
                              ( 6 ,  927) SOL. STATE COMM. 1,. . . . . . .  205, (1963).
                              ( 6 , 2276) SOL. STATE COMM. 2,. . . . . . .   73, (1964).
                              ( 6 , 2071) PHYS. REV. LETT. 14,. . . . . . .  554, (1965).
                              (10A,   17) SCOTTISH UNIV. SUMMER SCHOOL,. .  110, (1965).
                              ( 6 ,  619) SOL. STATE COMM. 3,. . . . . . .  245, (1965).
                              ( 5 ,  693) MATERIALS RES. BULL. VOL.2,. . .   69, (1967).
                              (10C,   52) METAL. SOC. CONF. LOS ANGELES, .  161, (1967).
                              ( 6 ,  616) PHYS. REV. LETT. VOL.18,. . . . .  858, (1967).
                              ( 6 ,  606) PHYS. REV. 155,. . . . . . . . .  619, (1967).
                              ( 6 ,  780) SOL. STATE COMM. 5,. . . . . . .  211, (1967).
                              ( 6 , 2190) SOL. STATE COMM. 5,. . . . . . .   79, (1967).
                              ( 6 , 1954) IAEA SYMP. COPENHAGEN, VOL.1,. .  209, (1968).
                              ( 6 ,  164) IAEA SYMP. COPENHAGEN, VOL.1,. .  253, (1968).
                              (23 ,   33) IAEA SYMP. COPENHAGEN, VOL.2,. .  259, (1968).
                              ( 6 , 1754) J. APP. PHYS. 39,. . . . . . . .  451, (1968).
                              ( 6 , 1852) PHYS. REV. LETT. 20,. . . . . . .  798, (1968).
                              ( 6 , 1764) CAN. J. PHYS. 47,. . . . . . . . 1117, (1969).
                              (17A,   37) IAEA SYMP. (INSTRUMENT.), VIENNA 105, (1969).
                              ( 6 ,   15) PHYS. LETT. 29A,. . . . . . . .   639, (1969).
                              ( 6 , 1541) PHYS. LETT. 29A,. . . . . . . .   694, (1969).
                              ( 6 , 1793) CAN. J. PHYS. 48,. . . . . . . . 1781, (1970).
                              ( 6 ,  578) CAN. J. PHYS. 49,. . . . . . . .  704, (1971).
                              ( 6 , 1531) PHYS. REV. B-3,. . . . . . . . . 2733, (1971).
                              ( 6 ,  201) BULL. AMER. PHYS. SOC. VOL.17, .  123, (1972).
                              ( 6 , 1944) CAN. J. PHYS. VOL.50,. . . . . . 2915, (1972).
                              ( 6 , 1772) IAEA SYMP. GRENOBLE,. . . . . .   73, (1972).
                              ( 6 ,  639) BULL. AMER. PHYS. SOC. VOL.18, .  112, (1973).
                              ( 6 , 1958) CAN. J. PHYS. VOL.51,. . . . . .  657, (1973).
                              ( 6 , 1536) PHYS. REV. B VOL.8,. . . . . . . 3928, (1973).
                              ( 6 ,  609) BULL. AMER. PHYS. SOC. VOL.19, .  320, (1974).
                              ( 6 , 1774) PHYS. REV. B VOL.10 NO. 2, . . . ,  (1974).

513  BROCKMAN.......(F.G.)..:  ( 5 , 1607) PHYS. REV. VOL.90,. . . . . . . 1013, (1953).

514  BRODSKY........(M.B.)..:  ( 5 , 2369) PHYS. REV. B VOL.9,. . . . . . . 1041, (1974).

515  BROOKS.........(M.S.S.):  ( 6 , 2139) J. OF PHYS. C VOL.1,. . . . . . 1596, (1968).
                              ( 6 , 2140) J. OF PHYS. C VOL.1,. . . . . . .  132, (1968).
                              (11A,   15) J. PHYS. -C- VOL.1,. . . . . . . 1279, (1968).
                              (11A,   22) J. PHYS. C VOL.7,. . . . . . . .  979, (1974).

516  BROOMHALL......(G.J.)..:  (24 ,   46) NUCL. INST. MET. 57,. . . . . .   82, (1967).

517  BROQUETAS-COLOMI(C.)....:  ( 5 ,  999) COLL. INTER. N.126 (GRENOBLE), .  102, (1963).

518  BROSCH.........(R.)....:  ( 9B,   31) ATOMKERNENERGIE VOL.13,. . . . .  183, (1968).

519  BROT...........(C.)....:  ( 6 , 1527) J. PHYSIQUE TOME 34,. . . . . .  473, (1973).
```

```
520  BROUT...........(R.).....!  (11A,    30) PHYS. LETT. 24A, . . . . . . . .    117, (1967).
                                 ( 6 ,    83) PHYS. REV. LETT. VOL.25, . . . . .  1423, (1970).
521  BROVCHENKO.......(V.G.)..!  (22B,    12) INSTRUM. EXP. TECH. NO.2,. . . . .   415, (1970).
522  BROVMAN.........(E.G.)..!   ( 6 ,  2186) SOV. PHYS. SOL. STATE 5, . . . . .    78, (1963).
                                 ( 6 ,  2072) SOV. PHYS. SOL. STATE 8, . . . . .  1120, (1966).
                                 ( 6 ,  2076) SOV. PHYS. JETP VOL.25,. . . . . .   365, (1967).
                                 (10B,   225) SOV. PHYS. SOL. STATE VOL.11,. . .   733, (1969).
523  BROWNE..........(H.N.)..!   ( 7A,    72) PHYS. REV. LETT. 16, . . . . . .    495, (1966).
524  BROWNE..........(J.0.)..!   (20A,     3) ACTA CRYST. 16,. . . . . . . . .    435, (1963).
                                 ( 6 ,  1507) PROC. PHYS. SOC. 86, . . . . . .    139, (1965).
525  BROWN...........(C.L.)..!   ( 6 ,  1946) NUCL. SCI. ENG. VOL.46,. . . . .    244, (1971).
526  BROWN...........(E.A.)..!   (11C,    25) PHYS. LETT. VOL.33A, . . . . . .    375, (1970).
527  BROWN...........(G.M.)..!   ( 5 ,  1488) COLL. INTER. N.126 (GRENOBLE), .     45, (1963).
                                 (17B,    30) COLL. INTER. N.126 (GRENOBLE), .     73, (1963).
                                 ( 5 ,  2506) SCIENCE VOL.141, . . . . . . . .    921, (1963).
                                 ( 5 ,  1487) ACTA CRYST. 20,. . . . . . . . .    220, (1966).
                                 ( 5 ,   733) ACTA CRYST. VOL.B25, . . . . . .    676, (1969).
                                 ( 5 ,  1662) ACTA CRYST. VOL.B27, . . . . . .     66, (1971).
                                 ( 5 ,   694) ACTA CRYST. VOL.B29, . . . . . .   2393, (1973).
                                 ( 5 ,  2505) ACTA CRYST. VOL.B29, . . . . . .    790, (1973).
                                 ( 5 ,   250) CRYST. STRUC. COMM. VOL.2 NO.1,.   107, (1973).
528  BROWN...........(I.0.)..!   ( 5 ,  1952) CAN. J. PHYS. 42,. . . . . . . .    229, (1964).
                                 ( 5 ,  1570) ACTA CRYST. VOL.A28, . . . . . .    663, (1972).
529  BROWN...........(J.R.)..!   ( 5 ,  1306) NUCL. SCI. ENG. VOL.33,. . . . .    265, (1968).
530  BROWN...........(J.S.)..!   (10B,   192) PHYS. STAT. SOL. VOL.39, . . . .    K75, (1970).
531  BROWN...........(L.H.)..!   (22B,    30) NUCL. INST. MET. 60, . . . . . .    237, (1968).
532  BROWN...........(N.).....!  ( 5 ,  2160) J. APPL. PHYS. VOL.32, . . . . .    375, (1961).
533  BROWN...........(P.J.)..!   ( 5 ,  1683) PHYS. REV. LETT. VOL.8,. . . . .    237, (1962).
                                 ( 5 ,  2190) PHYS. REV. VOL.125,. . . . . . .   1283, (1962).
                                 ( 6 ,  1460) J. APP. PHYS. 34,. . . . . . . .   1182, (1963).
                                 ( 6 ,  1513) J. APP. PHYS. 34,. . . . . . . .   1201, (1963).
                                 ( 5 ,   501) PHYS. REV. 129,. . . . . . . . .   1145, (1963).
                                 (12B,     5) BRIT. J. APPL. PHYS. VOL.15, . .   1529, (1964).
                                 ( 5 ,  1025) PHIL. MAG. VOL.10, . . . . . . .    713, (1964).
                                 ( 6 ,   846) PHYS. REV. VOL.136,. . . . . . .  A1641, (1964).
                                 ( 5 ,  1016) PROC. INT. CONF. (NOTTINGHAM), .    524, (1964).
                                 (17B,     7) ACTA CRYST. 18,. . . . . . . . .    398, (1965).
                                 ( 5 ,  1804) J. APP. PHYS. 36,. . . . . . . .   1095, (1965).
                                 ( 5 ,   744) PHYS. REV. 138,. . . . . . . . .   A153, (1965).
                                 ( 5 ,  1803) PHYS. REV. 140,. . . . . . . . .  A2139, (1965).
                                 ( 5 ,   539) PROC. PHYS. SOC. 85, . . . . . .   1185, (1965).
                                 ( 5 ,  1766) J. APP. PHYS. 37,. . . . . . . .   1053, (1966).
                                 ( 6 ,  1506) PROC. PHYS. SOC. 88, . . . . . .    333, (1966).
                                 ( 5 ,  1817) PROC. PHYS. SOC. 91, . . . . . .    332, (1967).
                                 ( 5 ,  1660) PROC. PHYS. SOC. 92, . . . . . .    125, (1967).
                                 (17B,    12) ACTA CRYST. A24,. . . . . . . . .   347, (1968).
                                 ( 9D,    26) HARWELL SUMMER SCHOOL, . . . . .    176, (1968).
                                 ( 5 ,  1818) J. APP. PHYS. 39,. . . . . . . .   1331, (1968).
                                 ( 5 ,  1160) J. PHYS.F: METAL PHYS. VOL.2,. .    358, (1972).
                                 ( 5 ,   446) J. PHYS. C VOL.6,. . . . . . . .   1405, (1973).
534  BROWN...........(R.A.)..!   (20B,   105) NUCL. INST. MET. VOL.109,. . . .    313, (1973).
535  BROWN...........(W.K.)..!   (23 ,    55) NUCL. INST. MET. 26, . . . . . .      1, (1964).
                                 (20C,    49) NUCL. INST. MET. 50, . . . . . .    181, (1967).
536  BRUCE...........(A.0.)..!   (10D,    18) J. PHYS. C VOL.6,. . . . . . . .   2422, (1973).
537  BRUESCH.........(P.).....!  ( 6 ,  1333) PHYS. REV. LETT. VOL.30, . . . .   1144, (1973).
538  BRUGGER.........(R.M.)..!   ( 5 ,   195) IAEA SYMPOSIUM VIENNA . . . . .    277, (1960).
                                 (20B,    33) IAEA SYMP. (PILE RESEARCH)VIENNA   433, (1960).
                                 (20B,    68) NUCL. INST. MET. VOL.8,. . . . .    203, (1960).
                                 ( 6 ,  2353) NUCL. SCI. ENGNG. VOL.5, . . . .     99, (1960).
                                 ( 5 ,  2869) NUCL. SCI. ENGNG. VOL.7, . . . .    193, (1960).
                                 (19A,    32) NUCL. INST. MET. VOL.12, . . . .    365, (1961).
                                 (23 ,    53) NUCL. INST. MET. VOL.12, . . . .     75, (1961).
                                 ( 6 ,   251) PHYS. REV. VOL.124,. . . . . . .    450, (1961).
                                 (20B,    74) NUCL. INST. MET. VOL.17, . . . .    169, (1962).
                                 ( 6 ,   311) PHYS. REV. VOL.125,. . . . . . .    933, (1962).
                                 ( 6 ,  2168) PHYS. REV. VOL.126,. . . . . . .     29, (1962).
                                 ( 6 ,   146) PHYS. REV. VOL.128,. . . . . . .    562, (1962).
                                 ( 6 ,   148) IAEA SYMP. BOMBAY, VOL.1,. . . .    379, (1964).
                                 ( 6 ,   265) PHYS. REV. VOL.136,. . . . . . .   A106, (1964).
                                 (17A,   100) UKAEA AERE REP. R 4562 (9 PP.),.         (1964).
                                 ( 6 ,  1116) J. CHEM. PHYSICS VOL.42, . . . .   1568, (1965).
                                 ( 6 ,   269) J. CHEM. PHYSICS VOL.42, . . . .    275, (1965).
                                 (21 ,    55) TH. NEUTRON SCATT./EGELSTAFF,. .     53, (1965).
                                 ( 6 ,  1058) NUCL. SCI. ENGNG. VOL.25,. . . .    248, (1966).
                                 ( 6 ,   231) PHYS. REV. VOL.148,. . . . . . .    163, (1966).
                                 (14B,   104) REACTOR PHYSICS, . . . . . . . .     91, (1967).
                                 ( 6 ,   275) J. CHEM. PHYSICS VOL.46, . . . .   2201, (1967).
                                 ( 6 ,   309) J. CHEM. PHYSICS VOL.47, . . . .    421, (1967).
                                 ( 6 ,   162) J. PHYS. CHEM. SOL. 28,. . . . .    249, (1967).
                                 ( 5 ,   209) PHYS. LETT. 24A, . . . . . . . .    714, (1967).
                                 ( 6 ,   153) IAEA SYMP. COPENHAGEN, VOL.1,. .    315, (1968).
                                 (17A,    34) IAEA SYMP. COPENHAGEN, VOL.2,. .    323, (1968).
                                 ( 5 ,   603) J. PHYS. CHEM. SOLIDS VOL.29,. .    435, (1968).
                                 ( 6 ,  1045) NUCL. SCI. ENGNG. VOL.33,. . . .    187, (1968).
                                 ( 6 ,   319) PHYS. REV. LETT. VOL.20, . . . .    983, (1968).
                                 (20B,   129) PHYS. TODAY VOL.21, NO.12, . . .     23, (1968).
                                 (23 ,    37) IAEA SYMP. (INSTRUMENT.), VIENNA   197, (1969).
                                 ( 6 ,   652) J. CHEM. PHYS. VOL.50, . . . . .   1030, (1969).
                                 ( 9C,    36) PROC/SYMP/CRYS. STRUC/HIGH PRESS   141, (1969).
                                 ( 6 ,  1939) J. CHEM. PHYS. VOL.52, . . . . .   4424, (1970).
                                 (20B,   114) NUCL. TECHNOL. VOL.15, . . . . .     14, (1972).
                                 ( 6 ,   313) J. CHEM. PHYS. VOL.58, . . . . .   1438, (1973).
539  BRUIN...........(C.).....!  ( 6 ,    74) PHYS. LETT. VOL.33A, . . . . . .    338, (1970).
540  BRUNEL..........(M.).....!  ( 9D,    56) PHYS. LETT. VOL.39A, . . . . . .    141, (1972).
541  BRUNHART........(G.).....!  ( 5 ,  2972) BULL. AMER. PHYS. SOC. VOL.17, .    125, (1972).
542  BRUNNER.........(J.).....!  ( 5 ,   380) NUKLEONIK VOL.11,. . . . . . . .    297, (1968).
543  BRUNO...........(R.).....!  ( 6 ,   613) CAN. J. PHYS. 49,. . . . . . . .   2496, (1971).
544  BRUNTON.........(G.).....!  ( 5 ,  1976) ACTA CRYST. VOL.B25, . . . . . .   2519, (1969).
```

```
545  BRUN............(T.O.)..:  ( 6 ,  764) J. APPL. PHYS. VOL.35,. . . . .    933, (1964).
                               ( 6 ,  524) PROC. INT. CONF. (NOTTINGHAM),. .   101, (1964).
                               ( 6 ,  753) PROC. INT. CONF. (NOTTINGHAM),. .    99, (1964).
                               ( 6 ,  523) SOL. STATE COMM. 2,. . . . . .      109, (1964).
                               ( 6 ,  756) PHYS. REV. VOL.139,. . . . . . . :A1866, (1965).
                               ( 9D, 1125) IAEA SYMP. COPENHAGEN, VOL.1,. .    339, (1968).
                               ( 9D,   37) J. CHEM. PHYSICS VOL.53,. . . . .  1387, (1970).
                               ( 6 , 2284) PHYS. REV. B-1,. . . . . . . . .   2430, (1970).
                               ( 6 , 2561) AIP CONF. PROC. VOL.5,. . . . . .  1376, (1971).
                               ( 6 , 2521) DISS. ABS. VOL.31,. . . . . . . .  5560B, (1971).
                               ( 5 , 2522) J. PHYSIQUE TOME 32 COL.1 VOL.I,   571, (1971).
                               ( 5 , 1764) J. PHYSIQUE TOME 32 COL.1 VOL.I,   577, (1971).
                               ( 6 , 1141) PHYS. REV. A VOL.3,. . . . . . .   1688, (1971).
                               ( 6 ,  838) PHYS. REV. LETT. VOL.28,. . . . .   746, (1972).
                               ( 5 , 2698) PHYS. REV. LETT. VOL.29,. . . . .  1172, (1972).
                               ( 5 , 2592) ACTA CRYST. VOL.A29,. . . . . . .   684, (1973).
                               ( 5 , 2697) PHYS. REV. B VOL.7,. . . . . . .   1988, (1973).
                               ( 5 , 2593) PHYS. REV. B VOL.8,. . . . . . .   3237, (1973).
                               ( 5 ,  840) PHYS. REV. B VOL.8,. . . . . . .    260, (1973).
                               ( 5 , 2594) PHYS. REV. B VOL.9,. . . . . . .   3003, (1974).
                               ( 5 , 2338) PHYS. REV. B VOL.9,. . . . . . .    248, (1974).

546  BRUSETTI........(R.)..:  ( 5 , 2189) PHYS. REV. LETT. VOL.32, . . . .  1257, (1974).

547  BRYANT..........(W.A.)..:  (19B,   29) REPORT KAPL-1657 (22 PP.),. . .  407, (1957).
                               ( 6 , 2352) CONF. NEUTRON THERMALIZ. VOL.I,   407, (1968).

548  BUCHANAN........(D.R.)..:  ( 5 ,   66) DISS. ABS. XXIV,. . . . . . .    90, (1964).
                               ( 5 ,   67) ACTA CRYST VOL.B24,. . . . . . .  954, (1968).

549  BUCHER..........(E.)..:  ( 6 , 1930) PHYS. REV. LETT. VOL.25,. . . .    752, (1970).
                               ( 6 , 1935) J. APPL. PHYS. 42,. . . . . . .   1746, (1971).
                               ( 6 , 2195) PHYS. REV. B VOL.4,. . . . . . .    718, (1971).
                               ( 6 , 1927) PHYS. REV. LETT. VOL.27,. . . .   1530, (1971).
                               ( 6 , 1932) IAEA SYMP. GRENOBLE,. . . . . .    543, (1972).
                               ( 5 , 2335) PHYS. REV. B VOL.6,. . . . . . .   2724, (1972).
                               ( 5 ,  838) PHYS. REV. LETT. VOL.28,. . . .    746, (1972).
                               ( 6 , 1953) PHYS. REV. B VOL.8,. . . . . . .   5345, (1973).

550  BUDINGER........(T.F.)..:  (24 ,   86) PHYS. MED. BIOL.(GB) VOL. 16,. .  439, (1971).

551  BUDZANOWSKI.....(A.)..:  ( 6 ,  415) ACTA PHYS. POLON. VOL.16,. . . .   335, (1957).

552  BUEHLER.........(E.)..:  ( 5 , 1582) J. PHYS. CHEM. SOLIDS VOL.34,. .   521, (1973).

553  BUEVOZ..........(J.L.)..:  (19B,    1) ACTA CRYST. (INTERACT.) VOL.A25, S256, (1969).
                               (22C,   66) REV. SCI. INSTRUM. VOL.45,. . .    341, (1974).

554  BUGAYONG........(R.R.)..:  ( 5 ,   23) ACTA CRYST. VOL.B28,. . . . . .  3214, (1972).

555  BUHL............(R.)..:  ( 5 , 2166) COMP. REND. 263,. . . . . . . .   3448, (1966).
                               ( 5 ,  472) J. APP. PHYS. 39,. . . . . . .     632, (1968).
                               ( 5 , 1047) J. APP. PHYS. 40,. . . . . . .    1120, (1969).
                               ( 5 , 1744) J. PHYS. CHEM. SOLIDS 30,. . . .   805, (1969).
                               ( 5 ,  471) J. PHYSIQUE TOME 31,. . . . . .    113, (1970).
                               ( 5 , 2941) PHYS. STAT. SOL. VOL.40,. . . .    171, (1970).
                               ( 5 , 1748) J. APPL. PHYS. 42,. . . . . . .   1615, (1971).
                               ( 5 , 1045) J. PHYSIQUE TOME 32 COL.1 VOL.I,   322, (1971).
                               ( 5 , 1742) J. PHYS. CHEM. SOLIDS VOL.32,.    2429, (1971).

556  BUHRER..........(W.)..:  ( 6 ,  604) SOL. STATE COMM. 4,. . . . . .    443, (1966).
                               ( 6 ,  765) HELV. PHYS. ACTA VOL.40,. . . .    378, (1967).
                               ( 6 ,  585) HELV. PHYS. ACTA VOL.41,. . . .    399, (1968).
                               ( 6 , 1282) PHYS. STAT. SOL. VOL.41,. . . .    789, (1970).
                               ( 6 ,  569) PHYS. STAT. SOL. B VOL.46,. . .    679, (1971).
                               ( 6 ,  466) IAEA SYMP. GRENOBLE,. . . . . .    563, (1972).
                               ( 6 ,  469) SOL. STATE COMM. VOL.11,. . . .     21, (1972).
                               ( 6 , 2197) INT. J. MAGN. VOL.4 NO.1,. . . .    63, (1973).
                               ( 6 ,  564) J. PHYS. C VOL.6,. . . . . . .    2931, (1973).
                               ( 6 , 2196) SOL. STATE COMM. VOL.12 NO.2,.     117, (1973).
                               ( 6 , 2146) SOL. STATE COMM. VOL.13,. . . .    881, (1973).

557  BUISSON.........(G.)..:  ( 5 ,  579) PROC. INT. CONF. (NOTTINGHAM),    275, (1964).
                               ( 5 ,  591) J. APPL. PHYS. VOL.37,. . . . .   1038, (1966).
                               ( 5 , 2908) COMP. REND. 264,. . . . . . . .   3978, (1967).
                               ( 5 ,  221) SOLID STATE COMM. 5,. . . . . .     25, (1967).
                               ( 5 ,  635) J. PHYS. CHEM. SOLIDS VOL. 31,.  1171, (1970).
                               ( 5 ,  359) J. PHYS. CHEM. SOLIDS VOL.32,.   1189, (1971).
                               ( 5 , 2903) PHYS. STAT. SOLIDI A VOL.16 NO.2   533, (1973).
                               ( 5 , 2584) PHYS. STAT. SOLIDI A VOL.17 NO.1   191, (1973).

558  BUKIN...........(V.I.)..:  ( 5 , 2050) SOV. PHYS. DOKLADY VOL.17, . . .  1131, (1973).

559  BULATOV.........(A.S.)..:  ( 5 , 1241) SOV. PHYS. J.E.T.P. 27,. . . . .   210, (1968).

560  BULLOUGH........(R.)..:  (10B,   81) J. PHYS.F: METAL PHYS. VOL.1,. .   554, (1971).

561  BULL............(R.)..:  ( 5 , 2167) ACTA CRYST. B25, . . . . . . .    2326, (1969).

562  BUNCE...........(L.J.)..:  (23 ,   71) REPORT AERE-R 6246 (25PP.),. . . . ,  (1970).

563  BUNGE...........(H.J.)..:  ( 5 ,  684) J. APPL. CRYST. VOL.4,. . . . .    303, (1971).
                               ( 5 ,  756) J. APPL. CRYST. VOL.5,. . . . .     27, (1972).

564  BURAS...........(B.)..:  ( 9B,   27) ARCH. HUTNICTWA VOL.4,. . . . .     81, (1959).
                               (10C,   58) NUKLEONIKA VOL.4,. . . . . . .    119, (1959).
                               (24 ,   27) KERNTECHNIK VOL.2,. . . . . . .   153, (1960).
                               (19A,   57) NUKLEONIKA VOL.8,. . . . . . .    259, (1963).
                               (19A,   58) NUKLEONIKA VOL.8,. . . . . . .     75, (1963).
                               (19A,   59) NUKLEONIKA VOL.9,. . . . . . .    523, (1964).
                               ( 5 , 2443) PHYS. STAT. SOL. 41,. . . . . .    349, (1964).
                               ( 9C,   34) PHYS. STAT. SOL. 11,. . . . . .    567, (1965).
                               (17A,   90) REPORT NO. 702/II/PS,. . . . . .   17, (1966).
                               (24 ,  101) RES. APPL./NUCL. PULSED SYSTEMS,   17, (1967).
                               (17A,   99) TRIESTE/INTERN. COURSE,. . . . .   443, (1967).
                               ( 9B,   68) NUKLEONIKA VOL.13, . . . . . .    1111, (1968).
                               (19A,   61) NUKLEONIKA (POL.) VOL.13,. . . .   591, (1968).
                               (19A,   62) NUKLEONIKA VOL.14,. . . . . . .    487, (1969).
                               (19A,   67) REPORT NO.1108/II/PS,. . . . . .    ., (1969).
                               (17C,   26) Z. KRIST. VOL.130,. . . . . . .    318, (1969).
                               (17A,   66) NUCL. INST. MET. 77,. . . . . .     13, (1970).
                               ( 9B,   17) ACTA CRYST. VOL.A28,. . . . . .    151, (1972).
                               (17A,   16) ACTA PHYS. POLON. VOL.A42,. . .    259, (1972).
                               (17A,   92) PHYS. STAT. SOLIDI A VOL.9,. . .   423, (1972).
                               (17A,   71) NUCL. INST. MET. VOL.106,. . . .   461, (1973).

565  BUREWICZ........(A.)..:  ( 5 , 1095) PHYS. STAT. SOL. 26,. . . . . .    429, (1968).

566  BURFORD.........(R.J.)..:  (13 ,   48) PHYS. REV. VOL.156,. . . . . . .  583, (1967).

567  BURGER..........(G.)..:  ( 5 , 2859) PHYS. LETT. 20,. . . . . . . .    470, (1966).

568  BURGMAN.........(J.O.)..:  ( 6 , 1082) PHYS. REV. 170,. . . . . . . .    808, (1968).
```

```
569  BURGY..........(M.T.)..!   ( 6 ,  767) PHYS. REV. VOL.75, . . . . . . .   565, (1949).
                              (12B,   37) PHYS. REV. VOL.76, : : : : : : :  1413, (1949).
                              (20A,   65) PHYS. REV. VOL.77, : : : : : : :   291, (1950).
                              ( 6 ,  769) PHYS. REV. VOL.80, : : : : : : :   481, (1950).
                              (12B,   39) PHYS. REV. VOL.80, : : : : : : :   953, (1950).
                              (20A,   66) PHYS. REV. VOL.81, : : : : : : :   498, (1951).
                              ( 5 , 1260) PHYS. REV. VOL.84. VOL.39, : : :  1160, (1951).
                              (12B,   53) REV. SCI. INSTRUM. VOL.39; . . . .  101, (1968).

570  BURKHOLDER......(H.R.)..!  (20C,   70) NUCL. INST. MET. VOL.84, . . . .   67, (1970).

571  BURLET.........(MME.P.)!   ( 5 , 1681) SOL. STATE COMM. VOL.7 NO.19,. . 1403, (1969).
                              ( 5 , 1680) SOL. STATE COMM. VOL.7 NO.2,. .   343, (1969).
                              ( 5 , 1028) SOL. STATE COMM. VOL. 9,. . . .  1633, (1971).
                              ( 5 , 1659) SOL. STATE COMM. VOL.14, . . . .   665, (1974).

572  BURLET.........(P.)....!   ( 5 , 1139) SOL. STATE COMM. 3,. . . . . .   335, (1965).
                              ( 5 ,  591) J. APPL. PHYS. VOL.37, : : : : :  1036, (1966).
                              ( 5 , 1674) COMP. REND. 264, . . . . . . .  3238, (1967).
                              ( 5 , 1675) SOL. STATE COMM. 5,. . . . . .   279, (1967).
                              ( 5 , 1672) J. APP. PHYS. VOL.39,. . . . .   590, (1968).
                              ( 5 , 1681) SOL. STATE COMM. VOL.7 NO.19,. . 1403, (1969).
                              ( 5 , 1680) SOL. STATE COMM. VOL.7 NO.2,. .   343, (1969).
                              ( 5 ,  166) J. PHYSIQUE TOME 31, . . . . .   401, (1970).
                              ( 5 ,  736) SOL. STATE COMM. VOL.8,. . . .   935, (1970).
                              (90 ,   12) AIP CONF. PROC.(USA) VOL.5,. .  1355, (1971).
                              ( 5 , 1028) SOL. STATE COMM. VOL. 9, . . .  1633, (1971).
                              ( 5 , 2567) SOL. STATE COMM. VOL. 9, . . .   435, (1971).
                              ( 5 , 1659) SOL. STATE COMM. VOL.14, . . .   665, (1974).

573  BURNS..........(J.H.)..!   ( 5 , 2871) SCIENCE VOL.139, . . . . . . .  1208, (1963).
                              ( 5 , 1997) ACTA CRYST. B24, . . . . . . .   230, (1968).

574  BUSCHERT........(R.C.)..!  ( 5 , 1974) J. CHEM. PHYSICS VOL.44, . . .  1758, (1966).
                              (14A,    6) ADV. IN PHYSICS VOL.16, . . . .   177, (1967).
                              ( 5 , 2431) J. CHEM. PHYSICS VOL.46, . . .   586, (1967).
                              ( 5 , 2457) J. PHYS. CHEM. SOL. 28,. . . .   423, (1967).

575  BUSCHOW.........(K.H.J.)!  ( 5 , 2635) PHYS. STAT. SOLIDI B VOL.57 NO.2 K155, (1973).

576  BUSCH..........(G.)....!   ( 6 ,  467) J. OF PHYS. C VOL.1 . . . . .   679, (1968).

577  BUSHEV.........(M.K.)..!   (9B,   110) VESTN. MOSK. UNIV. FIZ. ASTR. 12  710, (1971).
                              ( 8 ,  160) VESTN. MOSK. UNIV. FIZ. ASTRON.,  149, (1972).

578  BUSING.........(W.R.)..!   (17B,    1) ACTA CRYST. VOL.10,. . . . . .   180, (1957).
                              ( 5 ,  368) J. CHEM. PHYS. VOL.26, . . . .   563, (1957).
                              ( 5 ,   87) ACTA CRYST. VOL.11,. . . . . .   798, (1958).
                              (23 ,    7) COLL. INTER. N.126 (GRENOBLE),   71, (1963).
                              (17C,    2) ACTA CRYST. 17,. . . . . . . .   142, (1964).
                              ( 5 , 1283) J. CHEM. PHYSICS VOL.42, . . .  3054, (1965).
                              (17A,    6) ACTA CRYST. 22,. . . . . . . .   457, (1967).
                              ( 5 , 1599) ACTA CRYST. VOL.A25, . . . . .  S118, (1969).
                              ( 5 , 1918) ACTA CRYST. VOL.B25, . . . . .   572, (1969).
                              ( 5 ,   36) ACTA CRYST. VOL.B28, . . . . .  2454, (1972).

579  BUSSER.........(F.W.)..!   ( 6 , 1155) NUCL. PHYS. VOL.A129,. . . . .   666, (1969).

580  BUTLAND........(A.T.D.)!   ( 5 , 1291) REPORT AEEW-R 701 (31PP.), . . . . , (1970).

581  BUTLER.........(D.)....!   ( 6 ,  666) PROC. PHYS. SOC. 81, : : : : :   276, (1963).
                              (14B,   97) PROC. PHYS. SOC. 81, : : : : :   294, (1963).

582  BUTTERWORTH.....(I.)....!  (20B,  118) PHIL. MAG. (8) VOL.2,. . . . .   917, (1957).

583  BUTT...........(N.M.)..!   ( 5 , 1963) ACTA CRYST. VOL.A29, . . . . .   727, (1973).

584  BUYERS.........(W.J.L.)!   ( 6 , 1642) PHYS. REV. 153,. . . . . . .   923, (1967).
                              ( 6 , 1277) IAEA SYMP. COPENHAGEN, VOL.1,. .  267, (1968).
                              ( 6 , 1247) IAEA SYMP. COPENHAGEN, VOL.1,. .  281, (1968).
                              ( 6 , 1256) IAEA SYMP. COPENHAGEN, VOL.1,. .   43, (1968).
                              ( 6 ,  505) IAEA SYMP. COPENHAGEN, VOL.2,. .  123, (1968).
                              ( 6 , 1476) J. APP. PHYS. 39,. . . . . . .  1118, (1968).
                              ( 6 , 1253) PHYS. REV. 165,. . . . . . . .  1063, (1968).
                              ( 6 ,  509) PHYS. REV. 167,. . . . . . . .   510, (1968).
                              ( 6 , 1455) SOL. STATE COMM. 6,. . . . . .   145, (1968).
                              ( 6 , 1322) CAN. J. PHYS. VOL.47,. . . . .  1983, (1969).
                              ( 6 , 1475) J. APPL. PHYS. VOL.40, . . . .   991, (1969).
                              (19E,   16) J. PHYS. C, SER.2, VOL.2,. . .  2262, (1969).
                              ( 6 , 1639) PHYS. REV. LETT. VOL.23, . . .   525, (1969).
                              ( 6 ,  492) PHYS. REV. LETT. VOL.23, . . .    86, (1969).
                              ( 6 , 1244) PHYS. REV. VOL.180,. . . . . .   755, (1969).
                              ( 6 , 1481) SOL. STATE COMM. VOL.7,. . . .  1693, (1969).
                              ( 6 , 2994) SOL. STATE COMM. 7,. . . . . .   181, (1969).
                              ( 6 , 1449) J. APPL. PHYS. 41,. . . . . .   896, (1970).
                              ( 6 , 1320) J. PHYS. SOC. JAPAN VOL.28, SUP.  249, (1970).
                              ( 6 , 2080) J. PHYS. SOC. JAPAN VOL.28, SUP.  249, (1970).
                              ( 6 , 1632) PHYS. REV. B-1,. . . . . . . .  1833, (1970).
                              ( 6 , 1274) PHYS. REV. B-2,. . . . . . . .  4603, (1970).
                              ( 6 , 1323) J. PHYSIQUE TOME 32 COL.1 VOL.II 1184, (1971).
                              ( 6 , 1269) J. PHYS. -C, VOL. 4,. . . . .  2127, (1971).
                              (11A,   18) J. PHYS. -C, VOL. 4,. . . . .  2139, (1971).
                              ( 5 , 2056) PHYS. CAN. VOL.27, . . . . . .    68, (1971).
                              ( 6 ,  501) PHYS. REV. LETT. VOL.27, . . .  1442, (1971).
                              ( 6 , 1517) CAN. J. PHYS. VOL.50, . . . .  3069, (1972).
                              ( 6 , 1699) IAEA SYMP. GRENOBLE, . . . . .   399, (1972).
                              ( 6 , 2158) IAEA SYMP. GRENOBLE, . . . . .   553, (1972).
                              ( 6 , 1307) IAEA SYMP. GRENOBLE, . . . . .   581, (1972).
                              ( 6 , 1306) J. PHYS. C VOL.5,. . . . . . .  2611, (1972).
                              ( 6 , 1321) MAGN. AND MAG. MATERIALS-1971,  1315, (1972).
                              ( 8 ,  122) REV. MOD. PHYS. VOL.44,. . . .   406, (1972).
                              ( 6 , 1502) INT. J. MAGN. VOL.4 NO.1,. . .   167, (1973).
                              ( 6 , 1841) J. NONMET. VOL.1,. . . . . . .   159, (1973).
                              ( 6 , 1514) J. PHYS. C VOL.6,. . . . . . .  1933, (1973).
                              ( 6 ,  497) J. PHYS. C VOL.6,. . . . . . .  2997, (1973).
                              ( 6 , 1941) PHYS. REV. B VOL.9,. . . . . .  3797, (1974).

585  BUZANO.........(C.)....!   (20D,   19) NUCL. INST. MET. 47, . . . . .   320, (1967).
                              (15 ,   41) PHYS. REV. VOL.167, . . . . .    97, (1968).

586  BUZIN..........(V.N.)..!   ( 5 , 1911) SOV. PHYS. DOKLADY VOL.16, . . .    9, (1971).

587  BYKOV..........(V.N.)..!   (22A,   66) SOV. PHYS. CRYST. VOL.2, . . .   626, (1957).
                              ( 5 ,  131) SOV. PHYS. CRYST. VOL.3, . . .   308, (1958).
                              ( 5 ,  553) SOV. PHYS. DOKLADY VOL.4,. . .  1070, (1960).
                              (90 ,   75) SOV. PHYS. CRYST. VOL.9, . . .   535, (1965).
                              ( 5 ,  530) SOV. PHYS. J.E.T.P. VOL.22,. .   754, (1965).
                              (21 ,   52) SOV. PHYS. CRYST. VOL.12, . .   477, (1967).
                              ( 5 , 2967) SOV. PHYS. CRYST. VOL.14, . .   913, (1969).
                              ( 5 , 2969) SOV. PHYS. CRYST. VOL.15, . .   376, (1970).
                              ( 6 ,  638) SOV. PHYS. CRYST. VOL.15, . .   935, (1971).
                              ( 5 , 2427) SOV. PHYS. CRYST. VOL.17, . .   822, (1971).
                              ( 5 , 2899) SOV. PHYS. CRYST. VOL.17, . .   342, (1972).
                              (20B,  152) SOV. PHYS. CRYST. VOL.17,. . .   826, (1973).
```

588 BYLANDER........(E.G.)..: (6 , 1843) SOL. STATE COMM. 4,. . . . . . . 51, (1966).

589 BYSTRITSKII.....(V.M.)..: (22B, 13) INSTRUM. EXP. TECH. VOL.15,... 67, (1972).

590 BYSTROV.........(M.V.)..: (5 , 2892) SOV. PHYS. SOL. STATE VOL.14,. . 3037, (1973).

591 CABASSI.........(F.)....: (6 , 1910) MOL. PHYS. VOL.22, ... 241, (1971).

592 CABLE...........(J.W.)..: (5 , 1658) J. PHYS. RADIUM VOL.20,. . . . . 180, (1959).
(5 , 968) PHYS. REV. VOL.113,. . . . . . . 497, (1959).
(5 , 2386) IAEA SYMP. (PILE RESEARCH)VIENNA 379, (1960).
(5 , 2586) J. APPL. PHYS. VOL.31, SUPPL., . 358, (1960).
(5 , 573) PHYS. REV. VOL.118,. . . . . . . 950, (1960).
(5 , 807) J. APPL. PHYS. VOL.32, SUPPL., . 48S, (1961).
(5 , 852) J. APPL. PHYS. VOL.32, SUPPL., . 49S, (1961).
(5 , 567) J. PHYS. CHEM. SOLIDS VOL.19,. . 29, (1961).
(5 , 1866) PHYS. REV. VOL.121,. . . . . . . 74, (1961).
(5 , 2682) J. APPL. PHYS. VOL.33, SUPPL., . 1124, (1962).
(5 , 2301) J. APPL. PHYS. VOL.33, SUPPL., . 1340, (1962).
(5 , 2384) J. PHYS. SOC. JAPAN VOL.17 B-III 27, (1962).
(5 , 2606) J. PHYS. SOC. JAPAN VOL.17 B-III 32, (1962).
(5 , 2298) J. PHYS. SOC. JAPAN VOL.17 B-III 38, (1962).
(5 , 1720) PHYS. REV. VOL.125,. . . . . . . 1860, (1962).
(5 , 2683) PHYS. REV. VOL.126,. . . . . . . 1672, (1962).
(6 , 527) PHYS. REV. VOL.127,. . . . . . . 2080, (1962).
(5 , 977) PHYS. REV. VOL.127,. . . . . . . 714, (1962).
(5 , 2310) PHYS. REV. VOL.128,. . . . . . . 2118, (1962).
(5 , 2146) COLL. INTER. N.126 (GRENOBLE), . 36, (1963).
(5 , 1112) J. APP. PHYS. 34,. . . . . 1189, (1963).
(5 , 2610) J. APP. PHYS. 34,. . . . . 1335, (1963).
(5 , 2147) J. PHYS. CHEM. SOL. 24,. . . . . 1141, (1963).
(5 , 725) J. PHYS. CHEM. SOL. 24,. . . . . 1663, (1963).
(5 , 2540) PHYS. REV. VOL.131,. . . . . . . 922, (1963).
(6 , 2027) J. APPL. PHYS. VOL.35,. . . . . . 1041, (1964).
(5 , 1855) J. PHYS. CHEM. SOL. 25,. . . . . 1453, (1964).
(5 , 1391) PHYS. LETT. VOL.9,. . . . . . . 93, (1964).
(5 , 2337) PHYS. REV. LETT. VOL.12,. . . . 553, (1964).
(6 , 2662) PHYS. REV. VOL.136,. . . . . . . A240, (1964).
(6 , 2147) PROC. INT. CONF. (NOTTINGHAM), . 271, (1964).
(5 , 2599) J. APP. PHYS. 36,. . . . . 1096, (1965).
(5 , 2909) PHYS. REV. 138,. . . . . A1655, (1965).
(5 , 2101) PHYS. REV. 138,. . . . . A752, (1965).
(5 , 850) PHYS. REV. 140,. . . . . A1896, (1965).
(6 , 1853) PHYS. REV. 140,. . . . . A2003, (1965).
(5 , 1359) PHYS. REV. 151,. . . . . 414, (1966).
(6 , 1732) J. APP. PHYS. 38,. . . . . 1247, (1967).
(5 , 2634) J. APP. PHYS. 38,. . . . . 1384, (1967).
(9D, 53) METAL. SOC. CONF. LOS ANGELES, . 31, (1967).
(5 , 1358) PHYS. REV. 158,. . . . . 450, (1967).
(5 , 1209) PHYS. REV. 165,. . . . . 733, (1968).
(5 , 1214) J. APP. PHYS. VOL.40,. . . . . . 1003, (1969).
(5 , 2116) PHYS. REV. LETT. VOL.22,. . . . 1256, (1969).
(6 , 906) J. APPL. PHYS. 41,. . . . . 1182, (1970).
(5 , 2182) PHYS. REV. B-1,. . . . . 3809, (1970).
(5 , 516) PHYS. REV. B-2,. . . . . 176, (1970).
(6 , 907) PHYS. REV. LETT. VOL.24,. . . . 16, (1970).
(6 , 395) AIP CONF. PROC. VOL.5,. . . . . 1381, (1971).
(6 , 909) J. PHYSIQUE TOME 32 COL.1 VOL.I, 296, (1971).
(5 , 2153) J. PHYSIQUE TOME 32 COL.1 VOL.I, 67, (1971).
(5 , 2470) ACTA CRYST. VOL.A28 PART S-4,. . 197, (1972).
(5 , 1066) INT. J. MAGN. VOL.2 NO.2,. . . . 1, (1972).
(5 , 984) MAGN. AND MAG. MATERIALS-1971,. 1415, (1972).
(5 , 1205) PHYS. REV. B VOL.5,. . . . . . . 997, (1972).
(5 , 2612) A.I.P. CONF. PROC. NO.10 PART 2, 1554, (1973).
(5 , 2144) A.I.P. CONF. PROC. NO.10 PART 2, 1623, (1973).
(5 , 2134) PHYS. REV. B VOL.7,. . . . . . . 2005, (1973).

593 CABRERA.........(T.N.)..: (10B, 146) PHYS. REV. LETT. 22,. . . . . . 346, (1969).

594 CAGLIOTI........(G.)....: (20D, 12) NUCL. INSTRUM. VOL.3,. . . . . . 223, (1958).
(6 , 1801) IAEA SYMPOSIUM VIENNA,. . . . . 531, (1960).
(22A, 34) NUCL. INST. MET. VOL.9,. . . . . 195, (1960).
(6 , 193) IAEA SYMP. CHALK RIVER, VOL.1,. 259, (1962).
(17D, 11) J. PHYS. SOC. JAPAN VOL.17 B-II, 347, (1962).
(5 , 739) J. PHYS. SOC. JAPAN VOL.17 B-II, 348, (1962).
(17D, 13) NUCL. INST. MET. VOL.15,. . . . 155, (1962).
(23 , 65) NUOVO CIMENTO SUPPL. VOL.23,. . 17, (1962).
(9B, 69) NUOVO CIMENTO VOL.24,. . . . . . 1174, (1962).
(5 , 226) NUOVO CIMENTO VOL.24,. . . . . . 103, (1962).
(6 , 1810) PHYS. REV. VOL.128,. . . . . . . 1099, (1962).
(20D, 32) REV. SCI. INSTRUM. VOL.33,. . . 1103, (1962).
(6 , 2310) COPENHAGEN CONF.-LATT.DYNAMICS,. 17, (1963).
(6 , 2315) PHYS. REV. 132,. . . . . 683, (1963).
(17B, 3) ACTA CRYST. 17,. . . . . 1202, (1964).
(17A, 32) IAEA SYMPOSIUM BOMBAY, VOL.2,. . 505, (1964).
(6 , 2061) IAEA SYMP. BOMBAY, VOL.1,. . . . 117, (1964).
(10C, 75) PHYS. REV. VOL.136,. . . . . . . A1280, (1964).
(18A, 17) NUCL. INST. MET. 32,. . . . . . 181, (1965).
(17A, 83) NUOVO CIMENTO B VOL.43,. . . . . 375, (1966).
(5 , 1910) NUOVO CIMENTO B VOL.46,. . . . . 248, (1966).
(6 , 1733) J. APP. PHYS. 38,. . . . . 1245, (1967).
(5 , 2931) NUOVO CIMENTO B VOL.49,. . . . . 222, (1967).
(24 , 117) TRIESTE/INTERN. COURSE,. . . . . 539, (1967).
(10C, 15) C.N.E.N. SYMP. CASACCIA,. . . . 13, (1968).
(17A, 28) HARWELL SUMMER SCHOOL,. . . . . 14, (1968).
(6 , 2304) IAEA SYMP. COPENHAGEN, VOL.1,. . 373, (1968).
(6 , 829) J. APP. PHYS. 39,. . . . . 455, (1968).
(10A, 7) 2ND SUMMER SCHOOL S.S. PHYS.,. . 137, (1968).
(10C, 28) ENERGIA NUCLEARE VOL.16,. . . . 369, (1969).
(10C, 27) ENERGIA NUCLEARE (ITALY) VOL.16, 425, (1969).
(5 , 2928) NUOVO CIMENTO B VOL.60,. . . . . 48, (1969).
(6 , 2306) SOL. STATE COMM. VOL.8,. . . . . 367, (1970).
(10B, 41) IAEA SYMP. GRENOBLE, ... 149, (1972).

595 CALANDRA........(C.)....: (10B, 171) PHYS. REV. LETT. VOL.28,. . . . 1578, (1972).
(11B, 186) PHYS. REV. LETT. VOL.31,. . . . 1466, (1973).
(10B, 92) J. PHYS. F VOL.4,. . . . . 19, (1974).
(10B, 188) PHYS. REV. B VOL.9,. . . . . . . 1710, (1974).

596 CALDER..........(R.S.)..: (5 , 1563) J. PHYS. CHEM. SOLIDS VOL.23,. . 621, (1962).

597 CALDWELL........(R.L.)..: (20C, 94) REV. SCI. INSTRUM. VOL.33,. . . 866, (1962).

598 CALLAWAY........(J.)....: (8 , 49) LECT/ELEM. EXCIT./INTER. IN SOL. , (1968).
(11A, 31) PHYS. LETT. 27A,. . . . . 215, (1968).

599 CALLEN..........(E.)....: (5 , 1076) PHYS. REV. 150,. . . . . 367, (1966).

600 CAMBIAGHI.......(M.)....: (22C, 38) NUCL. INST. MET. 62,. . . . . . 233, (1968).

601 CAMPBELL........(I.A.)..: (5 , 928) PROC. PHYS. SOC. 89,. . . . . . 71, (1966).

602 CAMPOS..........(F.P.)..: (5 , 2664) J. APPL. PHYS. VOL.27,. . . . . 1040, (1956).
(5 , 2719) J. PHYS. CHEM. SOLIDS 29,. . . . 1702, (1968).
(5 , 2728) J. PHYS. CHEM. SOLIDS 30,. . . . 1642, (1969).

```
603  CANFIELD.........(E.H.)..:  ( 7A,    37) NUCL. PHYS. A VOL.A169,. . . . .    385, (1971).

604  CAPE.............(J.A.)..:  ( 5 , 1758) J. APP. PHYS. VOL.40,. . . . . .   1136, (1969).

605  CARABATOS.......(C.)....:  ( 6 ,   624) COMP. REND. 268, . . . . . . .    1658, (1969).
                                ( 6 ,   623) PHYS. REV. LETT. VOL.26,. . . .    770, (1971).
                                ( 6 ,   635) PHYS. STAT. SOL. VOL.44,. . . .    701, (1971).
                                (10B,    15) CAN. J. PHYS. 50,. . . . . . .     122, (1972).
                                ( 6 ,   627) PHYS. REV. LETT. VOL.28,. . . .    964, (1972).
                                ( 6 ,   622) SOL. STATE COMM. VOL.13,. . . .   1725, (1973).

606  CARBONI.........(F.)....:  ( 6 ,   387) PHYS. REV. B VOL.5,. . . . . .    2014, (1972).

607  CARBOTTE........(J.B.)..:  ( 6 ,    34) PHYS. LETT. 25A, . . . . . . .     685, (1967).

608  CARLOS..........(P.)....:  (21 ,    28) J. PHYSIQUE TOME 24, . . . . .      15A, (1963).
                                (20B,    62) J. PHYSIQUE TOME 24, . . . . .     359, (1963).
                                (12B,    22) J. PHYSIQUE TOME 24, . . . . .      89A, (1963).

609  CARLSON.........(A.D.)..:  ( 5 ,   926) NUCL. SCI. ENG. VOL.42,. . . .      28, (1970).

610  CARMICHAEL......(H.)....:  (22A,     8) CAN. JOURN. PHYS. 42,. . . . .    2443, (1964).

611  CARNEIRO........(K.)....:  ( 6 ,   960) IAEA SYMP. GRENOBLE, . . . . .     111, (1972).
                                ( 6 ,   967) PHYS. REV. LETT. VOL.30,. . . .    481, (1973).

612  CARON...........(L.G.)..:  ( 5 ,   364) J. PHYS. CHEM. SOL. 26,. . . .     927, (1965).
                                ( 5 ,   329) J. AMER. CERAM. SOC. VOL.49, . .    284, (1966).

613  CARPENTER.......(J.M.)..:  (18B,     3) IAEA SYMPOSIUM VIENNA, . . . . .    199, (1960).
                                ( 6 ,   305) BROOKHAVEN SYMPOSIUM., . . . . .    142, (1965).
                                (19A,     7) DISS. ABS. VOL.25,. . . . . . .   1131, (1965).
                                ( 7B,    25) J. CHEM. PHYSICS VOL.46,. . . .    465, (1967).
                                ( 6 ,   162) J. PHYS. CHEM. SOL. 28,. . . . .   249, (1967).
                                (22B,    29) NUCL. INST. MET. 47,. . . . . .    179, (1967).
                                ( 6 ,   412) IAEA SYMP. COPENHAGEN, VOL.1,. .    491, (1968).
                                ( 6 ,   243) IAEA SYMP. COPENHAGEN, VOL.2,. .    205, (1968).
                                ( 6 ,   413) J. CHEM. PHYS. VOL.51,. . . . .   5543, (1969).
                                (17C,    24) REV. SCI. INSTRUM. VOL.40,. . .    555, (1969).
                                (16 ,    33) NUCL. INST. MET. 85,. . . . . .    163, (1970).
                                ( 6 ,   995) J. CHEM. PHYSICS VOL.54,. . . .   5193, (1971).
                                ( 9B,    66) NUCL. INST. MET. 99,. . . . . .    453, (1972).
                                (20B,   134) REPORT ANL-8032 (104 PP.),. . .     . ,  (1973).
                                (22A,    48) NUCL. INST. MET. VOL.114,. . . .   417, (1974).
                                (17C,    25) REV. SCI. INSTRUM. VOL.45,. . .    572, (1974).

614  CARRARO.........(G.)....:  ( 5 , 2459) PHYSICS/NON-CRYSTALLINE SOLIDS,.   152, (1965).

615  CARRIVEAU.......(G.W.)..:  ( 6 , 1423) PHYS REV 174,. . . . . . . . .     953, (1968).

616  CARROLL-JR......(E.E.)..:  ( 6 , 2335) PHYS. REV. C VOL.2,. . . . . .    2390, (1970).
                                (17A,    68) NUCL. INST. MET. VOL.94,. . . .    173, (1971).

617  CARROLL.........(H.)....:  ( 7A,    62) PHYS. REV. VOL.60, . . . . . .     702, (1941).

618  CARTER..........(R.S.)..:  ( 5 ,   143) PHYS. REV. VOL.92,. . . . . .     716, (1953).
                                ( 6 , 2246) PHYS. REV. VOL.104,. . . . . .     271, (1956).
                                ( 6 ,    56) PHYS. REV. VOL.106,. . . . . .    1168, (1957).

619  CARVALHO........(F.)....:  ( 6 , 2228) IAEA SYMP. BOMBAY, VOL.1,. . .      99, (1964).
                                (19A,    36) NUCL. INST. MET. 49,. . . . . .    197, (1967).
                                ( 6 ,   193) NUCL. SCI. ENGNG. VOL.34,. . . .   224, (1968).

620  CASELLA.........(R.C.)..:  ( 6 , 1658) CONF. INTERN., RENNES, . . . . .    104, (1971).
                                ( 7B,    87) PHYS. REV. B VOL.6,. . . . . .    4533, (1972).
                                ( 6 , 1661) PHYS. REV. B VOL.10 NO.2,. . . .     . ,  (1974).

621  CASE............(K.M.)..:  ( 7B,    72) PHYS. REV. VOL.126,. . . . . .    1165, (1962).

622  CASSELS.........(J.M.)..:  ( 5 , 2258) NATURE VOL.161,. . . . . . . .     282, (1948).
                                ( 6 ,   773) PHYS. REV. VOL.74, . . . . . .     103, (1948).
                                ( 7B,   105) PROGR. NUCL. PHYS. VOL.1,. . . .    185, (1950).
                                ( 6 ,    45) PROC. ROY. SOC. A VOL.208, . . .    527, (1951).

623  CASSOLA.........(R.L.)..:  ( 7A,    28) DISS. ABS. VOL.27, . . . . . .   2825B, (1967).

624  CASTAGNO........(G.)....:  (22A,    49) NUCL. INST. MET. VOL.114,. . . .    21, (1974).

625  CASTELLIZ.......(L.)....:  ( 5 ,   902) CANAD. J. CHEM. VOL.43,. . . . .   2819, (1965).

626  CASWELL.........(R.S.)..:  (24 ,    99) RADIATION RESEARCH VOL.35, . . .     1, (1968).

627  CATICHA-ELLIS...(S.)....:  (17B,    15) ACTA CRYST. VOL.A25, . . . . .     666, (1969).

628  CAUVIN..........(B.)....:  ( 7A,    25) CONF/NUCL. STRUC. STUDY/NEUTRONS  144, (1972).

629  CAVALHEIRO......(R.)....:  ( 6 , 1818) J. PHYS. SOC. JAPAN VOL.34,. . .   1002, (1973).

630  CAVALLARI.......(F.)....:  (20C,    10) ENERGIA NUCLEARE VOL.16, . . . .    400, (1969).

631  CECH............(J.)....:  ( 5 ,   929) PHYS. LETT. 29A, . . . . . . .     679, (1969).
                                ( 9B,    72) PHYS. LETT. 37A,. . . . . . . .    403, (1971).
                                ( 5 , 2449) PHYS. STAT. SOLIDI A VOL.17 NO.1   163, (1973).

632  CELLI...........(V.)....:  (10B,   146) PHYS. REV. LETT. 22, . . . . .     346, (1969).

633  CERBONE.........(R.J.)..:  ( 5 ,   926) NUCL. SCI. ENG. VOL.42,. . . . .    28, (1970).

634  CEULEMANS.......(H.)....:  ( 5 , 2702) 2ND INTERNAT. CONF./NUCL. DATA,. . . , (1970).

635  CHADWICK........(B.M.)..:  ( 5 ,   649) Z. KRIST. VOL.134,. . . . . . .    308, (1971).
                                ( 5 ,   661) INORG. NUCL. CHEM. LETT. VOL.9,.  1025, (1973).

636  CHADWICK........(J.)....:  (25 ,     8) NATURE VOL.129, . . . . . . . .    312, (1932).
                                (25 ,    70) PROC. ROY. SOC. A VOL.136,. . .    692, (1932).
                                (25 ,    72) PROC. ROY. SOC. A VOL.142, . . .     1, (1933).

637  CHAIKOVSKII.....(V.G.)..:  (20C,    25) INSTRUM. EXP. TECH. VOL.15 PT.1,   351, (1972).

638  CHAKERA.........(A.)....:  ( 9B,    11) ACTA CRYST. VOL.A26, . . . . .     682, (1970).

639  CHAKRABARTI.....(D.J.)..:  ( 5 ,   125) AIP CONF. PROC. VOL.5, . . . .     497, (1971).

640  CHALUPA.........(B.)....:  (20B,    15) CZECH. J. PHYS. B VOL.13,. . . .    474, (1963).
                                (22A,    10) CZECH. J. PHYS. B VOL.16,. . . .    942, (1966).
                                ( 5 , 2376) PHYS. LETT. 28A, . . . . . . .     546, (1968).
                                ( 5 , 2371) PHYS. STAT. SOL. 29, . . . . . .    K51, (1968).
                                ( 5 , 2375) BRIT. J. APPL. PHYS. VOL.2,. . .   1041, (1969).
                                ( 9B,    37) CZECH. J. PHYS. B VOL.19,. . . .   1608, (1969).
                                ( 5 , 2379) NUCL. INST. MET. 67, . . . . . .    357, (1969).
                                ( 5 ,   929) PHYS. LETT. 29A, . . . . . . .     679, (1969).
                                ( 9B,    89) PHYS. STAT. SOL. A VOL.2,. . . .    211, (1970).
                                ( 5 , 2373) PHYS. STAT. SOL. VOL.42,. . . . .    895, (1970).
                                ( 5 , 2372) ACTA CRYST. VOL.A27,. . . . . .     410, (1971).
                                ( 5 , 2449) PHYS. STAT. SOLIDI A VOL.17 NO.1   163, (1973).
```

```
641  CHAMARD-BOIS....(R.)....:  ( 9D,    6) ACTA CRYST. VOL.A28, . . . . .     341, (1972).
                                ( 5 , 2416) SOL. STATE COMM. VOL.10, . . . . .  685, (1972).
                                ( 6 , 1232) SOL. STATE COMM. VOL.13, . . . . . 1549, (1973).

642  CHAMBERLAIN.....(O.)....:  ( 5 , 2419) PHYS. REV. VOL.77, . . . . . . .    305, (1950).

643  CHAMPIER........(G.)....:  ( 6 ,  205) C.R. ACAD. SCI. VOL.257, . . . .   1843, (1963).

644  CHANDRASEKHAR...(S.)....:  ( 9D,    2) ACTA CRYST. VOL.10, . . . . . . .   598, (1957).

645  CHANDROSS.......(R.J.)..:  ( 5 , 2841) J. PHYS. SOC. JAPAN VOL.17 B-III     16, (1962).

646  CHANG...........(Y.I.)..:  ( 6 , 1892) J. POLYM. SCI/A-2 POLYM. PHYS. 7    405, (1969).

647  CHAN............(H.H.)..:  ( 7A,   23) CAN. J. PHYS. 42, . . . . . . . .  1017, (1964).

648  CHAPPERT........(J.)....:  ( 5 , 1139) SOL. STATE COMM. 3, . . . . . . .    335, (1965).
                                ( 5 , 1092) BULL. SOC. FRANC. MIN. CRIST. 89    206, (1966).
                                ( 6 ,  591) J. APPL. PHYS. VOL.37, . . . . .   1038, (1966).
                                ( 5 , 1093) SOL. STATE COMM. 4, . . . . . . .    395, (1966).
                                ( 5 , 2571) SOL. STATE COMM. 5, . . . . . . .    293, (1967).
                                ( 5 ,  174) J. PHYSIQUE TOME 30, . . . . . .     566, (1969).

649  CHARAN-KHAN.....(O.)....:  ( 5 ,  481) DISS. ABS. XXVII, . . . . . . . .   2768, (1967).

650  CHARPIN.........(P.)....:  ( 5 ,   44) BULL. SOC. FRA. MINER. CRIST. 93      7, (1970).

651  CHASTEL.........(A.)....:  (20C,   75) NUCL. INST. MET. VOL.94, . . . .    493, (1971).

652  CHATTERJEE......(A.)....:  (22C,   33) NUCL. INST. MET. 49, . . . . . .    101, (1967).

653  CHATURVEDI......(D.K.)..:  ( 5 ,   94) J. PHYS. SOC. JAP. VOL.31, . . .   1162, (1971).
                                ( 6 , 1630) PHYS. REV. A VOL.9, . . . . . . .  2128, (1974).

654  CHAUSSY.........(J.)....:  ( 5 ,  874) J. PHYSIQUE TOME 31, . . . . . .    607, (1970).
                                ( 5 ,  642) SOL. STATE COMM. VOL.13, . . . .    905, (1973).

655  CHEETHAM........(A.K.)..:  ( 5 ,  378) SOL. STATE COMM. VOL.8, . . . . .    171, (1970).
                                ( 5 , 1079) J. PHYS. C VOL.4, . . . . . . . .  2160, (1971).
                                ( 5 ,  338) J. PHYS. C VOL.4, . . . . . . . .  3107, (1971).
                                ( 5 ,  405) J. PHYS. C VOL.5, . . . . . . . .   L35, (1972).
                                ( 5 , 1963) ACTA CRYST. VOL.A29, . . . . . .    727, (1973).
                                ( 9C,   10) CHEM. APPL. THERMAL NEUTRON SCAT    225, (1973).
                                ( 5 , 2655) NATURE (PHYS. SCI.) VOL.244, . .    139, (1973).

656  CHEKANOVA.......(S.S.)..:  (16 ,   82) SOV. AT. ENERGY VOL.26, . . . . .    334, (1969).

657  CHENAVAS........(J.)....:  ( 5 , 1017) SOL. STATE COMM. 3, . . . . . . .    117, (1965).

658  CHENG...........(V.K.C.):  (22A,    9) CHIN. J. PHYS. (TAIWAN) VOL.8, .     31, (1970).

659  CHENTSOV........(P.)....:  ( 6 , 2354) SOV. PHYS. USPEKHI VOL.1, . . . .    165, (1958).

660  CHEN............(C.W.)..:  ( 5 , 1131) J. APP. PHYS. 34, . . . . . . . .   1044, (1963).
                                ( 5 , 1125) PHYS. REV. VOL.134, . . . . . . . A1547, (1964).

661  CHEN............(S.H.)..:  ( 6 , 1519) SOL. STATE COMM. 2, . . . . . . .    233, (1964).
                                ( 6 , 2276) SOL. STATE COMM. 2, . . . . . . .     73, (1964).
                                ( 6 , 1516) THESIS (208 PP.), . . . . . . . .      , (1964).
                                ( 6 ,   81) PHYS. LETT. VOL.19, . . . . . . .    269, (1965).
                                ( 6 , 1706) PROC. PHYS. SOC. 89, . . . . . .    833, (1966).
                                ( 6 , 1954) IAEA SYMP. COPENHAGEN, VOL.1, .    209, (1968).
                                (10B,   48) J. CHEM. PHYSICS VOL.48, . . . .   4060, (1968).
                                ( 6 ,  961) IAEA SYMP. GRENOBLE, . . . . . .    445, (1972).
                                ( 6 ,  983) PHYS. REV. A VOL.8, . . . . . . .  3163, (1973).
                                ( 6 ,  889) PROPERTIES LIQUID METALS, . . . .    119, (1973).
                                (14B,  125) SPECTROSCOPY BIOL. CHEM., . . . .      1, (1974).

662  CHEPARIN........(V.P.)..:  ( 5 ,  176) SOV. PHYS. JETP LETT. VOL.7, . .    158, (1968).
                                ( 5 ,  167) SOV. PHYS. CRYST. VOL.18, . . .    393, (1973).

663  CHERIF..........(H.S.)..:  (14B,  129) THESIS (154PP.), . . . . . . . . .    , (1970).

664  CHERKASOV.......(A.P.)..:  ( 5 ,  167) SOV. PHYS. CRYST. VOL.18, . . .    393, (1973).

665  CHERNOPLEKOV....(N.A.)..:  ( 6 ,  328) SOV. AT. ENERGY VOL.14, . . . .    252, (1963).
                                ( 6 , 2239) SOV. PHYS. J.E.T.P. VOL.16, . .   1472, (1963).
                                ( 6 , 1750) SOV. PHYS. J.E.T.P. VOL.17, . .    584, (1963).
                                ( 6 , 2186) SOV. PHYS. SOL. STATE 5, . . . .     78, (1963).
                                ( 6 , 1427) SOV. PHYS. J.E.T.P. VOL.22, . .    315, (1966).
                                ( 6 , 2101) SOV. PHYS. SOL. STATE 8, . . . .   2156, (1966).
                                ( 6 , 2183) PHYS. STAT. SOL. 20, . . . . . .    767, (1967).
                                ( 6 , 1759) SOV. PHYS. J.E.T.P. VOL. 25, . .    436, (1967).
                                ( 6 , 2211) SOV. PHYS. SOL. STATE 9, . . . .   1366, (1967).
                                ( 5 , 2011) SOV. PHYS. SOL. STATE VOL.10, .   1076, (1968).
                                ( 5 , 2515) SOV. PHYS. SOL. STATE VOL.10, .    212, (1968).
                                ( 6 , 1680) SOV. PHYS. SOL. STATE VOL.11, .   2343, (1969).

666  CHERNOV.........(E.N.)..:  (20C,   21) INSTRUM. EXP. TECH. NO.1, . . .     77, (1970).

667  CHERNUKHIN......(V.L.)..:  (22B,   11) INSTRUM. EXP. TECH. NO.4, . . .    832, (1965).

668  CHERNYSHOV......(A.A.)..:  ( 5 ,  204) SOV. AT. ENERGY VOL.13, . . . .    852, (1962).
                                (16 ,   83) SOV. AT. ENERGY VOL.32, . . . .     33, (1972).
                                ( 6 ,  456) SOV. PHYS. JETP LETT. VOL.18, .    177, (1973).
                                ( 6 , 1408) SOV. PHYS. SOLID STATE VOL.15,   1309, (1974).

669  CHERTKOV........(A.A.)..:  ( 5 , 2011) SOV. PHYS. SOL. STATE VOL.10, .   1076, (1968).
                                ( 5 , 2515) SOV. PHYS. SOL. STATE VOL.10, .    212, (1968).
                                ( 6 , 1680) SOV. PHYS. SOL. STATE VOL.11, .   2343, (1969).
                                ( 5 , 2012) SOV. PHYS. CRYST. VOL.14, . . .    522, (1970).
                                ( 5 , 2837) SOV. PHYS. SOL. STATE VOL. 13,   2172, (1972).
                                ( 5 , 2835) SOV. PHYS. SOL. STATE VOL. 13,   2178, (1972).

670  CHERVONENKIS....(A.YA.).:  ( 5 , 2689) SOV. PHYS. J.E.T.P. VOL.31, . .    808, (1970).

671  CHERVYAKOV......(A.YU.).:  ( 5 , 2837) SOV. PHYS. SOL. STATE VOL. 13,   2172, (1972).
                                ( 5 , 2835) SOV. PHYS. SOL. STATE VOL. 13,   2178, (1972).

672  CHESSER.........(N.J.)..:  (20A,   44) NUCL. INST. MET. VOL.101, . . .    183, (1972).
                                ( 6 , 2311) THESIS (121 PP.), . . . . . . . .     , (1972).
                                (18A,    7) ACTA CRYST. VOL.A29, . . . . . .    160, (1973).
                                ( 6 , 2318) PHYS. REV. B VOL.9, . . . . . . .  4060, (1974).

673  CHEVALIER.......(R.)....:  ( 5 ,  893) COMP. REND. 262, . . . . . . . . 1707B, (1966).
                                ( 5 , 1747) SOL. STATE COMM. 5, . . . . . . .      7, (1967).
                                ( 5 , 2748) COMPTES RENDUS 267, SÉRIE B, .    518, (1968).
                                ( 5 , 1178) SOL. STATE COMM. VOL.6 NO.5, . .    317, (1968).
                                ( 5 , 2750) J. APP. PHYS. VOL.40, . . . . .   1131, (1969).
                                ( 5 ,  166) J. PHYSIQUE TOME 31, . . . . . .    401, (1970).
                                ( 9D,   12) AIP CONF. PROC.(USA) VOL.5, . .   1355, (1971).

674  CHEVEAU.........(L.)....:  (10B,  140) PHYS. REV. 169, . . . . . . . .    496, (1968).
```

```
675  CHEVRETON.......(M.)....:  ( 5 ,  615)  COLL. INTER. N.126 (GRENOBLE), .  158, (1963).
                                ( 5 , 2651)  COMP. REND. 263, . . . . . . . .  621B, (1966).
                                ( 5 , 1001)  COMP. REND. 264, . . . . . . . .  316B, (1967).
                                ( 5 , 2650)  BULL. SOC. FRANC. MIN. CRIST. 91   88, (1968).

676  CHEVYCHELOV.....(A.D.)..:  (10B,  118)  PHYS. MET. METALLOGR. VOL.33, .  939, (1972).

677  CHIA-HO-HUANG..........:  ( 6 , 1160)  THESIS (150 PP.), . . . . . . . . . . ,  (1972).

678  CHIA-WEI-WOO...........:  ( 6 , 1144)  PHYS. REV. A VOL.1, . . . . . . 1536, (1970).
                                ( 6 , 1129)  PHYS. REV. LETT. VOL.24, . . . . 1044, (1970).

679  CHICHERIN.......(A.G.)..:  ( 6 , 2239)  SOV. PHYS. J.E.T.P. VOL.16, . . 1472, (1963).
                                ( 6 , 1750)  SOV. PHYS. J.E.T.P. VOL.17, . .  584, (1963).
                                ( 6 , 2186)  SOV. PHYS. SOL. STATE 5, . . . .   78, (1963).

680  CHIDAMBARAM.....(R.)....:  ( 5 , 1425)  J. CHEM. PHYSICS VOL.41, . . . . 3616, (1964).
                                (22A,   20)  NUCL. INST. MET. 26, . . . . . .  340, (1964).
                                ( 5 , 1513)  ACTA CRYST. 20, . . . . . . . .  910, (1966).
                                ( 5 ,  251)  ACTA CRYST. 23, . . . . . . . .  107, (1967).
                                ( 5 , 1485)  ACTA CRYST. VOL.B24, . . . . . . 1176, (1968).
                                (23 ,   40)  INDIAN J. PURE APPL. PHYS. VOL.6  550, (1968).
                                ( 5 ,  158)  J. CHEM. PHYSICS VOL.48, . . . . 1883, (1968).
                                ( 5 ,  206)  ACTA CRYST. VOL.B25, . . . . . .  310, (1969).
                                ( 5 ,  733)  ACTA CRYST. VOL.B25, . . . . . .  676, (1969).
                                ( 5 , 1433)  ACTA CRYST. VOL.B26, . . . . . .  827, (1970).
                                ( 5 , 1503)  ACTA CRYST. VOL.B26, . . . . . .   77, (1970).
                                ( 5 ,   18)  ACTA CRYST. VOL.A28, PART S-4, .  193, (1972).
                                ( 5 ,   19)  ACTA CRYST. VOL.B28, . . . . . . 2514, (1972).
                                ( 5 ,   14)  ACTA CRYST. VOL.B28, . . . . . . 3000, (1972).
                                ( 5 ,   23)  ACTA CRYST. VOL.B28, . . . . . . 3214, (1972).
                                ( 5 ,   15)  ACTA CRYST. VOL.B29, . . . . . . 1167, (1973).
                                ( 5 ,  694)  ACTA CRYST. VOL.B29, . . . . . . 2393, (1973).
                                ( 5 ,   16)  ACTA CRYST. VOL.B30, . . . . . .  562, (1974).

681  CHIKOBAVA.......(V.S.)..:  (22A,   19)  INSTRUM. EXP. TECH. NO.2, . . .  306, (1964).
                                (22A,   21)  INSTRUM. EXP. TECH. NO.4, . . .  808, (1967).

682  CHILDS..........(P.E.)..:  ( 6 ,  441)  J. PHYS. C VOL.5, . . . . . . . 2677, (1972).

683  CHILD...........(H.R.)..:  ( 5 , 2839)  PHYS. REV. VOL.112, . . . . . . 1132, (1958).
                                ( 5 , 2386)  IAEA SYMP. (PILE RESEARCH)VIENNA  379, (1960).
                                ( 5 , 2586)  J. APPL. PHYS. VOL.31, SUPPL., .  358, (1960).
                                ( 5 , 1866)  PHYS. REV. VOL.121, . . . . . .   74, (1961).
                                ( 5 ,  390)  PHYS. REV. VOL.122, . . . . . . 1409, (1961).
                                ( 5 , 2384)  J. PHYS. SOC. JAPAN VOL.17 B-III   27, (1962).
                                ( 5 , 2310)  PHYS. REV. VOL.128, . . . . . . 2118, (1962).
                                ( 5 , 2610)  J. APP. PHYS. 34, . . . . . . . 1335, (1963).
                                ( 5 , 2540)  PHYS. REV. VOL.131, . . . . . .  922, (1963).
                                ( 6 , 1855)  J. PHYS. CHEM. SOL. 25, . . . . 1453, (1964).
                                ( 6 , 2147)  PROC. INT. CONF. (NOTTINGHAM), .  271, (1964).
                                ( 5 , 2599)  J. APP. PHYS. 36, . . . . . . .  106, (1965).
                                ( 5 , 2909)  PHYS. REV. 138, . . . . . . . . A1655, (1965).
                                ( 5 , 2602)  J. APP. PHYS. 37, . . . . . . . 1353, (1966).
                                ( 5 , 1212)  J. APP. PHYS. 38, . . . . . . . 1381, (1967).
                                ( 5 , 2634)  J. APP. PHYS. 38, . . . . . . . 1384, (1967).
                                ( 5 , 1358)  PHYS. REV. 158, . . . . . . . .  450, (1967).
                                ( 5 , 2605)  J. APP. PHYS. 39, . . . . . . . 1329, (1968).
                                ( 5 , 2603)  PHYS. REV. 174, . . . . . . . .  562, (1968).
                                ( 5 , 2922)  PHYS. REV. 176, . . . . . . . .  722, (1968).
                                ( 5 , 2840)  J. APPL. PHYS. VOL.40, . . . . . 1136, (1969).
                                ( 5 , 1214)  J. APP. PHYS. VOL.40, . . . . . 1003, (1969).
                                ( 5 , 2116)  PHYS. REV. LETT. VOL.22, . . . . 1256, (1969).
                                ( 6 ,  906)  J. APPL. PHYS. 41, . . . . . . . 1182, (1970).
                                ( 5 , 2182)  PHYS. REV. B-1, . . . . . . . . 3809, (1970).
                                ( 6 ,  907)  PHYS. REV. LETT. VOL.24, . . . .   16, (1970).
                                ( 5 ,  841)  J. PHYSIQUE TOME 32 COL.1 VOL.II 1128, (1971).
                                ( 6 ,  909)  J. PHYSIQUE TOME 32 COL.1 VOL.I,  296, (1971).
                                ( 5 , 2153)  J. PHYSIQUE TOME 32 COL.1 VOL.I,   67, (1971).
                                ( 5 , 2470)  ACTA CRYST. VOL.A28 PART S-4, .  197, (1972).
                                ( 5 , 1205)  PHYS. REV. B VOL.5, . . . . . .  997, (1972).
                                ( 5 , 2327)  A.I.P. CONF. PROC. NO.10 PART 2, 1319, (1973).
                                ( 5 , 2144)  A.I.P. CONF. PROC. NO.10 PART 2, 1623, (1973).

684  CHIPMAN.........(D.R.)..:  ( 9B,   10)  ACTA CRYST. VOL.A26, . . . . . .  447, (1970).

685  CHISHOLM........(A.)....:  (12A,   16)  NUCL. INST. MET. 75, . . . . . .  333, (1969).

686  CHOI............(C.S.)..:  ( 5 , 2275)  J. CHEM. PHYS. VOL.48, . . . . . 1397, (1968).
                                (20B,   84)  NUCL. INST. MET. 60, . . . . . .  182, (1968).
                                ( 5 ,  171)  ACTA CRYST. VOL.B25, . . . . . . 2638, (1969).
                                ( 5 , 2276)  ACTA CRYST. VOL.B25, . . . . . .  982, (1969).
                                ( 5 ,  759)  ACTA CRYST. VOL.B26, . . . . . . 1235, (1970).
                                ( 5 ,  292)  ACTA CRYST. VOL.B28, . . . . . .  193, (1972).

687  CHOL............(G.)....:  ( 5 , 1860)  J. APPL. CRYST. VOL.1 PT.2, . .  124, (1968).
                                ( 5 , 2942)  J. PHYSIQUE TOME 32 COL.1 VOL.I,  320, (1971).

688  CHOUDRY.........(A.)....:  (20D,   23)  NUCL. INST. MET. 68, . . . . . .  293, (1969).
                                (24 ,   66)  NUCL. INST. MET. 71, . . . . . .  221, (1969).

689  CHOWDHURY.......(M.R.)..:  ( 6 , 1868)  SOL. STATE COMM. VOL.13 NO.2, .  229, (1973).
                                ( 6 , 1381)  J. PHYS. C VOL.7, . . . . . . .  L99, (1974).

690  CHRIEN..........(R.E.)..:  (20B,   71)  NUCL. INST. MET. VOL.13, . . . .    1, (1961).
                                ( 6 ,   94)  PHYS. REV. VOL.125, . . . . . .  275, (1962).
                                (19A,   68)  REV. SCI. INSTRUM. VOL.35, . . . 1150, (1964).
                                (23 ,   59)  NUCL. INST. MET. 53, . . . . . .   93, (1967).
                                ( 5 , 2913)  PROC/CONF/NEUTRON C. S. + TECH.,  875, (1968).

691  CHRISTENSEN.....(A.N.)..:  ( 5 , 1410)  ACTA CHEM. SCAND. VOL.24, . . . 1662, (1970).
                                ( 5 ,  835)  SOL. STATE COMM. VOL.9, . . . .  925, (1971).
                                ( 5 , 1759)  SOL. STATE COMM. VOL.10, . . . .  609, (1972).
                                ( 5 ,  895)  SOL. STATE COMM. VOL.10, . . . .  765, (1972).
                                ( 5 , 2595)  J. SOLID STATE CHEM. VOL.9, . .  234, (1974).

692  CHRISTENSEN.....(C.J.)..:  (22C,   30)  NUCL. INST. MET. 33, . . . . . .  155, (1965).
                                (20D,   34)  REV. SCI. INSTRUM. VOL.36, . . .   48, (1965).

693  CHRIST..........(J.)....:  (20A,   54)  NUKLEONIK VOL.4, . . . . . . . .   23, (1962).
                                ( 6 ,  590)  PHYS. STAT. SOL. 7, . . . . . .  557, (1964).
                                ( 6 ,  674)  PHYS. LETT. 15, . . . . . . . .  231, (1965).
                                (10C,  137)  Z. ANGEW. PHYS. VOL.18, . . . .  295, (1965).
                                (20A,   82)  UKAEA AERE TRANS 1074 (20 PP.), .    ,  (1967).

694  CHUBASKOV.......(V.I.)..:  ( 5 , 1308)  SOV. PHYS. JETP. 27, . . . . . .   15, (1968).

695  CHUDLEY.........(C.T.)..:  (14B,   95)  PROC. PHYS. SOC. VOL.77, . . . .  353, (1961).
                                ( 5 , 1291)  REPORT AEEW-R 701 (31PP.), . . . . ,  (1970).

696  CHUDNOVSKII.....(E.M.)..:  (11A,   64)  SOV. PHYS. SOLID STATE VOL.14, . 1035, (1972).
                                (11B,   47)  SOV. PHYS. SOLID STATE VOL.14, .  387, (1972).
                                ( 8 ,  152)  SOV. PHYS. SOLID STATE VOL.14, . 1761, (1973).
```

```
697  CHUDNOVSKII.....(F.A.)..:  ( 6 ,  893) SOV. PHYS. SOL. STATE VOL.13,. . 1256, (1971).

698  CHUMBLEY.........(L.)....:  (18B,    8) J. SCI. INSTRUM. SER.2 VOL.1,. .  528, (1968).

699  CHUNG...........(C.H.)..:  (14A,   48) PHYS. REV. VOL.182,. . .. . . .  323, (1969).
                               ( 6 , 1819) PROPERTIES LIQUID METALS,. . . .  125, (1973).

700  CHUNG...........(M.K.)..:  ( 5 , 1510) ACTA CRYST. 23,. . . . . . . .  578, (1967).

701  CIALELLA........(C.M.)..:  (24 ,   62) NUCL. INST. MET. 60, . . . . .  269, (1968).

702  CIERJACKS.......(S.)....:  ( 5 , 1880) NUCL. PHYS. VOL.A166,. . . . .  443, (1971).
                               ( 5 , 1879) NUCL. PHYS. VOL.A166,. . . . .  461, (1971).

703  CISZEWSKI.......(R.)....:  ( 5 , 1709) PHYS. STAT. SOL. 3,. . . . . . 1999, (1963).
                               ( 5 , 1708) PHYS. STAT. SOL. 4,. . . . . .  199, (1964).
                               ( 5 , 1707) PHYS. STAT. SOL. 7,. . . . . . 1015, (1964).
                               ( 5 , 2796) PHYS. STAT. SOL. 11,. . . . . .  K85, (1965).
                               ( 5 , 2800) PHYS. STAT. SOL. 15,. . . . . .  515, (1966).
                               ( 6 ,  781) PHYS. LETT. 28A,. . . . . . .  389, (1968).
                               ( 6 ,  736) PHYS. LETT. 29A,. . . . . . .  513, (1969).
                               (13 ,   39) PHYS. LETT. 30A, A VOL.A39,. .   68, (1969).
                               ( 5 ,  923) ACTA PHYS. POL. A VOL.A39, . .  465, (1971).
                               ( 8 ,  104) PHYS. REV. B VOL.7,. . . . . . 4142, (1973).

704  CLAESSEN........(Y.)....:  ( 6 , 1679) SOL. STATE COMM. VOL.8,. . . . . 1321, (1970).

705  CLARKE..........(J.H.)..:  ( 5 ,  781) ACTA CRYST. VOL.A28 PART S-4,. .  188, (1972).

706  CLARK...........(A.E.)..:  ( 5 , 2920) J. APPL. PHYS. VOL.41, . . . . . 1192, (1970).

707  CLARK...........(B.C.)..:  ( 6 , 2230) PHYS. REV. VOL.134,. . . . . . .A1486, (1964).

708  CLARK...........(C.D.)..:  (20B,   65) J. SCI. INSTRUM. VOL.43,. . . .    1, (1966).
                               (20B,   36) IAEA SYMP. COPENHAGEN, VOL.2,. .  341, (1968).
                               ( 5 , 1220) CRYSTAL LATTICE DEFECTS VOL.2,.  105, (1971).

709  CLARK...........(G.E.)..:  (20C,   70) NUCL. INST. MET. VOL.84, . . . .   67, (1970).

710  CLARK...........(J.R.)..:  ( 5 ,  757) J. APPL. PHYS. VOL.23, . . . . . 1379, (1952).

711  CLARK...........(M.A.)..:  (23 ,   29) IAEA SYMP. (PILE RESEARCH)VIENNA   25, (1960).
                               (20B,   11) CAN. J. PHYS. VOL.39,. . . . . .    1, (1961).

712  CLAUDET.........(G.)....:  (22C,   66) REV. SCI. INSTRUM. VOL.45, . . .  341, (1974).

713  CLAYTON.........(E.)....:  ( 7A,   19) AUSTRAL. J. PHYS. VOL.23,. . . .  823, (1970).

714  CLAYTON.........(G.T.)..:  (22C,   50) REV. SCI. INSTRUM. VOL.31, . . . 1355, (1960).
                               ( 5 , 1518) DISS. ABSTR. VOL.21,. . . . . . 2759, (1961).
                               ( 5 , 1520) PHYS. REV. VOL.121,. . . . . . .  649, (1961).

715  CLEMENTEL.......(E.)....:  (23 ,   21) IAEA SYMP. (PILE RESEARCH)VIENNA  123, (1960).

716  CLEMENT.........(G.F.)..:  ( 5 ,  939) PHYS. REV. VOL.68, . . . . . . .  159, (1945).

717  CLEMENT.........(J.C.)..:  (22B,   35) NUCL. INST. MET. VOL.91, . . . .  541, (1971).

718  CLEWS...........(C.J.B.):  ( 5 ,  279) NATURE VOL.192,. . . . . . . .  154, (1961).
                               ( 5 , 2627) ACTA CRYST. VOL.B26,. . . . . .  693, (1970).

719  CLOSS...........(H.)....:  (10B,   87) J. PHYS. F VOL.3,. . . . . . .  L83, (1973).
                               ( 6 ,  586) J. PHYS. F VOL.3,. . . . . . . .  L1, (1973).
                               (10B,   77) J. PHYS. CHEM. SOLIDS VOL.35,. .  123, (1974).

720  CLOUD...........(H.)....:  ( 5 ,  592) PHYS. REV. 131,. . . . . . . . 1511, (1963).

721  CLOUD...........(W.H.)..:  ( 5 , 1807) PHYS. REV. VOL.120,. . . . . . 1969, (1960).
                               ( 5 , 1640) J. APPL. PHYS. VOL.33, SUPPL.; . 1356, (1962).

722  CLUNE...........(L.C.)..:  ( 6 , 2345) PHYS. REV. B-4,. . . . . . . . 2675, (1971).

723  COATES..........(M.S.)..:  (16C,   73) RENSSELAER POL. INST. SYMP.,. .  248, (1961).
                               (20C,   34) 2ND INTERNAT. CONF./NUCL. DATA,. . .  . , (1970).

724  COCEVA..........(C.)....:  (24 ,   61) NUCL. INST. MET. 58, . . . . .    1, (1968).

725  COCHRAN.........(L.W.)..:  (12B,   41) PHYS. REV. VOL.94, . . . . . .  374, (1954).

726  COCHRAN.........(W.)....:  ( 6 ,  933) PHYS. REV. LETT. VOL.2,. . . .  495, (1959).
                               ( 6 ,  934) PROC. ROY. SOC. A VOL.253, . . .  260, (1959).
                               ( 6 , 1649) PHYS. REV. VOL 119,. . . . . . .  980, (1960).
                               (10B,   28) IAEA SYMP. CHALK RIVER, VOL.1,.   59, (1962).
                               ( 5 , 1563) J. PHYS. CHEM. SOLIDS VOL.23,. .  621, (1962).
                               (10A,   10) PHYSICS OF SEMICONDUCTORS,. . .  467, (1962).
                               ( 6 , 1248) PHYS. REV. 131,. . . . . . . . 1025, (1963).
                               (10D,   26) PHYS. REV. 131,. . . . . . . . 1030, (1963).
                               ( 6 , 1625) PROC. ROY. SOC. 276,. . . . . .  308, (1963).
                               (10A,   13) REP. PROGR. PHYS. VOL.26,. . . .    1, (1963).
                               (10A,   15) SCI. PROGR. VOL.51,. . . . . .  424, (1963).
                               (10B,  208) SEMICONDUCTORS, BOMBAY, VOL.1,.   50, (1963).
                               (10A,   35) IAEA SYMPOSIUM BOMBAY, VOL.1,.     3, (1964).
                               (10A,   18) SOLID/LIQUID STATE PHYSICS,. . .   43, (1965).
                               ( 6 , 2078) PHYS. REV. LETT. VOL.17,. . . .  753, (1966).
                               ( 6 , 1842) PROC. ROY. SOC. 293,. . . . . .  433, (1966).
                               (10C,   12) ATLANTA SYMP. GEORGIA INST. TECH   17, (1967).
                               ( 6 , 1277) IAEA SYMP. COPENHAGEN, VOL.1,.   267, (1968).
                               (10B,   37) IAEA SYMP. COPENHAGEN, VOL.1,. .  275, (1968).
                               ( 5 , 1491) PHYS. STAT. SOL. 30,. . . . . .  K157, (1968).
                               (13 ,   36) 2ND SUMMER SCHOOL S.S. PHYS.,. .   93, (1968).
                               (17C,   13) ACTA CRYST. A25,. . . . . . . .   95, (1969).
                               (10C,    5) ADV. IN PHYS. VOL.18,. . . . . .  157, (1969).
                               ( 6 , 1274) PHYS. REV. B-2,. . . . . . . . 4607, (1970).
                               (10B,    3) ACTA CRYST. VOL.A27, . . . . . .  556, (1971).
                               ( 6 , 1336) FERROELECTRICS(GB) VOL.2,. . . .  261, (1972).
                               ( 6 , 1540) J. PHYSIQUE TOME 33 COL.2, . . .   59, (1972).

727  COCKING.........(S.J.)..:  (21 ,    8) IAEA SYMPOSIUM VIENNA, . . . . .  165, (1960).
                               ( 6 ,  227) IAEA SYMPOSIUM VIENNA, . . . . .  569, (1960).
                               ( 6 , 1049) IAEA SYMPOSIUM (VIENNA), . . . .  309, (1960).
                               (20B,   34) IAEA SYMP. CHALK RIVER, VOL.1,.  107, (1962).
                               ( 6 , 1598) IAEA SYMP. CHALK RIVER, VOL.1,.  227, (1962).
                               ( 6 , 2058) IAEA SYMP. CHALK RIVER, VOL.1,.  237, (1962).
                               ( 6 , 2175) IAEA SYMP. CHALK RIVER, VOL.2,.  265, (1962).
                               ( 6 , 1616) PROC. PHYS. SOC. VOL.80,. . . . 1201, (1962).
                               (22C,   25) J. SCI. INSTRUM. VOL.41, . . . .  376, (1964).
                               ( 6 , 2075) PHYS. LETT. 16,. . . . . . . .  130, (1965).
                               (20C,  111) TH. NEUTRON SCATT./EGELSTAFF,. .  141, (1965).
                               ( 6 , 1959) ADV. IN PHYSICS VOL.16,. . . . .  189, (1967).
                               ( 6 , 1798) IAEA SYMP. COPENHAGEN, VOL.1,. .  463, (1968).
                               ( 6 , 1802) J. OF PHYS. C VOL.1, . . . . . .  507, (1968).
                               (22C,   26) J. SCI. INSTRUM. SER.2 VOL.1,. .  367, (1968).
                               (14B,  109) REPORT AERE-R5867 (137 PP.),. . . .  . , (1968).
                               ( 6 , 1609) J. PHYS. C SER.2 VOL.2,. . . . . 2047, (1969).
```

728 CODDING-JR......(J.W.)..: (5 , 2325) NUCL. SCI. ENG. VOL.43,. . . . .   58, (1971).

729 CODDING.........(J.W.)..: (5 , 2248) NUCL. SCI. ENG. VOL.20,. . . . .   235, (1964).

730 COEY............(J.M.D.): (5 , 2189) PHYS. REV. LETT. VOL.32,. . . . 1257, (1974).

731 COFFINBERRY.....(A.S.)..: (5 , 2726) ACTA CRYST. VOL.5,. . . . . . .   394, (1952).

732 COHEN-ADDAD.....(C.).....: (5 ,   84) COLL. INTER. N.126 (GRENOBLE), .   54, (1963).
                                (5 ,   83) SOL. STATE COMM. 1,. . . . . . .   85, (1963).
                                (5 , 346) ACTA CRYST. VOL.A27,. . . . . .   68, (1971).
                                (5 , 2625) BU. SOC. FR. MIN. CRYSTALLOG. 94 172, (1971).

733 COHEN..........(E.)....:   (6 , 803) PHYS. REV. B VOL.5,. . . . . . . 2607, (1972).
                                (6 , 571) PHYS. REV. B VOL.6,. . . . . . . 1053, (1972).

734 COHEN..........(J.)....:   (5 , 618) J. PHYSIQUE TOME 29, ...   813, (1968).

735 COHEN..........(M.H.)..:   (10B,   38) IAEA SYMP. COPENHAGEN, VOL.1,. .   119, (1968).

736 COHEN..........(M.L.)..:   (10A, 176) PHYS. REV. LETT. VOL.29, ... 1593, (1972).

737 COHEN..........(M.)....:   (6 , 1200) PHYS. REV. VOL.118,. . . . . . .   27, (1960).

738 COHEN..........(S.S.)..:   (10C,   82) PHYS. STAT. SOL. VOL.50, ...   673, (1972).
                                (6 , 554) SOL. STATE COMM. VOL.11,. . . . 1269, (1972).

739 COHN...........(H.O.)..:   (5 , 2237) PROC/CONF/NEUTRON C. S. + TECH.,   851, (1968).

740 COING-BOYAT....(J.)....:   (5 , 500) PHYS. LETT. 3, ...   178, (1962).
                                (5 , 2789) J. APP. PHYS. 34,. . . . . 1333, (1963).
                                (5 , 453) BULL. SOC. FRANC. MIN. CRIST. 92 264, (1969).

741 COLELLA........(R.)....:   (9B,   16) ACTA CRYST. VOL.A28, ...   218, (1972).

742 COLE...........(F.E.)..:   (5 , 1313) DISS. ABS. XXVII,. . . . . . . .18508, (1967).

743 COLLIE.........(C.H.)..:   (24 ,   44) NUCL. INST. MET. 39, ...   350, (1966).

744 COLLING........(F.)....:   (21 ,   37) NUCL. INST. MET. 64, ...   52, (1968).

745 COLLINS........(B.M.)..:   (5 , 190) ACTA CRYST. VOL.B30, ...   816, (1974).

746 COLLINS........(M.F.)..:   (5 , 2253) PHIL. MAG. VOL.6,. . . . .   485, (1961).
                                (5 , 1062) J. PHYS. SOC. JAPAN VOL.17 B-III   19, (1962).
                                (5 , 1405) PROC. PHYS. SOC. VOL.80, ...   362, (1962).
                                (5 , 2107) ACTA CRYST. VOL.16,. . . . . . .   A126, (1963).
                                (18A,   9) BRIT. J. APPL. PHYS. VOL.14, . .   805, (1963).
                                (5 , 2131) J. APPL. PHYS. VOL.34, ... 1195, (1963).
                                (5 , 983) PHIL. MAG. 8,. . . . .   401, (1963).
                                (6 , 504) PROC. PHYS. SOC. 82, ...   633, (1963).
                                (5 , 1190) J. PHYSIQUE TOME 25, ...   596, (1964).
                                (6 , 1488) PROC. INT. CONF. (NOTTINGHAM), .   319, (1964).
                                (6 , 1655) J. APP. PHYS. 36,. . . . . 1092, (1965).
                                (5 , 930) PROC. PHYS. SOC. 86, ...   535, (1965).
                                (6 , 831) PROC. PHYS. SOC. 86, ...   973, (1965).
                                (6 , 1314) J. APP. PHYS. 37,. . . . . 1054, (1966).
                                (5 , 2224) PROC. PHYS. SOC. 89, ...   415, (1966).
                                (11C,   46) PROC. PHYS. SOC. 90, ... 1015, (1967).
                                (11C,   48) PROC. PHYS. SOC. 92, ...   390, (1967).
                                (8 ,   21) J. APPL. PHYS. VOL.39, ...   533, (1968).
                                (6 , 728) KYOTO CONF. STAT. MECHANICS, . .   169, (1968).
                                (10C,   51) LECT/ELEM. EXCIT./INTER. IN SOL. . ., (1968).
                                (6 , 752) PHYS. REV. LETT. 21, ...   99, (1968).
                                (6 , 220) CARBON., 7,. . . . .   663, (1969).
                                (6 , 734) J. APP. PHYS. VOL.40,. . . . . . 1442, (1969).
                                (6 , 1500) J. PHYS. C, SER.2, VOL.2,. . . .   46, (1969).
                                (6 , 738) J. PHYS. SOC. JAPAN (SUPPL.) 26,   169, (1969).
                                (6 , 794) PHYS. REV. VOL.179,. . . . . . .   417, (1969).
                                (6 , 1716) PHYS. REV. VOL.182,. . . . . . .   624, (1969).
                                (6 , 834) J. APPL. PHYS. 41, ... 1433, (1970).
                                (6 , 566) J. CHEM. PHYS. VOL.52, ... 1828, (1970).
                                (6 ,   4) J. CHEM. PHYS. VOL.52, ... 5740, (1970).
                                (6 ,   87) J. PHYS. C, SER.2, VOL.3,. . . .   249, (1970).
                                (11C,   39) PHYS. REV. B-2, ... 4552, (1970).
                                (5 , 1768) CAN. J. PHYS. 49,. . . . .   979, (1971).
                                (6 , 1485) J. PHYS. C: SOLID ST. PHYS. V-4, 1299, (1971).
                                (11C,   43) PHYS. REV. B VOL.4,. . . . . . . 1588, (1971).
                                (9D,   20) CAN. J. PHYS. VOL.50,. . . . . . 2991, (1972).
                                (5 , 264) J. CHEM. PHYS. VOL.57, ... 5007, (1972).
                                (6 , 1502) INT. J. MAGN. VOL.4 NO.1,. . . .   17, (1973).
                                (6 , 792) INT. J. MAGN. VOL.4, ...   205, (1973).
                                (13 ,   55) PHYS. REV. LETT. VOL.30 NO.17,   781, (1973).
                                (6 , 500) CAN. J. PHYS. VOL.52,. . . . . .   396, (1974).
                                (13 ,   25) J. PHYS. C VOL.7,. . . . .   573, (1974).

747 COLLOMB........(A.)....:   (5 , 2496) J. SOLID STATE CHEM. VOL.7,. . .   337, (1973).

748 COLOMINAS-BROQUE(C.)....: (5 , 456) BULL. SOC. FRANC. MIN. CRIST. 90 109, (1967).

749 COLOMINAS.......(C.)....: (5 , 707) PHYS. REV. 153,. . . . .   558, (1967).

750 COLWELL........(J.F.)..:   (12B,   10) IAEA SYMP. COPENHAGEN, VOL.2,. .   429, (1968).
                                (19A,   44) NUCL. INST. MET. 76, ...   135, (1969).
                                (19A,   47) NUCL. INST. MET. 77, ...   29, (1970).

751 COMES..........(R.)....:   (6 , 1343) PHYS. REV. B VOL.5,. . . . . . . 1866, (1972).
                                (10E,   28) PHYS. STAT. SOLIDI B VOL.56 NO.1 313, (1973).
                                (6 , 1334) PHYS. REV. LETT. VOL.32, ...   836, (1974).

752 COMLY..........(J.B.)..:   (6 , 1734) J. APP. PHYS. 38,. . . . . 1240, (1967).
                                (5 , 955) PROC. PHYS. SOC. VOL.92, ...   726, (1967).
                                (6 , 1715) J. OF PHYS. C VOL.1, ...   458, (1968).
                                (5 , 734) PHYS. REV. LETT. 22, ...   531, (1969).
                                (6 , 630) PHYS. REV. LETT. VOL.24, ...   598, (1970).
                                (5 , 2612) A.I.P. CONF. PROC. NO.10 PART 2, 1554, (1973).

753 COMPAGNER......(A.)....:   (6 ,   78) PHYS. LETT. VOL.28A, ...   642, (1969).

754 CONDON.........(E.U.)..:   (7B,   51) PHYS. REV. VOL.49, ...   229, (1936).

755 CONNOR.........(D.W.)..:   (6 , 2292) J. CHEM. PHYS. VOL.45, ... 3817, (1966).

756 CONNOR.........(D.)....:   (20B,   44) IEEE TRANS. NUCL. SCI. VOL.NS-13 311, (1966).
                                (20D,   35) REV. SCI. INSTRUM. VOL.40, . . .   49, (1968).

757 CONN...........(R.)....:   (6 , 219) NUCL. SCI. ENGNG. VOL.40,. . . .   17, (1970).

758 CONRAD.........(C.A.)..:   (5 , 2324) PROC/CONF/NEUTRON C. S. + TECH.,   687, (1968).
                                (5 , 2326) NUCL. SCI. ENG. VOL.52,. . . . .   310, (1973).

759 CONSTANTINESCU..(M.)....: (5 , 110) REV. ROUMAINE PHYS. VOL.9, . . .   121, (1964).

775 CORLISS.........(L.M.)..: (5 , 812) PHYS. REV. LETT. VOL.32, ... 544, (1974).

776 CORLISS.........(L.)....: (5 , 2516) C.R. ACAD. SCI. VOL.251, ... 1733, (1960).
                                (5 , 1442) ACTA CRYST. VOL.14,. . . . . . .   19, (1961).
                                (5 , 2018) J. PHYS. CHEM. SOLIDS VOL.21,. .   234, (1961).

777 CORNGOLD........(N.)....: (16 ,   22) IAEA SYMPOSIUM VIENNA, ...   607, (1960).
                                (16 ,   60) PHYS. REV. VOL.137,. . . . . . . .A1686, (1965).

778 CORTELLESSA.....(G.)....: (22C,   37) NUCL. INST. MET. 59, ...   117, (1968).

779 CORVI...........(F.)....: (24 ,   61) NUCL. INST. MET. 58, ...   1, (1968).

780 COSTA-PAIVA.....(M.M.)..: (5 , 1287) NUCL. SCI. ENG. VOL.45,. . . . .   308, (1971).

781 COTE............(R.E.)..: (5 , 1625) PHYS. REV. VOL.134,. . . . . . .B1047, (1964).
                                (5 , 422) PHYS. REV. VOL.134,. . . . . . .B1281, (1964).

782 COTTERILL.......(R.M.J.): (6 , 602) PHYS. LETT. 23,. . . . .   309, (1966).

783 COTTER..........(M.J.)..: (6 , 1837) DISS. ABSTR. VOL.23, ... 3944, (1963).

784 COTTIER.........(J.M.)..: (21 ,   5) ARCH. SCI. VOL.13, ...   375, (1960).

785 COTTON..........(J.P.)..: (5 , 2329) J. CHEM. PHYS. VOL.57, ...   290, (1972).
                                (5 , 2330) J. APPL. CRYST. VOL.7,. . . . .   189, (1974).
                                (6 , 1922) PHYS. REV. LETT. VOL.32,. . . . 1170, (1974).

786 COUCH...........(J.G.)..: (6 , 2345) PHYS. REV. B-4,. . . . . 2675, (1971).

787 COUDERCHON......(G.)....: (12B,   28) NUCL. INST. MET. VOL.95, ...   589, (1971).

788 COULSON.........(C.A.)..: (15 ,   11) HARWELL SUMMER SCHOOL, ...   68, (1968).

789 COULTHARD.......(M.A.)..: (10B,   67) J. PHYS. C, SER.2, VOL.4,. . . .   820, (1971).

790 COWAN...........(R.D.)..: (6 , 1190) PHYS. REV. VOL.113,. . . . . . . 1386, (1959).

791 COWLAM..........(N.)....: (5 , 546) J. OF PHYS.-C-, SER.2, VOL.2,. .   238, (1969).
                                (5 , 1630) J. OF PHYS.-C-, SER.2, VOL.3,. .   675, (1970).
                                (5 , 1732) J. PHYS. F VOL.3,. . . . .   6, (1973).

792 COWLEY..........(E.R.)..: (10D,   53) PROC. ROY. SOC. 287, ...   259, (1965).
                                (6 , 1652) PROC. ROY. SOC. 292, ...   209, (1966).
                                (6 , 2191) PROC. ROY. SOC. 300, ...   45, (1967).
                                (6 , 513) CAN. J. PHYS. 47,. . . . . 1381, (1969).
                                (6 , 2079) J. OF PHYS. C VOL.2, ... 1916, (1969).
                                (17C,   14) ACTA CRYST. VOL.A26, ...   539, (1970).
                                (10D,   17) J. PHYS. C: SOLID ST. PHYS. V-4,   988, (1971).
                                (10B, 170) PHYS. REV. B-3,. . . . . 2743, (1971).
                                (6 , 2166) CAN. J. PHYS. VOL.51,. . . . . .   843, (1973).
                                (6 , 434) PHYS. REV. B VOL.8,. . . . . . . 4795, (1973).
                                (6 , 1832) SOL. STATE COMM. VOL.14, ...   587, (1974).

793 COWLEY..........(R.A.)..: (10B,   28) IAEA SYMP. CHALK RIVER, VOL.1, .   59, (1962).
                                (6 , 2103) IAEA SYMP. CHALK RIVER, VOL.2, .   229, (1962).
                                (6 , 2102) PHYS. REV. LETT. VOL.9,. . . . .   159, (1962).
                                (10B, 203) PROC. ROYAL SOC. A VOL.268,. . .   109, (1962).
                                (10D,   2) ADVANCES IN PHYS. VOL.12,. . . .   421, (1963).
                                (6 , 1248) PHYS. REV. 131,. . . . . 1025, (1963).
                                (10D,   26) PHYS. REV. 131,. . . . . 1030, (1963).
                                (6 , 2089) IAEA SYMP. BOMBAY, VOL.1,. . . .   297, (1964).
                                (6 , 2215) IAEA SYMP. BOMBAY, VOL.1,. . . .   373, (1964).
                                (6 , 2092) PHYS. REV. VOL.134,. . . . . . . A981, (1964).
                                (6 , 2221) CAN. JOURN. PHYS. 43,. . . . . . 1397, (1965).
                                (6 , 1844) PHYS. REV. LETT. 14,. . . . . .   549, (1965).
                                (10D,   53) PROC. ROY. SOC. 287, ...   259, (1965).
                                (6 , 2078) PHYS. REV. LETT. VOL.17, ...   753, (1966).
                                (6 , 2220) PHYS. REV. LETT. 16, ...   683, (1966).
                                (6 , 1304) PHYS. REV. 147,. . . . .   577, (1966).
                                (6 , 1246) PHYS. REV. 150,. . . . .   487, (1966).
                                (6 , 2033) PROC. PHYS. SOC. 88,. . . . . .   463, (1966).
                                (6 , 1652) PROC. ROY. SOC. 292, ...   209, (1966).
                                (6 , 1842) PROC. ROY. SOC. 293, ...   433, (1966).
                                (6 , 493) PHYS. REV. LETT. 18, ...   162, (1967).
                                (10D, 217) PHYS. REV. 158,. . . . .   805, (1967).
                                (10D,   52) PROC. PHYS. SOC. VOL.90, ... 1127, (1967).
                                (6 , 496) CAN. J. PHYS. 46,. . . . . 1355, (1968).
                                (6 , 1243) IAEA SYMP. COPENHAGEN, VOL.1,. .   141, (1968).
                                (6 , 1277) IAEA SYMP. COPENHAGEN, VOL.1,. .   267, (1968).
                                (6 , 1247) IAEA SYMP. COPENHAGEN, VOL.1,. .   281, (1968).
                                (6 , 1124) IAEA SYMP. COPENHAGEN, VOL.1,. .   609, (1968).
                                (6 , 1256) IAEA SYMP. COPENHAGEN, VOL.1,. .   43, (1968).
                                (6 , 505) IAEA SYMP. COPENHAGEN, VOL.2,. .   123, (1968).
                                (23 ,   35) IAEA SYMP. COPENHAGEN, VOL.2,. .   281, (1968).
                                (6 , 2219) J. APP. PHYS. 39,. . . . . 1111, (1968).
                                (6 , 494) J. APP. PHYS. 39,. . . . . 1116, (1968).
                                (6 , 1476) J. APP. PHYS. 39,. . . . . 1118, (1968).
                                (6 , 1149) PHYS. REV. LETT. 21,. . . . . .   787, (1968).
                                (6 , 2222) PHYS. REV. 167,. . . . .   464, (1968).
                                (6 , 509) PHYS. REV. 167,. . . . .   510, (1968).
                                (14B, 108) REPORT AECL-3189 (11 PP.),. . .   145, (1968).
                                (6 , 1455) SOL. STATE COMM. 6,. . . . . . .   145, (1968).
                                (6 , 1322) CAN. J. PHYS. VOL.47,. . . . . . 1983, (1969).
                                (17A,   40) IAEA SYMP. (INSTRUMENT.), VIENNA   1, (1969).
                                (10E,   16) J. PHYS. C, SER.2, VOL.2,. . . . 2262, (1969).
                                (6 , 1639) PHYS. REV. LETT. VOL.23, ...   525, (1969).
                                (6 , 492) PHYS. REV. LETT. VOL.23, ...   86, (1969).
                                (6 , 1244) PHYS. REV. VOL.180,. . . . . . .   755, (1969).
                                (6 , 1481) SOL. STATE COMM. VOL.7,. . . . . 1693, (1969).
                                (6 , 2094) SOL. STATE COMM. 7,. . . . . . .   181, (1969).
                                (6 , 1449) J. APPL. PHYS. 41, ...   896, (1970).
                                (6 , 1320) J. PHYS. SOC. JAPAN VOL.28, SUP.   242, (1970).
                                (6 , 2080) J. PHYS. SOC. JAPAN VOL.28, SUP.   249, (1970).
                                (6 , 1659) J. PHYS. SOC. JAPAN VOL.28, SUP.   258, (1970).
                                (6 , 1660) J. PHYS. SOC. JAPAN VOL.28,. . . 1426, (1970).
                                (6 , 1274) PHYS. REV. B-2,. . . . . 4603, (1970).
                                (6 , 1130) PHYS. REV. LETT. VOL.24, ...   646, (1970).
                                (10D,   58) SOL. STATE COMM. VOL. 8, ...   923, (1970).
                                (6 , 1121) CAN. J. PHYS. 49,. . . . .   177, (1971).
                                (6 , 1323) J. PHYSIQUE TOME 32 COL.1 VOL.II 1184, (1971).
                                (6 , 1269) J. PHYS. -C-, VOL. 4,. . . . . . 2127, (1971).
                                (11A,   18) J. PHYS. -C-, VOL. 4,. . . . . . 2139, (1971).
                                (6 , 501) PHYS. REV. LETT. VOL.27, ... 1442, (1971).
                                (6 , 1307) IAEA SYMP. GRENOBLE, ...   581, (1972).
                                (10B,   40) IAEA SYMP. GRENOBLE, ...   813, (1972).
                                (6 , 2106) J. PHYSIQUE TOME 33 COL.2,. . .   135, (1972).
                                (13 ,   30) J. PHYSIQUE TOME 33, COL.2,. . .   7, (1972).
                                (6 , 1857) J. PHYS.F: METAL PHYS. VOL.2,. .   421, (1972).
                                (6 , 1321) MAGN. AND MAG. MATERIALS-1971, 1315, (1972).
                                (11A,   56) PROC. ROYAL SOC. EDINBURGH A-70,   75, (1972).
                                (8 , 122) REV. MOD. PHYS. VOL.44,. . . . .   406, (1972).
                                (6 , 1858) SOL. STATE COMM. VOL.11 NO.1,. .   271, (1972).
                                (10D,   18) J. PHYS. C VOL.6,. . . . . 2422, (1973).
                                (6 , 497) J. PHYS. C VOL.6,. . . . . 2997, (1973).
                                (13 ,   24) J. PHYS. C VOL.6,. . . . .   143, (1973).

793 COWLEY..........(R.A.)..‡ (14B, 113) REP. PROGR. PHYS. VOL.36,. . . . 1135, (1973).

794 COX.............(A.E.)..‡ (24 ,   44) NUCL. INST. MET. 39,. . . . . 350, (1966).

795 COX.............(D.E.)..‡ (5 , 1102) J. PHYS. CHEM. SOLIDS VOL.23,.   863, (1962).
                               (5 , 1186) J. PHYS. SOC. JAPAN VOL.17,..   1598, (1962).
                               (5 , 2564) J. APP. PHYS. 34,. . . . . .   1352, (1963).
                               (5 , 606) J. PHYS. CHEM. SOL. 24,. . . .   405, (1963).
                               (5 , 1809) PHYS. REV. 129,. . . . . . .   2008, (1963).
                               (5 , 1560) PHYS. REV. 132,. . . . . . .   1547, (1963).
                               (5 , 1000) J. APPL. PHYS. VOL.35,. . . .   954, (1964).
                               (5 , 622) PROC. INT. CONF. (NOTTINGHAM),   291, (1964).
                               (5 , 1031) PROC. INT. CONF. (NOTTINGHAM),   327, (1964).
                               (5 , 1802) ACTA CRYST. 19,. . . . . . .   854, (1965).
                               (5 , 1804) J. APP. PHYS. 36,. . . . . .   1095, (1965).
                               (5 , 448) PHYS. LETT. 17,. . . . . . .   103, (1965).
                               (5 , 744) PHYS. REV. 138,. . . . . . .   A153, (1965).
                               (5 , 2760) PHYS. REV. 140,. . . . . . .   A1448, (1965).
                               (5 , 1803) PHYS. REV. 140,. . . . . . .   A2139, (1965).
                               (5 , 450) J. APP. PHYS. 37,. . . . . .   1126, (1966).
                               (5 , 632) J. APP. PHYS. 37,. . . . . .   973, (1966).
                               (5 , 127) J. PHYS. CHEM. SOL. 27,. . .   413, (1966).
                               (5 , 1151) PHYS. REV. 147,. . . . . . .   415, (1966).
                               (5 , 161) J. APPL. PHYS. VOL.38,. . . .   1459, (1967).
                               (5 , 697) J. APPL. PHYS. VOL.38,. . . .   1461, (1967).
                               (5 , 2215) J. PHYS. CHEM. SOL. 28,. . .   1649, (1967).
                               (5 , 696) PHYS. REV. VOL.167,. . . . .   519, (1968).
                               (5 , 563) J. APPL. PHYS. VOL.40,. . . .   1124, (1969).
                               (5 , 743) PHYS. REV. VOL.181,. . . . .   936, (1969).
                               (5 , 360) PHYS. REV. VOL.188,. . . . .   930, (1969).
                               (5 , 2804) PHYS. REV. VOL.188,. . . . .   963, (1969).
                               (5 , 2387) ACTA CRYST. VOL.A26,. . . . .   377, (1970).
                               (5 , 172) J. APPL. PHYS. 41,. . . . . .   943, (1970).
                               (5 , 1620) J. PHYS. SOC. JAPAN VOL.28,..   615, (1970).
                               (5 , 670) SOL. STATE COMM. VOL.8,. . .   1001, (1970).
                               (5 , 650) ACTA CRYST. VOL.A27,. . . . .   494, (1971).
                               (5 , 276) ACTA CRYST. VOL.B27,. . . . .   26, (1971).
                               (5 , 2391) J. PHYSIQUE TOME 32 COL.1 VOL.II   892, (1971).
                               (5 , 2222) J. PHYSIQUE TOME 32 COL.1 VOL.I,   582, (1971).
                               (5 , 2396) PHYS. REV. B VOL.3,. . . . . .   3946, (1971).
                               (5 , 671) PHYS. REV. B VOL.4,. . . . . .   2209, (1971).
                               (90 ,   30) IEEE TRANS. MAGN. (USA) VOL.MAG-8 161, (1972).
                               (5 , 2395) MAGN. AND MAG. MATERIALS-1971,   436, (1972).
                               (5 , 2397) MAGN. AND MAG. MATERIALS-1971,   670, (1972).
                               (5 , 159) PHYS. REV. B VOL.6,. . . . . .   2677, (1972).
                               (5 , 2407) PHYS. REV. B VOL.6,. . . . . .   204, (1972).
                               (5 , 1898) SOL. STATE COMM. VOL.11,. . .   1011, (1972).
                               (5 , 2400) A.I.P. CONF. PROC. NO.10 PART 1,   659, (1973).
                               (5 , 162) A.I.P. CONF. PROC. NO.10 PART 1,   674, (1973).
                               (5 , 662) A.I.P. CONF. PROC. NO.10 PART 1,   684, (1973).
                               (5 , 672) PHYS. REV. B VOL.7,. . . . . .   2024, (1973).
                               (5 , 2227) PHYS. REV. B VOL.7,. . . . . .   3102, (1973).
                               (5 , 861) J. PHYS. CHEM. SOLIDS VOL.35,.   861, (1974).
                               (5 , 1901) PHYS. REV. B VOL.9 NO.11,. . .   . ., (1974).

796 COX.............(D.)....‡ (5 , 1746) AN. FIS. (SPAIN) VOL.66, ...   407, (1970).

797 COX.............(G.W.)..‡ (5 , 536) J. PHYS. CHEM. SOL. 27,. . . . .   1955, (1966).
                               (5 , 2421) ACTA CRYST. 23,. . . . .   574, (1967).
                               (5 , 1510) ACTA CRYST. 23,. . . . .   578, (1967).
                               (5 ,   3) ACTA CRYST. VOL.B25,. . . . . .   2437, (1969).

798 COX.............(S.A.)..‡ (17C,   19) NUCL. INST. MET. 56, ...   245, (1967).

799 CRACKNELL.......(A.P.)..‡ (6 , 1762) J. OF PHYS. C VOL.2, ...   24, (1969).
                               (6 , 507) PHYS. REV. 177,. . . . .   932, (1969).
                               (6 , 2216) PHYS. STAT. SOL. VOL.36,. . .   737, (1969).
                               (6 , 904) J. OF PHYS.-C-, SER.2, VOL.4,.   3215, (1971).
                               (11B,   12) J. PHYS. C VOL.6 NO.6,. . . . .   1054, (1973).
                               (11B,   11) J. PHYS. C VOL.6,. . . . . . .   2350, (1973).
                               (11A,   53) PHYS. STAT. SOLIDI B VOL.56 NO.1   157, (1973).
                               (11A,   24) J. PHYS. F VOL.4,. . . . . . .   466, (1974).

800 CRAIG...........(D.S.)..‡ (5 , 2366) CAN. J. PHYS. 42,. . . . .   2384, (1964).

801 CRAIG...........(R.S.)..‡ (5 , 393) PHYS. REV. B VOL.9,. . . . . . .   154, (1974).

802 CRANBERG........(L.)....‡ (20C,   46) NUCL. INST. MET. 57, ...   237, (1967).

803 CRANE...........(H.R.)..‡ (20B, 122) PHYS. REV. VOL.45, ...   507, (1934).

804 CRANGLE.........(J.)....‡ (5 , 2355) PROC. ROY. SOC. A VOL.272, . . .   387, (1963).

805 CRANSTON........(F.P.)..‡ (5 , 317) NUCL. PHYS. A VOL.A169,. . . . .   95, (1971).

806 CRAVEN..........(B.M.)..‡ (5 ,   37) ACTA CRYST. 17,. . . . .   415, (1964).
                               (5 , 773) ACTA CRYST. 20,. . . . .   214, (1966).
                               (5 , 305) ACTA CRYST. VOL.B25,. . . . . .   1970, (1969).
                               (5 ,   3) ACTA CRYST. VOL.B25,. . . . . .   2437, (1969).

807 CREER...........(J.G.)..‡ (5 , 1711) SOL. STATE COMM. VOL.8,. . . . .   1183, (1970).

808 CREMET..........(J.P.)..‡ (19B,   32) REV. PHYS. APPL. (FRANCE) VOL.4, 111, (1969).

809 CRETEN..........(W.L.)..‡ (20C,   6) APPL. SCI. RES. B VOL.10,. . . .   247, (1963).

810 CRIBIER.........(D.)....‡ (6 , 1468) C.R. ACAD. SCI. VOL.248, ...   1631, (1959).
                               (6 , 1231) C.R. ACAD. SCI. VOL.250, ...   2871, (1960).
                               (17A,   25) C.R. ACAD. SCI. VOL.251, ...   230, (1960).
                               (6 , 582) IAEA SYMPOSIUM VIENNA, ...   549, (1960).
                               (6 , 1034) IAEA SYMPOSIUM (VIENNA) . . . . .   347, (1960).
                               (6 , 963) IAEA SYMPOSIUM (VIENNA) . . . . .   411, (1960).
                               (6 , 595) J. PHYS. RADIUM VOL.21, ...   67, (1960).
                               (6 , 1067) J. PHYS. RADIUM VOL.21 . . . . .   69, (1960).
                               (6 , 436) IAEA SYMP. CHALK RIVER, VOL.2,.   225, (1962).
                               (6 , 1450) IAEA SYMP. CHALK RIVER, VOL.2,.   309, (1962).
                               (6 , 1757) IAEA SYMP. CHALK RIVER, VOL.2,.   317, (1962).
                               (6 , 1463) J. PHYS. RADIUM VOL.23,. . . . .   494, (1962).
                               (6 , 1744) J. PHYS. SOC. JAPAN VOL.17 9-III   67, (1962).
                               (5 , 336) PHYS. LETT. 1,. . . . .   187, (1962).
                               (10C,   20) COPENHAGEN CONF.-LATT.DYNAMICS,.   71, (1963).
                               (5 , 2000) PHYS. LETT. 9,. . . . .   106, (1964).
                               (5 , 1999) PROC. INT. CONF. (NOTTINGHAM),   285, (1964).
                               (8 , 117) PROG. LOW TEMP. PHYS. VOL.5,.   161, (1967).
                               (6 , 1414) IAEA SYMP. COPENHAGEN, VOL.1,.   289, (1968).
                               (5 , 2008) PHYS. REV. LETT. VOL.28,. . . .   1370, (1972).
                               (5 , 2004) J. PHYSIQUE TOME 34, ...   447, (1973).

811 CRISTU..........(M.)....‡ (23 ,   17) IAEA SYMP. (PILE RESEARCH)VIENNA 577, (1960).

812 CROAT...........(J.)....‡ (5 , 1400) J. CHEM. PHYSICS VOL.54, ...   1995, (1971).

813 CROMER..........(D.T.)..‡ (6 , 1640) ACTA CRYST. 19,. . . . .   224, (1965).
                               (5 , 651) ACTA CRYST. 21,. . . . .   383, (1966).
                               (5 , 716) J. CHEM. PHYSICS VOL.44, ...   1648, (1966).
                               (5 , 1950) ACTA CRYST. 22,. . . . .   182, (1967).

813 CROMER..........(D.T.)..: (5 , 1892) ACTA CRYST. 22,. . . . . 800, (1967).

814 CROUCH..........(M.F.)..: (5 , 2870) PHYS. REV. VOL.102,. . . . . . . 1321, (1956).

815 CSELIK..........(M.)....: (5 , 1785) PHYS. LETT. VOL.24A, ... 198, (1967).

816 CUBIOTTI........(G.)....: (6 , 2304) IAEA SYMP. COPENHAGEN, VOL.1,. . 373, (1968).
                             (6 , 2306) SOL. STATE COMM. VOL.8,. . . . . 367, (1970).
                             (14A,   31) NUOVO CIMENTO B VOL.13B NO.1,. . 185, (1973).

817 CUNNINGHAM......(R.M.)..: (6 , 550) PHYS. REV. B-2,. . . . . 4864, (1970).
                             (6 , 551) THESIS (139PP.),. . . . . , (1970).
                             (6 , 528) IAEA SYMP. GRENOBLE, ...   53, (1972).
                             (6 , 529) A.I.P. CONF. PROC. NO.10 PART 1,   41, (1973).

818 CURIEN..........(H.)....: (6 , 862) PHYS. STAT. SOL. 21, ... K163, (1967).

819 CURIE...........(I.)....: (25 ,   9) NATURE VOL.130,. . . . .   57, (1932).

820 CURRAT..........(R.)....: (20B, 110) NUCL. INST. MET. VOL.107,. . . .   21, (1973).

821 CURRIE..........(M.)....: (5 , 1924) J. CHEM. SOC. A (1967),. . . . . 1862, (1967).
                             (5 , 1452) J. CHEM. SOC. A (1970),. . . . . 1923, (1970).

822 CURRY...........(N.A.)..: (5 , 1958) ACTA CRYST. VOL.9, ...   82, (1956).
                             (5 , 2414) PROC. ROY. SOC. A VOL.235, . . . 552, (1956).
                             (5 , 1444) ACTA CRYST. VOL.10,. . . . . . . 524, (1957).
                             (5 , 1428) ACTA CRYST. VOL.12,. . . . . . . 674, (1959).
                             (5 , 1443) ACTA CRYST. VOL.13,. . . . . . . 717, (1960).
                             (5 , 2507) ACTA CRYST. VOL.13,. . . . . . .   10, (1960).
                             (5 , 745) PROC. ROY. SOC. A VOL.266, . . .   95, (1962).
                             (6 , 867) PHYS. LETT. 7, ... 177, (1963).
                             (5 , 270) PROC. ROY. SOC. A VOL.279, . . .   98, (1964).
                             (5 , 1094) PHIL. MAG. VOL.12, ... 221, (1965).
                             (5 , 2756) PROC. PHYS. SOC. 86, ... 1193, (1965).
                             (5 , 2801) PROC. PHYS. SOC. 89, ... 427, (1966).
                             (5 , 1924) J. CHEM. SOC. A (1967),. . . . . 1862, (1967).
                             (5 , 356) BULL. SOC. CHIM. FR. 1968, . . . 1748, (1968).
                             (5 , 355) J. CHEM. SOC. A (1971),. . . . . 3725, (1971).

823 CUTLER..........(P.H.)..: (10B, 211) SOL. STATE COMM. 7,. . . . . . . 295, (1969).
                             (10B, 164) PHYS. REV. B-3,. . . . . 2485, (1971).

824 CUTTLER.........(J.M.)..: (22A,   32) NUCL. INST. MET. 75, ... 309, (1969).

825 CVIKL...........(B.)....: (6 , 2041) PHIL. MAG. VOL.28, ... 1353, (1973).

826 CZACHOR.........(A.)....: (6 , 2309) COPENHAGEN CONF.-LATT.DYNAMICS,.   33, (1963).
                             (10B,   33) IAEA SYMPOSIUM BOMBAY, VOL.1,. . 181, (1964).
                             (6 , 165) IAEA SYMP. COPENHAGEN, VOL.1,. . 157, (1968).
                             (6 , 1854) J. PHYS.F: METAL PHYS. VOL.1,. . 339, (1971).
                             (6 , 2012) PHYS. STAT. SOL. VOL.44,. . . . K65, (1971).
                             (6 , 2013) IAEA SYMP. GRENOBLE, ...   61, (1972).
                             (6 , 178) ACTA PHYS. POLON. VOL.A43, . . .   37, (1973).

827 CZARNECKI.......(R.)....: (9B,   27) ARCH. HUTNICTWA VOL.4, ...   81, (1959).
                             (24 ,   27) KERNTECHNIK VOL.2, ... 153, (1960).

828 CZERLINSKY......(E.R.)..: (5 , 2888) HELV. PHYS. ACTA VOL.46, ... 429, (1973).

829 CZERLUNCZARKIEWI(B.)....: (6 , 1551) ACTA PHYS. POLON. VOL.25,. . . . 141, (1964).

830 DABBS...........(J.W.T.): (8 , 109) PROC....CONF. ON POL. TARGETS, . 235, (1971).

831 DACHS...........(H.)....: (5 , 1575) Z. KRIST. VOL.112, ...   60, (1959).
                             (20A,   83) Z. KRIST. VOL.115, ...   80, (1961).
                             (5 , 1762) J. PHYS. SOC. JAPAN VOL.17 B-II, 387, (1962).
                             (20B, 157) Z. KRIST. VOL.117, ... 135, (1962).
                             (5 , 1761) COLL. INTER. N.126 (GRENOBLE), . 139, (1963).
                             (5 , 1845) SOL. STATE COMM. 4,. . . . . . . 473, (1966).
                             (5 , 1847) Z. KRIST. VOL.125, ... 120, (1967).
                             (20B,   53) J. APPL. CRYST. VOL.3, ... 214, (1970).
                             (20B,   52) J. APPL. CRYST. VOL.3, ... 220, (1970).
                             (5 , 2398) SOL. STATE COMM. VOL.10, ... 1299, (1972).
                             (5 , 674) A.I.P. CONF. PROC. NO.10 PART 1, 664, (1973).
                             (5 , 1763) INT. J. MAGN. VOL.4 NO.1,. . . .   5, (1973).
                             (5 , 1846) SOL. STATE COMM. VOL.12 NO.8,. . 779, (1973).
                             (5 , 675) SOL. STATE COMM. VOL.14 NO.9,. . 841, (1974).

832 DADAKINA........(A.F.)..: (5 , 2242) UKRAINIAN PHYS. J. VOL.14, . . . 1968, (1969).

833 DAHLBORG........(U.)....: (20B,   7) ARK. FYS. VOL.16 PAPER 19, . . . 199, (1959).
                             (6 ,   42) ARK. FYS. VOL.17, PAPER 21,. . . 369, (1960).
                             (6 ,   52) IAEA SYMPOSIUM VIENNA, ... 587, (1960).
                             (6 , 946) IAEA SYMP. CHALK RIVER, VOL.1, . 317, (1962).
                             (6 , 1109) J. NUCL. ENERGY VOL.16,. . . . .   81, (1962).
                             (6 ,   54) IAEA SYMP. BOMBAY, VOL.2,. . . . 117, (1964).
                             (22A,   12) IAEA SYMP. BOMBAY, VOL.2,. . . . 487, (1964).
                             (14B,   62) PHYSICA VOL.30,. . . . . 1561, (1964).
                             (6 ,   38) ARK. FYS. VOL.31, PAPER 16,. . . 271, (1966).
                             (14B,   86) PHYS. REV. VOL.151,. . . . . . . 126, (1966).
                             (6 , 409) IAEA SYMP. COPENHAGEN, VOL.1,. . 581, (1968).
                             (6 , 1592) PHYSICA VOL.49,. . . . .   1, (1970).
                             (6 , 322) PHYSICA VOL.59,. . . . . 672, (1972).
                             (6 , 1568) SOL. STATE COMM. VOL.13 NO.5,. . 543, (1973).

834 DAITCH..........(P.B.)..: (16 ,   75) RENSSELAER POL. INST. SYMP., . . 187, (1961).

835 DALE............(D.H.)..: (5 , 2191) ACTA CRYST. 21,. . . . . 705, (1966).

836 DALLUGGE........(W.)....: (22C,   27) KERNTECHNIK VOL.10 NO.7, ... 371, (1968).

837 DALTON..........(A.W.)..: (20C,   76) NUCL. INST. MET. VOL.92, ... 221, (1971).

838 DAMASK..........(A.C.)..: (6 , 1871) J. CHEM. PHYSICS VOL.54, ... 2597, (1971).

839 DANELYAN........(L.S.)..: (20B,   47) INSTRUM. EXP. TECH. NO.3,. . . . 444, (1961).

840 DANERI..........(A.)....: (17C,   17) J. APPL. CRYST. VOL.3, ... 145, (1970).

841 DANFORD.........(M.D.)..: (5 , 1548) ANN. N.Y. ACAD. SCI. VOL.79, . . 762, (1960).

842 DANIELSON.......(G.C.)..: (6 , 1426) PHYS. REV. B VOL.5,. . . . . . . 2103, (1972).

843 DANIELS.........(E.)....: (5 , 191) J. CHEM. PHYSICS VOL.55, ... 1093, (1971).

844 DANIELS.........(W.B.)..: (5 , 1075) PHYS. LETT. VOL.22,. . . . . . . 407, (1966).
                             (5 , 1349) PHYS. REV. LETT. 18,. . . . . . 548, (1967).
                             (5 , 1356) SOLID STATE COMM. VOL. 5,. . . . 591, (1967).
                             (5 , 1356) PHYS. REV. VOL.165,. . . . . . . 688, (1968).
                             (6 , 1702) PHYS. REV. VOL.181,. . . . . . . 1251, (1969).
                             (5 , 1701) PHYS. REV. B VOL.6,. . . . . . . 4766, (1972).
                             (5 , 1901) PHYS. REV. B VOL.9 NO.11,. . . . . , (1974).

845   DANIEL..........(M.R.)..:   (6 ,   507) PHYS. REV. 177,. . . . .   932, (1969).

846   DANNER..........(H.R.)..:   (5 ,   187) PHYS. REV. VOL.100,. . . . . . .   745, (1955).
                                  (6 , 1032) IAEA SYMPOSIUM (VIENNA),. . . . .   363, (1960).
                                  (6 , 1068) PHYSICA VOL.27,. . . . .   373, (1961).
                                  (6 ,   948) IAEA SYMP. CHALK RIVER, VOL.1,.   273, (1962).
                                  (6 , 1894) IAEA SYMP. BOMBAY, VOL.2,. . . .   407, (1964).
                                  (6 , 1789) J. CHEM. PHYSICS VOL.40,. . . . 1417, (1964).
                                  (6 , 1919) J. CHEM. PHYSICS VOL.40,. . . . 1426, (1964).
                                  (6 , 1094) J. CHEM. PHYSICS VOL.40,. . . . 2670, (1964).
                                  (6 , 1213) J. CHEM. PHYSICS VOL.41,. . . . 3649, (1964).
                                  (6 ,   5) J. CHEM. PHYS. VOL.55,. . . . . 4156, (1971).
                                  (6 ,   324) J. CHEM. PHYS. VOL.59,. . . . . 2305, (1973).

847   DANNER..........(H.)....:   (5 ,   188) PHYS. REV. VOL.105,. . . . . . .   856, (1957).

848   DANYLENKO.......(V.M.)..:   (7B,   138) UKR. FIZ. ZH. (USSR) VOL.3,. . .   743, (1958).

849   DARBY...........(J.K.)..:   (6 , 2079) J. OF PHYS. C VOL.2,. . . . . . 1916, (1969).

850   DARGEL..........(L.)....:   (5 , 1574) ACTA PHYS. POLON. VOL.A43,. . .   673, (1973).
                                  (5 , 1032) ACTA PHYS. POLON. VOL.A44,. . .   587, (1973).

851   DARIER..........(P.)....:   (22B,   14) INTERNAT. COMPUTING SYMP.,. . .   291, (1974).

852   DARRES..........(M.)....:   (20A,   11) ANN. FAC. SCI. UNIV. TOULOUSE 3J   9, (1966).

853   DASANNACHARYA...(B.A.)..:   (6 ,   782) IAEA SYMPOSIUM VIENNA,. . . . .   555, (1960).
                                  (6 ,   69) IAEA SYMP. CHALK RIVER, VOL.1,.   189, (1962).
                                  (5 , 1174) J. PHYS. SOC. JAPAN VOL.17 B-III   41, (1962).
                                  (5 , 1173) J. PHYS. SOC. JAPAN VOL.17,. . .   247, (1962).
                                  (6 ,   927) SOL. STATE COMM. 1,. . . . . . .   205, (1963).
                                  (6 , 1499) IAEA SYMP. BOMBAY, VOL.1,. . . .   433, (1964).
                                  (6 ,   260) IAEA SYMP. BOMBAY, VOL.2,. . . .   157, (1964).
                                  (6 , 1497) PROC. INT. CONF. (NOTTINGHAM),.   322, (1964).
                                  (6 ,   86) PHYS. REV. VOL.137,. . . . . . . A417, (1965).
                                  (6 ,   376) PHYS. LETT. 23,. . . . .   226, (1966).
                                  (5 , 1906) PROC. PHYS. SOC. 89,. . . . . . .   379, (1966).
                                  (6 ,   274) CAN. JOURN. PHYS. 45,. . . . . . 3185, (1967).
                                  (6 ,   278) PHYS. REV. 156,. . . . .   196, (1967).
                                  (6 ,   377) PHYS. REV. 161,. . . . .   133, (1967).
                                  (6 ,   816) SOL. STATE COMM. 6,. . . . . . .   593, (1968).
                                  (6 , 1234) PHYS. STAT. SOL. 34, NO1,. . . .   279, (1969).
                                  (6 , 1496) SOL. STATE COMM. 7,. . . . . . .   123, (1969).
                                  (6 ,   118) SOL. STATE COMM. VOL.8,. . . . .   497, (1970).
                                  (6 , 1591) SOL. STATE COMM. VOL.8,. . . . .   889, (1970).
                                  (6 ,   134) CONF. INTERN., RENNES,. . . . .   140, (1971).
                                  (6 ,   291) J. OF PHYS.-C-, SER.2, VOL.4,. 2725, (1971).
                                  (6 , 1544) IAEA SYMP. GRENOBLE,. . . . . .   477, (1972).
                                  (5 ,   265) NUCL./S.S PHYS. SYMP. ABSTR.,.   , (1972).
                                  (6 ,   66) PHYS. STAT. SOL. VOL.50,. . . .   701, (1972).
                                  (17A,   74) NUCL. INST. MET. VOL.113,. . . .   15, (1973).

854   DASH............(J.G.)..:   (20B,   139) REV. SCI. INSTRUM. VOL.24,. . .   91, (1953).
                                  (5 , 1324) PHYS. REV. VOL.97,. . . . . . .   855, (1955).
                                  (5 , 1903) PHYS. REV. LETT. VOL.32,. . . .   724, (1974).

855   DAS-GUPTA.......(K.)....:   (20B,   145) REV. SCI. INSTRUM. VOL.32,. . .   602, (1961).

856   DATARS..........(W.R.)..:   (6 , 1242) PHYS. REV. B VOL.9,. . . . . . .   572, (1974).

857   DATT............(I.D.)..:   (5 ,   307) SOV. PHYS. CRYST. VOL.11,. . . .   177, (1966).
                                  (5 , 1550) SOV. PHYS. CRYST. VOL.13,. . . .   204, (1968).
                                  (5 , 1293) SOV. PHYS. CRYST. VOL.17, NO.2,   383, (1972).

858   DAUBERT.........(J.)....:   (6 ,   559) PHYS. LETT. VOL.32A,. . . . . .   437, (1970).
                                  (6 ,   560) IAEA SYMP. GRENOBLE,. . . . . .   85, (1972).

859   DAVIDOVIC.......(M.)....:   (20A,   37) NUCL. INST. MET. 65,. . . :. . .   233, (1968).
                                  (6 ,   3) CHEM. PHYS. LETT. VOL.17,. . . .   53, (1972).
                                  (6 ,   678) PHYS. LETT. VOL.42A NO.7,. . . .   509, (1973).

860   DAVIDSON........(G.R.)..:   (5 , 2395) MAGN. AND MAG. MATERIALS-1971,.   436, (1972).
                                  (5 , 2397) MAGN. AND MAG. MATERIALS-1971,.   670, (1972).
                                  (5 , 2369) PHYS. REV. B VOL.9,. . . . . . . 1041, (1974).

861   DAVIDSON........(W.L.)..:   (5 , 1972) PHYS. REV. VOL.73,. . . . . . .   842, (1948).
                                  (5 ,   788) PHYS. REV. VOL.75,. . . . . . . 1348, (1949).

862   DAVIES..........(F.)....:   (20B,   36) IAEA SYMP. COPENHAGEN, VOL.2,. .   341, (1968).

863   DAVISON.........(A.)....:   (5 , 1282) INORG. CHEM. VOL.8,. . . . . . 1928, (1969).

864   DAVIS-JR........(C.M.)..:   (5 ,   56) PHYS. REV. A VOL.8,. . . . . . .   368, (1973).

865   DAVIS...........(H.L.)..:   (6 , 1937) A.I.P. CONF. PROC. NO.10 PART 2, 1548, (1973).

866   DAVIS...........(J.)....:   (7B,   101) PROC. PHYS. SOC. 92,. . . . . .   889, (1967).

867   DAVIS...........(M.B.)..:   (20C,   75) NUCL. INST. MET. VOL.94,. . . .   493, (1971).
                                  (7A,   54) PHYS. LETT. B VOL.36B,. . . . .   560, (1971).

868   DAVIS...........(R.E.)..:   (20C,   36) NUCLEONICS VOL.17 APRIL,. . . .   116, (1959).

869   DAVIS...........(R.L.)..:   (6 ,   820) J. PHYS. SOC. JAPAN VOL.25,. . .   367, (1968).

870   DAVYDOV.........(L.N.)..:   (8 ,   154) SOV. PHYS. SOLID STATE VOL.14, 1774, (1973).

871   DAWSON..........(B.)....:   (17B,   60) PROC. ROY. SOC. 298,. . . . . .   255, (1967).
                                  (5 , 2771) PROC. ROY. SOC. 298,. . . . . .   289, (1967).
                                  (5 , 2444) PROC. ROY. SOC. 298,. . . . . .   307, (1967).
                                  (5 ,   163) HARWELL SUMMER SCHOOL,. . . . .   101, (1968).
                                  (17B,   21) ACTA CRYST. A25,. . . . .   12, (1969).

872   DAWSON..........(W.K.)..:   (19B,   15) NUCL. INST. MET. VOL.81,. . . .   301, (1970).

873   DAYAL...........(B.)....:   (6 , 2277) PHYS. REV. 143,. . . . .   443, (1966).
                                  (6 , 2316) PHYS. STAT. SOL. 13,. . . . . .   519, (1966).
                                  (6 ,   51) PHYS. STAT. SOL. 16,. . . . . .   513, (1966).
                                  (6 , 1751) PHYS. STAT. SOL. 19,. . . . . .   729, (1967).
                                  (10B,   191) PHYS. STAT. SOL. VOL.38,. . . .   141, (1970).
                                  (6 ,   922) SOL. STATE COMM. VOL.11 NO.6,.   775, (1972).
                                  (6 , 1970) SOL. STATE COMM. VOL.11,. . . .   185, (1972).
                                  (10B,   75) J. PHYS. C VOL.6,. . . . . 2943, (1973).

874   DAYANIDHI.......(P.K.)..:   (5 , 1906) PROC. PHYS. SOC. 89,. . . . . .   379, (1966).

875   DAY.............(D.H.)..:   (5 ,   223) J. OF PHYS.-C-, SER. 2, VOL.2,.   870, (1969).
                                  (19B,   14) NUCL. INST. MET. VOL.72,. . . .   237, (1969).
                                  (17A,   64) NUCL. INST. MET. 70,. . . . . .   164, (1969).
                                  (20B,   92) NUCL. INST. MET. 72,. . . . . .   237, (1969).
                                  (5 , 1705) ACTA CRYST. VOL.B26,. . . . . . 2079, (1970).
                                  (6 , 1046) J. CHEM. PHYS. VOL.55,. . . . . 2807, (1971).

876 DA-CUNHA-LIMA...(I.C.)..:   (6 , 1610) SOL. STATE COMM. VOL.11,. . . . . 1431; (1972):
                                 (10B, 109) PHYSICA VOL.72,. . . . . . . : : : : : 179; (1974):

877 DEAN............(P.J.)..:   (6 , 901) IAEA SYMP. COPENHAGEN, VOL.1,. .   301, (1968).

878 DEBEAUVAIS......(M.)....:   (20C,   86) REV. PHYS. APPL. (FRANCE) VOL.4, 259, (1969).

879 DEBETS..........(P.C.)..:   (5 , 2770) ACTA CRYST. 21,. . . . . 589, (1966).

880 DEBIESSE........(J.)....:   (20B,   12) C. R. ACAD. SCI. B TOME 270, . . 185, (1970).

881 DEBRAY..........(D.)....:   (5 , 412) CONF./RARE EARTHS/ACTINIDES, . . 218, (1971).
                                 (5 , 413) J. CHEM. PHYS. VOL.56,. . . . . . 4325, (1972).
                                 (5 , 1404) J. CHEM. PHYS. VOL.57,. . . . . . 2156, (1972).
                                 (5 , 2055) J. CHEM. PHYS. VOL.58,. . . . . . 1783, (1973).
                                 (5 , 2616) PHYS. STAT. SOLIDI A VOL.18 NO.1 227, (1973).

882 DECIU...........(N.)....:   (14B, 114) REV. ROUMAINE PHYS. VOL.18,. . . 997, (1973).

883 DECKER..........(D.L.)..:   (5 , 1074) PHYS REV 171,. . . . . 596, (1968).
                                 (5 , 487) BULL. AMER. PHYS. SOC. VOL.17, . 667, (1972).

884 DECKER..........(D.)....:   (6 , 1922) PHYS. REV. LETT. VOL.32, ... 1170, (1974).

885 DEGENKOLBE......(G.)....:   (23 ,   47) KERNTECHNIK VOL.15,. . . . . . . 450, (1973).

886 DEGTYAREV.......(YU.G.).:   (6 , 173) BU. ACAD. SCI. USSR PHYS. SR. 32 651, (1968).

887 DEJUREN.........(J.A.)..:   (16 ,   31) NUCLEAR SCI. ENGNG. VOL.24,. . . 417, (1966).

888 DEKHTYAR........(M.V.)..:   (5 , 2157) SOV. PHYS. SOL. STATE 5, ... 918, (1963).

889 DEKKER..........(A.L.)..:   (24 ,   11) DISS. ABS. VOL.30, ...3817B, (1970).

890 DELANEY.........(R.M.)..:   (5 , 2462) PHYS. REV. VOL.105,. . . . . . . 517, (1957).
                                 (10C,   24) DISS. ABSTR. VOL.19,. . . . . . . 2991, (1959).

891 DELAPALME.......(A.)....:   (5 , 2738) J. APPL. PHYS. VOL.33, SUPPL.,, 1123, (1962).
                                 (5 , 1837) J. PHYS. RADIUM VOL.23,. . . . . 477, (1962).
                                 (5 , 1128) J. PHYS. SOC. JAPAN VOL.17 B-III   53, (1962).
                                 (5 , 560) PHYS. LETT. 3,. . . . . 178, (1962).
                                 (5 , 1054) COLL. INTER. N.126 (GRENOBLE),. 121, (1963).
                                 (5 , 427) COLL. INTER. N.126 (GRENOBLE),. 186, (1963).
                                 (5 ,   84) COLL. INTER. N.126 (GRENOBLE),.   54, (1963).
                                 (5 , 1053) COMP. REND. 257,. . . . . 421, (1963).
                                 (5 , 2789) J. APP. PHYS. 34,. . . . . 1333, (1963).
                                 (5 , 435) SOL. STATE COMM. 1,. . . . . . .   81, (1963).
                                 (5 ,   83) SOL. STATE COMM. 1,. . . . . . .   85, (1963).
                                 (5 , 579) PROC. INT. CONF. (NOTTINGHAM),. 275, (1964).
                                 (5 , 591) J. APPL. PHYS. VOL.37,. . . . . . 1038, (1966).
                                 (5 , 1013) J. PHYS. CHEM. SOL. 28,. . . . . 1451, (1967).
                                 (5 , 1737) SOL. STATE COMM. 5,. . . . . . . 769, (1967).
                                 (6 , 819) J. APP. PHYS. 39,. . . . . 1332, (1968).
                                 (12A,   14) NUCL. INST. MET. 63,. . . . . . . 283, (1968).
                                 (12B,   1) ACTA CRYST. (INTERACT.) VOL.A25, S255, (1969).
                                 (6 , 1995) J. APPL. PHYS. VOL.40,. . . . . . 1278, (1969).
                                 (6 , 1985) PHYS. REV. LETT. VOL.23,. . . . . 1225, (1969).
                                 (6 , 1996) J. APPL. PHYS. VOL.41,. . . . . . 1384, (1970).
                                 (12B,   28) NUCL. INST. MET. VOL.95,. . . . . 589, (1971).
                                 (6 , 1978) PHYS. REV. B-4,. . . . . 3206, (1971).
                                 (17B,   48) NUCL. INST. MET. VOL.105,. . . . 533, (1972).

892 DELAPLANE.......(R.G.)..:   (5 ,   28) ACTA CRYST. VOL.B25, ... 2451, (1969).

893 DELAROCHE.......(J.P.)..:   (7A,   25) CONF/NUCL. STRUC. STUDY/NEUTRONS 144, (1972).

894 DELONNGI........(D.A.)..:   (5 , 2056) PHYS. CAN. VOL.27, ...   68, (1971).

895 DELPIERRE.......(P.)....:   (12A,   33) REV. PHYS. APPL. (FRANCE) VOL.4, 115; (1969).
                                 (12B,   50) REV. PHYS. APPL. (FRANCE) VOL.4; 254; (1969).

896 DELSANTO........(P.P.)..:   (20B, 108) NUCL. INST. MET. VOL.107,. . . .   37, (1973).

897 DELTOUR.........(J.)....:   (17B,   66) SOL. STATE COMM. VOL.8,. . . . . 1799, (1970).

898 DEMARCO.........(J.J.)..:   (17B,   5) ACTA CRYST. 19,. . . . .   68, (1965).

899 DEME............(SH.)...:   (20C,   26) INSTRUM. EXP. TECH. VOL.16 NO.2, 345, (1973).

900 DEMICHELIS......(F.)....:   (20D,   19) NUCL. INST. MET. 47,. . . . . . . 320, (1967).
                                 (20B,   85) NUCL. INST. MET. 59,. . . . . . . 136, (1968).
                                 (15 ,   41) PHYS. REV. VOL.167,. . . . . . . 139, (1968).
                                 (22A,   49) NUCL. INST. MET. VOL.114,. . . .   21, (1974).

901 DEMIDENKO.......(Z.A.)..:   (10C, 107) SOV. PHYS. SOL. STATE, 4,. . . . 2459, (1962).

902 DEMPSTER........(A.B.)..:   (10B,   58) J. CHEM. PHYSICS VOL.54,. . . . . 3600, (1971).

903 DENGEL..........(O.)....:   (6 , 674) PHYS. LETT. 15,. . . . . 231, (1965).

904 DENHAM..........(P.)....:   (10B, 205) PROC. ROY. SOC. A VOL.317, . . .   55, (1970).

905 DENIS...........(J.)....:   (5 , 2392) J. PHYSIQUE TOME 32 COL.1 VOL.II 611, (1971).

906 DENIS...........(P.)....:   (20D,   4) ARCH. SCI. VOL.12,. . . . . 676; (1959).
                                 (21 ,   5) ARCH. SCI. VOL.13,. . . . . 375; (1960).

907 DENNE...........(W.A.)..:   (5 , 356) BULL. SOC. CHIM. FR. 1968, . . . 1748, (1968).
                                 (5 , 354) J. CRYST. MOL. STRUCT.(GB) VOL.1 347, (1971).

908 DENT-GLASSER....(L.S.)..:   (5 , 1991) ACTA CRYST. VOL.B27, ... 2269, (1971).

909 DEOPURA.........(B.L.)..:   (6 , 421) MOL. PHYS. VOL.19,. . . . . 589, (1970).

910 DERYAGIN........(B.V.)..:   (24 , 112) SOV. PHYS. DOKLADY VOL.16, . . . 154, (1971).

911 DESAI...........(R.C.)..:   (16 ,   38) NUCL. SCI. ENGNG. VOL.24,. . . . 142, (1966).
                                 (6 ,   89) PHYS. REV. LETT. VOL.16,. . . . . 839, (1966).
                                 (7B,   49) PHYS. LETT. 25A,. . . . . 211, (1967).
                                 (6 ,   71) IAEA SYMP. COPENHAGEN, VOL.1,. . 545, (1968).
                                 (6 ,   93) PHYS. REV. VOL.166,. . . . . . . 129, (1968).
                                 (6 , 1607) PHYS. REV. VOL.180,. . . . . . . 308, (1969).

912 DESCLAUX........(J.P.)..:   (5 , 1208) INT. J. MAGN. VOL.3 NO.4,. . . . 311, (1972).
                                 (5 , 2593) PHYS. REV. B VOL.8,. . . . . . . 3237, (1973).

913 DESJARDINS......(J.S.)..:   (22A,   62) REV. SCI. INSTRUM. VOL.31,. . . 481, (1960).
                                 (20B,   54) J. APPL. CRYST. VOL.3,. . . . . . 361, (1970).
                                 (5 , 2528) PHYS. REV. B VOL.8,. . . . . . . 2595, (1973).

914 DEVANNEY........(J.A)...:   (24 ,   62) NUCL. INST. MET. 60, ... 269, (1968).

915 DEVINE..........(S.)....:   (10C,   47) J. PHYS. C, SER.2, VOL.4,. . . . 1091, (1971).

916 DEVKIN..........(B.V.)..: (6 , 2169) BU. ACAD. SCI. USSR PHYS. SR. 32 600, (1968).

917 DEVRIES.........(R.C.)..: (5 , 2267) J. APP. PHYS. 38,. . . . . 951, (1967).

918 DEWIT...........(G.A.)..: (23 ,   33) IAEA SYMP. COPENHAGEN, VOL.2,. . 259, (1968).
                               (6 , 1754) J. APP. PHYS. 39,. . . . . 451, (1968):
                               (6 , 1541) PHYS. LETT. 29A,. . . . . 694, (1969):

919 DEWOLFF.........(P.M.)..: (17A,   8) ACTA CRYST. VOL.A25, ... 206, (1969).

920 DE-AGOSTINO.....(E.)....: (23 ,   65) NUOVO CIMENTO SUPPL. VOL.23,. .   17, (1962).
                               (22B,   28) NUCL. INST. MET. 42, ...   29, (1966).

921 DE-BERGEVIN.....(F.)....: (5 , 2738) J. APPL. PHYS. VOL.33, SUPPL.,. 1123, (1962).
                               (5 , 1126) COMP. REND. 256, ... 1688, (1963).
                               (5 , 555) SOL. STATE COMM. VOL.5,. . . . . 955, (1967).
                               (9D,   56) PHYS. LETT. VOL.39A,. . . . . . 141, (1972).

922 DE-BOER.........(J.)....: (14A,   1) AARHUS SUMMER SCHOOL,. . . . . . 373, (1963).

923 DE-BROIN........(C.E.)..: (5 , 1278) ACTA CRYST. VOL.B28, ... 2530, (1972).

924 DE-CICCO........(P.D.)..: (6 , 747) PHYS REV. 162, ... 486, (1967).

925 DE-COMBARIEU....(A.)....: (5 , 1681) SOL. STATE COMM. VOL.7 NO.19,. . 1403, (1969).

926 DE-GENNES.......(P.G.)..: (8 ,   29) J. PHYS. CHEM. SOLIDS VOL.4, . . 223, (1958).
                               (8 ,   30) J. PHYS. CHEM. SOLIDS VOL.6, . .   43, (1958).
                               (5 , 978) J. PHYS. RADIUM VOL.19,. . . . . 617, (1958).
                               (14B,   61) PHYSICA VOL.25,. . . . . 825, (1959).
                               (14B,   27) IAEA SYMPOSIUM VIENNA, ... 239, (1960).
                               (10B, 209) SOL. STATE COMM. 1,. . . . . . . 132, (1963).
                               (14B,   65) PHYSICS VOL.3 NO.1,. . . . . . .   37, (1967).

927 DE-GIORGI.......(G.)....: (5 , 1236) J. NON-CRYST. SOLIDS VOL.11 NO.5 417, (1973).

928 DE-GRAAF........(L.A.)..: (6 , 650) BROOKHAVEN SYMPOSIUM,. . . . . . 138, (1965).
                               (22A,   58) REPORT 132-67-02,. . . . . , (1967).
                               (14B, 128) THESIS,. . . . . , (1967).
                               (6 , 340) PHYSICA VOL.40,. . . . . 497, (1969).
                               (6 , 321) PHYSICA VOL.48,. . . . .   79, (1970).
                               (5 , 2058) J. CHEM. PHYS. VOL.55, ... 4967, (1971).
                               (5 , 1993) J. CHEM. PHYS. VOL.55, ... 5363, (1971).
                               (6 , 575) IAEA SYMP. GRENOBLE, ... 247, (1972).
                               (6 , 2259) JULICH CONF/H IN METALS VOL.I, . 301, (1972).
                               (6 , 2257) J. CHEM. PHYS. VOL.56, ... 4574, (1972).
                               (6 , 1862) PHYS. REV. LETT. VOL.29, ... 1250, (1972).
                               (6 , 1665) J. CHEM. PHYS. VOL.58, ... 3439, (1973).
                               (6 , 2117) J. CHEM. PHYS. VOL.59, ... 6570, (1973).
                               (5 , 1328) PHYS. REV. A VOL.9, ... 448, (1974).
                               (6 , 1867) PHYS. REV. LETT. VOL.32, ... 1087, (1974).

929 DE-JONGE........(W.J.M.): (5 , 2097) PHYS. LETT. VOL.41A NO.5,. . . . 411, (1972).

930 DE-LANGE........(P.W.)..: (12B,   52) REV. SCI. INSTRUM. VOL.33, . . . 524, (1962).

931 DE-MEULENAER....(J.)....: (17C,   7) ACTA CRYST. 19,. . . . . 1014, (1965).

932 DE-MORTON.......(M.E.)..: (5 , 535) PHYS. REV. LETT. 11, ... 208, (1963).

933 DE-MUL..........(F.F.M.): (6 , 452) CHEM. PHYS. LETT. VOL.22,. . . . 476, (1973).

934 DE-MUL..........(F.F.)..: (6 , 649) PHYS. LETT. VOL.33A, ...   87, (1970).
                               (19A,   50) NUCL. INST. MET. VOL.93, ... 109, (1971).
                               (19A,   51) NUCL. INST. MET. VOL.98,. . . . .   53, (1972).
                               (6 , 651) PHYS. LETT. VOL.41A NO.3,. . . . 272, (1972).

935 DE-NOVION.......(C.)....: (5 , 2730) COMP. REND. 263, ... 457B, (1966).

936 DE-SALAMANCA....(M.E.)..: (20B,   6) AN. REAL SOC. ESPAN/FIS. VOL.63A 211, (1967).

937 DE-SANTIS.......(P.)....: (11A,   10) J. APPL. PHYS. 40, ... 2359, (1969).

938 DE-SAUSSURE.....(G.)....: (20B,   31) IAEA SYMP. (PILE RESEARCH)VIENNA 509, (1960).

939 DE-VRIES........(G.F.)..: (5 , 2559) J. PHYS. CHEM. SOL. 28,. . . . . 2143, (1967).
                               (5 , 2570) J. APPL. PHYS. VOL.39, ... 1364, (1968).

940 DE-WAMES........(R.E.)..: (16 ,   62) PHYS. REV. B VOL.7,. . . . . . . 917, (1973).

941 DE-WETTE........(F.W.)..: (6 , 602) PHYS. LETT. 23,. . . . . 309, (1966).
                               (10C,   77) PHYS. REV. 176,. . . . . 784, (1968).

942 DIACONESCU......(A.)....: (6 , 1106) REV. ROUMAINE PHYS. VOL.9, . . . 737, (1964).
                               (6 , 1524) PHYS. LETT. 22,. . . . . 558, (1966).
                               (6 , 167) IAEA SYMP. COPENHAGEN, VOL.1,. . 439, (1968).
                               (6 , 653) REV. ROUMAINE PHYS. VOL.15,. . . 783, (1970).

943 DIAMOND.........(D.J.)..: (16 ,   43) NUCL. SCI. ENGNG. VOL.40,. . . . 460, (1970).

944 DIAMOND.........(H.)....: (5 , 422) PHYS. REV. VOL.134,. . . . . . .B1281, (1964).

945 DIAZ-COLON......(F.A.)..: (5 , 2486) MATER. RES. BULL. VOL.8, ... 1183, (1973).

946 DIAZ............(L.M.)..: (21 ,   4) AN. REAL SOC. ESPAN/FIS. VOL.58A 119, (1962).
                               (20B,   6) AN. REAL SOC. ESPAN/FIS. VOL.63A 211, (1967).

947 DICKENS.........(B.)....: (5 , 1936) ACTA CRYST. VOL.B30, ... 1167, (1974).

948 DICKEY..........(J.M.)..: (10B, 143) PHYS. REV. VOL.188,. . . . . . . 1407, (1969).

949 DICKINSON.......(W.C.)..: (20B, 143) REV. SCI. INSTRUM. VOL.32, . . . 870, (1961).
                               (7A,   67) PHYS. REV. VOL.126,. . . . . . . 632, (1962).

950 DIERKS..........(H.)....: (5 , 150) ACTA CRYST. VOL.B27, ... 2003, (1971).

951 DIETERICH.......(W.)....: (6 , 2266) PHYS. LETT. VOL.35A,. . . . . . .   48, (1971).
                               (10C,   98) SOL. STATE COMM. VOL.12, ... 1191, (1973).

952 DIETRICH........(H.)....: (5 , 150) ACTA CRYST. VOL.B27, ... 2003, (1971).

953 DIETRICH........(O.W.)..: (6 , 1168) PHYS. REV. VOL.133,. . . . . . .B925, (1964).
                               (20A,   4) ACTA CRYST. 18,. . . . . 184, (1965).
                               (6 , 643) NBS MISC. PUB.273, ... 144, (1966).
                               (5 , 753) PHYS. REV. 153,. . . . . 706, (1967).
                               (5 , 752) PHYS. REV. 153,. . . . . 711, (1967).
                               (6 , 644) PHYS. REV. 153,. . . . . 717, (1967).
                               (5 , 2524) PHYS. REV. 162,. . . . . 315, (1967).
                               (6 , 2137) SOL. STATE COMM. 5,. . . . . . . 607, (1967).
                               (6 , 2126) IAEA SYMP. COPENHAGEN, VOL.2,. .   63, (1968).
                               (6 , 2134) J. APP. PHYS. 39,. . . . . 1229, (1968).
                               (18A,   18) NUCL. INST. MET. 61,. . . . . . . 296, (1968).
                               (6 , 1466) J. PHYS. C, SER.2, VOL.2,. . . . 1168, (1969).
                               (6 , 1467) J. PHYS. C, SER.2, VOL.2,. . . . 2022, (1969).
                               (6 , 2129) J. PHYS. SOC. JAPAN VOL.26,. . . 183, (1969).
                               (6 , 521) PHYS. REV. LETT. 22, ... 290, (1969).

953 DIETRICH........(O.W.)..: (6 , 2142) J. PHYS. C, SER.2, VOL.4,. . . . .   71, (1971).
                              (6 , 718) PHYS. REV. LETT. VOL.27,. . . . .   741, (1971).
                              (6 , 1165) IAEA SYMP. GRENOBLE,. . . . . . .   357, (1972).
                              (6 , 719) IAEA SYMP. GRENOBLE,. . . . . . .   619, (1972).
                              (6 , 720) MAGN. AND MAG. MATERIALS-1971,. 1251, (1972).
                              (6 , 1145) PHYS. REV. A 5,. . . . . 1377, (1972).

954 DILG............(W.)....: (5 , 438) Z. PHYSIK VOL.264,. . . . . . .   427, (1973).
                              (5 , 2831) Z. PHYSIK VOL.266,. . . . . . .   157, (1974).

955 DIMIC...........(V.)....: (6 , 1299) PHYS. LETT. VOL.26A, ...   8, (1967).
                              (5 , 372) PHYS. LETT. 25A,. . . . .   123, (1967).
                              (6 , 1383) PHYS. LETT. VOL.31A,. . . . . .   531, (1970).
                              (21 ,   25) J. PHYS. E VOL.4 NO.11,. . . . .   905, (1971).
                              (6 , 1392) PHYS. STAT. SOL. VOL.B-54,. . .   121, (1972).
                              (20B, 107) NUCL. INST. MET. VOL.107,. . . .   405, (1973).
                              (6 , 1579) PHYS. STAT. SOLIDI B VOL.59 NO.2 471, (1973).

956 DIMITRIJEVIC....(A.)....: (6 , 482) PHYSICA VOL.37,. . . . .   501, (1967).

957 DIMITRIJEVIC....(Z.)....: (6 , 879) COLL. INTER. N.126 (GRENOBLE),.   210, (1963).
                              (18A,   12) BULL/KIDRICH INST. NUCL. SCI. 15 115, (1964).
                              (6 , 835) IAEA SYMP. BOMBAY, VOL.1,. . . .   443, (1964).
                              (6 , 843) PHYS. STAT. SOL. 15, ...   119, (1966).
                              (6 , 862) PHYS. STAT. SOL. 21, ... K163, (1967).
                              (6 , 864) PHYS. STAT. SOL. 22, ...   K55, (1967).
                              (5 , 2940) PHYS. STAT. SOL. 32, ...   K91, (1969).
                              (5 , 1098) PHYS. STAT. SOL. A VOL.3,. . . . 1033, (1970).
                              (6 , 863) PHYS. STAT. SOL. 41, ... K103, (1970).
                              (6 , 1442) PHYSICA VOL.57,. . . . .   628, (1972).
                              (5 , 2169) PHYS. STAT. SOLIDI A VOL.9 NO.1   97, (1972).
                              (5 , 1697) PHYS. STAT. SOLIDI A VOL.19 NO.1 K13, (1973).

958 DINGLE..........(R.)....: (6 , 237) J. APPL. PHYS. 42, ... 1265, (1971).
                              (90,   62) PHYS. REV. LETT. VOL.26, ...   718, (1971).

959 DINITRIJEVIC....(L.)....: (6 , 545) PHYS. STAT. SOL. 32, ...   41, (1969).

960 DINITRIJEVIC....(Z.)....: (5 , 1095) PHYS. STAT. SOL. 26, ...   429, (1968).

961 DISDIER.........(F.)....: (22C,   66) REV. SCI. INSTRUM. VOL.45, . . .   341, (1974).

962 DIVATIA.........(A.S.)..: (20C,   51) NUCL. INST. MET. 55, ...   349, (1967).

963 DIXMIER.........(M.)....: (20C,   59) NUCL. INST. MET. 74, ...   123, (1969).

964 DIXON...........(D.R.)..: (6 , 1614) PROC. PHYS. SOC. 81, ...   973, (1963).
                              (12B,   3) ANN. OF PHYSICS VOL.74,. . . . .   250, (1972).

965 DI-BELLA........(R.)....: (5 , 471) J. PHYSIQUE TOME 31, ...   113, (1970).

966 DMITRIEV........(A.B.)..: (20C,   18) INSTRUM. EXP. TECH. NO.1,. . . .   87, (1966).

967 DMITRIEV........(R.P.)..: (12B,   64) SOV. PHYS. TECH. PHYS. VOL.40,. 1317, (1970).
                              (5 , 2084) SOV. PHYS. SOL. STATE VOL.15,.   919, (1973).

968 DMITRIEV........(V.B.)..: (5 , 2846) SOV. PHYS. DOKLADY VOL.15, . . .   776, (1971).

969 DOBRZYNSKI......(L.)....: (6 , 1305) IAEA SYMP. BOMBAY, VOL.2,. . . .   383, (1964).
                              (6 , 847) J. APP. PHYS. 39,. . . . . 1114, (1968).
                              (5 , 460) PHYS. STAT. SOL. VOL.38, ...   103, (1970).
                              (5 , 2798) PHYS. STAT. SOL. VOL.42, ...   K15, (1970).
                              (5 , 2203) PHYSICA VOL.51,. . . . .   627, (1971).
                              (17A,   69) NUCL. INST. MET. 99,. . . . . .   227, (1972).
                              (6 , 1780) PHYS. STAT. SOL. B VOL.54 NO.2,. K99, (1972).

970 DOEBBLER........(G.F.)..: (6 , 2010) J. CHEM. PHYS. VOL.50, ... 2140, (1969).

971 DOENITZ.........(F.D.)..: (5 , 525) PHYS. STAT. SOL. 3,. . . . . . . K249, (1963).

972 DOGONADZE.......(R.R.)..: (8 , 108) PHYS. STAT. SOL. VOL.49, ...   453, (1972).

973 DOHERTY.........(P.D.)..: (14B,   21) DISS. ABS. VOL.24, ... 4250, (1964).

974 DOIDZHASHVILI...(G.I.)..: (22A,   19) INSTRUM. EXP. TECH. NO.2,. . . .   306, (1964).

975 DOI.............(K.)....: (5 , 582) COLL. INTER. N.126 (GRENOBLE),.   38, (1963).
                              (5 , 2763) ACTA CRYST B24,. . . . . 1393, (1968).
                              (9B,   46) J. APPL. CRYST. VOL.4, ...   528, (1971).
                              (5 , 2439) J. PHYS. SOC. JAPAN VOL.31,. . .   954, (1971).
                              (5 , 1217) ACTA CRYST. VOL.A28 PART S-4,. .   220, (1972).
                              (5 , 2765) ACTA CRYST. VOL.B28, ...   785, (1972).
                              (8 , 1924) J. POLYM. SCI./POLYM. LETT. 11,.   377, (1973).
                              (9 , 1221) J. APPL. CRYST. VOL.7, ...   59, (1974).

976 DOLGINOV........(A.Z.)..: (12B,   62) SOV. PHYS. J.E.T.P. VOL.31,. . .   378, (1970).
                              (12B,   71) YADERNAYA FIZ. (USSR) VOL.12,. .   815, (1970).

977 DOLLEY..........(P.E.)..: (24 ,   47) NUCL. INST. MET. 59, ...   120, (1968).

978 DOLLING.........(G.)....: (6 , 2034) IAEA SYMPOSIUM VIENNA, ...   563, (1960).
                              (5 , 2253) PHIL. MAG. VOL.6,. . . . .   485, (1961).
                              (6 , 2035) IAEA SYMP. CHALK RIVER, VOL.2, .   37, (1962).
                              (6 , 206) PHYS. REV. VOL.128,. . . . . . . 1120, (1962).
                              (6 , 895) COPENHAGEN CONF.-LATT.DYNAMICS,.   19, (1963).
                              (6 , 897) PHYS. REV. 132,. . . . . 2410, (1963).
                              (6 , 916) IAEA SYMP. BOMBAY, VOL.1,. . . .   249, (1964).
                              (6 , 640) IAEA SYMP. BOMBAY, VOL.1,. . . .   343, (1964).
                              (6 , 2215) IAEA SYMP. BOMBAY, VOL.1,. . . .   373, (1964).
                              (10B, 113) PHYS. LETT. VOL.8,. . . . . . .   304, (1964).
                              (6 , 1731) PHYS. REV. VOL.136,. . . . . . . A1359, (1964).
                              (6 , 645) SOL. STATE COMM. 2,. . . . . . .   79, (1964).
                              (6 , 2221) CAN. JOURN. PHYS. 43,. . . . . . 1397, (1965).
                              (6 , 1844) PHYS. REV. LETT. 14,. . . . . .   549, (1965).
                              (6 , 642) PHYS. REV. VOL.138,. . . . . . . A1053, (1965).
                              (10A,   21) TH. NEUTRON SCATT./EGELSTAFF,.   193, (1965).
                              (6 , 2078) PHYS. REV. LETT. VOL.17, ...   753, (1966).
                              (6 , 2220) PHYS. REV. LETT. 16, ...   683, (1966).
                              (6 , 1304) PHYS. REV. 147,. . . . .   577, (1966).
                              (6 , 1246) PHYS. REV. 150,. . . . .   487, (1966).
                              (6 , 2033) PROC. PHYS. SOC. 88,. . . . . .   463, (1966).
                              (6 , 1842) PROC. ROY. SOC. 293,. . . . . .   433, (1966).
                              (20A,   36) NUCL. INST. MET. 49,. . . . . .   117, (1967).
                              (6 , 217) PHYS. REV. 158,. . . . .   805, (1967).
                              (5 , 1948) CAN. J. PHYS. 46,. . . . . 1727, (1968).
                              (6 , 1360) IAEA SYMP. COPENHAGEN, VOL.1,. .   149, (1968).
                              (6 , 611) IAEA SYMP. COPENHAGEN, VOL.1,. .   47, (1968).
                              (6 , 505) IAEA SYMP. COPENHAGEN, VOL.2,. .   123, (1968).
                              (6 , 2219) J. APP. PHYS. 39,. . . . . 1111, (1968).
                              (6 , 2222) PHYS. REV. 167,. . . . .   464, (1968).
                              (6 , 2509) PHYS. REV. 167,. . . . .   510, (1968).
                              (6 , 1373) PHYS. REV. 168,. . . . .   970, (1968).
                              (17A,   40) IAEA SYMP. (INSTRUMENT.), VIENNA   1, (1969).
                              (10B, 100) MOL. DYN. AND STRUCT. OF SOLIDS,   289, (1969).
                              (6 , 2094) SOL. STATE COMM. 7,. . . . . . .   181, (1969).
                              (6 , 2080) J. PHYS. SOC. JAPAN VOL.28, SUP.   249, (1970).

978 DOLLING.........(G.)....: (6 , 1659) J. PHYS. SOC. JAPAN VOL.28, SUP.   258, (1970).
                              (6 , 1660) J. PHYS. SOC. JAPAN VOL.28,   1426, (1970).
                              (6 , 1211) PROC. ROY. SOC. LOND. A.319,   269, (1970).
                              (6 , 425) BULL. AMER. PHYS. SOC. VOL.17,   291, (1972).
                              (6 , 423) IAEA SYMP. GRENOBLE,   207, (1972).
                              (6 , 1926) J. CHEM. PHYS. VOL.58,   158, (1973).
                              (6 , 1841) J. NONMET. VOL.1,   159, (1973).
                              (18A,   21) NUCL. INST. MET. VOL.106,   419, (1973).
                              (10B, 207) PROC. ROY. SOC. A VOL.333,   363, (1973).
                              (5 , 2272) CAN. J. PHYS. VOL.52,   748, (1974).
                              (6 , 1242) PHYS. REV. B VOL.9,   572, (1974).

979 DOMBROVSKI......(L.)....: (5 , 2954) SOV. PHYS. SOL. STATE VOL.13,   321, (1971).

980 DOMENICI........(M.)....: (6 , 2056) COLL. INTER. N.126 (GRENOBLE),   69, (1963).
                              (5 , 2459) PHYSICS/NON-CRYSTALLINE SOLIDS,   152, (1965).

981 DONAHUE.........(D.J.)..: (5 , 685) NUCL. SCI. ENGNG. VOL.7,   184, (1960).

982 DONALDSON.......(R.E.)..: (5 , 2236) PHYS. REV. VOL.138,   B1116, (1965).
                              (5 , 151) PHYS. REV. VOL.146,   660, (1966).
                              (5 , 766) PHYS. REV. VOL.174,   313, (1968).
                              (5 , 1244) PHYS. REV. C VOL.5,   1952, (1972).

983 DONIACH.........(S.)....: (11B,   28) PROC. PHYS. SOC. 91,   86, (1967).
                              (11A,   37) PHYS. REV. VOL.173,   603, (1968).

984 DONNAY..........(G.)....: (9D,   59) PHYS. REV. VOL.112,   1917, (1958).
                              (9D,   49) J. PHYS. SOC. JAPAN VOL.17 B-II,   332, (1962).

985 DONNAY..........(J.D.H.): (9D,   59) PHYS. REV. VOL.112,   1917, (1958).
                              (9D,   49) J. PHYS. SOC. JAPAN VOL.17 B-II,   332, (1962).
                              (9D,   18) ATLANTA SYMP/GEORGIA INST. TECH.   74, (1967).

986 DONNER..........(H.J.)..: (24 ,   64) NUCL. INST. MET. 70,   221, (1969).

987 DONOHUE.........(J.)....: (5 , 1335) PHYS. REV. VOL.114,   1009, (1959).

988 DONOVAN.........(J.L.)..: (6 , 1884) DISS. ABS. 25,   7333, (1965).
                              (6 , 1896) J. CHEM. PHYS. VOL.42,   4299, (1965).

989 DONZE...........(P.)....: (6 , 2196) SOL. STATE COMM. VOL.12 NO.2,   117, (1973).

990 DOOSE...........(C.)....: (23 ,   9) CRYOGENICS VOL.11,   107, (1971).

991 DORAIN..........(P.B.)..: (5 , 1476) J. PHYS. CHEM. SOLIDS 29,   881, (1968).

992 DORE............(J.C.)..: (6 , 1079) MOL. PHYS. VOL.24 NO.5,   1025, (1972).

993 DORING..........(W.)....: (8 , 161) Z. ANGEW. PHYS. VOL.20,   311, (1966).

994 DORNER..........(B.)....: (6 , 262) IAEA SYMP. BOMBAY, VOL.2,   291, (1964).
                              (6 , 266) PHYS. STAT. SOL. 5,   511, (1964).
                              (6 , 1833) PHYSICA VOL.31,   1537, (1965).
                              (6 , 271) PHYS. STAT. SOL. 18,   795, (1966).
                              (14A,   13) DISC. FARADAY SOC. NO.43,   160, (1967).
                              (6 , 422) IAEA SYMP. COPENHAGEN, VOL.1,   599, (1968).
                              (6 ,   97) PHYS. LETT. 28A,   433, (1968).
                              (14B,   34) IAEA SYMP. (INSTRUMENT.), VIENNA   19, (1969).
                              (5 , 253) PHYS. LETT. VOL.31A,   253, (1970).
                              (20A,   21) J. APPL. CRYST. VOL.4,   185, (1971).
                              (6 , 2155) PHYS. REV. LETT. VOL.26,   519, (1971).
                              (6 ,   95) PHYS. STAT. SOL. B VOL.43,   611, (1971).
                              (17D,   6) ACTA CRYST. VOL.A28,   319, (1972).
                              (6 , 2156) PHYS. REV. B VOL.6,   1950, (1972).
                              (20D,   3) ACTA CRYST. VOL.A29,   526, (1973).
                              (6 , 572) SOL. STATE COMM. VOL.12 NO.6,   537, (1973).
                              (20A,   22) J. APPL. CRYST. VOL.7,   38, (1974).

995 DOROFEEV........(YU.A.).: (5 , 988) BULL. ACAD. SCI. USSR VOL.30,   1007, (1966).

996 DOROSHENKO......(A.V.)..: (5 , 1828) PHYS. MET. METALLOGR. VOL.12,   119, (1961).
                              (5 , 2158) PHYS. MET. METALLOGR. VOL.15 N.6,   119, (1963).
                              (5 , 2159) PHYS. MET. METALLOGR. VOL.20 N.6,   48, (1965).
                              (5 , 2155) PHYS. STAT. SOL. VOL.16,   737, (1966).
                              (5 , 2649) PHYS. MET. METALLOGR. VOL.23 N.3,   168, (1967).

997 DORRIAN.........(J.F.)..: (5 , 224) DISS. ABS. VOL.32,   65878, (1972).

998 DOTSENKO........(B.B.)..: (7A,   74) PHYS. REV. VOL.178,   1647, (1969).

999 DOVER...........(C.B.)..: (7A,   36) NUCL. PHYS. A VOL.A177,   559, (1971).
                              (5 , 1880) NUCL. PHYS. VOL.A166,   443, (1971).
                              (5 , 1879) NUCL. PHYS. VOL.A166,   461, (1971).

1000 DOWNES..........(J.S.)..: (6 , 1547) PHYS. REV. LETT. VOL.17,   533, (1966).
                              (6 , 1553) MOL. SIEVES, PAP. CONF. 1967,   306, (1967).

1001 DOYAMA..........(M.)....: (6 , 602) PHYS. LETT. 23,   309, (1966).

1002 DO-AMARAL.......(L.Q.)..: (6 ,   59) J. NUCL. ENERGY VOL.26,   379, (1972).

1003 DO-DINH.........(C.)....: (5 , 155) BULL. SOC. FRANC. MIN. CRIST. 88   413, (1965).
                              (5 , 174) J. PHYSIQUE TOME 30,   566, (1969).
                              (5 , 166) J. PHYSIQUE TOME 31,   401, (1970).
                              (9D,   12) AIP CONF. PROC.(USA) VOL.5,   1355, (1971).
                              (5 , 2640) J. APPL. PHYS. VOL.44,   5096, (1973).

1004 DRABKIN.........(G.M.)..: (6 , 1749) SOV. PHYS. JETP LETT. VOL.2,   336, (1965).
                              (11B,   43) SOV. PHYS. JETP 20,   1548, (1965).
                              (5 , 2270) SOV. PHYS. SOL. STATE VOL.7,   997, (1965).
                              (5 , 2291) SOV. PHYS. SOL. STATE VOL.10,   511, (1968).
                              (11A,   63) SOV. PHYS. JETP LETT. VOL.9,   204, (1969).
                              (5 , 2065) SOV. PHYS. J.E.T.P. VOL.29,   261, (1969).
                              (12B,   64) SOV. PHYS. TECH. PHYS. VOL.40,   1317, (1970).
                              (12A,   43) SOV. PHYS. SOL. STATE VOL.13,   258, (1971).
                              (12B,   68) SOV. PHYS. JETP LETT. VOL.15,   324, (1972).
                              (6 , 1153) SOV. PHYS. TECH. PHYS. VOL.17,   180, (1972).

1005 DRAGHICESCU.....(P.)....: (12B,   31) PHYS. LETT. 12,   334, (1964).

1006 DREGICHESKU.....(M.)....: (20C,   87) REV. ROUMAINE PHYS. VOL.6,   207, (1961).

1007 DRESSELHAUS.....(G.)....: (6 , 2163) PROC/10TH INT. CONF/PHYS/SEMIC.,   338, (1970).
                              (6 , 2161) PHYS. REV. B VOL.4,   356, (1971).

1008 DRESSELHAUS.....(M.S.)..: (6 , 2163) PROC/10TH INT. CONF/PHYS/SEMIC.,   338, (1970).

1009 DREXEL..........(W.)....: (6 , 1942) IAEA SYMP. COPENHAGEN, VOL.1,   203, (1968).
                              (23 ,   32) IAEA SYMP. COPENHAGEN, VOL.2,   331, (1968).
                              (19A,   1) ACTA CRYST. (INTERACT.) VOL.A25, S250, (1969).
                              (6 ,   12) PHYS. LETT. VOL.28A,   531, (1969).
                              (6 ,   13) Z. PHYS. VOL.255,   281, (1972).
                              (6 , 1232) SOL. STATE COMM. VOL.13,   1549, (1973).
                              (6 , 709) PHYS. REV. B VOL.9 NO.11,   (1974).
                              (6 , 1334) PHYS. REV. LETT. VOL.32,   836, (1974).

1010 DRIGO...........(L.)...: (5 , 231) NUCL. PHYS. A VOL.A181,. . . . . 177, (1972).

1011 DROZDOV.........(S.I.).: (5 , 1258) J. NUCL. ENERGY VOL.4, ... 115, (1957).

1012 DROZHZHINOV.....(YU.N.).: (16 ,   85) SOV. PHYS. DOKLADY VOL.15, . . . 724, (1971).

1013 DROZ............(M.)...: (6 , 1154) HELV. PHYS. ACTA VOL.45, ...   34, (1972).

1014 DRUILHE.........(R.)...: (5 , 1672) J. APP. PHYS. VOL.39,. . . . . . 590, (1968).

1015 DRUSCHEL........(R.E.)..: (5 , 414) NUCL. SCI. ENG. VOL.37,. . . . . 228, (1969).
                                (5 , 1515) NUCL. SCI. ENG. VOL.47,. . . . . 371, (1972).

1016 DUBENKROPP......(H.)...: (6 , 1155) NUCL. PHYS. VOL.A129,. . . . . . 666, (1969).
                                (5 , 1880) NUCL. PHYS. VOL.A166,: . . . . . 443, (1971).
                                (5 , 1879) NUCL. PHYS. VOL.A166,. . . . . . 461, (1971).

1017 DUBININ.........(S.F.).: (5 , 2163) PHYS. STAT. SOLIDI A VOL.21 NO.1 K31, (1974).

1018 DUBOIS..........(D.M.).: (14A,   36) PHYSICA VOL.60,. . . . .   27, (1972).

1019 DUBROVSKAYA.....(L.B.).: (5 , 2646) SOV. PHYS. DOKLADY VOL.15, . . . 276, (1970).

1020 DUCKWORTH.......(J.A.K.): (5 , 1341) ACTA CRYST. VOL.A25, ... 482, (1969).
                                (5 , 1342) ACTA CRYST. VOL.A26, ... 263, (1970).

1021 DUCROS..........(P.)...: (5 ,   84) COLL. INTER. N.126 (GRENOBLE), .   54, (1963).
                                (5 ,   83) SOL. STATE COMM. 1,. . . . . . .   85, (1963).

1022 DUC.............(T.)...: (5 , 1713) SOL. STATE COMM. VOL.7,. . . . . 641, (1969).

1023 DUDERSTADT......(J.J.).: (16 ,   17) DISS. ABS. VOL.29, ...1120B, (1968).
                                (7A,   43) NUCL. SCI. ENG. VOL.45,. . . . . 167, (1971).

1024 DUFFEY..........(D.)...: (20B,   19) ENG. J. (CANADA),. . . . .   22, (1972).

1025 DUFFILL.........(C.)...: (20C,   11) IAEA SYMP. CHALK RIVER VOL.1, . 171, (1962).
                                (18B,   12) UKAEA AERE REPORT R 4895 (22 PP)   , (1965).
                                (5 , 2929) PHYS. LETT. 21,. . . . . 286, (1966).
                                (6 , 1957) CAN. J. PHYS. VOL.50,. . . . . . 3062, (1972).
                                (6 , 1777) IAEA SYMP. GRENOBLE,. . . . . . 669, (1972).

1026 DUGGAL..........(V.P.).: (24 ,   79) NUCL. SCI. ENGNG. VOL.6, ...   76, (1959).
                                (20A,   68) PROC. INDIAN ACAD. SCI. A VOL.53   59, (1961).
                                (20A,   78) REV. SCI. INSTRUM. VOL.33, . . .   49, (1962).

1027 DUKHOVSKAYA.....(E.L.).: (5 , 348) LATV. PSR...FIZ. TEHN. SER. NO.2 124, (1970).
                                (5 , 322) LATV. PSR...FIZ. TEHN. SER. NO.4 87, (1970).

1028 DUK-JOO-KIM............: (8 ,   42) J. PHYS. SOC. JAPAN VOL.18,. . . 1025, (1963).

1029 DULLOW..........(R.J.).: (22B,   15) J. APPL. CRYST. VOL.1, ... 272, (1968).

1030 DUMITRU.........(O.)...: (6 , 945) REV. ROUMAINE PHYS. VOL.18 NO.2, 135, (1973).

1031 DUNIN...........(S.Z.).: (11A,   35) PHYS. MET. METALLOGR. VOL.33 N.1   22, (1972).

1032 DUNITZ..........(J.D.).: (5 , 295) ACTA CRYST. VOL.B29, ... 2278, (1973).

1033 DUNLAP..........(B.D.).: (5 , 2216) A.I.P. CONF. PROC. NO.10 PART 1,   88, (1973).
                                (5 , 2217) INT. J. MAGN. VOL.4 NO.2,. . . .   99, (1973).
                                (5 , 2369) PHYS. REV. B VOL.9,. . . . . . . 1041, (1974).
                                (5 , 2212) PHYS. REV. B VOL.9,. . . . . . . 3766, (1974).

1034 DUNMUR..........(I.W.).: (5 , 1679) PROC. ROY. SOC. A VOL.241, . . . 223, (1957).

1035 DUNNING.........(J.R.).: (20A,   57) PHYS. REV. VOL.48,. . . . . . . 265, (1935).
                                (20B, 123) PHYS. REV. VOL.48,. . . . . . . 704, (1935).
                                (20A,   58) PHYS. REV. VOL.49,. . . . . . . 103, (1936).
                                (5 , 1255) PHYS. REV. VOL.52,. . . . . . . 1076, (1937).
                                (5 , 1250) PHYS. REV. VOL.54,. . . . . . . 266, (1938).
                                (25 ,   36) PHYS. REV. VOL.54,. . . . . . . 771, (1938).
                                (5 , 2136) PHYS. REV. VOL.58,. . . . . . . 1031, (1940).
                                (5 , 1249) PHYS. REV. VOL.69,. . . . . . . 236, (1946).
                                (22A,   55) PHYS. REV. VOL.71,. . . . . . .   65, (1947).
                                (5 , 2233) PHYS. REV. VOL.73,. . . . . . . 1399, (1948).
                                (6 , 975) PHYS. REV. VOL.73,. . . . . . . 733, (1948).
                                (6 , 598) PHYS. REV. VOL.73,. . . . . . . 963, (1948).

1036 DUNN............(F.E.).: (16 ,   50) NUCL. SCI. ENG. VOL.47,. . . . .   66, (1972).

1037 DURCANSKY.......(G.)...: (6 , 1723) J. PHYSIQUE TOME 32 COL.1 VOL.II 679, (1971).

1038 DURIF...........(A.)...: (5 ,   84) COLL. INTER. N.126 (GRENOBLE), .   54, (1963).
                                (5 ,   83) SOL. STATE COMM. 1,. . . . . . .   85, (1963).
                                (5 , 5211) ACTA CRYST. VOL.B26, ... 2036, (1970).

1039 DUTESCU.........(N.)...: (23 ,   77) STUD. CERCETARI FIZ. VOL.20, . . 1033, (1968).

1040 DUTKIEWICZ......(J.)...: (9B,   32) BU. ACAD. POLON. SCI...TECH. 16, 330, (1968).

1041 DUTTON..........(D.H.).: (6 , 1943) THESIS (68 PP.),. . . . .   , (1970).
                                (6 , 1944) CAN. J. PHYS. VOL.50,. . . . . . 2915, (1972).

1042 DVORAK..........(V.)...: (10B,   48) J. CHEM. PHYSICS VOL.48, ... 4060, (1968).

1043 DWIGHT..........(K.)...: (5 , 452) COLL. INTER. N.126 (GRENOBLE), . 104, (1963).
                                (5 , 517) PROC. INT. CONF. (NOTTINGHAM), . 538, (1964).
                                (5 , 1676) J. APP. PHYS. 36,. . . . . 1088, (1965).
                                (6 , 1441) J. APP. PHYS. 36,. . . . . 1090, (1965).
                                (5 , 1668) J. APP. PHYS. 37,. . . . . 962, (1966).
                                (5 , 1528) J. PHYS. CHEM. SOL. 28,. . . . . 549, (1967).

1044 DWIVEDI.........(A.M.).: (6 , 1912) NUCL./S.S. PHYS. SYMP. ABSTR.,. . . , (1972).

1045 DYCK............(W.)...: (5 , 2672) PHYS. STAT. SOLIDI B VOL.61 NO.1 241, (1974).

1046 DYER............(R.F.).: (22A,   28) J. SCI. INSTRUM. VOL.32, ... 256, (1955).
                                (22C,   24) J. SCI. INSTRUM. VOL.36, ... 419, (1959).
                                (23 ,   50) NATURE VOL.183,. . . . .   35, (1959).
                                (21 ,   7) IAEA SYMPOSIUM VIENNA,. . . . . 179, (1960).
                                (23 ,   81) UKAEA AERE REPORT 5259 (13 PP.), . . , (1966).
                                (19B,   37) UKAEA AERE REPORT 5259 (9PP.),. . . , (1966).
                                (18B,   8) J. SCI. INSTRUM. SER.2 VOL.1,. . 528, (1968).

1047 DYER............(R.H.).: (21 ,   48) REV. SCI. INSTRUM. 41, ...   61, (1970).

1048 DYMOND..........(R.R.).: (17A,   37) IAEA SYMP. (INSTRUMENT.), VIENNA 105, (1969).
                                (5 , 682) THESIS (86 PP.),. . . . . , (1970).

1049 DYNES...........(R.C.).: (6 ,   34) PHYS. LETT. 25A,. . . . . 685, (1967).
                                (10B, 144) PHYS. REV. VOL.187,. . . . . . . 821, (1969).
                                (10C,   16) CONF. INTERN., RENNES,. . . . . 150, (1971).

1050 DZHELEPOV.......(V.P.)..: (22B,   13) INSTRUM. EXP. TECH. VOL.15,. . .   67, (1972).
1051 DZYUB...........(I.P.)..: (10C, 114) SOV. PHYS. SOL. STATE, 6,. . . . 1469, (1964).
                                 (10C, 116) SOV. PHYS. SOL. STATE, 6,. . . . 2955, (1964).
                                 (14B, 136) UKR. FIZ. ZH. (USSR) VOL.9,. . . 684, (1964).
                                 (5 , 2229) SOV. AT. ENERGY VOL.18,. . . . . 585, (1965).
                                 (15 ,   55) SOV. PHYS. JETP. 22, ... 347, (1966).
                                 (6 , 878) UKR. FIZ. ZH. (USSR) VOL.13,. . 1682, (1968).
                                 (10E,   31) UKR. FIZ. ZH.(USSR) VOL.16,. . . 1985, (1971).
                                 (10C, 127) SOV. PHYS. SOLID STATE VOL.14, .   1, (1972).

1052 D≠ABRAMO........(G.)....: (13 ,   53) PHYS. REV. LETT. VOL.28, ...   22, (1972).
1053 D≠AGOSTINO......(O.)....: (16 ,   66) PROC. ROY. SOC. A VOL.149, . . . 522, (1935).
1054 D≠OULTREMONT....(P.)....: (17C,   22) NUCL. INST. MET. 77, ...   1, (1970).
1055 EASTWOOD........(T.A.)..: (22C,   9) CAN. JOURN. PHYS. 41,. . . . . . 1519, (1963).
1056 EAST............(L.V.)..: (20C,   63) NUCL. INST. MET. 72, ... 161, (1969).
1057 EBDON...........(F.R.)..: (21 ,   22) J. APPL. CRYST. VOL.4,. . . . .   254, (1971).
1058 ECKER...........(B.M.)..: (5 ,   95) NUCL. PHYS. VOL.A133,. . . . . . 410, (1969).
1059 ECONOMOU........(E.N.)..: (10B, 214) SOL. STATE COMM. VOL. 9,. . . . 1317, (1971).
                                 (11A,   46) PHYS. REV. LETT. VOL.28, ... 1206, (1972).
1060 EDDINE..........(M.N.)..: (5 , 595) COMP. REND. 269, ...   574, (1969).
1061 EDEN............(R.C.)..: (6 ,   2) DISC. FARADAY SOC. NO.43,. . . . 169, (1967).
1062 EDER............(O.J.)..: (6 ,   81) PHYS. LETT. VOL.19,. . . . . . . 269, (1965).
                                 (6 , 1706) PROC. PHYS. SOC. 89,. . . . . .   833, (1966).
                                 (6 , 300) IAEA SYMP. COPENHAGEN, VOL.2,. . 223, (1968).
                                 (22C,   65) REV. SCI. INSTRUM. 41, ... 1877, (1970).
                                 (14B,   4) ACTA PHYS. AUSTRIACA VOL.37, . . 305, (1973).
                                 (21 ,   26) J. PHYS. E VOL.6 NO.6,. . . . .   576, (1973).
                                 (22B,   36) NUCL. INST. MET. VOL.111,. . . . 375, (1973).
1063 EDGE............(R.D.)..: (24 ,   92) PHYS. REV. 145,. . . . . 1023, (1966).
1064 EDMONDS.........(J.W.)..: (5 , 275) DISC. FARADAY SOC. NO.48,. . . . 192, (1969).
1065 EDWARDS.........(D.M.)..: (11A,   16) J. PHYS. -C-, VOL. 2,. . . . . .   84, (1969).
1066 EDWARDS.........(T.R.)..: (6 , 169) DISS. ABS. VOL. 28,. . . . . . .11078, (1967).
1067 EFIMOV..........(B.V.)..: (20B,   47) INSTRUM. EXP. TECH. NO.3,. . . . 444, (1961).
1068 EGAMI...........(T.)....: (5 , 494) J. PHYS. C VOL.5,. . . . . L261, (1972).
                                 (11A,   22) J. PHYS. C VOL.7,. . . . . 979, (1974).
1069 EGELSTAFF.......(P.A.)..: (6 , 2238) NATURE VOL.168,. . . . .   290, (1951).
                                 (10C,   93) REPORT A.E.R.E. N/R 1164,. . . . . ., (1953).
                                 (20A,   73) REPORT A.E.R.E. N/R 1165,. . . . . ., (1953).
                                 (22A,   24) J. NUCL. ENERGY VOL.1,. . . . .   57, (1954).
                                 (20D,   11) J. SCI. INSTRUM. VOL.31,. . . .   207, (1954).
                                 (5 , 243) J. NUCL. ENERGY VOL.5,. . . . .   203, (1957).
                                 (20B, 118) PHIL. MAG. (8) VOL.2,. . . . . . 917, (1957).
                                 (6 , 1196) PROC. ROY. SOC. A VOL.242, . . . 374, (1957).
                                 (20A,   1) ACTA CRYST. VOL.11,. . . . . . . 228, (1958).
                                 (6 , 1835) NATURE VOL.181,. . . . .   643, (1958).
                                 (7A,   76) PROC. PHYS. SOC. VOL.71, ... 910, (1958).
                                 (14B,   12) BRIT. J. APPL. PHYS. VOL.10, . .   1, (1959).
                                 (21 ,   8) IAEA SYMPOSIUM VIENNA, ... 165, (1960).
                                 (6 , 227) IAEA SYMPOSIUM VIENNA, ... 569, (1960).
                                 (6 , 2244) IAEA SYMPOSIUM VIENNA, ... 581, (1960).
                                 (7B,   13) IAEA SYMPOSIUM VIENNA, ...   24, (1960).
                                 (6 , 1049) IAEA SYMPOSIUM (VIENNA). . . . . 309, (1960).
                                 (20B,   30) IAEA SYMP. (PILE RESEARCH)VIENNA 527, (1960).
                                 (20B,   20) IAEA SYMP. (PILE RESEARCH)VIENNA 549, (1960).
                                 (19A,   12) IAEA SYMP. (PILE RESEARCH)VIENNA 559, (1960).
                                 (24 , 100) REPORT AERE NP/GEN/13,. . . . . . ., (1960).
                                 (14B,   5) ADV. IN PHYSICS VOL.11,. . . . . 203, (1962).
                                 (20B,   34) IAEA SYMP. CHALK RIVER, VOL.1, . 107, (1962).
                                 (6 , 1796) IAEA SYMP. CHALK RIVER, VOL.1, . 203, (1962).
                                 (6 , 1113) IAEA SYMP. CHALK RIVER, VOL.1, . 343, (1962).
                                 (7B,   18) IAEA SYMP. CHALK RIVER, VOL.1, .   65, (1962).
                                 (6 , 1214) IAEA SYMP. CHALK RIVER, VOL.2, . 183, (1962).
                                 (16 ,   35) NUCL. SCI. ENGNG. VOL.12,. . . . 250, (1962).
                                 (17B,   52) NUCL. SCI. ENGNG. VOL.12,. . . . 260, (1962).
                                 (6 , 2237) PHYS. REV. VOL.127,. . . . . . . 1017, (1962).
                                 (14B, 134) UKAEA AERE REP. R4043 (28 PP.),. . ., (1962).
                                 (10C, 131) UKAEA AERE REP. R4101 (24 PP.),. . ., (1962).
                                 (6 , 208) PHYS. LETT. VOL.7,. . . . . . .   229, (1963).
                                 (20A,   72) REACTOR SCI. TECHNOL. VOL.17,. . 147, (1963).
                                 (22B,   7) AUTOMATIC/NUCLEAR DATA,. . . . .   83, (1964).
                                 (24 ,   19) IAEA SYMPOSIUM BOMBAY, VOL.2,. . 553, (1964).
                                 (6 , 969) PHYS. LETT. VOL.12,. . . . . . . 188, (1964).
                                 (14A,   6) BRIT. J. APPL. PHYS. VOL.16,. . 1219, (1965).
                                 (14A,   11) CONTEMP. PHYS. VOL.6,. . . . . . 274, (1965).
                                 (14A,   12) CONTEMP. PHYS. VOL.6,. . . . . . 453, (1965).
                                 (20C,   40) NUCL. INST. MET. 33,. . . . . .   181, (1965).
                                 (14B,   73) PHYS/NON-CRYSTALLINE SOLIDS,. .   127, (1965).
                                 (6 ,   81) PHYS. LETT. VOL.19,. . . . . . . 269, (1965).
                                 (6 , 2075) PHYS. LETT. 16,. . . . .   130, (1965).
                                 (14B, 117) SOLID/LIQUID STATE PHYSICS,. .   103, (1965).
                                 (5 , 749) PHIL. MAG. VOL.14, ... 961, (1966).
                                 (5 , 2929) PHYS. LETT. 21,. . . . .   286, (1966).
                                 (6 , 1547) PHYS. REV. LETT. VOL.17,. . . .   533, (1966).
                                 (6 , 1706) PROC. PHYS. SOC. 89,. . . . . .   833, (1966).
                                 (14A,   67) REP. PROGR. PHYS. VOL.29,. . . . 333, (1966).
                                 (14B,   7) ADV. IN PHYSICS VOL.16,. . . . . 147, (1967).
                                 (5 , 748) ADV. IN PHYSICS VOL.16,. . . . . 171, (1967).
                                 (14A,   14) DISC. FARADAY SOC. NO.43,. . . . 149, (1967).
                                 (6 , 1553) MOL. SIEVES, PAP. CONF. 1967,. . 306, (1967).
                                 (6 , 965) PROC. PHYS. SOC. 90, ... 681, (1967).
                                 (13 ,   61) REPORT AERE-R5627 (15 PP.),. .   . ., (1967).
                                 (14B,   19) C.N.E.N. SYMP. CASSACCIA,. . . .   5, (1968).
                                 (6 , 300) IAEA SYMP. COPENHAGEN, VOL.2,. . 223, (1968).
                                 (5 , 207) J. OF PHYS. C VOL.1,. . . . . . 1075, (1968).
                                 (6 , 1802) J. OF PHYS. C VOL.1,. . . . . .   507, (1968).
                                 (6 , 189) J. OF PHYS. C VOL.1,. . . . . . 1088, (1968).
                                 (6 , 1830) J. OF PHYS. C VOL.1,. . . . . .   519, (1968).
                                 (14B,   45) J. OF PHYS. C VOL.1,. . . . . .   784, (1968).
                                 (23 ,   61) NUCL. INST. MET. VOL.59,. . . . 245, (1968).
                                 (14B, 116) SCI. PROGR. VOL.56,. . . . . . . 223, (1968).
                                 (20B,   39) IAEA SYMP. (INSTRUMENT.), VIENNA 249, (1969).
                                 (6 , 1008) J. CHEM. PHYS. VOL.51,. . . . .   762, (1969).
                                 (15 ,   26) MOL. DYN. AND STRUCT. OF SOLIDS, 135, (1969).
                                 (5 , 2390) PHYS. LETT. VOL.29A,. . . . . .   296, (1969).
                                 (6 , 1521) PHYS. REV. 177,. . . . .   1111, (1969).
                                 (5 , 236) REPORT AERE-R 6052,. . . . ., (1969).
                                 (14A,   15) DISC. FARADAY SOC. NO.49,. . . . 193, (1970).
                                 (15 ,   19) J. CHEM. PHYS. VOL.53,. . . . . 2590, (1970).

1069 EGELSTAFF........(P.A.)...‡   (6 ,   899) J. PHYS. C, SER.2, VOL.3,. . . . . 1673, (1970).
                                   (23 ,   51) NATURE VOL.228,. . . . .   324, (1970).
                                   (6 , 1095) PROC. ROY. SOC. A319, NO.1537, .   189, (1970).
                                   (14A,   25) J. PHYS. C VOL.4,. . . . . 1453, (1971).
                                   (6 ,   939) MOL. PHYS. VOL.20, ...   881, (1971).
                                   (19B,   17) NUCL. INST. MET. 92, ...   51, (1971).
                                   (14A,   40) PHYS. LETT. VOL.37A, ...   321, (1971).
                                   (6 , 1957) CAN. J. PHYS. VOL.50,. . . . . . 3062, (1972).
                                   (14A,   20) IAEA SYMP. GRENOBLE, ...   383, (1972).
                                   (14A,   5) ANN. REV. PHYS. CHEM. VOL.24,. .   159, (1973).
                                   (5 , 1519) CAN. J. PHYS. VOL.51,. . . . . . 1965, (1973).
                                   (15 ,   30) MOL. PHYS. VOL.25, ... 1353, (1973).
                                   (6 , 1819) PROPERTIES LIQUID METALS,. . . .   125, (1973).
                                   (14A,   63) PROPERTIES LIQUID METALS,. . . .   13, (1973).
                                   (5 ,   316) CAN. J. PHYS. VOL.52,. . . . . .   241, (1974).
                                   (15 ,   58) SPECTROSCOPY BIOL. CHEM.,. . . .   269, (1974).

1070 EGERT...........(G.)....‡   (2JB,   53) J. APPL. CRYST. VOL.3, ...   214, (1970).

1071 EGGER...........(H.)....‡   (5 ,   92) PHYS LETT 27A, ...   695, (1968).
                                  (6 ,   97) PHYS. LETT. 28A, ...   433, (1968).
                                  (6 ,   95) PHYS. STAT. SOL. B VOL.43, . . .   611, (1971).

1072 EHRENREICH......(H.)....‡   (5 , 2087) J. APPL. PHYS. 37, ... 1449, (1966).
                                  (5 , 2088) PHYS. REV. VOL.152,. . . . . . .   505, (1966).

1073 EHRET...........(G.)....‡   (6 , 2228) IAEA SYMP. BOMBAY, VOL.1,. . . .   99, (1964).
                                  (6 , 1025) IAEA SYMP. BOMBAY, VOL.2,. . . .   167, (1964).
                                  (19A,   36) NUCL. INST. MET. 49, ...   197, (1967).
                                  (19A,   23) IAEA SYMP. (INSTRUMENT.), VIENNA   147, (1969).

1074 EIBSCHUTZ.......(M.)....‡   (5 ,   172) J. APPL. PHYS. 41, ...   943, (1970).
                                  (5 ,   165) J. PHYSIQUE TOME 32 COL.1 VOL.II   759, (1971).
                                  (5 , 2395) MAGN. AND MAG. MATERIALS-1971, .   436, (1972).
                                  (5 , 2397) MAGN. AND MAG. MATERIALS-1971, .   670, (1972).
                                  (5 ,   159) PHYS. REV. B VOL.6,. . . . . . . 2677, (1972).
                                  (5 ,   662) A.I.P. CONF. PROC. NO.10 PART 1,   684, (1973).

1075 EICHHORN........(F.)....‡   (5 , 2451) PHYS. STAT. SOL. VOL.23, ...   237, (1967).
                                  (5 , 1463) ACTA CRYST. A24, ...   237, (1968).
                                  (6 ,   919) PHYS. STAT. SOL. A VOL.4,. . . .   445, (1971).

1076 EILAND..........(H.M.)..‡   (5 , 2324) PROC/CONF/NEUTRON C. S. + TECH.,   687, (1968).
                                  (5 , 2326) NUCL. SCI. ENG. VOL.52,. . . . .   310, (1973).

1077 EIRIKSSON.......(V.R.)..‡   (5 , 1460) SOL. STATE COMM. VOL.11, ... 1261, (1972).

1078 EISBERG.........(R.M.)..‡   (11C,   33) PHYS. REV. VOL.98, ...   492, (1955).

1079 EISENHAUER......(C.M.)..‡   (6 , 2355) PHYS. REV. VOL.108,. . . . . . . 1091, (1957).
                                  (6 , 2240) PHYS. REV. VOL.109,. . . . . . . 1046, (1958).
                                  (6 ,   937) PHYS. REV. VOL.113,. . . . . . .   49, (1959).

1080 EISENRIEGLER....(E.)....‡   (5 , 2262) ACTA CRYST. (INTERACT.) VOL.A25, S211, (1969).
                                  (10C,   22) CRYSTAL LATTICE DEFECTS VOL.2, .   181, (1971).

1081 EKATOV..........(A.B.)..‡   (22B,   11) INSTRUM. EXP. TECH. NO.4,. . . .   832, (1965).

1082 EKSTEIN.........(H.)....‡   (8 ,   73) PHYS. REV. VOL.76, ... 1328, (1949).
                                  (8 ,   74) PHYS. REV. VOL.78, ...   731, (1950).

1083 ELAZIN..........(YU.P.).‡   (7A,   86) SOV. PHYS. JETP. 16, ... 1531, (1963).

1084 ELBRUS..........(D.T.)..‡   (20B,   137) REV. FAC. SCI/ISTANBUL C VOL.25,   40, (1960).

1085 ELCOMBE.........(M.M.)..‡   (6 , 1842) PROC. ROY. SOC. 293, ...   433, (1966).
                                  (6 , 2053) PROC. PHYS. SOC. 91, ...   946, (1967).
                                  (6 , 1840) PROC. ROY. SOC. 300, ...   210, (1967).
                                  (5 , 2429) ACTA CRYST. VOL.A24, ...   410, (1968).
                                  (6 , 2299) J. OF PHYS.-C-, SER. 2, VOL.2, . 1857, (1969).
                                  (6 ,   438) J. PHYS. C, SER.2, VOL.4,. . . .   492, (1971).
                                  (6 , 2085) J. PHYS. C VOL.5,. . . . . 2702, (1972).

1086 ELEMANS.........(J.B.A.)‡   (5 ,   823) J. PHYSIQUE TOME 32, ...   301, (1971).
                                  (5 , 2148) PHYSICA VOL.57,. . . . .   215, (1972).
                                  (5 , 2635) PHYS. STAT. SOLIDI B VOL.57 NO.2 K155, (1973).

1087 ELK.............(K.)....‡   (5 ,   952) ACTA PHYS. AUSTRALIA VOL.29, . .   342, (1969).

1088 ELLIOTT.........(N.)....‡   (5 , 1799) PHYS. REV. VOL.104,. . . . . . .   924, (1956).
                                  (5 , 1830) J. APPL. PHYS. VOL.29, ...   391, (1958).
                                  (9D,   59) PHYS. REV. VOL.112,. . . . . . . 1917, (1958).
                                  (5 , 1831) PHYS. REV. VOL.115,. . . . . . .   13, (1959).
                                  (5 ,   596) PHYS. REV. VOL.117,. . . . . . .   929, (1960).
                                  (5 , 1422) ACTA CRYST. VOL.14,. . . . . . . 1018, (1961).
                                  (5 , 1442) ACTA CRYST. VOL.14,. . . . . . .   19, (1961).
                                  (5 ,   624) PHYS. REV. VOL.122,. . . . . . . 1402, (1961).

1089 ELLIOTT.........(R.J.)..‡   (8 ,   115) PROC. ROY. SOC. A VOL.230, . . .   46, (1955).
                                  (8 ,   116) PROC. ROY. SOC. A VOL.235, . . .   289, (1956).
                                  (8 ,   120) REV. MOD. PHYS. VOL.30,. . . . .   75, (1958).
                                  (10C,   31) IAEA SYMPOSIUM VIENNA,. . . . .   61, (1960).
                                  (14B,   95) PROC. PHYS. SOC. VOL.77,. . . . .   353, (1961).
                                  (10C,   35) IAEA SYMPOSIUM BOMBAY, VOL.1,. .   231, (1964).
                                  (10E,   3) BROOKHAVEN SYMPOSIUM,. . . . . .   23, (1965).
                                  (10A,   16) SCOTTISH UNIV. SUMMER SCHOOL,. .   377, (1965).
                                  (6 ,   955) PROC. PHYS. SOC. 90, ...   671, (1967).
                                  (7B,   99) PROC. PHYS. SOC. 91, ...   903, (1967).
                                  (5 , 1838) PROC. ROY. SOC. 296, ...   161, (1967).
                                  (11A,   25) J. PHYSIQUE TOME 32 COL.1 VOL.II   585, (1971).
                                  (6 , 1306) J. PHYS. C VOL.5,. . . . . 2611, (1972).
                                  (6 , 1514) J. PHYS. C VOL.6,. . . . . 1933, (1973).

1090 ELLIOTT.........(R.)....‡   (10A,   6) 2ND SUMMER SCHOOL S.S. PHYS.,. .   241, (1968).

1091 ELLISON.........(R.D.)..‡   (5 , 1447) ACTA CRYST. 19,. . . . .   260, (1965).
                                  (5 , 1997) ACTA CRYST. B24, ...   230, (1968).
                                  (5 ,   7) ACTA CRYST. VOL.B27, ...   333, (1971).

1092 ELLIS...........(D.E.)..‡   (9D,   35) J. APPL. PHYS. 41, ...   937, (1970).

1093 ELLIS...........(P.J.)..‡   (5 , 2378) APPL. PHYS. LETT. VOL.10,. . . .   293, (1967).
                                  (22B,   15) J. APPL. CRYST. VOL.1, ...   272, (1968).

1094 ELMALEH.........(D.)....‡   (5 ,   834) J. PHYSIQUE TOME 32 COL.1 VOL.II   741, (1971).

1095 ELSASSER........(W.M.)..‡   (9C,   12) COMP. REND. TOME 202,. . . . . . 1029, (1936).

1096 ELWYN...........(A.J.)..‡   (5 ,   925) NUCL. PHYS. VOL.A123,. . . . . .   33, (1969).

1097 EL-ELA..........(M.A.)..‡   (5 , 2933) J. NUCL. ENERGY VOL.22,. . . . .   389, (1968).

1098 EL-KHOSHT.......(M.)....‡   (5 , 2965) ATOMKERNENERGIE VOL.22,. . . . .   87, (1973).

```
1099  EL-MINIAWY......(S.)....:  ( 5 , 2965) ATOMKERNENERGIE VOL.22,. . . . .    87, (1973).

1100  EL-SAFFAR.......(Z.M.)..:  ( 5 , 1662) ACTA CRYST. VOL.B27,. . . . . .    66, (1971).

1101  EL-VAKIL........(S.A.)..:  (24 ,  109) SOV. AT. ENERGY VOL.25,. . . . .   435, (1968).

1102  EL-WAKIL........(S.A.)..:  (24 ,    8) ATOMKERNENERGIE VOL.18,. . . . .   215, (1971).

1103  EMMONS..........(A.H.)..:  ( 5 ,  113) NUCL. APPL. TECHNOL. VOL.9,. . .   662, (1970).

1104  ENAYETULLAH.....(M.)....:  (22C,   21) J. NUCL. ENERGY VOL.24,. . . . .   385, (1970).

1105  ENDERBY.........(J.E.)..:  (14A,    2) ADV. IN PHYSICS VOL.14,. . . . .   453, (1965).
                                 ( 5 ,  749) PHIL. MAG. VOL.14,. . . . . . .   961, (1966).
                                 ( 5 , 2929) PHYS. LETT. 21,. . . . . . . .   286, (1966).
                                 ( 5 ,  748) ADV. IN PHYSICS VOL.16,. . . . .   171, (1967).
                                 ( 5 ,  207) J. OF PHYS. C VOL.1,. . . . . . 1075, (1968).
                                 (14B,   45) J. OF PHYS. C VOL.1,. . . . . .   784, (1968).
                                 ( 5 , 2391) PHYS. LETT. VOL.29A,. . . . . .   296, (1969).
                                 (14A,    3) ADV. STRUC. RES. DIFFR. METHOD 4  65, (1972).
                                 (14A,    9) CHEM. PHYS. LETT. VOL.21 NO.1,.   109, (1973).
                                 (14A,   64) PROPERTIES LIQUID METALS,. . . .    3, (1973).
                                 ( 5 , 2095) J. PHYS. C VOL.7,. . . . . . . . L111, (1974).

1106  ENDOH...........(Y.)....:  ( 5 ,  993) SOL. STATE COMM. 4,. . . . . .   657, (1966).
                                 ( 5 ,  575) J. PHYS. SOC. JAPAN VOL.22,. . . 1221, (1967).
                                 ( 5 , 1037) J. PHYS. SOC. JAPAN VOL.23,. . .  205, (1967).
                                 ( 5 , 1040) J. APP. PHYS. VOL.39,. . . . . . 1318, (1968).
                                 ( 5 ,  568) J. PHYS. SOC. JAPAN VOL.24,. . .  263, (1968).
                                 (22C,    1) ACTA CRYST. VOL. A25,. . . . . . S-65, (1969).
                                 ( 5 , 1834) SOL. STATE COMM. VOL.8,. . . . .   87, (1970).
                                 ( 6 ,  822) J. PHYSIQUE TOME 32 COL.1 VOL.II 1017, (1971).
                                 ( 5 , 1042) J. PHYS. SOC. JAPAN VOL.30,. . . 1614, (1971).
                                 ( 6 ,  112) SOL. STATE COMM. VOL.9,. . . . .  921, (1971).
                                 ( 6 , 1492) ACTA CRYST. VOL.A28 PART S-4,. .  208, (1972).
                                 ( 5 , 1498) SOL. STATE COMM. VOL.11,. . . . .  391, (1972).
                                 ( 5 ,  979) A.I.P. CONF. PROC. NO.10 PART 1,  98, (1973).
                                 ( 6 , 1345) BULL. AMER. PHYS. SOC. VOL.18,.   111, (1973).
                                 ( 6 ,  825) J. PHYS. SOC. JAPAN VOL.35,. . . 1616, (1973).
                                 ( 6 , 1347) PHYS. REV. B VOL.7,. . . . . . . 4670, (1973).
                                 ( 6 ,  826) SOL. STATE COMM. VOL.13,. . . . . 1179, (1974).
                                 ( 6 , 1350) PHYS. REV. B VOL.9,. . . . . . . 1797, (1974).
                                 ( 6 ,  625) PHYS. REV. LETT. VOL.32 NO.4,. .  170, (1974).

1107  ENGEBRETSEN.....(J.E.)..:  ( 5 ,  561) ACTA CHEM. SCAND. VOL.25,. . . . 1703, (1971).
                                 ( 5 , 1656) ACTA CHEM. SCAND. VOL.26,. . . .  175, (1972).
                                 ( 5 , 1655) ACTA CRYST. VOL.A28 PART S-4,. .  196, (1972).

1108  ENGELBRECHT.....(C.A.)..:  ( 7B,   41) NUCL. INST. MET. VOL.80,. . . .  187, (1970).

1109  ENGELSMAN.......(F.M.R.):  ( 5 , 1432) J. SOLID STATE CHEM. VOL.6,. . .  384, (1973).
                                 ( 5 , 1966) J. SOLID STATE CHEM. VOL.6,. . .  574, (1973).

1110  ENGLANDER.......(M.)....:  ( 5 , 2714) J. NUCL. MATERIALS VOL.2,. . . .   69, (1960).
                                 ( 5 , 2710) BULL. INFORM. SCI. TECH. NO.49,.   69, (1961).

1111  ENGLEHART.......(R.W.)..:  ( 9B,   63) NATURE VOL.229,. . . . . . . . .  111, (1971).

1112  ENGLE...........(W.W.)..:  (24 ,   69) NUCL. INST. MET. 72,. . . . . .  213, (1969).

1113  ENTIN...........(I.R.)..:  ( 5 , 2837) SOV. PHYS. SOL. STATE VOL. 13,. 2172, (1972).
                                 ( 5 , 2835) SOV. PHYS. SOL. STATE VOL. 13,. 2178, (1972).

1114  ENZ.............(C.P.)..:  (10D,   41) PHYS. REV. A VOL.5,. . . . . . . 1528, (1972).
                                 ( 6 , 2097) PHYS. REV. B VOL.6,. . . . . . . 4695, (1972).

1115  EPSTEIN.........(A.)....:  ( 5 , 1987) PHYS. REV. 174,. . . . . . . . .  560, (1968).
                                 ( 5 , 2041) PHYS. LETT. 29A,. . . . . . . .  659, (1969).
                                 ( 5 , 2403) PHYS. REV. B VOL.2,. . . . . . . 3703, (1970).
                                 ( 5 , 2401) PHYS. REV. LETT. VOL.25,. . . . 1713, (1970).

1116  EPSTEIN.........(S.T.)..:  ( 7A,   68) PHYS. REV. 129,. . . . . . . . . 1396, (1963).

1117  EREMEEV.........(I.P.)..:  ( 6 ,  456) SOV. PHYS. JETP LETT. VOL.18,. .  177, (1973).
                                 ( 6 , 1408) SOV. PHYS. SOLID STATE VOL.15,. 1309, (1974).

1118  ERGUN...........(S.)....:  (22B,   52) REV. SCI. INSTRUM. VOL.40,. . . 1144, (1969).

1119  ERICKSON........(J.O.)..:  (10C,   25) DISS. ABS. VOL.27,. . . . . . . 3224B, (1966).

1120  ERICKSON........(R.A.)..:  ( 5 , 2124) PHYS. REV. VOL.90,. . . . . . .  779, (1953).
                                 ( 5 ,  768) PHYS. REV. LETT. 15,. . . . . .  586, (1965).
                                 ( 5 ,  770) PHYS. REV. LETT. 16,. . . . . .  799, (1966).
                                 ( 5 ,  761) J. CHEM. PHYSICS VOL.48,. . . . 1273, (1968).
                                 ( 5 ,  760) J. CHEM. PHYSICS VOL.49,. . . . 1922, (1968).
                                 ( 5 ,  485) J. PHYS. CHEM. SOLIDS 29,. . . . 2087, (1968).
                                 ( 5 ,  423) PHYS. REV. B-1,. . . . . . . . . 2243, (1970).
                                 (22C,   64) REV. SCI. INSTRUM. 41,. . . . .  107, (1970).

1121  ERICSON.........(M.)....:  ( 6 ,  786) C.R. ACAD. SCI. VOL.246,. . . . 1018, (1958).
                                 ( 6 ,  978) J. PHYS. RADIUM VOL.19,. . . . .  617, (1958).
                                 ( 6 , 1468) C.R. ACAD. SCI. VOL.248,. . . . 1631, (1959).
                                 ( 6 ,  789) J. PHYS. RADIUM VOL.20,. . . . .  178, (1959).
                                 (13 ,   23) J. PHYS. CHEM. SOLIOS VOL.13,. .  235, (1960).

1122  ERICSSON........(T.)....:  ( 5 , 1105) J. PHYS. C VOL.7,. . . . . . . . 1344, (1974).

1123  ERLANOV.........(M.V.)..:  ( 9B,   33) BU. ACAD. SCI. USSR PHYS. SER.35 2406, (1971).

1124  ERMAKOV.........(O.N.)..:  (12B,   60) SOV. PHYS. CRYST. VOL.11,. . . .  597, (1967).

1125  ERMER...........(O.)....:  ( 5 ,  295) ACTA CRYST. VOL.B29,. . . . . . 2278, (1973).

1126  ERMOLOV.........(P.F.)..:  (22B,   13) INSTRUM. EXP. TECH. VOL.15,. . .   67, (1972).

1127  ERNST...........(G.)....:  ( 6 , 1647) IAEA SYMP. COPENHAGEN, VOL.1,. .  367, (1968).
                                 ( 6 , 1608) ACTA PHYS. AUSTRIACA VOL.33,. .    27, (1971).
                                 (18A,   20) NUCL. INST. MET. VOL.107,. . . .  461, (1973).
                                 (17A,   92) REV. SCI. INSTRUM. VOL.45,. . . .  526, (1974).

1128  ERNST...........(M.)....:  ( 5 ,  701) PHYS. STAT. SOL. A VOL.7,. . . .  469, (1971).
                                 ( 5 ,  700) PHYS. STAT. SOL. A VOL.7,. . . .  477, (1971).

1129  ERRINGTON.......(P.R.)..:  (23 ,   78) THESIS (106 PP.),. . . . . . . . .   , (1969).

1130  ESCHRIG.........(H.)....:  ( 6 , 1431) IAEA SYMP. GRENOBLE,. . . . . .   157, (1972).
                                 ( 6 , 1430) PHYS. STAT. SOL. VOL.49,. . . . .  807, (1972).
                                 (10C,   85) PHYS. STAT. SOLIDI B VOL.58 NO.2 K159, (1973).

1131  ESCH............(L.J.)..:  ( 6 , 1890) MOL. DYN. AND STRUCT. OF SOLIDS,  547, (1969).
                                 ( 6 , 1038) NUCL. SCI. ENG. VOL.46,. . . . .  223, (1971).

1132  ESPINOSA........(G.P.)..:  ( 5 , 1758) J. APP. PHYS. VOL.40,. . . . . . 1136, (1969).
```

```
1133  ESSAM...........(J.W.)..!  ( 8 ,   31) J. PHYS. C VOL.2,. . . . . . . .    317, (1969).

1134  EUGENIO.........(M.R.)..!  ( 5 , 2503) DISS. ABS. 24, . . . . . . . .   5316, (1964).

1135  EVANS...........(A.E.)..!  ( 5 , 2361) PHYS. REV. VOL.187,. . . . . . .  1506, (1969).

1136  EVANS...........(J.E.)..!  ( 5 ,  195) IAEA SYMPOSIUM VIENNA . . . . . .  277, (1960).
                                 (23 ,   27) IAEA SYMP. (PILE RESEARCH)VIENNA   93, (1960).
                                 ( 6 , 2353) NUCL. SCI. ENGNG. VOL.5,. . . . .  99, (1960).
                                 (23 ,   53) NUCL. INST. MET. VOL.12,. . . .    75, (1961).
                                 (19B,   28) RENSSELAER POL. INST. SYMP., . .   35, (1961).

1137  EVANS...........(M.T.)..!  ( 6 , 1990) J. PHYSIQUE TOME 32 COL.1 VOL.II  614, (1971).
                                 ( 6 , 1989) IAEA SYMP. GRENOBLE, . . . . . .  639, (1972).
                                 ( 6 ,  839) IAEA SYMP. GRENOBLE, . . . . . .  649, (1972).
                                 ( 6 , 1994) J. PHYS. C VOL.6,. . . . . . . .  495, (1973).
                                 ( 6 ,  849) SOL. STATE COMM. VOL.12 NO.8,. .  795, (1973).

1138  EVENSON.........(W.E.)..!  ( 9D,   61) PHYS. REV. 178,. . . . . . . . .  783, (1969).

1139  EWALD...........(R.)....!  (23 ,   13) ENERGIE NUCLEAIRE VOL.13,. . . .   15, (1971).

1140  FABIANI.........(F.)....!  (17A,   82) NUOVO CIMENTO SUPPL. VOL.3,. . .  187, (1965).

1141  FADEEVA.........(N.A.)..!  ( 5 , 2517) J. PHYSIQUE TOME 32 COL.1 VOL.I,  503, (1971).

1142  FADEEVA.........(N.V.)..!  ( 5 , 2494) 2ND INTERNAT. MEET./FERROELECTR.  219, (1969).
                                 ( 5 , 1959) SOV. PHYS. JETP LETT. VOL.16,. .  198, (1972).
                                 ( 5 , 2856) SOV. PHYS. CRYST. VOL.17,. . . . 1017, (1973).

1143  FADIN...........(V.P.)..!  ( 5 , 2164) SOV. PHYS. DOKLADY VOL.14, . . . 1119, (1970).

1144  FAEK............(M.K.)..!  ( 5 , 2323) SOV. PHYS. CRYST. VOL.14, 3, . .  447, (1969).
                                 ( 5 , 2192) SOV. PHYS. CRYST. VOL.16,. . . .  634, (1971).
                                 ( 5 , 2892) SOV. PHYS. SOL. STATE VOL.14,. . 3037, (1973).

1145  FAGOT...........(J.)....!  (23 ,   31) IAEA SYMP. COPENHAGEN, VOL.2,. .  353, (1968).

1146  FAIRAND.........(B.P.)..!  ( 6 ,  110) DISS. ABS. VOL.31, . . . . . . . 8678, (1970).

1147  FAIRCHILD.......(R.)....!  (22B,   35) NUCL. INST. MET. VOL.91, . . . .  541, (1971).

1148  FALKSTROM.......(E.)....!  (24 ,   59) NUCL. INST. MET. 66, . . . . . .  229, (1968).

1149  FANG............(J.H.)..!  ( 5 ,  511) ACTA CRYST. 17,. . . . . . . . .  240, (1964).
                                 ( 5 ,  508) J. PHYS. CHEM. SOL. 25,. . . . .  901, (1964).
                                 ( 5 , 2196) ACTA CRYST. 19,. . . . . . . . .  147, (1965).

1150  FARAGO..........(P.S.)..!  (22C,   28) NUCL. INST. MET. 30,. . . . . .   271, (1964).

1151  FARFALETTI-CASAL(F.)....!  (21D,   32) REV. SCI. INSTRUM. VOL.33, . . . 1103, (1962).

1152  FARNOUX.........(B.)....!  ( 6 ,  436) IAEA SYMP. CHALK RIVER, VOL.2, .  225, (1962).
                                 ( 5 ,  336) PHYS. LETT. 1, . . . . . . . . .  187, (1962).
                                 ( 5 , 2000) PHYS. LETT. 9,. . . . . . . . .   106, (1964).
                                 ( 5 , 1999) PROC. INT. CONF. (NOTTINGHAM), .  285, (1964).
                                 ( 8 ,  117) PROG. LOW TEMP. PHYS. VOL.5,. .   161, (1967).
                                 (23 ,   31) IAEA SYMP. COPENHAGEN, VOL.2,. .  353, (1968).
                                 ( 5 , 2329) J. CHEM. PHYS. VOL.57, . . . . .  290, (1972).
                                 ( 5 , 2330) J. APPL. CRYST. VOL.7, . . . . .  189, (1974).
                                 ( 6 , 1922) PHYS. REV. LETT. VOL.32, . . . . 1170, (1974).

1153  FARRELL.........(J.)....!  ( 5 , 2553) J. APP. PHYS. 36,. . . . . . . .  978, (1965).

1154  FARROW..........(R.F.C.)!  ( 6 ,  662) DISC. FARADAY SOC. NO.48,. . . .   78, (1969).

1155  FARZTDINOV......(M.M.)..!  ( 9A,   26) SOV. PHYS. USPEKHI VOL.7,. . . .  855, (1965).
                                 (11A,   34) PHYS. MET. METALLOGR. VOL.30, 5, 1064, (1970).
                                 (11A,   66) ZH. EKSPER. TEOR. FIZ.(USSR) 58,  918, (1970).

1156  FAYEK...........(M.K.)..!  ( 5 , 2950) PHYS. STAT. SOL. VOL.37, . . . .  843, (1970).
                                 ( 5 , 2946) PHYS. STAT. SOLIDI A VOL.4 NO.1,   53, (1971).
                                 ( 5 , 2951) PHYS. STAT. SOLIDI A VOL.7 NO.2,  K71, (1971).

1157  FEDERIGHI.......(F.D.)..!  ( 6 , 1028) IAEA SYMP. CHALK RIVER, VOL.1, .  389, (1962).

1158  FEDER...........(J.)....!  ( 6 ,   83) PHYS. REV. LETT. VOL.25, . . . . 1423, (1970).
                                 ( 6 , 2091) SOL. STATE COMM. VOL. 9, . . . . 1103, (1971).
                                 ( 6 , 2090) SOL. STATE COMM. VOL. 9, . . . . 1455, (1971).
                                 (13 ,   64) SOL. STATE COMM. VOL. 9, . . . . 2021, (1971).

1159  FEDOROVICH......(G.V.)..!  ( 7B,  133) SOV. PHYS. JETP. 26, . . . . . . 1139, (1968).

1160  FEDOROV.........(M.B.)..!  ( 7A,   97) UKR. FIZ. ZH. (USSR) VOL.16, . .  280, (1971).

1161  FEDRO...........(A.J.)..!  ( 5 ,  132) AIP CONF. PROC. VOL.5, . . . . . 1390, (1971).

1162  FEENBERG........(E.)....!  (14A,   50) PHYS. REV. A VOL.2,. . . . . . . 2158, (1970).

1163  FEIGL...........(B.)....!  (20C,   45) NUCL. INST. MET. 46, . . . . . .  153, (1966).
                                 (24 ,   48) NUCL. INST. MET. 61, . . . . . .  349, (1968).

1164  FEIKEMA.........(Y.D.)..!  ( 5 , 1280) ACTA CRYST. 20,. . . . . . . . .  765, (1966).

1165  FEIL............(D.)....!  ( 5 , 1923) ACTA CRYST. VOL.A25, . . . . . .  438, (1969).

1166  FEINLEIB........(J.)....!  ( 5 , 1668) J. APP. PHYS. 37,. . . . . . . .  962, (1966).

1167  FEINSTEIN.......(D.L.)..!  ( 6 , 1612) PHYS. REV. 158,. . . . . . . . .   97, (1967).

1168  FEINSTEIN.......(L.G.)..!  ( 5 , 2842) J. PHYS. CHEM. SOLIDS 29,. . . .  184, (1968).

1169  FEIN............(A.E.)..!  (1UE,   24) PHYS. REV. VOL.128,. . . . . . . 2589, (1962).

1170  FELCHER.........(G.P.)..!  (10C,   26) ENERGIA NUCLEARE VOL.7,. . . . .  772, (1960).
                                 ( 5 ,  951) COLL. INTER. N.126 (GRENOBLE), .  183, (1963).
                                 ( 5 , 2578) COLL. INTER. N.126 (GRENOBLE), .  190, (1963).
                                 ( 5 ,  725) J. PHYS. CHEM. SOL. 24,. . . . . 1663, (1963).
                                 ( 5 , 2577) PHYS. REV. 131,. . . . . . . . . 1518, (1963).
                                 ( 5 , 2294) PROC. INT. CONF. (NOTTINGHAM), .  288, (1964).
                                 ( 5 , 2691) J. APP. PHYS. 36,. . . . . . . . 1001, (1965).
                                 ( 5 , 1765) J. APP. PHYS. 37,. . . . . . . . 1056, (1966).
                                 ( 5 , 1619) PHYS. LETT 28A,. . . . . . . . .  267, (1968).
                                 ( 5 , 1641) J. APPL. PHYS. 41, . . . . . . .  939, (1970).
                                 ( 5 , 1617) PHYS. REV. B-2, . . . . . . . .   670, (1970).
                                 ( 5 , 2561) AIP CONF. PROC. VOL.5, . . . . . 1376, (1971).
                                 ( 5 , 1632) J. APPL. PHYS. 42, . . . . . . . 1621, (1971).
                                 ( 5 , 1764) J. PHYSIQUE TOME 32 COL.1 VOL.I,  577, (1971).
                                 ( 5 , 1104) PHYS. REV. B-3, . . . . . . . . 3046, (1971).
                                 ( 5 ,  838) PHYS. REV. LETT. VOL.28, . . . .  746, (1972).
                                 ( 5 ,  840) PHYS. REV. B VOL.8, . . . . . .   260, (1973).
                                 ( 9D,   69) SOL. STATE COMM. VOL.12, . . . . 1167, (1973).
                                 ( 5 , 2338) PHYS. REV. B VOL.9,. . . . . . .  248, (1974).
```

1171 FELDKAMP.........(L.A.)..:   (6 , 1880) IAEA SYMP. COPENHAGEN, VOL.2,. .   159, (1968).
                                   (6 , 1911) J. CHEM. PHYSICS VOL.48, ...   912, (1968).
                                   (6 , 1883) DISS. ABS. VOL.30, ... :23698, (1969).
                                   (6 , 1891) MOL. DYN. AND STRUCT. OF SOLIDS,   543, (1969).
                                   (6 , 2327) SOL. STATE COMM. 7 NO.21   1571, (1969).
                                   (8 , 2324) J. PHYS. CHEM. SOLIDS VOL.32,: :   1573, (1971).

1172 FELDMANN.........(G.)....:   (6 , 1600) IAEA SYMP. BOMBAY, VOL.2,. . . .   85, (1964).

1173 FELDMANN.........(K.)....:   (6 , 1936) PHYS. STAT. SOL.45,. . . . . . .   K105, (1971).
                                   (6 , 1431) IAEA SYMP. GRENOBLE, ...   157, (1972).
                                   (10C,   85) PHYS. STAT. SOLIDI B VOL.58 NO.2 K159, (1973).

1174 FELDMANN.........(W.L.)..:   (6 , 2192) PHYS. REV. 178,. . . . .   897, (1969).
                                   (6 , 2062) PHYS. REV. B VOL.3,. . . . . . .   4065, (1971).

1175 FELICI...........(M.)....:   (6 , 1508) PHYS. LETT. 30A, ...   310, (1969).

1176 FENDER..........(B.E.F.):   (5 , 2832) ACTA CRYST. VOL.B26, ...   1882, (1970).
                                   (5 , 2896) J. PHYS. CHEM. SOLIDS VOL.31,. .   2741, (1970).
                                   (5 , 378) SOL. STATE COMM. VOL.8,. . . . .   171, (1970).
                                   (5 ,   85) J. PHYS. C VOL.4,. . . . .   1279, (1971).
                                   (5 , 1079) J. PHYS. C VOL.4,. . . . .   2160, (1971).
                                   (5 , 338) J. PHYS. C VOL.4,. . . . .   3107, (1971).
                                   (5 , 2766) ACTA CRYST. VOL.B28, ...   3609, (1972).
                                   (5 , 156) ACTA CRYST. VOL.B28, ...   956, (1972).
                                   (6 , 441) J. PHYS. C VOL.5,. . . . .   2677, (1972).
                                   (5 , 175) J. PHYS. C VOL.5,. . . . .   2887, (1972).
                                   (5 , 405) J. PHYS. C VOL.5,. . . . .   L35, (1972).
                                   (9C,   9) CHEM. APPL. THERMAL NEUTRON SCAT   250, (1973).
                                   (5 , 1602) J. PHYS. C VOL.6,. . . . .   L333, (1973).
                                   (5 , 1751) J. PHYS. C VOL.6,. . . . .   1615, (1973).
                                   (5 , 2518) ACTA CRYST. VOL.B30, ...   809, (1974).
                                   (5 , 190) ACTA CRYST. VOL.B30, ...   816, (1974).
                                   (5 , 1006) J. PHYS. C VOL.7,. . . . .   783, (1974).
                                   (5 , 2979) J. PHYS. C VOL.7,. . . . .   1, (1974).

1177 FERER...........(M.)....:   (13C,   50) PHYS. REV. B VOL.4,. . . . . . .   3954, (1971).
                                   (11C,   42) PHYS. REV. B VOL.4,. . . . . . .   3964, (1971).

1178 FERGUSON-JR.....(G.A.)..:   (5 , 1088) PHYS. REV. VOL.112,. . . . . . .   1130, (1958).

1179 FERGUSON........(A.T.G.):   (20C,   39) NUCL. INST. MET. 34, ...   21, (1965).
                                   (24 ,   42) NUCL. INST. MET. 34, ...   29, (1965).

1180 FERGUSON........(G.A.)..:   (6 , 854) J. PHYS. CHEM. SOLIDS VOL.23,. .   117, (1962).
                                   (20A,   50) NUCL. SCI. ENGNG. VOL.14,. . . .   397, (1962).
                                   (5 , 2308) DISS. ABSTR. VOL.25, ...   3368, (1964).
                                   (5 , 2302) PHYS. REV. 137,. . . . .   A483, (1965).
                                   (6 , 2288) J. APP. PHYS. 37,. . . . .   1050, (1966).
                                   (6 , 2289) PHYS. REV. 156,. . . . .   632, (1967).
                                   (5 , 1234) J. AMER. CERAMIC SOC. VOL.53,. .   109, (1970).
                                   (6 , 1862) PHYS. REV. LETT. VOL.29,. . . .   1250, (1972).
                                   (6 , 1867) PHYS. REV. LETT. VOL.32, ...   1087, (1974).

1181 FERGUSON........(L.N.)..:   (11A,   43) PHYS. REV. B VOL.3,. . . . . . .   1025, (1971).

1182 FERMI...........(E.)....:   (16 ,   56) NUOVO CIMENTO VOL.12 SER.8,. .   201, (1935).
                                   (16 ,   66) PROC. ROY. SOC. A VOL.149, . . .   522, (1935).
                                   (7A,   58) PHYS. REV. VOL.50, ...   899, (1936).
                                   (7B, 111) RICERCA SCI. VOL.7, II,. . . . .   13, (1936).
                                   (16 ,   80) RICERCA SCI. VOL.7, I,. . . . .   393, (1936).
                                   (20D,   28) PHYS. REV. VOL.70, ...   815, (1946).
                                   (5 , 201) PHYS. REV. VOL.71, ...   589, (1947).
                                   (7B,   57) PHYS. REV. VOL.71, ...   666, (1947).
                                   (8 ,   70) PHYS. REV. VOL.72, ...   1139, (1947).
                                   (20A,   61) PHYS. REV. VOL.72, ...   193, (1947).
                                   (5 , 213) PHYS. REV. VOL.72, ...   408, (1947).
                                   (5 , 771) PHYS. REV. VOL.75, ...   578, (1949).

1183 FERNANDEZ.......(J.F.)..:   (6 , 1150) PHYS. LETT. 30A, ...   261, (1969).

1184 FERRARIS........(G.)....:   (5 , 352) ACTA CRYST. VOL.B27, ...   349, (1971).
                                   (5 , 1970) ACTA CRYST. VOL.B27, ...   354, (1971).
                                   (5 , 278) ACTA CRYST. VOL.A28 PART S-4,.   27, (1972).
                                   (5 , 351) ACTA CRYST. VOL.B28, ...   2430, (1972).
                                   (5 , 1310) ACTA CRYST. VOL.B28, ...   3572, (1972).
                                   (5 , 350) ACTA CRYST. VOL.B28, ...   209, (1972).
                                   (5 , 1951) Z. KRIST. VOL.135, ...   240, (1972).

1185 FERRARO.........(J.R.)..:   (6 , 1004) J. CHEM. PHYS. VOL.44, ...   2496, (1966).
                                   (6 , 2223) INORG. CHEM. VOL.6,. . . . . . .   346, (1967).

1186 FERREIRA........(L.G.)..:   (10B,   64) J. PHYS. C: SOLID ST. PHYS. V-4,   20, (1971).
                                   (10B,   65) J. PHYS. C: SOLID ST. PHYS. V-4,   3, (1971).

1187 FERRIER.........(R.P.)..:   (6 , 2234) PHYS. REV. LETT. 10, ...   295, (1963).

1188 FERTEL..........(G.E.F.):   (20A,   30) NATURE VOL.142,. . . . .   829, (1938).
                                   (20B, 132) PROC. ROY. SOC. A VOL.175, . . .   316, (1940).

1189 FERT............(A.R.)..:   (5 , 1029) SOL. STATE COMM. VOL.14 NO.2,.   187, (1974).

1190 FERZIGER........(J.H.)..:   (9B,   82) PHYS. REV. VOL.128,. . . . . . .   2188, (1962).
                                   (15 ,   40) PHYS. REV. VOL.138A, ...   791, (1965).
                                   (6 , 1612) PHYS. REV. 158,. . . . .   97, (1967).

1191 FETISOV.........(N.I.)..:   (6 , 2169) BU. ACAD. SCI. USSR PHYS. SR. 32 600, (1968).

1192 FIALA...........(W.)....:   (22C,   36) NUCL. INST. MET. 52, ...   15, (1967).
                                   (22A,   51) NUCL. TECHNOL. VOL.10,. . . . .   215, (1971).
                                   (20B, 109) NUCL. INST. MET. VOL.107,. . . .   33, (1973).
                                   (20B, 108) NUCL. INST. MET. VOL.107,. . . .   37, (1973).

1193 FIEDELDEY.......(H.)....:   (7A,   10) ANN. OF PHYS. VOL.16,. . . . . .   387, (1961).

1194 FIELD...........(G.R.)..:   (10B, 205) PROC. ROY. SOC. A VOL.317, . . .   55, (1970).

1195 FIH≠CHENKOV.....(V.V.)..:   (22B,   13) INSTRUM. EXP. TECH. VOL.15,. .   67, (1972).

1196 FILEVICH........(A.)....:   (20A,   49) NUCL. INST. MET. VOL.116,. . . .   615, (1974).

1197 FILIPPI.........(G.)....:   (23 ,   6) BULL. INF. SCI. TECH. NO.170,. .   61, (1972).

1198 FILIPPI.........(J.)....:   (5 , 994) J. PHYS. CHEM. SOL. 26,. . . . .   1727, (1965).

1199 FILLION.........(G.)....:   (5 , 2382) PHYS. STAT. SOLIDI A VOL.15 NO.2 613, (1973).

1200 FILYUSHKIN......(I.V.)..:   (20C,   21) INSTRUM. EXP. TECH. NO.1,. . . .   77, (1970).

1201 FINCH...........(E.D.)..:   (5 , 796) J. CHEM. PHYSICS VOL.49, ...   2514, (1968).
                                   (5 , 783) J. CHEM. PHYSICS VOL.49, ...   4361, (1968).
                                   (5 , 782) J. CHEM. PHYSICS VOL.49, ...   4365, (1968).
                                   (5 , 792) PHYS. ICE, ...   59, (1969).

1202 FINEGOLD........(L.)....‡   (22C,   63) REV. SCI. INSTRUM. 40, ...   1397, (1969).
                                   (6 , 2334) PHYS. STAT. SOL. VOL.42, ...   275, (1970).
                                   (6 ,   137) PHYS. REV. B VOL.4, ...   1390, (1971).

1203 FINGERLAND......(A.)....‡   (17B,   26) ACTA CRYST. VOL.A27, ...   280, (1971).

1204 FINHOLT.........(J.E.)..‡   (5 ,   797) J. CHEM. PHYS. VOL.59, ...   5114, (1973).

1205 FINKELSTEIN.....(R.J.)..‡   (10C,   66) PHYS. REV. VOL.72, ...   907, (1947).

1206 FINKEL..........(V.A.)..‡   (5 , 2609) SOV. PHYS. J.E.T.P. 26, ...   1086, (1968).

1207 FINK............(G.A.)..‡   (20A,   57) PHYS. REV. VOL.48, ...   265, (1935).
                                   (20B,   123) PHYS. REV. VOL.48, ...   704, (1935).
                                   (20A,   58) PHYS. REV. VOL.49, ...   103, (1936).
                                   (20B,   124) PHYS. REV. VOL.50, ...   738, (1936).

1208 FINK............(J.)....‡   (5 , 1242) NUCL. SCI. ENGNG. VOL.38, ...   180, (1969).

1209 FINLAYSON.......(T.R.)..‡   (6 , 2267) J. APPL. PHYS. VOL.39, ...   3501, (1968).

1210 FINNEMORE.......(D.K.)..‡   (5 , 1541) PHYS. REV. 176, ...   712, (1968).

1211 FIRK............(F.W.K.)‡   (5 , 2825) PROC. PHYS. SOC. 82, ...   477, (1963).
                                   (12A,   17) NUCL. INST. MET. VOL.98, ...   385, (1972).

1212 FISCHER.........(C.O.)..‡   (14B,   72) PHYS. LETT. VOL.30A, ...   393, (1969).
                                   (6 ,   400) BER. BUNSENGES. PHYS. CHEM. 74,   696, (1970).
                                   (6 ,   383) BER. BUNSENGES. PHYS. CHEM. 75,   361, (1971).

1213 FISCHER.........(P.)....‡   (5 ,   78) COLL. INTER. N.126 (GRENOBLE), .   23, (1963).
                                   (5 , 2953) Z. ANGEW. MATH. PHYS. VOL.16, . .   820, (1965).
                                   (5 , 2890) ACTA. CRYST. 21, ...   765, (1966).
                                   (5 , 2460) NATURWISSENSCHAFTEN VOL.53, . .   16, (1966).
                                   (5 , 1654) ACTA CHEM. SCAND. VOL.21, ...   1543, (1967).
                                   (5 , 2934) Z. KRIST. VOL.124, ...   275, (1967).
                                   (5 ,   914) PHYS. KONDENS. MAT. (GER.) VOL.9   249, (1969).
                                   (5 ,   776) J. CHEM. PHYSICS VOL.53, ...   1917, (1970).
                                   (5 ,   409) CONF./RARE EARTHS/ACTINIDES, . .   204, (1971).
                                   (5 , 2054) CONF./RARE EARTHS/ACTINIDES, . .   206, (1971).
                                   (5 ,   42) J. PHYS. CHEM. SOLIDS VOL.32, . .   1641, (1971).
                                   (5 ,   43) J. PHYS. CHEM. SOLIDS VOL.32, . .   543, (1971).
                                   (5 , 2888) HELV. PHYS. ACTA VOL.46, ...   429, (1973).
                                   (5 , 2048) J. PHYS. C VOL.6, ...   725, (1973).

1214 FISHELSON.......(Z.)....‡   (20C,   65) NUCL. INST. MET. 71, ...   292, (1969).

1215 FISHER-JR.......(C.P.)..‡   (5 , 1311) DISS. ABSTR. VOL.23, ...   967, (1962).

1216 FISHER..........(I.Z.)..‡   (14B,   121) SOV. PHYS. ACOUSTICS VOL.9, . .   349, (1964).
                                   (15 ,   62) UKR. FIZ. ZH. (USSR) VOL.9, . . .   349, (1964).
                                   (14A,   69) SOV. PHYS. J.E.T.P. VOL.32, . . .   729, (1971).

1217 FISHER..........(M.E.)..‡   (13 ,   48) PHYS. REV. VOL.156, ...   583, (1967).
                                   (13 ,   5) AIP CONF. PROC. VOL.5, ...   1250, (1971).
                                   (13 ,   52) PHYS. REV. B VOL.5, ...   2668, (1972).

1218 FISHER..........(T.R.)..‡   (5 ,   432) PHYS. REV. LETT. VOL.25, ...   117, (1970).
                                   (5 ,   430) REPORT HEPL-630 (14PP.), ... , (1970).

1219 FISHMAN.........(A.YA.).‡   (10C,   83) PHYS. STAT. SOL. B VOL.51 NO.2, .   853, (1972).

1220 FISK............(S.)....‡   (14A,   58) PHYS. REV. A VOL.5, ...   2519, (1972).

1221 FLANAGAN........(T.B.)..‡   (5 , 2306) PLATINUM METALS REV. 10, ...   20, (1966).

1222 FLANNERY........(M.R.)..‡   (7B,   100) PROC. PHYS. SOC. 92, ...   551, (1967).

1223 FLECK...........(C.M.)..‡   (24 ,   50) NUCL. INST. MET. 61, ...   45, (1968).
                                   (24 ,   51) NUCL. INST. MET. 61, ...   53, (1968).
                                   (24 ,   57) NUCL. INST. MET. 66, ...   141, (1968).
                                   (24 ,   58) NUCL. INST. MET. 66, ...   149, (1968).

1224 FLEEMAN.........(J.)....‡   (12B,   8) DISS. ABSTR. VOL.13, ...   248, (1953).

1225 FLEURY..........(P.A.)..‡   (6 , 1761) PHYS. REV. B-2, ...   1362, (1970).

1226 FLINN...........(P.A.)..‡   (5 , 1131) J. APP. PHYS. 34, ...   1044, (1963).
                                   (5 , 1560) PHYS. REV. 132, ...   1547, (1963).

1227 FLOTOW..........(H.E.)..‡   (6 , 2292) J. CHEM. PHYS. VOL.45, ...   3817, (1966).
                                   (6 , 2258) J. CHEM. PHYS. VOL.48, ...   3795, (1968).
                                   (6 , 2256) J. PHYS. CHEM. SOLIDS VOL.32, . .   41, (1971).
                                   (6 , 2259) JULICH CONF/H IN METALS VOL.I, .   301, (1972).
                                   (6 , 2257) J. CHEM. PHYS. VOL.56, ...   4574, (1972).
                                   (6 , 2117) J. CHEM. PHYS. VOL.59, ...   6570, (1973).
                                   (6 , 2119) PHYS. REV. B VOL.9 NO.12, ... , (1974).

1228 FLOWERS.........(R.H.)..‡   (5 , 2916) J. CHEM. PHYSICS VOL.52, ...   6430, (1970).

1229 FLOYD...........(E.R.)..‡   (10B,   152) PHYS. REV. B-2, ...   3947, (1970).

1230 FLUHARTY........(R.G.)..‡   (23 ,   27) IAEA SYMP. (PILE RESEARCH)VIENNA   93, (1960).
                                   (19B,   28) RENSSELAER POL. INST. SYMP., . .   35, (1961).
                                   (19A,   34) NUCL. INST. MET. 30, ...   293, (1964).

1231 FOGEDBY.........(H.C.)..‡   (11A,   49) PHYS. REV. B VOL.5, ...   1941, (1972).

1232 FOGLIO-PARA.....(A.)....‡   (20C,   52) NUCL. INST. MET. 65, ...   110, (1968).
                                   (24 ,   63) NUCL. INST. MET. 70, ...   52, (1969).

1233 FOLDY...........(L.L.)..‡   (24 ,   91) PHYS. REV. LETT. VOL.3, ...   105, (1959).

1234 FOOTE-JR........(H.L.)..‡   (22A,   61) REV. SCI. INSTRUM. VOL.27, . . .   26, (1956).
                                   (5 , 2712) PHYS. REV. VOL.109, ...   1641, (1958).
                                   (20B,   109) NUCL. INST. MET. VOL.107, ...   33, (1973).

1235 FOREMAN-JR......(D.W.)..‡   (5 ,   367) DISS. ABS. XXVII, ...   2459B, (1967).
                                   (5 ,   319) J. CHEM. PHYSICS VOL.48, ...   3037, (1968).

1236 FOREMAN.........(A.J.E.)‡   (10B,   202) PROC. PHYS. SOC. B. VOL.70, . .   1143, (1957).

1237 FORGACS.........(R.L.)..‡   (12B,   54) REV. SCI. INSTRUM. VOL.42, . . .   1007, (1971).

1238 FORMAN..........(L.)....‡   (24 ,   104) REV. SCI. INSTRUM. VOL.41, . . .   1900, (1970).

1239 FORRAT..........(F.)....‡   (5 , 2897) COMP. REND. TOME 243, ...   898, (1956).
                                   (5 , 2516) C.R. ACAD. SCI. VOL.251, ...   1733, (1960).
                                   (5 , 2018) J. PHYS. CHEM. SOLIDS VOL.21, . .   234, (1961).
                                   (5 , 2738) J. APPL. PHYS. VOL.33, SUPPL., .   1123, (1962).
                                   (5 , 1837) J. PHYS. RADIUM VOL.23, ...   477, (1962).

1240 FORSTER.........(R.H.)..‡   (5 , 1184) ACTA CRYST. VOL. 18, ...   857, (1965).
                                   (5 ,   732) J. PHYS. CHEM. SOLIDS 29, . . .   855, (1968).

```
1266  FRITSCH.........(G.)....:  (  5 , 1949) PHYS. LETT. VOL.47A NO.1,. . . .      91, (1974).

1267  FROHNER.........(F.H.)..:  (24 ,   53) NUCL. INST. MET. 62, . . . . . .      62, (1968).

1268  FROMAN..........(P.O.)..:  ( 9B,   29) ARK. FYSIK VOL.4,. . . . . . . .     183, (1952).
                                 ( 9B,   28) ARK. FYSIK VOL.4,. . . . . . . .     181, (1952).
                                 ( 7B,    3) ARK. FYSIK VOL.5,. . . . . . . .      53, (1952).
                                 ( 9B,   30) ARK. FYSIK VOL.5,. . . . . . . .     113, (1953).

1269  FRUCHART........(D.)....:  (  5 , 1702) SOL. STATE COMM. VOL.6 NO.5, . .     251, (1968).
                                 ( 9D,   12) AIP CONF. PROC.(USA) VOL.5,. . .    1355, (1971).
                                 (  5 ,  834) J. PHYSIQUE TOME 32 COL.1 VOL.II    741, (1971).
                                 (  5 , 1861) J. PHYSIQUE TOME 32 COL.1 VOL.II    876, (1971).
                                 (  5 , 1741) SOL. STATE COMM. VOL. 9,. . . .    1793, (1971).
                                 (  5 , 2948) J. SOLID STATE CHEM. VOL.8,. . .     182, (1973).

1270  FRUCHART........(R.)....:  (  5 , 1702) SOL. STATE COMM. VOL.6 NO.5, . .     251, (1968).
                                 ( 9D,   12) AIP CONF. PROC.(USA) VOL.5,. . .    1355, (1971).
                                 (  5 , 1791) INT. J. MAGN.(GB) VOL.1,. . . .     341, (1971).
                                 (  5 , 1861) J. PHYSIQUE TOME 32 COL.1 VOL.II    876, (1971).
                                 (  5 , 1770) J. PHYSIQUE TOME 32 COL.1 VOL.II    980, (1971).
                                 (  5 , 1741) SOL. STATE COMM. VOL. 9,. . . .    1793, (1971).
                                 (  5 , 1793) SOL. STATE COMM. VOL. 9,. . . .      27, (1971).

1271  FRYER...........(E.M.)..:  (  5 ,  940) PHYS. REV. VOL.70, . . . . . . .     235, (1946).

1272  FUESS...........(H.)....:  (  5 ,  898) CHEM. PHYS. LETT. VOL.2,NO.1,. .      47, (1968).
                                 (  5 , 1815) J. APP. PHYS. 39,. . . . . . . .     628, (1968).
                                 (  5 , 2195) Z. ANORG. ALLG. CHEM. BAND 358,.     125, (1968).
                                 (  5 ,  506) Z. ANGEW. PHYS. VOL.27,. . . . .     311, (1969).
                                 (  5 ,  521) ACTA CRYST. VOL.B26, . . . . . .    2036, (1970).
                                 (  5 , 1203) SOL. STATE COMM. VOL.8,. . . . .    1745, (1970).
                                 (  5 , 1202) SOL. STATE COMM. VOL.8,. . . . .    1751, (1970).
                                 (  5 , 1015) J. PHYSIQUE TOME 32 COL.1 VOL.I,    202, (1971).
                                 (  5 ,  844) PHYS. LETT. VOL.34A,. . . . . .      361, (1971).
                                 (  5 ,  837) SOL. STATE COMM. VOL. 9,. . . .     1949, (1971).
                                 (  5 , 2925) J. SOLID STATE CHEM. VOL.5,. . .      11, (1972).
                                 (  5 , 2960) ACTA CRYST. VOL.A29,. . . . . .       49, (1973).
                                 (  5 , 2189) PHYS. REV. LETT. VOL.32, . . . .    1257, (1974).

1273  FUJII...........(H.)....:  (  5 , 1116) J. PHYS. SOC. JAPAN VOL.35,. . .    1554, (1973).

1274  FUJII...........(Y.)....:  (  5 ,  774) ACTA CRYST. VOL.A28 PART S-4,. .     192, (1972).
                                 (  5 ,  775) J. PHYS. SOC. JAPAN VOL.35,. . .     842, (1973).
                                 (  6 ,  574) PHYS. REV. B VOL.9,. . . . . . .    4549, (1974).

1275  FUJINO..........(M.)....:  (  6 , 1039) ANNU./REACT. INST. KYOTO UNIV. 3    121, (1970).

1276  FUJITA..........(Y.)....:  (  6 , 1039) ANNU./REACT. INST. KYOTO UNIV. 3    121, (1970).
                                 (  6 ,  295) J. NUCL. SCI. TECHNOL. VOL.9,. .     374, (1972).

1277  FUKASE..........(M.)....:  (  5 ,  344) J. PHYS. SOC. JAPAN VOL.24,. . .     446, (1968).

1278  FUKETA..........(T.)....:  (24 ,   15) IAEA SYMP. (PILE RESEARCH)VIENNA    633, (1960).

1279  FUKUI...........(S.)....:  (  5 , 1980) J. PHYS. SOC. JAPAN VOL.23,. . .     461, (1967).

1280  FUKUI...........(Y.)....:  (15 ,   46) PROPERTIES LIQUID METALS,. . . .     143, (1973).

1281  FUKUSHIMA.......(Y.)....:  (  5 ,  228) PHYS. LETT. VOL.45A NO.4,. . . .     273, (1973).

1282  FULINSKI........(A.)....:  ( 7A,    7) ACTA PHYS. POLON. VOL.22,. . . .     399, (1962).
                                 (14B,    2) ACTA PHYS. POLON. VOL.26,. . . .      19, (1964).
                                 (  6 , 1060) PHYSICA VOL.36,. . . . . . . . .      35, (1967).
                                 (  6 ,  283) PHYS LETT 27A,. . . . . . . . .        9, (1968).
                                 (15 ,   32) PHYSICA VOL.48,. . . . . . . . .     126, (1970).

1283  FULLMER.........(L.O.)..:  (  5 ,  339) J. PHYS. CHEM. SOLIDS VOL.31,. .     793, (1970).

1284  FULLWOOD........(R.R.)..:  (  6 ,  136) DISS. ABS. 27, . . . . . . . . .    573B, (1966).
                                 (24 ,   45) NUCL. INST. MET. 52,. . . . . .      321, (1967).
                                 (24 ,   74) NUCL. INST. MET. 77,. . . . . .      245, (1970).

1285  FULMER..........(R.H.)..:  (  5 , 2964) NUCL. SCI. ENG. VOL.46,. . . . .     314, (1971).

1286  FURRER..........(A.)....:  (  6 ,  473) SOL. STATE COMM. 3,. . . . . . .     339, (1965).
                                 (  6 , 1729) SOL. STATE COMM. 4,. . . . . . .      99, (1966).
                                 (  6 ,  765) HELV. PHYS. ACTA VOL.40, . . . .     378, (1967).
                                 (  6 ,  464) IAEA SYMP. COPENHAGEN, VOL.2,. .     133, (1968).
                                 (  6 , 1800) PHYS. STAT. SOL. VOL.42, . . . .     821, (1970).
                                 (19A,    2) ACTA CRYST. VOL.A27, . . . . . .     461, (1971).
                                 (  5 , 2054) CONF./RARE EARTHS/ACTINIDES, . .     206, (1971).
                                 (19A,    3) ACTA CRYST. VOL.A28, . . . . . .     287, (1972).
                                 (  6 ,  466) IAEA SYMP. GRENOBLE, . . . . . .     563, (1972).
                                 (  6 , 1693) J. PHYS. C VOL.5,. . . . . . . .    2246, (1972).
                                 (  6 ,  469) SOL. STATE COMM. VOL.11,. . . .       21, (1972).
                                 (  6 , 2197) INT. J. MAGN. VOL.4, NO.1,. . .       63, (1973).
                                 (  6 , 2196) SOL. STATE COMM. VOL.12 NO.2,. .     117, (1973).
                                 (  6 ,  716) J. PHYS. C VOL.7,. . . . . . . .    1207, (1974).

1287  FURUKAWA........(K.)....:  (14B,  112) REP. PROGR. PHYS. VOL.25,. . . .     395, (1962).

1288  FURUSETH........(S.)....:  (  5 ,  498) ACTA CHEM. SCAND. VOL.21,. . . .     833, (1967).
                                 (  5 , 2193) ACTA CHEM. SCAND. VOL.23,. . . .    2325, (1969).

1289  FURUTA..........(Y.)....:  (22B,   54) REV. SCI. INSTRUM. VOL.41, . . .      11, (1970).

1290  FUSE............(T.)....:  (20C,   60) NUCL. INST. MET. 74,. . . . . .      322, (1969).

1291  FYKIN...........(L.E.)..:  (  5 , 1579) SOV. PHYS. DOKLADY VOL.8,. . . .     131, (1963).
                                 (  5 , 2665) DOKL. AKAD. NAUK SSSR VOL.194,.    1374, (1970).

1292  GABELA..........(F.)....:  (  5 , 1354) ACTA CRYST. VOL.A28 PART S-4,. .      60, (1972).

1293  GAHN............(U.)....:  (14A,   38) PHYS. LETT. VOL.31A, . . . . . .     237, (1970).

1294  GAJIC...........(B.)....:  (  5 , 1697) PHYS. STAT. SOLIDI A VOL.19 NO.1   K13, (1973).

1295  GALLOWAY........(R.B.)..:  (19A,   40) NUCL. INST. MET. 70, . . . . . .     262, (1969).

1296  GALOCIOVA.......(D.)....:  (22A,   10) CZECH. J. PHYS. B VOL.16,. . . .     942, (1966).
                                 (  5 , 2376) PHYS. LETT. 28A,. . . . . . . .      546, (1968).
                                 ( 9B,   37) CZECH. J. PHYS. B VOL.19,. . . .    1608, (1969).
                                 (  5 , 2379) NUCL. INST. MET. 67,. . . . . .      357, (1969).
                                 ( 9B,   89) PHYS. STAT. SOL. A VOL.2,. . . .     211, (1970).
                                 (  5 , 2372) ACTA CRYST. VOL.A27, . . . . . .     410, (1971).

1297  GALPERIN........(F.M.)..:  ( 9D,   63) PHYS. STAT. SOL. VOL.40, . . . .      59, (1970).

1298  GAMARI-SEALE....(H.)....:  (  5 , 1009) PHYS. STAT. SOL. 29, . . . . . .     323, (1968).

1299  GAMBETTA........(V.)....:  (  5 ,  273) PHYSICA VOL.32,. . . . . . . . .     119, (1966).
```

```
1300  GAMBINO.........(R.J.)..:  ( 5 , 1213) J. APPL. PHYS. VOL.40, . . . . .  1009, (1969).
                                 ( 6 ,  912) J. APPL. PHYS. 41,.: .: .: .: .:   933, (1970).
                                 ( 5 ,  840) PHYS. REV. B VOL.8,. : : : : : :   260, (1973).
1301  GAMEEL..........(Y.H.)..:  ( 6 , 2301) IAEA SYMP. COPENHAGEN, VOL.1,. .   195, (1968).
1302  GANGWANI........(G.S.)..:  ( 5 , 1288) NUCL. SCI. ENG. VOL.47,. . . . .   153, (1972).
1303  GARCIA..........(A.)....:  ( 5 , 1433) ACTA CRYST. VOL.B26, : : : : : :   827, (1970).
                                 ( 5 , 1994) ACTA CRYST. VOL.B27, : : : : : :   253, (1971).
1304  GARDNER.........(W.E.)..:  ( 5 ,  580) PROC. ROY. SOC. A VOL.246, . . .    78, (1958).
1305  GARELICK........(H.)....:  ( 8 ,   31) J. PHYS. C VOL.2,. . . . . . . .   317, (1969).
1306  GARLAND.........(J.W.)..:  ( 8 ,   25) J. APPL. PHYS. 42, . . . . . . .  1610, (1971).
1307  GARLAND.........(W.J.)..:  (24 ,   77) NUCL. INST. MET. VOL.104,. . . .   217, (1972).
1308  GARNSWORTHY.....(R.K)...:  (24 ,   46) NUCL. INST. MET. 57, . . . . . .    82, (1967).
1309  GARRETT-SR......(B.S.)..:  ( 5 , 1281) DISS. ABSTR. VOL.14, . . . . . .  1152, (1954).
1310  GARRETT.........(D.A.)..:  (24 ,  104) REV. SCI. INSTRUM. VOL.41, . . .  1900, (1970).
1311  GARRETT.........(R.)....:  (12A,   16) NUCL. INST. MET. 75, . . . . . .   333, (1969).
1312  GARRETT.........(W.)....:  ( 5 , 2980) J. SOLID STATE CHEM. VOL.7,. . .   448, (1973).
1313  GARWIN..........(E.L.)..:  (10D,   15) JOUR. APP. PHYS. 40, . . . . . .  2766, (1969).
                                 (10D,   16) JOUR. APP. PHYS. 40, . . . . . .  2776, (1969).
1314  GASKELL.........(T.)....:  ( 6 ,   82) J. PHYS. C VOL.5,. . . . . . . .  3279, (1972).
1315  GASSER..........(F.)....:  (22B,    6) AUTOMATIC/NUCLEAR DATA,. .: . . .   233, (1964).
                                 (22B,   57) Z. ANGEW. MATH. PHYS. VOL.15,. .   481, (1964).
1316  GAULT...........(J.D.)..:  (23 ,   11) DISS. ABS. VOL.31, . . . . . .  2897B, (1970).
                                 ( 6 ,  337) J. CHEM. PHYSICS VOL.53, : : : :  3417, (1970).
                                 ( 6 , 1080) J. CHEM. PHYS. VOL.56, : : : : :  3217, (1972).
                                 ( 6 ,  324) J. CHEM. PHYS. VOL.59, : : : : :  2305, (1973).
1317  GAUTIER.........(F.)....:  ( 6 , 1741) J. PHYS. CHEM. SOLIDS VOL.24,. .   387, (1963).
1318  GAVIN...........(W.R.)..:  ( 6 , 2074) PHYS. REV. LETT. VOL.14, . . . .    44, (1965).
1319  GAZIS...........(D.C.)..:  ( 6 , 2230) PHYS. REV. VOL.134,. . . . . . .A1486, (1964).
1320  GEHRING.........(K.A.)..:  ( 5 , 2122) PROC. PHYS. SOC. 85, . . . . . .   967, (1965).
1321  GEIGER..........(K.W.)..:  (22A,    7) CAN. JOURN. PHYS. 42,. . . . . .  1593, (1964).
                                 (20B,   66) METROLOGIA (GER.) VOL.4, . . . .     8, (1968).
1322  GEISLER.........(A.H.)..:  ( 5 , 1864) J. METALS VOL.8, . . . . . . . .  1259, (1956).
1323  GELARD..........(J.)....:  ( 5 , 1029) SOL. STATE COMM. VOL.14 NO.2,. .   187, (1974).
1324  GELBERG.........(D.)....:  (18A,   25) REV. ROUMAINE PHYS. VOL.10,. . .  1035, (1965).
                                 (20A,   35) NUCL. INST. MET. 40, . . . . . .    77, (1966).
                                 (20B,   89) NUCL. INST. MET. 69, : : : : : :   125, (1969).
1325  GELDART.........(D.J.W.):  (10B,   13) CAN. J. PHYS. 48,. . . . . . . .   183, (1970).
1326  GELLBERG........(D.)....:  (23 ,   77) STUD. CERCETARI FIZ. VOL.20, . .  1033, (1968).
1327  GELLER..........(R.)....:  ( 5 , 1441) CONF. ON LOW-TEMP. PHYS.(ABSTR.)    31, (1972).
                                 ( 6 , 1286) PHYS. STAT. SOL. B VOL.53 NO.1,.   227, (1972).
1328  GELLER..........(S.)....:  ( 5 , 1758) J. APP. PHYS. VOL.40, . . . . .  1136, (1969).
                                 ( 5 ,  339) J. PHYS. CHEM. SOLIDS VOL.31,. .   793, (1970).
1329  GEL≠D...........(P.V.)..:  ( 5 , 2646) SOV. PHYS. DOKLADY VOL.15, : : :   276, (1970).
                                 ( 5 ,  636) SOV. PHYS. SOL. STATE 13, : : :   644, (1971).
1330  GENIN...........(J.M.)..:  ( 6 ,  205) C.R. ACAD. SCI. VOL.257, . . . .  1843, (1963).
1331  GENIN...........(R.)....:  ( 5 , 2971) J. PHYS. RADIUM VOL.22,. : : : :   648, (1961).
                                 (21 ,   28) J. PHYSIQUE TOME 24, : : : : : :   15A, (1963).
                                 (12A,   22) J. PHYSIQUE TOME 24, : : : : : :   89A, (1963).
                                 ( 5 , 2060) J. PHYSIQUE TOME 24, : : : : : :    21, (1963).
1332  GENKIN..........(YA.E.).:  ( 6 , 1671) SOV. PHYS. SOL. STATE VOL. 13, .  1793, (1972).
1333  GEORGES.........(R.)....:  (12A,   14) NUCL. INST. MET. 63, . . . . . .   283, (1968).
1334  GERASIMOV.......(S.B.)..:  ( 7A,   87) SOV. PHYS. JETP. 16, . . . . . .  1321, (1963).
1335  GERASIMOV.......(V.F.)..:  (19B,   35) SOV. AT. ENERGY VOL.28,. .: : :   150, (1970).
                                 (20C,  109) SOV. AT. ENERGY VOL.32, NO.5,. .   416, (1972).
1336  GERJUOY.........(E.)....:  ( 9B,   76) PHYS. REV. VOL.76, . . . . . . .  1117, (1949).
1337  GERSCH..........(H.A.)..:  ( 6 ,  768) PHYS. REV. VOL.103, . . . . . .   525, (1956).
                                 ( 9D,   38) J. PHYS. CHEM. SOLIDS VOL.5, : :   180, (1958).
                                 (11C,   15) J. PHYS. -C- VOL. 1, : : : : : :  1563, (1968).
                                 (11C,   37) PHYS. REV. 167, : : : : : : : :   458, (1968).
                                 ( 6 , 1150) PHYS. LETT. 30A, : : : : : : : :   261, (1969).
                                 ( 6 , 1997) J. APPL. PHYS. VOL.41, : : : : :  1370, (1970).
                                 ( 8 ,   33) J. PHYS. CHEM. SOLIDS VOL.31,. .   255, (1970).
                                 ( 6 ,  801) AIP CONF. PROC. VOL.5, : : : : :  1334, (1971).
                                 ( 6 , 1203) PHYS. REV. A VOL.4,. : : : : : :   281, (1971).
                                 ( 6 , 1142) PHYS. REV. A VOL.5,. : : : : : :  1547, (1972).
                                 ( 7B,   88) PHYS. REV. A VOL.8,. : : : : : :   905, (1973).
                                 ( 6 , 1205) PHYS. REV. A VOL.9,. . . . . . .  2085, (1974).
1338  GERSON..........(R.)....:  ( 5 ,  215) J. PHYS. CHEM. SOLIDS VOL.32,. .  1315, (1971).
1339  GESI............(K.)....:  ( 6 , 1505) J. APPL. PHYS. 42, : : : : : : :  1374, (1971).
                                 ( 6 , 1317) PHYS. REV. B VOL.5,. : : : : : :  1933, (1972).
                                 ( 6 , 2104) J. PHYS. C VOL.6,. : : : : : : :  3021, (1973).
1340  GHATAK..........(A.K.)..:  ( 7B,   93) PHYS. STAT. SOL. VOL.49, . . . .   K75, (1972).
1341  GHEORGHIU.......(Z.)....:  (17C,    5) ACTA CRYST. 17, . . . . . . . .  1529, (1964).
                                 (23 ,   77) STUD. CERCETARI FIZ. VOL.20, : :  1033, (1968).
                                 (20B,   89) NUCL. INST. MET. 69, . . . . . .   125, (1969).
1342  GHOSE...........(A.M.)..:  (22C,   33) NUCL. INST. MET. 49, . . . . . .   101, (1967).
1343  GHOSE...........(A.)....:  ( 6 ,  937) PHYS. REV. VOL.113,. . . . . . .    49, (1959).
1344  GHOSH...........(R.E.)..:  ( 6 ,  565) IAEA SYMP. GRENOBLE, . .: . . .   345, (1972).
                                 ( 6 ,  304) J.C.S. FARADAY TRANS. II VOL.69,  275, (1973).
```

1377 GLASER..........(W.)....: (16 ,   54) NUKLEONIK VOL.11,. . . . .   282, (1968).
                                (19A,   1) ACTA CRYST. (INTERACT.) VOL.A25, S250, (1969).
                                (24 ,   12) ELECTRONIC STRUCTURES IN SOLIDS,   153, (1969).
                                (19A,   23) IAEA SYMP. (INSTRUMENT.), VIENNA   147, (1969).
                                (19A,   63) NUKLEONIK (GER.) VOL.12,. . . . . 153, (1969).
                                (6 ,   12) PHYS. LETT. VOL.28A,. . . . . . . 531, (1969).
                                (6 ,   289) PHYS. LETT. VOL.31A,. . . . . . . 158, (1970).
                                (6 , 1216) PHYS. REV. LETT. VOL.25,. . . . . 1611, (1970).
                                (6 ,   418) CONF. INTERN., RENNES,. . . . . . 145, (1971).
                                (6 , 1956) IAEA SYMP. GRENOBLE,. . . . . . . 435, (1972).
                                (6 ,   294) IAEA SYMP. GRENOBLE,. . . . . . . 841, (1972).
                                (6 , 1333) PHYS. REV. LETT. VOL.30,. . . . . 1144, (1973).
                                (6 , 1960) PROPERTIES LIQUID METALS,. . . .   111, (1973).
                                (6 , 1334) PHYS. REV. LETT. VOL.32,. . . . . 836, (1974).

1378 GLASSER.........(M.L.)..: (6 ,   845) PHYS. REV. 130,. . . . . 1783, (1963).

1379 GLASSGOLD.......(A.E.)..: (6 , 1156) BROOKHAVEN SYMPOSIUM,. . . . . .   69, (1965).

1380 GLASS...........(L.)....: (14B,   90) PHYS. REV. VOL.165,. . . . . . . 186, (1968).
                                (14A,   46) PHYS. REV. VOL.176,. . . . . . . 239, (1968).

1381 GLATTLI.........(H.)....: (8 ,   18) C.R. ACAD. SCI. B VOL.274,. . .   423, (1972).
                                (17A,   87) PHYS. REV. LETT. VOL.31,. . . .   776, (1973).

1382 GLATZ...........(A.C.)..: (5 ,   40) J. MATERIALS SCI. VOL.3,. . . .   498, (1968).

1383 GLAUBERMAN......(A.E.)..: (10C,   81) PHYS. STAT. SOL. 13,. . . . . .   233, (1966).

1384 GLAUBER.........(R.J.)..: (10C,   69) PHYS. REV. VOL.98,. . . . . . . 1692, (1955).
                                (15 ,   36) PHYS. REV. VOL.101,. . . . . . . 118, (1956).
                                (14B,   75) PHYS. REV. VOL.101,. . . . . . . 129, (1956).

1385 GLAVINA.........(C.)....: (19B,   15) NUCL. INST. MET. VOL.81,. . . .   301, (1970).

1386 GLEN............(G.L.)..: (5 , 2274) J. AMER. CHEM. SOC. VOL.85,. . . 3892, (1963).

1387 GLICKSTEIN......(S.S.)..: (6 , 2335) PHYS. REV. C VOL.2,. . . . . . . 2390, (1970).
                                (6 , 2336) PHYS. REV. C VOL.4,. . . . . . . 1818, (1971).

1388 GLINKA..........(C.J.)..: (5 , 2400) A.I.P. CONF. PROC. NO.10 PART 1, 659, (1973).

1389 GLUCK...........(P.)....: (14A,   52) PHYS. REV. A VOL.2,. . . . . . . 1095, (1970).

1390 GLYDE...........(H.R.)..: (10D,   58) SOL. STATE COMM. VOL. 8,. . . .   923, (1970).
                                (6 , 1122) CAN. J. PHYS. 49,. . . . . 2997, (1971).
                                (10D,   8) CRIT. REV. SOLID STATE SCI. 2,   181, (1971).
                                (6 , 1152) CAN. J. PHYS. VOL.50,. . . . . . 1152, (1972).
                                (10D,   44) PHYS. REV. B VOL.5,. . . . . . . 1206, (1972).

1391 GNEDIN..........(YU.N.).: (12B,   62) SOV. PHYS. J.E.T.P. VOL.31,. . . 378, (1970).
                                (12B,   71) YADERNAYA FIZ. (USSR) VOL.12,. . 815, (1970).

1392 GNIDAK..........(N.L.)..: (5 ,   379) UKR. FIZ. ZH.(USSR) VOL.17,. . .   38, (1972).

1393 GOBERT..........(G.)....: (20B,   63) J. PHYS. RADIUM VOL.19 SUPPL.4,   51A, (1958).
                                (20B,   61) J. PHYS. E VOL.6 NO.5,. . . . .   488, (1973).

1394 GOODARD.........(A.J.H.): (16 ,   42) NUCL. SCI. ENGNG. VOL.37,. . .   127, (1969).

1395 GODEL...........(J.B.)..: (21 ,   20) J. APPL. CRYST. VOL.2,. . . . .   109, (1969).

1396 GODET...........(M.)....: (6 , 2146) SOL. STATE COMM. VOL.13,. . . .   881, (1973).

1397 GOEBEL..........(H.)....: (5 ,   842) CONF./RARE EARTHS/ACTINIDES, . . 226, (1971).

1398 GOEDKOOP........(J.A.)..: (5 ,   718) ACTA CRYST. VOL.8,. . . . . . .   118, (1955).
                                (5 ,   194) ACTA CRYST. VOL.12,. . . . . . . 476, (1959).
                                (6 ,   65) IAEA SYMPOSIUM VIENNA,. . . . .   501, (1960).
                                (23 ,   30) IAEA SYMP. (PILE RESEARCH)VIENNA 361, (1960).
                                (6 ,   859) NATURE VOL.185,. . . . . 450, (1960).
                                (5 , 2304) PHYSICA VOL.26,. . . . . 744, (1960).
                                (5 ,   89) ACTA CRYST. VOL.14,. . . . . . . 223, (1961).
                                (5 ,   79) ACTA CRYST. 16,. . . . . 737, (1963).
                                (9C,   18) INTERN. SUMMER SCHOOL, MOL,. . . 534, (1963).
                                (24 ,   32) NED. TIJDSCHR. NATUURK. VOL.36,. 250, (1970).

1399 GOEL............(C.M.)..: (6 ,   599) PHYS. LETT. VOL.46A NO.3,. . . . 221, (1973).

1400 GOER............(D.A.)..: (5 ,   492) DISS. ABSTR. B VOL.34,. . . . . 2247, (1973).

1401 GOLAND..........(A.N.)..: (5 ,   691) PHYS. REV. VOL.111,. . . . . . . 261, (1958).
                                (20B,   140) REV. SCI. INSTRUM. VOL.30,. . .   269, (1959).
                                (11C,   36) PHYS. REV. 147,. . . . . 457, (1966).
                                (5 , 1962) PHYS. REV. 153,. . . . . 184, (1967).

1402 GOLDBERGER......(M.L.)..: (9B,   74) PHYS. REV. VOL.71,. . . . . . .   294, (1947).
                                (17B,   57) PHYS. REV. 132,. . . . . 2764, (1963).

1403 GOLDHABER.......(M.)....: (20A,   69) PROC. ROY. SOC. A VOL.162,. . .   127, (1937).

1404 GOLDMAN.........(D.T.)..: (6 , 1028) IAEA SYMP. CHALK RIVER, VOL.1,   389, (1962).

1405 GOLDMAN.........(J.E.)..: (5 , 1082) PHYS. REV. VOL.93,. . . . . . .   893, (1954).

1406 GOLDMAN.........(V.V.)..: (7B,   23) J. APPL. PHYS. 41,. . . . . . . 5138, (1970).
                                (6 , 1127) PHYS. REV. LETT. VOL.24,. . . . 1424, (1970).
                                (6 , 1698) DISS. ABS. VOL.31,. . . . . . . 4905B, (1971).
                                (10B,   166) PHYS. REV. B-4,. . . . . 567, (1971).

1407 GOLDSCHMIDT.....(G.H.)..: (5 , 1899) PHYS. REV. VOL.83,. . . . . . .   88, (1951).
                                (5 , 1900) PHYS. REV. VOL.86,. . . . . . .   797, (1952).

1408 GOLDSMITH.......(H.J.)..: (10B,   115) PHYS. LETT. 27A,. . . . .   523, (1968).

1409 GOLDSMITH.......(M.)....: (7A,   41) NUCL. SCI. ENGNG. VOL.34,. . .   93, (1968).
                                (6 , 2336) PHYS. REV. C VOL.4,. . . . . . . 1818, (1971).

1410 GOLDSTEIN.......(H.)....: (20A,   72) REACTOR SCI. TECHNOL. VOL.17,.   147, (1963).

1411 GOLDSTEIN.......(L.)....: (6 , 1182) PHYS. REV. VOL.77,. . . . . . .   319, (1950).
                                (6 , 1183) PHYS. REV. VOL.80,. . . . . . .   141, (1950).
                                (14A,   41) PHYS. REV. VOL.84,. . . . . . .   466, (1951).
                                (5 , 1324) PHYS. REV. VOL.97,. . . . . . .   855, (1955).
                                (5 , 1327) PHYS. REV. VOL.101,. . . . . . . 1235, (1956).

1412 GOLDSTEIN.......(M.)....: (6 , 1182) PHYS. REV. VOL.77,. . . . . . .   319, (1950).

1413 GOLD............(L.)....: (9B,   87) PHYS. STAT. SOL. VOL.5,. . . . .   K9, (1964).

1414 GOLIC...........(L.)....: (5 ,   4) ACTA CRYST. A28 PART S-4,. . . . 193, (1972).
                                (5 ,   30) MATER. RES. BULL. VOL.7,. . . . 1225, (1972).

```
1415  GOLIKOV.........(V.V.)..:  (20C,     14) INSTRUM. EXP. TECH. NO.2,. . .     243, (1963).
                                 (20A,     19) INSTRUM. EXP. TECH. NO.1,. . . .    58, (1966).
                                 (  5 ,  2255) UKR. FIZ. ZH.(USSR) VOL.15,. : :  1772, (1970).
                                 (20B,    153) SOV. PHYS. JETP VOL.37,. . . . .    41, (1973).

1416  GOLLNITZ........(H.)....:  (24 ,     68) NUCL. INST. MET. 74,. . . . . .    109, (1969).

1417  GOLOSOVSKII.....(I.V.)..:  (  5 ,  2880) PHYS. STAT. SOL. B VOL.53 NO.1;.  K37, (1972).
                                 (  5 ,  2051) PHYS. STAT. SOL. B VOL.56 NO.1;.   61, (1973).
                                 (  5 ,  2879) SOV. PHYS. JETP LETT. VOL.18,. .   47, (1973).
                                 (  5 ,   347) SOV. PHYS. SOL. STATE VOL.14,. . 2387, (1973).
                                 (  5 ,  1960) SOL. STATE COMM. VOL.14 NO.4,. .  309, (1974).

1418  GOLOVKIN........(V.S.)..:  (22A,     66) SOV. PHYS. CRYST. VOL.2,. . . .    626, (1957).
                                 (  5 ,  2140) SOV. PHYS. CRYST. VOL.3, . . . .   147, (1958).
                                 (  5 ,   553) SOV. PHYS. DOKLADY VOL.4,. . . .  1070, (1960).
                                 ( 90,     75) SOV. PHYS. CRYST. VOL.9,. . . .    535, (1965).
                                 (  5 ,   530) SOV. PHYS. J.E.T.P. VOL.22,. . .   754, (1965).
                                 (21 ,     52) SOV. PHYS. CRYST. VOL.12,. . . .   477, (1967).
                                 (  5 ,  2967) SOV. PHYS. CRYST. VOL.14,. . . .   913, (1969).
                                 (  5 ,  2969) SOV. PHYS. CRYST. VOL.15,. . . .   376, (1970).
                                 (  6 ,   638) SOV. PHYS. CRYST. VOL.15,. . . .   935, (1971).
                                 (20B,    152) SOV. PHYS. CRYST. VOL.17,. . . .   826, (1973).

1419  GOLUB...........(R.)....:  (17A,     67) NUCL. INST. MET. VOL.91, . . . .   205, (1971).
                                 (20B,    120) PHYS. LETT. VOL.38A,. . . . . .    177, (1972).

1420  GOMAN≠KOV.......(V.I.)..:  (22A,     67) SOV. PHYS. CRYST. VOL.5,. . . .    294, (1960).
                                 (23 ,     41) INSTRUM. EXP. TECH. NO.6,. . . .  902, (1961).
                                 (  5 ,  1828) PHYS. MET. METALLOGR. VOL.12,. .   119, (1961).
                                 (  5 ,  2108) PHYS. MET. METALLOGR. VOL.14,. .    25, (1962).
                                 (  5 ,   588) SOV. PHYS. CRYST. VOL.6,. . . .    628, (1962).
                                 (  5 ,   986) SOV. PHYS. CRYST. VOL.7,. . . .    637, (1963).
                                 (  5 ,   541) SOV. PHYS. CRYST. VOL.7,. . . .    639, (1963).
                                 (  5 ,   989) UKR. FIZ. ZH. (USSR) VOL.8,. . .   268, (1963).
                                 (  5 ,  1069) BULL. ACAD. SCI. USSR VOL.28,. .   354, (1964).
                                 (  5 ,  2132) SOV. PHYS. CRYST. VOL.10,. . . .   338, (1965).
                                 (  5 ,  2143) SOV. PHYS. DOKLADY VOL.15,. . .    874, (1971).

1421  GOMPF...........(F.)....:  (23 ,     32) IAEA SYMP. COPENHAGEN, VOL.2,.    331, (1968).
                                 (19A,     17) IAEA SYMP. COPENHAGEN, VOL.2,.    417, (1968).
                                 (19A,      1) ACTA CRYST. (INTERACT.) VOL.A25, S250, (1969).
                                 (19A,     23) IAEA SYMP. (INSTRUMENT.), VIENNA   147, (1969).
                                 (19A,     63) NUKLEONIK (GER.) VOL.12,. . . .    153, (1969).
                                 (  6 ,     12) PHYS. LETT. VOL.28A,. . . . . .    531, (1969).
                                 (  6 ,     33) IAEA SYMP. GRENOBLE,. . . . . .    137, (1972).

1422  GONDRAND........(J.C.)..:  (19B,     32) REV. PHYS. APPL. (FRANCE) VOL.4,  111, (1969).

1423  GONZALO.........(J.A.)..:  (  5 ,  2311) J. PHYS. SOC. JAPAN VOL.21,. .   1626, (1966).
                                 (  5 ,  1151) PHYS. REV. 147,. . . . . . . .     415, (1966).
                                 (  5 ,  1978) 2ND INTERNAT. MEET./FERROELECTR. . .,  (1969).
                                 (  5 ,  1746) AN. FIS. (SPAIN) VOL.66, . . . .   407, (1970).

1424  GOODENOUGH......(J.B.)..:  (  5 ,  2198) PHYS. REV. 164,. . . . . . . .     768, (1967).
                                 (  5 ,   683) SOL. STATE COMM. 5,. . . . . .     577, (1967).

1425  GOODINGS........(D.A.)..:  (  6 ,  2138) J. APP. PHYS. VOL.39,. . . . .     887, (1968).
                                 (  6 ,  2139) J. OF PHYS. C VOL.1,. . . . . .   1596, (1968).
                                 (  6 ,  2141) J. OF PHYS. C VOL.1,. . . . . .    125, (1968).
                                 (  6 ,  2140) J. OF PHYS. C VOL.1,. . . . . .    132, (1968).
                                 (11A,     15) J. PHYS. -C-, VOL. 1,. . . . .    1279, (1968).

1426  GOODMAN.........(B.)....:  (  7B,      4) ARK. FYSIK VOL.32, PAPER 31, . .   537, (1966).
                                 (10B,    135) PHYS. REV. 175,. . . . . . . .     963, (1968).
                                 (10D,     29) PHYS. REV. 178,. . . . . . . .    1171, (1969).

1427  GOODYEAR........(G.)....:  (  6 ,  2044) FARADAY SYMP. CHEM. SOC. NO.6,.   169, (1972).

1428  GORBACHEV.......(B.I.)..:  (  6 ,  1752) SOV. PHYS. SOL. STATE 7,. . . .  1547, (1965).
                                 (  6 ,  1889) UKRAINIAN PHYS. J. VOL.14, 9,. .  1525, (1969).
                                 (10C,    136) UKR. FIZ. ZH. (USSR) VOL.17,. .    789, (1972).
                                 (  7B,    139) UKR. FIZ. ZH. (USSR) VOL.17,. .     45, (1972).
                                 (  6 ,    785) UKR. FIZ. ZH. (USSR) VOL.18,. .   1384, (1973).
                                 (  6 ,    608) UKR. FIZ. ZH. (USSR) VOL.18,. .   1390, (1973).
                                 (  6 ,    784) UKR. FIZ. ZH. (USSR) VOL.18,. .   1528, (1973).
                                 (  6 ,   1748) UKR. FIZ. ZH. (USSR) VOL.18,. .    558, (1973).
                                 (  6 ,    568) UKR. FIZ. ZH. (USSR) VOL.18,. .     92, (1973).

1429  GORBACHEV.......(V.V.)..:  (  6 ,    601) SOV. PHYS. SOL. STATE 7,. . . .   296, (1965).
                                 (21 ,     53) SOV. PHYS. CRYST. VOL.12,. . . .   978, (1968).

1430  GORDEEV.........(G.P.)..:  (  5 ,    349) SOV. PHYS. SOL. STATE VOL.9, . .  1762, (1968).

1431  GORDON..........(G.)....:  (12B,     11) IAEA SYMP. COPENHAGEN, VOL.2,.    407, (1968).

1432  GORDON..........(J.)....:  (20C,      3) ACTA PHYS. HUNGAR. VOL.12, . . .   333, (1960).
                                 (22A,      5) ACTA PHYS. HUNGAR. VOL.15, . . .   203, (1963).
                                 (23 ,     60) NUCL. INST. MET. 55,. . . . . .    151, (1967).
                                 (  6 ,    726) IAEA SYMP. COPENHAGEN, VOL.2,.     55, (1968).
                                 (19B,     30) REPORT KFKI-07/1968 (14PP.),. .    . .,  (1968).
                                 (  6 ,    731) PHYS. LETT. VOL.32A,. . . . . .    341, (1970).

1433  GORDON..........(R.G.)..:  (  6 ,    997) CHEM. PHYS. LETTERS VOL.2, . . .   584, (1968).

1434  GORETZKI........(H.)....:  (  5 ,  2658) NATURWISSENSCHAFTEN VOL.54,. . .   163, (1967).
                                 (  5 ,  2644) PHYS. STAT. SOL. 20, . . . . . .  K141, (1967).

1435  GORLOV..........(G.V.)..:  (  7B,    134) SOV. PHYS. J. VOL.34,. . . . . .   138, (1970).

1436  GORODETSKY......(G.)....:  (  5 ,    882) PHYS. REV. B VOL.8,. . . . . .    3398, (1973).

1437  GORYACHEV.......(I.V.)..:  (20C,     20) INSTRUM. EXPER. TECH. NO.3,. . .   529, (1968).

1438  GORYUNOV........(A.F.)..:  (  5 ,   1304) J. NUCL. ENERGY VOL.4, . . . . .   109, (1957).

1439  GOSSMANN........(G.)....:  (  5 ,   1300) NUKLEONIK VOL.4, . . . . . . .     110, (1962).

1440  GOTOH...........(Y.)....:  (  5 ,    785) NUCL. SCI. ENG. VOL.45,. . . . .   126, (1971).

1441  GOTTARDI........(N.A.)..:  (20C,     52) NUCL. INST. MET. 65, . . . . . .   110, (1968).

1442  GOTTLIEB........(A.M.)..:  (  6 ,   1758) PHYS. REV. LETT. VOL.31,. . . .   1222, (1973).

1443  GOUDSMIT........(S.)....:  (17A,     86) PHYS. REV. VOL.50, . . . . . . .   570, (1936).

1444  GOULDING........(C.G.)..:  (22B,     35) NUCL. INST. MET. VOL.91, . . . .   541, (1971).

1445  GOULD...........(F.T.)..:  (22A,     50) NUCL. SCI. ENGNG. VOL.8,. . . .    453, (1960).
                                 (22A,     64) REV. SCI. INSTRUM. VOL.36; . . .   887, (1965).

1446  GOULD...........(H.)....:  (10B,    168) PHYS. REV. A VOL.3,. . . . . . .  1453, (1971).
```

```
1447  GOVOR...........(G.A.)..!   ( 5 , 1633) J. PHYSIQUE TOME 32 COL.1 VOL.II  987, (1971).

1448  GOVOR...........(G.G.)..!   ( 6 , 1436) PHYS. STAT. SOL. VOL.43, . . . .  K165, (1971).

1449  GOYAL...........(I.C.)..!   ( 7B,   84) PHYS. REV. 168,. . . . . . . . .   752, (1968).
                                 (16 ,   26) J. NUCL. ENERGY VOL.23,. . . . .   505, (1969).

1450  GOYAL...........(P.S.)..!   ( 6 , 1591) SOL. STATE COMM. VOL.8,. . . . .   889, (1970).
                                 ( 6 , 1311) J. PHYSIQUE TOME 32 COL.1 VOL.II   619, (1971).
                                 ( 6 , 1544) IAEA SYMP. GRENOBLE,. . . . . .    477, (1972).
                                 ( 6 ,   66) PHYS. STAT. SOL. VOL.50; . . . .   701, (1972).

1451  GOZANI..........(T.)....!   (24 ,   55) NUCL. INST. MET. 63, . . . . . .   307, (1968).

1452  GRABCEV.........(B.)....!   ( 6 ,  755) J. PHYS. CHEM. SOL. 28,. . . . .  1947, (1967).
                                 ( 6 ,  480) PHYS. LETT. 25A,. . . . . . . .    595, (1967).
                                 ( 6 ,  470) IAEA SYMP. COPENHAGEN, VOL.2,. .    75, (1968).
                                 ( 6 ,  759) J. APPL. PHYS. VOL.39, . . . . .   459, (1968).
                                 ( 6 ,  741) PHYS. LETT. VOL.26A,. . . . . .    392, (1968).
                                 ( 6 ,  805) ACTA CRYST. (INTERACT.) VOL.A25, S262, (1969).
                                 (18A,   22) NUCL. INST. MET. VOL.106,. . . .   349, (1973).
                                 ( 9B,   97) REPORT FN-43-1973 (31 PP.),. . .  , (1973).
                                 (17C,   15) ACTA CRYST. VOL.A30,. . . . . .    207, (1974).

1453  GRABCHEV........(B.)....!   (20A,   74) REV. ROUMAINE PHYS. VOL.8, . . .   277, (1963).
                                 ( 5 ,  437) REV. ROUMAINE PHYS. VOL.9,. . .    245, (1964).
                                 (17D,   15) REV. ROUMAINE PHYS. VOL.18,. . .   373, (1973).

1454  GRAF............(E.H.)..!   ( 6 , 1165) IAEA SYMP. GRENOBLE, . . . . . .   357, (1972).
                                 ( 6 , 1145) PHYS. REV. A VOL.5,. . . . . . .  1377, (1972).

1455  GRAF............(R.)....!   ( 5 ,   76) C.R. ACAD. SCI. B TOME 277,. . .   225, (1973).

1456  GRAHAM..........(A.J.)..!   ( 5 , 2394) CRYST. STRUCT. COMM. VOL.1 N.4,.   367, (1972).

1457  GRAHAM..........(K.F.)..!   (16 ,   33) NUCL. INST. MET. 85, . . . . . .   163, (1970).

1458  GRANT...........(D.F.)..!   (17B,   24) ACTA CRYST. B25, . . . . . . . .   374, (1969).

1459  GRANT...........(D.M.)..!   (17A,   34) IAEA SYMP. COPENHAGEN, VOL.2,. .   323, (1968).
                                 ( 6 ,  319) PHYS. REV. LETT. VOL.20,. . . .    983, (1968).
                                 ( 6 ,  652) J. CHEM. PHYS. VOL.50, . . . . .  1030, (1969).
                                 ( 6 , 1939) J. CHEM. PHYS. VOL.52, . . . . .  4424, (1970).
                                 ( 6 ,  313) J. CHEM. PHYS. VOL.58, . . . . .  1438, (1973).

1460  GRANT...........(P.J.)..!   ( 6 , 1037) PROC. ROY. SOC. 289, . . . . . .   342, (1966).

1461  GRANT...........(R.W.)..!   ( 5 , 1758) J. APP. PHYS. VOL.40,. . . . . .  1136, (1969).
                                 ( 5 ,  339) J. PHYS. CHEM. SOLIDS VOL.31,. .   793, (1970).

1462  GRASLUND........(C.)....!   ( 6 ,  322) PHYSICA VOL.59,. . . . . . . . .   672, (1972).

1463  GRASS...........(F.)....!   (20C,   45) NUCL. INST. MET. 46, . . . . . .   153, (1966).

1464  GRAY............(C.G.)..!   (14A,   40) PHYS. LETT. VOL.37A, . . . . . .   321, (1971).
                                 (15 ,   30) MOL. PHYS. VOL.25, . . . . . . .  1353, (1973).

1465  GREAVES.........(C.)....!   ( 5 , 2766) ACTA CRYST. VOL.B28, . . . . . .  3609, (1972).

1466  GREENBERGER.....(S.)....!   (22B,   32) NUCL. INST. MET. 75, . . . . . .   309, (1969).

1467  GREEN...........(J.L.)..!   ( 5 , 2363) J. NUCL. MATERIALS VOL.34, . . .   281, (1970).

1468  GREEN...........(J.M.)..!   ( 5 ,  924) NUCL. SCI. ENG. VOL.46,. . . . .   255, (1971).

1469  GREISS..........(H.B.)..!   (23 ,   47) KERNTECHNIK VOL.15,. . . . . . .   450, (1973).

1470  GRENIER.........(J.C.)..!   ( 5 , 2016) BULL. SOC. FRANC. MIN. CRIST. 88  345, (1965).

1471  GRENOT..........(M.)....!   ( 5 , 1608) J. PHYS. CHEM. SOLIDS VOL.28,. .  2441, (1967).

1472  GRENTHE.........(I.)....!   ( 5 , 2399) ACTA CRYST. VOL.B29, . . . . . .  2751, (1973).

1473  GRESS...........(M.E.)..!   ( 5 , 1511) DISS. ABSTR. B VOL.34, . . . . .  1958, (1973).

1474  GRIFFING........(G.W.)..!   ( 6 ,  255) IAEA SYMP. CHALK RIVER, VOL.1, .   435, (1962).
                                 ( 6 ,  257) PHYS. REV. VOL.127,. . . . . . .  1179, (1962).
                                 ( 6 ,  375) PHYS. REV. VOL.136,. . . . . . .  A988, (1964).
                                 ( 6 ,  268) BROOKHAVEN SYMPOSIUM,. . . . . .   149, (1965).
                                 (14B,   38) J. CHEM. PHYS. VOL.43, . . . . .  3328, (1965).
                                 ( 6 ,  231) PHYS. REV. VOL.148,. . . . . . .   163, (1966).

1475  GRIFFITHS.......(D.)....!   ( 5 , 1563) J. PHYS. CHEM. SOLIDS VOL.23,. .   621, (1962).

1476  GRIMES..........(S.M.)..!   ( 5 , 2438) THESIS (132PP.), . . . . . . . .   , (1969).

1477  GRIMM...........(H.)....!   ( 5 , 1456) PHYS. STAT. SOL. VOL.42, . . . .   207, (1970).
                                 ( 5 , 1459) REPORT JUL-696-FF, . . . . . . .     1, (1970).
                                 ( 5 , 1458) 1ST EUR. CONF./CONDENSED MATTER,   60, (1971).
                                 ( 5 , 1461) IAEA SYMP. GRENOBLE, . . . . . .   825, (1972).
                                 ( 6 , 1346) J. PHYS. CHEM. SOLIDS VOL.34,. .   255, (1973).

1478  GRISHANIN.......(E.I.)..!   ( 5 ,  379) UKR. FIZ. ZH.(USSR) VOL.17,. . .    38, (1972).

1479  GROENEWEGEN.....(P.P.M.)!   (10B,    5) ACTA CRYST. VOL.A28, . . . . . .   166, (1972).
                                 (10B,    4) ACTA CRYST. VOL.A28, . . . . . .   170, (1972).

1480  GROSESCU........(R.)....!   ( 6 , 1873) REV. ROUMAINE PHYS. VOL.18,. . .   313, (1973).

1481  GROSSHOG........(G.)....!   (22B,   38) NUCL. INST. MET. VOL.116,. . . .   561, (1974).

1482  GROS............(Y.)....!   ( 5 , 2942) J. PHYSIQUE TOME 32 COL.1 VOL.I,  320, (1971).

1483  GROTOWSKI.......(K.)....!   ( 6 ,  415) ACTA PHYS. POLON. VOL.16,. . . .   335, (1957).

1484  GROVES..........(D.J.)..!   ( 5 ,  151) PHYS. REV. VOL.146,. . . . . . .   660, (1966).
                                 ( 5 ,  766) PHYS. REV. VOL.174,. . . . . . .   313, (1968).

1485  GROVES..........(D.)....!   ( 5 , 2236) PHYS. REV. VOL.138,. . . . . . .B1116, (1965).

1486  GRUZIN..........(P.L.)..!   ( 6 ,  601) SOV. PHYS. SOL. STATE 7, . . . .   296, (1965).

1487  GSANGER.........(M.)....!   ( 6 ,   97) PHYS. LETT. 28A, . . . . . . . .   433, (1968).

1488  GSCHNEIDNER-JR..(K.A.)..!   ( 5 , 2707) ACTA CRYST. VOL.15,. . . . . . .   351, (1962).

1489  GUBBINS.........(K.E.)..!   (14A,   40) PHYS. LETT. VOL.37A, . . . . . .   321, (1971).
                                 (15 ,   30) MOL. PHYS. VOL.25, . . . . . . .  1353, (1973).

1490  GUDEL...........(H.U.)..!   ( 5 ,  776) J. CHEM. PHYSICS VOL.53, . . . .  1917, (1970).

1491  GUGGENHEIM......(H.J.)..!   ( 6 ,  814) J. APP. PHYS. 39,. . . . . . . .  1120, (1968).
                                 ( 6 , 1328) PHYS. REV. LETT. VOL.23, . . . .  1394, (1969).
                                 ( 5 , 1493) PHYS. REV. LETT. 22, . . . . . .   720, (1969).
                                 ( 5 ,  172) J. APPL. PHYS. 41, . . . . . . .   943, (1970).
                                 ( 5 , 1499) J. APPL. PHYS. 41; . . . . . . .   945, (1970).
```

```
1491  GUGGENHEIM.......(H.J.)...:  ( 5 , 1498) J. OF PHYS.-C., SER.2, VOL.3,.  .  1303, (1970).
                                   ( 6 , 1327) PHYS. REV. B-1,. . . . . . . .  2211, (1970).
                                   ( 6 , 1764) PHYS. REV. B-2,. . . . . . . .  1392, (1970).
                                   ( 6 ,  815) J. APPL. PHYS. VOL.42, . . . .  1376, (1971).
                                   ( 6 ,  811) J. OF PHYS-C. SER.2 VOL.44,. .   307, (1971).
                                   ( 5 ,  165) J. PHYSIQUE TOME 32 COL.1 VOL.II  759, (1971).
                                   ( 6 ,  812) IAEA SYMP. GRENOBLE,. . . . .    655, (1972).
                                   ( 5 , 2397) MAGN. AND MAG. MATERIALS-1971, :  670, (1972).
                                   ( 6 ,  817) PHYS. REV. B VOL.5,. . . . . .   154, (1972).
                                   ( 6 , 2000) PHYS. REV. B VOL.6,. . . . . .  2030, (1972).
                                   ( 5 ,  159) PHYS. REV. B VOL.6,. . . . . .  2677, (1972).
                                   ( 5 , 1483) PHYS. REV. B VOL.8,. . . . . .   304, (1973).
                                   ( 5 , 1583) PHYS. REV. LETT. VOL.32, . . .   610, (1974).
1492  GUILLEN..........(M.)....:  ( 5 , 1542) COMP. REND. 262, . . . . . . .  9628, (1966).
1493  GUINIER..........(A.)....:  ( 9B,   59) MOD. DIFF./IMAG. TECH./MAT. SCI,  521, (1970).
1494  GUNER............(Z.)....:  ( 6 , 2058) IAEA SYMP. CHALK RIVER, VOL.1,    237, (1962).
1495  GUNNING.........(J.E.)...:  (22A,   48) NUCL. INST. MET. VOL.114,. . .    417, (1974).
1496  GUNTHER..........(L.)....:  ( 9C,   32) PHYS. LETT. 29A, . . . . . . .    483, (1969).
1497  GUPTA...........(H.C.)...:  (10B,  160) PHYS. REV. B-2 METAL PHYS. VOL.1,  248, (1970).
                                   (10B,   85) J. PHYS.F: METAL PHYS. VOL.1,:    12, (1971).
                                   (10D,   50) PHYS. STAT. SOL. VOL.45,. . .    235, (1971).
                                   (10D,   48) PHYS. STAT. SOL. VOL.45 . . .    537, (1971).
                                   (10B,   76) J. PHYS. CHEM. SOLIDS VOL.34,.  1867, (1973).
                                   (10B,   98) J. PHYS. SOC. JAPAN VOL.34,. :  1006, (1973).
                                   ( 6 , 2018) PHYS. LETT. VOL.43A NO.4,. . .   365, (1973).
1498  GUPTA...........(J.B.)...:  ( 6 , 1804) INDIAN J. PHYS. VOL.42,. . . .    408, (1968).
1499  GUPTA...........(M.P.)...:  ( 5 , 1420) Z. KRIST. VOL.137, . . . . . .    173, (1973).
1500  GUPTA...........(N.P.)...:  (10B,   10) CAN. J. PHYS. 47,. . . . . . .    617, (1969).
                                   ( 6 , 1763) J. SOLID STATE CHEM. VOL.5,. :   477, (1972).
1501  GUPTA...........(R.K.)...:  (10B,   10) CAN. J. PHYS. 47,. . . . . . .    617, (1969).
                                   (10D,   46) PHYS. STAT. SOL. VOL.40, . . .   K73, (1970).
1502  GUPTA...........(R.P.)...:  ( 6 ,  776) PHYS. STAT. SOL. 12, . . . . .    305, (1965).
                                   (10B,   96) J. PHYS. SOC. JAPAN VOL.21,. .  2208, (1966).
                                   ( 6 , 2316) PHYS. STAT. SOL. 13, . . . . .   519, (1966).
                                   ( 6 , 1753) SOL. STATE COMM. 4,. . . . . .    83, (1966).
                                   ( 6 , 2202) REV. ROUMAINE PHYS. VOL.14,. .   247, (1969).
                                   (10B,  212) SOL. STATE COMM. VOL.8,. . . .   991, (1970).
                                   (10B,  167) PHYS. REV. LETT. VOL.26,. . .  1324, (1971).
                                   (10B,  189) PHYS. REV. B VOL.9,. . . . . .  2573, (1974).
1503  GUPTA...........(R.)....:  ( 6 ,   18) PHYS. REV. VOL.126,. . . . . .    933, (1962).
1504  GUPTA...........(S.C.)...:  ( 5 ,   16) ACTA CRYST. VOL.B30, . . . . .    562, (1974).
1505  GUPTA...........(V.D.)...:  ( 6 ,  421) MOL. PHYS. VOL.19,. . . . . .    589, (1970).
                                   ( 6 , 1912) NUCL./S.S. PHYS. SYMP. ABSTR.,  . . ,  (1972).
1506  GUREVICH........(A.G.)...:  ( 6 , 2291) SOV. PHYS. SOL. STATE VOL.10,.    511, (1968).
1507  GUREVICH........(I.I.)...:  (24 ,  111) SOV. PHYS. JETP VOL.14,. . . .    838, (1962).
1508  GUREWITZ........(E.)....:  ( 5 , 2403) PHYS. REV. B VOL.2,. . . . . .  3703, (1970).
                                   ( 5 , 2401) PHYS. REV. LETT. VOL.25; . . .  1713, (1970).
                                   ( 5 , 1440) PHYS. REV. B VOL.9,. . . . . .  1071, (1974).
1509  GURGENISHVILI...(G.E.)...:  (11C,   58) SOV. PHYS. J.E.T.P. VOL.29,. .  1089, (1969).
1510  GURMEN..........(E.)....:  ( 5 , 1413) PHYS. REV. 186,. . . . . . . .    705, (1969).
                                   ( 6 ,  827) J. APPL. PHYS. 41,. . . . . .  1363, (1970).
                                   ( 5 ,  191) J. CHEM. PHYSICS VOL.55,. . .  1093, (1971).
                                   ( 6 ,  528) IAEA SYMP. GRENOBLE, . . . . .    53, (1972).
                                   ( 6 ,  743) MAGN. AND MAG. MATERIALS-1971,  1340, (1972).
                                   ( 6 ,  529) A.I.P. CONF. PROC. NO.10 PART 1,   41, (1973).
                                   ( 5 , 1929) J. APPL. PHYS. VOL.45, . . . .  2021, (1974).
1511  GURSKAYA.........(A.V.)...:  ( 5 , 2011) SOV. PHYS. SOL. STATE VOL.10,:  1076, (1968).
                                   ( 5 , 2515) SOV. PHYS. SOL. STATE VOL.10,.   212, (1968).
1512  GUSTAVSSON......(B.)....:  (17B,   28) ARK. FYS. VOL.34, PAPER 40,. .    481, (1967).
1513  GUTTMAN.........(L.)....:  ( 5 ,  932) PHYS. REV. VOL.127, . . . . .  2052, (1962).
                                   (17A,   61) NUCL. INST. MET. VOL.25; . . .   188, (1963).
1514  GUYADER.........(J.)....:  ( 9C,   37) REVUE CHIM. MINER.(FRANCE) VOL.8  185, (1971).
                                   ( 5 , 1710) SOL. STATE COMM. VOL.11 NO.11,  1485, (1972).
1515  GUYOT...........(P.)....:  ( 5 ,   76) C.R. ACAD. SCI. B TOME 277,. .    225, (1973).
1516  GVOZDEV.........(M.M.)...:  (20C,   20) INSTRUM. EXPER. TECH. NO.3,. .    529, (1968).
1517  GWAN-TING.......(OU)....:  ( 5 ,  991) ACTA PHYS. SINICA VOL.21,. . .  1304, (1965).
1518  GYLDEN-HOUMANN..(J.C.)...:  ( 6 , 2127) IAEA SYMP. COPENHAGEN, VOL.2,.     29, (1968).
1519  HAAS............(C.)....:  ( 5 , 1810) J. PHYSIQUE TOME 32 COL.1 VOL.I,   78, (1971).
                                   ( 5 ,  619) J. PHYS. CHEM. SOLIDS VOL.32,:   581, (1971).
1520  HAAS............(F.X.)...:  ( 5 , 2237) PROC/CONF/NEUTRON C. S. + TECH.,  851, (1968).
1521  HAAS............(L.)....:  ( 6 , 2225) IAEA SYMP. CHALK RIVER, VOL.2,    155, (1962).
1522  HAAS............(M.)....:  ( 6 , 1843) SOL. STATE COMM. 4,. . . . . .     51, (1966).
1523  HAAS............(R.)....:  (20A,   76) REV. SCI. INSTRUM. VOL.30,. . .    17, (1959).
                                   ( 6 ,  166) IAEA SYMP. CHALK RIVER, VOL.1,   139, (1962).
                                   ( 6 , 2226) IAEA SYMP. CHALK RIVER, VOL.2;   145, (1962).
1524  HABENSCHUSS......(M.)....:  ( 5 ,  884) PHYS. LETT. VOL.45A NO.4,. . .    281, (1973).
1525  HACKNEY.........(R.)....:  ( 6 , 2205) PROC/CONF/FAST CRIT. EXP./ANALY.   524, (1967).
1526  HAGEN...........(S.)....:  ( 6 , 1960) PROPERTIES LIQUID METALS. . . .    111, (1973).
1527  HAGGSTROM.......(L.)....:  ( 5 , 1105) J. PHYS. C VOL.7,. . . . . . .   1344, (1974).
1528  HAGIHARA........(S.)....:  (21 ,   32) MITSUB. DENKI LAB. REP. VOL.3,    111, (1962).
1529  HAGIWARA........(S.)....:  (22B,   16) J. PHYS. SOC. JAPAN VOL.17 B-II,  358, (1962).
1530  HAHN............(H.)....:  (10E,   10) IAEA SYMP. CHALK RIVER, VOL.1;     37, (1962).
                                   (10C,   34) IAEA SYMPOSIUM BOMBAY, VOL.2,. .  279, (1964).
```

```
1531  HAHN...........(J.)....:   (22B,    45) REV. SCI. INSTRUM. VOL.31, . . .   490, (1960).

1532  HALG...........(G.)....:   ( 5 ,   776) J. CHEM. PHYSICS VOL.53, . . . . 1917, (1970).

1533  HALG...........(W.)....:   ( 5 ,    78) COLL. INTER. N.126 (GRENOBLE), .   23, (1963).
                                  (22B,     6) AUTOMATIC/NUCLEAR DATA, . . . .   233, (1964).
                                  (22B,    57) Z. ANGEW. MATH. PHYS. VOL.15,. .   481, (1964).
                                  ( 6 ,   473) SOL. STATE COMM. 3,. . . . . .   339, (1965).
                                  (22A,    73) Z. ANGEW. MATH. PHYS. VOL.16,. .   817, (1965).
                                  ( 5 ,  2953) Z. ANGEW. MATH. PHYS. VOL.16,. .   820, (1965).
                                  ( 5 ,  2890) ACTA. CRYST. 21,. . . . . . .   765, (1966).
                                  (19A,     9) HELV. PHYS. ACTA VOL.89, . . .   569, (1966).
                                  ( 6 ,  1729) SOL. STATE COMM. 4,. . . . . .    99, (1966).
                                  ( 5 ,  1654) ACTA CHEM. SCAND. VOL.21,. . . 1543, (1967).
                                  (24 ,    14) HELV. PHYS. ACTA VOL.41, . . .   868, (1968).
                                  ( 6 ,   464) IAEA SYMP. COPENHAGEN, VOL.2,.   133, (1968).
                                  ( 5 ,   914) PHYS. KONDENS. MAT. (GER.) VOL.9 249, (1969).
                                  ( 6 ,  1800) PHYS. STAT. SOL. VOL.42, . . .   821, (1970).
                                  ( 6 ,    68) SOL. STATE COMM. VOL.8,. . . .   165, (1970).
                                  ( 5 ,  2054) CONF./RARE EARTHS/ACTINIDES, .   206, (1971).
                                  ( 6 ,   569) PHYS. STAT. SOL. B VOL.46, . .   679, (1971).
                                  (19A,    72) Z. ANGEW. MATH. PHYS. VOL.22,.    62, (1971).
                                  ( 6 ,   466) IAEA SYMP. GRENOBLE, . . . . .   583, (1972).
                                  ( 5 ,  2888) HELV. PHYS. ACTA VOL.46, . . .   429, (1973).

1534  HALLMAN.........(E.O.)..:  ( 5 ,   693) MATERIALS RES. BULL. VOL.2,. .    69, (1967).
                                  (10C,    52) METAL. SOC. CONF. LOS ANGELES,   161, (1967).
                                  ( 6 ,   780) SOL. STATE COMM. 5,. . . . . .   211, (1967).
                                  (23 ,    33) IAEA SYMP. COPENHAGEN, VOL.2,.   259, (1968).
                                  ( 9 ,  1764) CAN. J. PHYS. 47,. . . . . . . 1117, (1969).
                                  ( 6 ,  1765) THESIS (183 PP.), . . . . . .     , (1969).

1535  HALLOCK.........(R.B.)..:  (14B,    93) PHYS. REV. A VOL.5,. . . . . .   320, (1972).
                                  ( 6 ,  1188) PHYS. REV. A VOL.8,. . . . . . 3244, (1973).

1536  HALL...........(E.L.)..:   ( 5 ,  1641) J. APPL. PHYS. 41,. . . . . . .   939, (1970).
                                  ( 5 ,  1642) DISS. ABS. VOL.31,. . . . . . .42788, (1971).
                                  ( 5 ,  1632) J. APPL. PHYS. 42,. . . . . .  1621, (1971).

1537  HALL...........(E.O.)..:   ( 5 ,  1184) ACTA CRYST. VOL. 18,. . . . .   857, (1965).
                                  ( 5 ,   726) J. PHYS. CHEM. SOLIDS 29,. . .   193, (1968).
                                  ( 5 ,   727) J. PHYS. CHEM. SOLIDS 29,. . .   201, (1968).

1538  HALL...........(J.W.)..:   (17B,    70) UKAEA AERE REPORT 5237 (13 PP.), . , (1966).
                                  (17B,    63) REPORT AERE-R5642 (8 PP.), . . . , (1967).
                                  (19B,    17) NUCL. INST. MET. 92, . . . . .    51, (1971).

1539  HALL...........(S.R.)..:   ( 5 ,    27) ACTA CRYST. 22,. . . . . . . .   216, (1967).

1540  HALL...........(T.)....:   ( 5 ,  1256) PHYS. REV. VOL.72, . . . . . . 1147, (1947).

1541  HALPERIN........(B.I.)..:  (13 ,    49) PHYS. REV. VOL.177,. . . . . .   952, (1969).
                                  (11A,    39) PHYS. REV. VOL.188,. . . . . .   898, (1969).
                                  ( 6 ,  2269) PHYS. REV. B-1, . . . . . . .   509, (1970).
                                  (11A,    44) PHYS. REV. B VOL.3,. . . . . .   961, (1971).

1542  HALPERIN........(J.)....:  ( 5 ,   414) NUCL. SCI. ENG. VOL.37,. . . .   228, (1969).
                                  ( 5 ,  1515) NUCL. SCI. ENG. VOL.47,. . . .   371, (1972).

1543  HALPERN.........(O.)....:  ( 8 ,    68) PHYS. REV. VOL.55, . . . . . .   898, (1939).
                                  (12B,    33) PHYS. REV. VOL.59, . . . . . .   960, (1941).
                                  ( 7B,    54) PHYS. REV. VOL.59, . . . . . .   981, (1941).
                                  ( 9B,    76) PHYS. REV. VOL.76, . . . . . . 1117, (1949).
                                  ( 8 ,    71) PHYS. REV. VOL.76, . . . . . . 1130, (1949).
                                  ( 7B,    60) PHYS. REV. VOL.76, . . . . . . 1811, (1949).
                                  ( 8 ,    77) PHYS. REV. VOL.88, . . . . . . 1003, (1952).
                                  ( 8 ,    78) PHYS. REV. VOL.88, . . . . . .   232, (1952).
                                  ( 6 ,  1784) PHYS. REV. VOL.90, . . . . . .   869, (1953).
                                  ( 7A,    67) PHYS. REV. VOL.126,. . . . . .   632, (1962).
                                  ( 8 ,    88) PHYS. REV. VOL.133,. . . . . .  B579, (1964).
                                  ( 8 ,    85) PHYS. REV. VOL.133,. . . . . .  B581, (1964).
                                  (12B,    45) PHYS. REV. VOL.134,. . . . . .  A126, (1964).

1544  HAMAGUCHI.......(Y.)....:  ( 5 ,  2746) J. PHYS. SOC. JAPAN VOL.17 B-III,  46, (1962).
                                  (23 ,    44) J. PHYS. SOC. JAPAN VOL.17 B-II,  354, (1962).
                                  ( 5 ,  1827) COLL. INTER. N.126 (GRENOBLE),   144, (1963).
                                  ( 5 ,   582) COLL. INTER. N.126 (GRENOBLE),    38, (1963).
                                  ( 5 ,   587) J. PHYS. SOC. JAPAN VOL.19,. . 1849, (1964).
                                  (20A,    26) J. PHYS. SOC. JAPAN VOL.19,. . 2280, (1964).
                                  ( 5 ,   639) PHYS. REV. 138,. . . . . . . . A737, (1965).
                                  ( 5 ,   641) PHYS. LETT. 24A,. . . . . . .   299, (1967).
                                  ( 6 ,   629) IAEA SYMP. COPENHAGEN, VOL.1,.   181, (1968).
                                  ( 6 ,   525) J. APP. PHYS. 39,. . . . . . . 1227, (1968).
                                  ( 6 ,   522) J. PHYS. SOC. JAPAN VOL.32,. .   394, (1972).

1545  HAMA...........(J.)....:   (15 ,    44) PROGR. THEOR. PHYS. VOL.46,. . 1666, (1971).
                                  (15 ,    45) PROGR. THEOR. PHYS. VOL.50,. . 1142, (1973).

1546  HAMERMESH.......(M.)....:  ( 7B,    54) PHYS. REV. VOL.59, . . . . . .   981, (1941).
                                  ( 8 ,    69) PHYS. REV. VOL.61, . . . . . .    17, (1942).
                                  (12B,    34) PHYS. REV. VOL.64, . . . . . .    47, (1943).
                                  ( 6 ,   660) PHYS. REV. VOL.69, . . . . . .   145, (1946).
                                  ( 5 ,  1253) PHYS. REV. VOL.71, . . . . . .   678, (1947).
                                  (12B,    35) PHYS. REV. VOL.74, . . . . . . 1285, (1948).
                                  ( 5 ,   846) PHYS. REV. VOL.85, . . . . . .   483, (1952).

1547  HAMILTON........(W.C.)..:  ( 5 ,  2277) ACTA CRYST. VOL.10,. . . . . .   103, (1957).
                                  ( 9B,     3) ACTA CRYST. VOL.10,. . . . . .   629, (1957).
                                  ( 9D,     3) ACTA CRYST. VOL.11,. . . . . .   585, (1958).
                                  ( 5 ,  1089) PHYS. REV. VOL.110,. . . . . . 1050, (1958).
                                  ( 5 ,   312) ACTA CRYST. VOL.14,. . . . . .    95, (1961).
                                  ( 5 ,  1946) J. PHYS. CHEM. VOL.65, . . . . 1453, (1961).
                                  ( 5 ,  1169) ACTA CRYST. VOL.15,. . . . . .   353, (1962).
                                  ( 5 ,  1266) ANN. REV. PHYS. CHEM. VOL.13,.    19, (1962).
                                  ( 5 ,  1264) J. PHYS. SOC. JAPAN VOL.17 B-II,  374, (1962).
                                  ( 5 ,  1276) J. PHYS. SOC. JAPAN VOL.17 B-II,  383, (1962).
                                  ( 5 ,  1275) ACTA CRYST. 16,. . . . . . . . 1209, (1963).
                                  (17C,     1) ACTA CRYST. 16,. . . . . . . .   609, (1963).
                                  ( 6 ,   416) INORG. CHEM. VOL.5,. . . . . . 2238, (1966).
                                  ( 5 ,  2240) J. CHEM. PHYSICS VOL.44, . . . 1748, (1966).
                                  ( 5 ,  1942) J. CHEM. PHYSICS VOL.44, . . . 2499, (1966).
                                  ( 5 ,  1937) J. CHEM. PHYSICS VOL.45, . . . 4498, (1966).
                                  ( 5 ,  1943) J. CHEM. PHYSICS VOL.45, . . .   408, (1966).
                                  ( 5 ,  1572) J. PHYS. CHEM. SOL. 27,. . . . 1013, (1966).
                                  (22B,    41) REP. NO. BNL-10238 (3 PP.),. .     , (1966).
                                  ( 9C,    43) TRANS. AMER. CRYST. ASSOC. VOL.2   53, (1966).
                                  ( 5 ,  2246) J. CHEM. PHYSICS VOL.47, . . . 1818, (1967).
                                  ( 5 ,  1581) J. PHYS. CHEM. SOL. 28,. . . . 1693, (1967).
                                  ( 5 ,   310) ACTA CRYST. VOL.B24, . . . . . 1147, (1968).
                                  ( 9C,    20) J. CHEM. EDUC. VOL.45,. . . . .   296, (1968).
                                  ( 5 ,  1448) J. MOLEC. STRUC. VOL.1,. . . .   283, (1968).
                                  ( 5 ,  1269) STRUC. CHEM. MOL. BIOL. . . . .   466, (1968).
                                  (17B,    20) ACTA CRYST. VOL.A25, . . . . .   194, (1969).
                                  ( 5 ,   275) DISC. FARADAY SOC. NO.48,. . .   192, (1969).
                                  ( 5 ,  1282) INORG. CHEM. VOL.8,. . . . . . 1928, (1969).
                                  (15 ,    25) MOL. CRYST. LIQUID CRYST. VOL.9,   11, (1969).
```

```
1547  HAMILTON........(W.C.)..:  ( 9C,     26) NAT. BUR. STAND., SPEC. PUB. 301   193, (1969).
                                 ( 5,     793) PHYS. ICE.. . . . . . . . . . .      44, (1969).
                                 (17A,     11) ACTA CRYST. VOL.A26, . . . . . .     48, (1970).
                                 (17A,     10) ACTA CRYST. VOL.A26, . . . . . .      7, (1970).
                                 (17A,     12) ACTA CRYST. VOL.A26, . . . . . .     71, (1970).
                                 ( 5,    1940) ACTA CRYST. VOL.B26, . . . . . .    536, (1970).
                                 ( 5,     415) J. PHARM. PHARMACOL. VOL.22, . .    724, (1970).
                                 ( 5,     294) ACTA CRYST. VOL.B27, . . . . . .   2393, (1971).
                                 ( 6,    2022) AMER. CRYSTALL. ASSOC., . . . .      27, (1971).
                                 ( 5,     229) AMER. MINERAL. VOL.56, . . . .      101, (1971).
                                 ( 5,    2093) J. CHEM. PHYS. VOL.54, . . . . .   3990, (1971).
                                 ( 5,    1292) J. CHEM. PHYS. VOL.55, . . . . .   1934, (1971).
                                 ( 5,       4) ACTA CRYST. A28 PART S-4, . . .     193, (1972).
                                 ( 5,     311) ACTA CRYST. VOL.B28, . . . . .     2083, (1972).
                                 ( 5,      13) ACTA CRYST. VOL.B28, . . . . .     3006, (1972).
                                 ( 5,      24) ACTA CRYST. VOL.B28, . . . . .     3207, (1972).
                                 ( 5,    2725) INORG. CHEM. VOL.11, . . . . .     3009, (1972).
                                 ( 5,      22) INT. J. PEPTIDE PROTEIN RES. 4, .   229, (1972).
                                 ( 5,      10) J. AMER. CHEM. SOC. VOL.94, . .    2657, (1972).
                                 ( 5,     981) J. CHEM. PHYS. VOL.56, . . . . .   3257, (1972).
                                 ( 5,      35) J. CHEM. PHYS. VOL.56, . . . . .   4433, (1972).
                                 ( 5,      20) J. CRYST. MOL. STRUC. VOL.2, . .    225, (1972).
                                 ( 5,      30) MATER. RES. BULL. VOL.7, . . . .   1225, (1972).
                                 ( 5,    2350) NATURE VOL.235, . . . . . . . .     328, (1972).
                                 ( 5,      21) ACTA CRYST. VOL.B29, . . . . .     2571, (1973).
                                 ( 5,     291) ACTA CRYST. VOL.B29, . . . . .      231, (1973).
                                 ( 5,      33) ACTA CRYST. VOL.B29, . . . . .      876, (1973).
                                 ( 5,      11) J.C.S. PERKIN TRANS. II, . . . .    133, (1973).
                                 ( 5,      25) J. CHEM. PHYS. VOL.58 NO.6, . .    2547, (1973).
                                 ( 5,     786) J. CHEM. PHYS. VOL.58, . . . . .    567, (1973).
                                 ( 5,    1594) J. CHEM. PHYS. VOL.59, . . . . .    915, (1973).
                                 ( 5,    1582) J. PHYS. CHEM. SOLIDS VOL.34, .     521, (1973).

1548  HAMMANN.........(J.)....:  ( 5,     883) J. PHYSIQUE TOME 29, . . . . . .    495, (1968).
                                 ( 5,    2045) PHYS. LETT. VOL.26A, . . . . . .    263, (1968).
                                 ( 5,    1366) ACTA CRYST. B25, . . . . . . .     1853, (1969).
                                 ( 5,     457) PHYSICA VOL.43, . . . . . . . .     277, (1969).

1549  HAMOUDA.........(I.)....:  (22A,     37) NUCL. INST. MET. 40, . . . . .      153, (1966).
                                 ( 5,    2510) J. NUCL. ENERGY VOL.21, . . . .     425, (1967).
                                 ( 5,    2933) J. NUCL. ENERGY VOL.22, . . . .     389, (1968).
                                 (20B,     87) NUCL. INST. MET. 60, . . . . .      349, (1968).
                                 ( 5,    2965) ATOMKERNENERGIE VOL.22, . . . .      87, (1973).
                                 (19A,     56) NUCL. INST. MET. VOL.116, . . .     509, (1974).

1550  HANDEL..........(P.H.)..:  ( 8,     163) Z. NATURFORSCH VOL.24A, . . . .    1646, (1969).
                                 ( 8,     100) PHYS. REV. LETT. VOL.28, . . .      596, (1972).
                                 ( 9B,     112) Z. PHYSIK BAND 252, . . . . . .      7, (1972).

1551  HANKE...........(W.)....:  (10A,      3) IAEA SYMP. GRENOBLE, . . . . .        3, (1972).

1552  HANLEY..........(P.)....:  (12B,     20) J. PHYS. E VOL.6 NO.8, . . . .      714, (1973).

1553  HANSEN..........(K.B.)..:  (22B,     25) NUCL. INST. MET. 41, . . . . .       61, (1966).
                                 (22C,     40) NUCL. INST. MET. 65, . . . . .      152, (1968).

1554  HANSEN..........(P.)....:  (11A,     65) Z. ANGEW PHYS. VOL.32, . . . .      104, (1971).

1555  HANSON..........(J.C.)..:  ( 5,    2504) ACTA CRYST. VOL.B29, . . . . .      797, (1973).

1556  HARADA..........(I.)....:  ( 6,    1494) J. PHYS. SOC. JAPAN VOL.36, . .     112, (1974).

1557  HARADA..........(J.)....:  ( 5,     184) ACTA CRYST. VOL.A26, . . . . .      336, (1970).
                                 ( 6,    1341) ACTA CRYST. VOL.A26, . . . . .      608, (1970).
                                 ( 6,    1340) PHYS. REV. P-1, . . . . . . . .    1227, (1970).
                                 ( 6,     124) PHYS. REV. B-2, . . . . . . . .    3651, (1970).
                                 ( 6,    1846) PHYS. REV. B-2, . . . . . . . .     155, (1970).
                                 ( 6,     129) PHYS. REV. B VOL.4, . . . . . .     155, (1971).
                                 ( 6,    2104) J. PHYS. C VOL.6, . . . . . . .    3021, (1973).

1558  HARDCASTLE......(K.I.)..:  (17B,     38) J. SCI. INSTRUM.(J. PHYS. E)S2,1   951, (1968).

1559  HARDELL.........(R.)....:  ( 5,     919) NUCL. PHYS. VOL.A123, . . . . .     215, (1969).
                                 ( 5,      96) PHYS. SCRIPTA VOL.1, . . . . . .     85, (1970).
                                 ( 5,    2068) PHYS. SCRIPTA(SWEDEN) VOL.4, . .     89, (1971).

1560  HARDY...........(J.R.)..:  ( 6,     200) IAEA SYMP. CHALK RIVER, VOL.2, .     49, (1962).
                                 (10B,     51) J. CHEM. PHYSICS VOL.48, . . . .   3173, (1968).
                                 (10B,     221) SOL. STATE COMM. VOL.14 NO.3, .     239, (1974).

1561  HARGROVE........(C.K.)..:  (22A,      7) CAN. JOURN. PHYS. 42, . . . . .    1593, (1964).

1562  HARIDASAN.......(T.M.)..:  ( 5,     660) PHYS LETT 28A, . . . . . . . .      301, (1968).

1563  HARIG...........(H.D.)..:  (23,      13) ENERGIE NUCLEAIRE VOL.13, . . .      15, (1971).

1564  HARKER..........(Y.D.)..:  ( 6,    1116) J. CHEM. PHYSICS VOL.42, . . .     1568, (1965).
                                 ( 6,     269) J. CHEM. PHYSICS VOL.42, . . . .    275, (1965).
                                 ( 6,     275) J. CHEM. PHYSICS VOL.46, . . . .   2201, (1967).

1565  HARLING.........(O.K.)..:  ( 6,    2342) BROOKHAVEN SYMPOSIUM, . . . . .      96, (1965).
                                 (20C,     42) NUCL. INST. MET. VOL.34, . . .      141, (1965).
                                 ( 6,    1061) PHYS. LETT. VOL.22, . . . . . .      15, (1966).
                                 (19B,     34) REV. SCI. INSTRUM. VOL.37, . . .    697, (1966).
                                 ( 6,    1108) BNWL-436, . . . . . . . . . . .           (1967).
                                 ( 6,     664) IAEA SYMP. COPENHAGEN, VOL.1, .     507, (1968).
                                 (23,      34) IAEA SYMP. COPENHAGEN, VOL.2, .     271, (1968).
                                 ( 6,     672) NUCL. SCI. ENGNG. VOL.33, . . .      41, (1968).
                                 ( 6,    1019) J. CHEM. PHYSICS VOL.50, . . . .   5279, (1969).
                                 ( 6,    1128) PHYS. REV. LETT. VOL.24, . . .     1046, (1970).
                                 (22B,     34) NUCL. INST. MET. VOL.96, . . . .    609, (1971).
                                 ( 6,    1172) PHYS. LETT. VOL.36A, . . . . . .    203, (1971).
                                 ( 6,    1146) PHYS. REV. A VOL.3, . . . . . .    1073, (1971).
                                 ( 6,    1169) PHYS. REV. A VOL.3, . . . . . .    1713, (1971).
                                 ( 6,    2345) PHYS. REV. B-4, . . . . . . . .    2675, (1971).
                                 (17B,     58) PHYS. REV. A VOL.7, . . . . . .    1748, (1973).

1566  HARMER..........(D.S.)..:  (24,      98) PROC/7TH/BIOMED/INSTRUM. SYMP...     93, (1969).

1567  HARMON..........(B.N.)..:  ( 5,    1210) A.I.P. CONF. PROC. NO.10 PART 2,  1309, (1973).

1568  HARMS...........(A.A.)..:  (24,      77) NUCL. INST. MET. VOL.104, . . .     217, (1972).

1569  HARMS...........(J.)....:  (12B,     12) IAEA SYMP. COPENHAGEN, VOL.2, .     387, (1968).
                                 (21,      39) NUCL. INST. MET. 58, . . . . .      261, (1968).

1570  HARP...........(G.D.)..:  (15,      42) PHYS. REV. A VOL.2, . . . . . .     975, (1970).

1571  HARRINGTON......(E.L.)..:  (17A,     86) PHYS. REV. VOL.50, . . . . . .      570, (1936).

1572  HARRISON........(R.J.)..:  (17C,      3) ACTA CRYST. 17, . . . . . . . .    325, (1964).

1573  HARRISON........(S.E.)..:  ( 5,    1610) PROC. I.E.E. VOL.104B SUP.4-7, .    217, (1957).
```

1574	HARRIS..........(A.B.)..:	(11A,	44)	PHYS. REV. B VOL.3,.	961,	(1971).
1575	HARRIS..........(D.H.C.):	(6 ,	208)	PHYS. LETT. VOL.7,.	220,	(1963).
		(20C,	40)	NUCL. INST. MET. 33,.	181,	(1965).
		(5 ,	236)	REPORT AERE-R 6052,.	., .	(1969).
		(14A,	45)	DISC. FARADAY SOC. NO.43,. . . .	193,	(1970).
		(23 ,	71)	REPORT AERE-R 6246 (25PP.),. . .	., .	(1970).
1576	HARRIS..........(D.)....:	(20B,	34)	IAEA SYMP. CHALK RIVER, VOL.1, .	107,	(1962).
		(20C,	11)	IAEA SYMP. CHALK RIVER, VOL.1, .	171,	(1962).
1577	HARRIS..........(M.J.)..:	(5 ,	1247)	PROC. PHYS. SOC., 85,.	79,	(1965).
1578	HARRIS..........(P.M.)..:	(5 ,	1891)	J. CHEM. PHYS VOL.35,.	1730,	(1961).
		(5 ,	768)	PHYS. REV. LETT. 15,.	586,	(1965).
		(5 ,	770)	PHYS. REV. LETT. 16,.	799,	(1966).
		(5 ,	67)	ACTA CRYST VOL.B24,.	954,	(1968).
		(5 ,	761)	J. CHEM. PHYSICS VOL.48,. . . .	1273,	(1968).
		(5 ,	760)	J. CHEM. PHYSICS VOL.49,. . . .	1922,	(1968).
1579	HARRYMAN........(M.B.M.):	(10B,	17)	CONF. INTERN., RENNES,.	209,	(1971).
1580	HARTLEY.........(R.H.)..:	(16 ,	39)	NUCL. SCI. ENGNG. VOL.28,. . . .	404,	(1967).
1581	HARTMANN........(W.M.)..:	(6 ,	955)	PROC. PHYS. SOC. 90,.	671,	(1967).
		(10B,	137)	PHYS. REV. 172,.	677,	(1968).
		(10D,	40)	PHYS. REV. LETT. VOL.28,. . . .	1261,	(1972).
1582	HARUNA..........(J.)....:	(5 ,	1725)	J. PHYS. SOC. JAPAN VOL.25,. . .	234,	(1968).
1583	HARVEY..........(A.R.)..:	(5 ,	2216)	A.I.P. CONF. PROC. NO.10 PART 1,	88,	(1973).
		(5 ,	2369)	PHYS. REV. B VOL.9,.	1041,	(1974).
		(5 ,	2212)	PHYS. REV. B VOL.9,.	3766,	(1974).
1584	HARVEY..........(J.)....:	(23 ,	28)	IAEA SYMP. (PILE RESEARCH)VIENNA	535,	(1960).
1585	HASEGAWA........(M.)....:	(15 ,	46)	PROPERTIES LIQUID METALS,. . . .	143,	(1973).
1586	HASER...........(R.)....:	(5 ,	1278)	ACTA CRYST. VOL.B28,.	2530,	(1972).
1587	HASE............(W.)....:	(20D,	36)	Z. KRIST. VOL.116,.	48,	(1961).
		(5 ,	2696)	PHYS. STAT. SOL. 2,.	K164,	(1962).
		(5 ,	832)	PHYS. STAT. SOL. 3,.	K446,	(1963).
		(5 ,	1409)	Z. KRIST. VOL.118,.	473,	(1963).
		(5 ,	2758)	PHYS. STAT. SOL. 21,.	K11,	(1967).
		(6 ,	808)	SOL. STATE COMM. VOL. 9,. . . .	353,	(1971).
1588	HASHIGUCHI......(R.)....:	(25 ,	65)	PROC. JAPAN ACAD. VOL.19,. . . .	26,	(1943).
		(5 ,	941)	PROC. PHYS. MATH. SOC. JAPAN 25,	495,	(1943).
		(5 ,	2463)	PROC. PHYS. MATH. SOC. JAPAN 25,	530,	(1943).
1589	HASHIMOTO.......(S.)....:	(5 ,	2977)	ACTA CRYST. VOL.A28 PART S-4,. .	99,	(1972).
		(5 ,	2978)	J. APPL. CRYST. VOL.7,.	67,	(1974).
1590	HASHIURA........(H.)....:	(5 ,	993)	SOL. STATE COMM. 4,.	657,	(1966).
1591	HASLER..........(H.G.)..:	(12A,	4)	ATOMKERNENERGIE VOL.7,.	170,	(1962).
1592	HASMAN..........(A.)....:	(6 ,	78)	PHYS. LETT. VOL.28A,.	642,	(1969).
		(14A,	34)	PHYSICA VOL.50,.	511,	(1970).
		(6 ,	74)	PHYS. LETT. VOL.33A,.	338,	(1970).
		(5 ,	106)	PHYS. LETT. VOL.34A,.	112,	(1971).
		(14B,	130)	THESIS,.	., .	(1971).
		(6 ,	90)	PHYSICA VOL.63 NO.3,.	499,	(1973).
1593	HASSELGREN......(A.)....:	(5 ,	919)	NUCL. PHYS. VOL.A123,.	215,	(1969).
		(5 ,	2068)	PHYS. SCRIPTA(SWEDEN) VOL.4,. .	89,	(1971).
1594	HASS............(M.)....:	(5 ,	1088)	PHYS. REV. VOL.112,.	1130,	(1958).
		(5 ,	1234)	J. AMER. CERAMIC SOC. VOL.53,. .	109,	(1970).
1595	HASTINGS........(J.M.)..:	(9B,	109)	TECHNIQUE ORG. CHEM. 1 PT.1, .	2361,	(1949).
		(20B,	126)	PHYS. REV. 83,.	863,	(1951).
		(5 ,	757)	J. APPL. PHYS. VOL.23,.	1379,	(1952).
		(5 ,	1607)	PHYS. REV. VOL.90,.	1013,	(1953).
		(5 ,	2944)	REV. MOD. PHYS. VOL.25,.	114,	(1953).
		(5 ,	1082)	PHYS. REV. VOL.93,.	893,	(1954).
		(11C,	31)	PHYS. REV. VOL.98,.	1721,	(1955).
		(5 ,	2945)	PHYS. REV. VOL.102,.	1460,	(1956).
		(5 ,	1694)	PHYS. REV. VOL.104,.	328,	(1956).
		(5 ,	1799)	PHYS. REV. VOL.104,.	924,	(1956).
		(5 ,	1830)	J. APPL. PHYS. VOL.29,.	364,	(1958).
		(9D,	59)	PHYS. REV. VOL.112,.	1917,	(1958).
		(5 ,	550)	PHYS. REV. LETT. VOL.3,.	211,	(1959).
		(5 ,	1831)	PHYS. REV. VOL.115,.	13,	(1959).
		(23 ,	18)	IAEA SYMP. (PILE RESEARCH)VIENNA	355,	(1960).
		(5 ,	596)	PHYS. REV. VOL.117,.	929,	(1960).
		(5 ,	624)	PHYS. REV. VOL.122,.	1402,	(1961).
		(5 ,	1671)	J. PHYS. SOC. JAPAN VOL.17 B-III	43,	(1962).
		(5 ,	1670)	PHYS. REV. VOL.126,.	556,	(1962).
		(5 ,	891)	COLL. INTER. N.126 (GRENOBLE), .	133,	(1963).
		(5 ,	1823)	J. APP. PHYS. 34,.	1192,	(1963).
		(5 ,	2582)	J. APP. PHYS. 35,.	1051,	(1964).
		(5 ,	2691)	J. APP. PHYS. 36,.	1001,	(1965).
		(5 ,	604)	J. APP. PHYS. 36,.	1099,	(1965).
		(5 ,	814)	PHYS. REV. 138,.	A176,	(1965).
		(5 ,	330)	J. APP. PHYS. 38,.	946,	(1967).
		(5 ,	375)	J. PHYS. CHEM. SOL. 28,.	1089,	(1967).
		(5 ,	331)	PHYS. REV. 160,.	408,	(1967).
		(6 ,	699)	J. APP. PHYS. 39,.	1232,	(1968).
		(5 ,	1822)	J. APP. PHYS. 39,.	461,	(1968).
		(5 ,	1179)	J. PHYS. CHEM. SOLIDS 29,. . . .	1359,	(1968).
		(5 ,	1351)	J. PHYS. CHEM. SOLIDS 29,. . . .	9,	(1968).
		(6 ,	1995)	J. APPL. PHYS. VOL.40,.	1278,	(1969).
		(6 ,	1985)	PHYS. REV. LETT. VOL.23,. . . .	1225,	(1969).
		(5 ,	816)	PHYS. REV. VOL.186,.	557,	(1969).
		(5 ,	813)	PHYS. REV. VOL.186,.	567,	(1969).
		(6 ,	1996)	J. APPL. PHYS. VOL.41,.	1384,	(1970).
		(5 ,	870)	J. APPL. PHYS. VOL.41,.	2326,	(1970).
		(5 ,	1600)	PHYS. REV. B-1,.	3116,	(1970).
		(6 ,	1981)	J. APPL. PHYS. 42,.	1378,	(1971).
		(6 ,	1978)	PHYS. REV. B-4,.	3206,	(1971).
		(6 ,	2441)	PHYS. REV. B VOL.4,.	2472,	(1971).
		(6 ,	1983)	PHYS. REV. LETT. VOL.26,. . . .	257,	(1971).
		(5 ,	2442)	PHYS. REV. LETT. VOL.27,. . . .	320,	(1971).
		(5 ,	183)	ACTA CRYST. VOL.B28,.	3429,	(1972).
		(6 ,	1998)	PHYS. REV. B VOL.8,.	1103,	(1973).
		(5 ,	812)	PHYS. REV. LETT. VOL.32,. . . .	544,	(1973).
1596	HASTINGS........(J.)....:	(5 ,	1422)	ACTA CRYST. VOL.14,.	1018,	(1961).
		(5 ,	1442)	ACTA CRYST. VOL.14,.	19,	(1961).
1597	HATHERLY........(M.)....:	(6 ,	830)	PROC. PHYS. SOC. 84,.	55,	(1964).

```
1598  HATTON .........(C.J.)..:  (22A,    8) CAN. JOURN. PHYS. 42,. . . . . .   2443, (1964).

1599  HAUG............(H.)....:  (14A,   56) PHYS. REV. A VOL.3,. . . . . . .    717, (1971).

1600  HAUMANN.........(J.)....:  (20B,   44) IEEE TRANS. NUCL. SCI. VOL.NS-13   311, (1966).

1601  HAUTECLER.......(S.)....:  (20B,   26) IAEA SYMPOSIUM VIENNA,. . . . .    453, (1960).
                                 ( 6 ,  254) IAEA SYMP. CHALK RIVER, VOL.1,.    423, (1962).
                                 ( 6 ,  252) IAEA SYMP. CHALK RIVER, VOL.2,.    281, (1962).
                                 ( 6 ,  258) Z. PHYSIK VOL.166,. . . . . . .    393, (1962).
                                 ( 6 ,  342) COLL. INTER. N.126 (GRENOBLE),.    224, (1963).
                                 ( 6 , 1755) COLL. INTER. N.126 (GRENOBLE),.    229, (1963).
                                 (17A,   48) INTERN. SUMMER SCHOOL, MOL,. .     580, (1963).
                                 ( 6 , 1756) BROOKHAVEN SYMPOSIUM., . . . .      83, (1965).
                                 ( 6 , 1742) PHYSICA VOL.34,. . . . . . . .     257, (1967).
                                 ( 6 ,  758) PHYSICA VOL.37,. . . . . . . .     603, (1967).
                                 (10B,   39) IAEA SYMP. COPENHAGEN, VOL.1,.      91, (1968).
                                 ( 6 , 2323) PHYS. LETT. VOL.31A,. . . . .        2, (1970).
                                 ( 6 , 2040) IAEA SYMP. GRENOBLE, . . . . .     489, (1972).
                                 ( 6 ,  297) PHYS. LETT. VOL.43A NO.2,. . .     189, (1973).

1602  HAVENS-JR.......(W.W.)..:  ( 5 , 1249) PHYS. REV. VOL.69,. . . . . . .    236, (1946).
                                 (20A,   60) PHYS. REV. VOL.70,. . . . . . .    136, (1946).
                                 ( 6 , 2112) PHYS. REV. VOL.71,. . . . . . .    165, (1947).
                                 ( 5 , 1407) PHYS. REV. VOL.71,. . . . . . .    174, (1947).
                                 (22A,   55) PHYS. REV. VOL.71,. . . . . . .     65, (1947).
                                 ( 5 , 2261) PHYS. REV. VOL.72,. . . . . . .    634, (1947).
                                 ( 5 , 2233) PHYS. REV. VOL.73,. . . . . . .   1399, (1948).
                                 ( 6 ,  975) PHYS. REV. VOL.73,. . . . . . .    733, (1948).
                                 ( 6 ,  598) PHYS. REV. VOL.73,. . . . . . .    963, (1948).
                                 (24 ,   34) NUCLEONICS VOL.5 DECEMBER,. .        4, (1949).
                                 ( 6 , 1465) PHYS. REV. VOL.75,. . . . . . .    895, (1949).
                                 (7B ,   40) NUCLEONICS VOL.6 NO.2,. . . .       66, (1950).
                                 (20A,   33) NUCLEONICS VOL.6 NO.4,. . . .       54, (1950).
                                 (24 ,  107) SCIENCE VOL.114, . . . . . . .     341, (1951).
                                 ( 5 , 2709) NUCL. SCI. ENGNG. VOL.6,. . .      433, (1959).
                                 (22A,   50) NUCL. SCI. ENGNG. VOL.8,. . .      453, (1960).
                                 ( 6 , 1578) PHYS. REV. LETT. VOL.5,. . . .     507, (1960).
                                 ( 5 , 2713) PHYS. REV. VOL.118,. . . . . .     799, (1960).
                                 ( 5 ,  146) PHYS. REV. VOL.119,. . . . . .    1291, (1960).
                                 (22A,   62) REV. SCI. INSTRUM. VOL.31,. .      481, (1960).
                                 (22B,   45) REV. SCI. INSTRUM. VOL.31,. .      490, (1960).
                                 ( 5 , 1265) J. CHEM. PHYS. VOL.35,. . . .     2265, (1961).
                                 ( 5 ,  141) NUCL. SCI. ENGNG. VOL.9,. . .       98, (1961).
                                 ( 6 ,   67) J. CHEM. PHYSICS VOL.37,. . .      234, (1962).
                                 ( 6 , 1550) NUCL. SCI. ENGNG. VOL.14,. . .     339, (1962).
                                 (22A,   64) REV. SCI. INSTRUM. VOL.36,. .      887, (1965).
                                 ( 6 , 1577) J. CHEM. PHYSICS VOL.48, . . .    4912, (1968).

1603  HAVLIN..........(S.)....:  ( 6 , 1157) PHYS. LETT. VOL.42A NO.2,. . .     133, (1972).

1604  HAYES...........(W.)....:  (10B,   74) J. PHYS. C VOL.6,. . . . . . .    1149, (1973).

1605  HAYTER..........(J.B.)..:  (12A,   20) J. PHYS. E VOL.6 NO.8, . . . .     714, (1973).

1606  HAYWOOD.........(B.C.G.):  (20B,   20) IAEA SYMP. (PILE RESEARCH)VIENNA  549, (1960).
                                 ( 6 ,  220) CARBON., 7,. . . . . . . . . .     663, (1969).
                                 ( 6 , 1500) J. PHYS. C, SER.2, VOL.2,. . .      46, (1969).
                                 ( 6 ,   87) J. PHYS. C, SER.2, VOL.3,. . .     249, (1970).
                                 ( 6 , 1435) J. PHYSIQUE TOME 32 COL.1 VOL.II 1186, (1971).
                                 ( 6 ,   98) J. PHYS. C, SER.2, VOL.4,. . .     910, (1971).
                                 (22A,   25) J. PHYS. E VOL.6 NO.6,. . . .      568, (1973).
                                 (20A,    6) ACTA CRYST. VOL.A30, . . . . .     448, (1974).

1607  HAYWOOD.........(B.C.)..:  ( 6 ,  191) IAEA SYMP. CHALK RIVER, VOL.1,    249, (1962).
                                 ( 6 , 1113) IAEA SYMP. CHALK RIVER, VOL.1,    343, (1962).
                                 ( 6 ,  199) IAEA SYMP. CHALK RIVER, VOL.2,    111, (1962).
                                 ( 6 , 2213) IAEA SYMP. CHALK RIVER, VOL.2,    213, (1962).
                                 ( 6 ,  969) PHYS. LETT. VOL.12,. . . . . .     188, (1964).
                                 ( 6 ,  679) UKAEA AERE REP. 4582 (13 PP.),   . . ., (1964).
                                 ( 6 , 1062) UKAEA AERE REP. 4484 (16 PP.),   . . ., (1964).
                                 ( 6 ,   81) PHYS. LETT. VOL.19,. . . . . .     269, (1965).
                                 ( 6 ,  965) PROC. PHYS. SOC. 90,. . . . .      681, (1967).
                                 ( 6 , 1077) CONF. NEUTRON THERMALIZ. VOL.I,.  361, (1968).
                                 ( 6 ,  566) J. CHEM. PHYS. VOL.52,. . . .     1828, (1970).
                                 ( 6 ,    4) J. CHEM. PHYS. VOL.52,. . . .     5740, (1970).
                                 (18A,    3) ACTA CRYST. VOL.A27,. . . . .      408, (1971).
                                 ( 6 , 1485) J. PHYS. C: SOLID ST. PHYS. V-4, 1299, (1971).

1608  HAY.............(H.J.)..:  (20A,    1) ACTA CRYST. VOL.11,. . . . . .     228, (1958).
                                 ( 5 , 2029) J. NUCL. ENERGY VOL.7,. . . .      199, (1958).
                                 (20B,   20) IAEA SYMP. (PILE RESEARCH)VIENNA  549, (1960).
                                 (19A,   10) IAEA SYMP. (PILE RESEARCH)VIENNA  559, (1960).
                                 (22A,   11) IAEA SYMP. (PILE RESEARCH)VIENNA  597, (1960).

1609  HAZELL..........(A.C.)..:  ( 5 ,  277) ACTA CRYST. VOL.B28, . . . . .    2977, (1972).

1610  HAZELL..........(R.G.)..:  ( 5 ,  284) ACTA CRYST. VOL.B28, . . . . .    1388, (1972).

1611  HEALD...........(P.T.)..:  (10B,  193) PHYS. STAT. SOL.46,. . . . . .     361, (1971).

1612  HEALEY..........(D.C.)..:  ( 5 ,  432) PHYS. REV. LETT. VOL.25,. . .      117, (1970).
                                 ( 5 ,  430) REPORT HEPL-630 (14PP.),. . .      . ., (1970).

1613  HEARD...........(C.R.T.):  (24 ,  100) REPORT AERE NP/GEN/13, . . . .     . ., (1960).
                                 (14A,   25) J. PHYS. C VOL.4,. . . . . . .    1453, (1971).

1614  HEARN...........(C.J.)..:  ( 5 ,  543) PROC. PHYS. SOC. VOL.8,. . . .     893, (1965).

1615  HEATON..........(L.)....:  ( 5 , 2654) ACTA CRYST. VOL.9, . . . . . .     607, (1956).
                                 ( 5 , 2669) J. APPL. PHYS. VOL.27,. . . .     1040, (1956).
                                 ( 5 , 2669) PHYS. REV. VOL.105,. . . . . .     216, (1957).
                                 (9B ,   51) J. APPL. PHYS. VOL.30,. . . .      1323, (1959).
                                 (22C,   50) REV. SCI. INSTRUM. VOL.31,. .      1355, (1960).
                                 ( 5 , 1593) J. CHEM. PHYS. VOL.34,. . . .       873, (1961).
                                 ( 5 , 1520) PHYS. REV. VOL.121,. . . . . .     649, (1961).
                                 (21 ,   43) REV. SCI. INSTRUM. VOL.32,. .      456, (1961).
                                 (20B,  146) REV. SCI. INSTRUM. VOL.34,. .      847, (1963).
                                 (22C,   53) REV. SCI. INSTRUM. VOL.34,. .       74, (1963).
                                 ( 5 , 1974) J. CHEM. PHYSICS VOL.44,. . .     1758, (1966).
                                 (14A,    6) ADV. IN PHYSICS VOL.16,. . .       177, (1967).
                                 ( 5 , 2431) J. CHEM. PHYSICS VOL.46,. . .      586, (1967).
                                 ( 5 , 2211) J. PHYS. CHEM. SOL. 28,. . . .    1651, (1967).
                                 ( 5 , 2457) J. PHYS. CHEM. SOL. 28,. . . .     423, (1967).
                                 ( 5 , 2719) J. PHYS. CHEM. SOLIDS 29,. . .    1702, (1968).
                                 ( 5 , 2799) J. PHYS. CHEM. SOLIDS 30,. . .     453, (1969).
                                 ( 5 , 2213) J. PHYS. CHEM. SOLIDS 30,. . .     733, (1969).
                                 (21 ,   21) J. APPL. CRYST. VOL.3,. . . .      289, (1970).

1616  HEBERT..........(D.)....:  (19B,   27) RC ACCAD. NAZ. LINCEI VOL.45,.    163, (1968).

1617  HEDLEY..........(I.G.)..:  ( 5 , 1097) J. OF PHYS. C VOL.1, . . . . .     179, (1968).
```

```
1618  HEER............(H.)....:  ( 5 , 2054)  CONF./RARE EARTHS/ACTINIDES, . .    206, (1971).
                                 ( 6 ,  466)  IAEA SYMP. GRENOBLE, . . . . . .    563, (1972).
                                 ( 6 , 2197)  INT. J. MAGN. VOL.4 NO.1,. . . .     63, (1973).
                                 ( 6 , 2196)  SOL. STATE COMM. VOL.12 NO.2,. .    117, (1973).
                                 ( 6 ,  716)  J. PHYS. C VOL.7,. . . . . . . .   1237, (1974).

1619  HEGER...........(G.)....:  ( 5 , 1441)  CONF. ON LOW-TEMP. PHYS.(ABSTR.)     31, (1972).
                                 ( 6 , 1286)  PHYS. STAT. SOL. B VOL.53 NO.1,.    227, (1972).
                                 ( 5 , 2398)  SOL. STATE COMM. VOL.10, . . . .   1299, (1972).
                                 ( 5 , 1988)  SOL. STATE COMM. VOL.11, . . . .   1119, (1972).
                                 ( 5 , 1986)  INT. J. MAGN. VOL.5, . . . . . .    119, (1973).
                                 ( 5 ,  289)  SOL. STATE COMM. VOL.12, . . . .   1157, (1973).
                                 ( 5 , 1484)  J. CRYST. GROWTH VOL.21, . . . .     82, (1974).

1620  HEIBERG.........(E.)....:  (17A,   63)  NUCL. INST. MET. 47, . . . . . .     89, (1967).
                                 (20C,   47)  NUCL. INST. MET. 52, . . . . . .    175, (1967).

1621  HEIDBREDER.......(E.)....:  (24 ,   68)  NUCL. INST. MET. 74, . . . . . .    109, (1969).

1622  HEIDEMANN........(A.)....:  (22A,   13)  IAEA SYMP. COPENHAGEN, VOL.2,. .    381, (1968).
                                 ( 6 , 2261)  Z. PHYS. VOL.238,. . . . . . . .    208, (1970).
                                 (22A,   42)  NUCL. INST. MET. 95, . . . . . .    435, (1971).
                                 ( 6 , 2262)  IAEA SYMP. GRENOBLE, . . . . . .    851, (1972).
                                 ( 6 , 2263)  PHYS. STAT. SOLIDI A VOL.16 NO.2 K139, (1973).
                                 ( 5 , 2498)  Z. PHYSIK VOL.258, . . . . . . .    429, (1973).
                                 ( 5 , 2447)  Z. PHYSIK VOL.263, . . . . . . .    291, (1973).
                                 ( 5 , 1949)  PHYS. LETT. VOL.47A NO.1,. . . .     91, (1974).

1623  HEIKER..........(D.)....:  (173,   39)  LATV. PSR...FIZ. TEHN. SER. NO.5  124, (1971).

1624  HEIKES..........(R.R.)..:  ( 5 , 1724)  PHYS. REV. VOL.125,. . . . . . .   1893, (1962).

1625  HEIMTZ..........(W.)....:  ( 9B,   96)  REPORT FMRB-37-71, . . . . . . . ,  . , (1971).

1626  HEINLOTH........(K.)....:  ( 5 , 1307)  IAEA SYMPOSIUM (VIENNA), . . . .    323, (1960).
                                 ( 5 , 1295)  Z. PHYSIK VOL.163, . . . . . . .    218, (1961).

1627  HEINRICH........(B.)....:  (13 ,    6)  AIP CONF. PROC. NO.10 PART 1,. .    822, (1973).

1628  HEINZELMANN.....(M.)....:  (20C,    7)  ATOMKERNENERGIE VOL.19 NO.4, . .    312, (1972).

1629  HELLEGOUARCH....(J.)....:  (22C,   75)  TECH. CEM (FRANCE) NO.84,. . . .     16, (1972).

1630  HELLER..........(P.)....:  ( 6 , 1446)  PHYS. REV. B-1,. . . . . . . . .   2304, (1970).
                                 ( 6 , 1451)  PHYS. REV. LETT. VOL.24, . . . .   1184, (1970).
                                 ( 6 , 1448)  J. APPL. PHYS. 42, . . . . . . .   1258, (1971).
                                 ( 6 , 1447)  PHYS. REV. B-4,. . . . . . . . .   2254, (1971).
                                 (13 ,   17)  DYN. ASPECTS OF CRITICAL PHEN. .     50, (1972).
                                 ( 6 , 1469)  A.I.P. CONF. PROC. NO.10 PART 1,     93, (1973).

1631  HELLER..........(R.B.)..:  ( 6 ,  767)  PHYS. REV. VOL.75, . . . . . . .    565, (1949).
                                 (12B,   39)  PHYS. REV. VOL.80, . . . . . . .    953, (1950).

1632  HENDERSON.......(B.)....:  ( 6 , 1422)  PROC. PHYS. SOC. 89, . . . . . .    153, (1966).

1633  HENDRICKS.......(R.W.)..:  ( 5 , 2002)  PHYS. STAT. SOLIDI B VOL.58 NO.2  633, (1973).

1634  HENDRICK........(L.D.)..:  (24 ,   92)  PHYS. REV. 145,. . . . . . . . .   1023, (1966).

1635  HENFREY.........(A.W.)..:  ( 5 , 2832)  ACTA CRYST. VOL.B26, . . . . . .   1882, (1970).

1636  HENNIG..........(K.)....:  ( 6 , 1936)  PHYS. STAT. SOL.45,. . . . . . .   K105, (1971).
                                 ( 6 , 1431)  IAEA SYMP. GRENOBLE, . . . . . .    157, (1972).
                                 (10C,   85)  PHYS. STAT. SOLIDI A VOL.58 NO.2 K159, (1973).

1637  HENNINGER.......(E.H.)..:  ( 5 , 1974)  J. CHEM. PHYSICS VOL.44, . . . .   1758, (1966).
                                 (14B,    6)  ADV. IN PHYSICS VOL.16,. . . . .    177, (1967).
                                 ( 5 , 2431)  J. CHEM. PHYSICS VOL.46, . . . .    586, (1967).
                                 ( 5 , 2457)  J. PHYS. CHEM. SOL. 28,. . . . .    423, (1967).

1638  HENNION.........(B.)....:  (23 ,   31)  IAEA SYMP. COPENHAGEN, VOL.2,. .    353, (1968).
                                 ( 6 ,  836)  SOL. STATE COMM. VOL.8,. . . . .   2141, (1970).
                                 ( 6 , 2331)  PHYS. LETT. VOL.36A, . . . . . .    376, (1971).
                                 ( 6 ,  623)  PHYS. REV. LETT. VOL.26, . . . .    770, (1971).
                                 ( 8 ,   13)  ANN. DE PHYSIQUE TOME 7 NO.4,. .    233, (1972).
                                 ( 6 , 1490)  IAEA SYMP. GRENOBLE, . . . . . .    631, (1972).
                                 ( 6 ,  627)  PHYS. REV. LETT. VOL.28, . . . .    964, (1972).
                                 ( 6 , 1486)  SOL. STATE COMM. VOL.10, . . . .    553, (1972).
                                 ( 6 ,  622)  SOL. STATE COMM. VOL.13, . . . .   1725, (1973).
                                 ( 6 ,  444)  SOL. STATE COMM. VOL.13, . . . .    919, (1973).

1639  HENRICH.........(E.)....:  ( 5 ,  289)  SOL. STATE COMM. VOL.12, . . . .   1157, (1973).

1640  HENRYSON........(H.)....:  ( 7A,   42)  NUCL. SCI. ENG. VOL.43,. . . . .    235, (1971).

1641  HENRY...........(A.W.)..:  ( 6 , 2045)  J. POLYMER SCI. A-2 VOL.7, . . .    433, (1969).

1642  HENSHAW.........(D.G.)..:  ( 5 , 1337)  PHYS. REV. VOL.91, . . . . . . .   1222, (1953).
                                 ( 5 , 2232)  PHYS. REV. VOL.92, . . . . . . .   1229, (1953).
                                 ( 5 , 1326)  PHYS. REV. VOL.100,. . . . . . .    994, (1955).
                                 ( 5 , 2779)  BULL. AMER. PHYS. SOC. SER.II 2,      9, (1957).
                                 ( 5 ,  109)  PHYS. REV. VOL.105,. . . . . . .    976, (1957).
                                 ( 6 , 1199)  PHYS. REV. LETT. VOL.1,. . . . .    127, (1958).
                                 ( 5 , 1330)  PHYS. REV. VOL.109,. . . . . . .    328, (1958).
                                 ( 5 , 2061)  PHYS. REV. VOL.111,. . . . . . .   1470, (1958).
                                 ( 5 , 1331)  PHYS. REV. VOL.119,. . . . . . .     14, (1960).
                                 ( 5 , 2235)  PHYS. REV. VOL.119,. . . . . . .     22, (1960).
                                 ( 5 , 1332)  PHYS. REV. VOL.119,. . . . . . .      9, (1960).
                                 ( 6 , 1176)  PHYS. REV. VOL.121,. . . . . . .   1266, (1961).

1643  HENSON..........(R.W.)..:  ( 6 ,  207)  PHIL. MAG. VOL.9,. . . . . . . .    659, (1964).

1644  HERAK...........(R.)....:  ( 5 , 2761)  ACTA CRYST. VOL.B25, . . . . . .   2505, (1969).
                                 ( 5 , 1354)  ACTA CRYST. VOL.A28 PART S-4,. .     60, (1972).
                                 ( 5 , 1474)  ACTA CRYST. VOL.B28, . . . . . .   3104, (1972).

1645  HERBSTEIN.......(F.H.)..:  ( 5 ,  969)  ACTA CRYST. 17,. . . . . . . . .   1331, (1964).

1646  HERDADE.........(S.B.)..:  ( 6 ,  688)  IAEA SYMP. COPENHAGEN, VOL.2,. .    197, (1968).
                                 (22A,   40)  NUCL. INST. MET. 63, . . . . . .     13, (1968).
                                 (14B,   41)  J. CHEM. PHYS. VOL.56, . . . . .   3118, (1972).
                                 ( 6 ,   59)  J. NUCL. ENERGY VOL.26,. . . . .    379, (1972).

1647  HERGET..........(P.)....:  ( 5 ,  681)  ACTA CRYST. (INTERACT.) VOL.A25, S218, (1969).

1648  HERPIN..........(A.G.)..:  ( 5 , 1703)  J. APP. PHYS. 37,. . . . . . . .    960, (1966).

1649  HERPIN..........(A.)....:  ( 5 , 2897)  COMP. REND. TOME 243,. . . . . .    898, (1956).
                                 ( 9B,   56)  J. PHYS. RADIUM VOL.18,. . . . .    649, (1957).
                                 ( 5 , 1647)  COMP. REND. TOME 246,. . . . . .   3170, (1958).
                                 ( 5 ,  978)  J. PHYS. RADIUM VOL.19,. . . . .    617, (1958).
                                 ( 5 , 1649)  COMP. REND. 249, . . . . . . . .   1334, (1959).
                                 ( 5 , 1648)  COMP. REND. 250, . . . . . . . .   1450, (1960).
                                 ( 5 , 1388)  COMP. REND. 251, . . . . . . . .   1359, (1960).
                                 (23 ,   23)  IAEA SYMP. (PILE RESEARCH)VIENNA   337, (1960).
                                 ( 5 , 1645)  J. PHYS. RADIUM VOL.21,. . . . .     67, (1960).
```

```
1649  HERPIN..........(A.)...:  ( 5 , 1646) J. PHYS. RADIUM VOL.22,. . . . . .    337, (1961).
                               ( 5 ,   86) COMP. REND. 259,. . . . . . . .     2416, (1964).
                               (12A,    7) COMP. REND. 264,. . . . . . . .      840, (1967).
                               ( 5 , 2392) J. PHYSIQUE TOME 32 COL.1 VOL.II     611, (1971).
                               ( 5 , 1361) J. PHYSIQUE TOME 34,. . . . . . .    423, (1973).

1650  HERPIN..........(P.)...:  ( 5 , 1445) COLL. INTER. N.126 (GRENOBLE),.      6C, (1963).
                               ( 5 , 1712) PHYS. STAT. SOL. VOL.44,. . . . .     71, (1971).

1651  HEWAT...........(A.W.).:  ( 6 , 2322) SOL. STATE COMM. VOL.8,. . . . .     187, (197C).
                               ( 5 , 1512) FERROELECT. VOL.4,. . . . . . .      153, (1972).
                               ( 6 , 1338) J. PHYSIQUE TOME 33 COL.2,. . .      133, (1972).
                               (10B,  195) PHYS. STAT. SOL. B VOL.53,. . . .     K33, (1972).
                               (21 ,   23) J. APPL. CRYST. VOL.6,. . . . . .     42, (1973).
                               ( 5 , 1489) J. PHYS. C VOL.6 NO.6,. . . . .     1074, (1973).
                               ( 5 , 1490) J. PHYS. C VOL.6,. . . . . . . .    2559, (1973).
                               ( 5 , 1902) NATURE VOL.246 NO.5428,. . . . .      90, (1973).
                               (10D,   54) REPORT U.K.A.E.A. R7350 (62 PP.) . . ,  (1973).
                               ( 6 , 1339) J. PHYS. C VOL.7,. . . . . . . .      15, (1974).

1652  HEYMAN..........(M.)...:  (12A,   33) REV. PHYS. APPL. (FRANCE) VOL.4,     115, (1969).
                               (12B,   50) REV. PHYS. APPL. (FRANCE) VOL.4,     254, (1969).

1653  HIBDON..........(C.T.).:  ( 5 , 1348) PHYS. REV. VOL.82,. . . . . . .      560, (1951).

1654  HICKMAN.........(B.S.).:  ( 6 ,  160) PHIL. MAG. VOL.8,. . . . . . . .      43, (1963).

1655  HICKS...........(T.J.).:  ( 6 , 15C7) PROC. PHYS. SOC. 86,. . . . . .      139, (1965).
                               ( 5 , 1773) J. OF PHYS. C VOL.1,. . . . . .     1683, (1968).
                               ( 5 , 2299) J. OF PHYS. C VOL.1,. . . . . .      528, (1968).
                               ( 6 ,  734) PHYS. REV. LETT. 22,. . . . . .      531, (1969).
                               ( 6 ,  631) PHYS. LETT. VOL.32A,. . . . . .      410, (197C).
                               ( 5 ,  516) PHYS. REV. B-2,. . . . . . . .      176, (1970).
                               ( 5 , 2089) J. PHYS. F VOL.3,. . . . . . . .     697, (1973).
                               ( 5 , 2115) PHYS. REV. B VOL.7,. . . . . .      218, (1973).

1656  HIGGINS.........(J.S.).:  ( 6 ,  362) MOLEC. PHYS. VOL.19,. . . . . .      645, (197C).
                               ( 6 ,  361) MOL. PHYS. VOL.21,. . . . . . .      721, (1971).
                               ( 6 , 2044) FARADAY SYMP. CHEM. SOC. NO.6,.      169, (1972).
                               ( 6 ,  371) IAEA SYMP. GRENOBLE,. . . . . .      461, (1972).
                               ( 6 , 1921) POLYMER VOL.13,. . . . . . . .       157, (1972).
                               (15 ,   51) REP. PROGR. PHYS. VOL.36,. . .      1073, (1973).
                               ( 6 , 2004) J.C.S. FARADAY TRANS. II VOL.70,     348, (1974).

1657  HIISMAKI........(P.)....:  ( 9B,    8) ACTA CRYST. VOL.A25,. . . . . .      378, (1969).
                               ( 9C,    6) ACTA POLYTECH. SCANDINAVIA NO.7J       6, (197C).
                               (19A,   24) IAEA SYMP. GRENOBLE,. . . . . .      803, (1972).

1658  HINKS...........(D.G.).:  ( 5 , 1423) J. CHEM. PHYS. VOL.56,. . . . .     3697, (1972).
                               ( 5 , 1957) J. CHEM. PHYS. VOL.58,. . . . .     2039, (1973).

1659  HIRABAYASHI.....(M.)...:  ( 5 ,  737) J. PHYS. SOC. JAPAN VOL.20,. .       381, (1965).
                               ( 5 , 2660) J. PHYS. SOC. JAPAN VOL.28,. .      1014, (197C).
                               ( 5 , 2977) ACTA CRYST. VOL.A28 PART S-4,.        99, (1972).
                               ( 5 , 1345) J. PHYS. SOC. JAPAN VOL.35,. .       473, (1973).
                               ( 5 , 2836) PHYS. STAT. SOLIDI A VOL.15 NO.1    267, (1973).
                               ( 5 , 2978) J. APPL. CRYST. VOL.7,. . . . .       67, (1974).

1660  HIRAGA..........(K.)...:  ( 5 , 2660) J. PHYS. SOC. JAPAN VOL.28,. .      1014, (197C).

1661  HIRAHARA........(E.)...:  ( 6 , 15C5) J. APPL. PHYS. 42,. . . . . . .     1374, (1971).

1662  HIRAI...........(Y.)...:  ( 6 , 1776) SOL. STATE COMM. VOL.13 NO.4,. .     495, (1973).

1663  HIRAKAWA........(K.)...:  ( 6 ,  830) PROC. PHYS. SOC. 84,. . . . . .       55, (1964).
                               ( 5 , 1434) PHYS. REV. VOL.188,. . . . . .       919, (1969).
                               ( 5 , 1482) ACTA CRYST. VOL.A28 PART S-4,.       175, (1972).
                               (21 ,   19) JAP. J. APPL. PHYS. VOL.11,. . .       82, (1972).
                               ( 5 , 1435) J. PHYS. SOC. JAPAN VOL.33,. .      1483, (1972).
                               ( 5 , 1480) J. PHYS. SOC. JAPAN VOL.33,. .       393, (1972).
                               ( 6 , 1273) J. PHYS. SOC. JAPAN VOL.35,. .      1328, (1973).
                               ( 6 , 1272) J. PHYS. SOC. JAPAN VOL.35,. .      1608, (1973).
                               ( 6 , 1266) J. PHYS. SOC. JAPAN VOL.35,. .      1795, (1973).
                               ( 6 , 1319) J. PHYS. SOC. JAPAN VOL.35,. .       617, (1973).
                               ( 5 , 1436) J. PHYS. SOC. JAPAN VOL.35,. .       722, (1973).
                               ( 6 , 1267) SOL. STATE COMM. VOL.14,. . . .      529, (1974).

1664  HIRONAKA........(K.)...:  (23 ,   48) MITSUBISHI DENHI ENG. NO.26,. .      10, (1970).

1665  HIRSCHFELDER....(J.O.).:  ( 7A,   70) PHYS. REV. 129,. . . . . . . .     1391, (1963).

1666  HIRSCH..........(P.B.).:  ( 7B,   45) PHIL. MAG. VOL.3,. . . . . . .       213, (1958).

1667  HITTERMAN.......(R.L.).:  ( 5 , 2705) PHYS. REV. VOL.129,. . . . . .       625, (1963).
                               ( 5 , 2639) ACTA CRYST. 20,. . . . . . . .       842, (1966).
                               ( 5 , 1439) ACTA CRYST. VOL.A26,. . . . . .       559, (197C).
                               (21 ,   21) J. APPL. CRYST. VOL.3,. . . . .       289, (197C).

1668  HNATOWICH.......(D.J.).:  (22C,    9) CAN. JOURN. PHYS. 41,. . . . . .     1519, (1963).

1669  HOCKENBURG......(R.W.).:  (24 ,   74) NUCL. INST. MET. 77,. . . . . .      245, (197C).
                               ( 5 , 2362) NUCL. PHYS. A VOL.164,. . . . .       34, (1971).

167C  HOCK-KEE-SIM............:  ( 6 , 1144) PHYS. REV. A VOL.1,. . . . . . .     1536, (197C).

1671  HODGES..........(L.)...:  ( 5 , 2087) J. APPL. PHYS. 37,. . . . . . .     1449, (1966).
                               ( 5 , 2088) PHYS. REV. VOL.152,. . . . . .       505, (1966).

1672  HODGKIN.........(D.C.).:  ( 5 , 2858) NATURE VOL.214,. . . . . . . .       130, (1967).

1673  HOEHNE..........(P.)...:  (22C,    6) ANN. PHYS. (GERMANY) VOL.7,. . .      50, (1961).

1674  HOFFMAN.........(C.M.).:  (20C,   75) NUCL. INST. MET. VOL.94,. . . .      493, (1971).

1675  HOFFMAN.........(W.)...:  ( 5 , 2460) NATURWISSENSCHAFTEN VOL.53,. . .      16, (1966).

1676  HOFFMAN.........(Z.)...:  (19B,   25) NUKLEONIKA VOL.8,. . . . . . .       695, (1963).

1677  HOFMANN.........(A.)...:  (24 ,   67) NUCL. INST. MET. 72,. . . . . .      3C1, (1969).

1678  HOGAN...........(W.S.).:  ( 7A,   73) PHYS. REV. VOL.177,. . . . . .      1706, (1969).

1679  HOGBERG.........(T.)...:  (10E,    2) ARK. FYS. VOL.34,PAPER 12,. . .      121, (1967).

168C  HOGE............(H.J.).:  ( 5 , 1255) PHYS. REV. VOL.52,. . . . . .      1076, (1937).
                               ( 5 , 1250) PHYS. REV. VOL.54,. . . . . .       266, (1938).

1681  HOGG............(S.)...:  ( 5 , 2959) ACTA CRYST. VOL.B25,. . . . . .     2254, (1969).

1682  HOHENBERG.......(P.C.).:  (13 ,   49) PHYS. REV. VOL.177,. . . . . .       952, (1969).
                               (11A,   39) PHYS. REV. VOL.188,. . . . . .       898, (1969).
                               (11A,   44) PHYS. REV. B VOL.3,. . . . . .       961, (1971).
```

```
1683  HOHENEMSER......(C.)....:  ( 6 , 1758) PHYS. REV. LETT. VOL.31, . . . . 1222, (1973).
1684  HOLAS...........(A.)....:  ( 6 , 2300) IAEA SYMP. BOMBAY, VOL.1,. . .    205, (1964).
                                 (19A,   61) NUKLEONIKA (POL.) VOL.13,. . . .  591, (1968).
                                 (19A,   60) NUKLEONIKA (POL.) VOL.13,. . . .  753, (1968).
                                 (22A,   31) NUCL. INST. MET. 69, . . . . .   173, (1969).
1685  HOLAS...........(J.)....:  (22A,   31) NUCL. INST. MET. 69, . . . . .   173, (1969).
1686  HOLDEN..........(T.M.)..:  ( 6 , 1859) PROC. PHYS. SOC. 89, . . . . .   119, (1966).
                                 ( 6 , 1734) J. APP. PHYS. 38, . . . . . .  1240, (1967).
                                 ( 6 ,  708) PHYS. REV. LETT. 19, . . . . .   908, (1967).
                                 ( 5 ,  955) PROC. PHYS. SOC. VOL.92, . . .   726, (1967).
                                 ( 6 , 1476) J. APP. PHYS. 39, . . . . . .  1118, (1968).
                                 ( 6 , 1225) J. APP. PHYS. 39, . . . . . .   457, (1968).
                                 ( 6 , 1715) J. OF PHYS. C VOL.1, . . . . .   458, (1968).
                                 ( 5 , 2299) J. OF PHYS. C VOL.1, . . . . .   528, (1968).
                                 ( 6 , 1455) SOL. STATE COMM. 6, . . . . .    145, (1968).
                                 ( 6 , 1322) CAN. J. PHYS. VOL.47, . . . .  1983, (1969).
                                 ( 6 , 1475) J. APPL. PHYS. VOL.40, . . . .   991, (1969).
                                 ( 6 , 1227) J. APP. PHYS. VOL.40, . . . .  1443, (1969).
                                 ( 6 , 1221) PHYS. REV. LETT. VOL.23, . . .  1693, (1969).
                                 ( 6 , 1481) SOL. STATE COMM. VOL.7, . . .  1693, (1969).
                                 ( 6 ,  704) COLLOQUE INTERNAT. NO.180 TOME 2  283, (1970).
                                 ( 6 , 1228) J. APPL. PHYS. 41, . . . . . . 1176, (1970).
                                 ( 6 , 1449) J. APPL. PHYS. 41, . . . . . .  896, (1970).
                                 ( 6 , 1223) METAL PHYS. VOL.3, . . . . . .  189, (1970).
                                 ( 6 ,  702) CAN. J. PHYS. 49, . . . . . .  2875, (1971).
                                 ( 6 ,  705) J. PHYSIQUE TOME 32 COL.1 VOL.II 1179, (1971).
                                 ( 6 , 1323) J. PHYSIQUE TOME 32 COL.1 VOL.II 1184, (1971).
                                 ( 6 , 1269) J. PHYS. -C-, VOL. 4, . . . .  2127, (1971).
                                 (11A,   18) J. PHYS. -C-, VOL. 4, . . . .  2139, (1971).
                                 ( 6 ,  501) PHYS. REV. LETT. VOL.27, . . . 1442, (1971).
                                 ( 6 , 1437) CAN. J. PHYS. 50, . . . . . .   687, (1972).
                                 ( 6 , 2158) IAEA SYMP. GRENOBLE, . . . . .  553, (1972).
                                 ( 6 , 1307) IAEA SYMP. GRENOBLE, . . . . .  581, (1972).
                                 ( 6 , 1321) MAGN. AND MAG. MATERIALS-1971, . 1315, (1972).
                                 ( 5 , 2272) CAN. J. PHYS. VOL.52, . . . .   748, (1974).
                                 ( 6 , 1941) PHYS. REV. B VOL.9, . . . . .  3797, (1974).
1687  HOLLEY-JR.......(C.E.)..:  ( 5 , 1604) ACTA CRYST. 16, . . . . . . .   352, (1963).
1688  HOLMES..........(D.E.)..:  ( 6 , 1355) J. CHEM. PHYS. VOL.54, . . . .  3979, (1971).
1689  HOLMES..........(L.M.)..:  ( 5 , 1583) PHYS. REV. LETT. VOL.32, . . .  610, (1974).
1690  HOLMES..........(L.)....:  ( 5 ,  172) J. APPL. PHYS. 41, . . . . . .  943, (1970).
                                 ( 5 ,  165) J. PHYSIQUE TOME 32 COL.1 VOL.II  759, (1971).
                                 ( 5 ,  159) PHYS. REV. B VOL.6, . . . . .  2677, (1972).
1691  HOLMQVIST.......(B.)....:  (17B,   28) ARK. FYS. VOL.34, PAPER 40, . .  481, (1967).
                                 ( 5 ,  527) 2ND INTERNAT. CONF./NUCL. DATA,. . . , (1970).
1692  HOLMRYD.........(S.)....:  (20B,    7) ARK. FYS. VOL.16 PAPER 19, . .  199, (1959).
                                 ( 6 ,   42) ARK. FYS. VOL.17, PAPER 21, . .  369, (1960).
                                 ( 6 ,   52) IAEA SYMPOSIUM VIENNA, . . . .   587, (1960).
                                 ( 6 ,  669) IAEA SYMPOSIUM (VIENNA), . . .   329, (1960).
                                 (19B,   10) NUCL. INST. MET. VOL.12, . . .   355, (1961).
                                 (23 ,   57) NUCL. INST. MET. 27, . . . . .    51, (1964).
                                 ( 6 ,  365) IAEA SYMP. COPENHAGEN, VOL.1,.   475, (1968).
                                 (20D,   35) REV. SCI. INSTRUM. 40, . . . .    49, (1968).
1693  HOLSETH.........(H.)....:  ( 5 , 1149) ACTA CHEM. SCAND. VOL.24, . . . 3309, (1970).
1694  HOLSOPPLE.......(H.L.)..:  (20C,   41) NUCL. INST. MET. VOL.33, . . .   194, (1965).
1695  HOLSTEIN........(T.)....:  (12B,   33) PHYS. REV. VOL.59, . . . . . .  960, (1941).
1696  HOLTZBERG.......(F.)....:  ( 5 ,  911) A.I.P. CONF. PROC. NO.10 PART 2, 1569, (1973).
1697  HOLTZMAN........(R.H.)..:  (12B,   36) PHYS. REV. VOL.73, . . . . . . 1277, (1948).
1698  HOLT............(G.)....:  ( 6 , 1214) IAEA SYMP. CHALK RIVER, VOL.2, .  183, (1962).
1699  HOLT............(R.J.)..:  (12A,   17) NUCL. INST. MET. VOL.98, . . .   385, (1972).
1700  HOLT............(S.L.)..:  ( 6 ,  237) J. APPL. PHYS. 42, . . . . . . 1265, (1971).
                                 ( 9D,   62) PHYS. REV. LETT. VOL.26, . . .   718, (1971).
                                 ( 6 ,  238) PHYS. REV. B VOL.5, . . . . .  1999, (1972).
                                 ( 6 ,  625) PHYS. REV. LETT. VOL.32 NO.4,.   170, (1974).
1701  HOLZL...........(K.)....:  (10B,  210) SOL. STATE COMM. 5, . . . . .    159, (1967).
1702  HONE............(D.)....:  ( 8 ,  102) PHYS. REV. B VOL.5, . . . . .  1915, (1972).
1703  HONZATKO........(J.)....:  (12A,    9) CZECH. J. PHYS. B VOL.22, . . .   38, (1972).
1704  HOPKINS.........(D.C.)..:  ( 5 , 1541) PHYS. REV. 176, . . . . . . .   712, (1968).
1705  HOPKINS.........(J.C.)..:  (12B,    3) ANN. OF PHYSICS VOL.74, . . . .  250, (1972).
1706  HORI............(T.)....:  ( 5 , 1858) J. PHYS. SOC. JAPAN VOL.19, . . 1255, (1964).
                                 ( 5 , 1719) J. PHYS. SOC. JAPAN VOL.19, . . 2078, (1964).
                                 ( 5 , 1856) J. PHYS. SOC. JAPAN VOL.19, . . 2082, (1964).
                                 ( 5 , 1857) J. PHYS. SOC. JAPAN VOL.21, . . 2080, (1966).
1707  HORL............(E.M.)..:  ( 6 , 1419) PHYS. REV. LETT. 20, . . . . .  209, (1968).
1708  HORNER..........(H.)....:  ( 6 , 1132) PHYS. REV. LETT. VOL.25, . . .  147, (1970).
                                 (10B,  216) SOL. STATE COMM. VOL. 9, . . .    79, (1971).
                                 (10D,   13) IAEA SYMP. GRENOBLE, . . . . .   119, (1972).
                                 (10D,   14) J. LOW TEMP. PHYS. VOL.8, . . .  511, (1972).
                                 ( 6 , 1161) PHYS. REV. LETT. VOL.29, . . .   556, (1972).
1709  HORNREICH.......(R.M.)..:  ( 5 ,  882) PHYS. REV. B VOL.8, . . . . .  3398, (1973).
1710  HORSEY..........(R.S.)..:  ( 5 , 2486) MATER. RES. BULL. VOL.8, . . . 1183, (1973).
1711  HORSLEY.........(A.)....:  (17B,   44) NUCL. INST. MET. 62, . . . . .   29, (1968).
1712  HORTON..........(G.K.)..:  ( 7B,   23) J. APPL. PHYS. 41, . . . . . . 5138, (1970).
                                 ( 6 , 1127) PHYS. REV. LETT. VOL.24, . . . 1424, (1970).
                                 (10B,  166) PHYS. REV. B-4, . . . . . . .   567, (1971).
1713  HOSEMANN........(R.)....:  (14A,   59) PHYS. REV. A VOL.6, . . . . .  2243, (1972).
1714  HOSHINO.........(S.)....:  (22B,   16) J. PHYS. SOC. JAPAN VOL.17 B-II,  358, (1962).
                                 ( 5 ,  576) J. PHYS. SOC. JAPAN VOL.20, . . 1729, (1965).
                                 ( 5 ,  737) J. PHYS. SOC. JAPAN VOL.20, . .  381, (1965).
                                 ( 5 ,  575) J. PHYS. SOC. JAPAN VOL.22, . . 1221, (1967).
                                 (22C,    1) ACTA CRYST. VOL. A25, . . . . . S-65, (1969).
                                 (21 ,   18) JAP. J. APPL. PHYS. VOL.10, . .  933, (1971).
                                 ( 5 ,  774) ACTA CRYST. VOL.A28 PART S-4,.   192, (1972).
                                 (20C,    1) ACTA CRYST. VOL.A28 PART S-4,.   250, (1972).
                                 ( 5 , 1273) J. PHYS. SOC. JAPAN VOL.32, . . 1019, (1972).
                                 ( 5 ,  775) J. PHYS. SOC. JAPAN VOL.35, . .  842, (1973).
```

1714 HOSHINO.........(S.)....: (6 ,   574) PHYS. REV. B VOL.9,. . . . . . . 4549, (1974).

1715 HOSKINS.........(B.F.)..: (5 ,   288) CHEM. COMMUNICATIONS,. . . . . .   554, (1969).

1716 HOSOYA..........(S.)....: (5 ,   524) PHYS. REV. LETT. VOL.29, ...   281, (1972).

1717 HOSSFELD........(F.)....: (6 , 1009) DISC. FARADAY SOC. NO.48,. . . .   69, (1969).
                                  (19A,   21) IAEA SYMP. (INSTRUMENT.), VIENNA   117, (1969).
                                  (17B,   31) CONF. INTERN. RENNES,. . . . . .   162, (1971).
                                  (17C,   18) J. APPL. CRYST. VOL.4, ...   210, (1971).
                                  (7B,   22) J. APPL. CRYST. VOL.4, ...   419, (1971).
                                  (17A,   43) IAEA SYMP. GRENOBLE,. . . . . .   747, (1972).

1718 HOUK............(T.L.)..: (5 , 1262) PHYS. REV. LETT. VOL.26, ... 1581, (1971).

1719 HOUMANN.........(J.C.G.): (6 , 2128) BROOKHAVEN SYMPOSIUM,. . . . . .   159, (1965).
                                  (6 , 2123) PHYS. REV. B-1,. . . . . 3943, (1970).

1720 .HOUMANN........(J.G.)..: (6 , 2136) PHYS. REV. LETT. 16, ...   737, (1966).
                                  (6 , 2150) PHYS. REV. LETT. 19, ...   312, (1967).
                                  (6 , 2153) J. APP. PHYS. 39,. . . . .   807, (1968).
                                  (6 , 1931) CONF./RARE EARTHS/ACTINIDES, . .   40, (1971).
                                  (6 , 1929) PHYS. REV. LETT. VOL.26, ... 1254, (1971).
                                  (6 , 1928) PHYS. REV. LETT. VOL.27, ...   223, (1971).
                                  (6 , 2132) IAEA SYMP. GRENOBLE, ...   603, (1972).
                                  (6 ,   812) IAEA SYMP. GRENOBLE, ...   655, (1972).
                                  (6 , 2145) PHYS. REV. LETT. VOL.31, ... 1585, (1973).

1721 HOUSTON.........(S.K.)..: (7A,   79) PROC. PHYS. SOC. 89, ...   341, (1966).

1722 HOWELLS.........(W.S.)..: (14A,   9) CHEM. PHYS. LETT. VOL.21 NO.1, .   109, (1973).
                                  (14A,   62) PROPERTIES LIQUID METALS,. . . .   43, (1973).
                                  (5 , 2095) J. PHYS. C VOL.7,. . . . . L111, (1974).

1723 HOWERTON........(R.F.)..: (24 ,   86) PHYS. MED. BIOL.(GB) VOL. 16,. .   439, (1971).

1724 HOWE............(R.A.)..: (14A,   9) CHEM. PHYS. LETT. VOL.21 NO.1, .   109, (1973).
                                  (5 , 2095) J. PHYS. C VOL.7,. . . . . L111, (1974).

1725 HOWSMAN.........(A.J.)..: (10B,   20) DISS. ABS. VOL.27, ... 2487B, (1966).

1726 HO..............(P.S.)..: (6 , 1626) PHYS. STAT. SOL. 23, ...   489, (1967).
                                  (10B,   139) PHYS. REV. 169,. . . . .   523, (1968).

1727 HRDLICKA........(Z.)....: (6 , 2052) NATURE (PHYS. SCI.) VOL.242, . .   109, (1973).

1728 HSU.............(F.S.L.): (5 ,   662) A.I.P. CONF. PROC. NO.10 PART 1,   684, (1973).

1729 HSU.............(W.S.)..: (22A,   9) CHIN. J. PHYS. (TAIWAN) VOL.8, .   31, (1970).

1730 HUANG...........(C.H.)..: (6 , 1145) PHYS. REV. A VOL.5,. . . . . . . 1377, (1972).

1731 HUBBARD.........(C.R.)..: (9B,   65) NUCL. INST. MET. VOL.94, ...   185, (1971).
                                  (9B,   15) ACTA CRYST. VOL.A28, ...   236, (1972).

1732 HUBBARD.........(H.W.)..: (5 ,   924) NUCL. SCI. ENG. VOL.46,. . . . .   255, (1971).

1733 HUBBARD.........(J.)....: (9D,   64) PROC. PHYS. SOC. 86, ...   561, (1965).
                                  (11A,   14) J. PHYS. -C- VOL. 1, ... 1650, (1968).
                                  (14A,   24) J. PHYS. C VOL.2,. . . . .   556, (1969).
                                  (11C,   40) PHYS. REV. B-1,. . . . . 3815, (1970).
                                  (8 ,   26) J. APPL. PHYS. 42, ... 1390, (1971).
                                  (8 ,   37) J. PHYS. C VOL.4,. . . . .   53, (1971).

1734 HUBER...........(D.L.)..: (11C,   26) PHYS. LETT. VOL.31A, ...   267, (1970).
                                  (6 , 1984) PHYS. REV. LETT. VOL.24, ...   111, (1970).

1735 HUBER...........(M.)....: (5 , 1608) J. PHYS. CHEM. SOLIDS VOL.28,. . 2441, (1967).
                                  (5 , 1757) J. PHYS. CHEM. SOLIDS VOL.33,. .   737, (1972).
                                  (5 ,   469) J. SOLID STATE CHEM. VOL.8,. . .   50, (1973).

1736 HUBER...........(U.)....: (20C,   50) NUCL. INST. MET. 53, ...   325, (1967).

1737 HUET............(J.L.)..: (23 ,   6) BULL. INF. SCI. TECH. NO.170,. .   61, (1972).

1738 HUGHES..........(D.J.)..: (12B,   36) PHYS. REV. VOL.73, ... 1277, (1948).
                                  (6 ,   767) PHYS. REV. VOL.75, ...   565, (1949).
                                  (12B,   37) PHYS. REV. VOL.76, ... 1413, (1949).
                                  (24 ,   37) NUCLEONICS VOL.6 NO.5, ...   38, (1950).
                                  (20A,   65) PHYS. REV. VOL.77, ...   291, (1950).
                                  (6 ,   769) PHYS. REV. VOL.80, ...   481, (1950).
                                  (12B,   39) PHYS. REV. VOL.80, ...   953, (1950).
                                  (20A,   66) PHYS. REV. VOL.81, ...   498, (1951).
                                  (5 , 1260) PHYS. REV. VOL.84, ... 1160, (1951).
                                  (6 ,   772) PHYS. REV. VOL.92, ...   202, (1953).
                                  (5 ,   143) PHYS. REV. VOL.92, ...   716, (1953).
                                  (6 , 2246) PHYS. REV. VOL.104,. . . . . . .   271, (1956).
                                  (23 ,   52) NUCL. INSTRUM. VOL.1,. . . . . .   92, (1957).
                                  (6 ,   56) PHYS. REV. VOL.106,. . . . . . . 1168, (1957).
                                  (6 , 2355) PHYS. REV. VOL.108,. . . . . . . 1091, (1957).
                                  (6 , 2240) PHYS. REV. VOL.109,. . . . . . . 1046, (1958).
                                  (6 , 2036) PHYS. REV. LETT. VOL.2,. . . . .   258, (1959).
                                  (6 , 1090) PHYS. REV. LETT. VOL.3,. . . . .   91, (1959).
                                  (6 ,   937) PHYS. REV. VOL.113,. . . . . . .   49, (1959).
                                  (6 , 1084) PHYS. REV. VOL.119,. . . . . . .   872, (1960).

1739 HUISZOON........(C.)....: (10B,   5) ACTA CRYST. VOL.A28, ...   166, (1972).
                                  (10B,   4) ACTA CRYST. VOL.A28, ...   170, (1972).

1740 HUKIN...........(D.)....: (6 , 1269) J. PHYS. -C-, VOL. 4,. . . . . . 2127, (1971).

1741 HULLER..........(A.)....: (6 ,   126) Z. PHYS. VOL.220,. . . . .   145, (1969).
                                  (6 ,   233) PHYS. REV. LETT. VOL.29, ...   266, (1972).

1742 HULL............(D.E.)..: (22C,   58) REV. SCI. INSTRUM. VOL.37, . . . 1543, (1966).

1743 HUNTER..........(R.J.)..: (6 , 1081) NATURE (PHYS. SCI.) VOL.230, . .   192, (1971).

1744 HUNT............(D.G.)..: (6 , 1950) CONF./RARE EARTHS/ACTINIDES, . .   46, (1971).

1745 HUNT............(G.J.)..: (20C,   34) 2ND INTERNAT. CONF./NUCL. DATA,. . , (1970).

1746 HURLEY..........(A.C.)..: (5 , 2771) PROC. ROY. SOC. 298, ...   289, (1967).

1747 HURRELL.........(J.P.)..: (6 ,   120) SOL. STATE COMM. VOL.8,. . . . .   463, (1970).

1748 HURST...........(D.G.)..: (5 , 2234) PHYS. REV. VOL.75, ... 1609, (1949).
                                  (23 ,   75) REV. SCI. INSTRUM. VOL.21, . . .   705, (1950).
                                  (5 ,   772) CAN. J. PHYS. VOL.29,. . . . . .   36, (1951).
                                  (5 , 1886) PHYS. REV. VOL.83, ... 1100, (1951).
                                  (6 ,   384) PHYS. REV. VOL.83, ...   840, (1951).
                                  (6 , 2055) PHYS. REV. VOL.83, ...   841, (1951).
                                  (5 , 1899) PHYS. REV. VOL.83, ...   88, (1951).
                                  (5 , 1900) PHYS. REV. VOL.86, ...   797, (1952).
                                  (5 , 1824) PHYS. REV. VOL.88, ...   542, (1952).
                                  (5 , 1337) PHYS. REV. VOL.91, ... 1222, (1953).

1748 HURST...........(D.G.)..: (5 , 2232) PHYS. REV. VOL.92,.: : : : : : : 1229, (1953).
                                (5 , 1326) PHYS. REV. VOL.100,.: : : : : : : 994, (1955).

1749 HURST...........(H.J.)..: (5 , 744) PHYS. REV. 138,. . . . . A153, (1965).
                                (5 , 1945) ACTA CRYST. VOL.B26; : : : : : : 2136, (1970).
                                (5 , 2786) ACTA CRYST. VOL.B27,. : : : : : 2018, (1971).

1750 HUSSAIN.........(M.)...: (20C,   80) NUCL. SCI. APPL. B(PAKIS.) VOL.5   43, (1969).
                                (22C,   21) J. NUCL. ENERGY VOL.24,. . . . . 385, (1970).

1751 HUSSAIN.........(S.)...: (22C,   21) J. NUCL. ENERGY VOL.24,. . . . . 385, (1970).

1752 HUSTON..........(J.L.)..: (5 , 2873) J. CHEM. PHYS. VOL.59,. . . . . 453, (1973).

1753 HUTCHENS........(R.D.)..: (5 , 2046) J. SOLID STATE CHEM. VOL.9,. . . 152, (1974).

1754 HUTCHINGS.......(M.T.)..: (5 , 1477) PPOC. PHYS. SOC. 91,. . . . . . 928, (1967).
                                (6 , 814) J. APP. PHYS. 39,. . . . . . . 1120, (1968).
                                (5 , 1434) PHYS. REV. VOL.188,. . . . . . 919, (1969).
                                (6 , 546) SOL. STATE COMM. 7, NO15,. . . 1043, (1969).
                                (5 , 1499) J. APPL. PHYS. 41,. . . . . . . 945, (1970).
                                (5 , 1498) J. OF PHYS.-C-, SER.2, VOL.3,. 1303, (1970).
                                (6 , 543) PHYSICA VOL.48,. . . . .   13, (1970).
                                (6 , 1761) PHYS. REV. B-2,. . . . . 1362, (1970).
                                (6 , 815) J. APPL. PHYS. VOL.42,. . . . . 1376, (1971).
                                (6 , 237) J. APPL. PHYS. 42,. . . . . . . 1265, (1971).
                                (6 , 811) J. OF PHYS.-C-, SER.2, VOL.4,. 307, (1971).
                                (6 , 1269) J. PHYS. -C-, VOL. 4,. . . . . 2127, (1971).
                                (11A,   18) J. PHYS. -C-, VOL. 4,. . . . . 2139, (1971).
                                (90 ,   62) PHYS. REV. LETT. VOL.26,. . . . 718, (1971).
                                (6 , 1769) SOL. STATE COMM. VOL.9,. . . . 1011, (1971).
                                (6 , 1777) IAEA SYMP. GRENOBLE,. . . . . . 669, (1972).
                                (6 , 238) PHYS. REV. B VOL.5,. . . . . . 1999, (1972).
                                (6 , 817) PHYS. REV. B VOL.5,. . . . . . 154, (1972).
                                (6 , 1770) PHYS. REV. B VOL.6,. . . . . . 3447, (1972).
                                (6 , 1778) A.I.P. CONF. PROC. NO.10 PART 2, 1403, (1973).
                                (6 , 489) J. PHYS. C VOL.6,. . . . . . . 3143, (1973).
                                (6 , 849) SOL. STATE COMM. VOL.12 NO.8,. 795, (1973).

1755 HUTCHINGS.......(T.)....: (5 , 2184) 1ST EUR. CONF./CONDENSED MATTER, 39, (1971).

1756 HUTCHINSON......(P.)....: (14A,   61) PROC. ROY. SOC. A VOL.282,. . . 283, (1964).
                                (17B,   71) UKAEA AERE REP.R5536 (41 PP.),. . . , (1967).

1757 HUTCHISON.......(E.A.)..: (6 , 1989) IAEA SYMP. GRENOBLE,. . . . . . 639, (1972).

1758 HVOSLEF.........(J.)....: (5 , 247) ACTA CRYST. VOL.11,. . . . . . 383, (1958).
                                (5 , 194) ACTA CRYST. VOL.12,: : : : : : 476, (1959).
                                (5 ,   12) ACTA CRYST. B24,. . . . . . . 1431, (1968).

1759 HWA-WEN-LAI.............: (6 , 1144) PHYS. REV. A VOL.1,. . . . . . 1536, (1970).

1760 HYAKUTAKE.......(M.)...: (19A,   46) NUCL. INST. MET. VOL.75,. . . .   32, (1969).
                                (19B,   7) JAP. J. APPL. PHYS. VOL.10,. . 1090, (1971).

1761 HYDER...........(M.L.)..: (5 , 421) J. INORG. NUCL. CHEM. VOL.33,. 1553, (1971).

1762 IBEL............(K.)...: (17A,   20) ATOMKERNENERGIE VOL.17,. . . . . 15, (1971).
                                (5 , 410) J. PHYS. C VOL.6,. . . . . . . 3465, (1973).
                                (5 , 411) J. APPL. CRYST. VOL.7,. . . . . 230, (1974).
                                (17A,   52) J. APPL. CRYST. VOL.7,. . . . .   96, (1974).

1763 IBERS...........(J.A.)..: (10D,   1) ACTA CRYST. VOL.12,. . . . . . 251, (1959).
                                (5 , 1276) J. PHYS. SOC. JAPAN VOL.17 B-II, 383, (1962).
                                (5 , 1275) ACTA CRYST. 16,. . . . . 1209, (1963).
                                (5 , 1971) COLL. INTER. N.126 (GRENOBLE),   50, (1963).
                                (5 , 1449) J. CHEM. PHYSICS VOL.40,. . . . 402, (1964).
                                (5 , 2240) J. CHEM. PHYSICS VOL.44,. . . . 1748, (1966).
                                (5 ,   28) ACTA CRYST. VOL.B25,. . . . . . 2451, (1969).
                                (5 , 1282) INORG. CHEM. VOL.8,. . . . . . 1928, (1969).

1764 IGAMBERDIEV.....(SH.G.).: (22A,   22) INSTRUM. EXP. TECH. 3,. . . . . 726, (1970).

1765 IGARASI.........(S.)...: (6 , 2207) 2ND INTERNAT. CONF./NUCL. DATA,. . . , (1970).

1766 IGNAT≠EV........(K.G.)..: (22A,   17) INSTRUM. EXP. TECH. NO.4,. . . . 533, (1959).

1767 IIJIMA..........(S.)...: (6 ,   46) J. NUCL. SCI. TECHNOL. VOL.3,. . 160, (1966).

1768 IIZUMI..........(M.)...: (5 , 246) J. NUCL. SCI. TECHNOL. VOL.5,. 649, (1968).
                                (6 , 437) J. PHYS. SOC. JAPAN VOL.33,: : : 647, (1972).
                                (9B,   43) JAP. J. APPL. PHYS. VOL.12,. . . 167, (1973).
                                (6 , 2104) J. PHYS. C VOL.6,. . . . . . . 3021, (1973).
                                (6 , 439) J. PHYS. SOC. JAPAN VOL.35,. . . 204, (1973).
                                (6 , 1924) J. POLYM. SCI./POLYM. LETT. 11, 377, (1973).

1769 IJIRI...........(H.)...: (19A,   46) NUCL. INST. MET. VOL.75,. . . .   32, (1969).

1770 IKEDA...........(H.)...: (5 , 1482) ACTA CRYST. VOL.A28 PART S-4,. 175, (1972).
                                (21 ,   19) JAP. J. APPL. PHYS. VOL.11,. . .   82, (1972).
                                (5 , 1435) J. PHYS. SOC. JAPAN VOL.33,. . 1483, (1972).
                                (5 , 1480) J. PHYS. SOC. JAPAN VOL.33,. . . 393, (1972).
                                (6 , 1273) J. PHYS. SOC. JAPAN VOL.35,. . 1328, (1973).
                                (6 , 1272) J. PHYS. SOC. JAPAN VOL.35,. . 1608, (1973).
                                (6 , 1266) J. PHYS. SOC. JAPAN VOL.35,. . 1795, (1973).
                                (6 , 1319) J. PHYS. SOC. JAPAN VOL.35,. . . 617, (1973).
                                (5 , 1436) J. PHYS. SOC. JAPAN VOL.35,. . . 722, (1973).
                                (6 , 1267) SOL. STATE COMM. VOL.14,. . . . 529, (1974).

1771 IKEDA...........(S.)...: (6 , 825) J. PHYS. SOC. JAPAN VOL.35,. . 1616, (1973).

1772 ILBERG..........(D.)...: (20A,   38) NUCL. INST. MET. 61,. . . . . .   93, (1968).

1773 ILIESCU.........(N.)...: (5 , 281) PHYS. LETT. 26A,. . . . .   72, (1967).
                                (6 , 316) REV. ROUMAINE PHYS. VOL.12,: : : 943, (1967).
                                (5 , 2319) REV. ROUMAINE PHYS. VOL.13,. . . 287, (1968).

1774 ILYINA..........(I.L.)..: (5 , 2219) PHYS. STAT. SOL. VOL.50,. . . . 385, (1972).

1775 IL-HWAN-SUH.............: (5 , 1433) ACTA CRYST. VOL.B26,. : : : : : 827, (1970).
                                (5 , 1994) ACTA CRYST. VOL.B27,. : : : : : 253, (1971).

1776 IMAI............(H.)...: (23 ,   48) MITSUBISHI DENHI ENG. NO.26,. .   10, (1970).

1777 IMBERT..........(P.)...: (5 , 343) COMP. REND. 263,. . . . .   98, (1966).

1778 IMRY............(J.)...: (6 , 1293) IAEA SYMP. BOMBAY, VOL.2,. . . . 325, (1964).

1779 IMRY............(Y.)...: (6 , 449) J. CHEM. PHYSICS VOL.43,. . . . 1864, (1965).
                                (6 , 1048) PHYS. LETT. 21,. . . . . 248, (1966).
                                (5 , 1462) SOL. STATE COMM. 5,. : : : : : : 41, (1967).
                                (9C,   32) PHYS. LETT. 29A,. . . . . 483, (1969).

```
1818  IWATA..........(Y.)....:  ( 5 , 2284) J. PHYS. SOC. JAPAN VOL.35,...    314, (1973).
1819  IWAVOV.........(V.V.)..:  ( 78,  133) SOV. PHYS. JETP. 26, .......    1139, (1968).
1820  IYENGAR........(K.V.)..:  (19B,   15) NUCL. INST. MET. VOL.81, ....    301, (1970).
1821  IYENGAR........(P.K.)..:  ( 6 ,  931) PHYS. REV. VOL.108,.........    894, (1957).
                                ( 6 , 1501) BULL. AMER. PHYS. SOC. VOL.3,:   195, (1958).
                                ( 8 ,  932) PHYS. REV. VOL.111,.........    747, (1958).
                                ( 5 , 2753) BULL. AMER. PHYS. SOC. SER.II 4, 184, (1959).
                                ( 6 ,  782) IAEA SYMPOSIUM VIENNA,......    555, (1960).
                                ( 6 , 1567) IAEA SYMP. CHALK RIVER, VOL.2,:  253, (1962).
                                ( 6 ,  143) IAEA SYMP. CHALK RIVER, VOL.2,    99, (1962).
                                ( 5 ,  174) J. PHYS. SOC. JAPAN VOL.17 B-III  41, (1962).
                                ( 5 , 1173) J. PHYS. SOC. JAPAN VOL.17,.    247, (1962).
                                ( 6 , 1402) COPENHAGEN CONF.-LATT.DYNAMICS,: 223, (1963).
                                (23 ,   10) CRYSTALLOGR./CRYSTAL PERFECTION, 279, (1963).
                                ( 6 , 1399) IAEA SYMP. BOMBAY, VOL.1,:       153, (1964).
                                ( 6 , 1499) IAEA SYMP. BOMBAY, VOL.2,:       433, (1964).
                                ( 5 , 1919) IAEA SYMP. BOMBAY, VOL.2,:       347, (1964).
                                (17A,   33) IAEA SYMP. BOMBAY, VOL.2,:       483, (1964).
                                (20D,   15) NUCL. INST. MET. 25,........    357, (1964).
                                ( 6 , 1497) PROC. INT. CONF. (NOTTINGHAM),:  322, (1964).
                                ( 6 , 1566) SOL. STATE COMM. 2,.........     17, (1964).
                                (23 ,    4) BROOKHAVEN SYMPOSIUM,.......    179, (1965).
                                (20D,   16) NUCL. INST. MET. 37,........    121, (1965).
                                (17A,   98) TH. NEUTRON SCATT./EGELSTAFF,:    97, (1965).
                                ( 6 , 1582) J. PHYS. CHEM. SOL. VOL.27,:    1103, (1966).
                                ( 6 , 2236) PHYSICA VOL.34,.............    384, (1967).
                                (10B,  228) TRIESTE-INTERN. COURSE,......    665, (1967).
                                ( 6 , 2301) IAEA SYMP. COPENHAGEN, VOL.1,:   195, (1968).
                                ( 6 ,  532) J. APPL. PHYS. VOL.39 NO.2,:    1113, (1968).
                                ( 6 ,  533) PHYS. LETT. VOL.26A,........    108, (1968).
                                ( 6 ,  816) SOL. STATE COMM. 6,.........    593, (1968).
                                ( 6 , 1234) PHYS. STAT. SOL. 34, NO1,:      279, (1969).
                                ( 6 , 1496) SOL. STATE COMM. 7,.........    123, (1969).
                                ( 6 ,  118) SOL. STATE COMM. VOL.8,.....    497, (1970).
                                ( 6 ,  134) CONF. INTERN., RENNES,......    140, (1971).
                                ( 6 ,  446) IAEA SYMP. GRENOBLE,........     95, (1972).
                                ( 6 ,   66) PHYS. STAT. SOL. VOL.50,....    701, (1972).
                                (17A,   74) NUCL. INST. MET. VOL.113,...     15, (1973).

1822  IYENGAR........(R.D.)..:  ( 6 , 2042) ADV. COLLOID INTERFACE SCI. 2,     1, (1968).
1823  IZUYAMA........(T.)....:  ( 8 ,   42) J. PHYS. SOC. JAPAN VOL.18,.    1025, (1963).
1824  IZYUMOV........(YU.A.).:  ( 7A,   56) PHYS. MET. METALLOG. VOL.11 NO.5 139, (1961).
                                ( 8 ,  133) SOV. PHYS. CRYST.7,.........    286, (1962).
                                (12A,   39) SOV. PHYS. JETP VOL.14,.....    1168, (1962).
                                ( 9D,   71) SOV. PHYS. JETP VOL.15,.....    1162, (1962).
                                ( 9D,   72) SOV. PHYS. SOL. STATE, 4,...    158, (1962).
                                ( 8 ,  138) SOV. PHYS. USPEKHI VOL.80,..    359, (1963).
                                ( 8 ,   62) PHYS. MET. METALLOG. VOL.22 NO.6   1, (1966).
                                (11B,   27) PROC. PHYS. SOC. 87,........    521, (1966).
                                (11B,   45) SOV. PHYS. JETP. 24,........    960, (1967).

1825  JACCARD........(S.)....:  ( 5 ,  764) NUCL. PHYS. A(NETHERLANDS) A182, 541, (1972).
                                ( 7A,   39) NUCL. PHYS. A VOL.182,......    541, (1972).
1826  JACKLE.........(J.)....:  (10D,   32) PHYS. REV. LETT. VOL.24,....    1101, (1970).
                                (10D,   25) PHYS. LETT. VOL.39A,........    327, (1972).
1827  JACKSON........(H.E.)..:  (20C,  101) REV. SCI. INSTRUM. VOL.36,..    419, (1965).
                                ( 6 , 1348) PHYS. REV. LETT. VOL.31,....    296, (1973).
                                ( 6 ,  609) BULL. AMER. PHYS. SOC. VOL.19,:  320, (1974).
1828  JACKSON........(H.W.)..:  ( 6 , 1206) PHYS. REV. A VOL.9 NO.6, ......    , (1974).
1829  JACKSON........(W.R.)..:  (20C,   51) NUCL. INST. MET. 55,........    349, (1967).
1830  JACOBI.........(N.)....:  ( 6 ,  426) J. CHEM. PHYS. VOL.58,......    3647, (1973).
1831  JACOBSON.......(A.J.)..:  ( 5 ,  156) ACTA CRYST. VOL.B28,........    956, (1972).
                                ( 5 ,  175) J. PHYS. C VOL.5,...........    2887, (1972).
                                ( 9A,   10) CHEM. APPL. THERMAL NEUTRON SCAT 270, (1973).
                                ( 5 , 1751) J. PHYS. C VOL.6,...........    1615, (1973).
                                ( 5 ,  190) ACTA CRYST. VOL.B30,........    816, (1974).
                                ( 5 , 1006) J. PHYS. C VOL.7,...........    783, (1974).
1832  JACOBSON.......(R.A.)..:  ( 9B,   65) NUCL. INST. MET. VOL.94,....    185, (1971).
                                ( 9B,   15) ACTA CRYST. VOL.A28,........    236, (1972).
1833  JACOBS.........(A.M.)..:  ( 9B,   63) NATURE VOL.229,.............    111, (1971).
1834  JACOBS.........(H.)....:  ( 5 , 1569) ATOMKERNENERGIE VOL.21,.....    275, (1973).
1835  JACOBS.........(I.S.)..:  ( 5 , 2612) A.I.P. CONF. PROC. NO.10 PART 2, 1554, (1973).
1836  JACROT.........(B.)....:  (23D,    6) C.R. ACAD. SCI. TOME 240,...    745, (1955).
                                ( 6 ,  786) C.R. ACAD. SCI. 246,........    1018, (1958).
                                (20B,   63) J. PHYS. RADIUM VOL.19 SUPPL.4,:  51A, (1958).
                                ( 5 ,  978) J. PHYS. RADIUM VOL.19,.....    617, (1958).
                                ( 6 , 1468) C.R. ACAD. SCI. 248,........    1631, (1959).
                                ( 6 ,  789) J. PHYS. RADIUM VOL.20,.....    178, (1959).
                                ( 6 , 1231) C.R. ACAD. SCI. 250,........    2871, (1960).
                                ( 6 ,  582) IAEA SYMPOSIUM VIENNA,......    549, (1960).
                                ( 6 , 1034) IAEA SYMPOSIUM (VIENNA),....    347, (1960).
                                ( 6 ,  963) IAEA SYMPOSIUM (VIENNA),....    411, (1960).
                                (20B,   24) IAEA SYMP. (PILE RESEARCH) VIENNA 393, (1960).
                                (13 ,   23) J. PHYS. CHEM. SOLIDS VOL.13,:   235, (1960).
                                ( 6 ,  595) J. PHYS. RADIUM VOL.21,.....     67, (1960).
                                ( 6 , 1067) J. PHYS. RADIUM VOL.21,.....    225, (1960).
                                ( 6 ,  436) IAEA SYMP. CHALK RIVER, VOL.2,:  225, (1962).
                                ( 6 , 1450) IAEA SYMP. CHALK RIVER, VOL.2,:  309, (1962).
                                ( 6 , 1757) IAEA SYMP. CHALK RIVER, VOL.2,:  317, (1962).
                                ( 6 , 1463) J. PHYS. RADIUM VOL.23,.....    494, (1962).
                                ( 6 , 1744) J. PHYS. SOC. JAPAN VOL.17 B-III  67, (1962).
                                ( 5 ,  336) PHYS. LETT. 1,..............    187, (1962).
                                (13 ,   20) INTERN. SUMMER SCHOOL, MOL,.    560, (1963).
                                ( 5 , 2000) PHYS. LETT. 9,..............    106, (1964).
                                ( 5 , 1999) PROC. INT. CONF. (NOTTINGHAM),:  285, (1964).
                                ( 8 ,  157) TH. NEUTRON SCATT./EGELSTAFF,:   251, (1965).
                                ( 8 ,  117) PROG. LOW TEMP. PHYS. VOL.5,:    161, (1967).
                                (13 ,   73) TRIESTE/INTERN. COURSE,......    483, (1967).
                                (20B,   40) IAEA SYMP. (INSTRUMENT.), VIENNA 225, (1969).
                                ( 9B,   58) LA RECHERCHE VOL.3,.........    325, (1972).

1837  JAIN...........(A.P.)..:  ( 6 ,   94) PHYS. REV. VOL.125,.........    275, (1962).
1838  JAIN...........(K.P.)..:  (11A,   14) J. PHYS. -C-, VOL. 1,.......    1650, (1968).
1839  JAIN...........(M.)....:  (10B,  180) PHYS. REV. B VOL.7,.........    4338, (1973).
```

1840 JAIN...........(S.C.)..§ (10B,   180) PHYS. REV. B VOL.7,........ 4338, (1973).
1841 JAISWAL........(V.K.)..§ (6 ,   875) PHYSICA VOL.44,..........   69, (1969).
1842 JAKOB..........(W.)....§ (6 ,   502) PHYSICA VOL.35,..........   441, (1967).
1843 JAMES..........(W.J.)..§ (5 , 2789) J. APP. PHYS. 34,........ 1333, (1963).
                                (5 , 1385) J. PHYSIQUE TOME 32 COL.1 VOL.II   670, (1971).
                                (5 ,   215) J. PHYS. CHEM. SOLIDS VOL.32,.. 1315, (1971).
1844 JAMES..........(W.)....§ (5 , 2882) C.R. ACAD. SCI. VOL.255,....   896, (1962).
1845 JAMISON........(W.E.)..§ (5 ,   561) ACTA CHEM. SCAND. VOL.25,.... 1703, (1971).
1846 JANDL..........(S.)....§ (6 ,   903) SOL. STATE COMM. VOL.13,.... 1555, (1973).
1847 JANIK..........(J.A.)..§ (5 , 1119) ACTA PHYS. POLON. VOL.11,....   146, (1951).
                                (5 ,   301) ACTA PHYS. POLON. VOL.12,....   45, (1953).
                                (8 ,   381) BULL. ACAD. POLON. SCI. III V.1,   45, (1953).
                                (5 ,   302) ACTA PHYS. POLON. VOL.13,....   167, (1954).
                                (5 ,   415) ACTA PHYS. POLON. VOL.16,....   335, (1957).
                                (5 ,   438) ACTA PHYS. POLON. VOL.17,....   483, (1958).
                                (6 ,   301) ACTA PHYS. POLON. VOL.17,....   489, (1958).
                                (5 , 1286) IAEA SYMPOSIUM VIENNA,......   297, (1960).
                                (6 ,   974) NUKLEONIKA VOL.5,..........   495, (1960).
                                (5 , 1908) PHYSICA VOL.26,..........   449, (1960).
                                (5 , 1315) ACTA PHYS. POLON. VOL.22,....   517, (1962).
                                (6 , 1552) IAEA SYMP. CHALK RIVER, VOL.1,   405, (1962).
                                (6 , 1549) PHYSICA VOL.29,..........   485, (1963).
                                (6 ,   115) ACTA PHYS. POLON. VOL.25,....   845, (1964).
                                (6 , 1099) IAEA SYMP. BOMBAY, VOL.2,...   243, (1964).
                                (14B,   43) J. CHIM. PHYS. VOL.61,......   97, (1964).
                                (6 ,   263) J. PHYS. CHEM. SOL. VOL.25,.. 1091, (1964).
                                (6 , 1573) PHYS. STAT. SOL. 9,........   905, (1965).
                                (15 ,   61) TH. NEUTRON SCATT./EGELSTAFF,.   413, (1965).
                                (15 ,   60) TH. NEUTRON SCATT./EGELSTAFF,.   453, (1965).
                                (6 ,   502) PHYSICA VOL.35,..........   441, (1967).
                                (6 ,   380) PHYSICA VOL.35,..........   451, (1967).
                                (6 , 1016) PHYSICA VOL.35,..........   457, (1967).
                                (6 , 2007) PHYSICA VOL.35,..........   469, (1967).
                                (24 ,   116) TRIESTE/INTERN. COURSE,....   577, (1967).
                                (6 , 1017) ACTA PHYS. POLON. VOL.33,....   419, (1968).
                                (6 , 1104) IAEA SYMP. COPENHAGEN VOL.II,.   143, (1968).
                                (6 ,   621) IAEA SYMP. COPENHAGEN, VOL.1,.   65, (1968).
                                (6 ,   285) DISCUSSIONS FARADAY SOC. NO.48,.   87, (1969).
                                (6 ,   286) PHYSICA VOL.41,..........   397, (1969).
                                (15 ,   32) PHYSICA VOL.48,..........   126, (1970).
                                (5 , 1921) PHYS. STAT. SOL. VOL.44,....   437, (1971).
                                (6 , 1384) PHYS. STAT. SOL. VOL.44,....   497, (1971).
                                (6 , 1388) SOV. PHYS. CRYST. VOL.15,... 1010, (1971).
                                (6 ,   116) FARADAY SYMP. CHEM. SOC. NO.6,   48, (1972).
                                (6 ,   114) IAEA SYMP. GRENOBLE,......   515, (1972).
                                (6 , 1387) ACTA PHYS. POLON. VOL.A44,..   731, (1973).
                                (6 , 1103) PHYSICA VOL.72,..........   168, (1974).
1848 JANIK..........(J.M.)..§ (6 ,   115) ACTA PHYS. POLON. VOL.25,....   845, (1964).
                                (6 ,   263) J. PHYS. CHEM. SOL. VOL.25,.. 1091, (1964).
                                (6 , 1573) PHYS. STAT. SOL. 9,........   905, (1965).
                                (6 ,   502) PHYSICA VOL.35,..........   441, (1967).
                                (6 ,   380) PHYSICA VOL.35,..........   451, (1967).
                                (6 , 1016) PHYSICA VOL.35,..........   457, (1967).
                                (6 , 1017) ACTA PHYS. POLON. VOL.33,....   419, (1968).
                                (6 , 1104) IAEA SYMP. COPENHAGEN VOL.II,.   143, (1968).
                                (6 ,   286) PHYSICA VOL.41,..........   397, (1969).
                                (5 , 1921) PHYS. STAT. SOL. VOL.44,....   437, (1971).
                                (6 , 1384) PHYS. STAT. SOL. VOL.44,....   497, (1971).
                                (6 , 1388) SOV. PHYS. CRYST. VOL.15,... 1010, (1971).
                                (6 ,   116) FARADAY SYMP. CHEM. SOC. NO.6,   48, (1972).
                                (6 ,   114) IAEA SYMP. GRENOBLE,......   515, (1972).
                                (6 , 1387) ACTA PHYS. POLON. VOL.A44,..   731, (1973).
                                (6 , 1103) PHYSICA VOL.72,..........   168, (1974).
1849 JANIK..........(J.)....§ (6 ,   860) J. PHYS. CHEM. SOLIDS VOL.9,..   153, (1959).
                                (5 ,   306) IAEA SYMPOSIUM VIENNA,......   293, (1960).
                                (5 , 1286) IAEA SYMPOSIUM VIENNA,......   297, (1960).
                                (90 ,   28) IAEA SYMP. (PILE RESEARCH)VIENNA,   415, (1960).
                                (6 ,   974) NUKLEONIKA VOL.5,..........   495, (1960).
                                (5 , 1908) PHYSICA VOL.26,..........   449, (1960).
                                (5 , 1315) ACTA PHYS. POLON. VOL.22,....   517, (1962).
                                (6 , 1552) IAEA SYMP. CHALK RIVER, VOL.1,   405, (1962).
                                (6 , 1101) ACTA PHYS. POLON. VOL.27,....   491, (1965).
                                (21 ,   40) NUKLEONIKA VOL.12,........   385, (1967).
1850 JANKOWSKA......(J.)....§ (6 , 1780) PHYS. STAT. SOL. B VOL.54 NO.2,   K99, (1972).
                                (6 ,   499) J. PHYS. F VOL.3,........ L173, (1973).
                                (6 ,   535) PHYS. STAT. SOLIDI B VOL.55 NO.1   K1, (1973).
                                (6 ,   536) PHYS. STAT. SOLIDI B VOL.57 NO.1   K1, (1973).
                                (6 , 1763) PHYS. STAT. SOLIDI B VOL.59 NO.2   K97, (1973).
1851 JANNINK........(G.)....§ (15 ,   23) J. POLYMER SCI./B POLYM. LETT. 7   865, (1969).
                                (6 ,   431) J. CHEM. PHYSICS VOL.55,.... 2384, (1971).
                                (15 ,   14) IAEA SYMP. GRENOBLE,......   277, (1972).
                                (5 , 2329) J. CHEM. PHYS. VOL.57,.....   290, (1972).
                                (5 , 2330) J. APPL. CRYST. VOL.7,.....   189, (1974).
                                (6 , 1922) PHYS. REV. LETT. VOL.32,.... 1170, (1974).
1852 JANOT..........(C.)....§ (6 ,   205) C.R. ACAD. SCI. VOL.257,.... 1843, (1963).
1853 JANSEN.........(R.)....§ (5 ,   615) COLL. INTER. N.126 (GRENOBLE),   158, (1963).
1854 JARRETT........(H.S.)..§ (5 , 1807) PHYS. REV. VOL.120,........ 1969, (1960).
1855 JASZCZAK.......(R.J.)..§ (20A,   80) REV. SCI. INSTRUM. VOL.42,...   240, (1971).
1856 JAUHO..........(P.)....§ (7B,   90) PHYS. STAT. SOL. VOL.42,....   757, (1970).
1857 JAY............(B.)....§ (6 , 1684) PHYS. LETT. VOL.43A NO.3,...   279, (1973).
1858 JEFFREY........(G.A.)..§ (5 ,   294) ACTA CRYST. VOL.B27,...... 2393, (1971).
1859 JEFFRIES.......(J.D.)..§ (9B,   38) DISS. ABSTR. B VOL.34,..... 1125, (1973).
1860 JEITSCHKO......(W.)....§ (5 , 1870) ACTA CRYST. 19,......... 1031, (1965).
1861 JELINEK........(G.E.)..§ (10B,   47) J. CHEM. PHYSICS VOL.40,....   531, (1964).
1862 JELLINEK.......(F.)....§ (5 , 1966) J. SOLID STATE CHEM. VOL.6,..   574, (1973).
1863 JENSEN.........(G.B.)..§ (5 , 1754) J. PHYS. C VOL.7,........   409, (1974).
1864 JENSEN.........(J.)....§ (6 , 2132) IAEA SYMP. GRENOBLE,......   603, (1972).
1865 JENSEN.........(L.H.)..§ (5 , 2504) ACTA CRYST. VOL.B29,......   797, (1973).

```
1866  JENSEN..........(S.J.)..:   ( 5 , 2404) ACTA CHEM. SCAND. VOL.24,. . . .  3422, (1970).
1867  JEROME..........(D.)....:   (11C,   50) SOL. STATE COMM. VOL.8,. . . . .  1793, (1970).
1868  JEWELL..........(R.W.)..:   (20C,   54) NUCL. INST. MET. 63,. . . . .      185, (1968).
1869  JEX.............(H.)....:   ( 6 , 1532) PHYS. LETT. VOL.34A,. 9,. . . . .   118, (1971).
                                  (10B,  213) SOL. STATE COMM. VOL. 9, . . . .  2057, (1971).
                                  ( 6 ,  921) IAEA SYMP. GRENOBLE,. . . . . .     29, (1972).
                                  (24 ,   76) NUCL. INST. MET. VOL.103,. . . .   229, (1972).
                                  ( 5 , 2672) PHYS. STAT. SOLIDI B VOL.61 NO.1   241, (1974).
1870  JOENK...........(R.J.)..:   ( 5 ,  917) J. APP. PHYS. 37,. . . . . . . .   981, (1966).
1871  JOGI............(S.)....:   (10B,   93) J. PHYS. F VOL.4,. . . . . . . .    11, (1974).
1872  JOHANNIN-GILLES.(A.)....:   ( 9D,   46) J. PHYS. RADIUM TOME 17, . . . .    72, (1956).
1873  JOHANSON........(E.W.)..:   (21 ,   43) REV. SCI. INSTRUM. VOL.32, . . .   456, (1961).
                                  (22B,   40) PROC/NAT. ELECTRONICS CONF. 21,.   816, (1965).
                                  (23 ,   39) IEEE TRANS NUCL. SCI. NS-13 NO.2    69, (1966).
1874  JOHANSSON.......(B.)....:   (11A,   33) PHYS. LETT. VOL.31A, . . . . . .   508, (1970).
1875  JOHANSSON.......(T.)....:   ( 5 , 2336) PHYS. REV. LETT. VOL.25, . . . .   524, (1970).
1876  JOHNSON.........(C.E.)..:   ( 5 , 1625) PHIL. MAG. VOL.10, . . . . . . .   713, (1964).
                                  ( 5 , 1097) J. OF PHYS. C VOL.1; : : : : : :   179, (1968).
                                  ( 5 , 1107) J. OF PHYS. C VOL.3; . . . . . .  1127, (1970).
1877  JOHNSON.........(C.H.)..:   ( 5 , 2237) PROC/CONF/NEUTRON C. S. + TECH.,   851, (1968).
1878  JOHNSON.........(C.K.)..:   (10B,   26) HARWELL SUMMER SCHOOL,. . . . .    132, (1968).
                                  ( 5 ,    7) ACTA CRYST. VOL.B27, . . . . . .   333, (1971).
1879  JOHNSON.........(D.A.G.):   (17A,   64) NUCL. INST. MET. 70, . . . . . .   164, (1969).
1880  JOHNSON.........(F.A.)..:   ( 6 , 2029) PROC. ROY. SOC. 281,. . . .; : :   274, (1964).
                                  (10B,  204) PROC. ROY. SOC. A VOL.317; . . .   279, (1970).
1881  JOHNSON.........(M.D.)..:   (14A,   61) PROC. ROY. SOC. A VOL.282, . . .   283, (1964).
1882  JOHNSON.........(M.H.)..:   ( 8 ,   68) PHYS. REV. VOL.55, : : : : : : :   898, (1939).
                                  ( 7B,   54) PHYS. REV. VOL.59, . . . . . . .   981, (1941).
1883  JOHNSON.........(M.W.)..:   ( 6 , 1961) PROPERTIES LIQUID METALS,. . . .    99, (1973).
1884  JOHNSON.........(P.W.)..:   (16 ,   42) NUCL. SCI. ENGNG. VOL.37,. . . .   127, (1969).
1885  JOHNSON.........(R.A.)..:   ( 6 , 1376) PHYS. REV. B-1,. . . . . . . . .  3510, (1970).
1886  JOHNSON.........(R.G.)..:   ( 6 , 1073) J. PHYS. SOC. JAPAN VOL.17 B-II,   370, (1962).
1887  JOHNSON.........(R.T.)..:   (24 ,   71) NUCL. INST. MET. 77, . . . . . .   189, (1970).
1888  JOHNSON.........(R.)....:   (10C,   36) IAEA SYMPOSIUM BOMBAY, VOL.1,. .    61, (1964).
1889  JOHNSON.........(W.H.)..:   ( 5 , 1516) NUCL. SCI. ENG.(USA) VOL.47, . .   151, (1972).
1890  JOHNSTON........(D.F.)..:   ( 9D,   65) PROC. PHYS. SOC. 88,. . . . . .     37, (1966).
                                  ( 9D,   42) J. PHYS. C, SER.2, VOL.2,. . . .  1151, (1969).
1891  JOHNSTON........(G.B.)..:   ( 6 ,  867) PHYS. LETT. 7, . . . . . . . . .   177, (1963).
                                  ( 5 , 1094) PHIL. MAG. VOL.12, . . . . . . .   221, (1965).
                                  ( 5 ,  726) J. PHYS. CHEM. SOLIDS 29,. . . .   193, (1968).
                                  ( 5 ,  727) J. PHYS. CHEM. SOLIDS 29,. . . .   201, (1968).
                                  ( 5 ,  732) J. PHYS. CHEM. SOLIDS 29,. . . .   855, (1968).
1892  JOHNSTON........(N.T.)..:   ( 5 ,  264) J. CHEM. PHYS. VOL.57, . . . . .  5007, (1972).
1893  JOHN............(W.)....:   (23 ,   26) IAEA SYMP. (PILE RESEARCH)VIENNA    83, (1960).
                                  (20C,   54) NUCL. INST. MET. 63,. . . . . .    185, (1968).
                                  (10C,   85) PHYS. STAT. SOLIDI A VOL.58 NO.2 K159, (1973).
1894  JOLIOT..........(F.)....:   (25 ,    9) NATURE VOL.130,. . . . . . . . .    57, (1932).
1895  JOLY............(R.)....:   (12B,   22) J. PHYSIQUE TOME 24, . . . . . .   89A, (1963).
1896  JONA............(F.)....:   ( 5 , 2289) PHYS. REV. VOL.105,. . . . . . .   849, (1957).
                                  ( 5 , 1975) J. PHYS. CHEM. SOL. 24,. . . . .  1341, (1963).
1897  JONES...........(D.D.)..:   ( 5 ,   17) ACTA CRYST. VOL.B30, . . . . . .  1220, (1974).
1898  JONES...........(D.W.)..:   ( 5 ,  356) BULL. SOC. CHIM. FR. 1968, . . .  1748, (1968).
                                  (22A,   38) NUCL. INST. MET. 62, . . . . . .    19, (1968).
                                  ( 5 ,  352) ACTA CRYST. VOL.B27, . . . . . .   349, (1971).
                                  ( 5 , 1970) ACTA CRYST. VOL.B27, . . . . . .   354, (1971).
                                  ( 5 ,  355) J. CHEM. SOC. A (1971),. . . . .  3725, (1971).
                                  ( 5 ,  354) J. CRYST. MOL. STRUCT.(GB) VOL.1   347, (1971).
                                  ( 5 ,  649) Z. KRIST. VOL.134,. . . . . . .    308, (1971).
                                  ( 5 ,  278) ACTA CRYST. VOL.A28 PART S-4,. .    27, (1972).
                                  ( 5 ,  351) ACTA CRYST. VOL.B28, . . . . . .  2430, (1972).
                                  ( 5 ,  350) ACTA CRYST. VOL.B28, . . . . . .   209, (1972).
                                  ( 5 , 1951) Z. KRIST. VOL.135, . . . . . . .   240, (1972).
                                  ( 5 ,  661) INORG. NUCL. CHEM. LETT. VOL.9,.  1025, (1973).
1899  JONES...........(H.D.)..:   (10B,   24) DISS. ABS. VOL.29, . . . . . . .  730B, (1968).
1900  JONES...........(I.R.)..:   (20D,   31) REV. SCI. INSTRUM. VOL.33, . . .  1399, (1962).
                                  (20D,   33) REV. SCI. INSTRUM. VOL.34, . . .    28, (1963).
1901  JONES...........(L.V.)..:   ( 6 ,    8) NATURE (PHYS. SCI.) VOL.241, . .    40, (1973).
1902  JONES...........(L.W.)..:   ( 7A,   54) PHYS. LETT. B VOL.36B, . . . . .   560, (1971).
1903  JONES...........(R.V.)..:   ( 5 , 1062) J. PHYS. SOC. JAPAN VOL.17 B-III   19, (1962).
1904  JONES...........(T.C.)..:   (22A,   26) J. PHYS. E VOL.6 NO.1, . . . . .    51, (1973).
1905  JONKER..........(C.C.)..:   ( 9C,   29) NED. TIJDSCHR. NATUURK. VOL.18,.   201, (1952).
                                  (24 ,   73) NUCL. INST. MET. 77, . . . . . .    55, (1970).
                                  (12A,   12) NED. TIJDSCHR. NATUURK. VOL.38,.   171, (1972).
1906  JONSSON.........(P.G.)..:   ( 5 , 1940) ACTA CRYST. VOL.B26, . . . . . .   536, (1970).
                                  ( 5 , 1927) ACTA CHEM. SCAND. VOL.25,. . . .  1729, (1971).
                                  ( 5 ,    1) ACTA CRYST. VOL.B27, . . . . . .   893, (1971).
                                  ( 5 , 1928) ACTA CHEM. SCAND. VOL.26,. . . .  1087, (1972).
                                  ( 5 ,    2) ACTA CHEM. SCAND. VOL.26,. . . .  1599, (1972).
                                  ( 5 ,    4) ACTA CRYST. A28 PART S-4,. . . .   193, (1972).
                                  ( 5 ,    8) ACTA CRYST. VOL.B28, . . . . . .  1827, (1972).
                                  ( 5 ,   35) J. CHEM. PHYS. VOL.56, . . . . .  4433, (1972).
1907  JORDAN..........(R.G.)..:   ( 5 , 1401) J. CHEM. PHYSICS VOL.53, . . . .  1455, (1970).
                                  ( 5 , 1400) J. CHEM. PHYSICS VOL.54, . . . .  1995, (1971).
```

```
1908   JOSHI...........(A.W.)..!   (119,    29) PROC. PHYS. SOC. 91, . . . . . .     97, (1967).

1909   JOSHI...........(S.K.)..!   ( 6 ,    18) PHYS. REV. VOL.126,. . . . . . .    933, (1962).
                                   (10B,    44) J. CHEM. PHYSICS VOL.39,. . . . .   2633, (1963).
                                   ( 6 ,   917) IAEA SYMP. BOMBAY, VOL.1,. . . . .   285, (1964).
                                   (10B,    46) J. CHEM. PHYSICS VOL.40,. . . . .    662, (1964).
                                   (10B,   103) PHYSICA VOL.30,. . . . . . . . .   2105, (1964).
                                   (10B,   145) PHYS. REV. VOL.187,. . . . . . .    808, (1969).
                                   (10B,   165) PHYS. REV. B-4,. . . . . . . . .   1770, (1971).
                                   (10B,   174) PHYS. REV. B VOL.5,. . . . . . .   2880, (1972).

1910   JOSHUA..........(S.J.)..!   ( 6 ,  1762) J. OF PHYS. C VOL.2,. . . . . .      24, (1969).
                                   ( 6 ,  2216) PHYS. STAT. SOL. VOL.36,. . . .     737, (1969).
                                   ( 5 ,   699) PHYS. STAT. SOL. VOL.38,. . . .     643, (1970).
                                   ( 6 ,  1487) PHYS. LETT. VOL.36A,. . . . . .     439, (1971).

1911   JOSLIN..........(C.W.)..!   (20C,    35) NUCL. ENG. INT. VOL.17,. . . . .    399, (1972).

1912   JOSWIG..........(G.)....!   (20A,    29) KERNTECHNIK VOL.14,. . . . . . .      9, (1972).

1913   JOUANIN.........(C.)....!   (10C,    59) NUOVO CIMENTO VOL.34,. . . . . .    293, (1964).

1914   JOUBERT.........(J.C.)..!   ( 5 ,   834) J. PHYSIQUE TOME 32 COL.1 VOL.II    741, (1971).
                                   ( 5 ,  2496) J. SOLID STATE CHEM. VOL.7,. . .    337, (1973).

1915   JOVIC...........(D.M.)..!   (22A,     6) BULL. INST. NUCL. SCI.(YUGO.) 11     59, (1961).

1916   JOVIC...........(D.)....!   ( 6 ,    54) IAEA SYMP. BOMBAY, VOL.2,. . . .    117, (1964).
                                   ( 6 ,   807) IAEA SYMP. COPENHAGEN, VOL.2,. .     37, (1968).
                                   ( 6 ,   806) PHYS. LETT. VOL.28A,. . . . . .     213, (1968).
                                   (13 ,    41) PHYS. LETT. VOL.33A,. . . . . .     287, (1970).
                                   ( 6 ,     3) CHEM. PHYS. LETT. VOL.17,. . . .     53, (1972).
                                   ( 6 ,   678) PHYS. LETT. VOL.42A NO.7,. . . .    509, (1973).

1917   JUDE............(R.J.)..!   ( 5 ,  2413) ACTA CRYST. VOL.A28 PART S-4,. .    193, (1972).

1918   JUDIN...........(V.M.)..!   ( 5 ,   218) PROC. INT. CONF. (NOTTINGHAM), .    354, (1964).

1919   JUHNKE..........(D.G.)..!   ( 5 ,   418) J. INORG. NUCL. CHEM. VOL.31,. .   3721, (1969).
                                   ( 5 ,   318) J. INORG. NUCL. CHEM. VOL.32,. .   2839, (1970).

1920   JUNA............(J.)....!   (20B,    14) CZECH. J. PHYS. VOL.8, . . . . .    592, (1958).

1921   JUNG............(H.H.)..!   (20B,    94) NUCL. INST. MET. 68,. . . . . .     353, (1969).
                                   (19B,     3) ATOMKERNENERGIE VOL.15,. . . . .     91, (1970).

1922   JUST............(W.)....!   ( 5 ,   410) J. PHYS. C VOL.6,. . . . . . . .   3465, (1973).
                                   ( 8 ,   104) PHYS. REV. B VOL.7,. . . . . . .   4142, (1973).
                                   ( 5 ,   411) J. APPL. CRYST. VOL.7,. . . . .     230, (1974).

1923   KACHARSKAYA.....(L.V.)..!   (14B,   121) SOV. PHYS. ACOUSTICS VOL.9,. . .    349, (1964).

1924   KACHHAVA........(C.M.)..!   ( 6 ,  1361) PHYS. STAT. SOL. VOL.B-54, . . .    K29, (1972).
                                   ( 6 ,  1808) J. PHYS. F VOL.3,. . . . . . . .     24, (1973).
                                   ( 6 ,    41) PHYSICA VOL.65,. . . . . . . . .     63, (1973).

1925   KADAR...........(G.)....!   ( 5 ,  1798) PHYS. LETT. 20,. . . . . . . . .    331, (1966).
                                   ( 5 ,  1788) J. APPL. PHYS. VOL.38,. . . . .    1265, (1967).
                                   ( 5 ,  1785) PHYS. LETT. VOL.24A,. . . . . .     198, (1967).
                                   ( 5 ,  1781) PHYS. LETT. VOL.25A,. . . . . .      56, (1967).
                                   ( 5 ,  1786) CENT. RES INST PHYS. KFKI NO.2,.     1, (1968).
                                   ( 5 ,  1783) J. APP. PHYS. 39,. . . . . . . .    538, (1968).
                                   ( 5 ,  1696) PHYS. LETT. VOL.26A,. . . . . .     556, (1968).
                                   ( 5 ,  1695) PHYS. REV. 171,. . . . . . . . .    574, (1968).
                                   ( 5 ,  1780) PHYS. LETT. VOL.29A,. . . . . .     340, (1969).
                                   ( 5 ,  1776) J. APPL. PHYS. 41,. . . . . . .     941, (1970).
                                   ( 5 ,  1700) SOL. STATE COMM. VOL.8,. . . . .   1653, (1970).
                                   ( 5 ,  1791) INT. J. MAGN.(GB) VOL.1, . . . .    341, (1971).
                                   ( 5 ,  1706) INT. J. MAGN. VOL.1,. . . . . .     143, (1971).
                                   ( 5 ,  1770) J. PHYSIQUE TOME 32 COL.1 VOL.II    980, (1971).
                                   ( 5 ,  1778) J. PHYS. CHEM. SOLIDS VOL.33,. .    212, (1972).
                                   ( 5 ,  1775) SOL. STATE COMM. VOL.10, . . . .   1195, (1972).
                                   ( 5 ,  1777) SOL. STATE COMM. VOL.11, . . . .    933, (1972).
                                   ( 5 ,   464) SOL. STATE COMM. VOL.12, . . . .    909, (1973).

1926   KADOMATSU.......(H.)....!   ( 5 ,  1116) J. PHYS. SOC. JAPAN VOL.35,. . .   1554, (1973).

1927   KADOMTSEVA......(A.M.)..!   ( 5 ,  1127) SOV. PHYS. J.E.T.P. VOL.19,. . .   1348, (1964).

1928   KAGANOV.........(M.I.)..!   ( 7B,   123) SOV. PHYS. JETP. 16, . . . . . .    109, (1963).

1929   KAGAN...........(YU.)...!   (10C,   111) SOV. PHYS. JETP 15,. . . . . . .    954, (1962).
                                   (10C,   113) SOV. PHYS. JETP. 17,. . . . . .     925, (1963).
                                   (10D,    62) SOV. PHYS. JETP 21,. . . . . . .    646, (1965).
                                   (10C,   118) SOV. PHYS. JETP. 20,. . . . . .    1340, (1965).
                                   ( 6 ,  2072) SOV. PHYS. SOL. STATE 8,. . . .    1120, (1966).
                                   ( 6 ,  2076) SOV. PHYS. JETP VOL.25,. . . . .    365, (1967).
                                   (10B,   225) SOV. PHYS. SOL. STATE VOL.11,. .    733, (1969).

1930   KAHANE..........(J.)....!   (12A,    33) REV. PHYS. APPL. (FRANCE) VOL.4,   115, (1969).
                                   (12B,    50) REV. PHYS. APPL. (FRANCE) VOL.4,   254, (1969).

1931   KAHAN...........(T.)....!   ( 7A,    82) REV. GEN. SCI. TOME 50,. . . . .     89, (1939).

1932   KAHN............(R.)....!   ( 6 ,   724) C.R. ACAD. SC. (PARIS), 267-B, .     61, (1968).
                                   ( 6 ,  1414) IAEA SYMP. COPENHAGEN, VOL.1,. .    289, (1968).
                                   ( 6 ,   740) J. PHYSIQUE TOME 32 COL.1 VOL.I,    523, (1971).
                                   ( 6 ,   779) J. PHYSIQUE TOME 32, . . . . . .    447, (1971).
                                   ( 5 ,  2004) J. PHYSIQUE TOME 34,. . . . . .     447, (1973).
                                   ( 5 ,  2007) SOL. STATE COMM. VOL.13, . . . .   1839, (1973).

1933   KAISER..........(B.)....!   ( 6 ,  1209) J. MOL. BIOL. VOL.41,. . . . . .    231, (1969).

1934   KAISER..........(W.)....!   (22B,    14) INTERNAT. COMPUTING SYMP., . . .    291, (1974).

1935   KAI.............(K.)....!   ( 5 ,  2482) JAP. J. APPL. PHYS. VOL.12,. . .   1119, (1973).

1936   KAJAMA..........(J.)....!   (22B,    24) NUCL. INST. MET. 40, . . . . . .    125, (1966).

1937   KAJFOSZ.........(J.)....!   (20B,    15) CZECH. J. PHYS. B VOL.13,. . . .    474, (1963).
                                   (12A,     9) CZECH. J. PHYS. B VOL.22,. . . .     38, (1972).

1938   KAKIHARA........(K.)....!   (23 ,    20) IAEA SYMP. (PILE RESEARCH)VIENNA    117, (1960).

1939   KAKINOKI........(J.)....!   ( 9C,     1) ACTA CRYST. 19,. . . . . . . . .    137, (1965).

1940   KALANOV.........(M.)....!   (17A,    46) INSTRUM. EXP. TECH.(USA) VOL.14,    733, (1971).
                                   ( 5 ,  1167) SOV. PHYS. JETP VOL.36,. . . . .   1170, (1973).

1941   KALDIS..........(E.)....!   ( 5 ,  2048) J. PHYS. C VOL.6,. . . . . . . .    725, (1973).

1942   KALEBIN.........(S.M.)..!   (20B,    48) INSTRUM. EXP. TECH. NO.4,. . . .    632, (1963).
                                   ( 5 ,  2628) ATOMNAYA ENERGIYA (USSR) VOL.24,   243, (1968).
```

1943 KALINICHENKO....(A.I.)..‡ (8 , 155) SOV. PHYS. SOLID STATE VOL.15, .   656, (1973).

1944 KALININ.........(V.P.)..‡ (5 , 2967) SOV. PHYS. CRYST. VOL.14,. . . .   913, (1969).
                                 (5 , 2969) SOV. PHYS. CRYST. VOL.15,. . . .   376, (1970).

1945 KALLEL..........(A.)....‡ (5 , 555) SOL. STATE COMM. VOL.5, ...   955, (1967).
                                 (5 ,   70) COMP. REND. 268, ...   455, (1969).
                                 (9D,   12) AIP CONF. PROC.(USA) VOL.5,. .   1355, (1971).
                                 (5 , 844) PHYS. LETT. VOL.34A, ...   361, (1971).
                                 (5 , 557) SOL. STATE COMM. VOL. 9,. . .   1699, (1971).
                                 (5 , 837) SOL. STATE COMM. VOL. 9, . . . 1949, (1971).
                                 (5 , 2925) J. SOLID STATE CHEM. VOL.5,. .   11, (1972).
                                 (5 , 562) SOL. STATE COMM. VOL.12 NO.7,.   665, (1973).
                                 (5 , 2189) PHYS. REV. LETT. VOL.32, . . . 1257, (1974).

1946 KALLI...........(H.)....‡ (16 ,   34) NUCL. INST. MET. VOL.101,. . .   263, (1972).
                                 (17A,   70) NUCL. INST. MET. VOL.108,. . .   11, (1973).
                                 (20B, 111) NUCL. INST. MET. VOL.113,. . .   461, (1973).

1947 KALUS...........(J.)....‡ (10C,   6) ANN. DE PHYSIQUE TOME 7 NO.4,.   247, (1972).
                                 (22C,   11) CONF. NEUTRON BEAMS, ...   , (1972).
                                 (18A,   30) Z. PHYSIK BAND 254,. . . . . .   149, (1972).
                                 (18A,   29) Z. PHYSIK BAND 254,. . . . . .   165, (1972).
                                 (20D,   3) ACTA CRYST. VOL.A29, ...   526, (1973).
                                 (20B,   61) J. PHYS. E VOL.6 NO.5, ...   488, (1973).
                                 (18A,   16) KERNTECHNIK VOL.15,. . . . . .   228, (1973).

1948 KAMAL...........(M.)....‡ (6 , 2241) DISS. ABSTR. B VOL.34, ...   354, (1973).
                                 (5 , 2528) PHYS. REV. B VOL.8,. . . . . . 2595, (1973).

1949 KAMB............(B.)....‡ (5 , 793) PHYS. ICE, ...   44, (1969).
                                 (5 , 1292) J. CHEM. PHYS. VOL.55, ... 1934, (1971).
                                 (5 , 786) J. CHEM. PHYS. VOL.58, ...   567, (1973).

1950 KAMINKER........(D.M.)..‡ (5 , 2271) BULL. ACAD. SCI. USSR VOL.28,.   350, (1964).

1951 KAMITAKAHARA....(W.A.)..‡ (6 ,   15) PHYS. LETT. 29A, ...   639, (1969).
                                 (6 , 614) CAN. J. PHYS. 49, ...   2291, (1971).
                                 (6 , 1772) IAEA SYMP. GRENOBLE, ...   73, (1972).
                                 (6 , 633) THESIS (246 PP.), ...   , (1972).
                                 (1JB, 187) PHYS. REV. B VOL.10 NO.2, . .   , (1974).
                                 (6 , 1774) PHYS. REV. B VOL.10 NO. 2, . .   , (1974).

1952 KAM.............(Z.)....‡ (10B, 127) PHYS. REV. 175,. . . . .   1156, (1968).
                                 (10B, 173) PHYS. REV. B VOL.5,. . . . . . 2887, (1972).

1953 KANDUSAR........(A.)....‡ (6 , 1579) PHYS. STAT. SOLIDI B VOL.59 NO.2 471, (1973).

1954 KANEKO..........(Y.)....‡ (6 , 1042) J. NUCL. SCI. TECHNOL. VOL.6,.   671, (1969).

1955 KANELLAKOPULOS..(B.)....‡ (5 , 289) SOL. STATE COMM. VOL.12, . . .   1157, (1973).

1956 KAPISHEV........(A.G.)..‡ (5 , 2494) 2ND INTERNAT. MEET./FERROELECTR. 219, (1969).

1957 KAPLAN..........(S.F.)..‡ (6 , 1441) J. APP. PHYS. 36,. . . . . . .   1090, (1965).
                                 (5 , 1715) ACTA CRYST B24, ...   1312, (1968).
                                 (5 , 1973) FERROELECTRICS,. . . . .   31, (1970).

1958 KAPLAN..........(T.A.)..‡ (11A,   36) PHYS. REV. VOL.109,. . . . . .   782, (1958).

1959 KAPLAN..........(T.)....‡ (6 , 1585) PHYS. REV. B VOL.9,. . . . . .   353, (1974).

1960 KAPPATSCH.......(A.)....‡ (9D,   43) J. PHYSIQUE TOME 31, ...   369, (1970).
                                 (5 , 2567) SOL. STATE COMM. VOL. 9, . . .   435, (1971).

1961 KAPULLA.........(H.)....‡ (5 , 1242) NUCL. SCI. ENGNG. VOL.38,. . .   180, (1969).
                                 (6 , 289) PHYS. LETT. VOL.31A, ...   158, (1970).
                                 (6 , 294) IAEA SYMP. GRENOBLE, ...   841, (1972).

1962 KARAS...........(W.)....‡ (9B,   91) PHYS. STAT. SOLIDI A VOL.5 NO.2,   397, (1971).

1963 KARAYIANIS......(N.)....‡ (11C,   25) PHYS. LETT. VOL.33A, ...   375, (1970).

1964 KARIMOV.........(I.)....‡ (6 , 2211) SOV. PHYS. SOL. STATE 9, . . .   1366, (1967).

1965 KARLEN..........(L.)....‡ (21 ,   34) NUCL. INST. MET. 56, ...   305, (1967).

1966 KARLE...........(J.)....‡ (17B,   9) ACTA CRYST. 20, ...   881, (1966).

1967 KARO............(A.M.)..‡ (10B,   51) J. CHEM. PHYSICS VOL.48, . . .   3173, (1968).

1968 KARPOVICH.......(I.R.)..‡ (22A,   67) SOV. PHYS. CRYST. VOL.5, . . .   294, (1960).

1969 KARSONO.........(L.)....‡ (5 , 1994) ACTA CRYST. VOL.B27, ...   253, (1971).

1970 KASATKIN........(S.N.)..‡ (23 ,   41) INSTRUM. EXP. TECH. NO.6,. . .   902, (1961).

1971 KASHCHEEV.......(V.N.)..‡ (10E,   29) SOV. PHYS. SOL. STATE VOL.3,.   1107, (1961).
                                 (8 ,   47) LATV. PSR ZIN. AKAD. VESTIS NO.3   39, (1962).
                                 (10E,   19) LATV. PSR ZIN. AKAD. VESTIS NO.5   67, (1962).
                                 (8 ,   46) LATV. PSR ZIN. AKAD. VESTIS NO.7   53, (1962).
                                 (8 ,   45) LATV. PSR ZIN. AKAD. VESTIS NO.9   65, (1962).
                                 (10C, 109) SOV. PHYS. SOL. STATE VOL.3,.   2311, (1962).
                                 (8 , 132) SOV. PHYS. SOL. STATE, 4,. . .   1054, (1962).
                                 (11B,   37) SOV. PHYS. SOL. STATE, 4,. . .   1494, (1962).
                                 (11A,   59) SOV. PHYS. SOL. STATE, 4,. . .   556, (1962).
                                 (8 , 136) SOV. PHYS. SOL. STATE, 5,. . .   635, (1963).
                                 (8 , 137) SOV. PHYS. SOL. STATE, 5,. . .   668, (1963).
                                 (11A,   2) ACTA PHYS. POLON. VOL.25,. . .   337, (1964).
                                 (11A,   1) ACTA PHYS. POLON. VOL.25,. . .   349, (1964).
                                 (11A,   3) ACTA PHYS. POLON. VOL.26,. . .   257, (1964).
                                 (8 ,   4) ACTA PHYS. POLON. VOL.26,. . .   11, (1964).
                                 (10E,   27) PHYS. STAT. SOL. VOL.4,. . . .   31, (1964).
                                 (13 ,   33) LATV. PSR ZIN. AKAD. VESTIS NO.2   22, (1966).
                                 (8 ,   48) LATV. PSR ZIN. AKAD. VESTIS NO.2   6, (1966).

1972 KASHIWASE.......(Y.)....‡ (10C,   49) J. PHYS. SOC. JAPAN VOL.21,. . . 1263, (1966).

1973 KASHUBA.........(I.E.)..‡ (5 , 389) UKR. FIZ. ZH.(USSR) VOL.16,. .   1735, (1971).

1974 KASMAN..........(YA.A.).‡ (6 , 1749) SOV. PHYS. JETP LETT. VOL.2,.   336, (1965).
                                 (11A,   63) SOV. PHYS. JETP LETT. VOL.9,. .   204, (1969).
                                 (5 , 2065) SOV. PHYS. J.E.T.P. VOL.29,. .   261, (1969).

1975 KASPER..........(J.S.)..‡ (5 , 997) ACTA MET. VOL.2, ...   456, (1954).
                                 (5 , 2956) PHYS. REV. VOL.95, ...   1408, (1954).
                                 (5 , 2843) ACTA CRYST. VOL.9, ...   289, (1956).
                                 (5 , 1624) PHYS. REV. VOL.101, ...   537, (1956).
                                 (5 , 1666) PHYS. REV. VOL.109, ... 1551, (1956).
                                 (5 , 2162) J. PHYS. CHEM. SOLIDS VOL.11,.   231, (1959).
                                 (5 , 932) PHYS. REV. VOL.127, ... 2052, (1962).
                                 (5 , 1068) J. PHYS. CHEM. SOL. 24,. . . .   529, (1963).
                                 (5 , 1634) ACTA CRYST. 17, ...   95, (1964).
                                 (5 , 1848) PROC. INT. CONF. (NOTTINGHAM), . 169, (1964).

1976	KASPER..........(R.G.)..:	(7A,	37)	NUCL. PHYS. A VOL.A169,.	385,	(1971).
1977	KATASE..........(A.)....:	(19B,	7)	JAP. J. APPL. PHYS. VOL.10,. . .	1090,	(1971).
1978	KATIYAR.........(R.S.)..:	(6 ,	1415)	J. OF PHYS.-C-, SER.2, VOL.3,. .	1693,	(1970).
		(6 ,	2181)	CONF. INTERN., RENNES,.	84,	(1971).
1979	KATORI..........(K.)....:	(12B,	23)	J. PHYS. SOC. JAPAN VOL.28,. . .	1116,	(1970).
1980	KATO............(N.)....:	(9B,	4)	ACTA CRYST. 16,.	276,	(1963).
1981	KATSUKI.........(A.)....:	(11A,	55)	PROC. ROY. SOC. VOL.295,. . . .	182,	(1966).
1982	KATSURAGI.......(H.)....:	(22B,	16)	J. PHYS. SOC. JAPAN VOL.17 B-II,	358,	(1962).
1983	KATSURAGI.......(S.)....:	(24 ,	81)	NUCL. SCI. ENGNG. VOL.40,. . . .	25,	(1970).
1984	KATSURAKI.......(H.)....:	(5 ,	723)	J. PHYS. SOC. JAPAN VOL.18,. . .	93,	(1963).
		(5 ,	965)	J. PHYS. SOC. JAPAN VOL.19,. . .	1988,	(1964).
		(5 ,	1024)	J. PHYS. SOC. JAPAN VOL.19,. . .	863,	(1964).
		(5 ,	1022)	REV. ELECT. COMMUN. LAB. VOL.12,	424,	(1964).
		(5 ,	963)	J. APP. PHYS. 36,.	1094,	(1965).
		(5 ,	964)	J. PHYS. SOC. JAPAN VOL.21,. . .	2238,	(1966).
		(5 ,	1041)	J. PHYS. SOC. JAPAN VOL.22,. . .	674,	(1967).
1985	KATZ............(M.J.)..:	(5 ,	99)	PHYS. REV. A VOL.7,.	2130,	(1973).
1986	KAWACHI.........(M.)....:	(5 ,	1050)	J. PHYS. SOC. JAPAN VOL.33,. . .	1296,	(1972).
1987	KAWAMINAMI......(M.)....:	(5 ,	1156)	J. PHYS. SOC. JAPAN VOL.22,. . .	924,	(1967).
		(5 ,	1153)	J. PHYS. SOC. JAPAN VOL.29,. . .	649,	(1970).
1988	KAWANO..........(S.)....:	(5 ,	2579)	ACTA CRYST. VOL.A28 PART S-4,. .	99,	(1972).
		(5 ,	2580)	J. PHYS. SOC. JAPAN VOL.35,. . .	303,	(1973).
1989	KAWASAKI........(K.)....:	(13 ,	60)	PROGR. THEOR. PHYS. VOL.25,. . .	723,	(1961).
1990	KAWASAKI........(S.)....:	(24 ,	54)	NUCL. INST. MET. 62, ...	311,	(1968).
1991	KAWASAKI........(T.)....:	(11C,	23)	KYOTO CONF. STAT. MECHANICS, . .	157,	(1968).
1992	KAWEEB..........(B.H.)..:	(7A,	75)	PHYS. STAT. SOL. 4,.	31,	(1964).
1993	KAY.............(M.I.)..:	(5 ,	1984)	ACTA CRYST. VOL.14,.	56,	(1961).
		(5 ,	2282)	ACTA CRYST. VOL.14,.	80,	(1961).
		(5 ,	1983)	ACTA CRYST. VOL.15,.	506,	(1962).
		(5 ,	1981)	J. PHYS. SOC. JAPAN VOL.17 B-II,	389,	(1962).
		(5 ,	742)	COLL. INTER. N.126 (GRENOBLE), .	18,	(1963).
		(5 ,	376)	J. CHEM. PHYSICS VOL.40, ...	504,	(1964).
		(5 ,	369)	NATURE VOL.204,.	1050,	(1964).
		(5 ,	6)	ACTA CRYST. 21,.	237,	(1966).
		(5 ,	651)	ACTA CRYST. 21,.	383,	(1966).
		(5 ,	716)	J. CHEM. PHYSICS VOL.44,. . . .	1648,	(1966).
		(5 ,	2311)	J. PHYS. SOC. JAPAN VOL.21,. . .	1626,	(1966).
		(5 ,	1950)	ACTA CRYST. 22,.	182,	(1967).
		(5 ,	1892)	ACTA CRYST. 22,.	800,	(1967).
		(5 ,	2481)	ACTA CRYST. 23,.	868,	(1967).
		(5 ,	1715)	ACTA CRYST B24,.	1312,	(1968).
		(5 ,	1978)	2ND INTERNAT. MEET./FERROELECTR.	.,	(1969).
		(5 ,	1973)	FERROELECTRICS,.	31,	(1970).
		(5 ,	276)	ACTA CRYST. VOL.B27,.	26,	(1971).
		(6 ,	2022)	AMER. CRYSTALL. ASSOC.,. . . .	27,	(1971).
		(5 ,	708)	ACTA CRYST. VOL.A28 PART S-4,. .	186,	(1972).
		(5 ,	1914)	FERROELECTRICS VOL.5,.	45,	(1973).
		(5 ,	2486)	MATER. RES. BULL. VOL.8, ...	1183,	(1973).
1994	KAY.............(R.E.)..:	(5 ,	1247)	PROC. PHYS. SOC., 85,.	79,	(1965).
1995	KAZAMA..........(N.)....:	(5 ,	589)	J. PHYS. SOC. JAPAN VOL.30,. . .	1319,	(1971).
		(5 ,	1805)	ACTA CRYST. VOL.A28 PART S-4,. .	196,	(1972).
		(5 ,	1148)	J. PHYS. SOC. JAPAN VOL.34,. . .	58,	(1973).
1996	KAZARNOVSKII....(M.V.)..:	(7B,	118)	SOV. PHYS. JETP. 16, ...	1624,	(1963).
		(9B,	107)	SOV. PHYS. JETP. 20, ...	94,	(1965).
1997	KAZENNA.........(N.)....:	(5 ,	1637)	J. APP. PHYS. VOL.40,.	1128,	(1969).
1998	KAZI............(A.H.)..:	(24 ,	17)	IAEA SYMP. (PILE RESEARCH)VIENNA	609,	(1960).
1999	KEARNEY.........(R.J.)..:	(6 ,	153)	IAEA SYMP. COPENHAGEN, VOL.1,. .	315,	(1968).
		(6 ,	1428)	J. PHYS. CHEM. SOLIDS VOL.31,. .	1085,	(1970).
2000	KEATING.........(D.T.)..:	(9B,	50)	J. APPL. PHYS. VOL.26,.	1041,	(1955).
		(5 ,	691)	PHYS. REV. VOL.111,.	261,	(1958).
		(5 ,	754)	PHYS. REV. 130,.	1726,	(1963).
		(5 ,	735)	PHYS. REV. 175,.	868,	(1968).
		(6 ,	632)	PHYS. LETT. 30A,.	206,	(1969).
		(10C,	46)	J. PHYS. CHEM. SOLIDS VOL.31,. .	1317,	(1970).
		(5 ,	1222)	PHYS. STAT. SOLIDI B VOL.59,. .	K59,	(1973).
		(9 ,	2339)	SOL. STATE COMM. VOL.13,. . . .	1465,	(1973).
		(5 ,	2975)	J. PHYS. CHEM. SOLIDS VOL.35,. .	879,	(1974).
		(5 ,	2450)	PHYS. REV. B VOL.9,.	2590,	(1974).
2001	KEATING.........(D.)....:	(5 ,	2441)	PHYS. REV. B VOL.4,.	2472,	(1971).
		(5 ,	2442)	PHYS. REV. LETT. VOL.27,. . . .	320,	(1971).
2002	KEATING.........(P.N.)..:	(10B,	134)	PHYS. REV. 175,.	1171,	(1968).
2003	KEATON-JR.......(P.W.)..:	(12B,	3)	ANN. OF PHYSICS VOL.74,.	250,	(1972).
2004	KEDEM...........(D.)....:	(20A,	46)	NUCL. INST. MET. VOL.102,. . . .	87,	(1972).
2005	KEFFER..........(F.)....:	(5 ,	1760)	PHYS. REV. VOL.108,.	637,	(1957).
2006	KEHR............(K.W.)..:	(10D,	32)	PHYS. REV. LETT. VOL.24,. . . .	1101,	(1970).
		(10D,	25)	PHYS. LETT. VOL.39A,.	327,	(1972).
2007	KEIDERLING......(T.A.)..:	(5 ,	2725)	INORG. CHEM. VOL.11, ...	3009,	(1972).
2008	KEINERT.........(J.)....:	(16 ,	4)	ATOMKERNENERGIE VOL.18,.	261,	(1971).
2009	KEIRIM-MARKUS...(I.B.)..:	(20C,	21)	INSTRUM. EXP. TECH. NO.1,. . . .	77,	(1970).
2010	KELAREV........(V.V.)..:	(5 ,	1120)	PHYS. MET. METALLOGR. VOL.17 N.5	136,	(1964).
		(5 ,	1115)	PHYS. STAT. SOL. 24,.	385,	(1967).
2011	KELKAR..........(V.M.)..:	(6 ,	1601)	IAEA SYMP. BOMBAY, VOL.1,. . . .	77,	(1964).
2012	KELLER..........(R.)....:	(5 ,	1884)	HELV. PHYS. ACTA VOL.19, ...	493,	(1946).
2013	KELLY...........(J.W.)..:	(5 ,	2744)	ACTA CRYST. VOL.B29 PART 1,. . .	7,	(1973).
2014	KEMOKLIDZE......(M.P.)..:	(9D,	74)	SOV. PHYS. JETP. 20, ...	1505,	(1965).
		(14A,	68)	SOV. PHYS. J.E.T.P. VOL.32,. . .	1183,	(1971).

2015 KENDRICK........(H.)....‡ (5 , 542) PHYS. REV. LETT. VOL.14, ... 1022, (1965).
                                 (5 , 537) J. APP. PHYS. 37,. . . . . 1260, (1966).
                                 (9A, 53) J. APP. PHYS. 37,. . . . . 2343, (1966).
                                 (5 , 577) J. APP. PHYS. 38,. . . . . 1243, (1967).
                                 (5 , 574) PHYS. REV. 153,. . . . .   624, (1967).
                                 (5 , 532) PHYS. REV. 155,. . . . .   528, (1967).
                                 (5 , 2848) J. APP. PHYS. 39,. . . . .   585, (1968).
                                 (12A, 15) NUCL. INST. MET. 68, ...   50, (1969).

2016 KENNARD........(C.H.L.)‡ (5 , 1437) CRYST. STRUCT. COMM. VOL.1 N.3,   185, (1972).
                                 (5 , 657) CRYST. STRUCT. COMM. VOL.1 N.3,   189, (1972).
                                 (5 , 280) CRYST. STRUCT. COMM. VOL.1 N.3,   193, (1972).
                                 (5 , 2394) CRYST. STRUCT. COMM. VOL.1 N.4,   367, (1972).
                                 (5 , 2674) CRYST. STRUCT. COMM. VOL.1 N.4,   371, (1972).

2017 KENNETT........(T.J.)..‡ (5 , 2484) NUCL. PHYS. VOL.A139, ...   625, (1969).

2018 KEPA...........(H.)....‡ (17A, 69) NUCL. INST. MET. 99, ...   227, (1972).

2019 KERLER.........(R.)....‡ (5 , 2859) PHYS. LETT. 20,. . . . .   470, (1966).

2020 KERNELL........(R.L.)..‡ (5 , 2237) PROC/CONF/NEUTRON C. S. + TECH.,   851, (1968).

2021 KERN...........(R.)....‡ (5 , 778) ACTA CRYST. B25, ... 1685, (1969).

2022 KERR...........(E.C.)..‡ (6 , 1202) PHYS. REV. LETT. VOL.1,. . . . .   9, (1958).
                                 (6 , 1191) PHYS. REV. VOL.113,. . . . . . . 1379, (1959).
                                 (12B,   3) ANN. OF PHYSICS VOL.74,. . . . .   250, (1972).

2023 KERR...........(W.C.)..‡ (6 , 1136) PHYS. REV. A VOL.2,. . . . . . . 2416, (1970).
                                 (6 , 1137) PHYS. REV. A VOL.4,. . . . . . . 2413, (1971).
                                 (6 , 1704) PHYS. REV. A VOL.7,. . . . . . . 1043, (1973).

2024 KESHARWANI......(K.M.)..‡ (6 , 615) NUCL./S.S. PHYS. SYMP. ABSTR., .   9, (1972).
                                 (6 , 553) PHYS. REV. B VOL.6,. . . . . . . 2178, (1972).
                                 (6 , 552) SOL. STATE COMM. VOL.11 NO.6, .   771, (1972).
                                 (6 , 618) PHYS. REV. B VOL.7,. . . . . . . 5153, (1973).

2025 KESTIGIAN.......(M.)...‡ (6 , 1988) J. OF PHYS. C VOL.1, ...   940, (1968).
                                 (5 , 2396) PHYS. REV. B VOL.3,. . . . . . . 3946, (1971).

2026 KETUDAT.........(S.)...‡ (12B, 53) REV. SCI. INSTRUM. VOL.39, . . .   101, (1968).

2027 KHABIBULLIN.....(B.M.)..‡ (6 , 462) FIZ. MET. METALLOV. VOL.10,. . .   825, (1960).
                                 (11C, 51) SOV. PHYS. JETP VOL.12,. . . . .   714, (1961).
                                 (11C, 53) SOV. PHYS. SOL. STATE, 4,. . . . 1339, (1962).
                                 (11C, 54) SOV. PHYS. SOL. STATE, 4,. . . . 1449, (1962).
                                 (11C, 52) SOV. PHYS. SOL. STATE, 4,. . . .   428, (1962).
                                 (11C, 56) SOV. PHYS. SOL. STATE VOL.4, . . 1339, (1963).
                                 (11C, 55) SOV. PHYS. SOL. STATE VOL.4, . . 1449, (1963).
                                 (10C, 125) SOV. PHYS. SOL. STATE VOL.13,. .   215, (1971).

2028 KHACHATURYAN....(A.G.)..‡ (7B, 131) SOV. PHYS. SOL. STATE VOL.9, . . 2040, (1968).

2029 KHALATNIKOV.....(I.M.)..‡ (6 , 1181) DOKL. AKAD. NAUK VOL.93, ...   799, (1953).

2030 KHALFINA........(A.A.)..‡ (11A, 66) ZH. EKSPER. TEOR. FIZ.(USSR) 58,   918, (1970).

2031 KHAMYANOV.......(L.P.)..‡ (20B, 131) PRIB. TEKH. EKSPER. VOL.2, . . .   36, (1964).

2032 KHANNA..........(F.C.)..‡ (6 , 1122) CAN. J. PHYS. 49,. . . . . 2997, (1971).
                                 (6 , 1152) CAN. J. PHYS. VOL.50,. . . . . . 1152, (1972).
                                 (6 , 1163) PHYS. REV. LETT. VOL.29, ...   549, (1972).

2033 KHAN...........(D.C.)..‡ (5 , 485) J. PHYS. CHEM. SOLIDS 29,. . . . 2087, (1968).
                                 (5 , 423) PHYS. REV. B-1,. . . . . 2243, (1970).
                                 (22C, 64) REV. SCI. INSTRUM. 41, ...   107, (1970).
                                 (8 , 36) J. PHYS. C VOL.4,. . . . . 3307, (1971).
                                 (5 , 482) PHYS. REV. B VOL.4,. . . . . . . 3901, (1971).
                                 (8 , 52) NUCL./S.S. PHYS. SYMP. ABSTR.,.   9, (1972).

2034 KHAN...........(Q.H.)..‡ (5 , 1382) J. PHYSIQUE TOME 32 COL.1 VOL.I,   362, (1971).

2035 KHARADZE........(G.A.)..‡ (8 , 141) SOV. PHYS. JETP. 19, ...   986, (1964).
                                 (8 , 110) PROC. INDIAN ACAD. SCI. VOL.61,   202, (1965).
                                 (11C, 58) SOV. PHYS. J.E.T.P. VOL.29,. . . 1089, (1969).

2036 KHATTAK........(C.P.)..‡ (5 , 2400) A.I.P. CONF. PROC. NO.10 PART 1,   659, (1973).
                                 (5 , 162) A.I.P. CONF. PROC. NO.10 PART 1,   674, (1973).

2037 KHEIER.........(D.M.)..‡ (22A, 33) LATV. PSR...FIZ. TEHN. SER. NO.3   64, (1970).

2038 KHODAKOVSKAYA...(R.YA.)..‡ (5 , 2466) PHYS. STAT. SOLIDI A VOL.18 NO.2   K91, (1973).
                                 (5 , 2467) J. APPL. CRYST. VOL.7, ...   207, (1974).

2039 KHOKHLOV.......(S.F.)..‡ (5 , 2480) PHYS. MET. METALLOGR. VOL.18 N.3   134, (1964).

2040 KHOLAS.........(A.)...‡ (10B, 225) SOV. PHYS. SOL. STATE VOL.11,. .   733, (1969).

2041 KHOTEEV........(N.V.)..‡ (21, 52) SOV. PHYS. CRYST. VOL.12,. . . .   477, (1967).

2042 KHRUSHCHEV......(B.I.)..‡ (22A, 20) INSTRUM. EXPER. TECH. NO. 6, . . 1294, (1967).
                                 (5 , 2256) PHYS. MET. METALLOGR. VOL.22 N.2   123, (1967).
                                 (23, 76) SOV. PHYS. JOURNAL VOL.10 NO.4,   63, (1967).
                                 (5 , 208) PHYS. MET. METALLOGR. VOL.27, 6, 1011, (1969).
                                 (22A, 22) INSTRUM. EXP. TECH. 3, ...   726, (1970).
                                 (5 , 2477) PHYS. MET. METALLOGR. VOL.29,. .   188, (1970).
                                 (5 , 2927) PHYS. MET. METALLOGR. VOL.31,. .   79, (1971).

2043 KHUBCHANDANI....(P.G.)..‡ (6 , 181) PHIL. MAG. VOL.3,. . . . .   798, (1958).
                                 (6 , 223) PHYS. REV. VOL.110,. . . . . . .   70, (1958).
                                 (5 , 1298) REACTOR SCI. VOL.11, ...   89, (1960).
                                 (10C, 57) NUCL. SCI. ENGNG. VOL.13,. . . .   40, (1962).
                                 (11C, 17) J. PHYS. -C-, VOL. 1,. . . . . . 1706, (1968).

2044 KHVASHCHEVSKA...(YA.)...‡ (20C, 81) NUKLEONIKA VOL.8,. . . . .   489, (1963).

2045 KIKUTA.........(S.)...‡ (5 , 2439) J. PHYS. SOC. JAPAN VOL.31,. . .   954, (1971).
                                 (9C, 33) PHYS. LETT. VOL.37A, ...   29, (1971).

2046 KILLEAN........(R.C.G.)..‡ (17B, 11) ACTA CRYST. 23,. . . . .   54, (1967).
                                 (17B, 24) ACTA CRYST. B25, ...   374, (1969).
                                 (17B, 23) ACTA CRYST. B25, ...   977, (1969).

2047 KIMURA.........(M.)....‡ (5 , 1355) PROC. IMP. ACAD. JAPAN VOL.15,   214, (1939).
                                 (5 , 1347) PROC. PHYS.-MATH. SOC. JAPAN 22,   391, (1940).
                                 (5 , 2461) PROC. JAPAN ACAD. VOL.19,. . . .   152, (1943).
                                 (25, 65) PROC. JAPAN ACAD. VOL.19,. . . .   26, (1943).
                                 (5 , 941) PROC. PHYS. MATH. SOC. JAPAN 25,   495, (1943).
                                 (5 , 2463) PROC. PHYS. MATH. SOC. JAPAN 25,   530, (1943).
                                 (22C, 43) NUCL. INST. MET. 71,. . . . . .   102, (1969).
                                 (6 , 1920) J. CHEM. PHYSICS VOL.55,. . . .   983, (1971).
                                 (23, 67) PULSED NEUTRON SOURCES/UTILIZ,.   9, (1971).
                                 (19A,   4) ACTA CRYST. VOL.A28 PART S-4,.   197, (1972).
                                 (5 , 2446) J. PHYS. SOC. JAPAN VOL.33,. . . 1493, (1972).
                                 (5 , 2482) JAP. J. APPL. PHYS. VOL.12,. . . 1119, (1973).

2080   KLEBAN..........(P.)....:   (14B,   68) PHYSICA VOL.70,. . . . .   105, (1973).

2081   KLEB............(R.)....:   (19B,   20) NUCL. INST. MET. VOL.106,. . . .   221, (1973).

2082   KLEINBERG.......(R.)....:   (5 , 2098) J. APPL. PHYS. VOL.38,. . . . . . 1453, (1967).
                                   (5 , 2096) J. CHEM. PHYSICS VOL.29, ... 4690, (1969).
                                   (5 , 449) J. CHEM. PHYSICS VOL.53, ... 2860, (1970).
                                   (7A,   34) J. PHYS. SOC. JAPAN VOL.28-4,. .   644, (1970).
                                   (5 , 708) ACTA CRYST. VOL.A28 PART S-4,. .   186, (1972).
                                   (5 , 1914) FERROELECTRICS VOL.5,. . . . . .   45, (1973).

2083   KLEINERT........(P.)....:   (5 , 1689) PHYS. STAT. SOL. VOL.8,. . . . .   271, (1965).

2084   KLEINER.........(W.H.)..:   (11C,   34) PHYS. REV. VOL.97,. . . . . . .   411, (1955).

2085   KLEINMAN........(L.)....:   (10B, 152) PHYS. REV. B-2,. . . . .   3947, (1970).

2086   KLEINSTUCK......(K.)....:   (20D,   36) Z. KRIST. VOL.116,. . . . . . .   48, (1961).
                                   (23 ,   49) MON. DEUTSCHEN AKAD. WISS. VOL.4   586, (1962).
                                   (9B,   85) PHYS. STAT. SOL. 2, ... K104, (1962).
                                   (5 , 2696) PHYS. STAT. SOL. 2, ... K164, (1965).
                                   (6 , 525) PHYS. STAT. SOL. 3,. . . . . .   K269, (1963).
                                   (5 , 1409) Z. KRIST. VOL.118,. . . . . . .   473, (1963).
                                   (9B,   71) PHYS. LETT. 8,. . . . .   241, (1964).
                                   (22C,   15) EXPER. TECH. PHYS. VOL.2,. . .   111, (1965).
                                   (9C,   31) PHYS. LETT. VOL.14,. . . . . .   374, (1965).
                                   (5 , 1689) PHYS. STAT. SOL. VOL.8,. . . .   271, (1965).
                                   (5 , 2451) PHYS. STAT. SOL. VOL.23,. . .   237, (1967).
                                   (5 , 1606) PHYS. STAT. SOL. A VOL.1,. . .   749, (1970).

2087   KLEIN...........(A.G.)..:   (5 , 2378) APPL. PHYS. LETT. VOL.10,. . .   293, (1967).

2088   KLEIN...........(B.M.)..:   (10B, 161) PHYS. REV. LETT. VOL.25, . . . 1014, (1970).

2089   KLEIN...........(M.L.)..:   (7B,   23) J. APPL. PHYS. 41,. . . . . .   5138, (1970).
                                   (6 , 655) PHYS. LETT. VOL.33A,. . . . .   253, (1970).
                                   (10D,   31) PHYS. REV. B-2, ...   4176, (1970).
                                   (6 , 1127) PHYS. REV. LETT. VOL.24,. . .   1424, (1970).
                                   (10D,   8) CRIT. REV. SOLID STATE SCI. 2,   181, (1971).
                                   (6 , 1344) PHYS. LETT. VOL.34A,. . . . .   415, (1971).
                                   (10D,   34) PHYS. REV. B-4,. . . . . . .   1983, (1971).
                                   (10B, 166) PHYS. REV. B-4,. . . . . . .   567, (1971).

2090   KLEIN...........(S.)....:   (20B,   12) C. R. ACAD. SCI. B TOME 270,   185, (1970).
                                   (5 , 477) SOL. STATE COMM. VOL.12 NO.2,.   113, (1973).
                                   (5 , 1177) ACTA CRYST. VOL.A30,. . . . .   380, (1974).

2091   KLEMENS.........(P.G.)..:   (10B,   43) J. APPL. PHYS. 40, ... 4696, (1969).
                                   (6 , 1186) PHYS. LETT. 30A,. . . . . . .   127, (1969).

2092   KLEY............(W.)....:   (6 , 2036) PHYS. REV. LETT. VOL.2, . . .   258, (1959).
                                   (6 , 1302) PHYS. REV. LETT. VOL.2,. . .   94, (1959).
                                   (6 , 1090) PHYS. REV. LETT. VOL.3,. . .   91, (1959).
                                   (6 , 1084) PHYS. REV. VOL.119,. . . . .   872, (1960).
                                   (6 , 166) IAEA SYMP. CHALK RIVER, VOL.1,   139, (1962).
                                   (6 , 2226) IAEA SYMP. CHALK RIVER, VOL.2,   145, (1962).
                                   (6 , 2225) IAEA SYMP. CHALK RIVER, VOL.2,   155, (1962).
                                   (6 , 1681) BROOKHAVEN SYMPOSIUM,. . . . .   105, (1965).
                                   (6 , 2254) PHYS. LETT. 14,. . . . . . .   100, (1965).
                                   (17A, 103) Z. NATURFORSCH. VOL.21 SUPPL., 1770, (1966).
                                   (6 , 2253) J. PHYSIQUE TOME 28 COL.1,. .   26, (1967).
                                   (6 , 1677) IAEA SYMP. COPENHAGEN, VOL.1,.   223, (1968).
                                   (17A,   81) NUKLEONIK VOL.10 NO.6,. . . .   287, (1968).
                                   (5 , 1470) PHYS. REV. VOL.172,. . . . .   576, (1968).
                                   (20B, 135) RESEARCH APPL./PULSED REACTORS,. . , (1970).

2093   KLINE...........(G.R.)..:   (5 , 552) PHYS. REV. LETT. VOL.31,. . .   1498, (1973).

2094   KLOCHIKHIN......(A.A.)..:   (11B,   41) SOV. PHYS. SOL. STATE, 7,. . . 2916, (1965).

2095   KLUSHIN.........(N.A.)..:   (6 , 893) SOV. PHYS. SOL. STATE VOL.13,. 1256, (1971).

2096   KLYUSHIN........(V.V.)..:   (5 , 1828) PHYS. MET. METALLOGR. VOL.12,.   119, (1961).
                                   (5 , 1120) PHYS. MET. METALLOGR. VOL.17 N.5   136, (1964).
                                   (5 , 1115) PHYS. STAT. SOL. 24,. . . . .   385, (1967).

2097   KNICKLE.........(H.N.)..:   (16 ,   18) DISS. ABS. VOL.30,. . . . . . .38198, (1970).

2098   KNITTER.........(H.H.)..:   (5 , 1545) PROC/CONF/NEUTRON C. S. + TECH.,   827, (1968).

2099   KNOLL...........(W.)....:   (5 , 746) PHYS. CHEM. LIQ. VOL.4,. . . .   39, (1973).

2100   KNOPF...........(K.)....:   (7B, 140) Z. NATURFORSCH. A VOL.26A, . .   391, (1971).
                                   (7A,   98) Z. NATURFORSCH. A VOL.27A, . .   901, (1972).

2101   KNOP............(O.)....:   (5 , 902) CANAD. J. CHEM. VOL.43,. . . . 2819, (1965).

2102   KNORR...........(K.)....:   (6 , 1232) SOL. STATE COMM. VOL.13,. . .   1549, (1973).
                                   (6 , 709) PHYS. REV. B VOL.9 NO.11,. . .   . , (1974).

2103   KNOTT...........(H.W.)..:   (5 , 2757) ACTA CRYST. VOL.11,. . . . . .   751, (1958).
                                   (5 , 2838) J. APPL. CRYST. VOL.6,. . . .   206, (1973).

2104   KNOWLES.........(J.W.)..:   (24 ,   1) ACTA CRYST. VOL.9,. . . . . .   61, (1956).
                                   (23 ,   29) IAEA SYMP. (PILE RESEARCH)VIENNA   25, (1960).

2105   KNOX............(K.)....:   (5 , 1508) COLL. INTER. N.126 (GRENOBLE),.   37, (1963).
                                   (5 , 1509) INORG. CHEM. VOL.3,. . . . . .   558, (1964).

2106   KOBAYASHI.......(H.)....:   (20C, 108) SCI/INST. PHYS. CHEM. RES. 54,   271, (1960).
                                   (5 , 666) PHYS. REV. B-2,. . . . . . .   1310, (1970).
                                   (6 , 570) PHYS. REV. B-2,. . . . . . .   4632, (1970).

2107   KOBAYASHI.......(S.)....:   (12B,   23) J. PHYS. SOC. JAPAN VOL.28,. . 1116, (1970).

2108   KOBLA...........(J.)....:   (9C,   24) KRISTALL TECH. VOL.4,. . . . .   135, (1969).

2109   KOCHALOV........(O.V.)..:   (5 , 1854) SOV. PHYS. SOL. STATE 8, . . .   215, (1966).

2110   KOCHAROV........(A.G.)..:   (5 , 2689) SOV. PHYS. J.E.T.P. VOL.31,. .   808, (1970).
                                   (5 , 2946) PHYS. STAT. SOLIDI A VOL.4 NO.1,   53, (1971).
                                   (5 , 2951) PHYS. STAT. SOLIDI A VOL.7 NO.2, K71, (1971).
                                   (5 , 2192) SOV. PHYS. CRYST. VOL.16,. . .   634, (1971).
                                   (5 , 2892) SOV. PHYS. SOL. STATE VOL.14,. 3037, (1973).

2111   KOCHERGIN.......(A.V.)..:   (24 , 112) SOV. PHYS. DOKLADY VOL.16,. . .   154, (1971).

2112   KOCHMARSKII.....(V.Z.)..:   (10E,   31) UKR. FIZ. ZH.(USSR) VOL.16,. .   1985, (1971).
                                   (10C, 127) SOV. PHYS. SOLID STATE VOL.14,   1, (1972).

2113   KOCINSKI........(J.)....:   (8 ,   2) ACTA PHYS. POLON. VOL.19,. . .   691, (1960).
                                   (8 ,   5) ACTA PHYS. POLON. VOL.30,. . .   591, (1966).
                                   (13 ,   38) PHYS. LETT. 25A,. . . . . . .   600, (1967).
                                   (13 ,   2) ACTA PHYS. POLON. VOL.33,. . .   13, (1968).
                                   (9D,   10) ACTA PHYS. POLON. VOL.35,. . .   61, (1969).
                                   (13 ,   4) ACTA PHYS. POLON. 36,. . . . .   697, (1969).

```
2113   KOCINSKI........(J.)....:   { 8 ,    57) PHYS. LETT. VOL.34A,            306, (1971):
                                   {13 ,    47) PHYS. LETT. VOL.41A, NO.3,:: : : 179, (1972):

2114   KOEHLER.........(T.R.)..:   (10B,   130) PHYS. REV. 165,. . . . . . . . . 942, (1968).
                                   ( 6 ,   655) PHYS. LETT. VOL.33A, . . . . . . 253, (1970).
                                   (10D,    30) PHYS. REV. B-1,. . : : : : : : : 4521, (1970).
                                   (10D,    34) PHYS. REV. B-4,. . : : : : : : : 1983, (1971).
                                   ( 6 ,    31) PHYS. REV. B VOL.3,. : : : : : : 3568, (1971).
                                   (10D,    38) PHYS. REV. B VOL.4,. : : : : : : 3971, (1971).
                                   ( 6 ,  1138) PHYS. REV. A VOL.5,. . . . . . . 2230, (1972).

2115   KOEHLER.........(W.C.)..:   ( 5 ,  2080) PHYS. REV. VOL.79,. . . . . . . . 395, (1950).
                                   ( 5 ,  1967) PHYS. REV. VOL.83,. . . . . . . . 700, (1951).
                                   ( 5 ,  1085) PHYS. REV. VOL.84,. . . . . . . . 912, (1951).
                                   ( 5 ,  2726) ACTA CRYST. VOL.5,. . . . . . . . 394, (1952).
                                   ( 5 ,   240) PHYS. REV. VOL.85,. . . . . . . . 491, (1952).
                                   ( 5 ,  2349) ACTA CRYST. VOL.6,. . . . . . . . 741, (1953).
                                   ( 5 ,  2471) PHYS. REV. VOL.91,. . . . . . . . 597, (1953).
                                   ( 6 ,  1938) PHYS. REV. VOL.92,. . . . . . . 1380, (1953).
                                   ( 5 ,  2854) PHYS. REV. VOL.95,. . . . . . . . 280, (1954).
                                   ( 5 ,  1538) PHYS. REV. VOL.100,. . . . . . . . 545, (1955).
                                   ( 5 ,   853) PHYS. REV. VOL.97,. . . . . . . 1177, (1955).
                                   ( 5 ,  1529) J. PHYS. CHEM. SOLIDS VOL.2,. . . 100, (1957).
                                   ( 9D,    38) J. PHYS. CHEM. SOLIDS VOL.5,. . . 180, (1958).
                                   ( 5 ,  1657) PHYS. REV. VOL.110,. . . . . . . . 638, (1958).
                                   ( 5 ,  1394) PHYS. REV. VOL.110,. . . . . . . .  37, (1958).
                                   ( 5 ,  2839) PHYS. REV. VOL.112,. . . . . . . 1132, (1958).
                                   ( 5 ,  1658) J. PHYS. RADIUM VOL.20,. . . . . . 180, (1959).
                                   ( 5 ,   968) PHYS. REV. VOL.113,. . . . . . . . 497, (1959).
                                   ( 5 ,  1388) COMP. REND. 251,. . . . . . . . . 1359, (1960).
                                   ( 5 ,  2386) IAEA SYMP. (PILE RESEARCH) VIENNA  379, (1960).
                                   ( 5 ,  2586) J. APPL. PHYS. VOL.31, SUPPL.,    358, (1960).
                                   ( 5 ,  2042) PHYS. REV. VOL.118,. . . . . . . .  58, (1960).
                                   ( 9D,     5) ACTA CRYST. VOL.14,. . . . . . . . 535, (1961).
                                   ( 9D,    33) J. APPL. PHYS. VOL.32, SUPPL.,    205, (1961).
                                   ( 5 ,   807) J. APPL. PHYS. VOL.32, SUPPL.,    485, (1961).
                                   ( 5 ,   852) J. APPL. PHYS. VOL.32, SUPPL.,    495, (1961).
                                   ( 5 ,   390) PHYS. REV. VOL.122,. . . . . . . 1409, (1961).
                                   ( 5 ,  2682) J. APPL. PHYS. VOL.33, SUPPL.,   1124, (1962).
                                   ( 5 ,  2301) J. APPL. PHYS. VOL.33, SUPPL.,   1340, (1962).
                                   ( 5 ,  2384) J. PHYS. SOC. JAPAN VOL.17 B-III   27, (1962).
                                   ( 5 ,  2606) J. PHYS. SOC. JAPAN VOL.17 B-III   32, (1962).
                                   ( 5 ,  2298) J. PHYS. SOC. JAPAN VOL.17 B-III   38, (1962).
                                   ( 5 ,  1720) PHYS. REV. VOL.125,. . . . . . . 1860, (1962).
                                   ( 5 ,  2683) PHYS. REV. VOL.126,. . . . . . . 1672, (1962).
                                   ( 6 ,   527) PHYS. REV. VOL.127,. . . . . . . 2080, (1962).
                                   ( 5 ,   977) PHYS. REV. VOL.127,. . . . . . . . 714, (1962).
                                   ( 5 ,  2310) PHYS. REV. VOL.128,. . . . . . . 2118, (1962).
                                   ( 5 ,  2578) COLL. INTER. N.126 (GRENOBLE),    190, (1963).
                                   ( 5 ,  2146) COLL. INTER. N.126 (GRENOBLE),     36, (1963).
                                   ( 5 ,  1112) J. APP. PHYS. 34,. . . . . . . . 1189, (1963).
                                   ( 5 ,  2610) J. APP. PHYS. 34,. . . . . . . . 1335, (1963).
                                   ( 5 ,  2147) J. PHYS. CHEM. SOL. 24,. . . . . 1141, (1963).
                                   ( 5 ,  2540) PHYS. REV. VOL.131,. . . . . . . . 922, (1963).
                                   ( 5 ,  2577) PHYS. REV. 131,. . . . . . . . . 1518, (1963).
                                   ( 5 ,  2027) J. APPL. PHYS. VOL.35,. . . . . . 1041, (1964).
                                   ( 6 ,  1855) J. PHYS. CHEM. SOL. 25,. . . . . 1453, (1964).
                                   ( 5 ,  1391) PHYS. LETT. VOL.9,. . . . . . . .  93, (1964).
                                   ( 5 ,  2337) PHYS. REV. LETT. VOL.12,. . . . . 553, (1964).
                                   ( 5 ,  2562) PHYS. REV. VOL.136,. . . . . . . A240, (1964).
                                   ( 6 ,  2147) PROC. INT. CONF. (NOTTINGHAM),    271, (1964).
                                   ( 5 ,   391) J. APP. PHYS. 36,. . . . . . . . 1078, (1965).
                                   ( 5 ,  2599) J. APP. PHYS. 36,. . . . . . . . 1096, (1965).
                                   ( 5 ,  2553) J. APP. PHYS. 36,. . . . . . . . . 978, (1965).
                                   ( 5 ,  2909) PHYS. REV. 138,. . . . . . . . . A1655, (1965).
                                   ( 5 ,   639) PHYS. REV. 138,. . . . . . . . . A737, (1965).
                                   ( 5 ,  2101) PHYS. REV. 138,. . . . . . . . . A755, (1965).
                                   ( 5 ,   850) PHYS. REV. 140,. . . . . . . . . A1896, (1965).
                                   ( 5 ,   529) J. APP. PHYS. 37,. . . . . . . . 1036, (1966).
                                   ( 5 ,   598) J. APP. PHYS. 37,. . . . . . . . 1259, (1966).
                                   ( 5 ,  2602) J. APP. PHYS. 37,. . . . . . . . 1353, (1966).
                                   ( 5 ,   597) PHYS. REV. VOL.151,. . . . . . . . 405, (1966).
                                   ( 5 ,  1359) PHYS. REV. VOL.151,. . . . . . . . 414, (1966).
                                   ( 8 ,    15) ATLANTA SYMP/GEORGIA INST. TECH.   53, (1967).
                                   ( 5 ,  1212) J. APP. PHYS. 38,. . . . . . . . 1381, (1967).
                                   ( 5 ,  2634) J. APP. PHYS. 38,. . . . . . . . 1384, (1967).
                                   ( 9D,    52) METAL. SOC. CONF. LOS ANGELES,     99, (1967).
                                   ( 5 ,  1358) PHYS. REV. 158,. . . . . . . . . . 450, (1967).
                                   (22R,     8) IAEA SYMP. COPENHAGEN, VOL.2,.    253, (1968).
                                   ( 5 ,  2605) J. APP. PHYS. 39,. . . . . . . . 1329, (1968).
                                   ( 6 ,  1374) PHYS. REV. LETT. 20,. . . . . . . 997, (1968).
                                   ( 5 ,  2603) PHYS. REV. 174,. . . . . . . . . . 562, (1968).
                                   (10B,   126) PHYS. REV. 175,. . . . . . . . . 1110, (1968).
                                   ( 5 ,  2922) PHYS. REV. 176,. . . . . . . . . . 722, (1968).
                                   ( 5 ,  2840) J. APPL. PHYS. VOL.40,. . . . . . 1136, (1969).
                                   ( 5 ,  2727) J. APP. PHYS. VOL.40,. . . . . . 1135, (1969).
                                   ( 6 ,  2179) J. APP. PHYS. VOL.40,. . . . . . 1445, (1969).
                                   ( 5 ,   426) J. APP. PHYS. VOL.40,. . . . . . 1454, (1969).
                                   ( 5 ,   434) PHYS. REV. 181,. . . . . . . . . . 883, (1969).
                                   ( 5 ,  1682) PHYS. REV. 181,. . . . . . . . . . 920, (1969).
                                   ( 6 ,   906) J. APPL. PHYS. 41,. . . . . . . 1182, (1970).
                                   ( 6 ,   907) PHYS. REV. LETT. VOL.24,. . . . .  16, (1970).
                                   ( 5 ,   395) AIP CONF. PROC. VOL.5,. . . . . 1381, (1971).
                                   ( 5 ,  2381) CONF./RARE EARTHS/ACTINIDES,. . .  25, (1971).
                                   ( 5 ,   825) INT. J. MAGN.(GB) VOL.2,. . . . . 389, (1971).
                                   ( 5 ,   841) J. PHYSIQUE TOME 32 COL.1 VOL.II 1128, (1971).
                                   ( 6 ,   909) J. PHYSIQUE TOME 32 COL.1 VOL.I,  296, (1971).
                                   ( 5 ,  1382) J. PHYSIQUE TOME 32 COL.1 VOL.I,  362, (1971).
                                   ( 5 ,  2470) ACTA CRYST. VOL.A28 PART S-4,     197, (1972).
                                   ( 5 ,  1205) PHYS. REV. B VOL.5,. . . . . . . . 997, (1972).
                                   ( 5 ,  2468) PHYS. REV. LETT. VOL.29,. . . . 1468, (1972).
                                   ( 5 ,  2472) A.I.P. CONF. PROC. NO.10 PART 2, 1314, (1973).
                                   ( 5 ,  2327) A.I.P. CONF. PROC. NO.10 PART 2, 1319, (1973).
                                   ( 5 ,  2612) A.I.P. CONF. PROC. NO.10 PART 2, 1554, (1973).

2116   KOENIG..........(S.H.)..:   ( 6 ,    27) COPENHAGEN CONF.-LATT.DYNAMICS,.   57, (1963).
                                   ( 6 ,   180) IBM J. RES. DEVELOP. VOL.8,. . .  234, (1964).
                                   ( 5 ,    99) PHYS. REV. A VOL.7,. . . . . . . 2130, (1973).

2117   KOESTER.........(L.)....:   ( 5 ,  1350) Z. PHYSIK VOL.182,. . . . . . . . 328, (1964).
                                   ( 5 ,  1257) Z. PHYSIK VOL.198,. . . . . . . . 187, (1967).
                                   ( 5 ,  2860) Z. PHYSIK VOL.219,. . . . . . . . 300, (1969).
                                   ( 7B,   140) Z. NATURFORSCH. A VOL.26A,. . . . 391, (1971).
                                   ( 7A,    98) Z. NATURFORSCH. A VOL.27A,. . . . 901, (1972).

2118   KOETZLE.........(T.F.)..:   ( 5 ,     4) ACTA CRYST. A28 PART S-4,. . . . . 193, (1972).
                                   ( 5 ,   311) ACTA CRYST. VOL.B28,. . . . . . . 2083, (1972).
                                   ( 5 ,    13) ACTA CRYST. VOL.B28,. . . . . . . 3006, (1972).
                                   ( 5 ,    24) ACTA CRYST. VOL.B28,. . . . . . . 3207, (1972).
                                   ( 5 ,    22) INT. J. PEPTIDE PROTEIN RES. 4,.  229, (1972).
                                   ( 5 ,    10) J. AMER. CHEM. SOC. VOL.94,. . . 2657, (1972).
                                   ( 5 ,   981) J. CHEM. PHYS. VOL.56,. . . . . 3257, (1972).
                                   ( 5 ,    20) J. CRYST. MOL. STRUC. VOL.2,. . . 225, (1972).
                                   ( 5 ,    30) MATER. RES. BULL. VOL.7,. . . . 1225, (1972).
                                   ( 5 ,  2350) NATURE VOL.235,. . . . . . . . . . 328, (1972).
                                   ( 5 ,    21) ACTA CRYST. VOL.B29,. . . . . . . 2571, (1973).
```

```
2118  KOETZLE.........(T.F.)..:  ( 5 ,   291) ACTA CRYST. VOL.B29, . . . . . . .   231, (1973).
                                 ( 5 ,    33) ACTA CRYST. VOL.B29, . . . . . . .   876, (1973).
                                 ( 5 ,    11) J.C.S. PERKIN TRANS. II, . . . . .   133, (1973).
                                 ( 5 ,    25) J. CHEM. PHYS. VOL.58 NO.6, . . .  2547, (1973).
                                 ( 5 ,  1594) J. CHEM. PHYS. VOL.59, . . . . . .   915, (1973).
                                 ( 5 ,    17) ACTA CRYST. VOL.B30, . . . . . . .  1220, (1974).
                                 ( 5 ,  1271) SPECTROSCOPY BIOL. CHEM., . . . .   177, (1974).

2119  KOGAN...........(V.S.)..:  (22C,    71) SOV. PHYS. CRYST. VOL.5, . . . . .   297, (1960).
                                 ( 5 ,  1277) SOV. PHYS. JETP VOL.13, . . . . .   718, (1961).
                                 ( 5 ,   780) J. PHYS. SOC. JAPAN VOL.17 B-II,   385, (1962).
                                 ( 5 ,  1241) SOV. PHYS. J.E.T.P. 27, . . . . .   210, (1968).

2120  KOHGI...........(M.)....:  ( 6 ,   872) PHYS. LETT. VOL.37A, . . . . . .   333, (1971).
                                 ( 6 ,  1492) ACTA CRYST. VOL.A28 PART S-4, . .   208, (1972).
                                 ( 6 ,  1498) SOL. STATE COMM. VOL.11, . . . . .   391, (1972).
                                 ( 6 ,  1494) J. PHYS. SOC. JAPAN VOL.36, . . .   112, (1974).

2121  KOHLER.........(W.)....:  (20A,    20) INT. J. APPL. RADIAT. ISOTOP. 22   529, (1971).

2122  KOHN...........(W.)....:  (10B,   182) PHYS. REV. B VOL.7, . . . . . . .  2285, (1973).

2123  KOHRA..........(K.)....:  ( 5 ,  2439) J. PHYS. SOC. JAPAN VOL.31, . . .   954, (1971).
                                 ( 9C,    33) PHYS. LETT. VOL.37A, . . . . . .    29, (1971).

2124  KOIZUMI........(H.)....:  ( 5 ,  2284) J. PHYS. SOC. JAPAN VOL.35, . . .   314, (1973).

2125  KOKKEDEE.......(J.J.)..:  (10E,    11) IAEA SYMP. CHALK RIVER, VOL.1, .    15, (1962).
                                 (10E,    21) PHYSICA VOL.28, . . . . . . . . .   374, (1962).
                                 (10E,    22) PHYSICA VOL.28, . . . . . . . . .   893, (1962).
                                 (10A,     9) NED. TIJDSCHRIFT NATUURKUNDE 29,    61, (1963).
                                 (10A,     8) NED. TIJDSCHRIFT NATUURKUNDE 29,    93, (1963).

2126  KOLENDA........(M.)....:  ( 5 ,   578) PHYS. STAT. SOLIDI B VOL.57 NO.2 K107, (1973).

2127  KOLLMAR........(A.)....:  ( 6 ,  2167) DISC. FARADAY SOC. NO.50, . . . .    74, (1970).
                                 ( 6 ,  1705) PHYS. LETT. VOL.45A, . . . . . .   479, (1973).
                                 (20A,    22) J. APPL. CRYST. VOL.7, . . . . .    38, (1974).

2128  KOLODZIEJCZYK...(A.)....:  ( 5 ,  2170) PHYS. STAT. SOLIDI A VOL.11 NO.1    57, (1972).
                                 ( 5 ,  2169) PHYS. STAT. SOLIDI A VOL.9 NO.1,    97, (1972).

2129  KOLOMIETS.......(V.M.)..:  ( 7A,    20) BU. ACAD. SCI. USSR PHYS. SER.34  2183, (1970).
                                 ( 7A,    21) BU. ACAD. SCI. USSR PHYS. SER.35  2349, (1971).

2130  KOLOS..........(W.)....:  ( 7A,     6) ACTA PHYS. POLON. VOL.13, . . . .    67, (1954).
                                 ( 6 ,   415) ACTA PHYS. POLON. VOL.16, . . . .   335, (1957).

2131  KOLOTYI........(V.V.)..:  (20B,    49) INSTRUM. EXP. TECH. NO.2, . . . .   347, (1963).
                                 ( 5 ,  2411) SOV. AT. ENERGY VOL.19, . . . . .  1162, (1965).
                                 ( 5 ,  1248) BULL. ACAD. SCI. USSR/PHYS. 31, .   334, (1967).

2132  KOMAROV........(L.I.)..:  (14B,   121) SOV. PHYS. ACOUSTICS VOL.9, . . .   349, (1964).
                                 (15 ,    62) UKR. FIZ. ZH. (USSR) VOL.9, . . .   349, (1964).

2133  KOMAROV........(V.E.)..:  ( 6 ,  1388) SOV. PHYS. CRYST. VOL.15, . . . .  1010, (1971).

2134  KOMOR..........(Z.)....:  (19B,    25) NUKLEONIKA VOL.8, . . . . . . . .   695, (1963).

2135  KOMOTO.........(T.)....:  ( 5 ,  1140) J. PHYS. SOC. JAPAN VOL.17 B-I,    249, (1962).
                                 ( 5 ,   614) COLL. INTER. N.126 (GRENOBLE), .   143, (1963).
                                 ( 5 ,  2186) J. APP. PHYS. 34, . . . . . . . .  1191, (1963).
                                 ( 5 ,  1989) J. APP. PHYS. 37, . . . . . . . .  1040, (1966).
                                 (22C,    59) REV. SCI. INSTRUM. VOL.37, . . .   435, (1966).
                                 (22C,    61) REV. SCI. INSTRUM. VOL.38, . . .   275, (1967).

2136  KOMURA.........(S.)....:  ( 5 ,  2746) J. PHYS. SOC. JAPAN VOL.17 B-III    46, (1962).
                                 (23 ,    44) J. PHYS. SOC. JAPAN VOL.17 B-II,   354, (1962).
                                 ( 5 ,   582) COLL. INTER. N.126 (GRENOBLE), .    38, (1963).
                                 ( 5 ,   640) J. PHYS. SOC. JAPAN VOL.20, . . .   103, (1965).
                                 ( 5 ,   641) PHYS. LETT. 24A, . . . . . . . .   299, (1967).
                                 ( 6 ,  1708) IAEA SYMP. COPENHAGEN, VOL.2, . .   101, (1968).
                                 (19A,    28) JAP. J. APPL. PHYS. VOL.9, . . .   866, (1970).
                                 ( 5 ,  1063) J. PHYS. SOC. JAPAN VOL.35, . . .   706, (1973).
                                 ( 5 ,  1065) J. APPL. CRYST. VOL.7, . . . . .   233, (1974).

2137  KOMURA.........(Y.)....:  ( 9C,     1) ACTA CRYST. 19, . . . . . . . . .   137, (1965).

2138  KONAKHOVICH.....(YU.YA.):  (22A,    57) PRIBORY TEKH. EKSPER. NO.3, . . .    26, (1959).
                                 ( 5 ,  2949) SOV. PHYS. CRYST. VOL.8, . . . .    18, (1963).

2139  KONAKHOVICH.....(YU.YU.):  ( 5 ,  1199) SOV. PHYS. CRYST. VOL.8, . . . .   626, (1964).

2140  KONCZOS........(G.)....:  ( 5 ,  1101) PHYS. LETT. 19, . . . . . . . . .   103, (1965).
                                 (20C,    33) MAGYAR FIZ. FOLYOIRAT (HUNG.) 16   279, (1968).

2141  KONDO..........(M.)....:  (20C,    29) J. NUCL. SCI. TECHNOL. JAP. 4, .    22, (1967).

2142  KONDRATENKO.....(P.S.)..:  (11A,    60) SOV. PHYS. J.E.T.P. 26, . . . . .   764, (1968).

2143  KONDRAT#EV......(L.G.)..:  (22B,    12) INSTRUM. EXP. TECH. NO.2, . . . .   415, (1970).

2144  KONIG..........(U.)....:  (22A,    29) KERNTECHNIK (GER.) VOL.8, . . . .   560, (1966).
                                 ( 5 ,  1049) Z. ANGEW. PHYS. VOL.22, . . . . .   103, (1967).
                                 ( 5 ,  1860) J. APPL. CRYST. VOL.1 PT.2, . . .   124, (1968).
                                 (22C,    27) KERNTECHNIK VOL.10 NO.7, . . . .   371, (1968).
                                 ( 5 ,  2942) J. PHYSIQUE TOME 32 COL.1 VOL.I,   320, (1971).

2145  KONIN..........(A.D.)..:  (20C,   110) SOV. PHYS. J.E.T.P. VOL.31, . . .    59, (1970).

2146  KONOBEEVSKII....(E.S.)..:  ( 6 ,   920) BU. ACAD. SCI. USSR PHYS. SR. 33  1754, (1969).

2147  KONONENKO.......(V.A.)..:  (20C,    84) PRIB. TEKH. EKSPER NO.1, . . . .    61, (1959).

2148  KONSTANTINOVIC..(J.M.)..:  (22A,     6) BULL. INST. NUCL. SCI.(YUGO.) 11    59, (1961).

2149  KONSTANTINOVIC..(J.)....:  ( 6 ,  1757) IAEA SYMP. CHALK RIVER, VOL.2, .   317, (1962).
                                 ( 6 ,   760) BULL/KIDRICH INST/NUCL. SCI. 17,   329, (1966).
                                 ( 6 ,   748) SOL. STATE COMM. 4, . . . . . . .   425, (1966).
                                 (20A,    37) NUCL. INST. MET. 65, . . . . . .   233, (1968).
                                 (17B,    66) SOL. STATE COMM. VOL.8, . . . . .  1799, (1970).
                                 ( 5 ,  2180) FIZIKA (YUGOSLAVIA) VOL.5, . . .   179, (1973).
                                 ( 5 ,  2174) SOL. STATE COMM. VOL.13, . . . .   249, (1973).

2150  KONTI..........(A.)....:  (10B,    57) J. CHEM. PHYSICS VOL.55, . . . .  3997, (1971).

2151  KOPECKY........(J.)....:  (20B,    15) CZECH. J. PHYS. B VOL.13, . . . .   474, (1963).
                                 (12A,    19) NUCL. PHYS. A VOL.A181, . . . . .   241, (1972).

2152  KOPPEL.........(J.U.)..:  ( 6 ,   141) NUCL. SCI. ENGNG. VOL.19, . . . .   367, (1964).
                                 ( 6 ,  1056) NUCL. SCI. ENGNG. VOL.19, . . . .   412, (1964).
                                 ( 6 ,  2313) PHYS. REV. VOL.134, . . . . . . . A1476, (1964).
                                 ( 6 ,   964) PHYS. REV. VOL.135, . . . . . . .  A603, (1964).
                                 (16 ,    53) NUKLEONIK VOL.7, . . . . . . . .   408, (1965).
                                 ( 6 ,   138) PHYS. LETT. 16, . . . . . . . . .   235, (1965).
                                 ( 6 ,   972) NUKLEONIK VOL.8, . . . . . . . .    40, (1966).
```

```
2188  KOZHEVNIKOV.....(D.A.)..:  (16 ,   82) SOV. AT. ENERGY VOL.26,. . . . .    334, (1969).

2189  KOZLOV..........(ZH.A.).:  ( 5 , 2255) UKR. FIZ. ZH.(USSR) VOL.15,. . . 1772, (1970).

2190  KOZUBOWSKI......(J.)....:  ( 6 , 2317) IAEA SYMP. CHALK RIVER, VOL.2,.     87, (1962).
                                 ( 6 ,  596) J. PHYS. CHEM. SOLIDS VOL.23,. . 1021, (1962).

2191  KRASNICKI.......(SZ.)...:  ( 5 ,  306) IAEA SYMPOSIUM VIENNA, . . . . .    293, (1960).
                                 ( 6 ,  542) IAEA SYMP. CHALK RIVER, VOL.2,.     327, (1962).
                                 ( 6 ,  879) COLL. INTER. N.126 (GRENOBLE),.    210, (1963).
                                 (18A,   12) BULL/KIDRICH INST. NUCL. SCI. 15   115, (1964).

2192  KRASNICKI.......(S.)....:  ( 5 ,  138) ACTA PHYS. POLON. VOL.17,. . . .    483, (1958).
                                 ( 6 ,  974) NUKLEONIKA VOL.5,. . . . . . . .    495, (1960).
                                 ( 6 ,  384) ACTA PHYS. POLON. VOL.21,. . . .    529, (1962).
                                 (23 ,   63) NUKLEONIKA VOL.7,. . . . . . . .    223, (1962).
                                 ( 6 ,  843) PHYS. STAT. SOL. 15,. . . . . .    119, (1966).
                                 ( 6 ,  482) PHYSICA VOL.37,. . . . . . . . .    501, (1967).
                                 ( 6 ,  862) PHYS. STAT. SOL. 21,. . . . . .   K163, (1967).
                                 ( 6 ,  864) PHYS. STAT. SOL. 22,. . . . . .    K55, (1967).
                                 ( 5 , 1095) PHYS. STAT. SOL. 26,. . . . . .    429, (1968).
                                 ( 6 ,  545) PHYS. STAT. SOL. 32,. . . . . .     41, (1969).
                                 ( 6 ,  863) PHYS. STAT. SOL. 41,. . . . . .   K103, (1970).

2193  KRASNYT.........(YU.P.).:  ( 5 , 2249) UKRAINIAN PHYS. J. VOL.14, . . .   1909, (1969).

2194  KRAYTOR.........(S.N.)..:  (17A,   95) SOV. AT. ENERGY(USA) VOL.32, . .     68, (1972).

2195  KREBS...........(K.H.)..:  ( 6 , 2226) IAEA SYMP. CHALK RIVER, VOL.2,.    145, (1962).
                                 ( 6 , 2225) IAEA SYMP. CHALK RIVER, VOL.2,.    155, (1962).
                                 ( 5 , 2730) COMP. REND. 263, . . . . . . .    457B, (1966).

2196  KREBS...........(K.)....:  ( 6 ,  348) J. CHEM. PHYS. VOL.40, . . . . .   3502, (1964).
                                 (10B,  112) PHYS. LETT. VOL.10, . . . . . .     12, (1964).
                                 ( 6 ,  449) J. CHEM. PHYSICS VOL.43,. . . .   1864, (1965).
                                 (10B,  123) PHYS. REV. VOL.138,. . . . . .    A143, (1965).
                                 (10B,  210) SOL. STATE COMM. 5,. . . . . .     159, (1967).
                                 (23 ,   36) IAEA SYMP. COPENHAGEN, VOL.2,.     289, (1968).

2197  KREISLER........(M.M.)..:  (20C,   75) NUCL. INST. MET. VOL.94, . . . .    493, (1971).

2198  KREISLER........(M.N.)..:  ( 7A,   54) PHYS. LETT. B VOL.36B, . . . . .    560, (1971).

2199  KREN............(E.)....:  (22C,   18) INSTRUM. EXP. TECH. NO.2,. . . .    285, (1961).
                                 (22A,    5) ACTA PHYS. HUNGAR. VOL.15, . . .    203, (1963).
                                 ( 5 , 1020) PHYS. LETT. 11,. . . . . . . .     215, (1964).
                                 ( 5 , 1130) PHYS. LETT. 9,. . . . . . . . .    297, (1964).
                                 ( 5 , 1123) PROC. INT. CONF. (NOTTINGHAM),.    158, (1964).
                                 ( 5 , 1101) PHYS. LETT. 19,. . . . . . . .     103, (1965).
                                 ( 5 , 1119) SOL. STATE COMM. 3,. . . . . .     371, (1965).
                                 ( 5 , 1798) PHYS. LETT. 20,. . . . . . . .     331, (1966).
                                 ( 5 , 1795) PHYS. LETT. 21,. . . . . . . .     383, (1966).
                                 ( 5 , 1772) PHYS. LETT. 22,. . . . . . . .     273, (1966).
                                 ( 5 , 1019) SOL. STATE COMM. 4,. . . . . .     255, (1966).
                                 ( 5 , 2354) SOL. STATE COMM. 4,. . . . . .      31, (1966).
                                 ( 5 , 1788) J. APPL. PHYS. VOL.38, . . . .    1265, (1967).
                                 ( 5 , 1785) PHYS. LETT. VOL.24A, . . . . .     198, (1967).
                                 ( 5 , 1781) PHYS. LETT. VOL.25A, . . . . .      56, (1967).
                                 ( 5 , 1786) CENT. RES INST PHYS. KFKI NO.2,.     1, (1968).
                                 ( 5 , 1783) J. APP. PHYS. 39,. . . . . . .     538, (1968).
                                 ( 5 , 1730) J. PHYS. CHEM. SOLIDS 29,. . .     101, (1968).
                                 ( 5 , 1696) PHYS. LETT. VOL.26A, . . . . .     556, (1968).
                                 ( 5 , 1695) PHYS. REV. 171,. . . . . . . .     574, (1968).
                                 ( 5 , 2883) J. APP. PHYS. 40,. . . . . . .    1454, (1969).
                                 ( 5 , 1780) PHYS. LETT. VOL.29A, . . . . .     340, (1969).
                                 ( 5 , 2881) PHYS. REV. VOL.186,. . . . . .     479, (1969).
                                 ( 5 , 1776) J. APPL. PHYS. 41,. . . . . . .     941, (1970).
                                 ( 5 , 1700) SOL. STATE COMM. VOL.8,. . . .    1653, (1970).
                                 ( 5 , 1791) INT. J. MAGN.(GB) VOL.1, . . .     341, (1971).
                                 ( 5 , 1706) INT. J. MAGN. VOL.1,. . . . .     143, (1971).
                                 ( 5 , 1770) J. PHYSIQUE TOME 32 COL.1 VOL.II  980, (1971).
                                 ( 5 , 1793) SOL. STATE COMM. VOL. 9, . . .      27, (1971).
                                 ( 5 , 1778) J. PHYS. CHEM. SOLIDS VOL.33,.     212, (1972).
                                 ( 5 , 1821) MAGN. AND MAG. MATERIALS-1971,.    513, (1972).
                                 ( 5 , 1775) SOL. STATE COMM. VOL.10, . . .    1195, (1972).
                                 ( 5 , 1777) SOL. STATE COMM. VOL.11,. . .      933, (1972).
                                 ( 5 , 1701) A.I.P. CONF. PROC. NO.10 PART 2, 1379, (1973).
                                 ( 5 , 2889) A.I.P. CONF. PROC. NO.10 PART 2, 1603, (1973).

2200  KRESS...........(W.)....:  (10D,   51) PHYS. STAT. SOL. VOL.49, . . . .    235, (1972).

2201  KREY............(U.)....:  (11A,   65) Z. ANGEW PHYS. VOL.32, . . . . .    104, (1971).

2202  KRIEGER.........(T.J.)..:  (10C,   71) PHYS. REV. VOL.121,. . . . . .    1388, (1961).

2203  KRIESSMAN.......(C.J.)..:  ( 5 , 1610) PROC. I.E.E. VOL.104B SUP.4-7,     217, (1957).

2204  KRIKORIAN.......(N.H.)..:  ( 5 , 2631) ACTA CRYST. VOL.B24, . . . . .    1121, (1968).
                                 ( 5 , 1524) ACTA CRYST. VOL.B24, . . . . .     459, (1968).
                                 ( 5 , 2754) J. APPL. PHYS. 41,. . . . . . .   5080, (1970).
                                 ( 5 , 2752) ACTA CRYST. VOL.B27,. . . . .    1067, (1971).

2205  KRISEMENT.......(O.)....:  ( 5 ,  566) Z. NATURFORSCH. VOL.15A, . . . .    880, (1960).
                                 ( 5 , 1162) Z. PHYS. VOL.174,. . . . . . .     472, (1963).

2206  KRISHNAN........(R.S.)..:  (24 ,   18) IAEA SYMPOSIUM BOMBAY, VOL.2,. .    537, (1964).

2207  KRIVOGLAZ.......(M.A.)..:  ( 7B,  115) SOV. PHYS. JETP VOL.5, . . . .    1115, (1957).
                                 ( 8 ,  129) SOV. PHYS. DOKLADY VOL.3,. . .      61, (1958).
                                 ( 7B,  116) SOV. PHYS. JETP VOL.7, . . . .     139, (1958).
                                 (13 ,   65) SOV. PHYS. JETP VOL.7,. . . . .    281, (1958).
                                 (10C,   65) PHYS. MET. METALLOGR. VOL.10 N.2     9, (1960).
                                 (13 ,   66) SOV. PHYS. CRYST. VOL.4, . . .     290, (1960).
                                 (10C,  105) SOV. PHYS. JETP VOL.13,. . . .     397, (1961).
                                 (10E,   29) SOV. PHYS. SOL. STATE VOL.3, .    1107, (1961).
                                 (10C,  110) SOV. PHYS. DOKLADY VOL.3,. . .    2015, (1962).
                                 (10C,  108) SOV. PHYS. JETP VOL.13,. . . .    1273, (1962).
                                 (10C,  109) SOV. PHYS. SOL. STATE VOL.3, .    2301, (1962).
                                 ( 7B,  122) SOV. PHYS. SOL. STATE VOL.5, .    2526, (1963).
                                 (10C,  132) UKR. FIZ. ZH. (USSR) VOL.8,. .      162, (1963).
                                 ( 8 ,   61) PHYS. MET. METALLOGR. VOL.20 N.2     1, (1965).
                                 (10C,  119) SOV. PHYS. SOL. STATE, 6,. . .    2618, (1965).
                                 (10D,   63) SOV. PHYS. SOL. STATE VOL.11,. .     69, (1969).

2208  KROEGER.........(H.R.)..:  (24 ,   33) NUCLEONICS VOL.5 OCTOBER,. . . .     51, (1949).

2209  KROHN-JR........(V.E.)..:  (20C,   38) NUCL. INST. MET. VOL.27,. . . .    351, (1964).
                                 (20C,   99) REV. SCI. INSTRUM. VOL.35,. . .     853, (1964).
                                 ( 5 , 2210) PHYS. REV. VOL.148,. . . . . .    1303, (1966).

2210  KROHN...........(V.E.)..:  ( 5 , 2870) PHYS. REV. VOL.102,. . . . . .    1321, (1956).

2211  KROH............(A.)....:  ( 6 , 1558) IAEA SYMP. CHALK RIVER, VOL.2, .    237, (1962).
```

2212 KRONER..........(E.)....‡ (10C, 138) Z. NATURFORSCH. VOL.14A, ... 74, (1959).

2213 KRONMUELLER.....(H.)....‡ (10C, 139) Z. PHYS. VOL.171,........ 291, (1963).

2214 KROO...........(N.)....‡
(6 , 70) IAEA SYMP. BOMBAY, VOL.2,.... 101, (1964).
(6 , 88) PHYS. REV. LETT. 12,........ 721, (1964).
(6 , 717) PHYS. LETT. 19,............ 15, (1965).
(6 , 824) PHYS. LETT. 24A,.......... 22, (1967).
(6 , 424) PHYS. LETT. 24A,.......... 57, (1967).
(6 , 430) IAEA SYMP. COPENHAGEN, VOL.1,. 615, (1968).
(6 , 1484) IAEA SYMP. COPENHAGEN, VOL.2,. 111, (1968).
(12B, 11) IAEA SYMP. COPENHAGEN, VOL.2,. 407, (1968).
(6 , 807) IAEA SYMP. COPENHAGEN, VOL.2,. 37, (1968).
(6 , 823) J. APP. PHYS. 39,.......... 453, (1968).
(6 , 806) PHYS. LETT. VOL.28A,....... 213, (1968).
(19B, 30) REPORT KFKI-07/1968 (14PP.),. , (1968).
(8 , 51) 2ND SUMMER SCHOOL S.S. PHYS.,. 321, (1968).
(17A, 36) IAEA SYMP. (INSTRUMENT.), VIENNA 211, (1969).
(6 , 183) IAEA SYMP. GRENOBLE, ... 595, (1972).
(19A, 27) IAEA SYMP. GRENOBLE; .... 763, (1972).
(19A, 25) IAEA SYMP. GRENOBLE; .... 787, (1972).
(6 , 62) PHYS. LETT. VOL.40A, ... 173, (1972).

2215 KROSBIE.........(K.L.)..‡ (6 , 2205) PROC/CONF/FAST CRIT. EXP./ANALY. 524, (1967).

2216 KROTENKO........(V.T.)..‡
(6 , 1904) SOV. AT. ENERGY VOL.20,..... 36, (1966).
(6 , 1889) UKRAINIAN PHYS. J. VOL.14, 9,. 1525, (1969).
(10C, 136) UKR. FIZ. ZH. (USSR) VOL.17,.. 789, (1972).
(7B, 139) UKR. FIZ. ZH. (USSR) VOL.17,.. 45, (1972).
(6 , 785) UKR. FIZ. ZH. (USSR) VOL.18,.. 1384, (1973).
(6 , 608) UKR. FIZ. ZH. (USSR) VOL.18,.. 1390, (1973).
(6 , 784) UKR. FIZ. ZH. (USSR) VOL.18,.. 1528, (1973).
(6 , 1748) UKR. FIZ. ZH. (USSR) VOL.18,.. 558, (1973).
(6 , 568) UKR. FIZ. ZH. (USSR) VOL.18, .. 92, (1973).

2217 KRUEGER.........(D.A.)..‡ (6 , 1984) PHYS. REV. LETT. VOL.24, ... 111, (1970).

2218 KRUEGER.........(H.H.A.)‡
(5 , 242) PHYS. REV. VOL.75,......... 1098, (1949).
(9B, 77) PHYS. REV. VOL.80,......... 507, (1950).

2219 KRUH............(R.F.)..‡ (5 , 2263) J. CHEM. PHYS. VOL.32,...... 241, (1960).

2220 KRUMHANSL.......(J.A.)..‡
(10A, 4) J. APPL. PHYS. VOL.33, SUPPL.,. 307, (1962).
(10B, 129) PHYS. REV. 166,........... 856, (1968).
(10C, 78) PHYS. REV. 175,........... 841, (1968).
(10C, 79) PHYS. REV. VOL.180,........ 729, (1969).

2221 KRUSE...........(W.A.)..‡ (5 , 2332) J. APPL. CRYST. VOL.7, ... 188, (1974).

2222 KRYUKOVA........(N.A.)..‡ (5 , 2856) SOV. PHYS. CRYST. VOL.17,... 1017, (1973).

2223 KRYVOHLAZ.......(M.O.)..‡ (10C, 134) UKR. FIZ. ZH. (USSR) VOL.9,... 1331, (1964).

2224 KSHNYAKINA......(A.N.)..‡
(5 , 2322) SOV. PHYS. SOL. STATE 5,.... 2425, (1963).
(5 , 307) SOV. PHYS. CRYST. VOL.11,.... 177, (1966).

2225 KUBEL...........(W.)....‡ (5 , 1032) ACTA PHYS. POLON. VOL.A44, ... 587, (1973).

2226 KUBO............(R.)....‡ (8 , 42) J. PHYS. SOC. JAPAN VOL.18,... 1025, (1963).

2227 KUCHERNYUK......(V.D.)..‡ (20C, 17) INSTRUM. EXP. TECH. NO.2,.... 305, (1965).

2228 KUCHER..........(T.I.)..‡
(6 , 202) SOV. PHYS. SOL. STATE VOL.4,.. 1747, (1962).
(6 , 218) SOV. PHYS. SOL. STATE VOL.8,.. 261, (1966).
(6 , 1648) SOV. PHYS. SOL. STATE VOL.12,. 423, (1970).
(6 , 1635) UKR. FIZ. ZH.(USSR) VOL.15,.. 1409, (1970).

2229 KUCHIN..........(V.M.)..‡ (5 , 1204) SOV. PHYS. J.E.T.P., 28, ... 649, (1969).

2230 KUDRYASHEV......(V.A.)..‡ (5 , 2879) SOV. PHYS. JETP LETT. VOL.18,.. 47, (1973).

2231 KUEHN...........(N.H.)..‡ (14B, 58) NUCL. SCI. ENGNG. VOL.50,.... 164, (1973).

2232 KUGLER..........(A.A.)..‡
(14A, 33) PHYS. CHEM. LIQUIDS VOL.3, ... 205, (1972).
(14A, 29) J. STAT. PHYS. VOL.8 NO.2, ... 107, (1973).

2233 KUICH...........(G.)....‡ (17A, 15) ACTA PHYS. AUSTRIACA VOL.20, ... 7, (1965).

2234 KUIJPERS........(F.A.)..‡
(5 , 2345) J. PHYSIQUE TOME 32 COL.1 VOL.II 657, (1971).
(5 , 2346) J. PHYS. CHEM. SOLIDS VOL.35,.. 301, (1974).

2235 KUIJPER.........(P.)....‡ (24 , 73) NUCL. INST. MET. 77, ... 55, (1970).

2236 KUKAVADZE.......(G.M.)..‡ (5 , 2628) ATOMNAYA ENERGIYA (USSR) VOL.24, 243, (1968).

2237 KUKHTEVICH......(V.I.)..‡ (20C, 20) INSTRUM. EXPER. TECH. NO.3,.... 529, (1968).

2238 KUKHTO..........(O.L.)..‡ (6 , 2101) SOV. PHYS. SOL. STATE 8, ... 2156, (1966).

2239 KUKLA...........(D.)....‡ (20A, 82) UKAEA AERE TRANS 1074 (20 PP.),. , (1967).

2240 KUMAR...........(A.)....‡
(10B, 63) J. PHYS. C: SOLID ST. PHYS. V-4, 1674, (1971).
(6 , 1365) J. PHYS. F VOL.3,......... 1308, (1973).

2241 KUMAR...........(D.)....‡ (11A, 44) PHYS. REV. B VOL.3,....... 961, (1971).

2242 KUNC............(K.)....‡
(6 , 2326) PHYS. STAT. SOL. VOL.41, ... 491, (1970).
(6 , 2331) PHYS. LETT. VOL.36A, ... 376, (1971).
(6 , 623) PHYS. REV. LETT. VOL.26, ... 770, (1971).

2243 KUNITOMI........(N.)....‡
(5 , 2746) J. PHYS. SOC. JAPAN VOL.17 B-III 46, (1962).
(23 , 44) J. PHYS. SOC. JAPAN VOL.17 B-II, 354, (1962).
(5 , 1827) COLL. INTER. N.126 (GRENOBLE),. 144, (1963).
(5 , 582) COLL. INTER. N.126 (GRENOBLE), . 38, (1963).
(5 , 587) J. PHYS. SOC. JAPAN VOL.19,... 1849, (1964).
(20A, 26) J. PHYS. SOC. JAPAN VOL.19,... 2280, (1964).
(20D, 8) JAPAN. J. APPL. PHYS. VOL.4,.. 911, (1965).
(5 , 71) J. PHYS. SOC. JAPAN VOL.20,... 1723, (1965).
(5 , 640) J. PHYS. SOC. JAPAN VOL.20,... 103, (1965).
(5 , 1023) J. PHYS. SOC. JAPAN VOL.21,... 1932, (1966).
(5 , 641) PHYS. LETT. 24A,.......... 299, (1967).
(6 , 525) J. APP. PHYS. 39,......... 1227, (1968).
(5 , 1620) J. PHYS. SOC. JAPAN VOL.28,... 615, (1970).
(5 , 120) J. PHYS. SOC. JAPAN VOL.29,... 978, (1970).
(6 , 872) PHYS. LETT. VOL.37A, ... 333, (1971).
(6 , 112) SOL. STATE COMM. VOL.9, ... 921, (1971).
(6 , 522) J. PHYS. SOC. JAPAN VOL.32,... 394, (1972).
(6 , 113) J. PHYS. SOC. JAPAN VOL.34,... 1197, (1973).
(6 , 1776) SOL. STATE COMM. VOL.13 NO.4,. 495, (1973).

2244 KUNNMANN........(W.)....‡
(5 , 330) J. APP. PHYS. 38,......... 946, (1967).
(5 , 375) J. PHYS. CHEM. SOL. 28,..... 1089, (1967).
(5 , 331) PHYS. REV. 160,.......... 408, (1967).
(5 , 1179) J. PHYS. CHEM. SOLIDS 29,.... 1359, (1968).
(6 , 718) PHYS. REV. LETT. VOL.27, ... 741, (1971).
(5 , 183) ACTA CRYST. VOL.B28,....... 3429, (1972).

2245	KUNSCH.........(B.)....:	(21 ,	26)	J. PHYS. E VOL.6 NO.6, ...	576, (1973).
2246	KUOPPAMAKI......(R.)...:	(6 ,	168)	IAEA SYMP. COPENHAGEN, VOL.1,. .	431, (1968).
2247	KUPER..........(C.G.)..:	(5 ,	1333)	NATURE VOL.183,.	1544, (1959).
2248	KURIYAMA........(I.)...:	(6 ,	1924)	J. POLYM. SCI./POLYM. LETT. 11,.	377, (1973).
2249	KURIYAMA........(M.)...:	(9B,	52)	J. APPL. PHYS. VOL.40, ...	1697, (1969);
		(17A,	9)	ACTA CRYST. VOL.A26,.	667, (1970).
2250	KURKIN..........(M.I.).:	(8 ,	64)	PHYS. MET. METALLOGR. VOL.34 N.2	13, (1972).
2251	KURZ...........(H.)....:	(9B,	113)	Z. PHYS. VOL.220,.	419, (1969).
2252	KUSCER.........(I.)....:	(7B,	86)	PHYS. REV. VOL.188,.	1445, (1969).
2253	KUSE...........(D.)....:	(6 ,	1333)	PHYS. REV. LETT. VOL.30, . . .	1144, (1973).
2254	KUSHWAHA........(S.S.).:	(10B,	153)	PHYS. REV. B-2, ...	3943, (1970);
		(10B,	82)	J. PHYS.F: METAL PHYS. VOL.1,. .	377, (1971);
		(10B,	63)	J. PHYS. C: SOLID ST. PHYS. V-4,	1674, (1971);
		(10B,	116)	PHYS. LETT. VOL.37A,.	193, (1971);
		(10D,	21)	NUCL./S.S. PHYS. SYMP. ABSTR.,. . . ,	(1972);
		(6 ,	2308)	PHYS. LETT. VOL.38A, ...	497, (1972);
		(10B,	91)	J. PHYS. F VOL.3,.	1531, (1973);
		(10B,	108)	PHYSICA VOL.64,.	625, (1973).
2255	KUTAITSEV.......(V.I.).:	(6 ,	2183)	PHYS. STAT. SOL. 20, ...	767, (1967).
2256	KUVALDIN........(B.V.).:	(5 ,	2861)	LATV. PSR...FIZ. TEHN. SER. 6, .	30, (1968);
		(5 ,	2898)	LATV. PSR...FIZ. TEHN. SER. 2, .	49, (1969).
2257	KUZINA..........(A.A.).:	(5 ,	349)	SOV. PHYS. SOL. STATE VOL.9, . .	1762, (1968).
2258	KUZNIETZ.......(M.)....:	(5 ,	2805)	SOL. STATE COMM. VOL.6,. . . .	877, (1968);
		(5 ,	2806)	J. APP. PHYS. VOL.40,.	1130, (1969);
		(5 ,	2728)	J. PHYS. CHEM. SOLIDS 30,. . .	1642, (1969);
		(5 ,	2804)	PHYS. REV. VOL.188,.	963, (1969);
		(5 ,	2755)	CONF./RARE EARTHS/ACTINIDES, . .	168, (1971);
		(5 ,	1878)	ACTA CRYST. VOL.A28, ...	655, (1972);
		(5 ,	2810)	J. PHYS. C VOL.5,.	3012, (1972);
		(5 ,	2816)	J. PHYS. C VOL.6,.	1652, (1973);
		(5 ,	840)	PHYS. REV. B VOL.8,.	260, (1973).
2259	KUZ#MINOV.......(YU.S.).:	(5 ,	2895)	SOV. PHYS. CRYST. VOL.7, . . .	765, (1963);
		(5 ,	2876)	SOV. PHYS. CRYST. VOL.7, . . .	767, (1963);
		(22A,	69)	SOV. PHYS. CRYST. VOL.8, . . .	234, (1963);
		(5 ,	2907)	SOV. PHYS. CRYST. VOL.8, . . .	15, (1963);
		(5 ,	1195)	SOV. PHYS. CRYST. VOL.8, . . .	537, (1964);
		(5 ,	2906)	SOV. PHYS. CRYST. VOL.9, . . .	159, (1964);
		(5 ,	1842)	SOV. PHYS. CRYST. VOL.12,. . .	811, (1968);
		(20C,	23)	INSTRUM. EXP. TECH. NO.2,. . .	399, (1970).
2260	KUZ#MIN........(N.N.)..:	(6 ,	868)	SOV. PHYS. JETP LETT. VOL.16,. .	6, (1972).
2261	KVICK..........(A.)....:	(5 ,	1928)	ACTA CHEM. SCAND. VOL.26,. . .	1087, (1972);
		(5 ,	4)	ACTA CRYST. A28 PART S-4,. . .	193, (1972);
		(5 ,	8)	ACTA CRYST. VOL.B28, ...	1827, (1972);
		(5 ,	9)	J. CHEM. PHYS. VOL.59, ...	3901, (1973).
2262	KVIRIKADZE......(A.G.).:	(11B,	42)	SOV. PHYS. SOL. STATE, 7,. . . .	891, (1965).
2263	KWON...........(T.H.)..:	(11C,	15)	J. PHYS. -C-, VOL. 1, ...	1563, (1968);
		(11C,	37)	PHYS. REV. 167, ...	458, (1968);
		(6 ,	1997)	J. APPL. PHYS. VOL.41, ...	1370, (1970);
		(8 ,	33)	J. PHYS. CHEM. SOLIDS VOL.31,. .	255, (1970);
		(6 ,	1986)	PHYS. REV. B VOL.5,.	4561, (1972).
2264	LACAZE.........(A.)....:	(6 ,	963)	IAEA SYMPOSIUM (VIENNA), . . .	411, (1960).
2265	LACINA.........(W.B.)..:	(7B,	85)	PHYS. REV. 168, ...	725, (1968).
2266	LAFOREST........(J.)...:	(5 ,	2604)	SOL. STATE COMM. VOL.8 NO.1,. .	23, (1970);
		(5 ,	2453)	J. PHYSIQUE TOME 32 COL.1 VOL.II	1133, (1971);
		(5 ,	2633)	IEEE TRANS. MAGN. VOL.9, . . .	217, (1973).
2267	LAGARDE........(V.)....:	(6 ,	1923)	DISC. FARADAY SOC. NO.48,. . .	15, (1969).
2268	LAIKHTMAN.......(B.D.).:	(10C,	112)	SOV. PHYS. SOL. STATE, 5,. . .	2222, (1963).
2269	LAJEUNESSE......(C.)...:	(6 ,	2352)	CONF. NEUTRON THERMALIZ. VOL.I,.	407, (1968);
		(6 ,	2208)	NUCL. SCI. ENG. VOL.47,. . . .	349, (1972).
2270	LAKATOS........(K.S.)..:	(10B,	22)	DISS. ABS. VOL.28, ...	2586B, (1967).
2271	LAKATOS........(K.)....:	(10C,	78)	PHYS. REV. 175,.	841, (1968);
		(10C,	79)	PHYS. REV. VOL.180,.	729, (1969).
2272	LALLEMENT.......(R.)...:	(5 ,	1373)	BULL. SOC. FRANC. MIN. CRIST. 89	216, (1966).
2273	LAL............(H.H.)..:	(10D,	47)	PHYS. STAT. SOL. VOL.38, . . .	K19, (1970);
		(10B,	68)	J. PHYS. C: SOLID ST. PHYS. V-5,	1038, (1972);
		(10B,	69)	J. PHYS. C: SOLID ST. PHYS. V-5,	543, (1972);
		(10B,	101)	NUCL./S.S. PHYS. SYMP. ABSTR.,. . . ,	(1972).
2274	LAMBERT-ANDRON..(B.)....:	(5 ,	2650)	BULL. SOC. FRANC. MIN. CRIST. 91	88, (1968);
		(5 ,	618)	J. PHYSIQUE TOME 29, ...	813, (1968);
		(5 ,	1154)	SOL. STATE COMM. VOL.7,. . . .	623, (1969);
		(5 ,	1155)	J. PHYSIQUE TOME 32 COL.1 VOL.II	985, (1971);
		(5 ,	1159)	J. PHYS. CHEM. SOLIDS VOL.33,. .	87, (1972);
		(5 ,	1965)	PHYS. STAT. SOLIDI A VOL.18 NO.1	209, (1973).
2275	LAMB-JR........(W.R.)..:	(25 ,	42)	PHYS. REV. VOL.55, ...	190, (1939).
2276	LAMB...........(R.C.)..:	(6 ,	29)	DISS. ABS. VOL.31, ...	2171B, (1970).
2277	LAMMERS........(B.)....:	(24 ,	70)	NUCL. INST. MET. 76, ...	103, (1969).
2278	LAM............(D.J.)..:	(5 ,	2214)	J. PHYSIQUE TOME 32 COL.1 VOL.II	917, (1971);
		(5 ,	2216)	A.I.P. CONF. PROC. NO.10 PART 1,	88, (1973);
		(5 ,	2212)	PHYS. REV. B VOL.9,.	3766, (1974).
2279	LANDAU.........(D.P.)..:	(5 ,	974)	A.I.P. CONF. PROC. NO.10 PART 2,	1163, (1973).
2280	LANDER.........(G.H.)..:	(5 ,	1817)	PROC. PHYS. SOC. 91, ...	332, (1967);
		(5 ,	1818)	J. APP. PHYS. 39,.	1331, (1968);
		(5 ,	2805)	SOL. STATE COMM. VOL.6,. . . .	877, (1968);
		(5 ,	2806)	J. APP. PHYS. VOL.40,.	1130, (1969);
		(5 ,	2728)	J. PHYS. CHEM. SOLIDS 30,. . .	1642, (1969);
		(5 ,	2213)	J. PHYS. CHEM. SOLIDS 30,. . .	733, (1969);
		(5 ,	2804)	PHYS. REV. VOL.188,.	963, (1969);
		(5 ,	2699)	ACTA CRYST. VOL.B26, ...	129, (1970);
		(90 ,	37)	J. CHEM. PHYSICS VOL.53, . . .	1387, (1970);
		(5 ,	2367)	ACTA CRYST. VOL.B27, ...	2284, (1971);
		(5 ,	2214)	J. PHYSIQUE TOME 32 COL.1 VOL.II	917, (1971).

2280 LANDER.........(G.H.)..: (5 , 2522) J. PHYSIQUE TOME 32 COL.1 VOL.I, 571, (1971).
(5 , 1764) J. PHYSIQUE TOME 32 COL.1 VOL.I, 577, (1971).
(5 , 2720) MAGN. AND MAG. MATERIALS-1971, 1371, (1972).
(5 , 2724) PHYS. REV. B VOL.6, 1880, (1972).
(5 , 2698) PHYS. REV. LETT. VOL.29, 1172, (1972).
(5 , 2592) ACTA CRYST. VOL.A29, 684, (1973).
(5 , 2216) A.I.P. CONF. PROC. NO.10 PART 1, 88, (1973).
(5 , 1819) A.I.P. CONF. PROC. NO.10 PART 2, 1138, (1973).
(5 , 2217) INT. J. MAGN. VOL.4 NO.2, 99, (1973).
(5 , 2697) PHYS. REV. B VOL.7, 1988, (1973).
(5 , 2593) PHYS. REV. B VOL.8, 3237, (1973).
(5 , 2369) PHYS. REV. B VOL.9, 1041, (1974).
(5 , 2594) PHYS. REV. B VOL.9, 3003, (1974).
(5 , 2212) PHYS. REV. B VOL.9, 3766, (1974).
(5 , 2338) PHYS. REV. B VOL.9, 248, (1974).

2281 LANDHEER........(D.)....: (6 , 1348) PHYS. REV. LETT. VOL.31, 296, (1973).

2282 LANDKAMMER......(F.J.)..: (17A, 104) Z. PHYSIK VOL.189, 113, (1966).

2283 LANDON..........(H.H.)..: (22A, 61) REV. SCI. INSTRUM. VOL.27, 26, (1956).

2284 LANDOVITZ.......(L.F.)..: (10B, 181) PHYS. REV. LETT. VOL.30, 1196, (1973).

2285 LANE............(R.O.)..: (5 , 233) PHYS. REV. VOL.188, 1618, (1969).

2286 LANGDON.........(F.)....: (22B, 43) REV. SCI. INSTRUM. VOL.30, 997, (1959).

2287 LANGSFORD.......(A.)....: (24 , 47) NUCL. INST. MET. 59, 120, (1968).

2288 LANG............(H.Y.)..: (5 , 1396) SOL. STATE COMM. VOL.10, 735, (1972).

2289 LANG............(N.O.)..: (5 , 2087) J. APPL. PHYS. 37, 1449, (1966).
(5 , 2088) PHYS. REV. VOL.152, 505, (1966).

2290 LANIESSE........(J.)....: (5 , 2714) J. NUCL. MATERIALS VOL.2, 69, (1960).
(5 , 2710) BULL. INFORM. SCI. TECH. NO.49, 69, (1961).

2291 LANNING.........(D.D.)..: (5 , 685) NUCL. SCI. ENGNG. VOL.7, 184, (1960).

2292 LANTER..........(R.J.)..: (20C, 106) REV. SCI. INSTRUM. 39, 1588, (1968).

2293 LAPLAZE.........(D.)....: (10B, 7) ANN. DE PHYSIQUE TOME 5, 77, (1970).

2294 LAROSE..........(A.)....: (6 , 634) THESIS (166 PP.), (1970).
(6 , 609) BULL. AMER. PHYS. SOC. VOL.19, 320, (1974).

2295 LARSEN..........(F.K.)..: (5 , 1410) ACTA CHEM. SCAND. VOL.24, 1662, (1970).
(5 , 287) ACTA CHEM. SCAND. VOL.24, 3248, (1970).
(5 , 1316) ACTA CHEM. SCAND. VOL.25, 1233, (1971).
(5 , 1473) ACTA CHEM. SCAND. VOL.25, 3859, (1971).
(5 , 277) ACTA CRYST. VOL.B28, 2977, (1972).

2296 LARSON..........(A.C.)..: (5 , 651) ACTA CRYST. 21, 383, (1966).
(5 , 1950) ACTA CRYST. 22, 182, (1967).
(17C, 9) ACTA CRYST. 23, 664, (1967).

2297 LARSSON.........(K.E.)..: (6 , 1192) PHYS. REV. VOL.108, 1346, (1957).
(23 , 43) J. NUCL. ENERGY VOL.6, 222, (1958).
(6 , 1198) PHYSICA VOL.24, SUPPL., 145, (1958).
(6 , 1201) PHYS. REV. VOL.112, 11, (1958).
(20B, 7) ARK. FYS. VOL.16 PAPER 19, 199, (1959).
(6 , 42) ARK. FYS. VOL.17 PAPER 21, 369, (1960).
(6 , 52) IAEA SYMPOSIUM VIENNA, 587, (1960).
(6 , 669) IAEA SYMPOSIUM (VIENNA), 329, (1960).
(19B, 10) NUCL. INST. MET. VOL.12, 355, (1961).
(6 , 946) IAEA SYMP. CHALK RIVER, VOL.1, 317, (1962).
(6 , 1109) J. NUCL. ENERGY VOL.16, 81, (1962).
(14B, 70) PHYS. LETT. 3, 145, (1962).
(6 , 70) IAEA SYMP. BOMBAY, VOL.2, 101, (1964).
(6 , 54) IAEA SYMP. BOMBAY, VOL.2, 117, (1964).
(22A, 12) IAEA SYMP. BOMBAY, VOL.2, 487, (1964).
(6 , 392) IAEA SYMP. BOMBAY, VOL.2, 3, (1964).
(23 , 57) NUCL. INST. MET. 27, 61, (1964).
(14B, 62) PHYSICA VOL.30, 1561, (1964).
(6 , 88) PHYS. REV. LETT. 12, 721, (1964).
(14B, 131) TH. NEUTRON SCATT./EGELSTAFF, 347, (1965).
(6 , 38) ARK. FYS. VOL.33, PAPER 16, 271, (1966).
(14B, 88) PHYS. REV. VOL.151, 117, (1966).
(14B, 86) PHYS. REV. VOL.151, 126, (1966).
(6 , 79) PHYS. REV. VOL.161, 102, (1967).
(6 , 26) IAEA SYMP. COPENHAGEN, VOL.1, 397, (1968).
(6 , 409) IAEA SYMP. COPENHAGEN, VOL.1, 581, (1968).
(14A, 45) PHYS. REV. VOL.167, 171, (1968).
(14B, 51) 2ND SUMMER SCHOOL S.S. PHYS., 43, (1968).
(6 , 1592) PHYSICA VOL.49, 1, (1970).
(15 , 4) BER. BUNSENGES. PHYS. CHEM. 75, 352, (1971).
(15 , 43) PHYS. REV. A VOL.3, 1006, (1971).
(15 , 9) FARADAY SYMP. CHEM. SOC. NO.6, 122, (1972).
(6 , 322) PHYSICA VOL.59, 672, (1972).
(15 , 20) J. CHEM. PHYS. VOL.59, 4612, (1973).
(6 , 103) PHYSICA VOL.72, 300, (1974).

2298 LASSIER.........(B.)....: (6 , 2022) AMER. CRYSTALL. ASSOC., 27, (1971).
(6 , 1527) J. PHYSIQUE TOME 34, 473, (1973).

2299 LATHAM..........(R.)....: (5 , 2258) NATURE VOL.161, 282, (1948).
(6 , 773) PHYS. REV. VOL.74, 103, (1948).

2300 LAURENCE........(J.L.)..: (17B, 24) ACTA CRYST. B25, 374, (1969).

2301 LAURENT.........(Y.)....: (5 , 2947) MATER. RES. BULL. VOL.8, 1049, (1973).

2302 LAURITSEN.......(C.C.)..: (20B, 122) PHYS. REV. VOL.45, 507, (1934).

2303 LAUT............(P.)....: (11B, 9) J. PHYS. CHEM. SOL. 28, 1357, (1967).

2304 LAU.............(H.Y.)..: (5 , 2840) J. APPL. PHYS. VOL.40, 1136, (1969).
(6 , 1995) J. APPL. PHYS. VOL.40, 1278, (1969).
(6 , 1985) PHYS. REV. LETT. VOL.23, 1225, (1969).
(6 , 1996) J. APPL. PHYS. VOL.41, 1384, (1970).
(6 , 1978) PHYS. REV. B-4, 3206, (1971).

2305 LAU.............(H.)....: (6 , 33) IAEA SYMP. GRENOBLE, 137, (1972).

2306 LAVEISSIERE.....(J.)....: (5 , 2743) BULL. SOC. FRANC. MIN. CRIST. 90, 304, (1967).
(5 , 2742) BULL. SOC. FRANC. MIN. CRIST. 90, 308, (1967).

2307 LAVI............(N.)....: (20A, 47) NUCL. INST. MET. VOL.107, 13, (1973).

2308 LAX.............(M.)....: (10C, 19) COPENHAGEN CONF.-LATT.DYNAMICS, 427, (1963).
(10C, 43) J. PHYS. CHEM. SOL. 25, 487, (1964).

```
2309  LAZAREV.........(B.G.)..:   (22C,   71)  SOV. PHYS. CRYST. VOL.5, ....    297, (1960).
                                  ( 5 , 1277)  SOV. PHYS. JETP VOL.13,....      718, (1961).
                                  ( 5 ,  780)  J. PHYS. SOC. JAPAN VOL.17 B-II, 385, (1962).

2310  LAZAROVICI......(C.)....:   (22B,   42)  REV. DE PHYSIQUE (BUCAREST) 4,.  327, (1959).
                                  (19B,   36)  STUD. CERCETARI FIZ. VOL.10,..    89, (1959).

2311  LAZEBNIK........(I.M.)..:   ( 6 , 2291)  SOV. PHYS. SOL. STATE VOL.10,.   511, (1968).

2312  LAZENBY.........(R.)....:   ( 5 , 2122)  PROC. PHYS. SOC. 85, ......      967, (1965).

2313  LA-PLACA........(S.J.)..:   ( 5 ,  375)  J. PHYS. CHEM. SOL. 28,....     1089, (1967).
                                  ( 5 ,  310)  ACTA CRYST. VOL.B24,....        1147, (1968).
                                  ( 5 , 1179)  J. PHYS. CHEM. SOLIDS 29,....   1359, (1968).
                                  ( 5 , 1282)  INORG. CHEM. VOL.8,....         1928, (1969).
                                  ( 5 ,  793)  PHYS. ICE,....                    44, (1969).
                                  ( 5 , 2093)  J. CHEM. PHYS. VOL.54,....      3990, (1971).
                                  ( 5 , 1292)  J. CHEM. PHYS. VOL.55,....      1934, (1971).
                                  ( 5 ,  183)  ACTA CRYST. VOL.B28,....        3429, (1972).
                                  ( 5 , 2725)  INORG. CHEM. VOL.11,....        3009, (1972).
                                  ( 5 ,  786)  J. CHEM. PHYS. VOL.58,....       567, (1973).
                                  ( 5 , 1582)  J. PHYS. CHEM. SOLIDS VOL.34,.   521, (1973).
                                  ( 5 , 2975)  J. PHYS. CHEM. SOLIDS VOL.35,.   879, (1974).

2314  LEACER..........(J.)....:   (10D,   15)  JOUR. APP. PHYS. 40, ......     2766, (1969).
                                  (13D,   16)  JOUR. APP. PHYS. 40, ......     2776, (1969).

2315  LEADBETTER......(A.J.)..:   (10B,   29)  IAEA SYMP. CHALK RIVER, VOL.1,    49, (1962).
                                  ( 6 , 2047)  J. CHEM. PHYSICS VOL.51,....     779, (1969).
                                  ( 6 ,  940)  DISC. FARADAY SOC. NO.50,....     62, (1970).
                                  ( 6 ,  151)  J. NON-CRYST. SOLIDS VOL.3,..    239, (1970).
                                  ( 5 , 1233)  AMORPHOUS MATERIALS, . . . . .   423, (1972).
                                  ( 6 , 2050)  CONF./PHONON SCATT. IN SOLIDS,   338, (1972).
                                  ( 6 ,  244)  IAEA SYMP. GRENOBLE, ......      231, (1972).
                                  ( 6 , 1948)  IAEA SYMP. GRENOBLE, ......      501, (1972).
                                  ( 5 , 1238)  J. NON-CRYST. SOLIDS VOL.7,..    141, (1972).
                                  ( 5 ,  203)  J. NON-CRYST. SOLIDS VOL.7,..    156, (1972).
                                  ( 5 , 1239)  J. NON-CRYST. SOLIDS VOL.7,..     23, (1972).
                                  ( 6 , 2108)  SYMP. CHEM. ORGANIC SOL. STATE,  ..., (1972).
                                  ( 5 , 1237)  CHEM. APPL. THERMAL NEUTRON SCAT 146, (1973).

2316  LEAKE...........(J.A.)..:   ( 6 , 1349)  PHYS. REV. LETT. 18, ......      548, (1967).
                                  ( 6 ,  128)  PHYS. REV. LETT. 19, ......      234, (1967).
                                  ( 5 , 2690)  SOL. STATE COMM. VOL.6,....       15, (1968).
                                  ( 6 , 1702)  PHYS. REV. VOL.181,....         1251, (1969).
                                  ( 6 , 1222)  SOL. STATE COMM. 7,....          535, (1969).

2317  LEAKE...........(J.W.)..:   (20C,   44)  NUCL. INST. MET. VOL.45,....     151, (1966).
                                  (24 ,   56)  NUCL. INST. MET. 63, ......      329, (1968).

2318  LEARY...........(J.A.)..:   ( 5 , 2363)  J. NUCL. MATERIALS VOL.34, ...   281, (1970).

2319  LEATH...........(P.L.)..:   (10B,   18)  DISS. ABS. VOL.27, ......      3247B, (1966).
                                  (10B,  138)  PHYS. REV. 171,....              725, (1968).
                                  (10B,  135)  PHYS. REV. 175,....              963, (1968).
                                  (10D,   37)  PHYS. REV. B VOL.3,....         44C4, (1971).

2320  LEBECH..........(B.)....:   ( 9C,   34)  PHYS. STAT. SOL. 11,....         567, (1965).
                                  ( 5 , 2575)  SOL. STATE COMM. VOL.6,....      761, (1968).
                                  (20A,   41)  NUCL. INST. MET. VOL.79,....      51, (1970).
                                  ( 5 , 2336)  PHYS. REV. LETT. VOL.25,....     524, (1970).
                                  ( 5 ,  409)  CONF./RARE EARTHS/ACTINIDES, .   204, (1971).
                                  ( 5 , 2024)  CONF./RARE EARTHS/ACTINIDES, .    43, (1971).
                                  ( 5 , 2023)  J. PHYSIQUE TOME 32 COL.1 VOL.I, 370, (1971).
                                  ( 5 ,  609)  J. PHYS. CHEM. SOLIDS VOL.33,.  1651, (1972).

2321  LEBEDEVA........(N.S.)..:   ( 7B,  114)  SOV. AT. ENERGY VOL.28,....      398, (1970).
                                  ( 7B,  134)  SOV. PHYS. J. VOL.34,....        138, (1970).
                                  ( 7A,   92)  SOV. PHYS. JETP LETT. VOL.14,..   91, (1971).

2322  LEBEDEV.........(A.I.)..:   ( 7A,   87)  SOV. PHYS. JETP 16, ......      1321, (1963).

2323  LECHNER.........(R.E.)..:   ( 6 , 1828)  PHYS. REV. LETT. 17,....        1259, (1966).
                                  (22C,   60)  REV. SCI. INSTRUM. VOL.37,....  1534, (1966).
                                  ( 5 ,  268)  CHEM. PHYS. LETTERS VOL.4,....   444, (1969).
                                  ( 6 ,  323)  SOL. STATE COMM. VOL.10,....    1247, (1972).

2324  LECIEJEWICZ.....(J.)....:   ( 5 , 2280)  ACTA CRYST. VOL.14,....         1304, (1961).
                                  ( 5 , 2862)  ACTA CRYST. VOL.14,....          200, (1961).
                                  ( 5 , 2283)  ACTA CRYST. VOL.14,....           66, (1961).
                                  ( 5 , 2624)  Z. KRIST. VOL.116,....          345, (1961).
                                  ( 5 , 2279)  NATURWISSENSCHAFTEN VOL.49,...  373, (1962).
                                  ( 5 , 1152)  COLL. INTER. N.126 (GRENOBLE),  150, (1963).
                                  (19A,   58)  NUKLEONIKA VOL.8,....            75, (1963).
                                  (19A,   59)  NUKLEONIKA VOL.9,....           523, (1964).
                                  ( 5 , 2443)  PHYS. STAT. SOL. 4,....         349, (1964).
                                  ( 9C,   34)  PHYS. STAT. SOL. 11,....        567, (1965).
                                  ( 5 , 2273)  Z. KRIST. VOL.121,....          158, (1965).
                                  ( 5 , 2800)  PHYS. STAT. SOL. 15,....        515, (1966).
                                  ( 5 , 2813)  PHYS. STAT. SOL. 22,....        517, (1967).
                                  ( 5 , 2716)  PHYS. STAT. SOL. 23,....        K123, (1967).
                                  ( 5 , 2717)  PHYS. STAT. SOL. 30,....        157, (1968).
                                  ( 5 , 2791)  PHYS. STAT. SOL. 30,....         61, (1968).
                                  ( 5 , 2792)  PHYS. STAT. SOL. 34,....        K157, (1969).
                                  ( 5 , 2950)  PHYS. STAT. SOL. VOL.37,....    843, (1970).
                                  ( 5 , 2722)  PHYS. STAT. SOL. VOL.38,....    K89, (1970).
                                  ( 5 , 2946)  PHYS. STAT. SOLIDI A VOL.4 NO.1, 53, (1971).
                                  ( 5 , 2807)  PHYS. STAT. SOLIDI A VOL.13 NO.1 K79, (1972).
                                  ( 5 , 2721)  PHYS. STAT. SOLIDI A VOL.13,.   657, (1972).
                                  ( 5 , 2723)  PHYS. STAT. SOLIDI A VOL.16 NO.2 K171, (1973).
                                  ( 5 , 2803)  PHYS. STAT. SOLIDI A VOL.19 NO.1 K89, (1973).
                                  ( 5 , 2817)  PHYS. STAT. SOLIDI A VOL.20,.   395, (1973).
                                  ( 5 ,  578)  PHYS. STAT. SOLIDI B VOL.57 NO.2 K107, (1973).

2325  LECOMTE.........(J.)....:   ( 5 , 2392)  J. PHYSIQUE TOME 32 COL.1 VOL.II 611, (1971).

2326  LECOMTE.........(M.)....:   ( 5 ,  237)  J. NUCL. MATERIALS VOL.7,....    92, (1962).
                                  ( 5 , 2673)  C.R. ACAD. SCI. B TOME 276 NO.14 579, (1973).

2327  LEE.............(K.)....:   ( 5 ,  667)  J. CHEM. PHYSICS VOL.37,....     697, (1962).

2328  LEE.............(M.H.)..:   ( 6 , 1135)  DISS. ABS. VOL.30, ......      2839B, (1969).

2329  LEE.............(R.R.)..:   ( 6 ,  135)  DISS. ABS. 27, ......         3625B, (1966).

2330  LEE.............(T.D.)..:   ( 6 , 1197)  PHYS. REV. LETT. VOL.2,....      284, (1959).

2331  LEFEVRE.........(Y.)....:   ( 6 ,  961)  IAEA SYMP. GRENOBLE, ......      445, (1972).
                                  ( 6 ,  983)  PHYS. REV. A VOL.8,....         3163, (1973).
                                  ( 6 ,  889)  PROPERTIES LIQUID METALS,....   119, (1973).

2332  LEFKOWITZ.......(I.)....:   ( 6 , 1302)  PHYS. REV. LETT. VOL.2,....       94, (1959).
                                  ( 6 ,  125)  IAEA SYMPOSIUM VIENNA, ......    601, (1960).
                                  ( 6 , 2080)  J. PHYS. SOC. JAPAN VOL.28, SUP. 249, (1970).
                                  (21 ,   48)  REV. SCI. INSTRUM. 41,....        61, (1970).
                                  ( 6 , 1871)  J. CHEM. PHYSICS VOL.54,....    2597, (1971).
```

2333 LEGGE...........(G.J.F.)..: (22A, 39) NUCL. INST. MET. 63, ... 157, (1968).

2334 LEGLER..........(E.)....: (21 , 35) NUCL. INST. MET. 50, ... 251, (1967).

2335 LEGRAND.........(E.)....: (5 , 1495) J. PHYS. RADIUM VOL.23, ... 474, (1962).
(5 , 564) PHYS. STAT. SOL. VOL.2, ... : K112, (1962).
(5 , 1494) PHYS. STAT. SOL. 2, ... 317, (1962).
(5 , 1492) COLL. INTER. N.126 (GRENOBLE), . 154, (1963).
(5 , 101) PHYSICA VOL.57, ... 191, (1972).
(5 , 98) PHYSICA VOL.62, ... 474, (1972).
(5 , 1650) PHYS. STAT. SOL. VOL.49, ... 589, (1972).
(5 , 1651) SOL. STATE COMM. VOL.10, ... 883, (1972).
(5 , 1718) PHYS. STAT. SOLIDI A VOL.15 NO.1 K37, (1973).

2336 LEHINAN.........(S.R.)..: (19A, 44) NUCL. INST. MET. 76, ... 135, (1969).
(19A, 47) NUCL. INST. MET. 77, ... 29, (1970).

2337 LEHMANN.........(A.M.)..: (5 , 1318) ACTA PHYS. POLON. VOL.35, . . . 245, (1969).
(5 , 1317) ACTA PHYS. POLON. A VOL.A37, . . 625, (1970).

2338 LEHMANN.........(H.W.)..: (5 , 704) J. PHYS. CHEM. SOL. 28, ... 897, (1967).

2339 LEHMANN.........(M.S.)..: (5 , 1410) ACTA CHEM. SCAND. VOL.24, . . . 1662, (1970).
(5 , 287) ACTA CHEM. SCAND. VOL.24, . . . 3248, (1970).
(5 , 2404) ACTA CHEM. SCAND. VOL.24, . . . 3422, (1970).
(5 , 1316) ACTA CHEM. SCAND. VOL.25, . . . 1233, (1971).
(5 , 1473) ACTA CHEM. SCAND. VOL.25, . . . 3859, (1971).
(5 , 91) ACTA CHEM. SCAND. VOL.26, . . . 1996, (1972).
(5 , 4) ACTA CRYST. A28 PART S-4, . . . 193, (1972).
(5 , 284) ACTA CRYST. VOL.B28, ... 1388, (1972).
(5 , 277) ACTA CRYST. VOL.B28, ... 2977, (1972).
(5 , 13) ACTA CRYST. VOL.B28, ... 3006, (1972).
(5 , 24) ACTA CRYST. VOL.B28, ... 3207, (1972).
(5 , 22) INT. J. PEPTIDE PROTEIN RES. 4, . 229, (1972).
(5 , 10) J. AMER. CHEM. SOC. VOL.94, . . . 2657, (1972).
(5 , 981) J. CHEM. PHYS. VOL.56, ... 3257, (1972).
(5 , 20) J. CRYST. MOL. STRUC. VOL.2, . . 225, (1972).
(5 , 30) MATER. RES. BULL. VOL.7, ... 1225, (1972).
(5 , 2350) NATURE VOL.235, ... 328, (1972).
(5 , 2626) ACTA CHEM. SCAND. VOL.27, . . . 85, (1973).
(5 , 21) ACTA CRYST. VOL.B29, ... 2571, (1973).
(5 , 291) ACTA CRYST. VOL.B29, ... 231, (1973).
(5 , 33) ACTA CRYST. VOL.B29, ... 876, (1973).
(5 , 11) J.C.S. PERKIN TRANS. II, ... 133, (1973).
(5 , 25) J. CHEM. PHYS. VOL.58 NO.6, . . . 2547, (1973).
(5 , 1594) J. CHEM. PHYS. VOL.59, ... 915, (1973).

2340 LEIGH...........(R.S.)..: (10B, 206) PROC. ROY. SOC. A VOL.320, . . . 505, (1971).

2341 LEMAIRE.........(R.)....: (5 , 2882) C.R. ACAD. SCI. VOL.255, ... 896, (1962).
(5 , 2789) J. APP. PHYS. 34, ... 1333, (1963).
(5 , 579) PROC. INT. CONF. (NOTTINGHAM), . 275, (1964).
(5 , 2037) J. PHYS. CHEM. SOL. 27, ... 1287, (1966).
(5 , 491) PHYS. REV. 141, ... 538, (1966).
(5 , 2554) J. PHYSIQUE TOME 28, ... 216, (1967).
(5 , 1365) COMPTES RENDUS, SERIE B, 266, . 994, (1968).
(5 , 857) PHYS LETT 27A, ... 541, (1968).
(5 , 2537) SOL. STATE COMM. VOL.6 NO.2, . . 115, (1968).
(5 , 2587) C. R. ACAD. SCI. B TOME 270, . . 1131, (1970).
(5 , 867) SOL. STATE COMM. VOL.8, ... 391, (1970).
(5 , 810) Z. ANGEW PHYS. VOL.32, ... 113, (1971).
(5 , 1591) C.R. ACAD. SCI. B TOME 274, . . 1166, (1972).
(5 , 1590) J. LESS-COMMON MET. VOL.29, . . 361, (1972).
(5 , 2611) SOL. STATE COMM. VOL.14 NO.9, . 877, (1974).

2342 LENK............(R.)....: (7B, 92) PHYS. STAT. SOL. VOL.46, ... 273, (1971).

2343 LEONARD-JR......(B.R.)..: (7A, 81) RENSSELAER POL. INST. SYMP., . . 3, (1961).
(6 , 1051) IAEA SYMP. CHALK RIVER, VOL.1, . 359, (1962).
(6 , 1052) IAEA SYMP. CHALK RIVER, VOL.1, . 373, (1962).
(6 , 2342) BROOKHAVEN SYMPOSIUM, ... 96, (1965).
(5 , 765) NUCL. TECHNOL. VOL.15, ... 49, (1972).

2344 LEONARD.........(A.)....: (9B, 82) PHYS. REV. VOL.128, ... 2188, (1962).

2345 LEONIDOVA.......(G.G.)..: (5 , 1911) SOV. PHYS. DOKLADY VOL.16, . . . 9, (1971).

2346 LEONI...........(F.)....: (5 , 1044) PHYS. REV. B VOL.6, ... 178, (1972).
(5 , 2226) PHYS. REV. B VOL.7, ... 3112, (1973).
(5 , 1064) NUOVO CIMENTO VOL.20B NO.1, . . . 1, (1974).

2347 LEPENDIN........(V.I.)..: (5 , 379) UKR. FIZ. ZH.(USSR) VOL.17, . . . 38, (1972).

2348 LEPETIT.........(J.)....: (20B, 12) C. R. ACAD. SCI. B TOME 270, . . 185, (1970).

2349 LERIBAUX........(H.R.)..: (14A, 54) PHYS. REV. A VOL.3, ... 1752, (1971).

2350 LEROY...........(J.L.)..: (19B, 27) RC ACCAD. NAZ. LINCEI VOL.45, . . 163, (1968).
(23 , 6) BULL. INF. SCI. TECH. NO.170, . . 61, (1972).

2351 LETHUILLIER.....(P.)....: (5 , 2382) PHYS. STAT. SOLIDI A VOL.15 NO.2 613, (1973).

2352 LETSIEVICH......(YA.)...: (5 , 193) SOV. PHYS. CRYST. VOL.15, ... 280, (1970).

2353 LEUNG...........(P.S.)..: (6 , 1021) J. CHEM. PHYSICS VOL.45, ... 1312, (1966).
(6 , 1577) J. CHEM. PHYSICS VOL.48, ... 4912, (1968).
(6 , 2010) J. CHEM. PHYS. VOL.50, ... 2140, (1969).
(6 , 1115) J. CHEM. PHYS. VOL.50, ... 4444, (1969).
(6 , 1041) J. PHYS. CHEM. VOL.74, ... 3696, (1970).
(6 , 1040) J. PHYS. CHEM. VOL.74, ... 3710, (1970).
(14B, 11) BER. BUNSENGES. PHYS. CHEM. 75, . 366, (1971).
(6 , 1906) J. POLYMER SCI. A-2 VOL.9, . . . 1219, (1971).
(24 , 24) J. CHEM. PHYS. VOL.57, ... 175, (1972).
(14B, 127) TECH. ELECTROCHEM. VOL.2, ... 173, (1973).

2354 LEUNG...........(T.T.)..: (5 , 235) ANNALS OF PHYS. VOL.75, ... 132, (1973).
(7A, 14) ANN. OF PHYS. VOL.75, ... 132, (1973).

2355 LEUTHAUSER......(K.D.)..: (16 , 6) ATOMKERNENERGIE VOL.18, ... 158, (1971).

2356 LEVOIK.........(V.A.)..: (22A, 66) SOV. PHYS. CRYST. VOL.2, ... 626, (1957).
(5 , 553) SOV. PHYS. DOKLADY VOL.4, ... 1070, (1960).
(90, 75) SOV. PHYS. CRYST. VOL.9, ... 535, (1965).
(5 , 530) SOV. PHYS. J.E.T.P. VOL.22, . . . 754, (1965).
(21 , 52) SOV. PHYS. CRYST. VOL.12, ... 477, (1967).
(5 , 2967) SOV. PHYS. CRYST. VOL.14, ... 913, (1969).
(5 , 2969) SOV. PHYS. CRYST. VOL.15, ... 376, (1970).
(6 , 638) SOV. PHYS. CRYST. VOL.15, ... 935, (1971).
(5 , 2427) SOV. PHYS. CRYST. VOL.17, ... 822, (1971).
(5 , 2899) SOV. PHYS. CRYST. VOL.17, ... 342, (1972).
(20B, 152) SOV. PHYS. CRYST. VOL.17, ... 826, (1973).

2357 LEVINSON........(L.M.)..: (5 , 1819) A.I.P. CONF. PROC. NO.10 PART 2, 1138, (1973).

```
2358  LEVIN..........(K.)....:  ( 5 , 2119) PHYS. REV. B VOL.9,. . . . . .  2354, (1974).

2359  LEVITIN........(R.Z.)..:  ( 5 , 1127) SOV. PHYS. J.E.T.P. VOL.19,. . . 1348, (1964).
                               ( 5 , 1078) SOL. STATE COMM. 7, NO22,. . . 1665, (1969).
                               ( 5 , 1072) SOV. PHYS. SOL. STATE VOL.13,. . .  44, (1971).

2360  LEVY...........(H.A.)..:  ( 9C,   21) J. CHEM. PHYS. VOL.19,. . . . . 1416, (1951).
                               ( 5 , 1895) PHYS. REV. VOL.83,. . . . . .  127C, (1951).
                               ( 5 , 1450) J. CHEM. PHYS. VOL.20,. . . . .  704, (1952).
                               ( 5 , 1920) PHYS. REV. VOL.86,. . . . . .   766, (1952).
                               ( 5 , 2830) PHYS. REV. VOL.87,. . . . . .   462, (1952).
                               ( 5 , 2318) ACTA CRYST. VOL.6,. . . . . .   661, (1953).
                               ( 5 , 1896) J. AMER. CHEM. SOC. VOL.75,. .  1536, (1953).
                               ( 5 , 1472) J. CHEM. PHYS. VOL.21,. . . . . 2084, (1953).
                               ( 5 , 1897) J. CHEM. PHYS. VOL.21,. . . . .  366, (1953).
                               ( 5 , 2430) J. PHYS. CHEM. VOL.57,. . . . .  535, (1953).
                               ( 5 , 1468) PHYS. REV. VOL.93,. . . . . .  1120, (1954).
                               (17B,    1) ACTA CRYST. VOL.10,. . . . . .  180, (1957).
                               ( 5 , 2819) ACTA CRYST. VOL.10,. . . . . .  319, (1957).
                               ( 5 ,  789) ACTA CRYST. VOL.10,. . . . . .   70, (1957).
                               ( 5 ,  368) J. CHEM. PHYS. VOL.26,. . . . .  563, (1957).
                               ( 5 ,   87) ACTA CRYST. VOL.11,. . . . . .  798, (1958).
                               ( 5 , 1502) J. CHEM. PHYS. VOL.29,. . . . .  948, (1958).
                               ( 5 , 1274) REV. MOD. PHYS. VOL.30,. . . .   101, (1958).
                               ( 5 , 1548) ANN. N.Y. ACAD. SCI. VOL.79,.   762, (196L).
                               ( 5 , 1922) ACTA CRYST. VOL.15,. . . . . .  1201, (1962).
                               ( 5 , 1468) COLL. INTER. N.126 (GRENOBLE),   45, (1963).
                               (23 ,    7) COLL. INTER. N.126 (GRENOBLE),   71, (1963).
                               (17B,   30) COLL. INTER. N.126 (GRENOBLE),   73, (1963).
                               ( 5 , 2871) SCIENCE VOL.139,. . . . . . .  120B, (1963).
                               ( 5 , 2506) SCIENCE VOL.141,. . . . . . .   921, (1963).
                               (17C,    2) ACTA CRYST. 17,. . . . . . . .  142, (1964).
                               ( 5 , 1447) ACTA CRYST. 19,. . . . . . . .  260, (1965).
                               ( 5 , 1283) J. CHEM. PHYSICS VOL.42,. . . . 3054, (1965).
                               (17A,    6) ACTA CRYST. 22,. . . . . . . .  457, (1967).
                               ( 5 , 1997) ACTA CRYST. B24,. . . . . . .   230, (1968).
                               ( 5 , 1578) J. CHEM. PHYSICS VOL.48,. . . . 5561, (1968).
                               ( 5 ,    ?) ACTA CRYST. VOL.B27,. . . . . .  333, (1971).
                               (22C,   19) J. APPL. CRYST. VOL.5 PT.6,. .   432, (1972).
                               ( 5 , 2505) ACTA CRYST. VOL.B29,. . . . . .  790, (1973).
                               ( 5 ,  250) CRYST. STRUC. COMM. VOL.2 NO.1,. 107, (1973).
                               ( 5 , 1549) J. CHEM. PHYS. VOL.58,. . . . . 5017, (1973).

2361  LEWIS..........(H.W.)..:  (17B,   57) PHYS. REV. 132,. . . . . . . . 2764, (1963).

2362  LE-CLERC.......(B.)....:  ( 5 , 2948) J. SOLID STATE CHEM. VOL.8,. .  182, (1973).

2363  LE-DANG-KHOI...........:  ( 5 , 2948) J. SOLID STATE CHEM. VOL.8,. .  182, (1973).

2364  LE-NEINDRE.....(B.)....:  ( 5 , 1328) PHYS. REV. A VOL.9,. . . . . .  448, (1974).

2365  LE-RAY.........(M.)....:  (10D,   22) PHYS. LETT. VOL.31A,. . . . .   563, (1970).

2366  LE-RIGOLEUR....(C.)....:  (23 ,    6) BULL. INF. SCI. TECH. NO.170,. .  61, (1972).

2367  LE-SOURNE......(M.)....:  (22B,   14) INTERNAT. COMPUTING SYMP.,. . .  291, (1974).

2368  LIBBY..........(W.F.)..:  (20B,  125) PHYS. REV. VOL.52,. . . . . . .  592, (1937).
                               ( 5 , 1252) PHYS. REV. VOL.55,. . . . . . .  339, (1939).

2369  LICHTENBERGER...(H.V.)..:  (20A,   64) PHYS. REV. VOL.72,. . . . . . .  585, (1947).

2370  LIDER..........(V.V.)..:  ( 5 , 1132) SOV. PHYS. CRYST. VOL.16,. . . .  935, (1971).

2371  LIEBERT........(L.)....:  ( 6 , 1334) PHYS. REV. LETT. VOL.32,. . . .  836, (1974).

2372  LIFOROV........(V.G.)..:  ( 6 , 1238) IAEA SYMP. COPENHAGEN, VOL.1,. .  483, (1968).

2373  LIGENZA........(S.)....:  ( 5 , 2951) PHYS. STAT. SOLIDI A VOL.7 NO.2, K71, (1971).
                               ( 5 , 2723) PHYS. STAT. SOLIDI A VOL.16 NO.2 K171, (1973).
                               ( 5 , 2803) PHYS. STAT. SOLIDI A VOL.19 NO.1 K89, (1973).
                               ( 5 , 2817) PHYS. STAT. SOLIDI A VOL.20,. .   395, (1973).
                               ( 5 , 2892) SOV. PHYS. SOL. STATE VOL.14,. . 3037, (1973).

2374  LIMINGA........(R.)....:  ( 5 , 1925) ACTA CHEM. SCAND. VOL.22,. . . .  719, (1968).
                               ( 5 , 1927) ACTA CHEM. SCAND. VOL.25,. . . . 1729, (1971).
                               ( 5 , 1928) ACTA CHEM. SCAND. VOL.26,. . . . 1087, (1972).
                               ( 5 , 1564) J. SOLID STATE CHEM. VOL.4,. . .  255, (1972).

2375  LINDGARD.......(P.A.)..:  (11B,    9) J. PHYS. CHEM. SOL. 28,. . . . . 1357, (1967).
                               ( 6 , 2137) SOL. STATE COMM. 5,. . . . . .   607, (1967).
                               ( 6 , 2125) IAEA SYMP. COPENHAGEN, VOL.2,. .   93, (1968).
                               ( 6 , 2134) J. APP. PHYS. 39,. . . . . . .  1229, (1968).
                               ( 6 , 1466) J. PHYS. C, SER.2, VOL.2,. . . . 1168, (1969).
                               ( 6 , 2151) PHYS. REV. LETT. VOL.25,. . . . 1451, (1970).

2376  LINDOW.........(J.T.)..:  (17B,   46) NUCL. INST. MET. VOL.85,. . . .  151, (1970).

2377  LINDQVIST......(O.)....:  ( 5 , 2626) ACTA CHEM. SCAND. VOL.27,. . . .   85, (1973).

2378  LINGGOATMODJO...(K.)....:  ( 5 , 1433) ACTA CRYST. VOL.B26,. . . . . .  827, (1970).

2379  LING...........(P.C.)..:  ( 5 , 2089) J. PHYS. F VOL.3,. . . . . . .  697, (1973).

2380  LINKOAHO.......(M.)....:  ( 5 ,   50) PHYS. SCRIPTA(SWEDEN) VOL.4,. .  125, (1971).

2381  LINNEBUR.......(E.J.)..:  (14A,   19) IAEA SYMP. GRENOBLE,. . . . . .  365, (1972).

2382  LINZ...........(A.)....:  ( 6 ,  128) PHYS. REV. LETT. 19,. . . . . .  234, (1967).
                               ( 6 ,  130) PHYS. REV. VOL.177,. . . . . .   848, (1969).
                               ( 6 , 1446) PHYS. REV. B-1,. . . . . . . .  2304, (1970).
                               ( 6 ,  124) PHYS. REV. B-2,. . . . . . . .  3651, (1970).
                               ( 6 , 1451) PHYS. REV. LETT. VOL.24,. . . . 1184, (1970).
                               ( 5 , 1479) SOL. STATE COMM. VOL.8,. . . . . 1941, (1970).
                               ( 6 , 1336) FERROELECTRICS(GB) VOL.2,. . . .  261, (1971).
                               ( 6 , 1448) J. APPL. PHYS. 42,. . . . . . .  1258, (1971).
                               ( 6 , 1447) PHYS. REV. B-4,. . . . . . . .  2254, (1971).
                               ( 6 , 1317) PHYS. REV. B VOL.5,. . . . . .  1933, (1972).
                               ( 6 , 1469) A.I.P. CONF. PROC. NO.10 PART 1,   93, (1973).

2383  LIN-SHI-CHIEN...........:  ( 5 , 1433) ACTA CRYST. VOL.B26,. . . . . .  827, (1970).
                               ( 5 , 1994) ACTA CRYST. VOL.B27,. . . . . .  253, (1971).

2384  LIPIN..........(YU.V.).:  ( 5 , 2861) LATV. PSR...FIZ. TEHN. SER. 6,   30, (1968).
                               ( 5 , 2898) LATV. PSR...FIZ. TEHN. SER. 2,   49, (1969).
                               ( 5 ,   75) LATV. PSR...FIZ. TEHN. SER. NO.1  88, (1970).
                               ( 5 ,  348) LATV. PSR...FIZ. TEHN. SER. NO.2 124, (1970).
                               ( 5 ,  322) LATV. PSR...FIZ. TEHN. SER. NO.4  87, (1970).

2385  LIPIN..........(YU.)...:  ( 5 , 2885) LATV. PSR...FIZ. TEHN. SER. NO.3  28, (1970).
                               ( 5 , 2919) LATV. PSR...FIZ. TEHN. SER. 6,   46, (1970).
                               (17B,   39) LATV. PSR...FIZ. TEHN. SER. NO.6 124, (1971).
                               ( 5 ,  357) LATV. PSR...FIZ. TEHN. SER. NO.6 120, (1971).
                               ( 5 , 1961) LATV. PSR...FIZ. TEHN. SER. NO.5 122, (1971).
```

```
2386  LIPPARD.........(S.J.)..!  ( 5 , 2725) INORG. CHEM. VOL.11, . . . . . . 3009, (1972).

2387  LIPPMANN........(B.A.)..!  ( 7A,    65) PHYS. REV. VOL.79, : : : : : : : 469, (1950).
                                ( 7A,    64) PHYS. REV. VOL.79, : : : : : : : 481, (1950).

2388  LIPPMANN........(G.)....!  ( 5 , 2005) Z. PHYSIK VOL.253, : : : : : : : 219, (1972).
                                ( 5 , 2003) PHYS. STAT. SOLIDI B VOL.57 NO.2  515, (1973).
                                ( 5 , 2002) PHYS. STAT. SOLIDI B VOL.58 NO.2  633, (1973).
                                ( 5 , 1065) J. APPL. CRYST. VOL.7, : : : : : 233, (1974).
                                ( 5 , 2010) J. APPL. CRYST. VOL.7, : : : : : 236, (1974).
                                ( 5 , 2618) J. LOW TEMP. PHYS. VOL.14, : : : 213, (1974).

2389  LIPSCHULTZ......(F.P.)..!  ( 6 , 1185) PHYS. REV. LETT. 19, . . . . . 1307, (1967).
                                ( 6 , 1186) PHYS. LETT. 30A, : : : : : : : 127, (1969).

2390  LISHER..........(E.J.)..!  ( 9D,     7) ACTA CRYST. VOL.A29, : : : : : 453, (1973).
                                ( 5 , 1105) J. PHYS. C VOL.7, : : : : : : 1344, (1974).

2391  LITCHINSKY......(D.)....!  ( 6 ,  940) DISC. FARADAY SOC. NO.50,. . . . 62, (1970).
                                ( 6 ,  244) IAEA SYMP. GRENOBLE, : : : : : 231, (1972).

2392  LITOVITZ........(T.A.)..!  (24 ,   20) IAEA SYMP. COPENHAGEN, VOL.1,. . 623, (1968).

2393  LITTLE..........(G.R.)..!  ( 5 , 1362) DISS. ABSTR. B VOL.34, . . . . 3992, (1974).

2394  LITVIN..........(D.B.)..!  ( 9D,    9) ACTA CRYST. VOL.A29, . . . . . 651, (1973).

2395  LITVIN..........(D.F.)..!  ( 5 , 2140) SOV. PHYS. CRYST. VOL.3, . . .  147, (1958).
                                ( 9B,   41) INSTRUM. EXP. TECH. NO.3, . . .  359, (1960).
                                (20B,  150) SOV. PHYS. CRYST. VOL.4, . . .  623, (1960).
                                ( 5 , 1067) J. PHYS. SOC. JAPAN VOL.17 B-III  49, (1962).
                                ( 5 , 2108) PHYS. MET. METALLOGR. VOL.14,. .  25, (1962).
                                ( 5 ,  987) SOV. PHYS. CRYST. VOL.6, . . .  443, (1962).
                                ( 5 ,  986) SOV. PHYS. CRYST. VOL.7, . . .  637, (1963).
                                ( 5 ,  541) SOV. PHYS. CRYST. VOL.7, . . .  639, (1963).
                                ( 5 ,  989) UKR. FIZ. ZH. (USSR) VOL.8, . .  268, (1963).
                                (22A,   19) INSTRUM. EXP. TECH. NO.2, . . .  306, (1964).
                                ( 5 ,  547) SOV. PHYS. DOKLADY VOL.9, . . .  388, (1964).
                                (22C,   74) SOV. PHYS. CRYST. VOL.11, . . .  322, (1966).
                                ( 5 , 2151) SOV. PHYS. DOKLADY VOL.16, . . .  486, (1971).

2396  LITVIN..........(L.F.)..!  ( 5 , 1627) SOV. PHYS. DOKLADY VOL.18, . . .  432, (1973).

2397  LITZMAN.........(O.)....!  (10B,   71) J. PHYS. C: SOLID ST. PHYS. V-5,  287, (1972).

2398  LIU.............(S.H.)..!  ( 6 ,  539) PHYS. REV. LETT. VOL.23, . . .  311, (1969).
                                ( 9D,   61) PHYS. REV. 178, : : : : : : : 783, (1969).
                                ( 6 ,  537) J. APPL. PHYS. 41, : : : : : 1365, (1970).
                                ( 8 ,   97) PHYS. REV. B VOL.2, : : : : : 2664, (1970).
                                ( 8 ,  101) PHYS. REV. B VOL.5, : : : : : 2719, (1972).
                                (10B,  177) PHYS. REV. LETT. VOL.29, . . . .  793, (1972).

2399  LIU.............(Y.C.)..!  (22A,   9) CHIN. J. PHYS. (TAIWAN) VOL.8, .  31, (1970).

2400  LIVESAY.........(B.R.)..!  ( 8 ,   22) J. APPL. PHYS. VOL.40, . . . . 1226, (1969).

2401  LIVINGSTON......(R.C.)..!  ( 6 , 1939) J. CHEM. PHYS. VOL.52, . . . . 4424, (1970).
                                ( 6 ,  575) IAEA SYMP. GRENOBLE, . . . . .  247, (1972).
                                ( 6 , 2259) JULICH CONF/H IN METALS VOL.I, .  301, (1972).
                                ( 6 ,  313) J. CHEM. PHYS. VOL.58, . . . . 1438, (1973).
                                ( 6 , 1665) J. CHEM. PHYS. VOL.58, . . . . 3439, (1973).
                                ( 6 , 2001) J. CHEM. PHYS. VOL.58, . . . . 5469, (1973).
                                ( 6 , 2117) J. CHEM. PHYS. VOL.59, . . . . 6570, (1973).
                                ( 6 , 2002) J. CHEM. PHYS. VOL.59, . . . . 6652, (1973).

2402  LI-ZHI..........(F.)....!  ( 8 ,    3) ACTA PHYS. SINICA VOL.19,. . . .  673, (1963).

2403  LI..............(F.)....!  (20A,   49) NUCL. INST. MET. VOL.116, . . .  615, (1974).

2404  LLOYD...........(P.)....!  ( 7A,   80) PROC. PHYS. SOC. 91, . . . . .  678, (1967).
                                (11C,   29) J. PHYS. F VOL.2,. : : : : : 1145, (1972).

2405  LLOYD...........(R.C.)..!  ( 6 , 1946) NUCL. SCI. ENG. VOL.46,. . . . .  244, (1971).

2406  LOFFLER.........(E.)....!  ( 5 ,  806) Z. PHYS. VOL.210, . . . . . .  265, (1968).
                                ( 5 ,  800) J. PHYS. CHEM. SOLIDS VOL. 30, : 2175, (1969).

2407  LOFFLER.........(U.)....!  ( 6 , 1960) PROPERTIES LIQUID METALS,. . . .  111, (1973).

2408  LOGAN...........(K.W.)..!  ( 6 , 1790) DISS. ABS. VOL.30, . . . . . .4740B, (1970).
                                ( 6 ,  337) J. CHEM. PHYSICS VOL.53, . . . 3417, (1970).
                                (10B,   54) J. CHEM. PHYSICS VOL.53, . . . 4645, (1970).
                                ( 6 , 1658) CONF. INTERN., RENNES, . . . .  104, (1971).
                                ( 6 , 1080) J. CHEM. PHYS. VOL.56, . . . . 3217, (1972).
                                ( 6 ,  324) J. CHEM. PHYS. VOL.59, . . . . 2305, (1973).

2409  LOGIUDICE.......(G.)....!  ( 6 ,  797) J. PHYS. CHEM. SOL. 28,. . . . .  467, (1967).

2410  LOLY............(P.D.)..!  (11A,   37) PHYS. REV. VOL.173,. . . . . . .  603, (1968).

2411  LOMER...........(W.M.)..!  (10B,  202) PROC. PHYS. SOC. B. VOL.70,. . 1143, (1957).
                                ( 8 ,  158) TH. NEUTRON SCATT./EGELSTAFF,. .    1, (1965).
                                ( 8 ,   16) CONTEMPORARY PHYS. VOL.7, . . .  278, (1966).
                                (11B,    3) CONTEMPORARY PHYS. VOL.7, . . .  401, (1966).
                                ( 9D,   27) HARWELL SUMMER SCHOOL, . . . . .  171, (1968).

2412  LONDON..........(H.)....!  (20B,  118) PHIL. MAG. (8) VOL.2,. . . . .  917, (1957).
                                ( 6 , 1196) PROC. ROY. SOC. A VOL.242, . . .  374, (1957).

2413  LONGO...........(G.)....!  (17A,   82) NUOVO CIMENTO SUPPL. VOL.3,. . .  187, (1965).

2414  LONGO...........(M.J.)..!  ( 7A,   54) PHYS. LETT. B VOL.36B, . . . . .  560, (1971).

2415  LONGSTER........(G.F.)..!  ( 6 , 1913) J. CHEM. PHYS. VOL.48, . . . . 5271, (1968).
                                ( 6 ,  358) MOL. PHYS. VOL.17, . . . . . . .    1, (1969).

2416  LONGWORTH.......(G.)....!  ( 5 , 2816) J. PHYS. C VOL.6,. . . . . . . 1652, (1973).

2417  LONG............(C.E.)..!  ( 5 , 2797) DISS. ABS. VOL.32, . . . . . . 5079, (1971).

2418  LONG............(C.)....!  ( 6 ,  722) PHYS. REV. LETT. VOL.28, . . . .  805, (1972).

2419  LONG............(E.A.)..!  (20B,  125) PHYS. REV. VOL.52, . . . . . .  592, (1937).
                                ( 5 , 1252) PHYS. REV. VOL.55, : : : : : : 339, (1939).

2420  LONG............(W.R.)..!  (20C,   13) IEEE TRANS. NUCL. SCI. VOL.S-16,  419, (1969).

2421  LONNGI..........(D.A.)..!  ( 6 , 1699) IAEA SYMP. GRENOBLE, . . . . . .  399, (1972).

2422  LONNGI..........(P.A.)..!  ( 5 , 2056) PHYS. CAN. VOL.27, . . . . . . .   68, (1971).
                                ( 6 , 1699) IAEA SYMP. GRENOBLE, : : : : : 399, (1972).

2423  LONSDALE........(K.)....!  ( 9A,   17) NATURE VOL.164,. . . . . . . . .  205, (1949).
                                ( 9A,   23) REP. PROGR. PHYS. VOL.16,. : : :   1, (1953).
```

```
2424  LOOPSTRA........(R.O.)..:   ( 5 ,  353) ACTA CRYST. VOL.15,. . . . . .     92, (1962).
                                 ( 5 , 2764) COLL. INTER. NO.126 (GRENOBLE),.    5, (1963).
                                 ( 5 , 2769) ACTA CRYST. 17,. . . . . . . .    651, (1964).
                                 ( 5 , 2868) ACTA CRYST. 21,. . . . . . . .    158, (1966).
                                 (22B,   27) NUCL. INST. MET. 44,. . . . . .   181, (1966).
                                 ( 5 , 2268) ACTA CRYST. 22,. . . . . . . .    744, (1967).
                                 ( 5 , 1481) PHYS. LETT. VOL.26A NO.11,. . .   526, (1968).
                                 ( 5 , 2867) ACTA CRYST. B25,. . . . . . .    1420, (1969).
                                 ( 5 , 2501) ACTA CRYST. B25,. . . . . . . .   787, (1969).
                                 ( 5 , 2762) ACTA CRYST. VOL.B26,. . . . . .   656, (1970).
                                 ( 5 , 2775) J. APPL. CRYST. VOL.3,. . . . .    94, (1970).
                                 ( 5 , 2345) J. PHYSIQUE TOME 32 COL.1 VOL.II 657, (1971).
                                 ( 5 , 2148) PHYSICA VOL.57,. . . . . . . .    215, (1972).
                                 ( 5 , 2346) J. PHYS. CHEM. SOLIDS VOL.35,.    301, (1974).

2425  LOPEZ...........(W.M.)..:   ( 6 , 1895) IAEA SYMPOSIUM (VIENNA), . . . .  613, (1960).

2426  LORCH...........(E.)....:   ( 5 , 1235) J. OF PHYS.-C-, SER. 2, VOL.2,   229, (1969).
                                 ( 5 , 2374) PHYS. CHEM. GLASSES VOL.10,. .    185, (1969).
                                 ( 5 , 2458) J. OF PHYS.-C-, SER.2, VOL.3,.   1314, (1970).

2427  LORTHIOIR.......(G.)....:   ( 5 , 1741) SOL. STATE COMM. VOL. 9,. . . .  1793, (1971).

2428  LOSACCO.........(F.J.)..:   ( 6 , 1783) J. CHEM. PHYSICS VOL.43,. . . .  3404, (1965).
                                 ( 6 , 1382) J. CHEM. PHYSICS VOL.44, . . . .   345, (1966).

2429  LOSHMANOV.......(A.A.)..:   (22A,   67) SOV. PHYS. CRYST. VOL.5,. . . .   294, (1960).
                                 (23 ,   41) INSTRUM. EXP. TECH. NO.6,. . .    902, (1961).
                                 ( 5 , 1828) PHYS. MET. METALLOGR. VOL.12,.    119, (1961).
                                 ( 5 , 2108) PHYS. MET. METALLOGR. VOL.14,.     25, (1962).
                                 ( 5 ,  588) SOV. PHYS. CRYST. VOL.6,. . . .   628, (1962).
                                 ( 5 ,  986) SOV. PHYS. CRYST. VOL.7,. . . .   637, (1963).
                                 ( 5 ,  541) SOV. PHYS. CRYST. VOL.7,. . . .   639, (1963).
                                 ( 5 , 1069) BULL. ACAD. SCI. USSR VOL.28,.    354, (1964).
                                 ( 6 , 1767) PHYS. MET. METALLOGR. VOL.18 N.2   22, (1964).
                                 ( 6 ,  540) SOV. PHYS. CRYST. VOL.9,. . . .   301, (1964).
                                 ( 5 , 2132) SOV. PHYS. CRYST. VOL.10,. . . .  338, (1965).
                                 ( 5 , 2689) SOV. PHYS. J.E.T.P. VOL.31,. .    808, (1970).
                                 ( 9C,   19) IZV. AKAD. NAUK...NEORG. MATER 8    1, (1972).
                                 ( 5 , 2466) PHYS. STAT. SOLIDI A VOL.18 NO.2  K91, (1973).
                                 ( 5 , 2467) J. APPL. CRYST. VOL.7,. . . . .   207, (1974).

2430  LOSHMANOV.......(A.)....:   ( 5 , 2951) PHYS. STAT. SOLIDI A VOL.7 NO.2,  K71, (1971).

2431  LOTGERING.......(F.K.)..:   ( 5 , 1003) SOL. STATE COMM. VOL.8,. . . . .  477, (1970).

2432  LOUDON..........(R.)....:   ( 6 , 2029) PROC. ROY. SOC. 281, . . . . . .  274, (1964).

2433  LOVCHIKOVA......(G.N.)..:   ( 6 , 2169) BU. ACAD. SCI. USSR PHYS. SR. 32  600, (1968).

2434  LOVELUCK........(J.M.)..:   ( 6 , 1530) PHYS. REV. B VOL.7,. . . . . . . 1644, (1973).

2435  LOVESEY.........(S.W.)..:   ( 8 ,  114) PROC. PHYS. SOC. VOL.89,. . . .   613, (1966).
                                 ( 5 , 2128) PROC. PHYS. SOC. VOL.89,. . . .   893, (1966).
                                 (11B,   30) PPOC. PHYS. SOC. 91,. . . . . .   658, (1967).
                                 (11B,   32) PROC. PHYS. SOC. 92,. . . . . .   845, (1967).
                                 (11A,   11) J. PHYS. -C- VOL. 1,. . . . . .   102, (1968).
                                 (11A,   12) J. PHYS. -C- VOL. 1,. . . . . .   118, (1968).
                                 (11A,   13) J. PHYS. -C- VOL.1 . . . . . .    408, (1968).
                                 (24 ,   10) COMMENTS SOLID STATE PHYS. VOL.2   88, (1969).
                                 ( 7B,   35) J. PHYS. -C- VOL.2,. . . . . .    981, (1969).
                                 ( 9D,   41) J. PHYS. -C- VOL.2,. . . . . .    470, (1969).
                                 (11A,   32) PHYS. LETT. VOL.30A,. . . . . .    84, (1969).
                                 ( 8 ,  119) REP. PROGR. PHYS. VOL.32,. . .    333, (1969).
                                 (14B,  115) RIV. DEL NUOVO CIMENTO VOL.1,.    155, (1969).
                                 ( 8 ,   34) J. PHYS. C VOL.3,. . . . . . .   1232, (1970).
                                 ( 8 ,   56) PHYS. LETT. VOL.31A,. . . . . .    67, (1970).
                                 (11B,   20) PHYS. LETT. 31A,. . . . . . . .    53, (1970).
                                 ( 8 ,   40) J. PHYSIQUE TOME 32 COL.1,. . .   573, (1971).
                                 (14A,   26) J. PHYS. C VOL.4,. . . . . . .   3057, (1971).
                                 ( 6 , 1710) PHYS. REV. B VOL.4,. . . . . .   3048, (1971).
                                 ( 6 ,  813) J. PHYS. C VOL.5,. . . . . . .   2769, (1972).
                                 (11A,   51) PHYS. REV. LETT. VOL.28,. . . .   614, (1972).
                                 (14A,   27) J. PHYS. C VOL.6,. . . . . . .   1856, (1973).
                                 ( 5 , 2851) J. PHYS. C VOL.6,. . . . . . .   3746, (1973).
                                 (11C,   19) J. PHYS. C VOL.6,. . . . . . .     79, (1973).

2436  LOVSETH.........(J.)....:   (15 ,   34) PHYS. NORVEG. VOL.1, . . . . . .  127, (1962).

2437  LOWDE...........(R.D.)..:   ( 9B,    1) ACTA CRYST. VOL.1, . . . . . .    303, (1948).
                                 (23C,   88) REV. SCI. INSTRUM. VOL.21,. . .    835, (1950).
                                 (20A,   31) NATURE VOL.167,. . . . . . . .    243, (1951).
                                 ( 5 ,  944) PROC. PHYS. SOC. A VOL.65,. . .    857, (1952).
                                 ( 6 ,  771) PROC. ROY. SOC. A VOL.221,. . .    206, (1954).
                                 ( 8 ,  115) PROC. ROY. SOC. A VOL.230,. . .     46, (1955).
                                 ( 9B,    2) ACTA CRYST. VOL.9,. . . . . . .   151, (1956).
                                 ( 6 ,  770) PROC. ROY. SOC. A VOL.235,. . .    305, (1956).
                                 ( 6 ,  790) REV. MOD. PHYS. VOL.30,. . . .     69, (1958).
                                 (20D,   57) J. NUCL. ENERGY A:REAC. SCI. 11    69, (1960).
                                 ( 6 ,  787) PHYS. REV. LETT. VOL.4,. . . .    452, (1960).
                                 ( 5 , 1563) J. PHYS. CHEM. SOLIDS VOL.23,.    621, (1962).
                                 ( 5 , 1062) J. PHYS. SOC. JAPAN VOL.17 B-III   19, (1962).
                                 ( 9D,   48) J. PHYS. SOC. JAPAN VOL.17 B-II   342, (1962).
                                 ( 6 ,  830) PROC. PHYS. SOC. 84,. . . . . .    55, (1964).
                                 (11B,    8) J. APPL. PHYS. 36,. . . . . . .   884, (1965).
                                 ( 6 ,  833) PHYS. REV. LETT. VOL.14,. . . .   698, (1965).
                                 ( 6 , 1732) J. APP. PHYS. 38,. . . . . . .   1247, (1967).
                                 ( 6 , 1724) PHYS. REV. LETT. VOL.18,. . . .  1136, (1967).
                                 (11C,    4) C.N.E.N. SYMP. CASACCIA,. . . .    73, (1968).
                                 ( 6 , 1708) IAEA SYMP. COPENHAGEN, VOL.2,.    101, (1968).
                                 ( 6 , 1747) J. APP. PHYS. VOL.39,. . . . .    449, (1968).
                                 (11B,   34) REPORTS ON PROGRESS IN PHYS. 31,  705, (1968).
                                 ( 6 , 1720) SOL. STATE COMM. VOL.6,. . . .    189, (1968).
                                 ( 6 , 1711) J. APP. PHYS. VOL.40,. . . . .   1142, (1969).
                                 ( 6 , 1725) ADVANCES IN PHYS. VOL.19 NO.82,   845, (1970).
                                 ( 6 , 1435) J. PHYSIQUE TOME 32 COL.1 VOL.II 1186, (1971).
                                 ( 6 , 1777) IAEA SYMP. GRENOBLE,. . . . . .   669, (1972).
                                 ( 6 , 1778) A.I.P. CONF. PROC. NO.10 PART 2, 1403, (1973).

2438  LOWNDES.........(R.P.)..:   (10D,   36) PHYS. REV. LETT. VOL.27, . . . . 1134, (1971).

2439  LOW.............(G.G.E.):   (21 ,    7) IAEA SYMPOSIUM VIENNA,. . . . .   179, (1960).
                                 ( 6 ,  485) PROC. PHYS. SOC. VOL.79,. . . .   473, (1962).
                                 ( 6 ,  763) PROC. PHYS. SOC. VOL.79,. . . .   479, (1962).
                                 ( 5 , 2107) ACTA CRYST. VOL.16,. . . . . .    A126, (1963).
                                 ( 5 , 2131) J. APP. PHYS. VOL.34,. . . . .   1195, (1963).
                                 ( 5 , 1190) J. PHYSIQUE TOME 25,. . . . . .   596, (1964).
                                 ( 5 ,  930) PROC. PHYS. SOC. 86,. . . . . .   535, (1965).
                                 ( 6 , 1859) PROC. PHYS. SOC. 89,. . . . . .   119, (1966).
                                 ( 6 , 1734) J. APP. PHYS. 38,. . . . . . .   1240, (1967).
                                 ( 5 ,  955) PROC. PHYS. SOC. VOL.92,. . . .   726, (1967).

2440  LOW.............(G.G.)..:   (11B,   50) UKAEA AERE REP. R 4299 (7 PP.),.   ., (1963).
                                 ( 7B,   21) IAEA SYMPOSIUM BOMBAY,. . . . .   413, (1964).
                                 ( 6 , 1473) IAEA SYMPOSIUM BOMBAY, VOL.1,.    453, (1964).
                                 ( 6 , 1462) J. APPL. PHYS. VOL.35,. . . . .   998, (1964).
```

```
2440  LOW..............(G.G.)..!   ( 5 , 1189) PROC. INT. CONF. (NOTTINGHAM), .  133, (1964).
                                   ( 8 ,  124) SOLID/LIQUID STATE PHYSICS,. : .    1, (1965).
                                   ( 8 ,  158) TH. NEUTRON SCATT./EGELSTAFF,. : .   1, (1965).
                                   (90 ,   81) TRIESTE/INTERN. COURSE,. . . : .  501, (1967).
                                   ( 5 , 1193) J. APP. PHYS. VOL.39, . . . . . . 1174, (1968).
                                   ( 6 , 1715) J. OF PHYS. C VOL.1, . : . . . .  458, (1968).
                                   ( 5 , 2299) J. OF PHYS. C VOL.1, . . . . . .  528, (1968).
                                   (90 ,   86) Z. ANGEW. PHYS. VOL.24,. . . . .  254, (1968).
                                   ( 5 ,  734) PHYS. REV. LETT. 22 . . . . . .  531, (1969).
                                   ( 5 , 2106) J. PHYSIQUE TOME 32 COL.1 VOL.I;  575, (1971).

2441  LOYALKA.........(S.K.)..!   ( 5 , 1929) J. APPL. PHYS. VOL.45, . . . . . 2021, (1974).

2442  LUARSABISHVILI..(N.N.)..!   ( 5 , 1627) SOV. PHYS. DOKLADY VOL.18, . . .  432, (1973).

2443  LUBAN...........(M.)....!   ( 6 , 1157) PHYS. LETT. VOL.42A NO.2,. . . .  133, (1972).

2444  LUBKIN..........(G.B.)..!   ( 6 , 1692) PHYSICS TODAY (MARCH), . . . . .   17, (1973).

2445  LUCARI..........(F.)....!   ( 5 , 1790) PHYS. REV. VOL.187,. . . . . . .  611, (1969).
                                   ( 5 ,  478) SOL. STATE COMM. VOL.8,. . . . .    1, (1970).

2446  LUCAS...........(B.M.)..!   ( 5 , 1486) J. PHYS. C VOL.6,. . . . . . . .  201, (1973).

2447  LUCAS...........(B.)....!   (20C,   57) NUCL. INST. MET. 72, . . . . . .  307, (1969).

2448  LUDEWIG.........(J.)....!   (17B,   19) ACTA CRYST. A25, . . . . . . . .  116, (1969).

2449  LUDI............(A.)....!   ( 5 ,  776) J. CHEM. PHYSICS VOL.53, . . . . 1917, (1970).

2450  LUDMAN..........(C.J.)..!   ( 6 ,  567) J. CHEM. PHYS. VOL.52, . : . . . 2730, (1970).
                                   ( 6 , 1000) J.C.S. FARADAY TRANS. II VOL.69, 1477, (1973).

2451  LUDSTECK........(A.)....!   ( 5 ,  234) ACTA CRYST. VOL.A28, . . . . . .   59, (1972).

2452  LUDWIG..........(W.)....!   (10B,   25) ERGEB. EXAKT. NATURWISS. VOL.35,    1, (1964).
                                   (10A,   20) SPRINGER TRACTS IN MOD. PHYS. 43  232, (1967).

2453  LUGSCHEIDER.....(W.)....!   ( 5 ,  903) PHYS. STAT. SOL. B VOL.46, . . .  597, (1971).

2454  LUK-YANOV.......(A.A.)..!   (24 ,  109) SOV. AT. ENERGY VOL.25,. . . . .  435, (1968).

2455  LUNDGREN........(J.O.)..!   ( 5 , 1309) J. CHEM. PHYS. VOL.58, . . . . .  788, (1973).

2456  LUNDGREN........(L.)....!   ( 5 , 1105) J. PHYS. C VOL.7,. . . . . . . . 1344, (1974).

2457  LUNEBURG........(R.K.)..!   ( 7B,   60) PHYS. REV. VOL.76, . . . . . . . 1811, (1949).

2458  LUNGU...........(A.M.)..!   ( 6 ,  755) J. PHYS. CHEM. SOL. 28,. . . . . 1947, (1967).
                                   ( 6 ,  480) PHYS. LETT. 25A,. . . . . . . .  595, (1967).
                                   ( 6 ,  470) IAEA SYMP. COPENHAGEN, VOL.2,. .   75, (1968).
                                   ( 6 ,  759) J. APPL. PHYS. VOL.39, . . . . .  459, (1968).
                                   ( 6 ,  741) PHYS. LETT. VOL.26A,. . . . . .  396, (1968).
                                   ( 6 ,  805) ACTA CRYST. (INTERACT.) VOL.A25, S262, (1969).
                                   (13 ,   62) REPORT FN-40/70 (6PP.),. . . . .    ., (1970).
                                   ( 6 ,  808) SOL. STATE COMM. VOL. 9,. . . . .  353, (1971).

2459  LURIE...........(N.A.)..!   ( 7B,   26) J. CHEM. PHYSICS VOL.46,. . . . .  352, (1967).
                                   ( 6 ,  243) IAEA SYMP. COPENHAGEN, VOL.2,. .  205, (1968).
                                   ( 7B,   29) J. CHEM. PHYSICS VOL.49, . . . .  890, (1968).
                                   ( 6 ,  995) J. CHEM. PHYSICS VOL.54, . . . . 5193, (1971).
                                   ( 6 ,    5) J. CHEM. PHYS. VOL.55, . . . . . 4156, (1971).
                                   ( 6 , 1469) A.I.P. CONF. PROC. NO.10 PART 1,   93, (1973).
                                   (17A,   73) NUCL. INST. MET. VOL.112,. . . .  571, (1973).
                                   ( 6 , 2279) PHYS. LETT. VOL.46A NO.5,. . . .  357, (1974).
                                   ( 6 , 2281) PHYS. REV. B VOL.9 NO.12,. . . .    ., (1974).
                                   ( 6 , 2282) PHYS. REV. B VOL.9,. . . . . . . 2661, (1974).

2460  LUSCHER.........(E.)....!   ( 5 ,   92) PHYS LETT 27A,. . . . . . . . .  695, (1968).
                                   ( 6 ,   97) PHYS. LETT. 28A,. . . . . . . .  433, (1968).
                                   ( 6 , 1346) J. PHYS. CHEM. SOLIDS VOL.34,. .  255, (1973).
                                   ( 5 , 1949) PHYS. LETT. VOL.47A NO.1,. . . .   91, (1974).

2461  LUSHCHIKOV......(V.I.)..!   (12B,   31) PHYS. LETT. 12,. . . . . . . . .  334, (1964).
                                   ( 7A,   85) SOV. J. NUCL. PHYS. VOL.10,1,. .   47, (1969).
                                   (12B,   55) SOV. J. NUCL. PHYS. VOL.10,. . . 1178, (1969).
                                   (20B,  153) SOV. PHYS. JETP VOL.37,. . . . .   41, (1973).

2462  LUTZ............(G.)....!   ( 9C,   11) CHEM.-ING.-TECH. VOL.32, . . . .  651, (1960).

2463  LUTZ............(U.A.)..!   ( 6 ,   68) SOL. STATE COMM. VOL.8,. . . . .  165, (1970).

2464  LYASHCHENKO.....(B.G.)..!   ( 5 , 2140) SOV. PHYS. CRYST. VOL.3, . . . :  147, (1958).
                                   ( 5 , 1067) J. PHYS. SOC. JAPAN VOL.17 B-III   49, (1962).
                                   ( 5 , 2108) PHYS. MET. METALLOGR. VOL.14,. .   25, (1962).
                                   ( 9B,  104) SOV. PHYS. CRYST. VOL.6,. . . .  404, (1962).
                                   ( 5 ,  987) SOV. PHYS. CRYST. VOL.6,. . . .  443, (1962).
                                   ( 5 ,  986) SOV. PHYS. CRYST. VOL.7,. . . .  637, (1963).
                                   ( 5 ,  541) SOV. PHYS. CRYST. VOL.7,. . . .  639, (1963).
                                   ( 5 ,  970) SOV. PHYS. CRYST. VOL.8,. . . .  300, (1963).
                                   ( 6 , 1750) SOV. PHYS. J.E.T.P. VOL.17,. . .  584, (1963).
                                   (22A,   19) INSTRUM. EXP. TECH. NO.2,. . . .  306, (1964).
                                   ( 5 , 1120) PHYS. MET. METALLOGR. VOL.17 N.5  136, (1964).
                                   ( 5 ,  988) BULL. ACAD. SCI. USSR VOL.30,. . 1007, (1966).

2465  LYCKLAMA........(H.)....!   ( 5 , 2484) NUCL. PHYS. VOL.A139,. . . . . .  625, (1969).

2466  LYDEN...........(E.F.X.)!   ( 6 , 2010) J. CHEM. PHYS. VOL.50, . . . . . 2140, (1969).

2467  LYNCH-JR........(J.E.)..!   ( 6 ,  363) IAEA SYMP. COPENHAGEN, VOL.2,. .  167, (1968).
                                   ( 6 , 1911) J. CHEM. PHYSICS VOL.48, . . . .  912, (1968).

2468  LYNN............(J.W.)..!   ( 5 , 2825) PROC. PHYS. SOC. 82, . . . . . .  477, (1963).
                                   ( 5 ,  984) MAGN. AND MAG. MATERIALS-1971; 1415, (1972).
                                   ( 6 ,  111) PHYS. REV. B VOL.8, . . . . . . 3493, (1973).
                                   ( 6 , 1739) PHYS. REV. LETT. VOL.30, . . . .  556, (1973).

2469  LYO.............(S.K.)..!   (11A,   47) PHYS. REV. LETT. VOL.28,. . . . 1192, (1972).

2470  LYUBIMTSEV......(V.A.)..!   ( 5 ,  168) SOV. PHYS. CRYST. VOL.16,. . . .  711, (1972).

2471  LYUBOSHITZ......(V.L.)..!   ( 7B,  128) SOV. PHYS. JETP. 21, . . . . . .  765, (1965).

2472  MAAYOUF.........(R.)....!   (22A,   37) NUCL. INST. MET. 40, . . . . . .  153, (1966).

2473  MACCHESNEY......(J.B.)..!   ( 5 ,  360) PHYS. REV. VOL.188,. . . . . . .  930, (1969).

2474  MACDONALD.......(A.C.)..!   ( 5 ,  385) ACTA CRYST. B25, . . . . . . . . 1804, (1969).
                                   ( 5 ,  386) ACTA CRYST. VOL.A25,. . . . . .  S76, (1969).

2475  MACDONALD.......(A.L.)..!   ( 9C,    4) ACTA CRYST. VOL.B27, . . . . . . 2289, (1971).

2476  MACDONALD.......(R.A.)..!   ( 7B,   36) J. PHYS.F: METAL PHYS. VOL.2,. .  209, (1972).
```

```
2477  MACKELLAR........(A.D.)...!  ( 7A,    38) NUCL. PHYS. A VOL.178, . . . . .    249, (1971).

2478  MACKINTOSH......(A.R.)...!  ( 6 ,   516) IAEA SYMP. BOMBAY, VOL.1,. . . .     95, (1964).
                                  ( 6 ,   524) PROC. INT. CONF. (NOTTINGHAM),.    101, (1964).
                                  ( 6 ,   523) SOL. STATE COMM. 2,. . . . . . .    105, (1964).
                                  ( 6 ,   555) PHYS. REV. LETT. 25,. . . . . .    627, (1965).
                                  ( 5 ,   638) SOL. STATE COMM. VOL.3,. . . . .   157, (1965).
                                  ( 5 ,   598) J. APP. PHYS. 37,. . . . . . . .  1259, (1966).
                                  ( 5 ,   597) PHYS. REV. VOL.151,. . . . . . .   405, (1966).
                                  ( 6 ,  2150) PHYS. REV. LETT. 19,. . . . . .    312, (1967).
                                  ( 6 ,  2153) J. APP. PHYS. 39,. . . . . . . .   807, (1968).
                                  (10C,    91) PROC/1ST INT. CONF/LOC. EXC/SOL.   721, (1968).
                                  ( 6 ,  2124) J. APPL. PHYS. 41,. . . . . . .   1174, (1970).
                                  ( 6 ,  2151) PHYS. REV. LETT. VOL.25,. . . .   1451, (1970).
                                  ( 5 ,  2336) PHYS. REV. LETT. VOL.25,. . . .    524, (1970).
                                  ( 6 ,  2132) IAEA SYMP. GRENOBLE, . . . . .     603, (1972).

2479  MACKLIN.........(R.L.)...!  (20C,    67) NUCL. INST. MET. 74,. . . . . .    224, (1969).
                                  (20A,    80) REV. SCI. INSTRUM. VOL.42, . . .   240, (1971).

2480  MACPHERSON......(R.W.)...!  ( 6 ,  1260) PHYS. REV. 182,. . . . . . . .     965, (1969).
                                  ( 6 ,  1259) PHYS. REV. B VOL.1,. . . . . .    4193, (1970).

2481  MACVEAN.........(C.R.)...!  (16 ,    14) DISS. ABS. VOL.25, . . . . . .    6014, (1965).

2482  MADARASZ........(Z.)....!   (20C,    33) MAGYAR FIZ. FOLYOIRAT (HUNG.) 16   279, (1968).

2483  MADAR...........(R.)....!   ( 5 ,  1861) J. PHYSIQUE TOME 32 COL.1 VOL.II   876, (1971).
                                  ( 5 ,  1741) SOL. STATE COMM. VOL. 9, . . . .  1793, (1971).

2484  MADHAV-RAO......(L.)....!   ( 6 ,   157) J. PHYS. CHEM. SOLIDS VOL.23,. .  1747, (1962).
                                  ( 5 ,  1999) PROC. INT. CONF. (NOTTINGHAM),.    285, (1964).
                                  ( 8 ,   117) PROG. LOW TEMP. PHYS. VOL.5,. .    161, (1967).
                                  ( 6 ,   816) SOL. STATE COMM. 6,. . . . . .     593, (1968).
                                  ( 6 ,  1496) SOL. STATE COMM. 7,. . . . . .     123, (1969).
                                  ( 5 ,  2677) J. PHYSIQUE TOME 32 COL.1 VOL.II   617, (1971).
                                  ( 6 ,  2082) J. PHYSIQUE TOME 32 COL.1 VOL.I,   318, (1971).
                                  ( 5 ,   522) NUCL./S.S. PHYS. SYMP. ABSTR.,.    , (1972).
                                  ( 6 ,  1482) PHYS. LETT. VOL.40A,. . . . . .    101, (1972).
                                  (11C,    45) PHYS. STAT. SOL. VOL.50,. . . .    737, (1972).

2485  MADHAV..........(L.)....!   (12A,    20) NUCL./S.S. PHYS. SYMP. ABSTR.,. . . . , (1972).

2486  MAELAND.........(A.J.)...!  ( 5 ,  2292) CAN. J. PHYS. 46,. . . . . . .     121, (1968).
                                  ( 5 ,   170) J. CHEM. PHYSICS VOL.48, . . . .  4660, (1968).
                                  ( 6 ,   123) J. CHEM. PHYSICS VOL.51, . . . .  2915, (1969).
                                  ( 6 ,   122) J. CHEM. PHYS. VOL.52, . . . . .  3952, (1970).
                                  ( 6 ,  1355) J. CHEM. PHYS. VOL.54, . . . . .  3979, (1971).

2487  MAELAND.........(A.)....!   ( 5 ,  2306) PLATINUM METALS REV. 10, . . . .    20, (1966).

2488  MAGLIC..........(R.)....!   ( 6 ,   879) COLL. INTER. N.126 (GRENOBLE), .   210, (1963).

2489  MAHENDRA........(A.)....!   ( 8 ,    36) J. PHYS. C VOL.4,. . . . . . .    3307, (1971).
                                  ( 5 ,   482) PHYS. REV. B VOL.4,. . . . . .    3901, (1971).
                                  ( 8 ,    52) NUCL./S.S. PHYS. SYMP. ABSTR.,. . . , (1972).

2490  MAHESH..........(P.S.)...!  ( 6 ,  2277) PHYS. REV. 143,. . . . . . . .     443, (1966).
                                  (10B,   198) PHYS. STAT. SOLIDI B VOL.59,. .    279, (1973).

2491  MAHLAV..........(M.)....!   (20A,    46) NUCL. INST. MET. VOL.102,. . . .    87, (1972).

2492  MAHONEY.........(J.V.)...!  ( 5 ,   393) PHYS. REV. B VOL.9,. . . . . .     154, (1974).

2493  MAIER-LEIBNITZ..(H.)....!   (23 ,    25) IAEA SYMP. (PILE RESEARCH)VIENNA    63, (1960).
                                  (22A,    72) Z. ANGEW. PHYS. VOL.14,. . . . .   738, (1962).
                                  (22A,    74) Z. PHYSIK VOL.167, . . . . . .     386, (1962).
                                  (20A,    24) J. NUCL. ENERGY VOL.17,. . . . .   217, (1963).
                                  (20A,    12) ANN. REV. OF NUCLEAR SCI. VOL.16   207, (1966).
                                  (17A,    78) NUKLEONIK VOL.8, . . . . . . .      61, (1966).
                                  ( 9B,    24) ANN. ACAD. SCI. FENNICAE A 267,.     3, (1967).
                                  (20D,    41) I.A.E.A. SYMP., STUDSVIK,. . . .    93, (1969).
                                  (18B,     5) IAEA SYMP. GRENOBLE, . . . . .     681, (1972).

2494  MAIER...........(G.)....!   ( 5 ,    78) COLL. INTER. N.126 (GRENOBLE), .    23, (1963).
                                  ( 5 ,  2460) NATURWISSENSCHAFTEN VOL.53,. . .    16, (1966).
                                  ( 6 ,  1737) PHYS. LETT. 24A, . . . . . . .     625, (1967).
                                  ( 6 ,  1714) PHYS. LETT. 29A, . . . . . . .      75, (1969).
                                  ( 5 ,  2648) Z. ANGEW. PHYS. VOL.27,. . . . .    73, (1969).
                                  ( 6 ,  1719) Z. PHYS. VOL.238,. . . . . . .     389, (1970).

2495  MAIER...........(U.)....!   ( 5 ,   972) PHYS. KOND. MATER. VOL.17, . . .    11, (1973).

2496  MAINO...........(G.)....!   (22B,    37) NUCL. INST. MET. VOL.115,. . . .   393, (1974).

2497  MAIR............(S.L.)..!   ( 6 ,  2084) PHYS. LETT. VOL.35A, . . . . .     286, (1971).

2498  MAISTRENKO......(A.N.)..!   (14B,   136) UKR. FIZ. ZH. (USSR) VOL.9,. .     684, (1964).
                                  ( 5 ,  2229) SOV. AT. ENERGY VOL.18,. . . .     585, (1965).
                                  (20C,    24) INSTRUM. EXP. TECH. VOL.14, NO.5 1343, (1971).

2499  MAITA...........(J.P.)...!  ( 5 ,   838) PHYS. REV. LETT. VOL.28, . . . .   746, (1972).
                                  ( 6 ,  1953) PHYS. REV. B VOL.8,. . . . . .    5345, (1973).

2500  MAKAROV.........(V.A.)...!  ( 5 ,  2856) SOV. PHYS. CRYST. VOL.17,. . . .  1017, (1973).

2501  MAKHMUDOV.......(Z.Z.)...!  ( 8 ,   150) SOV. PHYS. SOL. STATE VOL.12,. .  1596, (1971).

2502  MAKHURANE.......(P.M.)...!  ( 5 ,   122) J. PHYS. F VOL.3,. . . . . . .    1054, (1973).

2503  MAKOVETSKII.....(G.I.)...!  ( 5 ,  1829) SOV. PHYS. DOKLADY VOL.11, . . .   888, (1967).

2504  MAKOVSKY........(J.)....!   ( 5 ,  1987) PHYS. REV. 174,. . . . . . . .     560, (1968).
                                  ( 5 ,  2403) PHYS. REV. B VOL.2,. . . . . .    3703, (1970).
                                  ( 5 ,  1964) PHYS. REV. B VOL.2,. . . . . .     179, (1970).
                                  ( 5 ,  2401) PHYS. REV. LETT. VOL.25,. . . .   1713, (1970).
                                  ( 5 ,   664) BU. ISRAEL PHYS. SOC. 1971,. .      27, (1971).
                                  ( 5 ,  2676) PHYS. REV. B VOL.3,. . . . . .    2344, (1971).
                                  ( 5 ,   663) PHYS. REV. B VOL.3,. . . . . .    3873, (1971).
                                  ( 5 ,  2402) PHYS. REV. B VOL.3,. . . . . .     821, (1971).
                                  ( 5 ,  1933) SOL. STATE COMM. VOL. 9, . . . .   493, (1971).
                                  ( 6 ,   803) PHYS. REV. B VOL.5,. . . . . .    2607, (1972).
                                  ( 5 ,  1557) PHYS. REV. B VOL.5,. . . . . .    1968, (1972).
                                  ( 5 ,  1440) PHYS. REV. B VOL.9,. . . . . .    1071, (1974).

2505  MAKSIMYUK.......(P.A.)...!  ( 5 ,    75) LATV. PSR...FIZ. TEHN. SER. NO.1    88, (1970).

2506  MALEEV..........(S.V.)...!  ( 8 ,   128) SOV. PHYS. JETP VOL.6, . . . . .   776, (1958).
                                  (11B,    36) SOV. PHYS. JETP VOL.7, . . . .    1048, (1958).
                                  ( 7B,   117) SOV. PHYS. JETP VOL.7, . . . . .    89, (1958).
                                  ( 8 ,   131) SOV. PHYS. JETP VOL.12,. . . . .   995, (1961).
                                  ( 8 ,   130) SOV. PHYS. JETP VOL.13,. . . . .   860, (1961).
                                  (12A,    39) SOV. PHYS. JETP VOL.14,. . . . .  1168, (1962).
                                  ( 9D,    70) SOV. PHYS. SOL. STATE, 4, . . .   2533, (1962).
                                  (11B,    39) SOV. PHYS. SOL. STATE, 5,. . . .   858, (1963).
                                  (10C,   115) SOV. PHYS. SOL. STATE, 6,. . . .  2162, (1964).
```

2506   MALEEV..........(S.V.)..: (8 ,   142) SOV. PHYS. JETP LETT. VOL.2, . .   338, (1965).
                                (11B,   40) SOV. PHYS. JETP. 21, . ; . .   969, (1965).
                                (12A,   40) SOV. PHYS. SOL. STATE, 7,.: . 2477, (1965).
                                (9D,   76) SOV. PHYS. SOL. STATE, 8,. . . 1852, (1966).
                                (11B,   44) SOV. PHYS. SOL. STATE, 9,. . .   575, (1967).
                                (12B,   61) SOV. PHYS. JETP LETT. VOL.10,. .   345, (1969).
                                (12B,   63) SOV. PHYS. J.E.T.P. VOL.31,. .   111, (1970).
                                (13 ,   44) PHYS. LETT. VOL.37A, ...   406, (1971).
                                (12B,   67) SOV. PHYS. JETP VOL.35,.: . . .   222, (1972).

2507   MALETSKI........(K.H.)..: (5 , 2433) SOV. J. NUCL. PHYS. VOL.9, 6,. . 1119, (1969).

2508   MALIK...........(S.S.)..: (5 , 2972) BULL. AMER. PHYS. SOC. VOL.17, .   125, (1972).
                                (5 , 2528) PHYS. REV. B VOL.8,. . . . . 2595, (1973).

2509   MALINOSKI.......(F.)....: (8 ,   24) J. APPL. PHYS. 41, ...   1365, (1970).

2510   MALINOWSKI......(K.)....: (13 ,   47) PHYS. LETT. VOL.41A NO.2,. . .   175, (1972).

2511   MALISZEWSKI.....(E.F.)..: (6 , 2312) PHYS. LETT. 1,. . . . . .   338, (1962).
                                (6 , 1854) J. PHYS.-F: METAL PHYS. VOL.1,:   339, (1971).
                                (6 , 2012) PHYS. STAT. SOL. VOL.44, . . .   K65, (1971).

2512   MALISZEWSKI.....(E.)....: (20B,   27) IAEA SYMPOSIUM VIENNA, ...   447, (1960).
                                (20A,   34) NUCL. INST. MET. VOL.9, . . .   341, (1961).
                                (6 , 2317) IAEA SYMP. CHALK RIVER, VOL.2,.   87, (1962).
                                (6 , 2309) COPENHAGEN CONF.-LATT.DYNAMICS,.   33, (1963).
                                (21 ,   15) IAEA SYMP. COPENHAGEN, VOL.2,.   313, (1968).
                                (19A,   45) NUCL. INST. MET. 69, ...   173, (1969).
                                (6 , 2013) IAEA SYMP. GRENOBLE, ...   61, (1972).

2513   MALKOV..........(E.I.)..: (9B,   108) SOV. PHYS. JETP. 22, ... 1069, (1966).

2514   MALLETT.........(J.F.)..: (6 ,   830) PROC. PHYS. SOC. 84, ...   55, (1964).

2515   MALM............(W.C.)..: (6 ,   378) THESIS (173 PP.),. . . . . .   (1972).

2516   MALYSHEV........(E.K.)..: (20C,   18) INSTRUM. EXP. TECH. NO.1,. . .   87, (1966).

2517   MAL#TSEV........(E.I.)..: (5 , 2876) SOV. PHYS. CRYST. VOL.7, . . .   767, (1963).
                                (22A,   69) SOV. PHYS. CRYST. VOL.8, . . .   234, (1963).
                                (5 , 2271) BULL. ACAD. SCI. USSR VOL.28,. .   350, (1964).
                                (5 ,   218) PROC. INT. CONF. (NOTTINGHAM),.   354, (1964).
                                (5 , 2270) SOV. PHYS. SOL. STATE VOL.7, . .   997, (1965).
                                (6 , 2291) SOV. PHYS. SOL. STATE VOL.11,. .   511, (1968).
                                (5 , 2143) SOV. PHYS. DOKLADY VOL.15, . .   874, (1971).

2518   MANCINI.........(F.)....: (5 , 2022) PHYS. REV. B VOL.9,. . . . .   130, (1974).

2519   MANDELLI-BETTONI(M.)....: (20C,   52) NUCL. INST. MET. 65, ...   110, (1968).
                                (24 ,   63) NUCL. INST. MET. 70, ...   52, (1969).

2520   MANDUCHI........(C.)....: (20C,   66) NUCL. INST. MET. 67,. . . .   267, (1969).
                                (5 ,   231) NUCL. PHYS. A VOL.A181,. . .   177, (1972).

2521   MANDUCHI........(M.T.R.): (5 ,   231) NUCL. PHYS. A VOL.A181,. . . .   177, (1972).

2522   MANDZHAVIDZE....(A.G.)..: (5 , 2529) SOV. PHYS. SOL. STATE VOL.15,. . 1295, (1973).

2523   MANIAWSKI.......(F.)....: (6 ,   415) ACTA PHYS. POLON. VOL.16,. . .   335, (1957).
                                (6 ,   301) ACTA PHYS. POLON. VOL.17,. . .   489, (1958).
                                (5 , 1286) IAEA SYMPOSIUM VIENNA, ...   297, (1960).
                                (6 ,   974) NUKLEONIKA VOL.5,. . . . . .   495, (1960).
                                (6 , 1552) IAEA SYMP. CHALK RIVER, VOL.1,.   405, (1962).
                                (5 ,   269) NUKLEONIKA VOL.8,. . . 3,. . .   581, (1963).
                                (5 ,   460) PHYS. STAT. SOL. VOL.38, . . .   103, (1970).
                                (5 , 2203) PHYSICA VOL.51,. . . . . .   627, (1971).

2524   MANI............(K.K.)..: (6 , 1601) IAEA SYMP. BOMBAY, VOL.1,. . .   77, (1964).

2525   MANLEY..........(J.H.)..: (5 , 1255) PHYS. REV. VOL.52, ... 1076, (1937).
                                (5 , 1250) PHYS. REV. VOL.54, ...   266, (1938).

2526   MANNHART........(W.)....: (5 ,   438) Z. PHYSIK VOL.264, ...   427, (1973).
                                (5 , 2831) Z. PHYSIK VOL.266, ...   157, (1974).

2527   MANNHEIM........(P.P.)..: (10B,   141) PHYS. REV. 165,. . . . . . 1011, (1968).

2528   MANN............(D.)....: (22B,   35) NUCL. INST. MET. VOL.91, . . .   541, (1971).

2529   MANOHAR.........(C.)....: (11C,   17) J. PHYS. -C-, VOL. 1,. . . . 1706, (1968).
                                (11A,   50) PHYS. REV. B VOL.5,. . . . . 1993, (1972).
                                (6 , 1271) PHYS. REV. B VOL.7,. . . . . 1128, (1973).

2530   MANSMANN........(M)....: (5 , 1381) COLL. INTER. N.126 (GRENOBLE), .   30, (1963).

2531   MANSON..........(R.)....: (10B,   146) PHYS. REV. LETT. 22, ...   346, (1969).

2532   MANTE...........(A.J.H.): (10D,   57) SOL. STATE COMM. VOL.8,. . . . 1415, (1970).

2533   MARACCI.........(G.)....: (22C,   41) NUCL. INST. MET. 62, ...   57, (1968).

2534   MARADUDIN.......(A.A.)..: (10E,   24) PHYS. REV. VOL.128,. . . . . 2589, (1962).
                                (10B,   1) AARHUS SUMMER SCHOOL,. . . . .   424, (1963).
                                (10C,   35) IAEA SYMPOSIUM BOMBAY, VOL.1,.   231, (1964).
                                (10E,   26) PHYS. REV. VOL.135,. . . . . A1071, (1964).
                                (10B,   223) SOV. PHYS. SOL. STATE, 8,. . .   850, (1966).

2535   MARCHIK.........(I.I.)..: (6 , 2291) SOV. PHYS. SOL. STATE VOL.10,. .   511, (1968).
                                (12A,   43) SOV. PHYS. SOL. STATE VOL.13,. .   258, (1971).

2536   MARCH...........(N.H.)..: (14A,   61) PROC. ROY. SOC. A VOL.282, . .   283, (1964).
                                (14A,   2) ADV. IN PHYSICS VOL.14,. . . .   453, (1965).
                                (14A,   4) ANN. OF PHYS. VOL.81,. . . .   414, (1973).
                                (6 , 1961) PROPERTIES LIQUID METALS,. . .   99, (1973).

2537   MARCH...........(R.H.)..: (6 , 1629) PHYS. REV. VOL.128,. . . . . 1112, (1962).
                                (6 , 1621) PROC. PHYS. SOC. 79, ...   440, (1962).

2538   MARCINKOWSKI....(M.J.)..: (5 , 2160) J. APPL. PHYS. VOL.32, . . .   375, (1961).

2539   MARCUS..........(J.A.)..: (5 , 1619) PHYS LETT 28A, ...   267, (1968).
                                (5 , 1617) PHYS. REV. B-2,. . . . . .   670, (1970).

2540   MARESCHAL.......(J.)....: (5 ,   579) PROC. INT. CONF. (NOTTINGHAM), .   275, (1964).
                                (5 , 2560) IEEE TRANS. MAGNETICS MAG-2, . .   453, (1966).
                                (5 ,   591) J. APPL. PHYS. VOL.37,. . . . 1038, (1966).
                                (5 , 2559) J. PHYS. CHEM. SOL. 28,. . . . 2143, (1967).
                                (5 , 2571) SOL. STATE COMM. 5,. . . . .   293, (1967).
                                (5 , 2039) SOL. STATE COMM. 5,. . . . .   93, (1967).
                                (5 , 2539) Z. ANGEW. PHYS. VOL.23,. . . .   243, (1967).
                                (5 , 2570) J. APPL. PHYS. VOL.39,. . . . 1364, (1968).
                                (5 ,   824) J. PHYSIQUE TOME 29,. . . . .   67, (1968).
                                (5 , 1532) SOL. STATE COMM. VOL.6,. . . .   751, (1968).
                                (5 , 2385) J. PHYSIQUE TOME 30,. . . . .   967, (1969).
                                (5 , 1533) SOL. STATE COMM. VOL.7 NO.22,. . 1669, (1969).

```
2540  MARESCHAL.........(J.)....:  (  5,    874) J. PHYSIQUE TOME 31, . . . . . .    607, (1970).

2541  MARINESCU........(L.)....:  (20B,   138) REV. ROUMAINE PHYS. VOL.11,. . .    601, (1966).

2542  MARIS............(H.J.)..:  (10E,    23) PHYS. LETT. 17,. . . . . . .        228, (1965).
                                  (  6,  1131) PHYS. REV. LETT. VOL.25, . . . .    220, (1970).

2543  MARKOVIC.........(V.)....:  (  6,    879) COLL. INTER. N.126 (GRENOBLE), .    210, (1963).

2544  MARKOVNIN........(V.I.)..:  (  6,   1671) SOV. PHYS. SOL. STATE VOL. 13, .   1793, (1972).

2545  MARKOV...........(YU.YA.):  (20C,    22) INSTRUM. EXP. TECH. 4, . . . . .   1013, (1970).

2546  MARKWORTH........(A.J.)..:  (  8,    19) DISS. ABS. VOL.30, . . . . . . .  47558, (1970).

2547  MARK.............(H.)....:  (24,     17) IAEA SYMP. (PILE RESEARCH)VIENNA    609, (1960).
                                  ( 7A,    37) NUCL. PHYS. A VOL.A169,. . . . .    385, (1971).

2548  MARNEY...........(M.C.)..:  ( 9B,    75) PHYS. REV. VOL.73, . . . . . . .    527, (1948).

2549  MARSEGUERRA......(M.)....:  (23,     21) IAEA SYMP. (PILE RESEARCH)VIENNA    123, (1960).

2550  MARSHALL.........(G.D.)..:  (  6,    311) PHYS. REV. VOL.125,. . . . . . .    933, (1962).

2551  MARSHALL.........(J.)....:  (20A,    61) PHYS. REV. VOL.72, . . . . . . .    193, (1947).

2552  MARSHALL.........(L.)....:  (20D,    28) PHYS. REV. VOL.70, . . . . . . .    815, (1946).
                                  ( 7B,    57) PHYS. REV. VOL.71, . . . . . . .    666, (1947).
                                  (  8,    70) PHYS. REV. VOL.72, . . . . . . .   1139, (1947).
                                  (20A,    61) PHYS. REV. VOL.72, . . . . . . .    193, (1947).
                                  (  5,    213) PHYS. REV. VOL.72, . . . . . . .    408, (1947).
                                  (  5,    771) PHYS. REV. VOL.75, . . . . . . .    578, (1949).

2553  MARSHALL.........(R.)....:  (10B,    45) J. CHEM. PHYSICS VOL.41, . . . .   3158, (1964).

2554  MARSHALL.........(W.)....:  (  8,    112) PROC. PHYS. SOC. A VOL.67, . .      85, (1954).
                                  (  8,    54) NUOVO CIMENTO SUPPL. VOL.6,. .     1183, (1957).
                                  (  8,    120) REV. MOD. PHYS. VOL.30, . . .       75, (1958).
                                  ( 7B,    16) IAEA SYMPOSIUM VIENNA, . . . .       75, (1960).
                                  (  5,   1143) J. APPL. PHYS. VOL.31, SUPPL.,     356, (1960).
                                  (10C,    33) IAEA SYMPOSIUM BOMBAY, VOL.1,.     399, (1964).
                                  ( 9D,    79) TRANSITION METAL COMPOUNDS,. .       29, (1964).
                                  ( 9D,    64) PROC. PHYS. SOC. 86, . . . . .     561, (1965).
                                  (13,     35) NBS MISC. PUB.273, . . . . . .     135, (1966).
                                  (  8,    114) PROC. PHYS. SOC. VOL.89, . . .     613, (1966).
                                  (11C,    48) PROC. PHYS. SOC. 92, . . . . .     390, (1967).
                                  (  6,   2137) SOL. STATE COMM. 5, . . . . .      607, (1967).
                                  ( 9D,    81) TRIESTE/INTERN. COURSE,. . . .     501, (1967).
                                  (11A,     7) C.N.E.N. SYMP. CASACCIA, . . .     101, (1968).
                                  (  6,   2134) J. APP. PHYS. 39, . . . . . .     1229, (1968).
                                  ( 9D,    40) J. PHYS. -C-, VOL. 1,. . . . .     966, (1968).
                                  ( 9D,    39) J. PHYS. -C- VOL. 1, . . . . .      88, (1968).
                                  (11B,    34) REPORTS ON PROGRESS IN PHYS. 31,   705, (1968).
                                  (24,     19) COMMENTS SOLID STATE PHYS. VOL.2    88, (1969).
                                  (11A,    17) J. PHYS. C VOL.2, . . . . . . .    539, (1969).
                                  (14B,    115) RIV. DEL NUOVO CIMENTO VOL.1,. .   155, (1969).

2555  MARSILI..........(F.)....:  (23,     65) NUOVO CIMENTO SUPPL. VOL.23, . .     17, (1962).

2556  MARSONGKOHADI...........:  (  6,    446) IAEA SYMP. GRENOBLE, . . . . . .     95, (1972).

2557  MARTEL...........(P.)....:  (  6,    493) PHYS. REV. LETT. 18, . . . . .     162, (1967).
                                  (  6,    496) CAN. J. PHYS. 46, . . . . . .     1355, (1968).
                                  (  5,   1948) CAN. J. PHYS. 46, . . . . . .     1727, (1968).
                                  (  6,    494) J. APP. PHYS. 39, . . . . . .     1116, (1968).
                                  (  6,   1686) PHYS. REV. 171, . . . . . . .      727, (1968).
                                  (  6,    492) PHYS. REV. LETT. VOL.23, . . .      86, (1969).
                                  (21,     48) REV. SCI. INSTRUM. 41, . . . .      61, (1970).
                                  (  6,    425) BULL. AMER. PHYS. SOC. VOL.17,     291, (1972).
                                  (  6,   1437) CAN. J. PHYS. 50, . . . . . .      687, (1972).
                                  (  6,    423) IAEA SYMP. GRENOBLE, . . . . .     207, (1972).
                                  (  6,   1164) IAEA SYMP. GRENOBLE, . . . . .     359, (1972).
                                  (  6,   1672) IAEA SYMP. GRENOBLE, . . . . .      43, (1972).
                                  (  6,   1167) PHYS. REV. LETT. VOL.29, . . .     1148, (1972).
                                  (  6,    497) J. PHYS. C VOL.6, . . . . . .     2997, (1973).
                                  (  6,   1171) PHYS. LETT. VOL.43A NO.3,. . .     223, (1973).

2558  MARTINEZ.........(P.)....:  (20A,    49) NUCL. INST. MET. VOL.116,. . . .    615, (1974).

2559  MARTINHO.........(E.)....:  (  5,   1287) NUCL. SCI. ENG. VOL.45,. . . . .    308, (1971).

2560  MARTIN...........(D.G.)..:  (10C,    60) PHIL. MAG. VOL.5,. . . . . . .     1235, (1960).
                                  ( 9C,     7) APPL. MATERIALS RES. VOL.1,. .      160, (1962).
                                  ( 9C,    17) INTERN. SUMMER SCHOOL, MOL,. .      643, (1963).
                                  ( 9A,    54) J. PHYS. CHEM. SOL. 25, . . .      1005, (1964).
                                  (  6,    207) PHIL. MAG. VOL.9, . . . . . .      659, (1964).
                                  (  6,    158) PROC. BRIT. CERAM. SOC. NO.7,.      391, (1967).
                                  ( 7B,    137) UKAEA AERE REP.R 5479 (35 PP.),.        (1967).
                                  (  5,   1611) PROC. PHYS. SOC. SER.2, 1, . .     333C, (1968).

2561  MARTIN...........(F.D.)..:  (  5,   2237) PROC/CONF/NEUTRON C. S. + TECH.,    851, (1968).

2562  MARTIN...........(R.M.)..:  (10B,    38) IAEA SYMP. COPENHAGEN, VOL.1,. .    119, (1968).
                                  (  6,   2030) PHYS. REV LETT, 21, . . . . . .    536, (1968).

2563  MARTIN...........(R.)....:  (22B,    30) NUCL. INST. MET. 60, . . . . . .    237, (1968).

2564  MARTON...........(M.)....:  (  5,   1778) J. PHYS. CHEM. SOLIDS VOL.33, .     212, (1972).
                                  (  5,   1775) SOL. STATE COMM. VOL.10, . . .     1195, (1972).

2565  MARUKHIN.........(V.I.)..:  (17A,    46) INSTRUM. EXP. TECH.(USA) VOL.14,    733, (1971).

2566  MARX.............(D.)....:  (20A,    43) NUCL. INST. MET. VOL.94, . . .      533, (1971).
                                  (18A,    28) Z. ANGEW. PHYS.(GER.) VOL.32,.      296, (1971).

2567  MASAKI...........(N.)....:  (  5,   2763) ACTA CRYST B24, . . . . . . .     1393, (1968).
                                  ( 9B,    46) J. APPL. CRYST. VOL.4, . . . .      528, (1971).
                                  (  5,   1217) ACTA CRYST. VOL.A28 PART S-4,.      220, (1972).
                                  (  5,   2765) ACTA CRYST. VOL.B28, . . . . .      785, (1972).
                                  (  6,   1924) J. POLYM. SCI./POLYM. LETT. 11,.    377, (1973).
                                  (  5,   2780) J. APPL. CRYST. VOL.7, . . . .      247, (1974).

2568  MASLEN...........(E.N.)..:  (  5,    279) NATURE VOL.192,. . . . . . .        154, (1961).
                                  (  5,   2875) ACTA CRYST., 19, . . . . . .        679, (1965).
                                  (  5,   1913) ACTA CRYST. 22, . . . . . . .       134, (1967).
                                  (  5,    27) ACTA CRYST. 22, . . . . . . .        216, (1967).
                                  ( 9A,     8) AT. ENERGY AUST. VOL.12, . . .         2, (1969).
                                  (  5,   2627) ACTA CRYST. VOL.B26, . . . . .      693, (1970).

2569  MASLEN...........(V.W.)..:  (  5,   2771) PROC. ROY. SOC. 298, . . . . .      289, (1967).

2570  MASLIN...........(E.E.)..:  (22C,    42) NUCL. INST. MET. 71, . . . . . .     13, (1969).
```

```
2623  MCKENZIE........(D.R.)..:  (10B,    66) J. PHYS. C, SER.2, VOL.4,. . . .  2304, (1971).

2624  MCKEOWN.........(M.L.)..:  (22C,    48) REV. SCI. INSTRUM. VOL.25, . . .   699, (1954).

2625  MCLEAN..........(F.B.)..:  ( 8 ,   105) PHYS. REV. B VOL.7,. . . . . . .  1149, (1973).
                                 ( 8 ,   103) PHYS. REV. B VOL.7,. . . . . . .  5017, (1973).

2626  MCMILLAN........(W.L.)..:  ( 6 ,  2192) PHYS. REV. 178,. . . . . . . .     897, (1969).
                                 ( 6 ,  2062) PHYS. REV. B VOL.3,. . . . . . .  4065, (1971).

2627  MCMURRAY........(W.R.)..:  (24 ,    42) NUCL. INST. MET. 34, . . . . . .    29, (1965).

2628  MCMURRY.........(H.L.)..:  (14B,    57) NUCL. SCI. ENGNG. VOL.15,. . . .   429, (1963).
                                 (14B,    56) NUCL. SCI. ENGNG. VOL.15,. . . .   438, (1963).
                                 ( 6 ,   265) PHYS. REV. VOL.136,. . . . . . .  A106, (1964).
                                 ( 6 ,  1058) NUCL. SCI. ENGNG. VOL.25,. . . .   248, (1966).
                                 ( 6 ,  1939) J. CHEM. PHYS. VOL.52,. . . . .   4424, (1970).

2629  MCPHEE..........(R.C.)..:  (22C,    26) J. SCI. INSTRUM. SER.2 VOL.1,. .   367, (1968).

2630  MCREYNOLDS......(A.W.)..:  ( 5 ,  2829) PHYS. REV. VOL.83,. . . . . . .    171, (1951).
                                 ( 5 ,  2231) PHYS. REV. VOL.84,. . . . . . .    969, (1951).
                                 (20B,   127) PHYS. REV. VOL.88,. . . . . . .    958, (1952).
                                 ( 6 ,  1823) PHYS. REV. VOL.91,. . . . . . .   1368, (1953).
                                 ( 5 ,  1083) PHYS. REV. VOL.95,. . . . . . .   1161, (1954).
                                 ( 6 ,  2356) PHYS. REV. VOL.108,. . . . . . .  1092, (1957).
                                 (13 ,    31) J. PHYS. RADIUM VOL.20,. . . . .   175, (1959).
                                 ( 6 ,  1834) PHYS. REV. VOL.113,. . . . . . .   767, (1959).
                                 ( 5 ,  1261) PHYS. REV. VOL.113,. . . . . . .   806, (1959).
                                 (20B,    25) IAEA SYMPOSIUM VIENNA. . . . . .   421, (1960).
                                 ( 6 ,  1030) IAEA SYMPOSIUM VIENNA. . . . . .   511, (1960).
                                 (20B,    22) IAEA SYMP. (PILE RESEARCH)VIENNA  469, (1960).
                                 ( 6 ,   949) IAEA SYMP. CHALK RIVER, VOL.1, .   263, (1962).

2631  MCTAGUE.........(J.P.)..:  ( 6 ,   967) PHYS. REV. LETT. VOL.30, . . . .   481, (1973).

2632  MCWHAN..........(D.B.)..:  (22C,    67) REV. SCI. INSTRUM. VOL.45, . . .   643, (1974).

2633  MEADOWS.........(J.W.)..:  (19A,    43) NUCL. INST. MET. 75,. . . . . .    163, (1969).
                                 ( 5 ,  1546) NUCL. SCI. ENGNG. VOL.40,. . . .    12, (1970).

2634  MEAD............(S.W.)..:  ( 5 ,  1565) J. PHYS. CHEM. SOLIDS VOL.24,. .  1066, (1963).

2635  MEAD............(W.)...:   ( 5 ,  1143) J. APPL. PHYS. VOL.31, SUPPL.. .   356, (1960).
                                 ( 5 ,  1140) J. PHYS. SOC. JAPAN VOL.17 B-I,.   249, (1962).

2636  MEAKER..........(C.L.)..:  ( 6 ,  2112) PHYS. REV. VOL.71,. . . . . . .    165, (1947).

2637  MEDHI-ALI.......(S.)...:   ( 5 ,  1917) ACTA CRYST. VOL.18,. . . . . . .   567, (1965).

2638  MEDINA..........(R.)...:   (11B,    17) NUCL. INST. MET. 87, . . . . . .   125, (1970).
                                 ( 6 ,   871) SOL. STATE COMM. VOL. 9, . . . .   257, (1971).

2639  MEDRUD..........(R.C.)..:  ( 5 ,  2734) J. CHEM. PHYS. VOL.31, . . . . .   332, (1959).

2640  MEDVEDEV........(M.V.)..:  ( 8 ,    62) PHYS. MET. METALLOG. VOL.22 NO.6    1, (1966).
                                 (11B,    45) SOV. PHYS. JETP. 24, . . . . . .   960, (1967).

2641  MEHNER..........(J.)...:   (16 ,     5) ATOMKERNENERGIE VOL.17,. . . . .   229, (1971).

2642  MEHRINGER.......(W.)...:   ( 7A,    99) Z. PHYS. VOL.210 NO.5,. . . . .    434, (1968).

2643  MEIER...........(F.)...:   (22B,     6) AUTOMATIC/NUCLEAR DATA,. . . . .   233, (1964).
                                 (22B,    57) Z. ANGEW. MATH. PHYS. VOL.15,. .   481, (1964).
                                 (19A,     9) HELV. PHYS. ACTA VOL.89, . . . .   569, (1966).

2644  MEIER...........(P.F.)..:  ( 7B,    47) PHYS. KOND. MATER. (GER.) VOL.14   336, (1972).
                                 (13 ,    56) PHYS. REV. B VOL.8,. . . . . . .  4422, (1973).

2645  MEINHARDT.......(D.)...:   ( 5 ,   971) ARCH. EISENHUTTENW. VOL.30,. . .    51, (1959).
                                 ( 5 ,   566) Z. NATURFORSCH. VOL.15A, . . . .   880, (1960).
                                 ( 5 ,  1162) Z. PHYS. VOL.174,. . . . . . . .   472, (1963).

2646  MEISSNER........(J.A.)..:  ( 5 ,  2361) PHYS. REV. VOL.187,. . . . . . .  1506, (1969).

2647  MEISTER.........(H.)...:   (17A,    79) NUKLEONIK BAND 10, . . . . . . .   101, (1967).
                                 (17A,    80) NUKLEONIK BAND 10, . . . . . . .    97, (1967).
                                 ( 6 ,   632) PHYS. LETT. 30A,. . . . . . . .    206, (1969).
                                 ( 6 ,  1539) PHYS. REV. VOL.184,. . . . . . .   550, (1969).
                                 (17A,    65) NUCL. INST. MET. VOL.84, . . . .    61, (1970).
                                 (17A,    44) IAEA SYMP. GRENOBLE,. . . . . .    713, (1972).
                                 (20C,    26) NUCL. INST. MET. VOL.108,. . . .   107, (1973).

2648  MEKATA..........(M.)...:   ( 5 ,  1725) J. PHYS. SOC. JAPAN VOL.25,. . .   234, (1968).
                                 ( 5 ,  1722) J. PHYS. SOC. JAPAN VOL.31,. . .   301, (1971).
                                 ( 5 ,  1721) J. PHYS. SOC. JAPAN VOL.36,. . .   438, (1974).

2649  MELAMUD.........(M.)...:   ( 5 ,  1987) PHYS. REV. 174,. . . . . . . .     560, (1968).
                                 ( 5 ,  1964) PHYS. REV. B VOL.2,. . . . . . .   179, (1970).
                                 ( 5 ,   664) BU. ISRAEL PHYS. SOC. 1971,. . .    27, (1971).
                                 ( 5 ,  2676) PHYS. REV. B VOL.3,. . . . . . .  2344, (1971).
                                 ( 5 ,   663) PHYS. REV. B VOL.3,. . . . . . .  3873, (1971).
                                 ( 5 ,  2402) PHYS. REV. B VOL.3,. . . . . . .   821, (1971).

2650  MELKONIAN.......(E.)...:   ( 5 ,  2233) PHYS. REV. VOL.73, . . . . . . .  1399, (1948).
                                 ( 5 ,  2230) PHYS. REV. VOL.76, . . . . . . .  1750, (1949).
                                 (20B,    64) J. PHYS. SOC. JAPAN VOL.15,. . .   630, (1960).
                                 (22A,    50) NUCL. SCI. ENGNG. VOL.8,. . . .    453, (1960).
                                 (22A,    64) REV. SCI. INSTRUM. VOL.36, . . .   887, (1965).

2651  MELLOR..........(J.)...:   ( 6 ,   263) J. PHYS. CHEM. SOL. VOL.25,. . .  1091, (1964).
                                 ( 5 ,  1868) ACTA CRYST. 18,. . . . . . . . .    37, (1965).

2652  MELVEGER........(A.J.)..:  ( 6 ,  1289) J. CHEM. PHYS. VOL.56, . . . . .  2793, (1972).

2653  MELVIN..........(J.S.)..:  (10B,   125) PHYS. REV. 175,. . . . . . . .    1083, (1968).

2654  MENARDI.........(S.)...:   ( 6 ,   166) IAEA SYMP. CHALK RIVER, VOL.1, .   139, (1962).
                                 ( 5 ,   787) NUOVO CIMENTO B VOL.46,. . . . .     7, (1966).

2655  MENDELL.........(R.B.)..:  (20C,    91) REV. SCI. INSTRUM. VOL.30, . . .   442, (1959).

2656  MENEGHETTI......(D.)...:   ( 5 ,   242) PHYS. REV. VOL.75, . . . . . . .  1098, (1949).
                                 ( 5 ,  2958) PHYS. REV. VOL.75, . . . . . . .   975, (1949).
                                 ( 9B,    77) PHYS. REV. VOL.80, . . . . . . .   507, (1950).
                                 ( 5 ,   722) PHYS. REV. VOL.105,. . . . . . .   133, (1957).

2657  MENIL...........(F.)...:   ( 5 ,  1558) SOL. STATE COMM. VOL.10, . . . .   739, (1972).

2658  MENON...........(C.S.)..:  (10B,   218) SOL. STATE COMM. VOL.10, . . . .   179, (1972).
                                 ( 6 ,  2017) SOL. STATE COMM. VOL.12 NO.6,. .   527, (1973).

2659  MENYUK..........(N.)...:   ( 5 ,   452) COLL. INTER. N.126 (GRENOBLE), .   104, (1963).
                                 ( 5 ,   517) PROC. INT. CONF. (NOTTINGHAM), .   538, (1964).
                                 ( 5 ,  1676) J. APP. PHYS. 36,. . . . . . . .  1088, (1965).
                                 ( 6 ,  1441) J. APP. PHYS. 36,. . . . . . . .  1090, (1965).
```

```
2659  MENYUK..........(N.)....!  ( 5 ,  1668)  J. APP. PHYS. 37,. . . . . .     962, (1966).
                                 ( 5 ,  1528)  J. PHYS. CHEM. SOL. 28,. . . .   549, (1967).

2660  MENZINGER.......(F.)....!  ( 5 ,   951)  COLL. INTER. N.126 (GRENOBLE), .   180, (1963).
                                 ( 6 ,   474)  PHYS. REV. LETT. 10,. . . . . .    290, (1963).
                                 ( 5 ,  2294)  PROC. INT. CONF. (NOTTINGHAM), .   288, (1964).
                                 ( 6 ,   512)  PHYS. REV. 143,. . . . . . . .     365, (1966).
                                 ( 5 ,   495)  PHYS. LETT. 25A,. . . . . . . .    372, (1967).
                                 ( 6 ,  1992)  J. APP. PHYS. 39,. . . . . . .    1237, (1968).
                                 ( 5 ,  2652)  J. APP. PHYS. 39,. . . . . . .    2221, (1968).
                                 ( 6 ,   829)  J. APP. PHYS. 39,. . . . . . .     455, (1968).
                                 ( 6 ,  1508)  PHYS. LETT. 30A,. . . . . . . .    310, (1969).
                                 ( 6 ,   883)  PHYS. REV. VOL.178,. . . . . .     833, (1969).
                                 ( 5 ,   743)  PHYS. REV. VOL.181,. . . . . .     936, (1969).
                                 ( 5 ,  1790)  PHYS. REV. VOL.187,. . . . . .     611, (1969).
                                 (11B,   17)  NUCL. INST. MET. 87,. . . . . .    125, (1970).
                                 ( 5 ,   478)  SOL. STATE COMM. VOL.8,. . . .       1, (1970).
                                 ( 5 ,   479)  INT. J. MAGN. VOL.1 NO.2,. . .    183, (1971).
                                 ( 6 ,   799)  J. PHYSIQUE TOME 32 COL.I VOL.II 1188, (1971).
                                 ( 6 ,   870)  SOL. STATE COMM. VOL. 9,. . . .   1579, (1971).
                                 ( 6 ,   871)  SOL. STATE COMM. VOL. 9,. . . .    257, (1971).
                                 (11B,   35)  SOL. STATE COMM. VOL. 9,. . . .    417, (1971).
                                 ( 5 ,   463)  NUOVO CIMENTO VOL.10B,. . . . .    565, (1972).
                                 ( 5 ,  1789)  PHYS. REV. B VOL.5,. . . . . .    3778, (1972).
                                 (13 ,    53)  PHYS. REV. LETT. VOL.28,. . . .     22, (1972).
                                 ( 5 ,   424)  SOL. STATE COMM. VOL.10,. . . .    667, (1972).
                                 ( 5 ,   464)  SOL. STATE COMM. VOL.12,. . . .    909, (1973).
                                 ( 5 ,  1064)  NUOVO CIMENTO VOL.20B NO.1,. . .     1, (1974).

2661  MEN≠SHIKOV......(A.Z.)..!  ( 7A,   90)  SOV. PHYS. CRYST. VOL.15,. . .    252, (1970).
                                 ( 7A,   91)  SOV. PHYS. CRYST. VOL.15,. . .    383, (1970).
                                 ( 6 ,   868)  SOV. PHYS. JETP LETT. VOL.16,.      6, (1972).
                                 ( 5 ,  2133)  SOV. PHYS. JETP VOL.34,. . . .    163, (1972).
                                 ( 5 ,  1059)  SOV. PHYS. JETP VOL.34,. . . .    799, (1972).

2662  MEN≠............(A.N.)..!  ( 8 ,   133)  SOV. PHYS. CRYST. VOL.7, . . .    286, (1962).

2663  MERCIER.........(M.)....!  ( 5 ,  1849)  COLL. INTER. N.126 (GRENOBLE), .   126, (1963).
                                 ( 5 ,  1852)  PHYS. LETT. 5,. . . . . . . . .     27, (1963).
                                 ( 5 ,  1851)  PHYS. LETT. 7,. . . . . . . . .    110, (1963).
                                 ( 5 ,  1853)  PHYS. LETT. 18,. . . . . . . .     13, (1965).

2664  MERIEL..........(P.)....!  ( 5 ,  2897)  COMP. REND. TOME 243,. . . . .    898, (1956).
                                 ( 5 ,  1647)  COMP. REND. TOME 246,. . . . .   3170, (1958).
                                 ( 5 ,   978)  J. PHYS. RADIUM VOL.19,. . . .    617, (1958).
                                 ( 5 ,  1649)  COMP. REND. 249,. . . . . . .    1334, (1959).
                                 (17C,   16)  BULL. SOC. FRANC. MIN. CRIST. 83  125, (1960).
                                 (90 ,   19)  CAHIERS DE PHYS. VOL.113,. . . .    29, (1960).
                                 ( 5 ,  1648)  COMP. REND. 250,. . . . . . .    1450, (1960).
                                 ( 5 ,  1388)  COMP. REND. 251,. . . . . . .    1359, (1960).
                                 ( 5 ,  2714)  J. NUCL. MATERIALS VOL.2,. . .     69, (1960).
                                 ( 5 ,  1645)  J. PHYS. RADIUM VOL.21,. . . .     67, (1960).
                                 ( 5 ,  2710)  BULL. INFORM. SCI. TECH. NO.49,.   69, (1961).
                                 ( 5 ,  1646)  J. PHYS. RADIUM VOL.22,. . . .    337, (1961).
                                 ( 5 ,  1445)  COLL. INTER. N.126 (GRENOBLE), .    60, (1963).
                                 ( 5 ,    86)  COMP. REND. 259,. . . . . . .    2416, (1964).
                                 ( 5 ,  1373)  BULL. SOC. FRANC. MIN. CRIST. 89  216, (1966).
                                 ( 5 ,  2730)  COMP. REND. 263,. . . . . . .    4578, (1966).
                                 ( 5 ,   815)  J. PHYSIQUE VOL.29 NO.2-3,. . .    220, (1968).
                                 ( 5 ,    44)  BULL. SOC. FRA. MINER. CRIST. 93     7, (1970).
                                 ( 5 ,  2389)  C. R. ACAD. SCI. B TOME 270,.     560, (1970).
                                 ( 5 ,   412)  CONF./RARE EARTHS/ACTINIDES,. .    218, (1971).
                                 ( 5 ,  2392)  J. PHYSIQUE TOME 32 COL.1 VOL.II  611, (1971).
                                 (12B,    2)  ANN. DE PHYSIQUE TOME 7 NO.4,.    255, (1972).
                                 ( 8 ,   18)  C.R. ACAD. SCI. B VOL.274,. . .    423, (1972).
                                 ( 5 ,   413)  J. CHEM. PHYS. VOL.56,. . . . .   4325, (1972).
                                 ( 6 ,   722)  PHYS. REV. LETT. VOL.28,. . . .    805, (1972).
                                 ( 5 ,  1361)  J. PHYSIQUE TOME 34,. . . . . .    423, (1973).
                                 (17A,   87)  PHYS. REV. LETT. VOL.31,. . . .    776, (1973).
                                 ( 5 ,  1029)  SOL. STATE COMM. VOL.14 NO.2,.    187, (1974).

2665  MERKEL..........(A.)....!  ( 6 ,  1025)  IAEA SYMP. BOMBAY, VOL.2,. . . .   167, (1964).

2666  MERZAGORA.......(N.)....!  (20B,   76)  NUCL. INST. MET. 33, . . . . . .   229, (1965).

2667  MESERVE.........(R.A.)..!  (11A,   51)  PHYS. REV. LETT. VOL.28,. . . .    614, (1972).
                                 (11C,   19)  J. PHYS. C VOL.6,. . . . . . .     79, (1973).

2668  MESHKOV.........(N.V.)..!  (22B,   11)  INSTRUM. EXP. TECH. NO.4,. . . .   832, (1965).

2669  MESSIAH.........(A.M.L.)!  ( 6 ,   979)  PHYS. REV. VOL.84,. . . . . . .    204, (1951).

2670  METZBOWER.......(E.A.)..!  (10B,  149)  PHYS. REV. 177,. . . . . . . .    1139, (1969).

2671  MEYER...........(A.P.)..!  ( 5 ,  1647)  COMP. REND. TOME 246,. . . . .    3170, (1958).

2672  MEYER...........(W.)....!  (24 ,   64)  NUCL. INST. MET. 70, . . . . . .   221, (1969).

2673  MEYR............(H.)....!  (19A,   72)  Z. ANGEW. MATH. PHYS. VOL.22,.     62, (1971).

2674  MEZEI...........(F.)....!  ( 8 ,   12)  ANN. DE PHYSIQUE TOME 7 NO.4,.    269, (1972).
                                 (19A,   25)  IAEA SYMP. GRENOBLE, . . . . . .   787, (1972).
                                 (12B,   14)  IAEA SYMP. GRENOBLE, . . . . . .   797, (1972).
                                 (12B,   30)  NUCL. INST. MET. VOL.99,. . . .    613, (1972).
                                 (12B,   76)  Z. PHYS. VOL.255,. . . . . . .     146, (1972).

2675  MICAL...........(R.O.)..!  ( 6 ,  1324)  DISS. ABS. VOL.31,. . . . . .    3658B, (1970).
                                 ( 6 ,  1658)  CONF. INTERN., RENNES,. . . . .    104, (1971).
                                 ( 6 ,  1325)  PHYS. REV. B VOL.4,. . . . . .    4551, (1971).

2676  MICHAEL.........(P.)....!  ( 5 ,   261)  PHYS. REV. VOL.138,. . . . . .    A692, (1965).

2677  MICHALEC........(R.)....!  (20B,   15)  CZECH. J. PHYS. B VOL.13,. . .     474, (1963).
                                 (22A,   10)  CZECH. J. PHYS. B VOL.16,. . .     942, (1966).
                                 ( 5 ,  2376)  PHYS. LETT. 28A, . . . . . . .     546, (1968).
                                 ( 5 ,  2371)  PHYS. STAT. SOL. 29, . . . . .     K51, (1968).
                                 ( 5 ,  2375)  BRIT. J. APPL. PHYS. VOL.2,. .    1041, (1969).
                                 (99 ,   37)  CZECH. J. PHYS. B VOL.19,. . .    1608, (1969).
                                 (12A,    8)  CZECH. J. PHYS. B VOL.19,. . .     278, (1969).
                                 ( 5 ,  2379)  NUCL. INST. MET. 67, . . . . .     357, (1969).
                                 ( 5 ,   929)  PHYS. LETT. 29A, . . . . . . .     679, (1969).
                                 (99 ,   89)  PHYS. STAT. SOL. A VOL.2,. . .     211, (1970).
                                 ( 5 ,  2373)  PHYS. STAT. SOL. VOL.42,. . . .    895, (1970).
                                 ( 5 ,  2372)  ACTA CRYST. VOL.A27,. . . . .      410, (1971).
                                 (9B ,   72)  PHYS. LETT. VOL.37A, . . . . .     403, (1971).
                                 ( 6 ,  2052)  NATURE (PHYS. SCI.) VOL.242,.     103, (1973).
                                 ( 5 ,  2449)  PHYS. STAT. SOLIDI A VOL.17 NO.1  163, (1973).

2678  MICHELINI.......(M.)....!  (16 ,   19)  ENERGIA NUCLEARE (ITALY) VOL.16,   700, (1969).

2679  MICHEL..........(C.)....!  ( 5 ,  2908)  COMP. REND. 264, . . . . . . .    397B, (1967).
                                 ( 5 ,  1385)  J. PHYSIQUE TOME 32 COL.1 VOL.II  670, (1971).
                                 ( 5 ,   215)  J. PHYS. CHEM. SOLIDS VOL.32,.    1315, (1971).
```

```
2680  MICHEL..........(K.H.)..!    (  8 ,     98) PHYS. REV. LETT. VOL.26, . . . .  1568, (1971).
                                   ( 7B,     30) J. CHEM. PHYS. VOL.58, . . . . .  1143, (1973).

2681  MIGDAL..........(A.B.)..!    (11C,     12) J. EXPER. THEOR. PHYS. (USSR) 10     5, (1940).

2682  MIGDAL..........(A.)....!    (  8 ;     17) C.R. ACAD. SCI. URSS VOL.20, . .   551, (1938).
                                   (  6 ,    116) FARADAY SYMP. CHEM. SOC. NO.6, .    48, (1972).

2683  MIGON...........(K.)....!    (  5 ,   1032) ACTA PHYS. POLON. VOL.A44, . . .   587, (1973).

2684  MIILLER.........(A.P.)..!    (  6 ,   1852) PHYS. REV. LETT. 20, . . . . . .   798, (1968).
                                   (  6 ,   1851) THESIS (181 PP.), . . . . . . .         (1969).
                                   (  6 ,    578) CAN. J. PHYS. 49, . . . . . . .    704, (1971).
                                   (  6 ,   1944) CAN. J. PHYS. VOL.50, . . . . .  2915, (1972).

2685  MIKA............(K.)....!    (  6 ,    422) IAEA SYMP. COPENHAGEN, VOL.1,.    599, (1968).
                                   (  5 ,    420) J. OF PHYS.-C-, SER.2, VOL.4,.  3034, (1971).

2686  MIKHALKO........(V.O.)..!    (  5 ,     75) LATV. PSR...FIZ. TEHN. SER. NO.1    88, (1970).

2687  MIKKE...........(K.)....!    (  6 ,    668) IAEA SYMPOSIUM (VIENNA), . . .    351, (1960).
                                   (  6 ,   1558) IAEA SYMP. CHALK RIVER, VOL.2,    237, (1962).
                                   (  6 ,   1305) IAEA SYMP. BOMBAY, VOL.2, . .     383, (1964).
                                   ( 9C,     34) PHYS. STAT. SOL. 11, . . . . .    567, (1965).
                                   (20A,     41) NUCL. INST. MET. VOL.79, . . .     51, (1970).
                                   (  5 ,    609) J. PHYS. CHEM. SOLIDS VOL.33,.   1651, (1972).
                                   (  6 ,   1780) PHYS. STAT. SOL. B VOL.54 NO.2,.  K99, (1972).
                                   (  6 ,    499) J. PHYS. F VOL.3, . . . . . . .   L173, (1973).
                                   (  6 ,    535) PHYS. STAT. SOLIDI B VOL.55 NO.1    K1, (1973).
                                   (  6 ,    536) PHYS. STAT. SOLIDI B VOL.57 NO.1    K1, (1973).
                                   (  6 ,   1763) PHYS. STAT. SOLIDI B VOL.59 NO.2   K97, (1973).

2688  MIKULA..........(P.)....!    (  5 ,   2376) PHYS. LETT. 28A, . . . . . . .    546, (1968).
                                   (  5 ,   2373) PHYS. STAT. SOL. VOL.42, . . .    895, (1970).
                                   (  6 ,   2052) NATURE (PHYS. SCI.) VOL.242, .    109, (1973).
                                   (  5 ,   2449) PHYS. STAT. SOLIDI A VOL.17 NO.1   163, (1973).

2689  MIKULI..........(E.)....!    (  6 ,   1387) ACTA PHYS. POLON. VOL.A44, . . .   731, (1973).

2690  MILONER.........(D.F.R.)!    (22A,     48) NUCL. INST. MET. VOL.114, . . .    417, (1974).
                                   (17C,     25) REV. SCI. INSTRUM. VOL.45, . . .    572, (1974).

2691  MILFORD.........(F.J.)..!    (  6 ,    845) PHYS. REV. 130, . . . . . . . .  1783, (1963).

2692  MILLER..........(A.)....!    (  6 ,   1179) PHYS. REV. VOL.127, . . . . . .  1452, (1962).

2693  MILLER..........(L.G.)..!    (24 ,     52) NUCL. INST. MET. 61, . . . . .    245, (1968).
                                   (  5 ,   2362) NUCL. PHYS. A VOL.A164, . . . .     34, (1971).

2694  MILLER..........(P.H.)..!    (12B,     10) IAEA SYMP. COPENHAGEN, VOL.2,.    429, (1968).
                                   (19A,     44) NUCL. INST. MET. 76, . . . . .    135, (1969).
                                   (19A,     47) NUCL. INST. MET. 77, . . . . .     29, (1970).

2695  MILLER..........(R.C.)..!    (  5 ,   1102) J. PHYS. CHEM. SOLIDS VOL.23,. .   863, (1962).

2696  MILLER..........(T.G.)..!    (22C,     31) NUCL. INST. MET. 40, . . . . .     93, (1966).
                                   (12B,     26) NUCL. INST. MET. 48, . . . . .    154, (1967).

2697  MILLHOUSE.......(A.H.)..!    (  5 ,   2605) J. APP. PHYS. 39, . . . . . . .  1329, (1968).
                                   (  5 ,    825) INT. J. MAGN.(GB) VOL.2, . . .    389, (1971).
                                   (  5 ,   1367) SOL. STATE COMM. VOL.11 NO.5,.    707, (1972).
                                   (  5 ,    909) SOL. STATE COMM. VOL.13, . . .    339, (1973).

2698  MILLIGAN........(W.O.)..!    (  5 ,   2430) J. PHYS. CHEM. VOL.57, . . . . .   535, (1953).

2699  MILLINGTON......(A.J.)..!    (  6 ,   2298) J. OF PHYS.-C-, SER. 2, VOL.2,.  2366, (1969).
                                   (  6 ,   1605) J. PHYS.F: METAL PHYS. VOL.1,.    244, (1971).

2700  MILLS...........(D.L.)..!    (  5 ,   2119) PHYS. REV. B VOL.9, . . . . . .  2354, (1974).

2701  MILLS...........(R.L.)..!    (  5 ,    769) ACTA CRYST. VOL.A28 PART S-4,. .   188, (1972).

2702  MILLS...........(W.R.)..!    (20C,     94) REV. SCI. INSTRUM. VOL.33, . . .   866, (1962).

2703  MILL............(B.V.)..!    (  5 ,   1960) SOL. STATE COMM. VOL.14 NO.4,. .   309, (1974).

2704  MILOJEVIC.......(A.)....!    (  6 ,    862) PHYS. STAT. SOL. 21, . . . . . .  K163, (1967).

2705  MILOSEVIC.......(S.)....!    (  5 ,   2180) FIZIKA (YUGOSLAVIA) VOL.5, . . .   179, (1973).

2706  MILSTEIN........(F.)....!    (11A,     43) PHYS. REV. B VOL.3, . . . . . .  1025, (1971).

2707  MILTON..........(J.C.D.)!    (23 ,     29) IAEA SYMP. (PILE RESEARCH)VIENNA    25, (1960).

2708  MINAKAWA........(N.)....!    (20D,      8) JAPAN. J. APPL. PHYS. VOL.4, . .   911, (1965).
                                   ( 9B,     46) J. APPL. CRYST. VOL.4, . . . .    528, (1971).
                                   (  5 ,   2439) J. PHYS. SOC. JAPAN VOL.31, . .    954, (1971).
                                   (  5 ,   1217) ACTA CRYST. VOL.A28 PART S-4,.    220, (1972).
                                   (  6 ,   1924) J. POLYM. SCI./POLYM. LETT. 11,.   377, (1973).

2709  MINKIEWICZ......(V.J.)..!    (  6 ,   1733) J. APP. PHYS. 38, . . . . . . .  1245, (1967).
                                   (  6 ,   1185) PHYS. REV. LETT. 19, . . . . .   1307, (1967).
                                   (  6 ,    128) PHYS. REV. LETT. 19, . . . . .    234, (1967).
                                   (  6 ,   1738) PHYS. REV. 156, . . . . . . . .    623, (1967).
                                   (  6 ,   1342) PHYS. REV. 157, . . . . . . . .    396, (1967).
                                   (  6 ,    778) PHYS. REV. 162, . . . . . . . .    528, (1967).
                                   (  6 ,   1728) J. APP. PHYS. 39, . . . . . . .    383, (1968).
                                   (  5 ,   1476) J. PHYS. CHEM. SOLIDS 29, . . .    881, (1968).
                                   (  6 ,    728) KYOTO CONF. STAT. MECHANICS, .    169, (1968).
                                   (  6 ,    734) J. APP. PHYS. VOL.40, . . . . .   1442, (1969).
                                   (  6 ,   1310) J. PHYS. SOC. JAPAN VOL.26, . .    674, (1969).
                                   (  6 ,    738) J. PHYS. SOC. JAPAN (SUPPL.) 26,   169, (1969).
                                   (  6 ,    794) PHYS. REV. VOL.179, . . . . . .    417, (1969).
                                   (  6 ,   1716) PHYS. REV. VOL.182, . . . . . .    624, (1969).
                                   (  6 ,   1222) SOL. STATE COMM. 7, . . . . . .    535, (1969).
                                   (20D,     25) NUCL. INST. MET. VOL.89, . . .    109, (1970).
                                   (  5 ,    670) SOL. STATE COMM. VOL.8, . . . .   1001, (1970).
                                   (  5 ,   1479) SOL. STATE COMM. VOL.8, . . . .   1941, (1970).
                                   (  6 ,    120) SOL. STATE COMM. VOL.8, . . . .    463, (1970).
                                   (  5 ,    650) ACTA CRYST. VOL.A27, . . . . .    494, (1971).
                                   (13 ,     21) INT. J. MAGN. VOL.1, . . . . .    149, (1971).
                                   (  6 ,   1505) J. APPL. PHYS. 42, . . . . . .   1374, (1971).
                                   (  5 ,   2391) J. PHYSIQUE TOME 32 COL.1 VOL.II   892, (1971).
                                   (  6 ,    671) PHYS. REV. B VOL.4, . . . . . .  2209, (1971).
                                   (13 ,     16) DYN. ASPECTS OF CRITICAL PHEN.,.    69, (1972).
                                   (  5 ,   2395) MAGN. AND MAG. MATERIALS-1971,    436, (1972).
                                   (  6 ,   1143) PHYS. REV. A VOL.5, . . . . . .   1537, (1972).
                                   (  6 ,   1701) PHYS. REV. B VOL.6, . . . . . .   4766, (1972).
                                   (  6 ,   1162) PHYS. REV. LETT. VOL.29, . . .    552, (1972).
                                   (  6 ,   1509) SOL. STATE COMM. VOL.10, . . .    203, (1972).
                                   (  5 ,   2400) A.I.P. CONF. PROC. NO.10 PART 1,   659, (1973).
                                   (  6 ,   1189) PHYS. REV. A VOL.8, . . . . . .  1513, (1973).
```

```
2710  MINOR...........(W.)....:  (17A,    16) ACTA PHYS. POLON. VOL.A42, . . .      259, (1972).
                                 ( 9B,    92) PHYS. STAT. SOLIDI A VOL.9, . . .     423, (1972).

2711  MIRON...........(N.F.)..:  ( 5 , 2969) SOV. PHYS. CRYST. VOL.15,. .          376, (1970).
                                 ( 5 , 2427) SOV. PHYS. CRYST. VOL.17,. . : : :    822, (1971).
                                 ( 5 , 2899) SOV. PHYS. CRYST. VOL.17,. . . .      342, (1972).

2712  MISAWA..........(M.)....:  ( 5 , 2482) JAP. J. APPL. PHYS. VOL.12,. .        1119, (1973).
                                 ( 6 ,  228) PHYS. LETT. VOL.45A NO.4,. : : :      273, (1973).

2713  MISENTA.........(R.)....:  (22A,    41) NUCL. INST. MET. 81, . . . . .       317, (1970).
                                 ( 6 ,  953) SOL. STATE COMM. VOL.8,. : : :        373, (1970).
                                 ( 5 ,  767) Z. NATURFORSCH. A VOL.25,. . .        967, (1970).

2714  MISHRA..........(N.)....:  ( 6 , 1364) J. PHYS. F VOL.3,. . . . . . . .     1388, (1973).

2715  MISIUK..........(A.)....:  ( 5 , 2807) PHYS. STAT. SOLIDI A VOL.13 NO.1     K79, (1972).
                                 ( 5 , 2817) PHYS. STAT. SOLIDI A VOL.20, . .      395, (1973).

2716  MITANI..........(S.)....:  ( 9B,    57) J. PHYS. SOC. JAPAN VOL.23,. . .      460, (1967).
                                 ( 5 , 1980) J. PHYS. SOC. JAPAN VOL.23,. . .      461, (1967).
                                 ( 5 ,  468) J. PHYS. SOC. JAPAN VOL.35,. . .      426, (1973).

2717  MITCHELL........(A.C.G.):  ( 7A,    60) PHYS. REV. VOL.50, . . . . . . .      133, (1936).

2718  MITCHELL........(D.P.)..:  (20A,    57) PHYS. REV. VOL.48, . . . . . . .      265, (1935).
                                 (20A,   123) PHYS. REV. VOL.48, . . . . . . .      704, (1935).
                                 (20A,    58) PHYS. REV. VOL.49, . . . . . . .      103, (1936).
                                 ( 9B,    73) PHYS. REV. VOL.50, . . . . . . .      486, (1936).

2719  MITCHELL........(E.W.J.):  ( 5 , 2377) PHIL. MAG. VOL.3, . . . . . . . .     1280, (1958).
                                 ( 6 ,  200) IAEA SYMP. CHALK RIVER VOL.2, .        49, (1962).
                                 (20R,    65) J. SCI. INSTRUM. VOL.43, . . . .        1, (1966).
                                 (20B,    36) IAEA SYMP. COPENHAGEN VOL.2, . .      341, (1968).
                                 ( 5 , 1220) CRYSTAL LATTICE DEFECTS VOL.2, .      105, (1971).

2720  MITIN...........(A.V.)..:  (10C,   125) SOV. PHYS. SOL. STATE VOL.13,. .      215, (1971).

2721  MITOMI..........(Y.)....:  (23 ,    48) MITSUBISHI DENHI ENG. NO.26, . .       10, (1970).

2722  MITRA...........(N.R.)..:  ( 6 , 1364) J. PHYS. F VOL.3,. . . . . . . .     1388, (1973).

2723  MITRA...........(S.S.)..:  (10B,    45) J. CHEM. PHYSICS VOL.41, . . . .     3158, (1964).
                                 ( 6 , 2038) PHYS. REV. 178, . . . . . . . .      1349, (1969).
                                 (10B,   156) PHYS. REV. B-2, . . . . . . . .      2167, (1970).
                                 (10B,   158) PHYS. REV. B-2, . . . . . . . .       967, (1970).
                                 (10B,   215) SOL. STATE COMM. VOL. 9, . . . .      185, (1971).
                                 ( 6 ,  562) PHYS. REV. B VOL.7,. . . . . . .     4001, (1973).

2724  MITROFANOV......(N.)....:  ( 5 , 2317) PHYS. REV. VOL.125,. . . . . . .     1141, (1962).

2725  MITSUI..........(T.)....:  ( 5 , 1573) J. PHYS. CHEM. SOL. 24,. . . . .     1057, (1963).

2726  MIURA...........(T.)....:  (20C,    60) NUCL. INST. MET. 74, . . . . . .      322, (1969).

2727  MIYADAI.........(T.)....:  ( 5 , 2187) PHYS. LETT. VOL.44A NO.7,. . . .      529, (1973).

2728  MIYAGI..........(H.)....:  (15 ,    45) PROGR. THEOR. PHYS. VOL.50,. . .     1142, (1973).

2729  MIYAHARA........(S.)....:  ( 5 , 2187) PHYS. LETT. VOL.44A NO.7,. . . .      529, (1973).

2730  MIYAKAWA........(T.)....:  ( 9B,    52) J. APPL. PHYS. VOL.40, . . . . .     1697, (1969).
                                 (17A,     9) ACTA CRYST. VOL.A26, . . . . . .      667, (1970).

2731  MIYAKE..........(S.)....:  (22B,    16) J. PHYS. SOC. JAPAN VOL.17 B-II,     358, (1962).

2732  MIYAKOSHI.......(J.)....:  (22B,    54) REV. SCI. INSTRUM. VOL.41, . . .       11, (1970).

2733  MIYANO..........(K.)....:  ( 5 ,  847) J. PHYS. SOC. JAP. VOL.31, . . .     1304, (1971).

2734  MIYASHITA.......(K.)....:  (22B,    16) J. PHYS. SOC. JAPAN VOL.17 B-II,     358, (1962).
                                 (21 ,    32) MITSUB. DENKI LAB. REP. VOL.3, .      111, (1962).

2735  MIYAZAWA........(T.)....:  ( 7B,    27) J. CHEM. PHYSICS VOL.47, . . . .      337, (1967).
                                 ( 6 , 1902) REP. PROGR. POLYMER PHYS/JAP. 10      185, (1967).
                                 ( 6 , 1907) J. POLYMER SCI. B VOL.6, . . . .       83, (1968).
                                 ( 6 , 1879) J. CHEM. PHYSICS VOL.50,. . . . ,     915, (1969).

2736  MODRZEJEWSKI....(A.)....:  ( 9B,    27) ARCH. HUTNICTWA VOL.4, . . . . .       81, (1959).
                                 (24 ,    27) KERNTECHNIK VOL.2, . . . . . . .      153, (1960).
                                 ( 5 ,    48) PHYS. STAT. SOL. 5,. . . . . . .      K23, (1964).
                                 ( 9C,    24) KRISTALL TECH. VOL.4,. . . . . .      135, (1969).
                                 ( 5 ,  460) PHYS. STAT. SOL. VOL.38, . . . .      103, (1970).
                                 ( 6 ,  499) J. PHYS. F VOL.3,. . . . . . . .     L173, (1973).
                                 ( 6 , 1763) PHYS. STAT. SOLIDI B VOL.59 NO.2     K97, (1973).

2737  MOFFITT.........(R.D.)..:  (23 ,    61) NUCL. INST. MET. VOL.59, . . . .      245, (1968).

2738  MOGENSEN........(P.A.)..:  ( 6 , 2128) BROOKHAVEN SYMPOSIUM., . . . . .      159, (1965).

2739  MOGI............(M.)....:  (21 ,    32) MITSUB. DENKI LAB. REP. VOL.3, .      111, (1962).

2740  MOGLESTUE.......(K.T.)..:  (11A,     8) IAEA SYMP. COPENHAGEN, VOL.2,. .      117, (1968).
                                 ( 6 ,  900) PHYS REV 178,. . . . . . . . . .      781, (1969).

2741  MOGUTNOV........(B.M.)..:  ( 5 , 2103) SOV. PHYS. DOKLADY VOL.14, . . .      263, (1969).

2742  MOHLING.........(F.)....:  ( 6 , 1197) PHYS. REV. LETT. VOL.2,. . . . .      284, (1959).

2743  MOHOS...........(B.)....:  (17A,    85) PHYS. LETT. 31A, . . . . . . . .       78, (1970).

2744  MOISEIWITSCH....(B.L.)..:  ( 7A,    77) PROC. PHYS. SOC. 81, . . . . . .       35, (1963).
                                 ( 7A,    78) PROC. PHYS. SOC. 86, . . . . . .      363, (1965).
                                 ( 7A,    79) PROC. PHYS. SOC. 89, . . . . . .      341, (1966).

2745  MOLDASCHL.......(H.)....:  (12B,    12) IAEA SYMP. COPENHAGEN, VOL.2,. .      387, (1968).
                                 (21 ,    39) NUCL. INST. MET. 58, . . . . . .      261, (1968).

2746  MOLLARD.........(P.)....:  ( 5 , 2651) COMP. REND. 263, . . . . . . . .     621B, (1966).
                                 ( 5 ,  646) SOL. STATE COMM. 4,. . . . . . .      249, (1966).
                                 ( 5 ,  618) J. PHYSIQUE TOME 29, . . . . . .      813, (1968).
                                 ( 5 ,  642) SOL. STATE COMM. VOL.13, . . . .      905, (1973).

2747  MOLLERUD........(R.)....:  ( 5 , 2358) PHIL. MAG. VOL.11, . . . . . . .     1245, (1965).
                                 ( 5 , 2360) ACTA CHEM. SCAND. VOL.20, . . . .     2529, (1966).
                                 ( 5 , 2312) PHIL. MAG. VOL.16, . . . . . . .     1063, (1967).

2748  MOLLER..........(H.B.)..:  (20B,   144) REV. SCI. INSTRUM. VOL.32, . . .      654, (1961).
                                 (20D,    10) J. NUCL. ENERGY VOL.17,. . . . .      227, (1963).
                                 ( 6 ,  516) IAEA SYMP. BOMBAY, VOL.1, . . . .       95, (1964).
                                 ( 6 ,  524) PROC. INT. CONF. (NOTTINGHAM), .      101, (1964).
                                 ( 6 ,  553) SOL. STATE COMM. 2,. . . . . . .      109, (1964).
                                 ( 6 , 2128) BROOKHAVEN SYMPOSIUM., . . . . .      159, (1965).
                                 ( 6 ,  555) PHYS. REV. LETT. 15, . . . . . .      623, (1965).
                                 ( 6 ,  638) SOL. STATE COMM. VOL.3, . . . . .      137, (1965).
                                 ( 6 , 2136) PHYS. REV. LETT. 16, . . . . . .      737, (1966).
```

```
2748  MOLLER..........(H.B.)..:    ( 6 ,  2150) PHYS. REV. LETT. 19,. . . . .   312, (1967).
                                   ( 6 ,  2152) IAEA SYMP. COPENHAGEN, VOL.2,. :     3, (1968).
                                   ( 6 ,  2153) J. APP. PHYS. 39,. . . . . . .   807, (1968).
                                   (10C,    91) PROC/1ST INT. CONF/LOC. EXC/SOL. 721, (1968).
                                   (17O,     5) ACTA CRYST. VOL.A25, . . . . .   547, (1969).
                                   (17O,     9) IAEA SYMP. (INSTRUMENT.), VIENNA  49, (1969).
                                   ( 6 ,  2124) J. APPL. PHYS. 41,. . . . . .  1174, (1970).
                                   ( 6 ,  2151) PHYS. REV. LETT. VOL.25,. . .  1451, (1970).
                                   ( 5 ,  2336) PHYS. REV. LETT. VOL.25,. . .   524, (1970).
                                   ( 6 ,   654) PHYS. REV. B-3 . . . . . . .   4383, (1971).
                                   ( 6 ,  2132) IAEA SYMP. GRENOBLE,. . . . .   603, (1972).

2749  MOLLINARI.......(V.G.)..:    ( 6 ,   326) IAEA SYMP. CHALK RIVER, VOL.1, .  285, (1962).

2750  MOLODKIN........(V.B.)..:    (10C,   133) UKR. FIZ. ZH. (USSR) VOL.8,. .   256, (1963).

2751  MOMIN...........(S.N.)..:    (23 ,    40) INDIAN J. PURE APPL. PHYS. VOL.6  550, (1968).
                                   ( 5 ,   158) J. CHEM. PHYSICS VOL.48,. . . .  1883, (1968).

2752  MONAHAN.........(J.E.)..:    ( 5 ,   925) NUCL. PHYS. VOL.A123,. . . . .    33, (1969).
                                   ( 5 ,   233) PHYS. REV. VOL.188,. . . . . .  1618, (1969).

2753  MONAHAN.........(J.)..:      (24 ,    16) IAEA SYMP. (PILE RESEARCH)VIENNA  615, (1960).

2754  MONCTON.........(D.E.)..:    ( 5 ,  2981) SOL. STATE COMM. VOL.13, . . . 1779, (1973).

2755  MONTANO.........(P.A.)..:    ( 6 ,   571) PHYS. REV. B VOL.6,. . . . . .  1053, (1972).

2756  MONTMORY........(M.C.)..:    ( 5 ,   646) SOL. STATE COMM. 4,. . . . . .   249, (1966).
                                   ( 5 ,  1178) SOL. STATE COMM. VOL.6 NO.5, . .  317, (1968).
                                   ( 5 ,   634) SOL. STATE COMM. VOL.6 NO.5, . .  323, (1968).

2757  MONTROLL........(E.W.)..:    ( 6 ,   693) BROOKHAVEN SYMPOSIUM,. . . . .    57, (1965).

2758  MONTROSE........(C.J.)..:    (24 ,    20) IAEA SYMP. COPENHAGEN, VOL.1, .   623, (1968).

2759  MOOK............(H.A.)..:    ( 6 ,  1736) J. APP. PHYS. 37,. . . . . .   1034, (1966).
                                   ( 5 ,   935) PHYS. REV. LETT. VOL.16; . . .   184, (1966).
                                   ( 5 ,  2075) PHYS. REV. 148,. . . . . . . .   495, (1966).
                                   (12R,    19) J. APP. PHYS. 39,. . . . . . .   447, (1968).
                                   (19A,    22) IAEA SYMP. (INSTRUMENT.),VIENNA, 173, (1969).
                                   ( 6 ,  1712) J. APP. PHYS. VOL.40,. . . . .  1450, (1969).
                                   ( 6 ,  1226) J. APP. PHYS. VOL.40,. . . . .  1452, (1969).
                                   ( 6 ,   745) J. PHYSIQUE TOME 32 COL.1 VOL.II 1177, (1971).
                                   ( 6 ,  1937) A.I.P. CONF. PROC. NO.10 PART 2, 1548, (1973).
                                   ( 6 ,   742) PHYS. REV. B VOL.7,. . . . . .   336, (1973).
                                   ( 6 ,  1739) PHYS. REV. LETT. VOL.30,. . . .  556, (1973).
                                   ( 6 ,   500) CAN. J. PHYS. VOL.52,. . . . .   396, (1974).
                                   ( 5 ,    59) J. APPL. PHYS. VOL.45,. . . . .   43, (1974).
                                   (19B,    23) NUCL. INST. MET. VOL.116,. . .   205, (1974).
                                   ( 6 ,  1205) PHYS. REV. A VOL.9,. . . . . .  2085, (1974).
                                   ( 6 ,  1207) PHYS. REV. LETT. VOL.32, . . .  1167, (1974).

2760  MOON............(P.B.)..:    (25 ,    10) NATURE VOL.135,. . . . . . . .   964, (1935).
                                   (16 ,    67) PROC. ROY. SOC. A VOL.153, . .   476, (1936).
                                   (20A,    30) NATURE VOL.142,. . . . . . . .   829, (1938).
                                   (20B,   132) PROC. ROY. SOC. A VOL.175, . .   316, (1940).

2761  MOON............(R.M.)..:    (17C,     4) ACTA CRYST. 17,. . . . . . . .   805, (1964).
                                   ( 5 ,  2027) J. APPL. PHYS. VOL.35,. . . .   1041, (1964).
                                   ( 5 ,  2337) PHYS. REV. LETT. VOL.12,. . . .  553, (1964).
                                   ( 5 ,   436) PHYS. REV. 136,. . . . . . . .  A195, (1964).
                                   ( 6 ,  2147) PROC. INT. CONF. (NOTTINGHAM), . 271, (1964).
                                   ( 5 ,  2553) J. APP. PHYS. 36,. . . . . . .   978, (1965).
                                   ( 5 ,   529) J. APP. PHYS. 37,. . . . . . .  1036, (1966).
                                   ( 5 ,   598) J. APP. PHYS. 37,. . . . . . .  1259, (1966).
                                   ( 5 ,   597) PHYS. REV. VOL.151,. . . . . .   405, (1966).
                                   ( 5 ,  1212) J. APP. PHYS. 38,. . . . . . .  1381, (1967).
                                   ( 5 ,  2634) J. APP. PHYS. 38,. . . . . . .  1384, (1967).
                                   (22B,     8) IAEA SYMP. COPENHAGEN, VOL.2,. .  253, (1968).
                                   ( 6 ,  1374) PHYS. REV. LETT. 20, . . . . .   997, (1968).
                                   ( 5 ,  2922) PHYS. REV. 176,. . . . . . . .   722, (1968).
                                   ( 6 ,  2179) J. APP. PHYS. VOL.40,. . . . .  1445, (1969).
                                   ( 5 ,   426) J. APP. PHYS. VOL.40,. . . . .  1454, (1969).
                                   ( 5 ,   434) PHYS. REV. 181,. . . . . . . .   883, (1969).
                                   ( 5 ,  1682) PHYS. REV. 181,. . . . . . . .   920, (1969).
                                   ( 6 ,   906) J. APPL. PHYS. 41,. . . . . .   1182, (1970).
                                   ( 6 ,   907) PHYS. REV. LETT. VOL.24,. . . .   16, (1970).
                                   ( 5 ,  2847) PHYS. REV. LETT. VOL.25,. . . .  527, (1970).
                                   ( 9D,    31) INT. J. MAGN. VOL.1 NO.3,. . .   219, (1971).
                                   ( 6 ,   909) J. PHYSIQUE TOME 32 COL.1 VOL.I, 296, (1971).
                                   ( 5 ,  2470) ACTA CRYST. VOL.A28 PART S-4,. . 197, (1972).
                                   ( 5 ,  1205) PHYS. REV. B VOL.5,. . . . . .   997, (1972).
                                   ( 5 ,  2468) PHYS. REV. LETT. VOL.29,. . . . 1468, (1972).
                                   ( 9D,    78) TRANS. AMER. CRYST. ASSOC. VOL.8  59, (1972).
                                   ( 5 ,  2472) A.I.P. CONF. PROC. NO.10 PART 2, 1314, (1973).
                                   ( 5 ,  2327) A.I.P. CONF. PROC. NO.10 PART 2, 1319, (1973).

2762  MOON............(Y.S.)..:    ( 5 ,  1956) J. KOREAN PHYS. SOC. VOL.1,. .   108, (1968).

2763  MOORE...........(F.H.)..:    (23 ,     3) AT. ENERGY AUST. VOL.15,. . . .    2, (1972).
                                   ( 5 ,   280) CRYST. STRUCT. COMM. VOL.1 N.3,. 193, (1972).
                                   ( 5 ,   296) ACTA CRYST. VOL.B29, . . . . .  1903, (1973).

2764  MOORE...........(F.M.)..:    ( 5 ,  2858) NATURE VOL.214,. . . . . . . .   130, (1967).

2765  MOORE...........(J.A.)..:    (22A,    64) REV. SCI. INSTRUM. VOL.36, . .   887, (1965).

2766  MOORE...........(M.A.)..:    (13 ,    50) PHYS. REV. B VOL.4,. . . . . .  3954, (1971).

2767  MOORE...........(M.S.)..:    (23 ,    27) IAEA SYMP. (PILE RESEARCH)VIENNA  93, (1960).
                                   ( 5 ,  2715) NUCL. SCI. ENG. VOL.7,. . . . .  187, (1960).
                                   (19A,    34) NUCL. INST. MET. 30,. . . . . .  293, (1964).
                                   ( 5 ,  2362) NUCL. PHYS. A VOL.A164,. . . .    34, (1971).

2768  MOORE...........(S.O.)..:    ( 7A,    71) PHYS. REV. VOL.135,. . . . . .  B895, (1964).

2769  MOORE...........(W.E.)..:    ( 6 ,  1899) NUCL. SCI. ENGNG. VOL.25,. . .   300, (1966).
                                   ( 6 ,  1391) MOL. DYN. AND STRUCT. OF SOLIDS,  315, (1969).
                                   ( 6 ,  1690) MOL. DYN. AND STRUCT. OF SOLIDS,  547, (1969).
                                   ( 6 ,  1038) NUCL. SCI. ENG. VOL.46,. . . .   223, (1971).
                                   ( 6 ,  2208) NUCL. SCI. ENG. VOL.47,. . . .   349, (1972).
                                   ( 7B,    43) NUCL. SCI. ENG. VOL.48,. . . .   266, (1972).

2770  MOORHOUSE.......(R.G.)..:    ( 8 ,   111) PROC. PHYS. SOC. A VOL.64, . .  1097, (1951).

2771  MOOR............(M.A.)..:    (10B,    72) J. PHYS. C VOL.5,. . . . . . .  3168, (1972).

2772  MORALES-AMADO...(A.)....:    (14B,    28) IAEA SYMPOSIUM VIENNA, . . . .   251, (1960).

2773  MORASH..........(K.R.)..:    (17C,    10) ACTA CRYST. A24, . . . . . . .   160, (1968).

2774  MORDASHEV.......(V.M.)..:    ( 6 ,    21) SOV. AT. ENERGY VOL.28,. . . .   214, (1970).
```

2775 MOREAU..........(J.M.)..# (5 , 1532) SOL. STATE COMM. VOL.6,. . . . . .   751, (1968).
                                (5 , 1533) SOL. STATE COMM. VOL.7 NO.22,. . 1669, (1969).
                                (5 , 1385) J. PHYSIQUE TOME 32 COL.1 VOL.II   670, (1971).
                                (5 , 215) J. PHYS. CHEM. SOLIDS VOL.32,. . 1315, (1971).

2776 MORGAN..........(D.)....# (6 , 1487) PHYS. LETT. VOL.36A, ...   439, (1971).

2777 MORGAN..........(I.L.)..# (20C,   94) REV. SCI. INSTRUM. VOL.33, . . .   866, (1962).

2778 MORIN...........(P.)....# (6 , 709) PHYS. REV. B VOL.9 NO.11,. . . . . . . , (1974).

2779 MORITA..........(T.)....# (5 , 2302) PHYS. REV. 137,. . . . . A483, (1965).

2780 MORI............(H.)....# (13 ,   60) PROGR. THEOR. PHYS. VOL.25,. . .   723, (1961).
                                (13 ,   9) BROOKHAVEN SYMPOSIUM,. . . . . .   49, (1965).

2781 MORI............(N.)....# (8 ,   39) J. PHYSIQUE TOME 32 COL.1 VOL.II 812, (1971).

2782 MORI............(T.)....# (16 ,   63) PROC. JAP. ACAD. VOL.46, ...   944, (1970).

2783 MORLEY..........(G.L.)..# (10B,   21) DISS. ABS. VOL.28, ...47248, (1967).

2784 MOROSIN.........(B.)....# (5 , 1973) FERROELECTRICS,. . . . .   31, (1970).

2785 MOROZOV.........(V.M.)..# (7B, 114) SOV. AT. ENERGY VOL.28,. . . . .   398, (1970).
                                (7B, 134) SOV. PHYS. J. VOL.34,. . . . . .   138, (1970).
                                (7A,   92) SOV. PHYS. JETP LETT. VOL.14,. .   91, (1971).

2786 MORRISH.........(A.H.)..# (6 , 867) PHYS. LETT. 7, ...   177, (1963).
                                (5 , 1094) PHIL. MAG. VOL.12, ...   221, (1965).

2787 MORRISON........(C.A.)..# (11C,   25) PHYS. LETT. VOL.33A, ...   375, (1970).

2788 MORRISON........(J.A.)..# (10B,   29) IAEA SYMP. CHALK RIVER, VOL.1, .   49, (1962).

2789 MORRISON........(R.C.)..# (20C,   70) NUCL. INST. MET. VOL.84, ...   67, (1970).

2790 MORSE...........(P.L.R.)# (10B, 205) PROC. ROY. SOC. A VOL.317, . . .   55, (1970).

2791 MORTENSEN.......(G.A.)..# (16 ,   12) DISS. ABS. VOL.24, ... 5484, (1964).

2792 MORTON..........(G.A.)..# (5 , 1972) PHYS. REV. VOL.73, ...   842, (1948).

2793 MOSBURG-JR......(E.R.)..# (20A,   71) REACTOR SCI. TECHNOL. VOL.14,. .   25, (1961).

2794 MOSCHINI........(G.)....# (5 , 231) NUCL. PHYS. A VOL.A181,. . . . .   177, (1972).

2795 MOSCHINI........(N.G.)..# (20C,   66) NUCL. INST. MET. 67, ...   267, (1969).

2796 MOSHINSKY.......(M.)....# (10C,   95) REV. MEX. FIS. VOL.2,. . . . . .   137, (1953).
                                (10C,   96) REV. MEX. FIS. VOL.3,. . . . . .   1, (1954).

2797 MOSKALEV........(O.B.)..# (16 ,   85) SOV. PHYS. DOKLADY VOL.15, . . .   724, (1971).

2798 MOSKALEV........(S.S.)..# (5 , 2478) SOV. J. NUCL. PHYS. VOL.10, 1, .   18, (1969).
                                (19B,   35) SOV. AT. ENERGY VOL.29,. . . . .   150, (1970).
                                (20C, 109) SOV. AT. ENERGY VOL.32, NO.5,. .   416, (1972).

2799 MOSKOVKIN.......(V.N.)..# (6 , 1098) SOV. AT. ENERGY(USA) VOL.31,. .   459, (1971).
                                (6 , 228) SOV. AT. ENERGY VOL.31,. . . . .   459, (1971).

2800 MOSSBAUER.......(R.L.)..# (9D,   15) ANN. DE PHYSIQUE TOME 7 NO.4,. .   287, (1972).

2801 MOSS............(J.)....# (5 , 1160) J. PHYS.F# METAL PHYS. VOL.2,. .   358, (1972).

2802 MOSS............(S.C.)..# (5 , 1749) PHYS. REV. 147,. . . . .   418, (1966).
                                (5 , 735) PHYS. REV. 175,. . . . .   868, (1968).
                                (6 , 2339) SOL. STATE COMM. VOL.13, ... 1465, (1973).

2803 MOSTOLLER.......(M.)....# (10B, 178) PHYS. REV. B VOL.5,. . . . . . . 1260, (1972).
                                (6 , 1585) PHYS. REV. B VOL.9,. . . . . . .   353, (1974).

2804 MOSTOVOI........(YU.A.).# (12B,   17) INSTRUM. EXP. TECH. 3, ...   729, (1970).

2805 MOSTOVO.........(V.I.)..# (6 , 1098) SOV. AT. ENERGY(USA) VOL.31,. .   459, (1971).
                                (6 , 228) SOV. AT. ENERGY VOL.31,. . . . .   459, (1971).

2806 MOSZKOWSKI......(S.A.)..# (5 , 924) NUCL. SCI. ENG. VOL.46,. . . . .   255, (1971).

2807 MOSZYNSKI.......(M.)....# (19B,   11) NUCL. INST. MET. VOL.29, ...   241, (1964).

2808 MOTEGI..........(H.)....# (5 , 774) ACTA CRYST. VOL.A28 PART S-4,. .   192, (1972).
                                (5 , 1273) J. PHYS. SOC. JAPAN VOL.32,. . . 1019, (1972).
                                (5 , 775) J. PHYS. SOC. JAPAN VOL.35,. . .   842, (1973).

2809 MOTIZUKI........(K.)....# (6 , 511) REV. MOD. PHYS. VOL.30,. . . . .   89, (1958).
                                (6 , 1494) J. PHYS. SOC. JAPAN VOL.36,. . .   112, (1974).

2810 MOTOHASHI.......(H.)....# (20D,   8) JAPAN. J. APPL. PHYS. VOL.4,. .   911, (1965).
                                (9B,   46) J. APPL. CRYST. VOL.4, ...   528, (1971).
                                (5 , 1217) ACTA CRYST. VOL.A28 PART S-4,. .   220, (1972).
                                (6 , 1924) J. POLYM. SCI./POLYM. LETT. 11,.   377, (1973).

2811 MOTT............(W.E.)..# (16 ,   90) TRANS. AM. NUCL. SOC. VOL.13,. .   488, (1970).

2812 MOTZ............(L.)....# (7A,   61) PHYS. REV. VOL.58, ...   26, (1940).

2813 MOUNTAIN........(R.D.)..# (5 , 2059) J. CHEM. PHYS. VOL.57, ... 3987, (1972).
                                (14A,   23) J. CHEM. PHYS. VOL.57, ... 4346, (1972).
                                (14A,   57) PHYS. REV. A VOL.5,. . . . . . . 2629, (1972).
                                (5 , 1329) PHYS. REV. A VOL.9,. . . . . . .   435, (1974).

2814 MOUSSA..........(F.)....# (6 , 2331) PHYS. LETT. VOL.36A, ...   376, (1971).
                                (6 , 623) PHYS. REV. LETT. VOL.26, ...   770, (1971).
                                (8 ,   13) ANN. DE PHYSIQUE TOME 7 NO.4,. .   233, (1972).
                                (6 , 1490) IAEA SYMP. GRENOBLE, ...   631, (1972).
                                (6 , 627) PHYS. REV. LETT. VOL.28, ...   964, (1972).
                                (6 , 1486) SOL. STATE COMM. VOL.10, ...   553, (1972).
                                (6 , 622) SOL. STATE COMM. VOL.13, ... 1725, (1973).
                                (6 , 444) SOL. STATE COMM. VOL.13, ...   919, (1973).

2815 MOXON...........(M.C.)..# (5 , 2825) PROC. PHYS. SOC. 82, ...   477, (1963).
                                (5 , 139) J. NUCL. ENERGY VOL.24,. . . . .   85, (1970).
                                (5 , 1544) 2ND INTERNAT. CONF./NUCL. DATA,.   , (1970).

2816 MOYER...........(M.W.)..# (20A,   32) NATURE VOL.211,. . . . .   400, (1966).
                                (22C,   35) NUCL. INST. MET. 53, ...   299, (1967).

2817 MOZER...........(B.)....# (6 , 1771) IAEA SYMP. CHALK RIVER, VOL.2, .   167, (1962).
                                (6 , 1869) PHYS. REV. LETT. VOL.8,. . . . .   278, (1962).
                                (6 , 1709) COPENHAGEN CONF.-LATT. DYNAMICS,.   63, (1963).
                                (6 , 2247) PHYS. REV. VOL.152,. . . . . . .   535, (1966).
                                (6 , 2248) IAEA SYMP. COPENHAGEN, VOL.1,. .   55, (1968).
                                (5 , 735) PHYS. REV. 175, ...   868, (1968).
                                (6 , 632) PHYS. LETT. 30A, ...   206, (1969).
                                (5 , 2058) J. CHEM. PHYS. VOL.55, ... 4967, (1971).

```
2817  MOZER...........(B.)....:  ( 5 , 1328) PHYS. REV. A VOL.9,. . . . . . .   448, (1974).

2818  MRYGON..........(B.)....:  (13 ,   38) PHYS. LETT. 25A, . . . . . . . .   600, (1967).

2819  MUCKER..........(K.F.)..:  ( 5 ,  768) PHYS. REV. LETT. 15, . . . . . .   586, (1965).
                                 ( 5 ,  770) PHYS. REV. LETT. 16, . . . . . .   799, (1966).
                                 ( 5 ,  762) DISS. ABS. XXVII, . . . . . .  .38928, (1967).
                                 ( 5 ,  760) J. CHEM. PHYSICS VOL.49, . . . .  1922, (1968).

2820  MUECK...........(K.)....:  (20C,   28) J. NUCL. ENERGY VOL.27,. . . . .   677, (1973).

2821  MUEHLHAUSE......(C.O.)..:  ( 5 , 1348) PHYS. REV. VOL.82, . . . . . . .   560, (1951).

2822  MUELLER.........(M.H.)..:  ( 5 , 2757) ACTA CRYST. VOL.11,. . . . . . .   751, (1958).
                                 ( 9B,   51) J. APPL. PHYS. VOL.30, . . . . .  1323, (1959).
                                 ( 5 , 2662) PHYS. REV. VOL.118, . . . . . .    797, (1960).
                                 (21 ,   43) REV. SCI. INSTRUM. VOL.32, . . .   456, (1961).
                                 ( 5 , 2705) PHYS. REV. VOL.129, . . . . . .    625, (1963).
                                 (20B,  146) REV. SCI. INSTRUM. VOL.34, . . .   847, (1963).
                                 (22C,   53) REV. SCI. INSTRUM. VOL.34, . . .    74, (1963).
                                 ( 5 , 2787) ACTA CRYST. 19,. . . . . . . . .   536, (1965).
                                 ( 5 , 2639) ACTA CRYST. 20, . . . . . . . .    842, (1966).
                                 ( 5 , 2211) J. PHYS. CHEM. SOL. 28, . . . .   1651, (1967).
                                 ( 5 , 2799) J. PHYS. CHEM. SOLIDS 30, . . .    453, (1969).
                                 ( 5 , 2213) J. PHYS. CHEM. SOLIDS 30, . . .    733, (1969).
                                 ( 5 , 1439) ACTA CRYST. VOL.A26, . . . . . .   559, (1970).
                                 ( 5 , 2699) ACTA CRYST. VOL.B26, . . . . . .   129, (1970).
                                 (21 ,   21) J. APPL. CRYST. VOL.3, . . . . .   289, (1970).
                                 ( 5 , 2367) ACTA CRYST. VOL.B27, . . . . . .  2284, (1971).
                                 ( 5 ,  248) J. AMER. CHEM. SOC. VOL.93 . . .  5945, (1971).
                                 ( 5 , 2214) J. PHYSIQUE TOME 32 COL.1 VOL.II  917, (1971).
                                 ( 5 , 2720) MAGN. AND MAG. MATERIALS-1971, .  1371, (1972).
                                 ( 5 , 2724) PHYS. REV. B VOL.6, . . . . . .   1880, (1972).
                                 ( 5 , 2216) A.I.P. CONF. PROC. NO.10 PART 1,   88, (1973).
                                 ( 5 , 2217) INT. J. MAGN. VOL.4 NO.2, . . .    99, (1973).
                                 ( 5 , 2838) J. APPL. CRYST. VOL.6, . . . . .   206, (1973).
                                 ( 5 , 2369) PHYS. REV. B VOL.9, . . . . . .   1041, (1974).
                                 ( 5 , 2212) PHYS. REV. B VOL.9, . . . . . .   3766, (1974).

2823  MUETHER.........(H.R.)..:  ( 6 ,   72) IAEA SYMP. COPENHAGEN, VOL.1,. .   457, (1968).
                                 ( 6 ,   92) PHYS. LETT. VOL.26A, . . . . . .   152, (1968).

2824  MUGHABGHAB......(S.F.)..:  ( 5 , 2913) PROC/CONF/NEUTRON C. S. + TECH.,   875, (1968).

2825  MUHLESTEIN......(L.D.)..:  ( 6 ,  539) PHYS. REV. LETT. VOL.23, . . . .   311, (1969).
                                 ( 6 ,  537) J. APPL. PHYS. 41, . . . . . .    1365, (1970).
                                 ( 6 , 2284) PHYS. REV. B-1, . . . . . . . .   2430, (1970).
                                 ( 6 ,  550) PHYS. REV. B-2, . . . . . . . .   4864, (1970).
                                 ( 6 , 1658) CONF. INTERN., RENNES, . . . . .   104, (1971).
                                 ( 6 ,  515) PHYS. REV. B-4, . . . . . . . .    969, (1971).
                                 ( 6 ,  528) IAEA SYMP. GRENOBLE, . . . . . .    53, (1972).
                                 ( 6 ,  529) A.I.P. CONF. PROC. NO.10 PART 1,   41, (1973).
                                 ( 5 , 1929) J. APPL. PHYS. VOL.45, . . . . .  2021, (1974).

2826  MULLER-HARTMANN.(E.)....:  (11A,   67) Z. PHYSIK VOL.223, . . . . . . .   277, (1969).

2827  MULLER..........(K.A.)..:  ( 6 , 1352) PHYS. REV. VOL.183, . . . . . .    820, (1969).
                                 ( 6 , 1351) BULL. AMER. PHYS. SOC. VOL.18, .   111, (1973).
                                 ( 6 , 1353) PHYS. REV. B VOL.8, . . . . . .   1119, (1973).

2828  MULLER..........(K.D.)..:  (20C,   72) NUCL. INST. MET. VOL.83, . . . .   111, (1970).

2829  MULLNER.........(M.)....:  (24 ,   76) NUCL. INST. MET. VOL.103,. . . .   229, (1972).
                                 ( 5 , 2672) PHYS. STAT. SOLIDI B VOL.61 NO.1  241, (1974).

2830  MUNNO...........(F.J.)..:  (16 ,   45) NUCL. SCI. ENG. VOL.43,. . . . .   120, (1971).

2831  MUNN............(R.W.)..:  ( 5 , 2932) ACTA CRYST. 22,. . . . . . . . .   170, (1967).

2832  MURADYAN........(G.V.)..:  ( 5 , 2478) SOV. J. NUCL. PHYS. VOL.10, 1, .    18, (1969).

2833  MURASIK.........(A.)....:  ( 5 ,  138) ACTA PHYS. POLON. VOL.17,. . . .   483, (1958).
                                 ( 5 ,  306) IAEA SYMPOSIUM VIENNA, . . . . .   293, (1960).
                                 ( 6 ,  974) NUKLEONIKA VOL.5,. . . . . . . .   495, (1960).
                                 ( 6 ,  886) PHYSICS VOL.27, . . . . . . . .    883, (1961).
                                 ( 6 ,  542) IAEA SYMP. CHALK RIVER, VOL.2,.    327, (1962).
                                 ( 6 ,  885) J. PHYS. SOC. JAPAN VOL.17 B-III   69, (1962).
                                 ( 6 ,  626) ACTA PHYS. POLON. VOL.24,. . . .   249, (1963).
                                 ( 5 , 1230) COLL. INTER. N.126 (GRENOBLE), .    92, (1963).
                                 ( 5 , 1044) COLL. INTER. N.126 (GRENOBLE), .    98, (1963).
                                 ( 5 , 2796) PHYS. STAT. SOL. 10, . . . . . .  K85, (1965).
                                 ( 5 , 2813) PHYS. STAT. SOL. 22, . . . . . .   517, (1967).
                                 ( 5 , 2716) PHYS. STAT. SOL. 23, . . . . . .  K123, (1967).
                                 ( 5 , 2717) PHYS. STAT. SOL. 30, . . . . . .   157, (1968).
                                 ( 5 , 2791) PHYS. STAT. SOL. 30, . . . . . .    61, (1968).
                                 ( 5 , 2792) PHYS. STAT. SOL. 34, . . . . . .  K157, (1969).
                                 ( 5 , 2950) PHYS. STAT. SOL. VOL.37, . . . .   843, (1970).
                                 ( 5 , 2722) PHYS. STAT. SOL. VOL.38, . . . .   K89, (1970).
                                 ( 5 , 2946) PHYS. STAT. SOLIDI A VOL.4 NO.1,    53, (1971).
                                 ( 5 , 2181) PHYS. STAT. SOL. B VOL.43, . . .   125, (1971).
                                 ( 5 , 2723) PHYS. STAT. SOLIDI A VOL.16 NO.2 K171, (1973).
                                 ( 5 , 2803) PHYS. STAT. SOLIDI A VOL.19 NO.1  K89, (1973).
                                 ( 5 , 2817) PHYS. STAT. SOLIDI A VOL.20, . .   395, (1973).

2834  MURATA..........(T.)....:  ( 6 , 2207) 2ND INTERNAT. CONF./NUCL. DATA,. . . , (1970).

2835  MURPHEY.........(W.M.)..:  (20A,   71) REACTOR SCI. TECHNOL. VOL.14,. .    25, (1961).

2836  MURPHY..........(E.J.)..:  ( 7A,   60) PHYS. REV. VOL.50, . . . . . . .   133, (1936).
                                 ( 5 , 2708) PHYS. REV. VOL.55, . . . . . . .   793, (1939).

2837  MURRAY..........(D.O.)..:  ( 5 , 2179) DISS. ABSTR. VOL.23, . . . . . .   668, (1962).

2838  MURRAY..........(G.)....:  (11A,   17) J. PHYS. C VOL.2,. . . . . . . .   539, (1969).

2839  MURRAY..........(R.L.)..:  (14B,   58) NUCL. SCI. ENGNG. VOL.50,. . . .   164, (1973).

2840  MURTAZIN........(I.A.)..:  (10D,   66) SOV. PHYS. SOL. STATE VOL.13,. .   112, (1971).

2841  MURTHY..........(M.R.L.):  ( 5 , 1824) PHYS. LETT. VOL.15,. . . . . . .   225, (1965).
                                 ( 5 , 1018) SOL. STATE COMM. 3,. . . . . . .   113, (1965).
                                 (12B,   15) INDIAN J. PURE APPL. PHYS. VOL.7  546, (1969).
                                 ( 5 , 1628) J. APP. PHYS. VOL.40,. . . . . .  1870, (1969).
                                 ( 5 , 1663) J. PHYS. CHEM. SOLIDS 30, . . .    939, (1969).
                                 ( 5 , 2314) PHYS. STAT. SOLIDI A VOL.3,. . .   959, (1970).
                                 ( 5 ,  522) NUCL./S.S. PHYS. SYMP. ABSTR.,. . . , (1972).

2842  MURTHY..........(N.S.S.):  ( 6 ,  782) IAEA SYMPOSIUM VIENNA, . . . . .   555, (1960).
                                 ( 6 , 1499) IAEA SYMP. BOMBAY, VOL.1,. . . .   433, (1964).
                                 ( 6 , 1497) PROC. INT. CONF. (NOTTINGHAM), .   322, (1964).
                                 ( 5 , 1824) PHYS. LETT. VOL.15,. . . . . . .   225, (1965).
                                 ( 5 , 1018) SOL. STATE COMM. 3,. . . . . . .   113, (1965).
                                 ( 6 ,  532) J. APPL. PHYS. VOL.39 NO.2, . .   1113, (1968).
                                 ( 6 ,  533) PHYS. LETT. VOL.26A, . . . . . .   108, (1968).
                                 ( 6 ,  816) SOL. STATE COMM. 6, . . . . . .    593, (1968).
                                 (12B,   15) INDIAN J. PURE APPL. PHYS. VOL.7  546, (1969).
```

2842 MURTHY.........(N.S.S.)..‡ (5 , 1628) J. APP. PHYS. VOL.40,...:...: 1870, (1969).
(5 , 2142) J. PHYS. CHEM. SOLIDS VOL.30,: : 1941, (1969).
(5 , 1663) J. PHYS. CHEM. SOLIDS 30,.: : : 939, (1969).
(5 , 2955) PHYS. REV. VOL.181,. . . . . . 969, (1969).
(6 , 1496) SOL. STATE COMM. 7,. . . . . . 123, (1969).
(5 , 2314) PHYS. STAT. SOLIDI'A VOL.3,.: : 959, (1970).
(5 , 2677) J. PHYSIQUE TOME 32 COL:1 VOL:II 617, (1971).
(6 , 2082) J. PHYSIQUE TOME 32 COL:1 VOL:I, 318, (1971).
(6 , 1777) IAEA SYMP. GRENOBLE,. . . . . 669, (1972).
(5 , 2319) J. PHYS. CHEM. SOLIDS VOL:33,.: 759, (1972).
(5 , 522) NUCL./S.S. PHYS. SYMP. ABSTR.; . . . . . (1972).
(6 , 1482) PHYS. LETT. VOL.40A,. . . . . . 101, (1972).

2843 MURZIN.........(A.V.)..‡ (7A, 21) BU. ACAD. SCI. USSR PHYS. SER.35 2349, (1971).

2844 MUSAELYAN.......(R.M.)..‡ (6 , 920) BU. ACAD. SCI. USSR PHYS. SR. 33 1754, (1969).
(17A, 22) BU. ACAD. SCI. USSR PHYS. SR. 33 36, (1969).

2845 MUSCI..........(M.)....‡ (10C, 26) ENERGIA NUCLEARE VOL.7,. . . . . 772, (1960).
(20A, 17) ENERGIA NUCLEARE VOL.10,. . . . 173, (1963).

2846 MUTO...........(M.)....‡ (5 , 1982) J. PHYS. SOC. JAPAN VOL.35,. . . 628, (1973).

2847 MYERS..........(H.P.)..‡ (5 , 1859) CAN. J. PHYS. VOL.35,. . . . . . 313, (1957).

2848 MYERS..........(V.W.)..‡ (5 , 239) PHYS. REV. VOL.75,. . . . . . . 217, (1949).
(5 , 143) PHYS. REV. VOL.92,. . . . . . . 716, (1953).
(6 , 1869) PHYS. REV. LETT. VOL.8,. . . . . 278, (1962).
(6 , 1559) J. CHEM. PHYS. VOL.46,. . . . . 4034, (1967).
(6 , 2067) J. PHYS. CHEM. SOL. 28,. . . . . 2207, (1967).

2849 MYERS..........(W.R.)..‡ (6 , 1885) DISS. ABS. VOL.27,. . . . . . . 586B, (1966).
(6 , 1901) J. CHEM. PHYSICS VOL.49,. . . . 1043, (1968).
(9C, 36) PROC/SYMP/CRYS. STRUC/HIGH PRESS 141, (1969).

2850 MYERS..........(W.)....‡ (6 , 1888) BROOKHAVEN SYMPOSIUM,. . . . . . 126, (1965).
(6 , 1896) J. CHEM. PHYS. VOL.42,. . . . . 4299, (1965).
(10C, 61) PHYS. LETT. 16,. . . . . 225, (1965).
(6 , 1900) J. CHEM. PHYS. VOL.44,. . . . . 184, (1966).

2851 MYNATT.........(F.R.)..‡ (24 , 69) NUCL. INST. MET. 72, ... 213, (1969).

2852 NABAUER........(M.)....‡ (17A, 101) Z. ANGEW. PHYS. VOL.12,. . . . . 133, (1960).

2853 NAGAI..........(O.)....‡ (6 , 1464) PROGR. THEOR. PHYS. VOL.25,. . . 595, (1961).
(6 , 804) J. PHYS. SOC. JAPAN VOL.18,. . . 74, (1963).
(11A, 40) PHYS. REV. VOL.188,. . . . . . 821, (1969).

2854 NAGAMIYA.......(T.)....‡ (6 , 511) REV. MOD. PHYS. VOL.30,. . . . . 89, (1958).
(9D, 34) J. APPL. PHYS. VOL.33, SUPPL., : 1029, (1962).

2855 NAGATA.........(T.)....‡ (12B, 23) J. PHYS. SOC. JAPAN VOL.28,. . . 1116, (1970).

2856 NAGIB..........(M.N.)..‡ (5 , 433) ATOMKERNENERGIE VOL.11,. . . . . 286, (1966).

2857 NAGIB..........(M.)....‡ (5 , 1569) ATOMKERNENERGIE VOL.21,. . . . . 275, (1973).

2858 NAGY...........(E.)....‡ (5 , 1730) J. PHYS. CHEM. SOLIDS 29,. . . . 101, (1968).

2859 NAGY...........(G.)....‡ (19A, 27) IAEA SYMP. GRENOBLE, ... 763, (1972).

2860 NAGY...........(I.)....‡ (5 , 1730) J. PHYS. CHEM. SOLIDS 29,. . . . 101, (1968).

2861 NAHORNIAK......(V.)....‡ (6 , 1106) REV. ROUMAINE PHYS. VOL.9, . . . 737, (1964).
(6 , 1524) PHYS. LETT. 22,. . . . . 558, (1966).
(6 , 167) IAEA SYMP. COPENHAGEN, VOL.1,. : 439, (1968).
(6 , 653) REV. ROUMAINE PHYS. VOL.15,. . . 783, (1970).

2862 NAIDEN.........(E.P.)..‡ (5 , 1547) SOV. PHYS. J. NO.11,. . . . . . 88, (1968).
(5 , 1556) SOV. PHYS. SOL. STATE VOL.12,. : 770, (1970).
(5 , 1555) SOV. PHYS. J. VOL.12 NO.8, . . . 1018, (1972).

2863 NAITO..........(K.)....‡ (5 , 2759) J. PHYS. CHEM. SOLIDS VOL.32,. . 235, (1971).

2864 NAKAGAWA.......(M.)....‡ (24 , 81) NUCL. SCI. ENGNG. VOL.40,. . . . 25, (1970).

2865 NAKAGAWA.......(T.)....‡ (5 , 2491) J. PHYS. SOC. JAPAN VOL.24,. . . 219, (1968).

2866 NAKAGAWA.......(Y.)....‡ (6 , 1669) COPENHAGEN CONF.-LATT.DYNAMICS,. 39, (1963).
(6 , 1673) PHYS. REV. LETT. 11,. . . . . . 271, (1963).
(5 , 1858) J. PHYS. SOC. JAPAN VOL.19,. . : 1255, (1964).
(5 , 1719) J. PHYS. SOC. JAPAN VOL.19,. . : 2078, (1964).
(5 , 1856) J. PHYS. SOC. JAPAN VOL.19,. . : 2082, (1964).
(5 , 558) J. PHYS. SOC. JAPAN VOL.20,. . : 2244, (1965).
(5 , 1857) J. PHYS. SOC. JAPAN VOL.21,. . : 2080, (1966).

2867 NAKAHARA.......(Y.)....‡ (6 , 1043) J. NUCL. SCI. TECHNOL. VOL.5,. . 635, (1968).
(14B, 44) J. NUCL. SCI. TECHNOL. VOL.8,. : 1, (1971).

2868 NAKAI..........(Y.)....‡ (5 , 1620) J. PHYS. SOC. JAPAN VOL.28,. . : 615, (1970).
(5 , 120) J. PHYS. SOC. JAPAN VOL.29,. . : 978, (1970).
(6 , 872) PHYS. LETT. VOL.37A,. . . . . . 333, (1971).
(6 , 112) SOL. STATE COMM. VOL.9,. . . . : 921, (1971).
(6 , 541) J. PHYS. SOC. JAPAN VOL.33,. . : 1348, (1972).
(6 , 113) J. PHYS. SOC. JAPAN VOL.34,. . : 1197, (1973).

2869 NAKAJIMA.......(T.)....‡ (20C, 29) J. NUCL. SCI. TECHNOL. JAP. 4, . 22, (1967).
(20C, 48) NUCL. INST. MET. 53,. . . . . . 163, (1967).

2870 NAKAMURA.......(H.)....‡ (6 , 2207) 2ND INTERNAT. CONF./NUCL. DATA,. , (1970).

2871 NAKAMURA.......(T.)....‡ (15 , 44) PROGR. THEOR. PHYS. VOL.46,. . . 1666, (1971).

2872 NAKAMURA.......(Y.)....‡ (5 , 1070) J. PHYS. SOC. JAPAN VOL.27,. . . 1470, (1969).
(5 , 394) PHYS. STAT. SOL. VOL.44,. . . . K25, (1971).

2873 NAKAYAMA.......(K.)....‡ (9C, 33) PHYS. LETT. VOL.37A, ... 29, (1971).

2874 NAKAYAMA.......(S.)....‡ (5 , 2491) J. PHYS. SOC. JAPAN VOL.24,. . . 219, (1968).

2875 NAMIOT.........(V.A.)..‡ (16 , 86) SOV. PHYS. DOKLADY VOL.18, . . . 481, (1974).

2876 NAMJOSHI.......(K.V.)..‡ (10B, 156) PHYS. REV. B-2,. . . . . 2167, (1970).
(10B, 158) PHYS. REV. B-2,. . . . . 967, (1970).
(6 , 562) PHYS. REV. B VOL.7,. . . . . . 4001, (1973).

2877 NAMTALISHVILI...(M.I.)..‡ (5 , 1132) SOV. PHYS. CRYST. VOL.16,. . . . 935, (1971).
(5 , 168) SOV. PHYS. CRYST. VOL.16,. . . . 711, (1972).
(5 , 182) SOV. PHYS. JETP VOL.35,. . . . . 370, (1972).

2878 NANDINI........(R.)....‡ (5 , 660) PHYS LETT 28A, ... 301, (1968).

2879 NAPOLI.........(C.)....‡ (22C, 37) NUCL. INST. MET. 59, ... 117, (1968).

```
2880  NARATH..........(A.)....:  (  5 ,   450) J. APP. PHYS. 37,. . . . . . . .   1126, (1966).

2881  NARAY-SZABO.....(I.)....:  (  5 ,    47) ACTA CRYST. 19,. . . . . . . .    180, (1965).

2882  NARDELLI........(G.F.)..:  (14B,    87) PHYS. REV. VOL.148,. . . . . .    124, (1966).
                                 (14B,   103) REACTOR PHYSICS, . . . . . . .     73, (1966).
                                 (15 ,    49) REPORT EUR-4037.1 (50 PP.),. .    . . , (1968).
                                 (10B,    52) J. CHEM. PHYSICS VOL.48,. . . .   5242, (1969).

2883  NARTEN..........(A.H.)..:  (  5 ,  1197) J. CHEM. PHYS. VOL.56,. . . . .   1185, (1972).
                                 (  5 ,  1301) J. CHEM. PHYS. VOL.56,. . . . .   5681, (1972).
                                 (  5 ,  1549) J. CHEM. PHYS. VOL.58,. . . . .   5017, (1973).

2884  NASR-EDDINE.....(M.)....:  (  5 ,   594) SOL. STATE COMM. VOL. 9,. . . .    717, (1971).
                                 (  5 ,   562) SOL. STATE COMM. VOL.12 NO.7,.    665, (1973).
                                 (  5 ,   642) SOL. STATE COMM. VOL.13, . . .    905, (1973).

2885  NASUHOGLU.......(R.)....:  (  6 ,   247) J. CHEM. PHYS. VOL.32, . . . .    476, (1960).

2886  NATERA..........(M.G.)..:  (  5 ,   816) SOL. STATE COMM. 6,. . . . . .    593, (1968).
                                 (  5 ,  2142) J. PHYS. CHEM. SOLIDS VOL.30,.   1941, (1969).
                                 (  5 ,  2955) PHYS. REV. VOL.181,. . . . . .    969, (1969).
                                 (  5 ,  2314) PHYS. STAT. SOLIDI A VOL.3,. .    959, (1970).
                                 (  6 ,  2082) J. PHYSIQUE TOME 32 COL.1 VOL.I,  318, (1971).

2887  NATHANS.........(R.)....:  (  5 ,  1610) PROC. I.E.E. VOL.104B SUP.4-7,    217, (1957).
                                 (  5 ,   958) J. PHYS. CHEM. SOLIDS VOL.6, .     38, (1958).
                                 (  5 ,  1677) J. APPL. PHYS. VOL.30, SUPPL.,    280, (1959).
                                 (  5 ,  2083) J. PHYS. CHEM. SOLIDS VOL.10,.    138, (1959).
                                 (  5 ,  1187) J. PHYS. CHEM. SOLIDS VOL.13,.     35, (1959).
                                 (  5 ,   440) PHYS. REV. LETT. VOL.2,. . . .    254, (1959).
                                 (  5 ,  1688) PHYS. REV. VOL.116,. . . . . .    317, (1959).
                                 ( 9D,    32) J. APPL. PHYS. VOL.31, SUPPL.,    350, (1960).
                                 (  5 ,   957) J. APPL. PHYS. VOL.31, SUPPL.,    372, (1960).
                                 (  5 ,  1568) PHYS. REV. VOL.121,. . . . . .    707, (1961).
                                 (  5 ,   954) PHYS. REV. VOL.123,. . . . . .   1163, (1961).
                                 (  5 ,   933) J. PHYS. SOC. JAPAN VOL.17 B-III    7, (1962).
                                 (  5 ,  1683) PHYS. REV. LETT. VOL.8,. . . .    237, (1962).
                                 (  5 ,  1090) COLL. INTER. N.126 (GRENOBLE),    118, (1963).
                                 (  5 ,  1131) J. APP. PHYS. 34,. . . . . . .   1044, (1963).
                                 (  6 ,  1460) J. APP. PHYS. 34,. . . . . . .   1182, (1963).
                                 (  6 ,  1513) J. APP. PHYS. 34,. . . . . . .   1201, (1963).
                                 (  5 ,  1784) J. APP. PHYS. 34,. . . . . . .   1203, (1963).
                                 (  5 ,   915) J. PHYS. CHEM. SOL. 24,. . . .   1679, (1963).
                                 (  5 ,  1206) J. APPL. PHYS. VOL.35,. . . . .   1045, (1964).
                                 (  6 ,   809) J. PHYS. CHEM. SOLIDS VOL.25,.    183, (1964).
                                 (  5 ,  2983) PHYS. REV. LETT. VOL.12,. . . .  1094, (1964).
                                 (  5 ,  1125) PHYS. REV. VOL.134,. . . . . .  A1547, (1964).
                                 (  6 ,   846) PHYS. REV. VOL.136,. . . . . .  A1641, (1964).
                                 (  5 ,  1122) PROC. INT. CONF. (NOTTINGHAM),    223, (1964).
                                 (  5 ,  1031) PROC. INT. CONF. (NOTTINGHAM),    327, (1964).
                                 (  6 ,  1655) J. APP. PHYS. 36,. . . . . . .   1092, (1965).
                                 (  5 ,   604) J. APP. PHYS. 36,. . . . . . .   1099, (1965).
                                 (  6 ,   751) PHYS. REV. LETT. 15, . . . . .    146, (1965).
                                 (  5 ,   539) PROC. PHYS. SOC. 85, . . . . .   1185, (1965).
                                 (  6 ,  1315) J. APP. PHYS. 37,. . . . . . .   1041, (1966).
                                 (  6 ,   478) J. APP. PHYS. 37,. . . . . . .   1052, (1966).
                                 (17D,     1) ACTA CRYST. 23,. . . . . . . .    357, (1967).
                                 ( 9D,    17) ATLANTA SYMP/GEORGIA INST. TECH.   96, (1967).
                                 (  5 ,  2530) J. APPL. PHYS. VOL.38,. . . . .    969, (1967).
                                 (11B,    15) METAL. SOC. CONF. LOS ANGELES,    143, (1967).
                                 (  6 ,  1185) PHYS. REV. LETT. 19, . . . . .   1307, (1967).
                                 (  6 ,   841) PHYS. REV. 154,. . . . . . . .    508, (1967).
                                 (  6 ,  1738) PHYS. REV. 156,. . . . . . . .    623, (1967).
                                 (  6 ,  1342) PHYS. REV. 157,. . . . . . . .    396, (1967).
                                 (  6 ,  2130) PHYS. REV. 161,. . . . . . . .    499, (1967).
                                 (  6 ,   778) PHYS. REV. 162,. . . . . . . .    528, (1967).
                                 (17D,     2) ACTA CRYST. A24,. . . . . . . .    481, (1968).
                                 (17D,     3) ACTA CRYST. A24,. . . . . . . .    619, (1968).
                                 (13 ,    11) C.N.E.N. SYMP. CASACCIA, . . .    133, (1968).
                                 (  6 ,   699) J. APP. PHYS. 39,. . . . . . .   1232, (1968).
                                 (  6 ,  1992) J. APP. PHYS. 39,. . . . . . .   1237, (1968).
                                 (  5 ,  2652) J. APP. PHYS. 39,. . . . . . .   2221, (1968).
                                 (  6 ,  1728) J. APP. PHYS. 39,. . . . . . .    383, (1968).
                                 (  6 ,   829) J. APP. PHYS. 39,. . . . . . .    455, (1968).
                                 (  6 ,   728) KYOTO CONF. STAT. MECHANICS,     169, (1968).
                                 (  6 ,   752) PHYS. REV. LETT. 21, . . . . .     99, (1968).
                                 (  6 ,  1995) J. APPL. PHYS. VOL.40,. . . . .   1278, (1969).
                                 (  6 ,   734) J. APP. PHYS. VOL.40,. . . . .   1442, (1969).
                                 (  6 ,   738) J. PHYS. SOC. JAPAN (SUPPL.) 26,  169, (1969).
                                 (  6 ,  1985) PHYS. REV. LETT. VOL.23, . . .   1225, (1969).
                                 (  6 ,   794) PHYS. REV. VOL.179,. . . . . .    417, (1969).
                                 (  6 ,  1716) PHYS. REV. VOL.182,. . . . . .    624, (1969).
                                 (  6 ,   816) PHYS. REV. VOL.186,. . . . . .    557, (1969).
                                 (  5 ,   813) PHYS. REV. VOL.186,. . . . . .    567, (1969).
                                 ( 9C,     8) BER. BUNSENGES. PHYS. CHEM. 74,  1202, (1970).
                                 (  6 ,  1996) J. APPL. PHYS. VOL.41,. . . . .   1384, (1970).
                                 (  6 ,  1446) PHYS. REV. B-1,. . . . . . . .   2304, (1970).
                                 (  6 ,  1451) PHYS. REV. LETT. VOL.24, . . .   1184, (1970).
                                 (22C,     2) ACTA CRYST. VOL.A27,. . . . . .    284, (1971).
                                 (  6 ,  1448) J. APPL. PHYS. 42,. . . . . . .  1258, (1971).
                                 (  6 ,  1447) PHYS. REV. B-4,. . . . . . . .   2254, (1971).

2888  NATH............(N.)....:  (  6 ,  1804) INDIAN J. PHYS. VOL.42,. . . .    408, (1968).

2889  NATH............(R.)....:  (12A,    17) NUCL. INST. MET. VOL.98, . . .    385, (1972).

2890  NATION..........(R.)....:  (  6 ,  1026) IAEA SYMP. BOMBAY, VOL.2,. . .    141, (1964).

2891  NATKANIEC.......(I.)....:  (  6 ,  1379) PHYS. LETT. 24A, . . . . . . .    517, (1967).
                                 (  6 ,  1017) ACTA PHYS. POLON. VOL.33,. . .    419, (1968).
                                 (  6 ,  1104) IAEA SYMP. COPENHAGEN VOL.II,.    143, (1968).
                                 (  6 ,   621) IAEA SYMP. COPENHAGEN, VOL.1,.     65, (1968).
                                 (  6 ,  2003) IAEA SYMP. COPENHAGEN, VOL.2,.    237, (1968).
                                 (  6 ,   286) PHYSICA VOL.41,. . . . . . . .    397, (1969).
                                 (  6 ,  1936) PHYS. STAT. SOL.45,. . . . . .   K105, (1971).
                                 (  6 ,  1388) SOV. PHYS. CRYST. VOL.15,. . .   1010, (1971).

2892  NATOLI..........(C.R.)..:  (11B,    49) 1ST EUR. CONF./CONDENSED MATTER,   45, (1971).
                                 (  6 ,   491) PHYS. LETT. VOL.39A,. . . . .    105, (1972).
                                 (11A,    20) J. PHYS. C VOL.6,. . . . . . .    323, (1973).
                                 (  6 ,   495) J. PHYS. C VOL.6,. . . . . . .    370, (1973).
                                 (11A,    19) J. PHYS. C VOL.6,. . . . . . .    386, (1973).
                                 (11A,    23) J. PHYS. C VOL.7,. . . . . . .    947, (1974).

2893  NAUCIEL-BLOCH...(M.)....:  (11A,     4) ANN. DE PHYSIQUE TOME 5, . . .    139, (1970).
                                 (  5 ,   836) SOL. STATE COMM. VOL.8,. . . .   2141, (1970).

2894  NAUMANN.........(A.W.)..:  (  6 ,  1905) J. CHEM. PHYSICS VOL.45,. . .   3787, (1966).
                                 (  6 ,  1916) ADV. POLYMER SCI. VOL.5, . . .      1, (1967).
                                 (  6 ,  1115) J. CHEM. PHYS. VOL.50,. . . . .   4444, (1969).

2895  NAVALKAR........(M.B.)..:  (24 ,    49) NUCL. INST. MET. 61, . . . . .     37, (1968).
```

```
2896  NAVARRO.........(Q.O.)..!  (20D,    16) NUCL. INST. MET. 37, . . . . . .    121, (1965).
                                 ( 5 , 1433) ACTA CRYST. VOL.B26, . : . : . :    827, (1970).
                                 ( 5 , 1994) ACTA CRYST. B27, . . . : . : .      253, (1971).

2897  NEBE............(J.)....!  ( 5 , 1880) NUCL. PHYS. VOL.A166,. . . . . .    443, (1971).
                                 ( 5 , 1879) NUCL. PHYS. VOL.A166,. . . . . .    461, (1971).

2898  NECHAI..........(E.F.)..!  ( 5 , 1616) PHYS. STAT. SOL. VOL.36, . . . .    K25, (1969).

2899  NECHIPORUK......(V.V.)..!  ( 6 ,  218) SOV. PHYS. SOL. STATE VOL.8, . .    261, (1966).

2900  NEDLIN..........(G.M.)..!  ( 5 ,  218) PROC. INT. CONF. (NOTTINGHAM), .    354, (1964).

2901  NEELAKANDAN.....(K.)....!  ( 6 ,  583) PHYSICA VOL.53,. . . . . . . . .    628, (1971).
                                 (10B,  217) SOL. STATE COMM. VOL. 9, . . . .    167, (1971).
                                 (10B,  104) PHYSICA VOL.62,. . . . . . . . .    587, (1972).
                                 ( 6 , 1365) J. PHYS. F VOL.3,. . . . . . . .   1308, (1973).

2902  NEELAVATHI......(V.N.)..!  (24 ,   94) PHYS. REV. B VOL.3,. . . . . . .   2929, (1971).

2903  NEGOVETIC.......(I.)....!  ( 5 , 2180) FIZIKA (YUGOSLAVIA) VOL.5, . . .    179, (1973).
                                 ( 5 , 2174) SOL. STATE COMM. VOL.13, . . . .    249, (1973).

2904  NEIOHARDT.......(W.J.)..!  ( 5 ,  691) PHYS. REV. VOL.111, . . . . . .     261, (1958).

2905  NEILL...........(J.M.)..!  ( 5 , 1306) NUCL. SCI. ENG. VOL.33,. . . . .    265, (1968).
                                 ( 6 , 1946) NUCL. SCI. ENG. VOL.46,. . . . .    244, (1971).

2906  NEILSON.........(G.C.)..!  (19B,   15) NUCL. INST. MET. VOL.81, . . . .    301, (1970).

2907  NEISIG..........(K.E.)..!  (22B,   25) NUCL. INST. MET. 41, . . . . . .     61, (1966).

2908  NELIN...........(G.)....!  ( 5 , 2303) SOL. STATE COMM. 4,. . . . . . .    303, (1966).
                                 ( 6 , 1861) J. PHYS. CHEM. SOLIDS VOL.28,. .   2369, (1967).
                                 ( 6 ,  365) IAEA SYMP. COPENHAGEN, VOL.1,. .    475, (1968).
                                 ( 6 ,  915) PHYS. REV. B-3, . . . . . . . .     364, (1971).
                                 ( 5 , 2305) PHYS. STAT. SOL. B VOL.45, . . .    527, (1971).
                                 ( 6 ,  918) PHYS. REV. B VOL.5,. . . . . . .   3151, (1972).
                                 ( 6 ,  924) PHYS. REV. B VOL.6,. . . . . . .   3777, (1972).
                                 ( 6 ,  938) PHYS. REV. B VOL.9 NO.12,. . . . . . , (1974).

2909  NELKIN..........(M.)....!  ( 6 , 2356) PHYS. REV. VOL.108,. . . . . . .   1092, (1957).
                                 (16 ,   20) IAEA SYMPOSIUM VIENNA, . . . . .      3, (1960).
                                 ( 6 , 1086) PHYS. REV. VOL.119,. . . . . . .    741, (1960).
                                 ( 7B,   20) IAEA SYMPOSIUM BOMBAY, VOL.2,. .     35, (1964).
                                 (16 ,   38) NUCL. SCI. ENGNG. VOL.24,. . . .    142, (1966).
                                 ( 6 ,   89) PHYS. REV. LETT. VOL.16, . . . .    839, (1966).
                                 (14A,   43) PHYS. REV. VOL.164,. . . . . . .    222, (1967).
                                 (14B,   33) IAEA SYMP. COPENHAGEN, VOL.1,. .    535, (1968).
                                 (14A,   47) PHYS. REV. VOL.183,. . . . . . .    349, (1969).
                                 (14B,  110) REPORT AERE-R-6277 (16PP.),. . . . . , (1969).

2910  NELLIS..........(W.J.)..!  ( 5 , 2369) PHYS. REV. B VOL.9,. . . . . . .   1041, (1974).

2911  NELMES..........(R.J.)..!  ( 5 , 1460) SOL. STATE COMM. VOL.11, . . . .   1261, (1972).

2912  NELSON..........(R.D.)..!  (10D,   40) PHYS. REV. LETT. VOL.28, . . . .   1261, (1972).

2913  NEMARICH........(J.)....!  (11C,   25) PHYS. LETT. VOL.33A, . . . . . .    375, (1970).

2914  NEMIROVSKII.....(P.E.)..!  (24 ,  111) SOV. PHYS. JETP VOL.14,. . . . .    838, (1962).

2915  NEMJOSHI........(K.V.)..!  (10B,  215) SOL. STATE COMM. VOL. 9, . . . .    185, (1971).

2916  NEMNONOV........(S.A.)..!  ( 5 , 2649) PHYS. MET. METALLOGR. VOL.23 N.3    168, (1967).

2917  NEOV............(S.)....!  ( 5 , 1650) PHYS. STAT. SOL. VOL.49, . . . .    589, (1972).
                                 ( 5 , 1651) SOL. STATE COMM. VOL.10, . . . .    883, (1972).

2918  NERESON.........(N.G.)..!  ( 5 ,  907) PHYS. REV. VOL.124,. . . . . . .   1848, (1961).
                                 ( 5 ,  908) J. APPL. PHYS. VOL.33, SUPPL.,.   1135, (1962).
                                 ( 5 ,  912) PHYS. REV. VOL.127,. . . . . . .   2101, (1962).
                                 ( 5 , 2014) PHYS. REV. 131, . . . . . . . .    2098, (1963).
                                 ( 5 ,  906) PHYS. REV. 135, . . . . . . . .    A176, (1964).
                                 ( 5 , 2731) ACTA CRYST. 21,. . . . . . . . .    670, (1966).
                                 ( 5 , 2030) J. APP. PHYS. 37,. . . . . . . .   4575, (1966).
                                 ( 5 , 2339) J. APP. PHYS. 38,. . . . . . . .   1395, (1967).
                                 ( 5 ,  809) J. CHEM. PHYS. VOL.46, . . . . .   4041, (1967).
                                 ( 5 , 2631) ACTA CRYST. VOL.B24, . . . . . .   1121, (1968).
                                 ( 5 , 1524) ACTA CRYST. VOL.B24, . . . . . .    459, (1968).
                                 ( 5 ,  783) J. CHEM. PHYSICS VOL.49,. . . . .   4361, (1968).
                                 ( 5 , 2363) J. NUCL. MATERIALS VOL.34,. . . .    281, (1970).
                                 ( 5 ,  784) J. CHEM. PHYSICS VOL.55, . . . .    589, (1971).
                                 ( 5 ,  565) ACTA CRYST. VOL.B28, . . . . . .   3102, (1972).
                                 ( 5 ,  393) PHYS. REV. B VOL.9,. . . . . . .    154, (1974).

2919  NERESON.........(N.)....!  ( 5 ,  408) AIP CONF. PROC. VOL.5, . . . . .   1385, (1971).
                                 ( 5 , 2544) J. APPL. PHYS. VOL.42,. . . . . .   1625, (1971).
                                 ( 5 , 1652) A.I.P. CONF. PROC. NO.10 PART 1,    658, (1973).
                                 ( 5 , 1364) A.I.P. CONF. PROC. NO.10 PART 1,    669, (1973).
                                 ( 5 ,  855) J. APPL. PHYS. VOL.44,. . . . . .   4727, (1973).
                                 ( 5 , 2046) J. SOLID STATE CHEM. VOL.9,. . .    152, (1974).

2920  NERONOVA........(N.N.)..!  ( 90,   50) J. PHYS. SOC. JAPAN VOL.17 B-II,    330, (1962).
                                 ( 90,   49) J. PHYS. SOC. JAPAN VOL.17 B-II,    332, (1962).

2921  NERSESYAN.......(A.A.)..!  (11C,   58) SOV. PHYS. J.E.T.P. VOL.29,. . .   1089, (1969).

2922  NEVE-DE-MEVERGNI(M.)....!  ( 5 , 1907) PHYSICA VOL.30,. . . . . . . . .    237, (1964).
                                 (13B,   39) IAEA SYMP. COPENHAGEN, VOL.1,. .     91, (1968).

2923  NEWMAN..........(D.J.)..!  (10B,   79) J. PHYS. C VOL.7,. . . . . . . .   1443, (1974).

2924  NEWNHAM.........(R.E.)..!  ( 5 ,  511) ACTA CRYST. 17,. . . . . . . . .    240, (1964).
                                 ( 5 ,  508) J. PHYS. CHEM. SOL. 25,. . . . .    901, (1964).
                                 ( 5 , 2196) ACTA CRYST. 19,. . . . . . . . .    147, (1965).
                                 ( 5 , 1566) J. PHYS. CHEM. SOL. 26,. . . . .    445, (1965).
                                 ( 5 ,  364) J. PHYS. CHEM. SOL. 26,. . . . .    927, (1965).
                                 ( 5 ,  444) Z. KRIST. VOL.121, . . . . . . .    418, (1965).
                                 ( 5 ,  329) J. AMER. CERAM. SOC. VOL.49, . .    284, (1966).
                                 ( 5 , 1552) J. PHYS. CHEM. SOL. 27,. . . . .   1192, (1966).
                                 ( 5 , 1820) J. PHYS. CHEM. SOL 27,. . . . . .    655, (1966).
                                 ( 5 ,  443) PHYS. STAT. SOL. 16, . . . . . .    K17, (1966).
                                 ( 5 ,  222) Z. KRIST. VOL.123, . . . . . . .     73, (1967).
                                 ( 5 , 1567) ACTA CRYST. 22,. . . . . . . . .    344, (1967).
                                 ( 5 ,  563) J. APPL. PHYS. VOL.40, . . . . .   1124, (1969).
                                 ( 5 , 2486) MATER. RES. BULL. VOL.8, . . . .   1183, (1973).

2925  NEWNHAM.........(R.)....!  ( 5 , 2661) J. PHYS. CHEM. SOLIDS VOL.13,. .    166, (1960).
                                 ( 5 , 1178) SOL. STATE COMM. VOL.6 NO.5,. .     317, (1968).
                                 ( 5 ,  634) SOL. STATE COMM. VOL.6 NO.5, . .    323, (1968).

2926  NEWSTEAD........(C.M.)..!  ( 7A,   25) CONF/NUCL. STRUC. STUDY/NEUTRONS   144, (1972).
```

```
2927  NGO-VAN-HAI.............:  (20B,   12) C. R. ACAD. SCI. B TOME 270, . .   185, (1970).

2928  NGUYEN-VAN-GIAI.........:  ( 7A,   36) NUCL. PHYS. A VOL.A177,. . . . .   559, (1971).

2929  NGUYEN-VAN-NHUNG(A.)....:  ( 5 , 2453) J. PHYSIQUE TOME 32 COL.1 VOL.II 1133, (1971).

2930  NGUYEN-VAN-NHUNG(J.)....:  ( 5 , 2416) SOL. STATE COMM. VOL.10, . . . .   685, (1972).

2931  NGUYEN.........(D.H.)...:  (20A,   51) NUCL. TECHNOL. VOL.14, . . . . .   284, (1972).

2932  NG.............(S.C.)...:  ( 5 ,  693) MATERIALS RES. BULL. VOL.2,. . .    69, (1967).
                                 (10C,   52) METAL. SOC. CONF. LOS ANGELES, .   161, (1967).
                                 ( 6 , 2190) SOL. STATE COMM. 5,. . . . . . .    79, (1967).
                                 ( 6 ,  186) THESIS (137 PP.),. . . . . . . .     1, (1967).
                                 ( 6 ,  164) IAEA SYMP. COPENHAGEN, VOL.1,. .   253, (1968).

2933  NICHOLSON.......(K.P.)..:  (20C,    4) AERE REPORT N/R-1639 (7 PP.),. .    .., (1955).
                                 (23 ,   22) IAEA SYMP. (PILE RESEARCH)VIENNA   129, (1960).

2934  NICHOLS.........(H.)....:  ( 6 , 2068) J. ELECTROCHEM. SOC. VOL.112,. .   108, (1965).

2935  NICKLOW.........(R.M.)..:  ( 5 , 1500) J. APP. PHYS. 37,. . . . . . . .  1047, (1966).
                                 ( 6 ,   49) PHYS. REV. 143,. . . . . . . . .   487, (1966).
                                 ( 6 ,  600) PHYS. REV. 164,. . . . . . . . .   922, (1967).
                                 ( 6 , 1360) IAEA SYMP. COPENHAGEN, VOL.1,. .   169, (1968).
                                 ( 6 ,  611) IAEA SYMP. COPENHAGEN, VOL.1,. .    47, (1968).
                                 (22A,    8) IAEA SYMP. COPENHAGEN, VOL.2,. .   253, (1968).
                                 ( 6 ,  612) PHYS. REV. LETT. VOL.20, . . . .  1245, (1968).
                                 ( 6 , 1373) PHYS. REV. 168,. . . . . . . . .   970, (1968).
                                 ( 6 , 1712) J. APP. PHYS. VOL.40,. . . . . .  1450, (1969).
                                 ( 6 , 1226) J. APP. PHYS. VOL.40,. . . . . .  1452, (1969).
                                 ( 6 ,  906) J. APPL. PHYS. 41,. . . . . . .  1182, (1970).
                                 ( 6 , 2123) PHYS. REV. B-1,. . . . . . . . .  3943, (1970).
                                 ( 6 ,   16) PHYS. REV. B-1,. . . . . . . . .  4819, (1970).
                                 ( 6 ,  907) PHYS. REV. LETT. VOL.24, . . . .    16, (1970).
                                 ( 6 , 2294) AIP CONF. PROC. VOL.5,. . . . .  1450, (1971).
                                 ( 6 , 1264) CONF. INTERN., RENNES, . . . . .   144, (1971).
                                 ( 6 ,  694) J. APPL. PHYS. 42,. . . . . . .  1672, (1971).
                                 ( 6 ,  745) J. PHYSIQUE TOME 32 COL.1 VOL.II 1177, (1971).
                                 ( 6 , 1220) PHYS. REV. B-3,. . . . . . . . .  1229, (1971).
                                 ( 6 , 1240) PHYS. REV. B-3,. . . . . . . . .  1268, (1971).
                                 ( 6 , 2178) PHYS. REV. B-3,. . . . . . . . .  3457, (1971).
                                 ( 6 , 2039) PHYS. REV. B-4,. . . . . . . . .  2558, (1971).
                                 ( 6 ,  695) PHYS. REV. LETT. VOL.26, . . . .   140, (1971).
                                 ( 6 ,  703) PHYS. REV. LETT. VOL.27, . . . .   334, (1971).
                                 ( 6 , 1522) BULL. AMER. PHYS. SOC. VOL.17, .   292, (1972).
                                 ( 6 , 1265) IAEA SYMP. GRENOBLE, . . . . . .   103, (1972).
                                 ( 6 ,  696) IAEA SYMP. GRENOBLE, . . . . . .   611, (1972).
                                 ( 6 ,  697) MAGN. AND MAG. MATERIALS-1971, .  1446, (1972).
                                 ( 6 ,  204) PHYS. REV. B VOL.5,. . . . . . .  4951, (1972).
                                 ( 6 ,  639) BULL. AMER. PHYS. SOC. VOL.18, .   112, (1973).
                                 ( 6 ,  742) PHYS. REV. B VOL.7,. . . . . . .   336, (1973).
                                 ( 6 ,  111) PHYS. REV. B VOL.8,. . . . . . .  3493, (1973).
                                 ( 6 , 1739) PHYS. REV. LETT. VOL.30, . . . .   556, (1973).
                                 ( 6 ,  461) PHYS. REV. B VOL.10 NO.2,. . . .    .., (1974).

2936  NICOTERA........(E.)....:  (20B,  103) NUCL. INST. MET. VOL.104,. . . .   147, (1972).
                                 ( 5 , 1236) J. NON-CRYST. SOLIDS VOL.11 NO.5   417, (1973).

2937  NIEBERGALL......(F.)....:  ( 6 , 1155) NUCL. PHYS. VOL.A129,. . . . . .   666, (1969).

2938  NIELSEN.........(A.)....:  (20D,   34) REV. SCI. INSTRUM. VOL.36, . . .    48, (1965).

2939  NIELSEN.........(M.)....:  (17D,    5) ACTA CRYST. VOL.A25,. . . . . .   547, (1969).
                                 (17D,    9) IAEA SYMP. (INSTRUMENT.), VIENNA    49, (1969).
                                 ( 6 , 2124) J. APPL. PHYS. 41,. . . . . . .  1174, (1970).
                                 ( 6 , 2151) PHYS. REV. LETT. VOL.25, . . . .  1451, (1970).
                                 ( 5 , 2336) PHYS. REV. LETT. VOL.25, . . . .   524, (1970).
                                 ( 6 ,  654) PHYS. REV. B-3,. . . . . . . . .  4383, (1971).
                                 ( 6 ,  960) IAEA SYMP. GRENOBLE, . . . . . .   111, (1972).
                                 ( 6 ,  966) PHYS. REV. B VOL.7,. . . . . . .  1626, (1973).
                                 ( 6 ,  967) PHYS. REV. LETT. VOL.30, . . . .   481, (1973).

2940  NIELSEN.........(O.V.)..:  ( 5 , 1754) J. PHYS. C VOL.7,. . . . . . . .   409, (1974).

2941  NIELSEN.........(P.)....:  ( 6 ,  764) J. APPL. PHYS. VOL.35,. . . . .   933, (1964).
                                 ( 6 ,  524) PROC. INT. CONF. (NOTTINGHAM), .   101, (1964).
                                 ( 6 ,  753) PROC. INT. CONF. (NOTTINGHAM), .    99, (1964).
                                 ( 6 ,  756) PHYS. REV. VOL.139,. . . . . . . A1866, (1965).

2942  NIELSON.........(J.A.)..:  (13 ,   71) 1ST EUR. CONF./CONDENSED MATTER,    34, (1971).

2943  NIEMAN..........(H.)....:  (20A,   36) NUCL. INST. MET. 49, . . . . . .   117, (1967).

2944  NIEWIADOMSKI....(T.)....:  (19B,   24) NUKLEONIKA VOL.7,. . . . . . . .   231, (1962).

2945  NIFENECKER......(H.)....:  (24 ,   60) NUCL. INST. MET. 61, . . . . . .   198, (1968).

2946  NIGHIGORI.......(T.)....:  ( 7B,  106) PROGR. THEOR. PHYS. VOL.37,. . .  1051, (1967).

2947  NIILER..........(A.)....:  (12B,    3) ANN. OF PHYSICS VOL.74,. . . . .   250, (1972).

2948  NIIMURA.........(N.)....:  (21 ,   18) JAP. J. APPL. PHYS. VOL.10,. . .   933, (1971).
                                 ( 5 ,  774) ACTA CRYST. VOL.A28 PART S-4,. .   192, (1972).
                                 (19A,    4) ACTA CRYST. VOL.A28 PART S-4,. .   197, (1972).
                                 ( 5 , 1273) J. PHYS. SOC. JAPAN VOL.32,. . .  1019, (1972).
                                 ( 5 , 2446) J. PHYS. SOC. JAPAN VOL.33,. . .  1493, (1972).
                                 ( 6 ,  230) CHEM. PHYS. LETT. VOL.18,. . . .   366, (1973).
                                 ( 5 , 1982) J. PHYS. SOC. JAPAN VOL.35,. . .   628, (1973).
                                 ( 5 ,  775) J. PHYS. SOC. JAPAN VOL.35,. . .   842, (1973).

2949  NIIZEKI.........(N.)....:  ( 5 , 2284) J. PHYS. SOC. JAPAN VOL.35,. . .   314, (1973).

2950  NIJBOER........(B.R.A.):  (14B,   74) PHYS. REV. VOL.82, . . . . . . .   392, (1951).

2951  NIJBOER........(B.R.)..:  (10A,    9) NED. TIJDSCHRIFT NATUURKUNDE 29,    61, (1963).
                                 (10A,    8) NED. TIJDSCHRIFT NATUURKUNDE 29,    93, (1963).
                                 (14B,   64) PHYSICA VOL.32,. . . . . . . . .   415, (1966).

2952  NIKITIN.........(V.I.)..:  ( 5 , 1955) SOV. AT. ENERGY VOL.18,. . . . .   350, (1965).

2953  NIKLAUS.........(J.P.)..:  ( 6 , 2031) Z. PHYSIK VOL.190,. . . . . . .   295, (1966).
                                 ( 6 ,   37) Z. ANGEW. PHYS. VOL.24,NO.6, . .   313, (1968).

2954  NIKOTIN.........(O.P.)..:  ( 6 , 1466) J. PHYS. C, SER.2, VOL.2,. . . .  1168, (1969).
                                 (20B,   93) NUCL. INST. MET. 72, . . . . . .    77, (1969).

2955  NIKULIN.........(YU.M.).:  ( 5 , 2163) PHYS. STAT. SOLIDI A VOL.21 NO.1   K31, (1974).

2956  NIKULIN.........(Y.K.)..:  ( 6 ,   14) PHYS. LETT. VOL.36A, . . . . . .   337, (1971).
                                 (10B,  118) PHYS. MET. METALLOGR. VOL.33,. .   939, (1972).

2957  NILSSON.........(A.)....:  ( 5 , 1925) ACTA CHEM. SCAND. VOL.22,. . . .   719, (1968).
```

```
2958  NILSSON.........(G.)....:  ( 6 ,     23)  IAEA SYMP. BOMBAY, VOL.1,. . . . .    211, (1964).
                                 ( 6 ,     47)  PHYS. REV. LETT. 15,. . . . . . . .   634, (1965).
                                 ( 6 ,     50)  PHYS. REV. 145,. . . . . . . . . .    492, (1966).
                                 ( 6 ,   1827)  PHYS. REV. 162,. . . . . . . . . .    545, (1967).
                                 ( 6 ,   1812)  PHYS. REV. 162,. . . . . . . . . .    549, (1967).
                                 ( 6 ,   1822)  PHYS. REV. 163,. . . . . . . . . .    567, (1967).
                                 ( 6 ,    579)  IAEA SYMP. COPENHAGEN, VOL.1,. . .    187, (1968).
                                 (18B,     10)  REV. SCI. INSTRUM. VOL.39,. . . .     637, (1968).
                                 (21 ,     47)  REV. SCI. INSTRUM. VOL.40,. . . .     249, (1968).
                                 ( 6 ,    915)  PHYS. REV. B-3, . . . . . . . . .     364, (1971).
                                 ( 6 ,    918)  PHYS. REV. B VOL.5,. . . . . . .     3151, (1972).
                                 ( 6 ,    924)  PHYS. REV. B VOL.6,. . . . . . .     3777, (1972).
                                 ( 6 ,    594)  PHYS. REV. B VOL.7,. . . . . . .     2393, (1973).
                                 ( 6 ,    938)  PHYS. REV. B VOL.9 NO.12,. . . . .     . ., (1974).
                                 ( 6 ,    610)  PHYS. REV. B VOL.9,. . . . . . .     3278, (1974).

2959  NILSSON.........(S.)....:  (24 ,     59)  NUCL. INST. MET. 66, . . . . . .      229, (1968).

2960  NIMMO...........(J.K.)..:  ( 5 ,   1486)  J. PHYS. C VOL.6,. . . . . . . .      201, (1973).

2961  NIR-EL..........(Y.)....:  (20C,     55)  NUCL. INST. MET. 62, . . . . . .       43, (1968).

2962  NISHIGORI.......(T.)....:  (14B,    101)  PROGR. THEOR. PHYS. VOL.43,. . .     1423, (1970).
                                 (10C,     40)  J. NUCL. SCI. TECHNOL. VOL.8,. .      319, (1971).
                                 ( 79,     34)  J. NUCL. SCI. TECHNOL. VOL.8,. .      406, (1971).
                                 (16 ,     28)  J. NUCL. ENERGY VOL.27,. . . . .      171, (1973).

2963  NISHIMURA.......(K.)....:  ( 6 ,   2207)  2ND INTERNAT. CONF./NUCL. DATA,. . . ,      (1970).

2964  NISLE...........(R.G.)..:  (17B,     45)  NUCL. INST. MET. 72, . . . . . .      205, (1969).

2965  NITC............(V.V.)..:  (21 ,     15)  IAEA SYMP. COPENHAGEN, VOL.2,. .      313, (1968).

2966  NITC............(W.)....:  (19A,     59)  NUKLEONIKA VOL.9,. . . . . . . .      523, (1964).

2967  NITTS...........(V.V.)..:  ( 5 ,   2957)  SOV. PHYS. SOL. STATE 6, . . . .     1070, (1964).
                                 ( 5 ,   1078)  SOL. STATE COMM. 7, NO22,. . . .     1665, (1969).
                                 ( 5 ,   1072)  SOV. PHYS. SOL. STATE VOL.13,. .       44, (1971).
                                 ( 5 ,   1084)  SOV. PHYS. SOLID STATE VOL.15, .     1429, (1974).

2968  NITYANANDA......(R.)....:  ( 8 ,    126)  SOL. STATE COMM. VOL. 9, . . . .     1003, (1971).

2969  NIWA............(T.)....:  (20C,     29)  J. NUCL. SCI. TECHNOL. JAP. 4, .       22, (1967).

2970  NIX.............(F.C.)..:  ( 5 ,   2136)  PHYS. REV. VOL.58,. . . . . . .      1031, (1940).
                                 ( 5 ,    939)  PHYS. REV. VOL.68,. . . . . . .       159, (1945).

2971  NIZEL...........(S.)....:  ( 5 ,   1072)  SOV. PHYS. SOL. STATE VOL.13,. .       44, (1971).
                                 ( 5 ,   1084)  SOV. PHYS. SOLID STATE VOL.15, .     1429, (1974).

2972  NIZIOL..........(S.)....:  ( 5 ,   1078)  SOL. STATE COMM. 7, NO22,. . . .     1665, (1969).
                                 ( 5 ,   2137)  PHYS. STAT. SOLIDI A VOL.17 NO.2  555, (1973).
                                 ( 5 ,   2962)  PHYS. STAT. SOLIDI A VOL.18 NO.1  K11, (1973).

2973  NIZZOLI.........(F.)....:  (10B,    186)  PHYS. REV. LETT. VOL.31,. . . .      1466, (1973).
                                 (10B,     92)  J. PHYS. F VOL.4,. . . . . . . .       19, (1974).

2974  NOAKES..........(J.E.)..:  (13 ,      6)  AIP CONF. PROC. NO.10 PART 1,. .      822, (1973).

2975  NODA............(Y.)....:  ( 6 ,   1529)  PHYS. REV. B VOL.9,. . . . . . .     4429, (1974).

2976  NOMURA..........(S.)....:  ( 5 ,    508)  J. PHYS. CHEM. SOL. 25,. . . . .      901, (1964).
                                 ( 5 ,   2196)  ACTA CRYST. 19,. . . . . . . . .      147, (1965).
                                 ( 5 ,   1820)  J. PHYS. CHEM. SOL. 27,. . . . .      655, (1966).
                                 ( 5 ,   2491)  J. PHYS. SOC. JAPAN VOL.24,. . .      219, (1968).
                                 ( 5 ,   1050)  J. PHYS. SOC. JAPAN VOL.33,. . .     1296, (1972).

2977  NOOLANDI........(J.)....:  ( 6 ,    952)  SOL. STATE COMM. VOL. 9, . . . .     1809, (1971).

2978  NORMAND.........(J.M.)..:  ( 5 ,    918)  NUCL. PHYS. A VOL.A176,. . . . .      225, (1971).

2979  NORMAN..........(P.L.)..:  ( 5 ,   2844)  J. APPL. CRYST. VOL.6, . . . . .      240, (1973).

2980  NORTH...........(D.M.)..:  ( 5 ,    749)  PHIL. MAG. VOL.14, . . . . . . .      961, (1966).
                                 ( 5 ,   2929)  PHYS. LETT. 21,. . . . . . . . .      286, (1966).
                                 ( 5 ,    748)  ADV. IN PHYSICS VOL.16,. . . . .      171, (1967).
                                 ( 5 ,    207)  J. OF PHYS. C VOL.1, . . . . . .     1075, (1968).
                                 (14B,     45)  J. OF PHYS. C VOL.1, . . . . . .      784, (1968).

2981  NORVELL.........(J.C.)..:  ( 6 ,    699)  J. APP. PHYS. 39,. . . . . . . .     1232, (1968).
                                 ( 5 ,    816)  PHYS. REV. VOL.186,. . . . . . .      557, (1969).
                                 ( 5 ,    813)  PHYS. REV. VOL.186,. . . . . . .      567, (1969).
                                 ( 5 ,    751)  PHYS. REV. B-2,. . . . . . . . .      277, (1970).

2982  NOSKIN..........(V.A.)..:  ( 6 ,   1153)  SOV. PHYS. TECH. PHYS. VOL.17,.      180, (1972).

2983  NOTEA...........(A.)....:  (20C,     55)  NUCL. INST. MET. 62, . . . . . .       43, (1968).

2984  NOUET...........(J.)....:  ( 5 ,   2392)  J. PHYSIQUE TOME 32 COL.1 VOL.II  611, (1971).
                                 ( 5 ,   2673)  C.R. ACAD. SCI. B TOME 276 NO.14  579, (1973).

2985  NOVAK...........(L.I.)..:  ( 5 ,    988)  BULL. ACAD. SCI. USSR VOL.30,. .     1007, (1966).
                                 ( 5 ,   2164)  SOV. PHYS. DOKLADY VOL.14,. . .      1119, (1970).

2986  NOVIKOV.........(A.G.)..:  (20C,     18)  INSTRUM. EXP. TECH. NO.1,. . . .       87, (1966).
                                 ( 6 ,   2200)  J. CHEM. PHYS. VOL.60, . . . . .     2832, (1974).

2987  NOVOTNY.........(V.)....:  ( 5 ,   2272)  CAN. J. PHYS. VOL.52,. . . . . .      748, (1974).

2988  NOWIK...........(I.)....:  ( 5 ,   2216)  A.I.P. CONF. PROC. NO.10 PART 1,      88, (1973).
                                 ( 5 ,   2217)  INT. J. MAGN. VOL.4 NO.2,. . . .       99, (1973).

2989  NOZIERES........(P.)....:  ( 6 ,   1179)  PHYS. REV. VOL.127,. . . . . . .     1452, (1962).

2990  NOZIK...........(YU.E.).:  ( 5 ,   2898)  LATV. PSR...FIZ. TEHN. SER. 2, .       49, (1969).

2991  NOZIK...........(YU.Z.).:  ( 9B,    103)  SOV. PHYS. CRYST. VOL.6, . . . .      374, (1961).
                                 ( 5 ,   2269)  SOV. PHYS. DOKLADY VOL.6,. . . .      370, (1961).
                                 ( 5 ,   1693)  J. PHYS. SOC. JAPAN VOL.17 B-III       55, (1962).
                                 ( 5 ,   1692)  SOV. PHYS. CRYST. VOL.6, . . . .      744, (1962).
                                 (22A,     68)  SOV. PHYS. CRYST. VOL.7, . . . .       58, (1962).
                                 ( 5 ,   2861)  LATV. PSR...FIZ. TEHN. SER. 6, .       30, (1968).
                                 ( 5 ,     75)  LATV. PSR...FIZ. TEHN. SER. NO.1       88, (1970).
                                 ( 5 ,    348)  LATV. PSR...FIZ. TEHN. SER. NO.2      124, (1970).
                                 (22A,     33)  LATV. PSR...FIZ. TEHN. SER. NO.3       64, (1970).
                                 ( 5 ,    322)  LATV. PSR...FIZ. TEHN. SER. NO.4       87, (1970).
                                 ( 5 ,    323)  PHYS. STAT. SOLIDI A VOL.13, . .     K119, (1972).

2992  NOZIK...........(YU.)...:  ( 5 ,   2885)  LATV. PSR...FIZ. TEHN. SER. NO.3       28, (1970).
                                 ( 5 ,   2919)  LATV. PSR...FIZ. TEHN. SER. 6, .       46, (1970).
                                 (17B,     39)  LATV. PSR...FIZ. TEHN. SER. NO.5      124, (1971).
                                 ( 5 ,    357)  LATV. PSR...FIZ. TEHN. SER. NO.6      120, (1971).
                                 ( 5 ,   1961)  LATV. PSR...FIZ. TEHN. SER. NO.5      122, (1971).
```

2993 NUNES...........(A.C.)...:   (9C,   8) BER. BUNSENGES. PHYS. CHEM. 74,.   1202, (1970).
                                    (5,   1215) ACTA CRYST. VOL.A27,. . . . . . .   219, (1971).
                                    (22C,   2) ACTA CRYST. VOL.A27,. . : . : . :   284, (1971).
                                    (18A,   19) NUCL. INST. MET. VOL.95 . . . . .   445, (1971).
                                    (24,   4) ANN. REV. BIOPHYS. BIOENGNG. 1,.   529, (1972).
                                    (22A,   44) NUCL. INST. MET. VOL.108,. . . .   189, (1973).

2994 NUNES...........(A.)....:   (5,   2441) PHYS. REV. B VOL.4,. . . . . . .   2472, (1971).
                                    (5,   2442) PHYS. REV. LETT. VOL.27,. : . : :   320, (1971).

2995 NURZYNSKI.......(J.)....:   (24,   43) NUCL. INST. MET. 41,. . . . . .   89, (1966).

2996 NUSIMOVICI......(M.A.)..:   (6,   460) PHYS. REV. 156,. . . . .   925, (1967).
                                    (6,   154) COMP. REND. 268,. . . . .   755, (1969).
                                    (6,   2326) PHYS. STAT. SOL. VOL.41,. : . : :   491, (1970).

2997 NUTKINS.........(M.A.E.):   (6,   1611) PROC. PHYS. SOC. 86,. . . . . .   181, (1965).

2998 OBATA...........(Y.)....:   (8,   39) J. PHYSIQUE TOME 32 COL.1 VOL.II 812, (1971).

2999 OBERMAYER.......(H.A.)..:   (5,   1846) SOL. STATE COMM. VOL.12 NO.8,. .   779, (1973).

3000 OBERTEUFFER.....(J.A.)..:   (5,   1619) PHYS LETT 28A, ...   267, (1968).
                                    (5,   1618) DISS. ABS. VOL.3G,. . . . . . .47568, (1970).
                                    (5,   1617) PHYS. REV. B-2,. . . . .   670, (1970).
                                    (5,   2440) PHYS. REV. LETT. VOL.29,. . . .   871, (1972).

3001 OBER............(R.)....:   (5,   2330) J. APPL. CRYST. VOL.7,. . . . .   189, (1974).
                                    (6,   1922) PHYS. REV. LETT. VOL.32,. . . .   1170, (1974).

3002 OBINYAKOV.......(B.A.)..:   (12B,   17) INSTRUM. EXP. TECH. 3,. . . . .   729, (1970).

3003 ODAJIMA.........(A.)....:   (6,   1924) J. POLYM. SCI./POLYM. LETT. 11,.   377, (1973).

3004 ODIOT...........(S.)....:   (11C,   14) J. PHYS. CHEM. SOLIDS VOL.17,. .   117, (1960).

3005 OEHME...........(H.)....:   (6,   1618) NATURWISSENSCHAFTEN VOL.53,. . .   16, (1966).

3006 OESTERREICHER...(H.)....:   (5,   870) J. APPL. PHYS. VOL.41, ...   2326, (1970).
                                    (5,   2347) J. LESS-COMMON MET.(SWIZ.) 26,.   165, (1972).
                                    (5,   858) J. PHYS. CHEM. SOLIDS VOL.33,.   1031, (1972).
                                    (5,   869) J. LESS-COMMON METALS VOL.33,.   25, (1973).
                                    (5,   2555) J. PHYS. CHEM. SOLIDS VOL.34,.   1267, (1973).

3007 OGANESYAN.......(K.O.)..:   (22B,   13) INSTRUM. EXP. TECH. VOL.15,. . .   67, (1972).

3008 OGAWA...........(K.)....:   (22C,   32) NUCL. INST. MET. 42,. . . . . .   309, (1966).

3009 OGAWA...........(S.)....:   (5,   2978) J. APPL. CRYST. VOL.7,. . . . .   67, (1974).

3010 OGUCHI..........(T.)....:   (13,   32) J. PHYS. SOC. JAPAN VOL.21,. .   2178, (1966).

3011 OHARA...........(S.)....:   (5,   1063) J. PHYS. SOC. JAPAN VOL.35,. .   706, (1973).

3012 OHASHI..........(M.)....:   (5,   345) J. PHYS. SOC. JAPAN VOL.22,. .   939, (1967).
                                    (5,   1637) J. APP. PHYS. VOL.40,. . . . .   1128, (1969).

3013 OHKUBO..........(M.)....:   (20C,   53) NUCL. INST. MET. 65,. . . . . .   113, (1968).

3014 OHNO............(E.)....:   (21,   32) MITSUB. DENKI LAB. REP. VOL.3, .   111, (1962).

3015 OHNO............(H.)....:   (5,   993) SOL. STATE COMM. 4,. . . . . :   657, (1966).
                                    (5,   568) J. PHYS. SOC. JAPAN VOL.24,. :   263, (1968).

3016 OHNO............(Y.)....:   (23,   16) IAEA SYMP. (PILE RESEARCH)VIENNA 585, (1960).

3017 OHRLICH.........(R.)....:   (6,   1942) IAEA SYMP. COPENHAGEN, VOL.1,.   203, (1968).

3018 OHTANI..........(M.)....:   (5,   2063) J. PHYS. CHEM. SOLIDS VOL.35,.   585, (1974).

3019 OJANEN..........(M.)....:   (23,   8) COMMENTAT. PHYS.-MATH.(FIN.) 42, 276, (1972).

3020 OKADA...........(K.)....:   (5,   716) J. CHEM. PHYSICS VOL.44,. . . .   1648, (1966).

3021 OKAMOTO.........(K.)....:   (23,   16) IAEA SYMP. (PILE RESEARCH)VIENNA 585, (1960).

3022 OKAMOTO.........(T.)....:   (5,   1116) J. PHYS. SOC. JAPAN VOL.35,. .   1554, (1973).

3023 OKAYA...........(Y.)....:   (5,   6) ACTA CRYST. 21,. . . . .   237, (1966).
                                    (5,   276) ACTA CRYST. VOL.B27,. . . . . .   26, (1971).

3024 OKAZAKI.........(A.)....:   (11B,   50) UKAEA AERE REP. R 4299 (7 PP.),. . .   , (1963).
                                    (6,   1462) J. APPL. PHYS. VOL.35,. . . . .   998, (1964).
                                    (6,   1458) PHYS. LETT. 8,. . . . .   9, (1964).
                                    (6,   1453) PROC. INT. CONF. (NOTTINGHAM),.   92, (1964).
                                    (6,   1461) PROC. PHYS. SOC. 85,. . . . .   743, (1965).
                                    (5,   1156) J. PHYS. SOC. JAPAN VOL.22,. :   924, (1967).
                                    (6,   2191) PROC. ROY. SOC. 300,. . . . .   45, (1967).
                                    (5,   1153) J. PHYS. SOC. JAPAN VOL.29,. :   649, (1970).

3025 OKOROKOV........(A.I.)..:   (6,   1749) SOV. PHYS. JETP LETT. VOL.2,. .   336, (1965).
                                    (11A,   63) SOV. PHYS. JETP LETT. VOL.9,. .   204, (1969).
                                    (5,   2065) SOV. PHYS. J.E.T.P. VOL.29,. .   261, (1969).
                                    (12B,   68) SOV. PHYS. JETP LETT. VOL.15,.   324, (1972).

3026 OKUNEVA.........(N.M.)..:   (6,   48) SOV. PHYS. SOL. STATE 7,. . .   1138, (1965).
                                    (21,   51) SOV. PHYS. SOL. STATE 7,. . . .   1146, (1965).
                                    (6,   2023) SOV. PHYS. SOL. STATE 9,. . . .   955, (1967).
                                    (6,   2077) SOV. PHYS. SOL. STATE VOL.10,.   402, (1968).
                                    (6,   171) SOV. PHYS. SOL. STATE VOL.11,.   1615, (1969).
                                    (6,   893) SOV. PHYS. SOL. STATE VOL.13,.   1256, (1971).

3027 OLEJARCZYK......(W.)....:   (21,   40) NUKLEONIKA VOL.12,. . . . . . .   385, (1967).

3028 OLEJNIK.........(S.)....:   (6,   2043) SPEC. DISC. FARADAY SOC. NO.1, .   194, (1970).

3029 OLEKSA..........(S.)....:   (5,   2711) PHYS. REV. VOL.109,. . . . . .   1645, (1958).

3030 OLES............(A.)....:   (5,   953) ACTA PHYS. POLON. VOL.27,. . . .   343, (1965).
                                    (5,   1010) COMP. REND. 260,. . . . .   6075, (1965).
                                    (5,   2718) J. PHYSIQUE TOME 26,. . . . . .   561, (1965).
                                    (5,   1716) PHYS. STAT. SOL. 8,. . . . . .   K167, (1965).
                                    (5,   1012) ACTA PHYS. POLON. VOL.30,. . . .   125, (1966).
                                    (5,   1703) J. APP. PHYS. 37,. . . . . . .   960, (1966).
                                    (5,   1704) J. PHYSIQUE TOME 27,. . . . . .   51, (1966).
                                    (5,   1717) PHYS. STAT. SOL. 14,. . . . . .   K39, (1966).
                                    (9B,   68) NUKLEONIKA VOL.13,. . . . . . .   1111, (1968).
                                    (9B,   67) NUKLEONIKA VOL.13,. . . . . . .   171, (1968).
                                    (5,   2938) PHYS. STAT. SOL. A VOL.3,. . . .   569, (1970).
                                    (5,   1099) PHYS. STAT. SOL. 41,. . . . . .   173, (1970).
                                    (5,   2939) J. PHYSIQUE TOME 32 COL.1 VOL.I,   328, (1971).

3031 OLIVEI..........(A.)....:   (14B,   1) ACTA CRYST. VOL.A29, ...   692, (1973).

```
3032  OLIVI..........(L.)....!  ( 6 ,   953) SOL. STATE COMM. VOL.8,. . . . .   373, (1970).

3033  OLLIVIER.......(G.)....!  ( 5 ,  1533) SOL. STATE COMM. VOL.7 NO.22,. . 1669, (1969).
                                ( 5 ,   359) J. PHYS. CHEM. SOLIDS VOL.32,. . 1189, (1971).
                                ( 5 ,  1759) SOL. STATE COMM. VOL.10,. . . . .  609, (1972).
                                ( 6 ,   444) SOL. STATE COMM. VOL.13,. . . . .  919, (1973).

3034  OLOVSSON.......(I.)....!  ( 5 ,  1925) ACTA CHEM. SCAND. VOL.22,. . . .   719, (1968).

3035  OLSEN..........(C.E.)..!  ( 5 ,   908) J. APPL. PHYS. VOL.33, SUPPL.,. . 1135, (1962).
                                ( 5 ,   912) PHYS. REV. VOL.127,. . . . . . . 2101, (1962).
                                ( 5 ,   906) PHYS. REV. 135,. . . . . . . . . A176, (1964).
                                ( 5 ,  2760) PHYS. REV. 140,. . . . . . . . . A1448, (1965).
                                ( 5 ,  2030) J. APP. PHYS. 37,. . . . . . . . 4575, (1966).
                                ( 5 ,  2339) J. APP. PHYS. 38,. . . . . . . . 1395, (1967).
                                ( 5 ,   809) J. CHEM. PHYS. VOL.46,. . . . . . 4041, (1967).
                                ( 5 ,  2727) J. APP. PHYS. VOL.40,. . . . . . 1135, (1969).

3036  OLSEN..........(W.C.)..!  (20B,    71) NUCL. INST. MET. VOL.13,. . . .     1, (1961).

3037  OLSON..........(N.T.)..!  (20A,    48) NUCL. INST. MET. VOL.114,. . . .  143, (1974).

3038  OLSSON.........(L.G.)..!  ( 6 ,   103) PHYSICA VOL.72,. . . . . . . . .  300, (1974).

3039  OLTEANU........(I.)....!  (23 ,    72) REV. DE PHYSIQUE (BUCAREST) 5,    83, (1960).
                                (20A,    77) REV. SCI. INSTRUM. VOL.31,. . .   640, (1960).

3040  OMEL≠YANENKO...(M.N.)..!  (22B,    13) INSTRUM. EXP. TECH. VOL.15,. . .   67, (1972).

3041  OMICINI........(E.)....!  (17A,    75) NUCL. SCI. ENG. VOL.48,. . . . .  281, (1972).

3042  ONO............(I.)....!  (13 ,    32) J. PHYS. SOC. JAPAN VOL.21,. . . 2178, (1966).

3043  ONO............(M.)....!  ( 6 ,  1404) PHYS. LETT. VOL.41A NO.2,. . . .  141, (1972).
                                ( 6 ,  1363) J. PHYS. SOC. JAPAN VOL.34,. . .   26, (1973).

3044  OOSTERHUIS.....(W.T.)..!  ( 5 ,   979) A.I.P. CONF. PROC. NO.10 PART 1,   98, (1973).

3045  ORBAN..........(D.)....!  (17A,    47) INSTRUM. EXP. TECH. VOL.15,. . .   44, (1972).

3046  ORIA...........(M.)....!  (20B,    12) C. R. ACAD. SCI. B TOME 270,. .  185, (1970).

3047  ORLOV..........(N.F.)..!  (20C,   109) SOV. AT. ENERGY VOL.32, NO.5,. .  416, (1972).

3048  ORNSTEIN.......(L.S.)..!  ( 7B,    95) PROC. ACAD. SCI. AMSTERDAM 39,. 1049, (1936).
                                ( 7B,    97) PROC. ACAD. SCI. AMSTERDAM 39,.  810, (1936).
                                ( 7B,    96) PROC. ACAD. SCI. AMSTERDAM 39,.  904, (1936).

3049  ORTOLEVA.......(P.J.)..!  (14B,    33) IAEA SYMP. COPENHAGEN, VOL.1,. .  535, (1968).

3050  OSBORNE........(C.F.)..!  (11C,    20) J. PHYS. F VOL.2,. . . . . . . . 1145, (1972).

3051  OSBORN.........(R.K.)..!  ( 7A,    40) NUCL. SCI. ENGNG. VOL.3,. . . .    29, (1958).
                                (14B,    28) IAEA SYMPOSIUM VIENNA,. . . . .   251, (1960).
                                (15 ,    39) PHYS. REV. 130,. . . . . . . .  1860, (1963).
                                ( 6 ,  1053) PHYS. REV. 131,. . . . . . . .  2547, (1963).
                                (10E,    13) IAEA SYMPOSIUM BOMBAY, VOL.1,.   261, (1964).
                                (10E,    20) NUOVO CIMENTO VOL.38,. . . . .   175, (1965).

3052  OSGOOD.........(E.B.)..!  ( 6 ,  1143) PHYS. REV. A VOL.5,. . . . . .  1537, (1972).
                                ( 6 ,  1162) PHYS. REV. LETT. VOL.29,. . . .   552, (1972).
                                ( 6 ,  1189) PHYS. REV. A VOL.8,. . . . . .  1513, (1973).

3053  OSHIMA.........(K.)....!  ( 5 ,  2759) J. PHYS. CHEM. SOLIDS VOL.32,. .  235, (1971).

3054  OSKOTSKII......(V.S.)..!  (10C,   103) SOV. PHYS. SOL. STATE VOL.2,. .   647, (1960).
                                ( 7B,   120) SOV. PHYS. JETP. 17,. . . . . .   445, (1963).
                                (14B,   120) SOV. PHYS. SOL. STATE, 5,. . . .  789, (1963).
                                (10C,   122) SOV. PHYS. SOL. STATE, 9,. . . .  420, (1967).

3055  OSREDKAR.......(M.)....!  ( 5 ,   372) PHYS. LETT. 25A,. . . . . . . .   123, (1967).
                                ( 6 ,  1579) PHYS. STAT. SOLIDI B VOL.59 NO.2  471, (1973).

3056  OSTHELLER......(G.L.)..!  ( 6 ,   153) IAEA SYMP. COPENHAGEN, VOL.1,. .  315, (1968).

3057  OSTROWSKI......(G.E.)..!  (19B,    20) NUCL. INST. MET. VOL.106,. . . .  221, (1973).
                                ( 6 ,  1195) PHYS. REV. LETT. VOL.31,. . . .   510, (1973).

3058  OSTROWSKI......(G.)....!  (20B,    44) IEEE TRANS. NUCL. SCI. VOL.NS-13  311, (1966).
                                ( 6 ,    85) PHYS. REV. A VOL.6,. . . . . .  1107, (1972).

3059  OTNES..........(K.)....!  ( 6 ,  1192) PHYS. REV. VOL.108,. . . . . .  1346, (1957).
                                ( 6 ,  1198) PHYSICA VOL.24, SUPPL.,. . . . .  145, (1958).
                                ( 6 ,  1201) PHYS. REV. VOL.112,. . . . . . .   11, (1958).
                                (20B,     7) ARK. FYS. VOL.16 PAPER 19,. . .   199, (1959).
                                ( 6 ,   669) IAEA SYMPOSIUM (VIENNA),. . . .   329, (1960).
                                (19B,    10) NUCL. INST. MET. VOL.12,. . . .   355, (1961).
                                (20A,    16) IAEA SYMP. CHALK RIVER, VOL.1,    95, (1962).
                                ( 6 ,  1771) IAEA SYMP. CHALK RIVER, VOL.2,   167, (1962).
                                ( 6 ,  1275) IAEA SYMP. CHALK RIVER, VOL.2,   273, (1962).
                                ( 6 ,  1869) PHYS. REV. LETT. VOL.8,. . . . .  278, (1962).
                                ( 6 ,  1709) COPENHAGEN CONF.-LATT.DYNAMICS,.  63, (1963).
                                ( 6 ,  2247) PHYS. REV. VOL.152,. . . . . . .  535, (1966).
                                ( 5 ,  1962) PHYS. REV. 153,. . . . . . . . .  184, (1967).
                                ( 6 ,   285) DISCUSSIONS FARADAY SOC. NO.48,.  87, (1969).
                                (20B,    90) NUCL. INST. MET. 75,. . . . . .   197, (1969).
                                ( 6 ,   286) PHYSICA VOL.41,. . . . . . . . .  397, (1969).
                                ( 6 ,  2091) SOL. STATE COMM. VOL. 9,. . . . 1103, (1971).
                                ( 6 ,  2090) SOL. STATE COMM. VOL. 9,. . . . 1455, (1971).
                                ( 6 ,   114) IAEA SYMP. GRENOBLE,. . . . . .   515, (1972).
                                ( 6 ,   136) PHYS. NORV. VOL.6 NO.3-4,. . . .  205, (1972).
                                ( 5 ,   134) SOL. STATE COMM. VOL.11,. . . . 1365, (1972).

3060  OTOMO..........(N.)....!  ( 6 ,   295) J. NUCL. SCI. TECHNOL. VOL.9,. .  374, (1972).

3061  OTOMO..........(S.)....!  (24 ,    15) IAEA SYMP. (PILE RESEARCH)VIENNA  633, (1960).

3062  OTSUKI.........(H.)....!  ( 5 ,  1244) PHYS. REV. C VOL.5,. . . . . .  1952, (1972).

3063  OVCHAROV.......(V.P.)..!  (12B,    59) SOV. PHYS. CRYST. VOL.10,. . . .   76, (1965).

3064  OVCHINNIKOV....(A.A.)..!  (10D,    65) SOV. PHYS. J.E.T.P. VOL.30,. . .  147, (1970).

3065  OVERHAUSER.....(A.W.)..!  ( 5 ,   549) PHYS. REV. LETT. VOL.4,. . . . .  226, (1960).
                                ( 9B,    85) PHYS. REV. B-3,. . . . . . . .  3173, (1971).

3066  OVERSLUIZEN....(T.)....!  ( 5 ,  1336) PHYS. REV. LETT. VOL.32,. . . .   791, (1974).

3067  OVERTON-JR.....(W.C.)..!  (10D,    24) PHYS. LETT. VOL.37A,. . . . . .   287, (1971).

3068  OWSTON.........(P.G.)..!  ( 5 ,  1303) ADV. IN PHYS. VOL.7,. . . . . .   171, (1958).

3069  OYAMADA........(M.)....!  (22C,    43) NUCL. INST. MET. 71,. . . . . .   102, (1969).
```

```
3070  OZEROV..........(R.P.)..:   ( 9A,     27) USPEKHI FIZ. NAUK VOL.45,. . . .     481, (1951).
                                  ( 9D,     83) USPEKHI FIZ. NAUK VOL.47,. . . . .     445, (1952).
                                  (22A,     69) SOV. PHYS. CRYST. VOL.5,. . . . .     294, (1960).
                                  (22C,     71) SOV. PHYS. CRYST. VOL.5,. . . . .     297, (1960).
                                  (23,      41) INSTRUM. EXP. TECH. NO.6,. . . . .     902, (1961).
                                  ( 5 ,   1277) SOV. PHYS. JETP VOL.13,. . . . .      718, (1961).
                                  ( 5 ,    780) J. PHYS. SOC. JAPAN VOL.17 B-II,     385, (1962).
                                  ( 5 ,   1475) SOV. PHYS. CRYST. VOL.7,. . . . .     499, (1963).
                                  ( 5 ,    217) SOV. PHYS. DOKLADY VOL.7,. . . . .     742, (1963).
                                  ( 8 ,    135) SOV. PHYS. USPEKHI VOL.5,. . . . .     104, (1962).
                                  ( 8 ,   1579) SOV. PHYS. DOKLADY VOL.8,. . . . .     131, (1963).
                                  ( 5 ,   2322) SOV. PHYS. SOL. STATE 5,. . . . .    2425, (1963).
                                  ( 5 ,   1915) DOKL. AKAD. NAUK SSSR VOL.155,.     1416, (1964).
                                  ( 5 ,    307) SOV. PHYS. CRYST. VOL.11,. . . .      177, (1966).
                                  ( 8 ,    143) SOV. PHYS. CRYST. VOL.12,. . . .      199, (1967).
                                  ( 5 ,   1550) SOV. PHYS. CRYST. VOL.13,. . . .      204, (1968).
                                  ( 5 ,   1078) SOL. STATE COMM. 7, NO22,. . . .     1665, (1969).
                                  ( 8 ,   2288) SOV. PHYS. SOL. STATE. 11,. . . .    1433, (1969).
                                  ( 5 ,   2665) DOKL. AKAD. NAUK SSSR VOL.194,.     1374, (1970).
                                  ( 5 ,   2517) J. PHYSIQUE TOME 32 COL.1 VOL.I,     503, (1971).
                                  ( 6 ,   1388) SOV. PHYS. CRYST. VOL.15,. . . .     1010, (1971).
                                  ( 5 ,   1072) SOV. PHYS. SOL. STATE VOL.13,. .      44, (1971).
                                  ( 5 ,   1293) SOV. PHYS. CRYST. VOL.17, NO.2,.     383, (1972).
                                  ( 5 ,   1959) SOV. PHYS. JETP LETT. VOL.16,. .     198, (1972).
                                  ( 5 ,   2856) SOV. PHYS. CRYST. VOL.17,. . . .     1017, (1973).
                                  ( 5 ,   1084) SOV. PHYS. SOLID STATE VOL.15, .    1429, (1974).

3071  OZORA...........(A.)....:   ( 6 ,    230) CHEM. PHYS. LETT. VOL.18,. . . .      306, (1973).

3072  O≠BRIEN.........(D.D.)..:   ( 7A,     54) PHYS. LETT. B VOL.36B,. . . . .      560, (1971).

3073  O≠CONNELL.......(A.M.)..:   ( 5 ,   1913) ACTA CRYST. 22,. . . . . . . . .     134, (1967).

3074  O≠CONNOR........(B.H.)..:   ( 5 ,   2191) ACTA CRYST. 21,. . . . . . . . .     705, (1966).
                                  ( 5 ,   2228) ACTA CRYST. 22,. . . . . . . . .     927, (1967).
                                  ( 5 ,   2905) ACTA CRYST. B25,. . . . . . . . .    2140, (1969).
                                  ( 5 ,    296) ACTA CRYST. VOL.B29,. . . . . .      1903, (1973).

3075  O≠CONNOR........(D.A.)..:   (22A,      3) ACTA PHYS. POLON. VOL.19,. . . .     329, (1960).
                                  (20B,     69) NUCL. INST. MET. VOL.8,. . . . .     244, (1960).
                                  (20A,      2) ACTA CRYST. VOL.14,. . . . . . .     292, (1961).
                                  (10C,     89) PROC. PHYS. SOC. VOL.91,. . . . .    917, (1967).

3076  O≠CONNOR........(D.)....:   (22B,      4) ACTA PHYS. POLON. VOL.16,. . . .     293, (1957).
                                  (21 ,      1) ACTA PHYS. POLON. VOL.18,. . . .     265, (1959).
                                  (10C,     58) NUKLEONIKA VOL.4,. . . . . . . .     119, (1959).
                                  (19A,     12) IAEA SYMPOSIUM VIENNA,. . . . .      159, (1960).
                                  (20B,     27) IAEA SYMPOSIUM VIENNA,. . . . .      447, (1960).

3077  O≠DANIEL........(H.)....:   ( 5 ,   1871) ACTA CRYST. VOL. A25,. . . . . .    S119, (1969).

3078  O≠DELL..........(R.D.)..:   (12B,      9) DISS. ABS. VOL.26,. . . . . . .     1111, (1966).

3079  O≠DONNELL.......(F.R.)..:   (20B,    161) NUCL. INST. MET. VOL.102,. . . .     501, (1972).

3080  O≠KEEFE.........(T.J.)..:   ( 5 ,   1385) J. PHYSIQUE TOME 32 COL.1 VOL.II    670, (1971).

3081  O≠LAUGHLIN......(J.W.)..:   (20C,     70) NUCL. INST. MET. VOL.84,. . . .       67, (1970).

3082  O≠SULLIVAN......(W.)....:   ( 5 ,   1760) PHYS. REV. VOL.108,. . . . . . .     637, (1957).

3083  PACCARD.........(D.)....:   ( 5 ,   2587) C. R. ACAD. SCI. B TOME 270,. .     1131, (1970).
                                  ( 5 ,    867) SOL. STATE COMM. VOL.8,. . . . .     391, (1970).
                                  ( 5 ,   2183) SOL. STATE COMM. VOL.10,. . . .      989, (1972).

3084  PADLO...........(I.)....:   ( 6 ,   2317) IAEA SYMP. CHALK RIVER, VOL.2, .      87, (1962).
                                  ( 5 ,   2279) NATURWISSENSCHAFTEN VOL.49,. .       373, (1962).

3085  PADMANABHAN.....(V.M.)..:   ( 9C,     13) CRYSTALLOGR./CRYSTAL PERFECTION,    269, (1963).
                                  ( 5 ,   1917) ACTA CRYST. VOL.18,. . . . . . .     567, (1965).
                                  ( 5 ,   1571) ACTA CRYST. 22,. . . . . . . . .     532, (1967).
                                  ( 5 ,   1935) ACTA CRYST. 22,. . . . . . . . .     928, (1967).
                                  ( 5 ,   1510) ACTA CRYST. 23,. . . . . . . . .     578, (1967).
                                  ( 5 ,   1994) ACTA CRYST. VOL.B27,. . . . . .      253, (1971).

3086  PADUREANU.......(I.)....:   ( 6 ,    945) REV. ROUMAINE PHYS. VOL.18 NO.2,    135, (1973).
                                  (14B,    114) REV. ROUMAINE PHYS. VOL.18,. . .     997, (1973).
                                  ( 6 ,   2200) J. CHEM. PHYS. VOL.60,. . . . .     2832, (1974).

3087  PAGE............(D.I.)..:   ( 6 ,   2235) PROC. PHYS. SOC. 91,. . . . . .       76, (1967).
                                  ( 6 ,    680) UKAEA AERE REPORT 5408 (21 PP.),. . , (1967).
                                  ( 6 ,    211) UKAEA AERE REP.5574 (24 PP.),. . . , (1967).
                                  ( 6 ,   1077) CONF. NEUTRON THERMALIZ. VOL.I,.     361, (1968).
                                  ( 6 ,    197) IAEA SYMP. COPENHAGEN, VOL.1,. .     325, (1968).
                                  ( 5 ,   2390) PHYS. LETT. VOL.29A,. . . . . .      296, (1969).
                                  ( 6 ,   1095) PROC. ROY. SOC. A319, NO.1537,.      189, (1970).
                                  ( 5 ,    420) J. OF PHYS.-C-, SER.2, VOL.4,.      3034, (1971).
                                  (14A,     25) J. PHYS. C VOL.4,. . . . . . . .    1453, (1971).
                                  ( 6 ,    939) MOL. PHYS. VOL.20,. . . . . . .      881, (1971).
                                  ( 6 ,   1302) MOL. PHYS. VOL.21,. . . . . . .      901, (1971).
                                  ( 6 ,   1957) CAN. J. PHYS. VOL.50,. . . . . .    3063, (1972).
                                  ( 6 ,   1079) MOL. PHYS. VOL.24 NO.5,. . . .      1025, (1972).
                                  (14A,     18) CHEM. APPL. THERMAL NEUTRON SCAT    173, (1973).
                                  ( 5 ,   1196) J. PHYS. C VOL.6,. . . . . . . .     212, (1973).
                                  ( 6 ,   1961) PROPERTIES LIQUID METALS,. . . .      99, (1973).
                                  ( 6 ,   1063) WATER-A COMPREHENSIVE TREATISE,. . . , (1973).

3088  PAILTHORP.......(K.G.)..:   (24 ,     22) IEEE TRANS. NUCL. SCI. VOL.17, .     138, (1970).

3089  PAKHOMOV........(V.I.)..:   ( 5 ,   1475) SOV. PHYS. CRYST. VOL.7,. . . .      499, (1962).

3090  PAK-HWANG-O.............:   ( 5 ,   1072) SOV. PHYS. SOL. STATE VOL.13,. .      44, (1971).

3091  PALENIK.........(G.J.)..:   ( 5 ,   2513) J. CHEM. PHYS. VOL.41,. . . . .     3260, (1964).

3092  PALEVSKY........(H.)....:   ( 6 ,    772) PHYS. REV. VOL.92,. . . . . . .      202, (1953).
                                  ( 5 ,    143) PHYS. REV. VOL.92,. . . . . . .      716, (1953).
                                  (11C,     33) PHYS. REV. VOL.98,. . . . . . .      492, (1955).
                                  ( 6 ,   2246) PHYS. REV. VOL.104,. . . . . . .     271, (1956).
                                  (23 ,     52) NUCL. INSTRUM. VOL.1,. . . . . .      92, (1957).
                                  ( 6 ,     56) PHYS. REV. VOL.106,. . . . . . .    1168, (1957).
                                  ( 6 ,   2355) PHYS. REV. VOL.108,. . . . . . .    1091, (1957).
                                  ( 6 ,   1192) PHYS. REV. VOL.108,. . . . . . .    1346, (1957).
                                  (23 ,     43) J. NUCL. ENERGY VOL.6,. . . . .      222, (1958).
                                  ( 6 ,   2240) PHYS. REV. VOL.109,. . . . . . .    1046, (1958).
                                  ( 6 ,   1201) PHYS. REV. VOL.112,. . . . . . .      11, (1958).
                                  ( 6 ,   2036) PHYS. REV. LETT. VOL.2,. . . . .     258, (1959).
                                  ( 6 ,   1090) PHYS. REV. LETT. VOL.3,. . . . .      91, (1959).
                                  ( 6 ,    937) PHYS. REV. VOL.113,. . . . . . .      49, (1959).
                                  ( 5 ,   1319) IAEA SYMPOSIUM VIENNA,. . . . .      223, (1960).
                                  ( 5 ,   1285) IAEA SYMPOSIUM VIENNA,. . . . .      265, (1960).
                                  ( 6 ,   1084) PHYS. REV. VOL.119,. . . . . . .     872, (1960).
                                  (20B,     71) NUCL. INST. MET. VOL.13,. . . .        1, (1961).
                                  (29A,     16) IAEA SYMP. CHALK RIVER, VOL.1,.      95, (1962).
                                  ( 6 ,   1275) IAEA SYMP. CHALK RIVER, VOL.2,.     273, (1962).
                                  ( 6 ,   1583) J. PHYS. SOC. JAPAN VOL.17 B-II,    367, (1962).
```

```
3092  PALEVSKY........(H.)....:  ( 6 ,    94) PHYS. REV. VOL.125,. . . . . . .  275, (1962).
                                ( 6 , 1709) COPENHAGEN CONF.-LATT.DYNAMICS,.  63, (1963).
                                ( 6 , 1587) J. PHYS. CHEM. SOL. 24,. . . . .  617, (1963).
                                (17A,   30) IAEA SYMPOSIUM BOMBAY, VOL.2,. .  455, (1964).
                                ( 6 ,  263) J. PHYS. CHEM. SOL. VOL.25,. . . 1001, (1964).
                                (14B,   49) LIQUIDS: STRUC.,PROP., SOL: INT.  201, (1965).
                                (15 ,   21) J. CHIM. PHYS. VOL.63,. . . . .  157, (1966).
                                ( 6 ,   72) IAEA SYMP. COPENHAGEN, VOL.1,. .  457, (1968).
                                ( 6 ,   92) PHYS. LETT. VOL.26A,. . . . . .  152, (1968).

3093  PALEWSKI........(T.)....:  ( 5 , 2722) PHYS. STAT. SOL. VOL.38,. . . .  K89, (1970).

3094  PALLA...........(G.)....:  ( 7A,   55) PHYS. LETT. B VOL.35B,. . . . .  477, (1971).

3095  PALMER..........(B.J.)..:  ( 6 , 2280) J. PHYS. C VOL.6,. . . . . . . L313, (1973).

3096  PALMER..........(D.W.)..:  (20B,   65) J. SCI. INSTRUM. VOL.43,. . . .    1, (1966).

3097  PALMER..........(H.E.)..:  (24 ,   22) IEEE TRANS. NUCL. SCI. VOL.17, .  138, (1970).

3098  PALMGREN........(A.)....:  (16 ,   32) NUCL. INST. MET. 36,. . . . . .   88, (1965).
                                (23 ,   58) NUCL. INST. MET. 46,. . . . . .  266, (1966).
                                (20B,   78) NUCL. INST. MET. 46,. . . . . .   70, (1966).
                                ( 5 ,  381) ACTA POLYTECH. SCANDIN. SER.52,.    3, (1968).
                                ( 6 ,  168) IAEA SYMP. COPENHAGEN, VOL.1,. .  431, (1968).
                                ( 6 , 2349) J. NUCL. ENERGY VOL.25,. . . . .  189, (1971).

3099  PALM............(J.H.)..:  (17C,    6) ACTA CRYST. 17,. . . . . . . . 1326, (1964).

3100  PAL.............(L.)....:  ( 5 , 1130) PHYS. LETT. 9,. . . . . . . . .  297, (1964).
                                ( 5 , 1123) PROC. INT. CONF. (NOTTINGHAM),.  158, (1964).
                                ( 6 ,  717) PHYS. LETT. 19,. . . . . . . . .   15, (1965).
                                ( 5 , 1798) PHYS. LETT. 20,. . . . . . . . .  331, (1966).
                                ( 5 , 1788) J. APPL. PHYS. VOL.38,. . . . . 1265, (1967).
                                ( 5 , 1785) PHYS. LETT. VOL.24A,. . . . . .  198, (1967).
                                ( 5 , 1786) CENT. RES INST PHYS. KFKI NO.2,.    1, (1968).
                                (12B,   11) IAEA SYMP. COPENHAGEN, VOL.2,. .  407, (1968).
                                ( 6 ,  807) IAEA SYMP. COPENHAGEN, VOL.2,. .   37, (1968).
                                ( 6 ,  726) IAEA SYMP. COPENHAGEN, VOL.2,. .   55, (1968).
                                ( 6 ,  823) J. APP. PHYS. 39,. . . . . . . .  453, (1968).
                                ( 5 , 1793) J. APP. PHYS. 39,. . . . . . . .  538, (1968).
                                ( 5 , 1730) J. PHYS. CHEM. SOLIDS 29,. . . .  104, (1968).
                                ( 6 ,  806) PHYS. LETT. VOL.28A,. . . . . .  213, (1968).
                                ( 5 , 1695) PHYS. REV. 171,. . . . . . . . .  574, (1968).
                                ( 9D,   54) 2ND SUMMER SCHOOL S.S. PHYS.,. .  171, (1968).
                                ( 5 , 1776) J. APPL. PHYS. 41,. . . . . . .  941, (1970).
                                ( 5 , 1770) J. PHYSIQUE TOME 32 COL.1 VOL.II  980, (1971).

3101  PAL.............(S.)....:  ( 6 , 2273) PHYS. LETT. 19,. . . . . . . . .  105, (1965).
                                (10B,   96) J. PHYS. SOC. JAPAN VOL.21,. . . 2208, (1966).
                                ( 6 , 1753) SOL. STATE COMM. 4,. . . . . . .   83, (1966).
                                ( 6 , 2202) REV. ROUMAINE PHYS. VOL.14,. . .  247, (1969).
                                (10B,  150) PHYS. REV. B-2,. . . . . . . . . 4741, (1970).
                                (10B,   80) J. PHYS.F: METAL PHYS. VOL.1,. .  588, (1971).
                                (10B,   99) J. PHYS. SOC. JAPAN VOL.35,. . . 1487, (1973).

3102  PANASYUK........(I.S.)..:  (22A,   57) PRIBORY TEKH. EKSPER. NO.3,. . .   26, (1959).

3103  PANDEY..........(B.P.)..:  ( 6 ,  922) SOL. STATE COMM. VOL.11 NO.6,. .  775, (1972).
                                (10B,   75) J. PHYS. C VOL.6,. . . . . . . . 2943, (1973).

3104  PANDEY..........(R.N.)..:  ( 6 , 1970) SOL. STATE COMM. VOL.11,. . . .  185, (1972).

3105  PANIN...........(V.E.)..:  ( 5 , 2164) SOV. PHYS. DOKLADY VOL.14,. . . 1119, (1970).

3106  PANOVA..........(G.KH.).:  ( 6 , 2183) PHYS. STAT. SOL. 20,. . . . . .  767, (1967).

3107  PANTAZATOS......(P.)....:  ( 9B,   21) ACTA CRYST. VOL.A29,. . . . . .  577, (1973).

3108  PANT............(A.K.)..:  (17C,   14) ACTA CRYST. VOL.A26,. . . . . .  539, (1970).
                                ( 6 ,  434) PHYS. REV. B VOL.8,. . . . . . . 4795, (1973).

3109  PAN.............(S.S.)..:  ( 6 , 2344) NUCL. SCI. ENGNG. VOL.23,. . . .  194, (1965).
                                (17A,   62) NUCL. INST. MET. 42,. . . . . .  197, (1966).
                                ( 6 , 2352) CONF. NEUTRON THERMALIZ. VOL.I,.  407, (1968).
                                ( 6 , 1391) MOL. DYN. AND STRUCT. OF SOLIDS,  315, (1969).

3110  PAOLETTI........(A.)....:  (20D,   12) NUCL. INSTRUM. VOL.3,. . . . . .  223, (1958).
                                ( 5 ,  440) PHYS. REV. LETT. VOL.2,. . . . .  254, (1959).
                                (23 ,   21) IAEA SYMP. (PILE RESEARCH)VIENNA  123, (1960).
                                ( 5 , 2852) J. CHEM. PHYS. VOL.32,. . . . .  308, (1960).
                                (22A,   34) NUCL. INST. MET. VOL.9,. . . . .  195, (1960).
                                (23 ,   65) NUOVO CIMENTO SUPPL. VOL.23,. .   17, (1962).
                                ( 5 ,  951) COLL. INTER. N.126 (GRENOBLE),.  180, (1963).
                                ( 5 , 1736) J. APP. PHYS. 34,. . . . . . . . 1571, (1963).
                                ( 6 ,  474) PHYS. REV. LETT. 10,. . . . . .  290, (1963).
                                (12A,   13) NUCL. INST. MET. 26,. . . . . .  122, (1964).
                                ( 5 , 1163) NUOVO CIMENTO VOL.32,. . . . . .   29, (1964).
                                ( 5 ,  271) PHYSICA VOL.30,. . . . . . . . . 1647, (1964).
                                ( 5 , 2294) PROC. INT. CONF. (NOTTINGHAM),.  288, (1964).
                                ( 5 , 1735) J. APP. PHYS. 37,. . . . . . . . 3236, (1966).
                                ( 5 ,  273) PHYSICA VOL.32,. . . . . . . . .  119, (1966).
                                ( 6 ,  512) PHYS. REV. 143,. . . . . . . . .  365, (1966).
                                ( 5 , 1739) PHYS. LETT. 24A,. . . . . . . .  371, (1967).
                                ( 5 ,  495) PHYS. LETT. 25A,. . . . . . . .  372, (1967).
                                ( 9D,   80) TRIESTE/INTERN. COURSE,. . . . .  561, (1967).
                                ( 9D,   22) C.N.E.N. SYMP. CASACCIA,. . . .  149, (1968).
                                ( 6 ,  883) PHYS. REV. VOL.178,. . . . . . .  833, (1969).
                                ( 5 , 1790) PHYS. REV. VOL.187,. . . . . . .  611, (1969).
                                (23 ,   14) EVOLUTION OF PARTICLE PHYSICS, .  204, (1970).
                                (12A,   34) RIV. NUOVO CIMENTO VOL.2,. . . .  451, (1970).
                                ( 5 ,  479) INT. J. MAGN. VOL.1 NO.2,. . . .  183, (1971).
                                ( 6 ,  799) J. PHYSIQUE TOME 32 COL.1 VOL.II 1188, (1971).
                                ( 5 ,  463) NUOVO CIMENTO VOL.10B,. . . . .  565, (1972).

3111  PAPAMANTELLOS...(P.)....:  ( 5 , 1009) PHYS. STAT. SOL. 29,. . . . . .  323, (1968).

3112  PAPOULAR........(M.)....:  (13 ,   27) J. PHYSIQUE TOME 28 COL.1,. . .  140, (1967).

3113  PAPP............(R.)....:  (12B,   29) NUCL. INST. MET. 96,. . . . . .   29, (1971).

3114  PAPULOVA........(Z.G.)..:  ( 5 , 2957) SOV. PHYS. SOL. STATE 6,. . . . 1070, (1964).

3115  PARANOPE........(S.K.)..:  ( 5 ,  522) NUCL./S.S. PHYS. SYMP. ABSTR.,. . . , (1972).

3116  PARDAVI.........(M.)....:  ( 5 , 2889) A.I.P. CONF. PROC. NO.10 PART 2, 1603, (1973).

3117  PAREDES.........(M.C.)..:  (20B,    6) AN. REAL SOC. ESPAN/FIS. VOL.63A  211, (1967).

3118  PARETTE.........(G.)....:  ( 6 , 1757) IAEA SYMP. CHALK RIVER, VOL.2,.  317, (1962).
                                ( 6 , 1744) J. PHYS. SOC. JAPAN VOL.17 B-III   67, (1962).
                                ( 6 , 1459) J. APP. PHYS. 39,. . . . . . . . 1232, (1968).
                                ( 6 ,  740) J. PHYSIQUE TOME 32 COL.1 VOL.I,  525, (1971).
                                ( 6 ,  779) J. PHYSIQUE TOME 32,. . . . . .  447, (1971).
                                ( 8 ,   10) ANN. DE PHYSIQUE TOME 7 NO.4,. .  299, (1972).
                                (13 ,    7) ANN. DE PHYSIQUE TOME 7 NO.5,. .  313, (1972).
```

3118 PARETTE.........(G.)....: (5 , 2007) SOL. STATE COMM. VOL.13, ... 1839, (1973).

3119 PARFENOVA.......(N.N.)..: (5 , 2879) SOV. PHYS. JETP LETT. VOL.18,. .   47, (1973).

3120 PARFENOV........(V.A.)..: (6 , 1238) IAEA SYMP. COPENHAGEN, VOL.1,. .   483, (1968).
                                 (6 , 2200) J. CHEM. PHYS. VOL.60,. . . . . 2832, (1974).

3121 PARISOT.........(G.)....: (22C,   67) REV. SCI. INSTRUM. VOL.45, . . .   643, (1974).

3122 PARKER.........(J.B.)..: (24 ,   39) NUCL. INST. MET. 30, ...   77, (1964).
                                 (17B,   44) NUCL. INST. MET. 62, ...   29, (1968).

3123 PARKER.........(K.)....: (17B,   44) NUCL. INST. MET. 62, ...   29, (1968).

3124 PARKINSON......(T.F.)..: (20A,   32) NATURE VOL.211,. . . . . . .   400, (1966).
                                 (22C,   35) NUCL. INST. MET. 53, ...   299, (1967).
                                 (5 , 1929) J. APPL. PHYS. VOL.45, ... 2021, (1974).

3125 PARKS..........(D.E.)..: (6 , 326) IAEA SYMP. CHALK RIVER, VOL.1, .   285, (1962).

3126 PARKS..........(D.)....: (5 , 432) PHYS. REV. LETT. VOL.25, ...   117, (1970).
                                 (5 , 430) REPORT HEPL-630 (14PP.), ...   , (1970).

3127 PARLINSKI......(K.)....: (15 ,   1) ACTA PHYS. POLON. VOL.22, SUPP.,   179, (1962).
                                 (5 , 269) NUKLEONIKA VOL.8,. . . . . . .   581, (1963).
                                 (6 , 1573) PHYS. STAT. SOL. 9,. . . . . . .   905, (1965).
                                 (6 , 502) PHYSICA VOL.35,. . . . .   441, (1967).
                                 (6 , 380) PHYSICA VOL.35,. . . . .   451, (1967).
                                 (6 , 1016) PHYSICA VOL.35,. . . . .   457, (1967).
                                 (6 , 1596) PHYSICA VOL.35,. . . . .   465, (1967).
                                 (6 , 2007) PHYSICA VOL.35,. . . . .   469, (1967).
                                 (6 , 1379) PHYS. LETT. 24A,. . . . .   517, (1967).
                                 (6 , 1017) ACTA PHYS. POLON. VOL.33,. . .   419, (1968).
                                 (6 , 1104) IAEA SYMP. COPENHAGEN VOL.II,.   143, (1968).
                                 (6 , 621) IAEA SYMP. COPENHAGEN, VOL.1,.   65, (1968).
                                 (6 , 2003) IAEA SYMP. COPENHAGEN, VOL.2,.   237, (1968).
                                 (6 , 1569) ACTA PHYS. POLON. VOL.35,. . .   223, (1969).
                                 (6 , 286) PHYSICA VOL.41,. . . . .   397, (1969).
                                 (10B, 1388) SOV. PHYS. CRYST. VOL.15,. . .   1010, (1971).
                                 (10B,   6) ACTA PHYS. POLON. VOL.A43, . .   247, (1973).
                                 (6 , 1187) ACTA PHYS. POLON. VOL.A44, . .   829, (1973).

3128 PARRINELLO......(M.)....: (14A,   28) J. PHYS. C VOL.6,. . . . . . .   L254, (1973).

3129 PARRY..........(W.E.)..: (6 , 1180) ANN. PHYS. VOL.17, ...   301, (1962).

3130 PARSONS........(J.L.)..: (6 , 2100) SOL STAT. COMM. 5, ...   387, (1967).

3131 PARTHASARATHY...(R.)....: (5 , 311) ACTA CRYST. VOL.B28, ...   2083, (1972).

3132 PARTHE.........(E.)....: (5 , 1865) ACTA CRYST. 16,. . . . .   202, (1963).
                                 (5 , 1870) ACTA CRYST. 19,. . . . .   1031, (1965).
                                 (5 , 2537) SOL. STATE COMM. VOL.6 NO.2, .   115, (1968).

3133 PASCULESCU.....(D.)....: (17C,   5) ACTA CRYST. 17,. . . . .   1529, (1964).

3134 PASECHNIK......(M.V.)..: (5 , 2229) SOV. AT. ENERGY VOL.18,. . . .   585, (1965).
                                 (5 , 1248) BULL. ACAD. SCI. USSR/PHYS. 31,.   334, (1967).
                                 (5 , 2242) UKRAINIAN PHYS. J. VOL.14,. .   1968, (1969).
                                 (6 , 785) UKR. FIZ. ZH. (USSR) VOL.18,. . 1384, (1973).
                                 (6 , 608) UKR. FIZ. ZH. (USSR) VOL.18,. . 1390, (1973).
                                 (6 , 784) UKR. FIZ. ZH. (USSR) VOL.18,. . 1528, (1973).
                                 (6 , 1748) UKR. FIZ. ZH. (USSR) VOL.18,. .   558, (1973).
                                 (6 , 568) UKR. FIZ. ZH. (USSR) VOL.18,. .   92, (1973).

3135 PASICHNYK......(H.V.)..: (14B, 136) UKR. FIZ. ZH. (USSR) VOL.9,. .   684, (1964).

3136 PASKIN.........(A.)....: (17C,   3) ACTA CRYST. 17,. . . . .   325, (1964).
                                 (10B, 143) PHYS. REV. VOL.188,. . . . . .   1407, (1969).

3137 PASSARI........(L.)....: (5 , 951) COLL. INTER. N.126 (GRENOBLE),   180, (1963).
                                 (5 , 1163) NUOVO CIMENTO VOL.32,. . . . .   25, (1964).
                                 (5 , 1735) J. APP. PHYS. 37,. . . . . . .   3236, (1966).

3138 PASSELL........(L.)....: (20B, 143) REV. SCI. INSTRUM. VOL.32, . .   870, (1961).
                                 (7A,   67) PHYS. REV. VOL.126,. . . . . .   632, (1962).
                                 (20D,   10) J. NUCL. ENERGY VOL.17,. . . .   227, (1963).
                                 (6 , 764) J. APPL. PHYS. VOL.35,. . . . .   933, (1964).
                                 (6 , 753) PROC. INT. CONF. (NOTTINGHAM),   99, (1964).
                                 (5 , 2236) PHYS. REV. VOL.138,. . . . . . B1116, (1965).
                                 (6 , 756) PHYS. REV. VOL.139,. . . . . . A1866, (1965).
                                 (6 , 728) KYOTO CONF. STAT. MECHANICS, .   169, (1968).
                                 (6 , 752) PHYS. REV. LETT. 21,. . . . .   99, (1968).
                                 (6 , 734) J. APP. PHYS. VOL.40,. . . . .   1442, (1969).
                                 (6 , 738) J. PHYS. SOC. JAPAN (SUPPL.) 26,   169, (1969).
                                 (6 , 794) PHYS. REV. VOL.179,. . . . . .   417, (1969).
                                 (6 , 1930) PHYS. REV. LETT. VOL.25, . . .   752, (1970).
                                 (6 , 1935) J. APPL. PHYS. 42,. . . . . .   1746, (1971).
                                 (6 , 2195) PHYS. REV. B VOL.4,. . . . . .   718, (1971).
                                 (6 , 718) PHYS. REV. LETT. VOL.27, . . .   741, (1971).
                                 (6 , 1165) IAEA SYMP. GRENOBLE, ...   357, (1972).
                                 (6 , 719) IAEA SYMP. GRENOBLE, ...   619, (1972).
                                 (6 , 720) MAGN. AND MAG. MATERIALS-1971,   1251, (1972).
                                 (6 , 1145) PHYS. REV. A VOL.5,. . . . . .   1377, (1972).
                                 (6 , 1953) PHYS. REV. B VOL.8,. . . . . .   5345, (1973).
                                 (5 , 1903) PHYS. REV. LETT. VOL.32, . . .   724, (1974).
                                 (5 , 1336) PHYS. REV. LETT. VOL.32, . . .   791, (1974).

3139 PASTUSHENKO.....(S.N.)..: (6 , 784) UKR. FIZ. ZH. (USSR) VOL.18, .   1528, (1973).

3140 PATAUD.........(P.)....: (5 , 1533) SOL. STATE COMM. VOL.7 NO.22,.   1669, (1969).
                                 (5 , 1134) J. PHYSIQUE TOME 31, ...   803, (1970).

3141 PATHAK.........(K.N.)..: (6 , 1136) PHYS. REV. A VOL.2,. . . . . .   2416, (1970).
                                 (14A,   49) PHYS. REV. A VOL.2,. . . . . .   2427, (1970).
                                 (6 , 1137) PHYS. REV. A VOL.4,. . . . . .   2443, (1971).
                                 (14A,   31) NUOVO CIMENTO B VOL.13B NO.1,.   185, (1973).
                                 (6 , 1630) PHYS. REV. A VOL.9,. . . . . .   2128, (1974).

3142 PATON..........(M.G.)..: (5 , 2875) ACTA CRYST., 19,. . . . . . .   679, (1965).

3143 PATRIKEEV......(YU.B.)..: (5 , 1204) SOV. PHYS. J.E.T.P., 28, . . .   649, (1969).

3144 PATSCHEKE......(E.)....: (5 , 898) CHEM. PHYS. LETT. VOL.2,NO.1,.   47, (1968).
                                 (5 , 903) PHYS. STAT. SOL. B VOL.46,. . .   597, (1971).

3145 PATTENDEN......(N.J.)..: (21 ,   42) REPORT NP/R-2551 (26 PP.), . .   , (1957).
                                 (20A,   1) ACTA CRYST. VOL.11,. . . . . .   228, (1958).
                                 (23 ,   28) IAEA SYMP. (PILE RESEARCH)VIENNA 535, (1960).
                                 (22A,   11) IAEA SYMP. (PILE RESEARCH)VIENNA 597, (1960).
                                 (5 , 153) NUCL. PHYS. A VOL.A177,. . . .   393, (1971).

3146 PAULIKAS.......(A.P.)..: (5 , 2214) J. PHYSIQUE TOME 32 COL.1 VOL.II 917, (1971).

```
3147  PAULI...........(G.)....:    (23 ,    21)  IAEA SYMP. (PILE RESEARCH)VIENNA   123, (1960).
                                  (20B,    76)  NUCL. INST. MET. 33, . . . . . .   229, (1965).

3148  PAULI...........(R.)....:    ( 6 ,  1192)  PHYS. REV. VOL.108,. . . . . . .  1346, (1957).

3149  PAUL............(G.L.)..:    ( 6 ,  1277)  IAEA SYMP. COPENHAGEN, VOL.1,..    267, (1968).
                                  ( 5 ,  2320)  ACTA CRYST. VOL.B26, . . . . . .   925, (1970).
                                  ( 6 ,  1323)  J. PHYS. SOC. JAPAN VOL.28, SUP.   242, (1970).
                                  ( 6 ,  1274)  PHYS. REV. B-2 . . . . . . . . .  4603, (1970).
                                  ( 5 ,  1979)  ACTA CRYST. VOL.B28, . . . . . .  2700, (1972).

3150  PAUTHENET.......(R.)....:    ( 5 ,  2018)  J. PHYS. CHEM. SOLIDS VOL.21,..    234, (1961).
                                  ( 5 ,  2738)  J. APPL. PHYS. VOL.33, SUPPL.,   1123, (1962).
                                  ( 5 ,  1837)  J. PHYS. RADIUM VOL.23, . . . .    477, (1962).
                                  ( 5 ,  1849)  COLL. INTER. N.126 (GRENOBLE),     126, (1963).
                                  ( 5 ,   615)  COLL. INTER. N.126 (GRENOBLE),     158, (1963).
                                  ( 5 ,   427)  COLL. INTER. N.126 (GRENOBLE),     186, (1963).
                                  ( 5 ,  1230)  COLL. INTER. N.126 (GRENOBLE),      92, (1963).
                                  ( 5 ,  2790)  J. PHYS. CHEM. SOL. 24,. . . .     487, (1963).
                                  ( 5 ,   435)  SOL. STATE COMM. 1,. . . . . .      81, (1963).
                                  ( 5 ,  1853)  PHYS. LETT. 7, . . . . . . . .     110, (1963).
                                  ( 5 ,  1853)  PHYS. LETT. 18, . . . . . . . .     13, (1965).
                                  ( 5 ,   591)  J. APPL. PHYS. VOL.37,. . . .     1038, (1966).
                                  ( 5 ,  1365)  COMPTES RENDUS, SERIE B, 266,.     994, (1968).
                                  ( 5 ,   521)  ACTA CRYST. VOL.B26, . . . . .    2036, (1970).

3151  PAVLOVSKAYA.....(T.F.)..:    (22C,    12)  CRYOGENICS VOL.10, . . . . . . .   440, (1970).

3152  PAVLUSHKIN......(N.M.)..:    ( 5 ,  2466)  PHYS. STAT. SOLIDI A VOL.18 NO.2   K91, (1973).
                                  ( 5 ,  2467)  J. APPL. CRYST. VOL.7,. . . . .    207, (1974).

3153  PAWELCZYK.......(J.)....:    (23 ,    63)  NUKLEONIKA VOL.7,. . . . . . .     223, (1962).

3154  PAWLEY..........(G.S.)..:    ( 6 ,  2078)  PHYS. REV. LETT. VOL.17, . . .     753, (1966).
                                  ( 9C,    16)  HARWELL SUMMER SCHOOL, . . . .     161, (1968).
                                  ( 5 ,   256)  ACTA CRYST. B25, . . . . . . .    2009, (1969).
                                  ( 5 ,  1341)  ACTA CRYST. VOL.A25, . . . . .     482, (1969).
                                  ( 6 ,   419)  DISC. FARADAY SOC. NO.48,. . .     125, (1969).
                                  (18B,     7)  J. APPL. CRYST. VOL.2,. . . .       37, (1969).
                                  ( 6 ,  2079)  J. OF PHYS. C VOL.2, . . . . .    1916, (1969).
                                  ( 6 ,   343)  SOL. STATE COMM. 7,. . . . . .     385, (1969).
                                  ( 5 ,  1342)  ACTA CRYST. VOL.A26, . . . . .     263, (1970).
                                  ( 6 ,  2006)  CONF. INTERN., RENNES, . . . .     223, (1971).
                                  (17B,    67)  SOL. STATE COMM. VOL.9,. . . .    1353, (1971).
                                  ( 5 ,    91)  ACTA CHEM. SCAND. VOL.26,. . .    1996, (1972).
                                  ( 5 ,   284)  ACTA CRYST. VOL.B28, . . . . .    1388, (1972).
                                  ( 5 ,  2370)  ADV. STRUC. RES. DIFFR. METHOD 4     1, (1972).
                                  (15 ,    16)  IAEA SYMP. GRENOBLE, . . . . .     175, (1972).
                                  ( 5 ,  1272)  PROC. ROYAL SOC. EDINBURGH A-70,   225, (1972).
                                  (10B,    16)  CHEM. APPL. THERMAL NEUTRON SCAT    78, (1973).
                                  (10B,   207)  PROC. ROY. SOC. A VOL.333, . . .    363, (1973).

3155  PAYTON..........(D.N.)..:    (10B,   133)  PHYS. REV. 175,. . . . . . . .    1201, (1968).

3156  PEARCE..........(D.G.)..:    (20B,    35)  IAEA SYMP. CHALK RIVER, VOL.1,.     83, (1962).

3157  PEARSON.........(R.K.)..:    ( 5 ,   151)  PHYS. REV. VOL.146,. . . . . .     660, (1966).

3158  PEARSON.........(W.B.)..:    ( 5 ,  2358)  PHIL. MAG. VOL.11, . . . . . .    1245, (1965).
                                  ( 5 ,  2360)  ACTA CHEM. SCAND. VOL.20,. . .    2529, (1966).
                                  ( 5 ,  2312)  PHIL. MAG. VOL.16,. . . . . . .    1063, (1967).
                                  ( 5 ,  1412)  ACTA CHEM. SCAND. VOL.22,. . .    3039, (1968).

3159  PEASE...........(R.S.)..:    ( 5 ,  1467)  PROC. ROY. SOC. A VOL.220, . .     397, (1953).
                                  (20D,    11)  J. SCI. INSTRUM. VOL.31, . . .     207, (1954).
                                  ( 5 ,  1466)  PROC. ROY. SOC. A VOL.230, . .     359, (1955).

3160  PECKHAM.........(G.E.)..:    (18A,    10)  BRIT. J. APPL. PHYS. VOL.18, .     473, (1967).
                                  ( 6 ,  1421)  PROC. PHYS. SOC. 90, . . . . .     657, (1967).
                                  ( 6 ,  1420)  J. PHYS. C, SER.2, VOL.3,. . .    1026, (1970).
                                  ( 6 ,   445)  CONF. INTERN., RENNES, . . . .     171, (1971).
                                  (10C,    47)  J. PHYS. C, SER.2, VOL.4,. . .    1091, (1971).
                                  ( 6 ,   448)  J. PHYS. C, SER.2, VOL.4,. . .    2009, (1971).
                                  ( 6 ,  1381)  J. PHYS. C VOL.7,. . . . . . .     L99, (1974).

3161  PECKHAM.........(G.)....:    ( 6 ,  1416)  COPENHAGEN CONF.-LATT.DYNAMICS,     49, (1963).
                                  (18A,    27)  UKAEA AERE REP. 4380 (6 PP.),.   . . . , (1964).
                                  ( 6 ,   226)  SOL. STATE COMM. 5,. . . . . .     311, (1967).

3162  PECORA..........(R.)....:    (14B,    37)  J. CHEM. PHYS. VOL.42, . . . .    1863, (1965).

3163  PEDERSEN........(T.)....:    ( 5 ,   184)  ACTA CRYST. VOL.A26, . . . . .     336, (1970).

3164  PEGRAM..........(G.B.)..:    (20A,    57)  PHYS. REV. VOL.48, . . . . . .     265, (1935).
                                  (20B,   123)  PHYS. REV. VOL.48, . . . . . .     704, (1935).
                                  (20A,    58)  PHYS. REV. VOL.49, . . . . . .     103, (1936).

3165  PEISVAUX........(J.)....:    ( 6 ,   722)  PHYS. REV. LETT. VOL.28, . . .     805, (1972).

3166  PEKOSHEVSKI.....(E.)....:    ( 5 ,  2954)  SOV. PHYS. SOL. STATE VOL.13,.     321, (1971).

3167  PELAH...........(I.)....:    ( 6 ,  2355)  PHYS. REV. VOL.108,. . . . . .    1091, (1957).
                                  ( 6 ,  2240)  PHYS. REV. VOL.109,. . . . . .    1046, (1958).
                                  ( 6 ,  1302)  PHYS. REV. LETT. VOL.2,. . . .      94, (1959).
                                  ( 6 ,  1834)  PHYS. REV. VOL.113,. . . . . .     767, (1959).
                                  ( 6 ,   937)  PHYS. REV. VOL.113,. . . . . .      49, (1959).
                                  ( 6 ,   125)  IAEA SYMPOSIUM VIENNA, . . . .     601, (1960).
                                  ( 6 ,  2225)  IAEA SYMP. CHALK RIVER, VOL.2,     155, (1962).
                                  ( 6 ,  1293)  IAEA SYMP. BOMBAY, VOL.2,. . .     325, (1964).
                                  ( 6 ,   449)  J. CHEM. PHYSICS VOL.43, . . .    1864, (1965).
                                  ( 6 ,  1048)  PHYS. LETT. 21,. . . . . . . .     248, (1966).
                                  ( 5 ,  1462)  SOL. STATE COMM. 5, . . . . .      41, (1967).
                                  (20A,    46)  NUCL. INST. MET. VOL.102,. . .      87, (1972).

3168  PELETTI.........(J.)....:    ( 6 ,  2225)  IAEA SYMP. CHALK RIVER, VOL.2, .   155, (1962).

3169  PELIZZARI.......(C.A.)..:    (22A,    48)  NUCL. INST. MET. VOL.114,. . .     417, (1974).
                                  (17C,    25)  REV. SCI. INSTRUM. VOL.45,. . .    572, (1974).

3170  PELLEGRINI......(U.)....:    (23 ,    65)  NUOVO CIMENTO SUPPL. VOL.23, . .    17, (1962).

3171  PELLIONISZ......(P.)....:    (23 ,    60)  NUCL. INST. MET. 55, . . . . .     154, (1967).
                                  (12B,    11)  IAEA SYMP. COPENHAGEN, VOL.2,.     407, (1968).
                                  (20B,    10)  ATOMKERNENERGIE VOL.17,. . . .     277, (1971).
                                  (19A,    49)  NUCL. INST. MET. VOL.92,. . . .    125, (1971).
                                  (19A,    27)  IAEA SYMP. GRENOBLE, . . . . .     763, (1972).
                                  (19A,    25)  IAEA SYMP. GRENOBLE, . . . . .     787, (1972).
                                  (12B,    14)  IAEA SYMP. GRENOBLE, . . . . .     797, (1972).
                                  (12B,    30)  NUCL. INST. MET. VOL.99, . . .     613, (1972).

3172  PENN............(A.W.)..:    (10B,   115)  PHYS. LETT. 27A, . . . . . . .     523, (1968).

3173  PEPELYSHEV......(YU.N.)..:   (20C,    26)  INSTRUM. EXP. TECH. VOL.16 NO.2,  345, (1973).
```

```
3174  PEPINSKY........(R.)....:  ( 9C,    38) SCIENCE VOL.117, . VOL.25,. . . . .      1, (1953).
                                (22C,    48) REV. SCI. INSTRUM VOL.25,. . . . .     699, (1954).
                                ( 5,    187) PHYS. REV. VOL.100,. . . . . . . .     745, (1955).
                                ( 2,   2286) PHYS. REV. VOL.97, . . . . . . . .    1179, (1955).
                                ( 2,   2287) ACTA CRYST. VOL.9, . . . . . . . .     131, (1956).
                                ( 2,   2289) PHYS. REV. VOL.105,. . . . . . . .     849, (1957).
                                ( 5,    188) PHYS. REV. VOL.105,. . . . . . . .     856, (1957).
                                ( 2,   1930) ACTA CRYST. VOL.11,. . . . . . . .     505, (1958).
                                ( 2,   1270) REV. MOD. PHYS. VOL.30,. . . . . .     100, (1958).

3175  PEPPER..........(A.R.)..:  ( 5,   1773) J. OF PHYS. C VOL.1, . . . . . .    1683, (1968).

3176  PEPPER..........(D.E.)..:  ( 6,   1306) J. PHYS. C VOL.5,. . . . . . . .    2611, (1972).
                                ( 6,   1514) J. PHYS. C VOL.6,. . . . . . . .    1933, (1973).

3177  PEPY............(G.)....:  ( 6,   2331) PHYS. LETT. VOL.36A, . . . . . .     376, (1971).
                                ( 6,    183) IAEA SYMP. GRENOBLE, . . . . . .     595, (1972).
                                ( 6,   1490) IAEA SYMP. GRENOBLE, . . . . . .     631, (1972).
                                ( 6,   1486) SOL. STATE COMM. VOL.10, . . . .     553, (1972).
                                ( 6,    444) SOL. STATE COMM. VOL.13, . . . .     919, (1973).
                                ( 6,   1495) J. PHYS. CHEM. SOLIDS VOL.35,. .     433, (1974).

3178  PEREKALINA......(T.M.)..:  ( 5,    160) SOV. PHYS. J.E.T.P. VOL.23,. . .     395, (1966).
                                ( 5,   2487) SOV. PHYS. J.E.T.P. VOL.25,. . .     266, (1967).

3179  PERELYAEV.......(V.A.)..:  ( 5,   1543) SOV. PHYS. SOL. STATE VOL.15,. .    1079, (1973).

3180  PERETTI.........(J.)....:  (10B,   122) PHYS. REV. 171,. . . . . . . . .     665, (1963).
                                (10C,    59) NUOVO CIMENTO VOL.34,. . . . . .     293, (1964).
                                ( 6,   1681) BROOKHAVEN SYMPOSIUM,. . . . . .     105, (1965).
                                ( 6,   2254) PHYS. LETT. 14,. . . . . . . . .     100, (1965).
                                ( 6,   2253) J. PHYSIQUE TOME 28 COL.1, . . .      26, (1967).
                                ( 6,   2233) REPORT EUR-4216E (41PP.),. . . .       . , (1969).

3181  PEREZ...........(R.B.)..:  (20B,    31) IAEA SYMP. (PILE RESEARCH)VIENNA     509, (1960).
                                (16,     39) NUCL. SCI. ENGNG. VOL.28,. . . .     404, (1967).

3182  PERKINS.........(D.R.)..:  ( 5,     51) J. NUCL. ENERGY VOL.24,. . . . .     419, (1970).

3183  PERNET..........(M.)....:  ( 5,    453) BULL. SOC. FRANC. MIN. CRIST. 92    264, (1969).

3184  PERRIER-DE-LA-BA(R.)....:  (12B,    28) NUCL. INST. MET. VOL.95, . . . .     589, (1971).

3185  PERRIN..........(C.)....:  (19B,    32) REV. PHYS. APPL. (FRANCE) VOL.4,    111, (1969).

3186  PERRIN..........(G.)....:  (24,     60) NUCL. INST. MET. 61, . . . . . .     198, (1968).

3187  PERRIN..........(M.)....:  ( 5,   2166) COMP. REND. 263, . . . . . . . .    344B, (1966).
                                ( 5,    343) COMP. REND. 263, . . . . . . . .      9B, (1966).
                                ( 5,    472) J. APP. PHYS. 39,. . . . . . . .     632, (1968).
                                ( 5,   2167) ACTA CRYST. B25, . . . . . . . .    2326, (1969).
                                ( 5,   1047) J. APP. PHYS. 40,. . . . . . . .    1126, (1969).
                                ( 5,    471) J. PHYSIQUE TOME 31, . . . . . .     113, (1970).
                                ( 5,   2941) PHYS. STAT. SOL. VOL.40, . . . .     171, (1970).
                                ( 5,   1748) J. APPL. PHYS. 42, . . . . . . .    1615, (1971).
                                ( 5,   1045) J. PHYSIQUE TOME 32 COL.1 VOL.I,    322, (1971).
                                ( 5,   1742) J. PHYS. CHEM. SOLIDS VOL.32,. .    2429, (1971).

3188  PERRIN..........(P.)....:  (19B,    32) REV. PHYS. APPL. (FRANCE) VOL.4,    111, (1969).

3189  PERSHAN.........(P.S.)..:  ( 7B,    85) PHYS. REV. 168,. . . . . . . . .     725, (1968).

3190  PERSIANI........(P.J.)..:  ( 5,   2462) PHYS. REV. VOL.105,. . . . . . .     517, (1957).

3191  PERTHEL.........(R.)....:  ( 5,   1689) PHYS. STAT. SOL. VOL.8,. . . . .     271, (1965).

3192  PESHKIN.........(M.)....:  ( 7A,     8) AM. J. PHYS. VOL.39, . . . . . .     324, (1971).

3193  PETCH...........(H.E.)..:  ( 5,   1952) CAN. J. PHYS. 42,. . . . . . . .     229, (1964).
                                ( 5,   1453) CAN. J. PHYS. 48,. . . . . . . .    1091, (1970).

3194  PETERLIN........(T.M.)..:  (20A,    13) ATOMKERNENERGIE VOL.17,. . . . .      93, (1971).

3195  PETERSON........(D.T.)..:  ( 6,   2170) PHYS. REV. B VOL.8,. . . . . . .    1332, (1973).

3196  PETERSON........(E.R.)..:  ( 5,     45) DISS. ABS. XXV,. . . . . . . . .    5588, (1965).

3197  PETERSON........(K.C.)..:  ( 5,   1411) PHYS. REV. VOL.72, . . . . . . .     109, (1947).

3198  PETERSON........(S.W.)..:  ( 9C,    21) J. CHEM. PHYS. VOL.19, . . . . .    1416, (1951).
                                ( 5,   1895) PHYS. REV. VOL.83, . . . . . . .    1270, (1951).
                                ( 5,   1450) J. CHEM. PHYS. VOL.20, . . . . .     704, (1952).
                                ( 5,   1920) PHYS. REV. VOL.86, . . . . . . .     766, (1952).
                                ( 5,   2830) PHYS. REV. VOL.87, . . . . . . .     462, (1952).
                                ( 5,   2318) ACTA CRYST. VOL.6, . . . . . . .     661, (1953).
                                ( 5,   1896) J. AMER. CHEM. SOC. VOL.75,. . .    1536, (1953).
                                ( 5,   1472) J. CHEM. PHYS. VOL.21, . . . . .    2084, (1953).
                                ( 5,   1897) J. CHEM. PHYS. VOL.21, . . . . .     366, (1953).
                                ( 5,   2430) J. PHYS. CHEM. VOL.57, . . . . .     535, (1953).
                                ( 5,   1468) PHYS. REV. VOL.93, . . . . . . .    1120, (1954).
                                ( 5,   2819) ACTA CRYST. VOL.10,. . . . . . .     319, (1957).
                                ( 5,    789) ACTA CRYST. VOL.10,. . . . . . .      70, (1957).
                                ( 5,   1502) J. CHEM. PHYS. VOL.29, . . . . .     948, (1958).
                                ( 5,   1274) REV. MOD. PHYS. VOL.30,. . . . .     101, (1958).
                                ( 5,    387) PHYS. REV. LETT. VOL.6,. . . . .       7, (1961).
                                ( 5,   1580) J. PHYS. SOC. JAPAN VOL.17 B-II,    335, (1962).
                                ( 6,    458) COLL. INTER. N.126 (GRENOBLE), .     191, (1963).
                                ( 5,   2209) COLL. INTER. N.126 (GRENOBLE), .      27, (1963).
                                (23,      7) COLL. INTER. N.126 (GRENOBLE), .      71, (1963).
                                ( 5,   1935) ACTA CRYST. 22,. . . . . . . . .     928, (1967).
                                ( 5,   1578) J. CHEM. PHYSICS VOL.48,. . . . .    5561, (1968).
                                ( 5,     26) ACTA CRYST. (INTERACT.) VOL.A25,   S113, (1969).
                                ( 9C,     4) SPECTROS. INORG. CHEM.,. . . . .       1, (1971).
                                ( 5,   2873) J. CHEM. PHYS. VOL.59, . . . . .     453, (1973).

3199  PETER...........(H.)....:  ( 6,   1346) J. PHYS. CHEM. SOLIDS VOL.34,. .     255, (1973).

3200  PETIT...........(G.A.)..:  (17B,    43) NUCL. INST. MET. 44, . . . . . .     341, (1966).

3201  PETKOVSEK.......(J.)....:  ( 6,   1299) PHYS. LETT. VOL.26A, . . . . . .       8, (1967).
                                ( 5,    372) PHYS. LETT. 25A, . . . . . . . .     123, (1967).
                                (21,     25) J. PHYS. E VOL.4 NO.11,. . . . .     905, (1971).

3202  PETRASCU........(M.)....:  (20B,   138) REV. ROUMAINE PHYS. VOL.11,. . .     601, (1966).

3203  PETRIE..........(M.)....:  ( 5,   1946) J. PHYS. CHEM. VOL.65, . . . . .    1453, (1961).

3204  PETROVA.........(I.I.)..:  ( 5,    179) SOV. PHYS. SOL. STATE VOL.11,. .    2177, (1969).

3205  PETRUNIN........(V.F.)..:  ( 5,   2012) SOV. PHYS. CRYST. VOL.14,. . . .     522, (1970).
                                ( 5,   1595) ACTA CRYST. VOL.A28,. . . . . . .     473, (1972).

3206  PETRUN≠KIN......(V.A.)..:  ( 7A,    87) SOV. PHYS. JETP. 16, . . . . . .    1321, (1963).
```

```
3207  PETRZILKA........(V.)....:  (  98,    36) CZECH. J. PHYS. VOL.18,. . . . .  1111, (1968).
                                  (  98,    62) NATURE VOL.218,. ,. . . . . . .    80, (1968).
                                  (   5,  2371) PHYS. STAT. SOL. 29,. . . . . .   K51, (1968).
                                  (   5,  2375) BRIT. J. APPL. PHYS. VOL.2,. . . 1041, (1969).
                                  (   5,  2373) PHYS. STAT. SOL. VOL.42,. . . .   895, (1970).
                                  (   5,  2372) ACTA CRYST. VOL.A27,. . . . . .   410, (1971).
                                  (  98,    72) PHYS. LETT. VOL.37A,. . . . . .   403, (1971).
                                  ( 188,     6) JAD. ENERG. (CZECH.) VOL.18,. .   367, (1972).
                                  (   6,  2052) NATURE (PHYS. SCI.) VOL.242,. .   109, (1973).
                                  (   5,  2449) PHYS. STAT. SOLIDI A VOL.17 NO.1  163, (1973).

3208  PETZ............(J.I.)..:  (  5,  2263) J. CHEM. PHYS. VOL.32,. . . . .   241, (1960).

3209  PFAFFELHUBER....(E.)....:  ( 7B,     2) ANN. OF PHYS. VOL.45,. . . . . .  404, (1967).

3210  PFEIFFER........(K.)....:  (  6,  1670) PHYS. LETT. VOL.34A,. . . . . .    35, (1971).

3211  PHILIPPE........(M.)....:  ( 21,     5) ARCH. SCI. VOL.13,. . . . . . .   375, (1960).

3212  PHILIPPOT.......(E.)....:  (  5,  1465) REV. CHIM. MINER. VOL.9,. . . .   825, (1972).

3213  PHILLIPS........(J.C.)..:  ( 10B,   119) PHYS. REV. VOL.113,. : : : : : :  147, (1959).
                                  (   5,  2435) PHYS. LETT. VOL.37A,. . . . . .   434, (1971).

3214  PHILLIPS........(V.A.)..:  (  5,   738) TRANS. METALL. SOC. AIME VOL.236 1012, (1966).

3215  PHILLIPS........(W.C.)..:  (  9B,   102) REV. SCI. INSTRUM. VOL.33,. . .   126, (1962).
                                  (   5,  1108) PHYS. REV. 138,. . . . . . . . A1649, (1965).

3216  PICKART.........(S.J.)..:  (  5,  1610) PROC. I.E.E. VOL.104B SUP.4-7,    217, (1957).
                                  (   5,  2138) DISS. ABSTR. VOL.20,. . . . . .   333, (1959).
                                  (   5,  1677) J. APPL. PHYS. VOL.30, SUPPL.,    280, (1959).
                                  (   5,  1187) J. PHYS. CHEM. SOLIDS VOL.10,.     35, (1959).
                                  (   5,  2199) J. PHYS. SOC. JAPAN VOL.14,. .  1352, (1959).
                                  (   5,  1688) PHYS. REV. VOL.116,. . . . . .   317, (1959).
                                  (   5,   957) J. APPL. PHYS. VOL.31, SUPPL.,    372, (1960).
                                  (   5,  2852) J. CHEM. PHYS. VOL.32,. . . . .   308, (1960).
                                  (   5,  2661) J. PHYS. CHEM. SOLIDS VOL.13,.    166, (1960).
                                  (   5,  1568) PHYS. REV. VOL.121,. . . . . .   707, (1961).
                                  (   5,   954) PHYS. REV. VOL.123,. . . . . .  1163, (1961).
                                  (   5,   933) J. PHYS. SOC. JAPAN VOL.17 B-III    7, (1962).
                                  (   5,  1683) PHYS. REV. LETT. VOL.8,. . . .   237, (1962).
                                  ( 179,     2) ACTA CRYST. VOL.16,. . . . . .   174, (1963).
                                  (   5,  1090) COLL. INTER. N.126 (GRENOBLE),   118, (1963).
                                  (   5,  2405) COLL. INTER. N.126 (GRENOBLE),   141, (1963).
                                  (   6,  1460) J. APP. PHYS. 34,. . . . . . .  1182, (1963).
                                  (   6,  1784) J. APP. PHYS. 34,. . . . . . .  1203, (1963).
                                  (   5,  1686) J. PHYS. CHEM. SOL. 24,. . . .  1531, (1963).
                                  (   5,   915) J. PHYS. CHEM. SOL. 24,. . . .  1679, (1963).
                                  (   5,  1000) J. APPL. PHYS. VOL.35,. . . . .   954, (1964).
                                  (   6,   809) J. PHYS. CHEM. SOLIDS VOL.25,.    183, (1964).
                                  (   5,  2983) PHYS. REV. LETT. VOL.12,. . . .   444, (1964).
                                  (   6,   846) PHYS. REV. VOL.136,. . . . . . A1641, (1964).
                                  (   6,  1122) PROC. INT. CONF. (NOTTINGHAM),    223, (1964).
                                  (   6,  1735) J. APP. PHYS. 36,. . . . . . .  1076, (1965).
                                  (   6,   751) PHYS. REV. LETT. 15,. . . . . .   146, (1965).
                                  (   5,  1384) J. APP. PHYS. 37,. . . . . . .  1032, (1966).
                                  (   5,  1766) J. APP. PHYS. 37,. . . . . . .  1053, (1966).
                                  (   6,  1314) J. APP. PHYS. 37,. . . . . . .  1054, (1966).
                                  (   5,   917) J. APP. PHYS. 37,. . . . . . .   981, (1966).
                                  (   6,  1506) PROC. PHYS. SOC. 88,. . . . . .   333, (1966).
                                  (   6,  1733) J. APP. PHYS. 38,. . . . . . .  1245, (1967).
                                  (   6,  1738) PHYS. REV. 156,. . . . . . . .   623, (1967).
                                  (   6,  2130) PHYS. REV. 161,. . . . . . . .   499, (1967).
                                  (   5,  1992) J. APP. PHYS. 39,. . . . . . .  1237, (1968).
                                  (   5,  2652) J. APP. PHYS. 39,. . . . . . .  2221, (1968).
                                  (   6,   829) J. APP. PHYS. 39,. . . . . . .   455, (1968).
                                  (   5,   913) J. PHYS. CHEM. SOLIDS 29,. . .   414, (1968).
                                  (   5,  1213) J. APPL. PHYS. VOL.40,. . . . .  1009, (1969).
                                  (   5,  2920) J. APPL. PHYS. VOL.41,. . . . .  1192, (1970).
                                  (   6,   912) J. APPL. PHYS. 41,. . . . . . .   933, (1970).
                                  (   5,  2409) J. APPL. PHYS. VOL.42,. . . . .  1617, (1971).
                                  (   5,   848) MAGN. AND MAG. MATERIALS-1971,   1436, (1972).
                                  (   5,  2566) PHYS. REV. LETT. VOL.29,. . . .  1562, (1972).
                                  (   5,   911) A.I.P. CONF. PROC. NO.10 PART 2, 1569, (1973).
                                  (   5,  2565) PHYS. LETT. VOL.47A NO.1,. . .     73, (1974).

3217  PICK............(R.M.)..:  ( 10B,    38) IAEA SYMP. COPENHAGEN, VOL.1,.    119, (1968).

3218  PICOT...........(A.)....:  ( 19B,    33) REV. PHYS. APPL.(FRANCE) VOL.5,   693, (1970).

3219  PICOT...........(C.)....:  (  6,  1922) PHYS. REV. LETT. VOL.32,. . . .  1170, (1974).

3220  PIERRE..........(J.)....:  (  5,  2563) COMPTES RENDUS, SERIE B, 265,.   1169, (1967).
                                  (   5,  2382) PHYS. STAT. SOLIDI A VOL.15 NO.2  613, (1973).
                                  (   6,   709) PHYS. REV. B VOL.9 NO.11,. . .     , (1974).

3221  PIERROT.........(M.)....:  (  5,   778) ACTA CRYST. B25,. . . . . . . .  1685, (1969).
                                  (   5,  1278) ACTA CRYST. VOL.B28,. . . . . .  2530, (1972).

3222  PIESVAUX........(J.)....:  (  8,    18) C.R. ACAD. SCI. B VOL.274,. . .   423, (1972).
                                  (  17A,    87) PHYS. REV. LETT. VOL.31,. . . .   776, (1973).

3223  PIETRASS........(B.)....:  ( 10B,   194) PHYS. STAT. SOL. B VOL.53 NO.1,.  279, (1972).

3224  PIETRAS.........(E.)....:  (  9B,    68) NUKLEONIKA VOL.13,. . . . . . .  1111, (1968).

3225  PIGOTT..........(M.T.)..:  (  5,   958) J. PHYS. CHEM. SOLIDS VOL.6,. .     38, (1958).

3226  PIKEL≠NER.......(L.B.)..:  (  5,  2433) SOV. J. NUCL. PHYS. VOL.9, 6,.   1119, (1969).
                                  (  17A,    95) SOV. AT. ENERGY(USA) VOL.32,. .    68, (1972).

3227  PILCHER.........(E.)....:  ( 23,    57) NUCL. INST. MET. 27,. . . . . .    61, (1964).

3228  PILLIONISZ......(P.)....:  ( 19B,    30) REPORT KFKI-07/1968 (14PP.),. . . , (1968).

3229  PILLON..........(J.J.)..:  (  5,  1681) SOL. STATE COMM. VOL.7 NO.19,.   1403, (1969).

3230  PINDOR..........(A.)....:  (  6,   178) ACTA PHYS. POLON. VOL.A43,. . .    37, (1973).

3231  PINEDA..........(V.M.)..:  ( 20D,    16) NUCL. INST. MET. 37,. . . . . .   121, (1965).

3232  PINELLA.........(A.)....:  (  5,    40) J. MATERIALS SCI. VOL.3,. . . .   498, (1968).

3233  PINELLI.........(T.)....:  ( 22C,    38) NUCL. INST. MET. 62,. . . . . .   233, (1968).

3234  PINES...........(D.)....:  (  6,  1179) PHYS. REV. VOL.127,. . . . . .   1452, (1962).
                                  (   6,  1129) PHYS. REV. LETT. VOL.24,. . . .  1044, (1970).

3235  PINE............(A.S.)..:  (  6,  2161) PHYS. REV. B VOL.4,. . . . . . .   356, (1971).

3236  PINGS...........(C.J.)..:  ( 7B,    28) J. CHEM. PHYSICS VOL.48,. . . .  3016, (1968).
```

```
3237   PINKEVICH........(I.P.)..:   (10D,    63)  SOV. PHYS. SOL. STATE VOL.11,. .      69,  (1969).

3238   PINOT...........(M.)....:   ( 5 ,   998)  COLL. INTER. N.126 (GRENOBLE),. .     113,  (1963).
                                   ( 5 ,  1373)  BULL. SOC. FRANC. MIN. CRIST. 89      216,  (1966).
                                   ( 8 ,    18)  C.R. ACAD. SCI. B VOL.274,. . .       423,  (1972).
                                   ( 6 ,   722)  PHYS. REV. LETT. VOL.28, . . .        805,  (1972).
                                   (17A,    87)  PHYS. REV. LETT. VOL.31, . . . .      776,  (1973).

3239   PINTO...........(H.)....:   ( 5 ,  2893)  SOL. STATE COMM. VOL.8,. . . . .      597,  (1970).
                                   ( 5 ,  2676)  PHYS. REV. B VOL.3,. . . . . . .      2344, (1971).
                                   ( 5 ,   879)  PHYS. REV. B VOL.3,. . . . . . .      3861, (1971).
                                   ( 5 ,  2043)  ACTA CRYST. VOL.A28 PART S-4,. .      194,  (1972).
                                   ( 5 ,  2040)  SOL. STATE COMM. VOL.10, . . . .      663,  (1972).
                                   ( 5 ,  2044)  PHYS. REV. B VOL.7,. . . . . . .      3261, (1973).
                                   ( 5 ,   882)  PHYS. REV. B VOL.8,. . . . . . .      3398, (1973).

3240   PINTSCHOVIUS....(L.)....:   ( 6 ,  1333)  PHYS. REV. LETT. VOL.30, . . . .      1144, (1973).
                                   ( 6 ,  1334)  PHYS. REV. LETT. VOL.32, . . . .      836,  (1974).

3241   PIRIE...........(J.D.)..:   (10B,   125)  PHYS. REV. 175,. . . . . . . . .      1083, (1968).

3242   PIRILA..........(P.)....:   ( 7B,    90)  PHYS. STAT. SOL. VOL.42, . . . .      757,  (1970).

3243   PIRKMAJER.......(E.)....:   ( 6 ,  1299)  PHYS. LETT. VOL.26A, . . . . . .        8,  (1967).
                                   ( 6 ,   409)  IAEA SYMP. COPENHAGEN, VOL.1,. .      581,  (1968).
                                   ( 6 ,  1592)  PHYSICA VOL.49,. . . . . . . . .        1,  (1970).

3244   PIRLOGEA........(P.)....:   (20B,   136)  REV. DE PHYSIQUE VOL.6,. . . . .      411,  (1961).
                                   (20R,   142)  REV. SCI. INSTRUM. VOL.32, . . .      297,  (1961).
                                   (18B,     9)  REV. ROUMAINE PHYS. VOL.10,. . .      445,  (1965).

3245   PISANKO.........(ZH.I.).:   ( 5 ,  2411)  SOV. AT. ENERGY VOL.19,. . . . .      1162, (1965).

3246   PISERI..........(L.)....:   (10B,    50)  J. CHEM. PHYSICS VOL.48, . . . .      3561, (1968).
                                   (10R,    49)  J. CHEM. PHYSICS VOL.49, . . . .      3840, (1968).
                                   ( 6 ,  1910)  MOL. PHYS. VOL.22, . . . . . . .      241,  (1971).
                                   ( 6 ,   425)  BULL. AMER. PHYS. SOC. VOL.17, .      291,  (1972).
                                   ( 6 ,   423)  IAEA SYMP. GRENOBLE, . . . . . .      297,  (1972).
                                   (10B,    59)  J. CHEM. PHYS. VOL.56, . . . . .      1022, (1972).
                                   ( 6 ,  1926)  J. CHEM. PHYS. VOL.58, . . . . .      158,  (1973).
                                   (10B,    73)  J. PHYS. C VOL.6,. . . . . . . .      1521, (1973).

3247   PISTELLA........(F.)....:   (17A,    75)  NUCL. SCI. ENG. VOL.48,. . . . .      281,  (1972).

3248   PITAEVSKII......(L.P.)..:   (14A,    68)  SOV. PHYS. J.E.T.P. VOL.32,. . .      1183, (1971).

3249   PITKANEN........(P.H.)..:   ( 5 ,   230)  TRANS. AM. NUCL. SOC. VOL.13,. .      728,  (1970).

3250   PITZER..........(K.S.)..:   ( 5 ,  1254)  PHYS. REV. VOL.58, . . . . . . .      1003, (1940).

3251   PLACHENOVA......(E.L.)..:   ( 6 ,  2077)  SOV. PHYS. SOL. STATE VOL.10,. .      402,  (1968).
                                   ( 6 ,   171)  SOV. PHYS. SOL. STATE VOL.11,. .      1615, (1969).
                                   ( 6 ,   893)  SOV. PHYS. SOL. STATE VOL.13,. .      1256, (1971).

3252   PLACZEK.........(G.)....:   (16 ,    59)  PHYS. REV. VOL.69, . . . . . . .      423,  (1946).
                                   (14B,    74)  PHYS. REV. VOL.82, . . . . . . .      392,  (1951).
                                   ( 7B,    63)  PHYS. REV. VOL.86, . . . . . . .      377,  (1952).
                                   (10C,    68)  PHYS. REV. VOL.93, . . . . . . .      1207, (1954).
                                   ( 7B,    68)  PHYS. REV. VOL.93, . . . . . . .      895,  (1954).
                                   ( 7B,    44)  NUOVO CIMENTO VOL.1, . . . . . .      233,  (1955).
                                   ( 7B,    69)  PHYS. REV. VOL.105,. . . . . . .      1240, (1957).

3253   PLAEI...........(P.N.)..:   ( 5 ,  2628)  ATOMNAYA ENERGIYA (USSR) VOL.24,      243,  (1968).

3254   PLAKHTII........(V.P.)..:   ( 5 ,  2271)  BULL. ACAD. SCI. USSR VOL.28,. .      350,  (1964).
                                   ( 5 ,   218)  PROC. INT. CONF. (NOTTINGHAM), .      354,  (1964).
                                   ( 5 ,  2270)  SOV. PHYS. SOL. STATE VOL.7, . .      997,  (1965).
                                   ( 5 ,   214)  SOV. PHYS. SOL. STATE VOL.10,. .      754,  (1968).
                                   ( 5 ,   349)  SOV. PHYS. SOL. STATE VOL.9, . .      1762, (1968).
                                   ( 5 ,   730)  SOV. PHYS. SOL. STATE VOL.11,. .      672,  (1969).
                                   ( 5 ,  2880)  PHYS. STAT. SOL. B VOL.53 NO.1,      K37,  (1972).
                                   ( 5 ,  2051)  PHYS. STAT. SOL. B VOL.56 NO.1,       61,  (1973).
                                   ( 5 ,  2879)  SOV. PHYS. JETP LETT. VOL.18,. .       47,  (1973).
                                   ( 5 ,   347)  SOV. PHYS. SOL. STATE VOL.14,. .      2387, (1973).
                                   ( 5 ,  1960)  SOL. STATE COMM. VOL.14 NO.4,. .      309,  (1974).

3255   PLAKIDA.........(N.M.)..:   (10D,    23)  PHYS. LETT. VOL.32A, . . . . . .      134,  (1970).
                                   (10E,    30)  SOV. PHYS. SOL. STATE VOL.14,. .      2462, (1973).

3256   PLANT...........(J.S.)..:   ( 5 ,  1417)  Z. KRIST. VOL.126, . . . . . . .      460,  (1968).
                                   ( 5 ,  1339)  J. PHYS.F: METAL PHYS. VOL.1,. .      524,  (1971).
                                   ( 5 ,   130)  J. PHYS. F VOL.3,. . . . . . . .      2003, (1973).

3257   PLASKETT........(J.S.)..:   ( 7B,   104)  PROC. ROY. SOC. 301, . . . . . .      363,  (1967).

3258   PLATONOV........(V.I.)..:   (21 ,    53)  SOV. PHYS. CRYST. VOL.12,. . . .      978,  (1968).

3259   PLATTNER........(R.)....:   (20C,    46)  NUCL. INST. MET. 57, . . . . . .      237,  (1967).

3260   PLECHATY........(E.F.)..:   (24 ,    86)  PHYS. MED. BIOL.(GB) VOL. 16,. .      439,  (1971).

3261   PLESSER.........(T.)....:   ( 6 ,  1833)  PHYSICA VOL.31,. . . . . . . . .      1537, (1965).
                                   (14A,    13)  DISC. FARADAY SOC. NO.43,. . . .      160,  (1967).
                                   ( 6 ,  1297)  2ND SUMMER SCHOOL S.S. PHYS.,. .      120,  (1968).
                                   ( 5 ,  1456)  PHYS. STAT. SOL. VOL.42, . . . .      207,  (1970).

3262   PLICQUE.........(F.)....:   ( 5 ,  2392)  J. PHYSIQUE TOME 32 COL.1 VOL.II      611,  (1971).

3263   PLIHAL..........(M.)....:   (10B,   200)  PHYS. STAT. SOLIDI B VOL.58 NO.1      315,  (1973).

3264   PLUMIER.........(R.)....:   ( 5 ,  1841)  C. R. ACAD. SCI. VOL.255,. . . .      2244, (1962).
                                   ( 5 ,  1495)  J. PHYS. RADIUM VOL.23,. . . . .      474,  (1962).
                                   ( 5 ,   564)  PHYS. STAT. SOL. VOL. 2, . . . .      K112, (1962).
                                   ( 5 ,  1494)  PHYS. STAT. SOL. 2,. . . . . . .      317,  (1962).
                                   ( 5 ,  1615)  COMP. REND. 257, . . . . . . . .      3858, (1963).
                                   ( 5 ,  1497)  J. PHYSIQUE TOME 24, . . . . . .      741,  (1963).
                                   ( 5 ,  1496)  J. APPL. PHYS. VOL.35, . . . . .      950,  (1964).
                                   ( 5 ,  1839)  PROC. INT. CONF. (NOTTINGHAM), .      295,  (1964).
                                   ( 5 ,  2936)  COMP. REND. 260, . . . . . . . .      3348, (1965).
                                   ( 5 ,  1227)  COMP. REND. 263, . . . . . . . .      1738, (1966).
                                   ( 5 ,  2935)  J. APP. PHYS. 37,. . . . . . . .      964,  (1966).
                                   ( 5 ,  2937)  J. PHYSIQUE TOME 27, . . . . . .      213,  (1966).
                                   ( 5 ,  1225)  COMP. REND. 264, . . . . . . . .      2788, (1967).
                                   ( 5 ,   454)  COMP. REND. 265, . . . . . . . .      6728, (1967).
                                   ( 5 ,  1667)  COMP. REND. 265, . . . . . . . .      7268, (1967).
                                   ( 5 ,  1843)  COMP. REND. 267, . . . . . . . .      1057, (1968).
                                   ( 5 ,  1603)  COMP. REND. 267, . . . . . . . .       98,  (1968).
                                   ( 5 ,  1669)  J. APPL. PHYS. VOL.39, . . . . .      635,  (1968).
                                   ( 5 ,  1673)  COMP. REND. 268, . . . . . . . .      1549, (1969).
                                   ( 5 ,  1601)  COMP. REND. 268, . . . . . . . .      365,  (1969).
                                   ( 5 ,  2483)  REPORT CEA-R-3756 (148PP.),. . .        *,  (1969).
                                   ( 5 ,  1003)  SOL. STATE COMM. VOL.8,. . . . .      477,  (1970).
                                   ( 5 ,    46)  SOL. STATE COMM. VOL.9,. . . . .      413,  (1971).
                                   ( 5 ,   358)  SOL. STATE COMM. VOL.9,. . . . .      1723, (1971).
                                   ( 5 ,   341)  SOL. STATE COMM. VOL.10, . . . .        5,  (1972).
                                   ( 5 ,  1629)  SOL. STATE COMM. VOL.12 NO.2,. .      109,  (1973).
```

3301 POWELL..........(B.M.)..:   (6 , 1926) J. CHEM. PHYS. VOL.58,. . . . .   159, (1973).
                                   (10B,   207) PROC. ROY. SOC. A VOL.333, ... 363, (1973).
                                   (6 ,   903) SOL. STATE COMM. VOL.13, ... 1555, (1973).

3302 POWERS..........(P.N.)..:   (9B,   73) PHYS. REV. VOL.50, ...   486, (1936).
                                   (5 ,   943) PHYS. REV. VOL.54, ...   827, (1938).

3303 POWLES..........(J.G.)..:   (6 ,   661) IAEA SYMP. COPENHAGEN, VOL.1,. .   379, (1968).
                                   (6 ,   939) MOL. PHYS. VOL.20, ...   881, (1971).
                                   (5 , 1302) MOL. PHYS. VOL.21, ...   901, (1971).
                                   (6 , 1079) MOL. PHYS. VOL.24 NO.5,. . . . .   1025, (1972).
                                   (14B,   8) ADV. IN PHYSICS VOL.22,. . . . .   1, (1973).
                                   (14B,   17) CHEM. APPL. THERMAL NEUTRON SCAT   118, (1973).
                                   (15 ,   29) MOL. PHYS. VOL.26 NO.6,. . . . .   1325, (1973).

3304 POYRY...........(H.O.)..:   (6 , 1064) COMMENTAT. PHYS.-MATH.(FIN.) 42,   277, (1972).

3305 PRADDAUDE.......(H.C.)..:   (8 ,   38) J. PHYSIQUE TOME 32 COL.1 VOL.II   818, (1971).

3306 PRAGER..........(P.R.)..:   (9B,   14) ACTA CRYST. VOL.A27, ...   563, (1971).

3307 PRAGER..........(P.)....:   (5 , 2378) APPL. PHYS. LETT. VOL.10,. . . .   293, (1967).

3308 PRAKASH.........(A.)....:   (5 ,   793) PHYS. ICE, ...   44, (1969).
                                   (5 , 1292) J. CHEM. PHYS. VOL.55, ...   1934, (1971).
                                   (5 ,   786) J. CHEM. PHYS. VOL.58, ...   567, (1973).

3309 PRAKASH.........(S.)....:   (10B,   145) PHYS. REV. VOL.187,. . . . . . .   808, (1969).
                                   (6 ,   61) PHYSICA VOL.50,. . . . .   10, (1970).
                                   (10B,   165) PHYS. REV. B-4,. . . . .   1770, (1971).
                                   (10B,   106) PHYSICA VOL.58,. . . . .   71, (1972).
                                   (10B,   174) PHYS. REV. B VOL.5,. . . . . . .   2880, (1972).

3310 PRANDL..........(W.)....:   (5 ,   65) COLL. INTER. N.126 (GRENOBLE),   16, (1969).
                                   (5 ,   340) ACTA CRYST. VOL.A28 PART S-4,. .   194, (1972).
                                   (5 ,   342) SOL. STATE COMM. VOL.10, ...   529, (1972).
                                   (5 ,   332) SOL. STATE COMM. VOL.11 NO.5,. .   645, (1972).
                                   (5 ,   82) PHYS. STAT. SOLIDI B VOL.55 NO.2 K159, (1973).

3311 PRASAD..........(B.)....:   (10B,   86) J. PHYS.F: METAL PHYS. VOL.2,. .   247, (1972).
                                   (6 ,   28) PHYS. LETT. VOL.38A, ...   527, (1972).
                                   (6 , 1362) PHYS. REV. B VOL.6,. . . . . . .   2192, (1972).
                                   (10B,   197) PHYS. STAT. SOLIDI B VOL.56 NO.1   327, (1973).

3312 PRASAD..........(M.A.)..:   (24 ,   94) PHYS. REV. B VOL.3,. . . . . . .   2929, (1971).

3313 PRASAD..........(N.)....:   (5 , 1420) Z. KRIST. VOL.137, ...   173, (1973).

3314 PRASK...........(H.J.)..:   (6 , 1386) J. CHEM. PHYSICS VOL.45, ...   699, (1966).

3315 PRASK...........(H.)....:   (6 , 1894) IAEA SYMP. BOMBAY, VOL.2,. . . .   407, (1964).
                                   (6 , 1096) SURFACE SCI. VOL.2,. . . . . . .   261, (1964).
                                   (6 , 1097) J. CHEM. PHYSICS VOL.42, ...   1469, (1965).
                                   (6 , 1083) J. CHEM. PHYSICS VOL.45, ...   3284, (1966).
                                   (6 , 1586) J. CHEM. PHYSICS VOL.45, ...   401, (1966).
                                   (6 , 2042) ADV. COLLOID INTERFACE SCI. 2, .   1, (1968).
                                   (6 , 1050) IAEA SYMP. COPENHAGEN, VOL.1,. .   345, (1968).
                                   (6 , 1020) J. CHEM. PHYSICS VOL.48, ...   3367, (1968).
                                   (6 , 1923) DISC. FARADAY SOC. NO.48,. . . .   15, (1969).
                                   (6 , 1044) MOL. DYN. AND STRUCT. OF SOLIDS,   335, (1969).
                                   (6 ,   337) J. CHEM. PHYSICS VOL.53, ...   3417, (1970).
                                   (6 , 1325) PHYS. REV. B VOL.4,. . . . . . .   4551, (1971).
                                   (6 , 1656) J. CHEM. PHYS. VOL.56 NO.11, . .   5377, (1972).
                                   (6 , 1080) J. CHEM. PHYS. VOL.56, ...   3217, (1972).
                                   (6 , 1661) PHYS. REV. B VOL.10 NO.2,. . . .   . , (1974).

3316 PRATT-JR........(G.W.)..:   (10B,   64) J. PHYS. C: SOLID ST. PHYS. V-4,   20, (1971).

3317 PREDA...........(I.M.)..:   (6 ,   316) REV. ROUMAINE PHYS. VOL.12,. . .   943, (1967).

3318 PREISWERK.......(P.)....:   (5 ,   937) COMP. REND. TOME 203,. . . . . .   73, (1936).
                                   (7B,   37) J. PHYS. RADIUM SER.7 VOL.8, . .   29, (1937).

3319 PRELESNIK.......(B.)....:   (5 ,   653) ACTA CRYST. VOL.20,. . . . . . .   315, (1966).
                                   (5 , 2201) CZECH. J. PHYS. B VOL.19,. . . .   857, (1969).
                                   (5 , 1354) ACTA CRYST. VOL.A28 PART S-4,. .   60, (1972).
                                   (5 , 1474) ACTA CRYST. VOL.B28, ...   3104, (1972).

3320 PRESKITT........(C.A.)..:   (6 , 1946) NUCL. SCI. ENG. VOL.46,. . . . .   244, (1971).

3321 PRESSESKY.......(A.J.)..:   (23 ,   75) REV. SCI. INSTRUM. VOL.21, . . .   705, (1950).

3322 PRESS...........(W.)....:   (5 ,   253) PHYS. LETT. VOL.31A, ...   253, (1970).
                                   (5 ,   255) J. CHEM. PHYS. VOL.56, ...   2597, (1972).
                                   (6 ,   233) PHYS. REV. LETT. VOL.29, ...   266, (1972).
                                   (5 , 1893) ACTA CRYST. VOL.A29, ...   257, (1973).

3323 PREUSSNER.......(A.)....:   (23 ,   9) CRYOGENICS VOL.11, ...   107, (1971).

3324 PREVENDER.......(T.S.)..:   (6 , 2295) PHYS. REV. B VOL.6,. . . . . . .   4438, (1972).

3325 PREVOT..........(B.)....:   (6 ,   635) PHYS. STAT. SOL. VOL.44, ...   701, (1971).
                                   (10B,   15) CAN. J. PHYS. 50,. . . . .   122, (1972).
                                   (6 ,   627) PHYS. REV. LETT. VOL.28, ...   964, (1972).
                                   (6 ,   622) SOL. STATE COMM. VOL.13, ...   1725, (1973).

3326 PRICE...........(D.L.)..:   (6 , 2066) IAEA SYMP. BOMBAY, VOL.1,. . . .   109, (1964).
                                   (6 , 2073) PROC. ROY. SOC. 300, ...   25, (1967).
                                   (6 ,   632) PHYS. LETT. 30A, ...   206, (1969).
                                   (6 , 2063) SOL. STATE COMM. 7,. . . . . . .   1433, (1969).
                                   (19A,   48) NUCL. INST. MET. VOL.82, ...   208, (1970).
                                   (17A,   65) NUCL. INST. MET. VOL.84, ...   61, (1970).
                                   (10B,   155) PHYS. REV. B-2,. . . . .   2983, (1970).
                                   (6 , 1240) PHYS. REV. B-3,. . . . .   1268, (1971).
                                   (10B,   167) PHYS. REV. LETT. VOL.26, ...   1324, (1971).
                                   (5 , 1423) J. CHEM. PHYS. VOL.56, ...   3697, (1972).
                                   (5 , 1957) J. CHEM. PHYS. VOL.58, ...   2039, (1973).
                                   (19B,   20) NUCL. INST. MET. VOL.106,. . . .   221, (1973).
                                   (19A,   53) NUCL. INST. MET. VOL.107,. . . .   501, (1973).
                                   (6 , 1195) PHYS. REV. LETT. VOL.31, ...   510, (1973).
                                   (20B,   134) REPORT ANL-8032 (104 PP.), ... , (1973).
                                   (17B,   51) NUCL. INST. MET. VOL.114,. . . .   411, (1974).
                                   (6 ,   461) PHYS. REV. B VOL.10 NO.2,. . . .   2573, (1974).
                                   (10B,   189) PHYS. REV. B VOL.9,. . . . . . .   2573, (1974).
                                   (5 , 2338) PHYS. REV. B VOL.9,. . . . . . .   248, (1974).

3327 PRICE...........(J.A.)..:   (17B,   44) NUCL. INST. MET. 62, ...   29, (1968).

3328 PRIESMEYER......(H.G.)..:   (20B,   94) NUCL. INST. MET. 68, ...   353, (1969).

3329 PRINCE..........(A.)....:   (24 ,   95) PROC/CONF/NEUTRON C. S. + TECH.,   951, (1968).

3330 PRINCE..........(E.)....:   (5 ,   714) ACTA CRYST. VOL.9, ...   1025, (1956).
                                   (5 ,   465) PHYS. REV. VOL.102,. . . . . . .   674, (1956).
                                   (5 ,   702) ACTA CRYST. VOL.10,. . . . . . .   554, (1957).
                                   (22B,   44) REV. SCI. INSTRUM. VOL.30, . . .   581, (1959).

3330 PRINCE.........(E.)....: (5 , 2114) J. APPL. PHYS. VOL.32, SUPPL.,. 685, (1961).
                              (5 , 710) J. CHEM. PHYSICS VOL.36, . . .   50, (1962).
                              (5 , 1033) COLL. INTER. N.126 (GRENOBLE),   79, (1963).
                              (5 , 2891) J. APPL. PHYS. VOL.36, ... 1845, (1965).
                              (5 , 1551) J. APPL. PHYS. VOL.36, ... 165, (1965).
                              (5 , 1423) J. CHEM. PHYS. VOL.58, ... 3697, (1973).
                              (5 , 274) ACTA CRYST. VOL.B29, ... 184, (1973).
                              (5 , 1936) ACTA CRYST. VOL.B30, ... 1167, (1974).

3331 PRIVOROTSKII....(I.A.)..: (7B, 127) SOV. PHYS. JETP. 20, ... 1037, (1965).

3332 PROFIO.........(A.E.)..: (6 , 2205) PROC/CONF/FAST CRIT. EXP./ANALY. 524, (1967).

3333 PRONMAN........(I.M.)..: (24 , 113) SOV. PHYS. DOKLADY VOL.16, . . . 49, (1971).

3334 PROSEN.........(R.J.)..: (9D, 51) MAGNETIC MATERIALS DIGEST 1964,. 37, (1964).

3335 PROSPERI.......(D.)....: (23 , 21) IAEA SYMP. (PILE RESEARCH)VIENNA 123, (1960).

3336 PROTOPOPOV......(N.N.)..: (22A, 19) INSTRUM. EXP. TECH. NO.2,. . . . 306, (1964).

3337 PROTOPOPOV......(V.N.)..: (6 , 173) BU. ACAD. SCI. USSR PHYS. SR. 32 651, (1968).

3338 PRUSHINSKII.....(V.V.)..: (5 , 2164) SOV. PHYS. DOKLADY VOL.14, . . . 1119, (1970).

3339 PRYOR..........(A.W.)..: (6 , 160) PHIL. MAG. VOL.8,. . . . .   43, (1963).
                              (6 , 159) J. NUCLEAR MATERIALS VOL.14, . . 275, (1964).
                              (17B,   8) ACTA CRYST. 20,. . . . . 138, (1966).
                              (5 , 2776) ACTA CRYST. VOL.B24,. . . . . . 117, (1968).
                              (22B, 15) J. APPL. CRYST. VOL.1,. . . . . 272, (1968).
                              (6 , 2299) J. OF PHYS.-C-, SER. 2, VOL.2,. 1857, (1969).
                              (5 , 2818) ACTA CRYST. VOL.A26, ... 543, (1970).
                              (5 , 1909) ACTA CRYST. VOL.B26, ... 1487, (1970).
                              (10B, 66) J. PHYS. C, SER.2, VOL.4,. . . 2304, (1971).
                              (6 , 438) J. PHYS. C, SER.2, VOL.4,. . . 492, (1971).
                              (5 , 1979) ACTA CRYST. VOL.B28, ... 2700, (1972).

3340 PRZYSTAWA.......(J.)....: (5 , 2798) PHYS. STAT. SOL. VOL.42, ... K15, (1970).

3341 PUCHER.........(M.)....: (6 , 1107) NUKLEONIK VOL.10 NO.3,. . . . . 129, (1967).

3342 PUFF...........(R.D.)..: (6 , 1148) PHYS. REV. A VOL.1,. . . . . . 125, (1970).

3343 PUGMIRE........(R.J.)..: (6 , 652) J. CHEM. PHYS. VOL.50,. . . . . 1030, (1969).
                              (6 , 1939) J. CHEM. PHYS. VOL.52, ... 4424, (1970).
                              (6 , 313) J. CHEM. PHYS. VOL.58, ... 1438, (1973).

3344 PULL...........(I.C.)..: (7B, 107) REPORT AEEW-M730,. . . . .   , (1969).

3345 PURICA.........(I.I.)..: (16 , 87) STUD. CERCETARI FIZ.(RUMAN.) 22, 365, (1970).

3346 PUROHIT........(S.N.)..: (16 , 36) NUCL. SCI. ENGNG. VOL.13,. . . . 250, (1962).
                              (6 , 2352) CONF. NEUTRON THERMALIZ. VOL.1,. 407, (1968).

3347 PURWINS........(H.G.)..: (9D, 14) ANN. DE PHYSIQUE TOME 7 NO.5,. . 329, (1972).
                              (5 , 1367) SOL. STATE COMM. VOL.11 NO.5,. . 707, (1972).
                              (6 , 2197) INT. J. MAGN. VOL.4 NO.1,. . . .   63, (1973).
                              (6 , 2145) PHYS. REV. LETT. VOL.31, ... 1585, (1973).
                              (6 , 2196) SOL. STATE COMM. VOL.12 NO.2,. . 117, (1973).
                              (6 , 2146) SOL. STATE COMM. VOL.13, ... 881, (1973).
                              (6 , 716) J. PHYS. C VOL.7,. . . . . 1207, (1974).

3348 PUSHKAROV.......(K.I.)..: (11A, 61) SOV. PHYS. J.E.T.P. 27,. . . . . 307, (1968).

3349 PUSTOVOIT.......(V.I.)..: (10C, 128) SOV. PHYS. JETP VOL.35,. . . . . 399, (1972).

3350 PUTZKI.........(R.)....: (5 , 1880) NUCL. PHYS. VOL.A166,. . . . . . 443, (1971).
                              (5 , 1879) NUCL. PHYS. VOL.A166,. . . . . . 461, (1971).

3351 PUZEI..........(I.M.)..: (5 , 2140) SOV. PHYS. CRYST. VOL.3,. . . . 147, (1958).
                              (5 , 1067) J. PHYS. SOC. JAPAN VOL.17 B-III   49, (1962).
                              (5 , 986) SOV. PHYS. CRYST. VOL.7, ... 637, (1963).
                              (5 , 1069) BULL. ACAD. SCI. USSR VOL.28,. . 354, (1964).
                              (5 , 2132) SOV. PHYS. CRYST. VOL.10, . . . 338, (1965).
                              (5 , 2143) SOV. PHYS. DOKLADY VOL.15, . . . 874, (1971).

3352 PYNN...........(R.)....: (6 , 1397) IAEA SYMP. COPENHAGEN, VOL.1,. . 215, (1968).
                              (6 , 1597) J. APPL. PHYS. 42,. . . . . . . 4736, (1971).
                              (6 , 114) IAEA SYMP. GRENOBLE,. . . . . . 515, (1972).
                              (5 , 136) PHYS. NORV. VOL.6 NO.3-4,. . . . 205, (1972).
                              (6 , 1403) PROC. ROY. SOC. A 326 NO.1566,. 347, (1972).
                              (5 , 134) SOL. STATE COMM. VOL.11, ... 1365, (1972).
                              (5 , 137) J. PHYS. CHEM. SOLIDS VOL.34,. . 735, (1973).
                              (5 , 135) SOL. STATE COMM. VOL.12 NO.5,. . 409, (1973).

3353 PYTASZ.........(G.)....: (6 , 1384) PHYS. STAT. SOL. VOL.44, ... 497, (1971).
                              (6 , 1387) ACTA PHYS. POLON. VOL.A44, . . . 731, (1973).

3354 PYTTE..........(E.)....: (10D, 42) PHYS. REV. B VOL.5,. . . . . . . 3758, (1972).

3355 PYZHOVA........(Z.I.)..: (5 , 2628) ATOMNAYA ENERGIYA (USSR) VOL.24, 243, (1968).

3356 QUEREL.........(G.)....: (5 , 836) J. PHYS. CHEM. SOLIDS 29,. . . . 1001, (1968).

3357 QUEROZ-DO-AMARAL(L.)....: (14B, 86) PHYS. REV. VOL.151,. . . . . . . 126, (1966).

3358 QUEZEL-AMBRUNAZ.(S.)....: (5 , 221) SOLID STATE COMM. 5,. . . . . .   25, (1967).
                              (5 , 1729) BU. SOC. FRA. MIN. CRIST. VOL.91 339, (1968).
                              (9D, 43) J. PHYSIQUE TOME 31,. . . . . . 369, (1970).
                              (5 , 863) C. R. ACAD. SCI. B TOME 273,. . 619, (1971).
                              (5 , 2589) SOL. STATE COMM. VOL.11 NO.5,. . 605, (1972).

3359 QUEZEL.........(G.)....: (5 , 221) SOLID STATE COMM. 5,. . . . . .   25, (1967).
                              (5 , 453) BULL. SOC. FRANC. MIN. CRIST. 92 264, (1969).
                              (5 , 1396) SOL. STATE COMM. VOL.10, ... 735, (1972).
                              (5 , 2597) SOL. STATE COMM. VOL.12, ... 985, (1973).

3360 QUEZEL.........(S.)....: (5 , 835) SOL. STATE COMM. VOL.9, ... 925, (1971).
                              (5 , 895) SOL. STATE COMM. VOL.10, ... 765, (1972).
                              (5 , 2595) J. SOLID STATE CHEM. VOL.9,. . . 234, (1974).

3361 QUICKSALL.......(C.O.)..: (9B, 65) NUCL. INST. MET. VOL.94,. . . . 185, (1971).
                              (9B, 15) ACTA CRYST. VOL.A28,. . . . . . 236, (1972).

3362 QUITTNER.......(G.)....: (24 , 40) NUCL. INST. MET. 31,. . . . . .   61, (1964).
                              (6 , 1828) PHYS. REV. LETT. 17,. . . . . . 1259, (1966).
                              (21 , 35) NUCL. INST. MET. 50,. . . . . . 251, (1967).
                              (6 , 1647) IAEA SYMP. COPENHAGEN, VOL.1,. . 367, (1968).
                              (6 , 1825) PHYS. LETT. 28A,. . . . . 226, (1968).
                              (19B, 13) NUCL. INST. MET. 69, ... 290, (1969).
                              (18A,   4) ACTA CRYST. VOL.A27, ... 605, (1971).
                              (18A,   5) ACTA CRYST. VOL.A28, ...   47, (1972).
                              (20A,   9) ACTA PHYS. AUSTRIACA VOL.38,. . 282, (1973).
                              (18A, 20) NUCL. INST. MET. VOL.107,. . . . 461, (1973).

```
3363  QUIVY..........(P.)...:  (19B,   32) REV. PHYS. APPL. (FRANCE) VOL.4,   111, (1969).

3364  RABIDEAU........(S.W.)..:  ( 5 ,  796) J. CHEM. PHYSICS VOL.49, . . . . . 2514, (1968).
                                 ( 5 ,  783) J. CHEM. PHYSICS VOL.49, . . . . . 4361, (1968).
                                 ( 5 ,  782) J. CHEM. PHYSICS VOL.49, . . . . . 4365, (1968).
                                 ( 5 ,  792) PHYS. ICE, . . . . . . . . . . . .   59, (1969).
                                 ( 5 ,  784) J. CHEM. PHYSICS VOL.55, . . . . .  589, (1971).

3365  RABI...........(I.I.)..:  ( 5 , 2261) PHYS. REV. VOL.72, . . . . . . .    634, (1947).

3366  RABOY..........(S.)...:  (24 ,   16) IAEA SYMP. (PILE RESEARCH)VIENNA  615, (1960).

3367  RACCAH.........(P.M.)..:  ( 5 , 1528) J. PHYS. CHEM. SOL. 28,. . . . .  549, (1967).

3368  RACHWALSKA......(M.)...:  ( 6 , 1103) PHYSICA VOL.72,. . . . . . . . .  168, (1974).

3369  RADHAKRISHNAN...(N.K.)..:  ( 5 ,  522) NUCL./S.S. PHYS. SYMP. ABSTR.,. . . , (1972).

3370  RADHAKRISHNA....(P.)...:  ( 8 ,   12) ANN. DE PHYSIQUE TOME 7 NO.4,. .  269, (1972).

3371  RADKEVICH.......(I.A.)..:  (20B,  130) PRIB. TEKH. EKSPER. NO.2,. . . .    3, (1956).

3372  RADULESCU.......(C.)...:  (20B,  136) REV. DE PHYSIQUE VOL.6,. . . . .  411, (1961).

3373  RAE............(E.R.)..:  (22B,    7) AUTOMATIC/NUCLEAR DATA,. . . . .   83, (1964).
                                 (20C,   34) 2ND INTERNAT. CONF./NUCL. DATA,. . . , (1970).

3374  RAFFLE.........(J.F.)..:  (19A,   10) IAEA SYMP. (PILE RESEARCH)VIENNA  559, (1960).

3375  RAFIZADEH......(H.A.)..:  ( 6 , 1656) J. CHEM. PHYS. VOL.56 NO.11,. . . 5377, (1972).
                                 ( 6 , 1563) J. CHEM. PHYS. VOL.57, . . . . . 2291, (1972).
                                 ( 6 ,  210) PHYS. REV. B VOL.7,. . . . . . . 4527, (1973).

3376  RAHMAN.........(A.)...:  ( 5 , 1298) REACTOR SCI. VOL.11, . . . . . .   89, (1960).
                                 (15 ,   22) J. NUCL. ENERGY VOL.13,. . . . .  128, (1961).
                                 (14B,   79) PHYS. REV. VOL.122,. . . . . . .    9, (1961).
                                 (15 ,   48) REACTOR SCI. VOL.13,. . . . . .  128, (1961).
                                 ( 6 , 1022) IAEA SYMP. CHALK RIVER, VOL.1,   215, (1962).
                                 (14B,   81) PHYS. REV. VOL.126,. . . . . . .  986, (1962).
                                 (14B,   80) PHYS. REV. VOL.126,. . . . . . .  997, (1962).
                                 ( 7B,   73) PHYS. REV. 130,. . . . . . . . . 1334, (1963).
                                 ( 5 ,  100) J. CHEM. PHYS. VOL.42, . . . . . 3540, (1965).
                                 (14B,   64) PHYSICA VOL.32,. . . . . . . . .  415, (1966).
                                 (14A,   65) REACTOR PHYSICS, . . . . . . . .  123, (1966).
                                 ( 6 ,  104) IAEA SYMP. COPENHAGEN, VOL.1,. .  561, (1968).
                                 (10C,   77) PHYS. REV. 176,. . . . . . . . .  784, (1968).
                                 ( 6 , 1076) J. CHEM. PHYS. VOL.55, . . . . . 3336, (1971).
                                 (14A,   55) PHYS. REV. A VOL.4,. . . . . . . 1616, (1971).
                                 ( 5 ,  794) J. CHEM. PHYS. VOL.60, . . . . . 1545, (1974).
                                 ( 6 , 1965) PHYS. REV. A VOL.9,. . . . . . . 1667, (1974).
                                 ( 6 , 1963) PHYS. REV. LETT. VOL.32 NO.2,. .   52, (1974).

3377  RAHMAN.........(H.U.)..:  ( 5 , 2773) J. PHYS. CHEM. SOL. 27,. . . . . 1833, (1966).

3378  RAINEY.........(V.S.)..:  ( 6 ,   58) IAEA SYMP. CHALK RIVER, VOL.1,   413, (1962).
                                 ( 6 ,  265) PHYS. REV. VOL.136,. . . . . . . A106, (1964).
                                 ( 5 , 2929) PHYS. REV. LETT. 21, . . . . . .  286, (1966).
                                 ( 6 , 1547) PHYS. REV. LETT. VOL.17, . . . .  533, (1966).

3379  RAINFORD.......(B.D.)..:  ( 6 ,  814) J. APP. PHYS. 39,. . . . . . . . 1120, (1968).
                                 ( 6 ,  467) J. OF PHYS. C VOL.1, . . . . . .  679, (1968).
                                 ( 5 ,  734) PHYS. REV. LETT. 22, . . . . . .  531, (1969).
                                 ( 5 , 2316) PHYS. REV. LETT. VOL.24, . . . .  897, (1970).
                                 ( 6 , 1931) CONF./RARE EARTHS/ACTINIDES, . .   40, (1971).
                                 ( 5 , 2024) CONF./RARE EARTHS/ACTINIDES, . .   43, (1971).
                                 ( 6 ,  811) J. OF PHYS.-C-, SER.2, VOL.4,. .  307, (1971).
                                 ( 5 , 2023) J. PHYSIQUE TOME 32 COL.1 VOL.I,  370, (1971).
                                 ( 5 , 2106) J. PHYSIQUE TOME 32 COL.1 VOL.I,  575, (1971).
                                 ( 6 , 1929) PHYS. REV. LETT. VOL.26, . . . . 1254, (1971).
                                 ( 6 , 1928) PHYS. REV. LETT. VOL.27, . . . .  223, (1971).
                                 ( 6 ,  812) IAEA SYMP. GRENOBLE, . . . . . .  655, (1972).
                                 ( 5 , 2115) PHYS. REV. B VOL.7,. . . . . . .  218, (1973).

3380  RAINWATER......(L.J.)..:  ( 5 , 1249) PHYS. REV. VOL.69, . . . . . . .  236, (1946).
                                 (20A,   60) PHYS. REV. VOL.70, . . . . . . .  136, (1946).
                                 ( 6 , 2112) PHYS. REV. VOL.71, . . . . . . .  165, (1947).
                                 ( 5 , 1407) PHYS. REV. VOL.71, . . . . . . .  174, (1947).
                                 (22A,   55) PHYS. REV. VOL.71, . . . . . . .   65, (1947).
                                 ( 5 , 2261) PHYS. REV. VOL.72, . . . . . . .  634, (1947).
                                 ( 5 , 2233) PHYS. REV. VOL.73, . . . . . . . 1399, (1948).
                                 ( 6 ,  975) PHYS. REV. VOL.73, . . . . . . .  733, (1948).
                                 ( 6 ,  598) PHYS. REV. VOL.73, . . . . . . .  963, (1948).
                                 ( 6 , 1465) PHYS. REV. VOL.75, . . . . . . .  895, (1949).
                                 (22A,   62) REV. SCI. INSTRUM. VOL.31, . . .  481, (1960).

3381  RAJAGOPAL......(A.K.)..:  (16 ,   36) NUCL. SCI. ENGNG. VOL.13,. . . .  250, (1962).

3382  RAJAGOPAL......(H.)...:  ( 5 ,  158) J. CHEM. PHYSICS VOL.48, . . . . 1883, (1968).
                                 ( 5 ,   18) ACTA CRYST. VOL.A28 PART S-4,. .  193, (1972).
                                 ( 5 ,   19) ACTA CRYST. VOL.B28, . . . . . . 2514, (1972).

3383  RAJCA..........(A.)...:  (17A,   16) ACTA PHYS. POLON. VOL.A42, . . .  259, (1972).
                                 ( 9B,   92) PHYS. STAT. SOLIDI A VOL.9,. . .  423, (1972).
                                 ( 6 ,  178) ACTA PHYS. POLON. VOL.A43, . . .   37, (1973).

3384  RAJPUT.........(J.S.)..:  (10B,  153) PHYS. REV. B-2,. . . . . . . . . 3943, (1970).
                                 (10B,   82) J. PHYS.F: METAL PHYS. VOL.1,. .  377, (1971).
                                 ( 6 , 2308) PHYS. LETT. VOL.38A, . . . . . .  497, (1972).
                                 (10B,   91) J. PHYS. F VOL.3,. . . . . . . . 1531, (1973).

3385  RAKHECHA.......(V.C.)..:  ( 6 , 1482) PHYS. LETT. VOL.40A, . . . . . .  101, (1972).

3386  RAKHECNA.......(V.C.)..:  (12A,   20) NUCL./S.S. PHYS. SYMP. ABSTR.,. . . , (1972).

3387  RAKOTONDRAFARA..(H.)...:  (20A,   11) ANN. FAC. SCI. UNIV. TOULOUSE 30   9, (1966).

3388  RALAROSY.......(J.)...:  (20C,   86) REV. PHYS. APPL. (FRANCE) VOL.4,  259, (1969).

3389  RALPH..........(H.I.)..:  ( 6 , 2139) J. OF PHYS. C VOL.1, . . . . . . 1596, (1968).
                                 ( 6 , 2140) J. OF PHYS. C VOL.1, . . . . . .  132, (1968).

3390  RAMAKRISHNAN...(T.V.)..:  (11A,   28) NUCL./S.S. PHYS. SYMP. ABSTR.,. . . , (1972).

3391  RAMAKRISHNA....(D.V.S.):  (24 ,   49) NUCL. INST. MET. 61, . . . . . .   37, (1968).

3392  RAMANADHAM.....(M.)...:  ( 5 ,   14) ACTA CRYST. VOL.B28, . . . . . . 3000, (1972).
                                 ( 5 ,   15) ACTA CRYST. VOL.B29, . . . . . . 1167, (1973).

3393  RAMANNA........(R.)...:  ( 6 ,  155) IAEA SYMPOSIUM (VIENNA), . . . .  631, (1960).

3394  RAMASESHAN.....(S.)...:  ( 9C,   14) CURRENT SCI. (INDIA) VOL.35, . .   87, (1966).
                                 (17B,   14) ACTA CRYST. VOL.B24, . . . . . . 1701, (1968).
                                 (17B,   13) ACTA CRYST. VOL.B24, . . . . . .   35, (1968).
                                 ( 6 ,  459) ACTA CRYST. VOL.A26, . . . . . .  364, (1970).
                                 (10C,   94) REPORT NAL-TN-21 (14PP.),. . . .    , (1970).
                                 (10C,    3) ACTA CRYST. VOL.A27, . . . . . .  332, (1971).
```

3394 RAMASESHAN......(S.)....* (9C, 5) ACTA CRYST. VOL.A27, ... 563, (1971).
(14B, 47) J. PHYS. C VOL.4, ... 3029, (1971).
(8 , 126) SOL. STATE COMM. VOL. 9, ... 1003, (1971).

3395 RAMESH..........(T.G.)..* (10C, 3) ACTA CRYST. VOL.A27, ... 332, (1971).
(9C, 5) ACTA CRYST. VOL.A27, ... 569, (1971).
(14B, 47) J. PHYS. C VOL.4, ... 3029, (1971).

3396 RAMSEY..........(W.J.)..* (5 , 1565) J. PHYS. CHEM. SOLIDS VOL.24,.. 1066, (1963).

3397 RAM............(P.N.)..* (108, 163) PHYS. REV. B-4, ... 2774, (1971).

3398 RANDOLPH........(P.O.)..* (5 , 2869) NUCL. SCI. ENGNG. VOL.7, ... 193, (1960).
(6 , 251) PHYS. REV. VOL.124, ... 460, (1961).
(6 , 1606) PHYS. LETT. 3, ... 162, (1962).
(6 , 311) PHYS. REV. VOL.125, ... 933, (1962).
(6 , 146) PHYS. REV. VOL.128, ... 562, (1962).
(6 , 148) IAEA SYMP. BOMBAY, VOL.1,. . . . 379, (1964).
(6 , 1807) PHYS. REV. VOL.152, ... 99, (1966).
(6 , 1613) PHYS. REV. VOL.134, ... A1238, (1967).
(6 , 1603) IAEA SYMP. COPENHAGEN, VOL.1,. . 449, (1968).
(6 , 1901) J. CHEM. PHYSICS VOL.49, ... 1043, (1968).
(6 , 1811) PHYS. REV. LETT. 20, ... 531, (1968).
(6 , 85) PHYS. REV. A VOL.6, ... 1107, (1972).

3399 RANGANATHAN.....(S.)....* (14A, 43) PHYS. REV. VOL.164,.. . . . . 222, (1967).

3400 RANKOWITZ.......(S.)....* (19A, 68) REV. SCI. INSTRUM. VOL.35, . . 1150, (1964).

3401 RANNEV..........(N.B.)..* (5 , 1915) DOKL. AKAD. NAUK SSSR VOL.155, . 1415, (1964).

3402 RANNEV..........(N.V.)..* (5 , 1475) SOV. PHYS. CRYST. VOL.7, . . . 499, (1962).
(5 , 1579) SOV. PHYS. DOKLADY VOL.8,. . . 131, (1963).
(5 , 307) SOV. PHYS. CRYST. VOL.11,. . . 177, (1966).
(5 , 1550) SOV. PHYS. CRYST. VOL.13,. . . 204, (1968).

3403 RANNINGER.......(J.)....* (6 , 491) PHYS. LETT. VOL.39A, ... 105, (1972).
(11A, 20) J. PHYS. C VOL.6, ... 323, (1973).
(6 , 495) J. PHYS. C VOL.6, ... 370, (1973).
(11A, 19) J. PHYS. C VOL.6, ... 386, (1973).
(11A, 23) J. PHYS. C VOL.7, ... 947, (1974).

3404 RAO............(K.R.)..* (6 , 1816) PHYS. REV. LETT. VOL.7, ... 93, (1961).
(20A, 68) PROC. INDIAN ACAD. SCI. A VOL.53 59, (1961).
(6 , 1628) IAEA SYMP. CHALK RIVER, VOL.2,. 23, (1962).
(6 , 1413) IAEA SYMP. CHALK RIVER, VOL.2,. 99, (1962).
(6 , 2157) J. PHYS. SOC. JAPAN VOL.17 B-III 63, (1962).
(6 , 1810) PHYS. REV. VOL.128, ... 1099, (1962).
(5 , 2000) PHYS. LETT. 9, ... 106, (1964).
(6 , 86) PHYS. REV. VOL.137, ... A417, (1965).
(6 , 376) PHYS. LETT. 23, ... 226, (1966).
(5 , 1906) PROC. PHYS. SOC. 89, ... 379, (1966).
(6 , 274) CAN. JOURN. PHYS. 45, ... 3185, (1967).
(6 , 377) PHYS. REV. 161, ... 133, (1967).
(6 , 2301) IAEA SYMP. COPENHAGEN, VOL.1,. . 195, (1968).
(5 , 249) J. CHEM. PHYSICS VOL.48, . . . 2395, (1968).
(10B, 55) J. CHEM. PHYSICS VOL.53, . . . 4624, (1970).
(10B, 54) J. CHEM. PHYSICS VOL.53, . . . 4645, (1970).
(10B, 53) J. CHEM. PHYSICS VOL.53, . . . 4661, (1970).
(6 , 134) CONF. INTERN., RENNES, ... 140, (1971).
(6 , 291) J. OF PHYS.-C-, SER.2, VOL.4,. 2725, (1971).
(6 , 1325) PHYS. REV. B VOL.4, ... 4551, (1971).
(5 , 265) NUCL./S.S. PHYS. SYMP. ABSTR., . . . (1972).

3405 RAO............(L.M.)..* (6 , 532) J. APPL. PHYS. VOL.39 NO.2,. . 1113, (1968).
(6 , 533) PHYS. LETT. VOL.26A, ... 108, (1968).

3406 RAO............(P.V.S.)* (10B, 78) J. PHYS. CHEM. SOLIDS VOL.35,.. 669, (1974).

3407 RAO............(R.R.)..* (10D, 12) IAEA SYMPOSIUM BOMBAY, VOL.1,. 325, (1964).
(10B, 218) SOL. STATE COMM. VOL.10, . . . 179, (1972).
(6 , 2017) SOL. STATE COMM. VOL.12 NO.6,. 527, (1973).

3408 RAO............(V.U.S.)* (5 , 393) PHYS. REV. B VOL.9, ... 154, (1974).

3409 RAPACKI.........(H.)....* (23 , 63) NUKLEONIKA VOL.7,. . . . . . . 223, (1962).

3410 RAPEANU.........(S.)....* (5 , 281) PHYS. LETT. 26A, ... 72, (1967).
(6 , 316) REV. ROUMAINE PHYS. VOL.12,. . 943, (1967).
(6 , 393) PHYS. LETT. 28A, ... 31, (1968).
(5 , 2319) REV. ROUMAINE PHYS. VOL.13,. . 287, (1968).
(6 , 394) REV. ROUMAINE PHYS. VOL.13,. . 913, (1968).
(6 , 945) REV. ROUMAINE PHYS. VOL.18 NO.2, 135, (1973).
(6 , 1873) REV. ROUMAINE PHYS. VOL.18,. . 313, (1973).
(14B, 114) REV. ROUMAINE PHYS. VOL.18,. . 997, (1973).
(6 , 2200) J. CHEM. PHYS. VOL.60, ... 2832, (1974).

3411 RASETTI.........(F.)....* (16 , 56) NUOVO CIMENTO VOL.12 SER.8,. . 201, (1935).
(16 , 66) PROC. ROY. SOC. A VOL.149, . . 522, (1935).
(5 , 328) PHYS. REV. VOL.58, ... 321, (1940).

3412 RASETTI.........(M.)....* (20D, 19) NUCL. INST. MET. 47, ... 320, (1967).
(20B, 85) NUCL. INST. MET. 59, ... 136, (1968).
(15 , 41) PHYS. REV. VOL.167,. . . . . . 97, (1968).

3413 RASMUSSEN.......(N.C.)..* (24 , 17) IAEA SYMP. (PILE RESEARCH)VIENNA 609, (1960).

3414 RASMUSSEN.......(S.E.)..* (5 , 1410) ACTA CHEM. SCAND. VOL.24,. . . 1662, (1970).
(5 , 287) ACTA CHEM. SCAND. VOL.24,. . . 3248, (1970).

3415 RASTAS..........(A.)....* (16 , 41) NUCL. SCI. ENGNG. VOL.36,. . . 351, (1969).

3416 RATHMANN........(O.)....* (5 , 758) PHYS. REV. B VOL.9, ... 3921, (1974).

3417 RATKOVIC........(S.)....* (6 , 3) CHEM. PHYS. LETT. VOL.17,. . . 53, (1972).

3418 RATYNSKI........(W.)....* (12B, 25) NUCL. INST. MET. VOL.45, . . . 293, (1966).

3419 RAUBENHEIMER....(L.J.)..* (5 , 1500) J. APP. PHYS. 37, ... 1047, (1966).
(5 , 1212) J. APP. PHYS. 38, ... 1381, (1967).
(6 , 600) PHYS. REV. 164, ... 922, (1967).
(5 , 2922) PHYS. REV. 176, ... 722, (1968).

3420 RAUCH...........(H.)....* (17A, 15) ACTA PHYS. AUSTRIACA VOL.20, . 7, (1965).
(16 , 2) ATOMKERNENERGIE VOL.10, . . . 243, (1965).
(20C, 45) NUCL. INST. MET. 46, ... 153, (1966).
(17A, 19) ATOMKERNENERGIE VOL.12, . . . 395, (1967).
(22C, 36) NUCL. INST. MET. 52, ... 15, (1967).
(9B, 31) ATOMKERNENERGIE VOL.13, . . . 183, (1968).
(12B, 12) IAEA SYMP. COPENHAGEN, VOL.2,. 387, (1968).
(21 , 39) NUCL. INST. MET. 58, ... 261, (1968).
(24 , 48) NUCL. INST. MET. 61, ... 349, (1968).
(8 , 53) NUKLEONIK VOL.11, ... 113, (1968).
(5 , 806) Z. PHYS. VOL.210, ... 265, (1968).
(19B, 6) IAEA SYMP. (INSTRUMENT.), VIENNA 181, (1969).
(5 , 800) J. PHYS. CHEM. SOLIDS VOL. 30, . 2175, (1969).

3455 REMY...........(G.)....! (20C,   86) REV. PHYS. APPL. (FRANCE) VOL.4,   259, (1969).

3456 RENKER.........(B.)....! (6 ;   670) PHYS. LETT. 30A,. . . . .   493; (1969).
                                (6 ;   665) CONF. INTERN., RENNES,. . . . . .   167; (1971).
                                (6 ; 1333) PHYS. REV. LETT. VOL.30,. . . . . 1144; (1973).
                                (6 ; 1334) PHYS. REV. LETT. VOL.32,. . . . .   836; (1974).

3457 RENNINGER......(A.)....! (5 , 1749) PHYS. REV. 147,. . . . .   418, (1966).

3458 REULAND........(R.J.)..! (5 ,   421) J. INORG. NUCL. CHEM. VOL.33,. . 1553, (1971).

3459 REYNOLDS.......(P.A.)..! (6 ,   109) DISC. FARADAY SOC. NO.48,. . . .   131, (1969).
                                (10B,   17) CONF. INTERN., RENNES,. . . . . .   209, (1971).
                                (17B,   67) SOL. STATE COMM. VOL.9,. . . . . 1353, (1971).
                                (6 ,   235) IAEA SYMP. GRENOBLE,. . . . . . .   195, (1972).
                                (20A,   17) IAEA SYMP. GRENOBLE,. . . . . . .   733, (1972).
                                (10E,   15) J. CHEM. PHYS. VOL.56,. . . . . . 2928, (1972).
                                (6 ,   353) J. CHEM. SOC. FARADAY II VOL.68, 1434, (1972).
                                (6 ,   420) J. CHEM. PHYS. VOL.59,. . . . . . 2777, (1973).
                                (6 ,   374) J. CHEM. PHYS. VOL.60,. . . . . .   824, (1974).

3460 REZ............(I.S.)..! (5 , 1475) SOV. PHYS. CRYST. VOL.7,. . . .   499, (1962).

3461 RHYNE..........(J.J.)..! (5 ,   848) MAGN. AND MAG. MATERIALS-1971,. 1436, (1972).
                                (5 , 2566) PHYS. REV. LETT. VOL.29,. . . . . 1562, (1972).
                                (5 , 2565) PHYS. LETT. VOL.47A NO.1,. . . .   73, (1974).

3462 RIBAR..........(B.)....! (5 , 1354) ACTA CRYST. VOL.A28 PART S-4,. .   60, (1972).

3463 RIBON..........(P.)....! (20C,   57) NUCL. INST. MET. 72,. . . . . . .   307, (1969).

3464 RIBRAG.........(M.)....! (5 , 2971) J. PHYS. RADIUM VOL.22,. . . . .   648, (1961).

3465 RICCI..........(F.P.)..! (20D,   12) NUCL. INSTRUM. VOL.3,. . . . . .   223, (1958).
                                (22A,   34) NUCL. INST. MET. VOL.9,. . . . .   195, (1960).
                                (17D,   11) J. PHYS. SOC. JAPAN VOL.17 B-II,   347, (1962).
                                (5 ,   739) J. PHYS. SOC. JAPAN VOL.17 B-II,   348, (1962).
                                (17D,   13) NUCL. INST. MET. VOL.15,. . . . .   155, (1962).
                                (23 ,   65) NUOVO CIMENTO SUPPL. VOL.23,. .   17, (1962).
                                (5 ,   226) NUOVO CIMENTO VOL.24,. . . . . .   103, (1962).
                                (5 ,   951) COLL. INTER. N.126 (GRENOBLE),.   180, (1963).
                                (5 , 1736) J. APP. PHYS. 34,. . . . . 1571, (1963).
                                (12A,   13) NUCL. INST. MET. 26,. . . . . . .   125, (1964).
                                (5 , 1735) J. APP. PHYS. 37,. . . . . 3236, (1966).
                                (5 , 1739) PHYS. LETT. 24A,. . . . .   371, (1967).
                                (13 ,   53) PHYS. REV. LETT. VOL.28,. . . .   22, (1972).

3466 RICE...........(M.J.)..! (9D,   36) J. APPL. PHYS. VOL.41,. . . . .   919, (1970).
                                (6 , 1333) PHYS. REV. LETT. VOL.30,. . . . 1144, (1973).

3467 RICE...........(S.A.)..! (14B,   90) PHYS. REV. VOL.165,. . . . . . .   186, (1968).
                                (14A,   46) PHYS. REV. VOL.176,. . . . . . .   239, (1968).

3468 RICE...........(T.M.)..! (6 , 2269) PHYS. REV. B-1,. . . . .   509, (1970).

3469 RICHARDSON......(J.W.)..! (8 ,   96) PHYS. REV. LETT. VOL.25,. . . .   110, (1970).

347C RICHARDS.......(P.M.)..! (6 ,   387) PHYS. REV. B VOL.5,. . . . . . 2014, (1972).
                                (6 ,   625) PHYS. REV. LETT. VOL.32 NO.4,. .   170, (1974).

3471 RICHARD........(P.)....! (5 , 1465) REV. CHIM. MINER. VOL.9,. . . .   825, (1972).

3472 RICHTER........(G.)....! (19B,   2) ATOMKERNENERGIE VOL.9,. . . . .   307, (1964).
                                (19B,   3) ATOMKERNENERGIE VOL.15,. . . . .   91, (1970).

3473 RICHTER........(H.)....! (14B,   137) Z. NATURFORSCH. VOL.16A,. . . .   187, (1961).
                                (17A,   27) FORTSCHR. PHYS. VOL.14 NO.2,. .   73, (1966).
                                (6 , 1618) NATURWISSENSCHAFTEN VOL.53,. . .   16, (1966).

3474 RICHTER........(J.)....! (7A,   52) PHYS. LETT. 28A,. . . . .   376, (1968).
                                (5 ,   97) ANN. PHYS. (GER.) VOL.23,. . . .   49, (1969).

3475 RICKARDS.......(J.)....! (24 ,   103) REV. MEXICANA FIS. VOL.18,. . .   115, (1969).

3476 RIDGLEY........(D.H.)..! (5 , 1641) J. APPL. PHYS. 41,. . . . . . .   939, (1970).

3477 RIEDEL.........(E.)....! (13 ,   42) PHYS. LETT. VOL.32A,. . . . . .   273, (1970).
                                (13 ,   19) DYN. ASPECTS OF CRITICAL PHEN.,   19, (1972).

3478 RIEDER.........(K.H.)..! (6 , 1419) PHYS. REV. LETT. 20,. . . . . .   209, (1968).

3479 RIETSCHEL......(H.)....! (5 , 1243) NUCL. PHYS. VOL.A139,. . . . . .   100, (1969).
                                (5 , 1242) NUCL. SCI. ENGNG. VOL.38,. . . .   180, (1969).
                                (6 , 1333) PHYS. REV. LETT. VOL.30,. . . . 1144, (1973).
                                (6 , 1334) PHYS. REV. LETT. VOL.32,. . . .   836, (1974).

3480 RIETVELD.......(H.M.)..! (5 ,   279) NATURE VOL.192,. . . . .   154, (1961).
                                (5 , 2499) ACTA CRYST. 20,. . . . .   508, (1966).
                                (17B,   10) ACTA CRYST. 22,. . . . .   151, (1967).
                                (5 , 2867) ACTA CRYST. B25,. . . . . 1420, (1969).
                                (5 , 2501) ACTA CRYST. B25,. . . . .   787, (1969).
                                (17A,   50) J. APPL. CRYST. VOL.2,. . . . .   65, (1969).
                                (5 , 2627) ACTA CRYST. VOL.B26,. . . . . .   693, (1970).
                                (5 ,   43) J. PHYS. CHEM. SOLIDS VOL.32,. .   543, (1971).

3481 RIMAI..........(L.)....! (6 , 2100) SOL STAT. COMM. 5,. . . . . . .   387, (1967).

3482 RIMMER.........(D.E.)..! (9D,   24) HARWELL SUMMER SCHOOL,. . . . .   211, (1968).
                                (9D,   42) J. PHYS. C, SER.2, VOL.2,. . . . 1151, (1969).
                                (8 ,   119) REP. PROGR. PHYS. VOL.32,. . . .   333, (1969).

3483 RINALDI........(R.P.)..! (6 , 2006) CONF. INTERN., RENNES,. . . . .   223, (1971).

3484 RINGO..........(G.R.)..! (5 ,   242) PHYS. REV. VOL.75,. . . . . . . 1098, (1949).
                                (20A,   65) PHYS. REV. VOL.77,. . . . . . .   291, (1950).
                                (9B,   77) PHYS. REV. VOL.80,. . . . . . .   507, (1950).
                                (5 , 1348) PHYS. REV. VOL.82,. . . . . . .   560, (1951).
                                (5 , 1260) PHYS. REV. VOL.84,. . . . . . . 1160, (1951).
                                (5 ,   846) PHYS. REV. VOL.85,. . . . . . .   483, (1952).
                                (5 , 2870) PHYS. REV. VOL.102,. . . . . . . 1321, (1956).
                                (24 ,   16) IAEA SYMP. (PILE RESEARCH) VIENNA   615, (1960).
                                (6 ,   247) J. CHEM. PHYS. VOL.32,. . . . .   476, (1960).
                                (5 , 2210) PHYS. REV. VOL.148,. . . . . . . 1303, (1966).
                                (7A,   8) AM. J. PHYS. VOL.39,. . . . . .   324, (1971).

3485 RING...........(J.W.)..! (6 , 1008) J. CHEM. PHYS. VOL.51,. . . . .   762, (1969).

3486 RIPEANU........(S.)....! (5 ,   57) NUCL. SCI. ENGNG. VOL.12,. . . .   157, (1962).
                                (20A,   79) REV. SCI. INSTRUM. VOL.33,. . .   916, (1962).
                                (20A,   74) REV. ROUMAINE PHYS. VOL.8,. . .   277, (1963).
                                (14B,   86) PHYS. REV. VOL.151,. . . . . . .   126, (1966).
                                (6 ,   314) REV. ROUMAINE PHYS. VOL.11,. . .   265, (1966).

3487 RIPFEL..........(H.)....‡ (23 ,   32) IAEA SYMP. COPENHAGEN, VOL.2,. .   331, (1968).

3488 RISTE...........(T.)....‡ (5 , 1083) PHYS. REV. VOL.95, ...   1161, (1954).
                               (6 , 860) J. PHYS. CHEM. SOLIDS VOL.9,. .   153, (1959).
                               (13 ,   31) J. PHYS. RADIUM VOL.20,. . . .   175, (1959).
                               (90,   28) IAEA SYMP. (PILE RESEARCH)VIENNA   415, (1960).
                               (6 , 859) NATURE VOL.185,. . . . .   450, (1960).
                               (20B, 141) REV. SCI. INSTRUM. VOL.31,. .   214, (1960).
                               (6 , 853) J. PHYS. CHEM. SOLIDS VOL.17,.   308, (1961).
                               (6 , 852) J. PHYS. CHEM. SOLIDS VOL.17,.   318, (1961).
                               (5 , 1133) J. PHYS. CHEM. SOLIDS VOL.19,.   117, (1961).
                               (6 , 542) IAEA SYMP. CHALK RIVER, VOL.2,   327, (1962).
                               (8 ,   41) J. PHYS. SOC. JAPAN VOL.17 B-III   60, (1962).
                               (12A,   22) PHYS. LETT. 6, ...   47, (1963).
                               (6 , 487) PROC. INT. CONF. (NOTTINGHAM),   299, (1964).
                               (6 , 1735) J. APP. PHYS. 36, ...   1076, (1965).
                               (8 , 157) TH. NEUTRON SCATT./EGELSTAFF,.   251, (1965).
                               (6 , 847) J. APP. PHYS. 39, ...   1114, (1968).
                               (6 , 1727) J. APP. PHYS. 39,. . . . . . .   528, (1968).
                               (6 , 1374) PHYS. REV. LETT 20, ...   997, (1968).
                               (20B,   42) IAEA SYMP. (INSTRUMENT.), VIENNA   91, (1969).
                               (6 , 2179) J. APP. PHYS. VOL.40,. . . . . 1445, (1969).
                               (20B,   90) NUCL. INST. MET. 75, ...   197, (1969).
                               (20B, 121) PHYS. NORVEG. VOL.3,. . . . .   208, (1969).
                               (5 , 1682) PHYS. REV. 181,. . . . .   920, (1969).
                               (20B,   95) NUCL. INST. MET. VOL.86,. . .   1, (1970).
                               (6 , 2091) SOL. STATE COMM. VOL.9, . . .   1103, (1971).
                               (6 , 2090) SOL. STATE COMM. VOL.9, . . .   1455, (1971).
                               (6 , 1375) INT. J. MAGN. VOL.3 NO.4,. . .   349, (1972).
                               (5 , 136) PHYS. NORV. VOL.6 NO.3-4,. . .   205, (1972).
                               (6 , 1312) PHYS. REV. B VOL.6,. . . . . . 4332, (1972).
                               (5 , 134) SOL. STATE COMM. VOL.11,. . .   1365, (1972).
                               (5 , 135) SOL. STATE COMM. VOL.12 NO.5,.   409, (1973).

3489 RITCHIE.........(A.I.M.)‡ (6 , 156) J. NUCL. ENERGY(GB) VOL.26,. . .   27, (1972).

3490 RITCHIE.........(D.S.)..‡ (13 ,   5) AIP CONF. PROC. VOL.5, ... 1250, (1971).
                               (13 ,   52) PHYS. REV. B VOL.5,. . . . . . 2668, (1972).

3491 RITTER..........(H.L.)..‡ (5 , 2481) ACTA CRYST. 23,. . . . .   868, (1967).

3492 RIZOWYANETSKII..(D.R.)..‡ (8 ,   60) PHYS. MET. METALLOGR. VOL.20 N.2   27, (1965).

3493 RIZZI...........(G.)....‡ (5 , 2931) NUOVO CIMENTO B VOL.49,. . . .   222, (1967).
                               (6 , 2304) IAEA SYMP. COPENHAGEN, VOL.1,.   373, (1968).
                               (5 , 2928) NUOVO CIMENTO B VOL.60,. . . .   48, (1969).
                               (6 , 2306) SOL. STATE COMM. VOL.8,. . . .   367, (1970).
                               (10B,   41) IAEA SYMP. GRENOBLE, ...   149, (1972).

3494 ROBBINS.........(M.)....‡ (5 , 1051) J. PHYS. CHEM. SOL. 25,. . . .   717, (1964).
                               (5 , 704) J. PHYS. CHEM. SOL. 28,. . . .   897, (1967).
                               (5 , 706) J. APP. PHYS. 39,. . . . . . .   664, (1968).

3495 ROBB............(A.D.)..‡ (24 ,   75) NUCL. INST. MET. VOL.91,. . . .   13, (1971).

3496 ROBERTO.........(J.B.)..‡ (5 , 1222) PHYS. STAT. SOLIDI B VOL.59,.   K59, (1973).
                               (5 , 2450) PHYS. REV. B VOL.9,. . . . . . 2590, (1974).

3497 ROBERTSON.......(J.C.)..‡ (24 ,   41) NUCL. INST. MET. 35, ...   153, (1965).

3498 ROBERTSON.......(J.M.)..‡ (9C,   4) ACTA CRYST. B27, ...   2289, (1971).

3499 ROBERTS.........(B.W.)..‡ (5 , 1624) PHYS. REV. VOL.101,. . . . . .   537, (1956).
                               (5 , 219) PHYS. REV. VOL.104,. . . . . .   607, (1956).
                               (5 , 738) TRANS. METALL. SOC. AIME VOL.236 1012, (1966).

3500 ROBERTS.........(D.E.)..‡ (7B, 101) PROC. PHYS. SOC. 92, ...   889, (1967).

3501 ROBERTS.........(F.F.)..‡ (5 , 1605) ACTA CRYST. VOL.6, ...   57, (1953).

3502 ROBERTS.........(L.D.)..‡ (9B,   81) PHYS. REV. VOL.97, ...   889, (1955).

3503 ROBERT..........(A.)....‡ (20C,   85) REV. PHYS. APPL. VOL.1,. . . .   149, (1966).

3504 ROBILLARD.......(T.R.)..‡ (5 , 1348) PHYS. REV. VOL.82, ...   560, (1951).

3505 ROBINSON........(D.J.)..‡ (5 , 1437) CRYST. STRUCT. COMM. VOL.1 N.3,.   185, (1972).

3506 ROBINSON........(L.B.)..‡ (11A,   43) PHYS. REV. B VOL.3,. . . . . . 1025, (1971).

3507 ROBINSON........(P.D.)..‡ (7A,   70) PHYS. REV. 129,. . . . . 1391, (1963).
                               (7A,   68) PHYS. REV. 129,. . . . . 1396, (1963).

3508 ROBSON..........(J.M.)..‡ (20B,   11) CAN. J. PHYS. VOL.39,. . . . .   1, (1961).
                               (20C,   83) PHYS. LETT. B VOL.40B,. . . . .   537, (1972).

3509 ROCCA-VOLMERANGE(B.)....‡ (16 ,   51) NUCL. SCI. ENG. VOL.48,. . . .   10, (1972).

3510 ROCKENBAUER.....(G.)....‡ (24 ,   7) ATOMKERNENERGIE VOL.15,. . . .   275, (1970).

3511 RODBELL.........(D.S.)..‡ (6 , 548) J. APP. PHYS. 38,. . . . . . .   979, (1967).

3512 RODDA...........(J.L.)..‡ (20C,   67) NUCL. INST. MET. 74, ...   224, (1969).

3513 RODGERS.........(A.L.)..‡ (22B,   26) NUCL. INST. MET. 45, ...   245, (1966).
                               (20B,   36) IAEA SYMP. COPENHAGEN, VOL.2,.   341, (1968).

3514 RODRIGUES.......(C.)....‡ (6 ,   59) J. NUCL. ENERGY VOL.26,. . . .   379, (1972).

3515 RODRIGUEZ.......(C.)....‡ (22A,   40) NUCL. INST. MET. 63, ...   13, (1968).
                               (14B,   41) J. CHEM. PHYS. VOL.56, ... 3118, (1972).

3516 RODRIGUEZ.......(L.J.)..‡ (6 , 1142) PHYS. REV. A VOL.5,. . . . . . 1547, (1972).
                               (7B,   88) PHYS. REV. A VOL.8,. . . . . .   905, (1973).
                               (6 , 1208) DISS. ABSTR. B VOL.34,. . . . . 4560, (1974).
                               (6 , 1205) PHYS. REV. A VOL.9,. . . . . . 2085, (1974).

3517 ROESSLER........(B.)....‡ (14A,   59) PHYS. REV. A VOL.6,. . . . . . 2243, (1972).

3518 ROGALSKA........(G.)....‡ (6 , 267) ACTA PHYS. POLONICA VOL.27,. .   581, (1965).

3519 ROGALSKA........(Z.)....‡ (6 , 1552) IAEA SYMP. CHALK RIVER, VOL.1,   405, (1962).
                               (6 , 259) PHYSICA VOL.29,. . . . .   491, (1963).

3520 ROGERS..........(B.J.)..‡ (16 ,   73) RENSSELAER POL. INST. SYMP., . .   248, (1961).

3521 ROGERS..........(D.B.)..‡ (5 , 517) PROC. INT. CONF. (NOTTINGHAM), .   538, (1964).

3522 ROGERS..........(J.G.)..‡ (17C,   10) ACTA CRYST. A24, ...   160, (1968).

3523 ROHLOFF.........(F.)....‡ (20C,   7) ATOMKERNENERGIE VOL.19 NO.4, . .   312, (1972).

3524 ROLANDSON.......(S.)....‡ (6 , 1969) J. OF PHYS.-C., SER.2, VOL.3,. . 1013, (1970).
                               (10B, 157) PHYS. REV. B-2,. . . . . 2098, (1970).
                               (6 , 1976) PHYS. STAT. SOL. VOL.40,. . . .   749, (1970).
                               (6 , 1967) J. OF PHYS.-C., SER.2, VOL.4,.   958, (1971).

```
3524   ROLANDSON........(S.)....:   ( 6 ,    558)  PHYS. REV. B-4,........   4617, (1971).
                                    ( 6 ,    563)  PHYS. STAT. SOL. VOL.52 NO.2,:    643, (1972).
                                    ( 6 ,    594)  PHYS. REV. B VOL.7,.........   2393, (1973).
                                    ( 6 ,    610)  PHYS. REV. B VOL.9,:........   3278, (1974).

3525   ROMANAZZO........(M.)....:   ( 5 ,   1789)  PHYS. REV. B VOL.5,.......   3778, (1972).

3526   ROOF-JR..........(R.B.)..:   ( 5 ,   2707)  ACTA CRYST. VOL.15,.......    351, (1962).

3527   RORER............(D.C.)..:   ( 5 ,     95)  NUCL. PHYS. VOL.A133,......    410, (1969).
                                    ( 5 ,   2972)  BULL. AMER. PHYS. SOC. VOL.17, :    125, (1972).

3528   ROSCISZEWSKI.....(K.)....:   (14B,      3)  ACTA PHYS. POLON. VOL.A41, ...    549, (1972).

3529   ROSENBAUM........(M.)....:   ( 7B,     72)  PHYS. REV. VOL.126,........   1165, (1962).
                                    ( 7A,     26)  DISS. ABS. VOL.24,.........   2410, (1964).
                                    ( 7B,     79)  PHYS. REV. VOL.137,........   B271, (1965).

3530   ROSENBLUTH.......(M.)....:   ( 6 ,   2356)  PHYS. REV. VOL.108,.......   1092, (1957).

3531   ROSENSTOCK.......(H.B.)..:   (10B,    132)  PHYS. REV. 176,...........   1004, (1968).

3532   ROSEN............(J.L.)..:   (22A,     62)  REV. SCI. INSTRUM. VOL.31, ...    481, (1960).

3533   ROSE.............(J.L.)..:   ( 6 ,   2206)  CONF/NUCL. STRUC. STUDY/NEUTRONS   172, (1972).

3534   ROSOLOWSKI.......(J.)....:   ( 6 ,   2309)  COPENHAGEN CONF.-LATT.DYNAMICS,.    33, (1963).

3535   ROSSAT-MIGNOD...(J.)....:    ( 5 ,   2923)  SOL. STATE COMM. VOL. 7 NO.14,   1011, (1969).
                                    ( 9D,     12)  AIP CONF. PROC.(USA) VOL.5,..   1355, (1971).
                                    ( 5 ,   2197)  PHYS. STAT. SOL. VOL.47,....    239, (1971).
                                    ( 5 ,    829)  PHYS. STAT. SOL. VOL.47,....    247, (1971).
                                    ( 5 ,   2694)  PHYS. STAT. SOLIDI A VOL.14 NO.2   483, (1972).
                                    ( 5 ,    520)  PHYS. STAT. SOL. VOL.50,....    747, (1972).
                                    ( 5 ,   1396)  SOL. STATE COMM. VOL.10,....    735, (1972).
                                    ( 5 ,   2597)  SOL. STATE COMM. VOL.12,....    985, (1973).
                                    ( 6 ,   1232)  SOL. STATE COMM. VOL.13,....   1549, (1973).
                                    ( 6 ,    709)  PHYS. REV. B VOL.9 NO.11,...     ., (1974).

3536   ROSSEL...........(J.)....:   ( 5 ,   1884)  HELV. PHYS. ACTA VOL.19, ....    493, (1946).
                                    ( 6 ,   1525)  HELV. PHYS. ACTA VOL.20, ....    105, (1947).

3537   ROSSITTO.........(F.)....:   (20D,     83)  NUCL. INST. MET. 55,........    288, (1967).
                                    (20D,      1)  ACTA CRYST. VOL.A27,........    341, (1971).
                                    (10B,     41)  IAEA SYMP. GRENOBLE,........    149, (1972).
                                    (20D,      2)  ACTA CRYST. VOL.A29,........    440, (1973).

3538   ROSS.............(D.K.)..:   ( 6 ,   1055)  CONF. NEUTRON THERMALIZ. VOL.I,.    477, (1968).
                                    (15 ,     13)  IAEA SYMP. COPENHAGEN, VOL.2,.    299, (1968).
                                    (20B,     36)  IAEA SYMP. COPENHAGEN, VOL.2,.    341, (1968).
                                    ( 6 ,   1864)  J. PHYS. C. VOL.3,.........   2487, (1970).
                                    ( 6 ,   1950)  CONF./RARE EARTHS/ACTINIDES, ..     46, (1971).
                                    (20B,     99)  NUCL. INST. MET. VOL.91,....    159, (1971).
                                    ( 6 ,   1682)  BER. BUNSENGES. PHYS. CHEM. 76,.    782, (1972).
                                    ( 6 ,    236)  MOL. PHYS. VOL.24,.........    753, (1972).
                                    ( 6 ,    212)  J. PHYS. C VOL.6,..........   3525, (1973).
                                    (20B,    105)  NUCL. INST. MET. VOL.109,...    313, (1973).
                                    ( 6 ,   1868)  SOL. STATE COMM. VOL.13 NO.2,.    229, (1973).

3539   ROSS.............(F.K.)..:   ( 5 ,    415)  J. PHARM. PHARMACOL. VOL.22, ...    724, (1970).

3540   ROSS.............(H.H.)..:   (20C,     79)  NUCL. SCI. ENGNG. VOL.20,...     23, (1964).
                                    (20C,     41)  NUCL. INST. MET. VOL.33,....    194, (1965).

3541   ROSS.............(R.T.)..:   ( 6 ,   1381)  J. PHYS. C VOL.7,..........    L99, (1974).

3542   ROSTOKER.........(W.)....:   ( 6 ,   2068)  J. ELECTROCHEM. SOC. VOL.112,.    108, (1965).

3543   ROTHBAUER........(R.)....:   ( 5 ,   1871)  ACTA CRYST. VOL. A25,......   S119, (1969).
                                    (20B,    102)  NUCL. INST. MET. VOL.101,...    221, (1972).

3544   ROTHENBERG.......(M.S.)..:   ( 9C,     30)  PHYSICS TODAY (MARCH),.....     19, (1973).

3545   ROTHER...........(H.)....:   ( 6 ,   1013)  PHYSICA VOL.50,...........    380, (1970).

3546   ROTH.............(M.)....:   ( 5 ,     76)  C.R. ACAD. SCI. B TOME 277,..    225, (1973).
                                    ( 5 ,     77)  J. APPL. CRYST. VOL.7, .....    219, (1974).

3547   ROTH.............(S.)....:   ( 5 ,    410)  J. PHYS. C VOL.6,..........   3465, (1973).
                                    ( 5 ,    411)  J. APPL. CRYST. VOL.7, .....    230, (1974).

3548   ROTH.............(W.L.)..:   ( 5 ,   1352)  ACTA CRYST. VOL.9,.........    277, (1956).
                                    ( 5 ,   2177)  PHYS. REV. VOL.110,........   1333, (1958).
                                    ( 5 ,   2178)  PHYS. REV. VOL.111,........    772, (1958).
                                    ( 5 ,   1086)  ACTA CRYST. VOL.13,........    140, (1960).
                                    ( 5 ,   2173)  J. APPL. PHYS. VOL.31, SUPPL.,    352, (1960).
                                    ( 5 ,   2175)  J. APPL. PHYS. VOL.31,......   2000, (1960).
                                    ( 5 ,    962)  COLL. INTER. N.12E (GRENOBLE),     83, (1963).
                                    ( 5 ,    489)  J. PHYS. CHEM. SOL. 25,.....      ., (1964).
                                    ( 5 ,   2267)  J. APP. PHYS. 38,..........    951, (1967).

3549   ROTTERUD.........(H.)....:   ( 5 ,   1812)  ACTA CRYST. (INTERACT.) VOL.A25, S250, (1969).
                                    ( 5 ,   1811)  PHYS. NORVEG. VOL.3,.......    203, (1969).

3550   ROUAULT..........(A.)....:   ( 5 ,   2102)  MATER. RES. BULL. VOL.8, ....    229, (1973).

3551   ROUBEAU..........(P.)....:   ( 6 ,    963)  IAEA SYMPOSIUM (VIENNA), ....    411, (1960).

3552   ROUDAULT.........(R.)....:   ( 5 ,   1465)  REV. CHIM. MINER. VOL.9, ....    825, (1972).

3553   ROULT............(G.)....:   ( 5 ,   2738)  J. APPL. PHYS. VOL.33, SUPPL.,   1123, (1962).
                                    ( 5 ,    999)  COLL. INTER. N.126 (GRENOBLE),    102, (1963).
                                    ( 5 ,    615)  COLL. INTER. N.126 (GRENOBLE),    158, (1963).
                                    ( 5 ,   1044)  COLL. INTER. N.126 (GRENOBLE),     98, (1963).
                                    ( 5 ,   1126)  COMP. REND. 256,...........   1688, (1963).
                                    ( 5 ,   1855)  J. APPL. PHYS. VOL.35,.....    952, (1964).
                                    ( 5 ,    591)  J. APPL. PHYS. VOL.37,.....   1038, (1966).
                                    (20C,     85)  REV. PHYS. APPL. VOL.1,....    149, (1966).
                                    ( 5 ,   1747)  SOL. STATE COMM. 5,.........      7, (1967).
                                    ( 5 ,   2748)  COMPTES RENDUS 267 SERIE B,.    518, (1968).
                                    (19B,      1)  ACTA CRYST. (INTERACT.) VOL.A25, S256, (1969).
                                    ( 5 ,   2750)  J. APP. PHYS. VOL.40,.......   1131, (1969).
                                    ( 5 ,   1681)  SOL. STATE COMM. VOL.7 NO.19,.   1403, (1969).
                                    ( 5 ,     42)  J. PHYS. CHEM. SOLIDS VOL.32,.   1641, (1971).
                                    ( 5 ,    487)  BULL. AMER. PHYS. SOC. VOL.17,    667, (1972).
                                    (22C,     66)  REV. SCI. INSTRUM. VOL.45, ...    341, (1974).

3554   ROUSE............(K.D.)..:   (17C,     11)  ACTA CRYST. A24,...........    405, (1968).
                                    ( 5 ,    164)  ACTA CRYST. A24,...........    484, (1968).
                                    ( 5 ,   2776)  ACTA CRYST. VOL.B24,.......    117, (1968).
                                    ( 9B,     39)  HARWELL SUMMER SCHOOL,......      1, (1968).
                                    (17C,     12)  ACTA CRYST. A25,...........    615, (1969).
                                    ( 9B,     12)  ACTA CRYST. VOL.A26,.......    214, (1970).
                                    ( 9B,      9)  ACTA CRYST. VOL.A26,.......    457, (1970).
                                    ( 9B,     11)  ACTA CRYST. VOL.A26,.......    682, (1970).
                                    ( 5 ,    333)  ACTA CRYST. VOL.A27,.......    622, (1971).
```

```
3554   ROUSE..........(K.D.)..:   ( 5 , 1512) FERROELECT. VOL.4,. . . . .   153, (1972).
                                  ( 6 , 1338) J. PHYSIQUE TOME 33 COL.2, . :   133, (1972).
                                  ( 2 , 1460) SOL. STATE COMM. VOL.11,. . . .  1261, (1972).
                                  ( 5 ,  154) Z. KRISTALLOGR. VOL.135,. . . . .  316, (1972).
                                  ( 5 , 1428) ACTA CRYST. VOL.A29,. . . . . .   514, (1973).
                                  ( 5 , 2961) ACTA CRYST. VOL.A29, . . . . .    49, (1973).

3555   ROUSSEL.........(J.)....:  ( 5 ,  755) BULL. SOC. FRANC. MIN. CRIST. 85   57, (1962).

3556   ROUX...........(D.)....:   (20D,    4) ARCH. SCI. VOL.12, . . . . .   676, (1959).
                                  (21 ,    5) ARCH. SCI. VOL.13, . . . . .   375, (1960).

3557   ROWELL.........(J.M.)..:   (10B,  144) PHYS. REV. VOL.187,. . . . .   821, (1969).
                                  ( 6 , 2132) PHYS. REV. 178,. . . . . . .   897, (1969).
                                  (10C,   16) CONF. INTERN. RENNES, . . . .   150, (1971).
                                  ( 6 , 2062) PHYS. REV. B VOL.3,. . . . .  4065, (1971).

3558   ROWE...........(J.M.)..:   ( 6 , 2071) PHYS. REV. LETT. 14, . . . .   554, (1965).
                                  ( 6 ,  619) SOL. STATE COMM. 3,. . . . .   245, (1965).
                                  ( 6 , 2064) THESIS (98 PP.), . . . . . .  . , (1966).
                                  ( 6 ,  606) PHYS. REV. 155,. . . . . . .   619, (1967).
                                  ( 6 , 2070) PHYS. REV. 163,. . . . . . .   547, (1967).
                                  (23 ,   33) IAEA SYMP. COPENHAGEN, VOL.2,.   259, (1968).
                                  ( 5 ,  268) CHEM. PHYS. LETTERS VOL.4, . .   444, (1969).
                                  ( 6 , 2063) SOL. STATE COMM. 7,. . . . .  1433, (1969).
                                  ( 6 , 2256) J. PHYS. CHEM. SOLIDS VOL.32,.    41, (1971).
                                  ( 6 , 1240) PHYS. REV. B-3,. . . . . . .  1268, (1971).
                                  ( 6 ,   80) IAEA SYMP. GRENOBLE, . . . .   413, (1972).
                                  ( 6 , 2259) JULICH CONF/H IN METALS VOL.I,   301, (1972).
                                  ( 5 , 1423) J. CHEM. PHYS. VOL.56, . . .  3697, (1972).
                                  ( 6 , 2257) J. CHEM. PHYS. VOL.56, . . .  4574, (1972).
                                  ( 6 ,   85) PHYS. REV. A VOL.6,. . . . .  1107, (1972).
                                  ( 6 , 1862) PHYS. REV. LETT. VOL.29, . .  1250, (1972).
                                  ( 6 , 2250) SOL. STATE COMM. VOL.11, . .  1299, (1972).
                                  ( 5 , 1957) J. CHEM. PHYS. VOL.58, . . .  2039, (1973).
                                  ( 6 , 2001) J. CHEM. PHYS. VOL.58, . . .  5469, (1973).
                                  ( 6 , 2117) J. CHEM. PHYS. VOL.59, . . .  6570, (1973).
                                  ( 6 , 2002) J. CHEM. PHYS. VOL.59, . . .  6652, (1973).
                                  (19B,   20) NUCL. INST. MET. VOL.106,. .   221, (1973).
                                  (19A,   53) NUCL. INST. MET. VOL.107,. .   501, (1973).
                                  ( 6 , 1865) PHYS. REV. B VOL.8,. . . . .  6013, (1973).
                                  ( 6 , 1195) PHYS. REV. LETT. VOL.31,. .    510, (1973).
                                  (17R,   51) NUCL. INST. MET. VOL.114,. .   411, (1974).
                                  ( 6 , 1966) PHYS. REV. A VOL.9,. . . . .  1656, (1974).
                                  ( 6 ,  461) PHYS. REV. B VOL.10 NO.2,. .  . , (1974).
                                  ( 6 , 2119) PHYS. REV. B VOL.9 NO.12,. .  . , (1974).
                                  ( 6 , 1964) PHYS. REV. LETT. VOL.32 NO.2,.   49, (1974).
                                  ( 6 , 1867) PHYS. REV. LETT. VOL.32, . .  1087, (1974).

3559   ROYL...........(P.)....:   ( 7A,   16) ATOMKERNENERGIE VOL.18,. . . .   219, (1971).

3560   ROYSTON........(R.)....:   ( 6 , 1049) IAEA SYMPOSIUM (VIENNA), . . .   309, (1960).

3561   ROY............(A.P.)..:   ( 6 , 1567) IAEA SYMP. CHALK RIVER, VOL.2,.   253, (1962).
                                  ( 6 , 1413) IAEA SYMP. CHALK RIVER, VOL.2,.    99, (1962).
                                  ( 5 , 1174) J. PHYS. SOC. JAPAN VOL.17 B-III   41, (1962).
                                  ( 5 , 1173) J. PHYS. SOC. JAPAN VOL.17,. .   247, (1962).
                                  ( 6 , 1402) COPENHAGEN CONF.-LATT.DYNAMICS,.   223, (1963).
                                  ( 6 , 1399) IAEA SYMP. BOMBAY, VOL.1,. . .   153, (1964).
                                  ( 5 , 1919) IAEA SYMP. BOMBAY, VOL.2,. . .   347, (1964).
                                  ( 6 , 1582) J. PHYS. CHEM. SOL. VOL.27,. .  1103, (1966).
                                  ( 6 , 2236) PHYSICA VOL.34,. . . . . . .   384, (1967).
                                  ( 6 , 1793) CAN. J. PHYS. 48,. . . . . .  1781, (1970).
                                  ( 6 ,  187) THESIS (209 PP.),. . . . . .  . , (1970).
                                  ( 6 ,  198) CAN. J. PHYS. 49,. . . . . .   277, (1971).
                                  ( 6 ,  134) CONF. INTERN., RENNES,. . . .   140, (1971).

3562   ROZHDESTVENSKII.(F.A.).:   ( 5 , 2856) SOV. PHYS. CRYST. VOL.17,. . . .  1017, (1973).

3563   RUANE..........(T.F.)..:   ( 5 , 2964) NUCL. SCI. ENG. VOL.46,. . . .   314, (1971).

3564   RUBAN..........(V.A.)..:   (12B,   61) SOV. PHYS. JETP LETT. VOL.10,.   345, (1969).
                                  (12B,   63) SOV. J.E.T.P. VOL.31,. . . .   111, (1970).
                                  (12B,   67) SOV. PHYS. JETP VOL.35,. . . .   222, (1972).

3565   RUBIN..........(R.)....:   ( 6 , 2226) IAEA SYMP. CHALK RIVER, VOL.2,   145, (1962).
                                  ( 6 , 2225) IAEA SYMP. CHALK RIVER, VOL.2,   155, (1962).
                                  ( 6 , 1681) BROOKHAVEN SYMPOSIUM,. . . . .   105, (1965).
                                  ( 6 , 2254) PHYS. LETT. 14,. . . . . . .   100, (1965).
                                  ( 6 , 2253) J. PHYSIQUE TOME 28 COL.1, . .    26, (1967).
                                  ( 6 , 1677) IAEA SYMP. COPENHAGEN, VOL.1,.   223, (1968).
                                  ( 6 , 1679) SOL. STATE COMM. VOL.8,. . . .  1321, (1970).
                                  ( 6 , 1018) PHYS. REV. LETT. VOL.27,. . .  1576, (1971).
                                  ( 6 , 1683) PHYS. STAT. SOL. VOL.B-54, . .   295, (1972).
                                  ( 6 , 1684) PHYS. LETT. VOL.43A NO.3,. . .   279, (1973).

3566   RUBY...........(S.L.)..:   (20C,   36) NUCLEONICS VOL.17 APRIL,. . .   116, (1959).
                                  ( 5 , 1186) J. PHYS. SOC. JAPAN VOL.17,. .  1598, (1962).
                                  ( 5 , 1560) PHYS. REV. 132,. . . . . . .  1547, (1963).

3567   RUDERMAN.......(I.W.)..:   ( 6 , 1465) PHYS. REV. VOL.75, . . . . .   895, (1949).
                                  (11C,   29) PHYS. REV. VOL.76, . . . . .  1572, (1949).
                                  ( 7A,   63) PHYS. REV. VOL.77, . . . . .   575, (1950).

3568   RUDMAN.........(R.)....:   (21 ,   20) J. APPL. CRYST. VOL.2,. . . .   109, (1969).

3569   RUIJGROK.......(T.W.)..:   ( 8 ,   55) PHYSICA VOL.29,. . . . . . .   617, (1963).

3570   RUKOLAINE......(G.V.)..:   (20B,   48) INSTRUM. EXP. TECH. NO.4,. . .   632, (1967).
                                  ( 5 , 2628) ATOMNAYA ENERGIYA (USSR) VOL.24,   243, (1968).

3571   RUNCIMAN.......(W.A.)..:   ( 5 , 1428) ACTA CRYST. VOL.12,. . . . .   674, (1959).
                                  ( 5 , 2773) J. PHYS. CHEM. SOL. 27,. . . .  1833, (1966).

3572   RUNDLE.........(R.E.)..:   ( 5 , 2740) J. AMER. CHEM. SOC. VOL.73,. .  4172, (1951).
                                  ( 5 , 2968) ACTA CRYST. VOL.5,. . . . .    22, (1952).
                                  ( 5 ,  370) J. CHEM. PHYS. VOL.29, . . .  1306, (1958).
                                  ( 5 , 1996) J. CHEM. PHYS. VOL.32, . . .   627, (1960).
                                  ( 5 , 2239) COLL. INTER. N.126 (GRENOBLE),    63, (1963).

3573   RUNNINGER......(J.)....:   (11B,   49) 1ST EUR. CONF./CONDENSED MATTER,   45, (1971).

3574   RUNOV..........(V.V.)..:   (12B,   68) SOV. PHYS. JETP LETT. VOL.15,.   324, (1972).

3575   RUOFF..........(A.L.)..:   ( 6 , 1626) PHYS. STAT. SOL. 23, . . . . .   489, (1967).

3576   RUPAR..........(H.)....:   (24 ,    7) ATOMKERNENERGIE VOL.15,. . . .   275, (1970).

3577   RUPPERSBERG....(H.)....:   ( 5 , 1576) PHYS. LETT. VOL.46A NO.1,. . .    75, (1973).

3578   RUSH...........(J.J.)..:   ( 6 , 1578) PHYS. REV. LETT. VOL.5,. . . .   507, (1960).
                                  ( 5 , 1265) J. CHEM. PHYS. VOL.35,. . . .  2265, (1961).
                                  ( 6 ,   67) J. CHEM. PHYSICS VOL.37, . .   234, (1962).
                                  ( 6 , 1550) NUCL. SCI. ENGNG. VOL.14,. .   339, (1962).
                                  (15 ,    8) DISS. ABSTR. VOL.23,. . . . .  3151, (1963).
                                  ( 6 , 1285) IAEA SYMP. BOMBAY, VOL.2,. . .   333, (1964).
```

3578 RUSH...........(J.J.).: (6 , 1572) J. PHYS. CHEM. VOL.68,...... 2534, (1964).
(6 , 416) INORG. CHEM. VOL.5,.. . . . . 2238, (1966).
(5 , 1942) J. CHEM. PHYSICS VOL.44,. . . . 2499, (1966).
(6 , 351) J. CHEM. PHYSICS VOL.44,. . . . 2749, (1966).
(6 , 1021) J. CHEM. PHYSICS VOL.45,. . . . 1312, (1966).
(6 , 1788) J. CHEM. PHYS. VOL.44,. . . . . 1722, (1966).
(6 , 1004) J. CHEM. PHYS. VOL.44,. . . . . 2496, (1966).
(6 , 2292) J. CHEM. PHYS. VOL.45,. . . . . 3817, (1966).
(6 , 2223) INORG. CHEM. VOL.6,. . . . . .   346, (1967).
(6 , 359) J. CHEM. PHYS. VOL.46,. . . . . 2285, (1967).
(6 , 1396) J. CHEM. PHYS. VOL.47,. . . . . 3936, (1967).
(6 , 1561) J. CHEM. PHYS. VOL.47,. . . . . 4278, (1967).
(6 , 2258) J. CHEM. PHYS. VOL.48,. . . . . 3795, (1968).
(5 , 268) CHEM. PHYS. LETTERS VOL.4,. . .   444, (1969).
(5 , 275) DISC. FARADAY SOC. NO.48,. . . . 192, (1969).
(5 , 2245) J. CHEM. PHYS. VOL.54,. . . . . 1968, (1971).
(5 , 1993) J. CHEM. PHYS. VOL.55,. . . . . 5363, (1971).
(6 , 2256) J. PHYS. CHEM. SOLIDS VOL.32,. .   41, (1971).
(6 , 575) IAEA SYMP. GRENOBLE,. . . . . . 247, (1972).
(6 , 2259) JULICH CONF/H IN METALS VOL.I, . 301, (1972).
(6 , 1289) J. CHEM. PHYS. VOL.56,. . . . . 2793, (1972).
(5 , 1423) J. CHEM. PHYS. VOL.56,. . . . . 3697, (1972).
(6 , 2257) J. CHEM. PHYS. VOL.56,. . . . . 4574, (1972).
(24 ,   24) J. CHEM. PHYS. VOL.57,. . . . . 175, (1972).
(6 , 1862) PHYS. REV. LETT. VOL.29,. . . . 1250, (1972).
(5 , 274) ACTA CRYST. VOL.B29,. . . . . . 184, (1973).
(5 , 1957) J. CHEM. PHYS. VOL.58,. . . . . 2039, (1973).
(6 , 1665) J. CHEM. PHYS. VOL.58,. . . . . 3439, (1973).
(6 , 2001) J. CHEM. PHYS. VOL.58,. . . . . 5469, (1973).
(6 , 2117) J. CHEM. PHYS. VOL.59,. . . . . 6570, (1973).
(6 , 2002) J. CHEM. PHYS. VOL.59,. . . . . 6652, (1973).
(6 , 1865) PHYS. REV. B VOL.8,. . . . . . 6013, (1973).
(5 , 1936) ACTA CRYST. VOL.B30,. . . . . . 1167, (1974).
(6 , 2119) PHYS. REV. B VOL.9 NO.12,. . . . ., (1974).
(6 , 1867) PHYS. REV. LETT. VOL.32,. . . . 1087, (1974).

3579 RUSSELL-JR......(J.L.).: (6 , 2205) PROC/CONF/FAST CRIT. EXP./ANALY. 524, (1967).
(5 , 1306) NUCL. SCI. ENG. VOL.33,. . . . 265, (1968).

3580 RUSSELL.........(G.J.).: (6 , 1058) NUCL. SCI. ENGNG. VOL.25,. . . . 248, (1966).

3581 RUSSELL.........(M.C.B.): (22B,   30) NUCL. INST. MET. 60, ... 237, (1968).

3582 RUSTAD..........(B.M.).: (5 , 2709) NUCL. SCI. ENGNG. VOL.6,. . . . 433, (1959).
(20B,   64) J. PHYS. SOC. JAPAN VOL.15,. . . 630, (1960).
(22A,   50) NUCL. SCI. ENGNG. VOL.8,. . . . 453, (1960).
(5 , 2713) PHYS. REV. VOL.118,. . . . . . 799, (1960).
(5 , 146) PHYS. REV. VOL.119,. . . . . . 1291, (1960).
(5 , 141) NUCL. SCI. ENGNG. VOL.9,. . . .   98, (1961).
(22C,   30) NUCL. INST. MET. 33,. . . . . . 155, (1965).
(22A,   64) REV. SCI. INSTRUM. VOL.36,. . . 887, (1965).
(20D,   34) REV. SCI. INSTRUM. VOL.36,. . .   48, (1965).

3583 RUSTGI.........(M.L.).: (24 ,   89) PHYS. REV. VOL.114,. . . . . . . 830, (1959).

3584 RUSTICHELLI.....(F.)...: (22C,   41) NUCL. INST. MET. 62, ...   57, (1968).
(20B,   91) NUCL. INST. MET. 74, ... 219, (1969).
(20B,   96) NUCL. INST. MET. 83, ... 124, (1970).
(9B,   18) ACTA CRYST. VOL.A28 PART S-4,. 215, (1972).
(20B,   43) IAEA SYMP. GRENOBLE, ... 697, (1972).
(20B, 103) NUCL. INST. MET. VOL.104,. . . . 147, (1972).
(20B, 106) NUCL. INST. MET. VOL.107,2,. . . 429, (1973).
(9B,   70) NUOVO CIMENTO VOL.13B NO.2,. . 249, (1973).

3585 RUTA-WALA.......(K.)...: (6 , 886) PHYSICS VOL.27,. . . . . 883, (1961).
(6 , 542) IAEA SYMP. CHALK RIVER, VOL.2, 327, (1962).
(6 , 885) J. PHYS. SOC. JAPAN VOL.17 B-III 69, (1962).

3586 RUVALDS.........(J.)...: (10D,   39) PHYS. REV. B VOL.3,. . . . . . . 3556, (1971).
(10D,   60) SOL. STATE COMM. VOL. 9,. . . . 129, (1971).

3587 RUVINSKII.......(M.A.).: (10C,   81) PHYS. STAT. SOL. 13, ... 233, (1966).

3588 RYAZHSKAYA......(O.G.).: (20C,   22) INSTRUM. EXP. TECH. 4, ... 1013, (1970).

3589 RYVES...........(T.B.).: (24 ,   41) NUCL. INST. MET. 35,. . . . . . 153, (1965).
(5 ,   51) J. NUCL. ENERGY VOL.24,. . . . 419, (1970).
(7A,   32) J. NUCL. ENERGY VOL.24,. . . .   35, (1970).
(24 ,   25) J. NUCL. ENERGY VOL.25,. . . . 129, (1971).

3590 RYZHKOVSKII.....(V.M.).: (5 , 1639) SOV. PHYS. DOKLADY VOL.17, . . . 370, (1972).

3591 RZANY...........(A.)...: (5 , 303) NUKLEONIKA VOL.10, ... 201, (1965).

3592 RZANY...........(H.)...: (6 , 415) ACTA PHYS. POLON. VOL.16,. . . . 335, (1957).
(6 , 301) ACTA PHYS. POLON. VOL.17,. . . . 489, (1958).
(5 , 1286) IAEA SYMPOSIUM VIENNA, ... 297, (1960).
(6 , 974) NUKLEONIKA VOL.5,. . . . . 495, (1960).
(5 , 1315) ACTA PHYS. POLON. VOL.22,. . . . 517, (1962).
(6 , 1552) IAEA SYMP. CHALK RIVER, VOL.1, . 405, (1962).
(6 , 1549) PHYSICA VOL.29,. . . . . 485, (1963).
(6 , 1117) PHYSICA VOL.29,. . . . . 488, (1963).
(6 , 835) IAEA SYMP. BOMBAY, VOL.1,. . . . 443, (1964).
(6 , 843) PHYS. STAT. SOL. 15,. . . . . . 119, (1966).
(6 , 862) PHYS. STAT. SOL. 21,. . . . . . K163, (1967).
(5 , 1095) PHYS. STAT. SOL. 26,. . . . . . 429, (1968).
(6 , 545) PHYS. STAT. SOL. 32,. . . . . .   41, (1969).
(6 , 863) PHYS. STAT. SOL. 41,. . . . . . K103, (1970).
(6 , 1442) PHYSICA VOL.57,. . . . . 628, (1972).
(5 , 2170) PHYS. STAT. SOLIDI A VOL.11 NO.1 57, (1972).

3593 SAAM...........(W.F.).: (6 , 1159) BULL. AMER. PHYS. SOC. VOL.18, .   22, (1973).
(6 , 1194) PHYS. REV. A VOL.8,. . . . . . . 1048, (1973).

3594 SAASTAMOINEN....(J.)...: (16 ,   41) NUCL. SCI. ENGNG. VOL.36,. . . . 351, (1969).
(6 , 2349) J. NUCL. ENERGY VOL.25,. . . . . 189, (1971).

3595 SABINE.........(T.M.).: (5 , 2793) PROC. AUSTRAL. AT. ENERGY SYMP,. 168, (1958).
(23 ,   45) J. PHYS. SOC. JAPAN VOL.17 B-II, 352, (1962).
(20A,   3) ACTA CRYST. 16,. . . . . 435, (1963).
(22C,   7) AUSTRAL. J. PHYS. VOL.16,. . . . 272, (1963).
(6 , 160) PHIL. MAG. VOL.8,. . . . .   43, (1963).
(6 , 159) J. NUCLEAR MATERIALS VOL.14,. . 275, (1964).
(5 , 2410) ACTA CRYST. 19,. . . . . 205, (1965).
(5 , 773) ACTA CRYST. 20,. . . . . 214, (1966).
(5 , 536) J. PHYS. CHEM. SOL. 27,. . . . . 1955, (1966).
(5 , 2421) ACTA CRYST. 23,. . . . . 574, (1967).
(5 , 1510) ACTA CRYST. 23,. . . . . 574, (1967).
(5 , 2378) APPL. PHYS. LETT. VOL.10,. . . . 293, (1967).
(5 , 305) ACTA CRYST. VOL.B25,. . . . . . 1970, (1969).
(5 , 2955) ACTA CRYST. VOL.B25,. . . . . . 2254, (1969).
(5 ,   29) ACTA CRYST. VOL.B25,. . . . . . 2442, (1969).
(5 ,   29) ACTA CRYST. VOL.B25,. . . . . . 2442, (1969).
(5 ,   28) ACTA CRYST. VOL.B25,2,. . . . . 2451, (1969).
(20D,   9) J. APPL. CRYST. VOL.2,. . . . . 141, (1969).
(5 , 2320) ACTA CRYST. VOL.B26,. . . . . . 925, (1970).

```
3595  SABINE..........(T.M.)..:  ( 6 ,  1418) J. PHYS. C. VOL.3,. . . . . .      86, (1970).
                                 ( 5 ,  2781) ACTA CRYST. VOL.A28 PART S-4,.    176, (1972).
                                 ( 5 ,  1916) ACTA CRYST. VOL.B28,. . . . . .  3340, (1972).
                                 ( 5 ,   602) J. SOLID STATE CHEM. VOL.4,. . .  400, (1972).

3596  SABO............(P.)....:  (22C,    18) INSTRUM. EXP. TECH. NO.2,. . . .  285, (1961).

3597  SACCHETTI.......(F.)....:  ( 5 ,   479) INT. J. MAGN. VOL.1 NO.2,. . .    183, (1971).
                                 ( 6 ,   870) SOL. STATE COMM. VOL. 9,. . . .  1579, (1971).
                                 ( 5 ,  1789) PHYS. REV. B VOL.5,. . . . . .   3778, (1972).
                                 ( 5 ,   424) SOL. STATE COMM. VOL.10,. . . .   667, (1972).
                                 ( 5 ,  2226) PHYS. REV. B VOL.7,. . . . . .   3112, (1973).
                                 ( 5 ,  1064) NUOVO CIMENTO VOL.20B NO.1,. .      1, (1974).

3598  SACHS...........(R.G.)..:  (15 ,    35) PHYS. REV. VOL.60, . . . . . .     18, (1941).
                                 ( 5 ,   201) PHYS. REV. VOL.71, . . . . . .    589, (1947).

3599  SACLI...........(O.A.)..:  ( 5 ,   494) J. PHYS C VOL.5,. . . . . . .    L261, (1972).

3600  SADAGOPAN.......(V.)....:  ( 5 ,  1865) ACTA CRYST. 16,. . . . . . . .    202, (1963).
                                 ( 5 ,  1870) ACTA CRYST. 19,. . . . . . . .   1031, (1965).

3601  SADANA..........(V.N.)..:  ( 5 ,  2209) COLL. INTER. N.126 (GRENOBLE), .   27, (1963).

3602  SADIKOV.........(I.P.)..:  ( 5 ,   204) SOV. AT. ENERGY VOL.13,. . . .    852, (1962).
                                 (16 ,    83) SOV. AT. ENERGY VOL.32,. . . .     33, (1972).
                                 (10C,   126) SOV. PHYS. SOLID STATE VOL.13,   3012, (1972).
                                 ( 6 ,   456) SOV. PHYS. JETP LETT. VOL.18,.    177, (1973).
                                 ( 6 ,  1408) SOV. PHYS. SOLID STATE VOL.15,   1309, (1974).

3603  SAENZ...........(A.W.)..:  (12A,    10) IAEA SYMP. (PILE RESEARCH)VIENNA  423, (1960).
                                 (11B,     7) J. APPL. PHYS. VOL.31, SUPPL.,   108, (1960).
                                 ( 8 ,    82) PHYS. REV. VOL.119,. . . . . .   1542, (1960).
                                 ( 6 ,   854) J. PHYS. CHEM. SOLIDS VOL.23,.    117, (1962).
                                 (11B,    22) PHYS. REV. VOL.125,. . . . . .   1940, (1962).
                                 ( 6 ,  2288) J. APP. PHYS. 37,. . . . . . .   1050, (1966).
                                 ( 6 ,  2289) PHYS. REV. 156,. . . . . . . .    632, (1967).

3604  SAFFORD.........(G.J.)..:  ( 5 ,  2709) NUCL. SCI. ENGNG. VOL.6, . . .    433, (1959).
                                 ( 5 ,  2713) PHYS. REV. VOL.118,. . . . . .    799, (1960).
                                 ( 5 ,   146) PHYS. REV. VOL.119,. . . . . .   1291, (1960).
                                 ( 5 ,   141) NUCL. SCI. ENGNG. VOL.9, . . .     98, (1961).
                                 ( 6 ,  1550) NUCL. SCI. ENGNG. VOL.14,. . .    339, (1962).
                                 ( 6 ,  1646) J. CHEM. PHYS. VOL.39,. . . .    3135, (1963).
                                 ( 6 ,  1587) J. PHYS. CHEM. SOL. 24,. . . .    617, (1963).
                                 ( 6 ,  1425) J. PHYS. CHEM. SOL. 24,. . . .    771, (1963).
                                 ( 6 ,   992) IAEA SYMP. BOMBAY, VOL.2,. . .    393, (1964).
                                 ( 6 ,  1789) J. CHEM. PHYSICS VOL.40,. . .    1417, (1964).
                                 ( 6 ,  1919) J. CHEM. PHYSICS VOL.40,. . .    1426, (1964).
                                 ( 6 ,  1094) J. CHEM. PHYSICS VOL.40,. . .    2670, (1964).
                                 ( 6 ,  1213) J. CHEM. PHYSICS VOL.41,. . .    3649, (1964).
                                 ( 6 ,  1097) J. CHEM. PHYSICS VOL.42,. . .    1469, (1965).
                                 ( 6 ,  1783) J. CHEM. PHYSICS VOL.43,. . .    3404, (1965).
                                 ( 6 ,  1382) J. CHEM. PHYSICS VOL.44,. . .     345, (1966).
                                 ( 6 ,  1905) J. CHEM. PHYSICS VOL.45,. . .    3787, (1966).
                                 ( 6 ,  1916) ADV. POLYMER SCI. VOL.5,. . .       1, (1967).
                                 ( 6 ,  2310) J. CHEM. PHYS. VOL.50,. . . .    2140, (1969).
                                 ( 6 ,  1115) J. CHEM. PHYS. VOL.50,. . . .    4444, (1969).
                                 ( 6 ,  2045) J. POLYMER SCI. A-2 VOL.7,. .     433, (1969).
                                 ( 6 ,  1041) J. PHYS. CHEM. VOL.74, . . . .   3696, (1970).
                                 ( 6 ,  1040) J. PHYS. CHEM. VOL.74,. . . .    3710, (1970).
                                 (14B,    11) BER. BUNSENGES. PHYS. CHEM. 75,   366, (1971).
                                 ( 6 ,  1906) J. POLYMER SCI. A-2 VOL.9,. .    1219, (1971).
                                 (14B,   127) TECH. ELECTROCHEM. VOL.2,. . .    173, (1973).

3605  SAFIN...........(YU.A.).:  ( 6 ,  1098) SOV. AT. ENERGY(USA) VOL.31, .    459, (1971).
                                 ( 6 ,   228) SOV. AT. ENERGY VOL.31,. . . .    459, (1971).

3606  SAGAN...........(U.)....:  ( 5 ,  1315) ACTA PHYS. POLON. VOL.22,. . .    517, (1962).
                                 ( 6 ,  1552) IAEA SYMP. CHALK RIVER, VOL.1,    405, (1962).

3607  SAGET...........(G.)....:  (12B,    50) REV. PHYS. APPL. (FRANCE) VOL.4,  254, (1969).

3608  SAHNI...........(V.C.)..:  (10A,    14) REV. MOD. PHYS. VOL.42,. . . .    409, (1970).

3609  SAILOR..........(V.L.)..:  (22A,    61) REV. SCI. INSTRUM. VOL.27, . .     26, (1956).
                                 (20B,   144) REV. SCI. INSTRUM. VOL.32, . .    654, (1961).
                                 ( 5 ,  2820) PHYS. REV. B-3,. . . . . . . .    128, (1970).

3610  SAINT-JAMES.....(D.)....:  ( 6 ,   582) IAEA SYMPOSIUM VIENNA, . . . .    549, (1960).
                                 (11C,    14) J. PHYS. CHEM. SOLIDS VOL.17,.    117, (1960).
                                 ( 6 ,   595) J. PHYS. RADIUM VOL.21,. . . .     67, (1960).
                                 ( 6 ,  1951) C.R. ACAD. SCI. VOL.253,. . . .  2884, (1961).

3611  SAITO...........(Y.)....:  ( 5 ,   315) ACTA CRYST. 23,. . . . . . . .     64, (1967).

3612  SAITTA..........(L.)....:  (22A,    49) NUCL. INST. MET. VOL.114,. . .     21, (1974).

3613  SAKAMOTO........(M.)....:  ( 6 ,  1801) IAEA SYMPOSIUM VIENNA, . . . .    531, (1960).
                                 ( 6 ,   203) IAEA SYMPOSIUM (VIENNA), . . .    487, (1960).
                                 ( 5 ,  2746) J. PHYS. SOC. JAPAN VOL.17 B-III   46, (1962).
                                 (23 ,    44) J. PHYS. SOC. JAPAN VOL.17 B-II,  354, (1962).
                                 ( 6 ,  1070) J. PHYS. SOC. JAPAN VOL.17 B-II,  370, (1962).
                                 ( 6 ,   582) COLL. INTER. N.126 (GRENOBLE),     38, (1963).
                                 ( 6 ,  1120) J. PHYS. SOC. JAPAN VOL.19,. .    1862, (1964).
                                 (20A,    26) J. PHYS. SOC. JAPAN VOL.19,. .    2280, (1964).
                                 (20D,     8) JAPAN. J. APPL. PHYS. VOL.4,. .    911, (1965).
                                 ( 5 ,    71) J. PHYS. SOC. JAPAN VOL.20,. .    1723, (1965).
                                 ( 6 ,   629) IAEA SYMP. COPENHAGEN, VOL.1,.    181, (1968).
                                 ( 6 ,  1924) J. POLYM. SCI./POLYM. LETT. 11,   377, (1973).

3614  SAKAMOTO........(S.)....:  ( 6 ,  1042) J. NUCL. SCI. TECHNOL. VOL.6,.    671, (1969).

3615  SAKSONOV........(YU.G.).:  ( 5 ,  2949) SOV. PHYS. CRYST. VOL.8,. . . .     18, (1963).

3616  SAKURAI.........(J.)....:  ( 5 ,  1691) J. PHYS. SOC. JAPAN VOL.23,. .    1426, (1967).
                                 ( 6 ,   505) IAEA SYMP. COPENHAGEN, VOL.2,.    123, (1968).
                                 ( 6 ,   509) PHYS. REV. 167,. . . . . . . .    510, (1968).
                                 ( 6 ,  1659) J. PHYS. SOC. JAPAN VOL.28, SUP   258, (1970).
                                 ( 6 ,  1660) J. PHYS. SOC. JAPAN VOL.28,. .    1426, (1970).
                                 ( 6 ,  2284) PHYS. REV. B-1,. . . . . . . .    2430, (1970).

3617  SALAMATIN.......(I.M.)..:  ( 5 ,  2433) SOV. J. NUCL. PHYS. VOL.9, 6,.   1119, (1969).

3618  SALAMA..........(M.)....:  ( 5 ,  2510) J. NUCL. ENERGY VOL.21,. . . .    425, (1967).
                                 ( 5 ,  2933) J. NUCL. ENERGY VOL.22,. . . .    389, (1968).
                                 (20B,    87) NUCL. INST. MET. 60,. . . . .     349, (1968).
                                 ( 5 ,  2965) ATOMKERNENERGIE VOL.22,. . . .     87, (1973).
                                 (19A,    56) NUCL. INST. MET. VOL.116,. . .    509, (1974).

3619  SALGADO.........(J.)....:  ( 6 ,    33) IAEA SYMP. GRENOBLE, . . . . .    137, (1972).

3620  SALTER..........(L.S.)..:  (10B,    29) IAEA SYMP. CHALK RIVER, VOL.1,     49, (1962).
```

```
3621  SALZBERG........(J.B.)..:  ( 6 ,  1619) J. PHYS. F VOL.3,. . . . . . . .   L99, (1973).
                                 (10B,   102) NUOVO CIM. VOL.17B NO.1,. . . . .  166, (1973).
                                 ( 6 ,   592) PHYS. LETT. VOL.43A NO.5,. . . . .  429, (1973).

3622  SAL-NIKOV.......(O.A.)..:  ( 6 ,  2169) BU. ACAD. SCI. USSR PHYS. SR. 32  606, (1968).

3623  SAMARAS.........(D.)....:  ( 5 ,  2496) J. SOLID STATE CHEM. VOL.7,. . .  337, (1973).

3624  SAMOILOV........(B.N.)..:  ( 6 ,  2183) PHYS. STAT. SOL. 20, . . . . . . .  767, (1967).

3625  SAMOSVAT........(S.S.)..:  ( 5 ,  1308) SOV. PHYS. JETP. 27, . . . . . .   15, (1968).

3626  SAMPSON.........(T.E.)..:  ( 6 ,   412) IAEA SYMP. COPENHAGEN, VOL.1,..  491, (1968).
                                 ( 6 ,  1394) DISS. ABS.INT.,30,. . . . . . . :2380B, (1969).
                                 ( 6 ,   413) J. CHEM. PHYS. VOL.51, . . . . . : 5543, (1969).

3627  SAMS............(D.)....:  (24 ,    39) NUCL. INST. MET. 30, . . . . . .   77, (1964).

3628  SAMUELSEN.......(E.J.)..:  (12A,    22) PHYS. LETT. 6,. . . . . . . . . .   47, (1963).
                                 ( 6 ,   851) PHYSICA VOL.34,. . . . . . . . .  241, (1967).
                                 ( 6 ,   847) J. APP. PHYS. 39,. . . . . . . . 1114, (1968).
                                 ( 6 ,   547) PHYS. LETT. VOL.26A,. . . . . . .  160, (1968).
                                 ( 6 ,    64) PHYSICA VOL.43,. . . . . . . . .  353, (1969).
                                 ( 6 ,   544) PHYSICA VOL.45,. . . . . . . . .   12, (1969).
                                 ( 5 ,  1434) PHYS. REV. VOL.188,. . . . . . .  919, (1969).
                                 ( 6 ,   546) SOL. STATE COMM. 7, NO15,. . . . 1043, (1969).
                                 ( 6 ,   543) PHYSICA VOL.48,. . . . . . . . .   13, (1970).
                                 ( 6 ,   837) PHYS. STAT. SOL. VOL.42, . . . .  241, (1970).
                                 ( 5 ,  2222) J. PHYSIQUE TOME 32 COL.1 VOL.I,  582, (1971).
                                 ( 6 ,   531) PHYS. REV. B VOL.3,. . . . . . .  157, (1971).
                                 ( 6 ,  1769) SOL. STATE COMM. VOL. 9, . . . . 1011, (1971).
                                 ( 6 ,  2090) SOL. STATE COMM. VOL. 9, . . . . 1455, (1971).
                                 ( 6 ,  1770) PHYS. REV. B VOL.6,. . . . . . . 3447, (1972).
                                 ( 5 ,  2227) PHYS. REV. B VOL.7,. . . . . . . 3102, (1973).
                                 ( 6 ,  1972) PHYS. REV. LETT. VOL.31, . . . .  936, (1973).
                                 ( 5 ,  2393) J. PHYS. CHEM. SOLIDS VOL.35,. .  785, (1974).
                                 ( 5 ,  1091) J. PHYS. C VOL.7,. . . . . . . . L115, (1974).

3629  SANALAN.........(Y.)....:  ( 6 ,  1055) CONF. NEUTRON THERMALIZ. VOL.I,.  477, (1968).
                                 (20B,    36) IAEA SYMP. COPENHAGEN, VOL.2,.  341, (1968).

3630  SANBORN.........(S.M.)..:  ( 6 ,  1041) J. PHYS. CHEM. VOL.74, . . . . . 3696, (1970).
                                 ( 6 ,  1040) J. PHYS. CHEM. VOL.74, . . . . . 3710, (1970).

3631  SANDA...........(T.)....:  (2JB,    59) J. NUCL. SCI. TECHNOL. VOL.7,. .  381, (1970).

3632  SANDOR..........(E.)....:  ( 6 ,   662) DISC. FARADAY SOC. NO.48,. . . .   78, (1969).
                                 ( 5 ,   781) ACTA CRYST. VOL.A28 PART S-4,. .  188, (1972).

3633  SANDRONI........(S.)....:  ( 6 ,   348) J. CHEM. PHYS. VOL.40, . . . . . 3502, (1964).

3634  SANDS...........(D.E.)..:  ( 5 ,  1565) J. PHYS. CHEM. SOLIDS VOL.24,. . 1066, (1963).

3635  SANGER..........(M.G.)..:  ( 5 ,    92) PHYS LETT 27A, . . . . . . . . .  695, (1968).

3636  SANGER..........(P.L.)..:  ( 5 ,  2818) ACTA CRYST. VOL.A26, . . . . . .  543, (1970).

3637  SANGSTER........(M.J.L.):  ( 6 ,  1420) J. PHYS. C, SER.2, VOL.3,. . . . 1026, (1970).

3638  SANTORO.........(A.)....:  ( 5 ,   739) J. PHYS. SOC. JAPAN VOL.17 B-II,  348, (1962).
                                 (17A,     3) ACTA CRYST. 17,. . . . . . . . .  597, (1964).
                                 (17A,     5) ACTA CRYST. 22,. . . . . . . . .  331, (1967).

3639  SANTORO.........(R.P.)..:  ( 5 ,   511) ACTA CRYST. 17,. . . . . . . . .  240, (1964).
                                 ( 5 ,   508) J. PHYS. CHEM. SOL. 25,. . . . .  901, (1964).
                                 ( 5 ,  2196) ACTA CRYST. 19,. . . . . . . . .  147, (1965).
                                 ( 5 ,  1566) J. PHYS. CHEM. SOL. 26,. . . . .  445, (1965).
                                 ( 5 ,   364) J. PHYS. CHEM. SOL. 26,. . . . .  927, (1965).
                                 ( 5 ,   444) Z. KRIST. VOL.121, . . . . . . .  418, (1965).
                                 ( 5 ,   329) J. AMER. CERAM. SOC. VOL.49, . .  284, (1966).
                                 ( 5 ,  1552) J. PHYS. CHEM. SOL. 27,. . . . . 1192, (1966).
                                 ( 5 ,  1820) J. PHYS. CHEM. SOL. 27,. . . . .  655, (1966).
                                 ( 5 ,   443) PHYS. STAT. SOL. 16, . . . . . .  K17, (1966).
                                 ( 5 ,   222) Z. KRIST. VOL.123, . . . . . . .   73, (1966).
                                 ( 5 ,  1567) ACTA CRYST. 22,. . . . . . . . .  344, (1967).
                                 ( 5 ,   563) J. APPL. PHYS. VOL.40, . . . . . 1124, (1969).

3640  SAPORETTI.......(F.)....:  (17A,    82) NUOVO CIMENTO SUPPL. VOL.3,. . .  187, (1965).

3641  SARDUTOVICH.....(G.I.)..:  ( 5 ,  1320) SOV. PHYS. J.E.T.P. VOL.21,. . .  733, (1965).

3642  SARMA...........(B.K.)..:  (10C,    62) PHYS. LETT. VOL.44A NO.7,. . . .  519, (1973).
                                 (10C,    63) PHYS. LETT. VOL.45A, . . . . . .  481, (1973).
                                 (10C,    64) PHYS. LETT. VOL.46A NO.7,. . . .  471, (1974).

3643  SARMA...........(G.)....:  ( 6 ,  1468) C.R. ACAD. SCI. VOL.248, . . . . 1631, (1959).
                                 ( 6 ,   956) IAEA SYMPOSIUM (VIENNA), . . . .  397, (1960).
                                 ( 6 ,   973) J. PHYS. RADIUM VOL.21,. . . . .  783, (1960).
                                 (10B,    32) IAEA SYMPOSIUM BOMBAY, VOL.1,.  313, (1964).
                                 ( 6 ,   836) SOL. STATE COMM. VOL.8,. . . . . 2141, (1970).

3644  SARMA...........(N.)....:  ( 6 ,   155) IAEA SYMPOSIUM (VIENNA), . . . .  631, (1960).

3645  SARNESKI........(J.E.)..:  ( 5 ,   649) Z. KRIST. VOL.134, . . . . . . .  308, (1971).
                                 ( 5 ,   661) INORG. NUCL. CHEM. LETT. VOL.9,: 1025, (1973).

3646  SASAKI..........(K.)....:  ( 8 ,    39) J. PHYSIQUE TOME 32 COL.1 VOL.II  812, (1971).

3647  SASS............(R.L.)..:  ( 5 ,  1941) ACTA CRYST. VOL.13,. . . . . . .  320, (1960).
                                 ( 5 ,   624) PHYS. REV. VOL.122,. . . . . . . 1402, (1961).

3648  SASS............(S.L.)..:  ( 5 ,  2974) ACTA CRYST. VOL.A29, . . . . . .  594, (1973).

3649  SAS.............(W.H.)..:  (17A,     8) ACTA CRYST. VOL.A25, . . . . . .  206, (1969).

3650  SATO............(H.)....:  ( 5 ,   956) J. APPL. PHYS. VOL.29, . . . . .  515, (1958).
                                 ( 5 ,   960) PHYS. REV. VOL.114,. . . . . . . 1420, (1959).
                                 ( 5 ,   959) PHYS. REV. VOL.114,. . . . . . . 1427, (1959).
                                 ( 5 ,   127) J. PHYS. CHEM. SOL. 27,. . . . .  413, (1966).
                                 ( 5 ,   133) J. APP. PHYS. VOL.40,. . . . . . 1373, (1969).
                                 ( 5 ,   719) MAGN. AND MAG. MATERIALS-1971,   508, (1972).
                                 ( 9B,    20) ACTA CRYST. VOL.A29, . . . . . .  372, (1973).
                                 ( 5 ,   721) A.I.P. CONF. PROC. NO.10 PART 1,  679, (1973).

3651  SATO............(K.)....:  ( 5 ,   558) J. PHYS. SOC. JAPAN VOL.20,. . . 2244, (1965).
                                 ( 5 ,   737) J. PHYS. SOC. JAPAN VOL.20,. . .  381, (1965).
                                 ( 5 ,   468) J. PHYS. SOC. JAPAN VOL.35,. . .  426, (1973).

3652  SATO............(S.)....:  ( 5 ,  1183) J. PHYS. SOC. JAP. VOL.31, . . .  452, (1971).
                                 ( 5 ,   659) J. PHYS. SOC. JAPAN VOL.32,. . . 1670, (1972).

3653  SATSUK..........(V.V.)..:  ( 6 ,  1671) SOV. PHYS. SOL. STATE VOL. 13, . 1793, (1972).
```

3654 SAUER...........(G.E.)..‡ (6 , 2048) DISS. ABS. VOL.31, ...68398, (1971).

3655 SAUKOV..........(A.I.)..‡ (5 , 1955) SOV. AT. ENERGY VOL.18,. . . . . 350, (1965).

3656 SAUNDERSON......(D.H.)..‡ (22A,   11) IAEA SYMP. (PILE RESEARCH)VIENNA 597, (1960).
                                (6 ,   58) IAEA SYMP. CHALK RIVER, VOL.1, . 413, (1962).
                                (6 , 2175) IAEA SYMP. CHALK RIVER, VOL.2, . 265, (1962).
                                (6 , 200) IAEA SYMP. CHALK RIVER, VOL.2, . 49, (1962).
                                (6 , 209) UKAEA AERE MEM.1199 (5 PP.). . . . . , (1965).
                                (18B,   12) UKAEA AERE REPORT R 4895 (22 PP). . . , (1965).
                                (6 , 1977) PHYS. REV. LETT. 17, ... 530, (1966).
                                (18A,   10) BRIT. J. APPL. PHYS. VOL.18, . . 473, (1967).
                                (23 ,   61) NUCL. INST. MET. VOL.59, ... 245, (1968).
                                (6 , 834) J. APPL. PHYS.41, ... 1433, (1970).
                                (6 , 1420) J. PHYS. C, SER.2, VOL.3,. . . .1026, (1970).
                                (6 , 445) CONF. INTERN. RENNES, ... 171, (1971).
                                (6 , 448) J. PHYS. C, SER.2, VOL.4,. . . .2009, (1971).
                                (6 ,   98) J. PHYS. C, SER.2, VOL.4,. . . . 910, (1971).
                                (6 , 1989) IAEA SYMP. GRENOBLE, ... 639, (1972).
                                (6 , 1777) IAEA SYMP. GRENOBLE, ... 669, (1972).
                                (6 , 1778) A.I.P. CONF. PROC. NO.10 PART 2,1403, (1973).
                                (5 , 2280) J. PHYS. C VOL.6,. . . . . L313, (1973).
                                (6 , 1196) J. PHYS. C VOL.6,. . . . . 212, (1973).
                                (6 , 1381) J. PHYS. C VOL.7,. . . . . L99, (1974).

3657 SAUR............(H.D.)..‡ (16 ,   78) REPORT REPT-6-49 (69PP.),. . . . . . . , (1970).

3658 SAUTER..........(G.D.)..‡ (20B,   82) NUCL. INST. MET. 55,. . . . . . 141, (1967).

3659 SAVAGE..........(H.)....‡ (5 , 2565) PHYS. LETT. VOL.47A NO.1,. . . . 73, (1974).

3660 SAVCHENKO.......(M.A.)..‡ (11A,   61) SOV. PHYS. J.E.T.P. 27,. . . . . 307, (1968).

3661 SAVEL≠EV........(V.YA.).‡ (20C,   84) PRIB. TEKH. EKSPER NO.1,. . . . 61, (1959).

3662 SAWENKO.........(B.N.)..‡ (6 , 1936) PHYS. STAT. SOL.45,. . . . . . . K105, (1971).

3663 SAWICKI.........(A.)....‡ (19B,   25) NUKLEONIKA VOL.8,. . . . . 695, (1963).

3664 SAWYER..........(R.B.)..‡ (5 , 1411) PHYS. REV. VOL.72, ... 109, (1947).

3665 SAYASOV.........(YU.S.).‡ (7B, 125) SOV. PHYS. JETP. 18,. . . . . .1006, (1964).
                                (15 ,   57) SOV. PHYS. USPEKHI VOL.9,. . . . 670, (1967).
                                (5 , 1308) SOV. PHYS. JETP. 27, ... 15, (1968).

3666 SAYETAT.........(F.)....‡ (5 , 595) COMP. REND. 269, ... 574, (1969).
                                (5 , 2568) SOL. STATE COMM. VOL.8,. . . . . 239, (1970).
                                (5 , 844) PHYS. LETT. VOL.34A,. . . . . . 361, (1971).

3667 SCALETTAR.......(R.)....‡ (10E,   8) IAEA SYMPOSIUM VIENNA, ... 101, (1960).

3668 SCARPETTA.......(G.)....‡ (5 , 2022) PHYS. REV. B VOL.9,. . . . . . . 130, (1974).

3669 SCATTURIN.......(V.)....‡ (5 , 1442) ACTA CRYST. VOL.14,. . . . . . . 19, (1961).
                                (5 , 739) J. PHYS. SOC. JAPAN VOL.17 B-II, 348, (1962).

3670 SCHAERPF........(O.)....‡ (5 , 2079) INT. J. MAGN. VOL.5, ... 223, (1973).

3671 SCHAFER.........(W.)....‡ (5 , 842) CONF./RARE EARTHS/ACTINIDES, . . 226, (1971).
                                (5 , 843) J. OF PHYS.-C-, SER.2, VOL.4,. . 811, (1971).
                                (5 , 817) J. PHYS. C, SER.2, VOL.4,. . . .3224, (1971).
                                (5 , 1844) INT. J. MAGN. VOL.5, ... 175, (1973).

3672 SCHAFFER........(P.C.)..‡ (6 , 2010) J. CHEM. PHYS. VOL.50, ...2140, (1969).
                                (6 , 1115) J. CHEM. PHYS. VOL.50, ...4444, (1969).

3673 SCHARENBERG.....(W.)....‡ (5 , 747) J. PHYSIQUE TOME 32 COL.1 VOL.II 855, (1971).

3674 SCHAWB..........(C.)....‡ (6 , 627) PHYS. REV. LETT. VOL.28, ... 964, (1972).

3675 SCHEDLER........(E.)....‡ (20B,   61) J. PHYS. E VOL.6 NO.5, ... 488, (1973).

3676 SCHEEL..........(H.J.)..‡ (6 , 1351) BULL. AMER. PHYS. SOC. VOL.18, . 111, (1973).
                                (6 , 1353) PHYS. REV. B VOL.8,. . . . . . .1119, (1973).

3677 SCHELBERG.......(A.D.)..‡ (24 , 104) REV. SCI. INSTRUM. VOL.41, . . .1900, (1970).

3678 SCHELTEN........(J.)....‡ (20C,   72) NUCL. INST. MET. VOL.83, ... 111, (1970).
                                (17C,   18) J. APPL. CRYST. VOL.4, ... 210, (1971).
                                (23 ,   42) J. APPL. CRYST. VOL.4, ... 410, (1971).
                                (5 ,   90) J. APPL. CRYST. VOL.4, ... 511, (1971).
                                (5 , 701) PHYS. STAT. SOL. A VOL.7,. . . . 469, (1971).
                                (5 , 700) PHYS. STAT. SOL. A VOL.7,. . . . 477, (1971).
                                (5 , 1998) PHYS. STAT. SOL. VOL.48,. . . . 619, (1971).
                                (6 , 1210) J. BIOL. CHEM. VOL.247,. . . . .5436, (1972).
                                (20A,   28) KERNTECHNIK VOL.14,. . . . . . . 86, (1972).
                                (20A,   29) KERNTECHNIK VOL.14,. . . . . . .   9, (1972).
                                (5 , 2005) Z. PHYSIK VOL.253, ... 219, (1972).
                                (5 , 2003) PHYS. STAT. SOLIDI B VOL.57 NO.2 515, (1973).
                                (5 , 2002) PHYS. STAT. SOLIDI B VOL.58 NO.2 633, (1973).
                                (5 , 2332) J. APPL. CRYST. VOL.7, ... 188, (1974).
                                (5 , 2010) J. APPL. CRYST. VOL.7, ... 236, (1974).
                                (17A,   52) J. APPL. CRYST. VOL.7, ... 96, (1974).
                                (5 , 2618) J. LOW TEMP. PHYS. VOL.14, . . . 213, (1974).

3679 SCHENK..........(C.)....‡ (6 , 1294) PHYS LETT 27A, ... 582, (1968).
                                (5 , 1470) PHYS. REV. VOL.172,. . . . . . . 576, (1968).
                                (5 , 1471) ACTA CRYST. A25, ... 514, (1969).

3680 SCHENTER........(R.E.)..‡ (7A,   38) NUCL. PHYS. A VOL.178, ... 249, (1971).

3681 SCHERBER........(W.)....‡ (24 ,   67) NUCL. INST. MET. 72, ... 301, (1969).

3682 SCHERMER........(R.I.)..‡ (6 , 476) PHYS. REV. 130,. . . . . .1907, (1963).
                                (8 ,   91) PHYS. REV. 166,. . . . . 554, (1968).
                                (5 , 1336) PHYS. REV. LETT. VOL.32, ... 791, (1974).

3683 SCHERM..........(R.)....‡ (6 , 175) Z. NATURFORSCH. VOL.19A, ... 354, (1964).
                                (20A,   82) UKAEA AERE TRANS 1074 (20 PP.),. . . , (1967).
                                (22A,   71) UKAEA AERE TRANS 1080 (15 PP.),. . . , (1967).
                                (6 , 174) NUKLEONIC VOL.12,. . . . .   4, (1968).
                                (19A,   21) IAEA SYMP. (INSTRUMENT.), VIENNA 117, (1969).
                                (6 , 1209) J. MOL. BIOL. VOL.41,. . . . . . 231, (1969).
                                (19B,   16) NUCL. INST. MET. VOL.80,. . . . 69, (1970).
                                (7B,   1) ANN. DE PHYSIQUE TOME 7 NO.5,. . 349, (1972).

3684 SCHIER..........(W.A.)..‡ (24 ,   75) NUCL. INST. MET. VOL.91, ... 13, (1971).

3685 SCHILLER........(E.)....‡ (5 , 812) PHYS. REV. LETT. VOL.32, ... 544, (1974).

3686 SCHILLING.......(W.)....‡ (6 , 588) PHYS. STAT. SOL. 4,. . . . . . . 95, (1964).
                                (10C, 137) Z. ANGEW. PHYS. VOL.18,. . . . . 295, (1965).

```
3687  SCHINDLER.......(A.I.)..:  ( 5 , 2302)  PHYS. REV. 137,. . . . . . . ., .  A483, (1965).
                                 ( 5 , 2307)  ENGELHARD IND., TECH. BULL. 7, .   21, (1966).

3688  SCHITTENHELM....(C.)....:  ( 6 , 1304)  PHYS. REV. 147,. . . . . . . .    577, (1966).

3689  SCHLECHT........(P.)....:  ( 6 , 1210)  J. BIOL. CHEM. VOL.247,. . . . . 5436, (1972).

3690  SCHLEMPER.......(E.O.)..:  ( 5 , 1942)  J. CHEM. PHYSICS VOL.44, . . . . 2499, (1966).
                                 ( 5 , 1937)  J. CHEM. PHYSICS VOL.44, . . . . 4498, (1966).
                                 ( 5 , 1943)  J. CHEM. PHYSICS VOL.45, . . . .  408, (1966).
                                 ( 5 , 2093)  J. CHEM. PHYS. VOL.54, . . . . . 3990, (1971).

3691  SCHLENKER.......(M.)....:  ( 5 , 1166)  J. APPL. PHYS. VOL.44, . . . . . 4181, (1973).

3692  SCHLUP..........(W.A.)..:  ( 5 ,   93)  BROOKHAVEN SYMPOSIUM., . . . . .  183, (1965).

3693  SCHMATZ.........(W.)....:  ( 6 ,  588)  PHYS. STAT. SOL. 4,. . . . . . .   95, (1964).
                                 ( 6 ,  175)  Z. NATURFORSCH. VOL.19A, . . . .  354, (1964).
                                 ( 6 ,  674)  PHYS. LETT. 15,. . . . . . . . .  231, (1965).
                                 (10C,  137)  Z. ANGEW. PHYS. VOL.18,. . . . .  295, (1965).
                                 (20A,    5)  ACTA CRYST. 20,. . . . . . . . .  311, (1966).
                                 ( 6 , 2031)  Z. PHYSIK VOL.190,. . . . . . .   295, (1966).
                                 ( 6 , 1740)  BULL. SOC. FRANC. MIN. CRIST. 90  428, (1967).
                                 (20A,   82)  UKAEA AERE TRANS 1074 (20 PP.),.       (1967).
                                 ( 6 ,   37)  Z. ANGEW. PHYS. VOL.24,NO.6,. .   313, (1968).
                                 ( 5 , 2262)  ACTA CRYST. (INTERACT.) VOL.A25, S211, (1969).
                                 ( 5 ,  681)  ACTA CRYST. (INTERACT.) VOL.A25, S218, (1969).
                                 ( 6 , 1209)  J. MOL. BIOL. VOL.41,. . . . . .  231, (1969).
                                 (20C,   72)  NUCL. INST. MET. VOL.83,. . . .   111, (1970).
                                 (17A,   20)  ATOMKERNENERGIE VOL.17,. . . . .   15, (1971).
                                 (23 ,   92)  J. APPL. CRYST. VOL.4,. . . . .   410, (1971).
                                 ( 5 ,   90)  J. APPL. CRYST. VOL.4,. . . . .   511, (1971).
                                 ( 6 , 1723)  J. PHYSIQUE TOME 32 COL.1 VOL.II  679, (1971).
                                 ( 7B,   38)  KERNTECHNIK VOL.13,. . . . . . .  525, (1971).
                                 ( 5 ,  701)  PHYS. STAT. SOL. A VOL.7,. . . .  469, (1971).
                                 ( 5 ,  700)  PHYS. STAT. SOL. A VOL.7,. . . .  477, (1971).
                                 ( 5 , 1998)  PHYS. STAT. SOL. VOL.48,. . . .   619, (1971).
                                 ( 6 , 1210)  J. BIOL. CHEM. VOL.247,. . . . . 5436, (1972).
                                 (20A,   29)  KERNTECHNIK VOL.14,. . . . . . .    9, (1972).
                                 ( 5 , 2265)  PHYS. STAT. SOLIDI A VOL.20,. .   109, (1973).
                                 ( 5 , 2002)  PHYS. STAT. SOLIDI B VOL.58 NO.2  633, (1973).
                                 ( 5 , 1065)  J. APPL. CRYST. VOL.7,. . . . .   233, (1974).
                                 (17A,   52)  J. APPL. CRYST. VOL.7, . . . . .   96, (1974).

3694  SCHMEISSNER.....(F.)....:  (17A,  101)  Z. ANGEW. PHYS. VOL.12,. . . . .  133, (1960).

3695  SCHMELZ.........(H.)....:  ( 5 ,  185)  PHYS. STAT. SOL. 31, . . . . . .  121, (1969).

3696  SCHMIDT.........(H.H.)..:  (22A,   31)  KERNTECHNIK VOL.14,. . . . . . .   17, (1972).

3697  SCHMIDT.........(H.)....:  (16 ,    2)  ATOMKERNENERGIE VOL.10,. . . . .  243, (1965).

3698  SCHMIDT.........(U.)....:  ( 6 ,  685)  ATOMKERNENERGIE VOL.12,. . . . .  385, (1967).

3699  SCHMIDT.........(W.F.)..:  (24 ,  115)  TEK. TIDSKR. (SWEDEN) VOL.103, .   37, (1973).

3700  SCHMITT.........(H.W.)..:  ( 5 ,  144)  NUCL. PHYS. VOL.17,. . . . . . .  109, (1960).

3701  SCHMUCK.........(PH.)...:  ( 6 , 1825)  PHYS. LETT. 28A, . . . . . . . .  226, (1968).

3702  SCHMUNK.........(R.E.)..:  ( 5 , 2869)  NUCL. SCI. ENGNG. VOL.7, . . . .  193, (1960).
                                 (19A,   32)  NUCL. INST. MET. VOL.12, . . . .  365, (1961).
                                 ( 6 ,  251)  PHYS. REV. VOL.124,. . . . . . .  460, (1961).
                                 ( 6 ,  146)  PHYS. REV. VOL.128,. . . . . . .  562, (1962).
                                 ( 6 ,  148)  IAEA SYMP. BOMBAY, VOL.1,. . . .  379, (1964).
                                 ( 6 ,  140)  PHYS. REV. VOL.136,. . . . . . . A1303, (1964).
                                 ( 6 , 2074)  PHYS. REV. LETT. VOL.14, . . . .   44, (1965).
                                 ( 6 ,  147)  PHYS. REV. 149,. . . . . . . . .  450, (1966).
                                 ( 6 ,  153)  IAEA SYMP. COPENHAGEN, VOL.1,. .  315, (1968).
                                 (22C,   63)  REV. SCI. INSTRUM. 40, . . . . . 1397, (1969).
                                 ( 6 , 1428)  J. PHYS. CHEM. SOLIDS VOL.31,. . 1085, (1970).
                                 ( 6 , 1633)  J. PHYS. CHEM. SOLIDS VOL.31,. .  131, (1970).
                                 ( 6 , 2334)  PHYS. STAT. SOL. VOL.42,. . . .   275, (1970).
                                 ( 6 , 2187)  PHYS. REV. B-3, . . . . . . . . 4115, (1971).

3703  SCHNEIDER.......(C.S.)..:  ( 6 ,  783)  PHYS. REV. B VOL.3,. . . . . . .  830, (1971).
                                 ( 8 ,  104)  PHYS. REV. B VOL.7,. . . . . . . 4142, (1973).
                                 (22A,   65)  REV. SCI. INSTRUM. VOL.44, . . . 1594, (1973).

3704  SCHNEIDER.......(J.)....:  (17A,   54)  J. CRYST. GROWTH VOL.13, . . . .  247, (1972).

3705  SCHNEIDER.......(R.)....:  ( 6 , 1209)  J. MOL. BIOL. VOL.41,. . . . . .  231, (1969).

3706  SCHNEIDER.......(T.)....:  ( 6 ,  473)  SOL. STATE COMM. 3,. . . . . . .  339, (1965).
                                 (10B,  111)  PHYS. KONDENS. MATERIE VOL.5,. .  331, (1966).
                                 (10B,  110)  PHYS. KONDENS. MATERIE VOL.5,. .  364, (1966).
                                 ( 6 ,  604)  SOL. STATE COMM. 4,. . . . . . .  443, (1966).
                                 ( 6 ,  589)  SOL. STATE COMM. 4,. . . . . . .   79, (1966).
                                 ( 6 , 1729)  SOL. STATE COMM. 4,. . . . . . .   99, (1966).
                                 ( 6 ,  585)  HELV. PHYS. ACTA VOL.41, . . . .  399, (1968).
                                 ( 6 ,   24)  IAEA SYMP. COPENHAGEN, VOL.1,. .  101, (1968).
                                 ( 6 ,  464)  IAEA SYMP. COPENHAGEN, VOL.2,. .  133, (1968).
                                 (14A,   66)  REPORT AF-SSP-27,. . . . . . . .       (1968).
                                 ( 6 ,   83)  PHYS. REV. LETT. VOL.25, . . . . 1423, (1970).
                                 ( 8 ,  125)  SOL. STATE COMM. VOL.8,. . . . .  279, (1970).
                                 ( 8 ,   35)  J. PHYS. C VOL.4,. . . . . . . . 1168, (1971).
                                 (14A,   53)  PHYS. REV. A VOL.3,. . . . . . . 2145, (1971).
                                 (10D,   41)  PHYS. REV. A VOL.5,. . . . . . . 1528, (1972).
                                 (13 ,   54)  PHYS. REV. B VOL.7,. . . . . . .  201, (1973).
                                 (13 ,   56)  PHYS. REV. B VOL.8,. . . . . . . 4422, (1973).

3707  SCHNEIDER.......(W.)....:  ( 5 ,  982)  SOL. STATE COMM. VOL.13, . . . .  303, (1973).

3708  SCHNEPP.........(O.)....:  ( 6 ,  426)  J. CHEM. PHYS. VOL.58, . . . . . 3647, (1973).

3709  SCHOBINGER-PAPAM(P.)....:  ( 5 , 2048)  J. PHYS. C VOL.6,. . . . . . . .  725, (1973).

3710  SCHOENBORN......(B.P.)..:  ( 5 , 1874)  NATURE VOL.224,. . . . . . . . .  143, (1969).
                                 ( 9C,    8)  BER. BUNSENGES. PHYS. CHEM. '74, 1202, (1970).
                                 (22C,    2)  ACTA CRYST. VOL.A27, . . . . . .  284, (1971).
                                 ( 5 , 1875)  SYMP. QUANT. BIOL. VOL.36,. . .   569, (1971).
                                 (24 ,    4)  ANN. REV. BIOPHYS. BIOENGNG. 1,.  529, (1972).

3711  SCHOFIELD.......(P.)....:  ( 7B,   14)  IAEA SYMPOSIUM VIENNA, . . . . .   39, (1960).
                                 ( 7B,   71)  PHYS. REV. LETT. VOL.4,. . . . .  239, (1960).
                                 (17A,   52)  NUCL. SCI. ENGNG. VOL.12,. . . .  260, (1962).
                                 (14A,   11)  CONTEMP. PHYS. VOL.6,. . . . . .  274, (1965).
                                 (14A,   12)  CONTEMP. PHYS. VOL.6,. . . . . .  453, (1965).
                                 (17B,   71)  UKAEA AERE REP.R5536 (41 PP.),.        (1967).
                                 (14A,   18)  IAEA SYMP. COPENHAGEN, VOL.1,. .  573, (1968).

3712  SCHOITSCH.......(E.)....:  (22B,   36)  NUCL. INST. MET. VOL.111,. . . .  375, (1973).
```

3713   SCHOLZ.........(J.)....:   (5 , 2447) Z. PHYSIK VOL.263, ...   291, (1973).

3714   SCHOMMERS.......(W.)...:   (6 , 1960) PROPERTIES LIQUID METALS,. . . .   111, (1973).

3715   SCHOPF.........(H.G.)..:   (6 , 919) PHYS. STAT. SOL. A VOL.4,. . . .   445, (1971).

3716   SCHOTT.........(W.)....:   (5 , 1242) NUCL. SCI. ENGNG. VOL.38,. . . .   180, (1969).
                                  (6 , 959) Z. PHYS. VOL.231,. . . . .   243, (1970).

3717   SCHROCKE.......(H.)....:   (5 , 1846) SOL. STATE COMM. VOL.12 NO.8,. .   779, (1973).

3718   SCHRODER.......(H.)....:   (5 , 1606) PHYS. STAT. SOL. A VOL.1,. . . .   749, (1970).

3719   SCHRODER.......(U.)....:   (10B, 172) PHYS. REV. LETT. VOL.28, ...   600, (1972).

3720   SCHROEDER......(L.W.)..:   (5 , 2245) J. CHEM. PHYS. VOL.54, ... 1968, (1971).
                                  (5 , 1993) J. CHEM. PHYS. VOL.55, ... 5363, (1971).
                                  (6 , 1289) J. CHEM. PHYS. VOL.56, ... 2793, (1972).
                                  (5 , 274) ACTA CRYST. VOL.B29, ...   184, (1973).

3721   SCHUCH.........(A.F.)..:   (5 , 769) ACTA CRYST. VOL.A28 PART S-4,. .   188, (1972).

3722   SCHULHOF.......(M.P.)..:   (6 , 1452) DISS. ABS. VOL.31, ...36598, (1970).
                                  (8 ,   20) INT. J. MAGN. VOL.1, ...   45, (1970).
                                  (6 , 1446) PHYS. REV. B-1, ... 2304, (1970).
                                  (6 , 1451) PHYS. REV. LETT. VOL.24, ... 1184, (1970).
                                  (6 , 815) J. APPL. PHYS. VOL.42, ... 1376, (1971).
                                  (6 , 1448) J. APPL. PHYS. 42, ... 1258, (1971).
                                  (6 , 1447) PHYS. REV. B-4, ... 2254, (1971).
                                  (13 ,   18) DYN. ASPECTS OF CRITICAL PHEN.,.   32, (1972).
                                  (6 , 817) PHYS. REV. B VOL.5,. . . . . . .   154, (1972).

3723   SCHULTZ........(H.L.)..:   (12A,   17) NUCL. INST. MET. VOL.98, ...   385, (1972).

3724   SCHULZE........(G.E.R.):   (9B,   86) PHYS. STAT. SOL. 2,. . . . . . . K104, (1962).
                                  (9D,   83) WISS. Z. TECH. UNIV/DRESDEN 12, 1159, (1963).
                                  (9B,   71) PHYS. LETT. 8, ...   241, (1964).
                                  (9C,   31) PHYS. LETT. VOL.14,. . . . . . .   174, (1965).
                                  (6 , 919) PHYS. STAT. SOL. A VOL.4,. . . .   445, (1971).

3725   SCHUMACHER.....(H.)....:   (5 , 2262) ACTA CRYST. (INTERACT.) VOL.A25, S211, (1969).
                                  (5 , 2265) PHYS. STAT. SOLIDI A VOL.20,. .   109, (1973).

3726   SCHUMAN........(R.P.)..:   (5 , 2508) NUCL. SCI. ENGNG. VOL.33, . . .   16, (1968).
                                  (5 , 2675) PROC/CONF/NEUTRON C. S. + TECH.,   693, (1968).

3727   SCHUREN........(H.)....:   (20C,   7) ATOMKERNENERGIE VOL.19 NO.4, . .   312, (1972).

3728   SCHUSTER.......(H.G.)..:   (10B, 117) PHYS. LETT. VOL.46A NO.3,. . . .   163, (1973).

3729   SCHUSTER.......(H.)....:   (6 , 1687) PHYS. LETT. VOL.35A, ...   31, (1971).
                                  (6 , 2266) PHYS. LETT. VOL.35A, ...   48, (1971).

3730   SCHWABL........(F.)....:   (8 ,   98) PHYS. REV. LETT. VOL.26, ... 1568, (1971).

3731   SCHWAB.........(C.)....:   (6 , 623) PHYS. REV. LETT. VOL.26, ...   770, (1971).
                                  (6 , 622) SOL. STATE COMM. VOL.13, ... 1725, (1973).

3732   SCHWARA........(R.)....:   (21 ,   26) J. PHYS. E VOL.6 NO.6, ...   576, (1973).

3733   SCHWARTZ.......(B.B.)..:   (8 ,   90) PHYS. REV. LETT. 21, ... 1744, (1968).
                                  (8 ,   38) J. PHYSIQUE TOME 32 COL.1 VOL.II 818, (1971).

3734   SCHWARTZ.......(L.H.)..:   (19A,   35) NUCL. INST. MET. 42, ...   81, (1966).
                                  (5 , 1619) PHYS LETT 28A, ...   267, (1968).
                                  (5 , 1641) J. APPL. PHYS. 41, ...   939, (1970).
                                  (5 , 1617) PHYS. REV. B-2, ...   670, (1970).
                                  (5 , 1632) J. APPL. PHYS. 42, ... 1621, (1971).

3735   SCHWARZENBACH...(D.)....:   (5 ,   42) J. PHYS. CHEM. SOLIDS VOL.32,. . 1641, (1971).
                                  (5 ,   43) J. PHYS. CHEM. SOLIDS VOL.32,. .   543, (1971).

3736   SCHWARZER......(J.)....:   (22B,   36) NUCL. INST. MET. VOL.111,. . . .   375, (1973).

3737   SCHWEISS.......(P.)....:   (6 , 765) HELV. PHYS. ACTA VOL.40, ...   378, (1967).

3738   SCHWEIZER......(J.)....:   (5 , 2037) J. PHYS. CHEM. SOL. 27,. . . . . 1287, (1966).
                                  (5 , 491) PHYS. REV. 141,. . . . .   538, (1966).
                                  (5 , 2554) J. PHYSIQUE TOME 28, ...   216, (1967).
                                  (5 , 2344) PHYS. LETT. VOL.24A, ...   739, (1967).
                                  (12A,   14) NUCL. INST. MET. 63, ...   283, (1968).
                                  (12B,   1) ACTA CRYST. (INTERACT.) VOL.A25, S255, (1969).
                                  (5 , 2883) J. APP. PHYS. 40,. . . . . 1454, (1969).
                                  (5 , 2881) PHYS. REV. VOL.186,. . . . . . .   479, (1969).
                                  (12B,   28) NUCL. INST. MET. VOL.95,. . . . .   589, (1971).
                                  (5 , 2189) PHYS. REV. LETT. VOL.32, ... 1257, (1974).

3739   SCHWINGER......(J.)....:   (8 ,   67) PHYS. REV. VOL.51, ...   544, (1937).
                                  (5 , 1251) PHYS. REV. VOL.52, ...   286, (1937).
                                  (7A,   61) PHYS. REV. VOL.58, ...   26, (1940).
                                  (6 , 660) PHYS. REV. VOL.69, ...   145, (1946).
                                  (5 , 1253) PHYS. REV. VOL.71, ...   678, (1947).
                                  (7A,   65) PHYS. REV. VOL.79, ...   469, (1950).

3740   SCHWINK........(CH.)...:   (5 , 2079) INT. J. MAGN. VOL.5, ...   223, (1973).

3741   SCHWOB.........(P.)....:   (5 , 914) PHYS. KONDENS. MAT. (GER.) VOL.9 249, (1969).

3742   SCIESINSKA......(E.)....:   (6 , 1387) ACTA PHYS. POLON. VOL.A44, . . .   731, (1973).

3743   SCIESINSKI......(J.)....:   (5 , 1286) IAEA SYMPOSIUM VIENNA, ...   297, (1960).
                                  (6 , 974) NUKLEONIKA VOL.5,. . . . .   495, (1960).
                                  (6 , 1552) IAEA SYMP. CHALK RIVER, VOL.1,   405, (1962).
                                  (19B,   24) NUKLEONIKA VOL.7,. . . . .   231, (1962).
                                  (6 , 1549) PHYSICA VOL.29,. . . . .   485, (1963).
                                  (6 , 1117) PHYSICA VOL.29,. . . . .   488, (1963).
                                  (21 ,   34) NUCL. INST. MET. 56, ...   305, (1967).
                                  (6 , 1082) PHYS. REV. 170,. . . . .   808, (1968).
                                  (6 , 321) PHYSICA VOL.48, ...   79, (1970).
                                  (6 , 1387) ACTA PHYS. POLON. VOL.A44,. . . .   731, (1973).

3744   SCIUTI.........(S.)....:   (23 ,   21) IAEA SYMP. (PILE RESEARCH)VIENNA   123, (1960).
                                  (23 ,   14) EVOLUTION OF PARTICLE PHYSICS,.   204, (1970).

3745   SEAGRAVE.......(J.D.)..:   (12B,   3) ANN. OF PHYSICS VOL.74,. . . . .   250, (1972).

3746   SEAL...........(P.F.)..:   (5 , 443) PHYS. STAT. SOL. 16, ...   K17, (1966).

3747   SEARS..........(D.R.)..:   (5 , 1976) ACTA CRYST. VOL.B25, ... 2519, (1969).

3748   SEARS..........(V.F.)..:   (14B,   98) PROC. PHYS. SOC. 86, ...   965, (1965).
                                  (14B,   14) CAN. J. PHYS. 44,. . . . . 1279, (1966).
                                  (14B,   15) CAN. J. PHYS. 44,. . . . . 1299, (1966).
                                  (14B,   16) CAN. J. PHYS. 44,. . . . .   867, (1966).
                                  (11C,   1) CAN. J. PHYS. 45,. . . . . 2923, (1967).
                                  (6 , 273) CAN. J. PHYS. 45,. . . . .   237, (1967).

```
3748  SEARS...........(V.F.)..:  (11C,    49)  PROC. PHYS. SOC. 92, . . . . . .   837, (1967).
                                 ( 7A,    24)  CAN. J. PHYS. 48,;. : : : : : :   616, (1973).
                                 ( 6 ,  1140)  PHYS. REV. A VOL.1,. : : : : : :  1699, (1970).
                                 ( 6 ,  1699)  IAEA SYMP. GRENOBLE, . . . . . .   399, (1972).
                                 (14B,    92)  PHYS. REV. A VOL.5,. . . . . .    452, (1972).
                                 ( 6 ,  1163)  PHYS. REV. LETT. VOL.29, . . . .  549, (1972).
                                 ( 6 ,  1166)  SOL. STATE COMM. VOL.11,. . . .  1327, (1972).
                                 (18A,    21)  NUCL. INST. MET. VOL.106,. . . .  419, (1973).
                                 (14B,    94)  PHYS. REV. A VOL.7 NO.1,. : : : :  340, (1973).

3749  SEDAGHAT........(A.K.)..:  ( 6 ,   904)  J. OF PHYS.-C-, SER.2, VOL.4,.. 3215, (1971).
                                 (119,    11)  J. PHYS. C VOL.6,. . . . . . . . 2350, (1973).

3750  SEDLAKOVA.......(L.)....:  (19A,    45)  NUCL. INST. MET. 69, . . . . . .  173, (1969).
                                 ( 5 ,  2372)  ACTA CRYST. VOL.A27, . . . . . .  410, (1971).
                                 ( 9B,    72)  PHYS. LETT. VOL.37A, . . . . . .  493, (1971).
                                 ( 6 ,  2052)  NATURE (PHYS. SCI.) VOL.242, . .  109, (1973).
                                 ( 5 ,  2449)  PHYS. STAT. SOLIDI A VOL.17 NO.1  163, (1973).

3751  SEEGER..........(A.)....:  (10C,   138)  Z. NATURFORSCH. VOL.14A, . . . .   74, (1959).
                                 (10C,   139)  Z. PHYS. VOL.171,. . . . . . . .  291, (1963).

3752  SEEGER..........(R.J.)..:  ( 7B,    55)  PHYS. REV. VOL.62, . . . . . . .   37, (1942).

3753  SEEMANN.........(K.W.)..:  ( 6 ,  1899)  NUCL. SCI. ENGNG. VOL.25,. . . .  300, (1966).
                                 ( 5 ,  2324)  PROC/CONF/NEUTRON C.S. + TECH.,   687, (1968).
                                 ( 6 ,  1890)  MOL. DYN. AND STRUCT. OF SOLIDS;  547, (1969).
                                 ( 6 ,  1038)  NUCL. SCI. ENG. VOL.46,. . . . .  223, (1971).
                                 ( 5 ,  2326)  NUCL. SCI. ENG. VOL.52,. . . . .  310, (1973).

3754  SEGAL...........(D.J.)..:  ( 5 ,  1552)  J. PHYS. CHEM. SOL. 27,. . . . . 1192, (1966).
                                 ( 5 ,   222)  Z. KRIST. VOL.123, . . . . . . .   73, (1966).

3755  SEGAL...........(Y.)....:  (20A,    38)  NUCL. INST. MET. 61, . . . . . .   93, (1968).

3756  SEGRE...........(E.)....:  (20B,   123)  PHYS. REV. VOL.48, . . . . . . .  704, (1935).
                                 (16 ,    66)  PROC. ROY. SOC. A VOL.149, . . .  522, (1935).

3757  SEIDEN..........(J.)....:  (13 ,    13)  C.R.ACAD.SC.(PARIS);TOME 260-B,  457, (1965).
                                 (13 ,    14)  C.R.ACAD.SC.(PARIS);TOME 260-B,  833, (1965).

3758  SEIDL...........(E.)....:  ( 9B,    91)  PHYS. STAT. SOLIDI A VOL.5 NO.2,  397, (1971).
                                 ( 5 ,   802)  Z. ANGEW PHYS. VOL.32, . . . . .  109, (1971).

3759  SEIDL...........(F.G.P.):  (23 ,    52)  NUCL. INSTRUM. VOL.1,. . . . . .   92, (1957).

3760  SEIFFERT........(W.D.)..:  ( 5 ,   767)  Z. NATURFORSCH. A VOL.25,. . . .  967, (1970).

3761  SEITZ...........(E.)....:  ( 5 ,  1626)  SOLID STATE COMM. VOL.11,. . . .  179, (1972).
                                 ( 5 ,  2265)  PHYS. STAT. SOLIDI A VOL.20; . .  109, (1973).

3762  SEITZ...........(F.)....:  ( 9B,    74)  PHYS. REV. VOL.71, . . . . . . .  294, (1947).

3763  SEKIYA..........(T.)....:  (10C,    40)  J. NUCL. SCI. TECHNOL. VOL.8,. .  319, (1971).
                                 ( 7B,    34)  J. NUCL. SCI. TECHNOL. VOL.8,. .  406, (1971).

3764  SELENGUT........(D.S.)..:  (16 ,    76)  RENSSELAER POL. INST. SYMP., . .  162, (1961).

3765  SELF............(A.G.)..:  ( 5 ,  1732)  J. PHYS. F VOL.3,. . . . . . . .    6, (1973).

3766  SELTE...........(K.)....:  ( 5 ,  1412)  ACTA CHEM. SCAND. VOL.22,. . . . 3039, (1968).
                                 ( 5 ,   561)  ACTA CHEM. SCAND. VOL.25,. . . . 1703, (1971).
                                 ( 5 ,   966)  ACTA CHEM. SCAND. VOL.26,. . . . 3101, (1972).

3767  SEMENOV.........(V.)....:  ( 5 ,  1092)  BULL. SOC. FRANC. MIN. CRIST. 89  206, (1966).

3768  SEMUEAL.........(B.S.)..:  ( 6 ,   584)  ACTA PHYS. ACAD. SCI. HUNG. 30,.  231, (1971).

3769  SENATEUR........(J.P.)..:  ( 5 ,  2102)  MATER. RES. BULL. VOL.8, . . . .  229, (1973).

3770  SENET...........(L.)....:  (22C,    66)  REV. SCI. INSTRUM. VOL.45, . . .  341, (1974).

3771  SENE............(R.)....:  (12A,    33)  REV. PHYS. APPL. (FRANCE) VOL.4,  115, (1969).
                                 (12B,    50)  REV. PHYS. APPL. (FRANCE) VOL.4,  254, (1969).

3772  SENGUPTA........(S.)....:  (10B,   185)  PHYS. REV. B VOL.8,. . . . . . . 2982, (1973).

3773  SEN.............(K.K.)..:  (16 ,     7)  CAN. J. PHYS. 43,. . . . . . . .  432, (1965).

3774  SEQUEIRA........(A.)....:  ( 5 ,  1425)  J. CHEM. PHYSICS VOL.41, . . . . 3616, (1964).
                                 (22B,    20)  NUCL. INST. MET. 26, . . . . . .  340, (1964).
                                 ( 5 ,  1514)  ACTA CRYST. 18,. . . . . . . . .  291, (1965).
                                 ( 5 ,  1513)  ACTA CRYST. 20,. . . . . . . . .  910, (1966).
                                 ( 5 ,  2246)  J. CHEM. PHYSICS VOL.47, . . . . 1818, (1967).
                                 ( 5 ,  1581)  J. PHYS. CHEM. SOL. 28,. . . . . 1693, (1967).
                                 ( 5 ,  1485)  ACTA CRYST. VOL.B24, . . . . . . 1176, (1968).
                                 (23 ,    40)  INDIAN J. PURE APPL. PHYS. VOL.6  550, (1968).
                                 ( 5 ,  1448)  J. MOLEC. STRUC. VOL.1,. . . . .  283, (1968).
                                 ( 5 ,  1551)  MOL. DYN. AND STRUCT. OF SOLIDS;  249, (1969).
                                 ( 6 ,  1234)  PHYS. STAT. SOL. 34, NO1,. . . .  279, (1969).
                                 ( 5 ,  1433)  ACTA CRYST. VOL.B26, . . . . . .  827, (1970).
                                 ( 5 ,  1503)  ACTA CRYST. VOL.B26, . . . . . .   77, (1970).
                                 ( 6 ,   118)  SOL. STATE COMM. VOL.8,. . . . .  497, (1970).
                                 ( 5 ,    18)  ACTA CRYST. VOL.A28 PART S-4,. .  193, (1972).
                                 ( 5 ,    19)  ACTA CRYST. VOL.B28, . . . . . . 2514, (1972).
                                 ( 5 ,    23)  ACTA CRYST. VOL.B28, . . . . . . 3214, (1972).
                                 ( 5 ,    16)  ACTA CRYST. VOL.B30, . . . . . .  562, (1974).

3775  SERMENT.........(V.)....:  ( 5 ,   113)  NUCL. APPL. TECHNOL. VOL.9,. . .  662, (1970).

3776  SERVOMAA........(A.)....:  ( 5 ,    50)  PHYS. SCRIPTA(SWEDEN) VOL.4, . .  125, (1971).

3777  SEYBOLT.........(A.U.)..:  ( 5 ,  2850)  ACTA MET. VOL.1, . . . . . . . .  390, (1953).

3778  SEYMOUR.........(R.S.)..:  ( 5 ,  1909)  ACTA CRYST. VOL.B26, . . . . . . 1487, (1970).
                                 ( 5 ,  1894)  ACTA CRYST. VOL.A27, . . . . . .  348, (1971).

3779  SHACHAR.........(G.)....:  ( 5 ,  2893)  SOL. STATE COMM. VOL.8,. . . . .  597, (1970).
                                 ( 5 ,  2676)  PHYS. REV. B VOL.3,. . . . . . . 2344, (1971).
                                 ( 5 ,   879)  PHYS. REV. B VOL.3,. . . . . . . 3861, (1971).
                                 ( 5 ,  1933)  SOL. STATE COMM. VOL. 9, . . . .  493, (1971).
                                 ( 5 ,  1557)  PHYS. REV. B VOL.6,. . . . . . . 1968, (1972).
                                 ( 5 ,   882)  PHYS. REV. B VOL.8,. . . . . . . 3398, (1973).

3780  SHAFER..........(M.W.)..:  ( 5 ,   916)  J. APPL. PHYS. VOL.35, . . . . .  984, (1964).
                                 ( 5 ,   917)  J. APP. PHYS. 37,. . . . . . . .  981, (1966).

3781  SHAFFER.........(J.C.)..:  ( 5 ,   132)  AIP CONF. PROC. VOL.5, . . . . . 1390, (1971).

3782  SHAFRAN.........(S.)....:  ( 5 ,  1072)  SOV. PHYS. SOL. STATE VOL.13,. .   44, (1971).

3783  SHAHABUDDIN.....(A.M.)..:  (20C,    80)  NUCL. SCI. APPL. B(PAKIS.) VOL.5   43, (1969).
```

```
3784  SHAH...........(J.S.)..?  ( 5 ,  830) SOL. STATE COMM. VOL.11, . . . .  1709, (1972).
                               ( 5 , 2633) IEEE TRANS. MAGN. VOL.9, . . . .   217, (1973).

3785  SHAKED.........(H.)....?  ( 5 ,  337) PHYS. LETT. 25A, . . . . . . . .     9, (1967).
                               ( 5 , 1987) PHYS. REV. 174, . . . . . . . .    560, (1968).
                               ( 5 , 2041) PHYS. LETT. 29A, . . . . . . . .   659, (1969).
                               ( 5 , 1600) PHYS. REV. B-1, . . . . . . . .   3116, (1970).
                               ( 5 , 2403) PHYS. REV. B VOL.2, . . . . . .   3703, (1970).
                               ( 5 , 1964) PHYS. REV. B VOL.2, . . . . . .    179, (1970).
                               ( 5 , 2401) PHYS. REV. LETT. VOL.25, . . .   1713, (1970).
                               ( 5 , 2893) SOL. STATE COMM. VOL.8, . . .     597, (1970).
                               ( 5 ,  664) BU. ISRAEL PHYS. SOC. 1971, . .     27, (1971).
                               ( 5 , 2676) PHYS. REV. B VOL.3, . . . . . .   2344, (1971).
                               ( 5 ,  879) PHYS. REV. B VOL.3, . . . . . .   3861, (1971).
                               ( 5 ,  663) PHYS. REV. B VOL.3, . . . . . .   3873, (1971).
                               ( 5 , 2402) PHYS. REV. B VOL.3, . . . . . .    821, (1971).
                               ( 5 , 1933) SOL. STATE COMM. VOL. 9, . . .     493, (1971).
                               ( 5 , 2043) ACTA CRYST. VOL.A28 PART S-4, .    194, (1972).
                               ( 5 , 1557) PHYS. REV. B VOL.6, . . . . . .   1968, (1972).
                               ( 5 , 2040) SOL. STATE COMM. VOL.10, . . .     663, (1972).
                               ( 5 , 2044) PHYS. REV. B VOL.7, . . . . . .   3261, (1973).
                               ( 5 ,  882) PHYS. REV. B VOL.9, . . . . . .   3398, (1973).
                               ( 5 , 1440) PHYS. REV. B VOL.9, . . . . . .   1071, (1974).

3786  SHAKH-BUDAGOV...(A.L.)..?  ( 6 ,   48) SOV. PHYS. SOL. STATE 7, . . . .  1138, (1965).
                               (21 ,   51) SOV. PHYS. SOL. STATE 7, . . . .  1146, (1965).
                               ( 6 , 2023) SOV. PHYS. SOL. STATE 9, . . . .   955, (1967).

3787  SHALDERVAN......(P.I.)..?  (10C,  134) UKR. FIZ. ZH. (USSR) VOL.9, . . .  1331, (1964).
                               (10C,  119) SOV. PHYS. SOL. STATE, 6, . . .    2618, (1965).

3788  SHALEV.........(S.)....?  (20C,   65) NUCL. INST. MET. 71, . . . . .     292, (1969).
                               (22B,   32) NUCL. INST. MET. 75, . . . . .     309, (1969).

3789  SHAMBROOM.......(O.)....?  ( 5 , 1262) PHYS. REV. LETT. VOL.26, . . .   1581, (1971).

3790  SHAMENKOVICH....(S.S.)..?  ( 5 , 1457) SOV. PHYS. SOL. STATE VOL.10, . .   675, (1968).

3791  SHAM...........(L.J.)..?  ( 6 , 1623) PROC. ROY. SOC. 283, . . . . . .     33, (1965).
                               (10B,  142) PHYS. REV. VOL.188, . . . . . .   1431, (1969).
                               (10D,   35) PHYS. REV. LETT. VOL.27, . . .    1725, (1971).
                               (10B,  175) PHYS. REV. B VOL.6, . . . . . .   3581, (1972).
                               ( 6 , 1690) PHYS. REV. B VOL.6, . . . . . .   3584, (1972).

3792  SHANG-KENG-MA...........?  (10B,  168) PHYS. REV. A VOL.3, . . . . . . .  1453, (1971).

3793  SHANKAR........(J.)....?  ( 9C,   13) CRYSTALLOGR./CRYSTAL PERFECTION,  269, (1963).

3794  SHANKS.........(H.R.)..?  ( 6 , 1426) PHYS. REV. B VOL.5, . . . . . . .  2103, (1972).

3795  SHANN..........(R.T.)..?  (12A,   21) NUOVO CIMENTO A VOL.5A, . . . . .   591, (1971).

3796  SHAPIRO........(C.S.)..?  (16 ,   60) PHYS. REV. VOL.137, . . . . . . . .A1686, (1965).

3797  SHAPIRO........(F.L.)..?  (20A,   19) INSTRUM. EXP. TECH. NO.1, . . .      58, (1966).
                               (12A,   31) PROC/CONF/POL. TARGETS/ION SOUR.    339, (1967).
                               (12B,   55) SOV. J. NUCL. PHYS. VOL.10, . .    1178, (1969).
                               (20D,  153) SOV. PHYS. JETP VOL.37, . . . .      41, (1973).

3798  SHAPIRO........(F.)....?  (19A,   59) NUKLEONIKA VOL.9, . . . . . . . .   523, (1964).

3799  SHAPIRO........(J.N.)..?  (10B,  151) PHYS. REV. B-1, . . . . . . . . .  3982, (1970).

3800  SHAPIRO........(S.M.)..?  ( 5 , 2497) CONF. INTERN., RENNES, . . . . .    155, (1971).
                               (20A,   44) NUCL. INST. MET. VOL.101, . . .     183, (1972).
                               ( 6 , 1312) PHYS. REV. B VOL.6, . . . . . .    4332, (1972).
                               ( 6 , 1674) SOL. STATE COMM. VOL.13, . . . .   1893, (1973).

3801  SHARAN.........(B.)....?  ( 6 , 1365) J. PHYS. F VOL.3, . . . . . . . .   1308, (1973).

3802  SHARAPOV.......(E.I.)..?  ( 5 , 2433) SOV. J. NUCL. PHYS. VOL.9, 6, . .   1119, (1969).

3803  SHARIPOVA......(L.S.)..?  (22A,   20) INSTRUM. EXPER. TECH. NO. 6, . .   1294, (1967).
                               ( 5 , 2256) PHYS. MET. METALLOGR. VOL.22 N.2    123, (1967).
                               (23 ,   76) SOV. PHYS. JOURNAL VOL.10 NO.4,     63, (1967).
                               ( 5 , 2477) PHYS. MET. METALLOGR. VOL.29, .     188, (1970).

3804  SHARMA.........(K.C.)..?  ( 6 ,  917) IAEA SYMP. BOMBAY, VOL.1, . . . .   285, (1964).

3805  SHARMA.........(P.K.)..?  ( 5 , 2732) ACTA CRYST. 16, . . . . . . . .     322, (1963).
                               (10B,   44) J. CHEM. PHYSICS VOL.39, . . . .   2633, (1963).
                               (10B,   46) J. CHEM. PHYSICS VOL.40, . . . .    662, (1964).
                               ( 6 , 2273) PHYS. LETT. 19, . . . . . . . .     105, (1965).
                               ( 6 ,  776) PHYS. STAT. SOL. 12, . . . . . .    305, (1965).
                               ( 6 ,  875) PHYSICA VOL.44, . . . . . . . .      69, (1969).
                               ( 6 , 2202) REV. ROUMAINE PHYS. VOL.14, . .     247, (1969).
                               ( 6 ,  584) ACTA PHYS. ACAD. SCI. HUNG. 30, .   231, (1971).
                               (10B,  162) PHYS. REV. B-4, . . . . . . . .    4636, (1971).
                               (10B,  107) PHYSICA VOL.59, . . . . . . . .     109, (1972).

3806  SHARMA.........(R.R.)..?  (10C,   57) NUCL. SCI. ENGNG. VOL.13, . . . .    40, (1962).

3807  SHARP..........(R.I.)..?  (18B,   12) UKAEA AERE REPORT R 4895 (22 PP)         (1965).
                               (18A,   10) BRIT. J. APPL. PHYS. VOL.18, . .    473, (1967).
                               ( 6 , 1667) J. OF PHYS.-C-, SER. 2, VOL.2, .    421, (1969).
                               ( 6 , 1666) J. OF PHYS.-C-, SER. 2, VOL.2, .    432, (1969).
                               ( 6 , 2011) J. PHYS.F? METAL PHYS. VOL.1, .     570, (1971).

3808  SHARRAH........(P.C.)..?  ( 5 , 2257) J. CHEM. PHYS. VOL.21, . . . . .    228, (1953).
                               ( 5 , 2263) J. CHEM. PHYS. VOL.32, . . . . .    241, (1960).

3809  SHATLOVSKAYA....(N.S.)..?  (12B,   18) INSTRUM. EXP. TECH. (USA) VOL.14   618, (1971).

3810  SHAW...........(W.M.)..?  ( 6 ,  550) PHYS. REV. B-2, . . . . . . . .    4864, (1970).
                               ( 6 , 1658) CONF. INTERN., RENNES, . . . . .    104, (1971).
                               ( 6 ,  517) DISS. ABS. VOL.32, . . . . . . .  17958, (1971).
                               ( 6 ,  515) PHYS. REV. B-4, . . . . . . . .     969, (1971).
                               ( 5 , 1219) Z. NATURFORSCH. VOL.28A, . . . .    657, (1973).

3811  SHCHEBETOV......(A.F.)..?  ( 6 , 1153) SOV. PHYS. TECH. PHYS. VOL.17, .    180, (1972).

3812  SHCHEPKIN......(YU.G.).?  ( 5 , 2478) SOV. J. NUCL. PHYS. VOL.10, 1, .     18, (1969).

3813  SHCHERBAK......(V.I.)..?  ( 5 , 2967) SOV. PHYS. CRYST. VOL.14, . . .    913, (1969).
                               ( 5 , 2427) SOV. PHYS. CRYST. VOL.17, . . .     822, (1971).
                               ( 5 , 2899) SOV. PHYS. CRYST. VOL.17, . . .     342, (1972).

3814  SHCHURIK.......(A.G.)..?  ( 5 , 2846) SOV. PHYS. DOKLADY VOL.15, . . .    776, (1971).

3815  SHCHUR.........(I.A.)..?  ( 6 , 1635) UKR. FIZ. ZH.(USSR) VOL.15, . .    1409, (1970).

3816  SHECHTER.......(H.)....?  ( 6 ,  571) PHYS. REV. B VOL.6, . . . . . .    1053, (1972).
```

```
3817  SHEEN...........(E.M.)...:  (24 ,    22)  IEEE TRANS. NUCL. SCI. VOL.17, .    138, (1970).
3818  SHEFTER.........(E.)....:  ( 5 ,   415)  J. PHARM. PHARMACOL. VOL.22, ..    724, (1970).
3819  SHEINKER........(M.E.)..:  ( 5 ,   636)  SOV. PHYS. SOL. STATE VOL. 13, .    644, (1971).
3820  SHEKHTER........(A.)....:  (20C,    81)  NUKLEONIKA VOL.8,. . . . . . . .    489, (1963).
3821  SHERIF..........(H.)....:  ( 7A,    46)  PHYS. CAN. VOL.27, . . . . . . .     70, (1971).
3822  SHERMAN.........(R.H.)..:  (12B,     3)  ANN. OF PHYSICS VOL.74,. . . . .    250, (1972).
3823  SHERWOOD........(R.C.)..:  ( 5 ,   662)  A.I.P. CONF. PROC. NO.10 PART 1,    684, (1973).
3824  SHEVALEEVSKY....(O.P.)..:  ( 5 , 1960)  SOL. STATE COMM. VOL.14 NO.4,. .    309, (1974).
3825  SHIBANOV........(YU.A.).:  (12B,    71)  YADERNAYA FIZ. (USSR) VOL.12,. .    815, (1970).
3826  SHIBUYA.........(I.)....:  ( 99,    57)  J. PHYS. SOC. JAPAN VOL.23,. . .    460, (1967).
                                ( 5 , 1980)  J. PHYS. SOC. JAPAN VOL.23,. . .    461, (1967).
                                ( 5 , 1464)  ACTA CRYST. VOL.A28 PART S-4,. .    181, (1972).
                                ( 5 , 2285)  J. PHYS. SOC. JAPAN VOL.35,. . .   1269, (1973).
                                ( 5 , 2284)  J. PHYS. SOC. JAPAN VOL.35,. . .    314, (1973).
3827  SHIELDS.........(K.G.)..:  ( 5 ,   657)  CRYST. STRUCT. COMM. VOL.1 N.3,.    189, (1972).
                                ( 5 ,   280)  CRYST. STRUCT. COMM. VOL.1 N.3,.    193, (1972).
                                ( 5 , 2394)  CRYST. STRUCT. COMM. VOL.1 N.4,.    367, (1972).
                                ( 5 , 2674)  CRYST. STRUCT. COMM. VOL.1 N.4,.    371, (1972).
3828  SHIELDS.........(M.)....:  ( 6 , 2080)  J. PHYS. SOC. JAPAN VOL.28, SUP.    249, (1970).
3829  SHIGA...........(M.)....:  ( 5 , 1070)  J. PHYS. SOC. JAPAN VOL.27,. . .   1470, (1969).
3830  SHILES..........(E.)....:  ( 8 ,   102)  PHYS. REV. B VOL.5,. . . . . . .   1915, (1972).
3831  SHIL≠SHTEIN.....(S.SH.).:  ( 5 , 1204)  SOV. PHYS. J.E.T.P. 28, . . . .    649, (1969).
                                ( 5 , 2012)  SOV. PHYS. CRYST. VOL.14,. . . .    522, (1970).
                                (17A,    46)  INSTRUM. EXP. TECH.(USA) VOL.14,    733, (1971).
                                ( 5 , 2837)  SOV. PHYS. SOL. STATE VOL. 13,.   2172, (1972).
                                ( 5 , 2835)  SOV. PHYS. SOL. STATE VOL. 13,.   2178, (1972).
                                ( 5 , 1167)  SOV. PHYS. JETP VOL.36,. . . . .   1170, (1973).
3832  SHIMAOKA........(K.)....:  (22C,     1)  ACTA CRYST. VOL. A25,. . . . . .   S-65, (1969).
                                (21 ,    18)  JAP. J. APPL. PHYS. VOL.10,. . .    933, (1971).
                                ( 5 , 1273)  J. PHYS. SOC. JAPAN VOL.32,. . .   1019, (1972).
3833  SHIMCHAK........(G.F.)..:  (20C,    14)  INSTRUM. EXP. TECH. NO.2,. . . .    243, (1963).
3834  SHIMIZU.........(M.)....:  ( 6 ,   833)  PHYS. REV. LETT. VOL.14, . . . .    698, (1965).
                                ( 8 ,    44)  J. PHYS. SOC. JAPAN VOL.25,. . .   1001, (1968).
                                (11A,    27)  J. PHYS. SOC. JAPAN VOL.28,. . .    327, (1970).
3835  SHINJO..........(T.)....:  ( 5 , 1691)  J. PHYS. SOC. JAPAN VOL.23,. . .   1426, (1967).
3836  SHINOZAKI.......(S.S.)..:  ( 5 ,   133)  J. APP. PHYS. VOL.40,. . . . . .   1373, (1969).
3837  SHIOZAKI........(Y.)....:  ( 5 , 1573)  J. PHYS. CHEM. SOL. 24,. . . . .   1057, (1963).
3838  SHIRANE.........(G.)....:  ( 5 , 2286)  PHYS. REV. VOL.97, . . . . . . .   1179, (1955).
                                ( 5 , 2287)  ACTA CRYST. VOL.9, . . . . . . .    131, (1956).
                                ( 5 , 2289)  PHYS. REV. VOL.105,. . . . . . .    849, (1957).
                                ( 5 ,   188)  PHYS. REV. VOL.105,. . . . . . .    856, (1957).
                                ( 90,     4)  ACTA CRYST. VOL.12,. . . . . . .    282, (1959).
                                ( 5 , 2083)  J. PHYS. CHEM. SOLIDS VOL.10,. .    138, (1959).
                                ( 5 , 1187)  J. PHYS. CHEM. SOLIDS VOL.10,. .     35, (1959).
                                ( 5 , 2199)  J. PHYS. SOC. JAPAN VOL.14,. . .   1352, (1959).
                                ( 5 , 2661)  J. PHYS. CHEM. SOLIDS VOL.13,. .    166, (1960).
                                ( 5 , 1726)  PHYS. REV. VOL.119,. . . . . . .    122, (1960).
                                ( 5 , 1568)  PHYS. REV. VOL.121,. . . . . . .    707, (1961).
                                ( 5 , 1102)  J. PHYS. CHEM. SOLIDS VOL.23,. .    863, (1962).
                                ( 5 ,   545)  J. PHYS. SOC. JAPAN VOL.17 B-III     35, (1962).
                                ( 5 , 1186)  J. PHYS. SOC. JAPAN VOL.17,. . .   1598, (1962).
                                ( 5 , 1724)  PHYS. REV. VOL.125,. . . . . . .   1893, (1962).
                                ( 5 , 1131)  J. APP. PHYS. 34,. . . . . . . .   1044, (1963).
                                ( 5 , 2564)  J. APP. PHYS. 34,. . . . . . . .   1352, (1963).
                                ( 5 ,   606)  J. PHYS. CHEM. SOL. 24,. . . . .    405, (1963).
                                ( 5 , 1809)  PHYS. REV. 129,. . . . . . . . .   2008, (1963).
                                ( 5 , 1560)  PHYS. REV. 132,. . . . . . . . .   1547, (1963).
                                ( 5 , 1077)  BULL. ACAD. SCI. USSR VOL.28,. .    359, (1964).
                                ( 5 , 1000)  J. APPL. PHYS. VOL.35,. . . . . .    954, (1964).
                                ( 5 , 2983)  PHYS. REV. LETT. VOL.12, . . . .    444, (1964).
                                ( 5 , 1125)  PHYS. REV. VOL.134,. . . . . . .  A1547, (1964).
                                ( 5 , 1122)  PROC. INT. CONF. (NOTTINGHAM), .    223, (1964).
                                ( 5 ,   622)  PROC. INT. CONF. (NOTTINGHAM), .    291, (1964).
                                ( 6 , 1735)  J. APP. PHYS. 36,. . . . . . . .   1076, (1965).
                                ( 9 , 1804)  J. APP. PHYS. 36,. . . . . . . .   1095, (1965).
                                ( 5 ,   604)  J. APP. PHYS. 36,. . . . . . . .   1099, (1965).
                                ( 5 ,   448)  PHYS. LETT. 17,. . . . . . . . .    103, (1965).
                                ( 5 ,   698)  PHYS. LETT. 17,. . . . . . . . .     95, (1965).
                                ( 6 ,   751)  PHYS. REV. LETT. 15, . . . . . .    146, (1965).
                                ( 5 , 2760)  PHYS. REV. 140,. . . . . . . . .  A1448, (1965).
                                ( 5 , 1803)  PHYS. REV. 140,. . . . . . . . .  A2139, (1965).
                                ( 5 , 1384)  J. APP. PHYS. 37,. . . . . . . .   1032, (1966).
                                ( 6 ,   478)  J. APP. PHYS. 37,. . . . . . . .   1052, (1966).
                                ( 5 ,   450)  J. APP. PHYS. 37,. . . . . . . .   1126, (1966).
                                ( 5 ,   632)  J. APP. PHYS. 37,. . . . . . . .    973, (1966).
                                ( 5 ,   127)  J. PHYS. CHEM. SOL. 27,. . . . .    413, (1966).
                                ( 5 , 1075)  PHYS. LETT. VOL.22,. . . . . . .    407, (1966).
                                ( 5 , 1151)  PHYS. REV. 147,. . . . . . . . .    415, (1966).
                                ( 5 ,   161)  J. APPL. PHYS. VOL.38, . . . . .   1459, (1967).
                                ( 5 ,   697)  J. APPL. PHYS. VOL.38, . . . . .   1461, (1967).
                                ( 5 , 2530)  J. APPL. PHYS. VOL.38, . . . . .    969, (1967).
                                ( 6 , 1349)  PHYS. REV. LETT. 18, . . . . . .    548, (1967).
                                ( 6 , 1185)  PHYS. REV. LETT. 19, . . . . . .   1307, (1967).
                                ( 6 ,   128)  PHYS. REV. LETT. 19, . . . . . .    234, (1967).
                                ( 6 ,   841)  PHYS. REV. 154,. . . . . . . . .    508, (1967).
                                ( 6 , 1738)  PHYS. REV. 156,. . . . . . . . .    623, (1967).
                                ( 6 , 1342)  PHYS. REV. 157,. . . . . . . . .    396, (1967).
                                ( 6 , 2130)  PHYS. REV. 161,. . . . . . . . .    499, (1967).
                                ( 6 ,   778)  PHYS. REV. 162,. . . . . . . . .    528, (1967).
                                ( 5 , 1438)  SOLID STATE COMM. VOL. 5,. . . .    591, (1967).
                                (11B,     1)  C.N.E.N. SYMP. CASACCIA, . . . .    185, (1968).
                                ( 6 , 1728)  J. APP. PHYS. 39,. . . . . . . .    383, (1968).
                                ( 6 ,   829)  J. APP. PHYS. 39,. . . . . . . .    455, (1968).
                                ( 5 , 1476)  J. PHYS. CHEM. SOLIDS 29,. . . .    881, (1968).
                                ( 5 ,   528)  J. PHYS. SOC. JAPAN VOL.24,. . .    368, (1968).
                                ( 6 ,   778)  KYOTO CONF. STAT. MECHANICS, . .    169, (1968).
                                ( 6 ,   752)  PHYS. REV. LETT. 21, . . . . . .     99, (1968).
                                ( 5 , 1356)  PHYS. REV. VOL.165,. . . . . . .    688, (1968).
                                ( 5 ,   696)  PHYS. REV. VOL.167,. . . . . . .    519, (1968).
                                ( 5 , 2690)  SOL. STATE COMM. VOL.6,. . . . .     15, (1968).
                                ( 6 ,   734)  J. APP. PHYS. VOL.40,. . . . . .   1442, (1969).
                                ( 6 , 2095)  J. PHYS. SOC. JAPAN VOL.26,. . .    396, (1969).
                                ( 6 , 1310)  J. PHYS. SOC. JAPAN VOL.26,. . .    674, (1969).
                                ( 6 ,   738)  J. PHYS. SOC. JAPAN (SUPPL.) 26,    169, (1969).
                                ( 6 , 1328)  PHYS. REV. LETT. VOL.23, . . . .   1394, (1969).
```

```
3838  SHIRANE.........(G.)....:  ( 5 , 1493) PHYS. REV. LETT. 22, . . . . . . .     720, (1969).
                                 ( 6 ,  139) PHYS. REV. VOL.177, . . . . . . .      848, (1969).
                                 ( 6 , 2107) PHYS. REV. VOL.177, . . . . . . .      858, (1969).
                                 ( 6 ,  794) PHYS. REV. VOL.179, . . . . . . .      417, (1969).
                                 ( 6 , 1702) PHYS. REV. VOL.181, . . . . . . .     1251, (1969).
                                 ( 6 , 1718) PHYS. REV. VOL.182, . . . . . . .      624, (1969).
                                 ( 6 , 1352) PHYS. REV. VOL.183, . . . . . . .      820, (1969).
                                 ( 6 , 1539) PHYS. REV. VOL.184, . . . . . . .      550, (1969).
                                 ( 5 , 1714) PHYS. REV. VOL.188, . . . . . . .     1037, (1969).
                                 ( 5 , 1434) PHYS. REV. VOL.188, . . . . . . .      919, (1969).
                                 ( 5 ,  368) PHYS. REV. VOL.188, . . . . . . .      930, (1969).
                                 ( 6 ,  546) SOL. STATE COMM. 7, NO15, . . . .     1043, (1969).
                                 ( 6 , 1222) SOL. STATE COMM. 7, . . . . . . .      535, (1969).
                                 (10C,   54) 2ND INTERNAT. MEET./FERROELECTR.        5, (1969).
                                 ( 6 , 1341) ACTA CRYST. VOL.A26, . . . . . . .     6C8, (1970).
                                 ( 6 , 1331) J. APPL. PHYS. VOL.41, . . . . . .     1303, (1970).
                                 ( 5 , 1620) J. PHYS. SOC. JAPAN VOL.28, . . .       815, (1970).
                                 (20D,   25) NUCL. INST. MET. VOL.89, . . . . .      109, (1970).
                                 ( 6 ,  543) PHYSICA VOL.48, . . . . . . . . .        13, (1970).
                                 ( 6 , 1340) PHYS. REV. B-1, . . . . . . . . .     1227, (1970).
                                 ( 6 , 1327) PHYS. REV. B-1, . . . . . . . . .     2211, (1970).
                                 ( 6 , 1278) PHYS. REV. B-1, . . . . . . . . .      278, (1970).
                                 ( 6 , 2054) PHYS. REV. B-1, . . . . . . . . .      342, (1970).
                                 ( 5 ,  666) PHYS. REV. B-2, . . . . . . . . .     1310, (1970).
                                 ( 6 ,  124) PHYS. REV. B-2, . . . . . . . . .     3651, (1970).
                                 ( 6 ,  571) PHYS. REV. B-2, . . . . . . . . .     4632, (1970).
                                 ( 6 , 1846) PHYS. REV. B-2, . . . . . . . . .      155, (1970).
                                 ( 6 ,  837) PHYS. STAT. SOL. VOL.42, . . . . .      241, (1970).
                                 ( 5 ,  670) SOL. STATE COMM. VOL.8, . . . . .     1001, (1970).
                                 ( 5 , 1479) SOL. STATE COMM. VOL.8, . . . . .     1941, (1970).
                                 ( 5 , 2497) CONF. INTERN., RENNES, . . . . . .     155, (1971).
                                 ( 6 , 1336) FERROELECTRICS(GB) VOL.2, . . . .      261, (1971).
                                 ( 6 ,  237) J. APPL. PHYS. 42, . . . . . . .      1265, (1971).
                                 ( 6 ,  514) J. APPL. PHYS. 42, . . . . . . .      1666, (1971).
                                 ( 6 , 1329) J. PHYSIQUE TOME 32 COL.1 VOL.II    882, (1971).
                                 ( 5 , 2391) J. PHYSIQUE TOME 32 COL.1 VOL.II    892, (1971).
                                 (18A,   19) NUCL. INST. MET. VOL.95, . . . . .      445, (1971).
                                 ( 6 , 1330) PHYS. REV. B VOL.3, . . . . . . .     1736, (1971).
                                 ( 6 ,  531) PHYS. REV. B VOL.3, . . . . . . .      157, (1971).
                                 ( 5 , 2020) PHYS. REV. B VOL.4, . . . . . . .     2957, (1971).
                                 ( 5 ,  129) PHYS. REV. B VOL.4, . . . . . . .      155, (1971).
                                 ( 6 , 2155) PHYS. REV. LETT. VOL.26, . . . . .      519, (1971).
                                 ( 9D,   62) PHYS. REV. LETT. VOL.26, . . . . .      718, (1971).
                                 ( 6 , 1689) PHYS. REV. LETT. VOL.27, . . . . .     1803, (1971).
                                 ( 6 , 2091) SOL. STATE COMM. VOL. 9, . . . . .     1103, (1971).
                                 ( 6 , 2265) SOL. STATE COMM. VOL.17, . . . . .      397, (1971).
                                 ( 6 ,  201) BULL. AMER. PHYS. SOC. VOL.17, . .      123, (1972).
                                 ( 6 , 1143) PHYS. REV. A VOL.5, . . . . . . .     1537, (1972).
                                 ( 6 , 1343) PHYS. REV. B VOL.5, . . . . . . .     1866, (1972).
                                 ( 6 , 1317) PHYS. REV. B VOL.5, . . . . . . .     1933, (1972).
                                 ( 6 ,  238) PHYS. REV. B VOL.5, . . . . . . .     1999, (1972).
                                 ( 6 , 2156) PHYS. REV. B VOL.6, . . . . . . .     1950, (1972).
                                 ( 6 , 1312) PHYS. REV. B VOL.6, . . . . . . .     4332, (1972).
                                 ( 6 , 1701) PHYS. REV. B VOL.6, . . . . . . .     4766, (1972).
                                 ( 6 , 1162) PHYS. REV. LETT. VOL.29, . . . . .      552, (1972).
                                 ( 6 , 1469) A.I.P. CONF. PROC. NO.10 PART 1,       93, (1973).
                                 ( 6 , 1351) BULL. AMER. PHYS. SOC. VOL.18, . .      111, (1973).
                                 ( 6 , 1189) PHYS. REV. A VOL.8, . . . . . . .     1513, (1973).
                                 ( 6 , 1353) PHYS. REV. B VOL.8, . . . . . . .     1119, (1973).
                                 ( 6 , 1691) PHYS. REV. B VOL.8, . . . . . . .     1965, (1973).
                                 ( 5 , 1483) PHYS. REV. B VOL.8, . . . . . . .      304, (1973).
                                 ( 6 , 1688) PHYS. REV. LETT. VOL.30, . . . . .      214, (1973).
                                 ( 6 , 1934) PHYS. REV. LETT. VOL.31, . . . . .     1300, (1973).
                                 (13 ,   58) PHYS. TODAY SEPT., 1973, . . . . .       32, (1973).
                                 ( 6 ,  826) SOL. STATE COMM. VOL.13, . . . . .     1179, (1973).
                                 ( 6 , 1674) SOL. STATE COMM. VOL.13, . . . . .     1893, (1973).
                                 ( 6 , 2281) PHYS. REV. B VOL.9 NO.12, . . . .      304, (1974).
                                 ( 6 , 1350) PHYS. REV. B VOL.9, . . . . . . .     1797, (1974).
                                 ( 6 , 2282) PHYS. REV. B VOL.9, . . . . . . .     2661, (1974).
                                 ( 6 , 1529) PHYS. REV. B VOL.9, . . . . . . .     4429, (1974).
                                 ( 6 ,  574) PHYS. REV. B VOL.9, . . . . . . .     4549, (1974).
                                 ( 6 ,  625) PHYS. REV. LETT. VOL.32 NO.4, . .      170, (1974).

3839  SHI-JIE.........(GU.)...:  ( 8 ,    3) ACTA PHYS. SINICA VOL.19, . . . .      673, (1963).

3840  SHKATULA........(A.A.)..:  (20C,   14) INSTRUM. EXP. TECH. NO.2, . . . .      243, (1963).
                                 (20A,   19) INSTRUM. EXP. TECH. NO.1, . . . .       58, (1966).

3841  SHOEMAKER.......(C.B.)..:  ( 5 , 1868) ACTA CRYST. 18, . . . . . . . . .       37, (1965).

3842  SHOEMAKER.......(D.P.)..:  ( 5 , 2841) J. PHYS. SOC. JAPAN VOL.17 B-III       16, (1962).
                                 ( 5 , 1868) ACTA CRYST. 18, . . . . . . . . .       37, (1965).
                                 (22B,    3) ACTA CRYST. A24, . . . . . . . . .      136, (1968).
                                 ( 5 , 2842) J. PHYS. CHEM. SOLIDS 29, . . . .      184, (1968).

3843  SHOOK...........(D.)....:  (22A,   59) REPORT NASA-TM-X-52828 (11PP.), . . . ,  (1970).

3844  SHORE...........(F.J.)..:  (20A,   76) REV. SCI. INSTRUM. VOL.30, . . .       17, (1959).
                                 (20B,  144) REV. SCI. INSTRUM. VOL.32, . . .      654, (1961).

3845  SHRADER.........(E.F.)..:  (17B,   46) NUCL. INST. MET. VOL.85, . . . .      151, (1970).

3846  SHTRANIKH.......(I.V.)..:  ( 6 ,  920) BU. ACAD. SCI. USSR PHYS. SR. 33  1754, (1969).
                                 (17A,   22) BU. ACAD. SCI. USSR PHYS. SR. 33    36, (1969).

3847  SHTRIKMAN.......(S.)....:  ( 5 ,  337) PHYS. LETT. 25A, . . . . . . . .        9, (1967).
                                 ( 5 ,  879) PHYS. REV. B VOL.3, . . . . . . .     3861, (1971).
                                 ( 5 , 2402) PHYS. REV. B VOL.3, . . . . . . .      821, (1971).

3848  SHUKLA..........(M.M.)..:  (10B,  190) PHYS. STAT. SOL. 7, . . . . . . .      K11, (1964).
                                 ( 6 ,   51) PHYS. STAT. SOL. 16, . . . . . . .      513, (1966).
                                 ( 6 , 1751) PHYS. STAT. SOL. 19, . . . . . . .      729, (1967).
                                 ( 6 , 1610) SOL. STATE COMM. VOL.11, . . . . .     1431, (1972).
                                 (10B,   87) J. PHYS. F VOL.3, . . . . . . . .      L83, (1973).
                                 ( 6 , 1619) J. PHYS. F VOL.3, . . . . . . . .      L99, (1973).
                                 ( 6 ,  586) J. PHYS. F VOL.3, . . . . . . . .       L1, (1973).
                                 ( 6 , 1818) J. PHYS. SOC. JAPAN VOL.34, . . .     1002, (1973).
                                 (10B,  102) NUOVO CIM. VOL.17B NO.1, . . . . .      166, (1973).
                                 ( 6 ,  592) PHYS. LETT. VOL.43A NO.5, . . . .      429, (1973).
                                 (10B,   77) J. PHYS. CHEM. SOLIDS VOL.35, . .      123, (1974).
                                 (10B,  109) PHYSICA VOL.72, . . . . . . . . .      179, (1974).

3849  SHULL...........(C.G.)..:  ( 9A,   19) NUCLEONICS VOL.3 NO.1, . . . . .        8, (1948).
                                 ( 9A,   18) NUCLEONICS VOL.3 NO.2, . . . . .       17, (1948).
                                 ( 9B,   75) PHYS. REV. VOL.73, . . . . . . .      527, (1948).
                                 ( 5 , 1985) PHYS. REV. VOL.73, . . . . . . .      830, (1948).
                                 ( 5 , 1972) PHYS. REV. VOL.73, . . . . . . .      842, (1948).
                                 ( 9A,   24) SCIENCE VOL.108, . . . . . . . .       69, (1948).
                                 ( 5 , 2161) PHYS. REV. VOL.75, . . . . . . .     1008, (1949).
                                 ( 5 ,  788) PHYS. REV. VOL.75, . . . . . . .     1348, (1949).
                                 ( 5 , 1752) PHYS. REV. VOL.76, . . . . . . .     1256, (1949).
                                 ( 5 , 2080) PHYS. REV. VOL.79, . . . . . . .      395, (1950).
                                 ( 5 , 1081) PHYS. REV. VOL.81, . . . . . . .      483, (1951).
                                 ( 5 , 2260) PHYS. REV. VOL.81, . . . . . . .      527, (1951).
                                 (12B,   40) PHYS. REV. VOL.81, . . . . . . .      626, (1951).
                                 ( 5 , 1753) PHYS. REV. VOL.83, . . . . . . .      333, (1951).
```

3849 SHULL.........(C.G.)..: (5 , 1967) PHYS. REV. VOL.83, ... 760, (1951).
(5 , 1085) PHYS. REV. VOL.84, ... 912, (1951).
(5 , 2968) ACTA CRYST. VOL.5, ... 22, (1952).
(5 , 2828) REV. MOD. PHYS. VOL.25, ... 100, (1953).
(9B, 2135) PHYS. REV. VOL.97, ... 304, (1955).
(5 , 81) PHYS. REV. VOL.97, ... 889, (1955).
(5 , 942) PHYS. REV. VOL.103, ... 516, (1956).
(6 , 768) PHYS. REV. VOL.103, ... 525, (1956).
(9A, 25) SOL. STATE PHYS. VOL.2, ... 137, (1956).
(9A, 1806) J. PHYS. CHEM. SOLIDS VOL.2, . . 289, (1957).
(5 , 2309) J. PHYS. CHEM. SOLIDS VOL.3, . . 303, (1957).
(5 , 958) J. PHYS. CHEM. SOLIDS VOL.6, . . 38, (1958).
(5 , 2083) J. PHYS. CHEM. SOLIDS VOL.10, . 138, (1959).
(9D, 47) J. PHYS. RADIUM VOL.20, ... 169, (1959).
(5 , 2662) PHYS. REV. VOL.118, ... 797, (1960).
(5 , 934) J. PHYS. SOC. JAPAN VOL.17 B-III 1, (1962).
(20C, 30) J. PHYS. SOC. JAPAN VOL.17 B-II, 340, (1962).
(9B, 102) REV. SCI. INSTRUM. VOL.33, . . . 126, (1962).
(5 , 2826) PHYS. REV. LETT. VOL.10, ... 297, (1963).
(6 , 2234) PHYS. REV. LETT. 10, ... 295, (1963).
(17C, 4) ACTA CRYST. 17, ... 805, (1964).
(5 , 936) J. APP. PHYS. 35, ... 678, (1964).
(6 , 1736) J. APP. PHYS. 37, ... 1034, (1966).
(5 , 935) PHYS. REV. LETT. VOL.16, ... 184, (1966).
(5 , 2855) PHYS. REV. LETT. VOL.16, ... 513, (1966).
(8 , 14) ATLANTA SYMP/GEORGIA INST. TECH. 1, (1967).
(5 , 425) METAL. SOC. CONF. LOS ANGELES, . 15, (1967).
(24 , 93) PHYS. REV. VOL.153, ... 1415, (1967).
(17C, 10) ACTA CRYST. A24, ... 160, (1968).
(9B, 35) C.N.E.N. SYMP. CASACCIA, ... 197, (1968).
(5 , 2437) PHYS. REV. LETT. 21, ... 1585, (1968).
(6 , 471) PHYS. REV. VOL.185, ... 961, (1969).
(9B, 84) PHYS. REV. 179, ... 752, (1969).
(5 , 712) J. APPL. PHYS. VOL.41, ... 1146, (1970).
(6 , 783) PHYS. REV. B VOL.3, ... 830, (1971).
(5 , 711) PHYS. REV. B VOL.5, ... 1040, (1972).
(5 , 2440) PHYS. REV. LETT. VOL.29, ... 873, (1972).
(5 , 2448) J. APPL. CRYST. VOL.6, ... 257, (1973).
(5 , 1166) J. APPL. PHYS. VOL.44, ... 4181, (1973).
(8 , 104) PHYS. REV. B VOL.7, ... 4142, (1973).
(5 , 1219) Z. NATURFORSCH. VOL.28A, ... 657, (1973).

3850 SHVEIKIN........(G.P.)..: (5 , 1543) SOV. PHYS. SOL. STATE VOL.15, . 1079, (1973).

3851 SIAUD...........(E.)....: (5 , 827) J. PHYSIQUE TOME 32 COL.1 VOL.II 1126, (1971).

3852 SIDEY...........(G.R.)..: (6 , 87) J. PHYS. C, SER.2, VOL.3, . . . 249, (1970).

3853 SIDHU...........(S.S.)..: (5 , 2864) J. APPL. PHYS. VOL.19, ... 639, (1948).
(5 , 2958) PHYS. REV. VOL.75, ... 975, (1949).
(5 , 2654) ACTA CRYST. VOL.9, ... 607, (1956).
(5 , 2664) J. APPL. PHYS. VOL.27, ... 1040, (1956).
(5 , 722) PHYS. REV. VOL.105, ... 130, (1957).
(5 , 2669) PHYS. REV. VOL.105, ... 216, (1957).
(9B, 51) J. APPL. PHYS. VOL.30, ... 1323, (1959).
(22C, 53) REV. SCI. INSTRUM. VOL.34, . . . 74, (1963).
(5 , 1577) ACTA CRYST. 18, ... 906, (1965).
(5 , 2359) ACTA CRYST. 19, ... 413, (1965).
(5 , 2802) J. PHYS. CHEM. SOLIDS VOL.27, . 1197, (1966).
(5 , 2077) PHYS. REV. VOL.156, ... 1225, (1967).

3854 SIDOROV........(S.K.)..: (5 , 2159) PHYS. MET. METALLOGR. VOL.20 N.6 48, (1965).
(5 , 2155) PHYS. STAT. SOL. VOL.16, ... 737, (1966).
(5 , 2649) PHYS. MET. METALLOGR. VOL.23 N.3 168, (1967).
(5 , 1115) PHYS. STAT. SOL. 24, ... 385, (1967).
(6 , 868) SOV. PHYS. JETP LETT. VOL.16, . 6, (1972).
(5 , 2133) SOV. PHYS. JETP VOL.34, ... 163, (1972).
(5 , 1059) SOV. PHYS. JETP VOL.34, ... 799, (1972).
(5 , 2163) PHYS. STAT. SOLIDI A VOL.21 NO.1 K31, (1974).

3855 SIEGEL..........(S.)....: (5 , 2161) PHYS. REV. VOL.75, ... 1008, (1949).

3856 SIEKER..........(L.C.)..: (5 , 2504) ACTA CRYST. VOL.B29, ... 797, (1973).

3857 SIEMINSKI.......(M.)....: (12B, 27) NUCL. INST. MET. 64, ... 77, (1968).

3858 SIEWERT.........(C.E.)..: (16 , 49) NUCL. SCI. ENG. VOL.47, ... 156, (1972).

3859 SIGAEV..........(V.N.)..: (5 , 2466) PHYS. STAT. SOLIDI A VOL.18 NO.2 K91, (1973).
(5 , 2467) J. APPL. CRYST. VOL.7, ... 207, (1974).

3860 SIGMAR..........(D.J.)..: (7B, 19) IAEA SYMPOSIUM BOMBAY, VOL.2, . 59, (1964).
(7B, 6) BROOKHAVEN SYMPOSIUM, ... 175, (1965).

3861 SIGNARBIEUX.....(C.)....: (12B, 22) J. PHYSIQUE TOME 24, ... 89A, (1963).

3862 SIKKA...........(S.K.)..: (5 , 1425) J. CHEM. PHYSICS VOL.41, ... 3616, (1964).
(22B, 20) NUCL. INST. MET. 26, ... 340, (1964).
(5 , 251) ACTA CRYST. 23, ... 107, (1967).
(5 , 158) J. CHEM. PHYSICS VOL.48, ... 1883, (1968).
(17B, 18) ACTA CRYST. A25, ... 539, (1969).
(5 , 2473) ACTA CRYST. A25, ... 621, (1969).
(5 , 385) ACTA CRYST. B25, ... 1804, (1969).
(7A, 1) ACTA CRYST. VOL.A25, ... 396, (1969).
(5 , 386) ACTA CRYST. VOL.A25, ... 576, (1969).
(5 , 206) ACTA CRYST. VOL.B25, ... 310, (1969).
(17B, 25) ACTA CRYST. VOL.A26, ... 662, (1970).
(5 , 14) ACTA CRYST. VOL.B28, ... 3000, (1972).
(7A, 5) ACTA CRYST. VOL.A29, ... 211, (1973).
(5 , 15) ACTA CRYST. VOL.B29, ... 1167, (1973).

3863 SIKORSKA........(D.)....: (5 , 460) PHYS. STAT. SOL. VOL.38, ... 103, (1970).
(5 , 2203) PHYSICA VOL.51, ... 627, (1971).

3864 SILANT≠EV.......(N.A.)..: (12B, 62) SOV. PHYS. J.E.T.P. VOL.31, . . 378, (1970).
(12B, 71) YADERNAYA FIZ. (USSR) VOL.12, . 815, (1970).

3865 SILBERGLITT.....(R.)....: (6 , 531) PHYS. REV. B VOL.3, ... 157, (1971).
(11A, 45) PHYS. REV. B VOL.4, ... 2236, (1971).
(6 , 530) PHYS. REV. B VOL.4, ... 2280, (1971).
(6 , 2096) SOL. STATE COMM. VOL.11 NO.1, . 247, (1972).

3866 SILK...........(M.G.)..: (5 , 431) J. NUCL. ENERGY VOL.24, ... 43, (1970).

3867 SILVENNOINEN....(P.)....: (16 , 91) TRANSP. THEORY/STAT. PHYS. VOL.1 263, (1971).

3868 SILVERSTEIN.....(S.D.)..: (9D, 36) J. APPL. PHYS. VOL.41, ... 919, (1970).

3869 SILVER.........(E.G.)..: (20B, 31) IAEA SYMP. (PILE RESEARCH)VIENNA 509, (1960).

3870 SIMMONS........(M.)....: (5 , 1385) J. PHYSIQUE TOME 32 COL.1 VOL.II 670, (1971).

3871 SIMONS.........(G.G.)..: (24 , 64) NUCL. INST. MET. 70, ... 221, (1969).

3872 SIMON..........(F.T.)..: (6 , 1905) J. CHEM. PHYSICS VOL.45, ... 3787, (1966).

3873 SIMON..........(Y.)....: (5 , 2008) PHYS. REV. LETT. VOL.28, ... 1370, (1972).
                                 (5 , 2004) J. PHYSIQUE TOME 34, ... 447, (1973).

3874 SIMPSON........(A.W.)..: (5 , 494) J. PHYS. C VOL.5,. . . . . . . L261, (1972).

3875 SIMPSON........(F.B.)..: (5 , 2715) NUCL. SCI. ENG. VOL.7, ... 187, (1960).
                                 (5 , 2248) NUCL. SCI. ENG. VOL.20,. . . . 235, (1964).
                                 (5 , 2362) NUCL. PHYS. A VOL.A164,. . . .   34, (1971).
                                 (5 , 2325) NUCL. SCI. ENG. VOL.43,. . . .   58, (1971).

3876 SIMPSON........(O.D.)..: (5 , 2715) NUCL. SCI. ENG. VOL.7, ... 187, (1960).
                                 (19A,   34) NUCL. INST. MET. 30, ... 293, (1965).
                                 (24 ,   52) NUCL. INST. MET. 61, ... 245, (1968).
                                 (5 , 2508) NUCL. SCI. ENGNG. VOL.33,. . .   16, (1968).

3877 SIMSON.........(R.)....: (6 , 2031) Z. PHYSIK VOL.190, ... 295, (1966).
                                 (6 ,   37) Z. ANGEW. PHYS. VOL.24,NO.6, . 313, (1968).

3878 SIMS...........(G.H.E.): (5 , 418) J. INORG. NUCL. CHEM. VOL.31,. 3721, (1969).
                                 (5 , 318) J. INORG. NUCL. CHEM. VOL.32,. 2839, (1970).

3879 SINCLAIR.......(R.N.)..: (6 , 1801) IAEA SYMPOSIUM VIENNA, ... 531, (1960).
                                 (23 ,   29) IAEA SYMP. (PILE RESEARCH)VIENNA   25, (1960).
                                 (6 , 498) PHYS. REV. VOL.120,. . . . . . 1638, (1960).
                                 (6 , 131) IAEA SYMP. CHALK RIVER, VOL.2, 199, (1962).
                                 (6 , 2157) J. PHYS. SOC. JAPAN VOL.17 B-III   63, (1962).
                                 (5 , 223) J. OF PHYS.-C- SER. 2, VOL.2,   870, (1969).
                                 (19B,   14) NUCL. INST. MET. VOL.72, . . . 237, (1969).
                                 (17A,   64) NUCL. INST. MET. 70, ... 164, (1969).
                                 (20B,   92) NUCL. INST. MET. 72, ... 237, (1969).
                                 (5 , 236) REPORT AERE-R 6052,. . . . . .   , (1969).
                                 (5 , 1705) ACTA CRYST. VOL.B26, ... 2079, (1970).
                                 (6 , 1046) J. CHEM. PHYS. VOL.55, ... 2807, (1971).
                                 (19A,   55) NUCL. INST. MET. VOL.114,. . . 451, (1974).

3880 SINCLAIR.......(R.W.)..: (6 , 203) IAEA SYMPOSIUM (VIENNA), . . . 487, (1960).

3881 SINEPOL#SKII...(O.I.)..: (8 , 139) SOV. PHYS. SOL. STATE, 6,. . . 339, (1964).

3882 SINGER.........(J.)....: (5 , 2726) ACTA CRYST. VOL.5, ... 394, (1952).

3883 SINGHAL........(U.)....: (10B,   70) J. PHYS. C: SOLID ST. PHYS. V-5, 293, (1972).

3884 SINGH..........(A.K.)..: (17B,   14) ACTA CRYST. VOL.B24, ... 1701, (1968).
                                 (17B,   13) ACTA CRYST. VOL.B24, ...   35, (1968).

3885 SINGH..........(B.D.)..: (10B, 191) PHYS. STAT. SOL. VOL.38, . . . 141, (1970).

3886 SINGH..........(D.N.)..: (6 , 2242) PHYS. REV. VOL.116,. . . . . . 297, (1959).

3887 SINGH..........(M.P.)..: (10B,   95) J. PHYSIQUE TOME 35, ... 263, (1974).

3888 SINGH..........(N.)....: (10B, 103) PHYSICA VOL.30,. . . . . 2105, (1964).
                                 (10B, 162) PHYS. REV. B-4,. . . . . 4636, (1971).
                                 (10B, 107) PHYSICA VOL.59,. . . . . 109, (1972).

3889 SINGH..........(O.N.)..: (10D,   21) NUCL./S.S. PHYS. SYMP. ABSTR., . . , (1972).

3890 SINGH..........(P.P.)..: (20B,   71) NUCL. INST. MET. VOL.13, . . .   1, (1961).

3891 SINGH..........(R.B.)..: (10B,   99) J. PHYS. SOC. JAPAN VOL.35,. . 1487, (1973).

3892 SINGH..........(R.O.)..: (6 , 1912) NUCL./S.S. PHYS. SYMP. ABSTR., . . , (1972).

3893 SINGH..........(R.K.)..: (6 , 1303) INDIAN J. PURE APPL. PHYS. VOL.7 151, (1969).
                                 (10D,   45) PHYS. STAT. SOL. VOL.36, . . . 335, (1969).
                                 (10B,   62) J. PHYS. C: SOLID ST. PHYS. V-4, 2749, (1971).
                                 (6 , 508) PHYS. LETT. VOL.40A NO.4,. . . 291, (1972).
                                 (6 , 1417) PHYS. REV. B VOL.6,. . . . . . 1589, (1972).
                                 (6 ,   17) PHYS. STAT. SOL. B VOL.51 NO.1, 389, (1972).
                                 (6 , 1371) SOL. STATE COMM. VOL.11 NO.4,. 559, (1972).
                                 (6 , 447) SOL. STATE COMM. VOL.11 NO.4,. 567, (1972).
                                 (6 , 1489) SOL. STATE COMM. VOL.11, . . . 109, (1972).

3894 SINGH..........(R.P.)..: (6 , 1601) IAEA SYMP. BOMBAY, VOL.1,. . .   77, (1964).

3895 SINGH..........(S.N.)..: (6 ,   61) PHYSICA VOL.50,. . . . .   10, (1970).
                                 (10B, 106) PHYSICA VOL.58,. . . . .   71, (1972).

3896 SINGH..........(V.)....: (20A,   68) PROC. INDIAN ACAD. SCI. A VOL.53   59, (1961).

3897 SINGWI.........(K.S.)..: (10C,   90) PROC. ROY. SOC. A VOL.231, . . 293, (1955).
                                 (14B,   60) PHIL. MAG. (8) VOL.1,. . . . . 560, (1956).
                                 (6 , 221) PHYS. REV. VOL.106,. . . . . . 230, (1957).
                                 (6 , 163) J. NUCL. ENERGY VOL.7, ...   45, (1958).
                                 (6 , 223) PHYS. REV. VOL.110,. . . . . .   70, (1958).
                                 (7B, 112) SOL. STATE PHYS. VOL.8,. . . . 109, (1959).
                                 (6 , 1085) PHYS. REV. VOL.119,. . . . . . 863, (1960).
                                 (14B,   79) PHYS. REV. VOL.122,. . . . . .   9, (1961).
                                 (14B, 105) RENSSELAER POL. INST. SYMP.,.   145, (1961).
                                 (6 , 1022) IAEA SYMP. CHALK RIVER, VOL.1, 213, (1962).
                                 (10B,   30) IAEA SYMP. CHALK RIVER, VOL.1,   3, (1962).
                                 (14B,   70) PHYS. LETT. 3, ... 143, (1962).
                                 (14B,   81) PHYS. REV. VOL.126,. . . . . . 986, (1962).
                                 (14B,   80) PHYS. REV. VOL.126,. . . . . . 997, (1962).
                                 (6 , 1600) IAEA SYMP. BOMBAY, VOL.2,. . .   85, (1964).
                                 (14B,   82) PHYS. REV. VOL.136,. . . . . . A969, (1964).
                                 (6 ,   84) PHYSICA VOL.31,. . . . . 1257, (1965).
                                 (14B,   85) PHYS. REV. LETT. 15, ... 693, (1965).
                                 (14B, 118) SOLID/LIQUID STATE PHYSICS,. . 140, (1965).
                                 (6 , 1807) PHYS. REV. VOL.152,. . . . . .   99, (1966).
                                 (14B, 133) TRIESTE-INTERN. COURSE,. . . . 603, (1967).
                                 (14A,   44) PHYS. REV. LETT. 21, ... 881, (1968).
                                 (14A,   51) PHYS. REV. A VOL.1,. . . . . . 454, (1970).
                                 (6 , 1136) PHYS. REV. A VOL.2,. . . . . . 2416, (1970).
                                 (14A,   49) PHYS. REV. A VOL.2,. . . . . . 2427, (1970).
                                 (10B, 155) PHYS. REV. B-2,. . . . . 2983, (1970).
                                 (6 , 1137) PHYS. REV. A VOL.4,. . . . . . 2413, (1971).
                                 (14A,   31) NUOVO CIMENTO B VOL.13B NO.1,. 185, (1973).
                                 (6 , 1704) PHYS. REV. A VOL.7,. . . . . . 1043, (1973).

3898 SINHA..........(K.P.)..: (11B,   29) PROC. PHYS. SOC. 91, ...   97, (1967).
                                 (11B,   10) J. PHYS. -C-, VOL. 1,. . . . . 673, (1968).

3899 SINHA..........(S.K.)..: (6 , 580) COPENHAGEN CONF.-LATT.DYNAMICS,.   53, (1963).
                                 (15 ,   5) BROOKHAVEN SYMPOSIUM,. . . . . 169, (1965).
                                 (7B,   82) PHYS. REV. VOL.149,. . . . . .   1, (1966).
                                 (6 , 603) PHYS. REV. VOL.143,. . . . . . 422, (1966).
                                 (6 , 1125) IAEA SYMP. COPENHAGEN, VOL.1,. 339, (1968).
                                 (10D,   27) PHYS. REV. VOL.169,. . . . . . 477, (1968).
                                 (6 , 539) PHYS. REV. LETT. VOL.23, . . . 311, (1969).
                                 (10B, 148) PHYS. REV. VOL.177,. . . . . . 1256, (1969).
                                 (6 , 537) J. APPL. PHYS. 41, ... 1365, (1970).
                                 (6 , 2284) PHYS. REV. B-1,. . . . . 2430, (1970).

```
3899  SINHA...........(S.K.)...:  ( 6 ,  1141) PHYS. REV. A VOL.3,. . . . . . . .  1688, (1971):
                                  ( 6 ,  2015) PHYS. REV. B-4,. . . . . . . . . .  2398, (1971):
                                  (10B,   167) PHYS. REV. LETT. VOL.26,. . . . .  1324, (1971):
                                  (10B,  1156) IAEA SYMP. GRENOBLE,. . . . . . .   129, (1972):
                                  ( 6 ,  2295) PHYS. REV. B VOL.6A NO.4,. . . . .  4438, (1972):
                                  ( 5 ,   880) PHYS. LETT. VOL.45A NO.4,. . . . .   281, (1973):
                                  (16 ,    62) PHYS. REV. B VOL.7,. . . . . . . .   917, (1973):
                                  ( 6 ,  2170) PHYS. REV. B VOL.8,. . . . . . . .  1332, (1973):
                                  ( 5 ,   552) PHYS. REV. LETT. VOL.31,. . . . .  1498, (1973):
                                  (10B,   189) PHYS. REV. B VOL.9,. . . . . . . .  2573, (1974):
3900  SINRAM..........(K.)....:  ( 6 ,  1155) NUCL. PHYS. VOL.A129,. . . . . .    666, (1969).
3901  SIPPEL..........(O.)....:  ( 9B,    86) PHYS. STAT. SOL. 2,. . . . . . .   K104, (1962).
                                  ( 9B,    71) PHYS. LETT. 8,. . . . . . . . .    241, (1964).
                                  ( 9C,    31) PHYS. LETT. VOL.14,. . . . . . .    174, (1965).
                                  ( 5 ,  2451) PHYS. STAT. SOL. VOL.23,. . . .    237, (1967).
                                  ( 5 ,  1463) ACTA CRYST. A24,. . . . . . . .    237, (1968).
3902  SIROTA..........(N.N.)..:  ( 5 ,  1829) SOV. PHYS. DOKLADY VOL.11, . . .    888, (1967).
                                  ( 5 ,  1616) PHYS. STAT. SOL. VOL.36,. . . .    K25, (1969).
                                  ( 5 ,  1633) J. PHYSIQUE TOME 32 COL.1 VOL.II   987, (1971).
                                  ( 6 ,  1436) PHYS. STAT. SOL. VOL.43,. . . .   K165, (1971).
                                  ( 5 ,  1639) SOV. PHYS. DOKLADY VOL.17, . . .    370, (1972).
3903  SIVARDIERE......(J.)....:  ( 9D,    68) SOL. STATE COMM. 5,. . . . . . .    289, (1967).
                                  ( 5 ,  2571) SOL. STATE COMM. 5,. . . . . . .    293, (1967).
                                  ( 5 ,  2539) Z. ANGEW. PHYS. VOL.23,. . . .     243, (1967).
                                  ( 5 ,  2570) J. APPL. PHYS. VOL.39,. . . . .   1364, (1968).
                                  ( 5 ,  2385) J. PHYSIQUE TOME 30,. . . . . .    967, (1969).
                                  ( 5 ,  1533) SOL. STATE COMM. VOL.7 NO.22,.   1669, (1969).
                                  ( 9D,    43) J. PHYSIQUE TOME 31,. . . . . .    369, (1970).
                                  ( 5 ,   874) J. PHYSIQUE TOME 31,. . . . . .    607, (1970).
                                  ( 5 ,  1134) J. PHYSIQUE TOME 31,. . . . . .    803, (1970).
                                  ( 5 ,  2604) SOL. STATE COMM. VOL.8 NO.1,.      23, (1970).
                                  ( 5 ,   863) C. R. ACAD. SCI. B TOME 273,.     619, (1971).
                                  (11A,    45) PHYS. REV. B VOL.4,. . . . . .    2236, (1971).
                                  ( 5 ,  2567) SOL. STATE COMM. VOL. 9,. . . .    435, (1971).
3904  SIZOV...........(R.A.)..:  ( 9B,   106) SOV. PHYS. CRYST. VOL.10,. . . .    346, (1965).
                                  (22A,    70) SOV. PHYS. CRYST. VOL.9,. . . .     797, (1965).
                                  ( 5 ,   518) SOV. PHYS. DOKLADY VOL.11, . . .    379, (1966).
                                  ( 5 ,   160) SOV. PHYS. J.E.T.P. VOL.23,. . .    395, (1966).
                                  ( 5 ,  2488) SOV. PHYS. JETP LETT. VOL.6, . .    176, (1967).
                                  ( 5 ,  2487) SOV. PHYS. J.E.T.P. VOL.25,. . .    266, (1967).
                                  ( 5 ,   176) SOV. PHYS. JETP LETT. VOL.7, . .    158, (1968).
                                  ( 5 ,   180) SOV. PHYS. J.E.T.P. VOL.26,. . .    736, (1968).
                                  ( 5 ,   181) SOV. PHYS. SOL. STATE VOL.10,. .   2537, (1969).
                                  ( 5 ,   177) SOV. PHYS. J.E.T.P. VOL.33,. . .    737, (1971).
                                  ( 5 ,   178) SOV. PHYS. SOL. STATE VOL.12,. .   2316, (1971).
3905  SIZOV...........(V.A.)..:  ( 5 ,  2488) SOV. PHYS. JETP LETT. VOL.6, . .    176, (1967).
                                  ( 5 ,  2487) SOV. PHYS. J.E.T.P. VOL.25,. . .    266, (1967).
                                  ( 5 ,   180) SOV. PHYS. J.E.T.P. VOL.26,. . .    736, (1968).
                                  ( 5 ,   192) SOV. PHYS. SOL. STATE VOL.11,. .   2258, (1969).
                                  ( 5 ,   181) SOV. PHYS. SOL. STATE VOL.10,. .   2537, (1969).
3906  SJOLANDER.......(A.)....:  (10C,    11) ARK. FYS. VOL.13, PAPER 17,. . .    199, (1958).
                                  (10E,     1) ARK. FYS. VOL.13, PAPER 18,. . .    215, (1958).
                                  ( 6 ,  1085) PHYS. REV. VOL.119,. . . . . . .    863, (1960).
                                  (14B,    79) PHYS. REV. VOL.122,. . . . . . .      9, (1961).
                                  ( 6 ,  1022) IAEA SYMP. CHALK RIVER, VOL.1,.    215, (1962).
                                  (14B,    81) PHYS. REV. VOL.126,. . . . . . .    986, (1962).
                                  (14B,    80) PHYS. REV. VOL.126,. . . . . . .    997, (1962).
                                  (10B,     2) AARHUS SUMMER SCHOOL,. . . . . .     76, (1963).
                                  (10C,    36) IAEA SYMPOSIUM BOMBAY, VOL.1,.      61, (1964).
                                  (10D,     4) BROOKHAVEN SYMPOSIUM,. . . . . .     29, (1965).
                                  (14B,   132) TH. NEUTRON SCATT./EGELSTAFF,.     291, (1965).
                                  (14A,    10) C.N.E.N. SYMP. CASACCIA,. . . .    219, (1968).
                                  (10D,    20) 2ND SUMMER SCHOOL S.S. PHYS.,.       7, (1968).
3907  SJOSTRAND.......(N.G.)..:  (20A,    72) REACTOR SCI. TECHNOL. VOL.17,.     147, (1963).
3908  SKAARUP.........(P.)....:  (22C,    40) NUCL. INST. MET. 65, . . . . . .    152, (1968).
                                  (22B,    33) NUCL. INST. MET. 75, . . . . . .      1, (1969).
3909  SKALYO-JR.......(J.)....:  ( 6 ,  1328) PHYS. REV. LETT. VOL.23, . . . .   1394, (1969).
                                  ( 6 ,  1702) PHYS. REV. VOL.181,. . . . . . .   1251, (1969).
                                  ( 6 ,  1539) PHYS. REV. VOL.184,. . . . . . .    550, (1969).
                                  ( 5 ,  1714) PHYS. REV. VOL.188,. . . . . . .   1037, (1969).
                                  ( 6 ,  1331) J. APPL. PHYS. VOL.41,. . . . .    1303, (1970).
                                  ( 6 ,  1278) PHYS. REV. B-1,. . . . . . . . .    278, (1970).
                                  ( 5 ,   666) PHYS. REV. B-2,. . . . . . . . .   1310, (1970).
                                  ( 6 ,   570) PHYS. REV. B-2,. . . . . . . . .   4632, (1970).
                                  ( 6 ,  1329) J. PHYSIQUE TOME 32 COL.1 VOL.II   882, (1971).
                                  ( 6 ,  1330) PHYS. REV. B VOL.3,. . . . . . .   1736, (1971).
                                  ( 6 ,  1701) PHYS. REV. B VOL.6,. . . . . . .   4766, (1972).
                                  ( 6 ,  1345) BULL. AMER. PHYS. SOC. VOL.18,.    111, (1973).
                                  ( 6 ,  1346) J. PHYS. CHEM. SOLIDS VOL.34,.     255, (1973).
                                  (17A,    73) NUCL. INST. MET. VOL.112,. . . .    571, (1973).
                                  ( 6 ,  1347) PHYS. REV. B VOL.7,. . . . . . .   4670, (1973).
                                  ( 6 ,  2279) PHYS. LETT. VOL.46A NO.5,. . . .    357, (1974).
                                  ( 6 ,  2281) PHYS. REV. B VOL.9 NO.12,. . . .       , (1974).
                                  ( 6 ,  1350) PHYS. REV. B VOL.9,. . . . . . .   1797, (1974).
                                  ( 6 ,  2282) PHYS. REV. B VOL.9,. . . . . . .   2661, (1974).
3910  SKALYO..........(J.)....:  ( 5 ,   979) A.I.P. CONF. PROC. NO.10 PART 1,    98, (1973).
3911  SKEPPSTEDT......(O.)....:  ( 5 ,  2068) PHYS. SCRIPTA(SWEDEN) VOL.4, . .     89, (1971).
3912  SKOLD...........(K.)....:  ( 6 ,    70) IAEA SYMP. BOMBAY, VOL.2,. . . .    101, (1964).
                                  (22A,    12) IAEA SYMP. BOMBAY, VOL.2,. . . .    487, (1964).
                                  (23 ,    57) NUCL. INST. MET. 27, . . . . . .     61, (1964).
                                  ( 6 ,    88) PHYS. REV. LETT. 12, . . . . . .    721, (1964).
                                  ( 5 ,  2303) SOL. STATE COMM. 4,. . . . . . .    303, (1966).
                                  ( 6 ,  1861) J. PHYS. CHEM. SOLIDS VOL.28,.    2369, (1967).
                                  (21 ,    34) NUCL. INST. MET. 56, . . . . . .    305, (1967).
                                  ( 6 ,   106) PHYS. REV. LETT. 19, . . . . . .   1023, (1967).
                                  ( 6 ,    79) PHYS. REV. VOL.161,. . . . . . .    102, (1967).
                                  ( 6 ,   282) J. CHEM. PHYSICS VOL.49,. . . .    2443, (1968).
                                  (21 ,    36) NUCL. INST. MET. 63, . . . . . .    114, (1968).
                                  (19A,    38) NUCL. INST. MET. 63, . . . . . .    347, (1968).
                                  (14A,    44) PHYS. REV. LETT. 21, . . . . . .    881, (1968).
                                  ( 6 ,  1082) PHYS. REV. 170,. . . . . . . . .    808, (1968).
                                  ( 5 ,   268) CHEM. PHYS. LETTERS VOL.4, . . .    444, (1969).
                                  (19A,    48) NUCL. INST. MET. VOL.82,. . . .    208, (1970).
                                  (14A,    51) PHYS. REV. A VOL.1,. . . . . . .    454, (1970).
                                  ( 6 ,  2256) J. PHYS. CHEM. SOLIDS VOL.32,.      41, (1971).
                                  ( 6 ,    80) IAEA SYMP. GRENOBLE,. . . . . .    413, (1972).
                                  ( 6 ,    85) PHYS. REV. A VOL.6,. . . . . . .   1107, (1972).
                                  ( 6 ,  1568) SOL. STATE COMM. VOL.13 NO.5,.     543, (1973).
3913  SKYTTE-JENSEN...(B.)....:  (22C,    30) NUCL. INST. MET. 33, . . . . . .    155, (1965).
```

3914 SLACK...........(G.A.)..: (5 , 2173) J. APPL. PHYS. VOL.31, SUPPL., .   352, (1960).
3915 SLAGGIE.........(E.L.)..: (16 ,   40) NUCL. SCI. ENGNG. VOL.30,. . . .   199, (1967).
3916 SLAK............(J.)...:   (6 , 1579) PHYS. STAT. SOLIDI B VOL.59 NO.2 471, (1973).
3917 SLATER..........(J.C.)..: (10D,   56) REV. MOD. PHYS. VOL.30,. . . . .   197, (1958).
3918 SLAUGHTER.......(G.)...:   (23 ,   28) IAEA SYMP. (PILE RESEARCH)VIENNA 535, (1960).
3919 SLAVIK..........(F.)...:   (17A,   47) INSTRUM. EXP. TECH. VOL.15,. . .   44, (1972).
3920 SLEDZIEWSKA-BLOC(D.)....: (20A,   41) NUCL. INST. MET. VOL.79,. . . .   51, (1970).
                                 (17A,   16) ACTA PHYS. POLON. VOL.A42, . . .   259, (1972).
3921 SLEDZIEWSKA.....(D.)....: (6 , 2317) IAEA SYMP. CHALK RIVER, VOL.2,.   87, (1962).
                                 (6 , 2309) COPENHAGEN CONF.-LATT.DYNAMICS,.   33, (1963).
3922 SLOTNICK........(M.)...:   (8 ,   75) PHYS. REV. VOL.83, ...   1226, (1951).
3923 SLOVACEK........(R.E.)..: (5 , 2324) PROC/CONF/NEUTRON C. S. + TECH.,   687, (1968).
                                 (5 , 2326) NUCL. SCI. ENG. VOL.52,. . . . .   310, (1973).
3924 SLUCHEVSKAYA....(V.M.)..: (5 , 2874) BU. ACAD. SCI. USSR PHYS. SR. 32 579, (1968).
3925 SLUTSKY.........(L.J.)..: (10B,   47) J. CHEM. PHYSICS VOL.40, ...   531, (1964).
3926 SMART...........(J.S.)..: (5 , 1752) PHYS. REV. VOL.76, ...   1256, (1949).
3927 SMIRNOV.........(A.A.)..: (9C,   23) J. PHYS. (USSR) VOL.5, . . . , .   263, (1941).
                                 (10C, 102) SOV. PHYS. SOL. STATE VOL.1, . .   1277, (1959).
                                 (10C, 133) UKR. FIZ. ZH. (USSR) VOL.8,. . .   256, (1963).
                                 (8 ,   60) PHYS. MET. METALLOGR. VOL.20 N.2   27, (1965).
3928 SMIRNOV.........(A.I.)..: (6 , 457) SOV. PHYS. SOL. STATE 6, ...   76, (1964).
3929 SMIRNOV.........(L.S.)..: (5 , 2219) PHYS. STAT. SOL. VOL.50, ...   385, (1972).
3930 SMIRNOV.........(O.P.)..: (5 , 2879) SOV. PHYS. JETP LETT. VOL.18,. .   47, (1973).
3931 SMIRNOV.........(V.I.)..: (22B,   11) INSTRUM. EXP. TECH. NO.4,. . . .   832, (1965).
3932 SMIRNOV.........(V.P.)..: (5 , 2494) 2ND INTERNAT. MEET./FERROELECTR.   219, (1969).
                                 (5 , 2517) J. PHYSIQUE TOME 32 COL.1 VOL.I,   503, (1971).
                                 (5 , 2856) SOV. PHYS. CRYST. VOL.17,. . . .   1017, (1973).
3933 SMITH...........(A.J.S.): (20C,   75) NUCL. INST. MET. VOL.94, ...   493, (1971).
3934 SMITH...........(A.J.)..: (11A,   25) J. PHYSIQUE TOME 32 COL.1 VOL.II 585, (1971).
3935 SMITH...........(D.B.J.): (22B,   30) NUCL. INST. MET. 60, ...   237, (1968).
3936 SMITH...........(D.B.)..: (6 , 182) BULL. AMER. PHYS. SOC. VOL.12, .   689, (1967).
                                 (6 , 170) DISS. ABS. 29, ...   7368, (1968).
3937 SMITH...........(D.L.)..: (20A,   45) NUCL. INST. MET. VOL.102,. . . .   193, (1972).
3938 SMITH...........(F.A.)..: (5 , 126) J. PHYS. CHEM. SOL. 27,. . . . .   925, (1966).
                                 (5 , 2185) J. APPL. PHYS. VOL.40,. . . . .   1332, (1969).
                                 (5 , 1104) PHYS. REV. B-3,. . . . .   3046, (1971).
3939 SMITH...........(G.P.)..: (5 , 2257) J. CHEM. PHYS. VOL.21, ...   228, (1953).
3940 SMITH...............(H.G.)..: (5 , 387) PHYS. REV. LETT. VOL.6,. . . . .   7, (1961).
                                 (5 , 1922) ACTA CRYST. VOL.15,. . . . , .   1201, (1962).
                                 (5 , 1580) J. PHYS. SOC. JAPAN VOL.17 B-II,   335, (1962).
                                 (20C,   96) REV. SCI. INSTRUM. VOL.33,. . .   128, (1962).
                                 (6 , 458) COLL. INTER. N.126 (GRENOBLE),.   191, (1963).
                                 (23 ,   7) COLL. INTER. N.126 (GRENOBLE),.   71, (1963).
                                 (5 , 1500) J. APP. PHYS. 37,. . . . .   1047, (1966).
                                 (5 , 1506) J. APP. PHYS. 37,. . . . .   979, (1966).
                                 (5 , 1935) ACTA CRYST. 22, ...   928, (1967).
                                 (6 , 600) PHYS. REV. 164,. . . . .   922, (1967).
                                 (6 , 1360) IAEA SYMP. COPENHAGEN, VOL.1,. .   149, (1968).
                                 (6 , 611) IAEA SYMP. COPENHAGEN, VOL.1,. .   47, (1968).
                                 (22B,   8) IAEA SYMP. COPENHAGEN, VOL.2,. .   253, (1968).
                                 (5 , 1578) J. CHEM. PHYSICS VOL.48, ...   5561, (1968).
                                 (6 , 612) PHYS. REV. LETT. VOL.20, ...   1245, (1968).
                                 (6 , 1373) PHYS. REV. 168,. . . . .   970, (1968).
                                 (6 , 1226) J. APP. PHYS. VOL.40,. . . . . .   1452, (1969).
                                 (6 , 1562) PHYS. REV. VOL.181,. . . . . . .   1218, (1969).
                                 (6 , 906) J. APPL. PHYS. 41, ...   1182, (1970).
                                 (6 ,   16) PHYS. REV. B-1,. . . . .   4819, (1970).
                                 (6 , 907) PHYS. REV. LETT. VOL.24, ...   16, (1970).
                                 (6 , 1216) PHYS. REV. LETT. VOL.25, ...   1611, (1970).
                                 (6 , 1264) CONF. INTERN., RENNES,. . . . .   144, (1971).
                                 (6 , 418) CONF. INTERN., RENNES,. . . . .   145, (1971).
                                 (6 , 2178) PHYS. REV. B-3,. . . . .   3457, (1971).
                                 (6 , 2039) PHYS. REV. B-4,. . . . .   2558, (1971).
                                 (6 , 1522) BULL. AMER. PHYS. SOC. VOL.17,.   292, (1972).
                                 (6 , 1265) IAEA SYMP. GRENOBLE, ...   103, (1972).
                                 (6 , 453) IAEA SYMP. GRENOBLE, ...   219, (1972).
                                 (6 , 204) PHYS. REV. B VOL.5,. . . . . . .   4951, (1972).
                                 (6 , 561) PHYS. REV. B VOL.6,. . . . . . .   3956, (1972).
                                 (6 , 2114) PHYS. REV. LETT. VOL.29, ...   353, (1972).
                                 (6 , 111) PHYS. REV. B VOL.8,. . . . . . .   3493, (1973).
                                 (6 , 1865) PHYS. REV. B VOL.8,. . . . . . .   6013, (1973).
3941 SMITH...........(J.A.G.): (21 ,   29) J. SCI. INSTRUM. VOL.27, ...   330, (1950).
3942 SMITH...........(J.A.S.): (6 , 1000) J.C.S. FARADAY TRANS. II VOL.69, 1477, (1973).
                                 (6 ,   20) J.C.S. FARADAY TRANS. II VOL.69,   1, (1973).
                                 (6 ,   8) NATURE (PHYS. SCI.) VOL.241,. .   40, (1973).
3943 SMITH...........(J.F.)..: (6 , 2295) PHYS. REV. B VOL.6,. . . . . . .   4438, (1972).
3944 SMITH...........(J.H.)..: (5 , 1679) PROC. ROY. SOC. A VOL.241, . . .   223, (1957).
                                 (5 , 1773) J. OF PHYS. C VOL.1,. . . . . .   1683, (1968).
                                 (5 , 124) J. OF PHYS. C VOL.2,. . . . . .   356, (1969).
                                 (5 , 1622) J. OF PHYS. C VOL.2,. . . . . .   761, (1969).
                                 (5 , 1643) ACTA CRYST. VOL.A26,. . . . . .   379, (1970).
                                 (6 , 1444) J. APPL. PHYS. 41, ...   1857, (1970).
                                 (6 , 1443) J. PHYS. CHEM. SOLIDS VOL.31,.   485, (1970).
                                 (5 , 1740) J. PHYSIQUE TOME 32 COL.1 VOL.I,   70, (1971).
                                 (6 , 1445) J. PHYS.F: METAL PHYS. VOL.1,.   763, (1971).
                                 (6 , 1454) J. PHYS. C VOL.4, NO.13, . . .   L258, (1971).
                                 (6 , 1434) PHYS. STAT. SOL. VOL.47,. . . .   K11, (1971).
3945 SMITH...........(J.R.)..: (20B,   33) IAEA SYMP. (PILE RESEARCH)VIENNA 433, (1960).
                                 (20B, 113) NUCL. SCI. ENGNG. VOL.8,. . . .   393, (1960).
                                 (20C,   74) NUCL. INST. MET. VOL.96,. . . .   551, (1971).
3946 SMITH...........(P.N.)..: (6 , 1203) PHYS. REV. A VOL.4,. . . . . . .   281, (1971).
                                 (6 , 1142) PHYS. REV. A VOL.5,. . . . . . .   1547, (1972).

3947 SMITH...........(R.B.)..!   (6 , 1052) IAEA SYMP. CHALK RIVER, VOL.1, .   373, (1962).

3948 SMITH...........(R.R.)..!   (11C,   6) DISS. ABSTR. VOL.14, ...   162, (1954).

3949 SMITH...........(T.)....!   (10B,   125) PHYS. REV. 175,. . . . .   1083, (1968).
                                   (6 , 1632) PHYS. REV. B-1,. . . . .   1833, (1970).

3950 SMOLENSKY.......(G.A.)..!   (5 ,   218) PROC. INT. CONF. (NOTTINGHAM),.   354, (1964).

3951 SMUTS...........(J.)....!   (5 ,   969) ACTA CRYST. 17,. . . . .   1331, (1964).

3952 SNELL...........(A.H.)..!   (23 ,   24) IAEA SYMP. (PILE RESEARCH)VIENNA   49, (1960).

3953 SNODGRASS.......(F.W.)..!   (19B,   23) NUCL. INST. MET. VOL.116,. . . .   205, (1974).

3954 SNOW............(A.I.)..!   (5 ,   620) PHYS. REV. VOL.85, ...   365, (1952).

3955 SOBHANA.........(S.)....!   (7B,   93) PHYS. STAT. SOL. VOL.49, ...   K75, (1972).

3956 SODERLUND.......(B.)....!   (20C,   73) NUCL. INST. MET. 78, ...   173, (1970).

3957 SOKOLOFF........(J.B.)..!   (11A,   38) PHYS. REV. 173,. . . . .   617, (1968).
                                   (8 ,   93) PHYS. REV. VOL.185,. . . . . . .   770, (1969).
                                   (8 ,   92) PHYS. REV. VOL.185,. . . . . . .   783, (1969).
                                   (6 ,   518) PHYS. REV. VOL.187,. . . . . . .   584, (1969).
                                   (8 ,   94) PHYS. REV. 180,. . . . .   613, (1969).
                                   (6 ,   519) PHYS. REV. B VOL.3,. . . . . . .   2367, (1971).
                                   (6 , 1530) PHYS. REV. B VOL.7,. . . . . . .   1644, (1973).

3958 SOKOLOVSKII.....(V.V.)..!   (20B,   130) PRIB. TEKH. EKSPER. NO.2,. . . .   3, (1956).
                                   (20B,   48) INSTRUM. EXP. TECH. NO.4,. . . .   632, (1963).

3959 SOKOLOV.........(V.I.)..!   (5 , 1960) SOL. STATE COMM. VOL.14 NO.4,. .   309, (1974).

3960 SOLBRIG.........(H.)....!   (7B,   92) PHYS. STAT. SOL. VOL.46, ...   273, (1971).
                                   (7B,   91) PHYS. STAT. SOL. VOL.47, ...   143, (1971).
                                   (7B,   94) PHYS. STAT. SOL. VOL.51, ...   555, (1972).

3961 SOLOV≠EV........(S.P.)..!   (6 , 2101) SOV. PHYS. SOL. STATE 8, ...   2156, (1966).
                                   (5 , 2494) 2ND INTERNAT. MEET./FERROELECTR.   219, (1969).
                                   (5 , 2665) DOKL. AKAD. NAUK SSSR VOL.194,.   1374, (1970).
                                   (6 , 1388) SOV. PHYS. CRYST. VOL.15,. . . .   1010, (1971).

3962 SOLTAN..........(A.)....!   (20B,   122) PHYS. REV. VOL.45, ...   567, (1934).

3963 SOLT............(G.)....!   (7A,   50) PHYS. LETT. 6, ...   51, (1963).
                                   (7A,   51) PHYS. LETT. VOL.13,. . . . . . .   223, (1964).
                                   (6 , 1581) PHYSICA VOL.32,. . . . .   1571, (1966).
                                   (7A,   48) PHYSICA VOL.32,. . . . .   16, (1966).
                                   (15 ,   33) PHYS. LETT. 22,. . . . .   288, (1966).
                                   (6 ,   276) PHYSICA VOL.37,. . . . .   253, (1967).
                                   (6 ,   285) DISCUSSIONS FARADAY SOC. NO.48,.   87, (1969).

3964 SOLYOM..........(J.)....!   (5 , 1798) PHYS. LETT. 20,. . . . .   331, (1966).
                                   (5 , 1772) PHYS. LETT. 22,. . . . .   273, (1966).
                                   (5 , 1019) SOL. STATE COMM. 4,. . . . . . .   255, (1966).
                                   (5 , 1786) CENT. RES INST PHYS. KFKI NO.2,.   1, (1968).
                                   (5 , 1695) PHYS. REV. 171,. . . . .   574, (1968).
                                   (11B,   54) Z. PHYS. VOL.243,. . . . .   382, (1971).

3965 SOMANATHAN......(C.S.)..!   (5 , 1018) SOL. STATE COMM. 3,. . . . . . .   113, (1965).
                                   (12B,   15) INDIAN J. PURE APPL. PHYS. VOL.7   546, (1969).
                                   (5 , 1628) J. APP. PHYS. VOL.40,. . . . . .   1870, (1969).
                                   (5 , 1663) J. PHYS. CHEM. SOLIDS 30,. . . .   939, (1969).

3966 SOMENKOV........(V.A.)..!   (5 , 1199) SOV. PHYS. CRYST. VOL.8, ...   626, (1964).
                                   (6 , 1759) SOV. PHYS. J.E.T.P. VOL. 25, . .   436, (1967).
                                   (6 , 2211) SOV. PHYS. SOL. STATE 9, ...   1366, (1967).
                                   (5 , 2011) SOV. PHYS. SOL. STATE VOL.10,. .   1076, (1968).
                                   (5 , 2515) SOV. PHYS. SOL. STATE VOL.10,. .   212, (1968).
                                   (5 , 1204) SOV. PHYS. J.E.T.P., 28, ...   649, (1969).
                                   (6 , 1680) SOV. PHYS. SOL. STATE VOL.11,. .   2343, (1969).
                                   (5 , 2012) SOV. PHYS. CRYST. VOL.14,. . . .   522, (1970).
                                   (17A,   46) INSTRUM. EXP. TECH.(USA) VOL.14,   733, (1971).
                                   (5 , 2837) SOV. PHYS. SOL. STATE VOL. 13,.   2172, (1972).
                                   (5 , 2835) SOV. PHYS. SOL. STATE VOL. 13,.   2178, (1972).
                                   (5 , 1167) SOV. PHYS. JETP VOL.36,. . . . .   1170, (1973).

3967 SOMMERS-JR......(H.S.)..!   (20B,   139) REV. SCI. INSTRUM. VOL.24, . . .   91, (1953).
                                   (5 , 1324) PHYS. REV. VOL.97,. . . . . . .   855, (1955).
                                   (5 , 1327) PHYS. REV. VOL.101,. . . . . . .   1235, (1956).

3968 SOMMER..........(F.)....!   (14A,   38) PHYS. LETT. VOL.31A, ...   237, (1970).

3969 SONDERICKER.....(J.H.)..!   (20B,   140) REV. SCI. INSTRUM. VOL.30, . . .   269, (1959).

3970 SONI............(J.N.)..!   (20D,   16) NUCL. INST. MET. 37, ...   121, (1965).

3971 SONNTAG.........(W.)....!   (5 ,   684) J. APPL. CRYST. VOL.4, ...   303, (1971).

3972 SONODA..........(M.)....!   (19A,   46) NUCL. INST. MET. VOL.75,. . . .   32, (1969).
                                   (19B,   7) JAP. J. APPL. PHYS. VOL.10,. . .   1090, (1971).

3973 SOROKIN.........(L.M.)..!   (5 ,   970) SOV. PHYS. CRYST. VOL.8, ...   300, (1963).

3974 SOSNOVSKAYA.....(I.)....!   (5 , 2957) SOV. PHYS. SOL. STATE 6, ...   1070, (1964).

3975 SOSNOVSKII......(E.)....!   (5 , 2957) SOV. PHYS. SOL. STATE 6, ...   1070, (1964).

3976 SOSNOWSKA.......(I.)....!   (19A,   59) NUKLEONIKA VOL.9,. . . . .   523, (1964).
                                   (21 ,   15) IAEA SYMP. COPENHAGEN, VOL.2,. .   313, (1968).
                                   (9B,   68) NUKLEONIKA VOL.13, ...   1111, (1968).
                                   (9B,   67) NUKLEONIKA VOL.13, ...   171, (1968).
                                   (6 , 1018) PHYS. REV. LETT. VOL.27,. . . .   1576, (1971).
                                   (17A,   16) ACTA PHYS. POLON. VOL.A42,. . .   259, (1972).
                                   (5 , 1926) J. PHYSIQUE TOME 33 COL.2, . . .   83, (1972).

3977 SOSNOWSKI.......(J.J.)..!   (6 ,   596) J. PHYS. CHEM. SOLIDS VOL.23,. .   1021, (1962).

3978 SOSNOWSKI.......(J.)....!   (22A,   3) ACTA PHYS. POLON. VOL.19,. . . .   329, (1960).
                                   (20B,   32) IAEA SYMP. (PILE RESEARCH)VIENNA   445, (1960).
                                   (20A,   2) ACTA CRYST. VOL.14,. . . . . . .   292, (1961).
                                   (20B,   70) NUCL. INST. MET. VOL.10,. . . .   289, (1961).
                                   (6 , 2317) IAEA SYMP. CHALK RIVER, VOL.2,.   87, (1962).
                                   (19A,   59) NUKLEONIKA VOL.9,. . . . .   523, (1964).
                                   (6 ,   165) IAEA SYMP. COPENHAGEN, VOL.1,. .   157, (1968).
                                   (21 ,   15) IAEA SYMP. COPENHAGEN, VOL.2,. .   313, (1968).
                                   (6 , 1854) J. PHYS.F: METAL PHYS. VOL.1,. .   339, (1971).
                                   (6 , 2012) PHYS. STAT. SOL. VOL.44, ...   K65, (1971).
                                   (6 , 2013) IAEA SYMP. GRENOBLE, ...   61, (1972).
                                   (6 ,   178) ACTA PHYS. POLON. VOL.A43, . . .   37, (1973).

3979 SOTOFTE.........(I.)....!   (5 ,   287) ACTA CHEM. SCAND. VOL.24,. . . .   3248, (1970).
                                   (5 , 1316) ACTA CHEM. SCAND. VOL.25,. . . .   1233, (1971).

```
3980  SOUGI...........(M.)....:  ( 5 ; 1871) COMP. REND. 268, . . . . . . . 1549, (1969).
                                 ( 5 ; 1001) COMP. REND. 268, . . . . . . .  365, (1969).
                                 ( 5 ;  412) CONF./RARE EARTHS/ACTINIDES, .  218, (1971).
                                 ( 5 ; 1229) J. PHYSIQUE TOME 32 COL.1 VOL.II 853, (1971).
                                 ( 5 ; 1712) PHYS. STAT. SOL. VOL.44, . . .   71, (1971).
                                 ( 5 ;   46) SOL. STATE COMM. VOL.9, . . . .  413, (1971).
                                 ( 5 ;  413) J. CHEM. PHYS. VOL.56, . . . . 4325, (1972).
                                 ( 5 ; 1404) J. CHEM. PHYS. VOL.57, . . . . 2156, (1972).
                                 (22C;   69) REVUE PHYS. APPL. VOL.7, . . .   29, (1972).
                                 ( 5 ; 2055) J. CHEM. PHYS. VOL.58, . . . . 1783, (1973).

3981  SOULES..........(T.F.)..:  ( 8 ;   96) PHYS. REV. LETT. VOL.25, . . .  110, (1970).

3982  SPARKS..........(J.T.)..:  ( 5 ; 1143) J. APPL. PHYS. VOL.31, SUPPL.   356, (1960).
                                 ( 5 ; 1140) J. PHYS. SOC. JAPAN VOL.17 B-1,  249, (1962).
                                 ( 5 ;  614) COLL. INTER. N.126 (GRENOBLE),   143, (1963).
                                 ( 5 ; 2186) J. APP. PHYS. 34,. . . . . . . 1191, (1963).
                                 (22C; 1983) J. APP. PHYS. 37,. . . . . . . 1040, (1966).
                                 (22C;   59) REV. SCI. INSTRUM. VOL.37, . .  435, (1966).
                                 (22C;   61) REV. SCI. INSTRUM. VOL.38, . .  275, (1967).
                                 ( 5 ; 2185) J. APPL. PHYS. VOL.40, . . . . 1332, (1969).

3983  SPEAKMAN........(J.C.)..:  ( 5 ; 1924) J. CHEM. SOC. A (1967),. . . . 1862, (1967).
                                 ( 5 ; 1452) J. CHEM. SOC. A (1970),. . . . 1923, (1970).
                                 ( 9C;    4) ACTA CRYST. VOL.B27, . . . . . 2289, (1971).
                                 ( 5 ; 1268) CHEM. APPL.THERMAL NEUTRON SCAT, 201, (1973).

3984  SPEDDING........(F.H.)..:  ( 5 ; 1541) PHYS. REV. 176,. . . . . . . .  712, (1968).
                                 ( 5 ; 1401) J. CHEM. PHYSICS VOL.53, . . . 1455, (1970).
                                 ( 5 ; 1400) J. CHEM. PHYSICS VOL.54, . . . 1995, (1971).
                                 ( 6 ; 2015) PHYS. REV. B-4,. . . . . . . . 2398, (1971).
                                 ( 5 ;  884) PHYS. LETT. VOL.45A NO.4,. . .  281, (1973).

3985  SPEKTOR.........(E.Z.)..:  ( 5 ; 2480) PHYS. MET. METALLOGR. VOL.18 N.3 134, (1964).

3986  SPENCER.........(R.R.)..:  (20B;   33) IAEA SYMP. (PILE RESEARCH)VIENNA 433, (1962).
                                 (20B;  113) NUCL. SCI. ENGNG. VOL.8, . . .  393, (1960).

3987  SPIELBERG.......(D.H.)..:  ( 6 ; 1871) J. CHEM. PHYSICS VOL.54, . . . 2597, (1971).

3988  SPINRAD.........(R.J.)..:  (19A;   68) REV. SCI. INSTRUM. VOL.35, . . 1150, (1964).

3989  SPOL≠NIK........(Z.A.)..:  ( 8 ;  154) SOV. PHYS. SOLID STATE VOL.14, 1774, (1973).

3990  SPOONER.........(S.)....:  ( 6 ;  750) PHYS. REV. 142,. . . . . . . .  291, (1966).
                                 ( 8 ;   22) J. APPL. PHYS. VOL.40,. . . . . 1226, (1969).
                                 ( 6 ;  801) AIP CONF. PROC. VOL.5,. . . . . 1334, (1971).
                                 ( 5 ;  984) MAGN. AND MAG. MATERIALS-1971, 1415, (1972).
                                 ( 5 ;  974) A.I.P. CONF. PROC. NO.10 PART 2, 1163, (1973).

3991  SPOSITO.........(G.)....:  ( 6 ; 1147) PHYS. REV. A VOL.3,. . . . . .  820, (1971).

3992  SPOWART.........(A.R.)..:  (20C;   62) NUCL. INST. MET. 75, . . . . .   35, (1969).

3993  SPREVAK.........(D.)....:  ( 6 ;  349) NUCL. SCI. ENG. VOL.35,. . . .   80, (1969).
                                 ( 6 ; 1903) NUKLEONIK VOL.12,. . . . . . .   87, (1969).
                                 ( 5 ;  196) NUCL. SCI. ENG. VOL.42,. . . .  137, (1970).

3994  SPRINGER........(T.)....:  ( 5 ; 1307) IAEA SYMPOSIUM (VIENNA), . . .  323, (1960).
                                 ( 5 ; 2062) Z. NATURFORSCH. VOL.15A, . . .  828, (1960).
                                 ( 5 ; 1305) NUKLEONIK VOL.3,. . . . . . . .  110, (1961).
                                 ( 5 ;  790) Z. KRIST. VOL.116, . . . . . .  328, (1961).
                                 ( 5 ; 1299) Z. NATURFORSCH. VOL.16A,. . . .  112, (1961).
                                 ( 5 ; 1297) Z. PHYSIK VOL.164,. . . . . . .  111, (1961).
                                 (20A;   54) NUKLEONIK VOL.4,. . . . . . . .   23, (1962).
                                 (22A;   74) Z. PHYSIK VOL.157,. . . . . . .  386, (1962).
                                 (20A;   24) J. NUCL. ENERGY VOL.17,. . . .  217, (1963).
                                 (16 ;   52) NUKLEONIK VOL.6,. . . . . . . .   87, (1964).
                                 (10C;  137) Z. ANGEW. PHYS. VOL.18,. . . .  295, (1965).
                                 (20A;   12) ANN. REV. OF NUCLEAR SCI. VOL.16 207, (1966).
                                 ( 6 ; 1740) BULL. SOC. FRANC. MIN. CRIST. 90 428, (1967).
                                 ( 6 ; 1133) PHYS. STAT. SOL. VOL.23,. . .  K155, (1967).
                                 ( 6 ; 2021) IAEA SYMP. COPENHAGEN, VOL.1,.  245, (1968).
                                 (24 ;   12) ELECTRONIC STRUCTURES IN SOLIDS, 153, (1969).
                                 ( 6 ; 2020) PROC/INTERNAT. SYMP/PHYS/SE/TE,  299, (1969).
                                 ( 6 ; 2167) DISC. FARADAY SOC. NO.50,. . .   74, (1970).
                                 ( 5 ; 2013) J. PHYS. CHEM. SOLIDS VOL.31,. 2361, (1970).
                                 (17A;   20) ATOMKERNENERGIE VOL.17,. . . .   15, (1971).
                                 ( 6 ;  958) Z. NATURFORSCH. A VOL.26A NO.3,  575, (1971).
                                 (10E;   18) KOLL. Z. Z. POLYM. BAND 250, .  993, (1972).
                                 ( 6 ; 1705) PHYS. LETT. VOL.45A, . . . . .  479, (1973).
                                 (17A;   52) J. APPL. CRYST. VOL.7, . . . .   96, (1974).

3995  SQUIRES.........(G.L.)..:  ( 6 ; 1746) PROC. ROY. SOC. A VOL.212, . .  192, (1952).
                                 ( 5 ; 1263) PHYS. REV. VOL.90, . . . . . . 1125, (1953).
                                 ( 5 ; 2085) PROC. PHYS. SOC. A VOL.67, . .  248, (1954).
                                 ( 6 ;  981) PROC. ROY. SOC. A VOL.230, . .   19, (1955).
                                 (10C;   70) PHYS. REV. VOL.103,. . . . . .  304, (1956).
                                 ( 6 ;   25) IAEA SYMP. CHALK RIVER, VOL.2,   55, (1962).
                                 (10B;   27) IAEA SYMP. CHALK RIVER, VOL.2,   71, (1962).
                                 (10B;    9) ARK. FYS. VOL.25, PAPER 3,. . .   21, (1963).
                                 ( 6 ;  580) COPENHAGEN CONF.-LATT.DYNAMICS,   53, (1963).
                                 ( 6 ;   39) ARK. FYS. VOL.26 PAPER 17, . .  223, (1964).
                                 ( 6 ; 1411) BROOKHAVEN SYMPOSIUM,. . . . .   78, (1965).
                                 ( 6 ; 1410) PROC. PHYS. SOC. 88,. . . . . .  919, (1966).
                                 ( 6 ; 1397) IAEA SYMP. COPENHAGEN, VOL.1,.  215, (1968).
                                 ( 6 ; 2298) J. OF PHYS.-C, SER. 2, VOL.2, 2366, (1969).
                                 ( 6 ; 1605) J. PHYS.F: METAL PHYS. VOL.1,.  244, (1971).
                                 ( 6 ;  839) IAEA SYMP. GRENOBLE,. . . . . .  403, (1972).
                                 ( 6 ; 1403) PROC. ROY. SOC. A 326 NO.1566,  347, (1972).

3996  SRIKANTA........(S.)....:  ( 5 ; 1917) ACTA CRYST. VOL.18,. . . . . .  567, (1965).
                                 ( 5 ; 1485) ACTA CRYST. VOL.B24,. . . . . . 1176, (1968).
                                 ( 5 ; 1433) ACTA CRYST. VOL.B26,. . . . . .  827, (1970).
                                 ( 5 ; 1503) ACTA CRYST. VOL.B26,. . . . . .   77, (1970).

3997  SRINIVASAN......(B.S.)..:  ( 5 ; 1824) PHYS. LETT. VOL.15,. . . . . .  225, (1965).
                                 (12B;   15) INDIAN J. PURE APPL. PHYS. VOL.7 546, (1969).
                                 ( 5 ; 2142) J. PHYS. CHEM. SOLIDS VOL.30,. 1941, (1969).
                                 ( 5 ; 1663) J. PHYS. CHEM. SOLIDS 30,. . .  939, (1969).
                                 (12A;   20) NUCL./S.S. PHYS. SYMP. ABSTR.,       (1972).
                                 ( 6 ; 1482) PHYS. LETT. VOL.40A, . . . . .  101, (1972).

3998  SRINIVASAN......(G.)....:  (10D;   41) PHYS. REV. A VOL.5,. . . . . . 1528, (1972).

3999  SRINIVASAN......(R.)....:  (10D;   12) IAEA SYMPOSIUM BOMBAY, VOL.1,.  325, (1964).

4000  SRINIVASAN......(V.)....:  ( 5 ; 2022) PHYS. REV. B VOL.9,. . . . . .  130, (1974).

4001  SRIVASTAVA......(C.M.)..:  ( 5 ; 2955) PHYS. REV. VOL.181,. . . . . .  969, (1969).

4002  SRIVASTAVA......(P.L.)..:  (10B;   34) IAEA SYMPOSIUM BOMBAY, VOL.1,.   49, (1964).
                                 ( 6 ; 1364) J. PHYS. F VOL.3,. . . . . . . 1388, (1973).
```

```
4003  SRIVASTAVA......(R.S.)..:   (10B,    86) J. PHYS.F: METAL PHYS. VOL.2,..   247, (1972).
                                  ( 6 ,    28) PHYS. LETT. VOL.38A,. . . . . .   527, (1972).
                                  ( 6 ,  1362) PHYS. REV. B VOL.6,. . . . . .   2192, (1972).
                                  (10B,   197) PHYS. STAT. SOLIDI B VOL.56 NO.1  327, (1973).

4004  STACEY-JR.......(M.)..:     (16 ,    89) TRANS. AM. NUCL. SOC. VOL.13,. .  726, (1970).

4005  STAGER..........(C.V.)..:   ( 5 ,  1768) CAN. J. PHYS. 49,. . . . . . .   979, (1971).
                                  ( 5 ,   741) CAN. J. PHYS. VOL.50,. . . . .  3079, (1972).

4006  STALLARD........(J.M.)..:   ( 5 ,    49) DISS. ABS. VOL.32,. . . . . . .35538, (1971).
                                  ( 5 ,    56) PHYS. REV. A VOL.8,. . . . . .   368, (1973).

4007  STALLINGS.......(G.R.)..:   ( 5 ,   443) PHYS. STAT. SOL. 16, . . . . . .  K17, (1966).

4008  STAMENKOVICH....(S.S.)..:   (11B,    44) SOV. PHYS. SOL. STATE, 9,. . .   575, (1967).
                                  ( 7A,    89) SOV. PHYS. SOL. STATE, 10,. . .  675, (1968).

4009  STAMENKOVIC.....(S.)....:   ( 7B,    33) J. LOW TEMP. PHYS. VOL.9,. . .   475, (1972).
                                  ( 7B,    32) J. LOW TEMP. PHYS. VOL.9,. . .   485, (1972).
                                  ( 6 ,  1276) PHYS. STAT. SOL. VOL.49,. . .    277, (1972).

4010  STAMPER-JR......(J.F.)..:   ( 5 ,  1604) ACTA CRYST. 16,. . . . . . . .   352, (1963).

4011  STAMPFEL........(J.P.)..:   ( 5 ,   979) A.I.P. CONF. PROC. NO.10 PART 1,  98, (1973).

4012  STANCIC.........(V.)....:   ( 7A,    30) FIZIKA (YUGOSLAVIA) VOL.4, SUPL.  11, (1971).
                                  ( 5 ,  1290) RAD. ZAVODA FIZ. (YUGOSLAVIA),.  41, (1971).

4013  STANFORD........(C.P.)..:   (12B,    41) PHYS. REV. VOL.94,. . . . . . .  374, (1954).

4014  STARFELT........(N.)....:   (19A,    65) PROC./SEMINAR/LOW-EN. NUCL. PHYS 138, (1967).

4015  STARITSYN.......(V.E.)..:   (22A,    68) SOV. PHYS. CRYST. VOL.7,. . . .   58, (1962).
                                  (22A,    69) SOV. PHYS. CRYST. VOL.8,. . . .  234, (1963).

4016  STARK...........(D.S.)..:   (20B,   117) PATENT USA 3683190,. . . . . . . . ,  (1969).

4017  STARR-JR........(E.F.)..:   (22C,    13) DISS. ABSTR. B VOL.33 NO.11, . . 5435, (1973).

4018  STARSBINETS.....(S.S.)..:   ( 6 ,  2291) SOV. PHYS. SOL. STATE VOL.10,. .  511, (1968).

4019  STASSIS.........(C.)....:   ( 5 ,   712) J. APPL. PHYS. VOL.41,. . . . .  1146, (1970).
                                  ( 8 ,    95) PHYS. REV. LETT. VOL.24,. . . .  1415, (1970).
                                  ( 6 ,  1158) IAEA SYMP. GRENOBLE,. . . . . .   129, (1972).
                                  ( 5 ,   711) PHYS. REV. B VOL.5,. . . . . .   1040, (1972).
                                  ( 5 ,   884) PHYS. LETT. VOL.45A NO.4,. . .    281, (1973).
                                  ( 5 ,   552) PHYS. REV. LETT. VOL.31,. . . .  1498, (1973).

4020  STAUB...........(H.)....:   (12B,    34) PHYS. REV. VOL.64, . . . . . . .   47, (1943).

4021  STAVINSKII......(V.S.)..:   (17B,    69) SOV. J. NUCL. PHYS. VOL.12,. . .  960, (1970).

4022  STECHER-RASMUSSE(F.)....:   (20D,    10) J. NUCL. ENERGY VOL.17,. . . . .  227, (1963).
                                  (12B,    25) NUCL. INST. MET. VOL.45,. . . .  293, (1966).
                                  (12A,    19) NUCL. PHYS. A VOL.A181,. . . . .  241, (1972).

4023  STEDMAN.........(R.)....:   ( 6 ,  1192) PHYS. REV. VOL.108,. . . . . .   1346, (1957).
                                  (23 ,    43) J. NUCL. ENERGY VOL.6,. . . . .   222, (1958).
                                  (20B,     7) ARK. FYS. VOL.16 PAPER 19,. . .   199, (1959).
                                  (20C,    93) REV. SCI. INSTRUM. VOL.31,. . .  1156, (1960).
                                  ( 6 ,    23) IAEA SYMP. BOMBAY, VOL.1,. . .    211, (1964).
                                  ( 6 ,    47) PHYS. REV. LETT. 15,. . . . . .   634, (1965).
                                  ( 6 ,    50) PHYS. REV. 145,. . . . . . . .    492, (1966).
                                  ( 6 ,  1827) PHYS. REV. 162,. . . . . . . .    545, (1967).
                                  ( 6 ,  1812) PHYS. REV. 162,. . . . . . . .    549, (1967).
                                  ( 6 ,  1822) PHYS. REV. 163,. . . . . . . .    567, (1967).
                                  ( 6 ,  1634) IAEA SYMP. COPENHAGEN, VOL.1,. .  295, (1968).
                                  (18B,    10) REV. SCI. INSTRUM. VOL.39,. . .   637, (1968).
                                  (18A,    26) REV. SCI. INSTRUM. VOL.39,. . .   878, (1968).
                                  (21 ,    47) REV. SCI. INSTRUM. VOL.40,. . .   249, (1968).
                                  (17A,    39) IAEA SYMP. (INSTRUMENT.), VIENNA   37, (1969).
                                  ( 6 ,  1637) PHYS. REV. 178,. . . . . . . .   1496, (1969).
                                  ( 6 ,    22) PHYS. REV. B-2,. . . . . . . .   4743, (1970).
                                  ( 6 ,  2305) J. PHYS.F: METAL PHYS. VOL.1,. .  785, (1971).

4024  STEEB...........(S.)....:   (14B,   126) SPRINGER TRACTS IN MOD. PHYS. 47    1, (1968).
                                  ( 5 ,   746) PHYS. CHEM. LIQ. VOL.4,. . . . .   39, (1973).
                                  ( 5 ,   972) PHYS. KOND. MATER. VOL.17,. . .    11, (1973).

4025  STEELE..........(D.)....:   ( 5 ,   378) SOL. STATE COMM. VOL.8,. . . . .  171, (1970).
                                  ( 6 ,   441) J. PHYS. C VOL.5,. . . . . . .   2677, (1972).
                                  ( 5 ,  2979) J. PHYS. C VOL.7,. . . . . . . .    1, (1974).

4026  STEELE..........(W.A.)..:   (14B,    37) J. CHEM. PHYS. VOL.42, . . . . . 1863, (1965).

4027  STEFANON........(M.)....:   (24 ,    61) NUCL. INST. MET. 58, . . . . . .    1, (1968).

4028  STEFANOVSKII....(E.P.)..:   (11A,    62) SOV. PHYS. SOL. STATE VOL.11,. . 1566, (1969).

4029  STEHR...........(H.)....:   (20B,   157) Z. KRIST. VOL.117,. . . . . . .   135, (1962).
                                  (20B,   158) Z. KRIST. VOL.118,. . . . . . .   263, (1963).

4030  STEICHELE.......(E.)....:   (19B,    26) PHYS. LETT. VOL.44A NO.3,. . . .  165, (1973).
                                  ( 5 ,   438) Z. PHYSIK VOL.264,. . . . . . .   427, (1973).

4031  STEINBERGER.....(J.)....:   ( 8 ,    72) PHYS. REV. VOL.76, . . . . . . .  994, (1949).

4032  STEINER.........(M.)....:   ( 5 ,   669) SOL. STATE COMM. VOL.11, . . . . 1471, (1972).
                                  ( 5 ,   673) SOL. STATE COMM. VOL.11,. . . .    73, (1972).
                                  ( 5 ,   674) A.I.P. CONF. PROC. NO.10 PART 1,  664, (1973).
                                  ( 6 ,   573) INT. J. MAGN. VOL.5,. . . . . .    95, (1973).
                                  ( 6 ,   572) SOL. STATE COMM. VOL.12 NO.6,. .  537, (1973).
                                  ( 5 ,   675) SOL. STATE COMM. VOL.14 NO.9,. .  841, (1974).

4033  STEINITZ........(M.O.)..:   ( 5 ,  1819) A.I.P. CONF. PROC. NO.10 PART 2, 1138, (1973).

4034  STEINMAN........(D.K.)..:   ( 6 ,  1301) J. PHYS. CHEM. SOLIDS VOL.30,. .  449, (1969).
                                  ( 6 ,  2324) J. PHYS. CHEM. SOLIDS VOL.32,. . 1573, (1971).
                                  ( 6 ,   453) IAEA SYMP. GRENOBLE, . . . . . .  219, (1972).

4035  STEINSVOLL......(O.)....:   (12B,    52) REV. SCI. INSTRUM. VOL.33,. . .   524, (1962).
                                  (12A,    22) PHYS. LETT. 6,. . . . . . . . .    47, (1963).
                                  ( 6 ,   751) PHYS. REV. LETT. 15,. . . . . .   146, (1965).
                                  ( 6 ,   478) J. APP. PHYS. 37,. . . . . . .   1052, (1966).
                                  ( 6 ,   841) PHYS. REV. 154,. . . . . . . .    508, (1967).
                                  ( 6 ,  1738) PHYS. REV. 156,. . . . . . . .    623, (1967).
                                  ( 6 ,  2130) PHYS. REV. 161,. . . . . . . .    499, (1967).
                                  (12B,    13) IAEA SYMP. COPENHAGEN, VOL.2,. .  395, (1968).
                                  ( 6 ,   727) IAEA SYMP. COPENHAGEN, VOL.2,. .   45, (1968).
                                  (20B,   119) PHYS. LETT. VOL.26A, . . . . . .  469, (1968).
                                  ( 6 ,  1375) INT. J. MAGN. VOL.3 NO.4,. . . .  349, (1972).
                                  (19A,    54) NUCL. INST. MET. VOL.106,. . . .  453, (1973).
```

```
4036  STEIN..........(H.)....:  ( 6 ,  954) CONF. INTERN., RENNES, . . . . .   156, (1971).
                                ( 6 ,  957) J. CHEM. PHYS. VOL.57, . . . . .  1726, (1972).

4037  STEIN..........(R.)....:  (20C,   86) REV. PHYS. APPL. (FRANCE) VOL.4,  259, (1969).

4038  STELZER........(F.)....:  (23 ,    9) CRYOGENICS VOL.11, . . . . . . .  107, (1971).

4039  STEMPLE........(N.R.)..:  ( 5 ,    6) ACTA CRYST. 21,. . . . . . . . .  237, (1966).

4040  STEPANENKO.....(V.A.)..:  ( 5 , 1248) BULL. ACAD. SCI. USSR/PHYS. 31,   334, (1967).

4041  STEPANOV.......(A.V.)..:  ( 7B,  118) SOV. PHYS. JETP. 16, . . . . . . 1624, (1963).
                                ( 9B,  107) SOV. PHYS. JETP. 20, . . . . . .   94, (1965).

4042  STEPHENSON.....(T.E.)..:  (12B,   41) PHYS. REV. VOL.94, . . . . . . .  374, (1954).

4043  STERN..........(H.)....:  (10C,   31) IAEA SYMPOSIUM VIENNA, . . . . .   61, (1960).

4044  STEVENSON......(R.W.H.):  ( 6 , 1462) J. APPL. PHYS. VOL.35, . . . . .  998, (1964).
                                ( 6 , 1458) PHYS. LETT. 8,. . . . . . . . .     9, (1964).
                                ( 6 , 1453) PROC. INT. CONF. (NOTTINGHAM), .   92, (1964).
                                ( 6 , 1461) PROC. PHYS. SOC. 85, . . . . . .  743, (1965).
                                ( 6 , 1991) PROC. PHYS. SOC. 87, . . . . . .  501, (1966).
                                ( 6 ,  493) PHYS. REV. LETT. 18, . . . . . .  162, (1967).
                                ( 6 ,  496) CAN. J. PHYS. 46,. . . . . . . . 1355, (1968).
                                ( 6 ,  494) J. APP. PHYS. 39,. . . . . . . . 1116, (1968).
                                ( 6 , 1476) J. APP. PHYS. 39,. . . . . . . . 1118, (1968).
                                ( 6 , 1455) SOL. STATE COMM. 6, . . . . . .   145, (1968).
                                ( 6 , 1322) CAN. J. PHYS. VOL.47, . . . . . 1983, (1969).
                                ( 6 , 1475) J. APPL. PHYS. VOL.40, . . . . .  991, (1969).
                                ( 6 ,  492) PHYS. REV. LETT. VOL.23, . . . .   86, (1969).
                                ( 6 , 1449) J. APPL. PHYS. 41, . . . . . . .  896, (1970).
                                ( 6 , 1323) J. PHYSIQUE TOME 32 COL.1 VOL.II 1184, (1971).
                                ( 6 , 1269) J. PHYS. C VOL.4, . . . . . . . 2127, (1971).
                                ( 6 ,  501) PHYS. REV. LETT. VOL.27, . . . . 1442, (1971).
                                ( 6 , 1307) IAEA SYMP. GRENOBLE, . . . . . .  581, (1972).
                                ( 6 , 1321) MAGN. AND MAG. MATERIALS-1971, . 1315, (1972).
                                ( 6 ,  497) J. PHYS. C VOL.6,. . . . . . . . 2997, (1973).

4045  STEVENSON......(R.)....:  ( 6 ,  724) C.R. ACAD. SC. (PARIS), 267-B, .   61, (1968).

4046  STEWART........(A.T.)..:  ( 5 , 1263) PHYS. REV. VOL.90, . . . . . . . 1125, (1953).
                                ( 6 ,   43) PHYS. REV. VOL.100,. . . . . . .  756, (1955).
                                ( 6 ,  981) PROC. ROY. SOC. A VOL.230, . . .   19, (1955).
                                ( 6 ,   57) REV. MOD. PHYS. VOL.30,. . . . .  236, (1958).
                                ( 6 , 2243) REV. MOD. PHYS. VOL.30,. . . . .  250, (1958).
                                ( 6 , 1629) PHYS. REV. VOL.128,. . . . . . . 1112, (1962).

4047  STEWART........(R.J.)..:  ( 5 , 1220) CRYSTAL LATTICE DEFECTS VOL.2, .  105, (1971).

4048  STEYERL........(A.)....:  ( 7A,   53) PHYS. LETT. VOL.29B, . . . . . .   33, (1969).
                                (19B,   19) NUCL. INST. MET. VOL.101,. . . .  295, (1972).
                                (16 ,   92) Z. PHYSIK BAND 252,. . . . . . .  371, (1972).
                                (20A,   84) Z. PHYSIK BAND 254,. . . . . . .  169, (1972).
                                ( 5 ,  686) Z. PHYSIK VOL.250, . . . . . . .  166, (1972).
                                (19B,   22) NUCL. INST. MET. VOL.114,. . . .  198, (1974).

4049  STEYN..........(J.J.)..:  (20C,   59) NUCL. INST. MET. 74, . . . . . .  123, (1969).

4050  STICH..........(V.)....:  (12A,    8) CZECH. J. PHYS. B VOL.19,. . . .  278, (1969).

4051  STICKLER.......(J.J.)..:  ( 5 , 2198) PHYS. REV. 164,. . . . . . . . .  768, (1967).

4052  STIEGLITZ......(R.G.)..:  ( 5 ,  526) THESIS (231PP.), . . . . . . . . . , (1970).

4053  STILES.........(J.A.R.):  ( 5 ,  741) CAN. J. PHYS. VOL.50,. . . . . . 3079, (1972).

4054  STILLER........(H.H.)..:  ( 6 , 1032) IAEA SYMPOSIUM (VIENNA), . . VOL.1,  363, (1960).
                                ( 6 ,  422) IAEA SYMP. COPENHAGEN, VOL.1,. .  599, (1968).
                                (14B,   34) IAEA SYMP. (INSTRUMENT.), VIENNA   19, (1969).

4055  STILLER........(H.)....:  ( 6 , 1068) PHYSICA VOL.27,. . . . . . . . .  373, (1961).
                                ( 6 ,  254) IAEA SYMP. CHALK RIVER, VOL.1, .  423, (1962).
                                ( 6 ,  252) IAEA SYMP. CHALK RIVER, VOL.2, .  281, (1962).
                                ( 6 ,  258) Z. PHYSIK VOL.166, . . . . . . .  393, (1962).
                                (17A,   48) INTERN. SUMMER SCHOOL, MOL,. . .  580, (1963).
                                ( 6 , 1005) IAEA SYMP. BOMBAY, VOL.2,. . . .  179, (1964).
                                ( 6 ,  262) IAEA SYMP. BOMBAY, VOL.2,. . . .  291, (1964).
                                ( 6 , 1105) PHYSICA VOL.30,. . . . . . . . .  931, (1964).
                                ( 6 ,  266) PHYS. STAT. SOL. 5, . . . . . .   511, (1964).
                                ( 5 ,  252) NATURWISSENSCHAFTEN VOL.52,. . .  512, (1965).
                                ( 6 , 1833) PHYSICA VOL.31,. . . . . . . . . 1537, (1965).
                                ( 6 ,  271) PHYS. STAT. SOL. 18,. . . . . .   795, (1966).
                                (14A,   13) DISC. FARADAY SOC. NO.43,. . . .  160, (1967).
                                ( 6 ,  344) PHYS. STAT. SOL. 19, . . . . . .  781, (1967).
                                ( 6 ,  989) BER. BUNS. PHYS. CHEM. VOL.72, .   94, (1968).
                                ( 6 ,  345) PHYS. STAT. SOL. 27, . . . . . .  269, (1968).
                                ( 6 , 1297) 2ND SUMMER SCHOOL S.S. PHYS.,. .  120, (1968).
                                ( 6 , 1009) DISC. FARADAY SOC. NO.48,. . . .   69, (1969).
                                ( 5 , 1456) PHYS. STAT. SOL. VOL.42, . . . .  207, (1970).
                                ( 6 ,  954) CONF. INTERN., RENNES, . . . . .  156, (1971).
                                (23 ,    9) CRYOGENICS VOL.11, . . . . . . .  107, (1971).
                                ( 7B,   38) KERNTECHNIK VOL.13,. . . . . . .  525, (1971).
                                ( 6 ,  958) Z. NATURFORSCH. A VOL.26A NO.3,.  575, (1971).
                                ( 5 , 1458) 1ST EUR. CONF./CONDENSED MATTER,   60, (1971).
                                ( 5 , 1461) IAEA SYMP. GRENOBLE, . . . . . .  825, (1972).
                                ( 6 ,  957) J. CHEM. PHYS. VOL.57, . . . . . 1726, (1972).

4056  STILLINGER.....(F.H.)..:  ( 6 , 1076) J. CHEM. PHYS. VOL.55, . . . . . 3336, (1971).
                                ( 5 ,  794) J. CHEM. PHYS. VOL.60, . . . . . 1545, (1974).

4057  STIMLER........(M.)....:  (20C,   71) NUCL. INST. MET. VOL.87, . . . .  299, (1970).

4058  STINCHCOMBE....(R.B.)..:  (11B,   26) PHYS. REV. B VOL.9,. . . . . . . 3786, (1974).

4059  STIRLING.......(G.C.)..:  (19B,   38) UKAEA REP. AERE-R-6035 (8 PP.),.      , (1969).
                                ( 6 ,  566) J. CHEM. PHYS. VOL.52, . . . . . 1828, (1970).
                                ( 6 ,  567) J. CHEM. PHYS. VOL.52, . . . . . 2730, (1970).
                                (23 ,   71) REPORT AERE-R 6246 (25PP.),. . .      , (1970).
                                ( 6 , 2043) SPEC. DISC. FARADAY SOC. NO.1,.   194, (1970).
                                ( 6 , 1081) NATURE (PHYS. SCI.) VOL.230, . .  192, (1971).
                                ( 6 , 2199) FARADAY SYMP. CHEM. SOC. NO.6, .  135, (1972).
                                (17A,   24) CHEM. APPL. THERMAL NEUTRON SCAT   31, (1973).

4060  STIRLING.......(W.G.)..:  ( 6 , 2106) J. PHYSIQUE TOME 33 COL.2, . . .  135, (1972).
                                ( 6 , 1857) J. PHYS. F: METAL PHYS. VOL.2,. .  421, (1972).
                                ( 6 , 2099) J. PHYS. C VOL.5,. . . . . . . . 2711, (1972).
                                ( 6 , 1858) SOL. STATE COMM. VOL.11 NO.1,. .  271, (1972).

4061  STOCKMEYER.....(R.)....:  (20B,  116) NUKLEONIK VOL.4, . . . . . . . .  266, (1962).
                                (18A,   23) NUKLEONIK VOL.5, . . . . . . . .  121, (1963).
                                ( 6 ,  344) PHYS. STAT. SOL. 19, . . . . . .  781, (1967).
                                ( 6 ,  345) PHYS. STAT. SOL. 27, . . . . . .  269, (1968).
                                ( 6 ,   11) DISC. FARADAY SOC. NO.48, . . . .  156, (1969).
                                ( 6 ,  954) CONF. INTERN., RENNES, . . . . .  156, (1971).
                                ( 6 ,  957) J. CHEM. PHYS. VOL.57, . . . . . 1726, (1972).
```

```
4061  STOCKMEYER......(R.)...:   (19B,     9)  KERNTECHNIK VOL.14,. . . . . . .      13, (1972).

4062  STOCK...........(A.D.)..:  (17B,    38)  J. SCI. INSTRUM.(J. PHYS. E)S2,1    951, (1968).

4063  STOICHEFF.......(B.P.)..:  ( 6 ,  1348)  PHYS. REV. LETT. VOL.31,. . . .     296, (1973).

4064  STOKELY.........(J.R.)..:  ( 5 ,   414)  NUCL. SCI. ENG. VOL.37,. . . . .    228, (1969).

4065  STOKES..........(G.E.)..:  ( 5 ,  2508)  NUCL. SCI. ENGNG. VOL.33,. . . .     16, (1968).
                                 ( 5 ,  2675)  PROC/CONF/NEUTRON C. S. + TECH.,    693, (1968).

4066  STOLER..........(P.)....:  (22B,    35)  NUCL. INST. MET. VOL.91, . . . .    541, (1971).

4067  STOLL...........(E.)....:  ( 5 ,    78)  COLL. INTER. N.126 (GRENOBLE), .     23, (1963).
                                 (22A,    73)  Z. ANGEW. MATH. PHYS. VOL.16,. .    817, (1965).
                                 ( 5 ,  2953)  Z. ANGEW. MATH. PHYS. VOL.16,. .    820, (1965).
                                 ( 5 ,  2890)  ACTA. CRYST. 21, . . . . . . . .    765, (1966).
                                 (10B,   111)  PHYS. KONDENS. MATERIE VOL.5,. .    331, (1966).
                                 (10B,   110)  PHYS. KONDENS. MATERIE VOL.5,. .    364, (1966).
                                 ( 5 ,  1845)  SOL. STATE COMM. 4,. . . . . . .    473, (1966).
                                 ( 6 ,   589)  SOL. STATE COMM. 4,. . . . . . .     79, (1966).
                                 ( 5 ,  1654)  ACTA CHEM. SCAND. VOL.21,. . . .   1543, (1967).
                                 ( 5 ,  1847)  Z. KRIST. VOL.125, . . . . . . .    120, (1967).
                                 ( 6 ,   585)  HELV. PHYS. ACTA VOL.41, . . . .    399, (1968).
                                 ( 6 ,    24)  IAEA SYMP. COPENHAGEN, VOL.1,. .    101, (1968).
                                 (14A,    66)  REPORT AF-SSP-27,. . . . . . . .    . , (1968).

4068  STORMS..........(E.K.)..:  ( 5 ,  2933)  ACTA CRYST. 19,. . . . . . . . .      6, (1965).
                                 ( 5 ,  5651)  ACTA CRYST. VOL.B28; . . . . . .   3102, (1972).

4069  STORY...........(J.S.)..:  (24 ,    31)  2ND INTERNAT. CONF./NUCL. DATA,    . , (1970).

4070  STOUT...........(J.W.)..:  ( 5 ,  2840)  J. APPL. PHYS. VOL.40, . . . . .   1136, (1969).

4071  STRAKER.........(E.A.)..:  ( 6 ,   305)  BROOKHAVEN SYMPOSIUM.,. . . . .     142, (1965).
                                 ( 6 ,   306)  J. CHEM. PHYS. VOL.43, . . . . .   4134, (1965).
                                 ( 6 ,   307)  DISS. ABS. 27, . . . . . . . . .   507B, (1967).
                                 (24 ,    80)  NUCL. SCI. ENGNG. VOL.34,. . . .    114, (1968).

4072  STRAUSER........(W.A.)..:  ( 5 ,  1081)  PHYS. REV. VOL.81, . . . . . . .    483, (1951).
                                 ( 5 ,  1753)  PHYS. REV. VOL.83, . . . . . . .    333, (1951).

4073  STREETMAN.......(G.B.)..:  ( 6 ,  2353)  NUCL. SCI. ENGNG. VOL.5, . . . .     99, (1960).

4074  STREET..........(R.)....:  ( 5 ,   473)  PROC. ROY. SOC. A VOL.217, . . .    252, (1953).
                                 ( 5 ,  1636)  NATURE VOL.175,. . . . . . . . .    518, (1955).
                                 ( 5 ,  1679)  PROC. ROY. SOC. A VOL.241, . . .    223, (1957).
                                 ( 5 ,   128)  PROC. PHYS. SOC. VOL.72, . . . .    470, (1958).
                                 ( 5 ,   534)  PHYS. REV. LETT. 10, . . . . . .    210, (1963).
                                 ( 5 ,   540)  PROC. PHYS. SOC. 86, . . . . . .   1143, (1965).
                                 ( 5 ,   533)  PHYS. REV. 141,. . . . . . . . .    510, (1966).

4075  STRICOS.........(D.P.)..:  ( 5 ,  2964)  NUCL. SCI. ENG. VOL.46,. . . . .    314, (1971).

4076  STRINDEHAG......(O.)....:  (20C,    71)  NUCL. INST. MET. VOL.87, . . . .    299, (1970).
                                 (20C,    73)  NUCL. INST. MET. 78, . . . . . .    173, (1970).

4077  STRINGFELLOW....(M.W.)..:  ( 6 ,   830)  PROC. PHYS. SOC. 84, . . . . . .     55, (1964).
                                 ( 6 ,   833)  PHYS. REV. LETT. VOL.14; . . . .    698, (1965).
                                 ( 6 ,   818)  PROC. PHYS. SOC. 89, . . . . . .    419, (1966).
                                 (11B,    51)  UKAEA AERE REP. R4535 (22 PP.),.   . , (1966).
                                 ( 6 ,   869)  J. OF PHYS. C VOL.1, . . . . . .   1699, (1968).
                                 ( 6 ,   735)  J. OF PHYS. C VOL.1, . . . . . .    950, (1968).
                                 ( 6 ,  1227)  J. APP. PHYS. VOL.40,. . . . . .   1443, (1969).
                                 ( 6 ,  1221)  PHYS. REV. LETT. VOL.23, . . . .     81, (1969).
                                 ( 6 ,   704)  COLLOQUE INTERNAT. NO.180 TOME 2    283, (1970).
                                 ( 6 ,  1228)  J. APPL. PHYS. 41, . . . . . . .   1176, (1970).
                                 ( 6 ,  1223)  METAL PHYS. VOL.3, . . . . . . .    189, (1970).
                                 ( 6 ,  2316)  PHYS. REV. LETT. VOL.24, . . . .    897, (1970).
                                 ( 6 ,   702)  CAN. J. PHYS. 49,. . . . . . . .   2875, (1971).
                                 ( 6 ,   705)  J. PHYSIQUE TOME 32 COL.1 VOL.II   1179, (1971).
                                 ( 6 ,  1435)  J. PHYSIQUE TOME 32 COL.1 VOL.II   1186, (1971).
                                 ( 6 ,  1948)  IAEA SYMP. GRENOBLE, . . . . . .    501, (1972).
                                 ( 6 ,  1777)  IAEA SYMP. GRENOBLE, . . . . . .    669, (1972).
                                 ( 6 ,  1540)  J. PHYSIQUE TOME 33 COL.2, . . .     59, (1972).
                                 ( 6 ,  1857)  J. PHYS.F: METAL PHYS. VOL.2,. .    421, (1972).
                                 ( 6 ,  1778)  A.I.P. CONF. PROC. NO.10 PART 2,   1403, (1973).
                                 ( 6 ,   849)  SOL. STATE COMM. VOL.12 NO.8,. .    795, (1973).

4078  STRONG..........(K.A.)..:  ( 6 ,   251)  PHYS. REV. VOL.124,. . . . . . .    460, (1961).
                                 (20B,    74)  NUCL. INST. MET. VOL.17, . . . .    129, (1962).
                                 ( 6 ,   311)  PHYS. REV. VOL.125,. . . . . . .    933, (1962).
                                 ( 6 ,   146)  PHYS. REV. VOL.128,. . . . . . .    562, (1962).
                                 ( 6 ,  1116)  J. CHEM. PHYSICS VOL.42, . . . .   1568, (1965).
                                 ( 6 ,   309)  J. CHEM. PHYSICS VOL.47, . . . .    421, (1967).
                                 ( 6 ,   162)  J. PHYS. CHEM. SOL. 28, . . . .     249, (1967).
                                 (17A,    34)  IAEA SYMP. COPENHAGEN, VOL.2,. .    323, (1968).
                                 ( 6 ,   319)  PHYS. REV. LETT. VOL.20, . . . .    983, (1968).
                                 ( 6 ,   652)  J. CHEM. PHYS. VOL.50, . . . . .   1030, (1969).
                                 ( 6 ,  1939)  J. CHEM. PHYS. VOL.52, . . . . .   4424, (1970).
                                 ( 6 ,   313)  J. CHEM. PHYS. VOL.58, . . . . .   1438, (1973).

4079  STRUB...........(A.)....:  (20C,    43)  NUCL. INST. MET. 35, . . . . . .    203, (1965).

4080  STRUEBING.......(V.)....:  ( 5 ,   408)  AIP CONF. PROC. VOL.5, . . . . .   1385, (1971).

4081  STRUGAR.........(P.V.)..:  (22C,    22)  J. OPTIMIZ. THEORY APPLIC. VOL.5    301, (1970).

4082  STUART..........(R.)....:  ( 7B,    16)  IAEA SYMPOSIUM VIENNA, . . . . .     75, (1960).

4083  STUBBINS........(W.F.)..:  (20D,    21)  NUCL. INST. MET. 60, . . . . . .    246, (1968).

4084  STUBER.........(W.)....:   (21 ,    37)  NUCL. INST. MET. 64, . . . . . .     52, (1968).

4085  STUCKY..........(G.D.)..:  ( 5 ,   248)  J. AMER. CHEM. SOC. VOL.93,. . .   5945, (1971).

4086  STUDACH.........(T.)....:  (19A,    72)  Z. ANGEW. MATH. PHYS. VOL.22,. .     62, (1971).

4087  STUHRMANN.......(H.B.)..:  (15 ,    17)  J. APPL. CRYST. VOL.7, . . . . .    173, (1974).

4088  STUMP...........(N.)....:  ( 6 ,  1737)  PHYS. LETT. 24A, . . . . . . . .    625, (1967).
                                 ( 6 ,  1714)  PHYS. LETT. 29A, . . . . . . . .     75, (1969).
                                 ( 6 ,  1719)  Z. PHYS. VOL.238,. . . . . . . .    389, (1970).
                                 ( 6 ,  1683)  PHYS. STAT. SOL. VOL.B-54, . . .    295, (1972).
                                 ( 6 ,  1012)  PHYSICA VOL.65,. . . . . . . . .    109, (1973).

4089  STUPAK..........(A.)....:  (22A,    37)  NUCL. INST. MET. 40, . . . . . .    153, (1966).

4090  STURM...........(W.J.)..:  ( 5 ,   201)  PHYS. REV. VOL.71; . . . . . . .    589, (1947).
                                 (20A,    63)  PHYS. REV. VOL.71; . . . . . . .    757, (1947).

4091  SUBRAMANIAM.....(R.)....:  ( 8 ,   110)  PROC. INDIAN ACAD. SCI. VOL.61,.    202, (1965).
```

4092 SUBRAMANIAN.....(R.)....: (10C,   42) J. PHYS. CHEM. SOLIDS VOL.19,. .   173, (1961).

4093 SUBRAMANIAN.....(T.S.)..: (24 ,   94) PHYS. REV. B VOL.3,. . . . . . 2929, (1971).

4094 SUBRAMANYAM.....(K.N.)..: (5 , 2139) J. OF PHYS.-C-, SER.2, VOL.4,. 2266, (1971).

4095 SUCK...........(J.B.)..: (6 , 1956) IAEA SYMP. GRENOBLE,. . . . .   435, (1972).
                               (6 , 1960) PROPERTIES LIQUID METALS,. . . .   111, (1973).

4096 SUDNIK-HRYNKIEWI(M.)....: (21 ,   40) NUKLEONIKA VOL.12,. . . . . . .   385, (1967).
                               (6 ,   502) PHYSICA VOL.35,. . . . .   441, (1967).
                               (6 ,   380) PHYSICA VOL.35,. . . . .   451, (1967).
                               (6 , 1016) PHYSICA VOL.35,. . . . .   457, (1967).
                               (6 , 1596) PHYSICA VOL.35,. . . . .   465, (1967).
                               (6 , 2007) PHYSICA VOL.35,. . . . .   469, (1967).
                               (6 , 1379) PHYS. LETT. 24A,. . . . . . .   517, (1967).
                               (6 , 1017) ACTA PHYS. POLON. VOL.33,. . .   419, (1968).
                               (6 ,   621) IAEA SYMP. COPENHAGEN, VOL.1,.   65, (1968).
                               (6 , 2003) IAEA SYMP. COPENHAGEN, VOL.2,.   237, (1968).
                               (6 ,   286) PHYSICA VOL.41,. . . . .   397, (1969).
                               (6 , 1388) SOV. PHYS. CRYST. VOL.15,. . . 1010, (1971).

4097 SUGAWARA........(M.)....: (22C,   43) NUCL. INST. MET. 71,. . . . . .   102, (1969).

4098 SUGIMOTO........(A.)....: (23 ,   20) IAEA SYMP. (PILE RESEARCH)VIENNA 117, (1960).

4099 SUGIMOTO........(M.)....: (5 ,   344) J. PHYS. SOC. JAPAN VOL.24,. .   446, (1968).

4100 SUMITA..........(K.)....: (20B,   59) J. NUCL. SCI. TECHNOL. VOL.7,.   381, (1970).

4101 SUMMERFIELD.....(G.C.)..: (7B,   74) PHYS. REV. 131,. . . . . . . 1149, (1963).
                               (7A,   69) PHYS. REV. 131,. . . . . . . 1153, (1963).
                               (7A,   11) ANN. OF PHYS. VOL.26,. . . . .   72, (1964).
                               (6 , 1888) BROOKHAVEN SYMPOSIUM,. . . . .   126, (1965).
                               (6 , 1897) J. CHEM. PHYS. VOL.43,. . . . 1079, (1965).
                               (10C,   61) PHYS. LETT. 16,. . . . .   225, (1965).
                               (6 , 1900) J. CHEM. PHYS. VOL.44,. . . .   184, (1966).
                               (7B,   24) J. CHEM. PHYSICS VOL.47,. . . 4923, (1967).
                               (7B,   83) PHYS. REV. 159,. . . . .   1, (1967).
                               (16 ,   10) CONF. NEUTRON THERMALIZ. VOL.I,.   283, (1968).
                               (6 ,   363) IAEA SYMP. COPENHAGEN, VOL.2,.   167, (1968).
                               (6 , 1911) J. CHEM. PHYSICS VOL.48,. . .   912, (1968).
                               (7B,   29) J. CHEM. PHYSICS VOL.49,. . .   890, (1968).
                               (6 , 1301) J. PHYS. CHEM. SOLIDS VOL.30,.   449, (1969).
                               (6 , 1892) J. POLYM. SCI/A-2 POLYM. PHYS. 7   405, (1969).
                               (7B,   86) PHYS. REV. VOL.188,. . . . . . 1445, (1969).
                               (7A,   34) J. PHYS. SOC. JAPAN VOL.28,. .   644, (1970).
                               (15 ,   14) IAEA SYMP. GRENOBLE,. . . . .   277, (1972).
                               (5 , 2331) SPECTROSCOPY BIOL. CHEM.,. . .   323, (1974).

4102 SUMSION........(H.T.)..: (5 , 2850) ACTA MET. VOL.1,. . . . .   390, (1953).

4103 SUNAKAWA........(S.)....: (7B,   106) PROGR. THEOR. PHYS. VOL.37,. . 1051, (1967).

4104 SUN............(K.H.)..: (20C,   36) NUCLEONICS VOL.17 APRIL,. . . .   116, (1959).

4105 SURIS..........(R.A.)..: (9D,   70) SOV. PHYS. SOL. STATE, 4,. . . 2533, (1962).

4106 SURJADI........(A.J.)..: (5 , 1510) ACTA CRYST. 23,. . . . .   578, (1967).

4107 SURKOVA........(I.V.)..: (6 ,   920) BU. ACAD. SCI. USSR PHYS. SR. 33 1754, (1969).
                               (17A,   22) BU. ACAD. SCI. USSR PHYS. SR. 33   36, (1969).

4108 SUSKI..........(W.)....: (5 , 2791) PHYS. STAT. SOL. 30,. . . . .   61, (1968).
                               (5 , 2792) PHYS. STAT. SOL. 34. . . . . K157, (1969).
                               (5 , 2815) ACTA PHYS. POLON. VOL.A43,. . .   631, (1973).

4109 SUSMAN.........(S.)....: (5 , 1423) J. CHEM. PHYS. VOL.56,. . . . . 3697, (1972).
                               (5 , 1957) J. CHEM. PHYS. VOL.58,. . . . . 2039, (1973).

4110 SUSZKIN........(A.)....: (5 , 1246) ATOMKERNENERGIE VOL.19 NO.4,.   325, (1972).

4111 SUTARNO.................: (5 ,   902) CANAD. J. CHEM. VOL.43,. . . . 2819, (1965).

4112 SUTTON.........(J.D.)..: (5 , 2455) DISS. ABS. VOL.32,. . . . . . .41568, (1972).
                               (9B,   66) NUCL. INST. MET. 99,. . . . .   453, (1972).
                               (22A,   48) NUCL. INST. MET. VOL.114,. . .   417, (1974).

4113 SUTTON.........(R.B.)..: (5 , 1256) PHYS. REV. VOL.72,. . . . . . 1147, (1947).

4114 SUVAL≠SKI.......(YA.)..: (5 , 2954) SOV. PHYS. SOL. STATE VOL.13,.   321, (1971).

4115 SUZAWA.........(T.)....: (5 ,   394) PHYS. STAT. SOL. VOL.44,. . .   K25, (1971).

4116 SUZUKI.........(K.)....: (22B,   16) J. PHYS. SOC. JAPAN VOL.17 B-II, 358, (1962).
                               (5 ,   723) J. PHYS. SOC. JAPAN VOL.18,. .   93, (1963).
                               (5 , 1022) REV. ELECT. COMMUN. LAB. VOL.12, 424, (1964).
                               (5 ,   963) J. APP. PHYS. 36,. . . . . . 1094, (1965).
                               (5 , 2067) PHYS. STAT. SOL. VOL.39,. . .   181, (1970).
                               (5 ,   922) PHYS. STAT. SOL. VOL.39,. . .   669, (1970).
                               (5 , 2422) PHYS. LETT. VOL.35A,. . . . .   315, (1971).
                               (5 , 2423) PHYS. STAT. SOL. VOL.47,. . .   581, (1971).
                               (5 ,   228) PHYS. LETT. VOL.45A NO.4,. . .   273, (1973).
                               (14A,   60) PHYS. STAT. SOLIDI B VOL.57 NO.1 351, (1973).
                               (14B,   102) PROPERTIES LIQUID METALS,. . . .   37, (1973).
                               (5 ,   316) CAN. J. PHYS. VOL.52,. . . . .   241, (1974).
                               (5 , 2063) J. PHYS. CHEM. SOLIDS VOL.35,.   585, (1974).

4117 SUZUKI.........(T.)....: (5 , 1805) ACTA CRYST. VOL.A28 PART S-4,.   196, (1972).
                               (5 , 1048) J. PHYS. SOC. JAPAN VOL.34,. .   911, (1973).

4118 SUZUOKA........(T.)....: (5 , 1043) ACTA CRYST. A24,. . . . . . .   513, (1968).

4119 SVAB...........(E.)....: (5 , 2889) A.I.P. CONF. PROC. NO.10 PART 2, 1603, (1973).

4120 SVENSON........(A.C.)..: (6 , 1418) J. PHYS. C. VOL.3,. . . . . .   86, (1970).

4121 SVENSSON.......(E.C.)..: (6 , 2071) PHYS. REV. LETT. 14,. . . . .   554, (1965).
                               (6 ,   619) SOL. STATE COMM. 3,. . . . .   245, (1965).
                               (6 ,   616) PHYS. REV. LETT. VOL.18,. . .   858, (1967).
                               (6 ,   606) PHYS. REV. 155,. . . . .   619, (1967).
                               (6 ,   587) THESIS (190 PP.),. . . . . .   , (1967).
                               (6 , 1247) IAEA SYMP. COPENHAGEN, VOL.1,.   281, (1968).
                               (6 , 1253) PHYS. REV. 165,. . . . . 1063, (1968).
                               (6 , 1322) CAN. J. PHYS. VOL.47,. . . . . 1983, (1969).
                               (6 , 1639) PHYS. REV. LETT. VOL.23,. . .   525, (1969).
                               (6 ,   492) PHYS. REV. LETT. VOL.23,. . .   86, (1969).
                               (6 , 1481) SOL. STATE COMM. VOL.7,. . . . 1693, (1969).
                               (6 , 1449) J. APPL. PHYS. 41,. . . . . .   896, (1970).
                               (6 ,   614) CAN. J. PHYS. 49,. . . . . . 2291, (1971).
                               (6 , 1323) J. PHYSIQUE TOME 32 COL.1 VOL.II 1184, (1971).
                               (6 , 1269) J. PHYS. -C-, VOL. 4,. . . . . 2127, (1971).
                               (11A,   18) J. PHYS. -C-, VOL. 4,. . . . . 2139, (1971).
                               (6 ,   501) PHYS. REV. LETT. VOL.27,. . . 1442, (1971).
                               (6 , 1437) CAN. J. PHYS. 50,. . . . . . .   687, (1972).
                               (6 , 1164) IAEA SYMP. GRENOBLE,. . . . .   359, (1972).

4121 SVENSSON........(E.C.)..: (6 , 2158) IAEA SYMP. GRENOBLE, ...   553; (1972):
                                (6 , 1307) IAEA SYMP. GRENOBLE, ...   581; (1972):
                                (6 , 1321) MAGN. AND MAG. MATERIALS-1971; . . 1315; (1972):
                                (6 , 1167) PHYS. REV. LETT. VOL.29, ... 1148; (1972):
                                (6 , 1171) PHYS. LETT. VOL.43A NO.3, ...   223; (1973).

4122 SVOBODA.........(H.J.)..: (22B,   34) NUCL. INST. MET. VOL.96, ...   609; (1971).

4123 SWALIN..........(R.A.)..: (5 , 1864) J. METALS VOL.8, ...   1259; (1956).

4124 SWANSON.........(J.)...:   (5 ,   248) J. AMER. CHEM. SOC. VOL.93,. . .   5945; (1971).

4125 SWEENEY........(D.W.)..:   (6 , 1183) PHYS. REV. VOL.80, ...   141; (1950).

4126 SWEENEY.........(D.)...:   (6 , 1182) PHYS. REV. VOL.77, ...   319; (1950).

4127 SWENSON........(C.A.)..:   (6 , 1125) IAEA SYMP. COPENHAGEN, VOL.1,. .   339; (1968).

4128 SYKES...........(J.)...:   (14A,   37) PHYS. LETT. 28A, ...   208; (1968).

4129 SYMONDS........(J.L.)..:   (23 ,   22) IAEA SYMP. (PILE RESEARCH)VIENNA   129; (1960).

4130 SYONO...........(Y.)...:   (5 , 1183) J. PHYS. SOC. JAP. VOL.31, . . .   452; (1971).

4131 SYSHCHENKO.....(V.G.)..:   (11B,   46) SOV. PHYS. SOL. STATE VOL.13,. .   462; (1971).

4132 SZABO..........(F.P.)..:   (6 , 1055) CONF. NEUTRON THERMALIZ. VOL.I,.   477; (1968).

4133 SZABO...........(I.)...:   (23 ,   6) BULL. INF. SCI. TECH. NO.170,. .   61; (1972).

4134 SZABO...........(N.)...:   (14A,   66) REPORT AF-SSP-27,. . . . . , (1968).

4135 SZABO...........(P.)...:   (20D,   13) NUCL. INST. MET. VOL.5,. . . . .   184; (1959).
                                (20C,   3) ACTA PHYS. HUNGAR. VOL.12; . . .   333; (1960).
                                (20D,   14) NUCL. INST. MET. VOL.6,. . . . .   183; (1960).
                                (22A,   5) ACTA PHYS. HUNGAR. VOL.15, . . .   203; (1963).
                                (5 , 1020) PHYS. LETT. 11,. . . . .   215; (1964).
                                (5 , 1130) PHYS. LETT. 9,. . . . .   297; (1964).
                                (5 , 1123) PROC. INT. CONF. (NOTTINGHAM); .   158; (1964).
                                (5 ,   47) ACTA CRYST. 19,. . . . .   180; (1965).
                                (5 , 1101) PHYS. LETT. 19,. . . . .   103; (1965).
                                (5 , 1119) SOL. STATE COMM. 3,. . . . . . .   371; (1965).
                                (5 , 1798) PHYS. LETT. 20,. . . . .   331; (1966).
                                (5 , 2354) SOL. STATE COMM. 4,. . . . . . .   31; (1966).
                                (5 , 1788) J. APPL. PHYS. VOL.38,. . . . .   1265; (1967).
                                (5 , 1786) CENT. RES INST PHYS. KFKI NO.2,.   1; (1968).
                                (5 , 1783) J. APP. PHYS. 39,. . . . .   538; (1968).
                                (5 , 1730) J. PHYS. CHEM. SOLIDS 29,. . . .   101; (1968).
                                (5 , 1696) PHYS. LETT. VOL.26A,. . . . . . .   556; (1968).
                                (5 , 1695) PHYS. REV. 171,. . . . .   574; (1968).

4136 SZARRAS.........(S.)...:   (5 ,   48) PHYS. STAT. SOL. 5,. . . . . . .   K23; (1964).

4137 SZATULA.........(A.)...:   (6 ,   974) NUKLEONIKA VOL.5,. . . . .   495; (1960).

4138 SZELAG.........(J.)...:   (5 , 2169) PHYS. STAT. SOLIDI A VOL.9 NO.1,   97; (1972).

4139 SZENTIRMAY......(Z.)...:   (6 ,   183) IAEA SYMP. GRENOBLE, ...   595; (1972):
                                (6 ,   62) PHYS. LETT. VOL.40A, ...   173; (1972).

4140 SZIGETI.........(B.)...:   (10B,   206) PROC. ROY. SOC. A VOL.320, . . .   505; (1971).

4141 SZKATULA........(A.)...:   (6 ,   415) ACTA PHYS. POLON. VOL.16,. . . .   335; (1957).
                                (19B,   24) NUKLEONIKA VOL.7,. . . . .   231; (1962).
                                (23 ,   60) NUCL. INST. MET. 55; . . . . . .   151; (1967).
                                (21 ,   40) NUKLEONIKA VOL.12,. . . . .   385; (1967).
                                (6 , 1060) PHYSICA VOL.36,. . . . .   35; (1967).

4142 SZKATULA........(J.)...:   (5 , 1286) IAEA SYMPOSIUM VIENNA, ...   297; (1960).

4143 SZLAVIK........(F.)...:   (12B,   11) IAEA SYMP. COPENHAGEN, VOL.2,. .   407; (1968).
                                (19B,   30) REPORT KFKI-07/1968 (14PP.), . . , (1968).

4144 SZPUNAR........(J.)...:   (9B,   32) BU. ACAD. POLON. SCI...TECH. 16,   330; (1968).
                                (9B,   68) NUKLEONIKA VOL.13,. . . . . . .   1111; (1968).
                                (9B,   67) NUKLEONIKA VOL.13,. . . . .   171; (1968).
                                (23 ,   8) COMMENTAT. PHYS.-MATH.(FIN.) 42,   276; (1972).

4145 SZYMCZAK.......(R.)...:   (9D,   23) ELECTRON TECHNOL. (POL.) VOL.1,.   5; (1968).

4146 SZYTULA........(A.)...:   (5 , 1095) PHYS. STAT. SOL. 26, ...   429; (1968).
                                (5 , 2940) PHYS. STAT. SOL. 32, ...   K91; (1969).
                                (5 , 1098) PHYS. STAT. SOL. A VOL.3,. . . .   1033; (1970).
                                (5 , 1099) PHYS. STAT. SOL. 41, ...   173; (1970).
                                (5 , 2181) PHYS. STAT. SOL. B VOL.43,. . . .   125; (1971).
                                (5 , 2170) PHYS. STAT. SOLIDI A VOL.11 NO.1   57; (1972).
                                (5 , 2169) PHYS. STAT. SOLIDI A VOL.9 NO.1,   97; (1972).
                                (5 , 2815) ACTA PHYS. POLON. VOL.A43, . . .   631; (1973).
                                (5 , 2120) ACTA PHYS. POLON. VOL.A43, . . .   787; (1973).
                                (5 , 2200) ACTA PHYS. POLON. VOL.A44, . . .   147; (1973).
                                (5 , 1697) PHYS. STAT. SOLIDI A VOL.19 NO.1 K13; (1973).
                                (5 ,   578) PHYS. STAT. SOLIDI B VOL.57 NO.2 K107; (1973).

4147 TAESCHNER......(M.)...:   (22B,   14) INTERNAT. COMPUTING SYMP., . . .   291; (1974).

4148 TAGLAUER.......(E.)...:   (5 ,   678) PHYS. STAT. SOL. 29, ...   259; (1968).

4149 TAHIR-KHELI....(R.A.)..:   (6 , 1979) PHYS. REV. B-1,. . . . .   3178; (1970).

415C TAHIR-KHELI....(R.)...:   (8 ,   24) J. APPL. PHYS. 41, ...   1365; (1970).

4151 TAJIMA.........(K.)...:   (6 , 1492) ACTA CRYST. VOL.A28 PART S-4,. .   208; (1972):
                                (6 ,   826) SOL. STATE COMM. VOL.13, ...   1179; (1973).

4152 TAKADA.........(H.)...:   (5 ,   723) J. PHYS. SOC. JAPAN VOL.18,. . .   93; (1963).

4153 TAKAHASHI......(H.)...:   (10D,   28) PHYS. REV. 172,. . . . .   747; (1968):
                                (5 ,   232) NUCL. SCI. ENGNG. VOL.37,. . . .   198; (1969):
                                (5 ,   785) NUCL. SCI. ENG. VOL.45,. . . . .   126; (1971).

4154 TAKAHASHI......(S.)...:   (20C,   1) ACTA CRYST. VOL.A28 PART S-4,. .   250; (1972).

4155 TAKAKI.........(H.)...:   (5 , 1725) J. PHYS. SOC. JAPAN VOL.25,. . .   234; (1968):
                                (5 , 1722) J. PHYS. SOC. JAPAN VOL.31,. . .   301; (1971):
                                (5 , 1721) J. PHYS. SOC. JAPAN VOL.36,. . .   438; (1974).

4156 TAKANO.........(H.)...:   (24 ,   81) NUCL. SCI. ENGNG. VOL.40,. . . .   25; (1970).

4157 TAKATA.........(H.)...:   (5 ,   724) J. PHYS. SOC. JAPAN VOL.20,. . .   1743; (1965).

4158 TAKEDA.........(T.)...:   (5 ,   344) J. PHYS. SOC. JAPAN VOL.24,. . .   446; (1968):
                                (5 , 2492) J. PHYS. SOC. JAPAN VOL.26,. . .   1320; (1969):
                                (5 , 2493) J. PHYS. SOC. JAPAN VOL.33,. . .   967; (1972):
                                (5 , 2489) J. PHYS. SOC. JAPAN VOL.33,. . .   970; (1972):
                                (5 , 1063) J. PHYS. SOC. JAPAN VOL.35,. . .   706; (1973).

```
4159  TAKEDA..........(Y.)....:  (  5 , 1070) J. PHYS. SOC. JAPAN VOL.27,. .  1470, (1969).

4160  TAKEI...........(W.J.)..:  (  5 , 1726) PHYS. REV. VOL.119,. . . . . . .   122, (1960).
                                 (  5 , 1102) J. PHYS. CHEM. SOLIDS VOL.23,.    863, (1962).
                                 (  5 ,  545) J. PHYS. SOC. JAPAN VOL.17 B-III    35, (1962).
                                 (  5 , 1186) J. PHYS. SOC. JAPAN VOL.17,. .   1598, (1962).
                                 (  5 , 1724) PHYS. REV. VOL.125,. . . . . .   1893, (1962).
                                 (  5 , 2564) J. APP. PHYS. 34,. . . . . . .   1352, (1963).
                                 (  5 ,  606) J. PHYS. CHEM. SOL. 24,. . . .    405, (1963).
                                 (  5 , 1809) PHYS. REV. 129,. . . . . . . .   2008, (1963).
                                 (  5 , 1560) PHYS. REV. 132,. . . . . . . .   1547, (1963).
                                 (  5 ,   37) ACTA CRYST. 17,. . . . . . . .    415, (1964).
                                 (  5 ,  622) PROC. INT. CONF. (NOTTINGHAM),.   291, (1964).
                                 (  5 ,  632) J. APP. PHYS. 37,. . . . . . .    973, (1966).

4161  TAKEUCHI........(S.)....:  (  5 , 2067) PHYS. STAT. SOL. VOL.39,. . . .   181, (1970).
                                 (  5 ,  228) PHYS. LETT. VOL.45A NO.4,. . .    273, (1973).
                                 ( 15 ,   46) PROPERTIES LIQUID METALS,. . .    143, (1973).

4162  TAKIZAWA........(K.)....:  (  5 , 2187) PHYS. LETT. VOL.44A NO.7,. . . .   529, (1973).

4163  TALHOUK.........(S.J.)..:  (  5 ,  768) PHYS. REV. LETT. 15,. . . . .    586, (1965).
                                 (  5 ,  770) PHYS. REV. LETT. 16,. . . . .    799, (1966).
                                 (  5 ,  763) DISS. ABS. XXVII,. . . . . . .:1447B, (1967).
                                 (  5 ,  761) J. CHEM. PHYSICS VOL.48,. . .   1273, (1968).

4164  TALLEY..........(W.K.)..:  ( 16 ,   13) DISS. ABS. VOL.25, . . . . . .   1986, (1965).

4165  TALWAR..........(D.N.)..:  (  6 , 1241) SOL. STATE COMM. VOL.11, . . .   1691, (1972).

4166  TAMAKI..........(S.)....:  (  5 , 2067) PHYS. STAT. SOL. VOL.39, . . .    181, (1970).

4167  TAMURA..........(N.)....:  (  6 , 1924) J. POLYM. SCI./POLYM. LETT. 11,.  377, (1973).

4168  TANAKA..........(M.)....:  ( 15 ,   46) PROPERTIES LIQUID METALS,. . .    143, (1973).

4169  TANAKA..........(T.)....:  (  5 , 2302) PHYS. REV. 137,. . . . . . . .  A483, (1965).
                                 (11A ,   40) PHYS. REV. VOL.188,. . . . . .    821, (1969).

4170  TANI............(K.)....:  (10B ,  114) PHYS. LETT. 25A,. . . . . . .    400, (1967).
                                 (10C ,   50) J. PHYS. SOC. JAPAN VOL.29,. .    594, (1970).
                                 (10C ,   17) CONF./PHONON SCATT. IN SOLIDS,    314, (1972).
                                 (10E ,   17) J. PHYS. SOC. JAPAN VOL.36,. .    406, (1974).

4171  TANKEEV.........(A.P.)..:  (  8 ,   64) PHYS. MET. METALLOGR. VOL.34 N.2   13, (1972).

4172  TARAN...........(YU.V.).:  (12A ,   55) SOV. J. NUCL. PHYS. VOL.10,. .   1178, (1969).

4173  TARASENKO.......(V.V.)..:  (11A ,   61) SOV. PHYS. J.E.T.P. 27,. . . .    307, (1968).
                                 (  8 ,  153) SOV. PHYS. SOLID STATE VOL.14,.  2006, (1973).

4174  TARASOV.........(L.V.)..:  (10C ,  104) SOV. PHYS. SOLID STATE VOL.3,.   1039, (1961).

4175  TARASOV.........(V.A.)..:  (14B ,  119) SOV. AT. ENERGY VOL.18,. . . .    146, (1965).

4176  TARASYUK........(YU.A.).:  (  5 ,  728) INORG. MATER. VOL.8 NO.12, . .   1916, (1973).

4177  TARDIEU.........(A.)....:  (  5 , 1615) COMP. REND. 257, . . . . . . .   3858, (1963).

4178  TARINA..........(E.)....:  (23  ,   72) REV. DE PHYSIQUE (BUCAREST) 5, .   83, (1960).
                                 (20A ,   77) REV. SCI. INSTRUM. VOL.31,. .    640, (1960).
                                 (20B ,  142) REV. SCI. INSTRUM. VOL.32,. .    297, (1961).
                                 (18B ,    9) REV. ROUMAINE PHYS. VOL.10,. .    445, (1965).

4179  TARINA..........(V.)....:  (  6 ,  951) IAEA SYMP. CHALK RIVER, VOL.1,.   451, (1962).
                                 (  6 ,  303) IAEA SYMP. BOMBAY, VOL.2,. . .    421, (1964).
                                 (  6 ,  334) J. CHEM. PHYS. VOL.46,. . . .    3273, (1967).
                                 (  5 ,  263) PHYSICA VOL.33,. . . . . . . .    523, (1967).
                                 (  6 ,  341) IAEA SYMP. COPENHAGEN, VOL.1,.    501, (1968).
                                 ( 7A ,   84) REV. ROUMAINE PHYS. VOL.17,. .     57, (1972).

4180  TARNOCZI........(T.)....:  (  5 , 1123) PROC. INT. CONF. (NOTTINGHAM), .  158, (1964).
                                 (  5 , 2354) SOL. STATE COMM. 4,. . . . . .     31, (1966).
                                 (  5 , 1781) PHYS. LETT. VOL.25A,. . . . .     56, (1967).
                                 (  5 , 1786) CENT. RES INST PHYS. KFKI NO.2,     1, (1968).
                                 (  5 , 1783) J. APP. PHYS. 39,. . . . . . .    538, (1968).
                                 (  5 , 1695) PHYS. REV. 171,. . . . . . . .    574, (1968).

4181  TARTAGLIA.......(A.)....:  (22A ,   49) NUCL. INST. MET. VOL.114,. . .     21, (1974).

4182  TASSET..........(F.)....:  (  5 , 2883) J. APP. PHYS. 40,. . . . . . .   1454, (1969).
                                 (  5 , 2881) PHYS. REV. VOL.186,. . . . . .    479, (1969).

4183  TAUB............(H.)....:  (  5 , 1903) PHYS. REV. LETT. VOL.32, . . .    724, (1974).

4184  TAUPIN..........(D.)....:  (  6 ,  431) J. CHEM. PHYSICS VOL.55, . . .   2384, (1971).

4185  TAVENDALE.......(A.J.)..:  (20C ,   32) J. SCI. INSTRUM. VOL.39,. . .    124, (1962).

4186  TAWARA..........(H.)....:  (19A ,   46) NUCL. INST. MET. VOL.75,. . .     32, (1969).
                                 (19B ,    7) JAP. J. APPL. PHYS. VOL.10,. .   1090, (1971).

4187  TAYLOR..........(D.W.)..:  (  6 ,  942) PHYS. REV. VOL.156,. . . . . .   1017, (1967).
                                 (  6 , 1838) PROC. ROY. SOC. 296,. . . . .     161, (1967).
                                 (  6 ,  613) CAN. J. PHYS. 49,. . . . . . .   2496, (1971).
                                 (10B ,  187) PHYS. REV. B VOL.10 NO.2,. . .  . . , (1974).

4188  TAYLOR..........(J.C.)..:  (  5 , 2410) ACTA CRYST. 19,. . . . . . . .    205, (1965).
                                 (  5 , 2787) ACTA CRYST. 19,. . . . . . . .    536, (1965).
                                 (  5 , 2639) ACTA CRYST. 20,. . . . . . . .    842, (1966).
                                 (  5 , 2429) ACTA CRYST. VOL.A24,. . . . .     410, (1968).
                                 (  5 , 1439) ACTA CRYST. VOL.A26,. . . . .     559, (1970).
                                 (  5 , 1945) ACTA CRYST. VOL.B26,. . . . .    2136, (1970).
                                 (  5 , 2786) ACTA CRYST. VOL.B27,. . . . .    2018, (1971).
                                 (  5 , 2788) ACTA CRYST. VOL.B28,. . . . .    2995, (1972).
                                 (  5 , 1916) ACTA CRYST. VOL.B28,. . . . .    3340, (1972).
                                 (  5 , 2744) ACTA CRYST. VOL.B29 PART 1,.       7, (1973).
                                 (  5 , 2782) ACTA CRYST. VOL.B29,. . . . .    1073, (1973).
                                 (  5 , 2737) ACTA CRYST. VOL.B29,. . . . .    1942, (1973).
                                 (  5 , 2866) ACTA CRYST. VOL.B30,. . . . .    1216, (1974).
                                 (  5 ,  799) ACTA CRYST. VOL.B30,. . . . .     151, (1974).
                                 (  5 , 2784) ACTA CRYST. VOL.B30,. . . . .     169, (1974).
                                 (  5 , 2783) ACTA CRYST. VOL.B30,. . . . .     175, (1974).
                                 (  5 , 2123) ACTA CRYST. VOL.B30,. . . . .     554, (1974).

4189  TAYLOR..........(M.C.)..:  (20A ,   80) REV. SCI. INSTRUM. VOL.42,. . .   240, (1971).

4190  TAYLOR..........(P.L.)..:  (10B ,  121) PHYS. REV. 131,. . . . . . . .   1995, (1963).

4191  TAYLOR..........(R.I.)..:  (  5 , 2772) PHYS. LETT. 17,. . . . . . . .    188, (1965).
                                 (  5 , 2066) ACTA CRYST. VOL.A25,. . . . .     714, (1969).
                                 (  5 ,  378) SOL. STATE COMM. VOL.8,. . . .    171, (1970).
                                 (  5 , 1079) J. PHYS. C VOL.4,. . . . . . .   2160, (1971).
```

```
4192  TAYLOR..........(R.)....:  (10B,    13) CAN. J. PHYS. 48,. . . . . . . .     183, (1970).
                                 (10D,    44) PHYS. REV. B VOL.5,. . . . . . . : 1206, (1972).

4193  TAYLOR..........(T.I.)..:  (24 ,    34) NUCLEONICS VOL.5 DECEMBER, . . .      4, (1949).
                                 ( 6 ,  1465) PHYS. REV. VOL.75,. . . . . . . :   895, (1949).
                                 ( 7B,    40) NUCLEONICS VOL.6 NO.2;. . . . . :    66, (1950).
                                 (20A,    33) NUCLEONICS VOL.6 NO.4,. . . . . :    54, (1950).
                                 (24 ,   107) SCIENCE VOL.114,. . . . . . . . :   341, (1951).
                                 (22A,    50) NUCL. SCI. ENGNG. VOL.8,. . . . :   453, (1960).
                                 ( 6 ,  1578) PHYS. REV. LETT. VOL.5,. . . . :    507, (1960).
                                 ( 5 ,   146) PHYS. REV. VOL.119,. . . . . . : 1291, (1960).
                                 ( 5 ,  1265) J. CHEM. PHYS. VOL.35,. . . . . : 2265, (1961).
                                 ( 5 ,   141) NUCL. SCI. ENGNG. VOL.9,. . . .     98, (1961).
                                 ( 6 ,    67) J. CHEM. PHYSICS VOL.37,. . . . :   234, (1962).
                                 ( 6 ,  1550) NUCL. SCI. ENGNG. VOL.14,. . . :   339, (1962).
                                 ( 6 ,  1285) IAEA SYMP. BOMBAY, VOL.2,. . . :   333, (1964).
                                 ( 6 ,  1572) J. PHYS. CHEM. VOL.68,. . . . . : 2534, (1964).
                                 (22A,    64) REV. SCI. INSTRUM. VOL.36,. . . :   887, (1965).
                                 ( 6 ,   351) J. CHEM. PHYSICS VOL.44,. . . . : 2749, (1966).
                                 ( 6 ,  1021) J. CHEM. PHYSICS VOL.45,. . . . : 1312, (1966).
                                 ( 6 ,  1577) J. CHEM. PHYSICS VOL.48,. . . . : 4912, (1968).
                                 (24 ,    24) J. CHEM. PHYS. VOL.57,. . . . . :   175, (1972).

4194  TCHEOU..........(F.)....:  ( 5 ,  1203) SOL. STATE COMM. VOL.8,. . . . . : 1745, (1970).
                                 ( 5 ,  1202) SOL. STATE COMM. VOL.8,. . . . . : 1751, (1970).
                                 ( 5 ,  2568) SOL. STATE COMM. VOL.8,. . . . . :   239, (1970).
                                 (90,     12) AIP CONF. PROC.(USA) VOL.5,. . . : 1355, (1971).
                                 ( 5 ,  1015) J. PHYSIQUE TOME 32 COL.1 VOL.I,   202, (1971).
                                 ( 5 ,   844) PHYS. LETT. VOL.34A,. . . . . . :   361, (1971).
                                 ( 5 ,   837) SOL. STATE COMM. VOL.9,. . . . : 1949, (1971).
                                 ( 5 ,  2694) PHYS. STAT. SOLIDI A VOL.14 NO.2  483, (1972).

4195  TEBBLE..........(R.S.)..:  ( 5 ,  2315) PHIL. MAG. VOL.16,. . . . . . . :   347, (1967).
                                 ( 5 ,  2313) J. APP. PHYS. VOL.39,. . . . . . :   471, (1968).

4196  TEETERS.........(W.D.)..:  (24 ,    84) NUOVO CIMENTO A VOL.68,. . . . .   657, (1970).

4197  TEH.............(H.C.)..:  ( 6 ,  1541) PHYS. LETT. 29A,. . . . . . . . :   694, (1969).
                                 ( 6 ,  1531) PHYS. REV. B-3,. . . . . . . . : 2733, (1971).
                                 ( 6 ,  1533) THESIS (183 PP.),. . . . . . . .        (1971).
                                 ( 6 ,  1537) CAN. J. PHYS. VOL.50,. . . . . : 2807, (1972).
                                 ( 6 ,  1536) PHYS. REV. B VOL.8,. . . . . . : 3928, (1973).
                                 (13 ,    55) PHYS. REV. LETT. VOL.30 NO.17,.   781, (1973).
                                 ( 6 ,   500) CAN. J. PHYS. VOL.52,. . . . . .   396, (1974).

4198  TEICHNER........(R.)....:  ( 6 ,   870) SOL. STATE COMM. VOL. 9, . . . . 1579, (1971).

4199  TELLER..........(E.)....:  ( 5 ,  1251) PHYS. REV. VOL.52,. . . . . . . :   286, (1937).
                                 (15 ,    35) PHYS. REV. VOL.60,. . . . . . . :    18, (1941).
                                 ( 7B,    55) PHYS. REV. VOL.62;. . . . . . . :    37, (1942).

4200  TELLGREN........(R.)....:  ( 5 ,  1564) J. SOLID STATE CHEM. VOL.4,. . .   255, (1972).

4201  TEMME...........(F.P.)..:  ( 6 ,   565) IAEA SYMP. GRENOBLE,. . . . . . :   345, (1972).
                                 ( 6 ,  1296) J.C.S. FARADAY TRANS. II VOL.68,   350, (1972).
                                 ( 6 ,  1000) J.C.S. FARADAY TRANS. II VOL.69, 1477, (1973).
                                 ( 6 ,  1631) J.C.S. FARADAY TRANS. II VOL.69,   783, (1973).
                                 ( 6 ,    20) J.C.S. FARADAY TRANS. II VOL.69,     1, (1973).
                                 ( 6 ,  1002) J. CHEM. PHYS. VOL.59,. . . . . :   817, (1973).
                                 ( 6 ,     8) NATURE (PHYS. SCI.) VOL.241,. . :    40, (1973).

4202  TEMPELHOFF......(K.)....:  ( 6 ,  1936) PHYS. STAT. SOL.45,. . . . . . . K105, (1971).

4203  TEMPLETON.......(D.H.)..:  ( 5 ,   667) J. CHEM. PHYSICS VOL.37,. . . .   697, (1962).

4204  TENN............(J.S.)..:  ( 6 ,  1148) PHYS. REV. A VOL.1,. . . . . . :   125, (1970).
                                 ( 6 ,  1134) OISS. ABS. VOL.31,. . . . . . . :61738, (1971).

4205  TENZER..........(L.)....:  ( 5 ,  1930) ACTA CRYST. VOL.11,. . . . . . :   505, (1958).
                                 ( 5 ,  1133) J. PHYS. CHEM. SOLIDS VOL.19,. . :   117, (1961).

4206  TEPEL...........(J.W.)..:  (17B,    42) NUCL. INST. MET. VOL.40,. . . .   100, (1966).

4207  TEPLYUGOV.......(S.G.)..:  ( 5 ,  2163) PHYS. STAT. SOLIDI A VOL.21 NO.1  K31, (1974).

4208  TERRANI.........(M.)....:  (20B,    83) NUCL. INST. MET. 55,. . . . . . :   288, (1967).
                                 (20C,    10) ENERGIA NUCLEARE VOL.16,. . . . :   400, (1969).

4209  TERRY...........(A.L.)..:  ( 5 ,   494) J. PHYS. C VOL.5,. . . . . . . . L261, (1972).

4210  TERRY...........(S.H.)..:  (22C,    42) NUCL. INST. MET. 71,. . . . . . .    13, (1969).

4211  TESSLER.........(G.)....:  ( 6 ,  2335) PHYS. REV. C VOL.2,. . . . . . : 2390, (1970).
                                 ( 6 ,  2336) PHYS. REV. C VOL.4,. . . . . . : 1818, (1971).

4212  TEUTSCH.........(H.)....:  (23 ,    17) IAEA SYMP. (PILE RESEARCH)VIENNA  577, (1960).
                                 (23 ,    62) NUKLEONIK VOL.2,. . . . . . . . :    41, (1960).
                                 (20B,   115) NUKLEONIK VOL.3,. . . . . . . . :    15, (1961).
                                 (20B,   136) REV. DE PHYSIQUE VOL.6,. . . . . :   411, (1961).
                                 (19A,    33) NUCL. INST. MET. VOL.15,. . . . :   203, (1962).
                                 ( 5 ,  2254) STUD. CERCETARI FIZ. VOL.13,. . :   477, (1962).
                                 ( 6 ,  1106) REV. ROUMAINE PHYS. VOL.9,. . . :   737, (1964).
                                 ( 6 ,  1524) PHYS. LETT. 22,. . . . . . . . :   558, (1966).
                                 ( 6 ,   167) IAEA SYMP. COPENHAGEN, VOL.1,. . :   439, (1968).
                                 ( 6 ,   653) REV. ROUMAINE PHYS. VOL.15,. . . :   783, (1970).

4213  TEWARI..........(S.P.)..:  ( 6 ,   290) PHYS. LETT. VOL.33A,. . . . . . :   209, (1970).
                                 ( 5 ,  1905) NUCL./S.S. PHYS. SYMP. ABSTR.,. . . . . (1972).
                                 ( 5 ,  1288) NUCL. SCI. ENG. VOL.47,. . . . . :   153, (1972).
                                 (10C,    62) PHYS. LETT. VOL.44A NO.7,. . . . :   519, (1973).
                                 (10C,    63) PHYS. LETT. VOL.45A,. . . . . . :   481, (1973).
                                 (10C,    64) PHYS. LETT. VOL.46A NO.7,. . . . :   471, (1974).

4214  TEWARY..........(V.K.)..:  (10B,    81) J. PHYS.F: METAL PHYS. VOL.1,. . :   554, (1971).
                                 (10B,   206) PROC. ROY. SOC. A VOL.320,. . . . :   505, (1971).

4215  THALER..........(R.M.)..:  (24 ,    90) PHYS. REV. VOL.114,. . . . . . .   827, (1959).

4216  THAPER..........(C.L.)..:  (20A,    68) PROC. INDIAN ACAD. SCI. A VOL.53    59, (1961).
                                 (20A,    78) REV. SCI. INSTRUM. VOL.33,. . . :    49, (1962).
                                 ( 6 ,  2292) J. CHEM. PHYS. VOL.45,. . . . . : 3817, (1966).
                                 ( 6 ,  2247) PHYS. REV. VOL.152,. . . . . . :   535, (1966).
                                 ( 6 ,  2236) PHYSICA VOL.34,. . . . . . . . :   384, (1967).
                                 ( 6 ,  1234) PHYS. STAT. SOL. 34, NO1,. . . . :   279, (1969).
                                 ( 6 ,   118) SOL. STATE COMM. VOL.8,. . . . . :   497, (1970).
                                 ( 6 ,  1591) SOL. STATE COMM. VOL.8,. . . . . :   889, (1970).
                                 ( 6 ,   134) CONF. INTERN., RENNES,. . . . . :   140, (1971).
                                 ( 6 ,  1544) IAEA SYMP. GRENOBLE,. . . . . . :   477, (1972).
                                 ( 6 ,    66) PHYS. STAT. SOL. VOL.50,. . . . :   701, (1972).
                                 (17A,    74) NUCL. INST. MET. VOL.113,. . . .    15, (1973).

4217  THEWLIS.........(J.)....:  ( 9A,    21) PROC. ROY. SOC. A VOL.196, . . .    50, (1949).
```

```
4218  THIELENS........(G.)....:  (20C,    5)  APPL. SCI. RES. B VOL.7, . . . .     87, (1958).
                                 (20C,    6)  APPL. SCI. RES. B VOL.10,. . . .    247, (1963).
4219  THIELE..........(G.)....:  ( 5 , 2244)  J. LESS-COMMON METALS VOL.17,.      459, (1969).
4220  THIES...........(H.H.)..:  (20C,   58)  NUCL. INST. MET. 68, . . . . . .    277, (1969).
4221  THOLEN..........(A.)....:  (23 ,    9)  CRYOGENICS VOL.11, . . . . . . .    107, (1971).
4222  THOMAS..........(G.E.)..:  (20C,   92)  REV. SCI. INSTRUM. VOL.30, . . . 1135, (1959).
                                 (20C,   37)  NUCL. INST. MET. VOL.17, . . . .     97, (1962).
                                 ( 5 , 1625)  PHYS. REV. VOL.134,. . . . . . . B1047, (1964).
                                 (20C,  101)  REV. SCI. INSTRUM. VOL.36, . . .    419, (1965).
4223  THOMAS..........(H.)....:  ( 6 ,   83)  PHYS. REV. LETT. VOL.25, . . . . 1423, (1970).
4224  THOMAS..........(J.O.)..:  ( 5 ,    9)  J. CHEM. PHYS. VOL.59, . . . . . 3901, (1973).
4225  THOMAS..........(P.)....:  (20C,    8)  BULL. D'INSTRUM. SCI. TECH. 172,     55, (1972).
                                 (17A,   51)  J. APPL. CRYST. VOL.5, . . . . .    373, (1972).
                                 (20B,   56)  J. APPL. CRYST. VOL.5, . . . . .     78, (1972).
                                 ( 9B,   47)  J. APPL. CRYST. VOL.5, . . . . .     83, (1972).
4226  THOMPSON........(B.V.)..:  (10E,   25)  PHYS. REV. 131, . . . . . . . . . 1420, (1963).
4227  THOMPSON........(E.D.)..:  (11B,   23)  PHYS. REV. LETT.19,. . . . . . .    635, (1967).
                                 ( 6 , 1712)  J. APP. PHYS. VOL.40,. . . . . . 1450, (1969).
4228  THOMPSON........(M.C.)..:  ( 5 ,  421)  J. INORG. NUCL. CHEM. VOL.33,. . 1553, (1971).
4229  THOMPSON........(T.J.)..:  (17B,   49)  NUCLEONICS VOL.20, . . . . . . .    157, (1962).
4230  THOMSON.........(G.P.)..:  (20A,   30)  NATURE VOL.142,. . . . . . . . .    829, (1938).
                                 (20B,  132)  PROC. ROY. SOC. A VOL.175,. . .     316, (1940).
4231  THOREL..........(P.)....:  ( 5 , 2008)  PHYS. REV. LETT. VOL.28, . . . . 1370, (1972).
                                 ( 5 , 2004)  J. PHYSIQUE TOME 34, . . . . . .    447, (1973).
4232  THORNTON........(S.T.)..:  (20C,   74)  NUCL. INST. MET. VOL.96, . . . .    551, (1971).
4233  THORPE..........(M.F.)..:  ( 7B,   99)  PROC. PHYS. SOC. 91, . . . . . .    903, (1967).
                                 ( 6 , 1761)  PHYS. REV. B-2, . . . . . . . . . 1362, (1970).
4234  THORSON.........(I.M.)..:  ( 6 , 1049)  IAEA SYMPOSIUM (VIENNA), . . . .    309, (1960).
                                 ( 6 , 1113)  IAEA SYMP. CHALK RIVER, VOL.1, .    343, (1962).
                                 ( 6 ,  199)  IAEA SYMP. CHALK RIVER, VOL.2, .    111, (1962).
                                 ( 6 , 2213)  IAEA SYMP. CHALK RIVER, VOL.2, .    213, (1962).
                                 ( 6 , 1304)  PHYS. REV. 147,. . . . . . . . .    577, (1966).
4235  THROOP..........(G.J.)..:  (14A,   58)  PHYS. REV. A VOL.5,. . . . . . . 2519, (1972).
4236  TICHY...........(J.)....:  ( 5 , 2371)  PHYS. STAT. SOL. 29, . . . . . .    K51, (1968).
                                 ( 5 , 2375)  BRIT. J. APPL. PHYS. VOL.2,. . . 1041, (1969).
                                 ( 9B,   89)  PHYS. STAT. SOL. A VOL.2,. . . .    211, (1970).
4237  TICHY...........(K.)....:  ( 5 , 2201)  CZECH. J. PHYS. B VOL.19,. . . .    857, (1969).
4238  TIITTA..........(A.)....:  ( 6 ,  292)  PHYSICA VOL.54,. . . . . . . . .    393, (1971).
                                 ( 6 ,  296)  NUCL. INST. MET. VOL.103,. . . .    575, (1972).
4239  TIKHONOVA.......(E.A.)..:  (10C,  102)  SOV. PHYS. SOL. STATE VOL.1, . . 1277, (1959).
                                 (10C,  133)  UKR. FIZ. ZH. (USSR) VOL.8,. . .    256, (1963).
4240  TILFORD.........(C.R.)..:  ( 6 , 1125)  IAEA SYMP. COPENHAGEN, VOL.1,. .    339, (1968).
                                 ( 6 , 1141)  PHYS. REV. A VOL.3,. . . . . . . 1688, (1971).
4241  TILLACK.........(J.V.)..:  ( 5 ,  280)  CRYST. STRUCT. COMM. VOL.1 N.3,.    193, (1972).
4242  TILLMAN.........(J.R.)..:  (25 ,   10)  NATURE VOL.135,. . . . . . . . .    904, (1935).
                                 (16 ,   64)  PROC. PHYS. SOC. VOL.48,. . . .     642, (1936).
                                 (16 ,   67)  PROC. ROY. SOC. A VOL.153,. . .     476, (1936).
4243  TIMIS...........(P.)....:  (22B,   42)  REV. DE PHYSIQUE (BUCAREST) 4, .    327, (1959).
                                 (19B,   36)  STUD. CERCETARI FIZ. VOL.10,. .      89, (1959).
                                 (23 ,   62)  NUKLEONIK VOL.2, . . . . . . . .     41, (1960).
                                 (20B,  136)  REV. DE PHYSIQUE VOL.6,. . . . .    411, (1961).
                                 ( 5 , 2254)  STUD. CERCETARI FIZ. VOL.13, . .    477, (1962).
                                 ( 6 , 1106)  REV. ROUMAINE PHYS. VOL.9, . . .    737, (1964).
4244  TIMMESFELD......(K.H.)..:  (11A,   52)  PHYS. STAT. SOL. VOL.49, . . . .    199, (1972).
4245  TIMUSK..........(T.)....:  ( 6 , 1260)  PHYS. REV. 182,. . . . . . . . .    965, (1969).
                                 ( 6 , 1259)  PHYS. REV. B VOL.1,. . . . . . . 4193, (1970).
4246  TIPPE...........(A.)....:  (22C,   27)  KERNTECHNIK VOL.10 NO.7,. . . .     371, (1968).
                                 ( 5 ,  275)  DISC. FARADAY SOC. NO.48,. . . .    192, (1969).
                                 ( 5 ,  229)  AMER. MINERAL. VOL.56,. . . . .     101, (1971).
4247  TOBISCH.........(J.)....:  ( 5 , 2696)  PHYS. STAT. SOL. 2,. . . . . . .    K164, (1962).
                                 ( 5 , 1409)  Z. KRIST. VOL.118,. . . . . . .     473, (1963).
                                 ( 5 , 2758)  PHYS. STAT. SOL. 21, . . . . . .     K11, (1967).
                                 ( 5 ,  684)  J. APPL. CRYST. VOL.4, . . . . .    303, (1971).
                                 ( 5 ,  756)  J. APPL. CRYST. VOL.5, . . . . .     27, (1972).
4248  TOCCHETTI.......(D.)....:  (17A,   32)  IAEA SYMPOSIUM BOMBAY, VOL.2,. .    545, (1964).
                                 (18A,   17)  NUCL. INST. MET. 32, . . . . . .    181, (1965).
                                 ( 6 ,  797)  J. PHYS. CHEM. SOL. 28,. . . . .    467, (1967).
                                 ( 6 ,  775)  PHYS. LETT. 24A, . . . . . . . .    270, (1967).
4249  TODD............(M.C.J.):  (20B,   36)  IAEA SYMP. COPENHAGEN, VOL.2,. .    341, (1968).
4250  TODIREANU.......(S.)....:  (23 ,   72)  REV. DE PHYSIQUE (BUCAREST) 5,       83, (1960).
                                 (20A,   77)  REV. SCI. INSTRUM. VOL.31, . . .    640, (1960).
                                 ( 6 ,  951)  IAEA SYMP. CHALK RIVER, VOL.1,.     451, (1962).
                                 ( 5 ,   57)  NUCL. SCI. ENGNG. VOL.12, . . .     157, (1962).
                                 (20A,   79)  REV. SCI. INSTRUM. VOL.33, . . .    916, (1962).
                                 ( 6 ,  303)  IAEA SYMP. BOMBAY, VOL.2,. . . .    421, (1964).
                                 (18B,    9)  REV. ROUMAINE PHYS. VOL.10,. . .    445, (1965).
                                 ( 5 ,  263)  PHYSICA VOL.33,. . . . . . . . .    523, (1967).
                                 ( 5 ,  267)  PHYS. LETT. 24A,. . . . . . . .     544, (1967).
                                 ( 6 ,  287)  PHYS. LETT. 30A,. . . . . . . .     367, (1969).
                                 ( 6 ,  297)  PHYS. LETT. VOL.43A NO.2,. . . .    189, (1973).
4251  TODOROVIC.......(J.)....:  ( 6 ,  879)  COLL. INTER. N.126 (GRENOBLE),     210, (1963).
                                 ( 6 ,  835)  IAEA SYMP. BOMBAY, VOL.1,. . . .    443, (1964).
                                 ( 6 ,  843)  PHYS. STAT. SOL. 15, . . . . . .    119, (1966).
                                 ( 6 ,  482)  PHYSICA VOL.37,. . . . . . . . .    501, (1967).
                                 ( 6 ,  862)  PHYS. STAT. SOL. 21, . . . . . .    K163, (1967).
                                 ( 6 ,  864)  PHYS. STAT. SOL. 22, . . . . . .     K55, (1967).
                                 ( 5 , 1095)  PHYS. STAT. SOL. 26, . . . . . .    429, (1968).
                                 ( 5 , 2940)  PHYS. STAT. SOL. 32, . . . . . .     K91, (1969).
                                 ( 6 ,  545)  PHYS. STAT. SOL. 32, . . . . . .     41, (1969).
                                 ( 6 ,  863)  PHYS. STAT. SOL. 41, . . . . . .    K103, (1970).
                                 ( 6 , 1442)  PHYSICA VOL.57,. . . . . . . . .    628, (1972).
                                 ( 5 , 2170)  PHYS. STAT. SOLIDI A VOL.11 NO.1     57, (1972).
                                 ( 5 , 2169)  PHYS. STAT. SOLIDI A VOL.9 NO.1,     97, (1972).
```

4252 TOFIELD..........(B.C.)..‡   (5 , 2896) J. PHYS. CHEM. SOLIDS VOL.31,. . 2741, (1970).
                                  (5 ,   85) J. PHYS. C VOL.4,. . . . . 1279, (1971).
                                  (5 ,   156) ACTA CRYST. VOL.B28. : : : : : : 956, (1972).
                                  (5 ,   175) J. PHYS. C VOL.5,. . . . . 2889, (1972).
                                  (5 , 1751) J. PHYS. C VOL.6,. . . . . 1615, (1973).

4253 TOJO............(T.)....‡ (20C,   29) J. NUCL. SCI. TECHNOL. JAP. 4, .   22, (1967).
                                 (20C,   48) NUCL. INST. MET. 53, ...   163, (1967).

4254 TOKUNAGA........(M.)....‡ (9B,   57) J. PHYS. SOC. JAPAN VOL.23,. . .   460, (1967).
                                 (5 , 1980) J. PHYS. SOC. JAPAN VOL.23,. . .   461, (1967).

4255 TOLCHENOV.......(YU.M.)..‡ (20C,   25) INSTRUM. EXP. TECH. VOL.15 PT.1,   351, (1972).

4256 TOLK............(N.H.)..‡ (20B,   68) NUCL. INST. MET. VOL.8,. . . . .   203, (1960).

4257 TOLPYGO.........(K.B.)..‡ (10B,   196) PHYS. STAT. SOLIDI B VOL.56 NO.2   591, (1973).
                                 (10B,   227) SOV. PHYS. SOLID STATE VOL.14,.   2480, (1973).

4258 TOMASEVICH......(O.F.)..‡ (6 , 1648) SOV. PHYS. SOL. STATE VOL.12,. .   423, (1970).

4259 TOMIGOSHI.......(S.)....‡ (5 ,   560) J. PHYS. SOC. JAPAN VOL.32,. . .   958, (1972).

4260 TOMILIN.........(I.A.)..‡ (5 , 2846) SOV. PHYS. DOKLADY VOL.15, . . .   776, (1971).

4261 TOMIMITSU.......(H.)....‡ (5 , 1217) ACTA CRYST. VOL.A28 PART S-4,.   220, (1972).
                                 (5 , 1221) J. APPL. CRYST. VOL.7,. . . . .   59, (1974).

4262 TOMITA..........(K.)....‡ (13 ,   59) PROC. INT. CONF. (NOTTINGHAM), .   103, (1964).
                                 (11C,   23) KYOTO CONF. STAT. MECHANICS, . .   157, (1968).

4263 TOMIYOSHI.......(S.)....‡ (5 ,   344) J. PHYS. SOC. JAPAN VOL.24,. . .   446, (1968).
                                 (5 , 2492) J. PHYS. SOC. JAPAN VOL.26,. . . 1320, (1969).
                                 (19A,   4) ACTA CRYST. VOL.A28 PART S-4,.   197, (1972).
                                 (5 , 2482) JAP. J. APPL. PHYS. VOL.12,. . . 1119, (1973).
                                 (5 , 1148) J. PHYS. SOC. JAPAN VOL.34,. . .   58, (1973).

4264 TOMKOWICZ.......(Z.)....‡ (5 , 2200) ACTA PHYS. POLON. VOL.A44, . . .   147, (1973).

4265 TOMPA...........(H.)....‡ (17C,   7) ACTA CRYST. 19,. . . . . 1014, (1965).

4266 TOMPSON.........(C.W.)..‡ (6 ,   550) PHYS. REV. B-2,. . . . . 4864, (1970).

4267 TONDON..........(V.K.)..‡ (9D,   20) CAN. J. PHYS. VOL.50,. . . . . . 2991, (1972).
                                 (6 , 1502) INT. J. MAGN. VOL.4 NO.1,. . . .   17, (1973).

4268 TONOLINI........(F.)....‡ (20C,   9) ENERGIA NUCLEARE VOL.15, ...   311, (1968).
                                 (24 ,   13) ENERGIA NUCLEARE (ITALY) VOL.16,   65, (1969).

4269 TOPERVERG.......(B.P.)..‡ (12B,   65) SOV. PHYS. SOL. STATE VOL.12,. . 2445, (1971).

4270 TORBO...........(P.)....‡ (5 , 1146) ACTA CHEM. SCAND. VOL.21,. . . . 2841, (1967).

4271 TORNIELLI.......(G.)....‡ (20C,   66) NUCL. INST. MET. 67,. . . . . .   267, (1969).
                                 (5 ,   231) NUCL. PHYS. A VOL.A181,. . . . .   177, (1972).

4272 TORRIE..........(B.H.)..‡ (5 , 1952) CAN. J. PHYS. 42,. . . . .   229, (1964).
                                 (6 ,   830) PROC. PHYS. SOC. 84,. . . . . .   55, (1964).
                                 (6 ,   833) PHYS. REV. LETT. VOL.14,. . . .   698, (1965).
                                 (6 , 1456) PROC. PHYS. SOC. 89,. . . . . .   77, (1966).
                                 (6 ,   838) SOL. STATE COMM. VOL.5,. . . . .   715, (1967).

4273 TOSATTI.........(E.)....‡ (10B,   171) PHYS. REV. LETT. VOL.28, ... 1578, (1972).
                                 (10B,   188) PHYS. REV. B VOL.9,. . . . . . . 1710, (1974).

4274 TOSELLI.........(G.)....‡ (17C,   17) J. APPL. CRYST. VOL.3, ...   145, (1970).

4275 TOSIMA..........(S.)....‡ (11C,   22) J. PHYS. SOC. JAPAN VOL.16,. . .   241, (1961).

4276 TOSI............(M.P.)..‡ (14A,   44) PHYS. REV. LETT. 21, ...   881, (1968).
                                 (14A,   51) PHYS. REV. A VOL.1,. . . . . . .   454, (1970).
                                 (10B,   155) PHYS. REV. B-2,. . . . . 2983, (1970).
                                 (14A,   4) ANN. OF PHYS. VOL.81,. . . . . .   414, (1973).
                                 (14A,   28) J. PHYS. C VOL.6,. . . . . L254, (1973).
                                 (14A,   31) NUOVO CIMENTO B VOL.13B NO.1,.   185, (1973).

4277 TOTH............(G.)....‡ (10B,   122) PHYS. REV. 171,. . . . .   665, (1963).
                                 (6 , 2233) REPORT EUR-4216E (41PP.),. . . .   . , (1969).

4278 TOTH............(J.)....‡ (5 , 1123) PROC. INT. CONF. (NOTTINGHAM),   158, (1964).

4279 TOTH............(R.S.)..‡ (5 ,   127) J. PHYS. CHEM. SOL. 27,. . . . .   413, (1966).
                                 (5 ,   133) J. APP. PHYS. VOL.40,. . . . . . 1373, (1969).

4280 TOTIA...........(H.)....‡ (22B,   42) REV. DE PHYSIQUE (BUCAREST) 4,   327, (1959).
                                 (19B,   36) STUD. CERCETARI FIZ. VOL.10,. .   89, (1959).

4281 TOTIA...........(M.)....‡ (6 ,   755) J. PHYS. CHEM. SOL. 28,. . . . . 1947, (1967).
                                 (6 ,   480) PHYS. LETT. 25A,. . . . .   595, (1967).
                                 (6 ,   470) IAEA SYMP. COPENHAGEN, VOL.2,.   75, (1968).
                                 (6 ,   759) J. APPL. PHYS. VOL.39, ...   459, (1968).
                                 (6 ,   741) PHYS. LETT. VOL.26A,. . . . . .   396, (1968).
                                 (6 ,   805) ACTA CRYST. (INTERACT.) VOL.A25, S262, (1969).
                                 (13 ,   62) REPORT FN-40/70 (6PP.),. . . . .   . , (1970).
                                 (6 ,   808) SOL. STATE COMM. VOL. 9,. . . .   353, (1971).

4282 TOURAND.........(G.)....‡ (5 , 1522) PHYS. LETT. 29A,. . . . .   506, (1969).
                                 (5 , 2619) C. R. ACAD. SCI. B (FRANCE) 270, 179, (1970).
                                 (5 ,   677) J. PHYS. CHEM. SOLIDS VOL.31,.   549, (1970).
                                 (5 , 2620) J. PHYSIQUE (FRA.) VOL.32, . . .   813, (1971).
                                 (5 , 2434) J. PHYSIQUE TOME 34,. . . . . .   937, (1973).

4283 TOURNIER........(R.)....‡ (5 ,   994) J. PHYS. CHEM. SOL. 26,. . . . . 1727, (1965).

4284 TOWLE...........(J.H.)..‡ (24 ,   39) NUCL. INST. MET. 30,. . . . . .   77, (1964).
                                 (6 , 2285) NUCL. PHYS. VOL.A131,. . . . . .   561, (1969).

4285 TOYA............(T.)....‡ (10B,   31) IAEA SYMP. BOMBAY, VOL.1,. . . .   25, (1964).

4286 TRACY...........(C.A.)..‡ (8 ,   107) PHYS. REV. LETT. VOL.31, ... 1500, (1973).

4287 TRAIL...........(C.C.)..‡ (24 ,   16) IAEA SYMP. (PILE RESEARCH)VIENNA 615, (1960).

4288 TRAMMELL........(G.T.)..‡ (8 ,   79) PHYS. REV. VOL.92, ... 1387, (1953).

4289 TRAYLOR.........(J.G.)..‡ (6 , 1562) PHYS. REV. VOL.181,. . . . . . . 1218, (1969).
                                 (6 , 2180) DISS. ABS. VOL.32,. . . . . . . 1164B, (1971).
                                 (6 , 2178) PHYS. REV. B-3,. . . . . 3457, (1971).
                                 (6 , 1158) IAEA SYMP. GRENOBLE,. . . . . .   129, (1972).

4290 TREDGOLD........(R.H.)..‡ (5 ,   473) PROC. ROY. SOC. A VOL.217, . . .   252, (1953).

4291 TREGO...........(A.L.)..‡ (5 ,   638) SOL. STATE COMM. VOL.3,. . . . .   137, (1965).
                                 (5 ,   529) J. APP. PHYS. 37,. . . . . 1036, (1966).
                                 (5 ,   598) J. APP. PHYS. 37,. . . . . 1259, (1966).
                                 (5 ,   597) PHYS. REV. VOL.151,. . . . . . .   405, (1966).

4292 TREIMER.........(W.)....:   (22A,   53) PHYS. LETT. VOL.47A NO.5,. . . .   369, (1974).

4293 TREPADUS........(V.)....:   (5 ,   263) PHYSICA VOL.33,. . . . .   523, (1967).
                                   (6 , 123A) IAEA SYMP. COPENHAGEN, VOL.1,.   483, (1968).
                                   (15 ,   59) STUD. CERCET. FIZ. (RUMANIA) 23, 699, (1971).
                                   (6 ,   945) REV. ROUMAINE PHYS. VOL.18 NO.2, 135, (1973).
                                   (6 , 1873) REV. ROUMAINE PHYS. VOL.18,. . . 313, (1973).
                                   (6 , 2200) J. CHEM. PHYS. VOL.60,. . . . . 2832, (1974).

4294 TRESSAUD........(M.A.)..:   (5 , 1558) SOL. STATE COMM. VOL.10,. . . .   739, (1972).

4295 TRET-YAKOV......(A.G.)..:   (10C,   123) SOV. PHYS. J. NO.9,. . . . . .   31, (1968).

4296 TREUTING........(R.G.)..:   (5 ,   714) ACTA CRYST. VOL.9,. . . . . .   1025, (1956).

4297 TREVINO.........(S.F.)..:   (6 , 1894) IAEA SYMP. BOMBAY, VOL.2,. . .   407, (1964).
                                   (6 , 1586) J. CHEM. PHYSICS VOL.45,. . . .   4C1, (1966).
                                   (6 , 1877) J. CHEM. PHYSICS VOL.45,. . . .   757, (1966).
                                   (6 , 1908) J. MACROMOL. SCI. VOL.A1,. . .   723, (1967).
                                   (6 , 1050) IAEA SYMP. COPENHAGEN, VOL.1,.   345, (1968).
                                   (6 ,   337) J. CHEM. PHYSICS VOL.53,. . . . 3417, (1970).
                                   (10B,   55) J. CHEM. PHYSICS VOL.53,. . . . 4624, (1970).
                                   (10B,   54) J. CHEM. PHYSICS VOL.53,. . . . 4645, (1970).
                                   (10B,   53) J. CHEM. PHYSICS VOL.53,. . . . 4661, (1970).
                                   (6 , 1658) CONF. INTERN., RENNES,. . . .   104, (1971).
                                   (6 , 1325) PHYS. REV. B VOL.4,. . . . . . 4551, (1971).
                                   (6 , 1080) J. CHEM. PHYS. VOL.56,. . . . . 3217, (1972).
                                   (7B,   87) PHYS. REV. B VOL.6,. . . . . . 4533, (1972).
                                   (6 , 1661) PHYS. REV. B VOL.10 NO.2,. . .   . , (.).

4298 TREVINO.........(S.)....:   (6 , 1918) J. CHEM. PHYSICS VOL.45,. . . . 2700, (1966).
                                   (6 , 1923) DISC. FARADAY SOC. NO.48,. . .   15, (1969).

4299 TREYVAUD........(A.)....:   (6 , 2196) SOL. STATE COMM. VOL.12 NO.2,.   117, (1973).
                                   (6 ,   716) J. PHYS. C VOL.7,. . . . . . . 1207, (1974).

430C TRIFTSHAUSER....(W.)....:   (6 , 2031) Z. PHYSIK VOL.190,. . . . . .   295, (1966).

4301 TRIKHA..........(S.K.)..:   (16 ,   26) J. NUCL. ENERGY VOL.23,. . . .   505, (1969).
                                   (6 ,   192) NUCL. SCI. ENG. VOL.44,. . . .   444, (1971).

4302 TRIMBLE.........(G.D.)..:   (6 , 1946) NUCL. SCI. ENG. VOL.46,. . . .   244, (1971).

4303 TRINKAUS........(H.)....:   (9C,   44) Z. NATURFORSCH. VOL.28A,. . . .   980, (1973).

4304 TRIPATHI........(B.B.)..:   (6 ,   605) PHYS. LETT. VOL.29A,. . . . . .   313, (1969).
                                   (10B,   160) PHYS. REV. B-2,. . . . .   248, (1970).
                                   (10B,   85) J. PHYS.F: METAL PHYS. VOL.1,.   12, (1971).
                                   (10B,   84) J. PHYS.F: METAL PHYS. VOL.1,.   19, (1971).
                                   (11B,   97) J. PHYS. SOC. JAPAN VOL.31,. . 1639, (1971).
                                   (10D,   50) PHYS. STAT. SOL. VOL.45,. . . .   235, (1971).
                                   (10D,   48) PHYS. STAT. SOL. VOL.45,. . . .   537, (1971).
                                   (6 ,   744) J. PHYS. SOC. JAP. VOL.33,. . . 1207, (1972).
                                   (10B,   76) J. PHYS. CHEM. SOLIDS VOL.34,. 1867, (1973).
                                   (10B,   98) J. PHYS. SOC. JAPAN VOL.34,. . 1006, (1973).
                                   (6 , 2018) PHYS. LETT. VOL.43A NO.4,. . .   365, (1973).

43C5 TRIPIER.........(J.)....:   (20C,   86) REV. PHYS. APPL. (FRANCE) VOL.4, 259, (1969).

4306 TROCHON.........(J.)....:   (20C,   57) NUCL. INST. MET. 72,. . . . . .   307, (1969).

4307 TROC............(R.)....:   (5 , 2796) PHYS. STAT. SOL. 10,. . . . . .   K85, (1965).
                                   (5 , 2800) PHYS. STAT. SOL. 15,. . . . . .   515, (1966).
                                   (5 , 2813) PHYS. STAT. SOL. 22,. . . . . .   517, (1967).
                                   (5 , 2716) PHYS. STAT. SOL. 23,. . . . . . K123, (1967).
                                   (5 , 2717) PHYS. STAT. SOL. 30,. . . . . .   157, (1968).
                                   (5 , 2791) PHYS. STAT. SOL. 30,. . . . . .   61, (1968).
                                   (5 , 2722) PHYS. STAT. SOL. VOL.38,. . . .   K89, (197C).
                                   (5 , 2803) PHYS. STAT. SOLIDI A VOL.19 NO.1 K89, (1973).

4308 TROFIMOVA.......(N.A.)..:   (5 , 2411) SOV. AT. ENERGY VOL.19,. . . . . 1162, (1965).
                                   (5 , 2242) UKRAINIAN PHYS. J. VOL.14,. . . 1968, (1969).
                                   (5 ,   379) UKR. FIZ. ZH.(USSR) VOL.17,. . .   38, (1972).

4309 TROITSKAYA......(E.P.)..:   (10B,   227) SOV. PHYS. SOLID STATE VOL.14,. 2480, (1973).

4310 TROMP...........(R.L.)..:   (5 , 2325) NUCL. SCI. ENG. VOL.43,. . . . .   58, (1971).

4311 TROTIN..........(J.P.)..:   (6 ,   724) C.R. ACAD. SC. (PARIS), 267-B,.   61, (1968).
                                   (6 , 1414) IAEA SYMP. COPENHAGEN, VOL.1,.   289, (1968).

4312 TROTT...........(A.J.)..:   (10B,   193) PHYS. STAT. SOL.46,. . . . . .   361, (1971).

4313 TROUP...........(G.J.F.):   (5 , 1711) SOL. STATE COMM. VOL.8,. . . . . 1183, (1970).

4314 TRUEBLOOD.......(K.N.)..:   (15 ,   27) MOL. DYN. AND STRUCT. OF SOLIDS, 355, (1969).

4315 TRUKHANOV.......(G.YA.).:   (6 , 1098) SOV. AT. ENERGY(USA) VOL.31,. .   459, (1971).
                                   (6 ,   228) SOV. AT. ENERGY VOL.31,. . . .   459, (1971).

4316 TRUNOV..........(V.A.)..:   (12B,   61) SOV. PHYS. JETP LETT. VOL.10,. .   345, (1969).
                                   (12B,   64) SOV. PHYS. TECH. PHYS. VOL.40,. 1317, (1970).
                                   (6 , 1153) SOV. PHYS. TECH. PHYS. VOL.17,.   180, (1972).
                                   (5 , 2084) SOV. PHYS. SOL. STATE VOL.15,. .   919, (1973).

4317 TRUSTEDT........(W.D.)..:   (5 , 2420) Z. NATURFORSCH. VOL.26A,. . . .   400, (1971).

4318 TRYKOVA.........(V.I.)..:   (5 , 2874) BU. ACAD. SCI. USSR PHYS. SR. 32 579, (1968).

4319 TRYKOV..........(L.A.)..:   (20C,   20) INSTRUM. EXPER. TECH. NO.3,. . .   529, (1968).

4320 TSAKADZE........(D.S.)..:   (6 , 1636) JETP LETT. VOL.3,. . . . .   110, (1966).

4321 TSAREV..........(YU.N.).:   (6 ,   14) PHYS. LETT. VOL.36A,. . . . . .   337, (1971).
                                   (10B,   118) PHYS. MET. METALLOGR. VOL.33,.   939, (1972).

4322 TSIBULNIK.......(A.N.)..:   (6 , 1889) UKRAINIAN PHYS. J. VOL.14, 9,. . 1525, (1969).

4323 TSIRLIN.........(YU.A.).:   (20C,   15) INSTRUM. EXP. TECH. NO.4,. . . .   808, (1965).

4324 TSITSENOVSKAYA..(S.E.)..:   (5 ,   713) LATV. PSR...FIZ. TEHN SER. NO.3,   64, (1972).
                                   (5 ,   728) INORG. MATER. VOL.8 NO.12,. . . 1916, (1973).

4325 TSUCHIDA........(T.)....:   (5 ,   394) PHYS. STAT. SOL. VOL.44,. . . .   K25, (1971).

4326 TSUNODA.........(Y.)....:   (6 ,   525) J. APP. PHYS. 39,. . . . . . .   1227, (1968).
                                   (5 , 1370) J. CHEM. PHYSICS VOL.54,. . . . 3510, (1971).
                                   (6 ,   522) J. PHYS. SOC. JAPAN VOL.32,. .   394, (1972).
                                   (6 , 1776) SOL. STATE COMM. VOL.13 NO.4,.   495, (1973).

4327 TUBBS...........(N.)....:   (5 , 1315) ACTA PHYS. POLON. VOL.22,. . . .   517, (1962).
                                   (6 , 1552) IAEA SYMP. CHALK RIVER, VOL.1,.   405, (1962).

4328 TUBINO..........(R.)....¡   (10B,   56) J. CHEM. PHYSICS VOL.53, ... 1428, (1970).
                                 (10B,   59) J. CHEM. PHYS. VOL.56, ... 1022, (1972).
                                 (10B,   60) J. CHEM. PHYS. VOL.59; . . . . . 4578, (1973).

4329 TUCCIARONE......(A.)....¡   (6 , 1985) PHYS. REV. LETT. VOL.23, ... 1225, (1969).
                                 (6 , 883) PHYS. REV. VOL.178, ...   853, (1969).
                                 (6 , 1996) J. APPL. PHYS. VOL.41, ... 1384, (1970).
                                 (6 , 1981) J. APPL. PHYS. 42, 41, ... 1378, (1971).
                                 (6 , 1979) PHYS. REV. B-4, ...   3206, (1971).
                                 (6 , 1963) PHYS. REV. LETT. VOL.26, ...   257, (1971).
                                 (6 , 1998) PHYS. REV. B VOL.8, ...   1103, (1973).

4330 TUCKER-JR.......(C.W.)..¡   (5 , 2850) ACTA MET. VOL.1, ...   390, (1953).

4331 TUCKEY..........(G.S.G.)¡   (20B,   36) IAEA SYMP. COPENHAGEN, VOL.2,. . 341, (1968).

4332 TUDOS...........(F.)....¡   (17A,   85) PHYS. LETT. 31A, ...   78, (1970).

4333 TUNKELO.........(E.)....¡   (6 , 2036) PHYS. REV. LETT. VOL.2,. . . . . 258, (1959).
                                 (6 , 1302) PHYS. REV. LETT. VOL.2,. . . . .   94, (1959).
                                 (6 , 1090) PHYS. REV. LETT. VOL.3,. . . . .   91, (1959).
                                 (6 , 1084) PHYS. REV. VOL.119,. . . . . . . 872, (1960).
                                 (21 ,   3) ACTA POLYTECH. SCANDIN. PH 38, . .., (1966).
                                 (22B,   24) NUCL. INST. MET. 40, ... 125, (1966).
                                 (23 ,   58) NUCL. INST. MET. 46, ... 266, (1966).
                                 (20B,   78) NUCL. INST. MET. 46, ...   70, (1966).
                                 (6 , 168) IAEA SYMP. COPENHAGEN, VOL.1,. . 431, (1968).
                                 (6 , 2003) IAEA SYMP. COPENHAGEN, VOL.2,. . 237, (1968).
                                 (6 , 286) PHYSICA VOL.41,. . . . . 397, (1969).
                                 (6 , 292) PHYSICA VOL.54,. . . . . 393, (1971).
                                 (6 , 296) NUCL. INST. MET. VOL.103,. . . . 575, (1972).

4334 TUNNICLIFFE.....(P.R.)..¡   (23 ,   75) REV. SCI. INSTRUM. VOL.21, . . . 705, (195x).
                                 (20C,   89) REV. SCI. INSTRUM. VOL.21, . . . 734, (1950).

4335 TURANO..........(T.)....¡   (5 , 2528) PHYS. REV. B VOL.8,. . . . . . . 2595, (1973).

4336 TURBERFIELD.....(K.C.)..¡   (6 , 2244) IAEA SYMPOSIUM VIENNA, ... 581, (1960).
                                 (6 , 2237) PHYS. REV. VOL.127,. . . . . . . 1017, (1962).
                                 (6 , 1616) PROC. PHYS. SOC. VOL.80, ... 1201, (1962).
                                 (6 , 1821) PROC. PHYS. SOC. VOL.80, ...   395, (1962).
                                 (6 , 1462) J. APPL. PHYS. VOL.35, ...   998, (1964).
                                 (6 , 1458) PHYS. LETT. 8, ...   9, (1964).
                                 (6 , 1453) PROC. INT. CONF. (NOTTINGHAM), .   92, (1964).
                                 (6 , 1461) PROC. PHYS. SOC. 85, ... 743, (1965).
                                 (17B,   63) REPORT AERE-R5642 (8 PP.), ..., (1967).
                                 (5 , 717) SOL. STATE COMM. 5,. . . . . . . 887, (1967).
                                 (19A,   8) HARWELL SUMMER SCHOOL, ...   34, (1968).
                                 (6 , 467) J. OF PHYS. C VOL.1, ... 679, (1968).
                                 (20B, 133) REPORT AERE-R5647 (33 PP.),. . . ..., (1968).
                                 (6 , 1930) PHYS. REV. LETT. VOL.25, ... 752, (1970).
                                 (6 , 1935) J. APPL. PHYS. 42, ... 1746, (1971).
                                 (6 , 2195) PHYS. REV. B VOL.4,. . . . . . .   718, (1971).
                                 (6 , 1953) PHYS. REV. B VOL.8,. . . . . . . 5345, (1973).

4337 TURCHANINOV.....(A.M.)..¡   (5 , 2856) SOV. PHYS. CRYST. VOL.17,. . . . 1017, (1973).

4338 TURCHIN.........(V.F.)..¡   (10D,   61) SOV. PHYS. JETP VOL.6, ...   96, (1958).
                                 (10C, 100) SOV. PHYS. JETP VOL.7, ... 151, (1958).
                                 (14B, 119) SOV. AT. ENERGY VOL.18, ... 146, (1965).
                                 (17A,   94) SOV. AT. ENERGY VOL.19, ... 1387, (1965).

4339 TURINSKY........(P.J.)..¡   (7A,   43) NUCL. SCI. ENG. VOL.45,. . . . . 167, (1971).

4340 TURKEVICH.......(E.I.)..¡   (5 , 214) SOV. PHYS. SOL. STATE VOL.10,. . 754, (1968).
                                 (5 , 349) SOV. PHYS. SOL. STATE VOL.9,. . . 1762, (1968).

4341 TURNBULL........(A.)....¡   (6 , 244) IAEA SYMP. GRENOBLE, ... 231, (1972).
                                 (6 , 2108) SYMP. CHEM. ORGANIC SOL. STATE,. .., (1972).

4342 TURNER..........(J.E.)..¡   (24 ,   94) PHYS. REV. B VOL.3,. . . . . . . 2929, (1971).

4343 TURNER..........(R.E.)..¡   (7A,   47) PHYSICA VOL.27,. . . . . 260, (1961).
                                 (6 , 1180) ANN. PHYS. VOL.17, ... 301, (1962).
                                 (7A,   49) PHYS. LETT. 2, ... 266, (1962).

4344 TUROS...........(A.)....¡   (19B,   11) NUCL. INST. MET. VOL.29, ... 241, (1964).

4345 TUROV...........(E.A.)..¡   (11A,   34) PHYS. MET. METALLOGR. VOL.30, 5, 1064, (1970).
                                 (8 ,   64) PHYS. MET. METALLOGR. VOL.34 N.2   13, (1972).

4346 TUROWSKI........(M.)....¡   (5 , 2200) ACTA PHYS. POLON. VOL.A44, . . . 147, (1973).

4347 TUSTANOVSKII....(V.T.)..¡   (24 , 113) SOV. PHYS. DOKLADY VOL.16, . . .   49, (1971).

4348 TUTOV...........(A.G.)..¡   (5 , 218) PROC. INT. CONF. (NOTTINGHAM), . 354, (1964).

4349 TWISLETON.......(J.F.)..¡   (6 , 1893) IAEA SYMP. GRENOBLE, ... 301, (1972).
                                 (6 , 1925) POLYMER VOL.13 NO.1, ...   40, (1972).

4350 TZENG...........(H.S.)..¡   (22A,   9) CHIN. J. PHYS. (TAIWAN) VOL.8, .   31, (1970).

4351 UCHIDA..........(A.)....¡   (12B,   23) J. PHYS. SOC. JAPAN VOL.28,. . . 1116, (1970).

4352 UCHINO..........(K.)....¡   (5 , 2187) PHYS. LETT. VOL.44A NO.7,. . . . 529, (1973).

4353 UDOVENKO........(V.A.)..¡   (5 , 2151) SOV. PHYS. DOKLADY VOL.16, . . . 486, (1971).
                                 (5 , 1627) SOV. PHYS. DOKLADY VOL.18, . . . 432, (1973).

4354 UEDA............(R.)....¡   (5 , 1983) ACTA CRYST. VOL.15, ... 506, (1962).
                                 (5 , 1981) J. PHYS. SOC. JAPAN VOL.17 B-II, 389, (1962).

4355 UGODENKO........(A.A.)..¡   (5 , 1955) SOV. AT. ENERGY VOL.18, ... 350, (1965).

4356 ULKU............(D.)....¡   (5 , 1192) Z. KRIST. VOL.124, ... 192, (1967).

4357 ULLMAIER........(H.)....¡   (5 , 1998) PHYS. STAT. SOL. VOL.48, ... 619, (1971).
                                 (5 , 2005) Z. PHYSIK VOL.253, ... 219, (1972).
                                 (5 , 2618) J. LOW TEMP. PHYS. VOL.14, . . . 213, (1974).

4358 UL≠YANOV........(V.A.)..¡   (5 , 2084) SOV. PHYS. SOL. STATE VOL.15,. . 919, (1973).

4359 UMAKANTHA.......(N.)....¡   (6 , 787) PHYS. REV. LETT. VOL.4, ... 452, (1960).
                                 (6 , 157) J. PHYS. CHEM. SOLIDS VOL.23,. . 1747, (1962).

4360 UMEBAYASHI......(H.)....¡   (5 , 1038) J. PHYS. SOC. JAPAN VOL.21,. . . 1281, (1966).
                                 (5 , 1075) PHYS. LETT. VOL.22, ... 407, (1966).
                                 (5 , 697) J. APPL. PHYS. VOL.38, ... 1461, (1967).
                                 (5 , 1349) PHYS. REV. LETT. 18, ... 548, (1967).
                                 (5 , 1438) SOLID STATE COMM. VOL. 5,. . . . 591, (1967).
                                 (5 , 528) J. PHYS. SOC. JAPAN VOL.24,. . . 368, (1968).
                                 (5 , 1356) PHYS. REV. VOL.165,. . . . . . . 688, (1968).
                                 (5 , 696) PHYS. REV. VOL.167,. . . . . . . 519, (1968).
                                 (5 , 743) PHYS. REV. VOL.181,. . . . . . . 936, (1969).

4361 UMEZAWA.........(H.)....: (5 , 2022) PHYS. REV. B VOL.9,. . . . . . 130, (1974).

4362 UNGERER.........(H.)...: (5 , 2860) Z. PHYSIK VOL.219, ... 300, (1969).

4363 UPADHYAYA.......(J.C.)..: (10B, 90) J. PHYS. F VOL.3,. . . . . . . 1672, (1973).
                              (10B, 89) J. PHYS. F VOL.3,. . . . . . . 640, (1973).
                              (10B, 179) PHYS. REV. B VOL.8,. . . . . 593, (1973).
                              (10B, 220) SOL. STATE COMM. VOL.13, . . . 779, (1973).

4364 UPADHYAYA.......(K.S.)..: (6 , 508) PHYS. LETT. VOL.40A NO.4,. . . . 291, (1972).
                              (6 , 1417) PHYS. REV. B VOL.6,. . . . . 1589, (1972).
                              (6 , 17) PHYS. STAT. SOL. B VOL.51 NO.1,. 389, (1972).
                              (6 , 447) SOL. STATE COMM. VOL.11 NO.4,. . 567, (1972).
                              (6 , 1489) SOL. STATE COMM. VOL.11,. . . . 109, (1972).
                              (10B, 198) PHYS. STAT. SOLIDI B VOL.59, . . 279, (1973).

4365 URANO...........(T.)....: (6 , 113) J. PHYS. SOC. JAPAN VOL.34,. . . 1197, (1973).

4366 URBAN...........(S.)....: (6 , 116) FARADAY SYMP. CHEM. SOC. NO.6, . 48, (1972).

4367 URETSKY.........(J.L.)..: (9B, 25) ANN. OF PHYS. VOL.33,. . . . . . 400, (1965).

4368 URUSHADZE.......(G.G.)..: (5 , 2103) SOV. PHYS. DOKLADY VOL.14, . . . 263, (1969).
                              (5 , 2104) SOV. PHYS. DOKLADY VOL.14, . . . 84, (1969).
                              (5 , 2105) UKRAINIAN PHYS. VOL.15,. . . . . 133, (1970).

4369 URWANK..........(P.)....: (6 , 1430) PHYS. STAT. SOL. VOL.49,. . . . 807, (1972).

4370 URYU............(N.)....: (5 , 503) J. PHYS. SOC. JAPAN VOL.18,. . . 1641, (1963).

4371 USHA-DENIZ......(K.)....: (6 , 1499) IAEA SYMP. BOMBAY, VOL.1,. . . . 433, (1964).
                              (6 , 260) IAEA SYMP. BOMBAY, VOL.2,. . . . 157, (1964).
                              (5 , 1919) IAEA SYMP. BOMBAY, VOL.2,. . . . 347, (1964).
                              (6 , 1497) PROC. INT. CONF. (NOTTINGHAM),. 323, (1964).
                              (6 , 1566) SOL. STATE COMM. 2,. . . . . . . 17, (1964).
                              (6 , 1582) J. PHYS. CHEM. SOL. VOL.27,. . 1103, (1966).
                              (6 , 1459) J. APP. PHYS. 39,. . . . . . . 1232, (1968).
                              (6 , 431) J. CHEM. PHYSICS VOL.55,. . . . 2384, (1971).
                              (6 , 1311) J. PHYSIQUE TOME 32 COL.1 VOL.II 619, (1971).

4372 USHA............(K.)....: (6 , 1567) IAEA SYMP. CHALK RIVER, VOL.2, . 253, (1962).

4373 USPENSKII.......(L.N.)..: (20C, 21) INSTRUM. EXP. TECH. NO.1,. . . . 77, (1970).

4374 UTSURO..........(M.)....: (6 , 295) J. NUCL. SCI. TECHNOL. VOL.9,. . 374, (1972).

4375 VADACCHINO......(M.)....: (20B, 85) NUCL. INST. MET. 59,. . . . . . 136, (1968).
                              (5 , 1926) J. PHYSIQUE TOME 33 COL.2,. . . 83, (1972).

4376 VAGELATOS.......(N.)....: (6 , 2324) J. PHYS. CHEM. SOLIDS VOL.32,. . 1573, (1971).
                              (6 , 2330) J. CHEM. PHYS. VOL.60,. . . . . 3613, (1974).

4377 VALENTA.........(L.)....: (6 , 1743) ACTA PHYS. HUNGAR. VOL.15, . . . 29, (1962).

4378 VALENTINE.......(T.M.)..: (5 , 2905) ACTA CRYST. B25, ... 2140, (1969).

4379 VALKO...........(J.)....: (6 , 1075) NUKLEONIK VOL.12,. . . . . . . 237, (1969).

4380 VANCE...........(E.R.)..: (6 , 2267) J. APPL. PHYS. VOL.39,. . . . . 3501, (1968).
                              (6 , 820) J. PHYS. SOC. JAPAN VOL.25,. . . 367, (1968).
                              (5 , 1622) J. OF PHYS. C VOL.2, ... 761, (1969).
                              (6 , 1444) J. APPL. PHYS. 41,. . . . . . . 1857, (1970).
                              (6 , 1443) J. PHYS. CHEM. SOLIDS VOL.31,. . 485, (1970).
                              (6 , 1454) J. PHYS. C VOL.4, NO.13,. . . . L258, (1971).
                              (6 , 1434) PHYS. STAT. SOL. VOL.47,. . . . K11, (1971).

4381 VANDERVELDE.....(J.C.)..: (7A, 54) PHYS. LETT. B VOL.36B, ... 560, (1971).

4382 VANINBROUKX.....(R.)....: (20A, 20) INT. J. APPL. RADIAT. ISOTOP. 22 529, (1971).

4383 VANNESTE........(L.)....: (5 , 2820) PHYS. REV. B-3,. . . . . 128, (1970).

4384 VAN-BRUGGEN.....(C.F.)..: (5 , 1810) J. PHYSIQUE TOME 32 COL.1 VOL.I, 78, (1971).

4385 VAN-DER-MERWE...(P.)....: (22A, 39) NUCL. INST. MET. 63, ... 157, (1968).

4386 VAN-DER-WEY.....(R.)....: (12A, 12) NED. TIJDSCHR. NATUURK. VOL.38,. 171, (1972).

4387 VAN-DER-ZEE.....(J.J.)..: (5 , 280) CRYST. STRUCT. COMM. VOL.1 N.3,. 193, (1972).
                              (5 , 2394) CRYST. STRUCT. COMM. VOL.1 N.4,. 367, (1972).
                              (5 , 2674) CRYST. STRUCT. COMM. VOL.1 N.4,. 371, (1972).

4388 VAN-DE-VYVER....(R.E.)..: (5 , 153) NUCL. PHYS. A VOL.A177,. . . . . 393, (1971).

4389 VAN-DIJK........(C.)....: (18A, 11) BROOKHAVEN SYMPOSIUM,. . . . . . 163, (1965).
                              (20D, 18) NUCL. INST. MET. 51,. . . . . . 121, (1967).
                              (6 , 775) PHYS. LETT. 24A,. . . . . 270, (1967).
                              (6 , 796) PHYS. LETT. VOL.32A,. . . . . . 255, (1970).
                              (6 , 739) RCN-129 (122 PP.),. . . . . . . , (1970).

4390 VAN-DINGENEN....(W.)....: (20B, 26) IAEA SYMPOSIUM VIENNA,. . . . . 453, (1960).
                              (20B, 75) NUCL. INST. MET. VOL.16,. . . . 116, (1962).
                              (6 , 1775) PHYSICA VOL.28,. . . . . 917, (1962).
                              (6 , 342) COLL. INTER. N.126 (GRENOBLE),. 224, (1963).
                              (6 , 1755) COLL. INTER. N.126 (GRENOBLE),. 229, (1963).
                              (5 , 1907) PHYSICA VOL.30,. . . . . 237, (1964).
                              (6 , 1756) BROOKHAVEN SYMPOSIUM,. . . . . . 83, (1965).
                              (6 , 1742) PHYSICA VOL.34,. . . . . 257, (1967).
                              (6 , 758) PHYSICA VOL.37,. . . . . 603, (1967).

4391 VAN-DYJK........(C.)....: (6 , 725) IAEA SYMP. COPENHAGEN, VOL.1,. . 233, (1968).

4392 VAN-HOVE........(L.)....: (14B, 74) PHYS. REV. VOL.82,. . . . . . . 392, (1951).
                              (10C, 68) PHYS. REV. VOL.93,. . . . . . . 1207, (1954).
                              (8 , 81) PHYS. REV. VOL.93,. . . . . . . 268, (1954).
                              (8 , 80) PHYS. REV. VOL.95,. . . . . . . 1374, (1954).
                              (14A, 42) PHYS. REV. VOL.95,. . . . . . . 249, (1954).
                              (7B, 44) NUOVO CIMENTO VOL.1,. . . . . . 233, (1955).

4393 VAN-KRANENDONK..(J.)....: (8 , 121) REV. MOD. PHYS. VOL.30,. . . . . 1, (1958).

4394 VAN-LAAR........(B.)....: (5 , 2353) COLL. INTER. N.126 (GRENOBLE),. 176, (1963).
                              (5 , 579) PROC. INT. CONF. (NOTTINGHAM), . 275, (1964).
                              (5 , 490) PHYS. REV. 138, A,. . . . . . . 584, (1965).
                              (5 , 2037) J. PHYS. CHEM. SOL. 27,. . . . 1297, (1966).
                              (5 , 491) PHYS. REV. 141,. . . . . 538, (1966).
                              (5 , 616) PHYS. LETT. 25A,. . . . . 27, (1967).
                              (5 , 617) PHYS. REV. 156,. . . . . 654, (1967).
                              (5 , 1481) PHYS. LETT. VOL.26A NO.11,. . . 626, (1968).
                              (5 , 1157) ACTA CHEM. SCAND. VOL.24,. . . 2435, (1970).
                              (5 , 1810) J. PHYSIQUE TOME 32 COL.1 VOL.I, 78, (1971).
                              (5 , 823) J. PHYSIQUE TOME 32,. . . . . . 301, (1971).
                              (5 , 619) J. PHYS. CHEM. SOLIDS VOL.32,. . 581, (1971).
                              (5 , 655) ACTA CRYST. VOL.A28 PART S-4,. . 194, (1972).
                              (5 , 2148) PHYSICA VOL.57,. . . . . 215, (1972).
                              (5 , 2097) PHYS. LETT. VOL.41A NO.5,. . . . 411, (1972).

```
4394  VAN-LAAR........(B.)....!  ( 5 ,   656) PHYS. REV. B VOL.6,. . .,. : : : 2669, (1972).
                                 ( 5 ,  1432) J. SOLID STATE CHEM. VOL.6,. : : :  384, (1973).
                                 ( 5 ,  1966) J. SOLID STATE CHEM. VOL.6,. : : :  574, (1973).

4395  VAN-LEEUWEN.....(B.)....!  (12A,   12) NED. TIJDSCHR. NATUURK. VOL.38,.  171, (1972).

4396  VAN-LEEUWEN.....(J.M.J.)!  ( 7B,   20) IAEA SYMPOSIUM BOMBAY, VOL.2,. .,.    35, (1964).
                                 (16 ,   61) PHYS. REV. VOL.139,. . . . . . .A1138, (1965).

4397  VAN-LOEF........(J.J.)..!  ( 6 ,  1775) PHYSICA VOL.28,. . . . . . . . .  917, (1962).
                                 ( 5 ,  2078) PHYS. LETT. VOL.27A, : : : : : :   69, (1968).
                                 ( 6 ,   78) PHYS. LETT. VOL.28A, : : : : : :  642, (1969).
                                 (14A,   34) PHYSICA VOL.50,. . . . : : : : :  511, (1970).
                                 ( 6 ,   75) PHYS. LETT. VOL.35A,. : : : : : :  169, (1971).
                                 (14A,   35) PHYSICA VOL.62,. . . . : : : : :  345, (1972).

4398  VAN-NHUNG.......(N.)....!  ( 5 ,  2604) SOL. STATE COMM. VOL.8 NO.1, . .   23, (1970).

4399  VAN-TRICHT......(J.B.)..!  ( 6 ,   452) CHEM. PHYS. LETT. VOL.22,. . . .  476, (1973).

440C  VAN-UITERT......(L.G.)..!  ( 6 ,  1934) PHYS. REV. LETT. VOL.31, . . . . 1300, (1973).

4401  VAN-VLECK.......(J.H.)..!  ( 7A,   57) PHYS. REV. VOL.48, . . . . . . .  367, (1935).
                                 (11C,   27) PHYS. REV. VOL.55,. . . . . . .  924, (1939).
                                 ( 8 ,  121) REV. MOD. PHYS. VOL.30,. . . . .    1, (1958).

4402  VAN-VUCHT.......(J.H.N.)!  ( 5 ,   89) ACTA CRYST. VOL.14,. . . . . . .  223, (1961).

4403  VAN-VUCHT.......(J.H.)..!  ( 5 ,  2630) PHILIPS RES. REP. VOL.18,. . . .   35, (1963).

4404  VAN-ZEVENBERGEN.(F.)....!  ( 6 ,   78) PHYS. LETT. VOL.28A, . . . . . .  642, (1969).

4405  VARGA...........(B.B.)..!  (10B,   19) DISS. ABS. VOL.27, . . . . . . .2492B, (1966).

4406  VARSHNI.........(Y.P.)..!  (10B,  131) PHYS. REV. 174,. : : : : : : : :  766, (1968).
                                 (10R,   11) CAN. J. PHYS. 47,. : : : : : : :  451, (1969).

4407  VASILEV.........(E.A.)..!  ( 5 ,  1633) J. PHYSIQUE TOME 32 COL.1 VOL.II  987, (1971).

4408  VASILIU.........(V.)....!  (20B,  136) REV. DE PHYSIQUE VOL.6,. : : : :  411, (1961).
                                 (20A,   74) REV. ROUMAINE PHYS. VOL.8,. . . .  277, (1963).

4409  VASLOW..........(F.)....!  ( 5 ,  1549) J. CHEM. PHYS. VOL.58, . . . . . 5017, (1973).

441C  VASTEL..........(M.)....!  ( 5 ,  2703) PROC/CONF/NEUTRON C. S. + TECH., 1129, (1968).

4411  VAVILOVA........(V.V.)..!  ( 5 ,  2665) DOKL. AKAD. NAUK SSSR VOL.194, . 1374, (197C).

4412  VEALL...........(N.)....!  (20C,   31) J. SCI. INSTRUM. VOL.24, . . . .  331, (1947).

4413  VEDEKHIN........(A.F.)..!  (20C,   17) INSTRUM. EXP. TECH. NO.2,. . . .  305, (1965).

4414  VEEKIND.........(J.C.)..!  (24 ,   73) NUCL. INST. MET. 77, . . . . . .   55, (1970).

4415  VEHOFF..........(H.)....!  ( 5 ,  2079) INT. J. MAGN. VOL.5,. . . . . .  223, (1973).

4416  VEILLET.........(P.)....!  ( 5 ,  2948) J. SOLID STATE CHEM. VOL.8,. . .  182, (1973).

4417  VELAZQUEZ.......(R.)....!  (24 ,  103) REV. MEXICANA FIS. VOL.18, . . .  115, (1969).

4418  VENKATARAMAN....(G.)....!  ( 6 ,  1567) IAEA SYMP. CHALK RIVER, VOL.2,.  253, (1962).
                                 ( 6 ,  1413) IAEA SYMP. CHALK RIVER, VOL.2,.   99, (1962).
                                 ( 6 ,  1402) COPENHAGEN CONF.-LATT.DYNAMICS,.  223, (1963).
                                 ( 6 ,  1399) IAEA SYMP. BOMBAY, VOL.1,. . . .  153, (1964).
                                 ( 6 ,  1499) IAEA SYMP. BOMBAY, VOL.1,. . . .  433, (1964).
                                 ( 6 ,   260) IAEA SYMP. BOMBAY, VOL.2,. . . .  157, (1964).
                                 ( 5 ,  1919) IAEA SYMP. BOMBAY, VOL.2,. . . .  347, (1964).
                                 ( 6 ,  1497) PROC. INT. CONF. (NOTTINGHAM),.  322, (1964).
                                 ( 6 ,  1566) SOL. STATE COMM. 2, . . . . . .   17, (1964).
                                 (15 ,    5) BROOKHAVEN SYMPOSIUM,. . .,. . .  169, (1965).
                                 ( 6 ,  1582) J. PHYS. CHEM. SOL. VOL.27,. . . 1103, (1966).
                                 ( 6 ,   376) PHYS. LETT. 23,. . . . . . . . .  226, (1966).
                                 ( 7B,   82) PHYS. REV. VOL.149,. . . . . . .    1, (1966).
                                 ( 5 ,  1906) PROC. PHYS. SOC. 89, . . . . . .  379, (1966).
                                 ( 6 ,   274) CAN. JOURN. PHYS. 45,. . . . . . 3185, (1967).
                                 ( 6 ,   278) PHYS. REV. 156,. . . . . . . . .  196, (1967).
                                 ( 6 ,   377) PHYS. REV. 161,. . . . . . . . .  133, (1967).
                                 ( 6 ,  2301) IAEA SYMP. COPENHAGEN, VOL.1,.   195, (1968).
                                 ( 6 ,  1880) IAEA SYMP. COPENHAGEN, VOL.2,.   159, (1968).
                                 ( 6 ,   532) J. APPL. PHYS. VOL.39 NO.2,. . . 1113, (1968).
                                 ( 6 ,   533) PHYS. LETT. VOL.26A, . . . . . .  108, (1968).
                                 ( 6 ,  2327) SOL. STATE COMM. 7, NO.21,. . . 1571, (1969).
                                 (10A,   14) REV. MOD. PHYS. VOL.42,. . . . .  409, (197C).
                                 ( 6 ,  1591) SOL. STATE COMM. VOL.8,. . . . .  889, (197C).
                                 ( 6 ,  2324) J. PHYS. CHEM. SOLIDS VOL.32,. . 1573, (1971).
                                 (11A,   50) PHYS. REV. B VOL.5,. . . . . . . 1993, (1972).

4419  VENTURINI.......(G.)....!  (17A,   82) NUOVO CIMENTO SUPPL. VOL.3,. . .  187, (1965).

442C  VERBIST.........(J.J.)..!  ( 5 ,    4) ACTA CRYST. A28 PART S-4,. . . .  193, (1972).
                                 ( 5 ,   13) ACTA CRYST. VOL.B28, . . . . . . 3006, (1972).
                                 ( 5 ,   24) ACTA CRYST. VOL.B28, . . . . . . 3207, (1972).
                                 ( 5 ,   981) J. CHEM. PHYS. VOL.56,. . . . . 3257, (1972).
                                 ( 5 ,   30) MATER. RES. BULL. VOL.7,. . . . 1225, (1972).
                                 ( 5 ,  2350) NATURE VOL.235,. . . . . . . . .  328, (1972).
                                 ( 5 ,   11) J.C.S. PERKIN TRANS. II, . . . .  133, (1973).

4421  VERBLE..........(J.L.)..!  ( 6 ,  1378) PHYS REV 168,. . . . . . . . . .  980, (1968).

4422  VERDAN..........(G.)....!  ( 6 ,  1681) BROOKHAVEN SYMPOSIUM,. . . . . .  105, (1965).
                                 ( 6 ,  2254) PHYS. LETT. 14,. . . . . . . . .  100, (1965).
                                 ( 6 ,  2253) J. PHYSIQUE TOME 28 COL.1,. . .   26, (1967).
                                 ( 6 ,   962) PHYS. LETT. 25A,. . . . . . . .  435, (1967).
                                 (15 ,   12) HELV. PHYS. ACTA VOL.41,. . . .  533, (1968).
                                 ( 6 ,  1677) IAEA SYMP. COPENHAGEN, VOL.1,.   223, (1968).

4423  VERDIER.........(J.)....!  (23 ,   13) ENERGIE NUCLEAIRE VOL.13,. . . .   15, (1971).

4424  VERGHESE........(K.)....!  (20C,   61) NUCL. INST. MET. 74, . . . . . .  355, (1969).

4425  VERGNOUX........(A.M.)..!  (10B,    7) ANN. DE PHYSIQUE TOME 5, . . . .   77, (197C).

4426  VERHAEGHE.......(J.L.)..!  (20C,    6) APPL. SCI. RES. B VOL.10,. . . .  247, (1963).

4427  VERMA...........(G.S.)..!  (10B,   95) J. PHYSIQUE TOME 35, . . . . . .  263, (1974).

4428  VERMA...........(M.P.)..!  ( 6 ,  1303) INDIAN J. PURE APPL. PHYS. VOL.7  151, (1969).
                                 (100,   45) PHYS. STAT. SOL. VOL.36,. . . .  335, (1969).
                                 (100,   47) PHYS. STAT. SOL. VOL.38,. . . .  K19, (197C).
                                 (10B,   62) J. PHYS. C: SOLID ST. PHYS. V-4, 2749, (1971).
                                 (10B,   68) J. PHYS. C: SOLID ST. PHYS. V-5, 1038, (1972).
                                 (10B,   69) J. PHYS. C: SOLID ST. PHYS. V-5,  543, (1972).
                                 (10B,  101) NUCL./S.S. PHYS. SYMP. ABSTR.,. .     , (1972).
                                 (10B,   90) J. PHYS. F VOL.3,. . . . . . . . 1672, (1973).
                                 (10B,   89) J. PHYS. F VOL.3,. . . . . . . .  640, (1973).
                                 (10B,  184) PHYS. REV. B VOL.8,. . . . . . . 4880, (1973).
```

4428 VERMA..........(M.P.)..: (10B, 179) PHYS. REV. B VOL.8,. . . . . 593, (1973).
                           (10B, 220) SOL. STATE COMM. VOL.13,. . . . : 779, (1973).

4429 VERNON.........(L.W.)..: (5 , 2430) J. PHYS. CHEM. VOL.57,. . . . . 535, (1953).

4430 VERSCHUEREN.....(M.)....: (5 , 1492) COLL. INTER. N.126 (GRENOBLE), . 154, (1963).

4431 VERTEBNYI.......(V.P.)..: (20B, 49) INSTRUM. EXP. TECH. NO.2,. . . . 347, (1963).
                           (14B, 136) UKR. FIZ. ZH. (USSR) VOL.9,. . . 684, (1964).
                           (5 , 2229) SOV. AT. ENERGY VOL.18,. . . . . 885, (1965).
                           (5 , 2411) SOV. AT. ENERGY VOL.19,. . . . . 1162, (1965).
                           (5 , 1248) BULL. ACAD. SCI. USSR/PHYS. 31,. 334, (1967).
                           (5 , 2242) UKRAINIAN PHYS. J. VOL.14,. . . 1968, (1969).
                           (5 , 379) UKR. FIZ. ZH.(USSR) VOL.17,. . : 38, (1972).

4432 VETELINO.......(J.F.)..: (6 , 2038) PHYS. REV. 178,. . . . . 1349, (1969).
                           (10B, 156) PHYS. REV. B-2,. . . . . . . : 2167, (1970).
                           (10B, 158) PHYS. REV. B-2,. . . . . . . : 987, (1970).
                           (10B, 215) SOL. STATE COMM. VOL. 9,. . . : 185, (1971).
                           (6 , 562) PHYS. REV. B VOL.7,. . . . . : 4001, (1973).

4433 VEYRET.........(C.)....: (5 , 874) J. PHYSIQUE TOME 31, ... 607, (1970).

4434 VIEBAHN-HANSLER.(R.)....: (5 , 1988) SOL. STATE COMM. VOL.11, ... 1119, (1972).

4435 VIENNET........(R.)....: (5 , 764) NUCL. PHYS. A(NETHERLANDS) A182, 541, (1972).
                           (7A, 39) NUCL. PHYS. A VOL.182, ... 541, (1972).

4436 VIGREN.........(D.T.)..: (8 , 101) PHYS. REV. B VOL.5,. . . . . . 2719, (1972).

4437 VIJAYARAGHAVAN..(P.R.)..: (6 , 1567) IAEA SYMP. CHALK RIVER, VOL.2, . 253, (1962).
                           (6 , 1413) IAEA SYMP. CHALK RIVER, VOL.2, : 99, (1962).
                           (5 , 1174) J. PHYS. SOC. JAPAN VOL.17 B-III 41, (1962).
                           (5 , 1173) J. PHYS. SOC. JAPAN VOL.17,. . : 247, (1962).
                           (6 , 1402) COPENHAGEN CONF.-LATT.DYNAMICS, 223, (1963).
                           (6 , 1399) IAEA SYMP. BOMBAY, VOL.1,. . . . 153, (1964).
                           (5 , 1919) IAEA SYMP. BOMBAY, VOL.2,. . . . 347, (1964).
                           (6 , 1566) SOL. STATE COMM. 2,. . . . . . . 17, (1964).
                           (6 , 1582) J. PHYS. CHEM. SOL. VOL.27,. . . 1103, (1966).
                           (6 , 1360) IAEA SYMP. COPENHAGEN, VOL.1,. . 149, (1968).
                           (6 , 611) IAEA SYMP. COPENHAGEN, VOL.1,. . 47, (1968).
                           (6 , 612) PHYS. REV. LETT. VOL.20,. . . . 1245, (1968).
                           (6 , 1373) PHYS. REV. 168,. . . . . 970, (1968).
                           (6 , 16) PHYS. REV. B-1,. . . . . 4819, (1970).
                           (6 , 1220) PHYS. REV. B-3,. . . . . 1229, (1971).
                           (6 , 446) IAEA SYMP. GRENOBLE, ... 95, (1972).

4438 VILLAIN........(J.)....: (5 , 1649) COMP. REND. 249,. . . . . 1334, (1959).
                           (5 , 1645) J. PHYS. RADIUM VOL.21,. . . . . 67, (1960).
                           (13 , 12) COLL. INTER. NO.126 (GRENOBLE),. 194, (1963).
                           (13 , 26) J. PHYSIQUE TOME 24, ... 622, (1963).
                           (13 , 29) J. PHYSIQUE TOME 29, ... 488, (1968).
                           (13 , 28) J. PHYSIQUE TOME 29, ... 687, (1968).
                           (13 , 34) MAT. SCI. ENGNG,. . . . . , (1972).
                           (11A, 21) J. PHYS. C VOL.6,. . . . . L97, (1973).
                           (11A, 26) J. PHYSIQUE TOME 35, ... 27, (1974).

4439 VILLA..........(F.)....: (5 , 1236) J. NON-CRYST. SOLIDS VOL.11 NO.5 417, (1973).

4440 VILLA..........(S.)....: (22B, 37) NUCL. INST. MET. VOL.115,. . . . 393, (1974).

4441 VILLERS........(G.)....: (5 , 1672) J. APP. PHYS. VOL.39,. . . . . . 590, (1968).

4442 VINCENT........(D.H.)..: (6 , 305) BROOKHAVEN SYMPOSIUM., ... 142, (1965).

4443 VINCENT........(H.)....: (5 , 1832) SOL. STATE COMM. VOL.6 NO.5, . . 269, (1968).
                           (5 , 1713) SOL. STATE COMM. VOL.7,. . . . : 641, (1969).
                           (5 , 1027) J. PHYS. CHEM. SOLIDS VOL.34,. . 151, (1973).

4444 VINEYARD.......(G.H.)..: (7B, 65) PHYS. REV. VOL.85,. . . . . . . 633, (1952).
                           (5 , 1349) J. CHEM. PHYS. VOL.22, ... 1665, (1954).
                           (9B, 80) PHYS. REV. VOL.96,. . . . . . . 93, (1954).
                           (14B, 76) PHYS. REV. VOL.110,. . . . . . . 999, (1958).
                           (14B, 78) PHYS. REV. VOL.119,. . . . . . . 1150, (1960).
                           (14A, 7) BROOKHAVEN SYMPOSIUM., ... 16, (1965).

4445 VINHAS.........(L.A.)..: (22A, 40) NUCL. INST. MET. 63,. . . . . . 13, (1968).
                           (14B, 41) J. CHEM. PHYS. VOL.56, ... 3118, (1972).
                           (6 , 59) J. NUCL. ENERGY VOL.26,. . . . . 379, (1972).
                           (6 , 1684) PHYS. LETT. VOL.43A NO.3,. . . . 279, (1973).

4446 VINIT..........(A.)....: (22B, 14) INTERNAT. COMPUTING SYMP., . . . 291, (1974).

4447 VINNIK.........(M.A.)..: (5 , 179) SOV. PHYS. SOL. STATE VOL.11,. . 2177, (1969).

4448 VINOGRADOV......(S.I.)..: (22A, 66) SOV. PHYS. CRYST. VOL.2, ... 626, (1957).
                           (5 , 131) SOV. PHYS. CRYST. VOL.3, ... 308, (1958).
                           (5 , 553) SOV. PHYS. DOKLADY VOL.4,. . . : 1070, (1960).

4449 VINTAIKIN......(E.Z.)..: (6 , 1752) SOV. PHYS. SOL. STATE 7, ... 1547, (1965).
                           (6 , 601) SOV. PHYS. SOL. STATE 7,. . . . 296, (1965).
                           (5 , 2103) SOV. PHYS. DOKLADY VOL.14,. . . 263, (1969).
                           (5 , 2104) SOV. PHYS. DOKLADY VOL.14,. . . 84, (1969).
                           (5 , 2105) UKRAINIAN PHYS. VOL.15,. . . . 133, (1970).
                           (5 , 2846) SOV. PHYS. DOKLADY VOL.15,. . . 776, (1971).
                           (5 , 2151) SOV. PHYS. DOKLADY VOL.16,. . . 486, (1971).
                           (5 , 1627) SOV. PHYS. DOKLADY VOL.18,. . . 432, (1973).

4450 VIRJO..........(A.)....: (12B, 13) IAEA SYMP. COPENHAGEN, VOL.2,. . 395, (1968).
                           (19A, 39) NUCL. INST. MET. 63, ... 351, (1968).
                           (20B, 119) PHYS. LETT. VOL.26A,. . . . . . 469, (1968).
                           (19A, 41) NUCL. INST. MET. 73,. . . . . . 189, (1969).
                           (19A, 43) NUCL. INST. MET. 75,. . . . . . 77, (1969).

4451 VISSCHER.......(W.M.)..: (10B, 133) PHYS. REV. 175,. . . . . 1201, (1968).

4452 VISWANATHAN.....(S.)....: (14B, 60) PHIL. MAG. (8) VOL.1,. . . . . . 560, (1956).

4453 VISWANATHAN....(K.S.)..: (6 , 459) ACTA CRYST. VOL.A26,. . . . . . 364, (1970).
                           (10C, 94) REPORT NAL-TN-21 (14PP.),. . . . , (1970).

4454 VIVET..........(B.)....: (5 , 1999) PROC. INT. CONF. (NOTTINGHAM), . 285, (1964).

4455 VIZI...........(I.)....: (23 , 60) NUCL. INST. MET. 55,. . . . . . 151, (1967).
                           (12B, 11) IAEA SYMP. COPENHAGEN, VOL.2,. . 407, (1968).
                           (6 , 726) IAEA SYMP. COPENHAGEN, VOL.2,. . 55, (1968).
                           (19B, 30) REPORT KFKI-07/1968 (14PP.),. . , (1968).
                           (19A, 27) IAEA SYMP. GRENOBLE, ... 763, (1972).

4456 VLADIMIRSKII...(V.V.)..: (20B, 130) PRIB. TEKH. EKSPER. NO.2,. . . . 3, (1956).

4457 VLASOV.........(M.F.)..: (5 , 2411) SOV. AT. ENERGY VOL.19,. . . . . 1162, (1965).
                           (5 , 1248) BULL. ACAD. SCI. USSR/PHYS. 31,. 334, (1967).
                           (5 , 2242) UKRAINIAN PHYS. J. VOL.14,. . . 1968, (1969).
                           (5 , 379) UKR. FIZ. ZH.(USSR) VOL.17,. . : 38, (1972).

4458   VOGELSANG........(W.)....: (5 , 2802) J. PHYS. CHEM. SOLIDS VOL.27,. . 1197, (1966).

4459   VOGT............(E.)....: (5 , 1644) Z. ANGEW. PHYS. VOL.15,. . . . . . 371, (1963).

4460   VOGT............(O.)....: (6 , 467) J. OF PHYS. C VOL.1,. . . . . . . 679, (1968).
                                  (5 , 914) PHYS. KONDENS. MAT. (GER.) VOL.9 249, (1969).
                                  (6 , 2158) IAEA SYMP. GRENOBLE,. . . . . . . 553, (1972).
                                  (6 , 466) IAEA SYMP. GRENOBLE,. . . . . . . 563, (1972).
                                  (6 , 1693) J. PHYS. C VOL.5,. . . . . 2246, (1972).
                                  (5 , 2612) A.I.P. CONF. PROC. NO.10 PART 2, 1554, (1973).
                                  (5 , 2048) J. PHYS. C VOL.6,. . . . . 725, (1973).
                                  (5 , 2697) PHYS. REV. B VOL.7,. . . . . . . 1988, (1973).

4461   VOGT............(R.H.)..: (20A,   50) NUCL. SCI. ENGNG. VOL.14,. . . . 397, (1962).

4462   VOLKIN..........(H.C.)..: (15 ,   37) PHYS. REV. VOL.113,. . . . . . . 866, (1959).
                                  (15 ,   38) PHYS. REV. VOL.117,. . . . . . . 1029, (1960).

4463   VOLKL...........(J.)....: (5 , 2859) PHYS. LETT. 20,. . . . . 470, (1966).

4464   VOL≠GEMUT.......(A.A.)..: (24 , 113) SOV. PHYS. DOKLADY VOL.16,. . . 49, (1971).

4465   VONACH..........(H.)....: (5 , 686) Z. PHYSIK VOL.250,. . . . . . . 166, (1972).

4466   VONSOVSKY.......(S.V.)..: (9C,   23) J. PHYS. (USSR) VOL.5,. . . . . 263, (1941).

4467   VON-BALTZ.......(R.)....: (6 , 2341) ATOMKERNENERGIE VOL.16,. . . . . 201, (1970).

4468   VON-DREELE......(R.B.)..: (5 , 2655) NATURE (PHYS. SCI.) VOL.244,. . 139, (1973).

4469   VON-HALBAN......(H.)....: (5 , 937) COMP. REND. TOME 203,. . . . . . 73, (1936).
                                  (7B,   37) J. PHYS. RADIUM SER.7 VOL.8,. . 29, (1937).

4470   VON-HEIDENSTAM..(O.)....: (5 , 1057) ARK. KEMI(SWEDEN)VOL.28 PAPER23, 375, (1968).

4471   VON-JAN.........(R.)....: (19B,   16) NUCL. INST. MET. VOL.80,. . . . 69, (1970).

4472   VON-LOYEN.......(L.)....: (10C,   85) PHYS. STAT. SOLIDI B VOL.58 NO.2 K159, (1973).

4473   VON-WARTBURG....(W.)....: (5 , 914) PHYS. KONDENS. MAT. (GER.) VOL.9 249, (1969).

4474   VORA............(R.B.)..: (24 ,   94) PHYS. REV. B VOL.3,. . . . . . . 2929, (1971).

4475   VORDERWISCH.....(P.)....: (6 , 2040) IAEA SYMP. GRENOBLE,. . . . . . 489, (1972).

4476   VOROBEV.........(V.V.)..: (5 , 2609) SOV. PHYS. J.E.T.P. 26,. . . . . 1086, (1968).

4477   VORONA..........(P.N.)..: (20C,   24) INSTRUM. EXP. TECH. VOL.14, NO.5 1343, (1971).

4478   VOSKANYAN.......(R.A.)..: (5 , 1072) SOV. PHYS. SOL. STATE VOL.13,. . 44, (1971).

4479   VOSKANYAN.......(R.)....: (5 , 2487) SOV. PHYS. J.E.T.P. VOL.25,. . . 266, (1967).

4480   VOSS............(K.)....: (7A,   52) PHYS. LETT. 28A, ... 376, (1968).
                                  (5 ,   97) ANN. PHYS. (GER.) VOL.23,. . . . 49, (1969).

4481   VOS.............(A.)....: (5 ,   5) ACTA CRYST. VOL.B27, ... 146, (1971).

4482   VRZAL...........(J.)....: (5 , 2373) PHYS. STAT. SOL. VOL.42,. . . . 895, (1970).

4483   VUISTER.........(P.H.)..: (20C,   80) NUCL. SCI. APPL. B(PAKIS.) VOL.5 43, (1969).

4484   VUKOVICH........(S.)....: (6 , 1647) IAEA SYMP. COPENHAGEN, VOL.1,. . 367, (1968).

4485   VU-HONG-LAC.............: (20C,   8) BULL. D≠INSTRUM. SCI. TECH. 172, 55, (1972).

4486   VU-VAN-QUI..............: (5 , 1230) COLL. INTER. N.126 (GRENOBLE),. 92, (1963).
                                  (5 , 456) BULL. SOC. FRANC. MIN. CRIST. 90 109, (1967).
                                  (5 , 1713) SOL. STATE COMM. VOL.7,. . . . . 641, (1969).

4487   WADDINGTON......(T.C.)..: (6 , 567) J. CHEM. PHYS. VOL.52, ... 2730, (1970).
                                  (6 , 565) IAEA SYMP. GRENOBLE,. . . . . . . 345, (1972).
                                  (6 , 1296) J.C.S. FARADAY TRANS. II VOL.68, 350, (1972).
                                  (6 , 1000) J.C.S. FARADAY TRANS. II VOL.69, 1477, (1973).
                                  (6 , 304) J.C.S. FARADAY TRANS. II VOL.69, 275, (1973).
                                  (6 , 1631) J.C.S. FARADAY TRANS. II VOL.69, 783, (1973).
                                  (6 ,   20) J.C.S. FARADAY TRANS. II VOL.69,   1, (1973).
                                  (6 , 1002) J. CHEM. PHYS. VOL.59,. . . . . . 817, (1973).
                                  (6 ,   8) NATURE (PHYS. SCI.) VOL.241,. . 40, (1973).

4488   WADE............(B.O.)..: (5 , 431) J. NUCL. ENERGY VOL.24,. . . . . 43, (1970).

4489   WADE............(W.R.)..: (22C,   59) REV. SCI. INSTRUM. VOL.37,. . . 435, (1966).
                                  (22C,   61) REV. SCI. INSTRUM. VOL.38,. . . 275, (1967).

4490   WAEBER..........(W.B.)..: (6 , 894) J. PHYS. C, SER.2, VOL.2,. . . . 882, (1969).
                                  (6 , 892) J. PHYS. C, SER.2 VOL.2,. . . . . 903, (1969).
                                  (6 , 1435) J. PHYSIQUE TOME 32 COL.1 VOL.II 1186, (1971).

4491   WAGENFELD.......(H.)....: (5 , 2378) APPL. PHYS. LETT. VOL.10,. . . . 293, (1967).

4492   WAGNER..........(H.)....: (13 ,   43) PHYS. LETT. VOL.33A, ... 58, (1970).

4493   WAJIMA..........(J.T.)..: (20B,   64) J. PHYS. SOC. JAPAN VOL.15,. . . 630, (1960).

4494   WAKABAYASHI.....(N.)....: (6 , 539) PHYS. REV. LETT. VOL.23, ... 311, (1969).
                                  (6 , 2016) DISS. ABS. VOL.30,. . . . . . .5202B, (1970).
                                  (6 , 537) J. APPL. PHYS. 41,. . . . . . . 1365, (1970).
                                  (6 , 2294) AIP CONF. PROC. VOL.5,. . . . . 1450, (1971).
                                  (6 , 1264) CONF. INTERN., RENNES,. . . . . 144, (1971).
                                  (6 , 1220) PHYS. REV. B-3,. . . . . 1229, (1971).
                                  (6 , 2015) PHYS. REV. B-4,. . . . . 2398, (1971).
                                  (6 , 2039) PHYS. REV. B-4,. . . . . 2558, (1971).
                                  (6 , 695) PHYS. REV. LETT. VOL.26,. . . . 140, (1971).
                                  (6 , 703) PHYS. REV. LETT. VOL.27,. . . . 334, (1971).
                                  (6 , 1522) BULL. AMER. PHYS. SOC. VOL.17,. 292, (1972).
                                  (6 , 1265) IAEA SYMP. GRENOBLE,. . . . . . 103, (1972).
                                  (6 , 696) IAEA SYMP. GRENOBLE,. . . . . . 611, (1972).
                                  (6 , 697) MAGN. AND MAG. MATERIALS-1971; 1446, (1972).
                                  (6 , 1426) PHYS. REV. B VOL.5,. . . . . . . 2103, (1972).
                                  (6 , 204) PHYS. REV. B VOL.5,. . . . . . . 4951, (1972).
                                  (6 , 561) PHYS. REV. B VOL.6,. . . . . . . 3956, (1972).
                                  (6 , 943) PHYS. REV. B VOL.8,. . . . . . . 6015, (1973).

4495   WAKUTA..........(Y.)....: (6 , 1275) IAEA SYMP. CHALK RIVER, VOL.2,. 273, (1962).
                                  (19A,   46) NUCL. INST. MET. VOL.75,. . . . 32, (1969).
                                  (19B,   7) JAP. J. APPL. PHYS. VOL.10,. . . 1090, (1971).

4496   WALKER..........(A.)....: (6 , 2223) INORG. CHEM. VOL.6,. . . . . . . 346, (1967).

4497   WALKER..........(C.B.)..: (5 , 754) PHYS. REV. 130,. . . . . 1726, (1963).
                                  (6 , 1521) PHYS. REV. 177,. . . . . 1111, (1969).
                                  (9B,   10) ACTA CRYST. VOL.A26,. . . . . . 447, (1970).

4498 WALKER...........(E.)....: (5 , 1367) SOL. STATE COMM. VOL.11 NO.5,. .   707, (1972).
                                 (6 , 2197) INT. J. MAGN. VOL.4 NO.1,. . . . .   63, (1973).
                                 (6 , 2145) PHYS. REV. LETT. VOL.31,. . . . .   1585, (1973).
                                 (6 , 2196) SOL. STATE COMM. VOL.12 NO.2,. .   117, (1973).
                                 (6 , 2146) SOL. STATE COMM. VOL.13,. . . . .   881, (1973).
                                 (6 , 716) J. PHYS. C VOL.7,. . . . .   1207, (1974).

4499 WALKER...........(J.)....: (20P,   36) IAEA SYMP. COPENHAGEN, VOL.2,. .   341, (1968).

4500 WALKER...........(L.A.)..: (5 , 1487) ACTA CRYST. 20,. . . . .   220, (1966).

4501 WALLACE..........(D.C.)..: (6 ,   30) PHYS. REV. VOL.187,. . . . . . .   991, (1969).
                                 (10B, 147) PHYS. REV. 178,. . . . .   900, (1969).
                                 (10D,   30) PHYS. REV. B-1,. . . . .   4521, (1970).

4502 WALLACE..........(E.A.)..: (6 , 1540) J. PHYSIQUE TOME 33 COL.2, . . .   59, (1972).

4503 WALLACE..........(J.R.)..: (12B,   36) PHYS. REV. VOL.73,. . . . . . .   1277, (1948).
                                 (12B,   39) PHYS. REV. VOL.80,. . . . . . .   953, (1950).

4504 WALLACE..........(J.W.)..: (6 , 767) PHYS. REV. VOL.75,. . . . . . .   565, (1949).

4505 WALLACE..........(T.C.)..: (5 , 2833) ACTA CRYST. 19,. . . . .   6, (1965).
                                 (5 , 2731) ACTA CRYST. 21,. . . . .   670, (1966).
                                 (5 , 2019) ACTA CRYST. 21,. . . . .   843, (1966).
                                 (5 , 2631) ACTA CRYST. VOL.B24,. . . . . .   1121, (1968).
                                 (5 , 1524) ACTA CRYST. VOL.B24,. . . . . .   459, (1968).

4506 WALLACE..........(W.E.)..: (5 , 2514) J. CHEM. PHYS. VOL.35,. . . . .   2156, (1961).
                                 (5 , 1381) COLL. INTER. N.126 (GRENOBLE),.   30, (1963).
                                 (5 , 2564) J. APP. PHYS. 34,. . . . . . .   1352, (1963).
                                 (5 , 2512) J. CHEM. PHYSICS VOL.41,. . . .   3261, (1964).
                                 (5 , 2640) J. APPL. PHYS. VOL.44,. . . . .   5096, (1973).
                                 (5 , 2046) J. SOLID STATE CHEM. VOL.9,. .   152, (1974).
                                 (5 , 393) PHYS. REV. B VOL.9,. . . . . .   154, (1974).

4507 WALLER...........(I.)....: (9B,   29) ARK. FYSIK VOL.4,. . . . . . .   183, (1952).
                                 (10C,   21) CRYSTALLOGR./CRYST. PERFECTION,.   189, (1963).
                                 (10E,   14) IAEA SYMPOSIUM BOMBAY, VOL.1,..   225, (1964).
                                 (7B,   4) ARK. FYSIK VOL.32, PAPER 31,. .   537, (1966).

4508 WALLIS...........(D.E.)..: (18B,   8) J. SCI. INSTRUM. SER.2 VOL.1,.   528, (1968).

4509 WALLIS...........(R.F.)..: (6 , 2230) PHYS. REV. VOL.134,. . . . . .   A1486, (1964).
                                 (10B, 223) SOV. PHYS. SOL. STATE, 8,. . .   850, (1966).

4510 WALL............(T.)....: (6 , 1050) IAEA SYMP. COPENHAGEN, VOL.1,..   345, (1968).

4511 WALTER..........(R.K.)..: (12B,   3) ANN. OF PHYSICS VOL.74,. . . .   250, (1972).

4512 WALTON..........(J.R.)..: (5 , 1515) NUCL. SCI. ENG. VOL.47,. . . .   371, (1972).

4513 WALTON..........(R.B.)..: (6 , 1895) IAEA SYMPOSIUM (VIENNA), . . .   613, (1960).
                                 (20C,   63) NUCL. INST. MET. 72,. . . . .   161, (1969).

4514 WANG-SHOU.......(A.)....: (22A,   4) ACTA PHYS. SINICA VOL.17,. . .   222, (1961).

4515 WANG............(F.F.Y.): (5 , 2387) ACTA CRYST. VOL.A26, ...   377, (1970).
                                 (5 , 2396) PHYS. REV. B VOL.3,. . . . . .   3946, (1971).
                                 (5 , 162) A.I.P. CONF. PROC. NO.10 PART 1,   674, (1973).

4516 WANG............(S.P.)..: (20C,   30) J. PHYS. SOC. JAPAN VOL.17 B-II,   340, (1962).
                                 (9B, 102) REV. SCI. INSTRUM. VOL.33,. . .   126, (1962).

4517 WANIC...........(A.)....: (6 , 415) ACTA PHYS. POLON. VOL.16,. . .   335, (1957).
                                 (6 , 1554) ACTA PHYS. POLON. VOL.18,. . .   255, (1959).
                                 (5 , 1286) IAEA SYMPOSIUM VIENNA, ...   297, (1960).
                                 (90 ,   28) IAEA SYMP. (PILE RESEARCH)VIENNA   415, (1960).
                                 (6 , 974) NUKLEONIKA VOL.5,. . . . . . .   495, (1960).
                                 (5 , 1908) PHYSICA VOL.26,. . . . .   449, (1960).
                                 (20B, 141) REV. SCI. INSTRUM. VOL.31,. . .   214, (1960).
                                 (6 , 852) J. PHYS. CHEM. SOLIDS VOL.17,.   318, (1961).
                                 (6 , 886) PHYSICS VOL.27,. . . . .   883, (1961).
                                 (6 , 542) IAEA SYMP. CHALK RIVER, VOL.2,.   327, (1962).
                                 (6 , 885) J. PHYS. SOC. JAPAN VOL.17 B-III   69, (1962).
                                 (6 , 880) COLL. INTER. N.126 (GRENOBLE),.   203, (1963).
                                 (6 , 879) COLL. INTER. N.126 (GRENOBLE),.   210, (1963).
                                 (6 , 835) IAEA SYMP. BOMBAY, VOL.1,. . .   443, (1964).
                                 (11B,   18) NUKLEONIKA VOL.11,. . . . . . .   839, (1966).
                                 (6 , 843) PHYS. STAT. SOL. 15,. . . . . .   119, (1966).
                                 (6 , 482) PHYSICA VOL.37,. . . . .   501, (1967).
                                 (6 , 862) PHYS. STAT. SOL. 21,. . . . . .   K163, (1967).
                                 (6 , 864) PHYS. STAT. SOL. 22,. . . . . .   K55, (1967).
                                 (5 , 1096) PHYS. STAT. SOL. 26,. . . . . .   429, (1968).
                                 (11B,   16) 2ND SUMMER SCHOOL S.S. PHYS,..   211, (1968).
                                 (6 , 545) PHYS. STAT. SOL. 32,. . . . . .   41, (1969).
                                 (6 , 863) PHYS. STAT. SOL. 41,. . . . . .   K103, (1970).
                                 (5 , 1099) PHYS. STAT. SOL. 41,. . . . . .   173, (1970).
                                 (6 , 1375) INT. J. MAGN. VOL.3 NO.4,. . .   349, (1972).
                                 (6 , 1442) PHYSICA VOL.57,. . . . .   628, (1972).
                                 (5 , 2170) PHYS. STAT. SOLIDI A VOL.11 NO.1   57, (1972).
                                 (5 , 2169) PHYS. STAT. SOLIDI A VOL.9 NO.1,   97, (1972).

4518 WAPPLING........(R.)....: (5 , 1105) J. PHYS. C VOL.7,. . . . . . .   1344, (1974).

4519 WARMING.........(E.)....: (12B,   25) NUCL. INST. MET. VOL.45,. . .   293, (1966).
                                 (6 , 2011) J. PHYS.F: METAL PHYS. VOL.1,.   570, (1971).
                                 (6 , 839) IAEA SYMP. GRENOBLE, ...   649, (1972).
                                 (6 , 849) SOL. STATE COMM. VOL.12 NO.8,.   795, (1973).

4520 WARNAS..........(A.A.)..: (21 ,   38) NUCL. INST. MET. 65,. . . . .   125, (1968).

4521 WARREN..........(J.L.)..: (6 ,   27) COPENHAGEN CONF.-LATT.DYNAMICS,   57, (1963).
                                 (6 , 196) IAEA SYMP. BOMBAY, VOL.1,. . .   361, (1964).
                                 (6 , 180) IBM J. RES. DEVELOP. VOL.8,. .   234, (1964).
                                 (6 , 215) PHYS. REV. LETT. 13,. . . . . .   13, (1964).
                                 (6 , 194) BROOKHAVEN SYMPOSIUM,. . . . .   88, (1965).
                                 (6 , 182) BULL. AMER. PHYS. SOC. VOL.12,.   689, (1967).
                                 (6 , 217) PHYS. REV. 158,. . . . .   805, (1967).
                                 (6 , 901) IAEA SYMP. COPENHAGEN, VOL.1,.   301, (1968).
                                 (6 , 1378) PHYS REV 168,. . . . .   980, (1968).

4522 WASAKI..........(F.F.)..: (5 , 315) ACTA CRYST. 23,. . . . .   64, (1967).

4523 WASEDA..........(Y.)....: (5 , 2067) PHYS. STAT. SOL. VOL.39, . . .   181, (1970).
                                 (5 , 922) PHYS. STAT. SOL. VOL.39, . . .   669, (1970).
                                 (5 , 2422) PHYS. LETT. VOL.35A, ...   315, (1971).
                                 (5 , 2423) PHYS. STAT. SOL. VOL.47, . . .   581, (1971).
                                 (14A,   60) PHYS. STAT. SOLIDI B VOL.57 NO.1   351, (1973).
                                 (14B, 102) PROPERTIES LIQUID METALS,. . .   37, (1973).
                                 (5 , 2063) J. PHYS. CHEM. SOLIDS VOL.35,.   585, (1974).

4524 WASER...........(J.)....: (5 , 2318) ACTA CRYST. VOL.6,. . . . . . .   661, (1953).
                                 (7B,   28) J. CHEM. PHYSICS VOL.48,. . . .   3016, (1968).

4525 WASIUTYNSKI.....(T.)....: (6 , 1384) PHYS. STAT. SOL. VOL.44, ...   497, (1971).

4526 WATABE..........(M.)....: (15 ,   46) PROPERTIES LIQUID METALS,. . . .   143, (1973).

4527 WATANABE........(H.)....: (6 , 861) IAEA SYMP. CHALK RIVER, VOL.2, .   297, (1962).
                                 (6 , 842) PHYS. LETT. 1,. . . . .   189, (1962).
                                 (6 , 2296) PHYS. REV. VOL.128,. . . . . . .   67, (1962).
                                 (5 , 1164) J. PHYS. SOC. JAPAN VOL.18,. . .   995, (1963).
                                 (5 , 1719) J. PHYS. SOC. JAPAN VOL.19,. . .   2078, (1964).
                                 (5 , 558) J. PHYS. SOC. JAPAN VOL.20,. . .   2244, (1965).
                                 (5 , 1023) J. PHYS. SOC. JAPAN VOL.21,. . .   1932, (1966).
                                 (5 , 1172) J. PHYS. SOC. JAPAN VOL.22,. . .   1210, (1967).
                                 (5 , 345) J. PHYS. SOC. JAPAN VOL.22,. . .   939, (1967).
                                 (5 , 344) J. PHYS. SOC. JAPAN VOL.24,. . .   446, (1968).
                                 (5 , 1637) J. APP. PHYS. VOL.40,. . . . . .   1128, (1969).
                                 (5 , 2492) J. PHYS. SOC. JAPAN VOL.26,. . .   1320, (1969).
                                 (5 , 589) J. PHYS. SOC. JAPAN VOL.30,. . .   1319, (1971).
                                 (5 , 1805) ACTA CRYST. VOL.A28 PART S-4,.   196, (1972).
                                 (5 , 560) J. PHYS. SOC. JAPAN VOL.32,. . .   958, (1972).
                                 (5 , 2493) J. PHYS. SOC. JAPAN VOL.33,. . .   967, (1972).
                                 (5 , 2489) J. PHYS. SOC. JAPAN VOL.33,. . .   970, (1972).
                                 (5 , 1048) J. PHYS. SOC. JAPAN VOL.34,. . .   911, (1973).
                                 (5 , 1148) J. PHYS. SOC. JAPAN VOL.34,. . .   58, (1973).

4528 WATANABE........(N.)....: (6 , 1920) J. CHEM. PHYSICS VOL.55, ...   983, (1971).
                                 (23 ,   67) PULSED NEUTRON SOURCES/UTILIZ.,.   . ., (1971).
                                 (19A,   4) ACTA CRYST. VOL.A28 PART S-4,.   197, (1972).
                                 (6 , 230) CHEM. PHYS. LETT. VOL.18,. . . .   306, (1973).
                                 (5 , 2482) JAP. J. APPL. PHYS. VOL.12,. . .   1119, (1973).

4529 WATANABE........(T.)....: (5 , 2675) PROC/CONF/NEUTRON C. S. + TECH.,   693, (1968).
                                 (5 , 2617) NUCL. SCI. ENGNG. VOL.41,. . . .   188, (1970).

4530 WATERSTRAT......(R.M.)..: (5 , 2843) ACTA CRYST. VOL.9,. . . . . . .   289, (1956).
                                 (5 , 1666) PHYS. REV. VOL.109,. . . . . . .   1551, (1958).

4531 WATERS..........(J.R.)..: (19A,   31) NUCL. INST. MET. VOL.7,. . . . .   174, (1960).

4532 WATSON..........(B.P.)..: (173,   57) PHYS. REV. 132,. . . . .   2764, (1963).
                                 (100,   37) PHYS. REV. B VOL.3,. . . . . . .   4404, (1971).

4533 WATSON..........(R.E.)..: (8 ,   27) J. CHEM. PHYS. VOL.37, ...   1245, (1962).

4534 WATTENBERG......(A.)....: (5 , 846) PHYS. REV. VOL.85,. . . . . . .   483, (1952).

4535 WAUGH...........(J.L.T.): (6 , 895) COPENHAGEN CONF.-LATT.DYNAMICS,.   19, (1963).
                                 (6 , 897) PHYS. REV. 132,. . . . .   2410, (1963).

4536 WEBB............(F.J.)..: (20B, 118) PHIL. MAG. (8) VOL.2,. . . . . .   917, (1957).
                                 (20B,   34) IAEA SYMP. CHALK RIVER, VOL.1,.   107, (1962).
                                 (6 , 950) IAEA SYMP. CHALK RIVER, VOL.1,.   457, (1962).
                                 (20B,   35) IAEA SYMP. CHALK RIVER, VOL.1,.   83, (1962).
                                 (20B,   58) J. NUCL. ENERGY VOL.17,. . . . .   187, (1963).
                                 (6 , 969) PHYS. LETT. VOL.12,. . . . . . .   188, (1964).
                                 (6 , 2344) NUCL. SCI. ENGNG. VOL.23,. . . .   194, (1965).
                                 (6 ,   81) PHYS. LETT. VOL.19,. . . . . . .   269, (1965).
                                 (20C, 111) TH. NEUTRON SCATT./EGELSTAFF,.   141, (1965).
                                 (20B, 155) UKAEA AERE REP. R4263 (30 PP.),.   . ., (1965).
                                 (17A,   62) NUCL. INST. MET. 42,. . . . . .   197, (1966).
                                 (6 , 965) PROC. PHYS. SOC. 90,. . . . . .   681, (1967).
                                 (6 , 1548) PROC. PHYS. SOC. 92,. . . . . .   912, (1967).
                                 (20D,   24) NUCL. INST. MET. 69,. . . . . .   325, (1969).

4537 WEBER...........(A.H.)..: (5 ,   63) PHYS. REV. VOL.73,. . . . . . .   1385, (1948).
                                 (5 , 239) PHYS. REV. VOL.75,. . . . . . .   217, (1949).
                                 (9A,   20) NUCLEONICS VOL.7 NO.6,. . . . .   31, (1950).
                                 (5 , 2462) PHYS. REV. VOL.105,. . . . . . .   517, (1957).

4538 WEBER...........(G.)....: (12A,   4) ATOMKERNENERGIE VOL.7,. . . . .   170, (1962).
                                 (12A,   5) ATOMKERNENERGIE VOL.10,. . . . .   177, (1965).

4539 WEBER...........(H.W.)..: (6 , 1670) PHYS. LETT. VOL.34A,. . . . . .   35, (1971).
                                 (5 , 2003) PHYS. STAT. SOLIDI B VOL.57 NO.2   515, (1973).

4540 WEBER...........(W.)....: (10B, 172) PHYS. REV. LETT. VOL.28,. . . .   600, (1972).
                                 (10B, 183) PHYS. REV. B VOL.8,. . . . . . .   5082, (1973).

4541 WEBSTER.........(P.J.)..: (5 , 2315) PHIL. MAG. VOL.16,. . . . . . .   347, (1967).
                                 (5 , 2313) J. APP. PHYS. VOL.39,. . . . . .   2313, (1968).
                                 (5 , 467) J. PHYS. CHEM. SOLIDS VOL. 32,   1221, (1971).
                                 (5 , 509) J. PHYS. CHEM. SOLIDS VOL.34,.   1647, (1973).
                                 (5 , 515) J. PHYS. CHEM. SOLIDS VOL.35,.   1, (1974).

4542 WECKERMANN......(B.)....: (6 , 1294) PHYS LETT 27A,. . . . .   582, (1968).
                                 (5 , 1470) PHYS. REV. VOL.172,. . . . . . .   576, (1968).
                                 (5 , 1471) ACTA CRYST. A25, ...   514, (1969).
                                 (5 , 767) Z. NATURFORSCH. A VOL.25,. . .   967, (1970).
                                 (17A,   44) IAEA SYMP. GRENOBLE,. . . . . .   713, (1972).
                                 (20D,   26) NUCL. INST. MET. VOL.108,. . .   107, (1973).

4543 WEDEPOHL........(P.T.)..: (5 , 2377) PHIL. MAG. VOL.3,. . . . .   1280, (1958).

4544 WEDGWOOD........(A.)....: (5 , 236) REPORT AERE-R 6052,. . . . ., (1969).

4545 WEDGWOOD........(F.A.)..: (5 , 2855) PHYS. REV. LETT. VOL.16,. . . .   513, (1966).
                                 (24 ,   93) PHYS. REV. VOL.153,. . . . . . .   1415, (1967).
                                 (8 ,   34) J. PHYS. C VOL.3,. . . . .   1292, (1970).
                                 (5 , 2755) CONF./RARE EARTHS/ACTINIDES,.   168, (1971).
                                 (5 , 494) J. PHYS. C VOL.5,. . . . .   L261, (1972).
                                 (5 , 2809) J. PHYS. C VOL.5,. . . . .   2427, (1972).
                                 (5 , 2810) J. PHYS. C VOL.5,. . . . .   3012, (1972).
                                 (5 , 2816) J. PHYS. C VOL.6,. . . . .   1652, (1973).
                                 (5 , 2851) J. PHYS. C VOL.6,. . . . .   3746, (1973).

4546 WEGENER.........(W.)....: (6 , 2323) PHYS. LETT. VOL.31A,. . . . . .   2, (1970).

4547 WEGNER..........(F.)....: (13 ,   42) PHYS. LETT. VOL.32A,. . . . . .   273, (1970).
                                 (13 ,   19) DYN. ASPECTS OF CRITICAL PHEN.,.   19, (1972).

4548 WEHE............(D.)....: (6 , 2330) J. CHEM. PHYS. VOL.60,. . . . .   3613, (1974).

4549 WEIGEL..........(D.)....: (5 , 2278) ACTA CRYST. VOL.A26,. . . . . .   501, (1970).

4550 WEIJERMANS......(J.P.)..: (6 , 452) CHEM. PHYS. LETT. VOL.22,. . . .   476, (1973).

4551 WEIL............(J.W.)..: (10C,   23) CZECH. J. PHYS. B VOL.20,. . . .   950, (1970).
                                 (24 , 118) Z. PHYS. VOL.233,. . . . .   178, (1970).

4552 WEINERT.........(M.)....: (22A,   43) NUCL. INST. MET. VOL.108,. . . .   401, (1973).

4553 WEINMAN.........(J.A.)..: (24 ,   38) NUCL. INST. MET. 29, ...   181, (1963).

4554 WEINSTOCK.......(E.V.)..: (20D,   9) J. APPL. CRYST. VOL.2, ...   141, (1969).

4555 WEINSTOCK.......(R.)....: (7B,   56) PHYS. REV. VOL.65, ...   1, (1944).

4556 WEISSHAUPL......(H.A.)..: (24 ,   50) NUCL. INST. MET. 61, : : : : : :   45, (1968).
                                 (24 ,   51) NUCL. INST. MET. 61, : : : : : :   53, (1968).
                                 (24 ,   57) NUCL. INST. MET. 66, : : : : : :   141, (1968).
                                 (24 ,   58) NUCL. INST. MET. 66, : : : : : :   149, (1968).

4557 WEISS..........(K.)....: (14A,   56) PHYS. REV. A VOL.3, ...   717, (1971).

4558 WEISS..........(L.)....: (22C,   15) EXPER. TECH. PHYS. VOL.2,. . . . .   111, (1965).
                                 (6 ,   1431) IAEA SYMP. GRENOBLE, ...   157, (1972).

4559 WEISS..........(R.J.)..: (5 ,   2829) PHYS. REV. VOL.83, ...   171, (1951).
                                 (7B,   62) PHYS. REV. VOL.83, ...   379, (1951).
                                 (20B,   126) PHYS. REV. VOL.83, ...   863, (1951).
                                 (5 ,   757) J. APPL. PHYS. VOL.23, ...   1379, (1952).
                                 (9B,   79) PHYS. REV. VOL.86, ...   271, (1952).
                                 (9D,   2) ACTA CRYST. VOL.11, ...   598, (1957).
                                 (5 ,   2082) J. PHYS. CHEM. SOLIDS VOL.10,. .   147, (1959).
                                 (5 ,   550) PHYS. REV. LETT. VOL.3, ...   211, (1959).
                                 (6 ,   484) PHYS. REV. LETT. VOL.11, ...   264, (1963).
                                 (17B,   5) ACTA CRYST. 19, ...   68, (1965).

4560 WEIS...........(O.)....: (6 ,   63) PHYS. LETT. VOL.43A NO.2,. . . .   97, (1973).

4561 WEITZEL........(H.)....: (5 ,   1845) SOL. STATE COMM. 4, ...   473, (1966).
                                 (5 ,   1847) Z. KRIST. VOL.125, ...   120, (1967).
                                 (5 ,   2207) SOL. STATE COMM. VOL.8,. . . .   2071, (1970).
                                 (5 ,   1699) Z. KRIST. VOL.131, ...   289, (1970).
                                 (5 ,   1056) SOL. STATE COMM. VOL.11 NO.2,. .   313, (1972).
                                 (5 ,   477) SOL. STATE COMM. VOL.12 NO.2,. .   113, (1973).
                                 (5 ,   982) SOL. STATE COMM. VOL.13, ...   303, (1973).
                                 (5 ,   1177) ACTA CRYST. VOL.A30, ...   380, (1974).

4562 WELFORD........(P.J.)..: (5 ,   446) J. PHYS. C VOL.6, ...   1405, (1973).

4563 WELLS..........(P.)....: (5 ,   124) J. OF PHYS. C VOL.2, ...   356, (1969).
                                 (5 ,   1643) ACTA CRYST. VOL.A26, ...   379, (1970).
                                 (5 ,   1740) J. PHYSIQUE TOME 32 COL.1 VOL.I,   70, (1971).
                                 (6 ,   1445) J. PHYS.F: METAL PHYS. VOL.1,. .   763, (1971).

4564 WENDEL.........(H.)....: (6 ,   63) PHYS. LETT. VOL.43A NO.2,. . . .   97, (1973).

4565 WENTOWSKA......(K.)....: (13 ,   4) ACTA PHYS. POLON. 36, ...   697, (1969).

4566 WENZEL.........(R.G.)..: (6 ,   196) IAEA SYMP. BOMBAY, VOL.1,. . . .   361, (1964).
                                 (6 ,   180) IBM J. RES. DEVELOP. VOL.8,. . .   234, (1964).
                                 (6 ,   215) PHYS. REV. LETT. 13, ...   13, (1964).
                                 (5 ,   2833) ACTA CRYST. 19,. . . . .   6, (1965).
                                 (5 ,   2019) ACTA CRYST. 21,. . . . .   843, (1966).
                                 (6 ,   901) IAEA SYMP. COPENHAGEN, VOL.1,. .   301, (1968).
                                 (5 ,   783) J. CHEM. PHYSICS VOL.49, ...   4361, (1968).
                                 (5 ,   782) J. CHEM. PHYSICS VOL.49, ...   4365, (1968).
                                 (5 ,   784) J. CHEM. PHYSICS VOL.55, ...   589, (1971).
                                 (5 ,   99) PHYS. REV. A VOL.7, ...   2130, (1973).

4567 WENZL..........(H.)....: (5 ,   2859) PHYS. LETT. 20, ...   470, (1966).
                                 (6 ,   37) Z. ANGEW. PHYS. VOL.24,NO.6,. .   313, (1968).

4568 WERBER.........(K.)....: (14A,   38) PHYS. LETT. VOL.31A, ...   237, (1970).

4569 WERNER.........(P.E.)..: (17C,   8) ACTA CRYST. 20,. . . . .   407, (1966).

4570 WERNER.........(S.A.)..: (5 ,   542) PHYS. REV. LETT. VOL.14, ...   1022, (1965).
                                 (20B,   128) PHYS. REV. VOL.140, ...   A675, (1965).
                                 (5 ,   537) J. APP. PHYS. 37, ...   1260, (1966).
                                 (9B,   53) J. APP. PHYS. 37, ...   2343, (1966).
                                 (5 ,   2530) J. APPL. PHYS. VOL.38, ...   969, (1967).
                                 (5 ,   577) J. APP. PHYS. 38, ...   1243, (1967).
                                 (5 ,   556) METAL. SOC. CONF. LOS ANGELES,   59, (1967).
                                 (5 ,   574) PHYS. REV. 153, ...   624, (1967).
                                 (5 ,   532) PHYS. REV. 155, ...   528, (1967).
                                 (5 ,   2848) J. APP. PHYS. 39, ...   585, (1968).
                                 (5 ,   531) J. APP. PHYS. 39, ...   671, (1968).
                                 (5 ,   133) J. APP. PHYS. VOL.40, ...   1373, (1969).
                                 (6 ,   520) J. APP. PHYS. VOL.40, ...   1447, (1969).
                                 (12A,   15) NUCL. INST. MET. 68, ...   50, (1969).
                                 (5 ,   1413) PHYS. REV. 186, ...   705, (1969).
                                 (5 ,   1416) SOL. STATE COMM. VOL.7,. . . .   1681, (1969).
                                 (6 ,   827) J. APPL. PHYS. 41, ...   1363, (1970).
                                 (9B,   13) ACTA CRYST. VOL.A27, ...   665, (1971).
                                 (17A,   17) AMER. CRYSTALL. ASSOC.,. . . . .   65, (1971).
                                 (6 ,   1597) J. APPL. PHYS. 42, ...   4736, (1971).
                                 (6 ,   743) MAGN. AND MAG. MATERIALS-1971,   1340, (1972).
                                 (5 ,   719) MAGN. AND MAG. MATERIALS-1971,   508, (1972).
                                 (9B,   20) ACTA CRYST. VOL.A29, ...   372, (1973).
                                 (9B,   21) ACTA CRYST. VOL.A29, ...   577, (1973).
                                 (5 ,   721) A.I.P. CONF. PROC. NO.10 PART 1,   679, (1973).

4571 WERTHAMER......(N.R.)..: (10B,   128) PHYS. REV. 167,. . . . .   607, (1968).
                                 (10B,   126) PHYS. REV. 175,. . . . .   1110, (1968).
                                 (6 ,   1139) PHYS. REV. A VOL.2,. . . . . .   2050, (1970).
                                 (10D,   33) PHYS. REV. B VOL.1,. . . . . .   572, (1970).
                                 (9C,   39) SOL. STATE COMM. VOL.9, ...   2239, (1971).
                                 (6 ,   1138) PHYS. REV. A VOL.5,. . . . . .   2230, (1972).
                                 (10D,   43) PHYS. REV. B VOL.5,. . . . . .   285, (1972).
                                 (6 ,   1126) PHYS. REV. LETT. VOL.28, ...   1102, (1972).

4572 WESTCOTT.......(C.H.)..: (20B,   71) NUCL. INST. MET. VOL.13, ...   1, (1961).
                                 (5 ,   2366) CAN. J. PHYS. 42, ...   2384, (1964).

4573 WESTLAKE.......(D.G.)..: (5 ,   2838) J. APPL. CRYST. VOL.6, ...   206, (1973).

4574 WESTON.........(L.W.)..: (22B,   55) REV. SCI. INSTRUM. VOL.41, . . .   1539, (1970).

4575 WESTPHAL.......(G.P.)..: (20B,   100) NUCL. INST. MET. 98,. . . . . .   87, (1972).
                                 (20C,   77) NUCL. INST. MET. VOL.106,. . . .   279, (1973).

4576 WEST...........(R.E.)..: (6 ,   231) PHYS. REV. VOL.148,. . . . . . .   163, (1966).

4577 WETTENGEL......(H.)....: (5 ,   2498) Z. PHYSIK VOL.258, ...   429, (1973).

4578 WEYMOUTH.......(J.W.)..: (6 ,   22) PHYS. REV. B-2,. . . . .   4743, (1970).

4579 WHALEN.........(J.F.)..: (5 ,   1546) NUCL. SCI. ENGNG. VOL.40,. . . .   12, (1970).

4580 WHEELER........(C.V.)..: (22A,   51) NUCL. TECHNOL. VOL.10, ...   215, (1971).

4581 WHEELER........(D.A.)..: (9D,   48) J. PHYS. SOC. JAPAN VOL.17 B-II,   342, (1962).
                                 (6 ,   504) PROC. PHYS. SOC. 82,. . . . . .   633, (1963).
                                 (5 ,   732) J. PHYS. CHEM. SOLIDS 29,. . . .   855, (1968).
                                 (21 ,   22) J. APPL. CRYST. VOL.4,. . . . .   254, (1971).
                                 (5 ,   602) J. SOLID STATE CHEM. VOL.4,. . .   400, (1972).

4582 WHEELER.........(D.H.)..:   (5 ,   280) CRYST. STRUCT. COMM. VOL.1 N.3,.   193, (1972).

4583 WHEELER.........(R.G.)..:   (5 ,   1476) J. PHYS. CHEM. SOLIDS 29,. . . .   881, (1968).

4584 WHITAKER........(M.D.)..:   (7A,   60) PHYS. REV. VOL.50, ...   133, (1936).
                                  (25 ,   36) PHYS. REV. VOL.54, ...   771, (1938).
                                  (5 ,   2464) PHYS. REV. VOL.55, ...   1101, (1939).
                                  (5 ,   2708) PHYS. REV. VOL.55, ...   793, (1939).
                                  (7B,   53) PHYS. REV. VOL.57, ...   976, (1940).
                                  (11C,   28) PHYS. REV. VOL.60; . . . .   280, (1941).

4585 WHITEHEAD.......(C.D.)..:   (21 ,   29) J. SCI. INSTRUM. VOL.27, ...   330, (1950).

4586 WHITE...........(D.H.)..:   (20C,   54) NUCL. INST. MET. 63,. . . . . .   185, (1968).
                                  (5 ,   317) NUCL. PHYS. A VOL.A169,. . . . .   95, (1971).

4587 WHITE...........(D.)....:   (5 ,   768) PHYS. REV. LETT. 15, ...   586, (1965).
                                  (5 ,   770) PHYS. REV. LETT. 16, ...   799, (1966).
                                  (5 ,   761) J. CHEM. PHYSICS VOL.48, ...   1273, (1968).
                                  (5 ,   760) J. CHEM. PHYSICS VOL.49, ...   1922, (1968).

4588 WHITE...........(J.C.B.):   (5 ,   288) CHEM. COMMUNICATIONS,. . . . . .   554, (1969).
                                  (5 ,   290) TETRAHEDRON LETT. NO.59, ...   5219, (1969).

4589 WHITE...........(J.G.)..:   (5 ,   1051) J. PHYS. CHEM. SOL. 25,. . . . .   717, (1964).
                                  (5 ,   704) J. PHYS. CHEM. SOL. 28,. . . . .   897, (1967).
                                  (5 ,   706) J. APP. PHYS. 39,. . . . .   664, (1968).

4590 WHITE...........(J.W.)..:   (6 ,   1547) PHYS. REV. LETT. VOL.17, ...   533, (1966).
                                  (6 ,   2224) CHEM. COMMUN. 1967,. . . . . . .   74, (1967).
                                  (6 ,   2) DISC. FARADAY SOC. NO.43,. . . .   169, (1967).
                                  (6 ,   1553) MOL. SIEVES, PAP. CONF. 1967,. .   306, (1967).
                                  (10C,   29) EXCIT./MAGN./PHONONS/MOL. CRYST.   43, (1968).
                                  (6 ,   1913) J. CHEM. PHYS. VOL.48, ...   5271, (1968).
                                  (6 ,   109) DISC. FARADAY SOC. NO.48,. . . .   131, (1969).
                                  (6 ,   358) MOL. PHYS. VOL.17, ...   1, (1969).
                                  (6 ,   399) NAT. BUR. STAND. PUBL. NO.301,.   463, (1969).
                                  (6 ,   1001) J. CHEM. SOC. D 1970,. . . . . .   970, (1970).
                                  (6 ,   1014) J. CHEM. SOC. D 1970,. . . . . .   971, (1970).
                                  (6 ,   1914) J. MACROMOL. SCI. VOL.A4, NO.1,   1275, (1970).
                                  (6 ,   2043) SPEC. DISC. FARADAY SOC. NO.1,.   194, (1970).
                                  (6 ,   1087) BER. BUNSENGES. PHYS. CHEM. 75,   379, (1971).
                                  (10B,   17) CONF. INTERN., RENNES,. . . . .   209, (1971).
                                  (6 ,   1015) J. CHEM. SOC. SECTION A,. . . .   2843, (1971).
                                  (6 ,   1081) NATURE (PHYS. SCI.) VOL.230, . .   192, (1971).
                                  (17B,   67) SOL. STATE COMM. VOL.9,. . . . .   1353, (1971).
                                  (6 ,   2199) FARADAY SYMP. CHEM. SOC. NO.6,.   135, (1972).
                                  (6 ,   235) IAEA SYMP. GRENOBLE, ...   195, (1972).
                                  (6 ,   1893) IAEA SYMP. GRENOBLE, ...   301, (1972).
                                  (14B,   35) IAEA SYMP. GRENOBLE, ...   315, (1972).
                                  (10E,   15) J. CHEM. PHYS. VOL.56,. . . . .   2928, (1972).
                                  (6 ,   1335) J. CHEM. SOC. FARADAY II VOL.68, 1414, (1972).
                                  (6 ,   1235) J. CHEM. SOC. FARADAY II VOL.68, 1423, (1972).
                                  (6 ,   353) J. CHEM. SOC. FARADAY II VOL.68, 1434, (1972).
                                  (6 ,   1925) POLYMER VOL.13 NO.1, ...   40, (1972).
                                  (10C,   14) CHEM. APPL. THERMAL NEUTRON SCAT   49, (1973).
                                  (6 ,   1527) J. PHYSIQUE TOME 34, ...   473, (1973).
                                  (12B,   20) J. PHYS. E VOL.6 NO.8, ...   714, (1973).
                                  (6 ,   374) J. CHEM. PHYS. VOL.60, ...   824, (1974).

4591 WHITE...........(R.E.)..:   (12A,   16) NUCL. INST. MET. 75, ...   333, (1969).

4592 WHITTEMORE......(W.L.)..:   (6 ,   2356) PHYS. REV. VOL.108,. . . . . . .   1092, (1957).
                                  (6 ,   1834) PHYS. REV. VOL.113,. . . . . . .   767, (1959).
                                  (5 ,   1261) PHYS. REV. VOL.113,. . . . . . .   806, (1959).
                                  (20B,   25) IAEA SYMPOSIUM VIENNA, ...   421, (1960).
                                  (6 ,   1030) IAEA SYMPOSIUM VIENNA, ...   511, (1960).
                                  (6 ,   949) IAEA SYMP. CHALK RIVER, VOL.1, .   263, (1962).
                                  (6 ,   948) IAEA SYMP. CHALK RIVER, VOL.1,.   273, (1962).
                                  (6 ,   2347) IAEA SYMP. BOMBAY, VOL.2, . . .   305, (1964).
                                  (6 ,   970) NUCL. SCI. ENGNG. VOL.18,. . . .   182, (1964).
                                  (6 ,   1887) BROOKHAVEN SYMPOSIUM,. . . . . .   131, (1965).
                                  (6 ,   225) BROOKHAVEN SYMPOSIUM,. . . . . .   94, (1965).
                                  (6 ,   10) J. CHEM. PHYSICS VOL.44, ...   3127, (1966).
                                  (6 ,   1898) NUCL. SCI. ENGNG. VOL.24,. . . .   394, (1966).
                                  (20B,   148) REV. SCI. INSTRUM. VOL.37,. . .   742, (1966).
                                  (6 ,   7) IAEA SYMP. COPENHAGEN, VOL.2,.   175, (1968).
                                  (12B,   10) IAEA SYMP. COPENHAGEN, VOL.2,.   429, (1968).
                                  (6 ,   671) NUCL. SCI. ENGNG. VOL.33,. . . .   195, (1968).
                                  (6 ,   195) NUCL. SCI. ENGNG. VOL.33,. . .   31, (1968).

4593 WHITTLESTONE....(S.)....:   (6 ,   156) J. NUCL. ENERGY(GB) VOL.26,. . .   27, (1972).

4594 WHULER..........(A.)....:   (5 ,   1229) J. PHYSIQUE TOME 32 COL.1 VOL.II   853, (1971).
                                  (5 ,   1712) PHYS. STAT. SOL. VOL.44,. . . .   71, (1971).

4595 WICK............(G.C.)..:   (10C,   87) PHYS. ZEITS. VOL.38, ...   403, (1937).
                                  (10C,   86) PHYS. ZEITS. VOL.38, ...   689, (1937).
                                  (7B,   111) RICERCA SCI. VOL.8, I, ...   400, (1937).
                                  (8 ,   72) PHYS. REV. VOL.76, ...   994, (1949).
                                  (7B,   67) PHYS. REV. VOL.94, ...   1228, (1954).

4596 WIDDER..........(F.)....:   (5 ,   380) NUKLEONIK VOL.11,. . . . .   297, (1968).
                                  (5 ,   648) HELV. PHYS. ACTA VOL.45, ...   46, (1972).

4597 WIEDEMANN.......(W.)....:   (5 ,   2062) Z. NATURFORSCH. VOL.15A, ...   828, (1960).
                                  (5 ,   790) Z. KRIST. VOL.116, ...   328, (1961).
                                  (5 ,   1297) Z. PHYSIK VOL.164, ...   111, (1961).

4598 WIEDLING........(T.)....:   (17B,   28) ARK. FYS. VOL.34, PAPER 40,. . .   481, (1967).
                                  (5 ,   527) 2ND INTERNAT. CONF./NUCL. DATA,. . .   (1970).

4599 WIEGERS.........(G.A.)..:   (5 ,   1966) J. SOLID STATE CHEM. VOL.6,. . .   574, (1973).

4600 WIENER..........(E.)....:   (6 ,   1293) IAEA SYMP. BOMBAY, VOL.2,. . . .   325, (1964).
                                  (5 ,   1462) SOL. STATE COMM. 5,. . . . . . .   41, (1967).
                                  (5 ,   1470) PHYS. REV. VOL.172,. . . . . . .   576, (1968).
                                  (6 ,   827) J. APPL. PHYS. 41, ...   1363, (1970).

4601 WIESER..........(E.)....:   (5 ,   1689) PHYS. STAT. SOL. VOL.8,. . . . .   271, (1965).
                                  (5 ,   1606) PHYS. STAT. SOL. A VOL.1,. . . .   749, (1970).

4602 WIGGINS.........(P.F.)..:   (20B,   19) ENG. J. (CANADA),. . . . .   22, (1972).

4603 WIGNALL.........(G.D.)..:   (13 ,   61) REPORT AERE-R5627 (15 PP.),. . . .   (1967).
                                  (6 ,   189) J. OF PHYS. C VOL.1, ...   1088, (1968).
                                  (6 ,   1830) J. OF PHYS. C VOL.1, ...   519, (1968).
                                  (22C,   26) J. SCI. INSTRUM. SER.2 VOL.1,. .   367, (1968).
                                  (6 ,   899) J. PHYS. C, SER.2, VOL.3,. . . .   1673, (1970).

4604 WIGNER..........(E.)....:   (7A,   59) PHYS. REV. VOL.49, ...   519, (1936).

4605 WILDE...........(H.J.)..:   (5 ,   649) Z. KRIST. VOL.134, ...   308, (1971).
                                  (5 ,   661) INORG. NUCL. CHEM. LETT. VOL.9,   1025, (1973).

4606 WILHELMI........(G.)....‡ (19A,   23) IAEA SYMP. (INSTRUMENT.), VIENNA 147, (1969).
4607 WILHELMI........(Z.)....‡ (19B,   11) NUCL. INST. MET. VOL.29, ...   241, (1964).
                                 (12B,   27) NUCL. INST. MET. 64, ...   77, (1968).
4608 WILKENS.........(M.)....‡ (10C,   139) Z. PHYS. VOL.171,. . . . .   291, (1963).
4609 WILKINSON.......(C.)....‡ (17B,   7) ACTA CRYST. 18,. . . . .   398, (1965).
                                 (5 ,   539) PROC. PHYS. SOC. 85, ... 1185, (1965).
                                 (17B,   12) ACTA CRYST. A24,. . . . .   347, (1968).
                                 (9D,   55) PHIL. MAG. VOL.17,. . . . .   609, (1968).
                                 (5 , 1107) J. OF PHYS. C VOL.3, ... 1127, (1970).
                                 (9D,   8) ACTA CRYST. VOL.A29, ...   449, (1973).
                                 (9D,   7) ACTA CRYST. VOL.A29, ...   453, (1973).
                                 (5 , 1105) J. PHYS. C VOL.7,. . . . . 1344, (1974).
4610 WILKINSON.......(G.R.)..‡ (10B,   205) PROC. ROY. SOC. A VOL.317, . . .   55, (1970).
4611 WILKINSON.......(M.K.)..‡ (5 , 2828) REV. MOD. PHYS. VOL.25,. . . . .   100, (1953).
                                 (9A,   5) AMER. J. PHYS. VOL.22, ...   263, (1954).
                                 (5 , 2135) PHYS. REV. VOL.97, ...   304, (1955).
                                 (9B,   81) PHYS. REV. VOL.37, ...   889, (1955).
                                 (5 ,   942) PHYS. REV. VOL.103,. . . . . . .   516, (1956).
                                 (6 ,   768) PHYS. REV. VOL.103,. . . . . . .   525, (1956).
                                 (5 , 1806) J. PHYS. CHEM. SOLIDS VOL.2, . .   289, (1957).
                                 (5 , 2309) J. PHYS. CHEM. SOLIDS VOL.3, . .   303, (1957).
                                 (5 , 1657) PHYS. REV. VOL.110,. . . . . . .   638, (1958).
                                 (5 , 1394) PHYS. REV. VOL.110,. . . . . . .   37, (1958).
                                 (5 , 2839) PHYS. REV. VOL.112,. . . . . . . 1132, (1958).
                                 (5 , 1658) J. PHYS. RADIUM VOL.20,. . . . .   186, (1959).
                                 (5 ,   968) PHYS. REV. VOL.113,. . . . . . .   497, (1959).
                                 (5 , 2386) IAEA SYMP. (PILE RESEARCH)VIENNA 379, (1960).
                                 (5 , 2586) J. APPL. PHYS. VOL.31, SUPPL., .   358, (1960).
                                 (5 , 2662) PHYS. REV. VOL.118,. . . . . . .   797, (1960).
                                 (5 ,   573) PHYS. REV. VOL.118,. . . . . . .   950, (1960).
                                 (5 , 2042) PHYS. REV. VOL.118,. . . . . . .   58, (1960).
                                 (5 ,   807) J. APPL. PHYS. VOL.32, SUPPL., .   485, (1961).
                                 (5 ,   852) J. APPL. PHYS. VOL.32, SUPPL., .   495, (1961).
                                 (5 ,   567) J. PHYS. CHEM. SOLIDS VOL.19,. .   29, (1961).
                                 (5 , 1866) PHYS. REV. VOL.121,. . . . . . .   74, (1961).
                                 (5 ,   390) PHYS. REV. VOL.122,. . . . . . . 1409, (1961).
                                 (5 , 2682) J. APPL. PHYS. VOL.33, SUPPL., . 1124, (1962).
                                 (5 , 2301) J. APPL. PHYS. VOL.33, SUPPL., . 1340, (1962).
                                 (5 , 2384) J. PHYS. SOC. JAPAN VOL.17 B-III   27, (1962).
                                 (5 , 2606) J. PHYS. SOC. JAPAN VOL.17 B-III   32, (1962).
                                 (5 , 2298) J. PHYS. SOC. JAPAN VOL.17 B-III   38, (1962).
                                 (5 , 1720) PHYS. REV. VOL.125,. . . . . . . 1860, (1962).
                                 (5 , 2683) PHYS. REV. VOL.126,. . . . . . . 1672, (1962).
                                 (6 ,   527) PHYS. REV. VOL.127,. . . . . . . 2080, (1962).
                                 (5 ,   977) PHYS. REV. VOL.127,. . . . . . .   714, (1962).
                                 (5 ,   725) J. PHYS. CHEM. SOL. 24,. . . . . 1663, (1963).
                                 (5 , 2540) PHYS. REV. VOL.131,. . . . . . .   922, (1963).
                                 (5 ,   850) PHYS. REV. 140,. . . . . A1896, (1965).
                                 (5 , 1500) J. APP. PHYS. 37,. . . . . 1047, (1966).
                                 (5 , 1359) PHYS. REV. 151,. . . . .   414, (1966).
                                 (5 , 1358) PHYS. REV. 158,. . . . .   450, (1967).
                                 (6 ,   600) PHYS. REV. 164,. . . . .   922, (1967).
                                 (6 , 1360) IAEA SYMP. COPENHAGEN, VOL.1,. .   149, (1968).
                                 (6 ,   611) IAEA SYMP. COPENHAGEN, VOL.1,. .   47, (1968).
                                 (223,   8) IAEA SYMP. COPENHAGEN, VOL.2,. .   253, (1968).
                                 (12B,   19) J. APP. PHYS. 39,. . . . .   447, (1968).
                                 (6 ,   612) PHYS. REV. LETT. VOL.20, ... 1245, (1968).
                                 (6 , 1373) PHYS. REV. 168,. . . . .   970, (1968).
                                 (19A,   22) IAEA SYMP. (INSTRUMENT.),VIENNA, 173, (1969).
                                 (6 , 1712) J. APP. PHYS. VOL.40,. . . . . . 1450, (1969).
                                 (6 , 1226) J. APP. PHYS. VOL.40,. . . . . . 1452, (1969).
                                 (6 ,   16) PHYS. REV. B-1,. . . . . 4819, (1970).
                                 (6 , 2178) PHYS. REV. B-3,. . . . . 3457, (1971).
                                 (6 ,   695) PHYS. REV. LETT. VOL.26, ...   140, (1971).
                                 (6 ,   703) PHYS. REV. LETT. VOL.27, ...   334, (1971).
                                 (6 ,   561) PHYS. REV. B VOL.6,. . . . . . . 3956, (1972).
4612 WILLEE..........(CH.)...‡ (14A,   38) PHYS. LETT. VOL.31A, ...   237, (1970).
4613 WILLETT.........(R.D.)..‡ (5 , 2873) J. CHEM. PHYS. VOL.59, ...   453, (1973).
4614 WILLIAMSON......(T.J.)..‡ (16 ,   44) NUCL. SCI. ENG. VOL.42,. . . . .   97, (1970).
4615 WILLIAMS........(A.R.)..‡ (10B,   88) J. PHYS. F VOL.3,. . . . .   772, (1973).
4616 WILLIAMS........(D.E.)..‡ (5 , 1540) J. CHEM. PHYS. VOL.31, ...   329, (1959).
                                 (5 , 2550) J. CHEM. PHYS. VOL.35, ... 1960, (1961).
4617 WILLIAMS........(H.C.W.)‡ (10B,   72) J. PHYS. C VOL.5,. . . . . 3168, (1972).
4618 WILLIAMS........(H.J.)..‡ (5 , 1553) J. CHEM. PHYS. VOL.39, ... 2923, (1963).
4619 WILLIAMS........(J.M.)..‡ (5 , 1312) DISS. ABS. XXVII,. . . . . 1871B, (1967).
                                 (5 , 2211) J. PHYS. CHEM. SOL. 28,. . . . . 1651, (1967).
                                 (5 , 2719) J. PHYS. CHEM. SOLIDS 29,. . . . 1702, (1968).
                                 (5 ,   26) ACTA CRYST. (INTERACT.) VOL.A25, S113, (1969).
                                 (9C,   42) SPECTROS. INORG. CHEM.,. . . . .   1, (1971).
                                 (5 , 1309) J. CHEM. PHYS. VOL.58, ...   788, (1973).
                                 (5 ,   797) J. CHEM. PHYS. VOL.59, ... 5114, (1973).
4620 WILLIAMS........(L.J.)..‡ (5 , 1541) PHYS. REV. 176,. . . . .   712, (1968).
                                 (5 , 1536) DISS. ABS. VOL.31,. . . . . 879B, (1970).
4621 WILLIAMS........(M.M.R.)‡ (16 ,   65) PROC. PHYS. SOC. 85, ...   413, (1965).
                                 (7A,   33) J. NUCL. ENERGY VOL.25,. . . . .   489, (1971).
4622 WILLIAMS........(P.M.)..‡ (16 ,   45) NUCL. SCI. ENG. VOL.43,. . . . .   120, (1971).
4623 WILLIAMS........(P.P.)..‡ (5 , 1991) ACTA CRYST. VOL.B27, ... 2269, (1971).
4624 WILLIAMS........(W.G.)..‡ (20D,   29) REPORT RL-73-034 (68 PP.), ... , (1973).
4625 WILLIS..........(B.T.M.)‡ (22A,   1) ACTA CRYST. VOL.13,. . . . . . .   763, (1960).
                                 (17A,   29) IAEA SYMP. (PILE RESEARCH)VIENNA 455, (1960).
                                 (17A,   2) ACTA CRYST. VOL.14,. . . . . . .   90, (1961).
                                 (22C,   8) BRIT. J. APPL. PHYS. VOL.13, . .   547, (1962).
                                 (23 ,   46) J. SCI. INSTRUM. VOL.39, ...   590, (1962).
                                 (5 , 2794) COLL. INTER. N.126 (GRENOBLE), .   7, (1963).
                                 (5 , 2778) NATURE VOL.197,. . . . .   755, (1963).
                                 (22B,   18) NUCL. INST. MET. 24, ...   255, (1963).
                                 (5 , 2768) PROC. ROY. SOC. 274, ...   122, (1963).
                                 (5 , 2767) PROC. ROY. SOC. 274, ...   134, (1963).
                                 (22A,   63) REV. SCI. INSTRUM. VOL.34, . . .   224, (1963).
                                 (5 , 2795) PROC. BR. CERAMIC SOC. NO.1, . .   9, (1964).
                                 (5 ,   334) ACTA CRYST. 18,. . . . .   75, (1965).
                                 (5 , 2772) PHYS. LETT. 17,. . . . .   188, (1965).
                                 (5 , 2774) PAPER NO.IIE-1 (2 PP.),. . . . . . , (1966).
                                 (5 , 2858) NATURE VOL.214,. . . . .   130, (1967).
                                 (5 , 2444) PROC. ROY. SOC. 298, ...   307, (1967).
                                 (5 ,   164) ACTA CRYST. A24, ...   484, (1968).
                                 (5 , 2776) ACTA CRYST. VOL.B24, ...   117, (1968).

```
4625  WILLIS..........(B.T.M.):  (10D,     9) HARWELL SUMMER SCHOOL, . . . . .    124, (1968).
                                 (17B,    22) ACTA CRYST. A25, . . . . . . . .    277, (1969).
                                 ( 5 ,  1341) ACTA CRYST. VOL.A25, . . . . . .    482, (1969).
                                 ( 5 ,  1342) ACTA CRYST. VOL.A26, . . . . . .    263, (1970).
                                 ( 9C,     3) ACTA CRYST. VOL.A26, . . . . . .    396, (1970).
                                 ( 5 ,   378) SOL. STATE COMM. VOL.8, . . . .    171, (1970).
                                 ( 5 ,  1963) ACTA CRYST. VOL.A29, . . . . . .    727, (1973).

4626  WILLMANN........(G.)....:  (14A,    59) PHYS. REV. A VOL.6, . . . . . .   2243, (1972).

4627  WILL...........(G.)....:   ( 5 ,   915) J. PHYS. CHEM. SOL. 24, . . . .   1679, (1963).
                                 ( 5 ,  1206) J. APPL. PHYS. VOL.35, . . . .    1045, (1964).
                                 ( 5 ,  1031) PROC. INT. CONF. (NOTTINGHAM),     327, (1964).
                                 ( 5 ,  1802) ACTA CRYST. 19, . . . . . . . .    854, (1965).
                                 ( 5 ,  1804) J. APP. PHYS. 36, . . . . . . .   1095, (1965).
                                 ( 5 ,  1803) PHYS. REV. 140, . . . . . . . .  A2139, (1965).
                                 ( 5 ,   898) CHEM. PHYS. LETT. VOL.2,NO.1, .     47, (1968).
                                 ( 5 ,  1815) J. APP. PHYS. 39, . . . . . . .    628, (1968).
                                 ( 9D,    85) Z. ANGEW. PHYS. VOL.24, . . . .    260, (1968).
                                 ( 5 ,  2380) Z. ANGEW. PHYS. VOL.26, . . . .     67, (1968).
                                 ( 5 ,  2195) Z. ANORG. ALLG. CHEM. BAND 358,    125, (1968).
                                 ( 5 ,   860) Z. NATURFORSCH. VOL.23A, . . .     413, (1968).
                                 ( 5 ,   253) PHYS. LETT. VOL.31A, . . . . .     253, (1970).
                                 ( 5 ,  2091) Z. KRIST. VOL.131, . . . . . .     139, (1970).
                                 ( 5 ,  2910) Z. KRIST. VOL.131, . . . . . .     278, (1970).
                                 ( 5 ,   859) CONF./RARE EARTHS/ACTINIDES, .     196, (1971).
                                 ( 5 ,   842) CONF./RARE EARTHS/ACTINIDES, .     226, (1971).
                                 ( 5 ,   843) J. OF PHYS.-C-, SER.2, VOL.4,.     811, (1971).
                                 ( 5 ,   877) J. PHYSIQUE TOME 32 COL.1 VOL.II   675, (1971).
                                 ( 5 ,   747) J. PHYSIQUE TOME 32 COL.1 VOL.II   855, (1971).
                                 ( 5 ,   817) J. PHYS. C, SER.2, VOL.4, . .     3224, (1971).
                                 ( 9C,    28) NATURWISSENSCHAFTEN VOL.58, . .     444, (1971).
                                 ( 5 ,   876) PHYS. KOND. MATER.(GER.) VOL.13,   137, (1971).
                                 ( 5 ,   856) PHYS. LETT. VOL.36A, . . . . .      50, (1971).
                                 ( 5 ,   903) PHYS. STAT. SOL. B VOL.46, . .     597, (1971).
                                 ( 9D,    87) Z. ANGEW. PHYS. VOL.32, . . . .      1, (1971).
                                 ( 9D,    13) ANN. DE PHYSIQUE TOME 7 NO.5,.     371, (1972).
                                 ( 5 ,   862) INT. J. MAGN. VOL.3, . . . . .      87, (1972).
                                 ( 5 ,   904) Z. NATURFORSCH. A VOL.27A, . .     1581, (1972).
                                 ( 5 ,  1844) INT. J. MAGN. VOL.5, . . . . .     175, (1973).
                                 ( 5 ,  1816) INT. J. MAGN. VOL.5, . . . . .     197, (1973).
                                 ( 5 ,   861) J. PHYS. CHEM. SOLIDS VOL.35, .     861, (1974).

4628  WILSON.........(A.J.C.):   (17B,    59) PROC. PHYS. SOC. 85, . . . . .     807, (1965).

4629  WILSON.........(I.H.)..:   (20B,    65) J. SCI. INSTRUM. VOL.43, . . .       1, (1966).

4630  WILSON.........(P.W.)..:   ( 5 ,  2744) ACTA CRYST. VOL.B29 PART 1, . .      7, (1973).
                                 ( 5 ,  2782) ACTA CRYST. VOL.B29, . . . . .    1073, (1973).
                                 ( 5 ,  2737) ACTA CRYST. VOL.B29, . . . . .    1942, (1973).
                                 ( 5 ,  2866) ACTA CRYST. VOL.B30, . . . . .    1216, (1974).
                                 ( 5 ,   799) ACTA CRYST. VOL.B30, . . . . .     151, (1974).
                                 ( 5 ,  2784) ACTA CRYST. VOL.B30, . . . . .     169, (1974).
                                 ( 5 ,  2783) ACTA CRYST. VOL.B30, . . . . .     175, (1974).
                                 ( 5 ,  2123) ACTA CRYST. VOL.B30, . . . . .     554, (1974).

4631  WILSON.........(R.H.)..:   ( 5 ,  1634) ACTA CRYST. 17, . . . . . . .      95, (1964).

4632  WILSON.........(R.)....:   ( 5 ,  1262) PHYS. REV. LETT. VOL.26, . . .    1581, (1971).

4633  WILSON.........(S.A.)..:   ( 5 ,   270) PROC. ROY. SOC. A VOL.279, . .      98, (1964).
                                 ( 9B,    19) ACTA CRYST. VOL.A29, . . . . .      90, (1973).

4634  WILSON.........(V.C.)..:   ( 5 ,  2956) PHYS. REV. VOL.95, . . . . . .    1408, (1954).

4635  WILSON.........(W.D.)..:   ( 6 ,  1376) PHYS. REV. B-1, . . . . . . .     3510, (1970).

4636  WILTSHIRE......(M.C.K.):   (10B,    74) J. PHYS. C VOL.6, . . . . . . .   1149, (1973).

4637  WINDER.........(D.R.)..:   ( 6 ,  1633) J. PHYS. CHEM. SOLIDS VOL.31, .    131, (1970).

4638  WINDSOR........(C.G.)..:   ( 5 ,   814) PHYS. REV. 138, . . . . . . .    A176, (1965).
                                 ( 6 ,  1314) J. APP. PHYS. 37, . . . . . . .   1054, (1966).
                                 ( 6 ,  1991) PROC. PHYS. SOC. 87, . . . . .     501, (1966).
                                 ( 6 ,  1993) PROC. PHYS. SOC. 89, . . . . .     825, (1966).
                                 ( 6 ,  1732) J. APP. PHYS. 38, . . . . . . .   1247, (1967).
                                 ( 6 ,  1724) PHYS. REV. LETT. VOL.18, . . .     1136, (1967).
                                 (11C,    46) PROC. PHYS. SOC. 90, . . . . .     1015, (1967).
                                 ( 5 ,  1477) PROC. PHYS. SOC. 91, . . . . .      928, (1967).
                                 ( 6 ,  1708) IAEA SYMP. COPENHAGEN, VOL.2, .     101, (1968).
                                 ( 6 ,  1982) IAEA SYMP. COPENHAGEN, VOL.2, .      83, (1968).
                                 ( 6 ,  1747) J. APP. PHYS. VOL.39, . . . . .     449, (1968).
                                 ( 6 ,  1988) J. OF PHYS. C VOL.1, . . . . .      940, (1968).
                                 ( 6 ,  1720) SOL. STATE COMM. VOL.6, . . . .     189, (1968).
                                 ( 6 ,  1711) J. APP. PHYS. VOL.40, . . . . .    1142, (1969).
                                 ( 6 ,  1725) ADVANCES IN PHYS. VOL.19 NO.82,    813, (1970).
                                 ( 6 ,  2006) CONF. INTERN., RENNES, . . . .     223, (1971).
                                 ( 6 ,  1990) J. PHYSIQUE TOME 32 COL.1 VOL.II   614, (1971).
                                 ( 8 ,    40) J. PHYSIQUE TOME 32 COL.1, . .     573, (1971).
                                 ( 6 ,  1710) PHYS. REV. B VOL.4, . . . . . .   3048, (1971).
                                 ( 6 ,  1989) IAEA SYMP. GRENOBLE, . . . . .     639, (1972).
                                 ( 6 ,  1778) A.I.P. CONF. PROC. NO.10 PART 2   1403, (1973).
                                 ( 7B,     9) CHEM. APPL. THERMAL NEUTRON SCAT     1, (1973).
                                 ( 5 ,  1196) J. PHYS. C VOL.6, . . . . . . .     212, (1973).
                                 ( 6 ,  1994) J. PHYS. C VOL.6, . . . . . . .     495, (1973).

4639  WINFIELD.......(D.J.)..:   ( 6 ,   366) J. CHEM. PHYS. VOL.54, . . . .    3643, (1971).
                                 (2JB,    99) NUCL. INST. MET. VOL.91, . . .     159, (1971).
                                 ( 6 ,   236) MOL. PHYS. VOL.24, . . . . . .     753, (1972).
                                 ( 5 ,  1519) CAN. J. PHYS. VOL.51, . . . . .   1965, (1973).

4640  WINFIELD.......(D.)....:   (20C,    83) PHYS. LETT. B VOL.40B, . . . .     537, (1972).

4641  WINGFIELD......(B.R.)..:   ( 5 ,  2390) PHYS. LETT. VOL.29A, . . . . .     296, (1969).

4642  WINKWORTH......(R.)....:   (22B,    19) NUCL. INST. MET. 23, . . . . .     181, (1963).
                                 (22B,    22) NUCL. INST. MET. 25, . . . . .     288, (1964).

4643  WINSBERG.......(L.)....:   ( 5 ,   242) PHYS. REV. VOL.75, . . . . . .    1098, (1949).
                                 ( 5 ,  2958) PHYS. REV. VOL.75, . . . . . .     975, (1949).
                                 ( 9B,    77) PHYS. REV. VOL.80, . . . . . .     507, (1950).

4644  WINTENBERGER...(M.)....:   ( 9D,     6) ACTA CRYST. VOL.A28, . . . . .     341, (1972).
                                 ( 5 ,  2416) SOL. STATE COMM. VOL.10, . . .     685, (1972).
                                 ( 5 ,  1558) SOL. STATE COMM. VOL.10, . . .     739, (1972).
                                 ( 5 ,  1710) SOL. STATE COMM. VOL.11 NO.11,    1485, (1972).
                                 ( 5 ,  2947) MATER. RES. BULL. VOL.8, . . .    1949, (1973).
                                 ( 5 ,  1965) PHYS. STAT. SOLIDI A VOL.18 NO.1   209, (1973).

4645  WISHART........(L.P.)..:   (20C,    46) NUCL. INST. MET. 57, . . . . .     237, (1967).

4646  WITTEMAN.......(W.G.)..:   ( 5 ,  2731) ACTA CRYST. 21, . . . . . . .      670, (1966).
                                 (22C,    58) REV. SCI. INSTRUM. VOL.37, . .    1543, (1966).
```

4647 WODITSCH.........(P.)....: (5 , 2244) J. LESS-COMMON METALS VOL.17,. .   459, (1969).

4648 WOHLFARTH.......(E.P.)..: (11A,   55) PROC. ROY. SOC. VOL.295, ...   182, (1966).

4649 WOJTCZAK........(K.)....: (17A,   16) ACTA PHYS. POLON. VOL.A42, . . .   259, (1972).

4650 WOJTCZAK........(L.)....: (8 , 123) REV. ROUMAINE PHYS. VOL.13,. .   637, (1968).
                                 (8 ,   57) PHYS. LETT. VOL.34A,. . . . . .   306, (1971).

4651 WOLD............(A.)....: (5 , 452) COLL. INTER. N.126 (GRENOBLE),   104, (1963).
                                 (5 , 517) PROC. INT. CONF. (NOTTINGHAM),   538, (1964).
                                 (5 , 1676) J. APP. PHYS. 36,. . . . . . . 1088, (1965).
                                 (5 , 1668) J. APP. PHYS. 37,. . . . . . .   962, (1966).
                                 (5 , 1104) PHYS. REV. B-3,. . . . . 3046, (1971).

4652 WOLFERS.........(P.)....: (5 , 2747) J. PHYSIQUE TOME 32 COL.1 VOL.II 859, (1971).

4653 WOLFE...........(G.A.)..: (100,   29) PHYS. REV. 178,. . . . . 1171, (1969).

4654 WOLFE...........(R.A)...: (20D,   21) NUCL. INST. MET. 60, ...   246, (1968).

4655 WOLFE...........(R.W.)..: (5 , 2486) MATER. RES. BULL. VOL.8, . . . 1183, (1973).

4656 WOLF............(W.P.)..: (6 , 699) J. APP. PHYS. 39,. . . . . . . 1232, (1968).
                                 (5 , 816) PHYS. REV. VOL.186,. . . . . .   557, (1969).
                                 (5 , 813) PHYS. REV. VOL.186,. . . . . .   567, (1969).

4657 WOLLAN..........(E.O.)..: (5 , 1411) PHYS. REV. VOL.72,. . . . . .   109, (1947).
                                 (9A,   19) NUCLEONICS VOL.3 NO.1,. . . .   8, (1948).
                                 (9A,   18) NUCLEONICS VOL.3 NO.2,. . . .   17, (1948).
                                 (9B,   75) PHYS. REV. VOL.73,. . . . . .   527, (1948).
                                 (5 , 1985) PHYS. REV. VOL.73,. . . . . .   830, (1948).
                                 (5 , 1972) PHYS. REV. VOL.73,. . . . . .   842, (1948).
                                 (9A,   24) SCIENCE VOL.108,. . . . . . .   69, (1948).
                                 (5 , 788) PHYS. REV. VOL.75,. . . . . . 1348, (1949).
                                 (5 , 2080) PHYS. REV. VOL.79,. . . . . .   395, (1950).
                                 (5 , 1081) PHYS. REV. VOL.81,. . . . . .   483, (1951).
                                 (5 , 2260) PHYS. REV. VOL.81,. . . . . .   527, (1951).
                                 (5 , 1753) PHYS. REV. VOL.83,. . . . . .   333, (1951).
                                 (5 , 1967) PHYS. REV. VOL.83,. . . . . .   700, (1951).
                                 (5 , 1085) PHYS. REV. VOL.84,. . . . . .   912, (1951).
                                 (5 , 2968) ACTA CRYST. VOL.5,. . . . . .   22, (1952).
                                 (5 , 240) PHYS. REV. VOL.85,. . . . . .   491, (1952).
                                 (5 , 2349) ACTA CRYST. VOL.6,. . . . . .   741, (1953).
                                 (5 , 2471) PHYS. REV. VOL.91,. . . . . .   597, (1953).
                                 (6 , 1938) PHYS. REV. VOL.92,. . . . . . 1380, (1953).
                                 (5 , 2854) PHYS. REV. VOL.95,. . . . . .   280, (1954).
                                 (5 , 1538) PHYS. REV. VOL.100,. . . . . .   545, (1955).
                                 (5 , 853) PHYS. REV. VOL.97,. . . . . . 1177, (1955).
                                 (9A,   25) SOL. STATE PHYS. VOL.2,. . . .   137, (1956).
                                 (5 , 1529) J. PHYS. CHEM. SOLIDS VOL.2, .   100, (1957).
                                 (5 , 1657) PHYS. REV. VOL.110,. . . . . .   638, (1958).
                                 (5 , 1394) PHYS. REV. VOL.110,. . . . . .   37, (1958).
                                 (5 , 2839) PHYS. REV. VOL.112,. . . . . . 1132, (1958).
                                 (5 , 1658) J. PHYS. RADIUM VOL.20,. . . .   180, (1959).
                                 (5 , 968) PHYS. REV. VOL.113,. . . . . .   497, (1959).
                                 (5 , 2386) IAEA SYMP. (PILE RESEARCH)VIENNA 379, (1960).
                                 (5 , 2586) J. APPL. PHYS. VOL.31, SUPPL.,   358, (1960).
                                 (5 , 573) PHYS. REV. VOL.118,. . . . . .   950, (1960).
                                 (5 , 2042) PHYS. REV. VOL.118,. . . . . .   58, (1960).
                                 (5 , 807) J. APPL. PHYS. VOL.32, SUPPL.,   48S, (1961).
                                 (5 , 852) J. APPL. PHYS. VOL.32, SUPPL.,   49S, (1961).
                                 (5 , 567) J. PHYS. CHEM. SOLIDS VOL.19,.   29, (1961).
                                 (5 , 1866) PHYS. REV. VOL.121,. . . . . .   74, (1961).
                                 (5 , 390) PHYS. REV. VOL.122,. . . . . . 1409, (1961).
                                 (5 , 2682) J. APPL. PHYS. VOL.33, SUPPL., 1124, (1962).
                                 (5 , 2301) J. APPL. PHYS. VOL.33, SUPPL., 1340, (1962).
                                 (5 , 2384) J. PHYS. SOC. JAPAN VOL.17 B-III   27, (1962).
                                 (5 , 2606) J. PHYS. SOC. JAPAN VOL.17 B-III   32, (1962).
                                 (5 , 2298) J. PHYS. SOC. JAPAN VOL.17 B-III   38, (1962).
                                 (5 , 1720) PHYS. REV. VOL.125,. . . . . . 1860, (1962).
                                 (5 , 2683) PHYS. REV. VOL.126,. . . . . . 1672, (1962).
                                 (6 , 527) PHYS. REV. VOL.127,. . . . . . 2080, (1962).
                                 (5 , 977) PHYS. REV. VOL.127,. . . . . .   714, (1962).
                                 (5 , 2310) PHYS. REV. VOL.128,. . . . . . 2118, (1962).
                                 (5 , 2146) COLL. INTER. N.126 (GRENOBLE),   36, (1963).
                                 (5 , 1112) J. APP. PHYS. 34,. . . . . . . 1189, (1963).
                                 (5 , 2610) J. APP. PHYS. 34,. . . . . . . 1335, (1963).
                                 (5 , 2147) J. PHYS. CHEM. SOL. 24,. . . . 1141, (1963).
                                 (5 , 2540) PHYS. REV. VOL.131,. . . . . .   922, (1963).
                                 (6 , 1855) J. PHYS. CHEM. SOL. 25,. . . . 1453, (1964).
                                 (5 , 1391) PHYS. LETT. VOL.9,. . . . . .   93, (1964).
                                 (5 , 2337) PHYS. LETT. VOL.12,. . . . . .   553, (1964).
                                 (5 , 2562) PHYS. REV. VOL.136,. . . . . . A240, (1964).
                                 (6 , 2147) PROC. INT. CONF. (NOTTINGHAM),   271, (1964).
                                 (5 , 2909) PHYS. REV. 138,. . . . . A165, (1965).
                                 (5 , 639) PHYS. REV. 138,. . . . . A737, (1965).
                                 (5 , 2101) PHYS. REV. 138,. . . . . A755, (1965).
                                 (5 , 850) PHYS. REV. 140,. . . . . . . .A1896, (1965).
                                 (6 , 1163) PHYS. REV. 151,. . . . . . . .A2003, (1965).
                                 (5 , 1358) PHYS. REV. 151,. . . . .   414, (1966).
                                 (11A,   6) PHYS. REV. 158,. . . . .   450, (1967).
                                 (5 , 1209) C.N.E.N. SYMP. CASACCIA, . . .   235, (1968).
                                 (5 , 2116) PHYS. REV. 165,. . . . .   733, (1968).
                                 (5 , 1066) PHYS. REV. LETT. VOL.22,. . . . 1256, (1969).
                                 (5 , 2134) INT. J. MAGN. VOL.2 NO.2,. . .   1, (1972).
                                 (5 , 2134) PHYS. REV. B VOL.7,. . . . . . 2005, (1973).

4658 WOLLEY..........(E.O.)..: (20C,   36) NUCLEONICS VOL.17 APRIL, . . .   116, (1959).

4659 WOLSKI..........(W.)....: (5 , 1096) PHYS. STAT. SOL. 26, ...   429, (1968).

4660 WONG............(V.K.)..: (10B, 168) PHYS. REV. A VOL.3,. . . . . . 1453, (1971).

4661 WONN............(H.)....: (6 , 1430) PHYS. STAT. SOL. VOL.49, . . .   807, (1972).

4662 WOODLOCK........(A.)....: (5 , 290) TETRAHEDRON LETT. NO.59, . . . 5219, (1969).

4663 WOODS...........(A.D.B.): (6 , 1801) IAEA SYMPOSIUM VIENNA,. . . .   531, (1960).
                                 (6 , 203) IAEA SYMPOSIUM (VIENNA), . . .   487, (1960).
                                 (6 , 1649) PHYS. REV. VOL 119,. . . . . .   980, (1960).
                                 (6 , 1177) CAN. J. PHYS. VOL.39,. . . . . 1082, (1961).
                                 (6 , 1816) PHYS. REV. LETT. VOL.7,. . . .   93, (1961).
                                 (6 , 1176) PHYS. REV. VOL.121,. . . . . . 1266, (1961).
                                 (6 , 1628) IAEA SYMP. CHALK RIVER, VOL.2,   23, (1962).
                                 (6 , 1831) IAEA SYMP. CHALK RIVER, VOL.2,   3, (1962).
                                 (6 , 2157) J. PHYS. SOC. JAPAN VOL.17 B-III   63, (1962).
                                 (6 , 1810) PHYS. REV. VOL.128,. . . . . . 1099, (1962).
                                 (6 , 1629) PHYS. REV. VOL.128,. . . . . . 1112, (1962).
                                 (6 , 1621) PROC. PHYS. SOC. VOL.79,. . . .   440, (1962).
                                 (6 , 1669) COPENHAGEN CONF.-LATT.DYNAMICS,   39, (1963).
                                 (6 , 1673) PHYS. REV. LETT. 11,. . . . .   271, (1963).
                                 (6 , 1244) PHYS. REV. 131,. . . . . 1025, (1963).
                                 (100,   26) PHYS. REV. 131,. . . . . 1030, (1963).
                                 (6 , 1614) PROC. PHYS. SOC. 81,. . . . . .   973, (1963).
                                 (6 , 2215) IAEA SYMP. BOMBAY, VOL.1,. . .   373, (1964).

```
4663  WOODS..........(A.D.B.)..:  ( 6 , 1515) IAEA SYMP. BOMBAY, VOL.1,. . . .     87, (1964):
                                 ( 6 , 1123) IAEA SYMP. BOMBAY, VOL.2,. . . . .   191, (1964):
                                 ( 6 , 1731) PHYS. REV. VOL.136,. . . . . . .A1359, (1964):
                                 ( 6 , 2111) PHYS. REV. VOL.136,. . . . . . . A781, (1964):
                                 ( 6 , 1519) SOL. STATE COMM. 2,. . . . . . .    233, (1964):
                                 ( 6 , 2201) BROOKHAVEN SYMPOSION.. . . . . .      8, (1965):
                                 ( 6 , 2221) CAN. JOURN. PHYS. 43,. . . . . .  1397, (1965):
                                 ( 6 , 1204) PHYS. REV. LETT. 14,. . . . . .    355, (1965):
                                 ( 6 , 1685) PHYS. REV. LETT. 15,. . . . . .    778, (1965):
                                 (10A,   21) TH. NEUTRON SCATT./EGELSTAFF,. .    193, (1965):
                                 ( 6 , 1246) PHYS. REV. 150,. . . . . . . .     487, (1966):
                                 ( 6 , 1170) SYMP. ON QUANTUM FLUIDS,. . . .     242, (1966):
                                 ( 6 , 1732) J. APP. PHYS. 38,. . . . . . .    1247, (1967):
                                 ( 6 ,  708) PHYS. REV. LETT. 19,. . . . . .    908, (1967):
                                 (11C,    2) CAN. J. PHYS. 46,. . . . . . .    1499, (1968):
                                 ( 6 , 1124) IAEA SYMP. COPENHAGEN, VOL.1,.     609, (1968):
                                 (23 ,   35) IAEA SYMP. COPENHAGEN, VOL.2,.     281, (1968):
                                 ( 6 , 1225) J. APP. PHYS. 39,. . . . . . .     457, (1968):
                                 ( 6 , 1149) PHYS. REV. LETT. 21,. . . . . .    787, (1968):
                                 ( 6 , 1686) PHYS. REV. 171,. . . . . . . .     727, (1968):
                                 (17A,   40) IAEA SYMP. (INSTRUMENT.), VIENNA       1, (1969):
                                 ( 6 , 1227) J. APP. PHYS. VOL.40,. . . . .     1443, (1969):
                                 ( 6 , 1221) PHYS. REV. LETT. VOL.23,. . .        81, (1969):
                                 ( 6 ,  704) COLLOQUE INTERNAT. NO.180 TOME 2  283, (1970):
                                 ( 6 , 1228) J. APPL. PHYS. 41,. . . . . .     1176, (1970):
                                 ( 6 , 1223) METAL PHYS. VOL.3,. . . . . .      189, (1970):
                                 ( 6 , 1130) PHYS. REV. LETT. VOL.24,. . .      646, (1970):
                                 ( 6 ,  702) CAN. J. PHYS. 49,. . . . . . .    2875, (1971):
                                 ( 6 , 1121) CAN. J. PHYS. 49,. . . . . . .     177, (1971):
                                 ( 6 ,  705) J. PHYSIQUE TOME 32 COL.1 VOL.II 1179, (1971):
                                 ( 6 , 1517) CAN. J. PHYS. VOL.50,. . . . .    3069, (1972):
                                 ( 6 , 1164) IAEA SYMP. GRENOBLE,. . . . . .    359, (1972):
                                 ( 6 , 1672) IAEA SYMP. GRENOBLE,. . . . . .     43, (1972):
                                 ( 6 , 1167) PHYS. REV. LETT. VOL.29,. . .     1148, (1972):
                                 ( 6 , 1171) PHYS. LETT. VOL.43A NO.3,. . .     223, (1973):
                                 (14B,  113) REP. PROGR. PHYS. VOL.36,. . .     1135, (1973).

4664  WOOD...........(J.L.)..:  ( 6 , 1895) IAEA SYMPOSIUM (VIENNA),. . . .     613, (1960).

4665  WOOD...........(R.E.)..:  (22A,   61) REV. SCI. INSTRUM. VOL.27,. . .      26, (1956).

4666  WOOLF..........(W.E.)..:  ( 6 ,  769) PHYS. REV. VOL.80,. . . . . . .    481, (1950):
                                 (12B,   39) PHYS. REV. VOL.80,. . . . . . .    953, (1950).

4667  WOOSTER........(W.A.)..:  (21 ,   30) J. SCI. INSTRUM. VOL.40,. . . .     14, (1963).

4668  WORCESTER......(D.L.)..:  (22A,   25) J. PHYS. E VOL.6 NO.6,. . . . .    568, (1973).

4669  WORLTON........(T.G.)..:  ( 5 ,  600) DISS. ABS. 28,. . . . . . . . .3029B, (1967):
                                 ( 5 ,  209) PHYS. LETT. 24A,. . . . . . . .    714, (1968):
                                 ( 5 ,  603) J. PHYS. CHEM. SOLIDS VOL.29,.      435, (1968):
                                 ( 5 , 1074) PHYS REV 171,. . . . . . . . .     596, (1968):
                                 ( 9C,   36) PROC/SYMP/CRYS. STRUC/HIGH PRESS   141, (1969):
                                 ( 6 , 1428) J. PHYS. CHEM. SOLIDS VOL.31,.    1085, (1970):
                                 ( 6 , 2187) PHYS. REV. B-3,. . . . . . . .    4115, (1971):
                                 ( 5 ,  487) BULL. AMER. PHYS. SOC. VOL.17,     667, (1972).

4670  WORSHAM-JR.....(J.E.)..:  ( 5 , 2819) ACTA CRYST. VOL.10,. . . . . .      319, (1957):
                                 ( 5 , 2309) J. PHYS. CHEM. SOLIDS VOL.3,. .     303, (1957).

4671  WORSHAM........(J.E.)..:  ( 5 , 1918) ACTA CRYST. VOL.B25,. . . . . .     572, (1969).

4672  WORTIS.........(M.)....:  (13 ,   50) PHYS. REV. B VOL.4,. . . . . .     3954, (1971).

4673  WRAIGHT........(L.A.)..:  (20C,   11) IAEA SYMP. CHALK RIVER, VOL.1,     171, (1962):
                                 (20C,   40) NUCL. INST. MET. 33,. . . . . .    181, (1965):
                                 (22C,   42) NUCL. INST. MET. 71,. . . . . .     13, (1969).

4674  WREGE..........(D.E.)..:  ( 8 ,   22) J. APPL. PHYS. VOL.40,. . . . .    1226, (1969):
                                 ( 6 ,  801) AIP CONF. PROC. VOL.5,. . . . .   1334, (1971):
                                 ( 6 ,  802) THESIS,. . . . . . . . . . . .          (1971).

4675  WRIGHT.........(A.C.)..:  ( 6 ,  151) J. NON-CRYST. SOLIDS VOL.3,. .      239, (1970):
                                 ( 5 , 1233) AMORPHOUS MATERIALS,. . . . . .    423, (1972):
                                 ( 5 , 1238) J. NON-CRYST. SOLIDS VOL.7,. .      141, (1972):
                                 ( 5 ,  203) J. NON-CRYST. SOLIDS VOL.7,. .      156, (1972):
                                 ( 5 , 1239) J. NON-CRYST. SOLIDS VOL.7,. .       23, (1972):
                                 (19A,   55) NUCL. INST. MET. VOL.114,. . .     451, (1974).

4676  WRIGHT.........(C.J.)..:  ( 6 , 1001) J. CHEM. SOC. D 1970,. . . . .      970, (1970):
                                 ( 6 , 1014) J. CHEM. SOC. D 1970,. . . . .      971, (1970):
                                 ( 6 , 1015) J. CHEM. SOC. SECTION A,. . . .   2043, (1971):
                                 ( 6 , 1335) J. CHEM. SOC. FARADAY II VOL.68, 1414, (1972):
                                 ( 6 , 1235) J. CHEM. SOC. FARADAY II VOL.68, 1423, (1972):
                                 ( 6 ,  304) J.C.S. FARADAY TRANS. II VOL.69,  275, (1973):
                                 ( 6 , 2004) J.C.S. FARADAY TRANS. II VOL.70,  348, (1974).

4677  WROBEL.........(S.)....:  ( 6 ,  116) FARADAY SYMP. CHEM. SOC. NO.6,      48, (1972).

4678  WUTTIG.........(M.)....:  ( 5 , 2357) Z. ANGEW. MATH. PHYS. VOL.16,.     535, (1965).

4679  WU.............(C.S.)..:  ( 5 , 1249) PHYS. REV. VOL.69,. . . . . . .    236, (1946):
                                 ( 6 , 2112) PHYS. REV. VOL.71,. . . . . . .    165, (1947):
                                 ( 5 , 1407) PHYS. REV. VOL.71,. . . . . . .    174, (1947):
                                 (22A,   55) PHYS. REV. VOL.71,. . . . . . .     65, (1947):
                                 ( 6 ,  975) PHYS. REV. VOL.73,. . . . . . .    733, (1948):
                                 ( 6 ,  598) PHYS. REV. VOL.73,. . . . . . .    963, (1948).

4680  WYNCHANK.......(S.A.R.):  (24 ,   44) NUCL. INST. MET. 39,. . . . . .    350, (1966):
                                 (20B,   80) NUCL. INST. MET. 46,. . . . . .    141, (1966).

4681  WYNN-WILLIAMS..(C.E.)..:  (20A,   30) NATURE VOL.142,. . . . . . . .     829, (1938):
                                 (20B,  132) PROC. ROY. SOC. A VOL.175,. . .    316, (1940).

4682  YABLONSKII.....(D.A.)..:  ( 8 ,  153) SOV. PHYS. SOLID STATE VOL.14,    2006, (1973).

4683  YADAVA.........(V.S.)..:  ( 5 , 1994) ACTA CRYST. VOL.B27,. . . . . .     253, (1971).

4684  YAEGER.........(I.)....:  ( 5 ,  882) PHYS. REV. B VOL.8,. . . . . .     3398, (1973).

4685  YAGUD..........(A.Z.)..:  ( 6 , 1153) SOV. PHYS. TECH. PHYS. VOL.17,     180, (1972).

4686  YAI-PRAKASH...........:   ( 6 ,  584) ACTA PHYS. ACAD. SCI. HUNG. 30,    231, (1971).

4687  YAKEL..........(H.L.)..:  ( 5 , 1391) PHYS. LETT. VOL.9,. . . . . . .     93, (1964).

4688  YAKINTHOS......(J.)....:  ( 5 , 2197) PHYS. STAT. SOL. VOL.47,. . . .    239, (1971):
                                 ( 5 ,  829) PHYS. STAT. SOL. VOL.47,. . . .    247, (1971):
                                 ( 5 ,  520) PHYS. STAT. SOL. VOL.50,. . . .    747, (1972):
                                 ( 5 , 2183) SOL. STATE COMM. VOL.10,. . . .    989, (1972).

4689  YAKOVLEV.......(A.A.)..:  ( 5 , 1084) SOV. PHYS. SOLID STATE VOL.15,    1429, (1974).
```

4690 YAMADA..........(H.)....: (8 , 44) J. PHYS. SOC. JAPAN VOL.25,. . . 1001, (1968).
                              (11A, 27) J. PHYS. SOC. JAPAN VOL.28,. . . 327, (1970).

4691 YAMADA..........(T.)....: (5 , 576) J. PHYS. SOC. JAPAN VOL.20,. . . 1729, (1965).
                              (5 , 1620) J. PHYS. SOC. JAPAN VOL.28,. . . 615, (1970).

4692 YAMADA..........(Y.)....: (11C, 21) J. PHYS. SOC. JAPAN VOL.15,. . . 429, (1960).
                              (5 , 934) J. PHYS. SOC. JAPAN VOL.17 B-III 1, (1962).
                              (5 , 576) J. PHYS. SOC. JAPAN VOL.20,. . . 1729, (1965).
                              (6 , 2095) J. PHYS. SOC. JAPAN VOL.26,. . . 396, (1969).
                              (22C, 43) NUCL. INST. MET. 71,. . . . . 102, (1969).
                              (6 , 130) PHYS. REV. VOL.177,. . . . . . 848, (1969).
                              (6 , 2107) PHYS. REV. VOL.177,. . . . . . 858, (1969).
                              (5 , 12) J. PHYS. SOC. JAPAN VOL.29,. . . 978, (1970).
                              (6 , 1920) J. CHEM. PHYSICS VOL.55,. . . 983, (1971).
                              (6 , 1529) PHYS. REV. B VOL.9,. . . . . . 4429, (1974).
                              (6 , 574) PHYS. REV. B VOL.9,. . . . . . 4549, (1974).

4693 YAMAGISHI........(T.)....: (24 , 26) J. NUCL. SCI. TECHNOL.(JAP.) 8,. 470, (1971).
                              (16 , 30) J. NUCL. SCI. TECHNOL. VOL.8,. . 153, (1971).

4694 YAMAGUCHI........(H.)....: (5 , 345) J. PHYS. SOC. JAPAN VOL.22,. . . 939, (1967).
                              (5 , 560) J. PHYS. SOC. JAPAN VOL.32,. . . 958, (1972).

4695 YAMAGUCHI........(K.)....: (5 , 1172) J. PHYS. SOC. JAPAN VOL.22,. . . 1210, (1967).

4696 YAMAGUCHI........(S.)....: (5 , 2660) J. PHYS. SOC. JAPAN VOL.28,. . . 1014, (1970).
                              (5 , 2977) ACTA CRYST. VOL.A28 PART S-4,. . 99, (1972).
                              (5 , 1345) J. PHYS. SOC. JAPAN VOL.35,. . . 473, (1973).
                              (5 , 2978) J. APPL. CRYST. VOL.7,. . . . . 67, (1974).

4697 YAMAGUCHI........(Y.)....: (5 , 344) J. PHYS. SOC. JAPAN VOL.24,. . . 446, (1968).
                              (5 , 1637) J. APP. PHYS. VOL.40,. . . . . 1128, (1969).
                              (5 , 2492) J. PHYS. SOC. JAPAN VOL.26,. . . 1320, (1969).
                              (5 , 1805) ACTA CRYST. VOL.A28 PART S-4,. . 196, (1972).
                              (5 , 560) J. PHYS. SOC. JAPAN VOL.32,. . . 958, (1972).
                              (5 , 2493) J. PHYS. SOC. JAPAN VOL.33,. . . 967, (1972).
                              (5 , 2489) J. PHYS. SOC. JAPAN VOL.33,. . . 970, (1972).
                              (5 , 1048) J. PHYS. SOC. JAPAN VOL.34,. . . 911, (1973).
                              (5 , 1148) J. PHYS. SOC. JAPAN VOL.34,. . . 58, (1973).

4698 YAMAJI..........(A.)....: (20C, 60) NUCL. INST. MET. 74,. . . . . . 322, (1969).

4699 YAMAMOTO........(H.)....: (5 , 1164) J. PHYS. SOC. JAPAN VOL.18,. . . 995, (1963).
                              (5 , 2492) J. PHYS. SOC. JAPAN VOL.26,. . . 1320, (1969).
                              (5 , 1048) J. PHYS. SOC. JAPAN VOL.34,. . . 911, (1973).

4700 YAMAOKA.........(T.)....: (5 , 1722) J. PHYS. SOC. JAPAN VOL.31,. . . 301, (1971).
                              (5 , 1721) J. PHYS. SOC. JAPAN VOL.36,. . . 438, (1974).

4701 YAMASAKI........(S.)....: (7B, 106) PROGR. THEOR. PHYS. VOL.37,. . . 1051, (1967).

4702 YAMZIN..........(I.I.)..: (22C, 73) SOV. PHYS. CRYST. VOL.4, ... 397, (1960).
                              (9A, 103) SOV. PHYS. CRYST. VOL.6, ... 374, (1961).
                              (5 , 2269) SOV. PHYS. DOKLADY VOL.6,. . . . 373, (1961).
                              (5 , 1693) J. PHYS. SOC. JAPAN VOL.17 B-III 55, (1962).
                              (5 , 1692) SOV. PHYS. CRYST. VOL.6,. . . . 744, (1962).
                              (22A, 68) SOV. PHYS. CRYST. VOL.7,. . . . 769, (1963).
                              (5 , 2895) SOV. PHYS. CRYST. VOL.7,. . . . 767, (1963).
                              (5 , 2876) SOV. PHYS. CRYST. VOL.8,. . . . 234, (1963).
                              (22A, 69) SOV. PHYS. CRYST. VOL.8,. . . . 15, (1963).
                              (5 , 2907) SOV. PHYS. CRYST. VOL.10,. . . . 346, (1965).
                              (9B, 106) SOV. PHYS. CRYST. VOL.9,. . . . 797, (1965).
                              (22A, 70) SOV. PHYS. DOKLADY VOL.11,. . . 379, (1966).
                              (5 , 518) SOV. PHYS. J.E.T.P. VOL.23,. . . 395, (1966).
                              (5 , 160) SOV. PHYS. CRYST. VOL.11,. . . . 597, (1967).
                              (12B, 60) SOV. PHYS. JETP. LETT. VOL.6,. . 176, (1967).
                              (5 , 2488) SOV. PHYS. J.E.T.P. VOL.25,. . . 266, (1967).
                              (5 , 2487) SOV. PHYS. JETP. LETT. VOL.7,. . 158, (1968).
                              (5 , 176) SOV. PHYS. J.E.T.P. VOL.26,. . . 736, (1968).
                              (5 , 180) SOV. PHYS. CRYST. VOL.14, 3,. . 447, (1969).
                              (5C, 2323) SOV. PHYS. J.E.T.P. VOL.29,. . . 655, (1969).
                              (9C, 41) PHYS. STAT. SOL. VOL.37,. . . . 843, (1970).
                              (5 , 2950) SOV. PHYS. CRYST. VOL.15,. . . . 280, (1970).
                              (5 , 193) SOV. PHYS. J.E.T.P. VOL.31,. . . 808, (1970).
                              (5 , 2689) SOV. PHYS. CRYST. VOL.16,. . . . 634, (1971).
                              (5 , 2192) SOV. PHYS. CRYST. VOL.16,. . . . 935, (1971).
                              (5 , 1132) IZV. AKAD. NAUK...NEORG. MATER 8 1, (1972).
                              (9C, 19) SOV. PHYS. CRYST. VOL.16,. . . . 711, (1972).
                              (5 , 168) SOV. PHYS. JETP VOL.35,. . . . 370, (1972).
                              (5 , 182) PHYS. STAT. SOLIDI A VOL.18 NO.2 K91, (1973).
                              (5 , 2466) SOV. PHYS. CRYST. VOL.18,. . . . 393, (1973).
                              (5 , 167) J. APPL. CRYST. VOL.7,. . . . . 207, (1974).
                              (5 , 2467)

4703 YARNELL.........(J.L.)..: (6 , 1202) PHYS. REV. LETT. VOL.1,. . . . . 9, (1958).
                              (6 , 1191) PHYS. REV. VOL.113,. . . . . . 1379, (1959).
                              (6 , 1190) PHYS. REV. VOL.113,. . . . . . 1386, (1959).
                              (6 , 27) COPENHAGEN CONF.-LATT.DYNAMICS,. 57, (1963).
                              (6 , 196) IAEA SYMP. BOMBAY, VOL.1,. . . 361, (1964).
                              (6 , 180) IBM J. RES. DEVELOP. VOL.8,. . 234, (1964).
                              (6 , 215) PHYS. REV. LETT. 13,. . . . . 13, (1964).
                              (5 , 2833) ACTA CRYST. 19,. . . . . 6, (1965).
                              (5 , 2019) ACTA CRYST. 21,. . . . . 843, (1966).
                              (6 , 182) BULL. AMER. PHYS. SOC. VOL.12,. 689, (1967).
                              (6 , 217) PHYS. REV. 158,. . . . . 805, (1967).
                              (6 , 9C1) IAEA SYMP. COPENHAGEN, VOL.1,. . 301, (1968).
                              (6 , 1378) PHYS REV 168,. . . . . 980, (1968).
                              (5 , 769) ACTA CRYST. VOL.A28 PART S-4,. . 188, (1972).
                              (5 , 99) PHYS. REV. A VOL.7,. . . . . . 2130, (1973).

4704 YARNZIN.........(I.I.)..: (5 , 192) SOV. PHYS. SOL. STATE VOL.10,. . 2258, (1969).

4705 YASHIRO.........(T.)....: (5 , 1148) J. PHYS. SOC. JAPAN VOL.34,. . . 58, (1973).

4706 YASUDA..........(H.)....: (20B, 59) J. NUCL. SCI. TECHNOL. VOL.7,. . 381, (1970).

4707 YASUKAWA........(T.)....: (6 , 1920) J. CHEM. PHYSICS VOL.55, . . . 983, (1971).

4708 YAZVITSKII......(YU.S.).: (20C, 16) INSTRUM. EXP. TECH. NO.4,. . . 805, (1965).

4709 YEATER..........(M.L.)..: (19B, 29) REPORT KAPL-1657 (22 PP.),. . . . . . , (1957).
                              (17A, 62) NUCL. INST. MET. 42,. . . . . 197, (1966).
                              (6 , 1391) MOL. DYN. AND STRUCT. OF SOLIDS, 315, (1969).
                              (6 , 1890) MOL. DYN. AND STRUCT. OF SOLIDS, 547, (1969).
                              (6 , 1038) NUCL. SCI. ENG. VOL.46,. . . . 223, (1971).
                              (6 , 2208) NUCL. SCI. ENG. VOL.47,. . . . 349, (1972).
                              (7B, 43) NUCL. SCI. ENG. VOL.48,. . . . 266, (1972).

4710 YEATS...........(E.A.)..: (5 , 256) ACTA CRYST. B25,. . . . . 2009, (1969).
                              (6 , 343) SOL. STATE COMM. 7,. . . . . . 385, (1969).

4711 YELON...........(W.B.)..: (6 , 1336) FERROELECTRICS(GB) VOL.2,. . . 261, (1971).
                              (6 , 530) PHYS. REV. B VOL.4,. . . . . . 2280, (1971).
                              (6 , 803) PHYS. REV. B VOL.5,. . . . . . 2607, (1972).
                              (5 , 976) PHYS. REV. B VOL.5,. . . . . . 2615, (1972).

4711 YELON...........(W.B.)..: (5 , 2407) PHYS. REV. B VOL.6,. . . . . .   204, (1972).
                                 (5 , 1898) SOL. STATE COMM. VOL.11; . . . . . 1011, (1972):
                                 (8 , 1535) CONF. NATO SCHOOL,. . . . .   , (1973):
                                 (8 ,   672) PHYS. REV. B VOL.7,. . . . . . . 2024, (1973):
                                 (5 , 1901) PHYS. REV. B VOL.9 NO.11,. . . . . . , (1974).

4712 YERGIN..........(P.F.)..: (229,   35) NUCL. INST. MET. VOL.91, ...   541, (1971).

4713 YERICK..........(R.E.)..: (20C,   79) NUCL. SCI. ENGNG. VOL.20,. . . .   23, (1964).

4714 YERKESS.........(J.)....: (5 ,   352) ACTA CRYST. VOL.B27, ...   349, (1971).
                                 (5 , 1970) ACTA CRYST. VOL.B27, ...   354, (1971).
                                 (5 ,   649) Z. KRIST. VOL.134,. . . . . . .   308, (1971).
                                 (5 ,   278) ACTA CRYST. VOL.A28 PART S-4,.   27, (1972).
                                 (5 ,   351) ACTA CRYST. VOL.B28, ...   2430, (1972).
                                 (5 ,   350) ACTA CRYST. VOL.B28, ...   209, (1972).
                                 (5 , 1951) Z. KRIST. VOL.135,. . . . . . .   240, (1972).
                                 (5 ,   661) INORG. NUCL. CHEM. LETT. VOL.9, 1025, (1973).

4715 YESSIK..........(M.)....: (5 , 1767) PHIL. MAG. VOL.17, ...   623, (1968).
                                 (5 , 2745) J. APP. PHYS. VOL.40 . . . . . .   1133, (1969).
                                 (5 ,   719) MAGN. AND MAG. MATERIALS-1971,   508, (1972).
                                 (9B,   20) ACTA CRYST. VOL.A29, ...   372, (1973).
                                 (5 ,   721) A.I.P. CONF. PROC. NO.10 PART 1,   679, (1973).

4716 YIN-YUAN-LI.............: (90,   58) PHYS. REV. VOL.100,. . . . . . .   627, (1955).
                                 (90,   1) ACTA CRYST. VOL.9,. . . . . . .   738, (1956).

4717 YIP.............(S.)....: (15 ,   39) PHYS. REV. 130,. . . . .   1860, (1963).
                                 (6 , 1053) PHYS. REV. 131,. . . . .   2547, (1963).
                                 (14B,   22) DISS. ABS. VOL.24,. . . . . . .   685, (1964).
                                 (7B,   20) IAEA SYMPOSIUM BOMBAY, VOL.2,.   35, (1964).
                                 (7B,   80) PHYS. REV. VOL.139,. . . . . .   A1138, (1965).
                                 (14A,   21) J. CHEM. PHYSICS VOL.46,. . . .   1999, (1967).
                                 (7B,   49) PHYS. LETT. 25A,. . . . .   211, (1967).
                                 (6 ,   997) CHEM. PHYS. LETTERS VOL.2,. . .   584, (1968).
                                 (6 , 1050) IAEA SYMP. COPENHAGEN, VOL.1,.   345, (1968).
                                 (6 ,   71) IAEA SYMP. COPENHAGEN, VOL.1,.   545, (1968).
                                 (6 , 1020) J. CHEM. PHYSICS VOL.48,. . . .   3367, (1968).
                                 (6 ,   93) PHYS. REV. VOL.166,. . . . . .   129, (1968).
                                 (6 ,   998) PHYS. REV. VOL.171,. . . . . .   263, (1968).
                                 (6 , 1044) MOL. DYN. AND STRUCT. OF SOLIDS,   335, (1969).
                                 (15 ,   31) NUCL. SCI. ENGNG. VOL.37,. . . .   368, (1969).
                                 (6 , 1607) PHYS. REV. VOL.180,. . . . . .   308, (1969).
                                 (14A,   48) PHYS. REV. VOL.182,. . . . . .   323, (1969).
                                 (16 ,   43) NUCL. SCI. ENGNG. VOL.40,. . . .   460, (1970).
                                 (6 ,   291) J. OF PHYS.-C-, SER.2, VOL.4,.   2725, (1971).
                                 (6 ,   961) IAEA SYMP. GRENOBLE,. . . . . .   445, (1972).
                                 (6 , 1656) J. CHEM. PHYS. VOL.56 NO.11,.   5377, (1972).
                                 (6 , 1563) J. CHEM. PHYS. VOL.57, ...   2291, (1972).
                                 (6 ,   983) PHYS. REV. A VOL.8,. . . . . .   3163, (1973).
                                 (6 ,   889) PROPERTIES LIQUID METALS,. . .   119, (1973).
                                 (6 , 1819) PROPERTIES LIQUID METALS,. . .   125, (1973).
                                 (14B, 1244) SPECTROSCOPY BIOL. CHEM.,. . .   53, (1974).

4718 YODA............(O.)....: (6 , 1924) J. POLYM. SCI./POLYM. LETT. 11,.   377, (1973).

4719 YOON-IL-CHANG...........: (11B,   4) DISS. ABS. VOL.32, ...   17868, (1971).

4720 YOON............(B.G.)..: (5 , 1956) J. KOREAN PHYS. SOC. VOL.1,. . .   108, (1968).

4721 YORK............(E.J.)..: (9B,   11) ACTA CRYST. VOL.A26, ...   682, (1970).

4722 YOSHIE..........(T.)....: (22B,   16) J. PHYS. SOC. JAPAN VOL.17 B-II,   358, (1962).
                                 (21 ,   32) MITSUB. DENKI LAB. REP. VOL.3, .   111, (1962).

4723 YOSHII..........(S.)....: (5 , 1041) J. PHYS. SOC. JAPAN VOL.22,. . .   674, (1967).

4724 YOSHIMORI.......(A.)....: (5 , 1756) J. PHYS. SOC. JAPAN VOL.14,. . .   807, (1959).
                                 (6 , 1464) PROGR. THEOR. PHYS. VOL.25,. . .   595, (1961).

4725 YOSHIMURA.......(T.)....: (20C,   60) NUCL. INST. MET. 74,. . . . . .   322, (1969).

4726 YOSIDA..........(K.)....: (90,   66) PROGRESS/LOW TEMP. PHYS. VOL.4,.   265, (1964).

4727 YOST............(K.J.)..: (5 ,   230) TRANS. AM. NUCL. SOC. VOL.13,. .   728, (1970).

4728 YOUNG-JA-PARK...........: (5 ,   294) ACTA CRYST. VOL.B27, ...   2393, (1971).

4729 YOUNG...........(J.A.)..: (6 ,   141) NUCL. SCI. ENGNG. VOL.19,. . . .   367, (1964).
                                 (6 , 1056) NUCL. SCI. ENGNG. VOL.19,. . . .   412, (1964).
                                 (6 , 2313) PHYS. REV. VOL.134,. . . . . .   A1476, (1964).
                                 (6 ,   964) PHYS. REV. VOL.135,. . . . . .   A603, (1964).
                                 (16 ,   53) NUKLEONIK VOL.7,. . . . .   408, (1965).
                                 (6 ,   138) PHYS. LETT. 16,. . . . .   235, (1965).
                                 (6 ,   972) NUKLEONIK VOL.8,. . . . .   40, (1966).
                                 (16 ,   72) REACTOR PHYSICS,. . . . .   3, (1966).
                                 (6 , 1526) CONF. NEUTRON THERMALIZ. VOL.I,.   343, (1968).
                                 (6 , 2217) NUKLEONIK VOL.12,. . . . . . .   205, (1969).

4730 YOUNG...........(J.C.)..: (6 , 2205) PROC/CONF/FAST CRIT. EXP./ANALY.   524, (1967).
                                 (6 , 1946) NUCL. SCI. ENG. VOL.46,. . . . .   244, (1971).

4731 YOUNG...........(P.L.C.): (13 ,   15) DISS. ABS. VOL.32, ...   11668, (1971).

4732 YOUNG...........(R.A.)..: (5 ,   369) NATURE VOL.204,. . . . .   1050, (1964).

4733 YOUSSEF.........(S.I.)..: (5 , 2142) J. PHYS. CHEM. SOLIDS VOL.30,.   1941, (1969).
                                 (5 , 2955) PHYS. REV. VOL.181,. . . . . .   969, (1969).
                                 (6 , 2082) J. PHYSIQUE TOME 32 COL.1 VOL.I,   318, (1971).

4734 YUEN............(P.S.)..: (10B,   131) PHYS. REV. 174,. . . . .   766, (1968).

4735 YULMETEV........(R.M.)..: (14A,   32) OPT. SPECTROSC. VOL.35 NO.3, . .   342, (1973).

4736 YUL≠MET≠EV......(R.M.)..: (7A,   96) UKR. FIZ. ZH. (USSR) VOL.10, . .   1168, (1965).

4737 YUN-PEEL-LEE............: (22A,   23) J. KOREAN PHYS. SOC. VOL.4,. . .   47, (1971).

4738 ZABIDAROV.......(E.I.)..: (6 , 1749) SOV. PHYS. JETP LETT. VOL.2,. .   336, (1965).
                                 (11B,   43) SOV. PHYS. JETP 20,. . . . . .   1548, (1965).
                                 (11A,   63) SOV. PHYS. JETP LETT. VOL.9, . .   204, (1969).
                                 (5 , 2065) SOV. PHYS. J.E.T.P. VOL.29,. . .   261, (1969).

4739 ZABIYAKIN.......(G.I.)..: (22A,   18) INSTRUM. EXP. TECH. NO.6,. . . .   1176, (1964).

4740 ZABUSKY.........(N.J.)..: (10D,   19) KYOTO CONF. STAT. MECHANICS, . .   196, (1968).

4741 ZACCAI..........(G.)....: (5 , 1512) FERROELECT. VOL.4, ...   153, (1972).
                                 (6 , 1338) J. PHYSIQUE TOME 33 COL.2, . .   133, (1972).
                                 (6 , 1339) J. PHYS. C VOL.7,. . . . . . .   15, (1972).

4742 ZACHARIASEN.....(W.H.)..: (5 , 1604) ACTA CRYST. 16,. . . . .   352, (1963).
                                 (17B,   4) ACTA CRYST. 18,. . . . .   705, (1965).
                                 (5 , 2752) ACTA CRYST. VOL.B27, ...   1067, (1971).
                                 (5 , 2632) ACTA CRYST. VOL.B28, ...   1724, (1972).

```
4743   ZAFRIR..........(H.)....:   ( 5 , 1462) SOL. STATE COMM. 5,. . . . . . .    41, (1967).

4744   ZAHN............(C.T.)..:   (17A,    86) PHYS. REV. VOL.50, . . . . . . .   570, (1936).

4745   ZAITSEV.........(K.N.)..:   (20B,    51) INSTRUM. EXP. TECH. VOL.16 NO.2,  399, (1973).

4746   ZAJAC...........(ST.)...:   ( 6 , 1743) ACTA PHYS. HUNGAR. VOL.15, . . .   29, (1962).

4747   ZAKHAROV........(A.I.)..:   ( 5 , 1127) SOV. PHYS. J.E.T.P. VOL.19,. . . 1348, (1964).

4748   ZAK.............(J.)....:   ( 7B,    81) PHYS. REV. 151,. . . . . . . . .  464, (1966).

4749   ZALESSKII.......(A.V.)..:   ( 5 ,  160) SOV. PHYS. J.E.T.P. VOL.23,. . .  395, (1966).

4750   ZALKIN..........(A.)....:   ( 5 ,  667) J. CHEM. PHYSICS VOL.37, . . . .  697, (1962).

4751   ZAMIR...........(D.)....:   ( 6 , 2176) J. PHYS. CHEM. SOLIDS VOL.34,. .  725, (1973).

4752   ZAMRII..........(V.N.)..:   (22A,    18) INSTRUM. EXP. TECH. NO.6,. . . . 1176, (1964).

4753   ZANBERIS........(D.O.)..:   ( 5 , 2799) J. PHYS. CHEM. SOLIDS 30,. . . .  453, (1969).

4754   ZANDVELD........(P.)....:   (14A,    34) PHYSICA VOL.50,. . . . . . . .   511, (1970).
                                   ( 5 ,  106) PHYS. LETT. VOL.34A, . . . . . .  112, (1971).

4755   ZANIO...........(K.)....:   ( 6 ,  461) PHYS. REV. B VOL.10 NO.2,. . . . . . , (1974).

4756   ZANNONI.........(G.)....:   (20C,    66) NUCL. INST. MET. 67, . . . . . .  267, (1969).
                                   ( 5 ,  231) NUCL. PHYS. A VOL.A181,. . . . .  177, (1972).

4757   ZARALOVA.......(Z.K.)..:    ( 5 , 2628) ATOMNAYA ENERGIYA (USSR) VOL.24,  243, (1968).

4758   ZAREMBOVITCH....(A.)....:   ( 5 , 2392) J. PHYSIQUE TOME 32 COL.1 VOL.II  611, (1971).

4759   ZAROCHENTSEV....(E.V.)..:   ( 8 ,    63) PHYS. MET. METALLOG. VOL.22 NO.4    7, (1967).
                                   (12A,    44) UKRAINIAN PHYS. J. VOL.14, . . . 1867, (1969).

4760   ZASLAVSKII......(A.I.)..:   ( 5 ,  730) SOV. PHYS. SOL. STATE VOL.11,. .  672, (1969).

4761   ZATOVSKII.......(A.V.)..:   ( 5 , 2249) UKRAINIAN PHYS. J. VOL.14, . . . 1909, (1969).

4762   ZATSERKOVSKA....(R.A.)..:   ( 5 , 2242) UKRAINIAN PHYS. J. VOL.14, . . . 1968, (1969).

4763   ZATSERKOVSKII...(R.A.)..:   ( 5 ,  379) UKR. FIZ. ZH.(USSR) VOL.17,. . .   38, (1972).

4764   ZAUBERIS........(D.O.)..:   ( 5 , 2654) ACTA CRYST. VOL.9,. . . . . . .   607, (1956).
                                   ( 5 , 2664) J. APPL. PHYS. VOL.27,. . . . . 1040, (1956).
                                   ( 5 , 1577) ACTA CRYST. 18,. . . . . . . . .  906, (1965).
                                   ( 5 , 2359) ACTA CRYST. 19,. . . . . . . . .  413, (1965).

4765   ZAVADIL.........(V.)....:   (10B,    71) J. PHYS. C: SOLID ST. PHYS. V-5,  287, (1972).

4766   ZAVODINSKII.....(V.G.)..:   ( 6 , 1671) SOV. PHYS. SOL. STATE VOL. 13, . 1793, (1972).

4767   ZAWADOWSKI......(A.)....:   (10D,    60) SOL. STATE COMM. VOL. 9, . . . .  129, (1971).

4768   ZECH............(H.J.)..:   (15 ,    64) Z. NATURFORSCH. VOL.20A, . . . .  380, (1965).

4769   ZEIDLER.........(M.D.)..:   ( 6 ,     1) BER. BUNSENGES. PHYS. CHEM. 75,.  769, (1971).

4770   ZEILINGER.......(A.)....:   ( 5 ,  802) Z. ANGEW PHYS. VOL.32,. . . . .   109, (1971).
                                   ( 5 ,  801) ATOMKERNENERGIE VOL.19,. . . . .  167, (1972).

4771   ZEITNITZ........(B.)....:   ( 5 , 1880) NUCL. PHYS. VOL.A166,. . . . . .  443, (1971).
                                   ( 5 , 1879) NUCL. PHYS. VOL.A166,. . . . . .  461, (1971).

4772   ZEKOVIC.........(S.)....:   ( 6 , 1276) PHYS. STAT. SOL. VOL.49, . . . .  277, (1972).

4773   ZELENKA.........(J.)....:   ( 5 , 2371) PHYS. STAT. SOL. 29, . . . . . .  K51, (1968).
                                   ( 5 , 2375) BRIT. J. APPL. PHYS. VOL.2,. . . 1041, (1969).
                                   ( 9B,    89) PHYS. STAT. SOL. A VOL.2,. . . .  211, (1970).
                                   ( 5 , 2373) PHYS. STAT. SOL. VOL.42,. . . . .  895, (1970).
                                   ( 6 , 2052) NATURE (PHYS. SCI.) VOL.242, . .  109, (1973).

4774   ZELENYUK........(F.M.)..:   (20B,    51) INSTRUM. EXP. TECH. VOL.16 NO.2,  399, (1973).

4775   ZEMACH..........(A.C.)..:   (15 ,    36) PHYS. REV. VOL.101,. . . . . . .  118, (1956).
                                   (14R,    75) PHYS. REV. VOL.101,. . . . . . .  129, (1956).
                                   (14R,    77) PHYS. REV. VOL.109,. . . . . . . 1564, (1958).

4776   ZEMLYANOV.......(M.G.)..:   ( 6 ,  328) SOV. AT. ENERGY VOL.14,. . . . .  252, (1963).
                                   ( 6 , 2239) SOV. PHYS. J.E.T.P. VOL.16,. . . 1472, (1963).
                                   ( 6 , 1750) SOV. PHYS. J.E.T.P. VOL.17,. . .  584, (1963).
                                   ( 6 , 2186) SOV. PHYS. SOL. STATE 5,. . . .    78, (1963).
                                   ( 6 , 1427) SOV. PHYS. J.E.T.P. VOL.22,. . .  315, (1966).
                                   ( 6 , 2483) PHYS. STAT. SOL. 20,. . . . . .   767, (1967).
                                   ( 6 , 1759) SOV. PHYS. J.E.T.P. VOL. 25,. .   436, (1967).
                                   ( 6 , 2211) SOV. PHYS. SOL. STATE 9,. . . . 1366, (1967).
                                   ( 5 , 2011) SOV. PHYS. SOL. STATE VOL.10,. . 1076, (1968).
                                   ( 5 , 2515) SOV. PHYS. SOL. STATE VOL.10,. .  212, (1968).
                                   ( 6 , 1680) SOV. PHYS. SOL. STATE VOL.11,. . 2343, (1969).
                                   ( 5 , 1595) ACTA CRYST. VOL.A28,. . . . . .   473, (1972).

4777   ZENKEVICH.......(V.S.)..:   ( 5 , 2478) SOV. J. NUCL. PHYS. VOL.10, 1,.   18, (1969).
                                   (19B,    35) SOV. AT. ENERGY VOL.28,. . . . .  150, (1970).
                                   (20C,   109) SOV. AT. ENERGY VOL.32, NO.5,. .  416, (1972).

4778   ZERBI...........(G.)....:   ( 6 ,  348) J. CHEM. PHYS. VOL.40, . . . . . 3502, (1964).
                                   (10B,    50) J. CHEM. PHYSICS VOL.48, . . . . 3561, (1968).
                                   (10B,    49) J. CHEM. PHYSICS VOL.49, . . . . 3840, (1968).
                                   (10B,    56) J. CHEM. PHYSICS VOL.53, . . . . 1428, (1970).
                                   (10B,    58) J. CHEM. PHYSICS VOL.54, . . . . 3600, (1971).
                                   ( 6 , 1910) MOL. PHYS. VOL.22, . . . . . . .  241, (1971).
                                   (10B,    59) J. CHEM. PHYS. VOL.56, . . . . . 1022, (1972).
                                   (10B,    60) J. CHEM. PHYS. VOL.59, . . . . . 4578, (1973).

4779   ZETLYANOV.......(M.G.)..:   ( 6 , 2101) SOV. PHYS. SOL. STATE 8, . . . . 2156, (1966).

4780   ZEYHER..........(R.)....:   (10D,    49) PHYS. STAT. SOL. VOL.48, . . . .  711, (1971).

4781   ZGIERSKI........(M.)....:   ( 6 ,  283) PHYS LETT 27A, . . . . . . . . .    9, (1968).

4782   ZHARKOV.........(V.N.)..:   ( 6 , 1181) DOKL. AKAD. NAUK VOL.93, . . . .  799, (1953).

4783   ZHDANOV.........(G.S.)..:   (22C,    71) SOV. PHYS. CRYST. VOL.5, . . . .  297, (1960).
                                   ( 5 , 1277) SOV. PHYS. JETP VOL.13,. . . . .  718, (1961).
                                   ( 5 ,  780) J. PHYS. SOC. JAPAN VOL.17 B-II,  385, (1962).
                                   ( 5 , 1475) SOV. PHYS. CRYST. VOL.7,. . . . .  499, (1962).
                                   ( 5 ,  217) SOV. PHYS. DOKLADY VOL.7,. . . .  742, (1962).
                                   ( 8 ,  135) SOV. PHYS. USPEKHI VOL.5,. . . .  194, (1962).
                                   ( 5 , 1579) SOV. PHYS. DOKLADY VOL.8,. . . .  101, (1963).
                                   ( 5 , 2322) SOV. PHYS. SOL. STATE 5, . . . . 2425, (1963).
```

4784 ZHELUDEV........(I.S.)..: (5 , 160) SOV. PHYS. J.E.T.P. VOL.23,. . . 395, (1966).

4785 ZHERNOV.........(A.P.)..: (10D, 62) SOV. PHYS. JETP 21,. . . . . . . 646, (1965).
(10C, 118) SOV. PHYS. JETP. 20, ... : 1340, (1965).

4786 ZHEZHERUN.......(I.F.)..: (5 , 204) SOV. AT. ENERGY VOL.13,. . . . . 852, (1962).

4787 ZHIDKOV.........(L.G.)..: (10C, 123) SOV. PHYS. J. NO.9,. . . . . . . 31, (1968).

4788 ZHILYAKOV.......(S.M.)..: (5 , 1556) SOV. PHYS. SOL. STATE VOL.12,. . 770, (1970).

4789 ZHUKOV..........(G.P.)..: (22A, 18) INSTRUM. EXP. TECH. NO.6,. . . . 1176, (1964).

4790 ZHUKOV..........(G.)....: (19A, 27) IAEA SYMP. GRENOBLE, ... 763, (1972).

4791 ZHURAVLEV.......(B.E.)..: (22A, 18) INSTRUM. EXP. TECH. NO.6,. . . . 1176, (1964).

4792 ZIEBECK.........(K.R.A.): (5 , 509) J. PHYS. CHEM. SOLIDS VOL.34,. : 1647, (1973).
(5 , 515) J. PHYS. CHEM. SOLIDS VOL.35,. : 1, (1974).

4793 ZIELENIEWSKI....(R.)....: (21 , 40) NUKLEONIKA VOL.12, ... 385, (1967).

4794 ZIJP............(E.)....: (12A, 12) NED. TIJDSCHR. NATUURK. VOL.38,. 171, (1972).

4795 ZILBER..........(R.)....: (5 , 736) SOL. STATE COMM. VOL.8,. . . . . 935, (1970).

4796 ZILSEL..........(P.R.)..: (7B, 58) PHYS. REV. VOL.71, ... 232, (1947).

4797 ZIMMERMAN.......(R.L.)..: (23 , 52) NUCL. INSTRUM. VOL.1,. . . . . . 92, (1957).
(20B, 71) NUCL. INST. MET. VOL.13, ... 1, (1961).

4798 ZIMMER..........(G.J.)..: (5 , 1821) MAGN. AND MAG. MATERIALS-1971,. 513, (1972).
(5 , 1701) A.I.P. CONF. PROC. NO.10 PART 2, 1379, (1973).

4799 ZINN............(W.H.)..: (20A, 62) PHYS. REV. VOL.71, ... 752, (1947).

4800 ZINN............(W.)....: (5 , 903) PHYS. STAT. SOL. B VOL.46, . . . 597, (1971).

4801 ZINOV...........(V.G.)..: (20C, 110) SOV. PHYS. J.E.T.P. VOL.31,. . . 59, (1970).

4802 ZISCEWSKI.......(R.)....: (5 , 1774) PHYS. MET. METALLOGR. VOL.16 N.5 145, (1963).

4803 ZIVADINOVIC.....(M.S.)..: (5 , 653) ACTA CRYST. VOL.20,. . . . . . . 315, (1966).

4804 ZIVADINOVIC.....(M.)....: (5 , 194) ACTA CRYST. VOL.12,. . . . . . . 476, (1959).

4805 ZIVANOVIC.......(M.D.)..: (22A, 6) BULL. INST. NUCL. SCI.(YUGO.) 11 59, (1961).

4806 ZIVANOVIC.......(M.)....: (20A, 37) NUCL. INST. MET. 65,. . . . . . 233, (1968).
(6 , 3) CHEM. PHYS. LETT. VOL.17,. . . . 53, (1972).
(6 , 678) PHYS. LETT. VOL.42A NO.7,. . . . 509, (1973).

4807 ZOCCHI..........(M.)....: (17A, 3) ACTA CRYST. 17,. . . . . 597, (1964).
(17A, 5) ACTA CRYST. 22,. . . . . 331, (1967).

4808 ZOLOTUKHIN......(V.G.)..: (20B, 131) PRIB. TEKH. EKSPER. VOL.2, . . . 36, (1964).

4809 ZORIN...........(R.B.)..: (5 , 2856) SOV. PHYS. CRYST. VOL.17,. . . . 1017, (1973).

4810 ZSIGMOND........(G.)....: (20B, 79) NUCL. INST. MET. 45, ... 255, (1966).
(23 , 60) NUCL. INST. MET. 55, ... 151, (1967).
(19A, 27) IAEA SYMP. GRENOBLE, ... 763, (1972).

4811 ZSOLDOS.........(E.)....: (5 , 1770) J. PHYSIQUE TOME 32 COL.1 VOL.II 980, (1971).
(5 , 1793) SOL. STATE COMM. VOL.9,. . . . . 27, (1971).
(5 , 2889) A.I.P. CONF. PROC. NO.10 PART 2, 1603, (1973).

4812 ZUBKOV..........(V.G.)..: (5 , 2646) SOV. PHYS. DOKLADY VOL.15,. . . 276, (1970).
(5 , 636) SOV. PHYS. SOL. STATE VOL. 13,. 644, (1971).
(5 , 1543) SOV. PHYS. SOL. STATE VOL.15,. . 1079, (1973).

4813 ZUBOV...........(YU.G.).: (7A, 92) SOV. PHYS. JETP LETT. VOL.14,. . 91, (1971).

4814 ZUCCA...........(T.)....: (5 , 2459) PHYSICS/NON-CRYSTALLINE SOLIDS,. 152, (1965).

4815 ZUEV............(V.I.)..: (22C, 12) CRYOGENICS VOL.10, ... 440, (1970).

4816 ZUPRANSKI.......(P.)....: (12B, 27) NUCL. INST. MET. 64, ... 77, (1968).

4817 ZVEREV..........(G.M.)..: (6 , 457) SOV. PHYS. SOL. STATE 6, ... 76, (1964).

4818 ZWANZIG.........(R.)....: (14B, 84) PHYS. REV. VOL.133,. . . . . . . A50, (1964).
(14A, 55) PHYS. REV. A VOL.4,. . . . . . . 1616, (1971).

4819 ZWEIFEL.........(P.F.)..: (18B, 3) IAEA SYMPOSIUM VIENNA, ... 199, (1960).
(7B, 72) PHYS. REV. VOL.126,. . . . . . . 1165, (1962).
(7B, 74) PHYS. REV. 131,. . . . . 1149, (1963).
(7B, 79) PHYS. REV. VOL.137,. . . . . . . B271, (1965).
(7B, 24) J. CHEM. PHYSICS VOL.47, ... 4923, (1967).
(7B, 83) PHYS. REV. 159,. . . . . 1, (1967).
(16 , 10) CONF. NEUTRON THERMALIZ. VOL.I,. 283, (1968).

4820 ZYCH............(W.)....: (12B, 27) NUCL. INST. MET. 64, ... 77, (1968).

4821 ZYGMUNT.........(A.)....: (5 , 2813) PHYS. STAT. SOL. 22, ... 517, (1967).
(5 , 2716) PHYS. STAT. SOL. 23, ... K123, (1967).
(5 , 2721) PHYS. STAT. SOLIDI A VOL.13, . . 657, (1972).
(5 , 2723) PHYS. STAT. SOLIDI A VOL.16 NO.2 K171, (1973).

4822 ZYUGANOV........(A.N.).: (10C, 133) UKR. FIZ. ZH. (USSR) VOL.8,. . . 256, (1963).

INDEX